CALCULUS

Early Transcendentals

Fourth Edition

CALCULUS

Early Transcendentals
Fourth Edition

Dennis G. Zill
Loyola Marymount University

Warren S. Wright
Loyola Marymount University

JONES AND BARTLETT PUBLISHERS
Sudbury, Massachusetts
BOSTON   TORONTO   LONDON   SINGAPORE

World Headquarters
Jones and Bartlett Publishers
40 Tall Pine Drive
Sudbury, MA 01776
978-443-5000
info@jbpub.com
www.jbpub.com

Jones and Bartlett Publishers Canada
6339 Ormindale Way
Mississauga, Ontario L5V 1J2
Canada

Jones and Bartlett Publishers International
Barb House, Barb Mews
London W6 7PA
United Kingdom

Jones and Bartlett's books and products are available through most bookstores and online booksellers. To contact Jones and Bartlett Publishers directly, call 800-832-0034, fax 978-443-8000, or visit our website, www.jbpub.com.

Substantial discounts on bulk quantities of Jones and Bartlett's publications are available to corporations, professional associations, and other qualified organizations. For details and specific discount information, contact the special sales department at Jones and Bartlett via the above contact information or send an email to specialsales@jbpub.com.

Production Credits
Chief Executive Officer: Clayton Jones
Chief Operating Officer: Don W. Jones, Jr.
President, Higher Education and Professional Publishing: Robert W. Holland, Jr.
V.P., Sales: William J. Kane
V.P., Design and Production: Anne Spencer
V.P., Manufacturing and Inventory Control: Therese Connell
Publisher: David Pallai
Acquisitions Editor: Timothy Anderson
Editorial Assistant: Melissa Potter
Production Director: Amy Rose
Production Manager: Jennifer Bagdigian
Production Assistant: Ashlee Hazeltine
Senior Marketing Manager: Andrea DeFronzo
Associate Marketing Manager: Lindsay Ruggiero
Cover and Title Page Design: Kristin E. Parker
Photo Research and Permissions Manager: Kimberly Potvin
Composition: Aptara®, Inc.
Front Cover Image: © Cenk E. Tezel & Tunç Tezel
Back Cover Image: © Dennis Mammana, all rights reserved. Original image courtesy NASA/JPL-Caltech.
Printing and Binding: Courier Kendallville
Cover Printing: Courier Kendallville

Library of Congress Cataloging-in-Publication Data
Zill, Dennis G., 1940-
 Calculus: early transcendentals / Dennis G. Zill and Scott Wright.
  p. cm.
 Includes Index.
 ISBN-13: 978-0-7637-5995-7 (pbk.)
 ISBN-10: 0-7637-5995-3 (pbk.)
1. Calculus. I. Wright, Scott. II. Title.
QA303.Z523 2009
515—dc22
                    2008040467
6048
Printed in the United States of America
13 12 11 10 09 10 9 8 7 6 5 4 3 2 1

≡ To the Instructor

Philosophy

The **fourth edition** of *Calculus: Early Transcendentals* represents a substantial revision of the last edition. Although there is much new in this edition, I have striven to keep intact my original goal of compiling a calculus text that is not just a collection of definitions and theorems, skills and formulas to be memorized, and problems to be solved, but rather a book that communicates with its primary audience, the students. It is my hope that these changes make the text more relevant and interesting for both student and instructor.

Features in This Revision

Precalculus Material   The precalculus review material on the real number system, the Cartesian plane, lines, and circles that appeared in Appendix I in the last edition, has been moved to the *Student Resource Manual* (*SRM*).

Sections and Exercises   Most of the material has been updated and, in some instances, reorganized. Many sections and exercise sets have been completely rewritten. Many new problems, especially applications, problems dealing with graphing calculators and computers, conceptual problems, and project problems have been added to the exercise sets. For the most part, the added applications are "real life" in that they have been thoroughly researched using original sources. Problems dealing with interpretation of graphs have also been added. Moreover, there is an increased emphasis on the trigonometric functions, both in the examples and in the exercise sets throughout the text. There are over 7300 problems in this edition.

To aid in assignment of problems, each exercise set is clearly partitioned into groups of problems using headings such as *Fundamentals*, *Applications*, *Mathematical Models*, *Projects*, *Calculator/CAS Problems*, and so on. I think most of the heads are self-explanatory, and so problems listed under the heading *Think About It* deal with conceptual aspects of the material covered in that section and are suitable either for assignment or for a classroom discussion. No answers are supplied for these problems in the text. Some problems are labeled *Mathematical Classics* and reflect the fact that they have been around for a long time, appear in almost every calculus text, or have an interesting twist, while other problems labeled *A Bit of History* naturally evince some historical content.

Chapter 1 is a review of functions and, following the current prevailing fashion, functions are presented from the algebraic, graphical, numerical, or verbal points of view. In point of fact, the entire last section in Chapter 1 is entitled *From Words to Functions*. Because many students invariably encounter difficulties in solving related rate and applied optimization problems, I have included this section to give them a preview of how to set up, or construct, a function from a verbal description (with calculus content removed). Indeed, many of the problems in Section 1.7 reappear in a calculus context in Section 4.8.

Differential equations appear in two chapters in this text, Chapters 8 and 16. First-order equations are considered in Chapter 8 to benefit those students who encounter their applications in courses in physics and engineering. In Chapter 16, the solution and applications of higher-order differential equations are considered. Of course, Chapters 8 and 16 can be combined and covered as one unit at any point in the course after Chapter 4 has been covered. Proofs of some of the longer theorems are given in the appendix. Biographical sketches of some of the mathematicians who had a significant impact on the development of calculus appear at the end of appropriate sections under the rubric *Postscript—A Bit of History*.

Special Features Each chapter opens with its own table of contents and an introduction to the material covered in that chapter. The text ends with *Resource Pages*, which is a compact review of basic concepts from algebra, geometry, trigonometry, and calculus: the laws of exponents, factorization formulas, binomial expansions, Pascal's triangle, formulas from geometry, graphs and functions, trigonometric functions, exponential and logarithmic functions, and differentiation and integration formulas. Many of the topics summarized in the *Resource Pages* are discussed in greater depth in the *SRM*.

The popular feature introduced in the last edition called *Test Yourself* has been retained and expanded in this edition. The *Test Yourself* feature is a self-test consisting of 56 questions on four broad areas of precalculus mathematics. This test is intended to encourage students to review, perhaps on their own, some of the more essential prerequisite subjects, such as absolute values, the Cartesian plane, equations of lines, circles, and so on, that are used throughout the text. Answers to all questions in the test are given in the answer section.

Users of the previous three editions have been very receptive to the *Remarks* that often conclude a section. As a consequence their number has been increased and they have been renamed as *Notes from the Classroom*. These *Notes* are intended to be informal discussions that are aimed directly at the student. These discussions range from warnings about common algebraic, procedural, and notational errors; to misinterpretations of theorems; to advice; to questions asking the student to think about and possibly extend the ideas just presented.

Also, at the request of users, the number of marginal annotations and guidance annotations in the examples have been increased.

Student Resource Manual I believe that this manual can be of significant help to a student's success in a calculus course. Unlike the traditional student solutions manual, where a selected subset of the problems are worked out, the *SRM* is divided into four sections:

- ESSAYS • TOPICS IN PRECALCULUS • USE OF A CALCULATOR
- SELECTED SOLUTIONS

In the opening section of the *SRM*, Essays, there are two interesting essays, one on the history of calculus and the other on its modern day applications:

The Story of Calculus, by Roger Cooke, University of Vermont, and
Calculus and Mathematical Modeling by Fred S. Roberts, Rutgers University.

(*The Story of Calculus* also appears on page xxv of this book.) In Topics in Precalculus, selected topics from precalculus (such as sets, exponents, real numbers and inequalities, the Cartesian Plane, lines, matrices, and determinants) are reviewed because of their relevance to calculus. Because I feel that a mathematics textbook is not the appropriate place for a discussion of the use of technology, graphing calculator essentials are discussed in the section entitled Use of a Calculator. In Selected Solutions, a detailed solution of every odd problem in the exercise sets is given.

Figures, Definitions, Theorems A word about the numbering of the figures, definitions, and theorems is in order. Because of the great number of figures, definitions, and theorems in this text, I have switched to a double-decimal numeration system. For example, the interpretation of "Figure 1.2.3" is

Chapter Section of Chapter 1
↓ ↓

1.2.3 ← Third figure in Section 1.2

I feel that this type of numeration will make it easier to find, say, a theorem or figure when it is referred to in a later section or chapter. In addition, to better link a figure with the text, the *first* textual reference to each figure is done in the same font style and color as the figure number. For example, the first reference to the first figure in Section 7.5 is given as Figure 7.5.1 and all subsequent references are written in the traditional style of Figure 7.5.1. Also, in this revision each figure in the text has a brief explanatory caption.

Supplements

For Instructors

- An *Instructor's ToolKit* (*ITK*) offers instructors a computerized test bank that allows instructors to create customized tests and quizzes. The questions and answers are sorted by chapter and can be easily installed on a computer for accessibility. The *ITK* also includes *PowerPoint*® *Slides*, which feature all labeled figures as they appear in the text. This useful tool allows instructors to easily display and discuss figures and problems found within the text. The *ITK* is located on a secured website for qualified instructors only.

    http://www.jbpub.com/catalog/9780763759957/Instructor

- A *Complete Solutions Manual* (*CSM*) is also available on the website or on a CD. This manual contains detailed solutions of every problem in the text.
- *WebAssign*   Developed by instructors for instructors, WebAssign is the premier independent online teaching and learning environment, guiding over 3 million students through their academic careers since 1997. With WebAssign, you can:

    ○ Create and distribute algorithmic assignments using questions specific to your textbook
    ○ Grade, record, and analyze student responses and performance instantly
    ○ Offer more practice exercises, quizzes, and homework
    ○ Upload resources to share and communicate with your students seamlessly

    For more detailed information and to sign up for free faculty access, please visit:

    www.webassign.net

    Please contact your Jones and Bartlett Publisher's Representative for information on how students can purchase access to WebAssign bundled with this textbook.
    Designated instructor's materials are for qualified instructors only. Jones and Bartlett reserves the right to evaluate all requests.

For Students

- A *Student Resource Manual* (*SRM*) prepared by Jeffrey M. Gervasi, EdD, of Porterville College. This printed manual can be ordered bundled with the text at substantial savings compared to buying the text and the *SRM* separately.
- WebAssign access code card

≡ To the Student

You are enrolled in one of the most interesting courses in mathematics. Many years ago when I was a student in Calculus I, I was struck by the power and beauty of the material. It was unlike any mathematics that I had studied up to that point. It was fun, it was exciting, and it was a challenge. After teaching collegiate mathematics for many years, I have seen almost every type of student, from a budding genius who invented his own calculus, to students who struggled to master the most rudimentary mechanics of the subject. Over these years I have also witnessed a sad phenomenon: some students fail calculus, not because they find the subject matter impossibly difficult, but because they have weak algebra skills and an inadequate working knowledge of trigonometry. Calculus builds immediately on your prior knowledge and skills and there is much new ground to be covered. Consequently there is very little time to review basics in the formal classroom setting. So those of us who teach calculus must assume that you can factor, simplify and solve equations, solve inequalities, handle absolute values, use a calculator, apply the laws of exponents, find equations of lines, plot points, sketch basic graphs, and apply important logarithmic and trigonometric identities. The ability to do algebra and trigonometry, work with exponentials and logarithms, and sketch *by hand* basic graphs quickly and accurately are keys to success in a calculus course.

On page xxi there is a list of 56 questions in a section called "Test Yourself." This "test" is an opportunity for you to check your knowledge on some of the topics that will be assumed in this text. Relax, take your time, read and work every question, and then compare your answers with those given on page ANS-1. Regardless of your "score" on this test, you are strongly encouraged to review precalculus material either given in the *Student Resource Manual* or from a precalculus text.

A word to those students who have taken calculus in high school: please do not assume that you can get by with minimal effort because you recognize some of the topics in differential and integral calculus. A sense of familiarity of the subject combined with an attitude of complacency is often the downfall of some students.

Learning mathematics is not like learning how to ride a bicycle, that once learned, the ability sticks for a lifetime. Mathematics is more like learning another language or learning to play a musical instrument; it requires time, effort, and a lot of practice to develop and maintain proficiency. Even experienced musicians still practice the fundamental scales. So ultimately, you the student can learn mathematics (that is, make it stick) only through the hard work of doing mathematics. Although I have tried to make *most* details in the solution of an example clear to the reader, inevitably you will have to fill in some missing steps. You cannot read a text such as this as you would a novel; you must work your way through the text with pencil and paper at the ready.

In conclusion, I wish you the best of luck in this course.

≡ Acknowledgments

Compiling a textbook of this complexity is a monumental task. Besides the authors, many people put much time and energy into this project. First and foremost, I would like to express my appreciation to the editorial, production, and marketing staffs at Jones and Bartlett Publishers and to the following reviewers of this and previous editions who contributed many suggestions, valid criticisms, and even an occasional word of support:

Scott Wilde, *Baylor University*
Salvatore Anastasio, *SUNY, New Paltz*
Thomas Bengston, *Penn State University, Delaware County*
Steven Blasberg, *West Valley College*
Robert Brooks, *University of Utah*
Dietrich Burbulla, *University of Toronto*
David Burton, *Chabot College*
Maurice Chabot, *University of Southern Maine*
H. Edward Donley, *Indiana University of Pennsylvania*
John W. Dulin, *GMI Engineering & Management Institute*
Arthur Dull, *Diablo Valley College*
Hugh Easler, *College of William and Mary*
Jane Edgar, *Brevard Community College*
Joseph Egar, *Cleveland State University*
Patrick J. Enright, *Arapahoe Community College*
Peter Frisk, *Rock Valley College*
Shirley Goldman, *University of California at Davis*
Joan Golliday, *Santa Fe Community College*
David Green, Jr., *GMI Engineering & Management Institute*
Harvey Greenwald, *California Polytechnic State University*
Walter Gruber, *Mercy College of Detroit*
Dave Hallenbeck, *University of Delaware*
Noel Harbetson, *California State University at Fresno*
Bernard Harvey, *California State University, Long Beach*

Christopher E. Hee, *Eastern Michigan University*
Jean Holton, *Tidewater Community College*
Rahim G. Karimpour, *Southern Illinois University*
Martin Kotler, *Pace University*
Carlon A. Krantz, *Kean College of New Jersey*
George Kung, *University of Wisconsin at Stevens Point*
John C. Lawlor, *University of Vermont*
Timothy Loughlin, *New York Institute of Technology*
Antonio Magliaro, *Southern Connecticut Slate University*
Walter Fred Martens, *University of Alabama at Birmingham*
William E. Mastrocola, *Colgate University*
Jill McKenney, *Lane Community College*
Edward T. Migliore, *Monterey Peninsula College*
Carolyn Narasimhan, *DePaul University*
Harold Olson, *Diablo Valley College*
Gene Ortner, *Michigan Technological University*
Aubrey Owen, *Community College of Denver*
Marvin C. Papenfuss, *Loras College*
Don Poulson, *Mesa Community College*
Susan Prazak, *College of Charleston*
James J. Reynolds, *Pennsylvania State University, Beaver Campus*
Susan Richman, *Penn State University, Harrisburg*

Rodd Ross, *University of Toronto*
Donald E. Rossi, *De Anza College*
Lillian Seese, *St. Louis Community College at Meramec*
Donald Sherbert, *University of Illinois*
Nedra Shunk, *Santa Clara University*
Phil R. Smith, *American River College*
Joseph Stemple, *CUNY Queens College*

Margaret Suchow, *Adirondack Community College*
John Suvak, *Memorial University of Newfoundland*
George Szoke, *University of Akron*
Hubert Walczak, *College of St. Thomas*
Richard Werner, *Santa Rosa Junior College*
Loyd V. Wilcox, *Golden West College*
Jack Wilson, *University of North Carolina, Asheville*

I would also like to extend an extra-special thank you to the following individuals:

- Jeff Dodd, Jacksonville State University, for the project in Problem 37 of Exercises 8.3,
- John David Dionisio, Loyola Marymount University, and Brian and Melanie Fulton, High Point University, for providing the solutions for the *Complete Solutions Manual* and the *Student Resource Manual*,
- Roger Cooke, University of Vermont, and Fred S. Roberts, Rutgers University, for graciously taking time out their busy schedules to write and contribute the excellent calculus essays,
- Carol Wright, for her help in the final stages of manuscript preparation for this and other texts,
- David Pallai, Publisher, and Tim Anderson, Editor, for putting up with all my verbal venting of frustrations,
- Jennifer Bagdigian, Production Manager, for smoothly coordinating the phases of production and for her patience in putting up with my never-ending changes, and
- Irving Drooyan and Charles Carico, for starting it all.

Even with all the help just listed, the accuracy of every letter, word, symbol, equation, and figure contained in this final product is the responsibility of the author. I would be very grateful to have any errors or "typos" called to my attention. Corrections can be sent to Tim Anderson at

tanderson@jbpub.com.

In conclusion, I welcome my long-time colleague Warren Scott Wright at Loyola Marymount University, and author of many of the supplements that accompany my texts, as a coauthor of this text.

Dennis G. Zill

Warren S. Wright

Contents

3 | The Derivative   121

4 | Applications of the Derivative   191

5 | Integrals   267

6 Applications of the Integral **321**

7 Techniques of Integration **379**

8 First-Order Differential Equations **439**

9   Sequences and Series   475

10   Conics and Polar Coordinates   547

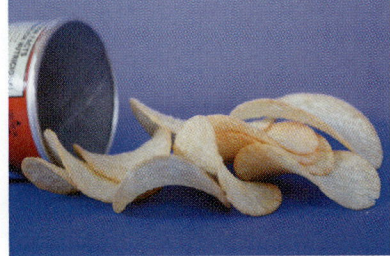

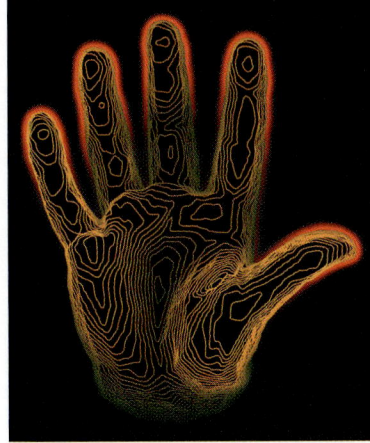

Front Cover: Total Solar Eclipse Analemma

This image is the world's first analemma (or *Tutulemma*, a term coined by the photographers based on the Turkish word for eclipse) photo that includes a total solar eclipse. This is a year-long image process, showing the Sun's motion in one frame. The photo was started in 2005 and ended in 2006 in Side, Turkey, about 500 km south of the photographer's home. Tunç Tezel, a leading amateur astronomer and night sky photographer, took the image with assistance from his brother, Cenk E. Tezel. Venus was also visible during the process, and can be seen toward the lower right portion of the photo. © Cenk E. Tezel & Tunç Tezel

Back Cover Inset: Martian Analemma

On Earth, the analemma is the figure-8 loop we get if we mark the Sun's position at the same time each day throughout the year. Its shape is determined by the tilt of our planet's axis and the variable speed at which Earth revolves around the Sun. If, however, we were to mark the Sun's position in the sky of Mars, we would discover a simpler, stretched, pear-shaped analemma. This digital illustration shows the late afternoon Sun that would have been seen from the Sagan Memorial Station once every 30 Martian days (sols) beginning on Sol 24 (July 29, 1997). Slightly less bright, the Martian Sun would appear only about one-third the size we see from Earth, while the Martian dust—responsible for Mars's pink sky—also scatters some blue light around the solar disk. This photo-illustration was created by first plotting the Martian analemma with Starry Night® Pro software (v6.2.3), and then using Adobe® Photoshop® CS2 (v9.0.2) to assemble it onto the correct celestial location on NASA's famous Presidential Panorama. © Dennis Mammana / dennismammana.com

Test Yourself

Answers to all questions are on page ANS-1.

In Preparation for Calculus

☰ Basic Mathematics

1. (True/False) $\sqrt{a^2 + b^2} = a + b$. _____

2. (True/False) For $a > 0$, $(a^{4/3})^{3/4} = a$. _____

3. (True/False) For $x \neq 0$, $x^{-3/2} = \dfrac{1}{x^{2/3}}$. _____

4. (True/False) $\dfrac{2^n}{4^n} = \dfrac{1}{2^n}$. _____

5. (Fill in the blank) In the expansion of $(1 - 2x)^3$ the coefficient of x^2 is _____.

6. Without the aid of a calculator, evaluate $(-27)^{5/3}$.

7. Write as one expression without negative exponents:
$$x^2 \frac{1}{2}(x^2 + 4)^{-1/2}2x + 2x\sqrt{x^2 + 4}.$$

8. Complete the square: $2x^2 + 6x + 5$.

9. Solve the equations:

   (a) $x^2 = 7x$ (b) $x^2 + 2x = 5$ (c) $\dfrac{1}{2x - 1} - \dfrac{1}{x} = 0$ (d) $x + \sqrt{x - 1} = 1$

10. Factor completely:
    (a) $10x^2 - 13x - 3$
    (b) $x^4 - 2x^3 - 15x^2$
    (c) $x^3 - 27$
    (d) $x^4 - 16$

☰ Real Numbers

11. (True/False) If $a < b$, then $a^2 < b^2$. _____

12. (True/False) $\sqrt{(-9)^2} = -9$. _____

13. (True/False) If $a < 0$, then $\dfrac{-a}{a} < 0$. _____

14. (Fill in the blanks) If $|3x| = 18$, then $x =$ _____ or $x =$ _____.

15. (Fill in the blank) If $a - 5$ is a negative number, then $|a - 5| =$ _____.

16. Which of the following real numbers are rational numbers?
    (a) 0.25   (b) 8.131313…   (c) π

    (d) $\dfrac{22}{7}$ (e) $\sqrt{16}$ (f) $\sqrt{2}$

    (g) 0   (h) -9 (i) $1\dfrac{1}{2}$

    (j) $\dfrac{\sqrt{5}}{\sqrt{2}}$ (k) $\dfrac{\sqrt{3}}{2}$ (l) $\dfrac{-2}{11}$

17. Match the given interval with the appropriate inequality.
    (i) $(2, 4]$ *(ii)* $[2, 4)$ *(iii)* $(2, 4)$ *(iv)* $[2, 4]$
    (a) $|x - 3| < 1$ (b) $|x - 3| \leq 1$ (c) $0 \leq x - 2 < 2$ (d) $1 < x - 1 \leq 3$

18. Express the interval $(-2, 2)$ as

    (a) an inequality and   (b) an inequality involving absolute values.

19. Sketch the graph of $(-\infty, -1] \cup [3, \infty)$ on the number line.

20. Find all real numbers x that satisfy the inequality $|3x - 1| > 7$. Write your solution using interval notation.

21. Solve the inequality $x^2 \geq -2x + 15$ and write your solution using interval notation.

22. Solve the inequality $x \leq 3 - \dfrac{6}{x + 2}$ and write your solution using interval notation.

☰ Cartesian Plane

23. (Fill in the blank) If (a, b) is a point in the third quadrant, then (a, b) is a point in the _____ quadrant.

24. (Fill in the blank) The midpoint of the line segment from $P_1(2, -5)$ to $P_2(8, -9)$ is _____.

25. (Fill in the blanks) If $(-2, 6)$ is the midpoint of the line segment from $P_1(x_1, 3)$ to $P_2(8, y_2)$, then $x_1 =$ _____ and $y_2 =$ _____.

26. (Fill in the blanks) The point $(1, 5)$ is on a graph. Give the coordinates of another point on the graph if the graph is:
(a) symmetric with respect to the x-axis. _____
(b) symmetric with respect to the y-axis. _____
(c) symmetric with respect to the origin. _____

27. (Fill in the blanks) The x- and y-intercepts of the graph of $|y| = 2x + 4$ are, respectively, _____ and _____.

28. In which quadrants of the Cartesian plane is the quotient x/y negative?

29. The y-coordinate of a point is 2. Find the x-coordinate of the point if the distance from the point to $(1, 3)$ is $\sqrt{26}$.

30. Find an equation of the circle for which $(-3, -4)$ and $(3, 4)$ are endpoints of a diameter.

31. If the points P_1, P_2, and P_3 are collinear as shown in FIGURE TY.1, then find an equation relating the distances $d(P_1, P_2)$, $d(P_2, P_3)$, and $d(P_1, P_3)$.

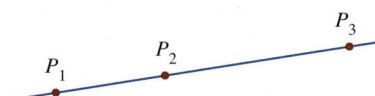

FIGURE TY.1   Graph for Problem 31

32. Which of the following equations best describes the circle given in FIGURE TY.2? The symbols a, b, c, d, and e stand for nonzero constants.
(a) $ax^2 + by^2 + cx + dy + e = 0$
(b) $ax^2 + ay^2 + cx + dy + e = 0$
(c) $ax^2 + ay^2 + cx + dy = 0$
(d) $ax^2 + ay^2 + c = 0$
(e) $ax^2 + ay^2 + cx + e = 0$

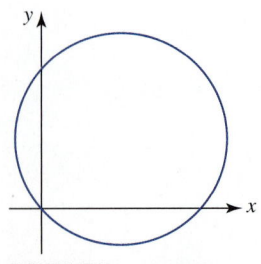

FIGURE TY.2   Graph for Problem 32

☰ Lines

33. (True/False) The lines $2x + 3y = 5$ and $-2x + 3y = 1$ are perpendicular. _____

34. (Fill in the blank) The lines $6x + 2y = 1$ and $kx - 9y = 5$ are parallel if $k =$ _____.

35. (Fill in the blank) A line with x-intercept $(-4, 0)$ and y-intercept $(0, 32)$ has slope _____.

36. (Fill in the blanks) The slope and the x- and y-intercepts of the line $2x - 3y + 18 = 0$ are, respectively, _____, _____, and _____.

37. (Fill in the blank) An equation of the line with slope -5 and y-intercept $(0, 3)$ is _____.

38. Find an equation of the line that passes through $(3, -8)$ and is parallel to the line $2x - y = -7$.

39. Find an equation of the line through the points $(-3, 4)$ and $(6, 1)$.

40. Find an equation of the line that passes through the origin and through the point of intersection of the graphs of $x + y = 1$ and $2x - y = 7$.

41. A tangent line to a circle at a point P on the circle is a line through P that is perpendicular to the line through P and the center of the circle. Find an equation of the tangent line L indicated in FIGURE TY.3.

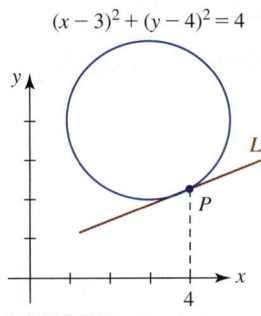

FIGURE TY.3 Graph for Problem 41

42. Match the given equation with the appropriate graph in FIGURE TY.4.

(*i*) $x + y - 1 = 0$ (*ii*) $x + y = 0$ (*iii*) $x - 1 = 0$

(*iv*) $y - 1 = 0$ (*v*) $10x + y - 10 = 0$ (*vi*) $-10x + y + 10 = 0$

(*vii*) $x + 10y - 10 = 0$ (*viii*) $-x + 10y - 10 = 0$

(a) 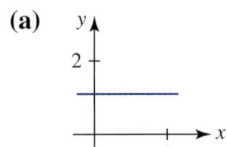   **(b)** 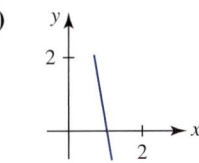   **(c)**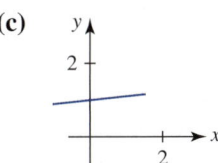

(d) 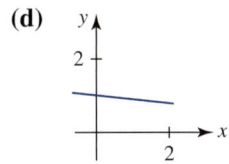   **(e)** 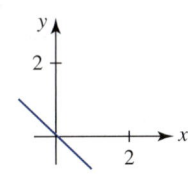   **(f)**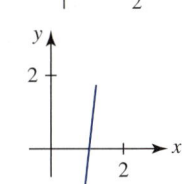

(g) 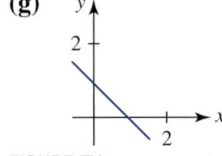   **(h)**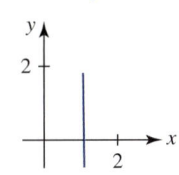

FIGURE TY.4 Graphs for Problem 42

☰ Trigonometry

43. (True/False) $1 + \sec^2\theta = \tan^2\theta$. _____

44. (True/False) $\sin(2t) = 2\sin t$. _____

45. (Fill in the blank) The angle 240 degrees is equivalent to _____ radians.

46. (Fill in the blank) The angle $\pi/12$ radians is equivalent to _____ degrees.

47. (Fill in the blank) If $\tan t = 0.23$, $\tan(t + \pi) =$ _____.

48. Find $\cos t$ if $\sin t = \frac{1}{3}$ and the terminal side of the angle t lies in the second quadrant.

49. Find the values of the six trigonometric functions of the angle θ given in FIGURE TY.5.

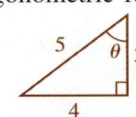

FIGURE TY.5 Triangle for Problem 49

50. Express the lengths b and c in FIGURE TY.6 in terms of the angle θ.

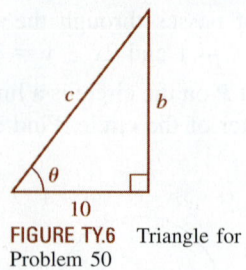

FIGURE TY.6   Triangle for
Problem 50

≡ Logarithms

51. Express the symbol k in the exponential statement $e^{(0.1)k} = 5$ as a logarithm.

52. Express the logarithmic statement $\log_{64} 4 = \frac{1}{3}$ as an equivalent exponential statement.

53. Express $\log_b 5 + 3\log_b 10 - \log_b 40$ as a single logarithm.

54. Use a calculator to evaluate $\dfrac{\log_{10} 13}{\log_{10} 3}$.

55. (Fill in the blank) $b^{3\log_b 10} = \underline{\qquad}$.

56. (True/False) $(\log_b x)(\log_b y) = \log_b(y^{\log_b x})$. _____

The Story of Calculus

by Roger Cooke

University of Vermont

Isaac Newton

Gottfried Leibniz

Calculus is generally considered to be a creation of the seventeenth-century European mathematicians, with the main work having been done by Isaac Newton (1642–1727) and Gottfried Wilhelm Leibniz (1646–1711). This traditional view is correct in broad outline. Any large-scale theory, however, is a mosaic whose tiles were laid over a long period of time; and in any living theory new tiles are continually being laid. The strongest statement the historian dares to make is that a pattern became apparent at a certain time and place. Such is the case with calculus. We can say with some confidence that the main outlines of the subject appeared in the seventeenth century and that the pattern was made much clearer by the work of Newton and Leibniz. However, many of the essential principles of calculus were discovered as early as the time of Archimedes (287–211 BCE), and some of these same discoveries were made independently in China and Japan. Moreover if you dig deeply into the problems and methods of calculus, you will soon find yourself pursuing problems that lead into the modern areas of analytic function theory, differential geometry, and functions of a real variable. To change the metaphor from art to transportation, we can think of calculus as a large railroad station, where passengers arriving from many different places all come together for a brief time before setting out again for a variety of destinations. In the present essay we shall try to look in both directions from that station, to the sources, and to the destinations. Let us begin by describing the station itself.

What Is Calculus?   Calculus is traditionally divided into two parts, called *differential calculus* and *integral calculus.* Differential calculus investigates the properties of the comparative rates of change of variables that are linked by equations. For example, a fundamental result of differential calculus is that if $y = x^n$, then the rate of change of y with respect to x is nx^{n-1}. It turns out that when we think intuitively about certain phenomena—the motion of bodies, changes in temperature, growth of populations, and many others—we are led to postulate certain relations between these variables and their rates of change. These relations are written down in a form known as *differential equations.* Thus the primary purpose of studying differential calculus is to understand what rates of change are and how to write down differential equations. Integral calculus provides methods of recovering the original variables knowing their rates of change. The technique for doing so is called *integration*, and the primary purpose of studying integral calculus is to learn how to *solve* the differential equations that are provided by differential calculus.

These goals are often masked in calculus books, where differential calculus is used to find the maximum and minimum values of certain variables, and integral calculus is used to compute lengths, areas, and volumes. There are two reasons for emphasizing these applications in a textbook. First, the full use of calculus involving differential equations involves some rather elaborate theory that must be introduced gradually; meanwhile, the student must be shown *some* use for the techniques that are being put forth. Second, such problems were the source of the ideas that led to calculus; the uses we now make of the subject arose only after it was discovered.

In describing the problems that led to calculus and the problems that can be solved using calculus, we still have not pointed out the fundamental techniques that make calculus so much more powerful a tool of analysis than mere algebra and geometry. These techniques involve the use of what was once called *infinitesimal analysis*. The constructions and formulas of high school geometry and algebra all have a finite character. For example, to construct the tangent to a circle or to bisect an angle, you perform a finite number of operations with straightedge and compass. Although Euclid knew considerably more geometry than is found in modern high school courses, he, too, confined himself mostly to finite processes. Only in the limited context of the theory of proportion does he allow the infinite into his geometry, and even there it is surrounded by so much logical caution that the proofs involved are extraordinarily cumbersome and hard to read. This same situation occurs in algebra. In order to solve a polynomial equation, you perform a finite number of operations of addition, sub-traction, multiplication, division, and root extraction. When the equation can be solved, the solution is expressed as a finite formula involving the coefficients.

These finite techniques, however, have a limited range of applicability. One cannot find the areas of most curved figures by a finite number of operations with straightedge and com-pass, nor can one solve most polynomial equations of degree five or higher using a finite number of algebraic operations. It was the desire to escape from the limitations of finite meth-ods that led to the creation of calculus. We shall now look at some of the early attempts to develop techniques for handling the more difficult problems of geometry, after which we shall summarize the process by which calculus was worked out and finally exhibit some of the harvest it has provided.

The Geometric Sources of the Calculus   One of the oldest mathematical problems is that of squaring the circle; that is, constructing a square equal in area to a given circle. It is now known that this problem cannot be solved by use of a finite number of applications of compass and straightedge. However, Archimedes discovered that if one could draw a spiral starting at the center of a circle that makes exactly one revolution before reaching the circle, then the tangent to that spiral at its point of intersection with the circle would form the hypotenuse of a right tri-angle with area exactly equal to the circle (see Figure 1). Thus if one could draw this spiral and its tangent, one could square the circle. Archimedes, however, was silent on the question of how one might draw this tangent.

We see here that one of the classical mathematical problems could be solved if only we could draw a certain curve and a tangent to it. This problem, and others like it, caused the purely mathematical problem of finding the tangent to a curve to become important. This

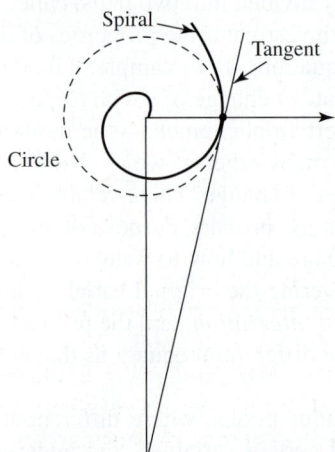

FIGURE 1   *The spiral of Archimedes.*   The tangent at the end of the first turn and the two axes form a triangle with area equal to the circle about the origin through the point of tangency.

problem is the main source of differential calculus. The "infinitesimal" trick that allows the problem to be solved is to think of the tangent as the line determined by two points on the curve "infinitely close" together. Another way of saying the same thing is that an "infinitely short" piece of any curve is straight. The trouble is that it is hard to be precise about the meaning of the phrases "infinitely close" and "infinitely short."

Little progress was made on this problem until the invention of analytic geometry in the seventeenth century by Pierre de Fermat (1601–1665) and René Descartes (1596–1650). Once a curve could be represented by an equation, it became possible to say with more confidence what was meant by "infinitely close" points, at least for polynomial equations such as $y = x^2$. With algebraic symbolism to represent points on the curve, it was possible to consider two points on the curve with x-coordinates x_0 and x_1, so that $x_1 - x_0$ is the distance between the x-coordinates. When the equation of the curve was written at each of these points and one of the two equations subtracted from the other, one side of the resulting equation contained the factor $x_1 - x_0$, which could therefore be divided out. Thus if $y_0 = x_0^2$ and $y_1 = x_1^2$, then $y_1 - y_0 = x_1^2 - x_0^2 = (x_1 - x_0) = (x_1 + x_0)$, and so $\dfrac{y_1 - y_0}{x_1 - x_0} = x_1 + x_0$. When $(x_1 = x_0)$, it follows that $y_1 = y_0$, and the expression $\dfrac{y_1 - y_0}{x_1 - x_0}$ has no meaning. However, the expression $x_1 + x_0$ has the perfectly definite value $2x_0$. Thus we can think of $2x_0$ as the ratio of the infinitely small difference in y, namely $y_1 - y_0$, to the infinitely small difference in x, namely $x_1 - x_0$, when the point (x_1, y_1) is infinitely close to the point (y_1, y_0) on the curve $y = x^2$. As you will learn in your study of calculus, this ratio gives enough information to draw the tangent line to the curve $y = x^2$.

The preceding argument is, except for small changes in notation, exactly the way Fermat found the tangent to a parabola. It was open to one logical objection, however: At one stage we divided both sides of an equation by $x_1 - x_0$, then at a later stage we decided that $x_1 - x_0 = 0$. Since division by zero is an illegal operation, we seem to be trying to eat our cake and have it, too. It took some time to find a convincing answer to this objection.

We have just seen that Archimedes was unable to solve the fundamental problem of differential calculus, drawing the tangent to a curve. Archimedes *was* able to solve some of the fundamental problems of integral calculus, however. In fact he found the volume of a sphere in an extremely ingenious way. He considered a cylinder containing a cone and a sphere and imagined this figure cut into infinitely thin slices. By looking at the areas of these sections of the cone, sphere, and cylinder, he was able to show how the cylinder would balance the cone and sphere if the figures were hung on opposite sides of a fulcrum. This balancing gave one relation among the three figures, and Archimedes already knew the volumes of the cone and cylinder; hence he was able to compute the volume of the sphere.

This argument illustrates the second infinitesimal technique that lies at the foundation of calculus: A volume can be regarded as a stack of plane figures, and an area can be regarded as a stack of line segments, in the sense that if every horizontal section of one region equals the same horizontal section of another region, then the two regions are equal. During the European Renaissance this principle came to be widely used under the name of the *method of indivisibles* for finding the areas and volumes of many figures. It is nowadays called *Cavalieri's principle* after Bonaventura Cavalieri (1598–1647), who used it to prove many of the elementary formulas that now make up integral calculus. Cavalieri's principle was also discovered in other lands where Euclid's work had never gone. The fifth-century Chinese mathematicians Zu Chongzhi and his son Zu Geng, for example, found the volume of a sphere using a technique very similar to Archimedes' method.

Thus we find mathematicians anticipating the integral calculus by using infinitesimal methods to find areas and volumes at a very early stage of geometry in both ancient Greece and China. Like the infinitesimal method of drawing tangents, however, this method of finding areas and volumes was open to objection. For example, the volume of each plane section of a figure is zero; how can a collection of zeros be put together to yield something that is not zero? Also, why doesn't the method work in one dimension? Consider the sections of a right triangle parallel to one of its legs. Each section intersects the hypotenuse and the other leg in congruent figures, namely one point each. Yet the hypotenuse and the other leg are not the same length. Objections like these were worrisome. The results obtained using these

Despite his notation and his arguments, which seem crude and inefficient today, the tremendous power of calculus shines through Newton's *Fluxions* in the solution of such difficult problems as finding the arc length of a curve. This "rectification" of a curve had been thought impossible, but Newton showed that one could find an infinite number of curves whose length could be expressed in finite terms.

Newton's approach to the calculus was algebraic, as we have just seen, and he inherited the fundamental theorem from Barrow. Leibniz, on the other hand, worked out the fundamental result on his own during the 1670s, and his approach was different from Newton's. Leibniz is considered the earliest pioneer of symbolic logic, and he had a much better appreciation than Newton of the importance of good symbolic notation. He invented the notation dx and dy that we still use today. For him dx was an abbreviation for "difference in x" and represented the difference between two infinitely close values of x. In other words, it expressed exactly what we had in mind above when we considered the infinitely small change $x_1 - x_0$. Leibniz thought of dx as an "infinitesimal" number, a number not zero, yet so small that no multiple of it could exceed any ordinary number. Not being zero, it could serve as the denominator in a fraction, and so dy/dx was the quotient of two infinitely small quantities. In this way he hoped to avoid the objections to the newly established method of finding tangents.

In the controversial technique of finding areas by adding up the sections, Leibniz also made a major contribution. Instead of thinking of an area [for example, the area under a curve $y = f(x)$] as a collection of line segments, he regarded it as the sum of the areas of "infinitely thin" rectangles of height $y = f(x)$ and infinitesimal base dx. Hence the difference between the area up to the point $x + dx$ and the area up to the point x was the infinitesimal difference in area $dA = f(x)\,dx$, and the total area was found by summing up these infinitesimal differences in area. Leibniz invented the elongated S (the integral sign $\int$) that is now universally used for expressing this summation process. Thus he would express the area under the curve $y = f(x)$ as $A = \int dA = \int f(x)\,dx$, and each part of this symbol expressed a simple and clear geometric idea.

With Leibniz's notation, Barrow's fundamental theorem of calculus merely says that the pair of equations

$$A = \int f(x)\,dx, \qquad dA = f(x)\,dx$$

are equivalent to each other. Because of what was just stated above, this equivalence is nearly obvious.

Both Newton and Leibniz had made huge advances in mathematics, and there was plenty of credit for both of them. It is unfortunate that the near coincidence of their work led to an acrimonious dispute over priority between their followers.

Some parts of the calculus, involving infinite series, had been invented in India in the fourteenth and fifteenth centuries. The late fifteenth-century Indian mathematician Jyesthadeva gave the series

$$\theta = r\left(\frac{\sin\theta}{\cos\theta} - \frac{\sin^3\theta}{3\cos^3\theta} + \frac{\sin^5\theta}{5\cos^5\theta} - \cdots\right)$$

for the length of an arc of a circle, proved this result, and explicitly stated that this series will converge only if θ is not larger than 45°. If we write $\theta = \arctan x$, and use the fact that $\dfrac{\sin\theta}{\cos\theta} = \tan\theta = x$, this series becomes the standard series for arctan x.

Likewise some infinite series were developed in Japan independently about the same time as in Europe. The Japanese mathematician Katahiro Takebe (1664–1739) found a series expansion equivalent to the series for the square of the arcsine function. He was considering the square of half the arc at height h in a circle of diameter d; this works out to be the function $f(h) = \left(\dfrac{d}{2}\arcsin\dfrac{h}{d}\right)^2$. Katahiro Takebe had no notation for the general term of a series, but he discovered patterns in the coefficients by computing the function geometrically at the particular value of $h = 0.000001$, $d = 10$ to a very large number of decimal places—more than fifty—and then using this extraordinary accuracy to refine the approximation by successively adding corrective terms. By proceeding in this way he was able to discern a pattern in the successive approximations, from which by extrapolation he was able to state the general term of the series:

$$f(h) = dh\left[1 + \sum_{n=1}^{\infty} \frac{2^{2n+1}(n!)^2}{(2n+2)!}\left(\frac{h}{d}\right)^n\right]$$

After Newton and Leibniz, there remained the problem of putting flesh on the skeleton these two geniuses had created. The majority of this work was completed by the Continental mathematicians, notably the circle around the Swiss mathematicians James and John Bernoulli ((1655–1705) and (1667–1748), respectively) and John Bernoulli's student the Marquis de l'Hôpital (1661–1704). These mathematicians and others worked out the familiar formulas for the derivatives and integrals of elementary functions that are found in textbooks today. The essential techniques of calculus were known by the early eighteenth century, and an eighteenth-century textbook such as Euler's *Introduction in analysin infinitorum* (1748), if translated into English, would look very much like a modern textbook.

The Legacy of the Calculus   Having looked at the sources of calculus and the procedure by which it was constructed, let us now examine briefly the results it produced.

The calculus scored an amazing number of triumphs in its first two centuries. Dozens of previously obscure physical phenomena involving heat, fluid flow, celestial mechanics, elasticity, light, electricity, and magnetism turned out to have measurable properties whose relations could be described as differential equations. Physics was forever committed to speaking the language of calculus.

By no means were all of the mathematical problems arising from physics solved, however. For example, the area under a curve whose equation involved the square root of a cubic polynomial could not be found in terms of familiar elementary functions. Such integrals arose frequently in both geometry and physics, and came to be known as *elliptic integrals* because the problem of finding the length could be understood only when the real variable x is replaced by a complex variable $z = x + iy$. The reworking of the calculus in terms of complex variables led to many new and fascinating discoveries, which eventually came to be codified as a new branch of mathematics called analytic function theory.

The proper definition of integration remained a problem for some time. Integrals arose out of the use of infinitesimal processes to find areas and volumes. Should the integral be defined as a "sum of infinitesimal differences," or should it be defined as the reverse of differentiation? What functions can be integrated? Many definitions of integral were proposed in the nineteenth century, and the elaboration of these ideas has led to the subject now known as real analysis.

While the applications of calculus have moved on to more and more triumphs in an unending stream for the last three hundred years, its foundations lay in an unsatisfactory state for the first half of this period. The root of the difficulty was the meaning to be attached to Leibniz's *dx*. What was this quantity? How could it be neither positive nor zero? If zero, it could not be used as a denominator; if positive, then the equations in which it occurred were not truly equations. Leibniz believed that infinitesimals were real things, that areas and volumes could be synthesized by "adding up" their sections, as Zu Chongzhi, Archimedes, and others had done. Newton was less confident of the validity of infinitesimal methods and tried to justify his arguments in ways that would meet Euclidean standards of rigor. In his *Principia Mathematica* he wrote:

> These Lemmas are premised to avoid the tediousness of deducing involved demonstrations *ad absurdum,* according to the method of the ancient geometers. For demonstrations are shorter by the method of indivisibles; but because the hypothesis of indivisibles seems somewhat harsh, and therefore that method is reckoned less geometrical, I chose rather to reduce the demonstrations of the following Propositions to the first and last sums and ratios of evanescent quantities, that is, to the limits of those sums and rations ... Therefore if hereafter I should happen to consider quantities as made up of particles, or should use little curved lines for right [straight] ones, I would not be understood to mean indivisibles, but evanescent divisible quantities ...

> ... For those ultimate ratios with which quantities vanish are not truly the ratios of ultimate quantities, but limits towards which the ratios of quantities decreasing without limit do always converge; and to which they approach nearer than by any given difference, but never go beyond, nor in effect attain to, till the quantities are diminished *in infinitum.*

In this passage Newton was claiming that the lack of rigor involved in using infinitesimal arguments could be compensated for by the use of limits. His formulation of this concept in the passage just quoted is not so clear as one might wish, however. This lack of clarity led the philosopher Berkeley to refer contemptuously to fluxions as "ghosts of departed quantities." The advances achieved in physics using calculus, however, were so outstanding that for more than a century no one bothered to supply the extra rigor Newton alluded to (and physicists still don't bother with it!). A completely rigorous and systematic presentation of the calculus came only in the nineteenth century.

After the work of Augustin-Louis Cauchy (1789–1856) and Karl Weierstrass (1815–1896), the received view was that infinitesimals are merely heuristic in nature, and students were subjected to a rigorous "epsilon-delta" approach to limits. Somewhat surprisingly, however, in the twentieth century it was shown by Abraham Robinson (1918–1974) that a logically consistent model of the real numbers can be developed in which there are actual infinitesimals, just as Leibniz had believed. This new approach, called "nonstandard analysis," does not seem to be supplanting the now-traditional presentation of calculus, however.

Exercises

1. The kind of spiral considered by Archimedes in now named after him. An Archimedean spiral is the locus of a point moving at a constant speed along a ray rotating with constant angular speed about a fixed point. If the linear speed along the ray (the *radial* component of its velocity) is v, the point will be at a distance vt from the center of rotation (assuming that is where it starts) at time t. Suppose the angular speed of rotation of the ray is ω (radians per unit time). Given a circle of radius R, and a radial speed of u, what must ω be in order for the spiral to reach the circle at the end of its first turn? *Ans.* $\left(\frac{2\pi v}{R}\right)$

   The point will have a *circumferential* velocity $r\omega = vt\,\omega$. According to a principle enunciated in Aristotle's *Mechanics,* the actual velocity of the particle will be directed along the diagonal of a parallelogram (a rectangle in this case) having the two components as sides. Use this principle to show how to construct the tangent to the spiral (it will be the line containing the diagonal of this rectangle). Verify that sides of this rectangle are in the ratio $1 : 2\pi$. See Figure 1.

2. Figure 2 illustrates how Archimedes found the relation between the volumes of a sphere, cone, and cylinder. The diameter AB is doubled, making $BC = AB$. When the figure is revolved about this line, the circle generates a sphere, the triangle DBG generates a cone, and the rectangle $DEFG$ generates a cylinder. Prove the following facts.

   (a) If B is used as a fulcrum, the cylinder has the center K of the circle as its center of gravity, and therefore could all be concentrated there without changing the torque about B.

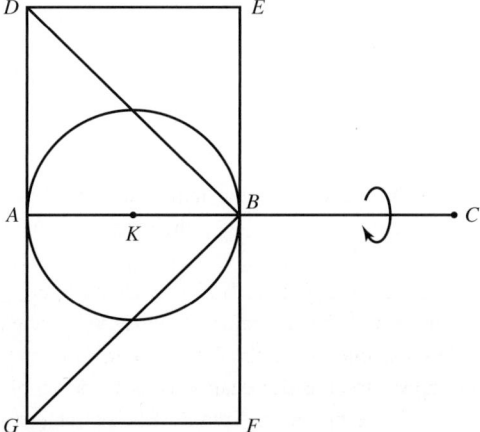

FIGURE 2   Section of Archimedes' sphere, cone, and cylinder

(b) Each section of the cylinder perpendicular to the line *AB,* remaining in its present position, would exactly balance the same section of the cone plus the section of the sphere if both of the latter were moved to the point *C.*

(c) Hence the cylinder concentrated at *K* would balance the cone and sphere concentrated at *C.*

(d) Therefore the cylinder equals twice the sum of the cone and the sphere.

(e) Since the cone is known to be one-third of the cylinder, it follows that the sphere must be one-sixth of it.

(f) Since the volume of the cylinder is $8\pi r^3$.

3. The method by which Zu Chongzhi and Zu Geng found the volume of a sphere is as follows: Imagine the sphere as a ball tightly stuck inside the intersection of two cylinders at right angles to each other. The solid formed by the intersection of the two cylinders (called a *double umbrella* in Chinese) and containing the ball is then tightly fitted inside a cube whose edge equals the diameter of the sphere.

From this description draw a section of the sphere within the double umbrella within the cube. Imagine this section being made parallel to the plane formed by the axes of the two cylinders and at a distance *h* below this plane. Verify the following facts.

(a) If the radius of the sphere is *r,* the circular section of it has diameter $2\sqrt{r^2 - h^2}$.

(b) Hence the square formed by the section of the double umbrella has area $4(r^2 - h^2)$, and so the area between the section of the cube and the section of the double umbrella is

$$4r^2 - 4(r^2 - h^2) = 4h^2.$$

(c) The corresponding section of a pyramid whose base is the bottom of the cube and whose vertex is at the center of the sphere (or cube) would also have area $4h^2$. Hence the volume between the double umbrella and the cube is exactly the volume of such a pyramid plus its mirror image above the central plane. Conclude that the region between the double umbrella and the cube is one-third of the cube.

(d) Therefore the double umbrella occupies two-thirds of the volume of the cube; that is, its volume is $\frac{16}{3}r^3$.

(e) Each circular section of the sphere is inscribed in the corresponding square section of the double umbrella. Hence the circular section is $\frac{\pi}{4}$ of the section of the double umbrella.

(f) Therefore the volume of the sphere is $\frac{\pi}{4}$ of the volume of the double umbrella; that is, $\frac{4}{3}\pi r^3$.

4. Give an "infinitesimal" argument that the area of a sphere is three times its volume divided by its radius by imagining the sphere to be a collection of "infinitely thin" pyramids with vertices all stuck together at the origin. [*Hint:* Use the fact that the volume of a pyramid is one-third the area of its base times its altitude. Archimedes says that this is the reasoning that led him to discover the area of a sphere.]

Answers to the Essay's exercise questions can be found in the *Student Resource Manual* and the *Complete Solutions Manual.*

Functions

In This Chapter Have you ever heard remarks such as "Success is a function of hard work" and "Demand is a function of price"? The word *function* is often used to suggest a relationship or a dependence of one quantity on another. As you may already know, in mathematics the notion of a function has a similar but slightly more specialized interpretation.

Calculus is mostly about functions. Thus it is appropriate that we begin its study with a chapter devoted to a review of this important concept.

1.1 Functions and Graphs

▌ Introduction   Using the objects and the persons around us, it is easy to make up a rule of correspondence that associates, or pairs, the members, or elements, of one set with the members of another set. For example, to each social security number there is a person, to each book there corresponds at least one author, to each state there is one governor, and so on. In mathematics we are interested in a special type of correspondence, a *single-valued correspondence*, called a **function**.

Definition 1.1.1   Function

A **function** from a set X to a set Y is a rule of correspondence that assigns to each element x in X exactly one element y in Y.

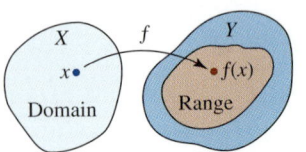

FIGURE 1.1.1 Domain and range of a function f

▌ Terminology   A function is usually denoted by a letter such as f, g, or h. We can then represent a function f from a set X to a set Y by the notation $f: X \rightarrow Y$. The set X is called the **domain** of f. The set of corresponding elements y in the set Y is called the **range** of the function. The unique element y in the range that corresponds to a selected element x in the domain X is called the **value** of the function at x, or the **image** of x, and is written $f(x)$. The latter symbol is read "f of x" or "f at x," and we write $y = f(x)$. It is also convenient at times to denote a function by $y = y(x)$. Notice in FIGURE 1.1.1 that the range of f need not be the entire set Y. Many instructors like to call an element x in the domain the *input* of the function and the corresponding element $f(x)$ in the range the *output* of the function. Since the value of y depends on the choice of x, y is called the **dependent variable**; x is called the **independent variable**. We will assume hereafter that the sets X and Y consist of real numbers; the function f is then called a **real-valued function of a real variable**.

Throughout the discussion and exercises of this text, functions are represented in several ways:

- *analytically*, that is, by a formula such as $f(x) = x^2$;
- *verbally*, that is, by a description in words;
- *numerically*, that is, by a table of numerical values; and
- *visually*, that is, by a graph.

EXAMPLE 1   Squaring Function

The rule for squaring a real number is given by the equation $f(x) = x^2$ or $y = x^2$. The values of f at $x = -5$ and $x = \sqrt{7}$ are obtained by replacing x, in turn, by the numbers -5 and $\sqrt{7}$:

$$f(-5) = (-5)^2 = 25 \quad \text{and} \quad f(\sqrt{7}) = (\sqrt{7})^2 = 7. \qquad ■$$

EXAMPLE 2   Student and Desk Correspondence

A natural correspondence occurs between a set of 20 students and a set of, say, 25 desks in a classroom when each student selects and sits in a different desk. If the set of 20 students is the set X and the set of 25 desks is the set Y, then this correspondence is a function from the set X to the set Y provided no student sits in two desks at the same time. The set of 20 desks actually occupied by the students constitutes the range of the function.   ■

Occasionally for emphasis we will write a function represented by a formula using parentheses in place of the symbol x. For example, we can write the squaring function $f(x) = x^2$ as

$$f(\ ) = (\ )^2. \qquad (1)$$

Thus, if we wish to evaluate (1) at, say, $3 + h$, where h represents a real number, we put $3 + h$ into the parentheses and carry out the appropriate algebra:

$$f(3 + h) = (3 + h)^2 = 9 + 6h + h^2.$$

Student/desk correspondence

See the *Resource Pages* for a review of binomial expansions.

If a function f is defined by means of a formula or an equation, then typically the domain of $y = f(x)$ is not expressly stated. We can usually deduce the domain of $y = f(x)$ either from the structure of the equation or from the context of the problem.

EXAMPLE 3   Domain and Range

In Example 1, since any real number x can be squared and the result x^2 is another real number, $f(x) = x^2$ is a function from R to R, that is, $f: R \to R$. In other words, the domain of f is the set R of real numbers. Using interval notation, we also write the domain as $(-\infty, \infty)$. Because $x^2 \geq 0$ for every real number x, it is easy to see that the range of f is the set of non-negative real numbers or $[0, \infty)$.   ■

■ **Domain of a Function**   As mentioned earlier, the domain of a function $y = f(x)$ that is defined by a formula is usually not specified. Unless stated or implied to the contrary, it is understood that

> • *The **domain** of a function f is the largest subset of the set of real numbers for which f(x) is a real number.*

This set is sometimes referred to as the **implicit domain** or **natural domain** of the function. For example, we cannot compute $f(0)$ for the **reciprocal function** $f(x) = 1/x$ since $1/0$ is not a real number. In this case we say that f is **undefined** at $x = 0$. Since every nonzero real number has a reciprocal, the domain of $f(x) = 1/x$ is the set of real numbers except 0. By the same reasoning, the function $g(x) = 1/(x^2 - 4)$ is not defined at either $x = -2$ or $x = 2$, and so its domain is the set of real numbers with -2 and 2 excluded. The **square root function** $h(x) = \sqrt{x}$ is not defined at $x = -1$ because $\sqrt{-1}$ is not a real number. In order for $h(x) = \sqrt{x}$ to be defined in the real number system we must require the **radicand**, in this case simply x, to be nonnegative. From the inequality $x \geq 0$ we see that the domain of the function h is the interval $[0, \infty)$. The domain of the **constant function** $f(x) = -1$ is the set of real numbers $(-\infty, \infty)$ and its range is the set consisting of the single number -1.

EXAMPLE 4   Domain and Range

Determine the domain and range of $f(x) = 4 + \sqrt{x - 3}$.

Solution   The radicand $x - 3$ must be nonnegative. By solving the inequality $x - 3 \geq 0$ we get $x \geq 3$, and so the domain of f is $[3, \infty)$. Now, since the symbol $\sqrt{}$ denotes the nonnegative square root of a number, $\sqrt{x - 3} \geq 0$ for $x \geq 3$ and consequently $4 + \sqrt{x - 3} \geq 4$. The smallest value of $f(x)$ occurs at $x = 3$ and is $f(3) = 4 + \sqrt{0} = 4$. Moreover, because $x - 3$ and $\sqrt{x - 3}$ increase as x increases, we conclude that $y \geq 4$. Consequently the range of f is $[4, \infty)$.   ■

EXAMPLE 5   Domains of Two Functions

Determine the domain of

(a) $f(x) = \sqrt{x^2 + 2x - 15}$ **(b)** $g(x) = \dfrac{5x}{x^2 - 3x - 4}$.

Solution

(a) As in Example 4, the expression under the radical symbol—the radicand—must be nonnegative, that is, the domain of f is the set of real numbers x for which $x^2 + 2x - 15 \geq 0$ or $(x - 3)(x + 5) \geq 0$. The solution set of the inequality $(-\infty, -5] \cup [3, \infty)$ is also the domain of f.

(b) A function that is given by a fractional expression is not defined at the x-values for which its denominator is equal to 0. Since the denominator of $g(x)$ factors, $x^2 - 3x - 4 = (x + 1)(x - 4)$, we see that $(x + 1)(x - 4) = 0$ for $x = -1$ and $x = 4$. These are the *only* numbers for which g is not defined. Hence, the domain of the function g is the set of real numbers with $x = -1$ and $x = 4$ excluded.   ■

◀ In precalculus a quadratic inequality such as $(x - 3)(x + 5) \geq 0$ is often solved by means of a sign chart.

Using interval notation, the domain of g in part (b) of Example 5 can be written as $(-\infty, -1) \cup (-1, 4) \cup (4, \infty)$. As an alternative to this ungainly union of disjoint intervals, this domain can also be written using set-builder notation as $\{x \mid x \neq -1 \text{ and } x \neq 4\}$.

■ **Graphs** A function is often used to describe phenomena in fields such as science, engineering, and business. In order to interpret and utilize data, it is useful to display this data in the form of a graph. In the **rectangular** or **Cartesian coordinate system**, the graph of a function f is the graph of the set of ordered pairs $(x, f(x))$, where x is in the domain of f. In the xy-plane an ordered pair $(x, f(x))$ is a point, so that the graph of a function is a set of points. If a function is defined by an equation $y = f(x)$, then the graph of f is the graph of the equation. To obtain points on the graph of an equation $y = f(x)$, we judiciously choose numbers $x_1, x_2, x_3, \ldots$ in its domain, compute $f(x_1), f(x_2), f(x_3), \ldots$, plot the corresponding points $(x_1, f(x_1)), (x_2, f(x_2)), (x_3, f(x_3)), \ldots$, and then connect these points with a smooth curve (if possible). See FIGURE 1.1.2. Keep in mind that

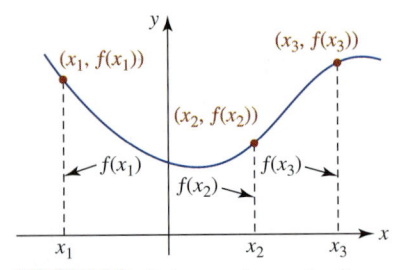

FIGURE 1.1.2 Points on the graph of an equation $y = f(x)$

- a value of x is a directed distance from the y-axis, and
- a function value $f(x)$ is a directed distance from the x-axis.

A word about the figures in this text is in order. With a few exceptions, it is usually impossible to display the complete graph of a function, and so we often display only the more important features of the graph. In FIGURE 1.1.3(a), notice that the graph goes down on its left and right sides. Unless indicated to the contrary, we may assume that there are no major surprises beyond what we have shown and the graph simply continues in the manner indicated. The graph in Figure 1.1.3(a) indicates the so-called **end behavior** or **global behavior** of the function. If a graph terminates at either its right or left end, we will indicate this by a dot when clarity demands it. We will use a solid dot to represent the fact that the end point is included on the graph and an open dot to signify that the end point is not included on the graph.

■ **Vertical Line Test** From the definition of a function we know that for each x in the domain of f there corresponds only one value $f(x)$ in the range. This means a vertical line that intersects the graph of a function $y = f(x)$ (this is equivalent to choosing an x) can do so in at most one point. Conversely, if *every* vertical line that intersects a graph of an equation does so in at most one point, then the graph is the graph of a function. The last statement is called the **vertical line test** for a function. On the other hand, if *some* vertical line intersects a graph of an equation more than once, then the graph is not that of a function. See Figures 1.1.3(a)–(c). When a vertical line intersects a graph in several points, the same number x corresponds to different values of y in contradiction to the definition of a function.

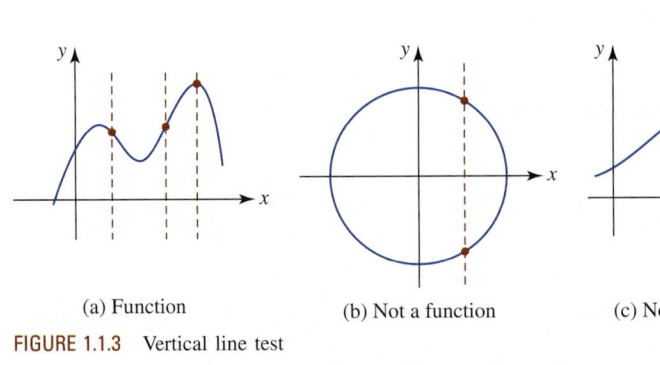

(a) Function   (b) Not a function   (c) Not a function

FIGURE 1.1.3 Vertical line test

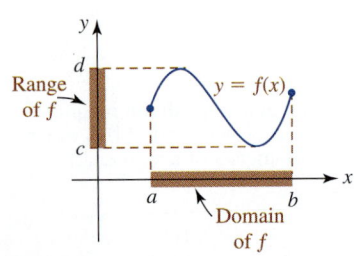

FIGURE 1.1.4 Domain and range interpreted graphically

If you have an accurate graph of a function $y = f(x)$, it is often possible to *see* the domain and range of f. In FIGURE 1.1.4 assume that the blue curve is the entire, or complete, graph of some function f. Then the domain of f is the interval $[a, b]$ on the x-axis and the range is the interval $[c, d]$ on the y-axis.

EXAMPLE 6   Example 4 Revisited

From the graph of $f(x) = 4 + \sqrt{x - 3}$ given in FIGURE 1.1.5, we can see that the domain and range of f are, respectively, $[3, \infty)$ and $[4, \infty)$. This agrees with the results in Example 4. ■

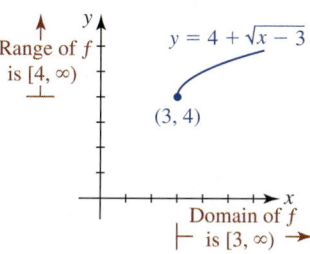

FIGURE 1.1.5   Graph of function f in Example 6

■ **Intercepts**   To graph a function defined by an equation $y = f(x)$, it is usually a good idea to first determine whether the graph of f has any intercepts. Recall that all points on the y-axis are of the form $(0, y)$. Thus, if 0 is the domain of a function f, the **y-intercept** is the point on the y-axis whose y-coordinate is $f(0)$; in other words, $(0, f(0))$. See FIGURE 1.1.6(a). Similarly, all points on the x-axis have the form $(x, 0)$. This means that to find the **x-intercepts** of the graph of $y = f(x)$, we determine the values of x that make $y = 0$. That is, we must solve the equation $f(x) = 0$ for x. A number c for which $f(c) = 0$ is referred to as either a **zero** of the function f or a **root** (or **solution**) of the equation $f(x) = 0$. The *real* zeros of a function f are the x-coordinates of the x-intercepts of the graph of f. In Figure 1.1.6(b), we have illustrated a function that has three zeros x_1, x_2, and x_3 because $f(x_1) = 0$, $f(x_2) = 0$, and $f(x_3) = 0$. The corresponding three x-intercepts are the points $(x_1, 0)$, $(x_2, 0)$, and $(x_3, 0)$. Of course, the graph of the function may have no intercepts. This is illustrated in Figure 1.1.5.

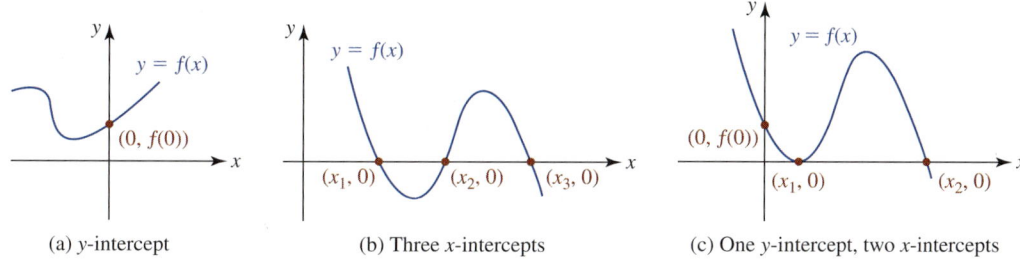

(a) y-intercept   (b) Three x-intercepts   (c) One y-intercept, two x-intercepts

FIGURE 1.1.6   Intercepts of the graph of a function f

A graph does not necessarily have to *cross* a coordinate axis at an intercept, a graph could simply touch, or be *tangent* to an axis. In Figure 1.1.6(c) the graph of $y = f(x)$ is tangent to the x-axis at $(x_1, 0)$.

EXAMPLE 7   Intercepts

Find, if possible, the x- and y-intercepts of the given function.

(a) $f(x) = x^2 + 2x - 2$ **(b)** $f(x) = \dfrac{x^2 - 2x - 3}{x}$

Solution

(a) Since 0 is in the domain of f, $f(0) = -2$ and so the y-intercept is the point $(0, -2)$. To obtain the x-intercepts we must determine whether f has any real zeros, that is, real solutions of the equation $f(x) = 0$. Since the left-hand side of the equation $x^2 + 2x - 2 = 0$ has no obvious factors, we use the quadratic formula to obtain $x = -1 \pm \sqrt{3}$. The x-intercepts are the points $(-1 - \sqrt{3}, 0)$ and $(-1 + \sqrt{3}, 0)$.

(b) Because 0 is not in the domain of f, the graph of f possesses no y-intercept. Now, since f is a fractional expression, the only way we can have $f(x) = 0$ is to have the numerator equal zero and the denominator not zero at the same number. Factoring the left-hand side of $x^2 - 2x - 3 = 0$ gives $(x + 1)(x - 3) = 0$. Therefore, the numbers -1 and 3 are the zeros of f. The x-intercepts are the points $(-1, 0)$ and $(3, 0)$. ■

■ **Piecewise-Defined Functions**   A function f may involve two or more expressions or formulas, with each formula defined on different parts of the domain of f. A function defined in this manner is called a **piecewise-defined function**. For example,

$$f(x) = \begin{cases} x^2, & x < 0 \\ x + 1, & x \geq 0 \end{cases}$$

is not two functions, but a single function in which the rule of correspondence is given in two pieces. In this case, one piece is used for the negative real numbers ($x < 0$) and the other part for the nonnegative numbers ($x \geq 0$); the domain of f is the union of the intervals $(-\infty, 0) \cup [0, \infty) = (-\infty, \infty)$. For example, since $-4 < 0$, the rule indicates that we square the number: $f(-4) = (-4)^2 = 16$; on the other hand, since $6 \geq 0$ we add 1 to the number: $f(6) = 6 + 1 = 7$.

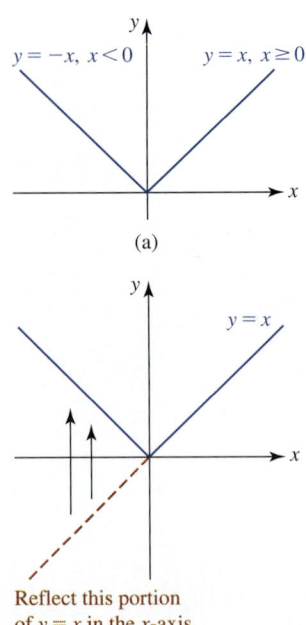

FIGURE 1.1.7 Graph of piecewise-defined function in Example 8

■ **EXAMPLE 8**   Graph of a Piecewise-Defined Function

Consider the piecewise-defined function

$$f(x) = \begin{cases} -1, & x < 0 \\ 0, & x = 0 \\ x + 1, & x > 0. \end{cases} \qquad (2)$$

Although the domain of f consists of all real numbers $(-\infty, \infty)$, each piece of the function is defined on a different part of this domain. We draw

- the horizontal line $y = -1$ for $x < 0$,
- the point $(0, 0)$ for $x = 0$, and
- the line $y = x + 1$ for $x > 0$.

The graph is given in FIGURE 1.1.7.   ■

■ **Semicircles**   As shown in Figure 1.1.3(b), a circle is not the graph of a function. Actually, an equation such as $x^2 + y^2 = 9$ defines (at least) two functions of x. If we solve this equation for y in terms of x, we get $y = \pm\sqrt{9 - x^2}$. Because of the single-valued convention of the $\sqrt{\ }$ sign, both equations $y = \sqrt{9 - x^2}$ and $y = -\sqrt{9 - x^2}$ define functions. The first equation defines an **upper semicircle** and the second defines a **lower semicircle**. From the graphs shown in FIGURE 1.1.8, the domain of $y = \sqrt{9 - x^2}$ is $[-3, 3]$ and the range is $[0, 3]$; the domain and range of $y = -\sqrt{9 - x^2}$ are $[-3, 3]$ and $[-3, 0]$, respectively.

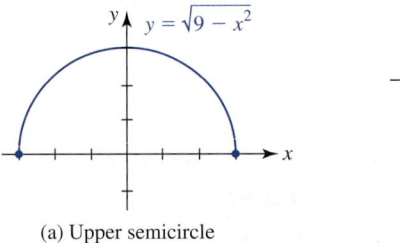

(a) Upper semicircle

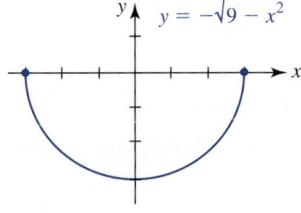

(b) Lower semicircle

FIGURE 1.1.8   These semicircles are graphs of functions

■ **Absolute-Value Function**   The function $f(x) = |x|$, called the **absolute-value function**, appears frequently in the discussion of subsequent chapters. The domain of f is the set of all real numbers $(-\infty, \infty)$ and its range is $[0, \infty)$. In other words, for any real number x, the function values $f(x)$ are nonnegative. For example,

$$f(3) = |3| = 3, \quad f(0) = |0| = 0, \quad f\left(-\frac{1}{2}\right) = \left|-\frac{1}{2}\right| = -\left(-\frac{1}{2}\right) = \frac{1}{2}.$$

By the definition of the absolute value of x, we see that f is a piecewise-defined function consisting of two pieces

$$f(x) = |x| = \begin{cases} -x, & \text{if } x < 0 \\ x, & \text{if } x \geq 0. \end{cases} \qquad (3)$$

FIGURE 1.1.9   Absolute-value function (3)

Its graph, shown in FIGURE 1.1.9(a), consists of two perpendicular half lines. Since $f(x) \geq 0$ for all x, another way of graphing (3) is simply to sketch the line $y = x$ and then reflect in the x-axis that portion of the line that is below the x-axis. See Figure 1.1.9(b).

▌ Greatest Integer Function   We consider next a piecewise-defined function f called the **greatest integer function**. This function, which has many notations, will be denoted here by $f(x) = \lfloor x \rfloor$ and is defined by the rule

$$\lfloor x \rfloor = n, \quad \text{where } n \text{ is an integer satisfying } n \le x < n + 1. \tag{4}$$

Translated into words, (4) means that

> • *The function value $f(x)$ is the greatest integer n that is less than or equal to x.*

For example,

$$f(-1.5) = -2,\ f(0.4) = 0,\ f(\pi) = 3,\ f(5) = 5,$$

and so on. The domain of f is the set of real numbers and consists of the union of an infinite number of disjoint intervals; in other words, $f(x) = \lfloor x \rfloor$ is a piecewise-defined function given by

$$f(x) = \lfloor x \rfloor = \begin{cases} \vdots \\ -2, & -2 \le x < -1 \\ -1, & -1 \le x < 0 \\ 0, & 0 \le x < 1 \\ 1, & 1 \le x < 2 \\ 2, & 2 \le x < 3 \\ \vdots \end{cases} \tag{5}$$

◀ The greatest integer function is also written as $f(x) = [\![x]\!]$.

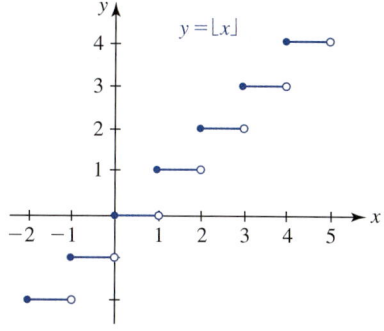

FIGURE 1.1.10   Greatest integer function

The range of f is the set of integers. The portion of the graph of f on the closed interval $[-2, 5]$ is given in FIGURE 1.1.10.

In computer science the greatest integer function is known as the **floor function**. A related function called the **ceiling function*** $g(x) = \lceil x \rceil$ is defined to be the least integer n that is greater than or equal to x. See Problems 57–59 in Exercises 1.1.

▌ A Mathematical Model   It is often desirable to describe the behavior of some real-life system or phenomenon, whether physical, sociological, or even economic, in mathematical terms. The mathematical description of a system or a phenomenon is called a **mathematical model** and can be as complicated as hundreds of simultaneous equations or as simple as a single function. We conclude this section with a real-world illustration of a piecewise-defined function called the *postage stamp function*. This function is similar to $f(x) = \lfloor x \rfloor$ in that both are examples of *step functions*; each function is constant on an interval and then jumps to another constant value on the next abutting interval.

As of this writing, the United States Postal Service (USPS) first-class mailing rates for a letter in a standard-size envelope depends on its weight in ounces:

$$\text{Postage} = \begin{cases} \$0.42, & 0 < \text{weight} \le 1 \text{ ounce} \\ \$0.59, & 1 < \text{weight} \le 2 \text{ ounces} \\ \$0.76, & 2 < \text{weight} \le 3 \text{ ounces} \\ \vdots \\ \$2.87, & 12 < \text{weight} \le 13 \text{ ounces}. \end{cases} \tag{6}$$

The rule in (6) is a function P consisting of 14 pieces (letters over 13 ounces are sent priority mail). A function value $P(w)$ is one of 14 constants; the constant changes depending on the weight w (in ounces) of the letter.† For example,

$$P(0.5) = \$0.42,\ P(1.7) = \$0.59,\ P(2.2) = \$0.76,\ P(2.9) = \$0.76, \text{ and } P(12.1) = \$2.87.$$

The domain of the function P is the union of the intervals:

$$(0, 1] \cup (1, 2] \cup (2, 3] \cup \cdots \cup (12, 13] = (0, 13].$$

---

*The floor and ceiling functions and their notation are due to the noted Canadian computer scientist Kenneth E. Iverson (1920–2004).

†Not shown in (6) is the fact that the postage of a letter whose weight falls in the interval (3, 4] is determined by whether its weight is in (3, 3.5] or in (3.5, 4]. This is the only interval subdivided in this manner.

$f(x)$ NOTES FROM THE CLASSROOM

When sketching the graph of a function, you should never resort to plotting a lot of points by hand. That is something a graphing calculator or a computer algebra system (CAS) does so well. On the other hand, you should not become dependent on a calculator to obtain a graph. Believe it or not, there are calculus instructors who do not allow the use of graphing calculators on quizzes or tests. Usually there is no objection to your using calculators or computers as an aid in checking homework problems, but in the classroom, instructors want to see the product of your own mind, namely, the ability to analyze. So you are strongly encouraged to develop your graphing skills to the point where you are able to quickly sketch by hand the graph of a function from a basic familiarity of types of functions and by plotting a minimum of well-chosen points.

Exercises 1.1   Answers to selected odd-numbered problems begin on page ANS-2.

≡ Fundamentals

In Problems 1–6, find the indicated function values.

1. If $f(x) = x^2 - 1$; $f(-5), f(-\sqrt{3}), f(3)$, and $f(6)$

2. If $f(x) = -2x^2 + x$; $f(-5), f\left(-\frac{1}{2}\right), f(2)$, and $f(7)$

3. If $f(x) = \sqrt{x + 1}$; $f(-1), f(0), f(3)$, and $f(5)$

4. If $f(x) = \sqrt{2x + 4}$; $f\left(-\frac{1}{2}\right), f\left(\frac{1}{2}\right), f\left(\frac{5}{2}\right)$, and $f(4)$

5. If $f(x) = \dfrac{3x}{x^2 + 1}$; $f(-1), f(0), f(1)$, and $f(\sqrt{2})$

6. If $f(x) = \dfrac{x^2}{x^3 - 2}$; $f(-\sqrt{2}), f(-1), f(0)$, and $f\left(\frac{1}{2}\right)$

In Problems 7 and 8, find

$$f(x), f(2a), f(a^2), f(-5x), f(2a + 1), f(x + h)$$

for the given function f and simplify as much as possible.

7. $f(\ ) = -2(\ )^2 + 3(\ )$

8. $f(\ ) = (\ )^3 - 2(\ )^2 + 20$

9. For what values of x is $f(x) = 6x^2 - 1$ equal to 23?

10. For what values of x is $f(x) = \sqrt{x - 4}$ equal to 4?

In Problems 11–26, find the domain of the given function f.

11. $f(x) = \sqrt{4x - 2}$

12. $f(x) = \sqrt{15 - 5x}$

13. $f(x) = \dfrac{10}{\sqrt{1 - x}}$

14. $f(x) = \dfrac{2x}{\sqrt{3x - 1}}$

15. $f(x) = \dfrac{2x - 5}{x(x - 3)}$

16. $f(x) = \dfrac{x}{x^2 - 1}$

17. $f(x) = \dfrac{1}{x^2 - 10x + 25}$

18. $f(x) = \dfrac{x + 1}{x^2 - 4x - 12}$

19. $f(x) = \dfrac{x}{x^2 - x + 1}$

20. $f(x) = \dfrac{x^2 - 9}{x^2 - 2x - 1}$

21. $f(x) = \sqrt{25 - x^2}$

22. $f(x) = \sqrt{x(4 - x)}$

23. $f(x) = \sqrt{x^2 - 5x}$

24. $f(x) = \sqrt{x^2 - 3x - 10}$

25. $f(x) = \sqrt{\dfrac{3 - x}{x + 2}}$

26. $f(x) = \sqrt{\dfrac{5 - x}{x}}$

In Problems 27–30, determine whether the graph in the figure is the graph of a function.

27.

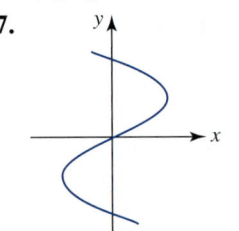

FIGURE 1.1.11   Graph for Problem 27

28.

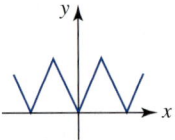

FIGURE 1.1.12   Graph for Problem 28

29.

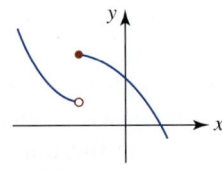

FIGURE 1.1.13   Graph for Problem 29

30.

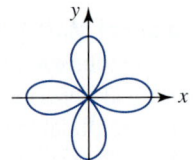

FIGURE 1.1.14   Graph for Problem 30

In Problems 31–34, use the graph of the function f given in the figure to find its domain and range.

31.

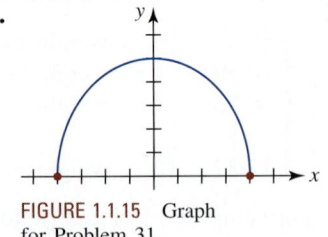

FIGURE 1.1.15   Graph for Problem 31

32.

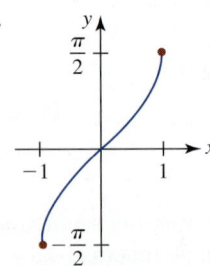

FIGURE 1.1.16   Graph for Problem 32

33.

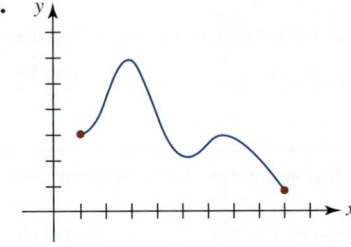

FIGURE 1.1.17   Graph for Problem 33

34.

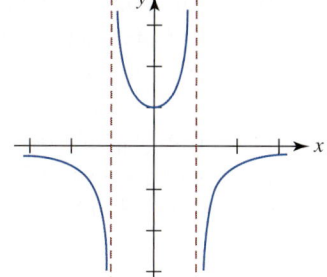

FIGURE 1.1.18   Graph for Problem 34

In Problems 35–44, find the x- and y-intercepts, if any, of the graph of the given function f. Do not graph.

35. $f(x) = \dfrac{1}{2}x - 4$

36. $f(x) = x^2 - 6x + 5$

37. $f(x) = 4(x - 2)^2 - 1$

38. $f(x) = (2x - 3)(x^2 + 8x + 16)$

39. $f(x) = x^3 - x^2 - 2x$

40. $f(x) = x^4 - 1$

41. $f(x) = \dfrac{x^2 + 4}{x^2 - 16}$

42. $f(x) = \dfrac{x(x + 1)(x - 6)}{x + 8}$

43. $f(x) = \dfrac{3}{2}\sqrt{4 - x^2}$

44. $f(x) = \dfrac{1}{2}\sqrt{x^2 - 2x - 3}$

In Problems 45 and 46, use the graph of the function f given in the figure to estimate the values of $f(-3), f(-2), f(-1), f(1),$ $f(2),$ and $f(3)$. Estimate the y-intercept.

45.

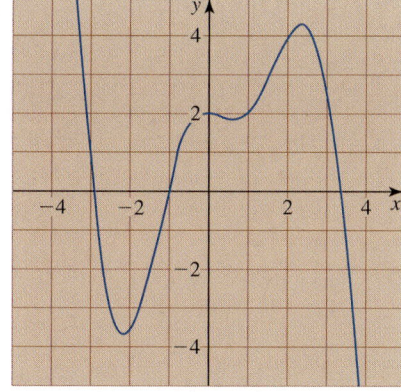

FIGURE 1.1.19   Graph for Problem 45

46.

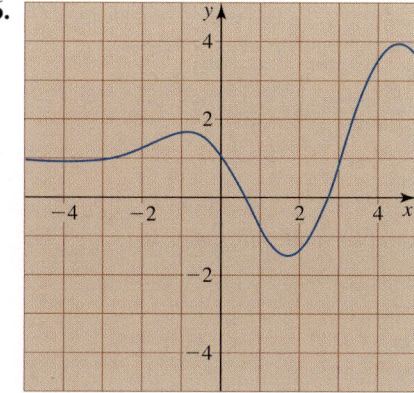

FIGURE 1.1.20   Graph for Problem 46

In Problems 47 and 48, use the graph of the function f given in the figure to estimate the values of $f(-2), f(-1.5), f(0.5),$ $f(1), f(2),$ and $f(3.2)$. Estimate the x-intercepts.

47.

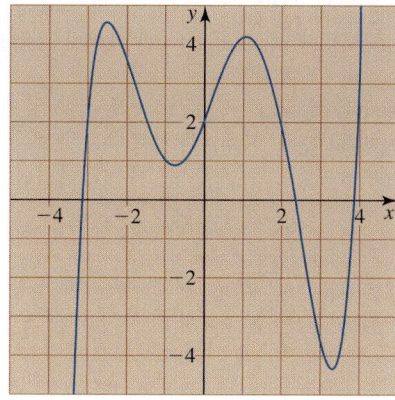

FIGURE 1.1.21   Graph for Problem 47

48.

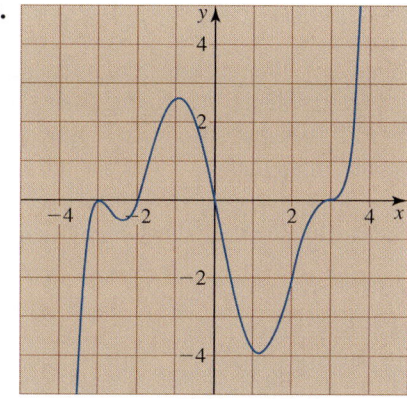

FIGURE 1.1.22   Graph for Problem 48

In Problems 49 and 50, find two functions $y = f_1(x)$ and $y = f_2(x)$ defined by the given equation. Find the domain of the functions f_1 and f_2.

49. $x = y^2 - 5$

50. $x^2 - 4y^2 = 16$

51. Some of the functions that you will encounter later on in this text will have the set of positive integers n as their domain. The **factorial function** $f(n) = n!$ is defined as the product of the first n positive integers, that is,

$$f(n) = n! = 1 \cdot 2 \cdot 3 \cdots (n - 1) \cdot n.$$

(a) Evaluate $f(2), f(3), f(5),$ and $f(7)$.

(b) Show that $f(n + 1) = f(n) \cdot (n + 1)$.

(c) Simplify $f(5)/f(4)$ and $f(7)/f(5)$.

(d) Simplify $f(n + 3)/f(n)$.

52. Another function of a positive integer n gives the sum of the first n squared positive integers:

$$S(n) = \frac{1}{6}n(n + 1)(2n + 1).$$

(a) Find the value of the sum $1^2 + 2^2 + \cdots + 99^2 + 100^2$.

(b) Find n such that $300 < S(n) < 400$. [*Hint*: Use a calculator.]

☰ Think About It

53. Determine an equation of a function $y = f(x)$ whose domain is
   (a) $[3, \infty)$ **(b)** $(3, \infty)$.

54. Determine an equation of a function $y = f(x)$ whose range is
   (a) $[3, \infty)$ **(b)** $(3, \infty)$.

55. From the graph of $f(x) = -x^2 + 2x + 3$ given in FIGURE 1.1.23 determine the range and domain of the function $g(x) = \sqrt{f(x)}$. Explain your reasoning in one or two sentences.

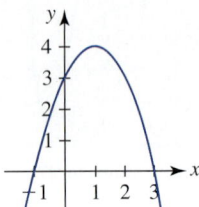

FIGURE 1.1.23   Graph for Problem 55

56. Let P denote any point $(x, f(x))$ on the graph of a function f. Suppose that the line segments PT and PS are perpendicular to the x- and y-axes, respectively. Let M_1, M_2, and M_3 be, in turn, the midpoints of PT, PS, and ST as shown in FIGURE 1.1.24. Find a function that describes the path of the points M_1. Repeat for the midpoints M_2 and M_3.

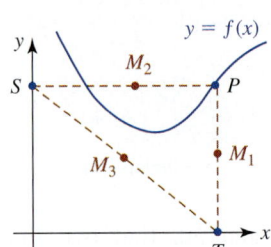

FIGURE 1.1.24   Graph for Problem 56

57. On page 7 we saw that the **ceiling function** $g(x) = \lceil x \rceil$ is defined to be the least integer n that is greater than or equal to x. Fill in the blanks.

$$g(x) = \lceil x \rceil = \begin{cases} \vdots & \\ \rule{1cm}{0.4pt}, & -3 < x \le -2 \\ \rule{1cm}{0.4pt}, & -2 < x \le -1 \\ \rule{1cm}{0.4pt}, & -1 < x \le 0 \\ \rule{1cm}{0.4pt}, & 0 < x \le 1 \\ \rule{1cm}{0.4pt}, & 1 < x \le 2 \\ \rule{1cm}{0.4pt}, & 2 < x \le 3 \\ \vdots & \end{cases}$$

58. Graph the ceiling function $g(x) = \lceil x \rceil$ defined in Problem 57.

59. The piecewise-defined function

$$\text{int}(x) = \begin{cases} \lfloor x \rfloor, & x \ge 0 \\ \lceil x \rceil, & x < 0 \end{cases}$$

is called the **integer function**. Graph int(x).

60. Discuss how to graph the function $f(x) = |x| + |x - 3|$. Carry out your ideas.

In Problems 61 and 62, describe in words how the graphs of the given functions differ.

61. $f(x) = \dfrac{x^2 - 9}{x - 3}$,

$$g(x) = \begin{cases} \dfrac{x^2 - 9}{x - 3}, & x \ne 3 \\ 4, & x = 3 \end{cases} \quad h(x) = \begin{cases} \dfrac{x^2 - 9}{x - 3}, & x \ne 3 \\ 6, & x = 3 \end{cases}$$

62. $f(x) = \dfrac{x^4 - 1}{x^2 - 1}$,

$$g(x) = \begin{cases} \dfrac{x^4 - 1}{x - 1}, & x \ne 1 \\ 0, & x = 1 \end{cases} \quad h(x) = \begin{cases} \dfrac{x^4 - 1}{x^2 - 1}, & x \ne 1 \\ 2, & x = 1 \end{cases}$$

1.2 Combining Functions

❚ Introduction   Two functions f and g can be combined in several ways to create new functions. In this section we will examine two such ways in which functions can be combined: through arithmetic operations, and through the operation of function composition.

❚ Power Functions   A function of the form

$$f(x) = x^n \tag{1}$$

is called a **power function**. In this section we consider n to be a rational real number. The domain of a power function depends on the power n. For example, for $n = 2$, $n = \frac{1}{2}$, and $n = -1$, respectively,

- the domain of $f(x) = x^2$ is the set R of real numbers or $(-\infty, \infty)$,
- the domain of $f(x) = x^{1/2} = \sqrt{x}$ is $[0, \infty)$,

- the domain of $f(x) = x^{-1} = \dfrac{1}{x}$ is the set R of real numbers except $x = 0$.

Simple power functions, or modified versions of these functions, occur so often in problems in calculus that you do not want to spend valuable time plotting their graphs. We suggest that you know (memorize) the short catalogue of graphs of power functions given in FIGURE 1.2.1. You should recognize that the graph in part (a) of Figure 1.2.1 is a **line** and the graph in part (b) is a **parabola**.

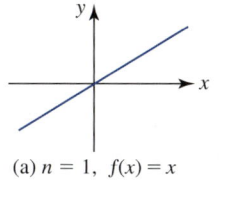

(a) $n = 1,\ f(x) = x$

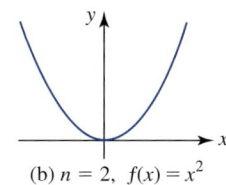

(b) $n = 2,\ f(x) = x^2$

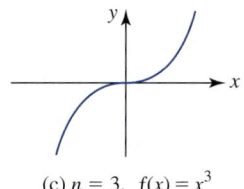

(c) $n = 3,\ f(x) = x^3$

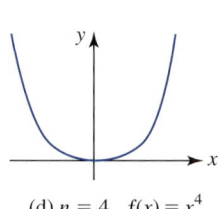

(d) $n = 4,\ f(x) = x^4$

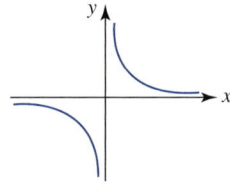

(e) $n = -1,\ f(x) = x^{-1} = \dfrac{1}{x}$

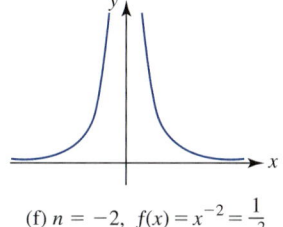

(f) $n = -2,\ f(x) = x^{-2} = \dfrac{1}{x^2}$

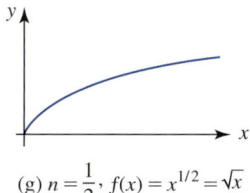

(g) $n = \dfrac{1}{2},\ f(x) = x^{1/2} = \sqrt{x}$

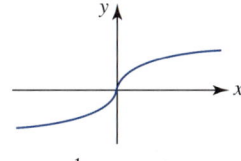

(h) $n = \dfrac{1}{3},\ f(x) = x^{1/3} = \sqrt[3]{x}$

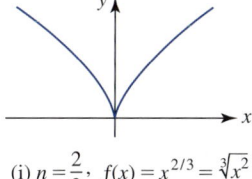

(i) $n = \dfrac{2}{3},\ f(x) = x^{2/3} = \sqrt[3]{x^2}$

FIGURE 1.2.1   Brief catalogue of graphs of power functions

■ **Arithmetic Combinations**   Two functions can be combined through the familiar four arithmetic operations of addition, subtraction, multiplication, and division.

Definition 1.2.1   Arithmetic Combinations

If f and g are two functions, then the **sum** $f + g$, the **difference** $f - g$, the **product** fg, and the **quotient** f/g are defined as follows:

$$(f + g)(x) = f(x) + g(x), \tag{2}$$

$$(f - g)(x) = f(x) - g(x), \tag{3}$$

$$(fg)(x) = f(x)g(x), \tag{4}$$

$$\left(\frac{f}{g}\right)(x) = \frac{f(x)}{g(x)}, \text{ provided } g(x) \neq 0. \tag{5}$$

■ **Domain of an Arithmetic Combination**   When combining two functions arithmetically it is necessary that both f and g be defined at a same number x. Hence the **domain** of the functions $f + g$, $f - g$, and fg is the set of real numbers that are *common* to both domains, that is, the domain is the *intersection* of the domain of f with the domain of g. In the case of the quotient f/g, the domain is also the intersection of the two domains, *but* we must also exclude any values of x for which the denominator $g(x)$ is zero. In other words, if the domain of f is the set X_1 and the domain of g is the set X_2, then the domain of $f + g, f - g$, and fg is $X_1 \cap X_2$, and the domain of f/g is the set $\{x \,|\, x \in X_1 \cap X_2,\ g(x) \neq 0\}$.

EXAMPLE 1   Sum of Two Power Functions

We have already seen that the domain of $f(x) = x^2$ is the set R of real numbers or $(-\infty, \infty)$ and that the domain of $g(x) = \sqrt{x}$ is $[0, \infty)$. Therefore, the domain of the sum

$$f(x) + g(x) = x^2 + \sqrt{x}$$

is the intersection of the two domains: $(-\infty, \infty) \cap [0, \infty) = [0, \infty)$. ∎

■ **Polynomial Functions**   Many of the functions that we work with in calculus are constructed by performing arithmetic operations on power functions. Of special interest are the power functions (1) where n is a nonnegative integer. For $n = 0, 1, 2, 3, \ldots$, the function $f(x) = x^n$ is called a **single-term polynomial function**. Using the arithmetic operations of addition, subtraction, and multiplication we can build polynomial functions with many terms. For example, if $f_1(x) = x^3, f_2(x) = x^2, f_3(x) = x,$ and $f_4(x) = 1$, then

$$f_1(x) - f_2(x) + f_3(x) + f_4(x) = x^3 - x^2 + x + 1.$$

In general, a **polynomial function** $y = f(x)$ is a function of the form

$$f(x) = a_n x^n + a_{n-1}x^{n-1} + \cdots + a_2 x^2 + a_1 x + a_0, \tag{6}$$

where n is a nonnegative integer and the coefficients $a_i, i = 0, 1, \ldots, n$ are real numbers. The **domain** of any polynomial function f is the set of all real numbers $(-\infty, \infty)$. The following functions are *not* polynomials:

not a nonnegative integer   not a nonnegative integer
↓   ↓

$$y = 5x^2 - 3x^{-1} \qquad \text{and} \qquad y = 2x^{1/2} - 4.$$

EXAMPLE 2   Sum, Difference, Product, and Quotient

Consider the polynomial functions $f(x) = x^2 + 4x$ and $g(x) = x^2 - 9$.

(a) From (2)–(4) of Definition 1.2.1 we can produce three new polynomial functions:

$$(f + g)(x) = f(x) + g(x) = (x^2 + 4x) + (x^2 - 9) = 2x^2 + 4x - 9,$$
$$(f - g)(x) = f(x) - g(x) = (x^2 + 4x) - (x^2 - 9) = 4x + 9,$$
$$(fg)(x) = f(x)g(x) = (x^2 + 4x)(x^2 - 9) = x^4 + 4x^3 - 9x^2 - 36x.$$

(b) Finally, from (5) of Definition 1.2.1,

$$\left(\frac{f}{g}\right)(x) = \frac{f(x)}{g(x)} = \frac{x^2 + 4x}{x^2 - 9}.$$ ∎

Notice in Example 2, since $g(-3) = 0$ and $g(3) = 0$, the domain of the quotient $(f/g)(x)$ is $(-\infty, \infty)$ with $x = 3$ and $x = -3$ excluded, in other words, the domain of $(f/g)(x)$ is the union of three intervals: $(-\infty, -3) \cup (-3, 3) \cup (3, \infty)$.

■ **Rational Functions**   The function in part (b) of Example 2 is an example of a rational functions. In general, a **rational function** $y = f(x)$ is a function of the form

Polynomial and rational functions ▶
will be discussed in greater detail in
Section 1.3.

$$f(x) = \frac{p(x)}{q(x)}, \tag{7}$$

where p and q are polynomial functions. For example, the functions

polynomial
↓

$$y = \frac{x}{x^2 + 5}, \quad y = \frac{x^3 - x + 7}{x + 3}, \quad y = \frac{1}{x},$$

↑
polynomial

are rational functions. The function

$$y = \frac{\sqrt{x}}{x^2 - 1} \quad \leftarrow \text{not a polynomial}$$

is not a rational function.

■ **Composition of Functions**   Another method of combining functions f and g is called **function composition**. To illustrate the idea, let us suppose that for a given x in the domain of g the function value $g(x)$ is a number in the domain of the function f. This means we are able to evaluate f at $g(x)$, in other words, $f(g(x))$. For example, suppose $f(x) = x^2$ and $g(x) = x + 2$. Then for $x = 1$, $g(1) = 3$, and since 3 is the domain of f, we can write $f(g(1)) = f(3) = 3^2 = 9$. Indeed, for these two particular functions it turns out that we can evaluate f at any function value $g(x)$, that is,

$$f(g(x)) = f(x + 2) = (x + 2)^2.$$

The resulting function, called the **composition of f and g**, is defined next.

Definition 1.2.2   Function Composition

If f and g are two functions, then the **composition of f and g**, denoted by $f \circ g$, is the function defined by

$$(f \circ g)(x) = f(g(x)). \qquad (8)$$

The **composition of g and f**, denoted by $g \circ f$, is the function defined by

$$(g \circ f)(x) = g(f(x)). \qquad (9)$$

EXAMPLE 3   Two Compositions

If $f(x) = x^2 + 3x$ and $g(x) = 2x^2 + 1$ find

(a) $(f \circ g)(x)$ and   (b) $(g \circ f)(x)$.

Solution

(a) For emphasis we replace x by the set of parentheses ( ) and write f in the form $f(x) = ( )^2 + 3( )$. Thus, to evaluate $(f \circ g)(x)$ we fill each set of parentheses with $g(x)$. We find

$$\begin{aligned}
(f \circ g)(x) = f(g(x)) &= f(2x^2 + 1) \\
&= (2x^2 + 1)^2 + 3(2x^2 + 1) \\
&= 4x^4 + 4x^2 + 1 + 3 \cdot 2x^2 + 3 \cdot 1 \\
&= 4x^4 + 10x^2 + 4.
\end{aligned}$$

(b) In this case write g in the form $g(x) = 2( )^2 + 1$. Then

$$\begin{aligned}
(g \circ f)(x) = g(f(x)) &= g(x^2 + 3x) \\
&= 2(x^2 + 3x)^2 + 1 \\
&= 2(x^4 + 6x^3 + 9x^2) + 1 \\
&= 2x^4 + 12x^3 + 18x^2 + 1.
\end{aligned}$$ ■

Parts (a) and (b) of Example 3 illustrate that function composition is not commutative. That is, in general

$$f \circ g \neq g \circ f.$$

EXAMPLE 4   Writing a Function as a Composition

Express $F(x) = \sqrt{6x^3 + 8}$ as the composition of two functions f and g.

Solution   If we define f and g as $f(x) = \sqrt{x}$ and $g(x) = 6x^3 + 8$, then

$$F(x) = (f \circ g)(x) = f(g(x)) = f(6x^3 + 8) = \sqrt{6x^3 + 8}.$$ ■

There are other solutions to Example 4. For instance, if the functions f and g are defined by $f(x) = \sqrt{6x + 8}$ and $g(x) = x^3$, then observe $(f \circ g)(x) = f(x^3) = \sqrt{6x^3 + 8}$.

■ Domain of a Composition To evaluate the composition $(f \circ g)(x) = f(g(x))$ the number $g(x)$ must be in the domain of f. For example, the domain of $f(x) = \sqrt{x}$ is $[0, \infty)$ and the domain of $g(x) = x - 2$ is the set of real numbers $(-\infty, \infty)$. Observe, we cannot evaluate $f(g(1))$ because $g(1) = -1$ and -1 is not in the domain of f. In order to substitute $g(x)$ into $f(x)$, $g(x)$ must satisfy the inequality that defines the domain of f, namely, $g(x) \geq 0$. This last inequality is the same as $x - 2 \geq 0$ or $x \geq 2$. The domain of the composition $f(g(x)) = \sqrt{g(x)} = \sqrt{x - 2}$ is $[2, \infty)$, which is only a portion of the original domain $(-\infty, \infty)$ of g. In general, the **domain of the composition** $f \circ g$ is the set of the numbers x in the domain of g such that $g(x)$ is in the domain of f.

For a constant $c > 0$, the functions defined by $y = f(x) + c$ and $y = f(x) - c$ are the *sum* and *difference* of the function $f(x)$ and the constant function $g(x) = c$. The function $y = cf(x)$ is the *product* of $f(x)$ and the constant function $g(x) = c$. The functions defined by $y = f(x + c)$, $y = f(x - c)$, and $y = f(cx)$ are *compositions* of $f(x)$ with the polynomial functions $g(x) = x + c$, $g(x) = x - c$, and $g(x) = cx$, respectively. As we see next, the graph of each of these is either a **rigid** or **nonrigid transformation** of the graph of $y = f(x)$.

■ Rigid Transformations A **rigid transformation** of a graph is one that changes only the *position* of the graph in the xy-plane but not its shape. For the graph of a function $y = f(x)$ we examine four kinds of shifts or translations.

> ### Translations
>
> Suppose $y = f(x)$ is a function and c is a positive constant. Then the graph of
>
> - $y = f(x) + c$ is the graph of f shifted vertically **up** c units,
> - $y = f(x) - c$ is the graph of f shifted vertically **down** c units,
> - $y = f(x + c)$ is the graph of f shifted horizontally to the **left** c units,
> - $y = f(x - c)$ is the graph of f shifted horizontally to the **right** c units.

Consider the graph of a function $y = f(x)$ given in FIGURE 1.2.2. Vertical and horizontal shifts of this graph are the graphs in red in parts (a)–(d) of FIGURE 1.2.3. If (x, y) is a point on the graph of $y = f(x)$ and the graph of f is shifted, say, upward by $c > 0$ units, then $(x, y + c)$ is a point on the new graph. In general, the x-coordinates do not change as a result of a vertical shift. See Figures 1.2.3(a) and 1.2.3(b). Similarly, in a horizontal shift the y-coordinates of points on the shifted graph are the same as on the original graph. See Figures 1.2.3(c) and 1.2.3(d).

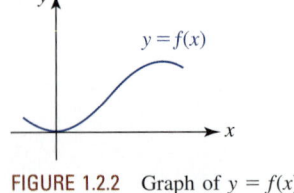

FIGURE 1.2.2   Graph of $y = f(x)$

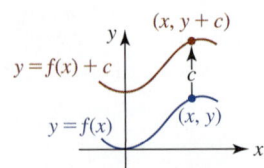

(a) Vertical shift up

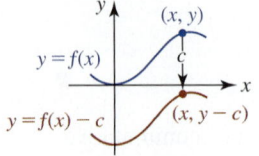

(b) Vertical shift down

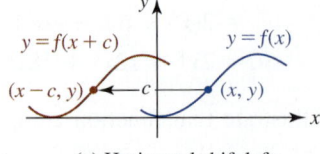

(c) Horizontal shift left

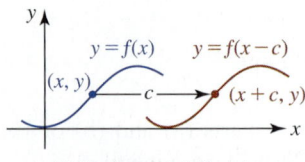

(d) Horizontal shift right

FIGURE 1.2.3   Vertical and horizontal shifts of the graph of $y = f(x)$ by an amount $c > 0$

EXAMPLE 5   Shifted Graphs

The graphs of $y = x^2 + 1$, $y = x^2 - 1$, $y = (x + 1)^2$, and $y = (x - 1)^2$ are obtained from the graph of $f(x) = x^2$ in FIGURE 1.2.4(a) by shifting this graph, in turn, 1 unit up (Figure 1.2.4(b)), 1 unit down (Figure 1.2.4(c)), 1 unit to the left (Figure 1.2.4(d)), and 1 unit to the right (Figure 1.2.4(e)).

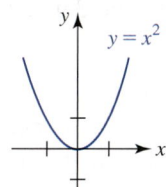

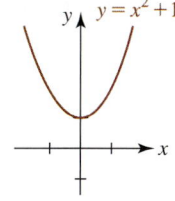

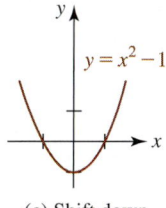

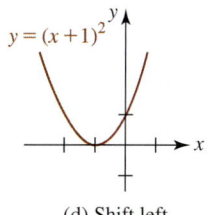

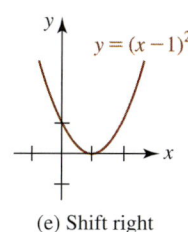

(a) Starting point   (b) Shift up   (c) Shift down   (d) Shift left   (e) Shift right

FIGURE 1.2.4 Shifted graphs in Example 5

■ **Combining Shifts**   In general, the graph of a function

$$y = f(x \pm c_1) \pm c_2, \tag{10}$$

◀ The order in which the shifts are done is irrelevant.

where c_1 and c_2 are positive constants, combines a horizontal shift (left or right) with a vertical shift (up or down). For example, the graph $y = (x + 1)^2 - 1$ is the graph of $f(x) = x^2$ shifted 1 unit to the left followed by a vertical shift 1 unit down. The graph is given in FIGURE 1.2.5.

Another way of rigidly transforming a graph of a function is by a **reflection** in a coordinate axis.

Reflections

Suppose $y = f(x)$ is a function. Then the graph of

- $y = -f(x)$ is the graph of f reflected in the **x-axis**,
- $y = f(-x)$ is the graph of f reflected in the **y-axis**.

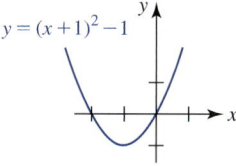

FIGURE 1.2.5 Graph obtained by a horizontal and vertical shift

In FIGURE 1.2.6(a) we have reproduced the graph of a function $y = f(x)$ given in Figure 1.2.2. The reflections of this graph in the x- and y-axes are illustrated in Figures 1.2.6(b) and 1.2.6(c). Each of these reflections is a **mirror image** of the graph of $y = f(x)$ in the respective coordinate axis.

Reflection or mirror image

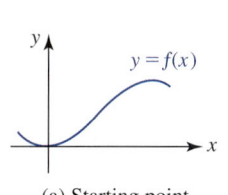

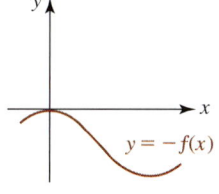

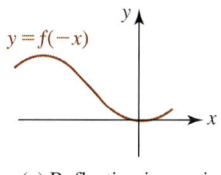

(a) Starting point   (b) Reflection in x-axis   (c) Reflection in y-axis

FIGURE 1.2.6 Reflections in the coordinate axes

EXAMPLE 6 Reflections

Graph

(a) $y = -\sqrt{x}$ **(b)** $y = \sqrt{-x}.$

Solution   The starting point is the graph of $f(x) = \sqrt{x}$ given in FIGURE 1.2.7(a).
  (a) The graph of $y = -\sqrt{x}$ is the reflection of the graph of $f(x) = \sqrt{x}$ in the x-axis. Observe in Figure 1.2.7(b) that since $(1, 1)$ is on the graph of f, the point $(1, -1)$ is on the graph of $y = -\sqrt{x}$.
  (b) The graph of $y = \sqrt{-x}$ is the reflection of the graph of $f(x) = \sqrt{x}$ in the y-axis. Observe in Figure 1.2.7(c) that since $(1, 1)$ is on the graph of f the point $(-1, 1)$ is on the graph of $y = \sqrt{-x}$. The function $y = \sqrt{-x}$ looks a little strange, but bear in mind that its domain is determined by the requirement that $-x \geq 0$, or equivalently $x \leq 0$, and so the reflected graph is defined on the interval $(-\infty, 0]$.

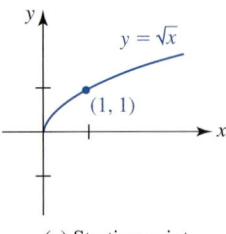

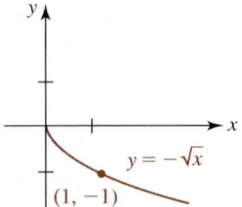

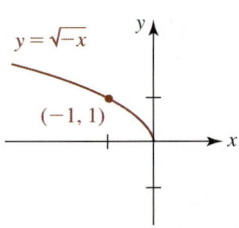

(a) Starting point   (b) Reflection in *x*-axis   (c) Reflection in *y*-axis

FIGURE 1.2.7 Graphs in Example 6

■ **Nonrigid Transformations** If a function *f* is multiplied by a constant $c > 0$, the shape of the graph is changed but retains, *roughly*, its original shape. The graph of $y = cf(x)$ is the graph of $y = f(x)$ distorted vertically; the graph of *f* is either stretched (or elongated) vertically or is compressed (or flattened) vertically depending on the value of *c*. Put another way, a vertical stretch is a stretch of the graph of $y = f(x)$ away from the *x*-axis, whereas a vertical compression is a squeezing of the graph of $y = f(x)$ toward the *x*-axis. The graph of the function $y = f(cx)$ is distorted horizontally, either by a stretch of the graph of $y = f(x)$ away from the *y*-axis or a squeezing of the graph of $y = f(x)$ toward the *y*-axis. Stretching or compressing a graph are examples of **nonrigid transformations**.

Stretches and Compressions

Suppose $y = f(x)$ is a function and *c* is a positive constant. Then the graph of

- $y = cf(x)$ is the graph of *f* **vertically stretched** by a factor of *c* if $c > 1$,
- $y = cf(x)$ is the graph of *f* **vertically compressed** by a factor of $1/c$ if $0 < c < 1$,
- $y = f(cx)$ is the graph of *f* **horizontally stretched** by a factor of $1/c$ if $0 < c < 1$,
- $y = f(cx)$ is the graph of *f* **horizontally compressed** by a factor of *c* if $c > 1$.

EXAMPLE 7 Two Compressions

Given $f(x) = x^2 - x$. Compare the graphs of

     (a) $y = \dfrac{1}{2}f(x)$ and   **(b)** $y = f(2x)$.

Solution The graph of the given polynomial function *f* is shown in FIGURE 1.2.8.
    (a) With the identification $c = \frac{1}{2}$, the graph of $y = \frac{1}{2}f(x)$ is the graph of *f* vertically compressed by a factor of 2. Of the three points shown on the graph of Figure 1.2.8(a), notice in Figure 1.2.8(b) the *y*-coordinates of the corresponding three points are one-half as large. The original graph is squeezed toward the *x*-axis.
    (b) With the identification $c = 2$, the graph of $y = f(2x)$ is the graph of *f* horizontally compressed by a factor of 2. Of the three points shown on the graph of Figure 1.2.8(a), in Figure 1.2.8(c) the *x*-coordinates of the corresponding three points are divided by 2. The original graph is squeezed toward the *y*-axis.

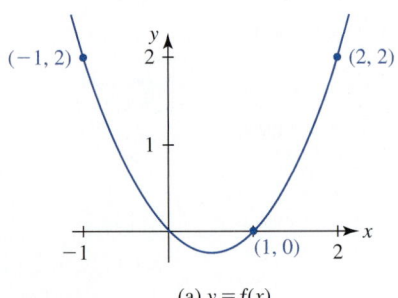

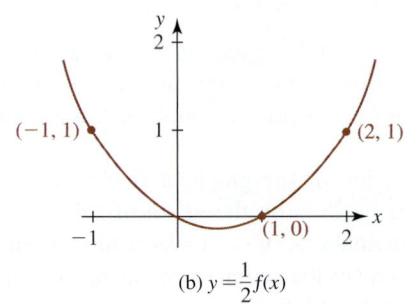

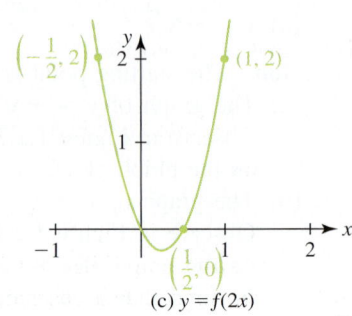

(a) $y = f(x)$ (b) $y = \frac{1}{2}f(x)$ (c) $y = f(2x)$

FIGURE 1.2.8 Graphs of functions in Example 7

The next example illustrates shifting, reflecting, and stretching of a graph.

EXAMPLE 8   Combining Transformations

Graph $y = 2 - 2\sqrt{x - 3}$.

Solution   You should recognize that the given function consists of four transformations of the basic function $f(x) = \sqrt{x}$:

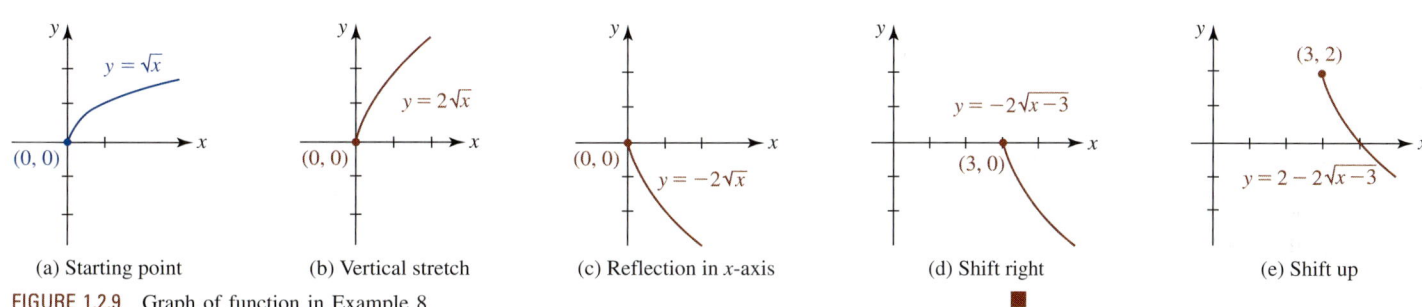

We start with the graph of $f(x) = \sqrt{x}$ in FIGURE 1.2.9(a). The four transformations are illustrated in Figures 1.2.9(b)–(e).

(a) Starting point	(b) Vertical stretch	(c) Reflection in x-axis	(d) Shift right	(e) Shift up

FIGURE 1.2.9   Graph of function in Example 8

▐ **Symmetry**   If the graph of a function is symmetric with respect to the y-axis, we say that f is an **even function**. A function whose graph is symmetric with respect to the origin is said to be an **odd function**. We have the following tests for symmetry.

Tests for Symmetry of the Graph of a Function

The graph of a function f with domain X is symmetric with respect to

- the **y-axis** if $f(-x) = f(x)$ for every x in X, or   (11)
- the **origin** if $f(-x) = -f(x)$ for every x in X.   (12)

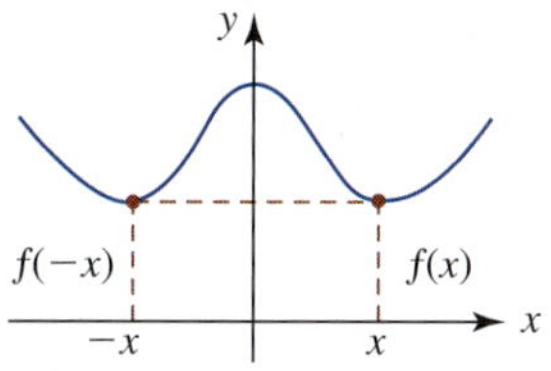

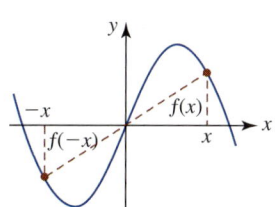

FIGURE 1.2.10   Even function; graph has y-axis symmetry

In FIGURE 1.2.10, observe that if f is an even function and

$$\overset{f(x)}{\underset{\downarrow}{}} \qquad\qquad \overset{f(-x)}{\underset{\downarrow}{}}$$
(x, y) is a point on its graph, then necessarily $(-x, y)$

is also a point on its graph. Similarly, we see in FIGURE 1.2.11 that if f is an odd function and

$$\overset{f(x)}{\underset{\downarrow}{}} \qquad\qquad \overset{f(-x) = -f(x)}{\underset{\downarrow}{}}$$
(x, y) is a point on its graph, then necessarily $(-x, -y)$

is a point on its graph.

FIGURE 1.2.11   Odd function; graph has origin symmetry

EXAMPLE 9   Odd and Even Functions

(a) $f(x) = x^3$ is an odd function since by (12),

$$f(-x) = (-x)^3 = (-1)^3 x^3 = -x^3 = -f(x).$$

Inspection of Figure 1.2.1(c) shows that the graph of f is symmetric with respect to the origin. For example, since $f(1) = 1$, $(1, 1)$ is a point on the graph of $y = x^3$. Because f is an odd function, $f(-1) = -f(1)$ implies $(-1, -1)$ is on the same graph.

(b) $f(x) = x^{2/3}$ is an even function since by (11) and the laws of exponents

$$\overset{\text{cube root of } -1 \text{ is } -1}{\downarrow}$$

$$f(-x) = (-x)^{2/3} = (-1)^{2/3}x^{2/3} = (\sqrt[3]{-1})^2 x^{2/3} = (-1)^2 x^{2/3} = x^{2/3} = f(x).$$

In Figure 1.2.1(i), we see that the graph of f is symmetric with respect to the y-axis. For example, $(8, 4)$ and $(-8, 4)$ are points on the graph of $y = x^{2/3}$.

(c) $f(x) = x^3 + 1$ is neither even nor odd. From

$$f(-x) = (-x)^3 + 1 = -x^3 + 1$$

we see that $f(-x) \neq f(x)$, and $f(-x) \neq -f(x)$. ∎

The graphs in Figure 1.2.1, with part (g) the only exception, possess either y-axis or origin symmetry. The functions in Figures 1.2.1(b), (d), (f), and (i) are even, whereas the functions in Figures 1.2.1(a), (c), (e), and (h) are odd.

Exercises 1.2   Answers to selected odd-numbered problems begin on page ANS-2.

≡ Fundamentals

In Problems 1–6, find $f + g, f - g, fg$, and f/g.

1. $f(x) = 2x + 5, g(x) = -4x + 8$

2. $f(x) = 5x^2, g(x) = 7x - 9$

3. $f(x) = \dfrac{x}{x + 1}, g(x) = \dfrac{1}{x}$

4. $f(x) = \dfrac{2x - 1}{x + 3}, g(x) = \dfrac{x - 3}{4x + 2}$

5. $f(x) = x^2 + 2x - 3, g(x) = x^2 + 3x - 4$

6. $f(x) = x^2, g(x) = \sqrt{x}$

In Problems 7–10, let $f(x) = \sqrt{x - 1}$ and $g(x) = \sqrt{2 - x}$. Find the domain of the given function.

7. $f + g$ **8.** fg **9.** f/g **10.** g/f

In Problems 11–16, find $f \circ g$ and $g \circ f$.

11. $f(x) = 3x - 2, g(x) = x + 6$

12. $f(x) = 4x + 1, g(x) = x^2$

13. $f(x) = x^2, g(x) = x^3 + x^2$

14. $f(x) = 2x + 4, g(x) = \dfrac{1}{2x + 4}$

15. $f(x) = \dfrac{3}{x}, g(x) = \dfrac{x}{x + 1}$

16. $f(x) = x^2 + \sqrt{x}, g(x) = x^2$

In Problems 17 and 18, let $f(x) = \sqrt{x - 3}$ and $g(x) = x^2 + 2$. Find the domain of the given function.

17. $f \circ g$ **18.** $g \circ f$

In Problems 19 and 20, let $f(x) = 5 - x^2$ and $g(x) = 2 - \sqrt{x}$. Find the domain of the given function.

19. $g \circ f$ **20.** $f \circ g$

In Problems 21 and 22, find $f \circ (2f)$ and $f \circ (1/f)$.

21. $f(x) = 2x^3$ **22.** $f(x) = \dfrac{1}{x - 1}$

The composition of three functions f, g, and h is the function

$$(f \circ g \circ h)(x) = f(g(h(x))).$$

In Problems 23 and 24, find $f \circ g \circ h$.

23. $f(x) = x^2 + 6, g(x) = 2x + 1, h(x) = 3x - 2$

24. $f(x) = \sqrt{x - 5}, g(x) = x^2 + 2, h(x) = \sqrt{2x + 1}$

In Problems 25 and 26, find a function g.

25. $f(x) = 2x - 5, (f \circ g)(x) = -4x + 13$

26. $f(x) = \sqrt{2x + 6}, (f \circ g)(x) = 4x^2$

In Problems 27 and 28, express the function F as a composition $f \circ g$ of two functions f and g.

27. $F(x) = 2x^4 - x^2$ **28.** $F(x) = \dfrac{1}{x^2 + 9}$

In Problems 29–36, the points $(-2, 1)$ and $(3, -4)$ are on the graph of the function $y = f(x)$. Find the corresponding points on the graph obtained by the given transformations.

29. the graph of f shifted up 2 units

30. the graph of f shifted down 5 units

31. the graph of f shifted to the left 6 units

32. the graph of f shifted to the right 1 unit

33. the graph of f shifted up 1 unit and to the left 4 units

34. the graph of f shifted down 3 units and to the right 5 units

35. the graph of f reflected in the y-axis

36. the graph of f reflected in the x-axis

In Problems 37–40, use the graph of the function $y = f(x)$ given in the figure to graph the following functions:

(a) $y = f(x) + 2$ (b) $y = f(x) - 2$
(c) $y = f(x + 2)$ (d) $y = f(x - 5)$
(e) $y = -f(x)$ (f) $y = f(-x)$

37.

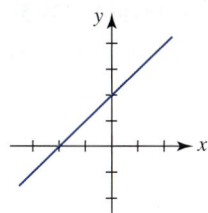

FIGURE 1.2.12 Graph for Problem 37

38.

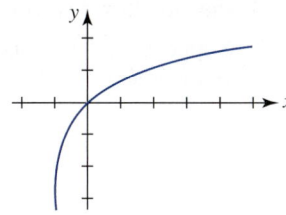

FIGURE 1.2.13 Graph for Problem 38

39.

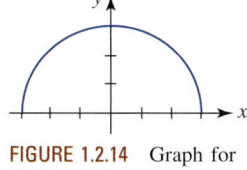

FIGURE 1.2.14 Graph for Problem 39

40.

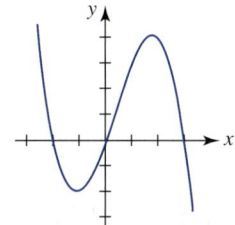

FIGURE 1.2.15 Graph for Problem 40

In Problems 41 and 42, use the graph of the function $y = f(x)$ given in the figure to graph the following functions:

(a) $y = f(x) + 1$ (b) $y = f(x) - 1$
(c) $y = f(x + \pi)$ (d) $y = f(x - \pi/2)$
(e) $y = -f(x)$ (f) $y = f(-x)$
(g) $y = 3f(x)$ (h) $y = -\frac{1}{2}f(x)$

41.

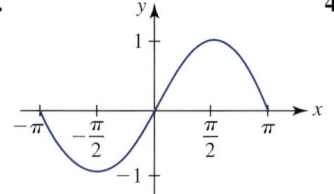

FIGURE 1.2.16 Graph for Problem 41

42.

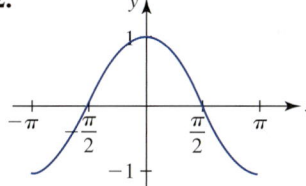

FIGURE 1.2.17 Graph for Problem 42

In Problems 43–46, find the equation of the final graph after the given transformations are applied to the graph of $y = f(x)$.

43. the graph of $f(x) = x^3$ shifted up 5 units and right 1 unit

44. the graph of $f(x) = x^{2/3}$ stretched vertically by a factor of 3 units, then shifted right 2 units

45. the graph of $f(x) = x^4$ reflected in the x-axis, then shifted left 7 units

46. the graph of $f(x) = \dfrac{1}{x}$ reflected in the y-axis, then shifted left 5 units and down 10 units

In Problems 47 and 48, complete the graph of the given function $y = f(x)$ if

(a) f is an even function   and   (b) f is an odd function.

47.

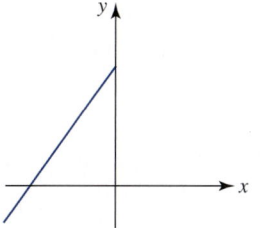

FIGURE 1.2.18 Graph for Problem 47

48.

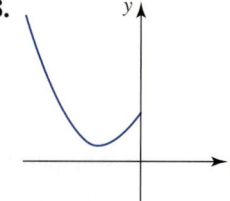

FIGURE 1.2.19 Graph for Problem 48

49. Fill in the table where f is an even function.

x	0	1	2	3	4
$f(x)$	-1	2	10	8	0
$g(x)$	2	-3	0	1	-4
$(f \circ g)(x)$					

50. Fill in the table where g is an odd function.

x	0	1	2	3	4
$f(x)$	-2	-3	0	-1	-4
$g(x)$	9	7	-6	-5	13
$(g \circ f)(x)$					

A Mathematical Classic In the mathematical analysis of circuits or of signals it is convenient to define a special function that is 0 (off) up to a certain number and then the number 1 (on) after that. The **Heaviside function**

$$U(x - a) = \begin{cases} 0, & x < a \\ 1, & x \geq a, \end{cases}$$

is named after the brilliant and controversial English electrical engineer and mathematician **Oliver Heaviside** (1850–1925). The function U is also known as the **unit step function**.

In Problems 51 and 52, sketch the given function. The function in Problem 52 is sometimes called the **boxcar function**.

51. $y = 2U(x - 1) + U(x - 2)$

52. $y = U\left(x + \frac{1}{2}\right) - U\left(x - \frac{1}{2}\right)$

53. Find an equation for the function f illustrated in FIGURE 1.2.20 in terms of $U(x - a)$.

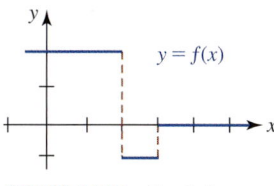

FIGURE 1.2.20 Graph for Problem 53

54. The Heaviside function $U(x - a)$ is frequently combined with other functions by addition and multiplication. Given that $f(x) = x^2$, compare the graphs of $y = f(x - 3)$ and $y = f(x - 3)U(x - 3)$.

In Problems 55 and 56, sketch the given function.

55. $y = (2x - 5)U(x - 1)$ **56.** $y = x - xU(x - 3)$

☰ Think About It

57. Determine whether $f \circ (g + h) = f \circ g + f \circ h$ is true or false.

58. Suppose $[-1, 1]$ is the domain of $f(x) = x^2$. What is the domain of $y = f(x - 2)$?

59. Explain why the graph of a function cannot be symmetric with respect to the x-axis.

60. What points, if any, on the graph of $y = f(x)$ remain fixed, that is, the same on the resulting graph after a vertical stretch or compression? After a reflection in the x-axis? After a reflection in the y-axis?

61. Suppose the domain of f is $(-\infty, \infty)$. What is the relationship between the graphs of $y = f(x)$ and $y = f(|x|)$?

62. Review the graphs of $y = x$ and $y = 1/x$ in Figure 1.2.1. Then discuss how to obtain the graph of $y = 1/f(x)$ from the graph of $y = f(x)$. Sketch the graph of $y = 1/f(x)$ for the function f whose graph is given in Figure 1.2.15.

63. Suppose $f(x) = x$ and $g(x) = \lfloor x \rfloor$ is the greatest integer or floor function. The difference of f and g is the function $\text{frac}(x) = x - \lfloor x \rfloor$ called the **fractional part of x**. Explain the name and then graph $\text{frac}(x)$.

64. Using the notion of a reflection of a graph in an axis, express the ceiling function $g(x) = \lceil x \rceil$ in terms of the floor function $f(x) = \lfloor x \rfloor$ (see pages 7 and 15).

1.3 Polynomial and Rational Functions

▍**Introduction** In this section we continue our review of polynomial functions and rational functions. Functions such as $y = 2x - 1$, $y = 5x^2 - 2x + 4$, and $y = x^3$, in which the variable x is raised to a *nonnegative integer power*, are examples of polynomial functions. In the preceding section we saw that a general **polynomial function** $y = f(x)$ has the form

$$f(x) = a_n x^n + a_{n-1} x^{n-1} + \cdots + a_2 x^2 + a_1 x + a_0, \tag{1}$$

where n is a nonnegative integer. A **rational function** is the quotient

$$f(x) = \frac{p(x)}{q(x)}, \tag{2}$$

where p and q are polynomial functions.

▍**Polynomial Functions** The constants $a_n, a_{n-1}, \ldots, a_1, a_0$ in (1) are called **coefficients**; the number a_n is called the **leading coefficient** and a_0 is called the **constant term** of the polynomial. The highest power of x in a polynomial is said to be its **degree**. So if $a_n \neq 0$, then we say that $f(x)$ in (1) has **degree n**. For example,

$$\overset{\text{degree 5}}{f(x) = 3x^5} - 4x^3 - 3\underset{\uparrow}{x} + \underset{\uparrow}{8}$$
$$\text{leading coefficient} \qquad \text{constant term}$$

is a polynomial function of degree 5.

Polynomials of degrees $n = 0$, $n = 1$, $n = 2$, and $n = 3$ are respectively,

$$f(x) = a, \qquad\qquad\qquad\qquad \textbf{constant function,}$$
$$f(x) = ax + b, \qquad\qquad\qquad \textbf{linear function,}$$
$$f(x) = ax^2 + bx + c, \qquad\qquad \textbf{quadratic function,}$$
$$f(x) = ax^3 + bx^2 + cx + d, \qquad \textbf{cubic function.}$$

The constant function $f(x) = 0$ is called the **zero polynomial**.

▍**Lines** You are undoubtedly familiar with the fact that the graphs of a constant function and a linear function are **lines**. Since the notion of a line plays an important role in the study of differential calculus, it is appropriate that we review equations of lines. There are three types of lines in the xy-plane: horizontal lines, vertical lines, and slant or oblique lines.

■ **Slope**   We begin with the recollection from plane geometry that through any two distinct points (x_1, y_1) and (x_2, y_2) in the plane there passes only one line L. If $x_1 \neq x_2$, then the number

$$m = \frac{y_2 - y_1}{x_2 - x_1} \qquad (3)$$

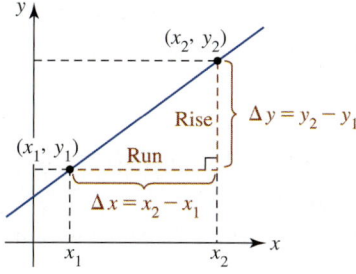

FIGURE 1.3.1   Slope of a line

is called the **slope** of the line determined by these two points. It is customary to denote the **change in y** or the **rise** of the line by $\Delta y = y_2 - y_1$ and the **change in x** or the **run** of the line by $\Delta x = x_2 - x_1$, so that (3) is written $m = \Delta y / \Delta x$. See FIGURE 1.3.1. As indicated in FIGURE 1.3.2, any pair of distinct points on a line with slope, say, $(x_1, y_1), (x_2, y_2)$, and $(x_3, y_3), (x_4, y_4)$, will determine the same slope. In other words, the slope of a line is independent of the choice of points on the line.

In FIGURE 1.3.3 we compare the graphs of lines with positive, negative, zero, and undefined slopes. In Figure 1.3.3(a) we see, reading the graph left to right, that a line with positive slope ($m > 0$) rises as x increases. Figure 1.3.3(b) shows that a line with negative slope ($m < 0$) falls as x increases. If (x_1, y_1) and (x_2, y_2) are points on a horizontal line, then $y_1 = y_2$ and so its rise is $\Delta y = y_2 - y_1 = 0$. Hence from (3) the slope is zero ($m = 0$). See Figure 1.3.3(c). If (x_1, y_1) and (x_2, y_2) are points on a vertical line, then $x_1 = x_2$ and so its run is $\Delta x = x_2 - x_1 = 0$. In this case we say that the slope of the line is **undefined** or that the line has no slope. See Figure 1.3.3(d). Only lines with slope are the graphs of functions.

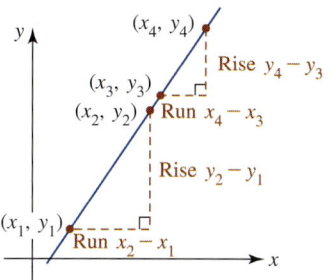

FIGURE 1.3.2   Similar triangles

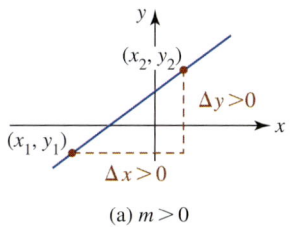

(a) $m > 0$

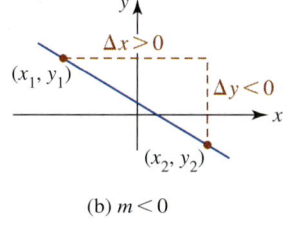

(b) $m < 0$

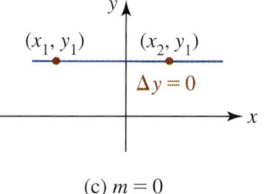

(c) $m = 0$

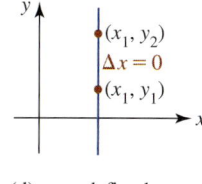

(d) m undefined

FIGURE 1.3.3   Lines with slope (a)–(c); line with no slope (d)

■ **Equations of Lines**   To find an equation of a line L with slope m let us suppose that (x_1, y_1) is on the line. If (x, y) represents any other point on L, then (3) gives

$$\frac{y - y_1}{x - x_1} = m.$$

Multiplying both sides of the last equality by $x - x_1$ gives an important equation. The **point–slope equation** of the line through (x_1, y_1) with slope m is

$$y - y_1 = m(x - x_1). \qquad (4)$$

Any line that is not vertical must cross the y-axis. If the y-intercept is $(0, b)$, then with $x_1 = 0, y_1 = b$, (4) gives $y - b = m(x - 0)$. The last equation simplifies to the **slope-intercept equation** of the line

$$y = mx + b. \qquad (5)$$

EXAMPLE 1   Point–Slope Equation

Find an equation of the line passing through the points $(4, 3)$ and $(-2, 5)$.

Solution   First we compute the slope of the line through the points. From (3),

$$m = \frac{5 - 3}{-2 - 4} = \frac{2}{-6} = -\frac{1}{3}.$$

The point–slope equation (4) then gives $y - 3 = -\frac{1}{3}(x - 4)$ or $y = -\frac{1}{3}x + \frac{13}{3}$.   ■

An equation of *any* line in the plane is a special case of the general **linear equation**

$$Ax + By + C = 0, \tag{6}$$

where A, B, and C are real constants. The characteristic that gives (6) its name *linear* is that the variables x and y appear only to the first power. The cases of special interest are

$$A = 0, B \neq 0, \text{ gives } y = -\frac{C}{B}, \tag{7}$$

$$A \neq 0, B = 0, \text{ gives } x = -\frac{C}{A}, \tag{8}$$

$$A \neq 0, B \neq 0, \text{ gives } y = -\frac{A}{B}x - \frac{C}{B}. \tag{9}$$

The first and the third of these three equations define functions. By relabeling $-C/B$ in (7) as a we get a constant function $y = a$. By relabeling $-A/B$ and $-C/B$ in (9) as a and b, respectively, we get the form of a linear function $f(x) = ax + b$, which, except for symbols, is the same as (5). By relabeling $-C/A$ in (8) as a we get the equation of a vertical line $x = a$, which is not a function.

◼ **Increasing–Decreasing Functions**   We have just seen in Figures 1.3.3(a) and 1.3.3(b) that if $a > 0$ (which, as we have just seen plays the part of m), the values of a linear function $f(x) = ax + b$ increase as x increases, whereas for $a < 0$, the values of $f(x)$ decrease as x increases. The notions of increasing and decreasing can be extended to *any* function. A function f is said to be

- **increasing** on an interval I if $f(x_1) < f(x_2)$, and $\tag{10}$
- **decreasing** on an interval I if $f(x_1) > f(x_2)$. $\tag{11}$

In FIGURE 1.3.4(a) the function f is increasing on the interval $[a, b]$, whereas f is decreasing on $[a, b]$ in Figure 1.3.4(b). A linear function $f(x) = ax + b$ increases on the interval $(-\infty, \infty)$ for $a > 0$ and decreases on the interval $(-\infty, \infty)$ for $a < 0$.

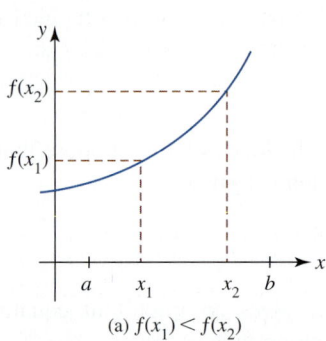
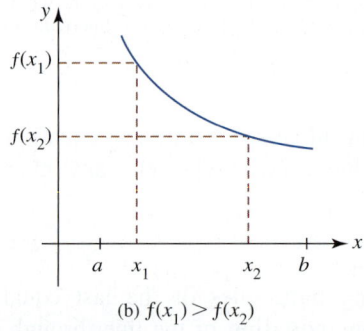

(a) $f(x_1) < f(x_2)$ (b) $f(x_1) > f(x_2)$

FIGURE 1.3.4   Increasing function in (a); decreasing function in (b)

◼ **Parallel and Perpendicular Lines**   If L_1 and L_2 are two distinct lines *with slope*, then necessarily L_1 and L_2 are either parallel or they intersect. If the lines intersect at a right angle, they are said to be perpendicular. We can determine whether two lines are parallel or are perpendicular by examining their slopes.

> This assumption means that L_1 and L_2 are nonvertical lines.

Parallel and Perpendicular Lines

Suppose L_1 and L_2 are lines with slopes m_1 and m_2, respectively. Then,

- L_1 is **parallel** to L_2 if and only if $m_1 = m_2$, and
- L_1 is **perpendicular** to L_2 if and only if $m_1 m_2 = -1$.

EXAMPLE 2   Parallel Lines

The linear equations $3x + y = 2$ and $6x + 2y = 15$ can be rewritten in the slope-intercept forms $y = -3x + 2$ and $y = -3x + \frac{15}{2}$, respectively. As noted in blue and red the slope of each line is -3. Therefore the lines are parallel. The graphs of these equations are shown in FIGURE 1.3.5. ◼

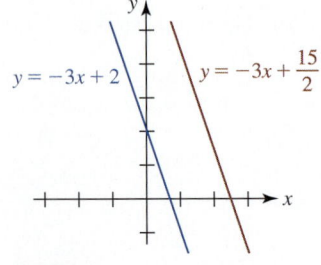

FIGURE 1.3.5   Parallel lines in Example 2

EXAMPLE 3 Perpendicular Lines

Find an equation of the line through $(0, -3)$ that is perpendicular to the graph of $4x - 3y + 6 = 0$.

Solution By solving for y, the given linear equation yields the slope-intercept form $y = \frac{4}{3}x + 2$. This line, whose graph is given in blue in FIGURE 1.3.6, has slope $\frac{4}{3}$. The slope of any line perpendicular to it is the negative reciprocal of $\frac{4}{3}$, namely, $-\frac{3}{4}$. Since $(0, -3)$ is the y-intercept of the required line, it follows from (5) that its equation is $y = -\frac{3}{4}x - 3$. The graph of the last equation is the red line in Figure 1.3.6. ∎

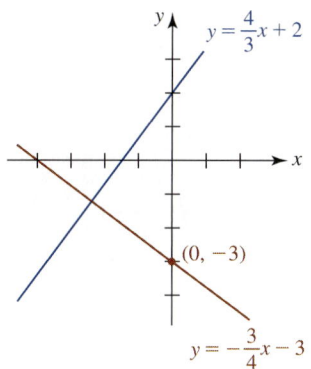

FIGURE 1.3.6 Perpendicular lines in Example 3

▍ Quadratic Functions The squaring function $y = x^2$ that we have seen in Sections 1.1 and 1.2 is a member of a family of functions called **quadratic functions**, that is, polynomial functions of the form $f(x) = ax^2 + bx + c$, where $a \neq 0$, b, and c are constants. The graphs of quadratic functions, called **parabolas**, are simply rigid and nonrigid transformations of the graph of $y = x^2$ shown in FIGURE 1.3.7.

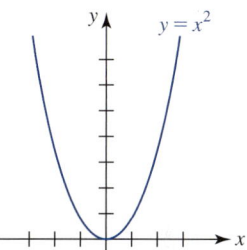

FIGURE 1.3.7 Graph of simplest parabola

▍ Vertex and Axis If the graph of a quadratic function opens upward $a > 0$ (or downward $a < 0$), the lowest (highest) point (h, k) on the parabola is called its **vertex**. All parabolas are symmetric with respect to a vertical line through the vertex (h, k). The line $x = h$ is called the **axis** of the parabola. See FIGURE 1.3.8.

▍ Standard Form The vertex (h, k) of a parabola can be determined by recasting the equation $f(x) = ax^2 + bx + c$ into the **standard form**

$$f(x) = a(x - h)^2 + k. \tag{12}$$

The form (12) is obtained from $f(x) = ax^2 + bx + c$ by completing the square in x. With the aid of differential calculus we will be able to find the vertex of a parabola without completing the square.

As the next example shows, a reasonable sketch of a parabola can be obtained by plotting the intercepts and the vertex. The form in (12) indicates that its graph is the graph of $y = ax^2$ shifted horizontally $|h|$ units and shifted vertically $|k|$ units.

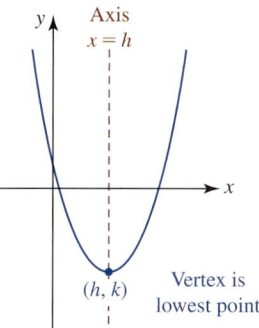

(a) $y = ax^2 + bx + c$, $a > 0$

EXAMPLE 4 Graph Using Intercepts and Vertex

Graph $f(x) = x^2 - 2x - 3$.

Solution Since $a = 1 > 0$ we know that the parabola will open upward. From $f(0) = -3$ we get the y-intercept $(0, -3)$. To see whether there are any x-intercepts we solve $x^2 - 2x - 3 = 0$ by factoring or by the quadratic formula. From $(x + 1)(x - 3) = 0$ we find the solutions $x = -1$ and $x = 3$. The x-intercepts are $(-1, 0)$ and $(3, 0)$. To locate the vertex we complete the square:

$$f(x) = (x^2 - 2x + 1) - 1 - 3 = (x^2 - 2x + 1) - 4.$$

Thus the standard form is $f(x) = (x - 1)^2 - 4$. By comparing the last equation with (12) we identify $h = 1$ and $k = -4$. We conclude that the vertex is $(1, -4)$. Using this information we draw a parabola through these four points as shown in FIGURE 1.3.9.

By finding the vertex of a parabola we automatically determine the range of the quadratic function. As Figure 1.3.9 clearly shows, the range of f is the interval $[-4, \infty)$ on the y-axis. Figure 1.3.9 also shows that f is decreasing on the interval $(-\infty, 1]$ but increasing on $[1, \infty)$. ∎

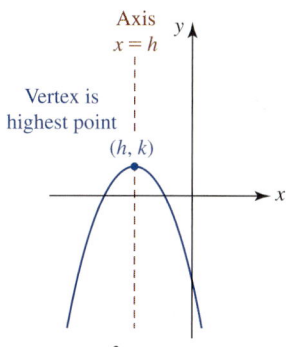

(b) $y = ax^2 + bx + c$, $a < 0$

FIGURE 1.3.8 Vertex and axis of a parabola

▍ Higher-Degree Polynomial Functions The graph of *every* linear function $f(x) = ax + b$ is a line and the graph of *every* quadratic function $f(x) = ax^2 + bx + c$ is a parabola. Such definitive descriptive statements cannot be made about the graph of a higher-degree polynomial function. What is the shape of the graph of a fifth-degree polynomial function? It turns out that the graph of a polynomial function of degree $n \geq 3$ can have several possible shapes. In general,

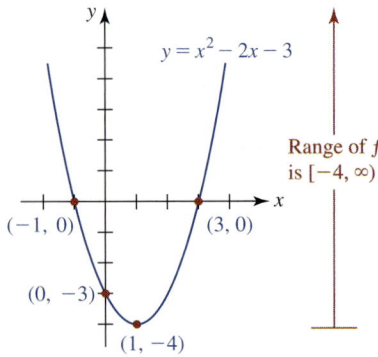

FIGURE 1.3.9 Parabola in Example 4

graphing a polynomial function f of degree $n \geq 3$ often demands the use of either calculus or a graphing utility. However, by keeping in mind shifting, end behavior, intercepts, and symmetry we can, in many instances, quickly sketch a reasonable graph of a higher-degree polynomial function while keeping point plotting to a minimum.

▪ **End Behavior** In rough terms, the **end behavior** of *any* function f is simply how f behaves for very large values of $|x|$. In the case of a polynomial function f of degree n, its graph resembles the graph of $y = a_n x^n$ for large values of $|x|$. To see why the graph of a polynomial function such as $f(x) = -2x^3 + 4x^2 + 5$ resembles the graph of the single-term polynomial $y = -2x^3$ when $|x|$ is large, let us factor out the highest power of x, that is, x^3:

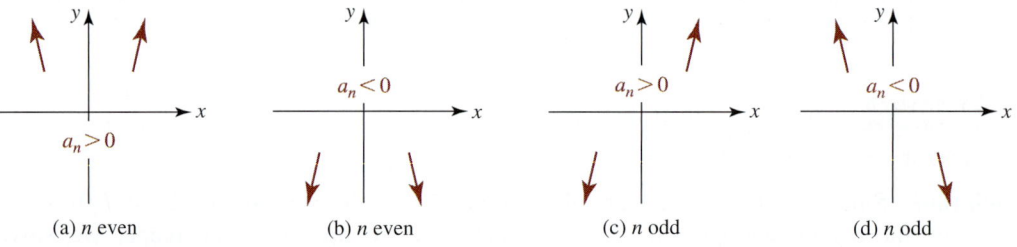

$$f(x) = x^3\left(-2 + \frac{4}{x} + \frac{5}{x^3}\right). \tag{13}$$

By letting $|x|$ increase without bound, both $4/x$ and $5/x^3$ can be made as close to 0 as we want. Thus, when $|x|$ is large, the values of the function f in (13) are closely approximated by the values of $y = -2x^3$. In general, there can be only four types of end behavior for polynomial functions. To interpret the arrows in FIGURE 1.3.10 let us examine the arrows in, say, Figure 1.3.10(c) where it is assumed that n is odd and $a_n > 0$. The position and direction of the left arrow (left arrow points down) indicates that as x becomes unbounded in the negative direction, the values $f(x)$ are decreasing. Stated another way, the graph is heading downward. Similarly, the position and direction of the right arrow (right arrow points up) indicates that as x becomes unbounded in the positive direction, the values $f(x)$ are increasing (the graph is heading upward). You can see the end behavior illustrated in Figures 1.3.10(a) and 1.3.10(c) in the graphs given in FIGURE 1.3.11 and FIGURE 1.3.12, respectively. The graphs of the functions $y = -x$, $y = -x^2$, $y = -x^3$, ..., $y = -x^8$ are the graphs in Figures 1.3.11 and 1.3.12 reflected in the x-axis, and so their end behavior is as shown in Figures 1.3.10(b) and 1.3.10(d).

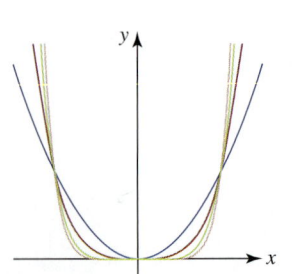

FIGURE 1.3.11 Graphs of $y = x^2$ (blue), $y = x^4$ (red), $y = x^6$ (green), $y = x^8$ (gold)

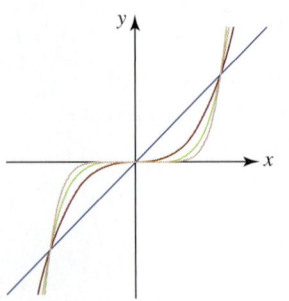

FIGURE 1.3.12 Graphs of $y = x$ (blue), $y = x^3$ (red), $y = x^5$ (green), $y = x^7$ (gold)

FIGURE 1.3.10 End behavior of a polynomial function f depends on its degree n and on the sign of its leading coefficient

▪ **Symmetry of Polynomial Functions** It is easy to tell by inspection those polynomial functions whose graphs possess **symmetry** with respect to either the y-axis or the origin. The words *even* and *odd* have special meaning for polynomial functions. The conditions $f(-x) = f(x)$ and $f(-x) = -f(x)$ hold for polynomial functions in which all the powers of x are even integers and odd integers, respectively. For example,

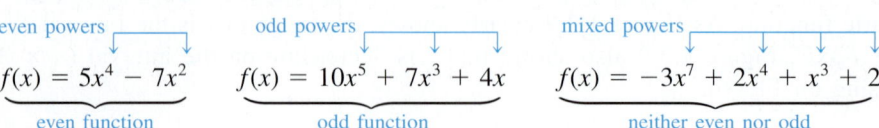

A function such as $f(x) = 3x^6 - x^4 + 6$ is an even function because all powers are even integers; the constant term 6 is actually $6x^0$, and 0 is an even nonnegative integer.

▪ **Intercepts of Polynomial Functions** The graph of every polynomial function f passes through the y-axis since $x = 0$ is in the domain of the function. The y-intercept is the point

$(0, f(0))$. The real **zeros** of a polynomial function are the x-coordinates of the **x-intercepts** of its graph. A number c is a zero of a polynomial function f of degree n if and only if $x - c$ is a factor of f, that is, $f(x) = (x - c)q(x)$, where $q(x)$ is a polynomial of degree $n - 1$. If $(x - c)^m$ is a factor of f, where $m > 1$ is a positive integer, and $(x - c)^{m+1}$ is *not* a factor of f, then c is said to be a **repeated zero**, or a **zero of multiplicity m**. When $m = 1$, c is called a **simple zero**. For example, $-\frac{1}{3}$ and $\frac{1}{2}$ are simple zeros of $f(x) = 6x^2 - x - 1$ since f can be written as $f(x) = 6\left(x + \frac{1}{3}\right)\left(x - \frac{1}{2}\right)$, whereas 5 is a repeated zero or a zero of multiplicity 2 for $f(x) = x^2 - 10x + 25 = (x - 5)^2$. The behavior of the graph of f at an x-intercept $(c, 0)$ depends on whether c is a simple zero, or a zero of multiplicity $m > 1$, where m is either an even or an odd integer. See FIGURE 1.3.13.

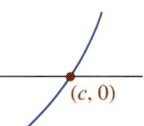

(a) Simple zero

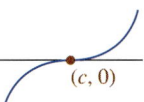

(b) Zero of odd multiplicity $m = 3, 5, \ldots$

x-intercepts of Polynomials

- If c is a simple zero, then the graph of f passes directly through the x-axis at $(c, 0)$. See Figure 1.3.13(a).
- If c is a zero of odd multiplicity $m = 3, 5, \ldots$, then the graph of f passes through the x-axis but is flattened at $(c, 0)$. See Figure 1.3.13(b).
- If c is a zero of even multiplicity $m = 2, 4, \ldots$, then the graph of f does not pass through the x-axis but is tangent to, or touches, the x-axis at $(c, 0)$. See Figure 1.3.13(c).

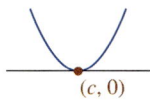

(c) Zero of even multiplicity $m = 2, 4, \ldots$

FIGURE 1.3.13 x-intercepts of a polynomial function f

In the case when c is either a simple zero or a zero of odd multiplicity, $f(x)$ changes sign at $(c, 0)$, whereas if c is a zero of even multiplicity, $f(x)$ does not change sign at $(c, 0)$. We note that depending on the sign of the leading coefficient of the polynomial, the graphs in Figure 1.3.13 could be reflected in the x-axis.

EXAMPLE 5   Graphs of Polynomial Functions

Graph

(a) $f(x) = x^3 - 9x$ **(b)** $g(x) = (1 - x)(x + 1)^2$ **(c)** $h(x) = -(x + 4)(x - 2)^3$.

Solution

(a) By ignoring all terms but the first, we see that the graph of f resembles the graph of $y = x^3$ for large $|x|$. This end behavior of f is shown in Figure 1.3.10(c). Since all the powers are odd integers, f is an odd function and its graph is symmetric with respect to the origin. Setting $f(x) = 0$ we see from

difference of two squares
$$\downarrow$$
$$x(x^2 - 9) = 0 \quad \text{or} \quad x(x - 3)(x + 3) = 0$$

that the zeros of f are $x = 0$ and $x = \pm 3$. Since these numbers are simple zeros the graph passes directly through x-intercepts at $(0, 0)$, $(-3, 0)$, and $(3, 0)$ as shown in FIGURE 1.3.14.

(b) Multiplying out, g is the same as $g(x) = -x^3 - x^2 + x + 1$ and so we see that the graph of g resembles the graph of $y = -x^3$ for large $|x|$, just the opposite of the end behavior of the function in part (a). Because there are both even and odd powers of x present, g is neither even nor odd; its graph possesses no y-axis or origin symmetry. Because -1 is a zero of multiplicity 2, the graph is tangent to the x-axis at $(-1, 0)$. Since 1 is a simple zero, the graph passes directly through the x-axis at $(1, 0)$. See FIGURE 1.3.15.

(c) Inspection of h shows that its graph resembles the graph of $y = -x^4$ for large $|x|$. This end behavior of h is shown in Figure 1.3.10(b). The function h is neither even nor odd. From the factored form of $h(x)$, we see that -4 is a simple zero and so the graph of h passes directly through the x-axis at $(-4, 0)$. Since 2 is a zero of multiplicity 3, its graph flattens as it passes through the x-intercept $(2, 0)$. See FIGURE 1.3.16.

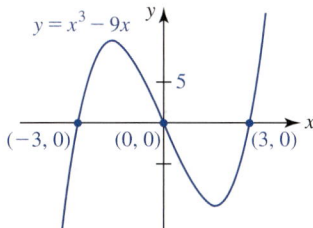

FIGURE 1.3.14   Graph of function in Example 5(a)

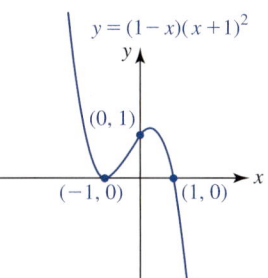

FIGURE 1.3.15   Graph of function in Example 5(b)

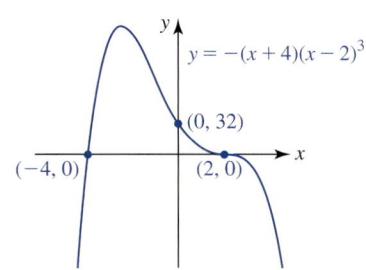

FIGURE 1.3.16   Graph of function in Example 5(c)

Rational Functions Graphing a rational function $f(x) = p(x)/q(x)$ is a little more complicated than graphing a polynomial function because in addition to paying attention to intercepts, symmetry, and shifting/reflecting/stretching of known graphs, you should also keep an eye on the domain of f and the degrees of $p(x)$ and $q(x)$. The last two items are important in determining whether a graph of a rational function possesses *asymptotes*.

Intercepts of Rational Functions The **y-intercept** of the graph of $f(x) = p(x)/q(x)$ is the point $(0, f(0))$, provided the number 0 is in the domain of f. For example, the graph of the rational function $f(x) = (1 - x)/x$ does not cross the y-axis since $f(0)$ is not defined. If the polynomials $p(x)$ and $q(x)$ have no common factors, then the **x-intercepts** of the graph of the rational function $f(x) = p(x)/q(x)$ are the points whose x-coordinates are the real zeros of the numerator $p(x)$. In other words, the only way we can have $f(x) = p(x)/q(x) = 0$ is to have $p(x) = 0$. Thus for $f(x) = (1 - x)/x$, $1 - x = 0$ gives $x = 1$ and so $(1, 0)$ is an x-intercept of the graph of f.

Asymptotes The graph of a rational function $f(x) = p(x)/q(x)$ can have asymptotes. For our purposes the asymptotes can be a horizontal line, a vertical line, or a slant line. On a practical level, vertical and horizontal asymptotes of the graph of a rational function f can be determined by inspection. So for the sake of discussion let us suppose that

$$f(x) = \frac{p(x)}{q(x)} = \frac{a_n x^n + a_{n-1}x^{n-1} + \cdots + a_1 x + a_0}{b_m x^m + b_{m-1}x^{m-1} + \cdots + b_1 x + b_0}, \quad a_n \neq 0,\ b_m \neq 0, \tag{14}$$

represents a general rational function. The degree of $p(x)$ is n and the degree of $q(x)$ is m.

Asymptotes of Graphs of Rational Functions

Suppose that the polynomial functions $p(x)$ and $q(x)$ in (14) *have no common factors.*

- If a is real zero of $q(x)$, then $x = a$ is a **vertical asymptote** for the graph of f.
- If $n = m$, then $y = a_n/b_m$ (the quotient of the leading coefficients) is a **horizontal asymptote** for the graph of f.
- If $n < m$, then $y = 0$ is a **horizontal asymptote** for the graph of f.
- If $n > m$, then the graph of f has *no* **horizontal asymptote**.
- If $n = m + 1$, then the quotient $y = mx + b$ of $p(x)$ and $q(x)$ is a **slant asymptote** for the graph of f.

We note from the foregoing bulleted list that horizontal and slant asymptotes are mutually exclusive. In other words, a graph of a rational function f cannot possess a slant asymptote *and* a horizontal asymptote.

EXAMPLE 6 Graphs of Rational Functions

Graph

(a) $f(x) = \dfrac{x}{1 - x^2}$ **(b)** $g(x) = \dfrac{x^2 - x - 6}{x - 5}$.

Solution

(a) We begin with the observation that the numerator $p(x) = x$ and denominator $q(x) = 1 - x^2$ of f have no common factors. Also, since $f(-x) = -f(x)$ the function f is odd. Therefore, its graph is symmetric with respect to the origin. Because $f(0) = 0$, the y-intercept is $(0, 0)$. Moreover, $p(x) = x = 0$ implies $x = 0$ and so the only intercept is $(0, 0)$. The zeros of the denominator $q(x) = 1 - x^2$ are ± 1. Therefore, the lines $x = -1$ and $x = 1$ are vertical asymptotes. Since the degree of the numerator x is 1 and the degree of the denominator $1 - x^2$ is 2 (and $1 < 2$) it follows that $y = 0$ is a horizontal asymptote for the graph of f. The graph consists of three distinct *branches*, one to the left of the line $x = -1$, one between the lines $x = -1$ and $x = 1$, and one to the right of the line $x = 1$. See FIGURE 1.3.17.

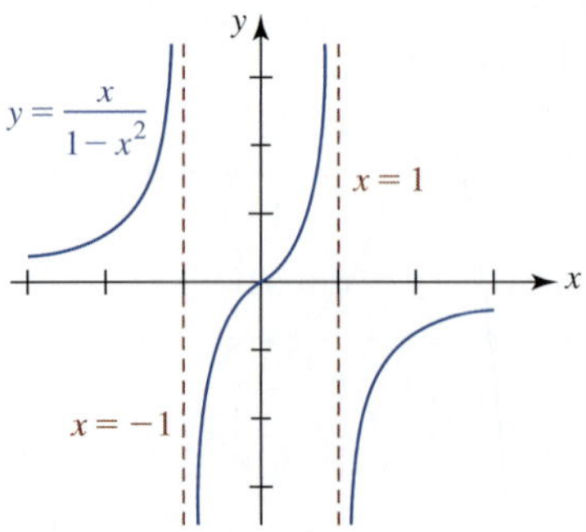

FIGURE 1.3.17 Graph of function in Example 6(a)

(b) Again, note that the numerator $p(x) = x^2 - x - 6$ and denominator $q(x) = x - 5$ of g have no common factors. Also, f is neither odd nor even. From $f(0) = \frac{6}{5}$ we get the y-intercept $\left(0, \frac{6}{5}\right)$. From $p(x) = x^2 - x - 6 = 0$ or $(x + 2)(x - 3) = 0$ we see that -2 and 3 are zeros of $p(x)$. The x-intercepts are $(-2, 0)$ and $(3, 0)$. The zero of $q(x) = x - 5$ is obviously 5 so that the line $x = 5$ is a vertical asymptote. Finally, from the fact that the degree of $p(x) = x^2 - x - 6$ (which is 2) *is exactly one greater than* the degree of $q(x) = x - 5$ (which is 1), the graph of $f(x)$ has a slant asymptote. To find it, we divide $p(x)$ by $q(x)$. By either long division or synthetic division, the result

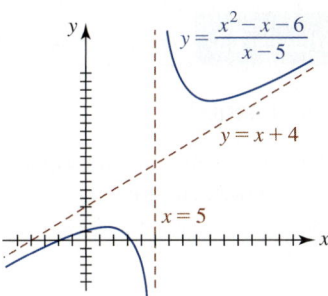

FIGURE 1.3.18   Graph of function in Example 6(b)

$$y = x + 4 \text{ is the slant asymptote}$$
$$\downarrow$$
$$\frac{x^2 - x - 6}{x - 5} = x + 4 + \frac{14}{x - 5}$$

shows that the slant asymptote is $y = x + 4$. The graph consists of two branches, one to the left of the line $x = 5$ and one to the right of the line $x = 5$. See FIGURE 1.3.18. ∎

■ **Postscript—Graph with a Hole**   We assumed throughout the discussion of asymptotes that the polynomial functions $p(x)$ and $q(x)$ in (14) had no common factors. We know that if $q(a) = 0$ ◄ and $p(x)$ and $q(x)$ have no common factors, then the line $x = a$ is necessarily a vertical asymptote for the graph of f. However, when $p(a) = 0$ *and* $q(a) = 0$, then $x = a$ *may not* be an asymptote; there may simply be a **hole** in the graph.

If $p(a) = 0$ and $q(a) = 0$, then by the Factor Theorem of algebra $x - a$ is a factor of both p and q.

EXAMPLE 7   Graph with a Hole

Graph the function $f(x) = \dfrac{x^2 - 2x - 3}{x^2 - 1}$.

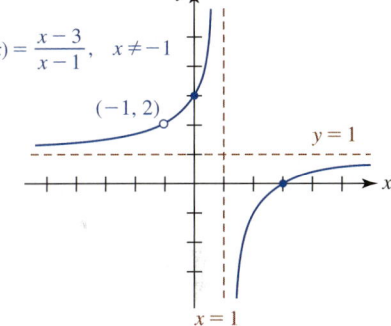

FIGURE 1.3.19   Graph of function in Example 7

Solution   Although the zeros of $x^2 - 1 = 0$ are ± 1, only $x = 1$ is a vertical asymptote. Note that the numerator $p(x)$ and denominator $q(x)$ have the common factor $x + 1$ which we cancel provided $x \neq -1$:

$$\text{equality is true for } x \neq -1$$
$$\downarrow$$
$$f(x) = \frac{(x + 1)(x - 3)}{(x + 1)(x - 1)} = \frac{x - 3}{x - 1}. \tag{15}$$

We graph $y = \dfrac{x - 3}{x - 1}$, $x \neq -1$, by observing that the y-intercept is $(0, 3)$, an x-intercept is $(3, 0)$, a vertical asymptote is $x = 1$, and a horizontal asymptote is $y = 1$. Although $x = -1$ is not a vertical asymptote, we represent the fact that f is not defined at that number by drawing an open circle or hole in the graph at the point corresponding to ◄ $(-1, 2)$. See FIGURE 1.3.19. ∎

The y-coordinate of the hole is the value of the reduced fraction (15) at $x = -1$.

$f(x)$ **NOTES FROM THE CLASSROOM**
....

In the last two sections we worked principally with polynomial functions. Polynomial functions are the fundamental building blocks of a class known as **algebraic functions**. In this section we saw that a rational function is the quotient of two polynomial functions. In general, an algebraic function f involves a finite number of additions, subtractions, multiplications, divisions, and roots of polynomial functions. Thus

$$y = 2x^2 - 5x, \quad y = \sqrt[3]{x^2}, \quad y = x^4 + \sqrt{x^2 + 5}, \quad \text{and} \quad y = \frac{\sqrt{x}}{x^3 - 2x^2 + 7}$$

are algebraic functions. Starting with the next section we consider functions that belong to a different class known as **transcendental functions**. A transcendental function f is defined to be one that is *not* algebraic. The six trigonometric functions and the exponential and logarithmic functions are examples of transcendental functions.

Exercises 1.3 Answers to selected odd-numbered problems begin on page ANS-3.

≡ Fundamentals

In Problems 1–6, find an equation of the line through $(1, 2)$ with the indicated slope.

1. $\dfrac{2}{3}$
2. $\dfrac{1}{10}$

3. 0
4. -2

5. -1
6. undefined

In Problems 7–10, find the slope and the x- and y-intercepts of the given line. Graph the line.

7. $3x - 4y + 12 = 0$
8. $\dfrac{1}{2}x - 3y = 3$

9. $2x - 3y = 9$
10. $-4x - 2y + 6 = 0$

In Problems 11–16, find an equation of the line that satisfies the given conditions.

11. through $(2, 3)$ and $(6, -5)$

12. through $(5, -6)$ and $(4, 0)$

13. through $(-2, 4)$ parallel to $3x + y - 5 = 0$

14. through $(5, -7)$ parallel to the y-axis

15. through $(2, 3)$ perpendicular to $x - 4y + 1 = 0$

16. through $(-5, -4)$ perpendicular to the line through $(1, 1)$ and $(3, 11)$

In Problems 17 and 18, find a linear function $f(x) = ax + b$ that satisfies both of the given conditions.

17. $f(-1) = 5, f(1) = 6$

18. $f(-1) = 1 + f(2), f(3) = 4f(1)$

In Problems 19 and 20, find an equation of the red line L shown in the given figure.

19.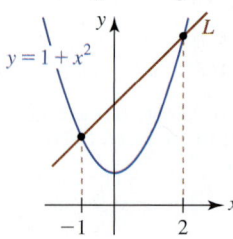
20.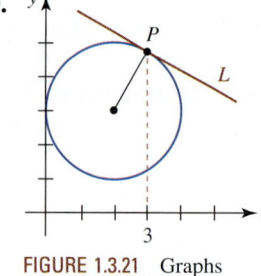

FIGURE 1.3.20 Graphs for Problem 19

FIGURE 1.3.21 Graphs for Problem 20

In Problems 21–26, consider the quadratic function f.
   (a) Find all intercepts of the graph of f.
   (b) Express the function f in standard form.
   (c) Find the vertex and axis of symmetry.
   (d) Sketch the graph of f.
   (e) What is the range of f?
   (f) On what interval is f increasing? Decreasing?

21. $f(x) = x(x + 5)$
22. $f(x) = -x^2 + 4x$

23. $f(x) = (3 - x)(x + 1)$
24. $f(x) = (x - 2)(x - 6)$

25. $f(x) = x^2 - 3x + 2$
26. $f(x) = -x^2 + 6x - 5$

In Problems 27–32, describe in words how the graph of the given function can be obtained from the graph of $y = x^2$ by rigid or nonrigid transformations.

27. $f(x) = (x - 10)^2$
28. $f(x) = (x + 6)^2$

29. $f(x) = -\dfrac{1}{3}(x + 4)^2 + 9$
30. $f(x) = 10(x - 2)^2 - 1$

31. $f(x) = (-x - 6)^2 - 4$
32. $f(x) = -(1 - x)^2 + 1$

In Problems 33–42, proceed as in Example 5 and sketch the graph of the given polynomial function f.

33. $f(x) = x^3 - 4x$
34. $f(x) = 9x - x^3$

35. $f(x) = -x^3 + x^2 + 6x$
36. $f(x) = x^3 + 7x^2 + 12x$

37. $f(x) = (x + 1)(x - 2)(x - 4)$

38. $f(x) = (2 - x)(x + 2)(x + 1)$

39. $f(x) = x^4 - 4x^3 + 3x^2$
40. $f(x) = x^2(x - 2)^2$

41. $f(x) = -x^4 + 2x^2 - 1$
42. $f(x) = x^5 - 4x^3$

In Problems 43–48, match the given graph with one of the polynomial functions in (a)–(f).

   (a) $f(x) = x^2(x - 1)^2$
   (b) $f(x) = -x^3(x - 1)$
   (c) $f(x) = x^3(x - 1)^3$
   (d) $f(x) = -x(x - 1)^3$
   (e) $f(x) = -x^2(x - 1)$
   (f) $f(x) = x^3(x - 1)^2$

43.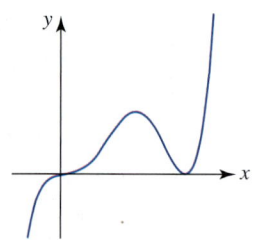

FIGURE 1.3.22 Graph for Problem 43

44.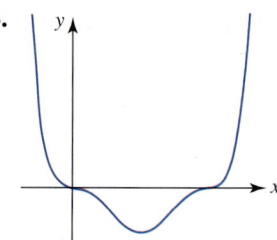

FIGURE 1.3.23 Graph for Problem 44

45.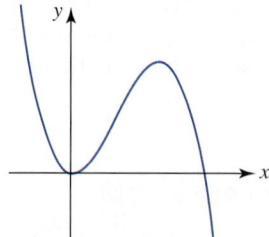

FIGURE 1.3.24 Graph for Problem 45

46.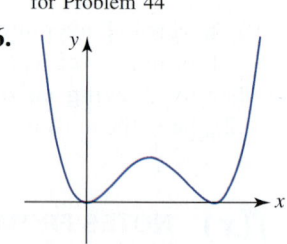

FIGURE 1.3.25 Graph for Problem 46

47.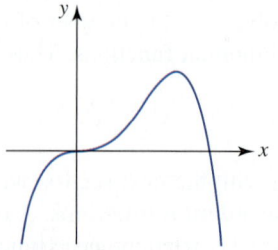

FIGURE 1.3.26 Graph for Problem 47

48.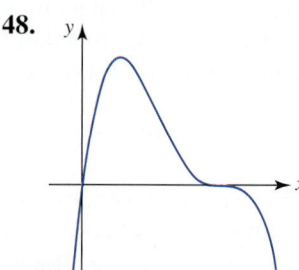

FIGURE 1.3.27 Graph for Problem 48

In Problems 49–62, find all asymptotes for the graph of the given rational function. Find x- and y-intercepts of the graph. Sketch the graph f.

49. $f(x) = \dfrac{4x - 9}{2x + 3}$

50. $f(x) = \dfrac{2x + 4}{x - 2}$

51. $f(x) = \dfrac{1}{(x - 1)^2}$

52. $f(x) = \dfrac{4}{(x + 2)^3}$

53. $f(x) = \dfrac{x}{x^2 - 1}$

54. $f(x) = \dfrac{x^2}{x^2 - 4}$

55. $f(x) = \dfrac{1 - x^2}{x^2}$

56. $f(x) = \dfrac{x(x - 5)}{x^2 - 9}$

57. $f(x) = \dfrac{x^2 - 9}{x}$

58. $f(x) = \dfrac{x^2 - 3x - 10}{x}$

59. $f(x) = \dfrac{x^2}{x + 2}$

60. $f(x) = \dfrac{x^2 - 2x}{x + 2}$

61. $f(x) = \dfrac{x^2 - 2x - 3}{x - 1}$

62. $f(x) = \dfrac{-(x - 1)^2}{x + 2}$

63. Determine whether the numbers -1 and 2 are in the range of the rational function $f(x) = \dfrac{2x - 1}{x + 4}$.

64. Determine the points where the graph of $f(x) = \dfrac{(x - 3)^2}{x^2 - 5x}$ crosses its horizontal asymptote.

≡ Mathematical Models

65. Related Temperatures The functional relationship between degrees Celsius T_C and degrees Fahrenheit T_F is linear. Express T_F as a function of T_C if $(0°C, 32°F)$ and $(60°C, 140°F)$ are on the graph of T_F. Show that $100°C$ is equivalent to the Fahrenheit boiling point $212°F$. See FIGURE 1.3.28.

66. Related Temperatures The functional relationship between degrees Celsius T_C and kelvin units T_K is linear. Express T_K as a function of T_C given that $(0°C, 273\text{ K})$ and $(27°C, 300\text{ K})$ are on the graph of T_K. Express the boiling point $100°C$ in kelvin units. Absolute zero is defined as 0 K. What is 0 K in degrees Celsius? Express T_K as a linear function of T_F. What is 0 K in degrees Fahrenheit? See Figure 1.3.28.

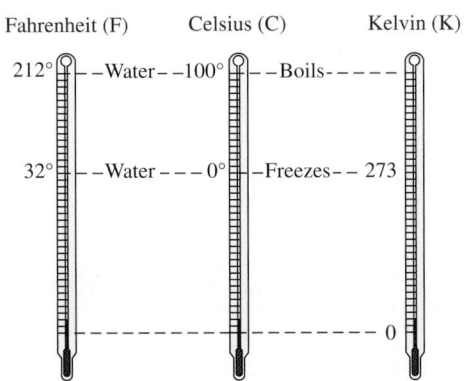

FIGURE 1.3.28   Thermometers in Problems 65 and 66

67. Simple Interest In simple interest, the amount A accrued over time is the linear function $A = P + Prt$, where P is the principal, t is measured in years, and r is the annual interest rate (expressed as a decimal). Compute A after 20 years if the principal is $P = 1000$, and the annual interest rate is 3.4%. At what time is $A = 2200$?

68. Linear Depreciation Straight line, or linear depreciation, consists of an item losing all its initial worth of A dollars over a period of n years by an amount A/n each year. If an item costing \$20,000 when new is depreciated linearly over 25 years, determine a linear function giving its value V after x years, where $0 \le x \le 25$. What is the value of the item after 10 years?

69. A ball is thrown upward from ground level with an initial velocity of 96 ft/s. The height of the ball from the ground is given by the quadratic function $s(t) = -16t^2 + 96t$. At what times is the ball on the ground? Graph s over the time interval for which $s(t) \ge 0$.

70. In Problem 69, at what times is the ball 80 ft above the ground? How high does the ball go?

≡ Think About It

71. Consider the linear function $f(x) = \frac{5}{2}x - 4$. If x is changed by 1 unit, how many units will y change? If x is changed by 2 units? If x is changed by n (n a positive integer) units?

72. Consider the interval $[x_1, x_2]$ and the linear function $f(x) = ax + b$, $a \ne 0$. Show that

$$f\left(\frac{x_1 + x_2}{2}\right) = \frac{f(x_1) + f(x_2)}{2},$$

and interpret this result geometrically for $a > 0$.

73. How would you find an equation of the line that is the perpendicular bisector of the line segment through $\left(\frac{1}{2}, 10\right)$ and $\left(\frac{3}{2}, 4\right)$?

74. Using only the concepts of this section, how would you prove or disprove that the triangle with vertices $(2, 3)$, $(-1, -3)$, and $(4, 2)$ is a right triangle?

1.4 Transcendental Functions

■ **Introduction** In the first two sections of this chapter we examined various properties and graphs of **algebraic functions**. For the next three sections we examine **transcendental functions**. Basically, a transcendental function *f* is one that is not algebraic. A transcendental function could be as simple as the power function $y = x^\pi$, where the power is an irrational number, but the familiar transcendental functions from precalculus mathematics are the trigonometric functions, the inverse trigonometric functions, and the exponential and logarithmic functions. In this section we review the six trigonometric functions and their graphs. In Section 1.5 we consider the inverse trigonometric functions and in Section 1.6 we review exponential and logarithmic functions.

For a review of the basics of unit-circle and right-triangle trigonometry see the *Student Resource Manual*. Also see the *Resource Pages* at the end of the text.

■ **Graphs of Sine and Cosine** Recall from precalculus mathematics that the trigonometric sine and cosine functions have **period 2π**:

$$\sin(x + 2\pi) = \sin x \qquad \text{and} \qquad \cos(x + 2\pi) = \cos x. \qquad (1)$$

The graph of *any* periodic function over an interval of length equal to its period is said to be one **cycle** of its graph. The graph of a periodic function is easily obtained by repeatedly drawing one cycle of its graph. FIGURE 1.4.1 shows one cycle of the graph of $f(x) = \sin x$ (in red); the graph of *f* on, say, the intervals $[-2\pi, 0]$ and $[2\pi, 4\pi]$ (in blue) is exactly the same as the graph on $[0, 2\pi]$. Because $f(-x) = \sin(-x) = -\sin x = -f(x)$ the sine function is an odd function and its graph is symmetric with respect to the origin.

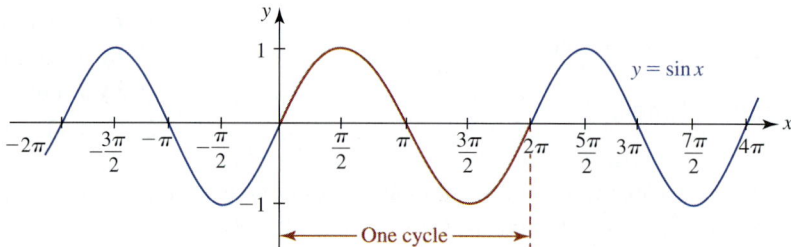

FIGURE 1.4.1 Graph of $y = \sin x$

FIGURE 1.4.2 shows one cycle (in red) of $g(x) = \cos x$ on $[0, 2\pi]$ along with the extension of that cycle (in blue) to the adjacent intervals $[-2\pi, 0]$ and $[2\pi, 4\pi]$. In contrast to the graph of $f(x) = \sin x$ where $f(0) = f(2\pi) = 0$, for the cosine function we have $g(0) = g(2\pi) = 1$. The cosine function is an even function: $g(-x) = \cos(-x) = \cos x = g(x)$, and so we see in Figure 1.4.2 its graph is symmetric with respect to the *y*-axis.

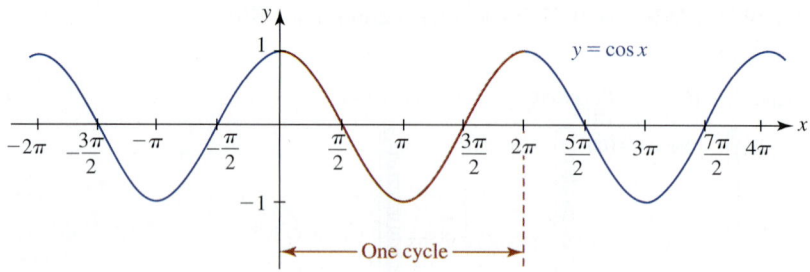

FIGURE 1.4.2 Graph of $y = \cos x$

The sine and cosine functions are defined for all real numbers *x*. Also, it is apparent in Figures 1.4.1 and 1.4.2 that

$$-1 \le \sin x \le 1 \qquad \text{and} \qquad -1 \le \cos x \le 1, \qquad (2)$$

or equivalently, $|\sin x| \le 1$ and $|\cos x| \le 1$. In other words,

- the domain of $\sin x$ and $\cos x$ is $(-\infty, \infty)$, and the range of $\sin x$ and $\cos x$ is $[-1, 1]$.

■ **Intercepts** In this and subsequent courses in mathematics it is important that you know the x-coordinates of the x-intercepts of the sine and cosine graphs, in other words, the zeros of $f(x) = \sin x$ and $g(x) = \cos x$. From the sine graph in Figure 1.4.1 we see that the zeros of the sine function, or the numbers for which $\sin x = 0$, are $x = 0, \pm\pi, \pm 2\pi, \pm 3\pi, \ldots$. These numbers are integer multiples of π. From the cosine graph in Figure 1.4.2 we see that $\cos x = 0$ when $x = \pm\pi/2, \pm 3\pi/2, \pm 5\pi/2, \ldots$. These numbers are odd-integer multiples of $\pi/2$.

If n represents an integer, then $2n + 1$ is an odd integer. Therefore the **zeros** of $f(x) = \sin x$ and $g(x) = \cos x$ can be written in a compact form:

- $\sin x = 0$ for $x = n\pi$, n an integer, $\qquad\qquad\qquad\qquad\qquad\qquad$ (3)

- $\cos x = 0$ for $x = (2n + 1)\dfrac{\pi}{2}$, n an integer. $\qquad\qquad\qquad\qquad$ (4)

Additional important numerical values of the sine and cosine functions on the interval $[0, \pi]$ are given in the table that follows.

x	0	$\dfrac{\pi}{6}$	$\dfrac{\pi}{4}$	$\dfrac{\pi}{3}$	$\dfrac{\pi}{2}$	$\dfrac{2\pi}{3}$	$\dfrac{3\pi}{4}$	$\dfrac{5\pi}{6}$	π	
$\sin x$	0	$\dfrac{1}{2}$	$\dfrac{\sqrt{2}}{2}$	$\dfrac{\sqrt{3}}{2}$	1	$\dfrac{\sqrt{3}}{2}$	$\dfrac{\sqrt{2}}{2}$	$\dfrac{1}{2}$	0	(5)
$\cos x$	1	$\dfrac{\sqrt{3}}{2}$	$\dfrac{\sqrt{2}}{2}$	$\dfrac{1}{2}$	0	$-\dfrac{1}{2}$	$-\dfrac{\sqrt{2}}{2}$	$-\dfrac{\sqrt{3}}{2}$	-1	

You should be able to discern the values $\sin x$ and $\cos x$ on $[\pi, 2\pi]$ from this table using the concept of the unit circle and a reference angle. Of course, outside the interval $[0, 2\pi]$ we can determine corresponding function values using periodicity.

■ **Other Trigonometric Functions** Four additional trigonometric functions are defined in terms of quotients or reciprocals of the sine and cosine functions. The **tangent**, **cotangent**, **secant**, and **cosecant** functions are defined, respectively, as

$$\tan x = \frac{\sin x}{\cos x}, \qquad \cot x = \frac{\cos x}{\sin x}, \qquad\qquad\qquad (6)$$

$$\sec x = \frac{1}{\cos x}, \qquad \csc x = \frac{1}{\sin x}. \qquad\qquad\qquad (7)$$

The domain of each function in (6) and (7) is the set of real numbers except those numbers for which the denominator is zero. From (4) we see

- the domain of $\tan x$ and of $\sec x$ is $\{x \mid x \neq (2n + 1)\pi/2, n = 0, \pm 1, \pm 2, \ldots\}$.

Similarly, from (3) it follows that

- the domain of $\cot x$ and of $\csc x$ is $\{x \mid x \neq n\pi, n = 0, \pm 1, \pm 2, \ldots\}$.

Moreover, from (2)

$$|\sec x| = \left|\frac{1}{\cos x}\right| = \frac{1}{|\cos x|} \geq 1 \qquad\qquad\qquad (8)$$

and

$$|\csc x| = \left|\frac{1}{\sin x}\right| = \frac{1}{|\sin x|} \geq 1. \qquad\qquad\qquad (9)$$

Recall that an absolute-value inequality such as (8) means $\sec x \geq 1$ *or* $\sec x \leq -1$. Hence the range of the secant and the cosecant functions is $(-\infty, -1] \cup [1, \infty)$. The tangent and cotangent functions have the same range: $(-\infty, \infty)$. Using (5) we can determine some numerical values of $\tan x, \cot x, \sec x$, and $\csc x$. For example,

$$\tan\frac{2\pi}{3} = \frac{\sin(2\pi/3)}{\cos(2\pi/3)} = \frac{\sqrt{3}/2}{-1/2} = -\sqrt{3}.$$

▌ **Graphs** The numbers that make the denominators of $\tan x$, $\cot x$, $\sec x$, and $\csc x$ equal to zero correspond to vertical asymptotes of their graphs. In view of (4), the vertical asymptotes of the graphs of $y = \tan x$ and $y = \sec x$ are $x = \pm\pi/2, \pm 3\pi/2, \pm 5\pi/2, \ldots$. On the other hand, from (3), the vertical asymptotes for the graphs of $y = \cot x$ and $y = \csc x$ are $x = 0$, $\pm\pi, \pm 2\pi, \pm 3\pi, \ldots$. These asymptotes are the red dashed lines in FIGURES 1.4.3–1.4.6.

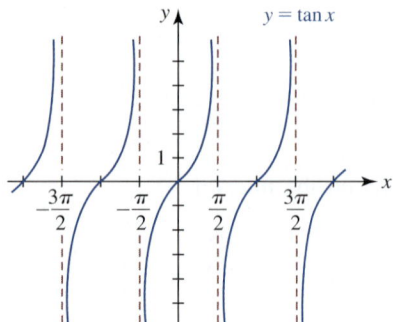

FIGURE 1.4.3 Graph of $y = \tan x$

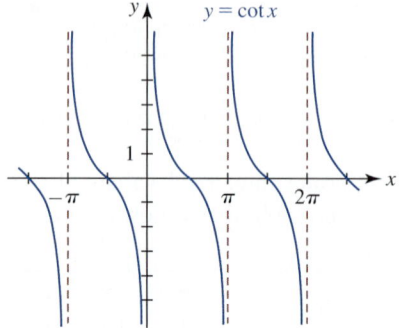

FIGURE 1.4.4 Graph of $y = \cot x$

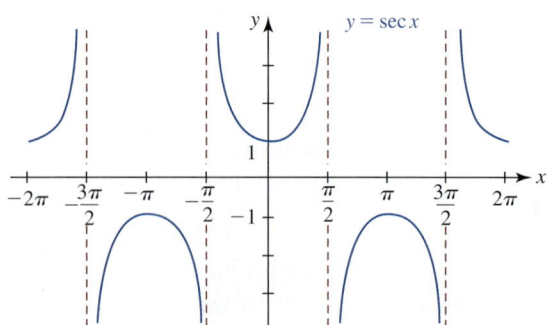

FIGURE 1.4.5 Graph of $y = \sec x$

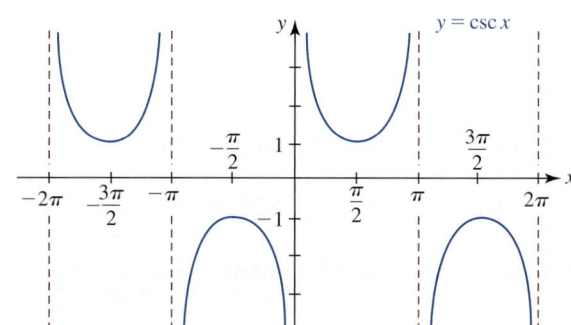

FIGURE 1.4.6 Graph of $y = \csc x$

Because the sine and cosine functions are 2π periodic, $\sec x$ and $\csc x$ are also 2π periodic. But it should be obvious from Figures 1.4.3 and 1.4.4 that tangent and cotangent are π periodic:

$$\tan(x + \pi) = \tan x \quad \text{and} \quad \cot(x + \pi) = \cot x. \tag{10}$$

Also, $\tan x$, $\cot x$, and $\csc x$ are odd functions; $\sec x$ is an even function.

▌ **Transformation and Graphs** We can obtain variations of the graphs of the trigonometric functions through rigid and nonrigid transformations. Graphs of functions of the form

$$y = D + A\sin(Bx + C) \quad \text{or} \quad y = D + A\cos(Bx + C), \tag{11}$$

where $A, B > 0, C,$ and D are real constants, represent shifts, compressions, and stretches of the basic sine and cosine graphs. For example,

$$\underset{\substack{\uparrow \\ \text{vertical shift}}}{\quad} \underset{\substack{\uparrow \\ \text{vertical stretch/compression/reflection}}}{\quad}$$
$$y = D + A\sin(Bx + C).$$
$$\underset{\substack{\downarrow \\ \text{horizontal stretch/compression} \\ \text{by changing period}}}{\quad} \underset{\substack{\downarrow \\ \text{horizontal shift}}}{\quad}$$

The number $|A|$ is called the **amplitude** of the functions or of their graphs. The amplitude of the basic functions $y = \sin x$ and $y = \cos x$ is $|A| = 1$. The **period** of each function in (11) is $2\pi/B, B > 0$, and the portion of the graph of each function in (11) over the interval $[0, 2\pi/B]$ is called one **cycle**.

EXAMPLE 1 Periods

(a) The period of $y = \sin 2x$ is $2\pi/2 = \pi$, and therefore one cycle of the graph is completed on the interval $[0, \pi]$.

(b) Before determining the period of $\sin(-\frac{1}{2}x)$ we must first rewrite the function as $\sin(-\frac{1}{2}x) = -\sin(\frac{1}{2}x)$ (the sine is an odd function). The period is now $2\pi/\frac{1}{2} = 4\pi$ and therefore one cycle of the graph is completed on the interval $[0, 4\pi]$. ∎

EXAMPLE 2 Vertically Transformed Graphs

Graph

(a) $y = -\dfrac{1}{2}\cos x$ **(b)** $y = 1 + 2\sin x$.

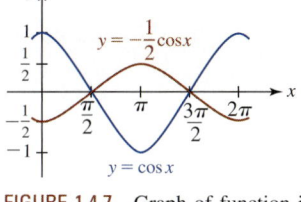

FIGURE 1.4.7 Graph of function in Example 2(a)

Solution

(a) The graph of $y = -\frac{1}{2}\cos x$ is the graph of $y = \cos x$ compressed vertically by a factor of 2 and the minus sign indicates that the graph is then reflected in the x-axis. With the identification $A = -\frac{1}{2}$ we see that the amplitude of the function is $|A| = |-\frac{1}{2}| = \frac{1}{2}$. The graph of $y = -\frac{1}{2}\cos x$ on the interval $[0, 2\pi]$ is shown in red in FIGURE 1.4.7.

(b) The graph of $y = 2\sin x$ is the graph of $y = \sin x$ stretched vertically by a factor of 2. The amplitude of the graph is $|A| = |2| = 2$. The graph of $y = 1 + 2\sin x$ is the graph of $y = 2\sin x$ shifted up 1 unit. See FIGURE 1.4.8. ∎

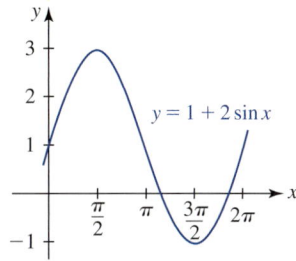

FIGURE 1.4.8 Graph of function in Example 2(b)

EXAMPLE 3 Horizontally Compressed Cosine Graph

Find the period of $y = \cos 4x$ and graph the function.

Solution With the identification that $B = 4$, we see that the period of $y = \cos 4x$ is $2\pi/4 = \pi/2$. We conclude that the graph of $y = \cos 4x$ is the graph of $y = \cos x$ compressed horizontally. To graph the function, we draw one cycle of the cosine graph with amplitude 1 on the interval $[0, \pi/2]$ and then use periodicity to extend the graph. FIGURE 1.4.9 shows four complete cycles of $y = \cos 4x$ (the basic cycle in red and the extended graph in blue) and one cycle of $y = \cos x$ (shown in green) on $[0, 2\pi]$. Notice that $y = \cos 4x$ attains its minimum at $x = \pi/4$ since $\cos 4(\pi/4) = \cos \pi = -1$ and its maximum at $x = \pi/2$ since $\cos 4(\pi/2) = \cos 2\pi = 1$. ∎

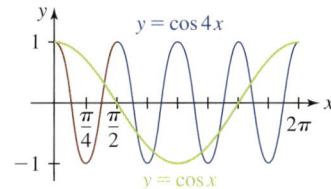

FIGURE 1.4.9 Graph of function in Example 3

From Section 1.2 we know that the graph of $y = \cos(x - \pi/2)$ is the basic cosine graph shifted to the right. In FIGURE 1.4.10 the graph of $y = \cos(x - \pi/2)$ (in red) on the interval $[0, 2\pi]$ is one cycle of $y = \cos x$ on the interval $[-\pi/2, 3\pi/2]$ (in blue) shifted horizontally $\pi/2$ units to the right. Similarly, the graphs of $y = \sin(x + \pi/2)$ and $y = \sin(x - \pi/2)$ are the basic sine graphs shifted $\pi/2$ units to the left and to the right, respectively. See FIGURE 1.4.11 and FIGURE 1.4.12.

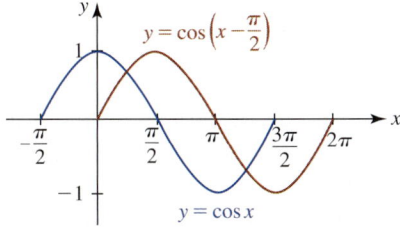

FIGURE 1.4.10 Horizontally shifted cosine graph

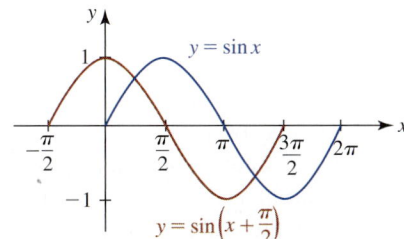

FIGURE 1.4.11 Horizontally shifted sine graph

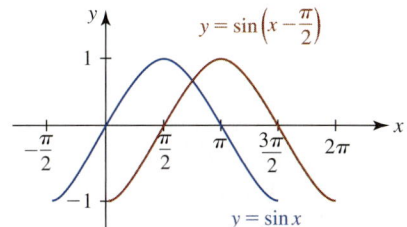

FIGURE 1.4.12 Horizontally shifted sine graph

By comparing the red graphs in Figures 1.4.10–1.4.12 with the graphs in Figures 1.4.1 and 1.4.2 we see that

- the cosine graph shifted $\pi/2$ units to the right is the sine graph,
- the sine graph shifted $\pi/2$ units to the left is the cosine graph, and
- the sine graph shifted $\pi/2$ units to the right is the cosine graph reflected in the x-axis.

In other words, we have graphically verified the identities

$$\cos\left(x - \frac{\pi}{2}\right) = \sin x, \qquad \sin\left(x + \frac{\pi}{2}\right) = \cos x, \qquad \text{and} \qquad \sin\left(x - \frac{\pi}{2}\right) = -\cos x. \quad (12)$$

Suppose $f(x) = A \sin Bx$, then

$$f\left(x + \frac{C}{B}\right) = A \sin B\left(x + \frac{C}{B}\right) = A \sin(Bx + C). \quad (13)$$

The result in (13) shows that the graph of $y = A \sin(Bx + C)$ can be obtained by shifting the graph of $f(x) = A \sin Bx$ horizontally a distance $|C|/B$. If $C < 0$, the shift is to the right, whereas if $C > 0$, the shift is to the left. The number $|C|/B$ is called the **phase shift** of the graphs of the functions in (3).

EXAMPLE 4   Horizontally Shifted Cosine Graph

The graph of $y = 10 \cos 4x$ is shifted $\pi/12$ units to the right. Find its equation.

Solution   By writing $f(x) = 10 \cos 4x$ and using (13), we find

$$f\left(x - \frac{\pi}{12}\right) = 10 \cos 4\left(x - \frac{\pi}{12}\right) \qquad \text{or} \qquad y = 10 \cos\left(4x - \frac{\pi}{3}\right).$$

In the last equation we would identify $C = -\pi/3$. The phase shift is $\pi/12$.   ∎

Note: As a practical matter the phase shift for either $y = A \sin(Bx + C)$ or $y = A \cos(Bx + C)$ can be obtained by factoring the number B from $Bx + C$. For example,

$$y = A \sin(Bx + C) = A \sin B\left(x + \frac{C}{B}\right).$$

EXAMPLE 5   Horizontally Shifted Graphs

Graph

    (a) $y = 3 \sin(2x - \pi/3)$ **(b)** $y = 2 \cos(\pi x + \pi)$.

Solution

(a) For purposes of comparison we will first graph $y = 3 \sin 2x$. The amplitude of $y = 3 \sin 2x$ is $|A| = 3$ and its period is $2\pi/2 = \pi$. Thus one cycle of $y = 3 \sin 2x$ is completed on the interval $[0, \pi]$. Then we extend this graph to the adjacent interval $[\pi, 2\pi]$ as shown in blue in FIGURE 1.4.13. Next, we rewrite $y = 3 \sin(2x - \pi/3)$ by factoring 2 from $2x - \pi/3$:

$$y = 3 \sin\left(2x - \frac{\pi}{3}\right) = 3 \sin 2\left(x - \frac{\pi}{6}\right).$$

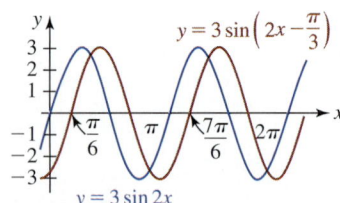

$y = 3 \sin\left(2x - \frac{\pi}{3}\right)$

$y = 3 \sin 2x$

FIGURE 1.4.13   Graph of function in Example 5(a)

From the form of the last expression we see that the phase shift is $\pi/6$. The graph of the given function, shown in red in Figure 1.4.13, is obtained by shifting the graph of $y = 3 \sin 2x$ (in blue) to the right $\pi/6$ units.

(b) The amplitude of $y = 2 \cos \pi x$ is $|A| = 2$ and the period is $2\pi/\pi = 2$. Thus one cycle of $y = 2 \cos \pi x$ is completed on the interval $[0, 2]$. In FIGURE 1.4.14 two cycles of the graph of $y = 2 \cos \pi x$ (in blue) are shown. The x-intercepts of this graph correspond to the values of x for which $\cos \pi x = 0$. By (4), this implies $\pi x = (2n + 1)\pi/2$ or $x = (2n + 1)/2$, n an integer. In other words, for $n = 0, -1, 1, -2, 2, -3, \ldots$ we get $x = \pm\frac{1}{2}, \pm\frac{3}{2}, \pm\frac{5}{2}$, and so on. Now by rewriting the given function as

$$y = 2 \cos \pi(x + 1)$$

we see that the phase shift is 1. The graph of $y = 2 \cos(\pi x + \pi)$ shown in red in Figure 1.4.14, is obtained by shifting the graph of $y = 2 \cos \pi x$ (in blue) to the left 1 unit. This means that the x-intercepts are the same for both graphs.   ∎

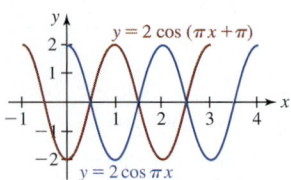

$y = 2 \cos(\pi x + \pi)$

$y = 2 \cos \pi x$

FIGURE 1.4.14   Graph of function in Example 5(b)

In applied mathematics, trigonometric functions serve as mathematical models for many periodic phenomena.

EXAMPLE 6   Alternating Current

A mathematical model for the current I (in amperes) in a wire of an alternating-current circuit is given by $I(t) = 30 \sin 120\pi t$, where t is time measured in seconds. Sketch one cycle of the graph. What is the maximum value of the current?

Solution   The graph has amplitude 30 and period $2\pi/120\pi = \frac{1}{60}$. Therefore, we sketch one cycle of the basic sine curve on the interval $\left[0, \frac{1}{60}\right]$, as shown in FIGURE 1.4.15. From the figure it is evident that the maximum value of the current is $I = 30$ amperes and occurs in the interval $\left[0, \frac{1}{60}\right]$ at $t = \frac{1}{240}$ since

$$I\left(\frac{1}{240}\right) = 30\sin\left(120\pi \cdot \frac{1}{240}\right) = 30 \sin \frac{\pi}{2} = 30.$$ ∎

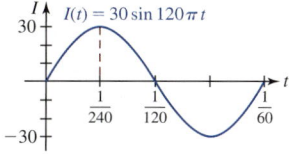

FIGURE 1.4.15   Graph of current in Example 6 shows that there are 60 cycles in 1 second

▮ **For Future Reference**   Trigonometric identities are used throughout calculus, especially in the study of integral calculus. For convenience of reference we list next some identities that are of particular importance.

Pythagorean Identities

$$\sin^2 x + \cos^2 x = 1 \qquad (14)$$

$$1 + \tan^2 x = \sec^2 x \qquad (15)$$

$$1 + \cot^2 x = \csc^2 x \qquad (16)$$

Sum and Difference Formulas

$$\sin(x_1 \pm x_2) = \sin x_1 \cos x_2 \pm \cos x_1 \sin x_2 \qquad (17)$$

$$\cos(x_1 \pm x_2) = \cos x_1 \cos x_2 \mp \sin x_1 \sin x_2 \qquad (18)$$

Double-Angle Formulas

$$\sin 2x = 2 \sin x \cos x \qquad (19)$$

$$\cos 2x = \cos^2 x - \sin^2 x \qquad (20)$$

Half-Angle Formulas

$$\sin^2 \frac{x}{2} = \frac{1}{2}(1 - \cos x) \qquad (21)$$

$$\cos^2 \frac{x}{2} = \frac{1}{2}(1 + \cos x) \qquad (22)$$

Additional identities can be found in the *Resource Pages* at the end of this text.

Exercises 1.4   Answers to selected odd-numbered problems begin on page ANS-5.

≡ Fundamentals

In Problems 1–6, use the techniques of shifting, stretching, compressing, and reflecting to sketch at least one cycle of the graph of the given function.

1. $y = \frac{1}{2} + \cos x$

2. $y = -1 + \cos x$

3. $y = 2 - \sin x$

4. $y = 3 + 3\sin x$

5. $y = -2 + 4\cos x$

6. $y = 1 - 2\sin x$

In Problems 7–14, find the amplitude and period of the given function. Sketch at least one cycle of the graph.

7. $y = 4 \sin \pi x$

8. $y = -5 \sin \frac{x}{2}$

9. $y = -3\cos 2\pi x$

10. $y = \frac{5}{2}\cos 4x$

11. $y = 2 - 4\sin x$

12. $y = 2 - 2\sin \pi x$

13. $y = 1 + \cos \frac{2x}{3}$

14. $y = -1 + \sin \frac{\pi x}{2}$

In Problems 15–18, the given figure shows one cycle of a sine or cosine graph. From the figure determine A and D and write an equation of the form $y = D + A\sin x$ or $y = D + A\cos x$ for the graph.

15.

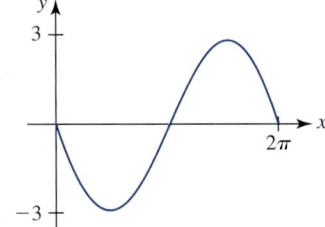

FIGURE 1.4.16 Graph for Problem 15

16.

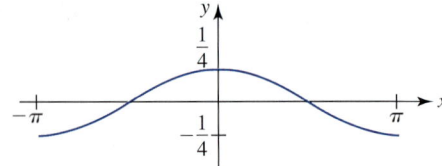

FIGURE 1.4.17 Graph for Problem 16

17.

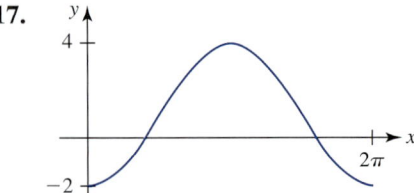

FIGURE 1.4.18 Graph for Problem 17

18.

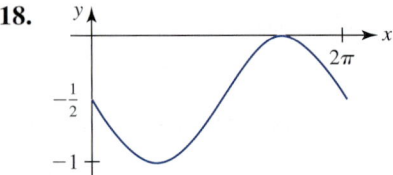

FIGURE 1.4.19 Graph for Problem 18

In Problems 19–24, the given figure shows one cycle of a sine or cosine graph. From the figure determine A and B and write an equation of the form $y = A\sin Bx$ or $y = A\cos Bx$ for the graph.

19.

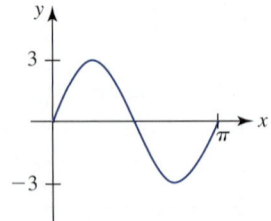

FIGURE 1.4.20 Graph for Problem 19

20.

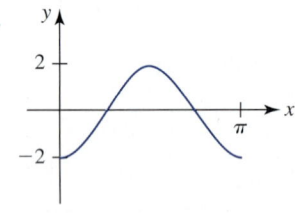

FIGURE 1.4.21 Graph for Problem 20

21.

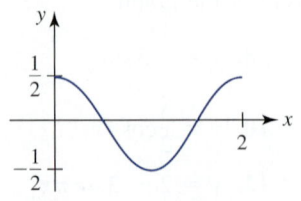

FIGURE 1.4.22 Graph for Problem 21

22.

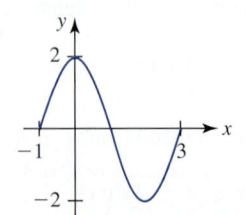

FIGURE 1.4.23 Graph for Problem 22

23.

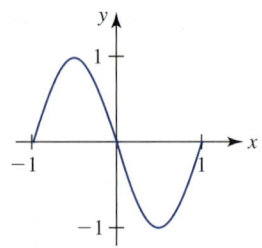

FIGURE 1.4.24 Graph for Problem 23

24.

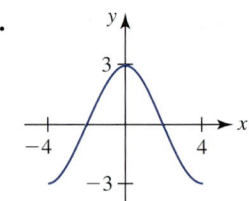

FIGURE 1.4.25 Graph for Problem 24

In Problems 25–34, find the amplitude, period, and phase shift of the given function. Sketch at least one cycle of the graph.

25. $y = \sin\left(x - \dfrac{\pi}{6}\right)$

26. $y = \sin\left(3x - \dfrac{\pi}{4}\right)$

27. $y = \cos\left(x + \dfrac{\pi}{4}\right)$

28. $y = -2\cos\left(2x - \dfrac{\pi}{6}\right)$

29. $y = 4\cos\left(2x - \dfrac{3\pi}{2}\right)$

30. $y = 3\sin\left(2x + \dfrac{\pi}{4}\right)$

31. $y = 3\sin\left(\dfrac{x}{2} - \dfrac{\pi}{3}\right)$

32. $y = -\cos\left(\dfrac{x}{2} - \pi\right)$

33. $y = -4\sin\left(\dfrac{\pi}{3}x - \dfrac{\pi}{3}\right)$

34. $y = 2\cos\left(-2\pi x - \dfrac{4\pi}{3}\right)$

In Problems 35 and 36, write an equation of the function whose graph is described in words.

35. The graph of $y = \sin \pi x$ is stretched vertically upward by a factor of 5 and is shifted $\frac{1}{2}$ unit to the right.

36. The graph of $y = 4\cos\dfrac{x}{2}$ is shifted downward 8 units and is shifted $2\pi/3$ units to the left.

In Problems 37 and 38, find the x-intercepts of the graph of the given function on the interval $[0, 2\pi]$. Then find all intercepts using periodicity.

37. $y = -1 + \sin x$

38. $y = 1 - 2\cos x$

In Problems 39–44, find the x-intercepts for the graph of the given function. Do not graph.

39. $y = \sin \pi x$

40. $y = -\cos 2x$

41. $y = 10\cos\dfrac{x}{2}$

42. $y = 3\sin(-5x)$

43. $y = \sin\left(x - \dfrac{\pi}{4}\right)$

44. $y = \cos(2x - \pi)$

In Problems 45–52, find the period, x-intercepts, and the vertical asymptotes of the given function. Sketch at least one cycle of the graph.

45. $y = \tan \pi x$

46. $y = \tan\dfrac{x}{2}$

47. $y = \cot 2x$

48. $y = -\cot\dfrac{\pi x}{3}$

49. $y = \tan\left(\dfrac{x}{2} - \dfrac{\pi}{4}\right)$

50. $y = \dfrac{1}{4}\cot\left(x - \dfrac{\pi}{2}\right)$

51. $y = -1 + \cot \pi x$

52. $y = \tan\left(x + \dfrac{5\pi}{6}\right)$

In Problems 53–56, find the period and the vertical asymptotes of the given function. Sketch at least one cycle of the graph.

53. $y = 3\csc \pi x$

54. $y = -2\csc\dfrac{x}{3}$

55. $y = \sec\left(3x - \dfrac{\pi}{2}\right)$

56. $y = \csc(4x + \pi)$

≡ **Mathematical Models**

57. Depth of Water The depth of water d at the entrance to a small harbor at time t is modeled by a function of the form

$$d(t) = D + A\sin B\left(t - \dfrac{\pi}{2}\right),$$

where A is one half the difference between the high- and low-tide depths, $2\pi/B, B > 0$, is the tidal period, and D is the average depth. Assume that the tidal period is 12 hours, the depth at high tide is 18 ft, and the depth at low tide is 6 ft. Sketch two cycles of the graph of d.

58. Fahrenheit Temperature Suppose that

$$T(t) = 50 + 10\sin\dfrac{\pi}{12}(t - 8), \quad 0 \le t \le 24$$

is a mathematical model of the Fahrenheit temperature at t hours after midnight on a certain day of the week.

(a) What is the temperature at 8 A.M.?
(b) At what time(s) does $T(t) = 60$?
(c) Sketch the graph of T.
(d) Find the maximum and minimum temperatures and the times at which they occur.

≡ **Calculator/CAS Problems**

59. Acceleration Due to Gravity Because of Earth's rotation, its shape is not spherical but bulges at the equator and is flattened at the poles. As a result, the acceleration due to gravity is not a constant 980 cm/s^2, but varies with latitude θ. Satellite studies have suggested that the acceleration due to gravity g is approximated by the mathematical model

$$g = 978.0309 + 5.18552\sin^2\theta - 0.00570\sin^2 2\theta.$$

Find g

(a) at the equator ($\theta = 0°$),
(b) at the north pole, and
(c) at 45° north latitude.

60. Putting the Shot The range of a shot put released from a height h above the ground with an initial velocity v_0 at an angle ϕ to the horizontal can be approximated by the mathematical model

$$R = \dfrac{v_0\cos\phi}{g}\left[v_0\sin\phi + \sqrt{v_0^2\sin^2\phi + 2gh}\,\right],$$

where g is the acceleration due to gravity. See FIGURE 1.4.26.

(a) If $v_0 = 13.7$ m/s, $\phi = 40°$, and $g = 9.81$ m/s^2, compare the ranges achieved for the release heights $h = 2.0$ m and $h = 2.4$ m.
(b) Explain why an increase in h yields an increase in the range R if the other parameters are held fixed.
(c) What does this imply about the advantage that height gives a shot-putter?

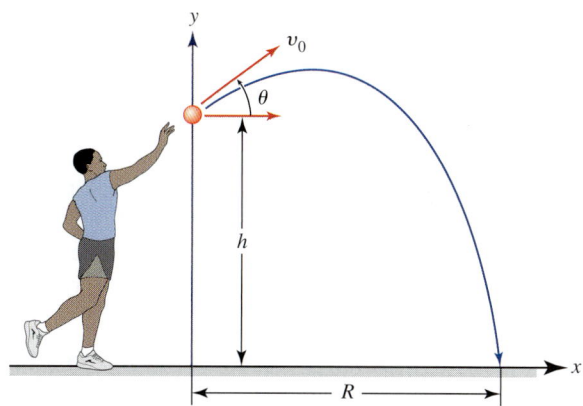

FIGURE 1.4.26 Projectile in Problem 60

≡ **Think About It**

61. The function $f(x) = \sin\frac{1}{2}x + \sin 2x$ is periodic. What is the period of f?

62. Discuss and then sketch the graphs of $y = |\sin x|$ and $y = |\cos x|$.

63. Discuss and then sketch the graphs of $y = |\sec x|$ and $y = |\csc x|$.

64. Can the given equation have any real-number solution x?

(a) $9\csc x = 1$ **(b)** $7 + 10\sec x = 0$
(c) $\sec x = -10.5$

In Problems 65 and 66, use the graphs of $y = \tan x$ and $y = \sec x$ to find numbers A and C for which the given equality is true.

65. $\cot x = A\tan(x + C)$ **66.** $\csc x = A\sec(x + C)$

1.5 Inverse Functions

▌ **Introduction** In Section 1.1 we saw that a function f is a rule of correspondence that assigns to each value x in its domain X, a single or unique value y in its range. This rule does not preclude having the same number y associated with several *different* values of x. For example, for $f(x) = -x^2 + 2x + 4$, the value $y = 4$ in the range of f occurs at either $x = 0$ or $x = 2$ in the domain of f. On the other hand, for the function $f(x) = 2x + 3$, the value $y = 4$ occurs only at

$x = \frac{1}{2}$. Indeed, for every value y in the range of $f(x) = 2x + 3$, there corresponds only one value of x in the domain. Functions of this last kind are given the special name **one-to-one**.

Definition 1.5.1   One-to-One Function

A function f is said to be **one-to-one** if each number in the range of f is associated with exactly one number in its domain X.

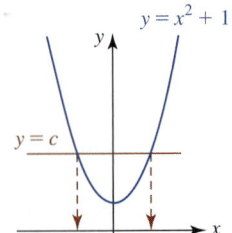

$y = x^2 + 1$

(a) Not one-to-one

■ **Horizontal Line Test**   Interpreted geometrically, Definition 1.5.1 means that a horizontal line (y = constant) can intersect the graph of a one-to-one function in at most one point. Furthermore, if *every* horizontal line that intersects the graph of a function does so in at most one point, then the function is necessarily one-to-one. A function is *not* one-to-one if *some* horizontal line intersects its graph more than once.

EXAMPLE 1   Horizontal Line Test

(a) The graph of the function $f(x) = x^2 + 1$ and a horizontal line $y = c$ intersecting the graph is shown in FIGURE 1.5.1(a). The figure clearly indicates that there are two numbers x_1 and x_2 in the domain of f for which $f(x_1) = f(x_2) = c$. Hence the function f is not one-to-one.

(b) Inspection of Figure 1.5.1(b) shows that for every horizontal line $y = c$ intersecting the graph of $f(x) = x^3$, there is only one number x_1 in the domain of f such that $f(x_1) = c$. The function f is one-to-one.   ■

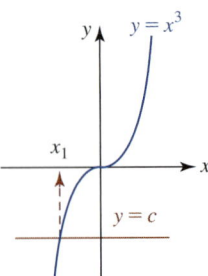

$y = x^3$

(b) One-to-one
FIGURE 1.5.1   Two types of functions in Example 1

■ **Inverse of a One-to-One Function**   Suppose f is a one-to-one function with domain X and range Y. Since every number y in Y corresponds to precisely one number x in X, the function f must actually determine a "reverse" function g whose domain is Y and range is X. As shown in FIGURE 1.5.2, f and g must satisfy

$$f(x) = y \qquad \text{and} \qquad g(y) = x. \tag{1}$$

The equations in (1) are actually the compositions of the functions f and g:

$$f(g(y)) = y \qquad \text{and} \qquad g(f(x)) = x. \tag{2}$$

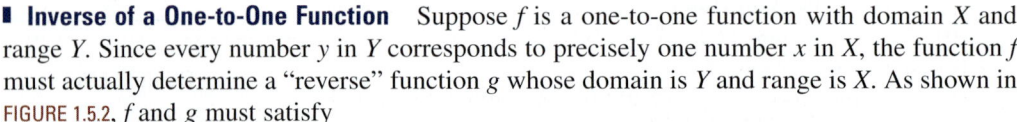

Domain of f Range of f

Range of g Domain of g
FIGURE 1.5.2   A function f and its inverse function g

The function g is called the **inverse** of f or the **inverse function** for f. Following convention that each domain element be denoted by the symbol x, the first equation in (2) is rewritten as $f(g(x)) = x$. We summarize the results given in (2).

Definition 1.5.2   Inverse Function

Let f be a one-to-one function with domain X and range Y. The **inverse** of f is the function g with domain Y and range X for which

$$f(g(x)) = x \text{ for every } x \text{ in } Y, \tag{3}$$

and

$$g(f(x)) = x \text{ for every } x \text{ in } X. \tag{4}$$

Of course, if a function f is not one-to-one, then it has no inverse function.

■ **Notation**   The inverse of a function f is usually written f^{-1} and is read "f inverse." This latter notation, although standard, is somewhat unfortunate. We hasten to point out that in the symbol $f^{-1}(x)$ the "-1" is *not* an exponent. In terms of the new notation, (3) and (4) become, respectively,

In (3) and (4), the symbol g is playing the part of the symbol f^{-1}. ▶

$$f(f^{-1}(x)) = x \qquad \text{and} \qquad f^{-1}(f(x)) = x. \tag{5}$$

■ **Properties**   Before we examine a method for finding the inverse of a one-to-one function f, let us list some important properties about f and its inverse f^{-1}.

Theorem 1.5.1   Properties of Inverse Functions

(i) The domain of f^{-1} = range of f.
(ii) The range of f^{-1} = domain of f.
(iii) An inverse function f^{-1} is one-to-one.
(iv) The inverse of f^{-1} is f.
(v) The inverse of f is unique.

■ **A Method for Finding f^{-1}**   If f^{-1} is the inverse of a one-to-one function $y = f(x)$, then from (1), $x = f^{-1}(y)$. Thus we need only do the following two things to find f^{-1}.

Guidelines for Finding the Inverse Function

Suppose $y = f(x)$ is a one-to-one function. Then to find f^{-1}:

- Solve $y = f(x)$ for the symbol x in terms of y (if possible). This gives $x = f^{-1}(y)$.
- Relabel the variable x as y and the variable y as x. This gives $y = f^{-1}(x)$.

Note: It is sometimes convenient to interchange the steps in the foregoing guidelines:

- Relabel x and y in the equation $y = f(x)$, and solve (if possible) $x = f(y)$ for y. This gives $y = f^{-1}(x)$.

EXAMPLE 2   Inverse of a Function

Find the inverse of $f(x) = x^3$.

Solution   In Example 1 we saw that this function was one-to-one. To begin, we rewrite the function as $y = x^3$. Solving for x then gives $x = y^{1/3}$. Next we relabel variables to obtain $y = x^{1/3}$. Thus $f^{-1}(x) = x^{1/3}$ or equivalently $f^{-1}(x) = \sqrt[3]{x}$. ■

Finding the inverse of a one-to-one function $y = f(x)$ is sometimes difficult and at times impossible. For example, FIGURE 1.5.3 suggests (and it can be shown) that the function $f(x) = x^3 + x + 3$ is one-to-one and so has an inverse f^{-1}. But solving the equation $y = x^3 + x + 3$ for x is difficult for everyone (including your instructor). Since f is a polynomial function its domain is $(-\infty, \infty)$ and, because its end behavior is that of $y = x^3$, the range of f is $(-\infty, \infty)$. Consequently the domain and range of f^{-1} are $(-\infty, \infty)$. Even though we do not know f^{-1} explicitly, it makes complete sense to talk about the values such as $f^{-1}(3)$ and $f^{-1}(5)$. In the case of $f^{-1}(3)$ note that $f(0) = 3$. This means that $f^{-1}(3) = 0$. Can you figure out the value of $f^{-1}(5)$?

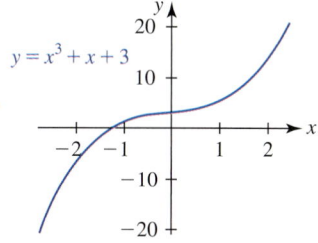

FIGURE 1.5.3   Graph suggests f is one-to-one

■ **Graphs of f and f^{-1}**   Suppose that (a, b) represents any point on the graph of a one-to-one function f. Then $f(a) = b$ and

$$f^{-1}(b) = f^{-1}(f(a)) = a$$

implies that (b, a) is a point on the graph of f^{-1}. As shown in FIGURE 1.5.4(a), the points (a, b) and (b, a) are reflections of each other in the line $y = x$. This means that the line $y = x$ is the perpendicular bisector of the line segment from (a, b) to (b, a). Because each point on one graph is the reflection of a corresponding point on the other graph, we see in Figure 1.5.4(b) that the graphs of f^{-1} and f are **reflections** of each other in the line $y = x$. We also say that the graphs of f^{-1} and f are **symmetric** with respect to the line $y = x$.

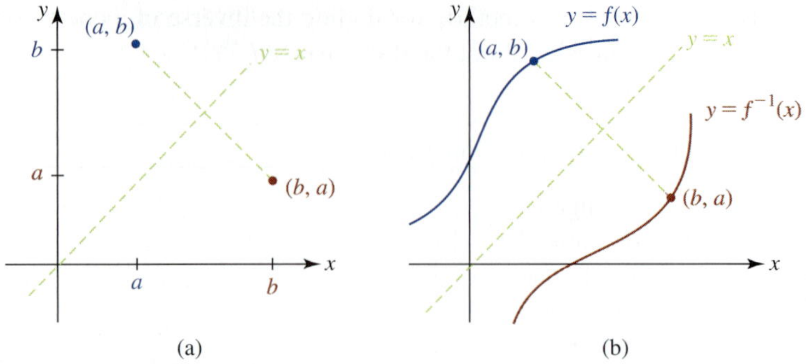

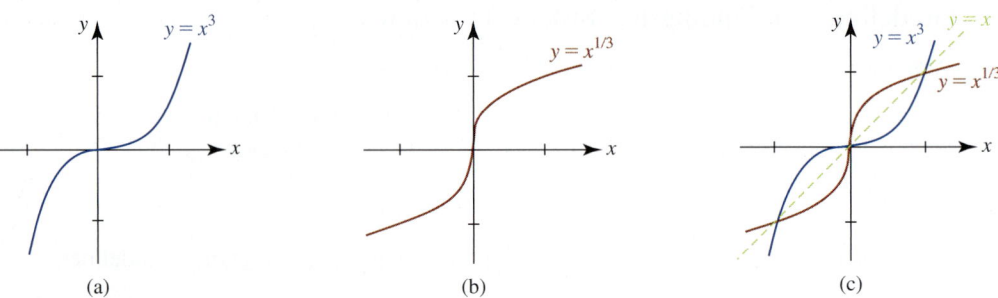

FIGURE 1.5.4 Graphs of f and f^{-1} are reflections in the line $y = x$

EXAMPLE 3   Graphs of f and f^{-1}

In Example 2 we saw that the inverse of $y = x^3$ is $y = x^{1/3}$. In FIGURES 1.5.5(a) and 1.5.5(b) we show the graphs of these functions; in Figure 1.5.5(c) the graphs are superimposed on the same coordinate system to illustrate that the graphs are reflections of each other in the line $y = x$.

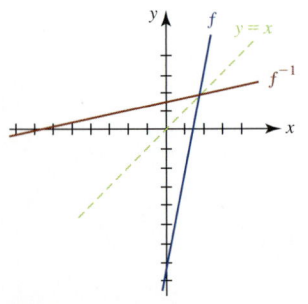

FIGURE 1.5.5   Graphs of f and f^{-1} in Example 3   ■

Every linear function $f(x) = ax + b$, $a \neq 0$, is one-to-one.

EXAMPLE 4   Inverse of a Function

Find the inverse of the linear function $f(x) = 5x - 7$.

Solution   Since the graph of $y = 5x - 7$ is a nonhorizontal line, it follows from the horizontal line test that f is a one-to-one function. To find f^{-1} solve $y = 5x - 7$ for x:

$$5x = y + 7 \qquad \text{implies} \qquad x = \frac{1}{5}y + \frac{7}{5}.$$

Relabeling variables in the last equation gives $y = \frac{1}{5}x + \frac{7}{5}$. Therefore $f^{-1}(x) = \frac{1}{5}x + \frac{7}{5}$. The graphs of f and f^{-1} are compared in FIGURE 1.5.6.   ■

FIGURE 1.5.6   Graphs of f and f^{-1} in Example 4

Every quadratic function $f(x) = ax^2 + bx + c$, $a \neq 0$, is *not* one-to-one.

▌ **Restricted Domains**   For a function f that is not one-to-one, it may be possible to restrict its domain in such a manner so that the new function consisting of f defined on this restricted domain is one-to-one and so has an inverse. In most cases we want to restrict the domain so that the new function retains its original range. The next example illustrates this concept.

EXAMPLE 5   Restricted Domain

In Example 1 we showed graphically that the quadratic function $f(x) = x^2 + 1$ is not one-to-one. The domain of f is $(-\infty, \infty)$, and as seen in FIGURE 1.5.7(a), the range of f is $[1, \infty)$. Now by defining $f(x) = x^2 + 1$ only on the interval $[0, \infty)$, we see two things in Figure 1.5.7(b): the range of f is preserved and $f(x) = x^2 + 1$ confined to the domain $[0, \infty)$ passes the horizontal line test, in other words, is one-to-one. The inverse of this new one-to-one function is obtained in the usual manner. Solving $y = x^2 + 1$ for x and relabeling variables gives

$$x = \pm\sqrt{y - 1} \qquad \text{and so} \qquad y = \pm\sqrt{x - 1}.$$

The appropriate algebraic sign in the last equation is determined from the fact that the domain and range of f^{-1} are $[1, \infty)$ and $[0, \infty)$, respectively. This forces us to choose $f^{-1}(x) = \sqrt{x - 1}$ as the inverse of f. See Figure 1.5.7(c).

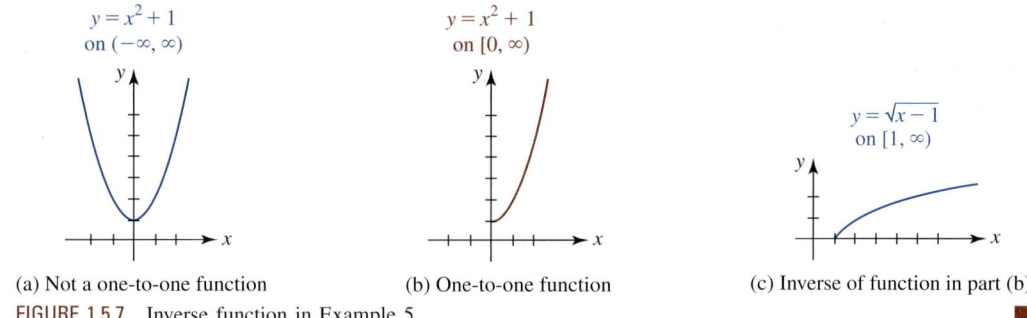

(a) Not a one-to-one function   (b) One-to-one function   (c) Inverse of function in part (b)

FIGURE 1.5.7   Inverse function in Example 5

■ **Inverse Trigonometric Functions**   Although none of the trigonometric functions are one-to-one, by suitably restricting each of their domains we can define six inverse trigonometric functions.

■ **Inverse Sine Function**   From FIGURE 1.5.8(a) we see that the function $y = \sin x$ on the closed interval $[-\pi/2, \pi/2]$ takes on all values in its range $[-1, 1]$. Notice that any horizontal line drawn to intersect the red portion of the graph can do so at most once. Thus the sine function on this restricted domain is one-to-one and has an inverse. There are two notations commonly used throughout mathematics to denote the inverse of the function shown in Figure 1.5.8(b):

$$\sin^{-1} x \quad \text{or} \quad \arcsin x,$$

and are read **inverse sine of x** and **arcsine of x**, respectively.

◀ The computer algebra system *Mathematica* uses the arcsine notation.

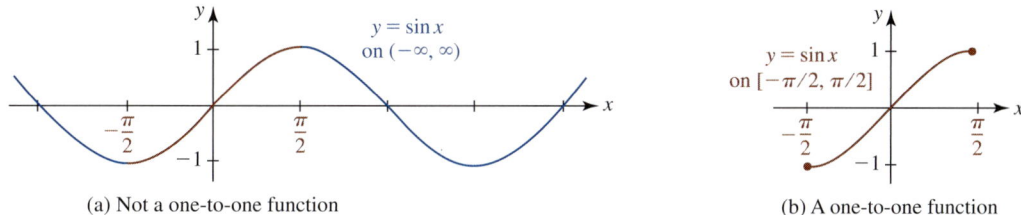

(a) Not a one-to-one function   (b) A one-to-one function

FIGURE 1.5.8   Restricting the domain of $y = \sin x$ to produce a one-to-one function

In FIGURE 1.5.9(a) we have reflected the portion of the graph of $y = \sin x$ on the interval $[-\pi/2, \pi/2]$ (the red graph in Figure 1.5.8(b)) in the line $y = x$ to obtain the graph of $y = \sin^{-1} x$ (in blue). For clarity, we have reproduced this blue graph in Figure 1.5.9(b). As this graph shows, the domain of the inverse sine function is $[-1, 1]$ and the range is $[-\pi/2, \pi/2]$.

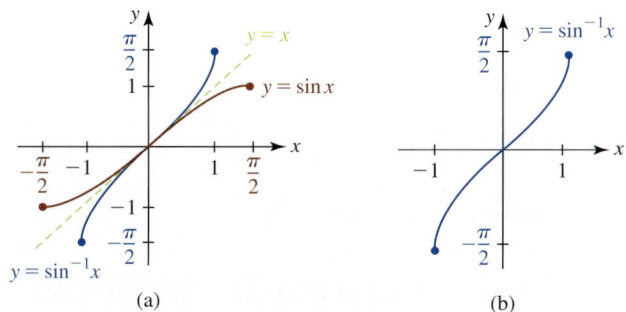

(a)   (b)

FIGURE 1.5.9   Graph of the inverse sine function is the blue curve

---

Definition 1.5.3   Inverse Sine Function

The **inverse sine function**, or **arcsine function**, is defined by

$$y = \sin^{-1}x \qquad \text{if and only if} \qquad x = \sin y, \tag{6}$$

where $-1 \le x \le 1$ and $-\pi/2 \le y \le \pi/2$.

In words:

- *The inverse sine of the number x is that number y (or radian-measured angle) between* $-\pi/2$ *and* $\pi/2$ *whose sine is x.*

The symbols $y = \arcsin x$ and $y = \sin^{-1}x$ are used interchangeably throughout mathematics and its applications and so we will alternate their use so that you become comfortable with both notations.

EXAMPLE 6   Evaluating the Inverse Sine Function

Find

(a) $\arcsin\dfrac{1}{2}$ (b) $\sin^{-1}\left(-\dfrac{1}{2}\right)$ and   (c) $\sin^{-1}(-1)$.

Solution

(a) If we let $y = \arcsin\frac{1}{2}$, then by (6) we must find the number y (or radian-measured angle) that satisfies $\sin y = \frac{1}{2}$ and $-\pi/2 \le y \le \pi/2$. Since $\sin(\pi/6) = \frac{1}{2}$, and $\pi/6$ satisfies the inequality $-\pi/2 \le y \le \pi/2$ it follows that

$$y = \frac{\pi}{6}.$$

(b) If we let $y = \sin^{-1}(-\frac{1}{2})$, then $\sin y = -\frac{1}{2}$. Since we must choose y such that $-\pi/2 \le y \le \pi/2$, we find that $y = -\pi/6$.

(c) Letting $y = \sin^{-1}(-1)$, we have that $\sin y = -1$ and $-\pi/2 \le y \le \pi/2$. Hence $y = -\pi/2$. ∎

Read this paragraph several times. ▶   In parts (b) and (c) of Example 6 we were careful to choose y so that $-\pi/2 \le y \le \pi/2$. For example, it is a common error to think that because $\sin(3\pi/2) = -1$, then necessarily $\sin^{-1}(-1)$ can be taken to be $3\pi/2$. Remember: If $y = \sin^{-1}x$, then y is subject to the restriction $-\pi/2 \le y \le \pi/2$, and $3\pi/2$ does not satisfy this inequality.

EXAMPLE 7   Evaluating a Composition

Without using a calculator, find $\tan\left(\sin^{-1}\frac{1}{4}\right)$.

Solution   We must find the tangent of the angle of t radians with sine equal to $\frac{1}{4}$, that is, $\tan t$ where $t = \sin^{-1}\frac{1}{4}$. The angle t is shown in **FIGURE 1.5.10**. Since

$$\tan t = \frac{\sin t}{\cos t} = \frac{1/4}{\cos t},$$

we want to determine the value of $\cos t$. From Figure 1.5.10 and the Pythagorean identity $\sin^2 t + \cos^2 t = 1$, we see that

$$\left(\frac{1}{4}\right)^2 + \cos^2 t = 1 \qquad \text{or} \qquad \cos t = \frac{\sqrt{15}}{4}.$$

Hence

$$\tan t = \frac{1/4}{\sqrt{15}/4} = \frac{1}{\sqrt{15}} = \frac{\sqrt{15}}{15},$$

and so

$$\tan\left(\sin^{-1}\frac{1}{4}\right) = \tan t = \frac{\sqrt{15}}{15}. \qquad ∎$$

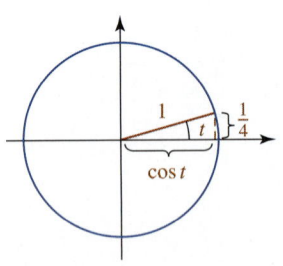

FIGURE 1.5.10   The angle $t = \sin^{-1}\frac{1}{4}$ in Example 7

The procedure that will be illustrated in Example 10 provides an alternative method for solving Example 7.

■ **Inverse Cosine Function** If we restrict the domain of the cosine function to the closed interval $[0, \pi]$, the resulting function is one-to-one and thus has an inverse. We denote this inverse by

$$\cos^{-1}x \quad \text{or} \quad \arccos x,$$

which gives us the following definition.

> **Definition 1.5.4** Inverse Cosine Function
>
> The **inverse cosine function**, or **arccosine function**, is defined by
>
> $$y = \cos^{-1}x \quad \text{if and only if} \quad x = \cos y, \tag{7}$$
>
> where $-1 \le x \le 1$ and $0 \le y \le \pi$.

The graphs shown in **FIGURE 1.5.11** illustrate how the function $y = \cos x$ restricted to the interval $[0, \pi]$ becomes a one-to-one function.

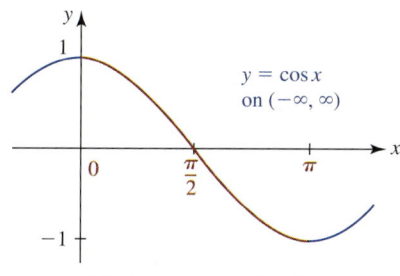

(a) Not a one-to-one function (b) A one-to-one function

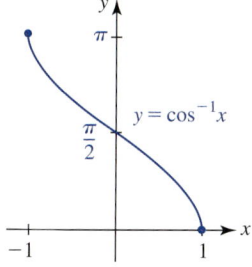

FIGURE 1.5.11 Restricting the domain of $y = \cos x$ to produce a one-to-one function

FIGURE 1.5.12 Graph of the inverse cosine function

By reflecting the graph of the one-to-one function in Figure 1.5.11(b) in the line $y = x$ we obtain the graph of $y = \cos^{-1}x$ shown in **FIGURE 1.5.12**. The figure clearly shows that the domain and range of $y = \cos^{-1}x$ are $[-1, 1]$ and $[0, \pi]$, respectively.

EXAMPLE 8 Evaluating the Inverse Cosine Function

Evaluate $\arccos(-\sqrt{3}/2)$.

Solution If $y = \arccos(-\sqrt{3}/2)$, then $\cos y = -\sqrt{3}/2$. The only number in $[0, \pi]$ for which this is true is $y = 5\pi/6$. That is,

$$\arccos\left(-\frac{\sqrt{3}}{2}\right) = \frac{5\pi}{6}.$$

■

EXAMPLE 9 Evaluating the Compositions of Functions

Write $\sin(\cos^{-1}x)$ as an algebraic expression in x.

Solution In **FIGURE 1.5.13** we have constructed an angle of t radians with cosine equal to x. Then $t = \cos^{-1}x$, or $x = \cos t$, where $0 \le t \le \pi$. Now to find $\sin(\cos^{-1}x) = \sin t$, we use the identity $\sin^2 t + \cos^2 t = 1$. Thus

$$\sin^2 t + x^2 = 1$$
$$\sin^2 t = 1 - x^2$$
$$\sin t = \sqrt{1 - x^2}$$
$$\sin(\cos^{-1}x) = \sqrt{1 - x^2}.$$

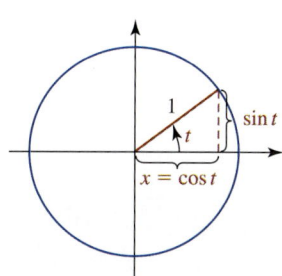

FIGURE 1.5.13 The angle $t = \cos^{-1}x$ in Example 9

We use the positive square root of $1 - x^2$, since the range of $\cos^{-1}x$ is $[0, \pi]$, and the sine of an angle t in the first or second quadrant is positive. ■

■ **Inverse Tangent Function** If we restrict the domain of $\tan x$ to the open interval $(-\pi/2, \pi/2)$, then the resulting function is one-to-one and thus has an inverse. This inverse is denoted by

$$\tan^{-1} x \quad \text{or} \quad \arctan x.$$

Definition 1.5.5 Arctangent Function

The **inverse tangent function**, or **arctangent function**, is defined by

$$y = \tan^{-1} x \quad \text{if and only if} \quad x = \tan y, \tag{8}$$

where $-\infty < x < \infty$ and $-\pi/2 < y < \pi/2$.

The graphs shown in FIGURE 1.5.14 illustrate how the function $y = \tan x$ restricted to the open interval $(-\pi/2, \pi/2)$ becomes a one-to-one function. By reflecting the graph of the one-to-one function in Figure 1.5.14(b) in the line $y = x$ we obtain the graph of $y = \tan^{-1} x$ shown in FIGURE 1.5.15. We see in the figure that the domain and range of $y = \tan^{-1} x$ are, in turn, the intervals $(-\infty, \infty)$ and $(-\pi/2, \pi/2)$. For example, $y = \tan^{-1}(-1) = -\pi/4$ since $-\pi/4$ is the only number in the interval $(-\pi/2, \pi/2)$ for which $\tan(-\pi/4) = -1$.

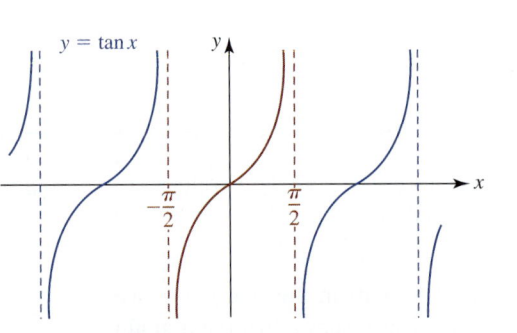
(a) Not a one-to-one function

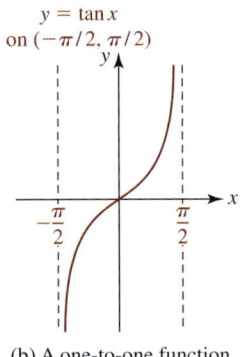
(b) A one-to-one function

FIGURE 1.5.14 Restricting the domain of $y = \tan x$ to produce a one-to-one function.

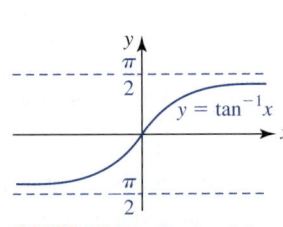
FIGURE 1.5.15 Graph of the inverse tangent function

EXAMPLE 10 Evaluating Compositions of Functions

Without using a calculator, find $\cos\left(\arctan\frac{2}{3}\right)$.

Solution If we let $y = \arctan\frac{2}{3}$, then $\tan y = \frac{2}{3}$. Using the right triangle in FIGURE 1.5.16 as an aid, we see that

$$\cos\left(\arctan\frac{2}{3}\right) = \cos y = \frac{3}{\sqrt{13}}. \qquad ■$$

FIGURE 1.5.16 Triangle in Example 10

■ **Properties of the Inverses** Recall from (5) that $f^{-1}(f(x)) = x$ and $f(f^{-1}(x)) = x$ hold for any function f and its inverse under suitable restrictions on x. Thus for the inverse trigonometric functions, we have the following properties.

Theorem 1.5.2 Properties of Inverse Trig Functions

(*i*) $\sin^{-1}(\sin x) = \arcsin(\sin x) = x$ if $-\pi/2 \le x \le \pi/2$
(*ii*) $\sin(\sin^{-1} x) = \sin(\arcsin x) = x$ if $-1 \le x \le 1$
(*iii*) $\cos^{-1}(\cos x) = \arccos(\cos x) = x$ if $0 \le x \le \pi$
(*iv*) $\cos(\cos^{-1} x) = \cos(\arccos x) = x$ if $-1 \le x \le 1$
(*v*) $\tan^{-1}(\tan x) = \arctan(\tan x) = x$ if $-\pi/2 < x < \pi/2$
(*vi*) $\tan(\tan^{-1} x) = \tan(\arctan x) = x$ if $-\infty < x < \infty$

EXAMPLE 11   Applying the Inverse Properties

Without using a calculator, evaluate

(a) $\cos\left(\cos^{-1}\frac{1}{3}\right)$ **(b)** $\tan^{-1}\left(\tan\frac{3\pi}{4}\right)$.

Solution

(a) By Theorem 1.5.2(*iv*), $\cos\left(\cos^{-1}\frac{1}{3}\right) = \frac{1}{3}$.

(b) In this case we *cannot* apply property (*v*), since $3\pi/4$ is not in the interval $(-\pi/2, \pi/2)$. If we first evaluate $\tan(3\pi/4) = -1$, then we have

$$\tan^{-1}\left(\tan\frac{3\pi}{4}\right) = \tan^{-1}(-1) = -\frac{\pi}{4}.$$

■

▌ Inverses of the Other Trigonometric Functions   With the domains suitably restricted the remaining trigonometric functions $y = \cot x$, $y = \sec x$, and $y = \csc x$ also have inverses.

Definition 1.5.6   Other Inverse Trig Functions

(*i*) $y = \cot^{-1}x$ if and only if $x = \cot y$, $-\infty < x < \infty$ and $0 < y < \pi$
(*ii*) $y = \sec^{-1}x$ if and only if $x = \sec y$, $|x| \geq 1$ and $0 \leq y \leq \pi, y \neq \pi/2$
(*iii*) $y = \csc^{-1}x$ if and only if $x = \csc y$, $|x| \geq 1$ and $-\pi/2 \leq y \leq \pi/2, y \neq 0$

The graphs of $y = \cot^{-1}x$, $y = \sec^{-1}x$, and $y = \csc^{-1}x$ as well as their domains and ranges are summarized in FIGURE 1.5.17.

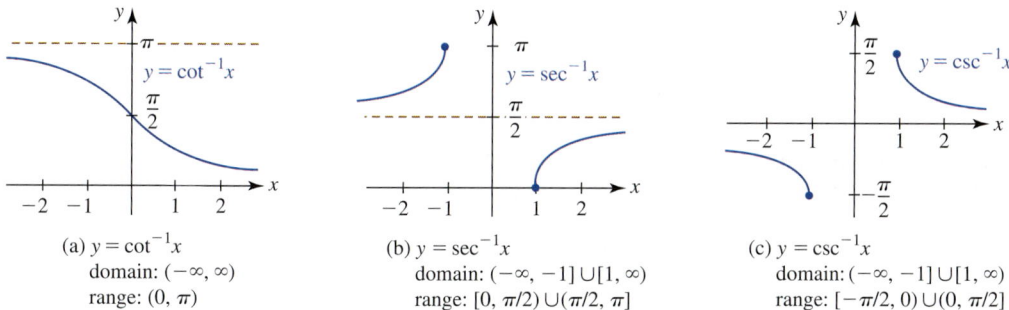

(a) $y = \cot^{-1}x$
domain: $(-\infty, \infty)$
range: $(0, \pi)$

(b) $y = \sec^{-1}x$
domain: $(-\infty, -1] \cup [1, \infty)$
range: $[0, \pi/2) \cup (\pi/2, \pi]$

(c) $y = \csc^{-1}x$
domain: $(-\infty, -1] \cup [1, \infty)$
range: $[-\pi/2, 0) \cup (0, \pi/2]$

FIGURE 1.5.17   Graphs of the inverse cotangent, inverse secant, and inverse cosecant functions

---

$f(x)$ **NOTES FROM THE CLASSROOM**

The ranges specified in Definitions 1.5.3, 1.5.4, 1.5.5, and 1.5.6(*i*) are universally agreed upon and grew out of the most logical and most convenient limitation of the original function. Thus, when we see arccos$x$ or $\tan^{-1}x$ in any context, we know that $0 \leq \arccos x \leq \pi$ and $-\pi/2 < \tan^{-1}x < \pi/2$. These conventions are the same as those used in calculators when the $\boxed{\sin^{-1}}$, $\boxed{\cos^{-1}}$, and $\boxed{\tan^{-1}}$ keys are employed. However, there has been no universal agreement on the ranges of either $y = \sec^{-1}x$ or $y = \csc^{-1}x$. The ranges specified in (*ii*) and (*iii*) in Definition 1.5.6 are gaining in popularity because these are the ranges employed in computer algebra systems such as *Mathematica* and *Maple*. But you should be aware that there are popular calculus texts that define the domain and range of $y = \sec^{-1}x$ to be

domain: $(-\infty, -1] \cup [1, \infty)$,   range: $[0, \pi/2) \cup [\pi, 3\pi/2)$,

and the domain and range of $y = \csc^{-1}x$ to be

domain: $(-\infty, -1] \cup [1, \infty)$,   range: $(0, \pi/2] \cup (\pi, 3\pi/2]$.

Exercises 1.5 Answers to selected odd-numbered problems begin on page ANS-6.

☰ Fundamentals

In Problems 1 and 2, reread the introduction to this section. Then explain why the given function f is not one-to-one.

1. $f(x) = 1 + x(x - 5)$

2. $f(x) = x^4 + 2x^2$

In Problems 3–8, determine whether the given function is one-to-one by examining its graph.

3. $f(x) = 5$

4. $f(x) = 6x - 9$

5. $f(x) = \frac{1}{3}x + 3$

6. $f(x) = |x + 1|$

7. $f(x) = x^3 - 8$

8. $f(x) = x^3 - 3x$

In Problems 9–12, the given function f is one-to-one. Find f^{-1}.

9. $f(x) = 3x^3 + 7$

10. $f(x) = \sqrt[3]{2x - 4}$

11. $f(x) = \frac{2 - x}{1 - x}$

12. $f(x) = 5 - \frac{2}{x}$

In Problems 13 and 14, verify that $f(f^{-1}(x)) = x$ and $f^{-1}(f(x)) = x$.

13. $f(x) = 5x - 10, f^{-1}(x) = \frac{1}{5}x + 2$

14. $f(x) = \frac{1}{x + 1}, f^{-1}(x) = \frac{1 - x}{x}$

In Problems 15–18, the given function f is one-to-one. Without finding the inverse, find the domain and range of f^{-1}.

15. $f(x) = \sqrt{x + 2}$

16. $f(x) = 3 + \sqrt{2x - 1}$

17. $f(x) = \frac{1}{x + 3}$

18. $f(x) = \frac{x - 1}{x - 4}$

In Problems 19 and 20, the given function f is one-to-one. Without finding the inverse, find the point on the graph of f^{-1} corresponding to the indicated value of x in the domain of f.

19. $f(x) = 2x^3 + 2x; \quad x = 2$

20. $f(x) = 8x - 3; \quad x = 5$

In Problems 21 and 22, the given function f is one-to-one. Without finding the inverse, find x in the domain of f^{-1} that satisfies the indicated equation.

21. $f(x) = x + \sqrt{x}; \quad f^{-1}(x) = 9$

22. $f(x) = \frac{4x}{x + 1}; \quad f^{-1}(x) = \frac{1}{2}$

In Problems 23 and 24, sketch the graph of f^{-1} from the graph of f.

23.

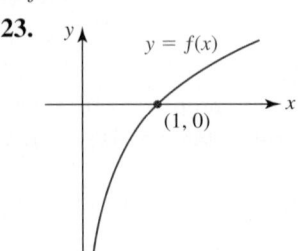

FIGURE 1.5.18 Graph for Problem 23

24.

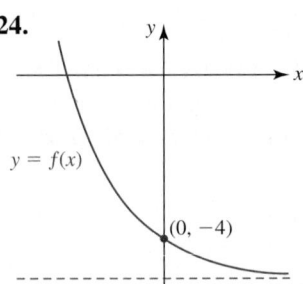

FIGURE 1.5.19 Graph for Problem 24

In Problems 25 and 26, sketch the graph of f from the graph of f^{-1}.

25.

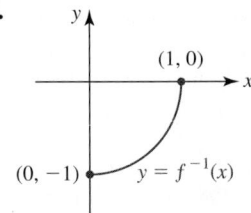

FIGURE 1.5.20 Graph for Problem 25

26.

FIGURE 1.5.21 Graph for Problem 26

In Problems 27–30, find an inverse function f^{-1} that has the same range as the given function by suitably restricting the domain of f.

27. $f(x) = (5 - 2x)^2$

28. $f(x) = 3x^2 + 9$

29. $f(x) = x^2 + 2x + 4$

30. $f(x) = -x^2 + 8x$

31. If the functions f and g have inverses, then it can be proved that

$$(f \circ g)^{-1} = g^{-1} \circ f^{-1}.$$

Verify this for $f(x) = x^3$ and $g(x) = 4x + 5$.

32. The equation $y = \sqrt[3]{x} - \sqrt[3]{y}$ defines a one-to-one function $y = f(x)$. Find $f^{-1}(x)$.

In Problems 33–44, obtain the exact value of the given expression. Do not use a calculator.

33. $\arccos\left(-\frac{\sqrt{2}}{2}\right)$

34. $\cos^{-1}\frac{1}{2}$

35. $\arctan(1)$

36. $\tan^{-1}\sqrt{3}$

37. $\cot^{-1}(-1)$

38. $\sec^{-1}(-1)$

39. $\arcsin\left(-\frac{\sqrt{3}}{2}\right)$

40. $\text{arccot}(-\sqrt{3})$

41. $\sin\left(\arctan\frac{4}{3}\right)$

42. $\cos\left(\sin^{-1}\frac{2}{5}\right)$

43. $\tan\left(\cot^{-1}\frac{1}{2}\right)$

44. $\csc\left(\tan^{-1}\frac{2}{3}\right)$

In Problems 45–48, evaluate the given expression by means of an appropriate trigonometric identity.

45. $\sin\left(2\sin^{-1}\frac{1}{3}\right)$ **46.** $\cos\left(2\cos^{-1}\frac{3}{4}\right)$

47. $\sin\left(\arcsin\frac{\sqrt{3}}{3} + \arccos\frac{2}{3}\right)$

48. $\cos(\tan^{-1}4 - \tan^{-1}3)$

In Problems 49–52, write the given expression as an algebraic quantity in x.

49. $\cos(\sin^{-1}x)$ **50.** $\tan(\sin^{-1}x)$
51. $\sec(\tan^{-1}x)$ **52.** $\sin(\sec^{-1}x), x \geq 1$

In Problems 53 and 54, graphically verify the identities by a reflection and a vertical shift.

53. $\sin^{-1}x + \cos^{-1}x = \frac{\pi}{2}$

54. $\text{arccot}\,x + \arctan x = \frac{\pi}{2}$

55. Prove that $\sec^{-1}x = \cos^{-1}(1/x)$ for $|x| \geq 1$.

56. Prove that $\csc^{-1}x = \sin^{-1}(1/x)$ for $|x| \geq 1$.

57. If $t = \sin^{-1}(-2/\sqrt{5})$, find the exact values of $\cos t$, $\tan t$, $\cot t$, $\sec t$, and $\csc t$.

58. If $\theta = \arctan\frac{1}{2}$, find the exact values of $\sin\theta$, $\cos\theta$, $\cot\theta$, $\sec\theta$, and $\csc\theta$.

≡ Calculator/CAS Problems

Most calculators do not have dedicated keys for $\csc^{-1}x$ and $\sec^{-1}x$. In Problems 59 and 60, use a calculator and the identities in Problems 55 and 56 to compute the given quantity.

59. (a) $\sec^{-1}(-\sqrt{2})$ (b) $\csc^{-1}2$
60. (a) $\sec^{-1}(3.5)$ (b) $\csc^{-1}(-1.25)$
61. Use a calculator to verify:

(a) $\tan(\tan^{-1}1.3) = 1.3$ and $\tan^{-1}(\tan 1.3) = 1.3$
(b) $\tan(\tan^{-1}5) = 5$ and $\tan^{-1}(\tan 5) = -1.2832$

Explain why $\tan^{-1}(\tan 5) \neq 5$.

62. Let $x = 1.7$ radians. Compare, if possible, the values of $\sin^{-1}(\sin x)$ and $\sin(\sin^{-1}x)$. Explain any differences.

≡ Applications

63. Consider a ladder of length L leaning against a house with a load at point P as shown in FIGURE 1.5.22. The angle β at which the ladder is at the verge of slipping is defined by

$$\frac{x}{L} = \frac{c}{1 + c^2}(c + \tan\beta),$$

where c is the coefficient of friction between the ladder and the ground.

(a) Find β when $c = 1$ and the load is at the top of the ladder.

(b) Find β when $c = 0.5$ and the load is $\frac{3}{4}$ of the way up the ladder.

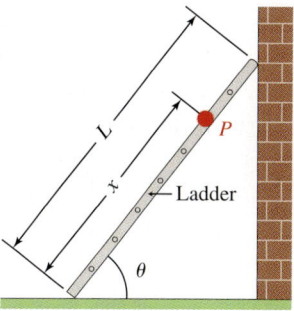

FIGURE 1.5.22 Ladder in Problem 63

64. An airplane flies west at a constant speed v_1 and a wind blows from the north at a constant speed v_2. The plane's course south of west is given by $\theta = \tan^{-1}(v_2/v_1)$. See FIGURE 1.5.23. Find the course of a plane flying west at 300 km/h if a wind from the north blows at 60 km/h.

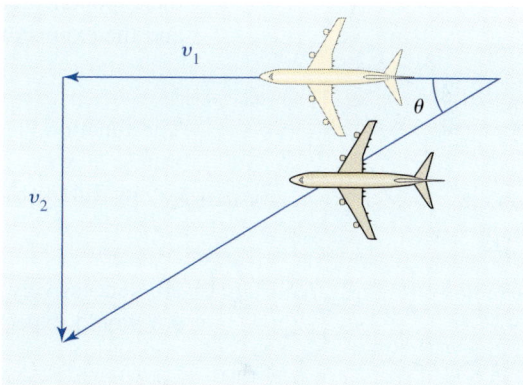

FIGURE 1.5.23 Plane in Problem 64

≡ Think About It

In Problems 65 and 66, use a calculator or CAS to obtain the graph of the given function where x is any real number. Explain why the graphs do not violate Theorems 1.5.2(i) and 1.5.2(iii).

65. $f(x) = \sin^{-1}(\sin x)$ **66.** $f(x) = \cos^{-1}(\cos x)$

67. Discuss: Can any periodic function be one-to-one?

68. How are the one-to-one functions $y = f(x)$ shown in FIGURES 1.5.24(a) and 1.5.24(b) related to the inverse functions $y = f^{-1}(x)$? Find at least three explicit functions with this property.

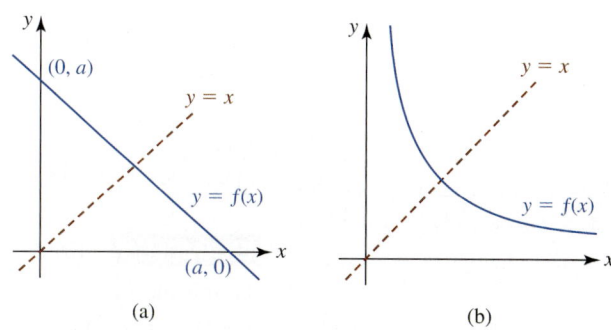

FIGURE 1.5.24 Graph for Problem 68

1.6 Exponential and Logarithmic Functions

■ Introduction In the preceding sections we considered functions such as $f(x) = x^2$, that is, a function with a variable base x and constant power or exponent 2. We now review functions such as $f(x) = 2^x$ having a constant base 2 and a variable exponent x.

Definition 1.6.1 Exponential Function

If $b > 0$ and $b \neq 1$, then an **exponential function** $y = f(x)$ is a function of the form

$$f(x) = b^x. \tag{1}$$

The number b is called the **base** and x is called the **exponent**.

In (1), the base b is restricted to positive numbers in order to guarantee that b^x is a real number. Also, $b = 1$ is of no interest since $f(x) = 1^x = 1$.

The **domain** of an exponential function f defined in (1) is the set of all real numbers $(-\infty, \infty)$.

■ Exponents Because the domain of an exponential function (1) is the set of real numbers, the exponent x can be either a rational or an irrational number. For example, if the base $b = 3$ and the exponent x is a *rational number*, for example, $x = \frac{1}{5}$ and $x = 1.4$, then

$$3^{1/5} = \sqrt[5]{3} \qquad \text{and} \qquad 3^{1.4} = 3^{14/10} = 3^{7/5} = \sqrt[5]{3^7}.$$

The function (1) is also defined for every *irrational number* x. The following procedure illustrates a way of defining a number such as $3^{\sqrt{2}}$. From the decimal representation $\sqrt{2} = 1.414213562\ldots$ we see that the rational numbers

$$1, 1.4, 1.41, 1.414, 1.4142, 1.41421, \ldots$$

are successively better approximations to $\sqrt{2}$. By using these rational numbers as exponents, we would expect that the numbers

$$3^1, 3^{1.4}, 3^{1.41}, 3^{1.414}, 3^{1.4142}, 3^{1.41421}, \ldots$$

are then successively better approximations to $3^{\sqrt{2}}$. In fact, this can be shown to be true with a precise definition of b^x for an irrational value of x. But on a practical level, we can use the $\boxed{y^x}$ key on a calculator to obtain the approximation 4.728804388 to $3^{\sqrt{2}}$.

One definition of b^x, for x irrational, is given by

$$b^x = \lim_{t \to x} b^t,$$

where t is rational. This is read "b^x is the **limit** of b^t as t approaches x". We will study limits in detail in Chapter 2.

■ Laws of Exponents Since b^x is defined for all real numbers x when $b > 0$, it can be proved that the **laws of exponents** hold for all real-number exponents. If $a > 0$, $b > 0$ and $x, x_1,$ and x_2 denote real numbers, then

(i) $b^{x_1} \cdot b^{x_2} = b^{x_1+x_2}$ $\qquad (ii)$ $\dfrac{b^{x_1}}{b^{x_2}} = b^{x_1-x_2}$ $\qquad (iii)$ $(b^{x_1})^{x_2} = b^{x_1 x_2}$

(iv) $\dfrac{1}{b^{x_2}} = b^{-x_2}$ $\qquad (v)$ $(ab)^x = a^x b^x$ $\qquad (vi)$ $\left(\dfrac{a}{b}\right)^x = \dfrac{a^x}{b^x}.$

■ Graphs We distinguish two types of graphs for (1) depending on whether the base b satisfies $b > 1$ or $0 < b < 1$. The next example illustrates the graphs of $f(x) = 3^x$ and $f(x) = \left(\frac{1}{3}\right)^x$. Before graphing, we can make some intuitive observations about both functions. Since the bases $b = 3$ and $b = \frac{1}{3}$ are positive, the values of 3^x and $\left(\frac{1}{3}\right)^x$ are positive for every real number x. Moreover, neither 3^x nor $\left(\frac{1}{3}\right)^x$ can be 0 for any x and so the graphs of $f(x) = 3^x$ and $f(x) = \left(\frac{1}{3}\right)^x$ have no x-intercepts. Also, $3^0 = 1$ and $\left(\frac{1}{3}\right)^0 = 1$ means that the graphs of $f(x) = 3^x$ and $f(x) = \left(\frac{1}{3}\right)^x$ have the same y-intercept $(0, 1)$.

EXAMPLE 1 Graphs of Exponential Functions

Graph the functions

(a) $f(x) = 3^x$, $\qquad$ **(b)** $f(x) = \left(\dfrac{1}{3}\right)^x.$

Solution

(a) We first construct a table of some function values corresponding to preselected values of x. As shown in FIGURE 1.6.1(a), we plot the corresponding points obtained from the table

x	-3	-2	-1	0	1	2
$f(x)$	$\dfrac{1}{27}$	$\dfrac{1}{9}$	$\dfrac{1}{3}$	1	3	9

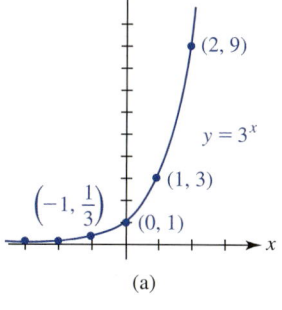

and connect them with a continuous curve. The graph shows that f is an increasing function on the interval $(-\infty, \infty)$.

(b) Proceeding as in part (a), we construct a table of some function values

x	-3	-2	-1	0	1	2
$f(x)$	27	9	3	1	$\dfrac{1}{3}$	$\dfrac{1}{9}$

corresponding to preselected values of x. Note, for example, by the laws of exponents $f(-2) = \left(\frac{1}{3}\right)^{-2} = (3^{-1})^{-2} = 3^2 = 9$. As shown in Figure 1.6.1(b), we plot the corresponding points obtained from the table and connect them with a continuous curve. In this case the graph shows that f is a decreasing function on the interval $(-\infty, \infty)$. ∎

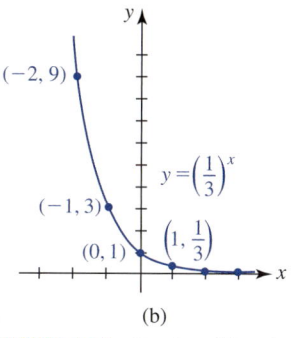

FIGURE 1.6.1   Graphs of functions in Example 1

Note: Exponential functions with bases satisfying $0 < b < 1$, such as $b = \frac{1}{3}$, are frequently written in an alternative manner. By writing $y = \left(\frac{1}{3}\right)^x$ as $y = (3^{-1})^x$ and using (iii) of the laws of exponents we see that $y = \left(\frac{1}{3}\right)^x$ is the same as $y = 3^{-x}$.

■ **Horizontal Asymptote**   FIGURE 1.6.2 illustrates the two general shapes that the graph of an exponential function $f(x) = b^x$ can have. But there is one more important aspect of all such graphs. Observe in Figure 1.6.2 that for $0 < b < 1$, the function values $f(x)$ approach 0 as x becomes unbounded in the positive direction (the red graph), and for $b > 1$, the function values $f(x)$ approach 0 as x becomes unbounded in the negative direction (the blue graph). In other words, the line $y = 0$ (the x-axis) is a **horizontal asymptote** for both types of exponential graphs.

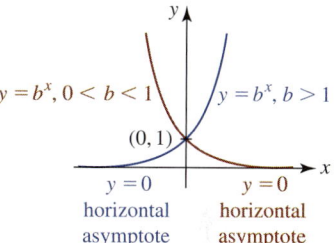

FIGURE 1.6.2 f increasing for $b > 1$; f decreasing for $0 < b < 1$

■ **Properties of an Exponential Function**   The following list summarizes some of the important properties of the exponential function f with base b. Reexamine the graphs in Figure 1.6.2 as you read this list.

- The domain of f is the set of real numbers, that is, $(-\infty, \infty)$.
- The range of f is the set of positive real numbers, that is, $(0, \infty)$.
- The y-intercept of f is $(0, 1)$. The graph of f has no x-intercept.
- The function f is increasing on the interval $(-\infty, \infty)$ for $b > 1$ and decreasing on the interval $(-\infty, \infty)$ for $0 < b < 1$.
- The x-axis, that is, $y = 0$, is a horizontal asymptote for the graph of f.
- The function f is one-to-one.

Although the graphs of $y = b^x$ in the case when $b > 1$ all share the same basic shape and all pass through the same point $(0, 1)$, there are subtle differences. The larger the base b the more steeply the graph rises as x increases. In FIGURE 1.6.3 we compare the graphs of $y = 5^x$, $y = 3^x$, $y = 2^x$, and $y = (1.2)^x$ in green, blue, gold, and red, respectively, on the same coordinate axes. We see from its graph that the values of $y = (1.2)^x$ increase slowly as x increases.

The fact that (1) is a one-to-one function follows from the horizontal line test discussed in Section 1.5.

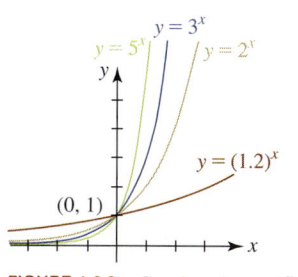

FIGURE 1.6.3   Graphs of $y = b^x$ for $b = 1.2, 2, 3, 5$

■ **The Number e**   Most every student of mathematics has heard of, and has likely worked with, the famous irrational number $\pi = 3.141592654\ldots$. In calculus and applied mathematics the irrational number,

$$e = 2.718281828459\ldots \qquad (2)$$

arguably plays a role more important than the number π. The usual definition of the number e is the number that the function $f(x) = (1 + 1/x)^x$ approaches as we let x become large without bound in the positive direction. If we let the arrow symbol $\rightarrow$ represent the word *approach*, then the fact that $f(x) \rightarrow e$ as $x \rightarrow \infty$ is evident in the table of numerical values of f

x	100	1000	10,000	100,000	1,000,000
$(1 + 1/x)^x$	2.704814	2.716924	2.718146	2.718268	2.718280

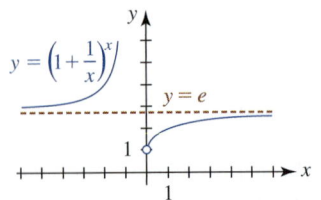

FIGURE 1.6.4 $y = e$ is a horizontal asymptote of the graph of f

and from the graph in FIGURE 1.6.4. In the figure the horizontal dashed red line $y = e$ is a horizontal asymptote for the graph of f. We also say that e is the *limit* of $f(x) = (1 + 1/x)^x$ as $x \rightarrow \infty$ and write

$$e = \lim_{x \to \infty}\left(1 + \frac{1}{x}\right)^x. \tag{3}$$

You will often see an alternative definition of the number e. If we let $h = 1/x$ in (3), then as $x \rightarrow \infty$ we have simultaneously $h \rightarrow 0$. Hence an equivalent form of (3) is

$$e = \lim_{h \to 0}(1 + h)^{1/h}. \tag{4}$$

■ **The Natural Exponential Function**   When the base in (1) is chosen to be $b = e$, the function $f(x) = e^x$ is called the **natural exponential function**. Since $b = e > 1$ and $b = 1/e < 1$, the graphs of $y = e^x$ and $y = e^{-x}$ are given in FIGURE 1.6.5. On the face of it, $f(x) = e^x$ possesses no noticeable graphical characteristic that distinguishes it from, say, the function $f(x) = 3^x$, and has no special properties other than the ones given in the bulleted list on page 49. Questions as to why $f(x) = e^x$ is a "natural" and frankly, the most important exponential function, will be answered in the chapters ahead and in your courses beyond calculus.

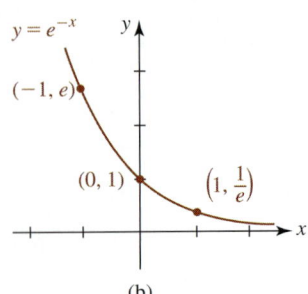

FIGURE 1.6.5 Natural exponential function in (a)

■ **Inverse of the Exponential Function**   Since an exponential function $y = b^x$ is one-to-one, we know that it has an inverse function. To find this inverse, we interchange the variables x and y to obtain $x = b^y$. This last formula defines y as a function of x:

- *y is that exponent of the base b that produces x.*

By replacing the word *exponent* with the word *logarithm*, we can rephrase the preceding line as:

- *y is that logarithm of the base b that produces x.*

This last line is abbreviated by the notation $y = \log_b x$ and is called the **logarithmic function**.

Definition 1.6.2   Logarithmic Function

The **logarithmic function** with base $b > 0, b \neq 1$, is defined by

$$y = \log_b x \qquad \text{if and only if} \qquad x = b^y. \tag{5}$$

For $b > 0$ there is no real number y for which b^y can either be 0 or negative. It then follows from $x = b^y$ that $x > 0$. In other words, the **domain** of a logarithmic function $y = \log_b x$ is the set of positive real numbers $(0, \infty)$.

For emphasis, all that is being said in the preceding sentences is:

- *The logarithmic expression $y = \log_b x$ and the exponential expression $x = b^y$ are equivalent,*

that is, they mean the same thing. As a consequence, within a specific context such as solving a problem, we can use whichever form happens to be more convenient. The following list illustrates several examples of equivalent logarithmic and exponential statements:

Logarithmic Form	Exponential Form
$\log_3 9 = 2$	$9 = 3^2$
$\log_8 2 = \frac{1}{3}$	$2 = 8^{1/3}$
$\log_{10} 0.001 = -3$	$0.001 = 10^{-3}$
$\log_b 5 = -1$	$5 = b^{-1}$

▌ **Graphs**   Because a logarithmic function is the inverse of an exponential function, we can obtain the graph of the former by reflecting the graph of the latter in the line $y = x$. As you inspect the two graphs in FIGURE 1.6.6, remember that the domain $(-\infty, \infty)$ and range $(0, \infty)$ of $y = b^x$ become, in turn, the range $(-\infty, \infty)$ and domain $(0, \infty)$ of $y = \log_b x$. Note that the y-intercept $(0, 1)$ for the exponential function (blue graph) becomes the x-intercept $(1, 0)$ for the logarithmic function (red graph). Also, when the exponential function is reflected in the line $y = x$, the horizontal asymptote $y = 0$ for the graph of $y = b^x$ becomes a vertical asymptote for the graph of $y = \log_b x$. In Figure 1.6.6 we see that for $b > 1$, $x = 0$, which is the equation of the y-axis, is a **vertical asymptote** for the graph of $y = \log_b x$.

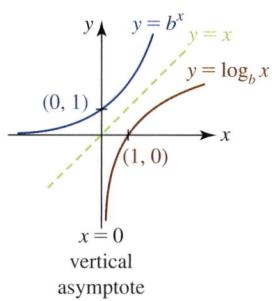

FIGURE 1.6.6   Graph of logarithmic function with base $b > 1$

▌ **Properties of a Logarithmic Function**   The following list summarizes some of the important properties of the logarithmic function $f(x) = \log_b x$:

- The domain of f is the set of positive real numbers, that is, $(0, \infty)$.
- The range of f is the set of real numbers, that is, $(-\infty, \infty)$.
- The x-intercept of f is $(1, 0)$. The graph of f has no y-intercept.
- The function f is increasing on the interval $(0, \infty)$ for $b > 1$ and decreasing on the interval $(0, \infty)$ for $0 < b < 1$.
- The y-axis, that is, $x = 0$, is a vertical asymptote for the graph of f.
- The function f is one-to-one.

We would like to call attention to the third entry in the foregoing list for special emphasis

$$\log_b 1 = 0 \qquad \text{since} \qquad b^0 = 1. \tag{6}$$

Also,

$$\log_b b = 1 \qquad \text{since} \qquad b^1 = b. \tag{7}$$

The result in (7) means that in addition to $(1, 0)$, the graph of any logarithmic function (5) with base b also contains the point $(b, 1)$. The equivalence of $y = \log_b x$ and $x = b^y$ also yields two sometimes-useful identities. By substituting $y = \log_b x$ into $x = b^y$, and then $x = b^y$ into $y = \log_b x$ gives

$$x = b^{\log_b x} \qquad \text{and} \qquad y = \log_b b^y. \tag{8}$$

For example, from (8), $2^{\log_2 10} = 10$ and $\log_3 3^7 = 7$.

▌ **Natural Logarithm**   Logarithms with base $b = 10$ are called **common logarithms** and logarithms with base $b = e$ are called **natural logarithms**. Furthermore, it is customary to write the natural logarithm $\log_e x$ as $\ln x$. The symbol "$\ln x$" is usually read phonetically as "ell-en of x." Since $b = e > 1$, the graph of $y = \ln x$ has the characteristic logarithmic shape shown in red in Figure 1.6.6. For base $b = e$, (5) becomes

$$y = \ln x \qquad \text{if and only if} \qquad x = e^y. \tag{9}$$

The analogues of (6) and (7) for the natural logarithm are

$$\ln 1 = 0 \qquad \text{since} \qquad e^0 = 1. \tag{10}$$

$$\ln e = 1 \qquad \text{since} \qquad e^1 = e. \tag{11}$$

The identities in (8) become

$$x = e^{\ln x} \qquad \text{and} \qquad y = \ln e^y. \tag{12}$$

For example, from (12), $e^{\ln 25} = 25$.

■ **Laws of Logarithms** The laws of exponents can be restated equivalently as the laws of logarithms. For example, if $M = b^{x_1}$ and $N = b^{x_2}$, then by (5), $x_1 = \log_b M$ and $x_2 = \log_b N$. By (*i*) of the laws of exponents, $MN = b^{x_1 + x_2}$. Expressed as a logarithm this is $x_1 + x_2 = \log_b MN$. Substituting for x_1 and x_2 gives $\log_b M + \log_b N = \log_b MN$. The remaining parts of the next theorem can be proved in the same manner.

Theorem 1.6.1 Laws of Logarithms

For any base $b > 0$, $b \neq 1$, and positive numbers M and N:

(*i*) $\log_b MN = \log_b M + \log_b N$

(*ii*) $\log_b \left(\dfrac{M}{N} \right) = \log_b M - \log_b N$

(*iii*) $\log_b M^c = c \log_b M$, for c any real number.

EXAMPLE 2 Laws of Logarithms

Simplify and write $\frac{1}{2} \ln 36 + 2 \ln 4$ as a single logarithm.

Solution By (*iii*) of the laws of logarithms we can write

$$\frac{1}{2} \ln 36 + 2 \ln 4 = \ln (36)^{1/2} + \ln 4^2 = \ln 6 + \ln 16.$$

Then by (*i*) of the laws of logarithms,

$$\frac{1}{2} \ln 36 + 2 \ln 4 = \ln 6 + \ln 16 = \ln (6 \cdot 16) = \ln 96.$$ ∎

EXAMPLE 3 Rewriting Logarithmic Expressions

Use the laws of logarithms to rewrite each expression and evaluate.

(a) $\ln \sqrt{e}$ (b) $\ln 5e$ (c) $\ln \dfrac{1}{e}$

Solution

(a) Since $\sqrt{e} = e^{1/2}$ we have from (*iii*) of the laws of logarithms:

$$\ln \sqrt{e} = \ln e^{1/2} = \frac{1}{2} \ln e = \frac{1}{2}. \qquad \leftarrow \text{from (11), } \ln e = 1$$

(b) From (*i*) of the laws of logarithms and a calculator:

$$\ln 5e = \ln 5 + \ln e = \ln 5 + 1 \approx 2.6094. \qquad \leftarrow \text{from (11), } \ln e = 1$$

(c) From (*ii*) of the laws of logarithms:

$$\ln \frac{1}{e} = \ln 1 - \ln e = 0 - 1 = -1. \qquad \leftarrow \text{from (10) and (11)}$$

Note that (*iii*) of the laws of logarithms can also be used here:

$$\ln \frac{1}{e} = \ln e^{-1} = (-1) \ln e = -1.$$ ∎

EXAMPLE 4 Solving Equations

(a) Solve the logarithmic equation $\ln 2 + \ln (4x - 1) = \ln (2x + 5)$ for x.
(b) Solve the exponential equation $e^{10k} = 7$ for k.

Solution

(a) By (*i*) of the laws of logarithms, the left-hand side of the equation can be written

$$\ln 2 + \ln(4x - 1) = \ln 2(4x - 1) = \ln(8x - 2).$$

The original equation is then

$$\ln(8x - 2) - \ln(2x + 5) = 0 \qquad \text{or} \qquad \ln \frac{8x - 2}{2x + 5} = 0.$$

It follows from (9) that

$$\frac{8x - 2}{2x + 5} = e^0 = 1 \qquad \text{or} \qquad 8x - 2 = 2x + 5.$$

From the last equation we find that $x = \frac{7}{6}$.

(b) We use (9) to rewrite the exponential expression $e^{10k} = 7$ as the logarithmic expression $10k = \ln 7$. Therefore, with the aid of a calculator

$$k = \frac{1}{10} \ln 7 \approx 0.1946.$$ ∎

■ **Change of Base**   The graph of $y = 2^x - 5$ is the graph $y = 2^x$ shifted down 5 units. As seen in FIGURE 1.6.7 the graph has an x-intercept. By setting $y = 0$, we see that x is the solution of the equation $2^x - 5 = 0$ or $2^x = 5$. Now a perfectly valid solution is $x = \log_2 5$. But from a computational viewpoint (that is, expressing x as a number), the last answer is not desirable because no calculator has a logarithmic function with base 2. We can compute the answer by changing $\log_2 5$ to the natural logarithm by simply taking the natural log of both sides of the exponential equation $2^x = 5$:

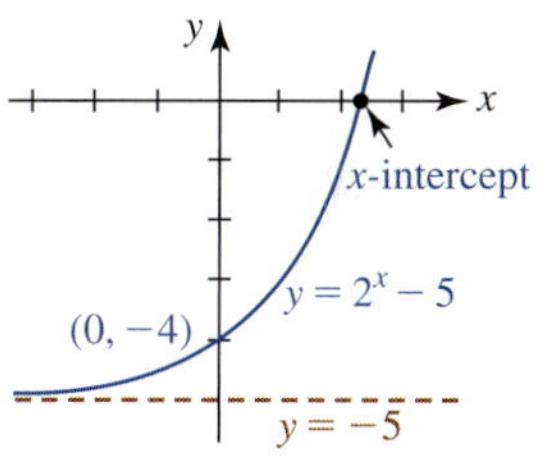

FIGURE 1.6.7 x-intercept of $y = 2^x - 5$

$$\ln 2^x = \ln 5$$
Note: We actually divide the logarithms here → $\quad x \ln 2 = \ln 5$
$$x = \frac{\ln 5}{\ln 2} \approx 2.3219.$$

By the way, since we started with $x = \log_2 5$, the last result also proves the equality $\log_2 5 = \frac{\ln 5}{\ln 2}$. The x-intercept of the graph is then $(\log_2 5, 0) = (\ln 5/\ln 2, 0) \approx (2.32, 0)$.

In general, to convert a logarithm with any base $b > 0$ to the natural logarithm, we first rewrite the logarithmic expression $x = \log_b N$ as an equivalent exponential expression $b^x = N$. Then take the natural logarithm of both sides of the last equality and solve the resulting equation $x \ln b = \ln N$ for x. This yields the general **change of base formula**:

$$\log_b N = \frac{\ln N}{\ln b}. \tag{13}$$

Exercises 1.6   Answers to selected odd-numbered problems begin on page ANS-6.

≡ **Fundamentals**

In Problems 1–6, sketch the graph of the given function f. Find the y-intercept and the horizontal asymptote of the graph.

1. $f(x) = \left(\frac{3}{4}\right)^x$

2. $f(x) = \left(\frac{4}{3}\right)^x$

3. $f(x) = -2^x$

4. $f(x) = -2^{-x}$

5. $f(x) = -5 + e^x$

6. $f(x) = 2 + e^{-x}$

In Problems 7–10, find an exponential function $f(x) = b^x$ such that the graph of f passes through the given point.

7. $(3, 216)$

8. $(-1, 5)$

9. $(-1, e^2)$

10. $(2, e)$

In Problems 11–14, use a graph to solve the given inequality for x.

11. $2^x > 16$

12. $e^x \leq 1$

13. $e^{x-2} < 1$

14. $\left(\dfrac{1}{2}\right)^x \geq 8$

In Problems 15 and 16, use $f(-x) = f(x)$ to demonstrate that the given function is even. Sketch the graph of f.

15. $f(x) = e^{x^2}$

16. $f(x) = e^{-|x|}$

In Problems 17 and 18, use the graphs obtained in Problems 15 and 16 as an aid in sketching the graph of the given function f.

17. $f(x) = 1 - e^{x^2}$

18. $f(x) = 2 + 3e^{-|x|}$

19. Show that $f(x) = 2^x + 2^{-x}$ is an even function. Sketch the graph of f.

20. Show that $f(x) = 2^x - 2^{-x}$ is an odd function. Sketch the graph of f.

In Problems 21 and 22, sketch the graph of the given piecewise-defined function f.

21. $f(x) \begin{cases} -e^x, & x < 0 \\ -e^{-x}, & x \geq 0 \end{cases}$

22. $f(x) \begin{cases} e^{-x}, & x \leq 0 \\ -e^x, & x > 0 \end{cases}$

In Problems 23–26, rewrite the given exponential expression as an equivalent logarithmic expression.

23. $4^{-1/2} = \dfrac{1}{2}$

24. $9^0 = 1$

25. $10^4 = 10{,}000$

26. $10^{0.3010} = 2$

In Problems 27–30, rewrite the given logarithmic expression as an equivalent exponential expression.

27. $\log_2 128 = 7$

28. $\log_5 \dfrac{1}{25} = -2$

29. $\log_{\sqrt{3}} 81 = 8$

30. $\log_{16} 2 = \dfrac{1}{4}$

In Problems 31 and 32, find a logarithmic function $f(x) = \log_b x$ such that the graph of f passes through the given point.

31. $(49, 2)$

32. $\left(4, \frac{1}{3}\right)$

In Problems 33–38, find the exact value of the given expression.

33. $\ln e^e$

34. $\ln(e^4 e^9)$

35. $10^{\log_{10} 6^2}$

36. $25^{\log_5 8}$

37. $e^{-\ln 7}$

38. $e^{\frac{1}{2}\ln \pi}$

In Problems 39–42, find the domain of the given function f. Find the x-intercept and the vertical asymptote of the graph. Sketch the graph of f.

39. $f(x) = -\ln x$

40. $f(x) = -1 + \ln x$

41. $f(x) = -\ln(x + 1)$

42. $f(x) = 1 + \ln(x - 2)$

In Problems 43 and 44, find the domain of the given function f.

43. $f(x) = \ln(9 - x^2)$

44. $f(x) = \ln(x^2 - 2x)$

45. Show that $f(x) = \ln|x|$ is an even function. Sketch the graph of f. Find the x-intercepts and the vertical asymptote of the graph.

46. Use the graph obtained in Problem 45 to sketch the graph of $y = \ln|x - 2|$. Find the x-intercepts and the vertical asymptote of the graph.

In Problems 47–50, use the laws of logarithms to rewrite the given expression as one logarithm.

47. $\ln(x^4 - 4) - \ln(x^2 + 2)$

48. $\ln\left(\dfrac{x}{y}\right) - 2\ln x^3 - 4\ln y$

49. $\ln 5 + \ln 5^2 + \ln 5^3 - \ln 5^6$

50. $5\ln 2 + 2\ln 3 - 3\ln 4$

In Problems 51–54, use the laws of logarithms so that $\ln y$ contains no products, quotients, or powers.

51. $y = \dfrac{x^{10}\sqrt{x^2 + 5}}{\sqrt[3]{8x^3 + 2}}$

52. $y = \sqrt{\dfrac{(2x + 1)(3x + 2)}{4x + 3}}$

53. $y = \dfrac{(x^3 - 3)^5(x^4 + 3x^2 + 1)^8}{\sqrt{x}(7x + 5)^9}$

54. $y = 64x^6\sqrt{x + 1}\sqrt[3]{x^2 + 2}$

In Problems 55 and 56, use the natural logarithm to find x in the domain of the given function for which f takes on the indicated value.

55. $f(x) = 6^x$; $f(x) = 51$

56. $f(x) = \left(\dfrac{1}{2}\right)^x$; $f(x) = 7$

In Problems 57–60, use the natural logarithm to solve for x.

57. $2^{x+5} = 9$

58. $4 \cdot 7^{2x} = 9$

59. $5^x = 2e^{x+1}$

60. $3^{2(x-1)} = 2^{x-3}$

In Problems 61 and 62, solve for x.

61. $\ln x + \ln(x - 2) = \ln 3$

62. $\ln 3 + \ln(2x - 1) = \ln 4 + \ln(x + 1)$

≡ Mathematical Models

63. Exponential Growth An exponential model for the number of bacteria in a culture at time t is given by $P(t) = P_0 e^{kt}$, where P_0 is the initial population and $k > 0$ is the growth constant.

(a) After 2 h the initial number of bacteria in a culture is observed to have doubled. Find an exponential growth model $P(t)$.

(b) According to the model in part (a), what is the number of bacteria present in the culture after 5 h?

(c) Find the time that it takes the culture to grow to 20 times its initial size.

64. Exponential Decay An exponential model for the amount of a radioactive substance remaining at time t is given by $A(t) = A_0 e^{kt}$, where A_0 is the initial amount and $k < 0$ is the decay constant.

(a) Initially 200 mg of a radioactive substance was present. After 6 h the mass had decreased by 3%. Construct an exponential model for the amount of the decaying substance remaining after t h.

(b) Find the amount remaining after 24 h.

(c) The time at which $A(t) = \frac{1}{2}A_0$ is called the **half-life** of the substance. What is the half-life of the substance in part (a)?

65. Logistic Growth   A student sick with a flu virus returns to an isolated college campus of 2000 students. The number of students infected with the flu t days after the student's return is predicted by the logistic function

$$P(t) = \frac{2000}{1 + 1999e^{-0.8905t}}.$$

(a) According to this mathematical model, how many students will be infected with the flu after 5 days?

(b) How long will it take for one-half of the student population to become infected?

(c) How many students does the model predict will become infected after a very long period of time?

(d) Sketch a graph of $P(t)$.

66. Newton's Law of Cooling   If an object or body is placed in a medium (such as air, water, etc.) that is held at a constant temperature T_m and if the initial temperature of the object is T_0, then Newton's law of cooling predicts that the temperature of the object at time t is given by

$$T(t) = T_m + (T_0 - T_m)e^{kt}, \, k < 0.$$

(a) A cake is removed from an oven where the temperature was $350°F$ into a kitchen where the temperature is a constant $75°F$. One minute later the temperature of the cake is measured to be $300°F$. What is the temperature of the cake after 6 minutes?

(b) At what time is the temperature of the cake $80°F$?

☰ Think About It

67. Discuss: How can the graphs of the given functions be obtained from the graph of $f(x) = \ln x$ by means of a rigid transformation (a shift or a reflection)?

(a) $y = \ln 5x$

(b) $y = \ln\dfrac{x}{4}$

(c) $y = \ln x^{-1}$

(d) $y = \ln(-x)$

68. (a) Use a graphing utility to obtain the graph of the function $f(x) = \ln(x + \sqrt{x^2 + 1})$.

(b) Show that f is an odd function, that is, $f(-x) = -f(x)$.

1.7   From Words to Functions

▐ **Introduction**   In Chapters 4 and 6 there will be several instances when you will be expected to translate the words that describe a *function* or an *equation* into mathematical symbols.

In this section we focus on problems that involve functions. We begin with a verbal description about the product of two numbers.

EXAMPLE 1   Product of Two Numbers

The sum of two nonnegative numbers is 5. Express the product of one and the square of the other as a function of one of the numbers.

Solution   We first represent the two numbers by the symbols x and y and recall that *nonnegative* means that $x \geq 0$ and $y \geq 0$. Using these symbols, the words "the sum . . . is 5" translates into the equation $x + y = 5$; this is *not* the function we are seeking. The word *product* in the second sentence suggests that we use the symbol P to denote the function we want. Now P is the product of one of the numbers, say, x and the square of the other, that is, y^2:

$$P = xy^2. \tag{1}$$

No, we are not finished because P is supposed to be a "function of *one* of the numbers." We now use the fact that the numbers x and y are related by $x + y = 5$. From this last equation we substitute $y = 5 - x$ into (1) to obtain the desired result:

$$P(x) = x(5 - x)^2. \tag{2} ■$$

Here is a symbolic diagram of the analysis of the problem given in Example 1:

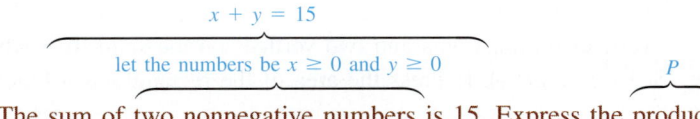

$$\tag{3}$$

Notice that the second sentence is vague about which number is squared. This implies that it really does not matter; (1) could also be written as $P = yx^2$. Also, we could have used $x = 5 - y$ in (1) to arrive at $P(y) = (5 - y)y^2$. In a calculus setting it would not have mattered whether we worked with $P(x)$ or with $P(y)$ because by finding *one* of the numbers we automatically find the other from the equation $x + y = 5$. This last equation is called a **constraint**. A constraint not only defines a relationship between the variables x and y but often puts a limitation on how x and y can vary. As we see in the next example, the constraint helps in determining the domain of the function.

EXAMPLE 2   Example 1 Continued

What is the domain of the function $P(x)$ in (2)?

Solution   Taken out of the context of the statement of the problem in Example 1, one would have to conclude that since

$$P(x) = x(5 - x)^2 = 25x - 10x^2 + x^3$$

is a polynomial function its domain is the set of real numbers $(-\infty, \infty)$. *But in the context* of the original problem, the numbers were to be nonnegative. From the requirement that $x \geq 0$ *and* $y = 5 - x \geq 0$ we get $x \geq 0$ and $x \leq 5$, which means that x must satisfy the simultaneous inequality $0 \leq x \leq 5$. Using interval notation, the domain of the product function P in (2) is the closed interval $[0, 5]$. ∎

If we allowed $x > 5$, then $y = 5 - x < 0$, contrary to the assumption that $y > 0$.

Often in problems that require words translated into a function, it is a good idea to sketch a curve or a picture and identify given quantities in your sketch. Keep your sketch simple.

EXAMPLE 3   Amount of Fencing

A rancher intends to mark off a rectangular plot of land that will have an area of 1000 m². The plot will be fenced and divided into two equal portions by an additional fence parallel to two sides. Express the amount of fence used as a function of the length of one side of the plot.

Solution   Your drawing should be a rectangle with a line drawn down its middle, similar to that given in FIGURE 1.7.1. As shown in the figure, let $x > 0$ be the length of the rectangular plot of land and let $y > 0$ denote its width. The function we seek is the "amount of fence." If the symbol F represents this amount, then the sum of the lengths of the *five* portions—two horizontal and three vertical—of the fence is

$$F = 2x + 3y. \tag{4}$$

But the fenced-in land is to have an area of 1000 m², and so x and y must be related by the constraint $xy = 1000$. From the last equation we get $y = 1000/x$, which can be used to eliminate y in (4). Thus, the amount of fence F as a function of the length x is $F(x) = 2x + 3(1000/x)$ or

$$F(x) = 2x + \frac{3000}{x}. \tag{5}$$

Since x represents a physical dimension that satisfies $xy = 1000$, we conclude that it is positive. But other than that, there is no restriction on x. Thus, unlike the previous example, the function (5) is not defined on a closed interval. The domain of $F(x)$ is the interval $(0, \infty)$. ∎

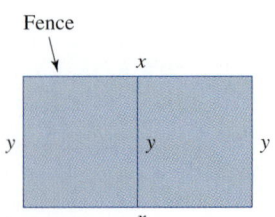

Fence

FIGURE 1.7.1   Rectangular plot of land in Example 3

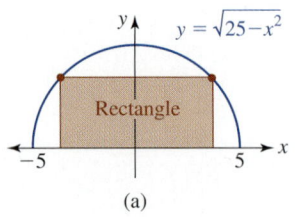

$y = \sqrt{25 - x^2}$

Rectangle

(a)

EXAMPLE 4   Area of a Rectangle

A rectangle has two vertices on the x-axis and two vertices on the semicircle whose equation is $y = \sqrt{25 - x^2}$. See FIGURE 1.7.2(a). Express the area of the rectangle as a function of x.

Solution   If $(x, y), x > 0, y > 0$, denotes the vertex of the rectangle on the circle in the first quadrant, then as shown in Figure 1.7.2(b) the area A is length × width, or

$$A = (2x) \times y = 2xy. \tag{6}$$

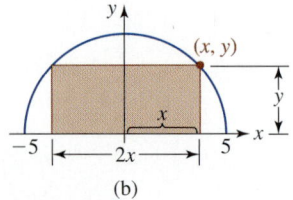

(x, y)

(b)

FIGURE 1.7.2   Rectangle in Example 4

The equation of the semicircle $y = \sqrt{25 - x^2}$ is the constraint in this problem. We use that equation to eliminate y in (6) and obtain the area of the rectangle as a function of x,

$$A(x) = 2x\sqrt{25 - x^2}. \tag{7}$$

The implicit domain of (7) is the closed interval $[-5, 5]$, but because we assumed that (x, y) was a point on the semicircle in the first quadrant we must have $x > 0$. Thus the domain of (7) is the interval $(0, 5)$. ∎

EXAMPLE 5 Distance

Express the distance from a point (x, y) in the first quadrant on the circle $x^2 + y^2 = 1$ to the point $(2, 4)$ as a function of x.

Solution Let (x, y) denote a point in the first quadrant on the circle, and let d represent the distance from (x, y) to $(2, 4)$. See FIGURE 1.7.3. Then from the distance formula,

$$d = \sqrt{(x - 2)^2 + (y - 4)^2} = \sqrt{x^2 + y^2 - 4x - 8y + 20}. \tag{8}$$

The constraint in this problem is the equation of the circle $x^2 + y^2 = 1$. From this equation we can immediately replace $x^2 + y^2$ in (8) by the number 1. Moreover, using the constraint to write $y = \sqrt{1 - x^2}$ allows us to eliminate the symbol y in (8). Thus the distance d as a function of x is:

$$d(x) = \sqrt{21 - 4x - 8\sqrt{1 - x^2}}. \tag{9}$$

Since (x, y) is a point on the circle in the first quadrant the variable x can range between 0 and 1, that is, the domain of the function in (9) is the open interval $(0, 1)$. ∎

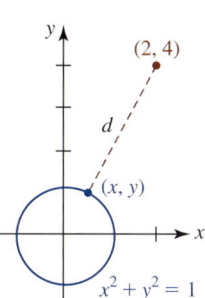

FIGURE 1.7.3 Distance d in Example 5

◀ A point on either the x-axis or the y-axis is not considered to be in any quadrant.

If a word problem involves triangles, you should study the problem carefully and determine whether the Pythagorean Theorem, similar triangles, or right-triangle trigonometry is applicable.

EXAMPLE 6 Length of a Shadow

A tree is planted 30 ft from the base of a streetlamp that is 25 ft tall. Express the length of the tree's shadow as a function of its height.

Solution As shown in FIGURE 1.7.4(a), we let h and s denote the height of the tree and the length of its shadow, respectively. Because the triangles shown Figure 1.7.4(b) are right triangles, we might think of using the Pythagorean Theorem. For this problem, however, the Pythagorean Theorem would lead us astray. The important thing to notice here is that triangles ABC and $AB'C'$ are similar. We then use the fact that the ratios of corresponding sides of similar triangles are equal to write

$$\frac{h}{s} = \frac{25}{s + 30} \qquad \text{or} \qquad (s + 30)h = 25s.$$

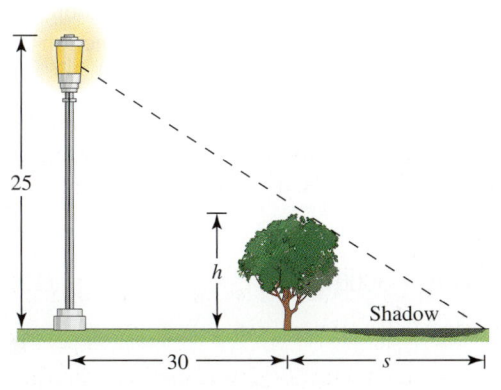

(a)

(b)

FIGURE 1.7.4 Streetlamp and tree in Example 6

By solving the last equation for s in terms of h, we obtain the rational function

$$s(h) = \frac{30h}{25 - h}. \tag{10}$$

It makes physical sense to take the domain of the function in (10) to be defined by $0 \le h < 25$. If $h > 25$, then $s(h)$ is negative, which makes no sense in the physical context of the problem. ∎

EXAMPLE 7   Length of a Ladder

A 10-ft wall stands 5 ft from a building. A ladder, supported by the wall, is to reach from the ground to the building as shown in FIGURE 1.7.5. Express the length of the ladder in terms of the distance x between the base of the wall and the base of the ladder.

Solution   Let L denote the length of the ladder. With the variables x and y defined in Figure 1.7.5, we see again that there are two right triangles; the larger triangle has three sides with lengths L, y, and $x + 5$ and the smaller triangle has two sides of lengths x and 10. The ladder is the hypotenuse of the larger right triangle, so by the Pythagorean Theorem,

$$L^2 = (x + 5)^2 + y^2. \tag{11}$$

The right triangles in Figure 1.7.5 are similar because they both contain a right angle and share the common acute angle the ladder makes with the ground. We again use the fact that the ratios of corresponding sides of similar triangles are equal. This enables us to write

$$\frac{y}{x + 5} = \frac{10}{x} \qquad \text{so that} \qquad y = \frac{10(x + 5)}{x}.$$

Using the last result, (11) becomes

$$L^2 = (x + 5)^2 + \left[\frac{10(x + 5)}{x}\right]^2$$

$$= (x + 5)^2\left[1 + \frac{100}{x^2}\right]$$

$$= (x + 5)^2\left[\frac{x^2 + 100}{x^2}\right].$$

Taking the square root gives L as a function of x,

$$L(x) = \frac{x + 5}{x}\sqrt{x^2 + 100}. \tag{12} \; ∎$$

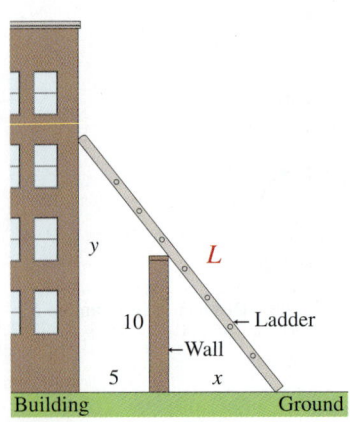

FIGURE 1.7.5   Ladder in Example 7

EXAMPLE 8   Distance

A plane flies at a constant height of 3000 ft over level ground away from an observer on the ground. Express the horizontal distance between the plane and the observer as a function of the angle of elevation of the plane measured by the observer.

Solution   As shown in FIGURE 1.7.6, let x be the horizontal distance between the plane and the observer, and let θ denote the angle of elevation. The triangle in the figure is a right triangle. Hence, from right-triangle trigonometry the side opposite θ is related to the side adjacent to θ by $\tan\theta = $ opp/adj. Therefore,

$$\tan\theta = \frac{3000}{x} \qquad \text{or} \qquad x(\theta) = 3000\cot\theta, \tag{13}$$

where $0 < \theta < \pi/2$. ∎

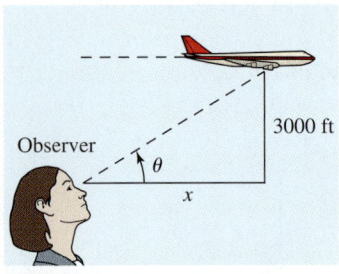

FIGURE 1.7.6   Plane in Example 8

Exercises 1.7   Answers to selected odd-numbered problems begin on page ANS-7.

≡ Fundamentals

In Problems 1–32, translate the words into an appropriate function. Give the domain of the function.

1. The product of two positive numbers is 50. Express their sum as a function of one of the numbers.

2. Express the sum of a nonzero number and its reciprocal as a function of the number.

3. The sum of two nonnegative numbers is 1. Express the sum of the square of one and twice the square of the other as a function of one of the numbers.

4. Let m and n be positive integers. The sum of two nonnegative numbers is S. Express the product of the mth power of one and the nth power of the other as a function of one of the numbers.

5. A rectangle has a perimeter of 200 in. Express the area of the rectangle as a function of the length of one of its sides.

6. A rectangle has an area of 400 in². Express the perimeter of the rectangle as a function of the length of one of its sides.

7. Express the area of the rectangle shaded in FIGURE 1.7.7 as a function of x.

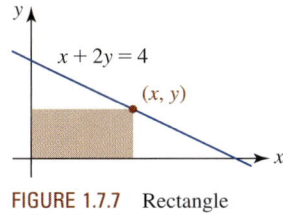

FIGURE 1.7.7 Rectangle in Problem 7

8. Express the length of the line segment containing the point $(2, 4)$ shown in FIGURE 1.7.8 as a function of x.

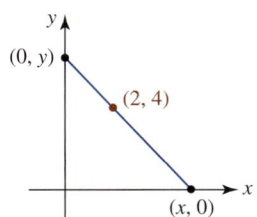

FIGURE 1.7.8 Line segment in Problem 8

9. Express the distance from a point (x, y) on the graph of $x + y = 1$ to the point $(2, 3)$ as a function of x.

10. Express the distance from a point (x, y) on the graph of $y = 4 - x^2$ to the point $(0, 1)$ as a function of x.

11. Express the perimeter of a square as a function of its area A.

12. Express the area of a circle as a function of its diameter d.

13. Express the diameter of a circle as a function of its circumference C.

14. Express the volume of a cube as a function of the area A of its base.

15. Express the area of an equilateral triangle as a function of its height h.

16. Express the area of an equilateral triangle as a function of the length s of one of its sides.

17. A wire of length x is bent into the shape of a circle. Express the area of the circle as a function of x.

18. A wire of length L is cut x units from one end. One piece of the wire is bent into a square and the other piece is bent into a circle. Express the sum of the areas as a function of x.

19. A rancher wishes to enclose a 1000-ft² rectangular corral using two different kinds of fence. Along two parallel sides, the fence costs $4 per foot; for the other two parallel sides, the fence costs $1.60 per foot. Express the total cost to enclose the corral as a function of the length of the sides with fence that costs $4 per foot.

20. The frame of a kite consists of six pieces of lightweight plastic. The outer frame of the kite consists of four precut pieces; two pieces of length 2 ft and two pieces of length 3 ft. Express the area of the kite as a function of x, where $2x$ is the length of the horizontal crossbar piece shown in FIGURE 1.7.9.

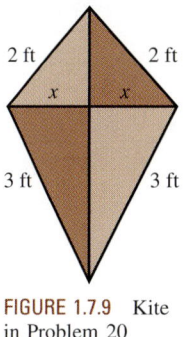

FIGURE 1.7.9 Kite in Problem 20

21. A company wants to construct an open rectangular box with a volume of 450 in³ so that the length of its base is three times its width. Express the surface area of the box as a function of its width.

22. A conical tank, with vertex down, has a radius of 5 ft and a height of 15 ft. See FIGURE 1.7.10. Water is pumped into the tank. Express the volume of the water as a function of its depth. [*Hint*: The volume of a cone is $V = \frac{1}{3}\pi r^2 h$.]

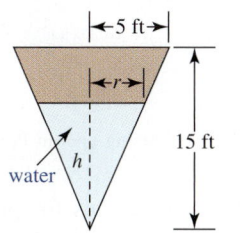

FIGURE 1.7.10 Conical tank in Problem 22

23. Car *A* passes point *O* heading east at a constant rate of 40 mi/h; car *B* passes the same point 1 hour later heading north at a constant rate of 60 mi/h. Express the distance between the cars as a function of time *t*, where *t* is measured starting when car *B* passes point *O*. See FIGURE 1.7.11.

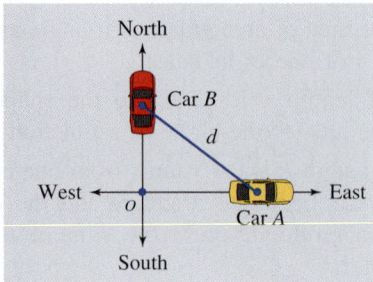

FIGURE 1.7.11   Cars in Problem 23

24. At time *t* = 0 (measured in hours), two planes with a vertical separation of 1 mi pass each other going in opposite directions. See FIGURE 1.7.12. The planes are flying horizontally at rates of 500 mi/h and 550 mi/h.

(a) Express the horizontal distance between them as a function of *t*. [*Hint*: Distance = rate × time.]

(b) Express the diagonal distance between them as a function of *t*.

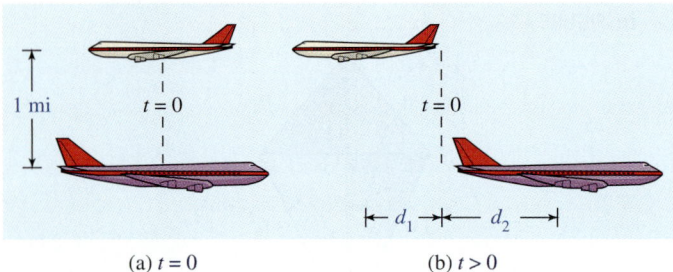

(a) *t* = 0   (b) *t* > 0

FIGURE 1.7.12   Planes in Problem 24

25. The swimming pool shown in FIGURE 1.7.13 is 3 ft deep at the shallow end, 8 ft deep at the deep end, 40 ft long, 30 ft wide, and the bottom is an inclined plane. Water is pumped into the pool. Express the volume of the water in the pool as a function of the height *h* of the water above the deep end. [*Hint*: The volume will be a piecewise-defined function with domain defined by $0 \le h \le 8$.]

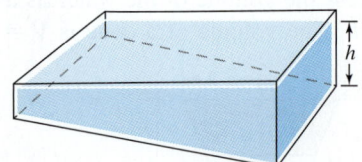

FIGURE 1.7.13   Swimming pool in Problem 25

26. U.S. Postal Service regulations for parcel post stipulate that the length plus girth (the perimeter of one end) of a package must not exceed 108 in. Express the volume

of the package as a function of the width *x* shown in FIGURE 1.7.14.

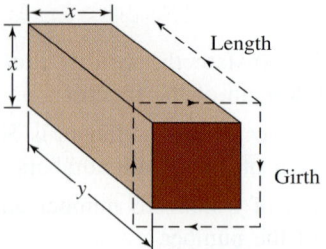

FIGURE 1.7.14   Package in Problem 26

27. Express the height of the balloon shown in FIGURE 1.7.15 as a function of its angle of elevation.

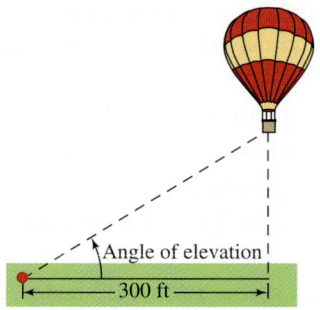

FIGURE 1.7.15   Balloon in Problem 27

28. A long sheet of metal 40 in. wide is made into a V-shaped trough by bending it in the middle along its length. Express the area of the triangular cross section of the trough as a function of the angle *θ* at the vertex of the V. See FIGURE 1.7.16.

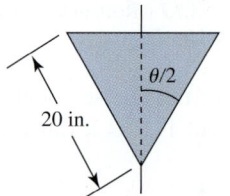

FIGURE 1.7.16   Triangular cross section in Problem 28

29. As shown in FIGURE 1.7.17, a plank is supported by a sawhorse so that one end rests on the ground and the other end rests against a building. Express the length *L* of the plank as a function of the indicated angle *θ*. [*Hint*: Use two right triangles.]

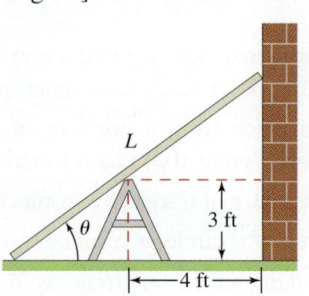

FIGURE 1.7.17   Plank in Problem 29

30. A farmer wishes to enclose a pasture in the form of a right triangle using 2000 ft of fencing on hand. See FIGURE 1.7.18. Express the area of the pasture as a function of the indicated angle θ. [*Hint*: Use the symbols in the figure to form $\cot\theta$ and $\csc\theta$.]

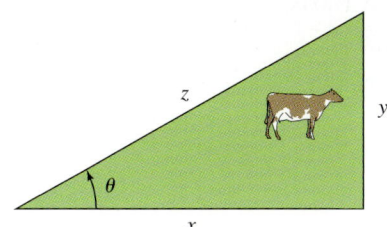

FIGURE 1.7.18   Pasture in Problem 30

31. A statue is placed on a pedestal as shown in FIGURE 1.7.19. Express the viewing angle θ as a function of the distance x from the pedestal.

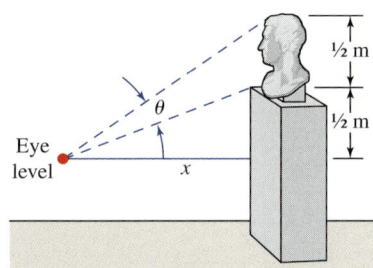

FIGURE 1.7.19   Statue in Problem 31

32. A woman on an island wishes to reach a point R on a straight shore on the mainland from a point P on the island. The point P is 9 mi from the shore and 15 mi from point R. See FIGURE 1.7.20. If the woman rows a boat at a rate of 3 mi/h to a point Q on land, then walks the rest of the way at a rate of 5 mi/h, express the total time it takes the woman to reach point R as a function of the indicated angle θ. [*Hint*: Distance = rate $\times$ time.]

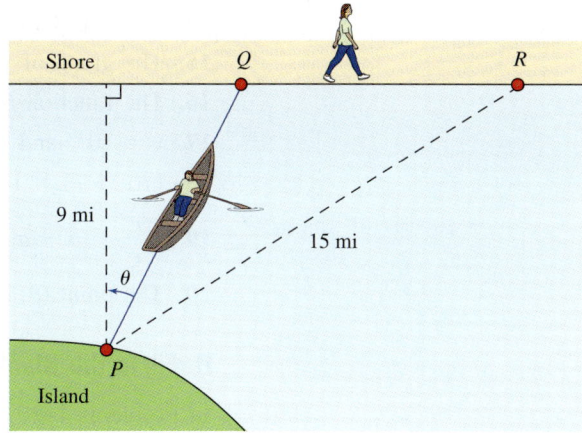

FIGURE 1.7.20   Woman rowing to shore in Problem 32

≡ Think About It

33. Suppose the height of the building in Example 7 is 60 ft. What is the domain of the function $L(x)$ given in (12)?

34. In an engineering text, the area of the octagon shown in FIGURE 1.7.21 is given as $A = 3.31r^2$. Show that this formula is actually an approximation to the area; that is, find the exact area A of the octagon as a function of r.

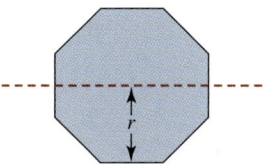

FIGURE 1.7.21   Octagon in Problem 34

Chapter 1 in Review

Answers to selected odd-numbered problems begin on page ANS-7.

A. True/False

In Problems 1–20, indicate whether the given statement is true or false.

1. If f is a function and $f(a) = f(b)$, then $a = b$. _____

2. The function $f(x) = x^5 - 4x^3 + 2$ is an odd function. _____

3. The graph of the function $f(x) = 5x^2\cos x$ is symmetric with respect to the y-axis. _____

4. The graph of $y = f(x + 3)$ is the graph of $y = f(x)$ shifted 3 units to the right. _____

5. The graph of the function $f(x) = \dfrac{1}{x - 1} + \dfrac{1}{x - 2}$ has no x-intercept. _____

6. An asymptote is a line that the graph of a function approaches but never crosses. _____

7. The graph of a function can have at most two horizontal asymptotes. _____

8. If $f(x) = p(x)/q(x)$ is a rational function and $q(a) = 0$, then the line $x = a$ is a vertical asymptote for the graph of f. _____

9. The function $y = -10\sec x$ has amplitude 10. _____

10. The range of the function $f(x) = 2 + \cos x$ is [1, 3]. _____

11. If $f(x) = 1 + x + 2e^x$ is one-to-one, then $f^{-1}(3) = 0.$ _____

12. If $\tan(5\pi/4) = -1$, then $\tan^{-1}(-1) = 5\pi/4.$ _____

13. No even function can be one-to-one. _____

14. A point of intersection of the graphs of f and f^{-1} must be on the line $y = x.$ _____

15. The graph of $y = \sec x$ does not cross the x-axis. _____

16. The function $f(x) = \sin^{-1}x$ is not periodic. _____

17. $y = 10^{-x}$ and $y = (0.1)^x$ are the same function. _____

18. $\ln(e + e) = 1 + \ln 2$ _____

19. $\ln\dfrac{e^b}{e^a} = b - a$ _____

20. The point $(b, 1)$ is on the graph of $f(x) = \log_b x.$ _____

B. Fill in the Blanks

In Problems 1–20, fill in the blanks.

1. The domain of the function $f(x) = \sqrt{x + 2}/x$ is _____.

2. If $f(x) = 4x^2 + 7$ and $g(x) = 2x + 3$, then $(f \circ g)(1) =$ _____, $(g \circ f)(1) =$ _____, and $(f \circ f)(1) =$ _____.

3. The vertex of the graph of the quadratic function $f(x) = x^2 + 16x + 70$ is _____.

4. The x-intercepts of the graph of $f(x) = x^2 + 2x - 35$ are _____.

5. The graph of the polynomial function $f(x) = x^3(x - 1)^2(x - 5)$ is tangent to the x-axis at _____ and passes through the x-axis at _____.

6. The range of the function $f(x) = 10/(x^2 + 1)$ is _____.

7. The y-intercept of the graph of $f(x) = (2x - 4)/(5 - x)$ is _____.

8. A rational function whose graph has the horizontal asymptote $y = 1$ and x-intercept $(3, 0)$ is $f(x) =$ _____.

9. The period of the function $y = 2\sin\dfrac{\pi}{3}x$ is _____.

10. The graph of the function $y = \sin(3x - \pi/4)$ is the graph of $f(x) = \sin 3x$ shifted _____ units to the _____.

11. $\sin^{-1}(\sin\pi) =$ _____.

12. If f is a one-to-one function such that $f^{-1}(3) = 1$, then a point on the graph of f is _____.

13. By rigid transformations, the point $(0, 1)$ on the graph of $y = e^x$ is moved to the point _____ on the graph of $y = 4 + e^{x-3}$.

14. $e^{3\ln 10} =$ _____.

15. If $3^x = 5$, then $x =$ _____.

16. If $3e^x = 4e^{-3x}$, then $x =$ _____.

17. If $\log_3 x = -2$, then $x =$ _____.

18. Written as an exponential statement, $\log_9 27 = 1.5$ is equivalent to _____.

19. The inverse of $y = e^x$ is _____.

20. If $f(x) = e^x - 3$, then $f(-\ln 2) =$ _____.

C. Exercises

1. Estimate the function value by referring to the graph of the function $y = f(x)$ in FIGURE 1.R.1.

(a) $f(-4)$ (b) $f(-3)$
(c) $f(-2)$ (d) $f(-1)$
(e) $f(0)$ (f) $f(1)$
(g) $f(1.5)$ (h) $f(2)$
(i) $f(3.5)$ (j) $f(4)$

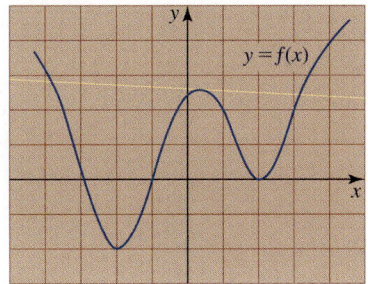

FIGURE 1.R.1 Graph for Problem 1

2. Given that

$$g(t)\begin{cases} t^2, & -1 < t \le 1 \\ 2t, & t \le -1 \text{ or } t > 1 \end{cases}$$

find for $0 < a < 1$:

(a) $g(1 + a)$ (b) $g(1 - a)$
(c) $g(1.5 - a)$ (d) $g(a)$
(e) $g(-a)$ (f) $g(2a)$

3. Determine whether the numbers 1, 5, and 8 are in the range of the function

$$f(x) = \begin{cases} 2x, & -2 \le x < 2 \\ 3, & x = 2 \\ x + 4, & x > 2. \end{cases}$$

4. Suppose $f(x) = \sqrt{x + 4}$, $g(x) = \sqrt{5 - x}$, and $h(x) = x^2$. Find the domain of each of the given functions.

(a) $f \circ h$ (b) $g \circ h$
(c) $f \circ f$ (d) $g \circ g$
(e) $f + g$ (f) f/g

In Problems 5 and 6, compute $\dfrac{f(x + h) - f(x)}{h}, h \ne 0$, and simplify.

5. $f(x) = -x^3 + 2x^2 - x + 5$ **6.** $f(x) = 1 + 2x - \dfrac{3}{x}$

In Problems 7–16, match the given rational function with one of the graphs (a)–(j).

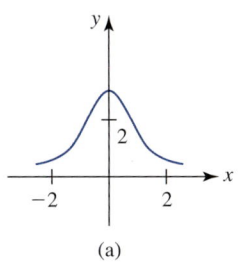

(a)

FIGURE 1.R.2

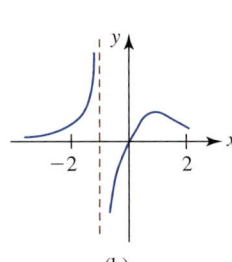

(b)

FIGURE 1.R.3

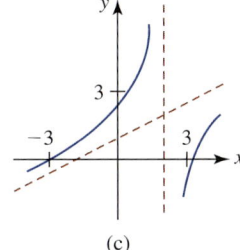

(c)

FIGURE 1.R.4

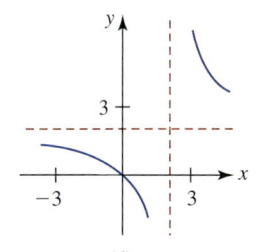

(d)

FIGURE 1.R.5

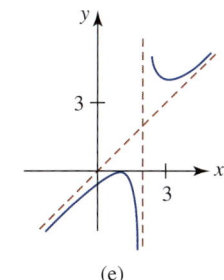

(e)

FIGURE 1.R.6

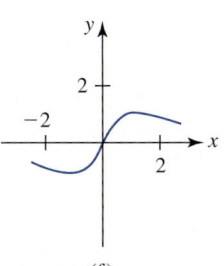

(f)

FIGURE 1.R.7

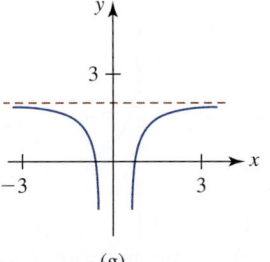

(g)

FIGURE 1.R.8

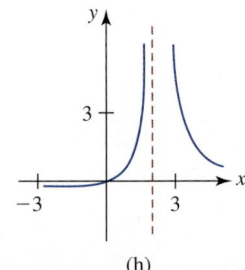

(h)

FIGURE 1.R.9

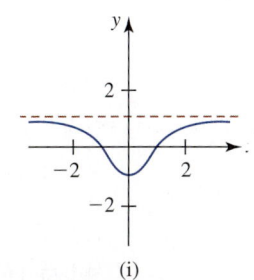

(i)

FIGURE 1.R.10

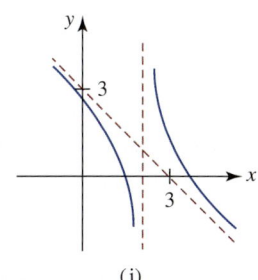

(j)

FIGURE 1.R.11

7. $f(x) = \dfrac{2x}{x^2 + 1}$

8. $f(x) = \dfrac{x^2 - 1}{x^2 + 1}$

9. $f(x) = \dfrac{2x}{x - 2}$

10. $f(x) = 2 - \dfrac{1}{x^2}$

11. $f(x) = \dfrac{x}{(x - 2)^2}$

12. $f(x) = \dfrac{(x - 1)^2}{x - 2}$

13. $f(x) = \dfrac{x^2 - 10}{2x - 4}$

14. $f(x) = \dfrac{-x^2 + 5x - 5}{x - 2}$

15. $f(x) = \dfrac{2x}{x^3 + 1}$

16. $f(x) = \dfrac{3}{x^2 + 1}$

In Problems 17 and 18, find the slope of the red line L in each figure.

17. $f(x) = 3^{-(x+1)}$

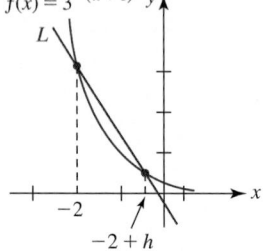

FIGURE 1.R.12   Graph for Problem 17

18.

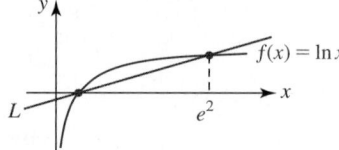

FIGURE 1.R.13   Graph for Problem 18

In Problems 19 and 20, suppose $2^t = a$ and $6^t = b$. Use the laws of exponents given in Section 1.6 to find the value of the given quantity.

19. (a) 12^t **(b)** 3^t **(c)** 6^{-t}

20. (a) 6^{3t} **(b)** $2^{-3t}2^{7t}$ **(c)** 18^t

21. Find a function $f(x) = ae^{kx}$ if $(0, 5)$ and $(6, 1)$ are points on the graph of f.

22. Find a function $f(x) = a10^{kx}$ if $f(3) = 8$ and $f(0) = \frac{1}{2}$.

23. Find a function $f(x) = a + b^x, 0 < b < 1$, if $f(1) = 5.5$ and the graph of f has a horizontal asymptote $y = 5$.

24. Find a function $f(x) = a + \log_3(x - c)$ if $f(11) = 10$ and the graph of f has a vertical asymptote $x = 2$.

In Problems 25–30, match each of the following functions with one of the given graphs.

   (a) $y = \ln(x - 2)$ **(b)** $y = 2 - \ln x$

   (c) $y = 2 + \ln(x + 2)$ **(d)** $y = -2 - \ln(x + 2)$

   (e) $y = -\ln(2x)$ **(f)** $y = 2 + \ln(-x + 2)$

25.

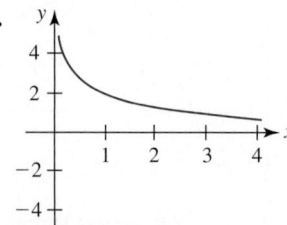

FIGURE 1.R.14   Graph for Problem 25

26.

FIGURE 1.R.15   Graph for Problem 26

27.

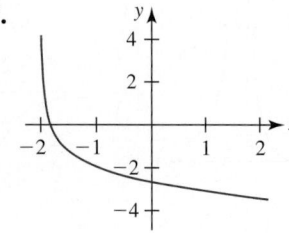

FIGURE 1.R.16   Graph for Problem 27

28.

FIGURE 1.R.17   Graph for Problem 28

29.

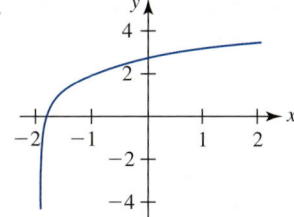

FIGURE 1.R.18 Graph for
Problem 29

30.

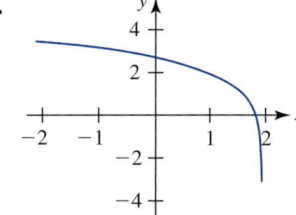

FIGURE 1.R.19 Graph for
Problem 30

31. The width of a rectangular box is three times its length and its height is two times its length.

(a) Express the volume V of the box as a function of its length l.
(b) As a function of its width w
(c) As a function of its height h

32. A closed box in the form of a cube is to be constructed from two different materials. The material for the sides costs 1 cent per square centimeter and the material for the top and bottom costs 2.5 cents per square centimeter. Express the total cost C of construction as a function of the length x of a side.

33. Express the volume V of the box shown in FIGURE 1.R.20 as a function of the indicated angle θ.

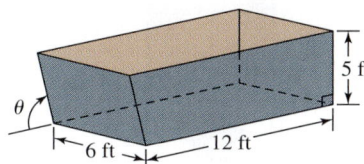

FIGURE 1.R.20   Box in Problem 33

34. Consider the circle of radius h with center (h, h) shown in FIGURE 1.R.21. Express the area A of the shaded region as a function of h.

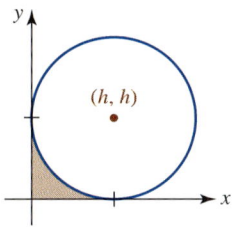

FIGURE 1.R.21   Circle in Problem 34

35. A gutter is to be made from a sheet of metal 30 cm wide by turning up the edges of width 10 cm along each side so that the sides make equal angles ϕ with the vertical. See FIGURE 1.R.22. Express the cross-sectional area of the gutter as a function of the angle ϕ.

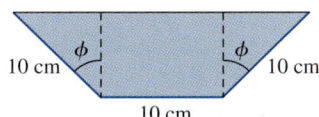

FIGURE 1.R.22   Gutter in Problem 35

36. A metal pipe is to be carried horizontally around a right-angled corner from a hallway 8 ft wide into a hallway that is 6 ft wide. See FIGURE 1.R.23. Express the length L of the pipe as a function of the angle θ shown in the figure.

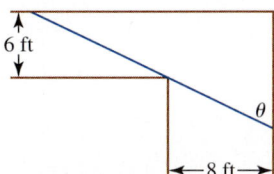

FIGURE 1.R.23   Pipe in Problem 36

37. FIGURE 1.R.24 shows a prism whose parallel faces are equilateral triangles. The rectangular base of the prism is perpendicular to the x-axis and is inscribed within the circle $x^2 + y^2 = 1$. Express the volume V of the prism as a function of x.

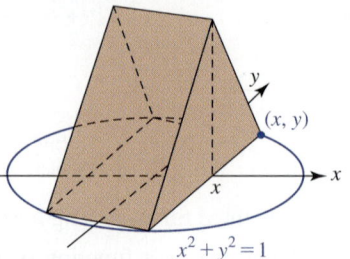

FIGURE 1.R.24   Prism in Problem 37

38. The container shown in FIGURE 1.R.25 consists of an inverted cone (open at its top) attached to the bottom of a right-circular cylinder (open at its top and bottom) of fixed radius R. The container has a fixed volume V. Express the total surface area S of the container as a function of the indicated angle θ. [*Hint*: The lateral surface area of a cone is given by $\pi R \sqrt{R^2 + h^2}$.]

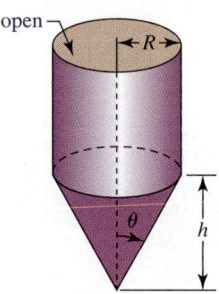

FIGURE 1.R.25   Container in Problem 38

Limit of a Function

In This Chapter   Many topics are included in a typical course in calculus. But the three most fundamental topics in this study are the concepts of *limit*, *derivative*, and *integral*. Each of these concepts deals with functions, which is why we began this text by first reviewing some important facts about functions and their graphs.

Historically, two problems are used to introduce the basic tenets of calculus. These are the *tangent line problem* and the *area problem*. We will see in this and the subsequent chapters that the solutions to both problems involve the limit concept.

2.1 Limits—An Informal Approach

▌ **Introduction**   The two broad areas of calculus known as *differential* and *integral calculus* are built on the foundation concept of a *limit*. In this section our approach to this important concept will be intuitive, concentrating on understanding *what* a limit is using numerical and graphical examples. In the next section, our approach will be analytical, that is, we will use algebraic methods to *compute* the value of a limit of a function.

▌ **Limit of a Function–Informal Approach**   Consider the function

$$f(x) = \frac{16 - x^2}{4 + x} \tag{1}$$

whose domain is the set of all real numbers except -4. Although f cannot be evaluated *at* -4 because substituting -4 for x results in the undefined quantity $0/0$, $f(x)$ can be calculated at any number x that is very *close* to -4. The two tables

x	-4.1	-4.01	-4.001
$f(x)$	8.1	8.01	8.001

x	-3.9	-3.99	-3.999
$f(x)$	7.9	7.99	7.999

$$\tag{2}$$

show that as x approaches -4 from either the left or right, the function values $f(x)$ appear to be approaching 8, in other words, when x is near -4, $f(x)$ is near 8. To interpret the numerical information in (1) graphically, observe that for every number $x \neq -4$, the function f can be simplified by cancellation:

$$f(x) = \frac{16 - x^2}{4 + x} = \frac{(4 + x)(4 - x)}{4 + x} = 4 - x.$$

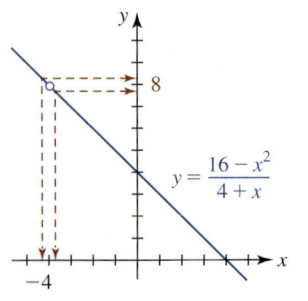

$$y = \frac{16 - x^2}{4 + x}$$

FIGURE 2.1.1   When x is near -4, $f(x)$ is near 8

As seen in **FIGURE 2.1.1**, the graph of f is essentially the graph of $y = 4 - x$ with the exception that the graph of f has a *hole* at the point that corresponds to $x = -4$. For x sufficiently close to -4, represented by the two arrowheads on the x-axis, the two arrowheads on the y-axis, representing function values $f(x)$, simultaneously get closer and closer to the number 8. Indeed, in view of the numerical results in (2), the arrowheads can be made as *close as we like* to the number 8. We say 8 is the **limit** of $f(x)$ as x approaches -4.

▌ **Informal Definition**   Suppose L denotes a finite number. The notion of $f(x)$ approaching L as x approaches a number a can be defined informally in the following manner.

- *If $f(x)$ can be made arbitrarily close to the number L by taking x sufficiently close to but different from the number a, from both the left and right sides of a, then the* **limit** *of $f(x)$ as x approaches a is L.*

▌ **Notation**   The discussion of the limit concept is facilitated by using a special notation. If we let the arrow symbol $\rightarrow$ represent the word *approach*, then the symbolism

$$x \rightarrow a^- \text{ indicates that } x \text{ approaches a number } a \text{ from the } \textbf{left},$$

that is, through numbers that are less than a, and

$$x \rightarrow a^+ \text{ signifies that } x \text{ approaches } a \text{ from the } \textbf{right},$$

that is, through numbers that are greater than a. Finally, the notation

$$x \rightarrow a \text{ signifies that } x \text{ approaches } a \text{ from } \textbf{both sides},$$

in other words, from the left and the right sides of a on a number line. In the left-hand table in (2) we are letting $x \rightarrow -4^-$ (for example, -4.001 is to the left of -4 on the number line), whereas in the right-hand table $x \rightarrow -4^+$.

▌ **One-Sided Limits**   In general, if a function $f(x)$ can be made arbitrarily close to a number L_1 by taking x sufficiently close to, but not equal to, a number a from the *left*, then we write

$$f(x) \rightarrow L_1 \text{ as } x \rightarrow a^- \quad \text{or} \quad \lim_{x \rightarrow a^-} f(x) = L_1. \tag{3}$$

The number L_1 is said to be the **left-hand limit of $f(x)$ as x approaches a**. Similarly, if $f(x)$ can be made arbitrarily close to a number L_2 by taking x sufficiently close to, but not equal to, a number a from the *right*, then L_2 is the **right-hand limit of $f(x)$ as x approaches a** and we write

$$f(x) \rightarrow L_2 \text{ as } x \rightarrow a^+ \qquad \text{or} \qquad \lim_{x \to a^+} f(x) = L_2. \tag{4}$$

The quantities in (3) and (4) are also referred to as **one-sided limits**.

■ **Two-Sided Limits** If both the left-hand limit $\lim_{x \to a^-} f(x)$ and the right-hand limit $\lim_{x \to a^+} f(x)$ exist and have a common value L,

$$\lim_{x \to a^-} f(x) = L \qquad \text{and} \qquad \lim_{x \to a^+} f(x) = L,$$

then we say that L is the **limit of $f(x)$ as x approaches a** and write

$$\lim_{x \to a} f(x) = L. \tag{5}$$

A limit such as (5) is said to be a **two-sided limit**. See FIGURE 2.1.2. Since the numerical tables in (2) suggest that

$$f(x) \rightarrow 8 \text{ as } x \rightarrow -4^- \qquad and \qquad f(x) \rightarrow 8 \text{ as } x \rightarrow -4^+, \tag{6}$$

we can replace the two symbolic statements in (6) by the statement

$$f(x) \rightarrow 8 \text{ as } x \rightarrow -4 \qquad \text{or equivalently} \qquad \lim_{x \to -4} \frac{16 - x^2}{4 + x} = 8. \tag{7}$$

■ **Existence and Nonexistence** Of course a limit (one-sided or two-sided) does not have to exist. But it is important that you keep firmly in mind:

- *The existence of a limit of a function f as x approaches a (from one side or from both sides), does not depend on whether f is defined at a but only on whether f is defined for x near the number a.*

For example, if the function in (1) is modified in the following manner

$$f(x) = \begin{cases} \dfrac{16 - x^2}{4 + x}, & x \neq -4 \\ 5, & x = -4, \end{cases}$$

then $f(-4)$ is defined and $f(-4) = 5$, but still $\lim_{x \to -4} \dfrac{16 - x^2}{4 + x} = 8$. See FIGURE 2.1.3. In general, the two-sided limit $\lim_{x \to a} f(x)$ **does not exist**

- if either of the one-sided limits $\lim_{x \to a^-} f(x)$ or $\lim_{x \to a^+} f(x)$ fails to exist, or
- if $\lim_{x \to a^-} f(x) = L_1$ and $\lim_{x \to a^+} f(x) = L_2$, but $L_1 \neq L_2$.

EXAMPLE 1 A Limit That Exists

The graph of the function $f(x) = -x^2 + 2x + 2$ is shown in FIGURE 2.1.4. As seen from the graph and the accompanying tables, it seems plausible that

$$\lim_{x \to 4^-} f(x) = -6 \qquad \text{and} \qquad \lim_{x \to 4^+} f(x) = -6$$

and consequently $\lim_{x \to 4} f(x) = -6$.

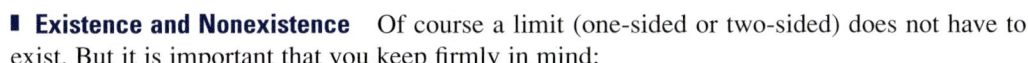

$x \to 4^-$	3.9	3.99	3.999
$f(x)$	-5.41000	-5.94010	-5.99400

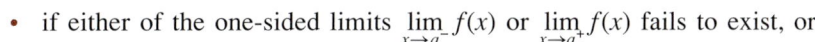

$x \to 4^+$	4.1	4.01	4.001
$f(x)$	-6.61000	-6.06010	-6.00600

Note that in Example 1 the given function is certainly defined at 4, but at no time did we substitute $x = 4$ into the function to find the value of $\lim_{x \to 4} f(x)$.

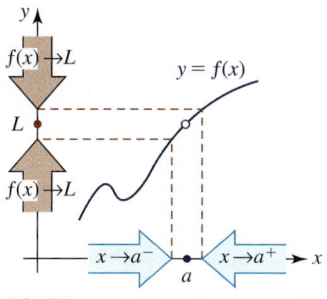

FIGURE 2.1.2 $f(x) \rightarrow L$ as $x \rightarrow a$ if and only if $f(x) \rightarrow L$ as $x \rightarrow a^-$ and $f(x) \rightarrow L$ as $x \rightarrow a^+$

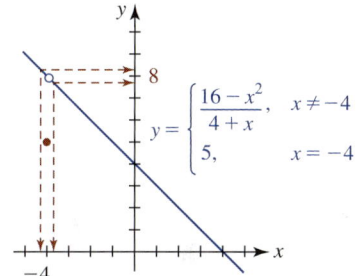

FIGURE 2.1.3 Whether f is defined at a or is not defined at a has no bearing on the existence of the limit of $f(x)$ as $x \rightarrow a$

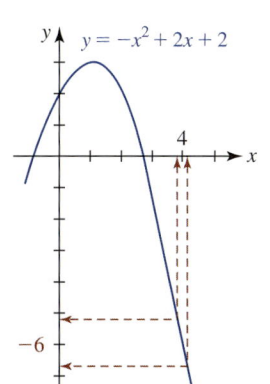

FIGURE 2.1.4 Graph of function in Example 1

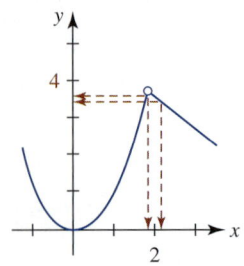

FIGURE 2.1.5 Graph of function in Example 2

EXAMPLE 2 A Limit That Exists

The graph of the piecewise-defined function

$$f(x) = \begin{cases} x^2, & x < 2 \\ -x + 6, & x > 2 \end{cases}$$

is given in FIGURE 2.1.5. Notice that $f(2)$ is not defined, but that is of no consequence when considering $\lim_{x \to 2} f(x)$. From the graph and the accompanying tables,

$x \to 2^-$	1.9	1.99	1.999
$f(x)$	3.61000	3.96010	3.99600

$x \to 2^+$	2.1	2.01	2.001
$f(x)$	3.90000	3.99000	3.99900

we see that when we make x close to 2, we can make $f(x)$ arbitrarily close to 4, and so

$$\lim_{x \to 2^-} f(x) = 4 \quad \text{and} \quad \lim_{x \to 2^+} f(x) = 4.$$

That is, $\lim_{x \to 2} f(x) = 4$. ∎

EXAMPLE 3 A Limit That Does Not Exist

The graph of the piecewise-defined function

$$f(x) = \begin{cases} x + 2, & x \le 5 \\ -x + 10, & x > 5 \end{cases}$$

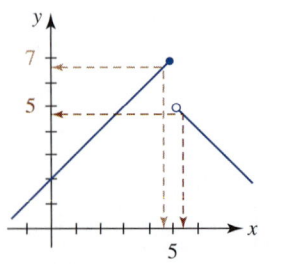

FIGURE 2.1.6 Graph of function in Example 3

is given in FIGURE 2.1.6. From the graph and the accompanying tables, it appears that as x approaches 5 through numbers less than 5 that $\lim_{x \to 5^-} f(x) = 7$. Then as x approaches 5 through numbers greater than 5 it appears that $\lim_{x \to 5^+} f(x) = 5$. But since

$$\lim_{x \to 5^-} f(x) \neq \lim_{x \to 5^+} f(x),$$

we conclude that $\lim_{x \to 5} f(x)$ does not exist.

$x \to 5^-$	4.9	4.99	4.999
$f(x)$	6.90000	6.99000	6.99900

$x \to 5^+$	5.1	5.01	5.001
$f(x)$	4.90000	4.99000	4.99900

∎

EXAMPLE 4 A Limit That Does Not Exist

The greatest integer function was discussed in Section 1.1.

▶ Recall, the **greatest integer function** or **floor function** $f(x) = \lfloor x \rfloor$ is defined to be the greatest integer that is less than or equal to x. The domain of f is the set of real numbers $(-\infty, \infty)$. From the graph in FIGURE 2.1.7 we see that $f(n)$ is defined for every integer n; nonetheless, for each integer n, $\lim_{x \to n} f(x)$ does not exist. For example, as x approaches, say, the number 3, the two one-sided limits exist but have different values:

$$\lim_{x \to 3^-} f(x) = 2 \quad \text{whereas} \quad \lim_{x \to 3^+} f(x) = 3. \quad (8)$$

In general, for an integer n,

$$\lim_{x \to n^-} f(x) = n - 1 \quad \text{whereas} \quad \lim_{x \to n^+} f(x) = n. \quad ∎$$

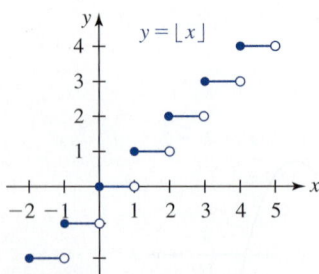

FIGURE 2.1.7 Graph of function in Example 4

EXAMPLE 5 A Right-Hand Limit

From FIGURE 2.1.8 it should be clear that $f(x) = \sqrt{x} \to 0$ as $x \to 0^+$, that is

$$\lim_{x \to 0^+} \sqrt{x} = 0.$$

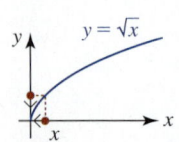

FIGURE 2.1.8 Graph of function in Example 5

It would be incorrect to write $\lim_{x \to 0} \sqrt{x} = 0$ since this notation carries with it the connotation that the limits from the left and from the right exist and are equal to 0. In this case $\lim_{x \to 0^-} \sqrt{x}$ does not exist since $f(x) = \sqrt{x}$ is not defined for $x < 0$. ∎

If $x = a$ is a vertical asymptote for the graph of $y = f(x)$, then $\lim\limits_{x \to a} f(x)$ will always fail to exist because the function values $f(x)$ must become unbounded from at least one side of the line $x = a$.

EXAMPLE 6   A Limit That Does Not Exist

A vertical asymptote always corresponds to an infinite break in the graph of a function f. In FIGURE 2.1.9 we see that the y-axis or $x = 0$ is a vertical asymptote for the graph of $f(x) = 1/x$. The tables

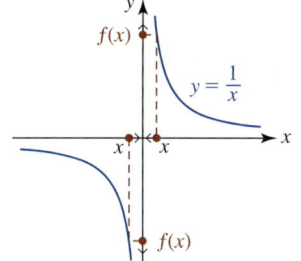

FIGURE 2.1.9   Graph of function in Example 6

$x \to 0^-$	-0.1	-0.01	-0.001
$f(x)$	-10	-100	-1000

$x \to 0^+$	0.1	0.01	0.001
$f(x)$	10	100	1000

clearly show that the function values $f(x)$ become unbounded in absolute value as we get close to 0. In other words, $f(x)$ is not approaching a real number as $x \to 0^-$ nor as $x \to 0^+$. Therefore, neither the left-hand nor the right-hand limit exists as x approaches 0. Thus we conclude that $\lim\limits_{x \to 0} f(x)$ does not exist.   ■

EXAMPLE 7   An Important Trigonometric Limit

To do the calculus of the trigonometric functions $\sin x$, $\cos x$, $\tan x$, and so on, it is important to realize that the variable x is either a real number or an angle measured in radians. With that in mind, consider the numerical values of $f(x) = (\sin x)/x$ as $x \to 0^+$ given in the table that follows.

$x \to 0^+$	0.1	0.01	0.001	0.0001
$f(x)$	0.99833416	0.99998333	0.99999983	0.99999999

It is easy to see that the same results given in the table hold as $x \to 0^-$. Because $\sin x$ is an odd function, for $x > 0$ and $-x < 0$ we have $\sin(-x) = -\sin x$ and as a consequence

$$f(-x) = \frac{\sin(-x)}{-x} = \frac{\sin x}{x} = f(x).$$

As can be seen in FIGURE 2.1.10, f is an even function. The table of numerical values as well as the graph of f strongly suggest the following result:

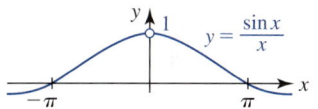

FIGURE 2.1.10   Graph of function in Example 7

$$\lim_{x \to 0} \frac{\sin x}{x} = 1. \qquad (9) \ \blacksquare$$

The limit in (9) is a very important result and will be used in Section 3.4. Another trigonometric limit that you are asked to verify as an exercise is given by

$$\lim_{x \to 0} \frac{1 - \cos x}{x} = 0. \qquad (10)$$

See Problem 43 in Exercises 2.1. Because of their importance, both (9) and (10) will be proven in Section 2.4.

▌ **An Indeterminate Form**   A limit of a quotient $f(x)/g(x)$, where both the numerator and the denominator approach 0 as $x \to a$, is said to have the **indeterminate form 0/0**. The limit (7) in our initial discussion has this indeterminate form. Many important limits, such as (9) and (10), and the limit

$$\lim_{h \to 0} \frac{f(x + h) - f(x)}{h},$$

which forms the backbone of differential calculus, also have the indeterminate form 0/0.

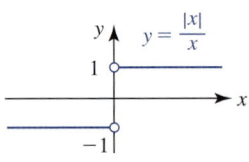

FIGURE 2.1.11   Graph of function in Example 8

EXAMPLE 8   An Indeterminate Form

The limit $\lim_{x\to0}|x|/x$ has the indeterminate form $0/0$, but unlike (7), (9), and (10) this limit fails to exist. To see why, let us examine the graph of the function $f(x) = |x|/x$. For

$$x \neq 0, \quad |x| = \begin{cases} x, & x > 0 \\ -x, & x < 0 \end{cases}$$ and so we recognize f as the piecewise-defined function

$$f(x) = \frac{|x|}{x} = \begin{cases} 1, & x > 0 \\ -1, & x < 0. \end{cases} \qquad (11)$$

From (11) and the graph of f in FIGURE 2.1.11 it should be apparent that both the left-hand and right-hand limits of f exist and

$$\lim_{x\to0^-} \frac{|x|}{x} = -1 \qquad \text{and} \qquad \lim_{x\to0^+} \frac{|x|}{x} = 1.$$

Because these one-sided limits are different, we conclude that $\lim_{x\to0}|x|/x$ does not exist. ∎

$\lim_{x\to a}$ NOTES FROM THE CLASSROOM

While graphs and tables of function values may be convincing for determining whether a limit does or does not exist, you are certainly aware that all calculators and computers work only with approximations and that graphs can be drawn inaccurately. A blind use of a calculator can also lead to a false conclusion. For example, $\lim_{x\to0} \sin(\pi/x)$ is known not to exist, but from the table of values

$x \to 0$	±0.1	±0.01	±0.001
$f(x)$	0	0	0

one would naturally conclude that $\lim_{x\to0} \sin(\pi/x) = 0$. On the other hand, the limit

$$\lim_{x\to0} \frac{\sqrt{x^2 + 4} - 2}{x^2} \qquad (12)$$

can be shown to exist and equals $\frac{1}{4}$. See Example 11 in Section 2.2. One calculator gives

$x \to 0$	±0.00001	±0.000001	±0.0000001
$f(x)$	0.200000	0.000000	0.000000

The problem in calculating (12) for x very close to 0 is that $\sqrt{x^2 + 4}$ is correspondingly very close to 2. When subtracting two numbers of nearly equal values on a calculator a loss of significant digits may occur due to round-off error.

Exercises 2.1   Answers to selected odd-numbered problems begin on page ANS-8.

≡ Fundamentals

In Problems 1–14, sketch the graph of the function to find the given limit, or state that it does not exist.

1. $\lim_{x\to2}(3x + 2)$

2. $\lim_{x\to2}(x^2 - 1)$

3. $\lim_{x\to0}\left(1 + \frac{1}{x}\right)$

4. $\lim_{x\to5}\sqrt{x - 1}$

5. $\lim_{x\to1}\frac{x^2 - 1}{x - 1}$

6. $\lim_{x\to0}\frac{x^2 - 3x}{x}$

7. $\lim_{x\to3}\frac{|x - 3|}{x - 3}$

8. $\lim_{x\to0}\frac{|x| - x}{x}$

9. $\lim_{x\to0}\frac{x^3}{x}$

10. $\lim_{x\to1}\frac{x^4 - 1}{x^2 - 1}$

11. $\lim_{x\to0}f(x)$ where $f(x) = \begin{cases} x + 3, & x < 0 \\ -x + 3, & x \geq 0 \end{cases}$

12. $\lim_{x\to2}f(x)$ where $f(x) = \begin{cases} x, & x < 2 \\ x + 1, & x \geq 2 \end{cases}$

13. $\lim_{x\to2}f(x)$ where $f(x) = \begin{cases} x^2 - 2x, & x < 2 \\ 1, & x = 2 \\ x^2 - 6x + 8, & x > 2 \end{cases}$

14. $\lim\limits_{x\to 0} f(x)$ where $f(x) = \begin{cases} x^2, & x < 0 \\ 2, & x = 0 \\ \sqrt{x} - 1, & x > 0 \end{cases}$

In Problems 15–18, use the given graph to find the value of each quantity, or state that it does not exist.

(a) $f(1)$ **(b)** $\lim\limits_{x\to 1^+} f(x)$ **(c)** $\lim\limits_{x\to 1^-} f(x)$ **(d)** $\lim\limits_{x\to 1} f(x)$

15.

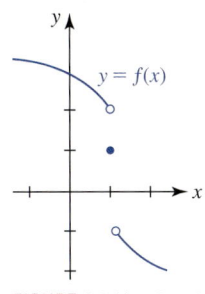

FIGURE 2.1.12   Graph for Problem 15

16.

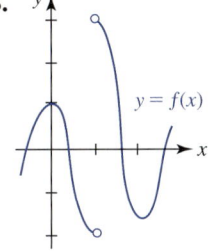

FIGURE 2.1.13   Graph for Problem 16

17.

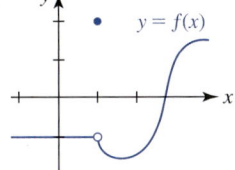

FIGURE 2.1.14   Graph for Problem 17

18.

FIGURE 2.1.15   Graph for Problem 18

In Problems 19–28, each limit has the value 0, but some of the notation is incorrect. If the notation is incorrect, give the correct statement.

19. $\lim\limits_{x\to 0} \sqrt[3]{x} = 0$ **20.** $\lim\limits_{x\to 0} \sqrt[4]{x} = 0$

21. $\lim\limits_{x\to 1} \sqrt{1 - x} = 0$ **22.** $\lim\limits_{x\to -2^+} \sqrt{x + 2} = 0$

23. $\lim\limits_{x\to 0^-} \lfloor x \rfloor = 0$ **24.** $\lim\limits_{x\to \frac{1}{2}} \lfloor x \rfloor = 0$

25. $\lim\limits_{x\to \pi} \sin x = 0$ **26.** $\lim\limits_{x\to 1} \cos^{-1} x = 0$

27. $\lim\limits_{x\to 3^+} \sqrt{9 - x^2} = 0$ **28.** $\lim\limits_{x\to 1} \ln x = 0$

In Problems 29 and 30, use the given graph to find each limit, or state that it does not exist.

29. (a) $\lim\limits_{x\to -4^+} f(x)$ **(b)** $\lim\limits_{x\to -2} f(x)$

   (c) $\lim\limits_{x\to 0} f(x)$ **(d)** $\lim\limits_{x\to 1} f(x)$

   (e) $\lim\limits_{x\to 3} f(x)$ **(f)** $\lim\limits_{x\to 4^-} f(x)$

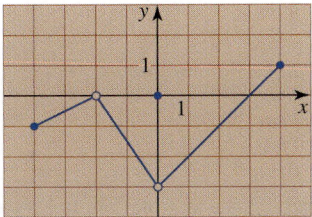

FIGURE 2.1.16   Graph for Problem 29

30. (a) $\lim\limits_{x\to -5} f(x)$ **(b)** $\lim\limits_{x\to -3^-} f(x)$

   (c) $\lim\limits_{x\to -3^+} f(x)$ **(d)** $\lim\limits_{x\to -3} f(x)$

   (e) $\lim\limits_{x\to 0} f(x)$ **(f)** $\lim\limits_{x\to 1} f(x)$

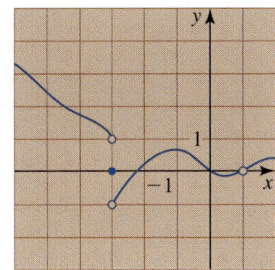

FIGURE 2.1.17   Graph for Problem 30

In Problems 31–34, sketch a graph of a function f with the given properties.

31. $f(-1) = 3, f(0) = -1, f(1) = 0, \lim\limits_{x\to 0} f(x)$ does not exist

32. $f(-2) = 3, \lim\limits_{x\to 0^-} f(x) = 2, \lim\limits_{x\to 0^+} f(x) = -1, f(1) = -2$

33. $f(0) = 1, \lim\limits_{x\to 1^-} f(x) = 3, \lim\limits_{x\to 1^+} f(x) = 3, f(1)$ is undefined, $f(3) = 0$

34. $f(-2) = 2, f(x) = 1, -1 \le x \le 1, \lim\limits_{x\to -1} f(x) = 1, \lim\limits_{x\to 1^-} f(x)$ does not exist, $f(2) = 3$

☰ Calculator/CAS Problems

In Problems 35–40, use a calculator or CAS to obtain the graph of the given function f on the interval $[-0.5, 0.5]$. Use the graph to conjecture the value of $\lim\limits_{x\to 0} f(x)$, or state that the limit does not exist.

35. $f(x) = \cos\dfrac{1}{x}$ **36.** $f(x) = x\cos\dfrac{1}{x}$

37. $f(x) = \dfrac{2 - \sqrt{4 + x}}{x}$

38. $f(x) = \dfrac{9}{x}\left[\sqrt{9 - x} - \sqrt{9 + x}\right]$

39. $f(x) = \dfrac{e^{-2x} - 1}{x}$ **40.** $f(x) = \dfrac{\ln|x|}{x}$

In Problems 41–50, proceed as in Examples 3, 6, and 7 and use a calculator to construct tables of function values. Conjecture the value of each limit, or state that it does not exist.

41. $\lim\limits_{x\to 1} \dfrac{6\sqrt{x} - 6\sqrt{2x - 1}}{x - 1}$ **42.** $\lim\limits_{x\to 1} \dfrac{\ln x}{x - 1}$

43. $\lim\limits_{x\to 0} \dfrac{1 - \cos x}{x}$ **44.** $\lim\limits_{x\to 0} \dfrac{1 - \cos x}{x^2}$

45. $\lim\limits_{x\to 0} \dfrac{x}{\sin 3x}$ **46.** $\lim\limits_{x\to 0} \dfrac{\tan x}{x}$

47. $\lim\limits_{x\to 4} \dfrac{\sqrt{x} - 2}{x - 4}$ **48.** $\lim\limits_{x\to 3} \left[\dfrac{6}{x^2 - 9} - \dfrac{6\sqrt{x - 2}}{x^2 - 9}\right]$

49. $\lim\limits_{x\to 1} \dfrac{x^4 + x - 2}{x - 1}$ **50.** $\lim\limits_{x\to -2} \dfrac{x^3 + 8}{x + 2}$

2.2 Limit Theorems

▌ Introduction The intention of the informal discussion in Section 2.1 was to give you an intuitive grasp of when a limit does or does not exist. However, it is neither desirable nor practical, in every instance, to reach a conclusion about the existence of a limit based on a graph or on a table of numerical values. We must be able to evaluate a limit, or discern its non-existence, in a somewhat mechanical fashion. The theorems that we shall consider in this section establish such a means. The proofs of some of these results are given in the *Appendix*.

The first theorem gives two basic results that will be used throughout the discussion of this section.

Theorem 2.2.1   Two Fundamental Limits

(*i*) $\lim\limits_{x \to a} c = c$, where c is a constant

(*ii*) $\lim\limits_{x \to a} x = a$

Although both parts of Theorem 2.2.1 require a formal proof, Theorem 2.2.1(*ii*) is almost tautological when stated in words:

- *The limit of x as x is approaching a is a.*

See the *Appendix* for a proof of Theorem 2.2.1(*i*).

EXAMPLE 1   Using Theorem 2.2.1

(**a**) From Theorem 2.2.1(*i*),

$$\lim_{x \to 2} 10 = 10 \qquad \text{and} \qquad \lim_{x \to 6} \pi = \pi.$$

(**b**) From Theorem 2.1.1(*ii*),

$$\lim_{x \to 2} x = 2 \qquad \text{and} \qquad \lim_{x \to 0} x = 0. \qquad ■$$

The limit of a constant multiple of a function f is the constant times the limit of f as x approaches a number a.

Theorem 2.2.2   Limit of a Constant Multiple

If c is a constant, then

$$\lim_{x \to a} c f(x) = c \lim_{x \to a} f(x).$$

We can now start using theorems in conjunction with each other.

EXAMPLE 2   Using Theorems 2.2.1 and 2.2.2

From Theorems 2.2.1 (*ii*) and 2.2.2,

(**a**) $\lim\limits_{x \to 8} 5x = 5 \lim\limits_{x \to 8} x = 5 \cdot 8 = 40$

(**b**) $\lim\limits_{x \to -2} \left(-\frac{3}{2} x \right) = -\frac{3}{2} \lim\limits_{x \to -2} x = \left(-\frac{3}{2} \right) \cdot (-2) = 3.$ ■

The next theorem is particularly important because it gives us a way of computing limits in an algebraic manner.

Theorem 2.2.3   Limit of a Sum, Product, and Quotient

Suppose a is a real number and $\lim_{x \to a} f(x)$ and $\lim_{x \to a} g(x)$ exist. If $\lim_{x \to a} f(x) = L_1$ and

$\lim_{x \to a} g(x) = L_2$, then

  (*i*) $\lim_{x \to a} [f(x) \pm g(x)] = \lim_{x \to a} f(x) \pm \lim_{x \to a} g(x) = L_1 \pm L_2,$

  (*ii*) $\lim_{x \to a} [f(x)g(x)] = \left(\lim_{x \to a} f(x)\right)\left(\lim_{x \to a} g(x)\right) = L_1 L_2,$ and

(*iii*) $\lim_{x \to a} \dfrac{f(x)}{g(x)} = \dfrac{\lim_{x \to a} f(x)}{\lim_{x \to a} g(x)} = \dfrac{L_1}{L_2}, \; L_2 \neq 0.$

Theorem 2.2.3 can be stated in words:

- *If both limits exist, then*
  - (*i*) *the limit of a sum is the sum of the limits,*
  - (*ii*) *the limit of a product is the product of the limits, and*
  - (*iii*) *the limit of a quotient is the quotient of the limits provided the limit of the denominator is not zero.*

Note: If all limits exist, then Theorem 2.2.3 is also applicable to one-sided limits, that is, the symbolism $x \to a$ in Theorem 2.2.3 can be replaced by either $x \to a^-$ or $x \to a^+$. Moreover, Theorem 2.2.3 extends to differences, sums, products, and quotients that involve more than two functions. See the *Appendix* for a proof of Theorem 2.2.3.

EXAMPLE 3   Using Theorem 2.2.3

Evaluate $\lim_{x \to 5} (10x + 7)$.

Solution   From Theorems 2.2.1 and 2.2.2, we know that $\lim_{x \to 5} 7$ and $\lim_{x \to 5} 10x$ exist. Hence, from Theorem 2.2.3(*i*),

$$\lim_{x \to 5} (10x + 7) = \lim_{x \to 5} 10x + \lim_{x \to 5} 7$$

$$= 10 \lim_{x \to 5} x + \lim_{x \to 5} 7$$

$$= 10 \cdot 5 + 7 = 57. \qquad \blacksquare$$

▌**Limit of a Power**   Theorem 2.2.3(*ii*) can be used to calculate the limit of a positive integer power of a function. For example, if $\lim_{x \to a} f(x) = L$, then from Theorem 2.2.3(*ii*) with $g(x) = f(x)$,

$$\lim_{x \to a} [f(x)]^2 = \lim_{x \to a} [f(x) \cdot f(x)] = \left(\lim_{x \to a} f(x)\right)\left(\lim_{x \to a} f(x)\right) = L^2.$$

By the same reasoning we can apply Theorem 2.2.3(*ii*) to the general case where $f(x)$ is a factor n times. This result is stated as the next theorem.

Theorem 2.2.4   Limit of a Power

Let $\lim_{x \to a} f(x) = L$ and n be a positive integer. Then

$$\lim_{x \to a} [f(x)]^n = \left[\lim_{x \to a} f(x)\right]^n = L^n.$$

For the special case $f(x) = x$, the result given in Theorem 2.2.4 yields

$$\lim_{x \to a} x^n = a^n. \qquad (1)$$

EXAMPLE 4 Using (1) and Theorem 2.2.3

Evaluate

(a) $\lim_{x\to 10} x^3$ **(b)** $\lim_{x\to 4} \dfrac{5}{x^2}$.

Solution

(a) From (1),

$$\lim_{x\to 10} x^3 = 10^3 = 1000.$$

(b) From Theorem 2.2.1 and (1) we know that $\lim_{x\to 4} 5 = 5$ and $\lim_{x\to 4} x^2 = 16 \neq 0$. Therefore by Theorem 2.2.3(*iii*),

$$\lim_{x\to 4} \frac{5}{x^2} = \frac{\lim_{x\to 4} 5}{\lim_{x\to 4} x^2} = \frac{5}{4^2} = \frac{5}{16}.$$ ∎

EXAMPLE 5 Using Theorem 2.2.3

Evaluate $\lim_{x\to 3}(x^2 - 5x + 6)$.

Solution In view of Theorem 2.2.1, Theorem 2.2.2, and (1) all limits exist. Therefore by Theorem 2.2.3(*i*),

$$\lim_{x\to 3}(x^2 - 5x + 6) = \lim_{x\to 3} x^2 - \lim_{x\to 3} 5x + \lim_{x\to 3} 6 = 3^2 - 5\cdot 3 + 6 = 0.$$ ∎

EXAMPLE 6 Using Theorems 2.2.3 and 2.2.4

Evaluate $\lim_{x\to 1}(3x - 1)^{10}$.

Solution First, we see from Theorem 2.2.3(*i*) that

$$\lim_{x\to 1}(3x - 1) = \lim_{x\to 1} 3x - \lim_{x\to 1} 1 = 2.$$

It then follows from Theorem 2.2.4 that

$$\lim_{x\to 1}(3x - 1)^{10} = \left[\lim_{x\to 1}(3x - 1)\right]^{10} = 2^{10} = 1024.$$ ∎

❚ **Limit of a Polynomial Function** Some limits can be evaluated by *direct substitution*. We can use (1) and Theorem 2.2.3(*i*) to compute the limit of a general polynomial function. If

$$f(x) = c_n x^n + c_{n-1} x^{n-1} + \cdots + c_1 x + c_0$$

is a polynomial function, then

$$\lim_{x\to a} f(x) = \lim_{x\to a}\left(c_n x^n + c_{n-1} x^{n-1} + \cdots + c_1 x + c_0\right)$$

$$= \lim_{x\to a} c_n x^n + \lim_{x\to a} c_{n-1} x^{n-1} + \cdots + \lim_{x\to a} c_1 x + \lim_{x\to a} c_0$$

$$= c_n a^n + c_{n-1} a^{n-1} + \cdots + c_1 a + c_0. \leftarrow \text{\small f is defined at $x = a$ and this limit is $f(a)$}$$

In other words, to evaluate a limit of a polynomial function f as x approaches a real number a, *we need only evaluate the function at $x = a$*:

$$\lim_{x\to a} f(x) = f(a). \tag{2}$$

A reexamination of Example 5 shows that $\lim_{x\to 3} f(x)$, where $f(x) = x^2 - 5x + 6$, is given by $f(3) = 0$.

Because a rational function f is a quotient of two polynomials $p(x)$ and $q(x)$, it follows from (2) and Theorem 2.2.3(*iii*) that a limit of a rational function $f(x) = p(x)/q(x)$ can also be found by evaluating f at $x = a$:

$$\lim_{x\to a} f(x) = \lim_{x\to a} \frac{p(x)}{q(x)} = \frac{p(a)}{q(a)}. \tag{3}$$

Of course we must add to (3) the all-important requirement that the limit of the denominator is not 0, that is, $q(a) \neq 0$.

EXAMPLE 7   Using (2) and (3)

Evaluate $\displaystyle\lim_{x \to -1} \frac{3x - 4}{8x^2 + 2x - 2}$.

Solution $f(x) = \dfrac{3x - 4}{8x^2 + 2x - 2}$ is a rational function and so if we identify the polynomials $p(x) = 3x - 4$ and $q(x) = 8x^2 + 2x - 2$, then from (2),

$$\lim_{x \to -1} p(x) = p(-1) = -7 \qquad \text{and} \qquad \lim_{x \to -1} q(x) = q(-1) = 4.$$

Since $q(-1) \neq 0$ it follows from (3) that

$$\lim_{x \to -1} \frac{3x - 4}{8x^2 + 2x - 2} = \frac{p(-1)}{q(-1)} = \frac{-7}{4} = -\frac{7}{4}. \qquad \blacksquare$$

You should not get the impression that we can *always* find a limit of a function by substituting the number *a directly into the function.*

EXAMPLE 8   Using Theorem 2.2.3

Evaluate $\displaystyle\lim_{x \to 1} \frac{x - 1}{x^2 + x - 2}$.

Solution   The function in this limit is rational, but if we substitute $x = 1$ into the function we see that this limit has the indeterminate form $0/0$. However, by simplifying *first*, we can then apply Theorem 2.2.3(*iii*):

$$\lim_{x \to 1} \frac{x - 1}{x^2 + x - 2} = \lim_{x \to 1} \frac{x - 1}{(x - 1)(x + 2)} \qquad \leftarrow \begin{array}{l} \text{cancellation is valid} \\ \text{provided that } x \neq 1 \end{array}$$

$$= \lim_{x \to 1} \frac{1}{x + 2}$$

$$= \frac{\displaystyle\lim_{x \to 1} 1}{\displaystyle\lim_{x \to 1}(x + 2)} = \frac{1}{3}. \qquad \blacksquare$$

▶ If a limit of a rational function has the indeterminate form $0/0$ as $x \to a$, then by the Factor Theorem of algebra $x - a$ must be a factor of both the numerator and the denominator. Factor those quantities and cancel the factor $x - a$.

Sometimes you can tell at a glance when *a limit does not exist.*

Theorem 2.2.5   A Limit That Does Not Exist

Let $\displaystyle\lim_{x \to a} f(x) = L_1 \neq 0$ and $\displaystyle\lim_{x \to a} g(x) = 0$. Then

$$\lim_{x \to a} \frac{f(x)}{g(x)}$$

does not exist.

PROOF   We will give an indirect proof of this result based on Theorem 2.2.3. Suppose $\displaystyle\lim_{x \to a} f(x) = L_1 \neq 0$ and $\displaystyle\lim_{x \to a} g(x) = 0$ and suppose further that $\displaystyle\lim_{x \to a}(f(x)/g(x))$ exists and equals L_2. Then

$$L_1 = \lim_{x \to a} f(x) = \lim_{x \to a}\left(g(x) \cdot \frac{f(x)}{g(x)}\right), \qquad g(x) \neq 0,$$

$$= \left(\lim_{x \to a} g(x)\right)\left(\lim_{x \to a} \frac{f(x)}{g(x)}\right) = 0 \cdot L_2 = 0.$$

By contradicting the assumption that $L_1 \neq 0$, we have proved the theorem. $\blacksquare$

EXAMPLE 9 Using Theorems 2.2.3 and 2.2.5

Evaluate

(a) $\displaystyle\lim_{x \to 5} \frac{x}{x - 5}$ (b) $\displaystyle\lim_{x \to 5} \frac{x^2 - 10x - 25}{x^2 - 4x - 5}$ (c) $\displaystyle\lim_{x \to 5} \frac{x - 5}{x^2 - 10x + 25}$.

Solution   Each function in the three parts of the example is rational.
   (a) Since the limit of the numerator x is 5, but the limit of the denominator $x - 5$ is 0, we conclude from Theorem 2.2.5 that the limit does not exist.
   (b) Substituting $x = 5$ makes both the numerator and denominator 0, and so the limit has the indeterminate form $0/0$. By the Factor Theorem of algebra, $x - 5$ is a factor of both the numerator and denominator. Hence,

$$\lim_{x \to 5} \frac{x^2 - 10x - 25}{x^2 - 4x - 5} = \lim_{x \to 5} \frac{(x - 5)^2}{(x - 5)(x + 1)} \quad \leftarrow \text{cancel the factor } x - 5$$

$$= \lim_{x \to 5} \frac{x - 5}{x + 1}$$

$$= \frac{0}{6} = 0. \quad \leftarrow \text{limit exists}$$

   (c) Again, the limit has the indeterminate form $0/0$. After factoring the denominator and canceling the factors we see from the algebra

$$\lim_{x \to 5} \frac{x - 5}{x^2 - 10x + 25} = \lim_{x \to 5} \frac{x - 5}{(x - 5)^2}$$

$$= \lim_{x \to 5} \frac{1}{x - 5}$$

that the limit does not exist since the limit of the numerator in the last expression is now 1 but the limit of the denominator is 0. ∎

❚ **Limit of a Root**   The limit of the nth root of a function is the nth root of the limit whenever the limit exists and has a real nth root. The next theorem summarizes this fact.

Theorem 2.2.6   Limit of a Root

Let $\lim_{x \to a} f(x) = L$ and n be a positive integer. Then

$$\lim_{x \to a} \sqrt[n]{f(x)} = \sqrt[n]{\lim_{x \to a} f(x)} = \sqrt[n]{L},$$

provided that $L \geq 0$ when n is even.

An immediate special case of Theorem 2.2.6 is

$$\lim_{x \to a} \sqrt[n]{x} = \sqrt[n]{a}, \tag{4}$$

provided $a \geq 0$ when n is even. For example, $\displaystyle\lim_{x \to 9} \sqrt{x} = \left[\lim_{x \to 9} x\right]^{1/2} = 9^{1/2} = 3$.

EXAMPLE 10 Using (4) and Theorem 2.2.3

Evaluate $\displaystyle\lim_{x \to -8} \frac{x - \sqrt[3]{x}}{2x + 10}$.

Solution   Since $\displaystyle\lim_{x \to -8} (2x + 10) = -6 \neq 0$, we see from Theorem 2.2.3(*iii*) and (4) that

$$\lim_{x \to -8} \frac{x - \sqrt[3]{x}}{2x + 10} = \frac{\displaystyle\lim_{x \to -8} x - \left[\lim_{x \to -8} x\right]^{1/3}}{\displaystyle\lim_{x \to -8} (2x + 10)} = \frac{-8 - (-8)^{1/3}}{-6} = \frac{-6}{-6} = 1. \quad ∎$$

When a limit of an algebraic function involving radicals has the indeterminate form $0/0$, rationalization of the numerator or the denominator may be something to try.

EXAMPLE 11   Rationalization of a Numerator

Evaluate $\displaystyle\lim_{x\to 0}\frac{\sqrt{x^2+4}-2}{x^2}$.

We have seen this limit in (12) in *Notes from the Classroom* at the end of Section 2.1.

Solution   Because $\displaystyle\lim_{x\to 0}\sqrt{x^2+4}=\sqrt{\lim_{x\to 0}(x^2+4)}=2$ we see by inspection that the given limit has the indeterminate form $0/0$. However, by rationalization of the numerator we obtain

$$\lim_{x\to 0}\frac{\sqrt{x^2+4}-2}{x^2}=\lim_{x\to 0}\frac{\sqrt{x^2+4}-2}{x^2}\cdot\frac{\sqrt{x^2+4}+2}{\sqrt{x^2+4}+2}$$

$$=\lim_{x\to 0}\frac{(x^2+4)-4}{x^2(\sqrt{x^2+4}+2)}$$

$$=\lim_{x\to 0}\frac{x^2}{x^2(\sqrt{x^2+4}+2)}\quad\leftarrow\text{cancel } x\text{'s}$$

$$=\lim_{x\to 0}\frac{1}{\sqrt{x^2+4}+2}.\quad\leftarrow\begin{array}{l}\text{this limit is no}\\\text{longer } 0/0\end{array}$$

We are now in a position to use Theorems 2.2.3 and 2.2.6:

$$\lim_{x\to 0}\frac{\sqrt{x^2+4}-2}{x^2}=\lim_{x\to 0}\frac{1}{\sqrt{x^2+4}+2}$$

$$=\frac{\displaystyle\lim_{x\to 0}1}{\sqrt{\displaystyle\lim_{x\to 0}(x^2+4)}+\displaystyle\lim_{x\to 0}2}$$

$$=\frac{1}{2+2}=\frac{1}{4}.\qquad\blacksquare$$

In case anyone is wondering whether there can be more than one limit of a function $f(x)$ as $x\to a$, we state the last theorem for the record.

Theorem 2.2.7   Existence Implies Uniqueness

If $\displaystyle\lim_{x\to a}f(x)$ exists, then it is unique.

$\displaystyle\lim_{x\to a}$ **NOTES FROM THE CLASSROOM**

In mathematics it is just as important to be aware of what a definition or a theorem does *not* say as what it says.

(*i*) Property (*i*) of Theorem 2.2.3 does not say that the limit of a sum is *always* the sum of the limits. For example, $\displaystyle\lim_{x\to 0}(1/x)$ does not exist, so

$$\lim_{x\to 0}\left[\frac{1}{x}-\frac{1}{x}\right]\neq\lim_{x\to 0}\frac{1}{x}-\lim_{x\to 0}\frac{1}{x}.$$

Nevertheless, since $1/x-1/x=0$ for $x\neq 0$, the limit of the difference exists

$$\lim_{x\to 0}\left[\frac{1}{x}-\frac{1}{x}\right]=\lim_{x\to 0}0=0.$$

(*ii*) Similarly, the limit of a product could exist and yet not be equal to the product of the limits. For example, $x/x=1$, for $x\neq 0$, and so

$$\lim_{x\to 0}\left(x\cdot\frac{1}{x}\right)=\lim_{x\to 0}1=1$$

but

$$\lim_{x\to 0}\left(x\cdot\frac{1}{x}\right)\neq\left(\lim_{x\to 0}x\right)\left(\lim_{x\to 0}\frac{1}{x}\right)$$

because $\displaystyle\lim_{x\to 0}(1/x)$ does not exist.

(*iii*) Theorem 2.2.5 does not say that the limit of a quotient fails to exist whenever the limit of the denominator is zero. Example 8 provides a counterexample to that interpretation. However, Theorem 2.2.5 states that a limit of a quotient does not exist whenever the limit of the denominator is zero *and* the limit of the numerator is not zero.

Exercises 2.2   Answers to selected odd-numbered problems begin on page ANS-8.

≡ Fundamentals

In Problems 1–52, find the given limit, or state that it does not exist.

1. $\displaystyle \lim_{x \to -4} 15$

2. $\displaystyle \lim_{x \to 0} \cos \pi$

3. $\displaystyle \lim_{x \to 3} (-4)x$

4. $\displaystyle \lim_{x \to 2} (3x - 9)$

5. $\displaystyle \lim_{x \to -2} x^2$

6. $\displaystyle \lim_{x \to 5} (-x^3)$

7. $\displaystyle \lim_{x \to -1} (x^3 - 4x + 1)$

8. $\displaystyle \lim_{x \to 6} (-5x^2 + 6x + 8)$

9. $\displaystyle \lim_{x \to 2} \frac{2x + 4}{x - 7}$

10. $\displaystyle \lim_{x \to 0} \frac{x + 5}{3x}$

11. $\displaystyle \lim_{t \to 1} (3t - 1)(5t^2 + 2)$

12. $\displaystyle \lim_{t \to -2} (t + 4)^2$

13. $\displaystyle \lim_{s \to 7} \frac{s^2 - 21}{s + 2}$

14. $\displaystyle \lim_{x \to 6} \frac{x^2 - 6x}{x^2 - 7x + 6}$

15. $\displaystyle \lim_{x \to -1} (x + x^2 + x^3)^{135}$

16. $\displaystyle \lim_{x \to 2} \frac{(3x - 4)^{40}}{(x^2 - 2)^{36}}$

17. $\displaystyle \lim_{x \to 6} \sqrt{2x - 5}$

18. $\displaystyle \lim_{x \to 8} (1 + \sqrt[3]{x})$

19. $\displaystyle \lim_{t \to 1} \frac{\sqrt{t}}{t^2 + t - 2}$

20. $\displaystyle \lim_{x \to 2} x^2 \sqrt{x^2 + 5x + 2}$

21. $\displaystyle \lim_{y \to -5} \frac{y^2 - 25}{y + 5}$

22. $\displaystyle \lim_{u \to 8} \frac{u^2 - 5u - 24}{u - 8}$

23. $\displaystyle \lim_{x \to 1} \frac{x^3 - 1}{x - 1}$

24. $\displaystyle \lim_{t \to -1} \frac{t^3 + 1}{t^2 - 1}$

25. $\displaystyle \lim_{x \to 10} \frac{(x - 2)(x + 5)}{(x - 8)}$

26. $\displaystyle \lim_{x \to -3} \frac{2x + 6}{4x^2 - 36}$

27. $\displaystyle \lim_{x \to 2} \frac{x^3 + 3x^2 - 10x}{x - 2}$

28. $\displaystyle \lim_{x \to 1.5} \frac{2x^2 + 3x - 9}{x - 1.5}$

29. $\displaystyle \lim_{t \to 1} \frac{t^3 - 2t + 1}{t^3 + t^2 - 2}$

30. $\displaystyle \lim_{x \to 0} x^3 (x^4 + 2x^3)^{-1}$

31. $\displaystyle \lim_{x \to 0^+} \frac{(x + 2)(x^5 - 1)^3}{(\sqrt{x} + 4)^2}$

32. $\displaystyle \lim_{x \to -2} x \sqrt{x + 4}\ \sqrt[3]{x - 6}$

33. $\displaystyle \lim_{x \to 0} \left[\frac{x^2 + 3x - 1}{x} + \frac{1}{x} \right]$

34. $\displaystyle \lim_{x \to 2} \left[\frac{1}{x - 2} - \frac{6}{x^2 + 2x - 8} \right]$

35. $\displaystyle \lim_{x \to 3^+} \frac{(x + 3)^2}{\sqrt{x - 3}}$

36. $\displaystyle \lim_{x \to 3} (x - 4)^{99}(x^2 - 7)^{10}$

37. $\displaystyle \lim_{x \to 10} \sqrt{\frac{10x}{2x + 5}}$

38. $\displaystyle \lim_{r \to 1} \frac{\sqrt{(r^2 + 3r - 2)^3}}{\sqrt[3]{(5r - 3)^2}}$

39. $\displaystyle \lim_{h \to 4} \sqrt{\frac{h}{h + 5}} \left(\frac{h^2 - 16}{h - 4} \right)^2$

40. $\displaystyle \lim_{t \to 2} (t + 2)^{3/2}(2t + 4)^{1/3}$

41. $\displaystyle \lim_{x \to 0^-} \sqrt[5]{\frac{x^3 - 64x}{x^2 + 2x}}$

42. $\displaystyle \lim_{x \to 1^+} \left(8x + \frac{2}{x} \right)^5$

43. $\displaystyle \lim_{t \to 1} (at^2 - bt)^2$

44. $\displaystyle \lim_{x \to -1} \sqrt{u^2 x^2 + 2xu + 1}$

45. $\displaystyle \lim_{h \to 0} \frac{(8 + h)^2 - 64}{h}$

46. $\displaystyle \lim_{h \to 0} \frac{1}{h}[(1 + h)^3 - 1]$

47. $\displaystyle \lim_{h \to 0} \frac{1}{h} \left(\frac{1}{x + h} - \frac{1}{x} \right)$

48. $\displaystyle \lim_{h \to 0} \frac{\sqrt{x + h} - \sqrt{x}}{h} \quad (x > 0)$

49. $\displaystyle \lim_{t \to 1} \frac{\sqrt{t} - 1}{t - 1}$

50. $\displaystyle \lim_{u \to 5} \frac{\sqrt{u + 4} - 3}{u - 5}$

51. $\displaystyle \lim_{v \to 0} \frac{\sqrt{25 + v} - 5}{\sqrt{1 + v} - 1}$

52. $\displaystyle \lim_{x \to 1} \frac{4 - \sqrt{x + 15}}{x^2 - 1}$

In Problems 53–60, assume that $\displaystyle \lim_{x \to a} f(x) = 4$ and $\displaystyle \lim_{x \to a} g(x) = 2$. Find the given limit, or state that it does not exist.

53. $\displaystyle \lim_{x \to a} [5f(x) + 6g(x)]$

54. $\displaystyle \lim_{x \to a} [f(x)]^3$

55. $\displaystyle \lim_{x \to a} \frac{1}{g(x)}$

56. $\displaystyle \lim_{x \to a} \sqrt{\frac{f(x)}{g(x)}}$

57. $\displaystyle \lim_{x \to a} \frac{f(x)}{f(x) - 2g(x)}$

58. $\displaystyle \lim_{x \to a} \frac{[f(x)]^2 - 4[g(x)]^2}{f(x) - 2g(x)}$

59. $\displaystyle \lim_{x \to a} x f(x) g(x)$

60. $\displaystyle \lim_{x \to a} \frac{6x + 3}{xf(x) + g(x)},\ a \neq -\frac{1}{2}$

≡ Think About It

In Problems 61 and 62, use the first result to find the limits in parts (a)–(c). Justify each step in your work citing the appropriate property of limits.

61. $\displaystyle \lim_{x \to 1} \frac{x^{100} - 1}{x - 1} = 100$

    (a) $\displaystyle \lim_{x \to 1} \frac{x^{100} - 1}{x^2 - 1}$ (b) $\displaystyle \lim_{x \to 1} \frac{x^{50} - 1}{x - 1}$ (c) $\displaystyle \lim_{x \to 1} \frac{(x^{100} - 1)^2}{(x - 1)^2}$

62. $\displaystyle \lim_{x \to 0} \frac{\sin x}{x} = 1$

    (a) $\displaystyle \lim_{x \to 0} \frac{2x}{\sin x}$ (b) $\displaystyle \lim_{x \to 0} \frac{1 - \cos^2 x}{x^2}$ (c) $\displaystyle \lim_{x \to 0} \frac{8x^2 - \sin x}{x}$

63. Using $\displaystyle \lim_{x \to 0} \frac{\sin x}{x} = 1$, show that $\displaystyle \lim_{x \to 0} \sin x = 0$.

64. If $\displaystyle \lim_{x \to 2} \frac{2f(x) - 5}{x + 3} = 4$, find $\displaystyle \lim_{x \to 2} f(x)$.

2.3 Continuity

Introduction In the discussion in Section 1.1 on graphing functions, we used the phrase "connect the points with a smooth curve." This phrase invokes the image of a graph that is a nice *continuous* curve—in other words, a curve with no breaks, gaps, or holes in it. Indeed, a continuous function is often described as one whose graph can be drawn without lifting pencil from paper.

In Section 2.2 we saw that the function value $f(a)$ played no part in determining the existence of $\lim_{x \to a} f(x)$. But we did see in Section 2.2 that limits as $x \to a$ of polynomial functions and certain rational functions could be found by simply evaluating the function *at $x = a$*. The reason we can do that in some instances is the fact that the function is *continuous* at a number a. In this section we will see that both the value $f(a)$ and the limit of f as x approaches a number a play major roles in defining the notion of continuity. Before giving the definition, we illustrate in FIGURE 2.3.1 some intuitive examples of graphs of functions that are *not* continuous at a.

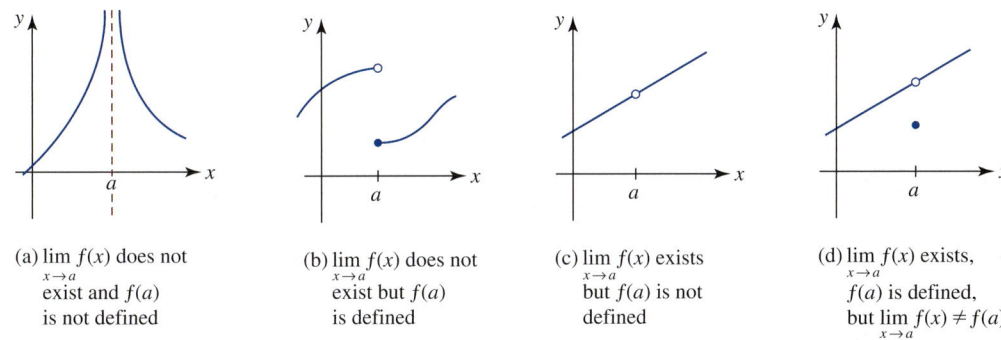

(a) $\lim_{x \to a} f(x)$ does not exist and $f(a)$ is not defined

(b) $\lim_{x \to a} f(x)$ does not exist but $f(a)$ is defined

(c) $\lim_{x \to a} f(x)$ exists but $f(a)$ is not defined

(d) $\lim_{x \to a} f(x)$ exists, $f(a)$ is defined, but $\lim_{x \to a} f(x) \neq f(a)$

FIGURE 2.3.1 Four examples of f *not* continuous at a

Continuity at a Number Figure 2.3.1 suggests the following threefold condition of continuity of a function f at a number a.

Definition 2.3.1   Continuity at a

A function f is said to be **continuous** at a number a if

(i) $f(a)$ is defined,   (ii) $\lim_{x \to a} f(x)$ exists, and   (iii) $\lim_{x \to a} f(x) = f(a)$.

If any one of the three conditions in Definition 2.3.1 fails, then f is said to be **discontinuous** at the number a.

EXAMPLE 1 Three Functions

Determine whether each of the functions is continuous at 1.

(a) $f(x) = \dfrac{x^3 - 1}{x - 1}$ **(b)** $g(x) = \begin{cases} \dfrac{x^3 - 1}{x - 1}, & x \neq 1 \\ 2, & x = 1 \end{cases}$ **(c)** $h(x) = \begin{cases} \dfrac{x^3 - 1}{x - 1}, & x \neq 1 \\ 3, & x = 1 \end{cases}$.

Solution

(a) f is discontinuous at 1 since substituting $x = 1$ into the function results in $0/0$. We say that $f(1)$ is not defined and so the first condition of continuity in Definition 2.3.1 is violated.

(b) Because g is defined at 1, that is, $g(1) = 2$, we next determine whether $\lim_{x \to 1} g(x)$ exists. From

$$\lim_{x \to 1} \frac{x^3 - 1}{x - 1} = \lim_{x \to 1} \frac{(x - 1)(x^2 + x + 1)}{x - 1} = \lim_{x \to 1}(x^2 + x + 1) = 3 \qquad (1)$$

◀ Recall from algebra that $a^3 - b^3 = (a - b)(a^2 + ab + b^2)$

we conclude $\lim_{x \to 1} g(x)$ exists and equals 3. Since this value is not the same as $g(1) = 2$, the second condition of Definition 2.3.1 is violated. The function g is discontinuous at 1.

(c) First, $h(1)$ is defined, in this case, $h(1) = 3$. Second, $\lim_{x \to 1} h(x) = 3$ from (1) of part (b). Third, we have $\lim_{x \to 1} h(x) = h(1) = 3$. Thus *all* three conditions in Definition 2.3.1 are satisfied and so the function h is continuous at 1.

The graphs of the three functions are compared in FIGURE 2.3.2.

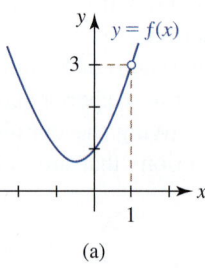
(a)

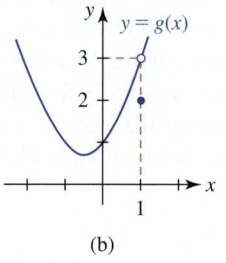
(b)

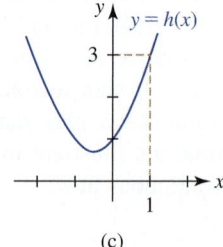
(c)

FIGURE 2.3.2   Graphs of functions in Example 1   ■

EXAMPLE 2   Piecewise-Defined Function

Determine whether the piecewise-defined function is continuous at 2.

$$f(x) = \begin{cases} x^2, & x < 2 \\ 5, & x = 2 \\ -x + 6, & x > 2. \end{cases}$$

Solution   First, observe that $f(2)$ is defined and equals 5. Next, we see from

$$\left.\begin{array}{l} \lim_{x \to 2^-} f(x) = \lim_{x \to 2^-} x^2 = 4 \\ \lim_{x \to 2^+} f(x) = \lim_{x \to 2^+} (-x + 6) = 4 \end{array}\right\} \text{ implies } \lim_{x \to 2} f(x) = 4$$

that the limit of f as $x \to 2$ exists. Finally, because $\lim_{x \to 2} f(x) \neq f(2) = 5$, it follows from (*iii*) of Definition 2.3.1 that f is discontinuous at 2. The graph of f is shown in FIGURE 2.3.3.   ■

<figure>
y
5
y = f(x)

2

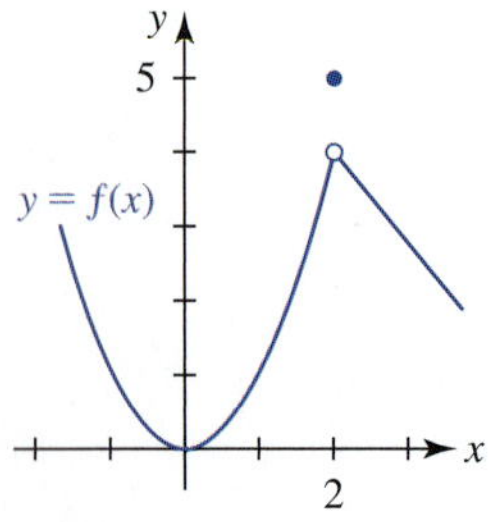

FIGURE 2.3.3   Graph of function in Example 2
</figure>

❚ Continuity on an Interval   We will now extend the notion of continuity at a number a to **continuity on an interval**.

Definition 2.3.2   Continuity on an Interval

A function f is continuous

(*i*) on an **open interval** (a, b) if it is continuous at every number in the interval; and
(*ii*) on a **closed interval** $[a, b]$ if it is continuous on (a, b) and, in addition,

$$\lim_{x \to a^+} f(x) = f(a) \qquad \text{and} \qquad \lim_{x \to b^-} f(x) = f(b).$$

If the right-hand limit condition $\lim_{x \to a^+} f(x) = f(a)$ given in (*ii*) of Definition 2.3.1 is satisfied, we say that **f is continuous from the right at a**; if $\lim_{x \to b^-} f(x) = f(b)$, then **$f$ is continuous from the left at b**.

Extensions of these concepts to intervals such as $[a, b)$, $(a, b]$, (a, ∞), $(-\infty, b)$, $(-\infty, \infty)$, $[a, \infty)$, and $(-\infty, b]$ are made in the expected manner. For example, f is continuous on $[1, 5)$ if it is continuous on the open interval $(1, 5)$ and continuous from the right at 1.

EXAMPLE 3   Continuity on an Interval

(a) As we see from FIGURE 2.3.4(a), $f(x) = 1/\sqrt{1 - x^2}$ is continuous on the open interval $(-1, 1)$ but is not continuous on the closed interval $[-1, 1]$, since neither $f(-1)$ nor $f(1)$ is defined.

(b) $f(x) = \sqrt{1 - x^2}$ is continuous on $[-1, 1]$. Observe from Figure 2.3.4(b) that

$$\lim_{x \to -1^+} f(x) = f(-1) = 0 \quad \text{and} \quad \lim_{x \to 1^-} f(x) = f(1) = 0.$$

(c) $f(x) = \sqrt{x - 1}$ is continuous on the unbounded interval $[1, \infty)$, because

$$\lim_{x \to a} f(x) = \sqrt{\lim_{x \to a}(x - 1)} = \sqrt{a - 1} = f(a),$$

for any real number a satisfying $a > 1$, and f is continuous from the right at 1 since

$$\lim_{x \to 1^+} \sqrt{x - 1} = f(1) = 0.$$

See Figure 2.3.4(c). ∎

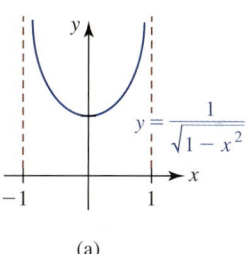

(a)

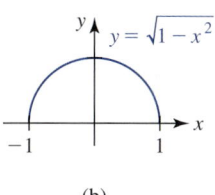

(b)

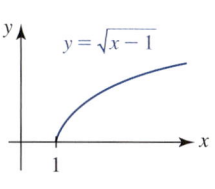

(c)

FIGURE 2.3.4   Graphs of functions in Example 3

A review of the graphs in Figures 1.4.1 and 1.4.2 shows that $y = \sin x$ and $y = \cos x$ are continuous on $(-\infty, \infty)$. Figures 1.4.3 and 1.4.5 show that $y = \tan x$ and $y = \sec x$ are discontinuous at $x = (2n + 1)\pi/2, n = 0, \pm 1, \pm 2, \ldots$, whereas Figures 1.4.4 and 1.4.6 show that $y = \cot x$ and $y = \csc x$ are discontinuous at $x = n\pi, n = 0, \pm 1, \pm 2, \ldots$. The inverse trigonometric functions $y = \sin^{-1} x$ and $y = \cos^{-1} x$ are continuous on the closed interval $[-1, 1]$. See Figures 1.5.9 and 1.5.12. The natural exponential function $y = e^x$ is continuous on $(-\infty, \infty)$, whereas the natural logarithmic function $y = \ln x$ is continuous on $(0, \infty)$. See Figures 1.6.5 and 1.6.6.

■ **Continuity of a Sum, Product, and Quotient**   When two functions f and g are continuous at a number a, then the combinations of functions formed by addition, multiplication, and division are also continuous at a. In the case of division f/g we must, of course, require that $g(a) \neq 0$.

Theorem 2.3.1   Continuity of a Sum, Product, and Quotient

If the functions f and g are continuous at a number a, then the sum $f + g$, the product fg, and the quotient f/g $(g(a) \neq 0)$ are continuous at $x = a$.

PROOF OF CONTINUITY OF THE PRODUCT *fg*   As a consequence of the assumption that the functions f and g are continuous at a number a, we can say that both functions are defined at $x = a$, the limits of both functions as x approaches a exist, and

$$\lim_{x \to a} f(x) = f(a) \quad \text{and} \quad \lim_{x \to a} g(x) = g(a).$$

Because the limits exist, we know that the limit of a product is the product of the limits:

$$\lim_{x \to a}(f(x)g(x)) = \left(\lim_{x \to a} f(x)\right)\left(\lim_{x \to a} g(x)\right) = f(a)g(a).$$

The proofs of the remaining parts of Theorem 2.3.1 are obtained in a similar manner. ∎

Since Definition 2.3.1 implies that $f(x) = x$ is continuous at any real number x, we see from successive applications of Theorem 2.3.1 that the functions $x, x^2, x^3, \ldots, x^n$ are also continuous for every x in the interval $(-\infty, \infty)$. Because a polynomial function is just a sum of powers of x, another application of Theorem 2.3.1 shows:

• *A polynomial function f is continuous on $(-\infty, \infty)$.*

Functions, such as polynomials and the sine and cosine, that are continuous for *all* real numbers, that is, on the interval $(-\infty, \infty)$, are said to be **continuous everywhere**. A function that is continuous everywhere is also just said to be **continuous**. Now, if $p(x)$ and $q(x)$ are polynomial functions, it also follows directly from Theorem 2.3.1 that:

- *A rational function $f(x) = p(x)/q(x)$ is continuous except at numbers at which the denominator $q(x)$ is zero.*

■ Terminology A discontinuity of a function f is often given a special name.

- If $x = a$ is a vertical asymptote for the graph of $y = f(x)$, then f is said to have an **infinite discontinuity** at a.

Figure 2.3.1(a) illustrates a function with an infinite discontinuity at a.

- If $\lim_{x \to a^-} f(x) = L_1$ and $\lim_{x \to a^+} f(x) = L_2$ and $L_1 \neq L_2$, then f is said to have a **finite discontinuity** or a **jump discontinuity** at a.

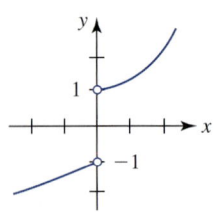

FIGURE 2.3.5 Jump discontinuity at $x = 0$

The function $y = f(x)$ given in FIGURE 2.3.5 has a jump discontinuity at 0, since $\lim_{x \to 0^-} f(x) = -1$ and $\lim_{x \to 0^+} f(x) = 1$. The greatest integer function $f(x) = \lfloor x \rfloor$ has a jump discontinuity at every integer value of x.

- If $\lim_{x \to a} f(x)$ exists but either f is not defined at $x = a$ or $f(a) \neq \lim_{x \to a} f(x)$, then f is said to have a **removable discontinuity** at a.

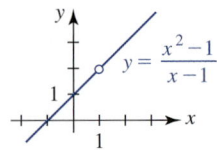

(a) Not continuous at 1

For example, the function $f(x) = (x^2 - 1)/(x - 1)$ is not defined at $x = 1$ but $\lim_{x \to 1} f(x) = 2$. By *defining* $f(1) = 2$, the new function

$$
f(x) = \begin{cases} \dfrac{x^2 - 1}{x - 1}, & x \neq 1 \\ 2, & x = 1 \end{cases}
$$

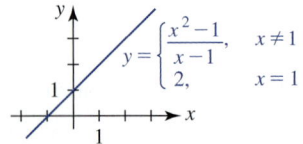

(b) Continuous at 1

FIGURE 2.3.6 Removable discontinuity at $x = 1$

is continuous everywhere. See FIGURE 2.3.6.

■ Continuity of f^{-1} The plausibility of the next theorem follows from the fact that the graph of an inverse function f^{-1} is a reflection of the graph of f in the line $y = x$.

Theorem 2.3.2 Continuity of an Inverse Function

If f is a continuous one-to-one function on an interval $[a, b]$, then f^{-1} is continuous on either $[f(a), f(b)]$ or $[f(b), f(a)]$.

The sine function, $f(x) = \sin x$, is continuous on $[-\pi/2, \pi/2]$ and, as noted previously, the inverse of f, $y = \sin^{-1} x$, is continuous on the closed interval $[f(-\pi/2), f(\pi/2)] = [-1, 1]$.

■ Limit of a Composite Function The next theorem tells us that if a function f is continuous, then the limit of the function is the function of the limit. The proof of Theorem 2.3.3 is given in the *Appendix*.

Theorem 2.3.3 Limit of a Composite Function

If $\lim_{x \to a} g(x) = L$ and f is continuous at L, then

$$
\lim_{x \to a} f(g(x)) = f\left(\lim_{x \to a} g(x)\right) = f(L).
$$

Theorem 2.3.3 is useful in proving other theorems. If the function g is continuous at a and f is continuous at $g(a)$, then we see that

$$\lim_{x \to a} f(g(x)) = f\left(\lim_{x \to a} g(x)\right) = f(g(a)).$$

We have just proved that the composite of two continuous functions is continuous.

Theorem 2.3.4 Continuity of a Composite Function

If g is continuous at a number a and f is continuous at $g(a)$, then the composite function $(f \circ g)(x) = f(g(x))$ is continuous at a.

EXAMPLE 4 Continuity of a Composite Function

$f(x) = \sqrt{x}$ is continuous on the interval $[0, \infty)$ and $g(x) = x^2 + 2$ is continuous on $(-\infty, \infty)$. But, since $g(x) \geq 0$ for all x, the composite function

$$(f \circ g)(x) = f(g(x)) = \sqrt{x^2 + 2}$$

is continuous everywhere. ∎

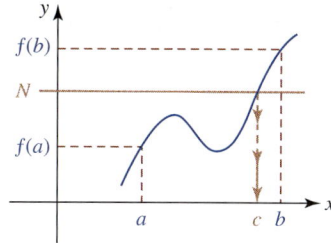

FIGURE 2.3.7 A continuous function f takes on all values between $f(a)$ and $f(b)$

If a function f is continuous on a closed interval $[a, b]$, then, as illustrated in FIGURE 2.3.7, f takes on all values between $f(a)$ and $f(b)$. Put another way, a continuous function f does not "skip" any values.

Theorem 2.3.5 Intermediate Value Theorem

If f denotes a function continuous on a closed interval $[a, b]$ for which $f(a) \neq f(b)$, and if N is any number between $f(a)$ and $f(b)$, then there exists at least one number c between a and b such that $f(c) = N$.

EXAMPLE 5 Consequence of Continuity

The polynomial function $f(x) = x^2 - x - 5$ is continuous on the interval $[-1, 4]$ and $f(-1) = -3, f(4) = 7$. For any number N for which $-3 \leq N \leq 7$, Theorem 2.3.5 guarantees that there is a solution to the equation $f(c) = N$, that is, $c^2 - c - 5 = N$ in $[-1, 4]$. Specifically, if we choose $N = 1$, then $c^2 - c - 5 = 1$ is equivalent to

$$c^2 - c - 6 = 0 \qquad \text{or} \qquad (c - 3)(c + 2) = 0.$$

Although the latter equation has two solutions, only the value $c = 3$ is between -1 and 4. ∎

The foregoing example suggests a corollary to the Intermediate Value Theorem.

- *If f satisfies the hypotheses of Theorem 2.3.5 and $f(a)$ and $f(b)$ have opposite algebraic signs, then there exists a number x between a and b for which $f(x) = 0$.*

This fact is often used in locating real zeros of a continuous function f. If the function values $f(a)$ and $f(b)$ have opposite signs, then by identifying $N = 0$, we can say that there is at least one number c in (a, b) for which $f(c) = 0$. In other words, if either $f(a) > 0, f(b) < 0$ or $f(a) < 0, f(b) > 0$, then $f(x)$ has at least one zero c in the interval (a, b). The plausibility of this conclusion is illustrated in FIGURE 2.3.8.

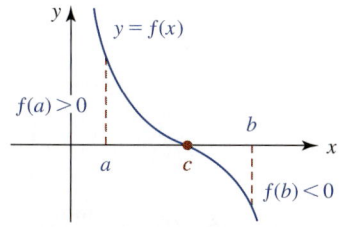

(a) One zero c in (a, b)

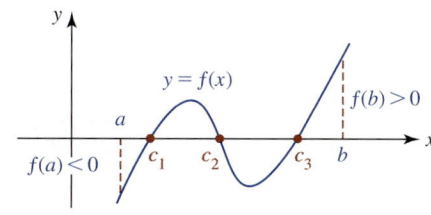

(b) Three zeros c_1, c_2, c_3 in (a, b)

FIGURE 2.3.8 Locating zeros of functions using the Intermediate Value Theorem

■ **Bisection Method**   As a direct consequence of the Intermediate Value Theorem, we can devise a means of approximating the zeros of a continuous function to any degree of accuracy. Suppose $y = f(x)$ is continuous on the closed interval $[a, b]$ such that $f(a)$ and $f(b)$ have opposite algebraic signs. Then, as we have just seen, f has a zero in $[a, b]$. Suppose we bisect the interval $[a, b]$ by finding its midpoint $m_1 = (a + b)/2$. If $f(m_1) = 0$, then m_1 is a zero of f and we proceed no further, but if $f(m_1) \neq 0$, then we can say that:

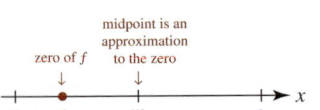

midpoint is an
approximation
zero of f to the zero

FIGURE 2.3.9   The number m_1 is an approximation to the number c

- If $f(a)$ and $f(m_1)$ have opposite algebraic signs, then f has a zero c in $[a, m_1]$.
- If $f(m_1)$ and $f(b)$ have opposite algebraic signs, then f has a zero c in $[m_1, b]$.

That is, if $f(m_1) \neq 0$, then f has a zero in an interval that is one-half the length of the original interval. See FIGURE 2.3.9. We now repeat the process by bisecting this new interval by finding its midpoint m_2. If m_2 is a zero of f, we stop, but if $f(m_2) \neq 0$, we have located a zero in an interval that is one-fourth the length of $[a, b]$. We continue this process of locating a zero of f in shorter and shorter intervals indefinitely. This method of approximating a zero of a continuous function by a sequence of midpoints is called the **bisection method**. Reinspection of Figure 2.3.9 shows that the error in an approximation to a zero in an interval is less than one-half the length of the interval.

EXAMPLE 6   Zeros of a Polynomial Function

(a) Show that the polynomial function $f(x) = x^6 - 3x - 1$ has a real zero in $[-1, 0]$ and in $[1, 2]$.

(b) Approximate the zero in $[1, 2]$ to two decimal places.

Solution

(a) Observe that $f(-1) = 3 > 0$ and $f(0) = -1 < 0$. This change in sign indicates that the graph of f must cross the x-axis at least once in the interval $[-1, 0]$. In other words, there is at least one zero of f in $[-1, 0]$.

Similarly, $f(1) = -3 < 0$ and $f(2) = 57 > 0$ implies that there is at least one zero of f in the interval $[1, 2]$.

(b) A first approximation to the zero in $[1, 2]$ is the midpoint of the interval:

$$m_1 = \frac{1 + 2}{2} = \frac{3}{2} = 1.5, \qquad \text{error} < \frac{1}{2}(2 - 1) = 0.5.$$

Now since $f(m_1) = f\left(\frac{3}{2}\right) > 0$ and $f(1) < 0$, we know that the zero lies in the interval $\left[1, \frac{3}{2}\right]$.

The second approximation is the midpoint of $\left[1, \frac{3}{2}\right]$:

$$m_2 = \frac{1 + \frac{3}{2}}{2} = \frac{5}{4} = 1.25, \qquad \text{error} < \frac{1}{2}\left(\frac{3}{2} - 1\right) = 0.25.$$

Since $f(m_2) = f\left(\frac{5}{4}\right) < 0$, the zero lies in the interval $\left[\frac{5}{4}, \frac{3}{2}\right]$.

The third approximation is the midpoint of $\left[\frac{5}{4}, \frac{3}{2}\right]$:

$$m_3 = \frac{\frac{5}{4} + \frac{3}{2}}{2} = \frac{11}{8} = 1.375, \qquad \text{error} < \frac{1}{2}\left(\frac{3}{2} - \frac{5}{4}\right) = 0.125.$$

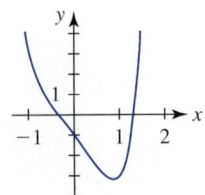

FIGURE 2.3.10   Graph of function in Example 6

If we wish the approximation to be accurate to *three* decimal places, we continue until the error becomes less than 0.0005, and so on.

▶   After eight calculations, we find that $m_8 = 1.300781$ with error less than 0.005. Hence, 1.30 is an approximation to the zero of f in $[1, 2]$ that is accurate to two decimal places. The graph of f is given in FIGURE 2.3.10.   ■

Exercises 2.3   Answers to selected odd-numbered problems begin on page ANS-8.

≡ **Fundamentals**

In Problems 1–12, determine the numbers, if any, at which the given function f is discontinuous.

1. $f(x) = x^3 - 4x^2 + 7$

2. $f(x) = \dfrac{x}{x^2 + 4}$

3. $f(x) = (x^2 - 9x + 18)^{-1}$

4. $f(x) = \dfrac{x^2 - 1}{x^4 - 1}$

5. $f(x) = \dfrac{x - 1}{\sin 2x}$

6. $f(x) = \dfrac{\tan x}{x + 3}$

7. $f(x) = \begin{cases} x, & x < 0 \\ x^2, & 0 \le x < 2 \\ x, & x > 2 \end{cases}$ 8. $f(x) = \begin{cases} \dfrac{|x|}{x}, & x \ne 0 \\ 1, & x = 0 \end{cases}$

9. $f(x) = \begin{cases} \dfrac{x^2 - 25}{x - 5}, & x \ne 5 \\ 10, & x = 5 \end{cases}$

10. $f(x) = \begin{cases} \dfrac{x - 1}{\sqrt{x} - 1}, & x \ne 1 \\ \dfrac{1}{2}, & x = 1 \end{cases}$

11. $f(x) = \dfrac{1}{2 + \ln x}$ 12. $f(x) = \dfrac{2}{e^x - e^{-x}}$

In Problems 13–24, determine whether the given function f is continuous on the indicated intervals.

13. $f(x) = x^2 + 1$
  (a) $[-1, 4]$ (b) $[5, \infty)$

14. $f(x) = \dfrac{1}{x}$
  (a) $(-\infty, \infty)$ (b) $(0, \infty)$

15. $f(x) = \dfrac{1}{\sqrt{x}}$
  (a) $(0, 4]$ (b) $[1, 9]$

16. $f(x) = \sqrt{x^2 - 9}$
  (a) $[-3, 3]$ (b) $[3, \infty)$

17. $f(x) = \tan x$
  (a) $[0, \pi]$ (b) $[-\pi/2, \pi/2]$

18. $f(x) = \csc x$
  (a) $(0, \pi)$ (b) $(2\pi, 3\pi)$

19. $f(x) = \dfrac{x}{x^3 + 8}$
  (a) $[-4, -3]$ (b) $(-\infty, \infty)$

20. $f(x) = \dfrac{1}{|x| - 4}$
  (a) $(-\infty, -1]$ (b) $[1, 6]$

21. $f(x) = \dfrac{x}{2 + \sec x}$
  (a) $(-\infty, \infty)$ (b) $[\pi/2, 3\pi/2]$

22. $f(x) = \sin\dfrac{1}{x}$
  (a) $[1/\pi, \infty)$ (b) $[-2/\pi, 2/\pi]$

23.

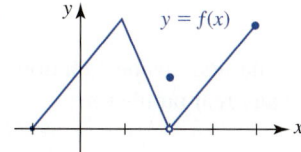

FIGURE 2.3.11   Graph for Problem 23
  (a) $[-1, 3]$ (b) $(2, 4]$

24.

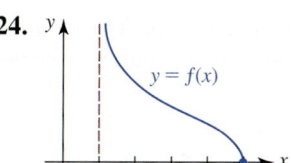

FIGURE 2.3.12   Graph for Problem 24
  (a) $[2, 4]$ (b) $[1, 5]$

In Problems 25–28, find values of m and n so that the given function f is continuous.

25. $f(x) = \begin{cases} mx, & x < 4 \\ x^2, & x \ge 4 \end{cases}$

26. $f(x) = \begin{cases} \dfrac{x^2 - 4}{x - 2}, & x \ne 2 \\ m, & x = 2 \end{cases}$

27. $f(x) = \begin{cases} mx, & x < 3 \\ n, & x = 3 \\ -2x + 9, & x > 3 \end{cases}$

28. $f(x) = \begin{cases} mx - n, & x < 1 \\ 5, & x = 1 \\ 2mx + n, & x > 1 \end{cases}$

In Problems 29 and 30, $\lfloor x \rfloor$ denotes the greatest integer not exceeding x. Sketch a graph to determine the points at which the given function is discontinuous.

29. $f(x) = \lfloor 2x - 1 \rfloor$ 30. $f(x) = \lfloor x \rfloor - x$

In Problems 31 and 32, determine whether the given function has a removable discontinuity at the given number a. If the discontinuity is removable, define a new function that is continuous at a.

31. $f(x) = \dfrac{x - 9}{\sqrt{x} - 3}$, $a = 9$ 32. $f(x) = \dfrac{x^4 - 1}{x^2 - 1}$, $a = 1$

In Problems 33–42, use Theorem 2.3.3 to find the given limit.

33. $\lim\limits_{x \to \pi/6} \sin(2x + \pi/3)$ 34. $\lim\limits_{x \to \pi^2} \cos\sqrt{x}$

35. $\lim\limits_{x \to \pi/2} \sin(\cos x)$ 36. $\lim\limits_{x \to \pi/2} (1 + \cos(\cos x))$

37. $\lim\limits_{t \to \pi} \cos\left(\dfrac{t^2 - \pi^2}{t - \pi}\right)$ 38. $\lim\limits_{t \to 0} \tan\left(\dfrac{\pi t}{t^2 + 3t}\right)$

39. $\lim\limits_{t \to \pi} \sqrt{t - \pi + \cos^2 t}$ 40. $\lim\limits_{t \to 1} (4t + \sin 2\pi t)^3$

41. $\lim\limits_{x \to -3} \sin^{-1}\left(\dfrac{x + 3}{x^2 + 4x + 3}\right)$ 42. $\lim\limits_{x \to \pi} e^{\cos 3x}$

In Problems 43 and 44, determine the interval(s) where $f \circ g$ is continuous.

43. $f(x) = \dfrac{1}{\sqrt{x} - 1}$, $g(x) = x + 4$

44. $f(x) = \dfrac{5x}{x - 1}$, $g(x) = (x - 2)^2$

In Problems 45–48, verify the Intermediate Value Theorem for f on the given interval. Find a number c in the interval for the indicated value of N.

45. $f(x) = x^2 - 2x$, $[1, 5]$; $N = 8$

46. $f(x) = x^2 + x + 1$, $[-2, 3]$; $N = 6$

47. $f(x) = x^3 - 2x + 1$, $[-2, 2]$; $N = 1$

48. $f(x) = \dfrac{10}{x^2 + 1}$, $[0, 1]$; $N = 8$

49. Given that $f(x) = x^5 + 2x - 7$, show that there is a number c such that $f(c) = 50$.

50. Given that f and g are continuous on $[a, b]$ such that $f(a) > g(a)$ and $f(b) < g(b)$, show that there is a number c in (a, b) such that $f(c) = g(c)$. [*Hint*: Consider the function $f - g$.]

In Problems 51–54, show that the given equation has a solution in the indicated interval.

51. $2x^7 = 1 - x$, $(0, 1)$

52. $\dfrac{x^2 + 1}{x + 3} + \dfrac{x^4 + 1}{x - 4} = 0$, $(-3, 4)$

53. $e^{-x} = \ln x$, $(1, 2)$

54. $\dfrac{\sin x}{x} = \dfrac{1}{2}$, $(\pi/2, \pi)$

≡ Calculator/CAS Problems

In Problems 55 and 56, use a calculator or CAS to obtain the graph of the given function. Use the bisection method to approximate, to an accuracy of two decimal places, the real zeros of f that you discover from the graph.

55. $f(x) = 3x^5 - 5x^3 - 1$ **56.** $f(x) = x^5 + x - 1$

57. Use the bisection method to approximate the value of c in Problem 49 to an accuracy of two decimal places.

58. Use the bisection method to approximate the solution in Problem 51 to an accuracy of two decimal places.

59. Use the bisection method to approximate the solution in Problem 52 to an accuracy of two decimal places.

60. Suppose a closed right-circular cylinder has a given volume V and surface area S (lateral side, top, and bottom).

(a) Show that the radius r of the cylinder must satisfy the equation $2\pi r^3 - Sr + 2V = 0$.

(b) Suppose $V = 3000$ ft^3 and $S = 1800$ ft^2. Use a calculator or CAS to obtain the graph of

$$f(r) = 2\pi r^3 - 1800r + 6000.$$

(c) Use the graph in part (b) and the bisection method to find the dimensions of the cylinder corresponding to the volume and surface area given in part (b). Use an accuracy of two decimal places.

≡ Think About It

61. Given that f and g are continuous at a number a, prove that $f + g$ is continuous at a.

62. Given that f and g are continuous at a number a and $g(a) \neq 0$, prove that f/g is continuous at a.

63. Let $f(x) = \lfloor x \rfloor$ be the greatest integer function and $g(x) = \cos x$. Determine the points at which $f \circ g$ is discontinuous.

64. Consider the functions

$$f(x) = |x| \quad \text{and} \quad g(x) = \begin{cases} x + 1, & x < 0 \\ x - 1, & x \geq 0. \end{cases}$$

Sketch the graphs of $f \circ g$ and $g \circ f$. Determine whether $f \circ g$ and $g \circ f$ are continuous at 0.

65. A Mathematical Classic The **Dirichlet function**

$$f(x) = \begin{cases} 1, & x \text{ rational} \\ 0, & x \text{ irrational} \end{cases}$$

is named after the German mathematician **Johann Peter Gustav Lejeune Dirichlet** (1805–1859). Dirichlet is responsible for the definition of a function as we know it today.

(a) Show that f is discontinuous at every real number a. In other words, f is a *nowhere continuous function*.

(b) What does the graph of f look like?

(c) If r is a positive rational number, show that f is r-periodic, that is, $f(x + r) = f(x)$.

2.4 Trigonometric Limits

▌ Introduction   In this section we examine limits that involve trigonometric functions. As the examples in this section will illustrate, computation of trigonometric limits entails both algebraic manipulations and knowledge of some basic trigonometric identities. We begin with some simple limit results that are consequences of continuity.

▌ Using Continuity   We saw in the preceding section that the sine and cosine functions are everywhere continuous. It follows from Definition 2.3.1 that for any real number a,

$$\lim_{x \to a} \sin x = \sin a, \tag{1}$$

$$\lim_{x \to a} \cos x = \cos a. \tag{2}$$

Similarly, for a number a in the domain of the given trigonometric function

$$\lim_{x \to a} \tan x = \tan a, \qquad \lim_{x \to a} \cot x = \cot a, \tag{3}$$

$$\lim_{x \to a} \sec x = \sec a, \qquad \lim_{x \to a} \csc x = \csc a. \tag{4}$$

EXAMPLE 1   Using (1) and (2)

From (1) and (2) we have

$$\lim_{x \to 0} \sin x = \sin 0 = 0 \quad \text{and} \quad \lim_{x \to 0} \cos x = \cos 0 = 1. \tag{5} \blacksquare$$

We will draw on the results in (5) in the following discussion on computing other trigonometric limits. But first, we consider a theorem that is particularly useful when working with trigonometric limits.

■ **Squeeze Theorem**   The next theorem has many names: **Squeeze Theorem**, **Pinching Theorem**, **Sandwiching Theorem**, **Squeeze Play Theorem**, and **Flyswatter Theorem** are just a few of them. As shown in FIGURE 2.4.1, if the graph of $f(x)$ is "squeezed" between the graphs of two other functions $g(x)$ and $h(x)$ for all x close to a, and if the functions g and h have a common limit L as $x \to a$, it stands to reason that f also approaches L as $x \to a$. The proof of Theorem 2.4.1 is given in the *Appendix*.

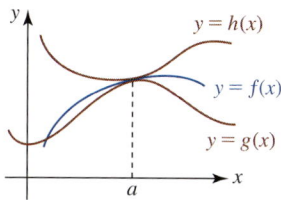

FIGURE 2.4.1   Graph of f squeezed between the graphs g and h

---

Theorem 2.4.1   Squeeze Theorem

Suppose f, g, and h are functions for which $g(x) \le f(x) \le h(x)$ for all x in an open interval that contains a number a, except possibly at a itself. If

$$\lim_{x \to a} g(x) = L \quad \text{and} \quad \lim_{x \to a} h(x) = L,$$

then $\lim_{x \to a} f(x) = L$.

◀ A colleague from Russia said this result was called the **Two Soldiers Theorem** when he was in school. Think about it.

Before applying Theorem 2.4.1, let us consider a trigonometric limit that does not exist.

EXAMPLE 2   A Limit That Does Not Exist

The limit $\lim_{x \to 0} \sin(1/x)$ does not exist. The function $f(x) = \sin(1/x)$ is odd but is not periodic. The graph f oscillates between -1 and 1 as $x \to 0$:

$$\sin \frac{1}{x} = \pm 1 \quad \text{for} \quad \frac{1}{x} = \frac{\pi}{2} + n\pi, \quad n = 0, \pm 1, \pm 2, \dots .$$

For example, $\sin(1/x) = 1$ for $n = 500$ or $x \approx 0.00064$, and $\sin(1/x) = -1$ for $n = 501$ or $x \approx 0.00063$. This means that near the origin the graph of f becomes so compressed that it appears to be one continuous smear of color. See FIGURE 2.4.2. $\blacksquare$

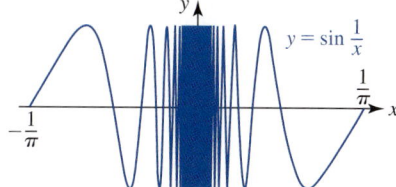

FIGURE 2.4.2   Graph of function in Example 2

EXAMPLE 3   Using the Squeeze Theorem

Find the limit $\lim_{x \to 0} x^2 \sin \frac{1}{x}$.

Solution   First observe that

$$\lim_{x \to 0} x^2 \sin \frac{1}{x} \ne \left(\lim_{x \to 0} x^2 \right) \left(\lim_{x \to 0} \sin \frac{1}{x} \right)$$

because we have just seen in Example 2 that $\lim_{x \to 0} \sin(1/x)$ does not exist. But for $x \ne 0$ we have $-1 \le \sin(1/x) \le 1$. Therefore,

$$-x^2 \le x^2 \sin \frac{1}{x} \le x^2.$$

Now if we make the identifications $g(x) = -x^2$ and $h(x) = x^2$, it follows from (1) of Section 2.2 that $\lim_{x \to 0} g(x) = 0$ and $\lim_{x \to 0} h(x) = 0$. Hence, from the Squeeze Theorem we conclude that

$$\lim_{x \to 0} x^2 \sin\frac{1}{x} = 0.$$

In FIGURE 2.4.3 note the small scale on the x- and y-axes.

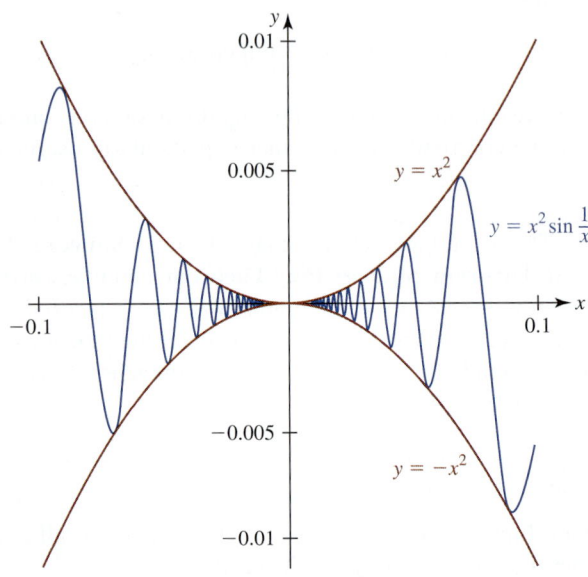

FIGURE 2.4.3   Graph of function in Example 3

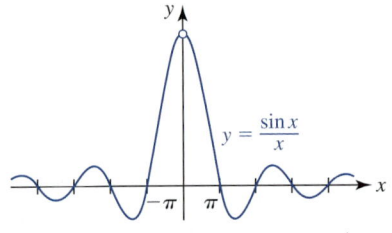

FIGURE 2.4.4   Graph of $f(x) = (\sin x)/x$

■ **An Important Trigonometric Limit**   Although the function $f(x) = (\sin x)/x$ is not defined at $x = 0$, the numerical table in Example 7 of Section 2.1 and the graph in FIGURE 2.4.4 suggests that $\lim_{x \to 0} (\sin x)/x$ exists. We are now able to prove this conjecture using the Squeeze Theorem.

Consider a circle centered at the origin O with radius 1. As shown in FIGURE 2.4.5(a), let the shaded region OPR be a sector of the circle with central angle t such that $0 < t < \pi/2$. We see from parts (b), (c), and (d) of Figure 2.4.5 that

$$\text{area of } \triangle OPR \le \text{area of sector } OPR \le \text{area of } \triangle OQR. \tag{6}$$

From Figure 2.4.5(b) the height of $\triangle OPR$ is $\overline{OP} \sin t = 1 \cdot \sin t = \sin t$, and so

$$\text{area of } \triangle OPR = \frac{1}{2}\overline{OR} \cdot (\text{height}) = \frac{1}{2} \cdot 1 \cdot \sin t = \frac{1}{2}\sin t. \tag{7}$$

From Figure 2.4.5(d), $\overline{QR}/\overline{OR} = \tan t$ or $\overline{QR} = \tan t$, so that

$$\text{area of } \triangle OQR = \frac{1}{2}\overline{OR} \cdot \overline{QR} = \frac{1}{2} \cdot 1 \cdot \tan t = \frac{1}{2}\tan t. \tag{8}$$

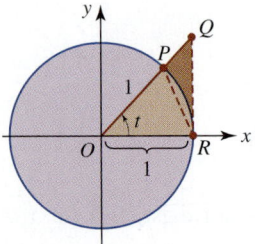
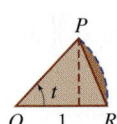
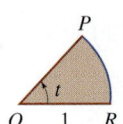
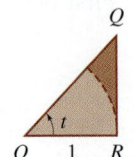

(a) Unit circle   (b) Triangle OPR (c) Sector OPR (d) Right triangle OQR

FIGURE 2.4.5   Unit circle along with two triangles and a circular sector

Finally, the area of a sector of a circle is $\frac{1}{2}r^2\theta$, where r is its radius and θ is the central angle measured in radians. Thus,

$$\text{area of sector } OPR = \frac{1}{2} \cdot 1^2 \cdot t = \frac{1}{2}t. \qquad (9)$$

Using (7), (8), and (9) in the inequality (6) gives

$$\frac{1}{2}\sin t < \frac{1}{2}t < \frac{1}{2}\tan t \qquad \text{or} \qquad 1 < \frac{t}{\sin t} < \frac{1}{\cos t}.$$

From the properties of inequalities, the last inequality can be written

$$\cos t < \frac{\sin t}{t} < 1.$$

We now let $t \to 0^+$ in the last result. Since $(\sin t)/t$ is "squeezed" between 1 and $\cos t$ (which we know from (5) is approaching 1), it follows from Theorem 2.4.1 that $(\sin t)/t \to 1$. While we have assumed $0 < t < \pi/2$, the same result holds for $t \to 0^-$ when $-\pi/2 < t < 0$. Using the symbol x in place of t, we summarize the result:

$$\lim_{x \to 0} \frac{\sin x}{x} = 1. \qquad (10)$$

As the following examples illustrate, the results in (1), (2), (3), and (10) are used often to compute other limits. Note that the limit (10) is the indeterminate form $0/0$.

EXAMPLE 4   Using (10)

Find the limit $\displaystyle\lim_{x \to 0} \frac{10x - 3\sin x}{x}$.

Solution   We rewrite the fractional expression as two fractions with the same denominator x:

$$\lim_{x \to 0} \frac{10x - 3\sin x}{x} = \lim_{x \to 0}\left[\frac{10x}{x} - \frac{3\sin x}{x}\right]$$

$$= \lim_{x \to 0}\frac{10x}{x} - 3\lim_{x \to 0}\frac{\sin x}{x} \quad \leftarrow \begin{array}{l}\text{since both limits exist, also cancel} \\ \text{the } x \text{ in the first expression}\end{array}$$

$$= \lim_{x \to 0} 10 - 3\lim_{x \to 0}\frac{\sin x}{x} \quad \leftarrow \text{ now use (10)}$$

$$= 10 - 3 \cdot 1$$

$$= 7. \qquad \blacksquare$$

EXAMPLE 5   Using the Double-Angle Formula

Find the limit $\displaystyle\lim_{x \to 0} \frac{\sin 2x}{x}$.

Solution   To evaluate the given limit we make use of the double-angle formula $\sin 2x = 2\sin x \cos x$ of Section 1.4, and the fact the limits exist:

$$\lim_{x \to 0} \frac{\sin 2x}{x} = \lim_{x \to 0} \frac{2\cos x \sin x}{x}$$

$$= 2\lim_{x \to 0}\left(\cos x \cdot \frac{\sin x}{x}\right)$$

$$= 2\left(\lim_{x \to 0}\cos x\right)\left(\lim_{x \to 0}\frac{\sin x}{x}\right).$$

From (5) and (10) we know that $\cos x \to 1$ and $(\sin x)/x \to 1$ as $x \to 0$, and so the preceding line becomes

$$\lim_{x \to 0} \frac{\sin 2x}{x} = 2 \cdot 1 \cdot 1 = 2. \qquad \blacksquare$$

EXAMPLE 6 Using (5) and (10)

Find the limit $\lim\limits_{x\to0}\dfrac{\tan x}{x}$.

Solution Using $\tan x = (\sin x)/\cos x$ and the fact that the limits exist we can write

$$\lim_{x\to0}\frac{\tan x}{x} = \lim_{x\to0}\frac{(\sin x)/\cos x}{x}$$

$$= \lim_{x\to0}\frac{1}{\cos x}\cdot\frac{\sin x}{x}$$

$$= \left(\lim_{x\to0}\frac{1}{\cos x}\right)\left(\lim_{x\to0}\frac{\sin x}{x}\right)$$

$$= \frac{1}{1}\cdot 1 = 1. \quad \leftarrow \text{ from (5) and (10)} \quad\blacksquare$$

■ **Using a Substitution** We are often interested in limits similar to that considered in Example 5. But if we wish to find, say, $\lim\limits_{x\to0}\dfrac{\sin 5x}{x}$ the procedure employed in Example 5 breaks down at a practical level since we do not have a readily available trigonometric identity for $\sin 5x$. There is an alternative procedure that allows us to quickly find $\lim\limits_{x\to0}\dfrac{\sin kx}{x}$, where $k \neq 0$ is any real constant, by simply changing the variable by means of a **substitution**. If we let $t = kx$, then $x = t/k$. Notice that as $x \to 0$ then necessarily $t \to 0$. Thus we can write

$$\lim_{x\to0}\frac{\sin kx}{x} = \lim_{t\to0}\frac{\sin t}{t/k} = \lim_{t\to0}\left(\frac{\sin t}{1}\cdot\frac{k}{t}\right) = k\,\overset{\substack{\text{this limit is 1 from (10)}\\\downarrow}}{\lim_{t\to0}\frac{\sin t}{t}} = k.$$

Thus we have proved the general result

$$\lim_{x\to0}\frac{\sin kx}{x} = k. \tag{11}$$

From (11), with $k = 2$, we get the same result $\lim\limits_{x\to0}\dfrac{\sin 2x}{x} = 2$ obtained in Example 5.

EXAMPLE 7 Using a Substitution

Find the limit $\lim\limits_{x\to1}\dfrac{\sin(x - 1)}{x^2 + 2x - 3}$.

Solution Before beginning observe that the limit has the indeterminate form $0/0$ as $x \to 1$. By factoring $x^2 + 2x - 3 = (x + 3)(x - 1)$ the given limit can be expressed as a limit of a product:

$$\lim_{x\to1}\frac{\sin(x - 1)}{x^2 + 2x - 3} = \lim_{x\to1}\frac{\sin(x - 1)}{(x + 3)(x - 1)} = \lim_{x\to1}\left[\frac{1}{x + 3}\cdot\frac{\sin(x - 1)}{x - 1}\right]. \tag{12}$$

Now if we let $t = x - 1$, we see that $x \to 1$ implies $t \to 0$. Therefore,

$$\lim_{x\to1}\frac{\sin(x - 1)}{x - 1} = \lim_{t\to0}\frac{\sin t}{t} = 1. \quad \leftarrow \text{ from (10)}$$

Returning to (12), we can write

$$\lim_{x\to1}\frac{\sin(x - 1)}{x^2 + 2x - 3} = \lim_{x\to1}\left[\frac{1}{x + 3}\cdot\frac{\sin(x - 1)}{x - 1}\right]$$

$$= \left(\lim_{x\to1}\frac{1}{x + 3}\right)\left(\lim_{x\to1}\frac{\sin(x - 1)}{x - 1}\right)$$

$$= \left(\lim_{x\to1}\frac{1}{x + 3}\right)\left(\lim_{t\to0}\frac{\sin t}{t}\right)$$

since both limits exist. Thus,

$$\lim_{x\to1}\frac{\sin(x-1)}{x^2+2x-3}=\left(\lim_{x\to1}\frac{1}{x+3}\right)\left(\lim_{t\to0}\frac{\sin t}{t}\right)=\frac{1}{4}\cdot1=\frac{1}{4}. \qquad ■$$

EXAMPLE 8   Using a Pythagorean Identity

Find the limit $\lim\limits_{x\to0}\dfrac{1-\cos x}{x}$.

Solution   To compute this limit we start with a bit of algebraic cleverness by multiplying the numerator and denominator by the conjugate factor of the numerator. Next we use the fundamental Pythagorean identity $\sin^2 x+\cos^2 x=1$ in the form $1-\cos^2 x=\sin^2 x$:

$$\lim_{x\to0}\frac{1-\cos x}{x}=\lim_{x\to0}\frac{1-\cos x}{x}\cdot\frac{1+\cos x}{1+\cos x}$$

$$=\lim_{x\to0}\frac{1-\cos^2 x}{x(1+\cos x)}$$

$$=\lim_{x\to0}\frac{\sin^2 x}{x(1+\cos x)}.$$

For the next step we resort back to algebra to rewrite the fractional expression as a product, then use the results in (5):

$$\lim_{x\to0}\frac{1-\cos x}{x}=\lim_{x\to0}\frac{\sin^2 x}{x(1+\cos x)}$$

$$=\lim_{x\to0}\left(\frac{\sin x}{x}\cdot\frac{\sin x}{1+\cos x}\right)$$

$$=\left(\lim_{x\to0}\frac{\sin x}{x}\right)\cdot\left(\lim_{x\to0}\frac{\sin x}{1+\cos x}\right).$$

Because $\lim\limits_{x\to0}(\sin x)/(1+\cos x)=0/2=0$ we have

$$\lim_{x\to0}\frac{1-\cos x}{x}=0. \qquad (13)\;■$$

Since the limit in (13) is equal to 0, we can write

$$\lim_{x\to0}\frac{1-\cos x}{x}=\lim_{x\to0}\frac{-(\cos x-1)}{x}=(-1)\lim_{x\to0}\frac{\cos x-1}{x}=0.$$

Dividing by -1 then gives another important trigonometric limit:

$$\lim_{x\to0}\frac{\cos x-1}{x}=0. \qquad (14)$$

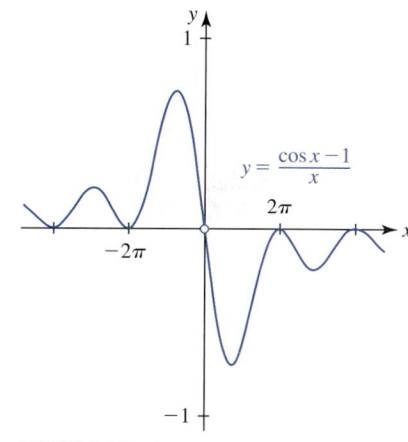

FIGURE 2.4.6   Graph of $f(x)=(\cos x-1)/x$

FIGURE 2.4.6 shows the graph of $f(x)=(\cos x-1)/x$. We will use the results in (10) and (14) in Exercises 2.7 and again in Section 3.4.

Exercises 2.4   Answers to selected odd-numbered problems begin on page ANS-8.

≡ **Fundamentals**

In Problems 1–36, find the given limit, or state that it does not exist.

1. $\lim\limits_{t\to0}\dfrac{\sin 3t}{2t}$

2. $\lim\limits_{t\to0}\dfrac{\sin(-4t)}{t}$

3. $\lim\limits_{x\to0}\dfrac{\sin x}{4+\cos x}$

4. $\lim\limits_{x\to0}\dfrac{1+\sin x}{1+\cos x}$

5. $\lim\limits_{x\to0}\dfrac{\cos 2x}{\cos 3x}$

6. $\lim\limits_{x\to0}\dfrac{\tan x}{3x}$

7. $\lim\limits_{t\to0}\dfrac{1}{t\sec t\csc 4t}$

8. $\lim\limits_{t\to0}5t\cot 2t$

9. $\lim\limits_{t\to0}\dfrac{2\sin^2 t}{t\cos^2 t}$

10. $\lim\limits_{t\to0}\dfrac{\sin^2(t/2)}{\sin t}$

11. $\lim\limits_{t\to0}\dfrac{\sin^2 6t}{t^2}$

12. $\lim\limits_{t\to0}\dfrac{t^3}{\sin^2 3t}$

13. $\lim\limits_{x\to1}\dfrac{\sin(x-1)}{2x-2}$

14. $\lim\limits_{x\to2\pi}\dfrac{x-2\pi}{\sin x}$

15. $\displaystyle\lim_{x\to 0}\frac{\cos x}{x}$

16. $\displaystyle\lim_{\theta\to\pi/2}\frac{1+\sin\theta}{\cos\theta}$

17. $\displaystyle\lim_{x\to 0}\frac{\cos(3x-\pi/2)}{x}$

18. $\displaystyle\lim_{x\to -2}\frac{\sin(5x+10)}{4x+8}$

19. $\displaystyle\lim_{t\to 0}\frac{\sin 3t}{\sin 7t}$

20. $\displaystyle\lim_{t\to 0}\sin 2t\,\csc 3t$

21. $\displaystyle\lim_{t\to 0^{+}}\frac{\sin t}{\sqrt{t}}$

22. $\displaystyle\lim_{t\to 0^{+}}\frac{1-\cos\sqrt{t}}{\sqrt{t}}$

23. $\displaystyle\lim_{t\to 0}\frac{t^{2}-5t\,\sin t}{t^{2}}$

24. $\displaystyle\lim_{t\to 0}\frac{\cos 4t}{\cos 8t}$

25. $\displaystyle\lim_{x\to 0^{+}}\frac{(x+2\sqrt{\sin x})^{2}}{x}$

26. $\displaystyle\lim_{x\to 0}\frac{(1-\cos x)^{2}}{x}$

27. $\displaystyle\lim_{x\to 0}\frac{\cos x-1}{\cos^{2}x-1}$

28. $\displaystyle\lim_{x\to 0}\frac{\sin x+\tan x}{x}$

29. $\displaystyle\lim_{x\to 0}\frac{\sin 5x^{2}}{x^{2}}$

30. $\displaystyle\lim_{t\to 0}\frac{t^{2}}{1-\cos t}$

31. $\displaystyle\lim_{x\to 2}\frac{\sin(x-2)}{x^{2}+2x-8}$

32. $\displaystyle\lim_{x\to 3}\frac{x^{2}-9}{\sin(x-3)}$

33. $\displaystyle\lim_{x\to 0}\frac{2\sin 4x+1-\cos x}{x}$

34. $\displaystyle\lim_{x\to 0}\frac{4x^{2}-2\sin x}{x}$

35. $\displaystyle\lim_{x\to\pi/4}\frac{1-\tan x}{\cos x-\sin x}$

36. $\displaystyle\lim_{x\to\pi/4}\frac{\cos 2x}{\cos x-\sin x}$

37. Suppose $f(x)=\sin x$. Use (10) and (14) of this section along with (17) of Section 1.4 to find the limit:
$$\lim_{h\to 0}\frac{f\left(\dfrac{\pi}{4}+h\right)-f\left(\dfrac{\pi}{4}\right)}{h}.$$

38. Suppose $f(x)=\cos x$. Use (10) and (14) of this section along with (18) of Section 1.4 to find the limit:
$$\lim_{h\to 0}\frac{f\left(\dfrac{\pi}{6}+h\right)-f\left(\dfrac{\pi}{6}\right)}{h}.$$

In Problems 39 and 40, use the Squeeze Theorem to establish the given limit.

39. $\displaystyle\lim_{x\to 0}x\sin\frac{1}{x}=0$

40. $\displaystyle\lim_{x\to 0}x^{2}\cos\frac{\pi}{x}=0$

41. Use the properties of limits given in Theorem 2.2.3 to show that

  (a) $\displaystyle\lim_{x\to 0}x^{3}\sin\frac{1}{x}=0$

  (b) $\displaystyle\lim_{x\to 0}x^{2}\sin^{2}\frac{1}{x}=0.$

42. If $|f(x)|\le B$ for all x in an interval containing 0, show that $\displaystyle\lim_{x\to 0}x^{2}f(x)=0$.

In Problems 43 and 44, use the Squeeze Theorem to evaluate the given limit.

43. $\displaystyle\lim_{x\to 2}f(x)$ where $2x-1\le f(x)\le x^{2}-2x+3, x\neq 2$

44. $\displaystyle\lim_{x\to 0}f(x)$ where $|f(x)-1|\le x^{2}, x\neq 0$

≡ **Think About It**

In Problems 45–48, use an appropriate substitution to find the given limit.

45. $\displaystyle\lim_{x\to\pi/4}\frac{\sin x-\cos x}{x-\pi/4}$

46. $\displaystyle\lim_{x\to\pi}\frac{x-\pi}{\tan 2x}$

47. $\displaystyle\lim_{x\to 1}\frac{\sin(\pi/x)}{x-1}$

48. $\displaystyle\lim_{x\to 2}\frac{\cos(\pi/x)}{x-2}$

49. Discuss: Is the function
$$f(x)=\begin{cases}\dfrac{\sin x}{x}, & x\neq 0\\[2mm] 1, & x=0\end{cases}$$
continuous at 0?

50. The existence of $\displaystyle\lim_{x\to 0}\frac{\sin x}{x}$ does not imply the existence of $\displaystyle\lim_{x\to 0}\frac{\sin|x|}{x}$. Explain why the second limit fails to exist.

2.5 Limits That Involve Infinity

■ **Introduction** In Sections 1.2 and 1.3 we considered some functions whose graphs possessed asymptotes. We will see in this section that vertical and horizontal asymptotes of a graph are defined in terms of limits involving the concept of *infinity*. Recall, the **infinity symbols,**

> Some texts use the symbol $+\infty$ and the words *plus infinity* instead of ∞ and *infinity*.

$-\infty$ ("minus infinity") and ∞ ("infinity"), are notational devices used to indicate, in turn, that a quantity becomes unbounded in the negative direction (in the Cartesian plane this means to the left for x and downward for y) and in the positive direction (to the right for x and upward for y).

Although the terminology and notation used when working with $\pm\infty$ is standard, it is nevertheless a bit unfortunate and can be confusing. So let us make it clear at the outset that we are going to consider two kinds of limits. First, we are going to examine

- *infinite limits.*

The words *infinite limit* always refer to *a limit that does not exist* because the function f exhibits unbounded behavior: $f(x)\to -\infty$ or $f(x)\to\infty$. Next, we will consider

- *limits at infinity.*

The words *at infinity* mean that we are trying to determine whether a function f possesses a limit when the variable x is allowed to become unbounded: $x \to -\infty$ or $x \to \infty$. Such limits may or may not exist.

Throughout the discussion, bear in mind that $-\infty$ and ∞ do not represent real numbers and should *never* be manipulated arithmetically like a number.

■ **Infinite Limits**   The limit of a function f will fail to exist as x approaches a number a whenever the function values increase or decrease without bound. The fact that the function values $f(x)$ increase without bound as x approaches a is denoted symbolically by

$$f(x) \to \infty \text{ as } x \to a \qquad \text{or} \qquad \lim_{x \to a} f(x) = \infty. \qquad (1)$$

If the function values decrease without bound as x approaches a, we write

$$f(x) \to -\infty \text{ as } x \to a \qquad \text{or} \qquad \lim_{x \to a} f(x) = -\infty. \qquad (2)$$

Recall, the use of the symbol $x \to a$ signifies that f exhibits the same behavior—in this instance, unbounded behavior—from both sides of the number a on the x-axis. For example, the notation in (1) indicates that

$$f(x) \to \infty \text{ as } x \to a^- \qquad and \qquad f(x) \to \infty \text{ as } x \to a^+.$$

See FIGURE 2.5.1.

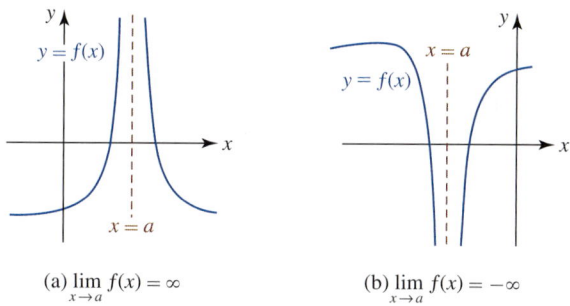

(a) $\lim_{x \to a} f(x) = \infty$ (b) $\lim_{x \to a} f(x) = -\infty$

FIGURE 2.5.1   Two types of infinite limits

Similarly, FIGURE 2.5.2 shows the unbounded behavior of a function f as x approaches a from one side. Note in Figure 2.5.2(c), we cannot describe the behavior of f near a using just one limit symbol.

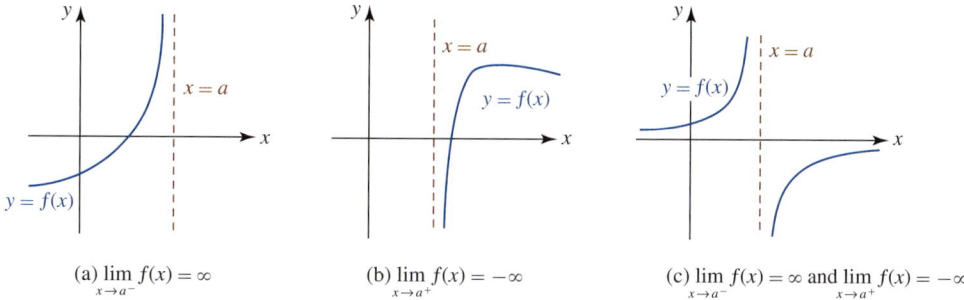

(a) $\lim_{x \to a^-} f(x) = \infty$ (b) $\lim_{x \to a^+} f(x) = -\infty$ (c) $\lim_{x \to a^-} f(x) = \infty$ and $\lim_{x \to a^+} f(x) = -\infty$

FIGURE 2.5.2   Three more types of infinite limits

In general, any limit of the six types

$$\lim_{x \to a^-} f(x) = -\infty, \qquad \lim_{x \to a^-} f(x) = \infty,$$

$$\lim_{x \to a^+} f(x) = -\infty, \qquad \lim_{x \to a^+} f(x) = \infty, \qquad (3)$$

$$\lim_{x \to a} f(x) = -\infty, \qquad \lim_{x \to a} f(x) = \infty,$$

is called an **infinite limit**. Again, in each case of (3) we are simply describing in a symbolic manner the behavior of a function f near the number a. *None of the limits in (3) exist.*

In Section 1.3 we reviewed how to identify a vertical asymptote for the graph of a rational function $f(x) = p(x)/q(x)$. We are now in a position to define a vertical asymptote of any function in terms of the limit concept.

> **Definition 2.5.1** Vertical Asymptote
>
> A line $x = a$ is said to be a **vertical asymptote** for the graph of a function f if at least one of the six statements in (3) is true.

See Figure 1.2.1. ▶ In the review of functions in Chapter 1 we saw that the graphs of rational functions often possess asymptotes. We saw that the graphs of the rational functions $y = 1/x$ and $y = 1/x^2$ were similar to the graphs in Figure 2.5.2(c) and Figure 2.5.1(a), respectively. The y-axis, that is, $x = 0$, is a vertical asymptote for each of these functions. The graphs of

$$y = \frac{1}{x - a} \qquad \text{and} \qquad y = \frac{1}{(x - a)^2} \tag{4}$$

are obtained by shifting the graphs of $y = 1/x$ and $y = 1/x^2$ horizontally $|a|$ units. As seen in FIGURE 2.5.3, $x = a$ is a vertical asymptote for the rational functions in (4). We have

$$\lim_{x \to a^-} \frac{1}{x - a} = -\infty \qquad \text{and} \qquad \lim_{x \to a^+} \frac{1}{x - a} = \infty \tag{5}$$

and

$$\lim_{x \to a} \frac{1}{(x - a)^2} = \infty. \tag{6}$$

The infinite limits in (5) and (6) are just special cases of the following general result:

$$\lim_{x \to a^-} \frac{1}{(x - a)^n} = -\infty \qquad \text{and} \qquad \lim_{x \to a^+} \frac{1}{(x - a)^n} = \infty, \tag{7}$$

for n an odd positive integer, and

$$\lim_{x \to a} \frac{1}{(x - a)^n} = \infty, \tag{8}$$

for n an even positive integer. As a consequence of (7) and (8), the graph of a rational function $y = 1/(x - a)^n$ either resembles the graph in Figure 2.5.3(a) for n odd or that in Figure 2.5.3(b) for n even.

For a general rational function $f(x) = p(x)/q(x)$, where p and q have no common factors, it should be clear from this discussion that when q contains a factor $(x - a)^n$, n a positive integer, then the shape of the graph near the vertical line $x = a$ must be either one of those shown in Figure 2.5.3 or its reflection in the x-axis.

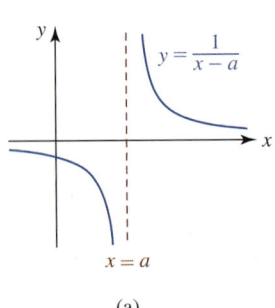

$y = \dfrac{1}{x - a}$

$x = a$

(a)

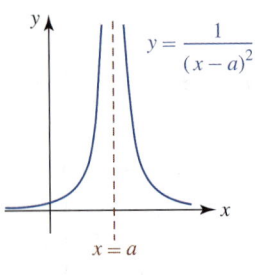

$y = \dfrac{1}{(x - a)^2}$

$x = a$

(b)

FIGURE 2.5.3 Graphs of functions in (4)

EXAMPLE 1 Vertical Asymptotes of a Rational Function

Inspection of the rational function

$$f(x) = \frac{x + 2}{x^2(x + 4)}$$

shows that $x = -4$ and $x = 0$ are vertical asymptotes for the graph of f. Since the denominator contains the factors $(x - (-4))^1$ and $(x - 0)^2$ we expect the graph of f near the line $x = -4$ to resemble Figure 2.5.3(a) or its reflection in the x-axis, and the graph near $x = 0$ to resemble Figure 2.5.3(b) or its reflection in the x-axis.

For x close to 0, from either side of 0, it is easily seen that $f(x) > 0$. But, for x close to -4, say $x = -4.1$ and $x = -3.9$, we have $f(x) > 0$ and $f(x) < 0$, respectively. Using the additional information that there is only a single x-intercept $(-2, 0)$, we obtain the graph of f in FIGURE 2.5.4. ■

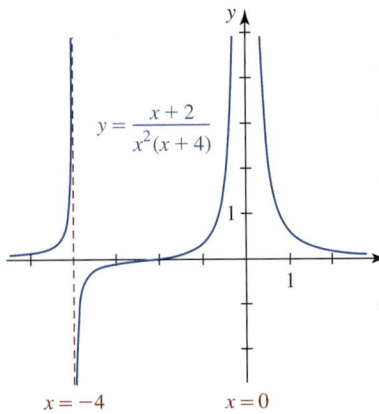

$y = \dfrac{x + 2}{x^2(x + 4)}$

$x = -4$ $\qquad$ $x = 0$

FIGURE 2.5.4 Graph of function in Example 1

EXAMPLE 2 One-Sided Limit

In Figure 1.6.6 we saw that the y-axis, or the line $x = 0$, is a vertical asymptote for the natural logarithmic function $f(x) = \ln x$ since

$$\lim_{x \to 0^+} \ln x = -\infty.$$

The graph of the logarithmic function $y = \ln(x + 3)$ is the graph of $f(x) = \ln x$ shifted 3 units to the left. Thus $x = -3$ is a vertical asymptote for the graph of $y = \ln(x + 3)$ since $\lim_{x \to -3^+} \ln(x + 3) = -\infty$. ∎

EXAMPLE 3   One-Sided Limit

Graph the function $f(x) = \dfrac{x}{\sqrt{x + 2}}$.

Solution   Inspection of f reveals that its domain is the interval $(-2, \infty)$ and the y-intercept is $(0, 0)$. From the accompanying table we conclude that f decreases

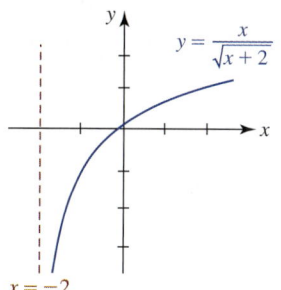

$x \to -2^+$	-1.9	-1.99	-1.999	-1.9999
$f(x)$	-6.01	-19.90	-63.21	-199.90

without bound as x approaches -2 from the right:

$$\lim_{x \to -2^+} f(x) = -\infty.$$

Hence, the line $x = -2$ is a vertical asymptote. The graph of f is given in FIGURE 2.5.5. ∎

FIGURE 2.5.5   Graph of function in Example 3

■ **Limits at Infinity**   If a function f approaches a constant value L as the independent variable x increases without bound ($x \to \infty$) or as x decreases ($x \to -\infty$) without bound, then we write

$$\lim_{x \to -\infty} f(x) = L \qquad \text{or} \qquad \lim_{x \to \infty} f(x) = L \tag{9}$$

and say that f possesses a **limit at infinity**. Here are all the possibilities for limits at infinity $\lim_{x \to -\infty} f(x)$ and $\lim_{x \to \infty} f(x)$:

- One limit exists but the other does not,
- Both $\lim_{x \to -\infty} f(x)$ and $\lim_{x \to \infty} f(x)$ exist and equal the same number,
- Both $\lim_{x \to -\infty} f(x)$ and $\lim_{x \to \infty} f(x)$ exist but are different numbers,
- Neither $\lim_{x \to -\infty} f(x)$ nor $\lim_{x \to \infty} f(x)$ exists.

If at least one of the limits exists, say, $\lim_{x \to \infty} f(x) = L$, then the graph of f can be made arbitrarily close to the line $y = L$ as x increases in the positive direction.

Definition 2.5.2   Horizontal Asymptote

A line $y = L$ is said to be a **horizontal asymptote** for the graph of a function f if at least one of the two statements in (9) is true.

In FIGURE 2.5.6 we have illustrated some typical horizontal asymptotes. We note, in conjunction with Figure 2.5.6(d) that, in general, the graph of a function can have at most *two* horizontal asymptotes but the graph of a *rational function* $f(x) = p(x)/q(x)$ can have at most *one*. If the graph of a rational function f possesses a horizontal asymptote $y = L$, then its end behavior is as shown in Figure 2.5.6(c), that is:

$$f(x) \to L \text{ as } x \to -\infty \qquad and \qquad f(x) \to L \text{ as } x \to \infty.$$

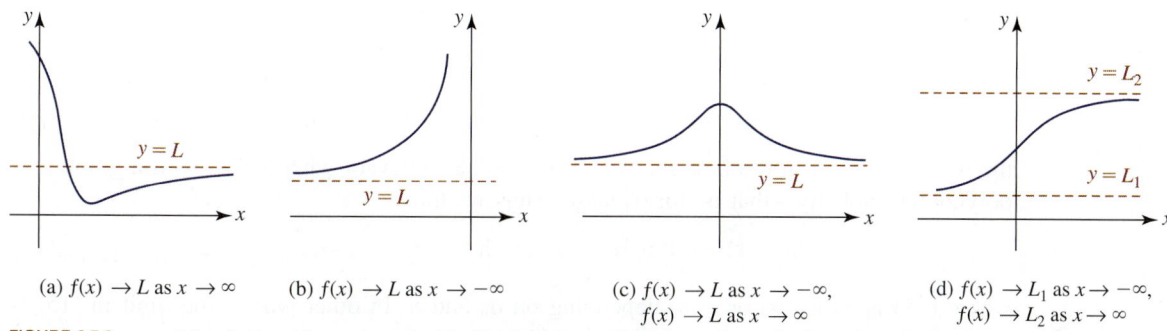

(a) $f(x) \to L$ as $x \to \infty$ (b) $f(x) \to L$ as $x \to -\infty$ (c) $f(x) \to L$ as $x \to -\infty$, $f(x) \to L$ as $x \to \infty$ (d) $f(x) \to L_1$ as $x \to -\infty$, $f(x) \to L_2$ as $x \to \infty$

FIGURE 2.5.6 $y = L$ is a horizontal asymptote in (a), (b), and (c); $y = L_1$ and $y = L_2$ are horizontal asymptotes in (d)

For example, if x becomes unbounded in either the positive or negative direction, the functions in (4) decrease to 0 and we write

$$\lim_{x \to -\infty} \frac{1}{x - a} = 0, \lim_{x \to \infty} \frac{1}{x - a} = 0 \quad \text{and} \quad \lim_{x \to -\infty} \frac{1}{(x - a)^2} = 0, \lim_{x \to \infty} \frac{1}{(x - a)^2} = 0. \quad (10)$$

In general, if r is a positive rational number and if $(x - a)^r$ is defined, then

These results are also true when $x - a$ is replaced by $a - x$, provided $(a - x)^r$ is defined.

$$\lim_{x \to -\infty} \frac{1}{(x - a)^r} = 0 \quad \text{and} \quad \lim_{x \to \infty} \frac{1}{(x - a)^r} = 0. \quad (11)$$

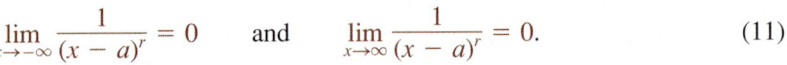

EXAMPLE 4 Horizontal and Vertical Asymptotes

The domain of the function $f(x) = \dfrac{4}{\sqrt{2 - x}}$ is the interval $(-\infty, 2)$. In view of (11) we can write

$$\lim_{x \to -\infty} \frac{4}{\sqrt{2 - x}} = 0.$$

Note that we cannot consider the limit of f as $x \to \infty$ because the function is not defined for $x \geq 2$. Nevertheless $y = 0$ is a horizontal asymptote. Now from infinite limit

$$\lim_{x \to 2^-} \frac{4}{\sqrt{2 - x}} = \infty$$

we conclude that $x = 2$ is a vertical asymptote for the graph of f. See **FIGURE 2.5.7**. ∎

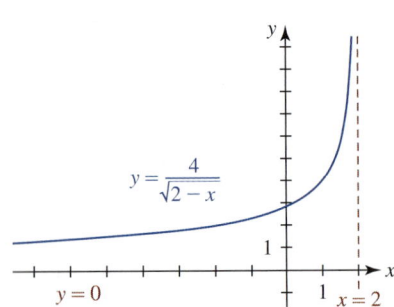

FIGURE 2.5.7 Graph of function in Example 4

In general, if $F(x) = f(x)/g(x)$, then the following table summarizes the limit results for the forms $\lim_{x \to a} F(x)$, $\lim_{x \to \infty} F(x)$, and $\lim_{x \to -\infty} F(x)$. The symbol L denotes a real number.

limit form: $x \to a, \infty, -\infty$	$\dfrac{L}{\pm\infty}$	$\dfrac{\pm\infty}{L}, L \neq 0$	$\dfrac{L}{0}, L \neq 0$
limit is:	0	infinite	infinite

(12)

Limits of the form $\lim_{x \to \infty} F(x) = \pm\infty$ or $\lim_{x \to -\infty} F(x) = \pm\infty$ are said to be **infinite limits at infinity**. Furthermore, the limit properties given in Theorem 2.2.3 hold by replacing the symbol a by ∞ or $-\infty$ provided the limits exist. For example,

$$\lim_{x \to \infty} f(x)g(x) = \left(\lim_{x \to \infty} f(x)\right)\left(\lim_{x \to \infty} g(x)\right) \quad \text{and} \quad \lim_{x \to \infty} \frac{f(x)}{g(x)} = \frac{\lim_{x \to \infty} f(x)}{\lim_{x \to \infty} g(x)}, \quad (13)$$

whenever $\lim_{x \to \infty} f(x)$ and $\lim_{x \to \infty} g(x)$ exist. In the case of the limit of a quotient we must also have $\lim_{x \to \infty} g(x) \neq 0$.

▌ **End Behavior** In Section 1.3 we saw that how a function f behaves when $|x|$ is very large is its **end behavior**. As already discussed, if $\lim_{x \to \infty} f(x) = L$, then the graph of f can be made arbitrarily close to the line $y = L$ for large positive values of x. The graph of a polynomial function,

$$f(x) = a_n x^n + a_{n-1} x^{n-1} + \cdots + a_2 x^2 + a_1 x + a_0,$$

resembles the graph of $y = a_n x^n$ for $|x|$ very large. In other words, for

$$f(x) = a_n x^n \boxed{+ a_{n-1} x^{n-1} + \cdots + a_1 x + a_0} \quad (14)$$

the terms enclosed in the blue rectangle in (14) are irrelevant when we look at a graph of a polynomial globally—that is, for $|x|$ large. Thus we have

$$\lim_{x \to \pm\infty} a_n x^n = \lim_{x \to \pm\infty} \left(a_n x^n + a_{n-1} x^{n-1} + \cdots + a_1 x + a_0\right), \quad (15)$$

where (15) is either ∞ or $-\infty$ depending on a_n and n. In other words, the limit in (15) is an example of an infinite limit at infinity.

EXAMPLE 5 Limit at Infinity

Evaluate $\lim\limits_{x\to\infty} \dfrac{-6x^4 + x^2 + 1}{2x^4 - x}$.

Solution We cannot apply the limit quotient law in (13) to the given function, since $\lim\limits_{x\to\infty}(-6x^4 + x^2 + 1) = -\infty$ and $\lim\limits_{x\to\infty}(2x^4 - x) = \infty$. However, by dividing the numerator and the denominator by x^4, we can write

$$\lim_{x\to\infty} \frac{-6x^4 + x^2 + 1}{2x^4 - x} = \lim_{x\to\infty} \frac{-6 + \left(\dfrac{1}{x^2}\right) + \left(\dfrac{1}{x^4}\right)}{2 - \left(\dfrac{1}{x^3}\right)}$$

$$= \frac{\lim\limits_{x\to\infty}\left[-6 + \left(\dfrac{1}{x^2}\right) + \left(\dfrac{1}{x^4}\right)\right]}{\lim\limits_{x\to\infty}\left[2 - \left(\dfrac{1}{x^3}\right)\right]} \quad \leftarrow \begin{array}{l}\text{\color{blue}Limit of the numerator}\\ \text{\color{blue}and denominator both}\\ \text{\color{blue}exist and the limit of}\\ \text{\color{blue}the denominator is not}\\ \text{\color{blue}zero}\end{array}$$

$$= \frac{-6 + 0 + 0}{2 - 0} = -3.$$

This means the line $y = -3$ is a horizontal asymptote for the graph of the function.

Alternative Solution In view of (14), we can discard all powers of x other than the highest:

discard terms in the blue boxes
$\downarrow$

$$\lim_{x\to\infty} \frac{-6x^4 \boxed{+\ x^2 + 1}}{2x^4 \boxed{-\ x}} = \lim_{x\to\infty} \frac{-6x^4}{2x^4} = \lim_{x\to\infty} \frac{-6}{2} = -3. \qquad ■$$

EXAMPLE 6 Infinite Limit at Infinity

Evaluate $\lim\limits_{x\to\infty} \dfrac{1 - x^3}{3x + 2}$.

Solution By (14),

$$\lim_{x\to\infty} \frac{1 - x^3}{3x + 2} = \lim_{x\to\infty} \frac{-x^3}{3x} = -\frac{1}{3}\lim_{x\to\infty} x^2 = -\infty.$$

In other words, the limit does not exist. ■

EXAMPLE 7 Graph of a Rational Function

Graph the function $f(x) = \dfrac{x^2}{1 - x^2}$.

Solution Inspection of the function f reveals that its graph is symmetric with respect to the y-axis, the y-intercept is $(0, 0)$, and the vertical asymptotes are $x = -1$ and $x = 1$. Now, from the limit

$$\lim_{x\to\infty} f(x) = \lim_{x\to\infty} \frac{x^2}{1 - x^2} = \lim_{x\to\infty} \frac{x^2}{-x^2} = -\lim_{x\to\infty} 1 = -1$$

we conclude that the line $y = -1$ is a horizontal asymptote. The graph of f is given in FIGURE 2.5.8. ■

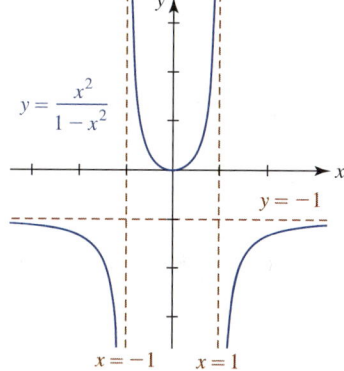

FIGURE 2.5.8 Graph of function in Example 7

Another limit law that holds true for limits at infinity is that the limit of an nth root of a function is the nth root of the limit, whenever the limit exists and the nth root is defined. In symbols, if $\lim\limits_{x\to\infty} g(x) = L$, then

$$\lim_{x\to\infty} \sqrt[n]{g(x)} = \sqrt[n]{\lim_{x\to\infty} g(x)} = \sqrt[n]{L}, \qquad (16)$$

provided $L \geq 0$ when n is even. The result also holds for $x \to -\infty$.

EXAMPLE 8 Limit of a Square Root

Evaluate $\displaystyle\lim_{x\to\infty}\sqrt{\dfrac{2x^3 - 5x^2 + 4x - 6}{6x^3 + 2x}}$.

Solution Because the limit of the rational function inside the radical exists and is positive, we can write

$$\lim_{x\to\infty}\sqrt{\frac{2x^3 - 5x^2 + 4x - 6}{6x^3 + 2x}} = \sqrt{\lim_{x\to\infty}\frac{2x^3 - 5x^2 + 4x - 6}{6x^3 + 2x}} = \sqrt{\lim_{x\to\infty}\frac{2x^3}{6x^3}} = \sqrt{\frac{1}{3}} = \frac{1}{\sqrt{3}}. \qquad \blacksquare$$

EXAMPLE 9 Graph with Two Horizontal Asymptotes

Determine whether the graph of $f(x) = \dfrac{5x}{\sqrt{x^2 + 4}}$ has any horizontal asymptotes.

Solution Since the function is not rational, we must investigate the limit of f as $x\to\infty$ and as $x\to-\infty$. First, recall from algebra that $\sqrt{x^2}$ is nonnegative, or more to the point,

$$\sqrt{x^2} = |x| = \begin{cases} x, & x \geq 0 \\ -x, & x < 0. \end{cases}$$

We then rewrite f as

$$f(x) = \frac{\dfrac{5x}{\sqrt{x^2}}}{\dfrac{\sqrt{x^2 + 4}}{\sqrt{x^2}}} = \frac{\dfrac{5x}{|x|}}{\dfrac{\sqrt{x^2 + 4}}{\sqrt{x^2}}} = \frac{\dfrac{5x}{|x|}}{\sqrt{1 + \dfrac{4}{x^2}}}.$$

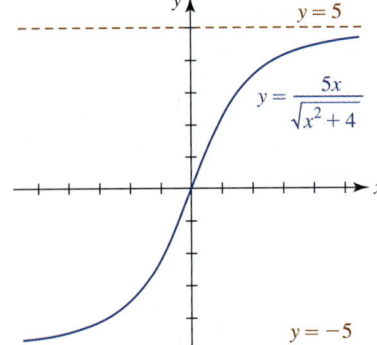

$y = 5$

$y = \dfrac{5x}{\sqrt{x^2 + 4}}$

$y = -5$

FIGURE 2.5.9 Graph of function in Example 9

The limits of f as $x\to\infty$ and as $x\to-\infty$ are, respectively,

$$\lim_{x\to\infty} f(x) = \lim_{x\to\infty}\frac{\dfrac{5x}{|x|}}{\sqrt{1 + \dfrac{4}{x^2}}} = \lim_{x\to\infty}\frac{\dfrac{5x}{x}}{\sqrt{1 + \dfrac{4}{x^2}}} = \frac{\displaystyle\lim_{x\to\infty} 5}{\sqrt{\displaystyle\lim_{x\to\infty}\left(1 + \dfrac{4}{x^2}\right)}} = \frac{5}{1} = 5,$$

and $\displaystyle\lim_{x\to-\infty} f(x) = \lim_{x\to-\infty}\frac{\dfrac{5x}{|x|}}{\sqrt{1 + \dfrac{4}{x^2}}} = \lim_{x\to-\infty}\frac{\dfrac{5x}{-x}}{\sqrt{1 + \dfrac{4}{x^2}}} = \frac{\displaystyle\lim_{x\to-\infty}(-5)}{\sqrt{\displaystyle\lim_{x\to-\infty}\left(1 + \dfrac{4}{x^2}\right)}} = \frac{-5}{1} = -5.$

Thus the graph of f has two horizontal asymptotes $y = 5$ and $y = -5$. The graph of f, which is similar to Figure 2.5.6(d), is given in **FIGURE 2.5.9**. $\qquad \blacksquare$

In the next example we see that the form of the given limit is $\infty - \infty$, but the limit exists and is *not* 0.

EXAMPLE 10 Using Rationalization

Evaluate $\displaystyle\lim_{x\to\infty}\left(x^2 - \sqrt{x^4 + 7x^2 + 1}\right)$.

Solution Because $f(x) = x^2 - \sqrt{x^4 + 7x^2 + 1}$ is an even function (verify that $f(-x) = f(x)$) with domain $(-\infty, \infty)$, if $\displaystyle\lim_{x\to\infty} f(x)$ exists it must be the same as $\displaystyle\lim_{x\to-\infty} f(x)$. We first rationalize the numerator:

$$\lim_{x\to\infty}\left(x^2 - \sqrt{x^4 + 7x^2 + 1}\right) = \lim_{x\to\infty}\frac{\left(x^2 - \sqrt{x^4 + 7x^2 + 1}\right)}{1}\cdot\left(\frac{x^2 + \sqrt{x^4 + 7x^2 + 1}}{x^2 + \sqrt{x^4 + 7x^2 + 1}}\right)$$

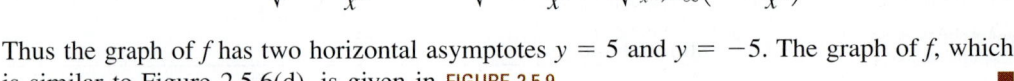

$$= \lim_{x\to\infty}\frac{x^4 - (x^4 + 7x^2 + 1)}{x^2 + \sqrt{x^4 + 7x^2 + 1}}$$

$$= \lim_{x\to\infty}\frac{-7x^2 - 1}{x^2 + \sqrt{x^4 + 7x^2 + 1}}.$$

Next, we divide the numerator and denominator by $\sqrt{x^4} = x^2$:

$$\lim_{x \to \infty} \frac{-7x^2 - 1}{x^2 + \sqrt{x^4 + 7x^2 + 1}} = \lim_{x \to \infty} \frac{\dfrac{-7x^2}{\sqrt{x^4}} - \dfrac{1}{\sqrt{x^4}}}{\dfrac{x^2 + \sqrt{x^4 + 7x^2 + 1}}{\sqrt{x^4}}}$$

$$= \lim_{x \to \infty} \frac{-7 - \dfrac{1}{x^2}}{1 + \sqrt{1 + \dfrac{7}{x^2} + \dfrac{1}{x^4}}}$$

$$= \frac{\displaystyle\lim_{x \to \infty}\left(-7 - \frac{1}{x^2}\right)}{\displaystyle\lim_{x \to \infty}1 + \sqrt{\lim_{x \to \infty}\left(1 + \frac{7}{x^2} + \frac{1}{x^4}\right)}}$$

$$= \frac{-7}{1 + 1} = -\frac{7}{2}.$$

With the help of a CAS, the graph of the function f is given in FIGURE 2.5.10. The line $y = -\frac{7}{2}$ is a horizontal asymptote. Note the symmetry of the graph with respect to the y-axis. ∎

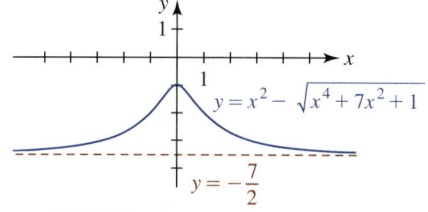

FIGURE 2.5.10 Graph of function in Example 10

When working with functions containing the natural exponential function, the following four limits merit special attention:

$$\lim_{x \to \infty} e^x = \infty, \quad \lim_{x \to -\infty} e^x = 0, \quad \lim_{x \to \infty} e^{-x} = 0, \quad \lim_{x \to -\infty} e^{-x} = \infty. \tag{17}$$

As discussed in Section 1.6 and verified by the second and third limit in (17), $y = 0$ is a horizontal asymptote for the graphs of $y = e^x$ and $y = e^{-x}$. See FIGURE 2.5.11.

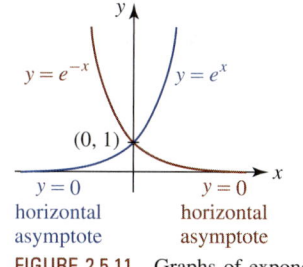

FIGURE 2.5.11 Graphs of exponential functions

EXAMPLE 11 Graph with Two Horizontal Asymptotes

Determine whether the graph of $f(x) = \dfrac{6}{1 + e^{-x}}$ has any horizontal asymptotes.

Solution Because f is not a rational function, we must examine $\lim_{x \to \infty} f(x)$ and $\lim_{x \to -\infty} f(x)$. First, in view of the third result given in (17) we can write

$$\lim_{x \to \infty} \frac{6}{1 + e^{-x}} = \frac{\displaystyle\lim_{x \to \infty} 6}{\displaystyle\lim_{x \to \infty}(1 + e^{-x})} = \frac{6}{1 + 0} = 6.$$

Thus $y = 6$ is a horizontal asymptote. Now, because $\lim_{x \to -\infty} e^{-x} = \infty$ it follows from the table in (12) that

$$\lim_{x \to -\infty} \frac{6}{1 + e^{-x}} = 0.$$

Therefore $y = 0$ is a horizontal asymptote. The graph of f is given in FIGURE 2.5.12. ∎

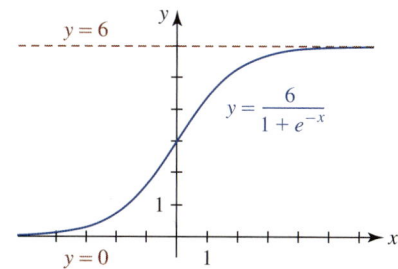

FIGURE 2.5.12 Graph of function in Example 11

■ **Composite Functions** Theorem 2.3.3, the limit of a composite function, holds when a is replaced by $-\infty$ or ∞ and the limit exists. For example, if $\lim_{x \to \infty} g(x) = L$ and f is continuous at L, then

$$\lim_{x \to \infty} f(g(x)) = f\left(\lim_{x \to \infty} g(x)\right) = f(L). \tag{18}$$

The limit result in (16) is just a special case of (18) when $f(x) = \sqrt[n]{x}$. The result in (18) also holds for $x \to -\infty$. Our last example illustrates (18) involving a limit at ∞.

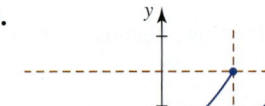

$y = \sin\frac{1}{x}$

$y = 0$

FIGURE 2.5.13 Graph of function in Example 12

EXAMPLE 12 A Trigonometric Function Revisited

In Example 2 of Section 2.4 we saw that $\lim_{x \to 0} \sin(1/x)$ does not exist. However, the limit at infinity, $\lim_{x \to \infty} \sin(1/x)$, exists. By (18) we can write

$$\lim_{x \to \infty} \sin\frac{1}{x} = \sin\left(\lim_{x \to \infty} \frac{1}{x}\right) = \sin 0 = 0.$$

As we see in **FIGURE 2.5.13**, $y = 0$ is a horizontal asymptote for the graph of $f(x) = \sin(1/x)$. You should compare this graph with that given in Figure 2.4.2. ∎

Exercises 2.5 Answers to selected odd-numbered problems begin on page ANS-8.

≡ **Fundamentals**

In Problems 1–24, express the given limit as a number, as $-\infty$, or as ∞.

1. $\lim_{x \to 5^-} \dfrac{1}{x - 5}$

2. $\lim_{x \to 6} \dfrac{4}{(x - 6)^2}$

3. $\lim_{x \to -4^+} \dfrac{2}{(x + 4)^3}$

4. $\lim_{x \to 2^-} \dfrac{10}{x^2 - 4}$

5. $\lim_{x \to 1} \dfrac{1}{(x - 1)^4}$

6. $\lim_{x \to 0^+} \dfrac{-1}{\sqrt{x}}$

7. $\lim_{x \to 0^+} \dfrac{2 + \sin x}{x}$

8. $\lim_{x \to \pi^+} \csc x$

9. $\lim_{x \to \infty} \dfrac{x^2 - 3x}{4x^2 + 5}$

10. $\lim_{x \to \infty} \dfrac{x^2}{1 + x^{-2}}$

11. $\lim_{x \to \infty} \left(5 - \dfrac{2}{x^4}\right)$

12. $\lim_{x \to -\infty} \left(\dfrac{6}{\sqrt[3]{x}} + \dfrac{1}{\sqrt[5]{x}}\right)$

13. $\lim_{x \to \infty} \dfrac{8 - \sqrt{x}}{1 + 4\sqrt{x}}$

14. $\lim_{x \to -\infty} \dfrac{1 + 7\sqrt[3]{x}}{2\sqrt[3]{x}}$

15. $\lim_{x \to \infty} \left(\dfrac{3x}{x + 2} - \dfrac{x - 1}{2x + 6}\right)$

16. $\lim_{x \to \infty} \left(\dfrac{x}{3x + 1}\right)\left(\dfrac{4x^2 + 1}{2x^2 + x}\right)^3$

17. $\lim_{x \to \infty} \sqrt{\dfrac{3x + 2}{6x - 8}}$

18. $\lim_{x \to -\infty} \sqrt[3]{\dfrac{2x - 1}{7 - 16x}}$

19. $\lim_{x \to \infty} \left(x - \sqrt{x^2 + 1}\right)$

20. $\lim_{x \to \infty} \left(\sqrt{x^2 + 5x} - x\right)$

21. $\lim_{x \to \infty} \cos\left(\dfrac{5}{x}\right)$

22. $\lim_{x \to -\infty} \sin\left(\dfrac{\pi x}{3 - 6x}\right)$

23. $\lim_{x \to -\infty} \sin^{-1}\left(\dfrac{x}{\sqrt{4x^2 + 1}}\right)$

24. $\lim_{x \to \infty} \ln\left(\dfrac{x}{x + 8}\right)$

In Problems 25–32, find $\lim_{x \to -\infty} f(x)$ and $\lim_{x \to \infty} f(x)$ for the given function f.

25. $f(x) = \dfrac{4x + 1}{\sqrt{x^2 + 1}}$

26. $f(x) = \dfrac{\sqrt{9x^2 + 6}}{5x - 1}$

27. $f(x) = \dfrac{2x + 1}{\sqrt{3x^2 + 1}}$

28. $f(x) = \dfrac{-5x^2 + 6x + 3}{\sqrt{x^4 + x^2 + 1}}$

29. $f(x) = \dfrac{e^x - e^{-x}}{e^x + e^{-x}}$

30. $f(x) = 1 + \dfrac{2e^{-x}}{e^x + e^{-x}}$

31. $f(x) = \dfrac{|x - 5|}{x - 5}$

32. $f(x) = \dfrac{|4x| + |x - 1|}{x}$

In Problems 33–42, find all vertical and horizontal asymptotes for the graph of the given function. Sketch the graph.

33. $f(x) = \dfrac{1}{x^2 + 1}$

34. $f(x) = \dfrac{x}{x^2 + 1}$

35. $f(x) = \dfrac{x^2}{x + 1}$

36. $f(x) = \dfrac{x^2 - x}{x^2 - 1}$

37. $f(x) = \dfrac{1}{x^2(x - 2)}$

38. $f(x) = \dfrac{4x^2}{x^2 + 4}$

39. $f(x) = \sqrt{\dfrac{x}{x - 1}}$

40. $f(x) = \dfrac{1 - \sqrt{x}}{\sqrt{x}}$

41. $f(x) = \dfrac{x - 2}{\sqrt{x^2 + 1}}$

42. $f(x) = \dfrac{x + 3}{\sqrt{x^2 - 1}}$

In Problems 43–46, use the given graph to find:

(a) $\lim_{x \to 2^-} f(x)$ (b) $\lim_{x \to 2^+} f(x)$

(c) $\lim_{x \to -\infty} f(x)$ (d) $\lim_{x \to \infty} f(x)$

43.

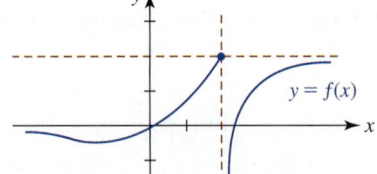

$y = f(x)$

FIGURE 2.5.14 Graph for Problem 43

44.

$y = f(x)$

FIGURE 2.5.15 Graph for Problem 44

45. $y = f(x)$

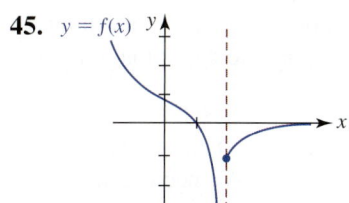

FIGURE 2.5.16   Graph for Problem 45

46.

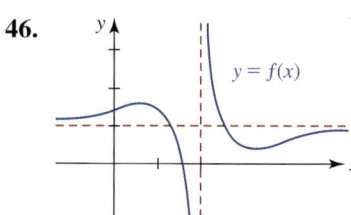

FIGURE 2.5.17   Graph for Problem 46

In Problems 47–50, sketch a graph of a function f that satisfies the given conditions.

47. $\lim\limits_{x \to 1^+} f(x) = -\infty,\ \lim\limits_{x \to 1^-} f(x) = -\infty,\ f(2) = 0,\ \lim\limits_{x \to \infty} f(x) = 0$

48. $f(0) = 1,\ \lim\limits_{x \to -\infty} f(x) = 3,\ \lim\limits_{x \to \infty} f(x) = -2$

49. $\lim\limits_{x \to 2} f(x) = \infty,\ \lim\limits_{x \to -\infty} f(x) = \infty,\ \lim\limits_{x \to \infty} f(x) = 1$

50. $\lim\limits_{x \to 1^-} f(x) = 2,\ \lim\limits_{x \to 1^+} f(x) = -\infty,\ f\left(\frac{3}{2}\right) = 0,\ f(3) = 0,$
$\lim\limits_{x \to -\infty} f(x) = 0,\ \lim\limits_{x \to \infty} f(x) = 0$

51. Use an appropriate substitution to evaluate

$$\lim_{x \to \infty} x \sin \frac{3}{x}.$$

52. According to Einstein's theory of relativity, the mass m of a body moving with velocity v is $m = m_0/\sqrt{1 - v^2/c^2}$, where m_0 is the initial mass and c is the speed of light. What happens to m as $v \to c^-$?

☰ Calculator/CAS Problems

In Problems 53 and 54, use a calculator or CAS to investigate the given limit. Conjecture its value.

53. $\lim\limits_{x \to \infty} x^2 \sin \dfrac{2}{x^2}$ **54.** $\lim\limits_{x \to \infty} \left(\cos \dfrac{1}{x} \right)^x$

55. Use a calculator or CAS to obtain the graph of $f(x) = (1 + x)^{1/x}$. Use the graph to conjecture the values of $f(x)$ as
(a) $x \to -1^+$, (b) $x \to 0$, and (c) $x \to \infty$.

56. (a) A regular n-gon is an n-sided polygon inscribed in a circle; the polygon is formed by n equally spaced points on the circle. Suppose the polygon shown in

FIGURE 2.5.18 represents a regular n-gon inscribed in a circle of radius r. Use trigonometry to show that the area $A(n)$ of the n-gon is given by

$$A(n) = \frac{n}{2} r^2 \sin\left(\frac{2\pi}{n} \right).$$

(b) It stands to reason that the area $A(n)$ approaches the area of the circle as the number of sides of the n-gon increases. Use a calculator to compute $A(100)$ and $A(1000)$.

(c) Let $x = 2\pi/n$ in $A(n)$ and note that as $n \to \infty$ then $x \to 0$. Use (10) of Section 2.4 to show that $\lim\limits_{n \to \infty} A(n) = \pi r^2$.

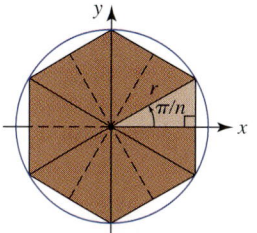

FIGURE 2.5.18   Inscribed n-gon for Problem 56

☰ Think About It

57. (a) Suppose $f(x) = x^2/(x + 1)$ and $g(x) = x - 1$. Show that

$$\lim_{x \to \pm\infty} [f(x) - g(x)] = 0.$$

(b) What does the result in part (a) indicate about the graphs of f and g where $|x|$ is large?

(c) If possible, give a name to the function g.

58. Very often students and even instructors will sketch vertically shifted graphs incorrectly. For example, the graphs of $y = x^2$ and $y = x^2 + 1$ are incorrectly drawn in FIGURE 2.5.19(a) but are correctly drawn in Figure 2.5.19(b). Demonstrate that Figure 2.5.19(b) is correct by showing that the horizontal distance between the two points P and Q shown in the figure approaches 0 as $x \to \infty$.

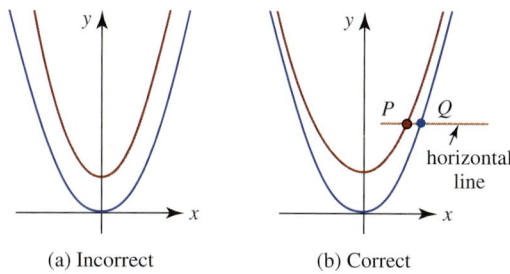

(a) Incorrect   (b) Correct
FIGURE 2.5.19   Graphs for Problem 58

2.6   Limits—A Formal Approach

▌ Introduction   In the discussion that follows we will consider an alternative approach to the notion of a limit that is based on analytical concepts rather than on intuitive concepts. A **proof** of the existence of a limit can never be based on one's ability to sketch graphs or on tables of numerical values. Although a good intuitive understanding of $\lim\limits_{x \to a} f(x)$ is sufficient for proceeding with the study of the calculus in this text, an intuitive understanding is admittedly too vague to be of any use in proving theorems. To give a rigorous demonstration of the existence of a limit, or to prove the important theorems of Section 2.2, we must start with a precise definition of a limit.

■ **Limit of a Function**   Let us try to prove that $\lim_{x \to 2}(2x + 6) = 10$ by elaborating on the following idea: "If $f(x) = 2x + 6$ can be made arbitrarily close to 10 by taking x sufficiently close to 2, from either side but different from 2, then $\lim_{x \to 2} f(x) = 10$." We need to make the concepts of *arbitrarily close* and *sufficiently close* precise. In order to set a standard of arbitrary closeness, let us demand that the distance between the numbers $f(x)$ and 10 be less than 0.1; that is,

$$|f(x) - 10| < 0.1 \qquad \text{or} \qquad 9.9 < f(x) < 10.1. \qquad (1)$$

Then, how close must x be to 2 to accomplish (1)? To find out, we can use ordinary algebra to rewrite the inequality

$$9.9 < 2x + 6 < 10.1$$

as $1.95 < x < 2.05$. Adding -2 across this simultaneous inequality then gives

$$-0.05 < x - 2 < 0.05.$$

Using absolute values and remembering that $x \neq 2$, we can write the last inequality as $0 < |x - 2| < 0.05$. Thus, for an "arbitrary closeness to 10" of 0.1, "sufficiently close to 2" means within 0.05. In other words, if x is a number different from 2 such that its distance from 2 satisfies $|x - 2| < 0.05$, then the distance of $f(x)$ from 10 is guaranteed to satisfy $|f(x) - 10| < 0.1$. Expressed in yet another way, when x is a number different from 2 but in the open interval (1.95, 2.05) on the x-axis, then $f(x)$ is in the interval (9.9, 10.1) on the y-axis.

Using the same example, let us try to generalize. Suppose ε (the Greek letter *epsilon*) denotes an arbitrary *positive number* that is our measure of arbitrary closeness to the number 10. If we demand that

$$|f(x) - 10| < \varepsilon \qquad \text{or} \qquad 10 - \varepsilon < f(x) < 10 + \varepsilon, \qquad (2)$$

then from $10 - \varepsilon < 2x + 6 < 10 + \varepsilon$ and algebra, we find

$$2 - \frac{\varepsilon}{2} < x < 2 + \frac{\varepsilon}{2} \qquad \text{or} \qquad -\frac{\varepsilon}{2} < x - 2 < \frac{\varepsilon}{2}. \qquad (3)$$

Again using absolute values and remembering that $x \neq 2$, we can write the last inequality in (3) as

$$0 < |x - 2| < \frac{\varepsilon}{2}. \qquad (4)$$

If we denote $\varepsilon/2$ by the new symbol δ (the Greek letter *delta*), (2) and (4) can be written as

$$|f(x) - 10| < \varepsilon \qquad \text{whenever} \qquad 0 < |x - 2| < \delta.$$

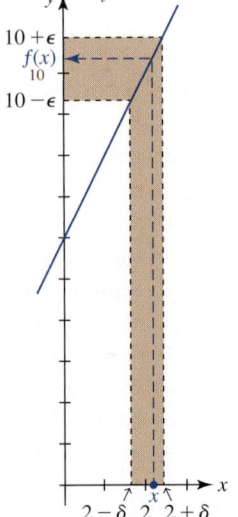

FIGURE 2.6.1 $f(x)$ is in $(10 - \varepsilon, 10 + \varepsilon)$ whenever x is in $(2 - \delta, 2 + \delta)$, $x \neq 2$

Thus, for a new value for ε, say $\varepsilon = 0.001$, $\delta = \varepsilon/2 = 0.0005$ tells us the corresponding closeness to 2. For any number x different from 2 in (1.9995, 2.0005),* we can be sure $f(x)$ is in (9.999, 10.001). See FIGURE 2.6.1.

■ **A Definition**   The foregoing discussion leads us to the so-called ε-δ **definition of a limit**.

Definition 2.6.1   Definition of a Limit

Suppose a function f is defined everywhere on an open interval, except possibly at a number a in the interval. Then

$$\lim_{x \to a} f(x) = L$$

means that for every $\varepsilon > 0$, there exists a number $\delta > 0$ such that

$$|f(x) - L| < \varepsilon \qquad \text{whenever} \qquad 0 < |x - a| < \delta.$$

---

*For this reason, we use $0 < |x - 2| < \delta$ rather than $|x - 2| < \delta$. Keep in mind when considering $\lim_{x \to 2} f(x)$, we do not care about f at 2.

Let $\lim_{x \to a} f(x) = L$ and suppose $\delta > 0$ is the number that "works" in the sense of Definition 2.6.1 for a given $\varepsilon > 0$. As shown in **FIGURE 2.6.2(a)**, every x in $(a - \delta, a + \delta)$, with the possible exception of a itself, will then have an image $f(x)$ in $(L - \varepsilon, L + \varepsilon)$. Furthermore, as in Figure 2.6.2(b), a choice $\delta_1 < \delta$ for the same ε also "works" in that every x not equal to a in $(a - \delta_1, a + \delta_1)$ gives $f(x)$ in $(L - \varepsilon, L + \varepsilon)$. However, Figure 2.6.2(c) shows that choosing a smaller $\varepsilon_1, 0 < \varepsilon_1 < \varepsilon$, will demand finding a new value of δ. Observe in Figure 2.6.2(c) that x is in $(a - \delta, a + \delta)$ but not in $(a - \delta_1, a + \delta_1)$, and so $f(x)$ is not necessarily in $(L - \varepsilon_1, L + \varepsilon_1)$.

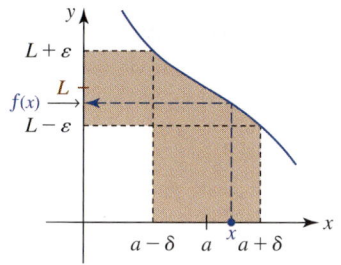

(a) A δ that works for a given ε

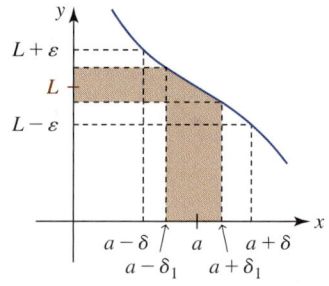

(b) A smaller δ_1 will also work for the same ε

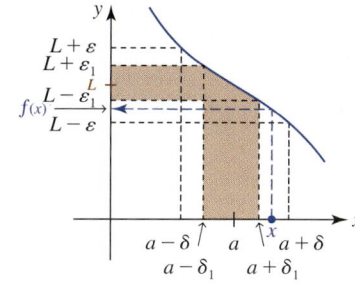

(c) A smaller ε_1 will require a $\delta_1 < \delta$. For x in $(a - \delta, a + \delta)$, $f(x)$ is not necessarily in $(L - \varepsilon_1, L + \varepsilon_1)$

FIGURE 2.6.2 $f(x)$ is in $(L - \varepsilon, L + \varepsilon)$ whenever x is in $(a - \delta, a + \delta), x \neq a$

EXAMPLE 1   Using Definition 2.6.1

Prove that $\lim_{x \to a} (5x + 2) = 17$.

Solution   For any arbitrary $\varepsilon > 0$, regardless how small, we wish to find a δ so that

$$|(5x + 2) - 17| < \varepsilon \qquad \text{whenever} \qquad 0 < |x - 3| < \delta.$$

To do this consider

$$|(5x + 2) - 17| = |5x - 15| = 5|x - 3|.$$

Thus, to make $|(5x + 2) - 17| = 5|x - 3| < \varepsilon$, we need only make $0 < |x - 3| < \varepsilon/5$; that is, choose $\delta = \varepsilon/5$.

Verification   If $0 < |x - 3| < \varepsilon/5$, then $5|x - 3| < \varepsilon$ implies

$$|5x - 15| < \varepsilon \qquad \text{or} \qquad |(5x + 2) - 17| < \varepsilon \qquad \text{or} \qquad |f(x) - 17| < \varepsilon. \qquad ■$$

EXAMPLE 2   Using Definition 2.6.1

Prove that $\lim_{x \to -4} \dfrac{16 - x^2}{4 + x} = 8$.

◀ We examined this limit in (1) and (2) of Section 2.1.

Solution   For $x \neq -4$,

$$\left| \frac{16 - x^2}{4 + x} - 8 \right| = |4 - x - 8| = |-x - 4| = |x + 4| = |x - (-4)|$$

Thus,

$$\left| \frac{16 - x^2}{4 + x} - 8 \right| = |x - (-4)| < \varepsilon$$

whenever we have $0 < |x - (-4)| < \varepsilon$; that is, choose $\delta = \varepsilon$. $\qquad ■$

EXAMPLE 3   A Limit That Does Not Exist

Consider the function

$$f(x) = \begin{cases} 0, & x \leq 1 \\ 2, & x > 1. \end{cases}$$

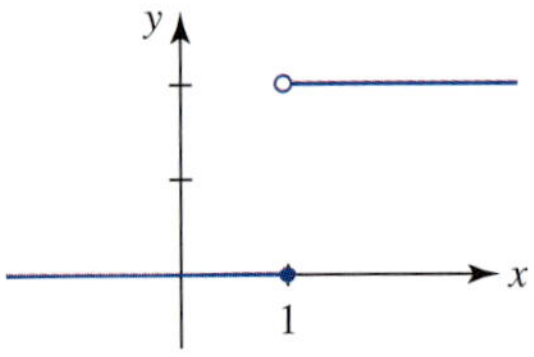

FIGURE 2.6.3 Limit of f does not exist as x approaches 1 in Example 3

We recognize in FIGURE 2.6.3 that f has a jump discontinuity at 1 and so $\lim_{x \to 1} f(x)$ does not exist. However, to *prove* this last fact, we shall proceed indirectly. Assume that the limit exists, namely, $\lim_{x \to 1} f(x) = L$. Then from Definition 2.6.1 we know that for the choice $\varepsilon = \frac{1}{2}$ there must exist a $\delta > 0$ so that

$$|f(x) - L| < \frac{1}{2} \qquad \text{whenever} \qquad 0 < |x - 1| < \delta.$$

Now to the right of 1, let us choose $x = 1 + \delta/2$. Since

$$0 < \left|1 + \frac{\delta}{2} - 1\right| = \left|\frac{\delta}{2}\right| < \delta$$

we must have

$$\left|f\left(1 + \frac{\delta}{2}\right) - L\right| = |2 - L| < \frac{1}{2}. \tag{5}$$

To the left of 1, choose $x = 1 - \delta/2$. But

$$0 < \left|1 - \frac{\delta}{2} - 1\right| = \left|-\frac{\delta}{2}\right| < \delta$$

implies

$$\left|f\left(1 - \frac{\delta}{2}\right) - L\right| = |0 - L| = |L| < \frac{1}{2}. \tag{6}$$

Solving the absolute-value inequalities (5) and (6) gives, respectively,

$$\frac{3}{2} < L < \frac{5}{2} \qquad \text{and} \qquad -\frac{1}{2} < L < \frac{1}{2}.$$

Since no number L can satisfy both of these inequalities, we conclude that $\lim_{x \to 1} f(x)$ does not exist. ∎

In the next example we consider the limit of a quadratic function. We shall see that finding the δ in this case requires a bit more ingenuity than in Examples 1 and 2.

EXAMPLE 4 Using Definition 2.6.1

We examined this limit in Example 1 ▶ of Section 2.1.

Prove that $\lim_{x \to 4} (-x^2 + 2x + 2) = -6$.

Solution For an arbitrary $\varepsilon > 0$ we must find a $\delta > 0$ so that

$$|-x^2 + 2x + 2 - (-6)| < \varepsilon \qquad \text{whenever} \qquad 0 < |x - 4| < \delta.$$

Now,

$$\begin{aligned}
|-x^2 + 2x + 2 - (-6)| &= |(-1)(x^2 - 2x - 8)| \\
&= |(x + 2)(x - 4)| \\
&= |x + 2||x - 4|. \tag{7}
\end{aligned}$$

In other words, we want to make $|x + 2||x - 4| < \varepsilon$. But since we have agreed to examine values of x *near* 4, let us consider only those values for which $|x - 4| < 1$. This last inequality gives $3 < x < 5$ or equivalently $5 < x + 2 < 7$. Consequently we can write $|x + 2| < 7$. Hence from (7),

$$0 < |x - 4| < 1 \qquad \text{implies} \qquad |-x^2 + 2x + 2 - (-6)| < 7|x - 4|.$$

If we now choose δ to be the minimum of the two numbers, 1 and $\varepsilon/7$, written $\delta = \min\{1, \varepsilon/7\}$ we have

$$0 < |x - 4| < \delta \qquad \text{implies} \qquad |-x^2 + 2x + 2 - (-6)| < 7|x - 4| < 7 \cdot \frac{\varepsilon}{7} = \varepsilon. \ ∎$$

The reasoning in Example 4 is subtle. Consequently it is worth a few minutes of your time to reread the discussion immediately following Definition 2.6.1, reexamine

Figure 2.3.2(b), and then think again about why $\delta = \min\{1, \varepsilon/7\}$ is the δ that "works" in the example. Remember, you can pick the ε arbitrarily; think about δ for, say, $\varepsilon = 8, \varepsilon = 6$, and $\varepsilon = 0.01$.

▌ **One-Sided Limits**   We state next the definitions of the **one-sided limits**, $\lim_{x \to a^-} f(x)$ and $\lim_{x \to a^+} f(x)$.

Definition 2.6.2   Left-Hand Limit

Suppose a function f is defined on an open interval (c, a). Then

$$\lim_{x \to a^-} f(x) = L$$

means for every $\varepsilon > 0$ there exists a $\delta > 0$ such that

$$|f(x) - L| < \varepsilon \qquad \text{whenever} \qquad a - \delta < x < a.$$

Definition 2.6.3   Right-Hand Limit

Suppose a function f is defined on an open interval (a, c). Then

$$\lim_{x \to a^+} f(x) = L$$

means for every $\varepsilon > 0$ there exists a $\delta > 0$ such that

$$|f(x) - L| < \varepsilon \qquad \text{whenever} \qquad a < x < a + \delta.$$

EXAMPLE 5   Using Definition 2.6.3

Prove that $\lim_{x \to 0^+} \sqrt{x} = 0$.

Solution   First, we can write

$$|\sqrt{x} - 0| = |\sqrt{x}| = \sqrt{x}.$$

Then, $|\sqrt{x} - 0| < \varepsilon$ whenever $0 < x < 0 + \varepsilon^2$. In other words, we choose $\delta = \varepsilon^2$.

Verification   If $0 < x < \varepsilon^2$, then $0 < \sqrt{x} < \varepsilon$ implies

$$|\sqrt{x}| < \varepsilon \qquad \text{or} \qquad |\sqrt{x} - 0| < \varepsilon. \qquad ■$$

▌ **Limits Involving Infinity**   The two concepts of **infinite limits**

$$f(x) \to \infty \ (\text{or} \ -\infty) \quad \text{as} \quad x \to a$$

and a **limit at infinity**

$$f(x) \to L \quad \text{as} \quad x \to \infty \ (\text{or} \ -\infty)$$

are formalized in the next two definitions.

Recall, an infinite limit is a limit that does not exist as $x \to a$.

Definition 2.6.4   Infinite Limits

(i) $\lim_{x \to a} f(x) = \infty$ means for each $M > 0$, there exists a $\delta > 0$ such that $f(x) > M$ whenever $0 < |x - a| < \delta$.

(ii) $\lim_{x \to a} f(x) = -\infty$ means for each $M < 0$, there exists a $\delta > 0$ such that $f(x) < M$ whenever $0 < |x - a| < \delta$.

Parts (*i*) and (*ii*) of Definition 2.6.4 are illustrated in FIGURE 2.6.4(a) and Figure 2.6.4(b), respectively. Recall, if $f(x) \to \infty$ (or $-\infty$) as $x \to a$, then $x = a$ is a vertical asymptote for the graph of f. In the case when $f(x) \to \infty$ as $x \to a$, then $f(x)$ can be made larger than any arbitrary positive number (that is, $f(x) > M$) by taking x sufficiently close to a (that is, $0 < |x - a| < \delta$).

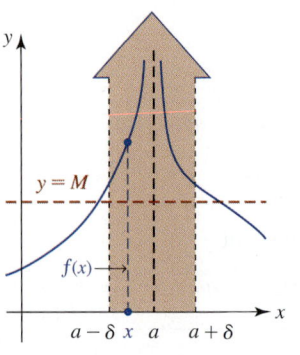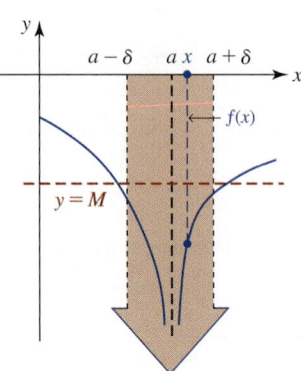

(a) For a given M, whenever
$a - \delta < x < a + \delta, x \neq a$,
then $f(x) > M$

(b) For a given M, whenever
$a - \delta < x < a + \delta, x \neq a$,
then $f(x) < M$

FIGURE 2.6.4 Infinite limits as $x \to a$

The four one-sided infinite limits

$$f(x) \to \infty \text{ as } x \to a^-, \qquad f(x) \to -\infty \text{ as } x \to a^-$$
$$f(x) \to \infty \text{ as } x \to a^+, \qquad f(x) \to -\infty \text{ as } x \to a^+$$

are defined in a manner analogous to that given in Definitions 2.6.2 and 2.6.3.

Definition 2.6.5 Limits at Infinity

(*i*) $\lim\limits_{x \to \infty} f(x) = L$ if for each $\varepsilon > 0$, there exists an $N > 0$ such that $|f(x) - L| < \varepsilon$ whenever $x > N$.

(*ii*) $\lim\limits_{x \to -\infty} f(x) = L$ if for each $\varepsilon > 0$, there exists an $N < 0$ such that $|f(x) - L| < \varepsilon$ whenever $x < N$.

Parts (*i*) and (*ii*) of Definition 2.6.5 are illustrated in FIGURE 2.6.5(a) and Figure 2.6.5(b), respectively. Recall, if $f(x) \to L$ as $x \to \infty$ (or $-\infty$), then $y = L$ is a horizontal asymptote for the graph of f. In the case when $f(x) \to L$ as $x \to \infty$, then the graph of f can be made arbitrarily close to the line $y = L$ (that is, $|f(x) - L| < \varepsilon$) by taking x sufficiently far out on the positive x-axis (that is, $x > N$).

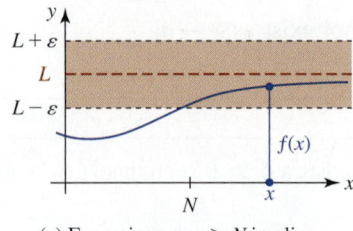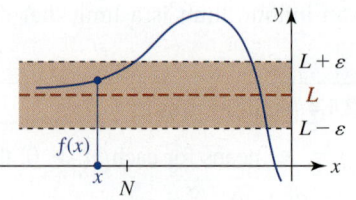

(a) For a given ε, $x > N$ implies
$L - \varepsilon < f(x) < L + \varepsilon$

(b) For a given ε, $x < N$ implies
$L - \varepsilon < f(x) < L + \varepsilon$

FIGURE 2.6.5 Limits at infinity

EXAMPLE 6   Using Definition 2.6.5(i)

Prove that $\lim\limits_{x\to\infty}\dfrac{3x}{x+1}=3$.

Solution   By Definition 2.6.5(i), for any $\varepsilon>0$, we must find a number $N>0$ such that

$$\left|\frac{3x}{x+1}-3\right|<\varepsilon \qquad \text{whenever} \qquad x>N.$$

Now, by considering $x>0$, we have

$$\left|\frac{3x}{x+1}-3\right|=\left|\frac{-3}{x+1}\right|=\frac{3}{x+1}<\frac{3}{x}<\varepsilon$$

whenever $x>3/\varepsilon$. Hence, choose $N=3/\varepsilon$. For example, if $\varepsilon=0.01$, then $N=3/(0.01)=300$ will guarantee that $|f(x)-3|<0.01$ whenever $x>300$.   ∎

▌ **Postscript—A Bit of History**   After this section you may agree with English philosopher, priest, historian, and scientist William Whewell (1794–1866), who wrote in 1858 that "A limit is a peculiar . . . conception." For many years after the invention of calculus in the seventeenth century, mathematicians argued and debated the nature of a limit. There was an awareness that intuition, graphs, and numerical examples of ratios of vanishing quantities provide at best a shaky foundation for such a fundamental concept. As you will see beginning in the next chapter, the limit concept plays a central role in calculus. The study of calculus went through several periods of increased mathematical rigor beginning with the French mathematician Augustin-Louis Cauchy and continuing later with the German mathematician Karl Wilhelm Weierstrass.

Cauchy

Augustin-Louis Cauchy (1789–1857) was born during an era of upheaval in French history. Cauchy was destined to initiate a revolution of his own in mathematics. For many contributions, but especially for his efforts in clarifying mathematical obscurities, his incessant demand for satisfactory definitions and rigorous proofs of theorems, Cauchy is often called "the father of modern analysis." A prolific writer whose output has been surpassed by only a few, Cauchy produced nearly 800 papers in astronomy, physics, and mathematics. But the same mind that was always open and inquiring in science and mathematics was also narrow and unquestioning in many other areas. Outspoken and arrogant, Cauchy's passionate stands on political and religious issues often alienated him from his colleagues.

Weierstrass

Karl Wilhelm Weierstrass (1815–1897) One of the foremost mathematical analysts of the nineteenth century never earned an academic degree! After majoring in law at the University of Bonn, but concentrating in fencing and beer drinking for four years, Weierstrass "graduated" to real life with no degree. In need of a job, Weierstrass passed a state examination and received a teaching certificate in 1841. During a period of 15 years as a secondary school teacher, his dormant mathematical genius blossomed. Although the quantity of his research publications was modest, especially when compared with that of Cauchy, the quality of these works so impressed the German mathematical community that he was awarded a doctorate, *honoris causa*, from the University of Königsberg and eventually was appointed a professor at the University of Berlin. While there, Weierstrass achieved worldwide recognition both as a mathematician and as a teacher of mathematics. One of his students was Sonja Kowalewski, the greatest female mathematician of the nineteenth century. It was Karl Wilhelm Weierstrass who was responsible for putting the concept of a limit on a firm foundation with the ε-δ definition.

Exercises 2.6 Answers to selected odd-numbered problems begin on page ANS-9.

≡ Fundamentals

In Problems 1–24, use Definitions 2.6.1, 2.6.2, or 2.6.3 to prove the given limit result.

1. $\lim_{x \to 5} 10 = 10$

2. $\lim_{x \to -2} \pi = \pi$

3. $\lim_{x \to 3} x = 3$

4. $\lim_{x \to 4} 2x = 8$

5. $\lim_{x \to -1} (x + 6) = 5$

6. $\lim_{x \to 0} (x - 4) = -4$

7. $\lim_{x \to 0} (3x + 7) = 7$

8. $\lim_{x \to 1} (9 - 6x) = 3$

9. $\lim_{x \to 2} \dfrac{2x - 3}{4} = \dfrac{1}{4}$

10. $\lim_{x \to 1/2} 8(2x + 5) = 48$

11. $\lim_{x \to -5} \dfrac{x^2 - 25}{x + 5} = -10$

12. $\lim_{x \to 3} \dfrac{x^2 - 7x + 12}{2x - 6} = -\dfrac{1}{2}$

13. $\lim_{x \to 0} \dfrac{8x^5 + 12x^4}{x^4} = 12$

14. $\lim_{x \to 1} \dfrac{2x^3 + 5x^2 - 2x - 5}{x^2 - 1} = 7$

15. $\lim_{x \to 0} x^2 = 0$

16. $\lim_{x \to 0} 8x^3 = 0$

17. $\lim_{x \to 0^+} \sqrt{5x} = 0$

18. $\lim_{x \to (1/2)^+} \sqrt{2x - 1} = 0$

19. $\lim_{x \to 0^-} f(x) = -1$, $f(x) = \begin{cases} 2x - 1, & x < 0 \\ 2x + 1, & x > 0 \end{cases}$

20. $\lim_{x \to 1^+} f(x) = 3$, $f(x) = \begin{cases} 0, & x \le 1 \\ 3, & x > 1 \end{cases}$

21. $\lim_{x \to 3} x^2 = 9$

22. $\lim_{x \to 2} (2x^2 + 4) = 12$

23. $\lim_{x \to 1} (x^2 - 2x + 4) = 3$

24. $\lim_{x \to 5} (x^2 + 2x) = 35$

25. For $a > 0$, use the identity.

$$|\sqrt{x} - \sqrt{a}| = |\sqrt{x} - \sqrt{a}| \cdot \frac{\sqrt{x} + \sqrt{a}}{\sqrt{x} + \sqrt{a}} = \frac{|x - a|}{\sqrt{x} + \sqrt{a}}$$

and the fact that $\sqrt{x} \ge 0$ to prove that $\lim_{x \to a} \sqrt{x} = \sqrt{a}$.

26. Prove that $\lim_{x \to 2} (1/x) = \frac{1}{2}$. [*Hint*: Consider only those numbers x for which $1 < x < 3$.]

In Problems 27–30, prove that $\lim_{x \to a} f(x)$ does not exist.

27. $f(x) = \begin{cases} 2, & x < 1 \\ 0, & x \ge 1 \end{cases}$; $a = 1$

28. $f(x) = \begin{cases} 1, & x \le 3 \\ -1, & x > 3 \end{cases}$; $a = 3$

29. $f(x) = \begin{cases} x, & x \le 0 \\ 2 - x, & x > 0 \end{cases}$; $a = 0$

30. $f(x) = \dfrac{1}{x}$; $a = 0$

In Problems 31–34, use Definition 2.6.5 to prove the given limit result.

31. $\lim_{x \to \infty} \dfrac{5x - 1}{2x + 1} = \dfrac{5}{2}$

32. $\lim_{x \to \infty} \dfrac{2x}{3x + 8} = \dfrac{2}{3}$

33. $\lim_{x \to -\infty} \dfrac{10x}{x - 3} = 10$

34. $\lim_{x \to -\infty} \dfrac{x^2}{x^2 + 3} = 1$

≡ Think About It

35. Prove that $\lim_{x \to 0} f(x) = 0$, where $f(x) = \begin{cases} x, & x \text{ rational} \\ 0, & x \text{ irrational}. \end{cases}$

2.7 The Tangent Line Problem

■ **Introduction** In a calculus course you will study many different things, but as mentioned in the introduction to Section 2.1, the subject "calculus" is roughly divided into two broad but related areas known as **differential calculus** and **integral calculus**. The discussion of each of these topics invariably begins with a motivating problem involving the graph of a function. Differential calculus is motivated by the problem

• *Find a tangent line to the graph of a function f,*

whereas integral calculus is motivated by the problem

• *Find the area under the graph of a function f.*

The first problem will be addressed in this section; the second problem will be discussed in Section 5.3.

■ **Tangent Line to a Graph** The word *tangent* stems from the Latin verb *tangere*, meaning "to touch." You might remember from the study of plane geometry that a tangent to a circle is a line L that intersects, or touches, the circle in exactly one point P. See **FIGURE 2.7.1**. It is not quite as easy to define a tangent line to the graph of a function f. The idea of *touching* carries over to the notion of a tangent line to the graph of a function, but the idea of *intersecting the graph in one point* does not carry over.

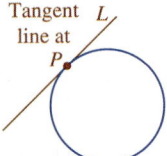

Tangent line at P

FIGURE 2.7.1 Tangent line L touches a circle at point P

Suppose $y = f(x)$ is a continuous function. If, as shown in FIGURE 2.7.2, f possesses a line L tangent to its graph at a point P, then what is an equation of this line? To answer this question, we need the coordinates of P and the slope m_{tan} of L. The coordinates of P pose no difficulty, since a point on the graph of a function f is obtained by specifying a value of x in the domain of f. The coordinates of the point of tangency at $x = a$ are then $(a, f(a))$. Therefore, the problem of finding a tangent line comes down to the problem of finding the slope m_{tan} of the line. As a means of *approximating* m_{tan}, we can readily find the slopes m_{sec} of *secant lines* (from the Latin verb *secare*, meaning "to cut") that pass through the point P and any other point Q on the graph. See FIGURE 2.7.3.

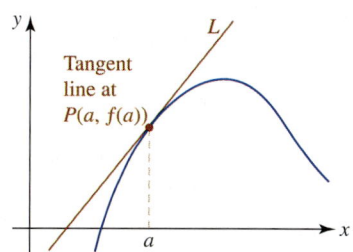

FIGURE 2.7.2   Tangent line L to a graph at point P

■ **Slope of Secant Lines**   If P has coordinates $(a, f(a))$ and if Q has coordinates $(a + h, f(a + h))$, then as shown in FIGURE 2.7.4, the slope of the secant line through P and Q is

$$m_{sec} = \frac{\text{change in } y}{\text{change in } x} = \frac{f(a + h) - f(a)}{(a + h) - a}$$

or

$$m_{sec} = \frac{f(a + h) - f(a)}{h}. \tag{1}$$

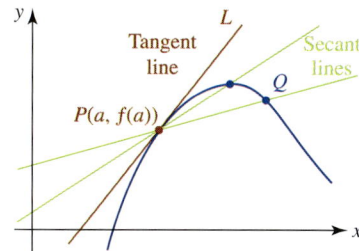

FIGURE 2.7.3   Slopes of secant lines approximate the slope m_{tan} of L

The expression on the right-hand side of the equality in (1) is called a **difference quotient**. When we let h take on values that are closer and closer to zero, that is, as $h \to 0$, then the points $Q(a + h, f(a + h))$ move along the curve closer and closer to the point $P(a, f(a))$. Intuitively, we expect the secant lines to approach the tangent line L, and that $m_{sec} \to m_{tan}$ as $h \to 0$. That is,

$$m_{tan} = \lim_{h \to 0} m_{sec}$$

provided this limit exists. We summarize this conclusion in an equivalent form of the limit using the difference quotient (1).

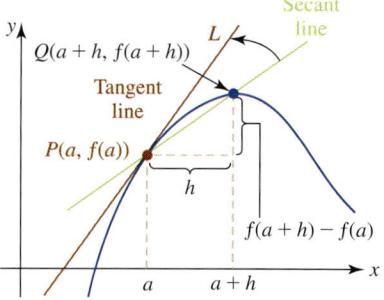

FIGURE 2.7.4   Secant lines swing into the tangent line L as $h \to 0$

Definition 2.7.1   Tangent Line with Slope

Let $y = f(x)$ be continuous at the number a. If the limit

$$m_{tan} = \lim_{h \to 0} \frac{f(a + h) - f(a)}{h} \tag{2}$$

exists, then the **tangent line** to the graph of f at $(a, f(a))$ is that line passing through the point $(a, f(a))$ with slope m_{tan}.

Just like many of the problems discussed earlier in this chapter, observe that the limit in (2) has the indeterminate form $0/0$ as $h \to 0$.

If the limit in (2) exists, the number m_{tan} is also called the **slope of the curve** $y = f(x)$ at $(a, f(a))$.

The computation of (2) is essentially a *four-step process*; three of these steps involve only precalculus mathematics: algebra and trigonometry. If the first three steps are done accurately, the fourth step, or the calculus step, *may* be the easiest part of the problem.

Guidelines for Computing (2)

(*i*) Evaluate $f(a)$ and $f(a + h)$.

(*ii*) Evaluate the difference $f(a + h) - f(a)$. Simplify.

(*iii*) Simplify the difference quotient

$$\frac{f(a + h) - f(a)}{h}.$$

(*iv*) Compute the limit of the difference quotient

$$\lim_{h \to 0} \frac{f(a + h) - f(a)}{h}.$$

The computation of the difference $f(a + h) - f(a)$ in step (*ii*) is in most instances the most important step. It is imperative that you simplify this step as much as possible. Here is Note ▶ a tip: In *many* problems involving the computation of (2) you will be able to factor h from the difference $f(a + h) - f(a)$.

EXAMPLE 1   The Four-Step Process

Find the slope of the tangent line to the graph of $y = x^2 + 2$ at $x = 1$.

Solution   We use the four-step procedure outlined above with the number 1 playing the part of the symbol a.

(*i*) The initial step is the computation of $f(1)$ and $f(1 + h)$. We have $f(1) = 1^2 + 2 = 3$, and

$$
\begin{aligned}
f(1 + h) &= (1 + h)^2 + 2 \\
&= (1 + 2h + h^2) + 2 \\
&= 3 + 2h + h^2.
\end{aligned}
$$

(*ii*) Next, from the result in the preceding step the difference is:

$$
\begin{aligned}
f(1 + h) - f(1) &= 3 + 2h + h^2 - 3 \\
&= 2h + h^2 \\
&= h(2 + h). \leftarrow \text{notice the factor of } h
\end{aligned}
$$

(*iii*) The computation of the difference quotient $\dfrac{f(1 + h) - f(1)}{h}$ is now straightforward. Again, we use the results from the preceding step:

$$
\frac{f(1 + h) - f(1)}{h} = \frac{h(2 + h)}{h} = 2 + h. \leftarrow \text{cancel the } h\text{'s}
$$

(*iv*) The last step is now easy. The limit in (2) is seen to be

$$
m_{\tan} = \lim_{h \to 0} \frac{f(1 + h) - f(1)}{h} = \lim_{h \to 0} (2 + h) = 2.
$$

The slope of the tangent line to the graph of $y = x^2 + 2$ at $(1, 3)$ is 2.   ■

EXAMPLE 2   Equation of Tangent Line

Find an equation of the tangent line whose slope was found in Example 1.

Solution   We know the point of tangency $(1, 3)$ and the slope $m_{\tan} = 2$, and so from the point–slope equation of a line we find

$$
y - 3 = 2(x - 1) \qquad \text{or} \qquad y = 2x + 1.
$$

Observe that the last equation is consistent with the x- and y-intercepts of the red line in FIGURE 2.7.5.   ■

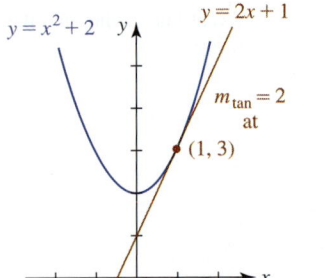

FIGURE 2.7.5   Tangent line in Example 2

EXAMPLE 3   Equation of Tangent Line

Find an equation of the tangent line to the graph of $f(x) = 2/x$ at $x = 2$.

Solution   We start by using (2) to find $m_{\tan}$ with a identified as 2. In the second of the four steps, we will have to combine two symbolic fractions by means of a common denominator.

(*i*) We have $f(2) = 2/2 = 1$ and $f(2 + h) = 2/(2 + h)$.

(*ii*) $f(2 + h) - f(2) = \dfrac{2}{2 + h} - 1$

$$
= \frac{2}{2 + h} - \frac{1}{1} \cdot \frac{2 + h}{2 + h} \leftarrow \text{a common denominator is } 2 + h
$$

$$
= \frac{2 - 2 - h}{2 + h}
$$

$$
= \frac{-h}{2 + h}. \leftarrow \text{here is the factor of } h
$$

(*iii*) The last result is to be divided by h or more precisely $\dfrac{h}{1}$. We invert and multiply by $\dfrac{1}{h}$:

$$\frac{f(2+h)-f(2)}{h} = \frac{\dfrac{-h}{2+h}}{\dfrac{h}{1}} = \frac{-h}{2+h}\cdot\frac{1}{h} = \frac{-1}{2+h}. \quad \leftarrow \text{cancel the } h\text{'s}$$

(*iv*) From (2) $m_{\tan}$ is

$$m_{\tan} = \lim_{h\to 0}\frac{f(2+h)-f(2)}{h} = \lim_{h\to 0}\frac{-1}{2+h} = -\frac{1}{2}.$$

From $f(2)=1$ the point of tangency is (2, 1) and the slope of the tangent line at (2, 1) is $m_{\tan} = -\frac{1}{2}$. From the point–slope equation of a line, the tangent line is

$$y - 1 = \frac{1}{2}(x-2) \qquad \text{or} \qquad y = -\frac{1}{2}x + 2.$$

The graphs of $y = 2/x$ and the tangent line at (2, 1) are shown in FIGURE 2.7.6. ■

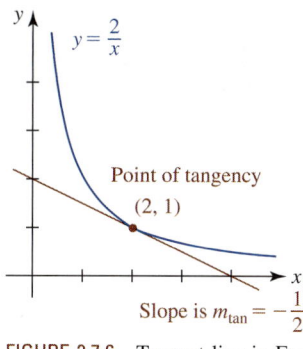

FIGURE 2.7.6 Tangent line in Example 3

EXAMPLE 4 Slope of Tangent Line

Find the slope of the tangent line to the graph of $f(x) = \sqrt{x-1}$ at $x = 5$.

Solution Replacing a by 5 in (2), we have:

(*i*) $f(5) = \sqrt{5-1} = \sqrt{4} = 2$, and

$$f(5+h) = \sqrt{5+h-1} = \sqrt{4+h}.$$

(*ii*) The difference is

$$f(5+h) - f(5) = \sqrt{4+h} - 2.$$

Because we expect to find a factor of h in this difference, we proceed to rationalize the numerator:

$$f(5+h)-f(5) = \frac{\sqrt{4+h}-2}{1}\cdot\frac{\sqrt{4+h}+2}{\sqrt{4+h}+2}$$

$$= \frac{(4+h)-4}{\sqrt{4+h}+2}$$

$$= \frac{h}{\sqrt{4+h}+2}. \qquad \leftarrow \text{here is the factor of } h$$

(*iii*) The difference quotient $\dfrac{f(5+h)-f(5)}{h}$ is then:

$$\frac{f(5+h)-f(5)}{h} = \frac{\dfrac{h}{\sqrt{4+h}+2}}{h}$$

$$= \frac{h}{h(\sqrt{4+h}+2)}$$

$$= \frac{1}{\sqrt{4+h}+2}.$$

(*iv*) The limit in (2) is

$$m_{\tan} = \lim_{h\to 0}\frac{f(5+h)-f(5)}{h} = \lim_{h\to 0}\frac{1}{\sqrt{4+h}+2} = \frac{1}{\sqrt{4}+2} = \frac{1}{4}.$$

The slope of the tangent line to the graph of $f(x) = \sqrt{x-1}$ at (5, 2) is $\frac{1}{4}$. ■

The result obtained in the next example should come as no surprise.

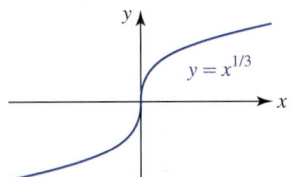

FIGURE 2.7.7 Vertical tangent in Example 6

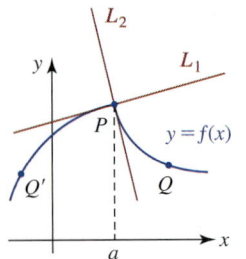

FIGURE 2.7.8 Tangent fails to exist at $(a, f(a))$

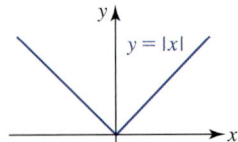

FIGURE 2.7.9 Function in Example 7

EXAMPLE 5 Tangent Line to a Line

For any linear function $y = mx + b$, the tangent line to its graph coincides with the line itself. Not unexpectedly then, the slope of the tangent line for any number $x = a$ is

$$m_{\tan} = \lim_{h \to 0} \frac{f(a + h) - f(a)}{h} = \lim_{h \to 0} \frac{m(a + h) + b - (ma + b)}{h} = \lim_{h \to 0} \frac{mh}{h} = \lim_{h \to 0} m = m. \quad \blacksquare$$

■ **Vertical Tangents** The limit in (2) can fail to exist for a function f at $x = a$ and yet there may be a tangent at the point $(a, f(a))$. The tangent line to a graph can be **vertical**, in which case its slope is undefined. We will consider the concept of vertical tangents in more detail in Section 3.1.

EXAMPLE 6 Vertical Tangent Line

Although we will not pursue the details at this time, it can be shown that the graph of $f(x) = x^{1/3}$ possesses a vertical tangent line at the origin. In FIGURE 2.7.7 we see that the y-axis, that is, the line $x = 0$, is tangent to the graph at the point $(0, 0)$. ■

■ **A Tangent May Not Exist** The graph of a function f that is continuous at a number a does not have to possess a tangent line at the point $(a, f(a))$. A tangent line will not exist whenever the graph of f has a sharp corner at $(a, f(a))$. FIGURE 2.7.8 indicates what can go wrong when the graph of a function f has a "corner." In this case f is continuous at a, but the secant lines through P and Q approach L_2 as $Q \to P$, and the secant lines through P and Q' approach a different line L_1 as $Q' \to P$. In other words, the limit in (2) fails to exist because the one-sided limits of the difference quotient (as $h \to 0^+$ and as $h \to 0^-$) are different.

EXAMPLE 7 Graph with a Corner

Show that the graph of $f(x) = |x|$ does not have a tangent at $(0, 0)$.

Solution The graph of the absolute-value function in FIGURE 2.7.9 has a corner at the origin. To prove that the graph of f does not possess a tangent line at the origin we must examine

$$\lim_{h \to 0} \frac{f(0 + h) - f(0)}{h} = \lim_{h \to 0} \frac{|0 + h| - |0|}{h} = \lim_{h \to 0} \frac{|h|}{h}.$$

From the definition of absolute value

$$|h| = \begin{cases} h, & h > 0 \\ -h, & h < 0 \end{cases}$$

we see that

$$\lim_{h \to 0^+} \frac{|h|}{h} = \lim_{h \to 0^+} \frac{h}{h} = 1 \qquad \text{whereas} \qquad \lim_{h \to 0^-} \frac{|h|}{h} = \lim_{h \to 0^-} \frac{-h}{h} = -1.$$

Since the right-hand and left-hand limits are not equal we conclude that the limit (2) does not exist. Even though the function $f(x) = |x|$ is continuous at $x = 0$, the graph of f possesses no tangent at $(0, 0)$. ■

■ **Average Rate of Change** In different contexts the difference quotient in (1) and (2), or slope of the secant line, is written in terms of alternative symbols. The symbol h in (1) and (2) is often written as Δx and the difference $f(a + \Delta x) - f(a)$ is denoted by Δy, that is, the difference quotient is

$$\frac{\text{change in } y}{\text{change in } x} = \frac{f(a + \Delta x) - f(a)}{(a + \Delta x) - a} = \frac{f(a + \Delta x) - f(a)}{\Delta x} = \frac{\Delta y}{\Delta x}. \tag{3}$$

Moreover, if $x_1 = a + \Delta x$, $x_0 = a$, then $\Delta x = x_1 - x_0$ and (3) is the same as

$$\frac{f(x_1) - f(x_0)}{x_1 - x_0} = \frac{\Delta y}{\Delta x}. \tag{4}$$

The slope $\Delta y / \Delta x$ of the secant line through the points $(x_0, f(x_0))$ and $(x_1, f(x_1))$ is called the **average rate of change of the function** f over the interval $[x_0, x_1]$. The limit $\lim_{\Delta x \to 0} \Delta y / \Delta x$ is then called the **instantaneous rate of change of the function** with respect to x at x_0.

Almost everyone has an intuitive notion of speed as a rate at which a distance is covered in a certain length of time. When, say, a bus travels 60 mi in 1 h, the *average speed* of the bus must have been 60 mi/h. Of course, it is difficult to maintain the rate of 60 mi/h for the entire trip because the bus slows down for towns and speeds up when it passes cars. In other words, the speed changes with time. If a bus company's schedule demands that the bus travel the 60 mi from one town to another in 1 h, the driver knows instinctively that he or she must compensate for speeds less than 60 mi/h by traveling at speeds greater than this at other points in the journey. Knowing that the average velocity is 60 mi/h does not, however, answer the question: What is the velocity of the bus at a particular instant?

∎ **Average Velocity**   In general, the **average velocity** or **average speed** of a moving object is defined by

$$v_{\text{ave}} = \frac{\text{change of distance}}{\text{change in time}}. \tag{5}$$

Consider a runner who finishes a 10-km race in an elapsed time of 1 h 15 min (1.25 h). The runner's average velocity or average speed for the race was

$$v_{\text{ave}} = \frac{10 - 0}{1.25 - 0} = 8 \text{ km/h}.$$

But suppose we now wish to determine the runner's *exact* velocity v at the instant the runner is one-half hour into the race. If the distance run in the time interval from 0 h to 0.5 h is measured to be 5 km, then

$$v_{\text{ave}} = \frac{5}{0.5} = 10 \text{ km/h}.$$

Again, this number is not a measure, or necessarily even a good indicator, of the instantaneous rate v at which the runner is moving 0.5 h into the race. If we determine that at 0.6 h the runner is 5.7 km from the starting line, then the average velocity from 0 h to 0.6 h is $v_{\text{ave}} = 5.7/0.6 = 9.5$ km/h. However, during the time interval from 0.5 h to 0.6 h,

$$v_{\text{ave}} = \frac{5.7 - 5}{0.6 - 0.5} = 7 \text{ km/h}.$$

The latter number is a more realistic measure of the rate v. See FIGURE 2.7.10. By "shrinking" the time interval between 0.5 h and the time that corresponds to a measured position close to 5 km, we expect to obtain even better approximations to the runner's velocity at time 0.5 h.

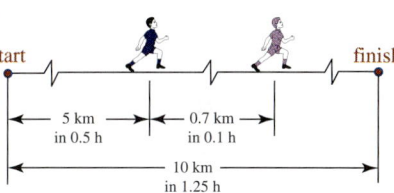

FIGURE 2.7.10   Runner in a 10-km race

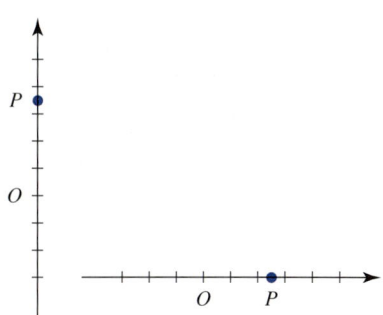

FIGURE 2.7.11   Coordinate lines

∎ **Rectilinear Motion**   To generalize the preceding discussion, let us suppose an object, or particle, at point P moves along either a vertical or horizontal coordinate line as shown in FIGURE 2.7.11. Furthermore, let the particle move in such a manner that its position, or coordinate, on the line is given by a function $s = s(t)$, where t represents time. The values of s are directed distances measured from O in units such as centimeters, meters, feet, or miles. When P is either to the right of or above O, we take $s > 0$, whereas $s < 0$ when P is either to the left of or below O. Motion in a straight line is called **rectilinear motion**.

If an object, such as a toy car moving on a horizontal coordinate line, is at point P at time t_0 and at point P' at time t_1, then the coordinates of the points, shown in FIGURE 2.7.12, are $s(t_0)$ and $s(t_1)$. By (4) the **average velocity** of the object in the time interval $[t_0, t_1]$ is

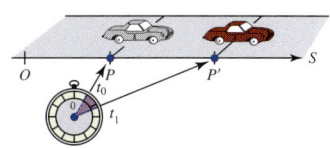

FIGURE 2.7.12   Position of toy car on a coordinate line at two times

$$v_{\text{ave}} = \frac{\text{change in position}}{\text{change in time}} = \frac{s(t_1) - s(t_0)}{t_1 - t_0}. \tag{6}$$

EXAMPLE 8   Average Velocity

The height s above ground of a ball dropped from the top of the St. Louis Gateway Arch is given by $s(t) = -16t^2 + 630$, where s is measured in feet and t in seconds. See FIGURE 2.7.13. Find the average velocity of the falling ball between the time the ball is released and the time it hits the ground.

Solution   The time at which the ball is released is determined from the equation $s(t) = 630$ or $-16t^2 + 630 = 630$. This gives $t = 0$ s. When the ball hits the ground then $s(t) = 0$ or

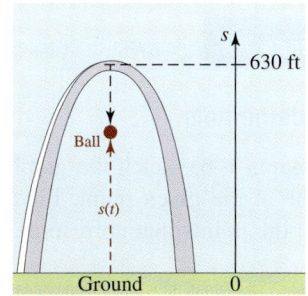

FIGURE 2.7.13   Falling ball in Example 8

$-16t^2 + 630 = 0$. The last equation gives $t = \sqrt{315/8} \approx 6.27$ s. Thus from (6) the average velocity in the time interval $\left[0, \sqrt{315/8}\right]$ is

$$v_{ave} = \frac{s\left(\sqrt{351/8}\right) - s(0)}{\sqrt{351/8} - 0} = \frac{0 - 630}{\sqrt{351/8} - 0} \approx -100.40 \text{ ft/s.} \qquad \blacksquare$$

If we let $t_1 = t_0 + \Delta t$, or $\Delta t = t_1 - t_0$, and $\Delta s = s(t_0 + \Delta t) - s(t_0)$, then (6) is equivalent to

$$v_{ave} = \frac{\Delta s}{\Delta t}. \qquad (7)$$

This suggests that the limit of (7) as $\Delta t \to 0$ gives the **instantaneous rate of change** of $s(t)$ at $t = t_0$ or the **instantaneous velocity**.

Definition 2.7.2 Instantaneous Velocity

Let $s = s(t)$ be a function that gives the position of an object moving in a straight line. Then the **instantaneous velocity** at time $t = t_0$ is

$$v(t_0) = \lim_{\Delta t \to 0} \frac{s(t_0 + \Delta t) - s(t_0)}{\Delta t} = \lim_{\Delta t \to 0} \frac{\Delta s}{\Delta t}, \qquad (8)$$

whenever the limit exists.

Note: Except for notation and interpretation, there is no mathematical difference between (2) and (8). Also, the word *instantaneous* is often dropped, and so one often speaks of the *rate of change* of a function or the *velocity* of a moving particle.

EXAMPLE 9 Example 8 Revisited

Find the instantaneous velocity of the falling ball in Example 8 at $t = 3$ s.

Solution We use the same four-step procedure as in the earlier examples with $s = s(t)$ given in Example 8.

(*i*) $s(3) = -16(9) + 630 = 486$. For any $\Delta t \neq 0$,
$$s(3 + \Delta t) = -16(3 + \Delta t)^2 + 630 = -16(\Delta t)^2 - 96\Delta t + 486.$$

(*ii*) $s(3 + \Delta t) - s(3) = [-16(\Delta t)^2 - 96\Delta t + 486] - 486$
$$= -16(\Delta t)^2 - 96\Delta t = \Delta t(-16\Delta t - 96)$$

(*iii*) $\dfrac{\Delta s}{\Delta t} = \dfrac{\Delta t(-16\Delta t - 96)}{\Delta t} = -16\Delta t - 96$

(*iv*) From (8),

$$v(3) = \lim_{\Delta t \to 0} \frac{\Delta s}{\Delta t} = \lim_{\Delta t \to 0}(-16\Delta t - 96) = -96 \text{ ft/s.} \qquad (9) \ \blacksquare$$

In Example 9, the number $s(3) = 486$ ft is the height of the ball above ground at 3 s. The minus sign in (9) is significant because the ball is moving opposite to the positive or upward direction.

Exercises 2.7   Answers to selected odd-numbered problems begin on page ANS-9.

≡ Fundamentals

In Problems 1–6, sketch the graph of the function and the tangent line at the given point. Find the slope of the secant line through the points that correspond to the indicated values of x.

1. $f(x) = -x^2 + 9, (2, 5); x = 2, x = 2.5$

2. $f(x) = x^2 + 4x, (0, 0); x = -\dfrac{1}{4}, x = 0$

3. $f(x) = x^3, (-2, -8); x = -2, x = -1$

4. $f(x) = 1/x, (1, 1); x = 0.9, x = 1$

5. $f(x) = \sin x, (\pi/2, 1); x = \pi/2, x = 2\pi/3$

6. $f(x) = \cos x, \left(-\pi/3, \tfrac{1}{2}\right); x = -\pi/2, x = -\pi/3$

In Problems 7–8, use (2) to find the slope of the tangent line to the graph of the function at the given value of x. Find an equation of the tangent line at the corresponding point.

7. $f(x) = x^2 - 6, x = 3$

8. $f(x) = -3x^2 + 10, x = -1$

9. $f(x) = x^2 - 3x, x = 1$

10. $f(x) = -x^2 + 5x - 3, x = -2$

11. $f(x) = -2x^3 + x, x = 2$ **12.** $f(x) = 8x^3 - 4, x = \dfrac{1}{2}$

13. $f(x) = \dfrac{1}{2x}, x = -1$ **14.** $f(x) = \dfrac{4}{x-1}, x = 2$

15. $f(x) = \dfrac{1}{(x-1)^2}, x = 0$ **16.** $f(x) = 4 - \dfrac{8}{x}, x = -1$

17. $f(x) = \sqrt{x}, x = 4$ **18.** $f(x) = \dfrac{1}{\sqrt{x}}, x = 1$

In Problems 19 and 20, use (2) to find the slope of the tangent line to the graph of the function at the given value of x. Find an equation of the tangent line at the corresponding point. Before starting, review the limits in (10) and (14) of Section 2.4 and the sum formulas (17) and (18) in Section 1.4.

19. $f(x) = \sin x, x = \pi/6$ **20.** $f(x) = \cos x, x = \pi/4$

In Problems 21 and 22, determine whether the line that passes through the red point is tangent to the graph of $f(x) = x^2$ at the blue point.

21.

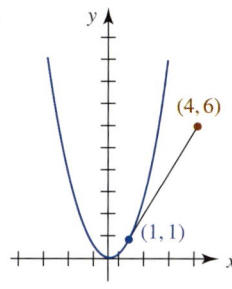

FIGURE 2.7.14 Graph for Problem 21

22.

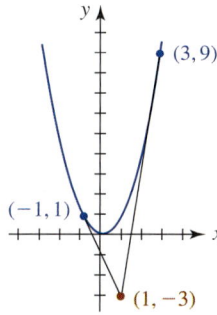

FIGURE 2.7.15 Graph for Problem 22

23. In FIGURE 2.7.16, the red line is tangent to the graph of $y = f(x)$ at the indicated point. Find an equation of the tangent line. What is the y-intercept of the tangent line?

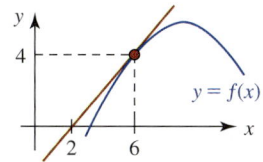

FIGURE 2.7.16 Graph for Problem 23

24. In FIGURE 2.7.17, the red line is tangent to the graph of $y = f(x)$ at the indicated point. Find $f(-5)$.

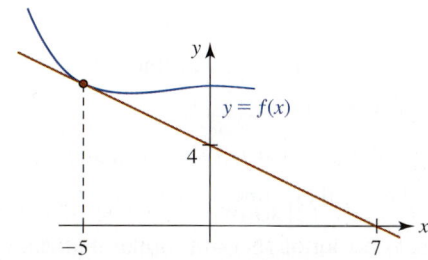

FIGURE 2.7.17 Graph for Problem 24

In Problems 25–28, use (2) to find a formula for m_{tan} at a general point $(x, f(x))$ on the graph of f. Use the formula for m_{tan} to determine the points where the tangent line to the graph is horizontal.

25. $f(x) = -x^2 + 6x + 1$ **26.** $f(x) = 2x^2 + 24x - 22$

27. $f(x) = x^3 - 3x$ **28.** $f(x) = -x^3 + x^2$

≡ Applications

29. A car travels the 290 mi between Los Angeles and Las Vegas in 5 h. What is its average velocity?

30. Two marks on a straight highway are $\frac{1}{2}$ mi apart. A highway patrol plane observes that a car traverses the distance between the marks in 40 s. Assuming a speed limit of 60 mi/h, will the car be stopped for speeding?

31. A jet airplane averages 920 km/h to fly the 3500 km between Hawaii and San Francisco. How many hours does the flight take?

32. A marathon race is run over a straight 26-mi course. The race begins at noon. At 1:30 P.M. a contestant passes the 10-mi mark and at 3:10 P.M. the contestant passes the 20-mi mark. What is the contestant's average running speed between 1:30 P.M. and 3:10 P.M.?

In Problems 33 and 34, the position of a particle moving on a horizontal coordinate line is given by the function. Use (8) to find the instantaneous velocity of the particle at the indicated time.

33. $s(t) = -4t^2 + 10t + 6, t = 3$ **34.** $s(t) = t^2 + \dfrac{1}{5t+1}, t = 0$

35. The height above ground of a ball dropped from an initial altitude of 122.5 m is given by $s(t) = -4.9t^2 + 122.5$, where s is measured in meters and t in seconds.

(a) What is the instantaneous velocity at $t = \frac{1}{2}$?

(b) At what time does the ball hit the ground?

(c) What is the impact velocity?

36. Ignoring air resistance, if an object is dropped from an initial height h, then its height above ground at time $t > 0$ is given by $s(t) = -\frac{1}{2}gt^2 + h$, where g is the acceleration of gravity.

(a) At what time does the object hit the ground?

(b) If $h = 100$ ft, compare the impact times for Earth ($g = 32$ ft/s^2), for Mars ($g = 12$ ft/s^2), and for the Moon ($g = 5.5$ ft/s^2).

(c) Use (8) to find a formula for the instantaneous velocity v at a general time t.

(d) Using the times found in part (b) and the formula found in part (c), find the corresponding impact velocities for Earth, Mars, and the Moon.

37. The height of a projectile shot from ground level is given by $s = -16t^2 + 256t$, where s is measured in feet and t in seconds.

(a) Determine the height of the projectile at $t = 2, t = 6, t = 9$, and $t = 10$.

(b) What is the average velocity of the projectile between $t = 2$ and $t = 5$?

(c) Show that the average velocity between $t = 7$ and $t = 9$ is zero. Interpret physically.

(d) At what time does the projectile hit the ground?

(e) Use (8) to find a formula for instantaneous velocity v at a general time t.

(f) Using the time found in part (d) and the formula found in part (e), find the corresponding impact velocity.

(g) What is the maximum height the projectile attains?

38. Suppose the graph shown in FIGURE 2.7.18 is that of position function $s = s(t)$ of a particle moving in a straight line, where s is measured in meters and t in seconds.

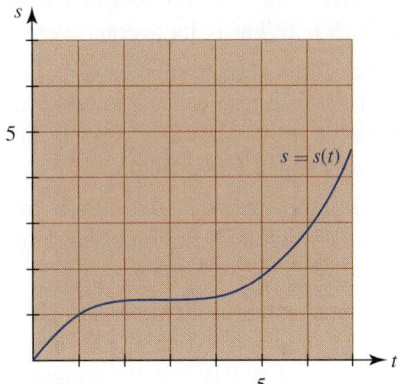

FIGURE 2.7.18   Graph for Problem 38

(a) Estimate the position of the particle at $t = 4$ and at $t = 6$.

(b) Estimate the average velocity of the particle between $t = 4$ and $t = 6$.

(c) Estimate the initial velocity of the particle—that is, its velocity at $t = 0$.

(d) Estimate a time at which the velocity of the particle is zero.

(e) Determine an interval on which the velocity of the particle is decreasing.

(f) Determine an interval on which the velocity of the particle is increasing.

≡ **Think About It**

39. Let $y = f(x)$ be an even function whose graph possesses a tangent line with slope m at $(a, f(a))$. Show that the slope of the tangent line at $(-a, f(a))$ is $-m$. [*Hint*: Explain why $f(-a + h) = f(a - h)$.]

40. Let $y = f(x)$ be an odd function whose graph possesses a tangent line with slope m at $(a, f(a))$. Show that the slope of the tangent line at $(-a, -f(a))$ is m.

41. Proceed as in Example 7 and show that there is no tangent line to graph of $f(x) = x^2 + |x|$ at $(0, 0)$.

Chapter 2 in Review

Answers to selected odd-numbered problems begin on page ANS-9.

A. True/False

In Problems 1–22, indicate whether the given statement is true or false.

1. $\lim_{x \to 2} \dfrac{x^3 - 8}{x - 2} = 12$ _____

2. $\lim_{x \to 5} \sqrt{x - 5} = 0$ _____

3. $\lim_{x \to 0} \dfrac{|x|}{x} = 1$ _____

4. $\lim_{x \to \infty} e^{2x - x^2} = \infty$ _____

5. $\lim_{x \to 0^+} \tan^{-1}\left(\dfrac{1}{x}\right)$ does not exist. _____

6. $\lim_{z \to 1} \dfrac{z^3 + 8z - 2}{z^2 + 9z - 10}$ does not exist. _____

7. If $\lim_{x \to a} f(x) = 3$ and $\lim_{x \to a} g(x) = 0$, then $\lim_{x \to a} f(x)/g(x)$ does not exist. _____

8. If $\lim_{x \to a} f(x)$ exists and $\lim_{x \to a} g(x)$ does not exist, then $\lim_{x \to a} f(x)g(x)$ does not exist. _____

9. If $\lim_{x \to a} f(x) = \infty$ and $\lim_{x \to a} g(x) = \infty$, then $\lim_{x \to a} f(x)/g(x) = 1$. _____

10. If $\lim_{x \to a} f(x) = \infty$ and $\lim_{x \to a} g(x) = \infty$, then $\lim_{x \to a} [f(x) - g(x)] = 0$. _____

11. If f is a polynomial function, then $\lim_{x \to \infty} f(x) = \infty$. _____

12. Every polynomial function is continuous on $(-\infty, \infty)$. _____

13. For $f(x) = x^5 + 3x - 1$ there exists a number c in $[-1, 1]$ such that $f(c) = 0$. _____

14. If f and g are continuous at the number 2, then f/g is continuous at 2. _____

15. The greatest integer function $f(x) = \lfloor x \rfloor$ is not continuous on the interval $[0, 1]$. _____

16. If $\lim_{x \to a^-} f(x)$ and $\lim_{x \to a^+} f(x)$ exist, then $\lim_{x \to a} f(x)$ exists. _____

17. If a function f is discontinuous at the number 3, then $f(3)$ is not defined. _____

18. If a function f is continuous at the number a, then $\lim_{x \to a} (x - a)f(x) = 0$. _____

19. If f is continuous and $f(a)f(b) < 0$, there is a root of $f(x) = 0$ in the interval $[a, b]$. _____

20. The function $f(x) = \begin{cases} \dfrac{x^2 - 6x + 5}{x - 5}, & x \neq 5 \\ 4, & x = 5 \end{cases}$ is discontinuous at 5. _____

21. The function $f(x) = \dfrac{\sqrt{x}}{x + 1}$ has a vertical asymptote at $x = -1$. _____

22. If $y = x - 2$ is a tangent line to the graph of a function $y = f(x)$ at $(3, f(3))$, then $f(3) = 1$. _____

B. Fill in the Blanks _____

In Problems 1–22, fill in the blanks.

1. $\lim\limits_{x \to 2}(3x^2 - 4x) =$ _____

2. $\lim\limits_{x \to 3}(5x^2)^0 =$ _____

3. $\lim\limits_{t \to \infty} \dfrac{2t - 1}{3 - 10t} =$ _____

4. $\lim\limits_{x \to -\infty} \dfrac{\sqrt{x^2 + 1}}{2x + 1} =$ _____

5. $\lim\limits_{t \to 1} \dfrac{1 - \cos^2(t - 1)}{t - 1} =$ _____

6. $\lim\limits_{x \to 0} \dfrac{\sin 3x}{5x} =$ _____

7. $\lim\limits_{x \to 0^+} e^{1/x} =$ _____

8. $\lim\limits_{x \to 0^-} e^{1/x} =$ _____

9. $\lim\limits_{x \to \infty} e^{1/x} =$ _____

10. $\lim\limits_{x \to -\infty} \dfrac{1 + 2e^x}{4 + e^x} =$ _____

11. $\lim\limits_{x \to -} \dfrac{1}{x - 3} = -\infty$

12. $\lim\limits_{x \to -}(5x + 2) = 22$

13. $\lim\limits_{x \to -} x^3 = -\infty$

14. $\lim\limits_{x \to -} \dfrac{1}{\sqrt{x}} = \infty$

15. If $f(x) = 2(x - 4)/|x - 4|$, $x \neq 4$, and $f(4) = 9$, then $\lim\limits_{x \to 4^-} f(x) =$ _____.

16. Suppose $x^2 - x^4/3 \le f(x) \le x^2$ for all x. Then $\lim\limits_{x \to 0} f(x)/x^2 =$ _____.

17. If f is continuous at a number a and $\lim\limits_{x \to a} f(x) = 10$, then $f(a) =$ _____.

18. If f is continuous at $x = 5$, $f(5) = 2$, and $\lim\limits_{x \to 5} g(x) = 10$, then $\lim\limits_{x \to 5}[g(x) - f(x)] =$ _____.

19. $f(x) = \begin{cases} \dfrac{2x - 1}{4x^2 - 1}, & x \neq \frac{1}{2} \\ 0.5, & x = \frac{1}{2} \end{cases}$ is _____ (continuous/discontinuous) at the number $\frac{1}{2}$.

20. The equation $e^{-x^2} = x^2 - 1$ has precisely _____ roots in the interval $(-\infty, \infty)$.

21. The function $f(x) = \dfrac{10}{x} + \dfrac{x^2 - 4}{x - 2}$ has a removable discontinuity at $x = 2$. To remove the discontinuity, $f(2)$ should be defined to be _____.

22. If $\lim\limits_{x \to -5} g(x) = -9$ and $f(x) = x^2$, then $\lim\limits_{x \to -5} f(g(x)) =$ _____.

C. Exercises _____

In Problems 1–4, sketch a graph of a function f that satisfies the given conditions.

1. $f(0) = 1, f(4) = 0, f(6) = 0,\ \lim\limits_{x \to 3^-} f(x) = 2,\ \lim\limits_{x \to 3^+} f(x) = \infty,\ \lim\limits_{x \to -\infty} f(x) = 0,\ \lim\limits_{x \to \infty} f(x) = 2$

2. $\lim\limits_{x \to -\infty} f(x) = 0, f(0) = 1,\ \lim\limits_{x \to 4^-} f(x) = \infty,\ \lim\limits_{x \to 4^+} f(x) = \infty, f(5) = 0,\ \lim\limits_{x \to \infty} f(x) = -1$

3. $\lim\limits_{x \to -\infty} f(x) = 2, f(-1) = 3, f(0) = 0, f(-x) = -f(x)$

4. $\lim\limits_{x \to \infty} f(x) = 0, f(0) = -3, f(1) = 0, f(-x) = f(x)$

In Problems 5–10, state which of the conditions (a)–(j) are applicable to the graph of $y = f(x)$.

(a) $f(a)$ is not defined   **(b)** $f(a) = L$ **(c)** f is continuous at $x = a$ **(d)** f is continuous on $[0, a]$ **(e)** $\lim\limits_{x \to a^+} f(x) = L$

(f) $\lim\limits_{x \to a} f(x) = L$ **(g)** $\lim\limits_{x \to a} |f(x)| = \infty$ **(h)** $\lim\limits_{x \to \infty} f(x) = L$ **(i)** $\lim\limits_{x \to \infty} f(x) = -\infty$ **(j)** $\lim\limits_{x \to \infty} f(x) = 0$

5.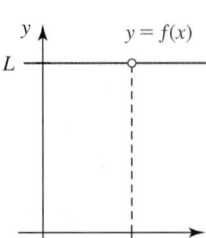

FIGURE 2.R.1   Graph for Problem 5

6.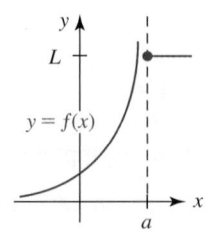

FIGURE 2.R.2   Graph for Problem 6

7.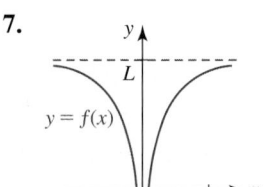

FIGURE 2.R.3   Graph for Problem 7

8.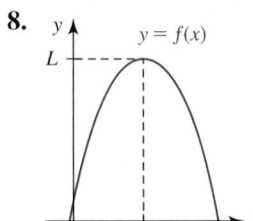

FIGURE 2.R.4   Graph for Problem 8

9.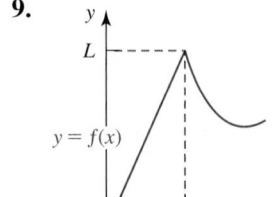

FIGURE 2.R.5   Graph for Problem 9

10.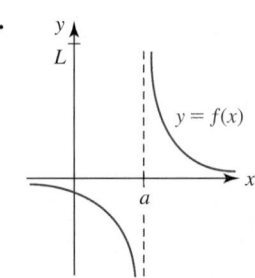

FIGURE 2.R.6   Graph for Problem 10

In Problems 11 and 12, sketch the graph of the given function. Determine the numbers, if any, at which f is discontinuous.

11. $f(x) = |x| + x$

12. $f(x) = \begin{cases} x + 1, & x < 2 \\ 3, & 2 < x < 4 \\ -x + 7, & x > 4 \end{cases}$

In Problems 13–16, determine intervals on which the given function is continuous.

13. $f(x) = \dfrac{x + 6}{x^3 - x}$

14. $f(x) = \dfrac{\sqrt{4 - x^2}}{x^2 - 4x + 3}$

15. $f(x) = \dfrac{x}{\sqrt{x^2 - 5}}$

16. $f(x) = \dfrac{\csc x}{\sqrt{x}}$

17. Find a number k so that

$$f(x) = \begin{cases} kx + 1, & x \le 3 \\ 2 - kx, & x > 3 \end{cases}$$

is continuous at the number 3.

18. Find numbers a and b so that

$$f(x) = \begin{cases} x + 4, & x \le 1 \\ ax + b, & 1 < x \le 3 \\ 3x - 8, & x > 3 \end{cases}$$

is continuous everywhere.

In Problems 19–22, find the slope of the tangent line to the graph of the function at the given value of x. Find an equation of the tangent line at the corresponding point.

19. $f(x) = -3x^2 + 16x + 12, \quad x = 2$

20. $f(x) = x^3 - x^2, \quad x = -1$

21. $f(x) = \dfrac{-1}{2x^2}, \quad x = \dfrac{1}{2}$

22. $f(x) = x + 4\sqrt{x}, \quad x = 4$

23. Find an equation of the line that is perpendicular to the tangent line at the point $(1, 2)$ on the graph of $f(x) = -4x^2 + 6x$.

24. Suppose $f(x) = 2x + 5$ and $\varepsilon = 0.01$. Find a $\delta > 0$ that will guarantee that $|f(x) - 7| < \varepsilon$ when $0 < |x - 1| < \delta$. What limit has been proved by finding δ?

Chapter 3

The Derivative

In This Chapter The word *calculus* is a diminutive form of the Latin word *calx*, which means "stone." In ancient civilizations small stones or pebbles were often used as a means of reckoning. Consequently, the word *calculus* can refer to any systematic method of computation. However, over the last several hundred years the connotation of the word *calculus* has evolved to mean that branch of mathematics concerned with the calculation and application of entities known as derivatives and integrals. Thus, the subject known as **calculus** has been divided into two rather broad but related areas: **differential calculus** and **integral calculus**.

In this chapter we will begin our study of differential calculus.

3.1 The Derivative

■ **Introduction** In the last section of Chapter 2 we saw that the tangent line to a graph of a function $y = f(x)$ is the line through a point $(a, f(a))$ with slope given by

Recall, m_{tan} is also called the slope of the curve at $(a, f(a))$.

$$m_{tan} = \lim_{h \to 0} \frac{f(a + h) - f(a)}{h}$$

whenever the limit exists. For many functions it is usually possible to obtain a general formula that gives the value of the slope of a tangent line. This is accomplished by computing

$$\lim_{h \to 0} \frac{f(x + h) - f(x)}{h} \tag{1}$$

for *any* x (for which the limit exists). We then substitute a value of x *after* the limit has been found.

■ **A Definition** The limit of the difference quotient in (1) defines a function—a function that is *derived* from the original function $y = f(x)$. This new function is called the **derivative function**, or simply the **derivative**, of f and is denoted by f'.

Definition 3.1.1 Derivative

The **derivative** of a function $y = f(x)$ at x is given by

$$f'(x) = \lim_{h \to 0} \frac{f(x + h) - f(x)}{h} \tag{2}$$

whenever the limit exists.

Let us now reconsider Examples 1 and 2 of Section 2.7.

EXAMPLE 1 A Derivative

Find the derivative of $f(x) = x^2 + 2$.

Solution As in the calculation of m_{tan} in Section 2.7, the process of finding the derivative $f'(x)$ consists of four steps:

(i) $f(x + h) = (x + h)^2 + 2 = x^2 + 2xh + h^2 + 2$

(ii) $f(x + h) - f(x) = [x^2 + 2xh + h^2 + 2] - x^2 - 2 = h(2x + h)$

(iii) $\dfrac{f(x + h) - f(x)}{h} = \dfrac{h(2x + h)}{h} = 2x + h \quad \leftarrow \text{cancel } h\text{'s}$

(iv) $\lim\limits_{h \to 0} \dfrac{f(x + h) - f(x)}{h} = \lim\limits_{h \to 0} [2x + h] = 2x.$

From step (iv) we see that the derivative of $f(x) = x^2 + 2$ is $f'(x) = 2x$. ■

Observe that the result $m_{tan} = 2$ obtained in Example 1 of Section 2.7 is obtained by evaluating the derivative $f'(x) = 2x$ at $x = 1$, that is, $f'(1) = 2$.

EXAMPLE 2 Value of the Derivative

For $f(x) = x^2 + 2$, find $f'(-2)$, $f'(0)$, $f'(\frac{1}{2})$, and $f'(1)$. Interpret.

Solution From Example 1 we know that the derivative is $f'(x) = 2x$. Hence,

at $x = -2,$ $\begin{cases} f(-2) = 6 & \leftarrow \text{point of tangency is } (-2, 6) \\ f'(-2) = -4 & \leftarrow \text{slope of tangent line at } (-2, 6) \text{ is } m = -4 \end{cases}$

at $x = 0,$ $\begin{cases} f(0) = 2 & \leftarrow \text{point of tangency is } (0, 2) \\ f'(0) = 0 & \leftarrow \text{slope of tangent line at } (0, 2) \text{ is } m = 0 \end{cases}$

$$\text{at } x = \tfrac{1}{2}, \quad \begin{cases} f\left(\tfrac{1}{2}\right) = \tfrac{9}{4} & \leftarrow \text{point of tangency is } \left(\tfrac{1}{2}, \tfrac{9}{4}\right) \\ f'\left(\tfrac{1}{2}\right) = 1 & \leftarrow \text{slope of tangent line at } \left(\tfrac{1}{2}, \tfrac{9}{4}\right) \text{ is } m = 1 \end{cases}$$

$$\text{at } x = 1, \quad \begin{cases} f(1) = 3 & \leftarrow \text{point of tangency is } (1, 3) \\ f'(1) = 2. & \leftarrow \text{slope of tangent line at } (1, 3) \text{ is } m = 2 \end{cases}$$

Recall that a horizontal line has 0 slope. So the fact that $f'(0) = 0$ means that the tangent line is horizontal at $(0, 2)$. ∎

By the way, if you trace back through the four-step process in Example 1, you will find that the derivative of $g(x) = x^2$ is also $g'(x) = 2x = f'(x)$. This makes intuitive sense; since the graph of $f(x) = x^2 + 2$ is a rigid vertical translation or shift of the graph of $g(x) = x^2$ for a given value of x, the points of tangency change but not the slope of the tangent line at the points. For example, at $x = 3$, $g'(3) = 6 = f'(3)$ but the points of tangency are $(3, g(3)) = (3, 9)$ and $(3, f(3)) = (3, 11)$.

EXAMPLE 3   A Derivative

Find the derivative of $f(x) = x^3$.

Solution To calculate $f(x + h)$, we use the Binomial Theorem.

(i) $f(x + h) = (x + h)^3 = x^3 + 3x^2h + 3xh^2 + h^3$

(ii) $f(x + h) - f(x) = [x^3 + 3x^2h + 3xh^2 + h^3] - x^3 = h(3x^2 + 3xh + h^2)$

(iii) $\dfrac{f(x + h) - f(x)}{h} = \dfrac{h[3x^2 + 3xh + h^2]}{h} = 3x^2 + 3xh + h^2$

(iv) $\displaystyle\lim_{h \to 0} \dfrac{f(x + h) - f(x)}{h} = \lim_{h \to 0} [3x^2 + 3xh + h^2] = 3x^2.$

◀ Recall from algebra that $(a + b)^3 = a^3 + 3a^2b + 3ab^2 + b^3$. Now replace a by x and b by h.

The derivative of $f(x) = x^3$ is $f'(x) = 3x^2$. ∎

EXAMPLE 4   Tangent Line

Find an equation of the tangent line to the graph of $f(x) = x^3$ at $x = \tfrac{1}{2}$.

Solution From Example 3 we have two functions $f(x) = x^3$ and $f'(x) = 3x^2$. As we saw in Example 2, when evaluated at the same number $x = \tfrac{1}{2}$ these functions give different information:

$$f\left(\tfrac{1}{2}\right) = \left(\tfrac{1}{2}\right)^3 = \tfrac{1}{8} \qquad \leftarrow \text{point of tangency is } \left(\tfrac{1}{2}, \tfrac{1}{8}\right)$$

$$f'\left(\tfrac{1}{2}\right) = 3\left(\tfrac{1}{2}\right)^2 = \tfrac{3}{4}. \qquad \leftarrow \text{slope of tangent at } \left(\tfrac{1}{2}, \tfrac{1}{8}\right) \text{ is } \tfrac{3}{4}$$

Thus, by the point–slope form of a line, an equation of the tangent line is given by

$$y - \frac{1}{8} = \frac{3}{4}\left(x - \frac{1}{2}\right) \qquad \text{or} \qquad y = \frac{3}{4}x - \frac{1}{4}.$$

The graph of the function and the tangent line are given in FIGURE 3.1.1. ∎

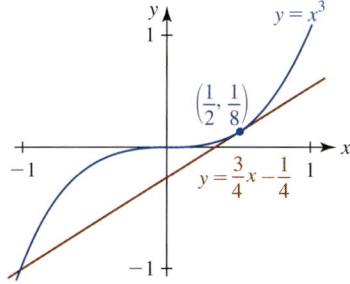

FIGURE 3.1.1   Tangent line in Example 4

EXAMPLE 5   A Derivative

Find the derivative of $f(x) = 1/x$.

Solution In this case you should be able to show that the difference is

$$f(x + h) - f(x) = \frac{1}{x + h} - \frac{1}{x} = \frac{-h}{(x + h)x}. \quad \leftarrow \begin{array}{l}\text{add fractions by using}\\ \text{a common denominator}\end{array}$$

Therefore,

$$\lim_{h \to 0} \frac{f(x + h) - f(x)}{h} = \lim_{h \to 0} \frac{-h}{h(x + h)x}$$

$$= \lim_{h \to 0} \frac{-1}{(x + h)x} = \frac{-1}{x^2}.$$

The derivative of $f(x) = 1/x$ is $f'(x) = -1/x^2$. ∎

▌ Notation The following is a list of some of the common **notations** used throughout mathematical literature to denote the derivative of a function:

$$f'(x), \quad \frac{dy}{dx}, \quad y', \quad Dy, \quad D_xy.$$

For a function such as $f(x) = x^2$, we write $f'(x) = 2x$; if the same function is written $y = x^2$, we then utilize $dy/dx = 2x$, $y' = 2x$, or $D_xy = 2x$. We will use the first three notations throughout this text. Of course other symbols are used in various applications. Thus, if $z = t^2$, then

$$\frac{dz}{dt} = 2t \quad \text{or} \quad z' = 2t.$$

The dy/dx notation has its origin in the derivative form of (3) of Section 2.7. Replacing h by Δx and denoting the difference $f(x + h) - f(x)$ by Δy in (2), the derivative is often defined as

$$\frac{dy}{dx} = \lim_{\Delta x \to 0} \frac{f(x + \Delta x) - f(x)}{\Delta x} = \lim_{\Delta x \to 0} \frac{\Delta y}{\Delta x}. \tag{3}$$

EXAMPLE 6 A Derivative Using (3)

Use (3) to find the derivative of $y = \sqrt{x}$.

Solution In the four-step procedure the important algebra takes place in the third step:

(i) $f(x + \Delta x) = \sqrt{x + \Delta x}$

(ii) $\Delta y = f(x + \Delta x) - f(x) = \sqrt{x + \Delta x} - \sqrt{x}$

(iii) $\dfrac{\Delta y}{\Delta x} = \dfrac{f(x + \Delta x) - f(x)}{\Delta x} = \dfrac{\sqrt{x + \Delta x} - \sqrt{x}}{\Delta x}$

$$= \frac{\sqrt{x + \Delta x} - \sqrt{x}}{\Delta x} \cdot \frac{\sqrt{x + \Delta x} + \sqrt{x}}{\sqrt{x + \Delta x} + \sqrt{x}} \quad \leftarrow \text{rationalization of numerator}$$

$$= \frac{x + \Delta x - x}{\Delta x(\sqrt{x + \Delta x} + \sqrt{x})}$$

$$= \frac{\Delta x}{\Delta x(\sqrt{x + \Delta x} + \sqrt{x})}$$

$$= \frac{1}{\sqrt{x + \Delta x} + \sqrt{x}}$$

(iv) $\displaystyle\lim_{\Delta x \to 0} \frac{\Delta y}{\Delta x} = \lim_{\Delta x \to 0} \frac{1}{\sqrt{x + \Delta x} + \sqrt{x}} = \frac{1}{\sqrt{x} + \sqrt{x}} = \frac{1}{2\sqrt{x}}.$

The derivative of $y = \sqrt{x}$ is $dy/dx = 1/(2\sqrt{x})$. ∎

▌ Value of a Derivative The **value** of the derivative at a number a is denoted by the symbols

$$f'(a), \quad \frac{dy}{dx}\bigg|_{x=a}, \quad y'(a), \quad D_xy\bigg|_{x=a}.$$

EXAMPLE 7 A Derivative

From Example 6, the value of the derivative of $y = \sqrt{x}$ at, say, $x = 9$ is written

$$\frac{dy}{dx}\bigg|_{x=9} = \frac{1}{2\sqrt{x}}\bigg|_{x=9} = \frac{1}{6}.$$

Alternatively, to avoid the clumsy vertical bar we can simply write $y'(9) = \frac{1}{6}$. ∎

▌ Differentiation Operators The process of finding or calculating a derivative is called **differentiation**. Thus differentiation is an operation that is performed on a function $y = f(x)$. The operation of differentiation of a function with respect to the variable x is represented by the symbols d/dx and D_x. These symbols are called **differentiation operators**. For instance, the results in Examples 1, 3, and 6 can be expressed, in turn, as

$$\frac{d}{dx}(x^2 + 2) = 2x, \quad \frac{d}{dx}x^3 = 3x^2, \quad \frac{d}{dx}\sqrt{x} = \frac{1}{2\sqrt{x}}.$$

The symbol

$$\frac{dy}{dx} \qquad \text{then means} \qquad \frac{d}{dx}y.$$

■ **Differentiability** If the limit in (2) exists for a given number x in the domain of f, the function f is said to be **differentiable** at x. If a function f is differentiable at every number x in the open intervals (a, b), $(-\infty, b)$, and (a, ∞), then f is **differentiable on the open interval**. If f is differentiable on $(-\infty, \infty)$, then f is said to be **differentiable everywhere**. A function f is **differentiable on a closed interval** $[a, b]$ when f is differentiable on the open interval (a, b), and

$$f'_+(a) = \lim_{h \to 0^+} \frac{f(a + h) - f(a)}{h}$$

$$f'_-(b) = \lim_{h \to 0^-} \frac{f(b + h) - f(b)}{h} \tag{4}$$

both exist. The limits in (4) are called **right-hand** and **left-hand derivatives**, respectively. A function is differentiable on $[a, \infty)$ when it is differentiable on (a, ∞) and has a right-hand derivative at a. A similar definition in terms of a left-hand derivative holds for differentiability on $(-\infty, b]$. Moreover, it can be shown:

- *A function f is differentiable at a number c in an interval (a, b) if and only if $f'_+(c) = f'_-(c)$.* (5)

■ **Horizontal Tangents** If $y = f(x)$ is continuous at a number a and $f'(a) = 0$, then the tangent line at $(a, f(a))$ is **horizontal**. In Examples 1 and 2 we saw that the value of derivative $f'(x) = 2x$ of the function $f(x) = x^2 + 2$ at $x = 0$ is $f'(0) = 0$. Thus, the tangent line to the graph is horizontal at $(0, f(0))$ or $(0, 0)$. It is left as an exercise (see Problem 7 in Exercises 3.1) to verify by Definition 3.1.1 that the derivative of the continuous function $f(x) = -x^2 + 4x + 1$ is $f'(x) = -2x + 4$. Observe in this latter case that $f'(x) = 0$ when $-2x + 4 = 0$ or $x = 2$. There is a horizontal tangent at the point $(2, f(2)) = (2, 5)$.

■ **Where f Fails to be Differentiable** A function f fails to have a derivative at $x = a$ if

(*i*) the function f is discontinuous at $x = a$, or
(*ii*) the graph of f has a corner at $(a, f(a))$.

In addition, since the derivative gives slope, f will fail to be differentiable

(*iii*) at a point $(a, f(a))$ at which the tangent line to the graph is vertical.

The domain of the derivative f', defined by (2), is the set of numbers x for which the limit exists. Thus the domain of f' is necessarily a subset of the domain of f.

EXAMPLE 8 Differentiability
_____

(**a**) The function $f(x) = x^2 + 2$ is differentiable for all real numbers x, that is, the domain of $f'(x) = 2x$ is $(-\infty, \infty)$.

(**b**) Because $f(x) = 1/x$ is discontinuous at $x = 0$, f is not differentiable at $x = 0$ and consequently not differentiable on any interval containing 0. ■

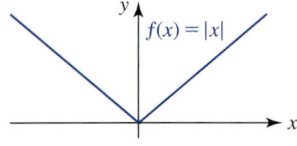

(a) Absolute-value function f

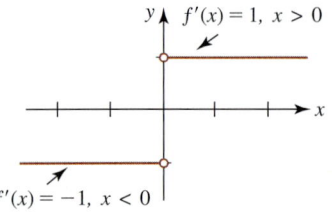

(b) Graph of the derivative f'

FIGURE 3.1.2 Graphs of f and f' in Example 9

EXAMPLE 9   Example 7 of Section 2.7 Revisited

In Example 7 of Section 2.7 we saw that the graph of $f(x) = |x|$ possesses no tangent at the origin $(0, 0)$. Thus $f(x) = |x|$ is not differentiable at $x = 0$. But $f(x) = |x|$ is differentiable on the open intervals $(0, \infty)$ and $(-\infty, 0)$. In Example 5 of Section 2.7, we proved that the derivative of a linear function $f(x) = mx + b$ is $f'(x) = m$. Hence, for $x > 0$ we have $f(x) = |x| = x$ and so $f'(x) = 1$. Also, for $x < 0, f(x) = |x| = -x$ and so $f'(x) = -1$. Since the derivative of f is a piecewise-defined function,

$$f'(x) = \begin{cases} 1, & x > 0 \\ -1, & x < 0, \end{cases}$$

we can graph it as we would any function. We see in FIGURE 3.1.2(b) that f' is discontinuous at $x = 0$.   ∎

In different symbols, what we have shown in Example 9 is that $f'_-(0) = -1$ and $f'_+(0) = 1$. Since $f'_-(0) \neq f'_+(0)$ it follows from (5) that f is not differentiable at 0.

■ **Vertical Tangents**   Let $y = f(x)$ be continuous at a number a. If $\lim_{x \to a} |f'(x)| = \infty$, then the graph of f is said to have a **vertical tangent** at $(a, f(a))$. The graphs of many functions with rational exponents possess vertical tangents.

In Example 6 of Section 2.7 we mentioned that the graph of $y = x^{1/3}$ possesses a vertical tangent line at $(0, 0)$. We verify this assertion in the next example.

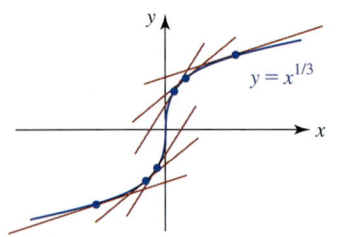

FIGURE 3.1.3 Tangent lines to the graph of the function in Example 10

EXAMPLE 10   Vertical Tangent

It is left as an exercise to prove that the derivative of $f(x) = x^{1/3}$ is given by

$$f'(x) = \frac{1}{3x^{2/3}}.$$

(See Problem 55 in Exercises 3.1.) Although f is continuous at 0, it is clear that f' is not defined at that number. In other words, f is not differentiable at $x = 0$. Moreover, because

$$\lim_{x \to 0^+} f'(x) = \infty \qquad \text{and} \qquad \lim_{x \to 0^-} f'(x) = \infty$$

we have $|f'(x)| \to \infty$ as $x \to 0$. This is sufficient to say that there is a tangent line at $(0, f(0))$ or $(0, 0)$ and that it is vertical. FIGURE 3.1.3 shows that the tangent lines to the graph on either side of the origin become steeper and steeper as $x \to 0$.   ∎

The graph of a function f can also have a vertical tangent at a point $(a, f(a))$ if f is differentiable only on one side of a, is continuous from the left (right) at a, and either $|f'(x)| \to \infty$ as $x \to a^-$ or $|f'(x)| \to \infty$ as $x \to a^+$.

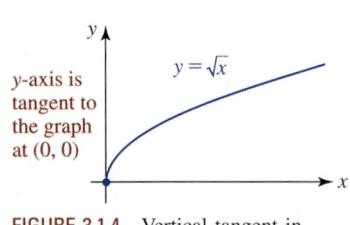

y-axis is tangent to the graph at $(0, 0)$

FIGURE 3.1.4 Vertical tangent in Example 11

EXAMPLE 11   One-Sided Vertical Tangent

The function $f(x) = \sqrt{x}$ is not differentiable on the interval $[0, \infty)$ because it is seen from the derivative $f'(x) = 1/(2\sqrt{x})$ that $f'_+(0)$ does not exist. The function $f(x) = \sqrt{x}$ is continuous on $[0, \infty)$ but differentiable on $(0, \infty)$. In addition, because f is continuous at 0 and $\lim_{x \to 0^+} f'(x) = \infty$, there is a one-sided vertical tangent at the origin $(0, 0)$. We see in FIGURE 3.1.4 that the vertical tangent is the y-axis.   ∎

Important ▶ The functions $f(x) = |x|$ and $f(x) = x^{1/3}$ are continuous everywhere. In particular, both functions are continuous at 0 but neither are differentiable at that number. In other words, continuity at a number a is not sufficient to guarantee that a function is differentiable at a. However, if a function f is differentiable at a, then f must be continuous at that number. We summarize this last fact in the next theorem.

Theorem 3.1.1   Differentiability Implies Continuity

If f is differentiable at a number a, then f is continuous at a.

PROOF To prove continuity of f at a number a it is sufficient to prove that $\lim\limits_{x \to a} f(x) = f(a)$ or equivalently that $\lim\limits_{x \to a}[f(x) - f(a)] = 0$. The hypothesis is that

$$f'(a) = \lim_{h \to 0} \frac{f(a + h) - f(a)}{h}$$

exists. If we let $x = a + h$, then as $h \to 0$ we have $x \to a$. Thus, the foregoing limit is equivalent to

$$f'(a) = \lim_{x \to a} \frac{f(x) - f(a)}{x - a}.$$

Then we can write

$$\lim_{x \to a} [f(x) - f(a)] = \lim_{x \to a} \frac{f(x) - f(a)}{x - a} \cdot (x - a) \qquad \leftarrow \text{multiplication by } \frac{x - a}{x - a} = 1$$

$$= \lim_{x \to a} \frac{f(x) - f(a)}{x - a} \cdot \lim_{x \to a} (x - a) \qquad \leftarrow \text{both limits exist}$$

$$= f'(a) \cdot 0 = 0. \qquad \blacksquare$$

▌ **Postscript—A Bit of History**   It is acknowledged that **Isaac Newton** (1642–1727), an English mathematician and physicist, was the first to set forth many of the basic principles of calculus in unpublished manuscripts on the *method of fluxions*, dated 1665. The word  *fluxion* originated from the concept of quantities that "flow"—that is, quantities that change at a certain rate. Newton used the dot notation $\dot{y}$ to represent a fluxion, or as we now know it: the derivative of a function. The symbol $\dot{y}$ never achieved overwhelming popularity among mathematicians and is used today primarily by physicists. For typographical reasons, the so-called "fly-speck

Newton   notation" has been superseded by the prime notation. Newton attained everlasting fame with the publication of his law of universal gravitation in his monumental treatise *Philosophiae Naturalis Principia Mathematica* in 1687. Newton was also the first to prove, using the calculus and his law of gravitation, Johannes Kepler's three empirical laws of planetary motion and was the first to prove that white light is composed of all colors. Newton was elected to Parliament, was appointed warden of the Royal Mint, and was knighted in 1705. Sir Isaac Newton said about his many accomplishments: "If I have seen farther than others, it is by standing on the shoulders of giants."

The German mathematician, lawyer, and philosopher **Gottfried Wilhelm Leibniz** (1646–1716) published a short version of his calculus in an article in a periodical journal in 1684. The dy/dx notation for a derivative of a function is due to Leibniz. In fact, it was Leibniz who introduced the word *function* into mathematical literature. But, since it was well known that Newton's manuscripts on the *method of fluxions* dated from 1665, Leibniz was accused of appropri-

Leibniz   ating his ideas from these unpublished works. Fueled by nationalistic prides, a controversy about who was the first to "invent" calculus raged for many years. Historians now agree that both Leibniz and Newton arrived at many of the major premises of calculus independent of each other. Leibniz and Newton are considered the "co-inventors" of the subject.

$\dfrac{d}{dx}$ NOTES FROM THE CLASSROOM

(*i*) In the preceding discussion, we saw that the derivative of a function is itself a function that gives the slope of a tangent line. The derivative is, however, *not* an equation of a tangent line. Also, to say that $y - y_0 = f'(x) \cdot (x - x_0)$ is an equation of the tangent at (x_0, y_0) is incorrect. Remember that $f'(x)$ must be evaluated at x_0 *before* it is used in the point–slope form. If f is differentiable at x_0, then an equation of the tangent line at (x_0, y_0) is $y - y_0 = f'(x_0) \cdot (x - x_0)$.

(*ii*) Although we have emphasized slopes in this section, do not forget the discussion on average rates of change and instantaneous rates of change in Section 2.7. The derivative $f'(x)$ is also the **instantaneous rate of change** of the function $y = f(x)$ with respect to the variable x. More will be said about rates in subsequent sections.

(*iii*) Mathematicians from the seventeenth to the nineteenth centuries believed that a continuous function *usually* possessed a derivative. (We have noted exceptions in this section.) In 1872 the German mathematician Karl Weierstrass conclusively destroyed this tenet by publishing an example of a function that was *everywhere continuous but nowhere differentiable*.

Exercises 3.1   Answers to selected odd-numbered problems begin on page ANS-10.

≡ Fundamentals

In Problems 1–20, use (2) of Definition 3.1.1 to find the derivative of the given function.

1. $f(x) = 10$

2. $f(x) = x - 1$

3. $f(x) = -3x + 5$

4. $f(x) = \pi x$

5. $f(x) = 3x^2$

6. $f(x) = -x^2 + 1$

7. $f(x) = -x^2 + 4x + 1$

8. $f(x) = \frac{1}{2}x^2 + 6x - 7$

9. $y = (x + 1)^2$

10. $f(x) = (2x - 5)^2$

11. $f(x) = x^3 + x$

12. $f(x) = 2x^3 + x^2$

13. $y = -x^3 + 15x^2 - x$

14. $y = 3x^4$

15. $y = \dfrac{2}{x + 1}$

16. $y = \dfrac{x}{x - 1}$

17. $y = \dfrac{2x + 3}{x + 4}$

18. $f(x) = \dfrac{1}{x} + \dfrac{1}{x^2}$

19. $f(x) = \dfrac{1}{\sqrt{x}}$

20. $f(x) = \sqrt{2x + 1}$

In Problems 21–24, use (2) of Definition 3.1.1 to find the derivative of the given function. Find an equation of the tangent line to the graph of the function at the indicated value of *x*.

21. $f(x) = 4x^2 + 7x; \quad x = -1$

22. $f(x) = \frac{1}{3}x^3 + 2x - 4; \quad x = 0$

23. $y = x - \dfrac{1}{x}; \quad x = 1$

24. $y = 2x + 1 + \dfrac{6}{x}; \quad x = 2$

In Problems 25–28, use (2) of Definition 3.1.1 to find the derivative of the given function. Find point(s) on the graph of the given function where the tangent line is horizontal.

25. $f(x) = x^2 + 8x + 10$

26. $f(x) = x(x - 5)$

27. $f(x) = x^3 - 3x$

28. $f(x) = x^3 - x^2 + 1$

In Problems 29–32, use (2) of Definition 3.1.1 to find the derivative of the given function. Find point(s) on the graph of the

given function where the tangent line is parallel to the given line.

29. $f(x) = \frac{1}{2}x^2 - 1; \quad 3x - y = 1$

30. $f(x) = x^2 - x; \quad -2x + y = 0$

31. $f(x) = -x^3 + 4; \quad 12x + y = 4$

32. $f(x) = 6\sqrt{x} + 2; \quad -x + y = 2$

In Problems 33 and 34, show that the given function is not differentiable at the indicated value of *x*.

33. $f(x) = \begin{cases} -x + 2, & x \le 2 \\ 2x - 4, & x > 2 \end{cases}; \quad x = 2$

34. $f(x) = \begin{cases} 3x, & x < 0 \\ -4x, & x \ge 0 \end{cases}; \quad x = 0$

In the proof of Theorem 3.1.1 we saw that an alternative formulation of the derivative of a function f at a is given by

$$f'(a) = \lim_{x \to a} \frac{f(x) - f(a)}{x - a}, \tag{6}$$

whenever the limit exists. In Problems 35–40, use (6) to compute $f'(a)$.

35. $f(x) = 10x^2 - 3$

36. $f(x) = x^2 - 3x - 1$

37. $f(x) = x^3 - 4x^2$

38. $f(x) = x^4$

39. $f(x) = \dfrac{4}{3 - x}$

40. $f(x) = \sqrt{x}$

41. Find an equation of the tangent line shown in red in FIGURE 3.1.5. What are $f(-3)$ and $f'(-3)$?

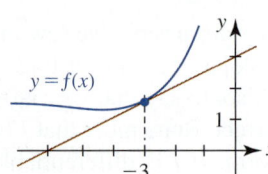

FIGURE 3.1.5   Graph for Problem 41

42. Find an equation of the tangent line shown in red in FIGURE 3.1.6. What is $f'(3)$? What is the y-intercept of the tangent line?

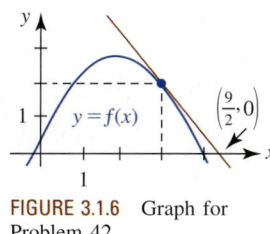

FIGURE 3.1.6 Graph for Problem 42

In Problems 43–48, sketch the graph of f' from the graph of f.

43.

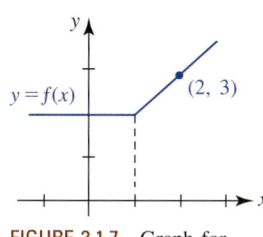

FIGURE 3.1.7 Graph for Problem 43

44.

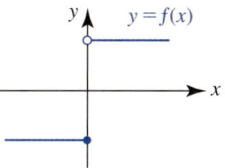

FIGURE 3.1.8 Graph for Problem 44

45.

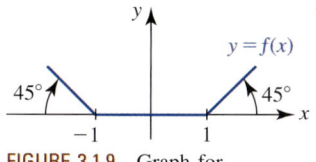

FIGURE 3.1.9 Graph for Problem 45

46.

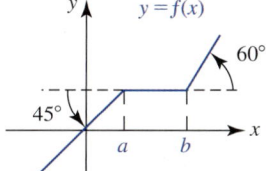

FIGURE 3.1.10 Graph for Problem 46

47.

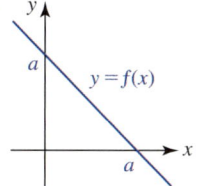

FIGURE 3.1.11 Graph for Problem 47

48.

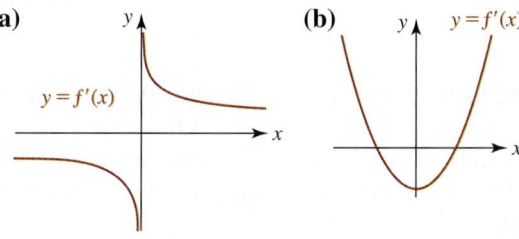

FIGURE 3.1.12 Graph for Problem 48

In Problems 49–54, match the graph of f with a graph of f' from (a)–(f).

(a)

(b)

(c)

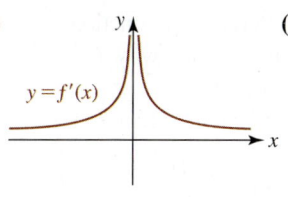

(d)

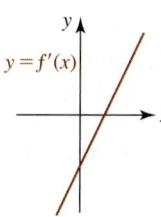

(e)

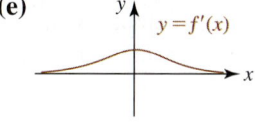

(f)

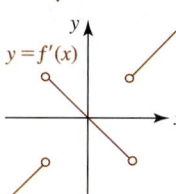

49.

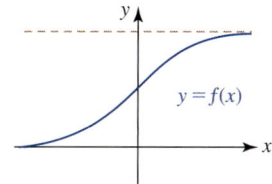

FIGURE 3.1.13 Graph for Problem 49

50.

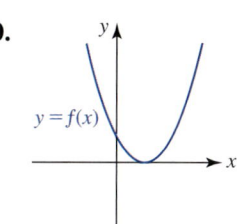

FIGURE 3.1.14 Graph for Problem 50

51.

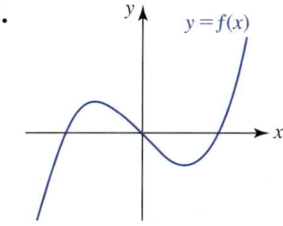

FIGURE 3.1.15 Graph for Problem 51

52.

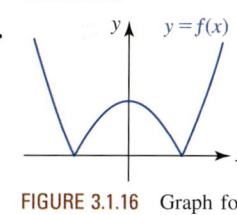

FIGURE 3.1.16 Graph for Problem 52

53.

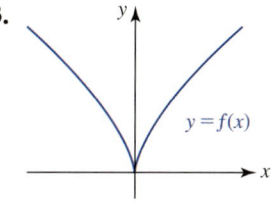

FIGURE 3.1.17 Graph for Problem 53

54.

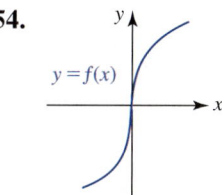

FIGURE 3.1.18 Graph for Problem 54

≡ Think About It

55. Use the alternative definition of the derivative (6) to find the derivative of $f(x) = x^{1/3}$.

[*Hint*: Note that $x - a = (x^{1/3})^3 - (a^{1/3})^3$.]

56. In Examples 10 and 11, we saw, respectively, that the functions $f(x) = x^{1/3}$ and $f(x) = \sqrt{x}$ possessed vertical tangents at the origin $(0, 0)$. Conjecture where the graphs of $y = (x - 4)^{1/3}$ and $y = \sqrt{x + 2}$ may have vertical tangents.

57. Suppose f is differentiable everywhere and has the three properties:

(*i*) $f(x_1 + x_2) = f(x_1)f(x_2)$, (*ii*) $f(0) = 1$, (*iii*) $f'(0) = 1$.

Use (2) of Definition 3.1.1 to show that $f'(x) = f(x)$ for all x.

58. (a) Suppose f is an even differentiable function on $(-\infty, \infty)$. On geometric grounds, explain why $f'(-x) = -f'(x)$; that is, f' is an odd function.

(b) Suppose f is an odd differentiable function on $(-\infty, \infty)$. On geometric grounds, explain why $f'(-x) = f'(x)$; that is, f' is an even function.

59. Suppose f is a differentiable function on $[a, b]$ such that $f(a) = 0$ and $f(b) = 0$. By experimenting with graphs discern whether the following statement is true or false: There is some number c in (a, b) such that $f'(c) = 0$.

60. Sketch graphs of various functions f that have the property $f'(x) > 0$ for all x in $[a, b]$. What do these functions have in common?

$\equiv$ **Calculator/CAS Problem**

61. Consider the function $f(x) = x^n + |x|$, where n is a positive integer. Use a calculator or CAS to obtain the graph of f for $n = 1, 2, 3, 4,$ and 5. Then use (2) to show that f is not differentiable at $x = 0$ for $n = 1, 2, 3, 4,$ and 5. Can you prove this for *any* positive integer n? What is $f'_-(0)$ and $f'_+(0)$ for $n > 1$?

3.2 Power and Sum Rules

❚ Introduction The definition of a derivative

$$f'(x) = \lim_{h \to 0} \frac{f(x + h) - f(x)}{h} \tag{1}$$

has the obvious drawback of being rather clumsy and tiresome to apply. To find the derivative of the polynomial function $f(x) = 6x^{100} + 4x^{35}$ using the above definition we would *only* have to juggle 137 terms in the binomial expansions of $(x + h)^{100}$ and $(x + h)^{35}$. There are more efficient ways of computing derivatives of a function than using the definition each time. In this section, and the sections that follow, we will see that there are shortcuts or general **rules** whereby derivatives of functions such as $f(x) = 6x^{100} + 4x^{35}$ can be obtained, literally, with just a flick of a pencil.

In the last section we saw that the derivatives of the power functions

$$f(x) = x^2, \quad f(x) = x^3, \quad f(x) = \frac{1}{x} = x^{-1}, \quad f(x) = \sqrt{x} = x^{1/2}$$

were, in turn,

See Examples 3, 5, and 6 in Section 3.1.

$$f'(x) = 2x, \quad f'(x) = 3x^2, \quad f'(x) = -\frac{1}{x^2} = -x^{-2}, \quad f'(x) = \frac{1}{2\sqrt{x}} = \frac{1}{2}x^{-1/2}.$$

If the right-hand sides of these four derivatives are written

$$2 \cdot x^{2-1}, \quad 3 \cdot x^{3-1}, \quad (-1) \cdot x^{-1-1}, \quad \frac{1}{2} \cdot x^{\frac{1}{2}-1},$$

we observe that each coefficient (indicated in red) corresponds with the original exponent of x in f and the new exponent of x in f' can be obtained from the old exponent (also indicated in red) by subtracting 1 from it. In other words, the pattern for the derivative of the general power function $f(x) = x^n$ appears to be

bring down exponent as a multiple
$$(\ ^{\downarrow})x^{(\ )-1}_{\uparrow}. \tag{2}$$
decrease exponent by 1

❚ Derivative of the Power Function The pattern illustrated in (2) does indeed hold for any real-number exponent n, and we will state it as a formal theorem, but at this point in the course we do not possess the necessary mathematical tools to prove its complete validity. We can, however, readily prove a special case of this power rule; the remaining parts of the proof will be given in the appropriate sections ahead.

Theorem 3.2.1   Power Rule

For any real number n,

$$\frac{d}{dx}x^n = nx^{n-1}. \qquad (3)$$

PROOF   We present the proof only in the case when n is a positive integer. To compute (1) for $f(x) = x^n$ we use the four-step method:

$$\overbrace{\text{general Binomial Theorem}}$$

$(i)\ f(x+h) = (x+h)^n = x^n + nx^{n-1}h + \frac{n(n-1)}{2!}x^{n-2}h^2 + \cdots + nxh^{n-1} + h^n$

◀ See the *Resource Pages* for a review of the Binomial Theorem.

$(ii)\ f(x+h) - f(x) = x^n + nx^{n-1}h + \frac{n(n-1)}{2!}x^{n-2}h^2 + \cdots + nxh^{n-1} + h^n - x^n$

$$= nx^{n-1}h + \frac{n(n-1)}{2!}x^{n-2}h^2 + \cdots + nxh^{n-1} + h^n$$

$$= h\left[nx^{n-1} + \frac{n(n-1)}{2!}x^{n-1}h + \cdots + nxh^{n-2} + h^{n-1}\right]$$

$(iii)\ \dfrac{f(x+h) - f(x)}{h} = \dfrac{h\left[nx^{n-1} + \dfrac{n(n-1)}{2!}x^{n-1}h + \cdots + nxh^{n-2} + h^{n-1}\right]}{h}$

$$= nx^{n-1} + \frac{n(n-1)}{2!}x^{n-1}h + \cdots + nxh^{n-2} + h^{n-1}$$

$(iv)\ f'(x) = \lim\limits_{h \to 0}\dfrac{f(x+h) - f(x)}{h}$

$$= \lim_{h \to 0}\left[nx^{n-1} + \underbrace{\frac{n(n-1)}{2!}x^{n-1}h + \cdots + nxh^{n-2} + h^{n-1}}\right] = nx^{n-1}. \qquad ■$$

$$\text{these terms} \to 0 \text{ as } h \to 0$$

EXAMPLE 1   Power Rule

Differentiate

(a) $y = x^7$ **(b)** $y = x$ **(c)** $y = x^{-2/3}$ **(d)** $y = x^{\sqrt{2}}$.

Solution   By the Power Rule (3),

(a) with $n = 7$: $\dfrac{dy}{dx} = 7x^{7-1} = 7x^6$,

(b) with $n = 1$: $\dfrac{dy}{dx} = 1x^{1-1} = x^0 = 1$,

(c) with $n = -\dfrac{2}{3}$: $\dfrac{dy}{dx} = \left(-\dfrac{2}{3}\right)x^{(-2/3)-1} = -\dfrac{2}{3}x^{-5/3} = -\dfrac{2}{3x^{5/3}}$,

(d) with $n = \sqrt{2}$: $\dfrac{dy}{dx} = \sqrt{2}x^{\sqrt{2}-1}$.   ■

Observe in part (b) of Example 1 that the result is consistent with the fact that the slope of the line $y = x$ is $m = 1$. See FIGURE 3.2.1.

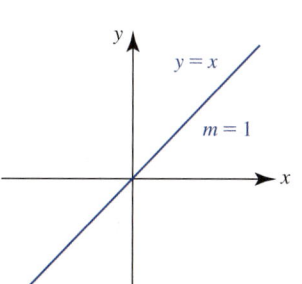

FIGURE 3.2.1   Slope of line $m = 1$ is consistent with $dy/dx = 1$

Theorem 3.2.2   Constant Function Rule

If $f(x) = c$ is a constant function, then $f'(x) = 0$.   (4)

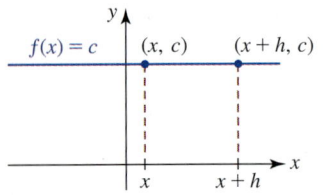

PROOF If $f(x) = c$ where c is any real number, then it follows that the difference is $f(x + h) - f(x) = c - c = 0$. Hence from (1),

$$f'(x) = \lim_{h \to 0} \frac{c - c}{h} = \lim_{h \to 0} 0 = 0. \qquad \blacksquare$$

Theorem 3.2.2 has an obvious geometric interpretation. As shown in FIGURE 3.2.2, the slope of the horizontal line $y = c$ is, of course, zero. Moreover, Theorem 3.2.2 agrees with (3) in the case when $x \neq 0$ and $n = 0$.

Theorem 3.2.3 Constant Multiple Rule

If c is any constant and f is differentiable at x, then cf is differentiable at x, and

$$\frac{d}{dx} cf(x) = cf'(x). \qquad (5)$$

PROOF Let $G(x) = cf(x)$. Then

$$G'(x) = \lim_{h \to 0} \frac{G(x + h) - G(x)}{h} = \lim_{h \to 0} \frac{cf(x + h) - cf(x)}{h}$$

$$= \lim_{h \to 0} c \left[\frac{f(x + h) - f(x)}{h} \right]$$

$$= c \lim_{h \to 0} \frac{f(x + h) - f(x)}{h} = cf'(x). \qquad \blacksquare$$

EXAMPLE 2 A Constant Multiple

Differentiate $y = 5x^4$.

Solution From (3) and (5),

$$\frac{dy}{dx} = 5 \frac{d}{dx} x^4 = 5(4x^3) = 20x^3. \qquad \blacksquare$$

Theorem 3.2.4 Sum and Difference Rules

If f and g are functions differentiable at x, then $f + g$ and $f - g$ are differentiable at x, and

$$\frac{d}{dx} [f(x) + g(x)] = f'(x) + g'(x), \qquad (6)$$

$$\frac{d}{dx} [f(x) - g(x)] = f'(x) - g'(x). \qquad (7)$$

PROOF OF (6) Let $G(x) = f(x) + g(x)$. Then

$$G'(x) = \lim_{h \to 0} \frac{G(x + h) - G(x)}{h} = \lim_{h \to 0} \frac{[f(x + h) + g(x + h)] - [f(x) + g(x)]}{h}$$

$$= \lim_{h \to 0} \frac{f(x + h) - f(x) + g(x + h) - g(x)}{h} \qquad \leftarrow \text{regrouping terms}$$

since limits exist, limit of a sum is → the sum of the limits

$$= \lim_{h \to 0} \frac{f(x + h) - f(x)}{h} + \lim_{h \to 0} \frac{g(x + h) - g(x)}{h}$$

$$= f'(x) + g'(x). \qquad \blacksquare$$

Theorem 3.2.4 holds for any finite sum of differentiable functions. For example, if f, g, and h are functions that are differentiable at x, then

$$\frac{d}{dx}[f(x) + g(x) + h(x)] = f'(x) + g'(x) + h'(x).$$

Since $f - g$ can be written as a sum, $f + (-g)$, there is no need to prove (7) since the result follows from a combination of (6) and (5). Hence, we can express Theorem 3.2.4 in words as:

- *The derivative of a sum is the sum of the derivatives.*

■ **Derivative of a Polynomial**   Because we now know how to differentiate powers of x and constant multiples of those powers we can easily differentiate sums of those constant multiples. The derivative of a polynomial function is particularly easy to obtain. For example, the derivative of the polynomial function $f(x) = 6x^{100} + 4x^{35}$, mentioned in the introduction to this section, is now readily seen to be $f'(x) = 600x^{99} + 140x^{34}$.

EXAMPLE 3   Polynomial with Six Terms

Differentiate $y = 4x^5 - \dfrac{1}{2}x^4 + 9x^3 + 10x^2 - 13x + 6$.

Solution   Using (3), (5), and (6), we obtain

$$\frac{dy}{dx} = 4\frac{d}{dx}x^5 - \frac{1}{2}\frac{d}{dx}x^4 + 9\frac{d}{dx}x^3 + 10\frac{d}{dx}x^2 - 13\frac{d}{dx}x + \frac{d}{dx}6.$$

Since $\dfrac{d}{dx}6 = 0$ by (4), we obtain

$$\frac{dy}{dx} = 4(5x^4) - \frac{1}{2}(4x^3) + 9(3x^2) + 10(2x) - 13(1) + 0$$

$$= 20x^4 - 2x^3 + 27x^2 + 20x - 13. \qquad ■$$

EXAMPLE 4   Tangent Line

Find an equation of a tangent line to the graph of $f(x) = 3x^4 + 2x^3 - 7x$ at the point corresponding to $x = -1$.

Solution   From the Sum Rule,

$$f'(x) = 3(4x^3) + 2(3x^2) - 7(1) = 12x^3 + 6x^2 - 7.$$

When evaluated at the same number $x = -1$ the functions f and f' give:

$$f(-1) = 8 \qquad \leftarrow \text{point of tangency is } (-1, 8)$$
$$f'(-1) = -13. \qquad \leftarrow \text{slope of tangent at } (-1, 8) \text{ is } -13$$

The point–slope form gives an equation of the tangent line

$$y - 8 = -13(x - (-1)) \qquad \text{or} \qquad y = -13x - 5. \qquad ■$$

■ **Rewriting a Function**   In some circumstances, in order to apply a rule of differentiation ◀ This discussion is worth remembering. efficiently it may be necessary to *rewrite* an expression in an alternative form. This alternative form is often the result of some algebraic manipulation or an application of the laws of exponents. For example, we can use (3) to differentiate the following expressions, but first we rewrite them using the laws of exponents

$$\boxed{\frac{4}{x^2}, \quad \frac{10}{\sqrt{x}}, \quad \sqrt{x^3}} \quad \rightarrow$$

rewrite square roots as powers	$\rightarrow \quad \dfrac{4}{x^2}, \quad \dfrac{10}{x^{1/2}}, \quad (x^3)^{1/2},$
then rewrite using negative exponents	$\rightarrow \quad 4x^{-2}, \quad 10x^{-1/2}, \quad x^{3/2},$
the derivative of each term using (3)	$\rightarrow \quad \boxed{-8x^{-3}, \quad -5x^{-3/2}, \quad \dfrac{3}{2}x^{1/2}}.$

A function such as $f(x) = (5x + 2)/x^2$ can be rewritten as two fractions

$$f(x) = \frac{5x + 2}{x^2} = \frac{5x}{x^2} + \frac{2}{x^2} = \frac{5}{x} + \frac{2}{x^2} = 5x^{-1} + 2x^{-2}.$$

From the last form of f it is now apparent that the derivative f' is

$$f'(x) = 5(-x^{-2}) + 2(-2x^{-3}) = -\frac{5}{x^2} - \frac{4}{x^3}.$$

EXAMPLE 5 Rewriting the Terms of a Function

Differentiate $y = 4\sqrt{x} + \dfrac{8}{x} - \dfrac{6}{\sqrt[3]{x}} + 10$.

Solution Before differentiating we rewrite the first three terms as powers of x:

$$y = 4x^{1/2} + 8x^{-1} - 6x^{-1/3} + 10.$$

Then

$$\frac{dy}{dx} = 4\frac{d}{dx}x^{1/2} + 8\frac{d}{dx}x^{-1} - 6\frac{d}{dx}x^{-1/3} + \frac{d}{dx}10.$$

By the Power Rule (3) and (4), we obtain

$$\frac{dy}{dx} = 4 \cdot \frac{1}{2}x^{-1/2} + 8 \cdot (-1)x^{-2} - 6 \cdot \left(-\frac{1}{3}\right)x^{-4/3} + 0$$

$$= \frac{2}{\sqrt{x}} - \frac{8}{x^2} + \frac{2}{x^{4/3}}. \qquad ■$$

EXAMPLE 6 Horizontal Tangents

Find the points on the graph of $f(x) = -x^3 + 3x^2 + 2$ where the tangent line is horizontal.

Solution At a point $(x, f(x))$ on the graph of f where the tangent is horizontal we must have $f'(x) = 0$. The derivative of f is $f'(x) = -3x^2 + 6x$ and the solutions of $f'(x) = -3x^2 + 6x = 0$ or $-3x(x - 2) = 0$ are $x = 0$ and $x = 2$. The corresponding points are then $(0, f(0)) = (0, 2)$ and $(2, f(2)) = (2, 6)$. See **FIGURE 3.2.3**. $\qquad ■$

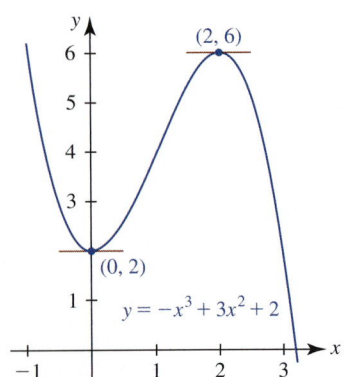

FIGURE 3.2.3 Graph of function in Example 6

■ **Normal Line** A **normal line** at a point P on a graph is one that is perpendicular to the tangent line at P.

EXAMPLE 7 Equation of a Normal Line

Find an equation of the normal line to the graph of $y = x^2$ at $x = 1$.

Solution Since $dy/dx = 2x$, we know that $m_{tan} = 2$ at $(1, 1)$. Thus the slope of the normal line shown in green in **FIGURE 3.2.4** is the negative reciprocal of the slope of the tangent line, that is, $m = -\frac{1}{2}$. By the point-slope form of a line, an equation of the normal line is then

$$y - 1 = -\frac{1}{2}(x - 1) \qquad \text{or} \qquad y = -\frac{1}{2}x + \frac{3}{2}. \qquad ■$$

FIGURE 3.2.4 Normal line in Example 7

EXAMPLE 8 Vertical Tangent

For the power function $f(x) = x^{2/3}$ the derivative is

$$f'(x) = \frac{2}{3}x^{-1/3} = \frac{2}{3x^{1/3}}.$$

Observe that $\lim\limits_{x \to 0^+} f(x) = \infty$ whereas $\lim\limits_{x \to 0^-} f(x) = -\infty$. Since f is continuous at $x = 0$ and $|f'(x)| \to \infty$ as $x \to 0$, we conclude that the y-axis is a vertical tangent at $(0, 0)$. This fact is apparent from the graph in **FIGURE 3.2.5**. $\qquad ■$

FIGURE 3.2.5 Graph of function in Example 8

■ **Cusp**   The graph of $f(x) = x^{2/3}$ in Example 8 is said to have a **cusp** at the origin. In general, the graph of a function $y = f(x)$ has a cusp at a point $(a, f(a))$ if f is continuous at a, $f'(x)$ has opposite signs on either side of a, and $|f'(x)| \to \infty$ as $x \to a$.

■ **Higher-Order Derivatives**   We have seen that the derivative $f'(x)$ is a function derived from $y = f(x)$. By differentiation of the first derivative, we obtain yet another function called the **second derivative**, which is denoted by $f''(x)$. In terms of the operation symbol d/dx, we define the second derivative with respect to x as the function obtained by differentiating $y = f(x)$ twice in succession:

$$\frac{d}{dx}\left(\frac{dy}{dx}\right).$$

The second derivative is commonly denoted by the symbols

$$f''(x), \quad y'', \quad \frac{d^2y}{dx^2}, \quad \frac{d^2}{dx^2}f(x), \quad D^2, \quad D_x^2.$$

EXAMPLE 9   Second Derivative

Find the second derivative of $y = \dfrac{1}{x^3}$.

Solution   We first simplify the function by rewriting it as $y = x^{-3}$. Then by the Power Rule (3), we have

$$\frac{dy}{dx} = -3x^{-4}.$$

The second derivative follows from differentiating the first derivative

$$\frac{d^2y}{dx^2} = \frac{d}{dx}(-3x^{-4}) = -3(-4x^{-5}) = 12x^{-5} = \frac{12}{x^5}. \qquad \blacksquare$$

Assuming that all derivatives exist, we can differentiate a function $y = f(x)$ as many times as we want. The **third derivative** is the derivative of the second derivative; the **fourth derivative** is the derivative of the third derivative; and so on. We denote the third and fourth derivatives by d^3y/dx^3 and d^4y/dx^4 and define them by

$$\frac{d^3y}{dx^3} = \frac{d}{dx}\left(\frac{d^2y}{dx^2}\right) \qquad \text{and} \qquad \frac{d^4y}{dx^4} = \frac{d}{dx}\left(\frac{d^3y}{dx^3}\right).$$

In general, if n is a positive integer, then the **nth derivative** is defined by

$$\frac{d^ny}{dx^n} = \frac{d}{dx}\left(\frac{d^{n-1}y}{dx^{n-1}}\right).$$

Other notations for the first n derivatives are

$$f'(x), \quad f''(x), \quad f'''(x), \quad f^{(4)}(x), \quad \dots, \quad f^{(n)}(x),$$
$$y', \quad y'', \quad y''', \quad y^{(4)}, \quad \dots, \quad y^{(n)},$$
$$\frac{d}{dx}f(x), \quad \frac{d^2}{dx^2}f(x), \quad \frac{d^3}{dx^3}f(x), \quad \frac{d^4}{dx^4}f(x), \quad \dots, \quad \frac{d^n}{dx^n}f(x),$$
$$D, \quad D^2, \quad D^3, \quad D^4, \quad \dots, \quad D^n,$$
$$D_x, \quad D_x^2, \quad D_x^3, \quad D_x^4, \quad \dots, \quad D_x^n.$$

Note that the "prime" notation is used to denote only the first three derivatives; after that we use the superscript $y^{(4)}$, $y^{(5)}$, and so on. The **value of the nth derivative** of a function $y = f(x)$ at a number a is denoted by

$$f^{(n)}(a), \qquad y^{(n)}(a), \qquad \text{and} \qquad \left.\frac{d^ny}{dx^n}\right|_{x=a}.$$

EXAMPLE 10 Fifth Derivative

Find the first five derivatives of $f(x) = 2x^4 - 6x^3 + 7x^2 + 5x$.

Solution We have

$$f'(x) = 8x^3 - 18x^2 + 14x + 5$$
$$f''(x) = 24x^2 - 36x + 14$$
$$f'''(x) = 48x - 36$$
$$f^{(4)}(x) = 48$$
$$f^{(5)}(x) = 0.$$

■

After reflecting a moment, you should be convinced that the $(n + 1)$st derivative of an nth-degree polynomial function is zero.

$\dfrac{d}{dx}$ **NOTES FROM THE CLASSROOM**

(*i*) In the different contexts of science, engineering, and business, functions are often expressed in variables other than x and y. Correspondingly we must adapt the derivative notation to the new symbols. For example,

Function	*Derivative*
$v(t) = 32t$	$v'(t) = \dfrac{dv}{dt} = 32$
$A(r) = \pi r^2$	$A'(r) = \dfrac{dA}{dr} = 2\pi r$
$r(\theta) = 4\theta^2 - 3\theta$	$r'(\theta) = \dfrac{dr}{d\theta} = 8\theta - 3$
$D(p) = 800 - 129p + p^2$	$D'(p) = \dfrac{dD}{dp} = -129 + 2p.$

(*ii*) You may be wondering what interpretation can be given to the higher-order derivatives. If we think in terms of graphs, then f'' gives the slope of tangent lines to the graph of the function f'; f''' gives the slope of the tangent lines to the graph of f'', and so on. In addition, if f is differentiable, then the first-derivative f' gives the instantaneous rate of change of f. Similarly, if f' is differentiable, then f'' gives the instantaneous rate of change of f'.

Exercises 3.2   Answers to selected odd-numbered problems begin on page ANS-10.

≡ **Fundamentals**

In Problems 1–8, find dy/dx.

1. $y = -18$

2. $y = \pi^6$

3. $y = x^9$

4. $y = 4x^{12}$

5. $y = 7x^2 - 4x$

6. $y = 6x^3 + 3x^2 - 10$

7. $y = 4\sqrt{x} - \dfrac{6}{\sqrt[3]{x^2}}$

8. $y = \dfrac{x - x^2}{\sqrt{x}}$

In Problems 9–16, find $f'(x)$. Simplify.

9. $f(x) = \dfrac{1}{5}x^5 - 3x^4 + 9x^2 + 1$

10. $f(x) = -\dfrac{2}{3}x^6 + 4x^5 - 13x^2 + 8x + 2$

11. $f(x) = x^3(4x^2 - 5x - 6)$

12. $f(x) = \dfrac{2x^5 + 3x^4 - x^3 + 2}{x^2}$

13. $f(x) = x^2(x^2 + 5)^2$

14. $f(x) = (x^3 + x^2)^3$

15. $f(x) = (4\sqrt{x} + 1)^2$

16. $f(x) = (9 + x)(9 - x)$

In Problems 17–20, find the derivative of the given function.

17. $h(u) = (4u)^3$

18. $p(t) = (2t)^{-4} - (2t^{-1})^2$

19. $g(r) = \dfrac{1}{r} + \dfrac{1}{r^2} + \dfrac{1}{r^3} + \dfrac{1}{r^4}$

20. $Q(t) = \dfrac{t^5 + 4t^2 - 3}{6}$

In Problems 21–24, find an equation of the tangent line to the graph of the given function at the indicated value of x.

21. $y = 2x^3 - 1;\quad x = -1$

22. $y = -x + \dfrac{8}{x};\quad x = 2$

23. $f(x) = \dfrac{4}{\sqrt{x}} + 2\sqrt{x};\quad x = 4$

24. $f(x) = -x^3 + 6x^2;\quad x = 1$

In Problems 25–28, find the point(s) on the graph of the given function at which the tangent line is horizontal.

25. $y = x^2 - 8x + 5$

26. $y = \frac{1}{3}x^3 - \frac{1}{2}x^2$

27. $f(x) = x^3 - 3x^2 - 9x + 2$

28. $f(x) = x^4 - 4x^3$

In Problems 29–32, find an equation of the normal line to the graph of the given function at the indicated value of x.

29. $y = -x^2 + 1;\quad x = 2$

30. $y = x^3;\quad x = 1$

31. $f(x) = \dfrac{1}{3}x^3 - 2x^2;\quad x = 4$

32. $f(x) = x^4 - x;\quad x = -1$

In Problems 33–38, find the second derivative of the given function.

33. $y = -x^2 + 3x - 7$

34. $y = 15x^2 - 24\sqrt{x}$

35. $y = (-4x + 9)^2$

36. $y = 2x^5 + 4x^3 - 6x^2$

37. $f(x) = 10x^{-2}$

38. $f(x) = x + \left(\dfrac{2}{x^2}\right)^3$

In Problems 39 and 40, find the indicated higher derivative.

39. $f(x) = 4x^6 + x^5 - x^3;\quad f^{(4)}(x)$

40. $y = x^4 - \dfrac{10}{x};\quad d^5y/dx^5$

In Problems 41 and 42, determine intervals for which $f'(x) > 0$ and intervals for which $f'(x) < 0$.

41. $f(x) = x^2 + 8x - 4$

42. $f(x) = x^3 - 3x^2 - 9x$

In Problems 43 and 44, find the point(s) on the graph of f at which $f''(x) = 0$.

43. $f(x) = x^3 + 12x^2 + 20x$

44. $f(x) = x^4 - 2x^3$

In Problems 45 and 46, determine intervals for which $f''(x) > 0$ and intervals for which $f''(x) < 0$.

45. $f(x) = (x - 1)^3$

46. $f(x) = x^3 + x^2$

An equation containing one or more derivatives of an unknown function $y(x)$ is called a **differential equation**. In Problems 47 and 48, show that the function satisfies the given differential equation.

47. $y = x^{-1} + x^4;\quad x^2y'' - 2xy' - 4y = 0$

48. $y = x + x^3 + 4;\quad x^2y'' - 3xy' + 3y = 12$

49. Find the point on the graph of $f(x) = 2x^2 - 3x + 6$ at which the slope of the tangent line is 5.

50. Find the point on the graph of $f(x) = x^2 - x$ at which the tangent line is $3x - 9y - 4 = 0$.

51. Find the point on the graph of $f(x) = x^2 - x$ at which the slope of the normal line is 2.

52. Find the point on the graph of $f(x) = \frac{1}{4}x^2 - 2x$ at which the tangent line is parallel to the line $3x - 2y + 1 = 0$.

53. Find an equation of the tangent line to the graph of $y = x^3 + 3x^2 - 4x + 1$ at the point where the value of the second derivative is zero.

54. Find an equation of the tangent line to the graph of $y = x^4$ at the point where the value of the third derivative is 12.

☰ Applications

55. The volume V of a sphere of radius r is $V = \frac{4}{3}\pi r^3$. Find the surface area S of the sphere if S is the instantaneous rate of change of the volume with respect to the radius.

56. According to the French physician Jean Louis Poiseuille (1799–1869) the velocity v of blood in an artery with a constant circular cross-section radius R is $v(r) = (P/4vl)(R^2 - r^2)$, where P, v, and l are constants. What is the velocity of blood at the value of r for which $v'(r) = 0$?

57. The potential energy of a spring-mass system when the spring is stretched a distance of x units is $U(x) = \frac{1}{2}kx^2$, where k is the spring constant. The force exerted on the mass is $F = -dU/dx$. Find the force if the spring constant is 30 N/m and the amount of stretch is $\frac{1}{2}$ m.

58. The height s above ground of a projectile at time t is given by

$$s(t) = \frac{1}{2}gt^2 + v_0t + s_0,$$

where g, v_0, and s_0 are constants. Find the instantaneous rate of change of s with respect to t at $t = 4$.

☰ Think About It

In Problems 59 and 60, the symbol n represents a positive integer. Find a formula for the given derivative.

59. $\dfrac{d^n}{dx^n} x^n$

60. $\dfrac{d^n}{dx^n} \dfrac{1}{x}$

61. From the graphs of f and g in FIGURE 3.2.6, determine which function is the derivative of the other. Explain your choice in words.

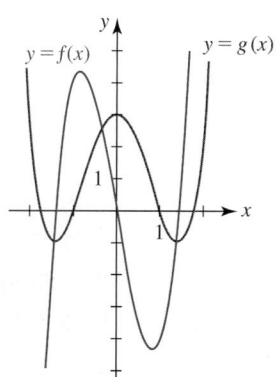

FIGURE 3.2.6 Graphs for Problem 61

62. From the graph of the function $y = f(x)$ given in FIGURE 3.2.7, sketch the graph of f'.

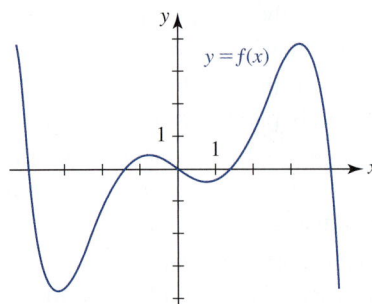

FIGURE 3.2.7   Graph for Problem 62

63. Find a quadratic function $f(x) = ax^2 + bx + c$ such that $f(-1) = -11$, $f'(-1) = 7$, and $f''(-1) = -4$.

64. The graphs of $y = f(x)$ and $y = g(x)$ are said to be **orthogonal** if the tangent lines to each graph are perpendicular at each point of intersection. Show that the graphs of $y = \frac{1}{8}x^2$ and $y = -\frac{1}{4}x^2 + 3$ are orthogonal.

65. Find the values of b and c so that the graph of $f(x) = x^2 + bx$ possesses the tangent line $y = 2x + c$ at $x = -3$.

66. Find an equation of the line(s) that passes through $\left(\frac{3}{2}, 1\right)$ and is tangent to the graph of $f(x) = x^2 + 2x + 2$.

67. Find the point(s) on the graph of $f(x) = x^2 - 5$ such that the tangent line at the point(s) has x-intercept $(-3, 0)$.

68. Find the point(s) on the graph of $f(x) = x^2$ such that the tangent line at the point(s) has y-intercept $(0, -2)$.

69. Explain why the graph of $f(x) = \frac{1}{5}x^5 + \frac{1}{3}x^3$ has no tangent line with slope -1.

70. Find coefficients A and B so that the function $y = Ax^2 + Bx$ satisfies the differential equation $2y'' + 3y' = x - 1$.

71. Find values of a and b such that the slope of the tangent to the graph of $f(x) = ax^2 + bx$ at $(1, 4)$ is -5.

72. Find the slopes of all the normal lines to the graph of $f(x) = x^2$ that pass through the point $(2, 4)$. [*Hint*: Draw a figure and note that *at* $(2, 4)$ there is only one normal line.]

73. Find a point on the graph of $f(x) = x^2 + x$ and a point on the graph of $g(x) = 2x^2 + 4x + 1$ at which the tangent lines are parallel.

74. Find a point on the graph of $f(x) = 3x^5 + 5x^3 + 2x$ at which the tangent has the least possible slope.

75. Find conditions on the coefficients a, b, and c so that the graph of the polynomial function

$$f(x) = ax^3 + bx^2 + cx + d$$

has exactly one horizontal tangent. Exactly two horizontal tangents. No horizontal tangents.

76. Let f be a differentiable function. If $f'(x) > 0$ for all x in the interval (a, b), sketch possible graphs of f on the interval. Describe in words the behavior of the graph of f on the interval. Repeat if $f'(x) < 0$ for all x in the interval (a, b).

77. Suppose f is a differentiable function such that $f'(x) - f(x) = 0$. Find $f^{(100)}(x)$.

78. The graphs of $y = x^2$ and $y = -x^2 + 2x - 3$ given in FIGURE 3.2.8 show that there are two lines L_1 and L_2 that are simultaneously tangent to both graphs. Find the points of tangency on both graphs. Find an equation of each tangent line.

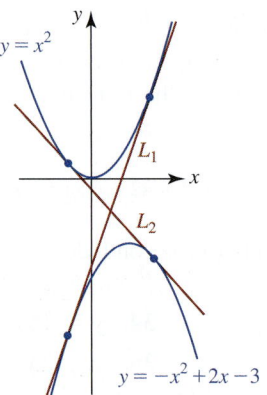

FIGURE 3.2.8   Graphs for Problem 78

≡ Calculator/CAS Problems

79. **(a)** Use a calculator or CAS to obtain the graph of $f(x) = x^4 - 4x^3 - 2x^2 + 12x - 2$.
(b) Evaluate $f''(x)$ at $x = -2$, $x = -1$, $x = 0$, $x = 1$, $x = 2$, $x = 3$, and $x = 4$.
(c) From the data in part (b), do you see any relationship between the shape of the graph of f and the algebraic signs of f''?

80. Use a calculator or CAS to obtain the graph of the given functions. By inspection of the graphs indicate where each function may not be differentiable. Find $f'(x)$ at all points where f is differentiable.
(a) $f(x) = |x^2 - 2x|$ **(b)** $f(x) = |x^3 - 1|$

3.3   Product and Quotient Rules

▌ Introduction   So far we know that the derivative of a constant function and a power of x are, in turn:

$$\frac{d}{dx}c = 0 \qquad \text{and} \qquad \frac{d}{dx}x^n = nx^{n-1}. \tag{1}$$

We also know that for differentiable functions f and g:

$$\frac{d}{dx}cf(x) = cf'(x) \quad \text{and} \quad \frac{d}{dx}[f(x) \pm g(x)] = f'(x) \pm g'(x). \tag{2}$$

Although the results in (1) and (2) allow us to quickly differentiate many algebraic functions (such as polynomials) neither (1) nor (2) are of immediate help in finding derivatives of functions such as $y = x^4\sqrt{x^2 + 4}$ or $y = x/(2x + 1)$. We need additional rules for differentiating products fg and quotients f/g.

■ **Product Rule** The rules of differentiation and the derivatives of functions ultimately stem from the definition of the derivative. The Sum Rule in (2), derived in the preceding section, follows from this definition and the fact that the limit of a sum is the sum of the limits whenever the limits exist. We also know that when the limits exist, the limit of a product is the product of the limits. Arguing by analogy, it would then seem plausible that the derivative of a product of two functions is the product of the derivatives. Regrettably, the Product Rule stated next is *not* that simple.

Theorem 3.3.1 Product Rule

If f and g are functions differentiable at x, then fg is differentiable at x, and

$$\frac{d}{dx}[f(x)g(x)] = f(x)g'(x) + g(x)f'(x). \tag{3}$$

PROOF Let $G(x) = f(x)g(x)$. Then by the definition of the derivative along with some algebraic manipulation:

$$G'(x) = \lim_{h \to 0} \frac{G(x + h) - G(x)}{h} = \lim_{h \to 0} \frac{f(x + h)g(x + h) - f(x)g(x)}{h}$$

$$= \lim_{h \to 0} \frac{f(x + h)g(x + h) \overbrace{- f(x + h)g(x) + f(x + h)g(x)}^{\text{zero}} - f(x)g(x)}{h}$$

$$= \lim_{h \to 0} \left[f(x + h)\frac{g(x + h) - g(x)}{h} + g(x)\frac{f(x + h) - f(x)}{h} \right]$$

$$= \lim_{h \to 0} f(x + h) \cdot \lim_{h \to 0} \frac{g(x + h) - g(x)}{h} + \lim_{h \to 0} g(x) \cdot \lim_{h \to 0} \frac{f(x + h) - f(x)}{h}.$$

Because f is differentiable at x, it is continuous there and so $\lim_{h \to 0} f(x + h) = f(x)$. Furthermore, $\lim_{h \to 0} g(x) = g(x)$. Hence the last equation becomes

$$G'(x) = f(x)g'(x) + g(x)f'(x). \qquad \blacksquare$$

The Product Rule is best memorized in words:

- *The first function times the derivative of the second plus the second function times the derivative of the first.*

EXAMPLE 1 Product Rule

Differentiate $y = (x^3 - 2x^2 + 3)(7x^2 - 4x)$.

Solution From the Product Rule (3),

$$\frac{dy}{dx} = \overbrace{(x^3 - 2x^2 + 3)}^{\text{first}} \cdot \overbrace{\frac{d}{dx}(7x^2 - 4x)}^{\substack{\text{derivative of}\\\text{second}}} + \overbrace{(7x^2 - 4x)}^{\text{second}} \cdot \overbrace{\frac{d}{dx}(x^3 - 2x^2 + 3)}^{\substack{\text{derivative of}\\\text{first}}}$$

$$= (x^3 - 2x^2 + 3)(14x - 4) + (7x^2 - 4x)(3x^2 - 4x)$$

$$= 35x^4 - 72x^3 + 24x^2 + 42x - 12.$$

Alternative Solution The two terms in the given function could be multiplied out to obtain a fifth-degree polynomial. The derivative can then be gotten using the Sum Rule. ∎

EXAMPLE 2 Tangent Line

Find an equation of the tangent line to the graph of $y = (1 + \sqrt{x})(x - 2)$ at $x = 4$.

Solution Before taking the derivative we rewrite $\sqrt{x}$ as $x^{1/2}$. Then from the Product Rule (3)

$$\frac{dy}{dx} = (1 + x^{1/2})\frac{d}{dx}(x - 2) + (x - 2)\frac{d}{dx}(1 + x^{1/2})$$

$$= (1 + x^{1/2}) \cdot 1 + (x - 2) \cdot \frac{1}{2}x^{-1/2}$$

$$= \frac{3x + 2\sqrt{x} - 2}{2\sqrt{x}}.$$

Evaluating the given function and its derivative at $x = 4$ gives:

$$y(4) = (1 + \sqrt{4})(4 - 2) = 6 \quad \leftarrow \text{point of tangency is } (4, 6)$$

$$\frac{dy}{dx}\Big|_{x=4} = \frac{12 + 2\sqrt{4} - 2}{2\sqrt{4}} = \frac{7}{2}. \quad \leftarrow \text{slope of the tangent at } (4, 6) \text{ is } \frac{7}{2}$$

By the point–slope form, the tangent line is

$$y - 6 = \frac{7}{2}(x - 4) \qquad \text{or} \qquad y = \frac{7}{2}x - 8. \qquad ∎$$

Although (3) is stated for only the product of two functions, it can be applied to functions with a greater number of factors. The idea is to group two (or more) functions and treat this grouping as one function. The next example illustrates the technique.

EXAMPLE 3 Product of Three Functions

Differentiate $y = (4x + 1)(2x^2 - x)(x^3 - 8x)$.

Solution We identify the first two factors as the "first function":

$$\frac{dy}{dx} = \overbrace{(4x + 1)(2x^2 - x)}^{\text{first}}\overbrace{\frac{d}{dx}(x^3 - 8x)}^{\substack{\text{derivative of} \\ \text{second}}} + \overbrace{(x^3 - 8x)}^{\text{second}}\overbrace{\frac{d}{dx}(4x + 1)(2x^2 - x)}^{\substack{\text{derivative of} \\ \text{first}}}.$$

Notice that to find the derivative of the first function, we must apply the Product Rule a second time:

$$\frac{dy}{dx} = (4x + 1)(2x^2 - x) \cdot (3x^2 - 8) + (x^3 - 8x) \cdot \overbrace{[(4x + 1)(4x - 1) + (2x^2 - x) \cdot 4]}^{\text{Product Rule again}}$$

$$= (4x + 1)(2x^2 - x)(3x^2 - 8) + (x^3 - 8x)(16x^2 - 1) + 4(x^3 - 8x)(2x^2 - x). \qquad ∎$$

■ **Quotient Rule** The derivative of the quotient of two functions f and g is given next.

Theorem 3.3.2 Quotient Rule

If f and g are functions differentiable at x and $g(x) \neq 0$, then f/g is differentiable at x, and

$$\frac{d}{dx}\left[\frac{f(x)}{g(x)}\right] = \frac{g(x)f'(x) - f(x)g'(x)}{[g(x)]^2}. \tag{4}$$

PROOF Let $G(x) = f(x)/g(x)$. Then

$$G'(x) = \lim_{h \to 0} \frac{G(x + h) - G(x)}{h} = \lim_{h \to 0} \frac{\dfrac{f(x + h)}{g(x + h)} - \dfrac{f(x)}{g(x)}}{h}$$

$$= \lim_{h \to 0} \frac{g(x)f(x + h) - f(x)g(x + h)}{hg(x + h)g(x)}$$

$$= \lim_{h \to 0} \frac{g(x)f(x + h) - \overbrace{g(x)f(x) + g(x)f(x)}^{\text{zero}} - f(x)g(x + h)}{hg(x + h)g(x)}$$

$$= \lim_{h \to 0} \frac{g(x)\dfrac{f(x + h) - f(x)}{h} - f(x)\dfrac{g(x + h) - g(x)}{h}}{g(x + h)g(x)}$$

$$= \frac{\displaystyle\lim_{h \to 0} g(x) \cdot \lim_{h \to 0} \frac{f(x + h) - f(x)}{h} - \lim_{h \to 0} f(x) \cdot \lim_{h \to 0} \frac{g(x + h) - g(x)}{h}}{\displaystyle\lim_{h \to 0} g(x + h) \cdot \lim_{h \to 0} g(x)}.$$

Since all limits are assumed to exist, the last line is the same as

$$G'(x) = \frac{g(x)f'(x) - f(x)g'(x)}{[g(x)]^2}. \qquad \blacksquare$$

In words, the Quotient Rule starts with the denominator:

- *The denominator times the derivative of the numerator minus the numerator times the derivative of the denominator all divided by the denominator squared.*

EXAMPLE 4 Quotient Rule

Differentiate $y = \dfrac{3x^2 - 1}{2x^3 + 5x^2 + 7}$.

Solution From the Quotient Rule (4),

$$\frac{dy}{dx} = \frac{\overbrace{(2x^3 + 5x^2 + 7)}^{\text{denominator}} \cdot \overbrace{\frac{d}{dx}(3x^2 - 1)}^{\substack{\text{derivative of} \\ \text{numerator}}} - \overbrace{(3x^2 - 1)}^{\text{numerator}} \cdot \overbrace{\frac{d}{dx}(2x^3 + 5x^2 + 7)}^{\substack{\text{derivative of} \\ \text{denominator}}}}{\underbrace{(2x^3 + 5x^2 + 7)^2}_{\substack{\text{square of} \\ \text{denominator}}}}$$

$$= \frac{(2x^3 + 5x^2 + 7) \cdot 6x - (3x^2 - 1) \cdot (6x^2 + 10x)}{(2x^3 + 5x^2 + 7)^2} \quad \leftarrow \text{multiply out numerator}$$

$$= \frac{-6x^4 + 6x^2 + 52x}{(2x^3 + 5x^2 + 7)^2}. \qquad \blacksquare$$

EXAMPLE 5 Quotient and Product Rule

Find the points on the graph of $y = \dfrac{(x^2 + 1)(2x^2 + 1)}{3x^2 + 1}$ where the tangent line is horizontal.

Solution We begin with the Quotient Rule and then use the Product Rule when differentiating the numerator:

$$\frac{dy}{dx} = \frac{(3x^2 + 1) \cdot \overbrace{\frac{d}{dx}[(x^2 + 1)(2x^2 + 1)]}^{\substack{\text{Product Rule} \\ \text{here}}} - (x^2 + 1)(2x^2 + 1) \cdot \frac{d}{dx}(3x^2 + 1)}{(3x^2 + 1)^2}$$

$$= \frac{(3x^2 + 1)[(x^2 + 1)4x + (2x^2 + 1)2x] - (x^2 + 1)(2x^2 + 1)6x}{(3x^2 + 1)^2} \quad \leftarrow \text{multiply out numerator}$$

$$= \frac{12x^5 + 8x^3}{(3x^2 + 1)^2}.$$

At a point where the tangent line is horizontal we must have $dy/dx = 0$. The derivative just found can only be 0 when the numerator satisfies

▶ Of course, values of x that make the numerator zero must *not* simultaneously make the denominator zero.

$$12x^5 + 8x^3 = 0 \quad \text{or} \quad x^3(12x^2 + 8) = 0. \tag{5}$$

In (5) because $12x^2 + 8 \neq 0$ for all real numbers x, we must have $x = 0$. Substituting this number into the function gives $y(0) = 1$. The tangent line is horizontal at the y-intercept $(0, 1)$. ∎

▮ **Postscript—Power Rule Revisited**   Remember in Section 3.2 we stated that the Power Rule, $(d/dx)x^n = nx^{n-1}$, is valid for all real number exponents n. We are now in a position to prove the rule when the exponent is a negative integer $-m$. Since, by definition, $x^{-m} = 1/x^m$, where m is a positive integer, we can obtain the derivative of x^{-m} by the Quotient Rule and the laws of exponents:

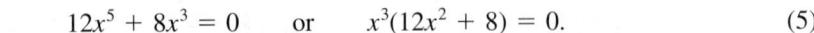

$$\frac{d}{dx}x^{-m} = \frac{d}{dx}\left(\frac{1}{x^m}\right) = \frac{x^m \cdot \frac{d}{dx}1 - 1 \cdot \frac{d}{dx}x^m}{(x^m)^2} = -\frac{\overset{\substack{\text{subtract exponents} \\ \downarrow}}{mx^{m-1}}}{x^{2m}} = -mx^{-m-1}.$$

$$\frac{d}{dx}$$ **NOTES FROM THE CLASSROOM**

(*i*) The Product and Quotient Rules will usually lead to expressions that demand simplification. If your answer to a problem does not look like the one in the text answer section, you may not have performed sufficient simplifications. Do not be content to simply carry through the mechanics of the various rules of differentiation; it is always a good idea to practice your algebraic skills.

(*ii*) The Quotient Rule is sometimes used when it is not required. Although we could use the Quotient Rule to differentiate functions such as

$$y = \frac{x^5}{6} \quad \text{and} \quad y = \frac{10}{x^3},$$

it is simpler (and faster) to rewrite the functions as $y = \frac{1}{6}x^5$ and $y = 10x^{-3}$ and then use the Constant Multiple and Power Rules:

$$\frac{dy}{dx} = \frac{1}{6}\frac{d}{dx}x^5 = \frac{5}{6}x^4 \quad \text{and} \quad \frac{dy}{dx} = 10\frac{d}{dx}x^{-3} = -30x^{-4}.$$

Exercises 3.3   Answers to selected odd-numbered problems begin on page ANS-10.

≡ **Fundamentals**

In Problems 1–10, find dy/dx.

1. $y = (x^2 - 7)(x^3 + 4x + 2)$

2. $y = (7x + 1)(x^4 - x^3 - 9x)$

3. $y = \left(4\sqrt{x} + \frac{1}{x}\right)\left(2x - \frac{6}{\sqrt[3]{x}}\right)$

4. $y = \left(x^2 - \frac{1}{x^2}\right)\left(x^3 + \frac{1}{x^3}\right)$

5. $y = \dfrac{10}{x^2 + 1}$

6. $y = \dfrac{5}{4x - 3}$

7. $y = \dfrac{3x + 1}{2x - 5}$

8. $y = \dfrac{2 - 3x}{7 - x}$

9. $y = (6x - 1)^2$

10. $y = (x^4 + 5x)^2$

In Problems 11–20, find $f'(x)$.

11. $f(x) = \left(\dfrac{1}{x} - \dfrac{4}{x^3}\right)(x^3 - 5x - 1)$

12. $f(x) = (x^2 - 1)\left(x^2 - 10x + \dfrac{2}{x^2}\right)$

13. $f(x) = \dfrac{x^2}{2x^2 + x + 1}$

14. $f(x) = \dfrac{x^2 - 10x + 2}{x(x^2 - 1)}$

15. $f(x) = (x + 1)(2x + 1)(3x + 1)$

16. $f(x) = (x^2 + 1)(x^3 - x)(3x^4 + 2x - 1)$

17. $f(x) = \dfrac{(2x + 1)(x - 5)}{3x + 2}$

18. $f(x) = \dfrac{x^5}{(x^2 + 1)(x^3 + 4)}$

19. $f(x) = (x^2 - 2x - 1)\left(\dfrac{x + 1}{x + 3}\right)$

20. $f(x) = (x + 1)\left(x + 1 - \dfrac{1}{x + 2}\right)$

In Problems 21–24, find an equation of the tangent line to the graph of the given function at the indicated value of x.

21. $y = \dfrac{x}{x - 1};\quad x = \dfrac{1}{2}$

22. $y = \dfrac{5x}{x^2 + 1};\quad x = 2$

23. $y = (2\sqrt{x} + x)(-2x^2 + 5x - 1);\quad x = 1$

24. $y = (2x^2 - 4)(x^3 + 5x + 3);\quad x = 0$

In Problems 25–28, find the point(s) on the graph of the given function at which the tangent line is horizontal.

25. $y = (x^2 - 4)(x^2 - 6)$

26. $y = x(x - 1)^2$

27. $y = \dfrac{x^2}{x^4 + 1}$

28. $y = \dfrac{1}{x^2 - 6x}$

In Problems 29 and 30, find the point(s) on the graph of the given function at which the tangent line has the indicated slope.

29. $y = \dfrac{x + 3}{x + 1};\quad m = -\dfrac{1}{8}$

30. $y = (x + 1)(2x + 5);\quad m = -3$

In Problems 31 and 32, find the point(s) on the graph of the given function at which the tangent line has the indicated property.

31. $y = \dfrac{x + 4}{x + 5};\quad$ perpendicular to $y = -x$

32. $y = \dfrac{x}{x + 1};\quad$ parallel to $y = \dfrac{1}{4}x - 1$

33. Find the value of k such that the tangent line to the graph of $f(x) = (k + x)/x^2$ has slope 5 at $x = 2$.

34. Show that the tangent to the graph of $f(x) = (x^2 + 14)/(x^2 + 9)$ at $x = 1$ is perpendicular to the tangent to the graph of $g(x) = (1 + x^2)(1 + 2x)$ at $x = 1$.

In Problems 35–40, f and g are differentiable functions. Find $F'(1)$ if $f(1) = 2$, $f'(1) = -3$, and $g(1) = 6$, $g'(1) = 2$.

35. $F(x) = 2f(x)g(x)$

36. $F(x) = x^2 f(x)g(x)$

37. $F(x) = \dfrac{2g(x)}{3f(x)}$

38. $F(x) = \dfrac{1 + 2f(x)}{x - g(x)}$

39. $F(x) = \left(\dfrac{4}{x} + f(x)\right)g(x)$

40. $F(x) = \dfrac{xf(x)}{g(x)}$

41. Suppose $F(x) = \sqrt{x}\,f(x)$, where f is a differentiable function. Find $F''(4)$ if $f(4) = -16$, $f'(4) = 2$, and $f''(4) = 3$.

42. Suppose $F(x) = xf(x) + xg(x)$, where f and g are differentiable functions. Find $F''(0)$ if $f'(0) = -1$ and $g'(0) = 6$.

43. Suppose $F(x) = f(x)/x$, where f is a differentiable function. Find $F''(x)$.

44. Suppose $F(x) = x^3 f(x)$, where f is a differentiable function. Find $F'''(x)$.

In Problems 45–48, determine intervals for which $f'(x) > 0$ and intervals for which $f'(x) < 0$.

45. $f(x) = \dfrac{5}{x^2 - 2x}$

46. $f(x) = \dfrac{x^2 + 3}{x + 1}$

47. $f(x) = (-2x + 6)(4x + 7)$

48. $f(x) = (x - 2)(4x^2 + 8x + 4)$

≡ Applications

49. The Law of Universal Gravitation states that the force F between two bodies of masses m_1 and m_2 separated by a distance r is $F = km_1m_2/r^2$, where k is constant. What is the instantaneous rate of change of F with respect to r when $r = \dfrac{1}{2}$ km?

50. The potential energy U between two atoms in a diatomic molecule is given by $U(x) = q_1/x^{12} - q_2/x^6$, where q_1 and q_2 are positive constants and x is the distance between the atoms. The force between the atoms is defined as $F(x) = -U'(x)$. Show that $F\left(\sqrt[6]{2q_1/q_2}\right) = 0$.

51. The **van der Waals equation of state** for an ideal gas is

$$\left(P + \dfrac{a}{V^2}\right)(V - b) = RT,$$

where P is pressure, V is volume per mole, R is the universal gas constant, T is temperature, and a and b are constants depending on the gas. Find dP/dV in the case where T is constant.

52. For a convex lens, the focal length f is related to the object distance p and the image distance q by the **lens equation**

$$\dfrac{1}{f} = \dfrac{1}{p} + \dfrac{1}{q}.$$

Find the instantaneous rate of change of q with respect to p in the case where f is constant. Explain the significance of the negative sign in your answer. What happens to q as p increases?

☰ Think About It

53. (a) Graph the rational function $f(x) = \dfrac{2}{x^2 + 1}$.

(b) Find all the points on the graph of f such that the normal lines pass through the origin.

54. Suppose $y = f(x)$ is a differentiable function.

(a) Find dy/dx for $y = [f(x)]^2$.

(b) Find dy/dx for $y = [f(x)]^3$.

(c) Conjecture a rule for finding the derivative of $y = [f(x)]^n$, where n is a positive integer.

(d) Use your conjecture in part (c) to find the derivative of $y = (x^2 + 2x - 6)^{500}$.

55. Suppose $y_1(x)$ satisfies the differential equation $y' + P(x)y = 0$, where P is a known function. Show that $y = u(x)y_1(x)$ satisfies the differential equation

$$y' + P(x)y = f(x)$$

whenever $u(x)$ satisfies $du/dx = f(x)/y_1(x)$.

3.4 Trigonometric Functions

▮ Introduction In this section we develop the derivatives of the six trigonometric functions. Once we have found the derivatives of $\sin x$ and $\cos x$ we can determine the derivatives of $\tan x$, $\cot x$, $\sec x$, and $\csc x$ using the Quotient Rule found in the preceding section. We will see immediately that the derivative of $\sin x$ utilizes the following two limit results

$$\lim_{x \to 0} \frac{\sin x}{x} = 1 \quad \text{and} \quad \lim_{x \to 0} \frac{\cos x - 1}{x} = 0 \tag{1}$$

found in Section 2.4.

▮ Derivatives of Sine and Cosine To find the derivative of $f(x) = \sin x$ we use the basic definition of the derivative

$$\frac{dy}{dx} = \lim_{h \to 0} \frac{f(x + h) - f(x)}{h} \tag{2}$$

and the four-step process introduced in Sections 2.7 and 3.1. In the first step we use the sum formula for the sine function,

$$\sin(x_1 + x_2) = \sin x_1 \cos x_2 + \cos x_1 \sin x_2, \tag{3}$$

but with x and h playing the parts of the symbols x_1 and x_2.

(*i*) $f(x + h) = \sin(x + h) = \sin x \cos h + \cos x \sin h$ ← from (3)

(*ii*) $f(x + h) - f(x) = \sin x \cos h + \cos x \sin h - \sin x$ ← factor $\sin x$ from first and third terms

$\qquad = \sin x(\cos h - 1) + \cos x \sin h$

As we see in the next line, we cannot cancel the h's in the difference quotient but we can rewrite the expression to make use of the limit results in (1).

(*iii*) $\dfrac{f(x + h) - f(x)}{h} = \dfrac{\sin x(\cos h - 1) + \cos x \sin h}{h}$

$\qquad\qquad\qquad = \sin x \cdot \dfrac{\cos h - 1}{h} + \cos x \cdot \dfrac{\sin h}{h}$

(*iv*) In this line, the symbol h plays the part of the symbol x in (1):

$$f'(x) = \lim_{h \to 0} \frac{f(x + h) - f(x)}{h} = \sin x \cdot \lim_{h \to 0} \frac{\cos h - 1}{h} + \cos x \cdot \lim_{h \to 0} \frac{\sin h}{h}.$$

From the limit results in (1), the last line is the same as

$$f'(x) = \lim_{h \to 0} \frac{f(x + h) - f(x)}{h} = \sin x \cdot 0 + \cos x \cdot 1 = \cos x.$$

Hence,

$$\frac{d}{dx} \sin x = \cos x. \tag{4}$$

In a similar manner it can be shown that

$$\frac{d}{dx}\cos x = -\sin x. \tag{5}$$

See Problem 50 in Exercises 3.4.

EXAMPLE 1   Equation of a Tangent Line

Find an equation of the tangent line to the graph of $f(x) = \sin x$ at $x = 4\pi/3$.

Solution   From (4) the derivative of $f(x) = \sin x$ is $f'(x) = \cos x$. When evaluated at the same number $x = 4\pi/3$ these functions give:

$$f\left(\frac{4\pi}{3}\right) = \sin\frac{4\pi}{3} = -\frac{\sqrt{3}}{2} \quad \leftarrow \text{point of tangency is } \left(\frac{4\pi}{3}, -\frac{\sqrt{3}}{2}\right)$$

$$f'\left(\frac{4\pi}{3}\right) = \cos\frac{4\pi}{3} = -\frac{1}{2}. \quad \leftarrow \text{slope of tangent at } \left(\frac{4\pi}{3}, -\frac{\sqrt{3}}{2}\right) \text{ is } -\frac{1}{2}$$

From the point–slope form of a line, an equation of the tangent line is

$$y + \frac{\sqrt{3}}{2} = -\frac{1}{2}\left(x - \frac{4\pi}{3}\right) \quad \text{or} \quad y = -\frac{1}{2}x + \frac{2\pi}{3} - \frac{\sqrt{3}}{2}.$$

The tangent line is shown in red in **FIGURE 3.4.1**.

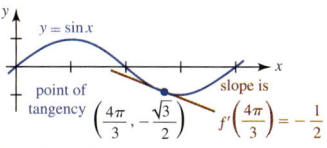

FIGURE 3.4.1   Tangent line in Example 1

▮ Other Trigonometric Functions   The results in (4) and (5) can be used in conjunction with the rules of differentiation to find the derivatives of the tangent, cotangent, secant, and cosecant functions.

To differentiate $\tan x = \sin x/\cos x$, we use the Quotient Rule:

$$\frac{d}{dx}\frac{\sin x}{\cos x} = \frac{\cos x \dfrac{d}{dx}\sin x - \sin x \dfrac{d}{dx}\cos x}{(\cos x)^2}$$

$$= \frac{\cos x (\cos x) - \sin x (-\sin x)}{(\cos x)^2} = \frac{\overbrace{\cos^2 x + \sin^2 x}^{\text{this equals } 1}}{\cos^2 x}.$$

Using the fundamental Pythagorean identity $\sin^2 x + \cos^2 x = 1$ and the fact that $1/\cos^2 x = (1/\cos x)^2 = \sec^2 x$, the last equation simplifies to

$$\frac{d}{dx}\tan x = \sec^2 x. \tag{6}$$

The derivative formula for the cotangent

$$\frac{d}{dx}\cot x = -\csc^2 x \tag{7}$$

is obtained in an analogous fashion and left as an exercise. See Problem 51 in Exercises 3.4.

Now $\sec x = 1/\cos x$. Therefore, we can use the Quotient Rule again to find the derivative of the secant function:

$$\frac{d}{dx}\frac{1}{\cos x} = \frac{\cos x \dfrac{d}{dx}1 - 1 \cdot \dfrac{d}{dx}\cos x}{(\cos x)^2}$$

$$= \frac{0 - (-\sin x)}{(\cos x)^2} = \frac{\sin x}{\cos^2 x}. \tag{8}$$

By writing

$$\frac{\sin x}{\cos^2 x} = \frac{1}{\cos x} \cdot \frac{\sin x}{\cos x} = \sec x \tan x$$

we can express (8) as

$$\frac{d}{dx}\sec x = \sec x \tan x. \tag{9}$$

The final result also follows immediately from the Quotient Rule:

$$\frac{d}{dx}\csc x = -\csc x \cot x. \tag{10}$$

See Problem 52 in Exercises 3.4.

EXAMPLE 2 Product Rule

Differentiate $y = x^2 \sin x$.

Solution The Product Rule along with (4) gives

$$\frac{dy}{dx} = x^2 \frac{d}{dx}\sin x + \sin x \frac{d}{dx}x^2$$

$$= x^2 \cos x + 2x \sin x. \qquad \blacksquare$$

EXAMPLE 3 Product Rule

Differentiate $y = \cos^2 x$.

Solution One way of differentiating this function is to recognize it as a product: $y = (\cos x)(\cos x)$. Then by the Product Rule and (5),

$$\frac{dy}{dx} = \cos x \frac{d}{dx}\cos x + \cos x \frac{d}{dx}\cos x$$

$$= \cos x(-\sin x) + (\cos x)(-\sin x)$$

$$= -2\sin x \cos x.$$

In the next section we will see that there is an alternative procedure for differentiating a power of a function. $\blacksquare$

EXAMPLE 4 Quotient Rule

Differentiate $y = \dfrac{\sin x}{2 + \sec x}$.

Solution By the Quotient Rule, (4), and (9),

$$\frac{dy}{dx} = \frac{(2 + \sec x)\dfrac{d}{dx}\sin x - \sin x \dfrac{d}{dx}(2 + \sec x)}{(2 + \sec x)^2}$$

$$= \frac{(2 + \sec x)\cos x - \sin x(\sec x \tan x)}{(2 + \sec x)^2} \quad \leftarrow \begin{array}{l} \sec x \cos x = 1 \text{ and} \\ \sin x(\sec x \tan x) = \sin^2 x/\cos^2 x \end{array}$$

$$= \frac{1 + 2\cos x - \tan^2 x}{(2 + \sec x)^2}. \qquad \blacksquare$$

EXAMPLE 5 Second Derivative

Find the second derivative of $f(x) = \sec x$.

Solution From (9) the first derivative is

$$f'(x) = \sec x \tan x.$$

To obtain the second derivative we must now use the Product Rule along with (6) and (9):

$$f''(x) = \sec x \frac{d}{dx}\tan x + \tan x \frac{d}{dx}\sec x$$

$$= \sec x(\sec^2 x) + \tan x(\sec x \tan x)$$

$$= \sec^3 x + \sec x \tan^2 x. \qquad \blacksquare$$

For future reference we summarize the derivative formulas introduced in this section.

Theorem 3.4.1 Derivatives of Trigonometric Functions

The derivatives of the six trigonometric functions are

$$\frac{d}{dx}\sin x = \cos x, \qquad\qquad \frac{d}{dx}\cos x = -\sin x, \qquad (11)$$

$$\frac{d}{dx}\tan x = \sec^2 x, \qquad\qquad \frac{d}{dx}\cot x = -\csc^2 x, \qquad (12)$$

$$\frac{d}{dx}\sec x = \sec x \tan x, \qquad\qquad \frac{d}{dx}\csc x = -\csc x \cot x. \qquad (13)$$

$\dfrac{d}{dx}$ **NOTES FROM THE CLASSROOM** ·······································

When working the problems in Exercises 3.4 you may not get the same answer as given in the answer section in the back of this book. This is because there are so many trigonometric identities that answers can often be expressed in a more compact form. For example, the answer in Example 3:

$$\frac{dy}{dx} = -2\sin x \cos x \quad \text{is the same as} \quad \frac{dy}{dx} = -\sin 2x$$

by the double-angle formula for the sine function. Try to resolve any differences between your answer and the given answer.

Exercises 3.4   Answers to selected odd-numbered problems begin on page ANS-10.

☰ Fundamentals

In Problems 1–12, find dy/dx.

1. $y = x^2 - \cos x$

2. $y = 4x^3 + x + 5\sin x$

3. $y = 1 + 7\sin x - \tan x$

4. $y = 3\cos x - 5\cot x$

5. $y = x\sin x$

6. $y = (4\sqrt{x} - 3\sqrt[3]{x})\cos x$

7. $y = (x^3 - 2)\tan x$

8. $y = \cos x \cot x$

9. $y = (x^2 + \sin x)\sec x$

10. $y = \csc x \tan x$

11. $y = \cos^2 x + \sin^2 x$

12. $y = x^3\cos x - x^3\sin x$

In Problems 13–22, find $f'(x)$.

13. $f(x) = (\csc x)^{-1}$

14. $f(x) = \dfrac{2}{\cos x \cot x}$

15. $f(x) = \dfrac{\cot x}{x + 1}$

16. $f(x) = \dfrac{x^2 - 6x}{1 + \cos x}$

17. $f(x) = \dfrac{x^2}{1 + 2\tan x}$

18. $f(x) = \dfrac{2 + \sin x}{x}$

19. $f(x) = \dfrac{\sin x}{1 + \cos x}$

20. $f(x) = \dfrac{1 + \csc x}{1 + \sec x}$

21. $f(x) = x^4\sin x \tan x$

22. $f(x) = \dfrac{1 + \sin x}{x\cos x}$

In Problems 23–26, find an equation of the tangent line to the graph of the given function at the indicated value of x.

23. $f(x) = \cos x; \quad x = \pi/3$

24. $f(x) = \tan x; \quad x = \pi$

25. $f(x) = \sec x; \quad x = \pi/6$

26. $f(x) = \csc x; \quad x = \pi/2$

In Problems 27–30, consider the graph of the given function on the interval $[0, 2\pi]$. Find the x-coordinates of the point(s) on the graph of the function where the tangent line is horizontal.

27. $f(x) = x + 2\cos x$

28. $f(x) = \dfrac{\sin x}{2 - \cos x}$

29. $f(x) = \dfrac{1}{x + \cos x}$

30. $f(x) = \sin x + \cos x$

In Problems 31–34, find an equation of the normal line to the graph of the given function at the indicated value of x.

31. $f(x) = \sin x; \quad x = 4\pi/3$

32. $f(x) = \tan^2 x; \quad x = \pi/4$

33. $f(x) = x\cos x; \quad x = \pi$

34. $f(x) = \dfrac{x}{1 + \sin x}; \quad x = \pi/2$

In Problems 35 and 36, find the derivative of the given function by first using an appropriate trigonometric identity.

35. $f(x) = \sin 2x$

36. $f(x) = \cos^2\dfrac{x}{2}$

In Problems 37–42, find the second derivative of the given function.

37. $f(x) = x\sin x$

38. $f(x) = 3x - x^2\cos x$

39. $f(x) = \dfrac{\sin x}{x}$

40. $f(x) = \dfrac{1}{1 + \cos x}$

41. $y = \csc x$

42. $y = \tan x$

In Problems 43 and 44, C_1 and C_2 are arbitrary real constants. Show that the function satisfies the given differential equation.

43. $y = C_1 \cos x + C_2 \sin x - \dfrac{1}{2}x \cos x; \quad y'' + y = \sin x$

44. $y = C_1 \dfrac{\cos x}{\sqrt{x}} + C_2 \dfrac{\sin x}{\sqrt{x}}; \quad x^2 y'' + xy' + \left(x^2 - \tfrac{1}{4}\right)y = 0$

≡ Applications

45. When the angle of elevation of the sun is θ, a telephone pole 40 ft high casts a shadow of length s as shown in FIGURE 3.4.2. Find the rate of change of s with respect to θ when $\theta = \pi/3$ radians. Explain the significance of the minus sign in the answer.

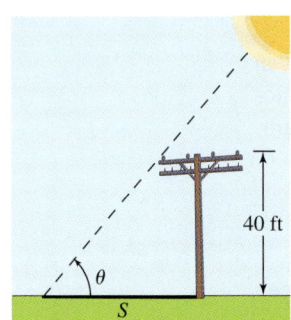

FIGURE 3.4.2   Shadow in Problem 45

46. The two ends of a 10-ft board are attached to perpendicular rails, as shown in FIGURE 3.4.3, so that point P is free to move vertically and point R is free to move horizontally.

    (a) Express the area A of triangle PQR as a function of the indicated angle θ.

    (b) Find the rate of change of A with respect to θ.

    (c) Initially the board rests flat on the horizontal rail. Suppose point R is then moved in the direction of point Q, thereby forcing point P to move up the vertical rail. Initially the area of the triangle is 0 ($\theta = 0$), but then it increases for a while as θ increases and then decreases as R approaches Q. When the board is vertical, the area of the triangle is again 0 ($\theta = \pi/2$). Graph the derivative $dA/d\theta$. Interpret this graph to find values of θ for which A is increasing and values of θ for which A is decreasing. Now verify your interpretation of the graph of the derivative by graphing $A(\theta)$.

    (d) Use the graphs in part (c) to find the value of θ for which the area of the triangle is the greatest.

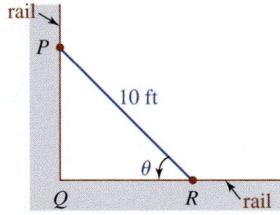

FIGURE 3.4.3   Board in Problem 46

≡ Think About It

47. (a) Find all positive integers n such that

$$\frac{d^n}{dx^n}\sin x = \sin x; \qquad \frac{d^n}{dx^n}\cos x = \cos x;$$

$$\frac{d^n}{dx^n}\cos x = \sin x; \qquad \frac{d^n}{dx^n}\sin x = \cos x.$$

    (b) Use the results in part (a) as an aid in finding

$$\frac{d^{21}}{dx^{21}}\sin x, \quad \frac{d^{30}}{dx^{30}}\sin x, \quad \frac{d^{40}}{dx^{40}}\cos x, \quad \text{and} \quad \frac{d^{67}}{dx^{67}}\cos x.$$

48. Find two distinct points P_1 and P_2 on the graph of $y = \cos x$ so that the tangent line at P_1 is perpendicular to the tangent line at P_2.

49. Find two distinct points P_1 and P_2 on the graph of $y = \sin x$ so that the tangent line at P_1 is parallel to the tangent line at P_2.

50. Use (1), (2), and the sum formula for the cosine to show that

$$\frac{d}{dx}\cos x = -\sin x.$$

51. Use (4) and (5) and the Quotient Rule to show that

$$\frac{d}{dx}\cot x = -\csc^2 x.$$

52. Use (4) and the Quotient Rule to show that

$$\frac{d}{dx}\csc x = -\csc x \cot x.$$

≡ Calculator/CAS Problems

In Problems 53 and 54, use a calculator or CAS to obtain the graph of the given function. By inspection of the graph indicate where the function may not be differentiable.

53. $f(x) = 0.5(\sin x + |\sin x|)$ **54.** $f(x) = |x + \sin x|$

55. As shown in FIGURE 3.4.4, a boy pulls a sled on which his little sister is seated. If the sled and girl weigh a total of 70 lb, and if the coefficient of sliding friction of snow-covered ground is 0.2, then the magnitude F of the force (measured in pounds) required to move the sled is

$$F = \frac{70(0.2)}{0.2\sin\theta + \cos\theta},$$

where θ is the angle the tow rope makes with the horizontal.

    (a) Use a calculator or CAS to obtain the graph of F on the interval $[-1, 1]$.

    (b) Find the derivative $dF/d\theta$.

    (c) Find the angle (in radians) for which $dF/d\theta = 0$.

    (d) Find the value of F corresponding to the angle found in part (c).

    (e) Use the graph in part (a) as an aid in interpreting the numbers found in parts (c) and (d).

FIGURE 3.4.4   Sled in Problem 55

3.5 Chain Rule

■ **Introduction**   As discussed in Section 3.2, the Power Rule

$$\frac{d}{dx}x^n = nx^{n-1}$$

is valid for all real number exponents n. In this section we see that a similar rule holds for the derivative of a power of a function $y = [g(x)]^n$. Before stating the formal result, let us consider an example when n is a positive integer.

Suppose we wish to differentiate

$$y = (x^5 + 1)^2. \tag{1}$$

By writing (1) as $y = (x^5 + 1) \cdot (x^5 + 1)$, we can find the derivative using the Product Rule:

$$\frac{d}{dx}(x^5 + 1)^2 = (x^5 + 1) \cdot \frac{d}{dx}(x^5 + 1) + (x^5 + 1) \cdot \frac{d}{dx}(x^5 + 1)$$

$$= (x^5 + 1) \cdot 5x^4 + (x^5 + 1) \cdot 5x^4$$

$$= 2(x^5 + 1) \cdot 5x^4. \tag{2}$$

Similarly, to differentiate the function $y = (x^5 + 1)^3$, we can write it as $y = (x^5 + 1)^2 \cdot (x^5 + 1)$ and use the Product Rule and the result given in (2):

$$\frac{d}{dx}(x^5 + 1)^3 = \frac{d}{dx}(x^5 + 1)^2 \cdot (x^5 + 1)$$

$$= (x^5 + 1)^2 \cdot \frac{d}{dx}(x^5 + 1) + (x^5 + 1) \cdot \overbrace{\frac{d}{dx}(x^5 + 1)^2}^{\text{we know this from (2)}}$$

$$= (x^5 + 1)^2 \cdot 5x^4 + (x^5 + 1) \cdot 2(x^5 + 1) \cdot 5x^4$$

$$= 3(x^5 + 1)^2 \cdot 5x^4. \tag{3}$$

In like manner, by writing $y = (x^5 + 1)^4$ as $y = (x^5 + 1)^3 \cdot (x^5 + 1)$ we can readily show by the Product Rule and (3) that

$$\frac{d}{dx}(x^5 + 1)^4 = 4(x^5 + 1)^3 \cdot 5x^4. \tag{4}$$

■ **Power Rule for Functions**   Inspection of (2), (3), and (4) reveals a pattern for differentiating a power of a function g. For example, in (4) we see

bring down exponent as a multiple
↓   ↓ derivative of function inside parentheses
$$4(x^5 + 1)^3 \cdot 5x^4$$
↑
decrease exponent by 1

For emphasis, if we denote a differentiable function by [ ], it appears that

$$\frac{d}{dx}[\ ]^n = n[\ ]^{n-1}\frac{d}{dx}[\ ].$$

The foregoing discussion suggests the result stated in the next theorem.

Theorem 3.5.1   Power Rule for Functions

If n is any real number and $u = g(x)$ is differentiable at x, then

$$\frac{d}{dx}[g(x)]^n = n[g(x)]^{n-1} \cdot g'(x), \tag{5}$$

or equivalently,

$$\frac{d}{dx}u^n = nu^{n-1} \cdot \frac{du}{dx}. \tag{6}$$

Theorem 3.5.1 is itself a special case of a more general theorem, called the **Chain Rule**, which will be presented after we consider some examples of this new power rule.

EXAMPLE 1 Power Rule for Functions

Differentiate $y = (4x^3 + 3x + 1)^7$.

Solution With the identification that $u = g(x) = 4x^3 + 3x + 1$, we see from (6) that

$$\frac{dy}{dx} = \overbrace{7}^{n}\underbrace{(4x^3 + 3x + 1)^6}_{u^{n-1}} \cdot \overbrace{\frac{d}{dx}(4x^3 + 3x + 1)}^{du/dx} = 7(4x^3 + 3x + 1)^6(12x^2 + 3).$$ ■

EXAMPLE 2 Power Rule for Functions

To differentiate $y = 1/(x^2 + 1)$, we could, of course, use the Quotient Rule. However, by rewriting the function as $y = (x^2 + 1)^{-1}$, it is also possible to use the Power Rule for Functions with $n = -1$:

$$\frac{dy}{dx} = (-1)(x^2 + 1)^{-2} \cdot \frac{d}{dx}(x^2 + 1) = (-1)(x^2 + 1)^{-2}\,2x = \frac{-2x}{(x^2 + 1)^2}.$$ ■

EXAMPLE 3 Power Rule for Functions

Differentiate $y = \dfrac{1}{(7x^5 - x^4 + 2)^{10}}$.

Solution Write the given function as $y = (7x^5 - x^4 + 2)^{-10}$. Identify $u = 7x^5 - x^4 + 2$, $n = -10$ and use the Power Rule (6):

$$\frac{dy}{dx} = -10(7x^5 - x^4 + 2)^{-11} \cdot \frac{d}{dx}(7x^5 - x^4 + 2) = \frac{-10(35x^4 - 4x^3)}{(7x^5 - x^4 + 2)^{11}}.$$ ■

EXAMPLE 4 Power Rule for Functions

Differentiate $y = \tan^3 x$.

Solution For emphasis, we first rewrite the function as $y = (\tan x)^3$ and then use (6) with $u = \tan x$ and $n = 3$:

$$\frac{dy}{dx} = 3(\tan x)^2 \cdot \frac{d}{dx}\tan x.$$

Recall from (6) of Section 3.4 that $(d/dx)\tan x = \sec^2 x$. Hence,

$$\frac{dy}{dx} = 3\tan^2 x \sec^2 x.$$ ■

EXAMPLE 5 Quotient Rule then Power Rule

Differentiate $y = \dfrac{(x^2 - 1)^3}{(5x + 1)^8}$.

Solution We start with the Quotient Rule followed by two applications of the Power Rule for Functions:

$$\frac{dy}{dx} = \frac{(5x + 1)^8 \cdot \overset{\text{Power Rule for Functions}}{\frac{d}{dx}(x^2 - 1)^3} - (x^2 - 1)^3 \cdot \frac{d}{dx}(5x + 1)^8}{(5x + 1)^{16}}$$

$$= \frac{(5x + 1)^8 \cdot 3(x^2 - 1)^2 \cdot 2x - (x^2 - 1)^3 \cdot 8(5x + 1)^7 \cdot 5}{(5x + 1)^{16}}$$

$$= \frac{6x(5x+1)^8(x^2-1)^2 - 40(5x+1)^7(x^2-1)^3}{(5x+1)^{16}}$$

$$= \frac{(x^2-1)^2(-10x^2+6x+40)}{(5x+1)^9}.$$

EXAMPLE 6 Power Rule then Quotient Rule

Differentiate $y = \sqrt{\dfrac{2x-3}{8x+1}}$.

Solution By rewriting the function as

$$y = \left(\frac{2x-3}{8x+1}\right)^{1/2} \quad \text{we can identify} \quad u = \frac{2x-3}{8x+1}$$

and $n = \frac{1}{2}$. Thus in order to compute du/dx in (6) we must use the Quotient Rule:

$$\frac{dy}{dx} = \frac{1}{2}\left(\frac{2x-3}{8x+1}\right)^{-1/2} \cdot \frac{d}{dx}\left(\frac{2x-3}{8x+1}\right)$$

$$= \frac{1}{2}\left(\frac{2x-3}{8x+1}\right)^{-1/2} \cdot \frac{(8x+1)\cdot 2 - (2x-3)\cdot 8}{(8x+1)^2}$$

$$= \frac{1}{2}\left(\frac{2x-3}{8x+1}\right)^{-1/2} \cdot \frac{26}{(8x+1)^2}.$$

Finally, we simplify using the laws of exponents:

$$\frac{dy}{dx} = \frac{13}{(2x-3)^{1/2}(8x+1)^{3/2}}.$$

■ Chain Rule A power of a function can be written as a composite function. If we identify $f(x) = x^n$ and $u = g(x)$, then $f(u) = f(g(x)) = [g(x)]^n$. The Chain Rule gives us a way of differentiating any composition $f \circ g$ of two differentiable functions f and g.

Theorem 3.5.2 Chain Rule

If the function f is differentiable at $u = g(x)$, and the function g is differentiable at x, then the composition $y = (f \circ g)(x) = f(g(x))$ is differentiable at x and

$$\frac{d}{dx}f(g(x)) = f'(g(x)) \cdot g'(x) \tag{7}$$

or equivalently,

$$\frac{dy}{dx} = \frac{dy}{du} \cdot \frac{du}{dx}. \tag{8}$$

PROOF FOR $\Delta u \neq 0$ In this partial proof it is convenient to use the form of the definition of the derivative given in (3) of Section 3.1. For $\Delta x \neq 0$,

$$\Delta u = g(x + \Delta x) - g(x) \tag{9}$$

or $g(x + \Delta x) = g(x) + \Delta u = u + \Delta u$. In addition,

$$\Delta y = f(u + \Delta u) - f(u) = f(g(x + \Delta x)) - f(g(x)).$$

When x and $x + \Delta x$ are in some open interval for which $\Delta u \neq 0$, we can write

$$\frac{\Delta y}{\Delta x} = \frac{\Delta y}{\Delta u} \cdot \frac{\Delta u}{\Delta x}.$$

Since g is assumed to be differentiable, it is continuous. Consequently, as $\Delta x \to 0$, $g(x + \Delta x) \to g(x)$, and so from (9) we see that $\Delta u \to 0$. Thus,

$$\lim_{\Delta x \to 0} \frac{\Delta y}{\Delta x} = \left(\lim_{\Delta x \to 0} \frac{\Delta y}{\Delta u} \right) \cdot \left(\lim_{\Delta x \to 0} \frac{\Delta u}{\Delta x} \right)$$

$$= \left(\lim_{\Delta u \to 0} \frac{\Delta y}{\Delta u} \right) \cdot \left(\lim_{\Delta x \to 0} \frac{\Delta u}{\Delta x} \right). \quad \leftarrow \text{note that } \Delta u \to 0 \text{ in the first term}$$

From the definition of the derivative, (3) of Section 3.1, it follows that

$$\frac{dy}{dx} = \frac{dy}{du} \cdot \frac{du}{dx}. \qquad \blacksquare$$

The assumption that $\Delta u \neq 0$ on some interval does not hold true for every differentiable function g. Although the result given in (7) remains valid when $\Delta u = 0$, the preceding proof does not.

It might help in the understanding of the derivative of a composition $y = f(g(x))$ to think of f as the *outside function* and $u = g(x)$ as the *inside function*. The derivative of $y = f(g(x)) = f(u)$ is then the *product of the derivative of the outside function* (evaluated at the inside function u) *and the derivative of the inside function* (evaluated at x):

derivative of outside function
↓

$$\frac{d}{dx} f(u) = f'(u) \cdot u'. \qquad (10)$$

↑
derivative of inside function

The result in (10) is written in various ways. Since $y = f(u)$, we have $f'(u) = dy/du$, and of course $u' = du/dx$. The product of the derivatives in (10) is the same as (8). On the other hand, if we replace the symbols u and u' in (10) by $g(x)$ and $g'(x)$ we obtain (7).

❚ Proof of the Power Rule for Functions

As noted previously, a power of a function can be written as a composition of $(f \circ g)(x)$ where the outside function is $y = f(x) = x^n$ and the inside function is $u = g(x)$. The derivative of the inside function $y = f(u) = u^n$ is $\dfrac{dy}{dx} = nu^{n-1}$ and the derivative of the outside function is $\dfrac{du}{dx}$. The product of these derivatives is then

$$\frac{dy}{dx} = \frac{dy}{du} \cdot \frac{du}{dx} = nu^{n-1} \frac{du}{dx} = n[g(x)]^{n-1} g'(x).$$

This is the Power Rule for Functions given in (5) and (6).

❚ Trigonometric Functions

We obtain the derivatives of the trigonometric functions composed with a differentiable function g as another direct consequence of the Chain Rule. For example, if $y = \sin u$, where $u = g(x)$, then the derivative of y with respect to the variable u is

$$\frac{dy}{du} = \cos u.$$

Hence, (8) gives

$$\frac{dy}{dx} = \frac{dy}{du} \cdot \frac{du}{dx} = \cos u \frac{du}{dx}$$

or equivalently,

$$\frac{d}{dx} \sin[\;] = \cos[\;] \frac{d}{dx}[\;].$$

Similarly, if $y = \tan u$ where $u = g(x)$, then $dy/du = \sec^2 u$ and so

$$\frac{dy}{dx} = \frac{dy}{du} \cdot \frac{du}{dx} = \sec^2 u \frac{du}{dx}.$$

We summarize the Chain Rule results for the six trigonometric functions.

Theorem 3.5.3   Derivatives of Trigonometric Functions

If $u = g(x)$ is a differentiable function, then

$$\frac{d}{dx}\sin u = \cos u \frac{du}{dx}, \qquad\qquad \frac{d}{dx}\cos u = -\sin u \frac{du}{dx}, \qquad (11)$$

$$\frac{d}{dx}\tan u = \sec^2 u \frac{du}{dx}, \qquad\qquad \frac{d}{dx}\cot u = -\csc^2 u \frac{du}{dx}, \qquad (12)$$

$$\frac{d}{dx}\sec u = \sec u \tan u \frac{du}{dx}, \qquad\qquad \frac{d}{dx}\csc u = -\csc u \cot u \frac{du}{dx}. \qquad (13)$$

EXAMPLE 7   Chain Rule

Differentiate $y = \cos 4x$.

Solution   The function is $\cos u$ with $u = 4x$. From the second formula in (11) of Theorem 3.5.3 the derivative is

$$\frac{dy}{dx} = \overbrace{-\sin 4x}^{\frac{dy}{du}} \cdot \overbrace{\frac{d}{dx}4x}^{\frac{du}{dx}} = -4\sin 4x. \qquad \blacksquare$$

EXAMPLE 8   Chain Rule

Differentiate $y = \tan(6x^2 + 1)$.

Solution   The function is $\tan u$ with $u = 6x^2 + 1$. From the first formula in (12) of Theorem 3.5.3 the derivative is

$$\frac{dy}{dx} = \overbrace{\sec^2(6x^2 + 1)}^{\sec^2 u} \cdot \overbrace{\frac{d}{dx}(6x^2 + 1)}^{\frac{du}{dx}} = 12x\sec^2(6x^2 + 1). \qquad \blacksquare$$

EXAMPLE 9   Product, Power, and Chain Rule

Differentiate $y = (9x^3 + 1)^2 \sin 5x$.

Solution   We first use the Product Rule:

$$\frac{dy}{dx} = (9x^3 + 1)^2 \cdot \frac{d}{dx}\sin 5x + \sin 5x \cdot \frac{d}{dx}(9x^3 + 1)^2$$

followed by the Power Rule (6) and the first formula in (11) of Theorem 3.5.3,

$$\frac{dy}{dx} = (9x^3 + 1)^2 \cdot \overset{\text{from (11)}}{\cos 5x} \cdot \frac{d}{dx}5x + \sin 5x \cdot \overset{\text{from (6)}}{2(9x^3 + 1)} \cdot \frac{d}{dx}(9x^3 + 1)$$

$$= (9x^3 + 1)^2 \cdot 5\cos 5x + \sin 5x \cdot 2(9x^3 + 1) \cdot 27x^2$$

$$= (9x^3 + 1)(45x^3\cos 5x + 5\cos 5x + 54x^2\sin 5x). \qquad \blacksquare$$

In Sections 3.2 and 3.3 we saw that even though the Sum and Product Rules were stated in terms of two functions f and g, they were applicable to any finite number of differentiable functions. So too, the Chain Rule is stated for the composition of two functions f and g but we can apply it to the composition of three (or more) differentiable functions. In the case of three functions f, g, and h, (7) becomes

$$\frac{d}{dx}f(g(h(x))) = f'(g(h(x))) \cdot \frac{d}{dx}g(h(x))$$

$$= f'(g(h(x))) \cdot g'(h(x)) \cdot h'(x).$$

EXAMPLE 10 Repeated Use of the Chain Rule

Differentiate $y = \cos^4(7x^3 + 6x - 1)$.

Solution For emphasis we first rewrite the given function as $y = [\cos(7x^3 + 6x - 1)]^4$. Observe that this function is the composition $(f \circ g \circ h)(x) = f(g(h(x)))$ where $f(x) = x^4$, $g(x) = \cos x$, and $h(x) = 7x^3 + 6x - 1$. We first apply the Chain Rule in the form of the Power Rule (6) followed by the second formula in (11):

$$\frac{dy}{dx} = 4[\cos(7x^3 + 6x - 1)]^3 \cdot \frac{d}{dx}\cos(7x^3 + 6x - 1) \qquad \leftarrow \begin{array}{l}\text{first Chain Rule:}\\ \text{differentiate the power}\end{array}$$

$$= 4\cos^3(7x^3 + 6x - 1) \cdot \left[-\sin(7x^3 + 6x - 1) \cdot \frac{d}{dx}(7x^3 + 6x - 1)\right] \leftarrow \begin{array}{l}\text{second Chain Rule:}\\ \text{differentiate the cosine}\end{array}$$

$$= -4(21x^2 + 6)\cos^3(7x^3 + 6x - 1)\sin(7x^3 + 6x - 1).$$

In the final example, the given function is a composition of four functions.

EXAMPLE 11 Repeated Use of the Chain Rule

Differentiate $y = \sin\left(\tan\sqrt{3x^2 + 4}\right)$.

Solution The function is $f(g(h(k(x))))$, where $f(x) = \sin x$, $g(x) = \tan x$, $h(x) = \sqrt{x}$, and $k(x) = 3x^2 + 4$. In this case we apply the Chain Rule three times in succession

$$\frac{dy}{dx} = \cos\left(\tan\sqrt{3x^2 + 4}\right) \cdot \frac{d}{dx}\tan\sqrt{3x^2 + 4} \qquad \leftarrow \begin{array}{l}\text{first Chain Rule:}\\ \text{differentiate the sine}\end{array}$$

$$= \cos\left(\tan\sqrt{3x^2 + 4}\right) \cdot \sec^2\sqrt{3x^2 + 4} \cdot \frac{d}{dx}\sqrt{3x^2 + 4} \qquad \leftarrow \begin{array}{l}\text{second Chain Rule:}\\ \text{differentiate the tangent}\end{array}$$

$$= \cos\left(\tan\sqrt{3x^2 + 4}\right) \cdot \sec^2\sqrt{3x^2 + 4} \cdot \frac{d}{dx}(3x^2 + 4)^{1/2} \qquad \leftarrow \text{rewrite power}$$

$$= \cos\left(\tan\sqrt{3x^2 + 4}\right) \cdot \sec^2\sqrt{3x^2 + 4} \cdot \frac{1}{2}(3x^2 + 4)^{-1/2} \cdot \frac{d}{dx}(3x^2 + 4) \quad \leftarrow \begin{array}{l}\text{third Chain Rule:}\\ \text{differentiate the}\\ \text{power}\end{array}$$

$$= \cos\left(\tan\sqrt{3x^2 + 4}\right) \cdot \sec^2\sqrt{3x^2 + 4} \cdot \frac{1}{2}(3x^2 + 4)^{-1/2} \cdot 6x \quad \leftarrow \text{simplify}$$

$$= \frac{3x\cos\left(\tan\sqrt{3x^2 + 4}\right) \cdot \sec^2\sqrt{3x^2 + 4}}{\sqrt{3x^2 + 4}}.$$

You should, of course, become so adept at applying the Chain Rule that you will not have to give a moment's thought as to the number of functions involved in the actual composition.

$\dfrac{d}{dx}$ **NOTES FROM THE CLASSROOM**

(*i*) Probably the most common mistake is to forget to carry out the second half of the Chain Rule, namely the derivative of the inside function. This is the du/dx part in

$$\frac{dy}{dx} = \frac{dy}{du}\frac{du}{dx}.$$

For instance, the derivative of $y = (1 - x)^{57}$ is not $dy/dx = 57(1 - x)^{56}$ since $57(1 - x)^{56}$ is only the dy/du part. It might help to consistently use the operation symbol d/dx:

$$\frac{d}{dx}(1 - x)^{57} = 57(1 - x)^{56} \cdot \frac{d}{dx}(1 - x) = 57(1 - x)^{56} \cdot (-1).$$

(*ii*) A less common but probably a worse mistake than the first is to differentiate inside the given function. A student wrote on an examination paper that the derivative of $y = \cos(x^2 + 1)$ was $dy/dx = -\sin(2x)$; that is, the derivative of the cosine is the negative of the sine and the derivative of $x^2 + 1$ is $2x$. Both observations are correct, but how they are put together is incorrect. Bear in mind that the derivative of the inside function is a multiple of the derivative of the outside function. Again, it might help to use the operation symbol d/dx. The correct derivative of $y = \cos(x^2 + 1)$ is the product of two derivatives.

$$\frac{dy}{dx} = -\sin(x^2 + 1) \cdot \frac{d}{dx}(x^2 + 1) = -2x\sin(x^2 + 1).$$

Exercises 3.5   Answers to selected odd-numbered problems begin on page ANS-11.

≡ Fundamentals

In Problems 1–20, find dy/dx.

1. $y = (-5x)^{30}$

2. $y = (3/x)^{14}$

3. $y = (2x^2 + x)^{200}$

4. $y = \left(x - \dfrac{1}{x^2}\right)^5$

5. $y = \dfrac{1}{(x^3 - 2x^2 + 7)^4}$

6. $y = \dfrac{10}{\sqrt{x^2 - 4x + 1}}$

7. $y = (3x - 1)^4(-2x + 9)^5$

8. $y = x^4(x^2 + 1)^6$

9. $y = \sin\sqrt{2x}$

10. $y = \sec x^2$

11. $y = \sqrt{\dfrac{x^2 - 1}{x^2 + 1}}$

12. $y = \dfrac{3x - 4}{(5x + 2)^3}$

13. $y = [x + (x^2 - 4)^3]^{10}$

14. $y = \left[\dfrac{1}{(x^3 - x + 1)^2}\right]^4$

15. $y = x(x^{-1} + x^{-2} + x^{-3})^{-4}$

16. $y = (2x + 1)^3\sqrt{3x^2 - 2x}$

17. $y = \sin(\pi x + 1)$

18. $y = -2\cos(-3x + 7)$

19. $y = \sin^3 5x$

20. $y = 4\cos^2\sqrt{x}$

In Problems 21–38, find $f'(x)$.

21. $f(x) = x^3\cos x^3$

22. $f(x) = \dfrac{\sin 5x}{\cos 6x}$

23. $f(x) = (2 + x\sin 3x)^{10}$

24. $f(x) = \dfrac{(1 - \cos 4x)^2}{(1 + \sin 5x)^3}$

25. $f(x) = \tan(1/x)$

26. $f(x) = x\cot(5/x^2)$

27. $f(x) = \sin 2x\cos 3x$

28. $f(x) = \sin^2 2x\cos^3 3x$

29. $f(x) = (\sec 4x + \tan 2x)^5$

30. $f(x) = \csc^2 2x - \csc 2x^2$

31. $f(x) = \sin(\sin 2x)$

32. $f(x) = \tan\left(\cos\dfrac{x}{2}\right)$

33. $f(x) = \cos\left(\sin\sqrt{2x + 5}\right)$

34. $f(x) = \tan(\tan x)$

35. $f(x) = \sin^3(4x^2 - 1)$

36. $f(x) = \sec(\tan^2 x^4)$

37. $f(x) = (1 + (1 + (1 + x^3)^4)^5)^6$

38. $f(x) = \left[x^2 - \left(1 + \dfrac{1}{x}\right)^{-4}\right]^2$

In Problems 39–42, find the slope of the tangent line to the graph of the given function at the indicated value of x.

39. $y = (x^2 + 2)^3$; $x = -1$

40. $y = \dfrac{1}{(3x + 1)^2}$; $x = 0$

41. $y = \sin 3x + 4x\cos 5x$; $x = \pi$

42. $y = 50x - \tan^3 2x$; $x = \pi/6$

In Problems 43–46, find an equation of the tangent line to the graph of the given function at the indicated value of x.

43. $y = \left(\dfrac{x}{x + 1}\right)^2$; $x = -\dfrac{1}{2}$

44. $y = x^2(x - 1)^3$; $x = 2$

45. $y = \tan 3x$; $x = \pi/4$

46. $y = (-1 + \cos 4x)^3$; $x = \pi/8$

In Problems 47 and 48, find an equation of the normal line to the graph of the given function at the indicated value of x.

47. $y = \sin\left(\dfrac{\pi}{6x}\right)\cos(\pi x^2)$; $x = \dfrac{1}{2}$

48. $y = \sin^3\dfrac{x}{3}$; $x = \pi$

In Problems 49–52, find the indicated derivative.

49. $f(x) = \sin\pi x$; $f'''(x)$

50. $y = \cos(2x + 1)$; d^5y/dx^5

51. $y = x\sin 5x$; d^3y/dx^3

52. $f(x) = \cos x^2$; $f''(x)$

53. Find the point(s) on the graph of $f(x) = x/(x^2 + 1)^2$ where the tangent line is horizontal. Does the graph of f have any vertical tangents?

54. Determine the values of t at which the instantaneous rate of change of $g(t) = \sin t + \frac{1}{2}\cos 2t$ is zero.

55. If $f(x) = \cos(x/3)$, what is the slope of the tangent line to the graph of f' at $x = 2\pi$?

56. If $f(x) = (1 - x)^4$, what is the slope of the tangent line to the graph of f'' at $x = 2$?

≡ Applications

57. The function $R = (v_0^2/g)\sin 2\theta$ gives the range of a projectile fired at an angle θ from the horizontal with an initial velocity v_0. If v_0 and g are constants, find those values of θ at which $dR/d\theta = 0$.

58. The volume of a spherical balloon of radius r is $V = \frac{4}{3}\pi r^3$. The radius is a function of time t and increases at a constant rate of 5 in/min. What is the instantaneous rate of change of V with respect to t?

59. Suppose a spherical balloon is being filled at a constant rate $dV/dt = 10$ in³/min. At what rate is its radius increasing when $r = 2$ in?

60. Consider a mass on a spring shown in FIGURE 3.5.1. In the absence of damping forces, the displacement (or directed distance) of the mass measured from a position called the **equilibrium position** is given by the function

$$x(t) = x_0\cos\omega t + \frac{v_0}{\omega}\sin\omega t,$$

where $\omega = \sqrt{k/m}$, k is the spring constant (an indicator of the stiffness of the spring), m is the mass (measured in slugs or kilograms), y_0 is the initial displacement of the mass (measured above or below the equilibrium position), v_0 is the initial velocity of the mass, and t is time measured in seconds.

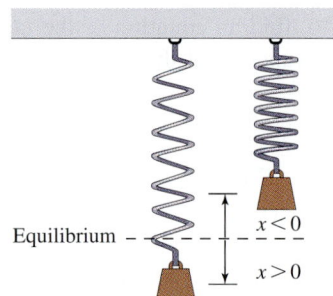

FIGURE 3.5.1 Mass on a spring in Problem 60

(a) Verify that $x(t)$ satisfies the differential equation

$$\frac{d^2x}{dt^2} + \omega^2 x = 0.$$

(b) Verify that $x(t)$ satisfies the initial conditions $x(0) = x_0$ and $x'(0) = v_0$.

≡ Think About It

61. Let F be a differentiable function. What is $\dfrac{d}{dx}F(3x)$?

62. Let G be a differentiable function. What is $\dfrac{d}{dx}[G(-x^2)]^2$?

63. Suppose $\dfrac{d}{du}f(u) = \dfrac{1}{u}$. What is $\dfrac{d}{dx}f(-10x + 7)$?

64. Suppose $\dfrac{d}{dx}f(x) = \dfrac{1}{1 + x^2}$. What is $\dfrac{d}{dx}f(x^3)$?

In Problems 65 and 66, the symbol n represents a positive integer. Find a formula for the given derivative.

65. $\dfrac{d^n}{dx^n}(1 + 2x)^{-1}$

66. $\dfrac{d^n}{dx^n}\sqrt{1 + 2x}$

67. Suppose $g(t) = h(f(t))$, where $f(1) = 3$, $f'(1) = 6$, and $h'(3) = -2$. What is $g'(1)$?

68. Suppose $g(1) = 2, g'(1) = 3, g''(1) - 1, f'(2) = 4$, and $f''(2) = 3$. What is $\dfrac{d^2}{dx^2}f(g(x))\Big|_{x=1}$?

69. Given that f is an odd differentiable function, use the Chain Rule to show that f' is an even function.

70. Given that f is an even differentiable function, use the Chain Rule to show that f' is an odd function.

3.6 Implicit Differentiation

❚ Introduction The graphs of many equations that we study in mathematics are not the graphs of functions. For example, the equation

$$x^2 + y^2 = 4 \tag{1}$$

describes a circle of radius 2 centered at the origin. Equation (1) is not a function, since for any choice of x satisfying $-2 < x < 2$ there corresponds two values of y. See FIGURE 3.6.1(a). Nevertheless, graphs of equations such as (1) can possess tangent lines at various points (x, y). Equation (1) defines *at least* two functions f and g on the interval $[-2, 2]$. Graphically, the obvious functions are the top half and the bottom half of the circle. To obtain formulas for these functions we solve $x^2 + y^2 = 4$ for y in terms of x:

$$y = f(x) = \sqrt{4 - x^2}, \quad \leftarrow \text{upper semicircle} \tag{2}$$

and $\qquad\qquad y = g(x) = -\sqrt{4 - x^2}. \quad \leftarrow \text{lower semicircle} \tag{3}$

See Figures 3.6.1(b) and (c). We can now find slopes of tangent lines for $-2 < x < 2$ by differentiating (2) and (3) by the Power Rule for Functions.

In this section we will see how the derivative dy/dx can be obtained for (1), as well as for more complicated equations $F(x, y) = 0$, without the necessity of solving the equation for the variable y.

■ **Explicit and Implicit Functions**   A function in which the dependent variable is expressed solely in terms of the independent variable x, namely, $y = f(x)$, is said to be an **explicit function**. For example, $y = \frac{1}{2}x^3 - 1$ is an explicit function. On the other hand, an equivalent equation $2y - x^3 + 2 = 0$ is said to define the function **implicitly**, or y is an **implicit function** of x. We have just seen that the equation $x^2 + y^2 = 4$ defines the two functions $f(x) = \sqrt{4 - x^2}$ and $g(x) = -\sqrt{4 - x^2}$ implicitly.

In general, if an equation $F(x, y) = 0$ defines a function f implicitly on some interval, then $F(x, f(x)) = 0$ is an identity on the interval. The graph of f is a portion or an arc (or all) of the graph of the equation $F(x, y) = 0$. In the case of the functions in (2) and (3), note that both equations

$$x^2 + [f(x)]^2 = 4 \quad \text{and} \quad x^2 + [g(x)]^2 = 4$$

are identities on the interval $[-2, 2]$.

The graph of the equation $x^3 + y^3 = 3xy$ shown in FIGURE 3.6.2(a) is a famous curve called the **Folium of Descartes**. With the aid of a CAS such as *Mathematica* or *Maple*, one of the implicit functions defined by $x^3 + y^3 = 3xy$ is found to be

$$y = \frac{2x}{\sqrt[3]{-4x^3 + 4\sqrt{x^6 - 4x^3}}} + \frac{1}{2}\sqrt[3]{-4x^3 + 4\sqrt{x^6 - 4x^3}}. \tag{4}$$

The graph of this function is the red arc shown in Figure 3.6.2(b). The graph of another implicit function defined by $x^3 + y^3 = 3xy$ is given in Figure 3.6.2(c).

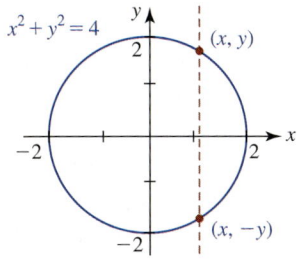

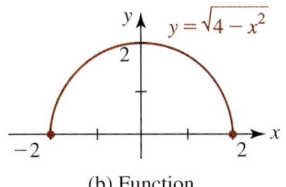

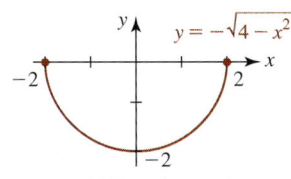

FIGURE 3.6.1   Equation $x^2 + y^2 = 4$ determines at least two functions

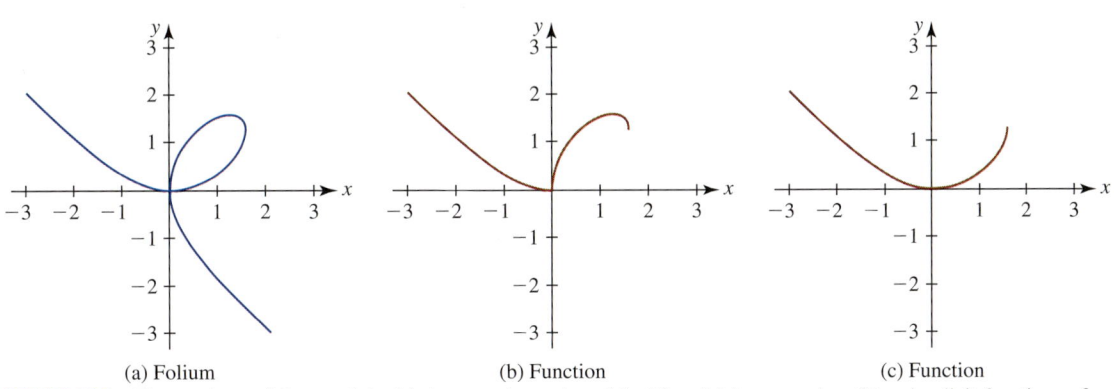

(a) Folium   (b) Function   (c) Function

FIGURE 3.6.2   The portions of the graph in (a) that are shown in red in (b) and (c) are graphs of two implicit functions of x

■ **Implicit Differentiation**   Do not jump to the conclusion from the preceding discussion that we can always solve an equation $F(x, y) = 0$ for an implicit function of x as we did in (2), (3), and (4). For example, solving an equation such as

$$x^4 + x^2y^3 - y^5 = 2x + y \tag{5}$$

for y in terms of x is more than an exercise in challenging algebra or a lesson in the use of the correct syntax of a CAS. It is *impossible*! Yet (5) may determine several implicit functions on a suitably restricted interval of the x-axis. Nevertheless, we *can* determine the derivative dy/dx by a process known as **implicit differentiation**. This process consists of differentiating both sides of an equation with respect to x, using the rules of differentiation, and then solving for dy/dx. Since we think of y as being determined by the given equation as a differentiable function of x, the Chain Rule, in the form of the Power Rule for Functions, gives the useful result

◀ Although we cannot solve certain equations for an explicit function, it still may be possible to graph the equation with the aid of a CAS. We can then *see* the functions as we did in Figure 3.6.2.

$$\frac{d}{dx}y^n = ny^{n-1}\frac{dy}{dx}, \tag{6}$$

where n is any real number. For example,

$$\frac{d}{dx}x^2 = 2x \qquad \text{whereas} \qquad \frac{d}{dx}y^2 = 2y\frac{dy}{dx}.$$

Similarly, if y is a function of x, then by the Product Rule,

$$\frac{d}{dx}xy = x\frac{d}{dx}y + y\frac{d}{dx}x = x\frac{dy}{dx} + y,$$

and by the Chain Rule,

$$\frac{d}{dx}\sin 5y = \cos 5y \cdot \frac{d}{dx}5y = 5\cos 5y\frac{dy}{dx}.$$

Guidelines for Implicit Differentiation

(*i*) Differentiate both sides of the equation with respect to x. Use the rules of differentiation and treat y as a differentiable function of x. For powers of the symbol y use (6).

(*ii*) Collect all terms involving dy/dx on the left-hand side of the differentiated equation. Move all other terms to the right-hand side of the equation.

(*iii*) Factor dy/dx from all terms containing this term. Then solve for dy/dx.

In the following examples we shall assume that the given equation determines at least one differentiable implicit function.

EXAMPLE 1   Using Implicit Differentiation

Find dy/dx if $x^2 + y^2 = 4$.

Solution   We differentiate both sides of the equation and then utilize (6):

use Power Rule (6) here
$$\frac{d}{dx}x^2 + \frac{d}{dx}y^2 = \frac{d}{dx}4$$

$$2x + 2y\frac{dy}{dx} = 0.$$

Solving for the derivative yields

$$\frac{dy}{dx} = -\frac{x}{y}. \qquad (7) \quad \blacksquare$$

As illustrated in (7) of Example 1, implicit differentiation usually yields a derivative that depends on both variables x and y. In our introductory discussion we saw that the equation $x^2 + y^2 = 4$ defines two differentiable implicit functions on the open interval $-2 < x < 2$. The symbolism $dy/dx = -x/y$ represents the derivative of either function on the interval. Note that this derivative clearly indicates that functions (2) and (3) are not differentiable at $x = -2$ and $x = 2$ since $y = 0$ for these values of x. In general, implicit differentiation yields the derivative of any differentiable implicit function defined by an equation $F(x, y) = 0$.

EXAMPLE 2   Slope of a Tangent Line

Find the slopes of the tangent lines to the graph of $x^2 + y^2 = 4$ at the points corresponding to $x = 1$.

Solution   Substituting $x = 1$ into the given equation gives $y^2 = 3$ or $y = \pm\sqrt{3}$. Hence, there are tangent lines at $\left(1, \sqrt{3}\right)$ and $\left(1, -\sqrt{3}\right)$. Although $\left(1, \sqrt{3}\right)$ and $\left(1, -\sqrt{3}\right)$ are points on the graphs of two different implicit functions, indicated by the different colors in FIGURE 3.6.3, (7) of Example 1 gives the correct slope at each number in the interval $(-2, 2)$. We have

$$\frac{dy}{dx}\bigg|_{(1,\sqrt{3})} = -\frac{1}{\sqrt{3}} \quad \text{and} \quad \frac{dy}{dx}\bigg|_{(1,-\sqrt{3})} = -\frac{1}{-\sqrt{3}} = \frac{1}{\sqrt{3}}. \quad \blacksquare$$

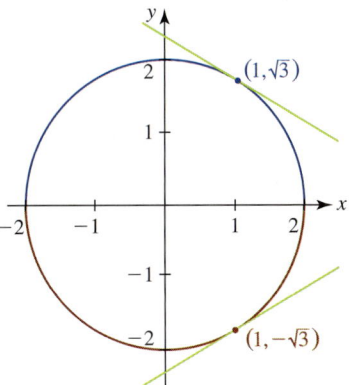

FIGURE 3.6.3 Tangent lines in Example 2 are shown in green

EXAMPLE 3 Using Implicit Differentiation

Find dy/dx if $x^4 + x^2y^3 - y^5 = 2x + 1$.

Solution In this case, we use (6) and the Product Rule:

$$\underset{\text{Product Rule here}}{\frac{d}{dx}x^4 + \frac{d}{dx}x^2y^3} - \underset{\text{Power Rule (6) here}}{\frac{d}{dx}y^5} = \frac{d}{dx}2x + \frac{d}{dx}1$$

$$4x^3 + x^2 \cdot 3y^2\frac{dy}{dx} + 2xy^3 - 5y^4\frac{dy}{dx} = 2 \quad \leftarrow \begin{array}{l}\text{factor } dy/dx \text{ from}\\ \text{second and fourth terms}\end{array}$$

$$(3x^2y^2 - 5y^4)\frac{dy}{dx} = 2 - 4x^3 - 2xy^3$$

$$\frac{dy}{dx} = \frac{2 - 4x^3 - 2xy^3}{3x^2y^2 - 5y^4}. \quad \blacksquare$$

▌ **Higher Derivatives** Through implicit differentiation we determine dy/dx. By differentiating dy/dx with respect to x we obtain the second derivative d^2y/dx^2. If the first derivative contains y, then d^2y/dx^2 will again contain the symbol dy/dx; we can eliminate that quantity by substituting its known value. The next example illustrates the method.

EXAMPLE 4 Second Derivative

Find d^2y/dx^2 if $x^2 + y^2 = 4$.

Solution From Example 1, we already know that the first derivative is $dy/dx = -x/y$. The second derivative is the derivative of dy/dx, and so by the Quotient Rule:

$$\frac{d^2y}{dx^2} = -\frac{d}{dx}\left(\frac{x}{y}\right) = -\frac{y \cdot 1 - x \cdot \overset{\text{substituting for } dy/dx}{\dfrac{dy}{dx}}}{y^2} = -\frac{y - x\left(-\dfrac{x}{y}\right)}{y^2} = -\frac{y^2 + x^2}{y^3}.$$

Noting that $x^2 + y^2 = 4$ permits us to rewrite the second derivative as

$$\frac{d^2y}{dx^2} = -\frac{4}{y^3}. \quad \blacksquare$$

EXAMPLE 5 Chain and Product Rules

Find dy/dx if $\sin y = y\cos 2x$.

Solution From the Chain Rule and Product Rule we obtain

$$\frac{d}{dx}\sin y = \frac{d}{dx}y\cos 2x$$

$$\cos y \cdot \frac{dy}{dx} = y(-\sin 2x \cdot 2) + \cos 2x \cdot \frac{dy}{dx}$$

$$(\cos y - \cos 2x)\frac{dy}{dx} = -2y\sin 2x$$

$$\frac{dy}{dx} = -\frac{2y\sin 2x}{\cos y - \cos 2x}. \quad \blacksquare$$

▮ Postscript—Power Rule Revisited So far we have proved the Power Rule $(d/dx)x^n = nx^{n-1}$ for all integer exponents n. Implicit differentiation provides a way of proving this rule when the exponent is a rational number p/q, where p and q are integers and $q \neq 0$. In the case $n = p/q$, the function

$$y = x^{p/q} \qquad \text{gives} \qquad y^q = x^p.$$

Now for $y \neq 0$, implicit differentiation

$$\frac{d}{dx} y^q = \frac{d}{dx} x^p \qquad \text{yields} \qquad qy^{q-1} \frac{dy}{dx} = px^{p-1}.$$

Solving the last equation for dy/dx and simplifying by the laws of exponents gives

$$\frac{dy}{dx} = \frac{p}{q} \frac{x^{p-1}}{y^{q-1}} = \frac{p}{q} \frac{x^{p-1}}{(x^{p/q})^{q-1}} = \frac{p}{q} \frac{x^{p-1}}{x^{p-p/q}} = \frac{p}{q} x^{p/q-1}.$$

Examination of the last result shows that it is (3) of Section 3.2 with $n = p/q$.

Exercises 3.6 Answers to selected odd-numbered problems begin on page ANS-11.

≡ Fundamentals

In Problems 1–4, assume that y is a differentiable function of x. Find the indicated derivative.

1. $\dfrac{d}{dx} x^2 y^4$

2. $\dfrac{d}{dx} \dfrac{x^2}{y^2}$

3. $\dfrac{d}{dx} \cos y^2$

4. $\dfrac{d}{dx} y \sin 3y$

In Problems 5–24, assume that the given equation defines at least one differentiable implicit function. Use implicit differentiation to find dy/dx.

5. $y^2 - 2y = x$

6. $4x^2 + y^2 = 8$

7. $xy^2 - x^2 + 4 = 0$

8. $(y - 1)^2 = 4(x + 2)$

9. $3y + \cos y = x^2$

10. $y^3 - 2y + 3x^3 = 4x + 1$

11. $x^3 y^2 = 2x^2 + y^2$

12. $x^5 - 6xy^3 + y^4 = 1$

13. $(x^2 + y^2)^6 = x^3 - y^3$

14. $y = (x - y)^2$

15. $y^{-3} x^6 + y^6 x^{-3} = 2x + 1$

16. $y^4 - y^2 = 10x - 3$

17. $(x - 1)^2 + (y + 4)^2 = 25$

18. $\dfrac{x + y}{x - y} = x$

19. $y^2 = \dfrac{x - 1}{x + 2}$

20. $\dfrac{x}{y^2} + \dfrac{y^2}{x} = 5$

21. $xy = \sin(x + y)$

22. $x + y = \cos(xy)$

23. $x = \sec y$

24. $x \sin y - y \cos x = 1$

In Problems 25 and 26, use implicit differentiation to find the indicated derivative.

25. $r^2 = \sin 2\theta; \quad dr/d\theta$

26. $\pi r^2 h = 100; \quad dh/dr$

In Problems 27 and 28, find dy/dx at the indicated point.

27. $xy^2 + 4y^3 + 3x = 0; \quad (1, -1)$

28. $y = \sin xy; \quad (\pi/2, 1)$

In Problems 29 and 30, find dy/dx at the points that correspond to the indicated number.

29. $2y^2 + 2xy - 1 = 0; \quad x = \dfrac{1}{2}$

30. $y^3 + 2x^2 = 11y; \quad y = 1$

In Problems 31–34, find an equation of the tangent line at the indicated point or number.

31. $x^4 + y^3 = 24; \quad (-2, 2)$

32. $\dfrac{1}{x} + \dfrac{1}{y} = 1; \quad x = 3$

33. $\tan y = x; \quad y = \pi/4$

34. $3y + \cos y = x^2; \quad (1, 0)$

In Problems 35 and 36, find the point(s) on the graph of the given equation where the tangent line is horizontal.

35. $x^2 - xy + y^2 = 3$

36. $y^2 = x^2 - 4x + 7$

37. Find the point(s) on the graph of $x^2 + y^2 = 25$ at which the slope of the tangent is $\frac{1}{2}$.

38. Find the point where the tangent lines to the graph of $x^2 + y^2 = 25$ at $(-3, 4)$ and $(-3, -4)$ intersect.

39. Find the point(s) on the graph of $y^3 = x^2$ at which the tangent line is perpendicular to the line $y + 3x - 5 = 0$.

40. Find the point(s) on the graph of $x^2 - xy + y^2 = 27$ at which the tangent line is parallel to the line $y = 5$.

In Problems 41–48, find d^2y/dx^2.

41. $4y^3 = 6x^2 + 1$

42. $xy^4 = 5$

43. $x^2 - y^2 = 25$

44. $x^2 + 4y^2 = 16$

45. $x + y = \sin y$

46. $y^2 - x^2 = \tan 2x$

47. $x^2 + 2xy - y^2 = 1$

48. $x^3 + y^3 = 27$

In Problems 49–52, first use implicit differentiation to find dy/dx. Then solve for y explicitly in terms of x and differentiate. Show that the two answers are equivalent.

49. $x^2 - y^2 = x$

50. $4x^2 + y^2 = 1$

51. $x^3 y = x + 1$

52. $y \sin x = x - 2y$

In Problems 53–56, determine an implicit function from the given equation such that its graph is the blue curve in the figure.

53. $(y - 1)^2 = x - 2$

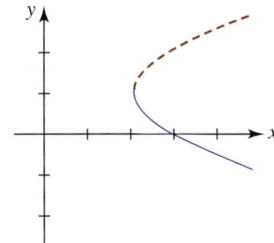

FIGURE 3.6.4 Graph for Problem 53

54. $x^2 + xy + y^2 = 4$

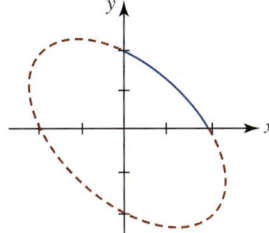

FIGURE 3.6.5 Graph for Problem 54

55. $x^2 + y^2 = 4$

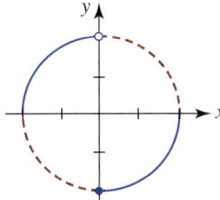

FIGURE 3.6.6 Graph for Problem 55

56. $y^2 = x^2(2 - x)$

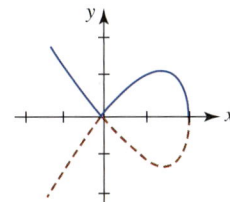

FIGURE 3.6.7 Graph for Problem 56

In Problems 57 and 58, assume that both x and y are differentiable functions of a variable t. Find dy/dt in terms of x, y, and dx/dt.

57. $x^2 + y^2 = 25$

58. $x^2 + xy + y^2 - y = 9$

59. The graph of the equation $x^3 + y^3 = 3xy$ is the Folium of Descartes given in Figure 3.6.2(a).

    (a) Find an equation of the tangent line at the point in the first quadrant where the Folium intersects the graph of $y = x$.

    (b) Find the point in the first quadrant at which the tangent line is horizontal.

60. The graph of $(x^2 + y^2)^2 = 4(x^2 - y^2)$ shown in FIGURE 3.6.8 is called a **lemniscate**.

    (a) Find the points on the graph that correspond to $x = 1$.

    (b) Find an equation of the tangent line to the graph at each point found in part (a).

    (c) Find the points on the graph at which the tangent is horizontal.

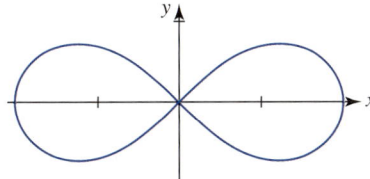

FIGURE 3.6.8 Lemniscate in Problem 60

In Problems 61 and 62, show that the graphs of the given equations are orthogonal at the indicated point of intersection. See Problem 64 in Exercises 3.2.

61. $y^2 = x^3$, $2x^2 + 3y^2 = 5$; $(1, 1)$

62. $y^3 + 3x^2y = 13$, $2x^2 - 2y^2 = 3x$; $(2, 1)$

If all the curves of one family of curves $G(x, y) = c_1$, c_1 a constant, intersect orthogonally all the curves of another family $H(x, y) = c_2$, c_2 a constant, then the families are said to be **orthogonal trajectories** of each other. In Problems 63 and 64, show that the families of curves are orthogonal trajectories of each other. Sketch the two families of curves.

63. $x^2 - y^2 = c_1$, $xy = c_2$

64. $x^2 + y^2 = c_1$, $y = c_2x$

≡ Applications

65. A woman drives toward a freeway sign as shown in FIGURE 3.6.9. Let θ be her viewing angle of the sign and let x be her distance (measured in feet) to that sign.

    (a) If her eye level is 4 ft from the surface of the road, show that

$$\tan\theta = \frac{4x}{x^2 + 252}.$$

    (b) Find the rate at which θ changes with respect to x.

    (c) At what distance is the rate in part (b) equal to zero?

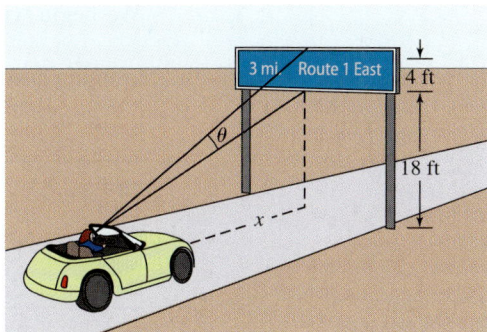

FIGURE 3.6.9 Car in Problem 65

66. A jet fighter "loops the loop" in a circle of radius 1 km as shown in FIGURE 3.6.10. Suppose a rectangular coordinate system is chosen so that the origin is at the center of the circular loop. The aircraft releases a missile that flies on a straight-line path that is tangent to the circle and hits a target on the ground whose coordinates are $(2, -2)$.

    (a) Determine the point on the circle where the missile was released.

    (b) If a missile is released at the point $\left(-\frac{1}{2}, -\frac{\sqrt{3}}{2}\right)$ on the circle, at what point does it hit the ground?

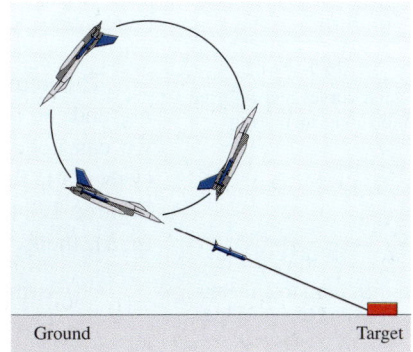

FIGURE 3.6.10 Jet fighter in Problem 66

≡ Think About It

67. The angle θ $(0 < \theta < \pi)$ between two curves is defined to be the angle between their tangent lines at the point P of intersection. If m_1 and m_2 are the slopes of the tangent lines at P, it can be shown that $\tan\theta = (m_1 - m_2)/(1 + m_1 m_2)$. Determine the angle between the graphs of $x^2 + y^2 + 4y = 6$ and $x^2 + 2x + y^2 = 4$ at $(1, 1)$.

68. Show that an equation of the tangent line to the ellipse $x^2/a^2 + y^2/b^2 = 1$ at the point (x_0, y_0) is given by

$$\frac{xx_0}{a^2} + \frac{yy_0}{b^2} = 1.$$

69. Consider the equation $x^2 + y^2 = 4$. Make up another implicit function $h(x)$ defined by this equation for $-2 \le x \le 2$ different from the ones given in (2), (3), and Problem 55.

70. For $-1 < x < 1$ and $-\pi/2 < y < \pi/2$, the equation $x = \sin y$ defines a differentiable implicit function.

 (a) Find dy/dx in terms of y.
 (b) Find dy/dx in terms of x.

3.7 Derivatives of Inverse Functions

▌ **Introduction**   In Section 1.5 we saw that the graphs of a one-to-one function f and its inverse f^{-1} are **reflections** of each other in the line $y = x$. As a consequence, if (a, b) is a point on the graph of f, then (b, a) is a point on the graph of f^{-1}. In this section we will also see that the slopes of tangent lines to the graph of a differentiable function f are related to the slopes of tangents to the graph of f^{-1}.

We begin with two theorems about the continuity of f and f^{-1}.

▌ **Continuity of f^{-1}**   Although we state the next two theorems without proof, their plausibility follows from the fact that the graph of f^{-1} is a reflection of the graph of f in the line $y = x$.

Theorem 3.7.1   Continuity of an Inverse Function

Let f be a continuous one-to-one function on its domain X. Then f^{-1} is continuous on its domain.

▌ **Increasing–Decreasing Functions**   Suppose $y = f(x)$ is a function defined on an interval I, and that x_1 and x_2 are any two numbers in the interval such that $x_1 < x_2$. Then from Section 1.3 and Figure 1.3.4 recall that f is said to be

- **increasing** on the interval if $f(x_1) < f(x_2)$, and   (1)
- **decreasing** on the interval if $f(x_1) > f(x_2)$.   (2)

The next two theorems establish a link between the notions of increasing/decreasing and the existence of an inverse function.

Theorem 3.7.2   Existence of an Inverse Function

Let f be a continuous function and increasing on an interval $[a, b]$. Then f^{-1} exists and is continuous and increasing on $[f(a), f(b)]$.

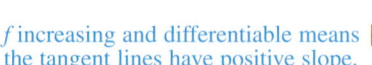

FIGURE 3.7.1 f (blue curve) and f^{-1} (red curve) are continuous and increasing

Theorem 3.7.2 also holds when the word *increasing* is replaced with the word *decreasing* and the interval in the conclusion is replaced by $[f(b), f(a)]$. See FIGURE 3.7.1. In addition, we can conclude from Theorem 3.7.2 that if f is continuous and increasing on an interval $(-\infty, \infty)$, then f^{-1} exists and is continuous and increasing on its domain. Inspection of Figures 1.3.4 and 3.7.1 also shows that if f in Theorem 3.7.2 is a differentiable function on (a, b), then:

▶ *f increasing and differentiable means the tangent lines have positive slope.*

- f is increasing on the interval $[a, b]$ if $f'(x) > 0$ on (a, b), and
- f is decreasing on the interval $[a, b]$ if $f'(x) < 0$ on (a, b).

We will prove these statements in the next chapter.

Theorem 3.7.3   Differentiability of an Inverse Function

Suppose f is a differentiable function on an open interval (a, b). If either $f'(x) > 0$ on the interval or $f'(x) < 0$ on the interval, then f is one-to-one. Moreover, f^{-1} is differentiable for all x in the range of f.

EXAMPLE 1   Existence of an Inverse

Prove that $f(x) = 5x^3 + 8x - 9$ has an inverse.

Solution   Since f is a polynomial function it is differentiable everywhere, that is, f is differentiable on the interval $(-\infty, \infty)$. Also, $f'(x) = 15x^2 + 8 > 0$ for all x implies that f is increasing on $(-\infty, \infty)$. It follows from Theorem 3.7.3 that f is one-to-one and hence f^{-1} exists.   ■

■ **Derivative of f^{-1}**   If f is differentiable on an interval I and is one-to-one on that interval, then for a in I the point (a, b) on the graph of f and the point (b, a) on the graph of f^{-1} are mirror images of each other in the line $y = x$. As we see next, the slopes of the tangent lines at (a, b) and (b, a) are also related.

EXAMPLE 2   Derivative of an Inverse

In Example 5 of Section 1.5 we showed that the inverse of the one-to-one function $f(x) = x^2 + 1, x \geq 0$ is $f^{-1}(x) = \sqrt{x - 1}$. At $x = 2$,

$$f(2) = 5 \qquad \text{and} \qquad f^{-1}(5) = 2.$$

Now from

$$f'(x) = 2x \qquad \text{and} \qquad (f^{-1})'(x) = \frac{1}{2\sqrt{x - 1}}$$

we see $f'(2) = 4$ and $(f^{-1})'(5) = \frac{1}{4}$. This shows that the slope of the tangent to the graph of f at $(2, 5)$ and the slope of the tangent to the graph of f^{-1} at $(5, 2)$ are reciprocals:

$$(f^{-1})'(5) = \frac{1}{f'(2)} \qquad \text{or} \qquad (f^{-1})'(5) = \frac{1}{f'(f^{-1}(5))}.$$

See FIGURE 3.7.2.

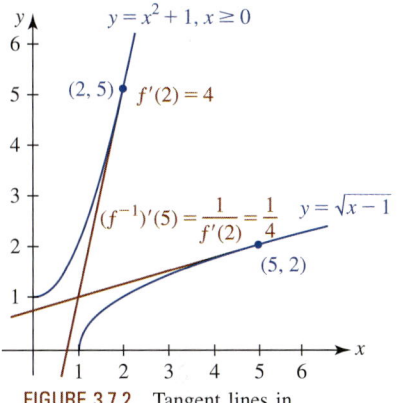

FIGURE 3.7.2   Tangent lines in Example 2

The next theorem shows that the result in Example 2 is no coincidence.

Theorem 3.7.4   Derivative of an Inverse Function

Suppose that f is differentiable on an interval I and $f'(x)$ is never zero on I. If f has an inverse f^{-1} on I, then f^{-1} is differentiable at a number x and

$$\frac{d}{dx} f^{-1}(x) = \frac{1}{f'(f^{-1}(x))}. \tag{3}$$

PROOF   As we have seen in (5) of Section 1.5, $f(f^{-1}(x)) = x$ for every x in the domain of f^{-1}. By implicit differentiation and the Chain Rule,

$$\frac{d}{dx} f(f^{-1}(x)) = \frac{d}{dx} x \qquad \text{or} \qquad f'(f^{-1}(x)) \cdot \frac{d}{dx} f^{-1}(x) = 1.$$

Solving the last equation for $\dfrac{d}{dx} f^{-1}(x)$ gives (3).   ■

Equation (3) clearly shows that to find the derivative function for f^{-1} we must know $f^{-1}(x)$ explicitly. For a one-to-one function $y = f(x)$ solving the equation $x = f(y)$ for y is sometimes difficult and often impossible. In this case it is convenient to rewrite (3) using different notation. Again by implicit differentiation

$$\frac{d}{dx} x = \frac{d}{dx} f(y) \qquad \text{gives} \qquad 1 = f'(y) \cdot \frac{dy}{dx}.$$

Solving the last equation for dy/dx and writing $dx/dy = f'(y)$ yields

$$\frac{dy}{dx} = \frac{1}{dx/dy}. \tag{4}$$

If (a, b) is a known point on the graph of f, the result in (4) enables us to evaluate the derivative of f^{-1} at (b, a) without an equation that defines $f^{-1}(x)$.

EXAMPLE 3 Derivative of an Inverse

It was pointed out in Example 1 that the polynomial function $f(x) = 5x^3 + 8x - 9$ is differentiable on $(-\infty, \infty)$ and hence continuous on the interval. Since the end behavior of f is that of the single-term polynomial function $y = 5x^3$ we can conclude that the range of f is also $(-\infty, \infty)$. Moreover, since $f'(x) = 15x^2 + 8 > 0$ for all x, f is increasing on its domain $(-\infty, \infty)$. Hence by Theorem 3.7.3, f has a differentiable inverse f^{-1} with domain $(-\infty, \infty)$. By interchanging x and y, the inverse is defined by the equation $x = 5y^3 + 8y - 9$, but solving this equation for y in terms of x is difficult (it requires the cubic formula). Nevertheless, using $dx/dy = 15y^2 + 8$, the derivative of the inverse function is given by (4):

$$\frac{dy}{dx} = \frac{1}{15y^2 + 8}. \tag{5}$$

For example, since $f(1) = 4$ we know that $f^{-1}(4) = 1$. Thus, the slope of the tangent line to the graph of f^{-1} at $(4, 1)$ is given by (5):

$$\left. \frac{dy}{dx} \right|_{x=4} = \left. \frac{1}{15y^2 + 8} \right|_{y=1} = \frac{1}{23}. \qquad \blacksquare$$

Read this paragraph a second time. ▶ In Example 3, the derivative of the inverse function can also be obtained directly from $x = 5y^3 + 8y - 9$ using implicit differentiation:

$$\frac{d}{dx} x = \frac{d}{dx}(5y^3 + 8y - 9) \qquad \text{gives} \qquad 1 = 15y^2 \frac{dy}{dx} + 8 \frac{dy}{dx}.$$

Solving the last equation for dy/dx gives (5). As a consequence of this observation implicit differentiation can be used to find the derivative of an inverse function with minimum effort. In the discussion that follows we will find the derivatives of the inverse trigonometric functions.

▮ **Derivatives of Inverse Trigonometric Functions** A review of Figures 1.5.15 and 1.5.17(a) reveals that the inverse tangent and inverse cotangent are differentiable for all x. However, the remaining four inverse trigonometric functions are not differentiable at either $x = -1$ or $x = 1$. We shall confine our attention to the derivations of the derivative formulas for the inverse sine, inverse tangent, and inverse secant and leave the others as exercises.

Inverse Sine: $y = \sin^{-1} x$ if and only if $x = \sin y$, where $-1 \le x \le 1$ and $-\pi/2 \le y \le \pi/2$. Therefore, implicit differentiation

$$\frac{d}{dx} x = \frac{d}{dx} \sin y \qquad \text{gives} \qquad 1 = \cos y \cdot \frac{dy}{dx}$$

and so

$$\frac{dy}{dx} = \frac{1}{\cos y}. \tag{6}$$

For the given restriction on the variable y, $\cos y \ge 0$ and so $\cos y = \sqrt{1 - \sin^2 y} = \sqrt{1 - x^2}$. By substituting this quantity in (6), we have shown that

$$\frac{d}{dx} \sin^{-1} x = \frac{1}{\sqrt{1 - x^2}}. \tag{7}$$

As predicted, note that (7) is not defined at $x = -1$ and $x = 1$. The inverse sine or arcsine function is differentiable on the open interval $(-1, 1)$.

Inverse Tangent: $y = \tan^{-1} x$ if and only if $x = \tan y$, where $-\infty < x < \infty$ and $-\pi/2 < y < \pi/2$. Thus,

$$\frac{d}{dx} x = \frac{d}{dx} \tan y \quad \text{gives} \quad 1 = \sec^2 y \cdot \frac{dy}{dx}$$

or

$$\frac{dy}{dx} = \frac{1}{\sec^2 y}. \tag{8}$$

In view of the identity $\sec^2 y = 1 + \tan^2 y = 1 + x^2$, (8) becomes

$$\frac{d}{dx} \tan^{-1} x = \frac{1}{1 + x^2}. \tag{9}$$

Inverse Secant: For $|x| > 1$ and $0 \le y < \pi/2$ or $\pi/2 < y \le \pi$,

$$y = \sec^{-1} x \quad \text{if and only if} \quad x = \sec y.$$

Differentiating the last equation implicitly gives

$$\frac{dy}{dx} = \frac{1}{\sec y \tan y}. \tag{10}$$

In view of the restrictions on y, we have $\tan y = \pm \sqrt{\sec^2 y - 1} = \pm \sqrt{x^2 - 1}$, $|x| > 1$. Hence, (10) becomes

$$\frac{d}{dx} \sec^{-1} x = \pm \frac{1}{x \sqrt{x^2 - 1}}. \tag{11}$$

We can get rid of the $\pm$ sign in (11) by observing in Figure 1.5.17(b) that the slope of the tangent line to the graph of $y = \sec^{-1} x$ is positive for $x < -1$ and positive for $x > 1$. Thus, (11) is equivalent to

$$\frac{d}{dx} \sec^{-1} x = \begin{cases} -\dfrac{1}{x\sqrt{x^2 - 1}}, & x < -1 \\[2ex] \dfrac{1}{x\sqrt{x^2 - 1}}, & x > 1. \end{cases} \tag{12}$$

The result in (12) can be rewritten in a compact form using the absolute value symbol:

$$\frac{d}{dx} \sec^{-1} x = \frac{1}{|x| \sqrt{x^2 - 1}}. \tag{13}$$

The derivative of the composition of an inverse trigonometric function with a differentiable function $u = g(x)$ is obtained from the Chain Rule.

Theorem 3.7.5   Inverse Trigonometric Functions

If $u = g(x)$ is a differentiable function, then

$$\frac{d}{dx} \sin^{-1} u = \frac{1}{\sqrt{1 - u^2}} \frac{du}{dx}, \qquad \frac{d}{dx} \cos^{-1} u = \frac{-1}{\sqrt{1 - u^2}} \frac{du}{dx}, \tag{14}$$

$$\frac{d}{dx} \tan^{-1} u = \frac{1}{1 + u^2} \frac{du}{dx}, \qquad \frac{d}{dx} \cot^{-1} u = \frac{-1}{1 + u^2} \frac{du}{dx}, \tag{15}$$

$$\frac{d}{dx} \sec^{-1} u = \frac{1}{|u| \sqrt{u^2 - 1}} \frac{du}{dx}, \qquad \frac{d}{dx} \csc^{-1} u = \frac{-1}{|u| \sqrt{u^2 - 1}} \frac{du}{dx}. \tag{16}$$

In the formulas in (14) we must have $|u| < 1$, whereas in the formulas in (16) we must have $|u| > 1$.

EXAMPLE 4 Derivative of Inverse Sine

Differentiate $y = \sin^{-1} 5x$.

Solution With $u = 5x$, we have from the first formula in (14),

$$\frac{dy}{dx} = \frac{1}{\sqrt{1 - (5x)^2}} \cdot \frac{d}{dx} 5x = \frac{5}{\sqrt{1 - 25x^2}}.$$ ■

EXAMPLE 5 Derivative of Inverse Tangent

Differentiate $y = \tan^{-1} \sqrt{2x + 1}$.

Solution With $u = \sqrt{2x + 1}$, we have from the first formula in (15),

$$\frac{dy}{dx} = \frac{1}{1 + (\sqrt{2x + 1})^2} \cdot \frac{d}{dx}(2x + 1)^{1/2}$$

$$= \frac{1}{1 + (2x + 1)} \cdot \frac{1}{2}(2x + 1)^{-1/2} \cdot 2$$

$$= \frac{1}{(2x + 2)\sqrt{2x + 1}}.$$ ■

EXAMPLE 6 Derivative of Inverse Secant

Differentiate $y = \sec^{-1} x^2$.

Solution For $x^2 > 1 > 0$, we have from the first formula in (16),

$$\frac{dy}{dx} = \frac{1}{|x^2|\sqrt{(x^2)^2 - 1}} \cdot \frac{d}{dx} x^2$$

$$= \frac{2x}{x^2\sqrt{x^4 - 1}} = \frac{2}{x\sqrt{x^4 - 1}}. \tag{17}$$

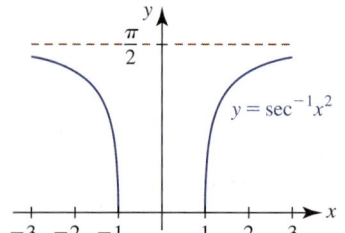

FIGURE 3.7.3 Graph of function in Example 6

With the aid of a graphing utility we obtain the graph of $y = \sec^{-1} x^2$ given in FIGURE 3.7.3. Notice that (17) gives positive slope for $x > 1$ and negative slope for $x < -1$. ■

EXAMPLE 7 Tangent Line

Find an equation of the tangent line to the graph of $f(x) = x^2 \cos^{-1} x$ at $x = -\frac{1}{2}$.

Solution By the Product Rule and the second formula in (14),

$$f'(x) = x^2 \left(\frac{-1}{\sqrt{1 - x^2}} \right) + 2x \cos^{-1} x.$$

Since $\cos^{-1}(-\frac{1}{2}) = 2\pi/3$, the two functions f and f' evaluated at $x = -\frac{1}{2}$ give:

$$f\left(-\frac{1}{2}\right) = \frac{\pi}{6} \qquad \leftarrow \text{point of tangency is } \left(-\frac{1}{2}, \frac{\pi}{6}\right)$$

$$f'\left(-\frac{1}{2}\right) = -\frac{1}{2\sqrt{3}} - \frac{2\pi}{3}. \leftarrow \text{slope of tangent at } \left(-\frac{1}{2}, \frac{\pi}{6}\right) \text{ is } -\frac{1}{2\sqrt{3}} - \frac{2\pi}{3}$$

By the point–slope form of a line, the unsimplified equation of the tangent line is

$$y - \frac{\pi}{6} = \left(-\frac{1}{2\sqrt{3}} - \frac{2\pi}{3}\right)\left(x + \frac{1}{2}\right).$$

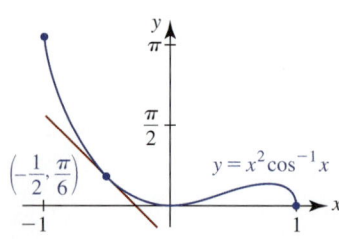

FIGURE 3.7.4 Tangent line in Example 7.

Since the domain of $\cos^{-1} x$ is the interval $[-1, 1]$ the domain of f is $[-1, 1]$. The corresponding range is $[0, \pi]$. FIGURE 3.7.4 was obtained with the aid of a graphing utility. ■

Exercises 3.7 Answers to selected odd-numbered problems begin on page ANS-11.

≣ Fundamentals

In Problems 1–4, without graphing determine whether the given function f has an inverse.

1. $f(x) = 10x^3 + 8x + 12$

2. $f(x) = -7x^5 - 6x^3 - 2x + 17$

3. $f(x) = x^3 + x^2 - 2x$

4. $f(x) = x^4 - 2x^2$

In Problems 5 and 6, use (3) to find the derivative of f^{-1} at the indicated point.

5. $f(x) = 2x^3 + 8$; $\left(f\left(\frac{1}{2}\right), \frac{1}{2}\right)$

6. $f(x) = -x^3 - 3x + 7$; $(f(-1), -1)$

In Problems 7 and 8, find f^{-1}. Use (3) to find $(f^{-1})'$ and then verify this result by direct differentiation of f^{-1}.

7. $f(x) = \dfrac{2x + 1}{x}$

8. $f(x) = (5x + 7)^3$

In Problems 9–12, without finding the inverse, find, at the indicated value of x, the corresponding point on the graph of f^{-1}. Then use (4) to find an equation of the tangent line at this point.

9. $y = \dfrac{1}{3}x^3 + x - 7$; $x = 3$

10. $y = \dfrac{2x + 1}{4x - 1}$; $x = 0$

11. $y = (x^5 + 1)^3$; $x = 1$

12. $y = 8 - 6\sqrt[3]{x + 2}$; $x = -3$

In Problems 13–32, find the derivative of the given function.

13. $y = \sin^{-1}(5x - 1)$

14. $y = \cos^{-1}\left(\dfrac{x + 1}{3}\right)$

15. $y = 4\cot^{-1}\dfrac{x}{2}$

16. $y = 2x - 10\sec^{-1}5x$

17. $y = 2\sqrt{x}\tan^{-1}\sqrt{x}$

18. $y = (\tan^{-1}x)(\cot^{-1}x)$

19. $y = \dfrac{\sin^{-1}2x}{\cos^{-1}2x}$

20. $y = \dfrac{\sin^{-1}x}{\sin x}$

21. $y = \dfrac{1}{\tan^{-1}x^2}$

22. $y = \dfrac{\sec^{-1}x}{x}$

23. $y = 2\sin^{-1}x + x\cos^{-1}x$

24. $y = \cot^{-1}x - \tan^{-1}\dfrac{x}{\sqrt{1 - x^2}}$

25. $y = \left(x^2 - 9\tan^{-1}\dfrac{x}{3}\right)^3$

26. $y = \sqrt{x - \cos^{-1}(x + 1)}$

27. $F(t) = \arctan\left(\dfrac{t - 1}{t + 1}\right)$

28. $g(t) = \arccos\sqrt{3t + 1}$

29. $f(x) = \arcsin(\cos 4x)$

30. $f(x) = \arctan\left(\dfrac{\sin x}{2}\right)$

31. $f(x) = \tan(\sin^{-1}x^2)$

32. $f(x) = \cos(x\sin^{-1}x)$

In Problems 33 and 34, use implicit differentiation to find dy/dx.

33. $\tan^{-1}y = x^2 + y^2$

34. $\sin^{-1}y - \cos^{-1}x = 1$

In Problems 35 and 36, show that $f'(x) = 0$. Interpret the result.

35. $f(x) = \sin^{-1}x + \cos^{-1}x$

36. $f(x) = \tan^{-1}x + \tan^{-1}(1/x)$.

In Problems 37 and 38, find the slope of the tangent line to the graph of the given function at the indicated value of x.

37. $y = \sin^{-1}\dfrac{x}{2}$; $x = 1$

38. $y = (\cos^{-1}x)^2$; $x = 1/\sqrt{2}$

In Problems 39 and 40, find an equation of the tangent line to the graph of the given function at the indicated value of x.

39. $f(x) = x\tan^{-1}x$; $x = 1$

40. $f(x) = \sin^{-1}(x - 1)$; $x = \dfrac{1}{2}$

41. Find the points on the graph of $f(x) = 5 - 2\sin x$, $0 \le x \le 2\pi$, at which the tangent line is parallel to the line $y = \sqrt{3}x + 1$.

42. Find all tangent lines to the graph of $f(x) = \arctan x$ that have slope $\frac{1}{4}$.

≣ Think About It

43. If f and $(f^{-1})'$ are differentiable, use (3) to find a formula for $(f^{-1})''(x)$.

3.8 Exponential Functions

▪ **Introduction** In Section 1.6 we saw that the exponential function $f(x) = b^x$, $b > 0$, $b \ne 1$, is defined for all real numbers, that is, the domain of f is $(-\infty, \infty)$. Inspection of Figure 1.6.2 shows that f is everywhere continuous. It turns out that an exponential function is also differentiable everywhere. In this section we develop the derivative of $f(x) = b^x$.

■ **Derivative of an Exponential Function** To find the derivative of an exponential function $f(x) = b^x$ we will use the definition of the derivative given in (2) of Definition 3.1.1. We first compute the difference quotient

$$\frac{f(x + h) - f(x)}{h} \tag{1}$$

in three steps. For the exponential function $f(x) = b^x$, we have

(*i*) $f(x + h) = b^{x+h} = b^x b^h$ ← laws of exponents

(*ii*) $f(x + h) - f(x) = b^{x+h} - b^x = b^x b^h - b^x = b^x(b^h - 1)$ ← laws of exponents and factoring

(*iii*) $\dfrac{f(x + h) - f(x)}{h} = \dfrac{b^x(b^h - 1)}{h} = b^x \cdot \dfrac{b^h - 1}{h}.$

In the fourth step, the calculus step, we let $h \to 0$ but analogous to the derivatives of $\sin x$ and $\cos x$ in Section 3.4, there is no apparent way of canceling the h in the difference quotient (*iii*). Nonetheless, the derivative of $f(x) = b^x$ is

$$f'(x) = \lim_{h \to 0} b^x \cdot \frac{b^h - 1}{h}. \tag{2}$$

Because b^x does not depend on the variable h, we can rewrite (2) as

$$f'(x) = b^x \cdot \lim_{h \to 0} \frac{b^h - 1}{h}. \tag{3}$$

Now here are the amazing results. The limit in (3),

$$\lim_{h \to 0} \frac{b^h - 1}{h}, \tag{4}$$

can be shown to exist for every positive base b. However, as one might expect, we will get a different answer for each base b. So for convenience let us denote the expression in (4) by the symbol $m(b)$. The derivative of $f(x) = b^x$ is then

$$f'(x) = b^x m(b). \tag{5}$$

You are asked to approximate the value of $m(b)$ in the four cases $b = 1.5, 2, 3,$ and 5 in Problems 57–60 of Exercises 3.8. For example, it can be shown that $m(10) \approx 2.302585\ldots$ and as a consequence if $f(x) = 10^x$, then

$$f'(x) = (2.302585\ldots)10^x. \tag{6}$$

We can get a better understanding of what $m(b)$ is by evaluating (5) at $x = 0$. Since $b^0 = 1$, we have $f'(0) = m(b)$. In other words, $m(b)$ is the slope of the tangent line to the graph of $f(x) = b^x$ at $x = 0$, that is, at the y-intercept $(0, 1)$. See FIGURE 3.8.1. Given that we have to calculate a different $m(b)$ for each base b, and that $m(b)$ is likely to be an "ugly" number as in (6), over time the following question arose *naturally*:

-   *Is there a base b for which m(b) = 1?*   (7)

■ **Derivative of the Natural Exponential Function** To answer the question posed in (7), we must return to the definitions of e given in Section 1.6. Specifically, (4) of Section 1.6,

$$e = \lim_{h \to 0} (1 + h)^{1/h} \tag{8}$$

provides the means for answering the question posed in (7). We know that on an intuitive level, the equality in (8) means that as h gets closer and closer to 0 then $(1 + h)^{1/h}$ can be made arbitrarily close to the number e. Thus for values of h near 0, we have the approximation $(1 + h)^{1/h} \approx e$ and so it follows that $1 + h \approx e^h$. The last expression written in the form

$$\frac{e^h - 1}{h} \approx 1 \tag{9}$$

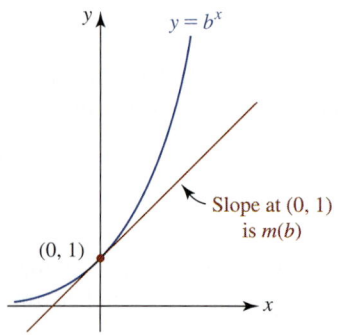

FIGURE 3.8.1 Find a base b so that the slope $m(b)$ of the tangent line at $(0, 1)$ is 1

suggests that

$$\lim_{h \to 0} \frac{e^h - 1}{h} = 1. \tag{10}$$

Since the left-hand side of (10) is $m(e)$ we have the answer to the question posed in (7):

- *The base b for which $m(b) = 1$ is $b = e$.* (11)

In addition, from (3) we have discovered a wonderfully simple result. The derivative of $f(x) = e^x$ is e^x. In summary,

$$\frac{d}{dx} e^x = e^x. \tag{12}$$

The result in (12) is the same as $f'(x) = f(x)$. Moreover, if $c \neq 0$ is a constant, then the only other nonzero function f in calculus whose derivative is equal to itself is $y = ce^x$ since by the Constant Multiple Rule of Section 3.2

$$\frac{dy}{dx} = \frac{d}{dx} ce^x = c \frac{d}{dx} e^x = ce^x = y.$$

■ **Derivative of $f(x) = b^x$—Revisited** In the preceding discussion we saw that $m(e) = 1$, but left unanswered the question of whether $m(b)$ has an exact value for each $b > 0$. It has. From the identity $e^{\ln b} = b, b > 0$, we can write any exponential function $f(x) = b^x$ in terms of the e base:

$$f(x) = b^x = (e^{\ln b})^x = e^{x(\ln b)}.$$

From the Chain Rule the derivative of b^x is

$$f'(x) = \frac{d}{dx} e^{x(\ln b)} = e^{x(\ln b)} \cdot \frac{d}{dx} x(\ln b) = e^{x(\ln b)}(\ln b).$$

Returning to $b^x = e^{x(\ln b)}$, the preceding line shows that

$$\frac{d}{dx} b^x = b^x(\ln b). \tag{13}$$

Matching the result in (5) with that in (13) we conclude that $m(b) = \ln b$. For example, the derivative of $f(x) = 10^x$ is $f'(x) = 10^x(\ln 10)$. Because $\ln 10 \approx 2.302585$ we see $f'(x) = 10^x(\ln 10)$ is the same as the result in (6).

The Chain Rule forms of the results in (12) and (13) are given next.

Theorem 3.8.1 Derivatives of Exponential Functions

If $u = g(x)$ is a differentiable function, then

$$\frac{d}{dx} e^u = e^u \frac{du}{dx}, \tag{14}$$

and

$$\frac{d}{dx} b^u = b^u(\ln b)\frac{du}{dx}. \tag{15}$$

EXAMPLE 1 Chain Rule

Differentiate

   (a) $y = e^{-x}$ **(b)** $y = e^{1/x^3}$ **(c)** $y = 8^{5x}$.

Solution

   (a) With $u = -x$ we have from (14),

$$\frac{dy}{dx} = e^{-x} \cdot \frac{d}{dx}(-x) = e^{-x}(-1) = -e^{-x}.$$

(b) By rewriting $u = 1/x^3$ as $u = x^{-3}$ we have from (14),

$$\frac{dy}{dx} = e^{1/x^3} \cdot \frac{d}{dx} x^{-3} = e^{1/x^3}(-3x^{-4}) = -3\frac{e^{1/x^3}}{x^4}.$$

(c) With $u = 5x$ we have from (15),

$$\frac{dy}{dx} = 8^{5x} \cdot (\ln 8) \cdot \frac{d}{dx} 5x = 5 \cdot 8^{5x}(\ln 8). \qquad \blacksquare$$

EXAMPLE 2　Product and Chain Rule

Find the points on the graph of $y = 3x^2 e^{-x^2}$ where the tangent line is horizontal.

Solution　We use the Product Rule along with (14):

$$\begin{aligned}
\frac{dy}{dx} &= 3x^2 \cdot \frac{d}{dx} e^{-x^2} + e^{-x^2} \cdot \frac{d}{dx} 3x^2 \\
&= 3x^2(-2xe^{-x^2}) + 6xe^{-x^2} \\
&= e^{-x^2}(-6x^3 + 6x).
\end{aligned}$$

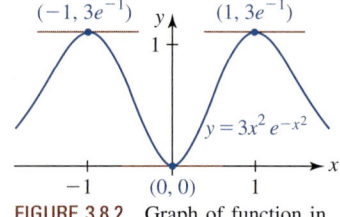

FIGURE 3.8.2　Graph of function in Example 2

Since $e^{-x^2} \neq 0$ for all real numbers x, $\dfrac{dy}{dx} = 0$ when $-6x^3 + 6x = 0$. Factoring the last equation gives $x(x + 1)(x - 1) = 0$ and so $x = 0$, $x = -1$, and $x = 1$. The corresponding points on the graph of the given function are then $(0, 0), (-1, 3e^{-1})$, and $(1, 3e^{-1})$. The graph of $y = 3x^2 e^{-x^2}$ along with the three tangent lines (in red) are shown in FIGURE 3.8.2. $\qquad \blacksquare$

In the next example we recall the fact that an exponential statement can be written in an equivalent logarithmic form. In particular, we use (9) of Section 1.6 in the form

$$y = e^x \qquad \text{if and only if} \qquad x = \ln y. \qquad (16)$$

EXAMPLE 3　Tangent Line Parallel to a Line

Find the point on the graph of $f(x) = 2e^{-x}$ at which the tangent line is parallel to $y = -4x - 2$.

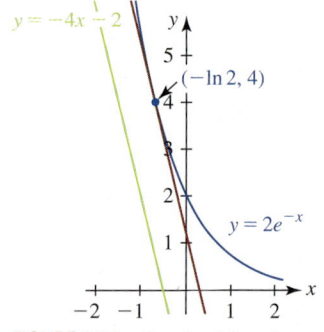

FIGURE 3.8.3　Graph of function and lines in Example 3

Solution　Let $(x_0, f(x_0)) = (x_0, 2e^{-x_0})$ be the unknown point on the graph of $f(x) = 2e^{-x}$ where the tangent line is parallel to $y = -4x - 2$. From the derivative $f'(x) = -2e^{-x}$ the slope of the tangent line at this point is then $f'(x_0) = -2e^{-x_0}$. Since $y = -4x - 2$ and the tangent line are parallel at that point, the slopes are equal:

$$f'(x_0) = -4 \qquad \text{or} \qquad -2e^{-x_0} = -4 \qquad \text{or} \qquad e^{-x_0} = 2.$$

From (16) the last equation gives $-x_0 = \ln 2$ or $x_0 = -\ln 2$. Hence, the point is $(-\ln 2, 2e^{\ln 2})$. Since $e^{\ln 2} = 2$, the point is $(-\ln 2, 4)$. In FIGURE 3.8.3 the given line is shown in green and the tangent line in red. $\qquad \blacksquare$

$\dfrac{d}{dx}$　**NOTES FROM THE CLASSROOM**

The numbers e and π are **transcendental** as well as irrational numbers. A transcendental number is one that is *not* a root of a polynomial equation with integer coefficients. For example, $\sqrt{2}$ is irrational but is not transcendental, since it is a root of the polynomial equation $x^2 - 2 = 0$. The number e was proved to be transcendental by the French mathematician **Charles Hermite** (1822–1901) in 1873, whereas π was proved to be transcendental nine years later by the **German mathematician** Ferdinand Lindemann (1852–1939). The latter proof showed conclusively that "squaring a circle" with a rule and a compass was impossible.

Exercises 3.8   Answers to selected odd-numbered problems begin on page ANS-11.

≡ Fundamentals

In Problems 1–26, find the derivative of the given function.

1. $y = e^{-x}$

2. $y = e^{2x+3}$

3. $y = e^{\sqrt{x}}$

4. $y = e^{\sin 10x}$

5. $y = 5^{2x}$

6. $y = 10^{-3x^2}$

7. $y = x^3 e^{4x}$

8. $y = e^{-x}\sin \pi x$

9. $f(x) = \dfrac{e^{-2x}}{x}$

10. $f(x) = \dfrac{xe^x}{x + e^x}$

11. $y = \sqrt{1 + e^{-5x}}$

12. $y = (e^{2x} - e^{-2x})^{10}$

13. $y = \dfrac{2}{e^{x/2} + e^{-x/2}}$

14. $y = \dfrac{e^x + e^{-x}}{e^x - e^{-x}}$

15. $y = \dfrac{e^{7x}}{e^{-x}}$

16. $y = e^{2x}e^{3x}e^{4x}$

17. $y = (e^3)^{x-1}$

18. $y = \left(\dfrac{1}{e^x}\right)^{100}$

19. $f(x) = e^{x^{1/3}} + (e^x)^{1/3}$

20. $f(x) = (2x + 1)^3 e^{-(1-x)^4}$

21. $f(x) = e^{-x}\tan e^x$

22. $f(x) = \sec e^{2x}$

23. $f(x) = e^{x\sqrt{x^2+1}}$

24. $y = e^{\frac{x+2}{x-2}}$

25. $y = e^{e^{x^2}}$

26. $y = e^x + e^{x+e^{-x}}$

27. Find an equation of the tangent line to the graph of $y = (e^x + 1)^2$ at $x = 0$.

28. Find the slope of the normal line to the graph of $y = (x - 1)e^{-x}$ at $x = 0$.

29. Find the point on the graph of $y = e^x$ at which the tangent line is parallel to $3x - y = 7$.

30. Find the point on the graph of $y = 5x + e^{2x}$ at which the tangent line is parallel to $y = 6x$.

In Problems 31 and 32, find the point(s) on the graph of the given function at which the tangent line is horizontal. Use a graphing utility to obtain the graph of each function.

31. $f(x) = e^{-x}\sin x$

32. $f(x) = (3 - x^2)e^{-x}$

In Problems 33–36, find the indicated higher derivative.

33. $y = e^{x^2}$; $\dfrac{d^3 y}{dx^3}$

34. $y = \dfrac{1}{1 + e^{-x}}$; $\dfrac{d^2 y}{dx^2}$

35. $y = \sin e^{2x}$; $\dfrac{d^2 y}{dx^2}$

36. $y = x^2 e^x$; $\dfrac{d^4 y}{dx^4}$

In Problems 37 and 38, C_1 and C_2 are arbitrary real constants. Show that the function satisfies the given differential equation.

37. $y = C_1 e^{-3x} + C_2 e^{2x}$; $y'' + y' - 6y = 0$

38. $y = C_1 e^{-x}\cos 2x + C_2 e^{-x}\sin 2x$; $y'' + 2y' + 5y = 0$

39. If C and k are real constants, show that the function $y = Ce^{kx}$ satisfies the differential equation $y' = ky$.

40. Use Problem 39 to find a function that satisfies the given conditions.

(a) $y' = -0.01y$ and $y(0) = 100$

(b) $\dfrac{dP}{dt} - 0.15P = 0$ and $P(0) = P_0$

In Problems 41–46, use implicit differentiation to find dy/dx.

41. $y = e^{x+y}$

42. $xy = e^y$

43. $y = \cos e^{xy}$

44. $y = e^{(x+y)^2}$

45. $x + y^2 = e^{x/y}$

46. $e^x + e^y = y$

47. (a) Sketch the graph of $f(x) = e^{-|x|}$.

(b) Find $f'(x)$.

(c) Sketch the graph of f'.

(d) Is the function differentiable at $x = 0$?

48. (a) Show that the function $f(x) = e^{\cos x}$ is periodic with period 2π.

(b) Find all points on the graph of f where the tangent is horizontal.

(c) Sketch the graph of f.

≡ Applications

49. The logistic function

$$P(t) = \frac{aP_0}{bP_0 + (a - bP_0)e^{-at}},$$

where a and b are positive constants, often serves as a mathematical model for an expanding but limited population.

(a) Show that $P(t)$ satisfies the differential equation

$$\frac{dP}{dt} = P(a - bP).$$

(b) The graph of $P(t)$ is called a **logistic curve** where $P(0) = P_0$ is the initial population. Consider the case when $a = 2, b = 1$, and $P_0 = 1$. Find horizontal asymptotes for the graph of $P(t)$ by determining the limits $\lim_{t \to -\infty} P(t)$ and $\lim_{t \to \infty} P(t)$.

(c) Graph $P(t)$.

(d) Find the value(s) of t for which $P''(t) = 0$.

50. The **Jenss mathematical model** (1937) represents one of the most accurate empirically devised formulas for predicting the height h (in centimeters) in terms of age t (in years) for preschool-age children (3 months to 6 years):

$$h(t) = 79.04 + 6.39t - e^{3.26 - 0.99t}.$$

(a) What height does this model predict for a 2-year-old?

(b) How fast is a 2-year-old increasing in height?

(c) Use a calculator or CAS to obtain the graph of h on the interval $\left[\frac{1}{4}, 6\right]$.

(d) Use the graph in part (c) to estimate the age of a preschool-age child who is 100 cm tall.

≡ Think About It

51. Show that the x-intercept of the tangent line to the graph of $y = e^{-x}$ at $x = x_0$ is one unit to the right of x_0.

52. How is the tangent line to the graph of $y = e^x$ at $x = 0$ related to the tangent line to the graph of $y = e^{-x}$ at $x = 0$?

53. Explain why there is no point on the graph of $y = e^x$ at which the tangent line is parallel to $2x + y = 1$.

54. Find all tangent lines to the graph of $f(x) = e^x$ that pass through the origin.

In Problems 55 and 56, the symbol n represents a positive integer. Find a formula for the given derivative.

55. $\dfrac{d^n}{dx^n}\sqrt{e^x}$

56. $\dfrac{d^n}{dx^n}xe^{-x}$

≡ Calculator/CAS Problems

In Problems 57–60, use a calculator to estimate the value $m(b) = \lim\limits_{h \to 0}\dfrac{b^h - 1}{h}$ for $b = 1.5$, $b = 2$, $b = 3$, and $b = 5$ by filling out the given table.

57.

$h \to 0$	0.1	0.01	0.001	0.0001	0.00001	0.000001
$\dfrac{(1.5)^h - 1}{h}$						

58.

$h \to 0$	0.1	0.01	0.001	0.0001	0.00001	0.000001
$\dfrac{2^h - 1}{h}$						

59.

$h \to 0$	0.1	0.01	0.001	0.0001	0.00001	0.000001
$\dfrac{3^h - 1}{h}$						

60.

$h \to 0$	0.1	0.01	0.001	0.0001	0.00001	0.000001
$\dfrac{5^h - 1}{h}$						

61. Use a calculator or CAS to obtain the graph of

$$f(x) = \begin{cases} e^{-1/x^2}, & x \neq 0 \\ 0, & x = 0. \end{cases}$$

Show that f is differentiable for all x. Compute $f'(0)$ using the definition of the derivative.

3.9 Logarithmic Functions

■ **Introduction** Because the inverse of the exponential function $y = b^x$ is the logarithmic function $y = \log_b x$ we can find the derivative of the latter function in three different ways: (3) of Section 3.7, implicit differentiation, or from the fundamental definition (2) of Section 3.1. We will demonstrate the last two methods.

■ **Derivative of the Natural Logarithm** We know from (9) of Section 1.6 that $y = \ln x$ is the same as $x = e^y$. By implicit differentiation, the Chain Rule, and (14) of Section 3.8,

$$\frac{d}{dx}x = \frac{d}{dx}e^y \qquad \text{gives} \qquad 1 = e^y\frac{dy}{dx}.$$

Therefore

$$\frac{dy}{dx} = \frac{1}{e^y}.$$

Replacing e^y by x, we get the following result:

Like the inverse trigonometric functions, the derivative of the inverse of the natural exponential function is an algebraic function.

$$\frac{d}{dx}\ln x = \frac{1}{x}. \tag{1}$$

■ **Derivative of $f(x) = \log_b x$** In precisely the same manner used to obtain (1), the derivative of $y = \log_b x$ can be gotten by differentiating $x = b^y$ implicitly:

$$\frac{d}{dx}x = \frac{d}{dx}b^y \qquad \text{gives} \qquad 1 = b^y(\ln b)\frac{dy}{dx}.$$

Therefore

$$\frac{dy}{dx} = \frac{1}{b^y(\ln b)}.$$

Replacing b^y by x gives

$$\frac{d}{dx}\log_b x = \frac{1}{x(\ln b)}. \tag{2}$$

Because $\ln e = 1$, (2) becomes (1) when $b = e$.

EXAMPLE 1   Product Rule

Differentiate $f(x) = x^2 \ln x$.

Solution   By the Product Rule and (1) we have

$$f'(x) = x^2 \cdot \frac{d}{dx}\ln x + (\ln x)\cdot\frac{d}{dx}x^2 = x^2 \cdot \frac{1}{x} + (\ln x)\cdot 2x$$

or

$$f'(x) = x + 2x\ln x. \qquad\blacksquare$$

EXAMPLE 2   Slope of a Tangent Line

Find the slope of the tangent to the graph of $y = \log_{10} x$ at $x = 2$.

Solution   By (2) the derivative of $y = \log_{10} x$ is

$$\frac{dy}{dx} = \frac{1}{x(\ln 10)}.$$

With the aid of a calculator, the slope of the tangent line at $(2, \log_{10} 2)$ is

$$\left.\frac{dy}{dx}\right|_{x=2} = \frac{1}{2\ln 10} \approx 0.2171. \qquad\blacksquare$$

We summarize the results in (1) and (2) in their Chain Rule forms.

Theorem 3.9.1   Derivatives of Logarithmic Functions

If $u = g(x)$ is a differentiable function, then

$$\frac{d}{dx}\ln u = \frac{1}{u}\frac{du}{dx}, \tag{3}$$

and

$$\frac{d}{dx}\log_b u = \frac{1}{u(\ln b)}\frac{du}{dx}. \tag{4}$$

EXAMPLE 3   Chain Rule

Differentiate

(a) $f(x) = \ln(\cos x)$ and   (b) $y = \ln(\ln x)$.

Solution

(a) By (3), with $u = \cos x$ we have

$$f'(x) = \frac{1}{\cos x}\cdot\frac{d}{dx}\cos x = \frac{1}{\cos x}\cdot(-\sin x)$$

or

$$f'(x) = -\tan x.$$

(b) Using (3) again, this time with $u = \ln x$, we get

$$\frac{dy}{dx} = \frac{1}{\ln x}\cdot\frac{d}{dx}\ln x = \frac{1}{\ln x}\cdot\frac{1}{x} = \frac{1}{x\ln x}. \qquad\blacksquare$$

EXAMPLE 4 Chain Rule

Differentiate $f(x) = \ln x^3$.

Solution Because x^3 must be positive it is understood that $x > 0$. Hence by (3), with $u = x^3$ we have

$$f'(x) = \frac{1}{x^3} \cdot \frac{d}{dx} x^3 = \frac{1}{x^3} \cdot (3x^2) = \frac{3}{x}.$$

Alternative Solution: From (*iii*) of the laws of logarithms (Theorem 1.6.1), $\ln N^c = c \ln N$ and so we can rewrite $y = \ln x^3$ as $y = 3 \ln x$ and then differentiate:

$$f(x) = 3 \frac{d}{dx} \ln x = 3 \cdot \frac{1}{x} = \frac{3}{x}. \qquad \blacksquare$$

Although the domain of the natural logarithm $y = \ln x$ is the set $(0, \infty)$, the domain of $y = \ln|x|$ extends to the set $(-\infty, 0) \cup (0, \infty)$. For the numbers in this last domain,

$$|x| = \begin{cases} x, & x > 0 \\ -x, & x < 0. \end{cases}$$

Therefore

$$\text{for } x > 0, \quad \frac{d}{dx} \ln x = \frac{1}{x}$$

$$\text{for } x < 0, \quad \frac{d}{dx} \ln(-x) = \frac{1}{-x} \cdot (-1) = \frac{1}{x}. \tag{5}$$

The derivatives in (5) prove that for $x \neq 0$,

$$\frac{d}{dx} \ln|x| = \frac{1}{x}. \tag{6}$$

The result in (6) then generalizes by the Chain Rule. For a differentiable function $u = g(x)$, $u \neq 0$,

$$\frac{d}{dx} \ln|u| = \frac{1}{u} \frac{du}{dx}. \tag{7}$$

EXAMPLE 5 Using (6)

Find the slope of the tangent line to the graph of $y = \ln|x|$ at $x = -2$ and at $x = 2$.

Solution Since (6) gives $dy/dx = 1/x$, we have

$$\left. \frac{dy}{dx} \right|_{x=-2} = -\frac{1}{2} \quad \text{and} \quad \left. \frac{dy}{dx} \right|_{x=2} = \frac{1}{2}. \tag{8}$$

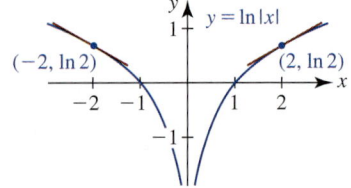

FIGURE 3.9.1 Graphs of tangent lines and function in Example 5

Because $\ln|-2| = \ln 2$, (8) gives, respectively, the slopes of the tangent lines at the points $(-2, \ln 2)$ and $(2, \ln 2)$. Observe in **FIGURE 3.9.1** that the graph of $y = \ln|x|$ is symmetric with respect to the y-axis; the tangent lines are shown in red. $\qquad \blacksquare$

EXAMPLE 6 Using (7)

Differentiate

(a) $y = \ln(2x - 3)$ and (b) $y = \ln|2x - 3|$.

Solution

(a) For $2x - 3 > 0$, or $x > \frac{3}{2}$, we have from (3),

$$\frac{dy}{dx} = \frac{1}{2x - 3} \cdot \frac{d}{dx}(2x - 3) = \frac{2}{2x - 3}. \tag{9}$$

(b) For $2x - 3 \neq 0$, or $x \neq \frac{3}{2}$, we have from (7),

$$\frac{dy}{dx} = \frac{1}{2x - 3} \cdot \frac{d}{dx}(2x - 3) = \frac{2}{2x - 3}. \tag{10}$$

Although (9) and (10) *appear* to be equal, they are definitely not the same function. The difference is simply that the domain of the derivative in (9) is the interval $(\frac{3}{2}, \infty)$, whereas the domain of the derivative in (10) is the set of real numbers except $x = \frac{3}{2}$.   ■

EXAMPLE 7   A Distinction

The functions $f(x) = \ln x^4$ and $g(x) = 4 \ln x$ are not the same. Since $x^4 > 0$ for all $x \neq 0$, the domain of f is the set of real numbers except $x = 0$. The domain of g is the interval $(0, \infty)$. Thus,

$$f'(x) = \frac{4}{x}, \quad x \neq 0 \qquad \text{whereas} \qquad g'(x) = \frac{4}{x}, \quad x > 0.$$ ■

EXAMPLE 8   Simplifying Before Differentiating

Differentiate $y = \ln \dfrac{x^{1/2}(2x + 7)^4}{(3x^2 + 1)^2}$.

Solution   Using the laws of logarithms given in Section 1.6, for $x > 0$ we can rewrite the right-hand side of the given function as

$$\begin{aligned}
y &= \ln x^{1/2}(2x + 7)^4 - \ln(3x^2 + 1)^2 && \leftarrow \ln(M/N) = \ln M - \ln N \\
&= \ln x^{1/2} + \ln(2x + 7)^4 - \ln(3x^2 + 1)^2 && \leftarrow \ln(MN) = \ln M + \ln N \\
&= \frac{1}{2} \ln x + 4 \ln(2x + 7) - 2 \ln(3x^2 + 1) && \leftarrow \ln N^c = c \ln N
\end{aligned}$$

so that $$\frac{dy}{dx} = \frac{1}{2} \cdot \frac{1}{x} + 4 \cdot \frac{1}{2x + 7} \cdot 2 - 2 \cdot \frac{1}{3x^2 + 1} \cdot 6x$$

or $$\frac{dy}{dx} = \frac{1}{2x} + \frac{8}{2x + 7} - \frac{12x}{3x^2 + 1}.$$ ■

■ **Logarithmic Differentiation**   Differentiation of a complicated function $y = f(x)$ that consists of products, quotients, and powers can be simplified by a technique known as **logarithmic differentiation**. The procedure consists of three steps.

> ## Guidelines for Logarithmic Differentiation
>
> (*i*) Take the natural logarithm of both sides of $y = f(x)$. Simplify the right-hand side of $\ln y = \ln f(x)$ as much as possible using the general properties of logarithms.
>
> (*ii*) Differentiate the simplified version of $\ln y = \ln f(x)$ implicitly:
>
> $$\frac{d}{dx} \ln y = \frac{d}{dx} \ln f(x).$$
>
> (*iii*) Since the derivative of the left-hand side is $\dfrac{1}{y} \dfrac{dy}{dx}$, multiply both sides by y and replace y by $f(x)$.

We know how to differentiate any function of the type

$$y = (\text{constant})^{\text{variable}} \qquad \text{and} \qquad y = (\text{variable})^{\text{constant}}.$$

For example,

$$\frac{d}{dx} \pi^x = \pi^x (\ln \pi) \qquad \text{and} \qquad \frac{d}{dx} x^\pi = \pi x^{\pi - 1}.$$

There are functions where both the base and the exponent are variable:

$$y = (\text{variable})^{\text{variable}}. \tag{11}$$

For example, $f(x) = (1 + 1/x)^x$ is a function of the type described in (11). Recall, in Section 1.6 we saw that $f(x) = (1 + 1/x)^x$ played an important role in the definition of the number e. Although we will not develop a general formula for the derivative of functions of the type given in (11), we can nonetheless obtain their derivatives through the process of logarithmic differentiation. The next example illustrates the method for finding dy/dx.

EXAMPLE 9   Logarithmic Differentiation

Differentiate $y = x^{\sqrt{x}}$, $x > 0$.

Solution   Taking the natural logarithm of both sides of the given equation and simplifying yields

$$\ln y = \ln x^{\sqrt{x}} = \sqrt{x}\, \ln x. \quad \leftarrow \text{property } (iii) \text{ of the laws of logarithms, Section 1.6}$$

Then we differentiate implicitly:

$$\frac{1}{y}\frac{dy}{dx} = \sqrt{x} \cdot \frac{1}{x} + \frac{1}{2}x^{-1/2} \cdot \ln x \quad \leftarrow \text{Product Rule}$$

$$\frac{dy}{dx} = y\left[\frac{1}{\sqrt{x}} + \frac{\ln x}{2\sqrt{x}}\right] \quad \leftarrow \text{now replace } y \text{ by } x^{\sqrt{x}}$$

$$= \frac{1}{2}x^{\sqrt{x}-\frac{1}{2}}(2 + \ln x). \quad \leftarrow \text{common denominator and laws of exponents}$$

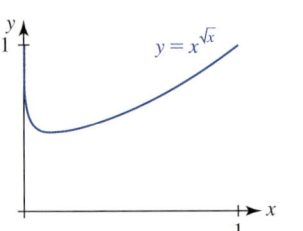

FIGURE 3.9.2   Graph of function in Example 9

We obtained the graph of $y = x^{\sqrt{x}}$ in **FIGURE 3.9.2** with the aid of a graphing utility. Note that the graph has a horizontal tangent at the point at which $dy/dx = 0$. Thus, the x-coordinate of the point of horizontal tangency is determined from $2 + \ln x = 0$ or $\ln x = -2$. The last equation gives $x = e^{-2}$. ∎

EXAMPLE 10   Logarithmic Differentiation

Find the derivative of $y = \dfrac{\sqrt[3]{x^4 + 6x^2}(8x + 3)^5}{(2x^2 + 7)^{2/3}}$.

Solution   Notice that the given function contains no logarithms. As such, we can find dy/dx using the ordinary application of the Quotient, Product, and Power Rules. This procedure, which is tedious, can be avoided by first taking the logarithm of both sides of the given equation, simplifying as we did in Example 9 by the laws of logarithms, and *then* differentiating implicitly. We take the natural logarithm of both sides of the given equation and simplify the right-hand side:

$$\ln y = \ln \frac{\sqrt[3]{x^4 + 6x^2}(8x + 3)^5}{(2x^2 + 7)^{2/3}}$$

$$= \ln \sqrt[3]{x^4 + 6x^2} + \ln (8x + 3)^5 - \ln (2x^2 + 7)^{2/3}$$

$$= \frac{1}{3}\ln (x^4 + 6x^2) + 5\ln (8x + 3) - \frac{2}{3}\ln (2x^2 + 7).$$

Differentiating the last line with respect to x gives

$$\frac{1}{y}\frac{dy}{dx} = \frac{1}{3} \cdot \frac{1}{x^4 + 6x^2} \cdot (4x^3 + 12x) + 5 \cdot \frac{1}{8x + 3} \cdot 8 - \frac{2}{3} \cdot \frac{1}{2x^2 + 7} \cdot 4x$$

$$\frac{dy}{dx} = y\left[\frac{4x^3 + 12x}{3(x^4 + 6x^2)} + \frac{40}{8x + 3} - \frac{8x}{3(2x^2 + 7)}\right] \quad \leftarrow \text{multiply both sides by } y$$

$$= \frac{\sqrt[3]{x^4 + 6x^2}(8x + 3)^5}{(2x^2 + 7)^{2/3}}\left[\frac{4x^3 + 12x}{3(x^4 + 6x^2)} + \frac{40}{8x + 3} - \frac{8x}{3(2x^2 + 7)}\right]. \quad \leftarrow \text{replace } y \text{ by the original expression} \quad ∎$$

■ **Postscript—Derivative of $f(x) = \log_b x$ Revisited**   As stated in the introduction to this section we can obtain the derivative of $f(x) = \log_b x$ using the definition of the derivative. From (2) of Section 3.1,

$$f'(x) = \lim_{h \to 0} \frac{\log_b(x + h) - \log_b x}{h}$$

$$= \lim_{h \to 0} \frac{1}{h} \log_b \frac{x + h}{x} \qquad \leftarrow \text{algebra and the laws of logarithms}$$

$$= \lim_{h \to 0} \frac{1}{h} \log_b \left(1 + \frac{h}{x} \right) \qquad \leftarrow \text{division of } x + h \text{ by } x$$

$$= \lim_{h \to 0} \frac{1}{x} \cdot \frac{x}{h} \log_b \left(1 + \frac{h}{x} \right) \qquad \leftarrow \text{multiplication by } x/x = 1$$

$$= \frac{1}{x} \lim_{h \to 0} \log_b \left(1 + \frac{h}{x} \right)^{x/h} \qquad \leftarrow \text{the laws of logarithms}$$

$$= \frac{1}{x} \log_b \left[\lim_{h \to 0} \left(1 + \frac{h}{x} \right)^{x/h} \right]. \qquad (12)$$

The last step, taking the limit inside the logarithmic function, is justified by invoking the continuity of the function on $(0, \infty)$ and assuming that the limit inside the brackets exists. If we let $t = h/x$ in the last equation, then since x is fixed, $h \to 0$ implies $t \to 0$. Consequently, we see from (4) of Section 1.6 that

$$\lim_{h \to 0} \left(1 + \frac{h}{x} \right)^{x/h} = \lim_{t \to 0} (1 + t)^{1/t} = e.$$

Hence the result in (12) shows that,

$$\frac{d}{dx} \log_b x = \frac{1}{x} \log_b e. \qquad (13)$$

◀ Those with sharp eyes and long memories will have noticed that (13) is not the same as (2). The results are equivalent, since by the change of base formula for logarithms $\log_b e = \ln e / \ln b = 1/\ln b$.

When the "natural" choice of $b = e$ is made, (13) becomes (1) since $\log_e e = \ln e = 1$.

■ **Postscript—Power Rule Revisited**   We are finally in a position to prove the Power Rule $(d/dx)x^n = nx^{n-1}$, (3) of Section 3.2, for all real number exponents n. Our demonstration uses the following fact: For $x > 0$, x^n is defined for all real numbers n. Then in view of the identity $x = e^{\ln x}$ we can write

$$x^n = (e^{\ln x})^n = e^{n \ln x}.$$

Thus, $\qquad \dfrac{d}{dx} x^n = \dfrac{d}{dx} e^{n \ln x} = e^{n \ln x} \dfrac{d}{dx} (n \ln x) = \dfrac{n}{x} e^{n \ln x}.$

Substituting $e^{n \ln x} = x^n$ in the last result completes the proof for $x > 0$,

$$\frac{d}{dx} x^n = \frac{n}{x} x^n = nx^{n-1}.$$

The last derivative formula is also valid for $x < 0$ when $n = p/q$ is a rational number and q is an odd integer.

Exercises 3.9   Answers to selected odd-numbered problems begin on page ANS-12.

≡ Fundamentals

In Problems 1–24, find the derivative of the given function.

1. $y = 10 \ln x$
2. $y = \ln 10x$
3. $y = \ln x^{1/2}$
4. $y = (\ln x)^{1/2}$
5. $y = \ln(x^4 + 3x^2 + 1)$
6. $y = \ln(x^2 + 1)^{20}$
7. $y = x^2 \ln x^3$
8. $y = x - \ln|5x + 1|$
9. $y = \dfrac{\ln x}{x}$
10. $y = x(\ln x)^2$
11. $y = \ln \dfrac{x}{x + 1}$
12. $y = \dfrac{\ln 4x}{\ln 2x}$

13. $y = -\ln|\cos x|$
14. $y = \dfrac{1}{3} \ln|\sin 3x|$
15. $y = \dfrac{1}{\ln x}$
16. $y = \ln \dfrac{1}{x}$
17. $f(x) = \ln(x \ln x)$
18. $f(x) = \ln(\ln(\ln x))$
19. $g(x) = \sqrt{\ln \sqrt{x}}$
20. $w(\theta) = \theta \sin(\ln 5\theta)$
21. $H(t) = \ln t^2 (3t^2 + 6)$
22. $G(t) = \ln \sqrt{5t + 1}(t^3 + 4)^6$
23. $f(x) = \ln \dfrac{(x + 1)(x + 2)}{x + 3}$
24. $f(x) = \ln \sqrt{\dfrac{(3x + 2)^5}{x^4 + 7}}$

25. Find an equation of the tangent line to the graph of $y = \ln x$ at $x = 1$.

26. Find an equation of the tangent line to the graph of $y = \ln(x^2 - 3)$ at $x = 2$.

27. Find the slope of the tangent to the graph of $y = \ln(e^{3x} + x)$ at $x = 0$.

28. Find the slope of the tangent to the graph of $y = \ln(xe^{-x^3})$ at $x = 1$.

29. Find the slope of the tangent to the graph of f' at the point where the slope of the tangent to the graph of $f(x) = \ln x^2$ is 4.

30. Determine the point on the graph of $y = \ln 2x$ at which the tangent line is perpendicular to $x + 4y = 1$.

In Problems 31 and 32, find the point(s) on the graph of the given function at which the tangent line is horizontal.

31. $f(x) = \dfrac{\ln x}{x}$

32. $f(x) = x^2 \ln x$

In Problems 33–36, find the indicated derivative and simplify as much as possible.

33. $\dfrac{d}{dx}\ln\left(x + \sqrt{x^2 - 1}\right)$

34. $\dfrac{d}{dx}\ln\left(\dfrac{1 + \sqrt{1 - x^2}}{x}\right)$

35. $\dfrac{d}{dx}\ln(\sec x + \tan x)$

36. $\dfrac{d}{dx}\ln(\csc x - \cot x)$

In Problems 37–40, find the indicated higher derivative.

37. $y = \ln x$; $\dfrac{d^3 y}{dx^3}$

38. $y = x\ln x$; $\dfrac{d^2 y}{dx^2}$

39. $y = (\ln|x|)^2$; $\dfrac{d^2 y}{dx^2}$

40. $y = \ln(5x - 3)$; $\dfrac{d^4 y}{dx^4}$

In Problems 41 and 42, C_1 and C_2 are arbitrary real constants. Show that the function satisfies the given differential equation for $x > 0$.

41. $y = C_1 x^{-1/2} + C_2 x^{-1/2}\ln x$; $4x^2 y'' + 8xy' + y = 0$

42. $y = C_1 x^{-1}\cos\left(\sqrt{2}\ln x\right) + C_2 x^{-1}\sin\left(\sqrt{2}\ln x\right)$; $x^2 y'' + 3xy' + 3y = 0$

In Problems 43–48, use implicit differentiation to find dy/dx.

43. $y^2 = \ln xy$

44. $y = \ln(x + y)$

45. $x + y^2 = \ln\dfrac{x}{y}$

46. $y = \ln xy^2$

47. $xy = \ln(x^2 + y^2)$

48. $x^2 + y^2 = \ln(x + y)^2$

In Problems 49–56, use logarithmic differentiation to find dy/dx.

49. $y = x^{\sin x}$

50. $y = (\ln|x|)^x$

51. $y = x(x - 1)^x$

52. $y = \dfrac{(x^2 + 1)^x}{x^2}$

53. $y = \dfrac{\sqrt{(2x + 1)(3x + 2)}}{4x + 3}$

54. $y = \dfrac{x^{10}\sqrt{x^2 + 5}}{\sqrt[3]{8x^2 + 2}}$

55. $y = \dfrac{(x^3 - 1)^5(x^4 + 3x^3)^4}{(7x + 5)^9}$

56. $y = x\sqrt{x + 1}\sqrt[3]{x^2 + 2}$

57. Find an equation of the tangent line to the graph of $y = x^{x+2}$ at $x = 1$.

58. Find an equation of the tangent line to the graph of $y = x(\ln x)^x$ at $x = e$.

In Problems 59 and 60, find the point on the graph of the given function at which the tangent line is horizontal. Use a graphing utility to obtain the graph of each function on the interval $[0.01, 1]$.

59. $y = x^x$

60. $y = x^{2x}$

≡ Think About It

61. Find the derivatives of
(a) $y = \tan x^x$ (b) $y = x^x e^{x^x}$ (c) $y = x^{x^x}$.

62. Find $d^2 y/dx^2$ for $y = \sqrt{x^x}$.

63. The function $f(x) = \ln|x|$ is not differentiable only at $x = 0$. The function $g(x) = |\ln x|$ is not differentiable at $x = 0$ and at one other value of $x > 0$. What is it?

64. Find a way to compute $\dfrac{d}{dx}\log_x e$.

≡ Calculator/CAS Problems

65. (a) Use a calculator or CAS to obtain the graph of $y = (\sin x)^{\ln x}$ on the interval $(0, 5\pi)$.
(b) Explain why there appears to be no graph on certain intervals. Identify the intervals.

66. (a) Use a calculator or CAS to obtain the graph of $y = |\cos x|^{\cos x}$ on the interval $[0, 5\pi]$.
(b) Determine, at least approximately, the values of x in the interval $[0, 5\pi]$ for which the tangent to the graph is horizontal.

67. Use a calculator or CAS to obtain the graph of $f(x) = x^3 - 12\ln x$. Then find the *exact* value of the least value of $f(x)$.

3.10 Hyperbolic Functions

▌ **Introduction** If you have ever toured the 630-ft-high Gateway Arch in St. Louis, Missouri, you may have asked the question, What is the shape of the arch? and received the rather cryptic reply: the shape of an inverted catenary. The word *catenary* stems from the Latin word *catena* and literally means "a hanging chain" (the Romans used the catena as a dog leash). It can be demonstrated that the shape assumed by a long flexible wire, chain, cable, or rope hanging under its own weight between two points is the shape of the graph of the function

$$f(x) = \frac{k}{2}(e^{cx} + e^{-cx}) \tag{1}$$

for appropriate choices of the constants c and k. The graph of any function of the form given in (1) is called a **catenary**.

The Gateway Arch in St. Louis, MO.

■ **Hyperbolic Functions** Combinations such as (1) involving the exponential functions e^x and e^{-x} occur so often in applied mathematics that they warrant special definitions.

Definition 3.10.1   Hyperbolic Sine and Cosine

For any real number x, the **hyperbolic sine** of x is

$$\sinh x = \frac{e^x - e^{-x}}{2} \tag{2}$$

and the **hyperbolic cosine** of x is

$$\cosh x = \frac{e^x + e^{-x}}{2}. \tag{3}$$

Because the domain of each of the exponential functions e^x and e^{-x} is the set of real numbers $(-\infty, \infty)$, the domain of $y = \sinh x$ and of $y = \cosh x$ is $(-\infty, \infty)$. From (2) and (3) of Definition 3.10.1 it is also apparent that

$$\sinh 0 = 0 \qquad \text{and} \qquad \cosh 0 = 1.$$

Analogous to the trigonometric functions $\tan x$, $\cot x$, $\sec x$, and $\csc x$ that are defined in terms of $\sin x$ and $\cos x$, we define four additional hyperbolic functions in terms of $\sinh x$ and $\cosh x$.

The shape of the St. Louis Gateway Arch is based on the mathematical model

$$y = A - B\cosh(Cx/L),$$

where $A = 693.8597$, $B = 68.7672$, $L = 299.2239$, $C = 3.0022$, and x and y are measured in feet. When $x = 0$, we get the approximate height of 630 ft.

Definition 3.10.2   Other Hyperbolic Functions

For a real number x, the **hyperbolic tangent** of x is

$$\tanh x = \frac{\sinh x}{\cosh x} = \frac{e^x - e^{-x}}{e^x + e^{-x}}, \tag{4}$$

the **hyperbolic cotangent** of x, $x \neq 0$, is

$$\coth x = \frac{\cosh x}{\sinh x} = \frac{e^x + e^{-x}}{e^x - e^{-x}}, \tag{5}$$

the **hyperbolic secant** of x is

$$\operatorname{sech} x = \frac{1}{\cosh x} = \frac{2}{e^x + e^{-x}}, \tag{6}$$

the **hyperbolic cosecant** of x, $x \neq 0$, is

$$\operatorname{csch} x = \frac{1}{\sinh x} = \frac{2}{e^x - e^{-x}}. \tag{7}$$

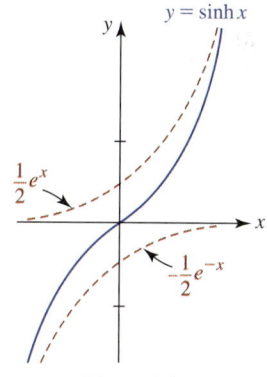

(a) $y = \sinh x$

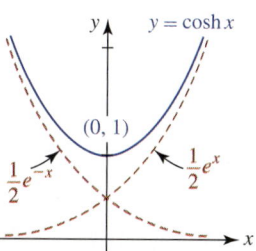

(b) $y = \cosh x$

FIGURE 3.10.1 Graphs of hyperbolic sine and cosine

■ **Graphs of Hyperbolic Functions** The graphs of the hyperbolic sine and hyperbolic cosine are given in FIGURE 3.10.1. Note the similarity of the graph in Figure 3.10.1(b) and the shape of the Gateway Arch in the photo at the beginning of this section. The graphs of the hyperbolic tangent, cotangent, secant, and cosecant are given in FIGURE 3.10.2. Note that $x = 0$ is a vertical asymptote of the graphs of $y = \coth x$ and $y = \operatorname{csch} x$.

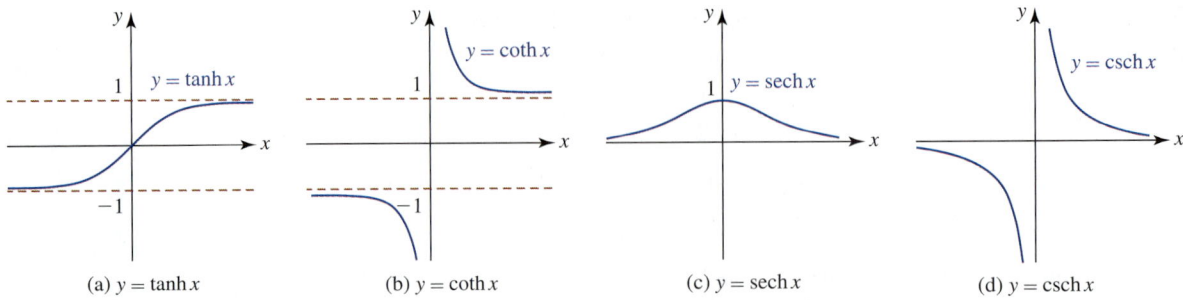

(a) $y = \tanh x$ (b) $y = \coth x$ (c) $y = \operatorname{sech} x$ (d) $y = \operatorname{csch} x$

FIGURE 3.10.2 Graphs of the hyperbolic tangent, cotangent, secant, and cosecant

■ **Identities**   Although the hyperbolic functions are not periodic, they possess many identities that are similar to those for the trigonometric functions. Notice that the graphs in Figure 3.10.1(a) and (b) are symmetric with respect to the origin and the y-axis, respectively. In other words, $y = \sinh x$ is an odd function and $y = \cosh x$ is an even function:

$$\sinh(-x) = -\sinh x, \tag{8}$$

$$\cosh(-x) = \cosh x. \tag{9}$$

In trigonometry a fundamental identity is $\cos^2 x + \sin^2 x = 1$. For hyperbolic functions the analogue of this identity is

$$\cosh^2 x - \sinh^2 x = 1. \tag{10}$$

To prove (10) we resort to (2) and (3) of Definition 3.10.1:

$$\cosh^2 x - \sinh^2 x = \left(\frac{e^x + e^{-x}}{2}\right)^2 - \left(\frac{e^x - e^{-x}}{2}\right)^2$$
$$= \frac{e^{2x} + 2 + e^{-2x}}{4} - \frac{e^{2x} - 2 + e^{-2x}}{4} = 1.$$

We summarize (8)–(10) along with eleven other identities in the theorem that follows.

Theorem 3.10.1   Hyperbolic Identities

$\sinh(-x) = -\sinh x$	$\sinh(x + y) = \sinh x \cosh y + \cosh x \sinh y$ $\qquad(11)$
$\cosh(-x) = \cosh x$	$\sinh(x - y) = \sinh x \cosh y - \cosh x \sinh y$ $\qquad(12)$
$\tanh(-x) = -\tanh x$	$\cosh(x + y) = \cosh x \cosh y + \sinh x \sinh y$ $\qquad(13)$
$\cosh^2 x - \sinh^2 x = 1$	$\cosh(x - y) = \cosh x \cosh y - \sinh x \sinh y$ $\qquad(14)$
$1 - \tanh^2 x = \operatorname{sech}^2 x$	$\sinh 2x = 2 \sinh x \cosh x$ $\qquad(15)$
$\coth^2 x - 1 = \operatorname{csch}^2 x$	$\cosh 2x = \cosh^2 x + \sinh^2 x$ $\qquad(16)$
$\sinh^2 x = \frac{1}{2}(-1 + \cosh 2x)$	$\cosh^2 x = \frac{1}{2}(1 + \cosh 2x)$ $\qquad(17)$

■ **Derivatives of Hyperbolic Functions**   The derivatives of the hyperbolic functions follow from (14) of Section 3.8 and the rules of differentiation; for example

$$\frac{d}{dx}\sinh x = \frac{d}{dx}\frac{e^x - e^{-x}}{2} = \frac{1}{2}\left[\frac{d}{dx}e^x - \frac{d}{dx}e^{-x}\right] = \frac{e^x + e^{-x}}{2}.$$

That is,

$$\frac{d}{dx}\sinh x = \cosh x. \tag{18}$$

Similarly, it should be apparent from the definition of the hyperbolic cosine in (3) that

$$\frac{d}{dx}\cosh x = \sinh x. \tag{19}$$

To differentiate, say, the hyperbolic tangent, we use the Quotient Rule and the definition given in (4):

$$\frac{d}{dx}\tanh x = \frac{d}{dx}\frac{\sinh x}{\cosh x}$$

$$= \frac{\cosh x \cdot \dfrac{d}{dx}\sinh x - \sinh x \cdot \dfrac{d}{dx}\cosh x}{\cosh^2 x}$$

$$= \frac{\cosh^2 x - \sinh^2 x}{\cosh^2 x} \leftarrow \text{this is equal to 1 by (10)}$$

$$= \frac{1}{\cosh^2 x}.$$

In other words,

$$\frac{d}{dx}\tanh x = \text{sech}^2 x. \tag{20}$$

The derivatives of the six hyperbolic functions in the most general case follow from the Chain Rule.

Theorem 3.10.2   Derivatives of Hyperbolic Functions

If $u = g(x)$ is a differentiable function, then

$$\frac{d}{dx}\sinh u = \cosh u \,\frac{du}{dx}, \qquad\qquad \frac{d}{dx}\cosh u = \sinh u \,\frac{du}{dx}, \tag{21}$$

$$\frac{d}{dx}\tanh u = \text{sech}^2 u \,\frac{du}{dx}, \qquad\qquad \frac{d}{dx}\coth u = -\text{csch}^2 u \,\frac{du}{dx}, \tag{22}$$

$$\frac{d}{dx}\text{sech}\, u = -\text{sech}\, u \tanh u \,\frac{du}{dx}, \qquad\qquad \frac{d}{dx}\text{csch}\, u = -\text{csch}\, u \coth u \,\frac{du}{dx}. \tag{23}$$

You should take careful note of the slight difference in the results in (21)–(23) and the analogous formulas for the trigonometric functions:

$$\frac{d}{dx}\cos x = -\sin x \qquad \text{whereas} \qquad \frac{d}{dx}\cosh x = \sinh x$$

$$\frac{d}{dx}\sec x = \sec x \tan x \qquad \text{whereas} \qquad \frac{d}{dx}\text{sech}\, x = -\text{sech}\, x \tanh x.$$

EXAMPLE 1   Chain Rule

Differentiate

(a) $y = \sinh\sqrt{2x + 1}$ (b) $y = \coth x^3$.

Solution

(a) From the first result in (21),

$$\frac{dy}{dx} = \cosh\sqrt{2x + 1} \cdot \frac{d}{dx}(2x + 1)^{1/2}$$

$$= \cosh\sqrt{2x + 1}\left(\frac{1}{2}(2x + 1)^{-1/2} \cdot 2\right)$$

$$= \frac{\cosh\sqrt{2x + 1}}{\sqrt{2x + 1}}.$$

(b) From the second result in (22),

$$\frac{dy}{dx} = -\text{csch}^2 x^3 \cdot \frac{d}{dx} x^3$$

$$= -\text{csch}^2 x^3 \cdot 3x^2.$$

■

EXAMPLE 2   Value of a Derivative

Evaluate the derivative of $y = \dfrac{3x}{4 + \cosh 2x}$ at $x = 0$.

Solution   From the Quotient Rule,

$$\frac{dy}{dx} = \frac{(4 + \cosh 2x) \cdot 3 - 3x(\sinh 2x \cdot 2)}{(4 + \cosh 2x)^2}.$$

Because $\sinh 0 = 0$ and $\cosh 0 = 1$, we have

$$\left.\frac{dy}{dx}\right|_{x=0} = \frac{15}{25} = \frac{3}{5}.$$

■

■ **Inverse Hyperbolic Functions**   Inspection of Figure 3.10.1(a) shows that $y = \sinh x$ is a one-to-one function. That is, for any real number y in the range $(-\infty, \infty)$ of the hyperbolic sine, there corresponds only one real number x in its domain $(-\infty, \infty)$. Hence, $y = \sinh x$ has an inverse function that is written $y = \sinh^{-1} x$. See FIGURE 3.10.3(a). As in our earlier discussion of the inverse trigonometric functions in Section 1.5, this later notation is equivalent to $x = \sinh y$. From Figure 3.10.2(a) it is also seen that $y = \tanh x$ with domain $(-\infty, \infty)$ and range $(-1, 1)$ is also one-to-one and has an inverse $y = \tanh^{-1} x$ with domain $(-1, 1)$ and range $(-\infty, \infty)$. See Figure 3.10.3(c). But from Figures 3.10.1(b) and 3.10.2(c) it is apparent that $y = \cosh x$ and $y = \text{sech } x$ are not one-to-one functions and so do not possess inverse functions unless their domains are suitably restricted. Inspection of Figure 3.10.1(b) shows that when the domain of $y = \cosh x$ is restricted to the interval $[0, \infty)$, the corresponding range is $[1, \infty)$. The inverse function $y = \cosh^{-1} x$ then has domain $[1, \infty)$ and range $[0, \infty)$. See Figure 3.10.3(b). The graphs of all the inverse hyperbolic functions along with their domains and ranges are summarized in Figure 3.10.3.

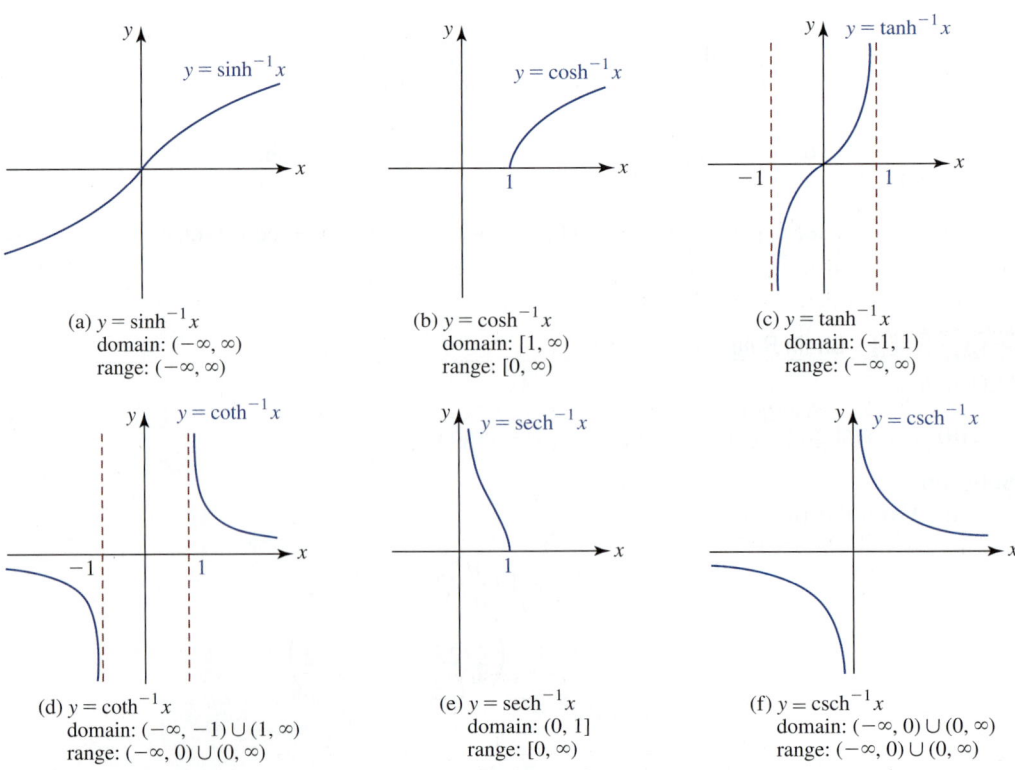

(a) $y = \sinh^{-1} x$
domain: $(-\infty, \infty)$
range: $(-\infty, \infty)$

(b) $y = \cosh^{-1} x$
domain: $[1, \infty)$
range: $[0, \infty)$

(c) $y = \tanh^{-1} x$
domain: $(-1, 1)$
range: $(-\infty, \infty)$

(d) $y = \coth^{-1} x$
domain: $(-\infty, -1) \cup (1, \infty)$
range: $(-\infty, 0) \cup (0, \infty)$

(e) $y = \text{sech}^{-1} x$
domain: $(0, 1]$
range: $[0, \infty)$

(f) $y = \text{csch}^{-1} x$
domain: $(-\infty, 0) \cup (0, \infty)$
range: $(-\infty, 0) \cup (0, \infty)$

FIGURE 3.10.3   Graphs of the inverses of the hyperbolic sine, cosine, tangent, cotangent, secant, and cosecant

■ Inverse Hyperbolic Functions as Logarithms Because all the hyperbolic functions are defined in terms of combinations of e^x, it should not come as any surprise to find that the inverse hyperbolic functions can be expressed in terms of the natural logarithm. For example, $y = \sinh^{-1}x$ is equivalent to $x = \sinh y$, so that

$$x = \frac{e^y - e^{-y}}{2} \quad \text{or} \quad 2x = \frac{e^{2y} - 1}{e^y} \quad \text{or} \quad e^{2y} - 2xe^y - 1 = 0.$$

Because the last equation is quadratic in e^y, the quadratic formula gives

$$e^y = \frac{2x \pm \sqrt{4x^2 + 4}}{2} = x \pm \sqrt{x^2 + 1}. \tag{24}$$

Now the solution corresponding to the minus sign in (24) must be rejected since $e^y > 0$ but $x - \sqrt{x^2 + 1} < 0$. Thus, we have

$$e^y = x + \sqrt{x^2 + 1} \quad \text{or} \quad y = \sinh^{-1}x = \ln(x + \sqrt{x^2 + 1}).$$

Similarly, for $y = \tanh^{-1}x$, $|x| < 1$,

$$x = \tanh y = \frac{e^y - e^{-y}}{e^y + e^{-y}}$$

gives

$$e^y(1 - x) = (1 + x)e^{-y}$$

$$e^{2y} = \frac{1 + x}{1 - x}$$

$$2y = \ln\left(\frac{1 + x}{1 - x}\right)$$

or

$$y = \tanh^{-1}x = \frac{1}{2}\ln\left(\frac{1 + x}{1 - x}\right).$$

We have proved two of the results in the next theorem.

Theorem 3.10.3 Logarithmic Identities

$$\sinh^{-1}x = \ln(x + \sqrt{x^2 + 1}) \qquad \cosh^{-1}x = \ln(x + \sqrt{x^2 - 1}), x \geq 1 \tag{25}$$

$$\tanh^{-1}x = \frac{1}{2}\ln\left(\frac{1 + x}{1 - x}\right), |x| < 1 \qquad \coth^{-1}x = \frac{1}{2}\ln\left(\frac{x + 1}{x - 1}\right), |x| > 1 \tag{26}$$

$$\text{sech}^{-1}x = \ln\left(\frac{1 + \sqrt{1 - x^2}}{x}\right), 0 < x \leq 1 \qquad \text{csch}^{-1}x = \ln\left(\frac{1}{x} + \frac{\sqrt{1 + x^2}}{|x|}\right), x \neq 0 \tag{27}$$

The foregoing identities are a convenient means for obtaining the numerical values of an inverse hyperbolic function. For example, with the aid of a calculator we see from the first result in (25) in Theorem 3.10.3 that when $x = 4$

$$\sinh^{-1}4 = \ln(4 + \sqrt{17}) \approx 2.0947.$$

■ Derivatives of Inverse Hyperbolic Functions To find the derivative of an inverse hyperbolic function, we can proceed in two different ways. For example, if

$$y = \sinh^{-1}x \quad \text{then} \quad x = \sinh y.$$

Using implicit differentiation, we can write

$$\frac{d}{dx}x = \frac{d}{dx}\sinh y$$

$$1 = \cosh y \frac{dy}{dx}.$$

Hence

$$\frac{dy}{dx} = \frac{1}{\cosh y} = \frac{1}{\sqrt{\sinh^2 y + 1}} = \frac{1}{\sqrt{x^2 + 1}}.$$

The foregoing result can be obtained in an alternative manner. We know from Theorem 3.10.3 that

$$y = \ln\left(x + \sqrt{x^2 + 1}\right).$$

Therefore, from the derivative of the logarithm, we obtain

$$\frac{dy}{dx} = \frac{1}{x + \sqrt{x^2 + 1}}\left(1 + \frac{1}{2}(x^2 + 1)^{-1/2} \cdot 2x\right) \quad \leftarrow \text{by (3) of Section 3.9}$$

$$= \frac{1}{x + \sqrt{x^2 + 1}} \cdot \frac{\sqrt{x^2 + 1} + x}{\sqrt{x^2 + 1}} = \frac{1}{\sqrt{x^2 + 1}}.$$

We have essentially proved the first entry in (28) in the next theorem.

Theorem 3.10.4 Derivatives of Inverse Hyperbolic Functions

If $u = g(x)$ is a differentiable function, then

$$\frac{d}{dx}\sinh^{-1}u = \frac{1}{\sqrt{u^2 + 1}}\frac{du}{dx}, \qquad \frac{d}{dx}\cosh^{-1}u = \frac{1}{\sqrt{u^2 - 1}}\frac{du}{dx}, u > 1, \quad (28)$$

$$\frac{d}{dx}\tanh^{-1}u = \frac{1}{1 - u^2}\frac{du}{dx}, \quad |u| < 1, \qquad \frac{d}{dx}\coth^{-1}u = \frac{1}{1 - u^2}\frac{du}{dx}, |u| > 1, \quad (29)$$

$$\frac{d}{dx}\operatorname{sech}^{-1}u = \frac{-1}{u\sqrt{1 - u^2}}\frac{du}{dx}, 0 < u < 1, \qquad \frac{d}{dx}\operatorname{csch}^{-1}u = \frac{-1}{|u|\sqrt{1 + u^2}}\frac{du}{dx}, u \neq 0. \quad (30)$$

EXAMPLE 3 Derivative of Inverse Hyperbolic Cosine

Differentiate $y = \cosh^{-1}(x^2 + 5)$.

Solution With $u = x^2 + 5$, we have from the second formula in (28),

$$\frac{dy}{dx} = \frac{1}{\sqrt{(x^2 + 5)^2 - 1}} \cdot \frac{d}{dx}(x^2 + 5) = \frac{2x}{\sqrt{x^4 + 10x^2 + 24}}. \qquad \blacksquare$$

EXAMPLE 4 Derivative of Inverse Hyperbolic Tangent

Differentiate $y = \tanh^{-1}4x$.

Solution With $u = 4x$, we have from the first formula in (29),

$$\frac{dy}{dx} = \frac{1}{1 - (4x)^2} \cdot \frac{d}{dx}4x = \frac{4}{1 - 16x^2}. \qquad \blacksquare$$

EXAMPLE 5 Product and Chain Rules

Differentiate $y = e^{x^2}\operatorname{sech}^{-1}x$.

Solution By the Product Rule and the first formula in (30), we have

$$\overset{\text{by first formula in (30)}}{\underset{\downarrow}{}} \qquad \overset{\text{by (14) of Section 3.8}}{\underset{\downarrow}{}}$$

$$\frac{dy}{dx} = e^{x^2}\left(\frac{-1}{x\sqrt{1 - x^2}}\right) + 2xe^{x^2}\operatorname{sech}^{-1}x$$

$$= -\frac{e^{x^2}}{x\sqrt{1 - x^2}} + 2xe^{x^2}\operatorname{sech}^{-1}x. \qquad \blacksquare$$

$\dfrac{d}{dx}$ **NOTES FROM THE CLASSROOM**

(a) hanging wires

(*i*) As mentioned in the introduction to this section, the graph of any function of the form $f(x) = k\cosh cx$, k and c constants, is called a **catenary**. The shape assumed by a flexible wire or heavy rope strung between two posts has the basic shape of a graph of a hyperbolic cosine. Furthermore, if two circular rings are held vertically and are not too far apart, then a soap film stretched between the rings will assume a surface having minimum area. The surface is a portion of a **catenoid**, which is the surface obtained by revolving a catenary about the *x*-axis. See FIGURE 3.10.4.

(*ii*) The similarity between trigonometric and hyperbolic functions extends beyond the derivative formulas and basic identities. If *t* is an angle measured in radians whose terminal side is *OP*, then the coordinates of *P* on a unit *circle* $x^2 + y^2 = 1$ are $(\cos t, \sin t)$. Now, the area of the shaded circular sector shown in FIGURE 3.10.5(a) is $A = \frac{1}{2}t$ and so $t = 2A$. In this manner, the *circular functions* $\cos t$ and $\sin t$ can be considered functions of the area *A*.

(b) soap film

FIGURE 3.10.4 Catenary in (a); catenoid in (b)

You might already know that the graph of the equation $x^2 - y^2 = 1$ is called a *hyperbola*. Because $\cosh t \geq 1$ and $\cosh^2 t - \sinh^2 t = 1$, it follows that the coordinates of a point *P* on the right-hand branch of the hyperbola are $(\cosh t, \sinh t)$. Furthermore, it can be shown that the area of the hyperbolic sector in Figure 3.10.5(b) is related to the number *t* by $t = 2A$. Whence we see the origin of the name *hyperbolic function*.

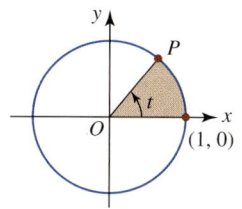

(a) circular sector

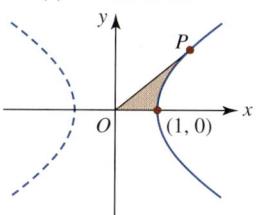

(b) hyperbolic sector

FIGURE 3.10.5 Circle in (a); hyperbola in (b)

Exercises 3.10 Answers to selected odd-numbered problems begin on page ANS-12.

☰ Fundamentals

1. If $\sinh x = -\frac{1}{2}$, find the values of the remaining hyperbolic functions.

2. If $\cosh x = 3$, find the values of the remaining hyperbolic functions.

In Problems 3–26, find the derivative of the given function.

3. $y = \cosh 10x$

4. $y = \operatorname{sech} 8x$

5. $y = \tanh \sqrt{x}$

6. $y = \operatorname{csch} \dfrac{1}{x}$

7. $y = \operatorname{sech}(3x - 1)^2$

8. $y = \sinh e^{x^2}$

9. $y = \coth(\cosh 3x)$

10. $y = \tanh(\sinh x^3)$

11. $y = \sinh 2x \cosh 3x$

12. $y = \operatorname{sech} x \coth 4x$

13. $y = x \cosh x^2$

14. $y = \dfrac{\sinh x}{x}$

15. $y = \sinh^3 x$

16. $y = \cosh^4 \sqrt{x}$

17. $f(x) = (x - \cosh x)^{2/3}$

18. $f(x) = \sqrt{4 + \tanh 6x}$

19. $f(x) = \ln(\cosh 4x)$

20. $f(x) = (\ln(\operatorname{sech} x))^2$

21. $f(x) = \dfrac{e^x}{1 + \cosh x}$

22. $f(x) = \dfrac{\ln x}{x^2 + \sinh x}$

23. $F(t) = e^{\sinh t}$

24. $H(t) = e^t e^{\operatorname{csch} t^2}$

25. $g(t) = \dfrac{\sin t}{1 + \sinh 2t}$

26. $w(t) = \dfrac{\tanh t}{(1 + \cosh t)^2}$

27. Find an equation of the tangent line to the graph of $y = \sinh 3x$ at $x = 0$.

28. Find an equation of the tangent line to the graph of $y = \cosh x$ at $x = 1$.

In Problems 29 and 30, find the point(s) on the graph of the given function at which the tangent is horizontal.

29. $f(x) = (x^2 - 2)\cosh x - 2x \sinh x$

30. $f(x) = \cos x \cosh x - \sin x \sinh x$

In Problems 31 and 32, find d^2y/dx^2 for the given function.

31. $y = \tanh x$

32. $y = \operatorname{sech} x$

In Problems 33 and 34, C_1, C_2, C_3, C_4 and k are arbitrary real constants. Show that the function satisfies the given differential equation.

33. $y = C_1 \cosh kx + C_2 \sinh kx$; $y'' - k^2 y = 0$

34. $y = C_1 \cos kx + C_2 \sin kx + C_3 \cosh kx + C_4 \sinh kx$; $y^{(4)} - k^4 y = 0$

In Problems 35–48, find the derivative of the given function.

35. $y = \sinh^{-1} 3x$

36. $y = \cosh^{-1} \dfrac{x}{2}$

37. $y = \tanh^{-1}(1 - x^2)$

38. $y = \coth^{-1} \dfrac{1}{x}$

39. $y = \coth^{-1}(\csc x)$

40. $y = \sinh^{-1}(\sin x)$

41. $y = x\sinh^{-1} x^3$

42. $y = x^2 \operatorname{csch}^{-1} x$

43. $y = \dfrac{\operatorname{sech}^{-1} x}{x}$

44. $y = \dfrac{\coth^{-1} e^{2x}}{e^{2x}}$

45. $y = \ln(\operatorname{sech}^{-1} x)$

46. $y = x\tanh^{-1} x + \ln\sqrt{1 - x^2}$

47. $y = (\cosh^{-1} 6x)^{1/2}$

48. $y = \dfrac{1}{(\tanh^{-1} 2x)^3}$

≡ Applications

49. (a) Assume that $k, m,$ and g are real constants. Show that the function

$$v(t) = \sqrt{\frac{mg}{k}} \tanh\left(\sqrt{\frac{kg}{m}}\, t\right)$$

satisfies the differential equation $m\dfrac{dv}{dt} = mg - kv^2.$

(b) The function v represents the velocity of a falling mass m when air resistance is taken to be proportional to the square of the instantaneous velocity. Find the limiting or **terminal velocity** $v_{\text{ter}} = \lim\limits_{t\to\infty} v(t)$ of the mass.

(c) Suppose a 80-kg skydiver delays opening the parachute until terminal velocity is attained. Determine the terminal velocity if it is known that $k = 0.25$ kg/m.

50. A woman, W, starting at the origin, moves in the direction of the positive x-axis pulling a boat along the curve C, called a **tractrix**, indicated in FIGURE 3.10.6. The boat, initially located on the y-axis at $(0, a)$, is pulled by a rope

of constant length a that is kept taut throughout the motion. An equation of the tractrix is given by

$$x = a\ln\left(\frac{a + \sqrt{a^2 - y^2}}{y}\right) - \sqrt{a^2 - y^2}.$$

(a) Rewrite this equation using a hyperbolic function.

(b) Use implicit differentiation to show that the equation of the tractrix satisfies the differential equation

$$\frac{dy}{dx} = -\frac{y}{\sqrt{a^2 - y^2}}.$$

(c) Interpret geometrically the differential equation in part (b).

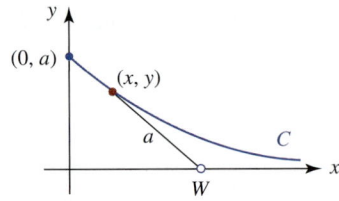

FIGURE 3.10.6 Tractrix in Problem 50

≡ Think About It

In Problems 51 and 52, find the exact numerical value of the given quantity.

51. $\cosh(\ln 4)$

52. $\sinh(\ln 0.5)$

In Problems 53 and 54, express the given quantity as a rational function of x.

53. $\sinh(\ln x)$

54. $\tanh(3\ln x)$

55. Show that for any positive integer n,

$$(\cosh x + \sinh x)^n = \cosh nx + \sinh nx.$$

Chapter 3 in Review

Answers to selected odd-numbered problems begin on page ANS-13.

A. True/False

In Problems 1–20, indicate whether the given statement is true or false.

1. If $y = f(x)$ is continuous at a number a, then there is a tangent line to the graph of f at $(a, f(a))$. _____

2. If f is differentiable at every real number x, then f is continuous everywhere. _____

3. If $y = f(x)$ has a tangent line at $(a, f(a))$, then f is necessarily differentiable at $x = a$. _____

4. The instantaneous rate of change of $y = f(x)$ with respect to x at x_0 is the slope of the tangent line to the graph at $(x_0, f(x_0))$. _____

5. At $x = -1$, the tangent line to the graph of $f(x) = x^3 - 3x^2 - 9x$ is parallel to the line $y = 2$. _____

6. The derivative of a product is the product of the derivatives. _____

7. A polynomial function has a tangent line at every point on its graph. _____

8. For $f(x) = -x^2 + 5x + 1$ an equation of the tangent line is $f'(x) = -2x + 5$.____

9. The function $f(x) = x/(x^2 + 9)$ is differentiable on the interval $[-3, 3]$.____

10. If $f'(x) = g'(x)$, then $f(x) = g(x)$. ____

11. If m is the slope of a tangent line to the graph of $f(x) = \sin x$, then $-1 \leq m \leq 1$. ____

12. For $y = \tan^{-1}x$, $dy/dx > 0$ for all x. ____

13. $\dfrac{d}{dx}\cos^{-1}x = -\sin^{-1}x$ ____

14. The function $f(x) = x^5 + x^3 + x$ has an inverse. ____

15. If $f'(x) < 0$ on the interval $[2, 8]$, then $f(3) > f(5)$. ____

16. If f is an increasing differentiable function on an interval, then $f'(x)$ is also increasing on the interval. ____

17. The only function for which $f'(x) = f(x)$ is $f(x) = e^x$. ____

18. $\dfrac{d}{dx}\ln|x| = \dfrac{1}{|x|}$ ____

19. $\dfrac{d}{dx}\cosh^2 x = \dfrac{d}{dx}\sinh^2 x$ ____

20. Every inverse hyperbolic function is a logarithm. ____

B. Fill in the Blanks

In Problems 1–20, fill in the blanks.

1. If $y = f(x)$ is a polynomial function of degree 3, then $\dfrac{d^4}{dx^4}f(x) = $ _____.

2. The slope of the tangent line to the graph of $y = \ln|x|$ at $x = -\dfrac{1}{2}$ is _____.

3. The slope of the normal line to the graph of $f(x) = \tan x$ at $x = \pi/3$ is _____.

4. $f(x) = \dfrac{x^{n+1}}{n+1}$, $n \neq -1$, then $f'(x) = $ _____.

5. An equation of the tangent line to the graph of $y = (x + 3)/(x - 2)$ at $x = 0$ is _____.

6. For $f(x) = 1/(1 - 3x)$ the instantaneous rate of change of f' with respect to x at $x = 0$ is _____.

7. If $f'(4) = 6$ and $g'(4) = 3$, then the slope of the tangent line to the graph of $y = 2f(x) - 5g(x)$ at $x = 4$ is _____.

8. If $f(2) = 1$, $f'(2) = 5$, $g(2) = 2$, and $g'(2) = -3$, then $\dfrac{d}{dx}\dfrac{x^2 f(x)}{g(x)}\bigg|_{x=2} = $ _____.

9. If $g(1) = 2$, $g'(1) = 3$, $g''(1) = -1$, $f'(2) = 4$, and $f''(2) = 3$, then $\dfrac{d^2}{dx^2}f(g(x))\bigg|_{x=1} = $ _____.

10. If $f'(x) = x^2$, then $\dfrac{d}{dx}f(x^3) = $ _____.

11. If F is a differentiable function, then $\dfrac{d^2}{dx^2}F(\sin 4x) = $ _____.

12. The function $f(x) = \cot x$ is not differentiable on the interval $[0, \pi]$ because _____.

13. The function

$$f(x) = \begin{cases} ax + b, & x \leq 3 \\ x^2, & x > 3 \end{cases}$$

is differentiable at $x = 3$ when $a = $ _____ and $b = $ _____.

14. If $f'(x) = \sec^2 2x$, then $f(x) = $ _____.

15. The tangent line to the graph of $f(x) = 5 - x + e^{x-1}$ is horizontal at the point _____.

16. $\dfrac{d}{dx} 2^x = $ _____.

17. $\dfrac{d}{dx} \log_{10} x = $ _____.

18. If $f(x) = \ln|2x - 4|$, the domain of $f'(x)$ is _____.

19. The graph of $y = \cosh x$ is called a _____.

20. $\cosh^{-1} 1 = $ _____.

C. Exercises

In Problems 1–28, find the derivative of the given function.

1. $f(x) = \dfrac{4x^{0.3}}{5x^{0.2}}$

2. $y = \dfrac{1}{x^3 + 4x^2 - 6x + 11}$

3. $F(t) = \left(t + \sqrt{t^2 + 1}\right)^{10}$

4. $h(\theta) = \theta^{1.5}(\theta^2 + 1)^{0.5}$

5. $y = \sqrt[4]{x^4 + 16}\,\sqrt[3]{x^3 + 8}$

6. $g(u) = \sqrt{\dfrac{6u - 1}{u + 7}}$

7. $y = \dfrac{\cos 4x}{4x + 1}$

8. $y = 10\cot 8x$

9. $f(x) = x^3 \sin^2 5x$

10. $y = \tan^2(\cos 2x)$

11. $y = \sin^{-1}\dfrac{3}{x}$

12. $y = \cos x \cos^{-1} x$

13. $y = (\cot^{-1} x)^{-1}$

14. $y = \text{arc}\sec(2x - 1)$

15. $y = 2\cos^{-1} x + 2x\sqrt{1 - x^2}$

16. $y = x^2 \tan^{-1}\sqrt{x^2 - 1}$

17. $y = xe^{-x} + e^{-x}$

18. $y = (e + e^2)^x$

19. $y = x^7 + 7^x + 7^\pi + e^{7x}$

20. $y = (e^x + 1)^{-e}$

21. $y = \ln\left(x\sqrt{4x - 1}\right)$

22. $y = (\ln\cos^2 x)^2$

23. $y = \sinh^{-1}(\sin^{-1} x)$

24. $y = (\tan^{-1} x)(\tanh^{-1} x)$

25. $y = xe^{x\cosh^{-1} x}$

26. $y = \sinh^{-1}\sqrt{x^2 - 1}$

27. $y = \sinh e^{x^3}$

28. $y = (\tanh 5x)^{-1}$

In Problems 29–34, find the indicated derivative.

29. $y = (3x + 1)^{5/2}$; $\dfrac{d^3 y}{dx^3}$

30. $y = \sin(x^3 - 2x)$; $\dfrac{d^2 y}{dx^2}$

31. $s = t^2 + \dfrac{1}{t^2}$; $\dfrac{d^4 s}{dt^4}$

32. $W = \dfrac{v - 1}{v + 1}$; $\dfrac{d^3 W}{dv^3}$

33. $y = e^{\sin 2x}$; $\dfrac{d^2 y}{dx^2}$

34. $f(x) = x^2 \ln x$; $f'''(x)$

35. First use the laws of logarithms to simplify

$$y = \ln\left|\dfrac{(x + 5)^4(2 - x)^3}{(x + 8)^{10}\sqrt[3]{6x + 4}}\right|,$$

and then find dy/dx.

36. Find dy/dx for $y = 5^{x^2} x^{\sin 2x}$.

37. Given that $y = x^3 + x$ is a one-to-one function, find the slope of the tangent line to the graph of the inverse function at $x = 1$.

38. Given that $f(x) = 8/(1 - x^3)$ is a one-to-one function, find f^{-1} and $(f^{-1})'$.

In Problems 39 and 40, find dy/dx.

39. $xy^2 = e^x - e^y$ **40.** $y = \ln(xy)$

41. Find an equation of a tangent line to the graph of $f(x) = x^3$ that is perpendicular to the line $y = -3x$.

42. Find the point(s) on the graph of $f(x) = \frac{1}{2}x^2 - 5x + 1$ at which
    (a) $f''(x) = f(x)$ and   **(b)** $f''(x) = f'(x)$.

43. Find equations for the lines through $(0, -9)$ that are tangent to the graph of $y = x^2$.

44. (a) Find the x-intercept of the tangent line to the graph of $y = x^2$ at $x = 1$.
    (b) Find an equation of the line with the same x-intercept that is perpendicular to the tangent line in part (a).
    (c) Find the point(s) where the line in part (b) intersects the graph of $y = x^2$.

45. Find the point on the graph of $f(x) = \sqrt{x}$ at which the tangent line is parallel to the secant line through $(1, f(1))$ and $(9, f(9))$.

46. If $f(x) = (1 + x)/x$, what is the slope of the tangent line to the graph of f'' at $x = 2$?

47. Find the x-coordinates of all points on the graph of $f(x) = 2\cos x + \cos 2x$, $0 \leq x \leq 2\pi$, at which the tangent line is horizontal.

48. Find the point on the graph of $y = \ln 2x$ such that the tangent line passes through the origin.

49. Suppose a series circuit contains a capacitor and a variable resistor. If the resistance at time t is given by $R = k_1 + k_2 t$, where k_1 and k_2 are positive known constants, then the charge $q(t)$ on the capacitor is given by

$$q(t) = E_0 C + (q_0 - E_0 C)\left(\frac{k_1}{k_1 + k_2 t}\right)^{1/Ck_2},$$

where C is a constant called the **capacitance** and $E(t) = E_0$ is the impressed voltage. Show that $q(t)$ satisfies the initial condition $q(0) = q_0$ and the differential equation

$$(k_1 + k_2 t)\frac{dq}{dt} + \frac{1}{C}q = E_0.$$

50. Assume that C_1 and C_2 are arbitrary real constants. Show that the function

$$y = C_1 x + C_2\left[\frac{x}{2}\ln\left(\frac{x-1}{x+1}\right) - 1\right]$$

satisfies the differential equation

$$(1 - x^2)y'' - 2xy' + 2y = 0.$$

In Problems 51 and 52, C_1, C_2, C_3, and C_4 are arbitrary real constants. Show that the function satisfies the given differential equation.

51. $y = C_1 e^{-x} + C_2 e^x + C_3 xe^{-x} + C_4 xe^x$; $y^{(4)} - 2y'' + y = 0$

52. $y = C_1\cos x + C_2\sin x + C_3 x\cos x + C_4 x\sin x$; $y^{(4)} + 2y'' + y = 0$

53. (a) Find the points on the graph of $y^3 - y + x^2 - 4 = 0$ corresponding to $x = 2$.
    (b) Find the slopes of the tangent lines at the points found in part (a).

54. Sketch the graph of f' from the graph of f given in **FIGURE 3.R.1**.

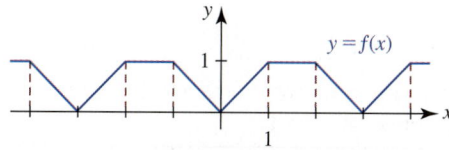

FIGURE 3.R.1   Graph for Problem 54

55. The graph of $x^{2/3} + y^{2/3} = 1$, shown in FIGURE 3.R.2, is called a **hypocycloid**.[*] Find equations of the tangent lines to the graph at the points corresponding to $x = \frac{1}{8}$.

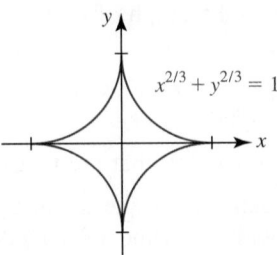

FIGURE 3.R.2   Hypocycloid in Problem 55

56. Find d^2y/dx^2 for the equation in Problem 55.

57. Suppose

$$f(x) = \begin{cases} x^2, & x \leq 0 \\ \sqrt{x}, & x > 0. \end{cases}$$

Find $f'(x)$ for $x \neq 0$. Use the definition of the derivative, (2) of Section 3.1, to determine whether $f'(0)$ exists.

[*]Go to the website **http://mathworld.wolfram.com/Hypocycloid.html** to see various kinds of hypocycloids and their properties.

Chapter 4

Applications of the Derivative

In This Chapter   The first and second derivatives of a function *f* can be used to determine the shape of its graph. If you imagine a graph of a function as a curve that rises and falls, then the high points and low points of the graph, or more precisely, the maximum and minimum values of the function, can be found using the derivative. As we have already seen, the derivative also gives a rate of change. We saw briefly in Section 2.7 that the rate of change with respect to time *t* of a function that gives the position of a moving object is the velocity of the object.

Finding the maximum and minimum values of a function along with the problem of finding rates of change are two of the central topics of study in this chapter.

4.1 Rectilinear Motion

■ **Introduction** In Section 2.7, the motion of an object in a straight line, either horizontally or vertically, was said to be **rectilinear motion**. A function $s = s(t)$ that gives the coordinate of the object on a horizontal or vertical line is called a **position function**. The variable t represents time and the function value $s(t)$ represents a directed distance, which is measured in centimeters, meters, feet, miles, and so on, from a reference point $s = 0$ on the line. Recall that on a horizontal scale, we take the positive s-direction to be to the right of $s = 0$, and on a vertical scale we take the positive s-direction to be upward.

EXAMPLE 1 Position of a Moving Particle

A particle moves on a horizontal line according to the position function $s(t) = -t^2 + 4t + 3$, where s is measured in centimeters and t in seconds. What is the position of the particle at 0, 2, and 6 seconds?

Solution Substitution into the position function gives

$$s(0) = 3, \quad s(2) = 7, \quad s(6) = -9.$$

As shown in FIGURE 4.1.1, $s(6) = -9 < 0$ means that the position of the particle is to the left of the reference point $s = 0$.

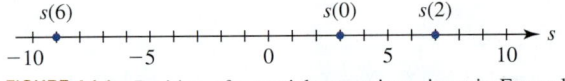

FIGURE 4.1.1 Position of a particle at various times in Example 1

■ **Velocity and Acceleration** If the **average velocity** of a body in motion over a time interval of length Δt is

$$\frac{\text{change in position}}{\text{change in time}} = \frac{s(t + \Delta t) - s(t)}{\Delta t},$$

then the instantaneous rate of change, or velocity of the body, is given by

$$v(t) = \lim_{\Delta t \to 0} \frac{s(t + \Delta t) - s(t)}{\Delta t}.$$

Thus, we have the following definition.

Definition 4.1.1 Velocity Function

If $s(t)$ is a position function of an object that moves rectilinearly, then its **velocity function** $v(t)$ at time t is

$$v(t) = \frac{ds}{dt}.$$

The **speed** of the object at time t is $|v(t)|$.

Velocity is measured in centimeters per second (cm/s), meters per second (m/s), feet per second (ft/s), kilometers per hour (km/h), miles per hour (mi/h), and so on.

We can also compute the rate of change of velocity.

Definition 4.1.2 Acceleration Function

If $v(t)$ is the velocity function of an object that moves rectilinearly, then its **acceleration function** $a(t)$ at time t is

$$a(t) = \frac{dv}{dt} = \frac{d^2s}{dt^2}.$$

Typical units for measuring acceleration are meters per second per second (m/s^2), feet per second per second (ft/s^2), miles per hour per hour (mi/h^2), and so on. Often we read units of acceleration literally as "meters per second squared."

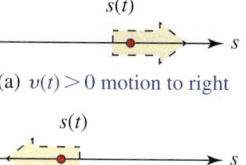

(a) $v(t) > 0$ motion to right

(b) $v(t) < 0$ motion to left
FIGURE 4.1.2   Significance of the sign of velocity function

■ **Significance of Algebraic Signs**   In Section 3.7 we saw that whenever the derivative of a function f is *positive* on an interval I, then f is *increasing* on I. Geometrically, the graph of an increasing function rises as x increases. Similarly, if the derivative of a function f is *negative* on I, then f is *decreasing,* which means its graph goes down as x increases. On a time interval for which $v(t) = s'(t) > 0$, we can say $s(t)$ is increasing. Thus the object is moving to the *right* on a horizontal line or moving *upward* on a vertical line. On the other hand, $v(t) = s'(t) < 0$ implies that $s(t)$ is decreasing and motion is to the *left* on a horizontal line or motion *downward* on a vertical line. See FIGURE 4.1.2. If $a(t) = v'(t) > 0$ on a time interval, then the velocity $v(t)$ of the object is *increasing,* whereas $a(t) = v'(t) < 0$ indicates that the velocity $v(t)$ of the object is *decreasing.* For example, an acceleration of -25 m/s^2 means that the velocity is decreasing by 25 m/s every second. Do not confuse the terms "velocity decreasing" and "velocity increasing" with the concepts of "slowing down" or "speeding up." For example, consider a stone that is dropped from the top of a tall building. The acceleration of gravity is a negative constant, -32 ft/s^2. The negative sign means that the velocity of the stone decreases starting from zero. When the stone hits the ground, its speed $|v(t)|$ is fairly large, but $v(t) < 0$. Specifically, an object that moves rectilinearly on, say, a horizontal line is *slowing down* when $v(t) > 0$ (motion to right) and $a(t) < 0$ (velocity decreasing), or when $v(t) < 0$ (motion to left) and $a(t) > 0$ (velocity increasing). Similarly, an object that moves rectilinearly on a horizontal line is *speeding up* when $v(t) > 0$ (motion to right) and $a(t) > 0$ (velocity increasing) or when $v(t) < 0$ (motion to left) and $a(t) < 0$ (velocity decreasing). In general:

An object that moves rectilinearly

- is **slowing down** *when its velocity and acceleration have opposite algebraic signs, and*
- is **speeding up** *when its velocity and acceleration have the same algebraic sign.*

Alternatively, an object is slowing down when its speed $|v(t)|$ is decreasing and speeding up when its speed is increasing.

EXAMPLE 2   Example 1 Revisited

In Example 1 the velocity and acceleration functions for the particle are, respectively,

$$v(t) = \frac{ds}{dt} = -2t + 4 \qquad \text{and} \qquad a(t) = \frac{dv}{dt} = -2.$$

At times 0, 2, and 6 s, the velocities are $v(0) = 4$ cm/s, $v(2) = 0$ cm/s, and $v(6) = -8$ cm/s, respectively. Since the acceleration is always negative, the velocity is always decreasing. Notice that $v(t) = 2(-t + 2) > 0$ for $t < 2$ and $v(t) = 2(-t + 2) < 0$ for $t > 2$. If the time t is allowed to be negative as well as positive, then the particle moves to the right for the time interval $(-\infty, 2)$ and moves to the left for the time interval $(2, \infty)$. The motion can be represented by the graph given in FIGURE 4.1.3(a). Since the motion actually takes place *on* the horizontal line, you should envision the movement of a point P that corresponds to the projection of a point on the graph onto the horizontal line. See Figure 4.1.3(b).

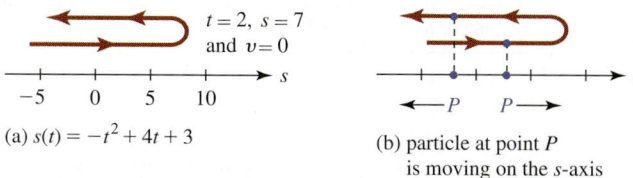

(a) $s(t) = -t^2 + 4t + 3$

(b) particle at point P is moving on the s-axis

FIGURE 4.1.3   Representation of the motion of the particle in Example 2

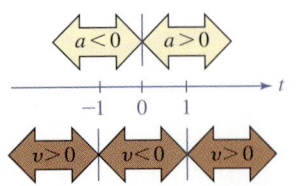

FIGURE 4.1.4 Signs of $v(t)$ and $a(t)$ in Example 3

EXAMPLE 3 Particle Slowing Down/Speeding Up

A particle moves on a horizontal line according to the position function $s(t) = \frac{1}{3}t^3 - t$. Determine the time intervals on which the particle is slowing down and the time intervals on which it is speeding up.

Solution The algebraic signs of the velocity and acceleration functions

$$v(t) = t^2 - 1 = (t + 1)(t - 1) \quad \text{and} \quad a(t) = 2t$$

are shown on the time scale in FIGURE 4.1.4. Since $v(t)$ and $a(t)$ have opposite signs on $(-\infty, -1)$ and $(0, 1)$, the particle is slowing down on these time intervals; $v(t)$ and $a(t)$ have the same algebraic sign on $(-1, 0)$ and $(1, \infty)$, so the particle is speeding up on these time intervals. ∎

In Example 2, you should verify that the particle is slowing down on the time interval $(-\infty, 2)$ and speeding up on the time interval $(2, \infty)$.

EXAMPLE 4 Motion of a Particle

An object moves on a horizontal line according to the position function $s(t) = t^4 - 18t^2 + 25$, where s is measured in centimeters and t in seconds. Use a graph to represent the motion during the time interval $[-4, 4]$.

Solution The velocity function is

$$v(t) = \frac{ds}{dt} = 4t^3 - 36t = 4t(t + 3)(t - 3)$$

and the acceleration function is

$$a(t) = \frac{d^2s}{dt^2} = 12t^2 - 36 = 12(t + \sqrt{3})(t - \sqrt{3}b).$$

Now, from the solutions of $v(t) = 0$, we can determine the time intervals for which $s(t)$ is increasing or decreasing. From the information given in the accompanying tables, we construct the graph shown in FIGURE 4.1.5. Inspection of the tables shows that the particle slows down on the time intervals $(-4, -3), (-\sqrt{3}, 0), (\sqrt{3}, 3)$ (shown in green in the figure) and speeds up on the time intervals $(-3, -\sqrt{3}), (0, \sqrt{3}), (3, 4)$ (shown in red in the figure).

Time Interval	Sign of $v(t)$	Direction of Motion
$(-4, -3)$	$-$	left
$(-3, 0)$	$+$	right
$(0, 3)$	$-$	left
$(3, 4)$	$+$	right

Time	Position	Velocity	Acceleration
-4	-7	-112	156
-3	-56	0	72
0	25	0	-36
3	-56	0	72
4	-7	112	156

Time Interval	Sign of $a(t)$	Velocity
$(-4, -\sqrt{3})$	$+$	increasing
$(-\sqrt{3}, \sqrt{3})$	$-$	decreasing
$(\sqrt{3}, 4)$	$+$	increasing

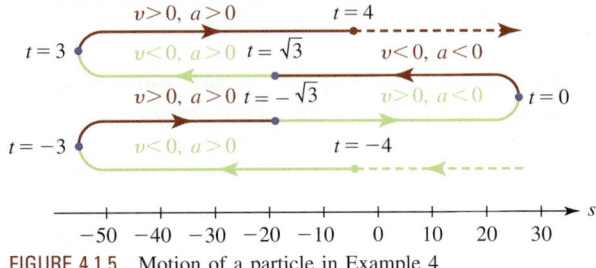

FIGURE 4.1.5 Motion of a particle in Example 4

Exercises 4.1 Answers to selected odd-numbered problems begin on page ANS-13.

☰ Fundamentals

In Problems 1–8, $s(t)$ is a position function of a particle that moves on a horizontal line. Find the position, velocity, speed, and acceleration of the particle at the indicated times.

1. $s(t) = 4t^2 - 6t + 1;$ $t = \dfrac{1}{2}, t = 3$

2. $s(t) = (2t - 6)^2;$ $t = 1, t = 4$

3. $s(t) = -t^3 + 3t^2 + t;$ $t = -2, t = 2$

4. $s(t) = t^4 - t^3 + t;$ $t = -1, t = 3$

5. $s(t) = t - \dfrac{1}{t};$ $t = \dfrac{1}{4}, t = 1$

6. $s(t) = \dfrac{t}{t + 2};$ $t = -1, t = 0$

7. $s(t) = t + \sin \pi t;$ $t = 1, t = \dfrac{3}{2}$

8. $s(t) = t \cos \pi t;$ $t = \dfrac{1}{2}, t = 1$

In Problems 9–12, $s(t)$ is a position function of a particle that moves on a horizontal line.

9. $s(t) = t^2 - 4t - 5$

   (a) What is the velocity of the particle when $s(t) = 0$?
   (b) What is the velocity of the particle when $s(t) = 7$?

10. $s(t) = t^2 + 6t + 10$

   (a) What is the position of the particle when $s(t) = v(t)$?
   (b) What is the velocity of the particle when $v(t) = -a(t)$?

11. $s(t) = t^3 - 4t$

   (a) What is the acceleration of the particle when $v(t) = 2$?
   (b) What is the position of the particle when $a(t) = 18$?
   (c) What is the velocity of the particle when $s(t) = 0$?

12. $s(t) = t^3 - 3t^2 + 8$

   (a) What is the position of the particle when $v(t) = 0$?
   (b) What is the position of the particle when $a(t) = 0$?
   (c) When is the particle slowing down? Speeding up?

In Problems 13 and 14, $s(t)$ is a position function of a particle that moves on a horizontal line. Determine the time intervals on which the particle is slowing down and the intervals on which it is speeding up.

13. $s(t) = t^3 - 27t$ 14. $s(t) = t^4 - t^3$

In Problems 15–20, $s(t)$ is a position function of a particle that moves on a horizontal line. Find the velocity and acceleration functions. Determine the time intervals on which the particle is slowing down and the intervals on which it is speeding up. Represent the motion during the indicated time interval with a graph.

15. $s(t) = t^2;$ $[-1, 3]$

16. $s(t) = t^3;$ $[-2, 2]$

17. $s(t) = t^2 - 4t - 2;$ $[-1, 5]$

18. $s(t) = (t + 3)(t - 1);$ $[-3, 1]$

19. $s(t) = 2t^3 - 6t^2;$ $[-2, 3]$

20. $s(t) = (t - 1)^2(t - 2);$ $[-2, 3]$

In Problems 21–28, $s(t)$ is a position function of a particle that moves on a horizontal line. Find the velocity and acceleration functions. Represent the motion during the indicated time interval with a graph.

21. $s(t) = 3t^4 - 8t^3;$ $[-1, 3]$

22. $s(t) = t^4 - 4t^3 - 8t^2 + 60;$ $[-2, 5]$

23. $s(t) = t - 4\sqrt{t};$ $[1, 9]$

24. $s(t) = 1 + \cos \pi t;$ $\left[-\frac{1}{2}, \frac{5}{2}\right]$

25. $s(t) = \sin \dfrac{\pi}{2} t;$ $[0, 4]$

26. $s(t) = \sin \pi t - \cos \pi t;$ $[0, 2]$

27. $s(t) = t^3 e^{-t};$ $[0, \infty)$

28. $s(t) = t^2 - 12 \ln(t + 1);$ $[0, \infty)$

29. The graph in the st-plane of a position function $s(t)$ of a particle moving rectilinearly is given in FIGURE 4.1.6. Complete the accompanying table by stating whether $v(t)$ and $a(t)$ are positive, negative, or zero. Give the time intervals on which the particle is slowing down and the intervals on which it is speeding up.

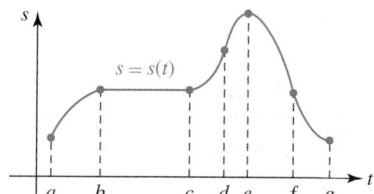

FIGURE 4.1.6   Graph for Problem 29

Interval	$v(t)$	$a(t)$
(a, b)		
(b, c)		
(c, d)		
(d, e)		
(e, f)		
(f, g)		

30. The graph of the velocity function v for a particle that moves on a horizontal line is given in FIGURE 4.1.7. Make a graph of a position function s with this velocity function.

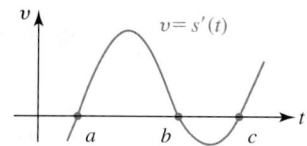

FIGURE 4.1.7   Graph for Problem 30

≡ Applications

31. The height (in feet) of a projectile shot vertically upward from a point 6 ft above ground level is given by $s(t) = -16t^2 + 48t + 6, 0 \le t \le T$, where T is the time the projectile hits the ground. See FIGURE 4.1.8.

(a) Determine the time interval for which $v > 0$ and the time interval for which $v < 0$.

(b) Find the maximum height attained by the projectile.

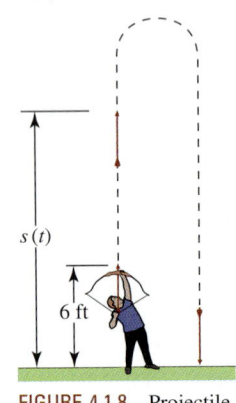

FIGURE 4.1.8 Projectile in Problem 31

32. A particle moves on a horizontal line according to the position function $s(t) = -t^2 + 10t - 20$, where s is measured in centimeters and t in seconds. Determine the total distance traveled by the particle during the time interval [1, 6].

In Problems 33 and 34, use the following information. When friction is ignored, the distance s (in feet) that a body moves down an inclined plane of inclination θ is given by $s(t) = 16t^2\sin\theta, [0, t_1]$, where $s(0) = 0, s(t_1) = L$, and t is measured in seconds. See FIGURE 4.1.9.

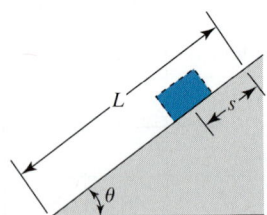

FIGURE 4.1.9 Inclined plane

33. An object is sliding down a 256-ft-long hill with an inclination of 30°. What are the velocity and acceleration of the object at the bottom of the hill?

34. An entry in a soapbox derby rolls down the hill shown in FIGURE 4.1.10. What are its velocity and acceleration at the bottom of the hill?

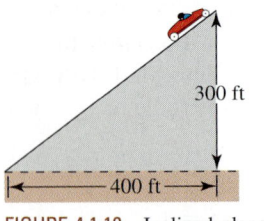

FIGURE 4.1.10 Inclined plane in Problem 34

35. A bucket, attached to a circular windlass by a rope, is permitted to fall in a straight line under the influence of gravity. If the rotational inertia of the windlass is ignored, then the distance the bucket falls is equal to the radian measure of the angle indicated in FIGURE 4.1.11—that is, $\theta = \frac{1}{2}gt^2$, where $g = 32$ ft/s^2 is the acceleration due to gravity. Find the rate at which the y-coordinate of a point P on the circumference of the windlass changes at $t = \sqrt{\pi}/4$ s. Interpret the result.

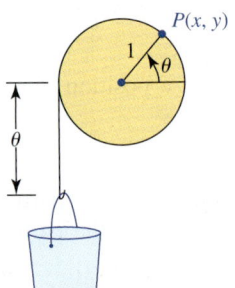

FIGURE 4.1.11 Bucket in Problem 35

36. In mechanics the force F that acts on a body is defined as the rate of change of its momentum: $F = (d/dt)(mv)$. When m is constant, we obtain from this definition the familiar formula known as Newton's Second Law $F = ma$, where the acceleration is $a = dv/dt$. According to Einstein's theory of relativity, when a particle of rest mass m_0 moves rectilinearly at a great velocity (such as in a linear accelerator), its mass varies with the velocity v according to the formula $m = m_0/\sqrt{1 - v^2/c^2}$, where c is the constant speed of light. Show that in the theory of relativity the force F acting on a particle is

$$F = \frac{m_0 a}{\sqrt{(1 - v^2/c^2)^3}},$$

where a is acceleration.

4.2 Related Rates

▌ Introduction In this section we are concerned with **related rates**. The derivative dy/dx of a function $y = f(x)$ is its instantaneous rate of change with respect to the variable x. In the preceding section we saw that when a function $s = s(t)$ describes the position of an object moving on a horizontal or vertical line, the time rate of change ds/dt is interpreted as the velocity of the object. In general, a time rate of change is the answer to the question: How *fast* is a quantity changing? For example, if V stands for volume that is changing in time, then dV/dt is the rate, or how fast, the volume is changing with respect to time t. A rate of, say, $dV/dt = 5$ ft³/s means that the volume is increasing 5 cubic feet each second. See FIGURE 4.2.1. Similarly, if a person is walking *toward* the street lamp shown in FIGURE 4.2.2 at a constant rate of 3 ft/s, then we know that $dx/dt = -3$ ft/s. On the other hand, if the person is walking *away* from the street lamp, then $dx/dt = 3$ ft/s. The negative and positive rates mean, of course, that the distance x from the person to the street lamp is decreasing (3 ft each second) and increasing (3 ft each second), respectively.

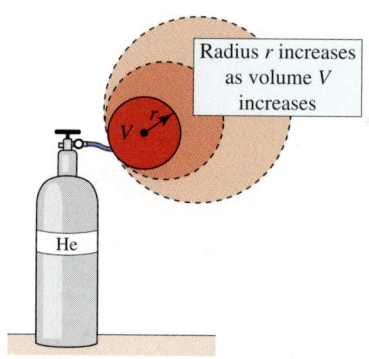

FIGURE 4.2.1 As a spherical balloon is filled with gas, its volume, radius, and surface area change with time

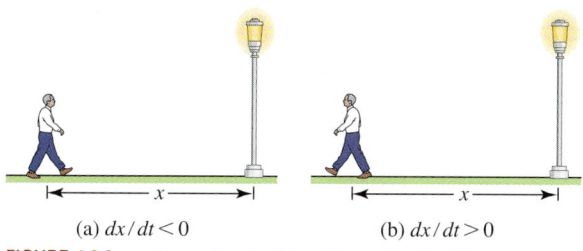

(a) $dx/dt < 0$ (b) $dx/dt > 0$

FIGURE 4.2.2 x decreasing in (a); x increasing in (b)

■ **Power Rule for Functions** Recall from (6) of Section 3.6 that if y denotes a function of x, then the Power Rule for Functions gives

$$\frac{d}{dx} y^n = n y^{n-1} \frac{dy}{dx}, \tag{1}$$

where n is a real number. Of course (1) is applicable to any function, say r, x, or z, that depends on the variable t:

$$\frac{d}{dt} r^n = n r^{n-1} \frac{dr}{dt}, \qquad \frac{d}{dt} x^n = n x^{n-1} \frac{dx}{dt}, \qquad \frac{d}{dt} z^n = n z^{n-1} \frac{dz}{dt}. \tag{2}$$

EXAMPLE 1 Using (2)

A spherical balloon is expanding with time. How is the rate at which the volume increases related to the rate at which the radius increases?

Solution At time t the volume V of a sphere is a function of the radius r, that is $V = \frac{4}{3}\pi r^3$. Thus, the related rates are obtained from the time derivative of this function. With the help of the first result in (2), we see that

$$\frac{dV}{dt} = \frac{4}{3}\pi \cdot \frac{d}{dt} r^3 = \frac{4}{3}\pi \left(3r^2 \frac{dr}{dt}\right)$$

is the same as

$$\underset{\text{related rates}}{\underbrace{\frac{dV}{dt} = 4\pi r^2 \frac{dr}{dt}}}.$$
■

Because the problems in this section will be stated in words, you must interpret these words in terms of mathematical symbols. The key to solving word problems is organization. Here are some suggestions.

Guidelines for Solving Related Problems

(*i*) Carefully read the problem several times. Draw a picture if possible.

(*ii*) Label with symbols all quantities that change with time.

(*iii*) Write down all the rates that are **given**. Using derivative notation, write down the rate that you want to **find**.

(*iv*) Set up an equation or a function that relates all the variables you have introduced.

(*v*) Differentiate the equation or function found in step (*iv*) with respect to time t. This step may require the use of implicit differentiation. The resulting equation after differentiation relates the rates at which the variables change with time.

EXAMPLE 2 Example 1 Revisited

Air is being pumped into a spherical balloon at a rate of 20 ft³/min. At what rate is the radius changing when the radius is 3 ft?

Solution As shown in Figure 4.2.1, we denote the radius of the balloon by r and its volume by V. Now, the interpretation of "air is being pumped ... at a rate of 20 ft³/min" and "at what rate is the radius changing when the radius is 3" are, in turn, the rate that we are

$$\textbf{Given: } \frac{dV}{dt} = 20 \text{ ft}^3/\text{min}$$

and the rate that we wish to

$$\textbf{Find: } \frac{dr}{dt}\bigg|_{r=3}.$$

Because we already know from Example 1 that

$$\frac{dV}{dt} = 4\pi r^2 \frac{dr}{dt}$$

we can substitute the constant rate $dV/dt = 20$, that is, $20 = 4\pi r^2 (dr/dt)$. Solving the last equation for dr/dt yields

$$\frac{dr}{dt} = \frac{20}{4\pi r^2} = \frac{5}{\pi r^2}.$$

Thus,

$$\frac{dr}{dt}\bigg|_{r=3} = \frac{5}{9\pi} \text{ ft/min} \approx 0.18 \text{ ft/min} \qquad \blacksquare$$

EXAMPLE 3 Using the Pythagorean Theorem

A woman jogging at a constant rate of 10 km/h crosses a point P heading north. Ten minutes later a man jogging at a constant rate of 9 km/h crosses the same point heading east. How fast is the distance between the joggers changing 20 min after the man crosses P?

Solution Let time t be measured in hours from the instant the man crosses point P. As shown in **FIGURE 4.2.3**, at $t > 0$ let the man M and woman W be located x and y km, respectively, from point P. Let z be the corresponding distance between the two joggers. Now, two rates are

$$\textbf{Given: } \frac{dx}{dt} = 9 \text{ km/h} \quad \text{and} \quad \frac{dy}{dt} = 10 \text{ km/h} \qquad (3)$$

and we wish to

$$\textbf{Find: } \frac{dz}{dt}\bigg|_{t=1/3} \quad \leftarrow 20 \text{ min} = \tfrac{1}{3}\text{h}$$

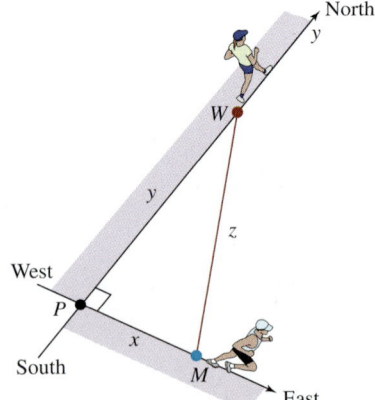

North

West

South

East

FIGURE 4.2.3 Joggers in Example 3

In Figure 4.2.3 we see that the triangle MPW is a right triangle and so from the Pythagorean Theorem, the variables x, y, and z are related by

$$z^2 = x^2 + y^2. \qquad (4)$$

Differentiating (4) with respect to t,

$$\frac{d}{dt} z^2 = \frac{d}{dt} x^2 + \frac{d}{dt} y^2 \quad \text{gives} \quad 2z\frac{dz}{dt} = 2x\frac{dx}{dt} + 2y\frac{dy}{dt}. \qquad (5)$$

Using the two rates given in (3), the last equation in (5) then yields

$$z\frac{dz}{dt} = 9x + 10y.$$

When $t = \tfrac{1}{3}$h we use *distance = rate × time* to obtain the distance the man has run: $x = 9 \cdot \left(\tfrac{1}{3}\right) = 3$ km. Because the woman has run $\tfrac{1}{6}$h (10 min) longer, the distance she has run is $y = 10 \cdot \left(\tfrac{1}{3} + \tfrac{1}{6}\right) = 5$ km. At $t = \tfrac{1}{3}$ h, it follows that $z = \sqrt{3^2 + 5^2} = \sqrt{34}$ km. Finally,

$$\sqrt{34}\frac{dz}{dt}\bigg|_{t=1/3} = 9 \cdot 3 + 10 \cdot 5 \quad \text{or} \quad \frac{dz}{dt}\bigg|_{t=1/3} = \frac{77}{\sqrt{34}} \approx 13.21 \text{ km/h}. \qquad \blacksquare$$

EXAMPLE 4 Using Trigonometry

A lighthouse is located on a small island 2 mi off a straight shore. The beacon of the lighthouse revolves at a constant rate of 6 deg/s. How fast is the light beam moving along the shore at a point 3 mi from a point on the shore closest to the lighthouse?

Solution We first introduce the variables θ and x as shown in **FIGURE 4.2.4**. In addition we change the information about θ to radian measure by recalling that $1°$ is equivalent to $\pi/180$ radians. Thus,

$$\textbf{Given: } \frac{d\theta}{dt} = 6 \cdot \frac{\pi}{180} = \frac{\pi}{30} \text{ rad/s} \qquad \textbf{Find: } \frac{dx}{dt}\Big|_{x=3}.$$

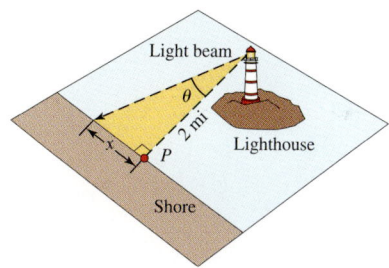

FIGURE 4.2.4 Lighthouse in Example 4

From right triangle trigonometry we see from the figure that

$$\frac{x}{2} = \tan\theta \qquad \text{or} \qquad x = 2\tan\theta.$$

Differentiating the last equation with respect to t and using the given rate yield

$$\frac{dx}{dt} = 2\sec^2\theta \cdot \frac{d\theta}{dt} = \frac{\pi}{15}\sec^2\theta. \quad \leftarrow \text{ Chain Rule: } \frac{dx}{dt} = \frac{dx}{d\theta}\frac{d\theta}{dt}$$

At the instant when $x = 3$, $\tan\theta = \frac{3}{2}$, so that from the trigonometric identity $1 + \tan^2\theta = \sec^2\theta$ we get $\sec^2\theta = \frac{13}{4}$. Hence,

$$\frac{dx}{dt}\Big|_{x=3} = \frac{\pi}{15} \cdot \frac{13}{4} = \frac{13\pi}{60} \text{ mi/s.} \qquad \blacksquare$$

In the next example we need to use the formula for the volume of a right circular cone of height H and base radius R:

$$V = \frac{\pi}{3}R^2H. \tag{6}$$

EXAMPLE 5 Using Similar Triangles

Sand flows from the top half of the conical hourglass shown in **FIGURE 4.2.5** to the bottom half at a constant rate of 4 cm³/s. Express the rate at which the height of the bottom pile increases in terms of the height of the sand.

Solution First, as suggested in Figure 4.2.5, let us make the assumption that the sand pile in the lower part of the hourglass has the shape of a frustum of a cone. At time $t > 0$, let V denote the volume of the sand pile, h its height, and r the radius of its top flat surface. So,

$$\textbf{Given: } \frac{dV}{dt} = 4 \text{ cm}^3/\text{s} \qquad \textbf{Find: } \frac{dh}{dt}.$$

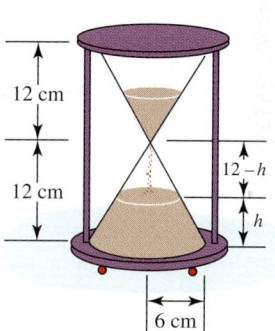

FIGURE 4.2.5 Hourglass in Example 5

We need to find the volume V of the sand pile at time $t > 0$. This can be done in the following way:

$$V = \textit{volume of complete lower cone } - \textit{ volume of cone that is not sand.}$$

Using Figure 4.2.5 and (6) with $R = 6$ and $H = 12$,

$$V = \frac{1}{3}\pi 6^2(12) - \frac{1}{3}\pi r^2(12 - h)$$

or

$$V = \pi\left(144 - 4r^2 + \frac{1}{3}r^2h\right). \tag{7}$$

We can eliminate the variable r from the last equation using similar triangles. As shown in **FIGURE 4.2.6**, the light red right triangle is similar to the blue right triangle and so the ratios of corresponding sides are equal:

$$\frac{12 - h}{r} = \frac{12}{6} \qquad \text{or} \qquad r = 6 - \frac{h}{2}.$$

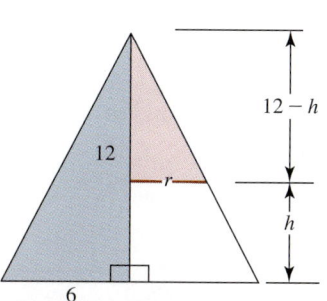

FIGURE 4.2.6 In cross section, the lower cone of the hour glass in Example 5 is a triangle

We substitute this last expression into (7) and simplify,

$$V = \pi\left(\frac{1}{12}h^3 - 3h^2 + 36h\right). \tag{8}$$

Differentiating (8) with respect to t gives

$$\frac{dV}{dt} = \pi\left(\frac{1}{4}h^2\frac{dh}{dt} - 6h\frac{dh}{dt} + 36\frac{dh}{dt}\right) = \pi\left(\frac{1}{4}h^2 - 6h + 36\right)\frac{dh}{dt}.$$

Finally, by using the given rate $dV/dt = 4$ we can solve for dh/dt:

$$\frac{dh}{dt} = \frac{16}{\pi(h - 12)^2}. \tag{9} \quad ■$$

Observe in (9) of Example 5 that the height of the lower sand pile in the hourglass increases fastest when the height h of the pile is close to 12 cm.

Exercises 4.2   Answers to selected odd-numbered problems begin on page ANS-14.

≡ Fundamentals

In the following problems, a solution may require a special formula with which you are not familiar. If necessary, consult the list of formulas given in the *Resource Pages*.

1. A cube is expanding with time. How is the rate at which the volume increases related to the rate at which the length of a side increases?

2. The volume of a rectangular box is $V = xyz$. Given that each side expands at a constant rate of 10 cm/min, find the rate at which the volume is expanding when $x = 1$ cm, $y = 2$ cm, and $z = 3$ cm.

3. A plate in the shape of an equilateral triangle expands with time. The length of a side increases at a constant rate of 2 cm/h. At what rate is the area increasing when a side is 8 cm?

4. In Problem 3, at what rate is the area increasing at the instant when the area is $\sqrt{75}$ cm²?

5. A rectangle expands with time. The diagonal of the rectangle increases at a rate of 1 in./h and the length increases at a rate of $\frac{1}{4}$ in./h. How fast is its width increasing when the width is 6 in. and the length is 8 in.?

6. The lengths of the sides of a cube increase at a rate of 5 cm/h. At what rate does the length of the diagonal of the cube increase?

7. A boat is sailing toward the vertical cliff shown in FIGURE 4.2.7. How are the rates at which x, s, and θ change related?

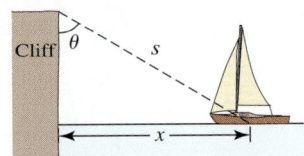

FIGURE 4.2.7   Boat in Problem 7

8. A bug crawls along the graph of $y = x^2 + 4x + 1$, where x and y are measured in centimeters. If the x-coordinate of the bug's position (x, y) changes at a constant rate of 3 cm/min, how fast is the y-coordinate changing when the bug is at the point $(2, 13)$? How fast is the y-coordinate changing when the bug is 6 cm above the x-axis?

9. A particle moves on the graph of $y^2 = x + 1$ so that $dx/dt = 4x + 4$. What is dy/dt when $x = 8$?

10. A particle in continuous motion moves on the graph of $4y = x^2 + x$. Find the point (x, y) on the graph at which the rate of change of the x-coordinate and the rate of change of the y-coordinate are the same.

11. The x-coordinate of the point P shown in FIGURE 4.2.8 increases at a rate of $\frac{1}{3}$ cm/h. How fast is the area of the right triangle OPA increasing when P has coordinates $(8, 2)$?

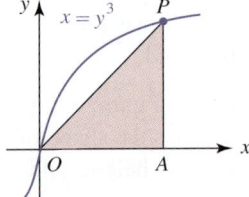

FIGURE 4.2.8   Triangle in Problem 11

12. A suitcase is carried up the conveyor belt shown in FIGURE 4.2.9 at a rate of 2 ft/s. How fast is the vertical distance of the suitcase from the bottom of the belt increasing?

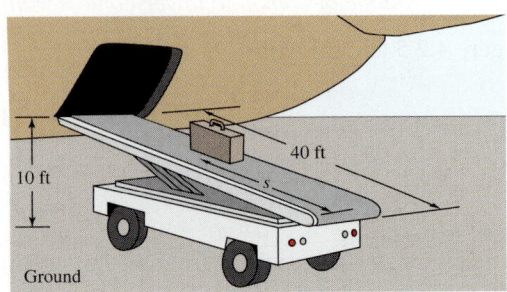

FIGURE 4.2.9   Conveyor belt in Problem 12

13. A 5-ft-tall person walks away from a 20-ft-tall streetlamp at a constant rate of 3 ft/s. See FIGURE 4.2.10.

(a) At what rate is the length of the person's shadow increasing?

(b) At what rate is the tip of the shadow moving away from the base of the streetlamp?

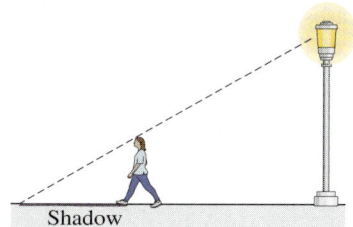

FIGURE 4.2.10   Shadow in Problem 13

14. A stone dropped into a still pond causes a circular wave. Assume that the radius of the wave expands at a constant rate of 2 ft/s.

(a) How fast does the diameter of the circular wave increase?

(b) How fast does the circumference of the circular wave increase?

(c) How fast does the area of the circular wave expand when the radius is 3 ft?

(d) How fast does the area of the circular wave expand when the area is 8π ft^2?

15. A 15-ft ladder is leaning against a wall of a house. The bottom of the ladder is pulled away from the base of the wall at a constant rate of 2 ft/min. At what rate is the top of the ladder sliding down the wall at the instant when the bottom of the ladder is 5 ft from the wall?

16. A 20-ft ladder is leaning against a wall of a house. The top of the ladder is sliding down the wall at a constant rate of $\frac{1}{2}$ ft/min. At what rate is the bottom of the ladder moving away from the wall at the instant when the top of the ladder is 18 ft above the ground?

17. Consider the ladder whose bottom is sliding away from the base of the vertical wall shown in FIGURE 4.2.11. Show that the rate at which θ_1 is increasing is the same as the rate at which θ_2 is decreasing.

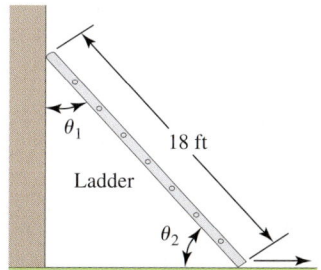

FIGURE 4.2.11   Ladder in Problem 17

18. A kite string is paid out at a constant rate of 3 ft/s. If the wind carries the kite horizontally at an altitude of 200 ft, how fast is the kite moving when 400 ft of string have been paid out?

19. Two tankers depart from the same floating oil terminal. One tanker sails east at noon at a rate of 10 knots. (1 knot = 1 nautical mi/h. A nautical mile is 6080 ft or 1.15 statute mi.) The other tanker sails north at 1:00 P.M.

at a rate of 15 knots. At what rate is the distance between the two ships changing at 2:00 P.M.?

20. At 8:00 A.M. ship S_1 is 20 km due north of ship S_2. Ship S_1 sails south at a rate of 9 km/h and ship S_2 sails west at a rate of 12 km/h. At 9:20 A.M., at what rate is the distance between the two ships changing?

21. A pulley is secured to the edge of a dock that is 15 ft above the surface of the water. A small boat is being pulled toward the dock by means of a rope on the pulley. The rope is attached to the bow of the boat 3 ft above the water line. See FIGURE 4.2.12. If the rope is pulled in at a constant rate of 1 ft/s, how fast does the boat approach the dock when it is 16 ft from the dock?

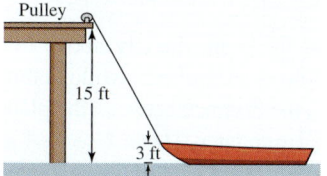

FIGURE 4.2.12   Boat and dock in Problem 21

22. A boat is being pulled toward a dock by means of a winch. The winch is located at the end of the dock and is 10 ft above the level at which the tow rope is attached to the bow of the boat. The rope is pulled in at a constant rate of 1 ft/s. Use an inverse trigonometric function to determine the rate at which the angle of elevation between the bow of the boat and the end of the dock is changing when 30 ft of tow rope is out.

23. A searchlight on a patrol boat that is situated $\frac{1}{2}$ km offshore follows a dune buggy that moves parallel to the water along a straight beach. The dune buggy is traveling at a constant rate of 15 km/h. Use an inverse trigonometric function to determine the rate at which the searchlight is rotating when the dune buggy is $\frac{1}{2}$ km from the point on the shore nearest the boat.

24. A baseball diamond is a square 90 ft on a side. See FIGURE 4.2.13. A player hits the ball and runs toward first base at a rate of 20 ft/s. At what rate is the distance from the runner to second base changing at the instant when the runner is 60 ft from home base? At what rate is the distance from the runner to third base changing at this same instant?

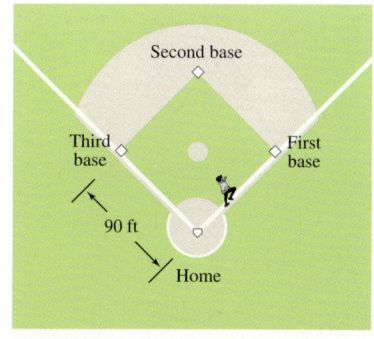

FIGURE 4.2.13   Baseball diamond in Problem 24

25. A plane flying parallel to level ground at a constant rate of 600 mi/h approaches a radar station. If the altitude of the plane is 2 mi, how fast is the distance between the plane and the radar station decreasing when the horizontal distance between them is 1.5 mi? See FIGURE 4.2.14.

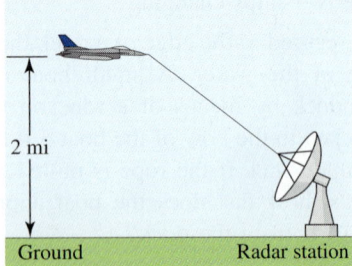

FIGURE 4.2.14   Plane in Problem 25

26. In Problem 25, at the point directly above the radar station, the plane goes into a 30° climb while retaining the same speed. How fast is the distance between the plane and the radar station increasing 1 min later? [*Hint*: Use the Law of Cosines.]

27. A plane at an altitude of 4 km passes directly over a tracking telescope on the ground. When the angle of elevation is 60°, it is observed that this angle is decreasing at a rate of 30 deg/min. How fast is the plane traveling?

28. A tracking camera, located 1200 ft from the point of launching, follows a vertically ascending hot-air balloon. At the instant that the angle of elevation θ of the camera is $\pi/6$ radians, the angle θ is increasing at the rate of 0.1 rad/min. See FIGURE 4.2.15. At what rate is the balloon rising at that instant?

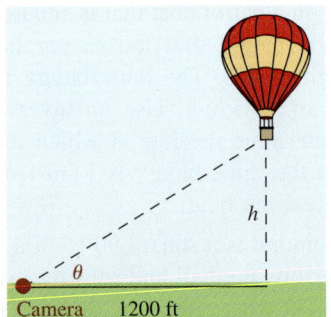

FIGURE 4.2.15   Balloon in Problem 28

29. A rocket is traveling at a constant rate of 1000 mi/h at an angle of 60° to the horizontal. See FIGURE 4.2.16.

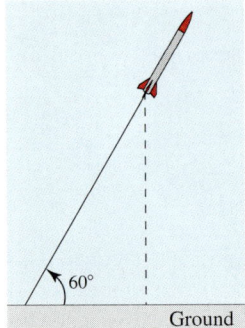

FIGURE 4.2.16   Rocket in Problem 29

(a) At what rate is its altitude increasing?
(b) What is the ground speed of the rocket?

30. A water tank in the shape of a right circular cylinder of diameter 40 ft is being drained so that the level of the water decreases at a constant rate of $\frac{3}{2}$ ft/min. How fast is the volume of the water decreasing?

31. An oil tank in the shape of a right circular cylinder of radius 8 m is being filled at a constant rate of 10 m³/min. How fast is the level of the oil rising?

32. As shown in FIGURE 4.2.17, a 5-ft-wide rectangular water tank is divided into two tanks by a partition that moves in the direction indicated at a rate of 1 in/min as water is pumped into the front tank at a rate of 1 ft³/min.

(a) At what rate is the level of the water changing when the volume of the water in the front tank is 40 ft³ and $x = 4$ ft?
(b) Is the level of the water rising or falling at that instant?

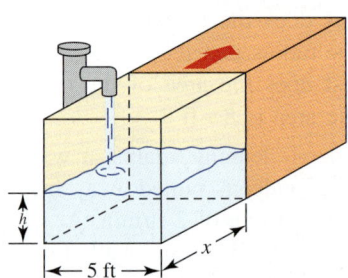

FIGURE 4.2.17   Tank in Problem 32

33. Water leaks out the bottom of the conical tank shown in FIGURE 4.2.18 at a constant rate of 1 ft³/min.

(a) At what rate is the level of the water changing when the water is 6 ft deep?
(b) At what rate is the radius of the water changing when the water is 6 ft deep?
(c) Assume the tank was full at $t = 0$. At what rate is the radius of the water changing at $t = 6$ min?

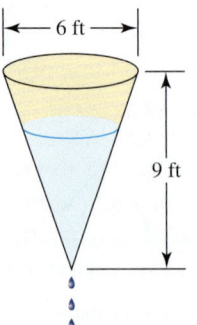

FIGURE 4.2.18   Tank in Problem 33

34. A water trough with vertical ends in the form of isosceles trapezoids has dimensions as shown in FIGURE 4.2.19. If water is pumped in at a constant rate of $\frac{1}{2}$ m³/s, how fast is the level of the water rising when the water is $\frac{1}{4}$ m deep?

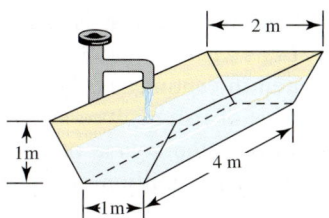

FIGURE 4.2.19   Tank in Problem 34

35. Each vertical end of a 20-ft-long water trough is an equilateral triangle with vertex down. Water is being pumped in at a constant rate of 4 ft³/min.

(a) How fast is the level h of the water rising when the water is 1 ft deep?

(b) If h_0 is the initial depth of water in the trough, show that

$$\frac{dh}{dt} = \frac{\sqrt{3}}{10}\left(h_0^2 + \frac{\sqrt{3}}{5}t\right)^{-1/2}.$$

[*Hint*: Consider the difference in volumes after t minutes.]

(c) If $h_0 = \frac{1}{2}$ ft and the height of the triangular end is 5 ft, determine the time when the trough is full. How fast is the level of the water rising when the trough is full?

36. The volume V between two concentric spheres is expanding. The radius of the outer sphere increases at a constant rate of 2 m/h, whereas the radius of the inner sphere decreases at a constant rate of $\frac{1}{2}$ m/h. At what rate is V changing when the outer radius is 3 m and the inner radius is 1 m?

37. Many spherical objects such as raindrops, snowballs, and mothballs evaporate at a rate proportional to their surface area. In this case show that the radius of the object decreases at a constant rate.

38. If the rate at which the volume of a sphere changes is constant, show that the rate at which its surface area changes is inversely proportional to the radius.

39. Assume that a cube of ice melts in such a manner that it always retains its cubical shape. If the volume of the cube decreases at a rate of $\frac{1}{4}$ in³/min, how fast is the surface area of the cube changing when the surface area is 54 in²?

40. The Ferris wheel shown in FIGURE 4.2.20 revolves counterclockwise once every 2 min. How fast is a passenger rising at the instant when she is 64 ft above the ground? How fast is she moving horizontally at the same instant?

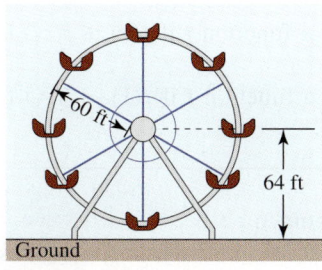

FIGURE 4.2.20   Ferris wheel in Problem 40

41. Suppose the Ferris wheel in Problem 40 is equipped with bidirectional colored spotlights fixed at various points on its circumference. Consider the spotlight located at point P in FIGURE 4.2.21. If the light beams are tangent to the wheel at point P, at what rate is the spot Q on the ground moving away from point R at the instant when $\theta = \pi/4$?

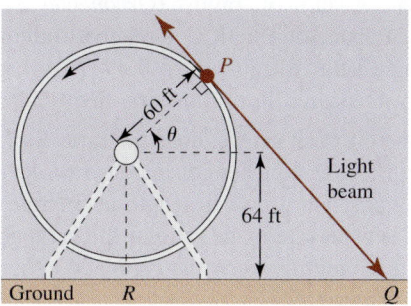

FIGURE 4.2.21   Ferris wheel in Problem 41

42. A diver jumps from a high platform with an initial downward velocity of 1 ft/s toward the center of a large circular tank of water. See FIGURE 4.2.22. From physics, the height of the diver above ground level is given by $s(t) = -16t^2 - t + 200$, where $t \geq 0$ is time measured in seconds.

(a) Express θ in terms of s using an inverse trigonometric function.

(b) Find the rate at which the angle θ subtended by the circular tank at the diver's eye is increasing at $t = 3$ s into the dive.

(c) What is the value of θ when the diver hits the water?

(d) What is the rate of change of θ when the diver hits the water?

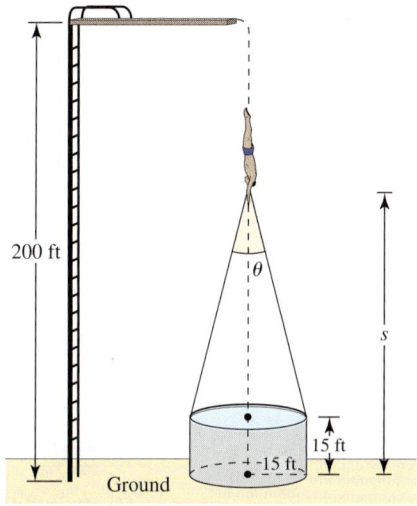

FIGURE 4.2.22   Diver in Problem 42

≡ Mathematical Models

43. Resistance   The total resistance R in a parallel circuit that contains two resistors of resistances R_1 and R_2 is given by $1/R = 1/R_1 + 1/R_2$. If each resistance changes with time t, then how are dR/dt, dR_1/dt, and dR_2/dt related?

44. Pressure In the adiabatic expansion of air, pressure P and volume V are related by $PV^{1.4} = k$, where k is a constant. At a certain instant the pressure is 100 lb/in^2 and the volume is 32 in^3. At what rate is the pressure changing at that instant if the volume is decreasing at a rate of 2 in^3/s?

45. Crayfish A study of crayfish (*Orconectes virilis*) indicates that the carapace of length C is related to the total length T according to the formula $C = 0.493T - 0.913$, where C and T are measured in millimeters. See FIGURE 4.2.23.

(a) As the crayfish grows, does the ratio R of the carapace length to the total length increase or decrease?

(b) If the crayfish grows in length at the rate of 1 mm per day, at what rate is the ratio of the carapace to the total length changing when the carapace is one-third of the total length?

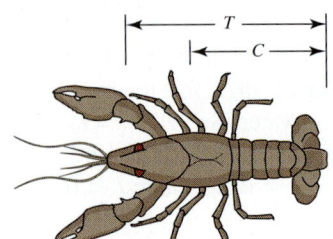

FIGURE 4.2.23 Crayfish in Problem 45

46. Brain Weight According to allometric studies, brain weight E in fish is related to body weight P by

$E = 0.007P^{2/3}$, and body weight is related to body length L by $P = 0.12L^{2.53}$, where E and P are measured in grams and L is measured in centimeters. Suppose that the length of a certain species of fish evolved at a constant rate from 10 cm to 18 cm over the course of 20 million years. At what rate, in grams per million years, was this species' brain growing when the fish was half its final body weight?

47. Momentum In physics the momentum p of a body of mass m that moves in a straight line with velocity v is given by $p = mv$. Suppose that an airplane of mass 10^5 kg flies in a straight line while ice builds up on the leading edges of its wings at a constant rate of 30 kg/h. See FIGURE 4.2.24.

(a) At what rate is the momentum of the airplane changing if it is flying at a constant rate of 800 km/h?

(b) At what rate is the momentum of the airplane changing at $t = 1$ h if at that instant its velocity is 750 km/h and is increasing at a rate of 20 km/h^2?

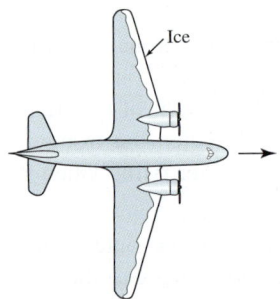

FIGURE 4.2.24 Airplane in Problem 47

4.3 Extrema of Functions

▌ Introduction We turn now to the problem of finding the maximum and minimum values of a function f on an interval I. We will see that by finding these **extrema** of f (if there are any) we can in many cases quickly sketch its graph. By finding the extrema of a function we will also be able to solve certain kinds of optimization problems. In this section we set forth some important definitions and show how to find the maximum and minimum values of a function f that is continuous on a closed interval.

▌ Absolute Extrema In FIGURE 4.3.1 we have illustrated the graph of the quadratic function $f(x) = x^2 - 3x + 4$. From this graph it should be apparent that the function value $f\left(\frac{3}{2}\right) = \frac{7}{4}$ is the y-coordinate of the vertex, and because the parabola opens upward, there is no number in the range of f smaller than $\frac{7}{4}$. We say that the extremum $f\left(\frac{3}{2}\right) = \frac{7}{4}$ is the absolute minimum of f. The concepts of an absolute maximum and an absolute minimum of a function are defined next.

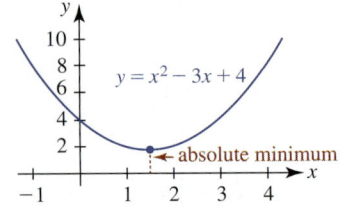

FIGURE 4.3.1 An absolute minimum of a function

Definition 4.3.1 Absolute Extrema

(i) A number $f(c_1)$ is an **absolute maximum** of a function f if $f(x) \leq f(c_1)$ for every x in the domain of f.

(ii) A number $f(c_1)$ is an **absolute minimum** of a function f if $f(x) \geq f(c_1)$ for every x in the domain of f.

Absolute extrema are also called **global extrema**.

From your experience of graphing functions it should be easy, in some cases, to see when a function possesses an absolute maximum or minimum. In general, a quadratic function

$f(x) = ax^2 + bx + c$ has either an absolute maximum or minimum. The function $f(x) = 4 - x^2$ has the absolute maximum $f(0) = 4$. A linear function $f(x) = ax + b$, $a \neq 0$, possesses no absolute extrema. The graphs of the familiar functions $y = 1/x$, $y = x^3$, $y = \tan x$, $y = e^x$, and $y = \ln x$ show that these functions do not have any absolute extrema. The trigonometric functions $y = \sin x$ and $y = \cos x$ possess both an absolute maximum and an absolute minimum.

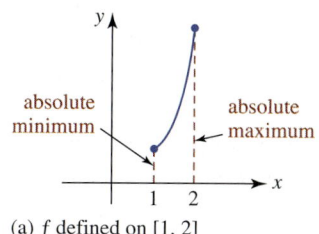

(a) f defined on [1, 2]

EXAMPLE 1 Absolute Extrema

For $f(x) = \sin x$, $f(\pi/2) = 1$ is its absolute maximum and $f(3\pi/2) = -1$ is its absolute minimum. By periodicity, the maximum and minimum values also occur at $x = \pi/2 + 2n\pi$ and $x = 3\pi/2 + 2n\pi$, $n = \pm 1, \pm 2, \ldots$, respectively. ∎

The interval on which the function is defined is very important in the consideration of extrema.

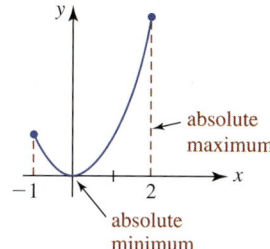

(b) f defined on [−1, 2]

FIGURE 4.3.2 Graphs of functions in Example 2

EXAMPLE 2 Functions Defined on a Closed Interval

(a) $f(x) = x^2$, defined only on the *closed* interval [1, 2], has the absolute maximum $f(2) = 4$ and the absolute minimum $f(1) = 1$. See FIGURE 4.3.2(a).

(b) On the other hand, if $f(x) = x^2$ is defined on the *open* interval (1, 2), then f has no absolute extrema. In this case, $f(1)$ and $f(2)$ are not defined.

(c) $f(x) = x^2$, defined on [−1, 2], has the absolute maximum $f(2) = 4$, but now the absolute minimum is $f(0) = 0$. See Figure 4.3.2(b).

(d) $f(x) = x^2$, defined on (−1, 2), has an absolute minimum $f(0) = 0$, but no absolute maximum. ∎

Parts (a) and (c) of Example 2 illustrate the following general result.

Theorem 4.3.1 Extreme Value Theorem

A function f continuous on a closed interval $[a, b]$ always has an absolute maximum and an absolute minimum on the interval.

In other words, when f is continuous on $[a, b]$, there are numbers $f(c_1)$ and $f(c_2)$ such that $f(c_1) \leq f(x) \leq f(c_2)$ for all x in $[a, b]$. The values $f(c_2)$ and $f(c_1)$ are the absolute maximum and the absolute minimum, respectively, on the closed interval $[a, b]$. See FIGURE 4.3.3.

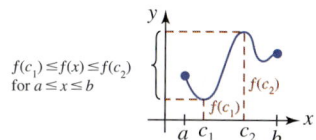

FIGURE 4.3.3 The function f has both an absolute maximum and an absolute minimum

■ Endpoint Extrema When an absolute extremum of a function occurs at an endpoint of an interval I, as in parts (a) and (c) of Example 2, we say it is an **endpoint extremum**. When I is not a closed interval, that is, when I is an interval such as $(a, b]$, $(-\infty, b]$, or $[a, \infty)$, then even when f is continuous there is no guarantee that an absolute extremum exists. See FIGURE 4.3.4.

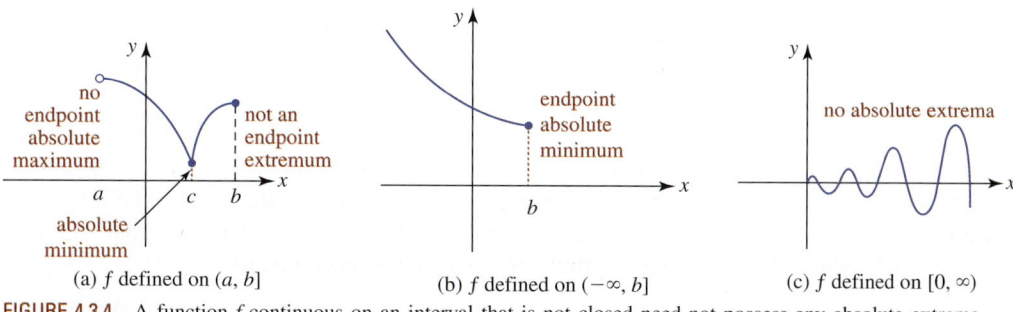

(a) f defined on $(a, b]$ (b) f defined on $(-\infty, b]$ (c) f defined on $[0, \infty)$

FIGURE 4.3.4 A function f continuous on an interval that is not closed need not possess any absolute extrema

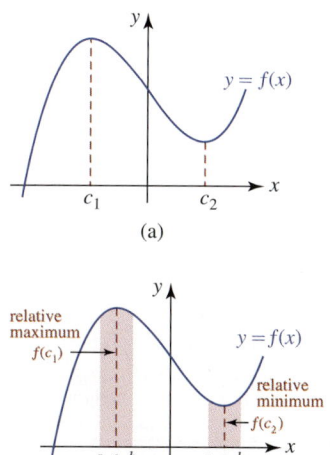

FIGURE 4.3.5 Relative maximum at c_1 and relative minimum at c_2

■ Relative Extrema In FIGURE 4.3.5(a) we have illustrated the graph of $f(x) = x^3 - 5x + 8$. Because the end behavior of f is that of $y = x^3$, $f(x) \to \infty$ as $x \to \infty$ and $f(x) \to -\infty$ as $x \to -\infty$. From this observation we can conclude that this polynomial function has no absolute extrema. However, suppose we focus our attention on values of x that are close to, or in a *neighborhood* of, the numbers c_1 and c_2. As shown in Figure 4.3.5(b), $f(c_1)$ is the largest or maximum value of the function f when compared with all other function values in the open interval (a_1, b_1); similarly $f(c_2)$ is the minimum value of f in the interval (a_2, b_2). These **relative**, or **local**, **extrema** are defined as follows.

Definition 4.3.2   Relative Extrema

  (*i*)   A number $f(c_1)$ is a **relative maximum** of a function f if $f(x) \leq f(c_1)$ for every x in some open interval that contains c_1.

  (*ii*)   A number $f(c_1)$ is a **relative minimum** of a function f if $f(x) \geq f(c_1)$ for every x in some open interval that contains c_1.

As a consequence of Definition 4.3.2, we can conclude that

  •   *Every absolute extremum, with the exception of an endpoint extremum, is also a relative extremum.*

An endpoint absolute extremum is precluded from being a relative extremum on the technicality that an open interval contained in the domain of the function cannot be found around an endpoint of the interval.

We have been leading up to an obvious question:

  •   *How do we find the extrema of a function?*

Even when we have graphs, for most functions the x-coordinate at which an extremum occurs is not apparent. With the aid of the zoom-in/zoom-out of a graphing utility we can search for, and, of course, approximate both the location and the value of an extremum. See FIGURE 4.3.6. Nevertheless it is desirable to be able to find the exact location and the exact value of an extremum.

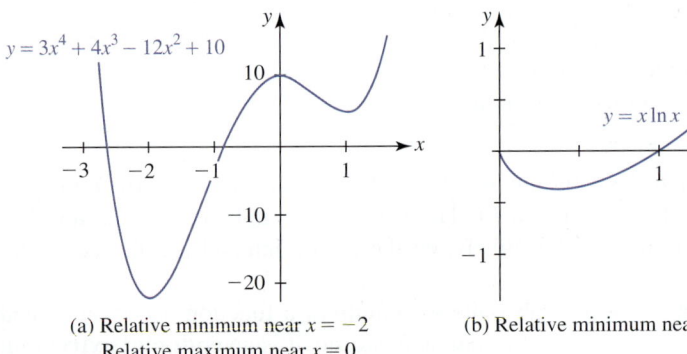

(a) Relative minimum near $x = -2$
Relative maximum near $x = 0$
Relative minimum near $x = 1$

(b) Relative minimum near $x = 0.4$

FIGURE 4.3.6   Approximate location of relative extrema

In Figure 4.3.6(a) we stated that a relative minimum occurs *near* $x = -2$. With the tools of a calculator or a CAS we can convince ourselves that this relative minimum is really an absolute or global minimum, but with the tools of calculus we can actually prove that this is the case.

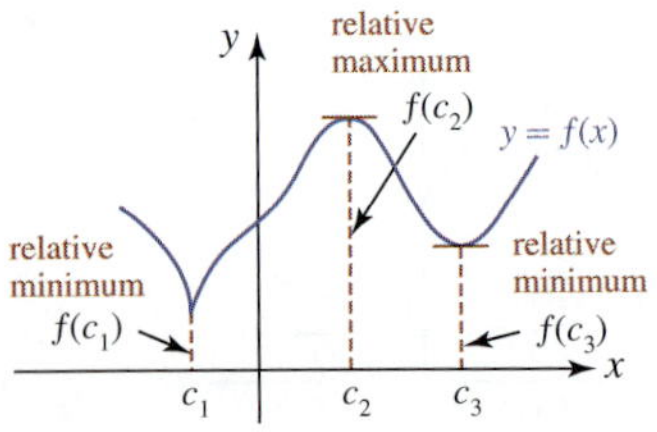

▌ **Critical Numbers**   An examination of FIGURE 4.3.7 along with Figures 4.3.5 and 4.3.6 suggest that if c is a number at which a function f has a relative extremum, then either the tangent is horizontal at the point corresponding to $x = c$ or is not differentiable at $x = c$. That is, either $f'(c) = 0$ or $f'(c)$ does not exist. Such a number c is given a special name.

FIGURE 4.3.7 f is not differentiable at c_1; f' is 0 at c_2 and c_3

Definition 4.3.3   Critical Number

A **critical number** of a function f is a number c in its domain for which $f'(c) = 0$ or $f'(c)$ does not exist.

In some texts a critical number $x = c$ is referred to as a **critical point**.

EXAMPLE 3 Finding Critical Numbers

Find the critical numbers of $f(x) = x \ln x$.

Solution By the Product Rule,

$$f'(x) = x \cdot \frac{1}{x} + 1 \cdot \ln x = 1 + \ln x.$$

The only solution of $f'(x) = 0$ or $\ln x = -1$ is $x = e^{-1}$. To two decimal places the critical number of f is $e^{-1} \approx 0.36$. ∎

EXAMPLE 4 Finding Critical Numbers

Find the critical numbers of $f(x) = 3x^4 + 4x^3 - 12x^2 + 10$.

Solution Differentiating f and factoring yield

$$f'(x) = 12x^3 + 12x^2 - 24x = 12x(x + 2)(x - 1).$$

Hence we see that $f'(x) = 0$ for $x = 0, x = -2$, and $x = 1$. The critical numbers of f are $0, -2$, and 1. ∎

EXAMPLE 5 Finding Critical Numbers

Find the critical numbers of $f(x) = (x + 4)^{2/3}$.

Solution By the Power Rule for Functions,

$$f'(x) = \frac{2}{3}(x + 4)^{-1/3} = \frac{2}{3(x + 4)^{1/3}}.$$

In this instance we see that $f'(x)$ does not exist when $x = -4$. Since -4 is in the domain of f, we conclude that it is a critical number. ∎

EXAMPLE 6 Finding Critical Numbers

Find the critical numbers of $f(x) = \dfrac{x^2}{x - 1}$.

Solution By the Quotient Rule, we find after simplifying,

$$f'(x) = \frac{x(x - 2)}{(x - 1)^2}.$$

Now $f'(x) = 0$ when the numerator of f is 0. The equation $x(x - 2) = 0$ gives $x = 0$ and $x = 2$. In addition, inspection of the denominator of f shows that $f'(x)$ does not exist when $x = 1$. However, examination of f reveals $x = 1$ is not in its domain, and so the only critical numbers are 0 and 2. ∎

> **Theorem 4.3.2** Relative Extrema Occur at Critical Numbers
>
> If a function f has a relative extremum at $x = c$, then c is a critical number.

PROOF Assume that $f(c)$ is a relative extremum.

(*i*) If $f'(c)$ does not exist, then c is a critical number by Definition 4.3.3.

(*ii*) If $f'(c)$ exists, there are three possibilities: $f'(c) > 0, f'(c) < 0$, or $f'(c) = 0$. For the sake of argument, let us further assume that $f(c)$ is a relative maximum. Hence, by Definition 4.3.2 there is some open interval that contains c in which

$$f(c + h) \leq f(c) \tag{1}$$

where the number h is sufficiently small in absolute value. The inequality in (1) then implies that

$$\frac{f(c + h) - f(c)}{h} \leq 0 \quad \text{for } h > 0 \qquad \text{and} \qquad \frac{f(c + h) - f(c)}{h} \geq 0 \quad \text{for } h < 0. \quad (2)$$

But since $\lim_{h \to 0} [f(c + h) - f(c)]/h$ exists and equals $f'(c)$, the inequalities in (2) show that $f'(c) \leq 0$ and $f'(c) \geq 0$, respectively. The only way this can happen is to have $f'(c) = 0$. The case when $f(c)$ is a relative minimum is proved in a similar manner. ∎

■ **Extrema of Functions Defined on a Closed Interval** We have seen that a function f that is continuous on a *closed* interval has both an absolute maximum and an absolute minimum. The next theorem tells us where these extrema occur.

Theorem 4.3.3 Finding Absolute Extrema

If f is continuous on a closed interval $[a, b]$, then an absolute extremum occurs either at an endpoint of the interval or at a critical number c in the open interval (a, b).

We summarize Theorem 4.3.3 in the following manner.

Guidelines for Finding Extrema on a Closed Interval

    (*i*) Evaluate f at the endpoints a and b of the interval $[a, b]$.
    (*ii*) Find all critical numbers $c_1, c_2, \ldots, c_n$ in the open interval (a, b).
    (*iii*) Evaluate f at all critical numbers.
    (*iv*) The largest and smallest values in the list

$$f(a), f(c_1), f(c_2), \ldots, f(c_n), f(b),$$

are the absolute maximum and the absolute minimum, respectively, of f on the interval $[a, b]$.

EXAMPLE 7 Finding Absolute Extrema

Find the absolute extrema of $f(x) = x^3 - 3x^2 - 24x + 2$ on the interval

    (a) $[-3, 1]$ **(b)** $[-3, 8]$.

Solution Because f is continuous, we need only evaluate f at the endpoints of each interval and at critical numbers within each open interval. From the derivative

$$f'(x) = 3x^2 - 6x - 24 = 3(x + 2)(x - 4)$$

we see that the critical numbers of the function f are -2 and 4.

    (a) From the data in the accompanying table it is evident that the absolute maximum of f on the interval $[-3, 1]$ is $f(-2) = 30$, and the absolute minimum is the endpoint extremum $f(1) = -24$.

On $[-3, 1]$			
x	-3	-2	1
$f(x)$	20	30	-24

    (b) On the interval $[-3, 8]$ we see from the table that $f(4) = -78$ is an absolute minimum and $f(8) = 130$ is an endpoint absolute maximum.

On $[-3, 8]$				
x	-3	-2	4	8
$f(x)$	20	30	-78	130

■

$f'(x)$ NOTES FROM THE CLASSROOM

(*i*) A function may, of course, assume its maximum and minimum values more than once on an interval. You should verify with the aid of a graphing utility that the function $f(x) = \sin x$ attains its maximum function value 1 fives times and its minimum function value -1 four times in the interval $[0, 9\pi]$.

(*ii*) The converse of Theorem 4.3.2 is not necessarily true. In other words:

A critical number of a function f need not correspond to a relative extremum.

Consider $f(x) = x^3$ and $g(x) = x^{1/3}$. The derivatives $f'(x) = 3x^2$ and $g'(x) = \frac{1}{3}x^{-2/3}$ show that 0 is a critical number of both functions. But from the graphs of f and g in FIGURE 4.3.8 we see that neither function possesses any extrema on the interval $(-\infty, \infty)$.

(*iii*) We have indicated how to find the absolute extrema of a function f that is continuous on a closed interval. In Sections 4.6 and 4.7 we will utilize the first and second derivatives to find the relative extrema of a function.

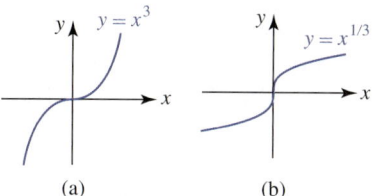

(a)   (b)

FIGURE 4.3.8   0 is a critical number for both functions, but neither has any extrema

---

Exercises 4.3   Answers to selected odd-numbered problems begin on page ANS-14.

≡ Fundamentals

In Problems 1–6, use the graph of the given function as an aid in determining any absolute extrema on the indicated intervals.

1. $f(x) = x - 4$
    (a) $[-1, 2]$ **(b)** $[3, 7]$ **(c)** $(2, 5)$ **(d)** $[1, 4]$

2. $f(x) = |x - 4|$
    (a) $[-1, 2]$ **(b)** $[3, 7]$ **(c)** $(2, 5)$ **(d)** $[1, 4]$

3. $f(x) = x^2 - 4x$
    (a) $[1, 4]$ **(b)** $[1, 3]$ **(c)** $(-1, 3)$ **(d)** $(4, 5]$

4. $f(x) = \sqrt{9 - x^2}$
    (a) $[-3, 3]$ **(b)** $(-3, 3)$ **(c)** $[0, 3)$ **(d)** $[-1, 1]$

5. $f(x) = \tan x$
    (a) $[-\pi/2, \pi/2]$ **(b)** $[-\pi/4, \pi/4]$
    (c) $[0, \pi/3]$ **(d)** $[0, \pi]$

6. $f(x) = 2\cos x$
    (a) $[-\pi, \pi]$ **(b)** $[-\pi/2, \pi/2]$
    (c) $[\pi/3, 2\pi/3]$ **(d)** $[-\pi/2, 3\pi/2]$

In Problems 7–22, find the critical numbers of the given function.

7. $f(x) = 2x^2 - 6x + 8$ **8.** $f(x) = x^3 + x - 2$

9. $f(x) = 2x^3 - 15x^2 - 36x$ **10.** $f(x) = x^4 - 4x^3 + 7$

11. $f(x) = (x - 2)^2(x - 1)$ **12.** $f(x) = x^2(x + 1)^3$

13. $f(x) = \dfrac{1 + x}{\sqrt{x}}$ **14.** $f(x) = \dfrac{x^2}{x^2 + 2}$

15. $f(x) = (4x - 3)^{1/3}$ **16.** $f(x) = x^{2/3} + x$

17. $f(x) = (x - 1)^2\sqrt[3]{x + 2}$ **18.** $f(x) = \dfrac{x + 4}{\sqrt[3]{x + 1}}$

19. $f(x) = -x + \sin x$ **20.** $f(x) = \cos 4x$

21. $f(x) = x^2 - 8\ln x$ **22.** $f(x) = e^{-x} + 2x$

In Problems 23–36, find the absolute extrema of the given function on the indicated interval.

23. $f(x) = -x^2 + 6x$; $[1, 4]$ **24.** $f(x) = (x - 1)^2$; $[2, 5]$

25. $f(x) = x^{2/3}$; $[-1, 8]$

26. $f(x) = x^{2/3}(x^2 - 1)$; $[-1, 1]$

27. $f(x) = x^3 - 6x^2 + 2$; $[-3, 2]$

28. $f(x) = -x^3 - x^2 + 5x$; $[-2, 2]$

29. $f(x) = x^3 - 3x^2 + 3x - 1$; $[-4, 3]$

30. $f(x) = x^4 + 4x^3 - 10$; $[0, 4]$

31. $f(x) = x^4(x - 1)^2$; $[-1, 2]$

32. $f(x) = \dfrac{\sqrt{x}}{x^2 + 1}$; $[\frac{1}{4}, \frac{1}{2}]$

33. $f(x) = 2\cos 2x - \cos 4x$; $[0, 2\pi]$

34. $f(x) = 1 + 5\sin 3x$; $[0, \pi/2]$

35. $f(x) = 3 + 2\sin^2 24x$; $[0, \pi]$

36. $f(x) = 2x - \tan x$; $[-1, 1.5]$

In Problems 37 and 38, find all critical numbers. Distinguish between absolute, endpoint absolute, and relative extrema.

37. $f(x) = x^2 - 2|x|$; $[-2, 3]$

38. $f(x) = \begin{cases} 4x + 12, & -5 \le x \le -2 \\ x^2, & -2 < x \le 1 \end{cases}$

39. Consider the continuous function f defined on $[a, b]$ shown in FIGURE 4.3.9. Given that c_1 through c_{10} are critical numbers:

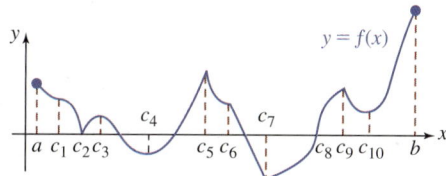

FIGURE 4.3.9   Graph for Problem 39

    (a) List critical numbers at which $f'(x) = 0$.
    (b) List critical numbers at which $f'(x)$ is not defined.
    (c) Distinguish between the absolute and endpoint absolute extrema.
    (d) Distinguish between the relative maxima and the relative minima.

40. Consider the function $f(x) = x + 1/x$. Show that the relative minimum is greater than the relative maximum.

≡ Applications

41. The height of a projectile launched from ground level is given by $s(t) = -16t^2 + 320t$, where t is measured in seconds and s in feet.

(a) $s(t)$ is defined only on the time interval $[0, 20]$. Why?
(b) Use the results of Theorem 4.3.3 to determine the maximum height attained by the projectile.

42. The French physician **Jean Louis Poiseuille** discovered that the velocity $v(r)$ (in cm/s) of blood flowing through an artery with circular cross-section of radius R is given by $v(r) = (P/4\nu l)(R^2 - r^2)$, where P, ν, and l are positive constants. See FIGURE 4.3.10.

(a) Determine a closed interval on which v is defined.
(b) Determine the maximum and minimum velocities of the blood.

Circular cross-section

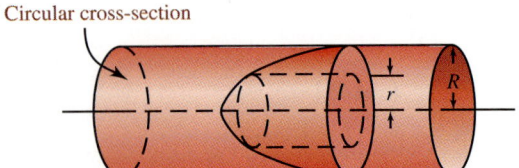

FIGURE 4.3.10 Artery in Problem 42

≡ Think About It

43. Draw a graph of a continuous function f that possesses no absolute extrema but has a relative maximum and a relative minimum that are the same value.

44. Give an example of a continuous function, defined on a closed interval $[a, b]$, for which the absolute maximum is the same as the absolute minimum.

45. Let $f(x) = \lfloor x \rfloor$ be the greatest integer function. Show that every value of x is a critical number.

46. Show that $f(x) = (ax + b)/(cx + d)$ has no critical numbers when $ad - bc \neq 0$. What happens when $ad - bc = 0$?

47. Let $f(x) = x^n$, where n is a positive integer. Determine the values of n for which f has a relative extremum.

48. Discuss: Why can a polynomial function of degree n have at most $n - 1$ critical numbers?

49. Suppose f is a continuous even function such that $f(a)$ is a relative minimum. What can be said about $f(-a)$?

50. Suppose f is a continuous odd function such that $f(a)$ is a relative maximum. What can be said about $f(-a)$?

51. Suppose f is an even function that is everywhere differentiable. Show that $x = 0$ is a critical number of f.

52. Suppose f is a differentiable function that possesses a single critical number c. If $k \neq 0$, find the critical numbers of:

(a) $k + f(x)$ **(b)** $kf(x)$ **(c)** $f(x + k)$ **(d)** $f(kx)$

≡ Calculator/CAS Problems

53. **(a)** Use a calculator or CAS to obtain the graph of $f(x) = -2\cos x + \cos 2x$.
(b) Find the critical numbers of f in the interval $[0, 2\pi]$.
(c) Find the absolute extrema of f in the interval $[0, 2\pi]$.

54. In the study of snow-crystal growth, the formula

$$I(t) = \frac{b}{\pi} + \frac{b}{2}\sin\omega t - \frac{2b}{3\pi}\cos 2\omega t$$

is a mathematical model for the daily variation in the intensity of solar radiation penetrating the surface of snow. Here t represents time measured in hours after sunrise ($t = 0$) and $\omega = 2\pi/24$.

(a) Use a calculator or CAS to obtain the graph of I on the interval $[0, 24]$. Use $b = 1$.
(b) Find the critical numbers of I in the interval $[0, 24]$.

4.4 Mean Value Theorem

▌ **Introduction** Suppose a function $y = f(x)$ is continuous and differentiable on a closed interval $[a, b]$ and that $f(a) = f(b) = 0$. These conditions mean that numbers a and b are the x-coordinates of x-intercepts of the graph of f. FIGURE 4.4.1(a) shows a typical graph of a function f satisfying these conditions. It seems plausible from Figure 4.4.1(b) that there must exist at least one number c in the open interval (a, b) corresponding to a point on the graph of f at which the tangent is horizontal. This observation leads to a result known as Rolle's Theorem. We will use Rolle's Theorem to prove the main result of this section: the Mean Value Theorem for derivatives.

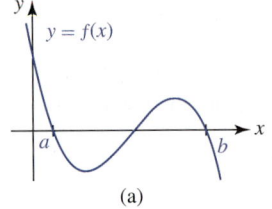

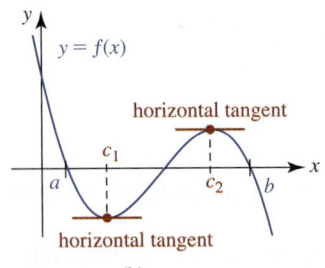

FIGURE 4.4.1 Two points where the tangent is horizontal

Theorem 4.4.1   Rolle's Theorem

Let f be a function that is continuous on $[a, b]$ and differentiable on (a, b). If $f(a) = f(b) = 0$, then there exists a number c in (a, b) such that $f'(c) = 0$.

PROOF   Either f is a constant function on the interval $[a, b]$ or it is not. If f is a constant function on $[a, b]$, then we must have $f'(c) = 0$ for every number c in (a, b). Now, if f is not a constant function on $[a, b]$, there must be some number x in (a, b) at which either

$f(x) > 0$ or $f(x) < 0$. Suppose $f(x) > 0$. Since f is continuous on $[a, b]$, we know from the Extreme Value Theorem that f attains an absolute maximum at some number c in $[a, b]$. But from $f(a) = f(b) = 0$ and $f(x) > 0$ for some x in (a, b), we conclude that the number c cannot be an endpoint of $[a, b]$. Consequently, c is in (a, b). Since f is differentiable on (a, b), it is differentiable at c. Hence, from Theorem 4.3.2, we have $f'(c) = 0$. The proof of the case when $f(x) < 0$ follows in a similar manner. ∎

EXAMPLE 1   Verifying Rolle's Theorem

Consider the function $f(x) = -x^3 + x$ defined on $[-1, 1]$. The graph of f is given in FIGURE 4.4.2. Since f is a polynomial function, it is continuous on $[-1, 1]$ and differentiable on $(-1, 1)$. Also, $f(-1) = f(1) = 0$. Thus, the hypotheses of Rolle's Theorem are satisfied. We conclude that there must be at least one number in $(-1, 1)$ for which $f'(x) = -3x^2 + 1$ is zero. To find this number, we solve $f'(c) = 0$ or $-3c^2 + 1 = 0$. The latter leads to *two* solutions in the interval: $c_1 = -\sqrt{3}/3 \approx -0.57$ and $c_2 = \sqrt{3}/3 \approx 0.57$. ∎

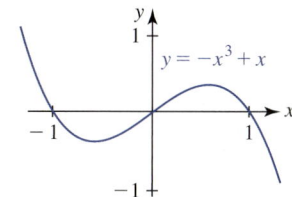

FIGURE 4.4.2   Graph of function in Example 1

In Example 1 notice that the given function f satisfies the hypotheses of Rolle's Theorem on $[0, 1]$ as well as on $[-1, 1]$. In the case of the interval $[0, 1]$, $f'(c) = -3c^2 + 1 = 0$ yields the single solution $c = \sqrt{3}/3$.

EXAMPLE 2   Verifying Rolle's Theorem

(a) The function $f(x) = x - 4x^{1/3}$, shown in FIGURE 4.4.3, is continuous on $[-8, 8]$ and satisfies $f(-8) = f(8) = 0$. But f is not differentiable on $(-8, 8)$, since there is a vertical tangent at the origin. Nevertheless, as the figure suggests, there are two numbers c_1 and c_2 in $(-8, 8)$ at which $f'(x) = 0$. You should verify that $f'(-8\sqrt{3}/9) = 0$ and $f'(8\sqrt{3}/9) = 0$. Bear in mind that the hypotheses of Rolle's Theorem are sufficient but not necessary conditions. In other words, if one or more of the three hypotheses: continuity on $[a, b]$, differentiability on (a, b), and $f(a) = f(b) = 0$ do not hold, the conclusion that there exists a number c in (a, b) such that $f'(c) = 0$ may or may not hold.

(b) Consider another function $g(x) = 1 - x^{2/3}$. This function is continuous on $[-1, 1]$ and $f(-1) = f(1) = 0$. But like the foregoing function f, g is not differentiable at $x = 0$ and so is not differentiable on the open interval $(-1, 1)$. In this case, however, there is no c in $(-1, 1)$ for which $f'(c) = 0$. See FIGURE 4.4.4. ∎

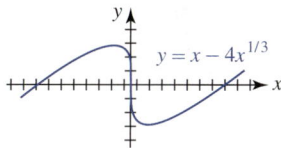

FIGURE 4.4.3   Graph of function f in Example 2

The conclusion of Rolle's Theorem also holds when the condition $f(a) = f(b) = 0$ is replaced with $f(a) = f(b)$. The plausibility of this fact is illustrated in FIGURE 4.4.5.

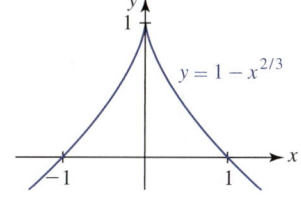

FIGURE 4.4.4   Graph of function g in Example 2

■ **Mean Value Theorem**   Rolle's Theorem is helpful in proving the next important result called the **Mean Value Theorem**. This theorem states that when a function f is continuous on $[a, b]$ and differentiable on (a, b) there must be at least one point on the graph at which the slope of the tangent line is the same as the slope of the secant line through the points $(a, f(a))$ and $(b, f(b))$. The word *mean* here refers to an average, that is, the value of the derivative at some point is the same as the average rate of change of the function on the interval.

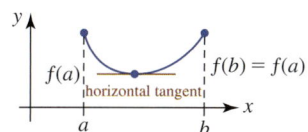

FIGURE 4.4.5   Rolle's Theorem holds when $f(a) = f(b)$

Theorem 4.4.2   Mean Value Theorem for Derivatives

Let f be a function that is continuous on $[a, b]$ and differentiable on (a, b). Then there exists a number c in (a, b) such that

$$f'(c) = \frac{f(b) - f(a)}{b - a}.$$

PROOF   As shown in FIGURE 4.4.6, let $d(x)$ denote the vertical distance between a point on the graph of $y = f(x)$ and the secant line through $(a, f(a))$ and $(b, f(b))$. Since the equation of the secant line is

$$y - f(b) = \frac{f(b) - f(a)}{b - a}(x - b)$$

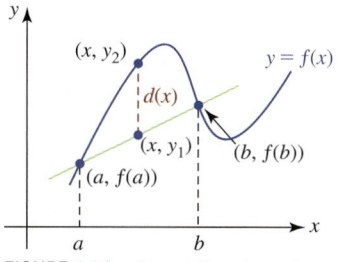

FIGURE 4.4.6   Secant line through $(a, f(a))$ and $(b, f(b))$

we have, as shown in the figure, $d(x) = y_2 - y_1$, or

$$d(x) = f(x) - \left[f(b) + \frac{f(b) - f(a)}{b - a}(x - b) \right].$$

Since $d(a) = d(b) = 0$ and $d(x)$ is continuous on $[a, b]$ and differentiable on (a, b), Rolle's Theorem implies there is some number c in (a, b) for which $d'(c) = 0$. Now

$$d'(x) = f'(x) - \frac{f(b) - f(a)}{b - a}$$

and so $d'(c) = 0$ is the same as

$$f'(c) = \frac{f(b) - f(a)}{b - a}. \qquad \blacksquare$$

As indicated in FIGURE 4.4.7, there may be more than one number c in (a, b) for which the tangent lines and secant line are parallel.

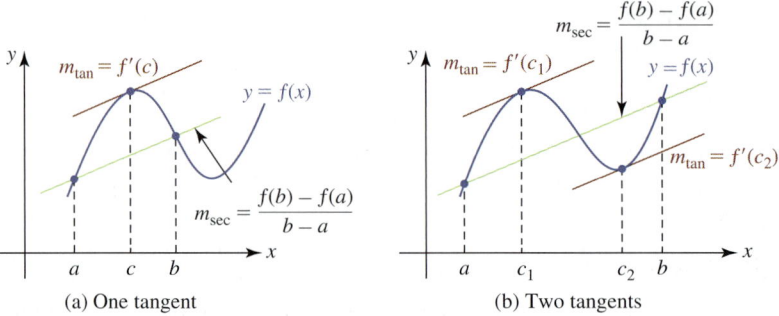

(a) One tangent (b) Two tangents

FIGURE 4.4.7 Tangents are parallel to secant line through $(a, f(a))$ and $(b, f(b))$

EXAMPLE 3 Verifying the Mean Value Theorem

Given the function $f(x) = x^3 - 12x$ defined on the closed interval $[-1, 3]$, does there exist a number c in the open interval $(-1, 3)$ that satisfies the conclusion of the Mean Value Theorem?

Solution Since f is a polynomial function, it is continuous on $[-1, 3]$ and differentiable on $(-1, 3)$. Now, $f(3) = -9, f(-1) = 11$,

$$f'(x) = 3x^2 - 12, \qquad \text{and} \qquad f'(c) = 3c^2 - 12.$$

Hence, we must have

$$\frac{f(3) - f(-1)}{3 - (-1)} = \frac{-20}{4} = 3c^2 - 12.$$

Thus, $3c^2 = 7$. Although the last equation has two solutions, the only solution in the interval $(-1, 3)$ is $c = \sqrt{7/3} \approx 1.53$. $\qquad \blacksquare$

The Mean Value Theorem is very useful in proving other theorems. Recall from Section 3.2 that if $f(x) = k$ is a constant function, then $f'(x) = 0$. The converse of this result is proved in the next theorem.

Theorem 4.4.3 Constant Function

If $f'(x) = 0$ for all x in an interval $[a, b]$, then $f(x)$ is a constant on the interval.

PROOF Let x_1 and x_2 be any numbers in $[a, b]$ such that $x_1 < x_2$. By the Mean Value Theorem, there is a number c in the interval (x_1, x_2) such that

$$\frac{f(x_2) - f(x_1)}{x_2 - x_1} = f'(c).$$

But $f'(c) = 0$ by hypothesis. Hence, $f(x_2) - f(x_1) = 0$ or $f(x_1) = f(x_2)$. Since x_1 and x_2 are arbitrarily chosen, the function f has the same value at all points in the interval. Thus, f is constant. $\qquad \blacksquare$

■ **Increasing and Decreasing Functions**   Suppose a function $y = f(x)$ is defined on an interval I and that x_1 and x_2 are any two numbers in the interval such that $x_1 < x_2$. We saw in Section 1.3 that f is **increasing** on I if $f(x_1) < f(x_2)$, and **decreasing** on I if $f(x_1) > f(x_2)$. See Figure 1.3.4. Intuitively, the graph of an increasing function *rises* as x increases (that is, the graph goes up when read left to right) and the graph of a decreasing function *falls* as x increases. For example, $y = e^x$ increases on $(-\infty, \infty)$ and $y = e^{-x}$ decreases on $(-\infty, \infty)$. Of course, a function f can be increasing on certain intervals and decreasing on different intervals. For example, $y = \sin x$ increases on $[-\pi/2, \pi/2]$ and decreases on $[\pi/2, 3\pi/2]$.

The graph in FIGURE 4.4.8 illustrates a function f that is increasing on the intervals $[b, c]$ and $[d, e]$ and decreasing on $[a, b]$, $[c, d]$, and $[e, h]$.

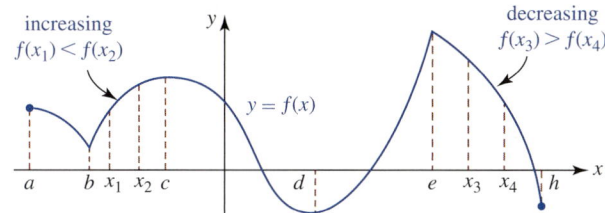

FIGURE 4.4.8   A function can increase on some intervals and decrease on others

The following theorem is a derivative test for increasing/decreasing.

Theorem 4.4.4   Test for Increasing/Decreasing

Let f be a function that is continuous on $[a, b]$ and differentiable on (a, b).

(*i*) If $f'(x) > 0$ for all x in (a, b), then f is increasing on $[a, b]$.
(*ii*) If $f'(x) < 0$ for all x in (a, b), then f is decreasing on $[a, b]$.

PROOF   (*i*) Let x_1 and x_2 be any two numbers in $[a, b]$ such that $x_1 < x_2$. By the Mean Value Theorem, there is a number c in the interval (x_1, x_2) such that

$$f'(c) = \frac{f(x_2) - f(x_1)}{x_2 - x_1}.$$

But $f'(c) > 0$ by hypothesis. Hence, $f(x_2) - f(x_1) > 0$ or $f(x_1) < f(x_2)$. Since x_1 and x_2 are arbitrarily chosen, it follows that f is increasing on $[a, b]$.
(*ii*) If $f'(c) < 0$, then $f(x_2) - f(x_1) < 0$ or $f(x_1) > f(x_2)$. Since x_1 and x_2 are arbitrarily chosen, it follows that f is decreasing on $[a, b]$.   ■

EXAMPLE 4   Derivative Test for Increasing/Decreasing

Determine the intervals on which $f(x) = x^3 - 3x^2 - 24x$ is increasing and the intervals on which f is decreasing.

Solution   The derivative is

$$f'(x) = 3x^2 - 6x - 24 = 3(x + 2)(x - 4).$$

To determine when $f'(x) > 0$ and $f'(x) < 0$ we must solve

$$(x + 2)(x - 4) > 0 \qquad \text{and} \qquad (x + 2)(x - 4) < 0,$$

respectively. One way of solving these inequalities is to examine the algebraic signs of the factors $(x + 2)$ and $(x - 4)$ on the intervals of the number line determined by the critical points -2 and 4: $(-\infty, -2]$, $[-2, 4]$, $[4, \infty)$. See FIGURE 4.4.9.

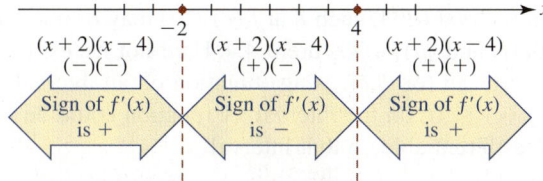

◀ In precalculus this procedure for solving nonlinear inequalities is called the *sign chart method*.

FIGURE 4.4.9   Signs of $f'(x)$ in three intervals in Example 4

The information garnered from Figure 4.4.9 is summarized in the accompanying table.

Interval	Sign of $f'(x)$	$y = f(x)$
$(-\infty, -2)$	$+$	increasing on $(-\infty, -2]$
$(-2, 4)$	$-$	decreasing on $[-2, 4]$
$(4, \infty)$	$+$	increasing on $[4, \infty)$

EXAMPLE 5 Derivative Test for Increasing/Decreasing

Determine the intervals on which $f(x) = \sqrt{x}\,e^{-x/2}$ is increasing and the intervals on which f is decreasing.

Solution First observe that the domain of f is defined by $x \geq 0$. Next, the derivative

$$f'(x) = x^{1/2}e^{-x/2}\left(-\frac{1}{2}\right) + \frac{1}{2}x^{-1/2}e^{-x/2} = \frac{e^{-x/2}}{2\sqrt{x}}(1 - x)$$

is zero at 1 and undefined at 0. Since 0 is in the domain of f and since $f'(x) \to \infty$ as $x \to 0^+$, we conclude that the graph of f has a vertical tangent (the y-axis) at $(0, 0)$. In addition, because $e^{-x/2}/2\sqrt{x} > 0$ for $x > 0$ we need only solve

$$1 - x > 0 \qquad \text{and} \qquad 1 - x < 0$$

to determine where $f'(x) > 0$ and $f'(x) < 0$, respectively. The results are given in the accompanying table.

Interval	Sign of $f'(x)$	$y = f(x)$
$(0, 1)$	$+$	increasing on $[0, 1]$
$(1, \infty)$	$-$	decreasing on $[1, \infty)$

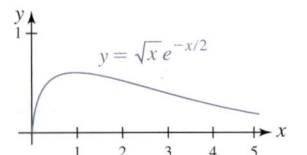

$y = \sqrt{x}\,e^{-x/2}$

FIGURE 4.4.10 Graph of function in Example 5

With the aid of a graphing utility we obtain the graph of f given in FIGURE 4.4.10.

If a function f is discontinuous at one or both endpoints of $[a, b]$, then $f'(x) > 0$ (or $f'(x) < 0$) on (a, b) implies f is increasing (or decreasing) on the open interval (a, b).

▌ **Postscript—A Bit of History** Michel Rolle (1652–1719), a French elementary school teacher, was deeply interested in mathematics and despite a very rudimentary education solved several theorems of note. But curiously Rolle did not prove the theorem that bears his name. Indeed, he was one of the early and vocal critics of the, then new, calculus. Rolle is also credited with inventing the symbolism $\sqrt[n]{x}$ to denote the nth root of a number x.

$f'(x)$ NOTES FROM THE CLASSROOM

(*i*) As mentioned previously, the hypotheses stated in Rolle's Theorem as well as the hypotheses of the Mean Value Theorem are sufficient but not necessary conditions. In Rolle's Theorem, for example, if one or more of the hypotheses: continuity on $[a, b]$, differentiability on (a, b), and $f(a) = f(b) = 0$ do not hold, the conclusion there exists a number c in the open interval (a, b) such that $f'(c) = 0$ may or may not hold.

(*ii*) The converses of parts (*i*) and (*ii*) of Theorem 4.4.4 are not necessarily true. In other words, when f is an increasing (or decreasing) function on an interval, it does not follow that $f'(x) > 0$ (or $f'(x) < 0$) on the interval. A function could be increasing on an interval and yet not be differentiable on that interval.

Exercises 4.4   Answers to selected odd-numbered problems begin on page ANS-14.

≡ Fundamentals

In Problems 1–10, determine whether the given function satisfies the hypotheses of Rolle's Theorem on the indicated interval. If so, find all values of c that satisfy the conclusion of the theorem.

1. $f(x) = x^2 - 4;$ $[-2, 2]$
2. $f(x) = x^2 - 6x + 5;$ $[1, 5]$
3. $f(x) = x^3 + 27;$ $[-3, -2]$
4. $f(x) = x^3 - 5x^2 + 4x;$ $[0, 4]$
5. $f(x) = x^3 + x^2;$ $[-1, 0]$
6. $f(x) = x(x - 1)^2;$ $[0, 1]$
7. $f(x) = \sin x;$ $[-\pi, 2\pi]$
8. $f(x) = \tan x;$ $[0, \pi]$
9. $f(x) = x^{2/3} - 1;$ $[-1, 1]$
10. $f(x) = x^{2/3} - 3x^{1/3} + 2;$ $[1, 8]$

In Problems 11 and 12, state why the function f whose graph is given does not satisfy the hypotheses of Rolle's Theorem on $[a, b]$.

11.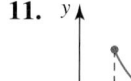

FIGURE 4.4.11   Graph for Problem 11

12.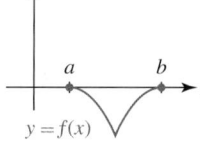

FIGURE 4.4.12   Graph for Problem 12

In Problems 13–22, determine whether the given function satisfies the hypotheses of the Mean Value Theorem on the indicated interval. If so, find all values of c that satisfy the conclusion of the theorem.

13. $f(x) = x^2;$ $[-1, 7]$
14. $f(x) = -x^2 + 8x - 6;$ $[2, 3]$
15. $f(x) = x^3 + x + 2;$ $[2, 5]$
16. $f(x) = x^4 - 2x^2;$ $[-3, 3]$
17. $f(x) = 1/x;$ $[-10, 10]$
18. $f(x) = x + \dfrac{1}{x};$ $[1, 5]$
19. $f(x) = 1 + \sqrt{x};$ $[0, 9]$
20. $f(x) = \sqrt{4x + 1};$ $[2, 6]$
21. $f(x) = \dfrac{x + 1}{x - 1};$ $[-2, -1]$
22. $f(x) = x^{1/3} - x;$ $[-8, 1]$

In Problems 23 and 24, state why the function f whose graph is given does not satisfy the hypotheses of the Mean Value Theorem on $[a, b]$.

23.

FIGURE 4.4.13   Graph for Problem 23

24.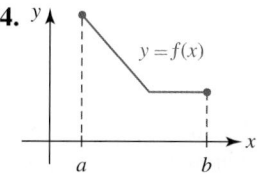

FIGURE 4.4.14   Graph for Problem 24

In Problems 25–46, determine the intervals on which the given function f is increasing and the intervals on which f is decreasing.

25. $f(x) = x^2 + 5$
26. $f(x) = x^3$
27. $f(x) = x^2 + 6x - 1$
28. $f(x) = -x^2 + 10x + 3$
29. $f(x) = x^3 - 3x^2$
30. $f(x) = \dfrac{1}{3}x^3 - x^2 - 8x + 1$
31. $f(x) = x^4 - 4x^3 + 9$
32. $f(x) = 4x^5 - 10x^4 + 2$
33. $f(x) = 1 - x^{1/3}$
34. $f(x) = x^{2/3} - 2x^{1/3}$
35. $f(x) = x + \dfrac{1}{x}$
36. $f(x) = \dfrac{1}{x} + \dfrac{1}{x^2}$
37. $f(x) = x\sqrt{8 - x^2}$
38. $f(x) = \dfrac{x + 1}{\sqrt{x^2 + 1}}$
39. $f(x) = \dfrac{5}{x^2 + 1}$
40. $f(x) = \dfrac{x^2}{x + 1}$
41. $f(x) = x(x - 3)^2$
42. $f(x) = (x^2 - 1)^3$
43. $f(x) = \sin x$
44. $f(x) = -x + \tan x$
45. $f(x) = x + e^{-x}$
46. $f(x) = x^2 e^{-x}$

In Problems 47 and 48, show, without graphing, that the given function has no relative extrema.

47. $f(x) = 4x^3 + x$
48. $f(x) = -x + \sqrt{2 - x}$

≡ Applications

49. A motorist enters a toll road and is given a stub stamped 1:15 P.M. Seventy miles down the road, when the motorist pays the toll at 2:15 P.M., he is also given a traffic ticket. Explain this by the Mean Value Theorem. Assume the speed limit is 65 mi/h.

50. In the mathematical analysis of the human cough, the trachea, or windpipe, is assumed to be a cylindrical tube. A mathematical model for the volume of air (in cm³/s) flowing through the trachea during its contraction is

$$V(r) = kr^4(r_0 - r), \quad r_0/2 \le r \le r_0,$$

where k is a positive constant and r_0 is its radius when there is no pressure difference at the ends of the tracheal tube. Determine an interval for which V is increasing and an interval for which V is decreasing. What radius will give the maximum volume flow of air?

≡ Think About It

51. Consider the function $f(x) = x^4 + x^3 - x - 1$. Use this function and Rolle's Theorem to show that the equation $4x^3 + 3x^2 - 1 = 0$ has at least one root in $[-1, 1]$.

52. Suppose the functions f and g are continuous on $[a, b]$ and differentiable on (a, b) such that $f'(x) > 0$ and $g'(x) > 0$ for all x in (a, b). Show that $f + g$ is an increasing function on $[a, b]$.

53. Suppose the functions f and g are continuous on $[a, b]$ and differentiable on (a, b) such that $f'(x) > 0$ and $g'(x) > 0$ for all x in (a, b). Give a condition on $f(x)$ and $g(x)$ that will guarantee that the product fg is increasing on $[a, b]$.

54. Show that the equation $ax^3 + bx + c = 0$, $a > 0$, $b > 0$, cannot have two real roots. [*Hint*: Consider the function $f(x) = ax^3 + bx + c$. Suppose there are two numbers r_1 and r_2 such that $f(r_1) = f(r_2) = 0$.]

55. Show that the equation $ax^2 + bx + c = 0$ has at most two real roots. [*Hint*: Consider the function $f(x) = ax^2 + bx + c$. Suppose there are three distinct numbers r_1, r_2, and r_3 such that $f(r_1) = f(r_2) = f(r_3) = 0$.]

56. For a quadratic polynomial function $f(x) = ax^2 + bx + c$ show that the value of x_3 that satisfies the conclusion of the Mean Value Theorem on any interval $[x_1, x_2]$ is $x_3 = (x_1 + x_2)/2$.

57. Suppose the graph of a polynomial function f has four distinct x-intercepts. Discuss: What is the minimum number of points at which a tangent line to the graph of f is horizontal?

58. As mentioned after Example 2, the hypothesis $f(a) = f(b) = 0$ in Rolle's Theorem can be replaced with the hypothesis $f(a) = f(b)$.

(a) Find an explicit function f defined on an interval $[a, b]$ such that f is continuous on the interval, differentiable on (a, b), and $f(a) = f(b)$.

(b) Find a number c for which $f'(c) = 0$.

59. Consider the function $f(x) = x \sin x$. Use f and Rolle's Theorem to show that the equation $\cot x = -1/x$ has a solution on the interval $(0, \pi)$.

≡ Calculator/CAS Problems

60. (a) Use a calculator or CAS to obtain the graph of $f(x) = x - 4x^{1/3}$.

(b) Verify that all but one of the hypotheses of Rolle's Theorem are satisfied on the interval $[-8, 8]$.

(c) Determine whether there exists a number c in $(-8, 8)$ for which $f'(c) = 0$.

In Problems 61 and 62, use a calculator to find a value of c that satisfies the conclusion of the Mean Value Theorem.

61. $f(x) = \cos 2x$; $\quad [0, \pi/4]$

62. $f(x) = 1 + \sin x$; $\quad [\pi/4, \pi/2]$

4.5 Limits Revisited—L'Hôpital's Rule

■ **Introduction** In Chapters 2 and 3, we saw how the concept of a limit leads to the notion of the derivative of a function. In this section, we turn the tables around. We will see how the derivative can be used to calculate certain limits with indeterminate forms.

■ **Terminology** Recall, in Chapter 2 we considered limits of quotients such as

$$\lim_{x \to 1} \frac{x^2 + 3x - 4}{x - 1} \quad \text{and} \quad \lim_{x \to \infty} \frac{2x^2 - x}{3x^2 + 1}. \tag{1}$$

The first limit in (1) has the indeterminate form $0/0$ at $x = 1$, whereas the second limit has the indeterminate form ∞/∞. In general, we say that the limit

$$\lim_{x \to a} \frac{f(x)}{g(x)}$$

has the **indeterminate form 0/0** at $x = a$ if

$$f(x) \to 0 \quad \text{and} \quad g(x) \to 0 \quad \text{as} \quad x \to a$$

and has the **indeterminate form ∞/∞** at $x = a$ if

$$|f(x)| \to \infty \quad \text{and} \quad |g(x)| \to \infty \quad \text{as} \quad x \to a.$$

The absolute value signs here mean that as x approaches a we could have, say,

$$f(x) \to \infty, \quad g(x) \to -\infty; \text{ or}$$
$$f(x) \to -\infty, \quad g(x) \to \infty; \text{ or}$$
$$f(x) \to -\infty, \quad g(x) \to -\infty,$$

and so on. A limit can also have an indeterminate form as

$$x \to a^-, \quad x \to a^+, \quad x \to -\infty, \quad \text{or} \quad x \to \infty.$$

Limits of the form

$$\frac{0}{k}, \qquad \frac{k}{0}, \qquad \frac{\infty}{k}, \qquad \text{and} \qquad \frac{k}{\infty},$$

◀ Note

where k is a *nonzero* constant, are *not* indeterminate forms. It is worth remembering that:

- *The value of a limit whose form is $0/k$ or k/∞ is 0.* $\hspace{2cm}$ (2)

- *A limit whose form is either $k/0$ or ∞/k does not exist.* $\hspace{1.5cm}$ (3)

In establishing whether limits of quotients such as those given in (1) exist, we resorted to the algebraic manipulations of factoring, canceling, and dividing. However, recall that the proof of $\lim\limits_{x\to 0}(\sin x)/x = 1$ used an elaborate geometric argument. But, algebra and geometric intuition fail miserably when confronted with a problem of the type

$$\lim_{x\to 0} \frac{\sin x}{e^x - e^{-x}},$$

which has the indeterminate form $0/0$. The next theorem will aid us in proving a rule that is extremely helpful in evaluating many limits that have an indeterminate form.

Theorem 4.5.1   Extended Mean Value Theorem

Let f and g be continuous on $[a, b]$ and differentiable on (a, b) and $g'(x) \neq 0$ for all x in (a, b). Then there exists a number c in (a, b) such that

$$\frac{f(b) - f(a)}{g(b) - g(a)} = \frac{f'(c)}{g'(c)}.$$

Observe that Theorem 4.5.1 reduces to the Mean Value Theorem when $g(x) = x$. A proof of this theorem, which is reminiscent of the proof of Theorem 4.4.2, will not be given.

The following rule is named after the French mathematician G.F.A. L'Hôpital.

Theorem 4.5.2   L'Hôpital's Rule

Suppose f and g are differentiable functions on an open interval containing the number a, except possibly at a itself, and that $g'(x) \neq 0$ for all x in the interval, except possibly at a. If $\lim\limits_{x\to a} f(x)/g(x)$ is an indeterminate form, and $\lim\limits_{x\to a} f'(x)/g'(x) = L$ or $\pm\infty$, then

$$\lim_{x\to a} \frac{f(x)}{g(x)} = \lim_{x\to a} \frac{f'(x)}{g'(x)}. \qquad (4)$$

PROOF OF THE CASE 0/0   Let the open interval be denoted by (r, s). Since we are assuming that

$$\lim_{x\to a} f(x) = 0 \qquad \text{and} \qquad \lim_{x\to a} g(x) = 0,$$

it can be further assumed that $f(a) = 0$ and $g(a) = 0$. It follows that f and g are continuous at a. Moreover, since f and g are differentiable, these functions are continuous on the open intervals (r, a) and (a, s). Consequently, f and g are continuous on the interval (r, s). Now, for any $x \neq a$ in the interval, Theorem 4.5.1 is applicable to either $[x, a]$ or $[a, x]$. In either case, there exists a number c between x and a such that

$$\frac{f(x) - f(a)}{g(x) - g(a)} = \frac{f(x)}{g(x)} = \frac{f'(c)}{g'(c)}.$$

Letting $x \to a$ implies $c \to a$, and so

$$\lim_{x\to a} \frac{f(x)}{g(x)} = \lim_{x\to a} \frac{f'(c)}{g'(c)} = \lim_{c\to a} \frac{f'(c)}{g'(c)} = \lim_{x\to a} \frac{f'(x)}{g'(x)}. \qquad \blacksquare$$

EXAMPLE 1   Indeterminate Form 0/0

Evaluate $\lim\limits_{x\to 0}\dfrac{\sin x}{x}$.

Solution   Since the given limit has the indeterminate form 0/0 at $x = 0$, it follows from (4) that we can write

$$\lim_{x\to 0}\frac{\sin x}{x}\overset{h}{=}\lim_{x\to 0}\frac{\dfrac{d}{dx}\sin x}{\dfrac{d}{dx}x}$$

$$=\lim_{x\to 0}\frac{\cos x}{1}=\frac{1}{1}=1.$$

The italic red *h* above the first equality indicates the two limits are equal as a result of an application of L'Hôpital's Rule.

EXAMPLE 2   Indeterminate Form 0/0

Evaluate $\lim\limits_{x\to 0}\dfrac{\sin x}{e^x - e^{-x}}$.

Solution   Since the given limit has the indeterminate form 0/0 at $x = 0$, we apply (4):

$$\lim_{x\to 0}\frac{\sin x}{e^x - e^{-x}}\overset{h}{=}\lim_{x\to 0}\frac{\dfrac{d}{dx}\sin x}{\dfrac{d}{dx}(e^x - e^{-x})}$$

$$=\lim_{x\to 0}\frac{\cos x}{e^x + e^{-x}}=\frac{1}{1+1}=\frac{1}{2}.$$

The result given in (4) remains valid when $x \to a$ is replaced by one-sided limits or by $x\to\infty$, $x\to -\infty$. The proof of the case $x\to\infty$ can be obtained by using the substitution $x = 1/t$ in $\lim\limits_{x\to\infty} f(x)/g(x)$ and noting that $x\to\infty$ is equivalent to $t\to 0^+$.

EXAMPLE 3   Indeterminate Form ∞/∞

Evaluate $\lim\limits_{x\to\infty}\dfrac{\ln x}{e^x}$.

Solution   The limit has the indeterminate form ∞/∞. Thus, from L'Hôpital's Rule we have

$$\lim_{x\to\infty}\frac{\ln x}{e^x}\overset{h}{=}\lim_{x\to\infty}\frac{1/x}{e^x}=\lim_{x\to\infty}\frac{1}{xe^x}.$$

In this latter limit, $xe^x\to\infty$ as $x\to\infty$, whereas 1 remains constant. Consequently by (2),

$$\lim_{x\to\infty}\frac{\ln x}{e^x}=\lim_{x\to\infty}\frac{1}{xe^x}=0$$

It may be necessary to apply L'Hôpital's Rule several times in the course of solving a problem.

EXAMPLE 4   Successive Applications of L'Hôpital's Rule

Evaluate $\lim\limits_{x\to\infty}\dfrac{6x^2 + 5x + 7}{4x^2 + 2x}$.

Solution   The indeterminate form is clearly ∞/∞, and so by (4),

$$\lim_{x\to\infty}\frac{6x^2 + 5x + 7}{4x^2 + 2x}\overset{h}{=}\lim_{x\to\infty}\frac{12x + 5}{8x + 2}.$$

Since the new limit still has the indeterminate form ∞/∞, we apply (4) a second time:

$$\lim_{x\to\infty}\frac{12x + 5}{8x + 2}\overset{h}{=}\lim_{x\to\infty}\frac{12}{8}=\frac{3}{2}.$$

We have shown that

$$\lim_{x\to\infty} \frac{6x^2 + 5x + 7}{4x^2 + 2x} = \frac{3}{2}$$ ∎

EXAMPLE 5   Successive Applications of L'Hôpital's Rule

Evaluate $\lim\limits_{x\to\infty} \dfrac{e^{3x}}{x^2}$.

Solution   The given limit and the limit obtained after one application of L'Hôpital's Rule have the indeterminate form ∞/∞:

$$\lim_{x\to\infty} \frac{e^{3x}}{x^2} \overset{h}{=} \lim_{x\to\infty} \frac{3e^{3x}}{2x} \overset{h}{=} \lim_{x\to\infty} \frac{9e^{3x}}{2}.$$

After the second application of (4), we observe that $e^{3x} \to \infty$ while the denominator remains constant. From this we conclude that

$$\lim_{x\to\infty} \frac{e^{3x}}{x^2} = \infty.$$

In other words, the given limit does not exist.   ∎

EXAMPLE 6   Successive Applications of L'Hôpital's Rule

Evaluate $\lim\limits_{x\to\infty} \dfrac{x^4}{e^{2x}}$.

Solution   We apply (4) four times:

$$\lim_{x\to\infty} \frac{x^4}{e^{2x}} \overset{h}{=} \lim_{x\to\infty} \frac{4x^3}{2e^{2x}} \quad (\infty/\infty)$$

$$\overset{h}{=} \lim_{x\to\infty} \frac{12x^2}{4e^{2x}} \quad (\infty/\infty)$$

$$\overset{h}{=} \lim_{x\to\infty} \frac{6x}{2e^{2x}} \quad (\infty/\infty)$$

$$\overset{h}{=} \lim_{x\to\infty} \frac{6}{4e^{2x}} = 0.$$ ∎

In successive applications of L'Hôpital's Rule, it is sometimes possible to change a limit from one indeterminate form to another, say, ∞/∞ to 0/0.

EXAMPLE 7   Indeterminate Form ∞/∞

Evaluate $\lim\limits_{t\to\pi/2^+} \dfrac{\tan t}{\tan 3t}$.

Solution   We observe that $\tan t \to -\infty$ and $\tan 3t \to -\infty$ as $t \to \pi/2^+$. Hence, from (4),

$$\lim_{t\to\pi/2^+} \frac{\tan t}{\tan 3t} \overset{h}{=} \lim_{t\to\pi/2^+} \frac{\sec^2 t}{3\sec^2 3t} \quad (\infty/\infty) \qquad \leftarrow \text{rewrite using } \sec t = 1/\cos t$$

$$= \lim_{t\to\pi/2^+} \frac{\cos^2 3t}{3\cos^2 t} \quad (0/0)$$

$$\overset{h}{=} \lim_{t\to\pi/2^+} \frac{2\cos 3t(-3\sin 3t)}{6\cos t(-\sin t)}$$

$$= \lim_{t\to\pi/2^+} \frac{2\sin 3t \cos 3t}{2\sin t \cos t} \qquad \leftarrow \text{rewrite using a double-angle formula on numerator and denominator}$$

$$= \lim_{t\to\pi/2^+} \frac{\sin 6t}{\sin 2t} \quad (0/0)$$

$$\overset{h}{=} \lim_{t\to\pi/2^+} \frac{6\cos 6t}{2\cos 2t} = \frac{-6}{-2} = 3.$$ ∎

EXAMPLE 8 One-Sided Limit

Evaluate $\lim\limits_{x \to 1^+} \dfrac{\ln x}{\sqrt{x-1}}$.

Solution The given limit has the indeterminate form 0/0 at $x = 1$. Hence, by L'Hôpital's Rule,

$$\lim_{x \to 1^+} \frac{\ln x}{\sqrt{x-1}} \overset{h}{=} \lim_{x \to 1^+} \frac{1/x}{\frac{1}{2}(x-1)^{-1/2}} = \lim_{x \to 1^+} \frac{2\sqrt{x-1}}{x} = \frac{0}{1} = 0.$$ ■

■ **Other Indeterminate Forms** There are five additional indeterminate forms:

$$\infty - \infty, \quad 0 \cdot \infty, \quad 0^0, \quad \infty^0, \quad \text{and} \quad 1^\infty. \tag{5}$$

By a combination of algebra and a little cleverness we can often convert one of these new limit forms to either 0/0 or ∞/∞.

■ **The Form $\infty - \infty$** The next example illustrates a limit that has the indeterminate form $\infty - \infty$. This example should destroy any unwarranted convictions that $\infty - \infty = 0$.

EXAMPLE 9 Indeterminate Form $\infty - \infty$

Evaluate $\lim\limits_{x \to 0^+} \left[\dfrac{3x+1}{\sin x} - \dfrac{1}{x} \right]$.

Solution We note that $(3x+1)/\sin x \to \infty$ and $1/x \to \infty$ as $x \to 0^+$. However, after writing the difference as a single fraction, we recognize the form 0/0:

$$\lim_{x \to 0^+} \left[\frac{3x+1}{\sin x} - \frac{1}{x} \right] = \lim_{x \to 0^+} \frac{3x^2 + x - \sin x}{x \sin x} \qquad \leftarrow \text{common denominator}$$

$$\overset{h}{=} \lim_{x \to 0^+} \frac{6x + 1 - \cos x}{x \cos x + \sin x}$$

$$\overset{h}{=} \lim_{x \to 0^+} \frac{6 + \sin x}{-x \sin x + 2 \cos x}$$

$$= \frac{6 + 0}{0 + 2} = 3.$$ ■

■ **The Form $0 \cdot \infty$** If

$$f(x) \to 0 \qquad \text{and} \qquad |g(x)| \to \infty \qquad \text{as} \quad x \to a,$$

then $\lim\limits_{x \to a} f(x)g(x)$ has the indeterminate form $0 \cdot \infty$. We can change a limit that has this form to one with the form 0/0 or ∞/∞ by writing, in turn,

$$f(x)g(x) = \frac{f(x)}{1/g(x)} \qquad \text{or} \qquad f(x)g(x) = \frac{g(x)}{1/f(x)}.$$

EXAMPLE 10 Indeterminate Form $0 \cdot \infty$

Evaluate $\lim\limits_{x \to \infty} x \sin \dfrac{1}{x}$.

Solution Since $1/x \to 0$, we have $\sin(1/x) \to 0$ as $x \to \infty$. Hence, the limit has the indeterminate form $0 \cdot \infty$. By writing

$$\lim_{x \to \infty} \frac{\sin(1/x)}{1/x}$$

we now have the form 0/0. Hence,

$$\lim_{x \to \infty} \frac{\sin(1/x)}{1/x} \overset{h}{=} \lim_{x \to \infty} \frac{(-x^{-2})\cos(1/x)}{(-x^{-2})}$$

$$= \lim_{x \to \infty} \cos \frac{1}{x} = 1.$$

In the last line we used the fact that $1/x \to 0$ as $x \to \infty$ and $\cos 0 = 1$. ■

■ **The Forms 0^0, ∞^0 and 1^∞**   Suppose $y = f(x)^{g(x)}$ tends toward 0^0, ∞^0, or 1^∞ as $x \to a$. By taking the natural logarithm of y:

$$\ln y = \ln f(x)^{g(x)} = g(x)\ln f(x)$$

we see that the right-hand side of

$$\lim_{x \to a} \ln y = \lim_{x \to a} g(x)\ln f(x)$$

has the form $0 \cdot \infty$. If it is assumed that $\lim_{x \to a} \ln y = \ln(\lim_{x \to a} y) = L$, then

$$\lim_{x \to a} y = e^L \qquad \text{or} \qquad \lim_{x \to a} f(x)^{g(x)} = e^L.$$

Of course, the procedure just outlined is applicable to limits involving

$$x \to a^-, \quad x \to a^+, \quad x \to \infty, \quad \text{or} \quad x \to -\infty.$$

EXAMPLE 11 Indeterminate Form 0^0

Evaluate $\lim_{x \to 0^+} x^{1/\ln x}$.

Solution Because $\ln x \to -\infty$ as $x \to 0^+$, it follows from (2) that $1/\ln x \to 0$. Thus the given limit has the indeterminate form 0^0. Now, if we set $y = x^{1/\ln x}$, then

$$\ln y = \frac{1}{\ln x} \ln x = 1.$$

Notice that we do not need L'Hôpital's Rule in this case, since

$$\lim_{x \to 0^+} \ln y = \lim_{x \to 0^+} 1 = 1 \qquad \text{or} \qquad \ln\left(\lim_{x \to 0^+} y\right) = 1.$$

Hence, $\lim_{x \to 0^+} y = e^1$ or equivalently $\lim_{x \to 0^+} x^{1/\ln x} = e$. ■

EXAMPLE 12 Indeterminate Form 1^∞

Evaluate $\lim_{x \to \infty}\left(1 - \frac{3}{x}\right)^{2x}$.

Solution Because $1 - 3/x \to 1$ as $x \to \infty$ the indeterminate form is 1^∞. If

$$y = \left(1 - \frac{3}{x}\right)^{2x} \qquad \text{then} \qquad \ln y = 2x \ln\left(1 - \frac{3}{x}\right).$$

Observe that the form of $\lim_{x \to \infty} 2x \ln(1 - 3/x)$ is $\infty \cdot 0$, whereas the form of

$$\lim_{x \to \infty} \frac{2 \ln\left(1 - \frac{3}{x}\right)}{\frac{1}{x}}$$

is $0/0$. Applying (4) to the latter limit and simplifying gives

$$\lim_{x \to \infty} 2 \frac{\ln(1 - 3/x)}{1/x} \overset{h}{=} \lim_{x \to \infty} 2 \frac{\frac{3/x^2}{(1 - 3/x)}}{-1/x^2} = \lim_{x \to \infty} \frac{-6}{(1 - 3/x)} = -6.$$

From $\lim_{x \to \infty} \ln y = \ln\left(\lim_{x \to \infty} y\right) = -6$ we conclude that $\lim_{x \to \infty} y = e^{-6}$ or

$$\lim_{x \to \infty}\left(1 - \frac{3}{x}\right)^{2x} = e^{-6}.$$ ■

■ **Postscript—A Bit of History** It is questionable whether the French mathematician **Marquis Guillaume François Antoine de L'Hôpital** (1661–1704) discovered the rule that bears his name. The result is probably due to Johann Bernoulli. However, L'Hôpital was the first to publish the rule in his text *Analyse des Infiniment Petits*. The book was published in 1696 and is considered to be the first textbook on calculus.

L'Hôpital

$f'(x)$ **NOTES FROM THE CLASSROOM**

(*i*) In the application of L'Hôpital's Rule, students will sometimes misinterpret

$$\lim_{x \to a} \frac{f'(x)}{g'(x)} \quad \text{as} \quad \lim_{x \to a} \frac{d}{dx}\frac{f(x)}{g(x)}.$$

Remember, L'Hôpital's Rule utilizes the *quotient of derivatives* and *not* the *derivative of the quotient*.

(*ii*) Inspect a problem before you leap to its solution. The limit $\lim_{x \to 0}(\cos x)/x$ is the form $1/0$ and, as a consequence, does not exist. Lack of mathematical forethought in writing

$$\lim_{x \to 0}\frac{\cos x}{x} = \lim_{x \to 0}\frac{-\sin x}{1} = 0$$

is an incorrect application of L'Hôpital's Rule. Of course, the "answer" has no significance.

(*iii*) L'Hôpital's Rule is not a cure-all for every indeterminate form. For example, $\lim_{x \to \infty} e^x/e^{x^2}$ is certainly of the form ∞/∞, but

$$\lim_{x \to \infty}\frac{e^x}{e^{x^2}} = \lim_{x \to \infty}\frac{e^x}{2xe^{x^2}}$$

is of no practical help.

Exercises 4.5   Answers to selected odd-numbered problems begin on page ANS-14.

≡ **Fundamentals**

In Problems 1–40, use L'Hôpital's Rule where appropriate to find the given limit, or state that it does not exist.

1. $\lim\limits_{x \to 0} \dfrac{\cos x - 1}{x}$

2. $\lim\limits_{t \to 3} \dfrac{t^3 - 27}{t - 3}$

3. $\lim\limits_{x \to 1} \dfrac{2x - 2}{\ln x}$

4. $\lim\limits_{x \to 0^+} \dfrac{\ln 2x}{\ln 3x}$

5. $\lim\limits_{x \to 0} \dfrac{e^{2x} - 1}{3x + x^2}$

6. $\lim\limits_{x \to 0} \dfrac{\tan x}{2x}$

7. $\lim\limits_{t \to \pi} \dfrac{5\sin^2 t}{1 + \cos t}$

8. $\lim\limits_{\theta \to 1} \dfrac{\theta^2 - 1}{e^{\theta^2} - e}$

9. $\lim\limits_{x \to 0} \dfrac{6 + 6x + 3x^2 - 6e^x}{x - \sin x}$

10. $\lim\limits_{x \to \infty} \dfrac{3x^2 - 4x^3}{5x + 7x^3}$

11. $\lim\limits_{x \to 0^+} \dfrac{\cot 2x}{\cot x}$

12. $\lim\limits_{x \to 0} \dfrac{\arcsin(x/6)}{\arctan(x/2)}$

13. $\lim\limits_{t \to 2} \dfrac{t^2 + 3t - 10}{t^3 - 2t^2 + t - 2}$

14. $\lim\limits_{r \to -1} \dfrac{r^3 - r^2 - 5r - 3}{(r + 1)^2}$

15. $\lim\limits_{x \to 0} \dfrac{x - \sin x}{x^3}$

16. $\lim\limits_{x \to 1} \dfrac{x^2 + 4}{x^2 + 1}$

17. $\lim\limits_{x \to 0} \dfrac{\cos 2x}{x^2}$

18. $\lim\limits_{x \to \infty} \dfrac{2e^{4x} + x}{e^{4x} + 3x}$

19. $\lim\limits_{x \to 1^+} \dfrac{\ln \sqrt{x}}{x - 1}$

20. $\lim\limits_{x \to \infty} \dfrac{\ln(3x^2 + 5)}{\ln(5x^2 + 1)}$

21. $\lim\limits_{x \to 2} \dfrac{e^{x^2} - e^{2x}}{x - 2}$

22. $\lim\limits_{x \to 0} \dfrac{4^x - 3^x}{x}$

23. $\lim\limits_{x \to \infty} \dfrac{x \ln x}{x^2 + 1}$

24. $\lim\limits_{t \to 0} \dfrac{1 - \cosh t}{t^2}$

25. $\lim\limits_{x \to 0} \dfrac{x - \tan^{-1} x}{x^3}$

26. $\lim\limits_{x \to 0} \dfrac{(\sin 2x)^2}{x^2}$

27. $\lim\limits_{x \to \infty} \dfrac{e^x}{x^4}$

28. $\lim\limits_{x \to \infty} \dfrac{e^{1/x}}{\sin(1/x)}$

29. $\lim\limits_{x \to 0} \dfrac{x - \tan^{-1} x}{x - \sin^{-1} x}$

30. $\lim\limits_{t \to 1} \dfrac{t^{1/3} - t^{1/2}}{t - 1}$

31. $\lim\limits_{u \to \pi/2} \dfrac{\ln(\sin u)}{(2u - \pi)^2}$

32. $\lim\limits_{\theta \to \pi/2} \dfrac{\tan \theta}{\ln(\cos \theta)}$

33. $\lim\limits_{x \to -\infty} \dfrac{1 + e^{-2x}}{1 - e^{-2x}}$

34. $\lim\limits_{x \to 0} \dfrac{e^x - x - 1}{2x^2}$

35. $\lim\limits_{r \to 0} \dfrac{r - \cos r}{r - \sin r}$

36. $\lim\limits_{t \to \pi} \dfrac{\csc 7t}{\csc 2t}$

37. $\lim\limits_{x \to 0^+} \dfrac{x^2}{\ln^2(1 + 3x)}$

38. $\lim\limits_{x \to 3} \left(\dfrac{\ln x - \ln 3}{x - 3}\right)^2$

39. $\lim\limits_{x \to 0} \dfrac{3x^2 + e^x - e^{-x} - 2\sin x}{x \sin x}$

40. $\lim\limits_{x \to 8} \dfrac{\sqrt{x + 1} - 3}{x^2 - 64}$

In Problems 41–74, identify the given limit as one of the indeterminate forms given in (5). Use L'Hôpital's Rule where appropriate to find the limit, or state that it does not exist.

41. $\lim\limits_{x \to 0}\left(\dfrac{1}{e^x - 1} - \dfrac{1}{x}\right)$

42. $\lim\limits_{x \to 0^+}(\cot x - \csc x)$

43. $\lim\limits_{x \to \infty} x(e^{1/x} - 1)$

44. $\lim\limits_{x \to 0^+} x \ln x$

45. $\lim\limits_{x \to 0^+} x^x$

46. $\lim\limits_{x \to 1^-} x^{1/(1-x)}$

47. $\lim\limits_{x \to 0}\left[\dfrac{1}{x} - \dfrac{1}{\sin x}\right]$

48. $\lim\limits_{x \to 0}\left[\dfrac{1}{x^2} - \dfrac{\cos 3x}{x^2}\right]$

49. $\lim\limits_{t \to 3}\left[\dfrac{\sqrt{t + 1}}{t^2 - 9} - \dfrac{2}{t^2 - 9}\right]$

50. $\lim\limits_{x \to 0^+}\left[\dfrac{1}{x} - \dfrac{1}{\ln(x + 1)}\right]$

51. $\lim\limits_{\theta \to 0} \theta \csc 4\theta$

52. $\lim\limits_{x \to \pi/2^-} (\sin^2 x)^{\tan x}$

53. $\lim\limits_{x \to \infty}(2 + e^x)^{e^{-x}}$

54. $\lim\limits_{x \to 0^-} (1 - e^x)^{x^2}$

55. $\lim\limits_{t \to \infty}\left(1 + \dfrac{3}{t}\right)^t$

56. $\lim\limits_{h \to 0}(1 + 2h)^{4/h}$

57. $\lim\limits_{x \to 0} x^{(1-\cos x)}$

58. $\lim\limits_{\theta \to 0}(\cos 2\theta)^{1/\theta^2}$

59. $\lim\limits_{x \to \infty} \dfrac{1}{x^2 \sin^2(2/x)}$

60. $\lim\limits_{x \to 1}(x^2 - 1)^{x^2}$

61. $\lim\limits_{x \to 1}\left[\dfrac{1}{x - 1} - \dfrac{5}{x^2 + 3x - 4}\right]$

62. $\lim\limits_{x \to 0}\left[\dfrac{1}{x^2} - \dfrac{1}{x}\right]$

63. $\lim\limits_{x \to \infty} x^5 e^{-x}$

64. $\lim\limits_{x \to \infty}(x + e^x)^{2/x}$

65. $\lim\limits_{x \to \infty} x\left(\dfrac{\pi}{2} - \arctan x\right)$

66. $\lim\limits_{t \to \pi/4}\left(t - \dfrac{\pi}{4}\right)\tan 2t$

67. $\lim\limits_{x \to \infty} x \tan\left(\dfrac{5}{x}\right)$

68. $\lim\limits_{x \to 0^+} x \ln(\sin x)$

69. $\lim\limits_{x \to -\infty}\left[\dfrac{1}{e^x} - x^2\right]$

70. $\lim\limits_{x \to 0}(1 + 5 \sin x)^{\cot x}$

71. $\lim\limits_{x \to \infty}\left(\dfrac{3x}{3x + 1}\right)^x$

72. $\lim\limits_{\theta \to \pi/2^-} (\sec^3 \theta - \tan^3 \theta)$

73. $\lim\limits_{x \to 0}(\sinh x)^{\tan x}$

74. $\lim\limits_{x \to 0^+} x^{(\ln x)^2}$

In Problems 75 and 76, find the given limit.

75. $\lim\limits_{x \to 0^+} \dfrac{1}{x} \ln\left(\dfrac{e^x - 1}{x}\right)$

76. $\lim\limits_{x \to \infty} \dfrac{1}{x} \ln\left(\dfrac{e^x - 1}{x}\right)$

≡ Calculator/CAS Problems

In Problems 77 and 78, use a calculator or CAS to obtain the graph of the given function for the value of n on the indicated interval. In each case conjecture the value of $\lim\limits_{x \to \infty} f(x)$.

77. $f(x) = \dfrac{e^x}{x^n}$; $n = 3$ on $[0, 15]$; $n = 4$ on $[0, 20]$;

$n = 5$ on $[0, 25]$

78. $f(x) = \dfrac{x^n}{e^x}$; $n = 3$ on $[0, 15]$; $n = 4$ on $[0, 15]$;

$n = 5$ on $[0, 20]$

In Problems 79 and 80, use $n! = 1 \cdot 2 \cdot 3 \cdots (n - 1) \cdot n$,

$$\dfrac{d^n}{dx^n} x^n = n!,$$

where n is a positive integer, and L'Hôpital's Rule to find the limit.

79. $\lim\limits_{x \to \infty} \dfrac{x^n}{e^x}$

80. $\lim\limits_{x \to \infty} \dfrac{e^x}{x^n}$

≡ Applications

81. Consider the circle shown in FIGURE 4.5.1.

(a) If the arc ABC is 5 in. long, express the area A of the shaded blue region as a function of the indicated angle θ. [*Hint:* The area of a circular sector is $\frac{1}{2}r^2\theta$ and the arc length on a circle is $r\theta$, where θ is measured in radians.]

(b) Evaluate $\lim\limits_{\theta \to 0} A(\theta)$.

(c) Evaluate $\lim\limits_{\theta \to 0} dA/d\theta$.

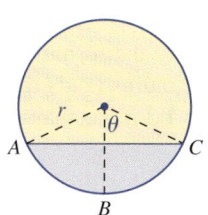

FIGURE 4.5.1 Circle in Problem 81

82. In the absence of damping forces, a mathematical model for the displacement $x(t)$ of a mass on a spring (see Problem 60 in Exercises 3.5) when the system is sinusoidally driven by an external force with amplitude F_0 and frequency $\gamma/2\pi$ is

$$x(t) = \dfrac{F_0}{\omega(\omega^2 - \gamma^2)}(-\gamma \sin \omega t + \omega \sin \gamma t), \quad \gamma \neq \omega,$$

where $\omega/2\pi$ is the frequency of free (undriven) vibrations of the system.

(a) When $\gamma = \omega$, the spring/mass system is said to be in **pure resonance**, and the displacement of the mass is defined by

$$x(t) = \lim\limits_{\gamma \to \omega} \dfrac{F_0}{\omega(\omega^2 - \gamma^2)}(-\gamma \sin \omega t + \omega \sin \gamma t).$$

Determine $x(t)$ by finding this limit.

(b) Use a graphing utility to examine the graph of $x(t)$ found in part (a) in the case where $F_0 = 2$, $\gamma = \omega = 1$. Describe the behavior of the spring/mass system in pure resonance as $t \to \infty$.

83. When an ideal gas expands from pressure p_1 and volume v_1 to pressure p_2 and volume v_2 such that $pv^\gamma = k$ (constant) throughout the expansion, if $\gamma \neq 1$, then the work done is given by

$$W = \dfrac{p_2 v_2 - p_1 v_1}{1 - \gamma}.$$

(a) Show that

$$W = p_1 v_1 \left[\frac{(v_2/v_1)^{1-\gamma} - 1}{1 - \gamma} \right].$$

(b) Find the work done in the case when $pv = k$ (constant) throughout the expansion by letting $\gamma \to 1$ in the expression in part (a).

84. The retina is most sensitive to photons that enter the eye near the center of the pupil, and less sensitive to light that enters near the edge of the pupil. (This phenomenon is known as the **Stiles–Crawford effect** of the first kind.) The percentage σ of photons that reach the photopigments is related to the pupil radius p (measured in mm) by the mathematical model

$$\sigma = \frac{1 - 10^{-0.05p^2}}{0.115p^2} \times 100.$$

See **FIGURE 4.5.2**.

(a) What percentage of photons reach the photopigments when $p = 2$ mm?

(b) What, according to the formula, is the limiting percentage as the pupil radius tends to zero? Can you explain why it seems to be more than 100%?

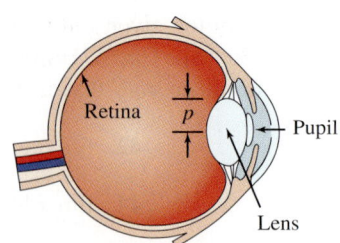

FIGURE 4.5.2 Eye in Problem 84

≡ **Think About It**

85. Suppose f has a second derivative. Evaluate

$$\lim_{h \to 0} \frac{f(x + h) - 2f(x) + f(x - h)}{h^2}.$$

86. (a) Use a calculator or CAS to obtain the graph of

$$f(x) = \frac{x \sin x}{x^2 + 1}.$$

(b) From the graph in part (a), conjecture the value of $\lim_{x \to \infty} f(x)$.

(c) Explain why L'Hôpital's Rule does not apply to $\lim_{x \to \infty} f(x)$.

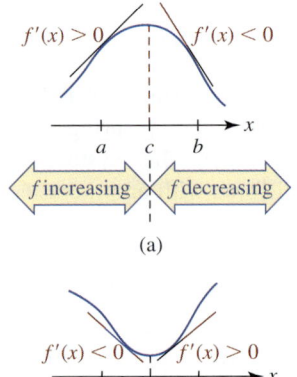

FIGURE 4.6.1 Relative maximum in (a); relative minimum in (b)

4.6 Graphing and the First Derivative

▌ **Introduction** Knowing that a function does, or does not, possess relative extrema is a great aid in drawing its graph. We saw in Section 4.3 (Theorem 4.3.2) that when a function has a relative extremum it must occur at a critical number. By finding the critical numbers of a function, we have a *list of candidates* for the x-coordinates of the points that correspond to relative extrema. We shall now combine the ideas of the earlier sections of this chapter to devise two tests for determining when a critical number actually is the x-coordinate of a relative extremum.

▌ **First Derivative Test** Suppose f is continuous on the closed interval $[a, b]$ and differentiable on an open interval (a, b) except possibly at a critical number c within the interval. If $f'(x) > 0$ for all x in (a, c) and $f'(x) < 0$ for all x in (c, b), then the graph of f on the interval (a, b) could be as shown in **FIGURE 4.6.1(a)**; that is, $f(c)$ is a relative maximum. On the other hand, when $f'(x) < 0$ for all x in (a, c) and $f'(x) > 0$ for all x in (c, b), then, as shown in Figure 4.6.1(b), $f(c)$ is a relative minimum. We have demonstrated two special cases of the next theorem.

Theorem 4.6.1 First Derivative Test

Let f be continuous on $[a, b]$ and differentiable on (a, b) except possibly at the critical number c.

(*i*) If $f'(x)$ changes from positive to negative at c, then $f(c)$ is a relative maximum.
(*ii*) If $f'(x)$ changes from negative to positive at c, then $f(c)$ is a relative minimum.
(*iii*) If $f'(x)$ has the same algebraic sign on each side of c, then $f(c)$ is not an extremum.

The conclusions of Theorem 4.6.1 can be summarized in one sentence:

- *A function f has a relative extremum at a critical number c where $f'(x)$ changes sign.*

FIGURE 4.6.2 illustrates what might be the case when $f'(c)$ *does not* change sign at a critical number c. In Figures 4.6.2(a) and 4.6.2(b) we have shown a horizontal tangent at $(c, f(c))$ and so $f'(c) = 0$ but $f(c)$ is neither a relative maximum nor a relative minimum. In

Figure 4.6.2(c) we have shown a vertical tangent at $(c, f(c))$ and so $f'(c)$ does not exist, but again $f(c)$ is not a relative extremum because $f'(c)$ does not change sign at the critical number c.

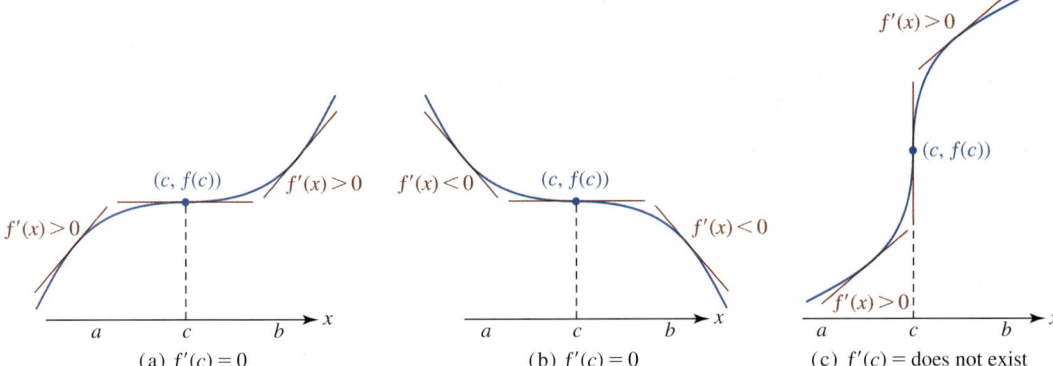

(a) $f'(c) = 0$ (b) $f'(c) = 0$ (c) $f'(c) =$ does not exist

FIGURE 4.6.2   No extremum because $f'(x)$ does not change sign at the critical number c

In the next five examples we will illustrate the usefulness of Theorem 4.6.1 in sketching a graph of a function f by hand. In addition to the calculus:

- Find the derivative of f and factor f' if possible.
- Find the critical numbers of f.
- Apply the First Derivative Test to each critical number.

it also pays to ask:

- What is the domain of f?
- Does the graph of f have any intercepts?   ← *x*-intercepts: Solve $f(x) = 0$
  y-intercept: Find $f(0)$
- Does the graph of f have any symmetry?   ← determine whether
  $f(-x) = f(x)$ or $f(-x) = -f(x)$
- Does the graph of f have any asymptotes?

The functions considered in Examples 1 and 2 are polynomials. Notice that these functions consist of both even and odd powers of x; this is sufficient to conclude that the graphs of these functions are not symmetric with respect to either the y-axis or the origin.

EXAMPLE 1   Polynomial Function of Degree 3

Graph $f(x) = x^3 - 3x^2 - 9x + 2$.

Solution   The first derivative

$$f'(x) = 3x^2 - 6x - 9 = 3(x + 1)(x - 3) \tag{1}$$

yields the critical numbers -1 and 3. Now the First Derivative Test is essentially the procedure used in finding the intervals on which f is either increasing or decreasing. We see in FIGURE 4.6.3(a) that $f'(x) > 0$ for $-\infty < x < -1$ and $f'(x) < 0$ for $-1 < x < 3$. In other words, $f'(x)$ changes from positive to negative at -1 and so it follows from part (*i*) of Theorem 4.6.1 that $f(-1) = 7$ is a relative maximum. Similarly, $f'(x) < 0$ for $-1 < x < 3$ and $f'(x) > 0$ for $3 < x < \infty$. Because $f'(x)$ changes from negative to positive at 3, part (*ii*) of Theorem 4.6.1 indicates that $f(3) = -25$ is a relative minimum. Now since $f(0) = 2$, the point $(0, 2)$ is the y-intercept for the graph of f. Furthermore, testing the equation $x^3 - 3x^2 - 9x + 2 = 0$ for rational roots reveals that $x = -2$ is a real root. Division by the factor $x + 2$ then gives $(x + 2)(x^2 - 5x + 1) = 0$. The quadratic formula applied to the quadratic factor reveals two additional real solutions:

◀ See the *SRM* for a brief review on how to find rational roots of polynomial equations.

$$\frac{1}{2}(5 - \sqrt{21}) \approx 0.21 \quad \text{and} \quad \frac{1}{2}(5 + \sqrt{21}) \approx 4.79.$$

The x-intercepts are then $(-2, 0)$, $\left(\frac{5}{2} - \frac{\sqrt{21}}{2}, 0\right)$, and $\left(\frac{5}{2} + \frac{\sqrt{21}}{2}, 0\right)$. Putting all this information together leads to the graph given in Figure 4.6.3(b).

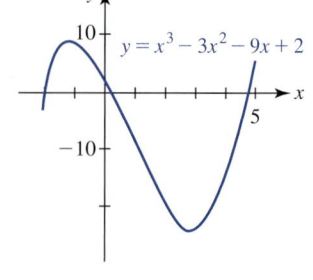

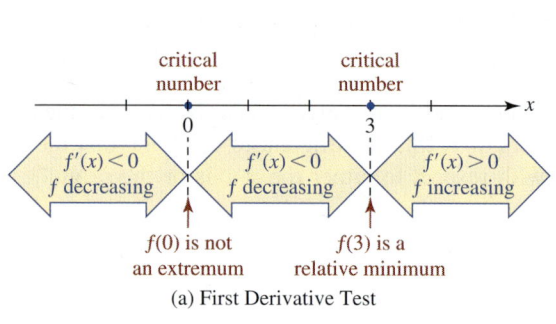

FIGURE 4.6.3 Graph of function in Example 1

EXAMPLE 2 Polynomial Function of Degree 4

Graph $f(x) = x^4 - 4x^3 + 10$.

Solution The derivative

$$f'(x) = 4x^3 - 12x^2 = 4x^2(x - 3)$$

shows that 0 and 3 are critical numbers. Now, as seen in FIGURE 4.6.4(a), f' has the same negative algebraic sign in the adjacent intervals $(-\infty, 0)$ and $(0, 3)$. Hence, $f(0) = 10$ is *not* an extremum. In this case $f'(0) = 0$ means there is only a horizontal tangent at the y-intercept $(0, f(0)) = (0, 10)$. However, it is evident from the First Derivative Test that $f(3) = -17$ is a relative minimum. Indeed, the information that f is decreasing on the left side and increasing on the right side of the critical number 3 (the graph of f cannot turn back down) allows us to conclude that $f(3) = -17$ is also an *absolute minimum*. Finally, we see that the graph of f has two x-intercepts. With the aid of a calculator or a CAS, the x-intercepts are approximately $(1.61, 0)$ and $(3.82, 0)$.

FIGURE 4.6.4 Graph of function in Example 2

EXAMPLE 3 Graph of a Rational Function

Graph $f(x) = \dfrac{x^2 - 3}{x^2 + 1}$.

Solution The following list summarizes some facts that can be discovered about the graph of this rational function f before actually graphing.

y-intercept: $f(0) = -3$, therefore the y-intercept is $(0, -3)$.
x-intercepts: $f(x) = 0$ when $x^2 - 3 = 0$. Thus, $x = -\sqrt{3}$ and $x = \sqrt{3}$. The x-intercepts are $(-\sqrt{3}, 0)$ and $(\sqrt{3}, 0)$.
Symmetry: y-axis, since $f(-x) = f(x)$.
Vertical asymptotes: None, since $x^2 + 1 \neq 0$ for all real numbers.
Horizontal asymptotes: Since the limit at infinity is the indeterminate form ∞/∞ we can use L'Hôpital's Rule to show

$$\lim_{x \to \infty} \frac{x^2 - 3}{x^2 + 1} \overset{h}{=} \lim_{x \to \infty} \frac{2x}{2x} = \lim_{x \to \infty} \frac{2}{2} = 1,$$

and so the line $y = 1$ is a horizontal asymptote.

Derivative: The Quotient Rule gives $f'(x) = \dfrac{8x}{(x^2 + 1)^2}$.

Critical numbers: $f'(x) = 0$ when $x = 0$. Therefore, 0 is the only critical number.

First Derivative Test: See FIGURE 4.6.5(a); $f(0) = -3$ is a relative minimum.

Graph: See Figure 4.6.5(b).

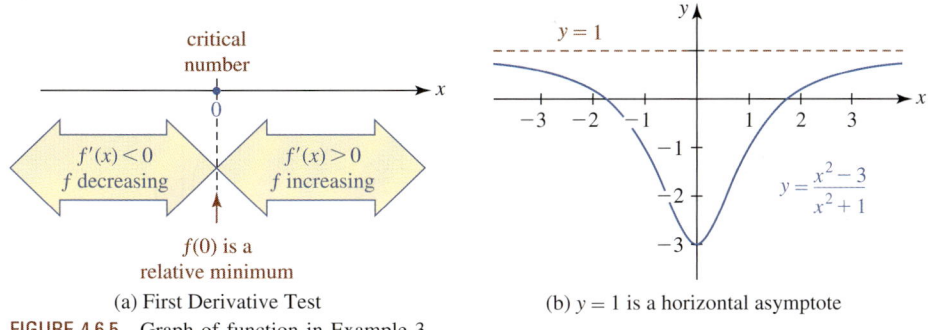

(a) First Derivative Test
(b) $y = 1$ is a horizontal asymptote

FIGURE 4.6.5   Graph of function in Example 3

EXAMPLE 4   Graph with a Vertical Asymptote

Graph $f(x) = x^2 + x - \ln|x|$.

Solution   First note that the domain of f is $(-\infty, 0) \cup (0, \infty)$. Then by setting the numerator of the derivative

$$f'(x) = 2x + 1 - \frac{1}{x} = \frac{2x^2 + x - 1}{x} = \frac{(2x - 1)(x + 1)}{x}$$

equal to zero we see that -1 and $\frac{1}{2}$ are critical numbers. Although f is not differentiable at $x = 0$, 0 is not a critical number since 0 is not in the domain of f. In fact, $x = 0$ is a vertical asymptote for $\ln|x|$ and is also a vertical asymptote for the graph of f. We put the critical numbers and 0 on the number line because the sign of the derivative to the left and right of 0 indicates the behavior of f. As seen in FIGURE 4.6.6(a), $f'(x) < 0$ for $-\infty < x < -1$ and $f'(x) > 0$ for $-1 < x < 0$. We conclude that $f(-1) = 0$ is a relative minimum (at the same time $f(-1) = 0$ shows that $x = -1$ is the x-coordinate of an x-intercept). Continuing, $f'(x) < 0$ for $0 < x < \frac{1}{2}$ and $f'(x) > 0$ for $\frac{1}{2} < x < \infty$ shows that $f(\frac{1}{2}) = \frac{3}{4} - \ln\frac{1}{2} \approx 1.44$ is another relative minimum.

As noted, f is not defined at $x = 0$ and so there is no y-intercept. Finally, there is no symmetry either with respect to the y-axis or with respect to the origin. The graph of the function f is given in Figure 4.6.6(b).

◀ Verify that $f(-x) \neq f(x)$ and $f(-x) \neq -f(x)$.

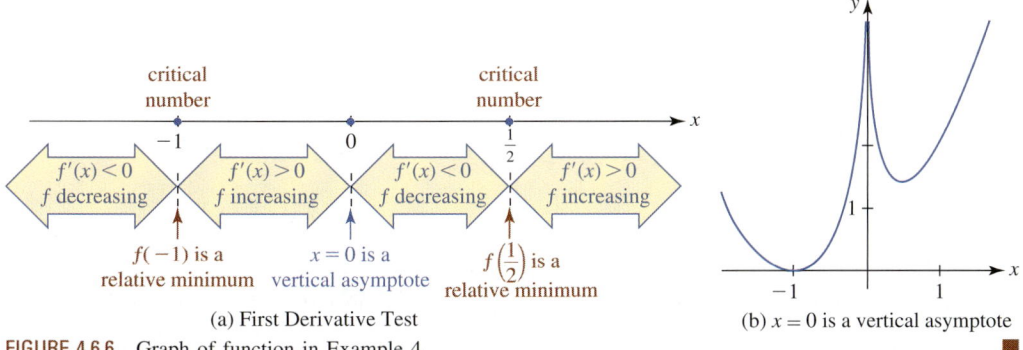

(a) First Derivative Test
(b) $x = 0$ is a vertical asymptote

FIGURE 4.6.6   Graph of function in Example 4

EXAMPLE 5   Graph with a Cusp

Graph $f(x) = -x^{5/3} + 5x^{2/3}$.

Solution   The derivative is

$$f'(x) = -\frac{5}{3}x^{2/3} + \frac{10}{3}x^{-1/3} = \frac{5}{3}\frac{(-x + 2)}{x^{1/3}}.$$

Notice that f' does not exist at 0 but 0 is in the domain of the function since $f(0) = 0$. The critical numbers are 0 and 2. The First Derivative Test, illustrated in FIGURE 4.6.7(a), shows that $f(0) = 0$ is a relative minimum and that $f(2) = -(2)^{5/3} + 5(2)^{2/3} \approx 4.76$ is a relative maximum.

Review the definition of a *cusp* in Section 3.2.

▶ Moreover, since $f'(x) \rightarrow \infty$ as $x \rightarrow 0^+$ and $f'(x) \rightarrow -\infty$ as $x \rightarrow 0^-$ there is a cusp at $(0, 0)$. Finally, by writing $f(x) = x^{2/3}(-x + 5)$, we see that $f(x) = 0$ at $x = 0$ and $x = 5$. The x-intercepts are the points $(0, 0)$ and $(5, 0)$. The graph of f is given in Figure 4.6.7(b).

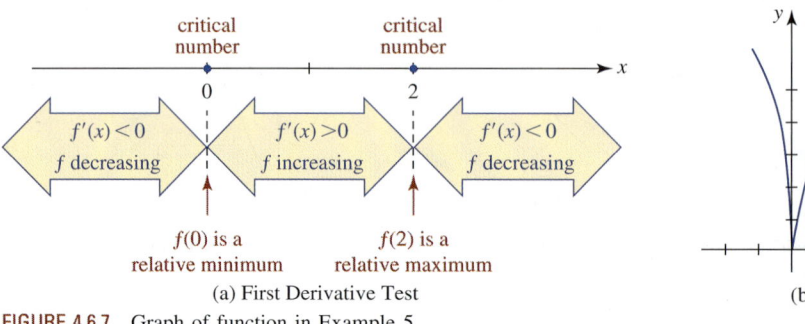

(a) First Derivative Test (b) Cusp at $(0, 0)$

FIGURE 4.6.7 Graph of function in Example 5

It is sometimes convenient to know in advance of graphing, or even with the bother of graphing, whether a relative extremum $f(c)$ is an *absolute* extremum. The next theorem helps a little. You should sketch some graphs and convince yourself of its plausibility.

Theorem 4.6.2 The Sole Critical Number Test

Suppose c is the only critical number of a function f within an interval I. If it is proved that $f(c)$ is a relative extremum, then $f(c)$ is an absolute extremum.

In Example 3, it was shown that $f(0) = 0$ is a relative minimum by the First Derivative Test. We could have also concluded immediately that this function value is an absolute minimum. This fact follows from Theorem 4.6.2 because 0 is the only critical number in the interval $(-\infty, \infty)$.

Exercises 4.6 Answers to selected odd-numbered problems begin on page ANS-15.

≣ **Fundamentals**

In Problems 1–32, use the First Derivative Test to find the relative extrema of the given function. Graph. Find intercepts when possible.

1. $f(x) = -x^2 + 2x + 1$ **2.** $f(x) = (x - 1)(x + 3)$

3. $f(x) = x^3 - 3x$ **4.** $f(x) = \frac{1}{3}x^3 - \frac{1}{2}x^2 + 1$

5. $f(x) = x(x - 2)^2$ **6.** $f(x) = -x^3 + 3x^2 + 9x - 1$

7. $f(x) = x^3 + x - 3$ **8.** $f(x) = x^3 + 3x^2 + 3x - 3$

9. $f(x) = x^4 + 4x$ **10.** $f(x) = (x^2 - 1)^2$

11. $f(x) = \frac{1}{4}x^4 + \frac{4}{3}x^3 + 2x^2$ **12.** $f(x) = 2x^4 - 16x^2 + 3$

13. $f(x) = -x^2(x - 3)^2$ **14.** $f(x) = -3x^4 + 8x^3 - 6x^2 - 2$

15. $f(x) = 4x^5 - 5x^4$ **16.** $f(x) = (x - 2)^2(x + 3)^3$

17. $f(x) = \frac{x^2 + 3}{x + 1}$ **18.** $f(x) = x + \frac{25}{x}$

19. $f(x) = \frac{1}{x} - \frac{1}{x^3}$ **20.** $f(x) = \frac{x^2}{x^2 - 4}$

21. $f(x) = \frac{10}{x^2 + 1}$ **22.** $f(x) = \frac{x^2}{x^4 + 1}$

23. $f(x) = (x^2 - 4)^{2/3}$ **24.** $f(x) = (x^2 - 1)^{1/3}$

25. $f(x) = x\sqrt{1 - x^2}$ **26.** $f(x) = x(x^2 - 5)^{1/3}$

27. $f(x) = x - 12x^{1/3}$ **28.** $f(x) = x^{4/3} + 32x^{1/3}$

29. $f(x) = x^3 - 24 \ln |x|$ **30.** $f(x) = \frac{\ln x}{x}$

31. $f(x) = (x + 3)^2 e^{-x}$ **32.** $f(x) = 8x^2 e^{-x^2}$

In Problems 33–36, sketch a graph of a function f whose derivative f' has the given graph.

33.

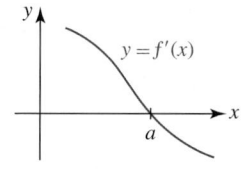

FIGURE 4.6.8   Graph for Problem 33

34.

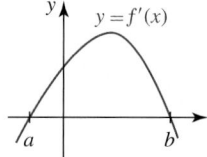

FIGURE 4.6.9   Graph for Problem 34

35.

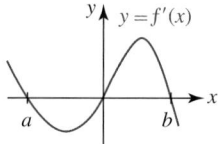

FIGURE 4.6.10   Graph for Problem 35

36.

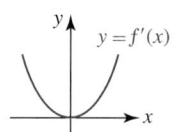

FIGURE 4.6.11   Graph for Problem 36

In Problems 37 and 38, sketch the graph of f' from the graph of f.

37.

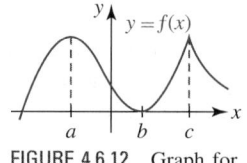

FIGURE 4.6.12   Graph for Problem 37

38.

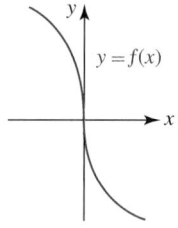

FIGURE 4.6.13   Graph for Problem 38

In Problems 39–42, sketch a graph of a function f that has the given properties.

39. $f(-1) = 0, f(0) = 1$
$f'(3)$ does not exist, $f'(5) = 0$
$f'(x) > 0, x < 3$ and $x > 5$
$f'(x) < 0, 3 < x < 5$

40. $f(0) = 0$
$f'(-1) = 0, f'(0) = 0, f'(1) = 0$
$f'(x) < 0, x < -1, -1 < x < 0$
$f'(x) > 0, 0 < x < 1, x > 1$

41. $f(-x) = f(x)$
$f(2) = 3$
$f'(x) < 0, 0 < x < 2$
$f'(x) > 0, x > 2$

42. $f(1) = -2, f(0) = -1$
$\lim\limits_{x \to 3} f(x) = \infty, f'(4) = 0$
$f'(x) < 0, x < 1$
$f'(x) < 0, x > 4$

In Problems 43 and 44, determine where the slope of the tangent to the graph of the given function has a relative maximum or a relative minimum.

43. $f(x) = x^3 + 6x^2 - x$ **44.** $f(x) = x^4 - 6x^2$

45. (a) From the graph of $g(x) = \sin 2x$ determine the intervals for which $g(x) > 0$ and the intervals for which $g(x) < 0$.

(b) Find the critical numbers of $f(x) = \sin^2 x$. Use the First Derivative Test and the information in part (a) to find the relative extrema of f.

(c) Sketch the graph of the function f in part (b).

46. (a) Find the critical numbers of $f(x) = x - \sin x$.

(b) Show that f has no relative extrema.

(c) Sketch the graph of f.

≡ Applications

47. The **arithmetic mean**, or **average**, of the n numbers $a_1, a_2, \ldots, a_n$ is given by

$$\bar{x} = \frac{a_1 + a_2 + \cdots + a_n}{n}.$$

(a) Show that $\bar{x}$ is a critical number of the function

$$f(x) = (x - a_1)^2 + (x - a_2)^2 + \cdots + (x - a_n)^2.$$

(b) Show that $f(\bar{x})$ is a relative minimum.

48. When sound passes from one medium to another, some of its energy can be lost because of a difference in the acoustic resistances of the two media. (Acoustic resistance is the product of density and elasticity.) The fraction of energy transmitted is given by

$$T(r) = \frac{4r}{(r + 1)^2},$$

where r is the ratio of the acoustic resistances of the two media.

(a) Show that $T(r) = T(1/r)$. Explain what this means physically.

(b) Use the First Derivative Test to find the relative extrema of T.

(c) Sketch the graph of the function T for $r \geq 0$.

≡ Think About It

49. Find values of a, b, and c such that $f(x) = ax^2 + bx + c$ has a relative maximum 6 at $x = 2$ and the graph of f has y-intercept 4.

50. Find values of a, b, c, and d such that $f(x) = ax^3 + bx^2 + cx + d$ has a relative minimum -3 at $x = 0$ and a relative maximum 4 at $x = 1$.

51. Suppose f is a differentiable function whose graph is symmetric about the y-axis. Prove that $f'(0) = 0$. Does f necessarily have a relative extremum at $x = 0$?

52. Let m and n denote positive integers. Show that $f(x) = x^m(x - 1)^n$ always has a relative minimum.

53. Suppose f and g are differentiable functions and have relative maxima at the same critical number c.

(a) Show that c is a critical number for the functions $f + g, f - g$, and fg.

(b) Does it follow that the functions $f + g, f - g$, and fg have relative maxima at c? Prove your assertions or give a counterexample.

4.7 Graphing and the Second Derivative

▪ **Introduction** In the discussion that follows, our goal is to relate the concept of the concavity of a graph with the second derivative of a function. The second derivative then provides us another way of testing to see whether a relative extremum of a function f occurs at a critical number.

▪ **Concavity** You probably have an *intuitive* idea of what is meant by concavity. FIGURES 4.7.1(a) and 4.7.1 (b) illustrate geometric shapes that are **concave up** and **concave down**, respectively. For example, the Gateway Arch in St. Louis is concave down; the cables between the vertical supports of the Golden Gate Bridge are concave up. Often a shape that is concave up is said to "hold water," whereas a shape that is concave down "spills water." However, the precise definition of concavity is given in terms of the derivative.

(a) "Holds water"

(b) "Spills water"
FIGURE 4.7.1 Concavity

Definition 4.7.1 Concavity

Let f be a differentiable function on an interval (a, b).

 (*i*) If f' is an increasing function on (a, b), then the graph of f is **concave up** on the interval.
(*ii*) If f' is a decreasing function on (a, b), then the graph of f is **concave down** on the interval.

In other words, if the slopes of the tangent lines to the graph of f increase (decrease) as x increases on (a, b), then the graph of f is concave up (down) on the interval. If the slopes increase (decrease) as x increases, then this means that the tangent lines are turning counterclockwise (clockwise) on the interval. The plausibility of Definition 4.7.1 is illustrated in FIGURE 4.7.2. An equivalent way of looking at concavity is also apparent in Figure 4.7.2. The graph of a function f is concave up (down) on an interval if the graph at any point lies above (below) the tangent lines.

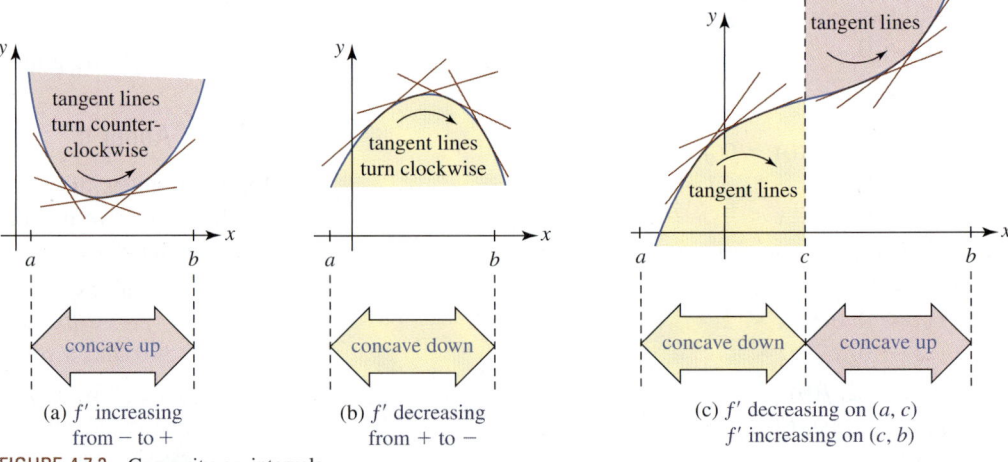

(a) f' increasing
from $-$ to $+$

(b) f' decreasing
from $+$ to $-$

(c) f' decreasing on (a, c)
f' increasing on (c, b)

FIGURE 4.7.2 Concavity on intervals

▪ **Concavity and the Second Derivative** In Theorem 4.4.4 of Section 4.4 we saw that the algebraic sign of the derivative of a function indicates when the function is increasing or decreasing on an interval. Specifically, if the function referred to in the preceding sentence is the derivative f', then we can conclude that the algebraic sign of the derivative of f', that is, f'', indicates when f' is either increasing or decreasing on an interval. For example, if $f''(x) > 0$ on (a, b), then f' increases on (a, b). In view of Definition 4.7.1, if f' increases on (a, b), then the graph of f is concave up on the interval. Therefore, we are led to the following test for concavity.

Theorem 4.7.1   Test for Concavity

Let f be a function for which f'' exists on (a, b).

 (i) If $f''(x) > 0$ for all x in (a, b), then the graph of f is concave up on (a, b).
(ii) If $f''(x) < 0$ for all x in (a, b), then the graph of f is concave down on (a, b).

EXAMPLE 1   Test for Concavity

Determine the intervals on which the graph of $f(x) = x^3 + \frac{9}{2}x^2$ is concave up and the intervals on which the graph is concave down.

Solution   From $f'(x) = 3x^2 + 9x$ we obtain

$$f''(x) = 6x + 9 = 6\left(x + \frac{3}{2}\right).$$

We see that $f''(x) < 0$ when $6\left(x + \frac{3}{2}\right) < 0$ or $x < -\frac{3}{2}$ and that $f''(x) > 0$ when $6\left(x + \frac{3}{2}\right) > 0$ or $x > -\frac{3}{2}$. It follows from Theorem 4.7.1 that the graph of f is concave down on the interval $\left(-\infty, -\frac{3}{2}\right)$ and concave up on $\left(-\frac{3}{2}, \infty\right)$. ∎

■ Point of Inflection   The graph of the function in Example 1 changes concavity at the point that corresponds to $x = -\frac{3}{2}$. As x increases through $-\frac{3}{2}$, the graph of f changes from concave down to concave up at the point $\left(-\frac{3}{2}, \frac{27}{4}\right)$. A point on the graph of a function where the concavity changes from up to down, or vice versa, is given a special name.

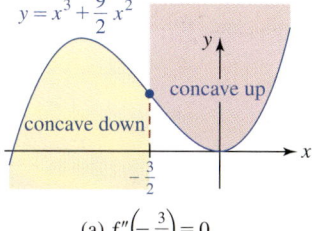
(a) $f''\left(-\frac{3}{2}\right) = 0$

Definition 4.7.2   Point of Inflection

Let f be continuous on an interval (a, b) containing the number c. A point $(c, f(c))$ is a **point of inflection** of the graph of f if there is a tangent line at $(c, f(c))$ and the graph changes concavity at this point.

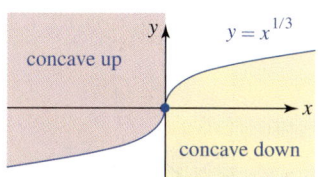
(b) $f''(x)$ does not exist at 0

FIGURE 4.7.3   Points of inflection

A reexamination of Example 1 shows that $f(x) = x^3 + \frac{9}{2}x^2$ is continuous at $-\frac{3}{2}$, has a tangent line at $\left(-\frac{3}{2}, \frac{27}{4}\right)$, and changes concavity at that point. Hence, $\left(-\frac{3}{2}, \frac{27}{4}\right)$ is a point of inflection. Also note that $f''\left(-\frac{3}{2}\right) = 0$. See **FIGURE 4.7.3(a)**. We also know that the function $f(x) = x^{1/3}$ is continuous at 0 and possesses a vertical tangent at $(0, 0)$ (see Example 10 of Section 3.1). From $f''(x) = -\frac{2}{9}x^{-5/3}$ it is seen that $f''(x) > 0$ for $x < 0$ and $f''(x) < 0$ for $x > 0$. Hence, $(0, 0)$ is a point of inflection. Note in this case $f''(x) = -\frac{2}{9}x^{-5/3}$ is not defined at $x = 0$. See Figure 4.7.3(b). We have illustrated two cases of the next theorem.

Theorem 4.7.2   Point of Inflection

If $(c, f(c))$ is a point of inflection for the graph of a function f, then either $f''(c) = 0$ or $f''(c)$ does not exist.

■ Second Derivative Test   If c is a critical number of a function $y = f(x)$, and, say, $f''(c) > 0$, then the graph of f is concave up on some interval (a, b) that contains c. Necessarily then, $f(c)$ is a relative minimum. Similarly, $f''(c) < 0$ at a critical value c implies $f(c)$ is a relative maximum. This so-called **Second Derivative Test** is illustrated in **FIGURE 4.7.4**.

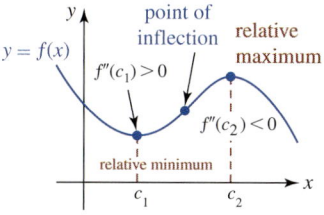

FIGURE 4.7.4   Second Derivative Test

Theorem 4.7.3   Second Derivative Test

Let f be a function for which f'' exists on an interval (a, b) that contains the critical number c.

  (i) If $f''(c) > 0$, then $f(c)$ is a relative minimum.
 (ii) If $f''(c) < 0$, then $f(c)$ is a relative maximum.
(iii) If $f''(c) = 0$, the test fails and $f(c)$ may or may not be a relative extremum. In this case, use the First Derivative Test.

At this point one might ask, Why do we need another test for relative extrema when we already have the First Derivative Test? If the function f under consideration is a polynomial, it is very easy to compute the second derivative. In using Theorem 4.7.3 we need only determine the algebraic sign of $f''(x)$ *at* the critical number. Contrast this with Theorem 4.6.1 where we must determine the sign of $f'(x)$ at numbers to the right and left of the critical number. If f' is not readily factored, the latter procedure may be somewhat difficult. On the other hand, it may be equally tedious to use Theorem 4.7.3 in the case of some functions that involve products, quotients, powers, and so on. Thus, Theorems 4.6.1 and 4.7.3 both have advantages and disadvantages.

EXAMPLE 2   Second Derivative Test

Graph $f(x) = 4x^4 - 4x^2$.

Solution   From $f(x) = 4x^2(x^2 - 1) = 4x^2(x + 1)(x - 1)$ we see that the graph of f has the intercepts $(-1, 0)$, $(0, 0)$, and $(1, 0)$. Furthermore, since f is a polynomial with only even powers, we conclude that its graph is symmetric with respect to the y-axis (even function). Now the first and second derivatives are

$$f'(x) = 16x^3 - 8x = 8x(\sqrt{2}x + 1)(\sqrt{2}x - 1)$$
$$f''(x) = 48x^2 - 8 = 8(\sqrt{6}x + 1)(\sqrt{6}x - 1).$$

From f' we see that the critical numbers of f are 0, $-\sqrt{2}/2$, and $\sqrt{2}/2$. The Second Derivative Test is summarized in the accompanying table.

x	Sign of $f''(x)$	$f(x)$	Conclusion
0	$-$	0	rel. max.
$\sqrt{2}/2$	$+$	-1	rel. min.
$-\sqrt{2}/2$	$+$	-1	rel. min.

Finally, from the factored form of f'' it is seen that $f''(x)$ changes signs at $x = -\sqrt{6}/6$ and at $x = \sqrt{6}/6$. Hence, the graph of f possesses two points of inflection: $\left(-\sqrt{6}/6, -\frac{5}{9}\right)$ and $\left(\sqrt{6}/6, -\frac{5}{9}\right)$. See **FIGURE 4.7.5**. ∎

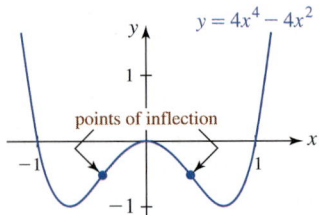

FIGURE 4.7.5   Graph of function in Example 2

EXAMPLE 3   Failure of the Second Derivative Test

Consider the simple function $f(x) = x^4 + 1$. From $f'(x) = 4x^3$ we see that 0 is a critical number. But from the second derivative $f''(x) = 12x^2$ we get $f''(0) = 0$. Thus the Second Derivative Test leads to no conclusion. However, from the first derivative $f'(x) = 4x^3$ we see:

$$f'(x) < 0 \quad \text{for} \quad x < 0 \qquad \text{and} \qquad f'(x) > 0 \quad \text{for} \quad x > 0.$$

This First Derivative Test indicates that $f(0) = 1$ is a relative minimum. **FIGURE 4.7.6** shows $f(0) = 1$ is actually an absolute minimum. ∎

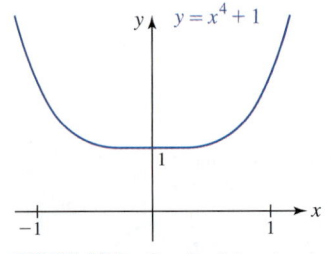

FIGURE 4.7.6   Graph of function in Example 3

EXAMPLE 4   Second Derivative Test

Graph $f(x) = 2\cos x - \cos 2x$.

Solution   Because $\cos x$ and $\cos 2x$ are even functions, the graph of f possesses symmetry with respect to the y-axis. Also, $f(0) = 1$ yields the y-intercept $(0, 1)$. Now the first and second derivatives are

$$f'(x) = -2\sin x + 2\sin 2x \qquad \text{and} \qquad f''(x) = -2\cos x + 4\cos 2x.$$

Using the trigonometric identity $\sin 2x = 2\sin x \cos x$, we can simplify the equation $f'(x) = 0$ to $\sin x(1 - 2\cos x) = 0$. The solutions of $\sin x = 0$ are $0, \pm\pi, \pm2\pi, \ldots$ and the solutions of $\cos x = \frac{1}{2}$ are $\pm\pi/3, \pm5\pi/3, \ldots$. But since f is 2π periodic (show this!), it suffices to consider only those critical numbers in $[0, 2\pi]$, namely, $0, \pi/3, \pi, 5\pi/3$, and 2π. The Second Derivative Test applied to these values is summarized in the accompanying table.

x	Sign of $f''(x)$	$f(x)$	Conclusion
0	$+$	1	rel. min.
$\pi/3$	$-$	$\frac{3}{2}$	rel. max.
π	$+$	-3	rel. min.
$5\pi/3$	$-$	$\frac{3}{2}$	rel. max.
2π	$+$	1	rel. min.

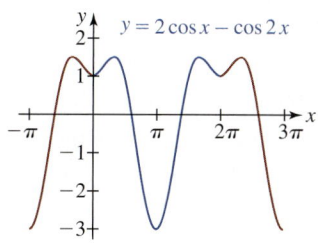

FIGURE 4.7.7 Graph of function in Example 4

The graph of f is the 2π-periodic extension of the blue portion shown in FIGURE 4.7.7 on the interval $[0, 2\pi]$. ■

$f'(x)$ NOTES FROM THE CLASSROOM

(*i*) If $(c, f(c))$ is a point of inflection, then either $f''(c) = 0$ or $f''(c)$ does not exist. The converse of this statement is not necessarily true. We cannot conclude, simply from the fact that when $f''(c) = 0$ or $f''(c)$ does not exist, that $(c, f(c))$ is a point of inflection. For example, in Example 3 we saw $f''(0) = 0$ for $f(x) = x^4 + 1$. But it is apparent from Figure 4.7.6 that $(0, f(0))$ is not a point of inflection. Also, for $f(x) = 1/x$, we see that $f''(x) = 2/x^3$ is undefined at $x = 0$ and that the graph of f changes concavity at $x = 0$:

$$f''(x) < 0 \quad \text{for} \quad x < 0 \quad \text{and} \quad f''(x) > 0 \quad \text{for} \quad x > 0.$$

However, $x = 0$ is not the x-coordinate of a point of inflection because f is not continuous at 0.

(*ii*) You should not think that the graph of a function *has to have* concavity. There are perfectly good differentiable functions whose graphs possess no concavity. See Problem 60 in Exercises 4.7.

(*iii*) You should be aware that textbooks disagree on the precise definition of a point of inflection. This is nothing to be concerned with but if interested, see Problem 65 in Exercises 4.7.

Exercises 4.7   Answers to selected odd-numbered problems begin on page ANS-16.

☰ Fundamentals

In Problems 1–12, use the second derivative to determine the intervals on which the graph of the given function is concave up and the intervals on which it is concave down.

1. $f(x) = -x^2 + 7x$

2. $f(x) = -(x + 2)^2 + 8$

3. $f(x) = -x^3 + 6x^2 + x - 1$

4. $f(x) = (x + 5)^3$

5. $f(x) = x(x - 4)^3$

6. $f(x) = 6x^4 + 2x^3 - 12x^2 + 3$

7. $f(x) = x^{1/3} + 2x$

8. $f(x) = x^{8/3} - 20x^{2/3}$

9. $f(x) = x + \dfrac{9}{x}$

10. $f(x) = \sqrt{x^2 + 10}$

11. $f(x) = \dfrac{1}{x^2 + 3}$

12. $f(x) = \dfrac{x - 1}{x + 2}$

In Problems 13–16, estimate from the graph of the given function f the intervals on which f' is increasing and the intervals on which f' is decreasing.

13.

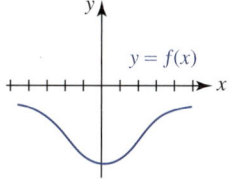

FIGURE 4.7.8 Graph for Problem 13

14.

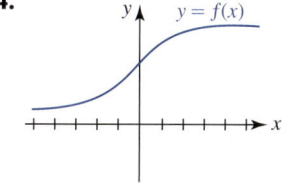

FIGURE 4.7.9 Graph for Problem 14

15.

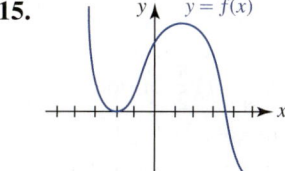

FIGURE 4.7.10 Graph for Problem 15

16.

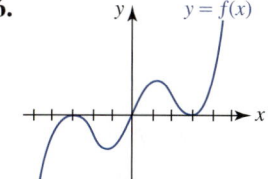

FIGURE 4.7.11 Graph for Problem 16

17. Show that the graph of $f(x) = \sec x$ is concave up on those intervals on which $\cos x > 0$, and concave down on those intervals on which $\cos x < 0$.

18. Show that the graph of $f(x) = \csc x$ is concave up on those intervals on which $\sin x > 0$, and concave down on those intervals on which $\sin x < 0$.

In Problems 19–26, use the second derivative to locate all points of inflection.

19. $f(x) = x^4 - 12x^2 + x - 1$ **20.** $f(x) = x^{5/3} + 4x$

21. $f(x) = \sin x$ **22.** $f(x) = \cos x$

23. $f(x) = x - \sin x$ **24.** $f(x) = \tan x$

25. $f(x) = x + xe^{-x}$ **26.** $f(x) = xe^{-x^2}$

In Problems 27–44, use the Second Derivative Test, when applicable, to find the relative extrema of the given function. Graph. Find intercepts and points of inflection when possible.

27. $f(x) = -(2x - 5)^2$ **28.** $f(x) = \dfrac{1}{3}x^3 - 2x^2 - 12x$

29. $f(x) = x^3 + 3x^2 + 3x + 1$ **30.** $f(x) = \dfrac{1}{4}x^4 - 2x^2$

31. $f(x) = 6x^5 - 10x^3$ **32.** $f(x) = x^3(x + 1)^2$

33. $f(x) = \dfrac{x}{x^2 + 2}$ **34.** $f(x) = x^2 + \dfrac{1}{x^2}$

35. $f(x) = \sqrt{9 - x^2}$ **36.** $f(x) = x\sqrt{x - 6}$

37. $f(x) = x^{1/3}(x + 1)$ **38.** $f(x) = x^{1/2} - \dfrac{1}{4}x$

39. $f(x) = \cos 3x, \; [0, 2\pi]$ **40.** $f(x) = 2 + \sin 2x, \; [0, 2\pi]$

41. $f(x) = \cos x + \sin x, \; [0, 2\pi]$

42. $f(x) = 2\sin x + \sin 2x, \; [0, 2\pi]$

43. $f(x) = 2x - x\ln x$ **44.** $f(x) = \ln(x^2 + 2)$

In Problems 45–48, determine whether the given function has a relative extremum at the indicated critical number.

45. $f(x) = \sin x \cos x; \quad \pi/4$ **46.** $f(x) = x\sin x; \quad 0$

47. $f(x) = \tan^2 x; \quad \pi$ **48.** $f(x) = (1 + \sin 4x)^3; \quad \pi/8$

In Problems 49–52, sketch a graph of a function f that has the given properties.

49. $f(-2) = 0, f(4) = 0$
$f'(3) = 0, f''(1) = 0, f''(2) = 0$
$f''(x) < 0, x < 1, x > 2$
$f''(x) > 0, 1 < x < 2$

50. $f(0) = 5, f(2) = 0$
$f'(2) = 0, f''(3)$ does not exist
$f''(x) > 0, x < 3$
$f''(x) < 0, x > 3$

51. $f(0) = -1, f(\pi/2) > 0$
$f'(x) \geq 0$ for all x
$f''(x) > 0, (2n - 1)\dfrac{\pi}{2} < x < (2n + 1)\dfrac{\pi}{2}, n$ even
$f''(x) < 0, (2n - 1)\dfrac{\pi}{2} < x < (2n + 1)\dfrac{\pi}{2}, n$ odd

52. $f(-x) = -f(x)$
vertical asymptote $x = 2, \displaystyle\lim_{x \to \infty} f(x) = 0$
$f''(x) < 0, 0 < x < 2$
$f''(x) > 0, x > 2$

≡ Think About It

53. Find values of a, b, and c such that the graph of $f(x) = ax^3 + bx^2 + cx$ passes through $(-1, 0)$ and has a point of inflection at $(1, 1)$.

54. Find values of a, b, and c such that the graph of $f(x) = ax^3 + bx^2 + cx$ has a horizontal tangent at the point of inflection $(1, 1)$.

55. Use the Second Derivative Test as an aid in graphing $f(x) = \sin(1/x)$. Observe that f is discontinuous at $x = 0$.

56. Show that the graph of a general polynomial function

$$f(x) = a_n x^n + a_{n-1} x^{n-1} + \cdots + a_1 x + a_0, a_n \neq 0$$

can have at most $n - 2$ points of inflection.

57. Let $f(x) = (x - x_0)^n$, where n is a positive integer.
   (a) Show that $(x_0, 0)$ is a point of inflection of the graph of f if n is an odd integer.
   (b) Show that $(x_0, 0)$ is not a point of inflection of the graph of f but corresponds to a relative minimum when n is an even integer.

58. Prove that the graph of a quadratic polynomial function $f(x) = ax^2 + bx + c, a \neq 0$, is concave upward on the x-axis when $a > 0$ and concave downward on the x-axis when $a < 0$.

59. Let f be a function for which f''' exists on an interval (a, b) that contains the number c. If $f''(c) = 0$ and $f'''(c) \neq 0$, what can be said about $(c, f(c))$?

60. Give an example of a differentiable function whose graph possesses no concavity. Do not think profound thoughts.

61. Prove or disprove: A point of inflection for a function f must occur at a critical value of f'.

62. Without graphing, explain why the graph of $f(x) = 10x^2 - x - 40 + e^x$ cannot have a point of inflection.

63. Prove or disprove: The function

$$f(x) = \begin{cases} 4x^2 - x, & x \leq 0 \\ -x^3, & x > 0 \end{cases}$$

has a point of inflection at $(0, 0)$.

64. Suppose f is a polynomial function of degree 3 and that c_1 and c_2 are distinct critical numbers.
   (a) Are $f(c_1)$ and $f(c_2)$ necessarily relative extrema of the function? Prove your assertion.
   (b) What do you think is the x-coordinate of the point of inflection for the graph of f? Prove your assertion.

≡ Project

65. Points of Inflection   Find other calculus texts and note how they define a point of inflection. Then do some Internet research on the definition of a point of inflection. Write a short paper comparing these definitions. Illustrate your paper with appropriate graphs.

4.8 Optimization

▌**Introduction** In science, engineering, and business one is often interested in the maximum and minimum values of functions; for example, a company is naturally interested in maximizing revenue while minimizing cost. The next time you go to a supermarket note that all cans containing, say, 15 oz of food (0.01566569 ft^3) have the same physical appearance. The fact that all cans of a specified volume have the same shape (same radius and height) is no coincidence, since there are specific dimensions that minimize the amount of metal used and, hence, minimize the cost of the construction of the can to a company. In the same vein, many of the so-called economy cars have appearances that are remarkably similar. This is not just a simple matter of one company copying the success of another company, but, rather, for a given volume engineers strive for a design that minimizes the amount of material used.

▌**Playing with Some Numbers** Let us begin with a simple problem:

> *Find two nonnegative numbers whose sum is 5 such that the product of one and the square of the other is as large as possible.* (1)

◀ At this point a review of Section 1.7 is strongly recommended.

In Example 1 of Section 1.7 we encountered the problem:

> *The sum of two nonnegative numbers is 5. Express the product of one and the square of the other as a function of one of the numbers.* (2)

A comparison of (1) and (2) shows that (2), where we are simply asked to set up a function, is embedded within the calculus problem (1). The calculus part of (1) requires that we find nonnegative numbers so that the product is a maximum. A review of Examples 1 and 2 of Section 1.7 indicates that the product described in (1) is

$$P = x(5 - x)^2 \quad \text{or} \quad P(x) = 25x - 10x^2 + x^3. \quad (3)$$

The domain of the function $P(x)$ in (3) is the interval $[0, 5]$. This fact came from combining the two inequalities $x \geq 0$ and $y = 5 - x \geq 0$ or the recognition that if x were allowed to be larger than 5, then y would be negative, contrary to our initial assumption. There are an infinite number of pairs of nonnegative real numbers (rational and irrational) that add up to 5. Note that we did not say nonnegative *integers*! For example

Numbers: x, y	Product: $P = xy^2$
$1, 4$	$P = 1 \cdot 4^2 = 16$
$2, 3$	$P = 2 \cdot 3^2 = 18$
$\dfrac{1}{2}, \dfrac{9}{2}$	$P = \dfrac{1}{2} \cdot \left(\dfrac{9}{2}\right)^2 = 10.125$
$\pi, 5 - \pi$	$P = \pi \cdot (5 - \pi)^2 \approx 10.85$

Pairs of numbers, such as $-1, 6$, that add up to 5 are rejected because both numbers must be nonnegative. Now how do we know when we have discovered the numbers x and y that give the largest, that is, the maximum or optimal, value of P? The answer lies in the realization that the domain of the function $P(x)$ is the closed interval $[0, 5]$. We know from Theorem 4.3.3 that the continuous function $P(x)$ has an absolute extremum either at an endpoint of the interval or at a critical number in the open interval $(0, 5)$. From (3) we see that $P'(x) = 25 - 20x + 3x^2 = (3x - 5)(x - 5)$ so that the only critical number in the open interval $(0, 5)$ is $\frac{5}{3}$. The function values $P(0) = 0$ and $P(5) = 0$ obviously represent the absolute minimum product, and so the absolute maximum product is $P\left(\frac{5}{3}\right) = \frac{5}{3}(5) - \left(\frac{5}{3}\right)^2 = \frac{500}{27} \approx 18.52$. In other words, the two numbers are $x = \frac{5}{3}$ and $y = 5 - \frac{5}{3} = \frac{10}{3}$.

▌**Terminology** In general, the function that describes the quantity we seek to optimize, by finding its maximum or its minimum value, is called the **objective function**. The function in (3) is the objective function for the problem given in (1). A relationship among the variables in an optimization problem, such as the equation $x + y = 5$ between the numbers x and y in the foregoing discussion, is called a **constraint**. The constraint allows us to eliminate one of the variables in the construction of the objective function, such as $P(x)$ in (3), as well as places a limitation on how the variables such as x and y can actually vary. We saw that the limitations $x \geq 0$ and $y = 5 - x \geq 0$ helped us to infer that the domain of the function $P(x)$ in (3) was the interval $[0, 5]$. You should be aware that the type of word problems considered in this section *may* or *may not* have a constraint.

■ **Suggestions**   In the examples and problems that follow either we will be *given* an objective function or we will have to translate the words into mathematical symbols and construct an objective function. These are the kinds of word problems that show off the power of calculus and provide one of many possible answers to the age-old question: What's it good for? While not guaranteeing anything, here are some suggestions to keep in mind when solving an optimization problem. First and foremost:

> *Develop a positive and analytical attitude. Try to be neat and organized.*

Guidelines for Solving Optimization Problems

(*i*) Read the problem slowly, then read it again.

(*ii*) Draw a picture when possible; keep it simple.

(*iii*) Introduce variables (in your picture, if there is one) and note any constraint between the variables.

(*iv*) Using all necessary variables, set up the objective function. If more than one variable is used, then employ the constraint to reduce the function to one variable.

(*v*) Note the interval on which the function is defined. Determine all critical numbers.

(*vi*) If the objective function is continuous and defined on a closed interval $[a, b]$, then test for endpoint extrema. If the desired extremum does not occur at an endpoint, it must occur at a critical number in the open interval (a, b).

(*vii*) If the objective function is defined on an interval that is not closed, then a derivative test should be used at each critical number in that interval.

In our first example, we examine a mathematical model that comes from physics.

EXAMPLE 1   Maximum Range

When air resistance is ignored, the horizontal range R of a projectile is given by

$$R(\theta) = \frac{v_0^2}{g} \sin 2\theta, \tag{4}$$

where v_0 is the constant initial velocity, g is the acceleration due to gravity, and θ is the angle of elevation or departure. Find the maximum range of the projectile.

Solution   As a physical model of the problem, let us imagine that the projectile is a cannonball. See FIGURE 4.8.1. For angles θ greater than $\pi/2$, the cannon shown in the figure would shoot backward. Thus, it makes physical sense to restrict the function in (4) to the closed interval $[0, \pi/2]$. From

$$\frac{dR}{d\theta} = \frac{v_0^2}{g} 2 \cos 2\theta$$

we see that $dR/d\theta = 0$ when $\cos 2\theta = 0$ or $2\theta = \pi/2$, and so the only critical number in the open interval $(0, \pi/2)$ is $\pi/4$. Evaluating the function at the endpoints and the critical number gives

$$R(0) = 0, \quad R(\pi/4) = \frac{v_0^2}{g}, \quad R(\pi/2) = 0.$$

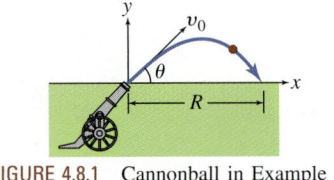

FIGURE 4.8.1   Cannonball in Example 1

Since $R(\theta)$ is continuous on the closed interval $[0, \pi/2]$, these values indicate that the minimum range is $R(0) = R(\pi/2) = 0$ and the maximum range is $R(\pi/4) = v_0^2/g$. In other words, to achieve maximum distance, the projectile should be launched at an angle of 45° to the horizontal. ■

If cannonballs in Example 1 are launched with the initial velocity v_0 but with varying angles of elevation θ different from 45° then their horizontal ranges are less than $R_{\max} = v_0^2/g$. Inspection of the function in (4) shows that the same horizontal range is attained for complementary angles such as 20° and 70°, and 30° and 60°. See **FIGURE 4.8.2**. If air resistance is taken into account, all projectiles will fall short of v_0^2/g, even when launched at an angle of elevation of 45°.

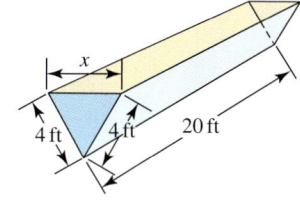

FIGURE 4.8.2 Same range for complementary angles

EXAMPLE 2   Maximum Volume

A 20-ft-long water trough has ends in the form of isosceles triangles with sides that are 4 ft long. Determine the dimension across the top of the triangular end so that the volume of the trough is a maximum. Find the maximum volume.

Solution   The trough with the unknown dimension x is shown in **FIGURE 4.8.3**. The volume V of the trough is

$$V = (\text{area of triangular end}) \times (\text{length}).$$

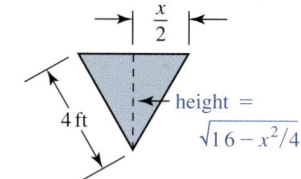

FIGURE 4.8.3   Water trough in Example 2

From **FIGURE 4.8.4** and the Pythagorean Theorem, the area of the triangular end as a function of x is $\frac{1}{2}x\sqrt{16 - x^2/4}$. Consequently, the volume of the trough as a function of x, the objective function, is

$$V(x) = 20 \cdot \left(\frac{1}{2}x\sqrt{16 - \frac{1}{4}x^2} \right) = 5x\sqrt{64 - x^2}.$$

The function $V(x)$ makes sense only on the closed interval $[0, 8]$. (Why?)
Taking the derivative and simplifying yield

$$V'(x) = -10\frac{x^2 - 32}{\sqrt{64 - x^2}}.$$

Although $V'(x) = 0$ for $x = \pm 4\sqrt{2}$, the only critical number in the open interval $(0, 8)$ is $4\sqrt{2}$. Since the function $V(x)$ is continuous on $[0, 8]$, we know from Theorem 4.3.3 that $V(0) = V(8) = 0$ must be its absolute minimum. The absolute maximum of $V(x)$ must then occur when the width across the top of the trough is $4\sqrt{2} \approx 5.66$ ft. The maximum volume is $V(4\sqrt{2}) = 160$ ft³. ■

FIGURE 4.8.4   Triangular end of trough in Example 2

Note: Often a problem can be solved in more than one way. In hindsight, you should verify that the solution of Example 2 is slightly "cleaner" if the dimension across the top of the end of the trough is labeled $2x$ rather than x. Indeed, as the next example shows, Example 2 can be solved using an entirely different variable.

EXAMPLE 3   Alternative Solution to Example 2

As shown in **FIGURE 4.8.5**, we let θ denote the angle between the vertical and one of the sides. From right-triangle trigonometry the height and base of the triangular end are $4\cos\theta$ and $8\sin\theta$, respectively. Expressed as a function of θ, V is $\left(\frac{1}{2} \cdot \text{base} \cdot \text{height} \right) \times (\text{length})$, or

$$
\begin{aligned}
V(\theta) &= \tfrac{1}{2}(4\cos\theta)(8\sin\theta) \cdot 20 \\
&= 320\sin\theta\cos\theta \\
&= 160(2\sin\theta\cos\theta) \\
&= 160\sin 2\theta, \qquad \leftarrow \text{double-angle formula}
\end{aligned}
$$

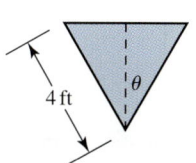

FIGURE 4.8.5   Triangular end of trough in Example 3

where $0 \le \theta \le \pi/2$. Proceeding as in Example 1, we find that the maximum value $V = 160$ ft³ occurs at $\theta = \pi/4$. The dimension across the top of the trough, or the base of the isosceles triangle, is $8\sin(\pi/4) = 4\sqrt{2}$ ft. ■

■ **Problems with Constraints**   It is often more convenient to set up a function in terms of two variables instead of one. In this case we need to find a relationship between these variables that can be used to eliminate one of the variables from the function under consideration. As discussed in conjunction with (1), this relationship is usually an equation called a **constraint**. The next two examples illustrate this concept.

EXAMPLE 4   Closest Point

Find the point in the first quadrant on the circle $x^2 + y^2 = 1$ that is closest to (2, 4).

Solution   Let (x, y), $x > 0, y > 0$, denote the point on the circle closest to the point (2, 4). See FIGURE 4.8.6.

As shown in the figure, the distance d between (x, y) and (2, 4) is

$$d = \sqrt{(x-2)^2 + (y-4)^2} \quad \text{or} \quad d^2 = (x-2)^2 + (y-4)^2.$$

Now the point that minimizes the square of the distance d^2 also minimizes the distance d. Let us write $D = d^2$. By expanding $(x-2)^2$ and $(y-4)^2$ and using the constraint $x^2 + y^2 = 1$ in the form $y = \sqrt{1 - x^2}$, we find

$$D(x) = x^2 - 4x + 4 + \overbrace{(1 - x^2)}^{y^2} - \overbrace{8\sqrt{1 - x^2}}^{y} + 16$$
$$= -4x - 8\sqrt{1 - x^2} + 21.$$

Because we have assumed x and y to be positive, the domain of the foregoing function is the open interval (0, 1). However, the solution of the problem will not be affected in any way by assuming that the domain is the closed interval [0, 1].

Differentiation gives

$$D'(x) = -4 - 4(1 - x^2)^{-1/2}(-2x) = \frac{-4\sqrt{1 - x^2} + 8x}{\sqrt{1 - x^2}}.$$

Now $D'(x) = 0$ only if $-4\sqrt{1 - x^2} + 8x = 0$ or $2x = \sqrt{1 - x^2}$. After squaring both sides and simplifying, we find that $\sqrt{5}/5$ is the only critical number in the interval (0, 1). Because $D(x)$ is continuous on [0, 1], we conclude from the function values

$$D(0) = 13, \quad D(\sqrt{5}/5) = 21 - 4\sqrt{5} \approx 12.06, \quad \text{and} \quad D(1) = 17$$

that D and, hence, the distance d are a minimum when $x = \sqrt{5}/5$. Using the constraint $x^2 + y^2 = 1$, we find correspondingly that $y = 2\sqrt{5}/5$. This means $(\sqrt{5}/5, 2\sqrt{5}/5)$ is the point on the circle closest to (2, 4). ∎

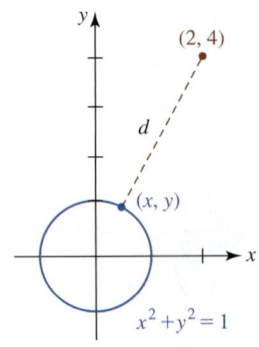

FIGURE 4.8.6   Circle and point in Example 4

EXAMPLE 5   Least Fencing

A rancher intends to mark off a rectangular plot of land that will have an area of 1500 m². The plot will be fenced and divided into two equal portions by an additional fence parallel to two sides. Find the dimensions of the land that require the least amount of fencing.

Solution   As shown in FIGURE 4.8.7, we let x and y denote the dimensions of the fenced-in land. The function we wish to minimize is the total amount of fence, that is, the sum of the lengths of the five portions of fence. If we denote this sum by the symbol L, we have

$$L = 2x + 3y. \tag{5}$$

Because the fenced-in land is to have an area of 1500 m², x and y must be related by the requirement that $xy = 1500$. We use this constraint in the form $y = 1500/x$ to eliminate y in (5) and write the objective function L as a function of x:

$$L(x) = 2x + \frac{4500}{x} \tag{6}$$

Since x represents a physical dimension that satisfies $xy = 1500$, we conclude that it is positive. But other than that, there is no restriction on x. Thus, unlike the prior examples, the

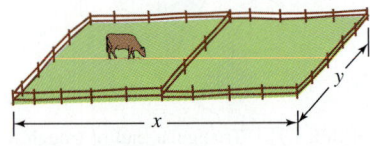

FIGURE 4.8.7   Rectangular plot of land in Example 5

function we are considering is not defined on a closed interval; $L(x)$ is defined on the unbounded interval $(0, \infty)$.

Setting the derivative

$$L'(x) = 2 - \frac{4500}{x^2}$$

equal to zero and solving for x, we find that the only critical number is $15\sqrt{10}$. Since the second derivative is easy to compute, we shall use the Second Derivative Test. From

$$L''(x) = \frac{9000}{x^3}$$

we observe that $L''(15\sqrt{10}) > 0$. It follows from Theorem 4.7.3 that $L(15\sqrt{10}) = 2(15\sqrt{10}) + 4500/(15\sqrt{10}) = 60\sqrt{10}$ m is the required minimum amount of fencing. Returning to the constraint $y = 1500/x$, we find the corresponding value of y is $10\sqrt{10}$. Therefore, the dimensions of the land should be $15\sqrt{10}$ m $\times$ $10\sqrt{10}$ m. ■

If an object is moving at a constant rate, then distance, rate, and time are related by *distance = rate × time*. We shall use this result in our last example in the form

$$\text{time} = \frac{\text{distance}}{\text{rate}}. \tag{7}$$

EXAMPLE 6 Minimum Time

A woman at point P on an island wishes to reach a village located at point S on a straight shore on the mainland. Point P is 9 mi from the closest point Q on the shore and the village at point S is 15 mi from point Q. See FIGURE 4.8.8. If the woman rows a boat at a rate of 3 mi/h to a point R on land, then walks the rest of the way to S at a rate of 5 mi/h, determine where she should land on shore in order to minimize the total time of travel.

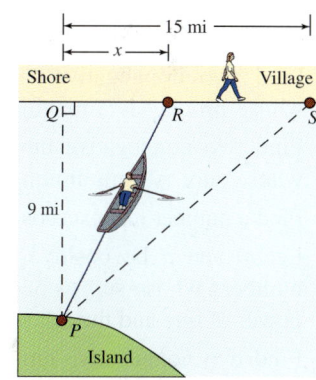

FIGURE 4.8.8 Traveling woman in Example 6

Solution As shown in the figure, if x denotes the distance from point Q on shore to the point R where she lands on shore, then by the Pythagorean Theorem, the distance she rows is $\sqrt{81 + x^2}$. The distance she walks is $15 - x$. By (7) the total time of the trip from P to S is

$$T = \text{time rowing} + \text{time walking} \quad \text{or} \quad T(x) = \frac{\sqrt{81 + x^2}}{3} + \frac{15 - x}{5}.$$

Since $0 \leq x \leq 15$, the function $T(x)$ is defined on the closed interval $[0, 15]$.

The derivative of T is

$$\frac{dT}{dx} = \frac{1}{6}(81 + x^2)^{-1/2}(2x) - \frac{1}{5} = \frac{x}{3\sqrt{81 + x^2}} - \frac{1}{5}.$$

We set this derivative equal to 0 and solve for x:

$$\frac{x}{3\sqrt{81 + x^2}} = \frac{1}{5}$$

$$\frac{x^2}{81 + x^2} = \frac{9}{25}$$

$$16x^2 = 729$$

$$x = \frac{27}{4}.$$

That is, $\frac{27}{4}$ is the only critical number in $[0, 15]$. Since $T(x)$ is continuous on the interval we see from the three function values

$$T(0) = 6\text{ h}, \quad T(\tfrac{27}{4}) = 5.4\text{ h}, \quad \text{and} \quad T(15) \approx 5.83\text{ h}$$

that the minimum total travel time occurs when $x = \frac{27}{4} = 6.75$. In other words, the woman lands the boat at point R, 6.75 mi from point Q, and then walks the remaining 8.25 mi to point S. ■

$L(x) = 2x + 4500/x,\ x > 0$

absolute
minimum

$L(15\sqrt{10})$

$15\sqrt{10}$

FIGURE 4.8.9 Graph of objective function in Example 5

$f'(x)$ NOTES FROM THE CLASSROOM

An observant reader may question at least two aspects of Example 5.

(*i*) Where did the assumption that the land be divided into two equal portions enter into the solution? In point of fact, it did not. What is important is that the dividing fence be parallel to the two ends. Ask yourself what $L(x)$ would be if this were *not* the case. However, the actual positioning of the dividing fence between the ends is irrelevant as long as it is parallel to them.

(*ii*) In an applied problem we are naturally interested in only absolute extrema. Therefore, another question might be: Since the function L in (6) is not defined on a closed interval and since the Second Derivative Test does not guarantee absolute extrema, how can we be certain that $L(15\sqrt{10})$ is an absolute minimum? When in doubt, we can always draw a graph. **FIGURE 4.8.9** answers the question for $L(x)$. Also, look again at Theorem 4.6.2 in Section 4.6. Because $15\sqrt{10}$ is the *only* critical number in the interval $(0, \infty)$ and because $L(15\sqrt{10})$ was proved to be a relative minimum, Theorem 4.6.2 guarantees that the function value $L(15\sqrt{10}) = 60\sqrt{10}$ is an absolute minimum.

Exercises 4.8 Answers to selected odd-numbered problems begin on page ANS-17.

≡ Fundamentals

1. Find two nonnegative numbers whose sum is 60 and whose product is a maximum.

2. Find two nonnegative numbers whose product is 50 and whose sum is a minimum.

3. Find a number that exceeds its square by the greatest amount.

4. Let m and n be positive integers. Find two nonnegative numbers whose sum is S such that the product of the mth power of one and the nth power of the other is a maximum.

5. Find two nonnegative numbers whose sum is 1 such that the sum of the square of one and twice the square of the other is a minimum.

6. Find the minimum value of the sum of a nonnegative number and its reciprocal.

7. Find the point(s) on the graph of $y^2 = 6x$ closest to $(5, 0)$. closest to $(3, 0)$.

8. Find the point on the graph of $x + y = 1$ closest to $(2, 3)$.

9. Determine the point on the graph of $y = x^3 - 4x^2$ at which the tangent line has minimum slope.

10. Determine the point on the graph of $y = 8x^2 + 1/x$ at which the tangent line has maximum slope.

In Problems 11 and 12, find the dimensions of the shaded region such that its area is a maximum.

11.

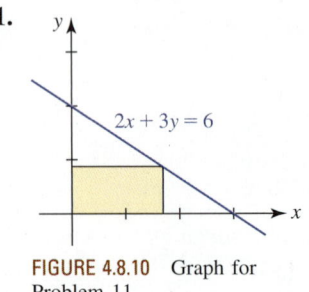

$2x + 3y = 6$

FIGURE 4.8.10 Graph for Problem 11

12.

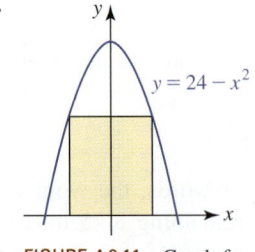

$y = 24 - x^2$

FIGURE 4.8.11 Graph for Problem 12

13. Find the vertices $(x, 0)$ and $(0, y)$ of the shaded triangular region in **FIGURE 4.8.12** so that its area is a minimum.

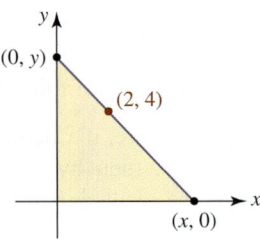

$(0, y)$

$(2, 4)$

$(x, 0)$

FIGURE 4.8.12 Graph for Problem 13

14. Find the maximum vertical distance d between the graphs of $y = x^2 - 1$ and $y = 1 - x$ for $-2 \le x \le 1$. See **FIGURE 4.8.13**.

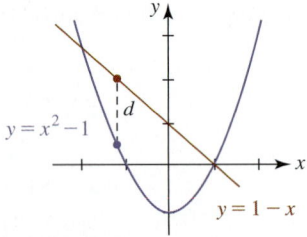

$y = x^2 - 1$

d

$y = 1 - x$

FIGURE 4.8.13 Graph for Problem 14

15. A rancher has 3000 ft of fencing on hand. Determine the dimensions of a rectangular corral that encloses a maximum area.

16. A rectangular plot of land will be fenced into three equal portions by two dividing fences parallel to two sides. See **FIGURE 4.8.14**. If the area to be enclosed is 4000 m², find the dimensions of the land that require the least amount of fence.

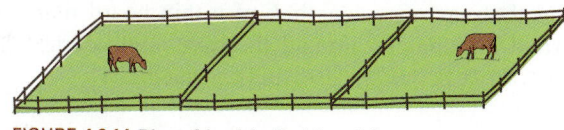

FIGURE 4.8.14 Plot of land in Problem 16

17. If the total fence to be used is 8000 m, find the dimensions of the enclosed land in Figure 4.8.14 that has the greatest area.

18. A rectangular yard is to be enclosed with a fence by attaching it to a house whose width is 40 ft. See FIGURE 4.8.15. The amount of fence to be used is 160 ft. Describe how the fence should be used so that the greatest area is enclosed.

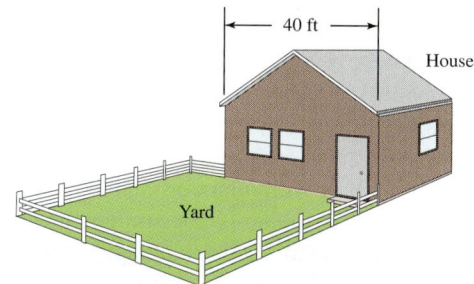

FIGURE 4.8.15   House and yard in Problem 18

19. Solve Problem 18 if the amount of fence to be used is 80 ft.

20. A rancher wishes to build a rectangular corral of 128,000 ft² with one side along a vertical cliff. The fencing along the cliff costs $1.50 per foot, whereas along the other three sides the fencing costs $2.50 per foot. Find the dimensions of the corral so that the cost of fencing is a minimum.

21. An open rectangular box is to be constructed with a square base and a volume of 32,000 cm³. Find the dimensions of the box that require the least amount of material.

22. In Problem 21, find the dimensions of a closed box that require the least amount of material.

23. A box, open at the top, is to be made from a square piece of cardboard by cutting a square out of each corner and turning up the sides. In FIGURE 4.8.16, the white squares are cut out and the cardboard is folded along the dashed lines. Given that the cardboard measures 40 cm on a side, find the dimensions of the box that will give the maximum volume. What is the maximum volume?

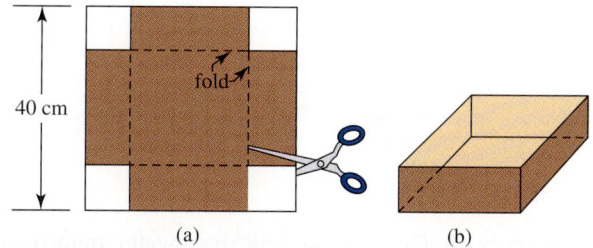

FIGURE 4.8.16   Open box in Problem 23

24. A box, open at the top, is to be made from a rectangular piece of cardboard that is 30 in. long and 20 in. wide. The box can hold itself together by cutting a square out of each corner, cutting on the interior solid lines, and then folding the cardboard on the dashed lines. See FIGURE 4.8.17. Express the volume of the box as a function of the indicated variable x. Find the dimensions of the box that give the maximum volume. What is the maximum volume?

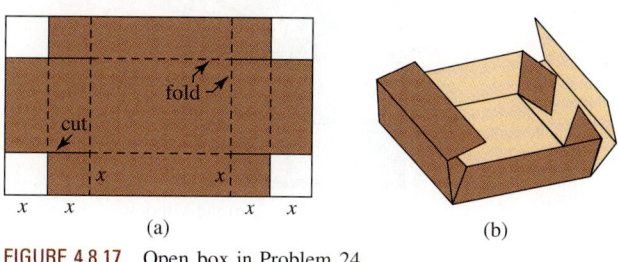

FIGURE 4.8.17   Open box in Problem 24

25. A gutter with a rectangular cross-section is made by bending up equal amounts from the ends of a 30-cm-wide piece of tin. What are the dimensions of the cross-section so that the volume is a maximum?

26. A gutter will be made so that its cross-section is an isosceles trapezoid with dimensions as indicated in FIGURE 4.8.18. Determine the value of θ so that the volume is a maximum.

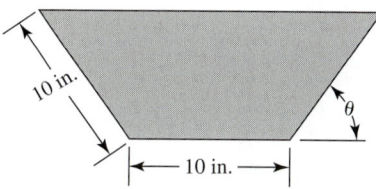

FIGURE 4.8.18   Gutter in Problem 26

27. Two flagpoles are secured by wires that are attached at a single point between the poles. See FIGURE 4.8.19. Where should the point be located to minimize the amount of wire used?

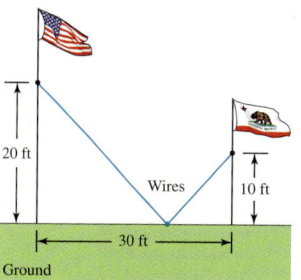

FIGURE 4.8.19   Flagpoles in Problem 27

28. The running track shown in FIGURE 4.8.20 is to consist of two parallel straight parts and two semicircular parts. The length of the track is to be 2 km. Find the design of the track so that the rectangular plot of land enclosed by the track is a maximum.

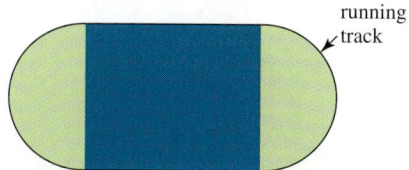

FIGURE 4.8.20   Running track in Problem 28

29. A Norman window consists of a rectangle surmounted by a semicircle. Find the dimensions of the window with largest area if its perimeter is 10 m. See FIGURE 4.8.21.

FIGURE 4.8.21 Norman window in Problem 29

30. Rework Problem 29 given that the rectangle is surmounted by an equilateral triangle.

31. A 10-ft wall stands 5 ft away from a building, as shown in FIGURE 4.8.22. Find the length L of the shortest ladder, supported by the wall, that reaches from the ground to the building.

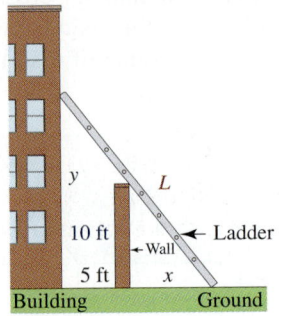

FIGURE 4.8.22 Ladder in Problem 31

32. U.S. Postal Service regulations state that a rectangular box sent by fourth-class mail must satisfy the requirement that its length plus its girth (perimeter of one end) must not exceed 108 in. Given that a box is to be constructed so that it has a square base, find the dimensions of the box that has a maximum volume. See FIGURE 4.8.23.

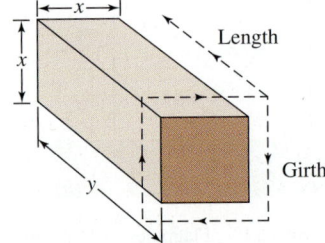

FIGURE 4.8.23 Box in Problem 32

33. Find the dimensions of the right circular cylinder with greatest volume that can be inscribed in a right circular cone of radius 8 in. and height 12 in. See FIGURE 4.8.24.

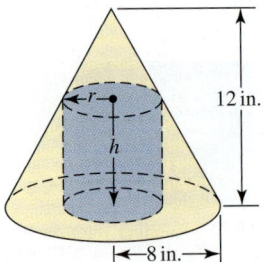

FIGURE 4.8.24 Inscribed cylinder in Problem 33

34. Find the maximum length L of a thin board than can be carried horizontally around the right-angle corner shown in FIGURE 4.8.25. [*Hint*: Use similar triangles.]

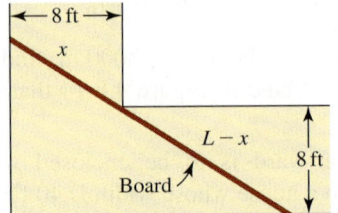

FIGURE 4.8.25 Board in Problem 34

35. A juice can is to be made in the form of a right circular cylinder and have a volume of 32 in^3. See FIGURE 4.8.26. Find the dimensions of the can so that the least amount of material is used in its construction. [*Hint*: Material = total surface area of can = area of top + area of bottom + area of lateral side. If the circular top and bottom covers are removed and the cylinder is cut straight up its side and flattened out, the result is the rectangle shown in Figure 4.8.26(c).]

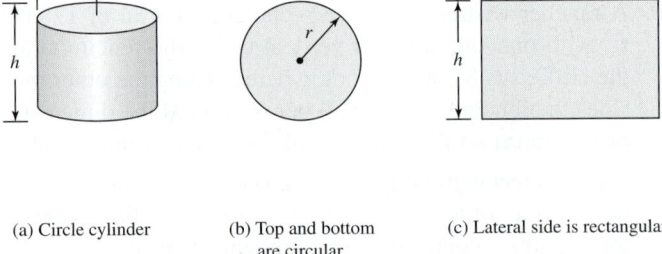

(a) Circle cylinder   (b) Top and bottom are circular   (c) Lateral side is rectangular

FIGURE 4.8.26 Juice can in Problem 35

36. In Problem 35, suppose that the circular top and bottom are cut from square sheets of metal as shown in FIGURE 4.8.27. If the metal cut from the corners of the square sheet is wasted, then find the dimensions of the can so that the least amount of material (including waste) is used in its construction.

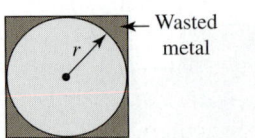

FIGURE 4.8.27 Top and bottom of can in Problem 36

37. Some birds fly more slowly over water than over land. A bird flies at constant rates of 6 km/h over water and 10 km/h over land. Use the information in FIGURE 4.8.28 to find the path the bird should take to minimize the total flying time between the shore of one island and its nest on the shore of another island. [*Hint*: Use *distance = rate × time*.]

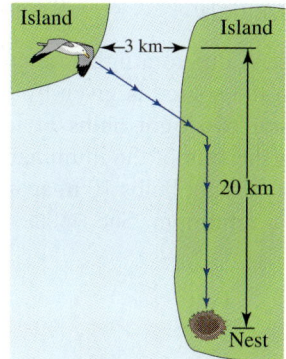

FIGURE 4.8.28   Bird in Problem 37

38. A pipeline is to be constructed from a refinery across a swamp to storage tanks. See FIGURE 4.8.29. The cost of construction is $25,000 per mile over the swamp and $20,000 per mile over land. How should the pipeline be made so that the cost of construction is a minimum?

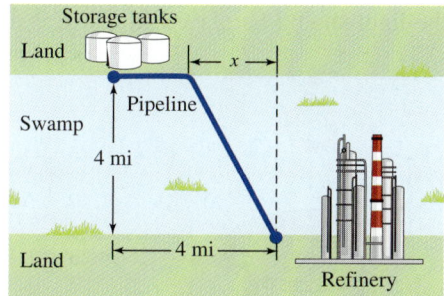

FIGURE 4.8.29   Pipeline in Problem 38

39. Rework Problem 38 given that the cost per mile across the swamp is twice the cost per mile over land.

40. At midnight ship A is 50 km north of ship B. Ship A is sailing south at 20 km/h and ship B is sailing west at 10 km/h. At what time will the distance between the ships be a minimum?

41. A container for transporting biohazardous waste is made of heavy plastic and is formed by adjoining two hemispheres to the ends of a right circular cylinder as shown in FIGURE 4.8.30. The total volume of the container is to be 30π ft^3. The cost per square foot of the plastic for the ends is one and a half times the cost per square foot of the plastic used in the cylindrical part. Find the dimensions of the container so that the cost of its construction is a minimum. [*Hint*: The volume of a sphere is $\frac{4}{3}\pi r^3$ and its surface area is $4\pi r^2$.]

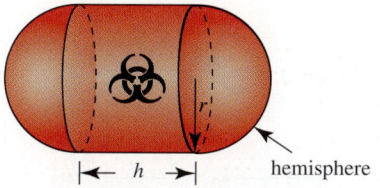

FIGURE 4.8.30   Container in Problem 41

42. A printed page will have 2-in. margins of white space on the sides and 1-in. margins of white space on the top and bottom. See FIGURE 4.8.31. The area of the printed portion is 32 in^2. Determine the dimensions of the page so that the least amount of paper is used.

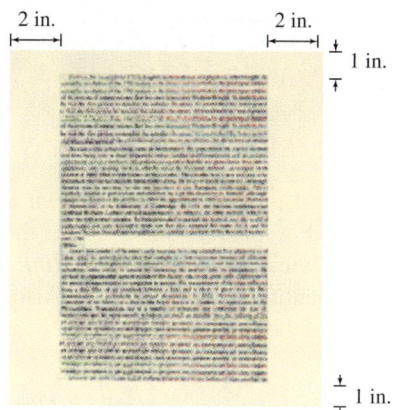

FIGURE 4.8.31   Printed page in Problem 42

43. One corner of an 8.5 in. × 11 in. piece of paper is folded over to the other edge of the paper as shown in FIGURE 4.8.32. Find the width x of the fold so that the length L of the crease is a minimum.

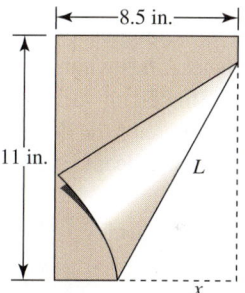

FIGURE 4.8.32   Piece of paper in Problem 43

44. The frame of a kite consists of six pieces of lightweight plastic. As shown in FIGURE 4.8.33, the outer frame of the kite consists of four precut pieces; two pieces of length 2 ft and two pieces of length 3 ft. The remaining crossbar pieces, labeled x in the figure, are to be cut to lengths so that the kite is as large as possible. Find these lengths.

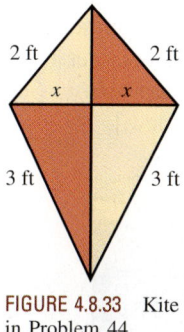

FIGURE 4.8.33   Kite in Problem 44

45. Find the dimensions of the rectangle of greatest area that can be circumscribed about a rectangle of length a and width b. See the red rectangle in FIGURE 4.8.34.

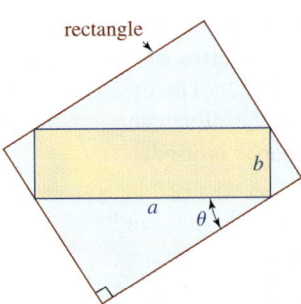

FIGURE 4.8.34 Rectangle in Problem 45

46. A statue is placed on a pedestal as shown in FIGURE 4.8.35. How far should a person stand from the pedestal to maximize the viewing angle θ? [*Hint*: Review the trigonometric identity for $\tan(\theta_2 - \theta_1)$. Also, it suffices to maximize $\tan\theta$ rather than θ. Why?]

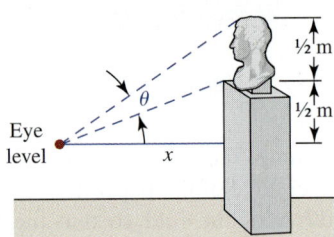

FIGURE 4.8.35 Statue in Problem 46

47. A cross-section of a rectangular wooden beam cut from a circular log of diameter d has width x and depth y. See FIGURE 4.8.36. The strength of the beam varies directly as the product of the width and the square of the depth. Find the dimensions of the cross-section of the beam of greatest strength.

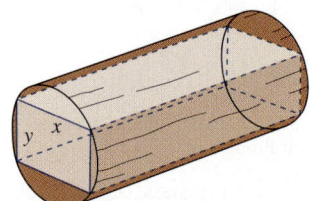

FIGURE 4.8.36 Log in Problem 47

48. The container shown in FIGURE 4.8.37 is to be constructed by attaching an inverted cone (open at its top) to the bottom of a right circular cylinder (open at its top and bottom) of radius 5 ft. The container is to have a volume of 100 ft^3. Find the value of the indicated angle so that the total surface area of the container is a minimum. What is the minimum surface area? [*Hint*: See Problem 38 in part C of Chapter 1 in Review.]

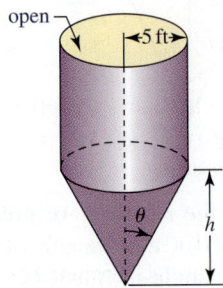

FIGURE 4.8.37 Container in Problem 48

Mathematical Models

49. The illuminance E due to a light source or intensity I at a distance r from the source is given by $E = I/r^2$. The total illuminance from two light bulbs of intensities $I_1 = 125$ and $I_2 = 216$ is the sum of the illuminances. Find the point P between the two light bulbs 10 m apart at which the total illuminance is a minimum. See FIGURE 4.8.38.

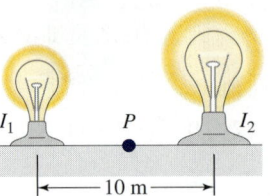

FIGURE 4.8.38 Light bulbs in Problem 49

50. The illuminance E at any point P on the edge of a circular table caused by a light placed directly above its center is given by $E = (i\cos\theta)/r^2$. See FIGURE 4.8.39. Given that the radius of the table is 1 m and $I = 100$, find the height at which the light should be placed so that E is a maximum.

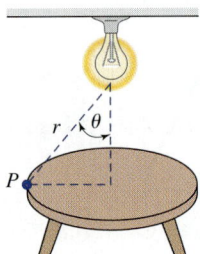

FIGURE 4.8.39 Light and table in Problem 50

51. Fermat's Principle in optics states that light travels from point A (in the xy-plane) in one medium to point B in another medium on a path that requires minimum time. Denote the speed of light in the medium that contains point A by c_1 and the speed of light in the medium that contains point B by c_2. Show that the time of travel from A to B is a minimum when the angles θ_1 and θ_2, shown in FIGURE 4.8.40, satisfy **Snell's law**:

$$\frac{\sin\theta_1}{c_1} = \frac{\sin\theta_2}{c_2}.$$

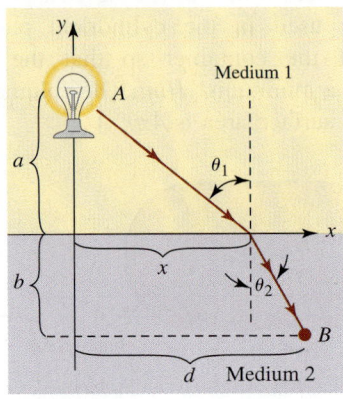

FIGURE 4.8.40 Two mediums in Problem 51

52. Blood is carried throughout the body by the vascular system, which consists of capillaries, veins, arterioles, and arteries. One consideration of the problem of minimizing the energy expended in moving the blood through the various organs is to find an optimum angle θ for *vascular branching* such that the total resistance to the blood along a path from a larger blood vessel to a smaller blood vessel is a minimum. See FIGURE 4.8.41. Use **Poiseuille's law**, which states that the resistance R of a blood vessel of length l and radius r is $R = kl/r^4$, where k is a constant, to show that the total resistance

$$R = k\left(\frac{x}{r_1^4}\right) + k\left(\frac{y}{r_2^4}\right)$$

along the path $P_1P_2P_3$ is a minimum when $\cos\theta = r_2^4/r_1^4$. [*Hint*: Express x and y in terms of θ and a.]

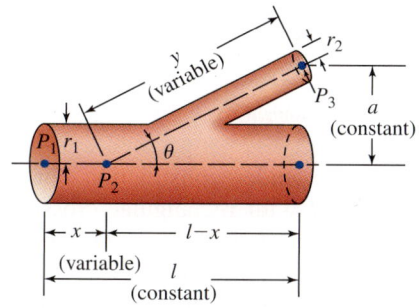

FIGURE 4.8.41 Vascular branching in Problem 52

53. The potential energy between two atoms in a diatomic molecule is given by $U(x) = 2/x^{12} - 1/x^6$. Find the minimum potential energy between the two atoms.

54. The height of a projectile launched with a constant initial velocity v_0 at an angle of elevation θ_0 is given by

$$y = (\tan\theta_0)x - \left(\frac{g}{2v_0^2\cos^2\theta_0}\right)x^2,$$ where x is its horizontal

displacement measured from the point of launch. Show that the maximum height attained by the projectile is $h = (v_0^2/2g)\sin^2\theta_0$.

55. A beam of length L is embedded in concrete walls as shown in FIGURE 4.8.42. When a constant load w_0 is uniformly distributed along its length, the deflection curve $y(x)$. for the beam is given by

$$y(x) = \frac{w_0 L^2}{24EI}x^2 - \frac{w_0 L}{12EI}x^3 + \frac{w_0}{24EI}x^4,$$

where E and I are constants. (E is **Young's modulus of elasticity** and I is a moment of inertia of a cross-section of the beam.) The deflection curve approximates the shape of the beam.

(a) Determine the maximum deflection of the beam.
(b) Sketch a graph of $y(x)$.

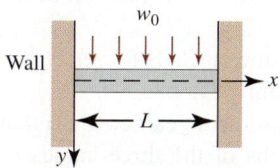

FIGURE 4.8.42 Beam in Problem 55

56. The relationship between the height h and the diameter d of a tree can be approximated by the quadratic expression $h = 137 + ad - bd^2$, where h and d are measured in centimeters, and a and b are positive parameters that depend on the type of tree. See FIGURE 4.8.43.

(a) Suppose a tree attains its maximum height of H centimeters at a diameter of D centimeters. Show that

$$h = 137 + 2\frac{H - 137}{D}d - \frac{H - 137}{D^2}d^2.$$

(b) Suppose a certain tree reaches its maximum possible height (according to the formula) of 15 m at a diameter of 0.8 m. What was the diameter of the tree when the tree was 10 m tall?

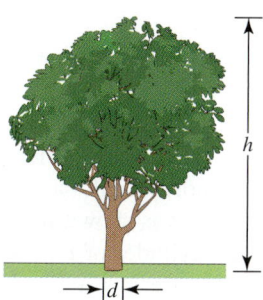

FIGURE 4.8.43 Tree in Problem 56

57. The long bones in mammals may be represented as hollow cylindrical tubes, filled with marrow, of outer radius R and inner radius r. Bones should be constructed to be lightweight yet capable of withstanding certain bending moments. In order to withstand a bending moment M, it can be shown that the mass m per unit length of the bone and marrow is given by

$$m = \pi\rho\left[\frac{M}{K(1 - x^4)}\right]^{2/3}\left(1 - \frac{1}{2}x^2\right),$$

where ρ is the density of the bone and K is a positive constant. If $x = r/R$, show that m is a minimum when $r = 0.63R$ (approximately).

58. The rate P (in mg carbon/m^3/h) at which photosynthesis takes place for a certain species of phytoplankton is related to the light intensity I (in 10^3 ft-candles) by the function

$$P = \frac{100I}{I^2 + I + 4}.$$

At what light intensity is P the largest?

≡ Think About It

59. A Mathematical Classic A person would like to cut a 1-m-long piece of wire into two pieces. One piece will be bent into the shape of a circle and the other into the shape of a square. How should the wire be cut so that the sum of the areas is a maximum?

60. In Problem 59, suppose one piece of wire will be bent into the shape of a circle and the other into the shape of an equilateral triangle. How should the wire be cut so that the sum of the areas is a minimum? A maximum?

61. A conical cup is made from a circular piece of paper of radius R by cutting out a circular sector and then joining the dashed edges shown in FIGURE 4.8.44.

(a) Determine the value of r indicated in Figure 4.8.44(b) so that the volume of the cup is a maximum.
(b) What is the maximum volume of the cup?
(c) Find the central angle θ of the circular sector so that the volume of the conical cup is a maximum.

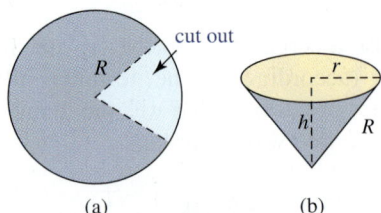

(a)   (b)

FIGURE 4.8.44   Conical cup in Problem 61

62. The lateral side of a cylinder is to be made from a rectangle of flimsy sheet plastic. Because the plastic material cannot support itself, a thin stiff wire is embedded in the material as shown in FIGURE 4.8.45(a). Find the dimensions of the cylinder of largest volume that can be constructed if the wire has a fixed length L. [*Hint*: There are two constraints in this problem. In Figure 4.8.45(b), the circumference of a circular end of the cylinder is y.]

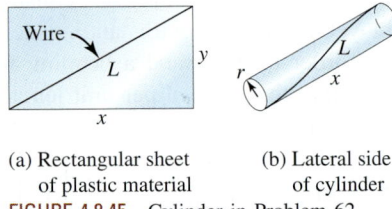

(a) Rectangular sheet   (b) Lateral side
of plastic material   of cylinder

FIGURE 4.8.45   Cylinder in Problem 62

63. In Problem 27, show that when the optimal amount of wire (the least amount) is used, then the angle θ_1 the wire to the left flagpole makes with the ground is the same as the angle θ_2 the wire to the right flagpole makes with the ground. See Figure 4.8.19.

64. Find an equation of the tangent line L to the graph of $y = 1 - x^2$ at $P(x_0, y_0)$ such that the triangle in the first quadrant bounded by the coordinate axes and L has minimum area. See FIGURE 4.8.46.

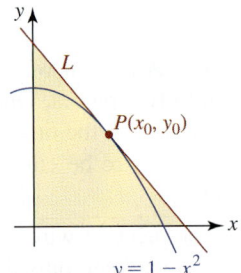

$y = 1 - x^2$

FIGURE 4.8.46   Triangle in Problem 64

≡ Calculator/CAS Problems

65. In a race a woman is required to swim from a floating dock A to the beach and, without stopping, swim from the beach out to another floating dock C. The distances are shown in FIGURE 4.8.47(a). She estimates that she can swim from dock A to the beach at a constant rate of 3 mi/h and out from the beach to dock C at a rate of 2 mi/h. Where should she touch the beach in order to minimize the total swimming time from A to C? Introduce an xy-coordinate system as shown in Figure 4.8.47(b). Use a CAS to find the critical numbers.

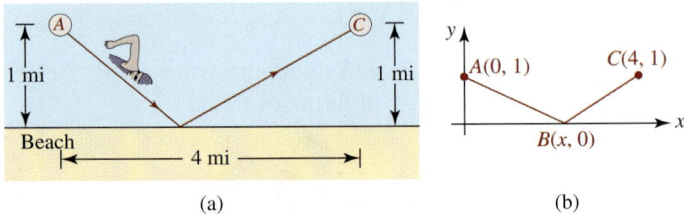

(a)   (b)

FIGURE 4.8.47   Swimmer in Problem 65

66. A two-story house under construction consists of two structures A and B with rectangular cross-sections and dimensions as indicated in FIGURE 4.8.48. The framing for structure B requires temporary wooden reinforcing buttresses from ground level that rest against structure A as shown.

(a) Express the length L of a buttress as a function of the indicated angle θ.
(b) Find $L'(\theta)$.
(c) Use a calculator or CAS to obtain the graph of $L'(\theta)$ on the interval $(0, \pi/2)$. Use this graph to show that L has only one critical number θ_c in $(0, \pi/2)$. Use this graph to determine the algebraic sign of $L'(\theta)$ for $0 < \theta < \theta_c$, and the algebraic sign of $L'(\theta)$ for $\theta_c < \theta < \pi/2$. What is your conclusion?
(d) Find the minimum length of a buttress.

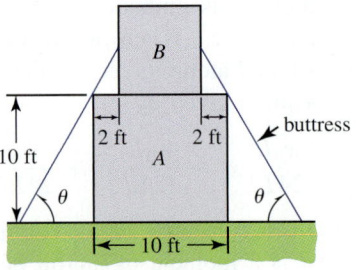

FIGURE 4.8.48   House in Problem 66

67. Consider the three cables shown in FIGURE 4.8.49.

(a) Express the total length L of the three cables shown in Figure 4.8.49(a) as a function of the length L of the cable AB.
(b) Use a calculator or CAS to verify that the graph of L has a minimum.
(c) Find the length of the cable AB so that the total length L of the lengths of the three cables is a minimum.

(d) Express the total length L of the three cables shown in Figure 4.8.49(b) as a function of the length of the cable AB.

(e) Use a calculator or CAS to verify that the graph of L has a minimum.

(f) Use the graph obtained in part (e) or a CAS as an aid in approximating the length of the cable AB that minimizes the function L obtained in part (d).

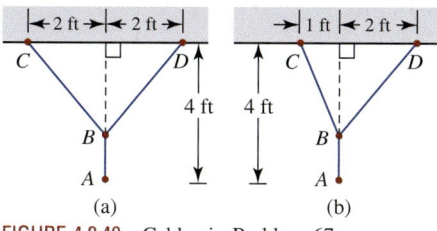

FIGURE 4.8.49 Cables in Problem 67

≡ **Project**

68. Frequency Interference When the Federal Aviation Administration (FAA) allocates numerous frequencies for an airport radio transmitter quite often nearby transmitters use the same frequencies. As a consequence, the FAA would like to minimize the interference between these transmitters. In FIGURE 4.8.50, the point (x_t, y_t) represents the location of a transmitter whose radio jurisdiction is indicated by the circle C of radius with center at the origin. A second transmitter is located at $(x_i, 0)$ as shown in the figure. In this problem you will develop and analyze a function to find the interference between two transmitters.

(a) The strength of the signal from a transmitter to a point is inversely proportional to the square of the distance between them. Assume that a point (x, y) is located on the upper portion of the circle C as shown in Figure 4.8.50. Express the primary strength of the signal at (x, y) from a transmitter at (x_t, y_t) as a function of x. Express the secondary strength at (x, y) from the transmitter at $(x_i, 0)$ as a function of x. Now define a

function $R(x)$ as a quotient of the primary signal strength to the secondary signal strength. $R(x)$ can be thought of as a *signal to noise ratio*. To guarantee that the interference remains small we need to show that the minimum signal to noise ratio is greater than the FAA's minimum threshold of -0.7.

(b) Suppose that $x_t = 760$ m, $y_t = -560$ m, $r = 1.1$ km, and $x_i = 12$ km. Use a CAS to simplify and then plot the graph of $R(x)$. Use the graph to estimate the domain and range of $R(x)$.

(c) Use the graph in part (b) to estimate the value of x where the minimum ratio R occurs. Estimate the value of R at that point. Does this value of R exceed the FAA's minimum threshold?

(d) Use a CAS to differentiate $R(x)$. Use a CAS to find the root of $R'(x) = 0$ and to compute the corresponding value of $R(x)$. Compare your answers here with the estimates in part (c).

(e) What is the point (x, y) on circle C?

(f) We assumed that the point (x, y) was in the top half plane when (x_t, y_t) was in the lower half plane. Explain why this assumption is correct.

(g) Use a CAS to find the value of x where the minimum interference occurs in terms of the symbols x_t, y_t, x_i, and r.

(h) Where is that point that minimizes the signal to noise ratio when the transmitter at (x_t, y_t) is on the x-axis? Give a convincing argument justifying your answer.

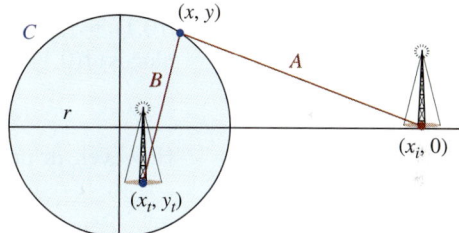

FIGURE 4.8.50 Radio transmitters in Problem 68

4.9 Linearization and Differentials

■ **Introduction** We started the discussion of the derivative with the problem of finding the tangent line to the graph of a function $y = f(x)$ at a point $(a, f(a))$. Intuitively, we would expect that a tangent line is very close to the graph of f whenever x is close to the number a. Put in other words, when x is in a small neighborhood of a the function values $f(x)$ are very close to the values of the y-coordinates on the tangent line. Thus, by finding an equation of the tangent line at $(a, f(a))$ we can use that equation to approximate $f(x)$.

An equation of the tangent line shown in red in FIGURE 4.9.1 is given by

$$y - f(a) = f'(a)(x - a) \qquad \text{or} \qquad y = f(a) + f'(a)(x - a). \tag{1}$$

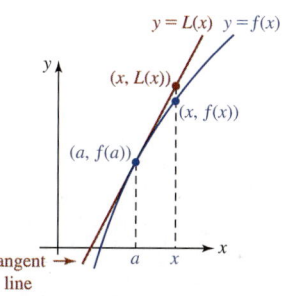

FIGURE 4.9.1 When x is close to a, the value $L(x)$ is close to $f(x)$

Using standard functional notation, let us write the last linear equation in (1) as $L(x) = f(a) + f'(a)(x - a)$. This linear function is given a special name.

Definition 4.9.1 Linearization

If a function $y = f(x)$ is differentiable at a number a, then the function

$$L(x) = f(a) + f'(a)(x - a) \qquad (2)$$

is said to be a **linearization** of f at a. For a number x near a, the approximation

$$f(x) \approx L(x) \qquad (3)$$

is called a **local linear approximation** of f at a.

There is no need to memorize (2); it is simply the point–slope form of the tangent line at $(a, f(a))$.

EXAMPLE 1 Linearization of $\sin x$

Find a linearization of $f(x) = \sin x$ at $a = 0$.

Solution Using $f(0) = 0, f'(x) = \cos x$, and $f'(0) = 1$ the tangent line to the graph of $f(x) = \sin x$ at $(0, 0)$ is $y - 0 = 1 \cdot (x - 0)$. Therefore, the linearization of $f(x) = \sin x$ at $a = 0$ is $L(x) = x$. As seen in **FIGURE 4.9.2** the graph of $f(x) = \sin x$ and its linearization at $a = 0$ are nearly indistinguishable near the origin. The local linear approximation $f(x) \approx L(x)$ of f at $a = 0$ is

$$\sin x \approx x. \qquad (4) \quad \blacksquare$$

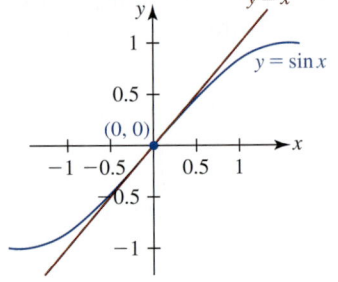

FIGURE 4.9.2 Graph of function and linearization in Example 1

▌ **Errors** Example 1 reemphasizes something you already know from trigonometry. The local linear approximation (4) shows that the sine of a small angle x (measured in radians) is approximately the same as the angle. For comparison, if we choose $x = 0.1$, then (4) indicates that $f(0.1) \approx L(0.1)$ or $\sin(0.1) \approx 0.1$. For comparison a calculator gives (rounded to five decimal places) $f(0.1) = \sin(0.1) = 0.09983$. Now an error in a calculation is defined by

$$\textbf{error = true value − approximate value.} \qquad (5)$$

However, in practice the

$$\textbf{relative error} = \frac{\textbf{error}}{\textbf{true value}} \qquad (6)$$

is usually more important than the error. Moreover, (relative error) $\cdot 100$ is called **percentage error**. Thus with the aid of a calculator, the percentage error in the approximation $f(0.1) \approx L(0.1)$ is roughly only 0.2%. Figure 4.9.2 clearly shows that as x moves away from 0, the accuracy of the approximation $\sin x \approx x$ diminishes. For example, for the number 0.9 a calculator gives $f(0.9) = \sin(0.9) = 0.78333$, whereas $L(0.9) = 0.9$. This time the percentage error is about 15%.

We have also seen the result of Example 1 presented in a slightly different manner in Section 2.4. If we divide the local linear approximation $\sin x \approx x$ by x, we get $\dfrac{\sin x}{x} \approx 1$ for values of x near 0. This leads us back to the important trigonometric limit $\lim\limits_{x \to 0} \dfrac{\sin x}{x} = 1$.

EXAMPLE 2 Linearization and Approximation

(a) Find a linearization of $f(x) = \sqrt{x + 1}$ at $a = 3$.
(b) Use a local linear approximation to approximate $\sqrt{3.95}$ and $\sqrt{4.01}$.

Solution

(a) By the Power Rule for Functions, the derivative of f is

$$f'(x) = \frac{1}{2}(x + 1)^{-1/2} = \frac{1}{2\sqrt{x + 1}}.$$

When evaluated at $a = 3$ the two functions give:

$$f(3) = \sqrt{4} = 2 \qquad \leftarrow \text{point of tangency is } (3, 2)$$

$$f'(3) = \frac{1}{2\sqrt{4}} = \frac{1}{4}. \qquad \leftarrow \text{slope of tangent at } (3, 2) \text{ is } \tfrac{1}{4}$$

Thus, by the point–slope form of a line the linearization of f at $a = 3$ is given by $y - 2 = \frac{1}{4}(x - 3)$ or

$$L(x) = 2 + \frac{1}{4}(x - 3). \tag{7}$$

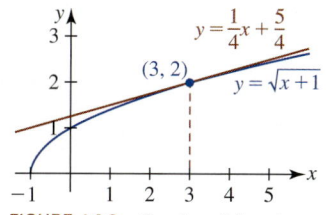

FIGURE 4.9.3   Graphs of function and linearization in Example 2

The graphs of f and L are given in FIGURE 4.9.3. Of course, L can be expressed in the slope–intercept form $L(x) = \frac{1}{4}x + \frac{5}{4}$ but for computational purposes the form given in (7) is more convenient.

(b) Using (7) from part (a), we have the local linear approximation $f(x) \approx L(x)$ or

$$\sqrt{x + 1} \approx 2 + \frac{1}{4}(x - 3), \tag{8}$$

whenever x is close to 3. Now, setting $x = 2.95$ and $x = 3.01$ in (8) gives, in turn, the approximations:

$$\overset{f(2.95)}{\overbrace{\sqrt{3.95}}} \approx \overset{L(2.95)}{\overbrace{2 + \frac{1}{4}(2.95 - 3)}} = 2 - \frac{0.05}{4} = 1.9875.$$

and

$$\overset{f(3.01)}{\overbrace{\sqrt{4.01}}} \approx \overset{L(3.01)}{\overbrace{2 + \frac{1}{4}(3.01 - 3)}} = 2 + \frac{0.01}{4} = 2.0025. \qquad \blacksquare$$

❙ Differentials   The fundamental idea of a linearization of a function was originally couched in the terminology of *differentials*. Suppose $y = f(x)$ is a differentiable function in an open interval containing the number a. If x_1 is a different number on the x-axis, then **increments** Δx and Δy are the differences

$$\Delta x = x_1 - a \qquad \text{and} \qquad \Delta y = f(x_1) - f(a).$$

But since $x_1 = a + \Delta x$, the **change in the function** is

$$\Delta y = f(a + \Delta x) - f(a).$$

For values of Δx that are close to 0, the difference quotient

$$\frac{f(a + \Delta x) - f(a)}{\Delta x} = \frac{\Delta y}{\Delta x}$$

is an approximation of the value of the derivative of f at a:

$$\frac{\Delta y}{\Delta x} \approx f'(a) \qquad \text{or} \qquad \Delta y \approx f'(a)\Delta x.$$

The quantities Δx and $f'(a)\Delta x$ are called **differentials** and are denoted by the symbols dx and dy, respectively. That is,

$$\Delta x = dx \qquad \text{and} \qquad dy = f'(a)\,dx.$$

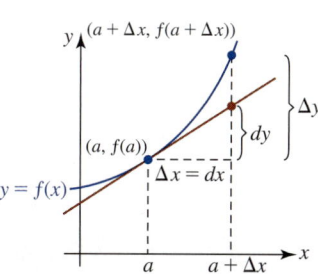

FIGURE 4.9.4   Geometric interpretations of dx, Δy, and dy

As shown in FIGURE 4.9.4, for a change dx in x the quantity $dy = f'(a)\,dx$ represents the **change in the linearization** (the *rise* in the tangent line at $(a, f(a))$*). And so when $dx \approx 0$, the change in the function Δy is approximately the same as the change in the linearization dy:

$$\Delta y \approx dy. \tag{9}$$

---

*For this reason, the Leibniz notation for the derivative dy/dx has the appearance of a quotient.

> **Definition 4.9.2** Differentials
>
> The **differential of the independent variable** x is the nonzero number Δx and is denoted by dx; that is,
>
> $$dx = \Delta x. \tag{10}$$
>
> If f is a differentiable function at x, then the **differential of the dependent variable** y is denoted by dy; that is,
>
> $$dy = f'(x)\Delta x = f'(x)\,dx. \tag{11}$$

EXAMPLE 3 Differentials

(a) Find Δy and dy for $f(x) = 5x^2 + 4x + 1$.
(b) Compare the values of Δy and dy for $x = 6$, $\Delta x = dx = 0.02$.

Solution

(a) $\Delta y = f(x + \Delta x) - f(x)$
$\quad = [5(x + \Delta x)^2 + 4(x + \Delta x) + 1] - [5x^2 + 4x + 1]$
$\quad = 10x\Delta x + 4\Delta x + 5(\Delta x)^2.$

Now, since $f'(x) = 10x + 4$, we have from (11) of Definition 4.9.2,

$$dy = (10x + 4)\,dx. \tag{12}$$

By rewriting Δy as $\Delta y = (10x + 4)\Delta x + 5(\Delta x)^2$ and using $dx = \Delta x$, observe that $dy = (10x + 4)\Delta x$ and $\Delta y = (10x + 4)\Delta x + 5(\Delta x)^2$ differ by the amount $5(\Delta x)^2$.

(b) When $x = 6$, $\Delta x = 0.02$:

$$\Delta y = 10(6)(0.02) + 4(0.02) + 5(0.02)^2 = 1.282$$

whereas $\qquad dy = (10(6) + 4)(0.02) = 1.28.$

The difference in answers is, of course, $5(\Delta x)^2 = 5(0.02)^2 = 0.002$. ∎

In Example 3 the value $\Delta y = 1.282$ is the *exact* amount by which the function $f(x) = 5x^2 + 4x + 1$ changes as x changes from 6 to 6.02. The differential $dy = 1.28$ represents an *approximation* of the amount by which the function changes. As shown in (9), for a small change or increment Δx in the independent variable, the corresponding change Δy in the dependent variable can be approximated by the differential dy.

▌ Linear Approximation Revisited Differentials can be used to approximate the value $f(x + \Delta x)$. From $\Delta y = f(x + \Delta x) - f(x)$, we get

$$f(x + \Delta x) = f(x) + \Delta y.$$

But in view of (9), for a small change in x we can write

$$f(x + \Delta x) \approx f(x) + dy.$$

With $dy = f'(x)\,dx = f'(x)\Delta x$ the preceding line is the same as

$$f(x + \Delta x) \approx f(x) + f'(x)\,dx. \tag{13}$$

We have already seen the formula in (13) in a different guise. If we let $x = a$ and $dx = \Delta x = x - a$, then (13) becomes

$$f(x) \approx f(a) + f'(a)(x - a). \tag{14}$$

The right-hand side of the equality in (14) is recognized as $L(x)$ and (13) becomes $f(x) \approx L(x)$ which is the result given in (3).

EXAMPLE 4 Approximation by Differentials

Use (13) to approximate $(2.01)^3$.

Solution First identify the function $f(x) = x^3$. We wish to calculate the approximate value of $f(x + \Delta x) = (x + \Delta x)^3$ when $x = 2$ and $\Delta x = 0.01$. Now from (11),

$$dy = 3x^2\,dx = 3x^2\Delta x.$$

Thus (13) gives

$$(x + \Delta x)^3 \approx x^3 + 3x^2\Delta x.$$

With $x = 2$ and $\Delta x = 0.01$, the preceding formula gives the approximation

$$(2.01)^3 \approx 2^3 + 3(2)^2(0.01) = 8.12.$$ ∎

EXAMPLE 5 Approximation by Differentials

A side of a cube is measured to be 30 cm with a possible error of ± 0.02 cm. What is the approximate maximum possible error in the volume of the cube?

Solution The volume of a cube is $V = x^3$, where x is the length of one side. If Δx represents the error in the length of one side, then the corresponding error in the volume is

$$\Delta V = (x + \Delta x)^3 - x^3.$$

To simplify matters, we utilize the differential $dV = 3x^2\,dx = 3x^2\Delta x$ as an approximation to ΔV. Thus, for $x = 30$ and $\Delta x = \pm 0.02$ the approximate maximum error is

$$dV = 3(30)^2(\pm 0.02) = \pm 54 \text{ cm}^3.$$ ∎

In Example 5, an error of about 54 cm^3 in the volume for an error of 0.02 cm in the length of a side seems considerable. But, observe, if the relative error (6) is $\Delta V/V$, then the *approximate* relative error is dV/V. When $x = 30$ and $V = (30)^3 = 27{,}000$ the approximate maximum relative error is $\pm 54/27{,}000 = \pm 1/500$, and the maximum percentage error is only about $\pm 0.2\%$.

■ **Rules for Differentials** The rules for differentiation considered in this chapter can be rephrased in terms of differentials; for example, if $u = f(x)$ and $v = g(x)$ and $y = f(x) + g(x)$, then $dy/dx = f'(x) + g'(x)$. Hence, $dy = [f'(x) + g'(x)]\,dx = f'(x)\,dx + g'(x)\,dx = du + dv$. We summarize the differential equivalents of the Sum, Product, and Quotient Rules:

$$d(u + v) = du + dv \tag{15}$$

$$d(uv) = u\,dv + v\,du \tag{16}$$

$$d(u/v) = \frac{v\,du - u\,dv}{v^2}. \tag{17}$$

As the next example shows, there is little need for memorizing (15), (16), and (17).

EXAMPLE 6 Differential of y

Find dy for $y = x^2\cos 3x$.

Solution To find the differential of a function, we can simply multiply its derivative by dx. Thus, by the Product Rule,

$$\frac{dy}{dx} = x^2(-\sin 3x \cdot 3) + \cos 3x(2x)$$

and so
$$dy = \left(\frac{dy}{dx}\right) \cdot dx = (-3x^2\sin 3x + 2x\cos 3x)\,dx. \tag{18}$$

Alternative Solution Applying (16) gives

$$dy = x^2 d(\cos 3x) + \cos 3x\, d(x^2)$$
$$= x^2(-\sin 3x \cdot 3\,dx) + \cos 3x(2x\,dx). \tag{19}$$

Factoring dx from (19) yields (18). ∎

Exercises 4.9   Answers to selected odd-numbered problems begin on page ANS-17.

≡ Fundamentals

In Problems 1–8, find a linearization of the given function at the indicated number.

1. $f(x) = \sqrt{x};\quad a = 9$

2. $f(x) = \dfrac{1}{x^2};\quad a = 1$

3. $f(x) = \tan x;\quad a = \pi/4$

4. $f(x) = \cos x;\quad a = \pi/2$

5. $f(x) = \ln x;\quad a = 1$

6. $f(x) = 5x + e^{x-2};\quad a = 2$

7. $f(x) = \sqrt{1 + x};\quad a = 3$

8. $f(x) = \dfrac{1}{\sqrt{3 + x}};\quad a = 6$

In Problems 9–16, use a linearization at $a = 0$ to establish the given local linear approximation.

9. $e^x \approx 1 + x$

10. $\tan x \approx x$

11. $(1 + x)^{10} \approx 1 + 10x$

12. $(1 + 2x)^{-3} \approx 1 - 6x$

13. $\sqrt{1 - x} \approx 1 - \dfrac{1}{2}x$

14. $\sqrt{x^2 + x + 4} \approx 2 + \dfrac{1}{4}x$

15. $\dfrac{1}{3 + x} \approx \dfrac{1}{3} - \dfrac{1}{9}x$

16. $\sqrt[3]{1 - 4x} \approx 1 - \dfrac{4}{3}x$

In Problems 17–20, use an appropriate result from Problems 1–8 to find an approximation of the given quantity.

17. $(1.01)^{-2}$ **18.** $\sqrt{9.05}$ **19.** $10.5 + e^{0.1}$ **20.** $\ln 0.98$

In Problems 21–24, use an appropriate result from Problems 9–16 to find an approximation of the given quantity.

21. $\dfrac{1}{(1.1)^3}$ **22.** $(1.02)^{10}$ **23.** $(0.88)^{1/3}$ **24.** $\sqrt{4.11}$

In Problems 25–32, use an appropriate function and local linear approximation to find an approximation of the given quantity.

25. $(1.8)^5$

26. $(7.9)^{2/3}$

27. $\dfrac{(0.9)^4}{(0.9) + 1}$

28. $(1.1)^3 + 6(1.1)^2$

29. $\cos\left(\dfrac{\pi}{2} - 0.4\right)$

30. $\sin 1°$

31. $\sin 33°$

32. $\tan\left(\dfrac{\pi}{4} + 0.1\right)$

In Problems 33 and 34, find a linearization $L(x)$ of f at the given value of a. Use $L(x)$ to approximate the indicated function value.

33. $a = 1;\quad f(1.04)$

34. $a = -2;\quad f(-1.98)$

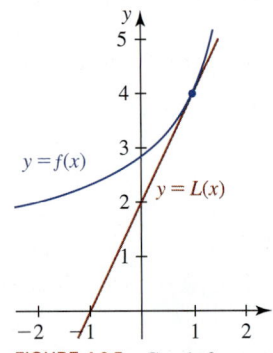

FIGURE 4.9.5   Graph for Problem 33

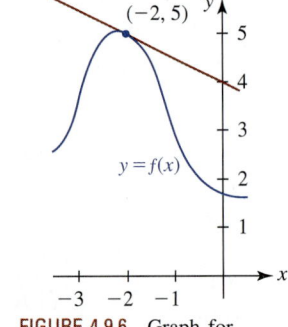

FIGURE 4.9.6   Graph for Problem 34

In Problems 35–42, find Δy and dy.

35. $y = x^2 + 1$

36. $y = 3x^2 - 5x + 6$

37. $y = (x + 1)^2$

38. $y = x^3$

39. $y = \dfrac{3x + 1}{x}$

40. $y = \dfrac{1}{x^2}$

41. $y = \sin x$

42. $y = -4\cos 2x$

In Problems 43 and 44, complete the following table for each function.

x	Δx	Δy	dy	$\Delta y - dy$
2	1			
2	0.5			
2	0.1			
2	0.01			

43. $y = 5x^2$ **44.** $y = 1/x$

45. Compute the approximate amount by which the function $f(x) = 4x^2 + 5x + 8$ changes as x changes from:

   (a) 4 to 4.03   (b) 3 to 2.9.

46. (a) Find an equation of the tangent line to the graph of $f(x) = x^3 + 3x^2$ at $x = 1$.
   (b) Find the y-coordinate of the point on the tangent line in part (a) that corresponds to $x = 1.02$.
   (c) Use (3) to find an approximation to $f(1.02)$. Compare your answer with that of part (b).

47. The area of a circle with radius r is $A = \pi r^2$.

   (a) Given that the radius of a circle changes from 4 cm to 5 cm, find the exact change in the area.
   (b) What is the approximate change in the area?

≡ Applications

48. According to Poiseuille, the resistance R of a blood vessel of length l and radius r is $R = kl/r^4$, where k is a constant. Given that l is constant, find the approximate change in R when r changes from 0.2 mm to 0.3 mm.

49. Many golf balls consist of a spherical cover over a solid core. Find the exact volume of the cover if its thickness is t and the radius of the core is r. [*Hint:* The volume of a sphere is $V = \frac{4}{3}\pi r^3$. Consider concentric spheres having radii r and $r + \Delta r$.] Use differentials to find an approximation to the volume of the cover. See **FIGURE 4.9.7**. Find an approximation to the volume of the cover if $r = 0.8$ in. and $t = 0.04$ in.

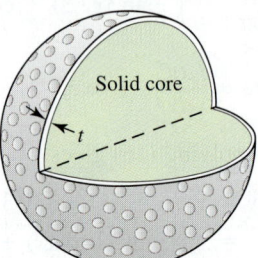

FIGURE 4.9.7   Golf ball in Problem 49

50. A hollow metal pipe is 1.5 m long. Find an approximation to the volume of the metal if the inner radius of the pipe is 2 cm and the thickness of the metal is 0.25 cm. See FIGURE 4.9.8.

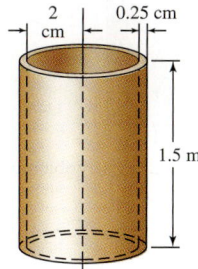

FIGURE 4.9.8   Pipe in Problem 50

51. The side of a square is measured to be 10 cm with a possible error of ± 0.3 cm. Use differentials to find an approximation to the maximum error in the area. Find the approximate relative error and the approximate percentage error.

52. An oil storage tank in the form of a circular cylinder has a height of 5 m. The radius is measured to be 8 m with a possible error of ± 0.25 m. Use differentials to estimate the maximum error in the volume. Find the approximate relative error and the approximate percentage error.

53. In the study of some adiabatic processes, the pressure P of a gas is related to the volume V that it occupies by $P = c/V^\gamma$, where c and γ are constants. Show that the approximate relative error in P is proportional to the approximate relative error in V.

54. The range R of a projectile with an initial velocity v_0 and angle of elevation θ is given by $R = (v_0^2/g)\sin 2\theta$, where g is the acceleration of gravity. If v_0 and θ are held constant, then show that the percentage error in the range is proportional to the percentage error in g.

55. Use the formula in Problem 54 to determine the range of a projectile when the initial velocity is 256 ft/s, the angle of elevation is $45°$, and the acceleration of gravity is 32 ft/s². What is the approximate change in the range of the projectile if the initial velocity is increased to 266 ft/s?

56. The acceleration due to gravity g is not constant but changes with altitude. For practical purposes, at the surface of the Earth g is taken to be 32 ft/s², 980 cm/s², or 9.8 m/s².

   (a) From the Law of Universal Gravitation, the force F between a body of mass m_1 and the Earth of mass m_2 is $F = km_1m_2/r^2$, where k is constant and r is the distance to the center of the Earth. Alternatively, Newton's second law of motion implies $F = m_1g$. Show that $g = km_2/r^2$.
   (b) Use the result from part (a) to show $dg/g = -2dr/r$.
   (c) Let $r = 6400$ km at the surface of the Earth. Use part (b) to show that the approximate value of g at an altitude of 16 km is 9.75 m/s².

57. The acceleration due to gravity g also changes with latitude. The International Geodesy Association has defined g (at sea level) as a function of latitude θ as follows:

$$g = 978.0318(1 + 53.024 \times 10^{-4}\sin^2\theta - 5.9 \times 10^{-6}\sin^2 2\theta),$$

where g is measured in cm/s².

   (a) According to this mathematical model, where is g a minimum? Where is g a maximum?
   (b) What is the value of g at latitude $60°$N?
   (c) What is the approximate change in g as θ changes from $60°$N to $61°$N? [*Hint*: Remember to use radian measure.]

58. The period (in seconds) of a simple pendulum of length L is $T = 2\pi\sqrt{L/g}$, where g is the acceleration due to gravity. Compute the exact change in the period if L is increased from 4 m to 5 m. Then use differentials to find an approximation to the change in the period. Assume $g = 9.8$ m/s².

59. In Problem 58, given that L is fixed at 4 m, find an approximation to the change in the period if the pendulum is moved to an altitude where $g = 9.75$ m/s².

60. Since license plates are all approximately the same size (12 in. across), a computerized optical sensor mounted at the front of car A could register the distance D to car B directly in front of car A by measuring the angle θ subtended by car B's license plate. See FIGURE 4.9.9.

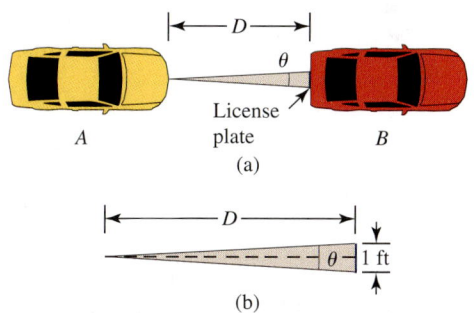

FIGURE 4.9.9   Cars in Problem 60

   (a) Express D as a function of the subtended angle θ.
   (b) Find the distance to the front car if the subtended angle θ is 30 min of an arc (that is, $\frac{1}{2}°$).
   (c) Suppose in part (b) that θ is decreasing at the rate of 2 min of arc per second, and that car A is traveling at a rate of 30 mi/h. At what rate is car B moving?
   (d) Show that the approximate relative error in measuring D is given by

$$\frac{dD}{D} = -\frac{d\theta}{\sin\theta},$$

where $d\theta$ is the approximate error (in radians) in measuring θ. What is the approximate relative error in D in part (b) if the subtended angle θ is measured with a possible error of ± 1 min of arc?

≡ Think About It

61. Suppose that the function $y = f(x)$ is differentiable at a number a. If a polynomial $p(x) = c_1x + c_0$ has the properties that $p(a) = f(a)$ and $p'(a) = f'(a)$, then show $p(x) = L(x)$, where L is defined in (2).

62. Without appealing to trigonometry, explain why for small values of x, $\cos x = 1$.

63. Suppose a function f and f' are differentiable at a number a and that $L(x)$ is a linearization of f at a. Discuss: If $f''(x) > 0$ for all x in some open interval containing a, will $L(x)$ overestimate or underestimate $f(x)$ for x near a?

64. Suppose $(c, f(c))$ is a point of inflection for the graph of $y = f(x)$ such that $f''(c) = 0$ and suppose further that $L(x)$ is a linearization of f at c. Describe what the graph of $y = f(x) - L(x)$ looks like in a neighborhood of c.

65. The area of a square with side of length x is $A = x^2$. Suppose, as shown in FIGURE 4.9.10, that each side of the square is increased by an amount Δx. In Figure 4.9.10 identify by color the areas ΔA, dA, and $\Delta A - dA$.

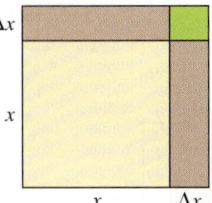

FIGURE 4.9.10   Square in Problem 65

4.10 Newton's Method

▌ Introduction   Finding the roots of certain kinds of equations was a problem that captivated mathematicians for centuries. The zeros of a *polynomial* function f of degree 4 or less—that is, the roots of the equation $f(x) = 0$, can always be found by means of an algebraic formula that expresses the unknown x in terms of the coefficients of f. For example, the polynomial equation of degree 2, $ax^2 + bx + c$, $a \neq 0$, can be solved by the quadratic formula. One of the major achievements in the history of mathematics was the proof in the nineteenth century that polynomial equations of degree greater than 4 cannot be solved by means of algebraic formulas, in other words, in terms of radicals. Thus, solving an algebraic equation such as

$$x^5 - 3x^2 + 4x - 6 = 0 \tag{1}$$

poses a quandary unless the fifth-degree polynomial $x^5 - 3x^2 + 4x - 6$ factors. Furthermore, in scientific analyses, one is often asked to find roots of transcendental equations such as

$$2x = \tan x. \tag{2}$$

In the case of problems such as (1) and (2) it is common practice to employ some method that yields an approximation or estimation of the roots. In this section we consider an approximation technique that makes use of the derivative of a function f, or more precisely, a tangent line to the graph of f. This new method is known as **Newton's Method** or the **Newton–Raphson Method**.

▌ Newton's Method   Suppose f is differentiable and suppose c represents an unknown real root of $f(x) = 0$; that is, $f(c) = 0$. Let x_0 denote a number that is chosen arbitrarily as a first guess to c. If $f(x_0) \neq 0$, compute $f'(x_0)$ and, as shown in FIGURE 4.10.1(a), construct a tangent to the graph of f at $(x_0, f(x_0))$. If we let $(x_1, 0)$ denote the x-intercept of the tangent line $y - f(x_0) = f'(x_0)(x - x_0)$, then the coordinates $x = x_1$ and $y = 0$ must satisfy this equation. By solving $0 - f(x_0) = f'(x_0)(x_1 - x_0)$ for x_1 we get

$$x_1 = x_0 - \frac{f(x_0)}{f'(x_0)}.$$

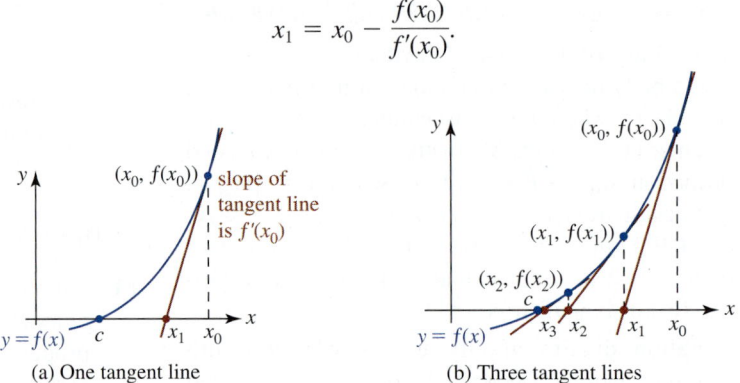

(a) One tangent line

(b) Three tangent lines

FIGURE 4.10.1   Successive x-coordinates of x-intercepts of tangent lines approximate the root c

Repeat the procedure at $(x_1, f(x_1))$ and let $(x_2, 0)$ be the x-intercept of the second tangent line $y - f(x_1) = f'(x_1)(x - x_1)$. From $0 - f(x_1) = f'(x_1)(x_2 - x_1)$ we find

$$x_2 = x_1 - \frac{f(x_1)}{f'(x_1)}.$$

Continuing in this fashion, we determine x_{n+1} from x_n using the formula

$$x_{n+1} = x_n - \frac{f(x_n)}{f'(x_n)}. \tag{3}$$

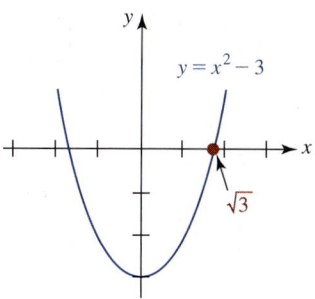

(a) x-coordinate of x-intercept

For $n = 0, 1, 2, \ldots$ formula (3) yields a sequence of approximations $x_1, x_2, x_3, \ldots$ to the root c. As suggested in Figure 4.10.1(b), if the terms in the sequence $x_1, x_2, x_3, \ldots$ become closer and closer to c as n increases without bound, that is, $x_n \to c$ as $n \to \infty$, we write $\lim_{x \to \infty} x_n = c$ and say that the sequence **converges** to c.

■ Graphical Analysis Before applying (3), it is a good idea to determine the existence and number of real roots of $f(x) = 0$ through graphical means. For example, the irrational number $\sqrt{3}$ can be interpreted as either

- a root of the quadratic equation $x^2 - 3 = 0$ and hence, a zero of the continuous function $f(x) = x^2 - 3$, or
- the x-coordinate of a point of intersection of the graphs of $y = x^2$ and $y = 3$.

Both interpretations are illustrated in FIGURE 4.10.2. Of course, another reason for a graph is to enable us to choose the initial guess x_0 so that it is close to the root c.

Although the actual computation of the number $\sqrt{3}$ is trivial on a calculator, its calculation serves nicely as an introduction to the use of Newton's Method.

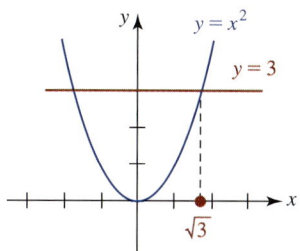

(b) x-coordinate of point of intersection of two graphs

FIGURE 4.10.2 Graphical location of $\sqrt{3}$

EXAMPLE 1 Using Newton's Method

Approximate $\sqrt{3}$ by Newton's Method.

Solution If we define $f(x) = x^2 - 3$, then $f'(x) = 2x$ and (3) becomes

$$x_{n+1} = x_n - \frac{x_n^2 - 3}{2x_n} \qquad \text{or} \qquad x_{n+1} = \frac{1}{2}\left(x_n + \frac{3}{x_n}\right). \tag{4}$$

From Figure 4.10.2 it seems reasonable to choose $x_0 = 1$ as an initial guess to the value of $\sqrt{3}$. We use (4) and display each calculation to eight decimal places:

$$x_1 = \frac{1}{2}\left(x_0 + \frac{3}{x_0}\right) = \frac{1}{2}(1 + 3) = 2$$

$$x_2 = \frac{1}{2}\left(x_1 + \frac{3}{x_1}\right) = \frac{1}{2}\left(2 + \frac{3}{2}\right) = 1.75$$

$$x_3 = \frac{1}{2}\left(x_2 + \frac{3}{x_2}\right) = \frac{1}{2}\left(\frac{7}{4} + \frac{12}{7}\right) \approx 1.73214286$$

$$x_4 = \frac{1}{2}\left(x_3 + \frac{3}{x_3}\right) \approx 1.73205081$$

$$x_5 = \frac{1}{2}\left(x_4 + \frac{3}{x_4}\right) \approx 1.73205081.$$

The process is continued until we obtain two consecutive approximations x_n and x_{n+1} that agree to the desired number of decimal places. Thus, if we are content with an eight decimal approximation to $\sqrt{3}$, we can stop with x_5 and conclude $\sqrt{3} \approx 1.73205081$. ■

EXAMPLE 2 Approximating a Root of an Equation

Use Newton's Method to approximate the real roots of $x^3 - x + 1 = 0$.

Solution Since the given equation is equivalent to $x^3 = x - 1$ we have graphed the functions $y = x^3$ and $y = x - 1$ in FIGURE 4.10.3. The figure should convince you that the original

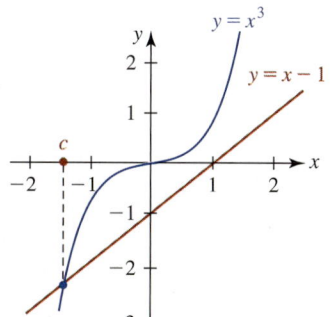

FIGURE 4.10.3 Graphs of functions in Example 2

equation has only one real root c, namely, the x-coordinate of the point of intersection of the two graphs. Now if $f(x) = x^3 - x + 1$, then $f'(x) = 3x^2 - 1$. Hence, (3) is

$$x_{n+1} = x_n - \frac{x_n^3 - x_n + 1}{3x_n^2 - 1} \quad \text{or} \quad x_{n+1} = \frac{2x_n^3 - 1}{3x_n^2 - 1}. \tag{5}$$

If we are interested in three and possibly four decimal place accuracy, we use (5) to compute $x_1, x_2, x_3, \ldots$ until two successive x_n in the sequence agree to four decimal places. Also, Figure 4.10.3 prompts us to make $x_0 = -1.5$ the initial guess. Consequently,

$$x_1 = \frac{2x_0^3 - 1}{3x_0^2 - 1} = \frac{2(-1.5)^3 - 1}{3(-1.5)^2 - 1} \approx -1.3478$$

$$x_2 = \frac{2x_1^3 - 1}{3x_1^2 - 1} \approx -1.3252$$

$$x_3 = \frac{2x_2^3 - 1}{3x_2^2 - 1} \approx -1.3247$$

$$x_4 = \frac{2x_3^3 - 1}{3x_3^2 - 1} \approx -1.3247.$$

Hence, the root of the given equation is approximately $c \approx -1.3247$. ∎

EXAMPLE 3 Approximating a Root of an Equation

Approximate the first positive root of $2x = \tan x$.

Solution FIGURE 4.10.4 shows that there are infinitely many points of intersection of the graphs of $y = 2x$ and $y = \tan x$. The first positive x-coordinate corresponding to a point of intersection is indicated by the letter c in the figure. With $f(x) = 2x - \tan x$ and $f'(x) = 2 - \sec^2 x$, (3) becomes

$$x_{n+1} = x_n - \frac{2x_n - \tan x_n}{2 - \sec^2 x_n}.$$

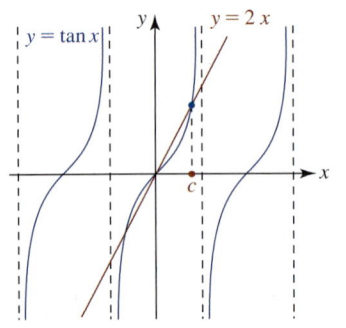

FIGURE 4.10.4 Graphs of functions in Example 3

If a calculator is used in the recursive use of the preceding formula, it is best to express the formula in terms of $\sin x$ and $\cos x$:

$$x_{n+1} = x_n - \frac{2x_n \cos^2 x_n - \sin x_n \cos x_n}{2 \cos^2 x_n - 1}. \tag{6}$$

Since the first vertical asymptote of $y = \tan x$ to the right of the y-axis is $x = \pi/2 \approx 1.57$, it appears from Figure 4.10.4 that the first positive root is near $x_0 = 1$. Using this initial guess, and setting our calculator in radian mode, (6) then yields

$$x_1 \approx 1.310478$$
$$x_2 \approx 1.223929$$
$$x_3 \approx 1.176051$$
$$x_4 \approx 1.165927$$
$$x_5 \approx 1.165562$$
$$x_6 \approx 1.165561$$
$$x_7 \approx 1.165561.$$

We conclude that the first positive root is approximately $c \approx 1.165561$. ∎

Example 3 illustrates the importance of the selection of the initial value x_0. You should verify that the choice $x_0 = \frac{1}{2}$ in (6) leads to a sequence of values $x_1, x_2, x_3, \ldots$ that converges to the one obvious root of $2x = \tan x$, namely, $c = 0$.

▌ **Postscript—A Bit of History** The problem of finding a formula that expresses the roots of a general nth degree *polynomial* equation $f(x) = 0$ in terms of its coefficients perplexed mathematicians for centuries. We know that in the case of a second-degree, or quadratic, polynomial function $f(x) = ax^2 + bx + c$ where the coefficients a, b, and c are real numbers, the roots c_1 and c_2 of the equation $ax^2 + bx + c = 0$ can be found using the quadratic formula.

The solution of finding roots of a general third-degree, or cubic, polynomial equation is usually attributed to the Italian mathematician **Nicolo Fontana** (1499–1557), also known as Tartaglia the "stammerer." Around 1540, the Italian mathematician **Lodovico Farrari** (1522–1565) discovered an algebraic formula for the roots of the general fourth-degree, or

quartic, polynomial equation. Since these formulas are complicated and difficult to use, they are rarely discussed in elementary courses.

For the next 284 years no one discovered any formula for roots of polynomial equations of degree 5 or greater. For good reason! In 1824, at age 22, the Norwegian mathematician **Niels Henrik Abel** (1802–1829), was the first to prove that it is *impossible* to find formulas for the roots of general polynomial equations of degrees $n \geq 5$ in terms of their coefficients.

Niels Henrik Abel

$f'(x)$ NOTES FROM THE CLASSROOM

There are problems with Newton's Method.

(*i*) We must compute $f'(x)$. Needless to say, the form of $f'(x)$ could be formidable when the equation $f(x) = 0$ is complicated.

(*ii*) If the root c of $f(x) = 0$ is near a value for which $f'(x) = 0$, then the denominator in (3) is approaching zero. This necessitates a computation of $f(x_n)$ and $f'(x_n)$ to a high degree of accuracy. A calculation of this kind requires a computer.

(*iii*) It is necessary to find an approximate location of a root $f(x) = 0$ before x_0 is chosen. Attendant to this are the usual difficulties in graphing. But, worse, the iteration of (3) *may not converge* for an imprudent or perhaps blindly chosen x_0. In FIGURE 4.10.5 we see that x_2 is undefined because $f'(x_1) = 0$.

(*iv*) Now some good news. The problems just discussed notwithstanding, the major advantage of Newton's Method is that when it converges to a root c of an equation $f(x) = 0$, it usually does so rather rapidly. It can be shown that under certain conditions Newton's Method converges quadratically, that is, the error at any step in the calculation is not greater than a constant multiple of the square of the error in the preceding step. Roughly this means that the number of places of accuracy can (but not necessarily) double with each step.

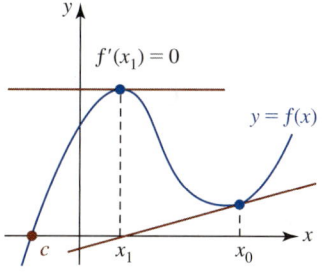

FIGURE 4.10.5 If $f'(x_1) = 0$, then x_2 is undefined

Exercises 4.10 Answers to selected odd-numbered problems begin on page ANS-17.

☰ Fundamentals

In Problems 1–6, determine graphically whether the given equation possesses any real roots.

1. $x^3 = -2 + \sin x$ **2.** $x^3 - 3x = x^2 - 1$

3. $x^4 + x^2 - 2x + 3 = 0$ **4.** $\cot x = x$

5. $e^{-x} = x + 2$ **6.** $e^{-x} - 2\cos x = 0$

In Problems 7–10, use Newton's Method to find an approximation for the given number.

7. $\sqrt{10}$ **8.** $1 + \sqrt{5}$

9. $\sqrt[3]{4}$ **10.** $\sqrt[5]{2}$

In Problems 11–16, use Newton's Method, if necessary, to find approximations to all real roots of the given equation.

11. $x^3 = -x + 1$ **12.** $x^3 - x^2 + 1 = 0$

13. $x^4 + x^2 - 3 = 0$ **14.** $x^4 = 2x + 1$

15. $x^2 = \sin x$ **16.** $x + \cos x = 0$

17. Find the smallest positive x-intercept of the graph of $f(x) = 3\cos x + 4\sin x$.

18. Consider the function $f(x) = x^5 + x^2$. Use Newton's Method to approximate the smallest positive number for which $f(x) = 4$.

☰ Applications

19. A cantilever beam 20 ft long with a load of 600 lb at its end is deflected by an amount $d = (60x^2 - x^3)/16{,}000$, where d is measured in inches and x in feet. See FIGURE 4.10.6. Use Newton's Method to approximate the value of x that corresponds to a deflection of 0.01 in.

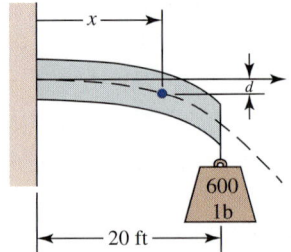

FIGURE 4.10.6   Beam in Problem 19

20. A vertical solid cylindrical column of fixed radius r that supports its own weight will eventually buckle when its height is increased. It can be proved that the maximum, or critical, height of such a column is $h_{cr} = kr^{2/3}$, where k is a constant and r is measured in meters. Use Newton's Method to approximate the diameter of a column for which $h_{cr} = 10$ m and $k = 35$.

21. A beam of light originating at point P in medium A, whose index of refraction is n_1, strikes the surface of medium B, whose index of refraction is n_2. It can be proved from Snell's Law that the beam is refracted tangent to the surface for the critical angle determined from $\sin\theta_c = n_2/n_1$, $0 < \theta_c < 90°$. For angles of incidence greater than the critical angle, all light is reflected internally to medium A. See FIGURE 4.10.7. If $n_2 = 1$ for air and $n_1 = 1.5$ for glass, use Newton's Method to approximate θ_c in radians.

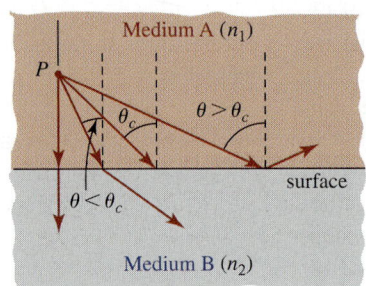

FIGURE 4.10.7   Refraction of light in Problem 21

22. For a suspension bridge, the length s of a cable between two vertical supports whose span is l (horizontal distance) is related to the sag d of the cable by

$$s = l + \frac{8d^2}{3l} - \frac{32d^4}{5l^3}.$$

See FIGURE 4.10.8. If $s = 404$ ft and $l = 400$ ft, use Newton's Method to approximate the sag. Round your answer to one decimal place.* [*Hint*: The root c satisfies $20 < c < 30$.]

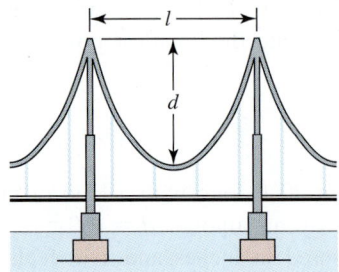

FIGURE 4.10.8   Suspension bridge in Problem 22

*The formula for s is itself only an approximation.

23. A rectangular block of steel is hollowed out, making a tub with a uniform thickness t. The dimensions of the tub are shown in FIGURE 4.10.9(a). For the tub to float in water, as shown in Figure 4.10.9(b), the weight of the water displaced must equal the weight of the tub (Archimedes' Principle). If the weight density of water is 62.4 lb/ft^3 and the weight density of the steel is 490 lb/ft^3, then

weight of water displaced = 62.4 × (volume of water displaced)

weight of tub = 490 × (volume of steel in tub).

(a) Show that t satisfies the equation

$$t^3 - 7t^2 + \frac{61}{4}t - \frac{1638}{1225} = 0.$$

(b) Use Newton's Method to approximate the largest positive root of the equation in part (a).

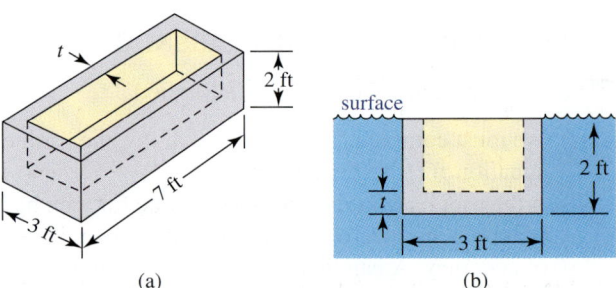

(a)   (b)

FIGURE 4.10.9   Floating tub in Problem 23

24. A flexible strip of metal 10 ft long is bent into the shape of a circular arc by securing the ends together by means of a cable that is 8 ft long. See FIGURE 4.10.10. Use Newton's Method to approximate the radius r of the circular arc.

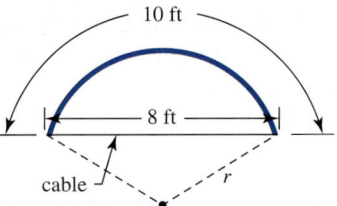

FIGURE 4.10.10   Bent metal strip in Problem 24

25. Two ends of a railroad track L feet long are pushed ℓ feet closer together so that the track bows upward in the arc of a circle of radius R. See FIGURE 4.10.11. The question is, what is the height h above ground of the highest point on the track?

(a) Use Figure 4.10.11 to show that

$$h = \frac{L(1 - \ell/L)^2\theta}{2\left(1 + \sqrt{1 - (1 - \ell/L)^2\theta^2}\right)},$$

where $\theta > 0$ satisfies $\sin\theta = (1 - \ell/L)\theta$. [*Hint*: In a circular sector, how are the arc length, the radius, and the central angle related?]

(b) If $L = 5280$ ft and $\ell = 1$ ft, use Newton's Method to approximate θ and then solve for the corresponding value of h.

(c) If ℓ/L and θ are very small, then $h \approx L\theta/4$ and $\sin\theta \approx \theta - \frac{1}{6}\theta^3$. Use these two approximations to show that $h \approx \sqrt{3\ell L/8}$. Use this formula with $L = 5280$ ft and $\ell = 1$ ft, and compare with the result in part (b).

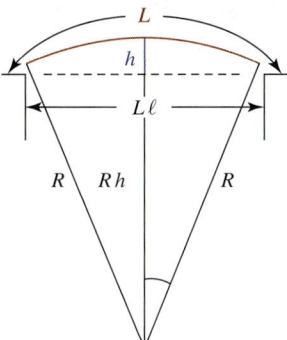

FIGURE 4.10.11 Bowed railroad track in Problem 25

26. At a foundry a metal sphere of radius 2 ft is recast in the form of a rod that is a right circular cylinder 15 ft long surmounted by a hemisphere at one end. The radius r of the hemisphere is the same as the base radius of the cylinder. Use Newton's Method to approximate r.

27. A round but unbalanced wheel of mass M and radius r is connected by a rope and frictionless pulleys to a mass m as shown in FIGURE 4.10.12. O is the center of the wheel and P is its center of mass. If it is released from rest, it can be shown that the angle θ at which the wheel first stops satisfies the equation

$$Mg\frac{r}{2}\sin\theta - mgr\theta = 0,$$

where g is the acceleration due to gravity. Use Newton's Method to approximate θ if the mass of the wheel is four times the mass m.

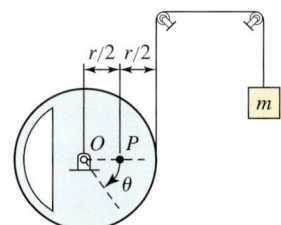

FIGURE 4.10.12 Unbalanced wheel in Problem 27

28. Two ladders of lengths $L_1 = 40$ ft and $L_2 = 30$ ft are placed against two vertical walls as shown in FIGURE 4.10.13. The height of the point where the ladders cross is $h = 10$ ft.

(a) Show that the indicated height x in the figure can be determined from the equation

$$x^4 - 2hx^3 + \left(L_1^2 - L_2^2\right)x^2 - 2h\left(L_1^2 - L_2^2\right)x + h^2\left(L_1^2 - L_2^2\right) = 0.$$

(b) Use Newton's Method to approximate the solution of the equation in part (a). Why does it make sense to choose $x_0 \geq 10$?

(c) Approximate the distance z between the two walls.

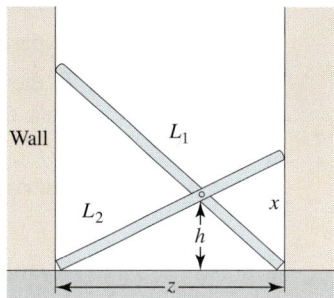

FIGURE 4.10.13 Ladders in Problem 28

≡ **Calculator/CAS Problems**

In Problems 29 and 30, use a calculator or CAS to obtain the graph of the given function. Use Newton's Method to approximate the roots of $f(x) = 0$ that you discover from the graph.

29. $f(x) = 2x^5 + 3x^4 - 7x^3 + 2x^2 + 8x - 8$

30. $f(x) = 4x^{12} + x^{11} - 4x^8 + 3x^3 + 2x^2 + x - 10$

31. (a) Use a calculator or CAS to obtain the graphs of $f(x) = 0.5x^3 - x$ and $g(x) = \cos x$ on the same coordinate axes.

(b) Use a calculator or CAS to obtain the graph of $y = f(x) - g(x)$, where f and g are as given in part (a).

(c) Use the graphs in part (a) or the graph in part (b) to determine the number of roots of the equation $0.5x^3 - x = \cos x$.

(d) Use Newton's Method to approximate the roots of the equation in part (c).

≡ **Think About It**

32. Let f be a differentiable function. Show that if $f(x_0) = -f(x_1)$ and $f'(x_0) = f'(x_1)$, then (3) implies $x_2 = x_0$.

33. For the piecewise-defined function

$$f(x) = \begin{cases} -\sqrt{4-x}, & x < 4 \\ \sqrt{x-4}, & x \geq 4 \end{cases}$$

observe that $f(4) = 0$. Show that for any choice of x_0, Newton's Method will fail to converge to the root. [*Hint*: See Problem 32.]

Chapter 4 in Review

Answers to selected odd-numbered problems begin on page ANS-17.

A. True/False

In Problems 1–20, indicate whether the given statement is true or false.

1. If f is increasing on an interval, then $f'(x) > 0$ on the interval. _____

2. A function f has an extremum at c when $f'(c) = 0$. _____

3. A particle moving rectilinearly slows down when the velocity $v(t)$ decreases. _____

4. If the position of a particle moving rectilinearly on a horizontal line is $s(t) = t^2 - 2t$, then the particle is speeding up for $t > 1$. _____

5. If $f''(x) < 0$ for all x in the interval (a, b), then the graph of f is concave down on the interval. _____

6. If $f''(c) = 0$, then $(c, f(c))$ is a point of inflection. _____

7. If $f(c)$ is a relative maximum, then $f'(c) = 0$ and $f'(x) > 0$ for $x < c$ and $f'(x) < 0$ for $x > c$. _____

8. If $f(c)$ is a relative minimum, then $f''(c) > 0$. _____

9. A function f that is continuous on a closed interval $[a, b]$ has both an absolute maximum and an absolute minimum on the interval. _____

10. Every absolute extremum is also a relative extremum. _____

11. If $c > 0$ is a constant and $f(x) = \frac{1}{3}x^3 - cx^2$, then $(c, f(c))$ is a point of inflection. _____

12. $x = 1$ is a critical number of the function $f(x) = \sqrt{x^2 - 2x}$. _____

13. If $f'(x) > 0$ and $g'(x) > 0$ on an interval I, then $f + g$ is increasing on I. _____

14. If $f'(x) > 0$ on an interval I, then $f''(x) > 0$ on I. _____

15. A limit of the form $\infty - \infty$ always has the value 0. _____

16. A limit of the form 1^∞ is always 1. _____

17. A limit of the form ∞/∞ is indeterminate. _____

18. A limit of the form $0/\infty$ is indeterminate. _____

19. If $\lim\limits_{x \to \infty} \dfrac{f(x)}{g(x)}$ and $\lim\limits_{x \to \infty} \dfrac{f'(x)}{g'(x)}$ are both of the form ∞/∞, then the first limit does not exist. _____

20. For an indeterminate form, L'Hôpital's Rule states that the limit of a quotient is the same as the derivative of the quotient. _____

B. Fill in the Blanks

In Problems 1–10, fill in the blanks.

1. For a particle moving rectilinearly, acceleration is the first derivative of _____.

2. The graph of a cubic polynomial can have at most _____ point(s) of inflection.

3. An example of a function $y = f(x)$ that is concave up on $(-\infty, 0)$, concave down on $(0, \infty)$, and increasing on $(-\infty, \infty)$ is _____.

4. Two nonnegative numbers whose sum is 8 such that the sum of their squares is a maximum are _____.

5. If f is continuous on $[a, b]$, differentiable on (a, b), and $f(a) = f(b) = 0$, then there exists some c in (a, b) such that $f'(c) = $ _____.

6. $\lim\limits_{x \to \infty} \dfrac{x^n}{e^x} = $ _____ for every integer n.

7. The sum of a positive number and its reciprocal is always greater than or equal to _____.

8. If $f(1) = 13$ and $f'(x) = 5^{x^2}$, then a linearization of f at $a = 1$ is _____ and $f(1.1) \approx $ _____.

9. If $y = x^2 - x$, then $\Delta y =$ _____.

10. If $y = x^3 e^{-x}$, then $dy =$ _____.

C. Exercises

In Problems 1–4, find the absolute extrema of the given function on the indicated interval.

1. $f(x) = x^3 - 75x + 150$; $[-3, 4]$

2. $f(x) = 4x^2 - \dfrac{1}{x}$; $\left[\frac{1}{4}, 1\right]$

3. $f(x) = \dfrac{x^2}{x + 4}$; $[-1, 3]$

4. $f(x) = (x^2 - 3x + 5)^{1/2}$; $[1, 3]$

5. Sketch a graph of a continuous function that has the properties:

$$f(0) = 1, \quad f(2) = 3$$
$$f'(0) = 0, \quad f'(2) \text{ does not exist}$$
$$f'(x) > 0, \quad x < 0$$
$$f'(x) > 0, \quad 0 < x < 2$$
$$f'(x) < 0, \quad x > 2.$$

6. Use the first and second derivatives as an aid in comparing the graphs of

$$y = x + \sin x \quad \text{and} \quad y = x + \sin 2x.$$

7. The position of a particle moving on a horizontal line is given by $s(t) = -t^3 + 6t^2$.

(a) Graph the motion on the time interval $[-1, 5]$.

(b) At what time is the velocity function a maximum?

(c) Does this time correspond to the maximum speed?

8. The height above ground of a projectile fired vertically is $s(t) = -4.9t^2 + 14.7t + 49$, where s is measured in meters and t in seconds.

(a) What is the maximum height attained by the projectile?

(b) At what speed does the projectile strike the ground?

9. Suppose f is a polynomial function with zeros of multiplicity 2 at $x = a$ and $x = b$; that is,

$$f(x) = (x - a)^2(x - b)^2 g(x)$$

where g is a polynomial function.

(a) Show that f' has at least three zeros in the closed interval $[a, b]$.

(b) If $g(x)$ is a constant, find the zeros of f' in $[a, b]$.

10. Show that the function $f(x) = x^{1/3}$ does not satisfy the hypothesis of the Mean Value Theorem on the interval $[-1, 8]$, yet a number c can be found in $(-1, 8)$ such that $f'(c) = [f(b) - f(a)]/(b - a)$. Explain.

In Problems 11–14, find the relative extrema of the given function f. Graph.

11. $f(x) = 2x^3 + 3x^2 - 36x$

12. $f(x) = x^5 - \dfrac{5}{3}x^3 + 2$

13. $f(x) = 4x - 6x^{2/3} + 2$

14. $f(x) = \dfrac{x^2 - 2x + 2}{x - 1}$

In Problems 15–18, find the relative extrema and the points of inflection of the given function f. Do not graph.

15. $f(x) = x^4 + 8x^3 + 18x^2$

16. $f(x) = x^6 - 3x^4 + 5$

17. $f(x) = 10 - (x - 3)^{1/3}$

18. $f(x) = x(x - 1)^{5/2}$

In Problems 19–24, match each figure with one or more of the following statements. On the interval corresponding to the portion of the graph of $y = f(x)$ shown:

(a) f has a positive first derivative.

(b) f has a negative second derivative.

(c) The graph of f has a point of inflection.

(d) f is differentiable.

(e) f has a relative extremum.

(f) The slopes of the tangent lines increase as x increases.

19.

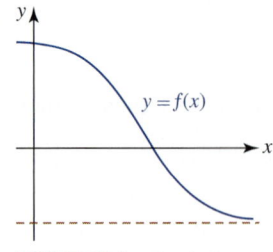

FIGURE 4.R.1   Graph for
Problem 19

20.

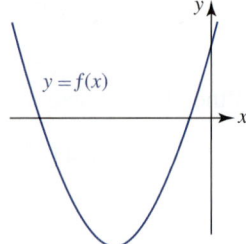

FIGURE 4.R.2   Graph for
Problem 20

21.

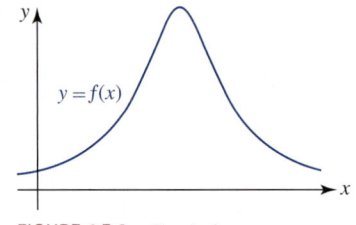

FIGURE 4.R.3   Graph for
Problem 21

22.

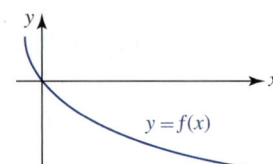

FIGURE 4.R.4   Graph for
Problem 22

23.

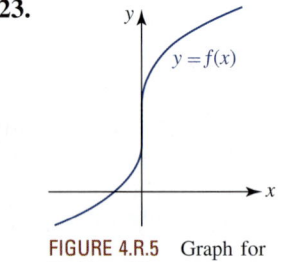

FIGURE 4.R.5   Graph for
Problem 23

24.

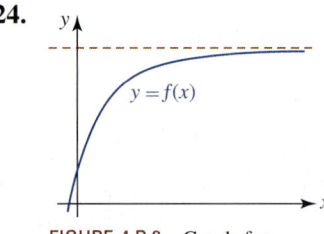

FIGURE 4.R.6   Graph for
Problem 24

25. Let a, b, and c be real numbers. Find the x-coordinate of the point of inflection for the graph of

$$f(x) = (x - a)(x - b)(x - c).$$

26. A triangle is expanding with time. The area of the triangle is increasing at a rate of 15 in²/min, whereas the length of its base is decreasing at a rate of $\frac{1}{2}$ in./min. At what rate is the altitude of the triangle changing when the altitude is 8 in. and the base is 6 in.?

27. A square is inscribed in a circle of radius r as shown in FIGURE 4.R.7. At what rate is the area of the square increasing at the instant the radius of the circle is 2 in. and increasing at a rate of 4 in./min?

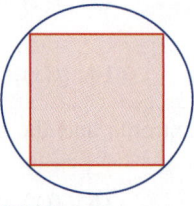

FIGURE 4.R.7   Circle
in Problem 27

28. Water drips into a hemispherical tank of radius 10 m at a rate of $\frac{1}{10}$ m³/min and drips out a hole in the bottom of the tank at a rate of $\frac{1}{5}$ m³/min. It can be shown that the volume of the water in the tank at t is $V = 10\pi h^2 - (\pi/3)h^3$. See FIGURE 4.R.8.

(a) Is the depth of the water increasing or decreasing?

(b) At what rate is the depth of the water changing when the depth is 5 m?

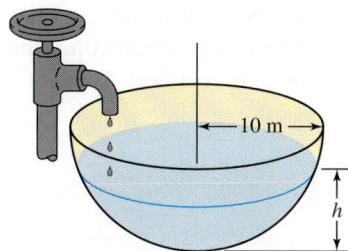

FIGURE 4.R.8   Tank in Problem 28

29. Two coils that carry the same current produce a magnetic field at point Q on the x-axis of strength

$$B = \frac{1}{2}\mu_0 r_0^2 I \left\{ \left[r_0^2 + \left(x + \frac{1}{2}r_0 \right)^2 \right]^{-3/2} + \left[r_0^2 + \left(x - \frac{1}{2}r_0 \right)^2 \right]^{-3/2} \right\},$$

where μ_0, r_0, and I are constants. See **FIGURE 4.R.9**. Show that the maximum value of B occurs at $x = 0$.

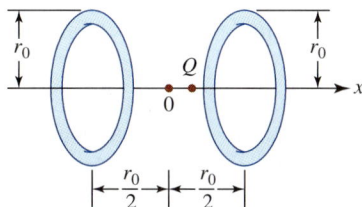

FIGURE 4.R.9   Coils in Problem 29

30. A battery with constant emf E and constant internal resistance r is wired in series with a resistor that has resistance R. The current in the circuit is then $I = E/(r + R)$. Find the value of R for which the power $P = RI^2$ dissipated in the external load is a maximum. This is called **impedance matching**.

31. When a hole is punched into the lateral side of a cylindrical tank full of water, the resulting stream hits the ground at a distance x ft from the base, where $x = 2\sqrt{y(h - y)}$. See **FIGURE 4.R.10**.

   (a) At what point should the hole be punched in the side of the tank so that the stream attains a maximum distance from the base?
   (b) What is the maximum distance?

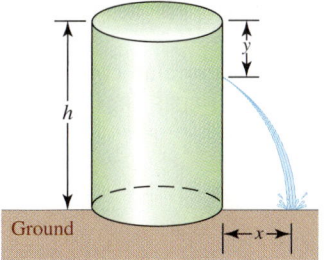

FIGURE 4.R.10   Leaking tank in Problem 31

32. The area of a circular sector of radius r and arc length s is $A = \frac{1}{2}rs$. See **FIGURE 4.R.11**. Find the maximum area of a sector enclosed by a perimeter of 60 cm.

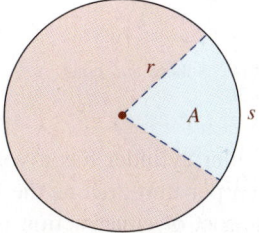

FIGURE 4.R.11   Circular sector in Problem 32

33. A pigpen, attached to a barn, is enclosed using fence on two sides, as shown in FIGURE 4.R.12. The amount of fence to be used is 585 ft. Find the values of x and y indicated in the figure so that the greatest area is enclosed. What is the greatest area?

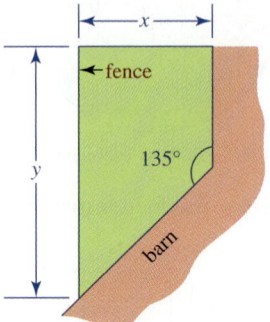

FIGURE 4.R.12 Pigpen in Problem 33

34. A rancher wants to use 100 m of fence to construct a diagonal fence connecting two existing walls that meet at a right angle. How should this be done so that the area enclosed by the walls and the fence is a maximum?

35. According to **Fermat's Principle**, a ray of light originating at point A and reflected from a plane surface to point B travels on a path requiring the least time. See FIGURE 4.R.13. Assume that the speed of light c as well as h_1, h_2, and d are constants. Show that the time is a minimum when $\tan\theta_1 = \tan\theta_2$. Since $0 < \theta_1 < \pi/2$ and $0 < \theta_2 < \pi/2$, it follows that $\theta_1 = \theta_2$. In other words, the angle of incidence equals the angle of reflection. [*Note*: Figure 4.R.13 is inaccurate on purpose.]

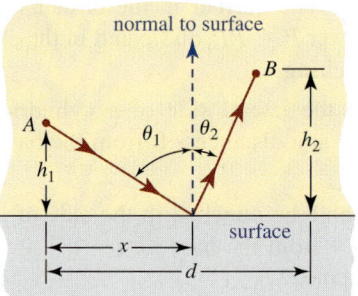

FIGURE 4.R.13 Reflected light rays in Problem 35

36. Determine the dimensions of a right circular cone having minimum volume V that circumscribes a sphere of radius r. See FIGURE 4.R.14. [*Hint*: Use similar triangles.]

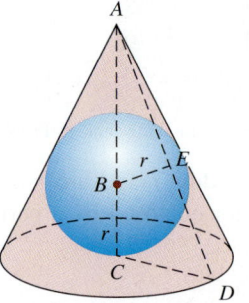

FIGURE 4.R.14 Sphere and cone in Problem 36

37. A container in the form of a right circular cylinder has a volume of 100 in^3. The top of the container costs three times as much per unit area as the bottom and the sides. Show that the dimension that gives the least cost of construction is a height that is four times the radius.

38. A box with a cover is to be made from a rectangular piece of cardboard 30 in. long and 15 in. wide by cutting a square out of each corner at one end of the cardboard and cutting a rectangle out of each corner at the other end. The cardboard is then folded on the dashed lines, as shown in FIGURE 4.R.15. Find the dimensions of the box that will give the maximum volume. What is the maximum volume?

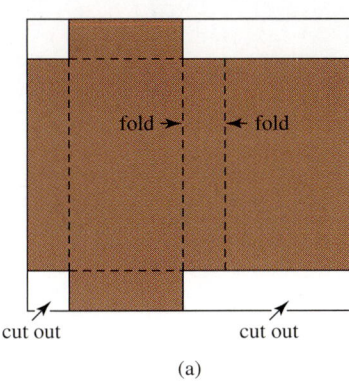

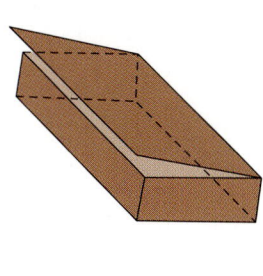

(a)   (b)

FIGURE 4.R.15   Box in Problem 38

In Problems 39–48, use L'Hôpital's Rule to find the limit.

39. $\displaystyle \lim_{x \to \sqrt{3}} \frac{\sqrt{3} - \tan(\pi/x^2)}{x - \sqrt{3}}$

40. $\displaystyle \lim_{\theta \to 0} \frac{10\theta - 5\sin 2\theta}{10\theta - 2\sin 5\theta}$

41. $\displaystyle \lim_{x \to \infty} x\left(\cos\frac{1}{x} - e^{2/x}\right)$

42. $\displaystyle \lim_{y \to 0} \left[\frac{1}{y} - \frac{1}{\ln(y+1)}\right]$

43. $\displaystyle \lim_{t \to 0} \frac{(\sin t)^2}{\sin t^2}$

44. $\displaystyle \lim_{x \to 0} \frac{\tan(5x)}{e^{3x/2} - e^{-x/2}}$

45. $\displaystyle \lim_{x \to 0^+} (3x)^{-1/\ln x}$

46. $\displaystyle \lim_{x \to 0} (2x + e^{3x})^{4/x}$

47. $\displaystyle \lim_{x \to \infty} \ln\left(\frac{x + e^{2x}}{1 + e^{4x}}\right)$

48. $\displaystyle \lim_{x \to 0^+} x(\ln x)^2$

In Problems 49 and 50, use Newton's Method to find the indicated root. Carry out the method until two successive approximations agree to four decimal places.

49. $x^3 - 4x + 2 = 0$, the largest positive root

50. $\left(\dfrac{\sin x}{x}\right)^2 = \dfrac{1}{2}$, the smallest positive root

Integrals

In This Chapter   In the last two chapters we have been concerned with the definition, properties, and applications of the derivative. We turn now from differential to integral calculus. Leibniz originally called this second of the two major divisions of calculus, *calculus summatorius.* In 1696, at the persuasion of the Swiss mathematician Johann Bernoulli, Leibniz changed its name to *calculus integralis.* As the original Latin words suggest, the notion of a *sum* plays an important role in the full development of the integral.

In Chapter 2 we saw that the tangent problem leads naturally to the derivative of a function. In the area problem, the motivational problem for integral calculus, we want to find the area bounded by the graph of a function and the *x*-axis. This problem leads to the concept of a *definite integral.*

5.1 The Indefinite Integral

■ Introduction In Chapters 3 and 4 we were concerned only with the basic problem:

- *Given a function f, find its derivative f′.*

In this chapter and in subsequent chapters of this text we shall see that an equally important problem is:

- *Given a function f, find a function F whose derivative is f.*

In other words, for a given function f, we now think of f as a derivative. We wish to find a function F whose derivative is f, that is, $F'(x) = f(x)$ for all x on some interval. Roughly put, we must do differentiation in reverse.

We begin with a definition.

Definition 5.1.1 Antiderivative

A function F is said to be an **antiderivative** of a function f on some interval I if $F'(x) = f(x)$ for all x in I.

EXAMPLE 1 An Antiderivative

An antiderivative of $f(x) = 2x$ is $F(x) = x^2$, since $F'(x) = 2x$. ■

There is always more than one antiderivative of a function. For instance, in the foregoing example, $F_1(x) = x^2 - 1$ and $F_2(x) = x^2 + 10$ are also antiderivatives of $f(x) = 2x$, since $F_1'(x) = F_2'(x) = 2x$.

We shall now prove that any antiderivative of f must be of the form $G(x) = F(x) + C$; that is, *two antiderivatives of the same function can differ by at most a constant.* Hence, $F(x) + C$ is *the most general antiderivative of* $f(x)$.

Theorem 5.1.1 Antiderivatives Differ by a Constant

If $G'(x) = F'(x)$ for all x in some interval $[a, b]$, then

$$G(x) = F(x) + C$$

for all x in the interval.

PROOF Suppose we define $g(x) = G(x) - F(x)$. Then, since $G'(x) = F'(x)$, it follows that $g'(x) = G'(x) - F'(x) = 0$ for all x in $[a, b]$. If x_1 and x_2 are any two numbers that satisfy $a \leq x_1 < x_2 \leq b$, it follows from the Mean Value Theorem (Theorem 4.4.2) that a number k exists in the open interval (x_1, x_2) for which

$$g'(k) = \frac{g(x_2) - g(x_1)}{x_2 - x_1} \qquad \text{or} \qquad g(x_2) - g(x_1) = g'(k)(x_2 - x_1).$$

But $g'(x) = 0$ for all x in $[a, b]$; in particular, $g'(k) = 0$. Hence, $g(x_2) - g(x_1) = 0$ or $g(x_2) = g(x_1)$. Now, by assumption, x_1 and x_2 are any two, but different, numbers in the interval. Since the function values $g(x_1)$ and $g(x_2)$ are the same, we must conclude that the function $g(x)$ is a constant C. Thus, $g(x) = C$ implies $G(x) - F(x) = C$ or $G(x) = F(x) + C$. ■

The notation $F(x) + C$ represents a *family of functions*; each member has a derivative equal to $f(x)$. Returning to Example 1, the most general antiderivative of $f(x) = 2x$ is the family $F(x) = x^2 + C$. As we see in FIGURE 5.1.1 the graph of an antiderivative of $f(x) = 2x$ is a vertical translation of the graph of x^2.

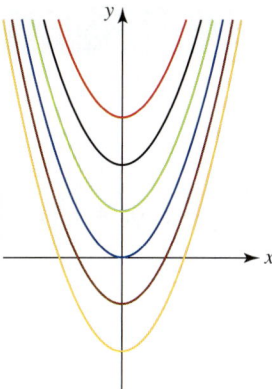

EXAMPLE 2   Most General Antiderivatives

(a) An antiderivative of $f(x) = 2x + 5$ is $F(x) = x^2 + 5x$ since $F'(x) = 2x + 5$. The most general antiderivative of $f(x) = 2x + 5$ is $F(x) = x^2 + 5x + C$.

(b) An antiderivative of $f(x) = \sec^2 x$ is $F(x) = \tan x$ since $F'(x) = \sec^2 x$. The most general antiderivative of $f(x) = \sec^2 x$ is $F(x) = \tan x + C$. ∎

FIGURE 5.1.1   Some members of the family of antiderivatives of $f(x) = 2x$

■ **Indefinite Integral Notation**   For convenience, let us introduce a notation for an antiderivative of a function. If $F'(x) = f(x)$, we shall represent the most general antiderivative of f by

$$\int f(x)\, dx = F(x) + C.$$

The symbol $\int$ was introduced by Leibniz and is called an **integral sign**. The notation $\int f(x)\, dx$ is called the **indefinite integral** of $f(x)$ with respect to x. The function $f(x)$ is called the **integrand**. The process of finding an antiderivative is called **antidifferentiation** or **integration**. The number C is called a **constant of integration**. Just as $\dfrac{d}{dx}(\ )$ denotes the operation of differentiation of $(\ )$ *with respect to x*, the symbolism $\int (\ )\, dx$ denotes the operation of integration of $(\ )$ *with respect to x*.

Differentiation and integration are fundamentally inverse operations. If $\int f(x)\, dx = F(x) + C$, then F is an antiderivative of f, that is, $F'(x) = f(x)$ and so

$$\int F'(x)\, dx = F(x) + C. \tag{1}$$

Moreover,
$$\frac{d}{dx}\int f(x)\, dx = \frac{d}{dx}(F(x) + C) = F'(x) = f(x) \tag{2}$$

In words, (1) and (2) are, respectively:

- *An antiderivative of the derivative of a function is that function plus a constant.*
- *The derivative of an antiderivative of a function is that function.*

From this it follows that whenever we obtain the derivative of a function, we get at the same time an integration formula. For example, in view of (1) if

$$\frac{d}{dx}\frac{x^{n+1}}{n+1} = x^n \qquad \text{then} \qquad \int \frac{d}{dx}\frac{x^{n+1}}{n+1}\, dx = \int x^n\, dx = \frac{x^{n+1}}{n+1} + C,$$

◀ This first result is valid only if $n \neq -1$.

$$\frac{d}{dx}\ln|x| = \frac{1}{x} \qquad \text{then} \qquad \int \frac{d}{dx}\ln|x|\, dx = \int \frac{1}{x}\, dx = \ln|x| + C,$$

$$\frac{d}{dx}\sin x = \cos x \qquad \text{then} \qquad \int \frac{d}{dx}\sin x\, dx = \int \cos x\, dx = \sin x + C,$$

$$\frac{d}{dx}\tan^{-1}x = \frac{1}{1+x^2} \qquad \text{then} \qquad \int \frac{d}{dx}\tan^{-1}x\, dx = \int \frac{1}{1+x^2}\, dx = \tan^{-1}x + C.$$

In this manner we can construct an integration formula from each derivative formula. TABLE 5.1.1 summarizes *some* of the important derivative formulas for the functions we have studied so far and their integration formula analogues.

TABLE 5.1.1

Differentiation Formula	Integration Formula	Differentiation Formula	Integration Formula
1. $\dfrac{d}{dx}x = 1$	$\displaystyle\int dx = x + C$	**10.** $\dfrac{d}{dx}\sin^{-1}x = \dfrac{1}{\sqrt{1-x^2}}$	$\displaystyle\int \dfrac{1}{\sqrt{1-x^2}}\,dx = \sin^{-1}x + C$
2. $\dfrac{d}{dx}\dfrac{x^{n+1}}{n+1} = x^n(n \neq -1)$	$\displaystyle\int x^n\,dx = \dfrac{x^{n+1}}{n+1} + C$	**11.** $\dfrac{d}{dx}\tan^{-1}x = \dfrac{1}{1+x^2}$	$\displaystyle\int \dfrac{1}{1+x^2}\,dx = \tan^{-1}x + C$
3. $\dfrac{d}{dx}\ln|x| = \dfrac{1}{x}$	$\displaystyle\int \dfrac{1}{x}\,dx = \ln|x| + C$	**12.** $\dfrac{d}{dx}\sec^{-1}x = \dfrac{1}{|x|\sqrt{x^2-1}}$	$\displaystyle\int \dfrac{1}{x\sqrt{x^2-1}}\,dx = \sec^{-1}|x| + C$
4. $\dfrac{d}{dx}\sin x = \cos x$	$\displaystyle\int \cos x\,dx = \sin x + C$	**13.** $\dfrac{d}{dx}b^x = b^x(\ln b),$ $(b > 0, b \neq 1)$	$\displaystyle\int b^x\,dx = \dfrac{b^x}{\ln b} + C$
5. $\dfrac{d}{dx}\cos x = -\sin x$	$\displaystyle\int \sin x\,dx = -\cos x + C$	**14.** $\dfrac{d}{dx}e^x = e^x$	$\displaystyle\int e^x\,dx = e^x + C$
6. $\dfrac{d}{dx}\tan x = \sec^2 x$	$\displaystyle\int \sec^2 x\,dx = \tan x + C$	**15.** $\dfrac{d}{dx}\sinh x = \cosh x$	$\displaystyle\int \cosh x\,dx = \sinh x + C$
7. $\dfrac{d}{dx}\cot x = -\csc^2 x$	$\displaystyle\int \csc^2 x\,dx = -\cot x + C$	**16.** $\dfrac{d}{dx}\cosh x = \sinh x$	$\displaystyle\int \sinh x\,dx = \cosh x + C$
8. $\dfrac{d}{dx}\sec x = \sec x \tan x$	$\displaystyle\int \sec x \tan x\,dx = \sec x + C$		
9. $\dfrac{d}{dx}\csc x = -\csc x \cot x$	$\displaystyle\int \csc x \cot x\,dx = -\csc x + C$		

With regard to entry 3 of Table 5.1.1, it is true that the derivative formulas

$$\frac{d}{dx}\ln x = \frac{1}{x}, \qquad \frac{d}{dx}\ln|x| = \frac{1}{x}, \qquad \frac{d}{dx}\frac{\log_b x}{\ln b} = \frac{1}{x}$$

mean that *an* antiderivative of $1/x = x^{-1}$ can be taken to be $\ln x, x > 0, \ln|x|, x \neq 0,$ or $\log_b x/\ln b, x > 0$. But we write

$$\int \frac{1}{x}\,dx = \ln|x| + C,$$

as the most general and useful result. Also note that only three formulas involving inverse trigonometric functions are given in Table 5.1.1. This is because, in indefinite integral form, the three remaining formulas are redundant. For example, from the derivatives

$$\frac{d}{dx}\sin^{-1}x = \frac{1}{\sqrt{1-x^2}} \qquad \text{and} \qquad \frac{d}{dx}\cos^{-1}x = \frac{-1}{\sqrt{1-x^2}}$$

it is seen that we could take either

$$\int \frac{1}{\sqrt{1-x^2}}\,dx = \sin^{-1}x + C \qquad \text{or} \qquad \int \frac{1}{\sqrt{1-x^2}}\,dx = -\cos^{-1}x + C.$$

Similar observations hold for the inverse cotangent and inverse cosecant.

EXAMPLE 3   An Important but Simple Antiderivative

The integration formula in entry 1 in Table 5.1.1 is included for emphasis:

$$\int dx = \int 1 \cdot dx = x + C \quad \text{because} \quad \frac{d}{dx}(x + C) = 1 + 0 = 1.$$

This result can also be obtained from integration formula 2 of Table 5.1.1 with $n = 0$. ∎

It is often necessary to rewrite an integrand $f(x)$ before carrying out the integration.

EXAMPLE 4   Rewriting the Integrand

Evaluate

(a) $\displaystyle\int \frac{1}{x^5}\, dx$ and   **(b)** $\displaystyle\int \sqrt{x}\, dx$.

Solution

(a) By rewriting $1/x^5$ as x^{-5} and identifying $n = -5$, we have from integration formula 2 of Table 5.1.1:

$$\int x^{-5}\, dx = \frac{x^{-5+1}}{-5+1} + C = -\frac{x^{-4}}{4} + C = -\frac{1}{4x^4} + C.$$

(b) We first rewrite the radical $\sqrt{x}$ as $x^{1/2}$ and then use integration formula 2 of Table 5.1.1 with $n = \frac{1}{2}$:

$$\int x^{1/2}\, dx = \frac{x^{3/2}}{3/2} + C = \frac{2}{3}x^{3/2} + C. \qquad\blacksquare$$

It should be kept in mind that the *results of integration can always be checked by differentiation*; for example, in part (b) of Example 4:

$$\frac{d}{dx}\left(\frac{2}{3}x^{3/2} + C\right) = \frac{2}{3}\cdot\frac{3}{2}x^{3/2-1} = x^{1/2} = \sqrt{x}.$$

Some properties of the indefinite integral are given in the next theorem.

Theorem 5.1.2   Properties of the Indefinite Integral

Let $F'(x) = f(x)$ and $G'(x) = g(x)$. Then

(i) $\displaystyle\int kf(x)\, dx = k\int f(x)\, dx = kF(x) + C$, where k is any constant,

(ii) $\displaystyle\int [\,f(x) \pm g(x)\,]\, dx = \int f(x)\, dx \pm \int g(x)\, dx = F(x) \pm G(x) + C.$

These properties follow immediately from the properties of the derivative. For example, (ii) is a consequence of the fact that the derivative of a sum is the sum of the derivatives.

Observe in Theorem 5.1.2(ii) that there is no reason to use two constants of integration, since

$$\int [\,f(x) \pm g(x)\,]\, dx = (F(x) + c_1) \pm (G(x) + C_2)$$
$$= F(x) \pm G(x) + (C_1 \pm C_2) = F(x) \pm G(x) + C,$$

where we have replaced $C_1 \pm C_2$ by the single constant C.

An indefinite integral of any finite sum of functions can be obtained by integrating each term.

EXAMPLE 5   Using Theorem 5.1.2

Evaluate $\displaystyle\int\left(4x - \frac{2}{x} + 5\sin x\right) dx$

Solution   From parts (i) and (ii) of Theorem 5.1.2 we can write this indefinite integral as three integrals:

$$\int\left(4x - \frac{2}{x} + 5\sin x\right) dx = 4\int x\, dx = 2\int\frac{1}{x}\, dx + 5\int \sin x\, dx.$$

In view of integration formulas 2, 3, and 5 in Table 5.1.1 we then have

$$\int \left(4x - \frac{2}{x} + 5\sin x\right) dx = 4 \cdot \frac{x^2}{2} - 2 \cdot \ln|x| + 5 \cdot (-\cos x) + C$$
$$= 2x^2 - 2\ln|x| - 5\cos x + C. \qquad \blacksquare$$

▪ **Using Division**   Putting an integrand in a more tractable form may sometimes entail division. The next two examples illustrate the idea.

EXAMPLE 6   Termwise Division

Evaluate $\displaystyle\int \frac{6x^3 - 5}{x} dx$.

If we read the common denominator concept

$$\frac{a}{c} + \frac{b}{c} = \frac{a+b}{c}$$

from right to left we are performing "termwise division."

▶ **Solution**   By termwise division, Theorem 5.1.2, and integration formulas 2 and 3 of Table 5.1.1:

$$\int \frac{6x^3 - 5}{x} dx = \int \left(\frac{6x^3}{x} - \frac{5}{x}\right) dx$$
$$= \int \left(6x^2 - \frac{5}{x}\right) dx = 6 \cdot \frac{x^3}{3} - 5 \cdot \ln|x| + C = 2x^3 - 5\ln|x| + C. \qquad \blacksquare$$

For the problem of evaluating $\int f(x)\,dx$, where $f(x) = p(x)/q(x)$ is a rational function, a working rule to keep in mind in this and subsequent sections is summarized next.

Integration of a Rational Function

Suppose $f(x) = p(x)/q(x)$ is a rational function. If the degree of the polynomial function $p(x)$ is greater than or equal to the degree of the degree of the polynomial function $q(x)$, use long division before integration, that is, write

$$\frac{p(x)}{q(x)} = \text{a polynomial} + \frac{r(x)}{q(x)},$$

where the degree of the polynomial $r(x)$ is less than the degree of $q(x)$.

EXAMPLE 7   Long Division

Evaluate $\displaystyle\int \frac{x^2}{1 + x^2} dx$.

Solution   Because the degree of the numerator of the integrand is equal to the degree of the denominator we perform long division:

$$\frac{x^2}{1 + x^2} = 1 - \frac{1}{1 + x^2}.$$

From (*ii*) of Theorem 5.1.2 and integration formulas 1 and 11 in Table 5.1.1 we obtain

$$\int \frac{x^2}{1 + x^2} dx = \int \left(1 - \frac{1}{1 + x^2}\right) dx = x - \tan^{-1}x + C. \qquad \blacksquare$$

▪ **Differential Equations**   In several exercise sets in Chapter 3 you were asked to verify that a given function satisfies a **differential equation**. Roughly, a differential equation is an equation that involves derivatives or the differential of an unknown function. Differential equations are classified by the **order** of the highest derivative appearing in the equation. The goal is to *solve* differential equations. A **first-order differential equation** of the form

$$\frac{dy}{dx} = g(x) \tag{3}$$

can be solved using indefinite integration. From (1) we see that

$$\int \left(\frac{dy}{dx}\right) dx = y.$$

Thus the solution of (3) is the most general antiderivative of g, that is,

$$y = \int g(x) \, dx. \tag{4}$$

EXAMPLE 8   Solving a Differential Equation

Find a function $y = f(x)$ whose graph passes through the point $(1, 2)$ that also satisfies the differential equation $dy/dx = 3x^2 - 3$.

Solution   From (3) and (4) it follows that if

$$\frac{dy}{dx} = 3x^2 - 3 \qquad \text{then} \qquad y = \int (3x^2 - 3) \, dx.$$

That is,

$$y = \int (3x^2 - 3) \, dx = 3 \cdot \frac{x^3}{3} - 3 \cdot x + C$$

or $y = x^3 - 3x + C$. Now when $x = 1$, $y = 2$, so that $2 = 1 - 3 + C$ or $C = 4$. Hence, $y = x^3 - 3x + 4$. Thus out of the family of antiderivatives of $3x^2 - 3$ shown in FIGURE 5.1.2, we see that there is only one whose graph (shown in red) passes through $(1, 2)$. ∎

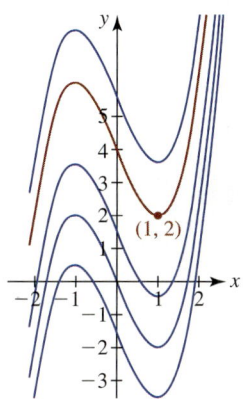

FIGURE 5.1.2   The red curve is the graph of the solution of the problem in Example 8

When solving a differential equation such as $dy/dx = 3x^2 - 3$ in Example 8, the specified side condition that the graph pass through $(1, 2)$, that is, $f(1) = 2$, is called an **initial condition**. It is common practice to write an initial condition such as this as $y(1) = 2$. The solution $y = x^3 - 3x + 4$ that was determined from the family of solutions $y = x^3 - 3x + C$ by the initial condition is called a **particular solution**. The problem of solving (3) subject to an initial condition,

$$\frac{dy}{dx} = g(x), \qquad y(x_0) = y_0$$

is called an **initial-value problem**.

We note that an *n*th-order differential equation of the form $d^n y/dx^n = g(x)$ can be solved by integrating the function $g(x)$ in succession n times. In this case the family of solutions will contain n constants of integration.

EXAMPLE 9   Solving a Differential Equation

Find a function $y = f(x)$ such that $\dfrac{d^2 y}{dx^2} = 1$.

Solution   We integrate the given differential equation in succession two times. The first integration gives

$$\frac{dy}{dx} = \int \frac{d^2 y}{dx^2} \, dx = \int 1 \cdot dx = x + C_1.$$

The second integration gives $y = f(x)$:

$$y = \int \frac{dy}{dx} \, dx = \int (x + C_1) \, dx = \frac{x^2}{2} + C_1 x + C_2. \qquad ∎$$

$\int$ **NOTES FROM THE CLASSROOM**

Students often have a more difficult time calculating antiderivatives than they do derivatives. Two words of advice. First, be very, very careful with your algebra—especially the laws of exponents. The second word of advice has been stated previously but it bears repeating: Keep firmly in mind that the *results of indefinite integration can always be checked.* On a quiz or test it is worth a few seconds of your valuable time to check your answer by taking its derivative. You can often do this in your head. For example,

$$\underset{\text{check by differentiation}}{\overset{\text{integration}}{\int x^2 \, dx = \frac{x^3}{3} + C}}$$

Exercises 5.1   Answers to selected odd-numbered problems begin on page ANS-18.

☰ Fundamentals

In Problems 1–30, evaluate the given indefinite integral.

1. $\int 3 \, dx$

2. $\int (\pi^2 - 1) \, dx$

3. $\int x^5 \, dx$

4. $\int 5x^{1/4} \, dx$

5. $\int \dfrac{1}{\sqrt[3]{x}} \, dx$

6. $\int \sqrt[3]{x^2} \, dx$

7. $\int (1 - t^{-0.52}) \, dt$

8. $\int 10w\sqrt{w} \, dw$

9. $\int (3x^2 + 2x - 1) \, dx$

10. $\int \left(2\sqrt{t} - t - \dfrac{9}{t^2}\right) dt$

11. $\int \sqrt{x}(x^2 - 2) \, dx$

12. $\int \left(\dfrac{5}{\sqrt[3]{s^2}} + \dfrac{2}{\sqrt{s^3}}\right) ds$

13. $\int (4x + 1)^2 \, dx$

14. $\int (\sqrt{x} - 1)^2 \, dx$

15. $\int (4w - 1)^3 \, dw$

16. $\int (5u - 1)(3u^3 + 2) \, du$

17. $\int \dfrac{r^2 - 10r + 4}{r^3} \, dr$

18. $\int \dfrac{(x + 1)^2}{\sqrt{x}} \, dx$

19. $\int \dfrac{x^{-1} - x^{-2} + x^{-3}}{x^2} \, dx$

20. $\int \dfrac{t^3 - 8t + 1}{(2t)^4} \, dt$

21. $\int (4\sin x - 1 + 8x^{-5}) \, dx$

22. $\int (-3\cos x + 4\sec^2 x) \, dx$

23. $\int \csc x(\csc x - \cot x) \, dx$

24. $\int \dfrac{\sin t}{\cos^2 t} \, dt$

25. $\int \dfrac{2 + 3\sin^2 x}{\sin^2 x} \, dx$

26. $\int \left(40 - \dfrac{2}{\sec\theta}\right) d\theta$

27. $\int (8x + 1 - 9e^x) \, dx$

28. $\int (15x^{-1} - 4\sinh x) \, dx$

29. $\int \dfrac{2x^3 - x^2 + 2x + 4}{1 + x^2} \, dx$

30. $\int \dfrac{x^6}{1 + x^2} \, dx$

In Problems 31 and 32, use a trigonometric identity to evaluate the given indefinite integral.

31. $\int \tan^2 x \, dx$

32. $\int \cos^2\dfrac{x}{2} \, dx$

In Problems 33–40, verify the given integration result by differentiation and the Chain Rule.

33. $\int \dfrac{1}{\sqrt{2x + 1}} \, dx = \sqrt{2x + 1} + C$

34. $\int (2x^2 - 4x)^9(x - 1) \, dx = \dfrac{1}{40}(2x^2 - 4x)^{10} + C$

35. $\int \cos 4x \, dx = \dfrac{1}{4}\sin 4x + C$

36. $\int \sin x \cos x \, dx = \dfrac{1}{2}\sin^2 x + C$

37. $\int x\sin x^2 \, dx = -\dfrac{1}{2}\cos x^2 + C$

38. $\int \dfrac{\cos x}{\sin^3 x} \, dx = -\dfrac{1}{2\sin^2 x} + C$

39. $\int \ln x \, dx = x\ln x - x + C$

40. $\int xe^x \, dx = xe^x - e^x + C$

In Problems 41 and 42, perform the indicated operations.

41. $\dfrac{d}{dx}\int (x^2 - 4x + 5) \, dx$

42. $\int \dfrac{d}{dx}(x^2 - 4x + 5) \, dx$

In Problems 43–48, solve the given differential equation.

43. $\dfrac{dy}{dx} = 6x^2 + 9$ **44.** $\dfrac{dy}{dx} = 10x + 3\sqrt{x}$

45. $\dfrac{dy}{dx} = \dfrac{1}{x^2}$ **46.** $\dfrac{dy}{dx} = \dfrac{(2+x)^2}{x^5}$

47. $\dfrac{dy}{dx} = 1 - 2x + \sin x$ **48.** $\dfrac{dy}{dx} = \dfrac{1}{\cos^2 x}$

49. Find a function $y = f(x)$ whose graph passes through the point $(2, 3)$ that also satisfies the differential equation $dy/dx = 2x - 1$.

50. Find a function $y = f(x)$ so that $dy/dx = 1/\sqrt{x}$ and $f(9) = 1$.

51. If $f''(x) = 2x$, find $f'(x)$ and $f(x)$.

52. Find a function f such that $f''(x) = 6, f'(-1) = 2,$ and $f(-1) = 0$.

53. Find a function f such that $f''(x) = 12x^2 + 2$ for which the slope of the tangent line to its graph at $(1, 1)$ is 3.

54. If $f^{(n)}(x) = 0$, what is f?

In Problems 55 and 56, the graph of a function f is shown in blue. Of the graphs of functions F, G, and H whose graphs are shown in black, green, and red, respectively, which function is the graph of an antiderivative of f? State your reasoning.

55.

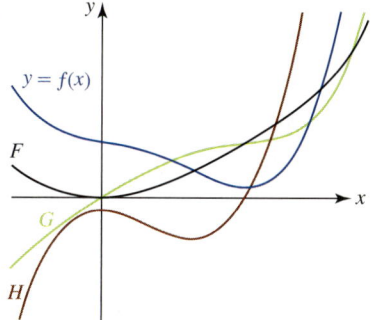

FIGURE 5.1.3   Graphs for Problem 55

56.

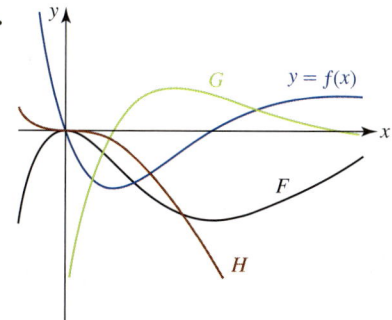

FIGURE 5.1.4   Graphs for Problem 56

≡ Applications

57. A bucket that contains liquid is rotating about a vertical axis at a constant angular velocity ω. The shape of the cross-section of the rotating liquid in the xy-plane is determined from

$$\frac{dy}{dx} = \frac{\omega^2}{g}\,x.$$

With coordinate axes as shown in FIGURE 5.1.5, find $y = f(x)$.

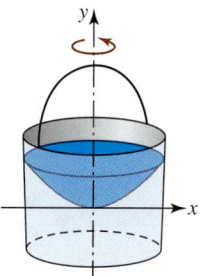

FIGURE 5.1.5   Bucket in Problem 57

58. The ends of a beam of length L rest on two supports as shown in FIGURE 5.1.6. With a uniform load on the beam, its shape (or elastic curve) is determined from

$$EIy'' = \frac{1}{2}qLx - \frac{1}{2}qx^2,$$

where E, I, and q are constants. Find $y = f(x)$ if $f(0) = 0$ and $f'(L/2) = 0$.

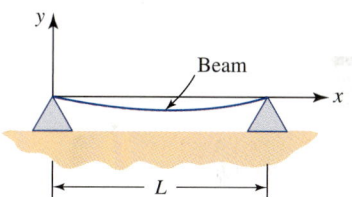

FIGURE 5.1.6   Beam in Problem 58

≡ Think About It

In Problems 59 and 60, determine f.

59. $\displaystyle\int f(x)\,dx = \ln|\ln x| + C$

60. $\displaystyle\int f(x)\,dx = x^2 e^x - 2xe^x + 2e^x + C$

61. Find a function f such that $f'(x) = x^2$ and $y = 4x + 7$ is a tangent line to the graph of f.

62. Simplify the expression $e^{4\int dx/x}$ as much as possible.

63. Determine which of the following two results is correct:

$$\int (x+1)^3\,dx = \frac{1}{4}(x+1)^4 + C$$

or

$$\int (x+1)^3\,dx = \frac{1}{4}x^4 + x^3 + \frac{3}{2}x^2 + x + C?$$

64. Given that $\dfrac{d}{dx}\sin \pi x = \pi \cos \pi x$. Find an antiderivative F of $\cos \pi x$ that has the property that $F\left(\frac{3}{2}\right) = 0$.

5.2 Integration by the *u*-Substitution

■ **Introduction** In the last section we discussed the fact that for each formula for the derivative of a function there is a corresponding antiderivative or indefinite integral formula. For example, by interpreting each of the functions

$$x^n \ (n \neq -1), \qquad x^{-1}, \qquad \text{and} \qquad \cos x$$

as a derivative, we find the corresponding "reverse of the derivative" is a family of antiderivatives:

$$\int x^n \, dx = \frac{x^{n+1}}{n+1} + C \quad (n \neq -1), \qquad \int \frac{1}{x} \, dx = \ln|x| + C, \qquad \int \cos x \, dx = \sin x + C. \quad (1)$$

Review Section 4.9 ▶ In the present exposition, we examine the "reverse of the Chain Rule." The concept of a **differential** of a function plays an important role in this discussion. Recall, if $u = g(x)$ is a differentiable function, then its differential is $du = g'(x) \, dx$.

We begin with an example.

■ **Power of a Function** If we wish to find a function F such that

$$\int (5x + 1)^{1/2} \, dx = F(x) + C,$$

we must have

$$F'(x) = (5x + 1)^{1/2}.$$

By reasoning "backward," we could argue that to obtain $(5x + 1)^{1/2}$ we must have differentiated $(5x + 1)^{3/2}$. It would then seem that we could proceed as in the first formula in (1)—namely, increase the power by 1 and divide by the new power:

$$\int (5x + 1)^{1/2} \, dx = \frac{(5x + 1)^{3/2}}{3/2} + C = \frac{2}{3}(5x + 1)^{3/2} + C. \quad (2)$$

Regrettably the "answer" in (2) does not check, since the Chain Rule, in the form of the Power Rule for Functions, gives

$$\frac{d}{dx}\left[\frac{2}{3}(5x + 1)^{3/2} + C\right] = \frac{2}{3} \cdot \frac{3}{2}(5x + 1)^{1/2} \cdot 5 = 5(5x + 1)^{1/2} \neq (5x + 1)^{1/2}. \quad (3)$$

To account for the missing factor of 5 in (2) we use Theorem 5.1.2(*i*) and a little bit of cleverness:

$$\int (5x + 1)^{1/2} \, dx = \int (5x + 1)^{1/2} \boxed{\frac{1}{5} \cdot 5} \, dx \quad \leftarrow \frac{5}{5} = 1$$

$$= \frac{1}{5} \int \boxed{(5x + 1)^{1/2} \, 5} \, dx \quad \leftarrow \text{derivative of } \frac{2}{3}(5x + 1)^{3/2}$$

$$= \frac{1}{5} \cdot \frac{2}{3}(5x + 1)^{3/2} + C \quad \leftarrow \text{from (3)}$$

$$= \frac{2}{15}(5x + 1)^{3/2} + C.$$

You should now verify by differentiation that the last function is indeed an antiderivative of $(5x + 1)^{1/2}$.

The key to evaluating indefinite integrals such as

$$\int (5x + 1)^{1/2} \, dx, \qquad \int \frac{x}{(4x^2 + 3)^6} \, dx, \qquad \text{and} \qquad \int \sin 10x \, dx \quad (4)$$

lies in the *recognition* that the integrands in (4),

$$(5x + 1)^{1/2}, \qquad \frac{x}{(4x^2 + 3)^6}, \qquad \text{and} \qquad \sin 10x$$

are the result of differentiating a composite function by the Chain Rule. In order to make this recognition it helps to make a substitution in an indefinite integral.

> **Theorem 5.2.1** u-Substitution Rule
>
> If $u = g(x)$ is a differentiable function whose range is an interval I, f is a function continuous on I, and F is an antiderivative of f on I, then
>
> $$\int f(g(x))g'(x)\,dx = \int f(u)\,du. \tag{5}$$

PROOF By the Chain Rule,

$$\frac{d}{dx}F(g(x)) = F'(g(x))g'(x)$$

and so from the definition of an antiderivative we have,

$$\int F'(g(x))g'(x)\,dx = F(g(x)) + C.$$

Because F is an antiderivative of f, that is, if $F' = f$, the preceding line becomes

$$\int f(g(x))g'(x)\,dx = F(g(x)) + C = F(u) + C = \int F'(u)\,du = \int f(u)\,du. \tag{6} \blacksquare$$

The interpretation of the result in (6) and its summary in (5) is subtle. In Section 5.1, the symbol dx was used simply as an indicator that we are integrating with respect to the variable x. In (6), we see that it is allowable to interpret dx and du as *differentials*.

■ **Using the u-Substitution** The basic idea is to be able to recognize an indefinite integral in a variable x (such as those given in (4)) that it is the reverse of the Chain Rule by converting it into a different indefinite integral in the variable u by means of the substitution $u = g(x)$. For convenience we list some guidelines for evaluating $\int f(g(x))g'(x)\,dx$ by carrying out a u-substitution.

> ### Guidelines for Using a u-Substitution
>
> (*i*) In the integral $\int f(g(x))g'(x)\,dx$ identify the functions $g(x)$ and $g'(x)\,dx$.
> (*ii*) Express the integral *entirely* in terms of the symbol u by substituting u and du for $g(x)$ and $g'(x)\,dx$, respectively. In your substitution there should be no x-variables left in the integral.
> (*iii*) Carry out the integration with respect to the variable u.
> (*iv*) Finally, resubstitute $g(x)$ for the symbol u.

■ **Indefinite Integral of a Power of a Function** The derivative of a power of a function was an important special case of the Chain Rule. Recall, if $F(x) = x^{n+1}/(n+1)$, n a real number, $n \neq -1$, and if $u = g(x)$ is a differentiable function, then

$$F(g(x)) = \frac{[g(x)]^{n+1}}{n+1} \qquad \text{and} \qquad \frac{d}{dx}F(g(x)) = [g(x)]^n g'(x).$$

Hence, Theorem 5.2.1 immediately implies

$$\int [g(x)]^n g'(x)\,dx = \frac{[g(x)]^{n+1}}{n+1} + C. \tag{7}$$

In terms of the substitutions

$$u = g(x) \qquad \text{and} \qquad du = g'(x)\,dx,$$

(7) can be summarized in the following manner:

$$\int u^n\,du = \frac{u^{n+1}}{n+1} + C, \qquad n \neq -1. \tag{8}$$

In the next example we evaluate the second of the three indefinite integrals in (4).

EXAMPLE 1 Using (8)

Evaluate $\displaystyle\int \frac{x}{(4x^2 + 3)^6}\, dx$.

Solution Let us rewrite the integral as

$$\int (4x^2 + 3)^{-6} x\, dx$$

and make the identifications

$$u = 4x^2 + 3 \qquad \text{and} \qquad du = 8x\, dx.$$

Now, to get the precise form $\int u^{-6}\, du$ we must adjust the integrand by multiplying and dividing by 8:

$$\int (4x^2 + 3)^{-6} x\, dx = \frac{1}{8}\int \overbrace{(4x^2 + 3)^{-6}}^{u^{-6}}\overbrace{(8x\, dx)}^{du} \quad \leftarrow \text{substitution}$$

$$= \frac{1}{8}\int u^{-6}\, du \qquad\qquad \leftarrow \text{now use (8)}$$

$$= \frac{1}{8}\cdot\frac{u^{-5}}{-5} + C$$

$$= -\frac{1}{40}(4x^2 + 3)^{-5} + C. \leftarrow \text{resubstitution}$$

Check by Differentiation: By the Power Rule for Functions,

$$\frac{d}{dx}\left[-\frac{1}{40}(4x^2 + 3)^{-5} + C \right] = \left(-\frac{1}{40}\right)(-5)(4x^2 + 3)^{-6}(8x) = \frac{x}{(4x^2 + 3)^6}. \quad \blacksquare$$

EXAMPLE 2 Using (8)

Evaluate $\displaystyle\int (2x - 5)^{11}\, dx$.

Solution If $u = 2x - 5$, then $du = 2\, dx$. We adjust the integral by multiplying and dividing by 2 to get the correct form of the differential du:

$$\int (2x - 5)^{11}\, dx = \frac{1}{2}\int \overbrace{(2x - 5)^{11}}^{u^{11}}\overbrace{(2\, dx)}^{du} \quad \leftarrow \text{substitution}$$

$$= \frac{1}{2}\int u^{11} du \qquad\qquad \leftarrow \text{now use (8)}$$

$$= \frac{1}{2}\cdot\frac{u^{12}}{12} + C$$

$$= \frac{1}{24}(2x - 5)^{12} + C. \quad \leftarrow \text{resubstitution} \quad \blacksquare$$

In Examples 1 and 2 we "fixed up" or adjusted the integrand by multiplying and dividing by a constant in order to obtain the appropriate du. This procedure works fine if you immediately recognize $g(x)$ in $\int f(g(x))g'(x)\, dx$ and that $g'(x)\, dx$ is simply missing an appropriate constant multiple. The next example illustrates a slightly different technique.

EXAMPLE 3 Using (8)

Evaluate $\displaystyle\int \cos^4 x \sin x\, dx$.

Solution For emphasis we rewrite the integrand as $\int (\cos x)^4 \sin x\, dx$. With the identification $u = \cos x$ we get $du = -\sin x\, dx$. Solving for the product $\sin x\, dx$ from the last differential we get $\sin x\, dx = -du$. Then

$$\int (\cos x)^4 \sin x \, dx = \int \overbrace{(\cos x)^4}^{u^4} \overbrace{(\sin x \, dx)}^{-du} \quad \leftarrow \text{ substitution}$$

$$= -\int u^4 \, du \qquad \leftarrow \text{ now use (8)}$$

$$= -\frac{u^5}{5} + C$$

$$= -\frac{1}{5}\cos^5 x + C. \qquad \leftarrow \text{ resubstitution}$$

You are again encouraged to differentiate the last result. ■

In the remaining examples in this section we will alternate between the methods employed in Examples 1 and 3.

On a practical level it not always obvious that we are dealing with an integral of the form $\int [g(x)]^n g'(x) \, dx$. As you work more and more problems you will see that integrals are not always what they seem to be on first inspection. For example, you should convince yourself using *u*-substitutions that the integral $\int \cos^2 x \, dx$ is *not* of the form $\int [g(x)]^n g'(x) \, dx$. In a more general sense, is it not always obvious in $\int f(g(x))g'(x) \, dx$ what functions should be chosen as *u* and *du*.

■ **Indefinite Integrals of Trigonometric Functions** If $u = g(x)$ is a differentiable function, then the differentiation formulas

$$\frac{d}{dx}\sin u = \cos u \frac{du}{dx} \qquad \text{and} \qquad \frac{d}{dx}(-\cos u) = \sin u \frac{du}{dx}$$

yield, in turn, the integration formulas

$$\int \cos u \frac{du}{dx} \, dx = \sin u + C \tag{9}$$

and

$$\int \sin u \frac{du}{dx} \, dx = -\cos u + C. \tag{10}$$

Since $du = g'(x) \, dx = \dfrac{du}{dx} \, dx$, (9) and (10) are, respectively, equivalent to

$$\int \cos u \, du = \sin u + C, \tag{11}$$

$$\int \sin u \, du = -\cos u + C. \tag{12}$$

EXAMPLE 4 Using (11)

Evaluate $\displaystyle\int \cos 2x \, dx$.

Solution If $u = 2x$, then $du = 2 \, dx$ and $dx = \dfrac{1}{2} \, du$. Accordingly, we write

$$\int \cos 2x \, dx = \int \cos \overbrace{2x}^{u} \overbrace{(dx)}^{\frac{1}{2}du} \quad \leftarrow \text{ substitution}$$

$$= \frac{1}{2} \int \cos u \, du \qquad \leftarrow \text{ now use (11)}$$

$$= \frac{1}{2}\sin u + C$$

$$= \frac{1}{2}\sin 2x + C. \quad \leftarrow \text{ resubstitution} \qquad ■$$

Integration formulas (8), (11), and (12) are the Chain Rule analogues of integration formulas 2, 4, and 5 in Table 5.1.1. In Table 5.2.1 that follows we summarize the Chain Rule analogues of the sixteen integration formulas in Table 5.1.1.

TABLE 5.2.1

Integration Formulas

1. $\int du = u + C$	**2.** $\int u^n \, du = \dfrac{u^{n+1}}{n+1} + C \quad (n \neq -1)$
3. $\int \dfrac{1}{u} \, du = \ln|u| + C$	**4.** $\int \cos u \, du = \sin u + C$
5. $\int \sin u \, du = -\cos u + C$	**6.** $\int \sec^2 u \, du = \tan u + C$
7. $\int \csc^2 du = -\cot u + C$	**8.** $\int \sec u \tan u \, du = \sec u + C$
9. $\int \csc u \cot u \, du = -\csc u + C$	**10.** $\int \dfrac{1}{\sqrt{1-u^2}} \, du = \sin^{-1} u + C$
11. $\int \dfrac{1}{1+u^2} \, du = \tan^{-1} u + C$	**12.** $\int \dfrac{1}{u\sqrt{u^2-1}} \, du = \sec^{-1}|u| + C$
13. $\int b^u \, du = \dfrac{b^u}{\ln b} + C$	**14.** $\int e^u \, du = e^u + C$
15. $\int \cosh u \, du = \sinh u + C$	**16.** $\int \sinh u \, du = \cosh u + C$

In other textbooks, formulas such as 3, 10, 11, and 12 in Table 5.2.1 are frequently written with the differential du as the numerator:

$$\int \frac{du}{u}, \quad \int \frac{du}{\sqrt{1-u^2}}, \quad \int \frac{du}{1+u^2}, \quad \int \frac{du}{u\sqrt{u^2-1}}.$$

But since we have found, over the years, that these latter formulas are often subject to misunderstanding in a classroom environment, we prefer the forms given in the table.

EXAMPLE 5 Using Table 5.2.1

Evaluate $\int \sec^2(1-4x) \, dx$.

Solution We recognize that the indefinite integral has the form of the integration formula 6 in Table 5.2.1. If $u = 1 - 4x$, then $du = -4 \, dx$. Adjusting the integrand to obtain the correct form of the differential requires multiplying and dividing by -4:

$$\int \sec^2(1-4x) \, dx = -\frac{1}{4} \int \sec^2(\overbrace{1-4x}^{u})(\overbrace{-4 \, dx}^{du})$$

$$= -\frac{1}{4} \int \sec^2 u \, du \quad \leftarrow \text{formula 6 in Table 5.2.1}$$

$$= -\frac{1}{4}\tan u + C$$

$$= -\frac{1}{4}\tan(1-4x) + C.$$

EXAMPLE 6   Using Table 5.2.1

Evaluate $\displaystyle\int \frac{x^2}{x^3 + 5}\,dx$.

Solution   If $u = x^3 + 5$, then $du = 3x^2\,dx$ and $x^2\,dx = \dfrac{1}{3}\,du$. Hence,

$$\int \frac{x^2}{x^3 + 5}\,dx = \int \frac{1}{x^3 + 5}\,(x^2\,dx)$$

$$= \frac{1}{3}\int \frac{1}{u}\,du$$

$$= \frac{1}{3}\ln|u| + C \qquad \leftarrow \text{formula 3 in Table 5.2.1}$$

$$= \frac{1}{3}\ln|x^3 + 5| + C. \qquad\qquad\blacksquare$$

EXAMPLE 7   Rewriting and Using Table 5.2.1

Evaluate $\displaystyle\int \frac{1}{1 + e^{-2x}}\,dx$.

Solution   The given integral does not look like any of the integration formulas in Table 5.2.1. However, if we multiply the numerator and denominator by e^{2x}, then we have

$$\int \frac{1}{1 + e^{-2x}}\,dx = \int \frac{e^{2x}}{e^{2x} + 1}\,dx.$$

If $u = e^{2x} + 1$, then $du = 2e^{2x}\,dx$ and so from formula 3 of Table 5.2.1,

$$\int \frac{1}{1 + e^{-2x}}\,dx = \frac{1}{2}\int \frac{1}{e^{2x} + 1}\,(2e^{2x}\,dx)$$

$$= \frac{1}{2}\int \frac{1}{u}\,du$$

$$= \frac{1}{2}\ln|u| + C$$

$$= \frac{1}{2}\ln(e^{2x} + 1) + C.$$

Note that the absolute value symbol can be dropped because $e^{2x} + 1 > 0$ for all values of x. $\blacksquare$

EXAMPLE 8   Using Table 5.2.1

Evaluate $\displaystyle\int e^{5x}\,dx$.

Solution   Let $u = 5x$ so that $du = 5\,dx$. Then

$$\int e^{5x}\,dx = \frac{1}{5}\int e^{5x}(5\,dx)$$

$$= \frac{1}{5}\int e^{u}\,du \qquad \leftarrow \text{formula 14 in Table 5.2.1}$$

$$= \frac{1}{5}e^{u} + C$$

$$= \frac{1}{5}e^{5x} + C. \qquad\qquad\blacksquare$$

EXAMPLE 9   Using Table 5.2.1

Evaluate $\displaystyle\int \frac{e^{4/x}}{x^2}\,dx$.

Solution   If we let $u = 4/x$, then $du = (-4/x^2)\,dx$ and $(1/x^2)\,dx = -\dfrac{1}{4}\,du$.

Again from formula 14 of Table 5.2.1 we see that

$$\int \frac{e^{4/x}}{x^2}\,dx = \int e^{4/x}\left(\frac{1}{x^2}\,dx\right)$$

$$= \int e^u\left(-\frac{1}{4}\,du\right)$$

$$= -\frac{1}{4}\int e^u\,du$$

$$= -\frac{1}{4}e^u + C$$

$$= -\frac{1}{4}e^{4/x} + C.\qquad\blacksquare$$

EXAMPLE 10 Using Table 5.2.1

Evaluate $\displaystyle\int \frac{(\tan^{-1}x)^2}{1 + x^2}\,dx$.

Solution Like Example 7, the given integral at first glance does not resemble any of the formulas in Table 5.2.1. But if we try the u-substitution with $u = \tan^{-1}x$ and $du = \dfrac{1}{1 + x^2}\,dx$, then

$$\int \frac{(\tan^{-1}x)^2}{1 + x^2}\,dx = \int \overbrace{(\tan^{-1}x)^2}^{u}\;\overbrace{\frac{1}{1 + x^2}}^{du}\,dx$$

$$= \int u^2\,du \quad\leftarrow \text{formula 2 in Table 5.2.1}$$

$$= \frac{u^3}{3} + C$$

$$= \frac{1}{3}(\tan^{-1}x)^3 + C.\qquad\blacksquare$$

EXAMPLE 11 Using Table 5.2.1

Evaluate $\displaystyle\int \frac{1}{\sqrt{100 - x^2}}\,dx$.

Solution By factoring 100 from the radical and identifying $u = \dfrac{1}{10}x$ and $du = \dfrac{1}{10}\,dx$, the result is obtained from formula 10 of Table 5.2.1:

$$\int \frac{1}{\sqrt{100 - x^2}}\,dx = \int \frac{1}{\sqrt{1 - \left(\dfrac{x}{10}\right)^2}}\left(\frac{1}{10}\,dx\right)$$

$$= \int \frac{1}{\sqrt{1 - u^2}}\,du$$

$$= \sin^{-1}u + C$$

$$= \sin^{-1}\frac{x}{10} + C.\qquad\blacksquare$$

▌ **Three Alternative Formulas** As a matter of convenience, integration formulas 10, 11, and 12 in Table 5.2.1 are extended in the following manner. For $a > 0$,

$$\int \frac{1}{\sqrt{a^2 - u^2}}\,du = \sin^{-1}\frac{u}{a} + C \qquad (13)$$

$$\int \frac{1}{a^2 + u^2}\,du = \frac{1}{a}\tan^{-1}\frac{u}{a} + C \qquad (14)$$

$$\int \frac{1}{u\sqrt{u^2 - a^2}}\,du = \frac{1}{a}\sec^{-1}\left|\frac{u}{a}\right| + C. \qquad (15)$$

For practice you are encouraged to verify these results by differentiation. Observe that the indefinite integral in Example 11 can be quickly evaluated by identifying $u = x$ and $a = 10$ in (13).

■ **Special Trigonometric Integrals** The integration formulas given next, which relate some trigonometric functions with the natural logarithm, occur often enough in practice to merit special attention:

$$\int \tan x \, dx = -\ln|\cos x| + C \qquad\qquad (16)$$

$$\int \cot x \, dx = \ln|\sin x| + C \qquad\qquad (17)$$

$$\int \sec x \, dx = \ln|\sec x + \tan x| + C \qquad\qquad (18)$$

$$\int \csc x \, dx = \ln|\csc x - \cot x| + C. \qquad\qquad (19)$$

◀ In tables of integral formulas you will often see (16) written as
$\int \tan x \, dx = \ln|\sec x| + C.$
By the properties of logarithms
$-\ln|\cos x| = \ln|\cos x|^{-1} = \ln|\sec x|.$

To obtain (16) we write

$$\int \tan x \, dx = \int \frac{\sin x}{\cos x} \, dx \qquad\qquad (20)$$

and identify $u = \cos x$, $du = -\sin x \, dx$ so that,

$$\int \tan x \, dx = \int \frac{\sin x}{\cos x} \, dx = -\int \frac{1}{\cos x}(-\sin x \, dx)$$

$$= -\int \frac{1}{u} \, du$$

$$= -\ln|u| + C$$

$$= -\ln|\cos x| + C.$$

To obtain (18) we write

$$\int \sec x \, dx = \int \sec x \frac{\sec x + \tan x}{\sec x + \tan x} \, dx$$

$$= \int \frac{\sec^2 x + \sec x \tan x}{\sec x + \tan x} \, dx.$$

If we let $u = \sec x + \tan x$, then $du = (\sec x \tan x + \sec^2 x) \, dx$ and so,

$$\int \sec x \, dx = \int \frac{1}{\sec x + \tan x}(\sec^2 x + \sec x \tan x) \, dx$$

$$= \int \frac{1}{u} \, du$$

$$= \ln|u| + C$$

$$= \ln|\sec x + \tan x| + C.$$

Also, each of the formulas (16)–(19) can be written in a general form:

$$\int \tan u \, dx = -\ln|\cos u| + C \qquad\qquad (21)$$

$$\int \cot u \, du = \ln|\sin u| + C \qquad\qquad (22)$$

$$\int \sec u \, dx = \ln|\sec u + \tan u| + C \qquad\qquad (23)$$

and

$$\int \csc u \, du = \ln|\csc u - \cot u| + C. \qquad\qquad (24)$$

Useful Identities When working with trigonometric functions it is often necessary to use a trigonometric identity to solve a problem. The half-angle formulas for the cosine and sine in the form

$$\cos^2 x = \frac{1}{2}(1 + \cos 2x) \qquad \text{and} \qquad \sin^2 x = \frac{1}{2}(1 - \cos 2x) \tag{25}$$

are particularly useful in problems that require antiderivatives of $\cos^2 x$ and $\sin^2 x$.

EXAMPLE 12 Using a Half-Angle Formula

Evaluate $\displaystyle\int \cos^2 x \, dx$.

Solution It should be verified that the integral is *not* of the form $\int u^2 \, du$. Now using the half-angle formula $\cos^2 x = \frac{1}{2}(1 + \cos 2x)$, we obtain

$$\int \cos^2 x \, dx = \int \frac{1}{2}(1 + \cos 2x) \, dx$$

$$= \frac{1}{2}\left[\int dx + \frac{1}{2}\int \cos 2x (2\, dx)\right] \quad \leftarrow \text{ see Example 4}$$

$$= \frac{1}{2}\left[x + \frac{1}{2}\sin 2x\right] + C$$

$$= \frac{1}{2}x + \frac{1}{4}\sin 2x + C. \qquad \blacksquare$$

Of course, the method illustrated in Example 12 works equally well in finding antiderivatives such as $\int \cos^2 5x \, dx$ and $\int \sin^2 \frac{1}{2}x \, dx$. With x replaced by $5x$ and then x replaced by $\frac{1}{2}x$, the formulas in (25) enable us to write, respectively,

$$\int \cos^2 5x \, dx = \int \frac{1}{2}(1 + \cos 10x) \, dx = \frac{1}{2}x + \frac{1}{20}\sin 10x + C$$

$$\int \sin^2 \frac{1}{2}x \, dx = \int \frac{1}{2}(1 - \cos x) \, dx = \frac{1}{2}x - \frac{1}{2}\sin x + C.$$

We will consider antiderivatives of more complicated powers of trigonometric functions in Section 7.4.

$\int$ **NOTES FROM THE CLASSROOM**

The following example illustrates a common, but *totally incorrect,* procedure for evaluating an indefinite integral. Because $2x/2x = 1$,

$$\int (4 + x^2)^{1/2} \, dx = \int (4 + x^2)^{1/2}\frac{2x}{2x} \, dx$$

$$= \frac{1}{2x}\int (4 + x^2)^{1/2}2x \, dx$$

$$= \frac{1}{2x}\int u^{1/2} \, du$$

$$= \frac{1}{2x} \cdot \frac{2}{3}(4 + x^2)^{3/2} + C.$$

You should verify that differentiation of the latter function does *not* yield $(4 + x^2)^{1/2}$. The mistake is in the first line of the "solution." Variables, in this case $2x$, cannot be brought outside an integral symbol. If $u = x^2 + 4$, then the integrand lacks the function $du = 2x \, dx$; in fact, there is no way of adjusting the problem to fit the form given in (8). With the "tools" we presently have on hand, the integral $\int (4 + x^2)^{1/2} \, dx$ simply cannot be evaluated.

Exercises 5.2   Answers to selected odd-numbered problems begin on page ANS-18.

≡ Fundamentals

In Problems 1–50, evaluate the given indefinite integral using an appropriate *u*-substitution.

1. $\displaystyle \int \sqrt{1 - 4x}\, dx$

2. $\displaystyle \int (8x + 2)^{1/3}\, dx$

3. $\displaystyle \int \frac{1}{(5x + 1)^3}\, dx$

4. $\displaystyle \int (7 - x)^{49}\, dx$

5. $\displaystyle \int x\sqrt{x^2 + 4}\, dx$

6. $\displaystyle \int \frac{t}{\sqrt[3]{t^2 + 9}}\, dt$

7. $\displaystyle \int \sin^5 3x \cos 3x\, dx$

8. $\displaystyle \int \sin 2\theta \cos^4 2\theta\, d\theta$

9. $\displaystyle \int \tan^2 2x \sec^2 2x\, dx$

10. $\displaystyle \int \sqrt{\tan x}\, \sec^2 x\, dx$

11. $\displaystyle \int \sin 4x\, dx$

12. $\displaystyle \int 5\cos\frac{x}{2}\, dx$

13. $\displaystyle \int \left(\sqrt{2t} - \cos 6t\right) dt$

14. $\displaystyle \int \sin(2 - 3x)\, dx$

15. $\displaystyle \int x\sin x^2\, dx$

16. $\displaystyle \int \frac{\cos(1/x)}{x^2}\, dx$

17. $\displaystyle \int x^2 \sec^2 x^3\, dx$

18. $\displaystyle \int \csc^2(0.1x)\, dx$

19. $\displaystyle \int \frac{\csc\sqrt{x}\cot\sqrt{x}}{\sqrt{x}}\, dx$

20. $\displaystyle \int \tan 5v \sec 5v\, dv$

21. $\displaystyle \int \frac{1}{7x + 3}\, dx$

22. $\displaystyle \int (5x + 6)^{-1}\, dx$

23. $\displaystyle \int \frac{x}{x^2 + 1}\, dx$

24. $\displaystyle \int \frac{x^2}{5x^3 + 8}\, dx$

25. $\displaystyle \int \frac{x}{x + 1}\, dx$

26. $\displaystyle \int \frac{(x + 3)^2}{x + 2}\, dx$

27. $\displaystyle \int \frac{1}{x\ln x}\, dx$

28. $\displaystyle \int \frac{1 - \sin\theta}{\theta + \cos\theta}\, d\theta$

29. $\displaystyle \int \frac{\sin(\ln x)}{x}\, dx$

30. $\displaystyle \int \frac{1}{x(\ln x)^2}\, dx$

31. $\displaystyle \int e^{10x}\, dx$

32. $\displaystyle \int \frac{1}{e^{4x}}\, dx$

33. $\displaystyle \int x^2 e^{-2x^3}\, dx$

34. $\displaystyle \int \frac{e^{1/x^3}}{x^4}\, dx$

35. $\displaystyle \int \frac{e^{-\sqrt{x}}}{\sqrt{x}}\, dx$

36. $\displaystyle \int \sqrt{e^x}\, dx$

37. $\displaystyle \int \frac{e^x - e^{-x}}{e^x + e^{-x}}\, dx$

38. $\displaystyle \int e^{3x}\sqrt{1 + 2e^{3x}}\, dx$

39. $\displaystyle \int \frac{1}{\sqrt{5 - x^2}}\, dx$

40. $\displaystyle \int \frac{1}{\sqrt{9 - 16x^2}}\, dx$

41. $\displaystyle \int \frac{1}{1 + 25x^2}\, dx$

42. $\displaystyle \int \frac{1}{2 + 9x^2}\, dx$

43. $\displaystyle \int \frac{e^x}{1 + e^{2x}}\, dx$

44. $\displaystyle \int \frac{\theta}{\sqrt{1 - \theta^4}}\, d\theta$

45. $\displaystyle \int \frac{2x - 3}{\sqrt{1 - x^2}}\, dx$

46. $\displaystyle \int \frac{x - 8}{x^2 + 2}\, dx$

47. $\displaystyle \int \frac{\tan^{-1}x}{1 + x^2}\, dx$

48. $\displaystyle \int \sqrt{\frac{\sin^{-1}x}{1 - x^2}}\, dx$

49. $\displaystyle \int \tan 5x\, dx$

50. $\displaystyle \int e^x \cot e^x\, dx$

In Problems 51–56, use the identities in (25) to evaluate the given indefinite integral.

51. $\displaystyle \int \sin^2 x\, dx$

52. $\displaystyle \int \cos^2 \pi x\, dx$

53. $\displaystyle \int \cos^2 4x\, dx$

54. $\displaystyle \int \sin^2\frac{3}{2}x\, dx$

55. $\displaystyle \int (3 - 2\sin x)^2\, dx$

56. $\displaystyle \int (1 + \cos 2x)^2\, dx$

In Problems 57 and 58, solve the given differential equation.

57. $\displaystyle \frac{dy}{dx} = \sqrt[3]{1 - x}$

58. $\displaystyle \frac{dy}{dx} = \frac{(1 - \tan x)^5}{\cos^2 x}$

59. Find a function $y = f(x)$ whose graph passes through the point $(\pi, -1)$ that also satisfies $dy/dx = 1 - 6\sin 3x$.

60. Find a function f such that $f''(x) = (1 + 2x)^5$, $f(0) = 0$, and $f'(0) = 0$.

61. Show that:

(a) $\displaystyle \int \sin x \cos x\, dx = \frac{1}{2}\sin^2 x + C_1$

(b) $\displaystyle \int \sin x \cos x\, dx = -\frac{1}{2}\cos^2 x + C_2$

(c) $\displaystyle \int \sin x \cos x\, dx = -\frac{1}{4}\cos 2x + C_3.$

62. In Problem 61:

(a) Verify that the derivative of each answer in parts (a), (b), and (c) is $\sin x \cos x$.

(b) By a trigonometric identity, show how the result in part (b) can be obtained from the answer in part (a).

(c) By adding the results in parts (a) and (b), obtain the result in part (c).

≡ Applications

63. Consider the plane pendulum, shown in **FIGURE 5.2.1**, that swings between points A and C. If B is midway between A and C, it can be shown that

$$\frac{dt}{ds} = \sqrt{\frac{L}{g(s_C^2 - s^2)}},$$

where g is the acceleration due to gravity.

(a) If $t(0) = 0$, then show that the time the pendulum takes to travel between points B and P is

$$t(s) = \sqrt{\frac{L}{g}} \sin^{-1}\left(\frac{s}{s_C}\right).$$

(b) Use the result in part (a) to determine the time of travel from B to C.

(c) Use (b) to determine the period T of the pendulum—that is, the time of an oscillation from A to C and back to A.

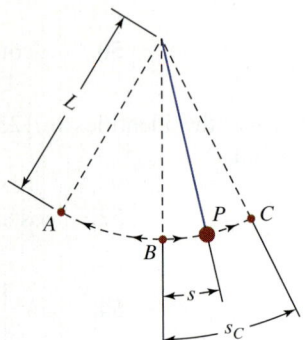

FIGURE 5.2.1 Pendulum in Problem 63

≡ **Think About It**

64. Find a function $y = f(x)$ for which $f(\pi/2) = 0$ and $\dfrac{dy}{dx} = \cos^3 x$. [*Hint*: $\cos^3 x = \cos^2 x \cos x$.]

In Problems 65 and 66, use the identities in (25) to evaluate the given indefinite integral.

65. $\displaystyle\int \cos^4 x \, dx$

66. $\displaystyle\int \sin^4 x \, dx$

In Problems 67 and 68, evaluate the given indefinite integral.

67. $\displaystyle\int \frac{1}{x\sqrt{x^4 - 16}} \, dx$

68. $\displaystyle\int \frac{e^{2x}}{e^x + 1} \, dx$

In Problems 69 and 70, evaluate the given indefinite integral.

69. $\displaystyle\int \frac{1}{1 - \cos x} \, dx$

70. $\displaystyle\int \frac{1}{1 + \sin 2x} \, dx$

In Problems 71–74, evaluate the given indefinite integral. Assume f is a differentiable function.

71. $\displaystyle\int f'(8x) \, dx$

72. $\displaystyle\int x f'(5x^2) \, dx$

73. $\displaystyle\int \sqrt{f(2x)} f'(2x) \, dx$

74. $\displaystyle\int \frac{f'(3x + 1)}{f(3x + 1)} \, dx$

75. Evaluate $\displaystyle\int f''(4x) \, dx$ if $f(x) = \sqrt{x^4 + 1}$.

76. Evaluate $\displaystyle\int \left\{ \int \sec^2 3x \, dx \right\} dx$.

5.3 The Area Problem

❙ **Introduction** As the derivative is motivated by the geometric problem of constructing a tangent to a curve, the historical problem leading to the definition of a definite integral is the problem of finding area. Specifically, we are interested in the following version of this problem:

• *Find the area A of a region bounded by the x-axis and the graph of a continuous nonnegative function $y = f(x)$ defined on an interval $[a, b]$.*

We shall call the area of this region the **area under the graph** of f on the interval $[a, b]$. The requirement that f be nonnegative on $[a, b]$ means that no portion of its graph on the interval is below the x-axis. See **FIGURE 5.3.1**.

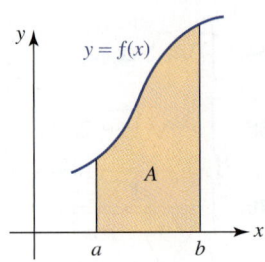

FIGURE 5.3.1 Area under the graph of f on $[a, b]$

Before pursuing the solution of the area problem we need to digress briefly to discuss a helpful notation for a sum of numbers such as

$$1 + 2 + 3 + \cdots + n \qquad \text{and} \qquad 1^2 + 2^2 + 3^2 + \cdots + n^2.$$

❙ Sigma Notation   Let a_k be a real number that depends on an integer k. We denote the sum $a_1 + a_2 + a_3 + \cdots + a_n$ by the symbol $\sum_{k=1}^{n} a_k$; that is,

$$\sum_{k=1}^{n} a_k = a_1 + a_2 + a_3 + \cdots + a_n. \tag{1}$$

Since Σ is the capital Greek letter *sigma*, (1) is called **sigma notation** or **summation notation**. The variable k is called the **index of summation**. Thus,

$$
\underset{\text{start with indicated}\atop\text{value of } k}{\overset{\text{end with this value of } k}{\underset{\text{symbol } \Sigma \text{ indicates}\atop\text{addition of } a_k}{\longrightarrow}\ \sum_{k=1}^{n} a_k}}
$$

is the sum of all numbers of the form a_k as k takes on the successive values $k = 1, k = 2, \ldots,$ end and concludes with $k = n$.

EXAMPLE 1   Using Sigma Notation

The sum of the first ten positive even integers

$$2 + 4 + 6 + \cdots + 18 + 20$$

can be written succinctly as $\sum_{k=1}^{10} 2k$. The sum of the first ten positive odd integers

$$1 + 3 + 5 + \cdots + 17 + 19$$

can be written $\sum_{k=1}^{10} (2k - 1)$.   ∎

The index of summation need not start at the value $k = 1$; for example,

$$\sum_{k=3}^{5} 2^k = 2^3 + 2^4 + 2^5 \qquad \text{and} \qquad \sum_{k=0}^{5} 2^k = 2^0 + 2^1 + 2^2 + 2^3 + 2^4 + 2^5.$$

Note that the sum of the first ten odd positive integers in Example 1 can also be written as $\sum_{k=0}^{9} (2k + 1)$. However, in a general discussion we shall always assume that the summation index starts at $k = 1$. This assumption is for convenience rather than necessity. The index of summation is often called a **dummy variable**, since the symbol itself is not important; it is the successive integer values of the index and the corresponding sum that are important. In general,

$$\sum_{k=1}^{n} a_k = \sum_{i=1}^{n} a_i = \sum_{j=1}^{n} a_j = \sum_{m=1}^{n} a_m.$$

For example,

$$\sum_{k=1}^{10} 4^k = \sum_{i=1}^{10} 4^i = \sum_{j=1}^{10} 4^j = 4^1 + 4^2 + 4^3 + \cdots + 4^{10}.$$

❙ Properties   The following is a list of some important properties of the sigma notation.

Theorem 5.3.1   Properties of Sigma Notation

For positive integers m and n,

(i) $\displaystyle\sum_{k=1}^{n} ca_k = c\sum_{k=1}^{n} a_k$, where c is any constant

(ii) $\displaystyle\sum_{k=1}^{n} (a_k \pm b_k) = \sum_{k=1}^{n} a_k \pm \sum_{k=1}^{n} b_k$

(iii) $\displaystyle\sum_{k=1}^{n} a_k = \sum_{k=1}^{m} a_k + \sum_{k=m+1}^{n} a_k, \; m < n.$

The proof of formula (*i*) is an immediate consequence of the distributive law. Of course, (*ii*) of Theorem 5.3.1 holds for the sum of three of more terms; for example

$$\sum_{k=1}^{n} (a_k + b_k + c_k) = \sum_{k=1}^{n} a_k + \sum_{k=1}^{n} b_k + \sum_{k=1}^{n} c_k.$$

■ **Special Summation Formulas**   For special kinds of indicated sums, particularly, sums involving positive integer powers of the summation index (such as sums of successive positive integers, successive squares, successive cubes, and so on) it is possible to find a formula that gives the actual numerical value of the sum. For purposes of this section we shall confine our attention to the following four formulas.

Theorem 5.3.2   Summation Formulas

For n a positive integer and c any constant,

(i) $\displaystyle\sum_{k=1}^{n} c = nc$

(ii) $\displaystyle\sum_{k=1}^{n} k = \frac{n(n+1)}{2}$

(iii) $\displaystyle\sum_{k=1}^{n} k^2 = \frac{n(n+1)(2n+1)}{6}$

(iv) $\displaystyle\sum_{k=1}^{n} k^3 = \frac{n^2(n+1)^2}{4}.$

Formulas (*i*) and (*ii*) can be easily justified. If c is a constant—that is, independent of the summation index k—then $\sum_{k=1}^{n} c$ means $c + c + c + \cdots + c$. Since there are n c's, we have $\sum_{k=1}^{n} c = n \cdot c$, which is (*i*) of Theorem 5.3.2. Now, the sum of the first n positive integers can be written as $\sum_{k=1}^{n} k$. If this sum is denoted by the letter S, then

$$S = 1 + 2 + 3 + \cdots + (n-2) + (n-1) + n. \tag{2}$$

Equivalently, $S = n + (n-1) + (n-2) + \cdots + 3 + 2 + 1. \tag{3}$

If we add (2) and (3) by adding corresponding first terms, second terms, and so on, then

$$2S = \underbrace{(n+1) + (n+1) + (n+1) + \cdots + (n+1)}_{n \text{ terms of } n+1} = n(n+1).$$

Solving for S gives $S = n(n+1)/2$, which is (*ii*). You should be able to derive formulas (*iii*) and (*iv*) with the hints supplied in Problems 55 and 56 in Exercises 5.3.

EXAMPLE 2   Using Summation Formulas

Find the numerical value of $\sum_{k=1}^{20} (k + 5)^2$.

Solution   By expanding $(k + 5)^2$ and using (*i*) and (*ii*) of Theorem 5.3.1, we can write

$$\sum_{k=1}^{20} (k + 5)^2 = \sum_{k=1}^{20} (k^2 + 10k + 25) \quad \leftarrow \text{squaring the binomial}$$

$$= \sum_{k=1}^{20} k^2 + 10 \sum_{k=1}^{20} k + \sum_{k=1}^{20} 25. \quad \leftarrow \text{(\textit{i}) and (\textit{ii}) of Theorem 5.3.1}$$

With the identification $n = 20$, it follows from summation formulas (*iii*), (*ii*), and (*i*) of Theorem 5.3.2, respectively, that

$$\sum_{k=1}^{20} (k + 5)^2 = \frac{20(21)(41)}{6} + 10\frac{20(21)}{2} + 20 \cdot 25 = 5470. \quad \blacksquare$$

Sigma notation and the foregoing summation formulas will be put to immediate use in the next discussion.

■ **Area of a Triangle**   Assume for the moment that we do not know a formula for calculating the area A of the right triangle given in FIGURE 5.3.2(a). By superimposing a rectangular coordinate system on the triangle, as shown in Figure 5.3.2(b), we see that the problem is the same as finding the area in the first quadrant bounded by the straight lines $y = (h/b)x$, $y = 0$ (the x-axis), and $x = b$. In other words, we wish to find the area under the graph of $y = (h/b)x$ on the interval $[0, b]$.

Using rectangles, FIGURE 5.3.3 indicates three different ways of *approximating* the area A. For convenience, let us pursue the procedure hinted at in Figure 5.3.3(b) in greater detail. We begin by dividing the interval $[0, b]$ into n subintervals of equal width $\Delta x = b/n$. If the right endpoint of each of these intervals is denoted by x_k^*, then

$$x_1^* = \Delta x = \frac{b}{n}$$

$$x_2^* = 2\Delta x = 2\left(\frac{b}{n}\right)$$

$$x_3^* = 3\Delta x = 3\left(\frac{b}{n}\right)$$

$$\vdots$$

$$x_n^* = n\Delta x = n\left(\frac{b}{n}\right) = b.$$

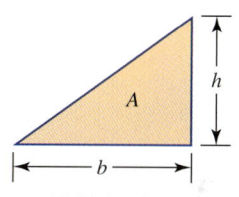

(a) Right triangle

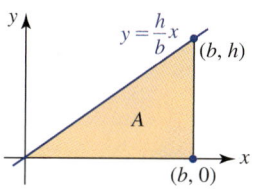

(b) Right triangle in a coordinate system

FIGURE 5.3.2   Find the area A of a right triangle

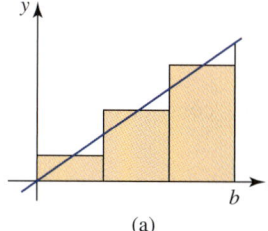

(a)

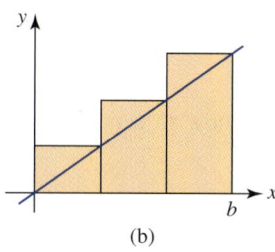

(b)

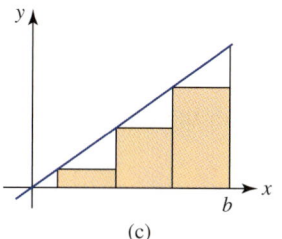

(c)

FIGURE 5.3.3   Approximating the area A using three rectangles

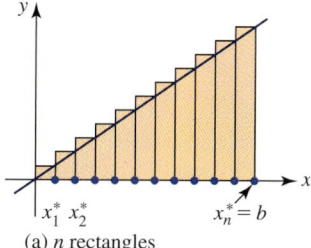

(a) n rectangles

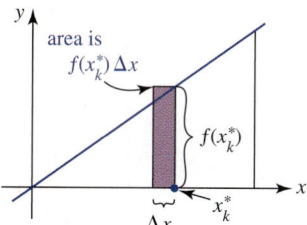

(b) Area of a general rectangle

FIGURE 5.3.4   Area A of the triangle is approximated by the sum of the areas of n rectangles

As shown in FIGURE 5.3.4(a), we now construct a rectangle of length $f(x_k^*)$ and width Δx on each of the n subintervals. Since the area of a rectangle is *length* $\times$ *width*, the area of each rectangle is $f(x_k^*)\Delta x$. See Figure 5.3.4(b). The sum of the areas of the n rectangles is an approximation to the number A. We write

$$A \approx f(x_1^*)\Delta x + f(x_2^*)\Delta x + \cdots + f(x_n^*)\Delta x,$$

or in sigma notation,

$$A \approx \sum_{k=1}^{n} f(x_k^*)\Delta x. \tag{4}$$

It seems plausible that we can reduce the error introduced by this method of approximation (the area of each rectangle is larger than the area under the graph on a subinterval $[x_{k-1}, x_k]$) by dividing the interval $[0, b]$ into finer subdivisions. In other words, we expect that a better approximation to A can be obtained by using more and more rectangles ($n \to \infty$) of decreasing widths ($\Delta x \to 0$). Now,

$$f(x) = \frac{h}{b}x, \quad x_k^* = k\left(\frac{b}{n}\right), \quad f(x_k^*) = \frac{h}{n} \cdot k, \quad \text{and} \quad \Delta x = \frac{b}{n},$$

so that with the aid of summation formula (*ii*) of Theorem 5.3.2, (4) becomes

$$A \approx \sum_{k=1}^{n} \left(\frac{h}{n} \cdot k\right)\frac{b}{n} = \frac{bh}{n^2} \sum_{k=1}^{n} k = \frac{bh}{n^2} \cdot \frac{n(n+1)}{2} = \frac{bh}{2}\left(1 + \frac{1}{n}\right). \tag{5}$$

Finally, by letting $n \to \infty$ on the right-hand side of (5), we obtain the familiar formula for the area of a triangle:

$$A = \frac{1}{2}bh \cdot \lim_{n \to \infty}\left(1 + \frac{1}{n}\right) = \frac{1}{2}bh.$$

■ **The General Problem** Now, let us turn from the preceding specific example to the general problem of finding the area A under the graph of a function $y = f(x)$ that is continuous on an interval $[a, b]$. As shown in FIGURE 5.3.5(a), we shall also assume that $f(x) \geq 0$ for all x in the interval $[a, b]$. As suggested in Figure 5.3.5(b), the area A can be approximated by adding the areas of n rectangles that are constructed on the interval. One possible procedure for determining A is summarized as follows:

- Divide the interval $[a, b]$ into n subintervals $[x_{k-1}, x_k]$, where

$$a = x_0 < x_1 < x_2 < \cdots < x_{n-1} < x_n = b,$$

so that each subinterval has the same width $\Delta x = (b - a)/n$. This collection of numbers is called a **regular partition** of the interval $[a, b]$.

- Choose a number x_k^* in each of the n subintervals $[x_{k-1}, x_k]$ and form the n products $f(x_k^*)\Delta x$. Since the area of a rectangle is length × width, $f(x_k^*)\Delta x$ is the area of the rectangle of length $f(x_k^*)$ and width Δx built up on the kth subinterval $[x_{k-1}, x_k]$. The n numbers $x_1^*, x_2^*, x_3^*, \ldots, x_n^*$ are called **sample points**.

- The sum of the areas of the n rectangles

$$\sum_{k=1}^{n} f(x_k^*)\Delta x = f(x_1^*)\Delta x + f(x_2^*)\Delta x + f(x_3^*)\Delta x + \cdots + f(x_n^*)\Delta x,$$

represents an approximation to the value of the area A under the graph of f on the interval $[a, b]$.

With these preliminaries, we are now in a position to define the concept of area under a graph.

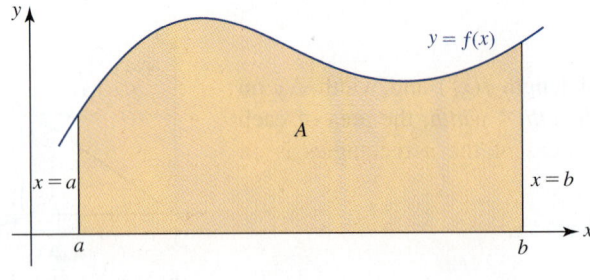

(a) Area A under the graph

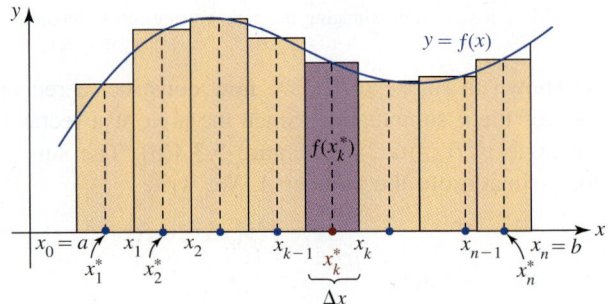

(b) n rectangles

FIGURE 5.3.5 Find the area A under the graph of f on the interval $[a, b]$

Definition 5.3.1 Area Under a Graph

Let f be continuous on $[a, b]$ and $f(x) \geq 0$ for all x in the interval. We define the **area A under the graph** of f on the interval to be

$$A = \lim_{n \to \infty} \sum_{k=1}^{n} f(x_k^*) \Delta x. \tag{6}$$

It can be proved that when f is *continuous*, the limit in (6) always exists regardless of the manner used to divide $[a, b]$ into subintervals; that is, the subintervals may or may not be taken of equal width, and the points x_k^* can be chosen quite arbitrarily in the subintervals $[x_{k-1}, x_k]$. However, if the subintervals are not of equal width, then a different kind of limiting process is necessary in (6). We must replace $n \to \infty$ with the requirement that the length of the widest subinterval approach zero.

■ **A Practical Form of (6)** To use (6), suppose we choose x_k^* as we did in the discussion of Figure 5.3.4; namely, let x_k^* be the **right endpoint** of each subinterval. Since the width of each of the n subintervals of equal width is $\Delta x = (b - a)/n$, we have

$$x_k^* = a + k\Delta x = a + k\frac{b - a}{n}.$$

Then for $k = 1, 2, \ldots, n$ we have

$$x_1^* = a + \Delta x = a + \frac{b - a}{n}$$

$$x_2^* = a + 2\Delta x = a + 2\left(\frac{b - a}{n}\right)$$

$$x_3^* = a + 3\Delta x = a + 3\left(\frac{b - a}{n}\right)$$

$$\vdots$$

$$x_n^* = a + n\Delta x = a + n\left(\frac{b - a}{n}\right) = b.$$

By substituting $a + k(b - a)/n$ for x_k^* and $(b - a)/n$ for Δx in (6), it follows that the area A is also given by

$$A = \lim_{n \to \infty} \sum_{k=1}^{n} f\left(a + k\frac{b - a}{n}\right) \cdot \frac{b - a}{n}. \tag{7}$$

We note that since $\Delta x = (b - a)/n$, $n \to \infty$ implies $\Delta x \to 0$.

EXAMPLE 3 Area Using (7)

Find the area A under the graph of $f(x) = x + 2$ on the interval $[0, 4]$.

Solution The area is bounded by the trapezoid indicated in FIGURE 5.3.6(a). By identifying $a = 0$ and $b = 4$, we find

$$\Delta x = \frac{4 - 0}{n} = \frac{4}{n}.$$

Thus, (7) becomes

$$A = \lim_{n \to \infty} \sum_{k=1}^{n} f\left(0 + k\frac{4}{n}\right)\frac{4}{n} = \lim_{n \to \infty} \frac{4}{n} \sum_{k=1}^{n} f\left(\frac{4k}{n}\right)$$

$$= \lim_{n \to \infty} \frac{4}{n} \sum_{k=1}^{n} \left(\frac{4k}{n} + 2\right)$$

$$= \lim_{n \to \infty} \frac{4}{n}\left[\frac{4}{n} \sum_{k=1}^{n} k + 2 \sum_{k=1}^{n} 1\right]. \quad \leftarrow \text{by properties } (i) \text{ and } (ii) \text{ of Theorem 5.3.1}$$

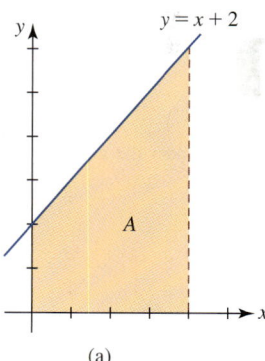

(a)

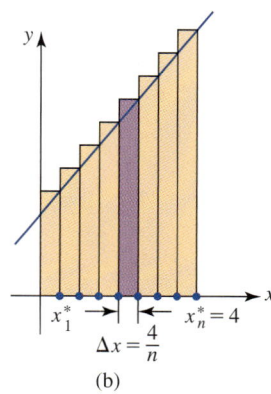

(b)

FIGURE 5.3.6 Area under the graph in Example 3

Now, by summation formulas (*ii*) and (*i*) of Theorem 5.3.2, we can write

$$A = \lim_{n \to \infty} \frac{4}{n} \left[\frac{4}{n} \cdot \frac{n(n+1)}{2} + 2n \right]$$

$$= \lim_{n \to \infty} \left[\frac{16}{2} \frac{n(n+1)}{n^2} + 8 \right] \quad \leftarrow \text{divide by } n^2$$

$$= \lim_{n \to \infty} \left[8 \left(1 + \frac{1}{n} \right) + 8 \right]$$

$$= 8 \lim_{n \to \infty} \left(1 + \frac{1}{n} \right) + 8 \lim_{n \to \infty} 1$$

$$= 8 + 8 = 16 \text{ square units.} \qquad \blacksquare$$

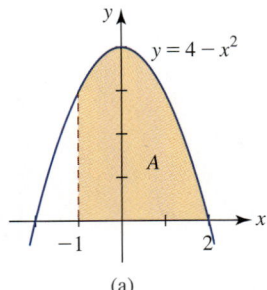

(a)

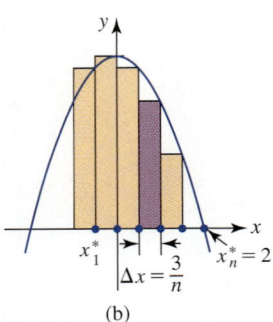

(b)

FIGURE 5.3.7 Area under the graph in Example 4

EXAMPLE 4   Area Using (7)

Find the area A under the graph of $f(x) = 4 - x^2$ on the interval $[-1, 2]$.

Solution   The area is indicated in FIGURE 5.3.7(a). Since $a = -1$ and $b = 2$, it follows that

$$\Delta x = \frac{2 - (-1)}{n} = \frac{3}{n}.$$

Let us review the steps leading up to (7). The width of each rectangle is given by $\Delta x = (2 - (-1))/n = 3/n$. Now, starting at $x = -1$, the right endpoint of each of the n subintervals is

$$x_1^* = -1 + \frac{3}{n}$$

$$x_2^* = -1 + 2 \left(\frac{3}{n} \right) = -1 + \frac{6}{n}$$

$$x_3^* = -1 + 3 \left(\frac{3}{n} \right) = -1 + \frac{9}{n}$$

$$\vdots$$

$$x_n^* = -1 + n \left(\frac{3}{n} \right) = 2.$$

The length of each rectangle is then

$$f(x_1^*) = f \left(-1 + \frac{3}{n} \right) = 4 - \left[-1 + \frac{3}{n} \right]^2$$

$$f(x_2^*) = f \left(-1 + \frac{6}{n} \right) = 4 - \left[-1 + \frac{6}{n} \right]^2$$

$$f(x_3^*) = f \left(-1 + \frac{9}{n} \right) = 4 - \left[-1 + \frac{9}{n} \right]^2$$

$$\vdots$$

$$f(x_n^*) = f \left(-1 + \frac{3n}{n} \right) = f(2) = 4 - (2)^2 = 0.$$

The area of the *k*th rectangle is *length* × *width*:

$$f(x_k^*) \frac{3}{n} = \left(4 - \left[-1 + k \frac{3}{n} \right]^2 \right) \frac{3}{n} = \left(3 + 6 \frac{k}{n} - 9 \frac{k^2}{n^2} \right) \frac{3}{n}.$$

Adding the areas of the n rectangles gives an approximation to the area under the graph on the interval: $A \approx \sum_{k=1}^{n} f(x_k^*)(3/n)$. As the number n of rectangles increases without bound, we obtain

$$A = \lim_{n \to \infty} \sum_{k=1}^{n} \left(3 + 6\frac{k}{n} - 9\frac{k^2}{n^2} \right) \frac{3}{n}$$

$$= \lim_{n \to \infty} \frac{3}{n} \sum_{k=1}^{n} \left(3 + 6\frac{k}{n} - 9\frac{k^2}{n^2} \right)$$

$$= \lim_{n \to \infty} \frac{3}{n} \left[\sum_{k=1}^{n} 3 + \frac{6}{n} \sum_{k=1}^{n} k - \frac{9}{n^2} \sum_{k=1}^{n} k^2 \right].$$

Using summation formulas (*i*), (*ii*), and (*iii*) of Theorem 5.3.2, we get

$$A = \lim_{n \to \infty} \frac{3}{n} \left[3n + \frac{6}{n} \cdot \frac{n(n+1)}{2} - \frac{9}{n^2} \cdot \frac{n(n+1)(2n+1)}{6} \right]$$

$$= \lim_{n \to \infty} \left[9 + 9\left(1 + \frac{1}{n} \right) - \frac{9}{2}\left(1 + \frac{1}{n} \right)\left(2 + \frac{1}{n} \right) \right]$$

$$= 9 + 9 - 9 = 9 \text{ square units.} \qquad \blacksquare$$

▮ Other Choices for x_k^* There is nothing special about choosing x_k^* to be the right endpoint of each subinterval. We reemphasize that x_k^* can be taken to be any convenient number in $[x_{k-1}, x_k]$. Had we chosen x_k^* to be the **left endpoint** of each subinterval, then

$$x_k^* = a + (k-1)\Delta x = a + (k-1)\frac{b-a}{n}, \quad k = 1, 2, \ldots, n,$$

and (7) would become

$$A = \lim_{n \to \infty} \sum_{k=1}^{n} f\left(a + (k-1)\frac{b-a}{n} \right) \cdot \frac{b-a}{n}. \qquad (8)$$

In Example 4 the corresponding rectangles would be as shown in FIGURE 5.3.8. In this case, we would have $x_k^* = -1 + (k-1)(3/n)$. In Problems 45 and 46 of Exercises 5.3 you will be asked to solve the area problem in Example 4 by choosing x_k^* to be first the left endpoint and then the midpoint of each subinterval $[x_{k-1}, x_k]$. By choosing x_k^* to be the midpoint of each $[x_{k-1}, x_k]$, then

$$x_k^* = a + \left(k - \frac{1}{2} \right)\Delta x, \quad k = 1, 2, \ldots, n. \qquad (9)$$

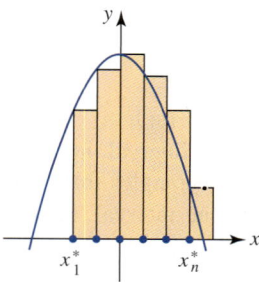

FIGURE 5.3.8 Rectangles using left endpoints of subintervals

Exercises 5.3 Answers to selected odd-numbered problems begin on page ANS-18.

≡ **Fundamentals**

In Problems 1–10, expand the indicated sum.

1. $\displaystyle\sum_{k=1}^{5} 3k$

2. $\displaystyle\sum_{k=1}^{5} (2k - 3)$

3. $\displaystyle\sum_{k=1}^{4} \frac{2^k}{k}$

4. $\displaystyle\sum_{k=1}^{4} \left(\frac{3}{10} \right)^k$

5. $\displaystyle\sum_{k=1}^{10} \frac{(-1)^k}{2k+5}$

6. $\displaystyle\sum_{k=1}^{10} \frac{(-1)^{k-1}}{k^2}$

7. $\displaystyle\sum_{j=2}^{5} (j^2 - 2j)$

8. $\displaystyle\sum_{m=0}^{4} (m+1)^2$

9. $\displaystyle\sum_{k=1}^{5} \cos k\pi$

10. $\displaystyle\sum_{k=1}^{5} \frac{\sin(k\pi/2)}{k}$

In Problems 11–20, write the given sum using sigma notation.

11. $3 + 5 + 7 + 9 + 11 + 13 + 15$

12. $2 + 4 + 8 + 16 + 32 + 64$

13. $1 + 4 + 7 + 10 + \cdots + 37$

14. $2 + 6 + 10 + 14 + \cdots + 38$

15. $1 - \dfrac{1}{2} + \dfrac{1}{3} - \dfrac{1}{4} + \dfrac{1}{5}$

16. $-\dfrac{1}{2} + \dfrac{2}{3} - \dfrac{3}{4} + \dfrac{4}{5} - \dfrac{5}{6}$

17. $6 + 6 + 6 + 6 + 6 + 6 + 6 + 6$

18. $1 + \sqrt{2} + \sqrt{3} + 2 + \sqrt{5} + \cdots + 3$

19. $\cos\dfrac{\pi}{p}x - \dfrac{1}{4}\cos\dfrac{2\pi}{p}x + \dfrac{1}{9}\cos\dfrac{3\pi}{p}x - \dfrac{1}{16}\cos\dfrac{4\pi}{p}x$

20. $f'(1)(x-1) - \dfrac{f''(1)}{3}(x-1)^2 + \dfrac{f'''(1)}{5}(x-1)^3$

$\qquad - \dfrac{f^{(4)}(1)}{7}(x-1)^4 + \dfrac{f^{(5)}(1)}{9}(x-1)^5$

In Problems 21–28, find the numerical value of the given sum.

21. $\displaystyle\sum_{k=1}^{20} 2k$

22. $\displaystyle\sum_{k=0}^{50} (-3k)$

23. $\displaystyle\sum_{k=1}^{10}(k+1)$

24. $\displaystyle\sum_{k=1}^{1000}(2k-1)$

25. $\displaystyle\sum_{k=1}^{6}(k^2+3)$

26. $\displaystyle\sum_{k=1}^{5}(6k^2-k)$

27. $\displaystyle\sum_{p=0}^{10}(p^3+4)$

28. $\displaystyle\sum_{i=1}^{10}(2i^3-5i+3)$

In Problems 29–42, use (7) and Theorem 5.3.2 to find the area under the graph of the given function on the indicated interval.

29. $f(x)=x,\quad[0,6]$

30. $f(x)=2x,\quad[1,3]$

31. $f(x)=2x+1,\quad[1,5]$

32. $f(x)=3x-6,\quad[2,4]$

33. $f(x)=x^2,\quad[0,2]$

34. $f(x)=x^2,\quad[-2,1]$

35. $f(x)=1-x^2,\quad[-1,1]$

36. $f(x)=2x^2+3,\quad[-3,-1]$

37. $f(x)=x^2+2x,\quad[1,2]$

38. $f(x)=(x-1)^2,\quad[0,2]$

39. $f(x)=x^3,\quad[0,1]$

40. $f(x)=x^3-3x^2+4,\quad[0,2]$

41. $f(x)=\begin{cases}2, & 0\le x<1\\ x+1, & 1\le x\le4\end{cases}$

42. $f(x)=\begin{cases}-x+1, & 0\le x<1\\ x+2, & 1\le x\le3\end{cases}$

43. Sketch the graph of $y=1/x$ on the interval $\left[\frac{1}{2},\frac{5}{2}\right]$. By dividing the interval into four subintervals of equal widths, construct rectangles that approximate the area A under the graph on the interval. First use the right endpoint of each subinterval, and then use the left endpoint.

44. Repeat Problem 43 for $y=\cos x$ on the interval $[-\pi/2,\pi/2]$.

45. Rework Example 4 by choosing x_k^* to be the left endpoint of each subinterval. See (8).

46. Rework Example 4 by choosing x_k^* to be the midpoint of each subinterval. See (9).

In Problems 47 and 48, sketch the region whose area A is given by the formula. Do not try to evaluate.

47. $A=\displaystyle\lim_{n\to\infty}\sum_{k=1}^{n}\sqrt{4-\dfrac{4k^2}{n^2}}\,\dfrac{2}{n}$

48. $A=\displaystyle\lim_{n\to\infty}\sum_{k=1}^{n}\left(\sin\dfrac{k\pi}{n}\right)\dfrac{\pi}{n}$

≡ Think About It

In Problems 49 and 50, write the given decimal number using sigma notation.

49. 0.11111111

50. 0.3737373737

51. Use summation formula (*iii*) of Theorem 5.3.2 to find the numerical value of $\sum_{k=21}^{60}k^2$.

52. Write the sum $8+7+8+9+10+11+12$ using sigma notation so that the index of summation starts with $k=0$. With $k=1$. With $k=2$.

53. Solve for $\bar{x}$: $\sum_{k=1}^{n}(x_k-\bar{x})^2=0$.

54. (a) Find the value of $\sum_{k=1}^{n}[f(k)-f(k-1)]$. A sum of this form is said to **telescope**.

(b) Use part (a) to find the numerical value of

$$\sum_{k=1}^{400}\left(\sqrt{k}-\sqrt{k-1}\right).$$

55. (a) Use part (a) of Problem 54 to show that

$$\sum_{k=1}^{n}[(k+1)^2-k^2]=-1+(n+1)^2=n^2+2n.$$

(b) Use the fact that $(k+1)^2-k^2=2k+1$ to show that

$$\sum_{k=1}^{n}[(k+1)^2-k^2]=n+2\sum_{k=1}^{n}k.$$

(c) Compare the results of parts (a) and (b) to derive summation formula (*iii*) of Theorem 5.3.2.

56. Show how the pattern illustrated in FIGURE 5.3.9 can be used to infer the summation formula (*iv*) of Theorem 5.3.2.

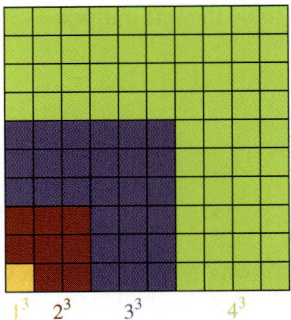

$1^3\qquad 2^3\qquad 3^3\qquad 4^3$

FIGURE 5.3.9 Array for Problem 56

57. Derive the formula for the area of the trapezoid given in FIGURE 5.3.10.

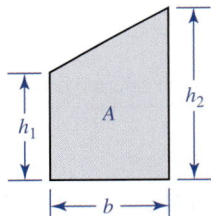

FIGURE 5.3.10 Trapezoid in Problem 57

58. In a supermarket, 136 cans are displayed in a triangular stack as shown in FIGURE 5.3.11. How many cans are there in the bottom row of the stack?

FIGURE 5.3.11 Stack of cans in Problem 58

59. Use (7) and the summation formula

$$\sum_{k=1}^{n} k^4 = \frac{n(n+1)(6n^3 + 9n^2 + n - 1)}{30}$$

to find the area under the graph of $f(x) = 16 - x^4$ on $[-2, 2]$.

60. Find the area under the graph of $y = \sqrt{x}$ on $[0, 1]$ by considering the area under the graph of $y = x^2$ on $[0, 1]$. Carry out your ideas.

61. Find the area under the graph of $y = \sqrt[3]{x}$ on $[0, 8]$ by considering the area under the graph of $y = x^3$ on $0 \le x \le 2$.

62. (a) Suppose $y = ax^2 + bx + c \ge 0$ on the interval $[0, x_0]$. Show that the area under the graph on $[0, x_0]$ is given by

$$A = a\frac{x_0^3}{3} + b\frac{x_0^2}{2} + cx_0.$$

(b) Use the result in part (a) to find the area under the graph of $y = 6x^2 + 2x + 1$ on the interval $[2, 5]$.

63. A summation formula for the sum of the n terms of a finite geometric sequence $a, ar, ar^2, \ldots, ar^{n-1}$ is given by

$$\sum_{k=1}^{n} ar^{k-1} = a\left(\frac{1 - r^n}{1 - r}\right).$$

Use this summation formula, (8) of this section, and L'Hôpital's Rule, to find the area under the graph of $y = e^x$ on $[0, 1]$.

64. A Bit of History Everyone knows in a beginning course in physics that the distance a body falls is proportional to the square of the elapsed time. **Galileo Galilei** (1564–1642) was the first to discover this fact. Galileo found that the distance a mass moves down an inclined plane in consecutive time intervals is proportional to a positive odd integer. Hence the total distance s that a mass moves in n seconds, with n a positive integer, is proportional to $1 + 3 + 5 + \cdots + 2n - 1$. Show that this is the same as saying that the total distance a mass moves down an inclined plane is proportional to the square of the elapsed time n.

5.4 The Definite Integral

■ Introduction In the previous section we saw that the area under the graph of a continuous nonnegative function f on an interval $[a, b]$ was defined as a limit of a sum. We see in this section that the same kind of limiting process leads to the notion of a **definite integral**.

Let $y = f(x)$ be a function defined on a closed interval $[a, b]$.
Consider the following four steps.

- Divide the interval $[a, b]$ into n subintervals $[x_{k-1}, x_k]$ of widths $\Delta x_k = x_k - x_{k-1}$, where

$$a = x_0 < x_1 < x_2 < \cdots < x_{n-1} < x_n = b. \tag{1}$$

The collection of numbers (1) is called a **partition** of the interval and is denoted by P.
- Let $\|P\|$ denote the largest number of the n subinterval widths $\Delta x_1, \Delta x_2, \ldots, \Delta x_n$. The number $\|P\|$ is called the **norm** of the partition P.
- Choose a number x_k^* in each subinterval $[x_{k-1}, x_k]$ as shown in FIGURE 5.4.1. The n numbers $x_1^*, x_2^*, x_3^*, \ldots, x_n^*$ are called **sample points** in these subintervals.
- Form the sum

$$\sum_{k=1}^{n} f(x_k^*)\Delta x_k. \tag{2}$$

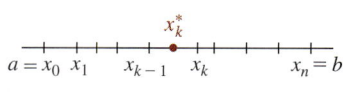

FIGURE 5.4.1 Sample point x_k^* in $[x_{k-1}, x_k]$

Sums of the kind given in (2) corresponding to various partitions of $[a, b]$ are known as **Riemann sums** and are named for the famous German mathematician, **Georg Friedrich Bernhard Riemann**.

Although the foregoing procedure looks very similar to the steps leading up to the definition of area under a graph given in Section 5.3, there are some important differences. Observe that a Riemann sum (2) does not require that f be either continuous or nonnegative on the interval $[a, b]$. Thus, (2) does not necessarily represent an approximation to the area under a graph. Keep in mind that "area under a graph" refers to *the area bounded between the graph of a continuous nonnegative function and the x-axis*. As shown in FIGURE 5.4.2, if $f(x) < 0$ for some x in $[a, b]$, a Riemann sum could contain terms $f(x_k^*)\Delta x_k$, where $f(x_{k1}^*) < 0$. In this case the products $f(x_k^*)\Delta x_k$ are numbers that are the negatives of the areas of rectangles drawn below the x-axis.

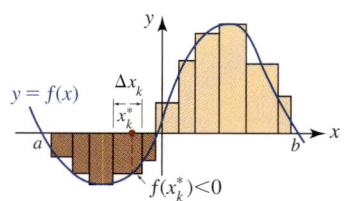

FIGURE 5.4.2 The function f is positive and negative on the interval $[a, b]$

EXAMPLE 1 A Riemann Sum

Compute the Riemann sum for $f(x) = x^2 - 4$ on $[-2, 3]$ with five subintervals determined by $x_0 = -2$, $x_1 = -\frac{1}{2}$, $x_2 = 0$, $x_3 = 1$, $x_4 = \frac{7}{4}$, $x_5 = 3$ and $x_1^* = -1$, $x_2^* = -\frac{1}{4}$, $x_3^* = \frac{1}{2}$, $x_4^* = \frac{3}{2}$, $x_5^* = \frac{5}{2}$. Find the norm of the partition.

Solution FIGURE 5.4.3 shows that the numbers x_k, $k = 0, 1, \ldots, 5$ determine five subintervals $\left[-2, -\frac{1}{2}\right], \left[-\frac{1}{2}, 0\right], [0, 1], \left[1, \frac{7}{4}\right]$, and $\left[\frac{7}{4}, 3\right]$ of the interval $[-2, 3]$ and a sample point x_k^* (in red) within each subinterval.

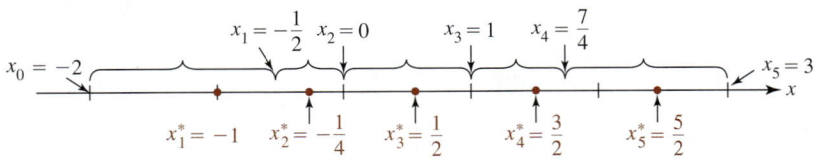

FIGURE 5.4.3 Five subintervals and sample points in Example 1

Now, evaluate the function f at each sample point and determine the width of each subinterval:

$$f(x_1^*) = f(-1) = -3, \qquad \Delta x_1 = x_1 - x_0 = -\frac{1}{2} - (-2) = \frac{3}{2}$$

$$f(x_2^*) = f\left(-\frac{1}{4}\right) = -\frac{63}{16}, \qquad \Delta x_2 = x_2 - x_1 = 0 - \left(-\frac{1}{2}\right) = \frac{1}{2}$$

$$f(x_3^*) = f\left(\frac{1}{2}\right) = -\frac{15}{4}, \qquad \Delta x_3 = x_3 - x_2 = 1 - 0 = 1$$

$$f(x_4^*) = f\left(\frac{3}{2}\right) = -\frac{7}{4}, \qquad \Delta x_4 = x_4 - x_3 = \frac{7}{4} - 1 = \frac{3}{4}$$

$$f(x_5^*) = f\left(\frac{5}{2}\right) = \frac{9}{4}, \qquad \Delta x_5 = x_5 - x_4 = 3 - \frac{7}{4} = \frac{5}{4}.$$

The **Riemann sum** for this partition and choice of sample points is then

$$f(x_1^*)\Delta x_1 + f(x_2^*)\Delta x_2 + f(x_3^*)\Delta x_3 + f(x_4^*)\Delta x_4 + f(x_5^*)\Delta x_5$$

$$= (-3)\left(\frac{3}{2}\right) + \left(-\frac{63}{16}\right)\left(\frac{1}{2}\right) + \left(-\frac{15}{4}\right)(1) + \left(-\frac{7}{4}\right)\left(\frac{3}{4}\right) + \left(\frac{9}{4}\right)\left(\frac{5}{4}\right) = -\frac{279}{32} \approx -8.72.$$

Inspection of the values of the five Δx_k shows that the norm of the partition is $\|P\| = \frac{3}{2}$. ∎

For a function f defined on an interval $[a, b]$, there are an infinite number of possible Riemann sums for a given partition P of the interval, since the numbers x_k^* can be chosen arbitrarily in each subinterval $[x_{k-1}, x_k]$.

EXAMPLE 2 Another Riemann Sum

Compute the Riemann sum for the function in Example 1 if the partition of $[-2, 3]$ is the same but the sample points are $x_1^* = -\frac{3}{2}$, $x_2^* = -\frac{1}{8}$, $x_3^* = \frac{3}{4}$, $x_4^* = \frac{3}{2}$, and $x_5^* = 2.1$.

Solution We need only compute f at the new sample points since the numbers Δx_k are the same as before:

$$f(x_1^*) = f\left(-\frac{3}{2}\right) = -\frac{7}{4}$$

$$f(x_2^*) = f\left(-\frac{1}{8}\right) = -\frac{255}{64}$$

$$f(x_3^*) = f\left(\frac{3}{4}\right) = -\frac{55}{16}$$

$$f(x_4^*) = f\left(\frac{3}{2}\right) = -\frac{7}{4}$$

$$f(x_5^*) = f(2.1) = 0.41.$$

The Riemann sum is now

$$f(x_1^*)\Delta x_1 + f(x_2^*)\Delta x_2 + f(x_3^*)\Delta x_3 + f(x_4^*)\Delta x_4 + f(x_5^*)\Delta x_5$$

$$= \left(-\frac{7}{4}\right)\left(\frac{3}{2}\right) + \left(-\frac{255}{64}\right)\left(\frac{1}{2}\right) + \left(-\frac{55}{16}\right)(1) + \left(-\frac{7}{4}\right)\left(\frac{3}{4}\right) + (0.41)\left(\frac{5}{4}\right) \approx -8.85. \quad \blacksquare$$

We are interested in a special kind of limit of (2). If the Riemann sums $\sum_{k=1}^{n} f(x_k^*)\Delta x_k$ are close to a number L for *every* partition P of $[a, b]$ for which the norm $\|P\|$ is close to zero, we then write

$$\lim_{\|P\|\to 0} \sum_{k=1}^{n} f(x_k^*)\Delta x_k = L \tag{3}$$

and say that L is the **definite integral** of f on the interval $[a, b]$. In the following definition we introduce a new symbol for the number L.

Definition 5.4.1 The Definite Integral

Let f be a function defined on a closed interval $[a, b]$. Then the **definite integral of f from a to b**, denoted by $\int_a^b f(x)\,dx$, is defined to be

$$\int_a^b f(x)\,dx = \lim_{\|P\|\to 0} \sum_{k=1}^{n} f(x_k^*)\Delta x_k. \tag{4}$$

If the limit in (4) exists, the function f is said to be **integrable** on the interval. The numbers a and b in the preceding definition are called the **lower** and **upper limits of integration**, respectively. The function f is called the **integrand**. The integral symbol $\int$, as used by Leibniz, is an elongated S for the word *sum*. Also, note that $\|P\|\to 0$ always implies that the number of subintervals n becomes infinite ($n\to\infty$). However, as shown in FIGURE 5.4.4, the fact that $n\to\infty$ does not necessarily imply $\|P\|\to 0$.

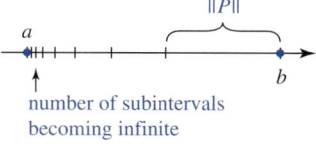

FIGURE 5.4.4 Infinite number of subintervals does not imply $\|P\|\to 0$.

■ **Integrability** In the next two theorems we state conditions that are sufficient for a function f to be integrable on an interval $[a, b]$. The proofs of these theorems will not be given.

Theorem 5.4.1 Continuity Implies Integrability

If f is continuous on the closed interval $[a, b]$, then $\int_a^b f(x)\,dx$ exists; that is, f is integrable on the interval.

There are functions that are defined for every value of x in $[a, b]$ for which the limit in (4) does not exist. Also, if the function f is not defined for all values of x in the interval, the definite integral *may* not exist; for example, later on we will see why an integral such as $\int_{-3}^{2}(1/x)\,dx$ does not exist. Notice that $y = 1/x$ is discontinuous at $x = 0$ and is unbounded on the interval. However, one should not conclude from this one example that when a function f has a discontinuity in $[a, b]$, $\int_a^b f(x)\,dx$ necessarily does not exist. Continuity of a function f on $[a, b]$ is *sufficient* but *not necessary* to guarantee the existence of $\int_a^b f(x)\,dx$. The set of functions continuous on $[a, b]$ is a subset of the set of functions that are integrable on the interval.

The next theorem gives another sufficient condition for integrability on $[a, b]$.

Theorem 5.4.2 Sufficient Conditions for Integrability

If a function f is bounded on the closed interval $[a, b]$, that is, if there exists a positive constant B such that $-B \leq f(x) \leq B$ for all x in the interval, and has a finite number of discontinuities in $[a, b]$, then f is integrable on the interval.

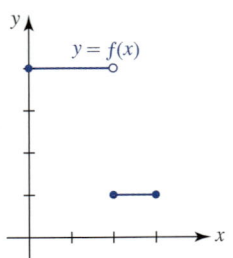

FIGURE 5.4.5 Definite integral of f on $[0, 3]$ exists

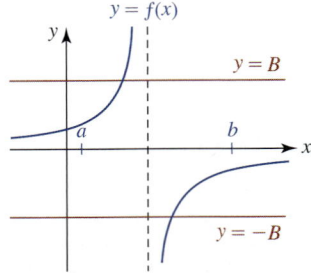

FIGURE 5.4.6 The function f is not bounded on $[a, b]$

When a function f is bounded, its complete graph must lie between two horizontal lines, $y = B$ and $y = -B$. In other words, $|f(x)| \le B$ for all x in $[a, b]$. The function

$$f(x) = \begin{cases} 4, & 0 \le x < 2 \\ 1, & 2 \le x \le 3 \end{cases}$$

shown in FIGURE 5.4.5 is discontinuous at $x = 2$ but is bounded on $[0, 3]$, since $|f(x)| \le 4$ for all x in $[0, 3]$. (For that matter, $1 \le f(x) \le 4$ for all x in $[0, 3]$ shows that f is bounded on the interval.) It follows from Theorem 5.4.2 that $\int_0^3 f(x)\, dx$ exists. FIGURE 5.4.6 shows the graph of a function f that is unbounded on an interval $[a, b]$. Regardless of how large the number B is chosen, the graph of f cannot be confined to the region between the horizontal lines, $y = B$ and $y = -B$.

■ **Regular Partition**   If it is known that a definite integral exists (say, the integrand f is continuous on $[a, b]$), then:

- *The limit in* (4) *exists for every possible way of partitioning* $[a, b]$ *and for every way of choosing* x_k^* *in the subintervals* $[x_{k-1}, x_k]$.

In particular, by choosing the subintervals of equal width and the sample points to be the right endpoints of the subintervals $[x_{k-1}, x_k]$, that is,

$$\Delta x = \frac{b - a}{n} \qquad \text{and} \qquad x_k^* = a + k\frac{b - a}{n}, \quad k = 1, 2, \ldots, n,$$

we can write (4) in the alternative manner

$$\int_a^b f(x)\, dx = \lim_{n \to \infty} \sum_{k=1}^n f\left(a + k\frac{b - a}{n}\right)\frac{b - a}{n}. \tag{5}$$

Recall from Section 5.3 that a partition P of $[a, b]$ in which the subintervals have the same width is called a **regular partition**.

■ **Area**   You might conclude that the formulations of $\int_a^b f(x)\, dx$ given in (4) and (5) are exactly the same as (6) and (7) of Section 5.3 for the general case of finding the area under the curve $y = f(x)$ on $[a, b]$. In a way this is correct; however, Definition 5.4.1 is a more general concept, since, as noted before, we are not requiring that f be continuous on $[a, b]$ or that $f(x) \ge 0$ on the interval. Thus, a *definite integral need not be area*. What then is a definite integral? For now, accept the fact that a definite integral is simply a real number. Contrast this with the indefinite integral, which is a function (or a family of functions). Is the area under the graph of a continuous nonnegative function a definite integral? The answer is *yes*.

Theorem 5.4.3   Area As a Definite Integral

If f is continuous on the closed interval $[a, b]$ and $f(x) \ge 0$ for all x in the interval, then the **area A under the graph** on $[a, b]$ is

$$A = \int_a^b f(x)\, dx. \tag{6}$$

EXAMPLE 3   Area as a Definite Integral

Consider the definite integral $\int_{-1}^1 \sqrt{1 - x^2}\, dx$. The integrand is continuous and nonnegative and so the definite integral represents the area under the graph of $f(x) = \sqrt{1 - x^2}$ on the interval $[-1, 1]$. Because the graph of the function f is the upper semicircle of $x^2 + y^2 = 1$, the area under the graph is the shaded region in FIGURE 5.4.7. From geometry we know that the area of a circle of radius r is πr^2, and so with $r = 1$ the area of the semicircle, and hence the value of the definite integral, is

$$\int_{-1}^1 \sqrt{1 - x^2}\, dx = \frac{1}{2}\pi(1)^2 = \frac{1}{2}\pi. \qquad \blacksquare$$

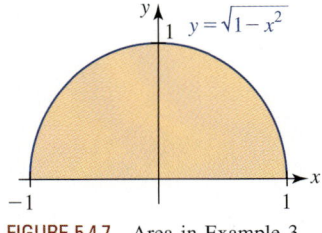

FIGURE 5.4.7   Area in Example 3

We shall return to the question of finding areas by means of the definite integral in Section 6.2.

EXAMPLE 4   Definite Integral Using (5)

Evaluate $\displaystyle\int_{-2}^{1} x^3\, dx$.

Solution   Since $f(x) = x^3$ is continuous on $[-2, 1]$, we know from Theorem 5.4.1 that the definite integral exists. We use a regular partition and the result given in (5). Choosing

$$\Delta x = \frac{1 - (-2)}{n} = \frac{3}{n} \qquad \text{and} \qquad x_k^* = -2 + k \cdot \frac{3}{n}$$

we have

$$f\left(-2 + \frac{3k}{n}\right) = \left(-2 + \frac{3k}{n}\right)^3 = -8 + 36\left(\frac{k}{n}\right) - 54\left(\frac{k^2}{n^2}\right) + 27\left(\frac{k^3}{n^3}\right).$$

It then follows from (5) and summation formulas (*i*), (*ii*), (*iii*), and (*iv*) of Theorem 5.3.2 that

$$\int_{-2}^{1} x^3\, dx = \lim_{n\to\infty} \sum_{k=1}^{n} f\left(-2 + \frac{3k}{n}\right)\frac{3}{n}$$

$$= \lim_{n\to\infty} \frac{3}{n} \sum_{k=1}^{n}\left[-8 + 36\left(\frac{k}{n}\right) - 54\left(\frac{k^2}{n^2}\right) + 27\left(\frac{k^3}{n^3}\right)\right]$$

$$= \lim_{n\to\infty} \frac{3}{n}\left[-8n + \frac{36}{n}\cdot\frac{n(n+1)}{2} - \frac{54}{n^2}\cdot\frac{n(n+1)(2n+1)}{6} + \frac{27}{n^3}\cdot\frac{n^2(n+1)^2}{4}\right]$$

$$= \lim_{n\to\infty}\left[-24 + 54\left(1 + \frac{1}{n}\right) - 27\left(1 + \frac{1}{n}\right)\left(2 + \frac{1}{n}\right) + \frac{81}{4}\left(1 + \frac{1}{n}\right)\left(1 + \frac{1}{n}\right)\right]$$

$$= -24 + 54 - 27(2) + \frac{81}{4} = -\frac{15}{4}.$$

FIGURE 5.4.8 shows that we are not considering area under the graph on $[-2, 1]$. ∎

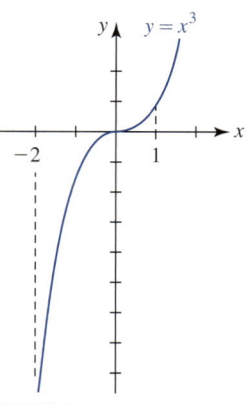

FIGURE 5.4.8 Graph of function in Example 4

EXAMPLE 5   Definite Integral Using (5)

The values of the Riemann sums in Examples 1 and 2 are approximations to the value of the definite integral $\int_{-2}^{3}(x^2 - 4)\, dx$. It is left as an exercise to show that (5) gives

$$\int_{-2}^{3}(x^2 - 4)\, dx = -\frac{25}{3} \approx -8.33.$$

See Problem 16 in Exercises 5.4. ∎

▍ **Properties of the Definite Integral**   We examine next some of the important properties of the definite integral defined in (4).

The following two definitions are useful when working with definite integrals.

Definition 5.4.2   Limits of Integration

(*i*) **Equality of Limits**   If a is in the domain of f, then

$$\int_{a}^{a} f(x)\, dx = 0. \tag{7}$$

(*ii*) **Reversing Limits**   If f is integrable on $[a, b]$, then

$$\int_{b}^{a} f(x)\, dx = -\int_{a}^{b} f(x)\, dx. \tag{8}$$

Definition 5.4.2(*i*) can be motivated by thinking that the area under the graph of f and above a single point a on the x-axis is zero.

In the definition of $\int_{a}^{b} f(x)\, dx$ it was assumed that $a < b$, and so the usual "direction" of definite integration is left to right. Part (*ii*) of Definition 5.4.2 states that reversing this direction, that is, interchanging the limits of integration, results in the negative of the integral.

EXAMPLE 6 Definition 5.4.2

By part (*i*) of Definition 5.4.2,

limits of integration → $\int_1^1 (x^3 + 3x)\,dx = 0.$ ← are the same

∎

EXAMPLE 7 Example 4 Revisited

In Example 4 we saw that $\int_{-2}^1 x^3\,dx = -\frac{15}{4}$. It follows from part (*ii*) of Definition 5.4.2 that

$$\int_1^{-2} x^3\,dx = -\int_{-2}^1 x^3\,dx = -\left(-\frac{15}{4}\right) = \frac{15}{4}.$$

∎

In the next theorem we list some of the basic properties of the definite integral. These properties are analogous to the properties of the sigma notation given in Theorem 5.3.1 as well as the properties of the indefinite integral, or antiderivative, discussed in Section 5.1.

Theorem 5.4.4 Properties of the Definite Integral

If *f* and *g* are integrable functions on the closed interval $[a, b]$, then

(*i*) $\displaystyle\int_a^b kf(x)\,dx = k\int_a^b f(x)\,dx$, where *k* is any constant

(*ii*) $\displaystyle\int_a^b [f(x) \pm g(x)]\,dx = \int_a^b f(x)\,dx + \int_a^b g(x)\,dx.$

Theorem 5.4.4(*ii*) extends to any finite sum of integrable functions on the interval $[a, b]$:

$$\int_a^b [f_1(x) + f_2(x) + \cdots + f_n(x)]\,dx = \int_a^b f_1(x)\,dx + \int_a^b f_2(x)\,dx + \cdots + \int_a^b f_n(x)\,dx.$$

The independent variable *x* in a definite integral is called a **dummy variable** of integration. The value of the integral does not depend on the symbol used. In other words,

$$\int_a^b f(x)\,dx = \int_a^b f(r)\,dr = \int_a^b f(s)\,ds = \int_a^b f(t)\,dt \qquad (9)$$

and so on.

EXAMPLE 8 Example 4 Revisited

From (9), it does not matter what symbol is used as the variable of integration:

$$\int_{-2}^1 x^3\,dx = \int_{-2}^1 r^3\,dr = \int_{-2}^1 s^3\,ds = \int_{-2}^1 t^3\,dt = -\frac{15}{4}.$$

∎

Theorem 5.4.5 Additive Interval Property

If *f* is an integrable function on a closed interval containing the numbers *a*, *b*, and *c*, then

$$\int_a^b f(x)\,dx = \int_a^c f(x)\,dx + \int_c^b f(x)\,dx. \qquad (10)$$

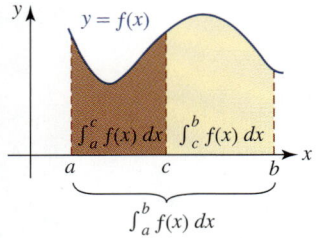

FIGURE 5.4.9 Areas are additive

It is easy to interpret the additive interval property given in Theorem 5.4.5 in the special case when *f* is continuous on $[a, b]$ and $f(x) \geq 0$ for all *x* in the interval. As seen in FIGURE 5.4.9, the area under the graph of *f* on $[a, c]$ plus the area under the graph on the adjacent interval $[c, b]$ is the same as the area under the graph on the entire interval $[a, b]$.

Note: The conclusion of Theorem 5.4.5 holds when *a*, *b*, and *c* are *any* three numbers in a closed interval. In other words, it is not necessary to have the order $a < c < b$ as shown in Figure 5.4.9. Moreover, the result in (10) extends to any finite number of numbers $a, b, c_1, c_2, \ldots, c_n$ in the interval. For example, for a closed interval containing the numbers $a, b, c_1,$ and c_2,

$$\int_a^b f(x)\,dx = \int_a^{c_1} f(x)\,dx + \int_{c_1}^{c_2} f(x)\,dx + \int_{c_2}^b f(x)\,dx.$$

For a given partition P of an interval $[a, b]$, it stands to reason that

$$\lim_{\|P\| \to 0} \sum_{k=1}^{n} \Delta x_k = b - a, \tag{11}$$

in other words, the limit $\lim_{\|P\| \to 0} \sum_{k=1}^{n} \Delta x_k$ is simply the width of the interval. As a consequence of (11), we have the following theorem.

Theorem 5.4.6 Definite Integral of a Constant

For any constant k,

$$\int_a^b k \, dx = k \int_a^b dx = k(b - a).$$

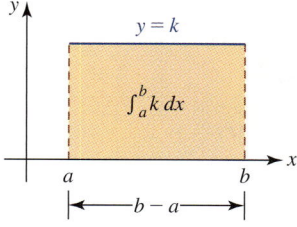

FIGURE 5.4.10 If $k > 0$, area under the graph is $k(b - a)$

If $k > 0$, then Theorem 5.4.6 implies that $\int_a^b k \, dx$ is simply the area of a rectangle of width $b - a$ and height k. See FIGURE 5.4.10.

EXAMPLE 9 Definite Integral of a Constant

From Theorem 5.4.6,

$$\int_2^8 5 \, dx = 5 \int_2^8 dx = 5(8 - 2) = 30.$$ ∎

EXAMPLE 10 Using Examples 4 and 9

Evaluate $\int_{-2}^{1} (x^3 + 5) \, dx$.

Solution From Theorem 5.4.4(*ii*) we can write the given integral as two integrals:

$$\int_{-2}^{1} (x^3 + 5) \, dx = \int_{-2}^{1} x^3 \, dx + \int_{-2}^{1} 5 \, dx.$$

Now, from Example 4 we know that $\int_{-2}^{1} x^3 \, dx = -\frac{15}{4}$ and with the help of Theorem 5.4.6 we see that $\int_{-2}^{1} 5 \, dx = 5[1 - (-2)] = 15$. Therefore,

$$\int_{-2}^{1} (x^3 + 5) \, dx = \left(-\frac{15}{4}\right) + 15 = \frac{45}{4}.$$ ∎

Finally, the following results are not surprising if you interpret the integrals as area.

Theorem 5.4.7 Comparison Properties

Let f and g be integrable functions on the closed interval $[a, b]$.
(*i*) If $f(x) \geq g(x)$ for all x in the interval, then

$$\int_a^b f(x) \, dx \geq \int_a^b g(x) \, dx.$$

(*ii*) If $m \leq f(x) \leq M$ for all x in the interval, then

$$m(b - a) \leq \int_a^b f(x) \, dx \leq M(b - a).$$

Properties (*i*) and (*ii*) of Theorem 5.4.7 are easily understood in terms of area. For (*i*) if we assume $f(x) \geq g(x) \geq 0$ for all x in $[a, b]$, then on the interval the area A_1 under the graph of f is greater than or equal to the area A_2 under the graph of g. Similarly, for (*ii*) if we assume that f is continuous and positive on the interval $[a, b]$, then by the Extreme Value Theorem,

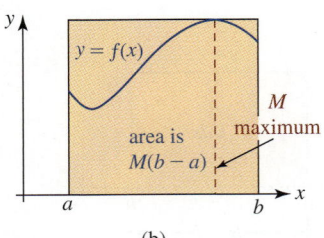

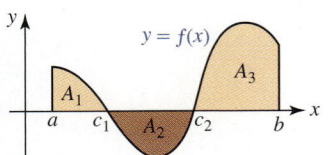

FIGURE 5.4.11 Motivation for part (*ii*) of Theorem 5.4.7

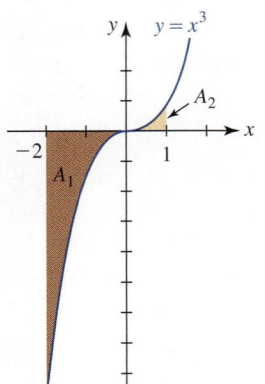

FIGURE 5.4.12 Definite integral of f on $[a, b]$ gives net signed area

f has an absolute minimum $m > 0$ and an absolute maximum $M > 0$ on the interval. The area under the graph $\int_a^b f(x)\,dx$ on the interval is then greater than or equal to the area $m(b - a)$ of the smaller rectangle shown in FIGURE 5.4.11(a) and less than or equal to the area $M(b - a)$ of the larger rectangle shown in Figure 5.4.11(b).

If we let $g(x) = 0$ in (*i*) of Theorem 5.4.7 and use the fact that $\int_a^b 0\,dx = 0$, we conclude:

- If $f(x) \geq 0$ on $[a, b]$, then $\int_a^b f(x)\,dx \geq 0$.   (12)

In like manner by choosing $f(x) = 0$ in (*i*), it follows that:

- If $g(x) \leq 0$ on $[a, b]$, then $\int_a^b g(x)\,dx \leq 0$.   (13)

▌ Net Signed Area Because the function f in FIGURE 5.4.12 takes on both positive and negative values on $[a, b]$ the definite integral $\int_a^b f(x)\,dx$ does not represent area under the graph of f on the interval. By Theorem 5.4.5, the additive interval property,

$$\int_a^b f(x)\,dx = \int_a^{c_1} f(x)\,dx + \int_{c_1}^{c_2} f(x)\,dx + \int_{c_2}^b f(x)\,dx. \qquad (14)$$

Because $f(x) \geq 0$ on $[a, c_1]$ and $[c_2, b]$ we have

$$\int_a^{c_1} f(x)\,dx = A_1 \quad \text{and} \quad \int_{c_2}^b f(x)\,dx = A_3,$$

where A_1 and A_3 denote the areas under the graph of f on the intervals $[a, c_1]$ and $[c_2, b]$, respectively. But since $f(x) \leq 0$ on $[c_1, c_2]$ we have in view of (13), $\int_{c_1}^{c_2} f(x)\,dx \leq 0$ and so $\int_{c_1}^{c_2} f(x)\,dx$ does not represent area. However, the value of $\int_{c_1}^{c_2} f(x)\,dx$ is the negative of the actual area A_2 bounded between the graph of f and the x-axis on the interval $[c_1, c_2]$. That is, $\int_{c_1}^{c_2} f(x)\,dx = -A_2$. Hence (14) is

$$\int_a^b f(x)\,dx = A_1 + (-A_2) + A_3 = A_1 - A_2 + A_3.$$

We see that the definite integral gives the **net signed area** between the graph of f and the x-axis on the interval $[a, b]$.

EXAMPLE 11   Net Signed Area

The result $\int_{-2}^1 x^3\,dx = -\frac{15}{4}$ obtained in Example 4 can be interpreted as the net signed area between the graph of $f(x) = x^3$ and the x-axis on $[-2, 1]$. Although the observation that

$$\int_{-2}^1 x^3\,dx = \int_{-2}^0 x^3\,dx + \int_0^1 x^3\,dx = -A_1 + A_2 = -\frac{15}{4}$$

does not give us the values of A_1 and A_2, the negative value is consistent with FIGURE 5.4.13 where it is obvious that the area A_1 is larger than A_2.   ■

▌ The Theory Let f be a function defined on $[a, b]$ and let L denote a real number. The intuitive concept that Riemann sums are close to L whenever the norm $\|P\|$ of a partition P is close to zero can be expressed in a precise manner using the ε-δ symbols introduced in Section 2.6. To say that f is integrable on $[a, b]$, we mean that for every real number $\varepsilon > 0$ there exists a real number $\delta > 0$ such that

$$\left| \sum_{k=1}^n f(x_k^*)\Delta x_k - L \right| < \varepsilon, \qquad (15)$$

whenever P is a partition of $[a, b]$ for which $\|P\| < \delta$ and the x_k^* are numbers in the subintervals $[x_{k-1}, x_k]$, $k = 1, 2, \ldots, n$. In other words,

$$\lim_{\|P\| \to 0} \sum_{k=1}^n f(x_k^*)\Delta x_k$$

exists and is equal to the number L.

FIGURE 5.4.13 Net signed area in Example 11

■ **Postscript—A Bit of History**   **Georg Friedrich Bernhard Riemann** (1826–1866) born in Hanover, Germany, in 1826, was the son of a Lutheran minister. Although a devout Christian,

Riemann was disinclined to follow his father's vocation and abandoned the study of theology at the University of Göttingen in favor of a course of studies in which his genius was obvious: mathematics. It is likely that the concept of Riemann sums grew out of a course on the definite integral that he had taken at the university; this concept reflects his attempt to give a precise mathematical meaning to the definite integral of Newton and Leibniz. After submitting his doctoral dissertation on the foundations of functions of a complex variable to the examining committee at the University of Göttingen, Karl Friedrich Gauss, the "prince of mathematicians," paid Riemann a very rare compliment: "The dissertation offers convincing evidence ... of a creative, active, truly mathematical mind ... of glorious fertile originality."

Riemann

Riemann, like so many other promising scholars of that time, possessed a fragile constitution. He died at age 39 of pleurisy. His original contributions to differential geometry, topology, non-Euclidean geometry, and his bold investigations into the nature of space, electricity, and magnetism, foreshadowed the work of Einstein in the next century.

$\int_a^b$ **NOTES FROM THE CLASSROOM**

The procedure outlined in (5) has limited utility as a practical means of computing a definite integral. In the next section we will introduce a theorem that enables us to find the number $\int_a^b f(x)\,dx$ in a much easier manner. This important theorem is the bridge between differential and integral calculus.

Exercises 5.4   Answers to selected odd-numbered problems begin on page ANS-19.

≡ Fundamentals

In Problems 1–6, compute the Riemann sum $\sum_{k=1}^{n} f(x_k^*)\Delta x_k$ for the given partition. Specify $\|P\|$.

1. $f(x) = 3x + 1$, $[0, 3]$, four subintervals; $x_0 = 0, x_1 = 1,$
$x_2 = \frac{5}{3}, x_3 = \frac{7}{3}, x_4 = 3; x_1^* = \frac{1}{2}, x_2^* = \frac{4}{3}, x_3^* = 2, x_4^* = \frac{8}{3}$

2. $f(x) = x - 4$, $[-2, 5]$, five subintervals; $x_0 = -2, x_1 = -1,$
$x_2 = -\frac{1}{2}, \ x_3 = \frac{1}{2}, \ x_4 = 3, \ x_5 = 5; \ x_1^* = -\frac{3}{2}, \ x_2^* = -\frac{1}{2},$
$x_3^* = 0, x_4^* = 2, x_5^* = 4$

3. $f(x) = x^2$, $[-1, 1]$, four subintervals: $x_0 = -1, x_1 = -\frac{1}{4},$
$x_2 = \frac{1}{4}, \ x_3 = \frac{3}{4}, \ x_4 = 1; \ x_1^* = -\frac{3}{4}, \ x_2^* = 0, \ x_3^* = \frac{1}{2},$
$x_4^* = \frac{7}{8}$

4. $f(x) = x^2 + 1$, $[1, 3]$, three subintervals; $x_0 = 1, x_1 = \frac{3}{2},$
$x_2 = \frac{5}{2}, x_3 = 3; x_1^* = \frac{5}{4}, x_2^* = \frac{7}{4}, x_3^* = 3$

5. $f(x) = \sin x$, $[0, 2\pi]$, three subintervals; $x_0 = 0, x_1 = \pi,$
$x_2 = 3\pi/2, x_3 = 2\pi; x_1^* = \pi/2, x_2^* = 7\pi/6, x_3^* = 7\pi/4$

6. $f(x) = \cos x$, $[-\pi/2, \pi/2]$, four subintervals; $x_0 = -\pi/2,$
$x_1 = -\pi/4, \ x_2 = 0, \ x_3 = \pi/3, \ x_4 = \pi/2; \ x_1^* = -\pi/3,$
$x_2^* = -\pi/6, x_3^* = \pi/4, x_4^* = \pi/3$

7. Given $f(x) = x - 2$ on $[0, 5]$, compute the Riemann sum using a partition with five subintervals of equal length. Let x_k^*, $k = 1, 2, \ldots, 5$, be the right endpoint of each subinterval.

8. Given $f(x) = x^2 - x + 1$ on $[0, 1]$, compute the Riemann sum using a partition with three subintervals of equal length. Let x_k^*, $k = 1, 2, 3$, be the left endpoint of each subinterval.

In Problems 9 and 10, let P be a partition of the indicated interval and x_k^* a number in the kth subinterval. Write the given sum as a definite integral on the indicated interval.

9. $\displaystyle\lim_{\|P\|\to 0} \sum_{k=1}^{n} \sqrt{9 + (x_k^*)^2}\,\Delta x_k;$ $[-2, 4]$

10. $\displaystyle\lim_{\|P\|\to 0} \sum_{k=1}^{n} (\tan x_k^*)\Delta x_k;$ $[0, \pi/4]$

In Problems 11 and 12, let P be a regular partition of the indicated interval and x_k^* the right endpoint of each subinterval. Write the given sum as a definite integral.

11. $\displaystyle\lim_{n\to\infty} \sum_{k=1}^{n} \left(1 + \frac{2k}{n}\right)\frac{2}{n};$ $[0, 2]$

12. $\displaystyle\lim_{n\to\infty} \sum_{k=1}^{n} \left(1 + \frac{3k}{n}\right)^3 \frac{3}{n};$ $[1, 4]$

In Problems 13–18, use (5) and the summation formulas in Theorem 5.3.2 to evaluate the given definite integral.

13. $\displaystyle\int_{-3}^{1} x\,dx$

14. $\displaystyle\int_{0}^{3} x\,dx$

15. $\displaystyle\int_{1}^{2} (x^2 - x)\,dx$

16. $\displaystyle\int_{-2}^{3} (x^2 - 4)\,dx$

17. $\int_0^1 (x^3 - 1)\, dx$

18. $\int_0^2 (3 - x^3)\, dx$

In Problems 19 and 20, proceed as in Problems 13–18 to obtain the given result.

19. $\int_a^b x\, dx = \frac{1}{2}(b^2 - a^2)$

20. $\int_a^b x^2\, dx = \frac{1}{3}(b^3 - a^3)$

21. Use Problem 19 to evaluate $\int_{-1}^3 x\, dx$.

22. Use Problem 20 to evaluate $\int_{-1}^3 x^2\, dx$.

In Problems 23 and 24, use Theorem 5.4.6 to evaluate the given definite integral.

23. $\int_3^6 4\, dx$

24. $\int_{-2}^5 (-2)\, dx$

In Problems 25–38, use Definition 5.4.2 and Theorems 5.4.4, 5.4.5, and 5.4.6 to evaluate the given definite integral. Use the results obtained in Problems 21 and 22 where appropriate.

25. $\int_4^{-2} \frac{1}{2}\, dx$

26. $\int_5^5 10x^4\, dx$

27. $-\int_3^{-1} 10x\, dx$

28. $\int_{-1}^3 (3x + 1)\, dx$

29. $\int_3^{-1} t^2\, dt$

30. $\int_{-1}^3 (3x^2 - 5)\, dx$

31. $\int_{-1}^3 (-3x^2 + 4x - 5)\, dx$

32. $\int_{-1}^3 6x(x - 1)\, dx$

33. $\int_{-1}^0 x^2\, dx + \int_0^3 x^2\, dx$

34. $\int_{-1}^{1.2} 2t\, dt - \int_3^{1.2} 2t\, dt$

35. $\int_0^4 x\, dx + \int_0^4 (9 - x)\, dx$

36. $\int_{-1}^0 t^2\, dt + \int_0^2 x^2\, dx + \int_2^3 u^2\, du$

37. $\int_0^3 x^3\, dx + \int_3^0 t^3\, dt$

38. $\int_{-1}^{-1} 5x\, dx - \int_3^{-1} (x - 4)\, dx$

In Problems 39–42, evaluate the definite integral using the given information.

39. $\int_2^5 f(x)\, dx$ if $\int_0^2 f(x)\, dx = 6$ and $\int_0^5 f(x)\, dx = 8.5$

40. $\int_1^3 f(x)\, dx$ if $\int_1^4 f(x)\, dx = 2.4$ and $\int_3^4 f(x)\, dx = -1.7$

41. $\int_{-1}^2 [2f(x) + g(x)]\, dx$ if

$\int_{-1}^2 f(x)\, dx = 3.4$ and $\int_{-1}^2 3g(x)\, dx = 12.6$

42. $\int_{-2}^2 g(x)\, dx$ if

$\int_2^{-2} f(x)\, dx = 14$ and $\int_{-2}^2 [f(x) - 5g(x)]\, dx = 24$

In Problems 43 and 44, evaluate the definite integrals

(a) $\int_a^b f(x)\, dx$ **(b)** $\int_b^c f(x)\, dx$ **(c)** $\int_c^d f(x)\, dx$

(d) $\int_a^c f(x)\, dx$ **(e)** $\int_b^d f(x)\, dx$ **(f)** $\int_a^d f(x)\, dx$

using the information in the given figure.

43.

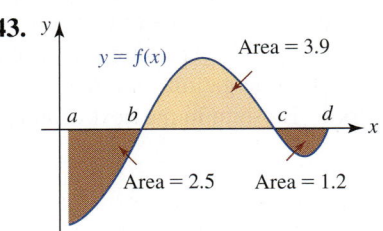

FIGURE 5.4.14 Graph for Problem 43

44.

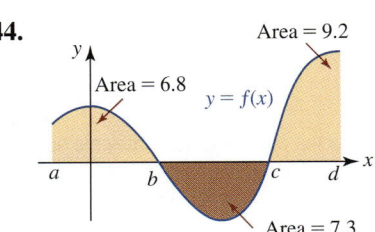

FIGURE 5.4.15 Graph for Problem 44

In Problems 45–48, the given integral represents the area under a graph on an interval. Sketch this region.

45. $\int_{-1}^1 (2x + 3)\, dx$

46. $\int_0^4 (-x^2 + 4x)\, dx$

47. $\int_{2\pi}^{3\pi} \sin x\, dx$

48. $\int_{-2}^0 \sqrt{x + 2}\, dx$

In Problems 49–52, the given integral represents the area under a graph on the interval. Use appropriate formulas from geometry to find the area.

49. $\int_{-2}^4 (x + 2)\, dx$

50. $\int_0^3 |x - 1|\, dx$

51. $\int_0^1 \sqrt{1 - x^2}\, dx$

52. $\int_{-3}^3 (2 + \sqrt{9 - x^2})\, dx$

In Problems 53–56, the given integral represents the net signed area between a graph and the x-axis on an interval. Sketch this region.

53. $\int_0^5 (-2x + 6)\, dx$

54. $\int_{-1}^2 (1 - x^2)\, dx$

55. $\int_{-1/2}^3 \frac{4x}{x + 1}\, dx$

56. $\int_0^{5\pi/2} \cos x\, dx$

In Problems 57–60, the given integral represents the net signed area between a graph and the x-axis on an interval. Use appropriate formulas from geometry to find the net signed area.

57. $\displaystyle\int_{-1}^{4} 2x \, dx$

58. $\displaystyle\int_{0}^{8} \left(\frac{1}{2}x - 2\right) dx$

59. $\displaystyle\int_{-1}^{1} \left(x - \sqrt{1 - x^2}\right) dx$

60. $\displaystyle\int_{-1}^{2} (1 - |x|) \, dx$

In Problems 61–64, the function f is defined to be

$$f(x) = \begin{cases} x, & x \le 3 \\ 3, & x > 3. \end{cases}$$

Use appropriate formulas from geometry to evaluate the given definite integral.

61. $\displaystyle\int_{-2}^{0} f(x) \, dx$

62. $\displaystyle\int_{-1}^{3} f(x) \, dx$

63. $\displaystyle\int_{-4}^{5} f(x) \, dx$

64. $\displaystyle\int_{0}^{10} f(x) \, dx$

In Problems 65–68, use Theorem 5.4.7 to establish the given inequality.

65. $\displaystyle\int_{-1}^{0} e^x \, dx \le \int_{-1}^{0} e^{-x} \, dx$

66. $\displaystyle\int_{0}^{\pi/4} (\cos x - \sin x) \, dx \ge 0$

67. $1 \le \displaystyle\int_{0}^{1} (x^3 + 1)^{1/2} \, dx \le 1.42$

68. $-2 \le \displaystyle\int_{0}^{2} (x^2 - 2x) \, dx \le 0$

In Problems 69 and 70, compare the given two integrals by means of an inequality symbol $\le$ or $\ge$.

69. $\displaystyle\int_{0}^{1} x^2 \, dx, \quad \int_{0}^{1} x^3 \, dx$

70. $\displaystyle\int_{0}^{1} \sqrt{4 + x^2} \, dx, \quad \int_{0}^{1} \sqrt{4 + x} \, dx$

≡ Think About It

71. If f is integrable on the interval $[a, b]$, then so is f^2. Explain why $\int_a^b f^2(x) \, dx \ge 0$.

72. Consider the function defined for all x in the interval $[-1, 1]$:

$$f(x) = \begin{cases} 0, & x \text{ rational} \\ 1, & x \text{ irrational.} \end{cases}$$

Show that f is not integrable on $[-1, 1]$, that is, $\int_{-1}^{1} f(x) \, dx$ does not exist. [*Hint*: The result in (11) may be useful.]

73. Evaluate the definite integral $\int_0^1 \sqrt{x} \, dx$ by using a partition of $[0, 1]$ in which the subintervals $[x_{k-1}, x_k]$ are defined by $[(k - 1)^2/n^2, k^2/n^2]$ and choosing x_k^* to be the right endpoint of each subinterval.

74. Evaluate the definite integral $\int_0^{\pi/2} \cos x \, dx$ by using a regular partition of $[0, \pi/2]$ and choosing x_k^* to be the midpoint of each subinterval $[x_{k-1}, x_k]$. Use the known results

$(i)\ \cos\theta + \cos 3\theta + \cdots + \cos(2n - 1)\theta = \dfrac{\sin 2n\theta}{2\sin\theta}$

$(ii)\ \displaystyle\lim_{n\to\infty} \frac{1}{n\sin(\pi/4n)} = \frac{4}{\pi}.$

5.5 Fundamental Theorem of Calculus

▌ **Introduction**   At the end of Section 5.4, we indicated that there is an easier way of evaluating a definite integral than by computing a limit of a sum. This "easier way" is by means of the so-called **Fundamental Theorem of Calculus**. In this section you will see that there are two forms of this important theorem; it is the first form presented next that enables us to evaluate many definite integrals.

▌ **Fundamental Theorem of Calculus—First Form**   In the next theorem we see that the concept of an antiderivative of a continuous function provides the bridge between differential calculus and integral calculus.

Theorem 5.5.1   Fundamental Theorem of Calculus—Antiderivative Form

If f is continuous on an interval $[a, b]$ and F is any antiderivative of f on the interval, then

$$\int_{a}^{b} f(x) \, dx = F(b) - F(a). \tag{1}$$

We will present two proofs of Theorem 5.5.1. In the proof given we use the basic premise that a definite integral is a limit of a sum. After we have proved the second form of the Fundamental Theorem of Calculus we will return to Theorem 5.5.1 and present an alternative proof.

▶ **PROOF** If F is an antiderivative of f, then by definition $F'(x) = f(x)$. Since F is differentiable on (a, b), the Mean Value Theorem (Theorem 4.4.2) guarantees that there exists an x_k^* in each subinterval (x_{k-1}, x_k) of the partition P:

$$a = x_0 < x_1 < x_2 < \cdots < x_{n-1} < x_n = b$$

such that

$$F(x_k) - F(x_{k-1}) = F'(x_k^*)(x_k - x_{k-1}) \quad \text{or} \quad F(x_k) - F(x_{k-1}) = f(x_k^*)\,\Delta x_k.$$

Now, for $k = 1, 2, 3, \ldots, n$ the last result gives

$$F(x_1) - F(a) = f(x_1^*)\,\Delta x_1$$
$$F(x_2) - F(x_1) = f(x_2^*)\,\Delta x_2$$
$$F(x_3) - F(x_2) = f(x_3^*)\,\Delta x_3$$
$$\vdots$$
$$F(b) - F(x_{n-1}) = f(x_n^*)\,\Delta x_n.$$

If we add the preceding columns,

$$[F(x_1) - F(a)] + [F(x_2) - F(x_1)] + \cdots + [F(b) - F(x_{n-1})] = \sum_{k=1}^{n} f(x_k^*)\,\Delta x_k$$

we see that all but the two terms in black type on the left-hand side of the equality add to 0 leaving

$$F(b) - F(a) = \sum_{k=1}^{n} f(x_k^*)\,\Delta x_k. \tag{2}$$

But $\lim\limits_{\|P\| \to 0} [F(b) - F(a)] = F(b) - F(a)$, and so the limit of (2) as $\|P\| \to 0$ is

$$F(b) - F(a) = \lim_{\|P\| \to 0} \sum_{k=1}^{n} f(x_k^*)\,\Delta x_k. \tag{3}$$

From Definition 5.4.1, the right-hand side of (3) is $\int_a^b f(x)\,dx$. ■

The difference $F(b) - F(a)$ in (1) is usually represented by the symbol $F(x)\big]_a^b$, that is,

$$\underbrace{\int_a^b f(x)\,dx}_{\substack{\text{definite} \\ \text{integral}}} = \underbrace{\left[\int f(x)\,dx \right]_a^b}_{\substack{\text{indefinite} \\ \text{integral}}} = F(x)\bigg]_a^b.$$

Since Theorem 5.5.1 indicates that F is *any* antiderivative of f, we may always choose the constant of integration C to be zero. Observe that if $C \neq 0$, then

$$(F(x) + C)\bigg]_a^b = (F(b) + C) - (F(a) + C) = F(b) - F(a) = F(x)\bigg]_a^b.$$

EXAMPLE 1 Using (1)

In Example 4 of Section 5.4 we resorted to the rather lengthy definition of the definite integral to show that $\int_{-2}^{1} x^3\,dx = -\frac{15}{4}$. Since $F(x) = \frac{1}{4}x^4$ is an antiderivative of $f(x) = x^3$, we now obtain immediately from (1)

$$\int_{-2}^{1} x^3\,dx = \frac{x^4}{4}\bigg]_{-2}^{1} = \frac{1}{4} - \frac{1}{4}(-2)^4 = \frac{1}{4} - \frac{16}{4} = -\frac{15}{4}. \quad ■$$

EXAMPLE 2 Using (1)

Evaluate $\displaystyle\int_1^3 x\,dx$.

Solution An antiderivative of $f(x) = x$ is $F(x) = \frac{1}{2}x^2$. Consequently (1) of Theorem 5.5.1 gives

$$\int_1^3 x\,dx = \frac{x^2}{2}\bigg]_1^3 = \frac{9}{2} - \frac{1}{2} = 4. \quad ■$$

EXAMPLE 3   Using (1)

Evaluate $\displaystyle\int_{-2}^{2} (3x^2 - x + 1)\, dx$.

Solution   We apply (*ii*) of Theorem 5.1.2 and integration formula 2 of Table 5.1.1 to each term of the integrand and then use the Fundamental Theorem:

$$\int_{-2}^{2} (3x^2 - x + 1)\, dx = \left(x^3 - \frac{x^2}{2} + x \right)\Bigg]_{-2}^{2}$$

$$= (8 - 2 + 2) - (-8 - 2 - 2) = 20.\ \blacksquare$$

EXAMPLE 4   Using (1)

Evaluate $\displaystyle\int_{\pi/6}^{\pi} \cos x\, dx$.

Solution   An antiderivative of $f(x) = \cos x$ is $F(x) = \sin x$. Therefore,

$$\int_{\pi/6}^{\pi} \cos x\, dx = \sin x\,\Bigg]_{\pi/6}^{\pi} = \sin \pi - \sin\frac{\pi}{6} = 0 - \frac{1}{2} = -\frac{1}{2}.\ \blacksquare$$

■ **Fundamental Theorem of Calculus—Second Form**   Suppose f is continuous on an interval $[a, b]$ and so we know that the integral $\int_a^b f(t)\, dt$ exists. For each x in the interval $[a, b]$, the definite integral

$$g(x) = \int_a^x f(t)\, dt \qquad (4)$$

◀ Keep in mind that a definite integral does not depend on the variable of integration t.

represents a single number. In this way, it is seen that (4) is a function with domain $[a, b]$. In **FIGURE 5.5.1** we have shown f to be a positive function on $[a, b]$, and so as x varies across the interval we can interpret $g(x)$ as the area under the graph on the interval $[a, x]$. In the second form of the Fundamental Theorem of Calculus we will show that $g(x)$ defined in (4) is a differentiable function.

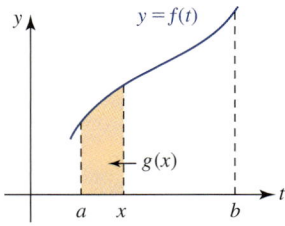

FIGURE 5.5.1 $g(x)$ as area

Theorem 5.5.2   Fundamental Theorem of Calculus—Derivative Form

Let f be continuous on $[a, b]$ and let x be any number in the interval. Then $g(x) = \int_a^x f(t)\, dt$ is continuous on $[a, b]$ and differentiable on (a, b) and

$$g'(x) = f(x). \qquad (5)$$

PROOF FOR $h > 0$ Let x and $x + h$ be in (a, b), where $h > 0$. From the definition of the derivative,

$$g'(x) = \lim_{h\to 0} \frac{g(x + h) - g(x)}{h}. \qquad (6)$$

Using the properties of the definite integral the difference $g(x + h) - g(x)$ can be written

$$g(x + h) - g(x) = \int_a^{x+h} f(t)\, dt - \int_a^x f(t)\, dt$$

$$= \int_a^{x+h} f(t)\, dt + \int_x^a f(t)\, dt \quad \leftarrow \text{by (8) of Section 5.4}$$

$$= \int_x^{x+h} f(t)\, dt. \qquad \leftarrow \text{by (10) of Section 5.4}$$

Hence (6) becomes

$$g'(x) = \lim_{h\to 0} \frac{1}{h} \int_x^{x+h} f(t)\, dt. \qquad (7)$$

Since f is continuous on the closed interval $[x, x + h]$, we know from the Extreme Value Theorem (Theorem 4.3.1) that f attains a minimum value m and a maximum value M on the interval. Since m and M are constant relative to the integration on the variable t it follows from Theorem 5.4.7(*ii*) that

$$\int_x^{x+h} m \, dt \leq \int_x^{x+h} f(t) \, dt \leq \int_x^{x+h} M \, dt. \tag{8}$$

With the aid of Theorem 5.5.1,

$$\int_x^{x+h} m \, dt = mt \Big]_x^{x+h} = m(x + h - x) = mh$$

and

$$\int_x^{x+h} M \, dt = Mt \Big]_x^{x+h} = M(x + h - x) = Mh.$$

Hence the inequality in (8) becomes

$$mh \leq \int_x^{x+h} f(t) \, dt \leq Mh \quad \text{or} \quad m \leq \frac{1}{h} \int_x^{x+h} f(t) \, dt \leq M. \tag{9}$$

Because f is continuous on $[x, x + h]$ it stands to reason that $\lim_{h \to 0^+} m = \lim_{h \to 0^+} M = f(x)$. Taking the limit of the second expression in (9) as $h \to 0^+$ gives

$$f(x) \leq \lim_{h \to 0^+} \frac{1}{h} \int_x^{x+h} f(t) \, dt \leq f(x).$$

This shows that $g'(x)$ exists and from $f(x) \leq g'(x) \leq f(x)$ we conclude that $g'(x) = f(x)$. Since g is differentiable, it is necessarily continuous. A similar argument holds for $h < 0$. ∎

An alternative, and more traditional, way of writing the result in (5) is

$$\frac{d}{dx} \int_a^x f(t) \, dt = f(x). \tag{10}$$

EXAMPLE 5 Using (10)

From (10),

(a) $\dfrac{d}{dx} \displaystyle\int_{-2}^x t^3 \, dt = x^3$ (b) $\dfrac{d}{dx} \displaystyle\int_1^x \sqrt{t^2 + 1} \, dt = \sqrt{x^2 + 1}.$ ∎

EXAMPLE 6 Chain Rule

Find $\dfrac{d}{dx} \displaystyle\int_\pi^{x^3} \cos t \, dt$.

Solution If we identify $g(x) = \int_\pi^x \cos t \, dt$, then the given integral is the composition $g(x^3)$. We carry out the differentiation using the Chain Rule with $u = x^3$:

$$\frac{d}{dx} \int_\pi^{x^3} \cos t \, dt = \frac{d}{du} \left(\int_\pi^u \cos t \, dt \right) \frac{du}{dx}$$

$$= \cos u \cdot \frac{du}{dx} = \cos x^3 \cdot 3x^2$$

$$= 3x^2 \cos x^3. \quad ∎$$

■ Alternative Proof of Theorem 5.5.1 It is worthwhile to examine yet another proof of Theorem 5.5.1 using Theorem 5.5.2. For a function f continuous on $[a, b]$, the statement $g'(x) = f(x)$ for $g(x) = \int_a^x f(t) \, dt$ means that $g(x)$ is an antiderivative of the integrand f. If F is any antiderivative of f, we know from Theorem 5.1.1 that $g(x) - F(x) = C$ or

$g(x) = F(x) + C$, where C is any arbitrary constant. Since $g(x) = \int_a^x f(t)\, dt$, it follows for any x in $[a, b]$ that

$$\int_a^x f(t)\, dt = F(x) + C. \tag{11}$$

If we substitute $x = a$ in (11), then

$$\int_a^a f(t)\, dt = F(a) + C$$

implies $C = -F(a)$, since $\int_a^a f(t)\, dt = 0$. Thus, (11) becomes

$$\int_a^x f(t)\, dt = F(x) - F(a).$$

Since the latter equation is valid at $x = b$, we find

$$\int_a^b f(t)\, dt = F(b) - F(a). \qquad \blacksquare$$

■ **Piecewise-Continuous Functions** A function f is said to be **piecewise continuous** on an interval $[a, b]$ if there are at most a finite number of points c_k, $k = 1, 2, \ldots, n$, $(c_{k-1} < c_k)$ at which f has a finite, or jump, discontinuity and f is continuous on each open interval (c_{k-1}, c_k). See FIGURE 5.5.2. If a function f is piecewise continuous on $[a, b]$, it is bounded on the interval, and hence by Theorem 5.4.2, f is integrable on $[a, b]$. A definite integral of a piecewise-continuous function on $[a, b]$ can be evaluated with the help of Theorem 5.4.5:

$$\int_a^b f(x)\, dx = \int_a^{c_1} f(x)\, dx + \int_{c_1}^{c_2} f(x)\, dx + \cdots + \int_{c_n}^b f(x)\, dx$$

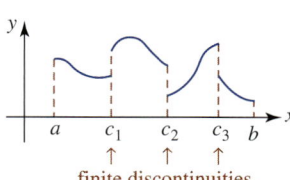

finite discontinuities

FIGURE 5.5.2 Piecewise-continuous function

and by simply treating the integrands of the definite integrals on the right side of the above equation as if they were continuous on the closed intervals $[a, c_1], [c_1, c_2], \ldots, [c_n, b]$.

EXAMPLE 7 Integrating a Piecewise-Continuous Function

Evaluate $\displaystyle\int_{-1}^4 f(x)\, dx$ where

$$f(x) = \begin{cases} x + 1, & -1 \le x < 0 \\ x, & 0 \le x < 2 \\ 3, & 2 \le x \le 4. \end{cases}$$

Solution The graph of the piecewise-continuous function f is given in FIGURE 5.5.3. Now, from the preceding discussion and the definition of f:

$$\int_{-1}^4 f(x)\, dx = \int_{-1}^0 f(x)\, dx + \int_0^2 f(x)\, dx + \int_2^4 f(x)\, dx$$

$$= \int_{-1}^0 (x + 1)\, dx + \int_0^2 x\, dx + \int_2^4 3\, dx$$

$$= \left(\frac{1}{2}x^2 + x \right) \Big]_{-1}^0 + \frac{1}{2}x^2 \Big]_0^2 + 3x \Big]_2^4 = \frac{17}{2}. \qquad \blacksquare$$

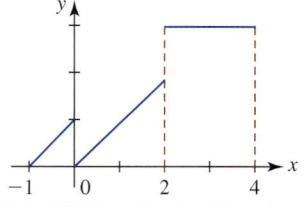

FIGURE 5.5.3 Graph of function in Example 7

EXAMPLE 8 Integrating a Piecewise-Continuous Function

Evaluate $\displaystyle\int_0^3 |x - 2|\, dx$.

Solution From the definition of absolute value,

$$|x - 2| = \begin{cases} x - 2 & \text{if } x - 2 \ge 0 \\ -(x - 2) & \text{if } x - 2 < 0 \end{cases} \quad \text{or} \quad |x - 2| = \begin{cases} x - 2 & \text{if } x \ge 2 \\ -x + 2 & \text{if } x < 2. \end{cases}$$

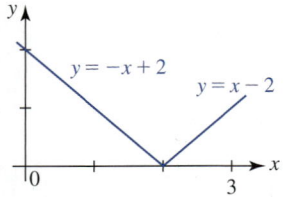

FIGURE 5.5.4   Graph of function in Example 8

The graph of $f(x) = |x - 2|$ is given in FIGURE 5.5.4. Now in view of (10) of Theorem 5.4.5 we can write

$$\int_0^3 |x - 2| \, dx = \int_0^2 |x - 2| \, dx + \int_2^3 |x - 2| \, dx$$

$$= \int_0^2 (-x + 2) \, dx + \int_2^3 (x - 2) \, dx$$

$$= \left(-\frac{1}{2}x^2 + 2x \right) \Big]_0^2 + \left(\frac{1}{2}x^2 - 2x \right) \Big]_2^3$$

$$= (-2 + 4) + \left(\frac{9}{2} - 6 \right) - (2 - 4) = \frac{5}{2}. \qquad \blacksquare$$

Substitution in a Definite Integral   Recall from Section 5.2 that we sometimes used a substitution as an aid in evaluating an indefinite integral of the form $\int f(g(x))g'(x) \, dx$. Care should be exercised when using a substitution in a definite integral $\int_a^b f(g(x))g'(x) \, dx$, since we can proceed in *two ways*.

Guidelines for Substituting in a Definite Integral

- Evaluate the indefinite integral $\int f(g(x))g'(x) \, dx$ by means of the substitution $u = g(x)$. Resubstitute $u = g(x)$ in the antiderivative and then apply the Fundamental Theorem of Calculus by using the original limits of integration $x = a$ and $x = b$.
- Alternatively, the resubstitution can be avoided by changing the limits of integration to correspond to the value of u at $x = a$ and u at $x = b$. The latter method, which is usually quicker, is summarized in the next theorem.

Theorem 5.5.3   Substitution in a Definite Integral

Let $u = g(x)$ be a function that has a continuous derivative on the interval $[a, b]$, and let f be a function that is continuous on the range of g. If $F'(u) = f(u)$ and $c = g(a)$, $d = g(b)$, then

$$\int_a^b f(g(x))g'(x) \, dx = \int_{g(a)}^{g(b)} f(u) \, du = F(d) - F(c). \qquad (12)$$

PROOF   If $u = g(x)$, then $du = g'(x) \, dx$. Therefore,

$$\int_a^b f(g(x))g'(x) \, dx = \int_{g(a)}^{g(b)} f(u) \frac{du}{dx} \, dx = \int_c^d f(u) \, du = F(u) \Big]_c^d = F(d) - F(c). \qquad \blacksquare$$

EXAMPLE 9   Substitution in a Definite Integral

Evaluate $\displaystyle\int_0^2 \sqrt{2x^2 + 1} \, x \, dx$.

Solution   We shall first illustrate the two procedures outlined in the guidelines preceding Theorem 5.5.3.

(a) To evaluate the indefinite integral $\int \sqrt{2x^2 + 1} \, x \, dx$, we use $u = 2x^2 + 1$ and $du = 4x \, dx$. Thus,

$$\int \sqrt{2x^2 + 1} \, x \, dx = \frac{1}{4} \int \sqrt{2x^2 + 1} \, (4x \, dx) \qquad \leftarrow \text{substitution}$$

$$= \frac{1}{4} \int u^{1/2} \, du$$

$$= \frac{1}{4} \frac{u^{3/2}}{3/2} + C$$

$$= \frac{1}{6} (2x^2 + 1)^{3/2} + C. \qquad \leftarrow \text{resubstitution}$$

Therefore, by Theorem 5.5.1,

$$\int_0^2 \sqrt{2x^2 + 1}\, x\, dx = \frac{1}{6}(2x^2 + 1)^{3/2}\Big]_0^2$$

$$= \frac{1}{6}[9^{3/2} - 1^{3/2}]$$

$$= \frac{1}{6}[27 - 1] = \frac{13}{3}.$$

(b) If $u = 2x^2 + 1$, then $x = 0$ implies $u = 1$, whereas $x = 2$ gives $u = 9$. Thus, by Theorem 5.5.3,

$$\underset{\substack{\downarrow \\ u\ \text{limits}}}{}$$

$$\int_0^2 \sqrt{2x^2 + 1}\, x\, dx = \frac{1}{4}\int_1^9 u^{1/2}\, du \ \underset{\substack{\text{integration with} \\ \text{respect to } u}}{\leftarrow}$$

$$= \frac{1}{4}\frac{u^{3/2}}{3/2}\Big]_1^9$$

$$= \frac{1}{6}[9^{3/2} - 1^{3/2}] = \frac{13}{3}. \qquad \blacksquare$$

When the graph of a function $y = f(x)$ is symmetric with respect to either the y-axis (even function) or the origin (odd function), then the definite integral of f on a symmetric interval $[-a, a]$, that is, $\int_{-a}^a f(x)\, dx$, can be evaluated by means of a "shortcut."

Theorem 5.5.4   Even Function Rule

If f is an even integrable function on $[-a, a]$, then

$$\int_{-a}^a f(x)\, dx = 2\int_0^a f(x)\, dx. \qquad (13)$$

We will prove the next theorem but leave the proof of Theorem 5.5.4 as an exercise.

Theorem 5.5.5   Odd Function Rule

If f is an odd integrable function on $[-a, a]$, then

$$\int_{-a}^a f(x)\, dx = 0. \qquad (14)$$

PROOF   Assume f is an odd function. By the additive interval property, Theorem 5.4.5, we have

$$\int_{-a}^a f(x)\, dx = \int_{-a}^0 f(x)\, dx + \int_0^a f(x)\, dx.$$

In the first integral on the right-hand side, let $x = -t$, so that $dx = -dt$, and when $x = -a$ and $x = 0$, then $t = a$ and $t = 0$:

$$\int_{-a}^a f(x)\, dx = \int_a^0 f(-t)(-dt) + \int_0^a f(x)\, dx \ \underset{}{\leftarrow f(-t) = -f(t),\ f\ \text{an odd function}}$$

$$= \int_a^0 f(t)\, dt + \int_0^a f(x)\, dx$$

$$= -\int_0^a f(t)\, dt + \int_0^a f(x)\, dx \qquad \leftarrow \text{by (8) of Section 5.4}$$

$$= -\int_0^a f(x)\, dx + \int_0^a f(x)\, dx \quad \leftarrow t \text{ was a ``dummy variable'' of integration}$$

$$= 0.$$ ∎

The point in Theorem 5.5.5 is this: When we integrate an odd integrable function f on a symmetric interval $[-a, a]$, there is no need to find an antiderivative of f; the value of the integral is always zero.

Geometric motivations for the results in Theorems 5.5.4 and 5.5.5 are given in FIGURE 5.5.5.

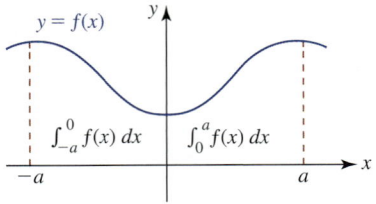
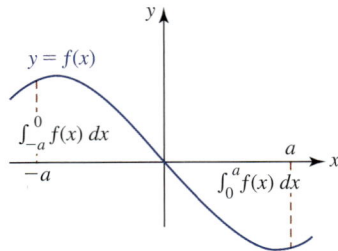

(a) Even function: Value of the definite integral on $[-a, 0]$ is the same as the value on $[0, a]$

(b) Odd function: Value of the definite integral on $[-a, 0]$ is the opposite of the value on $[0, a]$

FIGURE 5.5.5 Even Function Rule in (a); Odd Function Rule in (b)

EXAMPLE 10 Using the Even Function Rule

Evaluate $\displaystyle\int_{-1}^{1} (x^4 + x^2)\, dx$.

Solution The integrand $f(x) = x^4 + x^2$ is a polynomial function with all even powers, and so f is necessarily an even function. Since the interval of integration is the symmetric interval $[-1, 1]$, it follows from Theorem 5.5.4 that we can integrate on $[0, 1]$ and multiply the result by 2:

$$\int_{-1}^{1} (x^4 + x^2)\, dx = 2 \int_0^1 (x^4 + x^2)\, dx$$

$$= 2 \left(\frac{1}{5}x^5 + \frac{1}{3}x^3 \right) \Big]_0^1$$

$$= 2 \left(\frac{1}{5} + \frac{1}{3} \right) = \frac{16}{15}.$$ ∎

EXAMPLE 11 Using the Odd Function Rule

Evaluate $\displaystyle\int_{-\pi/2}^{\pi/2} \sin x\, dx$.

Solution In this case $f(x) = \sin x$ is an odd function on the symmetric interval $[-\pi/2, \pi/2]$. Thus, by Theorem 5.5.5 we have immediately

$$\int_{-\pi/2}^{\pi/2} \sin x\, dx = 0.$$ ∎

$\displaystyle\int_a^b$ **NOTES FROM THE CLASSROOM**

The antiderivative form of the Fundamental Theorem of Calculus is an extremely important and powerful tool for evaluating definite integrals. Why should we bother with a clumsy limit of a sum when the value of $\int_a^b f(x)\, dx$ can be found by computing $\int f(x)\, dx$ at the two numbers a and b? This is true up to a point—however, it is time to learn another fact of mathematical life. There are continuous functions for which the antiderivative $\int f(x)\, dx$

cannot be expressed in terms of *elementary functions*: sums, products, quotients, and powers of polynomial functions, trigonometric functions, inverse trigonometric functions, logarithmic, and exponential functions. The simple continuous function $f(x) = \sqrt{x^3 + 1}$ possesses no antiderivative that is an elementary function. Although, in view of Theorem 5.4.1 we can say that the definite integral $\int_0^1 \sqrt{x^3 + 1}\, dx$ exists, Theorem 5.5.1 provides no help in finding its value. The integral $\int_0^1 \sqrt{x^3 + 1}\, dx$ is called **nonelementary**. Nonelementary integrals are important and appear in many applications such as probability theory and optics. Here a few more nonelementary integrals:

$$\int \frac{\sin x}{x}\, dx, \quad \int \sin x^2\, dx, \quad \int_0^x e^{-t^2}\, dt, \quad \text{and} \quad \int \frac{e^x}{x}\, dx.$$

See Problems 71 and 72 in Exercises 5.5.

Exercises 5.5   Answers to selected odd-numbered problems begin on page ANS-19.

≡ Fundamentals

In Problems 1–42, use the Fundamental Theorem of Calculus given in Theorem 5.5.1 to evaluate the given definite integral.

1. $\int_3^7 dx$

2. $\int_2^{10} (-4)\, dx$

3. $\int_{-1}^2 (2x + 3)\, dx$

4. $\int_{-5}^4 t^2\, dt$

5. $\int_1^3 (6x^2 - 4x + 5)\, dx$

6. $\int_{-2}^1 (12x^5 - 36)\, dx$

7. $\int_0^{\pi/2} \sin x\, dx$

8. $\int_{-\pi/3}^{\pi/4} \cos\theta\, d\theta$

9. $\int_{\pi/4}^{\pi/2} \cos 3t\, dt$

10. $\int_{1/2}^1 \sin 2\pi x\, dx$

11. $\int_{1/2}^{3/4} \frac{1}{u^2}\, du$

12. $\int_{-3}^{-1} \frac{2}{x}\, dx$

13. $\int_{-1}^1 e^x\, dx$

14. $\int_0^2 (2x - 3e^x)\, dx$

15. $\int_0^2 x(1 - x)\, dx$

16. $\int_3^2 x(x - 2)(x + 2)\, dx$

17. $\int_{-1}^1 (7x^3 - 2x^2 + 5x - 4)\, dx$

18. $\int_{-3}^{-1} (x^2 - 4x + 8)\, dx$

19. $\int_1^4 \frac{x - 1}{\sqrt{x}}\, dx$

20. $\int_2^4 \frac{x^2 + 8}{x^2}\, dx$

21. $\int_1^{\sqrt{3}} \frac{1}{1 + x^2}\, dx$

22. $\int_0^{1/4} \frac{1}{\sqrt{1 - 4x^2}}\, dx$

23. $\int_{-4}^{12} \sqrt{z + 4}\, dz$

24. $\int_0^{7/2} (2x + 1)^{-1/3}\, dx$

25. $\int_0^3 \frac{x}{\sqrt{x^2 + 16}}\, dx$

26. $\int_{-2}^1 \frac{t}{(t^2 + 1)^2}\, dt$

27. $\int_{1/2}^1 \left(1 + \frac{1}{x}\right)^3 \frac{1}{x^2}\, dx$

28. $\int_1^4 \frac{\sqrt[3]{1 + 4\sqrt{x}}}{\sqrt{x}}\, dx$

29. $\int_0^1 \frac{x + 1}{\sqrt{x^2 + 2x + 3}}\, dx$

30. $\int_{-1}^1 \frac{u^3 + u}{(u^4 + 2u^2 + 1)^5}\, du$

31. $\int_0^{\pi/8} \sec^2 2x\, dx$

32. $\int_{\sqrt{\pi/4}}^{\sqrt{\pi/2}} x \csc x^2 \cot x^2\, dx$

33. $\int_{-1/2}^{3/2} (x - \cos \pi x)\, dx$

34. $\int_1^4 \frac{\cos \sqrt{x}}{2\sqrt{x}}\, dx$

35. $\int_0^{\pi/2} \sqrt{\cos x}\, \sin x\, dx$

36. $\int_{\pi/6}^{\pi/3} \sin x \cos x\, dx$

37. $\int_{\pi/6}^{\pi/2} \frac{1 + \cos\theta}{(\theta + \sin\theta)^2}\, d\theta$

38. $\int_{-\pi/4}^{\pi/4} (\sec x + \tan x)^2\, dx$

39. $\int_0^{3/4} \sin^2 \pi x\, dx$

40. $\int_{-\pi/2}^{\pi/2} \cos^2 x\, dx$

41. $\int_1^5 \frac{1}{1 + 2x}\, dx$

42. $\int_{-1}^1 \tan x\, dx$

In Problems 43–48, use the Fundamental Theorem of Calculus given in Theorem 5.5.2 to find the indicated derivative.

43. $\dfrac{d}{dx} \displaystyle\int_0^x te^t\, dt$

44. $\dfrac{d}{dx} \displaystyle\int_1^x \ln t\, dt$

45. $\dfrac{d}{dt} \displaystyle\int_2^t (3x^2 - 2x)^6\, dx$

46. $\dfrac{d}{dx} \displaystyle\int_x^9 \sqrt[3]{u^2 + 2}\, du$

47. $\dfrac{d}{dx} \displaystyle\int_3^{6x - 1} \sqrt{4t + 9}\, dt$

48. $\dfrac{d}{dx} \displaystyle\int_\pi^{\sqrt{x}} \sin t^2\, dt$

In Problems 49 and 50, use the Fundamental Theorem of Calculus given in Theorem 5.5.2 to find $F'(x)$. [*Hint*: Use two integrals.]

49. $F(x) = \displaystyle\int_{3x}^{x^2} \frac{1}{t^3 + 1}\, dt$

50. $F(x) = \displaystyle\int_{\sin x}^{5x} \sqrt{t^2 + 1}\, dt$

In Problems 51 and 52, verify the given result by first evaluating the definite integral and then differentiating.

51. $\dfrac{d}{dx}\displaystyle\int_1^x (6t^2 - 8t + 5)\, dt = 6x^2 - 8x + 5$

52. $\dfrac{d}{dt}\displaystyle\int_\pi^t \sin\dfrac{x}{3}\, dx = \sin\dfrac{t}{3}$

53. Consider the function $f(x) = \int_1^x \ln(2t + 1)\, dt$. Find the indicated function value.

(a) $f(1)$ (b) $f'(1)$
(c) $f''(1)$ (d) $f'''(1)$

54. Suppose $G(x) = \int_a^x f(t)\, dt$ and $G'(x) = f(x)$. Find the given expression.

(a) $G(x^2)$ (b) $\dfrac{d}{dx}G(x^2)$

(c) $G(x^3 + 2x)$ (d) $\dfrac{d}{dx}G(x^3 + 2x)$

In Problems 55 and 56, evaluate $\int_{-1}^2 f(x)\, dx$ for the given function f.

55. $f(x) = \begin{cases} -x, & x < 0 \\ x^2, & x \ge 0 \end{cases}$

56. $f(x) = \begin{cases} 2x + 3, & x \le 0 \\ 3, & x > 0 \end{cases}$

In Problems 57–60, evaluate the definite integral of the given piecewise-continuous function f.

57. $\displaystyle\int_0^3 f(x)\, dx$, where $f(x) = \begin{cases} 4, & 0 \le x < 2 \\ 1, & 2 \le x \le 3 \end{cases}$

58. $\displaystyle\int_0^\pi f(x)\, dx$, where $f(x) = \begin{cases} \sin x, & 0 \le x < \pi/2 \\ \cos x, & \pi/2 \le x \le \pi \end{cases}$

59. $\displaystyle\int_{-2}^2 f(x)\, dx$, where $f(x) = \begin{cases} x^2, & -2 \le x < -1 \\ 4, & -1 \le x < 1 \\ x^2, & 1 \le x \le 2 \end{cases}$

60. $\displaystyle\int_0^4 f(x)\, dx$, where $f(x) = \lfloor x \rfloor$ is the greatest integer function

In Problems 61–66, proceed as in Example 8 to evaluate the given definite integral.

61. $\displaystyle\int_{-3}^1 |x|\, dx$

62. $\displaystyle\int_0^4 |2x - 6|\, dx$

63. $\displaystyle\int_{-8}^3 \sqrt{|x| + 1}\, dx$

64. $\displaystyle\int_0^2 |x^2 - 1|\, dx$

65. $\displaystyle\int_{-\pi}^\pi |\sin x|\, dx$

66. $\displaystyle\int_0^\pi |\cos x|\, dx$

In Problems 67–70, proceed as in part (b) of Example 9 and evaluate the given definite integral using the indicated u-substitution.

67. $\displaystyle\int_{1/2}^e \dfrac{(\ln 2t)^5}{t}\, dt; \quad u = \ln 2t$

68. $\displaystyle\int_{\sqrt{2}/2}^1 \dfrac{1}{(\tan^{-1}x)(1 + x^2)}\, dx; \quad u = \tan^{-1}x$

69. $\displaystyle\int_0^1 \dfrac{e^{-2x}}{e^{-2x} + 1}\, dx; \quad u = e^{-2x} + 1$

70. $\displaystyle\int_0^{1/\sqrt{2}} \dfrac{x}{\sqrt{1 - x^4}}\, dx; \quad u = x^2$

≡ Applications

71. In applied mathematics some important functions are defined in terms of nonelementary integrals. One such special function is called the **error function** and is defined as

$$\text{erf}(x) = \dfrac{2}{\sqrt{\pi}}\int_0^x e^{-t^2}\, dt.$$

(a) Show that $\text{erf}(x)$ is an increasing function on the interval $(-\infty, \infty)$.
(b) Show that the function $y = e^{x^2}[1 + \sqrt{\pi}\,\text{erf}(x)]$ satisfies the differential equation

$$\dfrac{dy}{dx} - 2xy = 2,$$

and that $y(0) = 1$.

72. Another special function defined by a nonelementary integral is the **sine integral function**

$$\text{Si}(x) = \int_0^x \dfrac{\sin t}{t}\, dt.$$

The function $\text{Si}(x)$ has an infinite number of relative extrema.

(a) Find the first four critical numbers for $x > 0$. Use the second derivative test to determine whether these critical numbers correspond to a relative maximum or a relative minimum.
(b) Use a CAS to obtain the graph of $\text{Si}(x)$. [*Hint*: In *Mathematica* the sine integral function is denoted by SinIntegral[x].]

≡ Think About It

In Problems 73 and 74, let P be a partition of the indicated interval and x_k^* a number in the kth subinterval. Determine the value of the given limit.

73. $\displaystyle\lim_{\|P\| \to 0} \sum_{k=1}^n (2x_k^* + 5)\, \Delta x_k; \quad [-1, 3]$

74. $\displaystyle\lim_{\|P\| \to 0} \sum_{k=1}^n \cos\dfrac{x_k^*}{4}\, \Delta x_k; \quad [0, 2\pi]$

In Problems 75 and 76, let P be a regular partition of the indicated interval and x_k^* a number in the kth subinterval. Establish the given result.

75. $\displaystyle\lim_{n \to \infty} \dfrac{\pi}{n} \sum_{k=1}^n \sin x_k^* = 2; \quad [0, \pi]$

76. $\displaystyle\lim_{n \to \infty} \dfrac{2}{n} \sum_{k=1}^n x_k^* = 0; \quad [-1, 1]$

In Problems 77 and 78, evaluate the given definite integral.

77. $\int_{-1}^{2} \left\{ \int_{1}^{x} 12t^2 \, dt \right\} dx$ **78.** $\int_{0}^{\pi/2} \left\{ \int_{0}^{t} \sin x \, dx \right\} dt$

79. Prove the even function rule, Theorem 5.5.4.

80. Suppose f is an odd function that is defined on the interval $[-4, 4]$. Suppose further that f is differentiable on the interval, $f(-2) = 3.5$, has zeros at -3 and 3, and has critical numbers -2 and 2.

(a) What is $f(0)$?

(b) Sketch a rough graph of f.

(c) Suppose F is a function defined on $[-4, 4]$ by $F(x) = \int_{-3}^{x} f(t) \, dt$. Find $F(-3)$ and $F(3)$.

(d) Sketch a rough graph of F.

(e) Find the critical numbers and points of inflection of F.

81. Determine whether the following reasoning is correct:

$$\int_{-\pi/2}^{\pi/2} \sin^2 t \, dt = -\int_{-\pi/2}^{\pi/2} \sin t (-\sin t \, dt)$$

$$= -\int_{-\pi/2}^{\pi/2} \sqrt{1 - \cos^2 t} \, (-\sin t \, dt) \leftarrow \begin{cases} u = \cos t \\ du = -\sin t \, dt \end{cases}$$

$$= -\int_{0}^{0} \sqrt{1 - u^2} \, du = 0. \leftarrow \begin{cases} \text{Theorem 5.5.3} \\ \text{Definition 5.4.2}(i) \end{cases}$$

82. Compute the derivatives.

(a) $\dfrac{d}{dx} x \int_{1}^{2x} \sqrt{t^3 + 7} \, dt$ (b) $\dfrac{d}{dx} x \int_{1}^{4} \sqrt{t^3 + 7} \, dt$

≡ Calculator/CAS Problems

83. (a) Use a calculator or CAS to obtain the graphs of $f(x) = \cos^3 x$ and $g(x) = \sin^3 x$.

(b) Based on your interpretation of net signed area, use the graphs in part (a) to conjecture the values of $\int_{0}^{2\pi} \cos^3 x \, dx$ and $\int_{0}^{2\pi} \sin^3 x \, dx$.

≡ Projects

84. Integration by Darts In this problem we illustrate a method for approximating the area under a graph by "throwing darts." Suppose that we wish to find the area A under the graph of $f(x) = \cos^3(\pi x/2)$ on the interval $[0, 1]$; that is, we wish to approximate $A = \int_{0}^{1} \cos^3(\pi x/2) \, dx$.

If we throw, with no particular attempt at being skillful, a large number of darts, say N, at a 1×1 square target shown in FIGURE 5.5.6 and n darts strike the red-colored region under the graph of $f(x) = \cos^3(\pi x/2)$, then it can be shown that the probability of a dart striking the region is given by the ratio of two areas:

$$\frac{\text{area of region}}{\text{area of square}} = \frac{A}{1}.$$

Moreover, this theoretical probability is approximately the same as the empirical probability n/N:

$$\frac{A}{1} \approx \frac{n}{N} \qquad \text{or} \qquad A \approx \frac{n}{N}.$$

To simulate the throwing of darts at the target, use a CAS such as *Mathematica* and its random number function to generate a table of N ordered pairs (x, y), $0 < x < 1$, $0 < y < 1$.

(a) Let $N = 50$. Plot the points and the graph of f on the same set of coordinate axes. Use the figure to count the number of hits n. Construct at least 10 different tables of random points and plots. For each plot compute the ratio n/N.

(b) Repeat part (a) for $N = 100$.

(c) Use the CAS to find the exact value of area A and compare this value with the approximations obtained in parts (a) and (b).

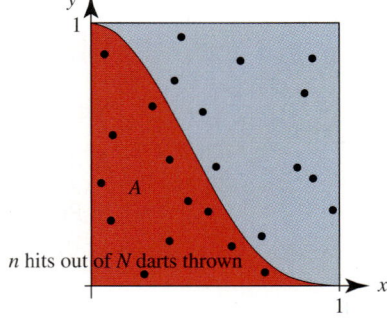

FIGURE 5.5.6 Target in Problem 84

85. Expanding Oil Spill A mathematical model that can be used to determine the time t required for an expanding oil spill to evaporate is given by the formula

$$\frac{RT}{Pv} = \int_{0}^{t} \frac{KA(u)}{V_0} \, du,$$

where $A(u)$ is the area of the spill at time u, RT/Pv is a dimensionless thermodynamic term, K is a mass transfer coefficient, and V_0 is the initial volume of the spill.

(a) Suppose the oil spill is expanding in the form of a circle whose initial radius is r_0. See FIGURE 5.5.7. If the radius r of the spill is increasing at a rate $dr/dt = C$ (in meters per second), solve for t in terms of the other symbols.

(b) Typical values for RT/Pv and K are 1.9×10^6 (for tridecane) and 0.01 mm/s, respectively. If $C = 0.01$ m/s², $r_0 = 100$ m, and $V_0 = 10,000$ m³, determine how long it will take for the oil to evaporate.

(c) Using the result in part (b), determine the final area of the oil spill.

Oil at time t

FIGURE 5.5.7 Circular oil spill in Problem 85

86. The Mercator Projection and the Integral of sec x
Roughly, a Mercator map is a representation of a three-dimensional global map onto a two-dimensional surface. See FIGURE 5.5.8. Find and study the article, "Mercator's World Map and the Calculus," Philip M. Tuchinsky, UMAP, Unit 206, Newton, MA, 1978. Write a short report summarizing the article and why **Gerhardus Mercator** (c. 1569) needed the value of the definite integral $\int_0^{\theta_0} \sec x \, dx$ to carry out his constructions.

(a) Globe

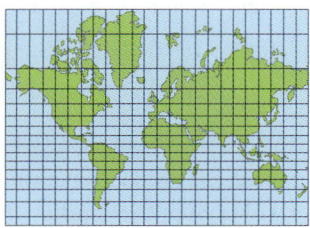
(b) Mercator map

FIGURE 5.5.8   Globe and Mercator Projection in Problem 86

Chapter 5 in Review

Answers to selected odd-numbered problems begin on page ANS-19.

A. True/False

In Problems 1–16, indicate whether the given statement is true or false.

1. If $f'(x) = 3x^2 + 2x$, then $f(x) = x^3 + x^2$. _____

2. $\displaystyle\sum_{k=2}^{6} (2k - 3) = \sum_{j=0}^{4} (2j + 1)$ _____

3. $\displaystyle\sum_{k=1}^{40} 5 = \sum_{k=1}^{20} 10$ _____

4. $\displaystyle\int_1^3 \sqrt{t^2 + 7} \, dt = -\int_3^1 \sqrt{t^2 + 7} \, dt$ _____

5. If f is continuous, then $\displaystyle\int_0^1 f(t) \, dt + \int_1^0 f(x)\,dx = 0$. _____

6. If f is integrable, then f is continuous. _____

7. $\displaystyle\int_0^1 (x - x^3) \, dx$ is the area under the graph of $y = x - x^3$ on the interval $[0, 1]$. _____

8. If $\displaystyle\int_a^b f(x) \, dx > 0$, then $\displaystyle\int_a^b f(x) \, dx$ is the area under the graph of f on $[a, b]$. _____

9. If P is a partition of $[a, b]$ into n subintervals, then $n \to \infty$ implies $\|P\| \to 0$. _____

10. If $F'(x) = 0$ for all x, then $F(x) = C$ for all x. _____

11. If f is an odd integrable function on $[-\pi, \pi]$, then $\displaystyle\int_{-\pi}^{\pi} f(x) \, dx = 0$. _____

12. $\displaystyle\int_{-1}^1 |x| \, dx = 2\int_0^1 x \, dx$ _____

13. $\displaystyle\int \sin x \, dx = \cos x + C$ _____

14. $\displaystyle\int x \cos x \, dx = x \sin x + \cos x + C$ _____

15. $\displaystyle\int_a^b f'(t) \, dt = f(b) - f(a)$ _____

16. The function $F(x) = \displaystyle\int_{-5}^{2x} (t + 4)e^{-t} \, dt$ is increasing on the interval $[-2, \infty)$. _____

B. Fill in the Blanks

In Problems 1–16, fill in the blanks.

1. If G is an antiderivative of a function f, then $G'(x) =$ _____.

2. $\displaystyle\int \frac{d}{dx} x^2 \, dx =$ _____.

3. If $\displaystyle\int f(x) \, dx = \frac{1}{2} (\ln x)^2 + C$, then $f(x) =$ _____.

4. The value of $\displaystyle\frac{d}{dx} \int_3^x \sqrt{t^2 + 5} \, dt$ at $x = 1$ is _____.

5. If g is differentiable, then $\displaystyle\frac{d}{dx} \int_{g(x)}^b f(t) \, dt =$ _____.

6. $\displaystyle\frac{d}{dx} \int_{5x}^{\sqrt{x}} e^{-t^2} \, dt =$ _____.

7. Using sigma notation, the sum $\dfrac{1}{3} + \dfrac{2}{5} + \dfrac{3}{7} + \dfrac{4}{9} + \dfrac{5}{11}$ can be expressed as _____.

8. The numerical value of $\displaystyle\sum_{k=1}^{15} (3k^2 - 2k)$ is _____.

9. If $u = t^2 + 1$, then the definite integral $\displaystyle\int_2^4 t(t^2 + 1)^{1/3} \, dt$ becomes $\dfrac{1}{2} \displaystyle\int_{\_}^{\_} u^{1/3} \, du$.

10. The area under the graph of $f(x) = 2x$ on the interval $[0, 2]$ is _____, and the net signed area between the graph of $f(x) = 2x$ and the x-axis on $[-1, 2]$ is _____.

11. If the interval $[1, 6]$ is partitioned into four subintervals determined by $x_0 = 1$, $x_1 = 2$, $x_2 = \dfrac{5}{2}$, $x_3 = 5$, and $x_4 = 6$, the norm of the partition is _____.

12. A partition of an interval $[a, b]$ in which all the subintervals have equal width is called a _____ partition.

13. If P is a partition of $[0, 4]$ and x_k^* is a number in the kth subinterval, then $\displaystyle\lim_{\|P\| \to 0} \sum_{k=1}^n \sqrt{x_k^*} \, \Delta x_k$ is the definition of the definite integral _____. By the Fundamental Theorem of Calculus, the value of this definite integral is _____.

14. If $\displaystyle\int_0^6 f(x) \, dx = 11$ and $\displaystyle\int_0^4 f(x) \, dx = 15$, then $\displaystyle\int_4^6 f(x) \, dx =$ _____.

15. $\displaystyle\int_{-1}^1 \left\{ \int_0^x e^{-t} \, dt \right\} dx =$ _____ and $\displaystyle\int_{-1}^1 \frac{d}{dx} \left\{ \int_0^x e^{-t} \, dt \right\} dx =$ _____.

16. For $t > 0$, the net signed area $\displaystyle\int_0^t (x^3 - x^2) \, dx = 0$ when $t =$ _____.

C. Exercises

In Problems 1–20, evaluate the given integral.

1. $\displaystyle\int_{-1}^1 (4x^3 - 6x^2 + 2x - 1) \, dx$

2. $\displaystyle\int_1^9 \frac{6}{\sqrt{x}} \, dx$

3. $\displaystyle\int (5t + 1)^{100} \, dt$

4. $\displaystyle\int w^2 \sqrt{3w^3 + 1} \, dw$

5. $\displaystyle\int_0^{\pi/4} (\sin 2x - 5 \cos 4x) \, dx$

6. $\displaystyle\int_{\pi^2/9}^{\pi^2} \frac{\sin \sqrt{z}}{\sqrt{z}} \, dz$

7. $\displaystyle\int_4^4 (-2x^2 + x^{1/2}) \, dx$

8. $\displaystyle\int_{-\pi/4}^{\pi/4} dx + \int_{-\pi/4}^{\pi/4} \tan^2 x \, dx$

9. $\displaystyle\int \cot^6 8x \csc^2 8x \, dx$

10. $\displaystyle\int \csc 3x \cot 3x \, dx$

11. $\displaystyle\int (4x^2 - 16x + 7)^4(x - 2) \, dx$

12. $\displaystyle\int (x^2 + 2x - 10)^{2/3}(5x + 5) \, dx$

13. $\displaystyle\int \frac{x^2 + 1}{\sqrt[3]{x^3 + 3x - 16}} \, dx$

14. $\displaystyle\int \frac{x^2 + 1}{x^3 + 3x - 16} \, dx$

15. $\displaystyle\int_0^4 \frac{x}{16 + x^2} \, dx$

16. $\displaystyle\int_0^4 \frac{1}{16 + x^2} \, dx$

17. $\displaystyle\int_0^2 \frac{1}{\sqrt{16 - x^2}} \, dx$

18. $\displaystyle\int_0^2 \frac{x}{\sqrt{16 - x^2}} \, dx$

19. $\displaystyle\int \tan 10x \, dx$

20. $\displaystyle\int \cot 10x \, dx$

21. Suppose $\displaystyle\int_0^5 f(x) \, dx = -3$ and $\displaystyle\int_0^7 f(x) \, dx = 2$. Evaluate $\displaystyle\int_5^7 f(x) \, dx$.

22. Suppose $\displaystyle\int_1^4 f(x) \, dx = 2$ and $\displaystyle\int_4^9 f(x) \, dx = -8$. Evaluate $\displaystyle\int_1^9 f(x) \, dx$.

In Problems 23–28, evaluate the given integral.

23. $\displaystyle\int_0^3 (1 + |x - 1|) \, dx$

24. $\displaystyle\int_0^1 \frac{d}{dt}\left[\frac{10t^4}{(2t^3 + 6t + 1)^2}\right] dt$

25. $\displaystyle\int_{\pi/2}^{\pi/2} \frac{\sin^{10} t}{16t^7 + 1} \, dt$

26. $\displaystyle\int_{-1}^1 t^5 \sin t^2 \, dt$

27. $\displaystyle\int_{-1}^1 \frac{1}{1 + 3x^2} \, dx$

28. $\displaystyle\int_{-2}^2 f(x) \, dx$, where $f(x) = \begin{cases} x^3, & x \le 0 \\ x^2, & 0 < x \le 1 \\ x, & x > 1 \end{cases}$

In Problems 29 and 30, find the given limit.

29. $\displaystyle\lim_{n\to\infty} \frac{1 + 2 + 3 + \cdots + n}{n^2}$

30. $\displaystyle\lim_{n\to\infty} \frac{1^2 + 2^2 + 3^2 + \cdots + n^2}{n^3}$

31. A bucket with dimensions (in feet) shown in FIGURE 5.R.1 is filled at a constant rate of $dV/dt = \frac{1}{4}$ ft³/min. At $t = 0$ the scale reads 31.2 lb. If water weighs 62.4 lb/ft³, what does the scale read at the end of 8 min? When is the bucket full? [*Hint*: See page RP-2 for the formula for the volume of a frustum of a cone. Also, ignore the weight of the bucket.]

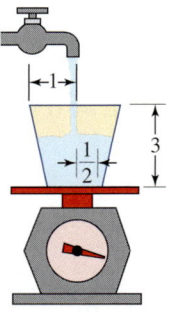

FIGURE 5.R.1 Bucket and scale in Problem 31

32. The **Tower of Hanoi** is a stack of circular disks, each larger than the one above it, set on a pole through holes in the disks' centers. See FIGURE 5.R.2. An ancient king once commanded that such a tower be built of gold disks to the following specifications: Each disk was to be one finger width thick, and the diameter of each disk was to be one finger width larger than the disk above it. The hole through the centers of the disks was to be one finger width in diameter, and the top disk was to be two finger widths in diameter. Assume that a finger width is 1.5 cm and gold weighs 19.3 g/cm³ and is valued at $14 per gram.

(a) Find a formula for the value of gold in the king's Tower of Hanoi if the tower has n disks.

(b) The usual number of gold disks in a Tower of Hanoi is 64. What is the value of the gold in this tower?

FIGURE 5.R.2 Tower of Hanoi in Problem 32

33. Consider the one-to-one function $f(x) = x^3 + x$ on the interval $[1, 2]$. See **FIGURE 5.R.3**. Without finding f^{-1}, determine the value of

$$\int_{f(1)}^{f(2)} f^{-1}(x)\,dx.$$

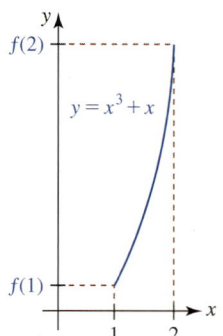

FIGURE 5.R.3 Graph for Problem 33

Applications of the Integral

In This Chapter   Although we return to the problem of finding areas by definite integration in Section 6.2, you will see in the subsequent sections of this chapter that a definite integral has many other interpretations besides area.

We begin the chapter with an application of the indefinite integral.

6.1 Rectilinear Motion Revisited

■ **Introduction** We began Chapter 4, *Applications of the Derivative*, with the notion of rectilinear motion. If $s = f(t)$ is the position function of an object moving rectilinearly—that is, in a straight line, then we know

$$\text{velocity} = v(t) = \frac{ds}{dt} \quad \text{and} \quad \text{acceleration} = a(t) = \frac{dv}{dt}.$$

As an immediate consequence of the definition of an antiderivative, the quantities s and v can be written as indefinite integrals

$$s(t) = \int v(t) \, dt \quad \text{and} \quad v(t) = \int a(t) \, dt. \tag{1}$$

By knowing the **initial position** $s(0)$ and the **initial velocity** $v(0)$, we can find specific values of the constants of integration used in (1).

Recall that when a body moves horizontally on a line, the positive direction is to the right. For motion in a vertical line, we take the positive direction to be upward. As shown in FIGURE 6.1.1, if an arrow is shot upward from ground level, then **initial conditions** are $s(0) = 0$, $v(0) > 0$, whereas if the arrow is shot downward from some initial height, say h meters off the ground, then the initial conditions are $s(0) = h$, $v(0) < 0$. A body that moves in a vertical line close to the surface of the earth, such as the arrow shot upward, is acted upon by the force of gravity. This force causes a body to accelerate. Near the surface of the earth the acceleration due to gravity, $a(t) = -g$, is assumed to be a constant. The magnitude g of this acceleration is approximately

$$32 \text{ ft/s}^2, \quad 9.8 \text{ m/s}^2, \quad \text{or} \quad 980 \text{ cm/s}^2.$$

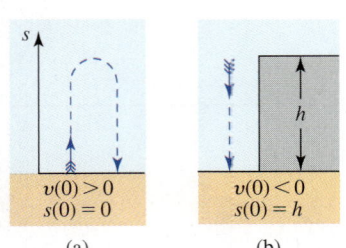

$v(0) > 0$
$s(0) = 0$
(a)

$v(0) < 0$
$s(0) = h$
(b)

FIGURE 6.1.1 Initial conditions

EXAMPLE 1 Projectile Motion

A projectile is shot vertically upward from ground level with an initial velocity of 49 m/s. What is its velocity at $t = 2$ s? What is the maximum height attained by the projectile? How long is the projectile in the air? What is its impact velocity?

Solution Starting with $a(t) = -9.8$ we obtain by indefinite integration,

$$v(t) = \int (-9.8) \, dt = -9.8t + C_1. \tag{2}$$

From the given initial condition $v(0) = 49$, we see that (2) implies $C_1 = 49$. Hence,

$$v(t) = -9.8t + 49,$$

and so $v(2) = -9.8(2) + 49 = 29.4$ m/s. Notice that $v(2) > 0$ implies the projectile is traveling upward.

Now, the height of the projectile, measured from ground level, is the indefinite integral of the velocity function,

$$s(t) = \int v(t) \, dt = \int (-9.8t + 49) \, dt = -4.9t^2 + 49t + C_2. \tag{3}$$

Since the projectile starts from ground level, $s(0) = 0$ and (3) gives $C_2 = 0$. Hence,

$$s(t) = -4.9t^2 + 49t. \tag{4}$$

When air resistance is ignored, the magnitude of the impact velocity (speed) is the same as the initial upward velocity from ground level. See Problem 32 in Exercises 6.1. This is not true when air resistance is taken into consideration.

When the projectile attains its maximum height, $v(t) = 0$. Solving $-9.8t + 49 = 0$ then gives $t = 5$. From (4) we find the corresponding height to be $s(5) = 122.5$ m.

Finally, to find the time that the projectile hits the ground, we solve $s(t) = 0$ or $-4.9t^2 + 49t = 0$. Writing the latter equation as $-4.9t(t - 10) = 0$, we see the projectile is in the air for 10 s. The impact velocity is $v(10) = -49$ m/s. ∎

EXAMPLE 2   Projectile Motion

A tennis ball is thrown vertically downward from a height of 54 ft with an initial velocity of 8 ft/s. What is its impact velocity if it hits a 6-ft-tall person on the head? See FIGURE 6.1.2.

Solution   In this case $a(t) = -32$, $s(0) = 54$, and, since the ball is thrown downward, $v(0) = -8$. Now

$$v(t) = \int (-32)\, dt = -32t + C_1.$$

Using the initial velocity $v(0) = -8$, we find $C_1 = -8$. Therefore,

$$v(t) = -32t - 8.$$

Continuing, we find

$$s(t) = \int (-32t - 8)\, dt = -16t^2 - 8t + C_2.$$

When $t = 0$, we know $s = 54$ and so the last equation implies $C_2 = 54$. Hence,

$$s(t) = -16t^2 - 8t + 54.$$

To determine the time that corresponds to $s = 6$ we solve

$$-16t^2 - 8t + 54 = 6.$$

Simplifying gives $-8(2t - 3)(t + 2) = 0$ and $t = \frac{3}{2}$. The velocity of the ball when it hits the person is then $v\left(\frac{3}{2}\right) = -56$ ft/s. ■

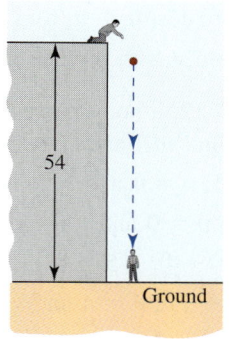

FIGURE 6.1.2   Thrown ball in Example 2

▌ Distance   The **total distance** an object travels rectilinearly in a time interval $[t_1, t_2]$ is given by the definite integral

$$\text{total distance} = \int_{t_1}^{t_2} |v(t)|\, dt. \qquad (5)$$

The absolute value is necessary in (5), since the object may be moving to the left and hence has negative velocity for some part of the time.

EXAMPLE 3   Distance Traveled

The position function of an object that moves on a coordinate line is $s(t) = t^2 - 6t$, where s is measured in centimeters and t in seconds. Find the distance traveled in the time interval $[0, 9]$.

Solution   The velocity function $v(t) = ds/dt = 2t - 6 = 2(t - 3)$ shows that the motion is as indicated in FIGURE 6.1.3; namely, $v < 0$ for $0 \le t < 3$ (motion to the left) and $v \ge 0$ for $3 \le t \le 9$ (motion to the right). Hence, from (5) the distance traveled is

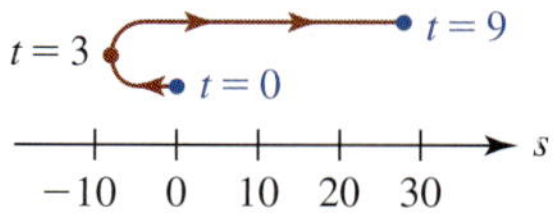

$$\int_0^9 |2t - 6|\, dt = \int_0^3 |2t - 6|\, dt + \int_3^9 |2t - 6|\, dt$$

$$= \int_0^3 -(2t - 6)\, dt + \int_3^9 (2t - 6)\, dt$$

$$= (-t^2 + 6t)\Big]_0^3 + (t^2 - 6t)\Big]_3^9 = 45 \text{ cm.}$$

FIGURE 6.1.3   Representation of the motion of the object in Example 3

Of course, the last result must be consistent with the number obtained by simply counting units in Figure 6.1.3 between $s(0)$ and $s(3)$, and between $s(3)$ and $s(9)$. ■

Exercises 6.1   Answers to selected odd-numbered problems begin on page ANS-20.

≡ Fundamentals

In Problems 1–6, a body moves in a straight line with velocity $v(t)$. Find the position function $s(t)$.

1. $v(t) = 6$; $s = 5$ when $t = 2$

2. $v(t) = 2t + 1$; $s = 0$ when $t = 1$

3. $v(t) = t^2 - 4t$; $s = 6$ when $t = 3$

4. $v(t) = \sqrt{4t + 5}$; $s = 2$ when $t = 1$

5. $v(t) = -10\cos(4t + \pi/6)$; $s = \frac{5}{4}$ when $t = 0$

6. $v(t) = 2\sin 3t$; $s = 0$ when $t = \pi$

In Problems 7–12, a body moves in a straight line with acceleration $a(t)$. Find $v(t)$ and $s(t)$.

7. $a(t) = -5$; $v = 4$ and $s = 2$ when $t = 1$

8. $a(t) = 6t$; $v = 0$ and $s = -5$ when $t = 2$

9. $a(t) = 3t^2 - 4t + 5$; $v = -3$ and $s = 10$ when $t = 0$

10. $a(t) = (t - 1)^2$; $v = 4$ and $s = 6$ when $t = 1$

11. $a(t) = 7t^{1/3} - 1$; $v = 50$ and $s = 0$ when $t = 8$

12. $a(t) = 100\cos 5t$; $v = -20$ and $s = 15$ when $t = \pi/2$

In Problems 13–18, an object moves in a straight line according to the given position function. If s is measured in centimeters, find the total distance traveled by the object in the indicated time interval.

13. $s(t) = t^2 - 2t$; [0, 5]

14. $s(t) = -t^2 + 4t + 7$; [0, 6]

15. $s(t) = t^3 - 3t^2 - 9t$; [0, 4]

16. $s(t) = t^4 - 32t^2$; [1, 5]

17. $s(t) = 6\sin\pi t$; [1, 3]

18. $s(t) = (t - 3)^2$; [2, 7]

≡ Applications

19. A driver of a car that is traveling in a straight line at a constant 60 mi/h takes his eyes off the road for 2 s. How many feet does the car move in this time?

20. A ball is dropped (released from rest) from a height of 144 ft. How long does it take for the ball to hit the ground? At what speed does it hit the ground?

21. An egg is dropped from the top of a building and hits the ground 4 s from release. How tall is the building?

22. A stone is dropped into a well and the splash is heard 2 s later. If the speed of sound in air is 1080 ft/s, find the depth of the well.

23. An arrow is projected vertically upward from ground level with an initial velocity of 24.5 m/s. How high does it rise?

24. How high would the arrow in Problem 23 rise on the planet Mars where $g = 3.6$ m/s²?

25. A golf ball is thrown vertically upward from the edge of the roof of a 384-ft-high building with an initial velocity of 32 ft/s. What is the maximum height attained by the ball? At what time does the ball hit the ground?

26. In Problem 25, what is the velocity of the golf ball as it passes an observer in a window that is 256 ft off the ground?

27. A person throws a marshmallow vertically downward with an initial velocity of 16 ft/s from a window that is 102 ft

off the ground. If the marshmallow hits a 6-ft-tall person on the head, what is the impact velocity?

28. The person hit on the head in Problem 27 climbs to the top of a 22-ft-high ladder and throws a stone vertically upward with an initial velocity of 96 ft/s. If the stone hits the culprit at the 102-ft level, what is the impact velocity?

≡ Think About It

29. In March 1979, the Voyager 1 space probe photographed an active volcanic eruption on Io, one of the moons of Jupiter. Find the ejection velocity of a rock from the volcano Loki if the rock attains a height of 200 km above the summit of the volcano. On Io the acceleration due to gravity is $g = 1.8$ m/s².

30. As shown in FIGURE 6.1.4, at a point 30 ft from a 25-ft-tall street lamp a ball is thrown vertically downward from a height of 25 ft with an initial velocity of 2 ft/s.

    (a) Find the rate at which the shadow of the ball is moving toward the base of the street lamp.

    (b) Find the rate at which the shadow of the ball is moving toward the base of the street lamp at $t = \frac{1}{2}$.

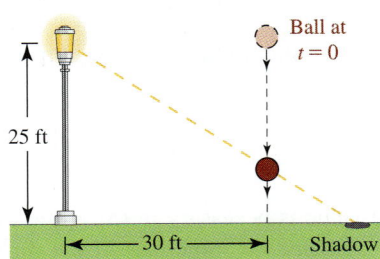

FIGURE 6.1.4   Street lamp in Problem 30

31. If a body is moving rectilinearly with a constant acceleration a and $v = v_0$ when $s = 0$, show that

$$v^2 = v_0^2 + 2as. \quad \left[\text{Hint: } \frac{dv}{dt} = \frac{dv}{ds}\frac{ds}{dt} = \frac{dv}{ds}v.\right]$$

32. Show that, when air resistance is ignored, a projectile shot vertically upward from ground level hits the ground again with a speed equal to the initial velocity v_0.

33. Suppose the acceleration due to gravity on a planet is one-half that on the Earth. Prove that a ball tossed vertically upward from the surface of the planet would attain a maximum height twice that on the Earth when the same initial velocity is used.

34. In Problem 33, suppose the initial velocity of the ball on the planet is v_0 and the initial velocity of the ball on the Earth is $2v_0$. Compare the maximum heights attained. Determine the initial velocity of the ball on the Earth (in terms of v_0) so that the maximum height attained is the same as on the planet.

6.2 Area Revisited

Introduction If a function f takes on both positive and negative values on $[a, b]$, then the definite integral $\int_a^b f(x)\,dx$ does not represent the area under the graph of f on the interval. As we saw in Section 5.4, we can interpret the value of $\int_a^b f(x)\,dx$ as the *net signed area* between the graph of f and the x-axis on the interval $[a, b]$. In this section we investigate two area problems:

- *Find the **total area** of a region bounded by the graph of f and the x-axis on an interval $[a, b]$.*
- *Find the **area of the region** bounded between two graphs on an interval $[a, b]$.*

We will see that the first problem is just a special case of the second problem.

Total Area Suppose the function $y = f(x)$ is continuous on the interval $[a, b]$ and that $f(x) < 0$ on $[a, c)$ and $f(x) \geq 0$ on $[c, b]$. The **total area** is the area of the region bounded by the graph of f, the x-axis, and the vertical lines $x = a$ and $x = b$. To find this area we employ the absolute value of the function $y = |f(x)|$, which is nonnegative for all x in $[a, b]$. Recall, $|f(x)|$ is defined in a piecewise manner. For the function f shown in FIGURE 6.2.1(a), $f(x) < 0$ on the interval $[a, c)$ and $f(x) \geq 0$ on the interval $[c, b]$. Thus,

$$|f(x)| = \begin{cases} -f(x), & \text{for } f(x) < 0 \\ f(x), & \text{for } f(x) \geq 0. \end{cases} \qquad (1)$$

As shown in Figure 6.2.1(b) the graph of $y = |f(x)|$ on the interval $[a, c)$ is obtained by reflecting that portion of the graph of $y = f(x)$ through the x-axis. On the interval $[c, b]$, where $f(x) \geq 0$, the graphs of $y = f(x)$ and $y = |f(x)|$ are the same. To find the total area $A = A_1 + A_2$ shown in Figure 6.2.1(b) we use the additive interval property of the definite integral along with (1):

◀ See Theorem 5.4.5.

$$\int_a^b |f(x)|\,dx = \int_a^c |f(x)|\,dx + \int_c^b |f(x)|\,dx$$

$$= \int_a^c (-f(x))\,dx + \int_c^b f(x)\,dx$$

$$= A_1 + A_2.$$

We summarize the ideas of the preceding discussion in the following definition.

Definition 6.2.1 Total Area

If $y = f(x)$ is continuous on $[a, b]$, then the **total area** A bounded by its graph and the x-axis on the interval is given by

$$A = \int_a^b |f(x)|\,dx. \qquad (2)$$

EXAMPLE 1 Total Area

Find the total area bounded by the graph of $y = x^3$ and the x-axis on $[-2, 1]$.

Solution From (2) we have

$$A = \int_{-2}^1 |x^3|\,dx.$$

In FIGURE 6.2.2 we have compared the graph of $y = x^3$ with the graph of $y = |x^3|$. Since $x^3 < 0$ for $x < 0$, we have on $[-2, 1]$,

$$|f(x)| = \begin{cases} -x^3, & -2 \leq x < 0 \\ x^3, & 0 \leq x \leq 1. \end{cases}$$

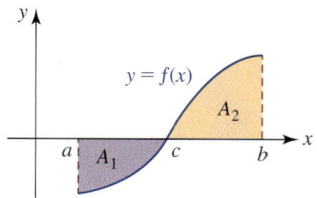

(a) The definite integral of f on $[a, b]$ is not area

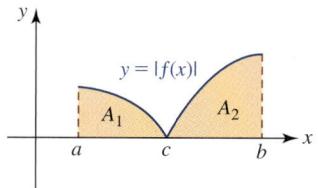

(b) The definite integral of $|f|$ on $[a, b]$ is area

FIGURE 6.2.1 Total area is $A = A_1 + A_2$

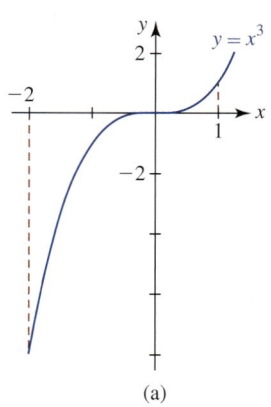

(a)

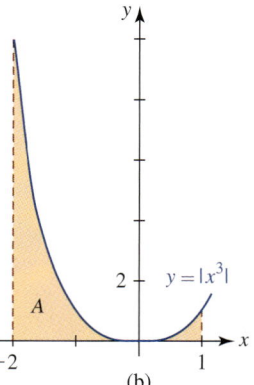

(b)

FIGURE 6.2.2 Graph of function and area in Example 1

Thus, by (2) of Definition 6.2.1 the desired area is

$$A = \int_{-2}^{1} |x^3| \, dx$$

$$= \int_{-2}^{0} |x^3| \, dx + \int_{0}^{1} |x^3| \, dx$$

$$= \int_{-2}^{0} (-x^3) \, dx + \int_{0}^{1} x^3 \, dx$$

$$= -\frac{1}{4}x^4 \Big]_{-2}^{0} + \frac{1}{4}x^4 \Big]_{0}^{1}$$

$$= 0 - \left(-\frac{16}{4}\right) + \frac{1}{4} - 0 = \frac{17}{4}. \qquad \blacksquare$$

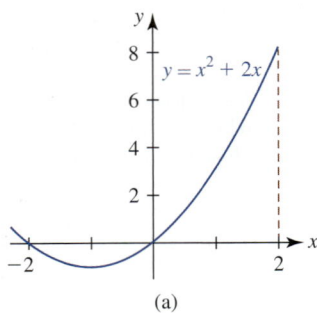

(a)

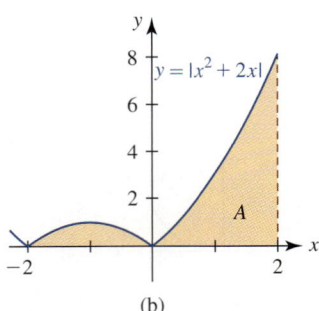

(b)

FIGURE 6.2.3 Graphs and area in Example 2

EXAMPLE 2 Total Area

Find the total area bounded by the graph of $y = x^2 + 2x$ and the x-axis on $[-2, 2]$.

Solution The graphs of $y = f(x)$ and $y = |f(x)|$ are given in FIGURE 6.2.3. Now, from Figure 6.2.3(a) we see that on $[-2, 2]$,

$$|f(x)| = \begin{cases} -(x^2 + 2x), & -2 \le x < 0 \\ x^2 + 2x, & 0 \le x \le 2. \end{cases}$$

Therefore, the total area bounded by the graph of f on the interval $[-2, 2]$ and the x-axis is

$$A = \int_{-2}^{2} |x^2 + 2x| \, dx$$

$$= \int_{-2}^{0} |x^2 + 2x| \, dx + \int_{0}^{2} |x^2 + 2x| \, dx$$

$$= \int_{-2}^{0} -(x^2 + 2x) \, dx + \int_{0}^{2} (x^2 + 2x) \, dx$$

$$= \left(-\frac{1}{3}x^3 - x^2\right)\Big]_{-2}^{0} + \left(\frac{1}{3}x^3 + x^2\right)\Big]_{0}^{2}$$

$$= 0 - \left(\frac{8}{3} - 4\right) + \left(\frac{8}{3} + 4\right) - 0 = 8. \qquad \blacksquare$$

▌ **Area Bounded by Two Graphs** The foregoing discussion is a special case of the more general problem of finding the **area of the region** bounded between the graphs of two functions f and g and the vertical lines $x = a$ and $x = b$. See FIGURE 6.2.4(a). The area *under* the graph of a continuous nonnegative function $y = f(x)$ on an interval $[a, b]$ can be interpreted as the area of

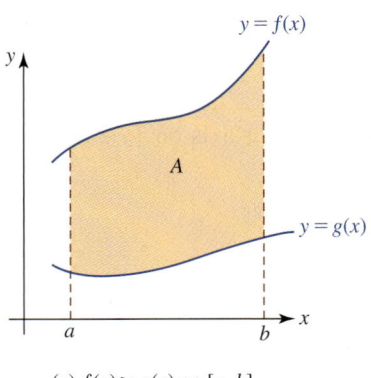

(a) $f(x) \ge g(x)$ on $[a, b]$

(b) Constructing n rectangles between two graphs

FIGURE 6.2.4 Area A bounded between two graphs

the region bounded by the graph of f and the graph of the function $y = 0$ (the x-axis) and the vertical lines $x = a$ and $x = b$.

■ **Building an Integral** Suppose $y = f(x)$ and $y = g(x)$ are continuous on $[a, b]$ and that $f(x) \geq g(x)$ for all x in the interval. Let P be a partition of the interval $[a, b]$ into n subintervals $[x_{k-1}, x_k]$. If we choose a sample point x_k^* in each subinterval, we can then construct n corresponding rectangles that have area

The assumption that $f(x) \geq g(x)$ on the interval means that the graphs of f and g can touch but do not cross each other.

$$A_k = [f(x_k^*) - g(x_k^*)]\Delta x_k.$$

See Figure 6.2.4(b). The area A of the region bounded by the two graphs on the interval $[a, b]$ is approximated by the Riemann sum

$$\sum_{k=1}^{n} A_k = \sum_{k=1}^{n} [f(x_k^*) - g(x_k^*)]\Delta x_k,$$

and this in turn suggests that the area is

$$A = \lim_{\|P\| \to 0} \sum_{k=1}^{n} [f(x_k^*) - g(x_k^*)]\Delta x_k.$$

Since f and g are continuous, so is $f - g$. Hence, the above limit exists and is, by definition, the definite integral

$$A = \int_a^b [f(x) - g(x)]\,dx. \tag{3}$$

Also, (3) applies to regions for which one or both of the functions f and g have negative values. See FIGURE 6.2.5. However, (3) is *not* valid on an interval $[a, b]$ where the graphs of f and g cross each other on the interval. Notice in FIGURE 6.2.6 that g is the upper graph on the intervals (a, c_1) and (c_2, b), whereas f is the upper graph on the interval (c_1, c_2). In the most general case, we have the following definition.

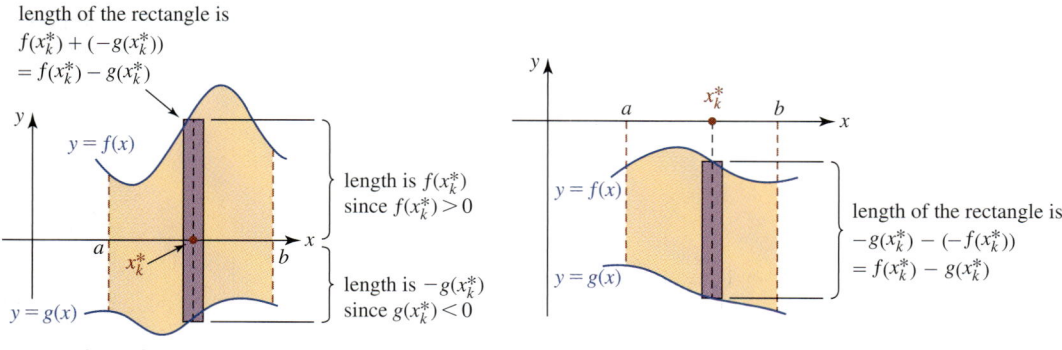

(a) $f(x) > 0$ and $g(x) < 0$ on $[a, b]$ (b) $f(x) < 0$ and $g(x) < 0$ on $[a, b]$

FIGURE 6.2.5 Graphs of f and g can be below the x-axis

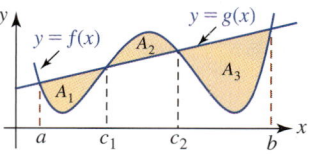

FIGURE 6.2.6 Graphs of f and g cross each other on $[a, b]$

Definition 6.2.2 Area Bounded by Two Graphs

If f and g are continuous functions on an interval $[a, b]$, then the **area A of the region bounded by their graphs** on the interval is given by

$$A = \int_a^b |f(x) - g(x)|\,dx. \tag{4}$$

Note that (4) reduces to (2) when $g(x) = 0$ for all x in $[a, b]$. Before using formulas (3) or (4), you are urged to sketch the necessary graphs. If the curves cross on the interval, then as we have seen in Figure 6.2.6 the relative position of the curves changes. In any event, on any subinterval of $[a, b]$ the appropriate integrand is always

(upper graph) − (lower graph).

As in (1), the absolute value of the integrand is given by

$$|f(x) - g(x)| = \begin{cases} -(f(x) - g(x)), & \text{for } f(x) - g(x) < 0 \\ f(x) - g(x), & \text{for } f(x) - g(x) \geq 0. \end{cases} \tag{5}$$

A more practical way of interpreting (5) is to draw the graphs of f and g accurately and visually determine that:

$$|f(x) - g(x)| = \begin{cases} g(x) - f(x), & \text{whenever } g \text{ is the upper graph} \\ f(x) - g(x), & \text{whenever } f \text{ is the upper graph.} \end{cases}$$

In Figure 6.2.6, the area A bounded by the graphs of f and g on $[a, b]$ is

$$A = \int_a^b |f(x) - g(x)|\, dx$$

$$= \int_a^{c_1} |f(x) - g(x)|\, dx + \int_{c_1}^{c_2} |f(x) - g(x)|\, dx + \int_{c_2}^b |f(x) - g(x)|\, dx$$

$$= \underbrace{\int_a^{c_1} [g(x) - f(x)]\, dx}_{g \text{ is the upper graph}} + \underbrace{\int_{c_1}^{c_2} [f(x) - g(x)]\, dx}_{f \text{ is the upper graph}} + \underbrace{\int_{c_2}^b [g(x) - f(x)]\, dx}_{g \text{ is the upper graph}}.$$

EXAMPLE 3   Area Bounded by Two Graphs

Find the area of the region bounded by the graphs of $y = \sqrt{x}$ and $y = x^2$.

Solution   As shown in FIGURE 6.2.7, the region in question is located in the first quadrant. Because 0 and 1 are the solutions of the equation $x^2 = \sqrt{x}$, the graphs intersect at the points $(0, 0)$ and $(1, 1)$. In other words, the region lies between the vertical lines $x = 0$ and $x = 1$. Since $y = \sqrt{x}$ is the upper graph on the interval $(0, 1)$, it follows that

$$A = \int_0^1 (\sqrt{x} - x^2)\, dx$$

$$= \left(\frac{2}{3} x^{3/2} - \frac{1}{3} x^3 \right) \Big]_0^1$$

$$= \frac{2}{3} - \frac{1}{3} - 0 = \frac{1}{3}.$$

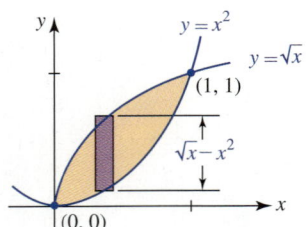

FIGURE 6.2.7   Area in Example 3

EXAMPLE 4   Area Bounded by Two Graphs

Find the area of the region bounded by the graphs of $y = x^2 + 2x$ and $y = -x + 4$ on the interval $[-4, 2]$.

Solution   Let us denote the given functions by

$$y_1 = x^2 + 2x \qquad \text{and} \qquad y_2 = -x + 4.$$

As FIGURE 6.2.8 shows, the graphs cross each other on the interval $[-4, 2]$.

To find the points of intersection we solve the equation $x^2 + 2x = -x + 4$ or $x^2 + 3x - 4 = 0$ and find that $x = -4$ and $x = 1$. The area in question is the sum of the areas $A = A_1 + A_2$:

$$A = \int_{-4}^2 |y_2 - y_1|\, dx = \int_{-4}^1 |y_2 - y_1|\, dx + \int_1^2 |y_2 - y_1|\, dx.$$

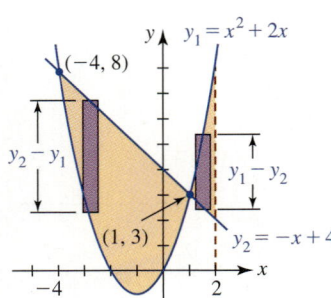

FIGURE 6.2.8   Area in Example 4

But since $y_2 = -x + 4$ is the upper graph on the interval $(-4, 1)$ and $y_1 = x^2 + 2x$ is the upper graph on the interval $(1, 2)$ we can write

$$A = \int_{-4}^{1} [(-x + 4) - (x^2 + 2x)]\, dx + \int_{1}^{2} [(x^2 + 2x) - (-x + 4)]\, dx$$

$$= \int_{-4}^{1} (-x^2 - 3x + 4)\, dx + \int_{1}^{2} (x^2 + 3x - 4)\, dx$$

$$= \left(-\frac{1}{3}x^3 - \frac{3}{2}x^2 + 4x\right)\Big]_{-4}^{1} + \left(\frac{1}{3}x^3 + \frac{3}{2}x^2 - 4x\right)\Big]_{1}^{2}$$

$$= \left(-\frac{1}{3} - \frac{3}{2} + 4\right) - \left(\frac{64}{3} - 24 - 16\right) + \left(\frac{8}{3} + 6 - 8\right) - \left(\frac{1}{3} + \frac{3}{2} - 4\right) = \frac{71}{3}. \quad \blacksquare$$

EXAMPLE 5   Area Bounded by Two Graphs

Find the area of the four regions bounded by the graphs of $y = \sin x$ and $y = \cos x$ shown in FIGURE 6.2.9.

Solution   There are an infinite number of such regions bounded by the graphs of $y = \sin x$ and $y = \cos x$ and the area of each region is the same. Therefore, we need only find the area of the region on the interval corresponding to the first two positive solutions of the equation $\sin x = \cos x$. By dividing by $\cos x$, a more useful form of the last equation is $\tan x = 1$. The first positive solution is $x = \tan^{-1} 1 = \pi/4$. Then since $\tan x$ is π-periodic, the next positive solution is $x = \pi + \pi/4 = 5\pi/4$. On the interval $(\pi/4, 5\pi/4)$, $y = \sin x$ is the upper graph so the area of the four regions is

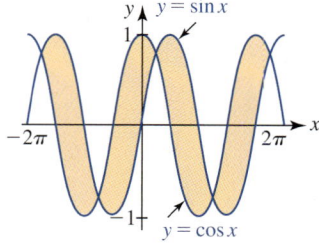

FIGURE 6.2.9   Each of the four regions has the same area in Example 5

$$A = 4\int_{\pi/4}^{5\pi/4} (\sin x - \cos x)\, dx$$

$$= 4(-\cos x - \sin x)\Big]_{\pi/4}^{5\pi/4}$$

$$= 4(2\sqrt{2}) = 8\sqrt{2}. \quad \blacksquare$$

In finding the area bounded by two graphs, it is not always convenient to integrate with respect to the variable x.

EXAMPLE 6   Area Bounded by Two Graphs

Find the area of the region bounded by the graphs of $y^2 = 1 - x$ and $2y = x + 2$.

Solution   We note that the equation $y^2 = 1 - x$ implicitly defines two functions, $y_2 = \sqrt{1 - x}$ and $y_1 = -\sqrt{1 - x}$ for $x \le 1$. If we define $y_3 = \frac{1}{2}x + 1$, we see from FIGURE 6.2.10 that the height of an element of area on the interval $(-8, 0)$ is $y_3 - y_1$, whereas the height of an element on the interval $(0, 1)$ is $y_2 - y_1$. Thus, if we integrate with respect to x, the desired area is the sum of

$$A_1 = \int_{-8}^{0} (y_3 - y_1)\, dx \qquad \text{and} \qquad A_2 = \int_{0}^{1} (y_2 - y_1)\, dx.$$

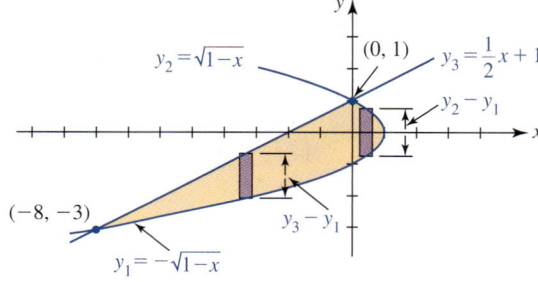

FIGURE 6.2.10   In Example 6 y_3 is the upper graph on the interval $(-8, 0)$; y_2 is the upper graph on the interval $(0, 1)$

Thus the area of the region is the sum of the areas $A = A_1 + A_2$, that is,

$$A = \int_{-8}^{0}\left[\left(\frac{1}{2}x + 1\right) - (-\sqrt{1-x})\right] dx + \int_{0}^{1}[\sqrt{1-x} - (-\sqrt{1-x})] \, dx$$

$$= \int_{-8}^{0}\left(\frac{1}{2}x + 1 + \sqrt{1-x}\right) dx + 2\int_{0}^{1}\sqrt{1-x} \, dx$$

$$= \left(\frac{1}{4}x^2 + x - \frac{2}{3}(1-x)^{3/2}\right)\Big|_{-8}^{0} - \frac{4}{3}(1-x)^{3/2}\Big|_{0}^{1}$$

$$= -\frac{2}{3}\cdot 1^{3/2} - \left(16 - 8 - \frac{2}{3}\cdot 9^{3/2}\right) - \frac{4}{3}\cdot 0 + \frac{4}{3}\cdot 1^{3/2} = \frac{32}{3}. \qquad \blacksquare$$

EXAMPLE 7 Alternative Solution to Example 6

The necessity of using two integrals in Example 6 to find the area is avoided by construct-ing horizontal rectangles and using y as the independent variable. If we define $x_2 = 1 - y^2$ and $x_1 = 2y - 2$, then, as shown in FIGURE 6.2.11, the area of a horizontal element is

$$A_k = [\text{right graph} - \text{left graph}] \cdot \text{width}.$$

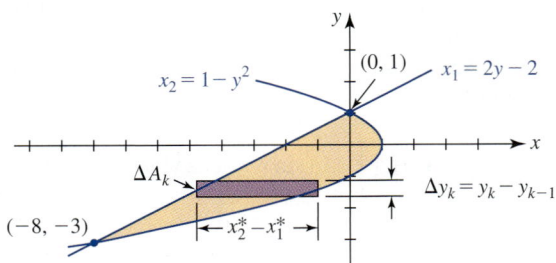

FIGURE 6.2.11 Using y as the variable of integration in Example 7

That is, $$A_k = [x_2^* - x_1^*]\Delta y_k,$$

where $\qquad x_2^* = 1 - (y_k^*)^2, \quad x_1^* = 2y_k^* - 2, \qquad$ and $\qquad \Delta y_k = y_k - y_{k-1}.$

Summing the rectangles in the positive y-direction leads to

$$A = \lim_{\|P\|\to 0}\sum_{k=1}^{n}[x_2^*(y) - x_1^*(y)] \, \Delta y_k,$$

where $\|P\|$ is the norm of a partition P of the interval on the y-axis defined by $-3 \le y \le 1$. In other words,

$$A = \int_{-3}^{1}(x_2 - x_1) \, dy,$$

where the lower limit -3 and the upper limit 1 are the y-coordinates of the points of inter-section $(-8, -3)$ and $(0, 1)$, respectively. Substituting for x_2 and x_1 then gives

$$A = \int_{-3}^{1}[(1 - y^2) - (2y - 2)] \, dy$$

$$= \int_{-3}^{1}(-y^2 - 2y + 3) \, dy$$

$$= \left(-\frac{1}{3}y^3 - y^2 + 3y\right)\Big|_{-3}^{1}$$

$$= \left(-\frac{1}{3} - 1 + 3\right) - \left(\frac{27}{3} - 9 - 9\right) = \frac{32}{3}. \qquad \blacksquare$$

$\int_a^b$ **NOTES FROM THE CLASSROOM**

As mentioned in the introduction, we are going to see different interpretations of the definite integral in this chapter. In each section you will see a variety of definite integrals derived under the paragraph heading *Building an Integral*. Before memorizing such integral formulas you should be aware that the derived result will usually not be applicable to every conceivable geometric or physical situation. For example, as we have seen in Example 7, to find the area of a region in the plane it may be more convenient to integrate with respect to y and you will have to build an entirely different integral. Rather than apply a formula blindly, you should try to understand the process and practice building integrals by analyzing the geometry of each problem.

Exercises 6.2   Answers to selected odd-numbered problems begin on page ANS-20.

≡ **Fundamentals**

In Problems 1–22, find the total area bounded by the graph of the given function and the x-axis on the indicated interval.

1. $y = x^2 - 1$; $[-1, 1]$ **2.** $y = x^2 - 1$; $[0, 2]$

3. $y = x^3$; $[-3, 0]$ **4.** $y = 1 - x^3$; $[0, 2]$

5. $y = x^2 - 3x$; $[0, 3]$

6. $y = -(x + 1)^2$; $[-1, 0]$ **7.** $y = x^3 - 6x$; $[-1, 1]$

8. $y = x^3 - 3x^2 + 2$; $[0, 2]$

9. $y = (x - 1)(x - 2)(x - 3)$; $[0, 3]$

10. $y = x(x + 1)(x - 1)$; $[-1, 1]$

11. $y = \dfrac{x^2 - 1}{x^2}$; $[\frac{1}{2}, 3]$ **12.** $y = \dfrac{x^2 - 1}{x^2}$; $[1, 2]$

13. $y = \sqrt{x} - 1$; $[0, 4]$ **14.** $y = 2 - \sqrt{x}$; $[0, 9]$

15. $y = \sqrt[3]{x}$; $[-2, 3]$ **16.** $y = 2 - \sqrt[3]{x}$; $[-1, 8]$

17. $y = \sin x$; $[-\pi, \pi]$

18. $y = 1 + \cos x$; $[0, 3\pi]$

19. $y = -1 + \sin x$; $[-3\pi/2, \pi/2]$

20. $y = \sec^2 x$; $[0, \pi/3]$

21. $y = \begin{cases} x, & -2 \le x < 0 \\ x^2, & 0 \le x \le 1 \end{cases}$; $[-2, 1]$

22. $y = \begin{cases} x + 2, & -3 \le x < 0 \\ 2 - x^2, & 0 \le x \le 2 \end{cases}$; $[-3, 2]$

In Problems 23–50, find the area of the region bounded by the graphs of the given functions.

23. $y = x, y = -2x, x = 3$ **24.** $y = x, y = 4x, x = 2$

25. $y = x^2, y = 4$ **26.** $y = x^2, y = x$

27. $y = x^3, y = 8, x = -1$

28. $y = x^3, y = \sqrt[3]{x}$, first quadrant

29. $y = 4(1 - x^2), y = 1 - x^2$

30. $y = 2(1 - x^2), y = x^2 - 1$

31. $y = x, y = 1/x^2, x = 3$

32. $y = x^2, y = 1/x^2, y = 9$, first quadrant

33. $y = -x^2 + 6, y = x^2 + 4x$ **34.** $y = x^2, y = -x^2 + 3x$

35. $y = x^{2/3}, y = 4$

36. $y = 1 - x^{2/3}, y = x^{2/3} - 1$

37. $y = x^2 - 2x - 3, y = 2x + 2$, on $[-1, 6]$

38. $y = -x^2 + 4x, y = \frac{3}{2}x$

39. $y = x^3, y = x + 6, y = -\frac{1}{2}x$

40. $x = y^2, x = 0, y = 1$

41. $x = -y, x = 2 - y^2$

42. $x = y^2, x = 6 - y^2$

43. $x = y^2 + 2y + 2, x = -y^2 - 2y + 2$

44. $x = y^2 - 6y + 1, x = -y^2 + 2y + 1$

45. $y = x^3 - x, y = x + 4, x = -1, x = 1$

46. $x = y^3 - y, x = 0$

47. $y = \cos x, y = \sin x, x = 0, x = \pi/2$

48. $y = 2 \sin x, y = -x, x = \pi/2$

49. $y = 4 \sin x, y = 2$, on $[\pi/6, 5\pi/6]$

50. $y = 2 \cos x, y = -\cos x$, on $[-\pi/2, \pi/2]$

In Problems 51 and 52, interpret the given definite integral as the area of a region bounded by the graphs of two functions. Sketch two regions that have the area given by the integral.

51. $\displaystyle\int_0^4 (\sqrt{x} + x)\, dx$ **52.** $\displaystyle\int_{-1}^2 \left(\frac{1}{2}x^2 + 3 - x\right) dx$

In Problems 53 and 54, interpret the given definite integral as the area of a region bounded by the graphs of two functions on an interval. Evaluate the given integral and sketch the region.

53. $\displaystyle\int_0^2 \left| \frac{3}{x + 1} - 4x \right| dx$ **54.** $\displaystyle\int_{-1}^1 |e^x - 2e^{-x}|\, dx$

In Problems 55–58, use the fact that the area of a circle of radius r is πr^2 to evaluate the given definite integral. Sketch a region whose area is given by the definite integral.

55. $\displaystyle\int_0^3 \sqrt{9 - x^2}\, dx$ **56.** $\displaystyle\int_{-5}^5 \sqrt{25 - x^2}\, dx$

57. $\displaystyle\int_{-2}^2 \left(1 + \sqrt{4 - x^2}\right) dx$

58. $\displaystyle\int_{-1}^1 \left(2x + 3 - \sqrt{1 - x^2}\right) dx$

59. Set up a definite integral that represents the area of an ellipse $x^2/a^2 + y^2/b^2 = 1$, $a > b > 0$. Use the idea used in Problems 55–58 to evaluate this definite integral.

60. Find the area of the triangle with vertices at $(1, 1)$, $(2, 4)$, and $(3, 2)$.

61. Consider the region bounded by the graphs of $y^2 = -x - 2$, $y = 2$, $y = -2$, and $y = 2(x - 1)$. Compute the area of the region by integrating with respect to x.

62. Compute the area of the region given in Problem 61 by integrating with respect to y.

63. Consider the region bounded by the graphs of $y = 2e^x - 1$, $y = e^x$, and $y = 2$ shown in FIGURE 6.2.12. Express the area of the region as definite integrals first using integration with respect to x and then using integration with respect to y. Choose one of these integral expressions to find the area.

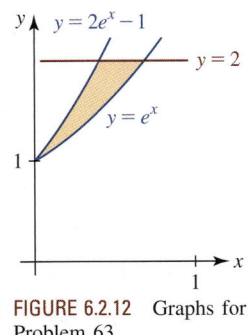

FIGURE 6.2.12 Graphs for Problem 63

≡ Calculator/CAS Problems

64. Use a calculator or CAS to approximate the x-coordinates of the points of intersection of the graphs shown in FIGURE 6.2.13. Find an approximate value of the area of the region.

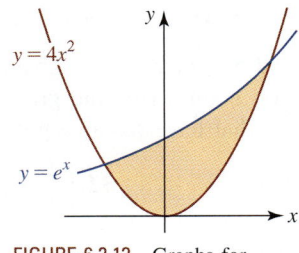

FIGURE 6.2.13 Graphs for Problem 64

≡ Think About It

65. The line segment between Q and R shown in FIGURE 6.2.14 is tangent to the graph of $y = 1/x$ at point P. Show that the area of triangle QOR is independent of the coordinates of P.

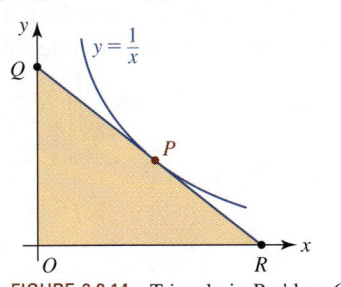

FIGURE 6.2.14 Triangle in Problem 65

66. A trapezoid is bounded by the graphs of $f(x) = Ax + B$, $x = a$, $x = b$, and $x = 0$. Show that the area of the trapezoid is $\dfrac{f(a) + f(b)}{2}(b - a)$.

67. Express the area of the shaded region shown in FIGURE 6.2.15 in terms of the number a. Try to be a little clever.

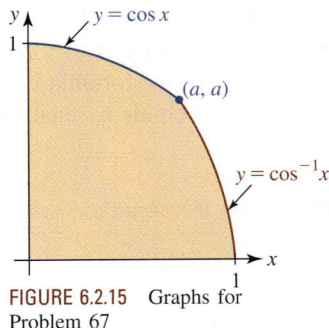

FIGURE 6.2.15 Graphs for Problem 67

68. Suppose the two swaths of paint shown in FIGURE 6.2.16 are done with one stroke using a paintbrush of width k, $k > 0$, over the interval $[a, b]$. In Figure 6.2.16(b) assume that the painted red region is parallel to the x-axis. Which swath has the greater area? Defend your answer with a solid mathematical demonstration. Can you formulate a general principle?

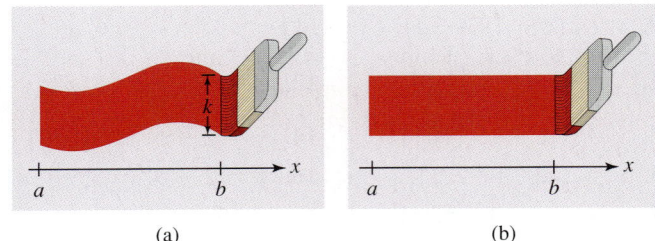

(a)   (b)

FIGURE 6.2.16 Swaths of paint in Problem 68

≡ Projects

69. The Larger Area The points A and B are on a line and the points C and D are on a line parallel to the first line. The points in FIGURE 6.2.17(a) form a rectangle $ABCD$. The points C and D are moved to the left as shown in Figure 6.2.17(b) in such a manner that $ABC'D'$ forms a parallelogram. Discuss: Which has the larger area, the rectangle $ABCD$ or the parallelogram $ABC'D'$?

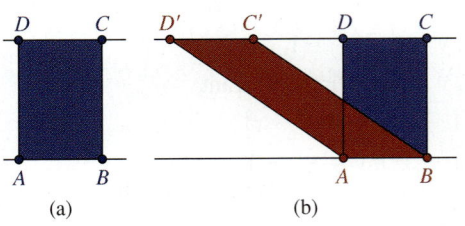

(a)   (b)

FIGURE 6.2.17 Rectangle and parallelogram in Problem 69

70. Cavalieri's Principle Write a short report on Cavalieri's Principle. Discuss Problems 68 and 69 in this report.

6.3 Volumes of Solids: Slicing Method

■ **Introduction**   The shape that undoubtedly springs to mind with the words **right cylinder** is the right *circular* cylinder—that is, the usual shape of a tin can. But a right cylinder need not be circular. From geometry, a **right cylinder** is defined as a solid bounded by two congruent plane regions, in parallel planes, and a lateral surface that is generated by a moving line segment that is perpendicular to both planes and whose ends are on the boundaries of the plane regions. When the regions are circles, we obtain the right circular cylinder. If the regions are rectangles, the cylinder is a rectangular parallelepiped. Common to all right cylinders, such as the five shown in FIGURE 6.3.1, is that their volume V is given by the formula:

$$V = B \cdot h, \tag{1}$$

where B denotes the area of a base (that is, the area of one of the plane regions) and h denotes the height of the cylinder (that is, the perpendicular distance between the plane regions).

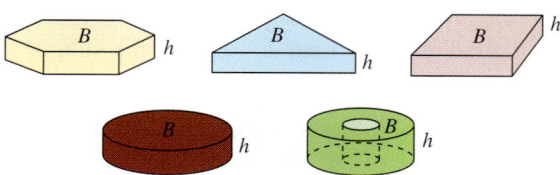

FIGURE 6.3.1   Five different right cylinders

In this section, we will show how the definite integral can be used to compute the volumes of certain kinds of solids, specifically solids with a known cross-sectional area. Formula (1) will be especially important in the discussion that follows.

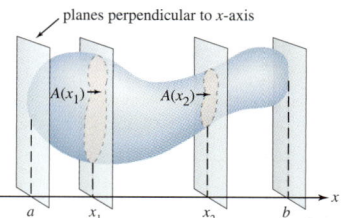

FIGURE 6.3.2   The regions or cross sections have known areas

■ **Slicing Method**   Suppose V is the volume of the solid shown in FIGURE 6.3.2 bounded by planes that are perpendicular to the x-axis at $x = a$ and $x = b$. Furthermore, suppose we know a continuous function $A(x)$ that gives the area of a cross-sectional region that is formed by *slicing* the solid by a plane perpendicular to the x-axis, in other words, a slice is the intersection of the solid and one plane. For example, for $a < x_1 < x_2 < b$ the areas of the cross sections shown in Figure 6.3.2 are $A(x_1)$ and $A(x_2)$. With this is mind, let us imagine slicing the solid into thin slabs by parallel planes (similar to slices of commercially baked bread) so that a slab has thickness or width Δx_k. By using right cylinders to approximate the volumes of these slabs, we can build a definite integral that gives the volume V of the solid.

A piece of bread is a slab formed by two slices

■ **Building an Integral**   Now think of slicing the solid into n slabs. If P is the partition

$$a = x_0 < x_1 < x_2 < \cdots < x_n = b$$

of the interval $[a, b]$ and x_k^* is a sample point in the kth subinterval $[x_{k-1}, x_k]$, then an approximation to the volume of the solid on this subinterval, or slab, is the volume V_k of the right cylinder, which is shown in the enlargement in FIGURE 6.3.3. The area B of the base of the right cylinder is the area $A(x_k^*)$ of the cross section and its height h is Δx_k and so by (1) its volume is

$$V_k = \text{area of base} \cdot \text{height} = A(x_k^*)(x_k - x_{k-1}) = A(x_k^*)\,\Delta x_k. \tag{2}$$

It follows that the Riemann sum of the volumes $V_k = A(x_k^*)\,\Delta x_k$ of the n right cylinders,

$$\sum_{k=1}^{n} V_k = \sum_{k=1}^{n} A(x_k^*)\,\Delta x_k,$$

is an approximation to the volume V of the solid on $[a, b]$. We use the definite integral

$$\lim_{\|P\| \to 0} \sum_{k=1}^{n} A(x_k^*)\,\Delta x_k = \int_a^b A(x)\,dx$$

as the definition of the volume V of the solid.

area of a cross section $A(x_k^*)$

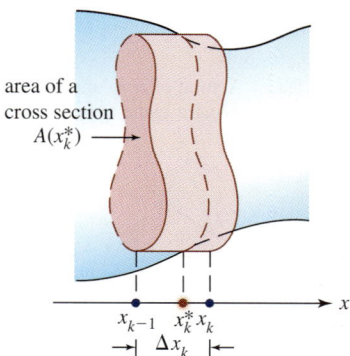

FIGURE 6.3.3   The volume of a right cylinder is an approximation to the volume of a slab

---

Definition 6.3.1   Volume by Slicing

Let V be the volume of a solid bounded by planes that are perpendicular to the x-axis at $x = a$ and $x = b$. If $A(x)$ is a continuous function that gives the area of a cross section of the solid formed by a plane perpendicular to the x-axis at any point in the interval $[a, b]$, then the volume of the solid is

$$V = \int_a^b A(x)\, dx. \qquad (3)$$

Bear in mind there is nothing special about the variable x in (3); depending on the geometry and the analysis of the problem we could just as well end up with an integral $\int_c^d A(y)\, dy$.

EXAMPLE 1   Solid with Square Cross Sections

For the solid in FIGURE 6.3.4(a), the cross sections perpendicular to a diameter of a circular base are squares. Given that the radius of the base is 4 ft, find the volume of the solid.

Solution   Let the x- and y-axes be as shown in Figure 6.3.4(a), namely, the origin is at the center of the circular base of the solid. In this figure a square cross section is shown perpendicular to the x-axis. Since the base of the solid is a circle we have $x^2 + y^2 = 4^2$. In Figure 6.3.4(b), the dashed line at x_k^* represents the cross section of the solid perpendicular to the x-axis in the subinterval $[x_{k-1}, x_k]$ in a partition of the interval $[-4, 4]$. From this we see that the length of one side of the square cross section is $2y_k^* = 2\sqrt{16 - (x_k^*)^2}$. Thus, the area of a square cross section is

$$A(x_k^*) = \left(2\sqrt{16 - (x_k^*)^2}\right)^2 = 64 - 4(x_k^*)^2.$$

The volume of the approximating right cylinder to the volume of the solid or slab on the subinterval $[x_{k-1}, x_k]$ is

$$V_k = A(x_k^*)\, \Delta x_k = (64 - 4(x_k^*)^2)\, \Delta x_k.$$

Forming the sum $\sum_{k=1}^n V_k$ and taking the limit as $\|P\| \to 0$ gives the definite integral

$$V = \int_{-4}^4 (64 - 4x^2)\, dx = 64x - \frac{4}{3}x^3 \Big]_{-4}^4 = \frac{512}{3} - \left(-\frac{512}{3}\right) = \frac{1024}{3}. \qquad \blacksquare$$

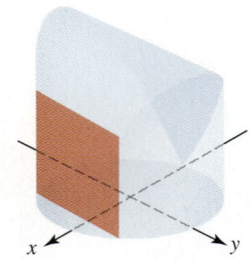

(a) Plane perpendicular to x-axis intersects solid in a square

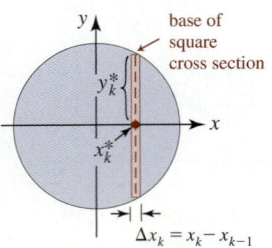

base of square cross section

$\Delta x_k = x_k - x_{k-1}$

(b) Circular base of solid

FIGURE 6.3.4   Solid in Example 1

■ **Solids of Revolution**   If a region R in the xy-plane is revolved about an axis L, it will generate a solid called a **solid of revolution**. See FIGURE 6.3.5.

■ **Disk Method**   As just discussed, we can find the volume V of a solid by means of a definite integral whenever we know a function $A(x)$ that gives the area of a cross-sectional region formed by passing a plane through the solid perpendicular to an axis. In the case of finding the volume of a solid of revolution, it is always possible to find $A(x)$; the axis in question is the axis of revolution L. We will see that by slicing the solid by two parallel planes perpendicular to the axis of revolution the volume of the resulting slabs of the solid can be approximated by right *circular* cylinders that are either disks or washers. We will next illustrate building a volume integral using disks.

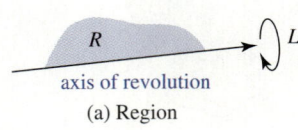

axis of revolution

(a) Region

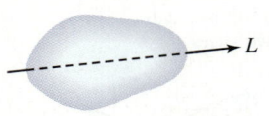

(b) Solid

FIGURE 6.3.5   A solid of revolution is formed by revolving a plane region R about an axis L

■ **Building an Integral**   Let R be the region bounded by the graph of a nonnegative continuous function $y = f(x)$, the x-axis, and the vertical lines $x = a$ and $x = b$, as shown in FIGURE 6.3.6. If this region is revolved about the x-axis, let us find the volume V of the resulting solid of revolution.

Let P be a partition of $[a, b]$ and let x_k^* be any number in the kth subinterval $[x_{k-1}, x_k]$ as shown in FIGURE 6.3.7(a). As the rectangular element of width Δx_k and height $f(x_k^*)$ is revolved about the x-axis, it generates a solid disk. Now the cross section of the solid determined by a plane cutting the surface at x_k^* is a circle of radius $r = f(x_k^*)$, and so the area of the cross-

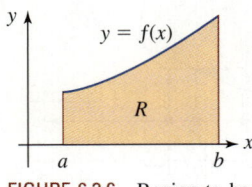

$y = f(x)$

R

FIGURE 6.3.6   Region to be revolved about the x-axis

sectional region is $A(x_k^*) = \pi[f(x_k^*)]^2$. The volume of the corresponding right-circular cylinder, or solid disk, of radius $r = f(x_k^*)$ and height $h = \Delta x_k$ is $\pi r^2 h$ or

$$V_k = A(x_k^*)\,\Delta x_k = \pi[f(x_k^*)]^2\,\Delta x_k.$$

The Riemann sum

$$\sum_{k=1}^{n} V_k = \sum_{k=1}^{n} A(x_k^*)\,\Delta x_k = \sum_{k=1}^{n} \pi[f(x_k^*)]^2\,\Delta x_k$$

represents an approximation to the volume of the solid shown in Figure 6.3.7(d). This suggests that the volume V of the solid of revolution is given by

$$V = \lim_{\|P\|\to 0} \sum_{k=1}^{n} \pi[f(x_k^*)]^2\,\Delta x_k$$

or
$$V = \int_a^b \pi[f(x)]^2\,dx. \tag{4}$$

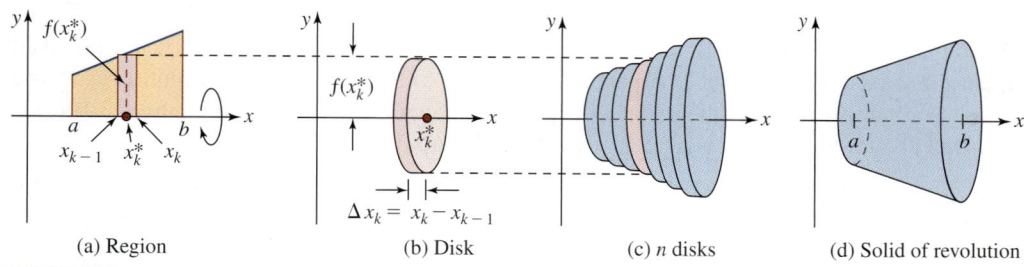

(a) Region   (b) Disk   (c) n disks   (d) Solid of revolution

FIGURE 6.3.7 Revolving the red rectangular element in (a) about the x-axis generates the red circular disk in (b)

If a region R is revolved about some other axis, then (4) may simply not be applicable to the problem of finding the volume of the resulting solid. Rather than apply a formula blindly, you should set up an appropriate integral by carefully analyzing the geometry of each problem. We will examine such a case in Example 6.

EXAMPLE 2 Disk Method

Find the volume V of the solid formed by revolving the region bounded by the graphs of $y = \sqrt{x}$, $y = 0$, and $x = 4$ about the x-axis.

Solution FIGURE 6.3.8(a) shows the region in question. Now, the area of a cross-sectional slice at x_k^* is

$$A(x_k^*) = \pi[f(x_k^*)]^2 = \pi[(x_k^*)^{1/2}]^2 = \pi x_k^*,$$

and so the volume of the corresponding disk shown in Figure 6.3.8(b) is

$$V_k = A(x_k^*)\,\Delta x_k = \pi x_k^*\,\Delta x_k.$$

Hence, the volume of the solid is

$$V = \pi \int_0^4 x\,dx = \pi \frac{1}{2}x^2 \Big]_0^4 = 8\pi.$$

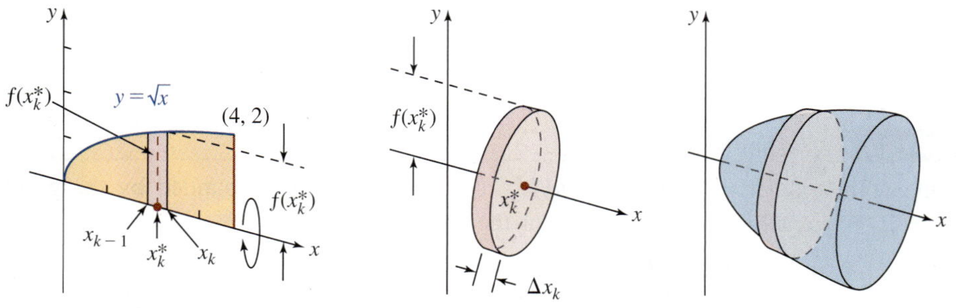

(a) Region   (b) Disk   (c) Solid of revolution

FIGURE 6.3.8 Region and solid of revolution in Example 2

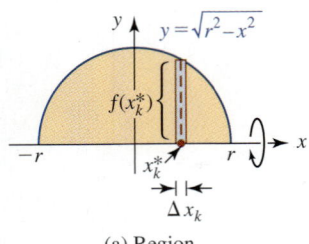

(a) Region

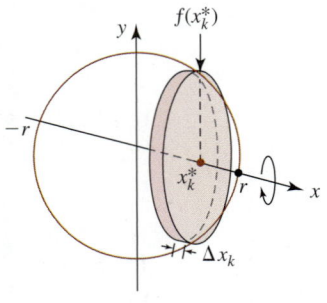

(b) Sphere

FIGURE 6.3.9 Semicircle and sphere in Example 3

EXAMPLE 3 Volume of a Sphere

Show that the volume V of a sphere of radius r is $V = \frac{4}{3}\pi r^3$.

Solution A sphere of radius r can be generated by revolving a semicircle $f(x) = \sqrt{r^2 - x^2}$ about the x-axis. From **FIGURE 6.3.9** we see that the area of a cross-sectional region of the solid perpendicular to the x-axis at x_k^* is

$$A(x_k^*) = \pi[f(x_k^*)]^2 = \pi(\sqrt{r^2 - (x_k^*)^2})^2 = \pi(r^2 - (x_k^*)^2)$$

and hence, the volume of one disk is

$$V_k = A(x_k^*)\,\Delta x_k = \pi(r^2 - (x_k^*)^2)\,\Delta x_k.$$

Using (4) we see that the volume of the sphere is

$$V = \int_{-r}^{r} \pi(r^2 - x^2)\,dx = \pi\left(r^2 x - \frac{1}{3}x^3\right)\Bigg]_{-r}^{r} = \pi\frac{2}{3}r^3 - \left(-\pi\frac{2}{3}r^3\right) = \frac{4}{3}\pi r^3. \qquad \blacksquare$$

▌ **Washer Method** Let the region R bounded by the graphs of the continuous functions $y = f(x)$, $y = g(x)$, and the lines $x = a$ and $x = b$, as shown in **FIGURE 6.3.10(a)**, be revolved about the x-axis. Then the slice perpendicular to the x-axis of the solid of revolution at x_k^* is a circular or annular ring. As the rectangular element of width Δx_k shown in Figure 6.3.10(a) is revolved about the x-axis, it generates a washer. The area of the ring is

$$A(x_k^*) = \text{area of circle} - \text{area of hole}$$
$$= \pi[f(x_k^*)]^2 - \pi[g(x_k^*)]^2 = \pi([f(x_k^*)]^2 - [g(x_k^*)]^2)$$

and the volume V_k of the representative washer shown in Figure 6.3.10(b) is

$$V_k = A(x_k^*)\,\Delta x_k = \pi([f(x_k^*)]^2 - [g(x_k^*)]^2)\,\Delta x_k.$$

Therefore, the volume of the solid is

$$V = \int_a^b \pi([f(x)]^2 - [g(x)]^2)\,dx. \qquad (5)$$

Observe that the integral (5) reduces to (4) when $g(x) = 0$.

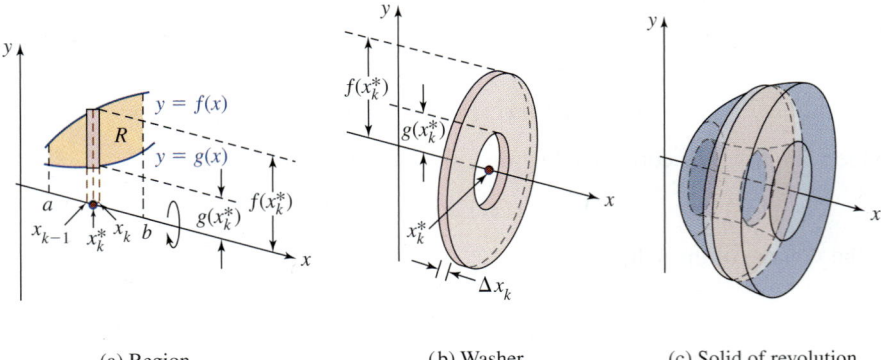

(a) Region   (b) Washer   (c) Solid of revolution

FIGURE 6.3.10 Revolving the red rectangular element in (a) about the x-axis generates the red circular washer in (b)

EXAMPLE 4 Washer Method

Find the volume V of the solid formed by revolving the region bounded by the graphs of $y = x + 2$, $y = x$, $x = 0$, and $x = 3$ about the x-axis.

Solution **FIGURE 6.3.11(a)** shows the region in question. Now, the area of a cross-sectional region of the solid corresponding to a plane perpendicular to the x-axis at x_k^* is

$$A(x_k^*) = \pi(x_k^* + 2)^2 - (x_k^*)^2 = \pi(4x_k^* + 4).$$

As seen in Figures 6.3.11(a) and (b), a vertical rectangular element of width Δx_k, when revolved about the x-axis, yields a washer having volume

$$V_k = A(x_k^*) \, \Delta x_k = \pi(4x_k^* + 4) \, \Delta x_k.$$

The usual summing and limiting process yields the definite integral for the volume V of the solid of revolution:

$$V = \pi \int_0^3 (4x + 4) \, dx = \pi(2x^2 + 4x) \Big]_0^3 = 30\pi.$$

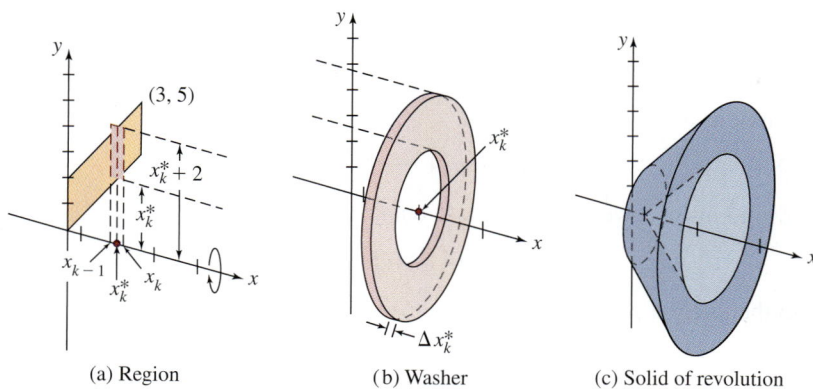

(a) Region   (b) Washer   (c) Solid of revolution

FIGURE 6.3.11 Region and solid of revolution in Example 4

EXAMPLE 5   Integration with Respect to y

Find the volume V of the solid formed by revolving the region bounded by the graphs of $y = \sqrt{x}$ and $y = x$ about the y-axis.

Solution   When the horizontal rectangular element in **FIGURE 6.3.12(a)** is revolved about the y-axis it generates a washer of width Δy_k. The area $A(y_k^*)$ of the annular ring at y_k^* is

$$A(y_k^*) = \text{area of circle} - \text{area of hole.} - 1$$

The radius of the circle and the radius of the hole are obtained by solving, in turn, $y = x$ and $y = \sqrt{x}$ for x in terms of y:

$$A(y_k^*) = \pi(y_k^*)^2 - \pi[(y_k^*)^2]^2 = \pi((y_k^*)^2 - (y_k^*)^4).$$

Thus, the volume of a washer is

$$V_k = A(y_k^*) \, \Delta y_k = \pi((y_k^*)^2 - (y_k^*)^4) \, \Delta y_k.$$

The usual summing of the V_k and taking the limit of that sum as $\|P\| \to 0$ leads to the definite integral for the volume of the solid:

$$V = \pi \int_0^1 (y^2 - y^4) \, dy = \pi\left(\frac{1}{3}y^3 - \frac{1}{5}y^5\right)\Big]_0^1 = \frac{2}{15}\pi.$$

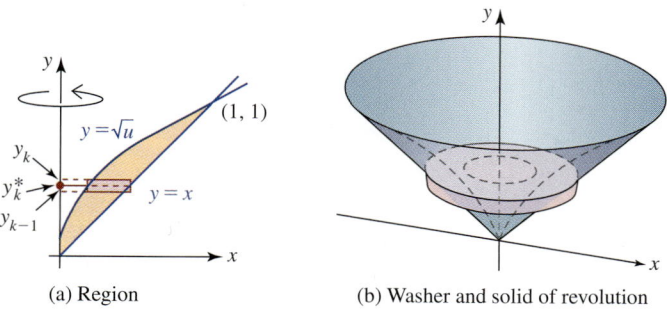

(a) Region   (b) Washer and solid of revolution

FIGURE 6.3.12 Region and solid of revolution in Example 5

∎ **Revolution about a Line**   The next example shows how to find the volume of a solid of revolution when a region is revolved about an axis that is not a coordinate axis.

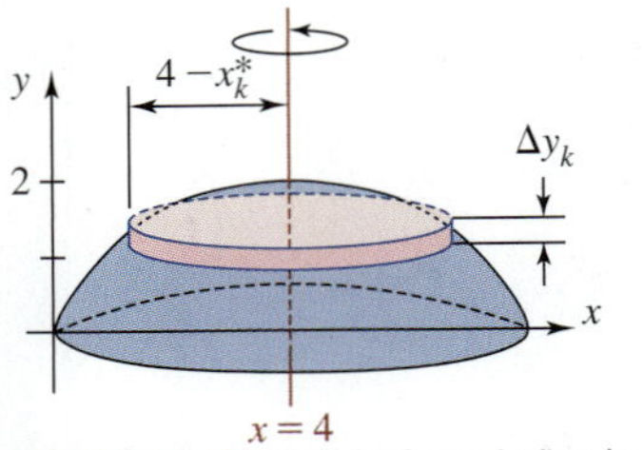

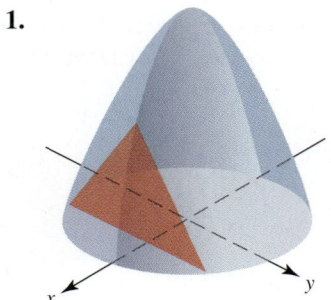

FIGURE 6.3.13   Solid of revolution in Example 6

EXAMPLE 6   Axis of Revolution not a Coordinate Axis

Find the volume V of the solid that is formed by revolving the region given in Example 2 about the line $x = 4$.

Solution   The domed-shaped solid of revolution is shown in FIGURE 6.3.13. From inspection of the figure we see that a horizontal rectangular element of width Δy_k that is perpendicular to the vertical line $x = 4$ generates a solid disk when revolved about that axis. The radius r of that disk is

$$r = (\text{right-most } x\text{-value}) - (\text{left-most } x\text{-value}) = 4 - x_k^*,$$

and so its volume is then

$$V_k = \pi(4 - x_k^*)^2\,\Delta y_k.$$

To express x in terms of y we use $y = \sqrt{x}$ to obtain $x_k^* = (y_k^*)^2$. Therefore,

$$V_k = \pi(4 - (y_k^*)^2)^2\,\Delta y_k.$$

This leads to the integral

$$
\begin{aligned}
V &= \pi \int_0^2 (4 - y^2)^2\,dy \\
&= \pi \int_0^2 (16 - 8y^2 + y^4)\,dy \\
&= \pi\left(16y - \frac{8}{3}y^3 + \frac{1}{5}y^5\right)\Bigg]_0^2 = \frac{256}{15}\pi.
\end{aligned}
$$

∎

Exercises 6.3   Answers to selected odd-numbered problems begin on page ANS-20.

≡ **Fundamentals**

In Problems 1 and 2, use the slicing method to find the volume of the solid if its cross sections perpendicular to a diameter of a circular base are as given. Assume that the radius of the base is 4.

1.

FIGURE 6.3.14   Cross sections are equilateral triangles

2.

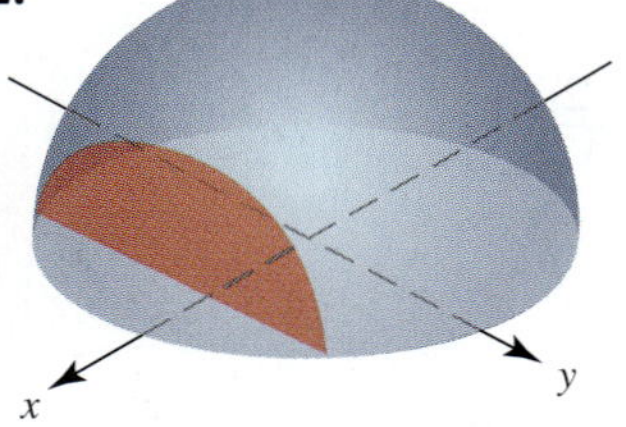

FIGURE 6.3.15   Cross sections are semicircles

3. The base of a solid is bounded by the curves $x = y^2$ and $x = 4$ in the xy-plane. The cross sections perpendicular to the x-axis are rectangles for which the height is four times the base. Find the volume of the solid.

4. The base of a solid is bounded by the curve $y = 4 - x^2$ and the x-axis. The cross sections perpendicular to the x-axis are equilateral triangles. Find the volume of the solid.

5. The base of a solid is an isosceles triangle whose base is 4 ft and height is 5 ft. The cross sections perpendicular to the altitude are semicircles. Find the volume of the solid.

6. A hole of radius 1 ft is drilled through the middle of the solid sphere of radius $r = 2$ ft. Find the volume of the remaining solid. See FIGURE 6.3.16.

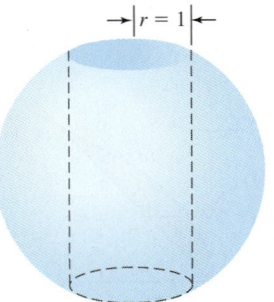

FIGURE 6.3.16   Hole through sphere in Problem 6

7. The base of a solid is a right isosceles triangle that is formed by the coordinate axes and the line $x + y = 3$. The cross sections perpendicular to the y-axis are squares. Find the volume of the solid.

8. Suppose the pyramid shown in FIGURE 6.3.17 has height h and a square base of area B. Show that the volume of the pyramid is given by $A = \frac{1}{3}hB$. [*Hint:* Let b denote the length of one side of the square base.]

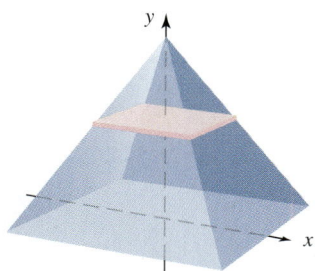

FIGURE 6.3.17 Pyramid in Problem 8

In Problems 9–14, refer to FIGURE 6.3.18. Use the disk or washer method to find the volume of the solid of revolution that is formed by revolving the given region about the indicated line.

9. R_1 about OC

10. R_1 about OA

11. R_2 about OA

12. R_2 about OC

13. R_1 about AB

14. R_2 about AB

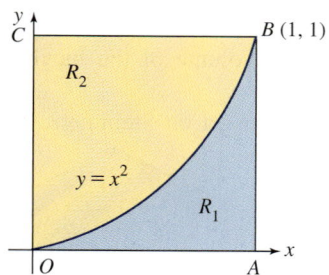

FIGURE 6.3.18 Regions for Problems 9–14

In Problems 15–40, use the disk or washer method to find the volume of the solid of revolution that is formed by revolving the region bounded by the graphs of the given equations about the indicated line or axis.

15. $y = 9 - x^2, y = 0$; x-axis

16. $y = x^2 + 1, x = 0, y = 5$; y-axis

17. $y = \dfrac{1}{x}, x = 1, y = \dfrac{1}{2}$; y-axis

18. $y = \dfrac{1}{x}, x = \dfrac{1}{2}, x = 3, y = 0$; x-axis

19. $y = (x - 2)^2, x = 0, y = 0$; x-axis

20. $y = (x + 1)^2, x = 0, y = 0$; y-axis

21. $y = 4 - x^2, y = 1 - \frac{1}{4}x^2$; x-axis

22. $y = 1 - x^2, y = x^2 - 1, x = 0$, first quadrant; y-axis

23. $y = x, y = x + 1, x = 0, y = 2$; y-axis

24. $x + y = 2, x = 0, y = 0, y = 1$; x-axis

25. $y = \sqrt{x - 1}, x = 5, y = 0$; $x = 5$

26. $x = y^2, x = 1$; $x = 1$

27. $y = x^{1/3}, x = 0, y = 1$; $y = 2$

28. $x = -y^2 + 2y, x = 0$; $x = 2$

29. $x^2 - y^2 = 16, x = 5$; y-axis

30. $y = x^2 - 6x + 9, y = 9 - \frac{1}{2}x^2$; x-axis

31. $x = y^2, y = x - 6$; y-axis

32. $y = x^3 + 1, x = 0, y = 9$; y-axis

33. $y = x^3 - x, y = 0$; x-axis

34. $y = x^3 + 1, x = 1, y = 0$; x-axis

35. $y = e^{-x}, x = 1, y = 1$; $y = 2$

36. $y = e^x, y = 1, x = 2$; x-axis

37. $y = |\cos x|, y = 0, 0 \le x \le 2\pi$; x-axis

38. $y = \sec x, x = -\pi/4, x = \pi/4, y = 0$; x-axis

39. $y = \tan x, y = 0, x = \pi/4$; x-axis

40. $y = \sin x, y = \cos x, x = 0$, first quadrant; x-axis

≡ Think About It

41. Reread Problems 68–70 in Exercises 6.2 on Cavalieri's Principle. Then show that the circular cylinders in FIGURE 6.3.19 have the same volume.

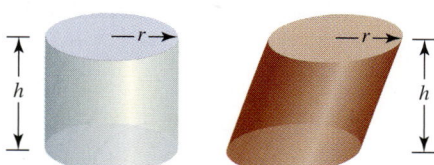

FIGURE 6.3.19 Cylinders in Problem 41

42. Consider the right circular cylinder of radius a shown in FIGURE 6.3.20. A plane inclined at an angle θ to the base of the cylinder passes through a diameter of the base. Find the volume of the resulting wedge cut from the cylinder when
(a) $\theta = 45°$ **(b)** $\theta = 60°$.

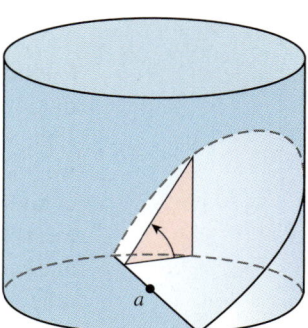

FIGURE 6.3.20 Cylinder and wedge in Problem 42

≡ Projects

43. For the Birds A mathematical model for the shape of an egg can be obtained by revolving the region bounded by the graphs of $y = 0$ and the function $f(x) = P(x)\sqrt{1 - x^2}$, where $P(x) = ax^3 + bx^2 + cx + d$ is a cubic polynomial, about the x-axis. For example, an egg of the Common Murre corresponds to $P(x) = -0.07x^3 - 0.02x^2 + 0.2x + 0.56$. FIGURE 6.3.21 shows the graph of f obtained with the aid of a CAS.

(a) Find a general formula for the volume V of an egg based on the mathematical model $f(x) = P(x)\sqrt{1 - x^2}$, where $P(x) = ax^3 + bx^2 + cx + d$. [*Hint*: This problem can be done by hand calculation but it is long and "messy". Use a CAS to carry out the integration.]

(b) Use the formula obtained in part (a) to estimate the volume of an egg of the Common Murre.

(c) An egg of the Red-throated Loon corresponds to $P(x) = -0.06x^3 + 0.04x^2 + 0.1x + 0.54$. Use a calculator or CAS to obtain the graph of f.

(d) Use part (a) to estimate the volume of an egg of the Red-throated Loon.

Common Murre eggs

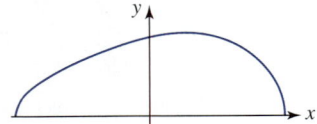

FIGURE 6.3.21 Model of the shape of the Common Murre egg in Problem 43

44. That Sinking Feeling A wooden spherical ball of radius r is floating on a pond of still water. Let h denote the depth that the ball will sink into the water. See FIGURE 6.3.22.

(a) Show that the volume of the submerged portion of the ball is given by $V = \pi r^2 h - \frac{1}{3}\pi h^3$.

(b) Suppose that the weight density of the ball is denoted by ρ_{ball} and the weight density of the water is ρ_{water} (measured in lb/ft^3). If $r = 3$ in. and $\rho_{ball} = 0.4\rho_{water}$, use Archimedes' Principle—the weight of the ball equals the weight of the water displaced—to determine the approximate depth h that the ball will sink. You will need a calculator or CAS to solve a cubic polynomial equation.

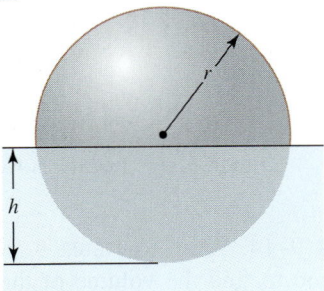

FIGURE 6.3.22 Floating wooden ball in Problem 44

45. Steinmetz Solids The solid formed by two intersecting circular cylinders of radius r whose axes intersect at a right angle is called a **bicylinder** and is a special case of Steinmetz solids. For clarity we have shown one-eighth of the solid in FIGURE 6.3.23.

(a) Find the total volume of the bicylinder illustrated in the figure.

(b) Write a short report on Steinmetz solids.

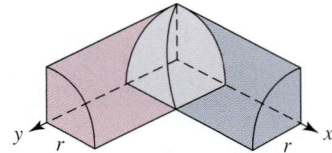

FIGURE 6.3.23 Intersecting right circular cylinders in Problem 45

6.4 Volumes of Solids: Shell Method

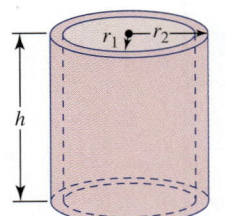

FIGURE 6.4.1 Cylindrical shell

■ **Introduction** In this section we continue the discussion of finding volumes of solids of revolution. But instead of using planes perpendicular to the axis of revolution to slice the solid into slabs whose volume can be approximated by right regular circular cylinders (disks or washers), we will develop a new method for finding volumes of solid of revolution that utilizes circular cylindrical shells. Before building an integral representing this **shell method** we need to find the volume of the general cylindrical shell shown in FIGURE 6.4.1. If, as shown in the figure, r_1 and r_2 denote, respectively, the inner and outer radii of the shell, and h is its height, then its volume is given by the difference

$$\text{volume of outer cylinder} - \text{volume of inner cylinder}$$
$$= \pi r_2^2 h - \pi r_1^2 h = \pi(r_2^2 - r_1^2)h = \pi(r_2 + r_1)(r_2 - r_1)h. \tag{1}$$

■ **Building an Integral** In Section 6.3 we saw that a rectangular element of area that is perpendicular to an axis of revolution will generate, when revolved, either a circular disk or a circular washer. However, if we were to revolve the rectangular element shown in FIGURE 6.4.2(a) about the y-axis, we generate a hollow shell as shown in Figure 6.4.2(b). To find the volume of the solid shown in Figure 6.4.2(c) we let P denote the arbitrary partition of the interval $[a, b]$:

$$a = x_0 < x_1 < x_2 < \cdots < x_{n-1} < x_n = b.$$

The partition P divides the interval into n subintervals $[x_{k-1}, x_k]$, $k = 1, 2, \ldots, n$, of width $\Delta x_k = x_k - x_{k-1}$. If we identify the outer radius as $r_2 = x_k$ and the inner radius as $r_1 = x_{k-1}$ and define $x_k^* = \frac{1}{2}(x_k + x_{k-1})$, then x_k^* is the midpoint of the subinterval $[x_{k-1}, x_k]$. With the further identification $h = f(x_k^*)$ it follows from (1) that the volume of the representative shell in Figure 6.4.2(b) be written as

$$V_k = \pi(x_k + x_{k-1})(x_k - x_{k-1})h$$

$$= 2\pi \frac{x_k + x_{k-1}}{2} h(x_k - x_{k-1})$$

or

$$V_k = 2\pi x_k^* f(x_k^*) \, \Delta x_k.$$

An approximation to the volume of the solid is given by the Riemann sum

$$\sum_{k=1}^{n} V_k = \sum_{k=1}^{n} 2\pi x_k^* f(x_k^*) \, \Delta x_k. \tag{2}$$

As the norm $\|P\|$ of the partition approaches zero, the limit of (2) is a definite integral that we use as the definition of the volume V of the solid:

$$V = 2\pi \int_a^b x f(x) \, dx. \tag{3}$$

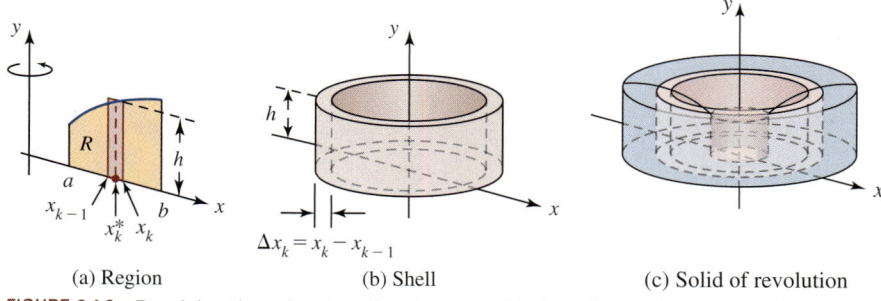

| (a) Region | (b) Shell | (c) Solid of revolution |

FIGURE 6.4.2 Revolving the red rectangular element in (a) about the y-axis generates the red shell in (b)

As mentioned in the *Notes from the Classroom* at the end of Section 6.2 it is not possible to derive an integral, in this case representing the volume of a solid of revolution, that "works" in every possible case. You are urged again not to memorize a particular formula such as (3). Try to understand the geometric interpretation of the component parts of the integrand. For example, $f(x)$, which represents the height of the rectangle in Figure 6.4.2, could be the difference $f(x) - g(x)$ if the rectangular element is between the graphs of two functions $y = f(x)$ and $y = g(x), f(x) \geq g(x)$. To set up an integral for a given problem without going through a lengthy analysis think of a shell as a circular tin can with its top and bottom removed. To find the volume of the shell, that is, the volume of the metal in the tin can analogy, imagine that the shell is cut straight down its lateral side and flattened out as illustrated in FIGURE 6.4.3(a) and (b). As Figure 6.4.3(c) shows, the volume of the shell is then the volume of a thin rectangular solid:

$$\text{volume} = (\text{length}) \cdot (\text{width}) \cdot (\text{thickness})$$

$$= (\text{circumference of the cylinder}) \cdot (\text{height}) \cdot (\text{thickness})$$

$$= 2\pi \, r h t. \tag{4}$$

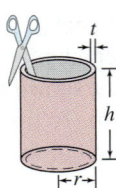

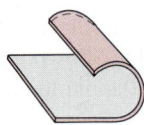
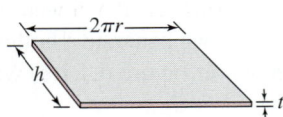

(a) Cut the shell down its side   (b) Flatten it out   (c) The result is a rectangular solid

FIGURE 6.4.3 Finding the volume of a shell

EXAMPLE 1   Using the Shell Method

Reread Example 5 in Section 6.3 before working through this example.

▶ Use the shell method to find the volume V of the solid formed by revolving the region bounded by the graphs of $y = \sqrt{x}$ and $y = x$ about the y-axis.

Solution   We solved this problem in Example 5 of Section 6.3. In that example we saw that using a horizontal rectangular element perpendicular to the y-axis of width Δy_k generated a washer when revolved about the y-axis. In contrast, a vertical rectangular element of width Δx_k revolved about the y-axis generates a shell. Using **FIGURE 6.4.4(a)** we make the identifications in (4) that $r = x_k^*$,

$$h = \text{uppergraph} - \text{lower graph} = \sqrt{x_k^*} - x_k^*,$$

and $t = \Delta x_k$. From the volume of the shell,

$$V_k = 2\pi x_k^* \left(\sqrt{x_k^*} - x_k^* \right) \Delta x_k = 2\pi \left((x_k^*)^{3/2} - (x_k^*)^2 \right) \Delta x_k.$$

we obtain the definite integral getting the volume of the solid:

$$V = 2\pi \int_0^1 (x^{3/2} - x^2)\, dx$$

$$= 2\pi \left(\frac{2}{5} x^{5/2} - \frac{1}{3} x^3 \right) \Bigg]_0^1 = \frac{2}{15}\pi.$$

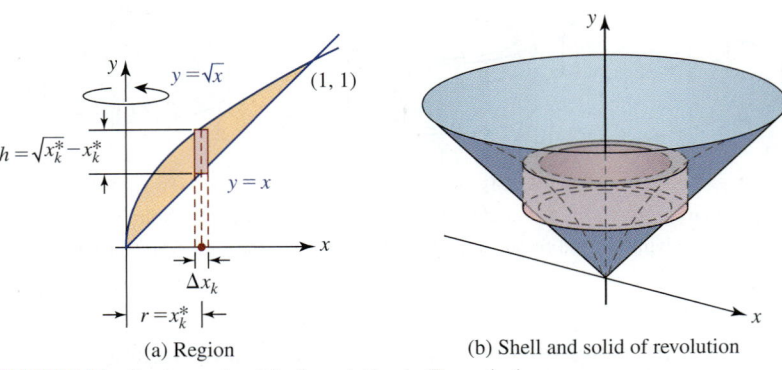

(a) Region   (b) Shell and solid of revolution

FIGURE 6.4.4 Region and solid of revolution in Example 1   ∎

It is not always convenient or even possible to use the disk or washer method discussed in the last section to find the volume of a solid of revolution.

EXAMPLE 2   Using the Shell Method

Find the volume V of the solid that is formed by revolving the region bounded by the graph of $y = \sin x^2$ and $y = 0, 0 \le x \le \sqrt{\pi}$ about the y-axis.

Solution   The graph of $y = \sin x^2$ on the indicated interval in **FIGURE 6.4.5** was obtained with the help of a CAS.

If we choose to use a horizontal rectangular element to revolve about the y-axis, a washer would be generated. To determine the inner and outer radii of the washer we would have to solve $y = \sin x^2$ for x in terms of y. While this simply leads to x^2 as an inverse sine, this poses the practical problem: We are not in a position to integrate an inverse trigonometric function at this time. Thus we switch our attention to a vertical rectangular element shown in Figure 6.4.5(a). When this element is revolved about the y-axis a shell with radius $r = x_k^*$, height $h = \sin(x_k^*)^2$, and thickness $t = \Delta x_k$ is generated. By (4) the volume of the shell is

$$V_k = 2\pi x_k^* \sin(x_k^*)^2 \, \Delta x_k.$$

Thus, by (3) we have

$$V = 2\pi \int_0^{\sqrt{\pi}} x \sin x^2 \, dx.$$

If we let $u = x^2$, then $du = 2x \, dx$ and $x \, dx = \frac{1}{2} \, du$. The u-limits of integration are determined from the fact that when $x = 0, u = 0$, and $x = \sqrt{\pi}, u = \pi$. Therefore, the volume of the solid of revolution shown in Figure 6.4.5(b) is

$$V = \pi \int_0^{\pi} \sin u \, du = -\pi \cos u \bigg]_0^{\pi} = \pi(1 + 1) = 2\pi.$$

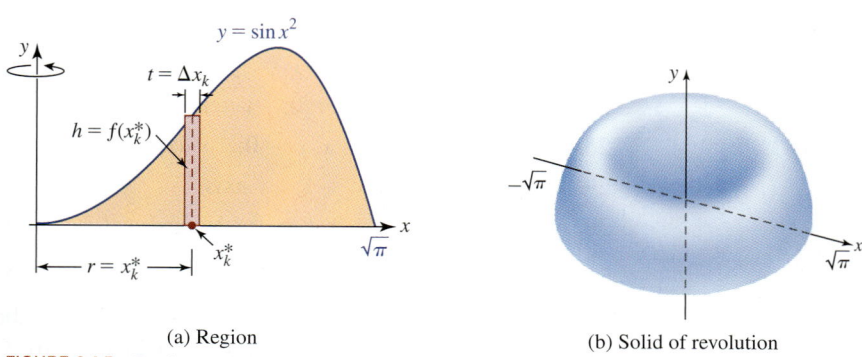

(a) Region   (b) Solid of revolution

FIGURE 6.4.5   Region and solid of revolution in Example 2

In the next example we illustrate the shell method when a region is revolved about a line that is not a coordinate axis.

EXAMPLE 3   Axis of Revolution not Coordinate Axis

Find the volume V of the solid that is formed by revolving the region bounded by the graphs of $x = y^2 - 2y$ and $x = 3$ about the line $y = 1$.

Solution   In this case a rectangular element of area that is perpendicular to a horizontal line and revolved about the line $y = 1$ would generate a disk. Since the radius of the disk is not measured from the x-axis but from the line $y = 1$, it would be necessary to solve $x = y^2 - 2y$ for y in terms of x. We can avoid this inconvenience by using horizontal elements of area, which then generate shells such as that shown in FIGURE 6.4.6(b). Note that when $x = 3$, the equation $3 = y^2 - 2y$ or equivalently $(y + 1)(y - 3) = 0$, has solutions -1 and 3. Thus, we need only partition the interval $[1, 3]$ on the y-axis. After making the identifications $r = y_k^* - 1, h = 3 - x_k^*$, and $t = \Delta y_k$, it follows from (4) that the volume of a shell is

$$\begin{aligned} V_k &= 2\pi(y_k^* - 1)(3 - x_k^*)\Delta y_k \\ &= 2\pi(y_k^* - 1)(3 - (y_k^*)^2 + 2y_k^*)\Delta y_k \\ &= 2\pi(-(y_k^*)^3 + 3(y_k^*)^2 + y_k^* - 3)\Delta y_k. \end{aligned}$$

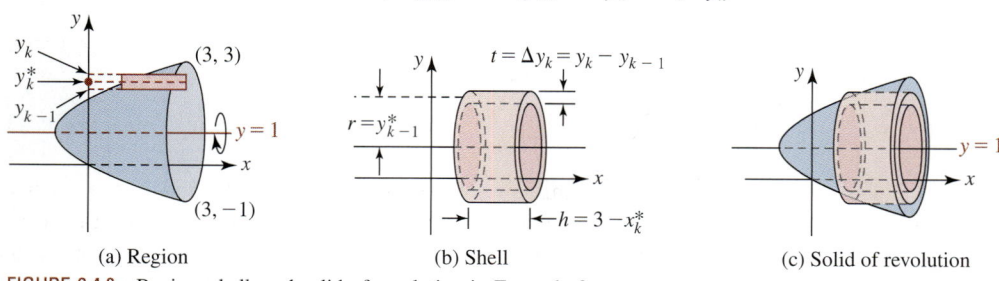

(a) Region   (b) Shell   (c) Solid of revolution

FIGURE 6.4.6   Region, shell, and solid of revolution in Example 3

From the last line we see that the volume of the solid is the definite integral

$$V = 2\pi \int_1^3 (-y^3 + 3y^2 + y - 3)\,dy$$

$$= 2\pi \left(-\frac{1}{4}y^4 + y^3 + \frac{1}{2}y^2 - 3y \right) \Big]_1^3$$

$$= 2\pi \left[\left(-\frac{81}{4} + 27 + \frac{9}{2} - 9 \right) - \left(-\frac{1}{4} + 1 + \frac{1}{2} - 3 \right) \right] = 8\pi. \qquad \blacksquare$$

Exercises 6.4   Answers to selected odd-numbered problems begin on page ANS-20.

≡ Fundamentals

In Problems 1–6, refer to FIGURE 6.4.7. Use the shell method to find the volume of the solid of revolution that is formed by revolving the given region about the indicated line.

1. R_1 about OC
2. R_1 about OA
3. R_2 about BC
4. R_2 about OA
5. R_1 about AB
6. R_2 about AB

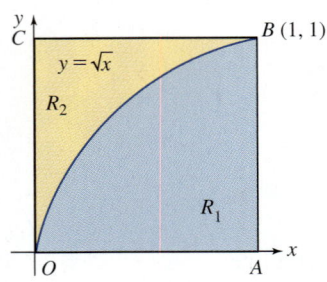

FIGURE 6.4.7   Regions for Problems 1–6

In Problems 7–30, use the shell method to find the volume of the solid of revolution that is formed by revolving the region bounded by the graphs of the given equations about the indicated line or axis.

7. $y = x, x = 0, y = 5$; x-axis
8. $y = 1 - x, x = 0, y = 0$; $y = -2$
9. $y = x^2, x = 0, y = 3$, first quadrant; x-axis
10. $y = x^2, x = 2, y = 0$; y-axis
11. $y = x^2, x = 1, y = 0$; $x = 3$
12. $y = x^2, y = 9$; x-axis
13. $y = x^2 + 4, x = 0, x = 2, y = 2$; y-axis
14. $y = x^2 - 5x + 4, y = 0$; y-axis
15. $y = (x - 1)^2, y = 1$; x-axis
16. $y = (x - 2)^2, y = 4$; $x = 4$
17. $y = x^{1/3}, x = 1, y = 0$; $y = -1$
18. $y = x^{1/3} + 1, y = -x + 1, x = 1$; $x = 1$
19. $y = x^2, y = x$; y-axis
20. $y = x^2, y = x$; $x = 2$
21. $y = -x^3 + 3x^2, y = 0$, first quadrant; y-axis
22. $y = x^3 - x, y = 0$, second quadrant; y-axis

23. $y = x^2 - 2, y = -x^2 + 2, x = 0$, second and third quadrants; y-axis
24. $y = x^2 - 4x, y = -x^2 + 4x$; $x = -1$
25. $x = y^2 - 5y, x = 0$; x-axis
26. $x = y^2 + 2, y = x - 4, y = 1$; x-axis
27. $y = x^3, y = x + 6, x = 0$; y-axis
28. $y = \sqrt{x}, y = \sqrt{1 - x}, y = 0$; x-axis
29. $y = \sin x^2, x = 0, y = 1$; y-axis
30. $y = e^{x^2}, y = 0, x = 0, x = 1$; y-axis

In Problems 31–36, the region in part (a) is revolved about the indicated axis generating the solid given in part (b). Choose between the disk, washer, or shell method to find the volume of the solid of revolution.

31.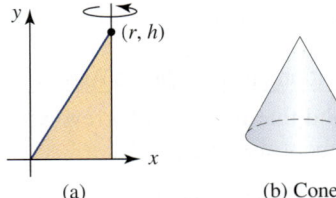

FIGURE 6.4.8   Region and solid for Problem 31

32.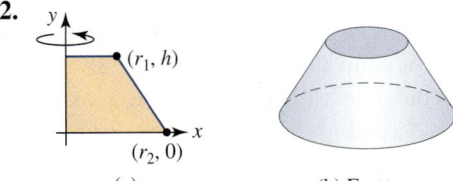

FIGURE 6.4.9   Region and solid for Problem 32

33.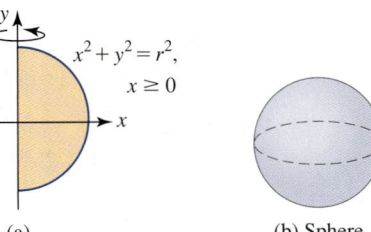

FIGURE 6.4.10   Region and solid for Problem 33

34.

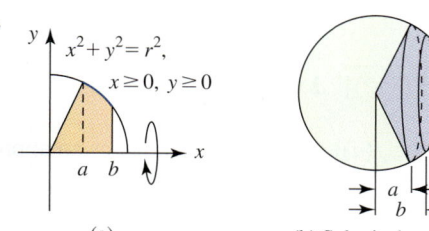

(a) (b) Spherical sector

FIGURE 6.4.11 Region and solid for Problem 34

35.

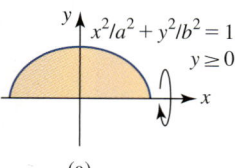 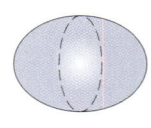

(a) (b) Prolate spheroid

FIGURE 6.4.12 Region and solid for Problem 35

36.

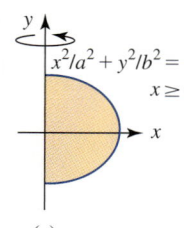 

(a) (b) Oblate spheroid

FIGURE 6.4.13 Region and solid for Problem 36

≡ Applications

37. A cylindrical bucket of radius r that contains a liquid is rotating about the y-axis with a constant angular velocity ω. It can be shown that the surface of the liquid has a parabolic cross-section given by $y = \omega^2 x^2/(2g)$, $-r \le x \le r$, where g is the acceleration due to gravity. Use the shell method to find the volume V of the liquid in the rotating bucket given that the height of the bucket is h. See FIGURE 6.4.14.

38. In Problem 37, determine the angular velocity ω for which the fluid will touch the bottom of the bucket. What is the corresponding volume V of the liquid?

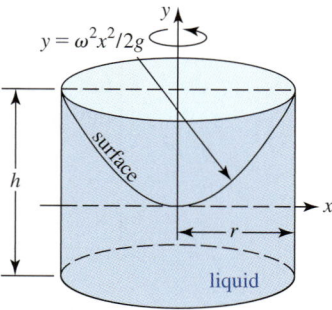

FIGURE 6.4.14 Bucket in Problems 37 and 38

6.5 Length of a Graph

■ **Introduction** If a function $y = f(x)$ has a continuous first derivative on an interval $[a, b]$, then its graph is said to be **smooth** and f is called a **smooth function**. As the name implies, a smooth graph has no sharp points. In the discussion that follows we will build an integral formula for the **length** L, or **arc length**, of a smooth graph on an interval $[a, b]$. See FIGURE 6.5.1.

■ **Building an Integral** Let f have a smooth graph on $[a, b]$ and let P denote an arbitrary partition of the interval:

$$a = x_0 < x_1 < x_2 < \cdots < x_{n-1} < x_n = b.$$

As usual, let the width of the kth subinterval be given by Δx_k and let $\|P\|$ be the width of the longest subinterval. As shown in 6.5.2(a), we can approximate the length of the graph on each subinterval $[x_{k-1}, x_k]$ by finding the length L_k of the chord between the points $(x_{k-1}, f(x_{k-1}))$ and $(x_k, f(x_k))$ for $k = 1, 2, \ldots, n$. From Figure 6.5.2(b), the Pythagorean Theorem gives the length L_k:

$$L_k = \sqrt{(\Delta x_k)^2 + (\Delta y_k)^2} = \sqrt{(x_k - x_{k-1})^2 + (f(x_k) - f(x_{k-1}))^2}. \quad (1)$$

By the Mean Value Theorem (Section 4.4), we know there exists a number x_k^* in each open subinterval (x_{k-1}, x_k) such that

$$\frac{f(x_k) - f(x_{k-1})}{x_k - x_{k-1}} = f'(x_k^*) \quad \text{or} \quad f(x_k) - f(x_{k-1}) = f'(x_k^*)(x_k - x_{k-1}).$$

Using the last equation, we replace $f(x_k) - f(x_{k-1})$ in (1) and simplify:

$$L_k = \sqrt{(x_k - x_{k-1})^2 + [f'(x_k^*)]^2(x_k - x_{k-1})^2}$$
$$= \sqrt{(x_k - x_{k-1})^2(1 + [f'(x_k^*)]^2)}$$
$$= \sqrt{1 + [f'(x_k^*)]^2}(x_k - x_{k-1})$$
$$= \sqrt{1 + [f'(x_k^*)]^2}\,\Delta x_k.$$

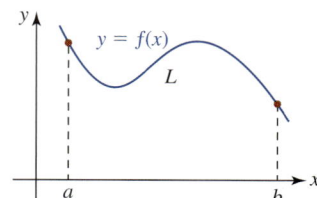

FIGURE 6.5.1 Find the length L of the graph of f on $[a, b]$

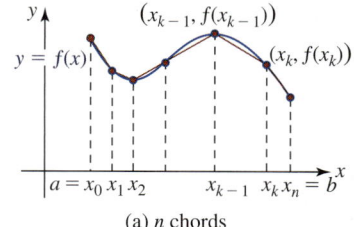

(a) n chords

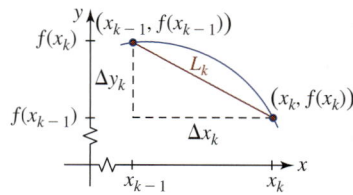

(b) Zoom in on chord on the kth subinterval

FIGURE 6.5.2 Approximating the length of a graph by summing the lengths of chords

The Riemann sum

$$\sum_{k=1}^{n} L_k = \sum_{k=1}^{n} \sqrt{1 + [f'(x_k^*)]^2}\, \Delta x_k$$

represents the length of the polygonal curve joining $(a, f(a))$ and $(b, f(b))$ and gives an approximation to the total length of the graph on $[a, b]$. As $\|P\| \to 0$, we obtain

$$\lim_{\|P\| \to 0} \sum_{k=1}^{n} \sqrt{1 + [f'(x_k^*)]^2}\, \Delta x_k = \int_a^b \sqrt{1 + [f'(x)]^2}\, dx. \tag{2}$$

The foregoing discussion prompts us to use (2) as the definition of the length of the graph on the interval.

Definition 6.5.1 Arc Length

Let f be a function for which f' is continuous on an interval $[a, b]$. Then the **length** L of the graph of $y = f(x)$ on the interval is given by

$$L = \int_a^b \sqrt{1 + [f'(x)]^2}\, dx. \tag{3}$$

The formula for arc length (3) is also written as

$$L = \int_a^b \sqrt{1 + \left(\frac{dy}{dx}\right)^2}\, dx. \tag{4}$$

A graph that has arc length is said to be **rectifiable**.

EXAMPLE 1 Length of a Curve

Find the length of the graph of $y = 4x^{3/2}$ from the origin $(0, 0)$ to the point $(1, 4)$.

Solution The graph of the function on the interval $[0, 1]$ is given in FIGURE 6.5.3. Now,

$$\frac{dy}{dx} = 6x^{1/2}$$

is continuous on the interval. Therefore, it follows from (4) that

$$L = \int_0^1 \sqrt{1 + [6x^{1/2}]^2}\, dx$$

$$= \int_0^1 (1 + 36x)^{1/2}\, dx$$

$$= \frac{1}{36} \int_0^1 (1 + 36x)^{1/2}(36\, dx)$$

$$= \frac{1}{54}(1 + 36x)^{3/2} \Big]_0^1 = \frac{1}{54}[37^{3/2} - 1] \approx 4.1493. \qquad \blacksquare$$

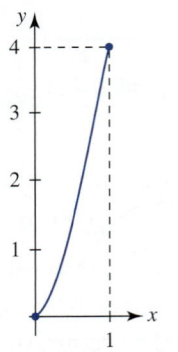

FIGURE 6.5.3 Graph of function in Example 1

■ Differential of Arc Length If C is a smooth curve defined by $y = f(x)$, then the arc length between an initial point $(a, f(a))$ and a variable point $(x, f(x))$, where $a \le x \le b$, is given by

$$s(x) = \int_a^x \sqrt{1 + [f'(t)]^2}\, dt, \tag{5}$$

where t represents a dummy variable of integration. The value of the integral in (5) obviously depends on x and so is called the **arc length function**. Then by (10) of Section 5.5, $ds/dx = \sqrt{1 + [f'(x)]^2}$ and, consequently,

$$ds = \sqrt{1 + [f'(x)]^2}\, dx. \tag{6}$$

The latter function is called the **differential of the arc length** and can be used to approximate lengths of curves. With $dy = f'(x)\,dx$, (6) can be written as

$$ds = \sqrt{(dx)^2 + (dy)^2} \quad \text{or} \quad (ds)^2 = (dx)^2 + (dy)^2. \tag{7}$$

FIGURE 6.5.4 shows that the differential ds can be interpreted as the hypotenuse of a right triangle with sides dx and dy.

If (3) is written $L = \int ds$ for brevity and the curve C is defined by $x = g(y)$, $c \le y \le d$, then the last expression in (7) can be used to solve for ds/dy:

$$\frac{ds}{dy} = \sqrt{1 + \left(\frac{dx}{dy}\right)^2} \quad \text{or} \quad ds = \sqrt{1 + \left(\frac{dx}{dy}\right)^2}\,dy.$$

Thus, the y-integration analogue of (4) is

$$L = \int_c^d \sqrt{1 + \left(\frac{dx}{dy}\right)^2}\,dy. \tag{8}$$

See Problems 17 and 18 in Exercises 6.5.

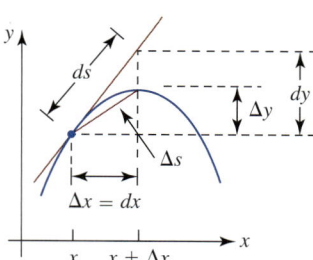

FIGURE 6.5.4 Geometric interpretation of the differential of the arc length

$\int_a^b$ NOTES FROM THE CLASSROOM

The integral in (3) often leads to problems in which specialized techniques of integration are necessary. See Chapter 7. But even with these subsequent procedures, it is not *always* possible to evaluate the indefinite integral $\int \sqrt{1 + [f'(x)]^2}\,dx$ in terms of the familiar elementary functions even for some of the simplest functions such as $y = x^2$. See Problem 45 in Exercises 7.8.

Exercises 6.5 Answers to selected odd-numbered problems begin on page ANS-20.

≡ Fundamentals

In Problems 1–12, find the length of the graph of the given function on the indicated interval. Use a calculator or CAS to obtain the graph.

1. $y = x$; $[-1, 1]$

2. $y = 2x + 1$; $[0, 3]$

3. $y = x^{3/2} + 4$; $[0, 1]$

4. $y = 3x^{2/3}$; $[1, 8]$

5. $y = \frac{2}{3}(x^2 + 1)^{3/2}$; $[1, 4]$

6. $(y + 1)^2 = 4(x + 1)^3$; $[-1, 0]$

7. $y = \frac{1}{3}x^{3/2} - x^{1/2}$; $[1, 4]$ **8.** $y = \frac{1}{6}x^3 + \frac{1}{2x}$; $[2, 4]$

9. $y = \frac{1}{4}x^4 + \frac{1}{8x^2}$; $[2, 3]$ **10.** $y = \frac{1}{5}x^5 + \frac{1}{12x^3}$; $[1, 2]$

11. $y = (4 - x^{2/3})^{3/2}$; $[1, 8]$

12. $y = \begin{cases} x - 2, & 2 \le x < 3 \\ (x - 2)^{2/3}, & 3 \le x < 10; \quad [2, 15] \\ \frac{1}{2}(x - 6)^{3/2}, & 10 \le x \le 15 \end{cases}$

In Problems 13–16, set up, but do not evaluate, an integral for the length of the graph of the given function on the indicated interval.

13. $y = x^2$; $[-1, 3]$ **14.** $y = 2\sqrt{x + 1}$; $[-1, 3]$

15. $y = \sin x$; $[0, \pi]$ **16.** $y = \tan x$; $[-\pi/4, \pi/4]$

In Problems 17 and 18, use (8) to find the length of the graph of the given equation on the indicated interval.

17. $x = 4 - y^{2/3}$; $[0, 8]$

18. $5x = y^{5/2} + 5y^{-1/2}$; $[4, 9]$

19. Consider the length of the graph of $x^{2/3} + y^{2/3} = 1$ in the first quadrant.

  (a) Show that the use of (3) leads to a discontinuous integrand.

  (b) By assuming that the Fundamental Theorem of Calculus can be used to evaluate the integral obtained in part (a), find the total length of the graph.

20. Set up, but make no attempt to evaluate, an integral that gives the total length of the ellipse $x^2/a^2 + y^2/b^2 = 1$, $a > b > 0$.

21. Given that the circumference of a circle of radius r is $2\pi r$, find the value of the integral

$$\int_0^1 \frac{1}{\sqrt{1 - x^2}}\,dx.$$

22. Use the differential of the arc length (6) to approximate the length of the graph of $y = \frac{1}{4}x^4$ from $(2, 4)$ to $(2.1, 4.862025)$. [*Hint:* Review (13) of Section 4.9.]

6.6 Area of a Surface of Revolution

▌ Introduction As we have seen in Sections 6.3 and 6.4, when the graph of a continuous function $y = f(x)$ on an interval $[a, b]$ is revolved about the x-axis, it generates a solid of revolution. In this section, we are interested in finding the area S of the corresponding surface—that is, a **surface of revolution** on $[a, b]$ as shown in FIGURE 6.6.1(b).

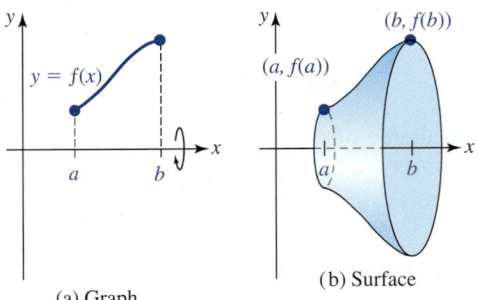

(a) Graph   (b) Surface

FIGURE 6.6.1 Surface of revolution

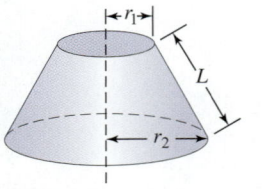

FIGURE 6.6.2 Frustum of a cone

▌ Building an Integral Before building a definite integral for the definition of the area of a surface of revolution, we first need the formula for the lateral area (top and bottom excluded) of a *frustum* of a right circular cone. See FIGURE 6.6.2. If r_1 and r_2 are the radii of the top and bottom and L is the slant height, then the lateral area is given by

$$\pi(r_1 + r_2)L. \tag{1}$$

See Problem 17 in Exercises 6.6. Now suppose $y = f(x)$ is a smooth function and $f(x) \geq 0$ on the interval $[a, b]$. Let P be a partition of the interval:

$$a = x_0 < x_1 < x_2 < \cdots < x_{n-1} < x_n = b.$$

Now, if we connect the points $(x_{k-1}, f(x_{k-1}))$ and $(x_k, f(x_k))$ shown in FIGURE 6.6.3(a) by a chord, we form a trapezoid. When revolved about the x-axis, this trapezoid generates a frustum of a cone with radii $f(x_{k-1})$ and $f(x_k)$. See Figure 6.6.3(b). As shown in cross-section in Figure 6.6.3(c), the slant height can be obtained from the Pythagorean Theorem:

$$\sqrt{(x_k - x_{k-1})^2 + (f(x_k) - f(x_{k-1}))^2}.$$

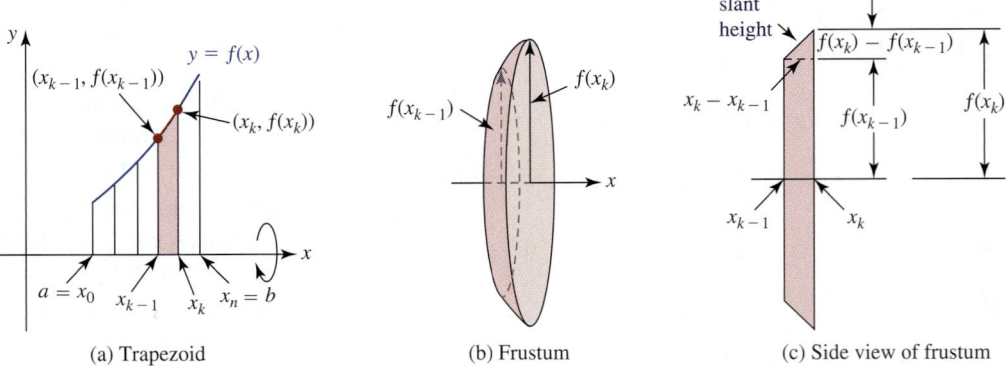

(a) Trapezoid   (b) Frustum   (c) Side view of frustum

FIGURE 6.6.3 Approximating area of surface of revolution by summing areas of frustums

Thus, from (1) the surface area of this element is

$$\begin{aligned}
S_k &= \pi[f(x_k) + f(x_{k-1})]\sqrt{(x_k - x_{k-1})^2 + (f(x_k) - f(x_{k-1}))^2} \\
&= \pi[f(x_k) + f(x_{k-1})]\sqrt{1 + \left(\frac{f(x_k) - f(x_{k-1})}{x_k - x_{k-1}}\right)^2}(x_k - x_{k-1}) \\
&= \pi[f(x_k) + f(x_{k-1})]\sqrt{1 + \left(\frac{f(x_k) - f(x_{k-1})}{x_k - x_{k-1}}\right)^2}\,\Delta x_k,
\end{aligned}$$

where $\Delta x_k = x_k - x_{k-1}$. This last quantity is an approximation to the actual area of the surface of revolution on the subinterval $[x_{k-1}, x_k]$.

Now, as in the discussion of arc length, we invoke the Mean Value Theorem for derivatives to assert that there exists an x_k^* in the open interval (x_{k-1}, x_k) such that

$$f'(x_k^*) = \frac{f(x_k) - f(x_{k-1})}{x_k - x_{k-1}}.$$

The Riemann sum

$$\sum_{k=1}^{n} S_k = \pi \sum_{k=1}^{n} [f(x_k) + f(x_{k-1})] \sqrt{1 + [f'(x_k^*)]^2} \, \Delta x_k$$

is an approximation to the area S on $[a, b]$. This suggests that the surface area S is given by the limit of the Riemann sum:

$$S = \lim_{\|P\| \to 0} \pi \sum_{k=1}^{n} [f(x_k) + f(x_{k-1})] \sqrt{1 + [f'(x_k^*)]^2} \, \Delta x_k. \tag{2}$$

Since we also expect $f(x_{k-1})$ and $f(x_k)$ to approach the common limit $f(x)$ as $\|P\| \to 0$, we have $f(x_k) + f(x_{k-1}) \to 2f(x)$.

The foregoing discussion prompts us to use (2) as the definition of the area of the surface of revolution on the interval.

Definition 6.6.1 Area of a Surface of Revolution

Let f be a function for which f' is continuous and $f(x) \geq 0$ for all x in the interval $[a, b]$. The **area** S of the surface that is obtained by revolving the graph of f on the interval about the x-axis is given by

$$S = 2\pi \int_a^b f(x) \sqrt{1 + [f'(x)]^2} \, dx. \tag{3}$$

EXAMPLE 1 Area of a Surface

Find the area S of the surface that is formed by revolving the graph of $y = \sqrt{x}$ on the interval $[1, 4]$ about the x-axis.

Solution We have $f(x) = x^{1/2}, f'(x) = \frac{1}{2}x^{-1/2} = 1/(2\sqrt{x})$, and from (3)

$$S = 2\pi \int_1^4 \sqrt{x} \sqrt{1 + \left(\frac{1}{2\sqrt{x}}\right)^2} \, dx$$

$$= 2\pi \int_1^4 \sqrt{x} \sqrt{1 + \frac{1}{4x}} \, dx$$

$$= 2\pi \int_1^4 \sqrt{x} \sqrt{\frac{4x + 1}{4x}} \, dx$$

$$= \pi \int_1^4 \sqrt{4x + 1} \, dx$$

$$= \frac{1}{4}\pi \int_1^4 (4x + 1)^{1/2}(4 \, dx) = \frac{1}{6}\pi(4x + 1)^{3/2} \Big|_1^4$$

$$= \frac{1}{6}\pi[17^{3/2} - 5^{3/2}] \approx 30.85.$$

See FIGURE 6.6.4.

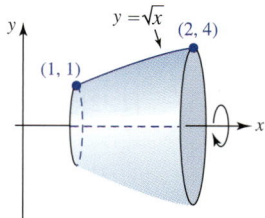

FIGURE 6.6.4 Surface of revolution about x-axis in Example 1

▌ Revolution about y-Axis It can be shown that if the graph of a continuous function $y = f(x)$ on $[a, b]$, $0 \le a < b$, is revolved about the y-axis, then the area S of the resulting surface of revolution is given by

$$S = 2\pi \int_a^b x\sqrt{1 + [f'(x)]^2}\, dx. \tag{4}$$

As in (3), we assume in (4) that $f'(x)$ is continuous on the interval $[a, b]$.

EXAMPLE 2   Area of a Surface

Find the area S of the surface formed by revolving the graph of $y = x^{1/3}$ on the interval $[0, 8]$ about the y-axis.

Solution   We have $f'(x) = \frac{1}{3}x^{-2/3}$, so that from (4) it follows that

$$S = 2\pi \int_0^8 x\sqrt{1 + \frac{1}{9}x^{-4/3}}\, dx$$

$$= 2\pi \int_0^8 x\sqrt{\frac{9x^{4/3} + 1}{9x^{4/3}}}\, dx$$

$$= \frac{2}{3}\pi \int_0^8 x^{1/3}\sqrt{9x^{4/3} + 1}\, dx.$$

Let us evaluate the last integral by reviewing the u-substitution method. If we let $u = 9x^{4/3} + 1$, then $du = 12x^{1/3}\, dx$, $dx = \frac{1}{12}x^{-1/3}\, du$, $x = 0$ implies $u = 1$, and $x = 8$ gives $u = 145$. Therefore,

$$S = \frac{1}{18}\pi \int_1^{145} u^{1/2}\, du = \frac{1}{27}\pi u^{3/2}\Big]_1^{145} = \frac{1}{27}\pi(145^{3/2} - 1^{3/2}) \approx 203.04.$$

See **FIGURE 6.6.5**.   ▮

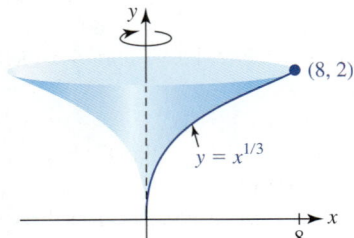

FIGURE 6.6.5   Surface of revolution about y-axis in Example 2

Exercises 6.6   Answers to selected odd-numbered problems begin on page ANS-20.

≡ Fundamentals

In Problems 1–10, find the area of the surface that is formed by revolving each graph on the given interval about the indicated axis.

1. $y = 2\sqrt{x}$, $[0, 8]$; x-axis

2. $y = \sqrt{x + 1}$, $[1, 5]$; x-axis

3. $y = x^3$, $[0, 1]$; x-axis

4. $y = x^{1/3}$, $[1, 8]$; y-axis

5. $y = x^2 + 1$, $[0, 3]$; y-axis

6. $y = 4 - x^2$, $[0, 2]$; y-axis

7. $y = 2x + 1$, $[2, 7]$; x-axis

8. $y = \sqrt{16 - x^2}$, $\left[0, \sqrt{7}\right]$; y-axis

9. $y = \frac{1}{4}x^4 + \frac{1}{8x^2}$, $[1, 2]$; y-axis

10. $y = \frac{1}{3}x^3 + \frac{1}{4x}$, $[1, 2]$; x-axis

11. (a) The shape of a dish antenna is a parabola revolved about its axis of symmetry and is called a **paraboloid**

of revolution. Find the surface area of an antenna of radius r and depth h obtained by revolving the graph of $f(x) = r\sqrt{1 - x/h}$ about the x-axis. See **FIGURE 6.6.6**.

(b) The depth of a dish antenna ranges from 10% to 20% of its radius. If the depth h of the antenna in part (a) is 10% of the radius, show that the surface area of the antenna is approximately the same as the area of a circle of radius r. What is the percentage error in this case?

Dish antennas are paraboloids of revolution

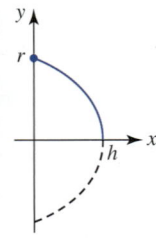

FIGURE 6.6.6   Graph of f in Problem 11

12. The surface formed by two parallel planes cutting a sphere of radius r is called a **spherical zone**. Find the area of the spherical zone shown in FIGURE 6.6.7.

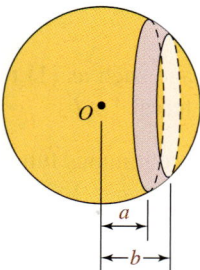

FIGURE 6.6.7   Spherical zone in Problem 12

13. The graph of $y = |x + 2|$ on $[-4, 2]$, shown in FIGURE 6.6.8, is revolved about the x-axis. Find the area S of the surface of revolution.

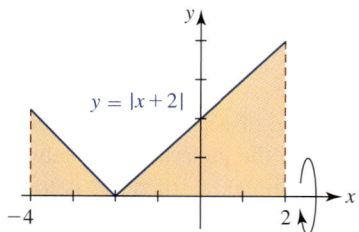

FIGURE 6.6.8   Graph of function in Problem 13

14. Find the area of the surface that is formed by revolving $x^{2/3} + y^{2/3} = a^{2/3}$, $[-a, a]$, about the x-axis.

☰ Think About It

15. Show that the lateral surface area of a right circular cone of radius r and slant height L is $\pi r L$. [*Hint*: A cone cut down its side and flattened forms a circular sector with area $\frac{1}{2}L^2\theta$.]

16. Use Problem 15 to show that the lateral surface area of a right circular cone of radius r and height h is given by $\pi r \sqrt{r^2 + h^2}$. Derive the same result using (3) or (4).

17. Use the result of Problem 15 to derive formula (1). [*Hint*: Consider a complete cone of radius r_2 and slant height L_2. Cut the conical top off. Similar triangles might help.]

18. Show that the surface area of a frustum of a right circular cone of radii r_1 and r_2 and height h is given by $\pi(r_1 + r_2)\sqrt{h^2 + (r_2 - r_1)^2}$.

19. Let $y = f(x)$ be a continuous nonnegative function on $[a, b]$ that has a continuous first derivative on the interval. Prove that if the graph of f is revolved around a horizontal line $y = L$, then the area S of the resulting surface of revolution is given by

$$S = 2\pi \int_a^b |f(x) - L| \sqrt{1 + [f'(x)]^2}\, dx.$$

20. Use the result of Problem 19 to find a definite integral that gives the area of the surface that is formed by revolving $y = x^{2/3}$, $[1, 8]$, about the line $y = 4$. Do not evaluate.

☰ Projects

21. **A View From Space**
    (a) From a spacecraft orbiting the Earth at a distance h from the surface, an astronaut can observe only a portion A_s of the Earth's total surface area A_e. See FIGURE 6.6.9(a). Find a formula for the fractional expression A_s/A_e as a function of h. In Figure 6.6.9(b) we have shown the Earth in cross-section as a circle with center C and radius R. Let the x- and y-axes be as shown and let the y-coordinates of the points B and E be y_B and $y_E = R$, respectively.
    (b) What percentage of the Earth's surface will an astronaut see from a height of 2000 km? Take the radius of the Earth to be $R = 6380$ km.
    (c) At what height h will an astronaut see one-fourth of the Earth's surface?
    (d) What is the limit of A_s/A_e as the height h increases without bound ($h \to \infty$)? Why does the answer make intuitive sense?
    (e) What percentage of the Earth's surface will an astronaut see from the Moon if $h = 3.76 \times 10^5$ km?

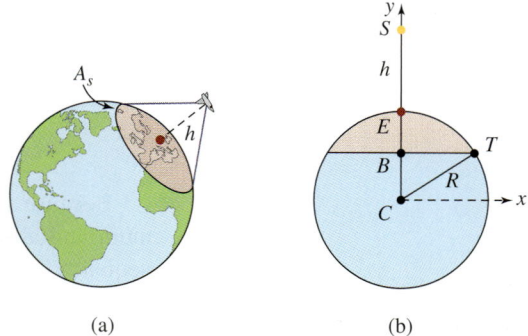

(a)   (b)
FIGURE 6.6.9   Portion of the Earth's surface in Problem 21

6.7 Average Value of a Function

▌ **Introduction**   Every student is aware of averages. If a student takes four examinations in a semester and scores 80%, 75%, 85%, and 92% on them, then his or her average score is

$$\frac{80 + 75 + 85 + 92}{4}$$

or 83%. In general, given n numbers $a_1, a_2, \ldots, a_n$, we say that their **arithmetic mean**, or **average**, is

$$\frac{a_1 + a_2 + \cdots + a_n}{n} = \frac{1}{n} \sum_{k=1}^{n} a_k. \tag{1}$$

In this section we shall extend the notion of a discrete average such as (1) to the average of *all* the values of a continuous function f defined over an interval $[a, b]$.

■ Average of Function Values Suppose now that we have a continuous function f defined on an interval $[a, b]$. For the arbitrarily chosen numbers $x_i, i = 1, 2, \ldots, n$ such that $a < x_1 < x_2 < \cdots < x_n < b$, then by (1) the average of the set of corresponding function values is

$$\frac{f(x_1) + f(x_2) + \cdots + f(x_n)}{n} = \frac{1}{n} \sum_{k=1}^{n} f(x_k). \tag{2}$$

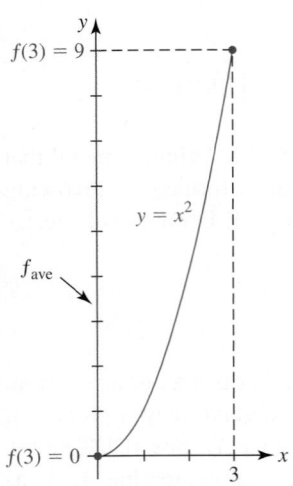

FIGURE 6.7.1 Find the average of all the numbers indicated in red on the y-axis

If we now consider the set of function values $f(x)$ that corresponds to all numbers x in an interval, it should be clear that we cannot use a discrete sum as in (1) since this set of function values is usually an uncountable set. For example, for $f(x) = x^2$ on $[0, 3]$, the values of the function range from a minimum of $f(0) = 0$ to a maximum of $f(3) = 9$. As indicated in FIGURE 6.7.1, we intuitively expect that there exists an average value f_{ave} such that $f(0) \le f_{\text{ave}} \le f(3)$.

■ Building an Integral Returning to the general case of a continuous function defined on a closed interval $[a, b]$, we let P be a regular partition of the interval into n subintervals of width $\Delta x = (b - a)/n$. If x_k^* is a number chosen in each subinterval, then the average

$$\frac{f(x_k^*) + f(x_k^*) + \cdots + f(x_n^*)}{n}$$

can be written as

$$\frac{f(x_k^*) + f(x_2^*) + \cdots + f(x_n^*)}{\dfrac{b - a}{\Delta x}} \tag{3}$$

since $n = (b - a)/\Delta x$. Rewriting (3) as

$$\frac{1}{b - a} \sum_{k=1}^{n} f(x_k^*) \, \Delta x$$

and taking the limit of this last expression as $\|P\| = \Delta x \to 0$, we obtain the definite integral

$$\frac{1}{b - a} \int_a^b f(x) \, dx. \tag{4}$$

Because we have assumed that f is continuous on $[a, b]$, let us denote its absolute minimum and absolute maximum on the interval by m and M, respectively. If we multiply the inequality

$$m \le f(x_k^*) \le M$$

by $\Delta x > 0$ and sum, we obtain

$$\sum_{k=1}^{n} m \, \Delta x \le \sum_{k=1}^{n} f(x_k^*) \, \Delta x \le \sum_{k=1}^{n} M \, \Delta x.$$

Because $\sum_{k=1}^{n} \Delta x = b - a$, the preceding inequality is equivalent to

$$(b - a)m \le \sum_{k=1}^{n} f(x_k^*) \, \Delta x \le (b - a)M.$$

And so as $\Delta x \to 0$, it follows that

$$(b - a)m \le \int_a^b f(x) \, dx \le (b - a)M.$$

From the last inequality we conclude that the number from (4) satisfies

$$m \le \frac{1}{b-a} \int_a^b f(x) \le M.$$

By the Intermediate Value Theorem, f takes on all values between m and M. Hence, the number given by (4) actually corresponds to a value of the function on the interval. This prompts us to state the following definition.

Definition 6.7.1 Average Value of a Function

Let $y = f(x)$ be continuous on $[a, b]$. The **average value** of f on the interval is the number

$$f_{\text{ave}} = \frac{1}{b-a} \int_a^b f(x)\, dx. \qquad (5)$$

Although we are interested primarily in continuous functions, Definition 6.7.1 is valid for any integrable function on the interval.

EXAMPLE 1 Finding an Average Value

Find the average value of $f(x) = x^2$ on $[0, 3]$.

Solution From (5) of Definition 6.7.1, we obtain

$$f_{\text{ave}} = \frac{1}{3-0} \int_0^3 x^2\, dx = \frac{1}{3}\left(\frac{1}{3}x^3\right)\Big]_0^3 = 3. \qquad \blacksquare$$

It is sometimes possible to determine the value of x in the interval that corresponds to the average value of a function.

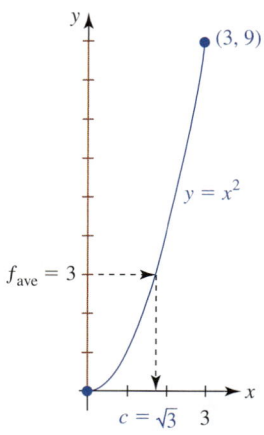

FIGURE 6.7.2 f_{ave} is the function value $f(\sqrt{3})$ in Example 1

EXAMPLE 2 Finding an x Corresponding to f_{ave}

Determine the value of x in the interval $[0, 3]$ that corresponds to the average value f_{ave} of the function $f(x) = x^2$.

Solution Since the function $f(x) = x^2$ is continuous on the closed interval $[0, 3]$, we know from the Intermediate Value Theorem that there exists a number c between 0 and 3 so that

$$f(c) = c^2 = f_{\text{ave}}.$$

But, from Example 1, we know $f_{\text{ave}} = 3$. Thus, the equation $c^2 = 3$ has solutions $c = \pm\sqrt{3}$. As shown in FIGURE 6.7.2, the only solution of this equation in $[0, 3]$ is $c = \sqrt{3}$. $\blacksquare$

■ **Mean Value Theorem for Definite Integrals** The following is an immediate consequence of the foregoing discussion. The result is called the Mean Value Theorem for Integrals.

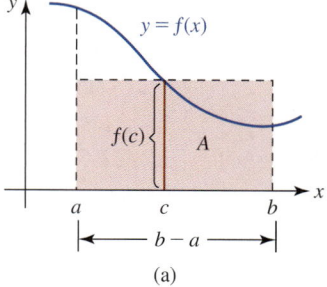

(a)

Theorem 6.7.1 Mean Value Theorem for Integrals

Let $y = f(x)$ be continuous on $[a, b]$. Then there exists a number c in the open interval (a, b) such that

$$f(c)(b-a) = \int_a^b f(x)\, dx. \qquad (6)$$

In the case when $f(x) \ge 0$ for all x in $[a, b]$, Theorem 6.7.1 is readily interpreted in terms of area. The result in (6) simply states that there is some number c in (a, b) for which the area A of a *rectangle* of height $f(c)$ and width $b - a$ shown in FIGURE 6.7.3(a) is the same as the *area A under the graph* indicated in Figure 6.7.3(b).

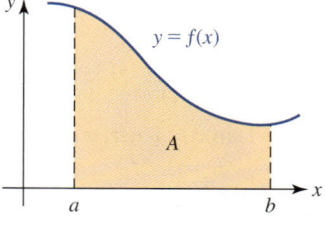

(b)

FIGURE 6.7.3 Area A of rectangle is the same as area under the graph on $[a, b]$

EXAMPLE 3 Finding an x Corresponding to f_{ave}

Find the height $f(c)$ of a rectangle so that the area A under the graph of $y = x^2 + 1$ on $[-2, 2]$ is the same as $f(c)[2 - (-2)] = 4f(c)$.

Solution This is basically the same type of problem as illustrated in Example 2. Now, the area under the graph shown in **FIGURE 6.7.4(a)** is

$$A = \int_{-2}^{2} (x^2 + 1)\, dx = \left(\frac{1}{3}x^3 + x\right)\Bigg]_{-2}^{2} = \frac{28}{3}.$$

Also, $4f(c) = 4(c^2 + 1)$, so that $4(c^2 + 1) = \frac{28}{3}$ implies $c^2 = \frac{4}{3}$. The solutions $c_1 = 2/\sqrt{3}$ and $c_2 = -2/\sqrt{3}$ are both in the interval $(-2, 2)$. For either number, we see that the height of the rectangle is $f(c_1) = f(c_2) = (\pm 2/\sqrt{3})^2 + 1 = \frac{7}{3}$. The area of the rectangle shown in Figure 6.7.4(b) is $f(c)[2 - (-2)] = \frac{7}{3} \cdot 4 = \frac{28}{3}$.

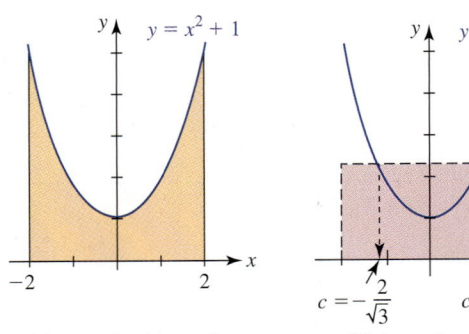

(a) Area under the graph (b) Area of rectangle

FIGURE 6.7.4 Area in (a) is the same as the area in (b) in Example 3

Exercises 6.7 Answers to selected odd-numbered problems begin on page ANS-21.

≡ Fundamentals

In Problems 1–20, find the average value f_{ave} of the given function on the indicated interval.

1. $f(x) = 4x$; $[-3, 1]$ **2.** $f(x) = 2x + 3$; $[-2, 5]$

3. $f(x) = x^2 + 10$; $[0, 2]$

4. $f(x) = 2x^3 - 3x^2 + 4x - 1$; $[-1, 1]$

5. $f(x) = 3x^2 - 4x$; $[-1, 3]$ **6.** $f(x) = (x + 1)^2$; $[0, 2]$

7. $f(x) = x^3$; $[-2, 2]$ **8.** $f(x) = x(3x - 1)^2$; $[0, 1]$

9. $f(x) = \sqrt{x}$; $[0, 9]$ **10.** $f(x) = \sqrt{5x + 1}$; $[0, 3]$

11. $f(x) = x\sqrt{x^2 + 16}$; $[0, 3]$ **12.** $f(x) = \left(1 + \frac{1}{x}\right)^{1/3}\frac{1}{x^2}$; $\left[\frac{1}{2}, 1\right]$

13. $f(x) = \frac{1}{x^3}$; $\left[\frac{1}{4}, \frac{1}{2}\right]$ **14.** $f(x) = x^{2/3} - x^{-2/3}$; $[1, 4]$

15. $f(x) = \frac{2}{(x + 1)^2}$; $[3, 5]$ **16.** $f(x) = \frac{(\sqrt{x} - 1)^3}{\sqrt{x}}$; $[4, 9]$

17. $f(x) = \sin x$; $[-\pi, \pi]$ **18.** $f(x) = \cos 2x$; $[0, \pi/4]$

19. $f(x) = \csc^2 x$; $[\pi/6, \pi/2]$ **20.** $f(x) = \frac{\sin \pi x}{\cos^2 \pi x}$; $\left[-\frac{1}{3}, \frac{1}{3}\right]$

In Problems 21 and 22, find a value of c in the given interval for which $f(c) = f_{ave}$.

21. $f(x) = x^2 + 2x$; $[-1, 1]$ **22.** $f(x) = \sqrt{x + 3}$; $[1, 6]$

23. The average value of a continuous nonnegative function $y = f(x)$ on the interval $[1, 5]$ is $f_{ave} = 3$. What is the area under the graph on the interval?

24. For $f(x) = 1 - \sqrt{x}$, find a value of b so that $f_{ave} = 0$ on $[0, b]$. Interpret geometrically.

≡ Applications

25. The function $T(t) = 100 + 3t - \frac{1}{2}t^2$ approximates the temperature at t hr past noon on a typical August day in Las Vegas. Find the average temperature between noon and 6 P.M.

26. A company determines that the revenue obtained after the sale of x units of a product is given by $R(x) = 50 + 4x + 3x^2$. Find the average revenue for sales $x = 1$ to $x = 5$. Compare the result with the average $\frac{1}{5}\sum_{k=1}^{5} R(k)$.

27. Let $s(t)$ denote the position of a particle on a horizontal axis as a function of time t. The average velocity $\bar{v}$ during the time interval $[t_1, t_2]$ is $\bar{v} = [s(t_2) - s(t_1)]/(t_2 - t_1)$. Use (5) to show that $v_{ave} = \bar{v}$. [*Hint*: Recall $ds/dt = v$.]

28. In the absence of damping, the position of a mass m on a freely vibrating spring is given by the function $x(t) = A\cos(\omega t + \phi)$, where A, ω, and ϕ are constants. The period of oscillation is $2\pi/\omega$. The potential energy of the system is $U(x) = \frac{1}{2}kx^2$, where k is the so-called spring constant. The kinetic energy of the system is $K = \frac{1}{2}mv^2$, where $v = dx/dt$. If $\omega^2 = k/m$, show that the average potential energy and the average kinetic energy over one period are the same and that each equals $\frac{1}{4}kA^2$.

29. In physics, the **Impulse-Momentum Theorem** states that the change in momentum of a body in a time interval $[t_0, t_1]$ is $mv_1 - mv_0 = (t_1 - t_0)\overline{F}$, where mv_0 is the initial momentum, mv_1 is the final momentum, and $\overline{F}$ is the average force acting on the body during the interval. Find the change in momentum of a pile driver dropped on a piling between times $t = 0$ and $t = t_1$ if

$$F(t) = k\left[1 - \left(\frac{2t}{t_1} - 1\right)^2\right],$$

where k is a constant.

30. In a small artery the velocity of blood (in cm/s) is given by $v(r) = (P/4\nu l)(R^2 - r^2), 0 \le r \le R$, where P is blood pressure, ν the viscosity of the blood, l the length of the artery, and R the radius of the artery. Find the average of $v(r)$ on the interval $[0, R]$.

≡ Think About It

31. If $y = f(x)$ is a continuous odd function, then what is f_{ave} on any interval $[-a, a]$?

32. For a linear function $f(x) = ax + b, a > 0, b > 0$, the average value of the function on $[x_1, x_2]$ is $f_{ave} = aX + b$, where X is some number in the interval. Conjecture the value of X. Prove your assertion.

33. If $y = f(x)$ is a differentiable function, find the average value of f' on the interval $[x, x + h]$, where $h > 0$.

34. Given that n is a positive integer and $a > 1$, show that the average value of $f(x) = (n + 1)x^n$ on the interval $[1, a]$ is $f_{ave} = a^n + a^{n-1} + \cdots + a + 1$.

35. Suppose $y = f(x)$ is a continuous function and f_{ave} is its average value on $[a, b]$. Explain: $\int_a^b [f(x) - f_{ave}] \, dx = 0$.

36. Let $f(x) = \lfloor x \rfloor$ be the greatest integer or floor function. Without integration, what is the average of f on $[0, 1]$? On $[0, 2]$? On $[0, 3]$? On $[0, 4]$? Conjecture the average value

of f on the interval $[0, n]$, where n is a positive integer. Prove your assertion.

37. As shown in FIGURE 6.7.5 a chord is drawn randomly between two points on a circle of radius $r = 1$. Discuss: What is the average length of the chords?

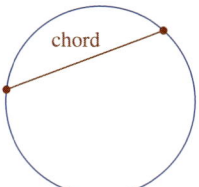

FIGURE 6.7.5 Circle in Problem 37

≡ Projects

38. **Human Limbs** The following formula is often used to approximate the surface area S of a human limb:

$$S \approx \text{average circumference} \times \text{length of limb}.$$

(a) As shown in FIGURE 6.7.6, a limb can be considered to be a solid of revolution. For many limbs, $f'(x)$ is small. If $|f'(x)| \le \varepsilon$ for $a \le x \le b$, show that

$$\int_a^b 2\pi f(x) \, dx \le S \le \sqrt{1 + \varepsilon^2} \int_a^b 2\pi f(x) \, dx.$$

(b) Show that $\overline{C}L \le S \le \sqrt{1 + \varepsilon^2}\, \overline{C}L$, where $\overline{C}$ is the average circumference of the limb over the interval $[a, b]$. Thus, the approximation formula stated above always underestimates S but does well when ε is small (such as for the forearm to wrist).

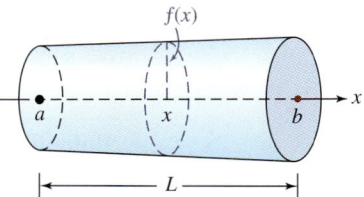

FIGURE 6.7.6 Model of a limb in Problem 38

6.8 Work

▌ Introduction In physics, when a *constant* force F moves an object a distance d in the same direction of the force, the **work** done is defined to be the product

$$W = Fd. \tag{1}$$

For example, if a 10-lb force moves an object 7 ft in the same direction as the force, then the work done is 70 ft-lb. In this section we shall see how to find the work done by a *variable* force.

Before examining work as a definite integral, let us review some important units.

Units Commonly used **units** of force, distance, and work are listed in the following table.

Quantity	Engineering system	SI	cgs
Force	pound (lb)	newton (N)	dyne
Distance	foot (ft)	meter (m)	centimeter (cm)
Work	foot-pound (ft-lb)	newton-meter (joule)	dyne-centimeter (erg)

Thus, if a force of 300 N moves an object 15 m, the work done is $W = 300 \cdot 15 = 4500$ N-m or 4500 joules. For comparison, and conversion of one unit to another, we note that

$$1 \text{ N} = 10^5 \text{ dynes} = 0.2247 \text{ lb}$$
$$1 \text{ ft-lb} = 1.356 \text{ joules} = 1.356 \times 10^7 \text{ ergs.}$$

So, for example, 70 ft-lb is equivalent to $70 \times 1.356 = 94.92$ joules, and 4500 joules is equivalent to $4500/1.356 = 3318.584$ ft-lb.

Building an Integral Now, if $F(x)$ represents a continuous variable force acting across an interval $[a, b]$, then the work is not simply a product as in (1). Suppose P is the partition

$$a = x_0 < x_1 < x_2 < \cdots < x_n = b$$

and Δx_k is the width of the kth subinterval $[x_{k-1}, x_k]$. Let x_k^* denote the sample point chosen arbitrarily in each subinterval. If the width of each $[x_{k-1}, x_k]$ is very small, then, since F is continuous, the function values $F(x)$ cannot vary much across the subinterval. Thus, we can reasonably consider the force that acts over $[x_{k-1}, x_k]$ as the constant $F(x_k^*)$ and the work done from x_{k-1} to x_k is given by the approximation

$$W_k = F(x_k^*) \, \Delta x_k.$$

An approximation to the total work done from a to b is then given by the Riemann sum

$$\sum_{k=1}^{n} W_k = F(x_1^*) \, \Delta x_1 + F(x_2^*) \, \Delta x_2 + \cdots + F(x_n^*) \, \Delta x_n = \sum_{k=1}^{n} F(x_k^*) \, \Delta x_k.$$

It is natural to assume that the work done by F over the interval is

$$W = \lim_{\|P\| \to 0} \sum_{k=1}^{n} F(x_k^*) \, \Delta x_k.$$

We summarize the foregoing discussion with the following definition.

Definition 6.8.1   Work

Let F be continuous on the interval $[a, b]$ and let $F(x)$ represent the force at a number x in the interval. Then the **work** W done by the force in moving an object from a to b is

$$W = \int_a^b F(x) \, dx. \qquad (2)$$

Note: If F is constant, $F(x) = k$ for all x in the interval, then (2) becomes $W = \int_a^b k \, dx = kx]_a^b = k(b - a)$, which is consistent with (1).

Spring Problems Hooke's Law states that, when a spring is stretched (or compressed) beyond its natural length, the restoring force exerted by the spring is directly proportional to the amount of elongation (or compression) x. Thus, in order to stretch a spring x units beyond its natural length, we need to apply the force

$$F(x) = kx, \qquad (3)$$

where k is a constant of proportionality called the **spring constant**. See FIGURE 6.8.1.

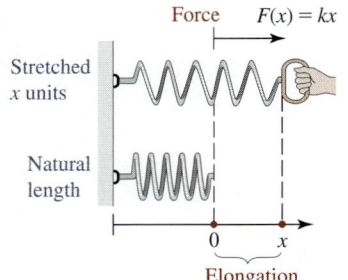

FIGURE 6.8.1   To stretch a spring x units a force $F(x) = kx$ is needed

EXAMPLE 1   Stretching a Spring

It takes a force of 130 N to stretch a spring 50 cm. Find the work done in stretching the spring 20 cm beyond its natural (unstretched) length.

Solution   When a force is measured in newtons, distances are commonly expressed in meters. Since $x = 50$ cm $= \frac{1}{2}$ m when $F = 130$ N, (3) becomes $130 = k\left(\frac{1}{2}\right)$, which implies the spring constant is $k = 260$ N/m. Thus, $F = 260x$. Now, 20 cm $= \frac{1}{5}$ m, so that the work done in stretching the spring by this amount is

$$W = \int_0^{1/5} 260x \, dx = 130x^2 \Big]_0^{1/5} = \frac{26}{5} = 5.2 \text{ joules.} \qquad \blacksquare$$

Note: Suppose the natural length of the spring in Example 1 is 40 cm. An equivalent way of stating the problem is: Find the work done in stretching the spring to a length of 60 cm. Since the elongation is $60 - 40 = 20$ cm $= \frac{1}{5}$ m, we still integrate $F = 260x$ on the interval $\left[0, \frac{1}{5}\right]$. However, if the problem were to find the work done in stretching the same spring from 50 cm to 60 cm, we would then integrate on the interval $\left[\frac{1}{10}, \frac{1}{5}\right]$. In this situation we are starting from a position where the spring is already stretched 10 cm $\left(\frac{1}{10}\text{ m}\right)$.

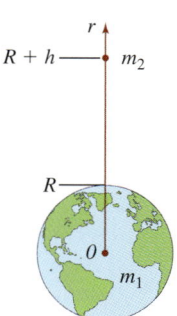

FIGURE 6.8.2   Lifting a mass m_2 to a height h

◼ **Work Done Against Gravity**   From the Universal Law of Gravitation, the force between a planet (or satellite) of mass m_1 and a body of mass m_2 is given by

$$F = k\frac{m_1 m_2}{r^2}, \qquad (4)$$

where k is a constant, called the **gravitational constant**, and r is the distance from the center of the planet to the mass m_2. See FIGURE 6.8.2. In lifting the mass m_2 off the surface of a planet of radius R to a height h, the work done can be obtained by using (4) in (2):

$$W = \int_R^{R+h} \frac{km_1 m_2}{r^2} \, dr = km_1 m_2\left(-\frac{1}{r}\right)\Big]_R^{R+h} = km_1 m_2\left(\frac{1}{R} - \frac{1}{R+h}\right). \qquad (5)$$

In SI units, $k = 6.67 \times 10^{-11}$ N·m²/kg². Some masses and values of R are given in the accompanying table.

Planets	m_1 (in kg)	R (in m)
Venus	4.9×10^{24}	6.2×10^6
Earth	6.0×10^{24}	6.4×10^6
Moon	7.3×10^{22}	1.7×10^6
Mars	6.4×10^{23}	3.3×10^6

EXAMPLE 2   Work Done in Lifting a Payload

The work done in lifting a 5000-kg payload from the surface of the Earth to a height of 30,000 m (0.03×10^6 m) follows from (5) and the preceding table:

$$W = (6.67 \times 10^{-11})(6.0 \times 10^{24})(5000)\left(\frac{1}{6.4 \times 10^6} - \frac{1}{6.43 \times 10^6}\right)$$

$$\approx 1.46 \times 10^9 \text{ joules.} \qquad \blacksquare$$

◼ **Pump Problems**   When a liquid that weighs ρ lb/ft³ is pumped from a tank, the work done in moving a fixed volume or layer of liquid d ft in a vertical direction is

$$W = \text{force} \cdot \text{distance} = (\text{weight per unit volume}) \cdot (\text{volume}) \cdot (\text{distance})$$

or

$$W = \underbrace{\rho \cdot (\text{volume})}_{\text{force}} \cdot d. \qquad (6)$$

In physics the quantity ρ is called the **weight density** of the fluid. For water, $\rho = 62.4$ lb/ft³ or 9800 N/m³.

In the next several examples we will use (6) to build the appropriate integral to find the work done in pumping water from a tank.

EXAMPLE 3 Work Done in Pumping Water

A hemispherical tank of radius 20 ft is filled with water to a 15-ft depth. Find the work done in pumping all the water to the top of the tank.

Solution As shown in FIGURE 6.8.3, we let the positive x-axis be directed *downward* and let the origin be at the center-top of the tank. Since the cross-section of the tank is a semicircle, x and y are related by $x^2 + y^2 = (20)^2$, $0 \leq x \leq 20$. Now suppose the interval $[5, 20]$, corresponding to the water on the x-axis, is partitioned into n subintervals $[x_{k-1}, x_k]$ of width Δx_k. Let x_k^* be any sample point in the kth subinterval and let W_k denote an approximation to the work done by the pump in lifting a circular layer of water of thickness Δx_k to the top of the tank. It follows from (6) that

$$W_k = \underbrace{[62.4\pi(y_k^*)^2\Delta x_k]}_{\text{force}} \cdot \underbrace{x_k^*,}_{\text{distance}}$$

where $(y_k^*)^2 = 400 - (x_k^*)^2$. Hence, the work done by the pump is approximated by the Riemann sum

$$\sum_{k=1}^{n} W_k = \sum_{k=1}^{n} 62.4\pi[400 - (x_k^*)^2]x_k^*\Delta x_k.$$

The work done in pumping all the water to the top of the tank is the limit of this last expression as $\|P\| \to 0$; that is,

$$W = \int_{5}^{20} 62.4\pi(400 - x^2)x\,dx = 62.4\pi\left(200x^2 - \frac{1}{4}x^4\right)\Big]_{5}^{20} \approx 6{,}891{,}869 \text{ ft-lb.}$$

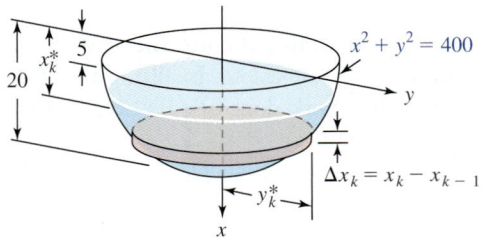

FIGURE 6.8.3 Hemispherical tank in Example 3

It is worth pursuing the analysis of Example 3 for the case where the positive x-axis is taken in the *upward* direction and the origin is at the center-bottom of the tank.

EXAMPLE 4 Alternative Solution to Example 3

With the axes as shown in FIGURE 6.8.4, we see that a circular layer of water must be lifted a distance of $20 - x_k^*$ ft. Since the center of the semicircle is at $(20, 0)$, x and y are now related by $(x - 20)^2 + y^2 = 400$. Hence,

$$W_k = \underbrace{(62.4\pi(y_k^*)^2\,\Delta x_k)}_{\text{force}} \cdot \underbrace{(20 - x_k^*)}_{\text{distance}}$$

$$= 62.4\pi[400 - (x - 20)^2](20 - x_k^*)\,\Delta x_k$$

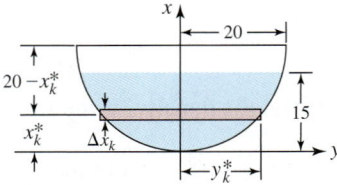

FIGURE 6.8.4 Hemispherical tank in Example 4

and so

$$W = 62.4\pi \int_0^{15} [400 - (x - 20)^2](20 - x)\,dx$$

$$= 62.4\pi \int_0^{15} (x^3 - 60x^2 + 800x)\,dx.$$

Note the new limits of integration; this is because the water shown in Figure 6.8.4 corresponds to the interval [0, 15] on the vertical x-axis. You should verify that the value of W in this case is the same as in Example 3. ∎

EXAMPLE 5   Example 3 Revisited

In Example 3, find the work done in pumping all the water to a point 10 ft above the hemispherical tank.

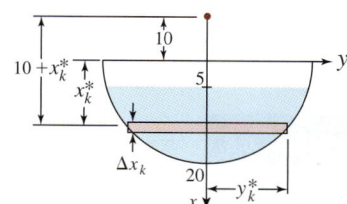

Solution   As in Figure 6.8.3, we position the positive x-axis downward. Now, from **FIGURE 6.8.5** we see

$$W_k = (62.4\pi(y_k^*)^2\,\Delta x_k) \cdot (10 + x_k^*)$$
$$= 62.4\pi[400 - (x_k^*)^2](10 + x_k^*)\,\Delta x_k.$$

FIGURE 6.8.5   Hemispherical tank in Example 5

Hence,

$$W = 62.4\pi \int_5^{20} (400 - x^2)(10 + x)\,dx$$

$$= 62.4\pi \int_5^{20} (-x^3 - 10x^2 + 400x + 4000)\,dx$$

$$= 62.4\pi\left(-\frac{1}{4}x^4 - \frac{10}{3}x^3 + 200x^2 + 4000x\right)\Bigg]_5^{20}$$

$$= 13{,}508{,}063 \text{ ft-lb.} \qquad \blacksquare$$

▌ **Cable Problems**   The next example illustrates the fact that when you are calculating the work done in lifting an object by means of a cable (heavy rope or chain), the weight of the cable must be taken into consideration.

EXAMPLE 6   Lifting an Elevator

A cable weighing 6 lb/ft is connected to a construction elevator weighing 1500 lb. Find the work done in lifting the elevator to a height of 500 ft.

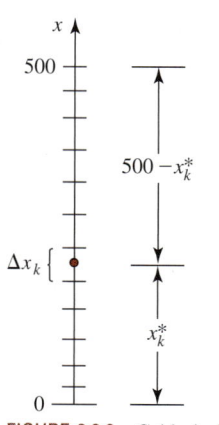

Solution   Since the weight of the elevator is a constant force, it follows from (1) that the work done in lifting the elevator a distance of 500 ft is simply

$$W_E = (1500) \cdot (500) = 750{,}000 \text{ ft-lb.}$$

The weight of the cable is the variable force. Let W_C denote the work done in lifting the cable. As shown in **FIGURE 6.8.6**, suppose the positive x-axis is directed upward and the interval [0, 500] is partitioned into n subintervals with lengths Δx_k. At a height of x_k^* ft off the ground, a segment of cable corresponding to the subinterval $[x_{k-1}, x_k]$ weighs $6\Delta x_k$ and must be pulled up an additional $500 - x_k^*$ ft. Hence, we can write

$$(W_C)_k = \underbrace{(6\,\Delta x_k)}_{\text{force}} \cdot \underbrace{(500 - x_k^*)}_{\text{distance}} = (3000 - 6x_k^*)\,\Delta x_k$$

FIGURE 6.8.6   Cable in Example 6

and so

$$W_C = \int_0^{500} (3000 - 6x)\,dx = (3000x - 3x^2)\Bigg]_0^{500} = 750{,}000 \text{ ft-lb.}$$

Thus, the total work done in lifting the elevator is

$$W = W_E + W_C = 1{,}500{,}000 \text{ ft-lb.} \qquad \blacksquare$$

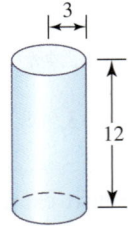

FIGURE 6.8.7 Elevator in Examples 6 and 7

EXAMPLE 7 Alternative Solution to Example 6

This is a slightly faster analysis of Example 6. As shown in FIGURE 6.8.7, when the elevator is at a height of x ft, it must be pulled up an additional $500 - x$ ft. The lifting force needed at that height is

$$\underbrace{1500}_{\substack{\text{weight of}\\\text{elevator}}} + \underbrace{6(500 - x)}_{\substack{\text{weight of}\\\text{cable}}} = 4500 - 6x.$$

Thus, by (2) the work done is

$$W = \int_0^{500} (4500 - 6x)\, dx = 1{,}500{,}000 \text{ ft-lb}. \qquad \blacksquare$$

Exercises 6.8   Answers to selected odd-numbered problems begin on page ANS-21.

☰ Fundamentals

1. Find the work done when a 55-lb force moves an object 20 yd in the same direction of the force.

2. A force of 100 N is applied to an object at an angle of $30°$ measured from the horizontal. If the object moves 8 m horizontally, find the work done by the force.

3. A mass that weighs 10 lb stretches a spring $\frac{1}{2}$ ft. How much will a mass that weighs 8 lb stretch the same spring?

4. A spring has a natural length of 0.5 m. A force of 50 N stretches the spring to a length of 0.6 m.

   (a) What force is needed to stretch the spring x m?

   (b) What force is required to stretch the spring to a length of 1 m?

   (c) How long is the spring when stretched by a force of 200 N?

5. In Problem 4:

   (a) Find the work done in stretching the spring 0.2 m.

   (b) Find the work done in stretching the spring from a length of 1 m to a length of 1.1 m.

6. A force of $F = \frac{3}{2}x$ lb is needed to stretch a 10-in. spring an additional x in.

   (a) Find the work done in stretching the spring to a length of 16 in.

   (b) Find the work done in stretching the spring 16 in.

7. A mass that weighs 10 lb is suspended from a 2-ft spring. The spring is stretched 8 in. and then the mass is removed.

   (a) Find the work done in stretching the spring to a length of 3 ft.

   (b) Find the work done in stretching the spring from a length of 4 ft to a length of 5 ft.

8. A 50-lb force compresses a 15-in.-long spring by 3 in. Find the work done in compressing the spring to a final length of 5 in.

9. Find the work done in lifting a mass of 10,000 kg from the surface of the Earth to a height of 500 km.

10. Find the work done in lifting a mass of 50,000 kg from the surface of the Moon to a height of 200 km.

11. A tank in the form of a right circular cylinder is filled with water. The dimensions of the tank (in feet) are shown in FIGURE 6.8.8. Find the work done in pumping all the water to the top of the tank.

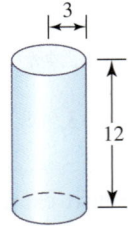

FIGURE 6.8.8 Cylindrical tank in Problem 11

12. A tank in the form of a right circular cone, vertex down, is filled with water to a depth of one-half its height. The dimensions of the tank (in feet) are shown in FIGURE 6.8.9. Find the work done in pumping all the water to the top of the tank. [*Hint*: Assume that the origin is at the vertex of the cone.]

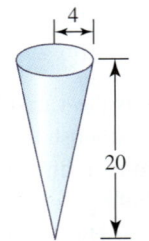

FIGURE 6.8.9 Conical tank in Problem 12

13. For the conical tank in Problem 12, find the work done in pumping all the water to a point 5 ft above the top of the tank.

14. Suppose the cylindrical tank in Problem 11 is horizontal. Find the work done in pumping all the water to a point 2 ft above the top of the tank. [*Hint*: See Problems 55–58 in Exercises 6.2.]

15. A tank has cross sections in the form of isosceles triangles, vertex down. The dimensions of the tank (in feet) are shown in FIGURE 6.8.10. Find the work done in filling the tank with water through a hole in its bottom by a pump located 5 ft below its vertex.

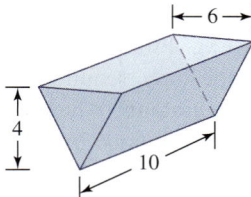

FIGURE 6.8.10 Tank with triangular cross sections in Problem 15

16. A horizontal vat with semicircular cross sections contains oil whose weight density is 80 lb/ft^3. The dimensions of the tank (in feet) are shown in FIGURE 6.8.11. If the depth of the oil is 3 ft, find the work done in pumping all the oil to the top of the tank.

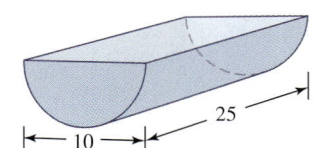

FIGURE 6.8.11 Semicircular vat in Problem 16

17. A 100-ft anchor chain, weighing 20 lb/ft, is hanging vertically over the side of a boat. How much work is performed by pulling in 40 ft of the chain?

18. A ship is anchored in 200 ft of water. In water the ship's anchor weighs 3000 lb and its anchor chain weighs 40 lb/ft. If the anchor chain hangs vertically, how much work is done in pulling in 100 ft of the chain?

19. A bucket of sand weighing 80 lb is lifted vertically by means of a rope and pulley to a height of 65 ft. Find the work done if

(a) the weight of the rope is negligible and
(b) the rope weighs $\frac{1}{2}$ lb/ft.

20. A bucket, initially containing 20 ft^3 of water, is lifted vertically from ground level. If the water leaks out at a rate of $\frac{1}{2}$ ft^3 per vertical foot, find the work done in lifting the bucket to a height at which it is empty.

21. The force of attraction between an electron and the nucleus of an atom is inversely proportional to the square of the distance separating them. If the initial distance between nucleus and electron is 1 unit, find the work done by an external force that moves the electron out to a distance four times the initial distance.

22. A rocket weighing 2,500,000 lb when fueled carries a 200,000-lb shuttle orbiter. Assume, in the early stages of the launch, that the rocket burns fuel at a rate of 100 lb/ft.

(a) Express the total weight of the system in terms of its altitude above the surface of the Earth. See FIGURE 6.8.12.
(b) Find the work done in lifting the system to an altitude of 1000 ft.

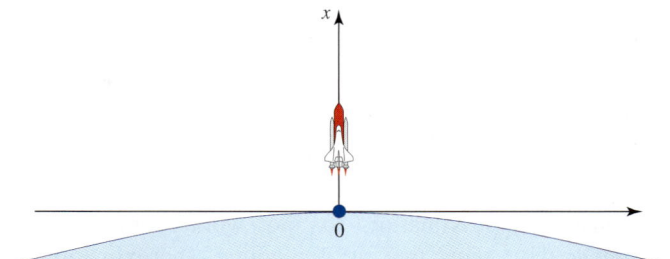

FIGURE 6.8.12 Rocket in Problem 22

23. In thermodynamics, if a gas enclosed in a cylinder expands against a piston so that the volume of the gas changes from v_1 to v_2, then the work done on the piston is given by $W = \int_{v_1}^{v_2} p\,dv$, where p is pressure (force per unit area). See FIGURE 6.8.13. In an adiabatic expansion of an ideal gas, pressure and volume are related by $pv^\gamma = k$, where γ and k are constants. Show that if $\gamma \neq 1$, then

$$W = \frac{p_2v_2 - p_1v_1}{1 - \gamma}$$

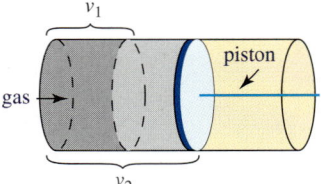

FIGURE 6.8.13 Piston in Problem 23

24. Show that when a body of weight mg is lifted vertically from a point y_1 to a point y_2, $y_2 > y_1$, the work done is the change in potential energy $W = mgy_2 - mgy_1$.

≡ **Think About It**

25. A person pushes against an immovable wall with a horizontal force of 75 lb. How much work is done?

26. The graph of a variable force F is given in FIGURE 6.8.14. Find the work done by the force in moving a particle from $x = 0$ to $x = 6$.

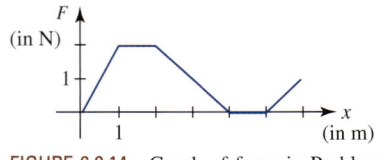

FIGURE 6.8.14 Graph of force in Problem 26

27. **A Bit of History—A Real Tall Story** In 1977 George Willig, known as the "human fly" or "spider man," scaled the outside of the south tower of the World Trade Center building in New York City to a height of 1350 ft in 3.5 h at a rate of 6.4 ft/min. At the time Willig weighed 165 lb. How much work did George do? (For his efforts, he was fined $1.10, 1 cent for each of the 110 stories of the building.)

28. A bucket containing water weighs 200 lb. As the bucket is lifted by a rope water leaks out of its bottom at a constant rate so that when it reaches a height of 10 ft it weighs 180 lb. Assume that the weight of the rope is negligible. Discuss: Explain why $\frac{200 + 180}{2} \cdot 10 = 1900$ ft-lb is a reasonable approximation to the work done. Without integration, show that the foregoing "approximation" is also the exact value of the work done.

29. As shown in FIGURE 6.8.15, a body of mass m is moved by a horizontal force F on a frictionless surface from a position at x_1 to a position at x_2. At these respective points, the body is moving at velocities v_1 and v_2, where $v_2 > v_1$. Show that the work done by the force is the increase in kinetic energy $W = \frac{1}{2}mv_2^2 - \frac{1}{2}mv_1^2$. [*Hint*: Use Newton's second law, $F = ma$, and express the acceleration a in terms of velocity v. Integrate with respect to time t and make a substitution.]

30. As shown in FIGURE 6.8.16, a bucket containing concrete that is suspended from a cable is pushed horizontally from the vertical by a construction worker. The length of the cable is 30 m and the combined mass m of the bucket and concrete is 550 kg. From the principles of physics it can be shown that the force required to move the bucket x m is given by $F = mg\tan\theta$, where g is the acceleration of gravity. Find the work done by the construction worker in pushing the bucket a horizontal distance of 3 m. [*Hint*: Use (2) and a substitution.]

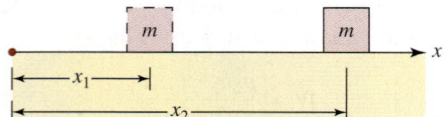

FIGURE 6.8.15 Mass in Problem 29

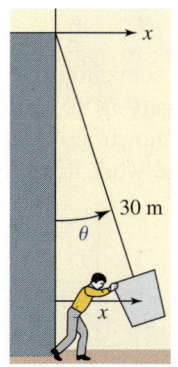

FIGURE 6.8.16 Bucket in Problem 30

6.9 Fluid Pressure and Force

■ **Introduction** Everyone has at one time experienced "ear popping" or even pain in the ear when descending in an airplane (or in an elevator) or when diving to the bottom of a swimming pool. These annoying ear sensations are due to a increase in the *pressure* exerted by air or water on mechanisms in the middle ear. Air and water are examples of fluids. In this section we will how the definite integral can be used to find the force exerted by a fluid.

A *fluid* includes liquids (such as water and oil) as well as gases (such as nitrogen).

■ **Force and Pressure** Suppose a *horizontal* flat plate is submerged below the surface of a fluid such as water. The force exerted on the plate by the fluid directly above it, called the **fluid force F**, is defined to be

$$F = \underbrace{(\text{force per unit area})}_{\text{fluid pressure } P} \cdot (\text{area of surface})$$

$$= PA. \tag{1}$$

If ρ denotes the weight density of the fluid (weight per unit volume) and A is the area of the horizontal plate submerged to a depth h, shown in FIGURE 6.9.1(a), then the **fluid pressure P** on the plate can be expressed in terms of ρ:

$$P = (\text{weight per unit volume}) \cdot (\text{depth}) = \rho h. \tag{2}$$

Therefore, the fluid force (1) is the same as

$$F = (\text{fluid pressure}) \cdot (\text{area of surface}) = \rho h A. \tag{3}$$

However, when a *vertical* plate is submerged, the fluid pressure and the fluid force on one side of the plate varies with the depth. See Figure 6.9.1(b). For example, the fluid pressure on a vertical dam is less at the top than at its base.

Before we begin, let us consider a simple example of pressure and force of a horizontally submerged plate.

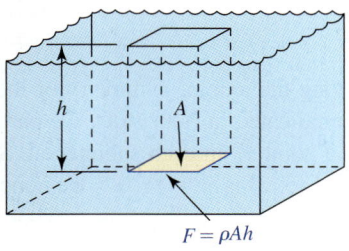
(a) Horizontal plate

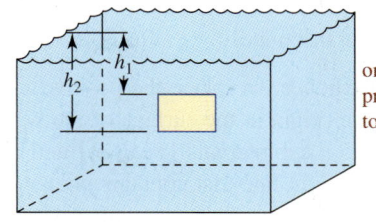
(b) Vertical plate

on a vertical plate pressure varies from top to bottom

FIGURE 6.9.1 Fluid pressure and force are constant on a horizontally submerged plate, but fluid pressure and force vary with the depth on a vertically submerged plate

EXAMPLE 1 Pressure and Force

A flat rectangular plate with dimensions 5 ft × 6 ft is submerged horizontally in water at a depth of 10 ft. Determine the pressure and the force exerted on the plate by the water above it.

Solution Recall that the weight density for water is 62.4 lb/ft^3. Hence, by (2) the fluid pressure is

$$P = \rho h = (62.4 \text{ lb/ft}^3) \cdot (10 \text{ ft}) = 624 \text{ lb/ft}^2.$$

Since the surface area of the plate is $A = 30$ ft^2, it follows from (3) that the fluid force on the plate is

$$F = PA = (\rho h)A = (624 \text{ lb/ft}^2) \cdot (30 \text{ ft}^2) = 18,720 \text{ lb}. \quad ■$$

To determine the total force F exerted by a fluid on one side of a vertically submerged flat surface, we employ one form of **Pascal's Principle**:

• *The pressure exerted by a fluid at a depth h is the same in all directions.*

Thus, if a large container with a flat bottom and vertical sidewalls is filled with water to a depth of 10 ft, the pressure of 624 lb/ft^2 at its bottom applies equally to the sidewalls. See **FIGURE 6.9.2**.

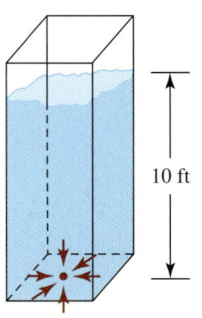

FIGURE 6.9.2 Pressure of 640 lb/ft^2 applies in all directions

■ **Building an Integral** Let the positive x-axis be directed downward with the origin at the surface of a fluid. Suppose a vertical flat plate, bounded by the horizontal lines $x = a$ and $x = b$, is submerged in the fluid as shown in **FIGURE 6.9.3(a)**. Let $w(x)$ be a function that denotes the width of the plate at any number x in $[a, b]$ and let P be any partition of the interval. If x_k^* is a sample point in the kth subinterval $[x_{k-1}, x_k]$, then from (3) with the identifications $h = x_k^*$ and $A = w(x_k^*)\,\Delta x_k$, the force F_k exerted by the fluid on the corresponding rectangular element is approximated by

$$F_k = \rho \cdot x_k^* \cdot w(x_k^*)\,\Delta x_k,$$

where, as before, ρ denotes the weight density of the fluid. Thus, an approximation to the fluid force on one side of the plate is given by the Riemann sum

$$\sum_{k=1}^{n} F_k = \sum_{k=1}^{n} \rho x_k^* w(x_k^*)\,\Delta x_k.$$

This suggests that the total fluid force on the plate is

$$F = \lim_{\|P\| \to 0} \sum_{k=1}^{n} \rho x_k^* w(x_k^*)\,\Delta x_k.$$

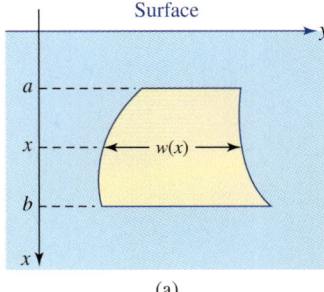
(a)

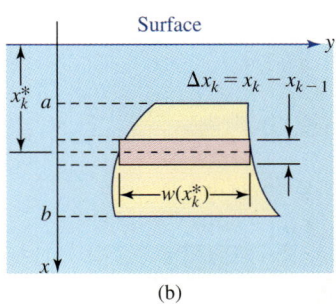
(b)

FIGURE 6.9.3 Submerged vertical plate with varying width $w(x)$ on $[a, b]$

Definition 6.9.1 Force Exerted by a Fluid

Let ρ be the weight density of a fluid and let $w(x)$ be a continuous function on $[a, b]$ that describes the width of a vertically submerged plate at a depth x. The **force F** exerted by the fluid on one side of the submerged plate is

$$F = \int_a^b \rho x w(x)\,dx. \qquad (4)$$

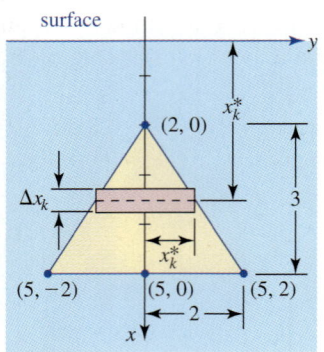

surface

(2, 0)

Δx_k

(5, −2) (5, 0) (5, 2)

FIGURE 6.9.4 Triangular plate in Example 2

EXAMPLE 2 Fluid Force

A plate in the shape of an isosceles triangle 3 ft high and 4 ft wide is submerged vertically in water, base downward, with the base 5 ft below the surface. Find the force exerted by the water on one side of the plate.

Solution For convenience, we place the positive x-axis along the axis of symmetry of the triangular plate with the origin at the surface of the water. As indicated in FIGURE 6.9.4, we partition the interval $[2, 5]$ into n subintervals $[x_{k-1}, x_k]$ and choose a point x_k^* in each subinterval. Since the equation of the straight line that contains points $(2, 0)$ and $(5, 2)$ is $y = \frac{2}{3}x - \frac{4}{3}$ we conclude by symmetry that the width of the rectangular element, shown in light red in Figure 6.9.4, is

$$2y_k^* = 2\left(\frac{2}{3}x_k^* - \frac{4}{3}\right).$$

Now $\rho = 62.4$ lb/ft^3 so that the fluid force on that portion of the plate that corresponds to the kth subinterval is approximated by

$$F_k = (62.4) \cdot x_k^* \cdot 2\left(\frac{2}{3}x_k^* - \frac{4}{3}\right) \Delta x_k.$$

Forming the sum $\sum_{k=1}^{n} F_k$ and taking the limit as $\|P\| \to 0$ give

$$F = \int_2^5 (62.4)2x\left(\frac{2}{3}x - \frac{4}{3}\right) dx$$

$$= (62.4)\frac{4}{3}\int_2^5 (x^2 - 2x)\, dx$$

$$= 83.2\left(\frac{1}{3}x^3 - x^2\right)\Big]_2^5$$

$$= (83.2) \cdot 18 = 1497.6 \text{ lb.} \qquad \blacksquare$$

In problems such as Example 2, the x- and y-axes are placed where convenient. If the y-axis is placed perpendicular to the x-axis at the top of the plate at the point $(2, 0)$, then the four points $(2, 0)$, $(5, -2)$, $(5, 0)$, and $(5, 2)$ in Figure 6.9.4 become $(0, 0)$, $(3, -2)$, $(3, 0)$, and $(3, 2)$, respectively. The equation of the straight line that contains the points $(0, 0)$ and $(3, 2)$ is $y = \frac{2}{3}x$. You should verify then that the force F exerted by the water against the plate is given the definite integral

$$F = (62.4)\frac{4}{3}\int_0^3 x(x + 2)\, dx.$$

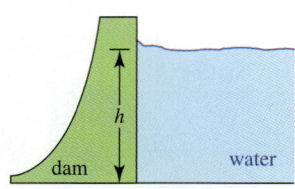

(a) Side view of dam and water

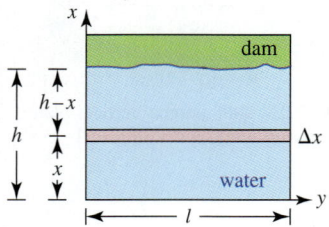

(b) Water against face of dam

FIGURE 6.9.5 Dam in Example 3

EXAMPLE 3 Force of Water Against a Dam

A dam has a vertical rectangular face. Find the force exerted by the water against the vertical face of the dam if the water is h ft deep and l ft wide. See FIGURE 6.9.5(a).

Solution For variety, let us take the positive x-axis pointing upward from the bottom of the rectangular face of the dam as shown in Figure 6.9.5(b). We then divide the interval $[0, h]$ into n subintervals. Suppressing the use of subscripts, the fluid force F_k on that rectangular portion of the plate that corresponds to the kth subinterval, shown in light red in Figure 6.9.5(b), is approximated by

$$F_k = (62.4) \cdot (h - x) \cdot (l\Delta x).$$

Here the depth is $h - x$ and the area of the rectangular element is $l\Delta x$. Summing these approximations and taking the limit as $\|P\| \to 0$ leads to

$$F = \int_0^h 62.4l(h - x)\, dx = \frac{1}{2}(62.4)lh^2. \qquad \blacksquare$$

In Example 3, if, say, the water is 100 ft deep and 300 ft wide, the fluid force on the face of the dam is then 93,600,000 lb.

≡ Fundamentals

1. Consider the tanks with flat circular bottoms shown in FIGURE 6.9.6. Each tank is full of water whose weight density is 62.4 lb/ft³. Find the pressure and force exerted by the water on the bottom of each tank.

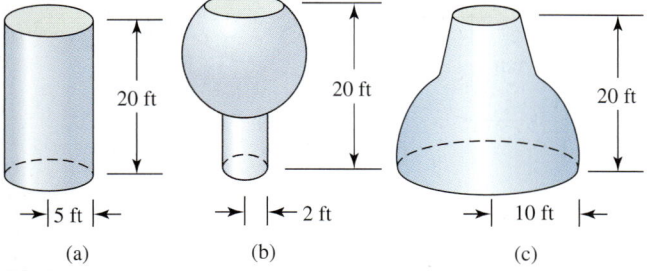

(a)   (b)   (c)
FIGURE 6.9.6   Tanks in Problem 1

2. The tanker shown in FIGURE 6.9.7 has a flat bottom and is filled with oil whose weight density is 55 lb/ft³. The tanker is 350 ft long.

(a) What is the pressure exerted on the bottom of the tanker by the oil?

(b) What is the pressure exerted on the bottom of the tanker by the water?

(c) What is the force exerted on the bottom of the tanker by the oil?

(d) What is the force exerted on the bottom of the tanker by the water?

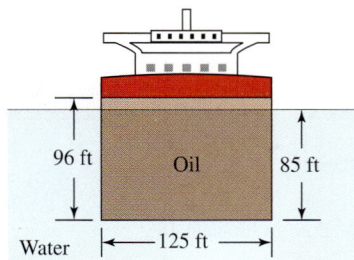

FIGURE 6.9.7   Tanker in Problem 2

3. A rectangular swimming pool in the form of a rectangular parallelepiped has dimensions of 30 ft × 15 ft × 9 ft.

(a) Find the pressure and force exerted on the flat bottom if the pool is filled with water to a depth of 8 ft. See FIGURE 6.9.8.

(b) Find the force exerted by the water on a vertical sidewall and on a vertical end.

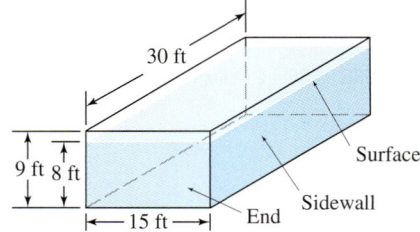

FIGURE 6.9.8   Swimming pool in Problem 3

4. A plate in the shape of an equilateral triangle $\sqrt{3}$ ft on a side is submerged vertically, base downward, with vertex

1 ft below the surface of the water. Find the force exerted by the water on one side of the plate.

5. Find the force on one side of the plate in Problem 4 if it is suspended with base upward 1 ft below the surface of the water.

6. A triangular plate is submerged vertically in water as shown in FIGURE 6.9.9. Find the force exerted by the water on one side of the plate.

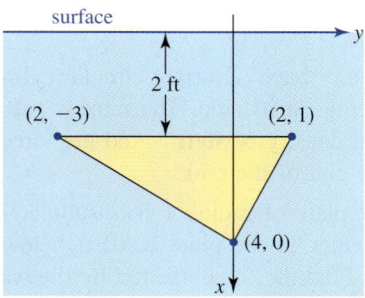

FIGURE 6.9.9   Triangular plate in Problem 6

7. Assuming the positive x-axis is downward, a plate bounded by the parabola $x = y^2$ and the line $x = 4$ is submerged vertically in oil that has weight density 50 lb/ft³. If the vertex of the parabola is at the surface, find the force exerted by the oil on one side of the plate.

8. Assuming the positive x-axis is downward, a plate bounded by the parabola $x = y^2$ and the line $y = -x + 2$ is submerged vertically in water. If the vertex of the parabola is at the surface, find the force exerted by the water on one side of the plate.

9. A full water trough has vertical ends in the form of trapezoids as shown in FIGURE 6.9.10. Find the force exerted by the water on one end of the trough.

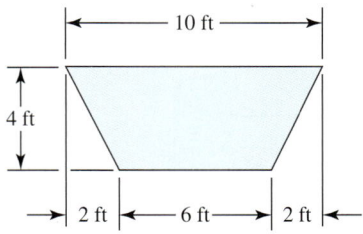

FIGURE 6.9.10   Water trough in Problem 9

10. A full water trough has vertical ends in the form shown in FIGURE 6.9.11. Find the force exerted by the water on one end of the trough.

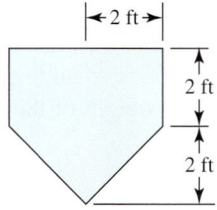

FIGURE 6.9.11   Water trough in Problem 10

11. A vertical end of a full swimming pool has the shape given in FIGURE 6.9.12. Find the force exerted by the water on this end of the pool.

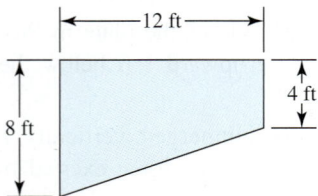

FIGURE 6.9.12   End of swimming pool in Problem 11

12. A tank in the shape of a right circular cylinder of diameter 10 ft is lying on its side. The tank is half full of oil that has weight density 60 lb/ft^3. Find the force exerted by the oil on one end of the tank.

13. A circular plate of radius 4 ft is submerged vertically so that the center of the plate is 10 ft below the surface of the water. Find the force exerted by the water on one side of the plate. [*Hint*: For simplicity, take the origin to be the center of the plate, positive *x*-axis downward. Also see Problems 55–58 in Exercises 6.2.]

14. A tank whose ends are in the form of an ellipse $x^2/4 + y^2/9 = 1$ is submerged in a liquid that has weight density ρ so that the end plates are vertical. Find the force exerted by the liquid on one end if its center is 10 ft below the surface of the liquid. [*Hint*: Proceed as in Problem 13 and use the fact that the area of an ellipse $x^2/a^2 + y^2/b^2 = 1$ is πab.]

15. A solid block in the shape of a cube 2 ft on a side is submerged in a large tank of water. The top of the block is horizontal and is 3 ft below the surface of the water. Find the total force on the block (six sides) that is caused by liquid pressure. See FIGURE 6.9.13.

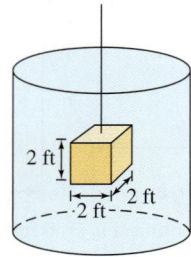

FIGURE 6.9.13   Submerged block in Problem 15

16. In Problem 15, what is the difference between the force on the bottom of the block and the force on the top of the block? This difference is the buoyant force of the water and, by Archimedes' Principle, is equal to the weight of the water displaced. What is the weight of the water displaced by the block?

≡ Think About It

17. Consider the rectangular swimming pool shown in FIGURE 6.9.14(a) whose ends are trapezoids. The pool is full of water. By taking the positive *x*-axis, as shown in Figure 6.9.14(b), find the force exerted by the water on the bottom of the pool. [*Hint*: Express the depth *d* in terms of *x*.]

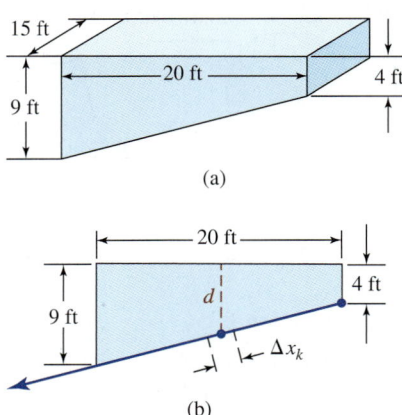

(a)

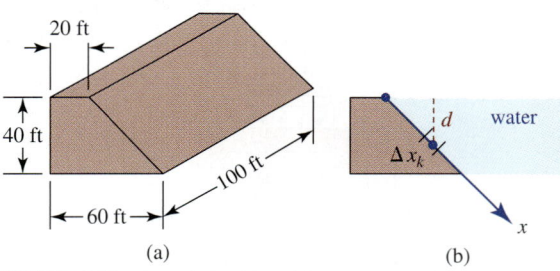

(b)

FIGURE 6.9.14   Swimming pool in Problem 17

18. An earthen dam is constructed with dimensions as shown in FIGURE 6.9.15(a). By taking the positive *x*-axis, as shown in Figure 6.9.15(b), find the force exerted by the water on the slanted wall of the dam.

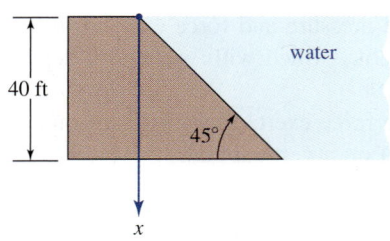

(a)   (b)

FIGURE 6.9.15   Dam in Problem 18

19. Analyze Problem 18 with the positive *x*-axis shown in FIGURE 6.9.16.

FIGURE 6.9.16   Orientation of *x*-axis in Problem 19

6.10 Centers of Mass and Centroids

■ **Introduction** In this section we consider another application from physics. We use the definite integral to find the mass and center of mass of rods and plane regions. We begin with a review of how to find the center of mass of one- and two-dimensional systems of n discrete or point masses.

■ **One-Dimensional Systems** If x denotes the directed distance from the origin O to a mass m, we say that the product mx is the **moment of the mass** about the origin. Some units are summarized in the following table.

Quantity	Engineering system	SI	cgs
Mass	slug	kilogram (kg)	gram (g)
Moment of mass	slug–feet	kilogram–meter	gram–centimeter

Now, for n point masses $m_1, m_2, \ldots, m_n$ at directed distances $x_1, x_2, \ldots, x_n$, respectively, from O, as in FIGURE 6.10.1, we say that

$$m = m_1 + m_2 + \cdots + m_n = \sum_{k=1}^{n} m_k$$

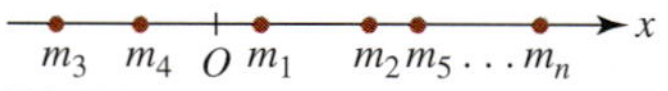

is the **total mass of the system**, and that

$$M_O = m_1 x_1 + m_2 x_2 + \cdots + m_n x_n = \sum_{k=1}^{n} m_k x_k$$

is the **moment of the system about the origin**. If $\sum_{k=1}^{n} m_k x_k = 0$, the system is said to be in **equilibrium**. See FIGURE 6.10.2. If the system of masses in Figure 6.10.1 is not in equilibrium, there is a point P with coordinate $\bar{x}$ such that

$$\sum_{k=1}^{n} m_k (x_k - \bar{x}) = 0 \qquad \text{or} \qquad \sum_{k=1}^{n} m_k x_k - \bar{x} \sum_{k=1}^{n} m_k = 0.$$

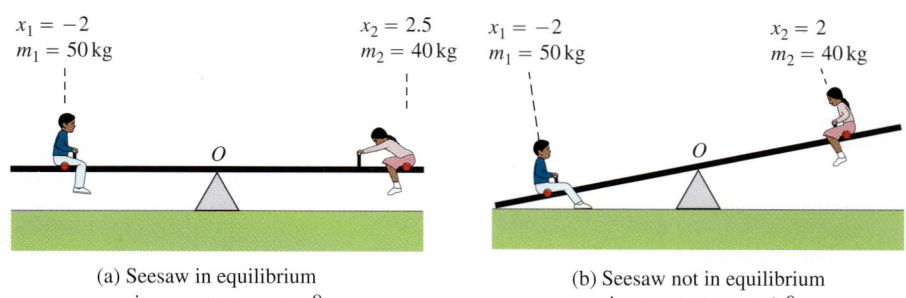

$x_1 = -2$
$m_1 = 50\,\text{kg}$

$x_2 = 2.5$
$m_2 = 40\,\text{kg}$

$x_1 = -2$
$m_1 = 50\,\text{kg}$

$x_2 = 2$
$m_2 = 40\,\text{kg}$

(a) Seesaw in equilibrium
since $m_1 x_1 + m_2 x_2 = 0$

(b) Seesaw not in equilibrium
since $m_1 x_1 + m_2 x_2 \neq 0$

FIGURE 6.10.2   Seesaw in equilibrium (a); not in equilibrium (b)

Solving for $\bar{x}$ gives

$$\bar{x} = \frac{M_O}{m} = \frac{\displaystyle\sum_{k=1}^{n} m_k x_k}{\displaystyle\sum_{k=1}^{n} m_k}. \tag{1}$$

The point with coordinate $\bar{x}$ is called the **center of mass** or the **center of gravity** of the system.
    Since (1) implies $\bar{x}\left(\sum_{k=1}^{n} m_k\right) = \sum_{k=1}^{n} m_k k_k$, it follows that $\bar{x}$ is the directed distance from the origin to a point at which the total mass of the system can be considered to be concentrated.

In the right margin, next to FIGURE 6.10.1:

$m_3 \quad m_4 \quad O \; m_1 \qquad m_2 m_5 \ldots m_n \to x$

FIGURE 6.10.1 n masses on the x-axes

◀ In a system in which the acceleration of gravity varies from mass to mass, the center of gravity is not the same as the center of mass.

Center of Mass of Three Objects

Three bodies of masses 4 kg, 6 kg, and 10 kg are located at $x_1 = -2$, $x_2 = 4$, and $x_3 = 9$, respectively. Distances are measured in meters. Find the center of mass.

Solution From (1),

$$\bar{x} = \frac{4 \cdot (-2) + 6 \cdot 4 + 10 \cdot 9}{4 + 6 + 10} = \frac{106}{20} = 5.3.$$

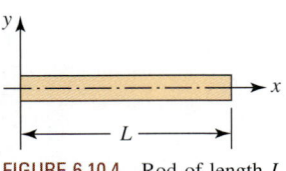

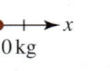

FIGURE 6.10.3 Center of mass of three point masses

FIGURE 6.10.3 shows that the center of mass $\bar{x}$ is 5.3 m to the right of the origin. ■

■ **Building an Integral** Now, let us consider the problem of finding the center of mass of a rod of length L that has a **variable linear density** ρ (mass/unit length measured in slugs/ft, kg/m, or g/cm). We assume that the rod coincides with the x-axis on the interval $[0, L]$, as shown in FIGURE 6.10.4, and that the density is a continuous function $\rho(x)$. After forming a partition P of the interval, we choose a point x_k^* in $[x_{k-1}, x_k]$. The number

$$m_k = \rho(x_k^*) \, \Delta x_k$$

is an approximation to the mass of that portion of the rod on the subinterval. Also, the moment of this element of mass about the origin is approximated by

$$(M_O)_k = x_k^* \rho(x_k^*) \, \Delta x_k.$$

Thus, we conclude that

$$m = \lim_{\|P\| \to 0} \sum_{k=1}^{n} \rho(x_k^*) \, \Delta x_k = \int_0^L \rho(x) \, dx$$

and

$$M_O = \lim_{\|P\| \to 0} \sum_{k=1}^{n} x_k^* \rho(x_k^*) \, \Delta x_k = \int_0^L x \rho(x) \, dx$$

are the **mass of the rod** and its **moment about the origin**, respectively. It then follows from $\bar{x} = M_O/m$ that the center of mass of the rod is given by

$$\bar{x} = \frac{\displaystyle\int_0^L x \rho(x) \, dx}{\displaystyle\int_0^L \rho(x) \, dx}. \tag{2}$$

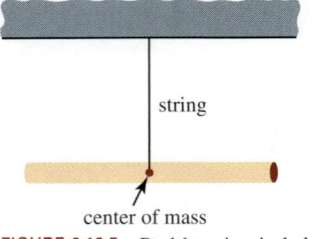

FIGURE 6.10.5 Rod hanging in balance

As shown in FIGURE 6.10.5, a rod suspended by a string attached to its center of mass would hang in perfect balance.

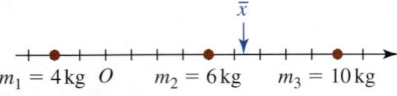

y

x

L

FIGURE 6.10.4 Rod of length L coinciding with x-axis

Center of Mass of a Rod

A 16-cm-long rod has a linear density, measured in g/cm, given by $\rho(x) = \sqrt{x}$, $0 \le x \le 16$. Find its center of mass.

Solution The mass of the rod in grams is

$$m = \int_0^{16} x^{1/2} \, dx = \frac{2}{3} x^{3/2} \Big]_0^{16} = \frac{128}{3}.$$

The moment about the origin (in g-cm) is

$$M_0 = \int_0^{16} x \cdot x^{1/2} \, dx = \frac{2}{5} x^{5/2} \Big]_0^{16} = \frac{2048}{5}.$$

From (2) we find

$$\bar{x} = \frac{2048/5}{128/3} = 9.6.$$

That is, the center of mass $\bar{x}$ of the rod is 9.6 cm from the left end of the rod that coincides with the origin. ■

■ **Two-Dimensional Systems** For n point masses located in the xy-plane, as indicated in FIGURE 6.10.6, the **center of mass of the system** is defined to be the point $(\bar{x}, \bar{y})$, where

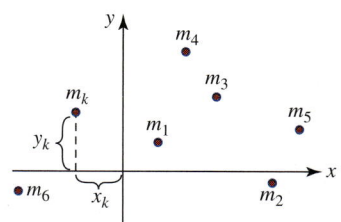

FIGURE 6.10.6 n masses in the xy-plane

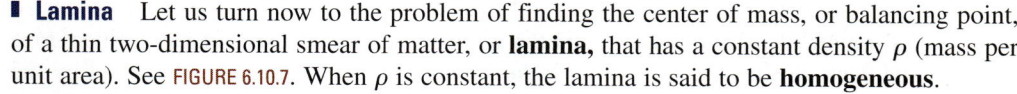

$$\bar{x} = \frac{M_y}{m} = \frac{\sum_{k=1}^{n} m_k x_k}{\sum_{k=1}^{n} m_k} = \frac{\text{moment of system about } y\text{-axis}}{\text{total mass}},$$

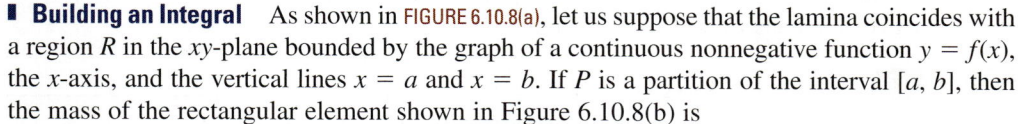

$$\bar{y} = \frac{M_x}{m} = \frac{\sum_{k=1}^{n} m_k y_k}{\sum_{k=1}^{n} m_k} = \frac{\text{moment of system about } x\text{-axis}}{\text{total mass}}.$$

■ **Lamina** Let us turn now to the problem of finding the center of mass, or balancing point, of a thin two-dimensional smear of matter, or **lamina,** that has a constant density ρ (mass per unit area). See FIGURE 6.10.7. When ρ is constant, the lamina is said to be **homogeneous.**

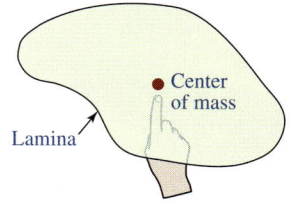

FIGURE 6.10.7 Center of mass of a lamina

■ **Building an Integral** As shown in FIGURE 6.10.8(a), let us suppose that the lamina coincides with a region R in the xy-plane bounded by the graph of a continuous nonnegative function $y = f(x)$, the x-axis, and the vertical lines $x = a$ and $x = b$. If P is a partition of the interval $[a, b]$, then the mass of the rectangular element shown in Figure 6.10.8(b) is

$$m_k = \rho \, \Delta A_k = \rho f(x_k^*) \, \Delta x_k,$$

where, in this case, we take x_k^* to be the midpoint of the subinterval $[x_{k-1}, x_k]$ and ρ is the constant density. The moment of this element about the y-axis is

$$(M_y)_k = x_k^* \, \Delta m_k = x_k^*(\rho \, \Delta A_k) = \rho x_k^* f(x_k^*) \, \Delta x_k.$$

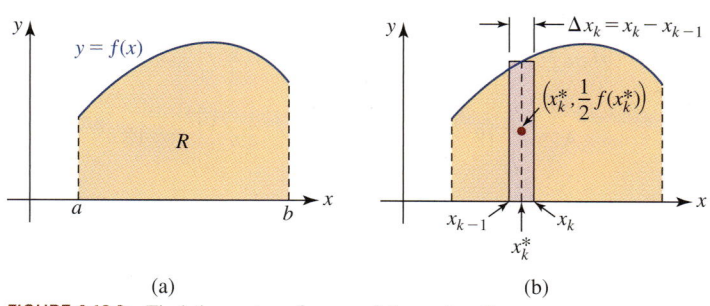

(a)　　　　　　　　(b)

FIGURE 6.10.8 Find the center of mass of the region R

Since the density is constant, the center of mass of the element is necessarily at its geometric center $(x_k^*, \frac{1}{2} f(x_k^*))$. Hence, the moment of the element about the x-axis is

$$(M_x)_k = \frac{1}{2} f(x_k^*)(\rho \, \Delta A_k) = \frac{1}{2} \rho [f(x_k^*)]^2 \, \Delta x_k.$$

We conclude that

$$m = \lim_{\|P\| \to 0} \sum_{k=1}^{n} \rho f(x_k^*) \, \Delta x_k = \int_a^b \rho f(x) \, dx,$$

$$M_y = \lim_{\|P\| \to 0} \sum_{k=1}^{n} \rho x_k^* f(x_k^*) \, \Delta x_k = \int_a^b \rho x f(x) \, dx,$$

and

$$M_x = \lim_{\|P\| \to 0} \frac{1}{2} \sum_{k=1}^{n} \rho [f(x_k^*)]^2 \Delta x_k = \frac{1}{2} \int_a^b \rho [f(x)]^2 \, dx.$$

Thus, the coordinates of the center of mass of the lamina are defined to be

$$\bar{x} = \frac{M_y}{m} = \frac{\displaystyle\int_a^b \rho x f(x) \, dx}{\displaystyle\int_a^b \rho f(x) \, dx}, \qquad \bar{y} = \frac{M_x}{m} = \frac{\displaystyle\frac{1}{2}\int_a^b \rho [f(x)]^2 \, dx}{\displaystyle\int_a^b \rho f(x) \, dx}. \qquad (3)$$

▌ Centroid   We note that the constant density ρ will cancel in equations (3) for $\bar{x}$ and $\bar{y}$, and that the denominator $\int_a^b f(x)\,dx$ is then the area A of the region R. In other words, the center of mass depends only on the shape of R:

$$\bar{x} = \frac{M_y}{A} = \frac{\displaystyle\int_a^b xf(x)\,dx}{\displaystyle\int_a^b f(x)\,dx}, \qquad \bar{y} = \frac{M_x}{A} = \frac{\dfrac{1}{2}\displaystyle\int_a^b [f(x)]^2\,dx}{\displaystyle\int_a^b f(x)\,dx}. \tag{4}$$

To emphasize the distinction, albeit minor, between the physical object, which is the homogeneous lamina, and the geometric object, which is the plane region R, we say the equations in (4) define the coordinates of the **centroid** of the region.

Note: It is important that you understand the result in (4), but it does not pay to memorize the integrals because we have assumed for the sake of discussion that R is bounded by the graph of a function f and the x-axis. R could just as well be the region bounded between the graphs of two functions f and g. See Example 5.

EXAMPLE 3   Centroid of a Region

Find the centroid of the region in the first quadrant bounded by the graph of $y = 9 - x^2$, the x-axis, and the y-axis.

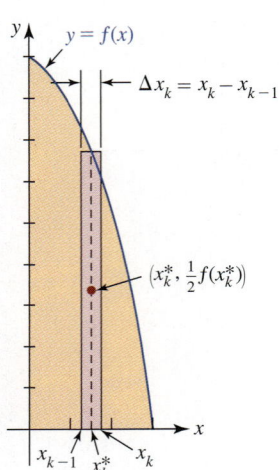

FIGURE 6.10.9   Region in Example 3

Solution   The region is shown in FIGURE 6.10.9. Now, if $f(x) = 9 - x^2$, then

$$A_k = f(x_k^*)\,\Delta x_k$$

$$(M_y)_k = xf(x_k^*)\,\Delta x_k$$

and

$$(M_x)_k = \frac{1}{2}f(x_k^*)(f(x_k^*)\,\Delta x_k) = \frac{1}{2}[f(x_k^*)]^2\,\Delta x_k \quad .$$

Hence,

$$A = \int_0^3 (9 - x^2)\,dx = \left(9x - \frac{1}{3}x^3\right)\Big]_0^3 = 18$$

$$M_y = \int_0^3 x(9 - x^2)\,dx = \left(\frac{9}{2}x^2 - \frac{1}{4}x^4\right)\Big]_0^3 = \frac{81}{4}$$

$$M_x = \frac{1}{2}\int_0^3 (9 - x^2)^2\,dx$$

$$= \frac{1}{2}\int_0^3 (81 - 18x^2 + x^4)\,dx$$

$$= \frac{1}{2}\left(81x - 6x^3 + \frac{1}{5}x^5\right)\Big]_0^3 = \frac{324}{5}.$$

It follows from (4) that the coordinates of the centroid are

$$\bar{x} = \frac{M_y}{A} = \frac{81/4}{18} = \frac{9}{8}, \qquad \bar{y} = \frac{M_x}{A} = \frac{324/5}{18} = \frac{54}{15}. \qquad ■$$

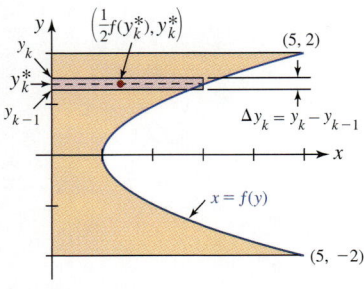

FIGURE 6.10.10   Region in Example 4

EXAMPLE 4   Integration with Respect to y

Find the centroid of the region bounded by the graphs of $x = y^2 + 1$, $x = 0$, $y = 2$, and $y = -2$.

Solution   The region is shown in FIGURE 6.10.10. Inspection of the figure suggests that we use horizontal rectangular elements. If $f(y) = y^2 + 1$, then

$$A_k = f(y_k^*)\,\Delta y_k$$

$$(M_x)_k = y_k^* f(y_k^*)\,\Delta y_k$$

$$(M_y)_k = \frac{1}{2}f(y_k^*)(f(y_k^*)\,\Delta y_k) = -\frac{1}{2}[f(y_k^*)]^2\,\Delta y_k$$

and so

$$A = \int_{-2}^{2} (y^2 + 1)\, dy = \left(\frac{1}{3}y^3 + y\right)\Big]_{-2}^{2} = \frac{28}{3},$$

$$M_x = \int_{-2}^{2} y(y^2 + 1)\, dy = \left(\frac{1}{4}y^4 + \frac{1}{2}y^2\right)\Big]_{-2}^{2} = 0,$$

$$M_y = \frac{1}{2}\int_{-2}^{2} (y^2 + 1)^2\, dy = \frac{1}{2}\int_{-2}^{2} (y^4 + 2y^2 + 1)\, dy$$

$$= \frac{1}{2}\left(\frac{1}{5}y^5 + \frac{2}{3}y^3 + y\right)\Big]_{-2}^{2} = \frac{206}{15}.$$

Thus, we have

$$\bar{x} = \frac{M_y}{A} = \frac{206/15}{28/3} = \frac{103}{70}, \quad \bar{y} = \frac{M_x}{A} = \frac{0}{28/3} = 0.$$

As we would expect, since the lamina is symmetric with respect to the x-axis, the centroid is on the axis of symmetry. We also note that the centroid is outside the region. ■

EXAMPLE 5 Region between Two Graphs

Find the centroid of the region bounded by the graphs of $y = -x^2 + 3$ and $y = x^2 - 2x - 1$.

Solution FIGURE 6.10.11 shows the region in question. We note that the points of intersection of the graphs are $(-1, 2)$ and $(2, -1)$. Now, if $f(x) = -x^2 + 3$ and $g(x) = x^2 - 2x - 1$, then the area of the region is

$$A = \int_{-1}^{2} [f(x) - g(x)]\, dx$$

$$= \int_{-1}^{2} (-2x^2 + 2x + 4)\, dx$$

$$= \left(-\frac{2}{3}x^3 + x^2 + 4x\right)\Big]_{-1}^{2} = 9.$$

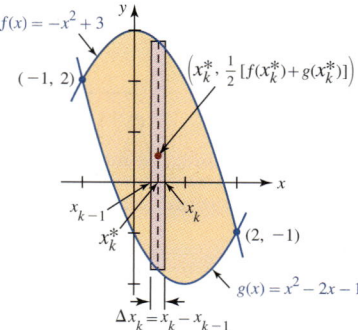

FIGURE 6.10.11 Region in Example 5

Since the coordinates of the midpoint of the indicated element are $(x_k^*, \frac{1}{2}[f(x_k^*) + g(x_k^*)])$, it follows that

$$M_y = \int_{-1}^{2} x[f(x) - g(x)]\, dx$$

$$= \int_{-1}^{2} (-2x^3 + 2x^2 + 4x)\, dx$$

$$= \left(-\frac{1}{2}x^4 + \frac{2}{3}x^3 + 2x^2\right)\Big]_{-1}^{2} = \frac{9}{2},$$

and

$$M_x = \frac{1}{2}\int_{-1}^{2} [f(x) + g(x)][f(x) - g(x)]\, dx$$

$$= \frac{1}{2}\int_{-1}^{2} ([f(x)]^2 - [g(x)]^2)\, dx$$

$$= \frac{1}{2}\int_{-1}^{2} [(-x^2 + 3)^2 - (x^2 - 2x - 1)^2]\, dx$$

$$= \frac{1}{2}\int_{-1}^{2} (4x^3 - 8x^2 - 4x + 8)\, dx$$

$$= \frac{1}{2}\left(x^4 - \frac{8}{3}x^3 - 2x^2 + 8x\right)\Big]_{-1}^{2} = \frac{9}{2}.$$

Thus, the coordinates of the centroid are

$$\bar{x} = \frac{M_y}{A} = \frac{9/2}{9} = \frac{1}{2}, \quad \bar{y} = \frac{M_x}{A} = \frac{9/2}{9} = \frac{1}{2}.$$

■

Exercises 6.10 Answers to selected odd-numbered problems begin on page ANS-21.

≡ **Fundamentals**

In Problems 1–4, find the center of mass of the given system of masses. The mass m_k is located on the x-axis at a point whose directed distance from the origin is x_k. Assume that mass is measured in grams and that distance is measured in centimeters.

1. $m_1 = 2, m_2 = 5; x_1 = 4, x_2 = -2$

2. $m_1 = 6, m_2 = 1, m_3 = 3; x_1 = -\frac{1}{2}, x_2 = -3, x_3 = 8$

3. $m_1 = 10, m_2 = 5, m_3 = 8, m_4 = 7; x_1 = -5, x_2 = 2,$
   $x_3 = 6, x_4 = -3$

4. $m_1 = 2, m_2 = \frac{3}{2}, m_3 = \frac{7}{2}, m_4 = \frac{1}{2}; x_1 = 9, x_2 = -4,$
   $x_3 = -6, x_4 = -10$

5. Two masses are placed at the ends of a uniform board of negligible mass, as shown in FIGURE 6.10.12. Where should a fulcrum be placed so that the system is in balance? [*Hint*: Although the origin can be placed anywhere, let us agree to choose it halfway between the masses.]

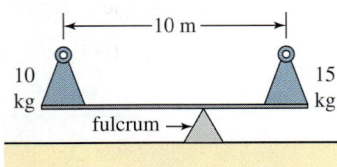

FIGURE 6.10.12 Masses in Problem 5

6. Find the center of mass of the three masses m_1, m_2, and m_3 located at the vertices of the equilateral triangle shown in FIGURE 6.10.13. [*Hint*: First find the center of mass of m_1 and m_2.]

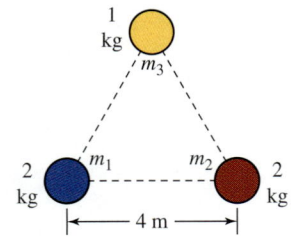

FIGURE 6.10.13 Masses in Problem 6

In Problems 7–14, a rod of linear density $\rho(x)$ kg/m coincides with the x-axis on the interval indicated. Find its center of mass.

7. $\rho(x) = 2x + 1;$ [0, 5]

8. $\rho(x) = -x^2 + 2x;$ [0, 2]

9. $\rho(x) = x^{1/3};$ [0, 1]

10. $\rho(x) = -x^2 + 1;$ [0, 1]

11. $\rho(x) = |x - 3|;$ [0, 4]

12. $\rho(x) = 1 + |x - 1|;$ [0, 3]

13. $\rho(x) = \begin{cases} x^2, & 0 \le x < 1 \\ 2 - x, & 1 \le x \le 2 \end{cases};$ [0, 2]

14. $\rho(x) = \begin{cases} x, & 0 \le x < 2 \\ 2 & 2 \le x \le 3 \end{cases};$ [0, 3]

15. The density of a 10-ft rod varies as the square of the distance from the left end. Find its center of mass if the density at its center is 12.5 slug/ft.

16. The linear density of a 3-m-long rod varies as the distance from the right end. Find the linear density at the center of the rod if its total mass is 6 kg.

In Problems 17–20, find the center of mass of the given system of masses. The mass m_k is located at the point P_k. Assume that mass is measured in grams and that distance is measured in centimeters.

17. $m_1 = 3, m_2 = 4; P_1 = (-2, 3), P_2 = (1, 2)$

18. $m_1 = 1, m_2 = 3, m_3 = 2; P_1 = (-4, 1), P_2 = (2, 2),$
    $P_3 = (5, -2)$

19. $m_1 = 4, m_2 = 8, m_3 = 10; P_1 = (1, 1), P_2 = (-5, 2),$
    $P_3 = (7, -6)$

20. $m_1 = 1, m_2 = \frac{1}{2}, m_3 = 4, m_4 = \frac{5}{2}; P_1 = (9, 3),$
    $P_2 = (-4, -6), P_3 = (\frac{3}{2}, -1), P_4 = (-2, 10)$

In Problems 21–38, find the centroid of the region bounded by the graphs of the given equations.

21. $y = 2x + 4, y = 0, x = 0, x = 2$

22. $y = x + 1, y = 0, x = 3$

23. $y = x^2, y = 0, x = 1$

24. $y = x^2 + 2, y = 0, x = -1, x = 2$

25. $y = x^3, y = 0, x = 3$

26. $y = x^3, y = 8, x = 0$

27. $y = \sqrt{x}, y = 0, x = 1, x = 4$

28. $x = y^2, x = 1$

29. $y = x^2, y - x = 2$

30. $y = x^2, y = \sqrt{x}$

31. $y = x^3, y = x^{1/3},$ first quadrant

32. $y = 4 - x^2, y = 0, x = 0,$ second quadrant

33. $y = 1/x^3, y = 0, x = 1, x = 3$

34. $y = x^2 - 2x + 1, y = -4x + 9$

35. $x = y^2 - 1, y = -1, y = 2, x = -2$

36. $y = x^2 - 4x + 6, y = 0, x = 0, x = 4$

37. $y = 4 - 4x^2, y = 1 - x^2$

38. $y^2 + x = 1, y + x = -1$

In Problems 39 and 40, use symmetry to locate $\bar{x}$ and integration to find $\bar{y}$ of the region bounded by the graphs of the given functions.

39. $y = 1 + \cos x, y = 1, -\pi/2 \leq x \leq \pi/2$

40. $y = 4\sin x, y = -\sin x, 0 \leq x \leq \pi$

☰ Think About It

41. A theorem due to **Pappus of Alexandria** (c. A.D. 350) states that:

> Let L be an axis in a plane and R a region in the same plane that does not intersect L. When R is revolved about L, the volume V of the resulting solid of revolution is equal to the area A of R times the length of the path traversed by the centroid of R.

(a) As shown in FIGURE 6.10.14, let the region R be bounded by the graphs of $y = f(x)$ and $y = g(x)$. Show that if R is revolved about the x-axis, then $V = (2\pi\bar{y})A$, where A is the area of the region.

(b) What do you think V is given by when the region R is revolved about the y-axis?

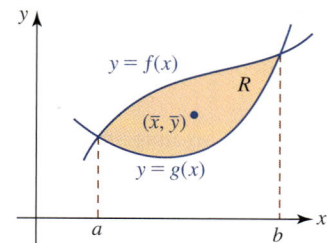

FIGURE 6.10.14 Region in Problem 41

42. Verify the Theorem of Pappus in Problem 41 by revolving the region bounded by $y = x^2 + 1, y = 1, x = 2$ about the x-axis.

43. Use the Theorem of Pappus in Problem 41 to find the volume of the torus shown in FIGURE 6.10.15.

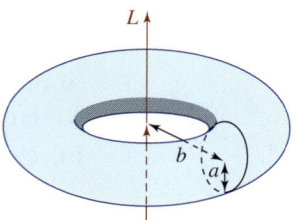

FIGURE 6.10.15 Torus in Problem 43

44. A rod of linear density $\rho(x)$ kg/m coincides with the x-axis on the interval [0, 6]. If $\rho(x) = x(6 - x) + 1$, where would one intuitively expect the center of mass to be? Prove your assertion.

45. Consider the triangular region R in FIGURE 6.10.16. Where do you think the centroid of the triangle is? Think geometrically.

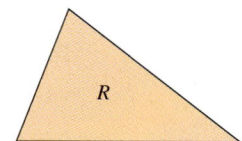

FIGURE 6.10.16 Triangular region in Problem 45

46. Without integration, determine the centroid of the region R shown in FIGURE 6.10.17.

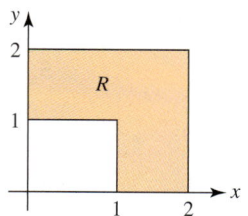

FIGURE 6.10.17 Region in Problem 46

Chapter 6 in Review

Answers to selected odd-numbered problems begin on page ANS-21.

A. True/False

In Problems 1–12, indicate whether the given statement is true or false.

1. When $\int_a^b f(x)\, dx > 0$, the integral gives the area under the graph of $y = f(x)$ on the interval [a, b]. _____

2. $\int_0^3 (x - 1)\, dx$ is the area under the graph of $y = x - 1$ on [0, 3]. _____

3. The integral $\int_a^b [f(x) - g(x)]\, dx$ gives the area between the graphs of the continuous functions f and g whenever $f(x) \geq g(x)$ for every x in [a, b]. _____

4. The disk and washer methods for finding volumes of solids of revolution are special cases of the slicing method. _____

5. The average value f_{ave} of a continuous function on an interval [a, b] is necessarily a number that satisfies $m \leq f_{ave} \leq M$, where m and M are the maximum and minimum values of f on the interval, respectively. _____

6. If f and g are continuous on [a, b], then the average value of $f + g$ is $(f + g)_{ave} = f_{ave} + g_{ave}$. _____

7. The center of mass of a pencil with a constant linear density ρ is at its geometric center. _____ .

8. The center of mass of a lamina that coincides with a plane region R is a point in R at which the lamina would hang in balance. _____

9. The pressure on the flat bottom of a swimming pool is the same as the horizontal pressure on the vertical sidewalls at the same depth. _____

10. Consider a circular tin can with radius 6 in. and a circular reservoir with radius 50 ft. If each has a flat bottom and is filled with water to a depth of 1 ft, then the liquid pressure on the bottom of the reservoir is greater than the pressure on the bottom of the tin can. _____

11. If $s(t)$ is the position function of a body that moves in a straight line, then $\int_{t_1}^{t_2} v(t)\,dt$ is the distance the body moves in the interval $[t_1, t_2]$. _____

12. In the absence of air resistance, when dropped simultaneously from the same height, a cannonball will hit the ground before a marshmallow. _____

B. Fill in the Blanks

In Problems 1–8, fill in the blanks.

1. The unit of work in the SI system of units is_____.

2. To warm up, a 200-lb jogger pushes against a tree for 5 min with a constant force of 60 lb and then runs 2 mi in 10 min. The total work done is _____.

3. The work done by a 100-lb constant force applied at an angle of 60° to the horizontal over a distance of 50 ft is _____.

4. If 80 N of force stretches a spring that is initially 1 m long into a spring that is 1.5 m long, then the spring will measure _____ m long when 100 N of force is applied.

5. The coordinates of the centroid of a region R are $(2, 5)$ and the moment of the region about the x-axis is 30. Hence, the area of R is _____ square units.

6. The weight density of water is _____ lb/ft^3.

7. The graph of a function with a continuous first derivative is said to be _____.

8. A ball dropped from a great height hits the ground in T seconds with a velocity v_{impact}. If its velocity function is $v(t) = -gt$, then the average velocity v_{ave} of the ball for $0 \le t \le T$ in terms of v_{impact} is _____.

C. Exercises

In Problems 1–8, set up the definite integral(s) for the area of the shaded region in each figure.

1.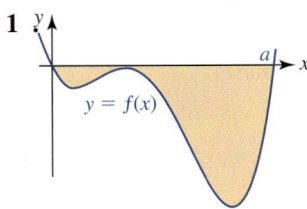

FIGURE 6.R.1   Graph for Problem 1

2.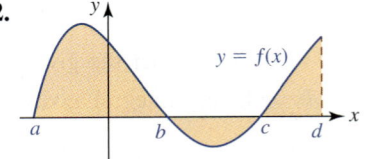

FIGURE 6.R.2   Graph for Problem 2

3.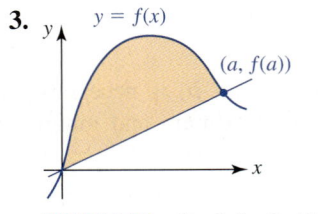

FIGURE 6.R.3   Graph for Problem 3

4.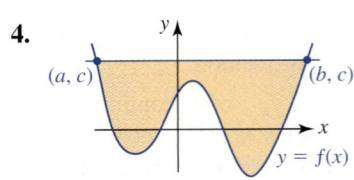

FIGURE 6.R.4   Graph for Problem 4

5.

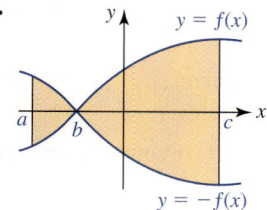

FIGURE 6.R.5   Graph for Problem 5

6.

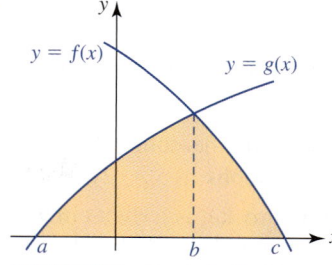

FIGURE 6.R.6   Graph for Problem 6

7.

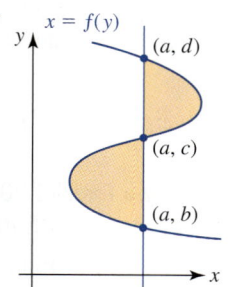

FIGURE 6.R.7   Graph for Problem 7

8.

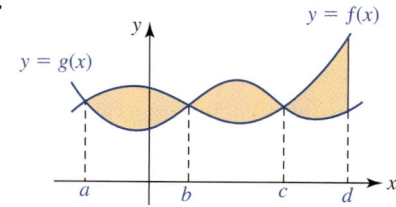

FIGURE 6.R.8   Graph for Problem 8

In Problems 9 and 10, use the definite integral to find the area of the shaded region in terms of a and b.

9.

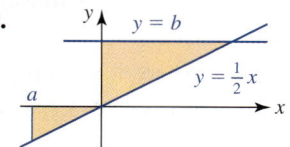

FIGURE 6.R.9   Graph for Problem 9

10.

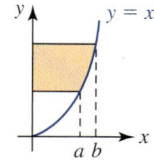

FIGURE 6.R.10   Graph for Problem 10

In Problems 11–16, consider the region R in FIGURE 6.R.11. Set up the definite integral(s) for the indicated quantity.

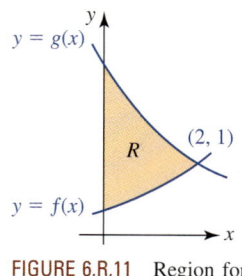

FIGURE 6.R.11   Region for
Problems 11–16

11. The centroid of the region

12. The volume of the solid of revolution that is formed by revolving R about the x-axis

13. The volume of the solid of revolution that is formed by revolving R about the y-axis

14. The volume of the solid of revolution that is formed by revolving R about the line $y = -1$

15. The volume of the solid of revolution that is formed by revolving R about the line $x = 2$

16. The volume of the solid with R as its base such that the cross sections of the solid parallel to the y-axis are squares

17. Find the area bounded by the graphs of $y = \sin x$ and $y = \sin 2x$ on the interval $[0, \pi]$.

18. Consider the region bounded by the graphs of $y = e^x$, $y = e^{-x}$, and $x = \ln 2$.

   (a) Find the area of the region.

   (b) Find the volume of the solid of revolution if the region is revolved about the x-axis.

19. Consider the region R bounded by the graphs of $x = y^2$ and $x = \sqrt{2}$. Use the slicing method to find the volume of the solid if the region R is its base and

    (a) cross sections of the solid perpendicular to the x-axis are squares,
    (b) cross sections of the solid perpendicular to the x-axis are circles.

20. Find the volume of the solid of revolution that is formed by revolving the region R bounded by the graphs of $x = 2y - y^2$ and $x = 0$ about the line $y = 3$.

21. A nose cone of a rocket is a right circular cone of height 8 ft and radius 10 ft. The lateral surface is to be covered with canvas except for a section of height 1 ft at the apex of the nose cone. Find the area of the canvas needed.

22. The area under the graph of a continuous nonnegative function $y = f(x)$ on the interval $[-3, 4]$ is 21 square units. What is the average value of the function on the interval?

23. Find the average value of $f(x) = x^{3/2} + x^{1/2}$ on $[1, 4]$.

24. Find a value of x in the interval $[0, 3]$ that corresponds to the average value of the function $f(x) = 2x - 1$.

25. A spring whose unstretched length is $\frac{1}{2}$ m is stretched to a length of 1 m by a force of 50 N. Find the work done in stretching the spring from a length of 1 m to a length of 1.5 m.

26. The work done in stretching a spring 6 in. beyond its natural length is 10 ft-lb. Find the spring constant.

27. A water tank, in the form of a cube that is 10 ft on a side, is filled with water. Find the work done in pumping all the water to a point 5 ft above the tank.

28. A bucket weighing 2 lb contains 30 lb of liquid. As the bucket is raised vertically at a rate of 1 ft/s, the liquid leaks out at a rate of $\frac{1}{4}$ lb/s. Find the work done in lifting the bucket a distance of 5 ft.

29. In Problem 28, find the work done in lifting the bucket to a point where it is empty.

30. In Problem 28, find the work done in lifting the leaking bucket a distance of 5 ft if the rope attached to the bucket weighs $\frac{1}{8}$ lb/ft.

31. A tank on top of a tower 15 ft high consists of a frustum of a cone surmounted by a right circular cylinder. The dimensions (in feet) are given in FIGURE 6.R.12. Find the work done in filling the tank with water from ground level.

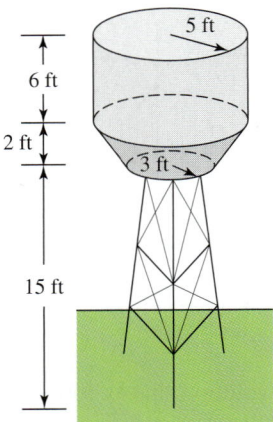

FIGURE 6.R.12 Tank in Problem 31

32. A rock is thrown vertically upward from the surface of the Moon with an initial velocity of 44 ft/s.

    (a) If the acceleration of gravity on the Moon is 5.5 ft/s^2, find the maximum height attained. Compare with the Earth.
    (b) On the way down, the rock hits a 6-ft-tall astronaut on the head. What is the impact velocity of the rock?

33. Find the length of the graph of $y = (x - 1)^{3/2}$ from $(1, 0)$ to $(5, 8)$.

34. The linear density of a 6-m-long rod is a linear function of the distance from its left end. The density in the middle of the rod is 11 kg/m and at the right end the density is 17 kg/m. Find the center of mass of the rod.

35. A flat plate, in the form of a quarter-circle, is submerged vertically in oil as shown in FIGURE 6.R.13. If the weight density of the oil is 800 kg/m³, find the force exerted by the oil on one side of the plate.

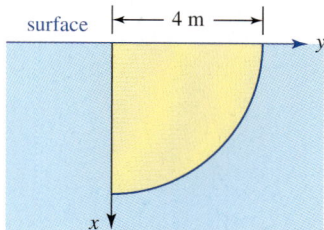

FIGURE 6.R.13   Submerged vertical plate
in Problem 35

36. A uniform metal bar of mass 4 kg and length 2m supports two masses, as shown in FIGURE 6.R.14. Where should the wire be attached to the bar so that the system hangs in balance?

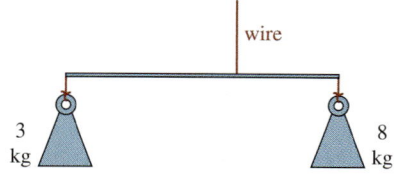

FIGURE 6.R.14   Masses in Problem 36

37. Three masses are suspended from uniform rigid bars of negligible mass as shown in FIGURE 6.R.15. Determine where the indicated wires should be attached so that the entire system hangs in balance.

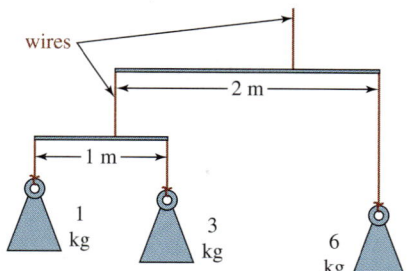

FIGURE 6.R.15   Masses in Problem 37

Techniques of Integration

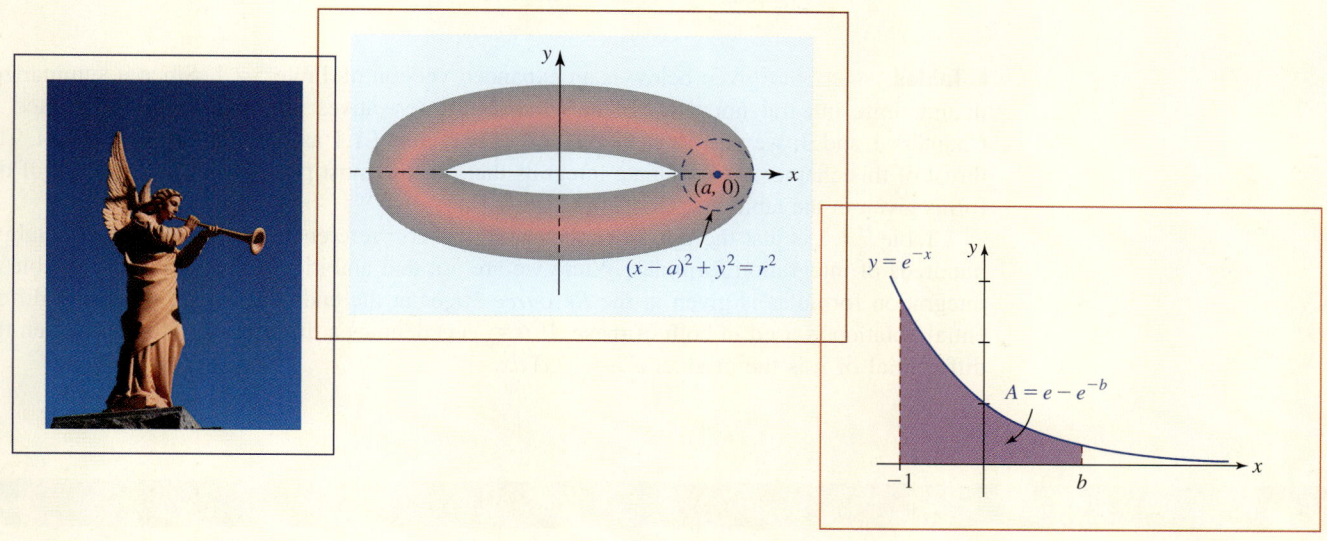

In This Chapter One often encounters an integral that cannot be categorized as a familiar form such as $\int u^n\, du$ or $\int e^u\, du$. For example, it is not possible to evaluate $\int x^2 \sqrt{x+1}\; dx$ by an immediate application of any one of the formulas listed on pages 380–381. However, by applying a **technique of integration**, an integral such as this can sometimes be reduced to one or more of these familiar forms.

7.1 Integration—Three Resources

■ **Introduction** In this chapter we are going to resume our study of antiderivatives begun in Chapter 5. In that earlier chapter we barely scratched the surface of how to obtain an antiderivative of a function f. Recall, an indefinite integral

$$\int f(x)\, dx = F(x) + C$$

is a family $F(x) + C$ of antiderivatives of the function f, that is, F is related to f by the fact that $F'(x) = f(x)$. In this manner, each time we devise a derivative of a specific function ($\sin x$, $\cos x$, e^x, $\ln x$, and so on) there corresponds an indefinite integral analogue. For example,

$$\frac{d}{dx}\cos x = -\sin x \quad \text{implies} \quad \int \sin x\, dx = -\cos x + C.$$

■ **Tables** TABLE 7.1.1 given below is an expanded version of Table 5.2.1. Since it summarizes in indefinite integral notation all the Chain-Rule derivatives of the functions discussed in Chapters 1 and 3, we will refer to the entries in Table 7.1.1 as *familiar* or *basic* forms. The thrust of this chapter is to evaluate integrals that, for the most part, do not fall into any of the forms given in the table.

Table 7.1.1 is just the tip of a rather large iceberg; reference handbooks often contained hundreds of integration formulas. While we are not that ambitious, a more extensive table of integration formulas is given in the *Resource Pages* at the end of this text. As usual, differential notation is used in both of these. If $u = g(x)$ denotes a differentiable function, then the differential of u is the product $du = g'(x)\, dx$.

TABLE 7.1.1

Integration Formulas

Constant Integrands

1. $\displaystyle\int du = u + C$

2. $\displaystyle\int k\, du = ku + C$

Integrands that are Powers

3. $\displaystyle\int u^n\, du = \frac{u^{n+1}}{n+1} + C \quad (n \neq -1)$

4. $\displaystyle\int u^{-1}\, du = \int \frac{1}{u}\, du = \ln|u| + C$

Exponential Integrands

5. $\displaystyle\int e^u\, du = e^u + C$

6. $\displaystyle\int a^u\, du = \frac{1}{\ln a}a^u + C$

Trigonometric Integrands

7. $\displaystyle\int \sin u\, du = -\cos u + C$

8. $\displaystyle\int \cos u\, du = \sin u + C$

9. $\displaystyle\int \sec^2 u\, du = \tan u + C$

10. $\displaystyle\int \csc^2 u\, du = -\cot u + C$

11. $\displaystyle\int \sec u \tan u\, du = \sec u + C$

12. $\displaystyle\int \csc u \cot u\, du = -\csc u + C$

13. $\displaystyle\int \tan u\, du = -\ln|\cos u| + C$

14. $\displaystyle\int \cot u\, du = \ln|\sin u| + C$

15. $\displaystyle\int \sec u\, du = \ln|\sec u + \tan u| + C$

16. $\displaystyle\int \csc u\, du = \ln|\csc u - \cot u| + C$

(continued)

Hyperbolic Integrands

17. $\displaystyle\int \sinh u \, du = \cosh u + C$

18. $\displaystyle\int \cosh u \, du = \sinh u + C$

19. $\displaystyle\int \text{sech}^2 u \, du = \tanh u + C$

20. $\displaystyle\int \text{csch}^2 u \, du = -\coth u + C$

21. $\displaystyle\int \text{sech}\, u \tanh u \, du = -\text{sech}\, u + C$

22. $\displaystyle\int \text{csch}\, u \coth u \, du = -\text{csch}\, u + C$

Algebraic Integrands

23. $\displaystyle\int \frac{1}{\sqrt{a^2 - u^2}} \, du = \sin^{-1}\frac{u}{a} + C$

24. $\displaystyle\int \frac{1}{a^2 + u^2} \, du = \frac{1}{a}\tan^{-1}\frac{u}{a} + C$

25. $\displaystyle\int \frac{1}{u\sqrt{u^2 - a^2}} \, du = \frac{1}{a}\sec^{-1}\left|\frac{u}{a}\right| + C$

26. $\displaystyle\int \frac{1}{\sqrt{a^2 + u^2}} \, du = \ln\left|u + \sqrt{u^2 + a^2}\right| + C$

27. $\displaystyle\int \frac{1}{\sqrt{u^2 - a^2}} \, du = \ln\left|u + \sqrt{u^2 - a^2}\right| + C$

28. $\displaystyle\int \frac{1}{a^2 - u^2} \, du = \frac{1}{2a}\ln\left|\frac{u + a}{u - a}\right| + C$

29. $\displaystyle\int \frac{1}{u^2 - a^2} \, du = \frac{1}{2a}\ln\left|\frac{u - a}{u + a}\right| + C$

30. $\displaystyle\int \frac{1}{u\sqrt{a^2 - u^2}} \, du = -\frac{1}{a}\ln\left|\frac{a + \sqrt{a^2 - u^2}}{u}\right| + C$

31. $\displaystyle\int \frac{1}{u\sqrt{a^2 + u^2}} \, du = -\frac{1}{a}\ln\left|\frac{a + \sqrt{a^2 + u^2}}{u}\right| + C$

Even though we have designated these integration formulas as *familiar* or *basic* you may not be *that* familiar with some of the formulas, especially 17–22 and 26–31. Because instructors sometimes give short shrift to the hyperbolic functions you are urged to review (or if need be, study for the first time) Section 3.10. Formulas 26–31, which resemble Formulas 23–25, are the indefinite integral forms of the differentiation formulas for the inverse hyperbolic functions combined with the fact that every inverse hyperbolic function is a natural logarithm. See page 183.

▮ **Techniques of Integration** In the sections that follow, the integrals that we are going to examine cannot be categorized as a single familiar form such as $\int u^n \, du$, $\int e^u \, du$, or $\int \sin u \, du$. Nevertheless, Table 7.1.1 is important; as we advance through this chapter we will, of necessity, frequently refer back to it. A wide variety of integrals can be evaluated by performing specific operations on the integrand—called a **technique of integration**—and reducing a given integral to *one or more* of the familiar forms in the table. For example, it is not possible to evaluate $\int \ln x \, dx$ by identifying it with any of the integration formulas in Table 7.1.1. However, we will see in Section 7.3 that by applying a technique of integration, $\int \ln x \, dx$ can be evaluated in a few seconds using the derivative of $\ln x$ along with Formula 1 in the table.

For purposes of review, you are urged to work the problems in Exercises 7.1. By an appropriate u-substitution, each problem can be matched with one of the formulas in Table 7.1.1.

Neither a table, regardless of how large it is, nor techniques of integration, regardless of how powerful they are, is a cure-all for all integration problems. While some integrals, such as $\int e^{x^2} \, dx$, completely defy tables and techniques of integration, others only *appear* to defy these resources. For example, the integral $\int e^{\sin x}\sin 2x \, dx$ does not appear in any tables but it *can* be evaluated by a technique of integration. The problem here is that it is not immediately obvious *which* technique can be applied. There will be times when you will be expected to give some thought to recasting an integrand into a form that is amenable to a technique of integration.

◀ It was pointed out in Section 5.5 (see pages 312–313) that a continuous function f may not have an antiderivative that is an elementary function.

▮ **Technology** A word about technology. If you have not worked with a computer algebra system (CAS) such as *Mathematica*, *Maple*, *Derive*, or *Axiom* you should rectify that deficiency in your background as quickly as possible. A computer algebra system is an extremely sophisticated program designed to perform a wide variety of symbolic mathematical operations such as ordinary algebra, matrix algebra, complex arithmetic, solving polynomial equations, approximating roots of equations, differentiation, integration, graphing equations in two and three dimensions, solving differential equations, manipulation of built-in special functions, and so on. If it is your goal to be a serious student of mathematics, science, or engineering, then an ideal aid to your lecture and laboratory classes (as well as your future career) would be a laptop computer equipped with a program such as *Mathematica, Maple,* or MATLAB. Also check the computer labs in your Mathematics and Engineering Departments; the computers therein are undoubtedly equipped with one or more of these programs. Some fundamental command syntax for performing calculus-related operations in *Mathematica* and *Maple* is given in the *Student Resource Manual* that accompanies this text.

As you become adept and comfortable using a CAS, you might be interested in exploring web resource sites such as

http://scienceworld.wolfram.com
http://mathworld.wolfram.com

Wolfram Research is the developer of the computer algebra system *Mathematica*.

Exercises 7.1   Answers to selected odd-numbered problems begin on page ANS-21.

☰ Fundamentals

In Problems 1–32, use a *u*-substitution and Table 7.1.1 to evaluate the given integral.

1. $\int 5^{-5x}\, dx$

2. $\int \dfrac{1}{\sqrt{x}\, e^{\sqrt{x}}}\, dx$

3. $\int \dfrac{\sin\sqrt{1+x}}{\sqrt{1+x}}\, dx$

4. $\int \dfrac{\cos e^{-x}}{e^x}\, dx$

5. $\int \dfrac{x}{\sqrt{25-4x^2}}\, dx$

6. $\int \dfrac{1}{\sqrt{25-4x^2}}\, dx$

7. $\int \dfrac{1}{x\sqrt{4x^2-25}}\, dx$

8. $\int \dfrac{1}{\sqrt{25+4x^2}}\, dx$

9. $\int \dfrac{1}{25+4x^2}\, dx$

10. $\int \dfrac{x}{25+4x^2}\, dx$

11. $\int \dfrac{1}{4x^2-25}\, dx$

12. $\int \dfrac{1}{x\sqrt{4x^2+25}}\, dx$

13. $\int \cot 10x\, dx$

14. $\int x\csc^2 x^2\, dx$

15. $\int \dfrac{6}{(3-5t)^{2.2}}\, dx$

16. $\int x^2\sqrt{(1-x^3)^5}\, dx$

17. $\int \sec 3x\, dx$

18. $\int 2\csc 2x\, dx$

19. $\int \dfrac{\sin^{-1}x}{\sqrt{1-x^2}}\, dx$

20. $\int \dfrac{1}{(1+x^2)\tan^{-1}x}\, dx$

21. $\int \dfrac{\sin x}{1+\cos^2 x}\, dx$

22. $\int \dfrac{\cos(\ln 9x)}{x}\, dx$

23. $\int \dfrac{x^3}{\cosh^2 x^4}\, dx$

24. $\int \tanh x\, dx$

25. $\int \tan 2x \sec 2x\, dx$

25. $\int \sin x \sin(\cos x)\, dx$

27. $\int \sin x \csc(\cos x)\cot(\cos x)\, dx$ **28.** $\int \cos x \csc^2(\sin x)\, dx$

29. $\int (1+\tan x)^2 \sec^2 x\, dx$

30. $\int \dfrac{1}{x(\ln x)^2}\, dx$

31. $\int \dfrac{e^{2x}}{1+e^{2x}}\, dx$

32. $\int \dfrac{e^x}{1+e^{2x}}\, dx$

7.2 Integration by Substitution

▎ Introduction   In this section we will extend the idea of the **u-substitution** introduced in Section 5.2. In Section 5.2 we basically used a *u*-substitution as an aid in recognizing that an integral was actually one of the familiar integration formulas such as $\int u^n\, du$, $\int du/u$, $\int e^u\, du$, and so on. For example, with the substitution $u = \ln x$ and $du = (1/x)\, dx$ we recognize that

$$\int \frac{(\ln x)^2}{x}\, dx \quad \text{is the same as} \quad \int u^2\, du.$$

You should verify that the integral $\int x^2\sqrt{2x+1}\, dx$ does not fit any *one* of the 31 integration formulas in Table 7.1.1. Nevertheless, with the aid of a substitution, the integral can be reduced to *several* cases of one of the formulas in Table 7.1.1.

The first example illustrates the general idea.

EXAMPLE 1 Using a *u*-Substitution

Evaluate $\int x^2 \sqrt{2x + 1} \, dx$.

Solution If we let $u = 2x + 1$, then the given integral can be recast entirely in terms of the variable u. To that end, observe that

$$x = \frac{1}{2}(u - 1), \qquad dx = \frac{1}{2} \, du,$$

$$x^2 = \frac{1}{4}(u - 1)^2 = \frac{1}{4}(u^2 - 2u + 1) \text{ and } \sqrt{2x + 1} = u^{1/2}.$$

Substituting these expressions into the given integral yields:

$$\int x^2 \sqrt{2x + 1} \, dx = \int \frac{1}{4}(u^2 - 2u + 1)u^{1/2} \frac{1}{2} \, du,$$

that is,

$$\int x^2 \sqrt{2x + 1} \, dx = \frac{1}{8} \int (u^2 - 2u + 1)u^{1/2} \, du$$

$$= \frac{1}{8} \int (u^{5/2} - 2u^{3/2} + u^{1/2}) \, du \qquad \leftarrow \text{three applications of formula 3 in Table 7.1.1}$$

$$= \frac{1}{8}\left[\left(\frac{2}{7}u^{7/2} - \frac{4}{5}u^{5/2} + \frac{2}{3}u^{3/2}\right) + C \right] \leftarrow \text{now resubstitute for } u$$

$$= \frac{1}{28}(2x + 1)^{7/2} - \frac{1}{10}(2x + 1)^{5/2} + \frac{1}{12}(2x + 1)^{3/2} + C.$$

You should verify that the derivative of the last line actually is $x^2 \sqrt{2x + 1}$. ∎

The choice of which, if any, substitution to use is not always obvious. Generally, if the integrand contains a power of a function, then it is a good idea to try to let u be that function *or* the power of the function itself. In Example 1, the alternative substitution $u = \sqrt{2x + 1}$ or $u^2 = 2x + 1$ leads to the different integral $\frac{1}{4}\int(1 - u^2)^2 u^2 \, du$. The latter can be evaluated by expanding the integrand and integrating each term.

EXAMPLE 2 Using a *u*-Substitution

Evaluate $\int \frac{1}{1 + \sqrt{x}} \, dx$.

Solution Let $u = \sqrt{x}$ so that $x = u^2$ and $dx = 2u \, du$. Then

$$\int \frac{1}{1 + \sqrt{x}} \, dx = \int \frac{1}{1 + u} 2u \, du$$

$$= \int \frac{2u}{1 + u} \, du \qquad \leftarrow \text{now use long division}$$

$$= \int \left(2 - \frac{2}{1 + u}\right) du \qquad \leftarrow \text{formulas 2 and 4 in Table 7.1.1}$$

$$= 2u - 2\ln|1 + u| + C \qquad \leftarrow \text{resubstitute for } u$$

$$= 2\sqrt{x} - 2\ln(1 + \sqrt{x}) + C. \qquad ∎$$

■ **Integrands Containing a Quadratic Expression** If an integrand contains a quadratic expression, $ax^2 + bx + c$, completion of the square may lead to an integral that can be expressed in terms of an inverse trigonometric function or an inverse hyperbolic function. Of course, more complicated integrals can yield other functions as well.

EXAMPLE 3 Completing the Square

Evaluate $\displaystyle\int \frac{x + 4}{x^2 + 6x + 18}\, dx$.

Solution After completing the square, the given integral can be written as

$$\int \frac{x + 4}{x^2 + 6x + 18}\, dx = \int \frac{x + 4}{(x + 3)^2 + 9}\, dx.$$

Now, if $u = x + 3$, then $x = u - 3$ and $dx = du$. Therefore,

$$\int \frac{x + 4}{x^2 + 6x + 18}\, dx = \int \frac{u + 1}{u^2 + 9}\, du \quad \leftarrow \text{termwise division}$$

$$= \int \frac{u}{u^2 + 9}\, du + \int \frac{1}{u^2 + 9}\, du$$

$$= \frac{1}{2}\int \frac{2u}{u^2 + 9}\, du + \int \frac{1}{u^2 + 9}\, du \quad \leftarrow \begin{array}{l}\text{formulas 4 and 24}\\ \text{in Table 7.1.1}\end{array}$$

$$= \frac{1}{2}\ln(u^2 + 9) + \frac{1}{3}\tan^{-1}\frac{u}{3} + C$$

$$= \frac{1}{2}\ln[(x + 3)^2 + 9] + \frac{1}{3}\tan^{-1}\frac{x + 3}{3} + C$$

$$= \frac{1}{2}\ln(x^2 + 6x + 18) + \frac{1}{3}\tan^{-1}\frac{x + 3}{3} + C. \qquad \blacksquare$$

The next example illustrates an algebraic substitution in a definite integral.

EXAMPLE 4 A Definite Integral

Evaluate $\displaystyle\int_0^2 \frac{6x + 1}{\sqrt[3]{3x + 2}}\, dx$

Solution If $u = 3x + 2$, then

$$x = \frac{1}{3}(u - 2), \qquad dx = \frac{1}{3}\, du,$$

$$6x + 1 = 2(u - 2) + 1 = 2u - 3 \text{ and } \sqrt[3]{3x + 2} = u^{1/3}.$$

Since we will change the variable of integration, we must convert the x-limits of integration to u-limits of integration. Observe when $x = 0$, $u = 2$, and when $x = 2$, $u = 8$. Therefore, the original integral becomes

$$\int_0^2 \frac{6x + 1}{\sqrt[3]{3x + 2}}\, dx = \int_2^8 \frac{2u - 3}{u^{1/3}}\frac{1}{3}\, du \quad \leftarrow \text{termwise division again}$$

$$= \int_2^8 \left(\frac{2}{3}u^{2/3} - u^{-1/3}\right) du$$

$$= \left(\frac{2}{5}u^{5/3} - \frac{3}{2}u^{2/3}\right)\Big]_2^8$$

$$= \left(\frac{2}{5}\cdot 2^5 - \frac{3}{2}\cdot 2^2\right) - \left(\frac{2}{5}\cdot 2^{5/3} - \frac{3}{2}\cdot 2^{2/3}\right)$$

$$= \frac{34}{5} - \frac{2}{5}\cdot 2^{5/3} + \frac{3}{2}\cdot 2^{2/3} \approx 7.9112. \qquad \blacksquare$$

You are encouraged to rework Example 4 again. The second time use the substitution $u = \sqrt[3]{3x + 2}$.

$\int$ **NOTES FROM THE CLASSROOM**

(*i*) When working the exercises throughout this chapter, do not be overly disturbed if you do not always obtain the same answer as given in the text. Different techniques applied to the same problem can lead to answers that look different. Remember, two antiderivatives of the same function can differ at most by a constant. Try to resolve any conflicts.

(*ii*) It might also prove helpful at this point to recall that integration of the quotient of two polynomial functions, $p(x)/q(x)$, usually begins with long division if the degree of $p(x)$ is greater than or equal to the degree of $q(x)$. See Example 2.

(*iii*) Look for problems that can be solved by previous methods.

Exercises 7.2   Answers to selected odd-numbered problems begin on page ANS-22.

≡ Fundamentals

In Problems 1–26, use a substitution to evaluate the given integral.

1. $\displaystyle\int x(x+1)^3\,dx$

2. $\displaystyle\int \frac{x^2-3}{(x+1)^3}\,dx$

3. $\displaystyle\int (2x+1)\sqrt{x-5}\,dx$

4. $\displaystyle\int (x^2-1)\sqrt{2x+1}\,dx$

5. $\displaystyle\int \frac{x}{\sqrt{x-1}}\,dx$

6. $\displaystyle\int \frac{x^2}{\sqrt{x+2}}\,dx$

7. $\displaystyle\int \frac{x+3}{(3x-4)^{3/2}}\,dx$

8. $\displaystyle\int (x^2+x)\sqrt[3]{x+7}\,dx$

9. $\displaystyle\int \frac{\sqrt{x}}{x+1}\,dx$

10. $\displaystyle\int \frac{t}{\sqrt{t}+1}\,dt$

11. $\displaystyle\int \frac{\sqrt{t}-3}{\sqrt{t}+1}\,dt$

12. $\displaystyle\int \frac{\sqrt{r}+3}{r+3}\,dr$

13. $\displaystyle\int \frac{x^3}{\sqrt[3]{x^2+1}}\,dx$

14. $\displaystyle\int \frac{x^5}{\sqrt[5]{x^2+4}}\,dx$

15. $\displaystyle\int \frac{x^2}{(x-1)^4}\,dx$

16. $\displaystyle\int \frac{2x+1}{(x+7)^2}\,dx$

17. $\displaystyle\int \sqrt{e^x-1}\,dx$

18. $\displaystyle\int \frac{1}{\sqrt{e^x-1}}\,dx$

19. $\displaystyle\int \sqrt{1-\sqrt{v}}\,dv$

20. $\displaystyle\int \frac{\sqrt{w}}{\sqrt{1-\sqrt{w}}}\,dw$

21. $\displaystyle\int \frac{\sqrt{1+\sqrt{t}}}{\sqrt{t}}\,dt$

22. $\displaystyle\int \sqrt{t}\sqrt{1+t\sqrt{t}}\,dt$

23. $\displaystyle\int \frac{2x+7}{x^2+2x+5}\,dx$

24. $\displaystyle\int \frac{6x-1}{4x^2+4x+10}\,dx$

25. $\displaystyle\int \frac{2x+5}{\sqrt{16-6x-x^2}}\,dx$

26. $\displaystyle\int \frac{4x-3}{\sqrt{11+10x-x^2}}\,dx$

In Problems 27 and 28, use the substitution $u=x^{1/6}$ to evaluate the integral.

27. $\displaystyle\int \frac{1}{\sqrt{x}-\sqrt[3]{x}}\,dx$

28. $\displaystyle\int \frac{\sqrt[6]{x}}{\sqrt[3]{x}+1}\,dx$

In Problems 29–40, use a substitution to evaluate the given definite integral.

29. $\displaystyle\int_0^1 x\sqrt{5x+4}\,dx$

30. $\displaystyle\int_{-1}^0 x\sqrt[3]{x+1}\,dx$

31. $\displaystyle\int_1^{16} \frac{1}{10+\sqrt{x}}\,dx$

32. $\displaystyle\int_4^9 \frac{\sqrt{x}-1}{\sqrt{x}+1}\,dx$

33. $\displaystyle\int_2^9 \frac{5x-6}{\sqrt[3]{x-1}}\,dx$

34. $\displaystyle\int_{-\sqrt{3}}^0 \frac{2x^3}{\sqrt{x^2+1}}\,dx$

35. $\displaystyle\int_0^1 (1-\sqrt{x})^{50}\,dx$

36. $\displaystyle\int_0^4 \frac{1}{(1+\sqrt{x})^3}\,dx$

37. $\displaystyle\int_1^8 \frac{1}{x^{1/3}+x^{2/3}}\,dx$

38. $\displaystyle\int_1^{64} \frac{x^{1/3}}{x^{2/3}+2}\,dx$

39. $\displaystyle\int_0^1 x^2(1-x)^5\,dx$

40. $\displaystyle\int_0^6 \frac{2x+5}{\sqrt{2x+4}}\,dx$

In Problems 41 and 42, use a substitution to establish the given result. Assume $x>0$.

41. $\displaystyle\int_1^{x^2} \frac{1}{t}\,dt = 2\int_1^x \frac{1}{t}\,dt$

42. $\displaystyle\int_1^{\sqrt{x}} \frac{1}{t}\,dt = \frac{1}{2}\int_1^x \frac{1}{t}\,dt$

≡ Review of Applications

43. Find the area under the graph of $y=\dfrac{1}{x^{1/3}+1}$ on the interval $[0, 1]$.

44. Find the area bounded by the graph of $y=x^3\sqrt{x+1}$ and the x-axis on the interval $[-1, 1]$.

45. Find the volume of the solid of revolution that is formed by revolving the region bounded by the graphs of $y=\dfrac{1}{\sqrt{x}+1}$, $x=0$, $x=4$, and $y=0$ about the y-axis.

46. Find the volume of the solid of revolution that is formed by revolving the region in Problem 45 about the x-axis.

47. Find the length of the graph of $y=\frac{4}{5}x^{5/4}$ on the interval $[0, 9]$.

48. **Bertalanffy's differential equation** is a mathematical model for the growth of an organism in which it is assumed that constructive metabolism (anabolism) of the organism proceeds at a rate proportional to the surface area, while destructive metabolism (catabolism) proceeds at a rate proportional to the volume. If it is also assumed that surface area is proportional to the two-thirds power of volume and that the organism's weight w is directly proportional to the volume, we can write Bertalanffy's equation as

$$\frac{dw}{dt} = Aw^{2/3} - Bw,$$

where A and B are positive parameters. One can conclude from this equation that the time it takes such an organism to grow from weight w_1 to weight w_2 is given by the definite integral

$$T = \int_{w_1}^{w_2} \frac{1}{Aw^{2/3} - Bw} \, dw.$$

Evaluate this integral. Find an upper limit on how large the organism can grow.

7.3 Integration by Parts

▌ **Introduction**   In this section we are going to develop an important formula that can often be used to integrate the product of two functions. To apply this formula we have to identify one of the functions in the product as a differential. Recall that if $v = g(x)$, then its differential is the function $dv = g'(x) \, dx$.

▌ **Integrating Products**   Since we wish to integrate a product it seems reasonable to begin with the Product Rule of differentiation. If $u = f(x)$ and $v = g(x)$ are differentiable functions, then the derivative of $f(x)g(x)$ is

$$\frac{d}{dx}[f(x)g(x)] = f(x)g'(x) + g(x)f'(x). \tag{1}$$

In turn, integration of both sides of (1),

$$\int \frac{d}{dx}[f(x)g(x)] \, dx = \int f(x)g'(x) \, dx + \int g(x)f'(x) \, dx$$

or

$$f(x)g(x) = \int f(x)g'(x) \, dx + \int g(x)f'(x) \, dx,$$

produces the formula

$$\int f(x)g'(x) \, dx = f(x)g(x) - \int g(x)f'(x) \, dx. \tag{2}$$

The formula in (2) is usually written in terms of the differentials $du = f'(x) \, dx$ and $dv = g'(x) \, dx$:

$$\int u \, dv = uv - \int v \, du. \tag{3}$$

The procedure defined by formula (3) is known as **integration by parts**. The essential idea behind (3) is to evaluate the integral $\int u \, dv$ by means of evaluating another, and it is hoped simpler, integral $\int v \, du$.

Guidelines for Integration by Parts

- The **first step** in the process of integrating by parts consists of choosing and integrating dv in the given integral. As a practical matter, the function dv is *usually* the most complicated factor in the product that can be integrated using one of the basic formulas given in Table 7.1.1.
- The **second step** is the differentiation of the remaining factor u in the given integral. We then form

differentiate

$$\int u \, dv = uv - \int v \, du.$$

integrate

- The **third step** is, of course, the evaluation of $\int v \, du$.

Integration problems can sometimes be done by several methods. In the first example, the integral can be evaluated by means of an algebraic substitution (Section 7.2) as well as by integration by parts.

EXAMPLE 1   Using (3)

Evaluate $\displaystyle\int \frac{x}{\sqrt{x+1}}\,dx$.

Solution   First, we write the integral as

$$\int x(x+1)^{-1/2}\,dx.$$

From this latter form we see that there are several possible choices for the function dv. Of the possible choices for dv,

$$dv = (x+1)^{-1/2}\,dx, \quad dv = x\,dx, \quad \text{or} \quad dv = dx,$$

we choose

$$dv = (x+1)^{-1/2}\,dx \quad \text{and} \quad u = x.$$

Then, by integration Formula 3 in Table 7.1.1, we find

$$v = \int (x+1)^{-1/2}\,dx = 2(x+1)^{1/2}.$$

◀ No constant of integration need be used when integrating dv.

Substituting $v = 2(x+1)^{1/2}$ and $du = dx$ into (3) then gives

$$\int \overset{u}{\overbrace{x}}\,\overset{dv}{\overbrace{(x+1)^{-1/2}\,dx}} = \overset{u}{\overbrace{x}} \cdot \overset{v}{\overbrace{2(x+1)^{1/2}}} - \int \overset{v}{\overbrace{2(x+1)^{1/2}}}\,\overset{du}{\overbrace{dx}}$$

$$= 2x(x+1)^{1/2} - 2 \cdot \frac{2}{3}(x+1)^{3/2} + C \;\leftarrow\; \text{we used formula 3 in Table 7.1.1}$$

$$= 2x(x+1)^{1/2} - \frac{4}{3}(x+1)^{3/2} + C.$$

Check by Differentiation   To verify the preceding result we use the Product Rule:

$$\frac{d}{dx}\left(2x(x+1)^{1/2} - \frac{4}{3}(x+1)^{3/2} + C\right) = 2x \cdot \frac{1}{2}(x+1)^{-1/2} + 2(x+1)^{1/2} - \frac{4}{3} \cdot \frac{3}{2}(x+1)^{1/2}$$

$$= x(x+1)^{-1/2} + 2(x+1)^{1/2} - 2(x+1)^{1/2}$$

$$= \frac{x}{\sqrt{x+1}}. \qquad\blacksquare$$

The key to making integration by parts work is to make the "right" choice for the function dv. In the guidelines given prior to Example 1 we stated that dv is usually the most complicated factor in the product that can be immediately integrated by a previously known formula. Yet this cannot be given as a firm rule. Realization that the "right" choice for dv has been made is often based on pragmatic hindsight: Is the second integral $\int v\,du$ less complicated than the first integral $\int u\,dv$? Can we evaluate this second integral? To see what happens when the "wrong" choice is made, let us consider Example 1 again, but this time we select

$$dv = x\,dx \qquad \text{and} \qquad u = (x+1)^{-1/2}$$

so that

$$v = \frac{1}{2}x^2 \qquad \text{and} \qquad du = -\frac{1}{2}(x+1)^{-3/2}\,dx.$$

Applying (3) in this instance gives

$$\int x(x+1)^{-1/2}\,dx = \frac{1}{2}x^2(x+1)^{-1/2} + \frac{1}{4}\int x^2(x+1)^{-3/2}\,dx.$$

The difficulty here is apparent; the second integral $\int v\, du$ is more complicated than the original $\int u\, dv$. The alternative selection $dv = dx$ also leads to an impasse.

EXAMPLE 2   Using (3)

Evaluate $\int x^3 \ln x\, dx$.

Solution   Again there are several possible choices for the function dv:

$$dv = \ln x\, dx, \quad dv = x^3\, dx, \quad \text{or} \quad dv = dx. \qquad (4)$$

Although the choice $dv = \ln x\, dx$ is undoubtedly the most complicated factor in the product $x^3 \ln x\, dx$, we reject this choice since it does not match any formula in Table 7.1.1. Of the remaining two functions in (4), the second is the more "complicated." So if we choose

$$dv = x^3\, dx \quad \text{and} \quad u = \ln x,$$

then

$$v = \frac{1}{4}x^4 \quad \text{and} \quad du = \frac{1}{x}\, dx.$$

Hence from (3),

$$\int x^3 \ln x\, dx = \overset{u}{\overbrace{\ln x}} \cdot \overset{v}{\overbrace{\frac{1}{4}x^4}} - \int \overset{v}{\overbrace{\frac{1}{4}x^4}} \cdot \overset{du}{\overbrace{\frac{1}{x}}}\, dx \qquad \leftarrow \text{simpify the integrand}$$

$$= \frac{1}{4}x^4 \ln x - \frac{1}{4}\int x^3\, dx \qquad \leftarrow \text{integrate } x^3$$

$$= \frac{1}{4}x^4 \ln x - \frac{1}{16}x^4 + C. \qquad \blacksquare$$

EXAMPLE 3   Using (3)

Evaluate $\int x\tan^{-1}x\, dx$.

Solution   The choice $dv = \tan^{-1}x\, dx$ is not a judicious one, since we cannot immediately integrate this function based on a previously known result. So we choose

$$dv = x\, dx \quad \text{and} \quad u = \tan^{-1}x$$

and find

$$v = \frac{1}{2}x^2 \quad \text{and} \quad du = \frac{1}{1+x^2}\, dx.$$

Therefore (3) gives

$$\int \overset{u}{\overbrace{(\tan^{-1}x)}}\, \overset{dv}{\overbrace{(x\, dx)}} = \overset{u}{\overbrace{(\tan^{-1}x)}}\overset{v}{\overbrace{\frac{1}{2}x^2}} - \int \overset{v}{\overbrace{\frac{1}{2}x^2}} \cdot \overset{du}{\overbrace{\frac{1}{1+x^2}}}\, dx. \quad \leftarrow \text{simplify the integrand}$$

To evaluate the indefinite integral $\int x^2\, dx/(1+x^2)$, we use long division (see Example 7 of Section 5.1). Hence,

$$\int x\tan^{-1}x\, dx = \frac{1}{2}x^2 \tan^{-1}x - \frac{1}{2}\int \left(1 - \frac{1}{1+x^2}\right) dx$$

$$= \frac{1}{2}x^2 \tan^{-1}x - \frac{1}{2}x + \frac{1}{2}\tan^{-1}x + C. \qquad \blacksquare$$

■ **Successive Integrations**   A problem may require integration by parts several times in succession. As a rule, integrals of the type

$$\int p(x)\sin kx\, dx, \quad \int p(x)\cos kx\, dx, \quad \text{and} \quad \int p(x)e^{kx}\, dx, \qquad (5)$$

where $p(x)$ is a polynomial of degree $n \geq 1$ and k a constant, will require integration by parts n times. Moreover, an integral such as

$$\int x^k (\ln x)^n \, dx, \tag{6}$$

where again n is a positive integer, will also require n applications of (3). The integral in Example 2 is of the form (6) with $k = 3$ and $n = 1$.

EXAMPLE 4   Using (3) Twice in Succession

Evaluate $\int x^2 \cos x \, dx$.

Solution   The integral $\int x^2 \cos x \, dx$ is the second of the three forms in (5) with $p(x) = x^2$ and $n = 2$. Consequently we apply (3) twice in succession. In the first integration we use

$$dv = \cos x \, dx \qquad \text{and} \qquad u = x^2$$

so that

$$v = \sin x \qquad \text{and} \qquad du = 2x \, dx.$$

Hence (3) becomes

requires integration by parts

$$\int x^2 \cos x \, dx = x^2 \sin x - 2 \int x \sin x \, dx. \tag{7}$$

The second integral in (7) is the first form in (5) and requires only one integration by parts since the degree of the polynomial $p(x) = x$ is $n = 1$. In this second integral we choose $dv = \sin x \, dx$ and $u = x$:

$$\int x^2 \cos x \, dx = x^2 \sin x - 2 \left[x(-\cos x) - \int (-\cos x) \, dx \right]$$

$$= x^2 \sin x + 2x \cos x - 2 \int \cos x \, dx \leftarrow \begin{array}{l} \text{formula 8} \\ \text{in Table 7.1.1} \end{array}$$

$$= x^2 \sin x + 2x \cos x - 2 \sin x + C. \tag{8} \blacksquare$$

The result in (8) can be obtained by a systematic shortcut. If we think of the integral in Example 4 as $\int f(x) g'(x) \, dx$ where $f(x) = x^2$ and $g'(x) = \cos x$, then we can display the derivatives and integrals in an array:

$f(x)$ and its derivatives		$g'(x)$ and its integrals
x^2	$+$	$\cos x$
$2x$	$-$	$\sin x$
2	$+$	$-\cos x$
0		$-\sin x$

We then form the products of the functions joined by the arrows and either add or subtract a product according to the algebraic sign indicated in blue:

$$\int x^2 \cos x \, dx = +x^2 (\sin x) - 2x(-\cos x) + 2(-\sin x) + C.$$

The last zero in the derivative column indicates that we need not integrate $g'(x)$ any further; the products from that point on are zero.

This technique for successive integrations by parts works on all integrals of the type shown in (5) and is called **tabular integration**. For an integral such as $\int x^4 e^{-2x} \, dx$ we would pick $f(x) = x^4$ and $g'(x) = e^{-2x}$. You should verify that tabular integration gives

$$\int x^4 e^{-2x} \, dx = +x^4 \left(-\frac{1}{2} e^{-2x} \right) - 4x^3 \left(\frac{1}{4} e^{-2x} \right) + 12x^2 \left(-\frac{1}{8} e^{-2x} \right) - 24x \left(\frac{1}{16} e^{-2x} \right) + 24 \left(-\frac{1}{32} e^{-2x} \right) + C$$

$$= -\frac{1}{2} x^4 e^{-2x} - x^3 e^{-2x} - \frac{3}{2} x^2 e^{-2x} - \frac{3}{2} x e^{-2x} - \frac{3}{4} e^{-2x} + C.$$

■ **Solving for Integrals** For certain integrals, one or more applications of integration by parts may result in a situation where the original integral occurs on the right-hand side. In this case the problem of evaluating that integral is completed by *solving* for the original integral. The next example illustrates the technique.

EXAMPLE 5 Solving for the Original Integral

Evaluate $\int \sec^3 x \, dx$.

Solution Inspection of the integral reveals no obvious choice for dv. However, by writing the integrand as the product $\sec^3 x = \sec x \cdot \sec^2 x$, we can identify

$$dv = \sec^2 x \, dx \qquad \text{and} \qquad u = \sec x$$

so that

$$v = \tan x \qquad \text{and} \qquad du = \sec x \tan x \, dx.$$

It follows from (3) that

$$\int \sec^3 x \, dx = \sec x \tan x - \int \tan^2 x \sec x \, dx$$

$$= \sec x \tan x - \int (\sec^2 x - 1) \sec x \, dx \quad \leftarrow \text{trig identity for } \tan^2 x$$

$$= \sec x \tan x + \int \sec x \, dx - \int \sec^3 x \, dx$$

$$= \sec x \tan x + \ln|\sec x + \tan x| - \int \sec^3 x \, dx.$$

See (18) in Section 5.2 for the evaluation of the integral $\int \sec x \, dx$. Also, see Formula 15 in Table 7.1.1.

We solve the last equation for $\int \sec^3 x \, dx$ and add a constant of integration:

$$2 \int \sec^3 x \, dx = \sec x \tan x + \ln|\sec x + \tan x|$$

and so

$$\int \sec^3 x \, dx = \frac{1}{2} \sec x \tan x + \frac{1}{2} \ln|\sec x + \tan x| + C. \qquad ■$$

Integrals of the type

$$\int e^{ax} \sin bx \, dx \qquad \text{and} \qquad \int e^{ax} \cos bx \, dx \qquad (9)$$

are important in certain aspects of applied mathematics. These integrals require two applications of integration by parts before recovering the original integral on the right-hand side.

EXAMPLE 6 Solving for the Original Integral

Evaluate $\int e^{2x} \cos 3x \, dx$.

Solution If we choose

$$dv = e^{2x} \, dx \qquad \text{and} \qquad u = \cos 3x,$$

then

$$v = \frac{1}{2} e^{2x} \qquad \text{and} \qquad du = -3 \sin 3x \, dx.$$

The integration by parts formula (3) then gives

$$\int e^{2x} \cos 3x \, dx = \frac{1}{2} e^{2x} \cos 3x + \frac{3}{2} \int e^{2x} \sin 3x \, dx.$$

We apply integration by parts again to the integral highlighted in color with $dv = e^{2x}\,dx$ and $u = \sin 3x$:

$$\int e^{2x}\cos 3x\,dx = \frac{1}{2}e^{2x}\cos 3x + \frac{3}{2}\left[\frac{1}{2}e^{2x}\sin 3x - \int \frac{1}{2}e^{2x}(3\cos 3x)\,dx\right]$$

$$= \frac{1}{2}e^{2x}\cos 3x + \frac{3}{4}e^{2x}\sin 3x - \frac{9}{4}\int e^{2x}\cos 3x\,dx.$$

Solving the last equation for $\int e^{2x}\cos 3x\,dx$ gives

$$\frac{13}{4}\int e^{2x}\cos 3x\,dx = \frac{1}{2}e^{2x}\cos 3x + \frac{3}{4}e^{2x}\sin 3x.$$

After dividing by $\frac{13}{4}$ and affixing a constant of integration we get

$$\int e^{2x}\cos 3x\,dx = \frac{2}{13}e^{2x}\cos 3x + \frac{3}{13}e^{2x}\sin 3x + C.$$ ■

In evaluating the integrals $\int e^{ax}\sin bx\,dx$ and $\int e^{ax}\cos bx\,dx$ it does not matter which of the functions are chosen as dv and u. In Example 6 we chose $dv = e^{2x}\,dx$ and $u = \cos 3x$; you are encouraged to rework this example using $dv = \cos 3x\,dx$ and $u = e^{2x}$.

■ **Definite Integrals**   A definite integral can be evaluated using integration by parts in the following manner:

$$\int_a^b f(x)g'(x)\,dx = f(x)g(x)\Big]_a^b - \int_a^b g(x)f'(x)\,dx.$$

For convenience, the foregoing equation is usually written as

$$\int_a^b u\,dv = uv\Big]_a^b - \int_a^b v\,du,\tag{10}$$

where it is understood the *limits of integration are values of* x and the integrations in the integrals are carried out with respect to the variable x.

EXAMPLE 7   Area Under the Graph

Find the area under the graph of $f(x) = \ln x$ on the interval $[1, e]$.

Solution   From FIGURE 7.3.1 we see that $f(x) \ge 0$ for all x in the interval. Hence, the area A is given by the definite integral

$$A = \int_1^e \ln x\,dx.$$

Choosing $\qquad dv = dx \qquad$ and $\qquad u = \ln x,$

then $\qquad v = x \qquad$ and $\qquad du = \frac{1}{x}\,dx.$

From (10) we have

$$A = x\ln x\Big]_1^e - \int_1^e x\cdot\frac{1}{x}\,dx$$

$$= x\ln x\Big]_1^e - \int_1^e dx$$

$$= x\ln x\Big]_1^e - x\Big]_1^e$$

$$= e\ln e - \ln 1 - e + 1 = 1.$$

Here we have used $\ln e = 1$ and $\ln 1 = 0$.   ■

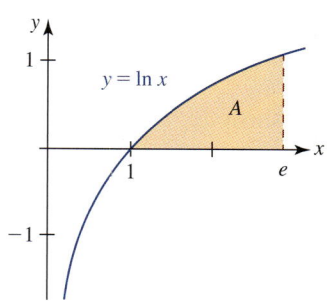

FIGURE 7.3.1   Area under graph in Example 7

≡ Fundamentals

In Problems 1–40, use integration by parts to evaluate the given integral.

1. $\displaystyle\int x\sqrt{x+3}\,dx$

2. $\displaystyle\int \frac{x}{\sqrt{2x-5}}\,dx$

3. $\displaystyle\int \ln 4x\,dx$

4. $\displaystyle\int \ln(x+1)\,dx$

5. $\displaystyle\int x\ln 2x\,dx$

6. $\displaystyle\int x^{1/2}\ln x\,dx$

7. $\displaystyle\int \frac{\ln x}{x^2}\,dx$

8. $\displaystyle\int \frac{\ln x}{\sqrt{x^3}}\,dx$

9. $\displaystyle\int (\ln t)^2\,dt$

10. $\displaystyle\int (t\ln t)^2\,dt$

11. $\displaystyle\int \sin^{-1}x\,dx$

12. $\displaystyle\int x^2\tan^{-1}x\,dx$

13. $\displaystyle\int xe^{3x}\,dx$

14. $\displaystyle\int x^2 e^{5x}\,dx$

15. $\displaystyle\int x^3 e^{-4x}\,dx$

16. $\displaystyle\int x^5 e^x\,dx$

17. $\displaystyle\int x^3 e^{x^2}\,dx$

18. $\displaystyle\int x^5 e^{2x^3}\,dx$

19. $\displaystyle\int t\cos 8t\,dt$

20. $\displaystyle\int x\sinh x\,dx$

21. $\displaystyle\int x^2\sin x\,dx$

22. $\displaystyle\int x^2\cos\frac{x}{2}\,dx$

23. $\displaystyle\int x^3\cos 3x\,dx$

24. $\displaystyle\int x^4\sin 2x\,dx$

25. $\displaystyle\int e^x\sin 4x\,dx$

26. $\displaystyle\int e^{-x}\cos 5x\,dx$

27. $\displaystyle\int e^{-2\theta}\cos\theta\,d\theta$

28. $\displaystyle\int e^{\alpha x}\sin\beta x\,dx$

29. $\displaystyle\int \theta\sec\theta\tan\theta\,d\theta$

30. $\displaystyle\int e^{2t}\cos e^t\,dt$

31. $\displaystyle\int \sin x\cos 2x\,dx$

32. $\displaystyle\int \cosh x\cosh 2x\,dx$

33. $\displaystyle\int x^3\sqrt{x^2+4}\,dx$

34. $\displaystyle\int \frac{t^5}{(t^3+1)^2}\,dt$

35. $\displaystyle\int \sin(\ln x)\,dx$

36. $\displaystyle\int \cos x\ln(\sin x)\,dx$

37. $\displaystyle\int \csc^3 x\,dx$

38. $\displaystyle\int x\sec^{-1}x\,dx$

39. $\displaystyle\int x\sec^2 x\,dx$

40. $\displaystyle\int x\tan^2 x\,dx$

In Problems 41–46, evaluate the given definite integral.

41. $\displaystyle\int_0^2 x\ln(x+1)\,dx$

42. $\displaystyle\int_0^1 \ln(x^2+1)\,dx$

43. $\displaystyle\int_2^4 xe^{-x/2}\,dx$

44. $\displaystyle\int_{-\pi}^{\pi} e^x\cos x\,dx$

45. $\displaystyle\int_0^1 \tan^{-1}x\,dx$

46. $\displaystyle\int_0^{\sqrt{2}/2} \cos^{-1}x\,dx$

≡ Review of Applications

47. Find the area under the graph of $y = 1 + \ln x$ on the interval $[e^{-1}, 3]$.

48. Find the area bounded by the graph of $y = \tan^{-1}x$ and the x-axis on the interval $[-1, 1]$.

49. The region in the first quadrant bounded by the graphs of $y = \ln x$, $x = 5$, and $y = 0$ is revolved about the x-axis. Find the volume of the solid of revolution.

50. The region in the first quadrant bounded by the graphs of $y = e^x$, $x = 0$, and $y = 3$ is revolved about the y-axis. Find the volume of the solid of revolution.

51. The region in the first quadrant bounded by the graphs of $y = \sin x$ and $y = 0, 0 \le x \le \pi$, is revolved about the y-axis. Find the volume of the solid of revolution.

52. Find the length of the graph of $y = \ln(\cos x)$ on the interval $[0, \pi/4]$.

53. Find the average value of $f(x) = \tan^{-1}(x/2)$ on the interval $[0, 2]$.

54. A body moves in a straight line with velocity $v(t) = e^{-t}\sin t$, where v is measured in cm/s. Find the position function $s(t)$ if it is known that $s = 0$ when $t = 0$.

55. A body moves in a straight line with acceleration $a(t) = te^{-t}$, where a is measured in cm/s². Find the velocity function $v(t)$ and the position function $s(t)$ if $v(0) = 1$ and $s(0) = -1$.

56. A water tank is formed by revolving the region bounded by the graphs of $y = \sin\pi x$ and $y = 0, 0 \le x \le 1$, about the x-axis, which is taken in the downward direction. The tank is filled to a depth of $\frac{1}{2}$ ft. Determine the work done in pumping all the water to the top of the tank.

57. Find the force caused by liquid pressure on one side of the vertical plate shown in FIGURE 7.3.2. Assume that the plate is submerged in water and that dimensions are in feet.

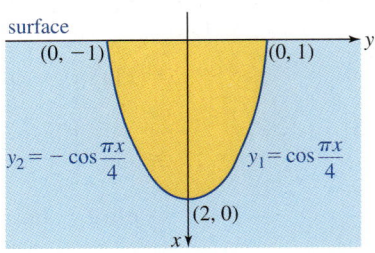

FIGURE 7.3.2 Submerged plate in Problem 57

58. Find the centroid of the region bounded by the graphs of $y = \sin x$, $y = 0$, and $x = \pi/2$.

In Problems 59–62, evaluate the given integral by first using a substitution followed by integration by parts.

59. $\displaystyle\int_1^4 \frac{\tan^{-1}\sqrt{x}}{\sqrt{x}}\, dx$

60. $\displaystyle\int xe^{\sqrt{x}}\, dx$

61. $\displaystyle\int \sin\sqrt{x+2}\, dx$

62. $\displaystyle\int_0^{\pi^2} \cos\sqrt{t}\, dt$

In Problems 63–66, use integration by parts to establish the given **reduction formula**.

63. $\displaystyle\int (\ln x)^n\, dx = x(\ln x)^n - n\int (\ln x)^{n-1}\, dx$

64. $\displaystyle\int \sin^n x\, dx = -\frac{\sin^{n-1}x\cos x}{n} + \frac{n-1}{n}\int \sin^{n-2}x\, dx$

65. $\displaystyle\int \cos^n x\, dx = \frac{\cos^{n-1}x\sin x}{n} + \frac{n-1}{n}\int \cos^{n-2}x\, dx$

66. $\displaystyle\int \sec^n x\, dx = \frac{\sec^{n-2}x\tan x}{n-1} + \frac{n-2}{n-1}\int \sec^{n-2}x\, dx$,

In Problems 67–70, use a reduction formula from Problems 63–66 to evaluate the given integral.

67. $\displaystyle\int \sin^3 x\, dx$

68. $\displaystyle\int \sec^4 x\, dx$

69. $\displaystyle\int \cos^3 10x\, dx$

70. $\displaystyle\int \cos^4 x\, dx$

71. Use Problem 64 to show that for $n \geq 2$,

$$\int_0^{\pi/2} \sin^n x\, dx = \frac{n-1}{n}\int_0^{\pi/2} \sin^{n-2}x\, dx.$$

72. Show how the repeated use of the formula in Problem 71 can be used to obtain the following results.

(a) $\displaystyle\int_0^{\pi/2} \sin^n x\, dx = \frac{\pi}{2}\cdot\frac{1\cdot3\cdot5\cdots(n-1)}{2\cdot4\cdot6\cdots n}$, n even and $n \geq 2$

(b) $\displaystyle\int_0^{\pi/2} \sin^n x\, dx = \frac{2\cdot4\cdot6\cdots(n-1)}{3\cdot5\cdot7\cdots n}$, n odd and $n \geq 3$

73. Use part (a) of Problem 72 to evaluate $\displaystyle\int_0^{\pi/2} \sin^8 x\, dx$.

74. Use part (b) of Problem 72 to evaluate $\displaystyle\int_0^{\pi/2} \sin^5 x\, dx$.

≡ **Think About It**

In Problems 75–82, the integration by parts is a bit more challenging. Evaluate the given integral.

75. $\displaystyle\int e^{2x}\tan^{-1}e^x\, dx$

76. $\displaystyle\int (\sin^{-1}x)^2\, dx$

77. $\displaystyle\int \frac{xe^x}{(x+1)^2}\, dx$

78. $\displaystyle\int \frac{x^2 e^x}{(x+2)^2}\, dx$

79. $\displaystyle\int xe^x \sin x\, dx$

80. $\displaystyle\int xe^{-x}\cos 2x\, dx$

81. $\displaystyle\int \ln\left(x + \sqrt{x^2+1}\right)\, dx$

82. $\displaystyle\int e^{\sin^{-1}x}\, dx$

≡ **Calculator/CAS Problems**

83. (a) Use a calculator or CAS to obtain the graph of $f(x) = 3 + 2\sin^2 x - 5\sin^4 x$.

(b) Find the area under the graph of the function given in part (a) on the interval $[0, 2\pi]$.

84. (a) Use a calculator or CAS to obtain the graphs of $y = x\sin x$ and $y = x\cos x$.

(b) Find the area of the region bounded by the graphs on the interval $[x_1, x_2]$, where x_1 and x_2 are the positive x-coordinates corresponding to the first and second points of intersection of the graphs for $x > 0$.

7.4 **Powers of Trigonometric Functions**

▮ **Introduction** In Section 5.2 we saw how to integrate $\sin^2 x$ and $\cos^2 x$. In this section we see how to integrate higher powers of $\sin x$ and $\cos x$, certain products of powers of $\sin x$ and $\cos x$, and products of powers of $\sec x$ and $\tan x$. The techniques illustrated in this section depend heavily on trigonometric identities.

▮ **Integrals of the Form $\int \sin^m x \cos^n x\, dx$** To evaluate integrals of the type

$$\int \sin^m x \cos^n x\, dx, \tag{1}$$

we distinguish two cases.

CASE I: *m* or *n* is an odd positive integer

Let us first assume that $m = 2k + 1$ in (1) is an odd positive integer. Then:

- Begin by splitting off the factor $\sin x$ from $\sin^{2k+1}x$, that is, write $\sin^{2k+1}x = \sin^{2k}x \sin x$, where $2k$ is now even.
- Use the basic Pythagorean identity $\sin^2 x = 1 - \cos^2 x$, to rewrite
$$\sin^{2k}x = (\sin^2 x)^k = (1 - \cos^2 x)^k.$$
- Expand the binomial $(1 - \cos^2 x)^k$.

In this manner we can express the integrand in (1) as a sum of powers of $\cos x$ times $\sin x$. The original integral can then be expressed as a sum of integrals, each having the recognizable form

$$\int \cos^r x \sin x \, dx = -\int \overbrace{(\cos x)^r}^{u^r} \overbrace{(-\sin x \, dx)}^{du} = -\int u^r \, du.$$

If $n = 2k + 1$ is an odd positive integer in (1), then the procedure is the same, except that we write $\cos^{2k+1}x = \cos^{2k}x \cos x$, use $\cos^2 x = 1 - \sin^2 x$, and write the integral as a sum of integrals of the form

$$\int \sin^r x \cos x \, dx = \int \overbrace{(\sin x)^r}^{u^r} \overbrace{(\cos x \, dx)}^{du} = \int u^r \, du.$$

We note that the exponent r need not be an integer.

EXAMPLE 1   Case I of Integral (1)

Evaluate $\int \sin^5 x \cos^2 x \, dx$.

Solution   We begin by writing the power of $\sin x$ as $\sin^5 x = \sin^4 x \sin x$:

$$\int \sin^5 x \cos^2 x \, dx = \int \cos^2 x \sin^4 x \sin x \, dx$$

$$= \int \cos^2 x (\sin^2 x)^2 \sin x \, dx \qquad \leftarrow \text{replace } \sin^2 x \text{ by } 1 - \cos^2 x$$

$$= \int \cos^2 x (1 - \cos^2 x)^2 \sin x \, dx$$

$$= \int \cos^2 x (1 - 2 \cos^2 x + \cos^4 x) \sin x \, dx \qquad \leftarrow \text{write as three integrals}$$

$$= -\int \overbrace{(\cos x)^2}^{u^2} \overbrace{(-\sin x \, dx)}^{du} + 2 \int \overbrace{(\cos x)^4}^{u^4} \overbrace{(-\sin x \, dx)}^{du} - \int \overbrace{(\cos x)^6}^{u^6} \overbrace{(-\sin x \, dx)}^{du}$$

$$= -\frac{1}{3} \cos^3 x + \frac{2}{5} \cos^5 x - \frac{1}{7} \cos^7 x + C. \qquad \blacksquare$$

EXAMPLE 2   Case I of Integral (1)

Evaluate $\int \sin^3 x \, dx$.

Solution   As in Example 1 we rewrite the power of $\sin x$ as $\sin^2 x \sin x$:

$$\int \sin^3 x \, dx = \int \sin^2 x \sin x \, dx$$

$$= \int (1 - \cos^2 x) \sin x \, dx$$

$$= \int \sin x \, dx + \int (\cos x)^2 (-\sin x \, dx)$$

$$= -\cos x + \frac{1}{3} \cos^3 x + C. \qquad \blacksquare$$

EXAMPLE 3 Case I of Integral (1)

Evaluate $\displaystyle\int \sin^4 x \cos^3 x \, dx$.

Solution This time we rewrite the power of $\cos x$ as $\cos^2 x \cos x$:

$$\int \sin^4 x \cos^3 x \, dx = \int \sin^4 x \cos^2 x \cos x \, dx$$

$$= \int \sin^4 x \, (1 - \sin^2 x) \cos x \, dx \qquad \leftarrow \text{write as two integrals}$$

$$= \int \overbrace{(\sin x)^4}^{u^4} \overbrace{(\cos x \, dx)}^{du} - \int \overbrace{(\sin x)^6}^{u^6} \overbrace{(\cos x \, dx)}^{du}$$

$$= \frac{1}{5} \sin^5 x - \frac{1}{7} \sin^7 x + C. \qquad \blacksquare$$

CASE II: m and n are both even nonnegative integers

When both m and n are even nonnegative integers, the evaluation of (1) relies heavily on the trigonometric identities

$$\sin x \cos x = \frac{1}{2} \sin 2x, \quad \sin^2 x = \frac{1}{2}(1 - \cos 2x), \quad \cos^2 x = \frac{1}{2}(1 + \cos 2x). \qquad (2)$$

We have already seen the last two identities in Section 5.2 as the useful forms for the half-angle formulas for the sine and cosine.

EXAMPLE 4 Using Identities (2) in Integral (1)

Evaluate $\displaystyle\int \sin^2 x \cos^2 x \, dx$.

Solution We will evaluate the integral in two different ways. We begin by using the second and third formulas in (2):

$$\int \sin^2 x \cos^2 x \, dx = \int \frac{1}{2}(1 - \cos 2x) \cdot \frac{1}{2}(1 + \cos 2x) \, dx$$

$$= \frac{1}{4} \int (1 - \cos^2 2x) \, dx$$

$$= \frac{1}{4} \int \left[1 - \frac{1}{2}(1 + \cos 4x) \right] dx \qquad \leftarrow \begin{array}{l} \text{third identity in (2)} \\ \text{with } x \text{ replaced by } 2x \end{array}$$

$$= \frac{1}{4} \int \left(\frac{1}{2} - \frac{1}{2} \cos 4x \right) dx$$

$$= \frac{1}{8} x - \frac{1}{32} \sin 4x + C.$$

Alternative Solution We now use the first formula in (2):

$$\int \sin^2 x \cos^2 x \, dx = \int (\sin x \cos x)^2 \, dx$$

$$= \int \left(\frac{1}{2} \sin 2x \right)^2 dx$$

$$= \frac{1}{4} \int \frac{1}{2} (1 - \cos 4x) \, dx.$$

The remainder of the solution is the same as before. $\qquad \blacksquare$

EXAMPLE 5 Using Identities (2) in Integral (1)

Evaluate $\int \cos^4 x \, dx$.

Solution We begin by rewriting $\cos^4 x$ as $(\cos^2 x)^2$ and then use the third identity in (2):

$$\int \cos^4 x \, dx = \int (\cos^2 x)^2 \, dx$$

$$= \int \left[\frac{1}{2}(1 + \cos 2x) \right]^2 dx$$

$$= \frac{1}{4} \int (1 + 2\cos 2x + \cos^2 2x) \, dx \quad \xleftarrow{\text{use the third formula in (2)}} \text{a second time}$$

$$= \frac{1}{4} \int \left[1 + 2\cos 2x + \frac{1}{2}(1 + \cos 4x) \right] dx$$

$$= \frac{1}{4} \int \left(\frac{3}{2} + 2\cos 2x + \frac{1}{2}\cos 4x \right) dx$$

$$= \frac{3}{8}x + \frac{1}{4}\sin 2x + \frac{1}{32}\sin 4x + C. \qquad \blacksquare$$

▌ Integrals of the Form $\int \tan^m x \sec^n x \, dx$ To evaluate an integral involving powers of secant and tangent,

$$\int \tan^m x \sec^n x \, dx, \tag{3}$$

we shall consider three cases. The procedure in the first two cases is similar to Case I for integral (1) in that we break off a factor from the product $\tan^m x \sec^n x$ to serve as part of the differential du.

CASE I: *m* is an odd positive integer

When $m = 2k + 1$ is an odd positive integer in (3), $2k$ is even. Then:

- Begin by splitting off the factor $\sec x \tan x$ from $\tan^{2k+1} x \sec^n x$, that is, write $\tan^{2k+1} x \sec^n x = \tan^{2k} x \sec^{n-1} x \sec x \tan x$, where $2k$ is now even.
- Use the identity $\tan^2 x = \sec^2 x - 1$ to rewrite

$$\tan^{2k} x = (\tan^2 x)^k = (\sec^2 x - 1)^k.$$

- Expand the binomial $(\sec^2 x - 1)^k$.

In this manner we can express the integrand in (3) as a sum of powers of $\sec x$ times $\sec x \tan x$. The original integral can then be expressed as a sum of integrals, each having the recognizable form

$$\int \overbrace{(\sec x)^r}^{u^r} \overbrace{(\sec x \tan x \, dx)}^{du} = \int u^r \, du.$$

EXAMPLE 6 Case I of Integral (3)

Evaluate $\int \tan^3 x \sec^7 x \, dx$.

Solution By writing $\tan^3 x \sec^7 x = \tan^2 x \sec^6 x \sec x \tan x$, the integral can be written as two integrals that we can evaluate:

$$\int \tan^3 x \sec^7 x \, dx = \int \tan^2 x \sec^6 x \sec x \tan x \, dx$$

$$= \int (\sec^2 x - 1) \sec^6 x \sec x \tan x \, dx$$

$$= \int \overbrace{(\sec x)^8}^{u^8} \overbrace{(\sec x \tan x \, dx)}^{du} - \int \overbrace{(\sec x)^6}^{u^6} \overbrace{(\sec x \tan x \, dx)}^{du}$$

$$= \frac{1}{9} \sec^9 x - \frac{1}{7} \sec^7 x + C.$$ ∎

CASE II: *n* is an even positive integer

Let $n = 2k$ represent an even positive integer in (3). Then:

- Begin by splitting the factor $\sec^2 x$ from $\sec^{2k} x \tan^m x$, that is, write $\sec^{2k} x \tan^m x = \sec^{2(k-1)} x \tan^m x \sec^2 x$.
- Use the identity $\sec^2 x = 1 + \tan^2 x$ to rewrite

$$\sec^{2(k-1)} x = (\sec^2 x)^{k-1} = (1 + \tan^2 x)^{k-1}.$$

- Expand the binomial $(1 + \tan^2 x)^{k-1}$.

In this manner we can express the integrand in (3) as a sum of powers of $\tan x$ times $\sec^2 x$. The original integral can then be expressed as a sum of integrals, each having the recognizable form

$$\int \overbrace{(\tan x)^r}^{u^r} \overbrace{(\sec^2 x \, dx)}^{du} = \int u^r \, du.$$

EXAMPLE 7   Case II of Integral (3)

Evaluate $\displaystyle \int \sqrt{\tan x} \, \sec^4 x \, dx$.

Solution   We rewrite the integrand using $\sec^4 x = \sec^2 x \sec^2 x$:

$$\int \sqrt{\tan x} \, \sec^4 x \, dx = \int (\tan x)^{1/2} \sec^2 x \sec^2 x \, dx$$

$$= \int (\tan x)^{1/2} (1 + \tan^2 x) \sec^2 x \, dx$$

$$= \int \overbrace{(\tan x)^{1/2}}^{u^{1/2}} \overbrace{(\sec^2 x \, dx)}^{du} + \int \overbrace{(\tan x)^{5/2}}^{u^{5/2}} \overbrace{(\sec^2 x \, dx)}^{du}$$

$$= \frac{2}{3} (\tan x)^{3/2} + \frac{2}{7} (\tan x)^{7/2} + C.$$ ∎

CASE III: *m* is even and *n* is odd

Finally, if *m* is an even positive integer and *n* is an odd positive integer, we write the integrand of (3) in terms of sec *x* and use integration by parts.

EXAMPLE 8   Case III of Integral (3)

Evaluate $\displaystyle \int \tan^2 x \, \sec x \, dx$.

Solution   By writing

$$\int \tan^2 x \, \sec x \, dx = \int (\sec^2 x - 1) \sec x \, dx$$

$$= \int \sec^3 x \, dx - \int \sec x \, dx$$

we encounter two integrals that were evaluated previously:

$$\int \sec^3 x \, dx = \frac{1}{2} \sec x \tan x + \frac{1}{2} \ln|\sec x + \tan x| + C_1, \qquad (4)$$

$$\int \sec x \, dx = \ln|\sec x + \tan x| + C_2. \qquad (5)$$

◀ For the result in (4), see Example 5 in Section 7.3. For (5), see (18) in Section 5.2.

Subtracting the results in (4) and (5) then yields the desired result:

$$\int \tan^2 x \sec x \, dx = \frac{1}{2} \sec x \tan x - \frac{1}{2} \ln |\sec x + \tan x| + C.$$ ∎

Integrals of the type

$$\int \cot^m x \csc^n x \, dx \tag{6}$$

are handled in a manner analogous to (3). In this case the identity $\csc^2 x = 1 + \cot^2 x$ is used.

Exercises 7.4   Answers to selected odd-numbered problems begin on page ANS-22.

≡ Fundamentals

In Problems 1–40, evaluate the given indefinite integral. Note that some of the integrals do not, strictly speaking, fall into any of the cases considered in this section. You should be able to evaluate these integrals by previous methods.

1. $\int (\sin x)^{1/2} \cos x \, dx$

2. $\int \cos^4 5x \sin 5x \, dx$

3. $\int \cos^3 x \, dx$

4. $\int \sin^3 4x \, dx$

5. $\int \sin^5 t \, dt$

6. $\int \cos^5 t \, dt$

7. $\int \sin^3 x \cos^3 x \, dx$

8. $\int \sin^5 2x \cos^2 2x \, dx$

9. $\int \sin^4 t \, dt$

10. $\int \cos^6 \theta \, d\theta$

11. $\int \sin^2 x \cos^4 x \, dx$

12. $\int \frac{\cos^3 x}{\sin^2 x} \, dx$

13. $\int \sin^4 x \cos^4 x \, dx$

14. $\int \sin^2 3x \cos^2 3x \, dx$

15. $\int \tan^3 2t \sec^4 2t \, dt$

16. $\int (2 - \sqrt{\tan x})^2 \sec^2 x \, dx$

17. $\int \tan^2 x \sec^3 x \, dx$

18. $\int \tan^2 3x \sec^2 3x \, dx$

19. $\int \tan^3 x (\sec x)^{-1/2} \, dx$

20. $\int \tan^3 \frac{x}{2} \sec^3 \frac{x}{2} \, dx$

21. $\int \tan^3 x \sec^5 x \, dx$

22. $\int \tan^5 x \sec x \, dx$

23. $\int \sec^5 x \, dx$

24. $\int \frac{1}{\cos^4 x} \, dx$

25. $\int \cos^2 x \cot x \, dx$

26. $\int \sin x \sec^7 x \, dx$

27. $\int \cot^{10} x \csc^4 x \, dx$

28. $\int (1 + \csc^2 t)^2 \, dt$

29. $\int \frac{\sec^4 (1 - t)}{\tan^8 (1 - t)} \, dt$

30. $\int \frac{\sin^3 \sqrt{t} \cos^2 \sqrt{t}}{\sqrt{t}} \, dt$

31. $\int (1 + \tan x)^2 \sec x \, dx$

32. $\int (\tan x + \cot x)^2 \, dx$

33. $\int \tan^4 x \, dx$

34. $\int \tan^5 x \, dx$

35. $\int \cot^3 t \, dt$

36. $\int \csc^5 t \, dt$

37. $\int (\tan^6 x - \tan^2 x) \, dx$

38. $\int \cot 2x \csc^{5/2} 2x \, dx$

39. $\int x \sin^3 x^2 \, dx$

40. $\int x \tan^8 (x^2) \sec^2 (x^2) \, dx$

In Problems 41–46, evaluate the given definite integral.

41. $\int_{\pi/3}^{\pi/2} \sin^3 \theta \sqrt{\cos \theta} \, d\theta$

42. $\int_0^{\pi/2} \sin^5 x \cos^5 x \, dx$

43. $\int_0^{\pi} \sin^3 2t \, dt$

44. $\int_{-\pi}^{\pi} \sin^4 x \cos^2 x \, dx$

45. $\int_0^{\pi/4} \tan y \sec^4 y \, dy$

46. $\int_0^{\pi/3} \tan x \sec^{3/2} x \, dx$

In Problems 47–52, use the trigonometric identities

$$\sin mx \cos nx = \frac{1}{2} [\sin(m + n)x + \sin(m - n)x]$$

$$\sin mx \sin nx = \frac{1}{2} [\cos(m - n)x - \cos(m + n)x]$$

$$\cos mx \cos nx = \frac{1}{2} [\cos(m - n)x + \cos(m + n)x]$$

to evaluate the given trigonometric integral.

47. $\int \sin x \cos 2x \, dx$

48. $\int \cos 3x \cos 5x \, dx$

49. $\displaystyle\int \sin 2x \sin 4x \, dx$

50. $\displaystyle\int \frac{5 - 3\sin 2x}{\sec 6x} \, dx$

51. $\displaystyle\int_0^{\pi/6} \cos 2x \cos x \, dx$

52. $\displaystyle\int_0^{\pi/2} \sin\frac{3}{2}x \sin\frac{1}{2}x \, dx$

53. Show that

$$\int_{-\pi}^{\pi} \sin mx \sin nx \, dx = \begin{cases} 0, & m \neq n \\ \pi, & m = n. \end{cases}$$

54. Evaluate $\displaystyle\int_{-\pi}^{\pi} \sin mx \cos nx \, dx$.

≡ **Review of Applications**

In Problems 55 and 56, the graphs of $f(x) = \sec^2(x/2)$ (Problem 55) and $f(x) = \sin^2 x$ (Problem 56) are given. Find the volume of the solid of revolution obtained by revolving the region R bounded by the graph of f on the interval $[-\pi/2, \pi/2]$ about the indicated axis.

55. the x-axis;

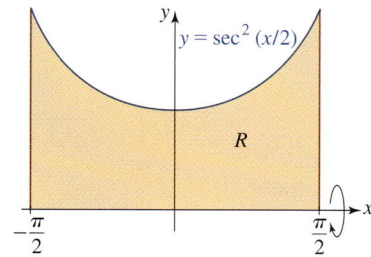

FIGURE 7.4.1 Region in Problem 55

56. the line $y = 1$;

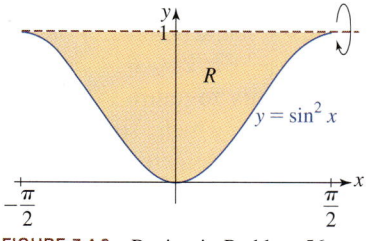

FIGURE 7.4.2 Region in Problem 56

57. Find the area of the region R bounded by the graphs of $y = \sin^3 x$ and $y = \cos^3 x$ on the interval $[-3\pi/4, \pi/4]$. See FIGURE 7.4.3.

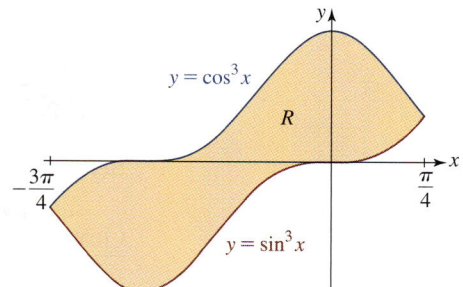

FIGURE 7.4.3 Graphs for Problem 57

58. The graph of the equation $r = |\sin 4\theta \sin\frac{1}{2}\theta|$, $0 \leq \theta \leq 2\pi$, encloses a region that is a mathematical model for the shape of a horse chestnut leaf. See FIGURE 7.4.4. We shall see in Chapter 10 that the area A bounded by this graph is given by $A = \frac{1}{2}\int_0^{2\pi} r^2 \, d\theta$. Find this area. [*Hint*: Use one of the identities given in the instructions for Problems 47–52.]

FIGURE 7.4.4 Region in Problem 58   Horse chestnut leaves

≡ **Calculator/CAS Problems**

59. Use a calculator or CAS to obtain the graphs of $y = \cos^3 x$, $y = \cos^5 x$, and $y = \cos^7 x$ on the interval $[0, \pi]$. Use the graphs to conjecture the values of the definite integrals

$$\int_0^{\pi} \cos^3 x \, dx, \quad \int_0^{\pi} \cos^5 x \, dx, \quad \text{and} \quad \int_0^{\pi} \cos^7 x \, dx.$$

60. In Problem 59, what do you think is the value of $\int_0^{\pi} \cos^n x \, dx$, where n is a positive odd integer? Prove your conjecture.

7.5 Trigonometric Substitutions

▌ **Introduction** When an integrand contains integer powers of x and integer powers of

$$\sqrt{a^2 - u^2}, \quad \sqrt{a^2 + u^2}, \quad \text{or} \quad \sqrt{u^2 - a^2}, a > 0 \tag{1}$$

we may be able to evaluate the integral by means of a trigonometric substitution. The three cases we shall consider in this section depend, in turn, on the fundamental Pythagorean identities written in the form:

$$1 - \sin^2\theta = \cos^2\theta, \quad 1 + \tan^2\theta = \sec^2\theta, \quad \text{and} \quad \sec^2\theta - 1 = \tan^2\theta.$$

The procedure for an indefinite integral is similar to the discussion in Sections 5.2 and 7.2:

- Make a substitution in an integral.
- After simplification, carry out the integration with respect to the new variable.
- Return to the original variable by resubstitution.

Before proceeding let us match the type of trigonometric substitution with the radicals in (1).

Guidelines for Trigonometric Substitutions

For integrands containing

- $\sqrt{a^2 - u^2}$, $a > 0$, let $u = a\sin\theta$, where $-\pi/2 \leq \theta \leq \pi/2$.
- $\sqrt{a^2 + u^2}$, $a > 0$, let $u = a\tan\theta$, where $-\pi/2 < \theta < \pi/2$.
- $\sqrt{u^2 - a^2}$, $a > 0$, let $u = a\sec\theta$, where $\begin{cases} 0 \leq \theta < \pi/2, & \text{if } u \geq a \\ \pi/2 < \theta \leq \pi, & \text{if } u \leq -a. \end{cases}$

See Section 1.5 for a review of inverse trigonometric functions. ▶

In each case, the restriction given on the variable θ is precisely the one that accompanies the corresponding inverse trigonometric function. In other words, we want to be able to write $\theta = \sin^{-1}(u/a)$, and so on. Moreover, with the aid of the foregoing identities each of these substitutions yields a perfect square. With the restriction on θ for the substitutions $u = a\sin\theta$ and $u = a\tan\theta$, the square root may be taken without recourse to absolute values. As we shall see, we have to be more careful using the substitution $u = a\sec\theta$.

- If $u = a\sin\theta$, where $-\pi/2 \leq \theta \leq \pi/2$, then

$$\sqrt{a^2 - u^2} = \sqrt{a^2 - a^2\sin^2\theta} = \sqrt{a^2(1 - \sin^2\theta)} = \sqrt{a^2\cos^2\theta} = a\cos\theta.$$

- If $u = a\tan\theta$, where $-\pi/2 < \theta < \pi/2$, then

$$\sqrt{a^2 + u^2} = \sqrt{a^2 + a^2\tan^2\theta} = \sqrt{a^2(1 + \tan^2\theta)} = \sqrt{a^2\sec^2\theta} = a\sec\theta.$$

- If $u = a\sec\theta$, where $0 \leq \theta < \pi/2$ or $\pi/2 < \theta \leq \pi$, then

$$\sqrt{u^2 - a^2} = \sqrt{a^2\sec^2\theta - a^2} = \sqrt{a^2(\sec^2\theta - 1)} = \sqrt{a^2\tan^2\theta} = a|\tan\theta|.$$

When an expression such as $\sqrt{a^2 - u^2}$ appears in the denominator of an integrand, there is the further restriction on the variable θ; in this case $-\pi/2 < \theta < \pi/2$.

After carrying out the integration in θ it is necessary to return to the original variable x. If we construct a reference right triangle, one in which $\sin\theta = u/a$, $\tan\theta = u/a$, or $\sec\theta = u/a$ as shown in FIGURE 7.5.1, then the other trigonometric functions can be readily expressed in terms of u.

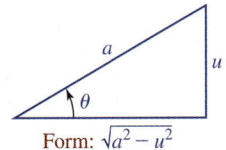
Form: $\sqrt{a^2 - u^2}$
Substitution: $u = a\sin\theta$

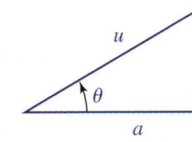
Form: $\sqrt{a^2 + u^2}$
Substitution: $u = a\tan\theta$

Form: $\sqrt{u^2 - a^2}$
Substitution: $u = a\sec\theta$

FIGURE 7.5.1 Reference right triangles used to express trigonometric functions in terms of algebraic expression in u and a

In the first two examples, the integrands contain the radical form $\sqrt{a^2 - u^2}$.

EXAMPLE 1　Using a Sine Substitution

Evaluate $\displaystyle\int \frac{x^2}{\sqrt{9 - x^2}}\, dx$.

Solution　Identifying $u = x$ and $a = 3$ leads to the substitutions

$$x = 3\sin\theta \quad \text{and} \quad dx = 3\cos\theta\, d\theta,$$

where $-\pi/2 < \theta < \pi/2$. The integral becomes

$$\int \frac{x^2}{\sqrt{9-x^2}}\,dx = \int \frac{9\sin^2\theta}{\sqrt{9-9\sin^2\theta}}(3\cos\theta\,d\theta) \quad \leftarrow \text{simplify}$$

$$= 9\int \sin^2\theta\,d\theta.$$

Recall, to evaluate this last trigonometric integral, we make use of the half-angle identity $\sin^2\theta = \frac{1}{2}(1 - \cos 2\theta)$:

$$\int \frac{x^2}{\sqrt{9-x^2}}\,dx = \frac{9}{2}\int (1 - \cos 2\theta)\,d\theta$$

$$= \frac{9}{2}\theta - \frac{9}{4}\sin 2\theta + C$$

$$= \frac{9}{2}\theta - \frac{9}{2}\sin\theta\cos\theta + C. \quad \leftarrow \text{double-angle formula}$$

In order to express this result back in terms of the variable x, we use $\sin\theta = x/3$ and $\theta = \sin^{-1}(x/3)$. Then from the reference right triangle in FIGURE 7.5.2 we see that $\cos\theta = \sqrt{9-x^2}/3$, and so

$$\int \frac{x^2}{\sqrt{9-x^2}}\,dx = \frac{9}{2}\sin^{-1}\left(\frac{x}{3}\right) - \frac{1}{2}x\sqrt{9-x^2} + C. \quad \blacksquare$$

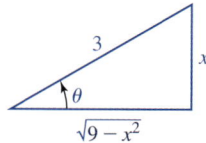

FIGURE 7.5.2 Right triangle in Example 1

EXAMPLE 2   Using a Sine Substitution

Evaluate $\displaystyle\int \frac{\sqrt{1-4x^2}}{x}\,dx$.

Solution   With the identifications $u = 2x$, $a = 1$ we let $2x = \sin\theta$, and so $x = \frac{1}{2}\sin\theta$, and $dx = \frac{1}{2}\cos\theta\,d\theta$. Then

$$\int \frac{\sqrt{1-4x^2}}{x}\,dx = \int \frac{\sqrt{1-\sin^2\theta}}{\frac{1}{2}\sin\theta}\left(\frac{1}{2}\cos\theta\,d\theta\right) \quad \leftarrow \text{simplify}$$

$$= \int \frac{\cos^2\theta}{\sin\theta}\,d\theta$$

$$= \int \frac{1-\sin^2\theta}{\sin\theta}\,d\theta \quad \leftarrow \text{use termwise division}$$

$$= \int (\csc\theta - \sin\theta)\,d\theta \quad \leftarrow \text{formulas 16 and 7 in Table 7.1.1}$$

$$= \ln|\csc\theta - \cot\theta| + \cos\theta + C.$$

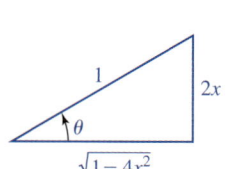

FIGURE 7.5.3 Right triangle in Example 2

In FIGURE 7.5.3 we have constructed a right triangle for which $\sin\theta = 2x$ and $\cos\theta = \sqrt{1-4x^2}$. Therefore, $\csc\theta = 1/\sin\theta = 1/(2x)$ and $\cot\theta = \cos\theta/\sin\theta = \sqrt{1-4x^2}/(2x)$. Hence, the result obtained by integrating with respect to θ can be written in terms of x as

$$\int \frac{\sqrt{1-4x^2}}{x}\,dx = \ln\left|\frac{1-\sqrt{1-4x^2}}{2x}\right| + \sqrt{1-4x^2} + C. \quad \blacksquare$$

As a refresher in differentiation skills, you are encouraged to verify periodically that the derivative of the antiderivative you have obtained is the integrand in the original integral. The derivative of the final answer in Example 2,

$$\frac{d}{dx}\left[\ln\left|\frac{1-\sqrt{1-4x^2}}{2x}\right| + \sqrt{1-4x^2} + C\right] = -\frac{-1+4x^2+\sqrt{1-4x^2}}{x(-1+\sqrt{1-4x^2})},$$

is on the face of it not the integrand in Example 2. Use algebra to resolve the difference between this result and the integrand $\sqrt{1-4x^2}/x$.

In the next two examples, the integrands contain an integer power of the radical form $\sqrt{a^2 + u^2}$.

EXAMPLE 3 Using a Tangent Substitution

Evaluate $\displaystyle\int \frac{1}{(4 + x^2)^{3/2}} \, dx$.

Solution Observe that the integrand is an integer power of $\sqrt{4 + x^2}$, since $(4 + x^2)^{3/2} = \left(\sqrt{4 + x^2}\right)^3$. Now, when $u = x$, $x = 2\tan\theta$ and $dx = 2\sec^2\theta \, d\theta$ we have $\sqrt{4 + x^2} = \sqrt{4\sec^2\theta} = 2\sec\theta$ and $(4 + x^2)^{3/2} = 8\sec^3\theta$. Thus,

$$\int \frac{1}{(4 + x^2)^{3/2}} \, dx = \int \frac{1}{8\sec^3\theta} (2\sec^2\theta \, d\theta)$$

$$= \frac{1}{4} \int \cos\theta \, d\theta$$

$$= \frac{1}{4}\sin\theta + C.$$

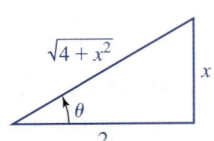

FIGURE 7.5.4 Right triangle in Example 3

From the triangle in **FIGURE 7.5.4**, we see that $\sin\theta = x/\sqrt{4 + x^2}$. Hence,

$$\int \frac{1}{(4 + x^2)^{3/2}} \, dx = \frac{1}{4} \frac{x}{\sqrt{4 + x^2}} + C.$$

$\blacksquare$

EXAMPLE 4 Arc Length

Find the length of the graph of $y = \frac{1}{2}x^2 + 3$ on the interval $[0, 1]$.

Solution Recall that the formula for arc length is $L = \displaystyle\int_a^b \sqrt{1 + [f'(x)]^2} \, dx$. Since $dy/dx = x$, we have

$$L = \int_0^1 \sqrt{1 + x^2} \, dx.$$

Now if $u = x$, then we substitute $x = \tan\theta$ and $dx = \sec^2\theta \, d\theta$. The θ-limits of integration in the resulting definite trigonometric integral are obtained from the x-limits in the original integral:

$$x = 0: \qquad \theta = \tan^{-1}0 = 0$$

and

$$x = 1: \qquad \theta = \tan^{-1}1 = \pi/4.$$

Therefore,

$$L = \int_0^{\pi/4} \sqrt{1 + \tan^2\theta} \, \sec^2\theta \, d\theta$$

$$= \int_0^{\pi/4} \sec^3\theta \, d\theta.$$

The indefinite integral of $\sec^3\theta$ was found in Example 5 of Section 7.3 using integration by parts:

$$L = \left(\frac{1}{2}\sec\theta\tan\theta + \frac{1}{2}\ln|\sec\theta + \tan\theta| \right)\Bigg]_0^{\pi/4}$$

$$= \frac{1}{2}\sec\frac{\pi}{4}\tan\frac{\pi}{4} + \frac{1}{2}\ln\left|\sec\frac{\pi}{4} + \tan\frac{\pi}{4}\right|$$

$$= \frac{1}{2}\sqrt{2} + \frac{1}{2}\ln(\sqrt{2} + 1) \approx 1.1478.$$

$\blacksquare$

In the next two examples, the integrands contain an integer power of the radical form $\sqrt{u^2 - a^2}$.

EXAMPLE 5 Using a Secant Substitution

Evaluate $\displaystyle\int \frac{\sqrt{x^2 - 16}}{x^4} \, dx$ assuming that $x > 4$.

Solution If we let $u = x$ and $x = 4\sec\theta$, $0 \le \theta < \pi/2$, and $dx = 4\sec\theta\tan\theta\, d\theta$, the integral becomes

$$\begin{aligned}\int \frac{\sqrt{x^2-16}}{x^4}\,dx &= \int \frac{\sqrt{16\sec^2\theta - 16}}{256\sec^4\theta}(4\sec\theta\tan\theta\, d\theta)\\ &= \frac{1}{16}\int \frac{\tan^2\theta}{\sec^3\theta}\,d\theta\\ &= \frac{1}{16}\int \frac{\sin^2\theta}{\cos^2\theta}\cos^3\theta\, d\theta\\ &= \frac{1}{16}\int (\sin\theta)^2(\cos\theta\, d\theta)\\ &= \frac{1}{48}\sin^3\theta + C.\end{aligned}$$

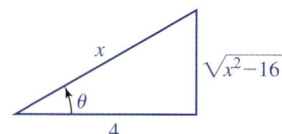

FIGURE 7.5.5 Right triangle in Example 5

The right triangle in FIGURE 7.5.5 was constructed so that $\sec\theta = x/4$ or $\cos\theta = 4/x$ and so we see that $\sin\theta = \sqrt{x^2-16}/x$. Because $\sin^3 x = \left(\sqrt{x^2-16}\right)^3/x^3$, it follows that

$$\int \frac{\sqrt{x^2-16}}{x^4}\,dx = \frac{1}{48}\frac{(x^2-16)^{3/2}}{x^3} + C.$$

■

EXAMPLE 6 A Definite Integral

Evaluate $\displaystyle\int_{-2}^{-1} \frac{\sqrt{x^2-1}}{x}\,dx.$

Solution We identify $u = x$ and $a = 1$ and use the substitutions $x = \sec\theta$ and $dx = \sec\theta\tan\theta\, d\theta$. In this case we must assume that $\pi/2 < \theta \le \pi$ since the interval of integration indicates that $x \le -a$, where $-a = -1$. See FIGURE 7.5.6 for a graph of the integrand $f(x) = \sqrt{x^2-1}/x$.

FIGURE 7.5.6 Graph of integrand in Example 6

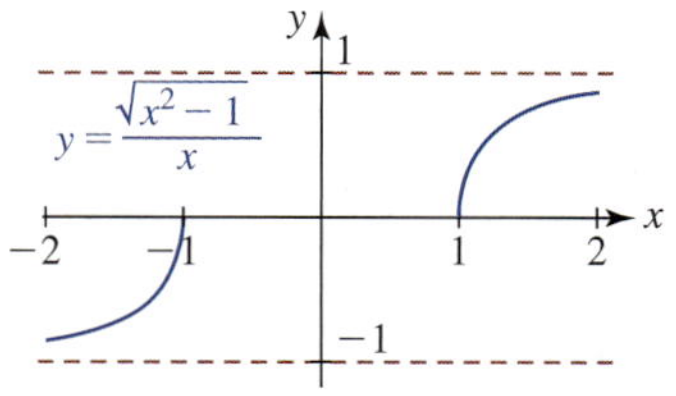

As in Example 4 we obtain the θ-limits of integration from the original x-limits of integration:

$$x = -2: \quad \theta = \sec^{-1}(-2) = \frac{2\pi}{3}$$
$$x = -1: \quad \theta = \sec^{-1}(-1) = \pi.$$

Therefore,
$$\begin{aligned}\int_{-2}^{-1}\frac{\sqrt{x^2-1}}{x}\,dx &= \int_{2\pi/3}^{\pi}\frac{\sqrt{\sec^2\theta - 1}}{\sec\theta}(\sec\theta\tan\theta\, d\theta)\\ &= \int_{2\pi/3}^{\pi}\sqrt{\tan^2\theta}(\tan\theta\, d\theta).\end{aligned}$$

Because $\pi/2 < \theta \le \pi$, $\tan\theta \le 0$, $\sqrt{\tan^2\theta} = |\tan\theta| = -\tan\theta$, the last integral becomes

$$\begin{aligned}\int_{-2}^{-1}\frac{\sqrt{x^2-1}}{x}\,dx &= \int_{2\pi/3}^{\pi}\sqrt{\tan^2\theta}(\tan\theta\, d\theta)\\ &= -\int_{2\pi/3}^{\pi}\tan^2\theta\, d\theta \quad \leftarrow \text{use trig identity}\\ &= -\int_{2\pi/3}^{\pi}(\sec^2\theta - 1)\, d\theta\\ &= -(\tan\theta - \theta)\Big]_{2\pi/3}^{\pi}\\ &= -\left[(0 - \pi) - \left(-\sqrt{3} - \frac{2\pi}{3}\right)\right]\\ &= \frac{\pi}{3} - \sqrt{3} \approx -0.6849.\end{aligned}$$

The negative answer stands to reason, since we see in Figure 7.5.6 that $f(x) \le 0$ on the interval $[-2, -1]$.

■

■ **Integrands Containing a Quadratic** A trigonometric substitution can also be used when an integrand contains an integer power of the square root of a quadratic expression $ax^2 + bx + c$. By completion of the square the radical can be expressed as one of the forms:

$$\sqrt{a^2 - u^2}, \quad \sqrt{a^2 + u^2}, \quad \text{or} \quad \sqrt{u^2 - a^2}.$$

If, say, an integrand contains an integer power of

$$\sqrt{x^2 + 4x + 7} = \sqrt{3 + (x + 2)^2},$$

we would identify $u = x + 2$, $a = \sqrt{3}$, and use the substitution $x + 2 = \sqrt{3}\tan\theta$.

EXAMPLE 7 Completing the Square

Evaluate $\displaystyle\int \frac{1}{(x^2 + 8x + 25)^{3/2}}\, dx$.

Solution By completing the square in x, we recognize that the integrand contains an integral power of $a^2 + u^2$,

$$\int \frac{1}{(x^2 + 8x + 25)^{3/2}}\, dx = \int \frac{1}{[9 + (x + 4)^2]^{3/2}}\, dx,$$

where $u = x + 4$ and $a = 3$. Using the substitutions $x + 4 = 3\tan\theta$ and $dx = 3\sec^2\theta\, d\theta$ we find

$$\int \frac{dx}{(x^2 + 8x + 25)^{3/2}} = \int \frac{3\sec^2\theta\, d\theta}{[9 + 9\tan^2\theta]^{3/2}}$$

$$= \frac{1}{9}\int \frac{\sec^2\theta}{\sec^3\theta}\, d\theta$$

$$= \frac{1}{9}\int \cos\theta\, d\theta$$

$$= \frac{1}{9}\sin\theta + C.$$

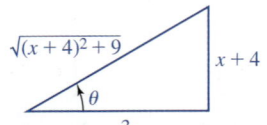

FIGURE 7.5.7 Right triangle in Example 7

Inspection of the triangle in **FIGURE 7.5.7** indicates how to express $\sin\theta$ in terms of x. It follows that

$$\int \frac{dx}{(x^2 + 8x + 25)^{3/2}} = \frac{1}{9}\frac{x + 4}{\sqrt{(x + 4)^2 + 9}} + C = \frac{x + 4}{9\sqrt{x^2 + 8x + 25}} + C. \qquad ■$$

∫ **NOTES FROM THE CLASSROOM**

Integrals of the form

$$\int \frac{1}{\sqrt{u^2 + a^2}}\, du, \quad \int \frac{1}{\sqrt{u^2 - a^2}}\, du, \quad \text{and} \quad \int \frac{1}{u^2 - a^2}\, du$$

can readily be evaluated by means of trigonometric substitutions. Inspection of Table 7.1.1 of integral formulas shows that each of these integrals is a logarithm. But those with long memories might recognize these are the indefinite integral forms of the differentiation formulas for three of the inverse hyperbolic functions:

$$\frac{d}{dx}\sinh^{-1}\left(\frac{u}{a}\right) = \frac{1}{\sqrt{u^2 + a^2}}\frac{du}{dx}, \qquad \frac{d}{dx}\cosh^{-1}\left(\frac{u}{a}\right) = \frac{1}{\sqrt{u^2 - a^2}}\frac{du}{dx},$$

$$\frac{d}{dx}\frac{1}{a}\tanh^{-1}\left(\frac{u}{a}\right) = \frac{1}{a^2 - u^2}\frac{du}{dx}.$$

Every inverse hyperbolic function can be expressed as a natural logarithm. See (25)–(27) in Section 3.10.

Exercises 7.5   Answers to selected odd-numbered problems begin on page ANS-23.

≡ **Fundamentals**

In Problems 1–38, evaluate the given indefinite integral by a trigonometric substitution where appropriate. You should be able to evaluate some of the integrals without a substitution.

1. $\displaystyle\int \frac{\sqrt{1-x^2}}{x^2}\,dx$

2. $\displaystyle\int \frac{x^3}{\sqrt{x^2-4}}\,dx$

3. $\displaystyle\int \frac{1}{\sqrt{x^2-36}}\,dx$

4. $\displaystyle\int \sqrt{3-x^2}\,dx$

5. $\displaystyle\int x\sqrt{x^2+7}\,dx$

6. $\displaystyle\int (1-x^2)^{3/2}\,dx$

7. $\displaystyle\int x^3\sqrt{1-x^2}\,dx$

8. $\displaystyle\int x^3\sqrt{x^2-1}\,dx$

9. $\displaystyle\int \frac{1}{(x^2-4)^{3/2}}\,dx$

10. $\displaystyle\int (9-x^2)^{3/2}\,dx$

11. $\displaystyle\int \sqrt{x^2+4}\,dx$

12. $\displaystyle\int \frac{x}{25+x^2}\,dx$

13. $\displaystyle\int \frac{1}{\sqrt{25-x^2}}\,dx$

14. $\displaystyle\int \frac{1}{x\sqrt{x^2-25}}\,dx$

15. $\displaystyle\int \frac{1}{x\sqrt{16-x^2}}\,dx$

16. $\displaystyle\int \frac{1}{x^2\sqrt{16-x^2}}\,dx$

17. $\displaystyle\int \frac{1}{x\sqrt{1+x^2}}\,dx$

18. $\displaystyle\int \frac{1}{x^2\sqrt{1+x^2}}\,dx$

19. $\displaystyle\int \frac{\sqrt{1-x^2}}{x^4}\,dx$

20. $\displaystyle\int \frac{\sqrt{x^2-1}}{x^4}\,dx$

21. $\displaystyle\int \frac{x^2}{(9-x^2)^{3/2}}\,dx$

22. $\displaystyle\int \frac{x^2}{(4+x^2)^{3/2}}\,dx$

23. $\displaystyle\int \frac{1}{(1+x^2)^2}\,dx$

24. $\displaystyle\int \frac{x^2}{(x^2-1)^2}\,dx$

25. $\displaystyle\int \frac{1}{(4+x^2)^{5/2}}\,dx$

26. $\displaystyle\int \frac{x^3}{(1-x^2)^{5/2}}\,dx$

27. $\displaystyle\int \frac{1}{\sqrt{x^2+2x+10}}\,dx$

28. $\displaystyle\int \frac{x}{\sqrt{4x-x^2}}\,dx$

29. $\displaystyle\int \frac{1}{(x^2+6x+13)^2}\,dx$

30. $\displaystyle\int \frac{1}{(11-10x-x^2)^2}\,dx$

31. $\displaystyle\int \frac{x-3}{(5-4x-x^2)^{3/2}}\,dx$

32. $\displaystyle\int \frac{1}{(x^2+2x)^{3/2}}\,dx$

33. $\displaystyle\int \frac{2x+4}{x^2+4x+13}\,dx$

34. $\displaystyle\int \frac{1}{4+(x-3)^2}\,dx$

35. $\displaystyle\int \frac{x^2}{x^2+16}\,dx$

36. $\displaystyle\int \frac{\sqrt{4-9x^2}}{x}\,dx$

37. $\displaystyle\int \sqrt{6x-x^2}\,dx$

38. $\displaystyle\int \frac{1}{\sqrt{6x-x^2}}\,dx$

In Problems 39–44, evaluate the given definite integral.

39. $\displaystyle\int_{-1}^{1} \sqrt{4-x^2}\,dx$

40. $\displaystyle\int_{-1}^{\sqrt{3}} \frac{x^2}{\sqrt{4-x^2}}\,dx$

41. $\displaystyle\int_{0}^{5} \frac{1}{(x^2+25)^{3/2}}\,dx$

42. $\displaystyle\int_{\sqrt{2}}^{2} \frac{1}{x^3\sqrt{x^2-1}}\,dx$

43. $\displaystyle\int_{1}^{6/5} \frac{16}{x^4\sqrt{4-x^2}}\,dx$

44. $\displaystyle\int_{0}^{1/2} x^3(1+x^2)^{-1/2}\,dx$

In Problems 45 and 46, use integration by parts followed by a trigonometric substitution.

45. $\displaystyle\int x^2\sin^{-1}x\,dx$

46. $\displaystyle\int x\cos^{-1}x\,dx$

≡ **Review of Applications**

47. Use a calculator or CAS to obtain the graph of $y=\dfrac{1}{x\sqrt{3+x^2}}$. Find the area under the graph on the interval $\left[1,\sqrt{3}\right]$.

48. Use a calculator or CAS to obtain the graph of $y=x^5\sqrt{1-x^2}$. Find the area under the graph on the interval $[0, 1]$.

49. Show that the area of a circle given by $x^2+y^2=a^2$ is πa^2.

50. Show that the area of an ellipse given by $a^2x^2+b^2y^2=a^2b^2$ is πab.

51. The region described in Problem 47 is revolved about the x-axis. Find the volume of the solid of revolution.

52. The region in the first quadrant bounded by the graphs of $y=\dfrac{4}{4+x^2}$, $x=2$, and $y=0$ is revolved about the x-axis. Find the volume of the solid of revolution.

53. The region in the first quadrant bounded by the graphs of $y=x\sqrt{4+x^2}$, $x=2$, and $y=0$ is revolved about the y-axis. Find the volume of the solid of revolution.

54. The region in the first quadrant bounded by the graphs of $y=\dfrac{x}{\sqrt{4-x^2}}$, $x=1$, and $y=0$ is revolved about the y-axis. Find the volume of the solid of revolution.

55. Find the length of the graph of $y=\ln x$ on the interval $\left[1,\sqrt{3}\right]$.

56. Find the length of the graph of $y=-\frac{1}{2}x^2+2x$ on the interval $[1, 2]$.

57. A woman, W, starting at the origin, moves in the direction of the positive y-axis pulling a mass along the curve C, called a **tractrix**, indicated in FIGURE 7.5.8. The mass, initially located on the x-axis at $(a, 0)$, is pulled by

a rope of constant length a that is kept taut throughout the motion.

(a) Show that the differential equation of the tractrix is

$$\frac{dy}{dx} = -\frac{\sqrt{a^2 - x^2}}{x}.$$

(b) Solve the equation in part (a). Assume that the initial point on the x-axis is $(10, 0)$ and the length of the rope is $a = 10$ ft.

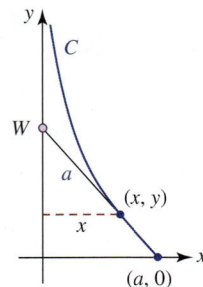

FIGURE 7.5.8 Tractrix in Problem 57

58. The region bounded by the graph of $(x - a)^2 + y^2 = r^2$, $r < a$, is revolved about the y-axis. Find the volume of the solid of revolution or **torus**. See FIGURE 7.5.9.

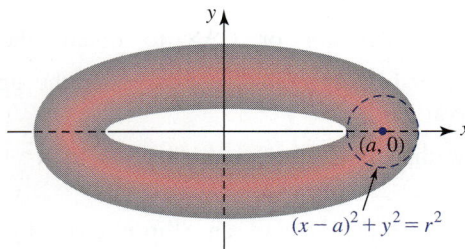

FIGURE 7.5.9 Torus in Problem 58

59. Find the fluid force on one side of the vertical plate shown in FIGURE 7.5.10. Assume that the plate is submerged in water and that dimensions are in feet.

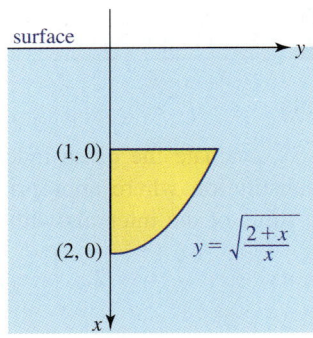

FIGURE 7.5.10 Submerged plate in Problem 59

60. Find the centroid of the region bounded by the graphs of

$$y = \frac{1}{\sqrt{1 + x^2}}, y = 0, x = 0, \text{ and } x = \sqrt{3}.$$

≣ **Think About It**

61. Evaluate the following integrals by an appropriate trigonometric substitution.

(a) $\displaystyle\int \frac{1}{\sqrt{e^{2x} - 1}} \, dx$ (b) $\displaystyle\int \sqrt{e^{2x} - 1} \, dx$

62. Find the area of the crescent-shaped region shown in yellow in FIGURE 7.5.11. The region, outside the circle of radius a but inside the circle of radius b, $a \neq b$, is called a **lune**.

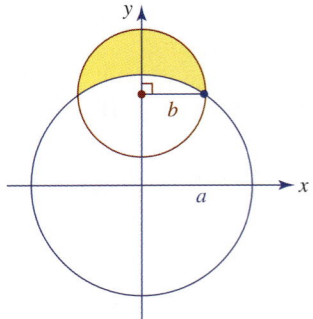

FIGURE 7.5.11 Lune in Problem 62

7.6 Partial Fractions

▌ **Introduction** When two rational functions, say, $g(x) = 2/(x + 5)$ and $h(x) = 1/(x + 1)$ are added, the terms are combined by means of a common denominator:

$$g(x) + h(x) = \frac{2}{x + 5} + \frac{1}{x + 1} = \frac{2}{x + 5}\left(\frac{x + 1}{x + 1}\right) + \frac{1}{x + 1}\left(\frac{x + 5}{x + 5}\right). \tag{1}$$

Adding numerators on the right-hand side of (1) yields the single rational function

$$f(x) = \frac{3x + 7}{(x + 5)(x + 1)}. \tag{2}$$

Now suppose that we are faced with the problem of integrating the function f. Of course, the solution is obvious: We use the equality of (1) and (2) to write

$$\int \frac{3x + 7}{(x + 5)(x + 1)} \, dx = \int \left[\frac{2}{x + 5} + \frac{1}{x + 1}\right] dx = 2\ln|x + 5| + \ln|x + 1| + C.$$

This example illustrates a procedure for integrating certain rational functions $f(x) = p(x)/q(x)$. This method consists of reversing the process illustrated in (1), in other words, starting with a rational function, such as (2), break it down into simpler component fractions $g(x) = 2/(x + 5)$ and $h(x) = 1/(x + 1)$ called **partial fractions**. We then evaluate the integral term-by-term.

■ **Partial Fractions** The algebraic process for breaking down a rational expression, such as (2), into partial fractions is known as **partial fraction decomposition**. For convenience we will assume that the rational function $f(x) = p(x)/q(x)$, $q(x) \neq 0$, is a **proper fraction** or **proper rational expression**, that is, the degree of $p(x)$ is less than the degree of $q(x)$. We will also assume that the polynomials $p(x)$ and $q(x)$ have no common factors.

In this section, we shall study four cases of partial fraction decomposition.

■ **Distinct Linear Factors** We state the following fact from algebra without proof. If the denominator $q(x)$ contains a product of n distinct linear factors,

$$(a_1x + b_1)(a_2x + b_2) \cdots (a_nx + b_n),$$

where the a_i and b_i, $i = 1, 2, \ldots, n$ are real numbers, then unique real constants $C_1, C_2, \ldots, C_n$ can be found such that the partial fraction decomposition of $f(x) = p(x)/q(x)$ contains the sum

$$\frac{C_1}{a_1x + b_1} + \frac{C_2}{a_2x + b_2} + \cdots + \frac{C_n}{a_nx + b_n}.$$

In other words, the assumed partial fraction decomposition for f contains one partial fraction for each of the linear factors $a_ix + b_i$.

EXAMPLE 1   Distinct Linear Factors

Evaluate $\displaystyle\int \frac{2x + 1}{(x - 1)(x + 3)}\, dx$.

Solution We make the assumption that the integrand can be written as

$$\frac{2x + 1}{(x - 1)(x + 3)} = \frac{A}{x - 1} + \frac{B}{x + 3}.$$

Combining the terms of the right-hand member of the equation over a common denominator gives

$$\frac{2x + 1}{(x - 1)(x + 3)} = \frac{A(x + 3) + B(x - 1)}{(x - 1)(x + 3)}.$$

Since the denominators are equal, the numerators of the two expressions must be identical:

$$2x + 1 = A(x + 3) + B(x - 1). \tag{3}$$

Since the last line is an identity, the coefficients of the powers of x are the same

$$2x + 1x^0 = (A + B)x + (3A - B)x^0$$

and therefore,

$$2 = A + B$$
$$1 = 3A - B. \tag{4}$$

By adding the two equations we get $3 = 4A$ and so we find that $A = \frac{3}{4}$. Substituting this value into either equation of (4) then yields $B = \frac{5}{4}$. Hence the desired partial fraction decomposition is

$$\frac{2x + 1}{(x - 1)(x + 3)} = \frac{\frac{3}{4}}{x - 1} + \frac{\frac{5}{4}}{x + 3}.$$

Therefore,

$$\int \frac{2x + 1}{(x - 1)(x + 3)} \, dx = \int \left[\frac{\frac{3}{4}}{x - 1} + \frac{\frac{5}{4}}{x + 3} \right] dx = \frac{3}{4} \ln|x - 1| + \frac{5}{4} \ln|x + 3| + C. \quad \blacksquare$$

■ **A Shortcut Worth Knowing** If the denominator contains, say, three linear factors, such as in $\dfrac{4x^2 - x + 1}{(x - 1)(x + 3)(x - 6)}$, then the partial fraction decomposition looks like this:

$$\frac{4x^2 - x + 1}{(x - 1)(x + 3)(x - 6)} = \frac{A}{x - 1} + \frac{B}{x + 3} + \frac{C}{x - 6}.$$

By following the same steps as in Example 1, we would find that the analogue of (4) is now three equations in the three unknowns A, B, and C. The point is this: The more linear factors in the denominator, the larger the system of equations we must solve. There is an algebraic procedure worth learning that can cut down on some of the algebra. To illustrate, let us return to the identity (3). Since the equality is true for every value of x, it holds for $x = 1$ and $x = -3$, *the zeros of the denominator*. Setting $x = 1$ in (3) gives $3 = 4A$, from which it follows immediately that $A = \frac{3}{4}$. Similarly, by setting $x = -3$ in (3), we obtain $-5 = (-4)B$ or $B = \frac{5}{4}$.

See the *Notes from the Classroom* at the end of this section for another quick method for determining the constants.

EXAMPLE 2   Area Under a Graph

Find the area A under the graph of $f(x) = \dfrac{1}{x(x + 1)}$ on the interval $\left[\frac{1}{2}, 2\right]$.

Solution   The area in question is shown in FIGURE 7.6.1. Since $f(x)$ is positive for all x in the interval, the area is the definite integral

$$A = \int_{1/2}^{2} \frac{1}{x(x + 1)} \, dx.$$

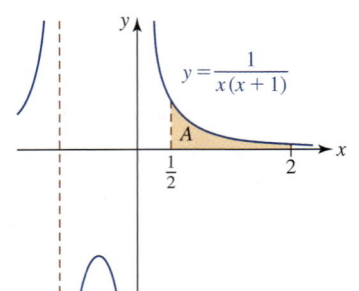

$y = \dfrac{1}{x(x + 1)}$

FIGURE 7.6.1 Area under graph in Example 2

Using partial fractions

$$\frac{1}{x(x + 1)} = \frac{A}{x} + \frac{B}{x + 1} = \frac{A(x + 1) + Bx}{x(x + 1)}$$

it follows that

$$1 = A(x + 1) + Bx. \tag{5}$$

Following the shortcut discussed prior to this example, we set, in turn, $x = 0$ and $x = -1$ in (5) and obtain $A = 1$ and $B = -1$. Therefore,

$$A = \int_{1/2}^{2} \left[\frac{1}{x} - \frac{1}{x + 1} \right] dx = \left(\ln|x| - \ln|x + 1| \right) \Big|_{1/2}^{2}$$

$$= \ln \left| \frac{x}{x + 1} \right| \Big|_{1/2}^{2} = \ln 2 \approx 0.6931. \quad \blacksquare$$

■ **Repeated Linear Factors** If the denominator of the rational function $f(x) = p(x)/q(x)$ contains a repeated linear factor $(ax + b)^n$, $n > 1$, a and b real numbers, then unique real constants $C_1, C_2, \ldots, C_n$ can be found such that the partial fraction decomposition of f contains the sum

$$\frac{C_1}{ax + b} + \frac{C_2}{(ax + b)^2} + \cdots + \frac{C_n}{(ax + b)^n}.$$

In other words, the assumed partial fraction decomposition for f contains a partial fraction for each power of $ax + b$.

EXAMPLE 3   Repeated Linear Factor

Evaluate $\displaystyle \int \frac{x^2 + 2x + 4}{(x + 1)^3} \, dx.$

Solution   The decomposition of the integrand contains a partial fraction for each of the three powers of $x + 1$:

$$\frac{x^2 + 2x + 4}{(x + 1)^3} = \frac{A}{x + 1} + \frac{B}{(x + 1)^2} + \frac{C}{(x + 1)^3}.$$

Equating numerators gives

$$x^2 + 2x + 4 = A(x + 1)^2 + B(x + 1) + C = Ax^2 + (2A + B)x + (A + B + C). \quad (6)$$

Note that setting $x = -1$ (the single zero of the denominator) in (6) yields only $C = 3$. But the coefficients of x^2 and x in (6) yield the system of equations

$$1 = A$$
$$2 = 2A + B.$$

From these equations we see that $A = 1$ and $B = 0$. Therefore,

$$\int \frac{x^2 + 2x + 4}{(x + 1)^3} \, dx = \int \left[\frac{1}{x + 1} + \frac{3}{(x + 1)^3} \right] dx$$

$$= \int \left[\frac{1}{x + 1} + 3(x + 1)^{-3} \right] dx$$

$$= \ln|x + 1| - \frac{3}{2}(x + 1)^{-2} + D. \qquad \blacksquare$$

When the denominator $q(x)$ contains distinct as well as repeated linear factors, we combine the two cases that we have just considered.

EXAMPLE 4   Repeated Factor and a Distinct Factor

Evaluate $\displaystyle\int \frac{6x - 1}{x^3(2x - 1)} \, dx.$

Solution   Since x is a repeated linear factor in the denominator of the integrand, the assumed partial fraction decomposition contains a partial fraction for each of the three powers of x and one partial fraction for the distinct linear factor $2x - 1$:

$$\frac{6x - 1}{x^3(2x - 1)} = \frac{A}{x} + \frac{B}{x^2} + \frac{C}{x^3} + \frac{D}{2x - 1}.$$

After putting the right-hand side over a common denominator we equate numerators:

$$6x - 1 = Ax^2(2x - 1) + Bx(2x - 1) + C(2x - 1) + Dx^3 \qquad (7)$$
$$= (2A + D)x^3 + (-A + 2B)x^2 + (-B + 2C)x - C. \qquad (8)$$

If we set $x = 0$ and $x = \frac{1}{2}$ in (7), we find $C = 1$ and $D = 16$, respectively. Now, by equating the coefficients of x^3 and x^2 in (8), we get

$$0 = 2A + D$$
$$0 = -A + 2B.$$

Since we know the value of D, the first equation yields $A = -\frac{1}{2}D = -8$. The second then gives $B = \frac{1}{2}A = -4$. Therefore,

$$\int \frac{6x - 1}{x^3(2x - 1)} \, dx = \int \left[-\frac{8}{x} - \frac{4}{x^2} + \frac{1}{x^3} + \frac{16}{2x - 1} \right] dx$$

$$= -8\ln|x| + 4x^{-1} - \frac{1}{2}x^{-2} + 8\ln|2x - 1| + E$$

$$= 8\ln\left| \frac{2x - 1}{x} \right| + 4x^{-1} - \frac{1}{2}x^{-2} + E. \qquad \blacksquare$$

The word *irreducible* means that the quadratic expression does not factor over the set of real numbers.

▶ **❚ Distinct Quadratic Factors** If the denominator of the rational function $f(x) = p(x)/q(x)$ contains a product of n distinct *irreducible* quadratic factors

$$(a_1x^2 + b_1x + c_1)(a_2x^2 + b_2x + c_2)\cdots(a_nx^2 + b_nx + c_n),$$

where the coefficients a_i, b_i, and c_i, $i = 1, 2, \ldots, n$ are real numbers, then unique real constants $A_1, A_2, \ldots, A_n, B_1, B_2, \ldots, B_n$ can be found such that the partial fraction decomposition for f contains the sum

$$\frac{A_1x + B_1}{a_1x^2 + b_1x + c_1} + \frac{A_2x + B_2}{a_2x^2 + b_2x + c_2} + \cdots + \frac{A_nx + B_n}{a_nx^2 + b_nx + c_n}.$$

Analogous to the case where $q(x)$ contains a product of distinct linear factors, the assumed partial fraction decomposition for f contains one partial fraction for each of the quadratic factors $a_ix^2 + b_ix + c_i$.

EXAMPLE 5 Repeated Linear and a Distinct Quadratic

Evaluate $\displaystyle\int \frac{x + 3}{x^4 + 9x^2}\,dx$.

Solution From $x^4 + 9x^2 = x^2(x^2 + 9)$, we see that the problem combines the irreducible quadratic factor $x^2 + 9$ with the repeated linear factor x. Accordingly, the partial fraction decomposition is

$$\frac{x + 3}{x^2(x^2 + 9)} = \frac{A}{x} + \frac{B}{x^2} + \frac{Cx + D}{x^2 + 9}.$$

Proceeding as usual, we find

$$x + 3 = Ax(x^2 + 9) + B(x^2 + 9) + (Cx + D)x^2 \tag{9}$$

$$= (A + C)x^3 + (B + D)x^2 + 9Ax + 9B. \tag{10}$$

Setting $x = 0$ in (9) yields $B = \frac{1}{3}$. Then (10) gives

$$0 = A + C$$
$$0 = B + D$$
$$1 = 9A.$$

From this system we get $A = \frac{1}{9}$, $C = -\frac{1}{9}$, and $D = -\frac{1}{3}$. This gives

$$\int \frac{x + 3}{x^2(x^2 + 9)}\,dx = \int \left[\frac{\frac{1}{9}}{x} + \frac{\frac{1}{3}}{x^2} + \frac{-\frac{1}{9}x - \frac{1}{3}}{x^2 + 9} \right] dx$$

$$= \int \left[\frac{\frac{1}{9}}{x} + \frac{\frac{1}{3}}{x^2} - \frac{1}{18}\frac{2x}{x^2 + 9} - \frac{1}{3}\frac{1}{x^2 + 9} \right] dx$$

$$= \frac{1}{9}\ln|x| - \frac{1}{3}x^{-1} - \frac{1}{18}\ln(x^2 + 9) - \frac{1}{9}\tan^{-1}\frac{x}{3} + E$$

$$= \frac{1}{18}\ln\frac{x^2}{x^2 + 9} - \frac{1}{3}x^{-1} - \frac{1}{9}\tan^{-1}\frac{x}{3} + E. \qquad ■$$

EXAMPLE 6 Distinct Quadratic Factors

Evaluate $\displaystyle\int \frac{4x}{(x^2 + 1)(x^2 + 2x + 3)}\,dx$.

Solution Since each quadratic factor in the denominator of the integrand is irreducible, we write

$$\frac{4x}{(x^2 + 1)(x^2 + 2x + 3)} = \frac{Ax + B}{x^2 + 1} + \frac{Cx + D}{x^2 + 2x + 3}$$

from which we find

$$4x = (Ax + B)(x^2 + 2x + 3) + (Cx + D)(x^2 + 1)$$
$$= (A + C)x^3 + (2A + B + D)x^2 + (3A + 2B + C)x + (3B + D).$$

Since the denominator of the integrand has no real zeros, we compare coefficients of powers of x:

$$0 = A + C$$
$$0 = 2A + B + D$$
$$4 = 3A + 2B + C$$
$$0 = 3B + D.$$

Solving the equations yields $A = 1$, $B = 1$, $C = -1$, and $D = -3$. Therefore,

$$\int \frac{4x}{(x^2 + 1)(x^2 + 2x + 3)} \, dx = \int \left[\frac{x + 1}{x^2 + 1} - \frac{x + 3}{x^2 + 2x + 3} \right] dx.$$

Now, the integral of each term still presents a slight challenge. For the first term in the integrand, we use termwise division to write

$$\frac{x + 1}{x^2 + 1} = \frac{1}{2} \frac{2x}{x^2 + 1} + \frac{1}{x^2 + 1}, \tag{11}$$

and then in the second term we complete the square:

$$\frac{x + 3}{x^2 + 2x + 3} = \frac{x + 1 + 2}{(x + 1)^2 + 2} = \frac{1}{2} \frac{2(x + 1)}{(x + 1)^2 + 2} + \frac{2}{(x + 1)^2 + 2}. \tag{12}$$

In the right-hand members of (11) and (12), we recognize that the integrals of the first and second terms are, respectively, of the forms $\int du/u$ and $\int du/(u^2 + a^2)$. Finally, we obtain

$$\int \frac{4x}{(x^2 + 1)(x^2 + 2x + 3)} \, dx$$

$$= \int \left[\frac{1}{2} \frac{2x}{x^2 + 1} + \frac{1}{x^2 + 1} - \frac{1}{2} \frac{2(x + 1)}{(x + 1)^2 + 2} - \frac{2}{(x + 1)^2 + (\sqrt{2})^2} \right] dx$$

$$= \frac{1}{2} \ln(x^2 + 1) + \tan^{-1}x - \frac{1}{2} \ln[(x + 1)^2 + 2] - \sqrt{2} \tan^{-1}\left(\frac{x + 1}{\sqrt{2}} \right) + E$$

$$= \frac{1}{2} \ln\left(\frac{x^2 + 1}{x^2 + 2x + 3} \right) + \tan^{-1}x - \sqrt{2} \tan^{-1}\left(\frac{x + 1}{\sqrt{2}} \right) + E. \qquad \blacksquare$$

❚ Repeated Quadratic Factors If the denominator of the rational function $f(x) = p(x)/q(x)$ contains a repeated irreducible quadratic factor $(ax^2 + bx + c)^n$, $n > 1$, where a, b, and c are real numbers, then unique real constants $A_1, A_2, \ldots, A_n, B_1, B_2, \ldots, B_n$ can be found so that the partial fraction decomposition of f contains the sum

$$\frac{A_1x + B_1}{ax^2 + bx + c} + \frac{A_2x + B_2}{(ax^2 + bx + c)^2} + \cdots + \frac{A_nx + B_n}{(ax^2 + bx + c)^n}.$$

That is, the assumed partial fraction decomposition for f contains a partial fraction for each power of $ax^2 + bx + c$.

❚ **EXAMPLE 7** Repeated Quadratic Factor

Evaluate $\displaystyle\int \frac{x^2}{(x^2 + 4)^2} \, dx$.

Solution The partial fraction decomposition of the integrand

$$\frac{x^2}{(x^2 + 4)^2} = \frac{Ax + B}{x^2 + 4} + \frac{Cx + D}{(x^2 + 4)^2}$$

leads to

$$x^2 = (Ax + B)(x^2 + 4) + Cx + D$$
$$= Ax^3 + Bx^2 + (4A + C)x + (4B + D)$$

and

$$0 = A$$
$$1 = B$$
$$0 = 4A + C$$
$$0 = 4B + D.$$

From this system we find $A = 0$, $B = 1$, $C = 0$, and $D = -4$. Consequently,

$$\int \frac{x^2}{(x^2 + 4)^2} \, dx = \int \left[\frac{1}{x^2 + 4} - \frac{4}{(x^2 + 4)^2} \right] dx.$$

The integral of the first term is an inverse tangent. However, to evaluate the integral of the second term, we employ the trigonometric substitution $x = 2 \tan \theta$:

$$\int \frac{1}{(x^2 + 4)^2} \, dx = \int \frac{2 \sec^2 \theta \, d\theta}{(4 \tan^2 \theta + 4)^2}$$

$$= \frac{1}{8} \int \frac{\sec^2 \theta}{\sec^4 \theta} \, d\theta = \frac{1}{8} \int \cos^2 \theta \, d\theta$$

$$= \frac{1}{16} \int (1 + \cos 2\theta) \, d\theta = \frac{1}{16} \left(\theta + \frac{1}{2} \sin 2\theta \right) \quad \leftarrow \text{use double-angle formula here}$$

$$= \frac{1}{16} (\theta + \sin \theta \cos \theta)$$

$$= \frac{1}{16} \left[\tan^{-1} \frac{x}{2} + \frac{x}{\sqrt{x^2 + 4}} \cdot \frac{2}{\sqrt{x^2 + 4}} \right]$$

$$= \frac{1}{16} \left[\tan^{-1} \frac{x}{2} + \frac{2x}{x^2 + 4} \right].$$

Therefore, the original integral is

$$\int \frac{x^2}{(x^2 + 4)^2} \, dx = \frac{1}{2} \tan^{-1} \frac{x}{2} - 4 \left[\frac{1}{16} \tan^{-1} \frac{x}{2} + \frac{1}{8} \frac{x}{x^2 + 4} \right] + E$$

$$= \frac{1}{4} \tan^{-1} \frac{x}{2} - \frac{1}{2} \frac{x}{x^2 + 4} + E. \qquad \blacksquare$$

Review Section 5.1.

▶ ▮ **Improper Fractions**   In each of the preceding examples the integrand $f(x) = p(x)/q(x)$ was a proper fraction. Recall, when $f(x) = p(x)/q(x)$ is an **improper fraction**, that is, the degree of $p(x)$ is greater than or equal to the degree of $q(x)$, we start with long division.

EXAMPLE 8   Integrand an Improper Fraction

Evaluate $\int \frac{x^3 - 2x}{x^2 + 3x + 2} \, dx$.

Solution   The integrand is recognized as an improper fraction and we divide the numerator by the denominator:

$$\frac{x^3 - 2x}{x^2 + 3x + 2} = x - 3 + \frac{5x + 6}{x^2 + 3x + 2}.$$

Now since the denominator factors as $x^2 + 3x + 2 = (x + 1)(x + 2)$, we decompose the remainder over the divisor into partial fractions:

See the *SRM* for the command syntax for doing partial fraction decomposition in *Mathematica* and *Maple*.

$$\frac{5x + 6}{x^2 + 3x + 2} = \frac{1}{x + 1} + \frac{4}{x + 2}.$$

With this information, evaluation of the integral is immediate:

$$\int \frac{x^3 - 2x}{x^2 + 3x + 2} \, dx = \int \left[x - 3 + \frac{1}{x + 1} + \frac{4}{x + 2} \right] dx$$

$$= \frac{1}{2} x^2 - 3x + \ln|x + 1| + 4 \ln|x + 2| + C. \qquad \blacksquare$$

$\int$ NOTES FROM THE CLASSROOM

There is another way, called the **cover-up method**, of determining the coefficients in a partial fraction decomposition in the special case when the denominator of the integrand is the product of distinct linear factors:

$$f(x) = \frac{p(x)}{(x - r_1)(x - r_2) \cdots (x - r_n)}.$$

Let us illustrate by means of a specific example. From the foregoing discussion we know there exists unique constants A, B, and C such that

$$\frac{x^2 + 4x - 1}{(x - 1)(x - 2)(x + 3)} = \frac{A}{x - 1} + \frac{B}{x - 2} + \frac{C}{x + 3}. \qquad (13)$$

Suppose we multiply both sides of this last expression by $x - 1$, simplify, and then set $x = 1$. Since the coefficients of B and C are zero, we get

$$\left. \frac{x^2 + 4x - 1}{(x - 2)(x + 3)} \right|_{x=1} = A \qquad \text{or} \qquad A = -1.$$

Written another way,

$$\left. \frac{x^2 + 4x - 1}{\boxed{(x - 1)}(x - 2)(x + 3)} \right|_{x=1} = A$$

where we have shaded or covered up the factor that canceled when the left side of (13) was multiplied by $x - 1$. We *do not evaluate this covered-up factor at $x = 1$*. Now to obtain B and C we simply evaluate the left-hand side of (13) while covering up, in turn, $x - 2$ and $x + 3$:

$$\left. \frac{x^2 + 4x - 1}{(x - 1)\boxed{(x - 2)}(x + 3)} \right|_{x=2} = B \qquad \text{or} \qquad B = \frac{11}{5}$$

$$\left. \frac{x^2 + 4x - 1}{(x - 1)(x - 2)\boxed{(x + 3)}} \right|_{x=3} = C \qquad \text{or} \qquad C = -\frac{1}{5}.$$

Thus, we obtain the decomposition

$$\frac{x^2 + 4x - 1}{(x - 1)(x - 2)(x + 3)} = \frac{-1}{x - 1} + \frac{\frac{11}{15}}{x - 2} + \frac{-\frac{1}{5}}{x + 3}.$$

Exercises 7.6   Answers to selected odd-numbered problems begin on page ANS-23.

≡ Fundamentals

In Problems 1–8, write out the appropriate form of the partial fraction decomposition of the given expression. Do not evaluate the coefficients.

1. $\dfrac{x - 1}{x^2 + x}$

2. $\dfrac{9x - 8}{(x - 3)(2x - 5)}$

3. $\dfrac{x^3}{(x - 1)(x + 2)^3}$

4. $\dfrac{2x^2 - 3}{x^3 + 6x^2}$

5. $\dfrac{4}{x^3(x^2 + 3)}$

6. $\dfrac{-x^2 + 3x + 7}{(x + 2)^2(x^2 + x + 1)}$

7. $\dfrac{2x^3 - x}{(x^2 + 9)^2}$

8. $\dfrac{3x^2 - x + 4}{x^4 + 2x^3 + x}$

In Problems 9–42, use partial fractions to evaluate the given integral.

9. $\displaystyle\int \frac{1}{x(x - 2)}\, dx$

10. $\displaystyle\int \frac{1}{x(2x + 3)}\, dx$

11. $\displaystyle\int \frac{x + 2}{2x^2 - x}\, dx$

12. $\displaystyle\int \frac{3x + 10}{x^2 + 2x}\, dx$

13. $\displaystyle\int \frac{x + 1}{x^2 - 16}\, dx$

14. $\displaystyle\int \frac{1}{4x^2 - 25}\, dx$

15. $\displaystyle\int \frac{x}{2x^2 + 5x + 2}\, dx$

16. $\displaystyle\int \frac{x + 5}{(x + 4)(x^2 - 1)}\, dx$

17. $\displaystyle\int \frac{x^2 + 2x - 6}{x^3 - x}\, dx$

18. $\displaystyle\int \frac{5x^2 - x + 1}{x^3 - 4x}\, dx$

19. $\int \dfrac{1}{(x + 1)(x + 2)(x + 3)}\, dx$ **20.** $\int \dfrac{1}{(4x^2 - 1)(x + 7)}\, dx$

21. $\int \dfrac{4t^2 + 3t - 1}{t^3 - t^2}\, dt$ **22.** $\int \dfrac{2x - 11}{x^3 + 2x^2}\, dx$

23. $\int \dfrac{1}{x^3 + 2x^2 + x}\, dx$ **24.** $\int \dfrac{t - 1}{t^4 + 6t^3 + 9t^2}\, dt$

25. $\int \dfrac{2x - 1}{(x + 1)^3}\, dx$ **26.** $\int \dfrac{1}{x^2(x^2 - 4)^2}\, dx$

27. $\int \dfrac{1}{(x^2 + 6x + 5)^2}\, dx$

28. $\int \dfrac{1}{(x^2 - x - 6)(x^2 - 2x - 8)}\, dx$

29. $\int \dfrac{x^4 + 2x^2 - x + 9}{x^5 + 2x^4}\, dx$ **30.** $\int \dfrac{5x - 1}{x(x - 3)^2(x + 2)^2}\, dx$

31. $\int \dfrac{x - 1}{x(x^2 + 1)}\, dx$ **32.** $\int \dfrac{1}{(x - 1)(x^2 + 3)}\, dx$

33. $\int \dfrac{x}{(x + 1)^2(x^2 + 1)}\, dx$ **34.** $\int \dfrac{x^2}{(x - 1)^3(x^2 + 4)}\, dx$

35. $\int \dfrac{1}{x^4 + 5x^2 + 4}\, dx$ **36.** $\int \dfrac{1}{x^4 + 13x^2 + 36}\, dx$

37. $\int \dfrac{1}{x^3 - 1}\, dx$ **38.** $\int \dfrac{81}{x^4 + 27x}\, dx$

39. $\int \dfrac{3x^2 - x + 1}{(x + 1)(x^2 + 2x + 2)}\, dx$

40. $\int \dfrac{4x + 12}{(x - 2)(x^2 + 4x + 8)}\, dx$

41. $\int \dfrac{x^2 - x + 4}{(x^2 + 4)^2}\, dx$ **42.** $\int \dfrac{1}{x^3(x^2 + 1)^2}\, dx$

In Problems 43 and 44, proceed as in Example 7 to evaluate the given integral.

43. $\int \dfrac{x^3 - 2x^2 + x - 3}{x^4 + 8x^2 + 16}\, dx$ **44.** $\int \dfrac{x^2}{(x^2 + 3)^2}\, dx$

In Problems 45 and 46, proceed as in Example 8 to evaluate the given integral.

45. $\int \dfrac{x^4 + 3x^2 + 4}{(x + 1)^2}\, dx$ **46.** $\int \dfrac{x^5 - 10x^3}{x^4 - 10x^2 + 9}\, dx$

In Problems 47–54, evaluate the given definite integral.

47. $\displaystyle\int_2^4 \dfrac{1}{x^2 - 6x + 5}\, dx$ **48.** $\displaystyle\int_0^1 \dfrac{1}{x^2 - 4}\, dx$

49. $\displaystyle\int_0^2 \dfrac{2x - 1}{(x + 3)^2}\, dx$ **50.** $\displaystyle\int_1^5 \dfrac{2x + 6}{x(x + 1)^2}\, dx$

51. $\displaystyle\int_0^1 \dfrac{1}{x^3 + x^2 + 2x + 2}\, dx$ **52.** $\displaystyle\int_0^1 \dfrac{x^2}{x^4 + 8x^2 + 16}\, dx$

53. $\displaystyle\int_{-1}^1 \dfrac{2x^3 + 5x}{x^4 + 5x^2 + 6}\, dx$ **54.** $\displaystyle\int_1^2 \dfrac{1}{x^5 + 4x^4 + 5x^3}\, dx$

In Problems 55–58, evaluate the given integral by first using the indicated substitution followed by partial fractions.

55. $\int \dfrac{\sqrt{1 - x^2}}{x^3}\, dx;\quad u^2 = 1 - x^2$

56. $\int \sqrt{\dfrac{x - 1}{x + 1}}\, dx;\quad u^2 = \dfrac{x - 1}{x + 1}$

57. $\int \dfrac{\sqrt[3]{x + 1}}{x}\, dx;\quad u^3 = x + 1$

58. $\int \dfrac{1}{\sqrt{x}(1 + \sqrt[3]{x})^2}\, dx;\quad u^6 = x$

≣ **Review of Applications**

In Problems 59 and 60, find the area under the graph of the given function on the indicated interval. If necessary, use a calculator or CAS to obtain the graph of the function.

59. $y = \dfrac{1}{x^2 + 2x - 3};\quad [2, 4]$

60. $y = \dfrac{x^3}{(x^2 + 1)(x^2 + 2)};\quad [0, 4]$

In Problems 61 and 62, find the area bounded by the graph of the given function and the x-axis on the indicated interval. If necessary, use a calculator or CAS to obtain the graph of the function.

61. $y = \dfrac{x}{(x + 2)(x + 3)};\quad [-1, 1]$

62. $y = \dfrac{3x^3}{x^3 - 8};\quad [-2, 1]$

In Problems 63–66, find the volume of the solid of revolution that is formed by revolving the region bounded in the first quadrant by the graphs of the given equations about the indicated axis. If necessary, use a calculator or CAS to obtain the graph of the given function.

63. $y = \dfrac{2}{x(x + 1)}, x = 1, x = 3, y = 0;\quad x$-axis

64. $y = \dfrac{1}{\sqrt{(x + 1)(x + 4)}}, x = 0, x = 2, y = 0;\quad x$-axis

65. $y = \dfrac{4}{(x + 1)^2}, x = 0, x = 1, y = 0;\quad y$-axis

66. $y = \dfrac{8}{(x^2 + 1)(x^2 + 4)}, x = 0, x = 1, y = 0;\quad y$-axis

≣ **Think About It**

In Problems 67–70, evaluate the given integral by first making a substitution followed by partial fraction decomposition.

67. $\int \dfrac{\cos x}{\sin^2 x + 3 \sin x + 2}\, dx$ **68.** $\int \dfrac{\sin x}{\cos^2 x - \cos^3 x}\, dx$

69. $\int \dfrac{e^t}{(e^t + 1)^2(e^t - 2)}\, dt$ **70.** $\int \dfrac{e^{2t}}{(e^t + 1)^3}\, dt$

71. Find the length of the graph of $y = e^x$ on the interval $[0, \ln 2]$. [*Hint*: Evaluate the integral by starting with a substitution.]

72. Explain why partial fraction decomposition would be either unnecessary or inappropriate for the given integral. Discuss how these integrals can be evaluated.

(a) $\displaystyle \int \frac{x^3}{(x^2 - 1)(x^2 + 1)}\, dx$ **(b)** $\displaystyle \int \frac{3x + 4}{x^2 + 4}\, dx$

(c) $\displaystyle \int \frac{x}{(x^2 + 5)^2}\, dx$ **(d)** $\displaystyle \int \frac{2x^3 + 5x}{x^4 + 5x^2 + 6}\, dx$

73. Although partial fraction decomposition could be used to evaluate

$$\int \frac{x^5}{(x - 1)^{10}(x + 1)^{10}}\, dx$$

it would require solving 20 equations in 20 unknowns. Evaluate the integral using an easier technique of integration.

74. Why could the answer to Problem 53 be obtained with *no work* whatsoever?

7.7 Improper Integrals

■ **Introduction** Up to now in our study of the definite integral $\int_a^b f(x)\, dx$, it was understood that

- the limits of integration were finite numbers, and
- the function f either was *continuous* on $[a, b]$ or, if discontinuous, was *bounded* on the interval.

When either of these two conditions is dropped, the resulting integral is said to be an **improper integral**. In the discussion that follows, we first consider integrals of functions that are defined and continuous on unbounded intervals, in other words,

- at least one of the limits of integration is ∞ or $-\infty$.

After that, we examine integrals over bounded intervals of functions that become unbounded on an interval. In the latter type of improper integral,

- the integrand f has an *infinite discontinuity* at some number in the interval of integration.

■ **Improper Integrals—Unbounded Intervals** If the integrand f is defined on an unbounded interval, then there are three possible **improper integrals** with infinite limits of integration. Their definitions are summarized as follows:

Definition 7.7.1 Unbounded Intervals

(*i*) If f is continuous on $[a, \infty)$, then

$$\int_a^\infty f(x)\, dx = \lim_{b \to \infty} \int_a^b f(x)\, dx. \tag{1}$$

(*ii*) If f is continuous on $(-\infty, b]$, then

$$\int_{-\infty}^b f(x)\, dx = \lim_{a \to -\infty} \int_a^b f(x)\, dx. \tag{2}$$

(*iii*) If f is continuous on $(-\infty, \infty)$, then

$$\int_{-\infty}^\infty f(x)\, dx = \int_{-\infty}^c f(x)\, dx + \int_c^\infty f(x)\, dx. \tag{3}$$

When the limits in (1) and (2) exist, the integrals are said to **converge**. If the limit fails to exist, the integral is said to **diverge**. In (3) the integral $\int_{-\infty}^\infty f(x)\, dx$ converges provided *both* $\int_{-\infty}^c f(x)\, dx$ and $\int_c^\infty f(x)\, dx$ converge. If either $\int_{-\infty}^c f(x)\, dx$ or $\int_c^\infty f(x)\, dx$ diverges, then the improper integral $\int_{-\infty}^\infty f(x)\, dx$ diverges.

EXAMPLE 1 Using (1)

Evaluate $\displaystyle\int_2^\infty \frac{1}{x^3}\,dx$.

Solution By (1),

$$\int_2^\infty \frac{1}{x^3}\,dx = \lim_{b\to\infty}\int_2^b x^{-3}\,dx = \lim_{b\to\infty}\frac{x^{-2}}{-2}\Big]_2^b = -\frac{1}{2}\lim_{b\to\infty}(b^{-2}-2^{-2}).$$

Since the limit $\displaystyle\lim_{b\to\infty} b^{-2} = \lim_{b\to\infty}(1/b^2)$ exists,

$$\lim_{b\to\infty}(b^{-2}-2^{-2}) = \lim_{b\to\infty}\left(\frac{1}{b^2}-\frac{1}{4}\right) = 0 - \frac{1}{4} = -\frac{1}{4},$$

the given integral converges, and

$$\int_2^\infty \frac{1}{x^3}\,dx = -\frac{1}{2}\left(-\frac{1}{4}\right) = \frac{1}{8}. \qquad \blacksquare$$

EXAMPLE 2 Using (1)

Evaluate $\displaystyle\int_1^\infty x^2\,dx$.

Solution By (1),

$$\int_1^\infty x^2\,dx = \lim_{b\to\infty}\int_1^b x^2\,dx = \lim_{b\to\infty}\frac{1}{3}x^3\Big]_1^b = \lim_{b\to\infty}\left(\frac{1}{3}b^3-\frac{1}{3}\right).$$

Since $\displaystyle\lim_{b\to\infty}\left(\frac{1}{3}b^3-\frac{1}{3}\right) = \infty$ we conclude that the integral diverges. $\qquad\blacksquare$

EXAMPLE 3 Using (3)

Evaluate $\displaystyle\int_{-\infty}^\infty x^2\,dx$.

Solution Since c can be chosen arbitrarily in (3), we pick $c = 1$ and write

$$\int_{-\infty}^\infty x^2\,dx = \int_{-\infty}^1 x^2\,dx + \int_1^\infty x^2\,dx.$$

But, in Example 2 we saw that $\int_1^\infty x^2\,dx$ diverges. This is sufficient to conclude that $\int_{-\infty}^\infty x^2\,dx$ also diverges. $\qquad\blacksquare$

▌ **Area** If $f(x) \geq 0$ for all x over $[a,\infty)$, $(-\infty, b]$, or $(-\infty,\infty)$, then each of the integrals in (1), (2), and (3) can be interpreted as area under the graph of f on the interval whenever the integral converges.

EXAMPLE 4 Area

Evaluate $\displaystyle\int_{-1}^\infty e^{-x}\,dx$. Interpret geometrically.

Solution By (1),

$$\int_{-1}^\infty e^{-x}\,dx = \lim_{b\to\infty}\int_{-1}^b e^{-x}\,dx = \lim_{b\to\infty}(-e^{-x})\Big]_{-1}^b = \lim_{b\to\infty}(e - e^{-b}).$$

Since $\displaystyle\lim_{b\to\infty}e^{-b} = 0$, $\displaystyle\lim_{b\to\infty}(e - e^{-b}) = e$ and so the given integral converges to e. In FIGURE 7.7.1(a) we see that the area under the graph of the positive function $f(x) = e^{-x}$ on the interval $[-1, b]$ is $e - e^{-b}$. But, by taking $b\to\infty$, $e^{-b}\to 0$, and hence, as shown in Figure 7.7.1(b), we can interpret $\int_{-1}^\infty e^{-x}\,dx = e$ as a measure of the area under the graph of f on $[-1,\infty)$. $\qquad\blacksquare$

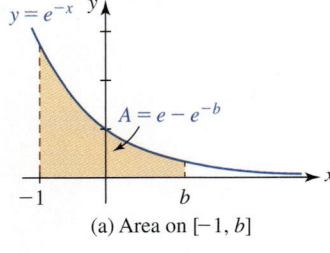

(a) Area on $[-1, b]$

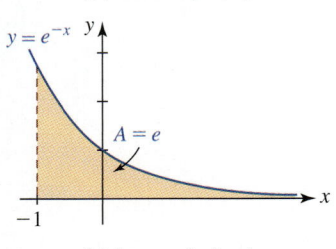

(b) Area on $[-1, \infty)$

FIGURE 7.7.1 Area under the graph in Example 4

EXAMPLE 5 Using (2)

Evaluate $\displaystyle\int_{-\infty}^{0} \cos x \, dx$.

Solution By (2),

$$\int_{-\infty}^{0} \cos x \, dx = \lim_{a \to -\infty} \int_{a}^{0} \cos x \, dx = \lim_{a \to -\infty} \sin x \bigg]_{a}^{0} = \lim_{a \to -\infty} (-\sin a).$$

Since $\sin a$ oscillates between -1 and 1, we conclude that $\displaystyle\lim_{a \to -\infty} (-\sin a)$ does not exist. Hence, $\displaystyle\int_{-\infty}^{0} \cos x \, dx$ diverges. ∎

EXAMPLE 6 Using (3)

Evaluate $\displaystyle\int_{-\infty}^{\infty} \frac{e^x}{e^x + 1} \, dx$.

Solution Choosing $c = 0$, we can write

$$\int_{-\infty}^{\infty} \frac{e^x}{e^x + 1} \, dx = \int_{-\infty}^{0} \frac{e^x}{e^x + 1} \, dx + \int_{0}^{\infty} \frac{e^x}{e^x + 1} \, dx = I_1 + I_2.$$

First, let us examine I_1:

$$I_1 = \lim_{a \to -\infty} \int_{a}^{0} \frac{e^x}{e^x + 1} \, dx = \lim_{a \to -\infty} \ln(e^x + 1) \bigg]_{a}^{0} = \lim_{a \to -\infty} [\ln 2 - \ln(e^a + 1)].$$

Now $e^a + 1 \to 1$ since $e^a \to 0$ as $a \to -\infty$. Therefore, $\ln(e^a + 1) \to \ln 1 = 0$ as $a \to -\infty$. Hence, $I_1 = \ln 2$.

Second, we have

$$I_2 = \lim_{b \to \infty} \int_{0}^{b} \frac{e^x}{e^x + 1} \, dx = \lim_{b \to \infty} \ln(e^x + 1) \bigg]_{0}^{b} = \lim_{b \to \infty} [\ln(e^b + 1) - \ln 2].$$

However, $e^b + 1 \to \infty$ as $b \to \infty$, so $\ln(e^b + 1) \to \infty$. Hence, I_2 diverges.

Because *both* I_1 and I_2 do not converge, it follows that the given integral is divergent. ∎

EXAMPLE 7 Using (3)

The improper integral $\displaystyle\int_{-\infty}^{\infty} \frac{1}{1 + x^2} \, dx$ converges because

$$\int_{-\infty}^{\infty} \frac{1}{1 + x^2} \, dx = \int_{-\infty}^{0} \frac{1}{1 + x^2} \, dx + \int_{0}^{\infty} \frac{1}{1 + x^2} \, dx = \frac{\pi}{2} + \frac{\pi}{2} = \pi.$$

The result follows from the facts that

$$\int_{-\infty}^{0} \frac{1}{1 + x^2} \, dx = \lim_{a \to -\infty} \int_{a}^{0} \frac{1}{1 + x^2} \, dx = -\lim_{a \to -\infty} \tan^{-1} a = -\left(-\frac{\pi}{2}\right) = \frac{\pi}{2}$$

$$\int_{0}^{\infty} \frac{1}{1 + x^2} \, dx = \lim_{b \to \infty} \int_{0}^{b} \frac{1}{1 + x^2} \, dx = \lim_{b \to \infty} \tan^{-1} b = \frac{\pi}{2}. \quad ∎$$

EXAMPLE 8 Work

In (5) of Section 6.8, we saw that the work done in lifting a mass m_2 off the surface of a planet of mass m_1 to a height h is given by

$$W = \int_{R}^{R+h} \frac{km_1 m_2}{r^2} \, dr,$$

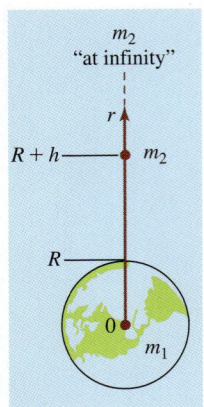

FIGURE 7.7.2 Mass m_2 lifted to "infinity" in Example 8

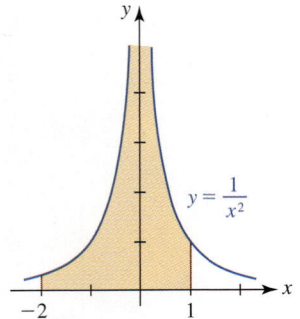

FIGURE 7.7.3 $x = 0$ is a vertical asymptote for the graph of $f(x) = 1/x^2$

where R is the radius of the planet. Hence, the amount of work done in lifting m_2 to an unlimited or "infinite distance" from the surface of the planet is

$$
\begin{aligned}
W &= \int_R^\infty \frac{km_1m_2}{r^2} \, dr \\
&= km_1m_2 \lim_{b\to\infty} \int_R^b r^{-2} \, dr \\
&= km_1m_2 \lim_{b\to\infty} \left[-\frac{1}{b} + \frac{1}{R} \right] \\
&= \frac{km_1m_2}{R}.
\end{aligned}
$$

See FIGURE 7.7.2. From the data in Example 2 of Section 6.8, it follows that the work done in lifting a payload of 5000 kg to an "infinite distance" from the surface of the Earth is

$$
W = \frac{(6.67 \times 10^{-11})(6.0 \times 10^{24})(5000)}{6.4 \times 10^6} \approx 3.13 \times 10^{11} \text{ joules.} \quad \blacksquare
$$

Recall, if f is continuous on $[a, b]$, then the definite integral $\int_a^b f(x)\,dx$ exists. Moreover, if $F'(x) = f(x)$, then $\int_a^b f(x)\,dx = F(b) - F(a)$. However, we cannot evaluate an integral such as

$$
\int_{-2}^1 \frac{1}{x^2} \, dx \tag{4}
$$

by the same procedure, since $f(x) = 1/x^2$ possesses an infinite discontinuity in $[-2, 1]$. See FIGURE 7.7.3. In other words, for the integral in (4), the "procedure"

$$
-x^{-1} \Big]_{-2}^1 = (-1) - \left(\frac{1}{2}\right) = -\frac{3}{2}
$$

is just meaningless scratchings on paper. Thus, we have another type of integral that demands special handling.

▌ Improper Integrals—Infinite Discontinuities An integral $\int_a^b f(x)\,dx$ is also said to be **improper** if f is unbounded on $[a, b]$—that is, if f has an infinite discontinuity at some number in the interval of integration. There are three possible **improper integrals** of this type. Their definitions are summarized in the next definition.

Definition 7.7.2 Infinite Discontinuities

(*i*) If f is continuous on $[a, b)$ and $|f(x)| \to \infty$ as $x \to b^-$, then

$$
\int_a^b f(x) \, dx = \lim_{t\to b^-} \int_a^t f(x) \, dx. \tag{5}
$$

(*ii*) If f is continuous on $(a, b]$ and $|f(x)| \to \infty$ as $x \to a^+$, then

$$
\int_a^b f(x) \, dx = \lim_{s\to a^+} \int_s^b f(x) \, dx. \tag{6}
$$

(*iii*) If $|f(x)| \to \infty$ as $x \to c$ for some c in (a, b) and f is continuous at all other numbers in $[a, b]$, then

$$
\int_a^b f(x) \, dx = \int_a^c f(x) \, dx + \int_c^b f(x) \, dx. \tag{7}
$$

When the limits in (5) and (6) exist, the integrals are said to **converge**. If the limit fails to exist, the integral is said to **diverge**. In (7) the integral $\int_a^b f(x)\,dx$ converges provided *both* $\int_a^c f(x)\,dx$ and $\int_c^b f(x)\,dx$ converge. If either $\int_a^c f(x)\,dx$ or $\int_c^b f(x)\,dx$ diverges, then $\int_a^b f(x)\,dx$ diverges.

EXAMPLE 9 Using (6)

Evaluate $\int_0^4 \dfrac{1}{\sqrt{x}} \, dx$.

Solution Observe that $f(x) = 1/\sqrt{x} \to \infty$ as $x \to 0^+$, that is, $x = 0$ is a vertical asymptote for the graph of f. Thus, by (6) of Definition 7.7.2,

$$\int_0^4 \frac{1}{\sqrt{x}} \, dx = \lim_{s \to 0^+} \int_s^4 x^{-1/2} \, dx = \lim_{s \to 0^+} 2x^{1/2} \Big]_s^4 = \lim_{s \to 0^+} [4 - 2s^{1/2}].$$

Since $\lim\limits_{s \to 0^+} s^{1/2} = 0$ we have $\lim\limits_{s \to 0^+} [4 - 2s^{1/2}] = 4$. Thus, the given integral converges and

$$\int_0^4 \frac{1}{\sqrt{x}} \, dx = 4.$$

As seen in FIGURE 7.7.4, the number 4 can be regarded as the area under the graph of f on the interval $[0, 4]$.

■

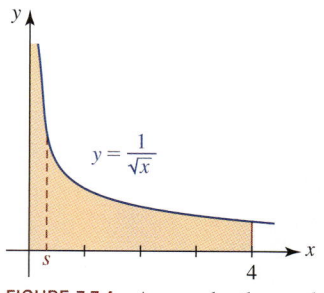

FIGURE 7.7.4 Area under the graph in Example 9

EXAMPLE 10 Using (6)

Evaluate $\int_0^e \ln x \, dx$.

Solution In this case we know $f(x) = \ln x \to -\infty$ as $x \to 0^+$. Using (6) and integration by parts gives

$$\int_0^e \ln x \, dx = \lim_{s \to 0^+} \int_s^e \ln x \, dx$$

$$= \lim_{s \to 0^+} (x \ln x - x) \Big]_s^e$$

$$= \lim_{s \to 0^+} [(e \ln e - e) - (s \ln s - s)] \quad \leftarrow \ln e = 1$$

$$= \lim_{s \to 0^+} s(1 - \ln s).$$

Now, the last limit has the indeterminate form $0 \cdot \infty$, but if it is written as

$$\lim_{s \to 0^+} \frac{1 - \ln s}{1/s},$$

we recognize the indeterminate form is now ∞ / ∞. Thus, by L'Hôpital's Rule,

$$\lim_{s \to 0^+} \frac{1 - \ln s}{1/s} \stackrel{h}{=} \lim_{s \to 0^+} \frac{-1/s}{-1/s^2} = \lim_{s \to 0^+} s = 0.$$

Therefore, the integral converges and $\int_0^e \ln x \, dx = 0$.

■

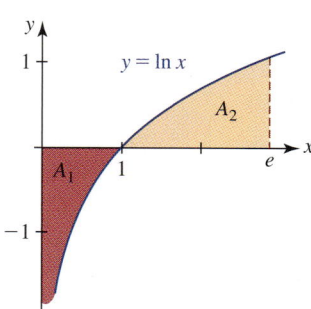

FIGURE 7.7.5 Net signed area in Example 10

The result $\int_0^e \ln x \, dx = 0$ in Example 10 indicates that the net signed area between the graph of $f(x) = \ln x$ and the x-axis on $[0, e]$ is 0. From FIGURE 7.7.5 we see that

$$\int_0^e \ln x \, dx = \int_0^1 \ln x \, dx + \int_1^e \ln x \, dx = -A_1 + A_2 = 0.$$

We saw in Example 7 in Section 7.3 that $\int_1^e \ln x \, dx = 1$, and so $A_1 = A_2 = 1$.

EXAMPLE 11 Using (7)

Evaluate $\int_1^5 \dfrac{1}{(x - 2)^{1/3}} \, dx$.

Solution In the interval $[1, 5]$ the integrand has an infinite discontinuity at 2. Consequently, from (7) we write

$$\int_1^5 \frac{1}{(x - 2)^{1/3}} \, dx = \int_1^2 (x - 2)^{-1/3} \, dx + \int_2^5 (x - 2)^{-1/3} \, dx = I_1 + I_2.$$

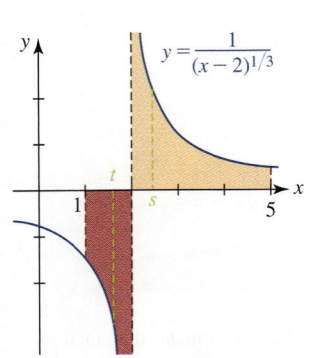

FIGURE 7.7.6 Graph of the integrand in Example 11

Now,

$$I_1 = \lim_{t \to 2^-} \int_1^t (x-2)^{-1/3}\, dx = \lim_{t \to 2^-} \frac{3}{2}(x-2)^{2/3}\Big]_1^t$$

$$= \frac{3}{2}\lim_{t \to 2^-}[(t-2)^{2/3}-1] = -\frac{3}{2}.$$

Similarly,

$$I_2 = \lim_{s \to 2^+} \int_s^5 (x-2)^{-1/3}\, dx = \lim_{s \to 2^+}\frac{3}{2}(x-2)^{2/3}\Big]_s^5$$

$$= \frac{3}{2}\lim_{s \to 2^+}[3^{2/3}-(s-2)^{2/3}] = \frac{3^{5/3}}{2}.$$

Since both I_1 and I_2 converge, the given integral converges and

$$\int_1^5 \frac{dx}{(x-2)^{1/3}} = -\frac{3}{2} + \frac{3^{5/3}}{2} \approx 1.62.$$

Note from **FIGURE 7.7.6** that this last number represents a net signed area on the interval $[1, 5]$. ∎

EXAMPLE 12   The Integral in (4) Revisited

Evaluate $\displaystyle\int_{-2}^1 \frac{1}{x^2}\, dx$.

Solution   This is the integral discussed in (4). Since, in the interval $[-2, 1]$ the integrand has an infinite discontinuity at 0, we write

$$\int_{-2}^1 \frac{1}{x^2}\, dx = \int_{-2}^0 \frac{1}{x^2}\, dx + \int_0^1 \frac{1}{x^2}\, dx = I_1 + I_2.$$

Now, the result

$$I_1 = \int_{-2}^0 \frac{1}{x^2}\, dx = \lim_{t \to 0^-} \int_{-2}^t x^{-2}\, dx = \lim_{t \to 0^-}(-x^{-1})\Big]_{-2}^t = \lim_{t \to 0^-}\left[-\frac{1}{t} - \frac{1}{2}\right] = \infty$$

indicates there is no need to evaluate $I_2 = \int_0^1 dx/x^2$. The integral $\int_{-2}^1 dx/x^2$ diverges.

$\displaystyle\int$ **NOTES FROM THE CLASSROOM**

(*i*) You should verify that $\int_{-\infty}^\infty x\, dx = \int_{-\infty}^0 x\, dx + \int_0^\infty x\, dx$ diverges since both $\int_{-\infty}^0 x\, dx$ and $\int_0^\infty x\, dx$ diverge. A common mistake when working with integrals with doubly infinite limits is to use one limit:

$$\int_{-\infty}^\infty x\, dx = \lim_{t \to \infty}\int_{-t}^t x\, dx = \lim_{t \to \infty}\frac{1}{2}x^2\Big]_{-t}^t = \frac{1}{2}\lim_{t \to \infty}[t^2 - t^2] = 0.$$

Of course, this "answer" is incorrect. Integrals of the type $\int_{-\infty}^\infty f(x)\, dx$ require the evaluation of *two independent* limits.

(*ii*) In our previous work we often wrote without thinking that an integral of a sum is the sum of the integrals:

$$\int_a^b [f(x) + g(x)]\, dx = \int_a^b f(x)\, dx + \int_a^b g(x)\, dx. \tag{8}$$

For improper integrals one should proceed with more caution. For example, the integral $\int_1^\infty \left[\frac{1}{x} - \frac{1}{x+1}\right] dx$ converges (see Problem 25 in Exercises 7.7), but

$$\int_1^\infty \left[\frac{1}{x} - \frac{1}{x+1}\right] dx \neq \int_1^\infty \frac{1}{x}\, dx - \int_1^\infty \frac{1}{x+1}\, dx.$$

The property in (8) remains valid for improper integrals whenever both integrals on the right-hand side of the equality converge.

(*iii*) From examples, problems, and graphs such as Figure 7.7.1, students are often left with the impression that $f(x) \to 0$ as $x \to \infty$ is a necessary condition for the integral $\int_a^\infty f(x)\,dx$ to converge. This is not so. Work Problem 70 when you get to Exercises 9.3.

(*iv*) It is possible for an integral to have infinite limits of integration *and* an integrand with an infinite discontinuity. To determine whether an integral such as

infinite limit $\to$

integrand discontinuous at $x = 1 \to$ $\displaystyle\int_1^\infty \frac{1}{x\sqrt{x^2 - 1}}\,dx$

converges, we break up the integration at some convenient point of continuity of the integrand, say, $x = 2$:

$$\int_1^\infty \frac{1}{x\sqrt{x^2 - 1}}\,dx = \int_1^2 \frac{1}{x\sqrt{x^2 - 1}}\,dx + \int_2^\infty \frac{1}{x\sqrt{x^2 - 1}}\,dx = I_1 + I_2. \qquad (9)$$

I_1 and I_2 are improper integrals; I_1 is of the type given in (6) and I_2 is of the type given in (1). If both I_1 and I_2 converge, then the original integral converges. See Problems 85 and 86 in Exercises 7.7.

(*v*) The integrand of $\int_a^b f(x)\,dx$ can also have infinite discontinuities at both $x = a$ and $x = b$. In this case the improper integral is defined in a manner analogous to (7). If an integrand f has an infinite discontinuity at several numbers in (a, b), then the improper integral is defined by a natural extension of (7). See Problems 87 and 88 in Exercises 7.7.

(*vi*) Sometimes strange things happen when working with improper integrals. It is possible to revolve a region with infinite area around an axis and the resulting volume of the solid of revolution can be finite. A very famous problem of this sort is given in Problem 89 of Exercises 7.7.

Exercises 7.7   Answers to selected odd-numbered problems begin on page ANS-24.

☰ Fundamentals

In Problems 1–30, evaluate the given improper integral or show that it diverges.

1. $\displaystyle\int_3^\infty \frac{1}{x^4}\,dx$

2. $\displaystyle\int_{-\infty}^{-1} \frac{1}{\sqrt[3]{x}}\,dx$

3. $\displaystyle\int_1^\infty \frac{1}{x^{0.99}}\,dx$

4. $\displaystyle\int_1^\infty \frac{1}{x^{1.01}}\,dx$

5. $\displaystyle\int_{-\infty}^3 e^{2x}\,dx$

6. $\displaystyle\int_{-\infty}^\infty e^{-x}\,dx$

7. $\displaystyle\int_1^\infty \frac{\ln x}{x}\,dx$

8. $\displaystyle\int_1^\infty \frac{\ln t}{t^2}\,dt$

9. $\displaystyle\int_e^\infty \frac{1}{x(\ln x)^3}\,dx$

10. $\displaystyle\int_e^\infty \ln x\,dx$

11. $\displaystyle\int_{-\infty}^\infty \frac{x}{(x^2 + 1)^{3/2}}\,dx$

12. $\displaystyle\int_{-\infty}^\infty \frac{x}{1 + x^2}\,dx$

13. $\displaystyle\int_{-\infty}^0 \frac{x}{(x^2 + 9)^2}\,dx$

14. $\displaystyle\int_5^\infty \frac{1}{\sqrt[4]{3x + 1}}\,dx$

15. $\displaystyle\int_2^\infty u e^{-u}\,du$

16. $\displaystyle\int_{-\infty}^3 \frac{x^3}{x^4 + 1}\,dx$

17. $\displaystyle\int_{2/\pi}^\infty \frac{\sin(1/x)}{x^2}\,dx$

18. $\displaystyle\int_{-\infty}^\infty t e^{-t^2}\,dt$

19. $\displaystyle\int_{-1}^\infty \frac{1}{x^2 + 2x + 2}\,dx$

20. $\displaystyle\int_{-\infty}^0 \frac{1}{x^2 + 2x + 3}\,dx$

21. $\displaystyle\int_0^\infty e^{-x}\sin x\,dx$

22. $\displaystyle\int_{-\infty}^0 e^x \cos 2x\,dx$

23. $\displaystyle\int_{1/2}^\infty \frac{x + 1}{x^3}\,dx$

24. $\displaystyle\int_0^\infty (e^{-x} - e^{-2x})^2\,dx$

25. $\displaystyle\int_1^\infty \left[\frac{1}{x} - \frac{1}{x + 1}\right]dx$

26. $\displaystyle\int_3^\infty \left[\frac{1}{x} + \frac{1}{x^2 + 9}\right]dx$

27. $\displaystyle\int_2^\infty \frac{1}{x^2 + 6x + 5}\,dx$

28. $\displaystyle\int_{-\infty}^0 \frac{1}{x^2 - 3x + 2}\,dx$

29. $\displaystyle\int_{-\infty}^{-2} \frac{x^2}{(x^3 + 1)^2}\,dx$

30. $\displaystyle\int_0^\infty \frac{1}{e^x + e^{-x}}\,dx$

In Problems 31–52, evaluate the given improper integral or show that it diverges.

31. $\displaystyle\int_0^5 \frac{1}{x}\,dx$

32. $\displaystyle\int_0^8 \frac{1}{x^{2/3}}\,dx$

33. $\displaystyle\int_0^1 \frac{1}{x^{0.99}}\,dx$

34. $\displaystyle\int_0^1 \frac{1}{x^{1.01}}\,dx$

35. $\displaystyle\int_0^2 \frac{1}{\sqrt{2 - x}}\,dx$

36. $\displaystyle\int_1^3 \frac{1}{(x - 1)^2}\,dx$

37. $\displaystyle\int_{-1}^{1} \frac{1}{x^{5/3}}\,dx$

38. $\displaystyle\int_{0}^{2} \frac{1}{\sqrt[3]{x-1}}\,dx$

39. $\displaystyle\int_{0}^{2} (x-1)^{-2/3}\,dx$

40. $\displaystyle\int_{0}^{27} \frac{e^{x^{1/3}}}{x^{2/3}}\,dx$

41. $\displaystyle\int_{0}^{1} x\ln x\,dx$

42. $\displaystyle\int_{1}^{e} \frac{1}{x\ln x}\,dx$

43. $\displaystyle\int_{0}^{\pi/2} \tan t\,dt$

44. $\displaystyle\int_{0}^{\pi/4} \frac{\sec^2\theta}{\sqrt{\tan\theta}}\,d\theta$

45. $\displaystyle\int_{0}^{\pi} \frac{\sin x}{1+\cos x}\,dx$

46. $\displaystyle\int_{0}^{\pi} \frac{\cos x}{\sqrt{1-\sin x}}\,dx$

47. $\displaystyle\int_{-1}^{0} \frac{x}{\sqrt{1+x}}\,dx$

48. $\displaystyle\int_{0}^{3} \frac{1}{x^2-1}\,dx$

49. $\displaystyle\int_{0}^{1} \frac{x^2}{\sqrt{1-x^2}}\,dx$

50. $\displaystyle\int_{0}^{2} \frac{e^w}{\sqrt{e^w-1}}\,dw$

51. $\displaystyle\int_{1}^{3} \frac{1}{\sqrt{3+2x-x^2}}\,dx$

52. $\displaystyle\int_{0}^{1} \left[\frac{1}{\sqrt{x}}+\frac{1}{\sqrt{1-x}}\right]dx$

In Problems 53 and 54, use a substitution to evaluate the given integral.

53. $\displaystyle\int_{12}^{\infty} \frac{1}{\sqrt{x}\,(x+4)}\,dx$

54. $\displaystyle\int_{1}^{\infty} \sqrt{x}\,e^{-\sqrt{x}}\,dx$

≡ Review of Applications

In Problems 55–58, find the area under the graph of the given function on the indicated interval.

55. $f(x) = \dfrac{1}{(2x+1)^2}; \quad [1,\infty)$

56. $f(x) = \dfrac{10}{x^2+25}; \quad (-\infty, 5]$

57. $f(x) = e^{-|x|}; \quad (-\infty, \infty)$

58. $f(x) = |x|^3 e^{-x^4}; \quad (-\infty, \infty)$

59. Find the area of the region that is bounded by the graphs of $y = 1/\sqrt{x-1}$ and $y = -1/\sqrt{x-1}$ on the interval $[1, 5]$.

60. Consider the region that is bounded by the graphs of $y = 1/\sqrt{x+2}$ and $y = 0$ on the interval $[-2, 1]$.
   (a) Show that the region has finite area.
   (b) Show that the solid of revolution that is formed by revolving the region around the x-axis has infinite volume.

61. Use a calculator or CAS to obtain the graphs of
$$y = \frac{1}{x} \quad \text{and} \quad y = \frac{1}{x(x^2+1)}.$$
Determine whether the area of the region that is bounded by these graphs on the interval $[0, 1]$ is finite.

62. Find the volume of the solid of revolution that is formed by revolving the region bounded by the graphs of $y = xe^{-x}$ and $y = 0$ on $[0, \infty)$ around the x-axis.

63. Find the work done against gravity in lifting a 10,000-kg payload to an infinite distance above the surface of the Moon. [*Hint:* Review page 357 of Section 6.8.]

64. The work done by an external force in moving a test charge q_0 radially from point A to point B in the electric field of a charge q is defined to be:
$$W = -\frac{qq_0}{4\pi e_0}\int_{r_A}^{r_B} \frac{1}{r^2}\,dr.$$
See FIGURE 7.7.7.
   (a) Show that $W = \dfrac{qq_0}{4\pi e_0}\left(\dfrac{1}{r_B} - \dfrac{1}{r_A}\right)$.
   (b) Find the work done in bringing the test charge in from an infinite distance to point B.

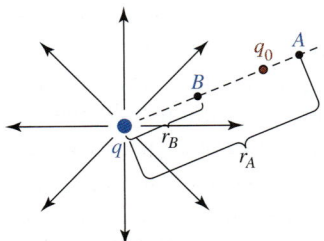

FIGURE 7.7.7 Charge in Problem 64

The **Laplace transform** of a function $y = f(x)$, defined by the integral
$$\mathscr{L}\{f(x)\} = \int_{0}^{\infty} e^{-st} f(t)\,dt,$$
is very useful in some areas of applied mathematics. In Problems 65–72, find the Laplace transform of the given function and state a restriction on s for which the integral converges.

65. $f(x) = 1$

66. $f(x) = x$

67. $f(x) = e^x$

68. $f(x) = e^{-5x}$

69. $f(x) = \sin x$

70. $f(x) = \cos 2x$

71. $f(x) = \begin{cases} 0, & 0 \le x < 1 \\ 1, & x \ge 1 \end{cases}$

72. $f(x) = \begin{cases} 0, & 0 \le x < 3 \\ e^{-x}, & x \ge 3 \end{cases}$

73. A **probability density function** is any nonnegative function f defined on an interval $[a, b]$ for which $\int_a^b f(x)\,dx = 1$. Verify that for $k > 0$,
$$f(x) = \begin{cases} 0, & x < 0 \\ ke^{-kx}, & x \ge 0 \end{cases}$$
is a probability density function on the interval $(-\infty, \infty)$.

74. Another integral in applied mathematics is the so-called **gamma function:**
$$\Gamma(\alpha) = \int_{0}^{\infty} t^{\alpha-1} e^{-t}\,dt, \quad x > 0.$$
   (a) Show that $\Gamma(\alpha + 1) = \alpha\Gamma(\alpha)$.
   (b) Use the result in part (a) to show that
$$\Gamma(n + 1) = 1 \cdot 2 \cdot 3 \cdots (n - 1) \cdot n = n!,$$
where the symbol $n!$ is read "n factorial." Because of this property, the gamma function is called the **generalized factorial function**.

≡ Think About It

In Problems 75–78, determine all values of k such that the given integral is convergent.

75. $\displaystyle\int_1^\infty \frac{1}{x^k}\,dx$

76. $\displaystyle\int_{-\infty}^1 x^{2k}\,dx$

77. $\displaystyle\int_0^\infty e^{kx}\,dx$

78. $\displaystyle\int_1^\infty \frac{(\ln x)^k}{x}\,dx$

The following is a **Comparison Test** for improper integrals. Suppose f and g are continuous and $0 \le f(x) \le g(x)$ for $x \ge a$. If $\int_a^\infty g(x)\,dx$ converges, then $\int_a^\infty f(x)\,dx$ also converges. In Problems 79–82, use this result to show that the given integral converges.

79. $\displaystyle\int_1^\infty \frac{\sin^2 x}{x^2}\,dx$

80. $\displaystyle\int_2^\infty \frac{1}{x^3 + 4}\,dx$

81. $\displaystyle\int_0^\infty \frac{1}{x + e^x}\,dx$

82. $\displaystyle\int_0^\infty e^{-x^2}\,dx$

In the Comparison Test for improper integrals, if the integral $\int_a^\infty f(x)\,dx$ diverges, then $\int_a^\infty g(x)\,dx$ is divergent. In Problems 83 and 84, use this result to show that the given integral diverges.

83. $\displaystyle\int_1^\infty \frac{1 + e^{-2x}}{\sqrt{x}}\,dx$

84. $\displaystyle\int_1^\infty e^{x^2}\,dx$

In Problems 85–88, determine whether the given integral converges or diverges.

85. $\displaystyle\int_1^\infty \frac{1}{x\sqrt{x^2 - 1}}\,dx$

86. $\displaystyle\int_{-\infty}^4 \frac{1}{(x - 1)^{2/3}}\,dx$

87. $\displaystyle\int_{-1}^1 \frac{1}{\sqrt{1 - x^2}}\,dx$

88. $\displaystyle\int_0^2 \frac{2x - 1}{\sqrt[3]{x^2 - x}}\,dx$

≡ Projects

89. A Mathematical Classic   The Italian mathematician and physicist **Evangelista Torricelli** (1608–1647) was the first to investigate the interesting properties of the region bounded by the graphs of $y = 1/x$ and $y = 0$ on the interval $[1, \infty)$.

(a) Show that the region has infinite area.

(b) Show, however, that the solid of revolution formed by revolving the region about the x-axis has finite volume. The solid shown in FIGURE 7.7.8 is called **Gabriel's horn** or **Torricelli's trumpet**. In some religious traditions

Gabriel is held to be the angel of judgment, the destroyer of Sodom and Gomorrah, and so he is frequently identified as the angel who blows the horn to announce the arrival of Judgment Day.

(c) Use (3) of Section 6.6 to show that the surface area S of the solid of revolution is given by

$$S = 2\pi \int_1^\infty \frac{\sqrt{x^4 + 1}}{x^3}\,dx.$$

Use the version of the Comparison Test given in Problems 83 and 84 to show that the surface area is infinite.

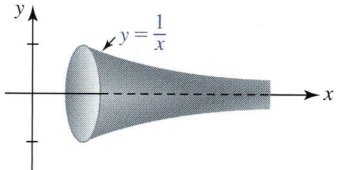

FIGURE 7.7.8   Gabriel's horn in Problem 89

90. A Bit of History—Return of the Plague   A study of the Bombay plague epidemic of 1905–06 found that the death rate for that epidemic could be approximated by the mathematical model

$$R = 890\,\text{sech}^2(0.2t - 3.4),$$

where R is the number of deaths per week and t is the time (in weeks) from the onset of the plague.

(a) What is the peak death rate, and when does it occur?

(b) Estimate the total number of deaths by computing the integral $\int_{-\infty}^\infty R_0(t)\,dt$.

(c) Show that more than 99% of the deaths occurred in the first 34 weeks of the epidemic; that is, compare $\int_0^{34} R(t)\,dt$ to the result in part (b).

(d) Suppose you want to use a "simpler" model for the death rate, of the form

$$R_0 = \frac{a}{t^2 - 2bt + c},$$

where $c > b^2$. You want this model to have the same peak death rate at the same time as the original model and you also want the total number of deaths, $\int_{-\infty}^\infty R_0(t)\,dt$, to be the same. Find coefficients a, b, and c that satisfy these requirements.

(e) For the model in part (d), show that less than 89% of the deaths occur in the first 34 weeks of the epidemic.

7.8 Approximate Integration

▌ Introduction   Life in mathematics would be extremely pleasant if the antiderivative of every continuous function could be expressed in terms of elementary functions such as polynomial, rational, exponential, or trigonometric functions. As discussed in the *Notes from the Classroom* in Section 5.5 this is not the case. Hence, Theorem 5.5.1 cannot be used to evaluate every definite integral. Sometimes the best we can hope for is an approximation of the value of a definite integral $\int_a^b f(x)\,dx$. In this concluding section of the chapter, we shall consider three such numerical or *approximate integration* procedures.

In the following discussion it will again be useful to interpret the definite integral $\int_a^b f(x)\,dx$ as the area under the graph of f on $[a, b]$. Although continuity of f is essential, there is no actual requirement that $f(x) \geq 0$ on the interval.

■ **Midpoint Rule** One way of approximating a definite integral is to proceed in the same manner as we did in the initial discussion about finding the area under a graph—namely, construct rectangular elements under the graph and add their areas. In particular, let us suppose that $y = f(x)$ is continuous on $[a, b]$ and that this interval is divided into n subintervals of equal length $\Delta x = (b - a)/n$. (Recall this is called a *regular partition*.) A simple, but fairly accurate, approximation rule consists of adding the areas of n rectangular elements whose lengths are calculated at the midpoint of each subinterval. See FIGURE 7.8.1(a).

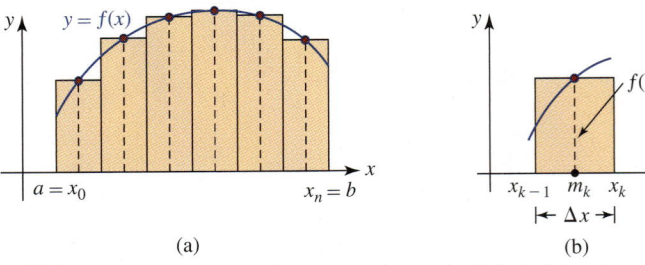

(a)   (b)

FIGURE 7.8.1   Using n rectangles to approximate the definite integral

Now, if $m_k = (x_{k-1} + x_k)/2$ is the midpoint of a subinterval $[x_{k-1}, x_k]$, then the area of the rectangular element shown in Figure 7.8.1(b) is

$$A_k = f(m_k)\,\Delta x = f\left(\frac{x_{k-1} + x_k}{2}\right)\Delta x.$$

Identifying $a = x_0$ and $b = x_n$ and summing the n areas, we obtain

$$\int_a^b f(x)\,dx \approx f\left(\frac{x_0 + x_1}{2}\right)\Delta x + f\left(\frac{x_1 + x_2}{2}\right)\Delta x + \cdots + f\left(\frac{x_{n-1} + x_n}{2}\right)\Delta x.$$

If we replace Δx by $(b - a)/n$, this midpoint approximation rule can be summarized as follows:

Definition 7.8.1   Midpoint Rule

The **Midpoint Rule** is the approximation $\displaystyle\int_a^b f(x)\,dx \approx M_n$, where

$$M_n = \frac{b - a}{n}\left[f\left(\frac{x_0 + x_1}{2}\right) + f\left(\frac{x_1 + x_2}{2}\right) + \cdots + f\left(\frac{x_{n-1} + x_n}{2}\right)\right]. \tag{1}$$

Since the function $f(x) = 1/x$ is continuous on any interval $[a, b]$ that does not include the origin, we know that $\int_a^b (1/x)\,dx$ exists. For the sake of the next example suspend your knowledge that $\ln|x|$ is an antiderivative of $1/x$.

EXAMPLE 1   Using (1)

Approximate $\displaystyle\int_1^2 (1/x)\,dx$ by the Midpoint Rule for $n = 1$, $n = 2$, and $n = 5$.

Solution   As shown in FIGURE 7.8.2(a), the case $n = 1$ is one rectangle in which $\Delta x = 1$. The midpoint of the interval is $m_1 = \frac{3}{2}$ and $f\left(\frac{3}{2}\right) = \frac{2}{3}$. Therefore, from (1),

$$M_1 = 1 \cdot \frac{2}{3} \approx 0.6666.$$

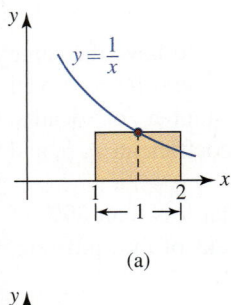

(a)

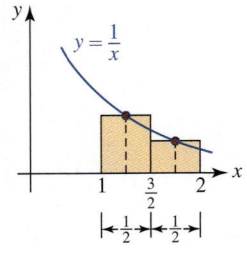

(b)

FIGURE 7.8.2   Rectangles in Example 1

When $n = 2$, Figure 7.8.2(b) shows $\Delta x = \frac{1}{2}$, $x_0 = 1$, $x_1 = 1 + \Delta x = \frac{3}{2}$, and $x_2 = 1 + 2\Delta x = 2$. The midpoints of intervals $[1, \frac{3}{2}]$, and $[\frac{3}{2}, 2]$ are, respectively, $m_1 = \frac{5}{4}$ and $m_2 = \frac{7}{4}$ and so $f(\frac{5}{4}) = \frac{4}{5}$ and $f(\frac{7}{4}) = \frac{4}{7}$. Hence, (1) gives

k	m_k	$f(m_k)$
1	$\frac{11}{10}$	$\frac{10}{11}$
2	$\frac{13}{10}$	$\frac{10}{13}$
3	$\frac{15}{10}$	$\frac{10}{15}$
4	$\frac{17}{10}$	$\frac{10}{17}$
5	$\frac{19}{10}$	$\frac{10}{19}$

$$M_2 = \frac{1}{2}\left[\frac{4}{5} + \frac{4}{7}\right] \approx 0.6857.$$

Finally, for $n = 5$, $\Delta x = \frac{1}{5}$, $x_0 = 1$, $x_1 = 1 + \Delta x = \frac{6}{5}$, $x_2 = 1 + 2\Delta x = \frac{7}{5}$, ..., $x_5 = 1 + 5\Delta x = 2$. The midpoints of the five subintervals $[1, \frac{6}{5}]$, $[\frac{6}{5}, \frac{7}{5}]$, $[\frac{7}{5}, \frac{8}{5}]$, $[\frac{8}{5}, \frac{9}{5}]$, $[\frac{9}{5}, 2]$ and the corresponding function values are shown in the accompanying table. The information in the table then gives

$$M_5 = \frac{1}{5}\left[\frac{10}{11} + \frac{10}{13} + \frac{10}{15} + \frac{10}{17} + \frac{10}{19}\right] \approx 0.6919.$$

In other words, $\int_1^2 (1/x)\, dx \approx M_5$ or $\int_1^2 (1/x)\, dx \approx 0.6919$. ∎

■ **Error in the Midpoint Rule** Suppose $I = \int_a^b f(x)\, dx$ and M_n is an approximation to I using n rectangles. We define the error in the method to be

$$E_n = |I - M_n|.$$

An upper bound for the error can be obtained by means of the next result. The proof is omitted.

> **Theorem 7.8.1** Error Bound for Midpoint Rule
>
> If there exists a number $M > 0$ such that $|f''(x)| \le M$ for all x in $[a, b]$, then
>
> $$E_n \le \frac{M(b-a)^3}{24n^2}. \tag{2}$$

Observe that this upper bound for the error E_n is inversely proportional to n^2. Hence, the accuracy in the method improves as we take more and more rectangles. For example, if the number of rectangles is doubled, the error E_{2n} is less than one-fourth the error bound for E_n. Thus, we see that $\lim_{n \to \infty} M_n = I$.

The next example illustrates how the error bound (2) can be utilized to determine the number of rectangles that will yield a prescribed accuracy.

EXAMPLE 2 Using (2)

Determine a value of n so that (1) will give an approximation to $\int_1^2 (1/x)\, dx$ that is accurate to two decimal places.

Solution The Midpoint Rule will be accurate to two decimal places for those values of n for which the upper bound $M(b-a)^3/24n^2$ for the error is strictly less than 0.005. For $f(x) = 1/x$, we have $f''(x) = 2/x^3$. Since f'' decreases on $[1, 2]$, it follows that $f''(x) \le f''(1) = 2$ for all x in the interval. Thus, with $M = 2$, $b - a = 1$, we want

◀ If we want accuracy to three decimal places, we use 0.0005, and so on.

$$\frac{2(1)^3}{24n^2} < 0.005 \qquad \text{or} \qquad n^2 > \frac{50}{3} \approx 16.67.$$

By taking $n \ge 5$ we obtain the desired accuracy. ∎

Example 2 indicates that the third approximation $\int_1^2 (1/x)\, dx \approx 0.6919$ obtained in Example 1 is accurate to two decimal places. By way of comparison, the exact value of the integral, using Theorem 5.5.1,

$$\int_1^2 \frac{1}{x}\, dx = \ln x \Big]_1^2 = \ln 2 - \ln 1 = \ln 2 \approx 0.6931$$

is correct to four decimal places. Thus, for $n = 5$ the error in the method E_n is approximately 0.0012.

■ **Trapezoidal Rule** A more popular method for approximating an integral is based on the plausibility that a better estimate of $\int_a^b f(x)\,dx$ can be obtained by adding the areas of trapezoids instead of the areas of rectangles. See **FIGURE 7.8.3(a)**. The area of the trapezoid shown in Figure 7.8.3(b) is $h(l_1 + l_2)/2$. Thus, the area A_k of the trapezoidal element shown in Figure 7.8.3(c) is

$$A_k = \Delta x \frac{f(x_{k-1}) + f(x_k)}{2}.$$

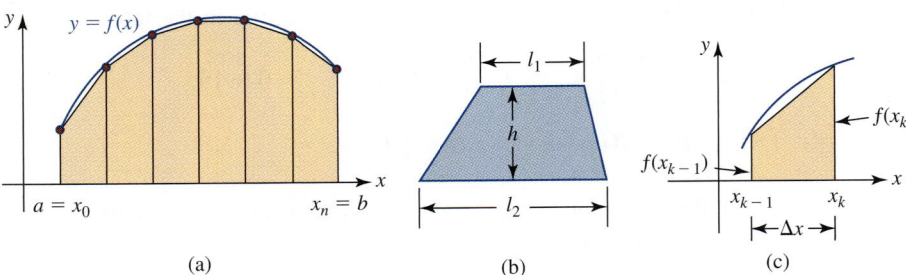

(a)   (b)   (c)

FIGURE 7.8.3 Using n trapezoids to approximate the definite integral

Thus, for a regular partition of the interval $[a, b]$ on which f is continuous, we obtain

$$\int_a^b f(x)\,dx \approx \Delta x \frac{f(x_0) + f(x_1)}{2} + \Delta x \frac{f(x_1) + f(x_2)}{2} + \cdots + \Delta x \frac{f(x_{n-1}) + f(x_n)}{2}.$$

We summarize this new approximation rule in the next definition after we combine like terms and substitute $\Delta x = (b - a)/n$.

Definition 7.8.2 Trapezoidal Rule

The **Trapezoidal Rule** is the approximation $\int_a^b f(x)\,dx \approx T_n$, where

$$T_n = \frac{b - a}{2n}[f(x_0) + 2f(x_1) + 2f(x_2) + \cdots + 2f(x_{n-1}) + f(x_n)]. \tag{3}$$

■ **Error in the Trapezoidal Rule** The error in the method for the Trapezoidal Rule is given by $E_n = |I - T_n|$, where $I = \int_a^b f(x)\,dx$. As the next theorem shows, the error bound for the Trapezoidal Rule is almost the same as that for the Midpoint Rule.

Theorem 7.8.2 Error Bound for Trapezoidal Rule

If there exists a number $M > 0$ such that $|f''(x)| \leq M$ for all x in $[a, b]$, then

$$E_n \leq \frac{M(b - a)^3}{12n^2}. \tag{4}$$

---

k	x_k	$f(x_k)$
0	1	1
1	$\frac{7}{6}$	$\frac{6}{7}$
2	$\frac{4}{3}$	$\frac{3}{4}$
3	$\frac{3}{2}$	$\frac{2}{3}$
4	$\frac{5}{3}$	$\frac{3}{5}$
5	$\frac{11}{6}$	$\frac{6}{11}$
6	2	$\frac{1}{2}$

EXAMPLE 3 Using (4) and (3)

Determine a value of n so that the Trapezoidal Rule will give an approximation to $\int_1^2 (1/x)\,dx$ that is accurate to two decimal places. Approximate the integral.

Solution Using the information in Example 2, we have immediately:

$$\frac{2(1)^3}{12n^2} < 0.005 \quad \text{or} \quad n^2 > \frac{100}{3} \approx 33.33.$$

In this case we take $n \geq 6$ to obtain the desired accuracy. Hence, $\Delta x = \frac{1}{6}$, $x_0 = 1$, $x_1 = 1 + \Delta x = \frac{7}{6}, \ldots, x_6 = 1 + 6\Delta x = 2$. With the information in the accompanying table (3), gives

$$T_6 = \frac{1}{12}\left[1 + 2\left(\frac{6}{7}\right) + 2\left(\frac{3}{4}\right) + 2\left(\frac{2}{3}\right) + 2\left(\frac{3}{5}\right) + 2\left(\frac{6}{11}\right) + \frac{1}{2}\right] \approx 0.6949. \quad \blacksquare$$

EXAMPLE 4   Using (4) and (3)

Approximate $\displaystyle\int_{1/2}^{1} \cos \sqrt{x}\, dx$ by the Trapezoidal Rule so that the error is less than 0.001.

Solution   The second derivative of $f(x) = \cos \sqrt{x}$ is

$$f''(x) = \frac{1}{4x}\left(\frac{\sin \sqrt{x}}{\sqrt{x}} - \cos \sqrt{x}\right).$$

For x in the interval $\left[\frac{1}{2}, 1\right]$ we have $0 < (\sin \sqrt{x})/\sqrt{x} \leq 1$ and $0 < \cos \sqrt{x} \leq 1$ and consequently $|f''(x)| \leq \frac{1}{4x}$. Therefore, on the interval, $|f''(x)| \leq \frac{1}{2}$. Thus, with $M = \frac{1}{2}$ and $b - a = \frac{1}{2}$, it follows from (4) that we want

$$\frac{\frac{1}{2}\left(\frac{1}{2}\right)^3}{12n^2} < 0.001 \qquad \text{or} \qquad n^2 > \frac{125}{24} \approx 5.21.$$

Hence, to obtain the desired accuracy it suffices to choose $n = 3$ and $\Delta x = \frac{1}{6}$. With the aid of a calculator to obtain the information in the accompanying table, we find the following approximation for $\int_{1/2}^{1} \cos \sqrt{x}\, dx$ from (3):

$$T_3 = \frac{1}{12}\left[\cos\sqrt{\frac{1}{2}} + 2\cos\sqrt{\frac{2}{3}} + 2\cos\sqrt{\frac{5}{6}} + \cos 1\right] \approx 0.3244. \quad \blacksquare$$

k	x_k	$f(x_k)$
0	$\frac{1}{2}$	0.7602
1	$\frac{2}{3}$	0.6848
2	$\frac{5}{6}$	0.6115
3	1	0.5403

Although not obvious from a figure, an improved method of approximating a definite integral $\int_a^b f(x)\, dx$ can be obtained by considering a series of parabolic arcs instead of a series of chords used in the Trapezoidal Rule. It can be proved, under certain conditions, that a parabolic arc passing through *three* specified points will "fit" the graph of f better than a single straight line. See **FIGURE 7.8.4**. By adding the areas under the parabolic arcs, we obtain an approximation to the integral.

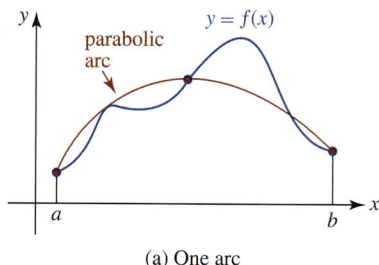

(a) One arc

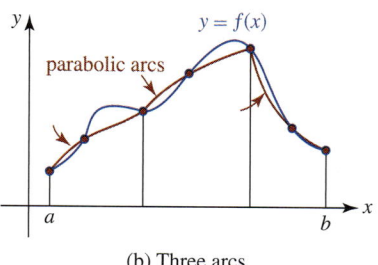

(b) Three arcs

FIGURE 7.8.4   Fitting a parabolic arc through three consecutive points on the graph of a function

To begin, let us find the area under an arc of a parabola that passes through three points $P_0(x_0, y_0)$, $P_1(x_1, y_1)$, and $P_2(x_2, y_2)$, where $x_0 < x_1 < x_2$ and $x_1 - x_0 = x_2 - x_1 = h$. As shown in **FIGURE 7.8.5**, this can be done by finding the area under the graph of $y = Ax^2 + Bx + C$ on the interval $[-h, h]$ so that P_0, P_1, and P_2 have coordinates $(-h, y_0)$, $(0, y_1)$, and (h, y_2), respectively. The interval $[-h, h]$ is chosen for simplicity; the area in question does not depend on the location of the y-axis. Using Theorem 5.5.1, the area is

$$\int_{-h}^{h} (Ax^2 + Bx + C)\, dx = \frac{h}{3}(2Ah^2 + 6C). \qquad (5)$$

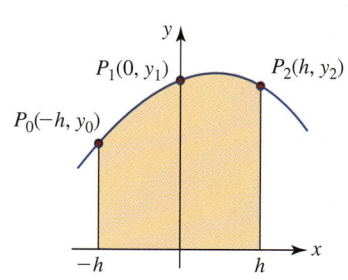

FIGURE 7.8.5   Area under a parabolic arc

But, since the graph is to pass through $(-h, y_0)$, $(0, y_1)$, and (h, y_2), we must have

$$y_0 = Ah^2 - Bh + C \tag{6}$$

$$y_1 = C \tag{7}$$

$$y_2 = Ah^2 + Bh + C. \tag{8}$$

By adding (6) and (8) and using (7), we find $2Ah^2 = y_0 + y_2 - 2y_1$. Thus, (5) can be expressed as

$$\text{area} = \frac{h}{3}(y_0 + 4y_1 + y_2). \tag{9}$$

▌ Simpson's Rule Now suppose that the interval $[a, b]$ is partitioned into n subintervals of equal width $\Delta x = (b - a)/n$, where n is *an even integer*. As shown in FIGURE 7.8.6, on each subinterval $[x_{k-2}, x_k]$ of width $2\Delta x$ we approximate the graph of f by an arc of a parabola through points P_{k-2}, P_{k-1}, and P_k on the graph that corresponds to the endpoints and midpoint of the subinterval. If A_k denotes the area under the parabola on $[x_{k-2}, x_k]$, it follows from (9) that

$$A_k = \frac{\Delta x}{3}[f(x_{k-2}) + 4f(x_{k-1}) + f(x_k)].$$

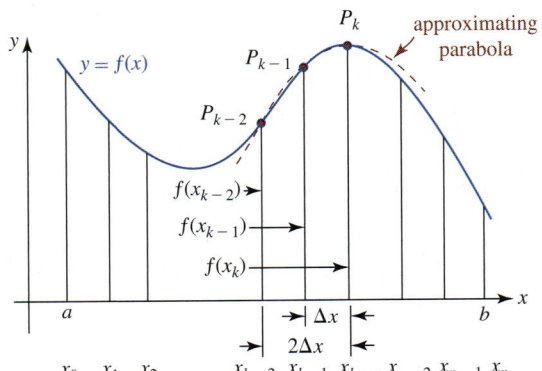

FIGURE 7.8.6 Approximating the function by a parabolic arc

Thus, summing all the A_k gives

$$\int_a^b f(x)\,dx \approx \frac{\Delta x}{3}[f(x_0) + 4f(x_1) + f(x_2)] + \frac{\Delta x}{3}[f(x_2) + 4f(x_3) + f(x_4)] + \cdots + \frac{\Delta x}{3}[f(x_{n-2}) + 4f(x_{n-1}) + f(x_n)].$$

This approximation rule, named after the English mathematician **Thomas Simpson** (1710–1761), is summarized in the following definition.

Definition 7.8.3 Simpson's Rule

Simpson's Rule is the approximation $\displaystyle\int_a^b f(x)\,dx \approx S_n$, where

$$S_n = \frac{b-a}{3n}[f(x_0) + 4f(x_1) + 2f(x_2) + 4f(x_3) + \cdots + 2f(x_{n-2}) + 4f(x_{n-1}) + f(x_n)]. \tag{10}$$

We note again that the integer n in (10) must be even, since each A_k represents the area under a parabolic arc on a subinterval of width $2\Delta x$.

▌ Error in Simpson's Rule If $I = \int_a^b f(x)\,dx$, the next theorem establishes an upper bound for the error in the method $E_n = |I - S_n|$ using an upper bound on the fourth derivative.

Theorem 7.8.3   Error Bound for Simpson's Rule

If there exists a number $M > 0$ such that $|f^{(4)}(x)| \le M$ for all x in $[a, b]$, then

$$E_n \le \frac{M(b-a)^5}{180n^4}. \tag{11}$$

EXAMPLE 5   Using (11)

Determine a value of n so that (10) will give an approximation to $\displaystyle\int_1^2 (1/x)\, dx$ that is accurate to two decimal places.

Solution For $f(x) = 1/x$, $f^{(4)}(x) = 24/x^5$ and on $[1, 2]$, $f^{(4)}(x) \le f^{(4)}(1) = 24$. Thus, with $M = 24$ it follows from (11) that

$$\frac{24(1)^5}{180n^4} < 0.005 \qquad \text{or} \qquad n^4 > \frac{80}{3} \approx 26.67$$

and so $n > 2.27$. Since n must be an even integer, it suffices to take $n \ge 4$. ∎

EXAMPLE 6   Using (10)

Approximate $\displaystyle\int_1^2 (1/x)\, dx$ by Simpson's Rule for $n = 4$.

Solution When $n = 4$, we have $\Delta x = \frac{1}{4}$. From (10) and the accompanying table we obtain

$$S_4 = \frac{1}{12}\left[1 + 4\left(\frac{4}{5}\right) + 2\left(\frac{2}{3}\right) + 4\left(\frac{4}{7}\right) + \frac{1}{2}\right] \approx 0.6933.$$ ∎

k	m_k	$f(m_k)$
0	1	1
1	$\frac{5}{4}$	$\frac{4}{5}$
2	$\frac{3}{2}$	$\frac{2}{3}$
3	$\frac{7}{4}$	$\frac{4}{7}$
4	2	$\frac{1}{2}$

In Example 6, keep in mind that even though we are using $n = 4$, the definite integral $\int_1^2 (1/x)\, dx$ is being approximated by the area under only two parabolic arcs. Recall that the Midpoint Rule gave $\int_1^2 (1/x)\, dx \approx 0.6919$ with $n = 5$, the Trapezoidal Rule gave $\int_1^2 (1/x)\, dx \approx 0.6949$ with $n = 6$, and 0.6931 is an approximation of the integral correct to four decimal places.

In some applications it may only be possible to obtain numerical values of a quantity $Q(x)$—say, by measurements or by experiment—at specific points in some interval $[a, b]$, and yet it may be necessary to have some idea of the value of the definite integral $\int_a^b Q(x)\, dx$. Even though Q is not defined by means of an explicit formula, we may still be able to apply the Trapezoidal Rule or Simpson's Rule to approximate the integral.

EXAMPLE 7   Area of a Plot of Land

Suppose we wish to find the area of an irregularly shaped piece of land that is bounded between a straight road and the shore of a lake. The boundaries of the land are indicated by the dashed lines in FIGURE 7.8.7(a). Suppose we divide the indicated 1-mi boundary along the road into, say, $n = 8$ subintervals and then, as shown in Figure 7.8.7(b), measure the perpendicular distances from the road to the shore of the lake. We are now in position to approximate the area of the land $A = \int_a^b f(x)\, dx$ by Simpson's Rule. With $b - a = 1$ mi $= 5280$ ft, $\Delta x = (b - a)/n = 5280/8 = 660$, and the identifications $f(x_0) = 83, \ldots, f(x_8) = 28$, (10) gives the following approximation for A:

$$S_8 = \frac{660}{3}[83 + 4(82) + 2(96) + 4(100) + 2(82) + 4(55) + 2(63) + 4(54) + 28]$$

$$= 386{,}540 \text{ ft}^2.$$

Using the fact that 1 acre $= 43,560 \text{ ft}^2$, we see that the land is approximately 8.9 acres.

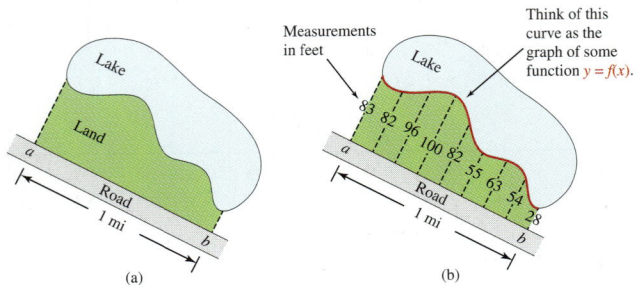

FIGURE 7.8.7   Lakeside land in Example 7

$\int_a^b$ **NOTES FROM THE CLASSROOM**

(*i*) The popularity of the Trapezoidal Rule notwithstanding, a direct comparison of the error bounds (2) and (4) shows that the Midpoint Rule is actually more accurate than the Trapezoidal Rule. Specifically, (2) suggests that in some cases the error in the Midpoint Rule can be one-half the error in the Trapezoidal Rule. See Problem 33 in Exercises 7.8.

(*ii*) Under some circumstances the rules considered in the foregoing discussion will give the *exact* value of an integral $\int_a^b f(x)\, dx$. The error bounds (2) and (4) indicate that M_n and T_n will yield the precise value whenever f is a linear function. See Problems 31, 32, and 35 in Exercises 7.8. Simpson's Rule will give the exact value of $\int_a^b f(x)\, dx$ whenever f is a linear, quadratic, or cubic polynomial function. See Problems 34 and 36 in Exercises 7.8.

(*iii*) In general, Simpson's Rule will give greater accuracy than either the Midpoint or the Trapezoidal Rule. So why should we even bother with these other rules? In some instances, the slightly simpler Midpoint and Trapezoidal Rules will yield accuracy that is sufficient for the purpose at hand. Furthermore, the requirement that n must be an even integer in Simpson's Rule may prevent its application to a given problem. Also, to find an error bound for Simpson's Rule, we must compute and then find an upper bound for the fourth derivative. The expression for $f^{(4)}(x)$ can, of course, be very complicated. The error bounds for the other two rules depend on the second derivative.

Exercises 7.8   Answers to selected odd-numbered problems begin on page ANS-24.

≡ **Fundamentals**

In Problems 1 and 2, compare the exact value of the integral with the approximation obtained from the Midpoint Rule for the indicated value of n.

1. $\displaystyle\int_1^4 (3x^2 + 2x)\, dx; \quad n = 3$ **2.** $\displaystyle\int_0^{\pi/6} \cos x\, dx; \quad n = 4$

In Problems 3 and 4, compare the exact value of the integral with the approximation obtained from the Trapezoidal Rule for the indicated value of n.

3. $\displaystyle\int_1^3 (x^3 + 1)\, dx; \quad n = 4$ **4.** $\displaystyle\int_0^2 \sqrt{x + 1}\, dx; \quad n = 6$

In Problems 5–12, use the Midpoint Rule and the Trapezoidal Rule to obtain an approximation to the given integral for the indicated value of n.

5. $\displaystyle\int_1^6 \frac{1}{x}\, dx; \quad n = 5$ **6.** $\displaystyle\int_0^2 \frac{1}{3x + 1}\, dx; \quad n = 4$

7. $\displaystyle\int_0^1 \sqrt{x^2 + 1}\, dx; \quad n = 10$ **8.** $\displaystyle\int_1^2 \frac{1}{\sqrt{x^3 + 1}}\, dx; \quad n = 5$

9. $\displaystyle\int_0^\pi \frac{\sin x}{x + \pi}\, dx; \quad n = 6$ **10.** $\displaystyle\int_0^{\pi/4} \tan x\, dx; \quad n = 3$

11. $\displaystyle\int_0^2 \cos x^2\, dx; \quad n = 6$

12. $\int_0^1 \frac{\sin x}{x}\, dx;\quad n = 5$ [*Hint:* Define $f(0) = 1$.]

In Problems 13 and 14, compare the exact value of the integral with the approximation obtained from Simpson's Rule for the indicated value of n.

13. $\int_0^4 \sqrt{2x + 1}\, dx;\quad n = 4$ **14.** $\int_0^{\pi/2} \sin^2 x\, dx;\quad n = 2$

In Problems 15–22, use Simpson's Rule to obtain an approximation to the given integral for the indicated value of n.

15. $\int_{1/2}^{5/2} \frac{1}{x}\, dx;\quad n = 4$ **16.** $\int_0^5 \frac{1}{x + 2}\, dx;\quad n = 6$

17. $\int_0^1 \frac{1}{1 + x^2}\, dx;\quad n = 4$ **18.** $\int_{-1}^1 \sqrt{x^2 + 1}\, dx;\quad n = 2$

19. $\int_0^\pi \frac{\sin x}{x + \pi}\, dx;\quad n = 6$ **20.** $\int_0^1 \cos \sqrt{x}\, dx;\quad n = 4$

21. $\int_2^4 \sqrt{x^3 + x}\, dx;\quad n = 4$ **22.** $\int_{\pi/4}^{\pi/2} \frac{1}{2 + \sin x}\, dx;\quad n = 2$

23. Determine the number of rectangles needed so that an approximation to $\int_{-1}^2 dx/(x + 3)$ is accurate to two decimal places.

24. Determine the number of trapezoids needed so that the error in an approximation to $\int_0^{1.5} \sin^2 x\, dx$ is less than 0.0001.

25. Use the Trapezoidal Rule so that an approximation to the area under the graph of $f(x) = 1/(1 + x^2)$ on $[0, 2]$ is accurate to two decimal places. [*Hint:* Examine $f'''(x)$.]

26. The domain of $f(x) = 10^x$ is the set of real numbers and $f(x) > 0$ for all x. Use the Trapezoidal Rule to approximate the area under the graph of f on $[-2, 2]$ with $n = 4$.

27. Using Simpson's Rule, determine n so that the error in approximating $\int_1^3 dx/x$ is less than 10^{-5}. Compare with the n needed in the Trapezoidal Rule to give the same accuracy.

28. Find an upper bound for the error in approximating $\int_0^3 dx/(2x + 1)$ by Simpson's Rule with $n = 6$.

In Problems 29 and 30, use the data given in the table and an appropriate rule to approximate the indicated definite integral.

29. $\int_{2.05}^{2.30} f(x)\, dx;$

x	2.05	2.10	2.15	2.20	2.25	2.30
$f(x)$	4.91	4.80	4.66	4.41	3.93	3.58

30. $\int_0^{1.20} f(x)\, dx;$

x	0.0	0.1	0.2	0.4	0.6	0.8	0.9	1.00	1.20
$f(x)$	-0.72	-0.55	-0.16	0.62	0.78	1.34	1.47	1.61	1.51

31. Compare the exact value of the integral $\int_0^4 (2x + 5)\, dx$ with the approximation obtained from the Midpoint Rule with $n = 2$ and $n = 4$.

32. Repeat Problem 31 using the Trapezoidal Rule.

33. (a) Find the exact value of the integral $I = \int_{-1}^1 (x^3 + x^2)\, dx$.
(b) Use the Midpoint Rule with $n = 8$ to find an approximation to I.
(c) Use the Trapezoidal Rule with $n = 8$ to find an approximation to I.
(d) Compare the errors $E_8 = |I - M_8|$ and $E_8 = |I - T_8|$.

34. Compare the exact value of the integral $\int_{-1}^3 (x^3 - x^2)\, dx$ with the approximations obtained from Simpson's Rule with $n = 2$ and $n = 4$.

35. Prove that the Trapezoidal Rule will give the exact value of $\int_a^b f(x)\, dx$ when $f(x) = c_1 x + c_0$, with c_0 and c_1 constants. Geometrically, why does this make sense?

36. Prove that Simpson's Rule will give the exact value of $\int_a^b f(x)\, dx$ where $f(x) = c_3 x^3 + c_2 x^2 + c_1 x + c_0$, with $c_0, c_1, c_2,$ and c_3 constants.

37. Use the data given in FIGURE 7.8.8 and Simpson's Rule to find an approximation to the area under the graph of the continuous function f on the interval $[1, 4]$.

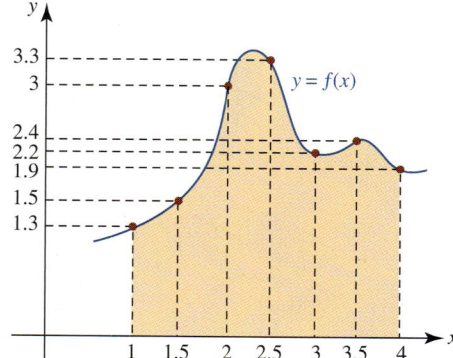

FIGURE 7.8.8   Graph for Problem 37

38. Use the Trapezoidal Rule with $n = 9$ to find an approximation to the area under the graph in FIGURE 7.8.9. Does the Trapezoidal Rule give the exact value of the area?

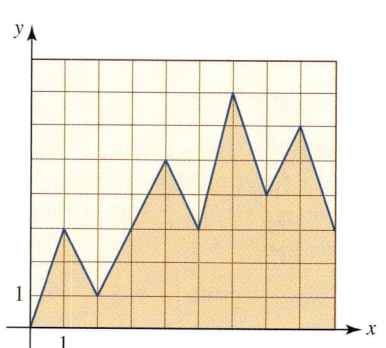

FIGURE 7.8.9   Graph for Problem 38

39. The large irregularly shaped fish pond shown in FIGURE 7.8.10 is filled with water to a uniform depth of 4 ft. Use Simpson's Rule to find an approximation to the number of gallons of water in the tank. Measurements are in feet; the vertical spacing between the horizontal measurements is 1.86 ft. There are 7.48 gal in 1 ft^3 of water.

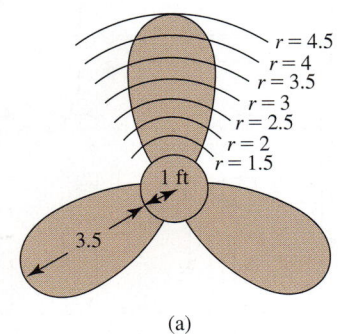

FIGURE 7.8.10   Pond in Problem 39

40. The moment of inertia I of a three-bladed ship's propeller whose dimensions are shown in **FIGURE 7.8.11(a)** is given by

$$I = \frac{3\rho\pi}{2g} + \frac{3\rho}{g}\int_1^{4.5} r^2 A\, dr,$$

where ρ is the density of the metal, g is the acceleration of gravity, and A is the area of a cross section of the propeller at a distance r ft from the center of the hub. If $\rho = 570$ lb/ft^3 for bronze, use the data in Figure 7.8.11(b) and the Trapezoidal Rule to find an approximation to I.

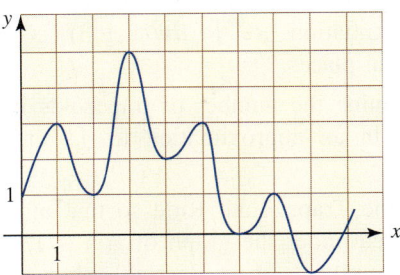

(a)

r (ft)	1	1.5	2	2.5	3	3.5	4	4.5
A (ft)	0.3	0.50	0.62	0.70	0.60	0.50	0.27	0

(b)

FIGURE 7.8.11   Propeller in Problem 40

≡ Calculator/CAS Problems

In Problems 41 and 42, use a calculator or CAS to obtain the graph of the given function. Use Simpson's Rule to approximate the area bounded by the graph of f and the x-axis on the indicated interval. Use $n = 10$.

41. $f(x) = \sqrt[5]{(5^{2.5} - |x|^{2.5})^2}$; $[-5, 5]$

42. $f(x) = 1 + |\sin x|^x$; $[0, 2\pi]$ [*Hint:* Use the graph to discern $f(0)$.]

43. (a) Show that the convergent integral $\displaystyle\int_1^\infty \frac{e^{1/x}}{x^{5/2}}\, dx$ can be written as $\int_0^1 t^{1/2} e^t\, dt$.
   (b) Use the result of part (a) and Simpson's Rule with $n = 4$ to find an approximation to the original improper integral.

44. Use (3) of Section 6.5 and Simpson's Rule with $n = 4$ to find an approximation to the length L of the graph of $y = \frac{1}{3}x^3 + 1$ from the point $(0, 1)$ to $\left(2, \frac{11}{3}\right)$.

45. Use (3) of Section 6.5 and the Trapezoidal Rule with $n = 10$ to find an approximation to the length L of the graph of $y = x^2$ from the origin $(0, 0)$ to the point $(1, 1)$.

46. Use (3) of Section 6.5 and Simpson's Rule with $n = 6$ to find an approximation to the length L of the graph of $y = \ln x$ on the interval $[1, 2]$.

47. Use (3) of Section 6.6 and the Midpoint Rule with $n = 5$ to find an approximation to the area S of the surface that is formed by revolving the graph of $y = \frac{1}{2}x^2$ on the interval $[0, 2]$ about the x-axis.

48. Use Simpson's Rule with $n = 6$ to find an approximation to the area S of the surface that is formed by revolving the graph of $x = y^2 + 1$ for $-1 \le y \le 1$ about the y-axis.

≡ Think About It

49. (a) Estimate the length L of the graph given in **FIGURE 7.8.12** on the interval $[1, 8]$.
   (b) Explain why using the Trapezoidal Rule with $n = 7$ is not particularly a good idea.

FIGURE 7.8.12   Graph for Problem 49

50. A Bit of History   The **logarithmic integral function**, $\text{Li}(x)$, is defined as the integral

$$\text{Li}(x) = \int_2^x \frac{1}{\ln t}\, dt$$

for $x > 2$. In 1896, the French mathematician **Jacques Hadamard** (1865–1963) and the Belgian mathematician **Charles-Jean de la Vallée Poussin** (1886–1962) independently proved the Prime Number Theorem, which that the number of prime numbers (2, 3, 5, 7, 11, etc.) less than or equal to x, denoted $\pi(x)$, can be approximated by the logarithmic integeral meaning that

$$\lim_{x\to\infty} \frac{\pi(x)}{\text{Li}(x)} = 1.$$

   (a) Show that $\pi(x)$ can also be approximated by the function $x/\ln x$ by using L'Hôpital's Rule and the Fundamental Theorem of Calculus to show that

$$\lim_{x\to\infty} \frac{\text{Li}(x)}{x/\ln x} = 1.$$

   Since there are an infinite number of primes, $\text{Li}(x) \to \infty$ as $x \to \infty$.
   (b) Use Simpson's Rule to approximate $\text{Li}(100)$. Compute $x/\ln x$ for $x = 100$. Compare these numbers with the actual number of prime numbers less than 100.

Chapter 7 in Review

Answers to selected odd-numbered problems begin on page ANS-24.

A. True/False

In Problems 1–20, indicate whether the given statement is true or false.

1. Under the change of variable $u = 2x + 3$, the integral $\int_1^5 \dfrac{4x}{\sqrt{2x + 3}}\, dx$ becomes $\int_5^{13}(u^{1/2} - 3u^{-1/2})\, du.$ _____

2. The trigonometric substitution $u = a\sec\theta$ is appropriate for integrals that contain $\sqrt{a^2 + u^2}.$ _____

3. The method of integration by parts is derived from the Product Rule for differentiation. _____

4. $\int_1^e 2x \ln x^2\, dx = e^2 + 1.$ _____

5. Partial fractions are not applicable to $\displaystyle\int \dfrac{1}{(x - 1)^3}\, dx.$ _____

6. A partial fraction decomposition of $x^2/(x + 1)^2$ can be found having the form $A/(x + 1) + B/(x + 1)^2$, where A and B are constants. _____

7. To evaluate $\displaystyle\int \dfrac{1}{(x^2 - 1)^2}\, dx$, we assume constants A, B, C, and D can be found such that

$$\dfrac{1}{(x^2 - 1)^2} = \dfrac{Ax + B}{x^2 - 1} + \dfrac{Cx + D}{(x^2 - 1)^2}.$$ _____

8. To evaluate $\int x^n e^x\, dx$, n a positive integer, integration by parts is used $n - 1$ times. _____

9. To evaluate $\displaystyle\int \dfrac{x}{\sqrt{9 - x^2}}\, dx$, it is necessary to use $x = 3\sin\theta.$ _____

10. When evaluated, the integral $\int \sin^3 x \cos^2 x\, dx$ can be expressed as a sum of powers of $\cos x.$ _____

11. If $\int_a^\infty f(x)\, dx$ and $\int_a^\infty g(x)\, dx$ converge, then $\int_a^\infty [f(x) + g(x)]\, dx$ converges. _____

12. If $\int_a^\infty [f(x) + g(x)]\, dx$ converges, then $\int_a^\infty f(x)\, dx$ converges. _____

13. If f is continuous for all x and $\int_{-\infty}^a f(x)\, dx$ diverges, then $\int_{-\infty}^\infty f(x)\, dx$ diverges. _____

14. The integral $\int_{-\infty}^\infty f(x)\, dx$ is defined by $\displaystyle\lim_{x\to\infty}\int_{-t}^t f(x)\, dx.$ _____

15. $\displaystyle\int_{\frac{1}{2}}^1 \dfrac{1}{1 + \ln x}\, dx$ is an improper integral. _____

16. $\int_{-1}^1 x^{-3}\, dx = 0.$ _____

17. $\int_0^4 x^{-0.999}\, dx$ converges. _____

18. $\int_1^\infty x^{-0.999}\, dx$ diverges. _____

19. $\displaystyle\int_2^\infty \left[\dfrac{e^x}{e^x + 1} - \dfrac{e^x}{e^x - 1} \right] dx$ diverges, since $\displaystyle\int_2^\infty \dfrac{e^x}{e^x + 1}\, dx$ diverges. _____

20. If a positive function f has an infinite discontinuity at a number in $[a, b]$, then the area under the graph on the interval is also infinite. _____

B. Fill in the Blanks

In Problems 1–6, fill in the blanks.

1. $\int_0^\infty e^{-5x}\, dx =$ _____.

2. If $\int_0^\infty e^{-x^2}\, dx = \sqrt{\pi}/2$, then $\int_{-\infty}^\infty e^{-x^2}\, dx =$ _____.

3. If $\int_0^\infty e^{-x^2}\, dx = \sqrt{\pi}/2$, then $\displaystyle\int_0^\infty \dfrac{e^{-x}}{\sqrt{x}}\, dx =$ _____.

4. The integral $\int_1^\infty x^p\, dx$ converges for $p <$ _____ and diverges for $p \geq$ _____.

5. $\int_0^x e^{-2t} \, dt = \int_x^\infty e^{-2t} \, dt$ for $x = $ _____.

6. $\int \sin x \ln (\sin x) \, dx = $ _____.

C. Exercises

In Problems 1–80, use the methods of this chapter, or previous chapters, to evaluate the given integral.

1. $\displaystyle\int \frac{1}{\sqrt{x} + 9} \, dx$

2. $\displaystyle\int e^{\sqrt{x+1}} \, dx$

3. $\displaystyle\int \frac{x}{\sqrt{x^2 + 4}} \, dx$

4. $\displaystyle\int \frac{1}{\sqrt{x^2 + 4}} \, dx$

5. $\displaystyle\int \frac{1}{(x^2 + 4)^3} \, dx$

6. $\displaystyle\int \frac{x^2}{x^2 + 4} \, dx$

7. $\displaystyle\int \frac{x^2 + 4}{x^2} \, dx$

8. $\displaystyle\int \frac{3x - 1}{x(x^2 - 4)} \, dx$

9. $\displaystyle\int \frac{x - 5}{x^2 + 4} \, dx$

10. $\displaystyle\int \frac{\sqrt[3]{x + 27}}{x} \, dx$

11. $\displaystyle\int \frac{(\ln x)^9}{x} \, dx$

12. $\displaystyle\int (\ln 3x)^2 \, dx$

13. $\displaystyle\int t \sin^{-1} t \, dt$

14. $\displaystyle\int \frac{\ln x}{(x - 1)^2} \, dx$

15. $\displaystyle\int (x + 1)^3(x - 2) \, dx$

16. $\displaystyle\int \frac{1}{(x + 1)^3(x - 2)} \, dx$

17. $\displaystyle\int \ln (x^2 + 4) \, dx$

18. $\displaystyle\int 8te^{2t^2} \, dt$

19. $\displaystyle\int \frac{1}{x^4 + 10x^3 + 25x^2} \, dx$

20. $\displaystyle\int \frac{1}{x^2 + 8x + 25} \, dx$

21. $\displaystyle\int \frac{x}{x^3 + 3x^2 - 9x - 27} \, dx$

22. $\displaystyle\int \frac{x + 1}{(x^2 - x)(x^2 + 3)} \, dx$

23. $\displaystyle\int \frac{\sin^2 t}{\cos^2 t} \, dt$

24. $\displaystyle\int \frac{\sin^3 \theta}{(\cos \theta)^{3/2}} \, d\theta$

25. $\displaystyle\int \tan^{10} x \sec^4 x \, dx$

26. $\displaystyle\int \frac{x \tan x}{\cos x} \, dx$

27. $\displaystyle\int y \cos y \, dy$

28. $\displaystyle\int x^2 \sin x^3 \, dx$

29. $\displaystyle\int (1 + \sin^2 t) \cos^3 t \, dt$

30. $\displaystyle\int \frac{\sec^3 \theta}{\tan \theta} \, d\theta$

31. $\displaystyle\int e^w(1 + e^w)^5 \, dw$

32. $\displaystyle\int (x - 1)e^{-x} \, dx$

33. $\displaystyle\int \cot^3 4x \, dx$

34. $\displaystyle\int (3 - \sec x)^2 \, dx$

35. $\displaystyle\int_0^{\pi/4} \cos^2 x \tan x \, dx$

36. $\displaystyle\int_0^{\pi/3} \sin^4 x \tan x \, dx$

37. $\displaystyle\int \frac{\sin x}{1 + \sin x} \, dx$

38. $\displaystyle\int \frac{\cos x}{1 + \sin x} \, dx$

39. $\displaystyle\int_0^1 \frac{1}{(x + 1)(x + 2)(x + 3)} \, dx$

40. $\displaystyle\int_{\ln 3}^{\ln 2} \sqrt{e^x + 1} \, dx$

41. $\displaystyle\int e^x \cos 3x \, dx$

42. $\displaystyle\int x(x - 5)^9 \, dx$

43. $\displaystyle\int \cos(\ln t)\, dt$

44. $\displaystyle\int \sec^2 x \ln(\tan x)\, dx$

45. $\displaystyle\int \cos\sqrt{x}\, dx$

46. $\displaystyle\int \frac{\cos\sqrt{x}}{\sqrt{x}}\, dx$

47. $\displaystyle\int \cos x \sin 2x\, dx$

48. $\displaystyle\int (\cos^2 x - \sin^2 x)\, dx$

49. $\displaystyle\int \sqrt{x^2 + 2x + 5}\, dx$

50. $\displaystyle\int \frac{1}{(8 - 2x - x^2)^{3/2}}\, dx$

51. $\displaystyle\int \tan^5 x \sec^3 x\, dx$

52. $\displaystyle\int \cos^4\frac{x}{2}\, dx$

53. $\displaystyle\int \frac{t^5}{1 + t^2}\, dt$

54. $\displaystyle\int \frac{1}{\sqrt{1 - x^2}}\, dx$

55. $\displaystyle\int \frac{5x^3 + x^2 + 6x + 1}{(x^2 + 1)^2}\, dx$

56. $\displaystyle\int \frac{\sqrt{x^2 + 9}}{x^2}\, dx$

57. $\displaystyle\int x\sin^2 x\, dx$

58. $\displaystyle\int (t + 1)^2 e^{3t}\, dt$

59. $\displaystyle\int e^{\sin x}\sin 2x\, dx$

60. $\displaystyle\int e^x \tan^2 e^x\, dx$

61. $\displaystyle\int_0^{\pi/6} \frac{\cos x}{\sqrt{1 + \sin x}}\, dx$

62. $\displaystyle\int_0^{\pi/2} \frac{1}{\sin x + \cos x}\, dx$

63. $\displaystyle\int \sinh^{-1} t\, dt$

64. $\displaystyle\int x\cot x^2\, dx$

65. $\displaystyle\int_3^8 \frac{1}{x\sqrt{x + 1}}\, dx$

66. $\displaystyle\int \frac{t + 3}{t^2 + 2t + 1}\, dt$

67. $\displaystyle\int \frac{\sec^4 3u}{\cot^{12} 3u}\, du$

68. $\displaystyle\int_0^2 x^5\sqrt{x^2 + 4}\, dx$

69. $\displaystyle\int \frac{3 + \sin x}{\cos^2 x}\, dx$

70. $\displaystyle\int \frac{\sin 2x}{5 + \cos^2 x}\, dx$

71. $\displaystyle\int x(1 + \ln x)^2\, dx$

72. $\displaystyle\int x\cos^2 x\, dx$

73. $\displaystyle\int e^x e^{e^x}\, dx$

74. $\displaystyle\int \frac{1}{\sqrt{x + 1} - \sqrt{x}}\, dx$

75. $\displaystyle\int \frac{2t}{1 + e^{t^2}}\, dt$

76. $\displaystyle\int \cos x \cos 2x\, dx$

77. $\displaystyle\int \frac{1}{\sqrt{1 - (5x + 2)^2}}\, dx$

78. $\displaystyle\int (\ln 2x)\ln x\, dx$

79. $\displaystyle\int \cos x \ln|\sin x|\, dx$

80. $\displaystyle\int \ln\!\left(\frac{x + 1}{x - 1}\right)\, dx$

In Problems 81–92, evaluate the given integral or show that it diverges.

81. $\displaystyle\int_0^3 x(x^2 - 9)^{-2/3}\, dx$

82. $\displaystyle\int_0^5 x(x^2 - 9)^{-2/3}\, dx$

83. $\displaystyle\int_{-\infty}^0 (x + 1)e^x\, dx$

84. $\displaystyle\int_0^\infty \frac{e^{2x}}{e^{4x} + 1}\, dx$

85. $\displaystyle\int_3^\infty \frac{1}{1 + 5x}\, dx$

86. $\displaystyle\int_0^\infty \frac{x}{(x^2 + 4)^2}\, dx$

87. $\displaystyle\int_0^e \ln\sqrt{x}\, dx$

88. $\displaystyle\int_0^{\pi/2} \frac{\sec^2 t}{\tan^3 t}\, dt$

89. $\displaystyle\int_0^{\pi/2} \frac{1}{1 - \cos x}\, dx$

90. $\displaystyle\int_0^{\infty} \frac{x}{x + 1}\, dx$

91. $\displaystyle\int_0^{1} \frac{1}{\sqrt{x}\, e^{\sqrt{x}}}\, dx$

92. $\displaystyle\int_0^{\infty} \frac{1}{\sqrt{x}\, e^{\sqrt{x}}}\, dx$

In Problems 93 and 94, prove the given result.

93. $\displaystyle\int_1^{\infty} \frac{\sqrt{x}}{(1 + x)^2}\, dx = \frac{\pi}{4} + \frac{1}{2}$

94. $\displaystyle\int_0^{\infty} \frac{1}{\sqrt{x}(x + 1)}\, dx = \pi$

In Problems 95 and 96, use the fact that $\displaystyle\int_0^{\infty} e^{t^2}\, dt = \lim_{x \to \infty} \int_0^{x} e^{t^2}\, dt = \infty$ to evaluate the given limit.

95. $\displaystyle\lim_{x \to \infty} \frac{x \displaystyle\int_0^{x} e^{t^2}\, dt}{e^{x^2}}$

96. $\displaystyle\lim_{x \to \infty} \frac{\displaystyle\int_0^{x} e^{t^2}\, dt}{x e^{x^2}}$

97. Find the area of the region that is bounded by the graphs of $y = e^{-x}$ and $y = e^{-3x}$ on $[0, \infty)$.

98. Consider the region that is bounded by the graphs of $y = 1/\sqrt[3]{1 - x}$ and $y = 0$ on the interval $[0, 1]$.

    (a) Find the area of the region.

    (b) Find the volume of the solid of revolution that is formed by revolving the region about the x-axis.

    (c) Find the volume of the solid of revolution that is formed by revolving the region about the line $x = 1$.

99. Consider the graph of $f(x) = (x^2 - 1)/(x^2 + 1)$ given in FIGURE 7.R.1.

    (a) Determine whether the region R_1, which is bounded between the graph of f and its horizontal asymptote, is finite.

    (b) Determine whether the regions R_2 and R_3 have finite areas.

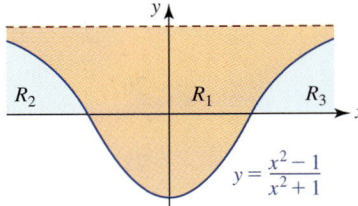

FIGURE 7.R.1 Graph for Problem 99

100. Use Newton's Method to find the number x^* for which the shaded region R in FIGURE 7.R.2 is 99% of the total area under the graph of $y = xe^{-x}$ on $[0, \infty)$.

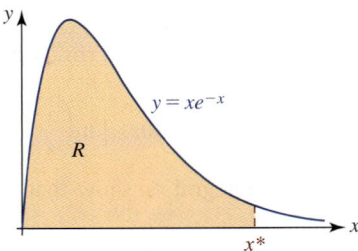

FIGURE 7.R.2 Graph for Problem 100

101. A continuous variable force $f(x)$ acts over the interval $[0, 1]$, where F is measured in Newtons and x in meters. It is determined empirically that

x (m)	0	0.2	0.4	0.6	0.8	1
$F(x)$ (N)	0	50	90	150	210	260

Use an appropriate numerical technique to approximate the work done over the interval.

102. The graph of a variable force F is given in FIGURE 7.R.3.

(a) Use rectangular elements of area to find an approximation to the work done by the force in moving a particle from $x = 1$ to $x = 5$.

(b) Use the Trapezoidal Rule to approximate the work done.

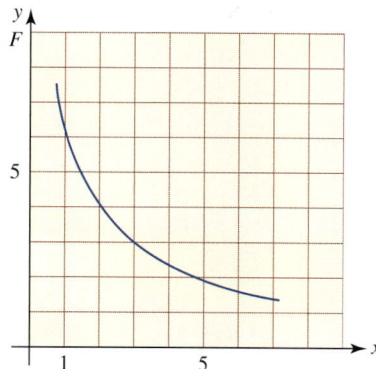

FIGURE 7.R.3   Graph for Problem 102

First-Order Differential Equations

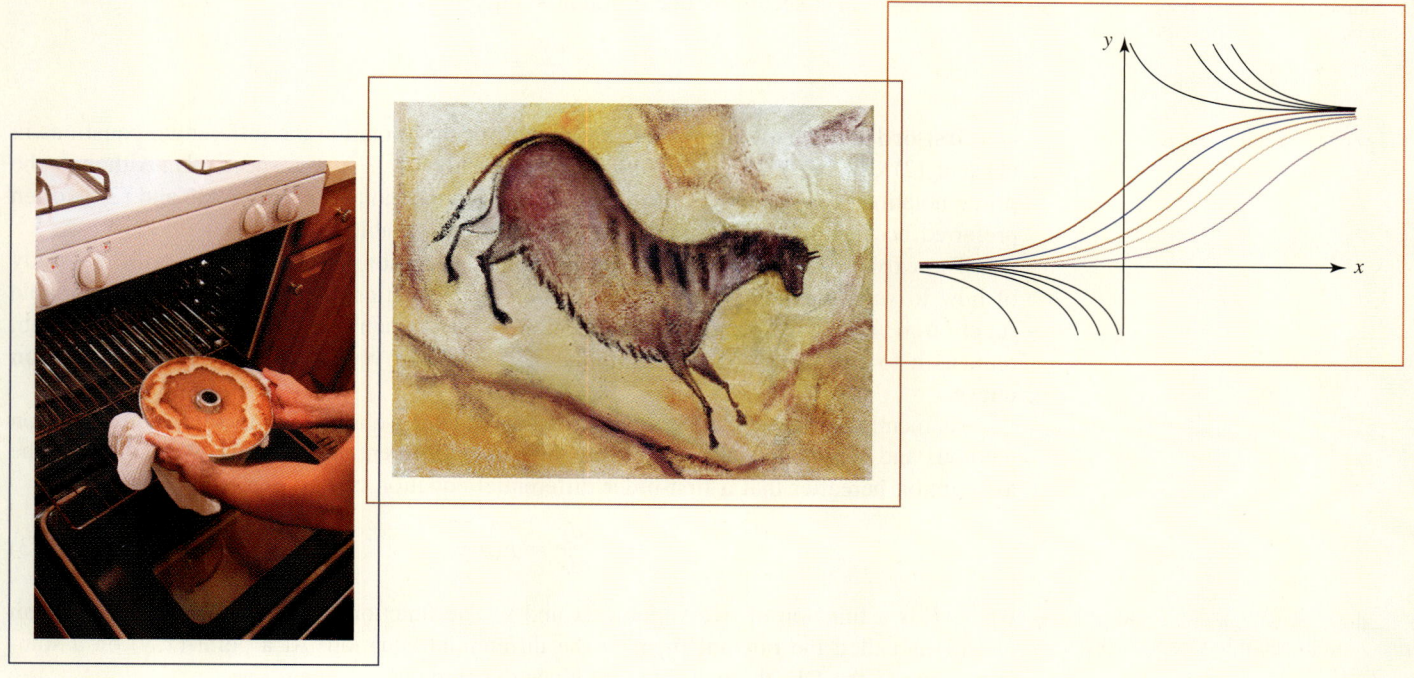

In This Chapter   We are going to study differential equations that have the form $dy/dx = F(x, y)$. These are called *first-order* differential equations. We will examine two solution methods and some applications of these equations. Higher-order differential equations are considered in Chapter 16.

8.1 Separable Equations

■ Introduction In several previous exercise sets you were asked to verify that a given function satisfies a **differential equation (DE)**. Roughly, a differential equation is an equation that involves an unknown function y and one or more of the derivatives of y. Differential equations are classified by the **order** of the highest derivative appearing in the equation. For example, the equation

highest-order derivative
$$\downarrow$$
$$\frac{d^2y}{dx^2} + 4\frac{dy}{dx} + 8y = 0 \tag{1}$$

▶ is an example of a **second-order** differential equation, whereas

$$\frac{dy}{dx} = -\frac{x}{y} \tag{2}$$

Of course, different symbols will often be used. For example,
$$\frac{d^2y}{dt^2} + 4y = \sin 2t$$
is also of a second-order DE.

is a **first-order** differential equation. Using the "prime" notation the differential equations in (1) and (2) can be written $y'' + 4y' + 8y = 0$ and $y' = -x/y$, respectively. Although the prime notation is easier to write and typeset, the Leibniz notation used in (1) and (2) is often preferred because it clearly displays the independent variable.

The exploration of the subject of differential equations usually begins with the study of how to *solve* them. A **solution** of a differential equation is a sufficiently differentiable function $y(x)$, defined explicitly or implicitly, that, when substituted into the equation, reduces it to an identity on some interval. The graph of $y(x)$ is naturally called a **solution curve**.

As mentioned in the chapter opener, in this chapter we are going to study some solution methods and some applications only for *first-order* differential equations. We will make the assumption hereafter that a first-order differential equation can be expressed as

$$\frac{dy}{dx} = F(x, y), \tag{3}$$

Functions of two variables will be discussed in detail in Chapter 13.

▶ where F is a function of two variables x and y. The function F is called the **slope function** and (3) is called the **normal form** of the differential equation. At a point (x, y) on a solution curve of the DE, the value $F(x, y)$ gives the slope of a tangent line.

■ A Definition We have already solved a simple kind of first-order differential equation in Section 5.1. Recall, the first-order DE,

$$\frac{dy}{dx} = g(x) \tag{4}$$

can be solved by finding the most general antiderivative of g; that is,

$$y = \int g(x)\, dx.$$

For example, a solution of the first-order DE

$$\frac{dy}{dx} = 2x + e^{-3x}$$

is given by

$$y = \int (2x + e^{-3x})\, dx = x^2 - \frac{1}{3}e^{-3x} + C.$$

Equations of the form in (4) are just a special case of a first-order differential equation $dy/dx = F(x, y)$, where the function F can be factored into a product of a function of x times a function of y.

> **Definition 8.1.1**   Separable Differential Equation
>
> A **separable first-order differential equation** is any equation $dy/dx = F(x, y)$ that can be put into the form
>
> $$\frac{dy}{dx} = g(x)f(y). \tag{5}$$

EXAMPLE 1   A Separable DE

The first-order DE

$$\frac{dy}{dx} = -\frac{x}{y} \tag{6}$$

is separable, since the right-hand side of the equality can be rewritten as the product of a function of x times a function of y:

$$\frac{dy}{dx} = \overset{g(x)}{-x} \cdot \overset{f(y)}{\frac{1}{y}}. \qquad\blacksquare$$

Notice that when $f(y) = 1$ in (5) we get (4). Analogous to differential equations of the form (4), we can also solve a separable DE by integration.

Before solving, we rewrite a separable equation in terms of differentials. For example, the equation in (6) can be written alternatively in the differential form

$$y\,dy = -x\,dx.$$

Similarly, by dividing by $f(y)$, (5) can be written as

$$p(y)\,dy = g(x)\,dx,$$

where for notational convenience we have written $p(y) = 1/f(y)$. Now if $y = \phi(x)$ denotes a solution of (5), we must have

$$p(\phi(x))\,\phi'(x) = g(x)$$

and therefore, by integration,

$$\int p(\phi(x))\,\phi'(x)\,dx = \int g(x)\,dx. \tag{7}$$

But $dy = \phi'(x)\,dx$, so (7) is the same as

$$\int p(y)\,dy = \int g(x)\,dx \qquad \text{or} \qquad H(y) = G(x) + C,$$

where $H(y)$ and $G(x)$ are antiderivatives of $p(y) = 1/f(y)$ and $g(x)$, respectively, and C is a constant. Since C is arbitrary, $H(y) = G(x) + C$ represents a **one-parameter family of solutions**. The parameter is the arbitrary constant C.

Note: There is no need to use two constants in the integration of a separable equation, because if we write $H(y) + C_1 = G(x) + C_2$, then the difference $C_2 - C_1$ can be replaced by a single constant C.

We summarize the discussion.

Guidelines for Solving a Separable DE

   (*i*) First, determine whether a first-order equation actually is separable. That is, can the DE be written in the form given in (5)?

(*ii*) If the DE is separable, then rewrite it in differential form:

$$p(y)\,dy = g(x)\,dx.$$

(*iii*) Integrate both sides of the differential form. Integrate the left-hand side with respect to y and the right-hand side with respect to x.

Before illustrating the above solution method, you should be aware that many first-order DEs are not separable. For example, neither of the differential equations

$$\frac{dy}{dx} = x^2 + y^2 \qquad \text{and} \qquad \frac{dy}{dx} = \sin(x + y)$$

are separable.

EXAMPLE 2   Solving a Separable DE

Solve $\dfrac{dy}{dx} = -\dfrac{x}{y}$.

Solution   Rewriting the given equation in differential form

$$y\,dy = -x\,dx$$

and integrating both sides give

$$\int y\,dy = -\int x\,dx \qquad \text{or} \qquad \frac{y^2}{2} = -\frac{x^2}{2} + C_1.$$

Thus, a one-parameter family of solutions is defined by $x^2 + y^2 = C^2$. Here we have chosen to replace the arbitrary constant $2C_1$ by C^2 because the equation $x^2 + y^2 = C^2$ represents a family of circles centered at the origin of radius $C > 0$. See FIGURE 8.1.1.

The solutions of the differential equation are the functions defined implicitly by the equation $x^2 + y^2 = C^2$, $C > 0$. ∎

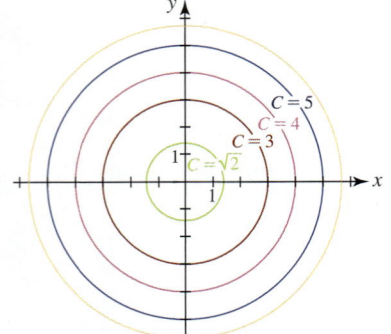

FIGURE 8.1.1   Family of circles in Example 2

▌ **Initial-Value Problem**   We are often interested in solving a first-order differential equation $dy/dx = F(x, y)$ subject to a prescribed side condition $y(x_0) = y_0$, where x_0 and y_0 are arbitrarily specified real numbers. The problem

$$\textit{Solve:} \qquad \frac{dy}{dx} = F(x, y)$$

$$\textit{Subject to:} \quad y(x_0) = y_0$$

is called an **initial-value problem** (**IVP**). The side condition $y(x_0) = y_0$ is called an **initial condition**. In geometric terms we are seeking at least one solution of the differential equation on an interval I containing x_0 so that a solution curve passes through the point (x_0, y_0). From a practical viewpoint, this often comes down to the problem of determining a specific value of the constant C in a family of solutions.

EXAMPLE 3   An Initial-Value Problem

Solve the initial-value problem $\dfrac{dy}{dx} = -\dfrac{x}{y}$, $y(4) = -3$.

Solution   From Example 2, a family of solutions for the given DE is $x^2 + y^2 = C^2$. When $x = 4$, then $y = -3$ so that $16 + 9 = C^2$ gives $C = 5$. Thus the IVP determines $x^2 + y^2 = 25$. Because of its simplicity we can solve the last equation for an explicit function or solution that satisfies the initial condition. Solving for y gives $y = \pm\sqrt{25 - x^2}$. Because the graph must be that of a function and the graph of this function must contain the point $(4, -3)$, we are forced to take the negative square root. In other words, the solution is $y = -\sqrt{25 - x^2}$ defined on the interval $(-5, 5)$. In Figure 8.1.1 the solution curve is the lower semicircle for the circle shown in blue. ∎

The simple first-order differential equation

$$\frac{dy}{dx} = ky, \tag{8}$$

where k is a constant, has many applications. The equation can be solved by separation of variables.

EXAMPLE 4   Solving a Separable DE

Solve $\dfrac{dy}{dx} = ky$, where $k \neq 0$ is a constant.

Solution   We write the differential equation as $\dfrac{1}{y} dy = k\, dx$. Integrating

$$\int \frac{1}{y}\, dy = k \int dx \qquad \text{gives} \qquad \ln|y| = kx + C_1.$$

Solving for y then yields

$$|y| = e^{kx + C_1} = e^{C_1} e^{kx} \qquad \text{or} \qquad y = \pm e^{kx + C_1} = \pm e^{C_1} e^{kx}.$$

By relabeling the constants $\pm e^{C_1}$ as C, a one-parameter family of solutions is given by $y = Ce^{kx}$. ∎

In order to solve separable differential equations it is obvious that a working knowledge of integration formulas and techniques is imperative. A review of Sections 7.1–7.3 and 7.6 is recommended.

EXAMPLE 5   Solving a Separable DE

Solve $(e^{2y} - y)\dfrac{dy}{dx} = e^y \sin x$.

Solution   By rewriting the equation as

$$\frac{e^{2y} - y}{e^y} \frac{dy}{dx} = \sin x \qquad \text{or} \qquad (e^y - ye^{-y})\, dy = \sin x\, dx$$

we see that the equation is separable. From the differential form of the equation,

$$\int (e^y - ye^{-y})\, dy = \int \sin x\, dx,$$

we see that integration by parts must be used to evaluate $\int ye^{-y}\, dy$. The result is

$$e^y + ye^{-y} + e^{-y} = \cos x + C.$$

The last equation defines a solution of the differential equation implicitly. Indeed, it is impossible to solve the last equation for y in terms of x. ∎

EXAMPLE 6   An Initial-Value Problem

Solve $\dfrac{dy}{dx} = y(1 - y)$, $y(0) = \dfrac{1}{3}$.

Solution   We rewrite the equation in differential form

$$\frac{1}{y(1 - y)} dy = dx \qquad \text{and integrate} \qquad \int \frac{1}{y(1 - y)}\, dy = \int dx.$$

Using partial fractions on the left-hand side of the equality gives

$$\int \left[\frac{1}{y} + \frac{1}{1 - y} \right] dy = \int dx$$

$$\ln|y| - \ln|1 - y| = x + C_1$$

$$\ln\left|\frac{y}{1-y}\right| = x + C_1$$

$$\left|\frac{y}{1-y}\right| = e^{x+C_1} = e^{C_1}e^x$$

$$\frac{y}{1-y} = C_2 e^x. \quad \leftarrow C_2 = \pm e^{C_1}$$

Solving for y gives

$$y = \frac{C_2 e^x}{1 + C_2 e^x} \quad \text{or} \quad y = \frac{1}{1 + Ce^{-x}}, \tag{9}$$

where we have replaced $1/C_2$ by C. Substituting $x = 0$ and $y = \frac{1}{3}$ into the last equation then yields $C = 2$. The solution of the initial-value problem is

$$y = \frac{1}{1 + 2e^{-x}}. \tag{10}$$

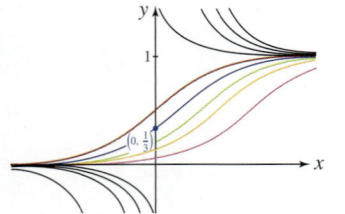

FIGURE 8.1.2 Family of solution curves in Example 6

Graphs of various members of the family of solutions given in (9) for various positive and negative values of C are illustrated in FIGURE 8.1.2. The graphs shown in color represent solutions of the DE that are defined on the interval $(-\infty, \infty)$. The graph of the solution given in (10) is the blue curve in the figure. ■

Exercises 8.1   Answers to selected odd-numbered problems begin on page ANS-25.

≡ Fundamentals

In Problems 1–20, solve the given differential equation by separation of variables.

1. $\dfrac{dy}{dx} = \sin 5x$

2. $\dfrac{dy}{dt} = (t + 1)^2$

3. $\dfrac{dy}{dx} = \dfrac{y^3}{x^2}$

4. $\dfrac{dy}{dx} = \dfrac{1}{5y^4}$

5. $\dfrac{dy}{dx} = \left(\dfrac{1 + x}{1 + y}\right)^2$

6. $\dfrac{dy}{dx} = \sqrt{xy}$

7. $\dfrac{dy}{dx} = \dfrac{1 + 5x^2}{x^2 \sin y}$

8. $\dfrac{dy}{dx} = y^3 \cos x$

9. $x\dfrac{dy}{dx} = 4y$

10. $\dfrac{dy}{dx} + 2xy = 0$

11. $\dfrac{dy}{dx} = e^{3x+2y}$

12. $e^x y \dfrac{dy}{dx} = e^{-y} + e^{-2x-y}$

13. $\left(\dfrac{y+1}{x}\right)^2 \dfrac{dy}{dx} = y \ln x$

14. $\dfrac{dy}{dx} = \left(\dfrac{2y+3}{4x+5}\right)^2$

15. $\dfrac{dN}{dt} + N = Nte^{t+2}$

16. $\dfrac{dQ}{dt} = k(Q - 70)$

17. $\dfrac{dP}{dt} = 5P - P^2$

18. $\dfrac{dX}{dt} = (10 - X)(50 - X)$

19. $\dfrac{dy}{dx} = \dfrac{xy + 3x - y - 3}{xy - 2x + 4y - 8}$

20. $\dfrac{dy}{dx} = \dfrac{xy + 2y - x - 2}{xy - 3y + x - 3}$

In Problems 21–26, solve the given initial-value problem.

21. $\dfrac{dy}{dx} = \dfrac{1}{(xy)^2}, \quad y(1) = 3$

22. $\dfrac{dy}{dx} = \dfrac{2x + \sec^2 x}{2y}, \quad y(0) = -2$

23. $\dfrac{dx}{dt} = 4(x^2 + 1), \quad x(\pi/4) = 1$

24. $\dfrac{dy}{dx} = \dfrac{y^2 - 1}{x^2 - 1}, \quad y(2) = 2$

25. $x^2 \dfrac{dy}{dx} = y - xy, \quad y(-1) = -1$

26. $\dfrac{dy}{dt} + 2y = 1, \quad y(0) = \dfrac{5}{2}$

In Problems 27 and 28, solve the given initial-value problem. Write the solution as an explicit *algebraic* function $y = f(x)$ (see *Notes From the Classroom* in Section 1.3). You may have to use a trigonometric identity.

27. $\sqrt{1 - x^2}\,\dfrac{dy}{dx} = \sqrt{1 - y^2}, \quad y(0) = \dfrac{\sqrt{3}}{2}$

28. $(1 + x^4)\dfrac{dy}{dx} + x + 4xy^2 = 0, \quad y(1) = 0$

In Problems 29–32, use the concept that $y = k$ on $(-\infty, \infty)$ is a constant function if and only if $dy/dx = 0$ to determine whether the given differential equation possesses constant solutions. Solve the given differential equation. Assume that k is a real number.

29. $x\dfrac{dy}{dx} + 6y = 18$

30. $2\dfrac{dy}{dx} = 5y + 40$

31. $\dfrac{dy}{dx} = y^2 - y - 20$

32. $x\dfrac{dy}{dx} = y^2 + 2y + 4$

In Problems 33 and 34, proceed as in Problems 29–32 to determine whether the given differential equation possesses constant solutions. Solve the given differential equation and then find a solution whose graph passes through the indicated point.

33. $x\dfrac{dy}{dx} = y^2 - y,$

    (a) $(0, 1)$ **(b)** $(0, 0)$ **(c)** $\left(\tfrac{1}{2}, \tfrac{1}{2}\right)$

34. $\dfrac{dy}{dx} = y^2 - 9,$

    (a) $(0, 0)$ **(b)** $(0, 3)$ **(c)** $\left(\tfrac{1}{3}, 1\right)$

≡ Think About It

35. Without solving, explain why the initial-value problem

$$\frac{dy}{dx} = \sqrt{y}, \quad y(x_0) = y_0$$

has no solution for $y_0 < 0$.

36. A solution of a differential equation that is not a member of the family of solutions of the equation is called a **singular solution**. Reexamine Problems 29, 31, 33, and 34 and find any singular solutions. In Example 6, what would be the solution of the IVP if the initial condition were changed to $y(0) = 1$?

37. In Example 3, it was stated that the solution $y = -\sqrt{25 - x^2}$ is defined on the open interval $(-5, 5)$. Why would it be incorrect to say that the solution is defined on the closed interval $[-5, 5]$?

8.2 Linear Equations

■ **Introduction** We continue our quest for solutions of first-order DEs by next examining linear equations. Linear differential equations are an especially "friendly" family of differential equations in that, given a linear equation, whether first-order or a higher-order kin, there is always a good possibility that we can find some sort of solution of the equation that we can look at. Nonlinear differential equations, especially equations of order greater than or equal to two, are often impossible to solve in terms of elementary functions.

The technique for solving a linear first-order equation, like a separable equation, consists of integration; but integration only after the original equation has been multiplied by a special function called an *integrating factor.*

■ **A Definition** We begin with the definition of a linear first-order equation. As you read the next definition bear in mind the following essential properties:

- In a linear differential equation the dependent variable and its derivative are of the first degree, that is, the power of each term involving the dependent variable is 1, and each coefficient depends at most only on the independent variable.

Definition 8.3.1   Linear Equation

A **linear first-order differential equation** is an equation $dy/dx = F(x, y)$ that can be put into the form

$$a_1(x)\frac{dy}{dx} + a_0(x)y = g(x). \tag{1}$$

The functions $a_1(x)$, $a_0(x)$, and $g(x)$ in (1) can of course be constants.

If a first-order DE is not linear, it is said to be **nonlinear**.

EXAMPLE 1   Linear/Nonlinear

(a) By direct comparison with (1) we see that the following differential equations

$$\frac{dy}{dx} + 3y = 6 \quad \text{and} \quad x\frac{dy}{dx} - 4y = x^6 e^x$$

are linear first-order equations.

(b) The following first-order equations are nonlinear:

$$\overset{\text{power not 1}}{\underset{\downarrow}{}}\qquad\qquad \overset{\text{coefficient depends on } y}{\underset{\downarrow}{}}$$

$$x\frac{dy}{dx} = y^2 \quad \text{and} \quad y\frac{dy}{dx} = 2y + \cos x.$$

■

It is important to note that not every first-order linear differential equation can be solved by the method of separation of variables. The linear equation

$$\frac{dy}{dx} + 2y = x$$

is not separable. Hence, we need a new procedure for solving linear equations.

■ **Standard Form** By dividing (1) by the lead coefficient $a_1(x)$, we obtain the more useful form of a linear equation:

$$\frac{dy}{dx} + P(x)y = f(x). \tag{2}$$

Equation (2) is called the **standard form** of a linear DE (1). We seek solutions of (2) on an interval I for which P and f are continuous. Equation (2) has the property that when multiplied by the function $e^{\int P(x)\,dx}$, the left-hand side of (2) becomes the derivative of the product $e^{\int P(x)\,dx}y$. To see this, observe that the Product Rule and Chain Rule give

By the Chain Rule:

$$\frac{d}{dx}e^{\int P(x)\,dx} = e^{\int P(x)\,dx}\frac{d}{dx}\int P(x)\,dx$$

$$= e^{\int P(x)\,dx}P(x)$$

▶

$$\frac{d}{dx}\big[e^{\int P(x)\,dx}y\big] = e^{\int P(x)\,dx}\frac{d}{dx}y + y\frac{d}{dx}e^{\int P(x)\,dx}$$

$$= e^{\int P(x)\,dx}\frac{dy}{dx} + e^{\int P(x)\,dx}P(x)y. \tag{3}$$

Thus, if we multiply both sides of (2) by $e^{\int P(x)\,dx}$, we get

$$\overbrace{e^{\int P(x)\,dx}\frac{dy}{dx} + e^{\int P(x)\,dx}P(x)y}^{\text{This is } \frac{d}{dx}[e^{\int P(x)\,dx}y]} = e^{\int P(x)\,dx}f(x).$$

By comparing the left-hand side of the last equation with the result in (3), it follows that the last equation is the same as

$$\frac{d}{dx}\big[e^{\int P(x)\,dx}y\big] = e^{\int P(x)\,dx}f(x). \tag{4}$$

The form of equation (4) is the key for solving linear first-order differential equations. We can simply integrate both sides of (4) with respect to x. The function $e^{\int P(x)\,dx}$ that makes this possible is called an **integrating factor** for the DE. The procedure is outlined next.

Guidelines for Solving Linear Equations

(*i*) Put the given differential equation into the standard form (2); that is, make the coefficient of dy/dx unity by division.

(*ii*) Identify $P(x)$ (the coefficient of y) and find the integrating factor

$$e^{\int P(x)\,dx}.$$

(*iii*) Multiply the equation obtained in step (*i*) by the integrating factor.

(*iv*) The left-hand side of the equation in step (*iii*) is the derivative of the integrating factor and the dependent variable:

$$\frac{d}{dx}\big[e^{\int P(x)\,dx}y\big] = e^{\int P(x)\,dx}f(x).$$

(*v*) Integrate both sides of the equation found in step (*iv*).

In Section 8.1 we solved the equation $dy/dx - ky = 0$ by separation of variables, but since the DE is linear, we can also solve it by the foregoing procedure.

EXAMPLE 2   Using an Integrating Factor

The linear differential equation

$$\frac{dy}{dx} - ky = 0,$$

k a constant, is already in standard form (2). By identifying $P(x) = -k$, the integrating factor is $e^{\int (-k)\,dx} = e^{-kx}$ and, after multiplying the equation by this factor, we see that

> We need not use a constant of integration in computing $e^{\int P(x)\,dx}$.

$$e^{-kx}\frac{dy}{dx} - ke^{-kx}y = 0 \cdot e^{-kx} \qquad \text{is the same as} \qquad \frac{d}{dx}[e^{-kx}y] = 0.$$

Integration of both sides of the last equation with respect to x,

$$\int \frac{d}{dx}[e^{-kx}y]\,dx = \int 0\,dx$$

gives $e^{-kx}y = C$. From this last expression we get the same family of solutions $y = Ce^{kx}$ as in Example 4 of Section 8.1. ∎

Recall, in the discussion of (2) we stated that we seek a solution of a linear equation on an interval I for which P and f are continuous. As you work your way through the next example, note that P and f are both continuous on the interval $(0, \infty)$.

EXAMPLE 3   Solving a Linear DE on an Interval

Solve $x\dfrac{dy}{dx} - 4y = x^6 e^x$.

Solution   By dividing by x we get the standard form

$$\frac{dy}{dx} - \frac{4}{x}y = x^5 e^x. \tag{5}$$

From this form we identify $P(x) = -4/x$ and $f(x) = x^5 e^x$ and observe that P and f are continuous for $x > 0$, that is, on $(0, \infty)$. Hence, the integrating factor is

we can use $\ln x$ instead of $\ln |x|$ since $x > 0$
$$\downarrow$$
$$e^{-4\int dx/x} = e^{-4\ln x} = e^{\ln x^{-4}} = x^{-4}.$$

> The identity $e^{\ln N} = N$ is useful in computing the integrating factor.

Now we multiply (5) by x^{-4},

$$x^{-4}\frac{dy}{dx} - 4x^{-5}y = xe^x \qquad \text{and obtain} \qquad \frac{d}{dx}[x^{-4}y] = xe^x.$$

Using integration by parts on the right-hand side of

$$\int \frac{d}{dx}[x^{-4}y]\,dx = \int xe^x\,dx$$

yields the solution defined on $(0, \infty)$:

$$x^{-4}y = xe^x - e^x + C \qquad \text{or} \qquad y = x^5 e^x - x^4 e^x + Cx^4. \qquad ∎$$

EXAMPLE 4   Solving a Linear DE on an Interval

Solve $(x^2 - 9)\dfrac{dy}{dx} + xy = 0$.

Solution   We write the equation in standard form

$$\frac{dy}{dx} + \frac{x}{x^2 - 9}y = 0 \tag{6}$$

and identify $P(x) = x/(x^2 - 9)$. Although P is continuous on $(-\infty, -3), (-3, 3)$ and on $(3, \infty)$, we shall solve the equation on the first and third intervals. On these intervals, the integrating factor is

$$e^{\int x\, dx/(x^2-9)} = e^{\frac{1}{2}\int 2x\, dx/(x^2-9)} = e^{\frac{1}{2}\ln(x^2-9)} = e^{\ln\sqrt{x^2-9}} = \sqrt{x^2 - 9}.$$

After multiplying the standard form (6) by this factor, we get

$$\frac{d}{dx}\left[\sqrt{x^2 - 9}\, y\right] = 0, \qquad \text{and integrating gives} \qquad \sqrt{x^2 - 9}\, y = C.$$

For either $(-\infty, -3)$ or $(3, \infty)$, the solution of the equation is $y = \dfrac{C}{\sqrt{x^2 - 9}}$. ∎

EXAMPLE 5   An Initial-Value Problem

Solve the initial-value problem $\dfrac{dy}{dx} + y = x,\ y(0) = 4$.

Solution   The equation is already in standard form, and $P(x) = 1$ and $f(x) = 1$ are continuous on $(-\infty, \infty)$. The integrating factor is $e^{\int dx} = e^x$, and so integrating

$$\frac{d}{dx}[e^x y] = xe^x$$

gives $e^x y = xe^x - e^x + C$. Solving this last equation for y yields the family of solutions

$$y = x - 1 + Ce^{-x}. \tag{7}$$

But from the initial condition we know that $y = 4$ when $x = 0$. Substituting these values in (7) implies $C = 5$. Hence, the solution of the problem on $(-\infty, \infty)$ is $y = x - 1 + 5e^{-x}$ and is the blue curve in FIGURE 8.2.1. ∎

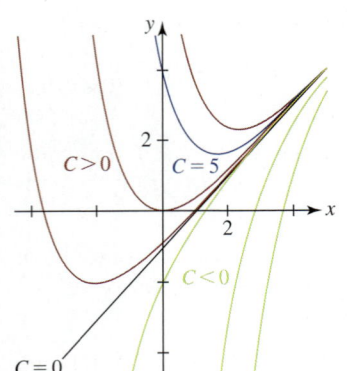

FIGURE 8.2.1 Family of solution curves in Example 5. Solution of the IVP is shown in blue.

It is interesting to observe that as x becomes unbounded in the positive direction, the graphs of *all* members of the family of solutions (7) for $C > 0$ or $C < 0$ are close to the graph of $y = x - 1$, which is shown in black in Figure 8.2.1. Indeed, $y = x - 1$ is the solution of the DE in Example 5 that corresponds to $C = 0$ in (7). This asymptotic behavior is attributable to the fact that the term Ce^{-x} in (7) becomes negligible for increasing values of x, that is, $e^{-x} \to 0$ as $x \to \infty$. We say that Ce^{-x} is a **transient term**. Although this behavior is not a characteristic of all families of solutions of linear equations (see Example 3), the notion of a transient is often important in applied problems.

$$\frac{dy}{dx} = F(x, y)$$ **NOTES FROM THE CLASSROOM**

If we solve (5) on an interval on which P and f are continuous, then it can be proved that a one-parameter family of solutions of the equation yields *all* solutions of the DE defined on the interval. In Example 3, the functions $P(x) = -4/x$ and $f(x) = x^5 e^x$ are continuous on the interval $(0, \infty)$. In this case, *every* solution of $dy/dx - (4/x)y = x^5 e^x$ on $(0, \infty)$ can be obtained from $y = x^5 e^x - x^4 e^x + Cx^4$ for appropriate choices of the constant C. For this reason, the family of solutions $y = x^5 e^x - x^4 e^x + Cx^4$ is called the **general solution** of the differential equation.

Exercises 8.2   Answers to selected odd-numbered problems begin on page ANS-25.

≡ Fundamentals

In Problems 1–22, solve the given linear differential equation.

1. $\dfrac{dy}{dx} = 4y$

2. $\dfrac{dy}{dx} + 2y = 0$

3. $2\dfrac{dy}{dx} + 10y = 1$

4. $x\dfrac{dy}{dx} + 2y = 3$

5. $\dfrac{dy}{dt} + y = e^{3t}$

6. $\dfrac{dy}{dt} = y + e^t$

7. $y' + 3x^2y = x^2$

8. $y' + 2xy = x^3$

9. $x^2\dfrac{dy}{dx} + xy = 1$

10. $(1 + x^2)\dfrac{dy}{dx} + xy = 2x$

11. $(1 + e^x)\dfrac{dy}{dx} + e^xy = 0$

12. $(1 - x^3)\dfrac{dy}{dx} = 3x^2y$

13. $x\dfrac{dy}{dx} - y = x^2\sin x$

14. $\dfrac{dy}{dx} + y = \cos(e^x)$

15. $\cos x\dfrac{dy}{dx} + (\sin x)y = 1$

16. $\sin x\dfrac{dy}{dx} + (\cos x)y = \sec^2x$

17. $\dfrac{dy}{dx} + (\cot x)y = 2\cos x$

18. $\dfrac{dr}{d\theta} + (\sec\theta)r = \cos\theta$

19. $(x + 2)^2\dfrac{dy}{dx} = 5 - 8y - 4xy$

20. $\dfrac{dP}{dt} + 2tP = P + 4t - 2$

21. $x^2\dfrac{dy}{dx} + x(x + 2)y = e^x$

22. $x\dfrac{dy}{dx} + (x + 1)y = e^{-x}\sin 2x$

In Problems 23–32, solve the given initial-value problem.

23. $\dfrac{dy}{dx} = x + y, \quad y(0) = -4$

24. $\dfrac{dy}{dx} = 2x - 3y, \quad y(0) = \tfrac{1}{3}$

25. $x\dfrac{dy}{dx} + y = e^x, \quad y(1) = 2$

26. $x\dfrac{dy}{dx} + y = 4x + 1, \quad y(1) = 8$

27. $x\dfrac{dy}{dx} - y = 2x^2, \quad y(5) = 1$

28. $x(x + 1)\dfrac{dy}{dx} + xy = 1, \quad y(1) = 10$

29. $(t + 1)\dfrac{dx}{dt} + x = \ln t, \quad x(1) = 10$

30. $y' + (\tan t)y = \cos^2 t, \quad y(0) = -1$

31. $L\dfrac{di}{dt} + Ri = E, i(0) = i_0, L, R,$ and E are constants

32. $\dfrac{dT}{dt} = k(T - T_m), T(0) = T_0, k, T_m,$ and T_0 are constants

≡ Calculator/CAS Problems

In Problems 33 and 34, before attempting to solve the given initial-value problem review Problems 71 and 72 in Exercises 5.5.

33. (a) Express the solution of the initial-value problem $y' - 2xy = 2, y(0) = 1$, in terms of the **error function**

$$\text{erf}(x) = \frac{2}{\sqrt{\pi}}\int_0^x e^{-t^2}\,dt.$$

(b) Use tables or a CAS to calculate $y(2)$. Use a CAS to graph the solution on the interval $(-\infty, \infty)$.

34. The **sine integral function** is defined by

$$\text{Si}(x) = \int_0^x \frac{\sin t}{t}\,dt.$$

(a) Show that the solution of the initial-value problem $x^3y' + 2x^2y = 10\sin x, y(1) = 0$ is

$$y = 10x^{-2}[\,\text{Si}(t) - \text{Si}(1)\,].$$

(b) Use tables or a CAS to calculate $y(2)$. Use a CAS to graph the solution on the interval $(0, \infty)$.

≡ Think About It

35. Find a continuous solution of the initial-value problem

$$\frac{dy}{dx} + y = f(x), \quad f(x) = \begin{cases} 1, & 0 \le x \le 1 \\ 0, & x > 1 \end{cases}, \quad y(0) = 0.$$

Graph f and the solution of the IVP. [*Hint:* Solve the problem in two parts and use continuity to match the parts of your solution.]

36. Explain why we do not have to use a constant of integration when computing an integrating factor $e^{\int P(x)\,dx}$ for a linear differential equation.

37. In Example 4 we solved the given differential equation on the intervals $(-\infty, -3)$ and $(3, \infty)$. Find a solution of the DE on the interval $(-3, 3)$.

38. Suppose $P(t)$ represents the population of some animal species present in an environment at time t. If the symbol $\propto$ means "proportional to," in words give a physical interpretation of the mathematical statement

$$\frac{dP}{dt} \propto P.$$

39. The following system of differential equations is encountered in the study of a special type of radioactive series of elements:

$$\frac{dx}{dt} = -\lambda_1 x$$

$$\frac{dy}{dt} = -\lambda_1 x - \lambda_2 y,$$

where λ_1 and λ_2 are constants. Solve the system subject to $x(0) = x_0, y(0) = y_0$.

40. The nonlinear first-order differential equation

$$\frac{dy}{dx} + \frac{1}{x}y = xy^2$$

is a member of a class of nonlinear DEs called **Bernoulli equations**.

(a) Use the substitution $y = u^{-1}$ to show that the given Bernoulli equation becomes

$$\frac{du}{dx} - \frac{1}{x}u = -x.$$

(b) Find a solution of the given Bernoulli equation by solving the DE in part (a).

41. The differential equation

$$\frac{dy}{dx} = -\frac{1}{x+y}$$

is neither separable nor linear in the variable y. Take the reciprocal of both sides of the equation. Can this new differential equation be solved?

42. Although the differential equation $y'' + y' = x$ is second order it can be solved using the method discussed in this section. Solve the equation by letting $Y = y'$.

43. (a) Find a one-parameter family of solutions of the linear equation

$$x\frac{dy}{dx} + 3y = 6x^2.$$

(b) Find the member of the family of solutions in part (a) that satisfies the initial condition $y(-1) = 2$. Give the interval over which this solution is valid.

(c) Find the member of the family of solutions in part (a) that satisfies the initial condition $y(1) = 2$. Give the interval over which this solution is valid.

(d) Find an initial condition so that the corresponding member of the family of solutions in part (a) passes through the origin and is valid on the interval $(-\infty, \infty)$.

8.3 Mathematical Models

▌ **Introduction**   So far our experience with first-order DEs has been limited to either solving them or verifying that a given function is a solution. But mathematics is a language as well as a tool. As you undoubtedly remember from algebra and Section 1.7, we translate words into mathematics when solving a "word problem." So too, we can interpret words, empirical laws, observations, or simply assumptions, into mathematical terms. When we try to describe something, let us call it a *system*, in mathematical terms we are constructing a *model* of that system. If something in the system changes with time, say, either growing or decreasing at a certain *rate*—and a rate of change is a derivative—then a **mathematical model** of the system may be a differential equation.

In this section we will consider a few simple mathematical models and their solutions.

▌ **Population Growth**   One of the earliest attempts to model human population growth by means of mathematics was by the English economist **Thomas Malthus** (1776–1834) in 1798. Basically, the idea of the Malthusian model is the assumption that the rate at which a population of a country *grows* or increases is proportional to the total population $P(t)$ of the country at time t. In other words, the more people there are at time t, the more there are going to be in the future. In mathematical terms this assumption can be expressed as

proportionality symbol
$$\downarrow$$
$$\frac{dP}{dt} \propto P \qquad \text{or} \qquad \frac{dP}{dt} = kP, \tag{1}$$

where k is a constant of proportionality. This simple model, which fails to take into account many factors (immigration and emigration, for example) that can influence human populations to either grow or decline, nevertheless turned out to be fairly accurate in predicting the population of the United States during the years 1790–1860. The differential equation given in (1) is still often used to model, over short intervals of time, the populations of bacteria or small animals.

The constant of proportionality k in (1) can be determined from the solution of the initial-value problem $dP/dt = kP$, $P(t_0) = P_0$ using a subsequent measurement of P at a time $t_1 > t_0$.

EXAMPLE 1   Bacterial Growth

A culture initially has P_0 number of bacteria. At $t = 1$ hr, the number of bacteria present is measured to be $\frac{3}{2}P_0$. If the rate of growth is proportional to the number of bacteria $P(t)$ present at time t, determine the time necessary for the number of bacteria to triple.

Solution   We first solve the differential equation in (1) subject to the initial condition $P(0) = P_0$. Then we use the empirical observation that $P(1) = \frac{3}{2}P_0$ to determine the constant of proportionality k. Now we have already see that the equation $dP/dt = kP$ is both separable and linear. From Example 2 of Section 8.2, with the symbols P and t, in turn, playing the parts of y and x, a family of solutions of the DE is

$$P(t) = Ce^{kt}.$$

At $t = 0$, it follows that $P_0 = Ce^0 = C$, so $P(t) = P_0e^{kt}$. At $t = 1$, we have $\frac{3}{2}P_0 = P_0e^k$ or $e^k = \frac{3}{2}$. From the last equation $k = \ln\frac{3}{2} = 0.4055$. Thus

$$P(t) = P_0e^{0.4055t}.$$

To find the time at which the number of bacteria has tripled, we solve $3P_0 = P_0e^{0.4055t}$ for t. It follows that $0.4055t = \ln 3$, so

$$t = \frac{\ln 3}{0.4055} \approx 2.71 \text{ h}.$$

See FIGURE 8.3.1. ■

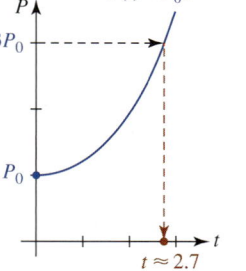

FIGURE 8.3.1 Graph of solution in Example 1

Note: We can write the function $P(t)$ obtained in the preceding example in an alternative form. From the laws of exponents,

$$P(t) = P_0e^{kt} = P_0(e^k)^t = P_0\left(\frac{2}{3}\right)^t,$$

since $e^k = \frac{3}{2}$. This latter solution provides a convenient method for computing $P(t)$ for small positive integral values of t; it also clearly shows the influence of the subsequent experimental observation at $t = 1$ on the solution for all time. We notice too that the actual number of bacteria present initially—that is, at time $t = 0$—is quite irrelevant in finding the time required to triple the number in the culture. The necessary time to triple, say, 100 or even 100,000 bacteria is still approximately 2.7 h.

■ **Radioactive Decay** The nucleus of an atom consists of combinations of protons and neutrons. Many of these combinations of protons and neutrons are unstable, that is, the atoms decay or transmute into the atoms of another substance. Such nuclei are said to be *radioactive*. For example, over time the highly radioactive radium, Ra-226, transmutes into the radioactive gas radon, Rn-222. To model the phenomenon of radioactive decay, it is assumed that the rate dA/dt at which the nuclei of a substance decays is proportional to the amount of the substance (more precisely, the number of nuclei) $A(t)$ remaining at time t:

$$\frac{dA}{dt} \propto A \qquad \text{or} \qquad \frac{dA}{dt} = kA. \tag{2}$$

The model (2) for decay also occurs in a biological setting, such as determining the time that it takes for 50% of a drug to be eliminated from a body by excretion or metabolism. The point of (1) and (2) is simply this:

- *A single differential equation can serve as a mathematical model for many different phenomena.*

Of course, since equations (1) and (2) are exactly the same, their solutions are exactly the same (namely, Ce^{kt}); the difference is only in the symbols and their interpretation. As shown in FIGURE 8.3.2, the exponential function e^{kt} increases as t increases for $k > 0$ and decreases as t increases for $k < 0$. Thus, problems describing growth (whether of animal populations, bacteria, or even capital) are characterized by a positive value of k, whereas problems involving decay yield a negative k value. Accordingly, we say that k is either a **growth constant** ($k > 0$) or a **decay constant** ($k < 0$).

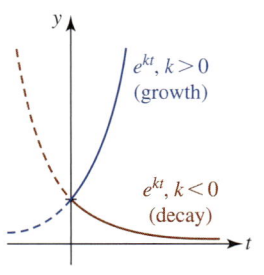

FIGURE 8.3.2 Exponential growth and decay

■ **Half-Life** In physics the **half-life** is a measure of the stability of a radioactive substance. The half-life is simply the time it takes for one-half of the atoms in an initial amount A_0 to disintegrate, or transmute, into the atoms of another element. In terms of the solution $A(t) = Ce^{kt}$ of (2), the half-life of a decaying element is the time t for which $A(t) = \frac{1}{2}A_0$. The longer the half-life of a substance, the more stable it is. For example, the half-life of highly radioactive radium, Ra-226, is about 1700 years. In 1700 years one-half of a given quantity of Ra-226 is transmuted into radon, Rn-222. The most commonly occurring uranium isotope, U-238, has a half-life of approximately 4,500,000,000 years. In about 4.5 billion years, one-half of a quantity of U-238 is transmuted into lead, Pb-206.

■ **Carbon Dating** In the 1940s the chemist Willard Libby devised a method of using radioactive carbon as a means of determining the approximate ages of fossils. The theory of **carbon dating** is based on the fact that the radioactive isotope carbon-14 is produced in the atmosphere by the action of cosmic radiation on nitrogen. The ratio of the amount of C-14 to ordinary carbon in the atmosphere appears to be a constant, and as a consequence the proportionate amount of the isotope present in all living organisms is the same as that in the atmosphere. When an organism dies, the absorption of C-14, by either breathing or eating, ceases. Thus, by comparing the proportionate amount of C-14 present, say, in a fossil with the constant ratio found in the atmosphere, it is possible to obtain a reasonable estimation of its age. The method is based on the knowledge that the half-life of the radioactive C-14 is approximately 5730 years. For his work Libby won the Nobel Prize for chemistry in 1960. Libby's method has been used to date wooden furniture in Egyptian tombs, the woven flax wrappings of the Dead Sea scrolls, and the controversial Shroud of Turin.

EXAMPLE 2 Dating a Fossil

A fossilized bone is found to contain $\frac{1}{1000}$ the original amount of C-14. Determine the age of the fossil.

Solution The starting point is the differential equation $dA/dt = kA$, where $A(t)$ is the amount of C-14 remaining at time t. If A_0 is the initial amount of C-14 in the bone, it follows as in Example 1 that

$$A(t) = A_0 e^{kt}.$$

We can use the fact that $A(5730) = \frac{1}{2}A_0$ to determine the decay constant k. Setting $t = 5730$ in $A(t)$ implies $\frac{1}{2}A_0 = A_0 e^{5730k}$ and so from $5730k = \ln\frac{1}{2} = -\ln 2$ we find that

$$k = -\frac{1}{5730}\ln 2 = -0.00012097.$$

Therefore, $A(t) = A_0 e^{-0.00012097t}$. Now the age of the fossil is determined from the equation $A(t) = \frac{1}{1000}A_0$. That is, $\frac{1}{1000}A_0 = A_0 e^{-0.00012097t}$ and so $-0.00012097t = \ln\frac{1}{1000} = -\ln 1000$ yields

$$t = \frac{\ln 1000}{0.00012097} \approx 57,103 \text{ years.} \qquad ■$$

The date found in Example 2 is really at the border of accuracy for this method. The usual carbon-14 technique is limited to about 9 half-lives of the isotope or about 50,000 years. One reason is that the chemical analysis needed to obtain an accurate measurement of the remaining C-14 becomes somewhat formidable around the point of $\frac{1}{1000}A_0$. Also, this analysis demands the destruction of a rather large sample of the specimen. If this measurement is accomplished indirectly, based on the actual radioactivity of the specimen, then it is very difficult to distinguish between the radiation from the fossil and the normal background radiation.

In recent developments geologists have shown that in some cases, dates determined by carbon dating may be off by as much as 3500 years. One conjecture for this possible error is the fact that carbon-14 levels in the air are known to vary with time. These same scientists have devised another dating technique based on the fact that living organisms ingest traces of uranium. By measuring the relative amounts of uranium and thorium (the isotope into which the uranium decays) and by knowing the half-lives of these elements, scientists can determine the age of a fossil. The advantage of this method is that it can date fossils up to 500,000 years; the disadvantage is that it is effective mostly on marine fossils. Another isotopic technique, using potassium-40 and argon-40, when applicable, can give dates of several million years. See Problem 37 in Exercises 8.3. Nonisotopic methods based on the use of amino acids are also sometimes possible.

■ **Cooling** Newton's law of cooling states that the rate at which the temperature $T(t)$ changes in a cooling body is proportional to the difference between the temperature in the body and the constant temperature T_m of the surrounding medium; that is,

$$\frac{dT}{dt} = k(T - T_m), \tag{3}$$

where k is a constant of proportionality.

EXAMPLE 3 Cooling Cake

When a cake is removed from a baking oven, its temperature is measured at 300°F. Three minutes later its temperature is 200°F. Determine the temperature of the cake at any time after leaving the oven if the room temperature is 70°F.

Solution We identify the temperature of the room (70°F) as T_m. To find the temperature of the cake at time t, we must solve the initial-value problem

$$\frac{dT}{dt} = k(T - 70), \quad T(0) = 300$$

and determine the value of k so that $T(3) = 200$. The DE is both separable and linear. Assuming $T > 70$, it follows by separation of variables that

$$\frac{1}{T - 70} dT = k\, dt$$

$$\int \frac{1}{T - 70} dT = \int k\, dt$$

$$\ln(T - 70) = kt + C_1$$

$$T - 70 = Ce^{kt} \quad \leftarrow C = e^{C_1}$$

$$T = 70 + Ce^{kt}.$$

When $t = 0$, $T = 300$ so that $300 = 70 + C$ gives $C = 230$ and therefore $T(t) = 70 + 230e^{kt}$. From $T(3) = 200$ we find $e^{3k} = \frac{13}{23}$ and so, to four decimal places, a calculator gives

$$k = \frac{1}{3}\ln\frac{13}{23} = -0.1902.$$

Thus, $T(t) = 70 + 230e^{-0.1902t}$. The graph of T along with some calculated values are given in FIGURE 8.3.3. ■

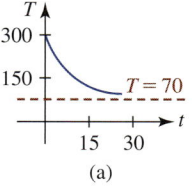

(a)

t (minutes)	$T(t)$
20.1	75°
21.3	74°
22.8	73°
24.9	72°
28.6	71°
32.3	70.5°

(b)

FIGURE 8.3.3 Graph of solution in Example 3

■ **Mixtures** The mixing of two fluids sometimes gives rise to a linear first-order differential equation. In the next example we consider the mixture of two salt solutions of different concentrations.

EXAMPLE 4 Mixing a Salt Solution

Initially 50 lb of salt is dissolved in a large tank holding 300 gal of water. A brine solution is pumped into the tank at a rate of 3 gal/min, and the well-stirred solution is then pumped out at the same rate. See FIGURE 8.3.4. If the concentration of the solution entering is 2 lb/gal, determine the amount of salt in the tank at time t. How much salt is present after 50 min? After a long time?

Solution Let $A(t)$ be the amount of salt (in pounds) in the tank at time t. For problems of this sort, the net rate at which $A(t)$ changes is given by

$$\frac{dA}{dt} = \left(\begin{array}{c}\text{rate of}\\\text{substance entering}\end{array}\right) - \left(\begin{array}{c}\text{rate of}\\\text{substance leaving}\end{array}\right) = R_1 - R_2. \tag{4}$$

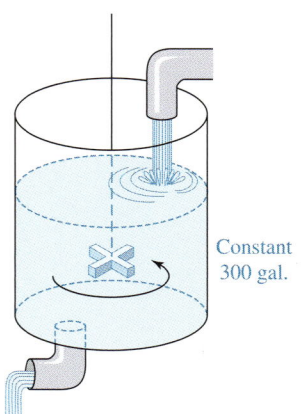

Constant 300 gal.

FIGURE 8.3.4 Mixing tank in Example 4

Now the rate at which the salt enters the tank is, in pounds per minute,

$$R_1 = (3 \text{ gal/min}) \cdot (2 \text{ lb/gal}) = 6 \text{ lb/min},$$

whereas the rate at which salt is leaving is

$$R_2 = (3 \text{ gal/min}) \cdot \left(\frac{A}{300} \text{ lb/gal} \right) = \frac{A}{100} \text{ lb/min}.$$

Thus, equation (4) becomes

$$\frac{dA}{dt} = 6 - \frac{A}{100} \qquad \text{or} \qquad \frac{dA}{dt} + \frac{1}{100}A = 6. \qquad (5)$$

We solve the last equation subject to the initial condition $A(0) = 50$.

Since the integrating factor is $e^{t/100}$, we can write (5) as

$$\frac{d}{dt}[e^{t/100}A] = 6e^{t/100}$$

and therefore $e^{t/100}A = 600e^{t/100} + C$ or $A = 600 + Ce^{-t/100}$. When $t = 0, A = 50$, so we find that $C = -550$. Finally, we obtain

$$A(t) = 600 - 550e^{-t/100}. \qquad (6)$$

At $t = 50$ we find $A(50) = 266.41$ lb. Also, as $t \to \infty$ it is seen from (6) and FIGURE 8.3.5 that $A \to 600$. Of course, this is what we would expect; over a long period of time the number of pounds of salt in the solution must be $(300 \text{ gal})(2 \text{ lb/gal}) = 600$ lb. ∎

In Example 4 we assumed that the rate at which the solution was pumped in was the same as the rate at which the solution was pumped out. However, this need not be the case; the mixed brine solution could be pumped out at a rate faster or slower than the rate at which the other solution is pumped in. For example, if the well-stirred solution is pumped out at the slower rate of 2 gal/min, then the solution is accumulating at a rate of $(3 - 2) \text{ gal/min} = 1 \text{ gal/min}$. After t min there are $300 + t$ gal of brine in the tank. The rate at which the salt is leaving is then

$$R_2 = (2 \text{ gal/min}) \cdot \left(\frac{A}{300 + t} \text{lb/gal} \right) = \frac{2A}{300 + t} \text{ lb/min}.$$

Equation (4) in this case becomes

$$\frac{dA}{dt} = 6 - \frac{2A}{300 + t} \qquad \text{or} \qquad \frac{dA}{dt} + \frac{2}{300 + t}A = 6. \qquad (7)$$

Inspection of the last equation shows that it is linear. We leave its solution as an exercise. See Problems 18–20 in Exercises 8.3.

■ **Newton's Second Law of Motion** To construct a mathematical model of the motion of a body moving in a force field, the usual starting point is Newton's second law of motion. Recall, **Newton's first law of motion** states that a body will either remain at rest or will continue to move with a constant velocity unless acted upon by an external force. In each of these two cases, this is equivalent to saying that when the sum of the forces $F = \Sigma F_k$—that is, the net or resultant force—acting on the body is zero, then the acceleration a of the body is zero. **Newton's second law of motion** indicates that when the net force acting on a body is *not* zero, then the net force is proportional to its acceleration a, or more precisely, $F = ma$, where m is the mass of the body.

■ **Falling Bodies and Air Resistance** It has been established empirically that when a body moves through a resistive medium such as air (or water), the retarding force due to the medium, called the **drag force**, acts in the direction opposite to that of the motion and is proportional to a power of the body's velocity, that is, kv^α. Here k is a constant of proportionality and α is constant in the range $1 \le \alpha \le 2$. Roughly, for slow speeds we take $\alpha = 1$. Now suppose a falling body of mass m encounters air resistance proportional to its instantaneous velocity v. If

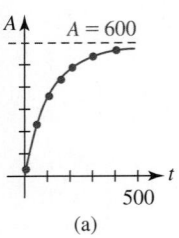

t (minutes)	$A(t)$
50	266.41
100	397.67
150	477.27
200	525.57
300	572.62
400	589.93

(b)

FIGURE 8.3.5 Graph of solution in Example 4

we take, in this circumstance, the positive direction to be oriented downward, then the net force acting on the mass is given by $mg - kv$ where the weight mg of the body is a force acting in the positive direction and air resistance is a force acting in the opposite or upward direction. See FIGURE 8.3.6. Now since v is related to acceleration a by $dv/dt = a$, Newton's second law becomes $F = ma = m\,dv/dt$. By equating the net force to this form of Newton's second law, we obtain a linear differential equation for the velocity v of the body at time t,

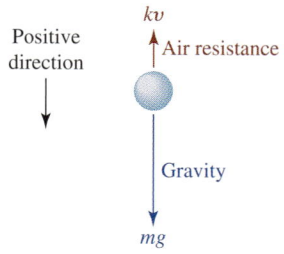

$$m\frac{dv}{dt} = mg - kv, \qquad (8)$$

FIGURE 8.3.6 Forces acting on falling body of mass m

where k is a positive constant of proportionality.

For high-speed motion—such as a skydiver who is falling before the parachute is opened—it is usually assumed that the air resistance is proportional to the square of the instantaneous velocity, in other words, $\alpha = 2$. If the positive direction is again taken to be downward, then a model for the velocity v of a falling body is given by the nonlinear differential equation

$$m\frac{dv}{dt} = mg - kv^2, \qquad (9)$$

where k is a positive constant of proportionality. See Problem 22 in Exercises 8.3.

EXAMPLE 5 Velocity of a Falling Body

Solve (8) subject to the initial condition $v(0) = v_0$.

Solution Equation (8) is linear and has the standard form

$$\frac{dv}{dt} + \frac{k}{m}v = g. \qquad (10)$$

Multiplying (10) by the integrating factor $e^{kt/m}$ enables us to write the equation as

$$\frac{d}{dt}\left[e^{kt/m}v\right] = g\,e^{kt/m}.$$

Integrating and solving for v then gives $v(t) = mg/k + Ce^{-kt/m}$. The initial condition $v(0) = v_0$ implies $C = v_0 - mg/k$ and so the velocity function for the falling body is

$$v(t) = \frac{mg}{k} + \left(v_0 - \frac{mg}{k}\right)e^{-kt/m}. \qquad (11) \; \blacksquare$$

Two observations are in order about the solution in Example 5. If we desire the position function $s(t)$ of the falling body, then it is a simple matter of integrating the equation

$$\frac{ds}{dt} = v(t),$$

where $v(t)$ is given in (11). See Problem 21 in Exercises 8.3. Also, because of the air resistance, the solution (11) clearly shows that the velocity of a body that falls a long distance does not increase indefinitely. Because the term $(v_0 - mg/k)e^{-kt/m}$ in (11) is *transient* (see page 448), we see that $v(t) \to mg/k$ as $t \to \infty$. This limiting value of the velocity $v_{\text{ter}} = mg/k$ is called the **terminal velocity** of the body. It is left as an exercise to find $v(t)$ and v_{ter} when the mathematical model for the velocity is given by (9). See Problem 22 in Exercises 8.3.

Exercises 8.3   Answers to selected odd-numbered problems begin on page ANS-25.

≡ Fundamentals

1. The population of a certain community is known to increase at a rate proportional to the number of people present at time t. If the population has doubled in 5 years, how long will it take to triple? To quadruple?

2. Suppose it is known that the population of the community in Problem 1 is 10,000 after 3 years. What was the initial population? What will the population be in 10 years?

3. The population of a town grows at a rate proportional to the population at time t. Its initial population of 500

increases by 15% in 10 years. What will the population be in 30 years?

4. The population of bacteria in a culture grows at a rate proportional to the number of bacteria present at time t. After 3 h it is observed that there are 400 bacteria present. After 10 h there are 2000 bacteria present. What was the initial number of bacteria?

5. The radioactive isotope of lead, Pb-209, decays at a rate proportional to the amount present at time t and has a half-life of 3.3 h. If 1 g of lead is present initially, how long will it take for 90% of the lead to decay?

6. Initially there were 100 mg of a radioactive substance present. After 6 h the mass decreased by 3%. If the rate of decay is proportional to the amount of the substance present at time t, find the amount remaining after 24 h.

7. Determine the half-life of the radioactive substance described in Problem 6.

8. Show that the half-life of a radioactive substance is, in general,

$$t = \frac{(t_2 - t_1)\ln 2}{\ln(A_1/A_2)},$$

where $A_1 = A(t_1)$ and $A_2 = A(t_2)$, $t_1 < t_2$.

9. When a vertical beam of light passes through a transparent substance, the rate at which its intensity I decreases is proportional to $I(t)$, where t represents the thickness of the medium (in feet). In clear seawater, the intensity 3 ft below the surface is 25% of the initial intensity I_0 of the incident beam. What is the intensity of the beam 15 ft below the surface?

10. When interest is compounded continuously, the amount of money increases at a rate proportional to the amount S present at time t: $dS/dt = rS$, where r is the annual rate of interest.

   (a) Find the amount of money accrued at the end of 5 years when $5000 is deposited in a savings account drawing $5\frac{3}{4}\%$ annual interest compounded continuously.

   (b) In how many years will the initial sum deposited be doubled?

   (c) Use a calculator to compare the number obtained in part (a) with the value

$$S = 5000\left(1 + \frac{0.0575}{4}\right)^{5(4)}.$$

   This value represents the amount accrued when interest is compounded quarterly.

11. In a piece of burned wood, or charcoal, it was found that 85.5% of the C-14 had decayed. Use the information in Example 2 to determine the approximate age of the wood. (It is precisely these data that archaeologists used to date prehistoric paintings in a cave in Lascoal, France.)

12. A thermometer is taken from an inside room to the outside where the air temperature is 5°F. After 1 min the thermometer reads 55°F, and after 5 min the reading is 30°F. What is the initial temperature of the room?

13. A thermometer is removed from a room where the air temperature is 70°F to the outside where the temperature is 10°F. After $\frac{1}{2}$ min the thermometer reads 50°F. What is the reading at $t = 1$ min? How long will it take for the thermometer to reach 15°F?

14. Equation (3) also holds when an object absorbs heat from the surrounding medium. If a small metal bar whose initial temperature is 20°C is dropped into a container of boiling water, how long will it take for the bar to reach 90°C if it is known that its temperature increased 2° in 1 s? How long will it take the bar to reach 98°C?

15. A tank contains 200 L of fluid in which 30 g of salt is dissolved. Brine containing 1 g of salt per liter is then pumped into the tank at a rate of 4 L/min; the well-mixed solution is pumped out at the same rate. Find the number of grams of salt $A(t)$ in the tank at time t.

16. Solve Problem 15 assuming pure water is pumped into the tank.

17. A large tank is filled with 500 gal of pure water. Brine containing 2 lb of salt per gallon is pumped into the tank at a rate of 5 gal/min. The well-mixed solution is pumped out at the same rate. Find the number of pounds of salt $A(t)$ in the tank at time t.

18. Solve the differential equation (7) subject to the initial condition $A(0) = 50$ lb.

19. Reread the discussion following Example 4. Then solve Problem 17 under the assumption that the solution is pumped out at a faster rate of 10 gal/min. Determine when the tank is empty.

20. A large tank is partially filled with 100 gal of fluid in which 10 lb of salt is dissolved. Brine containing $\frac{1}{2}$ lb of salt per gallon is pumped into the tank at a rate of 6 gal/min. The well-mixed solution is then pumped out at a slower rate of 4 gal/min. Find the number of pounds of salt in the tank after 30 min.

21. Reread the discussion following Example 5. Then find the position function $s(t)$ for the falling body in Example 5. Since the positive direction was assumed to be downward, assume that $s(0) = 0$.

22. Solve the differential equation (9) subject to $v(0) = v_0$. Express the velocity function $v(t)$ in terms of the hyperbolic tangent function. With the aid of Figure 3.10.2(a) determine the terminal velocity v_{ter} of a falling body.

≡ Additional Mathematical Models

23. The rate at which a drug disseminates into the bloodstream is governed by the differential equation

$$\frac{dX}{dt} = A - BX,$$

where A and B are positive constants. The function $X(t)$ describes the concentration of the drug in the bloodstream at time t. Find $X(t)$. What is the limiting value of $X(t)$ as $t \to \infty$? At what time is the concentration one-half this limiting value? Assume that $X(0) = 0$.

24. Suppose a cell is suspended in a solution containing a solute of constant concentration C_s. Suppose further that the cell has constant volume V and that the area of its permeable membrane is the constant A. By **Fick's law**, the rate of change of its mass m is directly proportional to the area A and the difference $C_s - C(t)$, where $C(t)$ is the concentration of the solute inside the cell at time t. Find $C(t)$ if $m = VC(t)$ and $C(0) = C_0$. See FIGURE 8.3.7.

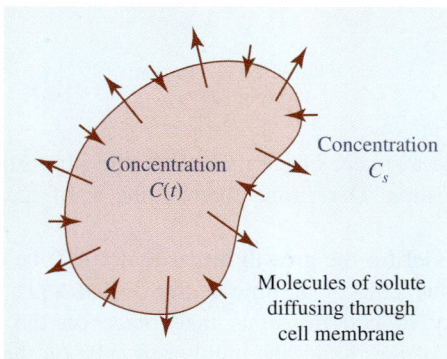

FIGURE 8.3.7   Cell in Problem 24

25. A heart pacemaker, shown in FIGURE 8.3.8, consists of a battery, a capacitor, and the heart as a resistor. When the switch S is at P, the capacitor charges; when S is at Q, the capacitor discharges, sending an electrical stimulus to the heart. During this time, the voltage E applied to the heart is given by the linear differential equation

$$\frac{dE}{dt} = -\frac{1}{RC}E, \quad t_1 < t < t_2,$$

where R and C are constants. Determine $E(t)$ if $E(t_1) = E_0$. (Of course, the opening and closing of the switch are periodic in time to simulate the natural heartbeat.)

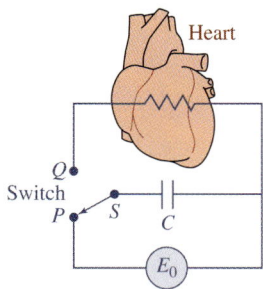

FIGURE 8.3.8   Heart pacemaker in Problem 25

26. In a series circuit that contains only a resistor and an inductor, Kirchhoff's second law states that the sum of the voltage drop across the inductor ($L(di/dt)$) and the voltage drop across the resistor (iR) is the same as the impressed voltage (E) on the circuit. See FIGURE 8.3.9. Thus, we obtain the differential equation for the current $i(t)$:

$$L\frac{di}{dt} + Ri = E,$$

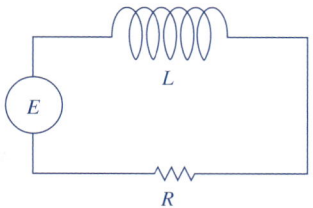

FIGURE 8.3.9   Series circuit in Problem 26

where L and R are constants known as the inductance and the resistance, respectively. Determine the current $i(t)$ if E is 12 volts, the inductance is $\frac{1}{2}$ henry, the resistance is 10 ohms, and $i(0) = 0$.

27. A 30-volt battery is connected to a series circuit in which the inductance is 0.1 henry and the resistance is 50 ohms. Find the current $i(t)$ if $i(0) = 0$. Determine the behavior of the current for large values of time. (See Problem 26.)

28. Suppose a water tank has the form of a right-circular cylinder. If water is allowed to drain under the influence of gravity through a hole in the bottom of the tank, then the height h of water at time t is given by the nonlinear differential equation

$$\frac{dh}{dt} = -c\frac{A_h}{A_w}\sqrt{2gh},$$

where A_w and A_h are the cross-sectional areas of the water and the hole, respectively, and c is a friction/contraction factor at the hole. See FIGURE 8.3.10.

(a) Solve the equation if the initial height of the water is 20 ft and $A_w = 50$ ft^2 and $A_h = \frac{1}{4}$ ft^2.
(b) If $c = 1$, at what time is the tank empty?
(c) How long would it take the tank to empty if the friction/contraction factor is $c = 0.6$?

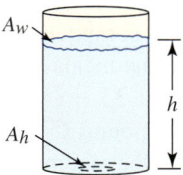

FIGURE 8.3.10   Tank in Problem 28

29. Around 1840, the Belgian mathematician–biologist P. F. Verhulst was concerned with mathematical models for predicting human population of countries. One of the equations he studied was

$$\frac{dP}{dt} = P(a - bP),$$

where $a > 0, b > 0$. This differential equation is now known as the **logistic equation**; the graph of a solution of the DE is known as a **logistic curve**. Show that a solution of this DE subject to the initial condition $P(0) = P_0$ is

$$P(t) = \frac{aP_0}{bP_0 + (a - bP_0)e^{-at}}.$$

30. The population $P(t)$ at time t in a suburb of a large city is modeled by the initial-value problem

$$\frac{dP}{dt} = P(10^{-1} - 10^{-7}P), \quad P(0) = 5000,$$

where t is measured in months. Find $P(t)$ and determine the limiting value of the population over a long period of time. At what time will the population be equal to one-half of this limiting value?

31. Suppose a student carrying a flu virus returns to an isolated college campus of 1000 students. If it is assumed that the rate at which the virus spreads is proportional not only to the number x of infected students but also to the number $1000 - x$ not infected, then a mathematical model for the number of infected students is

$$\frac{dx}{dt} = kx(1000 - x),$$

where $k > 0$ is a constant of proportionality, and t is time measured from the day the student returns to campus. If $x(0) = 1$ and if it is observed that $x(4) = 50$, then according to this model, how many students are infected after 6 days? Sketch a graph of the solution curve.

32. When two chemicals A and B are combined a compound C is formed. The resulting second-order reaction between the two chemicals is modeled by the differential equation

$$\frac{dX}{dt} = k(250 - X)(40 - X),$$

where $X(t)$ denotes the number of grams of the compound C present at time t.

(a) Determine $X(t)$ if it is known that $X(0) = 0$ g and $X(10) = 30$ g.

(b) How much of the compound C is present at 15 min?

(c) The amounts of chemicals A and B remaining at time t are $50 - \frac{1}{5}X$ and $32 - \frac{4}{5}X$, respectively. How many grams of the compound C is formed as $t \to \infty$? How many grams of the chemicals A and B remain as $t \to \infty$?

≡ Think About It

33. A rocket is shot vertically upward from the ground with an initial velocity v_0. See FIGURE 8.3.11. If the positive direction is taken to be upward, then in the absence of air resistance, the differential equation for the velocity v after fuel burnout is

$$v\frac{dv}{dy} = -\frac{k}{y^2},$$

where k is a positive constant.

(a) Solve the differential equation.

(b) If $k = gR^2$ and $g = 32$ ft/s^2, $R = 4000$ mi, use a calculator to show that the "escape velocity" of a rocket is approximately $v_0 = 25,000$ mi/h.

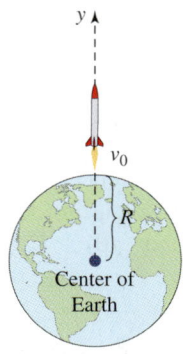

FIGURE 8.3.11 Rocket in Problem 33

34. Suppose a sphere of ice melts at a rate proportional to its surface area. Determine the volume V of the sphere at time t.

35. In a model for the growth of tissue, let $A(t)$ be the area of the tissue culture at time t. See FIGURE 8.3.12. Since the majority of cell divisions take place on the peripheral portion of the tissue, the number of cells on the periphery is proportional to $\sqrt{A(t)}$. If it is assumed that the rate of growth of the area is jointly proportional to $\sqrt{A(t)}$ and $M - A(t)$, then a mathematical model for A is given by

$$\frac{dA}{dt} = k\sqrt{A}(M - A),$$

where M is the final area of the tissue when growth is completed.

(a) Solve the differential equation by separation of variables. [*Hint*: Use a substitution as in Section 7.2 to carry out the integration with respect to A.]

(b) Find $\lim\limits_{t\to\infty} A(t)$.

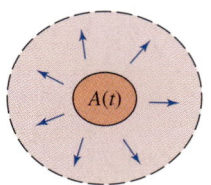

FIGURE 8.3.12 Tissue growth in Problem 35

≡ Projects

36. A Mathematical Classic—Time of Death The following problem occurs in almost all texts on differential equations.

A dead body was found within a closed room of a house where the temperature was a constant 70°F. A measurement of the core temperature of the body at the time of its discovery was found to be 85°F. A second measurement, one hour later, showed that the core temperature of the body was 80°F. Use the fact that if $t = 0$ corresponds to the time of death, then the core temperature of the body at that time was 98.6°F. Determine how many hours elapsed before the body was found. [*Hint:* Let $t_1 > 0$ denote the time that the body was discovered.]

37. Potassium/Argon Dating   The mineral potassium, whose chemical symbol is K, is the eighth most abundant element in the Earth's crust, making up about 2% of it by weight, and one of its naturally occurring isotopes, K-40, is radioactive. The radioactive decay of K-40 is more complex than that of carbon-14 because each of its atoms decays through one of two different nuclear decay reactions into one of two different substances: the mineral calcium-40 (Ca-40) or the gas argon-40 (Ar-40). Dating methods have been developed using both of these decay products. In each case, the age of a sample is calculated using the ratio of two numbers: the amount of the *parent* isotope K-40 in the sample and the amount of the *daughter isotope* (Ca-40 or Ar-40) in the sample that is **radiogenic**, in other words, the substance that originates from the decay of the parent isotope after the formation of the rock.

The amount of K-40 in a sample is easy to calculate. K-40 comprises 1.17% of naturally occurring potassium, and this small percentage is distributed quite uniformly, so that the mass of K-40 in the sample is just 1.17% of the total mass of potassium in the sample, which can be measured. But for several reasons it is complicated, and sometimes problematic, to determine how much of the Ca-40 in a sample is radiogenic. In contrast, when an igneous rock is formed by volcanic activity, all of the argon (and other) gas previously trapped in the rock is driven away by the intense heat. At the moment when the rock cools and solidifies, the gas trapped inside the rock has the same composition as the atmosphere. There are three stable isotopes of argon, and in the atmosphere they occur in the following relative abundances: 0.063% Ar-38, 0.337% Ar-36, and 99.60% Ar-40. Of these, just one, Ar-36, is not created radiogenically by the decay of any element, so any Ar-40 in excess of $99.60/(0.337) = 295.5$ times the amount of Ar-36 must be radiogenic. So the amount of radiogenic

Ar-40 in the sample can be determined from the amounts of Ar-38 and Ar-36 in the sample, which can be measured.

Assuming that we have a sample of rock for which the amount of K-40 and the amount of radiogenic Ar-40 have been determined, how can we calculate the age of the rock? Let $P(t)$ be the amount of K-40, $A(t)$ the amount of radiogenic Ar-40, and $C(t)$ the amount of radiogenic Ca-40 in the sample as functions of time t in years since the formation of the rock. Then a mathematical model for the decay of K-40 is the system of linear first-order differential equations

$$\frac{dA}{dt} = k_A P$$

$$\frac{dC}{dt} = k_C P$$

$$\frac{dP}{dt} = -(k_A + k_C)P,$$

where $k_A = 0.581 \times 10^{-10}$ and $k_C = 4.962 \times 10^{-10}$.

(a) Find a formula for $P(t)$. What is the half-life of K-40?

(b) Show that

$$A(t) = \frac{k_A}{k_A + k_C} P(t)(e^{(k_A + k_C)t} - 1).$$

(c) After a very long time (that is, let $t \to \infty$), what percentage of the K-40 originally present in the sample decays to Ar-40? What percentage decays to Ca-40?

(d) Show that the age t of the rock as a function of the present amounts $P(t)$ of K-40 and $A(t)$ of radiogenic Ar-40 in the sample is

$$t = \frac{1}{k_A + k_C} \ln\left[\frac{A(t)}{P(t)}\left(\frac{k_A + k_C}{k_A}\right) + 1\right].$$

(e) Suppose it is found that each gram of a rock sample contains 8.6×10^{-7} grams of radiogenic Ar-40 and 5.3×10^{-6} grams of K-40. How old is the rock?

8.4 Solution Curves without a Solution

▮ Introduction   Most differential equations cannot be solved. Perhaps the last sentence should be balanced by saying that *many* differential equations possess solutions, but the problem is finding them. When we say that a solution of a DE exists we do not mean that there also exists a method for finding it in the sense of being able to exhibit an exact solution, namely, a solution either given by an explicit function or as a function defined implicitly. It may be that the best we can do is to analyze a DE *qualitatively* or *numerically*.

In this section we shall examine two ways of analyzing a first-order DE qualitatively. We begin with a fundamental concept: A derivative dy/dx gives slope.

▮ A First-Order DE Defines Slope   Because a solution $y(x)$ of a first-order differential equation $dy/dx = F(x, y)$ is a differentiable function on some interval I, it must also be continuous on I. Thus, the corresponding solution curve on I has no breaks and must possess a tangent line at each point $(x, y(x))$. The slope of the tangent line at $(x, y(x))$ on a solution curve is the value of the first derivative dy/dx at this point, and this we know from the **slope function** F of the

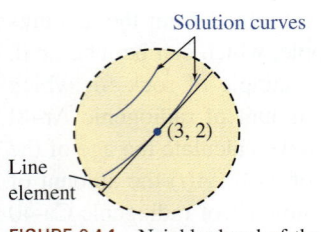

Solution curves

Line
element

FIGURE 8.4.1 Neighborhood of the
point (3, 2)

differential equation: $F(x, y(x))$. Now suppose that (x, y) represents any point in the xy-plane at which the function F is defined. The slope function F assigns a value $F(x, y)$ to the point; the value is the slope of a line. A short line segment, called a **line element**, is drawn through (x, y) with slope $F(x, y)$. For example, consider the equation $dy/dx = x - y$, where $F(x, y) = x - y$. At the point $(3, 2)$, for example, the slope of a line element is $F(3, 2) = 1$. As shown in FIGURE 8.4.1, a solution curve that passes through $(3, 2)$ does so tangent to the line element; a different solution curve that passes close to $(3, 2)$ will have a similar shape in a *small* neighborhood of the point.

■ **Direction Fields** Now suppose we systematically evaluate $F(x, y)$ over a rectangular grid of points in the xy-plane and draw a line element at each point where F is evaluated. The collection of all these line elements is called a **direction field** or a **slope field** of the differential equation $dy/dx = F(x, y)$. Visually, the direction field suggests the appearance or shape of a family of solution curves of the differential equation and consequently it may be possible to see certain qualitative aspects (for example, increasing, decreasing, and concavity) of a solution curve. A single solution curve that wends its way through the direction field must follow the flow pattern of the field; it is tangent to a line element when it intersects a point in the grid.

■ **Solution Curves without a Solution** Sketching a direction field by hand is straightforward, but time consuming; it is probably one of those tasks about which an argument can be made for doing it once in a lifetime, but it is overall most efficiently carried out by means of computer software. FIGURE 8.4.2 was obtained using a software direction field application with $dy/dx = x - y$ and a 5×5 rectangular region, where points in that region have a vertical and horizontal separation of $\frac{1}{2}$ unit, that is, at points (mh, nh), $h = \frac{1}{2}$, m and n integers such that $-10 \le m \le 10$, $-10 \le n \le 10$.

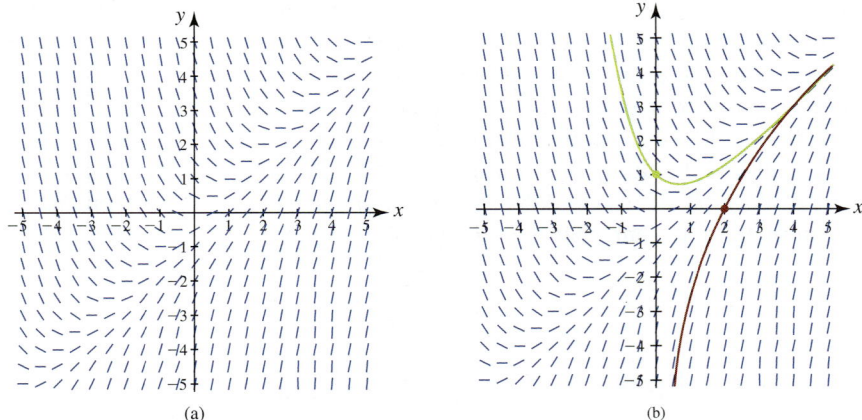

FIGURE 8.4.2 Direction field in (a); solution curves of DE superimposed on direction field in (b)

In Figure 8.4.2(a), notice that at any point along the line $y = x$, slopes are $F(x, x) = 0$ and so the line elements are horizontal. Moreover, the line $y = x$ splits the plane into two regions, above the line ($y > x$) the line elements have negative slope, whereas below the line ($y < x$) the line elements have positive slope. Reading left to right, imagine a solution curve starting at a point in the second quadrant, moving downward, becoming flat as it passes through the line $y = x$ and then moving upward into the first quadrant—in other words, its shape would be concave upward. We have seen the family of solutions of this DE in Example 5 of Section 8.2. You should compare the sample graphs in Figure 8.2.1 with the direction field in Figure 8.4.2(a). In Figure 8.4.2(b) we have given the two solution curves corresponding to the solutions of $dy/dx = x - y$ that pass through $(0, 1)$ (in green) and $(2, 0)$ (in red).

EXAMPLE 1   Direction Field

Sketch the direction field for $\dfrac{dy}{dx} = -\dfrac{x}{y}$.

Solution   This is the differential equation in Example 2 in Section 8.1.

(a) If we use a grid of points (x, y) with integer coordinates for $-5 \le x \le 5$, $-5 \le y \le 5$ then it is straight forward to compute the slopes of the line elements in the four quadrants by hand. For $x > 0, y > 0$ (first quadrant) the slopes $F(x, y) = -x/y$ are given in the following table.

$F(x, y)$	$y = 1$	$y = 2$	$y = 3$	$y = 4$	$y = 5$
$x = 1$	-1	$-\frac{1}{2}$	$-\frac{1}{3}$	$-\frac{1}{4}$	$-\frac{1}{5}$
$x = 2$	-2	-1	$-\frac{2}{3}$	$-\frac{1}{2}$	$-\frac{2}{5}$
$x = 3$	-3	$-\frac{3}{2}$	-1	$-\frac{3}{4}$	$-\frac{3}{5}$
$x = 4$	-4	-2	$-\frac{4}{3}$	-1	$-\frac{4}{5}$
$x = 5$	-5	$-\frac{5}{2}$	$-\frac{5}{3}$	$-\frac{5}{4}$	-1

For example, $F(3, 4) = -\frac{3}{4}$ is the slope of a line element at $(3, 4)$ and is given in red in the table at the intersection of the row labeled $x = 3$ and the column labeled $y = 4$.

Since the algebraic sign of the quotient x/y at (x, y), $x > 0, y > 0$ is the same as at (x, y), $x < 0, y < 0$ the slopes at the corresponding points in the third quadrant are the same as the slopes in the first quadrant. Similarly, it is easy to see that the slopes of the line segments in the second and fourth quadrants are the negatives of the slopes in the table. Drawing line elements through the points with the slopes determined from the table yields the direction field in FIGURE 8.4.3(a).

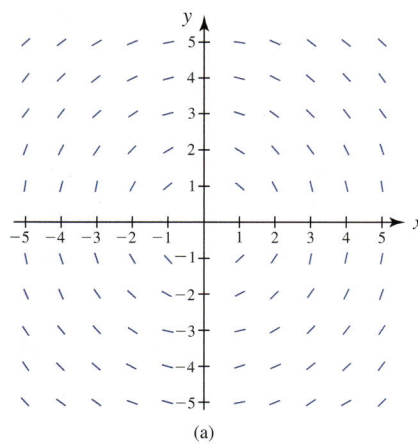 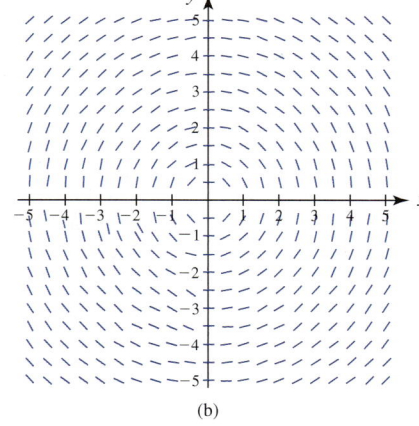 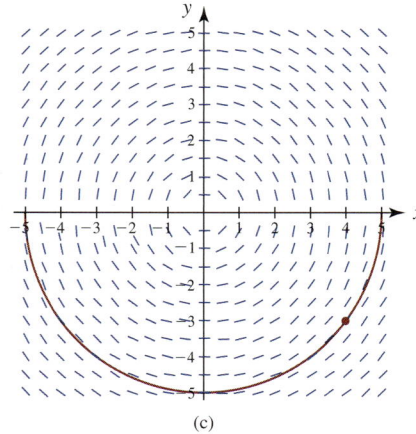

(a)   (b)   (c)

FIGURE 8.4.3   Direction fields in Example 1

(b) With the aid of a CAS and the grid of points again defined by (mh, nh), $h = \frac{1}{2}$, m and n integers, $-10 \le m \le 10, -10 \le n \le 10$, we get the direction field in Figure 8.4.3(b). Visually the flow of the field is circular. In Figure 8.4.3(c) we have superimposed the solution curve (in red) of the IVP,

◀ By using more gridpoints we get a better idea of the shapes of solution curves.

$$\frac{dy}{dx} = -\frac{x}{y}, \qquad y(4) = -3$$

obtained in Example 3 of Section 8.1 over the computer-generated direction field. ■

Of course the main purpose of constructing a direction field is to be able to obtain a rough sketch of a solution curve when it is impossible to solve a DE exactly.

EXAMPLE 2   Direction Field

Use a direction field to describe an approximate solution curve for the initial-value problem

$$\frac{dy}{dx} = \sin x + \cos y, \qquad y(-4) = 4.$$

Solution   We mark the initial point $(-4, 4)$ in the computer-generated direction field in FIGURE 8.4.4(a); moving to the left and to the right we try to draw a curve as long as possible that contains the initial point. When we move to the right of the initial point we see that the line segments almost immediately force a graph downward (roughly for $-3 < x < -0.5$), and then upward as a graph crosses the y-axis (roughly for $-0.5 < x < 2.5$) and finally followed by another downward movement (roughly for $x > 2.5$). The solution curve just described has the approximate shape shown in red in Figure 8.4.4(b).

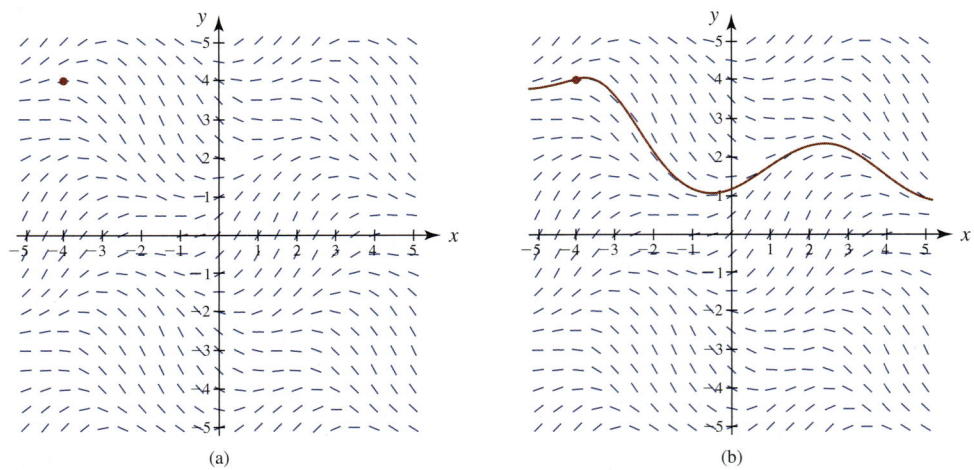

(a)   (b)

FIGURE 8.4.4   Direction field and approximate solution curve in Example 2   ■

■ **Phase Portraits and Stability**   Interpretation of the derivative dy/dx as a function that gives slope played the key role in the construction of direction fields. In the discussion that follows next, we will employ another telling property of the first derivative, namely, if $y(x)$ is a differentiable function, and if $dy/dx > 0$ (or $dy/dx < 0$) for all x in an interval I, then $y(x)$ is increasing (or decreasing) on I.

■ **Autonomous DEs**   A first-order differential equation in which the independent variable x does not appear explicitly, is said to be an **autonomous differential equation**. Hence, an autonomous first-order differential equation is one whose normal form is

$$\frac{dy}{dx} = f(y). \tag{1}$$

We shall assume throughout that f and its derivative f' are continuous functions of x on some interval I. The differential equations

$$\overset{f(y)}{\underset{\downarrow}{}} \qquad\qquad \overset{F(x, y)}{\underset{\downarrow}{}}$$

$$\frac{dy}{dx} = 1 + y^2 \quad \text{and} \quad \frac{dy}{dx} = 2xy$$

are autonomous and nonautonomous, respectively. Many first-order differential equations encountered in applications are autonomous and of the form (1). All but one of the mathematical models in Section 8.3 are autonomous; equation (7) of that section is nonautonomous. Of course different symbols in Section 8.3 are playing the part of x and y in the current discussion.

■ **Critical Points**   The zeros of the function f in (1) are of special interest. We say that a real number c is a **critical point** of the autonomous differential equation (1) if it is a zero of f, that is, $f(c) = 0$. Critical points are also called **equilibrium points** and **stationary points**. Moreover, substituting $y = c$ into (1) makes both sides of the equation zero, so we see that:

- *If c is a critical point of (1) then y(x) = c is a constant solution of the autonomous equation.*

A constant solution $y(x) = c$ of (1) is called an **equilibrium solution**, and:

- *Equilibrium solutions are the only constant solutions of (1).*

We can tell when a nonconstant solution $y(x)$ of (1) is increasing or decreasing by determining the algebraic sign of the derivative dy/dx; this we do by identifying the intervals over which $f(y)$ is positive or negative.

EXAMPLE 3   Autonomous First-Order DE

Inspection of the differential equation

$$\frac{dy}{dx} = y(a - by) \qquad (2)$$

◀ This is the DE in Problem 29, Exercises 8.3. Here the symbol y plays the part of P and x plays the part of t.

$a > 0, b > 0$, shows that it is autonomous. From $f(y) = y(a - by) = 0$ we also see that 0 and a/b are critical points of the equation. By putting these two numbers on a vertical number line, we divide the line into three intervals determined by the inequalities:

$$-\infty < x < 0, \qquad 0 < x < a/b, \qquad a/b < x < \infty.$$

The arrows on the line shown in FIGURE 8.4.5 indicate the algebraic sign of $f(y) = y(a - by)$ on these intervals, and whether a solution $y(x)$ is increasing or decreasing. The following table explains the figure.

Interval	Sign of $f(y)$	$y(x)$	Arrow
$(-\infty, 0)$	minus	decreasing	points down
$(0, a/b)$	plus	increasing	points up
$(a/b, \infty)$	minus	decreasing	points down

The equilibrium solutions of the DE are $y = 0$ and $y = a/b$.   ■

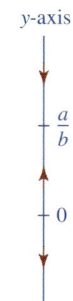

FIGURE 8.4.5   Two critical points determine three intervals in Example 3

Figure 8.4.5 is called a **one-dimensional phase portrait**, or simply a **phase portrait**, of the differential equation $dy/dx = y(a - by)$. The vertical line or y-axis is called a **phase line**. A phase portrait such as this can also be interpreted in terms of motion of a moving particle. If we imagine that $y(x)$ denotes the position of a particle at *time x* on a vertical line whose positive y-direction is upward, then the rate of change dy/dx represents the velocity of the particle. Positive velocity indicates motion upward and negative velocity indicates that the particle is moving downward. If a particle is placed at a critical point, then it must remain there for all time. Whence the origin of the alternative name *stationary point*.

▮ Solution Curves without the Solution   Without solving an autonomous differential equation, we can usually say a great deal about its solution curves. Relating this back to the first topic of this section, note that a direction field of an autonomous differential equation (1) is independent of x, so at any point on a line parallel to the x-axis all slopes are the same. Thus if $y(x)$ is a solution of (1), then any horizontal translation $y(x - k)$, k a constant, is also a solution. Since the function f in (1) is independent of the variable x, we may consider it defined for $-\infty < x < \infty$ or $0 \leq x < \infty$. Also, since f and its derivative f' are continuous functions of x on some interval I, it can be proved that in some horizontal strip or region R in the xy-plane corresponding to I, through any point (x_0, y_0) in R there passes only one solution curve of (1). See FIGURE 8.4.6(a). For the sake of discussion let us suppose that (1) possesses exactly two critical points c_1 and c_2 and that $c_1 < c_2$. The graphs of the equilibrium solutions $y(x) = c_1$ and $y(x) = c_2$ are horizontal lines, and these lines partition the region R into three subregions R_1, R_2, and R_3 as illustrated in Figure 8.4.6(b). Without proof, here are some conclusions that we can draw about a nonconstant solution $y(x)$ of (1):

- If (x_0, y_0) is in a subregion R_i, $i = 1, 2, 3$, and $y(x)$ is a solution whose graph passes through this point, then $y(x)$ remains in the subregion for all x. As illustrated in

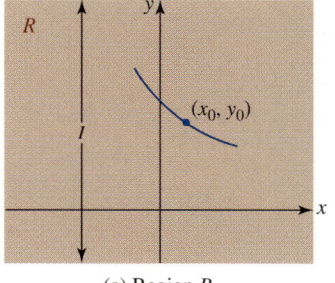

(a) Region R

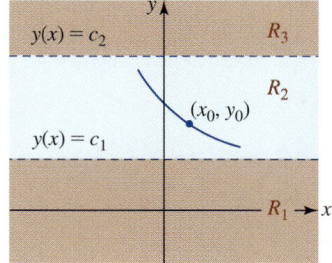

(b) Subregions R_1, R_2, and R_3 of R

FIGURE 8.4.6   Two equilibrium solutions determine three subregions in the plane

Figure 8.4.6(b) the solution $y(x)$ in R_2 is bounded below by c_1 and above by c_2, that is, $c_1 < y(x) < c_2$ for all x. The solution curve stays within R_2 for all x because the graph of a nonconstant solution of (1) cannot cross the graph of an equilibrium solution.

- By continuity of f we must then have either $f(y) > 0$ or $f(y) < 0$ for all y in a subregion R_i, $i = 1, 2, 3$. In other words, $f(y)$ cannot change signs in a subregion.

- Since $dy/dx = f(y(x))$ is either positive or negative in a subregion, a solution $y(x)$ is either increasing or decreasing in a subregion R_i, $i = 1, 2, 3$. Therefore $y(x)$ cannot be oscillatory, nor can it have a relative extremum (maximum or minimum).

- If $y(x)$ is *bounded above* by a critical point c_1 (as in subregion R_1 where $y(x) < c_1$ for all x), then the graph of $y(x)$ must approach the graph of the equilibrium solution $y(x) = c_1$ either as $x \to \infty$ or as $x \to -\infty$. If $y(x)$ is *bounded*, that is, bounded above and below by two consecutive critical points (as in subregion R_2 where $c_1 < y(x) < c_2$ for all x), then the graph of $y(x)$ must approach the graphs of the equilibrium solutions $y(x) = c_1$ and $y(x) = c_2$, one as $x \to \infty$ and the other as $x \to -\infty$. If $y(x)$ is *bounded below* by a critical point (as in subregion R_3 where $c_2 < y(x)$ for all x), then the graph of $y(x)$ must approach the graph of the equilibrium solution $y(x) = c_2$ either as $x \to \infty$ or as $x \to -\infty$.

With the foregoing facts in mind let us reexamine the differential equation in Example 3.

EXAMPLE 4   Example 3 Revisited

The three intervals determined on the y-axis or phase line by the critical points 0 and a/b now correspond in the xy-plane to three subregions defined by:

$$R_1: \quad -\infty < y < 0, \qquad R_2: \quad 0 < y < a/b, \qquad R_3: \quad a/b < y < \infty,$$

where $-\infty < x < \infty$. The phase portrait in Figure 8.4.5 tells us that $y(x)$ is decreasing in R_1, increasing in R_2, and decreasing in R_3. If $y(0) = y_0$ is an initial value, then in R_1, R_2, and R_3 we have, respectively:

   (*i*) For $y_0 < 0$, $y(x)$ is bounded above. Since $y(x)$ is decreasing, $y(x)$ decreases without bound for increasing x and so $y(x) \to 0$ as $x \to -\infty$. This means the negative x-axis, $y = 0$, is a horizontal asymptote for a solution curve.

   (*ii*) For $0 < y_0 < a/b$, $y(x)$ is bounded. Since $y(x)$ is increasing, $y(x) \to a/b$ as $x \to \infty$, and $y(x) \to 0$ as $x \to -\infty$. The two lines $y = 0$ and $y = a/b$ are horizontal asymptotes for any solution curve starting in this subregion.

   (*iii*) For $y_0 > a/b$, $y(x)$ is bounded below. Since $y(x)$ is decreasing, $y(x) \to a/b$ as $x \to \infty$. This means $y = a/b$ is a horizontal asymptote for a solution curve.

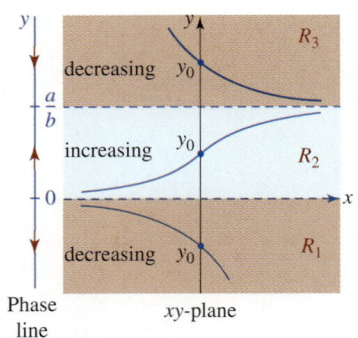

Phase line xy-plane

FIGURE 8.4.7 Phase portrait and solution curves in each of the three subregions in Example 4

In FIGURE 8.4.7, the original phase portrait is reproduced to the left of the xy-plane in which the subregions R_1, R_2, and R_3 are shaded. The graphs of the equilibrium solutions $y = a/b$ and $y = 0$ are shown in Figure 8.4.7 as dashed lines; the solid graphs represent typical graphs of $y(x)$ illustrating the three cases just discussed. ∎

In a subregion such as R_1 in Example 4, where $y(x)$ is decreasing and unbounded below, we must necessarily have $y(x) \to -\infty$. Do *not* interpret this last statement to mean $y(x) \to -\infty$ as $x \to \infty$; we could have $y(x) \to -\infty$ as $x \to a$, where $a > 0$ is a finite number that depends on the initial condition $y(x_0) = y_0$. Thinking in dynamic terms, $y(x)$ could "blow up" in finite time; or thinking graphically, $y(x)$ could have a vertical asymptote at $x = a > 0$. A similar remark holds true for the subregion R_3. The next example illustrates these concepts.

EXAMPLE 5   Solution Curves

The autonomous equation $dy/dx = (y - 1)^2$ possesses the single critical point 1 and hence has only one constant solution $y(x) = 1$. From the phase portrait in FIGURE 8.4.8(a), we conclude that a nonconstant solution $y(x)$ is an increasing function in the two subregions defined by $-\infty < y < 1$ and $1 < y < \infty$, where $-\infty < x < \infty$. For an initial condition $y(0) = y_0 < 1$, a solution $y(x)$ is increasing and bounded above by 1, so $y(x) \to 1$ as $x \to \infty$; for $y(0) = y_0 > 1$, a solution $y(x)$ is increasing and unbounded.

You should verify by separation of variables that a one-parameter family of solutions of the differential equation is $y(x) = 1 - 1/(x + C)$. For a given initial condition, say, $y(0) = -1 < 1$, we find $C = \frac{1}{2}$ and $y(x) = 1 - 1/(x + \frac{1}{2})$. Observe $x = -\frac{1}{2}$ is a vertical asymptote and $y(x) \to -\infty$ as $x \to -\frac{1}{2}^+$. See Figure 8.4.8(b). For a different initial condition $y(0) = 2 > 1$, we find $C = -1$ and $y(x) = 1 - 1/(x - 1)$. The last function has a vertical asymptote at $x = 1$ and we see in Figure 8.4.8(c) that $y(x) \to \infty$ as $x \to 1^-$. The solution curves are the portions of the graphs in Figures 8.4.8(b) and 8.4.8(c) shown in blue. As predicted by the phase portrait in Figure 8.4.8(a), for the solution curve in Figure 8.4.8(b), $y(x) \to 1$ as $x \to \infty$, whereas for the solution curve in Figure 8.4.8(c), $y(x) \to \infty$ as $x \to 1^-$.

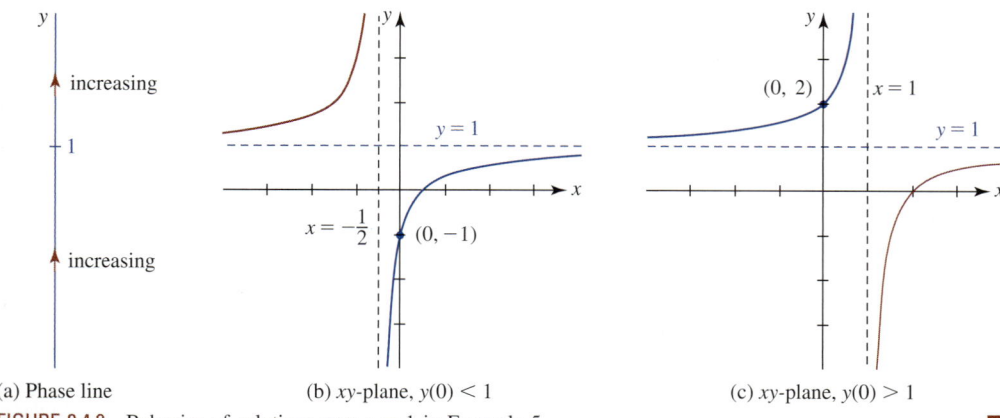

(a) Phase line   (b) xy-plane, $y(0) < 1$ (c) xy-plane, $y(0) > 1$

FIGURE 8.4.8 Behavior of solutions near $y = 1$ in Example 5

■ **Attractors and Repellers** Suppose $y(x)$ is a nonconstant solution of the autonomous differential equation (1) and that c is a critical point of the DE. There are basically three types of behavior $y(x)$ can exhibit near c. In FIGURE 8.4.9 we have placed c on four vertical phase lines. When both arrowheads on either side of the blue dot labeled c point toward c, as in Figure 8.4.9(a), all solutions $y(x)$ of (1) that start from an initial point (x_0, y_0) sufficiently near c exhibit the asymptotic behavior $\lim_{x \to \infty} y(x) = c$. For this reason the critical point c is said to be **asymptotically stable**. Using a physical analogy, a solution that starts near c is like a charged particle that, over time, is drawn to a particle of opposite charge, so c is often referred to as an **attractor**. When both arrowheads on either side of the blue dot labeled c point away from c, as in Figure 8.4.9(b), all solutions $y(x)$ of (1) that start from an initial point (x_0, y_0) move away from c as x increases. In this case the critical point c is said to be **unstable**. An unstable critical point is called a **repeller**, for obvious reasons. The critical point illustrated in Figures 8.4.9(c) and 8.4.9(d) is neither an attractor nor a repeller. But since c exhibits characteristics of both an attractor *and* a repeller—that is a solution starting from an initial point (x_0, y_0) sufficiently near c is attracted to c from one side and repelled from the other side—we say that the critical point c is **semi-stable**. In Example 3, the critical point a/b is asymptotically stable (an attractor) and the critical point 0 is unstable (a repeller). The critical point 1 in Example 5 is semi-stable.

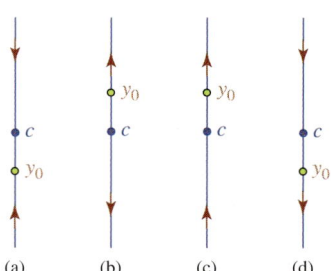

(a)   (b)   (c)   (d)
FIGURE 8.4.9 Critical point is: an attractor in (a), a repeller in (b), and semi-stable in (c) and (d)

Exercises 8.4   Answers to selected odd-numbered problems begin on page ANS-25.

≡ Fundamentals

In Problems 1–8, use the given computer-generated direction field to sketch an approximate solution curve for the indicated differential equation that passes through each of the given points.

1. $\dfrac{dy}{dx} = \dfrac{x}{y}$

   (a) $y(0) = 3$ **(b)** $y(3) = 3$

   (c) $y\left(-\frac{3}{2}\right) = 2$ **(d)** $y(-2) = -3$

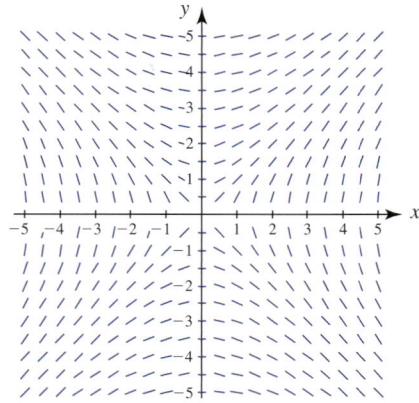

FIGURE 8.4.10 Direction field for Problem 1

2. $\dfrac{dy}{dx} = e^{-0.01xy^3}$

   (a) $y(-4) = 0$ **(b)** $y(3) = -2$

   (c) $y(0) = -2$ **(d)** $y(0) = 1$

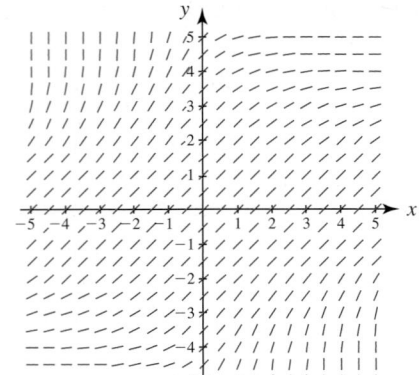

FIGURE 8.4.11 Direction field for Problem 2

3. $\dfrac{dy}{dx} = 1 - xy$

   (a) $y(0) = 0$ **(b)** $y(-1) = 0$

   (c) $y(2) = 0$ **(d)** $y(0) = -4$

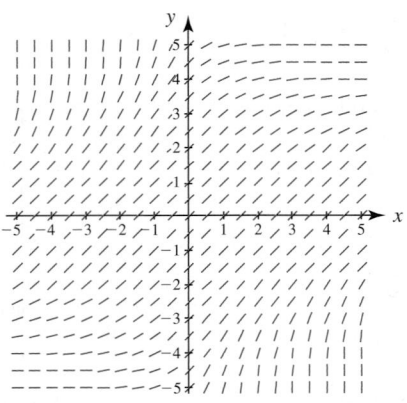

FIGURE 8.4.12 Direction field for Problem 3

4. $\dfrac{dy}{dx} = (\sin x)\cos y$

   (a) $y(0) = 0$ **(b)** $y(0) = 2$

   (c) $y(-3) = 0$ **(d)** $y(4) = 0$

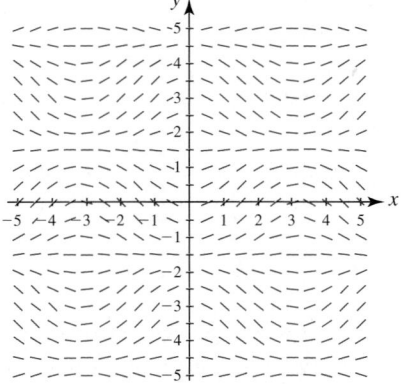

FIGURE 8.4.13 Direction field for Problem 4

5. $\dfrac{dy}{dx} = \dfrac{1}{2}x - \dfrac{1}{2}y^2$

   (a) $y(0) = -2$ **(b)** $y(0) = 2$

   (c) $y(-1) = 0$ **(d)** $y(-4) = 0$

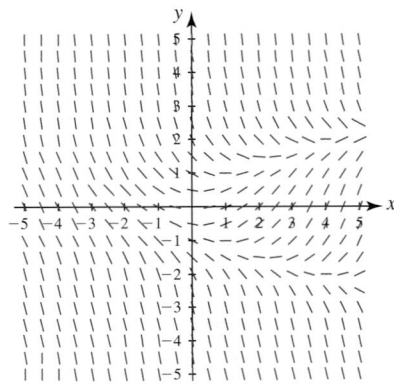

FIGURE 8.4.14 Direction field for Problem 5

6. $\dfrac{dy}{dx} = e^{-\sin y}$

   (a) $y(0) = 0$ **(b)** $y(0) = 2$

   (c) $y(-2) = 0$ **(d)** $y(4) = 0$

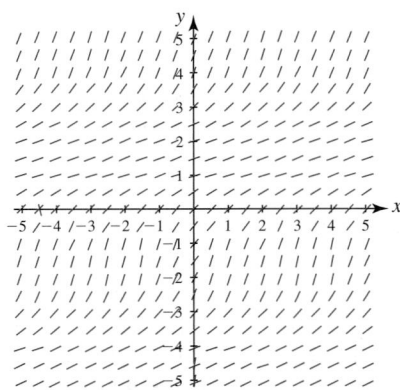

FIGURE 8.4.15 Direction field for Problem 6

7. $\dfrac{dy}{dx} = y\sin x$

   (a) $y(0) = 1$ **(b)** $y(-3) = -2$

   (c) $y(4) = 1$ **(d)** $y(2) = 2$

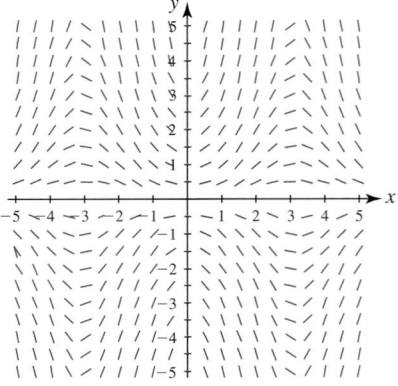

FIGURE 8.4.16 Direction field for Problem 7

8. $\dfrac{dy}{dx} = x^2 - y^2$

   (a) $y(-2) = 0$ **(b)** $y(0) = -2$

   (c) $y(0) = 0$ **(d)** $y(3) = 0$

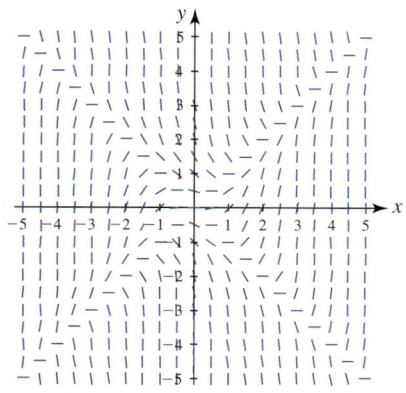

FIGURE 8.4.17   Direction field for Problem 8

In Problems 9–12, use computer software to obtain a direction field for the given differential equation. By hand, sketch an approximate solution curve that passes through each of the given points.

9. $\dfrac{dy}{dx} = \dfrac{1}{y}$ **10.** $\dfrac{dy}{dx} = x + y$

   (a) $y(0) = 3$ **(a)** $y(0) = -1$

   (b) $y(2) = -2$ **(b)** $y(3) = 0$

11. $\dfrac{dy}{dx} = \dfrac{1}{5}x^2 + y$ **12.** $\dfrac{dy}{dx} = y - 2\cos \pi x$

   (a) $y(0) = 1$ **(a)** $y(0) = 0$

   (b) $y(4) = 0$ **(b)** $y(2) = -4$

In Problems 13–20, find the critical points and phase portrait of the given autonomous first-order differential equation. Classify each critical point as asymptotically stable (attractor), unstable (repeller), or semi-stable.

13. $\dfrac{dy}{dx} = y^2 - 3y$ **14.** $\dfrac{dy}{dx} = y^2 - y^3$

15. $\dfrac{dy}{dx} = (y - 2)^2$ **16.** $\dfrac{dy}{dx} = 10 + 3y - y^2$

17. $\dfrac{dy}{dx} = y^2(4 - y^2)$ **18.** $\dfrac{dy}{dx} = y(2 - y)(4 - y)$

19. $\dfrac{dy}{dx} = y \ln(y + 2)$ **20.** $\dfrac{dy}{dx} = \dfrac{ye^y - 9y}{e^y}$

In Problems 21 and 22, consider the given autonomous first-order differential equation and the initial condition $y(0) = y_0$. By hand, sketch a graph of a typical solution $y(x)$ when y_0 has the values:

   (a) $y_0 > 1$ **(b)** $0 < y_0 < 1$

   (c) $-1 < y_0 < 0$ **(d)** $y_0 < -1$

21. $\dfrac{dy}{dx} = y - y^3$ **22.** $\dfrac{dy}{dx} = y^2 - y^4$

In Problems 23 and 24, consider the autonomous differential equation $dy/dx = f(y)$, where the graph of f is given. Use the graph to locate the critical points of each differential equation. Sketch a phase portrait of each equation. By hand, sketch typical solution curves in the subregions of the xy-plane determined by the graphs of the equilibrium solutions.

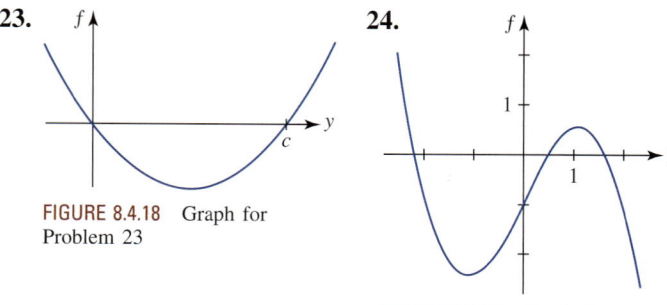

23.

FIGURE 8.4.18   Graph for Problem 23

24.

FIGURE 8.4.19   Graph for Problem 24

≡ Applications

25. We saw in Section 8.3 that the *linear* autonomous differential equation

$$m\dfrac{dv}{dt} = mg - kv$$

is a mathematical model for the velocity of a falling body when air resistance is taken into account. Using only a phase portrait, determine the terminal velocity of the falling body.

26. We also saw in Section 8.3 that the *nonlinear* autonomous differential equation

$$m\dfrac{dv}{dt} = mg - kv^2$$

is a mathematical model for the velocity of a falling body when air resistance is taken into account. Using only a phase portrait, determine the terminal velocity of the falling body.

27. In Problem 26 of Exercises 8.3, we saw that the current $i(t)$ in a series circuit is described by

$$L\dfrac{di}{dt} + Ri = E.$$

If the inductance L, resistance R, and impressed voltage E are positive constants, show that as $t \to \infty$ the current obeys Ohm's law that $E = iR$.

28. When two chemicals are combined, the rate at which a new compound is formed is governed by the differential equation

$$\dfrac{dX}{dt} = k(\alpha - X)(\beta - X),$$

where $k > 0$ is a constant of proportionality and $\beta > \alpha > 0$. Here $X(t)$ denotes the number of grams of the new compound formed in time t.

(a) Describe the behavior of X as $t \to \infty$.

(b) Consider the case when $\alpha = \beta$. What is the behavior of X as $t \to \infty$ if $X(0) < \alpha$? From the phase portrait of the differential equation, can you predict the behavior of X as $t \to \infty$ if $X(0) > \alpha$?

(c) Verify that an explicit solution of the differential equation in the case when $k = 1$ and $\alpha = \beta$ is $X(t) = \alpha - 1/(t + c)$. Find a solution satisfying $X(0) = \alpha/2$. Find a solution satisfying $X(0) = 2\alpha$. Graph these two solutions. Does the behavior of the solutions as $t \to \infty$ agree with your answers to part (b)?

≡ **Think About It**

29. For a differential equation $dy/dx = F(x, y)$, any member of the family of curves $F(x, y) = c$, where c is a constant, is called an **isocline** of the equation. In a direction field of

the DE $dy/dx = x^2 + y^2$, what is true about the line segments at points on the isocline $dy/dx = x^2 + y^2 = 1$? Identify the isoclines of the differential equation $dy/dx = x + y$.

30. For a differential equation $dy/dx = F(x, y)$, a curve in the plane defined by $F(x, y) = 0$ is called a **nullcline** of the equation. In a direction field of the DE $dy/dx = x^2 + y^2 - 1$, what is true about the line segments at points on a nullcline? Identify the nullclines of the differential equation $dy/dx = x^2 - y^2$ and indicate them in the direction field given in Figure 8.4.17.

31. The number 0 is a critical point of the autonomous differential equation $dy/dt = y^n$, where n is a positive integer. For what values of n is 0 asymptotically stable? Unstable? Repeat for the equation $dy/dx = -y^n$.

8.5 Euler's Method

■ **Introduction** We turn now from the visualization methods examined in the preceding section to a numerical method. By using the DE we are able to construct a simple procedure for obtaining approximations to the numerical values of the y-coordinates of points on a solution curve.

■ **Euler's Method** One of the simplest techniques for approximating a solution of a first-order initial-value problem

$$y' = F(x, y), \quad y(x_0) = y_0 \tag{1}$$

is known as **Euler's method**, or the **method of tangent lines**. This technique uses the fact that the derivative of a function $y(x)$ at a number x_0 determines a linearization of $y(x)$ at $x = x_0$:

$$L(x) = y_0 + y'(x_0)(x - x_0).$$

Recall from Section 4.9 that the linearization of $y(x)$ at x_0 is simply an equation of the tangent line to the graph of $y = y(x)$ at the point (x_0, y_0). We now let h be a positive increment on the x-axis, as shown in FIGURE 8.5.1. Then for $x_1 = x_0 + h$ we have

$$L(x_1) = y_0 + y'(x_0)(x_0 + h - x_0) = y_0 + hy_0',$$

where $y_0' = y'(x_0) = F(x_0, y_0)$. Letting $y_1 = L(x_1)$ we get

$$y_1 = y_0 + hF(x_0, y_0).$$

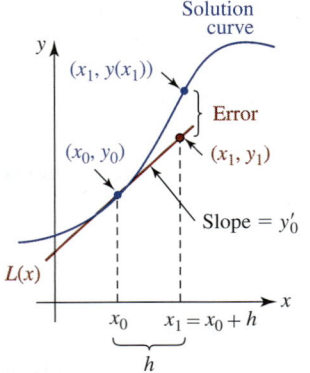

FIGURE 8.5.1 Approximating a point on a solution curve by a point on the tangent line

The point (x_1, y_1), which is seen in Figure 8.5.1 to be a point on the tangent line, is an approximation to the point $(x_1, y(x_1))$ on the actual solution curve; that is, $L(x_1) \approx y(x_1)$ or $y_1 \approx y(x_1)$ is an approximation of $y(x)$ at x_1. Of course, the accuracy of the approximation depends heavily on the size of the increment h. Usually we must choose this **step size** to be *reasonably small*. If we repeat the process, using (x_1, y_1) and the new slope $F(x_1, y_1)$ as the new starting point, then we obtain the approximation

$$y(x_2) = y(x_0 + 2h) = y(x_1 + h) \approx y_2 = y_1 + hF(x_1, y_1).$$

In general, it follows that

$$y_{n+1} = y_n + hF(x_n, y_n), \tag{2}$$

where $x_n = x_0 + nh$.

In the next example we apply Euler's method (2) to a differential equation for which we know an explicit solution; in this way we can compare the estimated values y_n with the true values $y(x_n)$.

EXAMPLE 1 Euler's Method

Consider the initial-value problem $y' = 0.2xy$, $y(1) = 1$. Use Euler's method to approximate $y(1.5)$ using first $h = 0.1$ and then $h = 0.05$.

Solution With the identification that $F(x, y) = 0.2xy$, (2) becomes

$$y_{n+1} = y_n + h(0.2x_ny_n).$$

Then for $x_0 = 1$, $y_0 = 1$, and $h = 0.1$ we find

$$y_1 = y_0 + (0.1)(0.2x_0y_0) = 1 + (0.1)[0.2(1)(1)] = 1.02,$$

which is an estimate to the value of $y(1.1)$. However, if we use $h = 0.05$, it takes *two* steps to reach $x = 1.1$. From

$$y_1 = 1 + (0.05)[0.2(1)(1)] = 1.01$$

$$y_2 = 1.01 + (0.05)[0.2(1.05)(1.01)] = 1.020605$$

we have $y_1 \approx y(1.05)$ and $y_2 \approx y(1.1)$. The remainder of the calculations were carried out using computer software. The results are summarized in TABLES 8.5.1 and 8.5.2. Each entry is rounded to four decimal places. Observe that it takes 5 steps with $h = 0.1$ and 10 steps with $h = 0.05$ to get to $x = 1.5$.

TABLE 8.5.1   Euler's Method with $h = 0.1$

x_n	y_n	True Value	Absolute Error	% Relative Error
1.00	1.0000	1.0000	0.0000	0.00
1.10	1.0200	1.0212	0.0012	0.12
1.20	1.0424	1.0450	0.0025	0.24
1.30	1.0675	1.0714	0.0040	0.37
1.40	1.0952	1.1008	0.0055	0.50
1.50	1.1259	1.1331	0.0073	0.64

TABLE 8.5.2   Euler's Method with $h = 0.05$

x_n	y_n	True Value	Absolute Error	% Relative Error
1.00	1.0000	1.0000	0.0000	0.00
1.05	1.0100	1.0103	0.0003	0.03
1.10	1.0206	1.0212	0.0006	0.06
1.15	1.0318	1.0328	0.0009	0.09
1.20	1.0437	1.0450	0.0013	0.12
1.25	1.0562	1.0579	0.0016	0.16
1.30	1.0694	1.0714	0.0020	0.19
1.35	1.0833	1.0857	0.0024	0.22
1.40	1.0980	1.1008	0.0028	0.25
1.45	1.1133	1.1166	0.0032	0.29
1.50	1.1295	1.1331	0.0037	0.32

In Example 1, the true values in the tables were calculated from the known solution $y = e^{0.1(x^2-1)}$. Also, **absolute error** is defined to be

◄ Verify this solution by solving the DE by separation of variables.

$$|true\ value - approximation|.$$

The **relative error** and **percentage relative error** are, in turn,

$$\frac{absolute\ error}{|true\ value|} \quad and \quad \frac{absolute\ error}{|true\ value|} \times 100.$$

It is apparent by comparing Tables 8.5.1 and 8.5.2 that the accuracy of the approximations improves as the step size h decreases. Also, we see that even though the percentage relative error is growing, it does not appear to be that bad. But you should not be deceived by one example. Watch what happens in the next example, when we simply change the coefficient of the right side of the differential equation from 0.2 to 2.

EXAMPLE 2 Comparison of Exact/Approximate Values

Use the Euler method to approximate $y(1.5)$ for the solution of the initial-value problem $y' = 2xy$, $y(1) = 1$.

Solution You should verify that the exact solution of the IVP is now $y = e^{x^2-1}$. Proceeding as in Example 1, we obtain the results shown in Tables 8.5.3 and 8.5.4.

In this case, with a step size $h = 0.1$, a 16% relative error in the calculation of the approximation to $y(1.5)$ is totally unacceptable. At the expense of doubling the number of calculations, a slight improvement in accuracy is obtained by halving the step size to $h = 0.05$.

TABLE 8.5.3 Euler's Method with $h = 0.1$

x_n	y_n	True Value	Absolute Error	% Relative Error
1.00	1.0000	1.0000	0.0000	0.00
1.10	1.2000	1.2337	0.0337	2.73
1.20	1.4640	1.5527	0.0887	5.71
1.30	1.8154	1.9937	0.1784	8.95
1.40	2.2874	2.6117	0.3244	12.42
1.50	2.9278	3.4904	0.5625	16.12

TABLE 8.5.4 Euler's Method with $h = 0.05$

x_n	y_n	True Value	Absolute Error	% Relative Error
1.00	1.0000	1.0000	0.0000	0.00
1.05	1.1000	1.1079	0.0079	0.72
1.10	1.2155	1.2337	0.0182	1.47
1.15	1.3492	1.3806	0.0314	2.27
1.20	1.5044	1.5527	0.0483	3.11
1.25	1.6849	1.7551	0.0702	4.00
1.30	1.8955	1.9937	0.0982	4.93
1.35	2.1419	2.2762	0.1343	5.90
1.40	2.4311	2.6117	0.1806	6.92
1.45	2.7714	3.0117	0.2403	7.98
1.50	3.1733	3.4904	0.3171	9.08

$$\frac{dy}{dx} = F(x, y) \quad \text{NOTES FROM THE CLASSROOM}$$

Euler's method is just one of many different ways a solution of a differential equation can be approximated. Although attractive in its simplicity, Euler's method is seldom used in serious calculations. We have introduced this topic simply to give you a first taste of numerical methods. You will delve into greater detail and examine methods that give significantly greater accuracy in a formal course in differential equations.

Exercises 8.5   Answers to selected odd-numbered problems begin on page ANS-26.

≡ **Fundamentals**

In Problems 1 and 2, use Euler's method (2) to obtain a four-decimal approximation to the indicated value. Carry out the recursion of (2) by hand, first using $h = 0.1$ and then $h = 0.05$.

1. $\dfrac{dy}{dx} = 2x - 3y + 1, y(1) = 5; \quad y(1.2)$

2. $\dfrac{dy}{dx} = x + y^2, y(0) = 0; \quad y(0.2)$

In Problems 3 and 4, use Euler's method to obtain a four-decimal approximation of the indicated value. First use $h = 0.1$ and then use $h = 0.05$. Find an explicit solution for each initial-value problem and then construct tables similar to Tables 8.5.1 and 8.5.2.

3. $y' = y, y(0) = 1; \quad y(1.0)$

4. $y' = 4x - 2y, y(0) = 2; \quad y(0.5)$

In Problems 5–10, use Euler's method to obtain a four-decimal approximation of the indicated value. First use $h = 0.1$ and then use $h = 0.05$.

5. $y' = e^{-y}, y(0) = 0; \quad y(0.5)$

6. $y' = x^2 + y^2, y(0) = 1; \quad y(0.5)$

7. $y' = (x - y)^2, y(0) = 0.5; \quad y(0.5)$

8. $y' = xy + \sqrt{y}, y(0) = 1; \quad y(0.5)$

9. $y' = xy^2 - \dfrac{y}{x}, y(1) = 1; \quad y(1.5)$

10. $y' = y - y^2, y(0) = 0.5; \quad y(0.5)$

Chapter 8 in Review

Answers to selected odd-numbered problems begin on page ANS-26.

A. True/False

In Problems 1–4, indicate whether the given statement is true or false.

1. The differential equation $dy/dx = x + xy$ is both separable and linear. _____
2. The differential equation $dy/dx = \sin y$ is nonlinear. _____
3. $y = 0$ is a solution of the initial-value problem $dy/dx = x^2 y$, $y(0) = 0$. _____
4. A solution of the differential equation $dy/dx = x^2 y^2 + 4$ is increasing on $(-\infty, \infty)$. _____

B. Fill in the Blanks

In Problems 1–8, fill in the blanks.

1. A one-parameter family of solutions of $dy/dx = 1 - 6x + 12e^{3x}$ is _____.
2. The order of the differential equation $(y'')^3 + y^4 = 1$ is _____.
3. An integrating factor for the linear equation $dy/dx - y = e^{3x}$ is _____.
4. In the direction field of the differential equation $dy/dx = x^2 - y^2$, the slope of a line element at $(2, 4)$ is _____.
5. The time that it takes for a substance decaying through radioactivity to go from its initial amount A_0 to $\frac{1}{2}A_0$ is called its _____.
6. If an initial population P_0 of bacteria doubles in 2 h, then the number of bacteria present after 32 h is _____.
7. If $P(t) = P_0 e^{0.16t}$ gives the population in an environment at time t, then $P(t)$ satisfies the initial-value problem _____.
8. Give an example of a first-order differential equation that is both separable and linear.

_____

C. Exercises

In Problems 1–10, solve the given differential equation.

1. $\sin x \dfrac{dy}{dx} + (\cos x)y = 0$

2. $\dfrac{dx}{dt} + x = e^{-t}\cos 2t$

3. $t\dfrac{dy}{dt} - 5y = t$

4. $\dfrac{y}{x^2}\dfrac{dy}{dx} + e^{2x^3 + y^2} = 0$

5. $(x^2 + 4)\dfrac{dy}{dx} = 2x - 8xy$

6. $y\sec^2 x\dfrac{dy}{dx} = y^2 + 1$

7. $\dfrac{dy}{dx} = 2x\sqrt{1 - y^2}$

8. $(e^x + e^{-x})\dfrac{dy}{dx} = y^2$

9. $y' - 2y = x(e^{3x} - e^{2x})$

10. $\dfrac{dy}{dx} + y = \sqrt{1 + e^x}$

In Problems 11–20, solve the given initial-value problems.

11. $\dfrac{dP}{dt} = 0.05P$, $P(0) = 1000$

12. $\dfrac{dA}{dt} = -0.015A$, $A(0) = 5$

13. $t\dfrac{dy}{dt} + y = t^4 \ln t$, $y(1) = 0$

14. $x\dfrac{dy}{dx} = 10y$, $y(1) = -3$

15. $\dfrac{dy}{dx} = 2y + y^2$, $y(0) = 3$

16. $\dfrac{dy}{dx} = y(10 - 2y)$, $y(0) = 7$

17. $\dfrac{dy}{dx} = 1 + y^2$, $y(\pi/3) = -1$

18. $x\dfrac{dy}{dx} = y^2 - 1$, $y(2) = 2$

19. $\dfrac{dy}{dx} = -8x^3 y^2$, $y(0) = \dfrac{1}{2}$

20. $\dfrac{dy}{dx} = e^{x-y}$, $y(0) = 1$

In Problems 21 and 22, find a function whose graph passes through the given point and has the indicated slope at a point (x, y) on its graph.

21. $(0, 2)$; $2x/3y^3$

22. $(0, 1)$; $x + y$

23. If P_0 is the initial population of a community, show that if the population P is modeled by $dP/dt = kP$, then

$$\left(\frac{P_1}{P_0}\right)^{t_2} = \left(\frac{P_2}{P_0}\right)^{t_1},$$

where $P_1 = P(t_1)$ and $P_2 = P(t_2)$, $t_1 < t_2$.

24. A metal bar is taken out of a furnace whose temperature is 150°C and put into a tank of water whose temperature is maintained at a constant 30°C. After $\frac{1}{4}$ h in the tank, the temperature of the bar is 90°C. What is the temperature of the bar in $\frac{1}{2}$ h? In 1 h?

25. When forgetfulness is taken into account, the rate at which a person can memorize a subject is given by the differential equation

$$\frac{dA}{dt} = k_1(M - A) - k_2A,$$

where k_1 and k_2 are positive constants, $A(t)$ is the amount of material memorized in time t, M is the total amount to be memorized, and $M - A$ is the amount remaining to be memorized.

(a) Solve for $A(t)$ if $A(0) = 0$.
(b) Find the limiting value of A as $t \to \infty$ and interpret the result.
(c) Graph the solution.

26. Suppose a series circuit contains a capacitor and a variable resistor. If the resistance at time t is given by $R = k_1 + k_2t$, where k_1 and k_2 are positive known constants, then the charge q on the capacitor is described by the first-order differential equation

$$(k_1 + k_2t)\frac{dq}{dt} + \frac{1}{C}q = E(t),$$

where C is a constant called the **capacitance** and $E(t)$ is the **impressed voltage**. Show that if $E(t) = E_0$ and $q(0) = q_0$ are constants, then

$$q(t) = E_0C + (q_0 - E_0C)\left(\frac{k_1}{k_1 + k_2t}\right)^{1/Ck_2}.$$

27. The differential equation $dP/dt = (k \cos t)P$, where k is a positive constant, is often used as a model of a population that undergoes yearly seasonal fluctuations.

(a) Solve for $P(t)$ if $P(0) = P_0$.
(b) Use a calculator or CAS to obtain the graph of the function found in part (a).

28. A projectile is shot vertically into the air with an initial velocity of v_0 ft/s. Assuming that air resistance is proportional to the square of the instantaneous velocity, the motion is described by the pair of differential equations:

$$m\frac{dv}{dt} = -mg - kv^2, \quad k > 0$$

positive y-axis up and origin at ground level so that $v = v_0$ at $y = 0$, and

$$m\frac{dv}{dt} = mg - kv^2, \quad k > 0$$

positive y-axis down and origin at the maximum height so that $v = 0$ at $y = h$. See FIGURE 8.R.1. The first and second equations describe the motion of the projectile when rising and falling, respectively. Prove that the impact velocity v_i is less than the initial velocity v_0. [*Hint*: By the Chain Rule, $dv/dt = v\,dv/dy$.]

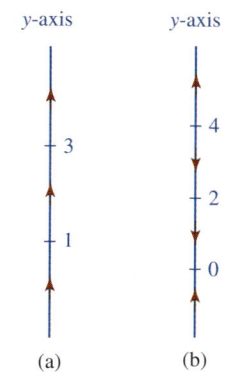

FIGURE 8.R.1 Initial and impact velocities in Problem 28

29. **(a)** Use a CAS to obtain the direction field for the differential equation $dy/dx = e^{-x} - 3y$ using a 3×3 rectangular grid with points (mh, nh), $h = 0.25$, $-12 \leq m \leq 12$, $-12 \leq n \leq 12$.
    (b) On the direction field, sketch by hand a solution curve that corresponds to each of the initial conditions: $y(0) = 1$, $y(-2) = 0$, $y(-1) = -2$.
    (c) Based on the direction field and the solution curves, form a conjecture about the behavior of all solutions $y(x)$ as $x \to \pm\infty$.

30. Construct an autonomous differential equation $dy/dx = f(y)$ whose phase portrait is consistent with part (a) of FIGURE 8.R.2. With part (b) of Figure 8.R.2.

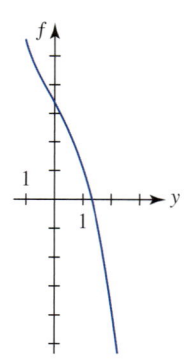

FIGURE 8.R.2 Phase portraits for Problem 30

31. Consider the autonomous differential equation $dy/dx = f(y)$, where

$$f(y) = -0.5y^3 - 1.7y + 3.4.$$

It is seen in FIGURE 8.R.3 that the function $f(y)$ has one zero. Without attempting to solve the differential equation for $y(x)$, estimate the value of $\lim_{x\to\infty} y(x)$.

FIGURE 8.R.3 Phase portrait for Problem 31

32. Use Euler's method with step size $h = 0.1$ to approximate $y(1.2)$, where $y(x)$ is the solution of the initial-value problem $dy/dx = 1 + x\sqrt{y}$, $y(1) = 9$.

33. Two curves are said to be **orthogonal** at a point if and only if their tangent lines L_1 and L_2 are perpendicular at the point of intersection. Show that the curves defined by $y = x^3$ and $x^2 + 3y^2 = 4$ are orthogonal at $(-1, -1)$ and $(1, 1)$.

34. When all the curves of one family of curves $F(x, y, C_1) = 0$ intersect orthogonally all the curves of another family $G(x, y, C_2) = 0$, then the families are said to be **orthogonal trajectories** of each another.

    (a) Find the differential equations of the families $xy = C_1$ and $y^2 - x^2 = C_2$. Show that the two families of curves are orthogonal trajectories of each other.

    (b) Sketch the graphs of some members of each family in part (a) on the same coordinate axis.

Sequences and Series

In This Chapter   Everyday experience gives one an intuitive feeling for the notion of a sequence. The words *sequence of events* or *sequence of numbers* suggest an arrangement whereby the events E or numbers n are set down in some order: $E_1, E_2, E_3, \ldots$ or $n_1, n_2, n_3, \ldots$.

Every student of mathematics is also familiar with the fact that any real number can be written as a decimal. For example, the rational number $\frac{1}{3} = 0.333\ldots$, where the mysterious three dots (an ellipsis) signify that the digit 3 repeats forever. This means that the decimal $0.333\ldots$ is an infinite sum or the *infinite series*

$$\frac{3}{10} + \frac{3}{100} + \frac{3}{1000} + \frac{3}{10,000} + \cdots .$$

In this chapter we will see that the concepts of sequence and infinite series are related.

9.1 Sequences

■ **Introduction** If the domain of a function f is the set of positive integers, then the elements $f(n)$ in the range can be arranged in an order corresponding to increasing values of n:

$$f(1), f(2), f(3), \ldots, f(n), \ldots$$

In the discussion that follows we consider only functions whose domain is the set of positive integers and whose range elements are real numbers.

EXAMPLE 1 Function with Domain the Positive Integers

If n is a positive integer, then the first several elements in the range of the function $f(n) = (1 + 1/n)^n$ are

$$f(1) = 2, \qquad f(2) = \frac{9}{4}, \qquad f(3) = \frac{64}{27}, \ldots.$$ ■

A function whose domain is the entire set of positive integers is given a special name.

Definition 9.1.1 Sequence

A **sequence** is a function whose domain is the set of positive integers.

Some texts use the words infinite *sequence. When the domain of the function is a finite subset of the set of positive integers, we get a* finite *sequence. All the sequences in this chapter will be infinite.*

■ **Notation and Terms** Instead of the customary function notation $f(n)$, a sequence is usually denoted by either $\{a_n\}$ or $\{a_n\}_{n=1}^{\infty}$. The integer n is sometimes called the **index** of a_n. The **terms** of the sequence are formed by letting the index n take on the values $1, 2, 3, \ldots$; the number a_1 is the *first term*, a_2 is the *second term*, and so on. The number a_n is called the *nth term* or the **general term** of the sequence. Thus, $\{a_n\}$ is equivalent to

$$a_1, a_2, a_3, \ldots, a_n, \ldots. \; \leftarrow \text{numbers in the range}$$
$$\uparrow \; \uparrow \; \uparrow \qquad \uparrow$$
$$1 \; \; 2 \; \; 3 \qquad n \qquad \leftarrow \text{numbers in the domain}$$

For example, the sequence defined in Example 1 would be written $\{(1 + 1/n)^n\}$.

In some circumstances it is convenient to take the first term of a sequence to be a_0 and the sequence is then

$$a_0, a_1, a_2, a_3, \ldots, a_n, \ldots.$$

EXAMPLE 2 Terms of a Sequence

Write out the first four terms of the sequences

(a) $\left\{\dfrac{1}{2^n}\right\}$ (b) $\{n^2 + n\}$ (c) $\{(-1)^n\}$.

Solution By substituting $n = 1, 2, 3, 4$ in the respective general term of each sequence, we obtain

(a) $\dfrac{1}{2}, \dfrac{1}{4}, \dfrac{1}{8}, \dfrac{1}{16}, \ldots$ (b) $2, 6, 12, 20, \ldots$ (c) $-1, 1, -1, 1, \ldots$ ■

■ **Convergent Sequence** For the sequence in part (a) of Example 2, we see that as the index n becomes progressively larger, the values $a_n = \frac{1}{2^n}$ do not increase without bound. Indeed, we see that as $n \to \infty$, the terms

$$\frac{1}{2}, \frac{1}{4}, \frac{1}{8}, \frac{1}{16}, \frac{1}{32}, \frac{1}{64}, \ldots$$

approach the limiting value 0. We say that the sequence $\{\frac{1}{2^n}\}$ **converges** to 0. In contrast, the terms of the sequences in parts (b) and (c) do not approach a limiting value as $n \to \infty$. In general we have the following definition.

Definition 9.1.2   Convergent Sequence

A sequence $\{a_n\}$ is said to **converge** to a real number L if for every $\varepsilon > 0$ there exists a positive integer N such that

$$|a_n - L| < \varepsilon \text{ whenever } n > N. \tag{1}$$

The number L is called the **limit** of the sequence.

◀ Compare this definition with the wording in Definition 2.6.5.

If a sequence $\{a_n\}$ converges, then its limit L is unique.

■ **Convergent Sequence**   If $\{a_n\}$ is a convergent sequence, (1) means that the terms a_n can be made arbitrarily close to L for n sufficiently large. We indicate that a sequence converges to a number L by writing

$$\lim_{n \to \infty} a_n = L.$$

When $\{a_n\}$ does not converge, that is, when $\lim_{n \to \infty} a_n$ does not exist, we say that the sequence **diverges**.

FIGURE 9.1.1 illustrates several ways in which a sequence $\{a_n\}$ can converge to a number L. Parts (a), (b), (c), and (d) of Figure 9.1.1 show that for four different convergent sequences $\{a_n\}$, *all but a finite* number of the terms a_n are in the interval $(L - \varepsilon, L + \varepsilon)$. The terms of the sequence $\{a_n\}$ that are in $(L - \varepsilon, L + \varepsilon)$ for $n > N$ are represented by red dots in the figure.

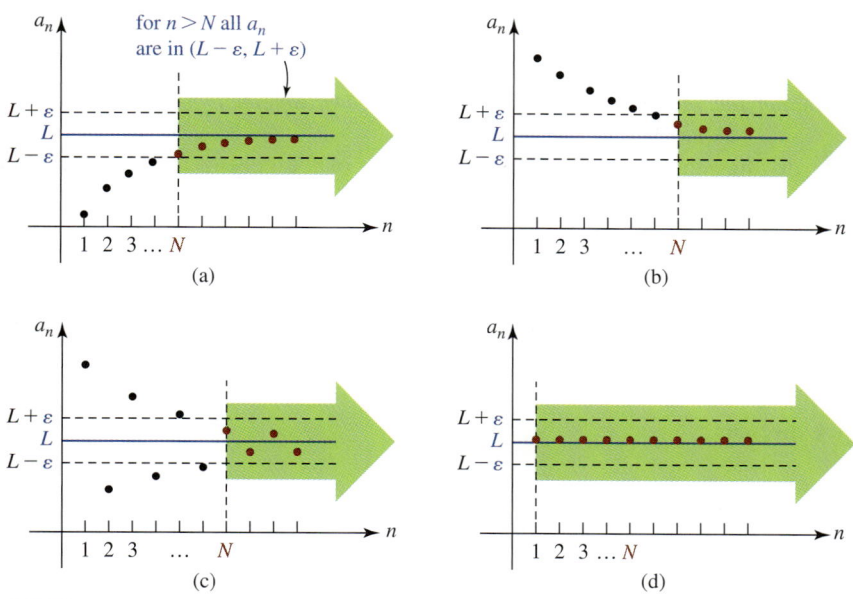

FIGURE 9.1.1   Four ways a sequence can converge to L

EXAMPLE 3   Convergent Sequence

Use Definition 9.1.2 to prove that the sequence $\{1/\sqrt{n}\}$ converges to 0.

Solution   Intuitively, we can see from the terms

$$1, \frac{1}{\sqrt{2}}, \frac{1}{\sqrt{3}}, \frac{1}{2}, \frac{1}{\sqrt{5}}, \ldots$$

that as the index n increases without bound the terms approach the limiting value 0. To prove convergence, we start by assuming that $\varepsilon > 0$ is given. Since the terms of the sequence are positive, the inequality $|a_n - 0| < \varepsilon$ is the same as

$$\frac{1}{\sqrt{n}} < \varepsilon.$$

This is equivalent to $\sqrt{n} > 1/\varepsilon$ or $n > 1/\varepsilon^2$. Hence, we need only choose N to be the first positive integer greater than or equal to $1/\varepsilon^2$. For instance, if we choose $\varepsilon = 0.01$, then $|1/\sqrt{n} - 0| = 1/\sqrt{n} < 0.01$ whenever $n > 10{,}000$. That is, we choose $N = 10{,}000$. ∎

In practice, to determine whether a sequence $\{a_n\}$ converges or diverges, we work directly with $\lim_{n\to\infty} a_n$ and proceed as we would in the examination of $\lim_{x\to\infty} f(x)$. If a_n either increases or decreases without bound as $n \to \infty$, then $\{a_n\}$ is necessarily divergent and we write, respectively,

$$\lim_{n\to\infty} a_n = \infty \qquad \text{or} \qquad \lim_{n\to\infty} a_n = -\infty. \tag{2}$$

In the first case in (2) we say that $\{a_n\}$ **diverges to infinity** and in the second, $\{a_n\}$ **diverges to negative infinity**. A sequence may diverge in a manner other than that given in (2). The next example illustrates two sequences; each diverges in a different way.

EXAMPLE 4 Divergent Sequences

    (a) The sequence $\{n^2 + n\}$ diverges to infinity, since $\lim_{n\to\infty}(n^2 + n) = \infty$.

    (b) The sequence $\{(-1)^n\}$ is divergent since $\lim_{n\to\infty}(-1)^n$ does not exist. The general term of the sequence does not approach a single constant as $n \to \infty$; as can be seen in part (c) of Example 2 the term $(-1)^n$ alternates between 1 and -1 as $n \to \infty$. ∎

EXAMPLE 5 Determining Convergence

Determine whether the sequence $\left\{\dfrac{3n(-1)^n}{n + 1}\right\}$ converges or diverges.

Solution By dividing the numerator and denominator of the general term by n we obtain

$$\lim_{n\to\infty}\frac{3n(-1)^n}{n + 1} = \lim_{n\to\infty}\frac{3(-1)^n}{1 + 1/n}.$$

Although $3/(1 + 1/n) \to 3$ as $n \to \infty$, the foregoing limit still does not exist. Because of the factor $(-1)^n$, we see that as $n \to \infty$,

$$a_n \to 3, \quad n \text{ even}, \qquad \text{and} \qquad a_n \to -3, \quad n \text{ odd}.$$

The sequence diverges. ∎

A sequence, such as those in part (b) of Example 4 and in Example 5, for which

$$\lim_{n\to\infty} a_{2n} = L \qquad \text{and} \qquad \lim_{n\to\infty} a_{2n+1} = -L,$$

$L \neq 0$, is said to **diverge by oscillation**.

■ **Constant Sequence** A sequence of constants

$$c, c, c, \ldots$$

is written $\{c\}$. Common sense tells us that this sequence converges and that its limit is c. See Figure 9.1.1(d). For example, the sequence $\{\pi\}$ converges to π.

In determining the limit of a sequence it is often useful to replace the discrete variable n by a continuous variable x. If f is a function such that $f(x) \to L$ as $x \to \infty$ and the values of f at the positive integers, $f(1), f(2), f(3), \ldots$, agree with the terms $a_1, a_2, a_3, \ldots$ of $\{a_n\}$, that is,

$$f(1) = a_1, \qquad f(2) = a_2, \qquad f(3) = a_3, \ldots,$$

then necessarily the sequence $\{a_n\}$ converges to the number L. The plausibility of this result is illustrated in FIGURE 9.1.2.

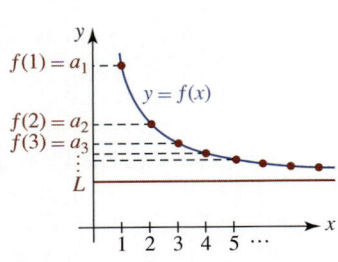

FIGURE 9.1.2 If $f(x) \to L$ as $x \to \infty$, then $f(n) = a_n \to L$ as $n \to \infty$

> **Theorem 9.1.1**   Limit of a Sequence
>
> Suppose $\{a_n\}$ is a sequence and f is a function such that $f(n) = a_n$ for $n \geq 1$. If
>
> $$\lim_{x \to \infty} f(x) = L \qquad \text{then} \qquad \lim_{n \to \infty} a_n = L. \qquad (3)$$

EXAMPLE 6   Using L'Hôpital's Rule

Show that the sequence $\{(n + 1)^{1/n}\}$ converges.

Solution   If we define $f(x) = (x + 1)^{1/x}$, then we recognize that $\lim_{x \to \infty} f(x)$ has the indeterminate form ∞^0 as $x \to \infty$. Hence, by L'Hôpital's Rule,

◀ See Section 4.5 for a review on how to handle the form ∞^0.

$$\lim_{x \to \infty} \ln f(x) = \lim_{x \to \infty} \frac{\ln(x + 1)}{x} \overset{h}{=} \lim_{x \to \infty} \frac{\dfrac{1}{x + 1}}{1} = \lim_{x \to \infty} \frac{1}{x + 1} = 0.$$

This shows $\lim_{x \to \infty} \ln f(x) = \ln\left[\lim_{x \to \infty} f(x)\right] = 0$ and that $\lim_{x \to \infty} f(x) = e^0 = 1$. Thus, by (3) we have $\lim_{n \to \infty} (n + 1)^{1/n} = e^0 = 1$. The sequence converges to 1. ∎

EXAMPLE 7   Convergent Sequence

Show that the sequence $\left\{ \dfrac{n(4n + 1)(5n + 3)}{6n^3 + 2} \right\}$ converges.

Solution   If $f(x) = \dfrac{x(4x + 1)(5x + 3)}{6x^3 + 2} = \dfrac{20x^3 + 17x^2 + 3x}{6x^3 + 2}$, then $\lim_{x \to \infty} f(x)$ has the indeterminate form ∞/∞. By L'Hôpital's Rule,

$$\lim_{x \to \infty} \frac{x(4x + 1)(5x + 3)}{6x^3 + 2} = \lim_{x \to \infty} \frac{20x^3 + 17x^2 + 3x}{6x^3 + 2}$$

$$\overset{h}{=} \lim_{x \to \infty} \frac{60x^2 + 34x + 3}{18x^2}$$

$$\overset{h}{=} \lim_{x \to \infty} \frac{120x + 34}{36x}$$

$$\overset{h}{=} \lim_{x \to \infty} \frac{120}{36} = \frac{10}{3}.$$

From (3) of Theorem 9.1.1, the given sequence converges to $\frac{10}{3}$. ∎

EXAMPLE 8   Determining Convergence

Determine whether the sequence $\left\{ \sqrt{\dfrac{n}{9n + 1}} \right\}$ converges.

Solution   It follows either by using L'Hôpital's Rule or by dividing the numerator and denominator by x that $x/(9x + 1) \to \frac{1}{9}$ as $x \to \infty$. Thus, we can write

$$\lim_{n \to \infty} \sqrt{\frac{n}{9n + 1}} = \sqrt{\lim_{n \to \infty} \frac{n}{9n + 1}} = \sqrt{\frac{1}{9}} = \frac{1}{3}.$$

The sequence converges to $\frac{1}{3}$. ∎

▪ **Properties**   The following **properties** of sequences are analogous to those given in Theorems 2.2.1, 2.2.2, and 2.2.3.

Theorem 9.1.2 Limit of a Sequence

Let $\{a_n\}$ and $\{b_n\}$ be convergent sequences. If $\lim\limits_{n\to\infty} a_n = L_1$ and $\lim\limits_{n\to\infty} b_n = L_2$, then

(i) $\lim\limits_{n\to\infty} c = c,\quad c$ a real number

(ii) $\lim\limits_{n\to\infty} ka_n = k\lim\limits_{n\to\infty} a_n = kL_1,\quad k$ a real number

(iii) $\lim\limits_{n\to\infty}(a_n + b_n) = \lim\limits_{n\to\infty} a_n + \lim\limits_{n\to\infty} b_n = L_1 + L_2$

(vi) $\lim\limits_{n\to\infty} a_n b_n = \lim\limits_{n\to\infty} a_n \cdot \lim\limits_{n\to\infty} b_n = L_1 \cdot L_2$

(v) $\lim\limits_{n\to\infty} \dfrac{a_n}{b_n} = \dfrac{\lim\limits_{n\to\infty} a_n}{\lim\limits_{n\to\infty} b_n} = \dfrac{L_1}{L_2},\quad L_2 \neq 0.$

EXAMPLE 9 Determining Convergence

Determine whether the sequence $\left\{\dfrac{2 - 3e^{-n}}{6 + 4e^{-n}}\right\}$ converges.

Solution Observe that $2 - 3e^{-n} \to 2$ and $6 + 4e^{-n} \to 6 \neq 0$ as $n \to \infty$. According to Theorem 9.1.2(v), we have

$$\lim_{n\to\infty} \frac{2 - 3e^{-n}}{6 + 4e^{-n}} = \frac{\lim\limits_{n\to\infty}(2 - 3e^{-n})}{\lim\limits_{n\to\infty}(6 + 4e^{-n})} = \frac{2}{6} = \frac{1}{3}.$$

The sequence converges to $\frac{1}{3}$. ∎

The first of the next two theorems should seem believable based on your knowledge of the behavior of the exponential function. Recall, for $0 < b < 1$, $b^x \to 0$ as $x \to \infty$, whereas for $b > 1$, $b^x \to \infty$ as $x \to \infty$.

Review Section 1.6, especially Figure 1.6.2.

Theorem 9.1.3 Sequences of the Form $\{r^n\}$

Suppose r is a nonzero constant. The sequence $\{r^n\}$ converges to 0 if $|r| < 1$ and diverges if $|r| > 1$.

Theorem 9.1.4 Sequences of the Form $\{1/n^r\}$

The sequence $\left\{\dfrac{1}{n^r}\right\}$ converges to 0 for r any positive rational number.

EXAMPLE 10 Applications of Theorems 9.1.3 and 9.1.4

(a) The sequence $\{e^{-n}\}$ converges to 0 by Theorem 9.1.3, since $e^{-n} = \left(\dfrac{1}{e}\right)^n$ and $r = 1/e < 1$.

(b) The sequence $\left\{\left(\dfrac{3}{2}\right)^n\right\}$ diverges by Theorem 9.1.3, since $r = \dfrac{3}{2} > 1$.

(c) The sequence $\left\{\dfrac{4}{n^{5/2}}\right\}$ converges to 0 by Theorem 9.1.2 *(ii)* and Theorem 9.1.4, since $r = \dfrac{5}{2}$ is a positive rational number. ∎

EXAMPLE 11 Determining Convergence

From Theorem 9.1.2(*iii*) and Theorem 9.1.4 we see that the sequence $\left\{10 + \dfrac{4}{n^{3/2}}\right\}$ converges to 10. ∎

Recursively Defined Sequence   As the following example indicates, a sequence can be defined by specifying the first term a_1 together with a rule for obtaining the subsequent terms from the preceding terms. In this case the sequence is said to be defined **recursively**. The defining rule is called a **recursion formula**. See Problems 59 and 60 in Exercises 9.1. Newton's Method, given in (3) in Section 4.10, is an example of a recursively defined sequence.

EXAMPLE 12   A Sequence Defined Recursively

Suppose a sequence is defined recursively by $a_{n+1} = 3a_n + 4$, where $a_1 = 2$. Then by letting $n = 1, 2, 3, \ldots$ we obtain

this number is given as 2
$\downarrow$

$$a_2 = 3a_1 + 4 = 3(2) + 4 = 10$$
$$a_3 = 3a_2 + 4 = 3(10) + 4 = 34$$
$$a_4 = 3a_3 + 4 = 3(34) + 4 = 106$$

and so on.   ∎

Squeeze Theorem   The following theorem is the sequence equivalent of Theorem 2.4.1.

Theorem 9.1.5   Squeeze Theorem

Suppose $\{a_n\}$, $\{b_n\}$, and $\{c_n\}$ are sequences such that

$$a_n \le c_n \le b_n$$

for all values of n larger than some index N (that is, $n > N$). If $\{a_n\}$ and $\{b_n\}$ converge to a common limit L, then $\{c_n\}$ also converges to L.

Factorial   Before presenting an example illustrating Theorem 9.1.5 we need to review a symbol that occurs frequently in this chapter. If n is a positive integer, the symbol $n!$, read "n factorial," is the product of the first n positive integers:

$$n! = 1 \cdot 2 \cdot 3 \cdots (n - 1) \cdot n. \qquad (4)$$

For example, $5! = 1 \cdot 2 \cdot 3 \cdot 4 \cdot 5 = 120$. An important property of the factorial is given by

$$n! = (n - 1)!n.$$

To see this, consider the case when $n = 6$:

$$6! = 1 \cdot 2 \cdot 3 \cdot 4 \cdot 5 \cdot 6 = \overbrace{(1 \cdot 2 \cdot 3 \cdot 4 \cdot 5)}^{5!} 6 = 5!6.$$

Stated in a slightly different manner, the property $n! = (n - 1)!n$ is equivalent to

$$(n + 1)! = n!(n + 1). \qquad (5)$$

One last point, for purposes of convenience and to ensure that the formula $n! = (n - 1)!n$ is valid when $n = 1$, we define $0! = 1$.

EXAMPLE 13   Determining Convergence

Determine whether the sequence $\left\{ \dfrac{2^n}{n!} \right\}$ converges.

Solution   The convergence or divergence of the given sequence is not obvious since $2^n \to \infty$ and $n! \to \infty$ as $n \to \infty$. Even though the limit form of $\lim\limits_{n \to \infty}(2^n/n!)$ is ∞/∞ we cannot use L'Hôpital's Rule since we have studied no function $f(x) = x!$. We can, however, use Theorem 9.1.5 by algebraically manipulating the general term of the sequence. In view of (4), the general term can be written

$$\frac{2^n}{n!} = \frac{\overbrace{2 \cdot 2 \cdot 2 \cdot 2 \cdots 2}^{n \text{ factor of } 2}}{1 \cdot 2 \cdot 3 \cdot 4 \cdots n} = \overbrace{\frac{2}{1} \cdot \frac{2}{2} \cdot \frac{2}{3} \cdot \frac{2}{4} \cdots \frac{2}{3}}^{n \text{ fractions}}$$

From the preceding line we obtain the inequality

$$0 \le \frac{2^n}{n!} = \overbrace{\frac{2}{1} \cdot \frac{2}{2} \cdot \frac{2}{3} \cdot \frac{2}{4} \cdots \frac{2}{n}}^{n \text{ fractions}} \le 2 \cdot 1 \cdot \overbrace{\frac{2}{3} \cdot \frac{2}{3} \cdots \frac{2}{3}}^{n-2 \text{ fractions}} = 2\left(\frac{2}{3}\right)^{n-2} \quad (6)$$

The $n - 2$ fractions of $\frac{2}{3}$ on the right-hand side of (6) results from the fact that after the second factor in the product of n fractions, 3 is the smallest denominator that makes $\frac{2}{3}$ larger than $\frac{2}{4}$, larger than $\frac{2}{5}$, and so on down to the last factor $\frac{2}{n}$. By the laws of exponents (6) is the same as

$$0 \le \frac{2^n}{n!} \le \frac{9}{2}\left(\frac{2}{3}\right)^n \quad \text{or} \quad a_n \le c_n \le b_n,$$

where we now identify the sequences $\{a_n\} = \{0\}$, $\{b_n\} = \{\frac{9}{2}(\frac{2}{3})^n\}$, and $\{c_n\} = \{2^n/n!\}$. The sequence $\{a_n\}$ is a sequence of 0's and so converges to 0. The sequence $\{b_n\} = \{\frac{9}{2}(\frac{2}{3})^n\}$ also converges to 0 by invoking Theorem 9.1.2(*ii*) and Theorem 9.1.3 with $r = \frac{2}{3} < 1$. Thus by Theorem 9.1.5, $\{c_n\} = \{2^n/n!\}$ must also converge to 0. ∎

The result $\lim\limits_{n \to \infty} \dfrac{2^n}{n!} = 0$ shows that $n!$ grows much faster than 2^n as $n \to \infty$. For example, for $n = 10$, $2^{10} = 1024$, whereas $10! = 3,628,800$

The sequence in the preceding example can also be defined recursively. For $n = 1$, $a_1 = 2^1/1! = 2$. Then by (5) and the laws of exponents,

$$a_{n+1} = \frac{2^{n+1}}{(n+1)!} = \frac{2 \cdot 2^n}{(n+1) \cdot n!} = \frac{2}{n+1} \cdot \overset{\text{this is } a_n}{\overbrace{\frac{2^n}{n!}}}.$$

Thus $\{2^n/n!\}$ is the same as

$$a_{n+1} = \frac{2}{n+1} a_n, \qquad a_1 = 2. \quad (7)$$

We can use the recursion formula in (7) as an alternative means of finding the limit L of the sequence $\{2^n/n!\}$. Since the sequence was shown to be convergent we have $\lim\limits_{n \to \infty} a_n = L$. This last statement is also equivalent to $\lim\limits_{n \to \infty} a_{n+1} = L$. By letting $n \to \infty$ in (7) and using the properties of limits we can write

$$\lim_{n \to \infty} a_{n+1} = \lim_{n \to \infty} \left(\frac{2}{n+1} a_n\right) = \left(\lim_{n \to \infty} \frac{2}{n+1}\right) \cdot \left(\lim_{n \to \infty} a_n\right). \quad (8)$$

From the last line we see that $L = 0 \cdot L$, which implies that the limit of the sequence is $L = 0$.

Our last theorem for this section is an immediate consequence of Theorem 9.1.5.

Theorem 9.1.6 Sequence of Absolute Values

If the sequence $\{|a_n|\}$ converges to 0, then $\{a_n\}$ converges to 0.

PROOF By the definition of absolute value, $|a_n| = a_n$ if $a_n \ge 0$ and $|a_n| = -a_n$ if $a_n < 0$. It follows that

$$-|a_n| \le a_n \le |a_n|. \quad (9)$$

By assumption $\{|a_n|\}$ converges to 0 and so $\lim\limits_{n \to \infty} |a_n| = 0$. From the inequality (9) and Theorem 9.1.5 we conclude that $\lim\limits_{n \to \infty} a_n = 0$. Therefore $\{a_n\}$ converges to 0. ∎

EXAMPLE 14 Using Theorem 9.1.6

The sequence $\left\{\dfrac{(-1)^n}{\sqrt{n}}\right\}$ converges to zero since we have already shown in Example 3 that the sequence of absolute values $\{|(-1)^n/\sqrt{n}|\} = \{1/\sqrt{n}\}$ converges to 0. ∎

Exercises 9.1   Answers to selected odd-numbered problems begin on page ANS-26.

≡ **Fundamentals**

In Problems 1–10, list the first four terms of the sequence whose general term is a_n.

1. $a_n = \dfrac{1}{2n + 1}$ **2.** $a_n = \dfrac{3}{4n - 2}$

3. $a_n = \dfrac{(-1)^n}{n}$ **4.** $a_n = \dfrac{(-1)^n n^2}{n + 1}$

5. $a_n = 10^n$ **6.** $a_n = 10^{-n}$

7. $a_n = 2n!$ **8.** $a_n = (2n)!$

9. $a_n = \displaystyle\sum_{k=1}^{n} \dfrac{1}{k}$ **10.** $a_n = \displaystyle\sum_{k=1}^{n} 2^{-k}$

In Problems 11–14, use Definition 9.1.2 to show that each sequence converges to the given number L.

11. $\left\{\dfrac{1}{n}\right\}$; $L = 0$ **12.** $\left\{\dfrac{1}{n^2}\right\}$; $L = 0$

13. $\left\{\dfrac{n}{n + 1}\right\}$; $L = 1$ **14.** $\left\{\dfrac{e^n + 1}{e^n}\right\}$; $L = 1$

In Problems 15–46, determine whether the given sequence converges. If the sequence converges, then find its limit.

15. $\left\{\dfrac{10}{\sqrt{n + 1}}\right\}$ **16.** $\left\{\dfrac{1}{n^{3/2}}\right\}$

17. $\left\{\dfrac{1}{5n + 6}\right\}$ **18.** $\left\{\dfrac{4}{2n + 7}\right\}$

19. $\left\{\dfrac{3n - 2}{6n + 1}\right\}$ **20.** $\left\{\dfrac{n}{1 - 2n}\right\}$

21. $\{20(-1)^{n+1}\}$ **22.** $\left\{\left(-\dfrac{1}{3}\right)^n\right\}$

23. $\left\{\dfrac{n^2 - 1}{2n}\right\}$ **24.** $\left\{\dfrac{7n}{n^2 + 1}\right\}$

25. $\{ne^{-n}\}$ **26.** $\{n^3 e^{-n}\}$

27. $\left\{\dfrac{\sqrt{n + 1}}{n}\right\}$ **28.** $\left\{\dfrac{n}{\sqrt{n + 1}}\right\}$

29. $\{\cos n\pi\}$ **30.** $\{\sin n\pi\}$

31. $\left\{\dfrac{\ln n}{n}\right\}$ **32.** $\left\{\dfrac{e^n}{\ln(n + 1)}\right\}$

33. $\left\{\dfrac{5 - 2^{-n}}{7 + 4^{-n}}\right\}$ **34.** $\left\{\dfrac{2^n}{3^n + 1}\right\}$

35. $\left\{\dfrac{e^n + 1}{e^n}\right\}$ **36.** $\left\{4 + \dfrac{3^n}{2^n}\right\}$

37. $\left\{n\sin\left(\dfrac{6}{n}\right)\right\}$ **38.** $\left\{\left(1 - \dfrac{5}{n}\right)^n\right\}$

39. $\left\{\dfrac{e^n - e^{-n}}{e^n + e^{-n}}\right\}$ **40.** $\left\{\dfrac{\pi}{4} - \arctan(n)\right\}$

41. $\{n^{2/(n+1)}\}$ **42.** $\{10^{(n+1)/n}\}$

43. $\left\{\ln\left(\dfrac{4n + 1}{3n - 1}\right)\right\}$ **44.** $\left\{\dfrac{\ln n}{\ln 3n}\right\}$

45. $\{\sqrt{n + 1} - \sqrt{n}\}$ **46.** $\{\sqrt{n}(\sqrt{n + 1} - \sqrt{n})\}$

In Problems 47–52, find a formula for the general term a_n of the sequence. Determine whether the given sequence converges. If the sequence converges, then find its limit.

47. $\dfrac{2}{1}, \dfrac{4}{3}, \dfrac{6}{5}, \dfrac{8}{7}, \ldots$

48. $1 + \dfrac{1}{2}, \dfrac{1}{2} + \dfrac{1}{3}, \dfrac{1}{3} + \dfrac{1}{4}, \dfrac{1}{4} + \dfrac{1}{5}, \ldots$

49. $3, -5, 7, -9, \ldots$

50. $-2, 2, -2, 2, \ldots$

51. $2, \dfrac{2}{3}, \dfrac{2}{9}, \dfrac{2}{27}, \ldots$

52. $\dfrac{1}{1 \cdot 4}, \dfrac{1}{2 \cdot 8}, \dfrac{1}{3 \cdot 16}, \dfrac{1}{4 \cdot 32}, \ldots.$

In Problems 53–56, for the given recursively defined sequence, write the next four terms after the indicated initial term(s).

53. $a_{n+1} = \dfrac{1}{2}a_n$, $a_1 = -1$

54. $a_{n+1} = 2a_n - 1$, $a_1 = 2$

55. $a_{n+1} = \dfrac{a_n}{a_{n-1}}$, $a_1 = 1$, $a_2 = 3$

56. $a_{n+1} = 2a_n - 3a_{n-1}$, $a_1 = 2$, $a_2 = 4$

In Problems 57 and 58, the recursively defined sequence is known to converge for a given initial value $a_1 > 0$. Assume $\lim_{n\to\infty} a_n = L$, and proceed as in (8) of this section to find the limit L of the sequence.

57. $a_{n+1} = \dfrac{1}{4}a_n + 6$ **58.** $a_{n+1} = \dfrac{1}{2}\left(a_n + \dfrac{5}{a_n}\right)$

In Problems 59 and 60, find a recursion formula that defines the given sequence.

59. $\left\{\dfrac{5^n}{n!}\right\}$

60. $\sqrt{3}, \sqrt{3 + \sqrt{3}}, \sqrt{3 + \sqrt{3 + \sqrt{3}}}, \ldots$

In Problems 61–64, use the Squeeze Theorem to establish convergence of the given sequence.

61. $\left\{\dfrac{\sin^2 n}{4^n}\right\}$ **62.** $\left\{\sqrt{16 + \dfrac{1}{n^2}}\right\}$

63. $\left\{\dfrac{\ln n}{n(n + 2)}\right\}$

64. $\left\{\dfrac{n!}{n^n}\right\}$ $\left[\text{Hint: } a_n = \dfrac{1}{n}\left(\dfrac{2}{n} \cdot \dfrac{3}{n} \cdot \dfrac{4}{n} \cdots \dfrac{n}{n}\right).\right]$

65. Show that for any real number x, the sequence $\{(1 + x/n)^n\}$ converges to e^x.

66. The sequence

$$\left\{ 1 + \frac{1}{2} + \frac{1}{3} + \cdots + \frac{1}{n} - \ln n \right\}$$

is known to converge to a number γ called **Euler's constant**. Calculate the first 10 terms of the sequence.

≡ Applications

67. A ball is dropped from an initial height of 15 ft onto a concrete slab. Each time it bounces, it reaches a height of $\frac{2}{3}$ its preceding height. What height does it reach on its third bounce? On its nth bounce? See FIGURE 9.1.3.

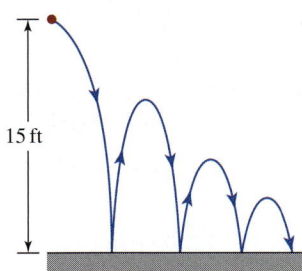

FIGURE 9.1.3 Bouncing ball in Problem 67

68. A ball, falling from a great height, travels 16 ft during the first second, 48 ft during the second, 80 ft during the third, and so on. How far does the ball travel during the sixth second?

69. A patient takes 15 mg of a drug each day. Suppose 80% of the drug accumulated is excreted each day by bodily functions. Write out the first six terms of the sequence $\{A_n\}$, where A_n is the amount of the drug present in the patient's body immediately after the nth dose.

70. One dollar is deposited in a savings account that pays an annual rate of interest r. If no money is withdrawn, what is the amount accrued in the account after the first, second, and third years?

71. Each person has two parents. Determine how many great-great-great-grandparents each person has.

72. The recursively defined sequence

$$p_{n+1} = 3p_n - \frac{p_n^2}{400}, \quad p_0 = 450,$$

is called a **discrete logistic equation**. Such a sequence is often used to model a population p_n in an environment; here p_0 is the initial population in the environment. Find the **carrying capacity** $K = \lim_{n \to \infty} p_n$ of the environment. Compute the next nine terms of the sequence and show that these terms oscillate around K.

≡ Think About It

73. Consider the sequence $\{a_n\}$ whose first four terms are

$$1, \quad 1 + \frac{1}{2}, \quad 1 + \frac{1}{2 + \frac{1}{2}}, \quad 1 + \frac{1}{2 + \frac{1}{2 + \frac{1}{2}}}, \dots.$$

(a) With $a_1 = 1$, find a recursion formula that defines the sequence.

(b) What are the fifth and sixth terms of the sequence?

(c) The sequence $\{a_n\}$ is known to converge. Find the limit of the sequence.

74. Conjecture the limit of the convergent sequence $\sqrt{3}, \sqrt{3\sqrt{3}}, \sqrt{3\sqrt{3\sqrt{3}}}, \dots.$

75. If the sequence $\{a_n\}$ converges, does the sequence $\{a_n^2\}$ diverge? Defend your answer with sound mathematics.

76. In FIGURE 9.1.4 the square shown in red is 1 unit on a side. A second blue square is constructed inside the first square by connecting the midpoints of the first one. A third green square is constructed by connecting the midpoints of the sides of the second square, and so on.

(a) Find a formula for the area A_n of the nth inscribed square.

(b) Consider the sequence $\{S_n\}$, where $S_n = A_1 + A_2 + \cdots + A_n$. Calculate the numerical values of the first 10 terms of this sequence.

(c) Make a conjecture about the convergence of $\{S_n\}$.

FIGURE 9.1.4 Embedded squares in Problem 76

≡ Projects

77. **A Mathematical Classic** Consider an equilateral triangle with sides of length 1 as shown in FIGURE 9.1.5(a). As shown in Figure 9.1.5(b), on each of the three sides of the first triangle, another equilateral triangle is constructed with sides of length $\frac{1}{3}$. As indicated in Figures 9.1.5(c) and 9.1.5(d), this construction is continued: equilateral triangles are constructed on the sides of each previously new triangle such that the length of the sides of the new triangles is $\frac{1}{3}$ the length of the sides of the previous triangle. Let the perimeter of the first figure be P_1, the perimeter of the second figure P_2, and so on.

(a) Find the values of P_1, P_2, P_3, and P_4.

(b) Find a formula for the perimeter P_n of the nth figure.

(c) What is $\lim_{n \to \infty} P_n$? The perimeter of the snowflake-like region obtained by letting $n \to \infty$ is called a **Koch snowflake curve** and was invented in 1904 by the Swedish mathematician **Helge von Koch** (1870–1924). The Koch curve plays a part in the theory of **fractals**.

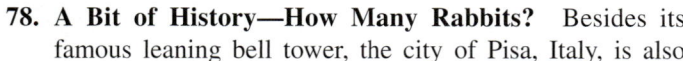

(a)   (b)

(c)   (d)

FIGURE 9.1.5 Snowflake regions in
Problem 77

78. A Bit of History—How Many Rabbits? Besides its
famous leaning bell tower, the city of Pisa, Italy, is also
noted as the birthplace of **Leonardo
Pisano**, aka **Leonardo Fibonacci**
(1170–1250). Fibonacci was the first in
Europe to introduce the Hindu–Arabic
place-valued decimal system and the
use of Arabic numerals. His book
Liber Abacci, published in 1202, is
basically a text on how to do arithmetic in this decimal
system. But in Chapter 12 of *Liber Abacci*, Fibonacci
poses and solves the following problem on the
reproduction of rabbits:

How many pairs of rabbits will be produced in a year begin-
ning with a single pair, if in every month each pair bears a
new pair which become productive from the second month
on?

Discern the pattern of the solution of this problem and
complete the following table.

	Start	After each month											
		1	2	3	4	5	6	7	8	9	10	11	12
Adult pairs	1	1	2	3	5	8	13	21					
Baby pairs	0	1	1	2	3	5	8	13					
Total pairs	1	2	3	5	8	13	21	34					

79. Write out five terms, after the initial two, of the
sequence defined recursively by $F_{n+1} = F_n + F_{n-1}$,
$F_1 = 1, F_2 = 1$. Reexamine Problem 78.

80. Golden Ratio If the recursion formula in Problem 79 is
divided by F_n, then

$$\frac{F_{n+1}}{F_n} = 1 + \frac{F_{n+1}}{F_n}.$$

If we define $a_n = F_{n+1}/F_n$, then the sequence $\{a_n\}$ is
defined recursively by

$$a_n = 1 + \frac{1}{a_{n-1}}, \quad a_1 = 1, n \geq 2.$$

The sequence $\{a_n\}$ is known to converge to the **golden
ratio** $\phi = \lim_{n \to \infty} a_n$.

(a) Find ϕ.
(b) Write a short report on the significance of the number ϕ.
Include in your report the relationship between the number ϕ and the shape of the multichambered nautilus shell.
See the photo in the Chapter 9 Opener on page 475.

9.2 Monotonic Sequences

▪ Introduction In the preceding section we showed that a sequence $\{a_n\}$ converged by
finding $\lim_{n \to \infty} a_n$. However, it is not always easy or even possible to determine whether a
sequence $\{a_n\}$ converges by seeking the exact value of $\lim_{n \to \infty} a_n$. For example, does the
sequence

$$\left\{ 1 + \frac{1}{2} + \frac{1}{3} + \cdots + \frac{1}{n} - \ln n \right\}$$

converge? It turns out that this sequence can be shown to converge, but not by using the basic
ideas of the last section. In this section we consider a special type of sequence that can be
proved convergent without finding the value of $\{a_n\}$.
We begin with a definition.

Definition 9.2.1 Monotonic Sequence

A sequence $\{a_n\}$ is said to be

(*i*) **increasing** if $a_{n+1} > a_n$ for all $n \geq 1$,
(*ii*) **nondecreasing** if $a_{n+1} \geq a_n$ for all $n \geq 1$,
(*iii*) **decreasing** if $a_{n+1} < a_n$ for all $n \geq 1$,
(*iv*) **nonincreasing** if $a_{n+1} \leq a_n$ for all $n \geq 1$.

If a sequence $\{a_n\}$ is one of the above types, then it is said to be **monotonic**.

In other words, sequences of the type

$$a_1 < a_2 < a_3 < \cdots < a_n < a_{n+1} < \cdots$$
$$a_1 > a_2 > a_3 > \cdots > a_n > a_{n+1} > \cdots,$$

are increasing and decreasing, respectively. Whereas

$$a_1 \leq a_2 \leq a_3 \leq \cdots \leq a_n \leq a_{n+1} \leq \cdots$$
$$a_1 \geq a_2 \geq a_3 \geq \cdots \geq a_n \geq a_{n+1} \geq \cdots,$$

are nondecreasing and nonincreasing sequences, respectively. The notions of *nondecreasing* and *nonincreasing* allow some adjacent terms in a sequence to be equal.

EXAMPLE 1 Monotonic/Not Monotonic

(**a**) The three sequences

$$4, 6, 8, 10, \ldots \qquad 1, \frac{1}{2}, \frac{1}{4}, \frac{1}{8}, \ldots \qquad \text{and} \qquad 5, 5, 4, 4, 4, 3, 3, 3, 3, \ldots$$

are monotonic. They are, respectively, increasing, decreasing, and nonincreasing.
(**b**) The sequence $-1, \frac{1}{2}, -\frac{1}{3}, \frac{1}{4}, -\frac{1}{5}, \ldots$ is not monotonic. ■

It is not always evident whether a sequence is increasing, decreasing, and so on. The following guidelines illustrate some of the ways that monotonicity can be demonstrated.

Guidelines for Demonstrating Monotonicity

(*i*) Form a **function** $f(x)$ such that $f(n) = a_n$. If $f'(x) > 0$, then $\{a_n\}$ is increasing. If $f'(x) < 0$, then $\{a_n\}$ is decreasing.
(*ii*) Form the **ratio** a_{n+1}/a_n where $a_n > 0$ for all n. If $a_{n+1}/a_n > 1$ for all n, then $\{a_n\}$ is increasing. If $a_{n+1}/a_n < 1$ for all n, then $\{a_n\}$ is decreasing.
(*iii*) Form the **difference** $a_{n+1} - a_n$. If $a_{n+1} - a_n > 0$ for all n, then $\{a_n\}$ is increasing. If $a_{n+1} - a_n < 0$ for all n, then $\{a_n\}$ is decreasing.

EXAMPLE 2 A Monotonic Sequence

Show that $\left\{ \dfrac{n}{e^n} \right\}$ is a monotonic sequence.

Solution If we define $f(x) = x/e^x$, then $f(n) = a_n$. Now,

$$f'(x) = \frac{1-x}{e^x} < 0$$

for $x > 1$ implies that f is decreasing on $[1, \infty)$. Thus it follows that

$$f(n+1) = a_{n+1} < f(n) = a_n.$$

By Definition 9.2.1 the given sequence is decreasing.

Alternative Solution   From the ratio

$$\frac{a_{n+1}}{a_n} = \frac{n+1}{e^{n+1}} \cdot \frac{e^n}{n} = \frac{n+1}{ne} = \frac{1}{e} + \frac{1}{ne} \le \frac{1}{e} + \frac{1}{e} = \frac{2}{e} < 1$$

we see that $a_{n+1} < a_n$ for $n \ge 1$. This shows the sequence is decreasing. ■

EXAMPLE 3   A Monotonic Sequence

The sequence $\left\{ \dfrac{2n+1}{n+1} \right\}$ or $\dfrac{3}{2}, \dfrac{5}{3}, \dfrac{7}{4}, \dfrac{9}{5}, \ldots$ *appears* to be increasing. From

$$a_{n+1} - a_n = \frac{2n+3}{n+2} - \frac{2n+1}{n+1} = \frac{1}{(n+2)(n+1)} > 0$$

we conclude $a_{n+1} > a_n$ for all $n \ge 1$. This proves the sequence is increasing. ■

Definition 9.2.2   Bounded Sequence

(*i*) A sequence $\{a_n\}$ is said to be **bounded above** if there is a positive number M such that $a_n \le M$ for all n.
(*ii*) A sequence $\{a_n\}$ is said to be **bounded below** if there is a positive number m such that $a_n \ge m$ for all n.
(*iii*) A sequence $\{a_n\}$ is said to be **bounded** if it is bounded above and bounded below.

Of course, if a sequence $\{a_n\}$ is not bounded, then it is said to be **unbounded**. An unbounded sequence is divergent. The Fibonacci sequence (see Problems 78 and 79 in Exercises 9.1)

$$1, 1, 2, 3, 5, 8, 13, 21, \ldots$$

is nondecreasing and is an example of an unbounded sequence.

   The sequence $1, \frac{1}{2}, \frac{1}{4}, \frac{1}{8}, \ldots$ in Example 1 is bounded since $0 \le a_n \le 1$ for all n. Any number smaller than a lower bound m of a sequence is also a lower bound and any number greater than an upper bound M is an upper bound; in other words the numbers m and M in Definition 9.2.2 are not unique. For the sequence $1, \frac{1}{2}, \frac{1}{4}, \frac{1}{8}, \ldots$ it is equally true that $-2 \le a_n \le 2$ for all $n \ge 1$.

EXAMPLE 4   A Bounded Sequence

The sequence $\left\{ \dfrac{2n+1}{n+1} \right\}$ is bounded above by 2, since the inequality

$$\frac{2n+1}{n+1} \le \frac{2n+2}{n+1} = \frac{2(n+1)}{n+1} = 2$$

shows that $a_n \le 2$ for $n \ge 1$. Moreover,

$$a_n = \frac{2n+1}{n+1} \ge 0$$

for $n \ge 1$ shows that the sequence is bounded below by 0. Thus, $0 \le a_n \le 2$ for all n implies that the sequence is bounded. ■

◀ Indeed, from Example 3 we see that the terms of the sequence are bounded below by the first term of the sequence.

   The next result will be useful in subsequent sections of this chapter.

Theorem 9.2.1   Sufficient Condition for Convergence

A bounded monotonic sequence $\{a_n\}$ converges.

PROOF We will prove the theorem in the case of a nondecreasing sequence. By assumption, $\{a_n\}$ is bounded and so $m \leq a_n \leq M$ for all n. In turn, this means the infinite set of terms

The existence of a *least upper bound,* that is, an upper bound that is smaller than all other upper bounds for the sequence, is one of the basic axioms in mathematics. It is called the **completeness property** of the real number system.

$S = \{a_1, a_2, a_3, \ldots, a_n, \ldots\}$ is bounded above and therefore has a least or smallest upper bound L. The sequence actually converges to L. For $\varepsilon > 0$ we know that $L - \varepsilon < L$, and consequently $L - \varepsilon$ is not an upper bound for S (there are no upper bounds smaller than the least upper bound). Hence, there exists a positive integer N such that $a_N > L - \varepsilon$. But, since $\{a_n\}$ is nondecreasing,

$$L - \varepsilon \leq a_N \leq a_{N+1} \leq a_{N+2} \leq a_{N+3} \leq \cdots \leq L + \varepsilon.$$

It follows that for $n > N$, $L - \varepsilon \leq a_n \leq L + \varepsilon$ or $|a_n - L| < \varepsilon$. From Definition 9.1.2 we conclude that $\lim\limits_{n \to \infty} a_n = L$. ∎

EXAMPLE 5 Bounded and Monotonic

The sequence $\left\{ \dfrac{2n + 1}{n + 1} \right\}$ was shown to be monotonic (Example 3) and bounded (Example 4). Hence, by Theorem 9.2.1 the sequence is convergent. ∎

EXAMPLE 6 Determining Convergence

Show that the sequence $\left\{ \dfrac{1 \cdot 3 \cdot 5 \cdots (2n - 1)}{2 \cdot 4 \cdot 6 \cdots (2n)} \right\}$ converges.

Solution First, the ratio

$$\frac{a_{n+1}}{a_n} = \frac{1 \cdot 3 \cdot 5 \cdots (2n - 1)(2n + 1)}{2 \cdot 4 \cdot 6 \cdots (2n)(2n + 2)} \cdot \frac{2 \cdot 4 \cdot 6 \cdots (2n)}{1 \cdot 3 \cdot 5 \cdots (2n - 1)} = \frac{2n + 1}{2n + 2} < 1$$

shows that $a_{n+1} < a_n$ for all n. The sequence is monotonic since it is decreasing. Next, from the inequality

Why is the product
$\dfrac{1}{2} \cdot \dfrac{3}{4} \cdot \dfrac{5}{6} \cdot \dfrac{7}{8} \cdots \dfrac{2n - 1}{2n}$ less than 1?

$$0 < \frac{1 \cdot 3 \cdot 5 \cdots (2n - 1)}{2 \cdot 4 \cdot 6 \cdots (2n)} = \frac{1}{2} \cdot \frac{3}{4} \cdot \frac{5}{6} \cdot \frac{7}{8} \cdots \frac{2n - 1}{2n} < 1$$

we see that the sequence is bounded. It follows from Theorem 9.2.1 that the sequence is convergent. ∎

Theorem 9.2.1 is handy in proving that a sequence $\{a_n\}$ converges, that is, $\lim\limits_{n \to \infty} a_n = L$, but the theorem does not provide us with the specific number L. But the next example shows how to determine L when the sequence is defined recursively.

EXAMPLE 7 Determining Convergence

Show that the sequence $\{a_n\}$ defined by the recursion formula $a_{n+1} = \frac{1}{4}a_n + 6$, $a_1 = 1$, converges.

This can be proved using a method called *mathematical induction.*

Solution First, the sequence $\{a_n\}$ is bounded. It can be proved that $a_n < 8$, for all n. This fact is suggested by calculating a_n for $n = 1, 2, 3, \ldots$

$$a_2 = \frac{1}{4}a_1 + 6 = \frac{1}{4}(1) + 6 = \frac{25}{4} = 6.25 < 8$$

$$a_3 = \frac{1}{4}a_2 + 6 = \frac{1}{4}\left(\frac{25}{4}\right) + 6 = \frac{121}{16} = 7.5625 < 8$$

$$a_4 = \frac{1}{4}a_3 + 6 = \frac{1}{4}\left(\frac{121}{16}\right) + 6 = \frac{505}{64} = 7.890625 < 8$$

$$\vdots$$

Because $a_n > 0$ for all n, we have $0 < a_n < 8$ for all n. Thus $\{a_n\}$ is bounded.

Next, we will show that the sequence $\{a_n\}$ is monotonic. Because $a_n < 8$ necessarily $\frac{3}{4}a_n < \frac{3}{4} \cdot 8 = 6$. Therefore, from the recursion formula,

$$a_{n+1} = \frac{1}{4}a_n + 6 > \frac{1}{4}a_n + \frac{3}{4}a_n = a_n.$$

This shows that $a_{n+1} > a_n$ for all n, and so the sequence is increasing.

Since $\{a_n\}$ is bounded and monotonic it follows from Theorem 9.2.1 that the sequence converges. Because we must have $\lim_{n\to\infty} a_n = L$ and $\lim_{n\to\infty} a_{n+1} = L$ the limit of the sequence can be determined from the recursion formula:

$$\lim_{n\to\infty} a_{n+1} = \lim_{n\to\infty}\left(\frac{1}{4}a_n + 6\right)$$

$$\lim_{n\to\infty} a_{n+1} = \frac{1}{4}\lim_{n\to\infty} a_n + 6$$

$$L = \frac{1}{4}L + 6.$$

By solving the last equation for L we find $\frac{3}{4}L = 6$ or $L = 8$. ∎

Σ NOTES FROM THE CLASSROOM

(*i*) Every convergent sequence $\{a_n\}$ is necessarily bounded. See Problem 31 in Exercises 9.2. But it does not follow that every bounded sequence is convergent. You will be asked to supply an example that illustrates this last statement in Problem 30 of Exercises 9.2.

(*ii*) Some sequences $\{a_n\}$ do not exhibit monotonic behavior until some point on in the sequence, that is, until the index satisfies $n \geq N$, where N is some positive integer. For example, the terms of the sequence $\{5^n/n!\}$ for $n = 1, 2, 3, 4, 5, 6, \ldots$ are:

$$5, \frac{25}{2}, \frac{125}{6}, \frac{625}{24}, \frac{625}{24}, \frac{3125}{144}, \ldots. \tag{1}$$

To see better what is happening in (1), let us approximate the terms using numbers rounded to two decimals:

$$5, 12.5, 20.83, 26.04, 26.04, 21.70, \ldots. \tag{2}$$

In (2) we see that the first four terms of $\{5^n/n!\}$ obviously increase, but starting with the *fourth* term the terms appear to turn to nonincreasing. This can be proven from a recursively defined version of the sequence. Proceeding as we did in obtaining the recurrence formula in (7) in Section 9.1, $\{5^n/n!\}$ is the same as $a_{n+1} = \frac{5}{n+1}a_n$, $a_1 = 5$. Since $\frac{5}{n+1} \leq 1$ for $n \geq 4$ we see that $a_{n+1} \leq a_n$, that is, $\{5^n/n!\}$ is nonincreasing only for $n \geq 4$. In like manner, it is easily shown that $\{100^n/n!\}$ eventually becomes nonincreasing only when $n \geq 99$. By taking the limit of the recursion formula as $n \to \infty$, as in Example 7, we can show that both $\{5^n/n!\}$ and $\{100^n/n!\}$ converge to 0.

Exercises 9.2   Answers to selected odd-numbered problems begin on page ANS-26.

☰ Fundamentals

In Problems 1–12, determine whether the given sequence is monotonic. If so, state whether it is increasing, decreasing, nondecreasing, or nonincreasing.

1. $\left\{\dfrac{n}{3n+1}\right\}$

2. $\left\{\dfrac{10+n}{n}\right\}$

3. $\{(-1)^n\sqrt{n}\}$

4. $\{(n-1)(n-2)\}$

5. $\left\{\dfrac{e^n}{n}\right\}$

6. $\left\{\dfrac{e^n}{n^5}\right\}$

7. $\left\{\dfrac{2^n}{n!}\right\}$

8. $\left\{\dfrac{2^{2n}(n!)^2}{(2n)!}\right\}$

9. $\left\{n + \dfrac{1}{n}\right\}$

10. $\{n^2 + (-1)^n n\}$

11. $\{(\sin 1)(\sin 2)\cdots(\sin n)\}$

12. $\left\{\ln\left(\dfrac{n+2}{n+1}\right)\right\}$

In Problems 13–24, use Theorem 9.2.1 to show that the given sequence converges.

13. $\left\{\dfrac{4n-1}{5n+2}\right\}$

14. $\left\{\dfrac{6-4n^2}{1+n^2}\right\}$

15. $\left\{\dfrac{3^n}{1+3^n}\right\}$

16. $\{n5^{-n}\}$

17. $\{e^{1/n}\}$

18. $\left\{\dfrac{n!}{n^n}\right\}$

19. $\left\{\dfrac{n!}{1 \cdot 3 \cdot 5 \cdots (2n-1)}\right\}$

20. $\left\{\dfrac{2 \cdot 4 \cdot 6 \cdots (2n)}{1 \cdot 3 \cdot 5 \cdots (2n+1)}\right\}$

21. $\{\tan^{-1}n\}$

22. $\left\{\dfrac{\ln(n+3)}{n+3}\right\}$

23. $(0.8), (0.8)^2, (0.8)^3, \ldots$

24. $\sqrt{3}, \sqrt{\sqrt{3}}, \sqrt{\sqrt{\sqrt{3}}}, \ldots$

In Problems 25 and 26, use Theorem 9.2.1 to show that the recursively defined sequence converges. Find the limit of the sequence.

25. $a_{n+1} = \dfrac{1}{2}a_n + 5, a_1 = 1$ **26.** $a_{n+1} = \sqrt{2 + a_n}, a_1 = 0$

27. Express

$$\sqrt{7}, \sqrt{7\sqrt{7}}, \sqrt{7\sqrt{7\sqrt{7}}}, \ldots$$

as a recursively defined sequence $\{a_n\}$. Use the fact that the sequence is bounded, $0 < a_n < 7$ for all n, to show that $\{a_n\}$ is increasing. Find the limit of the sequence.

28. Use Theorem 9.2.1 to show that the recursively defined sequence

$$a_{n+1} = \left(1 - \dfrac{1}{n^2}\right)a_n, \quad a_1 = 2, a_2 = 1, n \ge 2$$

is bounded and monotonic and hence converges. Explain why the recursion formula is no help in finding the limit of the sequence.

≡ **Applications**

29. Certain studies in fishery management hold that the size of an undisturbed fish population changes from one year to the next in accordance with the formula

$$p_{n+1} = \dfrac{bp_n}{a + p_n}, \quad n \ge 0,$$

where $p_n > 0$ is the population after n years, and a and b are positive parameters that depend on the species and its

environment. Suppose that a population size p_0 is introduced in year 0.

(a) Use the recursion formula to show that the only possible limit values for the sequence $\{p_n\}$ are 0 and $b - a$.

(b) Show that $p_{n+1} < (b/a)p_n$.

(c) Use the result in part (b) to show that if $a > b$, then the population dies out: that is, $\lim_{n \to \infty} p_n = 0$.

(d) Now assume $a < b$. Show that if $0 < p_0 < b - a$, then the sequence $\{p_n\}$ is increasing and bounded above by $b - a$. Show that if $0 < b - a < p_0$, then the sequence $\{p_n\}$ is decreasing and bounded below by $b - a$. Conclude that $\lim_{n \to \infty} p_n = b - a$ for any $p_0 > 0$.

[*Hint:* Examine $|b - a - p_{n+1}|$, which is the distance between p_{n+1} and $b - a$.]

≡ **Think About It**

30. Give an example of a bounded sequence that is not convergent.

31. Show that every convergent sequence $\{a_n\}$ is bounded. [*Hint:* Since $\{a_n\}$ is convergent, it follows from Definition 9.1.2 that there exists an N such that $|a_n - L| < 1$ whenever $n > N$.]

32. Show that $\{\int_1^n e^{-t^2}dt\}$ converges. [*Hint:* For $x > 1$, $e^{-x^2} \le e^{-x}$.]

33. A Mathematical Classic Prove that the sequence

$$\left\{1 + \dfrac{1}{2} + \dfrac{1}{3} + \cdots + \dfrac{1}{n} - \ln n\right\}$$

is bounded and monotonic and hence convergent. The limit of the sequence is denoted by γ and is called **Euler's constant** after the noted Swiss mathematician **Leonhard Euler** (1707–1783). From Problem 66 of Exercises 9.1, $\gamma \approx 0.5772\ldots$. [*Hint:* First prove the inequality

$$\dfrac{1}{2} + \dfrac{1}{3} + \cdots + \dfrac{1}{n-1} + \dfrac{1}{n} < \ln n < 1 + \dfrac{1}{2} + \dfrac{1}{3} + \cdots + \dfrac{1}{n-1}$$

by considering the area under the graph of $y = 1/x$ on the interval $[1, n]$.

9.3 Series

∎ **Introduction** The concept of a *series* is closely related to the concept of a *sequence*. If $\{a_n\}$ is the sequence $a_1, a_2, a_3, \ldots, a_n, \ldots$, then the sum of the terms

$$a_1 + a_2 + a_3 + \cdots + a_n + \cdots \tag{1}$$

is called an **infinite series** or simply a **series**. The $a_k, k = 1, 2, 3, \ldots$, are called the **terms** of the series and a_n is called the **general term**. We write (1) compactly using summation notation as

$$\sum_{k=1}^{\infty} a_k \quad \text{or for convenience} \quad \sum a_k.$$

The question we seek to answer in this and the next several sections is:

- *When does an infinite series of constants "add up" to a number?*

EXAMPLE 1   An Infinite Series

In the opening remarks to this chapter we noted that the decimal representation for the rational number $\frac{1}{3}$ is, in fact, an infinite series

$$0.333\cdots = \frac{3}{10} + \frac{3}{10^2} + \frac{3}{10^3} + \cdots = \sum_{k=1}^{\infty} \frac{3}{10^k}.$$ ∎

Intuitively, we expect that $\frac{1}{3}$ is the sum of the series $\sum_{k=1}^{\infty} \frac{3}{10^k}$. But, just as intuitively, we expect that an infinite series such as

$$100 + 1000 + 10{,}000 + 100{,}000 + \cdots$$

where the terms are becoming larger and larger, has no sum. In other words, we do not expect the latter series to "add up" or *converge* to any number. The concept of convergence of an infinite series is defined in terms of the convergence of a special kind of sequence.

■ **Sequence of Partial Sums**   Associated with every infinite series $\sum a_k$, there is a **sequence of partial sums** $\{S_n\}$ whose terms are defined by

$$S_1 = a_1$$
$$S_2 = a_1 + a_2$$
$$S_3 = a_1 + a_2 + a_3$$
$$\vdots$$
$$S_n = a_1 + a_2 + a_3 + \cdots + a_n$$
$$\vdots$$

The general term $S_n = a_1 + a_2 + \cdots + a_n = \sum_{k=1}^{n} a_k$ of this sequence is called the **nth partial sum** of the series.

EXAMPLE 2   An Infinite Series

The sequence of partial sums $\{S_n\}$ for the series $\sum_{k=1}^{\infty} \frac{3}{10^k}$ is

$$S_1 = \frac{3}{10} = 0.3$$

$$S_2 = \frac{3}{10} + \frac{3}{10^2} = 0.33$$

$$S_3 = \frac{3}{10} + \frac{3}{10^2} + \frac{3}{10^3} = 0.333$$

$$\vdots$$

$$S_n = \frac{3}{10} + \frac{3}{10^2} + \frac{3}{10^3} + \cdots + \frac{3}{10^n} = 0.\overset{n\,3's}{\overbrace{333\ldots 3}}$$

$$\vdots$$ ∎

In Example 2, when n is very large, S_n will give a good approximation to $\frac{1}{3}$, and so it seems reasonable to write

$$\frac{1}{3} = \lim_{n\to\infty} S_n = \lim_{n\to\infty} \sum_{k=1}^{n} \frac{3}{10^k} = \sum_{k=1}^{\infty} \frac{3}{10^k}.$$

This leads to the following definition.

Definition 9.3.1 Convergent Series

An infinite series $\sum_{k=1}^{\infty} a_k$ is said to be **convergent** if its sequence of partial sums $\{S_n\} = \{\sum_{k=1}^{n} a_k\}$ converges; that is

$$\lim_{n \to \infty} S_n = \lim_{n \to \infty} \sum_{k=1}^{n} a_k = S.$$

The number S is said to be the **sum** of the series. If $\lim_{n \to \infty} S_n$ does not exist, then the series is said to be **divergent**.

EXAMPLE 3 Using the Sequence of Partial Sums

Show that the series $\sum_{k=1}^{\infty} \dfrac{1}{(k+4)(k+5)}$ is convergent.

Solution By partial fractions the general term a_n of the series can be written as

$$a_n = \frac{1}{n+4} - \frac{1}{n+5}.$$

Thus, the nth partial sum of the series is

$$S_n = \left[\frac{1}{5} - \frac{1}{6}\right] + \left[\frac{1}{6} - \frac{1}{7}\right] + \left[\frac{1}{7} - \frac{1}{8}\right] + \cdots + \left[\frac{1}{n+3} - \frac{1}{n+4}\right] + \left[\frac{1}{n+4} - \frac{1}{n+5}\right]$$

$$= \frac{1}{5} - \frac{1}{6} + \frac{1}{6} - \frac{1}{7} + \frac{1}{7} - \frac{1}{8} + \cdots + \frac{1}{n+3} - \frac{1}{n+4} + \frac{1}{n+4} - \frac{1}{n+5}$$

$$= \frac{1}{5} - \frac{1}{n+5}.$$

From the last line we see that $\lim_{n \to \infty} 1/(n+5) = 0$, and so

$$\lim_{n \to \infty} S_n = \lim_{n \to \infty}\left[\frac{1}{5} - \frac{1}{n+5}\right] = \frac{1}{5} - 0 = \frac{1}{5}.$$

Hence, the series converges and we write

$$\sum_{k=1}^{\infty} \frac{1}{(k+4)(k+5)} = \frac{1}{5}. \qquad \blacksquare$$

■ **Telescoping Series** Because of the manner in which the general term of the sequence of partial sums "collapses" to two terms, the series in Example 3 is said to be a **telescoping series**. See Problems 11–14 in Exercises 9.3.

■ **Geometric Series** Another type of series that can be proven convergent or divergent directly from its sequence of partial sums has the form

$$a + ar + ar^2 + \cdots + ar^{n-1} + \cdots = \sum_{k=1}^{\infty} ar^{k-1}, \qquad (2)$$

where $a \neq 0$ and r are fixed real numbers. A series of the form (2) is called a **geometric series**. Note in (2) that each term after the first is obtained by multiplying the preceding term by r. The number r is called the **common ratio** and, as we see in the next theorem, its magnitude determines whether a geometric series converges or diverges.

Theorem 9.3.1 Sum of a Geometric Series

(*i*) If $|r| < 1$, then a geometric series converges and its sum is

$$\sum_{k=1}^{\infty} ar^{k-1} = \frac{a}{1-r}, \quad a \neq 0.$$

(*ii*) If $|r| \geq 1$, then a geometric series diverges.

PROOF The proof of Theorem 9.3.1 will be given in two parts. In each part we assume $a \neq 0$. We begin with the case that $|r| = 1$. For $r = 1$, the series is

$$\sum_{k=1}^{\infty} a = a + a + a + \cdots$$

and so the *n*th partial sum $S_n = \overbrace{a + a + \cdots + a}^{n \text{ } a\text{'s}}$ is simply $S_n = na$. In this case, $\lim_{n \to \infty} S_n = a \cdot \lim_{n \to \infty} n = \infty$. Thus the series diverges. For $r = -1$, the series is

$$\sum_{k=1}^{\infty} a(-1)^{k-1} = a + (-a) + a + (-a) + \cdots$$

and so the sequence of partial sums is

$$S_1, S_2, S_3, S_4, S_5, S_6, \ldots \quad \text{or} \quad a, 0, a, 0, a, 0, \ldots,$$

which is divergent.

Next we consider the case $|r| \neq 1$, which means that $|r| < 1$ or $|r| > 1$. Consider the general term of the sequence of partial sums of (2):

$$S_n = a + ar + ar^2 + \cdots + ar^{n-1}. \tag{3}$$

Multiplying both sides of (3) by *r* gives

$$rS_n = ar + ar^2 + ar^3 + \cdots + ar^n. \tag{4}$$

We then subtract (4) from (3) and solve for S_n:

$$S_n - rS_n = a - ar^n$$
$$(1 - r)S_n = a(1 - r^n)$$
$$S_n = \frac{a(1 - r^n)}{1 - r}, \quad r \neq 1. \tag{5}$$

Now, from Theorem 9.1.3 we know that $\lim_{n \to \infty} r^n = 0$ for $|r| < 1$. Consequently,

$$\lim_{n \to \infty} S_n = \lim_{n \to \infty} \frac{a(1 - r^n)}{1 - r} = \frac{a}{1 - r}, \quad |r| < 1.$$

If $|r| > 1$, then $\lim_{n \to \infty} r^n$ does not exist and so the limit of (5) fails to exist. ∎

EXAMPLE 4 Geometric Series

(a) In the geometric series

$$\sum_{k=1}^{\infty} \left(-\frac{1}{3}\right)^{k-1} = 1 - \frac{1}{3} + \frac{1}{9} - \frac{1}{27} + \cdots$$

we identify $a = 1$ and the common ratio $r = -\frac{1}{3}$. Since $|r| = \left|-\frac{1}{3}\right| = \frac{1}{3} < 1$, the series converges. From Theorem 9.3.1 the sum of the series is then

$$\sum_{k=1}^{\infty} \left(-\frac{1}{3}\right)^{k-1} = \frac{1}{1 - \left(-\frac{1}{3}\right)} = \frac{3}{4}.$$

(b) The common ratio in the geometric series

$$\sum_{k=1}^{\infty} 5\left(\frac{3}{2}\right)^{k-1} = 5 + \frac{15}{2} + \frac{45}{4} + \frac{135}{8} + \cdots$$

is $r = \frac{3}{2}$. The series diverges because $r = \frac{3}{2} > 1$. ∎

Every rational number p/q, where p and $q \neq 0$ are integers, can be expressed either as a terminating decimal or as a repeating decimal. Thus, the series $\sum_{k=1}^{\infty} \frac{3}{10^k}$ in Example 1 converges since it is a geometric series with $r = \frac{1}{10} < 1$. With $a = \frac{3}{10}$ we find

$$\sum_{k=1}^{\infty} \frac{3}{10^k} = \frac{\dfrac{3}{10}}{1 - \dfrac{1}{10}} = \frac{\dfrac{3}{10}}{\dfrac{9}{10}} = \frac{3}{9} = \frac{1}{3}.$$

In general:

- *Every repeating decimal is a convergent geometric series.*

EXAMPLE 5   Rational Number

Express the repeating decimal $0.121212\ldots$ as a quotient of integers.

Solution   We first write the given number as a geometric series

$$0.121212\ldots = \frac{12}{100} + \frac{12}{10,000} + \frac{12}{1,000,000} + \cdots$$

$$= \frac{12}{10^2} + \frac{12}{10^4} + \frac{12}{10^6} + \cdots$$

and make the identifications $a = \frac{12}{100}$ and $r = \frac{1}{10^2} = \frac{1}{100}$. By Theorem 9.3.1 the series converges since $r = \frac{1}{100} < 1$ and its sum is

$$0.121212\ldots = \frac{\dfrac{12}{100}}{1 - \dfrac{1}{100}} = \frac{\dfrac{12}{100}}{\dfrac{99}{100}} = \frac{12}{99} = \frac{4}{33}.$$ ∎

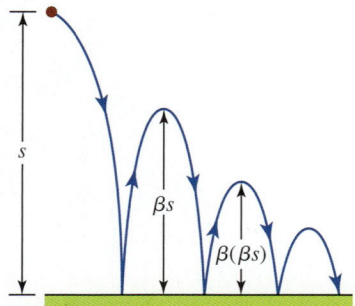

FIGURE 9.3.1   Bouncing ball in Example 6

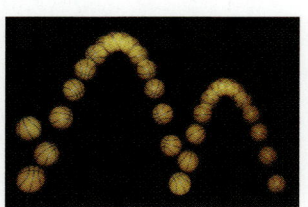

Stroboscopic photo of a bouncing basketball

EXAMPLE 6   Watch the Bouncing Ball

If a ball is dropped from a height of s ft above the ground, then the time t it takes to reach the ground is related to s by $s = \frac{1}{2}gt^2$. In other words, it takes the ball $t = \sqrt{2s/g}$ s to reach the ground. Suppose the ball always rebounds to a certain fixed fraction $\beta\, (0 < \beta < 1)$ of its prior height. Find a formula for the time T it takes for the ball to come to rest. See **FIGURE 9.3.1**.

Solution   The time to fall from a height of s ft to the ground is: $\sqrt{2s/g}$; the time to rise βs ft and then fall βs ft to the ground is: $2\sqrt{2\beta s/g}$; the time to rise $\beta(\beta s)$ ft and then fall $\beta(\beta s)$ ft to the ground is: $2\sqrt{2\beta^2 s/g}$; and so on. Thus, the total time T is given by the infinite series

$$T = \sqrt{2s/g} + 2\sqrt{2\beta s/g} + 2\sqrt{2\beta^2 s/g} + \cdots + 2\sqrt{2\beta^n s/g} + \cdots$$

$$= \sqrt{2s/g}\left[1 + 2\sum_{k=1}^{\infty}(\sqrt{\beta})^k\right].$$

Because $0 < \beta < 1$, the series $\sum_{k=1}^{\infty}(\sqrt{\beta})^k$ is a convergent geometric series with $a = \sqrt{\beta}$ and $r = \sqrt{\beta}$. Consequently, from Theorem 9.3.1

$$T = \sqrt{2s/g}\left[1 + 2\frac{\sqrt{\beta}}{1 - \sqrt{\beta}}\right] \quad \text{or} \quad T = \sqrt{2s/g}\left[\frac{1 + \sqrt{\beta}}{1 - \sqrt{\beta}}\right].$$ ∎

Harmonic Series One of the most famous series is also an example of a divergent series. The **harmonic series** is the sum of the reciprocals of the positive integers:

$$1 + \frac{1}{2} + \frac{1}{3} + \cdots + \frac{1}{n} + \cdots = \sum_{k=1}^{\infty} \frac{1}{k}. \tag{6}$$

Remember this series. It will be important in the subsequent sections of this chapter.

The general term of the sequence of partial sums for (6) is given by

$$S_n = 1 + \frac{1}{2} + \frac{1}{3} + \cdots + \frac{1}{n}.$$

Thus,

$$S_{2n} = 1 + \frac{1}{2} + \frac{1}{3} + \cdots + \frac{1}{n} + \frac{1}{n+1} + \frac{1}{n+2} + \cdots + \frac{1}{2n}$$

$$= S_n + \frac{1}{n+1} + \frac{1}{n+2} + \cdots + \frac{1}{2n}$$

$$\geq S_n + \underbrace{\frac{1}{2n} + \frac{1}{2n} + \cdots + \frac{1}{2n}}_{n \text{ terms of } \frac{1}{2n}} = S_n + n \cdot \frac{1}{2n} = S_n + \frac{1}{2}.$$

The inequality $S_{2n} \geq S_n + \frac{1}{2}$ implies that the sequence of partial sums for the harmonic series is unbounded. To see this, we observe that

$$S_2 \geq S_1 + \frac{1}{2} = 1 + \frac{1}{2} = \frac{3}{2}$$

$$S_4 \geq S_2 + \frac{1}{2} \geq \frac{3}{2} + \frac{1}{2} = 2$$

$$S_8 \geq S_4 + \frac{1}{2} \geq 2 + \frac{1}{2} = \frac{5}{2}$$

$$S_{16} \geq S_8 + \frac{1}{2} \geq \frac{5}{2} + \frac{1}{2} = 3$$

and so on. Hence, we conclude that the harmonic series is divergent.

A Consequence of Convergence If a_n and S_n are the general terms of a series and the corresponding sequence of partial sums, respectively, then from the subtraction

$$S_n - S_{n-1} = (a_1 + a_2 + \cdots + a_{n-1} + a_n) - (a_1 + a_2 + \cdots + a_{n-1}) = a_n$$

we see that $a_n = S_n - S_{n-1}$. Now, if the series $\sum a_k$ converges to a number S, we have $\lim_{n \to \infty} S_n = S$ and $\lim_{n \to \infty} S_{n-1} = S$. This implies that

$$\lim_{n \to \infty} a_n = \lim_{n \to \infty} (S_n - S_{n-1}) = S - S = 0.$$

We have established the next theorem.

Theorem 9.3.2 Necessary Condition for Convergence

If the series $\sum_{k=1}^{\infty} a_k$ converges, then $\lim_{n \to \infty} a_n = 0$.

Test for a Divergent Series Theorem 9.3.2 simply states that if an infinite series converges, it is necessary that the *n*th, or general, term of the series approach zero. Equivalently, we conclude:

- *If the* n*th term* a_n *of an infinite series **does not** approach zero as* $n \to \infty$, *then the series **does not** converge.*

We formalize this result as a test for divergence.

Theorem 9.3.3 nth Term Test for Divergence

If $\lim\limits_{n\to\infty} a_n \neq 0$, then the series $\sum_{k=1}^{\infty} a_k$ diverges.

Theorem 9.3.3 immediately corroborates part (*ii*) of the proof of Theorem 9.3.1, namely, a geometric series $\sum_{k=1}^{\infty} ar^{k-1}$, $a \neq 0$, diverges when $r = \pm 1$. For example, when $r = 1$, $\lim\limits_{n\to\infty} ar^{n-1} = \lim\limits_{n\to\infty} a \neq 0$.

EXAMPLE 7   Divergent Series

(a) Consider the series $\sum\limits_{k=1}^{\infty} \dfrac{4k-1}{5k+3}$. From

$$\lim_{n\to\infty} a_n = \lim_{n\to\infty} \frac{4n-1}{5n+3} = \lim_{n\to\infty} \frac{4 - \dfrac{1}{n}}{5 + \dfrac{3}{n}} = \frac{4}{5} \neq 0$$

it follows from Theorem 9.3.3 that the series diverges.

(b) Consider the series

$$\sum_{k=1}^{\infty} (-1)^{k-1} = 1 - 1 + 1 - 1 + \cdots .$$

Since $\lim\limits_{n\to\infty} a_n = \lim\limits_{n\to\infty} (-1)^{n-1}$ does not exist, we can assert that $\lim\limits_{n\to\infty} a_n \neq 0$. The series diverges by Theorem 9.3.3.? ■

You are encouraged to read (and remember) (*iii*) of *Notes From the Classroom* at this time. We state the following three theorems without proof.

Theorem 9.3.4   Constant Multiple of a Series

If c is any nonzero constant, then series $\sum_{k=1}^{\infty} a_k$ and $\sum_{k=1}^{\infty} ca_k$ both converge or both diverge.

Theorem 9.3.5   Sum of Two Convergent Series

If $\sum_{k=1}^{\infty} a_k$ and $\sum_{k=1}^{\infty} b_k$ converge to S_1 and S_2, respectively, then

(*i*) $\sum_{k=1}^{\infty} (a_k + b_k)$ converges to $S_1 + S_2$, and

(*ii*) $\sum_{k=1}^{\infty} (a_k - b_k)$ converges to $S_1 - S_2$.

Theorem 9.3.5 indicates that when $\sum_{k=1}^{\infty} a_k$ and $\sum_{k=1}^{\infty} b_k$ converge, then

$$\sum_{k=1}^{\infty} (a_k \pm b_k) = \sum_{k=1}^{\infty} a_k \pm \sum_{k=1}^{\infty} b_k.$$

Theorem 9.3.6   Sum of a Convergent and a Divergent Series

If $\sum_{k=1}^{\infty} a_k$ converges and $\sum_{k=1}^{\infty} b_k$ diverges, then $\sum_{k=1}^{\infty} (a_k + b_k)$ diverges.

EXAMPLE 8   Sum of Two Convergent Series

With the aid of Theorem 9.3.1, we see that the geometric series $\sum_{k=1}^{\infty}\left(\frac{1}{2}\right)^{k-1}$ and $\sum_{k=1}^{\infty}\left(\frac{1}{3}\right)^{k-1}$ converge to 2 and $\frac{3}{2}$, respectively. Hence, from Theorem 9.3.5, the series $\sum_{k=1}^{\infty}\left[\left(\frac{1}{2}\right)^{k-1} - \left(\frac{1}{3}\right)^{k-1}\right]$ converges and

$$\sum_{k=1}^{\infty}\left[\left(\frac{1}{2}\right)^{k-1} - \left(\frac{1}{3}\right)^{k-1}\right] = \sum_{k=1}^{\infty}\left(\frac{1}{2}\right)^{k-1} - \sum_{k=1}^{\infty}\left(\frac{1}{3}\right)^{k-1} = 2 - \frac{3}{2} = \frac{1}{2}.$$ ∎

EXAMPLE 9   Sum of Two Series

From Example 3 we know that $\sum_{k=1}^{\infty}\dfrac{1}{(k + 4)(k + 5)}$ converges. Since $\sum_{k=1}^{\infty}\dfrac{1}{k}$ is the divergent harmonic series, it follows from Theorem 9.3.6 that the series

$$\sum_{k=1}^{\infty}\left[\frac{1}{(k + 4)(k + 5)} + \frac{1}{k}\right]$$

diverges. ∎

Σ   NOTES FROM THE CLASSROOM

(*i*) The *n*th term of the sequence of partial sums of the harmonic series is often denoted by $H_n = \sum_{k=1}^{n}(1/k)$. The terms of the sequence $H_1 = 1$, $H_2 = \frac{3}{2}$, $H_3 = \frac{11}{6}, \ldots$ are called **harmonic numbers**. See Problem 71 in Exercises 9.3.

(*ii*) When written in terms of summation notation, a geometric series may not be immediately recognizable, or if it is, the values of a and r may not be apparent. For example, to see whether $\sum_{n=3}^{\infty}4\left(\frac{1}{2}\right)^{n+2}$ is a geometric series it is a good idea to write out two or three terms:

$$\sum_{n=3}^{\infty}4\left(\frac{1}{2}\right)^{n+2} = \overbrace{4\left(\frac{1}{2}\right)^{5}}^{a} + \overbrace{4\left(\frac{1}{4}\right)^{6}}^{ar} + \overbrace{4\left(\frac{1}{2}\right)^{7}}^{ar^2} + \cdots.$$

From the right side of the last equality, we can make the identifications $a = 4\left(\frac{1}{2}\right)^5$ and $r = \frac{1}{2} < 1$. Consequently, the sum of the series is $\dfrac{4\left(\frac{1}{2}\right)^5}{1 - \frac{1}{2}} = \dfrac{1}{4}$. If desired, although there is no real need to do this, we can express $\sum_{n=3}^{\infty}4\left(\frac{1}{2}\right)^{n+2}$ in the more familiar form $\sum_{k=1}^{\infty}ar^{k-1}$ by letting $k = n - 2$. The result is

$$\sum_{n=3}^{\infty}4\left(\frac{1}{2}\right)^{n+2} = \sum_{k=1}^{\infty}4\left(\frac{1}{2}\right)^{k+4} = \sum_{k=1}^{\infty}\overbrace{4\left(\frac{1}{2}\right)^{5}}^{a}\overbrace{\left(\frac{1}{2}\right)^{k-1}}^{r^{k-1}}.$$

(*iii*) Note carefully how Theorems 9.3.2 and 9.3.3 are stated. Specifically, Theorem 9.3.3 *does not* say if $\lim_{n\to\infty}a_n = 0$, then $\sum a_k$ converges. In other words, $\lim_{n\to\infty}a_n = 0$ is not *sufficient* to guarantee that $\sum a_k$ converges. In fact, if $\lim_{n\to\infty}a_n = 0$, the series may either converge or diverge. For example, in the harmonic series $\sum_{k=1}^{\infty}(1/k)$, $a_n = 1/n$ and $\lim_{n\to\infty}(1/n) = 0$, but the series diverges.

(*iv*) When determining convergence, it is possible, and sometimes convenient, to delete or ignore the first several terms of a series. In other words, the infinite series $\sum_{k=1}^{\infty} a_k$ and $\sum_{k=N}^{\infty} a_k$, $N > 1$, differ by at most a finite number of terms and are either both convergent or both divergent. Of course, deleting the first $N - 1$ terms of a convergent series usually does affect the sum of the series.

Exercises 9.3   Answers to selected odd-numbered problems begin on page ANS-27.

≡ Fundamentals

In Problems 1–10, write out the first four terms in each series.

1. $\displaystyle\sum_{k=1}^{\infty} \frac{2k+1}{k}$

2. $\displaystyle\sum_{k=1}^{\infty} \frac{2^k}{k}$

3. $\displaystyle\sum_{k=1}^{\infty} \frac{(-1)^{k-1}}{k(k+1)}$

4. $\displaystyle\sum_{k=1}^{\infty} \frac{(-1)^{k+1}}{k3^k}$

5. $\displaystyle\sum_{n=0}^{\infty} \frac{n+1}{n!}$

6. $\displaystyle\sum_{n=1}^{\infty} \frac{(2n)!}{n^2+1}$

7. $\displaystyle\sum_{m=1}^{\infty} \frac{2\cdot4\cdot6\cdots(2m)}{1\cdot3\cdot5\cdots(2m-1)}$

8. $\displaystyle\sum_{m=1}^{\infty} \frac{1\cdot3\cdot5\cdots(2m-1)}{m!}$

9. $\displaystyle\sum_{j=3}^{\infty} \frac{\cos j\pi}{2j+1}$

10. $\displaystyle\sum_{i=5}^{\infty} i\sin\frac{i\pi}{2}$

In Problems 11–14, proceed as in Example 3 to find the sum of the given telescoping series.

11. $\displaystyle\sum_{k=1}^{\infty} \frac{1}{k(k+1)}$

12. $\displaystyle\sum_{k=1}^{\infty} \frac{1}{(k+1)(k+2)}$

13. $\displaystyle\sum_{k=1}^{\infty} \frac{1}{4k^2-1}$

14. $\displaystyle\sum_{k=1}^{\infty} \frac{1}{k^2+7k+12}$

In Problems 15–24, determine whether the given geometric series converges or diverges. If convergent, find the sum of the series.

15. $\displaystyle\sum_{k=1}^{\infty} 3\left(\frac{1}{5}\right)^{k-1}$

16. $\displaystyle\sum_{k=1}^{\infty} 10\left(\frac{3}{4}\right)^{k-1}$

17. $\displaystyle\sum_{k=1}^{\infty} \frac{(-1)^{k-1}}{2^{k-1}}$

18. $\displaystyle\sum_{k=1}^{\infty} \pi^k\left(\frac{1}{3}\right)^{k-1}$

19. $\displaystyle\sum_{r=1}^{\infty} 5^r 4^{-r}$

20. $\displaystyle\sum_{s=1}^{\infty} (-3)^s 7^{-s}$

21. $\displaystyle\sum_{n=1}^{\infty} 1000(0.9)^n$

22. $\displaystyle\sum_{n=1}^{\infty} \frac{(1.1)^n}{1000}$

23. $\displaystyle\sum_{k=0}^{\infty} \frac{1}{(\sqrt{3}-\sqrt{2})^k}$

24. $\displaystyle\sum_{k=0}^{\infty} \left(\frac{\sqrt{5}}{1+\sqrt{5}}\right)^k$

In Problems 25–30, write each repeating decimal number as a quotient of integers.

25. $0.222\ldots$

26. $0.555\ldots$

27. $0.616161\ldots$

28. $0.393939\ldots$

29. $1.314314\ldots$

30. $0.5262626\ldots$

In Problems 31 and 32, find the sum of the given series.

31. $\displaystyle\sum_{k=1}^{\infty} \left[\left(\frac{1}{3}\right)^{k-1} + \left(\frac{1}{4}\right)^{k-1}\right]$

32. $\displaystyle\sum_{k=1}^{\infty} \frac{2^k-1}{4^k}$

In Problems 33–42, show that the given series is divergent.

33. $\displaystyle\sum_{k=1}^{\infty} 10$

34. $\displaystyle\sum_{k=1}^{\infty} (5k+1)$

35. $\displaystyle\sum_{k=1}^{\infty} \frac{k}{2k+1}$

36. $\displaystyle\sum_{k=1}^{\infty} \frac{k^2+1}{k^2+2k+3}$

37. $\displaystyle\sum_{k=1}^{\infty} (-1)^k$

38. $\displaystyle\sum_{k=1}^{\infty} \ln\left(\frac{k}{3k+1}\right)$

39. $\displaystyle\sum_{k=1}^{\infty} \frac{10}{k}$

40. $\displaystyle\sum_{k=1}^{\infty} \frac{1}{6k}$

41. $\displaystyle\sum_{k=1}^{\infty} \left[\frac{1}{2^{k-1}} + \frac{1}{k}\right]$

42. $\displaystyle\sum_{k=1}^{\infty} k\sin\frac{1}{k}$

In Problems 43–46, determine the values of x for which the given series converges.

43. $\displaystyle\sum_{k=1}^{\infty} \left(\frac{x}{2}\right)^{k-1}$

44. $\displaystyle\sum_{k=1}^{\infty} \left(\frac{1}{x}\right)^{k-1}$

45. $\displaystyle\sum_{k=1}^{\infty} (x+1)^k$

46. $\displaystyle\sum_{k=0}^{\infty} 2^k x^{2k}$

≡ Applications

47. A ball is dropped from an initial height of 15 ft onto a concrete slab. Each time the ball bounces, it reaches a height of $\frac{2}{3}$ its preceding height. Use geometric series to determine the distance the ball travels before it comes to rest.

48. In Problem 47 determine the time it takes for the ball to come to rest.

49. To eradicate agricultural pests (such as the Medfly), sterilized male flies are released into the general population at regular time intervals. Let N_0 be the number of flies released each day and let s be the proportion that survive a given day. Of the original N_0 sterilized males, $N_0 s^n$ will survive for n successive weeks. Hence, the total number of such males that survive n weeks after the program has begun is $N_0 + N_0 s + N_0 s^2 + \cdots + N_0 s^n$. What does this sum approach as $n \to \infty$? Suppose $s = 0.9$ and 10,000 sterilized males are needed to control the population in a certain area. Determine the number that should be released each day.

50. In some circumstances the amount of a drug that will accumulate in a patient's body after a long period of time is $A_0 + A_0 e^{-k} + A_0 e^{-2k} + \cdots$, where $k > 0$ is a constant and A_0 is the daily dose of the drug. Find the sum of the series.

51. A patient takes 15 mg of a drug each day. If 80% of the drug accumulated is excreted each day by bodily functions, how much of the drug will accumulate after a long period of time, that is, as $n \to \infty$? (Assume that the measurement of the accumulation is made immediately after each dose. See Problem 69 in Exercises 9.1.)

52. A force is applied to a particle, which moves in a straight line, in such a fashion that after each second the particle moves only one-half the distance it moved in the preceding second. If the particle moves 20 cm in the first second, how far will it move?

≡ **Think About It**

53. Suppose the sequence $\{a_n\}$ converges to a number $L \neq 0$. Explain why the series $\sum_{k=1}^{\infty} a_k$ diverges.

54. Determine whether the series

$$\frac{1}{1.1} + \frac{1}{1.11} + \frac{1}{1.111} + \cdots$$

converges or diverges.

55. Determine whether the sum of two divergent series is necessarily divergent.

56. Consider the series $\sum_{k=1}^{\infty} \frac{1}{k^2}$. Since $k^2 = k \cdot k$, the nth partial sum of the series is

$$S_n = \frac{1}{1 \cdot 1} + \frac{1}{2 \cdot 2} + \frac{1}{3 \cdot 3} + \cdots + \frac{1}{n \cdot n}.$$

Explain why the following inequalities are true and why they can be used to prove that the given series converges:

$$0 < S_n < 1 + \frac{1}{1 \cdot 2} + \frac{1}{2 \cdot 3} + \cdots + \frac{1}{(n-1) \cdot n}$$

or

$$0 < S_n < 1 + \left(\frac{1}{1} - \frac{1}{2}\right) + \left(\frac{1}{2} - \frac{1}{3}\right) + \cdots + \left(\frac{1}{n-1} - \frac{1}{n}\right).$$

57. Find the sum of the series

$$\frac{1+9}{25} + \frac{1+27}{125} + \frac{1+81}{625} + \cdots.$$

58. Find the sum of the series

$$\sum_{k=1}^{\infty} \left(\int_{k}^{k+1} x e^{-x} \, dx \right).$$

59. Find all values of x in $(-\pi/2, \pi/2)$ for which

$$\lim_{n \to \infty} \left(\frac{1}{1 - \tan x} - \sum_{k=0}^{n} \tan^k x \right) = 0.$$

60. Show that if $\lim_{n \to \infty} f(n+1) = L$, where L is a number, then

$$\sum_{k=1}^{\infty} [f(k+1) - f(k)] = L - f(1).$$

61. Determine whether $\sum_{n=1}^{\infty} \left(\sum_{k=1}^{n} \frac{1}{k} \right)$ converges or diverges.

62. Show that the series $\sum_{k=1}^{\infty} \frac{1}{\sqrt{k}}$ is divergent by showing that $S_n \geq \sqrt{n}$.

63. We saw that the harmonic series $\sum_{k=1}^{\infty} \frac{1}{k}$ diverges since the general term S_n of the sequence of partial sums can be made as large as we like by taking n to be sufficiently large ($S_n \to \infty$ as $n \to \infty$). Nevertheless, the harmonic series diverges *very slowly*.

   (a) Use the graph of $f(x) = 1/x$ for $x \geq 1$ to establish the inequality

   $$\ln(n+1) < 1 + \frac{1}{2} + \frac{1}{3} + \frac{1}{4} + \cdots + \frac{1}{n} < 1 + \ln n.$$

   (b) Use a calculator and the inequality in part (a) to estimate the value of n for which $S_n \geq 10$. Estimate the value of n for which $S_n \geq 100$.

64. In Problem 77 in Exercises 9.1 we considered the perimeters of the regions bounded by the Koch curves shown in Figure 9.1.5. In part (c) of the problem you should have shown that the perimeter of the limiting region is infinite. In this problem we consider the *areas* of the sequence of figures. Let the area of the first figure be A_1, the area of the second figure A_2, and so on.

   (a) Using the fact that the area of an equilateral triangle with sides of length s is $\frac{1}{4}\sqrt{3}s^2$, find the values of $A_1, A_2, A_3,$ and A_4.

   (b) Show that the area of the nth figure is

   $$A_n = \frac{1}{20}\sqrt{3}\left[8 - 3\left(\frac{4}{9}\right)^{n-1}\right].$$

   (c) What is $\lim_{n \to \infty} A_n$?

≡ **Projects**

65. **A Bit of History—Death by Bread** In 1972, an outbreak of methylmercury poisoning in Iraq resulted in 459 deaths

Homemade breads

among 6530 cases of poisoning admitted to hospitals. The outbreak was caused by the consumption of homemade bread prepared from wheat that had been treated with a methylmercury fungicide. The first symptoms of *parasthesia* (loss of sensation at the mouth, hands, and feet) begin to occur when the accumulated level of mercury reaches 25 mg. Symptoms of *ataxia* (loss of coordination in gait) begin to occur at 55 mg, *dysarthia* (slurred speech) at 90 mg, and deafness at 170 mg. Death becomes a possibility when the accumulated mercury level exceeds 200 mg. It was estimated that a typical loaf of bread made from the contaminated wheat contained

1.4 mg of mercury. It is also estimated that the body removes only about 0.9% of the accumulated mercury each day.

(a) Suppose that a person receives a dosage d of mercury each day, and that the body removes a fraction p of the accumulated mercury each day. Find a formula for L_n, the accumulated level after eating on the nth day, and a formula for the limiting level, $\lim_{n\to\infty} L_n$.

(b) Using $d = 1.4$ and $p = 0.009$, find the limiting level of mercury and determine on which days the various symptoms begin to occur.

(c) What would the daily dose have to be in order for death to become possible by the 100th day? (Use $p = 0.009$.)

66. **A Bit of History—Zeno's Paradox** The Greek philosopher **Zeno of Elea** (c. 490 BC) was a disciple of the pre-Socratic philosopher Permenides who held that change or motion was an illusion. Of the paradoxes Zeno advanced in support of this philosophy, the most famous is his argument that Achilles, known for his ability to run fast, could not overcome a moving tortoise. The usual form of the story goes something like this:

Achilles starts from point S and at exactly the same instant a tortoise starts from a point A in front of S. After a certain amount of time Achilles reaches the tortoise's starting point A, but during this time the tortoise has advanced to a new point B. During the time it takes Achilles to reach B, the tortoise has moved ahead again to a new point C. Continuing in this manner, forever, Achilles can never catch up to the tortoise.

See FIGURE 9.3.2. Use infinite series to resolve this apparent paradox. Make the assumption that each moves with a constant speed. It may help to make up reasonable values for the tortoise's head start and for the two speeds.

FIGURE 9.3.2 Achilles and the tortoise in Problem 66

67. **Prime Numbers** Write a short report in which you define a prime number. In the report include a proof on whether the series of the reciprocal of primes,

$$\sum_{n=1}^{\infty} \frac{1}{p_n} = \frac{1}{2} + \frac{1}{3} + \frac{1}{5} + \frac{1}{7} + \frac{1}{11} + \cdots$$

converges or diverges.

68. **Length of a Zigzag Path** In FIGURE 9.3.3(a), the blue triangle ABC is an isosceles right triangle. The line segment AP_1 is perpendicular to BC, the line segment P_1P_2 is perpendicular to AC, and so on. Find the length of the red zigzag path $AP_1P_2P_3\dots$.

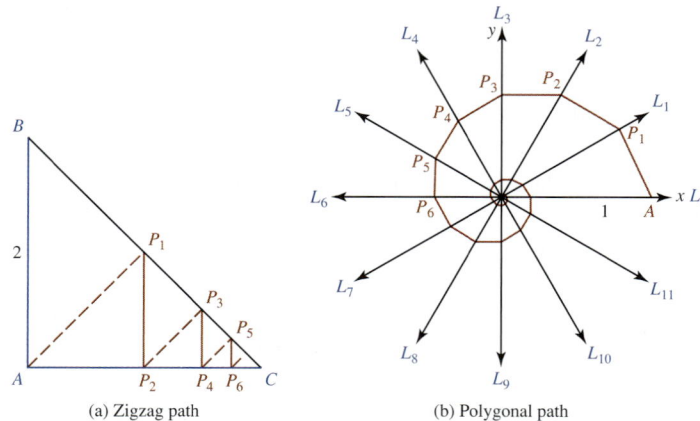

(a) Zigzag path   (b) Polygonal path

FIGURE 9.3.3 Zigzag and polygonal paths in Problems 68 and 69

69. **Length of a Polygonal Path** In Figure 9.3.3(b), there are twelve blue rays emanating from the origin and the angle between each pair of consecutive rays is 30°. The line segment AP_1 is perpendicular to ray L_1, the line segment P_1P_2 is perpendicular to ray L_2, and so on. Find the length of the red polygonal path $AP_1P_2P_3\dots$.

70. **An Improper Integral** At the end of Section 7.7 we left dangling the question of whether $f(x) \to 0$ as $x \to \infty$ is a necessary requirement for the convergence of an improper integral $\int_a^{\infty} f(x)\,dx$. Here is the answer. Observe that the function f whose graph is given in FIGURE 9.3.4 does *not* approach 0 as $x \to \infty$. Show that $\int_0^{\infty} f(x)\,dx$ converges.

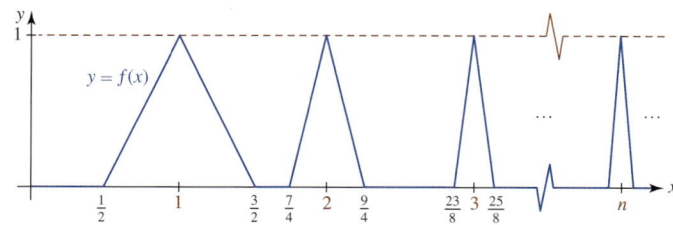

FIGURE 9.3.4 Graph for Problem 70

71. **A Stacking Problem** Take time out from doing your homework and perform an experiment. You will need a supply of n identical rectangular objects, let us say, the objects are books, but they could also be boards, playing cards, dominoes, and so on. Assume that the length of each book is L. Here is a rough statement of the problem:

How far can a stack of n books extend over the edge of a table without falling over?

Intuitively the stack should *not* fall provided its center of mass stays above the tabletop. Using the stacking rule illustrated in FIGURE 9.3.5, observe that the book shown in Figure 9.3.5(a) achieves its maximum overhang $d_1 = L/2$ when its center of mass is placed directly at the edge of the table.

(a) Compute the overhangs d_2, d_3, and d_4 from the edge of the table for the stack of books in Figures 9.3.5(b),

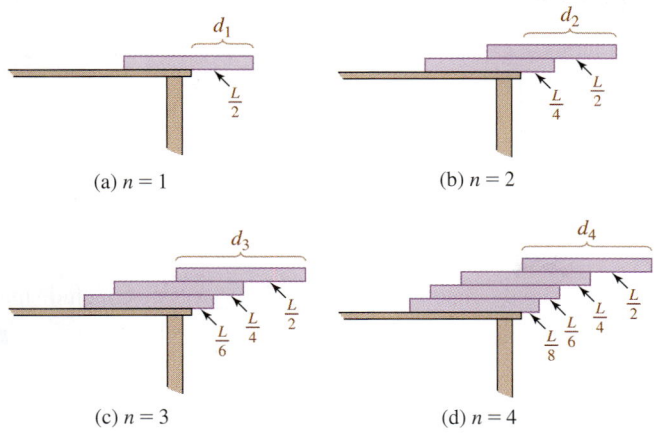

(a) $n = 1$ (b) $n = 2$

(c) $n = 3$ (d) $n = 4$

FIGURE 9.3.5 Method of stacking books in Problem 71

9.3.5(c), and 9.3.5(d), respectively. Then use (1) of Section 6.10 to show that the center of mass of each stack is at the edge of the table. [*Hint:* For n books put the x-axis along the horizontal tabletop with the origin O at the left edge of the first, or bottom, book in the stack.]

(b) What does the value of d_4 in part (a) indicate about the fourth, or top, book in the stack?

(c) Following the pattern of stacking indicated in Figure 9.3.5, for n books the overhang of the first book from the edge of the table would be $L/2n$, the overhang for the second book from the edge of the first book would be $L/2(n-1)$, the overhang for the third book from

the edge of the second book would be $L/2(n-2)$, and so on. Find a formula for d_n, the overhang of n books from the edge of the table. Show that the center of mass of the stack of n books is at the edge of the table.

(d) Use the formula d_n for the overhang found in part (c) and find the smallest value of n so that the overhang of n books stacked in the manner described in part (c) is greater than twice the length of one book.

(e) In theory, using the stacking rule in part (c), is there any limitation on the number of books in a stack?

72. A Mathematical Classic—The Trains and the Fly At a specified time two trains T_1 and T_2, 20 mi apart on the same track, start on a collision course at a rate of 10 mph. Suppose that at the precise instant the trains start a fly leaves the front of train T_1, flies at a rate of 20 mph in a straight line to the front of the engine of train T_2, then flies back to T_1 at 20 mph, then back to T_2, and so on. Use geometric series to find the total distance traversed by the fly when the trains collide (and the fly is squashed). Then use common sense to find the total distance the fly flies. See **FIGURE 9.3.6**.

FIGURE 9.3.6 Trains and fly in Problem 72

9.4 Integral Test

▌ **Introduction** Unless $\sum_{k=1}^{\infty} a_k$ is a telescoping series or a geometric series it is a difficult, if not futile, task to prove convergence or divergence directly from the sequence of partial sums. However, it is usually possible to determine whether a series converges or diverges by means of a *test* that utilizes only the terms of the series. In this and the next two sections we will examine five such tests that are applicable to infinite series of *positive* terms.

▌ **Integral Test** The first test that we shall consider relates the concepts of convergence and divergence of an improper integral to convergence and divergence of an infinite series.

Theorem 9.4.1   Integral Test

Suppose $\sum_{k=1}^{\infty} a_k$ is a series of positive terms and f is a continuous function that is nonnegative and decreasing on $[1, \infty)$ such that $f(k) = a_k$ for $k \geq 1$.

(*i*) If $\int_1^{\infty} f(x)\, dx$ converges, then $\sum_{k=1}^{\infty} a_k$ converges.

(*ii*) If $\int_1^{\infty} f(x)\, dx$ diverges, then $\sum_{k=1}^{\infty} a_k$ diverges.

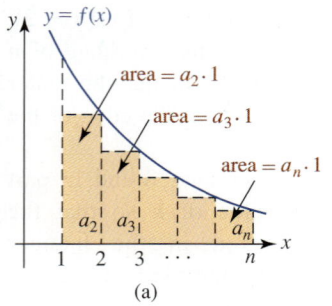

(a)

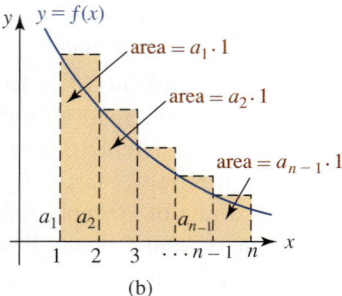

(b)

FIGURE 9.4.1 Rectangles in proof of Theorem 9.4.1

PROOF If the graph of f is given as in FIGURE 9.4.1, then by considering the areas of the rectangles shown in the figure, we see that

$$0 \le a_2 + a_3 + a_4 + \cdots + a_n \le \int_1^n f(x)\, dx \le a_1 + a_2 + a_3 + \cdots + a_{n-1}$$

or

$$S_n - a_1 \le \int_1^n f(x)\, dx \le S_{n-1}.$$

From the inequality $S_n - a_1 \le \int_1^n f(x)\, dx$, it is apparent that $\lim_{n\to\infty} S_n$ exists whenever $\lim_{n\to\infty} \int_1^n f(x)\, dx$ exists. On the other hand, from the inequality $S_{n-1} \ge \int_1^n f(x)\, dx$, we conclude that $\lim_{n\to\infty} S_{n-1}$ fails to exist whenever $\int_1^\infty f(x)\, dx$ diverges. ∎

EXAMPLE 1 Using the Integral Test

Test for convergence $\displaystyle\sum_{k=1}^{\infty} \frac{1}{1 + k^2}$.

Solution The function $f(x) = 1/(1 + x^2)$ is continuous, nonnegative, and decreasing for $x \ge 1$ such that $f(k) = a_k$ for $k \ge 1$. From

$$\int_1^\infty \frac{1}{1 + x^2}\, dx = \lim_{b\to\infty} \int_1^b \frac{1}{1 + x^2}\, dx$$

$$= \lim_{b\to\infty} \tan^{-1} x \bigg]_1^b$$

$$= \lim_{b\to\infty} \left(\tan^{-1} b - \tan^{-1} 1\right) \quad \leftarrow \tan^{-1} 1 = \pi/4$$

$$= \lim_{b\to\infty} \left(\tan^{-1} b - \frac{\pi}{4}\right) \quad \leftarrow \text{see Figure 1.5.15}$$

$$= \frac{\pi}{2} - \frac{\pi}{4} = \frac{\pi}{4}$$

we see that the improper integral is convergent. From Theorem 9.4.1(i) we conclude that the given series also converges. ∎

In the Integral Test, if the positive-term series is of the form $\sum_{k=N}^{\infty} a_k$, we then use

$$\int_N^\infty f(x)\, dx \quad \text{where } f(k) = a_k.$$

EXAMPLE 2 Using the Integral Test

Test for convergence $\displaystyle\sum_{k=3}^{\infty} \frac{\ln k}{k}$.

$f'(x) < 0$ on the interval $[3, \infty)$. ▶ **Solution** The function $f(x) = (\ln x)/x$ satisfies the hypotheses of the Integral Test on the interval $[3, \infty)$. Now,

$$\int_3^\infty \frac{\ln x}{x}\, dx = \lim_{b\to\infty} \int_3^b \frac{\ln x}{x}\, dx$$

$$= \lim_{b\to\infty} \frac{1}{2}(\ln x)^2 \bigg]_3^b$$

$$= \lim_{b\to\infty} \frac{1}{2}\left[(\ln b)^2 - (\ln 3)^2\right] = \infty$$

shows that the improper integral diverges. It follows from Theorem 9.4.1(ii) that the given series also diverges. ∎

p-Series The Integral Test is particularly useful on any series of the form

$$\sum_{k=1}^{\infty} \frac{1}{k^p} = 1 + \frac{1}{2^p} + \frac{1}{3^p} + \cdots, \tag{1}$$

where p is any fixed real number. The infinite series (1) is called the **p-series** or **hyperharmonic series**. The next theorem indicates the values of p for which the p-series converges (diverges).

Theorem 9.4.2 Convergence of p-Series

The p-series $\sum\limits_{k=1}^{\infty} \dfrac{1}{k^p}$ converges if $p > 1$ and diverges if $p \le 1$.

PROOF We distinguish four cases: $p > 1, p = 1, 0 < p < 1$, and $p \le 0$. In the first and third cases we use the Integral Test with $f(x) = 1/x^p = x^{-p}$.

(*i*) If $p > 1$, then $p - 1 > 0$ and so

$$\int_1^{\infty} x^{-p}\,dx = \lim_{b\to\infty} \frac{x^{-p+1}}{-p+1}\Big]_1^b = \frac{1}{1-p}\lim_{b\to\infty}\left[\frac{1}{b^{p-1}} - 1\right] = \frac{1}{1-p}[0-1] = \frac{1}{p-1}.$$

The p-series is convergent by Theorem 9.4.1(*i*).

(*ii*) If $p = 1$, then we recognize the p-series as the divergent harmonic series.

(*iii*) If $0 < p < 1$, then $-p + 1 > 0$ and so

$$\int_1^{\infty} x^{-p}dx = \lim_{b\to\infty} \frac{x^{-p+1}}{-p+1}\Big]_1^b = \frac{1}{1-p}\lim_{b\to\infty}[b^{-p+1} - 1] = \infty.$$

The p-series is divergent by Theorem 9.4.1(*ii*).

(*iv*) Finally, if $p \le 0$, then $-p \ge 0$ and so $\lim\limits_{n\to\infty}(1/n^p) = \lim\limits_{n\to\infty} n^{-p} \ne 0$. The p-series is divergent by the nth term test, Theorem 9.3.3. ∎

EXAMPLE 3 p-Series

(**a**) From Theorem 9.4.2, the p-series $\sum\limits_{k=1}^{\infty} \dfrac{1}{\sqrt{k}} = \sum\limits_{k=1}^{\infty} \dfrac{1}{k^{1/2}}$ diverges, since $p = \frac{1}{2} < 1$.

(**b**) From Theorem 9.4.2, the p-series $\sum\limits_{k=1}^{\infty} \dfrac{1}{k^2}$ converges, since $p = 2 > 1$. ∎

$\sum$ **NOTES FROM THE CLASSROOM**

(*i*) When applying the Integral Test, you should be aware that the value of the convergent improper integral $\int_1^{\infty} f(x)\,dx$ is not related to the actual sum of the corresponding infinite series. Thus, the series in Example 1 *does not* converge to $\pi/4$. See Problem 36 in Exercises 9.4.

(*ii*) The results of the Integral Test for $\sum_{k=n}^{\infty} a_k$ hold even if the continuous nonnegative function f does not begin to decrease until $x \ge N \ge n$. For the series $\sum_{k=1}^{\infty} (\ln k)/k$ the function $f(x) = (\ln x)/x$ decreases on the interval $[3, \infty)$. Nonetheless, in the Integral Test we may use $\int_1^{\infty} (\ln x\,dx)/x$.

Exercises 9.4   Answers to selected odd-numbered problems begin on page ANS-27.

☰ Fundamentals

In Problems 1–30, determine whether the given series converges or diverges. Use the Integral Test where appropriate.

1. $\sum\limits_{k=1}^{\infty} \dfrac{1}{k^{1.1}}$

2. $\sum\limits_{k=1}^{\infty} \dfrac{1}{k^{0.99}}$

3. $1 + \dfrac{1}{2\sqrt{2}} + \dfrac{1}{3\sqrt{3}} + \cdots$

4. $\dfrac{1}{100} + \dfrac{1}{100\sqrt{2}} + \dfrac{1}{100\sqrt{3}} + \cdots$

5. $\displaystyle\sum_{k=1}^{\infty} \frac{1}{2k+7}$

6. $\displaystyle\sum_{k=1}^{\infty} \frac{k}{3k+1}$

7. $\displaystyle\sum_{k=1}^{\infty} \frac{1}{1+5k^2}$

8. $\displaystyle\sum_{k=3}^{\infty} \frac{k}{k^2+5}$

9. $\displaystyle\sum_{k=1}^{\infty} ke^{-k^2}$

10. $\displaystyle\sum_{k=1}^{\infty} \frac{e^{1/k}}{k^2}$

11. $\displaystyle\sum_{k=1}^{\infty} \frac{k}{e^k}$

12. $\displaystyle\sum_{k=2}^{\infty} k^2 e^{-k}$

13. $\displaystyle\sum_{k=2}^{\infty} \frac{1}{k\ln k}$

14. $\displaystyle\sum_{k=2}^{\infty} \frac{k}{\ln k}$

15. $\displaystyle\sum_{k=2}^{\infty} \frac{10}{k(\ln k)^2}$

16. $\displaystyle\sum_{k=2}^{\infty} \frac{1}{k\sqrt{\ln k}}$

17. $\displaystyle\sum_{k=1}^{\infty} \frac{\arctan k}{1+k^2}$

18. $\displaystyle\sum_{k=1}^{\infty} \frac{k}{1+k^4}$

19. $\displaystyle\sum_{k=1}^{\infty} \frac{1}{\sqrt{1+k}}$

20. $\displaystyle\sum_{k=1}^{\infty} \frac{1}{\sqrt{1+k^2}}$

21. $\displaystyle\sum_{n=1}^{\infty} \frac{n}{(n^2+1)^3}$

22. $\displaystyle\sum_{n=2}^{\infty} \frac{1}{(4n+1)^{3/2}}$

23. $\displaystyle\sum_{k=1}^{\infty} k\sin\left(\frac{1}{k}\right)$

24. $\displaystyle\sum_{k=1}^{\infty} \ln\left(1+\frac{1}{3}k\right)$

25. $\displaystyle\sum_{k=1}^{\infty} \frac{1}{k(k+1)}$

26. $\displaystyle\sum_{k=1}^{\infty} \frac{2k+1}{k(k+1)}$

27. $\displaystyle\sum_{k=1}^{\infty} \frac{1}{(k+1)(k+2)}$

28. $\displaystyle\sum_{k=1}^{\infty} \frac{1}{k(k^2+1)}$

29. $\displaystyle\sum_{k=1}^{\infty} \frac{2}{e^k+e^{-k}}$

30. $\displaystyle\sum_{k=0}^{\infty} \frac{-1}{\sqrt{e^{3k}}}$

In Problems 31–34, without doing any work determine whether the given series converges or diverges. State your reasons.

31. $\displaystyle\sum_{k=1}^{\infty} \left(\frac{2}{k}+\frac{3}{k^2}\right)$

32. $\displaystyle\sum_{k=1}^{\infty} \left(5k^{-1.6}-10k^{-1.1}\right)$

33. $\displaystyle\sum_{k=1}^{\infty} \left(\frac{1}{k^2}+\frac{1}{2^k}\right)$

34. $\displaystyle\sum_{k=1}^{\infty} \frac{1+4\sqrt{k}}{k^2}$

In Problems 35 and 36, determine the values of p for which the given series converges.

35. $\displaystyle\sum_{k=2}^{\infty} \frac{1}{k(\ln k)^p}$

36. $\displaystyle\sum_{k=3}^{\infty} \frac{1}{k\ln k[\ln(\ln k)]^p}$

≡ **Think About It**

37. Determine the values of p for which the series

$$\sum_{k=2}^{\infty} k^p \ln k$$

is convergent.

38. Suppose that f is a continuous function that is positive and decreasing for $x \geq 1$ such that $f(k) = a_k$ for $k \geq 1$. Show that

$$\int_1^{n+1} f(x)\,dx \leq \sum_{k=1}^{n} a_k \leq a_1 + \int_1^{n} f(x)\,dx.$$

39. Show that

$$\frac{\pi}{4} \leq \sum_{k=1}^{\infty} \frac{1}{1+k^2} \leq \frac{1}{2}+\frac{\pi}{4}.$$

40. The harmonic series $\sum_{k=1}^{\infty}(1/k)$ was shown to be divergent because the sequence of partial sums diverges. Recall from page 495 that $S_n = \sum_{k=1}^{n}(1/k) \rightarrow \infty$ as $n \rightarrow \infty$.

(a) Use the result of Problem 38 to estimate the sum of the first 10 billion terms of the harmonic series.

(b) How many terms of the harmonic series are necessary to guarantee that $S_n \geq 100$?

41. Let S denote the sum of a positive-term series $\sum_{k=1}^{\infty} a_k$ and S_n the general term in its sequence of partial sums. Define the **remainder**, or the error made when S is approximated by S_n, to be

$$R_n = S - S_n = a_{n+1} + a_{n+2} + a_{n+3} + \cdots.$$

Suppose that f is a continuous function that is positive and decreasing for $x \geq 1$ such that $f(k) = a_k$ for $k \geq 1$ and that $\int_1^{\infty} f(x)\,dx$ converges. Show that

$$\int_{n+1}^{\infty} f(x)\,dx \leq R_n \leq \int_n^{\infty} f(x)\,dx.$$

42. The sum S of the convergent p-series $\sum_{k=1}^{\infty}(1/k^2)$ is known to be $\pi^2/6$. Use Problem 41 to determine n so that S_n will give an approximation to S that is accurate to three decimal places.

9.5 Comparison Tests

▌ **Introduction**   It is often possible to determine convergence or divergence of a positive-term series $\sum a_k$ by *comparing* its terms with the terms of a *test series* $\sum b_k$ that is known to be either convergent or divergent. In this section we will consider two comparison tests for convergence and divergence.

▌ **Direct Comparison Test**   The proof of the next test will utilize two important properties of sequences. Recall from Section 9.2 that if a sequence is bounded and monotonic it must converge. Also if the terms of a sequence become unbounded then it diverges. We apply these results to the sequence of partial sums of a series.

> **Theorem 9.5.1** Direct Comparison Test
>
> Suppose $\sum_{k=1}^{\infty} a_k$ and $\sum_{k=1}^{\infty} b_k$ are series of positive terms.
>
> (*i*) If $\sum_{k=1}^{\infty} b_k$ converges and $a_k \leq b_k$ for every positive integer k, then $\sum_{k=1}^{\infty} a_k$ converges.
>
> (*ii*) If $\sum_{k=1}^{\infty} b_k$ diverges and $a_k \geq b_k$ for every positive integer k, then $\sum_{k=1}^{\infty} a_k$ diverges.

PROOF Let $a_k > 0$ and $b_k > 0$ for $k = 1, 2, \ldots$ and let

$$S_n = a_1 + a_2 + \cdots + a_n \quad \text{and} \quad T_n = b_1 + b_2 + \cdots + b_n$$

be the general terms of the sequences of partial sums for $\sum a_k$ and $\sum b_k$, respectively.

(*i*) If $\sum b_k$ is a convergent series for which $a_k \leq b_k$, then $S_n \leq T_n$. Since $\lim_{n \to \infty} T_n$ exists, $\{S_n\}$ is a bounded increasing sequence and, hence, convergent by Theorem 9.2.1. Therefore, $\sum a_k$ is convergent.

(*ii*) If $\sum b_k$ diverges and $a_k > b_k$, then $S_n > T_n$. Since T_n increases without bound, so does S_n. Hence, $\sum a_k$ is divergent. ∎

In general, if $\sum c_k$ and $\sum d_k$ are two series for which $c_k \leq d_k$ for all k, we say that the series $\sum c_k$ is **dominated** by the series $\sum d_k$. Thus, for positive-term series, parts (*i*) and (*ii*) of Theorem 9.5.1 can be restated in the following manner:

- A series $\sum a_k$ is convergent if it is dominated by a convergent series $\sum b_k$.
- A series $\sum a_k$ diverges if it dominates a divergent series $\sum b_k$.

The next two examples illustrate the method. Of course, it goes without saying that to come up with a test series $\sum b_k$ you must be familiar with some series that converge and some that diverge.

◀ It might be a good idea at this point to review the notion of *p*-series in Section 9.4.

EXAMPLE 1 Using the Direct Comparison Test

Test for convergence $\displaystyle\sum_{k=1}^{\infty} \frac{k}{k^3 + 4}$.

Solution We observe that by decreasing the denominator in the general terms we obtain a larger fraction:

$$\frac{k}{k^3 + 4} \leq \frac{k}{k^3} = \frac{1}{k^2}.$$

Because the given series is dominated by the convergent *p*-series $\sum_{k=1}^{\infty}(1/k^2)$, it follows from Theorem 9.5.1(*i*) that the given series is also convergent. ∎

EXAMPLE 2 Using the Direct Comparison Test

Test for convergence $\displaystyle\sum_{k=1}^{\infty} \frac{\ln(k + 2)}{k}$.

Solution Since $\ln(k + 2) > 1$ for $k \geq 1$, we have

$$\frac{\ln(k + 2)}{k} > \frac{1}{k}.$$

In this case the given series has been shown to dominate the divergent harmonic series $\sum_{k=1}^{\infty}(1/k)$. Hence, by Theorem 9.5.1(*ii*) the given series diverges. ∎

■ **Limit Comparison Test** Another kind of comparison test involves taking the limit of the ratio of the general term of a series $\sum a_k$ to the general term of a test series $\sum b_k$ that is known to be convergent or divergent.

Theorem 9.5.2 Limit Comparison Test

Suppose $\sum_{k=1}^{\infty} a_k$ and $\sum_{k=1}^{\infty} b_k$ are series of positive terms. If

$$\lim_{n \to \infty} \frac{a_n}{b_n} = L,$$

where L is finite and $L > 0$, then the two series are either both convergent or both divergent.

PROOF Since $\lim_{n \to \infty} a_n/b_n = L > 0$, we can choose n so large, say $n \geq N$ for some positive integer N, that

$$\frac{1}{2}L \leq \frac{a_n}{b_n} \leq \frac{3}{2}L.$$

Since $a_n > 0$, the inequality implies that $a_n \leq \frac{3}{2}Lb_n$ for $n \geq N$. If $\sum_{k=1}^{\infty} b_k$ converges, it follows from the Direct Comparison Test that $\sum_{k=1}^{\infty} a_k$ and, therefore, $\sum_{k=1}^{\infty} a_k$ is convergent. Furthermore, since $\frac{1}{2}Lb_n \leq a_n$ for $n \geq N$, we see that if $\sum_{k=1}^{\infty} b_k$ diverges, then $\sum_{k=1}^{\infty} b_k$ and $\sum_{k=1}^{\infty} a_k$ diverge. ∎

The Limit Comparison Test is often applicable to series $\sum a_k$ for which the Direct Comparison Test is inconvenient.

EXAMPLE 3 Using the Limit Comparison Test

You should convince yourself that it is difficult to apply the Direct Comparison Test to the series $\sum_{k=1}^{\infty} \dfrac{1}{k^3 - 5k^2 + 1}$. However, we know that $\sum_{k=1}^{\infty} (1/k^3)$ is a convergent p-series ($p = 3 > 1$). Hence, with

$$a_n = \frac{1}{n^3 - 5n^2 + 1} \qquad \text{and} \qquad b_n = \frac{1}{n^3}$$

we have

$$\lim_{n \to \infty} \frac{a_n}{b_n} = \lim_{n \to \infty} \frac{n^3}{n^3 - 5n^2 + 1} = 1.$$

From Theorem 9.5.2, it follows that the given series converges. ∎

If the general term a_n of a series $\sum a_k$ is a quotient of either rational powers of n or roots of polynomials in n, it is possible to discern the general term b_n of a test series $\sum b_k$ by examining the "degree behavior" of a_n for large values of n. In other words, to find a candidate for b_n we need only examine the quotient of the *highest powers of n* in the numerator and denominator of a_n.

EXAMPLE 4 Using the Limit Comparison Test

Test for convergence $\displaystyle\sum_{k=1}^{\infty} \frac{k}{\sqrt[3]{8k^5 + 7}}$.

Solution For large values of n, the general term of the series $a_n = n/\sqrt[3]{8n^5 + 7}$ "behaves like" a constant multiple of

$$\frac{n}{\sqrt[3]{n^5}} = \frac{n}{n^{5/3}} = \frac{1}{n^{2/3}}.$$

Thus, we try the divergent p-series $\sum_{k=1}^{\infty} \dfrac{1}{k^{2/3}}$ as a test series:

$$\lim_{n\to\infty} \frac{a_n}{b_n} = \lim_{n\to\infty} \frac{\dfrac{n}{\sqrt[3]{8n^5 + 7}}}{\dfrac{1}{n^{2/3}}}$$

$$= \lim_{n\to\infty} \left(\frac{n^5}{8n^5 + 7}\right)^{1/3} = \left(\frac{1}{8}\right)^{1/3} = \frac{1}{2}.$$

Thus, from Theorem 9.5.2 the given series diverges. ∎

$\sum$ NOTES FROM THE CLASSROOM

(*i*) The hypotheses in the Direct Comparison Test can also be weakened, giving a stronger theorem. For a series with positive terms, it is only required that $a_k \leq b_k$ or $a_k \geq b_k$ for k sufficiently large and not for all positive integers.

(*ii*) In the application of the Direct Comparison Test, it is often easy to reach a point where the given series is dominated by a divergent series. For example,

$$\frac{1}{5^k + \sqrt{k}} \leq \frac{1}{\sqrt{k}}$$

is certainly true and $\sum_{k=1}^{\infty} \dfrac{1}{\sqrt{k}}$ diverges. This kind of reasoning proves nothing about the series $\sum_{k=1}^{\infty} \dfrac{1}{5^k + \sqrt{k}}$. Indeed, the last series converges. Why? Similarly, no conclusion can be reached by showing that a given series dominates a convergent series.

The following table summarizes the **Direct Comparison Test**. Let $\sum a_k$ be a series of positive terms and $\sum b_k$ a series that we know either converges or diverges (a test series).

Comparison of terms	Test Series $\sum b_k$	Conclusion about $\sum a_k$
$a_k \leq b_k$	converges	converges
$a_k \leq b_k$	diverges	none
$a_k \geq b_k$	diverges	diverges
$a_k \geq b_k$	converges	none

Exercises 9.5   Answers to selected odd-numbered problems begin on page ANS-27.

☰ Fundamentals

In Problems 1–14, use the Direct Comparison Test to determine whether the given series converges.

1. $\sum_{k=1}^{\infty} \dfrac{1}{(k+1)(k+2)}$

2. $\sum_{k=1}^{\infty} \dfrac{1}{k^2 + 5}$

3. $\sum_{k=2}^{\infty} \dfrac{1}{\sqrt{k} - 1}$

4. $\sum_{k=2}^{\infty} \dfrac{2k^2 + 1}{k^3 - k}$

5. $\sum_{k=2}^{\infty} \dfrac{1}{\ln k}$

6. $\sum_{k=3}^{\infty} \dfrac{\ln k}{k^5}$

7. $\sum_{k=1}^{\infty} \dfrac{1 + 3^k}{2^k}$

8. $\sum_{k=1}^{\infty} \dfrac{1 + 8^k}{3 + 10^k}$

9. $\sum_{k=1}^{\infty} \dfrac{2 + \sin k}{\sqrt[3]{k^4 + 1}}$

10. $\sum_{k=2}^{\infty} \dfrac{2k + 1}{k \ln k}$

11. $\displaystyle\sum_{j=1}^{\infty} \frac{j + e^{-j}}{5^j(j + 9)}$

12. $\displaystyle\sum_{i=1}^{\infty} \frac{ie^{-i}}{i + 1}$

13. $\displaystyle\sum_{k=1}^{\infty} \frac{\sqrt{k + 1} - \sqrt{k}}{k}$

14. $\dfrac{1}{1 \cdot 3} + \dfrac{1}{2 \cdot 9} + \dfrac{1}{3 \cdot 27} + \dfrac{1}{4 \cdot 81} + \cdots$

In Problems 15–28, use the Limit Comparison Test to determine whether the given series converges.

15. $\displaystyle\sum_{k=1}^{\infty} \frac{1}{2k + 7}$

16. $\displaystyle\sum_{k=1}^{\infty} \frac{1}{10 + \sqrt{k}}$

17. $\displaystyle\sum_{n=2}^{\infty} \frac{1}{n\sqrt{n^2 - 1}}$

18. $\displaystyle\sum_{n=1}^{\infty} \frac{1}{\sqrt{(n + 1)(n + 2)}}$

19. $\displaystyle\sum_{n=1}^{\infty} \frac{n^2 - n + 2}{3n^5 + n^2}$

20. $\displaystyle\sum_{n=2}^{\infty} \frac{n}{(4n + 1)^{3/2}}$

21. $\displaystyle\sum_{k=1}^{\infty} \frac{\sqrt{k + 1}}{\sqrt[3]{64k^9 + 40}}$

22. $\displaystyle\sum_{k=2}^{\infty} \frac{5k^2 - k}{2k^3 + 2k^2 - 8}$

23. $\displaystyle\sum_{k=2}^{\infty} \frac{k + \ln k}{k^3 + 2k - 1}$

24. $\displaystyle\sum_{k=1}^{\infty} \frac{10}{e^k - 2}$

25. $\displaystyle\sum_{k=1}^{\infty} \sin\!\left(\frac{1}{k}\right)$

26. $\displaystyle\sum_{k=1}^{\infty} \left(1 - \cos\!\left(\frac{1}{k}\right)\right)$

27. $\displaystyle\sum_{k=1}^{\infty} \left(\frac{1}{2} + \frac{1}{2k}\right)^k$

28. $\dfrac{1}{2 \cdot 3} + \dfrac{2}{3 \cdot 4} + \dfrac{3}{4 \cdot 5} + \dfrac{4}{5 \cdot 6} + \cdots$

In Problems 29–40, use any appropriate test to determine whether the given series converges.

29. $\displaystyle\sum_{k=1}^{\infty} \frac{k}{100\sqrt{k^2 + 1}}$

30. $\displaystyle\sum_{k=1}^{\infty} \frac{1}{k + \sqrt{k}}$

31. $\displaystyle\sum_{k=1}^{\infty} \ln\!\left(5 + \frac{k}{5}\right)$

32. $\displaystyle\sum_{k=1}^{\infty} \ln\!\left(1 + \frac{1}{3^k}\right)$

33. $\displaystyle\sum_{k=1}^{\infty} \frac{k}{(k^2 + 1)^2}$

34. $\displaystyle\sum_{k=2}^{\infty} \frac{k}{\sqrt{k - 1}\sqrt[3]{k^2 - 2}}$

35. $\displaystyle\sum_{k=1}^{\infty} \frac{1}{9 + \sin^2 k}$

36. $\displaystyle\sum_{k=1}^{\infty} \frac{3^k}{3^{2k} - 1}$

37. $\displaystyle\sum_{k=1}^{\infty} \frac{2}{2 + k2^k}$

38. $\displaystyle\sum_{k=1}^{\infty} \frac{2}{2 + k2^{-k}}$

39. $\displaystyle\sum_{k=2}^{\infty} \ln\!\left(1 + \frac{1}{k}\right)$

40. $\displaystyle\sum_{k=1}^{\infty} \frac{(0.9)^k}{k}$

≡ Think About It

41. Reread (*ii*) of *Notes from the Classroom* on page 507 and then discuss the reasons why the following statement is true:

If $a_k > 0$ for all k and Σa_k converges, then Σa_k^2 converges.

42. Suppose that p and q are polynomial functions with no common factors of degrees n and m, respectively, and that $p(x)/q(x) > 0$ for $x > 0$. Discuss: Under what conditions will the series $\sum_{k=1}^{\infty} p(k)/q(k)$ converge?

43. Discuss whether the following statement is true or false:

If $a_k < b_k$ for all k and Σb_k converges, then Σa_k converges.

44. Show that if the series Σa_k of positive terms converges, then $\Sigma \ln(1 + a_k)$ converges.

In Problems 45 and 46, determine whether the given series converges.

45. $\displaystyle\sum_{k=1}^{\infty} \frac{1}{k^{1+1/k}}$

46. $\displaystyle\sum_{k=1}^{\infty} \frac{1}{1 + 2 + 3 + \cdots + k}$

47. The decimal representation of a positive real number is an infinite series:

$$0.a_1a_2a_3a_4\ldots = \frac{a_1}{10} + \frac{a_2}{10^2} + \frac{a_3}{10^3} + \frac{a_4}{10^4} + \cdots,$$

where a_i represents one of the 10 nonnegative integers $0, 1, 2, \ldots, 9$. Prove that the series of the form

$$\frac{a_1}{10} + \frac{a_2}{10^2} + \frac{a_3}{10^3} + \frac{a_4}{10^4} + \cdots = \sum_{k=1}^{\infty} \frac{a_k}{10^k}$$

is always convergent.

≡ Project

48. How Big Is Infinity? The Integral Test can be used to verify that $\displaystyle\sum_{k=1}^{\infty} \frac{1}{k^{1.0001}}$ converges, whereas $\displaystyle\sum_{k=2}^{\infty} \frac{1}{k\ln k}$ diverges. However, with the aid of a CAS we see from the graphs of $y = 1/x^{1.0001}$ (in red) and $y = 1/(x\ln x)$ (in blue) in FIGURE 9.5.1 that

$$\frac{1}{k\ln k} < \frac{1}{k^{1.0001}}$$

for $2 \leq k \leq 15,000$. Indeed, the foregoing inequality is true for $2 \leq k \leq 99,999,999 \times 10^{99}$. So why doesn't $\displaystyle\sum_{k=2}^{\infty} \frac{1}{k\ln k}$ converge by the Direct Comparison Test?

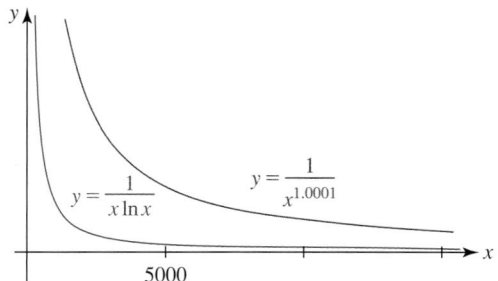

FIGURE 9.5.1 Graph for Problem 48

9.6 Ratio and Root Tests

▪ **Introduction** In this section, as in the last, the tests that we shall consider are applicable to infinite series of *positive terms*.

▪ **Ratio Test** The first of these tests employs a limit of the ratio of the $(n + 1)$st term to the nth term of the series. This test is especially useful when a_k involves factorials, kth powers of a constant, and sometimes, kth powers of k.

Theorem 9.6.1 Ratio Test

Suppose $\sum_{k=1}^{\infty} a_k$ is a series of positive terms such that

$$\lim_{n \to \infty} \frac{a_{n+1}}{a_n} = L.$$

(*i*) If $L < 1$, the series is convergent.
(*ii*) If $L > 1$, or if $L = \infty$, the series is divergent.
(*iii*) If $L = 1$, the test is inconclusive.

PROOF

(*i*) Let r be a positive number such that $0 \le L \le r \le 1$. For n sufficiently large, say $n \ge N$ for some positive integer N, $a_{n+1}/a_n < r$; that is $a_{n+1} < ra_n$, $n \ge N$. The last inequality implies

$$a_{N+1} < ra_N$$
$$a_{N+2} < ra_{N+1} < a_N r^2$$
$$a_{N+3} < ra_{N+2} < a_N r^3,$$

and so on. Thus, the series $\sum_{k=N+1}^{\infty} a_k$ converges by comparison with the convergent geometric series $\sum_{k=1}^{\infty} a_N r^k$. Since $\sum_{k=1}^{\infty} a_k$ differs from $\sum_{k=N+1}^{\infty} a_k$ by at most a finite number of terms, we conclude that the former series also converges.

(*ii*) Let r be a finite number such that $1 < r < L$. Then for n sufficiently large, say $n \ge N$ for some positive integer N, $a_{n+1}/a_n > r$ or $a_{n+1} > ra_n$. For $r > 1$ this last inequality implies $a_{n+1} > a_n$ and so $\lim_{n \to \infty} a_n \ne 0$. From Theorem 9.3.3 we conclude that $\sum_{k=1}^{\infty} a_k$ diverges. ▪

In the case when $L = 1$, we must apply another test to the series to determine its convergence or divergence.

EXAMPLE 1 Using the Ratio Test

Test for convergence $\displaystyle\sum_{k=1}^{\infty} \frac{5^k}{k!}$.

Solution We identify $a_n = 5^n/n!$ and so $a_{n+1} = 5^{n+1}/(n + 1)!$. We next form the quotient of a_{n+1} and a_n, simplify, and then take the limit as $n \to \infty$:

$$\lim_{n \to \infty} \frac{a_{n+1}}{a_n} = \lim_{n \to \infty} \frac{5^{n+1}}{(n + 1)!} \cdot \frac{n!}{5^n}$$

$$= \lim_{n \to \infty} 5 \frac{n!}{(n + 1)!}$$

$$= \lim_{n \to \infty} 5 \frac{n!}{n!(n + 1)}$$

$$= \lim_{n \to \infty} \frac{5}{n + 1} = 0.$$

◀ Review the properties of the factorial in Section 9.1. See (4) and (5) in that section.

Since $L = 0 < 1$, it follows from Theorem 9.6.1(*i*) that the series is convergent. ▪

EXAMPLE 2 Using the Ratio Test

Test for convergence $\displaystyle\sum_{k=1}^{\infty} \frac{k^k}{k!}$.

Solution In this case we have $a_n = n^n/n!$ and $a_{n+1} = (n+1)^{n+1}/(n+1)!$. Then

$$\lim_{n\to\infty}\frac{a_{n+1}}{a_n} = \lim_{n\to\infty}\frac{(n+1)^{n+1}}{(n+1)!} \cdot \frac{n!}{n^n}$$

$$= \lim_{n\to\infty}\frac{(n+1)^{n+1}}{n+1} \cdot \frac{1}{n^n}$$

$$= \lim_{n\to\infty}\left(\frac{n+1}{n}\right)^n$$

$$= \lim_{n\to\infty}\left(1 + \frac{1}{n}\right)^n = e. \quad \leftarrow \text{This limit is (3) of Section 1.6.}$$

Since $L = e > 1$, it follows from Theorem 9.6.1(*ii*) that the series is divergent. ∎

■ **Root Test** If the terms of a series $\sum a_k$ consist of only kth powers, then the following test, which involves taking the nth root of the nth term, may be applicable.

Theorem 9.6.2 Root Test

Suppose $\sum_{k=1}^{\infty} a_k$ is a series of positive terms such that

$$\lim_{n\to\infty} \sqrt[n]{a_n} = \lim_{n\to\infty} (a_n)^{1/n} = L.$$

(*i*) If $L < 1$, the series is convergent.
(*ii*) If $L > 1$, or if $L = \infty$, the series is divergent.
(*iii*) If $L = 1$, the test is inconclusive.

The proof of the Root Test is very similar to the proof of the Ratio Test and will not be given.

EXAMPLE 3 Using the Root Test

Test for convergence $\displaystyle\sum_{k=1}^{\infty}\left(\frac{5}{k}\right)^k$.

Solution We first identify $a_n = (5/n)^n$, and then compute the limit as $n \to \infty$ of the nth root of a_n:

$$\lim_{n\to\infty}\left[\left(\frac{5}{n}\right)^n\right]^{1/n} = \lim_{n\to\infty}\frac{5}{n} = 0.$$

Since $L = 0 < 1$, we conclude from Theorem 9.6.2(*i*) that the series converges. ∎

$\sum$ **NOTES FROM THE CLASSROOM**

(*i*) The Ratio Test will always give the inconclusive case when applied to a *p*-series. Try it on the series $\sum_{k=1}^{\infty} 1/k^2$ and see what happens.

(*ii*) The tests examined in this and the preceding two sections tell us when a series has a sum, but none of these tests gives so much as even a clue as to what the actual sum is. But knowing that a series converges, we can now add up five, a hundred, or a thousand terms on a computer to obtain an approximation of the sum.

Exercises 9.6   Answers to selected odd-numbered problems begin on page ANS-27.

≡ Fundamentals

In Problems 1–16, use the Ratio Test to determine whether the given series converges.

1. $\displaystyle\sum_{k=1}^{\infty} \frac{1}{k!}$

2. $\displaystyle\sum_{k=1}^{\infty} \frac{2^k}{k!}$

3. $\displaystyle\sum_{k=1}^{\infty} \frac{k!}{1000^k}$

4. $\displaystyle\sum_{k=1}^{\infty} k\left(\frac{2}{3}\right)^k$

5. $\displaystyle\sum_{j=1}^{\infty} \frac{j^{10}}{(1.1)^j}$

6. $\displaystyle\sum_{j=1}^{\infty} \frac{1}{j^5(0.99)^j}$

7. $\displaystyle\sum_{n=1}^{\infty} \frac{4^{n-1}}{n3^{n-2}}$

8. $\displaystyle\sum_{n=1}^{\infty} \frac{n^3 2^{n+3}}{7^{n-1}}$

9. $\displaystyle\sum_{k=1}^{\infty} \frac{k!}{(2k)!}$

10. $\displaystyle\sum_{k=1}^{\infty} \frac{(2k)!}{k!(2k)^k}$

11. $\displaystyle\sum_{k=1}^{\infty} \frac{99^k(k^3+1)}{k^2 10^{2k}}$

12. $\displaystyle\sum_{k=1}^{\infty} \frac{k!}{e^{k^2}}$

13. $\displaystyle\sum_{k=1}^{\infty} \frac{5^k}{k^k}$

14. $\displaystyle\sum_{k=1}^{\infty} \frac{k!3^k}{k^k}$

15. $\displaystyle\sum_{k=1}^{\infty} \frac{1 \cdot 3 \cdot 5 \cdots (2k-1)}{k!}$

16. $\displaystyle\sum_{k=1}^{\infty} \frac{k!}{2 \cdot 4 \cdot 6 \cdots (2k)}$

In Problems 17–24, use the Root Test to determine whether the given series converges.

17. $\displaystyle\sum_{k=1}^{\infty} \frac{1}{k^k}$

18. $\displaystyle\sum_{k=1}^{\infty} \left(\frac{ke}{k+1}\right)^k$

19. $\displaystyle\sum_{k=2}^{\infty} \left(\frac{k}{\ln k}\right)^k$

20. $\displaystyle\sum_{k=2}^{\infty} \frac{1}{(\ln k)^k}$

21. $\displaystyle\sum_{k=1}^{\infty} \left(\frac{k}{k+1}\right)^{k^2}$

22. $\displaystyle\sum_{k=1}^{\infty} \left(1-\frac{2}{k}\right)^{k^2}$

23. $\displaystyle\sum_{k=1}^{\infty} \frac{6^{2k+1}}{k^k}$

24. $\displaystyle\sum_{k=1}^{\infty} \frac{k^k}{e^{k+1}}$

In Problems 25–32, use any appropriate test to determine whether the given series converges.

25. $\displaystyle\sum_{k=1}^{\infty} \frac{k^2+k}{k^3+2k+1}$

26. $\displaystyle\sum_{k=1}^{\infty} \left(\frac{3k}{2k+1}\right)^k$

27. $\displaystyle\sum_{n=1}^{\infty} \frac{e^{1/n}}{n^2}$

28. $\displaystyle\sum_{n=1}^{\infty} \frac{n^2+n}{e^n}$

29. $\displaystyle\sum_{k=1}^{\infty} \frac{5^k k!}{(k+1)!}$

30. $\displaystyle\sum_{k=1}^{\infty} \frac{3}{2^k+k}$

31. $\displaystyle\sum_{k=0}^{\infty} \frac{2^k}{3^k+4^k}$

32. $\displaystyle \frac{1}{3}+\frac{2}{4}+\frac{3}{5}+\frac{4}{6}+\cdots$

In Problems 33 and 34, use the Ratio Test to determine the nonnegative values of p for which the given series converges.

33. $\displaystyle\sum_{k=1}^{\infty} kp^k$

34. $\displaystyle\sum_{k=1}^{\infty} k^2\left(\frac{2}{p}\right)^k$

In Problems 35 and 36, determine all real values of p for which the gives series converges.

35. $\displaystyle\sum_{k=1}^{\infty} \frac{k^p}{k!}$

36. $\displaystyle\sum_{k=2}^{\infty} \frac{\ln k}{k^p}$

37. In Problems 78 and 79 of Exercises 9.1 we saw that the Fibonacci sequence $\{F_n\}$,

$$1, 1, 2, 3, 5, 8, \ldots,$$

is defined by the recursion formula $F_{n+1} = F_n + F_{n-1}$, where $F_1 = 1, F_2 = 1$.

(a) Verify that the general term of the sequence is

$$F_n = \frac{1}{\sqrt{5}}\left(\frac{1+\sqrt{5}}{2}\right)^n - \frac{1}{\sqrt{5}}\left(\frac{1-\sqrt{5}}{2}\right)^n$$

by showing that this result satisfies the recursion formula.

(b) Use the general term in part (a) to calculate $F_1, F_2, F_3, F_4,$ and F_5.

38. Let F_n be the general term of the Fibonacci sequence given in Problem 37. Show that

$$\lim_{n\to\infty} \frac{F_{n+1}}{F_n} = \frac{1+\sqrt{5}}{2}.$$

39. Explain how the result in Problem 38 proves that the series

$$\frac{1}{1}+\frac{1}{1}+\frac{1}{2}+\frac{1}{3}+\frac{1}{5}+\frac{1}{8}+\cdots = \sum_{n=1}^{\infty} \frac{1}{F_n}$$

converges.

40. A Bit of History In 1985 William Gosper used the following identity to compute the first 17 million digits of π:

$$\frac{1}{\pi} = \frac{2\sqrt{2}}{9801} \sum_{n=0}^{\infty} (1103 + 26,390n) \frac{(4n)!}{(n!)^4(4 \cdot 99)^{4n}}.$$

This identity was discovered in 1920 by the Indian mathematician **Srinivasa Ramanujan** (1887–1920). Ramanujan was noted for his remarkable insights in handling exceedingly complex algebraic manipulations and calculations.

(a) Verify that the infinite series converges.

(b) How many correct decimal places of π does the first term of the series yield?

(c) How many correct decimal places of π do the first two terms of the series yield?

9.7 Alternating Series

▌ Introduction In the last three sections, we considered tests for convergence that were applicable only to series with positive terms. In the present discussion we consider series in which the terms alternate between positive and negative numbers, that is, series having either form

A geometric series such as

$$\sum_{k=1}^{\infty} \left(-\tfrac{1}{3}\right)^{k-1} = 1 - \tfrac{1}{3} + \tfrac{1}{9} - \tfrac{1}{27} + \cdots$$

is an alternating series. See Example 4 in Section 9.3.

$$a_1 - a_2 + a_3 - a_4 + \cdots + (-1)^{n+1}a_n + \cdots = \sum_{k=1}^{\infty} (-1)^{k+1}a_k \qquad (1)$$

or

$$-a_1 + a_2 - a_3 + a_4 - \cdots + (-1)^n a_n + \cdots = \sum_{k=1}^{\infty} (-1)^k a_k, \qquad (2)$$

where $a_k > 0$ for $k = 1, 2, 3, \ldots$. The series in (1) and (2) are said to be **alternating series**. We have already encountered a special type of alternating series in Section 9.3, but in this section we will examine properties of general alternating series and tests for their convergence. Because the series (2) is just a multiple of (1), we will confine our discussion to the latter series.

EXAMPLE 1 Alternating Series

The series

$$1 - \frac{1}{2} + \frac{1}{3} - \frac{1}{4} + \cdots = \sum_{k=1}^{\infty} \frac{(-1)^{k+1}}{k}$$

and

$$\frac{\ln 2}{4} - \frac{\ln 3}{8} + \frac{\ln 4}{16} - \frac{\ln 5}{32} + \cdots = \sum_{k=2}^{\infty} (-1)^k \frac{\ln k}{2^k}$$

are examples of alternating series. ▪

▌ Alternating Series Test The first series in Example 1, $1 - \frac{1}{2} + \frac{1}{3} - \frac{1}{4} + \cdots$, is called the **alternating harmonic series**. Although the harmonic series

$$\sum_{k=1}^{\infty} \frac{1}{k} = 1 + \frac{1}{2} + \frac{1}{3} + \frac{1}{4} + \cdots$$

is divergent, the introduction of positive and negative terms in the sequence of partial sums for the alternating harmonic series is sufficient to produce a convergent series. We will prove that $\displaystyle\sum_{k=1}^{\infty} \frac{(-1)^{k+1}}{k}$ converges by means of the next test.

> **Theorem 9.7.1** Alternating Series Test
>
> If $\displaystyle\lim_{n \to \infty} a_n = 0$ and $0 < a_{k+1} \le a_k$ for every positive integer k, then the alternating series $\sum_{k=1}^{\infty}(-1)^{k+1}a_k$ converges.

The condition $0 < a_{k+1} \le a_k$ means that
$$a_1 \ge a_2 \ge a_3 \ge \cdots \ge a_k \ge a_{k+1} \ge \cdots$$

PROOF Consider the partial sums that contain $2n$ terms:

$$S_{2n} = a_1 - a_2 + a_3 - a_4 + \cdots + a_{2n-1} - a_{2n} \qquad (3)$$
$$= (a_1 - a_2) + (a_3 - a_4) + \cdots + (a_{2n-1} - a_{2n}).$$

Since the assumption $0 < a_{k+1} \le a_k$ implies $a_k - a_{k+1} \ge 0$ for $k = 1, 2, 3, \ldots$ we have

$$S_2 \le S_4 \le S_6 \le \cdots \le S_{2n} \le \cdots.$$

Thus, the sequence $\{S_{2n}\}$, whose general term S_{2n} contains an even number of terms of the series, is a monotonic sequence. Rewriting (3) as

$$S_{2n} = a_1 - (a_2 - a_3) - \cdots - a_{2n}$$

shows that $S_{2n} < a_1$ for every positive integer n. Hence, $\{S_{2n}\}$ is bounded. By Theorem 9.2.1 it follows that $\{S_{2n}\}$ converges to a limit S. Now,

$$S_{2n+1} = S_{2n} + a_{2n+1}$$

implies that $\lim\limits_{n\to\infty} S_{2n+1} = \lim\limits_{n\to\infty} S_{2n} + \lim\limits_{n\to\infty} a_{2n+1} = S + 0 = S$. This shows that the sequence of partial sums $\{S_{2n+1}\}$, whose general term S_{2n+1} contains an odd number of terms, also converges to S. Because both $\{S_{2n}\}$ and $\{S_{2n+1}\}$ converge to S, we conclude that $\{S_n\}$ converges to S. ∎

EXAMPLE 2   Alternating Harmonic Series

Show that the alternating harmonic series $\sum\limits_{k=1}^{\infty} \dfrac{(-1)^{k+1}}{k}$ converges.

Solution   With the identification $a_n = 1/n$ we have immediately

$$\lim_{n\to\infty} a_n = \lim_{n\to\infty} \frac{1}{n} = 0.$$

Moreover, since

$$\frac{1}{k+1} \le \frac{1}{k}$$

for $k \ge 1$ we have $0 < a_{k+1} \le a_k$. If follows from Theorem 9.7.1 that the alternating harmonic series converges. ∎

EXAMPLE 3   Divergent Alternating Series

The alternating series $\sum\limits_{k=1}^{\infty} (-1)^{k+1} \dfrac{2k+1}{3k-1}$ diverges, since

$$\lim_{n\to\infty} a_n = \lim_{n\to\infty} \frac{2n+1}{3n-1} = \frac{2}{3}.$$

This last result indicates that

$$\lim_{n\to\infty} (-1)^{n+1} \frac{2n+1}{3n-1}$$

does not exist. Recall from Theorem 9.3.2 that it is necessary that the latter limit be 0 for the convergence of the series. ∎

Although showing that $a_{k+1} \le a_k$ may seem a straightforward task, this is often not the case.

EXAMPLE 4   Using the Alternating Series Test

Test for convergence $\sum\limits_{k=1}^{\infty} (-1)^{k+1} \dfrac{\sqrt{k}}{k+1}$.

Solution   In order to show that the terms of the series satisfy the condition $a_{k+1} \le a_k$, let us consider the function $f(x) = \sqrt{x}/(x+1)$ for which $f(k) = a_k$. From the derivative, we see that

$$f'(x) = -\frac{x-1}{2\sqrt{x}(x+1)^2} < 0 \quad \text{for} \quad x > 1,$$

and hence, the function f decreases for $x > 1$. Thus, $a_{k+1} \le a_k$ is true for $k \ge 1$. Moreover, L'Hôpital's Rule shows that

$$\lim_{x\to\infty} f(x) = 0 \qquad \text{and so} \qquad \lim_{n\to\infty} f(n) = \lim_{n\to\infty} a_n = 0.$$

Hence, the given series converges by the Alternating Series Test. ∎

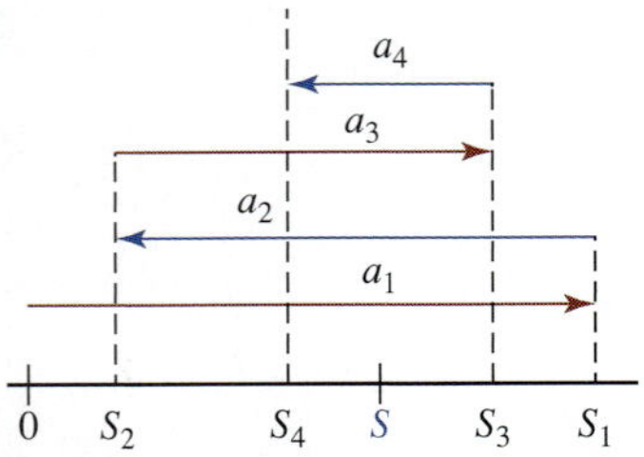

FIGURE 9.7.1 Partial sums on the number line

■ **Approximating the Sum of an Alternating Series** Suppose the alternating series $\sum_{k=1}^{\infty}(-1)^{k+1}a_k$ converges to a number S. The partial sums

$$S_1 = a_1, \quad S_2 = a_1 - a_2, \quad S_3 = a_1 - a_2 + a_3, \quad S_4 = a_1 - a_2 + a_3 - a_4, \ldots$$

can be represented on a number line as shown in FIGURE 9.7.1. The sequence $\{S_n\}$ converges in the manner illustrated in Figure 9.1.1(c); that is, the terms S_n get closer to S as $n \to \infty$ although they oscillate on either side of S. As indicated in Figure 9.7.1, the even-numbered partial sums are less than S and the odd-numbered partial sums are greater than S. Roughly, the even-numbered partial sums increase to the number S, and, in turn, the odd-numbered partial sums decrease to S. Because of this, *the sum S of the series must lie between consecutive partial sums S_n and S_{n+1}*:

$$S_n \leq S \leq S_{n+1}, \quad \text{if } n \text{ is even,} \tag{4}$$

and

$$S_{n+1} \leq S \leq S_n, \quad \text{if } n \text{ is odd.} \tag{5}$$

Now (4) yields $0 \leq S - S_n \leq S_{n+1} - S_n$ for n even, and (5) implies $0 \leq S_n - S \leq S_n - S_{n+1}$ for n odd. Thus, in either case $|S_n - S| \leq |S_{n+1} - S_n|$.

But $S_{n+1} - S_n = a_{n+1}$ for n even and $S_{n+1} - S_n = -a_{n+1}$ for n odd. Thus, $|S_n - S| \leq a_{n+1}$ for all n. We state this result as our next theorem.

Theorem 9.7.2 Error Bound for Alternating Series

Suppose the alternating series $\sum_{k=1}^{\infty}(-1)^{k+1}a_k$, $a_k > 0$, converges to a number S. If S_n is the nth partial sum of the series and $a_{k+1} \leq a_n$ for all k, then

$$|S_n - S| \leq a_{n+1}$$

for all n.

Theorem 9.7.2 is useful in approximating the sum of a convergent alternating series. It states that the **error** $|S_n - S|$ between the nth partial sum and the series is less than the absolute value of the $(n + 1)$st term of the series.

EXAMPLE 5 Approximating the Sum of a Series

Approximate the sum of the convergent series $\sum_{k=1}^{\infty} \frac{(-1)^{k+1}}{(2k)!}$ to four decimal places.

Solution First, we note that $a_n = 1/(2n)!$. Theorem 9.7.2 indicates that we must have

$$a_{n+1} = \frac{1}{(2n + 2)!} < 0.00005$$

in order to approximate the sum of the series to four decimal places. Now from

$$n = 1, \quad a_2 = \frac{1}{4!} \approx 0.041667$$

$$n = 2, \quad a_3 = \frac{1}{6!} \approx 0.001389$$

$$n = 3, \quad a_4 = \frac{1}{8!} \approx 0.000025 < 0.00005$$

we see that $|S_3 - S| \leq a_4 < 0.00005$. Therefore,

$$S_3 = \frac{1}{2!} - \frac{1}{4!} + \frac{1}{6!} \approx 0.4597$$

has the desired accuracy. ■

■ **Absolute and Conditional Convergence**   A series containing mixed signs such as

$$\frac{2}{3} + \left(\frac{2}{3}\right)^2 - \left(\frac{2}{3}\right)^3 - \left(\frac{2}{3}\right)^4 + \left(\frac{2}{3}\right)^5 + \left(\frac{2}{3}\right)^6 - - + + \cdots \tag{6}$$

is not strictly of the form given in (1) and so is not classified as an alternating series. Theorem 9.7.1 is not applicable to such a series. Nonetheless, we will see that the series (6) is convergent *because* the series of absolute values

$$\frac{2}{3} + \left(\frac{2}{3}\right)^2 + \left(\frac{2}{3}\right)^3 + \left(\frac{2}{3}\right)^4 + \left(\frac{2}{3}\right)^5 + \left(\frac{2}{3}\right)^6 + \cdots \tag{7}$$

◀ Peek ahead and read the two sentences immediately following Example 7.

is convergent (a geometric series with $r = \frac{2}{3} < 1$). The series (6) is an example of a series that is **absolutely convergent**.

In the next definition we are letting the symbol $\sum_{k=1}^{\infty} a_k$ represent *any* series—the terms a_k could alternate as in (1) or contain mixed signs—the signs can follow some rule (as in (6)) or not.

Definition 9.7.1   Absolute Convergence

A series $\sum_{k=1}^{\infty} a_k$ is said to be **absolutely convergent** if the series of absolute values $\sum_{k=1}^{\infty} |a_k|$ converges.

EXAMPLE 6   Absolute Convergence

The alternating series $\displaystyle\sum_{k=1}^{\infty} \frac{(-1)^{k+1}}{1 + k^2}$ is absolutely convergent, since the series of absolute values

$$\sum_{k=1}^{\infty} \left| \frac{(-1)^{k+1}}{1 + k^2} \right| = \sum_{k=1}^{\infty} \frac{1}{1 + k^2}$$

was shown to be convergent by the Integral Test in Example 1 of Section 9.4.   ■

Definition 9.7.2   Conditional Convergence

A series $\sum_{k=1}^{\infty} a_k$ is said to be **conditionally convergent** if $\sum_{k=1}^{\infty} a_k$ converges but the series of absolute values $\sum_{k=1}^{\infty} |a_k|$ diverges.

EXAMPLE 7   Conditional Convergence

In Example 2 we saw that the alternating harmonic series $\displaystyle\sum_{k=1}^{\infty} \frac{(-1)^{k+1}}{k}$ is convergent. But taking the absolute value of each term gives the divergent harmonic series $\displaystyle\sum_{k=1}^{\infty} \frac{1}{k}$. Thus, $\displaystyle\sum_{k=1}^{\infty} \frac{(-1)^{k+1}}{k}$ is conditionally convergent.   ■

The next result shows that every absolutely convergent series is also convergent. It is for this reason that the series in (6) converges.

Theorem 9.7.3   Absolute Convergence Implies Convergence

If $\sum_{k=1}^{\infty} |a_k|$ converges, then $\sum_{k=1}^{\infty} a_k$ converges.

PROOF If we define $c_k = a_k + |a_k|$, then $c_k \leq 2|a_k|$. Since $\sum |a_k|$ converges, it follows from the Comparison Test that $\sum c_k$ converges. Furthermore, $\sum(c_k - |a_k|)$ converges, since both $\sum c_k$ and $\sum |a_k|$ converge. But

$$\sum_{k=1}^{\infty} a_k = \sum_{k=1}^{\infty} (c_k - |a_k|).$$

Therefore, $\sum a_k$ converges. ∎

Note that $\sum |a_k|$ is a series of positive terms, and so the tests of the preceding section can be utilized to determine whether a series converges absolutely.

EXAMPLE 8   Absolute Convergence Implies Convergence

The series

$$\sum_{k=1}^{\infty} \frac{\sin k}{k^2} = \frac{\sin 1}{1} + \frac{\sin 2}{4} + \frac{\sin 3}{9} + \frac{\sin 4}{16} + \cdots$$

contains positive and negative terms since

$$\sin 1 > 0, \quad \sin 2 > 0, \quad \sin 3 > 0, \quad \sin 4 < 0, \quad \sin 5 < 0, \quad \sin 6 < 0,$$

and so on. From trigonometry we know that $|\sin k| \leq 1$ for all k. Therefore,

$$\left| \frac{\sin k}{k^2} \right| \leq \frac{1}{k^2}$$

for all k. By the Direct Comparison Test, Theorem 9.5.1, the series $\sum_{k=1}^{\infty} \left| \frac{\sin k}{k^2} \right|$ converges because it is dominated by the convergent p-series $\sum_{k=1}^{\infty} \frac{1}{k^2}$. Hence, $\sum_{k=1}^{\infty} \frac{\sin k}{k^2}$ is absolutely convergent, and so by Theorem 9.7.3 it converges. ∎

▮ Ratio and Root Tests   The following modified forms of the Ratio Test and the Root Test can be applied directly to an alternating series.

Theorem 9.7.4   Ratio Test

Suppose $\sum_{k=1}^{\infty} a_k$ is a series of nonzero terms such that:

$$\lim_{n \to \infty} \left| \frac{a_{n+1}}{a_n} \right| = L.$$

  (*i*) If $L < 1$, the series is absolutely convergent.
  (*ii*) If $L > 1$, or if $L = \infty$, the series is divergent.
  (*iii*) If $L = 1$, the test is inconclusive.

EXAMPLE 9   Using the Ratio Test

Test for convergence $\displaystyle\sum_{k=1}^{\infty} \frac{(-1)^{k+1} 2^{2k-1}}{k 3^k}$.

Solution   With $a_n = (-1)^{n+1} 2^{2n-1}/(n 3^n)$, we see that

$$\lim_{n \to \infty} \left| \frac{a_{n+1}}{a_n} \right| = \lim_{n \to \infty} \left| \frac{(-1)^{n+2} 2^{2n+1}}{(n+1) 3^{n+1}} \cdot \frac{n 3^n}{(-1)^{n+1} 2^{2n-1}} \right|$$

$$= \lim_{n \to \infty} \frac{4n}{3(n+1)} = \frac{4}{3}.$$

Since $L = \frac{4}{3} > 1$, we see from Theorem 9.7.4(*ii*) that the alternating series diverges. ▮

Theorem 9.7.5   Root Test

Suppose $\sum_{k=1}^{\infty} a_k$ is a series such that:

$$\lim_{n\to\infty} \sqrt[n]{|a_n|} = \lim_{n\to\infty} |a_n|^{1/n} = L.$$

(*i*) If $L < 1$, the series is absolutely convergent.
(*ii*) If $L > 1$, or if $L = \infty$, the series is divergent.
(*iii*) If $L = 1$, the test is inconclusive.

■ **Rearrangement of Terms**   When working with a *finite* series of terms such as

$$a_1 - a_2 + a_3 - a_4 + a_5 - a_6, \tag{8}$$

any rearrangement of the order of terms, such as

$$-a_2 + a_1 - a_4 + a_3 - a_6 + a_5$$

or

$$(a_1 - a_2) + (a_3 - a_4) + (a_5 - a_6)$$

has the same sum as the original (8). This kind of carefree manipulation of terms does not carry over to *infinite* series:

- *If the terms of a conditionally convergent series are written in a different order, the new series may diverge or converge to an entirely different number.*

Indeed, it can be proved that by a suitable rearrangement of its terms, a conditionally convergent series can be made to converge to any prescribed real number *r*.

In contrast, a rearrangement of the terms of an absolutely convergent series does not effects its sum:

- *If a series $\sum a_k$ is absolutely convergent, then the terms of the series can be rearranged in any manner and the resulting series will converge to the same number as the original series.*

For example, the geometric series $1 - \frac{1}{3} + \frac{1}{9} - \frac{1}{27} + \cdots$ is absolutely convergent and its sum is $\frac{3}{4}$. The rearrangement $-\frac{1}{3} + \frac{1}{1} - \frac{1}{27} + \frac{1}{9} - \cdots$ of the geometric series is *not* a geometric series, nevertheless the rearranged series converges and its sum is $\frac{3}{4}$. See Problems 53–56 in Exercises 9.7.

$\sum$ **NOTES FROM THE CLASSROOM**

(*i*) The conclusion of Theorem 9.7.1 remains true when the hypothesis "$a_{k+1} \leq a_k$ for every positive k" is replaced with the statement "$a_{k+1} \leq a_k$ for k sufficiently large." For the alternating series $\sum_{k=1}^{\infty} (-1)^{k+1}(\ln k)/k^{1/3}$, it is readily shown by the procedure used in Example 4 that $a_{k+1} \leq a_k$ for $k \geq 21$. Moreover, $\lim_{n\to\infty} a_n = 0$. Hence, the series converges by the Alternating Series Test.

(*ii*) If the series of absolute values $\sum |a_k|$ is found to be divergent, then no conclusion can be drawn concerning the convergence or divergence of the series $\sum a_k$.

Exercises 9.7   Answers to selected odd-numbered problems begin on page ANS-27.

≡ **Fundamentals**

In Problems 1–14, use the Alternating Series Test to determine whether the given series converges.

1. $\displaystyle\sum_{k=1}^{\infty} \frac{(-1)^{k+1}}{k+2}$

2. $\displaystyle\sum_{k=1}^{\infty} \frac{(-1)^{k-1}}{\sqrt{k}}$

3. $\displaystyle\sum_{k=1}^{\infty} (-1)^{k-1}\frac{k}{k+1}$

4. $\displaystyle\sum_{k=1}^{\infty} (-1)^{k}\frac{k}{k^2+1}$

5. $\displaystyle\sum_{k=1}^{\infty} (-1)^{k+1}\frac{k^2+2}{k^3}$

6. $\displaystyle\sum_{k=1}^{\infty} (-1)^{k+1}\frac{3k-1}{k+5}$

7. $\displaystyle\sum_{k=1}^{\infty}(-1)^{k+1}\left(\frac{1}{k}+\frac{1}{3^k}\right)$

8. $\displaystyle\sum_{k=1}^{\infty}(-1)^{k+1}\frac{k+1}{4^k}$

9. $\displaystyle\sum_{n=1}^{\infty}(-1)^{n+1}\frac{4\sqrt{n}}{2n+1}$

10. $\displaystyle\sum_{n=1}^{\infty}(-1)^{n-1}\frac{\sqrt[3]{n}}{n+1}$

11. $\displaystyle\sum_{n=2}^{\infty}(\cos n\pi)\frac{\sqrt{n+1}}{n+2}$

12. $\displaystyle\sum_{k=2}^{\infty}(-1)^{k}\frac{\sqrt{k^2+1}}{k^3}$

13. $\displaystyle\sum_{k=2}^{\infty}(-1)^{k}\frac{k}{\ln k}$

14. $\displaystyle\sum_{k=2}^{\infty}\frac{(-1)^{k}}{\ln k}$

In Problems 15–34, determine whether the given series is absolutely convergent, conditionally convergent, or divergent.

15. $\displaystyle\sum_{k=1}^{\infty}\frac{(-1)^{k+1}}{2k+1}$

16. $\displaystyle\sum_{k=1}^{\infty}\frac{(-1)^{k-1}}{\sqrt{k+5}}$

17. $\displaystyle\sum_{k=1}^{\infty}(-1)^{k+1}\left(\frac{2}{3}\right)^k$

18. $\displaystyle\sum_{k=1}^{\infty}(-1)^{k+1}\frac{2^{2k}}{3^k}$

19. $\displaystyle\sum_{k=1}^{\infty}(-1)^{k}\frac{k}{5^k}$

20. $\displaystyle\sum_{k=1}^{\infty}(-1)^{k}(k2^{-k})^2$

21. $\displaystyle\sum_{k=1}^{\infty}\frac{(-1)^{k}}{k!}$

22. $\displaystyle\sum_{k=1}^{\infty}(-1)^{k}\frac{(k!)^2}{(2k)!}$

23. $\displaystyle\sum_{k=1}^{\infty}(-1)^{k+1}\frac{k!}{100^k}$

24. $\displaystyle\sum_{k=1}^{\infty}(-1)^{k-1}\frac{5^{2k-3}}{10^{k+2}}$

25. $\displaystyle\sum_{k=1}^{\infty}(-1)^{k-1}\frac{k}{1+k^2}$

26. $\displaystyle\sum_{k=1}^{\infty}(-1)^{k+1}\frac{k}{1+k^4}$

27. $\displaystyle\sum_{k=1}^{\infty}\cos k\pi$

28. $\displaystyle\sum_{k=1}^{\infty}\frac{\sin\left(\frac{2k+1}{2}\pi\right)}{\sqrt{k+1}}$

29. $\displaystyle\sum_{k=1}^{\infty}(-1)^{k-1}\sin\left(\frac{1}{k}\right)$

30. $\displaystyle\sum_{k=1}^{\infty}\frac{(-1)^{k-1}}{k^2}\sin\left(\frac{1}{k}\right)$

31. $\displaystyle\sum_{k=1}^{\infty}(-1)^{k}\left[\frac{1}{k+1}-\frac{1}{k}\right]$

32. $\displaystyle\sum_{k=1}^{\infty}(-1)^{k}\left[\sqrt{k+1}-\sqrt{k}\right]$

33. $\displaystyle\sum_{k=1}^{\infty}(-1)^{k}\left(\frac{2k}{k+50}\right)^k$

34. $\displaystyle\sum_{k=1}^{\infty}(-1)^{k+1}\frac{6^{3k}}{k^k}$

In Problems 35 and 36, approximate the sum of the convergent series to the indicated number of decimal places.

35. $\displaystyle\sum_{k=1}^{\infty}\frac{(-1)^{k+1}}{(2k-1)!}$; five

36. $\displaystyle\sum_{k=1}^{\infty}\frac{(-1)^{k+1}}{k!}$; three

In Problems 37 and 38, find the smallest positive integer n so that S_n approximates the sum of the convergent series to the indicated number of decimal places.

37. $\displaystyle\sum_{k=1}^{\infty}\frac{(-1)^{k+1}}{k^3}$; two

38. $\displaystyle\sum_{k=1}^{\infty}\frac{(-1)^{k+1}}{\sqrt{k}}$; three

In Problems 39 and 40, approximate the sum of the convergent series so that the error is less than the indicated amount.

39. $1-\dfrac{1}{4^2}+\dfrac{1}{4^3}-\dfrac{1}{4^4}+\cdots$; 10^{-3}

40. $1-\dfrac{2}{5^2}+\dfrac{3}{5^3}-\dfrac{4}{5^4}+\cdots$; 10^{-4}

In Problems 41 and 42, estimate the error in using the indicated partial sum as an approximation to the sum of the convergent series.

41. $\displaystyle\sum_{k=1}^{\infty}\frac{(-1)^{k+1}}{k}$; S_{100}

42. $\displaystyle\sum_{k=1}^{\infty}\frac{(-1)^{k+1}}{k2^k}$; S_6

In Problems 43–48, state why the Alternating Series Test is not applicable to the given series. Determine whether the series converges.

43. $\displaystyle\sum_{k=1}^{\infty}\frac{\sin(k\pi/6)}{\sqrt{k^4+1}}$

44. $\displaystyle\sum_{k=1}^{\infty}\frac{100+(-1)^{k}2^{k}}{3^k}$

45. $1-\dfrac{1}{2}-\dfrac{1}{4}+\dfrac{1}{8}+\dfrac{1}{16}--++\cdots$

46. $\dfrac{1}{1}-\dfrac{1}{4}-\dfrac{1}{9}+\dfrac{1}{16}+\dfrac{1}{25}+\dfrac{1}{36}---+++\cdots$

47. $\dfrac{2}{1}-\dfrac{1}{1}+\dfrac{2}{2}-\dfrac{1}{2}+\dfrac{2}{3}-\dfrac{1}{3}+\dfrac{2}{4}-\dfrac{1}{4}+\cdots$

[*Hint*: Consider the partial sums S_{2n} for $n=1,2,3,\ldots$.]

48. $\dfrac{1}{2}+\dfrac{1}{2}-\dfrac{1}{3}-\dfrac{1}{3}-\dfrac{1}{3}+\dfrac{1}{4}+\dfrac{1}{4}+\dfrac{1}{4}+\dfrac{1}{4}-----\cdots$

In Problems 49–52, determine whether the given series converges.

49. $1-1+1-1+\cdots$

50. $(1-1)+(1-1)+(1-1)+\cdots$

51. $1+(-1+1)+(-1+1)+\cdots$

52. $1+(-1+1)+(-1+1-1)+\cdots$

≡ **Think About It**

53. Reread the discussion just prior to *Notes from the Classroom* in this section. Then explain why the following statement is true:

If a positive-term series $\sum a_k$ is convergent, then the terms of the series can be rearranged in any manner and the resulting series will converge to the same number as the original series.

54. Suppose S is the sum of the convergent alternating harmonic series $1-\frac{1}{2}+\frac{1}{3}-\frac{1}{4}+\frac{1}{5}-\frac{1}{6}+\cdots$.

Show that the rearrangement of the series

$$1-\dfrac{1}{2}-\dfrac{1}{4}+\dfrac{1}{3}-\dfrac{1}{6}-\dfrac{1}{8}+\dfrac{1}{5}-\dfrac{1}{10}-\dfrac{1}{12}+\dfrac{1}{7}-\dfrac{1}{14}\cdots$$

$$=\left(1-\dfrac{1}{2}\right)-\dfrac{1}{4}+\left(\dfrac{1}{3}-\dfrac{1}{6}\right)-\dfrac{1}{8}+\left(\dfrac{1}{5}-\dfrac{1}{10}\right)-\dfrac{1}{12}$$

$$+\left(\dfrac{1}{7}-\dfrac{1}{14}\right)-\cdots,$$

gives $\frac{1}{2}S=\frac{1}{2}-\frac{1}{4}+\frac{1}{6}-\frac{1}{8}+\cdots$.

55. Use $S=1-\frac{1}{2}+\frac{1}{3}-\frac{1}{4}+\frac{1}{5}-\frac{1}{6}+\cdots$ and the result of Problem 54 in the form

$$\dfrac{1}{2}S=0+\dfrac{1}{2}+0-\dfrac{1}{4}+0+\dfrac{1}{6}+0-\dfrac{1}{8}+\cdots$$

to show that the sum of another rearrangement of the terms of the alternating harmonic series is

$$\frac{3}{2}S = 1 + \frac{1}{3} - \frac{1}{2} + \frac{1}{5} + \frac{1}{7} - \frac{1}{4} + \cdots.$$

56. The series $1 - \frac{1}{3} + \frac{1}{9} - \frac{1}{27} + \cdots$ is an absolutely convergent geometric series. Show that its rearrangement $-\frac{1}{3} + \frac{1}{1} - \frac{1}{27} + \frac{1}{9} - \cdots$ is convergent. Try the Ratio Test and Root Test. [*Hint*: Examine $3^{k+(-1)^k}$, $k = 0, 1, 2, \ldots$.]

57. If $\sum a_k$ is absolutely convergent, prove that $\sum a_k^2$ converges. [*Hint*: For n sufficiently large, $|a_n| < 1$. Why?]

58. Give an example of a convergent series $\sum a_k$ for which $\sum a_k^2$ diverges.

59. Give an example of a convergent series $\sum a_k$ for which $\sum a_k^2$ converges.

60. Give an example of a divergent series $\sum a_k$ for which $\sum a_k^2$ converges.

61. Explain why the series

$$e^{-x}\sin x + e^{-2x}\sin 2x + e^{-3x}\sin 3x + \cdots$$

converges for every positive value of x.

9.8 Power Series

■ **Introduction** In applied mathematics it is common to work with infinite series of functions,

$$\sum_{k=0}^{\infty} c_k u_k(x) = c_0 u_0(x) + c_1 u_1(x) + c_2 u_2(x) + \cdots. \tag{1}$$

The coefficients c_k are constants depending on k and the functions $u_k(x)$ could be various kinds of polynomials or even the sine and cosine functions. When the variable x is specified, say, $x = 1$, then the series reduces to a series of constants. The convergence of a series such as (1) will, of course, depend on the variable x, the series usually converging for some values of x while diverging for other values. In this and the next section we consider infinite series (1) where the functions $u_k(x)$ are the polynomials $(x - a)^k$. We will study the properties of such series and show how to determine the values of x for which the series converge.

■ **Power Series** A series containing nonnegative integral powers of $(x - a)^k$,

$$\sum_{k=0}^{\infty} c_k(x - a)^k = c_0 + c_1(x - a) + c_2(x - a)^2 + \cdots + c_n(x - a)^n + \cdots, \tag{2}$$

is called a **power series in $x - a$**. The power series (2) is said to be **centered at a** or have **center a**. An important special case of (2), when $a = 0$,

$$\sum_{k=0}^{\infty} c_k x^k = c_0 + c_1 x + c_2 x^2 + \cdots + c_n x^n + \cdots, \tag{3}$$

is called a **power series in x**. The power series in (3) is centered at 0. A problem we face in this section is:

- *Find the values of x for which a power series converges.*

Observe that (2) and (3) converge to c_0 when $x = a$ and $x = 0$, respectively.

◀ It is conventient to define $(x - a)^0 = 1$ and $x^0 = 1$ even when $x = a$ and $x = 0$, respectively.

EXAMPLE 1 Power Series Centered at 0

The power series in x where the coefficients $c_k = 1$ for all k,

$$\sum_{k=0}^{\infty} x^k = 1 + x + x^2 + \cdots + x^n + \cdots,$$

is recognized as a geometric series with the common ratio $r = x$. By Theorem 9.3.1, the series converges for those values of x that satisfy $|x| < 1$ or $-1 < x < 1$. The series diverges for $|x| \geq 1$, that is, for $x \leq -1$ or $x \geq 1$. ∎

In general, the Ratio Test, as stated in Theorem 9.7.4, is especially helpful in finding the values of x for which a power series converges. The Root Test, in the form of Theorem 9.7.5, is also useful but to a lesser extent.

EXAMPLE 2 Interval of Convergence

Find the interval of convergence for $\displaystyle\sum_{k=0}^{\infty} \frac{x^k}{2^k(k+1)^2}$.

Solution With the identification that $a_n = x^n/(2^n(n+1)^2)$ we use the Ratio Test, Theorem 9.7.4,

$$\lim_{n\to\infty}\left|\frac{a_{n+1}}{a_n}\right| = \lim_{n\to\infty}\left|\frac{x^{n+1}}{2^{n+1}(n+2)^2}\cdot\frac{2^n(n+1)^2}{x^n}\right|$$

$$= \lim_{n\to\infty}\left(\frac{n+1}{n+2}\right)^2\frac{|x|}{2} \qquad\leftarrow\text{\small divide the numerator and}$$
$$\qquad\qquad\qquad\qquad\qquad\qquad\text{\small denominator of the first term by } n$$

$$= \frac{|x|}{2}\cdot\lim_{n\to\infty}\left(\frac{1+1/n}{1+2/n}\right)^2 = \frac{|x|}{2}.$$

From part (*i*) of Theorem 9.7.4, we have absolute convergence whenever this limit is strictly less than 1. Thus, the series is absolutely convergent for those values of x that satisfy $|x|/2 < 1$ or $|x| < 2$. Since the absolute-value inequality $|x| < 2$ is equivalent to $-2 < x < 2$, we see the given series will converge for any number x in the open interval $(-2, 2)$. However, if $|x|/2 = 1$, or $|x| = 2$, or when $x = 2$ or $x = -2$, then the Ratio Test gives no information. We must perform separate checks of the given series for convergence at these endpoints. Substituting 2 for x the series becomes

$$\sum_{k=1}^{\infty}\frac{1}{(k+1)^2},$$

which is convergent by direct comparison with the convergent p-series $\sum_{k=1}^{\infty}(1/k^2)$. Similarly, substituting -2 for x yields

$$\sum_{k=1}^{\infty}\frac{(-1)^k}{(k+1)^2},$$

which is convergent by the Alternating Series Test, Theorem 9.7.1. We conclude that the given series converges for all x in the closed interval $[-2, 2]$. The series diverges for $x < -2$ and $x > 2$, or equivalently, for $|x| > 2$. ∎

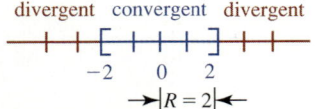

divergent convergent divergent

-2 0 2

$\rightarrow|R=2|\leftarrow$

FIGURE 9.8.1 The set of numbers x for which the series in Example 2 converges is shown in blue

■ **Interval of Convergence** In FIGURE 9.8.1 we have illustrated in blue the set $[-2, 2]$ of all real numbers x for which the series in Example 2 convergences and in red the set $(-\infty, -2) \cup (2, \infty)$ of numbers x for which the series diverges. The set of numbers for which the series converges is an interval centered at 0 (the center of the series). As shown in the figure, the radius of this interval is $R = 2$. In general, the set of *all* real numbers x for which a power series $\sum c_k(x - a)^k$ converges is said to be its **interval of convergence**. The center of the interval of convergence is the center a of the series. The radius R of the interval of convergence is called the **radius of convergence**.

The next theorem, given without proof, summarizes all the possible ways a power series can converge.

Theorem 9.8.1 Convergence of a Power Series

For a power series $\sum_{k=0}^{\infty} c_k(x - a)^k$ exactly one of the following is true:

(*i*) The series converges only at the *single number* $x = a$.

(*ii*) The series converges absolutely for *all real numbers* x.

(*iii*) The series converges absolutely for the *numbers* x *in a finite interval* $(a - R, a + R)$, $R > 0$, and diverges for numbers in the set $(-\infty, a - R) \cup (a + R, \infty)$. At an endpoint of the finite interval, $x = a - R$ or $x = a + R$, the series may converge absolutely, converge conditionally, or diverge.

Of course in (*ii*) and (*iii*), when the power series converges absolutely at a number x, we know, by Theorem 9.7.3, that it converges. In (*i*) of Theorem 9.8.1 the interval of convergence consists of the singleton set $\{a\}$ and we say that the series has **radius of convergence** $R = 0$. In (*ii*) of Theorem 9.8.1 the interval of convergence is $(-\infty, \infty)$ and the series has

radius of convergence $R = \infty$. Finally, in (*iii*) of Theorem 9.8.1, there are four possibilities for the interval of convergence with **radius of convergence $R > 0$:**

$$(a - R, a + R), \quad [a - R, a + R], \quad (a - R, a + R], \quad \text{or} \quad [a - R, a + R).$$

See **FIGURE 9.8.2.**

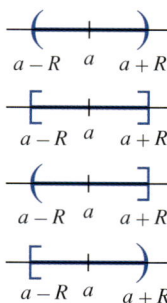

FIGURE 9.8.2   Possible finite intervals of convergence with $R > 0$

As in Example 1, if $R > 0$, we must handle the question of convergence at an endpoint $x = a \pm R$ by substituting these numbers into the given series and then either *recognizing* the resulting series as convergent or divergent or by *testing* the resulting series for convergence by an appropriate test other than the Ratio Test. Remember:

• *The Ratio Test is always inconclusive at an endpoint $x = a \pm R$.*

EXAMPLE 3   Interval of Convergence

Find the interval of convergence for $\displaystyle\sum_{k=0}^{\infty} \frac{x^k}{k!}$.

Solution   By the Ratio Test, Theorem 9.7.4, we have

$$\lim_{n \to \infty} \left| \frac{a_{n+1}}{a_n} \right| = \lim_{n \to \infty} \left| \frac{x^{n+1}}{(n+1)!} \cdot \frac{n!}{x^n} \right| = \lim_{n \to \infty} \frac{n!}{(n+1)!} |x| = \lim_{n \to \infty} \frac{|x|}{n+1}.$$

Since $\lim_{n \to \infty} |x|/(n+1) = 0$ for any choice of x, the series converges absolutely for every real number. Thus, the interval of convergence is $(-\infty, \infty)$ and the radius of convergence is $R = \infty$. ∎

EXAMPLE 4   Interval of Convergence

Find the interval of convergence for $\displaystyle\sum_{k=1}^{\infty} \frac{(x-5)^k}{k3^k}$.

Solution   By the Ratio Test, Theorem 9.7.4, we have

$$\lim_{n \to \infty} \left| \frac{a_{n+1}}{a_n} \right| = \lim_{n \to \infty} \left| \frac{(x-5)^{n+1}}{(n+1)3^{n+1}} \cdot \frac{n3^n}{(x-5)^n} \right|$$

$$= \lim_{n \to \infty} \left(\frac{n}{n+1} \right) \frac{|x-5|}{3}$$

$$= \lim_{n \to \infty} \left(\frac{1}{1+1/n} \right) \frac{|x-5|}{3} = \frac{|x-5|}{3}.$$

The series converges absolutely if $|x-5|/3 < 1$ or $|x-5| < 3$. This absolute-value inequality yields the open interval $(2, 8)$. At $x = 2$ and $x = 8$, the endpoints of the interval, we obtain, in turn,

$$\sum_{k=1}^{\infty} \frac{(-1)^k}{k} \quad \text{and} \quad \sum_{k=1}^{\infty} \frac{1}{k}.$$

The first series is a multiple of the alternating harmonic series and so is convergent, the second series is the divergent harmonic series. Consequently, the interval of convergence is $[2, 8)$. The radius of convergence is $R = 3$. The series diverges if $x < 2$ or $x \geq 8$. See **FIGURE 9.8.3.** ∎

The first series is

$$-1 + \tfrac{1}{2} - \tfrac{1}{3} + \cdots$$

or $\quad (-1)[1 - \tfrac{1}{2} + \tfrac{1}{3} + \cdots]$

◀ The series in the brackets is the convergent alternating harmonic series.

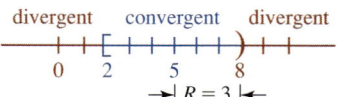

FIGURE 9.8.3   Interval of convergence (blue) in Example 4

EXAMPLE 5   Interval of Convergence

Find the interval of convergence for $\sum_{k=1}^{\infty} k!(x+10)^k$.

Solution   From the Ratio Test,

$$\lim_{n \to \infty} \left| \frac{a_{n+1}}{a_n} \right| = \lim_{n \to \infty} \left| \frac{(n+1)!(x+10)^{n+1}}{n!(x+10)^n} \right|$$

$$= \lim_{n \to \infty} (n+1)|x+10|$$

we see that the limit as $n \to \infty$ can only exist if $|x+10| = 0$, namely, when $x = -10$. Thus,

$$\lim_{n \to \infty} \left| \frac{a_{n+1}}{a_n} \right| = \begin{cases} \infty, & x \neq -10 \\ 0, & x = -10. \end{cases}$$

The series diverges for every real number x, *except $x = -10$.* At $x = -10$, we obtain a convergent series consisting of all zeros. The interval of convergence is the set $\{10\}$ and the radius of convergence is $R = 0$. ∎

Exercises 9.8 Answers to selected odd-numbered problems begin on page ANS-27.

≡ Fundamentals

In Problems 1–24, use the Ratio Test to find the interval and radius of convergence of the given power series.

1. $\displaystyle\sum_{k=1}^{\infty} \frac{(-1)^k}{k} x^k$

2. $\displaystyle\sum_{k=1}^{\infty} \frac{x^k}{k^2}$

3. $\displaystyle\sum_{k=1}^{\infty} \frac{2^k}{k} x^k$

4. $\displaystyle\sum_{k=0}^{\infty} \frac{5^k}{k!} x^k$

5. $\displaystyle\sum_{k=1}^{\infty} \frac{(x-3)^k}{k^3}$

6. $\displaystyle\sum_{k=1}^{\infty} \frac{(x+7)^k}{\sqrt{k}}$

7. $\displaystyle\sum_{k=1}^{\infty} \frac{(-1)^k}{10^k}(x-5)^k$

8. $\displaystyle\sum_{k=1}^{\infty} \frac{k}{(k+2)^2}(x-4)^k$

9. $\displaystyle\sum_{k=0}^{\infty} k! 2^k x^k$

10. $\displaystyle\sum_{k=0}^{\infty} \frac{k-1}{k^{2k}} x^k$

11. $\displaystyle\sum_{k=1}^{\infty} \frac{(3x-1)^k}{k^2+k}$

12. $\displaystyle\sum_{k=0}^{\infty} \frac{(4x-5)^k}{3^k}$

13. $\displaystyle\sum_{k=2}^{\infty} \frac{x^k}{\ln k}$

14. $\displaystyle\sum_{k=2}^{\infty} \frac{(-1)^k x^k}{k \ln k}$

15. $\displaystyle\sum_{k=1}^{\infty} \frac{k^2}{3^{2k}}(x+7)^k$

16. $\displaystyle\sum_{k=1}^{\infty} k^3 2^{4k}(x-1)^k$

17. $\displaystyle\sum_{k=1}^{\infty} \frac{2^{5k}}{5^{2k}} \left(\frac{x}{3}\right)^k$

18. $\displaystyle\sum_{k=1}^{\infty} \frac{1000^k}{k^k} x^k$

19. $\displaystyle\sum_{k=0}^{\infty} \frac{(-3)^k}{(k+1)(k+2)}(x-1)^k$

20. $\displaystyle\sum_{k=1}^{\infty} \frac{3^k}{(-2)^k k(k+1)}(x+5)^k$

21. $\displaystyle\sum_{k=1}^{\infty} \frac{(-1)^{k+1}}{(k!)^2} \left(\frac{x-2}{3}\right)^k$

22. $\displaystyle\sum_{k=0}^{\infty} \frac{(6-x)^{k+1}}{\sqrt{2k+1}}$

23. $\displaystyle\sum_{k=0}^{\infty} \frac{(-1)^k}{9^k} x^{2k+1}$

24. $\displaystyle\sum_{k=1}^{\infty} \frac{5^k}{(2k)!} x^{2k}$

In Problems 25-28, use the Root Test to find the interval and radius of convergence of the given power series.

25. $\displaystyle\sum_{k=2}^{\infty} \frac{x^k}{(\ln k)^k}$

26. $\displaystyle\sum_{k=1}^{\infty} (k+1)^k (x+1)^k$

27. $\displaystyle\sum_{k=1}^{\infty} \left(\frac{4}{3}\right)^k (x+3)^k$

28. $\displaystyle\sum_{k=1}^{\infty} \left(\frac{k}{k+1}\right)^{k^2} (x-e)^k$

In Problems 29 and 30, find the radius of convergence of the given power series.

29. $\displaystyle\sum_{k=1}^{\infty} \frac{k!}{1 \cdot 3 \cdot 5 \cdots (2k-1)} \left(\frac{x}{2}\right)^k$

30. $\displaystyle\sum_{k=2}^{\infty} \frac{1 \cdot 3 \cdot 5 \cdots (2k-3)}{3^k k!}(x-1)^k$

In Problems 31-38, the given series is not a power series. Nonetheless, find all values of x for which the given series converges.

31. $\displaystyle\sum_{k=1}^{\infty} \frac{1}{x^k}$

32. $\displaystyle\sum_{k=1}^{\infty} \frac{7^k}{x^{2k}}$

33. $\displaystyle\sum_{k=1}^{\infty} \left(\frac{x+1}{x}\right)^k$

34. $\displaystyle\sum_{k=1}^{\infty} \frac{1}{2^k} \left(\frac{x}{x+2}\right)^k$

35. $\displaystyle\sum_{k=0}^{\infty} \left(\frac{x^2+2}{6}\right)^{k^2}$

36. $\displaystyle\sum_{k=1}^{\infty} \frac{k!}{(kx)^k}$

37. $\displaystyle\sum_{k=0}^{\infty} e^{kx}$

38. $\displaystyle\sum_{k=0}^{\infty} k! e^{-kx^2}$

39. Find all values of x in $[0, 2\pi]$ for which $\displaystyle\sum_{k=1}^{\infty} \left(\frac{2}{\sqrt{3}}\right)^k \sin^k x$ converges.

40. Show that $\sum_{k=1}^{\infty} (\sin kx)/k^2$ converges for all real values of x.

≡ Calculator/CAS Problems

41. In Problems 71 and 72 of Exercises 5.5 we pointed out that some important functions in applied mathematics are defined in terms of nonelementary integrals. Some of these special functions of applied mathematics are also defined by infinite series. The power series

$$J_0(x) = \sum_{k=0}^{\infty} \frac{(-1)^k}{2^{2k}(k!)^2} x^{2k}$$

is called the **Bessel function of order 0**.

(a) The domain of the function $J_0(x)$ is its interval of convergence. Find the domain.

(b) The value of $J_0(x)$ is defined to be the sum of the series for x in its domain:

$$J_0(x) = \lim_{n \to \infty} S_n(x),$$

where $$S_n(x) = \sum_{k=0}^{n} \frac{(-1)^k}{2^{2k}(k!)^2} x^{2k}$$

is the general term of the sequence of partial sums. Use a calculator or CAS and graph the partial sums $S_0(x)$, $S_1(x)$, $S_2(x)$, $S_3(x)$, and $S_4(x)$.

(c) There are various kinds of Bessel functions of differing orders; $J_0(x)$ is a special case of a more general function $J_\nu(x)$ called the **Bessel function of the first kind of order ν.** Bessel functions are built-in functions in computer algebra systems such as *Mathematica* and *Maple*. Use a CAS to obtain the graph of $J_0(x)$ and compare it with the graphs of the partial sums in part (b). [*Hint*: In *Mathematica* $J_0(x)$ is denoted by BesselJ[0, x].]

9.9 Representing Functions by Power Series

Introduction   For each x in its interval of convergence a power series $\sum c_k(x - a)^k$ converges to a single number. For this reason, a power series is itself a function, which we denote as f, whose *domain* is its interval of convergence. Then for each x in the interval of convergence we define the corresponding element in the *range* of the function, the value $f(x)$, as the sum of the series:

$$f(x) = c_0 + c_1(x - a) + c_2(x - a)^2 + \cdots = \sum_{k=0}^{\infty} c_k(x - a)^k.$$

The next two theorems, which will be stated without proof, answer some of the fundamental questions about differentiability, integrability, and continuity of a function f defined by a power series.

Differentiation of a Power Series   The function f defined by a power series $\sum c_k(x - a)^k$ is differentiable.

Theorem 9.9.1   Differentiation of a Power Series

If $f(x) = \sum_{k=0}^{\infty} c_k(x - a)^k$ converges on an interval $(a - R, a + R)$ for which the radius of convergence R is either positive or ∞, then f is differentiable at each x in $(a - R, a + R)$, and

$$f'(x) = \sum_{k=1}^{\infty} k c_k(x - a)^{k-1}. \tag{1}$$

The radius of convergence R of (1) is the same as that of the original series.

The result in (1) simply states that a power series can be differentiated *term-by-term* as we would for a polynomial function:

$$f'(x) = \frac{d}{dx}c_0 + \frac{d}{dx}c_1(x - a) + \frac{d}{dx}c_2(x - a)^2 + \cdots + \frac{d}{dx}c_n(x - a)^n + \cdots$$

$$= c_1 + 2c_2(x - a) + 3c_3(x - a)^2 + \cdots + nc_n(x - a)^{n-1} + \cdots = \sum_{k=1}^{\infty} k c_k(x - a)^{k-1}. \tag{2}$$

Since (1) is a power series with radius of convergence R, we can apply Theorem 9.9.1 to f' defined in (2). That is, we can say f' is differentiable at each x in $(a - R, a + R)$ and f'' is given by

$$f''(x) = 2c_2 + 3 \cdot 2c_3(x - a) + \cdots + n(n - 1)c_n(x - a)^{n-2} + \cdots = \sum_{k=2}^{\infty} k(k - 1)c_k(x - a)^{k-2}.$$

Continuing in this manner, it follows that:

- *A function f defined by a power series on $(a - R, a + R)$, $R > 0$, or on $(-\infty, \infty)$ possesses derivatives of all orders in the interval.*

The radius of convergence R of each differentiated series is the same as that of the original series. Moreover, since differentiability implies continuity we also have the result:

- *A function f defined by a power series on $(a - R, a + R)$, $R > 0$, or on $(-\infty, \infty)$, is continuous at each x in the interval.*

Integration of a Power Series   As in (1), the process of integration of a power series can be carried out term-by-term:

$$\int f(x)\,dx = \int c_0(x - a)^0\,dx + \int c_1(x - a)\,dx + \int c_2(x - a)^2\,dx + \cdots + \int c_n(x - a)^n\,dx + \cdots$$

$$= c_0(x - a) + \frac{c_1}{2}(x - a)^2 + \frac{c_2}{3}(x - a)^3 + \cdots + \frac{c_n}{n + 1}(x - a)^{n+1} + \cdots + C$$

$$= \sum_{k=0}^{\infty} \frac{c_k}{k + 1}(x - a)^{k+1} + C.$$

This result is summarized in the next theorem.

> **Theorem 9.9.2**   Integration of a Power Series
>
> If $f(x) = \sum_{k=0}^{\infty} c_k(x - a)^k$ converges on an interval $(a - R, a + R)$ for which the radius of convergence R is either positive or ∞, then
>
> $$\int f(x)\, dx = \sum_{k=0}^{\infty} \frac{c_k}{k + 1}(x - a)^{k+1} + C. \tag{3}$$
>
> The radius of convergence R of (3) is the same as that of the original series.

Since the function $f(x) = \sum_{k=0}^{\infty} c_k(x - a)^k$ is continuous, its definite integral exists and is defined by

$$\int_{\alpha}^{\beta} f(x)\, dx = \sum_{k=0}^{\infty} c_k \left(\int_{\alpha}^{\beta} (x - a)^k\, dx \right)$$

for any numbers α and β in $(a - R, a + R)$, $R > 0$, or in $(-\infty, \infty)$ if $R = \infty$.

It is recommended that you read this ▶ In Theorems 9.9.1 and 9.9.2 it was stated that if the function $f(x) = \sum_{k=0}^{\infty} c_k(x - a)^k$ has
paragraph several times.   radius of convergence $R > 0$ or $R = \infty$, then the series obtained by forming $f'(x)$ and $\int f(x)\, dx$ have the same radius of convergence R. This does *not* mean that the power series defining $f(x)$, $f'(x)$, and $\int f(x)\, dx$ have the same intervals of convergence. This is not as bad as it sounds. If the radius of convergence of the series defining $f(x)$, $f'(x)$, and $\int f(x)\, dx$ is $R > 0$, then the intervals of convergence can differ only at the endpoints of the interval. As a rule, by differentiating a function defined by a power series with radius of convergence $R > 0$ we *may lose* convergence at an endpoint of the interval. By integrating a function defined by a power series with radius of convergence $R > 0$ we *may gain* convergence at an endpoint of the interval.

EXAMPLE 1   Interval of Convergence

For the function f defined by $f(x) = \sum_{k=1}^{\infty} \frac{x^k}{k}$, find the intervals of convergence of

(a) $f'(x)$ **(b)** $\int f(x)\, dx$.

Solution   It is readily shown from the Ratio Test that the interval of convergence of the power series that defines f is $[-1, 1)$.

(a) The derivative

$$f'(x) = \sum_{k=1}^{\infty} \frac{d}{dx} \frac{x^k}{k} = \sum_{k=1}^{\infty} x^{k-1} = 1 + x + x^2 + x^3 + \cdots \tag{4}$$

is recognized as a geometric series whose interval of convergence is $(-1, 1)$. The differentiated series (4) has lost convergence at the left endpoint of the interval of convergence for f.

(b) The integral of f is

$$\int f(x)\, dx = \sum_{k=1}^{\infty} \int \frac{x^k}{k}\, dx = \sum_{k=1}^{\infty} \frac{x^{k+1}}{k(k + 1)} + C. \tag{5}$$

At $x = -1$ and $x = 1$, the series in (5) become, respectively,

$$\sum_{k=1}^{\infty} \frac{(-1)^{k+1}}{k(k + 1)} \quad \text{and} \quad \sum_{k=1}^{\infty} \frac{1}{k(k + 1)}.$$

The first series converges by the ▶
Alternating Series Test; the second   Because both series converge, the interval of convergence of (5) is $[-1, 1]$. In this
converges by the Direct Comparison   instance, the integrated series (5) has gained convergence at the right endpoint of
Test (the series is dominated by the   the interval of convergence for f. ■
convergent p-series $\sum 1/k^2$.)

■ **Power Series Representation of a Function**   It is often possible to express a *known* or *given* function f (such as e^x or $\tan^{-1} x$) as the sum of a power series on some interval. In this case we then say that the series is a **power series representation of** f on the interval.

The next example is important because it leads to many other results.

EXAMPLE 2 Representing a Function by a Power Series

Find a power series representation of $\dfrac{1}{1-x}$ centered at 0.

Solution Recall that a geometric series converges to $a/(1-r)$ if $|r| < 1$:

$$\frac{a}{1-r} = a + ar + ar^2 + \cdots + ar^{n-1} + \cdots.$$

Identifying $a = 1$ and $r = x$, we see that

$$\frac{1}{1-x} = 1 + x + x^2 + x^3 + \cdots + x^n + \cdots = \sum_{k=0}^{\infty} x^k. \tag{6}$$

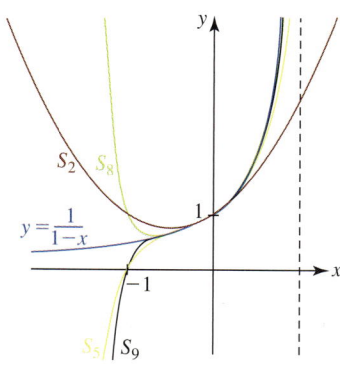

The series converges for $|x| < 1$. The interval of convergence is $(-1, 1)$. In **FIGURE 9.9.1** we have displayed the graph of $y = 1/(1-x)$ in blue along with the graphs of the partial sums $S_2(x)$, $S_5(x)$, $S_8(x)$, and $S_9(x)$ of the power series (6). When inspecting this figure, pay attention only to the interval $(-1, 1)$; the series does not represent the function outside this interval.

FIGURE 9.9.1 Graphs of partial sums in Example 2

By replacing x by $-x$ in (6), we obtain a power series representation for the function $1/(1 + x)$:

$$\frac{1}{1+x} = 1 - x + x^2 - x^3 + \cdots + (-1)^n x^n + \cdots = \sum_{k=0}^{\infty} (-1)^k x^k. \tag{7}$$

The series (7) converges for $|-x| < 1$ or $|x| < 1$. The interval of convergence is again $(-1, 1)$.

Many known functions can be represented by an infinite series by some sort of manipulation of the series in (6) and (7). For example, we could multiply the series by a power of x, we could replace x with another variable, or perhaps we could combine replacement of x by another variable with the process of integration (or differentiation), and so on.

EXAMPLE 3 Representing a Function by a Power Series

Find a power series representation of $\dfrac{1}{1 + 3x}$ centered at 0.

Solution By simply replacing the symbol x by $3x$ in (7) we get

$$\frac{1}{1+3x} = 1 - 3x + (3x)^2 - (3x)^3 + \cdots + (-1)^n (3x)^n + \cdots = \sum_{k=0}^{\infty} (-1)^k 3^k x^k.$$

This series converges when $|-3x| < 1$ or $|x| < \frac{1}{3}$. The interval of convergence is $\left(-\frac{1}{3}, \frac{1}{3}\right)$. ■

EXAMPLE 4 Representing a Function by a Power Series

Find a power series representation of $\dfrac{1}{5-x}$ centered at 0.

Solution By factoring 5 from the denominator,

$$\frac{1}{5-x} = \frac{1}{5\left(1 - \dfrac{x}{5}\right)} = \frac{1}{5} \cdot \frac{1}{1 - \dfrac{x}{5}},$$

we are in a position to use (6). Replacing the symbol x in (6) with $x/5$ we get

$$\frac{1}{5-x} = \frac{1}{5} \cdot \frac{1}{1 - \dfrac{x}{5}} = \frac{1}{5}\left[1 + \frac{x}{5} + \left(\frac{x}{5}\right)^2 + \left(\frac{x}{5}\right)^5 + \cdots\right]$$

or

$$\frac{1}{5-x} = \frac{1}{5}\sum_{k=0}^{\infty}\left(\frac{x}{5}\right)^k = \sum_{k=0}^{\infty} \frac{1}{5^{k+1}} x^k.$$

This series converges for $|x/5| < 1$ or $|x| < 5$. The interval of convergence is $(-5, 5)$. ■

With a little cleverness, the power series representations in (6) and (7) can often be used to find a power series representation of a function centered at a number a other than 0.

EXAMPLE 5 Power Series Centered at 3

Find a power series representation of $\dfrac{1}{1 + x}$ centered at 3.

Solution Since the center of the power is to be 3, we want the power series to contain only powers of $x - 3$. To that end, we subtract and add 3 in the denominator:

$$\frac{1}{1 + x} = \frac{1}{1 + x - 3 + 3} = \frac{1}{4 + (x - 3)}.$$

From this point on, we proceed as in Example 4, namely, we factor 4 from the denominator and use (7) with x replaced by $(x - 3)/4$:

$$\frac{1}{1 + x} = \frac{1}{4 + (x - 3)}$$

$$= \frac{1}{4} \cdot \frac{1}{1 + \dfrac{x - 3}{4}}$$

$$= \frac{1}{4}\left[1 - \frac{x - 3}{4} + \left(\frac{x - 3}{4}\right)^2 - \left(\frac{x - 3}{4}\right)^3 + \cdots\right]$$

or $\qquad \dfrac{1}{1 + x} = \dfrac{1}{4}\displaystyle\sum_{k=0}^{\infty}(-1)^k\left(\dfrac{x - 3}{4}\right)^k = \sum_{k=0}^{\infty}\dfrac{(-1)^k}{4^{k+1}}(x - 3)^k.$

This series converges for $|(x - 3)/4| < 1$ or $|x - 3| < 4$. Solving the last inequality shows that the interval of convergence is $(-1, 7)$. ∎

EXAMPLE 6 Differentiation of a Power Series

Term-by-term differentiation of (7) yields a power series representation of $1/(1 + x)^2$ on the interval $(-1, 1)$:

$$\frac{d}{dx}\frac{1}{1 + x} = \frac{d}{dx}1 - \frac{d}{dx}x + \frac{d}{dx}x^2 - \frac{d}{dx}x^3 + \cdots + (-1)^n\frac{d}{dx}x^n + \cdots$$

yields $\qquad \dfrac{-1}{(1 + x)^2} = -1 + 2x - 3x^2 + \cdots + (-1)^n n x^{n-1} + \cdots$ $\qquad \leftarrow$ multiply both sides by -1

or $\qquad \dfrac{1}{(1 + x)^2} = 1 - 2x + 3x^2 + \cdots + (-1)^{n+1}n x^{n-1} + \cdots = \displaystyle\sum_{k=1}^{\infty}(-1)^{k+1}k x^{k-1}.$ ∎

EXAMPLE 7 Integration of a Power Series

Find a power series representation of $\ln(1 + x)$ on $(-1, 1)$.

Solution We first introduce a dummy variable of integration by substituting $x = t$ in (7):

$$\frac{1}{1 + t} = 1 - t + t^2 - t^3 + \cdots + (-1)^n t^n + \cdots.$$

Then, for any x within the interval $(-1, 1)$,

$$\int_0^x \frac{1}{1 + t}\,dt = \int_0^x dt - \int_0^x t\,dt + \int_0^x t^2\,dt - \cdots + (-1)^n\int_0^x t^n\,dt + \cdots$$

$$= t\Big]_0^x - \frac{1}{2}t^2\Big]_0^x + \frac{1}{3}t^3\Big]_0^x - \cdots + (-1)^n\frac{1}{n + 1}t^{n+1}\Big]_0^x + \cdots$$

$$= x - \frac{x^2}{2} + \frac{x^3}{3} - \cdots + (-1)^n\frac{x^{n+1}}{n + 1} + \cdots.$$

But $\qquad \displaystyle\int_0^x \frac{1}{1 + t}\,dt = \ln(1 + t)\Big]_0^x = \ln(1 + x) - \ln 1 = \ln(1 + x)$

and so

$$\ln(1 + x) = x - \frac{x^2}{2} + \frac{x^3}{3} - \cdots + (-1)^n \frac{x^{n+1}}{n+1} + \cdots = \sum_{k=0}^{\infty} \frac{(-1)^k}{k+1} x^{k+1}. \qquad (8) \quad \blacksquare$$

Notice that the interval of convergence series in (8) is now $(-1, 1]$, that is, we have picked up convergence at $x = 1$. By setting $x = 1$ in (8), the series on the right-hand side of the equality is the convergent alternating harmonic series; on the left-hand side we get $\ln 2$. Thus, we have discovered the sum S of the alternating harmonic series:

$$\ln 2 = 1 - \frac{1}{2} + \frac{1}{3} - \frac{1}{4} + \cdots. \qquad (9)$$

EXAMPLE 8   Approximating a Value of ln x

Approximate $\ln(1.2)$ to four decimal places.

Solution   Substituting $x = 0.2$ in (8) gives

$$\ln(1.2) = 0.2 - \frac{(0.2)^2}{2} + \frac{(0.2)^3}{3} - \frac{(0.2)^4}{4} + \frac{(0.2)^5}{5} - \frac{(0.2)^6}{6} + \cdots \qquad (10)$$

$$= 0.2 - 0.02 + 0.00267 - 0.0004 + 0.000064 - 0.00001067 + \cdots$$

$$\approx 0.1823. \qquad (11) \quad \blacksquare$$

If the sum of the series (10) in Example 8 is denoted by S, then we know from Theorem 9.7.2 that $|S_n - S| \le a_{n+1}$. The number given in (11) is accurate to four decimal places, since, for the fifth partial sum of (10),

$$|S_5 - S| \le 0.00001067 < 0.00005.$$

▌ **Arithmetic of Power Series**   Two power series $f(x) = \sum b_k(x - a)^k$ and $g(x) = \sum c_k(x - a)^k$ can be combined by the arithmetic operations of addition, multiplication, and division. We can compute $f(x) + g(x)$ and $f(x)g(x)$ as in the addition and multiplication of two polynomials: We collect terms by like powers of $x - a$. At every point at which the power series defining f and g converge absolutely, the series

$$f(x) + g(x) = (b_0 + c_0) + (b_1 + c_1)(x - a) + (b_2 + c_2)(x - a)^2 + \cdots \qquad (12)$$

and $f(x)g(x) = b_0c_0 + (b_0c_1 + b_1c_0)(x - a) + (b_0c_2 + b_1c_1 + b_2c_0)(x - a)^2 + \cdots \qquad (13)$

converge absolutely. Similarly, for $c_0 \ne 0$ we can compute $f(x)/g(x)$ by long division:

$$\begin{array}{r}
\dfrac{b_0}{c_0} + \dfrac{b_1c_0 - b_0c_1}{c_0^2}(x - a) + \cdots \quad \leftarrow \text{quotient} \\[2mm]
c_0 + c_1(x - a) + \cdots \enclose{longdiv}{ b_0 + \qquad b_1(x - a) \qquad + \cdots} \\[2mm]
b_0 + \dfrac{b_0c_1}{c_0}(x - a) \qquad + \cdots \\[2mm]
\hline
0 + \dfrac{b_1c_0 - b_0c_1}{c_0}(x - a) + \cdots \\[2mm]
\vdots
\end{array} \qquad (14)$$

◀ Of course, do not memorize (12), (13), and (14); just carry out the algebra as you would for two polynomials.

The division is valid in *some* neighborhood of the center a of the two series.

We can sometimes use the arithmetic operations just illustrated along with previously known results to obtain a power series representation of a function.

EXAMPLE 9   Addition of Power Series

Find a power series representation of $\dfrac{4x}{x^2 + 2x - 3}$ centered at 0.

Solution   To start we decompose the function into partial fractions

$$\frac{4x}{x^2 + 2x - 3} = \frac{3}{3 + x} - \frac{1}{1 - x}.$$

We then factor 3 from the denominator of the first partial fraction and use (7) with x replaced by $x/3$:

$$\frac{3}{3+x} = \frac{1}{1+\dfrac{x}{3}} = 1 - \frac{x}{3} + \frac{x^2}{3^2} - \frac{x^3}{3^3} + \cdots = \sum_{k=0}^{\infty} \frac{(-1)^k}{3^k} x^k. \tag{15}$$

This series converges for $|x/3| < 1$ or $|x| < 3$. The interval of convergence for (15) is $(-3, 3)$. Now from (6) we know

$$\frac{1}{1-x} = 1 + x + x^2 + x^3 + \cdots = \sum_{k=0}^{\infty} x^k \tag{16}$$

converges for $|x| < 1$. The interval of convergence for (16) is $(-1, 1)$. Finally, the addition of (15) and (16) yields the following power series representation for the given function:

$$\frac{4x}{x^2 + 2x - 3} = \frac{3}{3+x} - \frac{1}{1-x} = -\frac{4}{3}x - \frac{8}{9}x^2 - \frac{28}{27}x^3 - \cdots = \sum_{k=1}^{\infty}\left(\frac{(-1)^k}{3^k} - 1\right)x^k. \tag{17}$$

The series (17) converges for all x common to (that is, the intersection of) the intervals $(-3, 3)$ and $(-1, 1)$, that is, for all x in $(-1, 1)$. ∎

The result in (17) can also be gotten by multiplying two power series.

EXAMPLE 10 Example 9 Revisited

If we rewrite the function in Example 9 as a product

$$\frac{4x}{x^2 + 2x - 3} = -\frac{4}{3}x \cdot \frac{1}{1+\dfrac{x}{3}} \cdot \frac{1}{1-x}$$

and then use (15) and (16), it follows that

$$\frac{4x}{x^2 + 2x - 3} = -\frac{4}{3}x \cdot \left(1 - \frac{x}{3} + \frac{x^2}{3^2} - \frac{x^3}{3^3} + \cdots\right) \cdot (1 + x + x^2 + x^3 + \cdots)$$

$$= -\frac{4}{3}x \cdot \left[1 + 1\left(1 - \frac{1}{3}\right)x + \left(1 - \frac{1}{3} + \frac{1}{3^2}\right)x^2 + \cdots\right]$$

$$= -\frac{4}{3}x - \frac{8}{9}x^2 - \frac{28}{27}x^3 - \cdots. ∎$$

Exercises 9.9   Answers to selected odd-numbered problems begin on page ANS-27.

≡ Fundamentals

In Problems 1–8, use (6) and (7) to find a power series representation, centered at 0, of the given function. Give the interval of convergence.

1. $\dfrac{1}{3-x}$ **2.** $\dfrac{1}{4+x}$

3. $\dfrac{1}{1+2x}$ **4.** $\dfrac{1}{5+2x}$

5. $\dfrac{1}{1+x^2}$ **6.** $\dfrac{x}{1+x^2}$

7. $\dfrac{1}{4+x^2}$ **8.** $\dfrac{4}{4-x^2}$

In Problems 9–14, use differentiation of an appropriate series from Problems 1–8 to find a power series representation, centered at 0, of the given function. Give the interval of convergence.

9. $\dfrac{1}{(3-x)^2}$ **10.** $\dfrac{1}{(1+2x)^2}$

11. $\dfrac{1}{(5+2x)^3}$ **12.** $\dfrac{1}{(4+x)^3}$

13. $\dfrac{x}{(1+x^2)^2}$ **14.** $\dfrac{1-x^2}{(1+x^2)^2}$

In Problems 15–20, use integration of an appropriate series from Problems 1–8 to find a power series representation, centered at 0, of the given function. Give the interval of convergence.

15. $\tan^{-1}x$ **16.** $\tan^{-1}(x/2)$

17. $\ln(1 + x^2)$ **18.** $\ln(5 + 2x)$

19. $\ln(4 + x)$ **20.** $\ln\left(\dfrac{3+x}{3-x}\right)$

In Problems 21–28, use (6), (7), or previous results to find a power series representation, centered at 0, of the given function. Give the interval of convergence.

21. $\dfrac{1-x}{1+2x}$

22. $\dfrac{3-x}{1-x}$

23. $\dfrac{x^2}{(1+x)^3}$

24. $\dfrac{x^3}{8+2x}$

25. $x\ln(1+x^2)$

26. $x^2\tan^{-1}x$

27. $\displaystyle\int_0^x \tan^{-1}t\,dt$

28. $\displaystyle\int_0^x \ln(1+t^2)\,dt$

In Problems 29–32, proceed as in Example 5 and find a power series representation, centered at the given number a, of the given function. Give the interval of convergence.

29. $\dfrac{1}{1-x}$; $a=6$

30. $\dfrac{1}{x}$; $a=-2$

31. $\dfrac{x}{2+x}$; $a=-1$

32. $\dfrac{x-2}{x-1}$; $a=2$

In Problems 33 and 34, proceed as in Example 9 and use partial fractions to find a power series representation, centered at 0, of the given function. Give the interval of convergence.

33. $\dfrac{7x}{x^2+x-12}$

34. $\dfrac{3}{x^2-x-2}$

In Problems 35 and 36, proceed as in Example 10 and use multiplication of power series to find the first four nonzero terms of a power series representation, centered at 0, for the given function.

35. $\dfrac{1}{(2-x)(1-x)}$

36. $\dfrac{x}{(1+2x)(1+x^2)}$

In Problems 37 and 38, find the domain of the given function.

37. $f(x)=\dfrac{x}{3}-\dfrac{x^2}{2\cdot3^2}+\dfrac{x^3}{3\cdot3^3}-\dfrac{x^4}{4\cdot3^4}+\cdots$

38. $f(x)=1+2x+\dfrac{4x^2}{1\cdot2}+\dfrac{8x^2}{1\cdot2\cdot3}+\cdots$

In Problems 39–44, use power series to approximate the given quantity to four decimal places.

39. $\ln(1.1)$

40. $\tan^{-1}(0.2)$

41. $\displaystyle\int_0^{1/2}\dfrac{1}{1+x^3}\,dx$

42. $\displaystyle\int_0^{1/3}\dfrac{x}{1+x^4}\,dx$

43. $\displaystyle\int_0^{0.3} x\tan^{-1}x\,dx$

44. $\displaystyle\int_0^{1/2}\tan^{-1}x^2\,dx$

45. Use Problem 15 to show that

$$\frac{\pi}{4}=1-\frac{1}{3}+\frac{1}{5}-\frac{1}{7}+\cdots.$$

46. The series in Problem 45 is known to converge very slowly. Show this by finding the smallest positive integer n so that S_n approximates $\pi/4$ to four decimal places.

In Problems 47 and 48, show that the function defined by the power series satisfies the given differential equation.

47. $y=\displaystyle\sum_{k=1}^{\infty}\dfrac{(-1)^{k+1}}{k}x^k$; $\quad(x+1)y''+y'=0$

48. $J_0(x)=\displaystyle\sum_{k=0}^{\infty}\dfrac{(-1)^k}{2^{2k}(k!)^2}x^{2k}$; $\quad xy''+y'+xy=0$

≡ Think About It

49. (a) If $f(x)=\displaystyle\sum_{k=0}^{\infty}\dfrac{x^k}{k!}$, then show that $f'(x)=f(x)$ for all x in $(-\infty,\infty)$.

(b) What function has the property that its first derivative equals the function? Conjecture what function is represented by the power series in part (a).

50. (a) If $f(x)=\displaystyle\sum_{k=0}^{\infty}\dfrac{(-1)^k}{(2k+1)!}x^{2k+1}$, then show that $f''(x)=-f(x)$ for all x in $(-\infty,\infty)$.

(b) What functions have the property that their second derivative equals the negative of the function? Conjecture what function is represented by the power series in part (a). Note that the powers of x in the power series are odd positive integers.

9.10 Taylor Series

▌ **Introduction** Suppose $\sum c_k(x-a)^k$ is a power series centered at a that has an interval of convergence with a nonzero radius of convergence R. Then, as we saw in the preceding section, within the interval of convergence a power series is a continuous function that possesses derivatives of all orders. We also touched on the idea of using a power series to *represent* a given function (such as $1/(1+x)$) on an interval. In this section we are going to expand upon the notion of representing a function by a power series. The basic problem is:

- *Suppose we are given a function f that possesses derivatives of all orders on an open interval I. Can we find a power series that **represents** f on I?*

In slightly different words: Can we **expand** an infinitely differentiable function (such as $f(x)=\sin x$, $f(x)=\cos x$, or $f(x)=e^x$) into a power series $\sum c_k(x-a)^k$ that converges to the correct function value $f(x)$ for all x in some open interval $(a-R,a+R)$, where R is either $R>0$ or $R=\infty$?

■ Taylor Series for a Function f Before answering the question in the last paragraph, let us simply make the *assumption* that an infinitely differentiable function f on an interval $(a - R, a + R)$ can be represented by a power series $\sum c_k(x - a)^k$ on that interval. It is then relatively easy to determine what the coefficients c_k must be. Repeated differentiation of

$$f(x) = c_0 + c_1(x - a) + c_2(x - a)^2 + c_3(x - a)^3 + \cdots + c_n(x - a)^n + \cdots \tag{1}$$

yields

$$f'(x) = c_1 + 2c_2(x - a) + 3c_3(x - a)^2 + \cdots \tag{2}$$

$$f''(x) = 2c_2 + 3 \cdot 2c_3(x - a) + \cdots \tag{3}$$

$$f'''(x) = 3 \cdot 2 \cdot 1c_3 + \cdots, \tag{4}$$

and so on. By evaluating (1), (2), (3), and (4) at $x = a$, we find that

$$f(a) = c_0, \quad f'(a) = 1!c_1, \quad f''(a) = 2!c_2, \quad \text{and} \quad f'''(a) = 3!c_3,$$

respectively. In general, we see that $f^{(n)}(a) = n!c_n$, or

$$c_n = \frac{f^{(n)}(a)}{n!}, n \geq 0. \tag{5}$$

When $n = 0$ we interpret the *zero*th derivative as $f(a)$ and $0! = 1$. Substituting (5) in (1) yields the results summarized in the next theorem.

Theorem 9.10.1 Form of a Power Series

If a function f possesses a power series representation $f(x) = \sum c_k(x - a)^k$ on an interval $(a - R, a + R)$, then the coefficients must be $c_k = f^{(k)}(a)/k!$.

In other words, if a function f has a power series representation centered at a then it must look like this:

$$f(x) = f(a) + \frac{f'(a)}{1!}(x - a) + \frac{f''(a)}{2!}(x - a)^2 + \frac{f'''(a)}{3!}(x - a)^3 + \cdots = \sum_{k=0}^{\infty} \frac{f^{(k)}(a)}{k!}(x - a)^k. \tag{6}$$

The series in (6) is called the **Taylor series of f at a**, or **centered at a**. The Taylor series centered at $a = 0$,

$$f(x) = f(0) + \frac{f'(0)}{1!}x + \frac{f''(0)}{2!}x^2 + \frac{f'''(0)}{3!}x^3 + \cdots = \sum_{k=0}^{\infty} \frac{f^{(k)}(0)}{k!}x^k \tag{7}$$

is called the **Maclaurin series of f**.

The question posed in the introduction can now be rephrased as:

- *Can we expand an infinitely differentiable function f into a Taylor series (6)?*

It would appear that the answer is yes—by simply calculating the coefficients as dictated by the formula (5). Unfortunately, the concept of expanding a given infinitely differentiable function f in a Taylor series is not that simple. You must bear in mind that (5) and (6) were obtained under the assumption that f was represented by a power series centered at a. If we do not know *a priori* that an infinitely differentiable function f has a power series representation, then we must look upon a power series obtained from either (6) or (7) as a *formal* result, in other words, a power series that is simply **generated** by the function f. We do not know whether the series generated in this manner converges or, even if it does, whether it converges to $f(x)$.

EXAMPLE 1 Taylor Series of ln x

Find the Taylor series of $f(x) = \ln x$ centered at $a = 1$. Find its interval of convergence.

Solution The function f, its derivatives, and their values at 1 are:

$$f(x) = \ln x \qquad\qquad f(1) = 0$$

$$f'(x) = \frac{1}{x} \qquad\qquad f'(1) = 1$$

$$f''(x) = -\frac{1}{x^2} \qquad\qquad f''(1) = -1$$

$$f'''(x) = \frac{1 \cdot 2}{x^3} \qquad\qquad f'''(1) = 2!$$

$$\vdots \qquad\qquad\qquad \vdots$$

$$f^{(n)}(x) = (-1)^{n-1}\frac{(n-1)!}{x^n} \qquad f^{(n)}(1) = (-1)^{n-1}(n-1)!$$

Since $(n-1)!/n! = 1/n$, $n \geq 1$, (6) yields

$$(x-1) - \frac{1}{2}(x-1)^2 + \frac{1}{3}(x-1)^3 - \cdots = \sum_{k=1}^{\infty} \frac{(-1)^{k-1}}{k}(x-1)^k. \qquad (8)$$

The Ratio Test,

$$\lim_{n\to\infty}\left|\frac{a_{n+1}}{a_n}\right| = \lim_{n\to\infty}\left|\frac{(-1)^n(x-1)^{n+1}}{n+1} \cdot \frac{n}{(-1)^{n-1}(x-1)^n}\right|$$

$$= \lim_{n\to\infty}\frac{n}{n+1}|x-1| = |x-1|,$$

shows that the series (8) converges for $|x-1| < 1$ or on the interval $(0, 2)$. At the endpoints $x = 0$ and $x = 2$, the series

$$-\sum_{k=1}^{\infty}\frac{1}{k} \qquad \text{and} \qquad \sum_{k=1}^{\infty}\frac{(-1)^{k-1}}{k}$$

are divergent and convergent, respectively. The interval of convergence for this series is $(0, 2]$. The radius of convergence is $R = 1$. ∎

Notice in Example 1 that we did not write the equality

$$\ln x = \sum_{k=1}^{\infty}\frac{(-1)^{k-1}}{k}(x-1)^k.$$

At this point it has not been established that the series given in (8) represents $\ln x$ on the interval $(0, 2]$.

■ Taylor's Theorem It is apparent from (5) that to have a Taylor series centered at a, it is necessary that a function f must possess derivatives of all orders that are defined at a. Thus, for example, $f(x) = \ln x$ does not possess a Maclaurin series, because $f(x) = \ln x$ and all its derivatives are undefined at 0. Moreover, it is important to note that even if a function f possesses derivatives of all orders and generates a Taylor series convergent on some interval, it is possible that the series does not represent f on the interval, that is, the series does not converge to $f(x)$ at every x in the interval. See Problem 63 in Exercises 9.10. The fundamental question of whether a Taylor series represents the function that generates it can be resolved by means of **Taylor's Theorem**.

Theorem 9.10.2 Taylor's Theorem

Let f be a function such that $f^{(n+1)}(x)$ exists for every x in an interval containing the number a. Then for all x in the interval,

$$f(x) = P_n(x) + R_n(x),$$

where
$$P_n(x) = f(a) + \frac{f'(a)}{1!}(x-a) + \cdots + \frac{f^{(n)}(a)}{n!}(x-a)^n \qquad (9)$$

(*continued*)

is called the nth-degree **Taylor polynomial of f at a**, and

There are several forms of the remainder. This form is due to the French mathematician **Joseph Louis Lagrange** (1736–1813).

$$R_n(x) = \frac{f^{(n+1)}(c)}{(n+1)!}(x-a)^{n+1} \qquad (10)$$

is called the **Lagrange form of the remainder**. The number c lies between a and x.

Since the proof of this theorem would deflect us from the main thrust of our discussion, it is given in the *Appendix*. The importance of Theorem 9.10.2 lies in the fact that the Taylor polynomials $P_n(x)$ are the partial sums of the Taylor series (6). The remainder is defined as

$$R_n(x) = f(x) - P_n(x) \qquad \text{and so} \qquad P_n(x) = f(x) - R_n(x). \qquad (11)$$

If $\lim_{n \to \infty} P_n(x) = f(x)$, then the function f is the sum of the Taylor series it generates. But from (11) we see that

$$\lim_{n \to \infty} P_n(x) = f(x) - \lim_{n \to \infty} R_n(x)$$

and so if we can somehow show that $R_n(x) \to 0$ as $n \to \infty$, then the sequence of partial sums converges to $f(x)$. We summarize the result.

Theorem 9.10.3 Convergence of a Taylor Series

Suppose f is a function that possesses derivatives of all orders on an interval centered at the number a. If

$$\lim_{n \to \infty} R_n(x) = 0$$

for every x in the interval, then the Taylor series generated by f converges to $f(x)$,

$$f(x) = \sum_{k=0}^{\infty} \frac{f^{(k)}(a)}{k!}(x-a)^k.$$

In practice, the proof that the remainder $R_n(x)$ approaches zero as $n \to \infty$ often depends on the fact that

$$\lim_{n \to \infty} \frac{|x|^n}{n!} = 0. \qquad (12)$$

This latter result follows from applying Theorem 9.3.2 to the series $\sum_{m=1}^{\infty} x^k/k!$, which is known to be absolutely convergent for all real numbers. (See Example 3 in Section 9.8.)

EXAMPLE 2 Example 1 Revisited

Prove that the series (8) represents $f(x) = \ln x$ on the interval $(0, 2]$.

Solution In the solution for Example 1 we saw that the nth derivative of $f(x) = \ln x$ is given by

$$f^{(n)}(x) = \frac{(-1)^{n-1}(n-1)!}{x^n}.$$

From $f^{(n+1)}(c) = \dfrac{(-1)^n n!}{c^{n+1}}$ we get from (10),

$$|R_n(x)| = \frac{|f^{(n+1)}(c)|}{(n+1)!}|x-1|^{n+1} = \left|\frac{(-1)^n n!}{c^{n+1}(n+1)!} \cdot (x-1)^{n+1}\right| = \frac{1}{n+1}\left|\frac{x-1}{c}\right|^{n+1},$$

where c is some number in the interval $(0, 2]$ between 1 and x.

If $1 \le x \le 2$, then $0 < x - 1 \le 1$. Since $1 < c < x$, we must have $0 < x - 1 \le 1 < c$ and, consequently, $(x-1)/c < 1$. Hence,

$$|R_n(x)| \le \frac{1}{n+1} \qquad \text{and} \qquad \lim_{n \to \infty} R_n(x) = 0.$$

In the case where $0 < x < 1$, we can also show that $\lim_{n \to \infty} R_n(x) = 0$. We omit the proof. Hence,

$$\ln x = (x - 1) - \frac{1}{2}(x - 1)^2 + \frac{1}{3}(x - 1)^3 - \cdots = \sum_{k=1}^{\infty} \frac{(-1)^{k-1}}{k}(x - 1)^k$$

for all values of x in the interval $(0, 2]$. ∎

EXAMPLE 3 Maclaurin Series Representation of cos x

Find the Maclaurin series of $f(x) = \cos x$. Prove that the Maclaurin series represents $\cos x$ for all x.

Solution We first find the Maclaurin series generated by $f(x) = \cos x$:

$$
\begin{array}{ll}
f(x) = \cos x & f(0) = 1 \\
f'(x) = -\sin x & f'(0) = 0 \\
f''(x) = -\cos x & f''(0) = -1 \\
f'''(x) = \sin x & f'''(0) = 0
\end{array}
$$

and so on. From (7) we obtain the power series

$$1 - \frac{x^2}{2!} + \frac{x^4}{4!} - \frac{x^6}{6!} + \cdots = \sum_{k=0}^{\infty} \frac{(-1)^k}{(2k)!} x^{2k}. \tag{13}$$

The Ratio Test shows that (13) converges absolutely for all real values of x, in other words, the interval of convergence is $(-\infty, \infty)$. Now in order to prove that $\cos x$ is represented by the series (13), we must show that $\lim_{n \to \infty} R_n(x) = 0$. To this end, we note that the derivatives of f satisfy

$$|f^{(n+1)}(x)| = \begin{cases} |\sin x|, & n \text{ even} \\ |\cos x|, & n \text{ odd.} \end{cases}$$

In either case, $|f^{(n+1)}(c)| \leq 1$ for any real number c, and so by (10),

$$|R_n(x)| = \frac{|f^{(n+1)}(c)|}{(n + 1)!}|x|^{n+1} \leq \frac{|x|^{n+1}}{(n + 1)!}.$$

In view of (12), we have for any fixed but arbitrary choice of x,

$$\lim_{n \to \infty} \frac{|x|^{n+1}}{(n + 1)!} = 0.$$

But $\lim_{n \to \infty} |R_n(x)| = 0$ implies that $\lim_{n \to \infty} R_n(x) = 0$. Therefore,

$$\cos x = 1 - \frac{x^2}{2!} + \frac{x^4}{4!} - \frac{x^6}{6!} + \cdots + (-1)^n \frac{x^{2n}}{(2n)!} + \cdots$$

is a valid representation of $\cos x$ for every real number x. ∎

EXAMPLE 4 Taylor Series Representation of sin x

Find the Taylor series of $f(x) = \sin x$ centered at $a = \pi/3$. Prove that the Taylor series represents $\sin x$ for all x.

Solution We have

$$
\begin{array}{ll}
f(x) = \sin x & f\left(\frac{\pi}{3}\right) = \frac{\sqrt{3}}{2} \\[2mm]
f'(x) = \cos x & f'\left(\frac{\pi}{3}\right) = \frac{1}{2} \\[2mm]
f''(x) = -\sin x & f''\left(\frac{\pi}{3}\right) = -\frac{\sqrt{3}}{2} \\[2mm]
f'''(x) = -\cos x & f'''\left(\frac{\pi}{3}\right) = -\frac{1}{2}
\end{array}
$$

and so on. Hence, the Taylor series centered at $\pi/3$ generated by $\sin x$ is

$$\frac{\sqrt{3}}{2} + \frac{1}{2 \cdot 1!}\left(x - \frac{\pi}{3}\right) - \frac{\sqrt{3}}{2 \cdot 2!}\left(x - \frac{\pi}{3}\right)^2 - \frac{1}{2 \cdot 3!}\left(x - \frac{\pi}{3}\right)^3 + \cdots. \quad (14)$$

Again, from the Ratio Test it follows that (14) converges absolutely for all real values of x, that is, its interval of convergence is $(-\infty, \infty)$. To show that

$$\sin x = \frac{\sqrt{3}}{2} + \frac{1}{2 \cdot 1!}\left(x - \frac{\pi}{3}\right) - \frac{\sqrt{3}}{2 \cdot 2!}\left(x - \frac{\pi}{3}\right)^2 - \frac{1}{2 \cdot 3!}\left(x - \frac{\pi}{3}\right)^3 + \cdots$$

for every real x, we note that, as in the preceding example, $|f^{(n+1)}(c)| \leq 1$. This implies that

$$|R_n(x)| \leq \frac{|x - \pi/3|^{n+1}}{(n+1)!}$$

from which we see, with the help of (12), that $\lim\limits_{n \to \infty} R_n(x) = 0$. ■

We summarize some important Maclaurin series representations and their intervals of convergence:

Maclaurin Series	**Interval of Convergence**	
$e^x = 1 + x + \dfrac{x^2}{2!} + \dfrac{x^3}{3!} + \cdots = \displaystyle\sum_{k=0}^{\infty} \dfrac{x^k}{k!}$	$(-\infty, \infty)$	(15)
$\cos x = 1 - \dfrac{x^2}{2!} + \dfrac{x^4}{4!} - \dfrac{x^6}{6!} + \cdots = \displaystyle\sum_{k=0}^{\infty} \dfrac{(-1)^k}{(2k)!}x^{2k}$	$(-\infty, \infty)$	(16)
$\sin x = x - \dfrac{x^3}{3!} + \dfrac{x^5}{5!} - \dfrac{x^7}{7!} + \cdots = \displaystyle\sum_{k=0}^{\infty} \dfrac{(-1)^k}{(2k+1)!}x^{2k+1}$	$(-\infty, \infty)$	(17)
$\tan^{-1}x = x - \dfrac{x^3}{3} + \dfrac{x^5}{5} - \dfrac{x^7}{7} + \cdots = \displaystyle\sum_{k=0}^{\infty} \dfrac{(-1)^k}{2k+1}x^{2k+1}$	$[-1, 1]$	(18)
$\cosh x = 1 + \dfrac{x^2}{2!} + \dfrac{x^4}{4!} + \dfrac{x^6}{6!} + \cdots = \displaystyle\sum_{k=0}^{\infty} \dfrac{x^{2k}}{(2k)!}$	$(-\infty, \infty)$	(19)
$\sinh x = x + \dfrac{x^3}{3!} + \dfrac{x^5}{5!} + \dfrac{x^7}{7!} + \cdots = \displaystyle\sum_{k=0}^{\infty} \dfrac{x^{2k+1}}{(2k+1)!}$	$(-\infty, \infty)$	(20)
$\ln(1+x) = x - \dfrac{x^2}{2} + \dfrac{x^3}{3} - \dfrac{x^4}{4} + \cdots = \displaystyle\sum_{k=0}^{\infty} \dfrac{(-1)^k}{k+1}x^{k+1}$	$[-1, 1]$	(21)

You are asked to demonstrate the validity of the representations (15), (17), (19), and (20) as exercises. See Problems 51–54 in Exercises 9.10.

Also, you are encouraged to look hard at the series given in (16)–(20). Then answer the question in Problem 61 of Exercise 9.10.

▌ **Some Graphs of Taylor Polynomials** In Example 3 we saw that the Taylor series of $f(x) = \cos x$ at $a = 0$ represents the function for all x, since $\lim\limits_{n \to \infty} R_n(x) = 0$. It is always of interest to see graphically how partial sums of the Taylor series, which are the Taylor polynomials defined in (9), converge to the function. In FIGURE 9.10.1(a) the graphs of the Taylor polynomials

$$P_0(x) = 1, \qquad P_2(x) = 1 - \frac{1}{2!}x^2, \qquad P_4(x) = 1 - \frac{1}{2!}x^2 + \frac{1}{4!}x^4,$$

and

$$P_{10}(x) = 1 - \frac{1}{2!}x^2 + \frac{1}{4!}x^4 - \frac{1}{6!}x^6 + \frac{1}{8!}x^8 - \frac{1}{10!}x^{10}$$

are compared with the graph of $f(x) = \cos x$ shown in blue.

A comparison of numerical values is given in Figure 9.10.1(b).

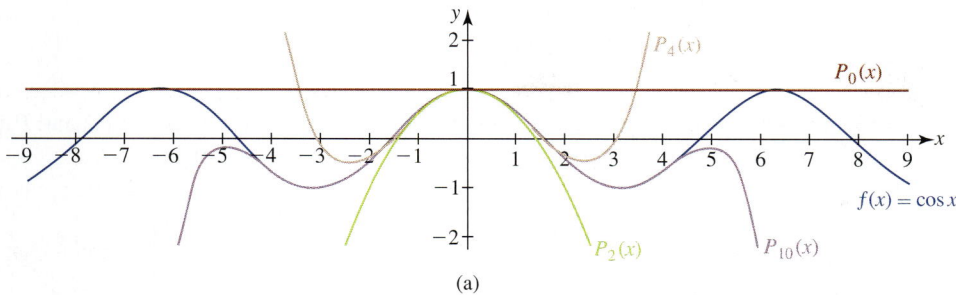

(a)

x	$P_2(x)$	$P_4(x)$	$P_{10}(x)$	$\cos x$
$\pi/6$	0.86292	0.86605	0.86603	0.86603
$\pi/4$	0.69157	0.70743	0.70711	0.70711
$\pi/3$	0.45169	0.50180	0.50000	0.5
$\pi/2$	-0.23370	0.01997	0.00000	0

(b)

FIGURE 9.10.1 Taylor polynomials P_0, P_2, P_4, and P_{10} for $\cos x$

■ **Approximations**   When the value of x is close to the center a ($x \approx a$) of a Taylor series the Taylor polynomial $P_n(x)$ of a function f at a can be used to approximate the function value $f(x)$. The error in this approximation is given by

$$|R_n(x)| = |f(x) - P_n(x)|.$$

EXAMPLE 5   Approximation Using a Taylor Polynomial

Approximate $e^{-0.2}$ by a Taylor polynomial $P_3(x)$. Determine the accuracy of the approximation.

Solution   Because the value $x = -0.2$ is close to zero, we use the Taylor polynomial of $f(x) = e^x$ at $a = 0$:

$$P_3(x) = f(0) + \frac{f'(0)}{1!}x + \frac{f''(0)}{1!}x^2 + \frac{f''(0)}{3!}x^3.$$

It follows from

$$f(x) = f'(x) = f''(x) = f'''(x) = e^x$$
$$f(0) = f'(0) = f''(0) = f'''(0) = 1$$

that

$$P_3(x) = 1 + x + \frac{1}{2}x^2 + \frac{1}{6}x^3.$$

This polynomial is the fourth partial sum of the series given in (15). Now,

$$P_3(-0.2) = 1 + (-0.2) + \frac{1}{2}(-0.2)^2 + \frac{1}{6}(-0.2)^3 \approx 0.8187$$

and so,

$$e^{-0.2} \approx 0.8187. \qquad (22)$$

Now, from (10) we can write

$$|R_3(x)| = \frac{e^c}{4!}|x|^4 < \frac{|x|^4}{4!}$$

since $-0.2 < c < 0$ and $e^c < 1$. The inequality,

$$|R_3(-0.2)| < \frac{|-0.2|^4}{24} < 0.0001$$

implies that the result in (22) is accurate to three decimal places.

In FIGURE 9.10.2 we have compared the graphs of the Taylor polynomials of $f(x) = e^x$ centered at $a = 0$:

$$P_1(x) = 1 + x, \qquad P_2(x) = 1 + x + \frac{1}{2}x^2, \qquad \text{and} \qquad P_3(x) = 1 + x + \frac{1}{2}x^2 + \frac{1}{6}x^3.$$

Notice in Figures 9.10.2(b) and 9.10.2(c) the graphs of the Taylor polynomials $P_2(x)$ and $P_3(x)$ are indistinguishable from the graph of $y = e^x$ in a small neighborhood of $x = 0.2$.

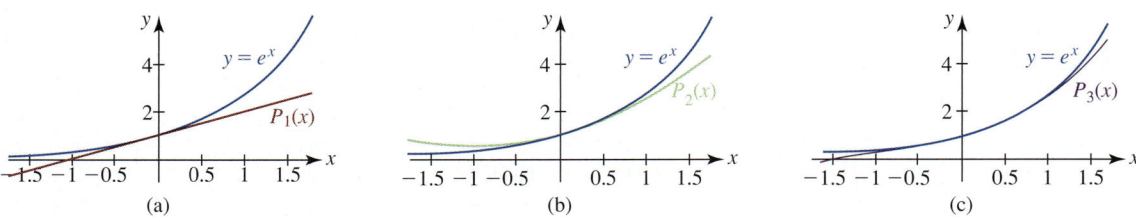

(a) (b) (c)

FIGURE 9.10.2 Graphs of Taylor polynomials in Example 5

In the *Notes from the Classroom* in Section 5.5 we introduced the notion of **nonelementary integrals**, namely, an integral such as $\int \sin x^2 \, dx$, where $\sin x^2$ does not possess an antiderivative in the form of an elementary function. Taylor series can be an aid when working with nonelementary integrals. For example, the Maclaurin series obtained by replacing x by x^2 in (17) converges for $-\infty < x < \infty$, and so by Theorem 9.9.2,

$$\int \sin x^2 \, dx = \int \left(x^2 - \frac{x^6}{3!} + \frac{x^{10}}{5!} - \frac{x^{14}}{7!} + \cdots \right) dx$$

$$= \frac{x^3}{3} - \frac{x^7}{7 \cdot 3!} + \frac{x^{11}}{11 \cdot 5!} - \frac{x^{15}}{15 \cdot 7!} + \cdots + C. \tag{23}$$

EXAMPLE 6 Approximation Using a Taylor Series

Approximate $\int_0^1 \sin x^2 \, dx$ to three decimal places.

Solution From (23) we see immediately that

$$\int_0^1 \sin x^2 \, dx = \frac{x^3}{3} - \frac{x^7}{7 \cdot 3!} + \frac{x^{11}}{11 \cdot 5!} - \frac{x^{15}}{15 \cdot 7!} + \cdots \Bigg]_0^1$$

$$= \frac{1}{3} - \frac{1}{7 \cdot 3!} + \frac{1}{11 \cdot 5!} - \frac{1}{15 \cdot 7!} + \cdots. \tag{24}$$

By the error-bound theorem for alternating series, Theorem 9.7.2, the fourth term in the series (24) satisfies

$$a_4 = \frac{1}{15 \cdot 7!} \approx 0.000013 < 0.0005.$$

Therefore, the approximation

$$\int_0^1 \sin x^2 \, dx \approx \frac{1}{3} - \frac{1}{7 \cdot 3!} + \frac{1}{11 \cdot 5!} \approx 0.3103$$

is accurate to three decimal places.

▌ **Limits** A power series representation of a function can sometimes be useful in computing limits. For example, in Section 2.4 we resorted to a subtle geometric argument to prove that $\lim\limits_{x \to 0} \dfrac{\sin x}{x} = 1$. But if we use (17) and division by x we see immediately that

limit of each of these terms is 0

$$\lim_{x \to 0} \frac{\sin x}{x} = \lim_{x \to 0} \frac{x - \dfrac{x^3}{3!} + \dfrac{x^5}{5!} - \cdots}{x} = \lim_{x \to 0} \left(1 - \overbrace{\frac{x^2}{3!} + \frac{x^4}{5!} - \cdots} \right) = 1.$$

EXAMPLE 7 Calculating a Limit

Evaluate $\lim\limits_{x \to 0} \dfrac{x - \tan^{-1}x}{x^3}$.

Solution Observe that the limit has the indeterminate form $0/0$. If you review Problem 25 in Exercises 4.5, you might recall evaluating this limit by L'Hôpital's Rule. But in view of (18), we can write

$$\lim_{x \to 0} \frac{x - \tan^{-1}x}{x^3} = \lim_{x \to 0} \frac{x - \left(x - \dfrac{x^3}{3} + \dfrac{x^5}{5} - \cdots\right)}{x^3} \quad \begin{array}{l} \text{also see Problem 15 in Exercises 9.9} \\ \leftarrow \text{for the power series representation of} \\ \tan^{-1}x \end{array}$$

$$= \lim_{x \to \infty} \frac{\dfrac{x^3}{3} - \dfrac{x^5}{5} + \cdots}{x^3} \quad \leftarrow \text{factor } x^3 \text{ from numerator and cancel}$$

$$= \lim_{x \to 0} \left(\frac{1}{3} - \frac{x^2}{5} + \cdots\right) = \frac{1}{3}. \qquad \blacksquare$$

■ **Using the Arithmetic of Power Series** In Section 9.9 we discussed the arithmetic of power series, that is, power series can basically be manipulated arithmetically like polynomials. In the case where the power series representations $f(x) = \sum b_k(x - a)^k$ and $g(x) = \sum c_k(x - a)^k$ converge on the same open interval $(a - R, a + R)$ for $R > 0$ or $(-\infty, \infty)$ for $R = \infty$, the power series representations for $f(x) + g(x)$ and $f(x)g(x)$ can be obtained, in turn, by adding the series and multiplying the series. The sum and product converge on the same interval. If we divide the power series of f by the power series of g, then the quotient represents $f(x)/g(x)$ in some neighborhood of a.

EXAMPLE 8 Maclaurin Series of tan x

Find the first three nonzero terms of the Maclaurin series of $f(x) = \tan x$.

Solution From (16) and (17) we can write

$$\tan x = \frac{\sin x}{\cos x} = \frac{x - \dfrac{x^3}{3!} + \dfrac{x^5}{5!} - \dfrac{x^7}{7!} + \cdots}{1 - \dfrac{x^2}{2!} + \dfrac{x^4}{4!} - \dfrac{x^6}{6!} + \cdots}$$

Then by long division

$$
\begin{array}{r}
x + \frac{1}{3}x^3 + \frac{2}{15}x^5 + \cdots \\
1 - \frac{1}{2}x^2 + \frac{1}{24}x^4 - \cdots \overline{\smash{\big)}\, x - \frac{1}{6}x^3 + \frac{1}{120}x^5 - \cdots} \\
\underline{x - \frac{1}{2}x^3 + \frac{1}{24}x^5 - \cdots} \\
\frac{1}{3}x^3 - \frac{1}{30}x^5 + \cdots \\
\underline{\frac{1}{3}x^3 - \frac{1}{6}x^5 + \cdots} \\
\frac{2}{15}x^5 + \cdots \\
\frac{2}{15}x^5 + \cdots \\
\vdots
\end{array}
$$

Hence, we have

$$\tan x = x + \frac{1}{3}x^3 + \frac{2}{15}x^5 + \cdots. \qquad \blacksquare$$

Of course, the last result could also be obtained using (7). See Problem 11 in Exercises 9.10. After working through Example 8 you are encouraged to read (*ii*) in the *Notes from the Classroom*.

■ **Taylor Polynomials—Redux**   In Section 4.9 we introduced the notion of a **local linear approximation** of f at a given by $f(x) \approx L(x)$, where

$$L(x) = f(a) + f'(a)(x - a). \tag{25}$$

This equation represents the tangent line to the graph of f at $x = a$. Because it is a linear polynomial, another appropriate symbol for (25) is

$$P_1(x) = f(a) + f'(a)(x - a). \tag{26}$$

The equation is now recognized as the first-degree Taylor polynomial of f at a. The idea behind (25) is that the tangent line can be used to approximate the value of $f(x)$ when x is in a small neighborhood of a. But, since most graphs have concavity and a tangent line does not, it makes sense to expect a polynomial of higher degree would provide a better approximation to $f(x)$ in the sense that its graph would stay near the graph of f over a larger interval containing a. Notice that (26) has the properties that P_1 and its first derivative agree with f and its first derivative at $x = a$:

$$P_1(a) = f(a) \qquad \text{and} \qquad P_1'(a) = f'(a).$$

If we want a quadratic polynomial function

$$P_2(x) = c_0 + c_1(x - a) + c_2(x - a)^2$$

to have the analogous properties, namely,

$$P_2(a) = f(a), \qquad P_2'(a) = f'(a), \qquad \text{and} \qquad P_2''(a) = f''(a),$$

then, following a procedure similar to (1)–(5), it is seen that P_2 must be

$P_n(x)$ is the polynomial of degree n defined in (9). ▶

$$P_2(x) = f(a) + \frac{f'(a)}{1!}(x - a) + \frac{f''(a)}{2!}(x - a)^2. \tag{27}$$

Graphically, this means that the graph of f and the graph of P_2 have the same tangent line and the same concavity at $x = a$. Of course, (27) is recognized as the second-degree Taylor polynomial. We say that $f(x) \approx P_2(x)$ is a **local quadratic approximation of f at a**. Continuing in this manner we then build up to $f(x) \approx P_n(x)$ which is a **local nth degree approximation of f at a**. With this discussion in mind, you should now look more closely at the graphs of $f(x) = \cos x$, P_0, P_2, P_4, and P_{10} near $x = 0$ in Figure 9.10.1(a) and the approximations in Figure 9.10.1(b). Also reinspect Figure 9.10.2.

■ **Postscript—A Bit of History**   Theorem 9.10.2 is named in honor of the English mathematician **Brook Taylor** (1685–1731), who published this result in 1715. However, the formula in (6) was discovered by Johann Bernoulli about 20 years earlier. The series in (7) is named after the Scottish mathematician and former student of Isaac Newton, **Colin Maclaurin** (1698–1746). It is not clear why Maclaurin's name is associated with this series.

∑   **NOTES FROM THE CLASSROOM**

(*i*) The Taylor series method of finding a power series for a function and then proving that the series represents the function has one big and obvious drawback. Obtaining a general expression for the nth derivative for most functions is nearly impossible. Thus, we are often limited to finding just the first few coefficients c_n.

(*ii*) It is easy to pass over the significance of the results in (6) and (7). Suppose we wish to find the Maclaurin series for $f(x) = 1/(2 - x)$. We can, of course, use (7)–and you are asked to do so in Problem 1 in Exercises 9.10. On the other hand, you should also recognize, from Examples 3–5 of Section 9.9, that a power series representation of f can be obtained utilizing geometric series. The point is:

   • *The representation is unique. Thus, on its interval of convergence, a power series representing a function, regardless of how it is obtained, **is** the Taylor or Maclaurin series of that function.*

Exercises 9.10 Answers to selected odd-numbered problems begin on page ANS-28.

≣ Fundamentals

In Problems 1–10, use (7) to find the Maclaurin series for the given function.

1. $f(x) = \dfrac{1}{2 - x}$

2. $f(x) = \dfrac{1}{1 + 5x}$

3. $f(x) = \ln(1 + x)$

4. $f(x) = \ln(1 + 2x)$

5. $f(x) = \sin x$

6. $f(x) = \cos 2x$

7. $f(x) = e^x$

8. $f(x) = e^{-x}$

9. $f(x) = \sinh x$

10. $f(x) = \cosh x$

In Problems 11 and 12, use (7) to find the first four nonzero terms of the Maclaurin series for the given function.

11. $f(x) = \tan x$

12. $f(x) = \sin^{-1} x$

In Problems 13–24, use (6) to find the Taylor series for the given function centered at the indicated value of a.

13. $f(x) = \dfrac{1}{1 + x}, \quad a = 4$

14. $f(x) = \sqrt{x}, \quad a = 1$

15. $f(x) = \dfrac{1}{x}, \quad a = 1$

16. $f(x) = \dfrac{1}{x}, \quad a = -5$

17. $f(x) = \sin x, \quad a = \pi/4$

18. $f(x) = \sin x, \quad a = \pi/2$

19. $f(x) = \cos x, \quad a = \pi/3$

20. $f(x) = \cos x, \quad a = \pi/6$

21. $f(x) = e^x, \quad a = 1$

22. $f(x) = e^{-2x}, \quad a = \dfrac{1}{2}$

23. $f(x) = \ln x, \quad a = 2$

24. $f(x) = \ln(x + 1), \quad a = 2$

In Problems 25–32, use previous results, methods, or problems to find the Maclaurin series for the given function.

25. $f(x) = e^{-x^2}$

26. $f(x) = x^2 e^{-3x}$

27. $f(x) = x \cos x$

28. $f(x) = \sin x^3$

29. $f(x) = \ln(1 - x)$

30. $f(x) = \ln\left(\dfrac{1 + x}{1 - x}\right)$

31. $f(x) = \sec^2 x$

32. $f(x) = \ln(\cos x)$

In Problems 33 and 34, use Maclaurin series as an aid in evaluating the given limit.

33. $\displaystyle\lim_{x \to 0} \dfrac{x^3}{x - \sin x}$

34. $\displaystyle\lim_{x \to 0} \dfrac{1 + x - e^x}{1 - \cos x}$

In Problems 35 and 36, use addition of the Maclaurin series for e^x and e^{-x} to find the Maclaurin series for the given function.

35. $f(x) = \cosh x$

36. $f(x) = \sinh x$

In Problems 37 and 38, use multiplication to find the first five nonzero terms of the Maclaurin series for the given function.

37. $f(x) = \dfrac{e^x}{1 - x}$

38. $f(x) = e^x \sin x$

In Problems 39 and 40, use division to find the first five nonzero terms of the Maclaurin series for the given function.

39. $f(x) = \dfrac{e^x}{\cos x}$

40. $f(x) = \sec x$

In Problems 41 and 42, establish the indicated value of the given definite integral.

41. $\displaystyle\int_0^1 e^{-x^2}\, dx = 1 - \dfrac{1}{3} + \dfrac{1}{10} - \dfrac{1}{42} + \cdots$

42. $\displaystyle\int_0^1 \dfrac{\sin x}{x}\, dx = 1 - \dfrac{1}{3 \cdot 3!} + \dfrac{1}{5 \cdot 5!} - \dfrac{1}{7 \cdot 7!} + \cdots$

In Problems 43–46, find the sum of the given series.

43. $1 - \dfrac{1}{3} + \dfrac{1}{5} - \dfrac{1}{7} + \cdots$

44. $\dfrac{1}{2!} - \dfrac{1}{3!} + \dfrac{1}{4!} - \dfrac{1}{5!} + \cdots$

45. $1 - \dfrac{\pi^2}{2!} + \dfrac{\pi^4}{4!} - \dfrac{\pi^6}{6!} + \cdots$

46. $\pi - \dfrac{\pi^3}{3!} + \dfrac{\pi^5}{5!} - \dfrac{\pi^7}{7!} + \cdots$

In Problems 47–50, approximate the given quantity using the Taylor polynomial $P_n(x)$ for the indicated values of n and a. Determine the accuracy of the approximation.

47. $\sin 46°, \quad n = 2, a = \pi/4$ [*Hint*: Convert $46°$ to radian measure.]

48. $\cos 29°, \quad n = 2, a = \pi/6$

49. $e^{0.3}, \quad n = 4, a = 0$

50. $\sinh(0.1), \quad n = 3, a = 0$

51. Prove that the series obtained in Problem 5 represents $\sin x$ for every real value of x.

52. Prove that the series obtained in Problem 7 represents e^x for every real value of x.

53. Prove that the series obtained in Problem 9 represents $\sinh x$ for every real value of x.

54. Prove that the series obtained in Problem 10 represents $\cosh x$ for every real value of x.

≣ Applications

55. In leveling a long roadway of length L, an allowance must be made for the curvature of the Earth.

(a) Show that the leveling correction y indicated in FIGURE 9.10.3 is $y = R\sec(L/R) - R$, where R is the radius of the Earth measured in miles.

(b) If $P_2(x)$ is the second-degree Taylor polynomial for $f(x) = \sec x$ at $a = 0$, use $\sec x \approx P_2(x)$ for x close to zero to show that the approximate leveling correction is $y \approx L^2/(2R)$.

(c) Find the number of inches of leveling correction needed for 1 mi of roadway. Use $R = 4000$ mi.

(d) If we use $\sec x \approx P_4(x)$, then show that the leveling correction is

$$y \approx \dfrac{L^2}{2R} + \dfrac{5L^4}{24R^3}.$$

Redo the calculation in part (c) using the last formula.

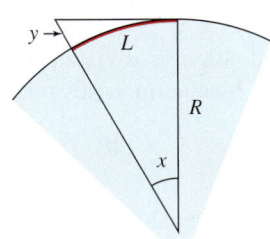

FIGURE 9.10.3 Earth in Problem 55

56. A wave of length L is traveling left to right across water of depth d (in feet), as illustrated in FIGURE 9.10.4. A mathematical model relating the speed v of the wave to L and d is

$$v = \sqrt{\frac{gL}{2\pi}\tanh\left(\frac{2\pi d}{L}\right)}.$$

(a) For deep water show that $v \approx \sqrt{gL/2\pi}$.
(b) Use (7) to find the first three nonzero terms of the Maclaurin series for $f(x) = \tanh x$. Show that when d/L is small, $v \approx \sqrt{gd}$. In other words, in shallow water the speed of a wave is independent of wave length.

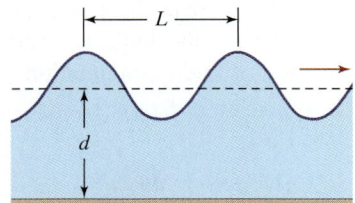

FIGURE 9.10.4 Wave in Problem 56

≡ **Think About It**

In Problems 57 and 58, find two ways, other than using (7), of finding the Maclaurin series representation of the given function.

57. $f(x) = \sin^2 x$ **58.** $f(x) = \sin x \cos x$

59. Without using (6), find the Taylor series for the function $f(x) = (x + 1)^2 e^x$ centered at $a = -1$. [*Hint:* $e^x = e^{x+1-1}$.]

60. Discuss: Does $f(x) = \cot x$ possess a Maclaurin series representation?

61. Explain why it stands to reason that the Maclaurin series (16) and (17) for $\cos x$ and $\sin x$ contain only even powers of x and only odd powers of x, respectively. Then reinspect the Maclaurin series in (18), (19), and (20) and make an observation.

62. Suppose it is desired to compute $f^{(10)}(0)$ for $f(x) = x^4\sin x^2$. Of course, one could use the brute force approach: Use the Product Rule and when the tenth derivative is (eventually) obtained set x equal to 0. Think of a more clever way of determining the value of this derivative.

≡ **Projects**

63. A Mathematical Classic The function

$$f(x) = \begin{cases} e^{-1/x^2}, & x \neq 0 \\ 0, & x = 0 \end{cases}$$

appears in almost every calculus text. The function f is continuous and possesses derivatives of all orders at every value of x.

(a) Use a calculator or CAS to obtain the graph of f.
(b) Use (7) to find the Maclaurin series for f. You will have to use the definition of the derivative to compute $f'(0), f''(0), \ldots$. For example,

$$f'(0) = \lim_{\Delta \to 0} \frac{f(0 + \Delta x) - f(0)}{\Delta x}.$$

It might help to use $t = \Delta x$ and to recall L'Hôpital's Rule. Show that the Maclaurin series for f converges for every x. Does the series represent the function f that generated it?

9.11 Binomial Series

▌**Introduction**   Most students of mathematics have a familiarity with binomial expansion in the two cases:

$$(1 + x)^2 = 1 + 2x + x^2$$
$$(1 + x)^3 = 1 + 3x + 3x^2 + x^3.$$

In general, if m is a positive integer, then

$$(1 + x)^m = 1 + mx + \frac{m(m - 1)}{2!}x^2 + \cdots + \frac{m(m - 1)(m - 2)\cdots(m - n + 1)}{n!}x^n \tag{1}$$
$$+ \cdots + mx^{m-1} + x^m.$$

The expansion of $(1 + x)^m$ in (1) is called the **Binomial Theorem**. Using summation notation, (1) is written

$$(1 + x)^m = \sum_{k=0}^{m} \binom{m}{k}x^k, \tag{2}$$

where the symbol $\begin{pmatrix} m \\ k \end{pmatrix}$ is defined as

for convenience this
term is defined to be 1 $(m - k + 1) = (m - (k - 1))$
↓   ↓

$$\begin{pmatrix} m \\ 0 \end{pmatrix} = 1, \quad k = 0 \qquad \text{and} \qquad \begin{pmatrix} m \\ k \end{pmatrix} = \frac{m(m - 1)(m - 2) \cdots (m - k + 1)}{k!}, \quad k \geq 1.$$

These numbers are called the **binomial coefficients**. For example, when $m = 3$, the four binomial coefficients are

$$\begin{pmatrix} 3 \\ 0 \end{pmatrix} = 1, \quad \begin{pmatrix} 3 \\ 1 \end{pmatrix} = \frac{3}{1} = 3, \quad \begin{pmatrix} 3 \\ 2 \end{pmatrix} = \frac{3(3 - 1)}{2} = 3, \quad \begin{pmatrix} 3 \\ 3 \end{pmatrix} = \frac{3(3 - 1)(3 - 2)}{6} = 1.$$

Although (2) has the appearance of a series, it is a finite sum consisting of $m + 1$ terms that ends with x^m. In this section we will see that when (1) is extended to powers m other than positive integers the result is an infinite series.

◄ The extension of the **Binomial Theorem** (m a positive integer) to **binomial series** (m fractional and negative real numbers) was first given by Isaac Newton in 1665.

■ **Binomial Series**   Now suppose $f(x) = (1 + x)^r$, where r represents any real number. From

$$\begin{array}{ll}
f(x) = (1 + x)^r & f(0) = 1 \\
f'(x) = r(1 + x)^{r-1} & f'(0) = r \\
f''(x) = r(r - 1)(1 + x)^{r-2} & f''(0) = r(r - 1) \\
f'''(x) = r(r - 1)(r - 2)(1 + x)^{r-3} & f'''(0) = r(r - 1)(r - 2) \\
\quad \vdots & \quad \vdots \\
f^{(n)}(x) = r(r - 1) \cdots (r - n + 1)(1 + x)^{r-n} & f^{(n)}(0) = r(r - 1) \cdots (r - n + 1)
\end{array}$$

we see that the Maclaurin series generated by f is

$$\sum_{k=0}^{\infty} \frac{f^{(k)}(0)}{k!} x^k = 1 + rx + \frac{r(r - 1)}{2!} x^2 + \frac{r(r - 1)(r - 2)}{3!} x^3 + \cdots + \frac{r(r - 1) \cdots (r - n + 1)}{n!} x^n + \cdots$$

$$= 1 + \sum_{k=1}^{\infty} \frac{r(r - 1) \cdots (r - k + 1)}{k!} x^k$$

$$= \sum_{k=0}^{\infty} \begin{pmatrix} r \\ k \end{pmatrix} x^k. \tag{3}$$

The power series given in (3) is called the **binomial series**. Note that (3) terminates only when r is a positive integer; in this case (3) reduces to (1). From the Ratio Test, the version given in Theorem 9.7.4,

$$\lim_{n \to \infty} \left| \frac{a_{n+1}}{a_n} \right| = \lim_{n \to \infty} \left| \frac{r(r - 1) \cdots (r - n + 1)(r - n) x^{n+1}}{(n + 1)!} \cdot \frac{n!}{r(r - 1) \cdots (r - n + 1) x^n} \right|$$

$$= \lim_{n \to \infty} \frac{|r - n|}{n + 1} |x|$$

$$= \lim_{n \to \infty} \frac{\left| \dfrac{r}{n} - 1 \right|}{1 + \dfrac{1}{n}} |x| = |x|$$

we conclude that the binomial series (3) converges for $|x| < 1$, or $-1 < x < 1$, and diverges for $|x| > 1$, that is, for $x > 1$ or $x < -1$. Convergence at the endpoints $x = \pm 1$ depends on the value of r.

Of course it is no big surprise to learn that the series (3) represents the function f that generated it. We state this as a formal theorem.

> **Theorem 9.11.1** Binomial Series
>
> If $|x| < 1$, then for any real number r,
>
> $$(1 + x)^r = \sum_{k=0}^{\infty} \binom{r}{k} x^k, \tag{4}$$
>
> where
>
> $$\binom{r}{0} = 1, k = 0, \quad \text{and} \quad \binom{r}{k} = \frac{r(r-1)(r-2)\cdots(r-k+1)}{k!}, \quad k \ge 1.$$

EXAMPLE 1 Representing a Function by a Binomial Series

Find a power series representation for $f(x) = \sqrt{1 + x}$.

Solution By rewriting f as $f(x) = (1 + x)^{1/2}$ we identify $r = \frac{1}{2}$. It then follows from (4) that for $|x| < 1$,

$$\sqrt{1 + x} = 1 + \binom{\frac{1}{2}}{1}x + \binom{\frac{1}{2}}{2}x^2 + \binom{\frac{1}{2}}{3}x^3 + \cdots + \binom{\frac{1}{2}}{n}x^n + \cdots$$

$$= 1 + \frac{1}{2}x + \frac{\frac{1}{2}(\frac{1}{2} - 1)}{2!}x^2 + \frac{\frac{1}{2}(\frac{1}{2} - 1)(\frac{1}{2} - 2)}{3!}x^3 + \cdots$$

$$+ \frac{\frac{1}{2}(\frac{1}{2} - 1)(\frac{1}{2} - 2)\cdots(\frac{1}{2} - n + 1)}{n!}x^n + \cdots$$

$$= 1 + \frac{1}{2}x - \frac{1}{2^2 2!}x^2 + \frac{1 \cdot 3}{2^3 3!}x^3 + \cdots + (-1)^{n-1}\frac{1 \cdot 3 \cdot 5 \cdots (2n - 3)}{2^n n!}x^n + \cdots.$$

The last line can be written using summation notation as

$$\sqrt{1 + x} = 1 + \frac{1}{2}x + \sum_{k=2}^{\infty}(-1)^{k-1}\frac{1 \cdot 3 \cdot 5 \cdots (2k - 3)}{2^k k!}x^k. \quad \blacksquare$$

Suppose the function in Example 1 had been $f(x) = \sqrt{4 + x}$. In order to get the binomial series representation of f we would have to rewrite the function in the form $(1 + x)^r$ by factoring the 4 out of the radical, that is,

$$f(x) = \sqrt{4 + x} = \sqrt{4}\left(1 + \frac{1}{4}x\right)^{1/2} = 2\left(1 + \frac{1}{4}x\right)^{1/2}.$$

We can now use (4) where the symbol x is replaced by $x/4$. The resulting series would then converge for $|x/4| < 1$ or $|x| < 4$.

EXAMPLE 2 A Formula from Physics

In Einstein's theory of relativity, the mass of a particle moving at a velocity v relative to an observer is given by

$$m = \frac{m_0}{\sqrt{1 - v^2/c^2}}, \tag{5}$$

where m_0 is the rest mass and c is the speed of light.

Many of the results from classical physics do not hold for particles, such as electrons, which may move close to the speed of light. Kinetic energy is no longer $K = \frac{1}{2}m_0 v^2$ but is

$$K = mc^2 - m_0 c^2. \tag{6}$$

If we identify $r = -\frac{1}{2}$ and $x = -v^2/c^2$ in (5), we have $|x| < 1$, since no particle can surpass the speed of light. Hence, (6) can be written:

$$K = \frac{m_0 c^2}{\sqrt{1 + x}} - m_0 c^2$$

$$= m_0 c^2\left[(1 + x)^{-1/2} - 1\right]$$

$$= m_0 c^2 \left[\left(1 - \frac{1}{2}x + \frac{3}{8}x^2 - \frac{5}{16}x^3 + \cdots \right) - 1 \right] \quad \leftarrow \text{now substitute for } x$$

$$= m_0 c^2 \left[\frac{1}{2}\left(\frac{v^2}{c^2}\right) + \frac{3}{8}\left(\frac{v^4}{c^4}\right) + \frac{5}{16}\left(\frac{v^6}{c^6}\right) + \cdots \right]. \tag{7}$$

In the everyday world where v is very much smaller than c, terms beyond the first in (7) are negligible. This leads to the well-known classical result

$$K \approx m_0 c^2 \left[\frac{1}{2}\left(\frac{v^2}{c^2}\right) \right] = \frac{1}{2}m_0 v^2. \qquad \blacksquare$$

Σ NOTES FROM THE CLASSROOM

As we come to the end of our discussion of infinite series you probably have a strong impression that divergent series are worthless. Not quite so. Mathematicians hate to see anything go to waste. Divergent series are used in a theory known as *asymptotic representations of functions*. It goes something like this; a divergent series of the form

$$a_0 + a_1/x + a_2/x^2 + \cdots$$

is an **asymptotic representation** of a function f if

$$\lim_{n\to\infty} x^n [f(x) - S_n(x)] = 0,$$

where $S_n(x)$ is the $(n+1)$st partial sum of the divergent series. Some important functions in applied mathematics are defined in this manner.

Exercises 9.11 Answers to selected odd-numbered problems begin on page ANS-28.

≡ Fundamentals

In Problems 1–10, use (4) to find the first four terms of a power series representation of the given function. Give the radius of convergence.

1. $f(x) = \sqrt[3]{1+x}$

2. $f(x) = \sqrt{1-x}$

3. $f(x) = \sqrt{9-x}$

4. $f(x) = \dfrac{1}{\sqrt{1+5x}}$

5. $f(x) = \dfrac{1}{\sqrt{1+x^2}}$

6. $f(x) = \dfrac{x}{\sqrt[3]{1-x^2}}$

7. $f(x) = (4+x)^{3/2}$

8. $f(x) = \dfrac{x}{\sqrt{(1+x)^5}}$

9. $f(x) = \dfrac{x}{(2+x)^2}$

10. $f(x) = x^2(1-x^2)^{-3}$

In Problems 11 and 12, explain why the error in the given approximation is less than the indicated amount. [*Hint:* Review Theorem 9.7.2.]

11. $(1+x)^{1/3} \approx 1 + \dfrac{x}{3}; \quad \dfrac{1}{9}x^2, x > 0$

12. $(1+x^2)^{-1/2} \approx 1 - \dfrac{x^2}{2} + \dfrac{3}{8}x^4; \quad \dfrac{5}{16}x^6$

13. Find a power series representation for $\sin^{-1}x$ using

$$\sin^{-1}x = \int_0^x \frac{1}{\sqrt{1-t^2}} \, dt.$$

14. (a) Show that the length of one-quarter of the ellipse $x^2/a^2 + y^2/b^2 = 1$ is given by $L = aE(k)$, where $E(k)$ is

$$E(k) = \int_0^{\pi/2} \sqrt{1 - k^2\sin^2\theta} \, d\theta$$

and $k^2 = (a^2 - b^2)/a^2 < 1$. This integral is called the **complete elliptic integral of the second kind**.

(b) Show that

$$L = a\frac{\pi}{2} - \frac{a}{2}\frac{\pi}{4}k^2 - \frac{a}{8}\frac{3\pi}{16}k^4 - \cdots.$$

15. In FIGURE 9.11.1 a hanging cable is supported at points A and B and carries a uniformly distributed load (such as the floor of a bridge). If $y = (4d/l^2)x^2$ is the equation of the cable, show that its length is given by

$$s = l + \frac{8d^2}{3l} - \frac{32d^4}{5l^3} + \cdots.$$

See Problem 22 in Exercises 4.10.

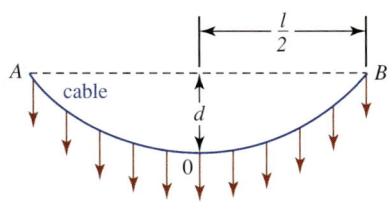

uniform load distributed horizontally

FIGURE 9.11.1 Hanging cable in Problem 15

16. Approximate the following integrals to three decimal places.

(a) $\int_0^{0.2} \sqrt{1 + x^3}\, dx$ **(b)** $\int_0^{1/2} \sqrt[3]{1 + x^4}\, dx$

17. By the law of cosines the potential at point A in **FIGURE 9.11.2** due to a unit charge at point B is $1/R = (1 - 2xr + r^2)^{-1/2}$, where $x = \cos\theta$. The expression $(1 - 2xr + r^2)^{-1/2}$ is said to be the **generating function** for the **Legendre polynomials** $P_k(x)$, since

$$(1 - 2xr + r^2)^{-1/2} = \sum_{k=0}^{\infty} P_k(x)r^k.$$

Use (4) to find $P_0(x)$, $P_1(x)$, and $P_2(x)$.

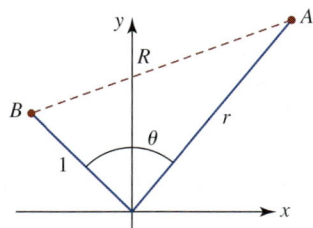

FIGURE 9.11.2 Unit charge at point B in Problem 17

18. (a) Suppose

$$f(x) = 1 + rx + \frac{r(r-1)}{2!}x^2 + \cdots$$
$$+ \frac{r(r-1)\cdots(r-n+1)}{n!}x^n + \cdots$$

for $|x| < 1$. Find $f'(x)$ and $xf'(x)$.

(b) Show that

$$(n+1)\frac{r(r-1)\cdots(r-n)}{(n+1)!} + n\frac{r(r-1)\cdots(r-n+1)}{n!}$$
$$= r\frac{r(r-1)\cdots(r-n+1)}{n!}.$$

(c) Show that $f'(x) + xf'(x) = rf(x)$.

(d) Solve the first-order differential equation

$$(1 + x)f'(x) = rf(x)$$

subject to $f(0) = 1$.

In Problems 19 and 20, use (4) to find a power series representation in $x - 1$ of the given function. [*Hint:* $1 + x = 2 + (x - 1)$.]

19. $f(x) = \sqrt{1 + x}$ **20.** $f(x) = (1 + x)^{-2}$

Chapter 9 in Review

Answers to selected odd-numbered problems begin on page ANS-28.

A. True/False

In Problems 1–30, indicate whether the given statement is true or false.

1. The sequence $\left\{\dfrac{(-1)^n n}{2n + 1}\right\}$ converges. _____

2. Every bounded sequence converges. _____

3. If a sequence is not monotonic, it is not convergent. _____

4. The sequence $\left\{\dfrac{10^n}{2^{n^2}}\right\}$ is not monotonic. _____

5. If $a_n \leq B$ for all n and $a_{n+1}/a_n \geq 1$ for all n, then $\{a_n\}$ converges. _____

6. $\lim\limits_{n\to\infty} \dfrac{|x|^n}{n!} = 0$ for every value of x. _____

7. If $\{a_n\}$ is a convergent sequence, then $\sum a_k$ always converges. _____

8. $0.999999\ldots = 1$ _____

9. If $\sum a_k = \frac{3}{2}$, then $a_n \to 0$ as $n \to \infty$. _____

10. If $a_n \to 0$ as $n \to \infty$, then $\sum a_k$ converges. _____

11. If $\sum a_k^2$ converges, then $\sum a_k$ converges. _____

12. If $\sum a_k$ converges and $\sum b_k$ diverges, then $\sum(a_k + b_k)$ diverges. _____

13. $\sum\limits_{k=1}^{\infty} \dfrac{1}{k^p}$ converges for $p = 1.0001$. _____

14. The series $\dfrac{2}{1} + \dfrac{2}{2} + \dfrac{2}{3} + \dfrac{2}{4} + \cdots$ diverges. _____

15. If $\sum |a_k|$ diverges, then $\sum a_k$ diverges. _____

16. If $\sum a_k$, $a_k > 0$, converges, then $\sum(-1)^{k+1}a_k$ converges. _____

17. If $\sum(-1)^{k+1}a_k$ converges absolutely, then $\sum(-1)^{k+1}\dfrac{a_k}{k}$ converges. _____

18. If $\sum b_k$ converges and $a_k \geq b_k$ for every positive integer k, then $\sum a_k$ converges. _____

19. If $\lim\limits_{n \to \infty} \left| \dfrac{a_{n+1}}{a_n} \right| = 1$, then $\sum a_k$ converges absolutely. _____

20. Every power series has a nonzero radius of convergence. _____

21. A power series converges absolutely at every number x in its interval of convergence. _____

22. A power series $\sum c_k x^k$ with an interval of convergence $[-R, R], R > 0$, is an infinitely differentiable function within $(-R, R)$. _____

23. If a power series $\sum c_k x^k$ converges for $-1 < x < 1$ and is convergent at $x = 1$, then the series must also converge at $x = -1$. _____

24. If the power series $\sum a_k x^k, a_k > 0$, has the interval of convergence $[-R, R), R > 0$, then the series converges conditionally, but not absolutely, at $x = -R$. _____

25. Since $\int_0^\infty e^{-x} \, dx = 1$, the series $\sum_{k=0}^{\infty} e^{-k}$ also converges to 1. _____

26. The series $1 + \dfrac{1}{2^2} - \dfrac{1}{3^2} - \dfrac{1}{4^2} + \dfrac{1}{5^2} + \dfrac{1}{6^2} - - + + \cdots$ converges. _____

27. $f(x) = \ln x$ cannot be represented by a Maclaurin series. _____

28. If the power series $\sum c_k(x - 4)^k$ diverges at $x = 7$, the series necessarily diverges at $x = 9$. _____

29. If the sequence $\{\sum_{k=1}^{n} a_k\}$ converges to 10, then $\sum_{k=1}^{\infty} a_k = 10$. _____

30. If $f(x) = \sum_{k=1}^{\infty} c_{2k-1} x^{2k-1}$ is the Maclaurin series of a function f, then $f^{(4)}(0) = 0$. _____

B. Fill in the Blanks

In Problems 1–12, fill in the blanks.

1. If $\{a_n\}$ converges to 4 and $\{b_n\}$ converges to 5, then $\{a_n b_n\}$ converges to _____, $\{a_n + b_n\}$ converges to _____, $\{a_n/b_n\}$ converges to _____, and $\{a_n^2\}$ converges to _____.

2. The sequence $\{\tan^{-1} n\}$ converges to _____.

3. To approximate the sum of the alternating series $\sum\limits_{k=1}^{\infty} \dfrac{(-1)^{k+1}}{10^k}$ to four decimal places, we need only use the _____ th partial sum.

4. The sum of the series $\sum\limits_{k=0}^{\infty} 4\left(\tfrac{2}{3}\right)^k$ is _____.

5. If n is an integer, $1 \leq n \leq 9$, then $0.nnn\ldots = $ _____ and so as a quotient of integers, $2.444444\ldots = $ _____.

6. The series $\sum_{k=1}^{\infty} [\tan^{-1} k - \tan^{-1}(k + 1)]$ converges to _____.

7. The power series $\sum\limits_{k=0}^{\infty} \dfrac{x^k}{k!}$ represents the function $f(x)$ _____ for all x.

8. The binomial series representation of $f(x) = (4 + x)^{1/2}$ has the radius of convergence _____.

9. The geometric series $\sum\limits_{k=1}^{\infty} \left(\dfrac{5}{x}\right)^{k-1}$ converges for the following values of x: _____.

10. If $e^x = \sum\limits_{k=0}^{\infty} \dfrac{x^k}{k!}$ for all real numbers x, then a power series for $e^{-x^3} = $ _____.

11. The interval of convergence of the power series $x - \dfrac{x^2}{2} + \dfrac{x^3}{3} - \dfrac{x^4}{4} + \cdots$ is _____.

12. If $\sum\limits_{k=1}^{n} a_k = 8 - 3\left(1 - \dfrac{1}{2^n}\right)$, then $\sum\limits_{k=1}^{\infty} a_k = $ _____.

C. Exercises

In Problems 1–12, determine whether the given series converges or diverges.

1. $\displaystyle\sum_{k=1}^{\infty} \frac{k}{(k^2+1)^2}$ **2.** $\displaystyle\sum_{k=1}^{\infty} \frac{1}{1+e^{-k}}$ **3.** $\displaystyle\sum_{k=1}^{\infty} \pi^{-k}$ **4.** $\displaystyle\sum_{k=0}^{\infty} \frac{1}{(\ln 2.5)^k}$

5. $\displaystyle\sum_{k=1}^{\infty} \frac{\sqrt{k}\ln k}{k^4+4}$ **6.** $\displaystyle\sum_{k=1}^{\infty} \frac{\sin k}{k^{3/2}}$ **7.** $\displaystyle\sum_{k=2}^{\infty} \frac{k}{\sqrt[3]{k^6-4k}}$ **8.** $\displaystyle\sum_{k=2}^{\infty} \frac{1}{k\sqrt{\ln k}}$

9. $\displaystyle\sum_{k=1}^{\infty} \frac{1+(-1)^k}{\sqrt{k}}$ **10.** $\displaystyle\sum_{k=1}^{\infty} \frac{(k^2)!}{(k!)^2}$ **11.** $\displaystyle\sum_{k=1}^{\infty} \frac{1}{3k^2+4k+6}$ **12.** $\displaystyle\sum_{k=1}^{\infty} \ln\left(\frac{3k}{k+1}\right)$

In Problems 13 and 14, find the sum of the given convergent series.

13. $\displaystyle\sum_{k=1}^{\infty} \frac{(-1)^{k-1}+3}{(1.01)^{k-1}}$ **14.** $\displaystyle\sum_{k=1}^{\infty} \frac{1}{k^2+11k+30}$

In Problems 15–18, find the interval of convergence of the given power series.

15. $\displaystyle\sum_{k=1}^{\infty} \frac{3^k}{k^3}x^k$ **16.** $\displaystyle\sum_{k=1}^{\infty} \frac{k}{4^k}(2x-1)^k$ **17.** $\displaystyle\sum_{k=1}^{\infty} k!(x+5)^k$ **18.** $\displaystyle\sum_{k=2}^{\infty} \frac{(2x)^k}{\ln k}$

19. Find the radius of convergence for the power series

$$\sum_{k=1}^{\infty} \frac{2\cdot 5\cdot 8\cdots(3k-1)}{3\cdot 7\cdot 11\cdots(4k-1)}x^k.$$

20. Find the values of x for which $\sum_{k=1}^{\infty}(\cos x)^k$ converges.

21. For $|\alpha|>1$, find the sum of the series

$$\frac{1}{\alpha}+\frac{1}{\alpha^2}+\frac{1}{\alpha^3}+\cdots.$$

22. Determine whether the following argument is valid. If

$$S = 1+2+4+8+\cdots, \qquad \text{then} \qquad 2S = 2+4+8+\cdots = S-1.$$

Solving $2S = S-1$ gives $S = -1$.

In Problems 23–26, find by any method, the first three nonzero terms of the Maclaurin series for the given function.

23. $f(x) = \dfrac{1}{\sqrt[3]{1+x^5}}$ **24.** $f(x) = \dfrac{x}{2-x}$ **25.** $f(x) = \sin x\cos x$ **26.** $f(x) = \displaystyle\int_0^x e^{t^2}\,dt$

27. Find the Taylor series for $f(x) = \cos x$ with center $a = \pi/2$.

28. Prove that the series in Problem 25 represents the function by showing that $R_n(x) \to 0$ as $n \to \infty$.

29. A large convention of free-spending mathematicians from out of town contributes \$3 million to the economy of the city of San Francisco. It is estimated that each resident of the city spends $\frac{2}{3}$ of his or her income in the city. Thus, of the amount of the money brought in by the convention, $3(\frac{2}{3}) = \$2$ million is spent by the people of San Francisco in the city. Of this last amount, $\frac{2}{3}$ is spent in the city, and so on. How much, in the long run, do the residents of San Francisco spend in their city as a result of the convention?

30. If P dollars are invested at an annual rate of interest r compounded annually, the return S after m years is $S = (1+r)^m$. The **Rule of 70**, which is often used by loan officers and stock analysts, says that the time required to double an investment earning an annual rate of interest r is approximately $70/(100r)$ years. For example, money invested at an annual rate of 5% takes approximately $79/100(0.05) = 14$ years to double.

  (a) Show that the true doubling time is $\ln 2/\ln(1+r)$.
  (b) Use the Maclaurin series for $\ln(1+r)$ to derive the Rule of 70.
  (c) Use the first three terms of the Maclaurin series for $\ln(1+r)$ to approximate that interest rate for which the Rule of 70 gives the true doubling time.

Conics and Polar Coordinates

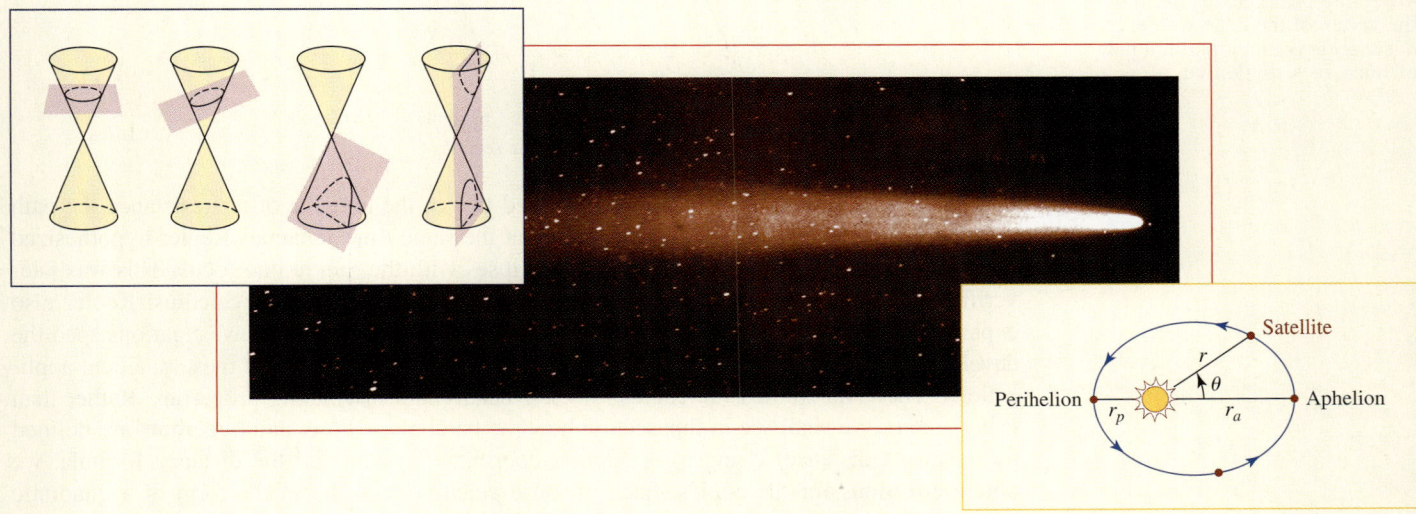

In This Chapter   A rectangular or Cartesian equation is not the only, and often not the most convenient, way of describing a curve in the plane. In this chapter we shall consider two additional means by which a curve can be represented. One of the two approaches utilizes an entirely new kind of coordinate system.

We begin by reviewing the notion of a conic section.

Hypatia

10.1 Conic Sections

■ **Introduction** **Hypatia** is the first woman in the history of mathematics about whom we have considerable knowledge. Born in 370 CE, in Alexandria, she was renowned as a mathematician and philosopher. Among her writings is *On the Conics of Apollonius*, which popularized Apollonius' (200 BCE) work on curves that can be obtained by intersecting a double-napped cone with a plane: the circle, parabola, ellipse, and hyperbola. See FIGURE 10.1.1. With the close of the Greek period, interest in conic sections waned; after Hypatia the study of these curves was neglected for over 1000 years.

> When the plane passes through the vertex of the cone we get a *degenerate conic*: a point, a pair of lines, or a single line.

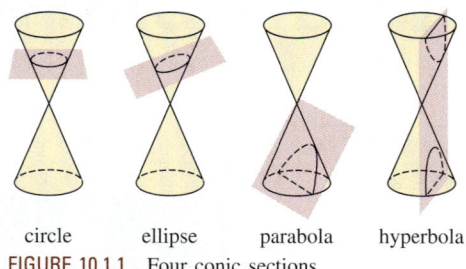

circle ⟶ ellipse ⟶ parabola ⟶ hyperbola

FIGURE 10.1.1 Four conic sections

In the seventeenth century, Galileo showed that in the absence of air resistance the path of a projectile follows a parabolic arc. At about the same time Johannes Kepler hypothesized that the orbits of planets about the Sun are ellipses with the Sun at one focus. This was later verified by Isaac Newton, using the methods of the newly developed calculus. Kepler also experimented with the reflecting properties of parabolic mirrors; these investigations sped the development of the reflecting telescope. The Greeks had known little of these practical applications. They had studied the conics for their beauty and fascinating properties. Rather than using a cone, we shall see in this section how the parabola, ellipse, and hyperbola are defined by means of distance. Using a rectangular coordinate system and the distance formula, we obtain equations for the conics. Each of these equations will be in the form of a quadratic equation in variables x and y:

$$Ax^2 + Bxy + Cy^2 + Dx + Ey + F = 0, \tag{1}$$

where A, B, C, D, E, and F are constants. The **standard form** of a circle with center (h, k) and radius r,

$$(x - h)^2 + (y - k)^2 = r^2, \tag{2}$$

is a special case of (1). Equation (2) is a direct result of the definition of a circle:

- A **circle** is defined to be the set of all points $P(x, y)$ in the coordinate plane that are a given fixed distance r, called the **radius**, from a given fixed point (h, k), called the **center**.

In a similar manner, we use the distance formula to obtain equations for the parabola, ellipse, and hyperbola.

The graph of a quadratic function $y = ax^2 + bx + c, a \neq 0$, is a parabola. However, not every parabola is the graph of a function of x. In general, a parabola is defined in the following manner.

Definition 10.1.1 Parabola

A **parabola** is the set of all points $P(x, y)$ in the plane that are equidistant from a fixed line L, called the **directrix**, and a fixed point F, called the **focus**.

The line through the focus perpendicular to the directrix is called the **axis** of the parabola. The point of intersection of the parabola and the axis is called the **vertex** of the parabola.

Equation of a Parabola To describe a parabola analytically, let us assume for the sake of discussion that the directrix L is the horizontal line $y = -p$ and that the focus is $F(0, p)$. Using Definition 10.1.1 and FIGURE 10.1.2, we see that $d(F, P) = d(P, Q)$ is the same as

$$\sqrt{x^2 + (y - p)^2} = y + p.$$

Squaring both sides and simplifying lead to

$$x^2 = 4py. \tag{3}$$

We say that (3) is the **standard form** for the equation of a parabola with focus $F(0, p)$ and directrix $y = -p$. In like manner, if the directrix and focus are $x = -p$ and $F(p, 0)$, respectively, we find that the standard form for the equation of the parabola is

$$y^2 = 4px. \tag{4}$$

Although we assumed that $p > 0$ in Figure 10.1.2, this, of course, need not be the case. FIGURE 10.1.3 summarizes information about equations (3) and (4).

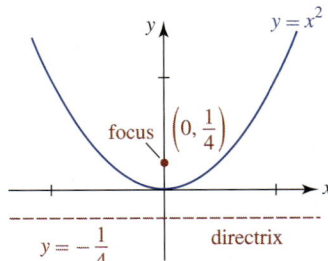

FIGURE 10.1.2 Parabola with vertex $(0, 0)$ and focus on the y-axis

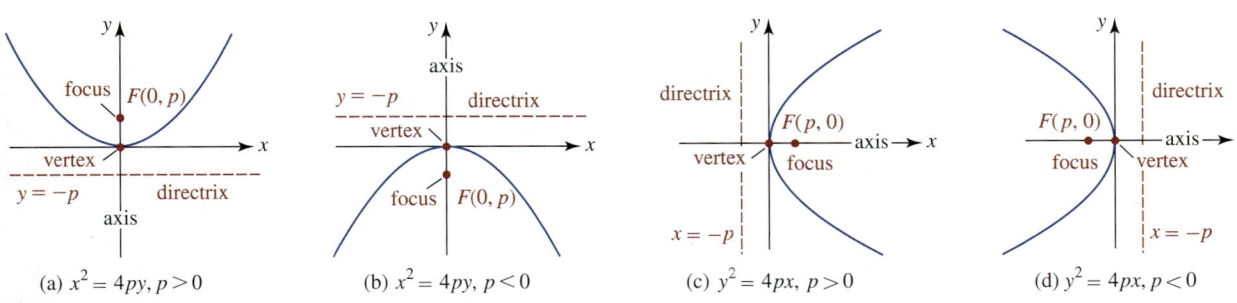

(a) $x^2 = 4py, p > 0$ (b) $x^2 = 4py, p < 0$ (c) $y^2 = 4px, p > 0$ (d) $y^2 = 4px, p < 0$

FIGURE 10.1.3 Pictorial summary of equations (3) and (4)

EXAMPLE 1   Focus and Directrix
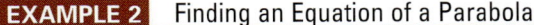

Find the focus and directrix of the parabola whose equation is $y = x^2$.

Solution Comparing the equation $y = x^2$ with (3) enables us to identify the coefficient of y, $4p = 1$ and so $p = \frac{1}{4}$. Hence, the focus of the parabola is $(0, \frac{1}{4})$ and its directrix is the horizontal line $y = -\frac{1}{4}$. The familiar graph, along with the focus and directrix, is given in FIGURE 10.1.4. ∎

By knowing the basic parabolic shape, all we need to know to sketch a *rough* graph of either equations (3) or (4) is the fact that the graph passes through its vertex (0, 0) and the direction in which the parabola opens. To add more accuracy to the graph it is convenient to use the number p determined by the standard-form equation to plot two additional points. Note that if we choose $y = p$ in (3), then $x^2 = 4p^2$ implies $x = \pm 2p$. Thus $(2p, p)$ and $(-2p, p)$ lie on the graph of $x^2 = 4py$. Similarly, the choice $x = p$ in (2) yields the points $(p, 2p)$ and $(p, -2p)$ on the graph of $y^2 = 4px$. The *line segment* through the focus with endpoints $(2p, p)$, $(-2p, p)$ for equations with standard form (3), and $(p, 2p)$, $(p, -2p)$ for equations with standard form (4) is called the **focal chord**. For example, in Figure 10.1.4, if we choose $y = \frac{1}{4}$, then $x^2 = \frac{1}{4}$ implies $x = \pm\frac{1}{2}$. Endpoints of the horizontal focal chord for $y = x^2$ are $(-\frac{1}{2}, \frac{1}{4})$ and $(\frac{1}{2}, \frac{1}{4})$.

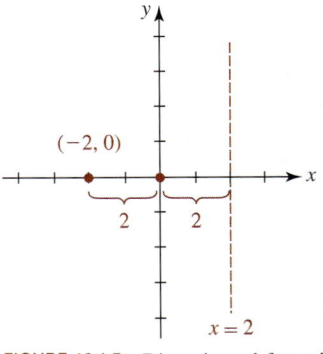

FIGURE 10.1.4 Graph of equation in Example 1

◀ Graphing tip for equations (3) and (4).

EXAMPLE 2   Finding an Equation of a Parabola

Find the equation in standard form of the parabola with directrix $x = 2$ and focus $(-2, 0)$. Graph.

Solution In FIGURE 10.1.5 we have graphed the directrix and the focus. We see from their placement that the equation we seek is of the form $y^2 = 4px$. Since $p = -2$, the parabola opens to the left and so

$$y^2 = 4(-2)x \qquad \text{or} \qquad y^2 = -8x.$$

As mentioned in the discussion preceding this example, if we substitute $x = p = -2$ into the equation $y^2 = -8x$ we can find two points on its graph. From $y^2 = -8(-2) = 16$ we get

FIGURE 10.1.5 Directrix and focus in Example 2

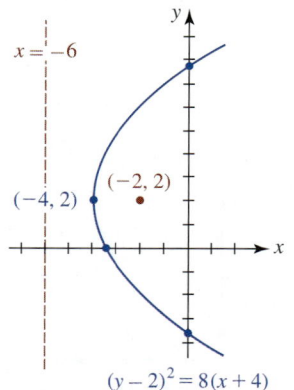

$y^2 = -8x$

$(-2, 4)$

$(-2, -4)$

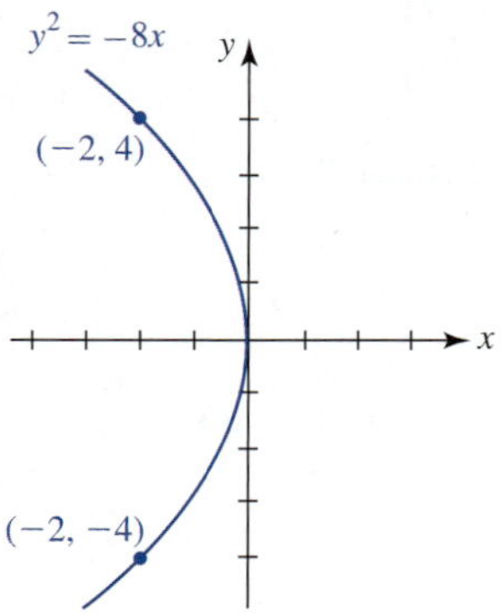

FIGURE 10.1.6 Graph of parabola in Example 2

$y = \pm4$. As shown in FIGURE 10.1.6, the graph passes through $(0, 0)$ as well as through the endpoints $(-2, -4)$ and $(-2, 4)$ of the focal chord. ∎

■ **Vertex Translated to (h, k)** In general, the **standard form** for the equation of a parabola with vertex (h, k) is given by either

$$(x - h)^2 = 4p(y - k) \tag{5}$$

or

$$(y - k)^2 = 4p(x - h). \tag{6}$$

The parabolas defined by these equations are identical in shape to the parabolas defined by equations (3) and (4) because equations (5) and (6) represent rigid transformations (shifts up, down, left, and right) of the graphs of (3) and (4). For example, the parabola $(x + 1)^2 = 8(y - 5)$ has vertex $(-1, 5)$. Its graph is the graph of $x^2 = 8y$ shifted horizontally one unit to the left followed by an upward vertical shift of five units.

For each of the equations, (3) and (4) or (5) and (6), the *distance* from the vertex to the focus, as well as the distance from the vertex to the directrix, is $|p|$.

EXAMPLE 3 Find Everything

Find the vertex, focus, axis, directrix, and graph of the parabola

$$y^2 - 4y - 8x - 28 = 0. \tag{7}$$

Solution In order to write the equation in one of the standard forms we complete the square in y:

$$y^2 - 4y + 4 = 8x + 28 + 4 \qquad \leftarrow \text{add 4 to both sides}$$
$$(y - 2)^2 = 8(x + 4).$$

Comparing the last equation with (6) we conclude that the vertex is $(-4, 2)$ and that $4p = 8$ or $p = 2$. From $p = 2 > 0$, the parabola opens to the right and the focus is 2 units to the right of the vertex at $(-2, 2)$. The directrix is the vertical line 2 units to the left of the vertex, $x = -6$. Knowing that the parabola opens to the right from the point $(-4, 2)$ also tells us that the graph has intercepts. To find the x-intercept we set $y = 0$ in (7) and find immediately that $x = -\frac{7}{2}$. The x-intercept is $\left(-\frac{7}{2}, 0\right)$. To find the y-intercepts we set $x = 0$ in (7) and find from the quadratic formula that $y = 2 \pm 4\sqrt{2}$ or $y \approx 7.66$ and $y \approx -3.66$. The y-intercepts are $(0, 2 - 4\sqrt{2})$ and $(0, 2 + 4\sqrt{2})$. Putting all this information together we get the graph in FIGURE 10.1.7. ∎

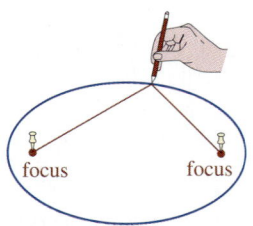

$x = -6$

$(-4, 2)$ $(-2, 2)$

$(y - 2)^2 = 8(x + 4)$

FIGURE 10.1.7 Graph of equation in Example 3

The ellipse is defined as follows.

Definition 10.1.2 Ellipse

An **ellipse** is the set of points $P(x, y)$ in the plane such that the sum of the distances between P and two fixed points F_1 and F_2 is a constant. The fixed points F_1 and F_2 are called **foci** (plural for **focus**). The midpoint of the line segment joining F_1 and F_2 is called the **center** of the ellipse.

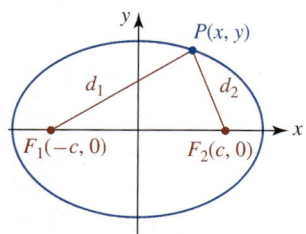

focus   focus

FIGURE 10.1.8 A way to draw an ellipse

If P is a point on the ellipse and $d_1 = d(F_1, P)$, $d_2 = d(F_2, P)$ are distances from the foci to P, then Definition 10.1.2 asserts that

$$d_1 + d_2 = k, \tag{8}$$

where $k > 0$ is a constant.

On a practical level, (8) can be used to sketch an ellipse. FIGURE 10.1.8 shows that if a string of length k is attached to a paper by two tacks, then an ellipse can be traced by inserting a pencil against the string and moving it in such a manner that the string remains taut.

■ **Equation of an Ellipse** For convenience, let us choose $k = 2a$ and put the foci on the x-axis with coordinates $F_1(-c, 0)$ and $F_2(c, 0)$. See FIGURE 10.1.9. It follows from (8) that

$P(x, y)$

d_1 d_2

$F_1(-c, 0)$ $F_2(c, 0)$

FIGURE 10.1.9 Ellipse with center $(0, 0)$ and foci on the x-axis

$$\sqrt{(x + c)^2 + y^2} + \sqrt{(x - c)^2 + y^2} = 2a. \tag{9}$$

Squaring (9), simplifying, and squaring again yields

$$(a^2 - c^2)x^2 + a^2y^2 = a^2(a^2 - c^2). \tag{10}$$

Referring to Figure 10.1.9, we see that the points F_1, F_2, and P form a triangle. Because the sum of the lengths of any two sides of a triangle is greater than the remaining side, we must have $2a > 2c$ or $a > c$. Hence, $a^2 - c^2 > 0$. When we let $b^2 = a^2 - c^2$, then (8) becomes $b^2x^2 + a^2y^2 = a^2b^2$. Dividing this last equation by a^2b^2 gives

$$\frac{x^2}{a^2} + \frac{y^2}{b^2} = 1. \tag{11}$$

Equation (11) is called the **standard form** of the equation of an ellipse centered at $(0, 0)$ with foci $(-c, 0)$ and $(c, 0)$, where c is defined by $b^2 = a^2 - c^2$ and $a > b > 0$.

If the foci are placed on the y-axis, then a repetition of the above analysis leads to

$$\frac{x^2}{b^2} + \frac{y^2}{a^2} = 1. \tag{12}$$

Equation (12) is called the **standard form** of the equation of an ellipse centered at $(0, 0)$ with foci $(0, -c)$ and $(0, c)$, where c is defined by $b^2 = a^2 - c^2$ and $a > b > 0$.

■ **Major and Minor Axes** The **major axis** of an ellipse is the line segment through its center, containing the foci, and with endpoints on the ellipse. For an ellipse with standard equation (11), the major axis is horizontal whereas, for (12) the major axis is vertical. The line segment through the center, perpendicular to the major axis, and with endpoints on the ellipse is called the **minor axis**. The two endpoints of the major axis are called the **vertices** of the ellipse. For (11) the vertices are the x-intercepts. Setting $y = 0$ in (11) gives $x = \pm a$. The vertices are then $(-a, 0)$ and $(a, 0)$. For (12) the vertices are the y-intercepts $(0, -a)$ and $(0, a)$. For equation (11), the endpoints of the minor axis are $(0, -b)$ and $(0, b)$; for (12) the endpoints are $(-b, 0)$ and $(b, 0)$. For either (11) or (12), the **length of the major axis** is $a - (-a) = 2a$; the length of the minor axis is $2b$. Since $a > b$, the major axis of an ellipse is always longer than its minor axis.

A summary of all this information for equations (11) and (12) is given in FIGURE 10.1.10.

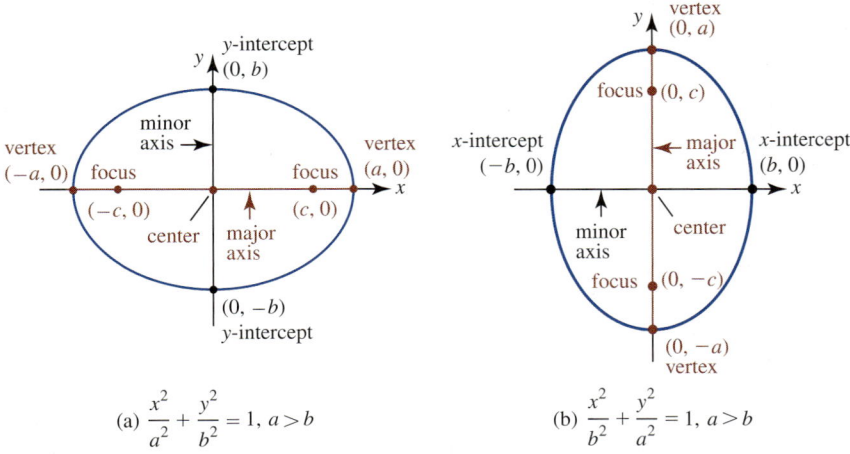

(a) $\dfrac{x^2}{a^2} + \dfrac{y^2}{b^2} = 1, a > b$

(b) $\dfrac{x^2}{b^2} + \dfrac{y^2}{a^2} = 1, a > b$

FIGURE 10.1.10   Pictorial summary of equations (11) and (12)

EXAMPLE 4   Vertices, Foci, Graph

Find the vertices and foci of the ellipse whose equation is $9x^2 + 3y^2 = 27$. Graph.

Solution   By dividing both sides of the equality by 27 the standard form of the equation is

$$\frac{x^2}{3} + \frac{y^2}{9} = 1.$$

We see that $9 > 3$ and so we identify the equation with (12). From $a^2 = 9$ and $b^2 = 3$, we see that $a = 3$ and $b = \sqrt{3}$. The major axis is vertical with endpoints or vertices $(0, -3)$

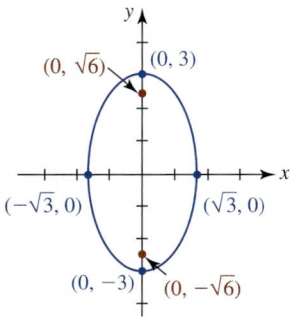

FIGURE 10.1.11 Ellipse in Example 4

and $(0, 3)$. The minor axis is horizontal with endpoints $(-\sqrt{3}, 0)$ and $(\sqrt{3}, 0)$. Of course, the vertices are also the y-intercepts and the endpoints of the minor axis are the x-intercepts. Now, to find the foci we use $b^2 = a^2 - c^2$ or $c^2 = a^2 - b^2$ to write $c = \sqrt{a^2 - b^2}$. With $a = 3, b = \sqrt{3}$, we get $c = \sqrt{6}$. Hence, the foci are on the y-axis at $(0, -\sqrt{6})$ and $(0, \sqrt{6})$. The graph is given in FIGURE 10.1.11. ∎

■ **Center Translated to (h, k)** When the center is at (h, k), the standard form for the equation of an ellipse is either

$$\frac{(x - h)^2}{a^2} + \frac{(y - k)^2}{b^2} = 1 \tag{13}$$

or

$$\frac{(x - h)^2}{b^2} + \frac{(y - k)^2}{a^2} = 1. \tag{14}$$

The ellipses defined by these equations are identical in shape to the ellipses defined by equations (11) and (12) since equations (13) and (14) represent rigid transformations of the graphs of (11) and (12). For example, the graph of the ellipse

$$\frac{(x - 1)^2}{9} + \frac{(y + 3)^2}{16} = 1$$

with center $(1, -3)$ is the graph of $x^2/9 + y^2/16 = 1$ shifted horizontally 1 unit to the right followed by a downward vertical shift of 3 units.

It is not a good idea to memorize formulas for the vertices and foci of an ellipse with center (h, k). Everything is the same as before, a, b, and c are positive, $a > b, a > c$, and $c^2 = a^2 - b^2$. You can locate vertices, foci, and endpoints of the minor axis using the fact that a is the distance from the center to a vertex, b is the distance from the center to an endpoint on the minor axis, and c is the distance from the center to a focus.

EXAMPLE 5 Find Everything

Find the vertices and foci of the ellipse $4x^2 + 16y^2 - 8x - 96y + 84 = 0$. Graph.

Solution To write the given equation in one of the standard forms (13) or (14) we complete the square in x and in y. To do this, recall we want the coefficients of the quadratic terms x^2 and y^2 to be 1. Factoring 4 from the x terms and 16 from the y terms gives

$$4(x^2 - 2x + 1) + 16(y^2 - 6y + 9) = -84 + 4 \cdot 1 + 16 \cdot 9$$

or $4(x - 1)^2 + 16(y - 3)^2 = 64$. The last equation yields the standard form

$$\frac{(x - 1)^2}{16} + \frac{(y - 3)^2}{4} = 1. \tag{15}$$

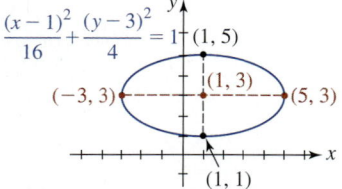

FIGURE 10.1.12 Ellipse in Example 5

In (15) we identify $a^2 = 16$ or $a = 4$, $b^2 = 4$ or $b = 2$, and $c^2 = a^2 - b^2 = 12$, or $c = 2\sqrt{3}$. The major axis is horizontal and lies on the horizontal line $y = 3$ passing through the center $(1, 3)$. This is the red horizontal dashed line segment in FIGURE 10.1.12. By measuring $a = 4$ units to the left and then to the right of the center along the line $y = 3$ we arrive at the vertices $(-3, 3)$ and $(5, 3)$. By measuring $b = 2$ units both down and up the vertical line $x = 1$ through the center we arrive at the endpoints $(1, 1)$ and $(1, 5)$ of the minor axis. The minor axis is the black dashed vertical line segment in Figure 10.1.12. Finally, by measuring $c = 2\sqrt{3}$ units to the left and right of the center along $y = 3$ we obtain the foci $(1 - 2\sqrt{3}, 3)$ and $(1 + 2\sqrt{3}, 3)$. ∎

The definition of a hyperbola is basically the same as the definition of the ellipse with only one exception: The word *sum* is replaced by the word *difference*.

Definition 10.1.3 Hyperbola

A **hyperbola** is the set of points $P(x, y)$ in the plane such that the difference of the distances between P and two fixed points F_1 and F_2 is constant. The fixed points F_1 and F_2 are called **foci** (plural for **focus**). The midpoint of the line segment joining points F_1 and F_2 is called the **center** of the hyperbola.

If P is a point on the hyperbola, then

$$|d_1 - d_2| = k, \tag{16}$$

where $d_1 = d(F_1, P)$ and $d_2 = d(F_2, P)$. Proceeding as for the ellipse, we place the foci on the x-axis at $F_1(-c, 0)$ and $F_2(c, 0)$ as shown in FIGURE 10.1.13 and choose the constant k to be $2a$ for algebraic convenience. As illustrated in the figure the graph of a hyperbola consists of two **branches**.

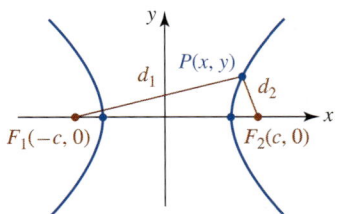

FIGURE 10.1.13 Hyperbola with center $(0, 0)$ and foci on the x-axis

■ **Hyperbola with Center (0, 0)** Applying the usual distance formula and algebra to (16) yields the **standard form** of the equation of a hyperbola centered at $(0, 0)$ with foci $(-c, 0)$ and $(c, 0)$,

$$\frac{x^2}{a^2} - \frac{y^2}{b^2} = 1 \tag{17}$$

When the foci lie on the y-axis, the **standard form** of the equation of a hyperbola centered at $(0, 0)$ with foci $(0, -c)$ and $(0, c)$ is

$$\frac{y^2}{a^2} - \frac{x^2}{b^2} = 1. \tag{18}$$

In both (17) and (18), c is defined by $b^2 = c^2 - a^2$ and $c > a$.

For the hyperbola (unlike the ellipse) bear in mind that in (17) and (18) there is no relationship between the relative sizes of a and b; rather, a^2 is always the denominator of the *positive term* and the intercepts *always* have $\pm a$ as a coordinate.

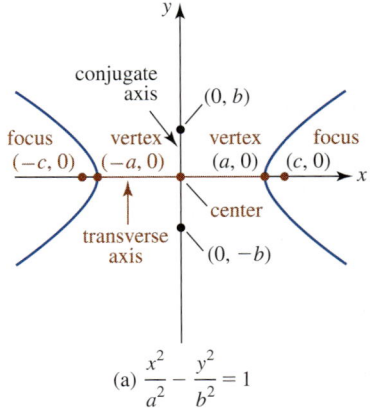

(a) $\dfrac{x^2}{a^2} - \dfrac{y^2}{b^2} = 1$

■ **Transverse and Conjugate Axes** The line segment with endpoints on the hyperbola and lying on the line through the foci is called the **transverse axis**; its endpoints are called the **vertices** of the hyperbola. For the hyperbola described by equation (17), the transverse axis lies on the x-axis. Therefore, the coordinates of the vertices are the x-intercepts. Setting $y = 0$ gives $x^2/a^2 = 1$, or $x = \pm a$. Thus, as shown in FIGURE 10.1.14 the vertices are $(-a, 0)$ and $(a, 0)$; the **length of the transverse axis** is $2a$. Notice that by setting $y = 0$ in (18), we get $-y^2/b^2 = 1$ or $y^2 = -b^2$, which has no real solutions. Hence the graph of any equation in that form has no y-intercepts. Nonetheless, the numbers $\pm b$ are important. The line segment through the center of the hyperbola perpendicular to the transverse axis and with endpoints $(0, -b)$ and $(0, b)$ is called the **conjugate axis**. Similarly, the graph of an equation in standard form (18) has no x-intercepts. The conjugate axis for (18) is the line segment with endpoints $(-b, 0)$ and $(b, 0)$.

This information for equations (17) and (18) is summarized in Figure 10.1.14.

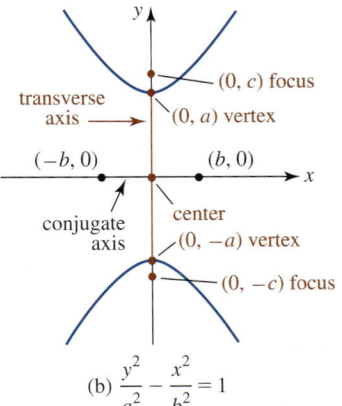

(b) $\dfrac{y^2}{a^2} - \dfrac{x^2}{b^2} = 1$

FIGURE 10.1.14 Pictorial summary of equations (17) and (18)

■ **Asymptotes** Every hyperbola possesses a pair of slant asymptotes that pass through its center. These asymptotes are indicative of end behavior, and as such are an invaluable aide in sketching the graph of a hyperbola. Solving (17) for y in terms of x gives

$$y = \pm \frac{b}{a} x \sqrt{1 - \frac{a^2}{x^2}}.$$

As $x \to -\infty$ or as $x \to \infty$, $a^2/x^2 \to 0$, and thus $\sqrt{1 - a^2/x^2} \to 1$. Therefore, for large values of $|x|$, points on the graph of the hyperbola are close to the points on the lines

$$y = \frac{b}{a} x \qquad \text{and} \qquad y = -\frac{b}{a} x. \tag{19}$$

By a similar analysis we find that the slant asymptotes for (18) are

$$y = \frac{a}{b} x \qquad \text{and} \qquad y = -\frac{a}{b} x. \tag{20}$$

Each pair of asymptotes intersect at the origin, which is the center of the hyperbola. Note, too, in FIGURE 10.1.15(a) that the asymptotes are simply the *extended diagonals* of a rectangle of width $2a$ (the length of the transverse axis) and height $2b$ (the length of the conjugate axis); in Figure 10.1.15(b) the asymptotes are the extended diagonals of a rectangle of width $2b$ and height $2a$.

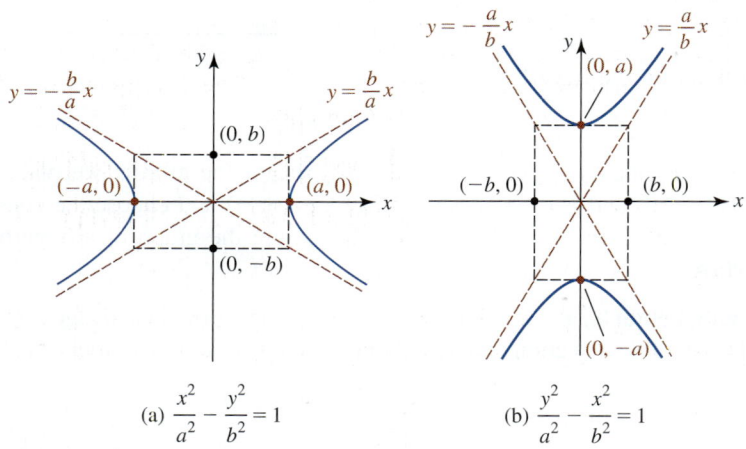

$$(a) \quad \frac{x^2}{a^2} - \frac{y^2}{b^2} = 1 \qquad\qquad (b) \quad \frac{y^2}{a^2} - \frac{x^2}{b^2} = 1$$

FIGURE 10.1.15 Hyperbolas (17) and (18) with slant asymptotes

We recommend that you *do not* memorize the equations in (19) and (20). There is an easy method for obtaining the asymptotes of a hyperbola. For example, since $y = \pm \dfrac{b}{a}x$ is equivalent to

$$\frac{x^2}{a^2} = \frac{y^2}{b^2} \qquad \text{or} \qquad \frac{x^2}{a^2} - \frac{y^2}{b^2} = 0. \tag{21}$$

Note that the last equation in (21) factors as the difference of two squares:

$$\left(\frac{x}{a} - \frac{y}{b}\right)\left(\frac{x}{a} + \frac{y}{b}\right) = 0.$$

▶ This is a mnemonic, or memory device. It has no geometric significance.

Setting each factor equal to zero and solving for y gives an equation of an asymptote. Equation (21) is simply the left-hand side of the standard form of the equation of a hyperbola given in (17). In like manner, to obtain the asymptotes for (18) just replace 1 by 0 in the standard form, factor $y^2/a^2 - x^2/b^2 = 0$, and solve for y.

EXAMPLE 6 Vertices, Foci, Asymptotes, Graph

Find the vertices, foci, and asymptotes of the hyperbola $9x^2 - 25y^2 = 225$. Graph.

Solution We first put the equation into standard form by dividing both sides of the equality by 225:

$$\frac{x^2}{25} - \frac{y^2}{9} = 1. \tag{22}$$

From this equation we see that $a^2 = 25$ and $b^2 = 9$, and so $a = 5$ and $b = 3$. Therefore, the vertices are $(-5, 0)$ and $(5, 0)$. Since $b^2 = c^2 - a^2$ implies $c^2 = a^2 + b^2$, we have $c^2 = 34$, and so the foci are $(-\sqrt{34}, 0)$ and $(\sqrt{34}, 0)$. To find the slant asymptotes we use the standard form (22) with 1 replaced by 0:

$$\frac{x^2}{25} - \frac{y^2}{9} = 0 \qquad \text{factors as} \qquad \left(\frac{x}{5} - \frac{y}{3}\right)\left(\frac{x}{5} + \frac{y}{3}\right) = 0.$$

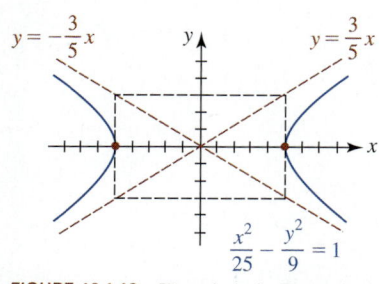

FIGURE 10.1.16 Hyperbola in Example 6

Setting each factor equal to zero and solving for y gives the asymptotes $y = \pm 3x/5$. We plot the vertices and graph the two lines through the origin. Both branches of the hyperbola must become arbitrarily close to the asymptotes as $x \to \pm\infty$. See FIGURE 10.1.16. ∎

■ **Center Translated to (h, k)** When the center of the hyperbola is (h, k) the **standard form** analogues of equations (17) and (18) are, in turn,

$$\frac{(x-h)^2}{a^2} - \frac{(y-k)^2}{b^2} = 1 \tag{23}$$

and

$$\frac{(y-k)^2}{a^2} - \frac{(x-h)^2}{b^2} = 1. \tag{24}$$

As in (17) and (18) the numbers a^2, b^2, and c^2 are related by $b^2 = c^2 - a^2$.

You can locate vertices and foci using the fact that a is the distance from the center to a vertex, and c is the distance from the center to a focus. The slant asymptotes for (23) can be obtained by factoring

$$\frac{(x-h)^2}{a^2} - \frac{(y-k)^2}{b^2} = 0 \quad \text{as} \quad \left(\frac{x-h}{a} - \frac{y-k}{b}\right)\left(\frac{x-h}{a} + \frac{y-k}{b}\right) = 0.$$

Similarly, the asymptotes for (24) can be obtained from factoring $\dfrac{(y-k)^2}{a^2} - \dfrac{(x-h)^2}{b^2} = 0$,

setting each factor equal to zero and solving for y in terms of x. As a check on your work, remember that (h, k) must be a point that lies on each asymptote.

EXAMPLE 7 Find Everything

Find the center, vertices, foci, and asymptotes of the hyperbola $4x^2 - y^2 - 8x - 4y - 4 = 0$. Graph.

Solution Before completing the square in x and y, we factor 4 from the two x-terms and factor -1 from the two y-terms so that the leading coefficient in each expression is 1. Then we have

$$4(x^2 - 2x + 1) - (y^2 + 4y + 4) = 4 + 4 \cdot 1 + (-1) \cdot 4$$

$$4(x - 1)^2 - (y + 2)^2 = 4$$

$$\frac{(x - 1)^2}{1} - \frac{(y + 2)^2}{4} = 1.$$

We see now that the center is $(1, -2)$. Since the term in the standard form involving x has the positive coefficient, the transverse axis is horizontal along the line $y = -2$, and we identify $a = 1$ and $b = 2$. The vertices are 1 unit to the left and to the right of the center at $(0, -2)$ and $(2, -2)$, respectively. From $b^2 = c^2 - a^2$ we have $c^2 = a^2 + b^2 = 5$, and so $c = \sqrt{5}$. Hence, the foci are $\sqrt{5}$ units to the left and the right of the center $(1, -2)$ at $(1 - \sqrt{5}, -2)$ and $(1 + \sqrt{5}, -2)$.

To find the asymptotes, we solve

$$\frac{(x - 1)^2}{1} - \frac{(y + 2)^2}{4} = 0 \quad \text{or} \quad \left(x - 1 - \frac{y + 2}{2}\right)\left(x - 1 + \frac{y + 2}{2}\right) = 0$$

for y. From $y + 2 = \pm 2(x - 1)$ we find that the asymptotes are $y = -2x$ and $y = 2x - 4$. Observe that by substituting $x = 1$, both equations give $y = -2$, which means that both lines pass through the center. We then locate the center, plot the vertices, and graph the asymptotes. As shown in FIGURE 10.1.17, the graph of the hyperbola passes through the vertices and becomes closer and closer to the asymptotes as $x \to \pm\infty$. ■

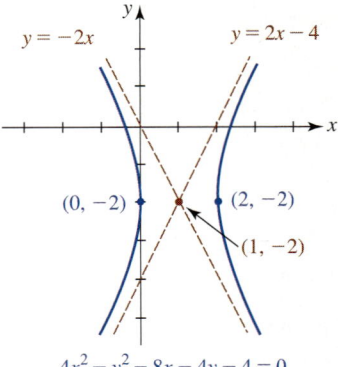

FIGURE 10.1.17 Hyperbola in Example 7

Eccentricity Associated with each conic section is a number e called its **eccentricity**.

Definition 10.1.4 Eccentricity

The **eccentricity** of an ellipse and a hyperbola is

$$e = \frac{c}{a}.$$

Of course, you must bear in mind that for an ellipse $c = \sqrt{a^2 - b^2}$ and for a hyperbola $c = \sqrt{a^2 + b^2}$. From the inequalities $0 < \sqrt{a^2 - b^2} < a$ and $0 < a < \sqrt{a^2 + b^2}$, we see, in turn, that

- the eccentricity of an ellipse satisfies $0 < e < 1$, and
- the eccentricity of a hyperbola satisfies $e > 1$.

The eccentricity of a parabola will be discussed in Section 10.7.

EXAMPLE 8 Eccentricity

Determine the eccentricity of
(a) the ellipse in Example 5,   (b) the hyperbola in Example 7.

Solution
(a) In the solution of Example 5 we found that $a = 4$ and $c = 2\sqrt{3}$. Hence, the eccentricity of the ellipse is $e = (2\sqrt{3})/4 = \sqrt{3}/2 \approx 0.87$.
(b) In Example 7 we found that $a = 1$ and $c = \sqrt{5}$. Hence, the eccentricity of the hyperbola is $e = \sqrt{5}/1 \approx 2.23$. ∎

Eccentricity is an indicator of the shape of an ellipse or a hyperbola. If $e \approx 0$, then $c = \sqrt{a^2 - b^2} \approx 0$ and consequently $a \approx b$. This means the ellipse is nearly circular. On the other hand, if $e \approx 1$, then $c = \sqrt{a^2 - b^2} \approx a$ and so $b \approx 0$. This means that each focus is close to a vertex and so the ellipse is elongated. See FIGURE 10.1.18. The shapes of a hyperbola in the two extreme cases $e \approx 1$ and e much bigger than 1 are illustrated in FIGURE 10.1.19.

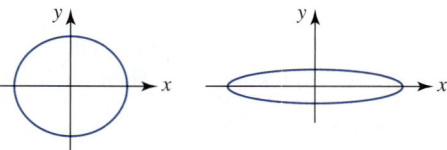

(a) e close to zero   (b) e close to 1
FIGURE 10.1.18 Effect of eccentricity on the shape of an ellipse

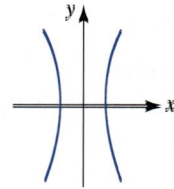

(a) e close to 1   (b) e much greater than 1
FIGURE 10.1.19 Effect of eccentricity on the shape of a hyperbola

■ **Applications** The parabola has many properties that make it suitable for certain applications. Reflecting surfaces are often designed to take advantage of a reflection property of parabolas. Such surfaces, called **paraboloids**, are three-dimensional and are formed by rotating a parabola about its axis. As illustrated in FIGURE 10.1.20, rays of light (or electronic signals) from a point source located at the focus of a parabolic reflecting surface will be reflected along lines parallel to the axis. This is the idea behind the design of searchlights, some flashlights, and on-location satellite dishes. Conversely, if the incoming rays of light are parallel to the axis of a parabola, they will be reflected off the surface along lines passing through the focus. Beams of light from a distant object such as a galaxy are essentially parallel, and so when these beams enter a reflecting telescope they are reflected by the parabolic mirror to the focus, where a camera is usually placed to capture the image over time. A parabolic home satellite dish operates on the same principle as the reflecting telescope; the digital signal from a TV satellite is captured at the focus of the dish by a receiver.

Ellipses have a reflection property analogous to the parabola. It can be shown that if a light or sound source is placed at one focus of an ellipse, then all rays or waves will be reflected off the ellipse to the other focus. See FIGURE 10.1.21. For example, if a ceiling is elliptical with two foci on (or near) the floor, but considerably distant from each other, then anyone whispering at one focus will be heard at the other. Some famous "whispering galleries"

reflecting surface

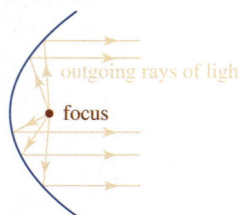

(a) Rays emitted at focus are reflected as parallel rays

reflecting surface

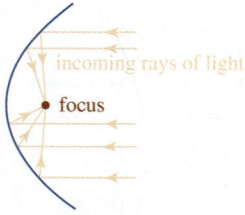

(b) Incoming rays reflected to focus
FIGURE 10.1.20 Parabolic reflecting surface

TV satellite dish

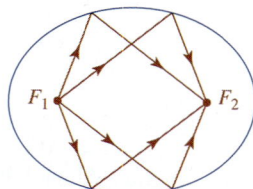

FIGURE 10.1.21 Reflection property of an ellipse

200 inch reflecting telescope at Mt. Palomar

are the Statuary Hall at the Capitol in Washington, DC, the Mormon Tabernacle in Salt Lake City, and St. Paul's Cathedral in London.

Using his Law of Universal Gravitation, Isaac Newton was the first to prove Kepler's first law of planetary motion: The orbit of each planet about the Sun is an ellipse with the Sun at one focus.

Statuary Hall in Washington, DC

EXAMPLE 9   Eccentricity of Earth's Orbit

The perihelion distance of the Earth (the least distance between the Earth and the Sun) is approximately 9.16×10^7 mi, and its aphelion distance (the greatest distance between the Earth and the Sun) is approximately 9.46×10^7 mi. What is the eccentricity of Earth's elliptical orbit?

Solution   Let us assume that the orbit of the Earth is as shown in FIGURE 10.1.22. From the figure we see that

$$a - c = 9.16 \times 10^7$$
$$a + c = 9.46 \times 10^7.$$

Solving this system of equations gives $a = 9.31 \times 10^7$ and $c = 0.15 \times 10^7$. Thus, the eccentricity $e = c/a$ is

$$e = \frac{0.15 \times 10^7}{9.31 \times 10^7} \approx 0.016.$$ ∎

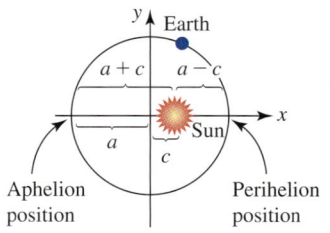

FIGURE 10.1.22   Graphical interpretation of data in Example 9

The orbits of seven of the planets have eccentricities less than 0.1 and, hence, the orbits are not far from circular. Mercury is an exception. Many of the asteroids and comets have highly eccentric orbits. The orbit of the asteroid Hildago is one of the most eccentric, with $e = 0.66$. Another notable case is the orbit of Comet Halley. See Problem 79 in Exercises 10.1.

The hyperbola has several important applications involving sounding techniques. In particular, several navigational systems utilize hyperbolas as follows. Two fixed radio transmitters at a known distance from each other transmit synchronized signals. The difference in reception times by a navigator determines the difference $2a$ of the distances from the navigator to the two transmitters. This information locates the navigator somewhere on the hyperbola with foci at the transmitters and fixed difference in distances from the foci equal to $2a$. By using two sets of signals obtained from a single master station paired with each of two second stations, the long-range navigation system LORAN locates a ship or plane at the intersection of two hyperbolas. See FIGURE 10.1.23.

There are many other applications of the hyperbola. As shown in FIGURE 10.1.24(a), a plane flying at a supersonic speed parallel to level ground leaves a hyperbolic sonic "footprint" on the ground. Like the parabola and ellipse, a hyperbola also possesses a reflecting property. The Cassegrain reflecting telescope shown in Figure 10.1.24(b) utilizes a convex hyperbolic secondary mirror to reflect a ray of light back through a hole to an eyepiece (or camera) behind the parabolic primary mirror. This telescope construction makes use of the fact that

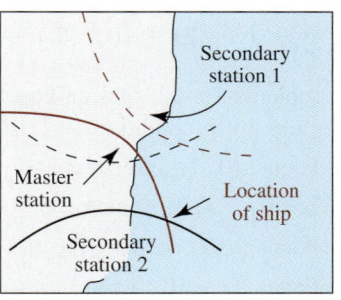

FIGURE 10.1.23   The idea behind LORAN

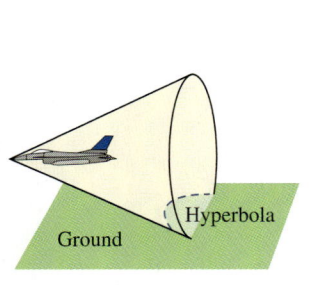

(a) Sonic footprint

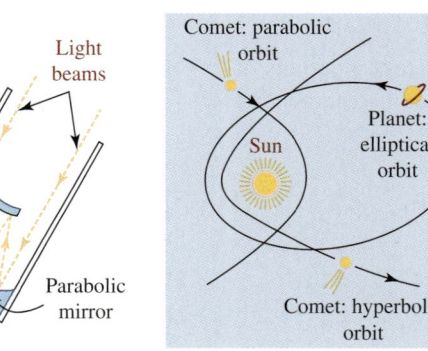

(b) Cassegrain telescope   (c) Orbits around the Sun

FIGURE 10.1.24   Applications of hyperbolas

a beam of light directed along a line through one focus of a hyperbolic mirror will be reflected on a line through the other focus.

Orbits of objects in the universe can be parabolic, elliptic, or hyperbolic. When an object passes close to the Sun (or a planet), it is not necessarily captured by the gravitational field of the larger body. Under certain conditions, the object picks up a fractional amount of orbital energy of this much larger body and the resulting "slingshot-effect" orbit of the object as it passes the Sun is hyperbolic. See Figure 10.1.24(c).

Exercises 10.1 Answers to selected odd-numbered problems begin on page ANS-29.

≡ Fundamentals

In Problems 1–14, find the vertex, focus, directrix, and axis of the given parabola. Graph the parabola.

1. $y^2 = 4x$

2. $y^2 = \dfrac{7}{2}x$

3. $x^2 = -16y$

4. $x^2 = \dfrac{1}{10}y$

5. $(y - 1)^2 = 16x$

6. $(y + 3)^2 = -8(x + 2)$

7. $(x + 5)^2 = -4(y + 1)$

8. $(x - 2)^2 + y = 0$

9. $y^2 + 12y - 4x + 16 = 0$

10. $x^2 + 6x + y + 11 = 0$

11. $x^2 + 5x - \dfrac{1}{4}y + 6 = 0$

12. $x^2 - 2x - 4y + 17 = 0$

13. $y^2 - 8y + 2x + 10 = 0$

14. $y^2 - 4y - 4x + 3 = 0$

In Problems 15–22, find an equation of the parabola that satisfies the given conditions.

15. Focus, (0, 7), directrix $y = -7$

16. Focus (−4, 0), directrix $x = 4$

17. Focus $\left(\frac{5}{2}, 0\right)$, vertex (0, 0)

18. Focus (0, −10), vertex (0, 0)

19. Focus (1, −7), directrix $x = -5$

20. Focus (2, 3), directrix $y = -3$

21. Vertex (0, 0), through (−2, 8), axis along the y-axis

22. Vertex (0, 0), through $\left(1, \frac{1}{4}\right)$, axis along the x-axis

In Problems 23 and 24, find the x- and y-intercepts of the given parabola.

23. $(y + 4)^2 = 4(x + 1)$

24. $(x - 1)^2 = -2(y - 1)$

In Problems 25–38, find the center, foci, vertices, endpoints of the minor axis, and eccentricity of the given ellipse. Graph the ellipse.

25. $x^2 + \dfrac{y^2}{16} = 1$

26. $\dfrac{x^2}{25} + \dfrac{y^2}{9} = 1$

27. $9x^2 + 16y^2 = 144$

28. $2x^2 + y^2 = 4$

29. $\dfrac{(x - 1)^2}{49} + \dfrac{(y - 3)^2}{36} = 1$

30. $\dfrac{(x + 1)^2}{25} + \dfrac{(y - 2)^2}{36} = 1$

31. $(x + 5)^2 + \dfrac{(y + 2)^2}{16} = 1$

32. $\dfrac{(x - 3)^2}{64} + \dfrac{(y + 4)^2}{81} = 1$

33. $4x^2 + \left(y + \frac{1}{2}\right)^2 = 4$

34. $36(x + 2)^2 + (y - 4)^2 = 72$

35. $25x^2 + 9y^2 - 100x + 18y - 116 = 0$

36. $9x^2 + 5y^2 + 18x - 10y - 31 = 0$

37. $x^2 + 3y^2 + 18y + 18 = 0$

38. $12x^2 + 4y^2 - 24x - 4y + 1 = 0$

In Problems 39–48, find an equation of the ellipse that satisfies the given conditions.

39. Vertices (±5, 0), foci (±3, 0)

40. Vertices (±9, 0), foci (±2, 0)

41. Vertices (−3, −3), (5, −3), endpoints of minor axis (1, −1), (1, −5)

42. Vertices (1, −6), (1, 2), endpoints of minor axis (−2, −2), (4, −2)

43. Foci $\left(\pm\sqrt{2}, 0\right)$, length of minor axis 6

44. Foci $\left(0, \pm\sqrt{5}\right)$, length of major axis 16

45. Foci (0, ±3), passing through $\left(-1, 2\sqrt{2}\right)$

46. Vertices (±5, 0), passing through $\left(\sqrt{5}, 4\right)$

47. Center (1, 3), one focus (1, 0), one vertex (1, −1)

48. Endpoints of major axis (2, 4), (13, 4), one focus (4, 4)

In Problems 49–62, find the center, foci, vertices, asymptotes, and eccentricity of the given hyperbola. Graph the hyperbola.

49. $\dfrac{x^2}{16} - \dfrac{y^2}{25} = 1$

50. $\dfrac{x^2}{4} - \dfrac{y^2}{4} = 1$

51. $y^2 - 5x^2 = 20$

52. $9x^2 - 16y^2 + 144 = 0$

53. $\dfrac{(x - 5)^2}{4} - \dfrac{(y + 1)^2}{49} = 1$

54. $\dfrac{(x + 2)^2}{10} - \dfrac{(y + 4)^2}{25} = 1$

55. $\dfrac{(y - 4)^2}{36} - x^2 = 1$

56. $\dfrac{\left(y - \frac{1}{4}\right)^2}{4} - \dfrac{(x + 3)^2}{9} = 1$

57. $25(x - 3)^2 - 5(y - 1)^2 = 125$

58. $10(x + 1)^2 - 2\left(y - \frac{1}{2}\right)^2 = 100$

59. $5x^2 - 6y^2 - 20x + 12y - 16 = 0$

60. $16x^2 - 25y^2 - 256x - 150y + 399 = 0$

61. $4x^2 - y^2 - 8x + 6y - 4 = 0$

62. $2y^2 - 9x^2 - 18x + 20y + 5 = 0$

In Problems 63–70, find an equation of the hyperbola that satisfies the given conditions.

63. Foci $(0, \pm 4)$, one vertex $(0, -2)$

64. Foci $(0, \pm 3)$, one vertex $\left(0, -\frac{3}{2}\right)$

65. Center $(1, -3)$, one focus $(1, -6)$, one vertex $(1, -5)$

66. Vertices $(2, 5)$, $(2, -1)$, one focus $(2, 7)$

67. Center $(-1, 3)$, one vertex $(-1, 4)$, passing through $\left(-5, 3 + \sqrt{5}\right)$

68. Center $(3, -5)$, one vertex $(3, -2)$, passing through $(1, -1)$

69. Center $(2, 4)$, one vertex $(2, 5)$, one asymptote $2y - x - 6 = 0$

70. Eccentricity $\sqrt{10}$, endpoints of conjugate axis $(-5, 4)$, $(-5, 10)$

☰ Applications

71. A large spotlight is designed so that a cross section through its axis is a parabola and the light source is at the focus. Find the position of the light source if the spotlight is 4 ft across at the opening and 2 ft deep.

72. A reflecting telescope has a parabolic mirror that is 20 ft across at the top and 4 ft deep at the center. Where should the eyepiece be located?

73. Suppose that two towers of a suspension bridge are 350 ft apart and the vertex of the parabolic cable is tangent to the road midway between the towers. If the cable is 1 ft above the road at a point 20 ft from the vertex, find the height of the towers above the road.

74. Two 75-ft towers of a suspension bridge with a parabolic cable are 250 ft apart. The vertex of the parabola is tangent to the road midway between the towers. Find the height of the cable above the roadway at a point 50 ft from one of the towers.

75. Assume that the water gushing from the end of a horizontal pipe follows a parabolic arc with vertex at the end of the pipe. The pipe is 20 m above the ground. At a point 2 m below the end of the pipe, the horizontal distance from the water to a vertical line through the end of the pipe is 4 m. See FIGURE 10.1.25. Where does the water strike the ground?

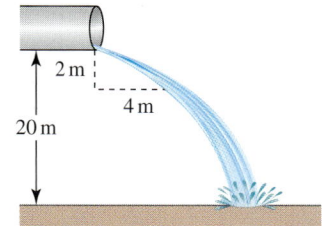

FIGURE 10.1.25 Pipe in Problem 75

76. A dart thrower releases a dart 5 ft above the ground. The dart is thrown horizontally and follows a parabolic path. It hits the ground $10\sqrt{10}$ ft from the dart thrower. At a distance of 10 ft from the dart thrower, how high should a bull's-eye be placed in order for the dart to hit it?

77. The orbit of the planet Mercury is an ellipse with the Sun at one focus. The length of the major axis of this orbit is 72 million mi and the length of the minor axis is 70.4 million mi. What is the least distance (perihelion) between Mercury and the Sun? What is the greatest distance (aphelion)?

78. What is the eccentricity of the orbit of Mercury in Problem 77?

79. The orbit of comet Halley is an ellipse whose major axis is 3.34×10^9 mi long, and whose minor axis is 8.5×10^8 mi long. What is the eccentricity of the comet's orbit?

80. A satellite orbits the Earth in an elliptical path with the center of the Earth at one focus. It has a minimum altitude of 200 mi and a maximum altitude of 1000 mi above the surface of the Earth. If the radius of the Earth is 4000 mi, what is an equation of the satellite's orbit?

81. A semielliptical archway has a vertical major axis. The base of the arch is 10 ft across and the highest part of the arch is 15 ft. Find the height of the arch above the point on the base of the arch 3 ft from the center.

82. Suppose that a room is constructed on a flat elliptical base by rotating a semiellipse 180° about its major axis. Then, by the reflection property of the ellipse, anything whispered at one focus will be distinctly heard at the other focus. If the height of the room is 16 ft and the length is 40 ft, find the location of the whispering and listening posts.

☰ Think About It

83. The graph of the ellipse $x^2/4 + (y - 1)^2/9 = 1$ is shifted 4 units to the right. What are the center, foci, vertices, and endpoints of the minor axis for the shifted graph?

84. The graph of the ellipse $(x - 1)^2/9 + (y - 4)^2 = 1$ is shifted 5 units to the left and 3 units up. What are the center, foci, vertices, and endpoints of the minor axis for the shifted graph?

85. The hyperbolas

$$\frac{x^2}{a^2} - \frac{y^2}{b^2} = 1 \quad \text{and} \quad \frac{y^2}{b^2} - \frac{x^2}{a^2} = 1$$

are said to be **conjugates** of each other.

(a) Find the equation of the hyperbola that is conjugate to

$$\frac{x^2}{25} - \frac{y^2}{144} = 1.$$

(b) Discuss how the graphs of conjugate hyperbolas are related.

86. A **rectangular hyperbola** is one for which the asymptotes are perpendicular.

(a) Show that $y^2 - x^2 + 5y + 3x = 1$ is a rectangular hyperbola.

(b) Which of the hyperbolas given in Problems 49–62 are rectangular?

87. It can be shown that a ray of light emanating from one focus of a hyperbola will be reflected back along the line from the opposite focus. See FIGURE 10.1.26. A light ray from the left focus of the hyperbola $x^2/16 - y^2/20 = 1$ strikes the hyperbola at $(-6, -5)$. Find an equation of the reflected ray.

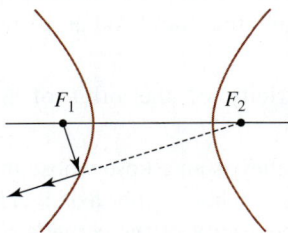

FIGURE 10.1.26 Reflecting property in Problem 87

88. An **oval** is an approximation to an ellipse consisting of arcs from symmetrically placed pairs of circles of different radii, each small circle being tangent to a large circle at two transition points as indicated in FIGURE 10.1.27. Architects in the Renaissance and Baroque periods commonly used ovals because they are simpler to construct than ellipses. In this problem, let the small circles be centered at $(\pm a, 0)$, $a > 0$, with radius r, and let the large circles be centered at $(0, \pm b)$, $b > 0$, with radius R. Also, let

$(\pm A, 0)$, $A > 0$, and $(0, \pm B)$, $B > 0$, be the points of intersection of the oval with the x-and y-axes.

(a) Express R in terms of a, b, and r.

(b) Show that $A > B$. This shows that the "major axis" of the oval is always in line with the centers of the small circles, and that the "minor axis" of the oval is always in line with the centers of the large circles. [*Hint*: Show that $A - B = a + b - \sqrt{a^2 + b^2}$.]

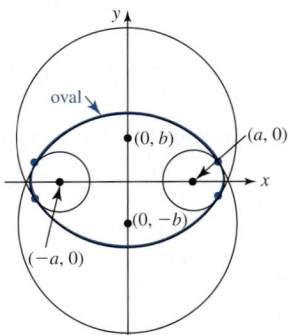

FIGURE 10.1.27 Oval in Problem 88

10.2 Parametric Equations

▌ Introduction A rectangular or Cartesian equation is not the only, and often not the most convenient, way of describing a curve in the coordinate plane. In this section we will consider a different way of representing a curve that is important in the many applications of calculus.

▌ Curvilinear Motion Let us begin with an example. The motion of a particle along a curve, in contrast to a straight line, is called **curvilinear motion**. If it is assumed that a golf ball is hit off the ground, perfectly straight (no hook or slice), and that its path stays in a coordinate plane, then its motion is governed by the fact that its acceleration in the x- and y-directions satisfies

$$a_x = 0, \qquad a_y = -g, \tag{1}$$

where g is the acceleration due to gravity and $a_x = d^2x/dt^2$, $a_y = d^2y/dt^2$. At $t = 0$ we take $x = 0$, $y = 0$, and the x- and y-components of the initial velocity v_0 to be

$$v_0 \cos \theta_0 \qquad \text{and} \qquad v_0 \sin \theta_0, \tag{2}$$

respectively. See FIGURE 10.2.1. Taking two antiderivatives of each equation in (1), we see from the initial conditions (2) that the x- and y-coordinates of the golf ball at time t are given by

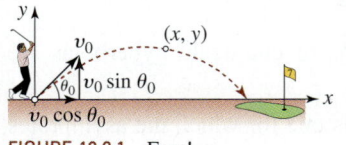

FIGURE 10.2.1 Fore!

$$x = (v_0 \cos \theta_0)t, \qquad y = -\frac{1}{2}gt^2 + (v_0 \sin \theta_0)t, \tag{3}$$

where θ_0 is the launch angle, v_0 is its initial velocity, and $g = 32 \text{ ft/s}^2$. These equations, which give the golf ball's position in the coordinate plane at time t, are said to be **parametric equations**. The third variable t in (3) is called a **parameter** and is restricted to some interval I, in this case I is defined by $0 \le t \le T$, where $t = 0$ gives the origin $(0, 0)$, and $t = T$ is the time the ball hits the ground.

The idea in (3), that is, representing x and y in an ordered pair (x, y) by functions of a third variable t, is used to *define* a curve.

Definition 10.2.1 Plane Curve

If f and g are continuous functions defined on a common interval I, then $x = f(t)$, $y = g(t)$ are called **parametric equations** and t is called a **parameter**. The set C of ordered pairs $(f(t), g(t))$ as t varies over I is called a **plane curve**.

It is also common practice to refer to the set of equations $x = f(t)$, $y = g(t)$, for t in I, as a **parameterization** for C. Hereafter, we will refer to a plane curve C as a **curve**, a **parametric curve**, or as a **parameterized curve**. The **graph** of a parametric curve C is the set of all points (x, y) in the coordinate plane corresponding to the ordered pairs $(f(t), g(t))$. For simplicity, we shall not belabor the distinction between a *parametric curve* and a *graph of a curve*.

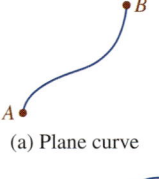

FIGURE 10.2.2 Curve in Example 1

EXAMPLE 1 Parametric Curve

Graph the curve C that has the parametric equations

$$x = t^2, \quad y = t^3, \quad -1 \le t \le 2.$$

Solution As shown in the accompanying table, for any choice of t in the interval $[-1, 2]$, we obtain a single ordered pair (x, y). By connecting the points with a curve, we obtain the graph in FIGURE 10.2.2.

t	-1	$-\dfrac{1}{2}$	0	$\dfrac{1}{2}$	1	$\dfrac{3}{2}$	2
x	1	$\dfrac{1}{4}$	0	$\dfrac{1}{4}$	1	$\dfrac{9}{4}$	4
y	-1	$-\dfrac{1}{8}$	0	$\dfrac{1}{8}$	1	$\dfrac{27}{8}$	2

∎

In Example 1, if we think in terms of motion and t as time, then as t increases from -1 to 2, a point P defined as (t^2, t^3) starts from $(1, -1)$, advances up the lower branch of the curve to the origin $(0, 0)$, passes to the upper branch, and finally stops at $(4, 8)$. In general, a parameter t need have no relation to time. As we plot points corresponding to *increasing values* of the parameter, a curve C is traced out by $(f(t), g(t))$ in a certain *direction* indicated by the arrowheads on the curve in Figure 10.2.2. This direction is called the **orientation** of the curve C.

When the interval I over which f and g are defined is a closed interval $[a, b]$, we say that $(f(a), g(a))$ is the **initial point** of the curve C and that $(f(b), g(b))$ is its **terminal point**. In Example 1, $(1, -1)$ and $(4, 8)$ are the initial and terminal points of C, respectively. If the terminal point is the same as the initial point, that is,

$$(f(a), g(a)) = (f(b), g(b)),$$

then C is a **closed curve**. If C is closed but does not cross itself, then it is called a **simple closed curve**. In FIGURE 10.2.3, A and B represent the initial and terminal points, respectively. The next example illustrates a simple closed curve.

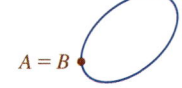

(a) Plane curve

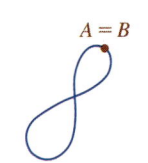

(b) Simple closed curve

(c) Closed but not simple

FIGURE 10.2.3 Some plane curves

EXAMPLE 2 A Parameterization of a Circle

Find a parameterization for the circle $x^2 + y^2 = a^2$.

Solution The circle has center at the origin and radius $a > 0$. If t represents the central angle, that is, an angle with vertex at the origin and initial side coinciding with the positive x-axis, then as shown in FIGURE 10.2.4 the equations

$$x = a\cos t, \quad y = a\sin t, \quad 0 \le t \le 2\pi \tag{4}$$

give every point P on the circle. For example, at $t = \pi/2$ we get $x = 0$ and $y = a$, in other words, the point is $(0, a)$. The initial point corresponds to $t = 0$ and is $(a, 0)$; the terminal point corresponds to $t = 2\pi$ and is also $(a, 0)$. Since the initial and terminal points are the same, this

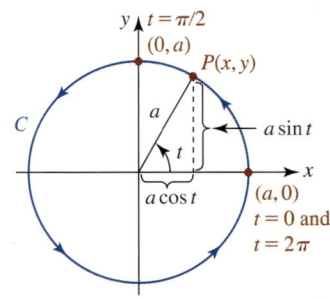

FIGURE 10.2.4 Circle in Example 2

proves the obvious, that the curve C defined by the parametric equations (4) is a closed curve. Note the orientation of C in Figure 10.2.4; as t increases from 0 to 2π, the point $P(x, y)$ traces out C in a counterclockwise direction. ∎

In Example 2, the upper *semicircle* $x^2 + y^2 = a^2, 0 \leq y \leq a$, is defined parametrically by restricting the parameter t to the interval $[0, \pi]$,

$$x = a\cos t, \quad y = a\sin t, \quad 0 \leq t \leq \pi.$$

Observe that when $t = \pi$, the terminal point is now $(-a, 0)$. On the other hand, if we wish to describe *two* complete counterclockwise revolutions around the circle, we again modify the parameter interval by writing

$$x = a\cos t, \quad y = a\sin t, \quad 0 \leq t \leq 4\pi.$$

■ **Eliminating the Parameter** Given a set of parametric equations, we sometimes desire to *eliminate* or *clear* the parameter to obtain a rectangular equation for the curve. To eliminate the parameter in (4), we simply square x and y and add the two equations:

$$x^2 + y^2 = a^2\cos^2 t + a^2\sin^2 t \quad \text{implies} \quad x^2 + y^2 = a^2$$

since $\sin^2 t + \cos^2 t = 1$. There is no unique way of eliminating a parameter.

EXAMPLE 3 Eliminating the Parameter

(a) From the first equation in (3) we have $t = x/(v_0\cos\theta_0)$. Substituting this into the second equation then gives

$$y = -\frac{g}{2(v_0\cos\theta_0)^2}x^2 + (\tan\theta_0)x.$$

Since v_0, θ_0, and g are constants, the last equation has the form $y = ax^2 + bx$ and so the trajectory of any projectile launched at the angle $0 < \theta_0 < \pi/2$ is a parabolic arc.

(b) In Example 1, we can eliminate the parameter from $x = t^2$, $y = t^3$ by solving the second equation for t in terms of y and then substituting in the first equation. We find

$$t = y^{1/3} \quad \text{and so} \quad x = (y^{1/3})^2 = y^{2/3}.$$

The curve shown in Figure 10.2.2 is only a portion of the graph of $x = y^{2/3}$. For $-1 \leq t \leq 2$ we have correspondingly $-1 \leq y \leq 8$. Thus, a rectangular equation for the curve in Example 1 is given by $x = y^{2/3}, -1 \leq y \leq 8$. ∎

A curve C can have more than one parameterization. For example, an alternative parameterization for the circle in Example 2 is

A curve C can have many different parameterizations.

$$x = a\cos 2t, \quad y = a\sin 2t, \quad 0 \leq t \leq \pi.$$

Note that the parameter interval is now $[0, \pi]$. We see that as t increases from 0 to π, the new angle $2t$ increases from 0 to 2π.

EXAMPLE 4 Alternative Parameterizations

Consider the curve C that has the parametric equations $x = t, y = 2t^2, -\infty < t < \infty$. We can eliminate the parameter by using $t = x$ and substituting in $y = 2t^2$. This gives the rectangular equation $y = 2x^2$, which we recognize as a parabola. Moreover, since $-\infty < t < \infty$ is equivalent to $-\infty < x < \infty$, the point $(t, 2t^2)$ traces out the complete parabola $y = 2x^2, -\infty < x < \infty$.

An alternative parameterization of C is given by $x = t^3/4, y = t^6/8, -\infty < t < \infty$. Using $t^3 = 4x$ and substituting in $y = t^6/8$ or $y = (t^3 \cdot t^3)/8$ gives $y = (4x)^2/8 = 2x^2$. Moreover, $-\infty < t < \infty$ implies $-\infty < t^3 < \infty$ and so $-\infty < x < \infty$. ∎

We note in Example 4 that a point on C need not correspond to the same value of the parameter in each set of parametric equations for C. For example, $(1, 2)$ is obtained for $t = 1$ in $x = t, y = 2t^2$, but $t = \sqrt[3]{4}$ yields $(1, 2)$ in $x = t^3/4, y = t^6/8$.

EXAMPLE 5 Example 4 Revisited

One has to be careful when working with parametric equations. Eliminating the parameter in $x = t^2, y = 2t^4, -\infty < t < \infty$, would seem to yield the same parabola $y = 2x^2$ as in Example 4. However, this is *not* the case because for any value of t, $t^2 \geq 0$ and so $x \geq 0$. In other words, the last set of equations is a parametric representation of only the right-hand branch of the parabola, that is, $y = 2x^2, 0 \leq x < \infty$. ◼

◀ You should proceed with caution when eliminating the parameter.

EXAMPLE 6 Eliminating the Parameter

Consider the curve C defined parametrically by

$$x = \sin t, \quad y = \cos 2t, \quad 0 \leq t \leq \pi/2.$$

Eliminate the parameter and obtain a rectangular equation for C.

Solution Using the double angle formula $\cos 2t = \cos^2 t - \sin^2 t$, we can write

$$\begin{aligned}
y &= \cos^2 t - \sin^2 t \\
&= (1 - \sin^2 t) - \sin^2 t \\
&= 1 - 2\sin^2 t \quad \leftarrow \text{substitute } \sin t = x \\
&= 1 - 2x^2.
\end{aligned}$$

Now the curve C described by the parametric equations does not consist of the complete parabola, that is, $y = 1 - 2x^2, -\infty < x < \infty$. See FIGURE 10.2.5(a). For $0 \leq t \leq \pi/2$ we have $0 \leq \sin t \leq 1$ and $-1 \leq \cos 2t \leq 1$. This means that C is only that portion of the parabola for which the coordinates of a point $P(x, y)$ satisfy $0 \leq x \leq 1$ and $-1 \leq y \leq 1$. The curve C, along with its orientation, is shown in Figure 10.2.5(b). A rectangular equation for C is $y = 1 - 2x^2$ with the restricted domain $0 \leq x \leq 1$. ◼

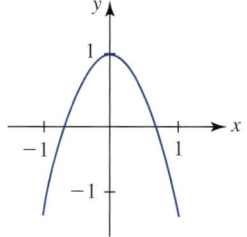

(a) $y = 1 - 2x^2, -\infty < x < \infty$

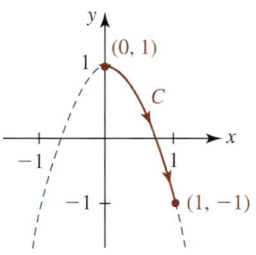

(b) $x = \sin t, y = \cos 2t,$
    $0 \leq t \leq \pi/2$

FIGURE 10.2.5 Curve C in Example 6

▮ **Intercepts** We can get intercepts of a curve C without finding its rectangular equation. For instance, in Example 6 we can find the x-intercept by finding the value of t in the parameter interval for which $y = 0$. The equation $\cos 2t = 0$ yields $2t = \pi/2$ so that $t = \pi/4$. The corresponding point at which C crosses the x-axis is $(\sqrt{2}, 0)$. Similarly, the y-intercept of C is found by solving $x = 0$ for t. From $\sin t = 0$ we immediately conclude $t = 0$ and so the y-intercept is $(0, 1)$.

▮ **Applications of Parametric Equations** Cycloidal curves were a popular topic of study by mathematicians in the seventeenth century. Suppose a point $P(x, y)$, marked on a circle of radius a, is at the origin when its diameter lies along the y-axis. As the circle rolls along the x-axis, the point P traces out a curve C that is called a **cycloid**. See FIGURE 10.2.6.

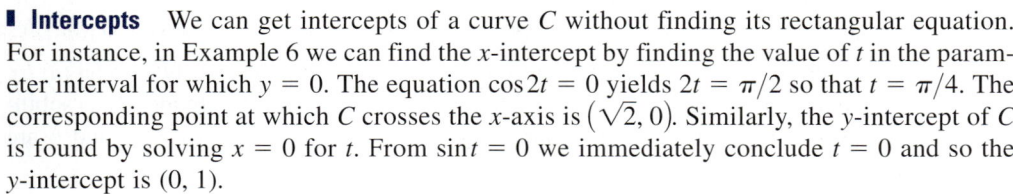

(a) Circle rolling on x-axis

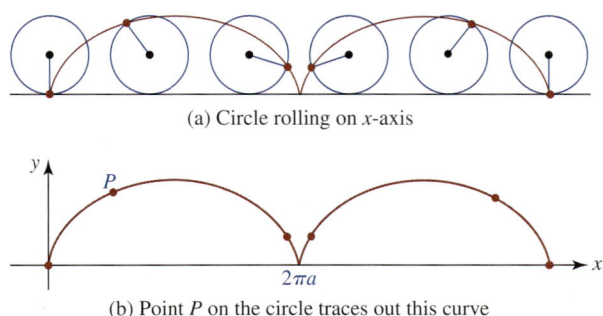

(b) Point P on the circle traces out this curve

FIGURE 10.2.6 Cycloid

Two problems were extensively studied in the seventeenth century. Consider a flexible (frictionless) wire fixed at points A and B and a bead free to slide down the wire starting at P. See FIGURE 10.2.7. Is there a particular shape of the wire so that, regardless of where the bead starts, the time to slide down the wire to B will be the same? Also, what would the shape of the wire be so that the bead slides from P to B in the shortest time? The so-called **tautochrone** (same time) and **brachistochrone** (least time) were shown to be an inverted half-arch of a cycloid.

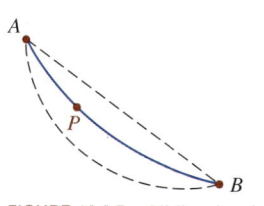

FIGURE 10.2.7 Sliding bead

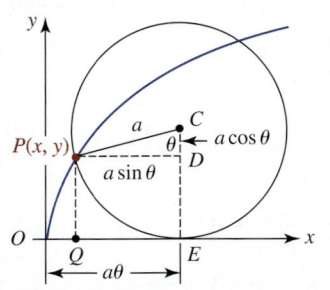

FIGURE 10.2.8 In Example 7 the angle θ is the parameter for the cycloid

EXAMPLE 7   Parameterization of a Cycloid

Find a parameterization for the cycloid shown in Figure 10.2.6(b).

Solution   A circle of radius a whose diameter initially lies along the y-axis rolls along the x-axis without slipping. We take as a parameter the angle θ (in radians) through which the circle has rotated. The point $P(x, y)$ starts at the origin, which corresponds to $\theta = 0$. As the circle rolls through an angle θ, its distance from the origin is the arc $PE = \overline{OE} = a\theta$. From FIGURE 10.2.8 we then see that the x-coordinate of P is

$$x = \overline{OE} - \overline{QE} = a\theta - a\sin\theta.$$

Now the y-coordinate of P is seen to be

$$y = \overline{CE} - \overline{CD} = a - a\cos\theta.$$

Hence, parametric equations for the cycloid are

$$x = a\theta - a\sin\theta, \quad y = a - a\cos\theta.$$

As shown in Figure 10.2.6(a), one arch of a cycloid is generated by one rotation of the circle and corresponds to the parameter interval $0 \le \theta \le 2\pi$. ■

▍ Parameterizations of Rectangular Curves   A curve C described by a continuous function $y = f(x)$ can always be parameterized by letting $x = t$. Parametric equations for C are then

$$x = t, \quad y = f(t). \tag{5}$$

For example, one cycle of the graph of the sine function $y = \sin x$ can be parameterized by $x = t, y = \sin t, 0 \le t \le 2\pi$.

▍ Smooth Curves   A curve C, given parametrically by

$$x = f(t), \quad y = g(t), \quad a \le t < b,$$

is said to be **smooth** if f' and g' are continuous on $[a, b]$ and not simultaneously zero on (a, b). A curve C is said to be **piecewise smooth** if the interval $[a, b]$ can be divided into subintervals such that C is smooth on each subinterval. The curves in Examples 2, 3, and 6 are smooth; the curves in Examples 1 and 7 are piecewise smooth.

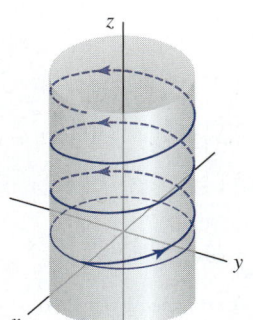

FIGURE 10.2.9   Circular helix

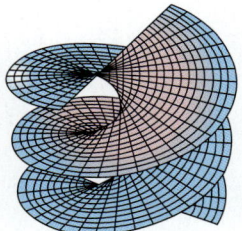

FIGURE 10.2.10   Circular Helicoid

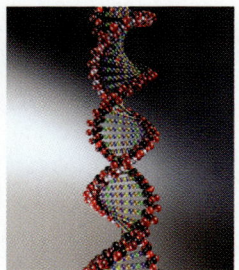

DNA is a double helix

Helical antenna

$\dfrac{d}{d\theta}$ **NOTES FROM THE CLASSROOM** ⋯⋯⋯⋯⋯⋯⋯⋯⋯⋯⋯⋯⋯⋯⋯⋯⋯⋯

In this section we have focused on **plane curves**, curves C defined parametrically in two dimensions. In the study of multivariable calculus you will see curves and surfaces in three dimensions that are defined by means of parametric equations. For example, a **space curve** C consists of a set of ordered triples $(f(t), g(t), h(t))$, where f, g, and h are defined on a common interval. Parametric equations for C are $x = f(t), y = g(t), z = h(t)$. For example, the **circular helix** such as shown in FIGURE 10.2.9 is a space curve whose parametric equations are

$$x = a\cos t, \quad y = a\cos t, \quad z = bt, \quad t \ge 0. \tag{6}$$

Surfaces in three dimensions can be represented by a set of parametric equations involving *two* parameters, $x = f(u, v), y = g(u, v), z = h(u, v)$. For example, the **circular helicoid** shown in FIGURE 10.2.10 arises from the study of minimal surfaces and is defined by the set of parametric equations similar to those in (6):

$$x = u\cos v, \quad y = u\sin v, \quad z = bv,$$

where b is a constant. The circular helicoid has a circular helix as its boundary. You might recognize the helicoid as the model for the rotating curved blade in machinery such as post hole diggers, ice augers, and snow blowers.

Exercises 10.2   Answers to selected odd-numbered problems begin on page ANS-31.

≡ Fundamentals

In Problems 1 and 2, fill in the table for the given set of parametric equations.

1. $x = 2t + 1, y = t^2 + t$

t	-3	-2	-1	0	1	2	3
x							
y							

2. $x = \cos t, y = \sin^2 t$

t	0	$\pi/6$	$\pi/4$	$\pi/3$	$\pi/2$	$5\pi/6$	$7\pi/4$
x							
y							

In Problems 3–10, graph the curve that has the given set of parametric equations.

3. $x = t - 1, y = 2t - 1; \quad -1 \le t \le 5$

4. $x = 3t, y = t^2 - 1; \quad -2 \le t \le 3$

5. $x = \sqrt{t}, y = 5 - t; \quad t \ge 0$

6. $x = 3 + 2\sin t, y = 4 + \sin t; \quad -\pi/2 \le t \le \pi/2$

7. $x = 4\cos t, y = 4\sin t; \quad -\pi/2 \le t \le \pi/2$

8. $x = t^3 + 1, y = t^2 - 1; \quad -2 \le t \le 2$

9. $x = e^t, y = e^{3t}; \quad 0 \le t \le \ln 2$

10. $x = -e^t, y = e^{-t}; \quad t \ge 0$

In Problems 11–16, eliminate the parameter from the given set of parametric equations and obtain a rectangular equation that has the same graph.

11. $x = t^2, y = t^4 + 3t^2 - 1$

12. $x = t^3 + t + 4, y = -2t^3 - 2t$

13. $x = -\cos 2t, y = \sin t; \quad -\pi/4 \le t \le \pi/4$

14. $x = e^t, y = \ln t; \quad t > 0$

15. $x = t^3, y = 3\ln t; \quad t > 0$

16. $x = \tan t, y = \sec t; \quad -\pi/2 < t < \pi/2$

In Problems 17–22, graphically show the difference between the given curves.

17. $y = x$ and $x = \sin t, y = \sin t$

18. $y = x^2$ and $x = -\sqrt{t}, y = t, t \ge 0$

19. $y = \frac{1}{4}x^2 - 1$ and $x = 2t, y = t^2 - 1, -1 \le t \le 2$

20. $y = -x^2$ and $x = e^t, y = -e^{2t}, t \ge 0$

21. $x^2 - y^2 = 1$ and $x = \cosh t, y = \sinh t$

22. $y = 2x - 2$ and $x = t^2 - 1, y = 2t^2 - 4$

In Problems 23–26, graphically show the difference between the given curves. Assume $a > 0, b > 0$.

23. $x = a\cos t, y = a\sin t, \quad 0 \le t \le \pi$
$x = a\sin t, y = a\cos t, \quad 0 \le t \le \pi$

24. $x = a\cos t, y = b\sin t, a > b, \quad \pi \le t \le 2\pi$
$x = a\sin t, y = b\cos t, a > b, \quad \pi \le t \le 2\pi$

25. $x = a\cos t, y = a\sin t, \quad -\pi/2 \le t \le \pi/2$
$x = a\cos 2t, y = a\sin 2t, \quad -\pi/2 \le t \le \pi/2$

26. $x = a\cos\dfrac{t}{2}, y = a\sin\dfrac{t}{2}, \quad 0 \le t \le \pi$
$x = a\cos\left(-\dfrac{t}{2}\right), y = a\sin\left(-\dfrac{t}{2}\right), \quad -\pi \le t \le 0$

In Problems 27 and 28, graph the curve that has the given parametric equations.

27. $x = 1 + 2\cosh t, y = 2 + 3\sinh t$

28. $x = -3 + 3\cos t, y = 5 + 5\sin t$

In Problems 29–34, determine whether the given set of parametric equations has the same graph as the rectangular equation $xy = 1$.

29. $x = \dfrac{1}{2t + 1}, y = 2t + 1$ **30.** $x = t^{1/2}, y = t^{-1/2}$

31. $x = \cos t, y = \sec t$

32. $x = t^2 + 1, y = (t^2 + 1)^{-1}$

33. $x = e^{-2t}, y = e^{2t}$ **34.** $x = t^3, y = t^{-3}$

≡ Applications

35. As shown in **FIGURE 10.2.11**, a piston is attached by means of a rod of length L to a circular crank mechanism of radius r. Parameterize the coordinates of the point P in terms of the angle ϕ.

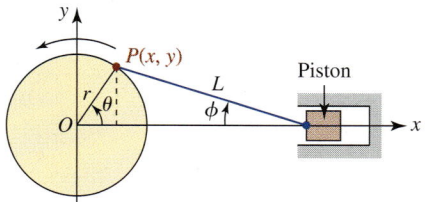

FIGURE 10.2.11   Crank mechanism in Problem 35

36. A point Q traces out a circular path of radius r and a point P moves in the manner shown in **FIGURE 10.2.12**. If R is constant, find parametric equations of the path traced by P. This curve is called an **epitrochoid**. (Those knowledgeable about automobiles might recognize the curve traced by P as the shape of the rotor housing in the rotary or Wankel engine.)

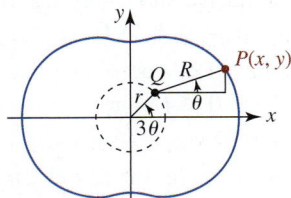

FIGURE 10.2.12 Epitrochoid in Problem 36

37. A circular spool wound with thread has its center at the origin. The radius of the spool is a. The end of the thread P, starting from $(a, 0)$, is unwound while the thread is kept taut. See FIGURE 10.2.13. Find parametric equations of the path traced by the point P if the thread PR is tangent to the circular spool at R. The curve is called an **involute** of a circle.

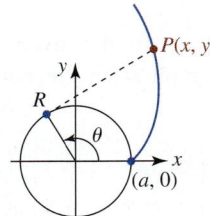

FIGURE 10.2.13 Involute of a circle in Problem 37

38. Imagine a small circle of radius a rolling around inside and on a larger circle of radius $b > a$. A point P on the smaller circle generates a curve called a **hypocycloid**. Use FIGURE 10.2.14 to show that parametric equations of a hypocycloid are

$$x = (b - a)\cos\theta + a\cos\frac{b - a}{a}\theta$$

$$y = (b - a)\sin\theta - a\sin\frac{b - a}{a}\theta.$$

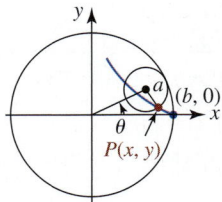

FIGURE 10.2.14 Hypocycloid in Problem 38

39. (a) Use the equations in Problem 38 to show that parametric equations of a **hypocycloid of four cusps** are

$$x = b\cos^3\theta, \quad y = b\sin^3\theta.$$

(b) Use a graphing utility to obtain the graph of the curve in part (a).
(c) Eliminate the parameter and obtain a rectangular equation for the hypocycloid of four cusps.

40. Use FIGURE 10.2.15 to show that parametric equations of an **epicycloid** are given by

$$x = (a + b)\cos\theta - a\cos\frac{a + b}{a}\theta$$

$$y = (a + b)\sin\theta - a\sin\frac{a + b}{a}\theta.$$

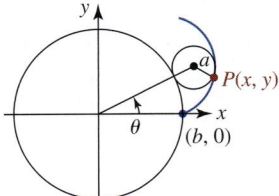

FIGURE 10.2.15 Epicycloid in Problem 40

41. (a) Use the equations in Problem 40 to show that parametric equations of an **epicycloid of three cusps** are

$$x = 4a\cos\theta - a\cos 4\theta, \quad y = 4a\sin\theta - a\sin 4\theta.$$

(b) Use a graphing utility to obtain the graph of the curve in part (a).

42. A Mathematical Classic

(a) Consider a circle of radius a, which is tangent to the x-axis at the origin O. Let B be a point on a horizontal line through $(0, 2a)$ and let the line segment OB cut the circle at point A. As shown in FIGURE 10.2.16, the projection of AB on the vertical gives the line segment BP. Find parametric equations of the path traced by the point P as A varies around the circle. The curve is called the **Witch of Agnesi**. No, the curve has nothing to do with witches and goblins. This curve, called *versoria*, which is Latin for a kind of rope, was included in a text on analytic geometry written in 1748 by the Italian mathematician **Maria Gaetana Agnesi** (1718–1799). This text proved to be so popular that it was soon translated into English. The translator confused *versoria* with the Italian word *versiera*, which means *female goblin*. In English, *female goblin* became a *witch*.

(b) In part (a) eliminate the parameter and show that the curve has the rectangular equation

$$y = 8a^3/(x^2 + 4a^2).$$

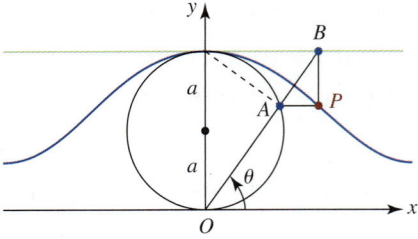

FIGURE 10.2.16 Witch of Agnesi in Problem 42

≣ Calculator/CAS Problems

In Problems 43–48, use a calculator or CAS to obtain the graph of the given set of parametric equations.

43. $x = 4\sin 2t$, $y = 2\sin t$; $0 \le t \le 2\pi$

44. $x = 6\cos 3t$, $y = 4\sin 2t$; $0 \le t \le 2\pi$

45. $x = 6\cos 4t$, $y = 4\sin t$; $0 \le t \le 2\pi$

46. $x = \cos t + t\sin t$, $y = \sin t - t\cos t$; $0 \le t \le 3\pi$

47. $x = t^3 - 4t + 1$, $y = t^4 - 4t^2$; $-5 \le t \le 5$

48. $x = t^5 - t + 1$, $y = t^3 + 2t - 1$; $-3 \le t \le 6$

≣ Think About It

49. Show that parametric equations for a line through (x_1, y_1) and (x_2, y_2) are

$$x = x_1 + (x_2 - x_1)t,\ y = y_1 + (y_2 - y_1)t,\ -\infty < t < \infty.$$

What do these equations represent when $0 \le t \le 1$?

50. (a) Use the result of Problem 49 to find parametric equations of the line through $(-2, 5)$ and $(4, 8)$.

(b) Eliminate the parameter in part (a) to obtain a rectangular equation for the line.

(c) Find parametric equations for the line segment with $(-2, 5)$ as the initial point and $(4, 8)$ as the terminal point.

51. A skier hits a mogul on a slope and is launched horizontally into the air with an initial velocity of 75 ft/s. As shown in **FIGURE 10.2.17** the slope falls away from the horizontal at an angle of 33°. Use the equations in (3) to determine how far down the slope she will land. [*Hint:* Observe the *x*- and *y*-axes in Figure 10.2.1 are in standard position (to the right and upward, respectively). In Figure 10.2.17 assume the origin is the point where the skier is launched into the air.]

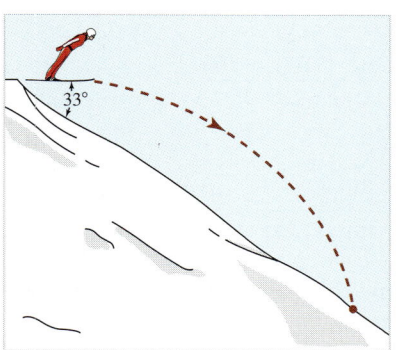

FIGURE 10.2.17 Skier in Problem 51

≣ Projects

52. Butterfly Curve The graph of the set of parametric equations

$$x = \sin t\left(e^{\cos t} - 2\cos 4t + \sin^5\frac{1}{12}t\right),$$

$$y = \cos t\left(e^{\cos t} - 2\cos 4t + \sin^5\frac{1}{12}t\right)$$

is said to be a **butterfly curve**. **FIGURE 10.2.18** consists of seven colored portions of the curve corresponding to different parameter intervals. Experiment with a CAS to determine these parameter intervals. Use the CAS to generate more colored portions and then combine all the colored curves on one set of coordinate axes.

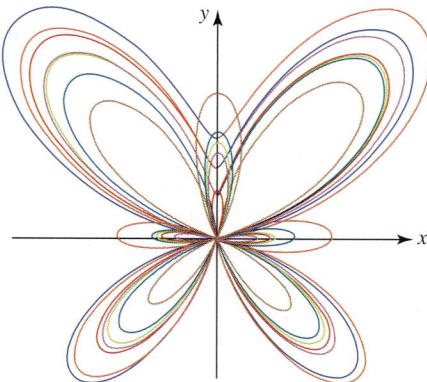

FIGURE 10.2.18 Butterfly curve in Problem 52

53. The curve in Figure 10.2.18 is one of two curves known as a butterfly curve. Write a short report that discusses both kinds of curves.

54. Bézier Curves Most computer graphing applications plot parametric equations in addition to graphs of functions. All graphic calculators can plot parametric equations by repeatedly calculating a point on the curve and then plotting it. In this project we introduce some special parametric curves called **Bézier curves**, which are commonly used in computer-aided design (CAD), in computer drawing programs, and in the mathematical representation of different fonts for many laser printers. A *cubic* Bézier curve is specified by four control points in the plane—for example,

$$P_0(p_0, q_0),\quad P_1(p_1, q_1),\quad P_2(p_2, q_2),\quad \text{and}\quad P_3(p_3, q_3).$$

The curve starts at the first point for the value $t = 0$, ends at the last point for $t = 1$, and roughly "heads toward" the middle points for parameter values between 0 and 1. Artists and engineering designers can move the control points to adjust the end locations and the shape of the parametric curve. The cubic Bézier curve for these four control points has the following parametric equations

$$x = p_0(1 - t)^3 + 3p_1(1 - t)^2 t + 3p_2(1 - t)t^2 + p_3 t^3$$

$$y = q_0(1 - t)^3 + 3q_1(1 - t)^2 t + 3q_2(1 - t)t^2 + q_3 t^3,$$

where $0 \le t \le 1$. Several Bézier curves can be pieced together continuously by making the last control point on one curve the first control point on the next curve. Equivalently, piecewise parametric equations can be constructed. For example, the next piece can be represented by

$$x = p_3(2 - t)^3 + 3p_4(2 - t)^2(t - 1)$$
$$+ 3p_5(2 - t)(t - 1)^2 + p_6(t - 1)^3$$

$$y = q_3(2 - t)^3 + 3q_4(2 - t)^2(t - 1)$$
$$+ 3q_5(2 - t)(t - 1)^2 + q_6(t - 1)^3,$$

where $1 \leq t \leq 2$.

In (a)–(f), use a graphing utility to obtain the graph of the piecewise continuous Bézier curve associated with the given control points.

(a) $P_0(5, 1)$, $P_1(1, 30)$, $P_2(50, 28)$, $P_3(55, 5)$
(b) $P_0(32, 1)$, $P_1(85, 25)$, $P_2(1, 30)$, $P_3(40, 3)$
(c) $P_0(10, 5)$, $P_1(16, 4)$, $P_2(25, 28)$, $P_3(30, 30)$,
$P_4(18, 1)$, $P_5(40, 18)$, $P_6(16, 20)$
(d) $P_0(55, 50)$, $P_1(45, 40)$, $P_2(38, 20)$, $P_3(50, 20)$,
$P_4(60, 20)$, $P_5(63, 35)$, $P_6(45, 32)$
(e) $P_0(30, 30)$, $P_1(40, 5)$, $P_2(12, 12)$, $P_3(45, 10)$,
$P_4(58, 10)$, $P_5(66, 31)$, $P_6(25, 30)$
(f) $P_0(48, 20)$, $P_1(20, 15)$, $P_2(20, 50)$, $P_3(48, 45)$,
$P_4(28, 47)$, $P_5(28, 18)$, $P_6(48, 20)$,
$P_7(48, 36)$, $P_8(52, 32)$, $P_9(40, 32)$

In (g)–(i), experiment with the locations for control points to obtain piecewise continuous Bézier curves approximating the indicated shape or object. Give the final control points chosen, and sketch the resulting parametric curve.

(g) Historically the letter "S" has been one of the most difficult to represent mathematically. Use two or three Bézier curve pieces to draw a letter "S" in some simple font style.
(h) The long cross-section of an egg is not quite an ellipse because one end is more pointed than the other. Use several Bézier curve pieces to represent an approximation for the shape of an egg.
(i) Give a curve approximating the shape of the Greek letter epsilon, ε, using as few pieces as possible.

Finish this project by writing a short report that discusses linear, quadratic, and nth degree Bézier curves. Include a discussion on the early history of Bézier curves; for example, what was the contribution of Pierre Bézier?

10.3 Calculus and Parametric Equations

■ **Introduction** As with graphs of functions $y = f(x)$, we can obtain useful information about a curve C defined parametrically by examining the derivative dy/dx.

■ **Slope** Let $x = f(t)$ and $y = g(t)$ be parametric equations of a smooth curve C. The **slope** of the tangent line at a point $P(x, y)$ on C is given by dy/dx. To calculate this derivative, we use the form of the derivative given in (3) of Section 3.1:

$$\frac{dy}{dx} = \lim_{\Delta x \to 0} \frac{\Delta y}{\Delta x}.$$

For an increment Δt, the increments in x and y are, respectively

$$\Delta x = f(t + \Delta t) - f(t) \qquad \text{and} \qquad \Delta y = g(t + \Delta t) - g(t)$$

and so

$$\frac{\Delta y}{\Delta x} = \frac{\dfrac{\Delta y}{\Delta t}}{\dfrac{\Delta x}{\Delta t}} = \frac{\dfrac{g(t + \Delta t) - g(t)}{\Delta t}}{\dfrac{f(t + \Delta t) - f(t)}{\Delta t}}.$$

Therefore,

$$\frac{dy}{dx} = \lim_{\Delta t \to 0} \frac{\Delta y}{\Delta x} = \frac{\lim_{\Delta t \to 0} \Delta y / \Delta t}{\lim_{\Delta t \to 0} \Delta x / \Delta t} = \frac{dy/dt}{dx/dt},$$

when the limit of the denominator is not zero. This parametric form of the derivative is summarized in the next theorem.

Theorem 10.3.1 Slope of a Tangent Line

If $x = f(t)$, $y = g(t)$ define a smooth curve C, then the **slope of the tangent line** at a point $P(x, y)$ on C is

$$\frac{dy}{dx} = \frac{dy/dt}{dx/dt} = \frac{g'(t)}{f'(t)}, \tag{1}$$

provided that $f'(t) \neq 0$.

EXAMPLE 1   Tangent Line

Find an equation of the tangent line to the curve $x = t^2 - 4t - 2$, $y = t^5 - 4t^3 - 1$ at the point corresponding to $t = 1$.

Solution   We first find the slope dy/dx of the tangent line. Since

$$\frac{dx}{dt} = 2t - 4 \qquad \text{and} \qquad \frac{dy}{dt} = 5t^4 - 12t^2$$

it follows from (1) that

$$\frac{dy}{dx} = \frac{dy/dt}{dx/dt} = \frac{5t^4 - 12t^2}{2t - 4}.$$

Thus, at $t = 1$ we have

$$\left.\frac{dy}{dx}\right|_{t=1} = \frac{-7}{-2} = \frac{7}{2}.$$

By substituting $t = 1$ back into the original parametric equations, we find the point of tangency to be $(-5, -4)$. Hence, an equation of the tangent line at that point is

$$y - (-4) = \frac{7}{2}(x - (-5)) \qquad \text{or} \qquad y = \frac{7}{2}x + \frac{27}{2}.$$

With the aid of a CAS we obtain the curve given in FIGURE 10.3.1.

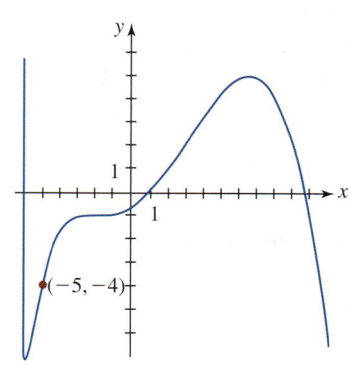

FIGURE 10.3.1   Curve in Example 1

■ Horizontal and Vertical Tangents   At a point (x, y) on a curve C at which $dy/dt = 0$ and $dx/dt \neq 0$, the tangent line is necessarily **horizontal** because $dy/dx = 0$ at that point. On the other hand, at a point at which $dx/dt = 0$ and $dy/dt \neq 0$, the tangent line is **vertical**. When both dy/dt and dx/dt are zero at a point, we can draw no immediate conclusion about the tangent line.

EXAMPLE 2   Graph of a Parametric Curve

Graph the curve that has the parametric equations $x = t^2 - 4$, $y = t^3 - 3t$.

Solution   *x-intercepts*: $y = 0$ implies $t(t^2 - 3) = 0$ at $t = 0, t = -\sqrt{3}$, and $t = \sqrt{3}$.
y-intercepts: $x = 0$ implies $t^2 - 4 = 0$ at $t = -2$ and $t = 2$.
Horizontal tangents: $\dfrac{dy}{dt} = 3t^2 - 3$; $\dfrac{dy}{dt} = 0$ implies $3(t^2 - 1) = 0$ at $t = -1$ and $t = 1$. Note that $dx/dt \neq 0$ at $t = -1$ and $t = 1$.
Vertical tangents: $\dfrac{dx}{dt} = 2t$; $\dfrac{dx}{dt} = 0$ implies $2t = 0$ at $t = 0$. Note that $dy/dt \neq 0$ at $t = 0$.

The points (x, y) on the curve corresponding to these values of the parameter are summarized in the accompanying table.

t	-2	$-\sqrt{3}$	-1	0	1	$\sqrt{3}$	2
x	0	-1	-3	-4	-3	-1	0
y	-2	0	2	0	-2	0	2

From the table we see that: the x-intercepts are $(-1, 0)$ and $(-4, 0)$, the y-intercepts are $(0, -2)$ and $(0, 2)$, the points of horizontal tangency are $(-3, 2)$ and $(-3, -2)$, the point of vertical tangency is $(-4, 0)$. A curve plotted through these points, consistent with the orientation and tangent information, is illustrated in FIGURE 10.3.2.

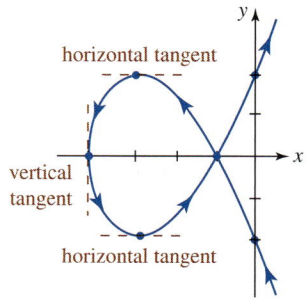

FIGURE 10.3.2   Curve in Example 2

The graph of a differentiable function $y = f(x)$ can have only one tangent line at a point on its graph. In contrast, since a curve C defined parametrically may not be the graph of a function, it is possible that such a curve may have more than one tangent line at a point.

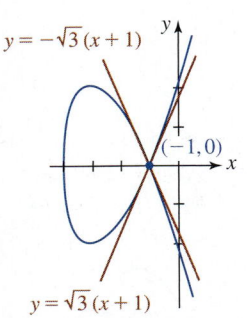

$y = -\sqrt{3}(x+1)$

$(-1, 0)$

$y = \sqrt{3}(x+1)$

FIGURE 10.3.3 Tangent lines in Example 3

EXAMPLE 3 Two Tangent Lines at a Point

In the table in Example 2 we see that for $t = -\sqrt{3}$ and $t = \sqrt{3}$ we get the single point $(-1, 0)$. As can be seen in Figure 10.3.2 this means the curve intersects itself at $(-1, 0)$. Now, from $x = t^2 - 4$, $y = t^3 - 3t$ we get

$$\frac{dy}{dx} = \frac{3t^2 - 3}{2t}$$

and

$$\left.\frac{dy}{dx}\right|_{t=-\sqrt{3}} = -\sqrt{3} \quad \text{and} \quad \left.\frac{dy}{dx}\right|_{t=\sqrt{3}} = \sqrt{3}.$$

Hence, we conclude that there are two tangent lines at $(-1, 0)$:

$$y = -\sqrt{3}(x + 1) \quad \text{and} \quad y = \sqrt{3}(x + 1).$$

See **FIGURE 10.3.3**. ∎

Higher-Order Derivatives Higher-order derivatives can be found in exactly the same manner as dy/dx. Suppose (1) is written as

$$\frac{d}{dx}(\ ) = \frac{d(\ )/dt}{dx/dt}. \tag{2}$$

If $y' = dy/dx$ is a differentiable function of t, it follows from (2) by replacing ( ) by y' that

$$\frac{d^2y}{dx^2} = \frac{d}{dx}y' = \frac{dy'/dt}{dx/dt}. \tag{3}$$

Similarly, if $y'' = d^2y/dx^2$ is a differentiable function of t, then the third derivative is

$$\frac{d^3y}{dx^3} = \frac{d}{dx}y'' = \frac{dy''/dt}{dx/dt}, \tag{4}$$

and so on.

EXAMPLE 4 Third Derivative

Find d^3y/dx^3 for the curve given by $x = 4t + 6$, $y = t^2 + t - 2$.

Solution To compute the third derivative we must first find the first and second derivatives. From (2) the first derivative is

$$\frac{dy}{dx} = \frac{dy/dt}{dx/dt} = \frac{2t + 1}{4} = y'.$$

Then using (3) and (4) we find the second and third derivatives:

$$\frac{d^2y}{dx^2} = \frac{dy'/dt}{dx/dt} = \frac{\frac{1}{2}}{4} = \frac{1}{8} = y''$$

$$\frac{d^3y}{dx^3} = \frac{dy''/dt}{dx/dt} = \frac{0}{4} = 0. \qquad ∎$$

Inspection of Example 4 shows that the curve has a horizontal tangent at $t = -\frac{1}{2}$ or $\left(4, -\frac{9}{4}\right)$. Furthermore, since $d^2y/dx^2 > 0$ for all t, the graph of the curve is concave upward at every point. Verify this by graphing the curve.

Length of a Curve In Section 6.5 we were able to find the length L of the graph of a smooth function $y = f(x)$ by means of a definite integral. We can now generalize the result given in (3) of that section to curves defined parametrically.

■ **Building an Integral** Suppose $x = f(t)$, $y = g(t)$, $a \leq t \leq b$, are parametric equations of a smooth curve C that does not intersect itself for $a < t < b$. If P is a partition of $[a, b]$ given by the numbers

$$a = t_0 < t_1 < t_2 < \cdots < t_{n-1} < t_n = b,$$

then, as shown in FIGURE 10.3.4, it seems reasonable that C can be approximated by a polygonal path through the points $Q_k(f(t_k), g(t_k))$, $k = 0, 1, \ldots, n$. Denoting the length of the line segment through Q_{k-1} and Q_k by L_k we write the approximate length of C as

$$\sum_{k=1}^{n} L_k, \tag{5}$$

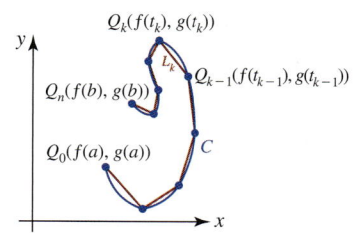

FIGURE 10.3.4 Approximating the length of C (blue) by the length of a polygonal path (red)

where

$$L_k = \sqrt{[f(t_k) - f(t_{k-1})]^2 + [g(t_k) - g(t_{k-1})]^2}.$$

Now, since f and g have continuous derivatives, the Mean Value Theorem (see Section 4.4) asserts that there exist numbers u_k^* and v_k^* in the interval (t_{k-1}, t_k) such that

$$f(t_k) - f(t_{k-1}) = f'(u_k^*)(t_k - t_{k-1}) = f'(u_k^*)\Delta t_k \tag{6}$$

and

$$g(t_k) - g(t_{k-1}) = g'(v_k^*)(t_k - t_{k-1}) = g'(v_k^*)\Delta t_k. \tag{7}$$

Using (6) and (7) in (5) and simplifying yield

$$\sum_{k=1}^{n} L_k = \sum_{k=1}^{n} \sqrt{[f'(u_k^*)]^2 + [g'(v_k^*)]^2} \Delta t_k. \tag{8}$$

By taking $\|P\| \to 0$ in (8), we obtain a formula for the length of a smooth curve. Notice that the limit of the sum in (8) is not the usual definition of a definite integral, since we are dealing with two numbers (u_k^* and v_k^*) rather than one in the interval (t_{k-1}, t_k). Nevertheless, it *can* be shown rigorously that the formula given in the next theorem results from (8) by taking $\|P\| \to 0$.

Theorem 10.3.2 Arc Length

If $x = f(t)$ and $y = g(t)$, $a \leq t \leq b$, define a smooth curve C that does not intersect itself for $a < t < b$, then the length L of C is

$$L = \int_a^b \sqrt{[f'(t)]^2 + [g'(t)]^2}\,dt = \int_a^b \sqrt{\left(\frac{dx}{dt}\right)^2 + \left(\frac{dy}{dt}\right)^2}\,dt. \tag{9}$$

Alternatively, (9) can be obtained using (1). If the curve defined by $x = f(t)$, $y = g(t)$, $a \leq t \leq b$, can also be represented by an explicit function $y = F(x)$, $x_0 \leq x \leq x_1$, then by changing variables of integration and using $f(a) = x_0$, $g(b) = x_1$, (3) of Section 6.5 becomes

$$L = \int_{x_0}^{x_1} \sqrt{1 + \left(\frac{dy}{dx}\right)^2}\,dx = \int_a^b \sqrt{1 + \left(\frac{f'(t)}{g'(t)}\right)^2}\,g'(t)\,dt = \int_a^b \sqrt{[f'(t)]^2 + [g'(t)]^2}\,dt.$$

EXAMPLE 5 Length of a Curve

Find the length of the curve given by $x = 4t$, $y = t^2$, $0 \leq t \leq 2$.

Solution: Since $f'(t) = 4$ and $g'(t) = 2t$, (9) gives

$$L = \int_0^2 \sqrt{16 + 4t^2}\,dt = 2\int_0^2 \sqrt{4 + t^2}\,dt.$$

With the trigonometric substitution $t = 2\tan\theta$, the last integral becomes

$$L = 8\int_0^{\pi/4} \sec^3\theta\,d\theta.$$

Integration by parts leads to (see Example 5, Section 7.3)

$$L = \left[4\sec\theta\tan\theta + 4\ln|\sec\theta + \tan\theta| \right]_0^{\pi/4} = 4\sqrt{2} + 4\ln\left(\sqrt{2} + 1\right) \approx 9.1823. \quad \blacksquare$$

Exercises 10.3 Answers to selected odd-numbered problems begin on page ANS-32.

≡ Fundamentals

In Problems 1–6, find the slope of the tangent line at the point corresponding to the indicated value of the parameter.

1. $x = t^3 - t^2, y = t^2 + 5t; \quad t = -1$

2. $x = 4/t, y = 2t^3 - t + 1; \quad t = 2$

3. $x = \sqrt{t^2 + 1}, y = t^4; \quad t = \sqrt{3}$

4. $x = e^{2t}, y = e^{-4t}; \quad t = \ln 2$

5. $x = \cos^2\theta, y = \sin\theta; \quad \theta = \pi/6$

6. $x = 2\theta - 2\sin\theta, y = 2 - 2\cos\theta; \quad \theta = \pi/4$

In Problems 7 and 8, find an equation of the tangent line to the given curve at the point corresponding to the indicated value of the parameter.

7. $x = t^3 + 3t, y = 6t^2 + 1; \quad t = -1$

8. $x = 2t + 4, y = t^2 + \ln t; \quad t = 1$

In Problems 9 and 10, find an equation of the tangent line to the given curve at the indicated point.

9. $x = t^2 + t, y = t^2; \quad (2, 4)$

10. $x = t^4 - 9, y = t^4 - t^2; \quad (0, 6)$

11. What is the slope of the tangent line to the curve given by $x = 4\sin 2t, y = 2\cos t, 0 \le t \le 2\pi$, at the point $(2\sqrt{3}, 1)$?

12. A curve C has parametric equations $x = t^2, y = t^3 + 1$. At what point on C is the tangent line given by $y + 3x - 5 = 0$?

13. A curve C has parametric equations $x = 2t - 5$, $y = t^2 - 4t + 3$. Find an equation of the tangent line to C that is parallel to the line $y = 3x + 1$.

14. Verify that the curve given by $x = -2/\pi + \cos\theta$, $y = -2\theta/\pi + \sin\theta, -\pi \le \theta \le \pi$, intersects itself. Find equations of tangent lines at the point of intersection.

In Problems 15–18, determine the points on the given curve at which the tangent line is either horizontal or vertical. Graph the curve.

15. $x = t^3 - t, y = t^2$ 16. $x = \frac{1}{8}t^3 + 1, y = t^2 - 2t$

17. $x = t - 1, y = t^3 - 3t^2$

18. $x = \sin t, y = \cos 3t, 0 \le t \le 2\pi$

In Problems 19–22, find dy/dx, d^2y/dx^2, and d^3y/dx^3.

19. $x = 3t^2, y = 6t^3$ 20. $x = \cos t, y = \sin t$

21. $x = e^{-t}, y = e^{2t} + e^{3t}$ 22. $x = \frac{1}{2}t^2 + t, y = \frac{1}{2}t^2 - t$

23. Use d^2y/dx^2 to determine the intervals of the parameter for which the curve in Problem 16 is concave upward and the intervals for which it is concave downward.

24. Use d^2y/dx^2 to determine whether the curve given by $x = 2t + 5, y = 2t^3 + 6t^2 + 4t$ has any points of inflection.

In Problems 25–30, find the length of the given curve.

25. $x = \frac{5}{3}t^3 + 2, y = 4t^3 + 6; \quad 0 \le t \le 2$

26. $x = \frac{1}{3}t^3, y = \frac{1}{2}t^2; \quad 0 \le t \le \sqrt{3}$

27. $x = e^t\sin t, y = e^t\cos t; \quad 0 \le t \le \pi$

28. One arch of the cycloid:

$$x = a(\theta - \sin\theta), y = a(1 - \cos\theta); \quad 0 \le \theta \le 2\pi$$

29. One arch of the hypocycloid of four cusps:

$$x = b\cos^3\theta, y = b\sin^3\theta; \quad 0 \le \theta \le \pi/2$$

30. One arch of the epicycloid of three cusps:

$$x = 4a\cos\theta - a\cos 4\theta, y = 4a\sin\theta - a\sin 4\theta, 0 \le \theta \le 2\pi/3$$

≡ Calculator/CAS Problems

31. Consider the curve $x = t^2 - 4t - 2, y = t^5 - 4t^3 - 1$ in Example 1.

    (a) Use a calculator to find an approximation of the y-coordinate of the y-intercept shown in Figure 10.3.1.

    (b) Use Newton's Method to approximate the x-coordinates of the three x-intercepts shown in Figure 10.3.1.

≡ Think About It

32. Let C be a curve described by $y = f(x)$, where F is a continuous nonnegative function on $x_1 \le x \le x_2$. Show that if C is given parametrically by $x = f(t), y = g(t), a \le t \le b$, f' and g continuous, then the **area** under the graph of C is $\int_a^b g(t)f'(t)dt$.

33. Use Problem 32 to show that the area under one arch of the cycloid in Figure 10.2.6(b) is three times the area of the circle.

10.4 Polar Coordinate System

■ **Introduction**   So far we have used the **rectangular** or **Cartesian coordinate system** to specify a point P or describe a curve C in the plane. We can regard this system as a grid of horizontal and vertical lines. The coordinates (a, b) of a point P are determined by the intersection of two lines: One line $x = a$ is perpendicular to the horizontal reference line called the x-axis, and the other $y = b$ is perpendicular to the vertical reference line called the y-axis. See FIGURE 10.4.1(a). Another system for locating points in the plane is the **polar coordinate system**.

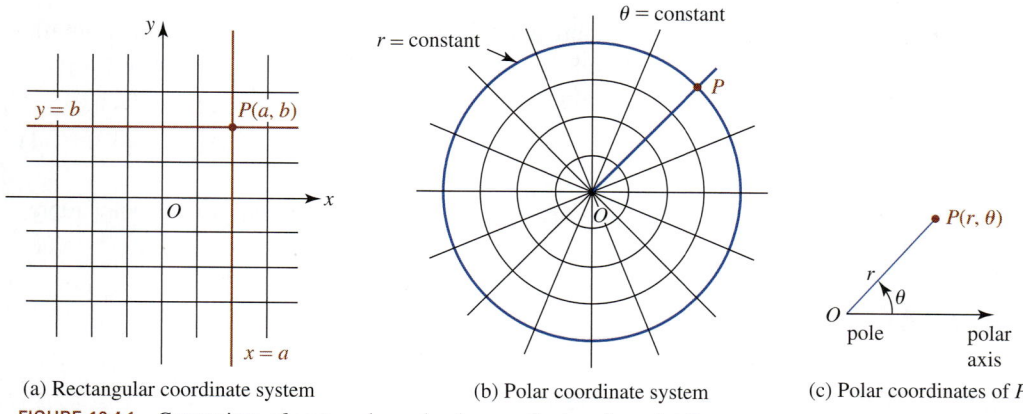

(a) Rectangular coordinate system   (b) Polar coordinate system   (c) Polar coordinates of P

FIGURE 10.4.1   Comparison of rectangular and polar coordinates of a point P

■ **Polar Coordinates**   To set up a **polar coordinate system**, we use a system of circles centered at a point O, called the **pole**, and straight lines or rays emanating from O. We take as a reference axis a horizontal half-line directed to the right of the pole and call it the **polar axis**. By specifying a directed (signed) distance r from O and an angle θ whose initial side is the polar axis and whose terminal side is the ray OP, we label the point P by (r, θ). We say that the ordered pair (r, θ) are the **polar coordinates** of P. See Figures 10.4.1(b) and 10.4.1(c).

Although the measure of the angle θ can be either in degrees or radians, in calculus radian measure is used almost exclusively. Consequently, we shall use only radian measure in this discussion.

In the polar coordinate system we adopt the following conventions.

Definition 10.4.1   Conventions in Polar Coordinates

(i) Angles $\theta > 0$ are measured counterclockwise from the polar axis, whereas angles $\theta < 0$ are measured clockwise.
(ii) To graph a point $(-r, \theta)$, where $-r < 0$, measure $|r|$ units along the ray $\theta + \pi$.
(iii) The coordinates of the pole O are $(0, \theta)$, where θ is any angle.

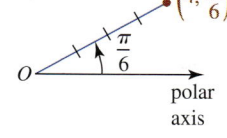

(a)

(b)

EXAMPLE 1   Plotting Polar Points

Plot the points whose polar coordinates are given.

   (a) $(4, \pi/6)$ **(b)** $(2, -\pi/4)$ **(c)** $(-3, 3\pi/4)$

Solution
   (a) Measure 4 units along the ray $\pi/6$ as shown in FIGURE 10.4.2(a).
   (b) Measure 2 units along the ray $-\pi/4$. See Figure 10.4.2(b).
   (c) Measure 3 units along the ray $3\pi/4 + \pi = 7\pi/4$. Equivalently, we can measure 3 units along the ray $3\pi/4$ extended *backward* through the pole. Note carefully in Figure 10.4.2(c) that the point $(-3, 3\pi/4)$ is not in the same quadrant as the terminal side of the given angle. ■

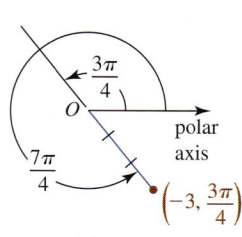

(c)

FIGURE 10.4.2   Points in polar coordinates in Example 1

In contrast to the rectangular coordinate system, the description of a point in polar coordinates is not unique. This is an immediate consequence of the fact that

$$(r, \theta) \quad \text{and} \quad (r, \theta + 2n\pi), \, n \text{ an integer},$$

are equivalent. To compound the problem, negative values of r can be used.

EXAMPLE 2 Equivalent Polar Points

The following coordinates are some alternative representations of the point $(2, \pi/6)$:

$$(2, 13\pi/6), \quad (2, -11\pi/6), \quad (-2, 7\pi/6), \quad (-2, -5\pi/6). \quad \blacksquare$$

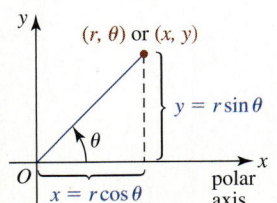

FIGURE 10.4.3 Relating polar and rectangular coordinates

■ **Conversion of Polar Coordinates to Rectangular** By superimposing a rectangular coordinate system on a polar coordinate system, as shown in FIGURE 10.4.3, we can convert a polar description of a point to rectangular coordinates by using

$$x = r\cos\theta, \quad y = r\sin\theta. \tag{1}$$

These conversion formulas hold true for any values of r and θ in an equivalent polar representation of (r, θ).

EXAMPLE 3 Polar to Rectangular

Convert $(2, \pi/6)$ in polar coordinates to rectangular coordinates.

Solution With $r = 2$, $\theta = \pi/6$, we have from (1),

$$x = 2\cos\frac{\pi}{6} = 2\left(\frac{\sqrt{3}}{2}\right) = \sqrt{3}$$

$$y = 2\sin\frac{\pi}{6} = 2\left(\frac{1}{2}\right) = 1.$$

Thus, $(2, \pi/6)$ is equivalent to $\left(\sqrt{3}, 1\right)$ in rectangular coordinates. ■

■ **Conversion of Rectangular Coordinates to Polar** It should be evident from Figure 10.4.3 that x, y, r, and θ are also related by

$$r^2 = x^2 + y^2, \quad \tan\theta = \frac{y}{x}. \tag{2}$$

The equations in (2) are used to convert the rectangular coordinates (x, y) to the polar coordinates (r, θ).

EXAMPLE 4 Rectangular to Polar

Convert $(-1, 1)$ in rectangular coordinates to polar coordinates.

Solution With $x = -1$, $y = 1$, we have from (2)

$$r^2 = 2 \quad \text{and} \quad \tan\theta = -1.$$

Now, $r^2 = 2$ or $r = \pm\sqrt{2}$, and two of many angles that satisfy $\tan\theta = -1$ are $3\pi/4$ and $7\pi/4$. From FIGURE 10.4.4 we see that two polar representations for $(-1, 1)$ are

$$\left(\sqrt{2}, 3\pi/4\right) \quad \text{and} \quad \left(-\sqrt{2}, 7\pi/4\right). \quad \blacksquare$$

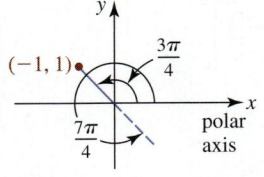

FIGURE 10.4.4 Point in Example 4

In Example 4, observe that we cannot pair just *any* angle θ and *any* value r that satisfy (2); these solutions must also be consistent with (1). Because the points $\left(-\sqrt{2}, 3\pi/4\right)$ and $\left(\sqrt{2}, 7\pi/4\right)$ lie in the fourth quadrant, they are not polar representations of the second-quadrant point $(-1, 1)$.

There are instances in calculus when a rectangular equation must be expressed as a polar equation $r = f(\theta)$. The next example shows how to do this using the conversion formulas in (1).

EXAMPLE 5 Rectangular Equation to Polar Equation

Find a polar equation that has the same graph as the circle $x^2 + y^2 = 8x$.

Solution Substituting $x = r\cos\theta$, $y = r\sin\theta$ into the given equation we find

$$r^2\cos^2\theta + r^2\sin^2\theta = 8r\cos\theta$$
$$r^2(\cos^2\theta + \sin^2\theta) = 8r\cos\theta \quad \leftarrow \cos^2\theta + \sin^2\theta = 1$$
$$r(r - 8\cos\theta) = 0.$$

The last equation implies that $r = 0$ or $r = 8\cos\theta$. Since $r = 0$ determines only the pole O, we conclude that a polar equation of the circle is $r = 8\cos\theta$. Note that the circle $x^2 + y^2 = 8x$ passes through the origin since $x = 0$ and $y = 0$ satisfy the equation. Relative to the polar equation $r = 8\cos\theta$ of the circle, the origin or pole corresponds to the polar coordinates $(0, \pi/2)$. ∎

EXAMPLE 6 Rectangular Equation to Polar Equation

Find a polar equation that has the same graph as the parabola $x^2 = 8(2 - y)$.

Solution We replace x and y in the given equation by $x = r\cos\theta$, $y = r\sin\theta$ and solve for r in terms of θ:

$$r^2\cos^2\theta = 8(2 - r\sin\theta)$$
$$r^2(1 - \sin^2\theta) = 16 - 8r\sin\theta$$
$$r^2 = r^2\sin^2\theta - 8r\sin\theta + 16 \quad \leftarrow \text{right side is a perfect square}$$
$$r^2 = (r\sin\theta - 4)^2$$
$$r = \pm(r\sin\theta - 4).$$

Solving for r gives two equations,

$$r = \frac{4}{1 + \sin\theta} \quad \text{or} \quad r = \frac{-4}{1 - \sin\theta}.$$

Now recall that, by convention (*ii*) of Definition 10.4.1, (r, θ) and $(-r, \theta + \pi)$ represent the same point. You should verify that if (r, θ) is replaced by $(-r, \theta + \pi)$ in the second of these two equations we obtain the first equation. In other words, the equations are equivalent and so we may simply take the polar equation of the parabola to be $r = 4/(1 + \sin\theta)$. ∎

In the last example we express a polar equation $r = f(\theta)$ as a rectangular equation using (1) and (2).

EXAMPLE 7 Polar Equation to Rectangular Equation

Find a rectangular equation that has the same graph as the polar equation $r^2 = 9\cos 2\theta$.

Solution First, we use the trigonometric identity for the cosine of a double angle:

$$r^2 = 9(\cos^2\theta - \sin^2\theta). \quad \leftarrow \cos 2\theta = \cos^2\theta - \sin^2\theta \qquad (3)$$

Now from (1) we can write $\cos\theta = x/r$ and $\sin\theta = y/r$, and from (2) we have $r^2 = x^2 + y^2$. Therefore,

$$\cos^2\theta = \frac{x^2}{r^2} = \frac{x^2}{x^2 + y^2} \quad \text{and} \quad \sin^2\theta = \frac{y^2}{r^2} = \frac{y^2}{x^2 + y^2}.$$

Substituting r^2, $\cos^2\theta$, and $\sin^2\theta$ into (3) then gives

$$x^2 + y^2 = 9\left(\frac{x^2}{x^2 + y^2} - \frac{y^2}{x^2 + y^2}\right) \quad \text{or} \quad (x^2 + y^2)^2 = 9(x^2 - y^2). \quad ∎$$

The next section will be devoted to graphing polar equations.

Exercises 10.4 Answers to selected odd-numbered problems begin on page ANS-32.

≡ Fundamentals

In Problems 1–6, plot the point with the given polar coordinates.

1. $(3, \pi)$ **2.** $(2, -\pi/2)$ **3.** $\left(-\frac{1}{2}, \pi/2\right)$

4. $(-1, \pi/6)$ **5.** $(-4, -\pi/6)$ **6.** $\left(\frac{2}{3}, 7\pi/4\right)$

In Problems 7–12, find alternative polar coordinates that satisfy

    (a) $r > 0, \theta < 0$ (b) $r > 0, \theta > 2\pi$
    (c) $r < 0, \theta > 0$ (d) $r < 0, \theta < 0$

for each point with the given polar coordinates.

7. $(2, 3\pi/4)$ **8.** $(5, \pi/2)$ **9.** $(4, \pi/3)$

10. $(3, \pi/4)$ **11.** $(1, \pi/6)$ **12.** $(3, 7\pi/6)$

In Problems 13–18, find the rectangular coordinates for each point with the given polar coordinates.

13. $\left(\frac{1}{2}, 2\pi/3\right)$ **14.** $(-1, 7\pi/4)$ **15.** $(-6, -\pi/3)$

16. $\left(\sqrt{2}, 11\pi/6\right)$ **17.** $(4, 5\pi/4)$ **18.** $(-5, \pi/2)$

In Problems 19–24, find polar coordinates that satisfy

    (a) $r > 0, -\pi < \theta \leq \pi$ (b) $r < 0, -\pi < \theta \leq \pi$

for each point with the given rectangular coordinates.

19. $(-2, -2)$ **20.** $(0, -4)$ **21.** $(1, -\sqrt{3})$

22. $(\sqrt{6}, \sqrt{2})$ **23.** $(7, 0)$ **24.** $(1, 2)$

In Problems 25–30, sketch the region on the plane that consists of points (r, θ) whose polar coordinates satisfy the given conditions.

25. $2 \leq r < 4, 0 \leq \theta \leq \pi$

26. $2 < r \leq 4$

27. $0 \leq r \leq 2, -\pi/2 \leq \theta \leq \pi/2$

28. $r \geq 0, \pi/4 < \theta < 3\pi/4$

29. $-1 \leq r \leq 1, 0 \leq \theta \leq \pi/2$

30. $-2 \leq r < 4, \pi/3 \leq \theta \leq \pi$

In Problems 31–40, find a polar equation that has the same graph as the given rectangular equation.

31. $y = 5$ **32.** $x + 1 = 0$

33. $y = 7x$ **34.** $3x + 8y + 6 = 0$

35. $y^2 = -4x + 4$ **36.** $x^2 - 12y - 36 = 0$

37. $x^2 + y^2 = 36$ **38.** $x^2 - y^2 = 1$

39. $x^2 + y^2 + x = \sqrt{x^2 + y^2}$ **40.** $x^3 + y^3 - xy = 0$

In Problems 41–52, find a rectangular equation that has the same graph as the given polar equation.

41. $r = 2 \sec\theta$ **42.** $r\cos\theta = -4$

43. $r = 6 \sin 2\theta$ **44.** $2r = \tan\theta$

45. $r^2 = 4 \sin 2\theta$ **46.** $r^2 \cos 2\theta = 16$

47. $r + 5 \sin\theta = 0$ **48.** $r = 2 + \cos\theta$

49. $r = \dfrac{2}{1 + 3\cos\theta}$ **50.** $r(4 - \sin\theta) = 10$

51. $r = \dfrac{5}{3\cos\theta + 8\sin\theta}$ **52.** $r = 3 + 3\sec\theta$

≡ Think About It

53. How would you express the distance d between two points (r_1, θ_1) and (r_2, θ_2) in terms of their polar coordinates?

54. You know how to find a rectangular equation of a line through two points with rectangular coordinates. How would you find a polar equation of a line through two points with polar coordinates (r_1, θ_1) and (r_2, θ_2)? Carry out your ideas by finding a polar equation of the line through $(3, 3\pi/4)$ and $(1, \pi/4)$. Find the polar coordinates of the x- and y-intercepts of the line.

55. In rectangular coordinates the x-intercepts of the graph of a function $y = f(x)$ are determined from the solutions of the equation $f(x) = 0$. In the next section we will graph polar equations $r = f(\theta)$. What is the significance of the solutions of the equation $f(\theta) = 0$?

10.5 Graphs of Polar Equations

▌ Introduction The graph of a polar equation $r = f(\theta)$ is the set of points P with *at least* one set of polar coordinates that satisfies the equation. Since it is most likely that your classroom does not have a polar coordinate grid, to facilitate graphing and discussion of graphs of a polar equation $r = f(\theta)$, we will, as in the preceding section, superimpose a rectangular coordinate over the polar coordinate system.

    We begin with some simple polar graphs.

EXAMPLE 1 A Circle Centered at Origin

Graph $r = 3$.

Solution   Since θ is not specified, the point $(3, \theta)$ lies on the graph of $r = 3$ for any value of θ and is 3 units from the origin. We see in FIGURE 10.5.1 that the graph is the circle of radius 3 centered at the origin.

Alternatively, we know from (2) of Section 10.4 that $r = \pm\sqrt{x^2 + y^2}$ so that $r = 3$ yields the familiar rectangular equation $x^2 + y^2 = 3^2$ of a circle of radius 3 centered at the origin. ■

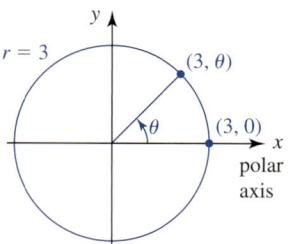

FIGURE 10.5.1   Circle in Example 1

■ **Circles Centered at the Origin**   In general, if a is any nonzero constant, the polar graph of

$$r = a \tag{1}$$

is a circle of radius $|a|$ with center at the origin.

EXAMPLE 2   A Line Through the Origin

Graph $\theta = \pi/4$.

Solution   Since r is not specified, the point $(r, \pi/4)$ lies on the graph for any value of r. If $r > 0$, then this point lies on the half-line in the first quadrant; if $r < 0$ then the point lies on the half-line in the third quadrant. For $r = 0$, the point $(0, \pi/4)$ is the pole or origin. Therefore, the polar graph of $\theta = \pi/4$ is the entire line through the origin that makes an angle of $\pi/4$ with the polar axis or positive x-axis. See FIGURE 10.5.2. ■

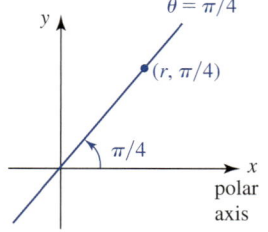

FIGURE 10.5.2   Line in Example 2

■ **Lines Through the Origin**   In general, if α is any nonzero real constant, the polar graph of

$$\theta = \alpha \tag{2}$$

is a line through the origin that makes an angle of α radians with the polar axis.

EXAMPLE 3   A Spiral

Graph $r = \theta$.

Solution   As $\theta \geq 0$ increases, r increases and the points (r, θ) wind around the pole in a counterclockwise manner. This is illustrated by the blue portion of the graph in FIGURE 10.5.3. The red portion of the graph is obtained by plotting points for $\theta < 0$.

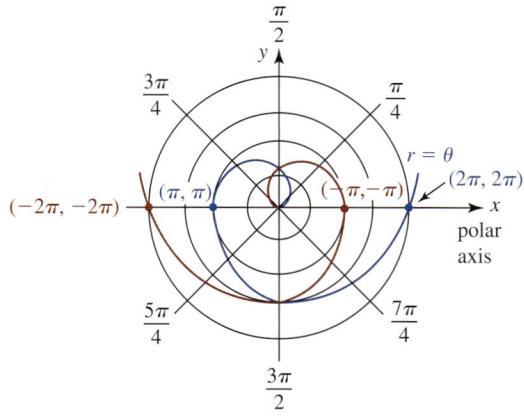

FIGURE 10.5.3   Graph of equation in Example 3   ■

■ **Spirals**   Many graphs in polar coordinates are given special names. The graph in Example 3 is a special case of

$$r = a\theta, \tag{3}$$

where a is a constant. A graph of this equation is called a **spiral of Archimedes**. A polar equation of the form

$$r = ae^{b\theta}, \tag{4}$$

is called a **logarithmic spiral**. The curve that describes the multichambered nautilus shell is an example of a logarithmic spiral. See Problems 31 and 32 in Exercises 10.5.

Half of multichambered Nautilus shell

Symmetries of a snowflake

In addition to basic point plotting, symmetry can often be utilized to graph a polar equation.

▌ **Symmetry**   As illustrated in FIGURE 10.5.4, a polar graph can have three types of symmetry. A polar graph is **symmetric with respect to the y-axis** if whenever (r, θ) is a point on the graph,

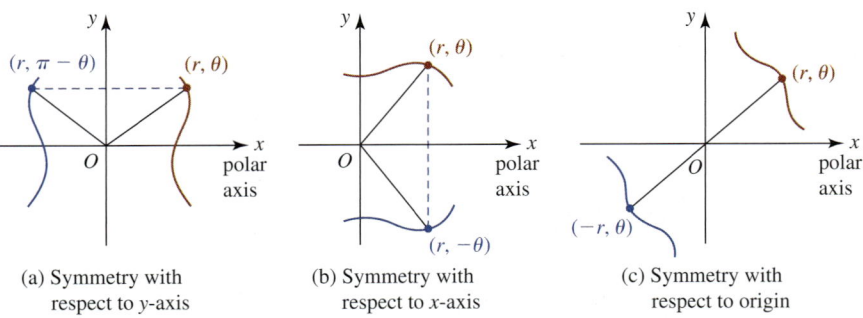

(a) Symmetry with respect to y-axis

(b) Symmetry with respect to x-axis

(c) Symmetry with respect to origin

FIGURE 10.5.4   Symmetries of a polar graph

$(r, \pi - \theta)$ is also a point on the graph. A polar graph is **symmetric with respect to the x-axis** if whenever (r, θ) is a point on the graph, $(r, -\theta)$ is also a point on the graph. Finally, a polar graph is **symmetric with respect to the origin** if whenever (r, θ) is on the graph, $(-r, \theta)$ is also a point on the graph.

We have the following tests for symmetries of a polar graph.

Tests for Symmetry of the Graph of a Polar Equation

The graph of a polar equation is symmetric with respect to:

- the **y-axis** if replacing (r, θ) by $(r, \pi - \theta)$ results in the same equation;   (5)
- the **x-axis** if replacing (r, θ) by $(r, -\theta)$ results in the same equation;   (6)
- the **origin** if replacing (r, θ) by $(-r, \theta)$ results in the same equation.   (7)

In rectangular coordinates the description of a point is unique. Hence, in rectangular coordinates if a test for a particular type of symmetry fails, then we can definitely say that the graph does not possess that symmetry.

▶ Because the polar description of a point is not unique, the graph of a polar equation may still have a particular type of symmetry even though the test for it may fail. For example, if replacing (r, θ) by $(r, -\theta)$ fails to give the original polar equation, the graph of that equation may still possess symmetry with respect to the x-axis. Therefore, if one of the replacement tests in (5)–(7) fails to give the same polar equation, the best we can say is "no conclusion."

EXAMPLE 4   Graphing a Polar Equation

Graph $r = 1 - \cos\theta$.

Solution   One way of graphing this equation is to plot a few well-chosen points corresponding to $0 \le \theta \le 2\pi$. As the following table shows

θ	0	$\pi/4$	$\pi/2$	$3\pi/4$	π	$5\pi/4$	$3\pi/2$	$7\pi/4$	2π
r	0	0.29	1	1.71	2	1.71	1	0.29	0

as θ advances from $\theta = 0$ to $\theta = \pi/2$, r increases from $r = 0$ (the origin) to $r = 1$. See

FIGURE 10.5.5(a). As θ advances from $\theta = \pi/2$ to $\theta = \pi$, r continues to increase from $r = 1$ to its maximum value of $r = 2$. See Figure 10.5.5(b). Then, for $\theta = \pi$ to $\theta = 3\pi/2$, r begins to decrease from $r = 2$ to $r = 1$. For $\theta = 3\pi/2$ to $\theta = 2\pi$, r continues to decrease and we end up again at the origin $r = 0$. See Figures 10.5.5(c) and (d).

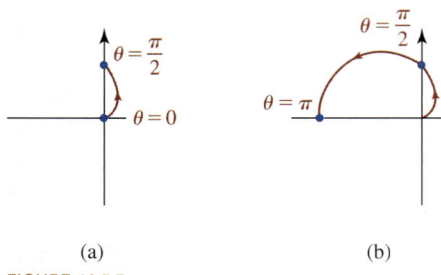 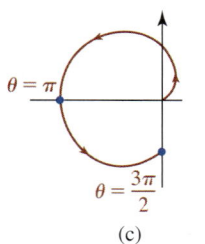 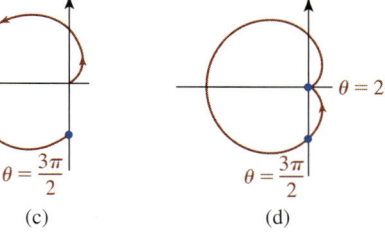

(a)   (b)   (c)   (d)

FIGURE 10.5.5   Graph of equation in Example 4

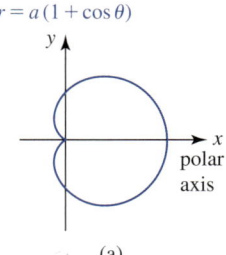

$r = a(1 + \cos\theta)$

(a)

By taking advantage of symmetry we could have simply plotted points for $0 \le \theta \le \pi$. From the trigonometric identity for the cosine function $\cos(-\theta) = \cos\theta$ it follows from (6) that the graph of $r = 1 - \cos\theta$ is symmetric with respect to the x-axis. We can obtain the complete graph of $r = 1 - \cos\theta$ by reflecting in the x-axis that portion of the graph given in Figure 10.5.5(b). ∎

■ **Cardioids**   The polar equation in Example 4 is a member of a family of equations that all have a "heart-shaped" graph that passes through the origin. A graph of any polar equation of the form

$$r = a \pm a\sin\theta \quad \text{or} \quad r = a \pm a\cos\theta \qquad (8)$$

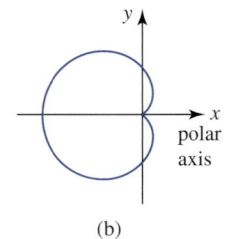

$r = a(1 - \cos\theta)$

(b)

is called a **cardioid**. The only difference in the graph of these four equations is their symmetry with respect to the y-axis ($r = a \pm a\sin\theta$) or symmetry with respect to the x-axis ($r = a \pm a\cos\theta$). In **FIGURE 10.5.6** we have assumed that $a > 0$.

By knowing the basic shape and orientation of a cardioid, you can obtain a quick and accurate graph by plotting the four points corresponding to $\theta = 0$, $\theta = \pi/2$, $\theta = \pi$, and $\theta = 3\pi/2$. The graphs of $r = a \pm a\sin\theta$ are symmetric with respect to the y-axis and the graphs of $r = a \pm a\cos\theta$ are symmetric with respect to the x-axis.

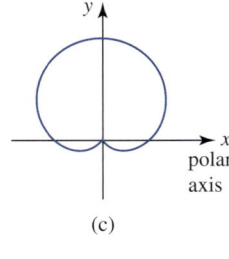

$r = a(1 + \sin\theta)$

(c)

■ **Limaçons**   Cardioids are special cases of polar curves known as **limaçons**:

$$r = a \pm b\sin\theta \quad \text{or} \quad r = a \pm b\cos\theta. \qquad (9)$$

The shape of a limaçon depends on the relative magnitudes of a and b. Let us assume that $a > 0$ and $b > 0$. For $0 < a/b < 1$, we get a **limaçon with an interior loop** as shown in **FIGURE 10.5.7(a)**. When $a = b$ or equivalently $a/b = 1$ we get a **cardioid**. For $1 < a/b < 2$, we get a **dimpled limaçon** as shown in Figure 10.5.7(b). For $a/b \ge 2$, the curve is called a **convex limaçon**. See Figure 10.5.7(c).

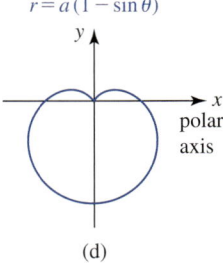

$r = a(1 - \sin\theta)$

(d)

FIGURE 10.5.6   Cardioids

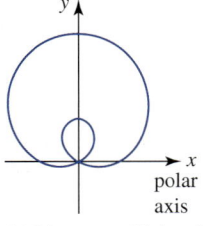 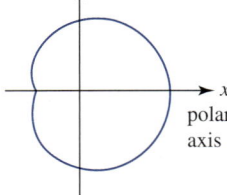 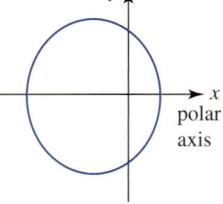

(a) Limaçon with interior loop   (b) Dimpled limaçon   (c) Convex limaçon

FIGURE 10.5.7   Three kinds of limaçons: for $0 < a/b < 1$ we get (a); for $1 < a/b < 2$ we get (b); for $a/b \ge 2$ we get (c)

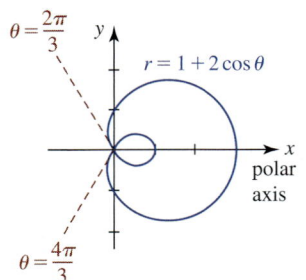

FIGURE 10.5.8 Graph of equation in Example 6

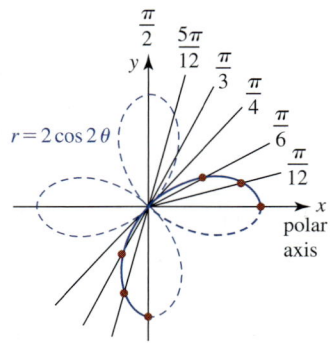

FIGURE 10.5.9 Graph of equation in Example 7

EXAMPLE 5 A Limaçon

The graph of $r = 3 - \sin\theta$ is a convex limaçon, since $a = 3$, $b = 1$, and $a/b = 3 > 2$. The graph of this equation is similar to that in Figure 10.5.7(c) except that the graph is symmetric with respect to the y-axis. ∎

EXAMPLE 6 A Limaçon

The graph of $r = 1 + 2\cos\theta$ is a limaçon with an interior loop, since $a = 1$, $b = 2$, and $a/b = \frac{1}{2} < 1$. For $\theta \geq 0$, notice in FIGURE 10.5.8 the limaçon starts at $\theta = 0$ or $(3, 0)$. The graph passes through the y-axis at $(1, \pi/2)$ and then enters the origin $(r = 0)$ for the first angle for which $r = 0$ or $1 + 2\cos\theta = 0$ or $\cos\theta = -\frac{1}{2}$. This implies that $\theta = 2\pi/3$. At $\theta = \pi$, the curve passes through $(-1, \pi)$. The remainder of the graph can then be completed using the fact that it is symmetric with respect to the x-axis. ∎

■ **Tangents to the Graph at the Origin** In Example 6, the lines $\theta = 2\pi/3$ and $\theta = 4\pi/3$ shown in red in Figure 10.5.8, where the graph of $r = 1 + 2\cos\theta$ enters and exits the origin, respectively, are actually tangent to the graph at the origin. In general, if $r = 0$ for $\theta = \theta_0$ and $dr/d\theta \neq 0$ when $\theta = \theta_0$, then the graph of $r = f(\theta)$ is tangent to the line $\theta = \theta_0$ at the origin. We will prove this in the next section.

EXAMPLE 7 A Rose Curve

Graph $r = 2\cos 2\theta$.

Solution Because

$$\cos 2(\pi - \theta) = \cos 2\theta \qquad \text{and} \qquad \cos(-2\theta) = \cos 2\theta$$

we conclude by (5) and (6) of the tests for symmetry that the graph is symmetric with respect to both the x- and the y-axes. A moment of reflection should convince you that we need only consider $0 \leq \theta \leq \pi/2$. Using the data in the following table, we see that the dashed portion of the graph given in FIGURE 10.5.9 is that completed by symmetry. The graph is called a **rose curve with four petals**.

θ	0	$\pi/12$	$\pi/6$	$\pi/4$	$\pi/3$	$5\pi/12$	$\pi/2$
r	2	$\sqrt{3}$	1	0	-1	$-\sqrt{3}$	-2

Note from the table that $r = 0$ and $dr/d\theta = -4\sin 2\theta \neq 0$ for $\theta = \pi/4$. Hence, the graph is tangent to the line $\theta = \pi/4$ at the origin. ∎

■ **Rose Curves** In general, if n is a positive integer, then the graphs of

$$r = a\sin n\theta \qquad \text{or} \qquad r = a\cos n\theta, \qquad n \geq 2 \tag{10}$$

are called **rose curves**, although as you can see in FIGURE 10.5.10 the curve looks more like a daisy. We note that the number of **petals** or **loops** of the curve is:

- n when n is odd, and
- $2n$ when n is even.

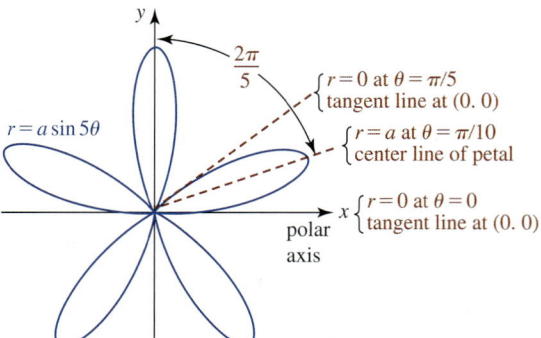

FIGURE 10.5.10 Rose curve with 5 petals

To graph a rose curve we can start by graphing one petal. To begin, we find an angle θ for which r is a maximum. This gives the *center line* of the petal. We then find corresponding values of θ for which the rose curve enters the origin ($r = 0$). To complete the graph we use the fact that the center lines of the petals are spaced $2\pi/n$ radians ($360/n$ degrees) apart if n is odd, and $2\pi/2n = \pi/n$ radians ($180/n$ degrees) apart if n is even. In Figure 10.5.10 we have drawn the graph of $r = a\sin 5\theta$, $a > 0$. The center line of the petal in the first quadrant is determined from the solution of

$$a = a\sin 5\theta \qquad \text{or} \qquad 1 = \sin 5\theta.$$

The last equation implies that $5\theta = \pi/2$ or $\theta = \pi/10$. The spacing between the center lines of the five petals is $2\pi/5$ radians ($72°$). Also, $r = 0$, or $\sin 5\theta = 0$, for $5\theta = 0$ and $5\theta = \pi$. Since $dr/d\theta = 5a\cos 5\theta \neq 0$ for $\theta = 0$ and $\theta = \pi/5$ the graph of the petal in the first quadrant is tangent to those lines at the origin.

In Example 5 in Section 10.4 we saw that the polar equation $r = 8\cos\theta$ is equivalent to the rectangular equation $x^2 + y^2 = 8x$. By completing the square in x in the rectangular equation, we recognize

$$(x - 4)^2 + y^2 = 16$$

as a circle of radius 4 centered at $(4, 0)$ on the x-axis. Polar equations such as $r = 8\cos\theta$ or $r = 8\sin\theta$ are circles and are also special cases of rose curves. See FIGURE 10.5.11.

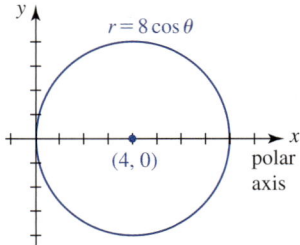

FIGURE 10.5.11   Graph of equation $r = 8\cos\theta$

■ **Circles with Centers on an Axis**   When $n = 1$ in (10) we get

$$r = a\sin\theta \qquad \text{or} \qquad r = a\cos\theta, \tag{11}$$

which are polar equations of circles passing through the origin with diameters $|a|$ and with centers $(a/2, 0)$ on the x-axis ($r = a\cos\theta$), or with centers $(0, a/2)$ on the y-axis ($r = a\sin\theta$). FIGURE 10.5.12 illustrates the graphs of the equations in (11) in the cases when $a > 0$ and $a < 0$.

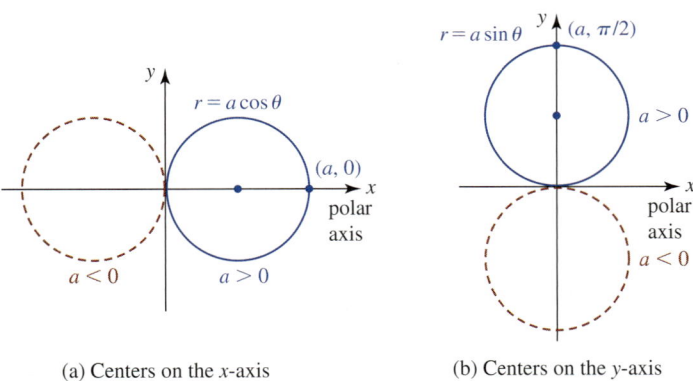

(a) Centers on the x-axis   (b) Centers on the y-axis

FIGURE 10.5.12   Circles through the origin with centers on an axis

■ **Lemniscates**   If n is a positive integer, the graphs of

$$r^2 = a\cos 2\theta \qquad \text{or} \qquad r^2 = a\sin 2\theta, \tag{12}$$

where $a > 0$, are called **lemniscates**. By (7) of the tests for symmetry you can see the graphs of both of the equations in (12) are symmetric with respect to the origin. Moreover, by (6) of the tests for symmetry, the graph of $r^2 = a\cos 2\theta$ is symmetric with respect to the x-axis. FIGURE 10.5.13 shows typical graphs of the equations $r^2 = a\cos 2\theta$ and $r^2 = a\sin 2\theta$, respectively.

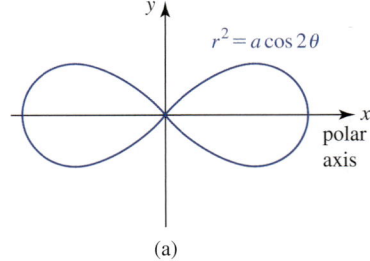

(a)

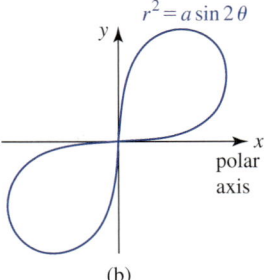

(b)

FIGURE 10.5.13   Lemniscates

■ **Points of Intersection**   In rectangular coordinates we can find the points (x, y) where the graphs of two functions $y = f(x)$ and $y = g(x)$ intersect by equating the y values. The real solutions of the equation $f(x) = g(x)$ correspond to *all* the x-coordinates of the points where the graphs intersect. In contrast, problems may arise in polar coordinates when we try the same method to determine where the graphs of two polar equations $r = f(\theta)$ and $r = g(\theta)$ intersect.

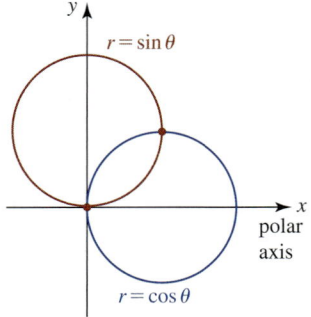

FIGURE 10.5.14 Intersecting circles in Example 8

See the identities in (18) of Section 1.4.

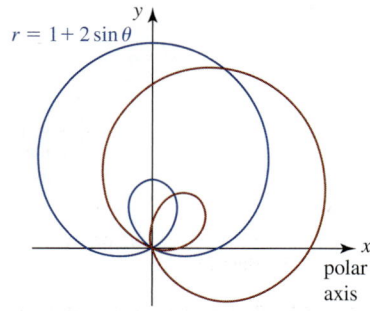

FIGURE 10.5.15 Graphs of polar equations in Example 9

EXAMPLE 8 Intersecting Circles

In FIGURE 10.5.14 we see that the circles $r = \sin\theta$ and $r = \cos\theta$ have two points of intersection. By equating the r values, the equation $\sin\theta = \cos\theta$ leads to $\theta = \pi/4$. Substituting this value into either equation yields $r = \sqrt{2}/2$. Thus, we have found only a single polar point $(\sqrt{2}/2, \pi/4)$ where the graphs intersect. From the figure, it is apparent that the graphs also intersect at the origin. But the problem here is that the origin or pole is $(0, \pi/2)$ on the graph of $r = \cos\theta$ but is $(0, 0)$ on the graph of $r = \sin\theta$. This situation is analogous to the curves reaching the same point at different times. ∎

■ **Rotation of Polar Graphs** In Section 1.2 we saw that if $y = f(x)$ is the rectangular equation of a function, then the graphs of $y = f(x - c)$ and $y = f(x + c)$, $c > 0$, are obtained by *shifting* the graph of f horizontally c units to the right and to the left, respectively. In contrast, if $r = f(\theta)$ is a polar equation, then the graphs of $f(\theta - \gamma)$ and $f(\theta + \gamma)$, where $\gamma > 0$, can be obtained by *rotating* the graph of f by an amount γ. Specifically:

- The graph of $r = f(\theta - \gamma)$ is the graph of $r = f(\theta)$ rotated *counterclockwise* about the origin by an amount γ.
- The graph of $r = f(\theta + \gamma)$ is the graph of $r = f(\theta)$ rotated *clockwise* about the origin by an amount γ.

For example, the graph of the cardioid $r = a(1 + \cos\theta)$ is shown in Figure 10.5.6(a). The graph of $r = a(1 + \cos(\theta - \pi/2))$ is the graph of $r = a(1 + \cos\theta)$ rotated counterclockwise about the origin by an amount $\pi/2$. Its graph then must be that given in Figure 10.5.6(c). This makes sense, because the sum formula of the cosine gives

$$r = a[1 + \cos(\theta - \pi/2)] = a[1 + \cos\theta\cos(\pi/2) + \sin\theta\sin(\pi/2)] = a(1 + \sin\theta).$$

Similarly, rotating $r = a(1 + \cos\theta)$ counterclockwise about the origin by an amount π gives

$$r = a[1 + \cos(\theta - \pi)] = a[1 + \cos\theta\cos\pi + \sin\theta\sin\pi] = a(1 - \cos\theta).$$

Now look again at Figure 10.5.13. From

$$r^2 = a\cos 2\left(\theta - \frac{\pi}{4}\right) = a\cos\left(2\theta - \frac{\pi}{2}\right) = a\sin 2\theta$$

we see that the graph of the lemniscate in Figure 10.5.13(b) is the graph in Figure 10.5.13(a) rotated counterclockwise about the origin by an amount $\pi/4$.

EXAMPLE 9 Rotated Polar Graph

Graph $r = 1 + 2\sin(\theta + \pi/4)$.

Solution The graph of the given equation is the graph of the limaçon $r = 1 + 2\sin\theta$ rotated clockwise about the origin by an amount $\pi/4$. In FIGURE 10.5.15 the blue graph is that of $r = 1 + 2\sin\theta$ and the red graph is the rotated graph. ∎

$$\frac{d}{d\theta}$$ **NOTES FROM THE CLASSROOM**

(i) The four-petal rose curve obtained in Example 7 is obtained by plotting r for θ-values satisfying $0 \le \theta < 2\pi$. See FIGURE 10.5.16. Do not assume this is true for every rose curve. Indeed, the five-petal rose curve discussed in Figure 10.5.10 is obtained using θ-values satisfying $0 \le \theta < \pi$. In general, a rose curve $r = a\sin n\theta$ or $r = a\cos n\theta$ is traced out exactly once for $0 \le \theta < 2\pi$ if n is even and once for $0 \le \theta < \pi$ if n is odd. Observations such as these will be important in the next section.

(ii) Example 8 illustrates one of several frustrating difficulties of working in polar coordinates:

> *A point can be on the graph of a polar equation even though its coordinates do not satisfy the equation.*

This is something that cannot happen in rectangular coordinates. For example, you should verify that $(2, \pi/2)$ is an alternative polar description of the point $(-2, 3\pi/2)$. Moreover, verify that $(-2, 3\pi/2)$ is a point on the graph of $r = 1 + 3\sin\theta$ by showing that the coordinates satisfy the equation. However, note that the alternative coordinates $(2, \pi/2)$ do not satisfy the equation.

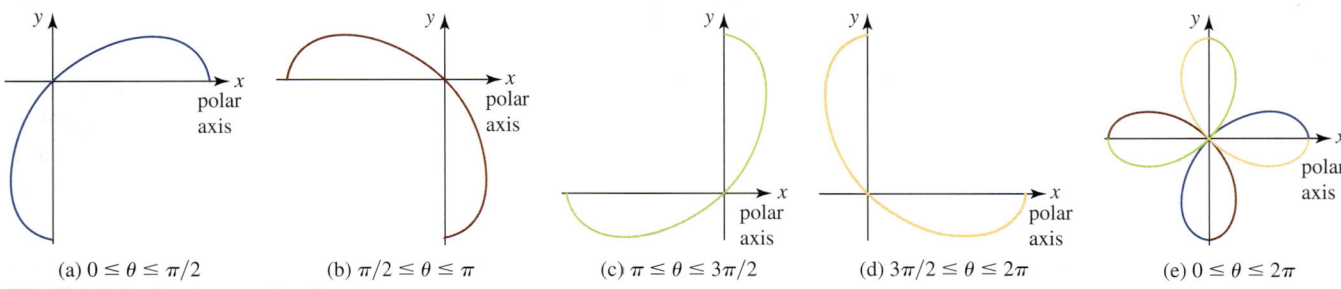

(a) $0 \le \theta \le \pi/2$ (b) $\pi/2 \le \theta \le \pi$ (c) $\pi \le \theta \le 3\pi/2$ (d) $3\pi/2 \le \theta \le 2\pi$ (e) $0 \le \theta \le 2\pi$

FIGURE 10.5.16 Plotting $r = 2\cos 2\theta$

Exercises 10.5   Answers to selected odd-numbered problems begin on page ANS-33.

≡ Fundamentals

In Problems 1–30, identify by name the graph of the given polar equation. Then sketch the graph of the equation.

1. $r = 6$

2. $r = -1$

3. $\theta = \pi/3$

4. $\theta = 5\pi/6$

5. $r = 2\theta, \theta \le 0$

6. $r = 3\theta, \theta \ge 0$

7. $r = 1 + \cos\theta$

8. $r = 5 - 5\sin\theta$

9. $r = 2(1 + \sin\theta)$

10. $2r = 1 - \cos\theta$

11. $r = 1 - 2\cos\theta$

12. $r = 2 + 4\sin\theta$

13. $r = 4 - 3\sin\theta$

14. $r = 3 + 2\cos\theta$

15. $r = 4 + \cos\theta$

16. $r = 4 - 2\sin\theta$

17. $r = \sin 2\theta$

18. $r = 3\sin 4\theta$

19. $r = 3\cos 3\theta$

20. $r = 2\sin 3\theta$

21. $r = \cos 5\theta$

22. $r = 2\sin 9\theta$

23. $r = 6\cos\theta$

24. $r = -2\cos\theta$

25. $r = -3\sin\theta$

26. $r = 5\sin\theta$

27. $r^2 = 4\sin 2\theta$

28. $r^2 = 4\cos 2\theta$

29. $r^2 = -25\cos 2\theta$

30. $r^2 = -9\sin 2\theta$

In Problems 31 and 32, the graph of the given equation is a spiral. Sketch its graph.

31. $r = 2^\theta, \theta \ge 0$ (logarithmic) **32.** $r\theta = \pi, \theta > 0$ (hyperbolic)

In Problems 33–38, find an equation of the given polar graph.

33.

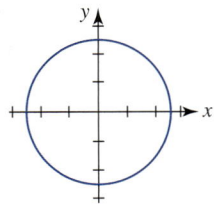

FIGURE 10.5.17 Graph for Problem 33

34.

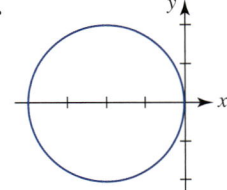

FIGURE 10.5.18 Graph for Problem 34

35.

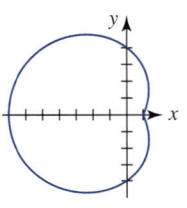

FIGURE 10.5.19 Graph for Problem 35

36.

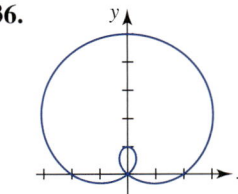

FIGURE 10.5.20 Graph for Problem 36

37.

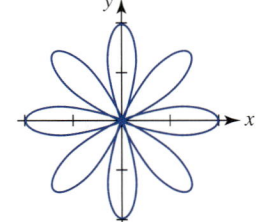

FIGURE 10.5.21 Graph for Problem 37

38.

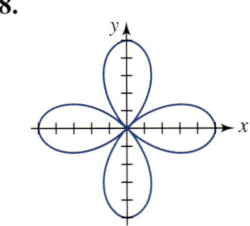

FIGURE 10.5.22 Graph for Problem 38

In Problems 39–42, find the points of intersection of the graphs of the given pair of polar equations.

39. $r = 2$
$r = 4\sin\theta$

40. $r = \sin\theta$
$r = \sin 2\theta$

41. $r = 1 - \cos\theta$
$r = 1 + \cos\theta$

42. $r = 3 - 3\cos\theta$
$r = 3\cos\theta$

In Problems 43 and 44, use the fact that $r = f(\theta)$ and $-r = f(\theta + \pi)$ describe the same curve as an aid in finding the points of intersection of the given pair of polar equations.

43. $r = 3$
$r = 6\sin 2\theta$

44. $r = \cos 2\theta$
$r = 1 + \cos\theta$

≡ Calculator/CAS Problems

45. Use a graphing utility to obtain the graph of the **bifolium** $r = 4\sin\theta \cos^2\theta$ and the circle $r = \sin\theta$ on the same axes. Find all points of intersection of the graphs.

46. Use a graphing utility to verify that the cardioid $r = 1 + \cos\theta$ and the lemniscate $r^2 = 4\cos\theta$ intersect at four points. Find these points of intersection of the graphs.

In Problems 47 and 48, the graphs of the equations (a)–(d) represent a rotation of the graph of the given equation. Try sketching these graphs by hand. If you have difficulties, then use a calculator or CAS.

47. $r = 1 + \sin\theta$
    (a) $r = 1 + \sin(\theta - \pi/2)$
    (b) $r = 1 + \sin(\theta + \pi/2)$
    (c) $r = 1 + \sin(\theta - \pi/6)$
    (d) $r = 1 + \sin(\theta + \pi/4)$

48. $r = 2 + 4\cos\theta$
    (a) $r = 2 + 4\cos(\theta + \pi/6)$
    (b) $r = 2 + 4\cos(\theta - 3\pi/2)$
    (c) $r = 2 + 4\cos(\theta + \pi)$
    (d) $r = 2 + 4\cos(\theta - \pi/8)$

In Problems 49–52, use a calculator or CAS, if necessary, to match the given graph with the appropriate polar equation in (a)–(d).

    (a) $r = 2\cos\dfrac{3\theta}{2}, \quad 0 \le \theta \le 4\pi$

    (b) $r = 2\cos\dfrac{\theta}{5}, \quad 0 \le \theta \le 5\pi$

    (c) $r = 2\sin\dfrac{\theta}{4}, \quad 0 \le \theta \le 8\pi$

    (d) $r = 2\sin\dfrac{\theta}{2}, \quad 0 \le \theta \le 4\pi$

49.

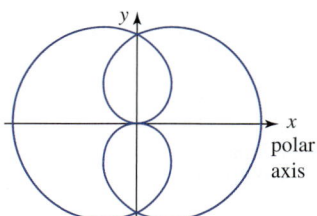

FIGURE 10.5.23 Graph for Problem 49

50.

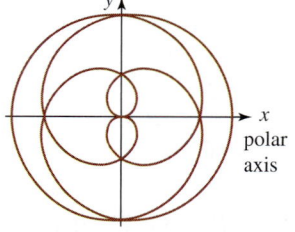

FIGURE 10.5.24 Graph for Problem 50

51.

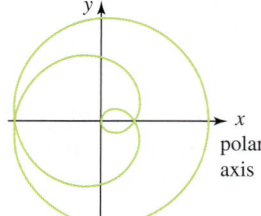

FIGURE 10.5.25 Graph for Problem 51

52.

FIGURE 10.5.26 Graph for Problem 52

53. Use a CAS to obtain graphs of the polar equation $r = a + \cos\theta$ for $a = 0, \frac{1}{4}, \frac{1}{2}, \frac{3}{4}, 1, \frac{5}{4}, \ldots, 3$.

54. Identify all the curves in Problem 53. What happens to the graphs as $a \to \infty$?

≡ Think About It

In Problems 55–58, identify the symmetries if the given pair of points are on the graph of $r = f(\theta)$.

55. $(r, \theta), (-r, \pi - \theta)$

56. $(r, \theta), (r, \theta + \pi)$

57. $(r, \theta), (-r, \theta + 2\pi)$

58. $(r, \theta), (-r, -\theta)$

In Problems 59 and 60, let $r = f(\theta)$ be a polar equation. Interpret the given result geometrically.

59. $f(-\theta) = f(\theta)$ (even function)

60. $f(-\theta) = -f(\theta)$ (odd function)

61. (a) What is the difference between the circles $r = -4$ and $r = 4$?
    (b) What is the difference between the lines through the origin $\theta = \pi/6$ and $\theta = 7\pi/6$?

62. A Bit of History The Italian **Galileo Galilei** (1564–1642) is remembered for his many discoveries and innovations in the fields of astronomy and physics. With a reflecting telescope of his own design he was the first to discover the moons of Jupiter. Through his observations of the planet Venus and sun spots, Galileo eventually came to support the controversial opinion of Nicolaus Copernicus that the planets revolved around the Sun. Galileo's empirical work on gravity predates the contributions of Isaac Newton. He was the first to perform scientific studies to determine the acceleration of gravity. By measuring the time it takes metal balls to roll down an inclined plane, Galileo was able to calculate the speed of each ball and from those observations concluded that the distance s a ball moved is related to time t by $s = \frac{1}{2}gt^2$, where g is the acceleration due to gravity.

    Suppose several metal balls are released simultaneously from a common point and allowed to slide down frictionless inclined planes at various angles, each ball accelerating because of gravity. See FIGURE 10.5.27. Show that at any instant, all the balls lie on a common circle whose topmost point is the point of release. Galileo was able to show this without the benefit of either rectangular or polar coordinates.

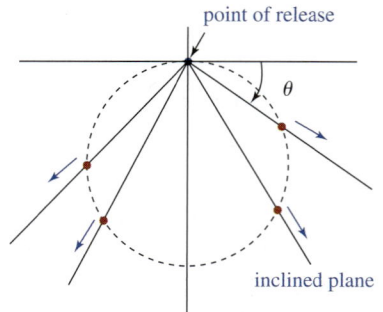

FIGURE 10.5.27 Inclined planes in Problem 62

10.6 Calculus in Polar Coordinates

Introduction In this section we will answer three standard calculus problems in the polar coordinate system.

- *What is the slope of a tangent line to a polar graph?*
- *What is the area bounded by a polar graph?*
- *What is the length of a polar graph?*

We begin with the tangent line problem.

Slope of a Tangent to a Polar Graph Somewhat surprisingly, the slope of a tangent line to the graph of a polar equation $r = f(\theta)$ is *not* the derivative $dr/d\theta = f'(\theta)$. The slope of a tangent line is still dy/dx. To find this latter derivative, we use $r = f(\theta)$ along with $x = r\cos\theta$, $y = r\sin\theta$ to write parametric equations of the curve:

$$x = f(\theta)\cos\theta, \quad y = f(\theta)\sin\theta. \tag{1}$$

Then from (1) of Section 10.3 and the Product Rule,

$$\frac{dy}{dx} = \frac{dy/d\theta}{dx/d\theta} = \frac{f(\theta)\cos\theta + f'(\theta)\sin\theta}{-f(\theta)\sin\theta + f'(\theta)\cos\theta}.$$

This result is summarized in the next theorem.

Theorem 10.6.1 Slope of Tangent Line

If f is a differentiable function of θ, then the **slope of the tangent line** to the graph of $r = f(\theta)$ at a point (r, θ) on the graph is

$$\frac{dy}{dx} = \frac{dy/d\theta}{dx/d\theta} = \frac{f(\theta)\cos\theta + f'(\theta)\sin\theta}{-f(\theta)\sin\theta + f'(\theta)\cos\theta}, \tag{2}$$

provided $dx/d\theta \neq 0$.

Formula (2) in Theorem 10.6.1 is presented "for the record"; do not memorize it. To find dy/dx in polar coordinates simply form the parametric equations $x = f(\theta)\cos\theta$, $y = f(\theta)\sin\theta$ and then use the parametric form of the derivative.

EXAMPLE 1 Slope

Find the slope of the tangent line to the graph of $r = 4\sin 3\theta$ at $\theta = \pi/6$.

Solution From the parametric equations $x = 4\sin 3\theta\cos\theta$, $y = 4\sin 3\theta\sin\theta$ we find

$$\frac{dy}{dx} = \frac{dy/d\theta}{dx/d\theta} = \frac{4\sin 3\theta\cos\theta + 12\cos 3\theta\sin\theta}{-4\sin 3\theta\sin\theta + 12\cos 3\theta\cos\theta}$$

and so

$$\left.\frac{dy}{dx}\right|_{\theta=\pi/6} = -\sqrt{3}.$$

The graph of the equation, which we recognize as a rose curve with three petals, and the tangent line are illustrated in FIGURE 10.6.1.

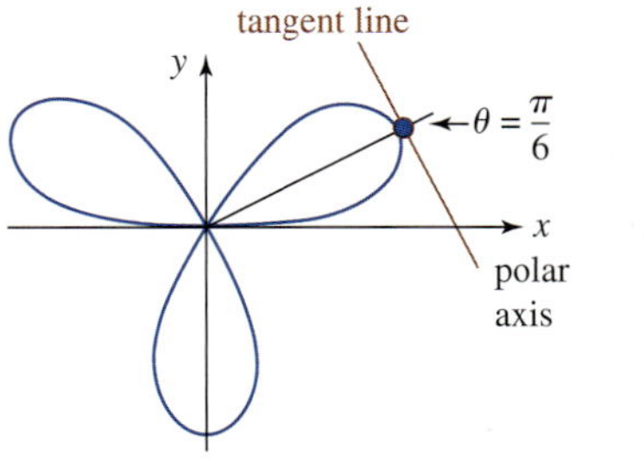

FIGURE 10.6.1 Tangent line in Example 1

EXAMPLE 2 Equation of Tangent Line

Find a rectangular equation of the tangent line in Example 1.

Solution At $\theta = \pi/6$ the parametric equations $x = 4\sin 3\theta\cos\theta$, $y = 4\sin 3\theta\sin\theta$ yield, respectively, $x = 2\sqrt{3}$ and $y = 2$. The rectangular coordinates of the point of tangency are $(2\sqrt{3}, 2)$. Using the slope found in Example 1, the point-slope form gives an equation of the red tangent line shown in Figure 10.6.1:

$$y - 2 = -\sqrt{3}(x - 2\sqrt{3}) \quad \text{or} \quad y = -\sqrt{3}x + 8.$$

We can get a polar equation of the line in Example 2 by replacing x and y in the rectangular equation by $x = r\cos\theta$, $y = r\sin\theta$ and solving for r:

$$r = \frac{8}{\sin\theta + \sqrt{3}\cos\theta}.$$

EXAMPLE 3 Horizontal and Vertical Tangents

Find the points on the graph of $r = 3 - 3\sin\theta$ at which the tangent line is horizontal and the points at which the tangent line is vertical.

Solution Recall from Section 10.3 that a horizontal tangent occurs at a point for which $dy/d\theta = 0$ and $dx/d\theta \neq 0$, whereas a vertical tangent occurs at a point for which $dx/d\theta = 0$ and $dy/d\theta \neq 0$. Now, from the parametric equations

$$x = (3 - 3\sin\theta)\cos\theta, \qquad y = (3 - 3\sin\theta)\sin\theta$$

we get

$$\frac{dx}{d\theta} = (3 - 3\sin\theta)(-\sin\theta) + \cos\theta(-3\cos\theta)$$

$$= -3\sin\theta + 3\sin^2\theta - 3\cos^2\theta$$

$$= -3 - 3\sin\theta + 6\sin^2\theta$$

$$= 3(2\sin\theta + 1)(\sin\theta - 1),$$

$$\frac{dy}{d\theta} = (3 - 3\sin\theta)\cos\theta + \sin\theta(-3\cos\theta)$$

$$= 3\cos\theta(1 - 2\sin\theta).$$

From these derivatives we see that:

$$\frac{dy}{d\theta} = 0 \left(\frac{dx}{d\theta} \neq 0\right) \text{ at } \theta = \frac{\pi}{6}, \theta = \frac{5\pi}{6}, \text{ and } \theta = \frac{3\pi}{2},$$

$$\frac{dx}{d\theta} = 0 \left(\frac{dy}{d\theta} \neq 0\right) \text{ at } \theta = \frac{7\pi}{6}, \text{ and } \theta = \frac{11\pi}{6}.$$

Thus, there are

horizontal tangents at: $\left(\frac{3}{2}, \pi/6\right), \left(\frac{3}{2}, 5\pi/6\right), (6, 3\pi/2),$
vertical tangents at: $\left(\frac{9}{2}, 7\pi/6\right), \left(\frac{9}{2}, 11\pi/6\right).$

These points, along with the tangent lines, are shown in FIGURE 10.6.2. ■

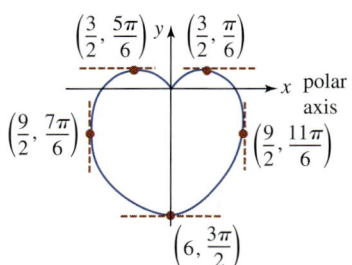

FIGURE 10.6.2 Horizontal and vertical tangent lines in Example 3

■ **Tangents to the Graph at the Origin** In the preceding section we stated that, in general, if $r = 0$ and $dr/d\theta = f'(\theta) \neq 0$ when $\theta = \theta_0$, then the graph of $r = f(\theta)$ is tangent to the line $\theta = \theta_0$ at the origin. This fact follows from (2). If $r = f(\theta)$ is a differentiable function of θ for which $f(\theta_0) = 0$ and $f'(\theta_0) \neq 0$, then at $\theta = \theta_0$, (2) gives

$$\frac{dy}{dx} = \frac{f(\theta_0)\cos\theta_0 + f'(\theta_0)\sin\theta_0}{-f(\theta_0)\sin\theta_0 + f'(\theta_0)\cos\theta_0} = \frac{f'(\theta_0)\sin\theta_0}{f'(\theta_0)\cos\theta_0} = \tan\theta_0.$$

In the last expression we recognize that $\tan\theta_0$ is the slope of the tangent line $\theta = \theta_0$.

Note in Example 3 that $r = 3 - 3\sin\theta = 0$ at $\theta = \pi/2$. But since both the numerator $dy/d\theta$ and the denominator $dx/d\theta$ in (2) are 0 at $\theta = \pi/2$ we do not draw any conclusion about the tangent line at the origin $(0, \pi/2)$.

■ **Area of Region** The problem of finding the area of a region bounded by polar graphs is not quite as straightforward as it was in Section 6.2. As we shall see in the ensuing discussion, in place of a rectangle we use a sector of a circle, such as shown in FIGURE 10.6.3. Since the area A of a circular sector is proportional to the central angle θ, measured in radians, and since the area of the complete circle is πr^2 we have

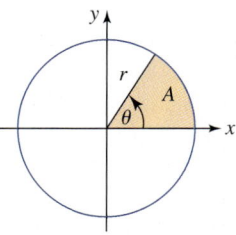

FIGURE 10.6.3 Area A of circular section

$$\frac{A}{\pi r^2} = \frac{\theta}{2\pi} \qquad \text{or} \qquad A = \frac{1}{2}r^2\theta. \tag{3}$$

■ **Building an Integral** Suppose $r = f(\theta)$ is a nonnegative continuous function on the interval $[\alpha, \beta]$, where $0 \le \alpha \le \beta < 2\pi$. To find the area A of the region shown in FIGURE 10.6.4(a) that is bounded by the graph of f and the rays $\theta = \alpha$ and $\theta = \beta$, we start by forming a partition P of $[\alpha, \beta]$:

$$\alpha = \theta_0 < \theta_1 < \theta_2 < \cdots < \theta_n = \beta.$$

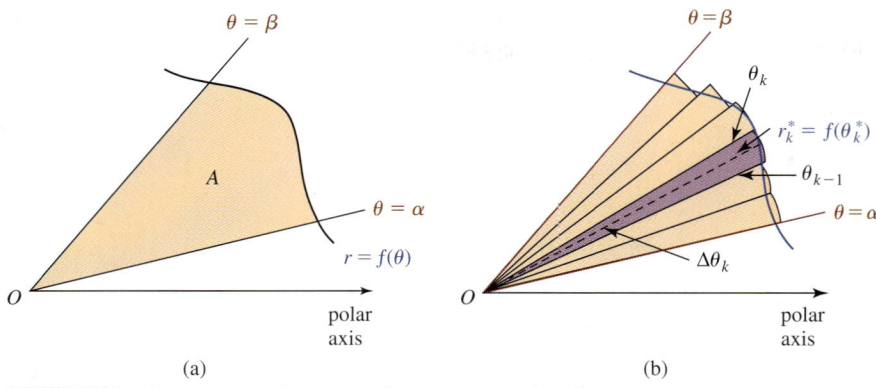

(a)   (b)

FIGURE 10.6.4   Area A of a region bounded by a polar graph and two rays

If θ_k^* denotes a sample point in the kth subinterval $[\theta_{k-1}, \theta_k]$, then by (3) the area of the circular sector of radius $r_k = f(\theta_k^*)$ indicated in Figure 10.6.4(b) is

$$A_k = \frac{1}{2}[f(\theta_k^*)]^2 \Delta\theta_k,$$

where $\Delta\theta_k = \theta_k - \theta_{k-1}$ is its central angle. In turn, the Riemann sum

$$\sum_{k=1}^{n} \frac{1}{2}[f(\theta_k^*)]^2 \Delta\theta_k$$

gives an approximation to A. The area A is then given by the limit as $\|P\| \to 0$:

$$A = \lim_{\|P\| \to 0} \sum_{k=1}^{n} \frac{1}{2}[f(\theta_k^*)]^2 \Delta\theta_k.$$

Theorem 10.6.2   Area in Polar Coordinates

If $r = f(\theta)$ is a nonnegative continuous function on $[\alpha, \beta]$, then the **area** bounded by its graph and the rays $\theta = \alpha$ and $\theta = \beta$ is given by

$$A = \int_{\alpha}^{\beta} \frac{1}{2}[f(\theta)]^2 \, d\theta = \frac{1}{2}\int_{\alpha}^{\beta} r^2 \, d\theta. \qquad (4)$$

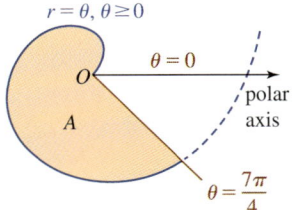

FIGURE 10.6.5   Area in Example 4

EXAMPLE 4   Area Bounded by a Spiral

Find the area of the region that is bounded by the spiral $r = \theta$, $\theta \ge 0$, between the rays $\theta = 0$ and $\theta = 7\pi/4$.

Solution   From (4), the area of the shaded region shown in FIGURE 10.6.5 is

$$A = \frac{1}{2}\int_{0}^{7\pi/4} \theta^2 \, d\theta = \frac{1}{6}\theta^3 \Big]_{0}^{7\pi/4} = \frac{343}{384}\pi^3 \approx 27.70. \qquad ■$$

EXAMPLE 5   Area Bounded by a Rose Curve

Find the area of one petal of the rose curve $r = 2\cos 5\theta$.

Solution   As shown in FIGURE 10.6.6 the rose curve has five petals. Because of symmetry we will find the area of one-half of one petal and multiply the result by 2. By setting $r = 0$ and solving

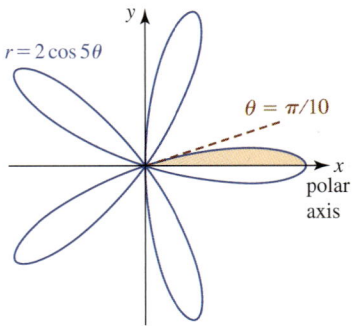

FIGURE 10.6.6   One-half of the area in Example 5

$2\cos 5\theta = 0$ we get $5\theta = \pi/2$ or $\theta = \pi/10$. In other words, the curve enters the origin tangent to the line $\theta = \pi/10$. From (4) the area A of the shaded half-petal in Figure 10.6.6 is then

$$
\begin{aligned}
A &= \frac{1}{2}\int_0^{\pi/10}(2\cos 5\theta)^2 d\theta \\
&= 2\int_0^{\pi/10}\cos^2 5\theta\, d\theta \\
&= 2\int_0^{\pi/10}\frac{1}{2}(1 + \cos 10\theta)\, d\theta \\
&= \theta + \frac{1}{10}\sin 10\theta\Big]_0^{\pi/10} \\
&= \frac{\pi}{10}.
\end{aligned}
$$

The half-angle formulas:

$$\cos^2\theta = \frac{1}{2}(1 + \cos 2\theta)$$

$$\sin^2\theta = \frac{1}{2}(1 - \cos 2\theta)$$

will be useful in this section.

The area of one petal is then $2(\pi/10) = \pi/5$. ∎

Of course, the area of each petal in Example 5 is the same and so the area enclosed by the complete five-petal rose curve is $5(\pi/5) = \pi$.

A word of caution. When working problems of the sort in Example 5 be careful with the limits of integration. Do not assume that the area enclosed by the complete five-petal rose curve can be obtained from (4) by integrating on the interval $[0, 2\pi]$. In other words, the area is not $\frac{1}{2}\int_0^{2\pi}(2\cos 5\theta)^2 d\theta$. This is because the complete curve is obtained for $0 \le \theta \le \pi$. See (*i*) in *Notes from the Classroom* in Section 10.5.

EXAMPLE 6   Area Bounded Between Two Graphs

Find the area of the region that is common to the interiors of the cardioid $r = 2 - 2\cos\theta$ and the limaçon $r = 2 + \cos\theta$.

Solution   Inspection of FIGURE 10.6.7 shows that we need two integrals. Solving the given equations simultaneously:

$$2 - 2\cos\theta = 2 + \cos\theta \qquad \text{or} \qquad \cos\theta = 0$$

yields $\theta = \pi/2$ so that a point of intersection is $(2, \pi/2)$. By symmetry, it follows that

$$
\begin{aligned}
A &= 2\left\{\frac{1}{2}\int_0^{\pi/2}(2 - 2\cos\theta)^2 d\theta + \frac{1}{2}\int_{\pi/2}^{\pi}(2 + \cos\theta)^2 d\theta\right\} \\
&= 4\int_0^{\pi/2}(1 - 2\cos\theta + \cos^2\theta)\, d\theta + \int_{\pi/2}^{\pi}(4 + 4\cos\theta + \cos^2\theta)\, d\theta \\
&= 4\int_0^{\pi/2}\left[1 - 2\cos\theta + \frac{1}{2}(1 + \cos 2\theta)\right] d\theta + \int_{\pi/2}^{\pi}\left[4 + 4\cos\theta + \frac{1}{2}(1 + \cos 2\theta)\right] d\theta \\
&= 4\left[\frac{3}{2}\theta - 2\sin\theta + \frac{1}{4}\sin 2\theta\right]_0^{\pi/2} + \left[\frac{9}{2}\theta + 4\sin\theta + \frac{1}{4}\sin 2\theta\right]_{\pi/2}^{\pi} \\
&= \frac{21}{4}\pi - 12 \approx 4.49.
\end{aligned}
$$

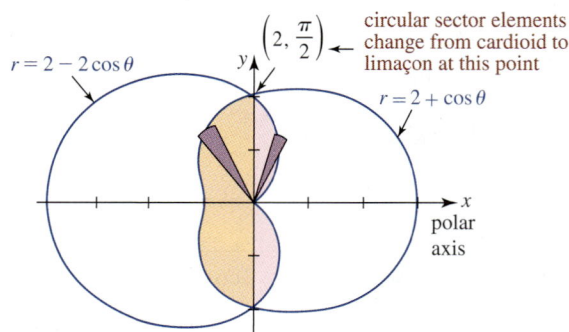

FIGURE 10.6.7   Area in Example 6

∎

■ **Area Bounded by Two Graphs** The area A of the region shown in FIGURE 10.6.8 can be found by subtracting areas. If f and g are continuous on $[\alpha, \beta]$ and $f(\theta) \geq g(\theta)$ on the interval, then the area bounded by the graphs of $r = f(\theta)$, $r = g(\theta)$, $\theta = \alpha$, and $\theta = \beta$ is

$$A = \frac{1}{2} \int_{\alpha}^{\beta} [f(\theta)]^2 \, d\theta - \frac{1}{2} \int_{\alpha}^{\beta} [g(\theta)]^2 \, d\theta.$$

Written as a single integral, the area is given by

$$A = \frac{1}{2} \int_{\alpha}^{\beta} ([f(\theta)]^2 - [g(\theta)]^2) \, d\theta. \tag{5}$$

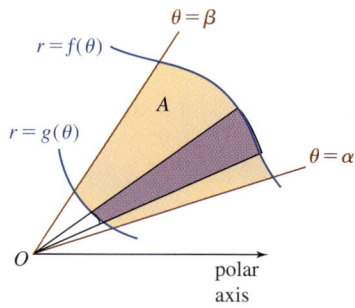

FIGURE 10.6.8 Area of region bounded between two graphs

EXAMPLE 7 Area Bounded by Two Graphs

Find the area of the region in the first quadrant that is outside the circle $r = 1$ and inside the rose curve $r = 2 \sin 2\theta$.

Solution Solving the two equations simultaneously:

$$1 = 2 \sin 2\theta \qquad \text{or} \qquad \sin 2\theta = \frac{1}{2}$$

implies that $2\theta = \pi/6$ and $2\theta = 5\pi/6$. Thus, two points of intersection in the first quadrant are $(1, \pi/12)$ and $(1, 5\pi/12)$. The area in question is shown in FIGURE 10.6.9. From (5),

$$A = \frac{1}{2} \int_{\pi/12}^{5\pi/12} [(2\sin 2\theta)^2 - 1^2] \, d\theta$$

$$= \frac{1}{2} \int_{\pi/12}^{5\pi/12} [4\sin^2 2\theta - 1] \, d\theta$$

$$= \frac{1}{2} \int_{\pi/12}^{5\pi/12} \left[4\left(\frac{1 - \cos 4\theta}{2} \right) - 1 \right] d\theta$$

$$= \frac{1}{2} \left[\theta - \frac{1}{2} \sin 4\theta \right]_{\pi/12}^{5\pi/12} = \frac{\pi}{6} + \frac{\sqrt{3}}{4} \approx 0.96. \qquad ■$$

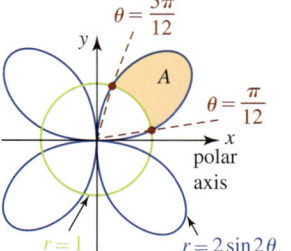

FIGURE 10.6.9 Area in Example 7

■ **Arc Length for Polar Graphs** We have seen that if $r = f(\theta)$ is the equation of a curve C in polar coordinates, then parametric equations of C are

$$x = f(\theta) \cos\theta, \quad y = f(\theta) \sin\theta, \quad \alpha \leq \theta \leq \beta.$$

If f has a continuous derivative, then it is a straightforward matter to derive a formula for arc length in polar coordinates. Since

$$\frac{dx}{d\theta} = f'(\theta) \cos\theta - f(\theta) \sin\theta, \quad \frac{dy}{d\theta} = f'(\theta) \sin\theta + f(\theta) \cos\theta,$$

straightforward algebra shows that

$$\left(\frac{dx}{d\theta} \right)^2 + \left(\frac{dy}{d\theta} \right)^2 = [f(\theta)]^2 + [f'(\theta)]^2 = r^2 + \left(\frac{dr}{d\theta} \right)^2.$$

The next result then follows from (9) of Section 10.3.

Theorem 10.6.3 Length of a Polar Graph

Let f be a function for which f' is continuous on an interval $[\alpha, \beta]$. Then the **length** L of the graph $r = f(\theta)$ on the interval is

$$L = \int_{\alpha}^{\beta} \sqrt{r^2 + \left(\frac{dr}{d\theta} \right)^2} \, d\theta. \tag{6}$$

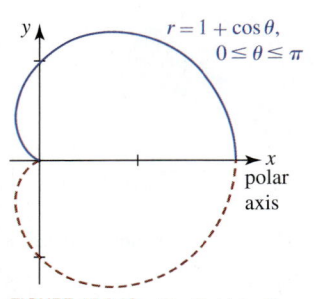

FIGURE 10.6.10 Cardioid in Example 8

EXAMPLE 8 Length of a Cardioid

Find the length of the cardioid $r = 1 + \cos\theta$ for $0 \le \theta \le \pi$.

Solution The graph of $r = 1 + \cos\theta$ for $0 \le \theta \le \pi$ is shown in blue in FIGURE 10.6.10. Now, $dr/d\theta = -\sin\theta$ so that

$$r^2 + \left(\frac{dr}{d\theta}\right)^2 = (1 + 2\cos\theta + \cos^2\theta) + \sin^2\theta = 2 + 2\cos\theta$$

and

$$\sqrt{r^2 + \left(\frac{dr}{d\theta}\right)^2} = \sqrt{2 + 2\cos\theta} = \sqrt{2}\sqrt{1 + \cos\theta}.$$

Hence, from (6) the length of the blue portion of the graph in Figure 10.6.10 is:

$$L = \sqrt{2}\int_0^\pi \sqrt{1 + \cos\theta}\, d\theta.$$

To evaluate this integral, we employ the half-angle formula for the cosine in the form $\cos^2(\theta/2) = \frac{1}{2}(1 + \cos\theta)$ or $1 + \cos\theta = 2\cos^2(\theta/2)$. The length of the graph for $0 \le \theta \le \pi$ is given by

$$L = 2\int_0^\pi \cos\frac{\theta}{2}\, d\theta = 4\sin\frac{\theta}{2}\Big]_0^\pi = 4\sin\frac{\pi}{2} = 4. \qquad \blacksquare$$

You are encouraged to read the following *Notes from the Classroom*.

$\dfrac{d}{d\theta}$ **NOTES FROM THE CLASSROOM**

It is easy to make an error in the limits of integration in the area and arc length integrals (4) and (6). In Example 8, we can use symmetry to see that the length of the complete cardioid, that is $r = 1 + \cos\theta$ for $0 \le \theta \le 2\pi$ is $2(4) = 8$ units, but yet this is *not* given by (6) by integrating on the interval $0 \le \theta \le 2\pi$:

$$L = 2\int_0^{2\pi} \cos(\theta/2)\, d\theta. \qquad (7)$$

Think about why an incorrect answer is obtained from (7) and then work Problems 45 and 46 in Exercises 10.6.

Exercises 10.6 Answers to selected odd-numbered problems begin on page ANS-33.

≡ **Fundamentals**

In Problems 1–6, find the slope of the tangent line at the indicated value of θ.

1. $r = \theta$; $\theta = \pi/2$
2. $r = 1/\theta$; $\theta = 3$
3. $r = 4 - 2\sin\theta$; $\theta = \pi/3$
4. $r = 1 - \cos\theta$; $\theta = 3\pi/4$
5. $r = \sin\theta$; $\theta = \pi/6$
6. $r = 10\cos\theta$; $\theta = \pi/4$

In Problems 7 and 8, find the points on the graph of the given equation at which the tangent line is horizontal and the points at which the tangent line is vertical.

7. $r = 2 + 2\cos\theta$
8. $r = 1 - \sin\theta$

In Problems 9 and 10, find a rectangular equation of the tangent line at the indicated point.

9. $r = 4\cos 3\theta$
10. $r = 1 + 2\cos\theta$

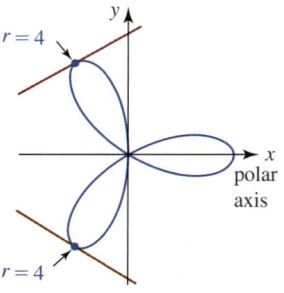

FIGURE 10.6.11 Graph for Problem 9

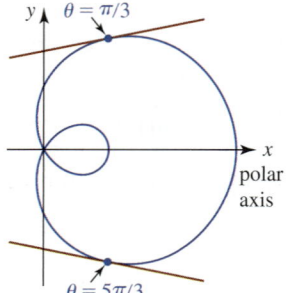

FIGURE 10.6.12 Graph for Problem 10

In Problems 11–16, find a polar equation of each tangent line to the polar graph at the origin.

11. $r = -2\sin\theta$

12. $r = 3\cos\theta$

13. $r = 1 + \sqrt{2}\sin\theta$

14. $r = 1 - 2\sin\theta$

15. $r = 2\cos 5\theta$

16. $r = 2\sin 2\theta$

In Problems 17–24, find the area of the region that is bounded by the graph of the given polar equation.

17. $r = 2\sin\theta$

18. $r = 10\cos\theta$

19. $r = 4 + 4\cos\theta$

20. $r = 1 - \sin\theta$

21. $r = 3 + 2\sin\theta$

22. $r = 2 + \cos\theta$

23. $r = 3\sin 2\theta$

24. $r = \cos 3\theta$

In Problems 25–30, find the area of the region that is bounded by the graph of the given polar equation and the indicated rays.

25. $r = 2\theta, \theta \geq 0, \theta = 0, \theta = 3\pi/2$

26. $r\theta = \pi, \theta > 0, \theta = \pi/2, \theta = \pi$

27. $r = e^\theta, \theta = 0, \theta = \pi$

28. $r = 10e^{-\theta}, \theta = 1, \theta = 2$

29. $r = \tan\theta, \theta = 0, \theta = \pi/4$

30. $r\sin\theta = 5, \theta = \pi/6, \theta = \pi/3$

In Problems 31 and 32, the graph is that of the polar equation $r = 1 + 2\cos\theta$. Find the area of the shaded region.

31.

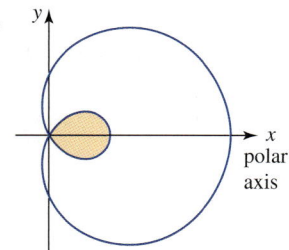

FIGURE 10.6.13 Region for Problem 31

32.

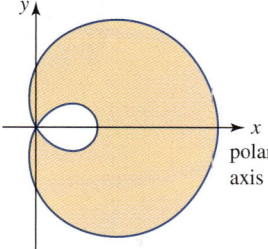

FIGURE 10.6.14 Region for Problem 32

In Problems 33–38, find the area of the described region.

33. Outside the circle $r = 1$ and inside the rose curve $r = 2\cos 3\theta$

34. Common to the interiors of the circles $r = \cos\theta$ and $r = \sin\theta$

35. Inside the circle $r = 5\sin\theta$ and outside the limaçon $r = 3 - \sin\theta$

36. Common to the interiors of the graphs of the equations in Problem 35

37. Inside the cardioid $r = 4 - 4\cos\theta$ and outside the circle $r = 6$

38. Common to the interiors of the graphs of the equations in Problem 37

In Problems 39–44, find the length of the curve for the indicated values of θ.

39. $r = 3, 0 \leq \theta \leq 2\pi$

40. $r = 6\cos\theta$, complete graph

41. $r = e^{\theta/2}, 0 \leq \theta \leq 4$

42. $r = \theta, 0 \leq \theta \leq 1$

43. $r = 3 - 3\cos\theta$, complete graph

44. $r = \sin^3(\theta/3), 0 \leq \theta \leq \pi$

≡ Think About It

45. Consider the lemniscate $r^2 = 9\cos 2\theta$.

　(a) Explain why the area of the region bounded by the graph is not given by the integral $\frac{1}{2}\int_0^{2\pi} 9\cos 2\theta \, d\theta$.

　(b) Using an appropriate integral, find the area of the region bounded by the graph.

46. In Example 8, explain why the length of the complete cardioid $r = 1 + \cos\theta, 0 \leq \theta \leq 2\pi$, is not given by the integral $2\int_0^{2\pi}\cos(\theta/2)\,d\theta$. Then reexamine Problem 43 and explain why there are no difficulties in integrating on the interval $[0, 2\pi]$.

47. Sketch the region common to the interiors of the graphs of $r = \sin 2\theta$ and $r = \cos 2\theta$. Find the area of this region.

48. The area of the region that is bounded by the graph of $r = 1 + \cos\theta$ is $3\pi/2$. What can you say about the areas bounded by the graphs of $r = 1 - \cos\theta$, $r = 1 + \sin\theta$, and $r = 1 - \sin\theta$? Justify your answer without calculating the areas using (4).

49. Is the area of the region bounded by the graph of $r = 2(1 + \cos\theta)$ equal to twice the area of the region bounded by the graph of $r = 1 + \cos\theta$?

50. Find the area of the shaded region in **FIGURE 10.6.15**. Each circle has radius 1.

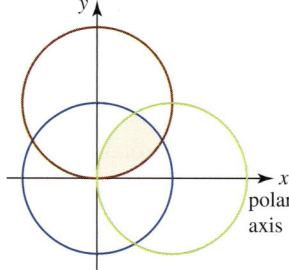

FIGURE 10.6.15 Intersecting circles in Problem 50

≡ Projects

51. Kepler's Second Law In polar coordinates the **angular momentum** of a moving particle of mass m is defined to be $L = mr^2 \, d\theta/dt$. Assume that the polar coordinates of a planet of mass m are (r_1, θ_1) and (r_2, θ_2) at times $t = a$ and $t = b, a < b$, respectively. Since the gravitational force acting on the planet is a central force, the angular momentum L of the planet is a constant. Use this fact to show that the area A swept out by r is $A = L(b - a)/2m$.

When the Sun is taken to be at the origin, this equation proves Kepler's second law of planetary motion:

- *A line joining a planet with the Sun sweeps out equal areas in equal time intervals.*

See FIGURE 10.6.16.

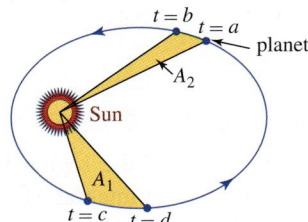

$A_1 = A_2$ when $b - a = d - c$

FIGURE 10.6.16   Orbit of planet in Problem 51

52. A Bit of History—For the Record   Prior to iPods, MP3 players, and CDs, music was obtained by playing a *record*. In the years 1960–1990 the popular format was the LP (long-playing) record that revolved on a record player at the rate of 33 revolutions per minute.* Although they can now be found in stores specializing in collectible objects, many of us still have collections of these large black vinyl 33-rpm records stored in boxes in the garage. Sound was encoded on these disks by mechanical means along a continuous groove. When a record was played, a needle started at a point near the outer edge of the disk and traversed the groove up to a point near its center. How long is the groove of a record? Suppose a record plays for 20 min at 33 revolutions per minute. As the record plays, the needle goes from an outer radius R_o to an inner radius R_i. See FIGURE 10.6.17. Assume that the groove of the record is a spiral that can be described by

---

*The 33s actually revolved at $33\frac{1}{3}$ revolutions per minute.

a polar equation of the form $r = R_o - k\theta$, where k is a constant and θ is measured in radians.

(a) Express k in terms of R_o, R_i, and N, where N is the number of revolutions completed by the record.

(b) Show that the length L of the record groove is given by

$$L = \frac{1}{k} \int_{R_i}^{R_o} \sqrt{k^2 + u^2} \, du.$$

(c) Use a binomial series to establish the approximation

$$\sqrt{k^2 + u^2} \approx u\left[1 + \frac{1}{2}\left(\frac{k}{u}\right)^2\right].$$

(d) In part (b), use the result obtained in part (c) to show that the length L of the record groove is given by the approximation

$$L \approx \pi N(R_i + R_o) + \frac{R_o - R_i}{4\pi N} \ln \frac{R_o}{R_i}.$$

(e) Use the result in part (d) to approximate the length L if $R_o = 6$ in. and $R_i = 2.5$ in.

(f) Use an appropriate substitution to evaluate the integral in part (b) using the specified values of R_o and R_i given in part (e). Compare this answer with that obtained in part (e).

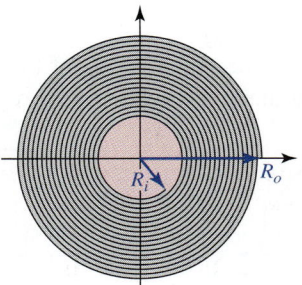

FIGURE 10.6.17   LP record in Problem 52

10.7 Conic Sections in Polar Coordinates

Introduction   In the first section of this chapter we derived equations for the parabola, ellipse, and hyperbola using the distance formula in rectangular coordinates. By using polar coordinates and the concept of eccentricity, we can now give one general definition of a conic section that encompasses all three curves.

Definition 10.7.1   Conic Section

Let L be a fixed line in the plane, and let F be a point not on the line. A **conic section** is the set of points P in the plane for which the distance from P to F divided by the distance from P to L is a constant.

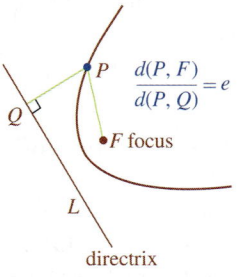

The fixed line L is called a **directrix**, and the point F is a **focus**. The fixed constant is the **eccentricity** e of the conic. As FIGURE 10.7.1 shows, the point P lies on the conic if and only if

$$\frac{d(P, F)}{d(P, Q)} = e, \tag{1}$$

where Q denotes the foot of the perpendicular from P to L. In (1), if

FIGURE 10.7.1   Geometric interpretation of (1)

- $e = 1$, the conic is a **parabola**,
- $0 < e < 1$, the conic is an **ellipse**, and
- $e > 1$, the conic is a **hyperbola**.

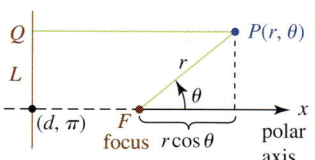

FIGURE 10.7.2   Polar coordinate interpretation of (2)

■ **Polar Equations of Conics**   Equation (1) is readily interpreted using polar coordinates. Suppose the focus F is placed at the pole and the directrix L is d units ($d > 0$) to the *left* of F perpendicular to the extended polar axis. We see from **FIGURE 10.7.2** that (1) written as $d(P, F) = ed(P, Q)$ is the same as

$$r = e(d + r\cos\theta) \qquad \text{or} \qquad r - er\cos\theta = ed. \qquad (2)$$

Solving for r yields

$$r = \frac{ed}{1 - e\cos\theta}. \qquad (3)$$

To see that (3) yields the familiar equations of the conics, let us superimpose a rectangular coordinate system on the polar coordinate system with origin at the pole and the positive x-axis coinciding with the polar axis. We then express the first equation in (2) in rectangular coordinates and simplify:

$$\pm\sqrt{x^2 + y^2} = ex + ed$$
$$x^2 + y^2 = e^2x^2 + 2e^2dx + e^2d^2$$
$$(1 - e^2)x^2 - 2e^2dx + y^2 = e^2d^2. \qquad (4)$$

Choosing $e = 1$, (4) becomes

$$-2dx + y^2 = d^2 \qquad \text{or} \qquad y^2 = 2d\left(x + \frac{d}{2}\right),$$

which is an equation in standard form of a parabola whose axis is the x-axis, vertex is at $(-d/2, 0)$ and, consistent with the placement of F, whose focus is at the origin.

It is a good exercise in algebra to show that (2) yields standard form equations of an ellipse in the case $0 < e < 1$ and a hyperbola in the case $e > 1$. See Problem 43 in Exercises 10.7. Thus, depending on the value of e, the polar equation (3) can have three possible graphs as shown in **FIGURE 10.7.3**.

If we place the directrix L to the *right* of the focus F at the origin in our derivation of the polar equation (3), then the resulting equation would be $r = ed/(1 + e\cos\theta)$. When the directrix L is chosen parallel to the polar axis, that is, horizontal, then the equation of the conic is found to be either $r = ed/(1 - e\sin\theta)$ (directrix below the origin) or $r = ed/(1 + e\sin\theta)$ (directrix above the origin). A summary of the preceding discussion is given in the next theorem.

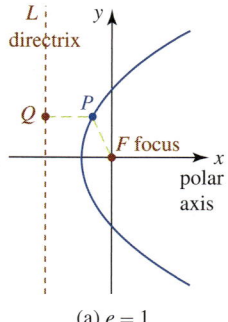

(a) $e = 1$

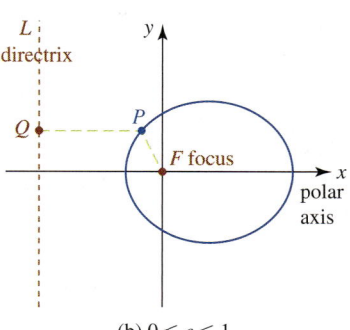

(b) $0 < e < 1$

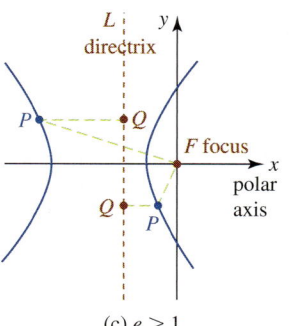

(c) $e > 1$

FIGURE 10.7.3   Graphs of equation (3); directrix L to the left of F

Theorem 10.7.1   Polar Equations of Conics

Any polar equation of the form

$$r = \frac{ed}{1 \pm e\cos\theta} \qquad (5)$$

or

$$r = \frac{ed}{1 \pm e\sin\theta} \qquad (6)$$

is a **conic section** with focus at the origin and directrix d units from the origin and either perpendicular (in the case of (5)) or parallel (in the case of (6)) to the x-axis. The conic is a parabola if $e = 1$, an ellipse if $0 < e < 1$, and a hyperbola if $e > 1$.

EXAMPLE 1 Identifying Conics

Identify each of the following conics

(a) $r = \dfrac{2}{1 - 2\sin\theta}$ (b) $r = \dfrac{3}{4 + \cos\theta}$.

Solution

(a) A term-by-term comparison of the given equation with the polar form $r = ed/(1 - e\sin\theta)$ enables us to make the identification $e = 2$. Hence the conic is a hyperbola.

(b) In order to identify the conic section, we divide the numerator and the denominator of the given equation by 4. This puts the equation into the form

$$r = \dfrac{\frac{3}{4}}{1 + \frac{1}{4}\cos\theta}.$$

Then by comparison with $r = ed/(1 + e\cos\theta)$ we see that $e = \frac{1}{4}$. Hence the conic is an ellipse. ∎

▌ Graphs A rough graph of a conic defined by (5) or (6) can be obtained by knowing the orientation of its axis, finding the x- and y-intercepts, and finding the vertices. In the cases of equations (5) and (6) we have, respectively:

- the two vertices of the **ellipse** or a **hyperbola** occur at $\theta = 0$ and $\theta = \pi$; the vertex of a **parabola** can occur at only one of the values: $\theta = 0$ or $\theta = \pi$.
- the two vertices of an **ellipse** or a **hyperbola** occur at $\theta = \pi/2$ and $\theta = 3\pi/2$; the vertex of a **parabola** can occur at only one of the values: $\theta = \pi/2$ or $\theta = 3\pi/2$.

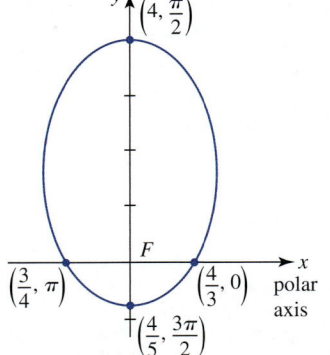

FIGURE 10.7.4 Graph of polar equation in Example 2

EXAMPLE 2 Graphing a Conic

Graph $r = \dfrac{4}{3 - 2\sin\theta}$.

Solution By writing the equation as $r = \dfrac{\frac{4}{3}}{1 - \frac{2}{3}\sin\theta}$ we see that the eccentricity is $e = \frac{2}{3}$ and so the conic is an ellipse. Moreover, because the equation is of the form given in (6), we know that the directrix is parallel to the x-axis. Now in view of the discussion preceding this example, we obtain:

$$vertices: \quad (4, \pi/2), \left(\tfrac{4}{5}, 3\pi/2\right)$$

$$x\text{-intercepts}: \quad \left(\tfrac{4}{3}, 0\right), \left(\tfrac{4}{3}, \pi\right).$$

As seen in **FIGURE 10.7.4** the major axis of the ellipse lies along the y-axis. ∎

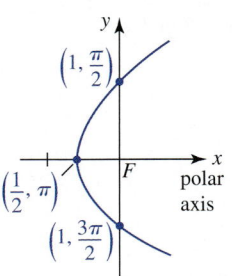

FIGURE 10.7.5 Graph of polar equation in Example 3

EXAMPLE 3 Graphing a Conic

Graph $r = \dfrac{1}{1 - \cos\theta}$.

Solution Inspection of the equation shows that it is of the form given in (5) with $e = 1$. Hence, the conic section is a parabola whose directrix is perpendicular to the x-axis. Since r is undefined at $\theta = 0$, the vertex of the parabola occurs at $\theta = \pi$:

$$vertex: \quad \left(\tfrac{1}{2}, \pi\right)$$

$$y\text{-intercepts}: \quad (1, \pi/2), (1, 3\pi/2).$$

As seen in **FIGURE 10.7.5** the axis of symmetry of the parabola lies along the x-axis. ∎

EXAMPLE 4 Graphing a Conic

Graph $r = \dfrac{2}{1 + 2\cos\theta}$.

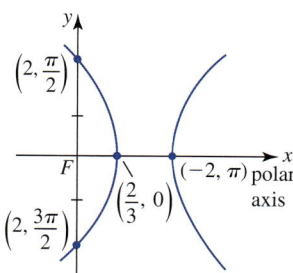

Solution From (5) we see that $e = 2$ and so the conic section is a hyperbola whose directrix is perpendicular to the x-axis. The vertices occur at $\theta = 0$ and at $\theta = \pi$:

$$\text{vertices:} \quad \left(\tfrac{2}{3}, 0\right), (-2, \pi)$$
$$\text{y-intercepts:} \quad (2, \pi/2), (2, 3\pi/2).$$

As seen in FIGURE 10.7.6 the transverse axis of the hyperbola lies along the x-axis. ∎

FIGURE 10.7.6 Graph of polar equation in Example 4

■ **Rotated Conics** We saw in Section 10.5 that graphs of $r = f(\theta - \gamma)$ and $r = f(\theta + \gamma)$, $\gamma > 0$, are rotations of the graph of the polar equation $r = f(\theta)$ about the origin by an amount γ. Thus,

$$\left.\begin{aligned} r &= \frac{ed}{1 \pm e\cos(\theta - \gamma)} \\ r &= \frac{ed}{1 \pm e\sin(\theta - \gamma)} \end{aligned}\right\} \begin{aligned} &\text{conics rotated} \\ &\text{counterclockwise} \\ &\text{about the origin} \end{aligned} \qquad \left.\begin{aligned} r &= \frac{ed}{1 \pm e\cos(\theta + \gamma)} \\ r &= \frac{ed}{1 \pm e\sin(\theta + \gamma)} \end{aligned}\right\} \begin{aligned} &\text{conics rotated} \\ &\text{clockwise} \\ &\text{about the origin} \end{aligned}$$

EXAMPLE 5 Rotated Conic

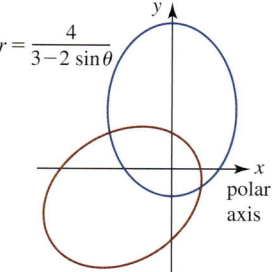

In Example 2 we saw that the graph of $r = \dfrac{4}{3 - 2\sin\theta}$ is an ellipse with major axis along the y-axis. This is the blue graph in FIGURE 10.7.7. The graph of $r = \dfrac{4}{3 - 2\sin(\theta - 2\pi/3)}$ is the red graph in Figure 10.7.7 and is a counter-clockwise rotation of the blue graph by the amount $2\pi/3$ (or 120°) about the origin. The major axis of the red graph lies along the line $\theta = 7\pi/6$. ∎

FIGURE 10.7.7 Graphs of polar equations in Example 5

■ **Applications** Equations of the type in (5) and (6) are well suited to describe a closed orbit of a satellite around the Sun (Earth or Moon) since such an orbit is an ellipse with the Sun (Earth or Moon) at one focus. Suppose that an equation of the orbit is given by $r = ed/(1 - e\cos\theta)$, $0 < e < 1$, and r_p is the value of r at perihelion (perigee or perilune) and r_a is the value of r at aphelion (apogee or apolune). These are the points in the orbit, occurring on the x-axis, at which the satellite is closest and farthest, respectively, from the Sun (Earth or Moon). See FIGURE 10.7.8. It is left as an exercise to show that the eccentricity e of the orbit is related to r_p and r_a by

$$e = \frac{r_a - r_p}{r_a + r_p}. \tag{7}$$

See Problem 44 in Exercises 10.7.

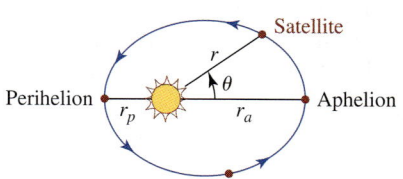

FIGURE 10.7.8 Orbit of satellite around the Sun

EXAMPLE 6 Finding a Polar Equation of an Orbit

Find a polar equation of the orbit of the planet Mercury around the Sun if $r_p = 2.85 \times 10^7$ mi and $r_a = 4.36 \times 10^7$ mi.

Solution From (7), the eccentricity of Mercury's orbit is

$$e = \frac{4.36 \times 10^7 - 2.85 \times 10^7}{4.36 \times 10^7 + 2.85 \times 10^7} = 0.21.$$

Hence,
$$r = \frac{0.21d}{1 - 0.21\cos\theta}.$$

Mercury is the closest planet to the Sun

All we need do now is to solve for the quantity $0.21d$. To do this we use the fact that aphelion occurs at $\theta = 0$:

$$4.36 \times 10^7 = \frac{0.21d}{1 - 0.21}.$$

Solving the last equation for the quantity $0.21d$ yields $0.21d = 3.44 \times 10^7$. Hence, a polar equation of Mercury's orbit is

$$r = \frac{3.44 \times 10^7}{1 - 0.21\cos\theta}. \qquad \blacksquare$$

Exercises 10.7 Answers to selected odd-numbered problems begin on page ANS-33.

☰ Fundamentals

In Problems 1–10, determine the eccentricity, identify the conic section, and sketch its graph.

1. $r = \dfrac{2}{1 + \cos\theta}$

2. $r = \dfrac{2}{2 - \sin\theta}$

3. $r = \dfrac{15}{4 - \cos\theta}$

4. $r = \dfrac{5}{2 - 2\sin\theta}$

5. $r = \dfrac{4}{1 + 2\sin\theta}$

6. $r = \dfrac{12}{6 + 2\sin\theta}$

7. $r = \dfrac{18}{3 + 6\cos\theta}$

8. $r = \dfrac{6\sec\theta}{\sec\theta - 1}$

9. $r = \dfrac{10}{5 + 4\sin\theta}$

10. $r = \dfrac{2}{2 + 5\cos\theta}$

In Problems 11–14, determine the eccentricity e of the given conic. Then convert the polar equation to a rectangular equation and verify that $e = c/a$.

11. $r = \dfrac{6}{1 + 2\sin\theta}$

12. $r = \dfrac{10}{2 - 3\cos\theta}$

13. $r = \dfrac{12}{3 - 2\cos\theta}$

14. $r = \dfrac{2\sqrt{3}}{\sqrt{3} + \sin\theta}$

In Problems 15–20, find a polar equation of the conic with focus at the origin that satisfies the given conditions.

15. $e = 1$, directrix $x = 3$

16. $e = \frac{3}{2}$, directrix $y = 2$

17. $e = \frac{2}{3}$, directrix $y = -2$

18. $e = \frac{1}{2}$, directrix $x = 4$

19. $e = 2$, directrix $x = 6$

20. $e = 1$, directrix $y = -2$

21. Find a polar equation of the conic in Problem 15 if the graph is rotated clockwise about the origin by an amount $2\pi/3$.

22. Find a polar equation of the conic in Problem 16 if the graph is rotated counterclockwise about the origin by an amount $\pi/6$.

In Problems 23–28, find a polar equation of the parabola with focus at the origin and the given vertex.

23. $\left(\frac{3}{2}, 3\pi/2\right)$

24. $(2, \pi)$

25. $\left(\frac{1}{2}, \pi\right)$

26. $(2, 0)$

27. $\left(\frac{1}{4}, 3\pi/2\right)$

28. $\left(\frac{3}{2}, \pi/2\right)$

In Problems 29–32, find the polar coordinates of the vertex or vertices of the given rotated conic.

29. $r = \dfrac{4}{1 + \cos(\theta - \pi/4)}$

30. $r = \dfrac{5}{3 + 2\cos(\theta - \pi/3)}$

31. $r = \dfrac{10}{2 - \sin(\theta + \pi/6)}$

32. $r = \dfrac{6}{1 + 2\sin(\theta + \pi/3)}$

☰ Applications

33. A communications satellite is 12,000 km above the Earth at its apogee. The eccentricity of its orbit is 0.2. Use (7) to find the perigee distance.

34. Find a polar equation $r = ed/(1 - e\cos\theta)$ of the orbit of the satellite in Problem 33.

35. Find a polar equation of the orbit of the Earth around the Sun if $r_p = 1.47 \times 10^8$ km and $r_a = 1.52 \times 10^8$ km.

36. **(a)** The eccentricity of the elliptical orbit of Comet Halley is 0.97 and the length of the major axis of its orbit is 3.34×10^9 mi. Find a polar equation of its orbit of the form $r = ed/(1 - e\cos\theta)$.

   (b) Use the equation in part (a) to obtain r_p and r_a for the orbit of Comet Halley.

☰ Calculator/CAS Problems

The orbital characteristics (eccentricity, perigee, and major axis) of a satellite near the Earth gradually degrade over time due to many small forces acting on the satellite other than the gravitational force of the Earth. These forces include atmospheric drag, the gravitational attractions of the Sun and the Moon, and magnetic forces. Approximately once a month tiny rockets are activated for a few seconds in order to "boost" the orbital characteristics back into the desired range. Rockets are turned on longer to a major change in the orbit of a satellite. The most fuel-efficient way to move from an inner orbit to an outer orbit, called a **Hohmann transfer**, is to add velocity in the direction of flight at the time the satellite reaches perigee on the inner orbit, follow the Hohmann transfer ellipse halfway around to its apogee, and add velocity again to achieve the outer orbit. A similar process (subtracting velocity at apogee on the outer orbit and subtracting velocity at perigee on the Hohmann transfer orbit) moves a satellite from an outer orbit to an inner orbit.

In Problems 37–40, use a calculator or CAS to superimpose the graphs of the given three polar equations on the same coordinate axes. Print out your result and use a colored pencil to trace out the Hohmann transfer.

37. Inner orbit $r = \dfrac{24}{1 + 0.2\cos\theta}$,

Hohmann transfer $r = \dfrac{32}{1 + 0.6\cos\theta}$,

outer orbit $r = \dfrac{56}{1 + 0.3\cos\theta}$

38. Inner orbit $r = \dfrac{5.5}{1 + 0.1\cos\theta}$,

Hohmann transfer $r = \dfrac{7.5}{1 + 0.5\cos\theta}$,

outer orbit $r = \dfrac{13.5}{1 + 0.1\cos\theta}$

39. Inner orbit $r = 9$,

Hohmann transfer $r = \dfrac{15.3}{1 + 0.7\cos\theta}$,

outer orbit $r = 51$

40. Inner orbit $r = \dfrac{73.5}{1 + 0.05\cos\theta}$,

Hohmann transfer $r = \dfrac{77}{1 + 0.1\cos\theta}$,

outer orbit $r = \dfrac{84.7}{1 + 0.01\cos\theta}$

In Problems 41 and 42, use a calculator or CAS to superimpose the graphs of the given two polar equations on the same coordinate axes.

41. $r = \dfrac{4}{4 + 3\cos\theta}$; $r = \dfrac{4}{4 + 3\cos(\theta - \pi/3)}$

42. $r = \dfrac{2}{1 - \sin\theta}$; $r = \dfrac{2}{1 - \sin(\theta + 3\pi/4)}$

≡ Think About It

43. Show that (2) yields standard form equations of an ellipse in the case $0 < e < 1$ and a hyperbola in the case $e > 1$.

44. Use the equation $r = ed/(1 - e\cos\theta)$ to derive the result in (7).

Chapter 10 in Review

Answers to selected odd-numbered problems begin on page ANS-34.

A. True/False

In Problems 1–26, indicate whether the given statement is true or false.

1. For a parabola, the distance from the vertex to the focus is the same as the distance from the vertex to the directrix. _____

2. The minor axis of an ellipse bisects the major axis. _____

3. The asymptotes of $x^2/a^2 - y^2/a^2 = 1$ are perpendicular. _____

4. The y-intercepts of the graph of $x^2/a^2 - y^2/b^2 = 1$ are $(0, b)$ and $(0, -b)$. _____

5. The point $(-2, 5)$ is on the ellipse $x^2/8 + y^2/50 = 1$. _____

6. The graphs of $y = x^2$ and $y^2 - x^2 = 1$ have at most two points in common. _____

7. If for all values of θ the points $(-r, \theta)$ and $(r, \theta + \pi)$ are on the graph of the polar equation $r = f(\theta)$, then the graph is symmetric with respect to the origin. _____

8. The graph of the curve $x = t^2$, $y = t^4 + 1$ is the same as the graph of $y = x^2 + 1$. _____

9. The graph of the curve $x = t^2 + t - 12$, $y = t^3 - 7t$ crosses the y-axis at $(0, 6)$. _____

10. $(3, \pi/6)$ and $(-3, -5\pi/6)$ are polar coordinates of the same point. _____

11. Rectangular coordinates of a point in the plane are unique. _____

12. The graph of the rose curve $r = 5\sin 6\theta$ has six "petals." _____

13. The point $(4, 3\pi/2)$ is not on the graph of $r = 4\cos 2\theta$, since its coordinates do not satisfy the equation. _____

14. The eccentricity of a parabola is $e = 1$. _____

15. The transverse axis of the hyperbola $r = 5/(2 + 3\cos\theta)$ lies along the x-axis. _____

16. The graph of the ellipse $r = 90/(15 - \sin\theta)$ is nearly circular. _____

17. The rectangular coordinates of the point $\left(-\sqrt{2}, 5\pi/4\right)$ in polar coordinates are $(1, 1)$. _____

18. The graph of the polar equation $r = -5\sec\theta$ is a line. _____

19. The terminal side of the angle θ is always in the same quadrant as the point (r, θ). _____

20. The slope of the tangent to the graph of $r = e^{\theta}$ at $\theta = \pi/2$ is -1. _____

21. The graphs of the cardioids $r = 3 + 3\cos\theta$ and $r = -3 + 3\cos\theta$ are the same. _____

22. The area bounded by $r = \cos 2\theta$ is $2\int_{-\pi/4}^{\pi/4}\cos^2 2\theta\, d\theta$. _____

23. The area bounded by $r = 2\sin 3\theta$ is $6\int_0^{\pi/3}\sin^2 3\theta\, d\theta$. _____

24. The area bounded by $r = 1 - 2\cos\theta$ is $\frac{1}{2}\int_0^{2\pi}(1 - 2\cos\theta)^2\, d\theta$. _____

25. The area bounded by $r^2 = 36\cos 2\theta$ is $18\int_0^{2\pi}\cos 2\theta\, d\theta$. _____

26. The θ-coordinate of a point of intersection of the graphs of the polar equations $r = f(\theta)$ and $r = g(\theta)$ must satisfy the equation $f(\theta) = g(\theta)$. _____

B. Fill in the Blanks

In Problems 1–22, fill in the blanks.

1. $y = 2x^2$, focus _____

2. $\dfrac{x^2}{4} - \dfrac{y^2}{12} = 1$, foci _____

3. $4x^2 + 5(y + 3)^2 = 20$, center _____

4. $25y^2 - 4x^2 = 100$, asymptotes _____

5. $8(y + 3) = (x - 1)^2$, directrix _____

6. $\dfrac{(x + 1)^2}{36} + \dfrac{(y + 7)^2}{16} = 1$, vertices _____

7. $x = y^2 + 4y - 6$, vertex _____

8. $x^2 - 2y^2 = 18$, length of conjugate axis _____

9. $(x - 4)^2 - (y + 1)^2 = 4$, endpoints of transverse axis _____

10. $\dfrac{(x - 3)^2}{7} + \dfrac{(y + 3/2)^2}{8} = 1$, equation of line containing major axis _____

11. $25x^2 + y^2 - 200x + 6y + 384 = 0$, center _____

12. $(x + 1)^2 + (y + 8)^2 = 100$, x-intercepts _____

13. $y^2 - (x - 2)^2 = 1$, y-intercepts _____

14. $y^2 - y + 3x = 3$, slope of tangent line at $(1, 1)$ _____

15. $x = t^3, y = 4t^3$, name of rectangular graph _____

16. $x = t^2 - 1, y = t^3 + t + 1$, y-intercepts _____

17. $r = -2\cos\theta$, name of polar graph _____

18. $r = 2 + \sin\theta$, name of polar graph _____

19. $r = \sin 3\theta$, tangents to the graph at the origin _____

20. $r = \dfrac{1}{2 + 5\sin\theta}$, eccentricity _____

21. $r = \dfrac{10}{1 - \sin\theta}$, focus_____ and vertex _____

22. $r = \dfrac{12}{2 + \cos\theta}$, center _____, foci _____, vertices_____

C. Exercises

1. Find an equation of the line that is normal to the graph of the curve $x = t - \sin t$, $y = 1 - \cos t, 0 \leq t \leq 2\pi$, at $t = \pi/3$.

2. Find the length of the curve given in Problem 1.

3. Find the points on the graph of the curve $x = t^2 + 4, y = t^3 - 9t^2 + 2$ at which the tangent line is parallel to $6x + y = 8$.

4. Find the points on the graph of the curve $x = t^2 + 1, y = 2t$ at which the tangent line passes through $(1, 5)$.

5. Consider the rectangular equation $y^2 = 4x^2(1 - x^2)$.

    (a) Explain why it is necessary that $|x| \le 1$.

    (b) If $x = \sin t$, then $|x| \le 1$. Find parametric equations that have the same graph as the given rectangular equation.

    (c) Using parametric equations, find the points on the graph of the rectangular equation at which the tangent is horizontal.

    (d) Sketch the graph of the rectangular equation.

6. Find the area of the region that is outside the circle $r = 4\cos\theta$ and inside the limaçon $r = 3 + \cos\theta$.

7. Find the area of the region that is common to the interiors of the circle $r = 3\sin\theta$ and the cardioid $r = 1 + \sin\theta$.

8. In polar coordinates, sketch the region whose area A is described by $A = \int_0^{\pi/2}(25 - 25\sin^2\theta)\, d\theta$.

9. Find **(a)** a rectangular equation and **(b)** a polar equation of the tangent line to the graph of $r = 2\sin 2\theta$ at $\theta = \pi/4$.

10. Determine the rectangular coordinates of the vertices of the ellipse whose polar equation is $r = 2/(2 - \sin\theta)$.

In Problems 11 and 12, find a rectangular equation that has the same graph as the given polar equation

11. $r = \cos\theta + \sin\theta$ **12.** $r = \sec\theta - 5\cos\theta$

In Problems 13 and 14, find a polar equation that has the same graph as the given rectangular equation

13. $2xy = 5$ **14.** $(x^2 + y^2 - 2x)^2 = 9(x^2 + y^2)$

15. Find a polar equation for the set of points that are equidistant from the origin (pole) and the line $r = -\sec\theta$.

16. Find a polar equation of the hyperbola with focus at the origin, vertices (in rectangular coordinates) $\left(0, -\frac{4}{3}\right)$ and $(0, -4)$, and eccentricity 2.

In Problems 17 and 18, write an equation of the given polar graph.

17.

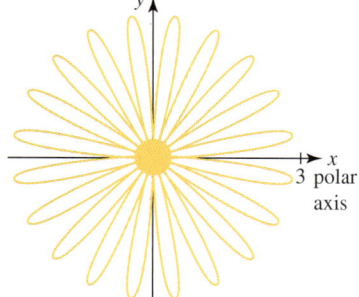

18.

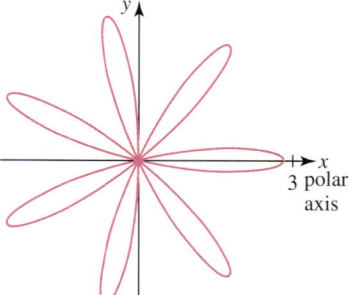

FIGURE 10.R.1   Graph for Problem 17   FIGURE 10.R.2   Graph for Problem 18

19. Find an equation of the hyperbola that has asymptotes $3y = 5x$ and $3y = -5x$ and vertices $(0, 10)$ and $(0, -10)$.

20. Find a rectangular equation of the tangent line to the graph of $r = 1/(1 + \cos\theta)$ at $\theta = \pi/2$.

21. The folium of Descartes, first discussed in Section 3.6, has the rectangular equation $x^3 + y^3 = 3axy$, where $a > 0$ is a constant. Use the substitution $y = tx$ to find parametric equations for the curve. See FIGURE 10.R.3.

22. Use the parametric equations found in Problem 21 to find the points on the folium of Descartes where the tangent line is horizontal. See Figure 10.R.3.

23. **(a)** Find a polar equation for the folium of Descartes in Problem 21.

    (b) Use the polar equation to find the area of the shaded loop in the first quadrant in Figure 10.R.3. [*Hint*: Let $u = \tan\theta$.]

24. Use the parametric equations found in Problem 21 to show that the folium of Descartes has the slant asymptote $x + y + a = 0$. This is the red dashed line in Figure 10.R.3. [*Hint*: Consider what happens to x, y, and $x + y$ as $r \to -1$.]

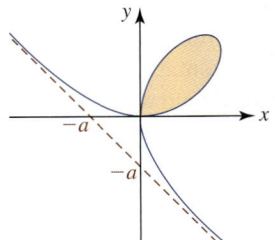

FIGURE 10.R.3   Graph for Problems 21–24

25. The graph of $r = 2\sin(\theta/3)$ given in FIGURE 10.R.4 resembles a limaçon with an interior loop. Find the area of the interior loop.

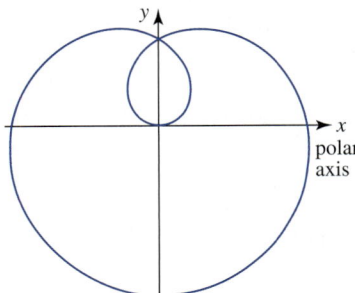

FIGURE 10.R.4   Graph for Problem 25

26. Find the area of the shaded region in FIGURE 10.R.5. Each circle has radius 1.

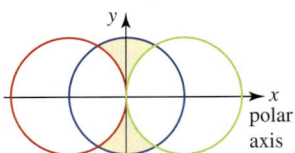

FIGURE 10.R.5   Graph for Problem 26

In Problems 27 and 28, the graph of the given polar equation is rotated about the origin by the indicated amount.

(a) Find a polar equation of the new graph.
(b) Find a rectangular equation for the new graph.

27. $r = 2\cos\theta$;   counterclockwise, $\pi/4$

28. $r = 1/(1 + \cos\theta)$;   clockwise, $\pi/6$

29. A satellite revolves around the planet Jupiter in an elliptical orbit with the center of the planet at one focus. The length of the major axis of the orbit is 10^9 m and the length of the minor axis is 6×10^8 m. Find the minimum distance between the satellite and the center of Jupiter. What is the maximum distance?

30. Find the width w of each petal of the rose curve $r = \cos 2\theta$. One petal is shown in FIGURE 10.R.6.

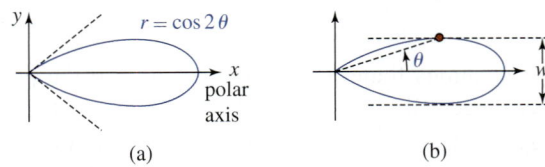

(a)   (b)

FIGURE 10.R.6   Graph for Problem 30

Vectors and 3-Space

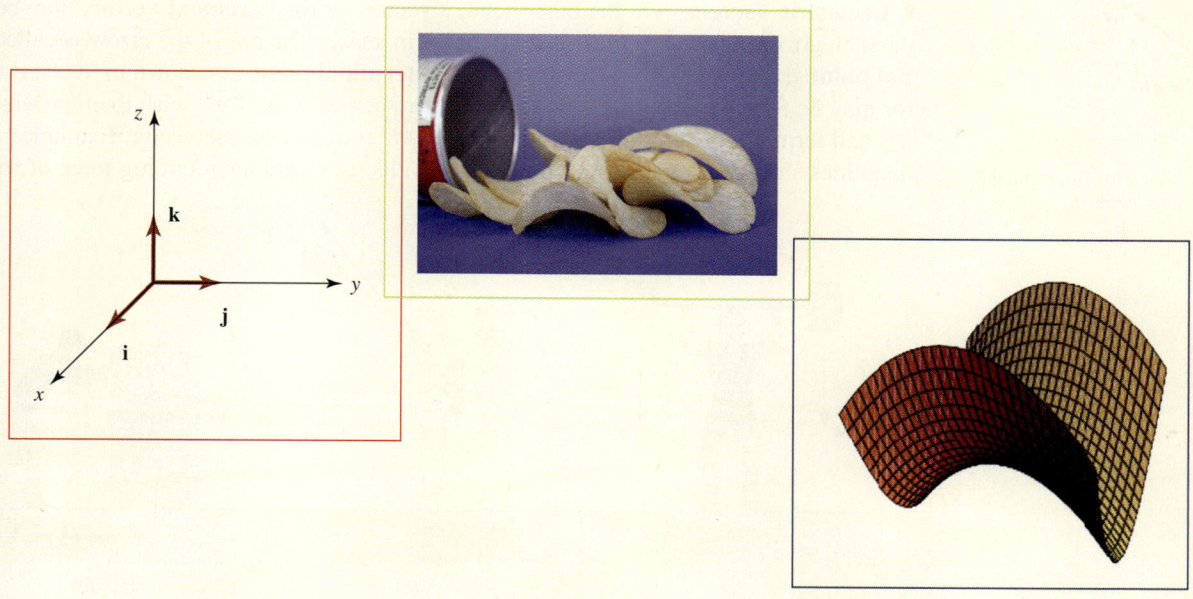

In This Chapter Until now we have carried out most of our endeavors in calculus in the flatland of the two-dimensional Cartesian plane or 2-space. For the next several chapters, we will be primarily interested in examining mathematical life in three dimensions or 3-space. We begin with an examination of vectors in two- and three-dimensions.

11.1 Vectors in 2-Space

▌ Introduction Up to this point we have, for the most part, concentrated on the study of functions of a single variable whose graphs exist in a two-dimensional plane. In this section we begin our study of multivariable calculus with an introduction to vectors in 2-space. In subsequent sections and chapters we will focus mainly on vectors and functions defined in 3-space.

▌ Scalars In science, mathematics, and engineering we distinguish two important quantities: *scalars* and *vectors*. A **scalar** is simply a real number and is generally represented by a lower-case italicized letter, such as a, k, or x. Scalars may be used to represent magnitudes and may have specific units attached to them; for example, 80 ft or 20°C.

▌ Geometric Vectors On the other hand, a **vector**, or **displacement vector**, may be thought of as an arrow connecting two points A and B in space. The *tail* of the arrow is called the **initial point** and the *tip* of the arrow is called the **terminal point**. As shown in FIGURE 11.1.1, a vector may be represented using a boldfaced letter such as **v** or, if we wish to emphasize the initial and terminal points A and B, we can use $\overrightarrow{AB}$ to represent the vector. Examples of vector quantities shown in FIGURE 11.1.2 are weight **w**, velocity **v**, and the retarding force of friction $\mathbf{F}_f$.

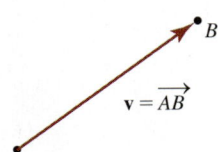

FIGURE 11.1.1 A vector from initial point A to terminal point B

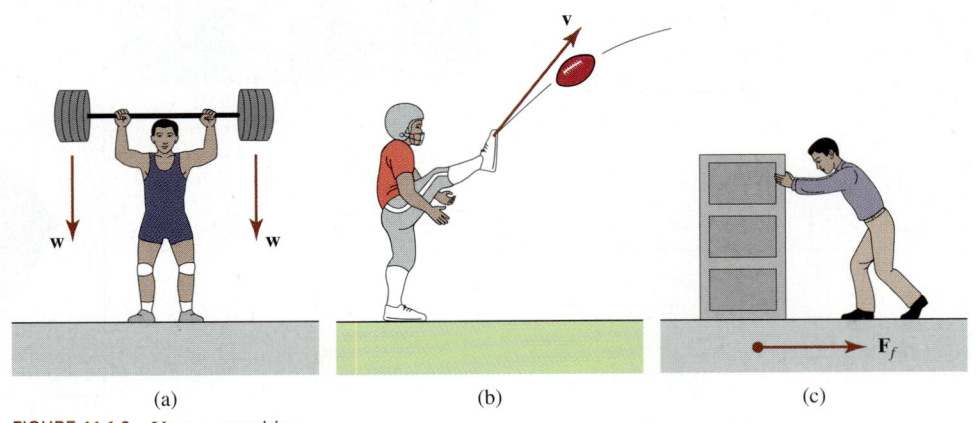

(a)　　　　(b)　　　　(c)
FIGURE 11.1.2　Vector quantities

▌ Notation and Terminology The distance between the initial and terminal points of a vector $\overrightarrow{AB}$ is called the **length**, **magnitude**, or **norm** of the vector and is denoted by $|\overrightarrow{AB}|$. Two vectors that have the same magnitude and same direction are said to be **equal**. Thus, in FIGURE 11.1.3, we have $\overrightarrow{AB} = \overrightarrow{CD}$. The **negative** of a vector $\overrightarrow{AB}$, written $-\overrightarrow{AB}$, is a vector that has the same magnitude as $\overrightarrow{AB}$ but is opposite in direction. If $k \neq 0$ is a scalar, the **scalar multiple** of a vector, $k\overrightarrow{AB}$, is a vector that is $|k|$ times as long as $\overrightarrow{AB}$. If $k > 0$, then $k\overrightarrow{AB}$ has the same direction as the vector $\overrightarrow{AB}$; if $k < 0$, then $k\overrightarrow{AB}$ has the direction opposite to that of $\overrightarrow{AB}$. When $k = 0$, we say $0\overrightarrow{AB} = \mathbf{0}$ is the **zero vector**. Two vectors are **parallel** if and only if they are nonzero scalar multiples of each other. See FIGURE 11.1.4.

The question of what is the direction of 0 is usually answered by saying that the zero vector can be assigned any direction. More to the point, **0** is needed in order to have a vector algebra.

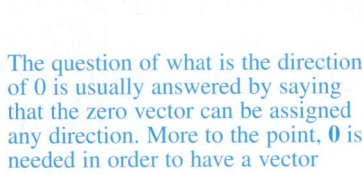

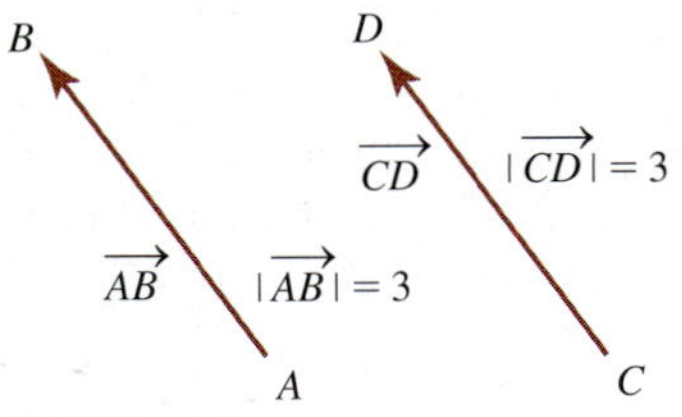

FIGURE 11.1.3　Equal vectors

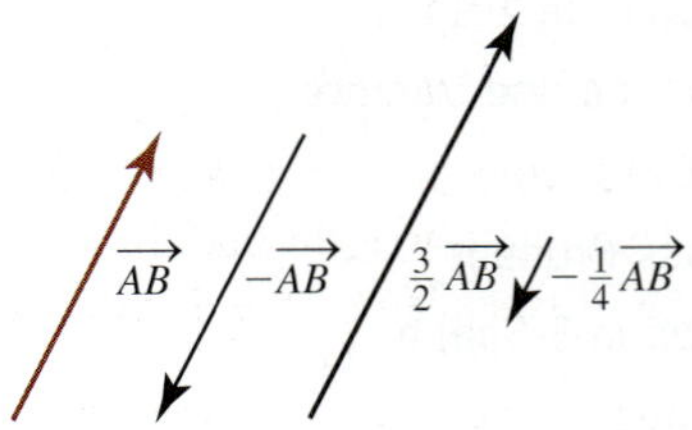

FIGURE 11.1.4　Parallel vectors

▌ Addition and Subtraction Two vectors can be considered as having a common initial point, such as A in FIGURE 11.1.5(a). Thus, if nonparallel vectors $\overrightarrow{AB}$ and $\overrightarrow{AC}$ are the sides of a parallelogram in Figure 11.1.5(b), we say the vector that is the main diagonal, or $\overrightarrow{AD}$, is the **sum** of $\overrightarrow{AB}$ and $\overrightarrow{AC}$. We write

$$\overrightarrow{AD} = \overrightarrow{AB} + \overrightarrow{AC}.$$

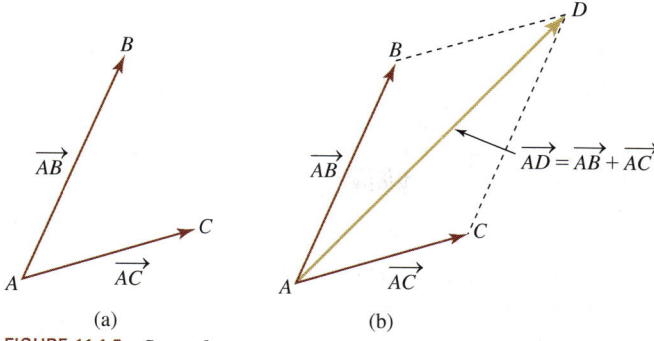

(a)

(b)

FIGURE 11.1.5   Sum of two vectors

In science and engineering, if two vectors represent forces, then their sum is called the **resultant force**.

The **difference** of two vectors $\overrightarrow{AB}$ and $\overrightarrow{AC}$ is defined by

$$\overrightarrow{AB} - \overrightarrow{AC} = \overrightarrow{AB} + (-\overrightarrow{AC}).$$

As seen in FIGURE 11.1.6(a), the difference $\overrightarrow{AB} - \overrightarrow{AC}$ can be interpreted as the main diagonal of the parallelogram with sides $\overrightarrow{AB}$ and $-\overrightarrow{AC}$. However, as shown in Figure 11.1.6(b), we can also interpret the same vector difference as the third side of a triangle with sides $\overrightarrow{AB}$ and $\overrightarrow{AC}$. In this second interpretation, observe that the vector difference $\overrightarrow{CB} = \overrightarrow{AB} - \overrightarrow{AC}$ points toward the terminal point of the vector *from* which we are subtracting the second vector. If $\overrightarrow{AB} = \overrightarrow{AC}$, then $\overrightarrow{AB} - \overrightarrow{AC} = 0$.

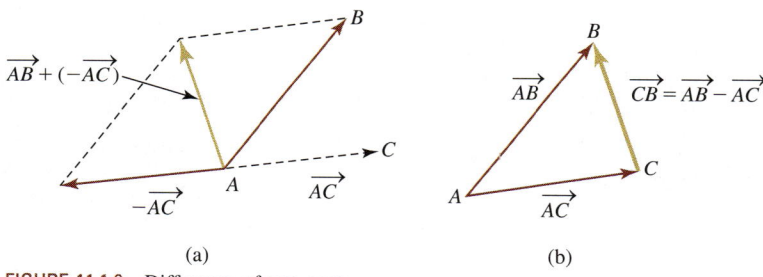

(a)

(b)

FIGURE 11.1.6   Difference of two vectors

■ **Vectors in a Coordinate Plane**   To describe a vector analytically, let us suppose for the remainder of this section that the vectors we are considering lie in a two-dimensional coordinate plane or **2-space**. The vector shown in FIGURE 11.1.7, whose initial point is the origin O and whose terminal point is $P(x_1, y_1)$, is called the **position vector** of the point P and is written

$$\overrightarrow{OP} = \langle x_1, y_1 \rangle.$$

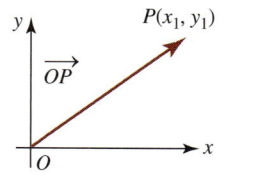

FIGURE 11.1.7   Position vector

■ **Components**   In general, any vector in 2-space can be identified with a unique position vector $\mathbf{a} = \langle a_1, a_2 \rangle$. The numbers a_1 and a_2 are said to be the **components** of the position vector $\mathbf{a}$.

EXAMPLE 1   Position Vector

The displacement from the initial point $P_1(x, y)$ to the terminal point $P_2(x + 4, y + 3)$ in FIGURE 11.1.8(a) is 4 units to the right and 3 units up. As seen in Figure 11.1.8(b), the position vector of $\mathbf{a} = \langle 4, 3 \rangle$ is equivalent to the displacement vector $\overrightarrow{P_1P_2}$ from $P_1(x, y)$ to $P_2(x + 4, y + 3)$.

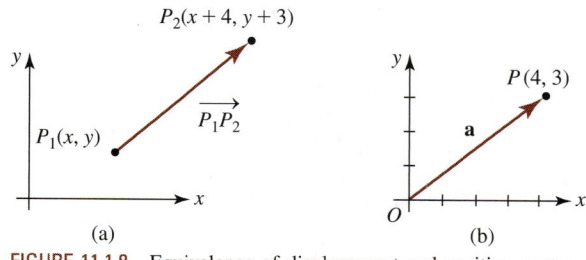

(a)

(b)

FIGURE 11.1.8   Equivalence of displacement and position vectors

We have already given geometric definitions of addition, scalar multiplication, and equality of vectors. We now give the equivalent algebraic definitions using the component form of vectors.

Definition 11.1.1 Component Arithmetic

Let $\mathbf{a} = \langle a_1, a_2 \rangle$ and $\mathbf{b} = \langle b_1, b_2 \rangle$ be vectors in 2-space.

(*i*) Addition: $\mathbf{a} + \mathbf{b} = \langle a_1 + b_1, a_2 + b_2 \rangle$	(1)
(*ii*) Scalar multiplication: $k\mathbf{a} = \langle ka_1, ka_2 \rangle$	(2)
(*iii*) Equality: $\mathbf{a} = \mathbf{b}$ if and only if $a_1 = b_1, a_2 = b_2$	(3)

■ **Subtraction** Using (2), we define the **negative** of a vector $\mathbf{b}$ by

$$-\mathbf{b} = (-1)\mathbf{b} = \langle -b_1, -b_2 \rangle.$$

We can then define the **subtraction**, or the difference, of two vectors as

$$\mathbf{a} - \mathbf{b} = \mathbf{a} + (-\mathbf{b}) = \langle a_1 - b_1, a_2 - b_2 \rangle. \tag{4}$$

In **FIGURE 11.1.9(a)**, we see the sum of two vectors $\overrightarrow{OP_1}$ and $\overrightarrow{OP_2}$ illustrated. In Figure 11.1.9(b) the vector $\overrightarrow{P_1P_2}$, with initial point P_1 and terminal point P_2, is the difference of position vectors

$$\overrightarrow{P_1P_2} = \overrightarrow{OP_2} - \overrightarrow{OP_1} = \langle x_2 - x_1, y_2 - y_1 \rangle.$$

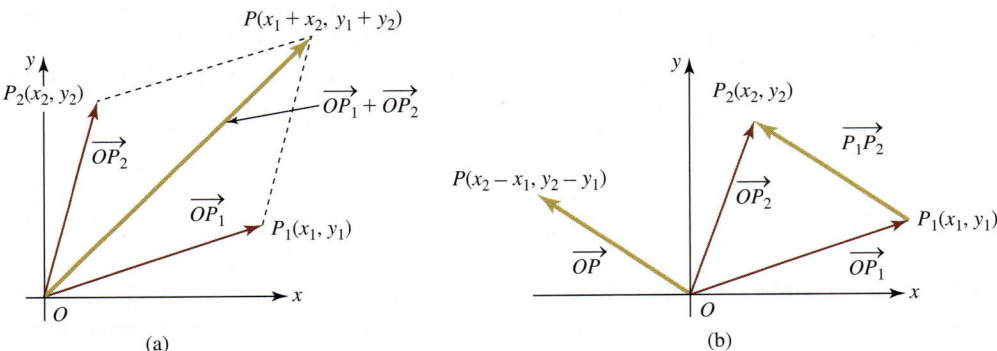

FIGURE 11.1.9 Subtraction of vectors

As shown in Figure 11.1.9(b), the vector $\overrightarrow{P_1P_2}$ can be drawn either starting from the terminal point of $\overrightarrow{OP_1}$ and ending at the terminal point of $\overrightarrow{OP_2}$, or as the position vector $\overrightarrow{OP}$ whose terminal point has coordinates $(x_2 - x_1, y_2 - y_1)$. Remember, $\overrightarrow{OP}$ and $\overrightarrow{P_1P_2}$ are considered equal because they have the same magnitude and direction.

EXAMPLE 2 Sum and Difference of Vectors

If $\mathbf{a} = \langle 1, 4 \rangle$ and $\mathbf{b} = \langle -6, 3 \rangle$, find

(a) $\mathbf{a} + \mathbf{b}$,   (b) $\mathbf{a} - \mathbf{b}$,   and   (c) $2\mathbf{a} + 3\mathbf{b}$.

Solution We use (1), (2), and (4).

(a) $\mathbf{a} + \mathbf{b} = \langle 1 + (-6), 4 + 3 \rangle = \langle -5, 7 \rangle$
(b) $\mathbf{a} - \mathbf{b} = \langle 1 - (-6), 4 - 3 \rangle = \langle 7, 1 \rangle$
(c) $2\mathbf{a} + 3\mathbf{b} = \langle 2, 8 \rangle + \langle -18, 9 \rangle = \langle -16, 17 \rangle$ ■

■ **Properties** The component form of a vector can be used to verify each of the following properties.

Theorem 11.1.1 Properties of Vector Arithmetic

(*i*) $\mathbf{a} + \mathbf{b} = \mathbf{b} + \mathbf{a}$ ← commutative law
(*ii*) $\mathbf{a} + (\mathbf{b} + \mathbf{c}) = (\mathbf{a} + \mathbf{b}) + \mathbf{c}$ ← associative law
(*iii*) $\mathbf{a} + \mathbf{0} = \mathbf{a}$ ← additive identity
(*iv*) $\mathbf{a} + (-\mathbf{a}) = \mathbf{0}$ ← additive inverse
(*v*) $k(\mathbf{a} + \mathbf{b}) = k\mathbf{a} + k\mathbf{b}$, k a scalar
(*vi*) $(k_1 + k_2)\mathbf{a} = k_1\mathbf{a} + k_2\mathbf{a}$, k_1 and k_2 scalars
(*vii*) $(k_1)(k_2\mathbf{a}) = (k_1 k_2)\mathbf{a}$, k_1 and k_2 scalars
(*viii*) $1\mathbf{a} = \mathbf{a}$
(*ix*) $0\mathbf{a} = \mathbf{0}$

The **zero vector 0** in properties (*iii*), (*iv*), and (*ix*) is defined as

$$\mathbf{0} = \langle 0, 0 \rangle.$$

■ **Magnitude** Motivated by the Pythagorean Theorem and FIGURE 11.1.10, we define the **magnitude**, **length**, or **norm** of a vector $\mathbf{a} = \langle a_1, a_2 \rangle$ to be

$$|\mathbf{a}| = \sqrt{a_1^2 + a_2^2}.$$

Clearly, $|\mathbf{a}| \geq 0$ for any vector $\mathbf{a}$, and $|\mathbf{a}| = 0$ if and only if $\mathbf{a} = \mathbf{0}$. For example, if $\mathbf{a} = \langle 6, -2 \rangle$, then

$$|\mathbf{a}| = \sqrt{6^2 + (-2)^2} = \sqrt{40} = 2\sqrt{10}.$$

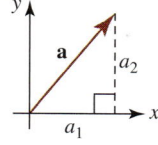

FIGURE 11.1.10 Magnitude of a vector

■ **Unit Vectors** A vector that has magnitude 1 is called a **unit vector**. We can obtain a unit vector $\mathbf{u}$ in the same direction as a nonzero vector $\mathbf{a}$ by multiplying $\mathbf{a}$ by the positive scalar $k = 1/|\mathbf{a}|$ (reciprocal of its magnitude). In this case we say that $\mathbf{u} = (1/|\mathbf{a}|)\mathbf{a}$ is the **normalization** of the vector $\mathbf{a}$. The normalization of the vector $\mathbf{a}$ is a unit vector because

$$|\mathbf{u}| = \left| \frac{1}{|\mathbf{a}|} \mathbf{a} \right| = \frac{1}{|\mathbf{a}|} |\mathbf{a}| = 1.$$

Note: It is often convenient to write the scalar multiple $\mathbf{u} = (1/|\mathbf{a}|)\mathbf{a}$ as

$$\mathbf{u} = \frac{\mathbf{a}}{|\mathbf{a}|}.$$

EXAMPLE 3 Unit Vector

Given $\mathbf{v} = \langle 2, -1 \rangle$, form a unit vector

    (a) in the same direction as $\mathbf{v}$ and   **(b)** in the opposite direction of $\mathbf{v}$.

Solution First, we find the magnitude of the vector $\mathbf{v}$:

$$|\mathbf{v}| = \sqrt{4 + (-1)^2} = \sqrt{5}.$$

(a) A unit vector in the same direction as $\mathbf{v}$ is then

$$\mathbf{u} = \frac{1}{\sqrt{5}} \mathbf{v} = \frac{1}{\sqrt{5}} \langle 2, -1 \rangle = \left\langle \frac{2}{\sqrt{5}}, \frac{-1}{\sqrt{5}} \right\rangle.$$

(b) A unit vector in the opposite direction of $\mathbf{v}$ is the negative of $\mathbf{u}$:

$$-\mathbf{u} = \left\langle -\frac{2}{\sqrt{5}}, \frac{1}{\sqrt{5}} \right\rangle. \qquad ■$$

If $\mathbf{a}$ and $\mathbf{b}$ are vectors and c_1 and c_2 are scalars, then the expression $c_1\mathbf{a} + c_2\mathbf{b}$ is called a **linear combination** of $\mathbf{a}$ and $\mathbf{b}$. As we shall see next, any vector in 2-space can be written as a linear combination of two special vectors.

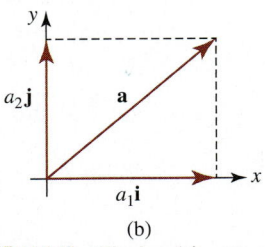

FIGURE 11.1.11 The **i** and **j** vectors in component form

The i, j Vectors In view of (1) and (2), any vector $\mathbf{a} = \langle a_1, a_2 \rangle$ can be written as a sum:

$$\langle a_1, a_2 \rangle = \langle a_1, 0 \rangle + \langle 0, a_2 \rangle = a_1 \langle 1, 0 \rangle + a_2 \langle 0, 1 \rangle. \tag{5}$$

The unit vectors $\langle 1, 0 \rangle$ and $\langle 0, 1 \rangle$ are usually given the special symbols **i** and **j**, respectively. See FIGURE 11.1.11(a). Thus, if

$$\mathbf{i} = \langle 1, 0 \rangle \quad \text{and} \quad \mathbf{j} = \langle 0, 1 \rangle,$$

then (5) becomes $\quad \mathbf{a} = a_1\mathbf{i} + a_2\mathbf{j}. \tag{6}$

Since any vector **a** can be written uniquely as a linear combination of **i** and **j**, these unit vectors are referred to as the **standard basis** for the system of two-dimensional vectors. If $\mathbf{a} = a_1\mathbf{i} + a_2\mathbf{j}$ is a position vector, then Figure 11.1.11(b) shows that **a** is the sum of the vectors $a_1\mathbf{i}$ and $a_2\mathbf{j}$, which have the origin as a common initial point and which lie on the x- and y-axes, respectively. The scalar a_1 is called the **horizontal component** of **a** and the scalar a_2 is called the **vertical component** of **a**.

EXAMPLE 4 Various Vector Forms

(a) $\langle 4, 7 \rangle = 4\mathbf{i} + 7\mathbf{j}$
(b) $(2\mathbf{i} - 5\mathbf{j}) + (8\mathbf{i} + 13\mathbf{j}) = 10\mathbf{i} + 8\mathbf{j}$
(c) $|\mathbf{i} + \mathbf{j}| = \sqrt{2}$
(d) $10(3\mathbf{i} - \mathbf{j}) = 30\mathbf{i} - 10\mathbf{j}$
(e) $\mathbf{a} = 6\mathbf{i} + 4\mathbf{j}$ and $\mathbf{b} = 9\mathbf{i} + 6\mathbf{j}$ are parallel, since **b** is a scalar multiple of **a**. In this case $\mathbf{b} = \frac{3}{2}\mathbf{a}$. ∎

EXAMPLE 5 Graphs of the Sum and Difference

Let $\mathbf{a} = 4\mathbf{i} + 2\mathbf{j}$ and $\mathbf{b} = -2\mathbf{i} + 5\mathbf{j}$. Graph the vectors $\mathbf{a} + \mathbf{b}$ and $\mathbf{a} - \mathbf{b}$.

Solution From (1) and (4), we have respectively

$$\mathbf{a} + \mathbf{b} = 2\mathbf{i} + 7\mathbf{j} \quad \text{and} \quad \mathbf{a} - \mathbf{b} = 6\mathbf{i} - 3\mathbf{j}.$$

The graphs of these two vectors in the xy-plane are given in FIGURE 11.1.12.

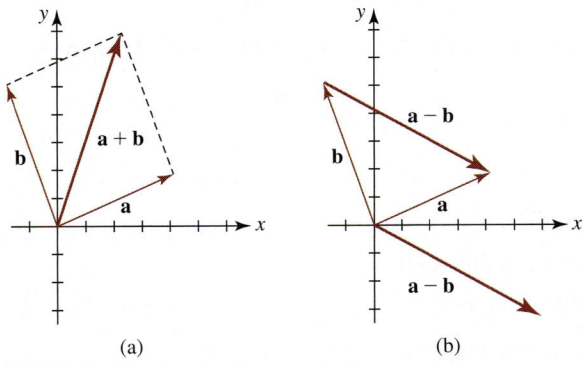

FIGURE 11.1.12 Graphs of vectors in Example 5 ∎

Exercises 11.1 Answers to selected odd-numbered problems begin on page ANS-35.

☰ Fundamentals

In Problems 1–8, find
(a) 3a, (b) a + b, (c) a − b,
(d) |a + b|, and (e) |a − b|.

1. $\mathbf{a} = 2\mathbf{i} + 4\mathbf{j}, \mathbf{b} = -\mathbf{i} + 4\mathbf{j}$ **2.** $\mathbf{a} = \langle 1, 1 \rangle, \mathbf{b} = \langle 2, 3 \rangle$

3. $\mathbf{a} = \langle 4, 0 \rangle, \mathbf{b} = \langle 0, -5 \rangle$

4. $\mathbf{a} = \dfrac{1}{6}\mathbf{i} - \dfrac{1}{6}\mathbf{j}, \mathbf{b} = \dfrac{1}{2}\mathbf{i} + \dfrac{5}{6}\mathbf{j}$

5. $\mathbf{a} = -3\mathbf{i} + 2\mathbf{j}, \mathbf{b} = 7\mathbf{j}$ **6.** $\mathbf{a} = \langle 1, 3 \rangle, \mathbf{b} = -5\mathbf{a}$

7. $\mathbf{a} = -\mathbf{b}, \mathbf{b} = -2\mathbf{i} - 9\mathbf{j}$ **8.** $\mathbf{a} = \langle 7, 10 \rangle, \mathbf{b} = \langle 1, 2 \rangle$

In Problems 9–14, find
(a) 4a − 2b and (b) −3a − 5b.

9. $\mathbf{a} = \langle 1, -3 \rangle, \mathbf{b} = \langle -1, 1 \rangle$ **10.** $\mathbf{a} = \mathbf{i} + \mathbf{j}, \mathbf{b} = 3\mathbf{i} - 2\mathbf{j}$

11. $\mathbf{a} = \mathbf{i} - \mathbf{j}, \mathbf{b} = -3\mathbf{i} + 4\mathbf{j}$ **12.** $\mathbf{a} = \langle 2, 0 \rangle, \mathbf{b} = \langle 0, -3 \rangle$

13. $\mathbf{a} = \langle 4, 10 \rangle, \mathbf{b} = -2\langle 1, 3 \rangle$

14. $\mathbf{a} = \langle 3, 1 \rangle + \langle -1, 2 \rangle, \mathbf{b} = \langle 6, 5 \rangle - \langle 1, 2 \rangle$

In Problems 15–18, find the vector $\overrightarrow{P_1P_2}$. Graph $\overrightarrow{P_1P_2}$ and its corresponding position vector.

15. $P_1(3, 2), P_2(5, 7)$ **16.** $P_1(-2, -1), P_2(4, -5)$

17. $P_1(3, 3), P_2(5, 5)$ **18.** $P_1(0, 3), P_2(2, 0)$

19. Find the terminal point of the vector $\overrightarrow{P_1P_2} = 4\mathbf{i} + 8\mathbf{j}$ if its initial point is $(-3, 10)$.

20. Find the initial point of the vector $\overrightarrow{P_1P_2} = \langle -5, -1 \rangle$ if its terminal point is $(4, 7)$.

21. Determine which of the following vectors are parallel to $\mathbf{a} = 4\mathbf{i} + 6\mathbf{j}$.

     (a) $-4\mathbf{i} - 6\mathbf{j}$ **(b)** $-\mathbf{i} - \dfrac{3}{2}\mathbf{j}$

     (c) $10\mathbf{i} + 15\mathbf{j}$ **(d)** $2(\mathbf{i} - \mathbf{j}) - 3\left(\dfrac{1}{2}\mathbf{i} - \dfrac{5}{12}\mathbf{j}\right)$

     (e) $8\mathbf{i} + 12\mathbf{j}$ **(f)** $(5\mathbf{i} + \mathbf{j}) - (7\mathbf{i} + 4\mathbf{j})$

22. Determine a scalar c so that $\mathbf{a} = 3\mathbf{i} + c\mathbf{j}$ and $\mathbf{b} = -\mathbf{i} + 9\mathbf{j}$ are parallel.

In Problems 23 and 24, find $\mathbf{a} + (\mathbf{b} + \mathbf{c})$ for the given vectors.

23. $\mathbf{a} = \langle 5, 1 \rangle, \mathbf{b} = \langle -2, 4 \rangle, \mathbf{c} = \langle 3, 10 \rangle$

24. $\mathbf{a} = \langle 1, 1 \rangle, \mathbf{b} = \langle 4, 3 \rangle, \mathbf{c} = \langle 0, -2 \rangle$

In Problems 25–28, find a unit vector

     (a) in the same direction as $\mathbf{a}$, and

     (b) in the opposite direction of $\mathbf{a}$.

25. $\mathbf{a} = \langle 2, 2 \rangle$ **26.** $\mathbf{a} = \langle -3, 4 \rangle$

27. $\mathbf{a} = \langle 0, -5 \rangle$ **28.** $\mathbf{a} = \langle 1, -\sqrt{3} \rangle$

In Problems 29 and 30, normalize the given vector when $\mathbf{a} = \langle 2, 8 \rangle$ and $\mathbf{b} = \langle 3, 4 \rangle$.

29. $\mathbf{a} + \mathbf{b}$ **30.** $2\mathbf{a} - 3\mathbf{b}$

In Problems 31 and 32, find a vector $\mathbf{b}$ that is parallel to the given vector $\mathbf{a}$ and has the indicated magnitude.

31. $\mathbf{a} = 3\mathbf{i} + 7\mathbf{j}, |\mathbf{b}| = 2$ **32.** $\mathbf{a} = \dfrac{1}{2}\mathbf{i} - \dfrac{1}{2}\mathbf{j}, |\mathbf{b}| = 3$

33. Find a vector in the opposite direction of $\mathbf{a} = \langle 4, 10 \rangle$ but $\frac{3}{4}$ as long.

34. Given that $\mathbf{a} = \langle 1, 1 \rangle$ and $\mathbf{b} = \langle -1, 0 \rangle$, find a vector in the same direction as $\mathbf{a} + \mathbf{b}$ but 5 times as long.

In Problems 35 and 36, use the given figure to draw the indicated vector.

35. $3\mathbf{b} - \mathbf{a}$ **36.** $\mathbf{a} + (\mathbf{b} + \mathbf{c})$

FIGURE 11.1.13 Vectors for Problem 35

FIGURE 11.1.14 Vectors for Problem 36

In Problems 37 and 38, express the vector $\mathbf{x}$ in terms of vectors $\mathbf{a}$ and $\mathbf{b}$.

37.   **38.**

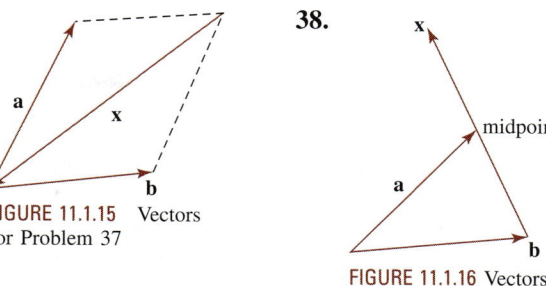

FIGURE 11.1.15 Vectors for Problem 37

FIGURE 11.1.16 Vectors for Problem 38

In Problems 39 and 40, use the given figure to prove the given result.

39. $\mathbf{a} + \mathbf{b} + \mathbf{c} = \mathbf{0}$ **40.** $\mathbf{a} + \mathbf{b} + \mathbf{c} + \mathbf{d} = \mathbf{0}$

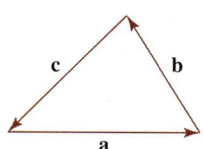

FIGURE 11.1.17 Vectors for Problem 39

FIGURE 11.1.18 Vectors for Problem 40

In Problems 41 and 42, express the vector $\mathbf{a} = 2\mathbf{i} + 3\mathbf{j}$ as a linear combination of the given vectors $\mathbf{b}$ and $\mathbf{c}$.

41. $\mathbf{b} = \mathbf{i} + \mathbf{j}, \mathbf{c} = \mathbf{i} - \mathbf{j}$

42. $\mathbf{b} = -2\mathbf{i} + 4\mathbf{j}, \mathbf{c} = 5\mathbf{i} + 7\mathbf{j}$

A vector is said to be tangent to a curve at a point if it is parallel to the tangent line at the point. In Problems 43 and 44, find a unit tangent vector to the given curve at the indicated point.

43. $y = \dfrac{1}{4}x^2 + 1; \quad (2, 2)$

44. $y = -x^2 + 3x; \quad (0, 0)$

45. Let, $P_1, P_2,$ and P_3 be distinct points such that $\mathbf{a} = \overrightarrow{P_1P_2}$, $\mathbf{b} = \overrightarrow{P_2P_3}$, and $\mathbf{a} + \mathbf{b} = \overrightarrow{P_1P_3}$.

     (a) What is the relation of $|\mathbf{a} + \mathbf{b}|$ to $|\mathbf{a}| + |\mathbf{b}|$?

     (b) Under what condition is $|\mathbf{a} + \mathbf{b}| = |\mathbf{a}| + |\mathbf{b}|$?

≡ Applications

46. An electric charge Q is uniformly distributed along the y-axis between $y = -a$ and $y = a$. See FIGURE 11.1.19. The total force exerted on the charge q on the x-axis by the charge Q is $\mathbf{F} = F_x\mathbf{i} + F_y\mathbf{j}$, where

$$F_x = \frac{qQ}{4\pi\varepsilon_0} \int_{-a}^{a} \frac{L}{2a(L^2 + y^2)^{3/2}} \, dy$$

and

$$F_y = -\frac{qQ}{4\pi\varepsilon_0} \int_{-a}^{a} \frac{y}{2a(L^2 + y^2)^{3/2}} \, dy.$$

Determine **F**.

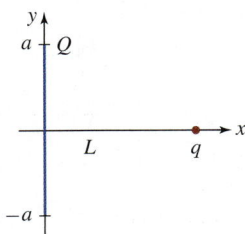

FIGURE 11.1.19 Electric charge in Problem 46

47. When walking, a person's foot strikes the ground with a force **F** at an angle θ from the vertical. In FIGURE 11.1.20, the vector **F** is resolved into vector components $\mathbf{F}_g$, which is parallel to the ground, and $\mathbf{F}_n$, which is perpendicular to the ground. In order that the foot does not slip, the force $\mathbf{F}_g$ must be offset by the opposing force $\mathbf{F}_f$ of friction; that is, $\mathbf{F}_f = -\mathbf{F}_g$.

(a) Use the fact that $|\mathbf{F}_f| = \mu|\mathbf{F}_n|$, where μ is the coefficient of friction, to show that $\tan\theta = \mu$. The foot will not slip for angles less than or equal to θ.

(b) Given that $\mu = 0.6$ for a rubber heel striking an asphalt sidewalk, find the "no-slip" angle.

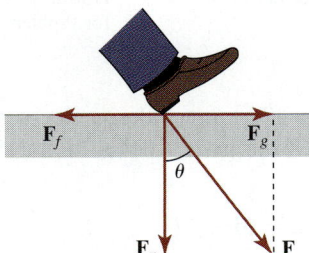

FIGURE 11.1.20 Vectors for Problem 47

48. A 200-lb traffic light supported by two cables hangs in equilibrium. As shown in FIGURE 11.1.21(b), let the weight of the light be represented by **w** and the forces in the two cables by $\mathbf{F}_1$ and $\mathbf{F}_2$. From Figure 11.1.21(c), we see that a condition of equilibrium is

$$\mathbf{w} + \mathbf{F}_1 + \mathbf{F}_2 = \mathbf{0}. \qquad (7)$$

See Problem 39. If

$$\mathbf{w} = -200\mathbf{j}$$
$$\mathbf{F}_1 = (|\mathbf{F}_1|\cos 20°)\mathbf{i} + (|\mathbf{F}_1|\sin 20°)\mathbf{j}$$
$$\mathbf{F}_2 = -(|\mathbf{F}_2|\cos 15°)\mathbf{i} + (|\mathbf{F}_2|\sin 15°)\mathbf{j},$$

use (7) to determine the magnitudes of $\mathbf{F}_1$ and $\mathbf{F}_2$. [*Hint*: Reread (*iii*) of Definition 11.1.1.]

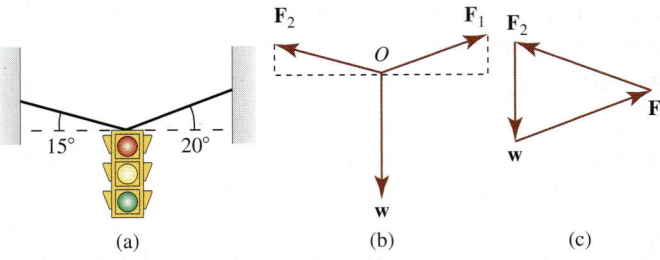

FIGURE 11.1.21 Traffic light in Problem 48

49. Water rushing from a fire hose exerts a horizontal force $\mathbf{F}_1$ of magnitude 200 lb. See FIGURE 11.1.22. What is the magnitude of the force $\mathbf{F}_3$ that a firefighter must exert to hold the hose at an angle of 45° from the horizontal?

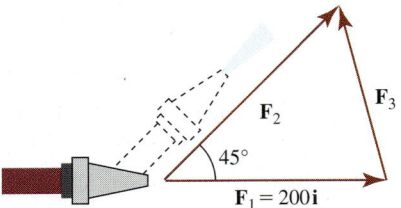

FIGURE 11.1.22 Vectors for Problem 49

50. An airplane starts from an airport located at the origin O and flies 150 mi in the direction 20° north of east to city A. From A the airplane then flies 200 mi in the direction 23° west of north to city B. From B the airplane flies 240 mi in the direction 10° south of west to city C. Express the location of city C as a vector **r** as shown in FIGURE 11.1.23. Find the distance from O to C.

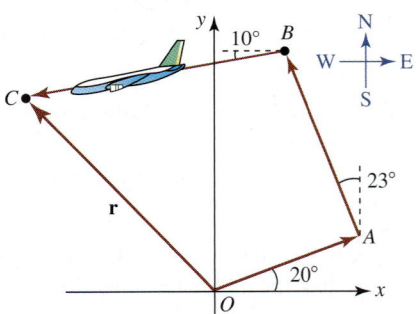

FIGURE 11.1.23 Vectors for Problem 50

≡ Think About It

51. Using vectors, show that the diagonals of a parallelogram bisect each other. [*Hint*: Let M be the midpoint of one diagonal and N the midpoint of the other.]

52. Using vectors, show that the line segment between the midpoints of two sides of a triangle is parallel to the third side and half as long.

11.2 3-Space and Vectors

❙ Introduction In the plane, or 2-space, one way of describing the position of a point P is to assign to it coordinates relative to two mutually orthogonal axes called the *x*- and *y*-axes. If P is the point of intersection of the line $x = a$ (perpendicular to the *x*-axis) and the line $y = b$ (perpendicular to the *y*-axis), then the **ordered pair** (a, b) is said to be the rectangular or Cartesian coordinates of the point. See **FIGURE 11.2.1**. In this section we extend this method of representation to three-dimensional space and then consider vectors in 3-space.

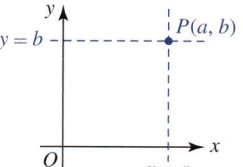

FIGURE 11.2.1   Point in 2-space

■ **Rectangular Coordinate System in 3-Space**   In three dimensions, or **3-space**, a rectangular coordinate system is constructed using three mutually perpendicular axes. The point at which these axes intersect is called the **origin O**. These axes, shown in **FIGURE 11.2.2(a)**, are labeled in accordance with the so-called **right-hand rule**: If the fingers of the right hand, pointing in the direction of the positive x-axis, are curled toward the positive y-axis, then the thumb will point in the direction of a new axis perpendicular to the plane of the x- and y-axes. This new axis is labeled the z-axis. The dashed lines in Figure 11.2.2(a), represent the negative axes. Now, if

◀ If the x- and y-axes are interchanged in Figure 11.2.2(a), the coordinate system is said to be **left-handed**.

$$x = a, \qquad y = b, \qquad z = c$$

are planes perpendicular to the x-axis, y-axis, and z-axis, respectively, the point P at which these planes intersect can be represented by an **ordered triple** of numbers (a, b, c) said to be the **rectangular** or **Cartesian coordinates** of the point. The numbers a, b, and c are, in turn, called the x-, y-, and z-coordinates of $P(a, b, c)$. See Figure 11.2.2(b).

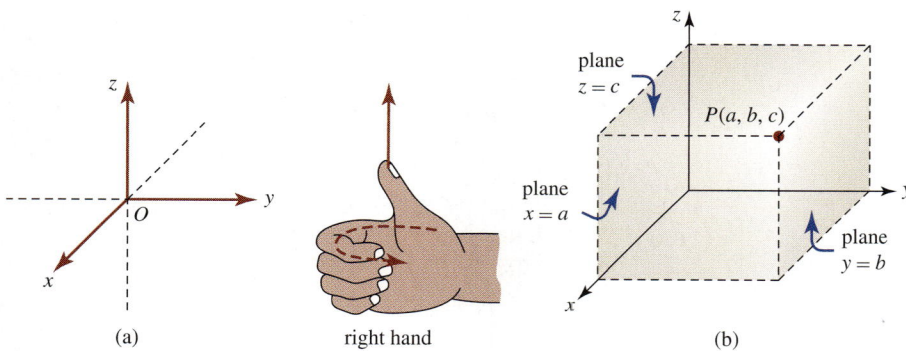

(a)   right hand   (b)

FIGURE 11.2.2   The right-hand rule and a point in 3-space

■ **Octants**   Each pair of coordinate axes determines a **coordinate plane**. As shown in **FIGURE 11.2.3**, the x- and y-axes determine the xy-plane, the x- and z-axes determine the xz-plane, and so on. The coordinate planes divide 3-space into eight parts known as **octants**. The octant in which all three coordinates of a point are *positive* is called the **first octant**. There is no agreement for naming the other seven octants.

The following table summarizes the coordinates of a point either on a coordinate axis or in a coordinate plane. As seen in the table, we can also describe, say, the xy-plane by the simple equation $z = 0$. Similarly, the xz-plane is $y = 0$ and the yz-plane is $x = 0$.

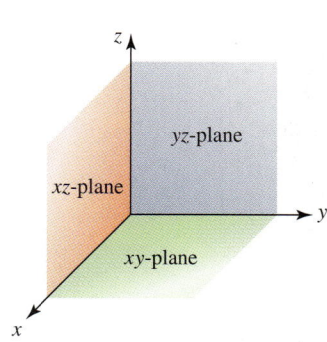

FIGURE 11.2.3   Coordinate planes

Axes	Coordinates	Plane	Coordinates
x	$(a, 0, 0)$	xy	$(a, b, 0)$
y	$(0, b, 0)$	xz	$(a, 0, c)$
z	$(0, 0, c)$	yz	$(0, b, c)$

EXAMPLE 1   Graphing Points in 3-Space

Graph the points $(4, 5, 6)$, $(3, -3, -1)$, and $(-2, -2, 0)$.

Solution   Of the three points shown in **FIGURE 11.2.4**, only $(4, 5, 6)$ is in the first octant. The point $(-2, -2, 0)$ is in the xy-plane.

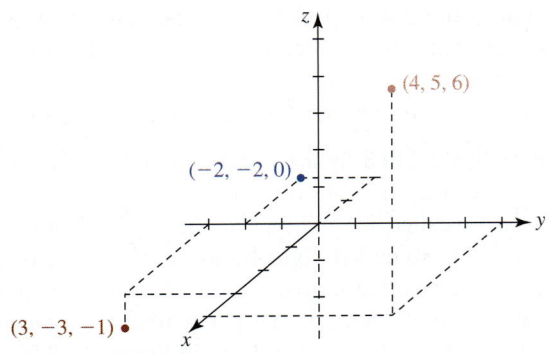

FIGURE 11.2.4 Points in Example 1

■ **Distance Formula** To find the **distance** between two points $P_1(x_1, y_1, z_1)$ and $P_2(x_2, y_2, z_2)$ in 3-space, let us first consider their projections onto the xy-plane. As seen in FIGURE 11.2.5, the distance between $(x_1, y_1, 0)$ and $(x_2, y_2, 0)$ follows from the usual distance formula in the plane and is $\sqrt{(x_2 - x_1)^2 + (y_2 - y_1)^2}$. Hence, from the Pythagorean Theorem applied to the right triangle $P_1P_3P_2$, we have

$$[d(P_1, P_2)]^2 = \left[\sqrt{(x_2 - x_1)^2 + (y_2 - y_1)^2}\right]^2 + |z_2 - z_1|^2$$

or
$$d(P_1, P_2) = \sqrt{(x_2 - x_1)^2 + (y_2 - y_1)^2 + (z_2 - z_1)^2}. \tag{1}$$

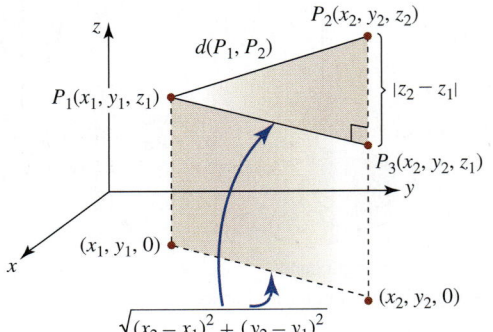

FIGURE 11.2.5 Distance between two points in 3-space

EXAMPLE 2 Distance Between Points in 3-Space

Find the distance between $(2, -3, 6)$ and $(-1, -7, 4)$.

Solution From (1),

$$d = \sqrt{(2 - (-1))^2 + (-3 - (-7))^2 + (6 - 4)^2} = \sqrt{29}.$$

■ **Midpoint Formula** The distance formula can be used to show that the coordinates of the **midpoint of the line segment** in 3-space connecting the distinct points $P_1(x_1, y_1, z_1)$ and $P_2(x_2, y_2, z_2)$ are

$$\left(\frac{x_1 + x_2}{2}, \frac{y_1 + y_2}{2}, \frac{z_1 + z_2}{2}\right). \tag{2}$$

See Problem 64 in Exercises 11.2.

EXAMPLE 3 Midpoint in 3-Space

Find the coordinates of the midpoint of the line segment between the two points in Example 2.

Solution From (2) we obtain

$$\left(\frac{2 + (-1)}{2}, \frac{-3 + (-7)}{2}, \frac{6 + 4}{2}\right) \qquad \text{or} \qquad \left(\frac{1}{2}, -5, 5\right).$$

■ **Vectors in 3-Space**   A **vector a** in 3-space is any ordered triple of real numbers

$$\mathbf{a} = \langle a_1, a_2, a_3 \rangle,$$

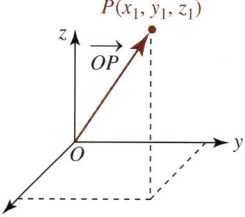

FIGURE 11.2.6   A vector in 3-space

where a_1, a_2, and a_3 are the **components** of the vector. The **position vector** of a point $P_1(x_1, y_1, z_1)$ in 3-space is the vector $\overrightarrow{OP} = \langle x_1, y_1, z_1 \rangle$ whose initial point is the origin O and whose terminal point is P. See FIGURE 11.2.6.

The component definitions of addition, subtraction, scalar multiplication, and so on, are natural generalizations of those given for vectors in 2-space.

Definition 11.2.1   Component Arithmetic

Let $\mathbf{a} = \langle a_1, a_2, a_3 \rangle$ and $\mathbf{b} = \langle b_1, b_2, b_3 \rangle$ be vectors in 3-space.

(*i*) Addition: $\mathbf{a} + \mathbf{b} = \langle a_1 + b_1, a_2 + b_2, a_3 + b_3 \rangle$
(*ii*) Scalar multiplication: $k\mathbf{a} = \langle ka_1, ka_2, ka_3 \rangle$
(*iii*) Equality: $\mathbf{a} = \mathbf{b}$ if and only if $a_1 = b_1, a_2 = b_2, a_3 = b_3$
(*iv*) Negative: $-\mathbf{b} = (-1)\mathbf{b} = \langle -b_1, -b_2, -b_3 \rangle$
(*v*) Subtraction: $\mathbf{a} - \mathbf{b} = \mathbf{a} + (-\mathbf{b}) = \langle a_1 - b_1, a_2 - b_2, a_3 - b_3 \rangle$
(*vi*) Zero vector: $\mathbf{0} = \langle 0, 0, 0 \rangle$
(*vii*) Magnitude: $|\mathbf{a}| = \sqrt{a_1^2 + a_2^2 + a_3^2}$

If $\overrightarrow{OP_1}$ and $\overrightarrow{OP_2}$ are the position vectors of the points $P_1(x_1, y_1, z_1)$, and $P_2(x_2, y_2, z_2)$, then the vector $\overrightarrow{P_1P_2}$ is given by

$$\overrightarrow{P_1P_2} = \overrightarrow{OP_2} - \overrightarrow{OP_1} = \langle x_2 - x_1, y_2 - y_1, z_2 - z_1 \rangle. \tag{3}$$

As in 2-space, $\overrightarrow{P_1P_2}$ can be drawn either as a vector whose initial point is P_1 and whose terminal point is P_2 or as a position vector $\overrightarrow{OP}$ with terminal point $P = (x_2 - x_1, y_2 - y_1, z_2 - z_1)$. See FIGURE 11.2.7.

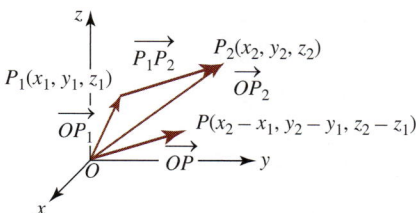

FIGURE 11.2.7   A vector connecting two points in 3-space

EXAMPLE 4   Vector Between Two Points

Find the vector $\overrightarrow{P_1P_2}$ if the points P_1 and P_2 are given by $P_1 = (4, 6, -2)$ and $P_2 = (1, 8, 3)$.

Solution   If the position vectors of the points are $\overrightarrow{OP_1} = \langle 4, 6, -2 \rangle$ and $\overrightarrow{OP_2} = \langle 1, 8, 3 \rangle$, then from (3) we have

$$\overrightarrow{P_1P_2} = \overrightarrow{OP_2} - \overrightarrow{OP_1} = \langle 1 - 4, 8 - 6, 3 - (-2) \rangle = \langle -3, 2, 5 \rangle. \qquad \blacksquare$$

EXAMPLE 5   A Unit Vector

Find a unit vector in the direction of $\mathbf{a} = \langle -2, 3, 6 \rangle$.

Solution   Since a unit vector has length 1, we first find the magnitude of $\mathbf{a}$ and then use the fact that $\mathbf{a}/|\mathbf{a}|$ is a unit vector in the direction of $\mathbf{a}$. The magnitude of $\mathbf{a}$ is

$$|\mathbf{a}| = \sqrt{(-2)^2 + 3^2 + 6^2} = \sqrt{49} = 7.$$

A unit vector in the direction of $\mathbf{a}$ is

$$\frac{\mathbf{a}}{|\mathbf{a}|} = \frac{1}{7}\langle -2, 3, 6 \rangle = \left\langle -\frac{2}{7}, \frac{3}{7}, \frac{6}{7} \right\rangle. \qquad \blacksquare$$

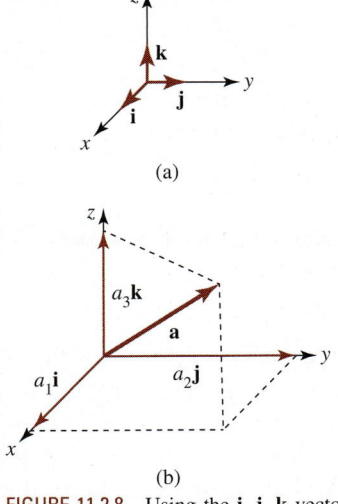

FIGURE 11.2.8 Using the **i**, **j**, **k** vectors to represent a position vector **a**

The i, j, k Vectors We saw in the preceding section that the set of two unit vectors $\mathbf{i} = \langle 1, 0 \rangle$ and $\mathbf{j} = \langle 0, 1 \rangle$ constitute a basis for the system of two-dimensional vectors. That is, any vector **a** in 2-space can be written as a linear combination of **i** and **j**: $a = a_1\mathbf{i} + a_2\mathbf{j}$. Likewise any vector $a = \langle a_1, a_2, a_3 \rangle$ in 3-space can be expressed as a linear combination of the unit vectors

$$\mathbf{i} = \langle 1, 0, 0 \rangle, \quad \mathbf{j} = \langle 0, 1, 0 \rangle, \quad \mathbf{k} = \langle 0, 0, 1 \rangle.$$

To see this we use (*i*) and (*ii*) of Definition 11.2.1 to write

$$\langle a_1, a_2, a_3 \rangle = \langle a_1, 0, 0 \rangle + \langle 0, a_2, 0 \rangle + \langle 0, 0, a_3 \rangle$$
$$= a_1\langle 1, 0, 0 \rangle + a_2\langle 0, 1, 0 \rangle + a_3\langle 0, 0, 1 \rangle,$$

that is, $\qquad \mathbf{a} = a_1\mathbf{i} + a_2\mathbf{j} + a_3\mathbf{k}.$

The vectors **i**, **j**, and **k** illustrated in FIGURE 11.2.8(a) are called the **standard basis** for the system of three-dimensional vectors. In Figure 11.2.8(b) we see that a position vector $\mathbf{a} = a_1\mathbf{i} + a_2\mathbf{j} + a_3\mathbf{k}$ is the sum of the vectors $a_1\mathbf{i}$, $a_2\mathbf{j}$, and $a_3\mathbf{k}$, which lie along the coordinate axes and have the origin as a common initial point.

EXAMPLE 6 Using the **i, j, k** Vectors

The vector $\mathbf{a} = \langle 7, -5, 13 \rangle$ is the same as $\mathbf{a} = 7\mathbf{i} - 5\mathbf{j} + 13\mathbf{k}$. ∎

When the third dimension is taken into consideration, any vector in the *xy*-plane is equivalently described as a three-dimensional vector that lies in the coordinate plane $z = 0$. Although the vectors $\langle a_1, a_2 \rangle$ and $\langle a_1, a_2, 0 \rangle$ are technically not equal, we shall ignore the distinction. That is why, for example, we denoted $\langle 1, 0 \rangle$ and $\langle 1, 0, 0 \rangle$ by the same symbol **i**. A vector in either the *yz*-plane or the *xz*-plane must also have one zero component. In the *yz*-plane a vector $\mathbf{b} = \langle 0, b_2, b_3 \rangle$ is written $\mathbf{b} = b_2\mathbf{j} + b_3\mathbf{k}$.

EXAMPLE 7 Vectors in the Coordinate Planes

(a) The vector $\mathbf{a} = 5\mathbf{i} + 3\mathbf{k} = 5\mathbf{i} + 0\mathbf{j} + 3\mathbf{k}$ lies in the *xz*-plane and can also be written as $\mathbf{a} = \langle 5, 0, 3 \rangle$.

(b) $|5\mathbf{i} + 3\mathbf{k}| = \sqrt{5^2 + 0^2 + 3^2} = \sqrt{25 + 9} = \sqrt{34}$ ∎

EXAMPLE 8 Combining Vectors

If $\mathbf{a} = 3\mathbf{i} - 4\mathbf{j} + 8\mathbf{k}$ and $\mathbf{b} = \mathbf{i} - 4\mathbf{k}$, find $5\mathbf{a} - 2\mathbf{b}$.

Solution By writing $5\mathbf{a} = 15\mathbf{i} - 20\mathbf{j} + 40\mathbf{k}$ and $2\mathbf{b} = 2\mathbf{i} + 0\mathbf{j} - 8\mathbf{k}$ we get

$$5\mathbf{a} - 2\mathbf{b} = (15\mathbf{i} - 20\mathbf{j} + 40\mathbf{k}) - (2\mathbf{i} + 0\mathbf{j} - 8\mathbf{k})$$
$$= 13\mathbf{i} - 20\mathbf{j} + 48\mathbf{k}.$$ ∎

Exercises 11.2 Answers to selected odd-numbered problems begin on page ANS-35.

≡ **Fundamentals**

In Problems 1–6, graph the given point. Use the same coordinate axes.

1. $(1, 1, 5)$ **2.** $(0, 0, 4)$

3. $(3, 4, 0)$ **4.** $(6, 0, 0)$

5. $(6, -2, 0)$ **6.** $(5, -4, 3)$

In Problems 7–10, describe geometrically all points $P(x, y, z)$ whose coordinates satisfy the given condition.

7. $z = 5$ **8.** $x = 1$

9. $x = 2, y = 3$ **10.** $x = 4, y = -1, z = 7$

11. Give the coordinates of the vertices of the rectangular parallelepiped whose sides are the coordinate planes and the planes $x = 2, y = 5, z = 8$.

12. In FIGURE 11.2.9, two vertices are shown of a rectangular parallelepiped having sides parallel to the coordinate planes. Find the coordinates of the remaining six vertices.

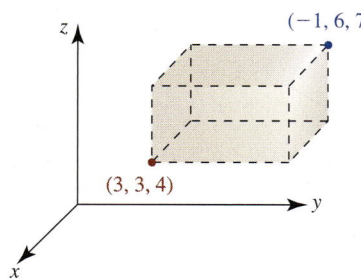

FIGURE 11.2.9 Parallelepiped in Problem 12

13. Consider the point $P(-2, 5, 4)$.
  (a) If lines are drawn from P perpendicular to the coordinate planes, what are the coordinates of the point at the base of each perpendicular?
  (b) If a line is drawn from P to the plane $z = -2$, what are the coordinates of the point at the base of the perpendicular?
  (c) Find the point in the plane $x = 3$ that is closest to P.

14. Determine an equation of a plane parallel to a coordinate plane that contains the given pair of points.
  (a) $(3, 4, -5), (-2, 8, -5)$
  (b) $(1, -1, 1), (1, -1, -1)$
  (c) $(-2, 1, 2), (2, 4, 2)$

In Problems 15–20, describe the set of points $P(x, y, z)$ in 3-space whose coordinates satisfy the given equation.

15. $xyz = 0$
16. $x^2 + y^2 + z^2 = 0$
17. $(x + 1)^2 + (y - 2)^2 + (z + 3)^2 = 0$
18. $(x - 2)(z - 8) = 0$
19. $z^2 - 25 = 0$
20. $x = y = z$

In Problems 21 and 22, find the distance between the given points.

21. $(3, -1, 2), (6, 4, 8)$
22. $(-1, -3, 5), (0, 4, 3)$

23. Find the distance from the point $(7, -3, -4)$ to
  (a) the yz-plane and **(b)** the x-axis.

24. Find the distance from the point $(-6, 2, -3)$ to
  (a) the xz-plane and **(b)** the origin.

In Problems 25–28, the given three points form a triangle. Determine which triangles are isosceles and which are right triangles.

25. $(0, 0, 0), (3, 6, -6), (2, 1, 2)$
26. $(0, 0, 0), (1, 2, 4), (3, 2, 2\sqrt{2})$
27. $(1, 2, 3), (4, 1, 3), (4, 6, 4)$
28. $(1, 1, -1), (1, 1, 1), (0, -1, 1)$

In Problems 29–32, use the distance formula to determine whether the given points are collinear.

29. $P_1(1, 2, 0), P_2(-2, -2, -3), P_3(7, 10, 6)$
30. $P_1(1, 2, -1), P_2(0, 3, 2), P_3(1, 1, -3)$

31. $P_1(1, 0, 4), P_2(-4, -3, 5), P_3(-7, -4, 8)$
32. $P_1(2, 3, 2), P_2(1, 4, 4), P_3(5, 0, -4)$

In Problems 33 and 34, solve for the unknown.

33. $P_1(x, 2, 3), P_2(2, 1, 1); \quad d(P_1, P_2) = \sqrt{21}$
34. $P_1(x, x, 1), P_2(0, 3, 5); \quad d(P_1, P_2) = 5$

In Problems 35 and 36, find the coordinates of the midpoint of the line segment between the given points.

35. $\left(1, 3, \frac{1}{2}\right), \left(7, -2, \frac{5}{2}\right)$ **36.** $(0, 5, -8), (4, 1, -6)$

37. The coordinates of the midpoint of the line segment between $P_1(x_1, y_1, z_1)$ and $P_2(2, 3, 6)$ are $(-1, -4, 8)$. Find the coordinates of P_1.

38. Let P_3 be the midpoint of the line segment between $P_1(-3, 4, 1)$ and $P_2(-5, 8, 3)$. Find the coordinates of the midpoint of the line segment
  (a) between P_1 and P_3 and **(b)** between P_3 and P_2.

In Problems 39–42, express the vector $\overrightarrow{P_1 P_2}$ in component form.

39. $P_1(3, 4, 5), P_2(0, -2, 6)$ **40.** $P_1(-2, 4, 0), P_2\left(6, \frac{3}{4}, 8\right)$
41. $P_1(0, -1, 0), P_2(2, 0, 1)$ **42.** $P_1\left(\frac{1}{2}, \frac{3}{4}, 5\right), P_2\left(-\frac{5}{2}, -\frac{9}{4}, 12\right)$

In Problems 43–46, sketch the given vector.

43. $\langle -3, 5, -2 \rangle$ **44.** $\langle 2, 0, 4 \rangle$
45. $\mathbf{i} + 2\mathbf{j} - 3\mathbf{k}$ **46.** $-4\mathbf{i} + 4\mathbf{j} + 2\mathbf{k}$

In Problems 47–50, determine the axis or plane in which the given vector lies.

47. $\langle 7, -3, 0 \rangle$ **48.** $\langle 0, 2, 0 \rangle$
49. $4\mathbf{k}$ **50.** $-2\mathbf{j} + 5\mathbf{k}$

In Problems 51–58, $\mathbf{a} = \langle 1, -3, 2 \rangle$, $\mathbf{b} = \langle -1, 1, 1 \rangle$, and $\mathbf{c} = \langle 2, 6, 9 \rangle$. Find the indicated vector or scalar.

51. $\mathbf{a} + (\mathbf{b} + \mathbf{c})$ **52.** $2\mathbf{a} - (\mathbf{b} - \mathbf{c})$
53. $\mathbf{b} + 2(\mathbf{a} - 3\mathbf{c})$ **54.** $4(\mathbf{a} + 2\mathbf{c}) - 6\mathbf{b}$
55. $|\mathbf{a} + \mathbf{c}|$ **56.** $|\mathbf{c}| |2\mathbf{b}|$
57. $\left| \dfrac{\mathbf{a}}{|\mathbf{a}|} \right| + 5 \left| \dfrac{\mathbf{b}}{|\mathbf{b}|} \right|$ **58.** $|\mathbf{b}|\mathbf{a} + |\mathbf{a}|\mathbf{b}$

59. Find a unit vector in the opposite direction of $\mathbf{a} = \langle 10, -5, 10 \rangle$.

60. Find a unit vector in the same direction as $\mathbf{a} = \mathbf{i} - 3\mathbf{j} + 2\mathbf{k}$.

61. Find a vector $\mathbf{b}$ that is four times as long as $\mathbf{a} = \mathbf{i} - \mathbf{j} + \mathbf{k}$ in the same direction as $\mathbf{a}$.

62. Find a vector $\mathbf{b}$ for which $|\mathbf{b}| = \frac{1}{2}$ that is parallel to $\mathbf{a} = \langle -6, 3, -2 \rangle$ but has the opposite direction.

≡ Think About It

63. Using the vectors $\mathbf{a}$ and $\mathbf{b}$ shown in FIGURE 11.2.10, sketch the "average vector" $\frac{1}{2}(\mathbf{a} + \mathbf{b})$.

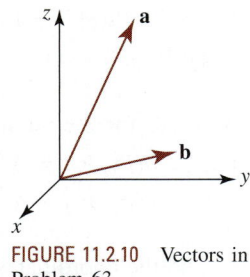

FIGURE 11.2.10 Vectors in Problem 63

64. Use the distance formula to prove that

$$M\left(\frac{x_1 + x_2}{2}, \frac{y_1 + y_2}{2}, \frac{z_1 + z_2}{2}\right)$$

is the midpoint of the line segment between $P_1(x_1, y_1, z_1)$ and $P_2(x_2, y_2, z_2)$. [*Hint*: Show that

$$d(P_1, M) = d(M, P_2) \text{ and } d(P_1, P_2) = d(P_1, M) + d(M, P_2).]$$

≡ Projects

65. As shown in FIGURE 11.2.11(a), a spacecraft can perform rotations called **pitch**, **roll**, and **yaw** about three distinct axes. To describe the coordinates of a point P we use two coordinate systems: a fixed three-dimensional Cartesian coordinate system in which the coordinates of P are (x, y, z) and a spacecraft coordinate system that moves with the particular rotation. In Figure 11.2.11(b) we have illustrated a yaw—that is, a rotation around the z-axis (which is perpendicular to the plane of the page). When the spacecraft performs a pitch, roll, and yaw *in sequence* through the angles α, β, and γ, respectively, the final coordinates of the point P in the spacecraft system (x_S, y_S, z_S) are obtained from the sequence of transformations:

$$x_P = x \qquad\qquad x_R = x_P\cos\beta - z_P\sin\beta$$
$$y_P = y\cos\alpha + z\sin\alpha \qquad y_R = y_P$$
$$z_P = -y\sin\alpha + z\cos\alpha, \qquad z_R = x_P\sin\beta + z_P\cos\beta,$$
$$x_S = x_R\cos\gamma + y_R\sin\gamma$$
$$y_S = -x_R\sin\gamma + y_R\cos\gamma$$
$$z_S = z_R.$$

Suppose the coordinates of a point are $(1, 1, 1)$ in the fixed coordinate system. Determine the coordinates of the point in the spacecraft system if the spacecraft performs

a pitch, roll, and yaw in sequence through the angles $\alpha = 30°$, $\beta = 45°$, $\gamma = 60°$.

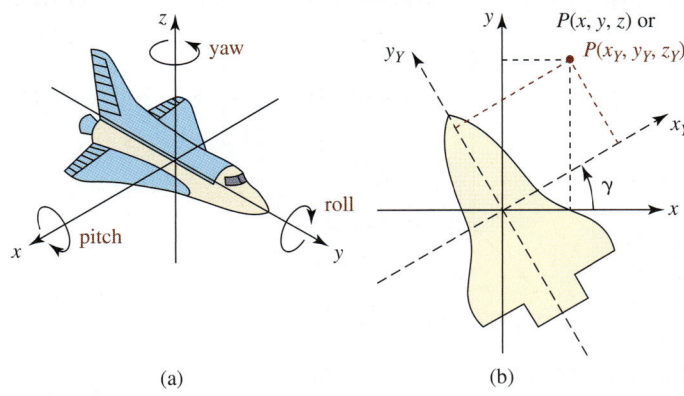

(a)   (b)

FIGURE 11.2.11 Spacecraft in Problem 65

66. (*In order to work this problem, you should learn about, or be familiar with, matrix multiplication.*)

(a) Each system of equations in Problem 65 can be written as a matrix equation. For example, the last system is

$$\begin{bmatrix} x_S \\ y_S \\ z_S \end{bmatrix} = \mathbf{M}_Y \begin{bmatrix} x_R \\ y_R \\ z_R \end{bmatrix},$$

where $\mathbf{M}_Y = \begin{bmatrix} \cos\gamma & \sin\gamma & 0 \\ -\sin\gamma & \cos\gamma & 0 \\ 0 & 0 & 1 \end{bmatrix}$. Identify the matrices $\mathbf{M}_P$

and $\mathbf{M}_R$ and write the first two systems as

$$\begin{bmatrix} x_P \\ y_P \\ z_P \end{bmatrix} = \mathbf{M}_P \begin{bmatrix} x \\ y \\ z \end{bmatrix} \quad \text{and} \quad \begin{bmatrix} x_R \\ y_R \\ z_R \end{bmatrix} = \mathbf{M}_R \begin{bmatrix} x_P \\ y_P \\ z_P \end{bmatrix}.$$

(b) Verify that the final coordinates (x_S, y_S, z_S) in the spacecraft system after a pitch, roll, and yaw are obtained from

$$\begin{bmatrix} x_S \\ y_S \\ z_S \end{bmatrix} = \mathbf{M}_Y \mathbf{M}_R \mathbf{M}_P \begin{bmatrix} x \\ y \\ z \end{bmatrix}.$$

(c) With $(x, y, z) = (1, 1, 1)$ and $\alpha = 30°$, $\beta = 45°$, $\gamma = 60°$, carry out the indicated matrix multiplication in part (b) and verify that your answer is the same as in Problem 65.

11.3 Dot Product

▌ Introduction In this and the following section, we shall consider two kinds of products between vectors that originated in the study of mechanics and electricity and magnetism. The first of these products, known as the **dot product**, is studied in this section.

▌ Component Form of the Dot Product The dot product, defined next, is also known as the **inner product**, or **scalar product**. The dot product of two vectors $\mathbf{a}$ and $\mathbf{b}$ is denoted by $\mathbf{a} \cdot \mathbf{b}$ and is a real number, or scalar, defined in terms of the components of the vectors.

Definition 11.3.1 Dot Product of Two Vectors

In 2-space the **dot product** of two vectors $\mathbf{a} = \langle a_1, a_2 \rangle$ and $\mathbf{b} = \langle b_1, b_2 \rangle$ is

$$\mathbf{a} \cdot \mathbf{b} = a_1 b_1 + a_2 b_2. \qquad (1)$$

In 3-space the **dot product** of two vectors $\mathbf{a} = \langle a_1, a_2, a_3 \rangle$ and $\mathbf{b} = \langle b_1, b_2, b_3 \rangle$ is

$$\mathbf{a} \cdot \mathbf{b} = a_1 b_1 + a_2 b_2 + a_3 b_3. \qquad (2)$$

EXAMPLE 1 Dot Product Using (2)

If $\mathbf{a} = 10\mathbf{i} + 2\mathbf{j} - 6\mathbf{k}$ and $\mathbf{b} = -\frac{1}{2}\mathbf{i} + 4\mathbf{j} - 3\mathbf{k}$, then it follows from (2) that

$$\mathbf{a} \cdot \mathbf{b} = (10)\left(-\frac{1}{2}\right) + (2)(4) + (-6)(-3) = 21. \qquad \blacksquare$$

EXAMPLE 2 Dot Products of the Basis Vectors

Since $\mathbf{i} = \langle 1, 0, 0 \rangle$, $\mathbf{j} = \langle 0, 1, 0 \rangle$, and $\mathbf{k} = \langle 0, 0, 1 \rangle$, we see from (2) that

$$\mathbf{i} \cdot \mathbf{j} = \mathbf{j} \cdot \mathbf{i} = 0, \quad \mathbf{j} \cdot \mathbf{k} = \mathbf{k} \cdot \mathbf{j} = 0, \quad \text{and} \quad \mathbf{k} \cdot \mathbf{i} = \mathbf{i} \cdot \mathbf{k} = 0. \qquad (3)$$

Similarly, by (2)

$$\mathbf{i} \cdot \mathbf{i} = 1, \quad \mathbf{j} \cdot \mathbf{j} = 1, \quad \text{and} \quad \mathbf{k} \cdot \mathbf{k} = 1. \qquad (4) \ \blacksquare$$

▌ **Properties** The dot product possesses the following properties.

Theorem 11.3.1 Properties of the Dot Product

(*i*) $\mathbf{a} \cdot \mathbf{b} = 0$ if $\mathbf{a} = \mathbf{0}$ or $\mathbf{b} = \mathbf{0}$
(*ii*) $\mathbf{a} \cdot \mathbf{b} = \mathbf{b} \cdot \mathbf{a}$ ← commutative law
(*iii*) $\mathbf{a} \cdot (\mathbf{b} + \mathbf{c}) = \mathbf{a} \cdot \mathbf{b} + \mathbf{a} \cdot \mathbf{c}$ ← distributive law
(*iv*) $\mathbf{a} \cdot (k\mathbf{b}) = (k\mathbf{a}) \cdot \mathbf{b} = k(\mathbf{a} \cdot \mathbf{b})$, k a scalar
(*v*) $\mathbf{a} \cdot \mathbf{a} \geq 0$
(*vi*) $\mathbf{a} \cdot \mathbf{a} = |\mathbf{a}|^2$

PROOF We prove parts (*iii*) and (*vi*). The remaining proofs are left for the student. See Problem 53 in Exercises 11.3. To prove part (*iii*) we let $\mathbf{a} = \langle a_1, a_2, a_3 \rangle$, $\mathbf{b} = \langle b_1, b_2, b_3 \rangle$, and $\mathbf{c} = \langle c_1, c_2, c_3 \rangle$. Then

$$\begin{aligned}
\mathbf{a} \cdot (\mathbf{b} + \mathbf{c}) &= \langle a_1, a_2, a_3 \rangle \cdot \left(\langle b_1, b_2, b_3 \rangle + \langle c_1, c_2, c_3 \rangle \right) \\
&= \langle a_1, a_2, a_3 \rangle \cdot \langle b_1 + c_1, b_2 + c_2, b_3 + c_3 \rangle \\
&= a_1(b_1 + c_1) + a_2(b_2 + c_2) + a_3(b_3 + c_3) \\
&= a_1 b_1 + a_1 c_1 + a_2 b_2 + a_2 c_2 + a_3 b_3 + a_3 c_3 \qquad \leftarrow \left\{ \begin{array}{l} \text{since multiplication} \\ \text{of real numbers is} \\ \text{distributive over addition} \end{array} \right. \\
&= (a_1 b_1 + a_2 b_2 + a_3 b_3) + (a_1 c_1 + a_2 c_2 + a_3 c_3) \\
&= \mathbf{a} \cdot \mathbf{b} + \mathbf{a} \cdot \mathbf{c}.
\end{aligned}$$

To prove part (*vi*) we note that

$$\mathbf{a} \cdot \mathbf{a} = \langle a_1, a_2, a_3 \rangle \cdot \langle a_1, a_2, a_3 \rangle = a_1^2 + a_2^2 + a_3^2 = |\mathbf{a}|^2. \qquad \blacksquare$$

▌ **Alternative Form** The dot product of two vectors can also be expressed in terms of the lengths of the vectors and the angle between them.

Theorem 11.3.2 Alternative Form of the Dot Product
The dot product of two vectors **a** and **b** is

$$\mathbf{a} \cdot \mathbf{b} = |\mathbf{a}||\mathbf{b}|\cos\theta, \qquad (5)$$

where θ is the angle between the vectors such that $0 \leq \theta \leq \pi$.

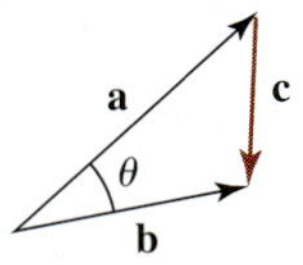

This more geometric form is what is generally used as the definition of the dot product in a physics course.

PROOF Suppose θ is the angle between the vectors $\mathbf{a} = a_1\mathbf{i} + a_2\mathbf{j} + a_3\mathbf{k}$ and $\mathbf{b} = b_1\mathbf{i} + b_2\mathbf{j} + b_3\mathbf{k}$. Then the vector

$$\mathbf{c} = \mathbf{b} - \mathbf{a} = (b_1 - a_1)\mathbf{i} + (b_2 - a_2)\mathbf{j} + (b_3 - a_3)\mathbf{k}$$

is the third side of the triangle indicated in FIGURE 11.3.1. By the law of cosines we can write

$$|\mathbf{c}|^2 = |\mathbf{b}|^2 + |\mathbf{a}|^2 - 2|\mathbf{a}||\mathbf{b}|\cos\theta \quad \text{or} \quad |\mathbf{a}||\mathbf{b}|\cos\theta = \tfrac{1}{2}(|\mathbf{b}|^2 + |\mathbf{a}|^2 - |\mathbf{c}|^2). \qquad (6)$$

Using
$$|\mathbf{a}|^2 = a_1^2 + a_2^2 + a_3^2, \qquad |\mathbf{b}|^2 = b_1^2 + b_2^2 + b_3^2,$$

and
$$|\mathbf{c}|^2 = |\mathbf{b} - \mathbf{a}|^2 = (b_1 - a_1)^2 + (b_2 - a_2)^2 + (b_3 - a_3)^2,$$

we can simplify the right side of the equation in (6) to $a_1b_1 + a_2b_2 + a_3b_3$. Since this is the definition of the dot product, we see that $|\mathbf{a}||\mathbf{b}|\cos\theta = \mathbf{a} \cdot \mathbf{b}$. ∎

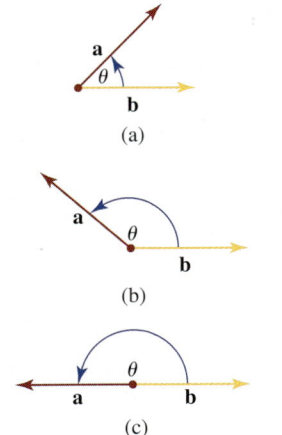

FIGURE 11.3.1 The vector **c** in the proof of Theorem 11.3.2

FIGURE 11.3.2 The angle θ in the dot product

Angle Between Vectors FIGURE 11.3.2 illustrates three cases of the angle θ in (5). If the vectors **a** and **b** are not parallel, then θ is the *smaller* of the two possible angles between them. Solving for $\cos\theta$ in (5) and then using the definition of the dot product in (2) we have a formula for the cosine of the angle between two vectors:

$$\cos\theta = \frac{\mathbf{a} \cdot \mathbf{b}}{|\mathbf{a}||\mathbf{b}|} = \frac{a_1b_1 + a_2b_2 + a_3b_3}{|\mathbf{a}||\mathbf{b}|}. \qquad (7)$$

EXAMPLE 3 Angle Between Two Vectors

Find the angle between $\mathbf{a} = 2\mathbf{i} + 3\mathbf{j} + \mathbf{k}$ and $\mathbf{b} = -\mathbf{i} + 5\mathbf{j} + \mathbf{k}$.

Solution We have $|\mathbf{a}| = \sqrt{14}$, $|\mathbf{b}| = \sqrt{27}$, and $\mathbf{a} \cdot \mathbf{b} = 14$. Hence, (7) gives

$$\cos\theta = \frac{14}{\sqrt{14}\sqrt{27}} = \frac{1}{9}\sqrt{42},$$

and so $\theta = \cos^{-1}(\sqrt{42}/9) \approx 0.77$ radian or $\theta \approx 44.9°$. ∎

Orthogonal Vectors If **a** and **b** are nonzero vectors, then Theorem 11.3.2 implies that

(*i*) $\mathbf{a} \cdot \mathbf{b} > 0$ if and only if θ is acute,
(*ii*) $\mathbf{a} \cdot \mathbf{b} < 0$ if any only if θ is obtuse, and
(*iii*) $\mathbf{a} \cdot \mathbf{b} = 0$ is and only if $\cos\theta = 0$.

But in the last case, the only number in $[0, 2\pi]$ for which $\cos\theta = 0$ is $\theta = \pi/2$. When $\theta = \pi/2$, we say that the vectors are **orthogonal** or **perpendicular**. Thus, we are led to the following result.

The words *orthogonal* and *perpendicular* are used interchangeably. As a general rule we will use *orthogonal* when referring to vectors and *perpendicular* when a line or plane is involved.

Theorem 11.3.3 Criterion for Orthogonal Vectors
Two nonzero vectors **a** and **b** are orthogonal if and only if $\mathbf{a} \cdot \mathbf{b} = 0$.

Since $\mathbf{0} \cdot \mathbf{b} = 0$ for every vector **b**, the zero vector is regarded to be orthogonal to every vector.

EXAMPLE 4 Orthogonal Vectors

If $\mathbf{a} = -3\mathbf{i} - \mathbf{j} + 4\mathbf{k}$ and $\mathbf{b} = 2\mathbf{i} + 14\mathbf{j} + 5\mathbf{k}$, then

$$\mathbf{a} \cdot \mathbf{b} = (-3)(2) + (-1)(14) + (4)(5) = 0.$$

From Theorem 11.3.3, we conclude that $\mathbf{a}$ and $\mathbf{b}$ are orthogonal. ∎

▌ Direction Cosines For a nonzero vector $\mathbf{a} = a_1\mathbf{i} + a_2\mathbf{j} + a_3\mathbf{k}$ in 3-space, the angles α, β, and γ between $\mathbf{a}$ and the unit vectors $\mathbf{i}$, $\mathbf{j}$, and $\mathbf{k}$, respectively, are called **direction angles** of $\mathbf{a}$. See FIGURE 11.3.3. Now, by (7),

$$\cos\alpha = \frac{\mathbf{a} \cdot \mathbf{i}}{|\mathbf{a}||\mathbf{i}|}, \quad \cos\beta = \frac{\mathbf{a} \cdot \mathbf{j}}{|\mathbf{a}||\mathbf{j}|}, \quad \cos\gamma = \frac{\mathbf{a} \cdot \mathbf{k}}{|\mathbf{a}||\mathbf{k}|},$$

which simplify to

$$\cos\alpha = \frac{a_1}{|\mathbf{a}|}, \quad \cos\beta = \frac{a_2}{|\mathbf{a}|}, \quad \cos\gamma = \frac{a_3}{|\mathbf{a}|}.$$

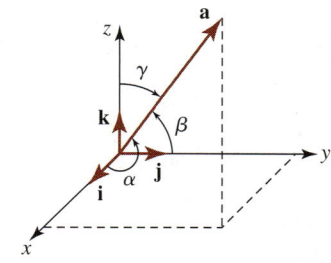

FIGURE 11.3.3 The direction angles of a vector

We say that $\cos\alpha$, $\cos\beta$, $\cos\gamma$ are the **direction cosines** of $\mathbf{a}$. The direction cosines of a nonzero vector $\mathbf{a}$ are simply the components of the unit vector $\mathbf{a}/|\mathbf{a}|$:

$$\frac{\mathbf{a}}{|\mathbf{a}|} = \frac{a_1}{|\mathbf{a}|}\mathbf{i} + \frac{a_2}{|\mathbf{a}|}\mathbf{j} + \frac{a_3}{|\mathbf{a}|}\mathbf{k}$$
$$= (\cos\alpha)\mathbf{i} + (\cos\beta)\mathbf{j} + (\cos\gamma)\mathbf{k}.$$

Since the magnitude of $\mathbf{a}/|\mathbf{a}|$ is 1, it follows from the last equation that

$$\cos^2\alpha + \cos^2\beta + \cos^2\gamma = 1.$$

EXAMPLE 5 Direction Cosines and Direction Angles

Find the direction cosines and direction angles of the vector $\mathbf{a} = 2\mathbf{i} + 5\mathbf{j} + 4\mathbf{k}$.

Solution From $|\mathbf{a}| = \sqrt{2^2 + 5^2 + 4^2} = \sqrt{45} = 3\sqrt{5}$, we see that the direction cosines are

$$\cos\alpha = \frac{2}{3\sqrt{5}}, \quad \cos\beta = \frac{5}{3\sqrt{5}}, \quad \cos\gamma = \frac{4}{3\sqrt{5}}.$$

The direction angles are

$$\alpha = \cos^{-1}\left(\frac{2}{3\sqrt{5}}\right) \approx 1.27 \text{ radians} \quad \text{or} \quad \alpha \approx 72.7°$$

$$\beta = \cos^{-1}\left(\frac{5}{3\sqrt{5}}\right) \approx 0.73 \text{ radian} \quad \text{or} \quad \beta \approx 41.8°$$

and

$$\gamma = \cos^{-1}\left(\frac{4}{3\sqrt{5}}\right) \approx 0.93 \text{ radian} \quad \text{or} \quad \gamma \approx 53.4°.$$ ∎

Observe in Example 5 that

$$\cos^2\alpha + \cos^2\beta + \cos^2\gamma = \frac{4}{45} + \frac{25}{45} + \frac{16}{45} = 1.$$

▌ Component of a on b Using the distributive law together with (3) and (4) enables us to express the components of a vector $\mathbf{a} = a_1\mathbf{i} + a_2\mathbf{j} + a_3\mathbf{k}$ in terms of the dot product:

$$a_1 = \mathbf{a} \cdot \mathbf{i}, \quad a_2 = \mathbf{a} \cdot \mathbf{j}, \quad a_3 = \mathbf{a} \cdot \mathbf{k}. \tag{8}$$

Symbolically, we write the components of $\mathbf{a}$ as

$$\text{comp}_\mathbf{i}\mathbf{a} = \mathbf{a} \cdot \mathbf{i}, \quad \text{comp}_\mathbf{j}\mathbf{a} = \mathbf{a} \cdot \mathbf{j}, \quad \text{comp}_\mathbf{k}\mathbf{a} = \mathbf{a} \cdot \mathbf{k}. \tag{9}$$

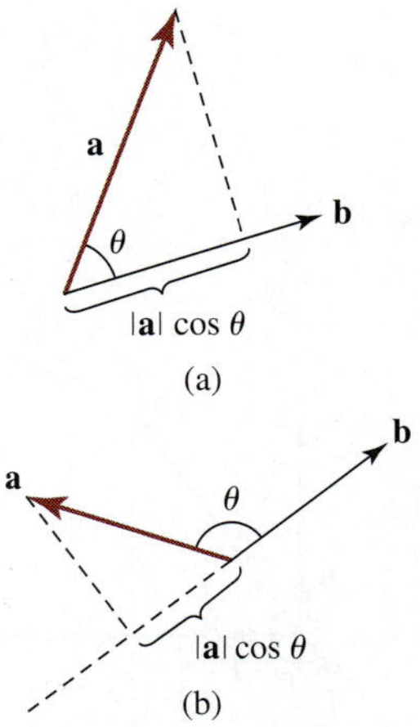

We shall now see that the procedure indicated in (9) carries over to finding the **component of a on a vector b**. Note that in either of the two cases shown in FIGURE 11.3.4,

$$\text{comp}_b \mathbf{a} = |\mathbf{a}| \cos \theta. \tag{10}$$

In Figure 11.3.4(b), $\text{comp}_b \mathbf{a} < 0$, since $\pi/2 < \theta \le \pi$. Now, by writing (10) as

$$\text{comp}_b \mathbf{a} = \frac{|\mathbf{a}||\mathbf{b}|\cos\theta}{|\mathbf{b}|} = \frac{\mathbf{a}\cdot\mathbf{b}}{|\mathbf{b}|},$$

we see that

$$\text{comp}_b \mathbf{a} = \mathbf{a} \cdot \left(\frac{\mathbf{b}}{|\mathbf{b}|}\right). \tag{11}$$

In other words:

- *To find the component of vector **a** on vector **b**, we dot **a** with a unit vector in the direction of **b**.*

EXAMPLE 6 Component of a Vector on Another

Let $\mathbf{a} = 2\mathbf{i} + 3\mathbf{j} - 4\mathbf{k}$ and $\mathbf{b} = \mathbf{i} + \mathbf{j} + 2\mathbf{k}$. Find

(a) $\text{comp}_b \mathbf{a}$ and (b) $\text{comp}_a \mathbf{b}$.

Solution

(a) We first form a unit vector in the direction of **b**:

$$|\mathbf{b}| = \sqrt{6} \quad \text{so} \quad \frac{\mathbf{b}}{|\mathbf{b}|} = \frac{1}{\sqrt{6}}(\mathbf{i} + \mathbf{j} + 2\mathbf{k}).$$

Then from (11) we have

$$\text{comp}_b \mathbf{a} = (2\mathbf{i} + 3\mathbf{j} - 4\mathbf{k}) \cdot \frac{1}{\sqrt{6}}(\mathbf{i} + \mathbf{j} + 2\mathbf{k}) = -\frac{3}{\sqrt{6}}.$$

(b) By modifying (11) accordingly, we have

$$\text{comp}_a \mathbf{b} = \mathbf{b} \cdot \left(\frac{\mathbf{a}}{|\mathbf{a}|}\right).$$

Then

$$|\mathbf{a}| = \sqrt{29} \quad \text{so} \quad \frac{\mathbf{a}}{|\mathbf{a}|} = \frac{1}{\sqrt{29}}(2\mathbf{i} + 3\mathbf{j} - 4\mathbf{k}),$$

and

$$\text{comp}_a \mathbf{b} = (\mathbf{i} + \mathbf{j} + 2\mathbf{k}) \cdot \frac{1}{\sqrt{29}}(2\mathbf{i} + 3\mathbf{j} - 4\mathbf{k}) = -\frac{3}{\sqrt{29}}. \qquad \blacksquare$$

Projection of a onto b As illustrated in FIGURE 11.3.5(a), the projection of a vector **a** in any of the directions determined by **i**, **j**, **k** is simply the *vector* formed by multiplying the component of **a** in the specified direction with a unit vector in that direction; for example,

$$\text{proj}_i \mathbf{a} = (\text{comp}_i \mathbf{a})\mathbf{i} = (\mathbf{a} \cdot \mathbf{i})\mathbf{i} = a_1 \mathbf{i},$$

and so on. Figure 11.3.5(b) shows the general case of the **projection of a onto b**:

$$\text{proj}_b \mathbf{a} = (\text{comp}_b \mathbf{a})\left(\frac{\mathbf{b}}{|\mathbf{b}|}\right). \tag{12}$$

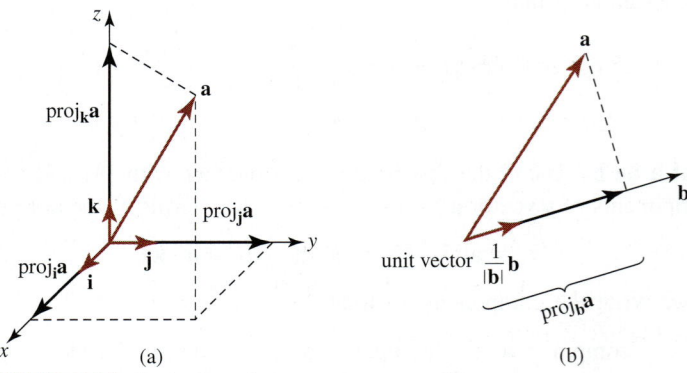

FIGURE 11.3.5 Projection of vector **a** onto vector **b**

FIGURE 11.3.4 Component of vector **a** on vector **b**

EXAMPLE 7   Projection of **a** onto **b**

Find the projection of $\mathbf{a} = 4\mathbf{i} + \mathbf{j}$ onto the vector $\mathbf{b} = 2\mathbf{i} + 3\mathbf{j}$. Graph.

Solution   First, we find the component of **a** on **b**. Since $|\mathbf{b}| = \sqrt{13}$, we find from (11),

$$\text{comp}_{\mathbf{b}}\mathbf{a} = (4\mathbf{i} + \mathbf{j}) \cdot \frac{1}{\sqrt{13}}(2\mathbf{i} + 3\mathbf{j}) = \frac{11}{\sqrt{13}}.$$

Thus, from (12),

$$\text{proj}_{\mathbf{b}}\mathbf{a} = \left(\frac{11}{\sqrt{13}}\right)\left(\frac{1}{\sqrt{13}}\right)(2\mathbf{i} + 3\mathbf{j}) = \frac{22}{13}\mathbf{i} + \frac{33}{13}\mathbf{j}.$$ ■

The graph of this vector is shown in FIGURE 11.3.6.

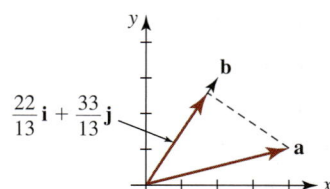

FIGURE 11.3.6   Projection of **a** onto **b**

▌ **Projection of a Orthogonal to b**   As seen in FIGURE 11.3.7, the vectors **a** and $\text{proj}_{\mathbf{b}}\mathbf{a}$ are the hypotenuse and one leg of a right triangle, respectively. The second leg of the triangle is then

$$\mathbf{a} - \text{proj}_{\mathbf{b}}\mathbf{a}.$$

This is a vector that is orthogonal to **b** and is called the **projection of a orthogonal to b**.

EXAMPLE 8   Projection of **a** Orthogonal to **b**

Let $\mathbf{a} = 3\mathbf{i} - \mathbf{j} + 5\mathbf{k}$ and $\mathbf{b} = 2\mathbf{i} + \mathbf{j} + 2\mathbf{k}$. Find the projection of **a** orthogonal to **b**.

Solution   We first find the projection of **a** onto **b**. Since $|\mathbf{b}| = 3$, we have by (11) that

$$\text{comp}_{\mathbf{b}}\mathbf{a} = (3\mathbf{i} - \mathbf{j} + 5\mathbf{k}) \cdot \frac{1}{3}(2\mathbf{i} + \mathbf{j} + 2\mathbf{k}) = 5,$$

so, using (12),

$$\text{proj}_{\mathbf{b}}\mathbf{a} = (5)\left(\frac{1}{3}\right)(2\mathbf{i} + \mathbf{j} + 2\mathbf{k}) = \frac{10}{3}\mathbf{i} + \frac{5}{3}\mathbf{j} + \frac{10}{3}\mathbf{k}.$$

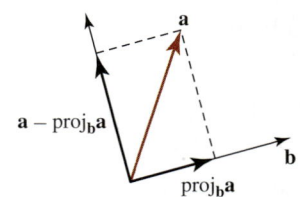

FIGURE 11.3.7   The vector $\mathbf{a} - \text{proj}_{\mathbf{b}}\mathbf{a}$ is orthogonal to **b**

Then, the projection of **a** orthogonal to **b** is

$$\mathbf{a} - \text{proj}_{\mathbf{b}}\mathbf{a} = (3\mathbf{i} - \mathbf{j} + 5\mathbf{k}) - \left(\frac{10}{3}\mathbf{i} + \frac{5}{3}\mathbf{j} + \frac{10}{3}\mathbf{k}\right) = -\frac{1}{3}\mathbf{i} - \frac{8}{3}\mathbf{j} + \frac{5}{3}\mathbf{k}.$$ ■

▌ **Physical Interpretation of the Dot Product**   In Section 6.8 we saw that when a constant force of magnitude F moves an object a distance d in the same direction of the force, the work done is simply

$$W = Fd. \tag{13}$$

However, if a constant force **F** applied to a body acts at an angle θ to the direction of motion, then the work done by **F** is defined to be the product of the component of **F** in the direction of the displacement and the distance $|\mathbf{d}|$ that the body moves:

$$W = (|\mathbf{F}|\cos\theta)|\mathbf{d}| = |\mathbf{F}||\mathbf{d}|\cos\theta.$$

See FIGURE 11.3.8. It follows from Theorem 11.3.2 that if **F** causes a displacement **d** of a body, then the work done is

$$W = \mathbf{F} \cdot \mathbf{d}. \tag{14}$$

Note that (14) reduces to (13) when $\theta = 0$.

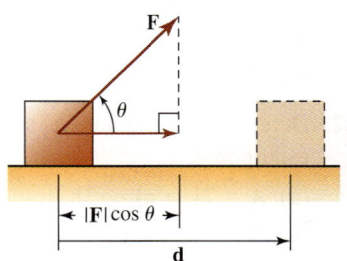

FIGURE 11.3.8   Work done by a force acting at an angle θ to the direction of motion

EXAMPLE 9   Work Done by a Force at an Angle

Find the work done by a constant force $\mathbf{F} = 2\mathbf{i} + 4\mathbf{j}$ on a block that moves from $P_1(1, 1)$ to $P_2(4, 6)$. Assume that $|\mathbf{F}|$ is measured in pounds and $|\mathbf{d}|$ is measured in feet.

Solution   The displacement of the block is given by

$$\mathbf{d} = \overrightarrow{P_1P_2} = \overrightarrow{OP_2} - \overrightarrow{OP_1} = 3\mathbf{i} + 5\mathbf{j}$$

It follows from (14) that the work done is

$$W = (2\mathbf{i} + 4\mathbf{j}) \cdot (3\mathbf{i} + 5\mathbf{j}) = 26 \text{ ft-lb.}$$ ■

Exercises 11.3 Answers to selected odd-numbered problems begin on page ANS-36.

≡ Fundamentals

In Problems 1–12, $\mathbf{a} = 2\mathbf{i} - 3\mathbf{j} + 4\mathbf{k}$, $\mathbf{b} = -\mathbf{i} + 2\mathbf{j} + 5\mathbf{k}$, and $\mathbf{c} = 3\mathbf{i} + 6\mathbf{j} - \mathbf{k}$. Find the indicated vector or scalar.

1. $\mathbf{a} \cdot \mathbf{b}$
2. $\mathbf{b} \cdot \mathbf{c}$
3. $\mathbf{a} \cdot \mathbf{c}$
4. $\mathbf{a} \cdot (\mathbf{b} + \mathbf{c})$
5. $\mathbf{a} \cdot (4\mathbf{b})$
6. $\mathbf{b} \cdot (\mathbf{a} - \mathbf{c})$
7. $\mathbf{a} \cdot \mathbf{a}$
8. $(2\mathbf{b}) \cdot (3\mathbf{c})$
9. $\mathbf{a} \cdot (\mathbf{a} + \mathbf{b} + \mathbf{c})$
10. $(2\mathbf{a}) \cdot (\mathbf{a} - 2\mathbf{b})$
11. $\left(\dfrac{\mathbf{a} \cdot \mathbf{b}}{\mathbf{b} \cdot \mathbf{b}}\right)\mathbf{b}$
12. $(\mathbf{c} \cdot \mathbf{b})\mathbf{a}$

In Problems 13–16, find $\mathbf{a} \cdot \mathbf{b}$ if the smaller angle between $\mathbf{a}$ and $\mathbf{b}$ is as given.

13. $|\mathbf{a}| = 10$, $|\mathbf{b}| = 5$, $\theta = \pi/4$
14. $|\mathbf{a}| = 6$, $|\mathbf{b}| = 12$, $\theta = \pi/6$
15. $|\mathbf{a}| = 2$, $|\mathbf{b}| = 3$, $\theta = 2\pi/3$
16. $|\mathbf{a}| = 4$, $|\mathbf{b}| = 1$, $\theta = 5\pi/6$

In Problems 17–20, find the angle θ between the given vectors.

17. $\mathbf{a} = 3\mathbf{i} - \mathbf{k}$, $\mathbf{b} = 2\mathbf{i} + 2\mathbf{k}$
18. $\mathbf{a} = 2\mathbf{i} + \mathbf{j}$, $\mathbf{b} = -3\mathbf{i} - 4\mathbf{j}$
19. $\mathbf{a} = \langle 2, 4, 0\rangle$, $\mathbf{b} = \langle -1, -1, 4\rangle$
20. $\mathbf{a} = \langle \frac{1}{2}, \frac{1}{2}, \frac{3}{2}\rangle$, $\mathbf{b} = \langle 2, -4, 6\rangle$

21. Determine which pairs of the following vectors are orthogonal.
    (a) $\langle 2, 0, 1\rangle$
    (b) $3\mathbf{i} + 2\mathbf{j} - \mathbf{k}$
    (c) $2\mathbf{i} - \mathbf{j} - \mathbf{k}$
    (d) $\mathbf{i} - 4\mathbf{j} + 6\mathbf{k}$
    (e) $\langle 1, -1, 1\rangle$
    (f) $\langle -4, 3, 8\rangle$

22. Determine a scalar c so that the given vectors are orthogonal.
    (a) $\mathbf{a} = 2\mathbf{i} - c\mathbf{j} + 3\mathbf{k}$, $\mathbf{b} = 3\mathbf{i} + 2\mathbf{j} + 4\mathbf{k}$
    (b) $\mathbf{a} = \langle c, \frac{1}{2}, c\rangle$, $\mathbf{b} = \langle -3, 4, c\rangle$

23. Find a vector $\mathbf{v} = \langle x_1, y_1, 1\rangle$ that is orthogonal to both $\mathbf{a} = \langle 3, 1, -1\rangle$ and $\mathbf{b} = \langle -3, 2, 2\rangle$.

24. A **rhombus** is an oblique-angled parallelogram with all four sides equal. Use the dot product to show that the diagonals of a rhombus are perpendicular.

25. Verify that the vector

$$\mathbf{c} = \mathbf{b} - \frac{\mathbf{a} \cdot \mathbf{b}}{|\mathbf{a}|^2}\mathbf{a}$$

is orthogonal to the vector $\mathbf{a}$.

26. Determine a scalar c so that the angle between $\mathbf{a} = \mathbf{i} + c\mathbf{j}$ and $\mathbf{b} = \mathbf{i} + \mathbf{j}$ is 45°.

In Problems 27–30, find the direction cosines and direction angles of the given vector.

27. $\mathbf{a} = \mathbf{i} + 2\mathbf{j} + 3\mathbf{k}$
28. $\mathbf{a} = 6\mathbf{i} + 6\mathbf{j} - 3\mathbf{k}$
29. $\mathbf{a} = \langle 1, 0, -\sqrt{3}\rangle$
30. $\mathbf{a} = \langle 5, 7, 2\rangle$

31. Find the angle between the diagonal $\overrightarrow{AD}$ of the cube shown in FIGURE 11.3.9 and the edge $\overrightarrow{AB}$. Find the angle between the diagonal $\overrightarrow{AD}$ and the diagonal $\overrightarrow{AC}$.

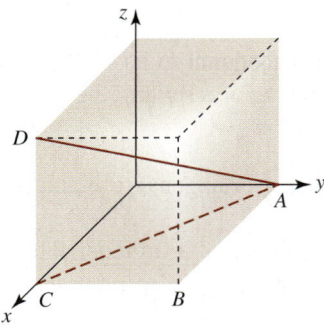

FIGURE 11.3.9 Cube in Problem 31

32. An airplane is 4 km high, 5 km south, and 7 km east of an airport. See FIGURE 11.3.10. Find the direction angles of the plane.

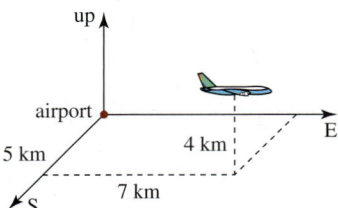

FIGURE 11.3.10 Airplane in Problem 32

In Problems 33–36, $\mathbf{a} = \mathbf{i} - \mathbf{j} + 3\mathbf{k}$ and $\mathbf{b} = 2\mathbf{i} + 6\mathbf{j} + 3\mathbf{k}$. Find the indicated number.

33. $\text{comp}_\mathbf{b}\mathbf{a}$
34. $\text{comp}_\mathbf{a}\mathbf{b}$
35. $\text{comp}_\mathbf{a}(\mathbf{b} - \mathbf{a})$
36. $\text{comp}_{2\mathbf{b}}(\mathbf{a} + \mathbf{b})$

In Problems 37 and 38, find the component of the given vector in the direction from the origin to the indicated point.

37. $\mathbf{a} = 4\mathbf{i} + 6\mathbf{j}$; $P(3, 10)$
38. $\mathbf{a} = \langle 2, 1, -1\rangle$; $P(1, -1, 1)$

In Problems 39–42, find (a) $\text{proj}_\mathbf{b}\mathbf{a}$ and (b) the projection of $\mathbf{a}$ orthogonal to $\mathbf{b}$.

39. $\mathbf{a} = -5\mathbf{i} + 5\mathbf{j}$, $\mathbf{b} = -3\mathbf{i} + 4\mathbf{j}$
40. $\mathbf{a} = 4\mathbf{i} + 2\mathbf{j}$, $\mathbf{b} = -3\mathbf{i} + \mathbf{j}$
41. $\mathbf{a} = \langle -1, -2, 7\rangle$, $\mathbf{b} = \langle 6, -3, -2\rangle$
42. $\mathbf{a} = \langle 1, 1, 1\rangle$, $\mathbf{b} = \langle -2, 2, -1\rangle$

In Problems 43 and 44, $\mathbf{a} = 4\mathbf{i} + 3\mathbf{j}$ and $\mathbf{b} = -\mathbf{i} + \mathbf{j}$. Find the indicated vector.

43. $\text{proj}_{(\mathbf{a} + \mathbf{b})}\mathbf{a}$
44. projection of $\mathbf{b}$ orthogonal to $\mathbf{a} - \mathbf{b}$

☰ Applications

45. A sled is pulled horizontally over ice by a rope attached to its front. A 20-lb force acting at an angle of 60° with the horizontal moves the sled 100 ft. Find the work done.

46. A train is pushed along a straight track by a force of 3000 lb acting at a 45° angle to the direction of motion. Find the work done in moving the train 400 ft.

47. Find the work done by a constant force $\mathbf{F} = 4\mathbf{i} + 3\mathbf{j} + 5\mathbf{k}$ that moves an object from $P_1(3, 1, -2)$ to $P_2(2, 4, 6)$. Assume that $|\mathbf{F}|$ is measured in newtons and $|\mathbf{d}|$ is measured in meters.

48. A block with weight $\mathbf{w}$ is pulled along a frictionless horizontal surface by a constant force $\mathbf{F}$ of magnitude 30 newtons in the direction given by a vector $\mathbf{d}$. See FIGURE 11.3.11. Assume $|\mathbf{d}|$ is measured in meters.

(a) What is the work done by the weight $\mathbf{w}$?
(b) What is the work done by the force $\mathbf{F}$ if $\mathbf{d} = 4\mathbf{i} + 3\mathbf{j}$?

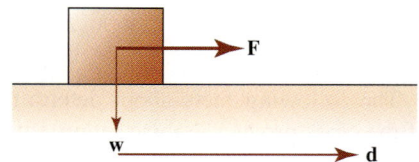

FIGURE 11.3.11   Block in Problem 48

49. A constant force $\mathbf{F}$ of magnitude 3 lb is applied to the block shown in FIGURE 11.3.12. $\mathbf{F}$ has the same direction as the vector $\mathbf{a} = 3\mathbf{i} + 4\mathbf{j}$. Find the work done in the direction of motion if the block moves from $P_1(3, 1)$ to $P_2(9, 3)$. Assume distance is measured in feet.

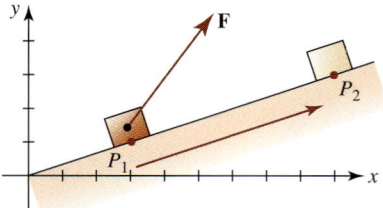

FIGURE 11.3.12   Block in Problem 49

50. The methane molecule CH_4 consists of four hydrogen atoms surrounding a single carbon atom. As shown in FIGURE 11.3.13, hydrogen atoms are located at the vertices of a regular tetrahedron. The distance between the center of a hydrogen atom and the center of a carbon atom is 1.10 angstroms (1 angstrom = 10^{-10} m) and the hydrogen–carbon–hydrogen bond angle is $\theta = 109.5°$. Using vector methods only, find the distance between two hydrogen atoms.

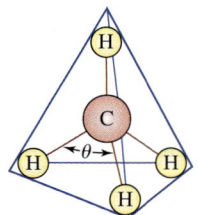

FIGURE 11.3.13   Atoms in the methane molecule in Problem 50

☰ Think About It

51. Show that if two nonzero vectors $\mathbf{a}$ and $\mathbf{b}$ are orthogonal, then their direction cosines satisfy

$$\cos\alpha_1 \cos\alpha_2 + \cos\beta_1 \cos\beta_2 + \cos\gamma_1 \cos\gamma_2 = 0.$$

52. Determine a unit vector whose direction angles, relative to the three coordinate axes, are equal.

53. Use the definition of the dot product to prove parts (*i*), (*ii*), (*iv*), and (*v*) of Theorem 11.3.1.

54. Use the dot product to prove the **Cauchy-Schwarz inequality**: $|\mathbf{a} \cdot \mathbf{b}| \le |\mathbf{a}||\mathbf{b}|$.

55. Use the dot product to prove the **triangle inequality**: $|\mathbf{a} + \mathbf{b}| \le |\mathbf{a}| + |\mathbf{b}|$. [*Hint*: Consider property (*vi*) in Theorem 11.3.1.]

56. Prove that the vector $\mathbf{n} = a\mathbf{i} + b\mathbf{j}$ is perpendicular to the line whose equation is $ax + by + c = 0$. [*Hint*: Let $P_1(x_1, y_1)$ and $P_2(x_2, y_2)$ be distinct points on the line.]

57. Use the result of Problem 56 and FIGURE 11.3.14 to show that the distance d from a point $P_1(x_1, y_1)$ to a line $ax + by + c = 0$ is $d = |ax_1 + by_1 + c|/\sqrt{a^2 + b^2}$.

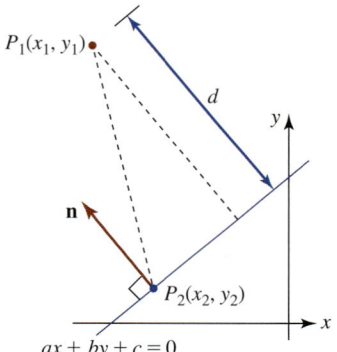

FIGURE 11.3.14   Distance from a point to a line in Problem 57

☰ Projects

58. Light from a source at point $S(a, b)$ is reflected by a spherical mirror of radius 1, centered at the origin, to an observer located at point $O(c, d)$ as shown in FIGURE 11.3.15. The point of reflection $P(x, y)$ from a spherical mirror lies in the plane determined by the source, the observer, and the center of the sphere. (The analysis of spherical mirrors occurs, among other places, in the study of radar design.)

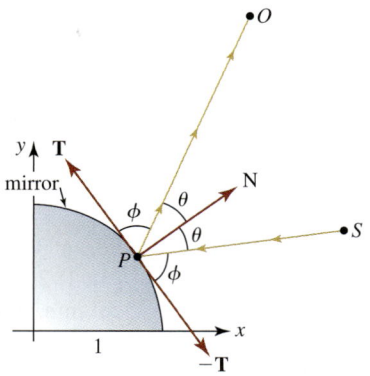

FIGURE 11.3.15   Mirror in Problem 58

(a) Use Theorem 11.3.2 twice, once with the angle θ and once with the angle ϕ, to show that the coordinates of the point of reflection $P(x, y)$ satisfy the equation

$$\frac{ax + by - 1}{ay - bx} = \frac{cx + dy - 1}{dx - cy}.$$

[*Hint*: As shown in the figure, let $\mathbf{N}$ and $\mathbf{T}$ denote, respectively, a unit normal vector and a unit tangent to the circle at $P(x, y)$. If $\mathbf{N} = x\mathbf{i} + y\mathbf{j}$, what is $\mathbf{T}$ in terms of x and y?]

(b) Let $a = 2$, $b = 0$, $c = 0$, and $d = 3$. Use the relationship $x^2 + y^2 = 1$ to show that the x-coordinate of the point of reflection is a root of a fourth-degree polynomial equation.

(c) Use Newton's Method or a CAS to find the point of reflection in part (b). You may have to consider all four roots of the equation in part (b) to find the one that corresponds to a solution of the equation in part (a).

11.4 Cross Product

■ **Introduction** The dot product, introduced in the preceding section, works in both two- and three-dimensional spaces and results in a *number*. On the other hand, the **cross product**, introduced in this section, is only defined for vectors in 3-space and results in another *vector* in 3-space.

■ **Second- and Third-Order Determinants** The following facts about determinants will be important in the definition and discussion of the cross product in this section.

Review of Determinants

The definition of a **second-order determinant** is the number

$$\begin{vmatrix} a_1 & a_2 \\ b_1 & b_2 \end{vmatrix} = a_1 b_2 - a_2 b_1.$$

A **third-order determinant** is defined in terms of three second-order determinants as follows:

$$\begin{vmatrix} a_1 & a_2 & a_3 \\ b_1 & b_2 & b_3 \\ c_1 & c_2 & c_3 \end{vmatrix} = a_1 \begin{vmatrix} b_2 & b_3 \\ c_2 & c_3 \end{vmatrix} - a_2 \begin{vmatrix} b_1 & b_3 \\ c_1 & c_3 \end{vmatrix} + a_3 \begin{vmatrix} b_1 & b_2 \\ c_1 & c_2 \end{vmatrix}.$$

This is called **expanding the determinant by cofactors** of the first row.

Even though a determinant is a *number* it is convenient to think of it as a square array. Thus, second-order and third-order determinants are referred to as 2×2 and 3×3 **determinants**, respectively. There are higher-order determinants, but since we will not encounter them in the remaining chapters of this text their definitions will not be given.

Read as a "two by two" determinant. ▶ To find the value of a 2×2 determinant we compute the products of the numbers on the two diagonals and subtract:

$$\begin{vmatrix} a_1 & a_2 \\ b_1 & b_2 \end{vmatrix} = a_1 b_2 - a_2 b_1.$$

For a 3×3 determinant, the **cofactor** of an entry a_{1j} in the 1st row and jth column, $j = 1, 2, 3$, is $(-1)^{1+j}$ times that 2×2 determinant formed by deleting the 1st row and jth column. The cofactors of a_1, a_2, and a_3 are, respectively,

$$\begin{vmatrix} b_2 & b_3 \\ c_2 & c_3 \end{vmatrix}, \quad -\begin{vmatrix} b_1 & b_3 \\ c_1 & c_3 \end{vmatrix}, \quad \text{and} \quad \begin{vmatrix} b_1 & b_2 \\ c_1 & c_2 \end{vmatrix}.$$

Thus:

$$\begin{vmatrix} a_1 & a_2 & a_3 \\ b_1 & b_2 & b_3 \\ c_1 & c_2 & c_3 \end{vmatrix} = a_1 \begin{vmatrix} a_1 & a_2 & a_3 \\ b_1 & b_2 & b_3 \\ c_1 & c_2 & c_3 \end{vmatrix} - a_2 \begin{vmatrix} a_1 & a_2 & a_3 \\ b_1 & b_2 & b_3 \\ c_1 & c_2 & c_3 \end{vmatrix} + a_3 \begin{vmatrix} a_1 & a_2 & a_3 \\ b_1 & b_2 & b_3 \\ c_1 & c_2 & c_3 \end{vmatrix}$$

$$= a_1 \begin{vmatrix} b_2 & b_3 \\ c_2 & c_3 \end{vmatrix} - a_2 \begin{vmatrix} b_1 & b_3 \\ c_1 & c_3 \end{vmatrix} + a_3 \begin{vmatrix} b_1 & b_2 \\ c_1 & c_2 \end{vmatrix}.$$

EXAMPLE 1 A 2 × 2 Determinant

$$\begin{vmatrix} -4 & -2 \\ 5 & 3 \end{vmatrix} = (-4)3 - (-2)5 = -2$$ ∎

EXAMPLE 2 A 3 × 3 Determinant

$$\begin{vmatrix} 8 & 5 & 4 \\ 2 & 4 & 6 \\ -1 & 2 & 3 \end{vmatrix} = 8 \begin{vmatrix} 4 & 6 \\ 2 & 3 \end{vmatrix} - 5 \begin{vmatrix} 2 & 6 \\ -1 & 3 \end{vmatrix} + 4 \begin{vmatrix} 2 & 4 \\ -1 & 2 \end{vmatrix} = 8(0) - 5(12) + 4(8) = -28$$ ∎

The following properties will be useful in the discussion that follows.

Three Properties of Determinants

(*i*) If every entry in a row (or column) of a determinant is 0, then the value of the determinant is zero.

(*ii*) If two rows (or columns) of a determinant are equal, then the value of the determinant is zero.

(*iii*) When two rows (or columns) of a determinant are interchanged, the resulting determinant is the negative of the original determinant.

■ **Component Form of the Cross Product**   As we did in the discussion of the dot product, we define the cross product of two vectors **a** and **b** in terms of the components of the vectors.

Definition 11.4.1 Cross Product of Two Vectors

The **cross product** of two vectors $\mathbf{a} = \langle a_1, a_2, a_3 \rangle$ and $\mathbf{b} = \langle b_1, b_2, b_3 \rangle$ is the vector

$$\mathbf{a} \times \mathbf{b} = (a_2 b_3 - a_3 b_2)\mathbf{i} - (a_1 b_3 - a_3 b_1)\mathbf{j} + (a_1 b_2 - a_2 b_1)\mathbf{k}. \tag{1}$$

The coefficients of the basis vectors in (1) are recognized as 2 × 2 determinants, so (1) can be written as

$$\mathbf{a} \times \mathbf{b} = \begin{vmatrix} a_2 & a_3 \\ b_2 & b_3 \end{vmatrix}\mathbf{i} - \begin{vmatrix} a_1 & a_3 \\ b_1 & b_3 \end{vmatrix}\mathbf{j} + \begin{vmatrix} a_1 & a_2 \\ b_1 & b_2 \end{vmatrix}\mathbf{k}.$$

This representation, in turn, suggests that the cross product can be written as a 3 × 3 determinant:

$$\mathbf{a} \times \mathbf{b} = \begin{vmatrix} \mathbf{i} & \mathbf{j} & \mathbf{k} \\ a_1 & a_2 & a_3 \\ b_1 & b_2 & b_3 \end{vmatrix}. \tag{2}$$

Technically the expression on the right-hand side of the equality in (2) is *not* a determinant, because its entries are not all scalars. Nevertheless, the "determinant" in (2) is used simply as a way of remembering the component definition of the cross product given in (1).

EXAMPLE 3   The Cross Product

Let $\mathbf{a} = 4\mathbf{i} - 2\mathbf{j} + 5\mathbf{k}$ and $\mathbf{b} = 3\mathbf{i} + \mathbf{j} - \mathbf{k}$. Find $\mathbf{a} \times \mathbf{b}$.

Solution   We use (2) and expand the determinant using cofactors of the first row:

$$\mathbf{a} \times \mathbf{b} = \begin{vmatrix} \mathbf{i} & \mathbf{j} & \mathbf{k} \\ 4 & -2 & 5 \\ 3 & 1 & -1 \end{vmatrix} = \begin{vmatrix} -2 & 5 \\ 1 & -1 \end{vmatrix}\mathbf{i} - \begin{vmatrix} 4 & 5 \\ 3 & -1 \end{vmatrix}\mathbf{j} + \begin{vmatrix} 4 & -2 \\ 3 & 1 \end{vmatrix}\mathbf{k}$$

$$= -3\mathbf{i} + 19\mathbf{j} + 10\mathbf{k}. \qquad \blacksquare$$

EXAMPLE 4   Cross Products of the Basis Vectors

Since $\mathbf{i} = \langle 1, 0, 0 \rangle$, $\mathbf{j} = \langle 0, 1, 0 \rangle$, and $\mathbf{k} = \langle 0, 0, 1 \rangle$, we see from (2) or the second property of determinants that

$$\mathbf{i} \times \mathbf{i} = \mathbf{0}, \quad \mathbf{j} \times \mathbf{j} = \mathbf{0}, \quad \text{and} \quad \mathbf{k} \times \mathbf{k} = \mathbf{0}. \qquad (3)$$

Also by (2)

$$\mathbf{i} \times \mathbf{j} = \mathbf{k}, \quad \mathbf{j} \times \mathbf{k} = \mathbf{i}, \quad \mathbf{k} \times \mathbf{i} = \mathbf{j},$$
$$\mathbf{j} \times \mathbf{i} = -\mathbf{k}, \quad \mathbf{k} \times \mathbf{j} = -\mathbf{i}, \quad \mathbf{i} \times \mathbf{k} = -\mathbf{j}. \qquad (4)$$
$$\qquad \blacksquare$$

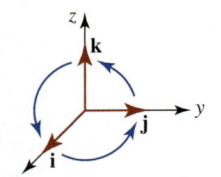

FIGURE 11.4.1   A mnemonic for cross products involving $\mathbf{i}$, $\mathbf{j}$, and $\mathbf{k}$

The cross products in (4) can be obtained using the circular mnemonic shown in FIGURE 11.4.1.

▮ **Properties**   The next theorem summarizes some of the important properties of the cross product.

Theorem 11.4.1   Properties of the Cross Product

(*i*) $\mathbf{a} \times \mathbf{b} = \mathbf{0}$ if $\mathbf{a} = \mathbf{0}$ or $\mathbf{b} = \mathbf{0}$

(*ii*) $\mathbf{a} \times \mathbf{b} = -\mathbf{b} \times \mathbf{a}$

(*iii*) $\mathbf{a} \times (\mathbf{b} + \mathbf{c}) = (\mathbf{a} \times \mathbf{b}) + (\mathbf{a} \times \mathbf{c}) \leftarrow$ distributive law

(*iv*) $(\mathbf{a} + \mathbf{b}) \times \mathbf{c} = (\mathbf{a} \times \mathbf{c}) + (\mathbf{b} \times \mathbf{c}) \leftarrow$ distributive law

(*v*) $\mathbf{a} \times (k\mathbf{b}) = (k\mathbf{a}) \times \mathbf{b} = k(\mathbf{a} \times \mathbf{b})$, k a scalar

(*vi*) $\mathbf{a} \times \mathbf{a} = \mathbf{0}$

(*vii*) $\mathbf{a} \cdot (\mathbf{a} \times \mathbf{b}) = 0$

(*viii*) $\mathbf{b} \cdot (\mathbf{a} \times \mathbf{b}) = 0$

Note in part (*i*) of Theorem 14.4.1 that the cross product is not commutative. As a consequence of this non-commutative property there are two distributive laws in parts (*iii*) and (*iv*) of the theorem.

PROOF   Parts (*i*), (*ii*), and (*vi*) follow directly from the three properties of determinants given above. We prove part (*iii*) and leave the remaining proofs for the student. See Problem 60 in Exercises 11.4. To prove part (*iii*) we let $\mathbf{a} = \langle a_1, a_2, a_3 \rangle$, $\mathbf{b} = \langle b_1, b_2, b_3 \rangle$, and $\mathbf{c} = \langle c_1, c_2, c_3 \rangle$. Then

$$\mathbf{a} \times (\mathbf{b} + \mathbf{c}) = \begin{vmatrix} a_2 & a_3 \\ b_2 + c_2 & b_3 + c_3 \end{vmatrix}\mathbf{i} - \begin{vmatrix} a_1 & a_3 \\ b_1 + c_1 & b_3 + c_3 \end{vmatrix}\mathbf{j} + \begin{vmatrix} a_1 & a_2 \\ b_1 + c_1 & b_2 + c_2 \end{vmatrix}\mathbf{k}$$

$$= [(a_2 b_3 + a_2 c_3) - (a_3 b_2 + a_3 c_2)]\mathbf{i} - [(a_1 b_3 + a_1 c_3) - (a_3 b_1 + a_3 c_1)]\mathbf{j}$$
$$\quad + [(a_1 b_2 + a_1 c_2) - (a_2 b_1 + a_2 c_1)]\mathbf{k}$$

$$= [(a_2 b_3 - a_3 b_2)\mathbf{i} - (a_1 b_3 - a_3 b_1)\mathbf{j} + (a_1 b_2 - a_2 b_1)\mathbf{k}]$$
$$\quad + [(a_2 c_3 - a_3 c_2)\mathbf{i} - (a_1 c_3 - a_3 c_1)\mathbf{j} + (a_1 c_2 - a_2 c_1)\mathbf{k}]$$

$$= (\mathbf{a} \times \mathbf{b}) + (\mathbf{a} \times \mathbf{c}). \qquad \blacksquare$$

■ Parallel Vectors We saw in Section 11.1 that two nonzero vectors are parallel if and only if one is a nonzero scalar multiple of the other. Thus, two vectors are parallel if they have the forms **a** and $k\mathbf{a}$, where **a** is any vector. By properties (v) and (vi) in Theorem 11.4.1, the cross product of parallel vectors must be **0**. This is stated formally in the next theorem.

Theorem 11.4.2 Criterion for Parallel Vectors

Two nonzero vectors **a** and **b** are parallel if and only if $\mathbf{a} \times \mathbf{b} = \mathbf{0}$.

EXAMPLE 5 Parallel Vectors

Determine if $\mathbf{a} = 2\mathbf{i} + \mathbf{j} - \mathbf{k}$ and $\mathbf{b} = -6\mathbf{i} - 3\mathbf{j} + 3\mathbf{k}$ are parallel vectors.

Solution From the cross product

$$\mathbf{a} \times \mathbf{b} = \begin{vmatrix} \mathbf{i} & \mathbf{j} & \mathbf{k} \\ 2 & 1 & -1 \\ -6 & -3 & 3 \end{vmatrix} = \begin{vmatrix} 1 & -1 \\ -3 & 3 \end{vmatrix}\mathbf{i} - \begin{vmatrix} 2 & -1 \\ -6 & 3 \end{vmatrix}\mathbf{j} + \begin{vmatrix} 2 & 1 \\ -6 & -3 \end{vmatrix}\mathbf{k}$$

$$= 0\mathbf{i} - 0\mathbf{j} + 0\mathbf{k} = \mathbf{0}$$

and Theorem 11.4.2 we conclude that **a** and **b** are parallel vectors. ■

■ Right-Hand Rule An alternative characterization of the cross product uses the **right-hand rule**. As seen in FIGURE 11.4.2(a), if the fingers of the right hand point along the vector **a** and then curl toward the vector **b**, the thumb will give the direction of $\mathbf{a} \times \mathbf{b}$. In Figure 11.4.1(b), the right-hand rule shows the direction of $\mathbf{b} \times \mathbf{a}$.

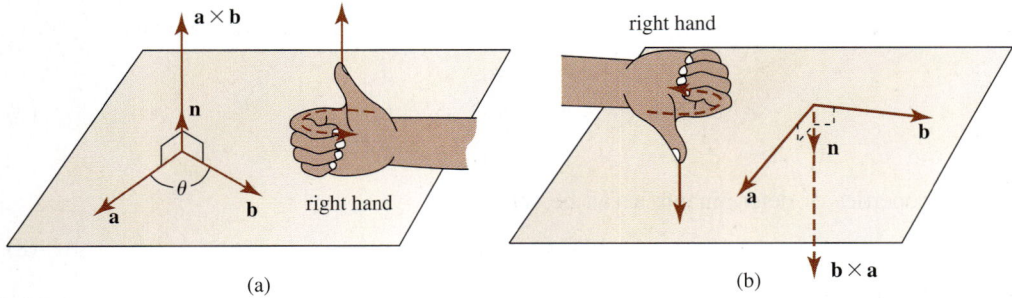

FIGURE 11.4.2 The right-hand rule

Theorem 11.4.3 Alternative Form of the Cross Product

Let **a** and **b** be two nonzero vectors that are not parallel to each other. Then the cross product of **a** and **b** is

$$\mathbf{a} \times \mathbf{b} = (|\mathbf{a}||\mathbf{b}|\sin\theta)\mathbf{n}, \tag{5}$$

where θ is the angle between the vectors such that $0 \leq \theta \leq \pi$ and **n** is a unit vector perpendicular to the plane of **a** and **b** with direction given by the right-hand rule.

PROOF We see from properties (vii) and (viii) of Theorem 11.4.1 that both **a** and **b** are perpendicular to $\mathbf{a} \times \mathbf{b}$. Thus, the direction of $\mathbf{a} \times \mathbf{b}$ is perpendicular to the plane of **a** and **b**, and it can be shown that the right-hand rule determines the appropriate direction. It remains to show that the magnitude of $\mathbf{a} \times \mathbf{b}$ is given by

$$|\mathbf{a} \times \mathbf{b}| = |\mathbf{a}||\mathbf{b}|\sin\theta. \tag{6}$$

We separately compute the squares of the left- and right-hand sides of this equation using the component forms of **a** and **b**:

$$|\mathbf{a} \times \mathbf{b}|^2 = (a_2b_3 - a_3b_2)^2 + (a_1b_3 - a_3b_1)^2 + (a_1b_2 - a_2b_1)^2$$
$$= a_2^2b_3^2 - 2a_2b_3a_3b_2 + a_3^2b_2^2 + a_1^2b_3^2 - 2a_1b_3a_3b_1 + a_3^2b_1^2$$
$$+ a_1^2b_2^2 - 2a_1b_2a_2b_1 + a_2^2b_1^2$$

$$(|\mathbf{a}||\mathbf{b}|\sin\theta)^2 = |\mathbf{a}|^2|\mathbf{b}|^2\sin^2\theta = |\mathbf{a}|^2|\mathbf{b}|^2(1 - \cos^2\theta)$$
$$= |\mathbf{a}|^2|\mathbf{b}|^2 - |\mathbf{a}|^2|\mathbf{b}|^2\cos^2\theta = |\mathbf{a}|^2|\mathbf{b}|^2 - (\mathbf{a} \cdot \mathbf{b})^2$$
$$= (a_1^2 + a_2^2 + a_3^2)(b_1^2 + b_2^2 + b_3^2) - (a_1b_1 + a_2b_2 + a_3b_3)^2$$
$$= a_2^2b_3^2 - 2a_2b_2a_3b_3 + a_3^2b_2^2 + a_1^2b_3^2 - 2a_1b_1a_3b_3 + a_3^2b_1^2$$
$$+ a_1^2b_2^2 - 2a_1b_1a_2b_2 + a_2^2b_1^2.$$

Since both sides are equal to the same quantity, they must be equal to each other, so

$$|\mathbf{a} \times \mathbf{b}|^2 = (|\mathbf{a}||\mathbf{b}|\sin\theta)^2.$$

Finally, taking the square root of both sides and using the fact that $\sqrt{\sin^2\theta} = \sin\theta$ since $\sin\theta \geq 0$ for $0 \leq \theta \leq \pi$, we have $|\mathbf{a} \times \mathbf{b}| = |\mathbf{a}||\mathbf{b}|\sin\theta$. ∎

Combining Theorems 11.4.2 and 11.4.3 we see that for *any* pair of vectors **a** and **b**,

$$\mathbf{a} \times \mathbf{b} = (|\mathbf{a}||\mathbf{b}|\sin\theta)\mathbf{n}.$$

▶ This more geometric form is generally used as the definition of the cross product on a physics course.

■ Special Products The **scalar triple product** of vectors **a**, **b**, and **c** is $\mathbf{a} \cdot (\mathbf{b} \times \mathbf{c})$. Using the component forms of the definitions of the dot and cross products, we have

$$\mathbf{a} \cdot (\mathbf{b} \times \mathbf{c}) = (a_1\mathbf{i} + a_2\mathbf{j} + a_3\mathbf{k}) \cdot \left[\begin{vmatrix} b_2 & b_3 \\ c_2 & c_3 \end{vmatrix}\mathbf{i} - \begin{vmatrix} b_1 & b_3 \\ c_1 & c_3 \end{vmatrix}\mathbf{j} + \begin{vmatrix} b_1 & b_2 \\ c_1 & c_2 \end{vmatrix}\mathbf{k} \right]$$

$$= a_1\begin{vmatrix} b_2 & b_3 \\ c_2 & c_3 \end{vmatrix} - a_2\begin{vmatrix} b_1 & b_3 \\ c_1 & c_3 \end{vmatrix} + a_3\begin{vmatrix} b_1 & b_2 \\ c_1 & c_2 \end{vmatrix}.$$

Thus, we see that the scalar triple product can be written as a 3×3 determinant:

$$\mathbf{a} \cdot (\mathbf{b} \times \mathbf{c}) = \begin{vmatrix} a_1 & a_2 & a_3 \\ b_1 & b_2 & b_3 \\ c_1 & c_2 & c_3 \end{vmatrix}. \tag{7}$$

Using properties of determinants it can be shown that

$$\mathbf{a} \cdot (\mathbf{b} \times \mathbf{c}) = (\mathbf{a} \times \mathbf{b}) \cdot \mathbf{c}. \tag{8}$$

See Problem 61 in Exercises 11.4.

The **vector triple product** of three vectors **a**, **b**, and **c** is

$$\mathbf{a} \times (\mathbf{b} \times \mathbf{c}).$$

The vector triple product is related to the dot product by

$$\mathbf{a} \times (\mathbf{b} \times \mathbf{c}) = (\mathbf{a} \cdot \mathbf{c})\mathbf{b} - (\mathbf{a} \cdot \mathbf{b})\mathbf{c}. \tag{9}$$

See Problem 62 in Exercises 11.4.

■ Areas Two nonzero and nonparallel vectors **a** and **b** can be considered to be the sides of a parallelogram. The **area A of a parallelogram** is

$$A = (\text{base})(\text{altitude}).$$

From FIGURE 11.4.3(a), we see that $A = |\mathbf{b}|(|\mathbf{a}|\sin\theta) = |\mathbf{a}||\mathbf{b}|\sin\theta$

or

$$A = |\mathbf{a} \times \mathbf{b}|. \tag{10}$$

Likewise from Figure 11.4.3(b), we see that the **area of a triangle** with sides **a** and **b** is

$$A = \frac{1}{2}|\mathbf{a} \times \mathbf{b}|. \tag{11}$$

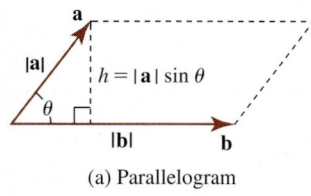

(a) Parallelogram

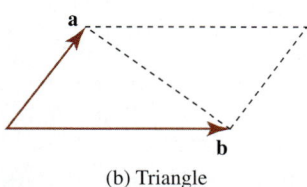

(b) Triangle

FIGURE 11.4.3 The area of a parallelogram

EXAMPLE 6 Area of a Triangle

Find the area of the triangle determined by the points $P_1(1, 1, 1)$, $P_2(2, 3, 4)$, and $P_3(3, 0, -1)$.

Solution The vectors $\overrightarrow{P_1P_2}$ and $\overrightarrow{P_2P_3}$ can be taken as two sides of the triangle. Since

$$\overrightarrow{P_1P_2} = \mathbf{i} + 2\mathbf{j} + 3\mathbf{k} \qquad \text{and} \qquad \overrightarrow{P_2P_3} = \mathbf{i} - 3\mathbf{j} - 5\mathbf{k}$$

we have

$$\overrightarrow{P_1P_2} \times \overrightarrow{P_2P_3} = \begin{vmatrix} \mathbf{i} & \mathbf{j} & \mathbf{k} \\ 1 & 2 & 3 \\ 1 & -3 & -5 \end{vmatrix} = \begin{vmatrix} 2 & 3 \\ -3 & -5 \end{vmatrix}\mathbf{i} - \begin{vmatrix} 1 & 3 \\ 1 & -5 \end{vmatrix}\mathbf{j} + \begin{vmatrix} 1 & 2 \\ 1 & -3 \end{vmatrix}\mathbf{k}$$
$$= -\mathbf{i} + 8\mathbf{j} - 5\mathbf{k}.$$

From (11) we see that the area is

$$A = \frac{1}{2}|-\mathbf{i} + 8\mathbf{j} - 5\mathbf{k}| = \frac{3}{2}\sqrt{10}. \qquad \blacksquare$$

▌ Volume of a Parallelepiped If the vectors $\mathbf{a}$, $\mathbf{b}$, and $\mathbf{c}$ do not lie in the same plane, then the volume of the parallelepiped with edges $\mathbf{a}$, $\mathbf{b}$, and $\mathbf{c}$ shown in FIGURE 11.4.4 is

$$V = (\text{area of base})(\text{height})$$
$$= |\mathbf{b} \times \mathbf{c}||\text{comp}_{\mathbf{b} \times \mathbf{c}}\mathbf{a}|$$
$$= |\mathbf{b} \times \mathbf{c}|\left|\mathbf{a} \cdot \left(\frac{1}{|\mathbf{b} \times \mathbf{c}|}\mathbf{b} \times \mathbf{c}\right)\right|$$

or

$$V = |\mathbf{a} \cdot (\mathbf{b} \times \mathbf{c})|. \qquad (12)$$

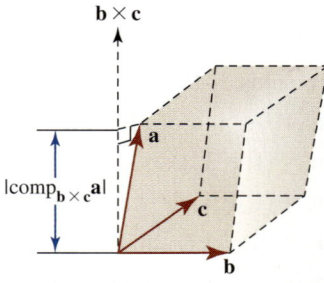

FIGURE 11.4.4 Parallelepiped formed by three vectors

Thus, the volume of a parallelepiped determined by three vectors is the absolute value of the scalar triple product of the vectors.

▌ Coplanar Vectors Vectors that lie in the same plane are said to be **coplanar**. We have just seen that if the vectors $\mathbf{a}$, $\mathbf{b}$, and $\mathbf{c}$ are not coplanar, then necessarily $\mathbf{a} \cdot (\mathbf{b} \times \mathbf{c}) \neq 0$, since the volume of a parallelepiped with edges $\mathbf{a}$, $\mathbf{b}$, and $\mathbf{c}$ has nonzero volume. Equivalently stated, this means that if $\mathbf{a} \cdot (\mathbf{b} \times \mathbf{c}) = 0$, then the vectors $\mathbf{a}$, $\mathbf{b}$, and $\mathbf{c}$ are coplanar. Since the converse of this last statement is also true (see Problem 64 in Exercises 11.4), we have

$$\mathbf{a} \cdot (\mathbf{b} \times \mathbf{c}) = 0 \qquad \textit{if and only if} \qquad \mathbf{a}, \mathbf{b}, \textit{and } \mathbf{c} \textit{ are coplanar.}$$

▌ Physical Interpretation of the Cross Product In physics a force $\mathbf{F}$ acting at the end of a position vector $\mathbf{r}$, as shown in FIGURE 11.4.5, is said to produce a **torque** $\boldsymbol{\tau}$ defined by $\boldsymbol{\tau} = \mathbf{r} \times \mathbf{F}$. For example, if $|\mathbf{F}| = 20$ N, $|\mathbf{r}| = 3.5$ m, and $\theta = 30°$, then from (6),

$$|\boldsymbol{\tau}| = (3.5)(20)\sin 30° = 35 \text{ N-m.}$$

If $\mathbf{F}$ and $\mathbf{r}$ are in the plane of the page, the right-hand rule implies that the direction of $\boldsymbol{\tau}$ is outward from, and perpendicular to, the page (toward the reader).

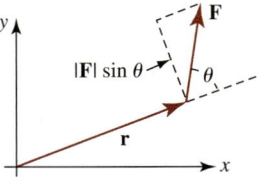

FIGURE 11.4.5 A force acting at the end of a vector

As we see in FIGURE 11.4.6, when a force $\mathbf{F}$ is applied to a wrench, the magnitude of the torque $\boldsymbol{\tau}$ is a measure of the turning effect about the pivot point P and the vector $\boldsymbol{\tau}$ is directed along the axis of the bolt. In this case $\boldsymbol{\tau}$ points inward from the page.

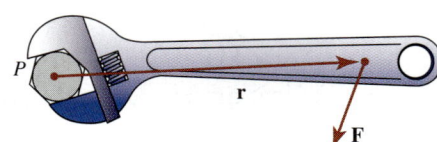

FIGURE 11.4.6 A wrench applying torque to a bolt

$\overrightarrow{OP}$ **NOTES FROM THE CLASSROOM**

When working with vectors, one should be careful not to mix the dot and cross symbols, that is, $\cdot$ and $\times$, with the symbols for ordinary multiplication, and to be especially careful in the use, or lack of use, of parentheses. For example, if a, b, and c are real numbers, then the product abc is well-defined because

$$abc = a(bc) = (ab)c.$$

On the other hand, the expression $\mathbf{a} \times \mathbf{b} \times \mathbf{c}$ is not well-defined because

$$\mathbf{a} \times (\mathbf{b} \times \mathbf{c}) \neq (\mathbf{a} \times \mathbf{b}) \times \mathbf{c}.$$

See Problem 59 in Exercises 11.4. Other expressions, such as $\mathbf{a} \cdot \mathbf{b} \cdot \mathbf{c}$, are not meaningful, even if parentheses are included. Why?

Exercises 11.4 Answers to selected odd-numbered problems begin on page ANS-36.

≡ Fundamentals

In Problems 1–10, find $\mathbf{a} \times \mathbf{b}$.

1. $\mathbf{a} = \mathbf{i} - \mathbf{j}, \mathbf{b} = 3\mathbf{j} + 5\mathbf{k}$ **2.** $\mathbf{a} = 2\mathbf{i} + \mathbf{j}, \mathbf{b} = 4\mathbf{i} - \mathbf{k}$

3. $\mathbf{a} = \langle 1, -3, 1 \rangle, \mathbf{b} = \langle 2,0,4 \rangle$

4. $\mathbf{a} = \langle 1, 1, 1 \rangle, \mathbf{b} = \langle -5, 2, 3 \rangle$

5. $\mathbf{a} = 2\mathbf{i} - \mathbf{j} + 2\mathbf{k}, \mathbf{b} = -\mathbf{i} + 3\mathbf{j} - \mathbf{k}$

6. $\mathbf{a} = 4\mathbf{i} + \mathbf{j} - 5\mathbf{k}, \mathbf{b} = 2\mathbf{i} + 3\mathbf{j} - \mathbf{k}$

7. $\mathbf{a} = \langle \frac{1}{2}, 0, \frac{1}{2} \rangle, \mathbf{b} = \langle 4, 6, 0 \rangle$ **8.** $\mathbf{a} = \langle 0, 5, 0 \rangle, \mathbf{b} = \langle 2, -3, 4 \rangle$

9. $\mathbf{a} = \langle 2, 2, -4 \rangle, \mathbf{b} = \langle -3, -3, 6 \rangle$

10. $\mathbf{a} = \langle 8, 1, -6 \rangle, \mathbf{b} = \langle 1, -2, 10 \rangle$

In Problems 11 and 12, find $\overrightarrow{P_1 P_2} \times \overrightarrow{P_1 P_3}$.

11. $P_1(2, 1, 3), P_2(0, 3, -1), P_3(-1, 2, 4)$

12. $P_1(0, 0, 1), P_2(0, 1, 2), P_3(1, 2, 3)$

In Problems 13 and 14, find a nonzero vector that is perpendicular to both $\mathbf{a}$ and $\mathbf{b}$.

13. $\mathbf{a} = 2\mathbf{i} + 7\mathbf{j} - 4\mathbf{k}, \mathbf{b} = \mathbf{i} + \mathbf{j} - \mathbf{k}$

14. $\mathbf{a} = \langle -1, -2, 4 \rangle, \mathbf{b} = \langle 4, -1, 0 \rangle$

In Problems 15 and 16, verify that $\mathbf{a} \cdot (\mathbf{a} \times \mathbf{b}) = 0$ and $\mathbf{b} \cdot (\mathbf{a} \times \mathbf{b}) = 0$.

15. $\mathbf{a} = \langle 5, -2, 1 \rangle, \mathbf{b} = \langle 2, 0, -7 \rangle$

16. $\mathbf{a} = \frac{1}{2}\mathbf{i} - \frac{1}{4}\mathbf{j} - 4\mathbf{k}, \mathbf{b} = 2\mathbf{i} - 2\mathbf{j} + 6\mathbf{k}$

In Problems 17 and 18,

(a) calculate $\mathbf{b} \times \mathbf{c}$ followed by $\mathbf{a} \times (\mathbf{b} \times \mathbf{c})$.

(b) Verify the results in part (a) by (9) of this section.

17. $\mathbf{a} = \mathbf{i} - \mathbf{j} + 2\mathbf{k}$ **18.** $\mathbf{a} = 3\mathbf{i} - 4\mathbf{k}$
  $\mathbf{b} = 2\mathbf{i} + \mathbf{j} + \mathbf{k}$ $\mathbf{b} = \mathbf{i} + 2\mathbf{j} - \mathbf{k}$
  $\mathbf{c} = 3\mathbf{i} + \mathbf{j} + \mathbf{k}$ $\mathbf{c} = -\mathbf{i} + 5\mathbf{j} + 8\mathbf{k}$

In Problems 19–36, find the indicated scalar or vector *without* using (2), (7), or (9).

19. $(2\mathbf{i}) \times \mathbf{j}$ **20.** $\mathbf{i} \times (-3\mathbf{k})$

21. $\mathbf{k} \times (2\mathbf{i} - \mathbf{j})$ **22.** $\mathbf{i} \times (\mathbf{j} \times \mathbf{k})$

23. $[(2\mathbf{k}) \times (3\mathbf{j})] \times (4\mathbf{j})$ **24.** $(2\mathbf{i} - \mathbf{j} + 5\mathbf{k}) \times \mathbf{i}$

25. $(\mathbf{i} + \mathbf{j}) \times (\mathbf{i} + 5\mathbf{k})$ **26.** $\mathbf{i} \times \mathbf{k} - 2(\mathbf{j} \times \mathbf{i})$

27. $\mathbf{k} \cdot (\mathbf{j} \times \mathbf{k})$ **28.** $\mathbf{i} \cdot [\mathbf{j} \times (-\mathbf{k})]$

29. $|4\mathbf{j} - 5(\mathbf{i} \times \mathbf{j})|$ **30.** $(\mathbf{i} \times \mathbf{j}) \cdot (3\mathbf{j} \times \mathbf{i})$

31. $\mathbf{i} \times (\mathbf{i} \times \mathbf{j})$ **32.** $(\mathbf{i} \times \mathbf{j}) \times \mathbf{i}$

33. $(\mathbf{i} \times \mathbf{i}) \times \mathbf{j}$ **34.** $(\mathbf{i} \cdot \mathbf{i})(\mathbf{i} \times \mathbf{j})$

35. $2\mathbf{j} \cdot [\mathbf{i} \times (\mathbf{j} - 3\mathbf{k})]$ **36.** $(\mathbf{i} \times \mathbf{k}) \times (\mathbf{j} \times \mathbf{i})$

In Problems 37–44, $\mathbf{a} \times \mathbf{b} = 4\mathbf{i} - 3\mathbf{j} + 6\mathbf{k}$ and $\mathbf{c} = 2\mathbf{i} + 4\mathbf{j} - \mathbf{k}$. Find the indicated scalar or vector.

37. $\mathbf{a} \times (3\mathbf{b})$ **38.** $\mathbf{b} \times \mathbf{a}$

39. $(-\mathbf{a}) \times \mathbf{b}$ **40.** $|\mathbf{a} \times \mathbf{b}|$

41. $(\mathbf{a} \times \mathbf{b}) \times \mathbf{c}$ **42.** $(\mathbf{a} \times \mathbf{b}) \cdot \mathbf{c}$

43. $\mathbf{a} \cdot (\mathbf{b} \times \mathbf{c})$ **44.** $(4\mathbf{a}) \cdot (\mathbf{b} \times \mathbf{c})$

In Problems 45 and 46,

(a) verify that the given quadrilateral is a parallelogram and

(b) find the area of the parallelogram.

45.

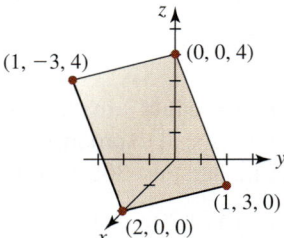

FIGURE 11.4.7 Parallelogram in Problem 45

46.

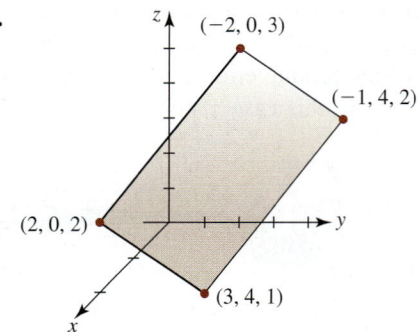

FIGURE 11.4.8 Parallelogram in Problem 46

In Problems 47–50, find the area of the triangle determined by the given points.

47. $P_1(1, 1, 1)$, $P_2(1, 2, 1)$, $P_3(1, 1, 2)$

48. $P_1(0, 0, 0)$, $P_2(0, 1, 2)$, $P_3(2, 2, 0)$

49. $P_1(1, 2, 4)$, $P_2(1, -1, 3)$, $P_3(-1, -1, 2)$

50. $P_1(1, 0, 3)$, $P_2(0, 0, 6)$, $P_3(2, 4, 5)$

In Problems 51 and 52, find the volume of the parallelepiped for which the given vectors are three edges.

51. $\mathbf{a} = \mathbf{i} + \mathbf{j}$, $\mathbf{b} = -\mathbf{i} + 4\mathbf{j}$, $\mathbf{c} = 2\mathbf{i} + 2\mathbf{j} + 2\mathbf{k}$

52. $\mathbf{a} = 3\mathbf{i} + \mathbf{j} + \mathbf{k}$, $\mathbf{b} = \mathbf{i} + 4\mathbf{j} + \mathbf{k}$, $\mathbf{c} = \mathbf{i} + \mathbf{j} + 5\mathbf{k}$

In Problems 53 and 54, determine whether the indicated vectors are coplanar.

53. $\mathbf{a} = 4\mathbf{i} + 6\mathbf{j}$, $\mathbf{b} = -2\mathbf{i} + 6\mathbf{j} - 6\mathbf{k}$, $\mathbf{c} = \frac{5}{2}\mathbf{i} + 3\mathbf{j} + \frac{1}{2}\mathbf{k}$

54. $\mathbf{a} = \mathbf{i} + 2\mathbf{j} - 4\mathbf{k}$, $\mathbf{b} = -2\mathbf{i} + \mathbf{j} + \mathbf{k}$, $\mathbf{c} = \frac{3}{2}\mathbf{j} - 2\mathbf{k}$

In Problems 55 and 56, determine whether the indicated four points lie in the same plane.

55. $P_1(1, 1, -2)$, $P_2(4, 0, -3)$, $P_3(1, -5, 10)$, $P_4(-7, 2, 4)$

56. $P_1(2, -1, 4)$, $P_2(-1, 2, 3)$, $P_3(0, 4, -3)$, $P_4(4, -2, 2)$

57. As shown in FIGURE 11.4.9, the vector $\mathbf{a}$ lies in the xy-plane and the vector $\mathbf{b}$ lies along the positive z-axis. Their magnitudes are $|\mathbf{a}| = 6.4$ and $|\mathbf{b}| = 5$.

(a) Use (5) to find $|\mathbf{a} \times \mathbf{b}|$.

(b) Use the right-hand rule to find the direction of $\mathbf{a} \times \mathbf{b}$.

(c) Use part (b) to express $\mathbf{a} \times \mathbf{b}$ in terms of the unit vectors $\mathbf{i}$, $\mathbf{j}$, and $\mathbf{k}$.

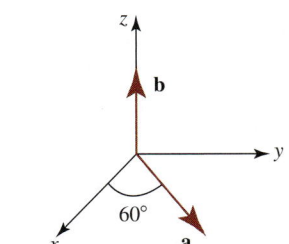

FIGURE 11.4.9 Vectors in Problem 57

58. Two vectors $\mathbf{a}$ and $\mathbf{b}$ lie in the xz-plane so that the angle between them is 120°. If $|\mathbf{a}| = \sqrt{27}$ and $|\mathbf{b}| = 8$, find all possible values of $\mathbf{a} \times \mathbf{b}$.

≡ Think About It

59. If $\mathbf{a} = \langle 1, 2, 3 \rangle$, $\mathbf{b} = \langle 4, 5, 6 \rangle$, and $\mathbf{c} = \langle 7, 8, 3 \rangle$, show that $\mathbf{a} \times (\mathbf{b} \times \mathbf{c}) \neq (\mathbf{a} \times \mathbf{b}) \times \mathbf{c}$.

60. Prove parts (iv), (v), (vii), and $(viii)$ of Theorem 11.4.1.

61. Prove $\mathbf{a} \cdot (\mathbf{b} \times \mathbf{c}) = (\mathbf{a} \times \mathbf{b}) \cdot \mathbf{c}$.

62. Prove $\mathbf{a} \times (\mathbf{b} \times \mathbf{c}) = (\mathbf{a} \cdot \mathbf{c})\mathbf{b} - (\mathbf{a} \cdot \mathbf{b})\mathbf{c}$.

63. Prove $\mathbf{a} \times (\mathbf{b} \times \mathbf{c}) + \mathbf{b} \times (\mathbf{c} \times \mathbf{a}) + \mathbf{c} \times (\mathbf{a} \times \mathbf{b}) = \mathbf{0}$.

64. Prove that if $\mathbf{a}$, $\mathbf{b}$, and $\mathbf{c}$ are coplanar, then $\mathbf{a} \cdot (\mathbf{b} \times \mathbf{c}) = 0$.

≡ Projects

65. A three-dimensional lattice is a collection of integer combinations of three noncoplanar basis vectors $\mathbf{a}$, $\mathbf{b}$, and $\mathbf{c}$. In crystallography, a lattice can specify the locations of atoms in a crystal. X-ray diffraction studies of crystals use the "reciprocal lattice," which has basis vectors

$$\mathbf{A} = \frac{\mathbf{b} \times \mathbf{c}}{\mathbf{a} \cdot (\mathbf{b} \times \mathbf{c})}, \quad \mathbf{B} = \frac{\mathbf{c} \times \mathbf{a}}{\mathbf{b} \cdot (\mathbf{c} \times \mathbf{a})}, \quad \mathbf{C} = \frac{\mathbf{a} \times \mathbf{b}}{\mathbf{c} \cdot (\mathbf{a} \times \mathbf{b})}.$$

(a) A certain lattice has basis vectors $\mathbf{a} = \mathbf{i}$, $\mathbf{b} = \mathbf{j}$, and $\mathbf{c} = \frac{1}{2}(\mathbf{i} + \mathbf{j} + \mathbf{k})$. Find basis vectors for the reciprocal lattice.

(b) The unit cell of the reciprocal lattice is the parallelepiped with edges $\mathbf{A}$, $\mathbf{B}$, and $\mathbf{C}$, while the unit cell of the original lattice is the parallelepiped with edges $\mathbf{a}$, $\mathbf{b}$, and $\mathbf{c}$. Show that the volume of the unit cell of the reciprocal lattice is the reciprocal of the volume of the unit cell of the original lattice. [*Hint*: Start with $\mathbf{B} \times \mathbf{C}$ and use (9).]

11.5 Lines in 3-Space

▮ Introduction In Section 1.3 we saw that the key to writing an equation of a line in the plane is the notion of slope. The slope of a line (or its angle of inclination) gives a line a direction. A line in the plane is determined by specifying either a point and a slope or any two distinct points. Basically the same is true in 3-space.

We see next that vector concepts are an important aid in obtaining an equation of a line in space.

▮ Vector Equation A line in space is determined by specifying a point $P_0(x_0, y_0, z_0)$ and a nonzero vector $\mathbf{v}$. Through the point P_0 there passes only one line L parallel to the given vector. Let us assume that $P(x, y, z)$ is *any* point on the line. If $\mathbf{r} = \overrightarrow{OP}$ and $\mathbf{r}_0 = \overrightarrow{OP_0}$ are position vectors of P and P_0, then because $\mathbf{r} - \mathbf{r}_0$ is parallel to the vector $\mathbf{v}$ there exists a scalar t such that $\mathbf{r} - \mathbf{r}_0 = t\mathbf{v}$. This gives us a **vector equation**

$$\mathbf{r} = \mathbf{r}_0 + t\mathbf{v} \tag{1}$$

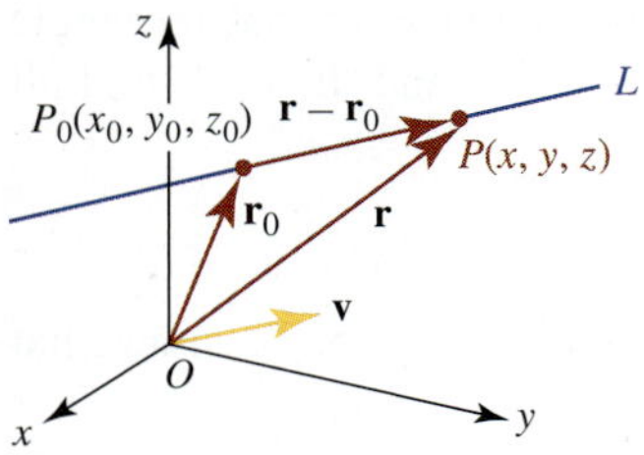

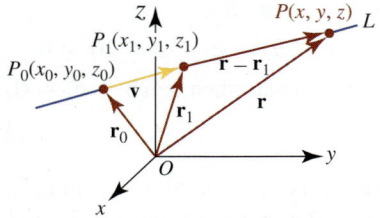

FIGURE 11.5.1 Line through P_0 parallel to **v**

FIGURE 11.5.2 Line through P_0 and P_1

of the line L. Using components, $\mathbf{r} = \langle x, y, z \rangle$, $\mathbf{r}_0 = \langle x_0, y_0, z_0 \rangle$, and $\mathbf{v} = \langle a, b, c \rangle$ we see that (1) is the same as

$$\langle x, y, z \rangle = \langle x_0 + at, y_0 + bt, z_0 + ct \rangle. \tag{2}$$

The scalar t is called a **parameter** and the nonzero vector **v** is called a **direction vector**; the components a, b, and c of the direction vector **v** are called **direction numbers** for the line L. For each real number t the vector **r** in (1) is the position vector of a point on L and so we can envision the line as being traced out in space by the moving arrowhead of **r**. See FIGURE 11.5.1.

Any two distinct points $P_0(x_0, y_0, z_0)$ and $P_1(x_1, y_1, z_1)$ in 3-space determine only one line L between them. If $\mathbf{r} = \overrightarrow{OP}$, $\mathbf{r}_0 = \overrightarrow{OP_0}$, and $\mathbf{r}_1 = \overrightarrow{OP_1}$ are position vectors, we see in FIGURE 11.5.2 that the vector $\mathbf{v} = \mathbf{r}_1 - \mathbf{r}_0$ is parallel to vector $\mathbf{r} - \mathbf{r}_1$. Thus, $\mathbf{r} - \mathbf{r}_1 = t(\mathbf{r}_1 - \mathbf{r}_0)$ or $\mathbf{r} = \mathbf{r}_1 + t(\mathbf{r}_1 - \mathbf{r}_0)$. Because $\mathbf{r} - \mathbf{r}_0$ is also parallel to **v** an alternative vector equation for the line is $\mathbf{r} - \mathbf{r}_0 = t(\mathbf{r}_1 - \mathbf{r}_0)$ or

$$\mathbf{r} = \mathbf{r}_0 + t(\mathbf{r}_1 - \mathbf{r}_0). \tag{3}$$

If we write $\mathbf{v} = \mathbf{r}_1 - \mathbf{r}_0 = \langle x_1 - x_0, y_1 - y_0, z_1 - z_0 \rangle = \langle a, b, c \rangle$ we see that (3) is the same as (1). Indeed, $\mathbf{r} = \mathbf{r}_0 + t(-\mathbf{v})$ and $\mathbf{r} = \mathbf{r}_0 + t(k\mathbf{v})$, k a nonzero scalar, are also equations for L.

EXAMPLE 1　Vector Equation of a Line

Find a vector equation for the line through $(4, 6, -3)$ and parallel to $\mathbf{v} = 5\mathbf{i} - 10\mathbf{j} + 2\mathbf{k}$.

Solution　With the identifications $x_0 = 4$, $y_0 = 6$, $z_0 = -3$, $a = 5$, $b = -10$, and $c = 2$ we obtain from (2) a vector equation of the line:

$$\langle x, y, z \rangle = \langle 4, 6, -3 \rangle + t\langle 5, -10, 2 \rangle \quad \text{or} \quad \langle x, y, z \rangle = \langle 4 + 5t, 6 - 10t, -3 + 2t \rangle. \quad \blacksquare$$

EXAMPLE 2　Vector Equation of a Line

Find a vector equation for the line through $(2, -1, 8)$ and $(5, 6, -3)$.

Solution　If we label the points as $P_0(2, -1, 8)$ and $P_1(5, 6, -3)$, then a direction vector for the line through P_0 and P_1 is

$$\mathbf{v} = \overrightarrow{P_0P_1} = \overrightarrow{OP_1} - \overrightarrow{OP_0} = \langle 5 - 2, 6 - (-1), -3 - 8 \rangle = \langle 3, 7, -11 \rangle.$$

From (3) a vector equation of the line is

$$\langle x, y, z \rangle = \langle 2, -1, 8 \rangle + t\langle 3, 7, -11 \rangle.$$

This is one of many possible vector equations of the line. For example, two alternative equations are

$$\langle x, y, z \rangle = \langle 5, 6, -3 \rangle + t\langle 3, 7, -11 \rangle$$
$$\langle x, y, z \rangle = \langle 5, 6, -3 \rangle + t\langle -3, -7, 11 \rangle. \quad \blacksquare$$

❚ Parametric Equations　By equating components in (2) we obtain

$$x = x_0 + at, \quad y = y_0 + bt, \quad z = z_0 + ct. \tag{4}$$

The equations in (4) are called **parametric equations** for the line through P_0. The entire line L, the line that extends indefinitely in both directions, is obtained by allowing the parameter t to increase from $-\infty$ to ∞, in other words, the parameter interval is $(-\infty, \infty)$. If the parameter t is restricted to a closed interval $[t_0, t_1]$, then as t increases (4) defines a **line segment** that starts at the point corresponding to t_0 and ends at the point corresponding to t_1.

EXAMPLE 3　Parametric Equations of a Line

Find parametric equations for the line
(a) through $(5, 2, 4)$ parallel to $\mathbf{v} = 4\mathbf{i} + 7\mathbf{j} - 9\mathbf{k}$, and **(b)** through $(-1, 0, 1)$ and $(2, -1, 6)$.

Solution

(a) With the identifications $x_0 = 5$, $y_0 = 2$, $z_0 = 4$, $a = 4$, $b = 7$, and $c = -9$, we see from (4) that parametric equations of the line are

$$x = 5 + 4t, y = 2 + 7t, z = 4 - 9t.$$

(b) Proceeding as in Example 2, a direction vector for the line is

$$\mathbf{v} = \langle 2, -1, 6 \rangle - \langle -1, 0, 1 \rangle = \langle 3, -1, 5 \rangle.$$

With direction numbers $a = 3$, $b = -1$, and $c = 5$, (4) gives

$$x = -1 + 3t, y = -t, z = 1 + 5t. \qquad \blacksquare$$

If we limit the parameter interval in part (a) of Example 3 to, say, $-1 \le t \le 0$, then

$$x = 5 + 4t, \quad y = 2 + 7t, \quad z = 4 - 9t, \quad -1 \le t \le 0$$

are parametric equations of the line segment starting at the point $(1, -5, 13)$ and ending at $(5, 2, 4)$.

EXAMPLE 4 Example 1 Revisited

Find the point where the line in Example 1 intersects the xy-plane.

Solution Equating components in the vector equation $\langle x, y, z \rangle = \langle 4 + 5t, 6 - 10t, -3 + 2t \rangle$ yields parametric equations of the line:

$$x = 4 + 5t, \quad y = 6 - 10t, \quad z = -3 + 2t.$$

Since an equation for the xy-plane is $z = 0$ we solve $z = -3 + 2t = 0$ for t. Substituting $t = \frac{3}{2}$ in the remaining two equations then gives $x = 4 + 5\left(\frac{3}{2}\right) = \frac{23}{2}$ and $y = 6 - 10\left(\frac{3}{2}\right) = -9$. The point of intersection in the z-plane is then $\left(\frac{23}{2}, -9, 0\right)$. $\qquad \blacksquare$

■ **Symmetric Equations** From (4) observe that we can eliminate the parameter by writing

$$t = \frac{x - x_0}{a} = \frac{y - y_0}{b} = \frac{z - z_0}{c}$$

provided that each of the three direction numbers a, b, and c is nonzero. The resulting equations

$$\frac{x - x_0}{a} = \frac{y - y_0}{b} = \frac{z - z_0}{c} \tag{5}$$

are said to be **symmetric equations** for the line through P_0.

If one of the direction numbers a, b, or c is zero, we use the remaining two equations to eliminate the parameter t. For example, if $a = 0$, $b \ne 0$, $c \ne 0$, then (4) yields

$$x = x_0 \quad \text{and} \quad t = \frac{y - y_0}{b} = \frac{z - z_0}{c}.$$

In this case,

$$x = x_0, \quad \frac{y - y_0}{b} = \frac{z - z_0}{c} \tag{6}$$

are symmetric equations for the line. Since $x = x_0$ is an equation of a vertical plane perpendicular to the x-axis, the line described by (6) lies in that plane.

EXAMPLE 5 Example 3 Revisited

Find symmetric equations of the line found in part (a) of Example 3.

Solution From the identifications given in the solution of Example 3 we can write immediately from (5) that

$$\frac{x - 5}{4} = \frac{y - 2}{7} = \frac{z - 4}{-9}. \qquad \blacksquare$$

EXAMPLE 6 Symmetric Equations

Find symmetric equations for the line through the points $(5, 3, 1)$ and $(2, 1, 1)$.

Solution Define $a = 5 - 2 = 3$, $b = 3 - 1 = 2$, and $c = 1 - 1 = 0$. From the preceding discussion it follows that symmetric equations for the line are

$$\frac{x - 5}{3} = \frac{y - 3}{2}, \quad z = 1.$$

In other words, the symmetric equations describe a line in the plane $z = 1$. ∎

▌ Perpendicular and Parallel Lines The following definition gives a way of using the direction vectors of two lines to determine whether the lines are perpendicular or parallel.

Definition 11.5.1 Perpendicular and Parallel Lines

Two lines L_1 and L_2 with direction vectors $\mathbf{v}_1$ and $\mathbf{v}_2$, respectively, are

(i) **perpendicular** if $\mathbf{v}_1 \cdot \mathbf{v}_2 = 0$, and
(ii) **parallel** if $\mathbf{v}_2 = k\mathbf{v}_1$, for some nonzero scalar k.

EXAMPLE 7 Perpendicular Lines

Determine whether the lines

$$L_1: \quad x = -6 - t, \quad y = 20 + 3t, \quad z = 1 + 2t$$
$$L_2: \quad x = 5 + 2s, \quad y = -9 - 4s, \quad z = 1 + 7s$$

are perpendicular.

Solution Reading off the coefficients of the parameters t and s, we see that

$$\mathbf{v}_1 = -\mathbf{i} + 3\mathbf{j} + 2\mathbf{k} \quad \text{and} \quad \mathbf{v}_2 = 2\mathbf{i} - 4\mathbf{j} + 7\mathbf{k}$$

are the direction vectors for L_1 and L_2, respectively. Because $\mathbf{v}_1 \cdot \mathbf{v}_2 = -2 - 12 + 14 = 0$ we conclude that the lines are perpendicular. ∎

EXAMPLE 8 Parallel Lines

Direction vectors for the lines

$$L_1: \quad x = 4 - 2t, \quad x = 1 + 4t, \quad z = 3 + 10t$$
$$L_2: \quad x = s, \quad y = 6 - 2s, \quad z = \frac{1}{2} - 5s$$

are $\mathbf{v}_1 = -2\mathbf{i} + 4\mathbf{j} + 10\mathbf{k}$ and $\mathbf{v}_2 = \mathbf{i} - 2\mathbf{j} - 5\mathbf{k}$. Because $\mathbf{v}_1 = -2\mathbf{v}_2$ (or $\mathbf{v}_2 = -\frac{1}{2}\mathbf{v}_1$), we conclude that the lines are parallel. ∎

Notice that (i) of Definition 11.5.1 does not demand that the two lines intersect in order to be perpendicular. FIGURE 11.5.3 shows two perpendicular lines L_1 and L_2 that do not intersect. In other words, L_1 can be perpendicular to a plane containing L_2.

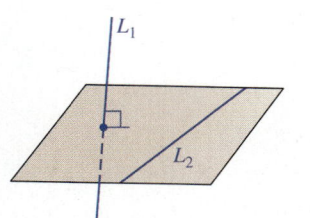

FIGURE 11.5.3 Perpendicular lines

EXAMPLE 9 Example 7 Revisited

Determine whether the lines L_1 and L_2 in Example 7 intersect.

Solution Since a point (x, y, z) of intersection is common to both lines, we must have

$$\left. \begin{array}{r} -6 - t = 5 + 2s \\ 20 + 3t = -9 - 4s \\ 1 + 2t = 1 + 7s \end{array} \right\} \quad \text{or} \quad \left\{ \begin{array}{r} 2s + t = -11 \\ 4s + 3t = -29 \\ -7s + 2t = 0 \end{array} \right. \tag{7}$$

We now solve any *two* of the equations in (7) simultaneously and use the remaining equation as a check. Choosing the first and third, we find from the system of equations

$$2s + t = -11$$
$$-7s + 2t = 0$$

that $s = -2$ and $t = -7$. Substitution of these values in the second equation in (7) yields the identity $-8 - 21 = -29$. Thus, L_1 and L_2 intersect. To find the point of intersection, we use, say, $s = -2$:

$$x = 5 + 2(-2) = 1, \quad y = -9 - 4(-2) = -1, \quad z = 1 + 7(-2) = -13.$$

The point of intersection is $(1, -1, -13)$. ∎

In Example 9, had the remaining equation not been satisfied when the values $s = -2$ and $t = -7$ were substituted, then the three equations would not be satisfied simultaneously and so the lines would not intersect. Two lines L_1 and L_2 in 3-space that do not intersect and are not parallel are called **skew lines**. As shown in FIGURE 11.5.4, skew lines lie in parallel planes.

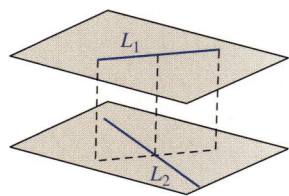

FIGURE 11.5.4   Skew lines

Exercises 11.5   Answers to selected odd-numbered problems begin on page ANS-36.

≡ Fundamentals

In Problems 1–4, find a vector equation for the line through the point parallel to the given vector.

1. $(4, 6, -7)$, $\mathbf{v} = \langle 3, \frac{1}{2}, -\frac{3}{2} \rangle$

2. $(1, 8, -2)$, $\mathbf{v} = -7\mathbf{i} - 8\mathbf{j}$

3. $(0, 0, 0)$, $\mathbf{v} = 5\mathbf{i} + 9\mathbf{j} + 4\mathbf{k}$

4. $(0, -3, 10)$, $\mathbf{v} = \langle 12, -5, -6 \rangle$

In Problems 5–10, find a vector equation for the line through the given points.

5. $(1, 2, 1)$, $(3, 5, -2)$

6. $(0, 4, 5)$, $(-2, 6, 3)$

7. $\left(\frac{1}{2}, -\frac{1}{2}, 1\right)$, $\left(-\frac{3}{2}, \frac{5}{2}, -\frac{1}{2}\right)$

8. $(10, 2, -10)$, $(5, -3, 5)$

9. $(1, 1, -1)$, $(-4, 1, -1)$

10. $(3, 2, 1)$, $\left(\frac{5}{2}, 1, -2\right)$

In Problems 11–16, find parametric equations for the line through the given points.

11. $(2, 3, 5)$, $(6, -1, 8)$

12. $(2, 0, 0)$, $(0, 4, 9)$

13. $(1, 0, 0)$, $(3, -2, -7)$

14. $(0, 0, 5)$, $(-2, 4, 0)$

15. $\left(4, \frac{1}{2}, \frac{1}{3}\right)$, $\left(-6, -\frac{1}{4}, \frac{1}{6}\right)$

16. $(-3, 7, 9)$, $(4, -8, -1)$

In Problems 17–22, find symmetric equations for the line through the given points.

17. $(1, 4, -9)$, $(10, 14, -2)$

18. $\left(\frac{2}{3}, 0, -\frac{1}{4}\right)$, $\left(1, 3, \frac{1}{4}\right)$

19. $(4, 2, 1)$, $(-7, 2, 5)$

20. $(-5, -2, -4)$, $(1, 1, 2)$

21. $(5, 10, -2)$, $(5, 1, -14)$

22. $\left(\frac{5}{6}, -\frac{1}{4}, \frac{1}{5}\right)$, $\left(\frac{1}{3}, \frac{3}{8}, -\frac{1}{10}\right)$

23. Find parametric equations for the line through $(6, 4, -2)$ that is parallel to the line $x/2 = (1 - y)/3 = (z - 5)/6$.

24. Find symmetric equations for the line through $(4, -11, -7)$ that is parallel to the line $x = 2 + 5t$, $y = -1 + \frac{1}{3}t$, $z = 9 - 2t$.

25. Find parametric equations for the line through $(2, -2, 15)$ that is parallel to the xz-plane and the xy-plane.

26. Find parametric equations for the line through $(1, 2, 8)$ that is
  (a) parallel to the y-axis and
  (b) perpendicular to the xy-plane.

In Problems 27 and 28, show that the lines L_1 and L_2 are the same.

27. L_1: $\mathbf{r} = t\langle 1, 1, 1 \rangle$
  L_2: $\mathbf{r} = \langle 6, 6, 6 \rangle + t\langle -3, -3, -3 \rangle$

28. L_1: $x = 2 + 3t, y = -5 + 6t, z = 4 - 9t$
  L_2: $x = 5 - t, y = 1 - 2t, z = -5 + 3t$

29. Given that the lines L_1 and L_2 defined by the parametric equations

  L_1: $x = 3 + 2t, y = 4 - t, z = -1 + 6t$
  L_2: $x = 5 - s, y = 3 + \frac{1}{2}s, z = 5 - 3s$

  are the same.

  (a) Find a value of t such that $(-7, 9, -31)$ is a point on L_1.
  (b) Find a value of s such that $(-7, 9, -31)$ is a point on L_2.

30. Determine which of the following lines are perpendicular and which are parallel.

  (a) $\mathbf{r} = \langle 1, 0, 2 \rangle + t\langle 9, -12, 6 \rangle$
  (b) $x = 1 + 9t, y = 12t, z = 2 - 6t$
  (c) $x = 2t, y = -3t, z = 4t$
  (d) $x = 5 + t, y = 4t, z = 3 + \frac{5}{2}t$
  (e) $x = 1 + t, y = \frac{3}{2}t, z = 2 - \frac{3}{2}t$
  (f) $\dfrac{x + 1}{-3} = \dfrac{y + 6}{4} = \dfrac{z - 3}{-2}$

In Problems 31 and 32, determine the points of intersection of the given line and the three coordinate planes.

31. $x = 4 - 2t, y = 1 + 2t, z = 9 + 3t$

32. $\dfrac{x-1}{2} = \dfrac{y+2}{3} = \dfrac{z-4}{2}$

In Problems 33–36, determine whether the lines L_1 and L_2 intersect. If so, find the point of intersection.

33. L_1: $x = 4 + t, y = 5 + t, z = -1 + 2t$
L_2: $x = 6 + 2s, y = 11 + 4s, z = -3 + s$

34. L_1: $x = 1 + t, y = 2 - t, z = 3t$
L_2: $x = 2 - s, y = 1 + s, z = 6s$

35. L_1: $x = 2 - t, y = 3 + t, z = 1 + t$
L_2: $x = 4 + s, y = 1 + s, z = 1 - s$

36. L_1: $x = 3 - t, y = 2 + t, z = 8 + 2t$
L_2: $x = 2 + 2s, y = -2 + 3s, z = -2 + 8s$

In Problems 37 and 38, determine whether the given points lie on the same line.

37. $(4, 3, -5), (10, 15, -11), (-1, -7, 0)$

38. $(1, 6, 6), (-11, 10, -2), (-2, 7, 5)$

39. Find parametric equations for the line segment joining the points $(2, 5, 9)$ and $(6, -1, 3)$.

40. Find parametric equations for the line segment joining the midpoints of the given line segments.

$$x = 1 + 2t, y = 2 - t, z = 4 - 3t, 1 \le t \le 2$$
$$x = -2 + 4t, y = 6 + t, z = 5 + 6t, -1 \le t \le 1$$

In Problems 41 and 42, find the angle between the given lines L_1 and L_2. The angle between two lines is the angle between their direction vectors $\mathbf{v}_1$ and $\mathbf{v}_2$.

41. L_1: $x = 4 - t, y = 3 + 2t, z = -2t$
L_2: $x = 5 + 2s, y = 1 + 3s, z = 5 - 6s$

42. L_1: $\dfrac{x-1}{2} = \dfrac{y+5}{7} = \dfrac{z-1}{-1}$
L_2: $\dfrac{x+3}{-2} = y - 9 = \dfrac{z}{4}$

In Problems 43 and 44, the lines L_1 and L_2 lie in the same plane. Find parametric equations for the line through the indicated point that is perpendicular to this plane.

43. L_1: $x = 3 + t, y = -2 + t, z = 9 + t$
L_2: $x = 1 - 2s, y = 5 + s, z = -2 - 5s$; $(4, 1, 6)$

44. L_1: $\dfrac{x-1}{3} = \dfrac{y+1}{2} = \dfrac{z}{4}$
L_2: $\dfrac{x+4}{6} = \dfrac{y-6}{4} = \dfrac{z-10}{8}$; $(1, -1, 0)$

In Problems 45 and 46, show that L_1 and L_2 are skew lines.

45. L_1: $x = -3 + t, y = 7 + 3t, z = 5 + 2t$
L_2: $x = 4 + s, y = 8 - 2s, z = 10 - 4s$

46. L_1: $x = 6 + 2t, y = 6t, z = -8 + 10t$
L_2: $x = 7 + 8s, y = 4 - 4s, z = 3 - 24s$

☰ Think About It

47. Suppose L_1 and L_2 are skew lines. Let L_1 and L_2 be points on line L_1 and let P_3 and P_4 be points on line L_2. Use the vector $\overrightarrow{P_1P_3}$, shown in FIGURE 11.5.5, to show that the shortest distance d between L_1 and L_2 (and hence the shortest distance between the planes) is

$$d = \frac{|\overrightarrow{P_1P_3} \cdot (\overrightarrow{P_1P_2} \times \overrightarrow{P_3P_4})|}{|\overrightarrow{P_1P_2} \times \overrightarrow{P_3P_4}|}.$$

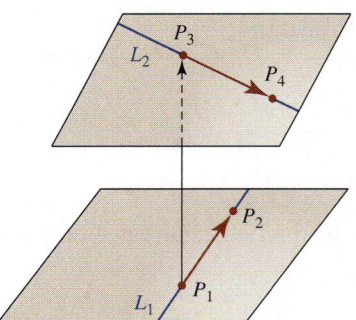

FIGURE 11.5.5 Distance between two skew lines in Problem 47

48. Using the result in Problem 47, find the distance between the skew lines in Problem 45.

11.6 Planes

▌ Introduction In this section we use vector methods to obtain equations of planes.

▌ Vector Equation FIGURE 11.6.1(a) illustrates the fact that there are an infinite number of planes $S_1, S_2, S_3, \ldots$ that pass through a given point $P_0(x_0, y_0, z_0)$. However, as shown in Figure 11.6.1(b), if a point P_0 and a nonzero vector $\mathbf{n}$ are specified, there is only *one* plane S containing P_0 with $\mathbf{n}$ **normal**, or perpendicular, to the plane. Moreover, if $P(x, y, z)$ represents any point on the plane, and $\mathbf{r} = \overrightarrow{OP}$, $\mathbf{r}_0 = \overrightarrow{OP_0}$, then as shown in Figure 11.6.1(c), $\mathbf{r} - \mathbf{r}_0$ lies in the plane S. It follows that a **vector equation** of the plane is

$$\mathbf{n} \cdot (\mathbf{r} - \mathbf{r}_0) = 0. \tag{1}$$

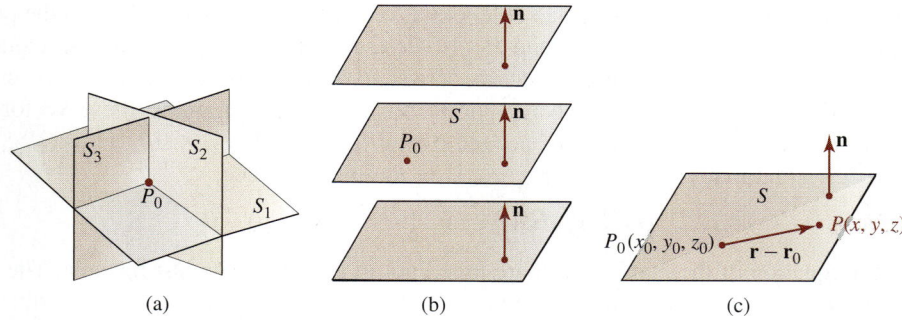

FIGURE 11.6.1 A point P_0 and a vector $\mathbf{n}$ determine a plane

❚ **Rectangular Equation**   Specifically, if the normal vector is $\mathbf{n} = a\mathbf{i} + b\mathbf{j} + c\mathbf{k}$, then (1) yields a **rectangular** or **Cartesian equation** of the plane containing $P_0(x_0, y_0, z_0)$:

$$a(x - x_0) + b(y - y_0) + c(z - z_0) = 0. \tag{2}$$

Equation (2) is called the **point-normal** form of the equation of a plane.

EXAMPLE 1   Equation of a Plane

Find an equation of the plane that contains the point $(4, -1, 3)$ and is perpendicular to the vector $\mathbf{n} = 2\mathbf{i} + 8\mathbf{j} - 5\mathbf{k}$.

Solution   It follows immediately from (2) with $x_0 = 4$, $y_0 = -1$, $z_0 = 3$, $a = 2$, $b = 8$, $c = -5$ that

$$2(x - 4) + 8(y + 1) - 5(z - 3) = 0 \quad \text{or} \quad 2x + 8y - 5z + 15 = 0. \quad ∎$$

The equation in (2) can always be written as $ax + by + cz + d = 0$ by identifying $d = -ax_0 - by_0 - cz_0$. Conversely, we shall now prove that a **linear equation**

$$ax + by + cz + d = 0, \tag{3}$$

a, b, c not all zero, is a plane.

Theorem 11.6.1   Plane and Normal Vector

The graph of a linear equation $ax + by + cz + d = 0$, a, b, c not all zero, is a plane with normal vector $\mathbf{n} = a\mathbf{i} + b\mathbf{j} + c\mathbf{k}$.

PROOF   Suppose x_0, y_0, and z_0 are numbers that satisfy the given equation. Then, $ax_0 + by_0 + cz_0 + d = 0$ implies that $d = -ax_0 - by_0 - cz_0$. Replacing this value of d in the original equation gives, after simplifying,

$$a(x - x_0) + b(y - y_0) + c(z - z_0) = 0$$

or, in terms of vectors,

$$[a\mathbf{i} + b\mathbf{j} + c\mathbf{k}] \cdot [(x - x_0)\mathbf{i} + (y - y_0)\mathbf{j} + (z - z_0)\mathbf{k}] = 0.$$

This last equation implies that $a\mathbf{i} + b\mathbf{j} + c\mathbf{k}$ is normal to the plane containing the point (x_0, y_0, z_0) and the vector

$$(x - x_0)\mathbf{i} + (y - y_0)\mathbf{j} + (z - z_0)\mathbf{k}. \quad ∎$$

EXAMPLE 2   Normal Vector to a Plane

By reading off the coefficients of x, y, and z in the linear equation $3x - 4y + 10z - 8 = 0$ we obtain a normal vector

$$\mathbf{n} = 3\mathbf{i} - 4\mathbf{j} + 10\mathbf{k}$$

to the plane.   ∎

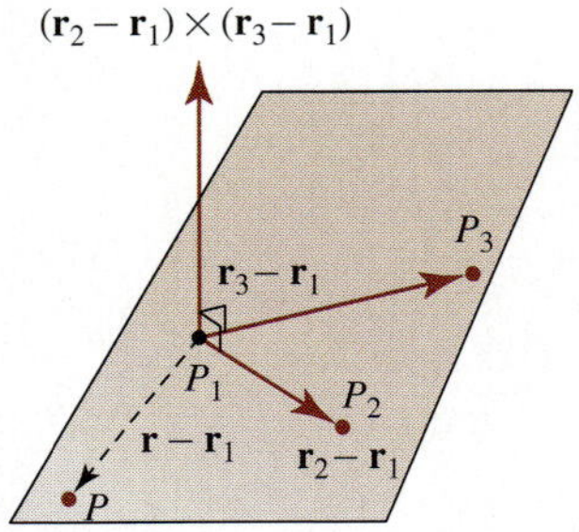

$(\mathbf{r}_2 - \mathbf{r}_1) \times (\mathbf{r}_3 - \mathbf{r}_1)$

FIGURE 11.6.2 Plane determined by three noncollinear points

Of course, a nonzero scalar multiple of a normal vector $\mathbf{n}$ is still perpendicular to the plane.

Three noncollinear points P_1, P_2, and P_3 also determine a plane S. To obtain an equation of the plane, we need only form two vectors between two pairs of points. As shown in FIGURE 11.6.2, their cross product is a vector normal to the plane containing these vectors. If $P(x, y, z)$ represents any point on the plane, and $\mathbf{r} = \overrightarrow{OP}$, $\mathbf{r}_1 = \overrightarrow{OP_1}$, $\mathbf{r}_2 = \overrightarrow{OP_2}$, $\mathbf{r}_3 = \overrightarrow{OP_3}$, then $\mathbf{r} - \mathbf{r}_1$ (or, for that matter, $\mathbf{r} - \mathbf{r}_2$ or $\mathbf{r} - \mathbf{r}_3$) is in the plane. Hence,

$$[(\mathbf{r}_2 - \mathbf{r}_1) \times (\mathbf{r}_3 - \mathbf{r}_1)] \cdot (\mathbf{r} - \mathbf{r}_1) = 0 \qquad (4)$$

is a vector equation of the plane S. You are urged not to memorize the last formula. The procedure is the same as (1) with the exception that the vector normal to the plane is obtained by means of the cross product.

EXAMPLE 3 Equation of a Plane

Find an equation of the plane that contains $(1, 0, -1)$, $(3, 1, 4)$, and $(2, -2, 0)$.

Solution We need three vectors. Pairing the points on the left as shown yields the vectors on the right. The order in which we subtract is irrelevant.

$$\left.\begin{array}{c} (1, 0, -1) \\ (3, 1, 4) \end{array}\right\} \mathbf{u} = 2\mathbf{i} + \mathbf{j} + 5\mathbf{k}$$

$$\left.\begin{array}{c} (3, 1, 4) \\ (2, -2, 0) \end{array}\right\} \mathbf{v} = \mathbf{i} + 3\mathbf{j} + 4\mathbf{k}$$

$$\left.\begin{array}{c} (2, -2, 0) \\ (x, y, z) \end{array}\right\} \mathbf{w} = (x - 2)\mathbf{i} + (y + 2)\mathbf{j} + z\mathbf{k}$$

Now,

$$\mathbf{u} \times \mathbf{v} = \begin{vmatrix} \mathbf{i} & \mathbf{j} & \mathbf{k} \\ 2 & 1 & 5 \\ 1 & 3 & 4 \end{vmatrix} = -11\mathbf{i} - 3\mathbf{j} + 5\mathbf{k}$$

is a vector normal to the plane containing the given points. Hence from (1), a vector equation of the plane is $(\mathbf{u} \times \mathbf{v}) \cdot \mathbf{w} = 0$. The latter equation yields

$$-11(x - 2) - 3(y + 2) + 5z = 0 \qquad \text{or} \qquad -11x - 3y + 5z + 16 = 0. \qquad \blacksquare$$

▮ **Perpendicular and Parallel Planes** FIGURE 11.6.3 illustrates the plausibility of the following definition about **perpendicular** and **parallel** planes.

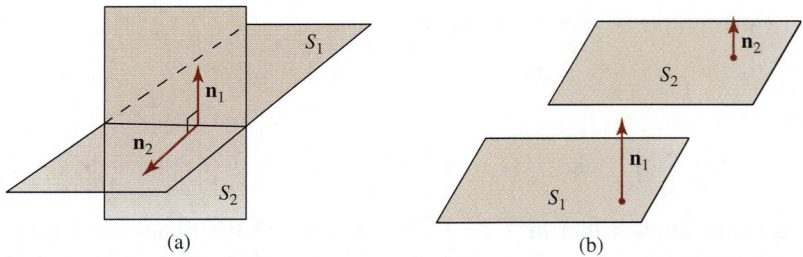

(a) (b)

FIGURE 11.6.3 Perpendicular planes (a); parallel planes (b)

Definition 11.6.1 Perpendicular and Parallel Planes

Two planes S_1 and S_2 with normal vectors $\mathbf{n}_1$ and $\mathbf{n}_2$, respectively, are

  (*i*) **perpendicular** if $\mathbf{n}_1 \cdot \mathbf{n}_2 = 0$, and

 (*ii*) **parallel** if $\mathbf{n}_2 = k\mathbf{n}_1$, for some nonzero scalar k.

EXAMPLE 4 Parallel Planes

The three planes given by

$$S_1: \quad 2x - 4y + 8z = 7$$
$$S_2: \quad x - 2y + 4z = 0$$
$$S_3: \quad -3x + 6y - 12z = 1$$

are parallel, since their respective normal vectors

$$\mathbf{n}_1 = 2\mathbf{i} - 4\mathbf{j} + 8\mathbf{k}$$
$$\mathbf{n}_2 = \mathbf{i} - 2\mathbf{j} + 4\mathbf{k} = \frac{1}{2}\mathbf{n}_1$$
$$\mathbf{n}_3 = -3\mathbf{i} + 6\mathbf{j} - 12\mathbf{k} = -\frac{3}{2}\mathbf{n}_1$$

are parallel. ∎

▌ Graphs The following lists are some guidelines for sketching the graph of a plane.

> ### Guidelines for Graphing a Plane
>
> - The graphs of each of the equations $x = x_0$, $y = y_0$, $z = z_0$, where x_0, y_0, and z_0 are constants, are planes perpendicular to the x-, y-, and z-axes, respectively.
> - To graph a linear equation $ax + by + cz + d = 0$, find the x-, y-, and z-intercepts or, if necessary, find the trace of the plane in each coordinate plane.

A **trace** of a plane in a coordinate plane is the line of intersection of the plane with a coordinate plane.

EXAMPLE 5 Graph

Graph the equation $2x + 3y + 6z = 18$.

Solution Setting:

$$y = 0, z = 0 \text{ gives } x = 9$$
$$x = 0, z = 0 \text{ gives } y = 6$$
$$x = 0, y = 0 \text{ gives } z = 3.$$

As shown in FIGURE 11.6.4, we use the x-, y-, and z-intercepts $(9, 0, 0)$, $(0, 6, 0)$, and $(0, 0, 3)$ to draw the graph of the plane in the first octant. ∎

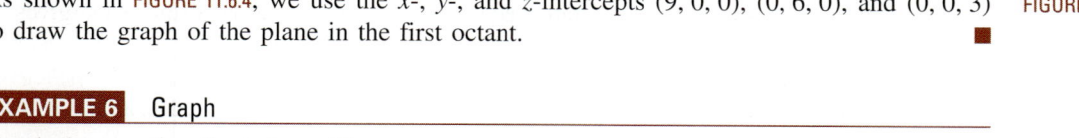

FIGURE 11.6.4 Plane in Example 5

EXAMPLE 6 Graph

Graph the equation $6x + 4y = 12$.

Solution In two dimensions the graph of the equation is a line with x-intercept $(2, 0)$ and y-intercept $(0, 3)$. However, in three dimensions this line is the trace of a plane in the xy-coordinate plane. Since z is not specified, it can be any real number. In other words, (x, y, z) is a point on the plane provided that x and y are related by the given equation. As shown in FIGURE 11.6.5, the graph is a plane parallel to the z-axis. ∎

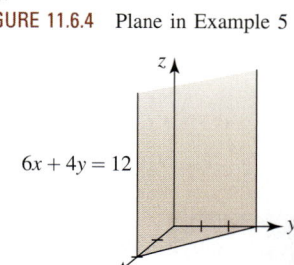

FIGURE 11.6.5 Plane in Example 6

EXAMPLE 7 Graph

Graph the equation $x + y - z = 0$.

Solution First observe that the plane passes through the origin $(0, 0, 0)$. Now, the trace of the plane in the xz-plane $(y = 0)$ is $z = x$, whereas its trace in the yz-plane $(x = 0)$ is $z = y$. Drawing these two lines leads to the graph given in FIGURE 11.6.6. ∎

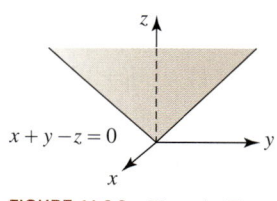

FIGURE 11.6.6 Plane in Example 7

Two planes S_1 and S_2 that are not parallel must intersect in a line L. See FIGURE 11.6.7. Example 8 illustrates one way of finding parametric equations for the line of intersection. In Example 9 we see how to find a point of intersection (x_0, y_0, z_0) of a plane S and a line L. See FIGURE 11.6.8.

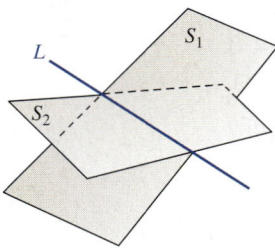

FIGURE 11.6.7 Two intersecting planes

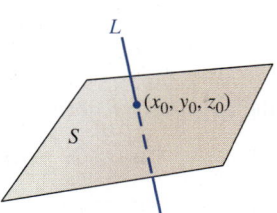

FIGURE 11.6.8 Intersection of a line and a plane

EXAMPLE 8 Line of Intersection

Find parametric equations for the line of intersection of $x - y + 2z = 1$ and $x + y + z = 3$.

Solution In a system of two equations and three unknowns, we choose one variable arbitrarily, say $z = t$, and solve for x and y from

$$x - y = 1 - 2t$$
$$x + y = 3 - t.$$

Solving the system then gives

$$x = 2 - \frac{3}{2}t, \quad y = 1 + \frac{1}{2}t, \quad z = t.$$

These are parametric equations for the line L of intersection of the given planes. The line is shown in red, the plane $x - y + 2z = 1$ is blue, and the plane $x + y + z = 3$ is purple in FIGURE 11.6.9. ■

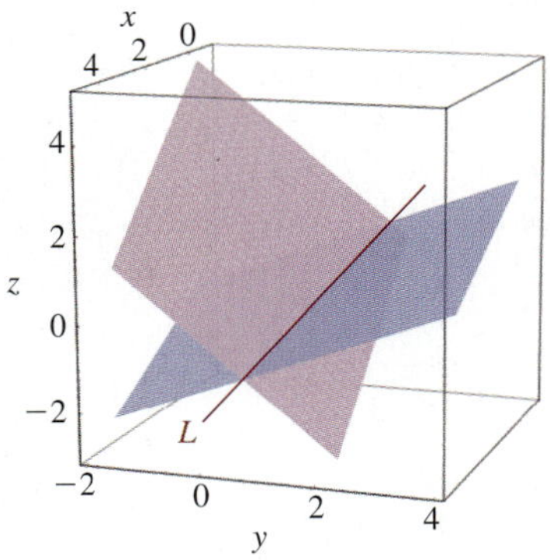

FIGURE 11.6.9 Line L of intersection of two planes in Example 8

The line in Example 8 can be obtained in another way. See Problem 52 in Exercises 11.6.

EXAMPLE 9 Point of Intersection

Find the point of intersection of the plane $3x - 2y + z = -5$ and the line $x = 1 + t$, $y = -2 + 2t, z = 4t$.

Solution If (x_0, y_0, z_0) denotes the point of intersection, then we must have

$$3x_0 - 2y_0 + z_0 = -5 \quad \text{and} \quad x_0 = 1 + t_0, y_0 = -2 + 2t_0, z_0 = 4t_0$$

for some number t_0. Substituting the latter equations into the equation of the plane gives

$$3(1 + t_0) - 2(-2 + 2t_0) + 4t_0 = -5 \quad \text{or} \quad t_0 = -4.$$

From the parametric equations for the line we then obtain $x_0 = -3$, $y_0 = -10$, and $z_0 = -16$. The point of intersection is $(-3, -10, -16)$. ■

Exercises 11.6 Answers to selected odd-numbered problems begin on page ANS-36.

≡ **Fundamentals**

In Problems 1–6, find an equation of the plane that contains the given point and is perpendicular to the indicated vector.

1. $(5, 1, 3)$; $2\mathbf{i} - 3\mathbf{j} + 4\mathbf{k}$ **2.** $(1, 2, 5)$; $4\mathbf{i} - 2\mathbf{j}$

3. $(6, 10, -7)$; $-5\mathbf{i} + 3\mathbf{k}$ **4.** $(0, 0, 0)$; $6\mathbf{i} - \mathbf{j} + 3\mathbf{k}$

5. $\left(\frac{1}{2}, \frac{3}{4}, -\frac{1}{2}\right)$; $6\mathbf{i} + 8\mathbf{j} - 4\mathbf{k}$ **6.** $(-1, 1, 0)$; $-\mathbf{i} + \mathbf{j} - \mathbf{k}$

In Problems 7–12, find, if possible, an equation of a plane that contains the given points.

7. $(3, 5, 2), (2, 3, 1), (-1, -1, 4)$

8. $(0, 1, 0), (0, 1, 1), (1, 3, -1)$

9. $(0, 0, 0), (1, 1, 1), (3, 2, -1)$

10. $(0, 0, 3), (0, -1, 0), (0, 0, 6)$

11. $(1, 2, -1), (4, 3, 1), (7, 4, 3)$

12. $(2, 1, 2), (4, 1, 0), (5, 0, -5)$

In Problems 13–22, find an equation of the plane that satisfies the given conditions.

13. Contains $(2, 3, -5)$ and is parallel to $x + y - 4z = 1$

14. Contains the origin and is parallel to $5x - y + z = 6$

15. Contains $(3, 6, 12)$ and is parallel to the xy-plane

16. Contains $(-7, -5, 18)$ and is perpendicular to the y-axis

17. Contains the lines $x = 1 + 3t$, $y = 1 - t$, $z = 2 + t$; $x = 4 + 4s$, $y = 2s$, $z = 3 + s$

18. Contains the lines $\dfrac{x - 1}{2} = \dfrac{y + 1}{-1} = \dfrac{z - 5}{6}$; $\mathbf{r} = \langle 1, -1, 5 \rangle + t\langle 1, 1, -3 \rangle$

19. Contains the parallel lines $x = 1 + t$, $y = 1 + 2t$, $z = 3 + t$; $x = 3 + s$, $y = 2s$, $z = -2 + s$

20. Contains the point $(4, 0, -6)$ and the line $x = 3t$, $y = 2t$, $z = -2t$

21. Contains $(2, 4, 8)$ and is perpendicular to the line $x = 10 - 3t$, $y = 5 + t$, $z = 6 - \frac{1}{2}t$

22. Contains $(1, 1, 1)$ and is perpendicular to the line through $(2, 6, -3)$ and $(1, 0, -2)$

23. Determine which of the following planes are perpendicular and which are parallel.
 (a) $2x - y + 3z = 1$ **(b)** $x + 2y + 2z = 9$
 (c) $x + y - \frac{3}{2}z = 2$ **(d)** $-5x + 2y + 4z = 0$
 (e) $-8x - 8y + 12z = 1$ **(f)** $-2x + y - 3z = 5$

24. Find parametric equations for the line that contains $(-4, 1, 7)$ and is perpendicular to the plane $-7x + 2y + 3z = 1$.

25. Determine which of the following planes are perpendicular to the line $x = 4 - 6t$, $y = 1 + 9t$, $z = 2 + 3t$.
 (a) $4x + y + 2z = 1$ **(b)** $2x - 3y + z = 4$
 (c) $10x - 15y - 5z = 2$ **(d)** $-4x + 6y + 2z = 9$

26. Determine which of the following planes are parallel to the line $(1 - x)/2 = (y + 2)/4 = z - 5$.
 (a) $x - y + 3z = 1$ **(b)** $6x - 3y = 1$
 (c) $x - 2y + 5z = 0$ **(d)** $-2x + y - 2z = 7$

In Problems 27–30, find parametric equations for the line of intersection of the given planes.

27. $5x - 4y - 9z = 8$ **28.** $x + 2y - z = 2$
 $x + 4y + 3z = 4$ $3x - y + 2z = 1$

29. $4x - 2y - z = 1$ **30.** $2x - 5y + z = 0$
 $x + y + 2z = 1$ $y = 0$

In Problems 31–34, find the point of intersection of the given plane and line.

31. $2x - 3y + 2z = -7$; $x = 1 + 2t$, $y = 2 - t$, $z = -3t$

32. $x + y + 4z = 12$; $x = 3 - 2t$, $y = 1 + 6t$, $z = 2 - \frac{1}{2}t$

33. $x + y - z = 8$; $x = 1$, $y = 2$, $z = 1 + t$

34. $x - 3y + 2z = 0$; $x = 4 + t$, $y = 2 + t$, $z = 1 + 5t$

In Problems 35 and 36, find parametric equations for the line through the indicated point that is parallel to the given planes.

35. $x + y - 4z = 2$, $2x - y + z = 10$; $(5, 6, -12)$

36. $2x + z = 0$, $-x + 3y + z = 1$; $(-3, 5, -1)$

In Problems 37 and 38, find an equation of the plane that contains the given line and that is perpendicular to the indicated plane.

37. $x = 4 + 3t$, $y = -t$, $z = 1 + 5t$; $x + y + z = 7$

38. $\dfrac{2 - x}{3} = \dfrac{y + 2}{5} = \dfrac{z - 8}{2}$; $2x - 4y - z + 16 = 0$

In Problems 39–44, graph the given equation.

39. $5x + 2y + z = 10$ **40.** $3x + 2z = 9$

41. $-y - 3z + 6 = 0$ **42.** $3x + 4y - 2z - 12 = 0$

43. $-x + 2y + z = 4$ **44.** $3x - y - 6 = 0$

45. Show that the line $x = -2t$, $y = t$, $z = -t$ is
 (a) parallel to but above the plane $x + y - z = 1$,
 (b) parallel to but below the plane $-3x - 4y + 2z = 8$.

46. Let $P_0(x_0, y_0, z_0)$ be a point on the plane $ax + by + cz + d = 0$ and let $\mathbf{n}$ be a normal vector to the plane. See FIGURE 11.6.10. Show that if $P_1(x_1, y_1, z_1)$ is any point not on the plane, then the **distance** D **from a point to a plane** is given by
$$D = \frac{|ax_1 + by_1 + cz_1 + d|}{\sqrt{a^2 + b^2 + c^2}}.$$

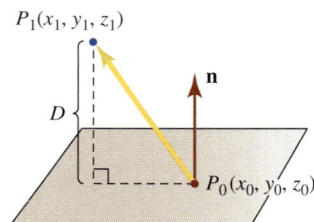

FIGURE 11.6.10 Distance between a point and a plane in Problem 46

47. Use the result of Problem 46 to find the distance from the point $(2, 1, 4)$ to the plane $x - 3y + z - 6 = 0$.

48. **(a)** Show that the planes $x - 2y + 3z = 3$, and $-4x + 8y - 12z = 7$ are parallel.
 (b) Find the distance between the planes in part (a).

As shown in FIGURE 11.6.11, the **angle between two planes** is defined to be the acute angle between their normal vectors. In Problems 49 and 50, find the angles between the given planes.

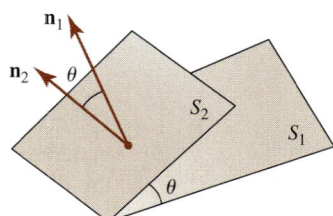

FIGURE 11.6.11 Angle between two planes in Problems 49 and 50

49. $x - 3y + 2z = 14$, $-x + y + z = 10$

50. $2x + 6y + 3z = 13$, $4x - 2y + 4z = -7$

☰ Think About It

51. If you have ever sat at a four-legged table that rocks, you might consider replacing it with a three-legged table. Why?

52. Reread Example 8. Find parametric equations for the line L of intersection of the two planes using the fact that L lies in both planes and so must be perpendicular to the normal vector of each plane. If you obtain an answer that differs from the equations in Example 8, show that the answers are equivalent.

53. (a) Find an equation of the plane whose points are equidistant from $(1, -2, 3)$ and $(2, 5, -1)$.

(b) Find the distance between the plane and the points given in part (a).

11.7 Cylinders and Spheres

■ **Introduction** In 2-space the graph of the equation $x^2 + y^2 = 1$ is a circle centered at the origin in the xy-plane. However, in 3-space we can interpret the graph of the set

$$\{(x, y, z) \,|\, x^2 + y^2 = 1, z \text{ arbitrary}\}$$

as a **surface** that is the right circular cylinder shown in FIGURE 11.7.1(b).

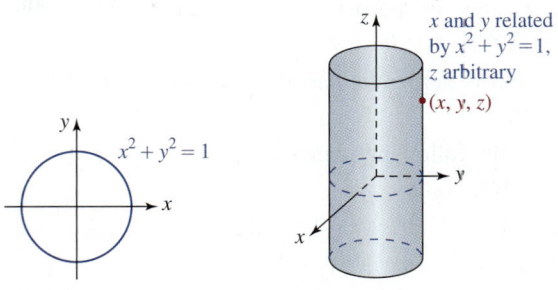

(a) Circle in 2-space   (b) Circular cylinder in 3-space

FIGURE 11.7.1   Interpretation of an equation of a circle in 2- and 3-space

Similarly, we have already seen in Section 11.6 that the graph of an equation such as $y + 2z = 2$ is a line in 2-space (the yz-plane), but in 3-space the graph of the set

$$\{(x, y, z) \,|\, y + 2z = 2, x \text{ arbitrary}\}$$

is the plane perpendicular to the yz-plane shown in FIGURE 11.7.2(b). Surfaces such as these are given a special name.

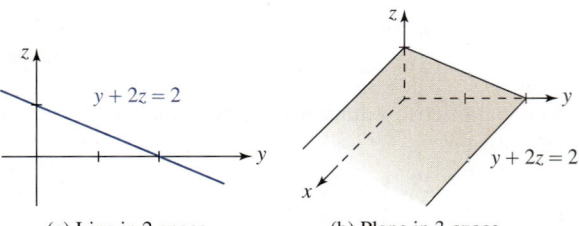

(a) Line in 2-space   (b) Plane in 3-space

FIGURE 11.7.2   Interpretation of an equation of a line in 2- and 3-space

■ **Cylinder** The surfaces illustrated in Figures 11.7.1 and 11.7.2 are called **cylinders**. We use the term *cylinder* in a more general sense than that of a right circular cylinder. Specifically, if C is a curve in a plane and L is a line not parallel to the plane, then the set of all points (x, y, z) generated by a moving line traversing C parallel to L is called a **cylinder**. The curve C is called the **directrix** of the cylinder. See FIGURE 11.7.3.

Thus, an equation of a curve in a coordinate plane, when considered in three dimensions, is an equation of a cylinder perpendicular to that coordinate plane.

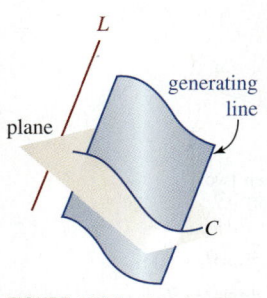

FIGURE 11.7.3   Moving line on C parallel to L generates a cylinder

- *If the graphs of $f(x, y) = c_1$, $g(y, z) = c_2$, and $h(x, z) = c_3$ are curves in the 2-space of their respective coordinate planes, then their graphs in 3-space are surfaces called cylinders. A cylinder is generated by a moving line that traverses the curve parallel to the coordinate axis that is represented by the variable missing in its equation.*

FIGURE 11.7.4 shows a curve C defined by $f(x, y) = c_1$ in the xy-plane and a collection of red lines called **rulings** that represent various positions of a generating line that is traversing C while moving parallel to the z-axis.

In the following example, we compare the graph of an equation in a coordinate plane with its interpretation as a cylinder in 3-space (FIGURES 11.7.5–11.7.8). As in Figure 11.7.2(b), we shall show only a portion of the cylinder.

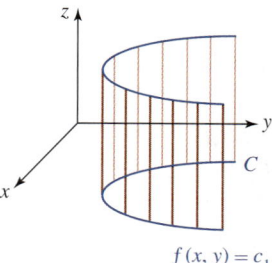

$f(x, y) = c_1$

FIGURE 11.7.4 Rulings of the cylinder $f(x, y) = c_1$

EXAMPLE 1 Cylinders

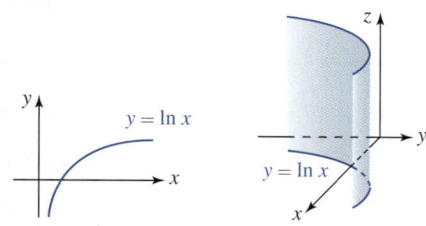

(a) Logarithmic curve (b) Logarithmic cylinder

FIGURE 11.7.5 Cylinder with rulings parallel to the z-axis

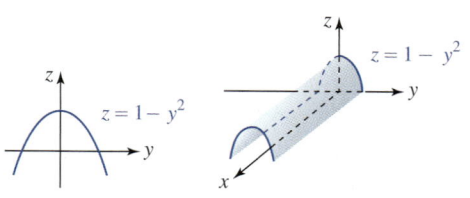

(a) Parabola (b) Parabolic cylinder

FIGURE 11.7.6 Cylinder with rulings parallel to the x-axis

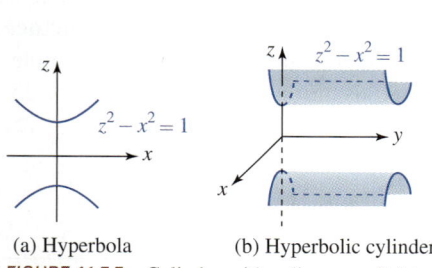

(a) Hyperbola (b) Hyperbolic cylinder

FIGURE 11.7.7 Cylinder with rulings parallel to the y-axis

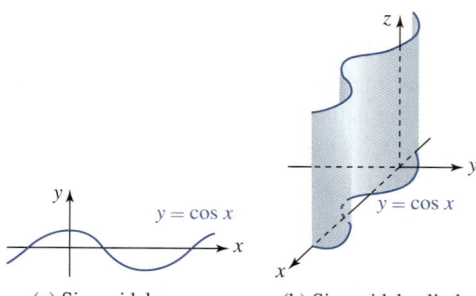

(a) Sinusoidal curve (b) Sinusoidal cylinder

FIGURE 11.7.8 Cylinder with rulings parallel to the z-axis ∎

❚ Spheres Like a circle, a sphere can be defined by means of the distance formula.

Definition 11.7.1 Sphere

A **sphere** is the set of all points $P(x, y, z)$ in 3-space that are equidistant from a fixed point called the **center**.

If r denotes the fixed distance, or **radius** of the sphere, and if the center is $P_1(a, b, c)$, then a point $P(x, y, z)$ is on the sphere if and only if $[d(P_1, P)]^2 = r^2$, or

$$(x - a)^2 + (y - b)^2 + (z - c)^2 = r^2. \tag{1}$$

EXAMPLE 2 Sphere

Graph $x^2 + y^2 + z^2 = 25$.

Solution We identify $a = 0, b = 0, c = 0$, and $r^2 = 25 = 5^2$ in (1), and so the graph of $x^2 + y^2 + z^2 = 25$ is a sphere of radius 5 whose center is at the origin. The graph of the equation is given in FIGURE 11.7.9. ∎

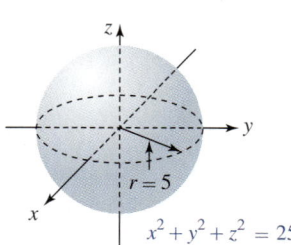

$x^2 + y^2 + z^2 = 25$

FIGURE 11.7.9 Sphere in Example 2

EXAMPLE 3 Sphere

Graph $(x - 5)^2 + (y - 7)^2 + (z - 6)^2 = 9$.

Solution In this case we identify $a = 5, b = 7, c = 6$, and $r^2 = 9$. From (1) we see that the graph of $(x - 5)^2 + (y - 7)^2 + (z - 6)^2 = 3^2$ is a sphere with center $(5, 7, 6)$ and radius 3. Its graph lies entirely in the first octant and is shown in FIGURE 11.7.10. ∎

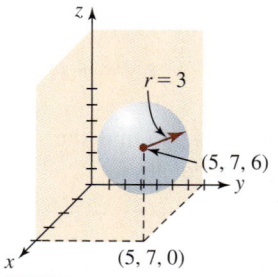

$(5, 7, 6)$

$(5, 7, 0)$

FIGURE 11.7.10 Sphere in Example 3

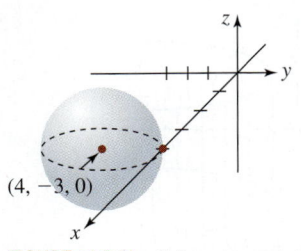

FIGURE 11.7.11 Sphere tangent to plane $y = 0$ in Example 4

EXAMPLE 4 Equation of a Sphere

Find an equation of the sphere whose center is $(4, -3, 0)$ that is tangent to the xz-plane.

Solution The perpendicular distance from the given point to the xz-plane $(y = 0)$, and hence the radius of the sphere, is the absolute value of the y-coordinate, $|-3| = 3$. Thus, an equation of the sphere is

$$(x - 4)^2 + (y + 3)^2 + z^2 = 3^2.$$

See FIGURE 11.7.11. ∎

EXAMPLE 5 Center and Radius

Find the center and radius of the sphere whose equation is

$$16x^2 + 16y^2 + 16z^2 - 16x + 8y - 32z + 16 = 0.$$

Solution Dividing by 16 and completing the square in x, y, and z yield

$$\left(x - \frac{1}{2}\right)^2 + \left(y + \frac{1}{4}\right)^2 + (z - 1)^2 = \frac{5}{16}.$$

The center and radius of the sphere are $\left(\frac{1}{2}, -\frac{1}{4}, 1\right)$ and $\frac{1}{4}\sqrt{5}$, respectively. ∎

■ **Trace of a Surface** We saw in Section 11.6 that the trace of a plane in a coordinate plane is the line of intersection of the plane with the coordinate plane. In general, a **trace of a surface** in *any* plane is the curve formed by the intersection of the surface and the plane. For example, in Figure 11.7.9 the trace of the sphere in the xy-plane $(z = 0)$ is the dashed circle $x^2 + y^2 = 25$. In the xz- and yz-planes, the traces of the sphere are the circles $x^2 + z^2 = 25$ and $y^2 + z^2 = 25$, respectively.

Exercises 11.7 Answers to selected odd-numbered problems begin on page ANS-37.

≡ **Fundamentals**

In Problems 1–16, sketch the graph of the given cylinder.

1. $y = x$ **2.** $z = -y$

3. $y = x^2$ **4.** $x^2 + z^2 = 25$

5. $y^2 + z^2 = 9$ **6.** $z = y^2$

7. $z = e^{-x}$ **8.** $z = 1 - e^y$

9. $y^2 - x^2 = 4$ **10.** $z = \cosh y$

11. $4x^2 + y^2 = 36$ **12.** $x = 1 - y^2$

13. $z = \sin x$ **14.** $y = \dfrac{1}{x^2}$

15. $yz = 1$ **16.** $z = x^3 - 3x$

In Problems 17–20, sketch the graph of the given equation.

17. $x^2 + y^2 + z^2 = 9$

18. $x^2 + y^2 + (z - 3)^2 = 16$

19. $(x - 1)^2 + (y - 1)^2 + (z - 1)^2 = 1$

20. $(x + 3)^2 + (y + 4)^2 + (z - 5)^2 = 4$

In Problems 21–24, find the center and radius of the sphere with the given equation.

21. $x^2 + y^2 + z^2 + 8x - 6y - 4z - 7 = 0$

22. $4x^2 + 4y^2 + 4z^2 + 4x - 12z + 9 = 0$

23. $x^2 + y^2 + z^2 - 16z = 0$

24. $x^2 + y^2 + z^2 - x + y = 0$

In Problems 25–32, find an equation of a sphere that satisfies the given conditions.

25. Center $(-1, 4, 6)$; radius $\sqrt{3}$

26. Center $(0, -3, 0)$; diameter $\frac{5}{2}$

27. Center $(1, 1, 4)$; tangent to the xy-plane

28. Center $(5, 2, -2)$; tangent to the yz-plane

29. Center on the positive y-axis; radius 2; tangent to $x^2 + y^2 + z^2 = 36$

30. Center on the line $x = 2t, y = 3t, z = 6t, t > 0$, at a distance 21 units from the origin; radius 5

31. Diameter has endpoints $(0, -4, 7)$ and $(2, 12, -3)$

32. Center $(-3, 1, 2)$; passing through the origin

In Problems 33–38, describe geometrically all points $P(x, y, z)$ whose coordinates satisfy the given condition(s).

33. $x^2 + y^2 + (z - 1)^2 = 4, 1 \le z \le 3$

34. $x^2 + y^2 + (z - 1)^2 = 4, z = 2$

35. $x^2 + y^2 + z^2 \ge 1$

36. $0 < (x - 1)^2 + (y - 2)^2 + (z - 3)^2 < 1$

37. $1 \le x^2 + y^2 + z^2 \le 9$

38. $1 \le x^2 + y^2 + z^2 \le 9, z \le 0$

11.8 Quadric Surfaces

▪ Introduction The equation of the sphere given in (1) of Section 11.7 is just a particular case of the general second-degree equation in three variables

$$Ax^2 + By^2 + Cz^2 + Dxy + Eyz + Fxz + Gx + Hy + Iz + J = 0, \qquad (1)$$

where $A, B, C, \ldots, J$ are constants. The graph of a second-degree equation of form (1) that describes a real set of points is said to be a **quadric surface**. For example, both the elliptical cylinder $x^2/4 + y^2/9 = 1$ and the parabolic cylinder $z = y^2$ are quadric surfaces. We conclude this chapter by considering the six additional quadric surfaces: the **ellipsoid**, **elliptic cone**, **elliptic paraboloid**, **hyperbolic paraboloid**, **hyperboloid of one sheet**, and the **hyperboloid of two sheets**.

▪ Ellipsoid The graph of any equation of the form

$$\frac{x^2}{a^2} + \frac{y^2}{b^2} + \frac{z^2}{c^2} = 1, \quad a > 0, b > 0, c > 0, \qquad (2)$$

is called an **ellipsoid**. When $a = b = c$, (2) is the equation of a sphere centered at the origin. For $|y_0| < b$, the equation

$$\frac{x^2}{a^2} + \frac{z^2}{c^2} = 1 - \frac{y_0^2}{b^2}$$

represents a family of ellipses (or circles if $a = c$) parallel to the xz-plane that are formed by slicing the surface by planes $y = y_0$. By choosing, in turn, $x = x_0$ and $z = z_0$, we would find that slices of the surface are ellipses (or circles) parallel to the yz- and xy-planes, respectively. FIGURE 11.8.1 summarizes a typical graph of an ellipsoid along with the traces of the surface in the three coordinate planes.

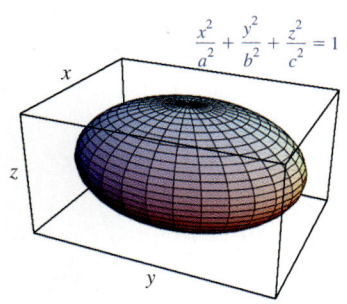

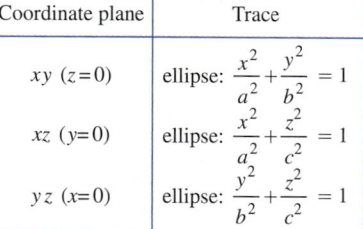

Coordinate plane	Trace
xy $(z=0)$	ellipse: $\dfrac{x^2}{a^2} + \dfrac{y^2}{b^2} = 1$
xz $(y=0)$	ellipse: $\dfrac{x^2}{a^2} + \dfrac{z^2}{c^2} = 1$
yz $(x=0)$	ellipse: $\dfrac{y^2}{b^2} + \dfrac{z^2}{c^2} = 1$

(a) *Mathematica* generated graph

(b) Traces of surface in coordinate planes

FIGURE 11.8.1 Ellipsoid

▪ Elliptic Cone The graph of an equation of the form

$$\frac{x^2}{a^2} + \frac{y^2}{b^2} = \frac{z^2}{c^2}, \quad a > 0, b > 0, c > 0, \qquad (3)$$

is called an **elliptic cone** (or circular if cone $a = b$). For arbitrary z_0, planes parallel to the xy-plane slice the surface in ellipses whose equations are

$$\frac{x^2}{a^2} + \frac{y^2}{b^2} = \frac{z_0^2}{c^2}.$$

FIGURE 11.8.2 summarizes a typical graph of an elliptic cone along with the traces in the coordinate planes.

As we see next, there are two kinds of paraboloids: elliptic and hyperbolic.

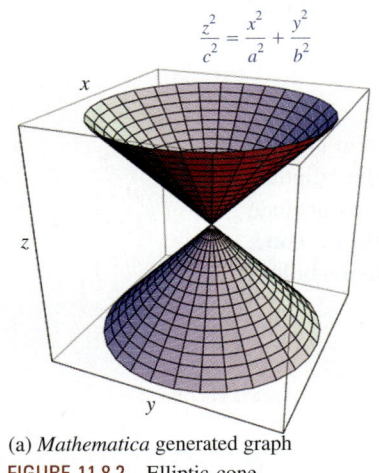

$$\frac{z^2}{c^2} = \frac{x^2}{a^2} + \frac{y^2}{b^2}$$

Coordinate plane	Trace
xy $(z=0)$	point: $(0, 0)$
xz $(y=0)$	lines: $z = \pm \dfrac{c}{a} x$
yz $(x=0)$	lines: $z = \pm \dfrac{c}{b} y$

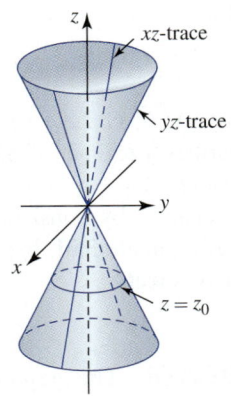

(a) *Mathematica* generated graph

FIGURE 11.8.2 Elliptic cone

(b) Traces of surface in coordinate planes

■ **Elliptic Paraboloid**   The graph of an equation of the form

$$cz = \frac{x^2}{a^2} + \frac{y^2}{b^2}, \quad a > 0, b > 0, \tag{4}$$

is called an **elliptic paraboloid**. In FIGURE 11.8.3(b) we see that for $c > 0$, planes $z = z_0 > 0$, parallel to the xy-plane, slice the surface in ellipses whose equations are

$$\frac{x^2}{a^2} + \frac{y^2}{b^2} = cz_0.$$

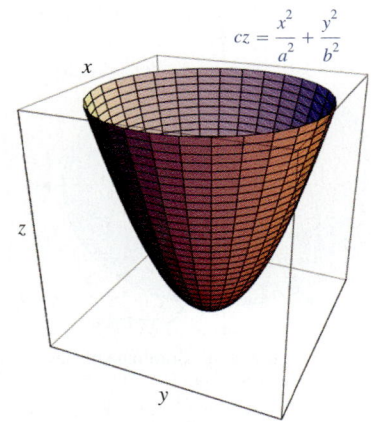

$$cz = \frac{x^2}{a^2} + \frac{y^2}{b^2}$$

Coordinate plane	Trace
xy $(z=0)$	point: $(0, 0)$
xz $(y=0)$	parabola: $cz = \dfrac{x^2}{a^2}$
yz $(x=0)$	parabola: $cz = \dfrac{y^2}{b^2}$

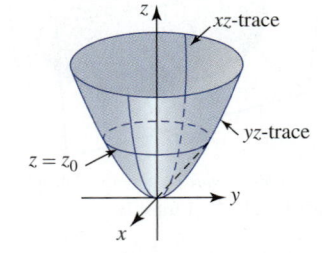

(a) *Mathematica* generated graph

FIGURE 11.8.3 Elliptic paraboloid

(b) Traces of surface in coordinate planes

■ **Hyperbolic Paraboloid**   The graph of an equation of the form

$$cz = \frac{y^2}{a^2} + \frac{x^2}{b^2}, \quad a > 0, b > 0, \tag{5}$$

is known as a **hyperbolic paraboloid**. Note that for $c > 0$, planes $z = z_0$, parallel to the xy-plane, cut the surface in hyperbolas whose equations are

$$\frac{y^2}{a^2} - \frac{x^2}{b^2} = cz_0.$$

The characteristic saddle shape of a hyperbolic paraboloid is shown in FIGURE 11.8.4.

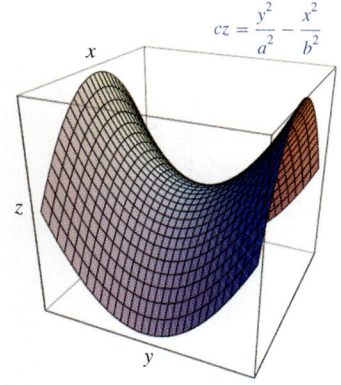

$$cz = \frac{y^2}{a^2} - \frac{x^2}{b^2}$$

Coordinate plane	Trace
xy $(z=0)$	lines: $y = \pm \dfrac{a}{b} x$
xz $(y=0)$	parabola: $cz = -\dfrac{x^2}{b^2}$
yz $(x=0)$	parabola: $cz = \dfrac{y^2}{a^2}$

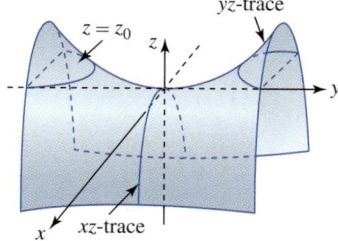

(a) *Mathematica* generated graph

FIGURE 11.8.4 Hyperbolic paraboloid

(b) Traces of surface in coordinate planes

There are two kinds of hyperboloids: of one sheet and of two sheets.

■ Hyperboloid of One Sheet   The graph of an equation of the form

$$\frac{x^2}{a^2} + \frac{y^2}{b^2} - \frac{z^2}{c^2} = 1, \quad a > 0, b > 0, c > 0, \tag{6}$$

is called a **hyperboloid of one sheet**. In this case, a plane $z = z_0$, parallel to the xy-plane, slices the surface into elliptical (or circular if $a = b$) cross sections. The equations of these ellipses are

$$\frac{x^2}{a^2} + \frac{y^2}{b^2} = 1 + \frac{z_0^2}{c^2}.$$

The smallest ellipse, $z_0 = 0$, corresponds to the trace in the xy-plane. A summary of the traces and a typical graph of (6) are given in FIGURE 11.8.5.

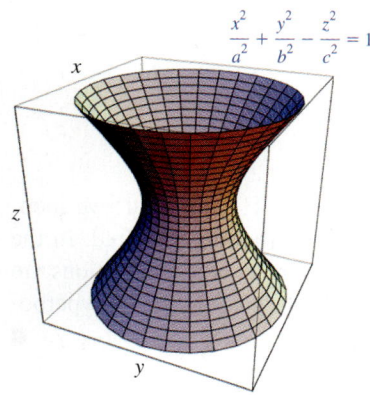

$$\frac{x^2}{a^2} + \frac{y^2}{b^2} - \frac{z^2}{c^2} = 1$$

Coordinate plane	Trace
xy $(z=0)$	ellipse: $\dfrac{x^2}{a^2} + \dfrac{y^2}{b^2} = 1$
xz $(y=0)$	hyperbola: $\dfrac{x^2}{a^2} - \dfrac{z^2}{c^2} = 1$
yz $(x=0)$	hyperbola: $\dfrac{y^2}{b^2} - \dfrac{z^2}{c^2} = 1$

(a) *Mathematica* generated graph

FIGURE 11.8.5 Hyperboloid of one sheet

(b) Traces of surface in coordinate planes

■ Hyperboloid of Two Sheets   As seen in FIGURE 11.8.6, a graph of

$$-\frac{x^2}{a^2} - \frac{y^2}{b^2} + \frac{z^2}{c^2} = 1, \quad a > 0, b > 0, c > 0, \tag{7}$$

is appropriately called a **hyperboloid of two sheets**. For $|z_0| > c$, the equation $x^2/a^2 + y^2/b^2 = z_0^2/c^2 - 1$ describes the elliptical curve of intersection of the surface with the plane $z = z_0$.

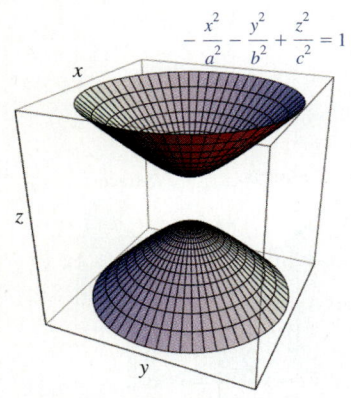

$$-\frac{x^2}{a^2} - \frac{y^2}{b^2} + \frac{z^2}{c^2} = 1$$

Coordinate plane	Trace
xy $(z=0)$	no locus
xz $(y=0)$	hyperbola: $-\dfrac{x^2}{a^2} + \dfrac{z^2}{c^2} = 1$
yz $(x=0)$	hyperbola: $-\dfrac{y^2}{b^2} + \dfrac{z^2}{c^2} = 1$

(a) *Mathematica* generated graph

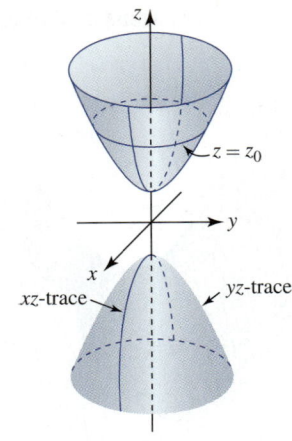

(b) Traces of surface in coordinate planes

FIGURE 11.8.6 Hyperboloid of two sheets

■ **Variation of the Equations** Interchanging the position of the variables in equations (2)–(7) does not change the basic nature of a surface, but *does* change the orientation of the surface in space. For example, graphs of the equations

$$\frac{x^2}{a^2} - \frac{y^2}{b^2} + \frac{z^2}{c^2} = 1 \qquad \text{and} \qquad -\frac{x^2}{a^2} + \frac{y^2}{b^2} + \frac{z^2}{c^2} = 1 \tag{8}$$

are still hyperboloids of one sheet. Similarly, the two minus signs in (7) that characterize hyperboloids of two sheets can occur anywhere in the equation. Similarly,

$$\frac{x^2}{a^2} + \frac{z^2}{b^2} = cy \qquad \text{and} \qquad \frac{y^2}{a^2} + \frac{z^2}{b^2} = cx \tag{9}$$

are paraboloids. Graphs of equations of the form

$$\frac{x^2}{a^2} - \frac{z^2}{b^2} = cy \qquad \text{and} \qquad \frac{y^2}{a^2} - \frac{z^2}{b^2} = cx \tag{10}$$

are hyperbolic paraboloids.

EXAMPLE 1 Quadric Surfaces

Identify

   (a) $y = x^2 + z^2$ and   **(b)** $y = x^2 - z^2$.

Compare the graphs.

Solution From the first equations in (9) and (10) with $a = 1$, $b = 1$, and $c = 1$, we identify the graph of (a) as a paraboloid and the graph of (b) as a hyperbolic paraboloid. In the case of equation (a), a plane $y = y_0$, $y_0 > 0$, slices the surface in circles whose equations are $y_0 = x^2 + z^2$. On the other hand, a plane $y = y_0$ slices the graph of equation (b) in hyperbolas $y_0 = x^2 - z^2$. The graphs are compared in FIGURE 11.8.7. ■

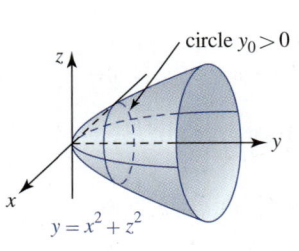

FIGURE 11.8.7 Surfaces in Example 1

EXAMPLE 2 Quadric Surfaces

Identify

(a) $2x^2 - 4y^2 + z^2 = 0$ and (b) $-2x^2 + 4y^2 + z^2 = -36$.

Solution

(a) From $\frac{1}{2}x^2 + \frac{1}{4}z^2 = y^2$, we identify the graph as an elliptic cone.

(b) From $\frac{1}{18}x^2 - \frac{1}{9}y^2 - \frac{1}{36}z^2 = 1$, we identify the graph as a hyperboloid of two sheets. ∎

■ **Origin at (h, k, l)** When the origin $(0, 0, 0)$ is translated to (h, k, l), the equations of the quadric surfaces become

$$\frac{(x-h)^2}{a^2} + \frac{(y-k)^2}{b^2} + \frac{(z-l)^2}{c^2} = 1 \quad \leftarrow \text{ellipsoid}$$

$$c(z-l) = \frac{(x-h)^2}{a^2} + \frac{(y-k)^2}{b^2} \quad \leftarrow \text{paraboloid}$$

$$\frac{(x-h)^2}{a^2} + \frac{(y-k)^2}{b^2} - \frac{(z-l)^2}{c^2} = 1 \quad \leftarrow \text{hyperboloid of one sheet}$$

and so on. You may have to complete the square to put a second-degree equation into one of these forms.

EXAMPLE 3 Paraboloid

Graph $z = 4 - x^2 - y^2$.

Solution By writing the equation as

$$-(z - 4) = x^2 + y^2$$

we recognize the equation of a paraboloid. The minus sign in front of the term on the left side of the equality indicates that the graph of the paraboloid opens downward from $(0, 0, 4)$. See FIGURE 11.8.8. ∎

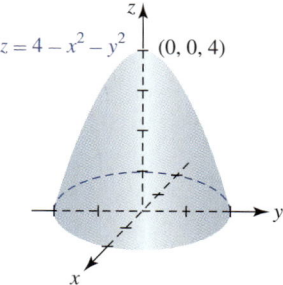

FIGURE 11.8.8 Paraboloid in Example 3

■ **Surfaces of Revolution** In Sections 6.3 and 6.4 we saw that a surface S could be generated by revolving a plane curve C about an axis. In the discussion that follows we shall find equations of **surfaces of revolution** when C is a curve in a coordinate plane and the axis of revolution is a coordinate axis.

For the sake of discussion, let us suppose that $f(y, z) = 0$ is an equation of a curve C in the yz-plane and that C is revolved about the z-axis generating a surface S. Let us also suppose for the moment that the y- and z-coordinates of points on C are nonnegative. If (x, y, z) denotes a general point on S that results from revolving the point $(0, y_0, z)$ on C, then we see from FIGURE 11.8.9 that the distance from (x, y, z) to $(0, 0, z)$ is the same as the distance from $(0, y_0, z)$ to $(0, 0, z)$; that is, $y_0 = \sqrt{x^2 + y^2}$. From the fact that $f(y_0, z) = 0$ we arrive at an equation for S:

$$f(\sqrt{x^2 + y^2}, z) = 0. \tag{11}$$

A curve in a coordinate plane can, of course, be revolved about each coordinate axis. If the curve C in the yz-plane defined by $f(y, z) = 0$ is now revolved about the y-axis, it can be shown that an equation of the resulting surface of revolution is

$$f(y, \sqrt{x^2 + z^2}) = 0. \tag{12}$$

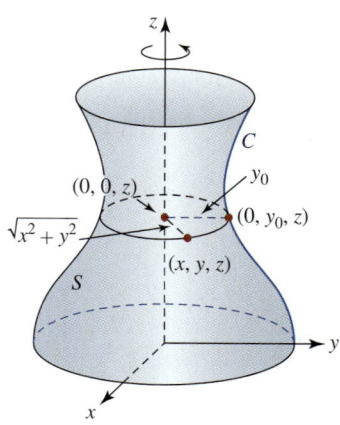

FIGURE 11.8.9 Surface S of revolution

Finally, we note that if there are points $(0, y, z)$ on C for which the y- or z-coordinates are negative, then we replace $\sqrt{x^2 + y^2}$ in (11) by $\pm\sqrt{x^2 + y^2}$ and $\sqrt{x^2 + z^2}$ in (12) by $\pm\sqrt{x^2 + z^2}$.

Equations of surfaces of revolution generated when a curve in the xy- or xz-plane is revolved about a coordinate axis are analogous to (11) and (12). As the following table shows, an equation of a surface generated by revolving a curve in a coordinate plane about the

$$\left.\begin{array}{l} x = \text{axis} \\ y = \text{axis} \\ z = \text{axis} \end{array}\right\} \quad \text{involves the term} \quad \left\{\begin{array}{l} \sqrt{y^2 + z^2} \\ \sqrt{x^2 + z^2} \\ \sqrt{x^2 + y^2}. \end{array}\right.$$

Equation of Curve C	Axis of Revolution	Equation of Surface S
$f(x, y) = 0$	x-axis	$f(x, \pm\sqrt{y^2 + z^2}) = 0$
	y-axis	$f(\pm\sqrt{x^2 + z^2}, y) = 0$
$f(x, z) = 0$	x-axis	$f(x, \pm\sqrt{y^2 + z^2}) = 0$
	z-axis	$f(\pm\sqrt{x^2 + y^2}, z) = 0$
$f(y, z) = 0$	y-axis	$f(y, \pm\sqrt{x^2 + z^2}) = 0$
	z-axis	$f(\pm\sqrt{x^2 + y^2}, z) = 0$

EXAMPLE 4   Paraboloid of Revolution

(a) In Example 1, the equation $y = x^2 + z^2$ can be written as

$$y = (\pm\sqrt{x^2 + z^2})^2.$$

Hence, from the preceding table we see that the surface is generated by revolving either the parabola $y = x^2$ or the parabola $y = z^2$ about the y-axis. The surface shown in Figure 11.8.7(a) is called a **paraboloid of revolution**.

(b) In Example 3, the equation $-(z - 4) = x^2 + y^2$ can be written as

$$-(z - 4) = (\pm\sqrt{x^2 + y^2})^2.$$

The surface is also a paraboloid of revolution. In this case the surface is generated by revolving either the parabola $-(z - 4) = x^2$ or the parabola $-(z - 4) = y^2$ about the z-axis. ∎

EXAMPLE 5   Ellipsoid of Revolution

The graph of $4x^2 + y^2 = 16$ is revolved about the x-axis. Find an equation of the surface of revolution.

Solution   The given equation has the form $f(x, y) = 0$. Since the axis of revolution is the x-axis, we see from the table that an equation of the surface of revolution can be found by replacing y by $\pm\sqrt{y^2 + z^2}$. It follows that

$$4x^2 + (\pm\sqrt{y^2 + z^2})^2 = 16 \quad \text{or} \quad 4x^2 + y^2 + z^2 = 16.$$

The surface is called an **ellipsoid of revolution**. ∎

EXAMPLE 6   Cone

The graph of $z = y, y \geq 0$, is revolved about the z-axis. Find an equation of the surface of revolution.

Solution   Since there are no points on the graph of $z = y, y \geq 0$, with a negative y-coordinate, we obtain an equation for the surface of revolution by substituting $\sqrt{x^2 + y^2}$ for y:

$$z = \sqrt{x^2 + y^2}. \tag{13} ∎$$

Observe that (13) is not the same as $z^2 = x^2 + y^2$. Technically the graph of (3) is a **double-napped** cone or a complete cone; the portions of the cone above and below the vertex are called **nappes**. If we solve (3) for z in terms of x and y, we obtain equations of

single-napped cones. For example, by solving $z^2 = x^2 + y^2$ we obtain $z = \sqrt{x^2 + y^2}$ and $z = -\sqrt{x^2 + y^2}$ that are, in turn, equations of the upper nappe and lower nappe of the cone. Thus, the graph of (13) is the single-napped cone in FIGURE 11.8.10(a).

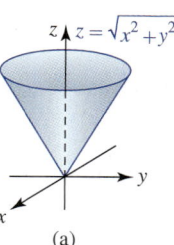
(a)

 NOTES FROM THE CLASSROOM

(*i*) The graph of $z = xy$ is also a hyperbolic paraboloid. In fact it can be shown that the surface $z = xy$ is congruent to the hyperbolic paraboloid $z = \frac{1}{2}x^2 - \frac{1}{2}y^2$ by means of a counterclockwise rotation of the *x*- and *y*-axes through an angle of $\pi/4$ radians about the *z*-axis.

(*ii*) The hyperboloid and hyperbolic paraboloid are often encountered in architectural engineering. The shape of the Kobe Port Tower in Kobe, Japan, is a hyperboloid of one sheet. Over the years the surface shown in FIGURE 11.8.11 was used in roof designs for houses and even gas stations. The most famous was the Catalano house built in Raleigh, North Carolina in 1954. Go to www.trianglemodernisthouses.com/catalano.htm for detailed photos. Finally, if you look carefully at potato chips, especially the uniform-size Pringles potato crisps, you see the shape of a hyperbolic paraboloid.

(*iii*) The surface of the mirror in a reflecting telescope is ground to the shape of a paraboloid of revolution.

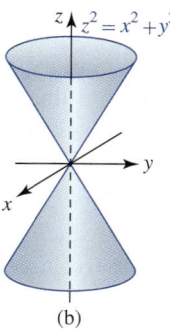
(b)
FIGURE 11.8.10 Single-napped cone in (a); double-napped cone in (b)

Kobe Port Tower, Kobe, Japan

Catalano house, 1954

Pringles potato crisps

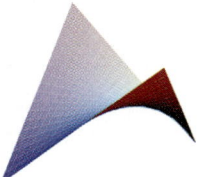
FIGURE 11.8.11 Surface $z = xy$

Exercises 11.8 Answers to selected odd-numbered problems begin on page ANS-37.

≡ **Fundamentals**

In Problems 1–14, identify and graph the quadric surface.

1. $x^2 + y^2 = z$
2. $-x^2 + y^2 = z^2$
3. $9x^2 + 36y^2 + 4z^2 = 36$
4. $x^2 + y^2 - z^2 = -4$
5. $36x^2 - y^2 + 9z^2 = 144$
6. $4x^2 + 4y^2 + z^2 = 100$
7. $y^2 + 5z^2 = x^2$
8. $-9x^2 + 16y^2 = 144z$
9. $y = 4x^2 - z^2$
10. $9z + x^2 + y^2 = 0$
11. $x^2 - y^2 - z^2 = 4$
12. $-x^2 + 9y^2 + z^2 = 9$
13. $y^2 + 4z^2 = x$
14. $x^2 + y^2 - z^2 = 1$

In Problems 15–18, graph the quadric surface.

15. $z = 3 + x^2 + y^2$
16. $y + x^2 + 4z^2 = 4$
17. $(x - 4)^2 + (y - 6)^2 - z^2 = 1$
18. $5x^2 + (y - 5)^2 + 5z^2 = 25$

In Problems 19–22, the given equation is an equation of a surface of revolution obtained by revolving a curve *C* in a coordinate plane about a coordinate axis. Find an equation for *C* and identify the axis of revolution.

19. $x^2 + y^2 + z^2 = 1$
20. $-9x^2 + 4y^2 + 4z^2 = 36$
21. $y = e^{x^2 + z}$
22. $x^2 + y^2 = \sin^2 z$

In Problems 23–30, the graph of the given equation is revolved about the indicated axis. Find an equation of the surface of revolution.

23. $y = 2x$; *y*-axis
24. $y = \sqrt{z}$; *y*-axis
25. $z = 9 - x^2, x \geq 0$; *x*-axis
26. $z = 1 + y^2, y \geq 0$; *z*-axis
27. $x^2 - z^2 = 4$; *x*-axis
28. $3x^2 + 4z^2 = 12$; *z*-axis
29. $z = \ln y$; *z*-axis
30. $xy = 1$; *x*-axis

31. Which of the surfaces in Problems 1–14 are surfaces of revolution? Identify the axis of revolution for each surface.

32. Sketch a graph of the equation in Problem 22 for $0 \leq z \leq 2\pi$.

In Problems 33 and 34, compare the graphs of the given equations.

33. $z + 2 = -\sqrt{x^2 + y^2}, (z + 2)^2 = x^2 + y^2$

34. $y - 1 = \sqrt{x^2 + z^2}, (y - 1)^2 = x^2 + z^2$

35. Consider the paraboloid

$$z - c = -\left(\frac{x^2}{a^2} + \frac{y^2}{b^2}\right), \quad c > 0.$$

(a) The area of an ellipse $x^2/A^2 + y^2/B^2 = 1$ is πAB. Use this fact to express the area of a cross section perpendicular to the z-axis as a function of z, $z \leq c$.

(b) Use the slicing method (see Section 6.3) to find the volume of the solid bounded by the paraboloid and the xy-plane.

36. (a) Use the slicing method as in part (b) of Problem 35 to find the volume of the ellipsoid

$$\frac{x^2}{a^2} + \frac{y^2}{b^2} + \frac{z^2}{c^2} = 1.$$

(b) What does your answer in part (a) become when $a = b = c$?

37. Determine the points where the line

$$\frac{x - 2}{2} = \frac{y + 2}{-3} = \frac{z - 6}{3/2}$$

intersects the ellipsoid $x^2/9 + y^2/36 + z^2/81 = 1$.

≡ Projects

38. Conic Sections Redux In the introduction to Section 10.1 we informally defined a conic section (circle, ellipse, parabola, and hyperbola) as the curve of intersection of a plane and a double-napped cone. With the newly acquired knowledge of equations of planes and cones you are in a position to actually prove the foregoing statement. For simplicity let us consider a single-napped cone with the equation $z = \sqrt{x^2 + y^2}$. It is fairly easy to see that a plane $z = a, a > 0$ parallel to the xy-plane cuts the cone in a circle. Substituting $z = a$ into the equation of the cone gives, after simplifying, $x^2 + y^2 = a^2$. This last equation is a circle of radius a and is the equation of the projection onto the xy-plane of the curve of intersection of the plane with the cone. Now suppose a plane defined by $z = b - (b/a)x$ passes through the cone as shown in FIGURE 11.8.12(a). Investigate how to demonstrate that the

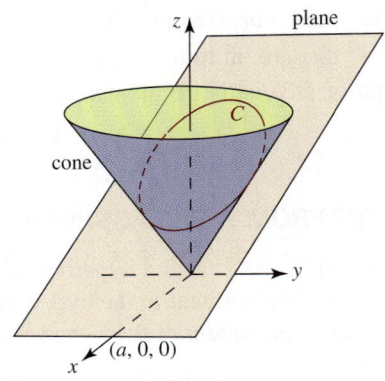

(a) C lies in the plane of intersection with the cone

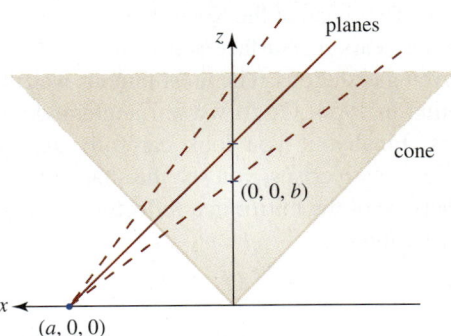

(b) Cross-sectional view

FIGURE 11.8.12 Intersection of planes with single-napped cone

curve C of intersection is either a parabola, ellipse, or a hyperbola. Consider cases as suggested by the various positions of the plane shown in Figure 11.8.12(b). You are probably going to have to dig a little deeper than you initially think.

39. Spheroids

(a) Write a short paper discussing under what conditions the equation

$$\frac{x^2}{a^2} + \frac{y^2}{b^2} + \frac{z^2}{c^2} = 1$$

describes an **oblate spheroid** and a **prolate spheroid**.

(b) Relate these two surfaces to the concept of a surface of revolution.

(c) The planet Earth is an example of an oblate spheroid. Compare the polar radius of the Earth with its equatorial radius.

(d) Give an example of a prolate spheroid.

Chapter 11 in Review

Answers to selected odd-numbered problems begin on page ANS-38.

A. True/False

In Problems 1–20, indicate whether the given statement is true or false.

1. The vectors $\langle -4, -6, 10 \rangle$ and $\langle -10, -15, 25 \rangle$ are parallel._____

2. In 3-space any three distinct points determine a plane. _____

3. The line $x = 1 + 5t, y = 1 - 2t, z = 4 + t$ and the plane $2x + 3y - 4z = 1$ are perpendicular. _____

4. Nonzero vectors **a** and **b** are parallel if $\mathbf{a} \times \mathbf{b} = \mathbf{0}$. _____

5. If $\mathbf{a} \cdot \mathbf{b} < 0$, the angle between **a** and **b** is obtuse. _____

6. If **a** is a unit vector, then $\mathbf{a} \cdot \mathbf{a} = 1$. _____

7. A line in 3-space can have many different vector equations. _____

8. The terminal point of the vector $\mathbf{a} - \mathbf{b}$ is at the terminal point of **a**. _____

9. $(\mathbf{a} \times \mathbf{b}) \cdot \mathbf{c} = \mathbf{a} \cdot (\mathbf{b} \times \mathbf{c})$ _____

10. If **a**, **b**, **c**, and **d** are nonzero coplanar vectors, then $(\mathbf{a} \times \mathbf{b}) \times (\mathbf{c} \times \mathbf{d}) = \mathbf{0}$. _____

11. The planes $x + 2y - z = 5$ and $-2x - 4y + 2z = 1$ are parallel. _____

12. Two perpendicular lines L_1 and L_2 intersect. _____

13. The surface $z = x^2$ is a cylinder. _____

14. The trace of the ellipsoid $x^2/9 + y^2/2 + z^2/2 = 1$ in the yz-plane is a circle. _____

15. The four points $(0, 1, 2), (1, -1, 1), (3, 2, 6), (2, 1, 2)$ lie in the same plane. _____

16. In general, for nonzero vectors **a** and **b** in 3-space, $\mathbf{a} \times \mathbf{b} \neq \mathbf{b} \times \mathbf{a}$. _____

17. The trace of the surface $x^2 + 9y^2 + z^2 = 1$ in the yz-plane is a circle. _____

18. $x^2 + 9y^2 + z^2 = 1$ is a surface of revolution. _____

19. If **a** and **b** are nonzero orthogonal vectors, then $|\mathbf{a} \times \mathbf{b}| = |\mathbf{a}||\mathbf{b}|$. _____

20. If **a** is a nonzero vector and $\mathbf{a} \cdot \mathbf{b} = \mathbf{a} \cdot \mathbf{c} = 0$, then $\mathbf{b} = \mathbf{c}$. _____

B. Fill in the Blanks

In Problems 1–24, fill in the blanks.

1. The sum of $3\mathbf{i} + 4\mathbf{j} + 5\mathbf{k}$ and $6\mathbf{i} - 2\mathbf{j} - 3\mathbf{k}$ is _____.

2. If $\mathbf{a} \cdot \mathbf{b} = 0$, the nonzero vectors **a** and **b** are _____.

3. $(-\mathbf{k}) \times (5\mathbf{j}) =$ _____

4. $\mathbf{i} \cdot (\mathbf{i} \times \mathbf{j}) =$ _____

5. $|-12\mathbf{i} + 4\mathbf{j} + 6\mathbf{k}| =$ _____

6. $\mathbf{k} \times (\mathbf{i} + 2\mathbf{j} - 5\mathbf{k}) =$ _____

7. $\begin{vmatrix} 2 & -5 \\ 4 & 3 \end{vmatrix} =$ _____

8. $\begin{vmatrix} \mathbf{i} & \mathbf{j} & \mathbf{k} \\ 2 & 1 & 5 \\ 0 & 4 & -1 \end{vmatrix} =$ _____

9. A vector that is normal to the plane $-6x + y - 7z + 10 = 0$ is _____.

10. The largest sphere with center $(4, 3, 7)$ whose interior lies entirely in the first octant has radius $r =$ _____.

11. The point of intersection of the line $x - 1 = (y + 2)/3 = (z + 1)/2$ and the plane $x + 2y - z = 13$ is _____.

12. A unit vector that has the opposite direction of $\mathbf{a} = 4\mathbf{i} + 3\mathbf{j} - 5\mathbf{k}$ is _____.

13. If $\overrightarrow{P_1 P_2} = \langle 3, 5, -4 \rangle$ and P_1 has coordinates $(2, 1, 7)$, then the coordinates of P_2 are _____.

14. The midpoint of the line segment between $P_1(4, 3, 10)$ and $P_2(6, -2, -5)$ has coordinates _____.

15. If $|\mathbf{a}| = 7.2, |\mathbf{b}| = 10$, and the angle between **a** and **b** is $135°$, then $\mathbf{a} \cdot \mathbf{b} =$ _____.

16. If $\mathbf{a} = \langle 3, 1, 0 \rangle, \mathbf{b} = \langle -1, 2, 1 \rangle$ and $\mathbf{c} = \langle 0, -2, 2 \rangle$, then $\mathbf{a} \cdot (2\mathbf{b} + 4\mathbf{c}) =$ _____.

17. The x-, y-, and z-intercepts of the plane $2x - 3y + 4z = 24$ are, respectively, _____.

18. The angle θ between the vectors $\mathbf{a} = \mathbf{i} + \mathbf{j}$ and $\mathbf{b} = \mathbf{i} - \mathbf{k}$ is _____.

19. The area of a triangle with two sides given by $\mathbf{a} = \langle 1, 3, -1 \rangle$ and $\mathbf{b} = \langle 2, -1, 2 \rangle$ is _____.

20. An equation of a sphere with center $(-5, 7, -9)$ and radius $\sqrt{6}$ is _____.

21. The distance from the plane $y = -5$ to the point $(4, -3, 1)$ is _____.

22. The vectors $\langle 1, 3, c \rangle$ and $\langle -2, -6, 5 \rangle$ are parallel for $c =$ _____ and orthogonal for $c =$ _____.

23. The surface $x^2 + 2y^2 + 2z^2 - 4y - 12z = 0$ is a(n) _____.

24. The trace of the surface $y = x^2 - z^2$ in the plane $z = 1$ is a(n) _____.

C. Exercises

1. Find a unit vector that is perpendicular to both $\mathbf{a} = \mathbf{i} + \mathbf{j}$ and $\mathbf{b} = \mathbf{i} - 2\mathbf{i} + \mathbf{k}$.

2. Find the direction cosines and direction angles of the vector $\mathbf{a} = \frac{1}{2}\mathbf{i} + \frac{1}{2}\mathbf{j} - \frac{1}{4}\mathbf{k}$.

In Problems 3–6, let $\mathbf{a} = \langle 1, 2, -2 \rangle$ and $\mathbf{b} = \langle 4, 3, 0 \rangle$. Find the indicated number or vector.

3. $\text{comp}_{\mathbf{b}}\mathbf{a}$ 4. $\text{proj}_{\mathbf{a}}\mathbf{b}$ 5. $\text{proj}_{\mathbf{b}}2\mathbf{a}$ 6. projection of $\mathbf{a} - \mathbf{b}$ orthogonal to $\mathbf{b}$

In Problems 7–12, identify the surface whose equation is given.

7. $x^2 + 4y^2 = 16$ 8. $y + 2x^2 + 4z^2 = 0$ 9. $x^2 + 4y^2 - z^2 = -9$

10. $x^2 + y^2 + z^2 = 10z$ 11. $9z - x^2 + y^2 = 0$ 12. $2x - 3y = 6$

13. Find an equation of the surface of revolution obtained by revolving the graph of $x^2 - y^2 = 1$ about the y-axis. About the x-axis. Identify the surface in each case.

14. A surface of revolution has an equation $y = 1 + \sqrt{x^2 + z^2}$. Find an equation of a curve C in a coordinate plane that, when revolved about a coordinate axis, generates the surface.

15. Let $\mathbf{r}$ be the position vector of a variable point $P(x, y, z)$ in space and let $\mathbf{a}$ be a constant vector. Determine the surface described by the following vector equations:
    (a) $(\mathbf{r} - \mathbf{a}) \cdot \mathbf{r} = 0$ (b) $(\mathbf{r} - \mathbf{a}) \cdot \mathbf{a} = 0$.

16. Use the dot product to determine whether the points $(4, 2, -2)$, $(2, 4, -3)$, and $(6, 7, -5)$ are vertices of a right triangle.

17. Find symmetric equations for the line through the point $(7, 3, -5)$ that is parallel to $(x - 3)/4 = (y + 4)/(-2) = (z - 9)/6$.

18. Find parametric equations for the line through the point $(5, -9, 3)$ that is perpendicular to the plane $8x + 3y - 4z = 13$.

19. Show that the lines $x = 1 - 2t, y = 3t, z = 1 + t$ and $x = 1 + 2s, y = -4 + s$, $z = -1 + s$ intersect and are perpendicular.

20. Find an equation of the plane containing the points $(0, 0, 0)$, $(2, 3, 1)$, $(1, 0, 2)$.

21. Find an equation of the plane containing the lines $x = t, y = 4t, z = -2t$ and $x = 1 + t$, $y = 1 + 4t, z = 3 - 2t$.

22. Find an equation of the plane containing $(1, 7, -1)$ that is perpendicular to the line of intersection of $-x + y - 8z = 4$ and $3x - y + 2z = 0$.

23. Find an equation of the plane containing $(1, -1, 2)$ that is parallel to the vectors $\mathbf{i} - 2\mathbf{j}$ and $2\mathbf{i} + 3\mathbf{k}$.

24. Find an equation of a sphere for which the line segment $x = 4 + 2t, y = 7 + 3t$, $z = 8 + 6t, -1 \leq t \leq 0$, is a diameter.

25. Show that the three vectors $\mathbf{a} = 3\mathbf{i} + 5\mathbf{j} + 2\mathbf{k}, \mathbf{b} = 3\mathbf{i} + 4\mathbf{j} + \mathbf{k}$, and $\mathbf{c} = 4\mathbf{i} + 5\mathbf{j} + \mathbf{k}$ are coplanar.

26. Consider the right triangle whose sides are the vectors $\mathbf{a}$, $\mathbf{b}$, and $\mathbf{c}$ shown in FIGURE 11.R.1. Show that the midpoint M of the hypotenuse is equidistant from all three vertices of the triangle.

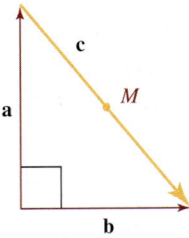

FIGURE 11.R.1   Triangle in Problem 26

27. **(a)** The force **F** acting on a particle of charge q moving with velocity **v** through a magnetic field **B** is given by $\mathbf{F} = q(\mathbf{v} \times \mathbf{B})$. Find **F** if **v** acts along the positive y-axis and **B** acts along the positive x-axis. Assume $|\mathbf{v}| = v$ and $|\mathbf{B}| = B$.

   (b) The angular momentum **L** of a particle of mass m moving with a linear velocity **v** in a circle of radius **r** is given by $\mathbf{L} = m(\mathbf{r} \times \mathbf{v})$, where **r** is perpendicular to **v**. Use vector methods to solve for **v** in terms of **L**, **r**, and m.

28. A constant force of 10 N in the direction of $\mathbf{a} = \mathbf{i} + \mathbf{j}$ moves a block on a frictionless surface from $P_1(4, 1, 0)$ to $P_2(7, 4, 0)$. Suppose distance is measured in meters. Find the work done.

29. In Problem 28 find the work done in moving the block between the same points if another constant force of 50 N in the direction of $\mathbf{b} = \mathbf{i}$ acts simultaneously with the original force.

30. A uniform ball of weight 50 lb is supported by two frictionless planes as shown in FIGURE 11.R.2. Let the force exerted by the supporting plane S_1 on the ball be $\mathbf{F}_1$ and the force exerted by the plane S_2 be $\mathbf{F}_2$. Since the ball is held in equilibrium, we must have $\mathbf{w} + \mathbf{F}_1 + \mathbf{F}_2 = \mathbf{0}$, where $\mathbf{w} = -50\mathbf{j}$. Find the magnitudes of the forces $\mathbf{F}_1$ and $\mathbf{F}_2$. [*Hint*: Assume the forces $\mathbf{F}_1$ and $\mathbf{F}_2$ are normal to the planes S_1 and S_2, respectively, and act along lines through the center C of the ball. Place the origin of a two-dimensional coordinate system at C.]

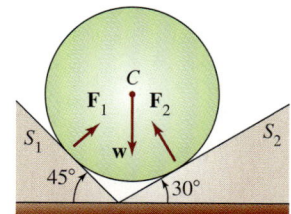

FIGURE 11.R.2 Ball in Problem 30

Vector-Valued Functions

$$\mathbf{r}(t_0) = \langle x(t_0), y(t_0), z(t_0) \rangle$$

In This Chapter A curve in the plane as well as a curve C in 3-space can be defined by means of parametric equations. Using the functions in a set of parametric equations as components, we can construct a vector-valued function whose values are position vectors of points on the curve C. In this chapter we will consider the calculus and applications of these vector functions.

12.1 Vector Functions

❚ Introduction   We saw in Section 10.2 that a curve C in the xy-plane can be parameterized by two equations

$$x = f(t), \quad y = g(t), \quad a \leq t \leq b. \tag{1}$$

It is often convenient in science and engineering to introduce a vector $\mathbf{r}$ with the functions f and g as components:

$$\mathbf{r}(t) = \langle f(t), g(t) \rangle = f(t)\mathbf{i} + g(t)\mathbf{j}, \tag{2}$$

where $\mathbf{i} = \langle 1, 0 \rangle$ and $\mathbf{j} = \langle 0, 1 \rangle$. In this section we study the analogues of (1) and (2) in three dimensions.

❚ Vector-Valued Functions   A curve C in three-dimensional space, or a **space curve**, is parameterized by three equations

$$x = f(t), \quad y = g(t), \quad z = h(t), \quad a \leq t \leq b. \tag{3}$$

As in Section 10.2, the **orientation** of C corresponds to *increasing values* of the parameter t. Using the functions in (3) as components, the 3-space counterpart of (2) is

$$\mathbf{r}(t) = \langle f(t), g(t), h(t) \rangle = f(t)\mathbf{i} + g(t)\mathbf{j} + h(t)\mathbf{k}, \tag{4}$$

where $\mathbf{i} = \langle 1, 0, 0 \rangle$, $\mathbf{j} = \langle 0, 1, 0 \rangle$, and $\mathbf{k} = \langle 0, 0, 1 \rangle$. We say that $\mathbf{r}$ in (2) and (4) is a **vector-valued function** or simply a **vector function**. As shown in FIGURE 12.1.1, for a given number t_0, the vector $\mathbf{r}(t_0)$ is the *position vector* of a point P on the curve C. In other words, as t varies, we can envision the curve C being traced out by the moving arrowhead of $\mathbf{r}(t)$.

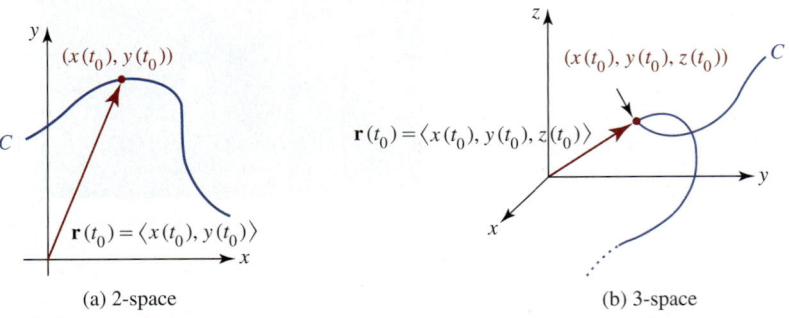

| (a) 2-space | (b) 3-space |

FIGURE 12.1.1   Vector functions in 2- and 3-space

❚ Lines   We have already seen an example of parametric equations as well as the vector function of a space curve in Section 11.5 where we discussed the line in 3-space. Recall, parametric equations of a line L that passes through a point $P_0(x_0, y_0, z_0)$ in space and is parallel to a vector $\mathbf{v} = \langle a, b, c \rangle$, $\mathbf{v} \neq \mathbf{0}$, are

$$x = x_0 + at, \quad y = y_0 + bt, \quad z = z_0 + ct, \quad -\infty < t < \infty.$$

These equations result from the fact that the vectors $\mathbf{r} - \mathbf{r}_0$ and $\mathbf{v}$ are parallel so that $\mathbf{r} - \mathbf{r}_0$ is a scalar multiple of $\mathbf{v}$, that is, $\mathbf{r} - \mathbf{r}_0 = t\mathbf{v}$. Hence a vector function of the line L is given by $\mathbf{r}(t) = \mathbf{r}_0 + t\mathbf{v}$. The last equation can be expressed in the alternative forms

$$\mathbf{r}(t) = \langle x_0 + at, y_0 + bt, z_0 + ct \rangle$$

and

$$\mathbf{r}(t) = (x_0 + at)\mathbf{i} + (y_0 + bt)\mathbf{j} + (z_0 + ct)\mathbf{k}.$$

If $\mathbf{r}_0 = \langle x_0, y_0, z_0 \rangle$ and $\mathbf{r}_1 = \langle x_1, y_1, z_1 \rangle$ are the position vectors of two distinct points P_0 and P_1, then we can take $\mathbf{v} = \mathbf{r}_1 - \mathbf{r}_0 = \langle x_1 - x_0, y_1 - y_0, z_1 - z_0 \rangle$. A vector function of the line through the two points is $\mathbf{r}(t) = \mathbf{r}_0 + t(\mathbf{r}_1 - \mathbf{r}_0)$ or

$$\mathbf{r}(t) = (1 - t)\mathbf{r}_0 + t\mathbf{r}_1. \tag{5}$$

If the parameter interval is a closed interval $[a, b]$, then the vector function (5) traces out the **line segment** between the points defined by $\mathbf{r}(a)$ and $\mathbf{r}(b)$. In particular, if $0 \le t \le 1$ and $\mathbf{r} = (1 - t)\mathbf{r}_0 + t\mathbf{r}_1$, then the orientation is such that $\mathbf{r}(t)$ traces out the line segment from the point P_0 to the point P_1.

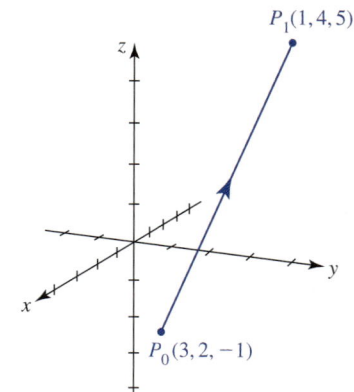

EXAMPLE 1 Graph of a Vector Function

Find a vector function of the line segment from the point $P_0(3, 2, -1)$ to the point $P_1(1, 4, 5)$.

Solution The position vectors corresponding to the given points are $\mathbf{r}_0 = \langle 3, 2, -1 \rangle$ and $\mathbf{r}_1 = \langle 1, 4, 5 \rangle$. Thus, from (5) a vector function for the line segment is

$$\mathbf{r}(t) = (1 - t)\langle 3, 2, -1 \rangle + t\langle 1, 4, 5 \rangle$$

or

$$\mathbf{r}(t) = \langle 3 - 2t, 2 + 2t, -1 + 6t \rangle,$$

where $0 \le t \le 1$. The graph of the vector equation is given in FIGURE 12.1.2. ∎

FIGURE 12.1.2 Line segment in Example 1

EXAMPLE 2 Graph of a Vector Function

Graph the curve C traced by the vector function

$$\mathbf{r}(t) = 2\cos t\,\mathbf{i} + 2\sin t\,\mathbf{j} + t\mathbf{k}, \quad t \ge 0.$$

Solution The parametric equations of the curve C are $x = 2\cos t$, $y = 2\sin t$, $z = t$. By eliminating the parameter t from the first two equations,

$$x^2 + y^2 = (2\cos t)^2 + (2\sin t)^2 = 2^2,$$

we see that a point on the curve lies on the circular cylinder $x^2 + y^2 = 4$. As seen in FIGURE 12.1.3 and the accompanying table, as the value of t increases, the curve winds upward in a cylindrical spiral or a circular helix.

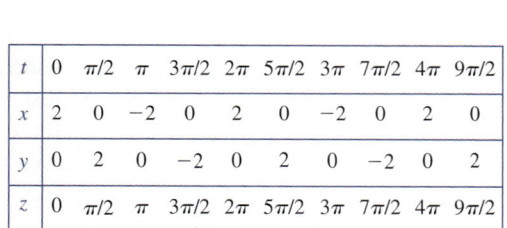

t	0	$\pi/2$	π	$3\pi/2$	2π	$5\pi/2$	3π	$7\pi/2$	4π	$9\pi/2$
x	2	0	-2	0	2	0	-2	0	2	0
y	0	2	0	-2	0	2	0	-2	0	2
z	0	$\pi/2$	π	$3\pi/2$	2π	$5\pi/2$	3π	$7\pi/2$	4π	$9\pi/2$

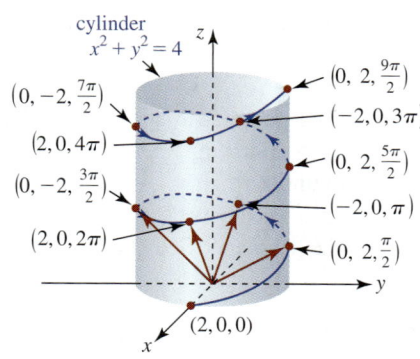

FIGURE 12.1.3 Graph of vector function in Example 2 ∎

Helical Curves The curve in Example 2 is one of several types of space curves known as **helical curves**. In general, a vector function of the form

$$\mathbf{r}(t) = a\cos kt\,\mathbf{i} + a\sin kt\,\mathbf{j} + ct\mathbf{k}, \tag{6}$$

describes a **circular helix**. The number $2\pi c/k$ is called the **pitch** of a helix. A circular helix is just a special case of the vector function

◄ The helix defined by (6) winds upward along the z-axis. The pitch is the vertical separation of the loops of the helix.

$$\mathbf{r}(t) = a\cos kt\,\mathbf{i} + b\sin kt\,\mathbf{j} + ct\mathbf{k}, \tag{7}$$

which describes an **elliptical helix** when $a \ne b$. The curve defined by

$$\mathbf{r}(t) = at\cos kt\,\mathbf{i} + bt\sin kt\,\mathbf{j} + ct\mathbf{k}, \tag{8}$$

is a called a **conical helix**. Finally, a curve given by

$$\mathbf{r}(t) = a\sin kt\cos t\,\mathbf{i} + a\sin kt\sin t\,\mathbf{j} + a\cos kt\,\mathbf{k}, \tag{9}$$

is called a **spherical helix**. In (6)–(9) we assume that a, b, c, and k are positive constants.

EXAMPLE 3 Helical Curves

(a) If we interchange, say, the y and z components of the vector function (7) we obtain an elliptical helix that winds sideways along the y-axis. For example, with the help of a CAS, the graph of elliptical helix

$$\mathbf{r}(t) = 4\cos t\,\mathbf{i} + t\mathbf{j} + 2\sin t\,\mathbf{k}$$

is shown in FIGURE 12.1.4(a).

(b) Figure 12.1.4(b) shows the graph of

$$\mathbf{r}(t) = t\cos t\,\mathbf{i} + t\sin t\,\mathbf{j} + t\mathbf{k}$$

and illustrates why a vector function of the form given in (8) defines a conical helix. For greater clarity, we have chosen to suppress the default box that surrounds the *Mathematica* 3D-output.

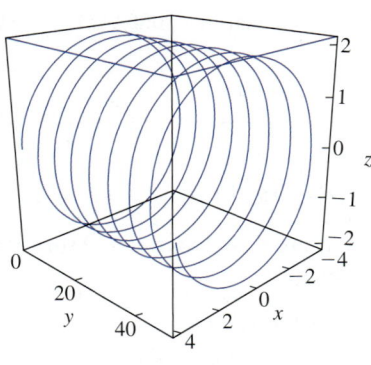

(a) Elliptical helix   (b) Conical helix

FIGURE 12.1.4 Helical curves in Example 3

EXAMPLE 4 Graph of a Vector Function

Graph the curve traced by the vector function $\mathbf{r}(t) = 2\cos t\,\mathbf{i} + 2\sin t\,\mathbf{j} + 3\mathbf{k}$.

Solution The parametric equations of the curve are the components of the vector function $x = 2\cos t$, $y = 2\sin t$, $z = 3$. As in Example 1, we see that a point on the curve must also lie on the cylinder $x^2 + y^2 = 4$. However, since the z-coordinate of any point has the constant value $z = 3$, the vector function $\mathbf{r}(t)$ traces out a circle in a plane 3 units above and parallel to the xy-plane. See FIGURE 12.1.5.

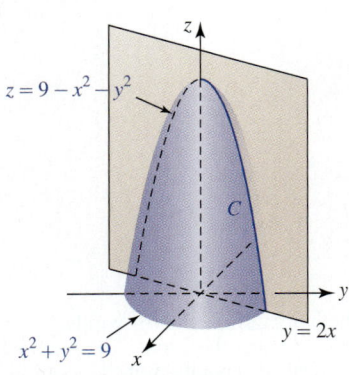

$x^2 + y^2 = 4, z = 3$

FIGURE 12.1.5 Circle in a plane in Example 4

EXAMPLE 5 Curve of Intersection of Two Surfaces

Find the vector function that describes the curve C of intersection of the plane $y = 2x$ and the paraboloid $z = 9 - x^2 - y^2$.

Solution We first parameterize the curve C of intersection by letting $x = t$. It follows that $y = 2t$ and $z = 9 - t^2 - (2t)^2 = 9 - 5t^2$. From the parametric equations

$$x = t, \quad y = 2t, \quad z = 9 - 5t^2, \quad -\infty < t < \infty,$$

we see that a vector function describing the trace of the paraboloid in the plane $y = 2x$ is given by

$$\mathbf{r}(t) = t\mathbf{i} + 2t\mathbf{j} + (9 - 5t^2)\mathbf{k}.$$

See FIGURE 12.1.6.

$z = 9 - x^2 - y^2$

$x^2 + y^2 = 9$

$y = 2x$

FIGURE 12.1.6 Curve C of intersection in Example 5

EXAMPLE 6 Curve of Intersection of Two Cylinders

Find the vector function that describes the curve C of intersection of the cylinders $y = x^2$ and $z = x^3$.

Solution In 2-space the graph of $y = x^2$ is a parabola in the xy-plane and so in 3-space is a parabolic cylinder whose rulings are perpendicular to the xy-plane, that is, parallel to the

z-axis. See FIGURE 12.1.7(a). On the other hand, $z = x^3$ can be interpreted as a cubic cylinder whose rulings are perpendicular to the xz-plane, that is, parallel to the y-axis. See Figure 12.1.7(b). As in Example 5, if we let $x = t$, then $y = t^2$ and $z = t^3$. A vector function describing the curve C of intersection of the two cylinders is then

$$\mathbf{r}(t) = t\mathbf{i} + t^2\mathbf{j} + t^3\mathbf{k}, \tag{10}$$

where $-\infty < t < \infty$.

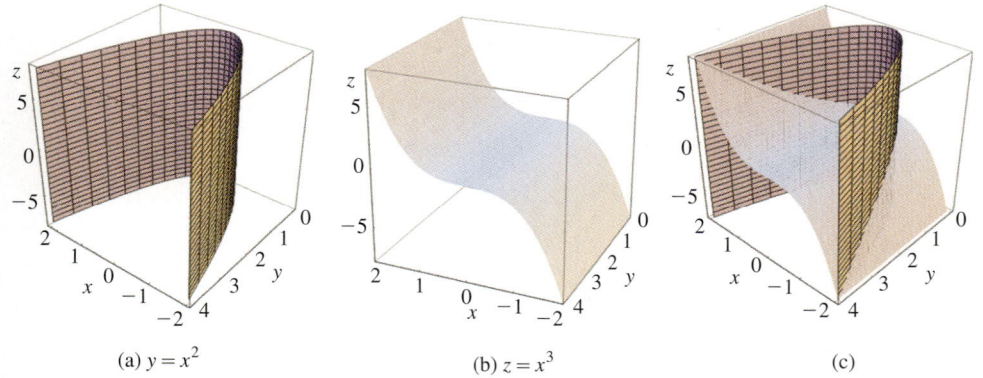

(a) $y = x^2$ (b) $z = x^3$ (c)

FIGURE 12.1.7 (a) and (b) two cylinders; (c) curve C of intersection in Example 6

The curve C defined by the vector function (10) is called a **twisted cubic**. With the aid of a CAS we have graphed $\mathbf{r}(t) = t\mathbf{i} + t^2\mathbf{j} + t^3\mathbf{k}$ in FIGURE 12.1.8. Parts (a) and (b) of the figure show two different perspectives, or viewpoints, of the curve C of intersection of the cylinders $y = x^2$ and $z = x^3$. In Figure 12.1.8(c) we see the cubic nature of C by using a viewpoint that is toward the xz-plane. The twisted cubic has various properties of interest to mathematicians and so it is often studied in advanced courses in algebraic geometry.

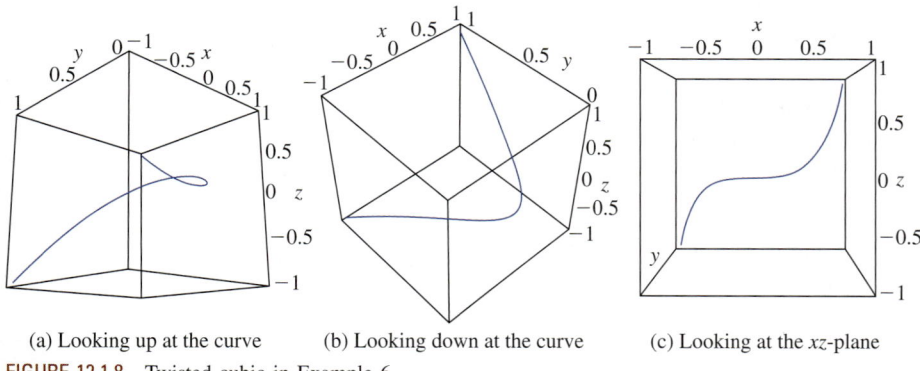

(a) Looking up at the curve   (b) Looking down at the curve   (c) Looking at the xz-plane

FIGURE 12.1.8 Twisted cubic in Example 6

Exercises 12.1   Answers to selected odd-numbered problems begin on page ANS-38.

≡ Fundamentals

In Problems 1–4, find the domain of the given vector function.

1. $\mathbf{r}(t) = \sqrt{t^2 - 9}\,\mathbf{i} + 3t\mathbf{j}$

2. $\mathbf{r}(t) = (t + 1)\mathbf{i} + \ln(1 - t^2)\mathbf{j}$

3. $\mathbf{r}(t) = t\mathbf{i} + \frac{1}{2}t^2\mathbf{j} - \sin^{-1}t\mathbf{k}$

4. $\mathbf{r}(t) = e^{-t}\mathbf{i} + \cos t\mathbf{j} + \sin 2t\mathbf{k}$

In Problems 5–8, write the given parametric equations as a vector function $\mathbf{r}(t)$.

5. $x = \sin \pi t,\ y = \cos \pi t,\ z = -\cos^2 \pi t$

6. $x = \cos^2 t,\ y = 2\sin^2 t,\ z = t^2$

7. $x = e^{-t},\ y = e^{2t},\ z = e^{3t}$

8. $x = -16t^2,\ y = 50t,\ z = 10$

In Problems 9–12, write the given vector function $\mathbf{r}(t)$ as parametric equations.

9. $\mathbf{r}(t) = t^2\mathbf{i} + \sin t\mathbf{j} + \cos t\mathbf{k}$

10. $\mathbf{r}(t) = t\sin t(\mathbf{i} + \mathbf{k})$

11. $\mathbf{r}(t) = \ln t\mathbf{i} + (1 + t)\mathbf{j} + t^3\mathbf{k}$

12. $\mathbf{r}(t) = 5\sin t\sin 3t\mathbf{i} + 5\cos 3t\mathbf{j} + 5\cos t\sin 3t\mathbf{k}$

In Problems 13–22, graph the curve traced by the given vector function.

13. $\mathbf{r}(t) = 2\sin t\mathbf{i} + 4\cos t\mathbf{j} + t\mathbf{k}, \quad t \geq 0$

14. $\mathbf{r}(t) = t\mathbf{i} + \cos t\mathbf{j} + \sin t\mathbf{k}, \quad t \geq 0$

15. $\mathbf{r}(t) = t\mathbf{i} + 2t\mathbf{j} + \cos t\mathbf{k}, \quad t \geq 0$

16. $\mathbf{r}(t) = 4\mathbf{i} + 2\cos t\mathbf{j} + 3\sin t\mathbf{k}$

17. $\mathbf{r}(t) = \langle e^t, e^{2t} \rangle$

18. $\mathbf{r}(t) = \cosh t\mathbf{i} + 3\sinh t\mathbf{j}$

19. $\mathbf{r}(t) = \langle \sqrt{2}\sin t, \sqrt{2}\sin t, 2\cos t \rangle, \quad 0 \leq t \leq \pi/2$

20. $\mathbf{r}(t) = t\mathbf{i} + t^3\mathbf{j} + t\mathbf{k}$

21. $\mathbf{r}(t) = e^t\cos t\mathbf{i} + e^t\sin t\mathbf{j} + e^t\mathbf{k}$

22. $\mathbf{r}(t) = \langle t\cos t, t\sin t, t^2 \rangle$

In Problems 23 and 24, graph the line whose vector function is given.

23. $\mathbf{r}(t) = (4 - 4t)\mathbf{i} + (2 - 2t)\mathbf{j} + 3t\mathbf{k}$

24. $\mathbf{r}(t) = (2 + 3t)\mathbf{i} + (3 + 2t)\mathbf{j} + 5t\mathbf{k}$

25. Find a vector function for the line segment in 2-space with orientation such that $\mathbf{r}(t)$ traces out the line from the point $(4, 0)$ to $(0, 3)$. Sketch the line segment.

26. Find a vector function for the line segment in 3-space with orientation such that $\mathbf{r}(t)$ traces out the line from the point $(1, 1, 1)$ to $(0, 0, 0)$. Sketch the line segment.

In Problems 27–32, find the vector function $\mathbf{r}(t)$ that describes the curve C of intersection between the given surfaces. Sketch the curve C. Use the indicated parameter.

27. $z = x^2 + y^2, y = x; \quad x = t$

28. $x^2 + y^2 - z^2 = 1, y = 2x; \quad x = t$

29. $x^2 + y^2 = 9, z = 9 - x^2; \quad x = 3\cos t$

30. $z = x^2 + y^2, z = 1; \quad x = \sin t$

31. $x + y + z = 1, y = x; \quad x = t$

32. $3x - 2y + z = 6, x = 1; \quad y = t$

In Problems 33–36, match the given graph with one of the vector functions in (a)–(d).

(a) $\mathbf{r}(t) = t\mathbf{i} + \cos 3t\mathbf{j} + \sin 3t\mathbf{k}$

(b) $\mathbf{r}(t) = \sin 6t\mathbf{i} + t\mathbf{j} + t\mathbf{k}$

(c) $\mathbf{r}(t) = \cos t\mathbf{i} + \sin t\mathbf{j} + (1 - \sin t)\mathbf{k}$

(d) $\mathbf{r}(t) = \cos^3 t\mathbf{i} + \sin^3 t\mathbf{j} + 5\mathbf{k}$

33.

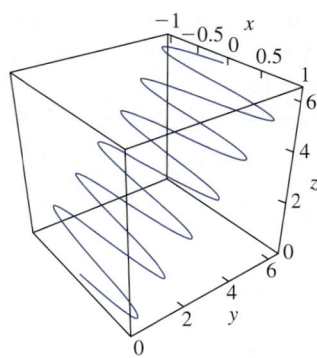

FIGURE 12.1.9 Graph for Problem 33

34.

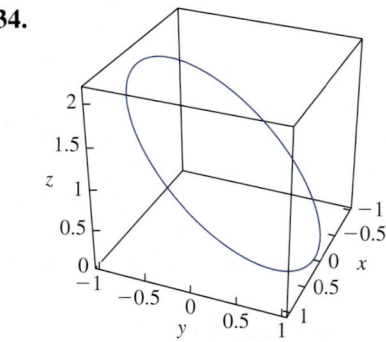

FIGURE 12.1.10 Graph for Problem 34

35.

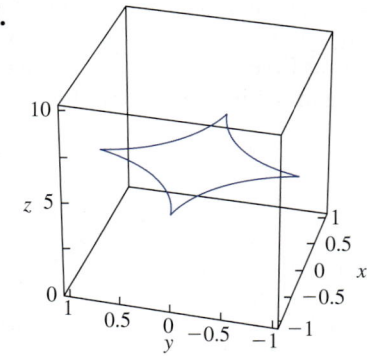

FIGURE 12.1.11 Graph for Problem 35

36.

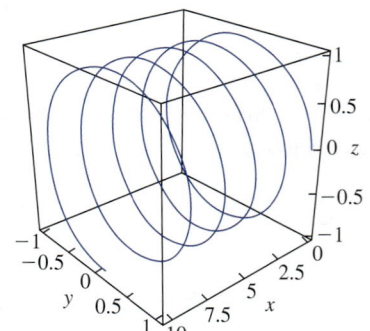

FIGURE 12.1.12 Graph for Problem 36

37. Show that points on a conical helix

$$\mathbf{r}(t) = at\cos t\mathbf{i} + bt\sin t\mathbf{j} + ct\mathbf{k},$$

$a > 0, b > 0, c > 0,$ lie on an elliptic cone whose equation is

$$\frac{z^2}{c^2} = \frac{x^2}{a^2} + \frac{y^2}{b^2}.$$

38. A variation of the conical helix in Problem 37 is given by

$$\mathbf{r}(t) = t\mathbf{i} + t\cos t\mathbf{j} + t\sin t\mathbf{k}.$$

(a) Before graphing $\mathbf{r}(t)$ discuss the orientation of the curve.

(b) Use a CAS to graph $\mathbf{r}(t)$. Experiment with the parameter interval and viewpoint of the curve.

39. The vector function

$$\mathbf{r}(t) = ae^{kt}\cos t\mathbf{i} + be^{kt}\sin t\mathbf{j} + ce^{kt}\mathbf{k},$$

$a > 0, b > 0, c > 0, k > 0$ also describes a conical helix. Show that points on this conical helix lie on an elliptic cone whose equation is given in Problem 37.

40. A special case of the curve in Problem 39 is given by

$$\mathbf{r}(t) = \frac{1}{2}e^{0.05t}\cos t\mathbf{i} + \frac{1}{2}e^{0.05t}\sin t\mathbf{j} + e^{0.05t}\mathbf{k}.$$

(a) Use a CAS to graph $\mathbf{r}(t)$ for $-30 \le t \le 30$.

(b) Reexamine Figure 12.1.4(b). Then discuss the basic geometric difference between the conical helix in Problem 37 and that given in Problem 39.

41. Show that points on a spherical helix

$$\mathbf{r}(t) = a\sin kt\cos t\mathbf{i} + a\sin kt\sin t\mathbf{j} + a\cos kt\mathbf{k}$$

lie on a sphere of radius $a > 0$.

42. A special case of the curve in Problem 41 is given by

$$\mathbf{r}(t) = \sin kt\cos t\mathbf{i} + \sin kt\sin t\mathbf{j} + \cos kt\mathbf{k}.$$

Use a CAS to graph $\mathbf{r}(t)$ for $k = 1, 2, 3, 4, 10, 20,$ and $0 \le t \le 2\pi$. Experiment with different viewpoints of the graphs.

43. (a) Use a CAS to superimpose the graphs of the cylinders $z = 4 - x^2$ and $z = 4 - y^2$ on the same coordinate axes.

(b) Find the vector functions describing the two curves of intersection of the cylinders.

(c) Use a CAS to plot both curves in part (b). Superimpose the curves on the same coordinate axes. Experiment with the viewpoint until the visualization of the graphs makes sense.

44. Suppose $\mathbf{r}(t)$ is a nonconstant vector function that defines a curve C with the property $|\mathbf{r}(t)| = a$, where $a > 0$ is a constant. Geometrically describe the curve C.

≡ Calculator/CAS Problems

45. Use a CAS to graph the vector function

$$\mathbf{r}(t) = (10 + \sin 20t)\cos t\mathbf{i} + (10 + \sin 20t)\sin t\mathbf{j} + \cos 2t\mathbf{k}$$

for $0 \le t \le 2\pi$. Experiment with different viewpoints of the graph. Discuss why the curve is called a **toroidal spiral**.

46. Use a CAS to graph the vector function

$$\mathbf{r}(t) = \cos(\arctan kt)\cos t\mathbf{i} + \cos(\arctan kt)\sin t\mathbf{j} - \sin(\arctan kt)\mathbf{k}$$

for $-10\pi \le t \le 10\pi$ and $k = 0.1, 0.2, 0.3$. Experiment with different viewpoints of the graph. The curve is called a **spherical spiral**.

In Problems 47 and 48, use a CAS to graph the given vector function for the indicated values of k. Experiment with different viewpoints of the graph.

47. $\mathbf{r}(t) = \sin kt\sin t\mathbf{i} + \sin kt\cos t\mathbf{j} + \sin t\mathbf{k};$ $k = 2, 4$

48. $\mathbf{r}(t) = \sin t\mathbf{i} + \cos t\mathbf{j} + \ln(kt)\sin t\mathbf{k};$ $k = \frac{1}{10}, 1$

12.2 Calculus of Vector Functions

▌**Introduction**   In this section we consider the calculus of vector-valued functions, in other words, limits, derivatives, and integrals of vector function. Because the concepts are similar to those discussed in Section 10.3, a review of that section is recommended.

▌**Limits and Continuity**   The fundamental notion of the **limit** of a vector function $\mathbf{r}(t) = \langle f(t), g(t), h(t) \rangle$ is defined in terms of the limits of the component functions.

Definition 12.2.1   Limit of a Vector Function

If $\lim\limits_{t \to a} f(t)$, $\lim\limits_{t \to a} g(t)$, and $\lim\limits_{t \to a} h(t)$ exist, then

$$\lim_{t \to a}\mathbf{r}(t) = \left\langle \lim_{t \to a}f(t), \lim_{t \to a}g(t), \lim_{t \to a}h(t)\right\rangle. \tag{1}$$

The symbol $t \to a$ in Definition 12.2.1 can, of course, be replaced by $t \to a^+$, $t \to a^-$, $t \to \infty$, or $t \to -\infty$.

As an immediate consequence of Definition 12.2.1, we have the following result.

Theorem 12.2.1 Limit Properties

Suppose a is a real number and $\lim_{t \to a} \mathbf{r}_1(t)$ and $\lim_{t \to a} \mathbf{r}_2(t)$ exist. If $\lim_{t \to a} \mathbf{r}_1(t) = \mathbf{L}_1$ and $\lim_{t \to a} \mathbf{r}_2(t) = \mathbf{L}_2$, then

(*i*) $\lim_{t \to a} c\mathbf{r}_1(t) = c\mathbf{L}_1$, $\quad c$ a scalar

(*ii*) $\lim_{t \to a} [\mathbf{r}_1(t) + \mathbf{r}_2(t)] = \mathbf{L}_1 + \mathbf{L}_2$

(*iii*) $\lim_{t \to a} \mathbf{r}_1(t) \cdot \mathbf{r}_2(t) = \mathbf{L}_1 \cdot \mathbf{L}_2$.

Definition 12.2.2 Continuity

A vector function $\mathbf{r}$ is **continuous** at a number a if

(*i*) $\mathbf{r}(a)$ is defined, $\qquad$ (*ii*) $\lim_{t \to a} \mathbf{r}(t)$ exists, and $\qquad$ (*iii*) $\lim_{t \to a} \mathbf{r}(t) = \mathbf{r}(a)$.

Equivalently the vector function $\mathbf{r}(t) = \langle f(t), g(t), h(t) \rangle$ is continuous at a number a if and only if the component functions f, g, and h are continuous at a. For brevity, we often say that a vector function $\mathbf{r}(t)$ is continuous at a number a if

$$\lim_{t \to a} \mathbf{r}(t) = \mathbf{r}(a). \tag{2}$$

By writing (2) it is assumed that conditions (*i*) and (*ii*) of Definition 12.2.2 hold at a number a.

■ **Derivative of a Vector Function** The definition of the derivative $\mathbf{r}'(t)$ of a vector function $\mathbf{r}(t)$ is the vector equivalent of Definition 3.1.1. In the next definition we assume that h represents a nonzero real number.

Definition 12.2.3 Derivative of a Vector Function

The **derivative** of a vector function $\mathbf{r}$ is

$$\mathbf{r}'(t) = \lim_{h \to 0} \frac{\mathbf{r}(t + h) - \mathbf{r}(t)}{h} \tag{3}$$

for all t for which the limit exists.

The derivative of $\mathbf{r}$ is also written $d\mathbf{r}/dt$. The next theorem shows that on a practical level, the derivative of a vector function is obtained by simply differentiating its component functions.

Theorem 12.2.2 Differentiation

If the component functions f, g, and h are differentiable, then the derivative of the vector function $\mathbf{r}(t)$ is given by

$$\mathbf{r}'(t) = \langle f'(t), g'(t), h'(t) \rangle. \tag{4}$$

PROOF From (3) we have

$$\mathbf{r}'(t) = \lim_{h \to 0} \frac{1}{h}[\langle f(t+h), g(t+h), h(t+h)\rangle - \langle f(t), g(t), h(t)\rangle]$$

$$= \lim_{h \to 0}\left\langle \frac{f(t+h) - f(t)}{h}, \frac{g(t+h) - g(t)}{h}, \frac{h(t+h) - h(t)}{h}\right\rangle$$

$$= \left\langle \lim_{h \to 0}\frac{f(t+h) - f(t)}{h}, \lim_{h \to 0}\frac{g(t+h) - g(t)}{h}, \lim_{h \to 0}\frac{h(t+h) - h(t)}{h}\right\rangle$$

$$= \langle f'(t), g'(t), h'(t)\rangle. \qquad \blacksquare$$

▌ Smooth Curves When the component functions of a vector function **r** have continuous first derivatives and $\mathbf{r}'(t) \neq \mathbf{0}$ for all t in an open interval (a, b), then **r** is said to be a **smooth function** and the curve C traced by **r** is called a **smooth curve**.

▌ Geometric Interpretation of r′(t) If the vector $\mathbf{r}'(t)$ exists and is not **0** at a point P on the curve C defined by a vector function $\mathbf{r}(t)$, then the derivative $\mathbf{r}'(t)$ is defined to be the **tangent vector** to the curve at P. The justification for this is similar to the discussion leading up to Definition 2.7.1 in Section 2.7. As can be seen in FIGURES 12.2.1(a) and (b), for $h > 0$ the vector $\mathbf{r}(t+h) - \mathbf{r}(t)$ and the scalar multiple

$$\frac{1}{h}[\mathbf{r}(t+h) - \mathbf{r}(t)] = \frac{\mathbf{r}(t+h) - \mathbf{r}(t)}{h}$$

are parallel. Assuming that the limit

$$\lim_{h \to 0}\frac{\mathbf{r}(t+h) - \mathbf{r}(t)}{h}$$

exists, then the vectors $\mathbf{r}(t)$ and $\mathbf{r}(t+h)$ become closer and closer as $h \to 0$. As suggested by Figures 12.2.1(b) and (c), the limiting position of the vector $[\mathbf{r}(t+h) - \mathbf{r}(t)]/h$ is a vector on the tangent line at P. We also define the **tangent line** to be the line through P that is parallel to the vector $\mathbf{r}'(t)$.

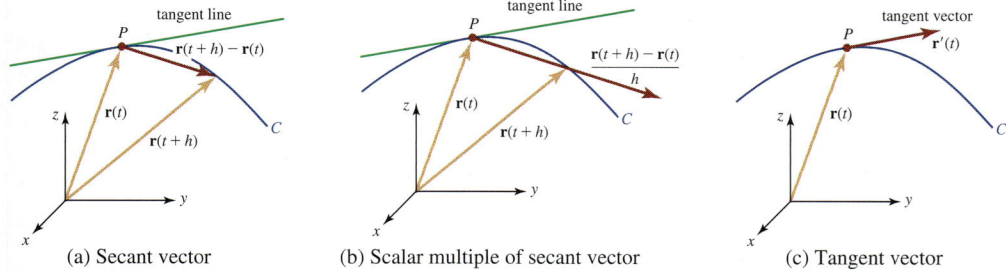

(a) Secant vector (b) Scalar multiple of secant vector (c) Tangent vector

FIGURE 12.2.1 Tangent vector at P on a curve C

EXAMPLE 1 **The Vector r′(t)**

Consider the curve C in 2-space that is traced by a point P whose position is given by $\mathbf{r}(t) = \cos 2t\mathbf{i} + \sin t\mathbf{j}, -\pi/2 \leq t \leq \pi/2$. Find the derivative $\mathbf{r}'(t)$ and graph the vectors $\mathbf{r}'(0)$ and $\mathbf{r}'(\pi/6)$.

Solution The curve C is smooth because the component functions of $\mathbf{r}(t) = \cos 2t\mathbf{i} + \sin t\mathbf{j}$ have continuous derivatives and $\mathbf{r}(t) \neq \mathbf{0}$ on the open interval $(-\pi/2, \pi/2)$. From (4),

$$\mathbf{r}'(t) = -2\sin 2t\mathbf{i} + \cos t\mathbf{j}.$$

Hence, $\qquad \mathbf{r}'(0) = \mathbf{j} \qquad$ and $\qquad \mathbf{r}'(\pi/6) = -\sqrt{3}\mathbf{i} + \frac{1}{2}\sqrt{3}\mathbf{j}.$

To graph C we first eliminate the parameter from the parametric equations $x = \cos 2t$, $y = \sin t$:

$$x = \cos 2t = \cos^2 t - \sin^2 t = 1 - 2\sin^2 t = 1 - 2y^2.$$

Since $-\pi/2 \leq t \leq \pi/2$, we see that the curve C is the portion of the parabola $x = 1 - 2y^2$ on the interval defined by $-1 \leq x \leq 1$. The vectors $\mathbf{r}'(0)$ and $\mathbf{r}'(\pi/6)$ are drawn tangent to the curve C at $(1, 0)$ and $\left(\frac{1}{2}, \frac{1}{2}\right)$, respectively. See FIGURE 12.2.2. $\blacksquare$

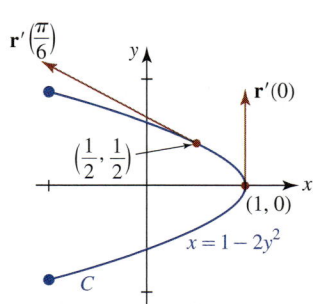

FIGURE 12.2.2 Curve C and vectors in Example 1

EXAMPLE 2 **Parametric Equations**

Find parametric equations of the tangent line to the curve C whose parametric equations are $x = t^2$, $y = t^2 - t$, $z = -7t$ at the point corresponding to $t = 3$.

Solution The vector function that gives the position of a point P on the curve is given by $\mathbf{r}(t) = t^2\mathbf{i} + (t^2 - t)\mathbf{j} - 7t\mathbf{k}$. Now,

$$\mathbf{r}'(t) = 2t\mathbf{i} + (2t - 1)\mathbf{j} - 7\mathbf{k} \quad \text{and so} \quad \mathbf{r}'(3) = 6\mathbf{i} + 5\mathbf{j} - 7\mathbf{k}.$$

The vector $\mathbf{r}'(3)$ is tangent to C at the point P whose position vector is

$$\mathbf{r}(3) = 9\mathbf{i} + 6\mathbf{j} - 21\mathbf{k},$$

that is, at the point $P(9, 6, -21)$. Using the components of $\mathbf{r}'(3)$, we see that parametric equations of the tangent line are

$$x = 9 + 6t, y = 6 + 5t, z = -21 - 7t. \qquad \blacksquare$$

■ **Higher-Order Derivatives** Higher-order derivatives of a vector function are also obtained by differentiating its components. In the case of the **second derivative**, we have

$$\mathbf{r}''(t) = \langle f''(t), g''(t), h''(t) \rangle = f''(t)\mathbf{i} + g''(t)\mathbf{j} + h''(t)\mathbf{k}. \qquad (5)$$

EXAMPLE 3 Vectors $\mathbf{r}'(t)$ and $\mathbf{r}''(t)$

If $\mathbf{r}(t) = (t^3 - 2t^2)\mathbf{i} + 4t\mathbf{j} + e^{-t}\mathbf{k}$, then

$$\mathbf{r}'(t) = (3t^2 - 4t)\mathbf{i} + 4\mathbf{j} - e^{-t}\mathbf{k}$$

and

$$\mathbf{r}''(t) = (6t - 4)\mathbf{i} + e^{-t}\mathbf{k}. \qquad \blacksquare$$

In the following theorem we list some rules of differentiation for vector functions.

Theorem 12.2.3 Rules of Differentiation

Let $\mathbf{r}$, $\mathbf{r}_1$, and $\mathbf{r}_2$ be differentiable vector functions and $f(t)$ a differentiable scalar function.

(*i*) $\dfrac{d}{dt}[\mathbf{r}_1(t) + \mathbf{r}_2(t)] = \mathbf{r}_1'(t) + \mathbf{r}_2'(t)$

(*ii*) $\dfrac{d}{dt}[f(t)\mathbf{r}(t)] = f(t)\mathbf{r}'(t) + f'(t)\mathbf{r}(t)$

(*iii*) $\dfrac{d}{dt}[\mathbf{r}(f(t))] = \mathbf{r}'(f(t))f'(t)$ (Chain Rule)

(*iv*) $\dfrac{d}{dt}[\mathbf{r}_1(t) \cdot \mathbf{r}_2(t)] = \mathbf{r}_1(t) \cdot \mathbf{r}_2'(t) + \mathbf{r}_1'(t) \cdot \mathbf{r}_2(t)$

(*v*) $\dfrac{d}{dt}[\mathbf{r}_1(t) \times \mathbf{r}_2(t)] = \mathbf{r}_1(t) \times \mathbf{r}_2'(t) + \mathbf{r}_1'(t) \times \mathbf{r}_2(t)$

PROOF OF (*iv*) If $\mathbf{r}_1(t) = \langle f_1(t), g_1(t), h_1(t) \rangle$ and $\mathbf{r}_2(t) = \langle f_2(t), g_2(t), h_2(t) \rangle$, then by (2) of Section 11.3 the dot product is the scalar function

$$\mathbf{r}_1(t) \cdot \mathbf{r}_2(t) = f_1(t)f_2(t) + g_1(t)g_2(t) + h_1(t)h_2(t).$$

After using the Product Rule we gather the terms in red and the terms shown in blue:

$$\frac{d}{dt}\mathbf{r}_1(t) \cdot \mathbf{r}_2(t) = \frac{d}{dt}f_1(t)f_2(t) + \frac{d}{dt}g_1(t)g_2(t) + \frac{d}{dt}h_1(t)h_2(t)$$

$$= f_1(t)f_2'(t) + f_1'(t)f_2(t) + g_1(t)g_2'(t) + g_1'(t)g_2(t) + h_1(t)h_2'(t) + h_1'(t)h_2(t)$$

$$= \langle f_1(t), g_1(t), h_1(t) \rangle \cdot \langle f_2'(t), g_2'(t), h_2'(t) \rangle + \langle f_1'(t), g_1'(t), h_1'(t) \rangle \cdot \langle f_2(t), g_2(t), h_2(t) \rangle$$

$$= \mathbf{r}_1(t) \cdot \mathbf{r}_2'(t) + \mathbf{r}_1'(t) \cdot \mathbf{r}_2(t). \qquad \blacksquare$$

Note: Since the cross product of two vectors is not commutative, the order in which $\mathbf{r}_1$ and $\mathbf{r}_2$ appear in part (*v*) of Theorem 12.2.3 must be strictly observed. Of course in (*iv*) and (*v*) we can carry out the dot product and cross product first and then differentiate the resulting scalar or vector function.

▌ Integrals of Vector Functions If $\mathbf{r}(t) = f(t)\mathbf{i} + g(t)\mathbf{j} + h(t)\mathbf{k}$ is a continuous vector function on an interval $[a, b]$, then the indefinite integral of $\mathbf{r}$ is defined by

$$\int \mathbf{r}(t)dt = \left[\int f(t)\,dt\right]\mathbf{i} + \left[\int g(t)\,dt\right]\mathbf{j} + \left[\int h(t)\,dt\right]\mathbf{k}.$$

The indefinite integral of $\mathbf{r}$ is another vector $\mathbf{R} + \mathbf{C}$, where $\mathbf{C}$ is a *constant vector*, such that $\mathbf{R}'(t) = \mathbf{r}(t)$. Because of the continuity of the component function f, g, and h, the definite integral of $\mathbf{r}(t)$ on $[a, b]$ can be defined as

$$\int_a^b \mathbf{r}(t)dt = \lim_{n\to\infty} \sum_{k=1}^n \mathbf{r}(t_k^*)\Delta t$$

$$= \left[\lim_{n\to\infty} \sum_{k=1}^n f(t_k^*)\Delta t\right]\mathbf{i} + \left[\lim_{n\to\infty} \sum_{k=1}^n g(t_k^*)\Delta t\right]\mathbf{j} + \left[\lim_{n\to\infty} \sum_{k=1}^n h(t_k^*)\Delta t\right]\mathbf{k}.$$

In other words,

$$\int_a^b \mathbf{r}(t)dt = \left[\int_a^b f(t)\,dt\right]\mathbf{i} + \left[\int_a^b g(t)\,dt\right]\mathbf{j} + \left[\int_a^b h(t)\,dt\right]\mathbf{k}.$$

The Fundamental Theorem of Calculus, extended to vector functions, is

$$\int_a^b \mathbf{r}(t)dt = \mathbf{R}(t)\,\Big]_a^b = \mathbf{R}(b) - \mathbf{R}(a),$$

where $\mathbf{R}$ is a vector function such that $\mathbf{R}'(t) = \mathbf{r}(t)$.

EXAMPLE 4   Integrals

(a) If $\mathbf{r}(t) = 6t^2\mathbf{i} + 4e^{-2t}\mathbf{j} + 8\cos 4t\,\mathbf{k}$, then

$$\int \mathbf{r}(t)dt = \left[\int 6t^2 dt\right]\mathbf{i} + \left[\int 4e^{-2t}dt\right]\mathbf{j} + \left[\int 8\cos 4t\,dt\right]\mathbf{k}$$

$$= [2t^3 + c_1]\mathbf{i} + [-2e^{-2t} + c_2]\mathbf{j} + [2\sin 4t + c_3]\mathbf{k}$$

$$= 2t^3\mathbf{i} - 2e^{-2t}\mathbf{j} + 2\sin 4t\,\mathbf{k} + \mathbf{C},$$

where $\mathbf{C} = c_1\mathbf{i} + c_2\mathbf{j} + c_3\mathbf{k}$. The components c_1, c_2, and c_3 of the last vector are arbitrary real constants.

(b) If $\mathbf{r}(t) = (4t - 3)\mathbf{i} + 12t^2\mathbf{j} + \dfrac{2}{1 + t^2}\mathbf{k}$, then

$$\int_{-1}^1 \mathbf{r}(t)dt = (2t^2 - 3t)\mathbf{i} + 4t^3\mathbf{j} + 2\tan^{-1}t\,\mathbf{k}\,\Big]_{-1}^1$$

$$= \left(-\mathbf{i} + 4\mathbf{j} + 2\cdot\frac{\pi}{4}\mathbf{k}\right) - \left(5\mathbf{i} - 4\mathbf{j} - 2\cdot\frac{\pi}{4}\mathbf{k}\right)$$

$$= -6\mathbf{i} + 8\mathbf{j} + \pi\mathbf{k}. \qquad\blacksquare$$

▌ Length of a Space Curve In Section 10.3 we saw that the arc length formula for a smooth curve C in 2-space defined by the parametric equations $x = f(t)$, $y = g(t)$, $a \leq t \leq b$, is

$$L = \int_a^b \sqrt{[f'(t)]^2 + [g'(t)]^2}\,dt = \int_a^b \sqrt{\left(\frac{dx}{dt}\right)^2 + \left(\frac{dy}{dt}\right)^2}\,dt.$$

Similarly, if C is a smooth curve in three-dimensional space defined by the parametric equations

$$x = f(t), \quad y = g(t), \quad z = h(t), \quad a \leq t \leq b,$$

then as we did in Section 10.3 we can build a definite integral using a polygonal path, as shown in FIGURE 12.2.3, to arrive at the definite integral

$$L = \int_a^b \sqrt{[f'(t)]^2 + [g'(t)]^2 + [h'(t)]^2}\,dt = \int_a^b \sqrt{\left(\frac{dx}{dt}\right)^2 + \left(\frac{dy}{dt}\right)^2 + \left(\frac{dz}{dt}\right)^2}\,dt \qquad (6)$$

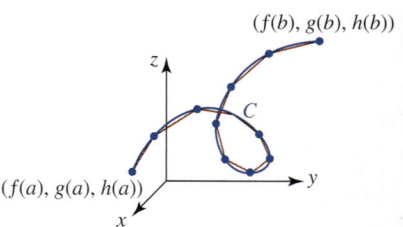

FIGURE 12.2.3 Approximating the length of C (blue) by the length of a polygonal path (red)

that defines the **length** L of the curve between the points $(f(a), g(a), h(a))$ and $(f(b), g(b), h(b))$. If the curve C is traced out by a smooth vector-valued function $\mathbf{r}(t)$, then its length between the initial point at $t = a$ and the terminal point at $t = b$ can be expressed in terms of the magnitude of $\mathbf{r}'(t)$:

$$L = \int_a^b |\mathbf{r}'(t)| \, dt. \tag{7}$$

In (7), $|\mathbf{r}'(t)|$ is either

$$|\mathbf{r}'(t)| = \sqrt{[f'(t)]^2 + [g'(t)]^2} \qquad \text{or} \qquad |\mathbf{r}'(t)| = \sqrt{[f'(t)]^2 + [g'(t)]^2 + [h'(t)]^2}$$

depending on whether C is in 2-space or 3-space, respectively.

▮ Arc Length Function   The definite integral

Review (5) in Section 6.5.

$$s(t) = \int_a^t |\mathbf{r}'(u)| \, du \tag{8}$$

is called the **arc length function** for the curve C. In (8) the symbol u is a dummy variable of integration. The function $s(t)$ represents the length of C between the points on the curve defined by the position vectors $\mathbf{r}(a)$ and $\mathbf{r}(t)$. Often it is useful to parameterize a smooth curve C in the plane or in space in terms of the arc length s. By evaluating (8) we express s as a function of the parameter t. If we can solve that equation for t in terms of s, then we can express $\mathbf{r}(t) = \langle f(t), g(t) \rangle$ or $\mathbf{r}(t) = \langle f(t), g(t), h(t) \rangle$ as

$$\mathbf{r}(s) = \langle x(s), y(s) \rangle \qquad \text{or} \qquad \mathbf{r}(s) = \langle x(s), y(s), z(s) \rangle.$$

The next example illustrates the procedure for finding an **arc length parameterization** $\mathbf{r}(s)$ for a curve C.

EXAMPLE 5   An Arc Length Parameterization

Find an arc length parameterization of the circular helix of Example 2 of Section 12.1:

$$\mathbf{r}(t) = 2\cos t \, \mathbf{i} + 2\sin t \, \mathbf{j} + t \, \mathbf{k}.$$

Solution   From $\mathbf{r}'(t) = -2\sin t \, \mathbf{i} + 2\cos t \, \mathbf{j} + \mathbf{k}$ we find $|\mathbf{r}'(t)| = \sqrt{5}$. It follows from (8) that the length of the curve starting at $\mathbf{r}(0)$ to an arbitrary point defined by $\mathbf{r}(t)$ is

$$s = \int_0^t \sqrt{5} \, du = \sqrt{5} \, u \Big]_0^t = \sqrt{5} \, t.$$

Solving $s = \sqrt{5} \, t$ for t gives $t = s/\sqrt{5}$. By substituting for t in $\mathbf{r}(t)$ we obtain a vector function of the helix as a function of arc length:

$$\mathbf{r}(s) = 2\cos \frac{s}{\sqrt{5}} \mathbf{i} + 2\sin \frac{s}{\sqrt{5}} \mathbf{j} + \frac{s}{\sqrt{5}} \mathbf{k}. \tag{9}$$

Parametric equations of the helix are then

$$x = 2\cos \frac{s}{\sqrt{5}}, \quad y = 2\sin \frac{s}{\sqrt{5}}, \quad z = \frac{s}{\sqrt{5}}. \qquad \blacksquare$$

Note that the derivative of the vector function (9) with respect to arc length s is

It is particularly easy to find an arc length parameterization of a line $\mathbf{r}(t) = \mathbf{r}_0 + t\mathbf{v}$. See Problem 49 in Exercises 12.2.

$$\mathbf{r}'(s) = -\frac{2}{\sqrt{5}} \sin \frac{s}{\sqrt{5}} \mathbf{i} + \frac{2}{\sqrt{5}} \cos \frac{s}{\sqrt{5}} \mathbf{j} + \frac{1}{\sqrt{5}} \mathbf{k}$$

and its magnitude is

$$|\mathbf{r}'(s)| = \sqrt{\frac{4}{5} \sin^2 \frac{s}{\sqrt{5}} + \frac{4}{5} \cos^2 \frac{s}{\sqrt{5}} + \frac{1}{5}} = \sqrt{\frac{5}{5}} = 1.$$

The fact that $|\mathbf{r}'(s)| = 1$ indicates that $\mathbf{r}'(s)$ is a unit vector. This is no coincidence. As we have seen, the derivative of a vector function $\mathbf{r}(t)$ with respect to the parameter t is a tangent vector to the curve C traced by $\mathbf{r}$. However, if the curve C is parameterized in terms of arc length s, then:

  • *The derivative $\mathbf{r}'(s)$ is a unit tangent vector.* (10)

To see why this is so, recall that the derivative form of the Fundamental Theorem of Calculus, Theorem 5.5.2, shows that the derivative of (8) with respect to t is

$$\frac{ds}{dt} = |\mathbf{r}'(t)|.$$ (11)

But if the curve C is described by an arc length parameterization $\mathbf{r}(s)$, then (8) shows that the length s of the curve from $\mathbf{r}(0)$ to $\mathbf{r}(s)$ is

$$s = \int_0^s |\mathbf{r}'(u)| \, du.$$ (12)

Because $\dfrac{d}{ds} s = 1$, the derivative of (12) with respect to s is

$$\frac{d}{ds} s = |\mathbf{r}'(s)| \qquad \text{or} \qquad |\mathbf{r}'(s)| = 1.$$

We will see why (10) is important in the next section.

Exercises 12.2   Answers to selected odd-numbered problems begin on page ANS-39.

≡ Fundamentals

In Problems 1–4, evaluate the given limit or state that it does not exist.

1. $\lim\limits_{t \to 2} [t^3 \mathbf{i} + t^4 \mathbf{j} + t^5 \mathbf{k}]$

2. $\lim\limits_{t \to 0^+} \left[\dfrac{\sin 2t}{t} \mathbf{i} + (t-2)^5 \mathbf{j} + t \ln t \, \mathbf{k} \right]$

3. $\lim\limits_{t \to 1} \left\langle \dfrac{t^2 - 1}{t - 1}, \dfrac{5t - 1}{t + 1}, \dfrac{2e^{t-1} - 2}{t - 1} \right\rangle$

4. $\lim\limits_{t \to \infty} \left\langle \dfrac{e^{2t}}{2e^{2t} + t}, \dfrac{e^{-t}}{2e^{-t} + 5}, \tan^{-1} t \right\rangle$

In Problems 5 and 6, suppose that

$$\lim_{t \to a} \mathbf{r}_1(t) = \mathbf{i} - 2\mathbf{j} + \mathbf{k} \quad \text{and} \quad \lim_{t \to a} \mathbf{r}_2(t) = 2\mathbf{i} + 5\mathbf{j} + 7\mathbf{k}.$$

Find the given limit.

5. $\lim\limits_{t \to a} [-4\mathbf{r}_1(t) + 3\mathbf{r}_2(t)]$

6. $\lim\limits_{t \to a} \mathbf{r}_1(t) \cdot \mathbf{r}_2(t)$

In Problems 7 and 8, determine whether the given vector function is continuous at $t = 1$.

7. $\mathbf{r}(t) = (t^2 - 2t)\mathbf{i} + \dfrac{1}{t + 1}\mathbf{j} + \ln(t - 1)\mathbf{k}$

8. $\mathbf{r}(t) = \sin \pi t \, \mathbf{i} + \tan \pi t \, \mathbf{j} + \cos \pi t \, \mathbf{k}$

In Problems 9 and 10, find the indicated two vectors for the given vector function.

9. $\mathbf{r}(t) = (3t - 1)\mathbf{i} + 4t^2\mathbf{j} + (5t^2 - t)\mathbf{k}$; $\mathbf{r}'(1), \dfrac{\mathbf{r}(1.1) - \mathbf{r}(1)}{0.1}$

10. $\mathbf{r}(t) = \dfrac{1}{1 + 5t}\mathbf{i} + (3t^2 + t)\mathbf{j} + (1 - t)^3\mathbf{k}$; $\mathbf{r}'(0), \dfrac{\mathbf{r}(0.05) - \mathbf{r}(0)}{0.05}$

In Problems 11–14, find $\mathbf{r}'(t)$ and $\mathbf{r}''(t)$ for the given vector function.

11. $\mathbf{r}(t) = \ln t \, \mathbf{i} + \dfrac{1}{t} \mathbf{j}, \quad t > 0$

12. $\mathbf{r}(t) = \langle t \cos t - \sin t, t + \cos t \rangle$

13. $\mathbf{r}(t) = \langle te^{2t}, t^3, 4t^2 - t \rangle$

14. $\mathbf{r}(t) = t^2\mathbf{i} + t^3\mathbf{j} + \tan^{-1} t \, \mathbf{k}$

In Problems 15–18, graph the curve C that is described by $\mathbf{r}(t)$ and graph $\mathbf{r}'(t)$ at the point corresponding to the indicated value of t.

15. $\mathbf{r}(t) = 2\cos t \, \mathbf{i} + 6\sin t \, \mathbf{j}; \quad t = \pi/6$

16. $\mathbf{r}(t) = t^3\mathbf{i} + t^2\mathbf{j}; \quad t = -1$

17. $\mathbf{r}(t) = 2\mathbf{i} + t\mathbf{j} + \dfrac{4}{1 + t^2}\mathbf{k}; \quad t = 1$

18. $\mathbf{r}(t) = 3\cos t \, \mathbf{i} + 3\sin t \, \mathbf{j} + 2t\mathbf{k}; \quad t = \pi/4$

In Problems 19 and 20, find parametric equations of the tangent line to the given curve at the point corresponding to the indicated value of t.

19. $x = t, y = \dfrac{1}{2}t^2, z = \dfrac{1}{3}t^3; \quad t = 2$

20. $x = t^3 - t, y = \dfrac{6t}{t + 1}, z = (2t + 1)^2; \quad t = 1$

In Problems 21 and 22, find a unit tangent vector to the given curve at the point corresponding to the indicated value of t. Find parametric equations of the tangent line at this point.

21. $\mathbf{r}(t) = te^t\mathbf{i} + (t^2 + 2t)\mathbf{j} + (t^3 - t)\mathbf{k}; \quad t = 0$

22. $\mathbf{r}(t) = (1 + \sin 3t)\mathbf{i} + \tan 2t \, \mathbf{j} + t\mathbf{k}; \quad t = \pi$

In Problems 23 and 24, find a vector function of the tangent line to the given curve at the point corresponding to the indicated value of t.

23. $\mathbf{r}(t) = \langle \cos t, \sin t, t \rangle; \quad t = \pi/3$

24. $\mathbf{r}(t) = \langle 6e^{-t/2}, e^{2t}, e^{3t} \rangle; \quad t = 0$

In Problems 25–30, find the indicated derivative. Assume that all vector functions are differentiable.

25. $\dfrac{d}{dt}[\mathbf{r}(t) \times \mathbf{r}'(t)]$

26. $\dfrac{d}{dt}[\mathbf{r}(t) \cdot (t\mathbf{r}(t))]$

27. $\dfrac{d}{dt}[\mathbf{r}(t) \cdot (\mathbf{r}'(t) \times \mathbf{r}''(t))]$

28. $\dfrac{d}{dt}[\mathbf{r}_1(t) \times (\mathbf{r}_2(t) \times \mathbf{r}_3(t))]$

29. $\dfrac{d}{dt}[\mathbf{r}_1(2t) + \mathbf{r}_2(1/t)]$

30. $\dfrac{d}{dt}[t^3\mathbf{r}(t^2)]$

In Problems 31–34, evaluate the given integral.

31. $\displaystyle\int_{-1}^{2} (t\mathbf{i} + 3t^2\mathbf{j} + 4t^3\mathbf{k})\,dt$

32. $\displaystyle\int_{0}^{4} (\sqrt{2t+1}\,\mathbf{i} - \sqrt{t}\,\mathbf{j} + \sin \pi t\mathbf{k})\,dt$

33. $\displaystyle\int (te^t\mathbf{i} - e^{-2t}\mathbf{j} + te^{t^2}\mathbf{k})\,dt$

34. $\displaystyle\int \dfrac{1}{1+t^2}(\mathbf{i} + t\mathbf{j} + t^2\mathbf{k})\,dt$

In Problems 35–38, find a vector function $\mathbf{r}(t)$ that satisfies the indicated conditions.

35. $\mathbf{r}'(t) = 6\mathbf{i} + 6t\mathbf{j} + 3t^2\mathbf{k}; \quad \mathbf{r}(0) = \mathbf{i} - 2\mathbf{j} + \mathbf{k}$

36. $\mathbf{r}'(t) = t\sin t^2\mathbf{i} - \cos 2t\mathbf{j}; \quad \mathbf{r}(0) = \frac{3}{2}\mathbf{i}$

37. $\mathbf{r}''(t) = 12t\mathbf{i} - 3t^{-1/2}\mathbf{j} + 2\mathbf{k}; \quad \mathbf{r}'(1) = \mathbf{j}, \mathbf{r}(1) = 2\mathbf{i} - \mathbf{k}$

38. $\mathbf{r}''(t) = \sec^2 t\mathbf{i} + \cos t\mathbf{j} - \sin t\mathbf{k};$
$\mathbf{r}'(0) = \mathbf{i} + \mathbf{j} + \mathbf{k}, \mathbf{r}(0) = -\mathbf{j} + 5\mathbf{k}$

In Problems 39–42, find the length of the curve traced by the given vector function on the indicated interval.

39. $\mathbf{r}(t) = a\cos t\mathbf{i} + a\sin t\mathbf{j} + ct\mathbf{k}; \quad 0 \le t \le 2\pi$

40. $\mathbf{r}(t) = t\mathbf{i} + t\cos t\mathbf{j} + t\sin t\mathbf{k}; \quad 0 \le t \le \pi$

41. $\mathbf{r}(t) = e^t\cos 2t\mathbf{i} + e^t\sin 2t\mathbf{j} + e^t\mathbf{k}; \quad 0 \le t \le 3\pi$

42. $\mathbf{r}(t) = 3t\mathbf{i} + \sqrt{3}t^2\mathbf{j} + \frac{2}{3}t^3\mathbf{k}; \quad 0 \le t \le 1$

In Problems 43–46, use (8) and integration from $u = 0$ to $u = t$ to find an arc length parameterization $\mathbf{r}(s)$ for the given curve. Verify that $\mathbf{r}'(s)$ is a unit vector.

43. $\mathbf{r}(t) = 9\sin t\mathbf{i} + 9\cos t\mathbf{j}$

44. $\mathbf{r}(t) = 5\cos t\mathbf{i} + 12t\mathbf{j} + 5\sin t\mathbf{k}$

45. $\mathbf{r}(t) = (1 + 2t)\mathbf{i} + (5 - 3t)\mathbf{j} + (2 + 4t)\mathbf{k}$

46. $\mathbf{r}(t) = e^t\cos t\mathbf{i} + e^t\sin t\mathbf{j} + \mathbf{k}$

≡ Think About It

47. Suppose $\mathbf{r}$ is a differentiable vector function for which $|\mathbf{r}(t)| = c$ for all t. Show that the tangent vector $\mathbf{r}'(t)$ is perpendicular to the position vector $\mathbf{r}(t)$ for all t.

48. If $\mathbf{v}$ is a constant vector and $\mathbf{r}(t)$ is integrable on $[a, b]$, prove that $\int_a^b \mathbf{v} \cdot \mathbf{r}(t)\,dt = \mathbf{v} \cdot \int_a^b \mathbf{r}(t)\,dt$.

49. Suppose $\mathbf{r}(t) = \mathbf{r}_0 + t\mathbf{v}$ is the vector equation of a line, where $\mathbf{r}_0$ and $\mathbf{v}$ are constant vectors. Use the arc length function $s = \int_0^t |\mathbf{r}'(u)|\,du$ to show that an arc length parameterization of the line is given by $\mathbf{r}(s) = \mathbf{r}_0 + s\dfrac{\mathbf{v}}{|\mathbf{v}|}$. Show that $\mathbf{r}'(s)$ is a unit vector. In other words, to obtain an arc length parameterization of a line we only need to normalize the vector $\mathbf{v}$.

50. Use the results of Problem 49 to find an arc length parameterization for each of the following lines.

(a) $\mathbf{r}(t) = \langle 1 + 3t, 2 - 4t \rangle = \langle 1, 2 \rangle + t\langle 3, -4 \rangle$

(b) $\mathbf{r}(t) = \langle 1 + t, 1 + 2t, 10 - t \rangle$

12.3 Motion on a Curve

▌ Introduction Suppose a particle or body moves along a curve C so that its position at time t is given by the vector-valued function

$$\mathbf{r}(t) = f(t)\mathbf{i} + g(t)\mathbf{j} + h(t)\mathbf{k}.$$

We can describe the velocity and acceleration of the particle in terms of derivatives of $\mathbf{r}(t)$.

▌ Velocity and Acceleration If f, g, and h have second derivatives, then the vectors

$$\mathbf{v}(t) = \mathbf{r}'(t) = f'(t)\mathbf{i} + g'(t)\mathbf{j} + h'(t)\mathbf{k} \tag{1}$$

$$\mathbf{a}(t) = \mathbf{r}''(t) = f''(t)\mathbf{i} + g''(t)\mathbf{j} + h''(t)\mathbf{k} \tag{2}$$

are called the **velocity** and **acceleration** of the particle, respectively. The scalar function

$$|\mathbf{v}(t)| = |\mathbf{r}'(t)| = \sqrt{[f'(t)]^2 + [g'(t)]^2 + [h'(t)]^2} \tag{3}$$

is the **speed** of the particle. Speed is related to arc length. From (7) of Section 12.2 we saw that if a curve C is traced out by a smooth vector-valued function $\mathbf{r}(t)$, then its length between the initial point at $t = a$ and terminal point at $t = b$ is given by $L = \int_a^b |\mathbf{r}'(t)|\, dt$. In view of (1) and (3) this is the same as

$$L = \int_a^b |\mathbf{v}(t)|\, dt. \qquad (4)$$

If $P(x_1, y_1, z_1)$ is the position of the particle on the curve C at time t_1, then in view of the discussion in Section 12.2 on the geometric interpretation of $\mathbf{r}'(t)$ we conclude that

- $\mathbf{v}(t_1)$ *is tangent to the curve C at P.*

Similar remarks hold for curves traced by the vector function $\mathbf{r}(t) = f(t)\mathbf{i} + g(t)\mathbf{j}$.

EXAMPLE 1 Graph of Velocity and Acceleration

The position of a moving particle is given by $\mathbf{r}(t) = t^2\mathbf{i} + t\mathbf{j} + \frac{5}{2}t\mathbf{k}$. Graph the curve C defined by $\mathbf{r}(t)$ and the vectors $\mathbf{v}(2)$ and $\mathbf{a}(2)$.

Solution Since $x = t^2$, $y = t$, the path of the particle is above the parabola $x = y^2$ that lies in the xy-plane. When $t = 2$, the position vector $\mathbf{r}(2) = 4\mathbf{i} + 2\mathbf{j} + 5\mathbf{k}$ indicates that the particle is at the point $P(4, 2, 5)$ on C. Now,

$$\mathbf{v}(t) = \mathbf{r}'(t) = 2t\mathbf{i} + \mathbf{j} + \frac{5}{2}\mathbf{k} \quad \text{and} \quad \mathbf{a}(t) = \mathbf{r}''(t) = 2\mathbf{i}$$

so that

$$\mathbf{v}(2) = 4\mathbf{i} + \mathbf{j} + \frac{5}{2}\mathbf{k} \quad \text{and} \quad \mathbf{a}(2) = 2\mathbf{i}.$$

These vectors are shown in FIGURE 12.3.1. ∎

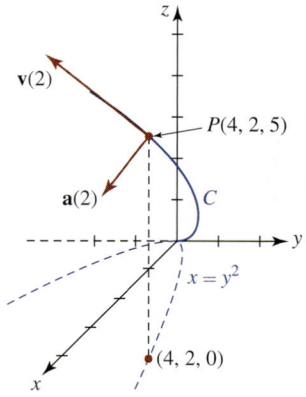

FIGURE 12.3.1 Velocity and acceleration vectors in Example 1

If a particle moves with a constant speed c, then its acceleration vector is perpendicular to the velocity vector $\mathbf{v}$. To see this, note that

$$|\mathbf{v}|^2 = c^2 \qquad \text{or} \qquad \mathbf{v} \cdot \mathbf{v} = c^2.$$

We differentiate both sides with respect to t, and obtain with the aid of Theorem 12.2.3(*iv*)

$$\frac{d}{dt}(\mathbf{v} \cdot \mathbf{v}) = \mathbf{v} \cdot \frac{d\mathbf{v}}{dt} + \frac{d\mathbf{v}}{dt} \cdot \mathbf{v} = 2\mathbf{v} \cdot \frac{d\mathbf{v}}{dt} = 0.$$

Thus,

$$\frac{d\mathbf{v}}{dt} \cdot \mathbf{v} = 0 \qquad \text{or} \qquad \mathbf{a}(t) \cdot \mathbf{v}(t) = 0 \quad \text{for all } t. \qquad (5)$$

EXAMPLE 2 Graph of Velocity and Acceleration

Suppose the vector function in Example 4 of Section 12.1 represents the position of a particle moving in a circular orbit. Graph the velocity and acceleration vectors at $t = \pi/4$.

Solution The vector-valued function

$$\mathbf{r}(t) = 2\cos t\, \mathbf{i} + 2\sin t\, \mathbf{j} + 3\mathbf{k}$$

is the position vector of a particle moving in a circular orbit of radius 2 in the plane $z = 3$. When $t = \pi/4$, the particle is at the point $P(\sqrt{2}, \sqrt{2}, 3)$. Now,

$$\mathbf{v}(t) = \mathbf{r}'(t) = -2\sin t\, \mathbf{i} + 2\cos t\, \mathbf{j}$$

and

$$\mathbf{a}(t) = \mathbf{r}''(t) = -2\cos t\, \mathbf{i} - 2\sin t\, \mathbf{j}.$$

Since the speed $|\mathbf{v}(t)| = 2$ is constant for all time t, it follows from (5) that $\mathbf{a}(t)$ is perpendicular to $\mathbf{v}(t)$. (Verify this.) As shown in FIGURE 12.3.2, the vectors

$$\mathbf{v}\!\left(\frac{\pi}{4}\right) = -2\sin\frac{\pi}{4}\mathbf{i} + 2\cos\frac{\pi}{4}\mathbf{j} = -\sqrt{2}\mathbf{i} + \sqrt{2}\mathbf{j}$$

and

$$\mathbf{a}\!\left(\frac{\pi}{4}\right) = -2\cos\frac{\pi}{4}\mathbf{i} - 2\sin\frac{\pi}{4}\mathbf{j} = -\sqrt{2}\mathbf{i} - \sqrt{2}\mathbf{j}$$

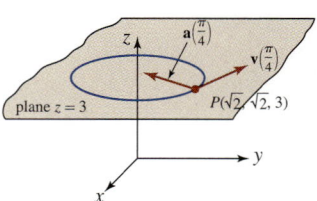

FIGURE 12.3.2 Velocity and acceleration vectors in Example 2

are drawn at the point P. The vector $\mathbf{v}(\pi/4)$ is tangent to the circular path whereas $\mathbf{a}(\pi/4)$ points along a radius toward the center of the circle. ∎

Centripetal Acceleration For circular motion in the plane, described by $\mathbf{r}(t) = r_0\cos\omega t\,\mathbf{i} + r_0\sin\omega t\,\mathbf{j}$, r_0 and ω constants, it is evident that $\mathbf{r}'' = -\omega^2\mathbf{r}$. This means that the acceleration vector $\mathbf{a}(t) = \mathbf{r}''(t)$ points in the direction opposite to that of the position vector $\mathbf{r}(t)$. We then say $\mathbf{a}(t)$ is **centripetal acceleration**. See FIGURE 12.3.3. If $v = |\mathbf{v}(t)|$ and $a = |\mathbf{a}(t)|$, we leave it as an exercise to show that $a = v^2/r_0$. See Problem 17 in Exercises 12.3.

Curvilinear Motion in the Plane Many important applications of vector functions occur in describing curvilinear motion in a plane. For example, planetary and projectile motions take place in a plane. In analyzing the motion of short-range ballistic projectiles, we begin with the acceleration of gravity written in vector form

▶ The projectile is shot or hurled rather than self-propelled. In the analysis of long-range ballistic motion, the curvature of the Earth must be taken into consideration.

$$\mathbf{a}(t) = -g\mathbf{j}.$$

If, as shown in FIGURE 12.3.4, a projectile is launched with an initial velocity $\mathbf{v}_0 = v_0\cos\theta\,\mathbf{i} + v_0\sin\theta\,\mathbf{j}$ from an initial height $\mathbf{s}_0 = s_0\mathbf{j}$, then

$$\mathbf{v}(t) = \int(-g\mathbf{j})dt = -gt\mathbf{j} + \mathbf{C}_1,$$

where $\mathbf{v}(0) = \mathbf{v}_0$ implies that $\mathbf{C}_1 = \mathbf{v}_0$. Therefore,

$$\mathbf{v}(t) = (v_0\cos\theta)\mathbf{i} + (-gt + v_0\sin\theta)\mathbf{j}.$$

Integrating again and using $\mathbf{r}(0) = \mathbf{s}_0$ yield

$$\mathbf{r}(t) = (v_0\cos\theta)t\mathbf{i} + \left[-\frac{1}{2}gt^2 + (v_0\sin\theta)t + s_0\right]\mathbf{j}.$$

Hence, parametric equations for the trajectory of the projectile are

$$x(t) = (v_0\cos\theta)t, \quad y(t) = -\frac{1}{2}gt^2 + (v_0\sin\theta)t + s_0. \tag{6}$$

See (3) of Section 10.2.

We are naturally interested in finding the maximum altitude, or height, H and the maximum horizontal distance, or range, R attained by a projectile. As shown in FIGURE 12.3.5, these quantities are the maximum values of $y(t)$ and $x(t)$, respectively. To compute these values we find the times t_1 and $t_2 > 0$ for which $y'(t_1) = 0$ and $y(t_2) = 0$, respectively. Then

$$H = y_{\max} = y(t_1) \quad \text{and} \quad R = x_{\max} = x(t_2). \tag{7}$$

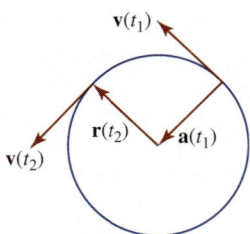

FIGURE 12.3.3 Centripetal acceleration vector $\mathbf{a}$

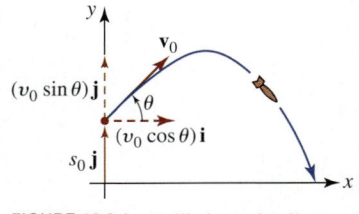

FIGURE 12.3.4 Ballistic projectile

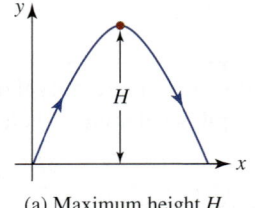

(a) Maximum height H (b) Range R
FIGURE 12.3.5 Maximum height and range of a projectile

EXAMPLE 3 Projectile Motion

A shell is fired from ground level with an initial speed of 768 ft/s at an angle of elevation of 30°. Find

    (a) the vector function and parametric equations of the shell's trajectory,
    (b) the maximum altitude attained,
    (c) the range of the shell, and
    (d) the speed at impact.

Solution
    (a) In terms of vectors, the initial position of the projectile is $\mathbf{s}_0 = \mathbf{0}$ and its initial velocity is

$$\mathbf{v}_0 = (768\cos30°)\mathbf{i} + (768\sin30°)\mathbf{j} = 384\sqrt{3}\,\mathbf{i} + 384\,\mathbf{j}. \tag{8}$$

Integrating $\mathbf{a}(t) = -32\mathbf{j}$ and using (8) give

$$\mathbf{v}(t) = (384\sqrt{3})\mathbf{i} + (-32t + 384)\mathbf{j}. \tag{9}$$

Integrating (9) and using $\mathbf{s}_0 = \mathbf{0}$ yield the vector function

$$\mathbf{r}(t) = (384\sqrt{3}t)\mathbf{i} + (-16t^2 + 384t)\mathbf{j}.$$

Hence, the parametric equations of the shell's trajectory are

$$x(t) = 384\sqrt{3}t, \quad y(t) = -16t^2 + 384t. \tag{10}$$

(b) From (10) we see that $dy/dt = 0$ when

$$-32t + 384 = 0 \quad \text{or} \quad t = 12.$$

Thus, from the first part of (7) the maximum height H attained by the shell is

$$H = y(12) = -16(12)^2 + 384(12) = 2304 \text{ ft.}$$

(c) From (6) we see that $y(t) = 0$ when

$$-16t(t - 24) = 0 \quad \text{or} \quad t = 0, t = 24.$$

From the second part of (7) the range R of the shell is

$$R = x(24) = 384\sqrt{3}(24) \approx 15,963 \text{ ft.}$$

(d) From (9) we obtain the impact speed of the shell:

$$|\mathbf{v}(24)| = \sqrt{(384\sqrt{3})^2 + (-384)^2} = 768 \text{ ft/s.} \quad \blacksquare$$

$r(t)$ NOTES FROM THE CLASSROOM

On page 667 we saw that the rate of change of arc length dL/dt is the same as the speed $|\mathbf{v}(t)| = |\mathbf{r}'(t)|$. However, as we will see in the next section, it does *not* follow that the *scalar acceleration* d^2L/dt^2 is the same as $|\mathbf{a}(t)| = |\mathbf{r}''(t)|$. See Problem 18 in Exercises 12.3.

Exercises 12.3 Answers to selected odd-numbered problems begin on page ANS-39.

☰ Fundamentals

In Problems 1–8, $\mathbf{r}(t)$ is the position vector of a moving particle. Graph the curve and the velocity and acceleration vectors at the indicated time. Find the speed at that time.

1. $\mathbf{r}(t) = t^2\mathbf{i} + \frac{1}{4}t^4\mathbf{j}; \quad t = 1$

2. $\mathbf{r}(t) = t^2\mathbf{i} + \frac{1}{t^2}\mathbf{j}; \quad t = 1$

3. $\mathbf{r}(t) = -\cosh 2t\mathbf{i} + \sinh 2t\mathbf{j}; \quad t = 0$

4. $\mathbf{r}(t) = 2\cos t\mathbf{i} + (1 + \sin t)\mathbf{j}; \quad t = \pi/3$

5. $\mathbf{r}(t) = 2\mathbf{i} + (t - 1)^2\mathbf{j} + t\mathbf{k}; \quad t = 2$

6. $\mathbf{r}(t) = t\mathbf{i} + t\mathbf{j} + t^3\mathbf{k}; \quad t = 2$

7. $\mathbf{r}(t) = t\mathbf{i} + t^2\mathbf{j} + t^3\mathbf{k}; \quad t = 1$

8. $\mathbf{r}(t) = t\mathbf{i} + t^3\mathbf{j} + t\mathbf{k}; \quad t = 1$

9. Suppose $\mathbf{r}(t) = t^2\mathbf{i} + (t^3 - 2t)\mathbf{j} + (t^2 - 5t)\mathbf{k}$ is the position vector of a moving particle.

  (a) At what points does the particle pass through the xy-plane?

  (b) What are its velocity and acceleration at the points in part (a)?

10. Suppose a particle moves in space so that $\mathbf{a}(t) = \mathbf{0}$ for all time t. Describe its path.

11. A shell is fired from ground level with an initial speed of 480 ft/s at an angle of elevation of 30°. Find:

  (a) a vector function and parametric equations of the shell's trajectory,

  (b) the maximum altitude attained,

  (c) the range of the shell, and

  (d) the speed at impact.

12. Rework Problem 11 if the shell is fired with the same initial speed and the same angle of elevation but from a cliff 1600 ft high.

13. A car is pushed off an 81-ft-high sheer seaside cliff with a speed of 4 ft/s. Find the speed at which the car hits the water.

14. A small projectile is launched from ground level with an initial speed of 98 m/s. Find the possible angles of elevation so that its range is 490 m.

15. A football quarterback throws a 100-yd "bomb" at an angle of 45° from the horizontal. What is the initial speed of the football at the point of release?

16. A quarterback throws a football with the same initial speed at an angle of 60° from the horizontal and then at an angle of 30° from the horizontal. Show that the range of the football is the same in each case. Generalize this result for any release angle $0 < \theta < \pi/2$.

17. Suppose that $\mathbf{r}(t) = r_0\cos\omega t\,\mathbf{i} + r_0\sin\omega t\,\mathbf{j}$ is the position vector of an object that is moving in a circle of radius r_0 in the xy-plane. If $|\mathbf{v}(t)| = v$, show that the magnitude of the centripetal acceleration is $a = |\mathbf{a}(t)| = v^2/r_0$.

18. The motion of a particle in 3-space is described by the vector function

$$\mathbf{r}(t) = b\cos t\,\mathbf{i} + b\sin t\,\mathbf{j} + ct\,\mathbf{k}, \quad t \geq 0.$$

(a) Compute $|\mathbf{v}(t)|$.
(b) Compute the arc length function $s(t) = \int_0^t |\mathbf{v}(u)|\,du$ and verify that ds/dt is the same as the result of part (a).
(c) Verify that $d^2s/dt^2 \neq |\mathbf{a}(t)|$.

≡ Applications

19. A projectile is fired from a cannon directly at a target that is dropped from rest simultaneously as the cannon is fired. Show that the projectile will strike the target in midair. See FIGURE 12.3.6. [*Hint*: Assume that the origin is at the muzzle of the cannon and that the angle of elevation is θ. If $\mathbf{r}_p$ and $\mathbf{r}_t$ are position vectors of the projectile and target, respectively, is there a time at which $\mathbf{r}_p = \mathbf{r}_t$?]

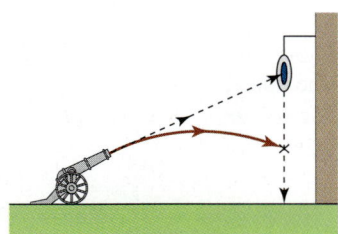

FIGURE 12.3.6 Cannon and target in Problem 19

20. To supply the victims of a natural disaster, sturdy equipment and food/medicine supply packs are simply dropped from planes that fly horizontally at a slow speed and a low altitude. A supply plane flies horizontally over a target at an altitude of 1024 ft at a constant speed of 180 mi/h. Use (2) to determine the horizontal distance a supply pack travels relative to the point from which it was dropped. At what line-of-sight angle α should the supply pack be released in order to hit the target indicated in FIGURE 12.3.7?

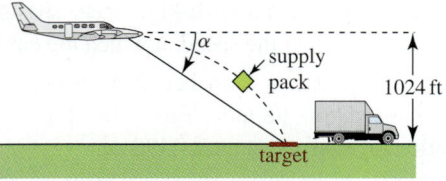

FIGURE 12.3.7 Supply plane in Problem 20

21. The **effective weight** w_e of a body of mass m at the equator of the Earth is defined by $w_e = mg - ma$, where a is the magnitude of the centripetal acceleration given in Problem 17. Determine the effective weight of a 192-lb person if the radius of the Earth is 4000 mi, $g = 32$ ft/s², and $v = 1530$ ft/s.

22. Consider a bicyclist riding on a flat circular track of radius r_0. If m is the combined mass of the rider and bicycle, fill in the blanks in FIGURE 12.3.8. [*Hint*: Use Problem 17 and *force = mass × acceleration*. Assume that the positive directions are upward and to the left.] The **resultant** vector $\mathbf{U}$ gives the direction the bicyclist must be tipped to avoid falling. Find the angle ϕ from the vertical at which the bicyclist must be tipped if her speed is 44 ft/s and the radius of the track is 60 ft.

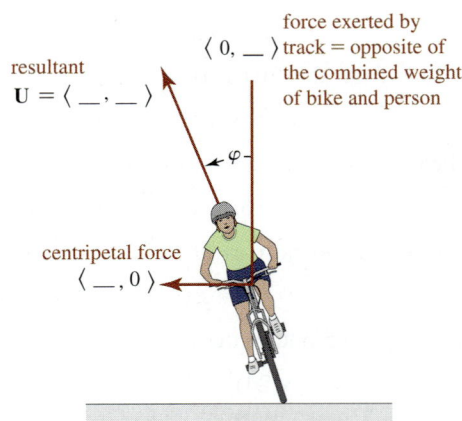

FIGURE 12.3.8 Bicyclist in Problem 22

23. Use the results given in (6) to show that the trajectory of a ballistic projectile is parabolic.

24. A projectile is launched with an initial speed v_0 from ground level at an angle of elevation θ. Use (6) to show that the maximum height and range of the projectile are

$$H = \frac{v_0^2\sin^2\theta}{2g} \quad \text{and} \quad R = \frac{v_0^2\sin 2\theta}{g},$$

respectively.

25. The velocity of a particle moving in a fluid is described by means of a **velocity field** $\mathbf{v} = v_1\mathbf{i} + v_2\mathbf{j} + v_3\mathbf{k}$, where the components v_1, v_2, and v_3 are functions of x, y, z, and time t. If the velocity of the particle is $\mathbf{v}(t) = 6t^2x\mathbf{i} - 4ty^2\mathbf{j} + 2t(z + 1)\mathbf{k}$, find $\mathbf{r}(t)$. [*Hint*: Use separation of variables. See Section 8.1 or Section 16.1]

26. Suppose m is the mass of a moving particle. Newton's second law of motion can be written in vector form as

$$\mathbf{F} = m\mathbf{a} = \frac{d}{dt}(m\mathbf{v}) = \frac{d\mathbf{p}}{dt},$$

where $\mathbf{p} = m\mathbf{v}$ is called **linear momentum**. The **angular momentum** of the particle with respect to the origin is defined to be $\mathbf{L} = \mathbf{r} \times \mathbf{p}$, where $\mathbf{r}$ is its position vector. If the torque of the particle about the origin is $\boldsymbol{\tau} = \mathbf{r} \times \mathbf{F} = \mathbf{r} \times d\mathbf{p}/dt$, show that $\boldsymbol{\tau}$ is the time rate of change of angular momentum.

27. Suppose the Sun is located at the origin. The gravitational force $\mathbf{F}$ exerted on a planet of mass m by the Sun of mass M is

$$\mathbf{F} = -k\frac{Mm}{r^2}\mathbf{u}.$$

F is a **central force**—that is, a force directed along the position vector **r** of the planet. Here k is the gravitational constant (see page 369), $r = |\mathbf{r}|$, $\mathbf{u} = (1/r)\mathbf{r}$ is a unit vector in the direction of **r**, and the minus sign indicates that **F** is an attractive force—that is, a force directed toward the Sun. See FIGURE 12.3.9.

(a) Use Problem 26 to show that the torque acting on the planet due to this central force is **0**.

(b) Explain why the angular momentum **L** of a planet is constant.

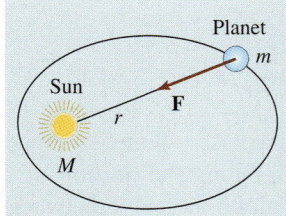

FIGURE 12.3.9 Central force vector **F** in Problem 27

≡ **Think About It**

28. A cannon launches a cannon ball horizontally as shown in FIGURE 12.3.10.

(a) The more gunpowder that is used, the greater the initial velocity $\mathbf{v}_0$ of the cannon ball and the farther it goes. Using sound mathematics explain why.

(b) If air resistance is ignored, explain why the cannon ball always reaches the ground in the same time regardless of the value of the initial velocity $\mathbf{v}_0 > 0$.

(c) If the cannon ball is simply dropped from the height s_0 shown in Figure 12.3.10, show that the time it hits the ground is the same as the time in part (b).

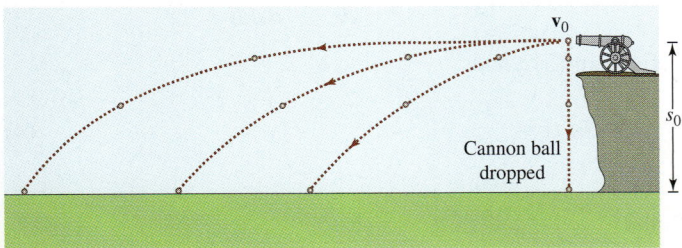

FIGURE 12.3.10 Cannon in Problem 28

≡ **Projects**

29. In this project you are going to use the properties in Sections 11.4 and 12.1 to prove **Kepler's first law of planetary motion**:

- *The orbit of a planet is an ellipse with the Sun at one focus.*

We assume that the Sun is of mass M and is located at the origin, **r** is the position vector of a body of mass m moving under the gravitational attraction of the Sun, and $\mathbf{u} = (1/r)\mathbf{r}$ is a unit vector in the direction of **r**.

(a) Use Problem 27 and Newton's second law of motion $\mathbf{F} = m\mathbf{a}$ to show that

$$\frac{d^2\mathbf{r}}{dt^2} = -\frac{kM}{r^2}\mathbf{u}.$$

(b) Use part (a) to show that $\mathbf{r} \times \mathbf{r}'' = \mathbf{0}$.

(c) Use part (b) to show that $\dfrac{d}{dt}(\mathbf{r} \times \mathbf{v}) = \mathbf{0}$.

(d) It follows from part (c) that $\mathbf{r} \times \mathbf{v} = \mathbf{c}$, where **c** is a constant vector. Show that $\mathbf{c} = r^2(\mathbf{u} \times \mathbf{u}')$.

(e) Show that $\dfrac{d}{dt}(\mathbf{u} \cdot \mathbf{u}) = 0$ and consequently $\mathbf{u} \cdot \mathbf{u}' = 0$.

(f) Use parts (a), (d), and (e) to show that

$$\frac{d}{dt}(\mathbf{v} \times \mathbf{c}) = kM\frac{d\mathbf{u}}{dt}.$$

(g) After integrating the result in part (f) with respect to t, it follows that $\mathbf{v} \times \mathbf{c} = kM\mathbf{u} + \mathbf{d}$, where **d** is another constant vector. Dot both sides of this last expression by the vector $\mathbf{r} = r\mathbf{u}$ and use Problem 61 in Exercises 11.4 to show that

$$r = \frac{c^2/kM}{1 + (d/kM)\cos\theta},$$

where $c = |\mathbf{c}|$, $d = |\mathbf{d}|$, and θ is the angle between **d** and **r**.

(h) Explain why the result in part (g) proves Kepler's first law.

(i) At perihelion (see page 595) the vectors **r** and **v** are perpendicular and have magnitudes r_0 and v_0, respectively. Use this information and parts (d) and (g) to show that $c = r_0v_0$ and $d = r_0v_0^2 - kM$.

12.4 Curvature and Acceleration

■ **Introduction** Let C be a smooth curve in 2- or 3-space that is traced out by a vector-valued function $\mathbf{r}(t)$. In this section we shall consider the acceleration vector $\mathbf{a}(t) = \mathbf{r}''(t)$, introduced in the last section, in greater detail. But before doing this, we need to examine a scalar quantity called the **curvature** of a curve.

▌ **Curvature** If $\mathbf{r}(t)$ defines a curve C, then we know that $\mathbf{r}'(t)$ is a tangent vector at a point P on C. As a consequence

$$\mathbf{T}(t) = \frac{\mathbf{r}'(t)}{|\mathbf{r}'(t)|} \tag{1}$$

is a **unit tangent**. But recall from the end of Section 12.2 that if C is parameterized by arc length s, then a unit tangent to the curve is also given by $d\mathbf{r}/ds$. As we saw in (11) of Section 12.3, the quantity $|\mathbf{r}'(t)|$ in (1) is related to the arc length function s by $ds/dt = |\mathbf{r}'(t)|$. Since the curve C is smooth, we know from page 667 that $ds/dt > 0$. Hence, by the Chain Rule,

$$\frac{d\mathbf{r}}{dt} = \frac{d\mathbf{r}}{ds}\frac{ds}{dt}$$

and so

$$\frac{d\mathbf{r}}{ds} = \frac{d\mathbf{r}/dt}{ds/dt} = \frac{\mathbf{r}'(t)}{|\mathbf{r}'(t)|} = \mathbf{T}(t). \tag{2}$$

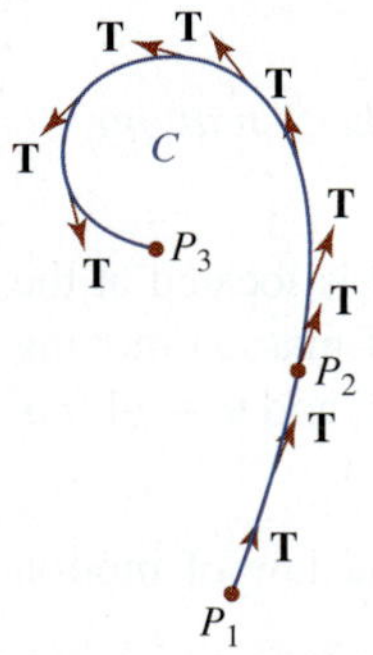

Now suppose C is as shown in FIGURE 12.4.1. As s increases, $\mathbf{T}$ moves along C changing direction but not length (it is always of unit length). Along the portion of the curve between P_1 and P_2 the vector $\mathbf{T}$ varies little in direction; along the curve between P_2 and P_3, where C obviously bends more sharply, the change in the direction of the tangent $\mathbf{T}$ is more pronounced. We use the *rate* at which the unit vector $\mathbf{T}$ changes direction with respect to arc length as an indicator of the *curvature* of a smooth curve C.

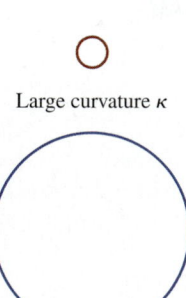
FIGURE 12.4.1 The tangent vector changes with respect to arc length

---

Definition 12.4.1 Curvature

Let $\mathbf{r}(t)$ be a vector function defining a smooth curve C. If s is the arc length parameter and $\mathbf{T} = d\mathbf{r}/ds$ is the unit tangent vector, then the **curvature** of C at a point P is defined to be

$$\kappa = \left|\frac{d\mathbf{T}}{ds}\right|. \tag{3}$$

The symbol κ in (3) is the Greek letter kappa. Now, since curves are often not parameterized by arc length, it is convenient to express (3) in terms of a general parameter t. Using the Chain Rule again, we can write

$$\frac{d\mathbf{T}}{dt} = \frac{d\mathbf{T}}{ds}\frac{ds}{dt} \quad \text{and consequently} \quad \frac{d\mathbf{T}}{ds} = \frac{d\mathbf{T}/dt}{ds/dt}.$$

In other words, the curvature defined in (3) yields

$$\kappa(t) = \frac{|\mathbf{T}'(t)|}{|\mathbf{r}'(t)|}. \tag{4}$$

EXAMPLE 1 Curvature of a Circle

Find the curvature of a circle of radius a.

Solution A circle can be described by the vector function $\mathbf{r}(t) = a\cos t\,\mathbf{i} + a\sin t\,\mathbf{j}$. Now from $\mathbf{r}'(t) = -a\sin t\,\mathbf{i} + a\cos t\,\mathbf{j}$ and $|\mathbf{r}'(t)| = a$, we get

$$\mathbf{T}(t) = \frac{\mathbf{r}'(t)}{|\mathbf{r}'(t)|} = -\sin t\,\mathbf{i} + \cos t\,\mathbf{j} \quad \text{and} \quad \mathbf{T}'(t) = -\cos t\,\mathbf{i} - \sin t\,\mathbf{j}.$$

Hence, from (4) the curvature is

$$\kappa(t) = \frac{|\mathbf{T}'(t)|}{|\mathbf{r}'(t)|} = \frac{\sqrt{\cos^2 t + \sin^2 t}}{a} = \frac{1}{a}. \tag{5}$$

The result in (5) shows that the curvature at a point on a circle is the reciprocal of the radius of the circle and indicates a fact that is in keeping with our intuition: A circle with a small radius curves more than one with a large radius. See FIGURE 12.4.2. ▀

Large curvature κ

Small curvature κ

FIGURE 12.4.2 Curvature of a circle in Example 1

■ **Tangential and Normal Components of Acceleration**   Suppose a particle moves in 2- or 3-space on a smooth curve C described by the vector function $\mathbf{r}(t)$. Then the velocity of the particle on C is $\mathbf{v}(t) = \mathbf{r}'(t)$, whereas its speed is $ds/dt = v = |\mathbf{v}(t)|$. Thus, (1) implies $\mathbf{v}(t) = v\mathbf{T}(t)$. Differentiating this last expression with respect to t gives acceleration:

$$\mathbf{a}(t) = v\frac{d\mathbf{T}}{dt} + \frac{dv}{dt}\mathbf{T}. \qquad (6)$$

Furthermore, with the help of Theorem 12.2.1(*iii*), it follows from differentiation of $\mathbf{T} \cdot \mathbf{T} = 1$ that $\mathbf{T} \cdot d\mathbf{T}/dt = 0$. Hence, at a point P on C the vectors $\mathbf{T}$ and $d\mathbf{T}/dt$ are orthogonal. If $|d\mathbf{T}/dt| \neq 0$, then the vector

$$\mathbf{N}(t) = \frac{\mathbf{T}'(t)}{|\mathbf{T}'(t)|} \qquad (7)$$

is a unit normal to the curve C at P with direction given by $d\mathbf{T}/dt$. The vector $\mathbf{N}$ is called the **principal normal vector**, or simply, the **unit normal**. But, since curvature is $\kappa(t) = |\mathbf{T}'(t)|/v$, it follows from (7) that $d\mathbf{T}/dt = \kappa v\mathbf{N}$. Thus, (6) becomes

$$\mathbf{a}(t) = \kappa v^2\mathbf{N} + \frac{dv}{dt}\mathbf{T}. \qquad (8)$$

By writing (8) as

$$\mathbf{a}(t) = a_N\mathbf{N} + a_T\mathbf{T} \qquad (9)$$

we see that the acceleration vector $\mathbf{a}$ of the moving particle is the sum of two orthogonal vectors $a_N\mathbf{N}$ and $a_T\mathbf{T}$. See FIGURE 12.4.3. The scalar functions

$$a_T = dv/dt \qquad \text{and} \qquad a_N = \kappa v^2$$

are called the **tangential** and **normal components of the acceleration**, respectively. Note that the tangential component of the acceleration results from a change in the *magnitude* of the velocity $\mathbf{v}$, whereas the normal component of the acceleration results from a change in the *direction* of $\mathbf{v}$.

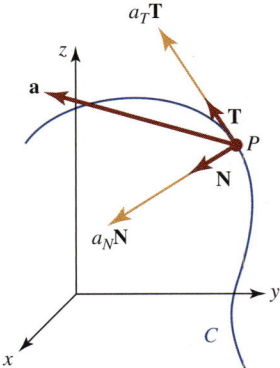

FIGURE 12.4.3   Components of acceleration vector

■ **The Binormal**   A third unit vector defined by the cross product

$$\mathbf{B}(t) = \mathbf{T}(t) \times \mathbf{N}(t) \qquad (10)$$

is called the **binormal vector**. The three unit vectors $\mathbf{T}$, $\mathbf{N}$, and $\mathbf{B}$ form a right-handed set of mutually orthogonal vectors called the **moving trihedral**. The plane of $\mathbf{T}$ and $\mathbf{N}$ is called the **osculating plane**, the plane of $\mathbf{N}$ and $\mathbf{B}$ is said to be the **normal plane**, and the plane of $\mathbf{T}$ and $\mathbf{B}$ is the **rectifying plane**. See FIGURE 12.4.4.

◀ Literally, the words "osculating plane" mean the "kissing plane."

The three mutually orthogonal unit vectors $\mathbf{T}$, $\mathbf{N}$, $\mathbf{B}$ can be thought of as a movable right-handed coordinate system since

$$\mathbf{B}(t) = \mathbf{T}(t) \times \mathbf{N}(t), \quad \mathbf{N}(t) = \mathbf{B}(t) \times \mathbf{T}(t), \quad \mathbf{T}(t) = \mathbf{N}(t) \times \mathbf{B}(t).$$

This movable coordinate system is referred to as the **TNB-frame**.

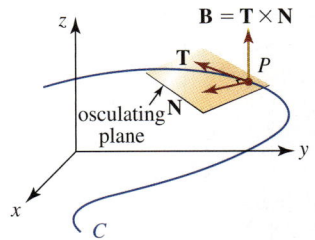

FIGURE 12.4.4   Moving trihedral and osculating plane

EXAMPLE 2   Finding **T**, **N**, and **B**

In 3-space the position of a moving particle is given by the vector function $\mathbf{r}(t) = 2\cos t\mathbf{i} + 2\sin t\mathbf{j} + 3t\mathbf{k}$. Find the vectors $\mathbf{T}(t)$, $\mathbf{N}(t)$, and $\mathbf{B}(t)$. Find the curvature $\kappa(t)$.

Solution   Since $\mathbf{r}'(t) = -2\sin t\mathbf{i} + 2\cos t\mathbf{j} + 3\mathbf{k}$, $|\mathbf{r}'(t)| = \sqrt{13}$, and so from (1) we see that a unit tangent is

$$\mathbf{T}(t) = \frac{\mathbf{r}'(t)}{|\mathbf{r}'(t)|} = -\frac{2}{\sqrt{13}}\sin t\mathbf{i} + \frac{2}{\sqrt{13}}\cos t\mathbf{j} + \frac{3}{\sqrt{13}}\mathbf{k}.$$

Next, we have

$$\mathbf{T}'(t) = -\frac{2}{\sqrt{13}}\cos t\mathbf{i} - \frac{2}{\sqrt{13}}\sin t\mathbf{j} \quad \text{and} \quad |\mathbf{T}'(t)| = \frac{2}{\sqrt{13}}.$$

Hence, (7) gives the principal normal

$$\mathbf{N}(t) = -\cos t\,\mathbf{i} - \sin t\,\mathbf{j}.$$

Thus, from (10) the binormal is

$$\mathbf{B}(t) = \mathbf{T}(t) \times \mathbf{N}(t) = \begin{vmatrix} \mathbf{i} & \mathbf{j} & \mathbf{k} \\ -\dfrac{2}{\sqrt{13}}\sin t & \dfrac{2}{\sqrt{13}}\cos t & \dfrac{3}{\sqrt{13}} \\ -\cos t & -\sin t & 0 \end{vmatrix} \qquad (11)$$

$$= \frac{3}{\sqrt{13}}\sin t\,\mathbf{i} - \frac{3}{\sqrt{13}}\cos t\,\mathbf{j} + \frac{2}{\sqrt{13}}\mathbf{k}.$$

Finally, using $|\mathbf{T}'(t)| = 2/\sqrt{13}$ and $|\mathbf{r}'(t)| = \sqrt{13}$, we find from (4) that the curvature at any point is the constant

$$\kappa(t) = \frac{2/\sqrt{13}}{\sqrt{13}} = \frac{2}{13}. \qquad \blacksquare$$

The fact that the curvature $\kappa(t)$ in Example 2 is constant is not surprising, because the curve defined by $\mathbf{r}(t)$ is a circular helix.

EXAMPLE 3   Osculating, Normal, Rectifying Planes

At the point corresponding to $t = \pi/2$ on the circular helix in Example 2, find an equation of

    (a) the osculating plane,
    (b) the normal plane, and
    (c) the rectifying plane.

Solution   From $\mathbf{r}(\pi/2) = \langle 0, 2, 3\pi/2 \rangle$ the point in question is $(0, 2, 3\pi/2)$.
    (a) From (11) a normal vector to the osculating plane at P is

$$\mathbf{B}(\pi/2) = \mathbf{T}(\pi/2) \times \mathbf{N}(\pi/2) = \frac{3}{\sqrt{13}}\mathbf{i} + \frac{2}{\sqrt{13}}\mathbf{k}.$$

To find an equation of a plane we do not require a *unit* normal, so in lieu of $\mathbf{B}(\pi/2)$ it is a bit simpler to use $\langle 3, 0, 2 \rangle$. From (2) of Section 11.6 an equation of the osculating plane is

$$3(x - 0) + 0(y - 2) + 2\left(z - \frac{3\pi}{2}\right) = 0 \qquad \text{or} \qquad 3x + 2z = 3\pi.$$

    (b) At the point P, the vector $\mathbf{T}(\pi/2) = \frac{1}{\sqrt{13}}\langle -2, 0, 3 \rangle$ or $\langle -2, 0, 3 \rangle$ is normal to the plane containing $\mathbf{N}(\pi/2)$ and $\mathbf{B}(\pi/2)$. Hence an equation of the normal plane is

$$-2(x - 0) + 0(y - 2) + 3\left(z - \frac{3\pi}{2}\right) = 0 \qquad \text{or} \qquad -4x + 6z = 9\pi.$$

    (c) Finally, at the point P, the vector $\mathbf{N}(\pi/2) = \langle 0, -1, 0 \rangle$ is normal to the plane containing $\mathbf{T}(\pi/2)$ and $\mathbf{B}(\pi/2)$. An equation of the rectifying plane is

$$0(x - 0) + (-1)(y - 2) + 0\left(z - \frac{3\pi}{2}\right) = 0 \qquad \text{or} \qquad y = 2. \qquad \blacksquare$$

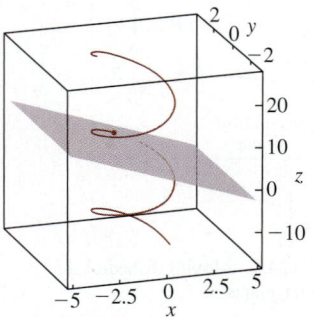

FIGURE 12.4.5 Helix and osculating plane in Example 3

Portions of the helix and the osculating plane in Example 3 are shown in FIGURE 12.4.5. The point $(0, 2, 3\pi/2)$ is indicated in the figure by the red dot.

■ **Formulas for a_T, a_N, and Curvature**   By first dotting, and then crossing, the vector $\mathbf{v} = v\mathbf{T}$ with the acceleration vector (9), it is possible to obtain explicit formulas involving $\mathbf{r}$, $\mathbf{r}'$, and $\mathbf{r}''$ for the tangential and normal components of the acceleration and the curvature. Observe that

$$\mathbf{v} \cdot \mathbf{a} = a_N(\underbrace{v\mathbf{T} \cdot \mathbf{N}}_{0}) + a_T(\underbrace{v\mathbf{T} \cdot \mathbf{T}}_{1}) = a_T v$$

yields the tangential component of acceleration:

$$a_T = \frac{dv}{dt} = \frac{\mathbf{v} \cdot \mathbf{a}}{|\mathbf{v}|} = \frac{\mathbf{r}'(t) \cdot \mathbf{r}''(t)}{|\mathbf{r}'(t)|}. \tag{12}$$

On the other hand,

$$\mathbf{v} \times \mathbf{a} = a_N(v\underbrace{\mathbf{T} \times \mathbf{N}}_{\mathbf{B}}) + a_T(v\underbrace{\mathbf{T} \times \mathbf{T}}_{\mathbf{0}}) = a_N v\mathbf{B}.$$

Since $|\mathbf{B}| = 1$, it follows that the normal component of acceleration is

$$a_N = \kappa v^2 = \frac{|\mathbf{v} \times \mathbf{a}|}{|\mathbf{v}|} = \frac{|\mathbf{r}'(t) \times \mathbf{r}''(t)|}{|\mathbf{r}'(t)|}. \tag{13}$$

Solving (13) for the curvature κ gives

$$\kappa(t) = \frac{|\mathbf{v} \times \mathbf{a}|}{|\mathbf{v}|^3} = \frac{|\mathbf{r}'(t) \times \mathbf{r}''(t)|}{|\mathbf{r}'(t)|^3}. \tag{14}$$

EXAMPLE 4 Finding a_T, a_N, and κ

The curve traced by $\mathbf{r}(t) = t\mathbf{i} + \frac{1}{2}t^2\mathbf{j} + \frac{1}{3}t^3\mathbf{k}$ is a variation of the twisted cubic discussed in Section 12.1. If $\mathbf{r}(t)$ is the position vector of a particle moving on a curve C, find the tangential and normal components of the acceleration at any point on C. Find the curvature.

Solution From

$$\mathbf{v}(t) = \mathbf{r}'(t) = \mathbf{i} + t\mathbf{j} + t^2\mathbf{k}$$

$$\mathbf{a}(t) = \mathbf{r}''(t) = \mathbf{j} + 2t\mathbf{k}$$

we find $\mathbf{v} \cdot \mathbf{a} = t + 2t^3$ and $|\mathbf{v}| = \sqrt{1 + t^2 + t^4}$. Hence, from (12) we get

$$a_T = \frac{dv}{dt} = \frac{t + 2t^3}{\sqrt{1 + t^2 + t^4}}.$$

Now,

$$\mathbf{v} \times \mathbf{a} = \begin{vmatrix} \mathbf{i} & \mathbf{j} & \mathbf{k} \\ 1 & t & t^2 \\ 0 & 1 & 2t \end{vmatrix} = t^2\mathbf{i} - 2t\mathbf{j} + \mathbf{k}$$

and $|\mathbf{v} \times \mathbf{a}| = \sqrt{t^4 + 4t^2 + 1}$. Thus, (13) gives

$$a_N = \kappa v^2 = \frac{\sqrt{t^4 + 4t^2 + 1}}{\sqrt{1 + t^2 + t^4}}.$$

Finally, from (14) we find that the curvature of the twisted cubic is given by

$$\kappa(t) = \frac{(t^4 + 4t^2 + 1)^{1/2}}{(1 + t^2 + t^4)^{3/2}}. \qquad \blacksquare$$

■ **Radius of Curvature** The reciprocal of the curvature, $\rho = 1/\kappa$, is called the **radius of curvature**. The radius of curvature at a point P on a curve C is the radius of a circle that "fits" the curve there better than any other circle. The circle at P is called the **circle of curvature** and its center is the **center of curvature**. The circle of curvature has the same tangent line at P as the curve C, and its center lies on the concave side of C. For example, a car moving on a curved track, as shown in FIGURE 12.4.6, can, at any instant, be thought to be moving on a circle of radius ρ. Hence, the normal component of its acceleration $a_N = kv^2$ must be the same as the magnitude of its centripetal acceleration $a = v^2/\rho$. Therefore, $\kappa = 1/\rho$ and $\rho = 1/\kappa$. By knowing the radius of curvature, it is possible to determine the speed v at which a car can negotiate a banked curve without skidding. (This is essentially the idea in Problem 22 in Exercises 12.3.)

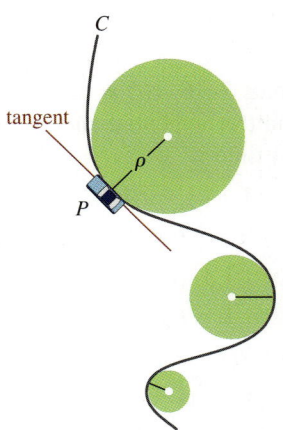

FIGURE 12.4.6 Circle and radius of curvature

$\mathbf{r}(t)$ **NOTES FROM THE CLASSROOM**

By writing (6) as

$$\mathbf{a}(t) = \frac{ds}{dt}\frac{d\mathbf{T}}{dt} + \frac{d^2s}{dt^2}\mathbf{T}$$

we observe that the so-called scalar acceleration d^2s/dt^2, referred to in the *Notes from the Classroom* in Section 12.3, is now seen to be the tangential component a_T of the acceleration $\mathbf{a}(t)$.

Exercises 12.4 Answers to selected odd-numbered problems begin on page ANS-40.

≡ **Fundamentals**

In Problems 1 and 2, for the given position function, find the unit tangent $\mathbf{T}(t)$.

1. $\mathbf{r}(t) = (t\cos t - \sin t)\mathbf{i} + (t\sin t + \cos t)\mathbf{j} + t^2\mathbf{k},\ t > 0$

2. $\mathbf{r}(t) = e^t\cos t\,\mathbf{i} + e^t\sin t\,\mathbf{j} + \sqrt{2}e^t\mathbf{k}$

3. Use the procedure outlined in Example 2 to find $\mathbf{T}(t), \mathbf{N}(t), \mathbf{B}(t),$ and $\kappa(t)$ for motion on a general circular helix that is described by $\mathbf{r}(t) = a\cos t\,\mathbf{i} + a\sin t\,\mathbf{j} + ct\mathbf{k}$.

4. Use the procedure outlined in Example 2 to show on the twisted cubic of Example 4 that at $t = 1$:

$$\mathbf{T}(1) = \frac{1}{\sqrt{3}}(\mathbf{i} + \mathbf{j} + \mathbf{k}), \qquad \mathbf{N}(1) = -\frac{1}{\sqrt{2}}(\mathbf{i} - \mathbf{k}),$$

$$\mathbf{B}(1) = -\frac{1}{\sqrt{6}}(-\mathbf{i} + 2\mathbf{j} - \mathbf{k}), \quad \kappa(1) = \frac{\sqrt{2}}{3}.$$

In Problems 5 and 6, find an equation of

(a) the osculating plane,
(b) the normal plane, and
(c) the rectifying plane to the given space curve at the point that corresponds to the indicated value of t.

5. The circular helix in Example 2; $\ t = \pi/4$

6. The twisted cubic in Example 4; $\ t = 1$

In Problems 7–16, $\mathbf{r}(t)$ is the position vector of a moving particle. Find the tangential and normal components of the acceleration at time t.

7. $\mathbf{r}(t) = \mathbf{i} + t\mathbf{j} + t^2\mathbf{k}$

8. $\mathbf{r}(t) = 3\cos t\,\mathbf{i} + 2\sin t\,\mathbf{j} + t\mathbf{k}$

9. $\mathbf{r}(t) = t^2\mathbf{i} + (t^2 - 1)\mathbf{j} + 2t^2\mathbf{k}$

10. $\mathbf{r}(t) = t^2\mathbf{i} - t^3\mathbf{j} + t^4\mathbf{k}$

11. $\mathbf{r}(t) = 2t\mathbf{i} + t^2\mathbf{j}$

12. $\mathbf{r}(t) = \tan^{-1}t\,\mathbf{i} + \frac{1}{2}\ln(1 + t^2)\mathbf{j}$

13. $\mathbf{r}(t) = 5\cos t\,\mathbf{i} + 5\sin t\,\mathbf{j}$

14. $\mathbf{r}(t) = \cosh t\,\mathbf{i} + \sinh t\,\mathbf{j}$

15. $\mathbf{r}(t) = e^{-t}(\mathbf{i} + \mathbf{j} + \mathbf{k})$

16. $\mathbf{r}(t) = t\mathbf{i} + (2t - 1)\mathbf{j} + (4t + 2)\mathbf{k}$

17. Find the curvature of an elliptical helix that is described by the vector function $\mathbf{r}(t) = a\cos t\,\mathbf{i} + b\sin t\,\mathbf{j} + ct\mathbf{k},$ $a > 0, b > 0, c > 0$.

18. (a) Find the curvature of an elliptical orbit that is described by the vector function $\mathbf{r}(t) = a\cos t\,\mathbf{i} + b\sin t\,\mathbf{j} + c\mathbf{k},$ $a > 0, b > 0, c > 0$.

(b) Show that when $a = b$, the curvature of a circular orbit is the constant $\kappa = 1/a$.

19. Show that the curvature of a straight line is the constant $\kappa = 0$. [*Hint*: Use (1) of Section 11.5.]

20. Find the curvature of the cycloid that is described by
$\mathbf{r}(t) = a(t - \sin t)\mathbf{i} + a(1 - \cos t)\mathbf{j}, a > 0$ at $t = \pi$.

21. Let C be a plane curve traced by $\mathbf{r}(t) = f(t)\mathbf{i} + g(t)\mathbf{j}$, where f and g have second derivatives. Show that the curvature at a point is given by

$$\kappa = \frac{|f'(t)g''(t) - g'(t)f''(t)|}{\left([f'(t)]^2 + [g'(t)]^2\right)^{3/2}}.$$

22. Show that if $y = f(x)$, the formula for curvature κ in Problem 21 reduces to

$$\kappa = \frac{|F''(x)|}{[1 + (F'(x))^2]^{3/2}}.$$

In Problems 23 and 24, use the result of Problem 22 to find the curvature and radius of curvature of the curve at the indicated points. Decide at which point the curve is "sharper."

23. $y = x^2;\ \ (0, 0), (1, 1)$

24. $y = x^3;\ \ (-1, -1), \left(\frac{1}{2}, \frac{1}{8}\right)$

25. Sketch the graph of the curvature $y = \kappa(x)$ for the parabola in Problem 23. Determine the behavior of $y = \kappa(x)$ as $x \to \pm\infty$. In words, describe this behavior in geometric terms.

≡ Calculator/CAS Problems

26. In Example 4 we showed that the curvature for $\mathbf{r}(t) = t\mathbf{i} + \frac{1}{2}t^2\mathbf{j} + \frac{1}{3}t^3\mathbf{k}$ is given by

$$\kappa(t) = \frac{(t^4 + 4t^2 + 1)^{1/2}}{(1 + t^2 + t^4)^{3/2}}.$$

   (a) Use a CAS to obtain the graph of $y = \kappa(t)$ for $-3 \le t \le 3$.

   (b) Use a CAS to obtain $\kappa'(t)$ and the critical numbers of the function $y = \kappa(t)$.

   (c) Find the maximum value of $y = \kappa(t)$ and the approximate corresponding points on the curve traced by $\mathbf{r}(t)$.

≡ Think About It

27. Assume that $(c, F(c))$ is a point of inflection for the graph of $y = F(x)$ and that F'' exists for all x in some interval containing c. Discuss the curvature near $(c, F(c))$.

28. Show that $|\mathbf{a}(t)|^2 = a_N^2 + a_T^2$.

Chapter 12 in Review

Answers to selected odd-numbered problems begin on page ANS-40.

A. True/False

In Problems 1–10, indicate whether the given statement is true or false.

1. A particle whose position vector is $\mathbf{r}(t) = \cos t\mathbf{i} + \cos t\mathbf{j} + \sqrt{2}\sin t\mathbf{k}$ moves with constant speed. _____

2. A circle has constant curvature. _____

3. The binormal vector is perpendicular to the osculating plane. _____

4. If $\mathbf{r}(t)$ is the position vector of a moving particle, then the velocity vector $\mathbf{v}(t) = \mathbf{r}'(t)$ and the acceleration vector $\mathbf{a}(t) = \mathbf{r}''(t)$ are orthogonal. _____

5. If s is the arc length of a curve C, then the magnitude of the velocity of a particle moving on C is ds/dt. _____

6. If s is the arc length of a curve C, then the magnitude of the acceleration of a particle on C is d^2s/dt^2. _____

7. If the binormal is defined by $\mathbf{B} = \mathbf{T} \times \mathbf{N}$, then the principal normal is $\mathbf{N} = \mathbf{B} \times \mathbf{T}$.

_____

8. If $\lim_{t \to a} \mathbf{r}_1(t) = 2\mathbf{i} + \mathbf{j}$ and $\lim_{t \to a} \mathbf{r}_2(t) = -\mathbf{i} + 2\mathbf{j}$, then $\lim_{t \to a} \mathbf{r}_1(t) \cdot \mathbf{r}_2(t) = 0$. _____

9. If $\mathbf{r}_1(t)$ and $\mathbf{r}_2(t)$ are integrable, then $\int_a^b [\mathbf{r}_1(t) \cdot \mathbf{r}_2(t)]\,dt = \left[\int_a^b \mathbf{r}_1(t)\,dt\right] \cdot \left[\int_a^b \mathbf{r}_2(t)\,dt\right]$. _____

10. If $\mathbf{r}(t)$ is differentiable, then $\frac{d}{dt}|\mathbf{r}(t)|^2 = 2\mathbf{r}(t) \cdot \frac{d\mathbf{r}}{dt}$. _____

B. Fill in the Blanks

In Problems 1–10, fill in the blanks.

1. The path of a moving particle whose position vector is $\mathbf{r}(t) = (t^2 + 1)\mathbf{i} + 4\mathbf{j} + t^4\mathbf{k}$ lies in the plane _____.

2. The curvature of a straight line is $\kappa = $ _____.

For the vector function $\mathbf{r}(t) = \langle t, t^2, \frac{1}{3}t^3 \rangle$,

3. $\mathbf{r}'(1) = $ _____,   **4.** $\mathbf{r}''(1) = $ _____,

5. $\kappa(1) = $ _____,   **6.** $\mathbf{T}(1) = $ _____,

7. $\mathbf{N}(1) = $ _____,   **8.** $\mathbf{B}(1) = $ _____,

and at the point corresponding to $t = 1$ an equation of the

9. normal plane is _____, and an equation of the

10. osculating plane is _____.

C. Exercises

1. Find the length of the curve that is traced by the vector function

$$\mathbf{r}(t) = \sin t\,\mathbf{i} + (1 - \cos t)\mathbf{j} + t\mathbf{k},\ 0 \le t \le \pi.$$

2. The position vector of a moving particle is given by $\mathbf{r}(t) = 5t\mathbf{i} + (1 + t)\mathbf{j} + 7t\mathbf{k}$. Given that the particle starts at the point corresponding to $t = 0$, find the distance the particle travels to the point corresponding to $t = 3$. At what point will the particle have traveled $80\sqrt{3}$ units along the curve?

3. Find parametric equations for the tangent line to the curve traced out by

$$\mathbf{r}(t) = -3t^2\mathbf{i} + 4\sqrt{t + 1}\,\mathbf{j} + (t - 2)\mathbf{k}$$

at the point corresponding to $t = 3$.

4. Sketch the curve traced by $\mathbf{r}(t) = t\cos t\,\mathbf{i} + t\sin t\,\mathbf{j} + t\mathbf{k}$.

5. Sketch the curve traced by $\mathbf{r}(t) = \cosh t\,\mathbf{i} + \sinh t\,\mathbf{j} + t\mathbf{k}$.

6. Given that

$$\mathbf{r}_1(t) = t^2\mathbf{i} + 2t\mathbf{j} + t^3\mathbf{k} \quad \text{and} \quad \mathbf{r}_2(t) = -t\mathbf{i} + t^2\mathbf{j} + (t^2 + 1)\mathbf{k},$$

calculate the derivative $\dfrac{d}{dt}[\mathbf{r}_1(t) \times \mathbf{r}_2(t)]$ in two different ways.

7. Given that

$$\mathbf{r}_1(t) = \cos t\,\mathbf{i} - \sin t\,\mathbf{j} + 4t^3\mathbf{k} \quad \text{and} \quad \mathbf{r}_2(t) = t^2\mathbf{i} + \sin t\,\mathbf{j} + e^{2t}\mathbf{k},$$

calculate $\dfrac{d}{dt}[\mathbf{r}_1(t) \cdot \mathbf{r}_2(t)]$ in two different ways.

8. Given that $\mathbf{r}_1$, $\mathbf{r}_2$, and $\mathbf{r}_3$ are differentiable, find $\dfrac{d}{dt}[\mathbf{r}_1(t) \cdot (\mathbf{r}_2(t) \times \mathbf{r}_3(t))]$.

9. A particle of mass m is acted on by a continuous force of magnitude 2, which is directed parallel to the positive y-axis. If the particle starts with an initial velocity $\mathbf{v}(0) = \mathbf{i} + \mathbf{j} + \mathbf{k}$ from $(1, 1, 0)$, find the position vector of the particle and the parametric equations of its path. [*Hint*: $\mathbf{F} = m\mathbf{a}$.]

10. The position vector of a moving particle is $\mathbf{r}(t) = t\mathbf{i} + (1 - t^3)\mathbf{j}$.
    (a) Sketch the path of the particle.
    (b) Sketch the velocity and acceleration vectors at $t = 1$.
    (c) Find the speed at $t = 1$.

11. Find the velocity and acceleration of a particle whose position vector is $\mathbf{r}(t) = 6t\mathbf{i} + t\mathbf{j} + t^2\mathbf{k}$ as it passes through the plane $-x + y + z = -4$.

12. The velocity of a moving particle is $\mathbf{v}(t) = -10t\mathbf{i} + (3t^2 - 4t)\mathbf{j} + \mathbf{k}$. If the particle starts at $t = 0$ at $(1, 2, 3)$, what is its position at $t = 2$?

13. The acceleration of a moving particle is $\mathbf{a}(t) = \sqrt{2}\sin t\,\mathbf{i} + \sqrt{2}\cos t\,\mathbf{j}$. Given that the velocity and position of the particle at $t = \pi/4$ are $\mathbf{v}(\pi/4) = -\mathbf{i} + \mathbf{j} + \mathbf{k}$ and $\mathbf{r}(\pi/4) = \mathbf{i} + 2\mathbf{j} + (\pi/4)\mathbf{k}$, respectively, what is the position of the particle at $t = 3\pi/4$?

14. Given that $\mathbf{r}(t) = \frac{1}{2}t^2\mathbf{i} + \frac{1}{3}t^3\mathbf{j} - \frac{1}{2}t^2\mathbf{k}$ is the position vector of a moving particle, find the tangential and normal components of the acceleration at time t. Find the curvature.

15. Suppose that the vector function of Problem 5 is the position vector of a moving particle. Find the vectors $\mathbf{T}$, $\mathbf{N}$, and $\mathbf{B}$ at $t = 1$. Find the curvature at that point.

Partial Derivatives

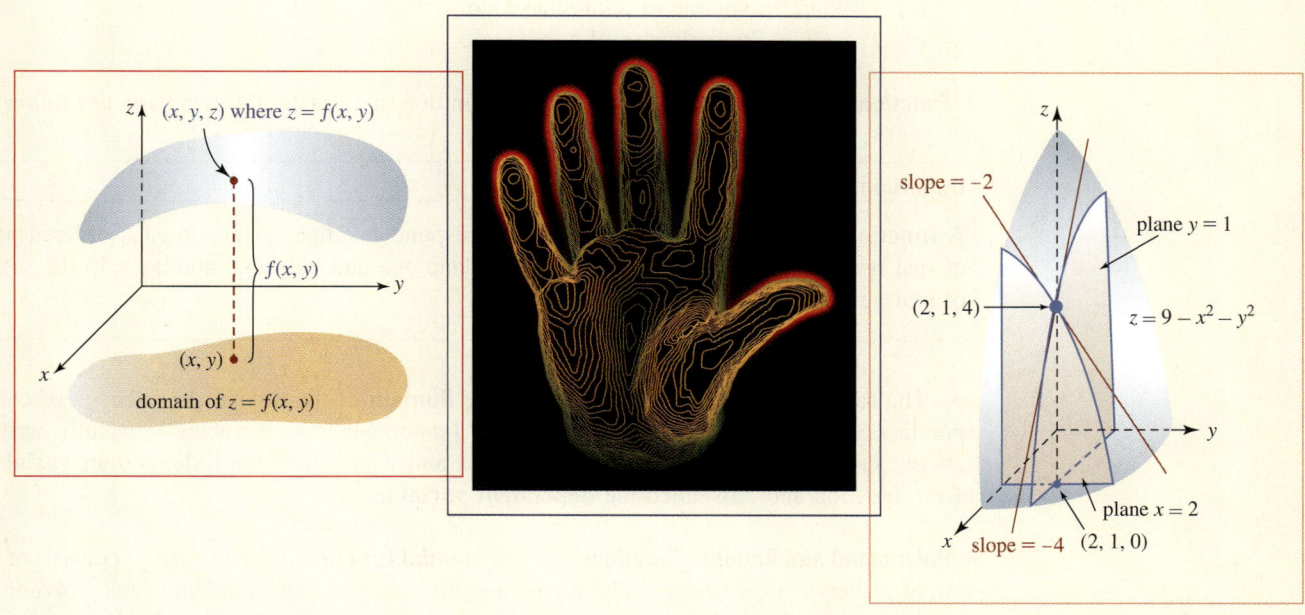

In This Chapter Up to this point in our study of calculus, we have considered only functions of a single variable. Previously considered concepts for functions of a single variable, such as limits, tangents, maxima and minima, integrals, and so on, extend to functions of two or more variables as well. This chapter is devoted primarily to the differential calculus of multivariable functions.

13.1 Functions of Several Variables

▌ **Introduction** Recall that a function of one variable $y = f(x)$ is a rule of correspondence that assigns to each element x in a subset X of the real numbers, called the *domain* of f, one and only one real number y in another set of real numbers Y. The set $\{y \mid y = f(x), x \text{ in } X\}$ is called the *range* of f. In this chapter we consider the calculus of functions that are, for the most part, functions of two variables. You are probably already aware of the existence of functions of two or more variables.

EXAMPLE 1 Some Functions of Two Variables

    (a) $A = xy$, area of a rectangle
    (b) $V = \pi r^2 h$, volume of a circular cylinder
    (c) $V = \frac{1}{3}\pi r^2 h$, volume of a circular cone
    (d) $P = 2x + 2y$, perimeter of a rectangle   ■

▌ **Functions of Two Variables** The formal definition of a function of two variables follows.

> **Definition 13.1.1** Function of Two Variables
>
> A **function of two variables** is a rule of correspondence that assigns to each ordered pair of real numbers (x, y) in a subset of the xy-plane one and only one number z in the set R of real numbers.

The set of ordered pairs (x, y) is called the **domain** of the function and the set of corresponding values of z is called the **range**. A function of two variables is usually written $z = f(x, y)$ and read "f of x, y." The variables x and y are called the **independent variables** of the function and z is called the **dependent variable**.

▌ **Polynomial and Rational Functions** A **polynomial function** of two variables consists of the sum of powers $x^m y^n$, where m and n are nonnegative integers. The quotient of two polynomial functions is called a **rational function**. For example,

    Polynomial Functions:

$$f(x, y) = xy - 5x^2 + 9 \quad \text{and} \quad f(x, y) = 3xy^2 - 5x^2y + x^3$$

    Rational Functions:

$$f(x, y) = \frac{1}{xy - 3y} \quad \text{and} \quad f(x, y) = \frac{x^4 y^2}{x^2 y + y^5 + 2x}.$$

The domain of a polynomial function is the entire xy-plane. The domain of a rational function is the xy-plane except those ordered pairs (x, y) for which the denominator is zero. For example, the domain of the rational function $f(x, y) = 4/(6 - x^2 - y^2)$ consists of the xy-plane except those points (x, y) that lie on the circle $6 - x^2 - y^2 = 0$ or $x^2 + y^2 = 6$.

EXAMPLE 2 Domain of a Function of Two Variables

    (a) Given that $f(x, y) = 4 + \sqrt{x^2 - y^2}$, find $f(1, 0)$, $f(5, 3)$, and $f(4, -2)$.
    (b) Sketch the domain of the function.

Solution
    (a) $f(1, 0) = 4 + \sqrt{1 - 0} = 5$

$$f(5, 3) = 4 + \sqrt{25 - 9} = 4 + \sqrt{16} = 8$$
$$f(4, -2) = 4 + \sqrt{16 - (-2)^2} = 4 + \sqrt{12} = 4 + 2\sqrt{3}$$

    (b) The domain of f consists of all ordered pairs (x, y) for which $x^2 - y^2 \geq 0$ or $(x - y)(x + y) \geq 0$. As shown in **FIGURE 13.1.1**, the domain consists of all points on the lines $y = x$ and $y = -x$ and in the shaded regions between them.   ■

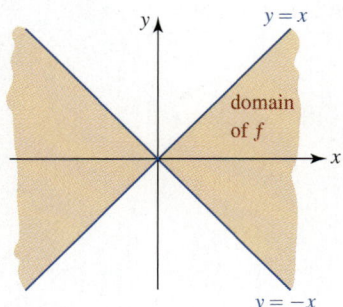

FIGURE 13.1.1 Domain of f in Example 2

EXAMPLE 3 Functions of Two Variables

(a) An equation of a plane $ax + by + cz = d$, $c \neq 0$, describes a function when written as

$$z = \frac{d}{c} - \frac{a}{c}x - \frac{b}{c}y \qquad \text{or} \qquad f(x, y) = \frac{d}{c} - \frac{a}{c}x - \frac{b}{c}y.$$

Since z is a polynomial in x and y, the domain of the function consists of the entire xy-plane.

(b) A mathematical model for the area S of the surface of a human body is a function of its weight w and height h:

$$S(w, h) = 0.1091w^{0.425}h^{0.725}.$$

■ **Graphs** The **graph** of a function $z = f(x, y)$ is a *surface* in 3-space. See FIGURE 13.1.2. In FIGURE 13.1.3 the surface is the graph of the polynomial function $z = 2x^2 - 2y^2 + 2$.

◄ Recall, the graph of this polynomial function is a hyperbolic paraboloid.

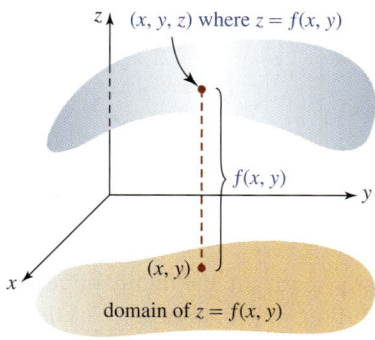

FIGURE 13.1.2 Graph of a function of x and y is a surface

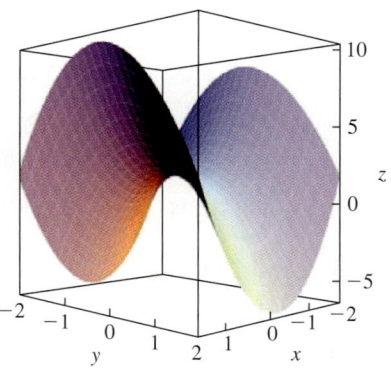

FIGURE 13.1.3 Graph of a polynomial function

EXAMPLE 4 Domain of a Function of Two Variables

From the discussion of quadric surfaces in Section 11.8 you should recognize the graph of the polynomial function $f(x, y) = x^2 + 9y^2$ as an elliptic paraboloid. Since f is defined for every ordered pair of real numbers, its domain is the entire xy-plane. From the fact that $x^2 \geq 0$ and $y^2 \geq 0$, we can argue to the fact that the range of f is defined by the inequality $z \geq 0$. ■

EXAMPLE 5 Domain of a Function of Two Variables

In Section 11.7 we saw that $x^2 + y^2 + z^2 = 9$ is a sphere centered at the origin of radius 3. Solving for z, and taking the nonnegative square root, gives the function

$$z = \sqrt{9 - x^2 - y^2} \qquad \text{or} \qquad f(x, y) = \sqrt{9 - x^2 - y^2}.$$

The graph of f is the upper hemisphere shown in FIGURE 13.1.4. The domain of the function is the set of ordered pairs (x, y) where the coordinates satisfy

$$9 - x^2 - y^2 \geq 0 \qquad \text{or} \qquad x^2 + y^2 \leq 9.$$

That is, the domain of f consists of the circle $x^2 + y^2 = 9$ and its interior. Inspection of Figure 13.1.4 shows that the range of the function is the interval $[0, 3]$ on the z-axis. ■

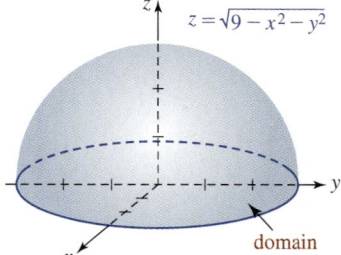

FIGURE 13.1.4 Hemisphere in Example 5

In science, one often encounters the words **isothermal**, **equipotential**, and **isobaric**. The prefix *iso* comes from the Greek word *isos*, which means *equal* or *the same*. Thus, these terms apply to lines or curves on which the temperature, potential, or barometric pressure is *constant*.

EXAMPLE 6 Potential Function

The electrostatic potential at a point $P(x, y)$ in the plane due to a unit point charge at the origin is given by $U = 1/\sqrt{x^2 + y^2}$. If the potential is a constant, say $U = c$, where c is a positive constant, then

$$\frac{1}{\sqrt{x^2 + y^2}} = c \qquad \text{or} \qquad x^2 + y^2 = \frac{1}{c^2}.$$

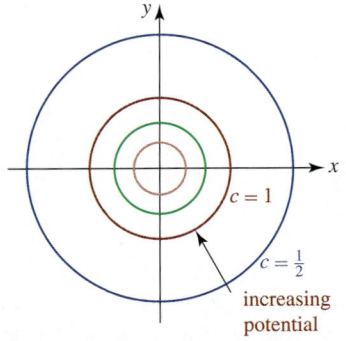

FIGURE 13.1.5 Equipotential curves in Example 6

Thus, as shown in FIGURE 13.1.5, the curves of equipotential are concentric circles surrounding the charge. Note that in Figure 13.1.5 we can get a feeling for the behavior of the function U, specifically where it is increasing (or decreasing), by observing the direction of increasing c. ∎

■ **Level Curves** In general, if a function of two variables is given by $z = f(x, y)$, then the curves defined by $f(x, y) = c$, for suitable c, are called the **level curves** of f. The word *level* arises from the fact that we can interpret $f(x, y) = c$ as the projection onto the xy-plane of the curve of intersection, or **trace**, of $z = f(x, y)$ and the (horizontal or level) plane $z = c$. See FIGURE 13.1.6.

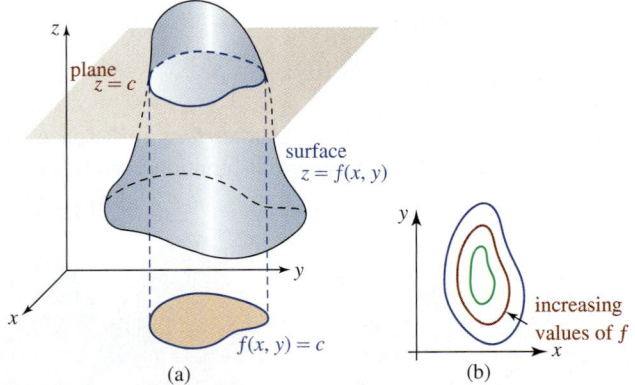

FIGURE 13.1.6 Surface in (a) and level curves in (b)

EXAMPLE 7 Level Curves

The level curves of the polynomial function $f(x, y) = y^2 - x^2$ are the family of curves defined by $y^2 - x^2 = c$. As shown in FIGURE 13.1.7, when $c > 0$ or $c < 0$, a member of this family of curves is a hyperbola. For $c = 0$, we obtain the lines $y = x$ and $y = -x$.

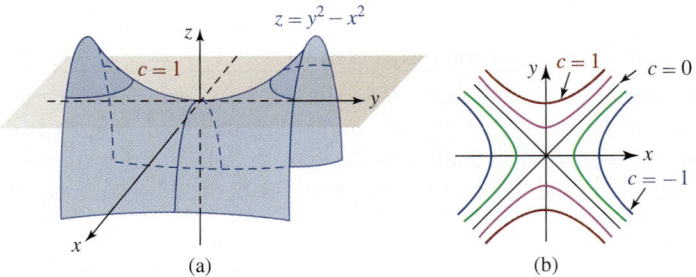

FIGURE 13.1.7 Surface and level curves in Example 7 ∎

In most instances the task of graphing level curves of a function of two variables $z = f(x, y)$ is formidable. A CAS was used to generate the surfaces and corresponding level curves in FIGURE 13.1.8 and FIGURE 13.1.9.

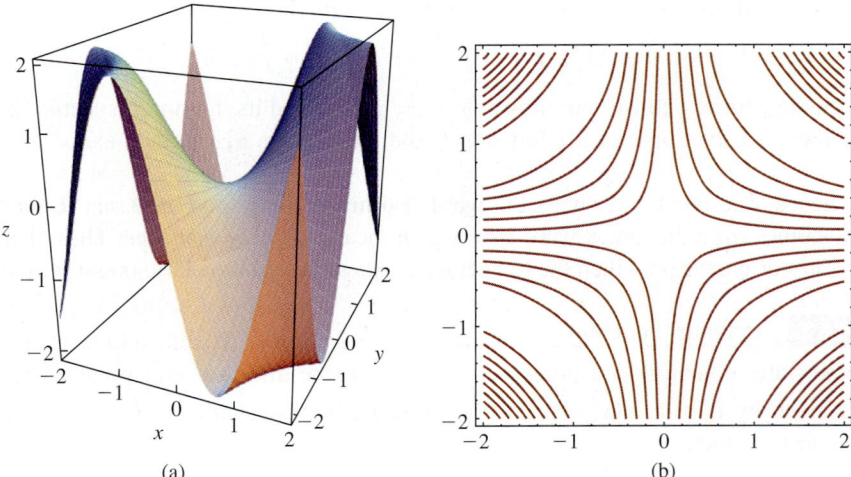

FIGURE 13.1.8 Graph of $f(x, y) = 2\sin xy$ in (a); level curves in (b)

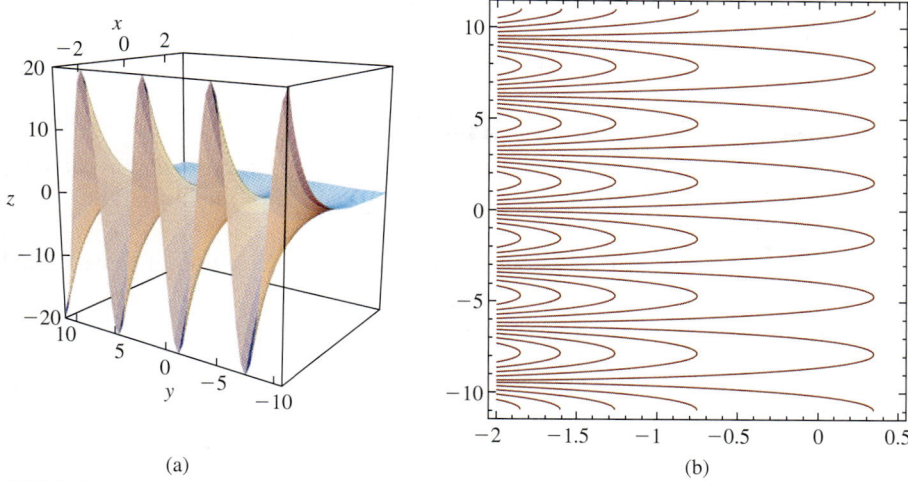

(a)   (b)

FIGURE 13.1.9   Graph of $f(x, y) = e^{-x} \sin y$ in (a); level curves in (b)

The level curves of a function f are also called **contour lines**. On a practical level, **contour maps** are often used to display curves of equal elevation. In FIGURE 13.1.10, we see that a contour map illustrates the various segments of a hill that have a given altitude. This is the idea of the contours in FIGURE 13.1.11,* which show the thickness of volcanic ash surrounding the volcano El Chichon. El Chichon, in the state of Chiapas, Mexico, erupted on March 28 and April 4, 1982.

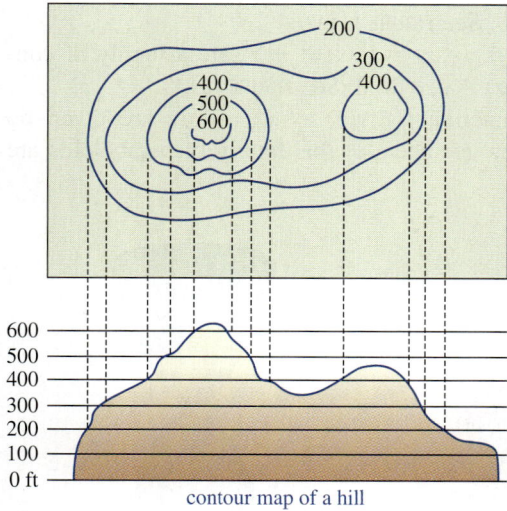

contour map of a hill

FIGURE 13.1.10   Contour map

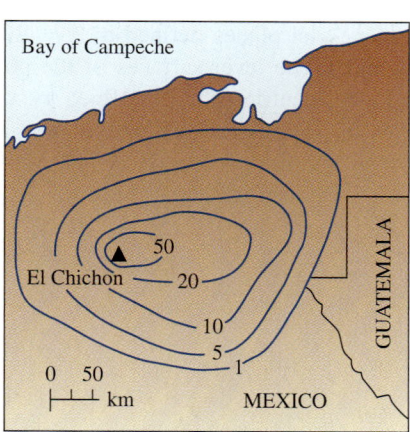

thickness (in mm) of rain-compacted volcanic ash surrounding El Chichon

FIGURE 13.1.11   Contour map showing the depth of ash around a volcano

■ **Functions of Three or More Variables**   The definitions of functions of three or more variables are simply generalizations of Definition 13.1.1. For example, a **function of three variables** is a rule of correspondence that assigns to each ordered triple of real numbers (x, y, z) in a subset of 3-space, one and only one number w in the set R of real numbers. A function of three variables is usually denoted by $w = f(x, y, z)$ or $w = F(x, y, z)$. A **polynomial function** of three variables consists of the sum of powers $x^m y^n z^k$, where m, n, and k are nonnegative integers. The quotient of two polynomial functions is called a **rational function**.

For example, the volume V and surface area S of a rectangular box are polynomial functions of three variables:

$$V = xyz \qquad \text{and} \qquad S = 2xy + 2xz + 2yz.$$

*Adapted, with permission, from *National Geographic* magazine.

Poiseuille's law states that the discharge rate, or rate of flow, of a viscous fluid (such as blood) through a tube (such as an artery) is

$$Q = k\frac{R^4}{L}(p_1 - p_2),$$

where k is a constant, R is the radius of the tube, L is its length, and p_1 and p_2 are the pressures at the ends of the tube. This is an example of a function *of four* variables.

Note: Since it would take four dimensions, we cannot graph a function of three variables.

EXAMPLE 8   Domain of a Function of Four Variables

The domain of the rational function of three variables

$$f(x, y, z) = \frac{2x + 3y + z}{4 - x^2 - y^2 - z^2}$$

is the set of points (x, y, z) that satisfy $x^2 + y^2 + z^2 \neq 4$. In other words, the domain of f is all of 3-space *except* the points that lie on the surface of a sphere of radius 2 centered at the origin. ∎

An unfortunate, but standard, choice of words, since *level surfaces* are usually not level. ▶ ∎ **Level Surfaces**   For a function of three variables, $w = f(x, y, z)$, the surfaces defined by $f(x, y, z) = c$, where c is a constant, are called **level surfaces** for the function f.

EXAMPLE 9   Some Level Surfaces

(a) The level surfaces of the polynomial $f(x, y, z) = x - 2y + 3z$ are a family of parallel planes defined by $x - 2y + 3z = c$. See FIGURE 13.1.12.
(b) The level surfaces of the polynomial $f(x, y, z) = x^2 + y^2 + z^2$ are a family of concentric spheres defined by $x^2 + y^2 + z^2 = c, c > 0$. See FIGURE 13.1.13.
(c) The level surfaces of the rational function $f(x, y, z) = (x^2 + y^2)/z$ are given by $(x^2 + y^2)/z = c$ or $x^2 + y^2 = cz$. A few members of this family of paraboloids are given in FIGURE 13.1.14.

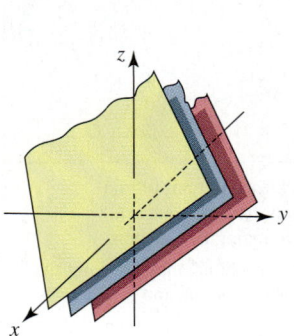

FIGURE 13.1.12   Level surfaces in (a) of Example 9

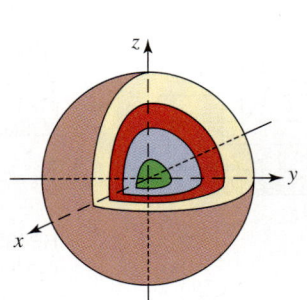

FIGURE 13.1.13   Level surfaces in (b) of Example 9

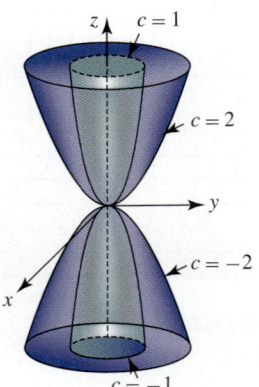

FIGURE 13.1.14   Level surfaces in (c) of Example 9 ∎

Exercises 13.1   Answers to selected odd-numbered problems begin on page ANS-40.

≡ **Fundamentals**

In Problems 1–10, find the domain of the given function.

1. $f(x, y) = \dfrac{xy}{x^2 + y^2}$

2. $f(x, y) = (x^2 - 9y^2)^{-2}$

3. $f(x, y) = \dfrac{y^2}{y + x^2}$

4. $f(x, y) = x^2 - y^2\sqrt{4 + y}$

5. $f(s, t) = s^3 - 2t^2 + 8st$

6. $f(u, v) = \dfrac{u}{\ln(u^2 + v^2)}$

7. $g(r, s) = e^{2r}\sqrt{s^2 - 1}$

8. $g(\theta, \phi) = \dfrac{\tan\theta + \tan\phi}{1 - \tan\theta\tan\phi}$

9. $H(u, v, w) = \sqrt{u^2 + v^2 + w^2 - 16}$

10. $f(x, y, z) = \dfrac{\sqrt{25 - x^2 - y^2}}{z - 5}$

In Problems 11–18, match the set of points given in the figure with the domain of one of the functions in (a)–(h).

(a) $f(x, y) = \sqrt{y - x^2}$ **(b)** $f(x, y) = \ln(x - y^2)$

(c) $f(x, y) = \sqrt{x} + \sqrt{y - x}$ **(d)** $f(x, y) = \sqrt{\dfrac{x}{y} - 1}$

(e) $f(x, y) = \sqrt{xy}$ **(f)** $f(x, y) = \sin^{-1}(xy)$

(g) $f(x, y) = \dfrac{x^4 + y^4}{xy}$ **(h)** $f(x, y) = \dfrac{\sqrt{x^2 + y^2 - 1}}{y - x}$

11.

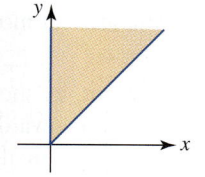

FIGURE 13.1.15 Graph for Problem 11

12.

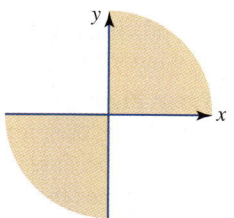

FIGURE 13.1.16 Graph for Problem 12

13.

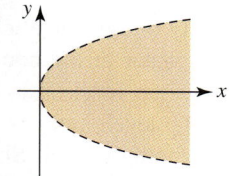

FIGURE 13.1.17 Graph for Problem 13

14.

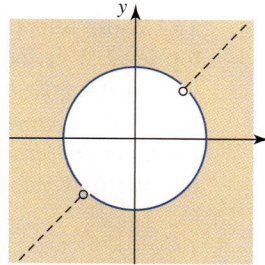

FIGURE 13.1.18 Graph for Problem 14

15.

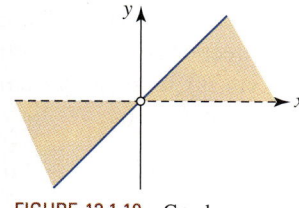

FIGURE 13.1.19 Graph for Problem 15

16.

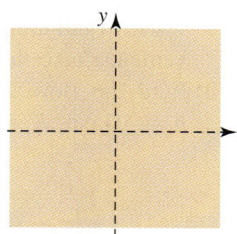

FIGURE 13.1.20 Graph for Problem 16

17.

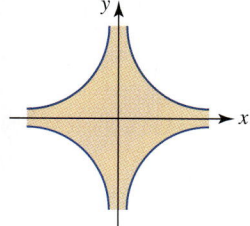

FIGURE 13.1.21 Graph for Problem 17

18.

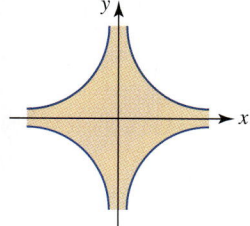

FIGURE 13.1.22 Graph for Problem 18

In Problems 19–22, sketch the domain of the given function.

19. $f(x, y) = \sqrt{x} - \sqrt{y}$

20. $f(x, y) = \sqrt{(1 - x^2)(y^2 - 4)}$

21. $f(x, y) = \sqrt{\ln(y - x + 1)}$

22. $f(x, y) = e^{\sqrt{xy + 1}}$

In Problems 23–26, find the range of the given function.

23. $f(x, y) = 10 + x^2 + 2y^2$ **24.** $f(x, y) = x + y$

25. $f(x, y, z) = \sin(x + 2y + 3z)$ **26.** $f(x, y, z) = 7 - e^{xyz}$

In Problems 27–30, evaluate the given function at the indicated points.

27. $f(x, y) = \displaystyle\int_x^y (2t - 1)\,dt;$ $(2, 4), (-1, 1)$

28. $f(x, y) = \ln\dfrac{x^2}{x^2 + y^2};$ $(3, 0), (5, -5)$

29. $f(x, y, z) = (x + 2y + 3z)^2;$ $(-1, 1, -1), (2, 3, -2)$

30. $F(x, y, z) = \dfrac{1}{x^2} + \dfrac{1}{y^2} + \dfrac{1}{z^2};$ $(\sqrt{3}, \sqrt{2}, \sqrt{6}), (\tfrac{1}{4}, \tfrac{1}{5}, \tfrac{1}{3})$

In Problems 31–36, describe the graph of the given function.

31. $z = x$ **32.** $z = y^2$

33. $z = \sqrt{x^2 + y^2}$ **34.** $z = \sqrt{1 + x^2 + y^2}$

35. $z = \sqrt{36 - x^2 - 3y^2}$ **36.** $z = -\sqrt{16 - x^2 - y^2}$

In Problems 37–42, sketch some of the level curves associated with the given function.

37. $f(x, y) = x + 2y$ **38.** $f(x, y) = y^2 - x$

39. $f(x, y) = \sqrt{x^2 - y^2 - 1}$ **40.** $f(x, y) = \sqrt{36 - 4x^2 - 9y^2}$

41. $f(x, y) = e^{y - x^2}$ **42.** $f(x, y) = \tan^{-1}(y - x)$

In Problems 43–46, describe the level surfaces but do not graph.

43. $f(x, y, z) = \tfrac{1}{9}x^2 + \tfrac{1}{4}z^2$

44. $f(x, y, z) = (x - 1)^2 + (y - 2)^2 + (z - 3)^2$

45. $f(x, y, z) = x^2 + 3y^2 + 6z^2$ **46.** $G(x, y, z) = 4y - 2z + 1$

47. Graph some of the level surfaces associated with $f(x, y, z) = x^2 + y^2 - z^2$ for $c = 0$, $c > 0$, and $c < 0$.

48. Given that

$$f(x, y, z) = \frac{x^2}{16} + \frac{y^2}{4} + \frac{z^2}{9},$$

find the x-, y-, and z-intercepts of the level surface that passes through $(-4, 2, -3)$.

≡ Applications

49. The temperature, pressure, and volume of an enclosed ideal gas are related by $T = 0.01PV$, where T, P, and V are measured in kelvins, atmospheres, and liters, respectively. Sketch the isotherms $T = 300$ K, 400 K, and 600 K.

50. Express the height of a rectangular box with a square bottom as a function of the volume and the length of one side of the box.

51. A soda can is constructed with a tin lateral side and an aluminum top and bottom. Given that the cost is 1.8 cents per square unit for the top, 1 cent per square unit for the bottom, and 2.3 cents per square unit for the side, determine the cost function $C(r, h)$, where r is the radius of the can and h is its height.

52. A closed rectangular box is to be constructed from 500 cm^2 of cardboard. Express the volume V as a function of the length x and width y.

53. As shown in FIGURE 13.1.23, a conical cap rests on top of a circular cylinder. If the height of the cap is two-thirds the height of the cylinder, express the volume of the solid as a function of the indicated variables.

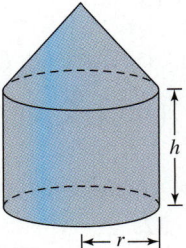

FIGURE 13.1.23   Conical capped cylinder in Problem 53

54. Often a tissue sample is an obliquely cut cylinder as shown in FIGURE 13.1.24. Express the thickness t of the cut as a function of x, y, and z.

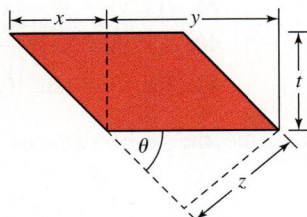

FIGURE 13.1.24   Tissue sample in Problem 54

55. In medicine, formulas for surface area (see Example 3(b)) are often used to calibrate drug doses, since it is assumed that a drug dose D and surface area S are directly proportional. The following simple function can be used to obtain a quick estimate of the body surface area of a human: $S = 2ht$, where h is the height (in cm) and t is the maximum thigh circumference (in cm). Estimate the surface area of a 156-cm-tall person with a maximum thigh circumference of 50 cm. Estimate your own surface area.

 Projects

56. Wind Chill Factor   During his investigation of the winter of 1941 in the Antarctic, Dr. Paul A. Siple devised

the following mathematical model for defining the wind chill factor:

$$H(v, T) = (10\sqrt{v} - v + 10.5)(33 - T),$$

where H is measured in kcal/m²h, v is wind velocity in m/s, and T is temperature in degrees Celsius. An example of this index is: 1000 = very cold, 1200 = bitterly cold, and 1400 = exposed flesh freezes. Determine the wind chill factor at $-6.67°C$ ($20°F$) with a wind velocity of 20 m/s (45 mi/h). Write a short report that defines wind chill precisely. Find at least one other mathematical model for wind chill.

57. Water Flow   When water flows from a spigot, as shown in FIGURE 13.1.25(a), it contracts as it accelerates downward. It does this because the flow rate Q, which is defined as velocity times the cross-sectional area of the water column, must be constant at each level. In this problem assume that the cross-sections of the fluid column are circular.

(a) Consider the column of water shown in Figure 13.1.25(b). Suppose v is the velocity of the water at the top level, V is the velocity of the water at the bottom level a distance h units below the top level, R is the radius of the cross-section at the top level, and r is the radius of the cross-section at the bottom level. Show that the flow rate Q as a function of r and R is

$$Q = \frac{\pi r^2 R^2 \sqrt{2gh}}{\sqrt{R^4 - r^4}},$$

where g is the acceleration due to gravity. [*Hint*: Start by expressing the time t it takes a cross-section of water to fall a distance h in terms of u and V. For convenience take the positive direction to be downward.]

(b) Find the flow rate Q (in cm³/s) if $g = 980 \text{ cm/s}^2$, $h = 10$ cm, $R = 1$ cm, and $r = 0.2$ cm.

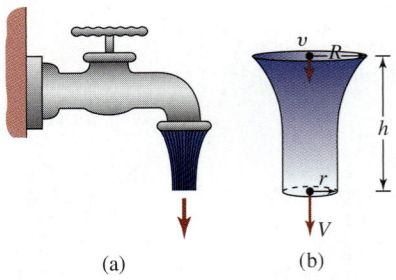

FIGURE 13.1.25   Water flowing from a spigot in Problem 57

13.2 Limits and Continuity

▌ **Introduction**   For functions of one variable, in many instances we were able to make a judgment about the existence of $\lim_{x \to a} f(x)$ from the graph of $y = f(x)$. Also, we utilized the fact that $\lim_{x \to a} f(x)$ exists if and only if $\lim_{x \to a^-} f(x)$ and $\lim_{x \to a^+} f(x)$ exist and are equal to the same number L, in which case, $\lim_{x \to a} f(x) = L$. In this section we will see that the situation is more demanding in the consideration of limits of functions of two variables.

■ **Terminology**   Before proceeding with the discussion on limits we need to introduce some terminology about sets that will be used in this as well as in the sections and chapters that follow. The set in 2-space

$$\{(x, y) \mid (x - x_0)^2 + (y - y_0)^2 < \delta^2\} \tag{1}$$

consists of all points *interior to*, but *not on*, a circle with center (x_0, y_0) and radius $\delta > 0$. The set (1) is called an **open disk**. On the other hand, the set

$$\{(x, y) \mid (x - x_0)^2 + (y - y_0)^2 \le \delta^2\} \tag{2}$$

is a **closed disk**. A closed disk includes all points *interior to* and *on* a circle with center (x_0, y_0) and radius $\delta > 0$. See FIGURE 13.2.1(a). If R is some region of the xy-plane, then a point (a, b) is said to be an **interior point** of R if there is *some* open disk centered at (a, b) that contains only points of R. In contrast, we say that (a, b) is a **boundary point** of R if the interior of *every* open disk centered at (a, b) contains both points in R and points not in R. The region R is said to be **open** if it contains no boundary points and **closed** if it contains all its boundary points. See Figure 13.2.1(b). A region R is said to be **bounded** if it can be contained in a sufficiently large rectangle in the plane. Figure 13.2.1(c) illustrates a bounded region; the first quadrant illustrated in Figure 13.2.1(d) is an example of an **unbounded** region. These concepts carry over naturally to 3-space. For example, the analogue of an open disk is an **open ball**. An open ball consists of all points *interior to*, but *not on*, a sphere with center (x_0, y_0) and radius $\delta > 0$:

$$\{(x, y, z) \mid (x - x_0)^2 + (y - y_0)^2 + (z - z_0) < \delta^2\}. \tag{3}$$

A region in 3-space is bounded if it can be contained in a sufficiently large rectangular box.

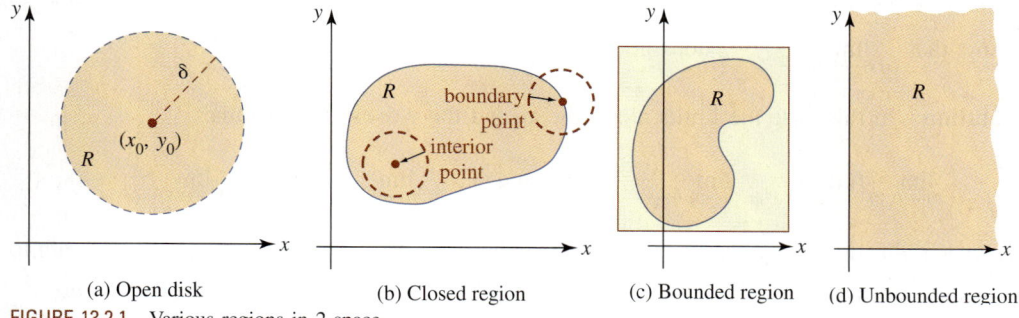

(a) Open disk   (b) Closed region   (c) Bounded region   (d) Unbounded region

FIGURE 13.2.1   Various regions in 2-space

■ **Limits of Functions of Two Variables**   To analyze a limit by sketching the graph of $z = f(x, y)$ is not convenient or even routinely possible for most functions of two variables. Intuitively, f has a limit at a point (a, b) if the function values $f(x, y)$ are approaching a number L as (x, y) approaches (a, b). We write $f(x, y) \to L$ as $(x, y) \to (a, b)$, or

$$\lim_{(x, y) \to (a, b)} f(x, y) = L.$$

To be a little more precise, f has a limit L at a point (a, b) if the points in space $(x, y, f(x, y))$ can be made arbitrarily close to (a, b, L) whenever (x, y) is close enough to (a, b).

The notion of (x, y) "approaching" a point (a, b) is not as simple as it is for functions of one variable where $x \to a$ means that x can approach a only from the left and from the right. In the xy-plane, there are an infinite number of ways of approaching a point (a, b). As shown in FIGURE 13.2.2, in order that $\lim_{(x, y) \to (a, b)} f(x, y)$ exist, we now require that f approach the same number L along *every* possible curve or **path** through (a, b). Put in a negative way:

- *If $f(x, y)$ does not approach the same number L for two different paths to (a, b), then $\lim_{(x, y) \to (a, b)} f(x, y)$ does not exist.* (4)

In the discussion of $\lim_{(x, y) \to (a, b)} f(x, y)$ that follows we shall assume that the function f is defined at every point (x, y) in an open disk centered at (a, b) but not necessarily *at* (a, b) itself.

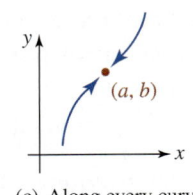

(a) Along horizontal
and vertical lines
through (a, b)

(b) Along every
straight line
through (a, b)

(c) Along every curve
through (a, b)

FIGURE 13.2.2 Three of many ways of approaching the point (a, b)

EXAMPLE 1   A Limit That Does Not Exist

Show that $\displaystyle\lim_{(x, y)\to(0, 0)} \frac{x^2 - 3y^2}{x^2 + 2y^2}$ does not exist.

Solution   The function $f(x, y) = (x^2 - 3y^2)/(x^2 + 2y^2)$ is defined everywhere except at $(0, 0)$. As illustrated in Figure 13.2.2(a), two ways of approaching $(0, 0)$ are along the x-axis ($y = 0$) and along the y-axis ($x = 0$). On $y = 0$ we have

$$\lim_{(x, 0)\to(0, 0)} f(x, 0) = \lim_{(x, 0)\to(0, 0)} \frac{x^2 - 0}{x^2 + 0} = 1$$

whereas on $x = 0$,

$$\lim_{(0, y)\to(0, 0)} f(0, y) = \lim_{(0, y)\to(0, 0)} \frac{0 - 3y^2}{0 + 2y^2} = -\frac{3}{2}.$$

In view of (4), we conclude that the limit does not exist.   ∎

EXAMPLE 2   A Limit That Does Not Exist

Show that $\displaystyle\lim_{(x, y)\to(0, 0)} \frac{xy}{x^2 + y^2}$ does not exist.

Solution   In this case the limits along the x- and the y-axes are the same:

$$\lim_{(x, 0)\to(0, 0)} f(x, 0) = \lim_{(x, 0)\to(0, 0)} \frac{0}{x^2} = 0 \quad \text{and} \quad \lim_{(0, y)\to(0, 0)} f(0, y) = \lim_{(0, y)\to(0, 0)} \frac{0}{y^2} = 0.$$

However, this does *not* mean $\displaystyle\lim_{(x, y)\to(0, 0)} f(x, y)$ exists, since we have not examined *every* path to $(0, 0)$. As illustrated in 13.2.2(b), we now try any line through the origin given by $y = mx$:

$$\lim_{(x, y)\to(0, 0)} f(x, y) = \lim_{(x, y)\to(0, 0)} \frac{mx^2}{x^2 + m^2x^2} = \frac{m}{1 + m^2}.$$

Since $\displaystyle\lim_{(x, y)\to(0, 0)} f(x, y)$ depends on the slope m of the line on which we approach the origin, we conclude that the limit does not exist. For example, on $y = x$ and on $y = 2x$, we have, respectively,

$$f(x, x) = \frac{x^2}{x^2 + x^2} \quad \text{and} \quad \lim_{(x, y)\to(0, 0)} f(x, x) = \lim_{(x, y)\to(0, 0)} \frac{x^2}{x^2 + x^2} = \frac{1}{2},$$

$$f(x, 2x) = \frac{2x^2}{x^2 + 4x^2} \quad \text{and} \quad \lim_{(x, y)\to(0, 0)} f(x, 2x) = \lim_{(x, y)\to(0, 0)} \frac{2x^2}{x^2 + 4x^2} = \frac{2}{5}.$$

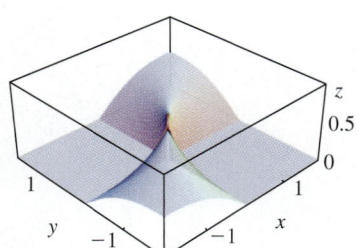

FIGURE 13.2.3 Graph of function in Example 2

A computer-generated graph of the surface is given in FIGURE 13.2.3. If you bear in mind that the origin is at the center of the box, it should be clear why different paths to $(0, 0)$ yield different values of the limit.   ∎

EXAMPLE 3   A Limit That Does Not Exist

Show that $\displaystyle\lim_{(x, y)\to(0, 0)} \frac{x^3y}{x^6 + y^2}$ does not exist.

Solution   Let $f(x, y) = x^3y/(x^6 + y^2)$. You are encouraged to show that along the x-axis, the y-axis, any line $y = mx, m \neq 0$ through $(0, 0)$, and along any parabola $y = ax^2, a \neq 0$, through $(0, 0)$, $\lim\limits_{(x,\,y)\to(0,\,0)} f(x, y) = 0$. Although this certainly constitutes an infinite number of paths to the origin, the limit *still* does not exist, since on $y = x^3$:

$$\lim_{(x,\,y)\to(0,\,0)} f(x, y) = \lim_{(x,\,y)\to(0,\,0)} f(x, x^3) = \lim_{(x,\,y)\to(0,\,0)} \frac{x^6}{x^6 + x^6} = \lim_{(x,\,y)\to(0,\,0)} \frac{x^6}{2x^6} = \frac{1}{2}. \qquad \blacksquare$$

❚ **Properties of Limits** In the following two theorems we list the properties of limits for functions of two variables. These theorems are the two-variable counterparts of Theorems 2.2.1, 2.2.2, and 2.2.3.

Theorem 13.2.1   Three Fundamental Limits

(*i*) $\lim\limits_{(x,\,y)\to(a,\,b)} c = c,$ c a constant

(*ii*) $\lim\limits_{(x,\,y)\to(a,\,b)} x = a$ and $\lim\limits_{(x,\,y)\to(a,\,b)} y = b$

(*iii*) $\lim\limits_{(x,\,y)\to(a,\,b)} cf(x, y) = c \lim\limits_{(x,\,y)\to(a,\,b)} f(x, y)$

Theorem 13.2.2   Limit of a Sum, Product, Quotient

Suppose (a, b) is a point in the xy-plane and that $\lim\limits_{(x,\,y)\to(a,\,b)} f(x, y)$ and $\lim\limits_{(x,\,y)\to(a,\,b)} g(x, y)$ exist. If $\lim\limits_{(x,\,y)\to(a,\,b)} f(x, y) = L_1$ and $\lim\limits_{(x,\,y)\to(a,\,b)} g(x, y) = L_2$, then

(*i*) $\lim\limits_{(x,\,y)\to(a,\,b)} [f(x, y) \pm g(x, y)] = L_1 \pm L_2,$

(*ii*) $\lim\limits_{(x,\,y)\to(a,\,b)} f(x, y)g(x, y) = L_1 L_2,$ and

(*iii*) $\lim\limits_{(x,\,y)\to(a,\,b)} \dfrac{f(x, y)}{g(x, y)} = \dfrac{L_1}{L_2},$ $L_2 \neq 0.$

EXAMPLE 4   Limit of a Sum

Evaluate $\lim\limits_{(x,\,y)\to(2,\,3)} (x + y^2).$

Solution   From (*ii*) of Theorem 13.2.1 we first note that

$$\lim_{(x,\,y)\to(2,\,3)} x = 2 \qquad \text{and} \qquad \lim_{(x,\,y)\to(2,\,3)} y = 3.$$

Then from parts (*i*) and (*ii*) of Theorem 13.2.2 we know that the limit of a sum is the sum of the limits and the limit of a product is the product of the limits whenever the limits exist:

$$\lim_{(x,\,y)\to(2,\,3)} (x + y^2) = \lim_{(x,\,y)\to(2,\,3)} x + \lim_{(x,\,y)\to(2,\,3)} y^2$$

$$= \lim_{(x,\,y)\to(2,\,3)} x + \left(\lim_{(x,\,y)\to(2,\,3)} y \right)\left(\lim_{(x,\,y)\to(2,\,3)} y \right)$$

$$= 2 + 3 \cdot 3 = 11. \qquad \blacksquare$$

❚ **Use of Polar Coordinates** In some cases polar coordinates can be helpful in evaluating a limit of the form $\lim\limits_{(x,\,y)\to(0,\,0)} f(x, y)$. If $x = r\cos\theta$, $y = r\sin\theta$ and $r^2 = x^2 + y^2$, then $(x, y) \to (0, 0)$ if and only if $r \to 0$.

EXAMPLE 5   Using Polar Coordinates

Evaluate $\lim\limits_{(x,\,y)\to(0,\,0)} \dfrac{10xy^2}{x^2 + y^2}.$

Solution   Substituting $x = r\cos\theta$, $y = r\sin\theta$ in the function gives

$$\frac{10xy^2}{x^2 + y^2} = \frac{10r^3 \cos\theta \sin^2\theta}{r^2} = 10r\cos\theta \sin^2\theta.$$

Since $\lim\limits_{r \to 0} r\cos\theta\sin^2\theta = 0$, we conclude that

$$\lim_{(x,\,y)\to(0,\,0)} \frac{10xy^2}{x^2 + y^2} = 0.$$

We will examine the limit in Example 5 again in Example 8.

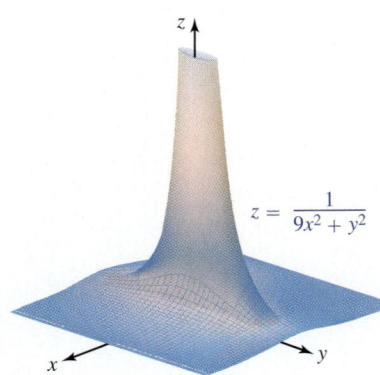

$z = \dfrac{1}{9x^2 + y^2}$

FIGURE 13.2.4 Function has an infinite discontinuity at $(0, 0)$

■ **Continuity** A function $z = f(x, y)$ is **continuous** at (a, b) if $f(a, b)$ is defined, $\lim\limits_{(x,\,y)\to(a,\,b)} f(x, y)$ exists, and the limit is the same as the function value $f(a, b)$; that is,

$$\lim_{(x,\,y)\to(a,\,b)} f(x, y) = f(a, b). \tag{5}$$

If f is not continuous at (a, b), it is said to be **discontinuous**. The graph of a continuous function is a surface with no breaks. From the graph of the function $f(x, y) = 1/(9x^2 + y^2)$ in FIGURE 13.2.4 we see that f has infinite discontinuity at $(0, 0)$, that is, $f(x, y) \to \infty$ as $(x, y) \to (0, 0)$. A function $z = f(x, y)$ is **continuous on a region R** of the xy-plane if f is continuous at every point in R. The **sum** and **product** of two continuous functions are continuous. The **quotient** of two continuous functions is continuous, except at points where the denominator is zero. Also, if g is a function of two variables continuous at (a, b) and if F is a function of one variable continuous at $g(a, b)$, then the **composition** $f(x, y) = F(g(x, y))$ is continuous at (a, b).

EXAMPLE 6 Discontinuous Function at (0, 0)

The function $f(x, y) = \dfrac{x^4 - y^4}{x^2 + y^2}$ is discontinuous at $(0, 0)$, since $f(0, 0)$ is not defined. However, as we see in the next example, f has a removable discontinuity at $(0, 0)$. ■

EXAMPLE 7 Continuous Function at (0, 0)

The function f defined by

$$f(x, y) = \begin{cases} \dfrac{x^4 - y^4}{x^2 + y^2}, & (x, y) \neq (0, 0) \\ 0, & (x, y) = (0, 0) \end{cases}$$

is continuous at $(0, 0)$, since $f(0, 0) = 0$ and

$$\lim_{(x,\,y)\to(0,\,0)} \frac{x^4 - y^4}{x^2 + y^2} = \lim_{(x,\,y)\to(0,\,0)} \frac{(x^2 + y^2)(x^2 - y^2)}{x^2 + y^2} = \lim_{(x,\,y)\to(0,\,0)} (x^2 - y^2) = 0^2 - 0^2 = 0.$$

Hence, we see that $\lim\limits_{(x,\,y)\to(0,\,0)} f(x, y) = f(0, 0)$.

With the aid of a CAS we see in FIGURE 13.2.5 two different perspectives (ViewPoint in *Mathematica*) of the surface defined by $z = f(x, y)$. Note the orientation of the x- and y-axes parts (a) and (b) of Figure 13.2.5.

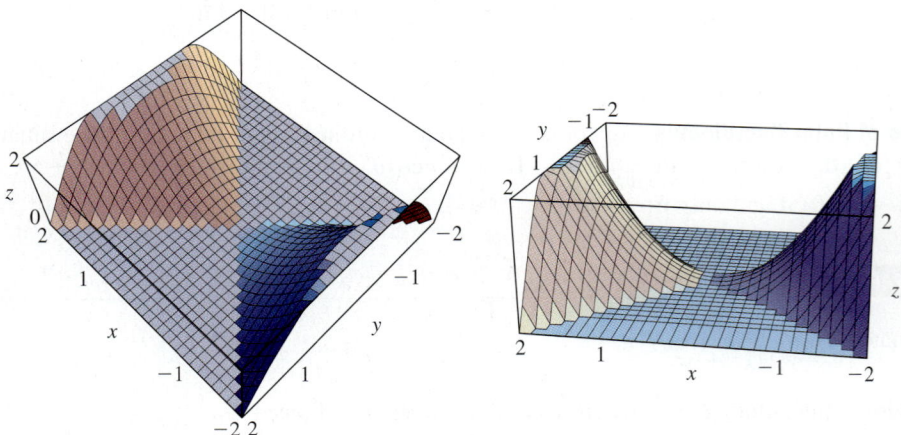

(a) Looking down on the surface   (b) Looking slightly down and facing x-axis

FIGURE 13.2.5 Graph of function in Example 7

■ **Polynomial and Rational Functions**   In Section 13.1 we saw that a **polynomial function** of two variables consists of the sum of powers $x^m y^n$, where m and n are nonnegative integers, and that the quotient of two polynomial functions is called a **rational function**. Polynomial functions, such as $f(x, y) = xy$, are continuous throughout the entire xy-plane. Rational functions are continuous except at points where the denominator is zero. For example, the rational function $f(x, y) = xy/(y - x)$ is continuous except at points on the line $y = x$. In FIGURE 13.2.6 we have illustrated the graphs of three functions that are discontinuous at points on a curve. In parts (a) and (c) of Figure 13.2.6 the given rational function is discontinuous at all points on the curve obtained by setting the denominator equal to 0. In Figure 13.2.6(b) the logarithmic function is discontinuous where $x^2 + y^2 - 4 = 0$, that is, on the circle $x^2 + y^2 = 4$.

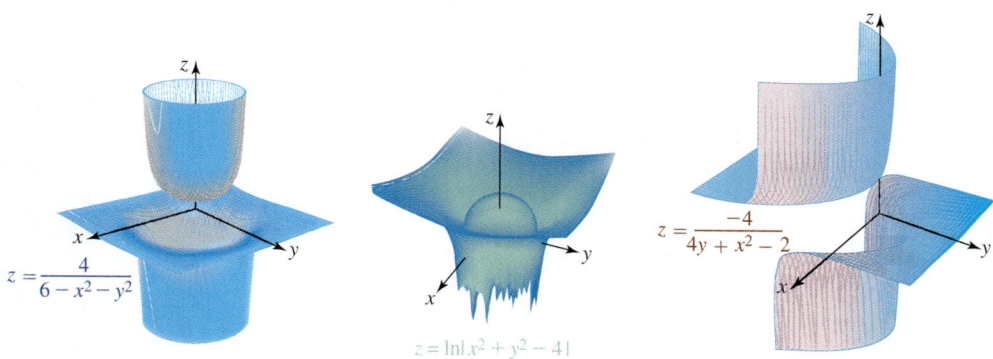

(a) Discontinuous on $x^2 + y^2 = 6$ (b) Discontinuous on $x^2 + y^2 = 4$ (c) Discontinuous on $y = -\frac{1}{4}x^2 + \frac{1}{2}$

FIGURE 13.2.6   Three discontinuous functions

■ **Functions of Three or More Variables**   The notions of limit and continuity for functions of three of more variables are natural extensions of those just considered. For example, a function of three variables $w = f(x, y, z)$ is continuous at (a, b, c) if

$$\lim_{(x, y, z) \to (a, b, c)} f(x, y, z) = f(a, b, c).$$

The polynomial function in three variables $f(x, y, z) = xy^2 z^3$ is continuous throughout 3-space. The rational function

$$f(x, y, z) = \frac{xy^2}{x^2 + y^2 + (z - 1)^2}$$

is continuous except at the single point $(0, 0, 1)$. The rational function

$$f(x, y, z) = \frac{x + 3y}{2x + 5y + z}$$

is continuous except at points (x, y, z) on the plane $2x + 5y + z = 0$.

■ **Formal Definition of a Limit**   The foregoing discussion leads us to the formal definition of a limit of a function $z = f(x, y)$ at a point (a, b). This **ε-δ definition** is analogous to Definition 2.6.1.

Definition 13.2.1   Definition of a Limit

Suppose a function f of two variables is defined at every point (x, y) in an open disk centered at (a, b), except possibly at (a, b). Then

$$\lim_{(x, y) \to (a, b)} f(x, y) = L$$

means that for every $\varepsilon > 0$, there exists a number $\delta > 0$ such that

$$|f(x, y) - L| < \varepsilon \quad \text{whenever} \quad 0 < \sqrt{(x - a)^2 + (y - a)^2} < \delta.$$

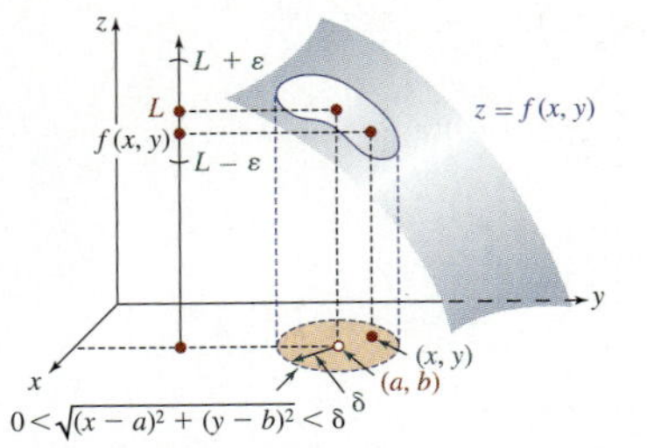

FIGURE 13.2.7 When $(x, y) \neq (a, b)$ is in the open disk, $f(x, y)$ is in the interval $(L - \varepsilon, L + \varepsilon)$

As illustrated in FIGURE 13.2.7, when f has a limit at (a, b), for a given $\varepsilon > 0$, regardless how small, we can find an open disk of radius δ centered at (a, b) so that $L - \varepsilon < f(x, y) < L + \varepsilon$ for every point $(x, y) \neq (a, b)$ within the disk. The open disk with radius $\delta > 0$ and its center (a, b) deleted is defined by the inequality

$$0 < \sqrt{(x - a)^2 + (y - a)^2} < \delta.$$

As mentioned previously, the values of f are close to L whenever (x, y) is close to (a, b). The concept of "close enough" is defined by the number δ.

EXAMPLE 8 Example 5 Revisited

Prove that $\displaystyle \lim_{(x, y) \to (0, 0)} \frac{10xy^2}{x^2 + y^2} = 0$.

Solution From Definition 13.2.1, if $\varepsilon > 0$ is given, we want to find a number $\delta > 0$ such that

$$\left| \frac{10xy^2}{x^2 + y^2} - 0 \right| < \varepsilon \qquad \text{whenever} \qquad 0 < \sqrt{x^2 + y^2} < \delta.$$

The last line is the same as

$$\frac{10|x|y^2}{x^2 + y^2} < \varepsilon \qquad \text{whenever} \qquad 0 < \sqrt{x^2 + y^2} < \delta.$$

Because $x^2 \geq 0$, we can write $y^2 \leq x^2 + y^2$ and

$$\frac{y^2}{x^2 + y^2} \leq 1.$$

Thus, $\displaystyle \frac{10|x|y^2}{x^2 + y^2} = 10|x| \cdot \frac{y^2}{x^2 + y^2} \leq 10|x| = 10\sqrt{x^2} \leq 10\sqrt{x^2 + y^2}$.

So if we choose $\delta = \varepsilon/10$, we have

$$\left| \frac{10xy^2}{x^2 + y^2} - 0 \right| \leq 10\sqrt{x^2 + y^2} \leq 10 \cdot \frac{\varepsilon}{10} = \varepsilon.$$

By Definition 13.2.1, this shows

$$\lim_{(x, y) \to (0, 0)} \frac{xy^2}{x^2 + y^2} = 0. \qquad \blacksquare$$

Exercises 13.2 Answers to selected odd-numbered problems begin on page ANS-41.

≡ Fundamentals

In Problems 1–30, evaluate the given limit, if it exists.

1. $\displaystyle \lim_{(x, y) \to (5, -1)} (x^2 + y^2)$

2. $\displaystyle \lim_{(x, y) \to (2, 1)} \frac{x^2 - y}{x - y}$

3. $\displaystyle \lim_{(x, y) \to (0, 0)} \frac{5x^2 + y^2}{x^2 + y^2}$

4. $\displaystyle \lim_{(x, y) \to (1, 2)} \frac{4x^2 + y^2}{16x^4 + y^4}$

5. $\displaystyle \lim_{(x, y) \to (1, 1)} \frac{4 - x^2 - y^2}{x^2 + y^2}$

6. $\displaystyle \lim_{(x, y) \to (0, 0)} \frac{2x^2 - y}{x^2 + 2y^2}$

7. $\displaystyle \lim_{(x, y) \to (0, 0)} \frac{x^2 y}{x^4 + y^2}$

8. $\displaystyle \lim_{(x, y) \to (0, 0)} \frac{6xy^2}{x^2 + y^4}$

9. $\displaystyle \lim_{(x, y) \to (1, 2)} x^3 y^2 (x + y)^3$

10. $\displaystyle \lim_{(x, y) \to (2, 3)} \frac{xy}{x^2 - y^2}$

11. $\displaystyle \lim_{(x, y) \to (0, 0)} \frac{e^{xy}}{x + y + 1}$

12. $\displaystyle \lim_{(x, y) \to (0, 0)} \frac{\sin xy}{x^2 + y^2}$

13. $\displaystyle \lim_{(x, y) \to (2, 2)} \frac{xy}{x^3 + y^2}$

14. $\displaystyle \lim_{(x, y) \to (\pi, \pi/4)} \cos(3x + y)$

15. $\displaystyle \lim_{(x, y) \to (0, 0)} \frac{x^2 - 3y + 1}{x + 5y - 3}$

16. $\displaystyle \lim_{(x, y) \to (0, 0)} \frac{x^2 y^2}{x^4 + 5y^4}$

17. $\displaystyle \lim_{(x, y) \to (4, 3)} xy^2 \left(\frac{x + 2y}{x - y} \right)$

18. $\displaystyle \lim_{(x, y) \to (1, 0)} \frac{x^2 y}{x^3 + y^3}$

19. $\displaystyle \lim_{(x, y) \to (1, 1)} \frac{xy - x - y + 1}{x^2 + y^2 - 2x - 2y + 2}$

20. $\displaystyle \lim_{(x, y) \to (0, 3)} \frac{xy - 3y}{x^2 + y^2 - 6y + 9}$

21. $\displaystyle\lim_{(x,\,y)\to(0,\,0)} \frac{x^3y + xy^3 - 3x^2 - 3y^2}{x^2 + y^2}$

22. $\displaystyle\lim_{(x,\,y)\to(-2,\,2)} \frac{y^3 + 2x^3}{x + 5xy^2}$

23. $\displaystyle\lim_{(x,\,y)\to(1,\,1)} \ln(2x^2 - y^2)$ **24.** $\displaystyle\lim_{(x,\,y)\to(1,\,2)} \frac{\sin^{-1}(x/y)}{\cos^{-1}(x - y)}$

25. $\displaystyle\lim_{(x,\,y)\to(0,\,0)} \frac{(x^2 - y^2)^2}{x^2 + y^2}$ **26.** $\displaystyle\lim_{(x,\,y)\to(0,\,0)} \frac{\sin(3x^2 + 3y^2)}{x^2 + y^2}$

27. $\displaystyle\lim_{(x,\,y)\to(0,\,0)} \frac{6xy}{\sqrt{x^2 + y^2}}$ **28.** $\displaystyle\lim_{(x,\,y)\to(0,\,0)} \frac{x^2 - y^2}{\sqrt{x^2 + y^2}}$

29. $\displaystyle\lim_{(x,\,y)\to(0,\,0)} \frac{x^3}{x^2 + y^2}$ **30.** $\displaystyle\lim_{(x,\,y)\to(0,\,0)} \frac{x^3 + y^3}{x^2 + y^2}$

In Problems 31–34, determine where the given function is continuous.

31. $f(x, y) = \sqrt{x}\cos\sqrt{x + y}$ **32.** $f(x, y) = y^2 e^{1/xy}$

33. $f(x, y) = \tan\dfrac{x}{y}$

34. $f(x, y) = \ln(4x^2 + 9y^2 + 36)$

In Problems 35 and 36, determine whether the given function is continuous on the indicated sets in the xy-plane.

35. $f(x, y) = \begin{cases} x + y, & x \geq 2 \\ 0, & x < 2 \end{cases}$

    (a) $x^2 + y^2 < 1$ **(b)** $x \geq 0$ **(c)** $y > x$

36. $f(x, y) = \dfrac{xy}{\sqrt{x^2 + y^2 - 25}}$

    (a) $y \geq 3$ **(b)** $|x| + |y| < 1$ **(c)** $(x - 2)^2 + y^2 < 1$

37. Determine whether the function f defined by

$$f(x, y) = \begin{cases} \dfrac{6x^2 y^3}{(x^2 + y^2)^2}, & (x, y) \neq (0, 0) \\ 0, & (x, y) = (0, 0) \end{cases}$$

is continuous at $(0, 0)$.

38. Show that

$$f(x, y) = \begin{cases} \dfrac{xy}{2x^2 + 2y^2}, & (x, y) \neq (0, 0) \\ 0, & (x, y) = (0, 0) \end{cases}$$

is continuous in each variable separately at $(0, 0)$; that is, that $f(x, 0)$ and $f(0, y)$ are continuous at $x = 0$ and $y = 0$, respectively. Show, however, that f is not continuous at $(0, 0)$.

≡ **Think About It**

In Problems 39 and 40, use Definition 13.2.1 to prove the given result; that is, find δ for an arbitrary $\varepsilon > 0$.

39. $\displaystyle\lim_{(x,\,y)\to(0,\,0)} \frac{3x^2 y}{2x^2 + 2y^2} = 0$ **40.** $\displaystyle\lim_{(x,\,y)\to(0,\,0)} \frac{x^2 y^2}{x^2 + y^2} = 0$

41. Determine whether there are any points at which the function

$$f(x, y) = \begin{cases} \dfrac{x^3 - y^3}{x - y}, & y \neq x \\ 3x^2, & y = x \end{cases}$$

is discontinuous.

42. Use Definition 13.2.1 to prove that $\displaystyle\lim_{(x,\,y)\to(a,\,b)} y = b$.

13.3 Partial Derivatives

▌ Introduction The derivative of a function of **one variable** $y = f(x)$ is given by the limit of a difference quotient

$$\frac{dy}{dx} = \lim_{h\to 0} \frac{f(x + h) - f(x)}{h}.$$

In exactly the same manner, we can define a first-order derivative of a function of **two variables** $z = f(x, y)$ with respect to *each* variable.

Definition 13.3.1 First-Order Partial Derivatives

If $z = f(x, y)$ is a function of two variables, then the **partial derivative with respect to x** at a point (x, y) is

$$\frac{\partial z}{\partial x} = \lim_{h\to 0} \frac{f(x + h, y) - f(x, y)}{h} \tag{1}$$

and the **partial derivative with respect to y** is

$$\frac{\partial z}{\partial y} = \lim_{h\to 0} \frac{f(x, y + h) - f(x, y)}{h} \tag{2}$$

provided each limit exists.

■ **Computing a Partial Derivative** In (1) observe that the variable y does not change in the limiting process; in other words, y is held fixed. Similarly, in the limit definition (2) the variable x is held fixed. The two first-order partial derivatives (1) and (2) then represent the *rates of change* of f with respect to x and y, respectively. On a practical level we have the following simple guidelines.

Guidelines for Partial Differentiation

By *rules of ordinary differentiation*, we mean the rules developed in Chapter 3: Constant Multiple, Sum, Product, Quotient, Power, and Chain Rules.

- To compute $\partial z / \partial x$, use the laws of ordinary differentiation while treating y as a constant.
- To compute $\partial z / \partial y$, use the laws of ordinary differentiation while treating x as a constant.

EXAMPLE 1 Partial Derivatives

If $z = 4x^3 y^2 - 4x^2 + y^6 + 1$, find

(a) $\dfrac{\partial z}{\partial x}$ and   (b) $\dfrac{\partial z}{\partial y}$.

Solution

(a) We differentiate z with respect to x while holding y fixed and treating constants in the usual manner:

$$\frac{\partial z}{\partial x} = (12x^2)y^2 - 8x + 0 + 0 = 12x^2 y^2 - 8x.$$

(b) Now by treating x as a constant, we obtain

$$\frac{\partial z}{\partial y} = 4x^3(2y) - 0 + 6y^5 + 0 = 8x^3 y + 6y^5.$$ ∎

■ **Alternative Symbols** The partial derivatives $\partial z / \partial x$ and $\partial z / \partial y$ are often represented by alternative symbols. If $z = f(x, y)$, then

$$\frac{\partial z}{\partial x} = \frac{\partial f}{\partial x} = z_x = f_x \quad \text{and} \quad \frac{\partial z}{\partial y} = \frac{\partial f}{\partial y} = z_y = f_y.$$

Symbols such as $\partial / \partial x$ and $\partial / \partial y$ are called **partial differentiation operators** and denote the *operation* of taking a partial derivative, in this case with respect to x and y, respectively. For example,

$$\frac{\partial}{\partial x}(x^2 - y^2) = \frac{\partial}{\partial x}x^2 - \frac{\partial}{\partial x}y^2 = 2x - 0 = 2x$$

and

$$\frac{\partial}{\partial y}e^{x^4 y^5} = e^{x^4 y^5} \cdot \frac{\partial}{\partial y}x^4 y^5 = e^{x^4 y^5}x^4 \cdot \frac{\partial}{\partial y}y^5 = e^{x^4 y^5}x^4(5y^4) = 5x^4 y^4 e^{x^4 y^5}.$$

The **value** of a partial derivative at a point (x_0, y_0) is written in various ways. For example, the partial derivative of $z = f(x, y)$ with respect to x at (x_0, y_0) is written as

$$\left.\frac{\partial z}{\partial x}\right|_{(x_0, y_0)}, \quad \left.\frac{\partial z}{\partial x}\right|_{x=x_0,\, y=y_0}, \quad \frac{\partial z}{\partial x}(x_0, y_0), \quad \text{or} \quad f_x(x_0, y_0).$$

EXAMPLE 2 Using the Product Rule

If $f(x, y) = x^5 y^{10} \cos(xy^2)$, find f_y.

Solution When x is held fixed, observe that

$$f(x, y) = x^5 \overbrace{y^{10} \cos(xy^2)}^{\substack{\text{product of two} \\ \text{functions of } y}}.$$

Hence, by the Product and Chain Rules the partial derivative of f with respect to y is,

$$f_y(x, y) = x^5 [y^{10}(-\sin(xy^2)) \cdot 2xy + 10y^9 \cdot \cos(xy^2)]$$
$$= -2x^6 y^{11} \sin(xy^2) + 10x^5 y^9 \cos(xy^2).$$ ∎

EXAMPLE 3 A Rate of Change

The function $S = 0.1091w^{0.425}h^{0.725}$ relates the surface area (in square feet) of a person's body as a function of weight w (in pounds) and height h (in inches). Find $\partial S/\partial w$ when $w = 150$ and $h = 72$. Interpret.

Solution The partial derivative of S with respect to w,

$$\frac{\partial S}{\partial w} = (0.1091)(0.425)w^{-0.575}h^{0.725},$$

evaluated at $(150, 72)$ is

$$\left.\frac{\partial S}{\partial w}\right|_{(150, 72)} = (0.1091)(0.425)(150)^{-0.575}(72)^{0.725} \approx 0.058.$$

The partial derivative $\partial S/\partial w$ is the rate at which the surface area of a person of *fixed* height h, such as an adult, changes with respect to weight w. Since the units for the derivative are ft²/lb and $\partial S/\partial w > 0$, we see that a gain of 1 lb, while h is fixed at 72, results in an *increase* in the area of the skin of approximately $0.058 \approx \frac{1}{17}$ ft². ∎

■ **Geometric Interpretation** As seen in FIGURE 13.3.1(a), when y is constant, say $y = b$, the trace of the surface $z = f(x, y)$ in the plane $y = b$ is the blue curve C. If we define the slope of the secant through the points $P(a, b, f(a, b))$ and $R(a + h, b, f(a + h, b))$ as

$$\frac{f(a + h, b) - f(a, b)}{(a + h) - a} = \frac{f(a + h, b) - f(a, b)}{h}$$

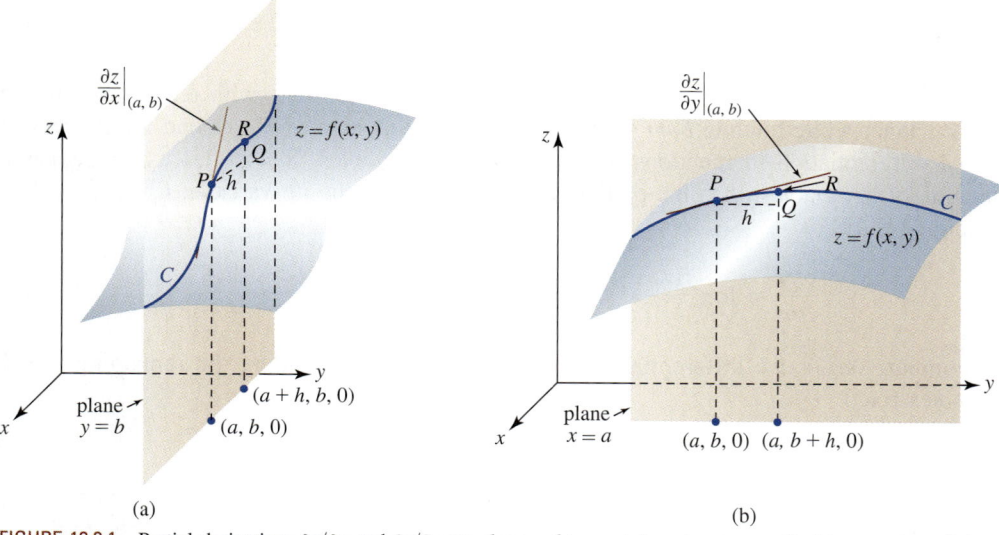

(a)   (b)

FIGURE 13.3.1 Partial derivatives $\partial z/\partial x$ and $\partial z/\partial y$ are slopes of tangent lines to a curve C of intersection of the surface and a plane parallel to the x- or y-axes.

we have
$$\frac{\partial z}{\partial x}\bigg|_{(a,\,b)} = \lim_{h \to 0} \frac{f(a + h, b) - f(a, b)}{h}.$$

In other words, we can interpret $\partial z / \partial x$ as the slope of the tangent line at the point P (for which the limit exists) on the curve C of intersection of the surface $z = f(x, y)$ and the plane $y = b$. In turn, an inspection of Figure 13.3.1(b) reveals that $\partial z / \partial y$ is the slope of the tangent line at the point P on the curve C of intersection between the surface $z = f(x, y)$ and the plane $x = a$.

EXAMPLE 4 Slopes of Tangent Lines

For $z = 9 - x^2 - y^2$, find the slope of the tangent line at $(2, 1, 4)$ in

(a) the plane $x = 2$ and   (b) the plane $y = 1$.

Solution

(a) By specifying the plane $x = 2$, we are holding all values of x constant. Hence, we compute the partial derivative of z with respect to y:
$$\frac{\partial z}{\partial y} = -2y.$$

At $(2, 1, 4)$ the slope is $\dfrac{\partial z}{\partial y}\bigg|_{(2,\,1)} = -2.$

(b) In the plane $y = 1$, y is constant and so we find the partial derivative of z with respect to x:
$$\frac{\partial z}{\partial x} = -2x.$$

At $(2, 1, 4)$ the slope is $\dfrac{\partial z}{\partial x}\bigg|_{(2,\,1)} = -4.$

See FIGURE 13.3.2.

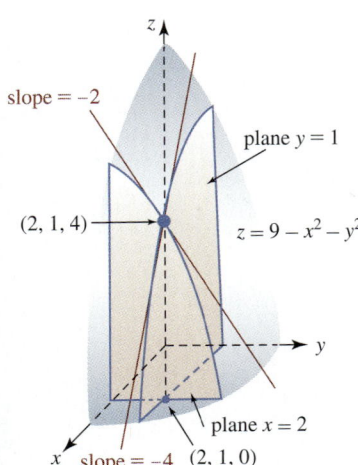

slope = −2

plane $y = 1$

$(2, 1, 4)$

$z = 9 - x^2 - y^2$

plane $x = 2$

x slope = −4 $(2, 1, 0)$

FIGURE 13.3.2 Slope of tangent lines in Example 4

If $z = f(x, y)$, then the values of the partial derivatives $\partial z / \partial x$ and $\partial z / \partial y$ at a point $(a, b, f(a, b))$ are also referred to as the **slopes of the surface** in the x- and y-directions, respectively.

■ **Functions of Three or More Variables**   The rates of change of a function of three variables $w = f(x, y, z)$ in the x, y, and z directions are the partial derivatives $\partial w / \partial x$, $\partial w / \partial y$, and $\partial w / \partial z$, respectively. The partial derivative of f with respect to z is defined to be
$$\frac{\partial w}{\partial z} = \lim_{h \to 0} \frac{f(x, y, z + h) - f(x, y, z)}{h}, \tag{3}$$

whenever the limit exists. To compute, say, $\partial w / \partial x$, we differentiate with respect to x in the usual manner while holding *both* y and z constant. In this manner we extend the process of partial differentiation to functions of any number of variables. If $u = f(x_1, x_2, \ldots, x_n)$ is a function of n variables, then the partial of f with respect to the ith variable, $i = 1, 2, \ldots, n$, is defined to be
$$\frac{\partial u}{\partial x_i} = \lim_{h \to 0} \frac{f(x_1, x_2, \ldots, x_i + h, \ldots x_n) - f(x_1, x_2, \ldots x_n)}{h}. \tag{4}$$

To compute $\partial u / \partial x_i$ we differentiate with respect to x_i while holding the remaining $n - 1$ variables fixed.

EXAMPLE 5 Using the Quotient Rule

If $w = \dfrac{x^2 - z^2}{y^2 + z^2}$, find $\dfrac{\partial w}{\partial z}$.

Solution   We use the Quotient Rule while holding both x and y constant:

$$\frac{\partial w}{\partial z} = \frac{(y^2 + z^2)(-2z) - (x^2 - z^2)2z}{(y^2 + z^2)^2} = -\frac{2z(x^2 + y^2)}{(y^2 + z^2)^2}.$$ ■

EXAMPLE 6   Three Partial Derivatives

If $f(x, y, t) = e^{-3\pi t}\cos 4x \sin 6y$, then the partial derivatives with respect to x, y, and t are, in turn,

$$f_x(x, y, t) = -4e^{-3\pi t}\sin 4x \sin 6y,$$
$$f_y(x, y, t) = 6e^{-3\pi t}\cos 4x \cos 6y,$$

and
$$f_t(x, y, t) = -3\pi e^{-3\pi t}\cos 4x \sin 6y.$$ ■

■ **Higher-Order and Mixed Derivatives**   For a function of two variables $z = f(x, y)$, the partial derivatives $\partial z/\partial x$ and $\partial z/\partial y$ are themselves functions of x and y. Consequently, we can compute **second**, and higher, **partial derivatives**. Indeed, we can find the partial derivative of $\partial z/\partial x$ with respect to y, and the partial derivative of $\partial z/\partial y$ with respect to x. The latter types of partial derivatives are called **mixed partial derivatives**. In summary, the second, third, and mixed partial derivatives of $z = f(x, y)$ are defined by:

Second-order partial derivatives:

$$\frac{\partial^2 z}{\partial x^2} = \frac{\partial}{\partial x}\left(\frac{\partial z}{\partial x}\right) \quad \text{and} \quad \frac{\partial^2 z}{\partial y^2} = \frac{\partial}{\partial y}\left(\frac{\partial z}{\partial y}\right)$$

Third-order partial derivatives:

$$\frac{\partial^3 z}{\partial x^3} = \frac{\partial}{\partial x}\left(\frac{\partial^2 z}{\partial x^2}\right) \quad \text{and} \quad \frac{\partial^3 z}{\partial y^3} = \frac{\partial}{\partial y}\left(\frac{\partial^2 z}{\partial y^2}\right)$$

Mixed second-order partial derivatives:

$$\frac{\partial^2 z}{\partial x\, \partial y} = \frac{\partial}{\partial x}\left(\frac{\partial z}{\partial y}\right) \quad \text{and} \quad \frac{\partial^2 z}{\partial y\, \partial x} = \frac{\partial}{\partial y}\left(\frac{\partial z}{\partial x}\right).$$

differentiate ↑ first with respect to y differentiate ↑ first with respect to x

Observe in the summary that there are four second partial derivatives. How many third-order partial derivatives of $z = f(x, y)$ are there? Higher-order partial derivatives for $z = f(x, y)$ and for functions of three or more variables are defined in a similar manner.

■ **Alternative Symbols**   The second- and third-order partial derivatives are also denoted by f_{xx}, f_{yy}, f_{xxx}, and so on. The subscript notation for mixed second partial derivatives is f_{xy} or f_{yx}.

Note   The order of the symbols in the subscripts for the mixed partials is just the opposite of the order of the symbols when partial differentiation operator notation is used:

$$f_{xy} = (f_x)_y = \frac{\partial}{\partial y}\left(\frac{\partial z}{\partial x}\right) = \frac{\partial^2 z}{\partial y\, \partial x}$$

and
$$f_{yx} = (f_y)_x = \frac{\partial}{\partial x}\left(\frac{\partial z}{\partial y}\right) = \frac{\partial^2 z}{\partial x\, \partial y}.$$

■ **Equality of Mixed Partials**   Although we shall not prove it, the next theorem states that under certain conditions the order in which a mixed second partial derivative is done is irrelevant; that is, the mixed partial derivatives f_{xy} and f_{yx} are equal.

> **Theorem 13.3.2** Equality of Mixed Partials
>
> Let f be a function of two variables. If the partial derivatives f_x, f_y, f_{xy}, and f_{yx} are continuous on some open disk, then
>
> $$f_{xy} = f_{yx}$$
>
> at each point on the disk.

See Problem 68 in Exercises 13.3.

EXAMPLE 7 Second-Order Partial Derivatives

If $z = x^2y^2 - y^3 + 3x^4 + 5$, find

(a) $\dfrac{\partial^2 z}{\partial x^2}, \dfrac{\partial^3 z}{\partial x^3}$ (b) $\dfrac{\partial^2 z}{\partial y^2}, \dfrac{\partial^3 z}{\partial y^3}$, and (c) $\dfrac{\partial^2 z}{\partial x\,\partial y}$.

Solution From the first partial derivatives

$$\frac{\partial z}{\partial x} = 2xy^2 + 12x^3 \quad \text{and} \quad \frac{\partial z}{\partial y} = 2x^2y - 3y^2$$

we get:

(a) $\dfrac{\partial^2 z}{\partial x^2} = \dfrac{\partial}{\partial x}\left(\dfrac{\partial z}{\partial x}\right) = 2y^2 + 36x^2 \quad$ and $\quad \dfrac{\partial^3 z}{\partial x^3} = \dfrac{\partial}{\partial x}\left(\dfrac{\partial^2 z}{\partial x^2}\right) = 72x,$

(b) $\dfrac{\partial^2 z}{\partial y^2} = \dfrac{\partial}{\partial y}\left(\dfrac{\partial z}{\partial y}\right) = 2x^2 - 6y \quad$ and $\quad \dfrac{\partial^3 z}{\partial y^3} = \dfrac{\partial}{\partial y}\left(\dfrac{\partial^2 z}{\partial y^2}\right) = -6,$

(c) $\dfrac{\partial^2 z}{\partial x\,\partial y} = \dfrac{\partial}{\partial x}\left(\dfrac{\partial z}{\partial y}\right) = 4xy.$

You should also verify that $\dfrac{\partial^2 z}{\partial y\,\partial x} = \dfrac{\partial}{\partial y}\left(\dfrac{\partial z}{\partial x}\right) = 4xy.$ ∎

If f is a function of two variables and has continuous first-, second-, and third-order partial derivatives on some open disk, then the mixed third-order derivatives are equal; that is,

$$f_{xyy} = f_{yxy} = f_{yyx} \quad \text{and} \quad f_{yxx} = f_{xyx} = f_{xxy}.$$

Similar remarks hold for functions of three or more variables. For example, if f is a function of three variables x, y, and z that possesses continuous partial derivatives of any order in some open ball, then the mixed partials such as $f_{xyz} = f_{zyx} = f_{yxz}$ are equal at each point in the ball.

EXAMPLE 8 Mixed Third-Order Partial Derivatives

If $f(x, y, z) = \sqrt{x^2 + y^4 + z^6}$, find f_{yzz}.

Solution f_{yzz} is a mixed third-order partial derivative. First we find the partial derivative with respect to y by the Power Rule for functions:

$$f_y = \frac{1}{2}(x^2 + y^4 + z^6)^{-1/2}4y^3 = 2y^3(x^2 + y^4 + z^6)^{-1/2}.$$

The partial derivative with respect to z of the function in the last line is then

$$f_{yz} = (f_y)_z = 2y^3\left(-\frac{1}{2}\right)(x^2 + y^4 + z^6)^{-3/2} \cdot 6z^5$$

$$= -6y^3z^5(x^2 + y^4 + z^6)^{-3/2}.$$

Finally, by the Product Rule,

$$f_{yzz} = (f_{yz})_z = -6y^3z^5\left(-\frac{3}{2}\right)(x^2 + y^4 + z^6)^{-5/2} \cdot 6z^5 - 30y^3z^4(x^2 + y^4 + z^6)^{-3/2}$$

$$= y^3z^4(x^2 + y^4 + z^6)^{-5/2}(24z^6 - 30x^2 - 30y^4).$$

You are encouraged to compute f_{zzy} and f_{zyz} and verify on any open disk not containing the origin that $f_{yzz} = f_{zzy} = f_{zyz}$. ■

▌ **Implicit Partial Differentiation** Implicit partial differentiation is carried out in the same manner as in Section 3.6.

EXAMPLE 9 Implicit Partial Derivative

Assume that the equation $z^2 = x^2 + xy^2z$ defines z implicitly as a function of x and y. Find $\partial z/\partial x$ and $\partial z/\partial y$.

Solution By holding y constant,

$$\frac{\partial}{\partial x}z^2 = \frac{\partial}{\partial x}(x^2 + xy^2z) \qquad \text{implies} \qquad \frac{\partial}{\partial x}z^2 = \frac{\partial}{\partial x}x^2 + y^2\frac{\partial}{\partial x}xz.$$

By the Power Rule for functions along with the Product Rule:

$$2z\frac{\partial z}{\partial x} = 2x + y^2\left(x\frac{\partial z}{\partial x} + z\right).$$

We then solve the last equation for $\partial z/\partial x$:

$$\frac{\partial z}{\partial x} = \frac{2x + y^2z}{2z - xy^2}.$$

Now by holding x constant,

$$\frac{\partial}{\partial y}z^2 = \frac{\partial}{\partial y}(x^2 + xy^2z) \qquad \text{implies} \qquad 2z\frac{\partial z}{\partial y} = x\left(y^2\frac{\partial z}{\partial y} + 2yz\right).$$

Solving for $\partial z/\partial y$ yields

$$\frac{\partial z}{\partial x} = \frac{2xyz}{2z - xy^2}. \qquad ■$$

Exercises 13.3 Answers to selected odd-numbered problems begin on page ANS-41.

≡ **Fundamentals**

In Problems 1–4, use Definition 13.3.1 to compute $\partial z/\partial x$ and $\partial z/\partial y$ for the given function.

1. $z = 7x + 8y^2$

2. $z = xy$

3. $z = 3x^2y + 4xy^2$

4. $z = \dfrac{x}{x + y}$

In Problems 5–24, find the first partial derivatives of the given function.

5. $z = x^2 - xy^2 + 4y^5$

6. $z = -x^3 + 6x^2y^3 + 5y^2$

7. $z = 5x^4y^3 - x^2y^6 + 6x^5 - 4y$

8. $z = \tan(x^3y^2)$

9. $z = \dfrac{4\sqrt{x}}{3y^2 + 1}$

10. $z = 4x^3 - 5x^2 + 8x$

11. $z = (x^3 - y^2)^{-1}$

12. $z = (-x^4 + 7y^2 + 3y)^6$

13. $z = \cos^2 5x + \sin^2 5y$

14. $z = e^{x^2 \tan^{-1} y^2}$

15. $f(x, y) = xe^{x^3y}$

16. $f(\theta, \phi) = \phi^2\sin\dfrac{\theta}{\phi}$

17. $f(x, y) = \dfrac{3x - y}{x + 2y}$

18. $f(x, y) = \dfrac{xy}{(x^2 - y^2)^2}$

19. $g(u, v) = \ln(4u^2 + 5v^3)$

20. $h(r, s) = \dfrac{\sqrt{r}}{s} - \dfrac{\sqrt{s}}{r}$

21. $w = 2\sqrt{xy} - ye^{y/z}$

22. $w = xy \ln xz$

23. $f(u, v, x, t) = u^2w^2 - uv^3 + vw\cos(ut^2) + (2x^2t)^4$

24. $G(p, q, r, s) = (p^2q^3)e^{2r^4s^5}$

In Problems 25 and 26, suppose $z = 4x^3y^4$.

25. Find the slope of the tangent line at $(1, -1, 4)$ in the plane $x = 1$.

26. Find the slope of the tangent line at $(1, -1, 4)$ in the plane $y = -1$.

In Problems 27 and 28, suppose $f(x, y) = \dfrac{18xy}{x + y}$.

27. Find parametric equations for the tangent line at $(-1, 4, -24)$ in the plane $x = -1$.

28. Find symmetric equations for the tangent line at $(-1, 4, -24)$ in the plane $y = 4$.

In Problems 29 and 30, suppose $z = \sqrt{9 - x^2 - y^2}$.

29. At what rate is z changing with respect to x in the plane $y = 2$ at the point $(2, 2, 1)$?

30. At what rate is z changing with respect to y in the plane $x = \sqrt{2}$ at the point $(\sqrt{2}, \sqrt{3}, 2)$?

In Problems 31–38, find the indicated partial derivative.

31. $z = e^{xy}$; $\dfrac{\partial^2 z}{\partial x^2}$
32. $z = x^4 y^{-2}$; $\dfrac{\partial^3 z}{\partial y^3}$

33. $f(x, y) = 5x^2 y^2 - 2xy^3$; f_{xy}
34. $f(p, q) = \ln\dfrac{p + q}{q^2}$; f_{qp}

35. $w = u^2 v^3 t^3$; w_{tuv}
36. $w = \dfrac{\cos(u^2 v)}{t^3}$; w_{vvt}

37. $F(r, \theta) = e^{r^2}\cos\theta$; $F_{r\theta r}$
38. $H(s, t) = \dfrac{s + t}{s - t}$; H_{tts}

In Problems 39 and 40, verify that $\dfrac{\partial^2 z}{\partial x\, \partial y} = \dfrac{\partial^2 z}{\partial y\, \partial x}$.

39. $z = x^6 - 5x^4 y^3 + 4xy^2$
40. $z = \tan^{-1}(2xy)$

In Problems 41 and 42, verify that the indicated partial derivatives are equal.

41. $w = u^3 v^4 - 4u^2 v^2 t^3 + v^2 t$; $w_{uvt}, w_{tvu}, w_{vut}$
42. $F(\eta, \xi, \tau) = (\eta^3 + \xi^2 + \tau)^2$; $F_{\eta\xi\eta}, F_{\xi\eta\eta}, F_{\eta\eta\xi}$

In Problems 43–46, suppose the given equation defines z as a function of the remaining two variables. Use implicit differentiation to find the first partial derivatives.

43. $x^2 + y^2 + z^2 = 25$
44. $z^2 = x^2 + y^2 z$

45. $z^2 + u^2 v^3 - uvz = 0$
46. $se^z - e^{st} + 4s^3 t = z$

47. The area A of a parallelogram with base x and height $y\sin\theta$ is $A = xy\sin\theta$. Find all first partial derivatives.

48. The volume of the frustum of a cone shown in FIGURE 13.3.3 is $V = \frac{1}{3}\pi h(r^2 + rR + R^2)$. Find all first partial derivatives.

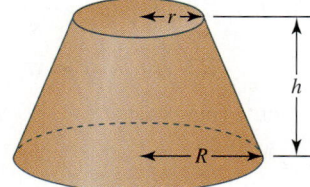

FIGURE 13.3.3 Frustum of a cone in Problem 48

≡ Applications

In Problems 49 and 50, verify that the given temperature distribution satisfies **Laplace's equation in two dimensions**

$$\frac{\partial^2 u}{\partial x^2} + \frac{\partial^2 u}{\partial y^2} = 0. \tag{5}$$

A solution $u(x, y)$ of Laplace's equation (5) can be interpreted as the time-independent temperature distribution throughout a thin two-dimensional plate. See FIGURE 13.3.4.

49. $u(x, y) = (\cosh 2\pi y + \sinh 2\pi y)\sin 2\pi x$
50. $u(x, y) = e^{-(n\pi x/L)}\sin(n\pi y/L)$, n and L constants

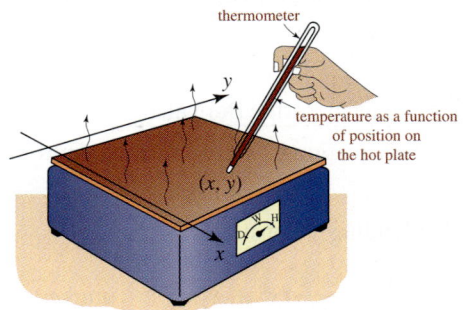

thermometer

temperature as a function of position on the hot plate

(x, y)

FIGURE 13.3.4 Hot plate in Problems 49 and 50

In Problems 51 and 52, verify that the given function satisfies Laplace's equation (5).

51. $u(x, y) = \ln(x^2 + y^2)$
52. $u(x, y) = \tan^{-1}\dfrac{y}{x}$

In Problems 53 and 54, verify that the given function satisfies **Laplace's equation in three dimensions**

$$\frac{\partial^2 u}{\partial x^2} + \frac{\partial^2 u}{\partial y^2} + \frac{\partial^2 u}{\partial z^2} = 0. \tag{6}$$

53. $u(x, y, z) = \dfrac{1}{\sqrt{x^2 + y^2 + z^2}}$

54. $u(x, y, z) = e^{\sqrt{m^2 + n^2}\, x}\cos my\, \sin nz$

In Problems 55 and 56, verify that the given function satisfies the **one-dimensional wave equation**

$$a^2\frac{\partial^2 u}{\partial x^2} = \frac{\partial^2 u}{\partial t^2}. \tag{7}$$

The wave equation (7) occurs in problems involving vibrational phenomena.

55. $u(x, t) = \cos at\, \sin x$
56. $u(x, t) = \cos(x + at) + \sin(x - at)$

57. The molecular concentration $C(x, t)$ of a liquid is given by $C(x, t) = t^{-1/2}e^{-x^2/kt}$. Verify that this function satisfies the **one-dimensional diffusion equation**

$$\frac{k}{4}\frac{\partial^2 C}{\partial x^2} = \frac{\partial C}{\partial t}.$$

58. The pressure P exerted by an enclosed ideal gas is given by $P = k(T/V)$, where k is a constant. T is temperature and V is volume. Find:

(a) the rate of change of P with respect to V,
(b) the rate of change of V with respect to T, and
(c) the rate of change of T with respect to P.

59. The vertical displacement of a long string fastened at the origin but falling under its own weight is given by

$$u(x, t) = \begin{cases} -\dfrac{g}{2a^2}(2axt - x^2), & 0 \le x \le at \\ -\frac{1}{2}gt^2, & x > at. \end{cases}$$

See **FIGURE 13.3.5**.

(a) Find $\partial u/\partial t$. Interpret for $x > at$.
(b) Find $\partial u/\partial x$. Interpret for $x > at$.

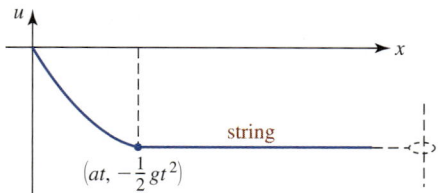

FIGURE 13.3.5 Falling string in Problem 59

60. For the skin-area function $S = 0.1091w^{0.425}h^{0.725}$ discussed in Example 3 find $\partial S/\partial h$ at $w = 60, h = 36$. If a girl grows in height from 36 to 37 in., while her weight is fixed at 60 lb, what is the approximate increase in the area of skin?

≡ **Think About It**

61. Formulate a limit definition that is analogous to Definition 13.3.1 for the second-order partial derivatives

(a) $\dfrac{\partial^2 z}{\partial x^2}$ (b) $\dfrac{\partial^2 z}{\partial y^2}$ (c) $\dfrac{\partial^2 z}{\partial x \, \partial y}$

62. Find a function $z = f(x, y)$ such that

$$\frac{\partial z}{\partial x} = 2xy^3 + 2y + \frac{1}{x} \quad \text{and} \quad \frac{\partial z}{\partial y} = 3x^2y^2 + 2x + 1.$$

63. Can a function $z = f(x, y)$, with partial derivatives continuous on an open set, be found such that

$$\frac{\partial z}{\partial x} = x^2 + y^2 \quad \text{and} \quad \frac{\partial z}{\partial y} = x^2 - y^2?$$

64. (a) Suppose the function $w = f(x, y, z)$ has continuous third-order partial derivatives. How many different third-order partial derivatives are there?
(b) Suppose the function $z = f(x, y)$ has continuous nth-order partial derivatives. How many different nth-order partial derivatives are there?

65. (a) Suppose $z = f(x, y)$ has the property that $\partial z/\partial x = 0$ and $\partial z/\partial y = 0$ for all (x, y). What can you say about the form of f?
(b) Suppose $z = f(x, y)$ has continuous second-order partial derivatives and $\partial^2 z/\partial x \, \partial y = 0$. What can you say about the form of f?

66. Some level curves of a function $z = f(x, y)$ are shown in FIGURE 13.3.6. Use these level curves to conjecture the algebraic signs of the partial derivatives $\partial z/\partial x$ and $\partial z/\partial y$ at the point indicated in red in the figure.

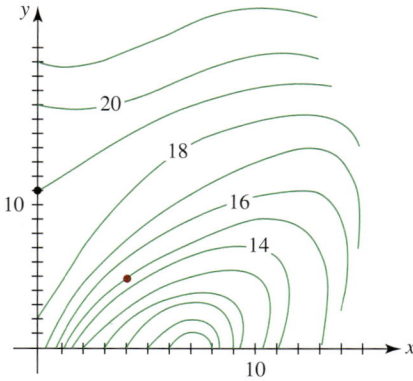

FIGURE 13.3.6 Level curves in Problem 66

67. A Mathematical Classic   A function $z = f(x, y)$ may not be continuous at a point but still may have partial derivatives at that point. The function

$$f(x, y) = \begin{cases} \dfrac{xy}{2x^2 + 2y^2}, & (x, y) \ne (0, 0) \\ 0, & (x, y) = (0, 0) \end{cases}$$

is not continuous at $(0, 0)$. (See Problem 38 in Exercises 13.2.) Use (1) and (2) in Definition 13.3.1 to show that

$$\frac{\partial z}{\partial x}\bigg|_{(0, 0)} = 0 \quad \text{and} \quad \frac{\partial z}{\partial y}\bigg|_{(0, 0)} = 0.$$

68. A Mathematical Classic   Consider the function $z = f(x, y)$ defined by

$$f(x, y) = \begin{cases} \dfrac{xy(y^2 - x^2)}{x^2 + y^2}, & (x, y) \ne (0, 0) \\ 0, & (x, y) = (0, 0). \end{cases}$$

(a) Compute $\dfrac{\partial z}{\partial x}\bigg|_{(0, y)}$ and $\dfrac{\partial z}{\partial y}\bigg|_{(x, 0)}$.

(b) Show that $\dfrac{\partial^2 z}{\partial y \, \partial x}\bigg|_{(0, 0)} \ne \dfrac{\partial^2 z}{\partial x \, \partial y}\bigg|_{(0, 0)}$.

13.4 Linearization and Differentials

▌ **Introduction**   In Section 4.9 we saw that a linearization $L(x)$ of a function of a single variable $y = f(x)$ at a number x_0 is given by $L(x) = f(x_0) + f'(x_0)(x - x_0)$. This equation can be used to approximate the function values $f(x)$ in a neighborhood of x_0, that is, $L(x) \approx f(x)$ for values of x near x_0. In like manner we can define a linearization $L(x, y)$ of a function of two variables at a point (x_0, y_0). In the case of a function of a single variable we assumed that $y = f(x)$ was differentiable at x_0, that is,

$$f'(x_0) = \lim_{\Delta x \to 0} \frac{f(x_0 + \Delta x) - f(x_0)}{\Delta x} \tag{1}$$

exists. Recall too, that if f is differentiable at x_0, it is also continuous at that number. Mimicking the assumption in (1), we wish $z = f(x, y)$ to be differentiable at a point (x_0, y_0). Although we have considered what it means for $z = f(x, y)$ to possess *partial derivatives* at a point, we have not as yet formulated a definition of *differentiability* of a function of two variables f at a point.

■ **Increment of the Dependent Variable** The definition of differentiability of a function of any number of independent variables depends not on the notion of a difference quotient as in (1), but rather on the notion of an *increment* of the dependent variable. Recall, for a function of one variable $y = f(x)$ the increment in the dependent variable is given by

$$\Delta y = f(x + \Delta x) - f(x).$$

Analogously, for a function of two variables $z = f(x, y)$, we define the **increment of the dependent variable** z as

$$\Delta z = f(x + \Delta x, y + \Delta y) - f(x, y). \tag{2}$$

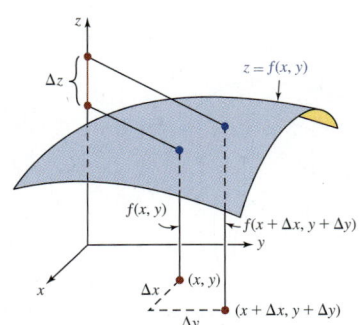

FIGURE 13.4.1 shows that Δz gives the amount of change in the function as (x, y) changes to $(x + \Delta x, y + \Delta y)$.

FIGURE 13.4.1 Increment in z

EXAMPLE 1 Finding Δz

Find Δz for the polynomial function $z = x^2 - xy$. What is the change in the function from $(1, 1)$ to $(1.2, 0.7)$?

Solution From (2),

$$\Delta z = [(x + \Delta x)^2 - (x + \Delta x)(y + \Delta y)] - (x^2 - xy)$$
$$= (2x - y)\Delta x - x\Delta y + (\Delta x)^2 - \Delta x \Delta y. \tag{3}$$

With $x = 1$, $y = 1$, $\Delta x = 0.2$, and $\Delta y = -0.3$,

$$\Delta z = (1)(0.2) - (1)(-0.3) + (0.2)^2 - (0.2)(-0.3) = 0.6. \qquad ■$$

■ **A Fundamental Increment Formula** A brief reinspection of the increment Δz in (3) shows that in the first two terms the coefficients of Δx and Δy are $\partial z / \partial x$ and $\partial z / \partial y$, respectively. The important theorem that follows shows that this is no accident.

Theorem 13.4.1 An Increment Formula

Let $z = f(x, y)$ have continuous partial derivatives $f_x(x, y)$ and $f_y(x, y)$ in an open rectangular region that is defined by $a < x < b, c < y < d$. If (x, y) is any point in this region, then there exist ε_1 and ε_2, which are functions of Δx and Δy, such that

$$\Delta z = f_x(x, y)\Delta x + f_y(x, y)\Delta y + \varepsilon_1 \Delta x + \varepsilon_2 \Delta y, \tag{4}$$

where $\varepsilon_1 \to 0$ and $\varepsilon_2 \to 0$ when $\Delta x \to 0$ and $\Delta y \to 0$.

PROOF By adding and subtracting $f(x, y + \Delta y)$ in (2), we have

$$\Delta z = [f(x + \Delta x, y + \Delta y) - f(x, y + \Delta y)] + [f(x, y + \Delta y) - f(x, y)].$$

Applying the Mean Value Theorem (Theorem 4.4.2) to each set of brackets then gives

$$\Delta z = f_x(x_0, y + \Delta y)\Delta x + f_y(x, y_0)\Delta y, \tag{5}$$

where, as shown in FIGURE 13.4.2, $x < x_0 < x + \Delta x$ and $y < y_0 < y + \Delta y$. Now, define

$$\varepsilon_1 = f_x(x_0, y + \Delta y) - f_x(x, y) \qquad \text{and} \qquad \varepsilon_2 = f_y(x, y_0) - f_y(x, y). \tag{6}$$

As $\Delta x \to 0$ and $\Delta y \to 0$, then, as shown in the figure, $P_2 \to P_1$ and $P_3 \to P_1$. Since f_x and f_y are assumed continuous in the region, we have

$$\lim_{(\Delta x, \Delta y) \to (0, 0)} \varepsilon_1 = 0 \qquad \text{and} \qquad \lim_{(\Delta x, \Delta y) \to (0, 0)} \varepsilon_2 = 0.$$

Solving (6) for $f_x(x_0, y + \Delta y)$ and $f_y(x, y_0)$ and substituting in (5) gives (4). ∎

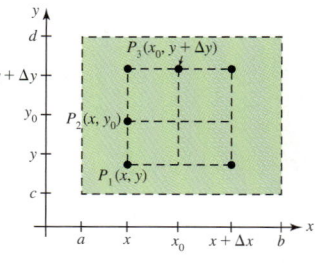

FIGURE 13.4.2 Rectangular region in Theorem 13.4.1

▌ Differentiability—Functions of Two Variables

We are now in a position to define differentiability of a function $z = f(x, y)$ at a point.

Definition 13.4.1 Differentiable Function

A function $z = f(x, y)$ is **differentiable** at (x_0, y_0) if the increment Δz can be written as

$$\Delta z = f_x(x_0, y_0)\Delta x + f_y(x_0, y_0)\Delta y + \varepsilon_1 \Delta x + \varepsilon_2 \Delta y,$$

where both ε_1 and $\varepsilon_2 \to 0$ as $(\Delta x, \Delta y) \to (0, 0)$.

If the function $z = f(x, y)$ is differentiable at each point in a region R of the xy-plane, then f is said to be **differentiable on R**. If f is differentiable on the region consisting of the entire xy-plane, then f is said to be **differentiable everywhere**.

It is interesting to note that the partial derivatives f_x and f_y may exist at a point (x_0, y_0) and yet f may not be differentiable at that point. Of course, if f_x and f_y fail to exist at a point (x_0, y_0), then f is not differentiable there. The following theorem gives us sufficient conditions under which the existence of the partial derivatives implies differentiability.

Theorem 13.4.2 Sufficient Condition for Differentiability

If the first partial derivatives f_x and f_y are continuous at every point in an open region R, then $z = f(x, y)$ is differentiable on R.

The next theorem is the analogue of Theorem 3.1.1; it states that if $z = f(x, y)$ is differentiable at a point, then it is continuous at the point.

Theorem 13.4.3 Differentiability Implies Continuity

If $z = f(x, y)$ is differentiable at a point (x_0, y_0), then f is continuous at (x_0, y_0).

PROOF Suppose f is differentiable at a point (x_0, y_0) and that

$$\Delta z = f(x_0 + \Delta x, y_0 + \Delta y) - f(x_0, y_0).$$

Using this expression in (4) gives

$$f(x_0 + \Delta x, y_0 + \Delta y) - f(x_0, y_0) = f_x(x_0, y_0)\Delta x + f_y(x_0, y_0)\Delta y + \varepsilon_1 \Delta x + \varepsilon_2 \Delta y.$$

As $(\Delta x, \Delta y) \to (0, 0)$ it follows from the last line that

$$\lim_{(\Delta x, \Delta y) \to (0, 0)} [f(x_0 + \Delta x, y_0 + \Delta y) - f(x_0, y_0)] = 0 \quad \text{or} \quad \lim_{(\Delta x, \Delta y) \to (0, 0)} f(x_0 + \Delta x, y_0 + \Delta y) = f(x_0, y_0).$$

If we let $x = x_0 + \Delta x$, $y = y_0 + \Delta y$, then the last result is equivalent to

$$\lim_{(x, y) \to (x_0, y_0)} f(x, y) = f(x_0, y_0).$$

By (5) of Section 13.2, f is continuous at (x_0, y_0). ∎

EXAMPLE 2 Differentiability

If (3) of Example 1 is written as

$$\Delta z = \overbrace{(2x - y)}^{f_x}\Delta x + \overbrace{(-x)}^{f_y}\Delta y + \overbrace{(\Delta x)}^{\varepsilon_1}(\Delta x) + \overbrace{(-\Delta x)}^{\varepsilon_2}\Delta y,$$

we can make the identifications $\varepsilon_1 = \Delta x$ and $\varepsilon_2 = -\Delta x$. Since $\varepsilon_1 \to 0$ and $\varepsilon_2 \to 0$ as $(\Delta x, \Delta y) \to (0, 0)$, the function $z = x^2 - xy$ is differentiable at every point in the xy-plane. ∎

As noted in Example 2 the given function is a polynomial. Any polynomial function of two or more variables is differentiable everywhere.

■ **Linearization** If $z = f(x, y)$ is differentiable at (x_0, y_0) and (x, y) is a point very close to (x_0, y_0), it follows from Definition 13.4.1 that since $\Delta x = x - x_0, \Delta y = y - y_0$ are both close to zero so are $\varepsilon_1 \Delta x$ and $\varepsilon_2 \Delta y$. In view of (4) this means

$$f(x_0 + \Delta x, y_0 + \Delta y) - f(x_0, y_0) \approx f_x(x_0, y_0)\Delta x + f_y(x_0, y_0)\Delta y.$$

Using $x = x_0 + \Delta x, y = y_0 + \Delta y$ the last line is the same as

$$f(x, y) \approx f(x_0, y_0) + f_x(x_0, y_0)\Delta x + f_y(x_0, y_0)\Delta y.$$

This prompts us to define the linearization of f at (x_0, y_0) in the following manner.

Definition 13.4.2 Linearization

If a function $z = f(x, y)$ is differentiable at a point (x_0, y_0), then the function

$$L(x, y) = f(x_0, y_0) + f_x(x_0, y_0)(x - x_0) + f_y(x_0, y_0)(y - y_0) \qquad (7)$$

is said to be a **linearization** of f at (x_0, y_0). For a point (x, y) near (x_0, y_0), the approximation

$$f(x, y) \approx L(x, y) \qquad (8)$$

is called a **local linear approximation** of f at (x_0, y_0).

EXAMPLE 3 Linearization

Find a linearization of $f(x, y) = \sqrt{x^2 + y^2}$ at $(4, 3)$.

Solution The first partial derivatives of f are

$$f_x(x, y) = \frac{x}{\sqrt{x^2 + y^2}} \qquad \text{and} \qquad f_y(x, y) = \frac{y}{\sqrt{x^2 + y^2}}.$$

Using the values $f(4, 3) = 5, f_x(4, 3) = \frac{4}{5}$, and $f_y(4, 3) = \frac{3}{5}$, it follows from (7) that a linearization of f at $(4, 3)$ is

$$L(x, y) = 5 + \frac{4}{5}(x - 4) + \frac{3}{5}(y - 3). \qquad (9)$$

The last equation is equivalent to $L(x, y) = \frac{4}{5}x + \frac{3}{5}y$ but for computational purposes (9) is more convenient. ∎

EXAMPLE 4 Local Linear Approximation

Use a local linear approximation to approximate $\sqrt{(4.01)^2 + (2.98)^2}$.

Solution First observe that we are asking for an approximation of the function value $f(4.01, 2.98)$, where $f(x, y) = \sqrt{x^2 + y^2}$. Because the point $(4.01, 2.98)$ is reasonably close to the point $(4, 3)$ we can use the linearization in (9) to form the local linear approximation $f(x, y) \approx L(x, y)$. From

$$L(4.01, 2.98) = 5 + \frac{4}{5}(4.01 - 4) + \frac{3}{5}(2.98 - 3) = 4.996$$

it follows that the desired approximation is $f(4.01, 2.98) \approx L(4.01, 2.98)$ or

$$\sqrt{(4.01)^2 + (2.98)^2} \approx 4.996. \qquad ∎$$

Suppose we let $z = L(x, y)$ and rewrite (7) as

$$f_x(x_0, y_0)(x - x_0) + f_y(x_0, y_0)(y - y_0) - (z - f(x_0, y_0)) = 0. \qquad (10)$$

Matching (10) term-by-term with (2) of Section 11.6 shows that a linearization of a function $z = f(x, y)$ at (x_0, y_0) is an equation of a plane.

▌ Tangent Plane   In Section 4.9 we saw that the linearization $L(x) = f(x_0) + f'(x_0)(x - x_0)$ of a function f of a single variable at a number x_0 is nothing more than an equation of the tangent line to the graph of $y = f(x)$ at $(x_0, f(x_0))$. In three-dimensions the analogue of a *tangent line* to a curve is a *tangent plane* to a surface. We will see in Section 13.7 that the linearization formula $z = L(x, y)$ in (7) is an equation of the tangent plane to the graph of $z = f(x, y)$ at the point $(x_0, y_0, f(x_0, y_0))$.

▌ Differentials   Recall too, for a function f of a single independent variable there are two differentials, $\Delta x = dx$ and $dy = f'(x)\,dx$. The differential dx is simply the change in the independent variable x. The differential dy is the change in the linearization $L(x)$; at a number x_0 we have

$$\begin{aligned}
\Delta L &= L(x_0 + \Delta x) - L(x_0) \\
&= [f(x_0) + f'(x_0)\Delta x] - [f(x_0) + f'(x_0) \cdot 0] \\
&= f'(x_0)\,dx = dy.
\end{aligned}$$

In the case of a function f of two independent variables we naturally have three differentials. The changes in the independent variables x and y are dx and dy; the change in the linearization $L(x, y)$ is denoted by dz. At a point (x_0, y_0) the change in the linearization is

$$\begin{aligned}
\Delta L &= L(x_0 + \Delta x, y_0 + \Delta y) - L(x_0, y_0) \\
&= f(x_0, y_0) + f_x(x_0, y_0)(x_0 + \Delta x - x_0) + f_y(x_0, y_0)(y_0 + \Delta y - y_0) - f(x_0, y_0) \\
&= f_x(x_0, y_0)\Delta x + f_y(x_0, y_0)\Delta y. \qquad (11)
\end{aligned}$$

Using the result in (11) we next define the differential dz of a function f at an arbitrary point in the xy-plane. If (x, y) denotes the point, then a nearby point is $(x + \Delta x, y + \Delta y)$ or $(x + dx, y + dy)$. The differential dz is commonly called the **total differential** of the function.

Definition 13.4.3   Differentials

Let $z = f(x, y)$ be a function for which the first partial derivatives f_x and f_y exist. Then the **differentials of x and y** are $dx = \Delta x$ and $dy = \Delta y$. The **differential of z**,

$$dz = f_x(x, y)\,dx + f_y(x, y)\,dy = \frac{\partial z}{\partial x}\,dx + \frac{\partial z}{\partial y}\,dy, \qquad (12)$$

is also called the **total differential of z**.

EXAMPLE 5   Total Differential

If $z = x^2 - xy$, then

$$\frac{\partial z}{\partial x} = 2x - y \quad \text{and} \quad \frac{\partial z}{\partial y} = -x.$$

From (12) the total differential of the function is

$$dz = (2x - y)\,dx - x\,dy. \qquad ∎$$

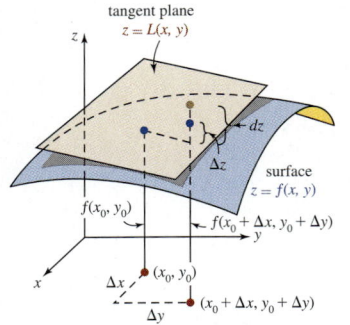

FIGURE 13.4.3 Geometric interpretations of dx, dy, Δz, and dz

It follows immediately from (4) of Theorem 13.4.1 that when f_x and f_y are continuous and when Δx and Δy are close to 0, then dz is an approximation for Δz, that is

$$dz \approx \Delta z. \tag{13}$$

FIGURE 13.4.3 is the three-dimensional version of Figure 4.9.4. The points in blue are the same points as shown in Figure 13.4.1 and are on the surface. The plane is tangent to the surface at $(x_0, y_0, f(x_0, y_0))$ and the point marked in brown is a point on the tangent plane.

EXAMPLE 6 Comparison of Δz and dz

In Example 1 we saw that the function $z = x^2 - xy$ changed by the exact amount $\Delta z = 0.6$ when we moved from the point $(1, 1)$ to $(1.2, 0.7)$. With the identifications $x = 1$, $y = 1$, $dx = 0.2$, and $dy = -0.3$, we see from (12) and (13) and the result in Example 5 that the change Δz of the function can be approximated by the change in the linearization:

$$dz = (1)(0.2) - (1)(-0.3) = 0.5. \qquad \blacksquare$$

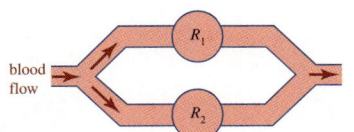

FIGURE 13.4.4 Blood flow through two resistances in Example 7

EXAMPLE 7 An Approximation of an Error

The human cardiovascular system is similar to electrical series and parallel circuits. For example, when blood flows through two resistances in parallel, as shown in **FIGURE 13.4.4**, then the equivalent resistance R of the network is

$$\frac{1}{R} = \frac{1}{R_1} + \frac{1}{R_2} \qquad \text{or} \qquad R = \frac{R_1 R_2}{R_1 + R_2}.$$

If the percentage errors in measuring R_1 and R_2 are $\pm 0.2\%$ and $\pm 0.6\%$, respectively, find the approximate maximum percentage error in R.

Solution We have $\Delta R_1 = \pm 0.002 R_1$ and $\Delta R_2 = \pm 0.006 R_2$. Now,

$$dR = \frac{R_2^2}{(R_1 + R_2)^2} dR_1 + \frac{R_1^2}{(R_1 + R_2)^2} dR_2,$$

and so

$$|\Delta R| \approx |dR| \leq \left| \frac{R_2^2}{(R_1 + R_2)^2}(\pm 0.002 R_1) \right| + \left| \frac{R_1^2}{(R_1 + R_2)^2}(\pm 0.006 R_2) \right|$$

$$= R \left[\frac{0.002 R_2}{R_1 + R_2} + \frac{0.006 R_1}{R_1 + R_2} \right]$$

$$\leq R \left[\frac{0.006 R_2}{R_1 + R_2} + \frac{0.006 R_1}{R_1 + R_2} \right] = (0.006)R.$$

Thus the maximum relative error is given by the approximation $|dR|/R \approx 0.006$; therefore, the maximum percentage error is approximately 0.6%. $\qquad \blacksquare$

■ **Functions of Three Variables** Definitions 13.4.1, 13.4.2, and 13.4.3 and Theorems 13.4.1, 13.4.2, and 13.4.3 generalize in the expected manner to functions of three or more variables. Here are some important points. If $w = f(x, y, z)$, then the **increment** Δw is given by

$$\Delta w = f(x + \Delta x, y + \Delta y, z + \Delta z) - f(x, y, z). \tag{14}$$

Then f is **differentiable** at a point (x_0, y_0, z_0) if Δw can be written

$$\Delta w = f_x \Delta x + f_y \Delta y + f_z \Delta z + \varepsilon_1 \Delta x + \varepsilon_2 \Delta y + \varepsilon_3 \Delta z, \tag{15}$$

where ε_1, ε_2, and $\varepsilon_3 \to 0$ as Δx, Δy, and $\Delta z \to 0$. If f is differentiable at (x_0, y_0, z_0), then the **linearization** of f is defined to be

$$L(x, y, z) = f(x_0, y_0, z_0) + f_x(x_0, y_0, z_0)(x - x_0) + f_y(x_0, y_0, z_0)(y - y_0) + f_z(x_0, y_0, z_0)(z - z_0). \tag{16}$$

Finally, the **total differential** of f is

$$dw = \frac{\partial w}{\partial x}\,dx + \frac{\partial w}{\partial y}\,dy + \frac{\partial w}{\partial z}\,dz. \tag{17}$$

EXAMPLE 8 Total Differential—Function of Three Variables

If $w = x^2 + 2y^3 + 3z^4$, then the three first partial derivatives are

$$\frac{\partial w}{\partial x} = 2x, \quad \frac{\partial w}{\partial y} = 6y^2, \quad \text{and} \quad \frac{\partial w}{\partial z} = 12z^3.$$

By (17) the total differential is

$$dw = 2x\,dx + 6y^2\,dy + 12z^3\,dz. \qquad \blacksquare$$

$\dfrac{\partial z}{\partial x}$ NOTES FROM THE CLASSROOM

(i) Since $dy \approx \Delta y$ whenever $f'(x)$ exists and Δx is close to 0, it seems reasonable to expect that $dz = f_x(x, y)\Delta x + f_y(x, y)\Delta y$ will be a good approximation to Δz when Δx and Δy are both close to 0. But life is not so simple for functions of several variables. The guarantee that $dz \approx \Delta z$ for increments close to 0 comes from the continuity of and not simply the existence of the partial derivatives $f_x(x, y)$ and $f_y(x, y)$.

(ii) When you work Problems 27–30 in Exercises 13.4 you will discover that the functions ε_1 and ε_2 introduced in (4) of Theorem 13.4.1 are not unique.

Exercises 13.4 Answers to selected odd-numbered problems begin on page ANS-41.

≡ Fundamentals

In Problems 1–6, find a linearization of the given function at the indicated point.

1. $f(x, y) = 4xy^2 - 2x^3y;\quad (1, 1)$

2. $f(x, y) = \sqrt{x^3y};\quad (2, 2)$

3. $f(x, y) = x\sqrt{x^2 + y^2};\quad (8, 15)$

4. $f(x, y) = 3\sin x\cos y;\quad (\pi/4, 3\pi/4)$

5. $f(x, y) = \ln(x^2 + y^3);\quad (-1, 1)$

6. $f(x, y) = e^{-2y}\sin 3x;\quad (0, \pi/3)$

In Problems 7–10, use a local linear approximation to approximate the given quantity.

7. $\sqrt{102} + \sqrt[4]{80}$

8. $\sqrt{\dfrac{35}{63}}$

9. $f(1.95, 2.01)$ for $f(x, y) = (x^2 + y^2)^2$

10. $f(0.52, 2.96)$ for $f(x, y) = \cos \pi xy$

In Problems 11–22, find the total differential of the given function.

11. $z = x^2\sin 4y$

12. $z = xe^{x^2 - y^2}$

13. $z = \sqrt{2x^2 - 4y^3}$

14. $z = (5x^3y + 4y^5)^3$

15. $f(s, t) = \dfrac{2s - t}{s + 3t}$

16. $g(r, \theta) = r^2\cos 3\theta$

17. $w = x^2y^4z^{-5}$

18. $w = e^{-z^2}\cos(x^2 + y^4)$

19. $F(r, s, t) = r^3 + s^{-2} - 4t^{1/2}$

20. $G(\rho, \theta, \phi) = \rho\sin\phi\cos\theta$

21. $w = \ln\left(\dfrac{uv}{st}\right)$

22. $w = \sqrt{u^2 + s^2t^2 - v^2}$

In Problems 23–26, compare the values of Δz and dz for the given function as (x, y) varies from the first to the second point.

23. $z = 3x + 4y + 8;\quad (2, 4), (2.2, 3.9)$

24. $z = 2x^2y + 5y + 8;\quad (0, 0), (0.2, -0.1)$

25. $z = (x + y)^2;\quad (3, 1), (3.1, 0.8)$

26. $z = x^2 + x^2y^2 + 2;\quad (1, 1), (0.9, 1.1)$

In Problems 27–30, find functions ε_1 and ε_2 from Δz as defined in (4) of Theorem 13.4.1.

27. $z = 5x^2 + 3y - xy$

28. $z = 10y^2 + 3x - x^2$

29. $z = x^2y^2$

30. $z = x^3 - y^3$

≡ Applications

31. When blood flows through three resistances R_1, R_2, R_3 in parallel, the equivalent resistance R of the network is

$$\frac{1}{R} = \frac{1}{R_1} + \frac{1}{R_2} + \frac{1}{R_3}.$$

Given that the percentage error in measuring each resistance is $\pm 0.9\%$, find the approximate maximum percentage error in R.

32. The pressure P of an enclosed ideal gas is given by $P = k(T/V)$, where V is volume, T is temperature, and k is a constant. Given that the percentage errors in measuring T and V are at most 0.6% and 0.8%, respectively, find the approximate maximum percentage error in P.

33. The tension T in the string of the yo-yo shown in FIGURE 13.4.5 is

$$T = mg\frac{R}{2r^2 + R^2},$$

where mg is its constant weight. Find the approximate change in the tension if R and r are increased from 4 cm and 0.8 cm to 4.1 cm and 0.9 cm, respectively. Does the tension increase or decrease?

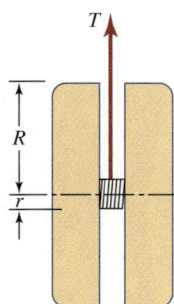

FIGURE 13.4.5 Yo-yo in Problem 33

34. Find the approximate increase in the volume of a right circular cylinder if its height is increased from 10 to 10.5 cm and its radius is increased from 5 to 5.3 cm. What is the approximate new volume?

35. If the length, width, and height of a closed rectangular box are increased by 2%, 5%, and 8%, respectively, what is the approximate percentage increase in volume?

36. In Problem 35 if the original length, width, and height are 3 ft, 1 ft, and 2 ft, respectively, what is the approximate increase in surface area of the box? What is the approximate new surface area?

37. The function $S = 0.1091w^{0.425}h^{0.725}$ gives the surface area of a person's body in terms of weight w and height h. If the error in the measurement of w is at most 3% and the error in the measurement of h is at most 5%, what is the approximate maximum percentage error in the measurement of S?

38. The impedance Z of the series circuit shown in FIGURE 13.4.6 is $Z = \sqrt{R^2 + X^2}$, where R is resistance, $X = 1000L - 1/(1000C)$ is net reactance, L is inductance, and C is capacitance. If the values of R, L, and C given in the figure are increased to 425 ohms, 0.45 henry, and 11.1×10^{-5} farad, respectively, what is the approximate change in the impedance of the circuit? What is the approximate new impedance?

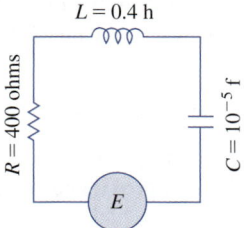

FIGURE 13.4.6 Series circuit in Problem 38

≡ Think About It

39. **(a)** Give a definition for the linearization of a function of three variables $w = f(x, y, z)$.
    (b) Use linearization to find an approximation for $\sqrt{(9.1)^2 + (11.75)^2 + (19.98)^2}$.

40. In Problem 67 in Exercises 13.3 we saw that for

$$f(x, y) = \begin{cases} \dfrac{xy}{2x^2 + 2y^2}, & (x, y) \neq (0, 0) \\ 0, & (x, y) = (0, 0) \end{cases}$$

both $\partial z/\partial x$ and $\partial z/\partial y$ exist at $(0, 0)$. Explain why f is not differentiable at $(0, 0)$.

41. **(a)** Give an intuitive explanation why $f(x, y) = \sqrt{x^2 + y^2}$ is not differentiable at $(0, 0)$.
    (b) Now prove that f is not differentiable at $(0, 0)$.

42. The length of the sides of the red rectangular box shown in FIGURE 13.4.7 are x, y, and z. Let the volume of the red box be V. When the sides of the box are increased by the amounts $\Delta x, \Delta y$, and Δz we obtain the rectangular box shown in the figure that is outlined in blue. Draw or trace Figure 13.4.7 on a piece of paper. Identify by different colors the quantities $\Delta x, \Delta y, \Delta z, \Delta V, dV$, and $\Delta V - dV$.

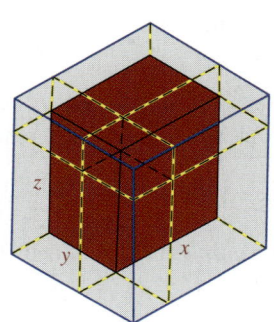

FIGURE 13.4.7 Box in Problem 42

≡ Projects

43. **Robotic Arm** A two-dimensional robot arm whose shoulder is fixed at the origin keeps track of its position by means of a shoulder angle θ and an elbow angle ϕ as shown in FIGURE 13.4.8. The shoulder angle is measured counterclockwise from the x-axis, and the elbow angle is measured counterclockwise from the upper arm to the lower arm, which are of length L and l, respectively.

    (a) The location of the elbow joint is given by (x_e, y_e), where

$$x_e = L\cos\theta, \quad y_e = L\sin\theta.$$

    Find corresponding formulas for the location (x_h, y_h) of the hand.

    (b) Show that the total differentials of x_h and y_h can be written as

$$dx_h = -y_h\, d\theta + (y_e - y_h)\, d\phi$$
$$dy_h = x_h\, d\theta + (x_e - x_h)\, d\phi.$$

    (c) Suppose that $L = l$ and that the arm is to be positioned so as to reach the point (L, L). Suppose also that the error in measuring each of the angles θ and ϕ is at most $\pm 1°$. Find the approximate maximum error in the x-coordinate of the hand's location for each of the two possible positions.

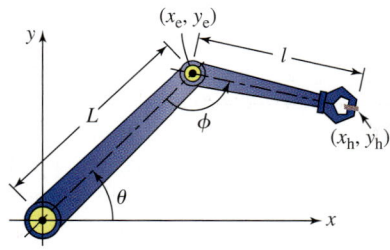

FIGURE 13.4.8 Robotic arm in Problem 43

44. Projectile Motion A projectile is fired at an angle θ with velocity v across a chasm of width D toward a vertical cliff wall that is essentially infinite in both height and depth. See **FIGURE 13.4.9**.

(a) If the projectile is subject only to the force of gravity, show that the height H at which the projectile strikes the cliff wall as a function of the variables v and θ is given by

$$H = D \tan\theta - \frac{1}{2}g\frac{D^2}{v^2}\sec^2\theta.$$

[*Hint*: See Section 10.2.]

(b) Find the total differential of H.
(c) Suppose that $D = 100$ ft, $g = 32$ ft/s^2, $v = 100$ ft/s, and $\theta = 45°$. Find H.
(d) Suppose, for the data in part (c), that the error in measuring v is at most ± 1 ft/s and that the error in measuring θ is at most $\pm 1°$. Find the approximate maximum error in H.
(e) By allowing D to vary, H can also be considered a function of three variables. Find the total differential of H. Using the data from parts (c) and (d) and assuming that the error in measuring D is at most ± 2 ft/s, find the approximate maximum error in H.

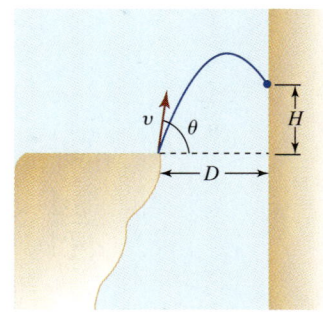

FIGURE 13.4.9 Chasm in Problem 44

13.5 Chain Rule

❚ **Introduction** The Chain Rule for functions of a single variable states that if $y = f(x)$ is a differentiable function of x, and $x = g(t)$ is a differentiable function of t, then the derivative of the composite function is

$$\frac{dy}{dt} = \frac{dy}{dx}\frac{dx}{dt}.$$

In this section we extend the Chain Rule to functions of several variables.

❚ **Chain Rule for Ordinary Derivatives** If $z = f(x, y)$ and x and y are functions of a single variable t, then the next theorem indicates how to compute the ordinary derivative dz/dt.

Theorem 13.5.1 Chain Rule

Suppose $z = f(x, y)$ is differentiable at (x, y) and $x = g(t)$ and $y = h(t)$ are differentiable functions at t. Then $z = f(g(t), h(t))$ is a differentiable function of t and

$$\frac{dz}{dt} = \frac{\partial z}{\partial x}\frac{dx}{dt} + \frac{\partial z}{\partial y}\frac{dy}{dt}. \qquad (1)$$

EXAMPLE 1 Chain Rule

If $z = x^3y - y^4$ and $x = 2t^2$, $y = 5t^2 - 6t$, find dz/dt at $t = 1$.

Solution From (1),

$$\frac{dz}{dt} = \frac{\partial z}{\partial x}\frac{dx}{dt} + \frac{\partial z}{\partial y}\frac{dy}{dt}$$
$$= (3x^2y)(4t) + (x^3 - 4y^3)(10t - 6).$$

Now, at $t = 1$, $x(1) = 2$ and $y(1) = -1$ so

$$\left.\frac{dz}{dt}\right|_{t=1} = (3 \cdot 4 \cdot (-1)) \cdot 4 + (8 + 4) \cdot 4 = 0. \qquad ■$$

Although there is no need to do it, we can also find the derivative dz/dt in Example 1 by substituting the functions $x = 2t^2$, $y = 5t^2 - 6t$ into $z = x^3y - y^4$ and then differentiating the resulting function of a single variable $z = 8t^6(5t^2 - 6t) - (5t^2 - 6t)^4$ with respect to t.

EXAMPLE 2 Related Rates

In Example 3 of Section 13.3 we saw that the function $S(w, h) = 0.1091w^{0.425}h^{0.725}$ relates the surface area (in square feet) of a person's body as a function of weight w (in pounds) and height h (in inches). Find the rate at which S changes when $dw/dt = 10$ lb/yr, $dh/dt = 2.3$ in/yr, $w = 100$ lb, and $h = 60$ in.

Solution With the symbols w and h playing the parts of x and y it follows from (1) the rate of change of S with respect to time t is

$$\frac{dS}{dt} = \frac{\partial S}{\partial w}\frac{dw}{dt} + \frac{\partial S}{\partial h}\frac{dh}{dt}$$

$$= (0.1091)(0.425)w^{-0.575}h^{0.725}\frac{dw}{dt} + (0.1091)(0.725)w^{0.425}h^{-0.275}\frac{dh}{dt}.$$

When $dw/dt = 10$, $dh/dt = 2.3$, $w = 100$, and $h = 60$ the value of the derivative is

$$\frac{dS}{dt}\bigg|_{(100,\,60)} = (0.1091)(0.425)(100)^{-0.575}(60)^{0.725} \cdot (10) + (0.1091)(0.725)(100)^{0.425}(60)^{-0.275} \cdot (2.3)$$

$$\approx 1.057.$$

Because $dS/dt > 0$ the person's surface is increasing at a rate of approximately 1.057 ft^2 per year. ∎

■ **Chain Rule for Partial Derivatives** For a composite function of two variables $z = f(x, y)$, where $x = g(u, v)$ and $y = h(u, v)$, we would naturally expect *two* formulas analogous to (1), since $z = f(g(u, v), h(u, v))$ and so we can compute both $\partial z/\partial u$ and $\partial z/\partial v$. The Chain Rule for functions of two variables is summarized in the next theorem.

Theorem 13.5.2   Chain Rule

If $z = f(x, y)$ is differentiable and $x = g(u, v)$ and $y = h(u, v)$ have continuous first partial derivatives, then

$$\frac{\partial z}{\partial u} = \frac{\partial z}{\partial x}\frac{\partial x}{\partial u} + \frac{\partial z}{\partial y}\frac{\partial y}{\partial u} \qquad \text{and} \qquad \frac{\partial z}{\partial v} = \frac{\partial z}{\partial x}\frac{\partial x}{\partial v} + \frac{\partial z}{\partial y}\frac{\partial y}{\partial v}. \qquad (2)$$

PROOF We prove the second of the results in (2). If $\Delta u = 0$, then

$$\Delta z = f(g(u, v + \Delta v), h(u, v + \Delta v)) - f(g(u, v), h(u, v))$$

Now, if

$$\Delta x = g(u, v + \Delta v) - g(u, v) \qquad \text{and} \qquad \Delta y = h(u, v + \Delta v) - h(u, v),$$

then

$$g(u, v + \Delta v) = x + \Delta x \qquad \text{and} \qquad h(u, v + \Delta v) = y + \Delta y.$$

Hence, Δz can be written as

$$\Delta z = f(x + \Delta x, y + \Delta y) - f(x, y).$$

Since f is differentiable, it follows from the increment formula (4) of Section 13.4 that Δz can be written

$$\Delta z = \frac{\partial z}{\partial x}\Delta x + \frac{\partial z}{\partial y}\Delta y + \varepsilon_1\Delta x + \varepsilon_2\Delta y,$$

where, recall, ε_1 and ε_2 are functions of Δx and Δy with the property that $\lim\limits_{(\Delta u, \Delta v) \to (0, 0)} \varepsilon_1 = 0$ and $\lim\limits_{(\Delta u, \Delta v) \to (0, 0)} \varepsilon_2 = 0$. Since ε_1 and ε_2 are not uniquely defined functions, a pair of functions can always be found for which $\varepsilon_1(0, 0) = 0$, $\varepsilon_2(0, 0) = 0$. Hence, ε_1 and ε_2 are continuous at $(0, 0)$. Therefore,

$$\frac{\Delta z}{\Delta v} = \frac{\partial z}{\partial x}\frac{\Delta x}{\Delta v} + \frac{\partial z}{\partial y}\frac{\Delta y}{\Delta v} + \varepsilon_1\frac{\Delta x}{\Delta v} + \varepsilon_2\frac{\Delta y}{\Delta v}.$$

Now, taking the limit of the last line as $\Delta v \to 0$ gives

$$\frac{\partial z}{\partial v} = \frac{\partial z}{\partial x}\frac{\partial x}{\partial v} + \frac{\partial z}{\partial y}\frac{\partial y}{\partial v} + 0 \cdot \frac{\partial x}{\partial v} + 0 \cdot \frac{\partial y}{\partial v} = \frac{\partial z}{\partial x}\frac{\partial x}{\partial v} + \frac{\partial z}{\partial y}\frac{\partial y}{\partial v}$$

since Δx and Δy both approach zero as $\Delta v \to 0$. ∎

EXAMPLE 3 Chain Rule

If $z = x^2 - y^3$ and $x = e^{2u - 3v}$, $y = \sin(u^2 - v^2)$, find $\partial z/\partial u$ and $\partial z/\partial v$.

Solution Because

$$\frac{\partial z}{\partial x} = 2x, \qquad\qquad \frac{\partial z}{\partial y} = -3y^2,$$

$$\frac{\partial x}{\partial u} = 2e^{2u - 3v}, \qquad \frac{\partial y}{\partial u} = 2u\cos(u^2 - v^2),$$

$$\frac{\partial x}{\partial v} = -3e^{2u - 3v}, \qquad \frac{\partial y}{\partial v} = -2v\cos(u^2 - v^2),$$

we see from (2) that $\partial z/\partial u$ and $\partial z/\partial v$ are, in turn,

$$\frac{\partial z}{\partial u} = 2x(2e^{2u - 3v}) - 3y^2[2u\cos(u^2 - v^2)] = 4xe^{2u-3v} - 6uy^2\cos(u^2 - v^2)$$

$$\frac{\partial z}{\partial v} = 2x(-3e^{2u-3v}) - 3y^2[-2v\cos(u^2 - v^2)] = -6xe^{2u-3v} + 6vy^2\cos(u^2 - v^2). \quad ∎$$

Of course, as in Example 1, we could substitute the expressions for x and y in the original function and then find the partial derivatives $\partial z/\partial u$ and $\partial z/\partial v$ directly. But there is no particular advantage gained by doing this.

❚ Generalizations The results given in (1) and (2) immediately generalize to any number of variables. If $z = f(x_1, x_2, \ldots, x_n)$ is differentiable at $(x_1, x_2, \ldots, x_n)$, and if the x_i, $i = 1, \ldots, n$, are differentiable functions of a single variable t, then (1) of Theorem 13.5.1 becomes

$$\frac{dz}{dt} = \frac{\partial z}{\partial x_1}\frac{dx_1}{dt} + \frac{\partial z}{\partial x_2}\frac{dx_2}{dt} + \cdots + \frac{\partial z}{\partial x_n}\frac{dx_n}{dt}. \tag{3}$$

Similarly, if $z = f(x_1, x_2, \ldots, x_n)$ and each of the variables $x_1, x_2, x_3, \ldots, x_n$ are functions of k variables $u_1, u_2, u_3, \ldots, u_k$, then under the same assumptions as in Theorem 13.5.2, we have

$$\frac{\partial z}{\partial u_i} = \frac{\partial z}{\partial x_1}\frac{\partial x_1}{\partial u_i} + \frac{\partial z}{\partial x_2}\frac{\partial x_2}{\partial u_i} + \cdots + \frac{\partial z}{\partial x_n}\frac{\partial x_n}{\partial u_i}, \tag{4}$$

where $i = 1, 2, \ldots, k$.

❚ Tree Diagrams The results in (1) and (2) can be memorized in terms of **tree diagrams**. The dots in FIGURE 13.5.1(a) indicate that z depends on x and y; x and y depend, in turn, on u and v. To compute $\partial z/\partial u$ for example, we read the diagram vertically downward starting from z and following the two blue paths leading to x and y. We then follow the blue paths leading to u, multiply the partial derivatives on each path, and then add the products. To compute $\partial z/\partial v$ we start on the two blue paths but then branch at x and y to the red paths to get to v, multiply the partial derivatives on each segment, and then add the products. The result in (1) is represented by the tree diagram in Figure 13.5.1(b). There is only one branch stemming from x and from y since these variables depend only on a single variable t.

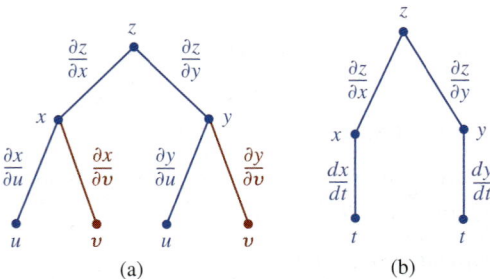

FIGURE 13.5.1 Tree diagrams: (a) for (2) and (b) for (1)

We use tree diagrams in the next three examples to illustrate special cases of (3) and (4).

EXAMPLE 4 Chain Rule

If $r = x^2 + y^5 z^3$ and $x = uve^{2s}$, $y = u^2 - v^2 s$, $z = \sin(uvs^2)$, find **(a)** $\partial r/\partial u$ and **(b)** $\partial r/\partial s$.

Solution Here r is a function of three variables x, y, and z, and each of these variables is itself a function of three variables u, v, and s. To construct a tree diagram we draw three blue paths from r to three points labeled x, y, and z. Then since x, y, and z depend on three variables, we draw three paths (blue, red, and green) stemming from the points x, y, and z to the points u, v, and s. On each of these twelve segments we indicate the appropriate partial derivative. See FIGURE 13.5.2. To compute $\partial r/\partial u$ we follow the three blue polygonal paths starting at r all the way down to u in the tree diagram. We form the products of the partial derivatives indicated on each segment of the three blue polygonal paths to u and add:

$$\frac{\partial r}{\partial u} = \frac{\partial r}{\partial x}\frac{\partial x}{\partial u} + \frac{\partial r}{\partial y}\frac{\partial y}{\partial u} + \frac{\partial r}{\partial z}\frac{\partial z}{\partial u}$$
$$= 2x(ve^{2s}) + 5y^4 z^3(2u) + 3y^5 z^2(vs^2 \cos(uvs^2)).$$

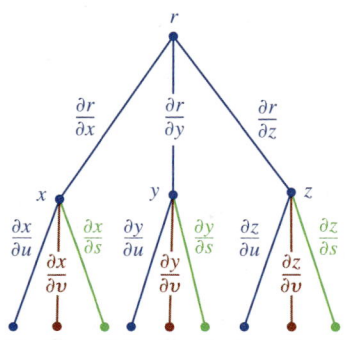

FIGURE 13.5.2 Tree diagram for Example 4

Now to compute $\partial r/\partial s$ we start from r on the three blue polygonal paths in Figure 13.5.2 and then branch to the green paths at x, y, and z to get to s. Adding the products of the partial derivative on each segment of the three polygonal paths leading to s gives

$$\frac{\partial r}{\partial s} = \frac{\partial r}{\partial x}\frac{\partial x}{\partial s} + \frac{\partial r}{\partial y}\frac{\partial y}{\partial s} + \frac{\partial r}{\partial z}\frac{\partial z}{\partial s}$$
$$= 2x(2uve^{2s}) + 5y^4 z^3(-v^2) + 3y^5 z^2(2uvs \cos(uvs^2)). \quad \blacksquare$$

EXAMPLE 5 Chain Rule

Suppose $w = f(x, y, z)$ is a differentiable function of x, y, and z and $x = g(u, v)$, $y = h(u, v)$, and $z = k(u, v)$ are differentiable functions of u and v. Construct a tree diagram to compute $\partial w/\partial u$ and $\partial w/\partial v$.

Solution Since f is a function of three variables x, y, and z and these are functions of two variables u and v, the tree diagram is as shown in FIGURE 13.5.3. The partial derivatives are then

$$\frac{\partial w}{\partial u} = \frac{\partial w}{\partial x}\frac{\partial x}{\partial u} + \frac{\partial w}{\partial y}\frac{\partial y}{\partial u} + \frac{\partial w}{\partial z}\frac{\partial z}{\partial u}$$

$$\frac{\partial w}{\partial v} = \frac{\partial w}{\partial x}\frac{\partial x}{\partial v} + \frac{\partial w}{\partial y}\frac{\partial y}{\partial v} + \frac{\partial w}{\partial z}\frac{\partial z}{\partial v}. \quad \blacksquare$$

FIGURE 13.5.3 Tree diagram for Example 5

EXAMPLE 6 Chain Rule

If $z = u^2v^3w^4$ and $u = t^2$, $v = 5t - 8$, $w = t^3 + t$, find dz/dt.

Solution In this case the tree diagram in FIGURE 13.5.4 indicates that

$$\frac{dz}{dt} = \frac{\partial z}{\partial u}\frac{du}{dt} + \frac{\partial z}{\partial v}\frac{dv}{dt} + \frac{\partial z}{\partial w}\frac{dw}{dt}$$

$$= 2uv^3w^4(2t) + 3u^2v^2w^4(5) + 4u^2v^3w^3(3t^2 + 1). \quad \blacksquare$$

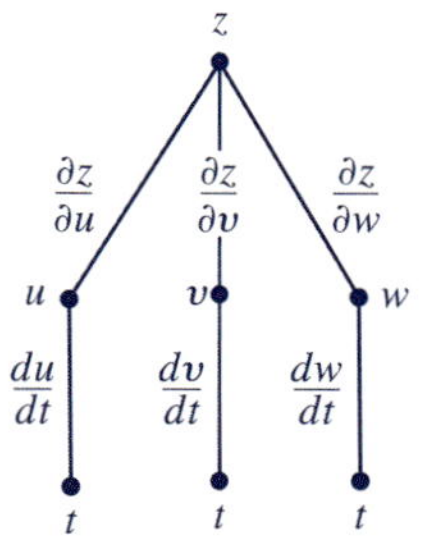

FIGURE 13.5.4 Tree diagram for Example 6

■ **Implicit Differentiation** If the equation $F(x, y) = 0$ defines a function $y = f(x)$ implicitly, then $F(x, f(x)) = 0$ for all x in the domain of f. Recall from Section 3.6 that we found the derivative dy/dx by a process called *implicit differentiation*. The derivative dy/dx can also be determined from the Chain Rule. If we assume $w = F(x, y)$ and $y = f(x)$ are differentiable functions, then from (1) we have

$$\frac{dw}{dx} = F_x(x, y)\frac{dx}{dx} + F_y(x, y)\frac{dy}{dx}. \quad (5)$$

Since $w = F(x, y) = 0$ and $dx/dx = 1$, (5) implies

$$F_x(x, y) + F_y(x, y)\frac{dy}{dx} = 0 \quad \text{or} \quad \frac{dy}{dx} = -\frac{F_x(x, y)}{F_y(x, y)},$$

provided $F_y(x, y) \neq 0$.

Moreover, if $F(x, y, z) = 0$ implicitly defines a function $z = f(x, y)$, then $F(x, y, f(x, y)) = 0$ for all (x, y) in the domain of f. If $w = F(x, y, z)$ is a differentiable function and $z = f(x, y)$ is differentiable in x and y, then (3) yields

$$\frac{\partial w}{\partial x} = F_x(x, y, z)\frac{\partial x}{\partial x} + F_y(x, y, z)\frac{\partial y}{\partial x} + F_z(x, y, z)\frac{\partial z}{\partial x}. \quad (6)$$

Since $w = F(x, y, z) = 0$, $\partial x/\partial x = 1$, and $\partial y/\partial x = 0$, (6) gives

$$F_x(x, y, z) + F_z(x, y, z)\frac{\partial z}{\partial x} = 0 \quad \text{or} \quad \frac{\partial z}{\partial x} = -\frac{F_x(x, y, z)}{F_z(x, y, z)},$$

provided $F_z(x, y, z) \neq 0$. The partial derivative $\partial z/\partial y$ can be obtained in a similar manner. We summarize these results in the following theorem.

Theorem 13.5.3 Implicit Differentiation

(*i*) If $w = F(x, y)$ is differentiable and $y = f(x)$ is a differentiable function of x defined implicitly by $F(x, y) = 0$, then

$$\frac{dy}{dx} = -\frac{F_x(x, y)}{F_y(x, y)}, \quad (7)$$

where $F_y(x, y) \neq 0$.

(*ii*) If $w = F(x, y, z)$ is differentiable and $z = f(x, y)$ is a differentiable function of x and y defined implicitly by $F(x, y, z) = 0$, then

$$\frac{\partial z}{\partial x} = -\frac{F_x(x, y, z)}{F_z(x, y, z)} \quad \text{and} \quad \frac{\partial z}{\partial y} = -\frac{F_y(x, y, z)}{F_z(x, y, z)}, \quad (8)$$

where $F_z(x, y, z) \neq 0$.

EXAMPLE 7 Implicit Differentiation

(**a**) Find dy/dx if $x^2 - 4xy - 3y^2 = 10$.
(**b**) Find $\partial z/\partial y$ if $x^2y - 5xy^2 = 2yz - 4z^3$.

Solution

(a) Let $F(x, y) = x^2 - 4xy - 3y^2 - 10$. Then y is defined as a function of x by $F(x, y) = 0$. Now $F_x = 2x - 4y$ and $F_y = -4x - 6y$, and so by (7) of Theorem 13.5.3 we have

$$\frac{dy}{dx} = -\frac{F_x(x, y)}{F_y(x, y)} = -\frac{2x - 4y}{-4x - 6y} = \frac{x - 2y}{2x + 3y}.$$

You are encouraged to verify this result by the procedure of Section 3.6.

(b) Let $F(x, y, z) = x^2y - 5xy^2 - 2yz + 4z^3$. Then z is defined as a function of x and y by $F(x, y, z) = 0$. Since $F_y = x^2 - 10xy - 2z$ and $F_z = -2y + 12z^2$, it follows from (8) in Theorem 13.5.3 that

$$\frac{\partial z}{\partial y} = -\frac{F_y(x, y, z)}{F_z(x, y, z)} = -\frac{x^2 - 10xy - 2z}{-2y + 12z^2} = \frac{x^2 - 10xy - 2z}{2y - 12z^2}. \qquad \blacksquare$$

Exercises 13.5 Answers to selected odd-numbered problems begin on page ANS-41.

≡ **Fundamentals**

In Problems 1–6, find the indicated derivative.

1. $z = \ln(x^2 + y^2)$; $x = t^2, y = t^{-2}$; $\dfrac{dz}{dt}$

2. $z = x^3y - xy^4$; $x = e^{5t}, y = \sec 5t$; $\dfrac{dz}{dt}$

3. $z = \cos(3x + 4y)$; $x = 2t + \dfrac{\pi}{2}, y = -t - \dfrac{\pi}{4}$; $\dfrac{dz}{dt}\Big|_{t=\pi}$

4. $z = e^{xy}$; $x = \dfrac{4}{2t + 1}, y = 3t + 5$; $\dfrac{dz}{dt}\Big|_{t=0}$

5. $p = \dfrac{r}{2s + t}$; $r = u^2, s = \dfrac{1}{u^2}, t = \sqrt{u}$; $\dfrac{dp}{du}$

6. $r = \dfrac{xy^2}{z^3}$; $x = \cos s, y = \sin s, z = \tan s$; $\dfrac{dr}{ds}$

In Problems 7–16, find the indicated partial derivatives.

7. $z = e^{xy^2}$; $x = u^3, y = u - v^2$; $\dfrac{\partial z}{\partial u}, \dfrac{\partial z}{\partial v}$

8. $z = x^2 \cos 4y$; $x = u^2v^3, y = u^3 + v^3$; $\dfrac{\partial z}{\partial u}, \dfrac{\partial z}{\partial v}$

9. $z = 4x - 5y^2$; $x = u^4 - 8v^3, y = (2u - v)^2$; $\dfrac{\partial z}{\partial u}, \dfrac{\partial z}{\partial v}$

10. $z = \dfrac{x - y}{x + y}$; $x = \dfrac{u}{v}, y = \dfrac{v^2}{u}$; $\dfrac{\partial z}{\partial u}, \dfrac{\partial z}{\partial v}$

11. $w = (u^2 + v^2)^{3/2}$; $u = e^{-t}\sin\theta, v = e^{-t}\cos\theta$; $\dfrac{\partial w}{\partial t}, \dfrac{\partial w}{\partial \theta}$

12. $w = \tan^{-1}\sqrt{uv}$; $u = r^2 - s^2, v = r^2s^2$; $\dfrac{\partial w}{\partial r}, \dfrac{\partial w}{\partial s}$

13. $R = rs^2t^4$; $r = ue^{v^2}, s = ve^{-u^2}, t = e^{u^2v^2}$; $\dfrac{\partial R}{\partial u}, \dfrac{\partial R}{\partial v}$

14. $Q = \ln(pqr)$; $p = t^2\sin^{-1}x, q = \dfrac{x}{t^2}, r = \tan^{-1}\dfrac{x}{t}$; $\dfrac{\partial Q}{\partial x}, \dfrac{\partial Q}{\partial t}$

15. $w = \sqrt{x^2 + y^2}$; $x = \ln(rs + tu)$, $y = \dfrac{t}{u}\cosh rs$; $\dfrac{\partial w}{\partial t}, \dfrac{\partial w}{\partial r}, \dfrac{\partial w}{\partial u}$

16. $s = p^2 + q^2 - r^2 + 4t$; $p = \phi e^{3\theta}, q = \cos(\phi + \theta)$, $r = \phi\theta^2, t = 2\phi + 8\theta$; $\dfrac{\partial s}{\partial \phi}, \dfrac{\partial s}{\partial \theta}$

In Problems 17–20, find dy/dx by two methods:
(a) implicit differentiation and
(b) Theorem 13.5.3(i).

17. $x^3 - 2x^2y^2 + y = 1$ **18.** $x + 2y^2 = e^y$

19. $y = \sin xy$ **20.** $(x + y)^{2/3} = xy$

In Problems 21–24, use Theorem 13.5.3(ii) to find $\partial z/\partial x$ and $\partial z/\partial y$.

21. $x^2 + y^2 - z^2 = 1$ **22.** $x^{2/3} + y^{2/3} + z^{2/3} = a^{2/3}$

23. $xy^2z^3 + x^2 - y^2 = 5z^2$ **24.** $z = \ln(xyz)$

25. If F and G have second partial derivatives, show that $u(x, t) = F(x + at) + G(x - at)$ satisfies the **wave equation**

$$a^2\frac{\partial^2 u}{\partial x^2} = \frac{\partial^2 u}{\partial t^2}.$$

26. Let $\eta = x + at$ and $\xi = x - at$. Show that the wave equation in Problem 25 becomes

$$\frac{\partial^2 u}{\partial \eta \partial \xi} = 0,$$

where $u = f(\eta, \xi)$.

27. If $u = f(x, y)$ and $x = r\cos\theta$, $y = r\sin\theta$, show that **Laplace's equation** $\partial^2 u/\partial x^2 + \partial^2 u/\partial y^2 = 0$ becomes

$$\frac{\partial^2 u}{\partial r^2} + \frac{1}{r}\frac{\partial u}{\partial r} + \frac{1}{r^2}\frac{\partial^2 u}{\partial \theta^2} = 0.$$

28. If $z = f(u)$ is a differentiable function of one variable and $u = g(x, y)$ possesses first partial derivatives, then what are $\partial z/\partial x$ and $\partial z/\partial y$?

29. Use the result of Problem 28 to show that for any differentiable function f, $z = f(y/x)$ satisfies the partial differential equation $x\partial z/\partial x + y\partial z/\partial y = 0$.

30. If $u = f(r)$ and $r = \sqrt{x^2 + y^2}$, show that Laplace's equation $\partial^2 u/\partial x^2 + \partial^2 u/\partial y^2 = 0$ becomes

$$\frac{d^2 u}{dr^2} + \frac{1}{r}\frac{du}{dr} = 0.$$

31. The **error function** defined by $\text{erf}(x) = (2/\sqrt{\pi})\int_0^x e^{-v^2}dv$ is important in applied mathematics. Show that $u(x, t) = A + B\,\text{erf}(x/\sqrt{4kt})$, A and B constants, satisfies the one-dimensional diffusion equation

$$k\frac{\partial^2 u}{\partial x^2} = \frac{\partial u}{\partial t}.$$

≣ Applications

32. The voltage across a conductor is increasing at a rate of 2 volts/min and the resistance is decreasing at a rate of 1 ohm/min. Use $I = E/R$ and the Chain Rule to find the rate at which the current passing through the conductor is changing when $R = 50$ ohms and $E = 60$ volts.

33. The length of the side labeled x of the triangle in FIGURE 13.5.5 increases at a rate of 0.3 cm/s, the side labeled y increases at a rate of 0.5 cm/s, and the included angle θ increases at a rate of 0.1 rad/s. Use the Chain Rule to find the rate at which the area of the triangle is changing at the instant $x = 10$ cm, $y = 8$ cm, and $\theta = \pi/6$.

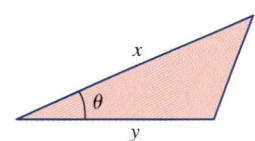

FIGURE 13.5.5 Triangle in Problem 33

34. Van der Waals equation of state for the real gas CO_2 is

$$P = \frac{0.08T}{V - 0.0427} - \frac{3.6}{V^2}.$$

If dT/dt and dV/dt are rates at which the temperature and volume change, respectively, use the Chain Rule to find dP/dt.

35. A very young child grows at a rate of 2 in/yr and gains weight at a rate of 4.2 lb/yr. Use $S = 0.1091w^{0.425}h^{0.725}$ and the Chain Rule to find the rate at which the surface area of the child is changing when the child weighs 25 lb and is 29 in. tall.

36. A particle moves in 3-space so that its coordinates at any time are $x = 4\cos t$, $y = 4\sin t$, $z = 5t$, $t \geq 0$. Use the Chain Rule to find the rate at which its distance

$$w = \sqrt{x^2 + y^2 + z^2}$$

from the origin is changing at $t = 5\pi/2$ seconds.

37. The equation of state for a thermodynamic system is $F(P, V, T) = 0$, where P, V, and T are pressure, volume, and temperature, respectively. If the equation defines V as a function of P and T, and also defines T as a function of V and P, show that

$$\frac{\partial V}{\partial T} = -\frac{\dfrac{\partial F}{\partial T}}{\dfrac{\partial F}{\partial V}} = -\frac{1}{\dfrac{\partial T}{\partial V}}.$$

38. Two coast guard ships (denoted by A and B in FIGURE 13.5.6), located a distance 500 yd apart, spot a suspect ship C at relative bearings θ and ϕ as shown in the figure.

(a) Use the law of sines to express the distance r from A to C in terms of θ and ϕ.

(b) How far is C from A when $\theta = 62°$ and $\phi = 75°$?

(c) Suppose that at the moment specified in part (b), the angle θ is increasing at the rate of 5° per minute, while ϕ is decreasing at the rate of 10° per minute. Is the distance from C to A increasing or decreasing? At what rate?

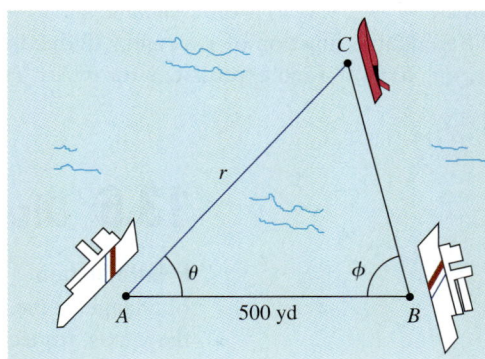

FIGURE 13.5.6 Ships in Problem 38

39. A **Helmholtz resonator** is any container with a neck and an opening (such as a jug or a beer bottle). When air is blown across the opening, the resonator produces a characteristic sound whose frequency, in cycles per second, is

$$f = \frac{c}{2\pi}\sqrt{\frac{A}{lV}},$$

where A is the cross-sectional area of the opening, l is the length of the neck, V is the volume of the container (not counting the neck), and c is the speed of sound (approximately 330 m/s). See FIGURE 13.5.7.

(a) What frequency sound will a bottle make if it has a circular opening 2 cm in diameter, a neck 6 cm long, and a volume of 100 cm³? [*Hint:* Be sure to convert c to cm/s.]

(b) Suppose the volume of the bottle in part (a) is decreasing at a rate of 10 cm³/s, while its neck is lengthening at the rate of 1 cm/s. At the instant specified in part (a) (that is, $V = 100$, $l = 6$) is the frequency increasing or decreasing?

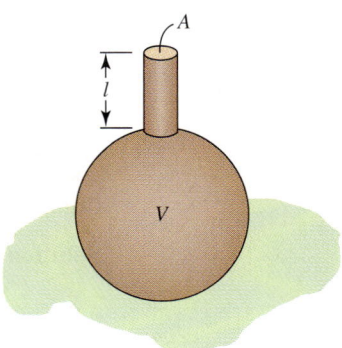

FIGURE 13.5.7 Container in Problem 39

≡ **Think About It**

40. (a) Suppose $w = F(x, y, z)$ and $y = g(x), z = h(x)$. Sketch an appropriate tree diagram and find an expression for dw/dx.

(b) Suppose $w = xy^2 - 2yz + x$ and $y = \ln x, z = e^x$. Use the Chain Rule to find dw/dx.

41. Suppose $z = F(u, v, w)$, where $u = F(t_1, t_2, t_3, t_4)$, $v = g(t_1, t_2, t_3, t_4)$, and $w = h(t_1, t_2, t_3, t_4)$. Sketch an appropriate tree diagram and find expressions for the partial derivatives $\partial z/\partial t_2$ and $\partial z/\partial t_4$.

42. Suppose $w = F(x, y, z, u)$ is differentiable and $u = f(x, y, z)$ is a differentiable function of x, y, and z defined implicitly by $f(x, y, z, u) = 0$. Find expressions for $\partial u/\partial x, \partial u/\partial y$, and $\partial u/\partial z$.

43. Use the results of Problem 42 to find $\partial u/\partial x, \partial u/\partial y$, and $\partial u/\partial z$ if u is a differentiable function of x, y, and z defined implicitly by $-xyz + x^2yu + 2xy^3u - u^4 = 8$.

44. (a) A function f is said to be **homogeneous of degree n** if $f(\lambda x, \lambda y) = \lambda^n f(x, y)$. If f has first partial derivatives, show that

$$x\frac{\partial f}{\partial x} + y\frac{\partial f}{\partial y} = nf.$$

(b) Verify that $f(x, y) = 4x^2y^3 - 3xy^4 + x^5$ is a homogeneous function of degree 5.

(c) Verify that the function in part (b) satisfies the differential equation in part (a).

(d) Reexamine Problem 29. Conjecture whether $z = f(y/x)$ is homogeneous.

13.6 Directional Derivative

■ **Introduction** In Section 13.3 we saw that the partial derivatives $\partial z/\partial x$ and $\partial z/\partial y$ are the rates of change of the function $z = f(x, y)$ in the directions that are either parallel to the x-axis or to the y-axis, respectively. In the present section we will generalize the notion of partial derivatives by showing how to find the rate of change of f in an arbitrary direction. To do this, it is convenient to introduce a new vector-valued function whose components are partial derivatives.

■ **The Gradient of a Function** When the vector **differential operator**

$$\nabla = \mathbf{i}\frac{\partial}{\partial x} + \mathbf{j}\frac{\partial}{\partial y} \qquad \text{or} \qquad \nabla = \mathbf{i}\frac{\partial}{\partial x} + \mathbf{j}\frac{\partial}{\partial y} + \mathbf{k}\frac{\partial}{\partial z}$$

is applied to a function $z = f(x, y)$ or $w = f(x, y, z)$, we obtain a very useful vector-valued function.

Definition 13.6.1 Gradients

(i) Suppose f is a function of two variables x and y whose partial derivatives f_x and f_y exist. Then the **gradient of f** is defined to be

$$\nabla f(x, y) = \frac{\partial f}{\partial x}\mathbf{i} + \frac{\partial f}{\partial y}\mathbf{j}. \tag{1}$$

(ii) Suppose f is a function of three variables x, y, and z whose partial derivatives f_x, f_y, and f_z exist. Then the **gradient of f** is defined to be

$$\nabla f(x, y, z) = \frac{\partial f}{\partial x}\mathbf{i} + \frac{\partial f}{\partial y}\mathbf{j} + \frac{\partial f}{\partial z}\mathbf{k}. \tag{2}$$

The symbol ∇ is an inverted capital Greek delta, is called *del* or *nabla*. The symbol ∇f is usually read "grad f."

EXAMPLE 1 Gradient of a Function of Two Variables

Compute $\nabla f(x, y)$ for $f(x, y) = 5y - x^3y^2$.

Solution From (1),

$$\nabla f(x, y) = \frac{\partial}{\partial x}(5y - x^3y^2)\mathbf{i} + \frac{\partial}{\partial y}(5y - x^3y^2)\mathbf{j}$$

$$= -3x^2y^2\mathbf{i} + (5 - 2x^3y)\mathbf{j}. \qquad \blacksquare$$

EXAMPLE 2 Gradient of a Function of Three Variables

If $f(x, y, z) = xy^2 + 3x^2 - z^3$, find $\nabla f(x, y, z)$ at $(2, -1, 4)$.

Solution From (2),

$$\nabla f(x, y, z) = (y^2 + 6x)\mathbf{i} + 2xy\mathbf{j} - 3z^2\mathbf{k}$$

and so

$$\nabla f(2, -1, 4) = 13\mathbf{i} - 4\mathbf{j} - 48\mathbf{k}.$$ ∎

The gradient of a function f has many applications. We see next that ∇f plays an important role in the generalization of the partial derivative concept.

▌ A Generalization of Partial Differentiation Recall the partial derivatives $\partial z/\partial x$ and $\partial z/\partial y$ give the slope of a tangent line to the trace, or curve of intersection, of the surface given by $z = f(x, y)$ and vertical planes that are, respectively, parallel to the x- and y-coordinate axes. Equivalently, $\partial z/\partial x$ is the rate of change of the function f in the direction given by the vector $\mathbf{i}$, and $\partial z/\partial y$ is the rate of change of $z = f(x, y)$ in the $\mathbf{j}$-direction. There is no reason to confine our attention to just two directions; we can find the rate of change of a differentiable function in *any* direction. See FIGURE 13.6.1. Suppose Δx and Δy denote increments in x and y, respectively, and that $\mathbf{u} = \cos\theta\mathbf{i} + \sin\theta\mathbf{j}$ is a *unit* vector in the xy-plane that makes an angle θ with the positive x-axis and that is parallel to the vector $\mathbf{v}$ from $(x, y, 0)$ to $(x + \Delta x, y + \Delta y, 0)$. If $h = \sqrt{(\Delta x)^2 + (\Delta y)^2} > 0$, then $\mathbf{v} = h\mathbf{u}$. Furthermore, let the plane perpendicular to the xy-plane that contains these points slice the surface $z = f(x, y)$ in a curve C. We ask:

- *What is the slope of the tangent line to C at a point P with coordinates $(x, y, f(x, y))$ in the direction given by* $\mathbf{v}$?

See FIGURE 13.6.2.

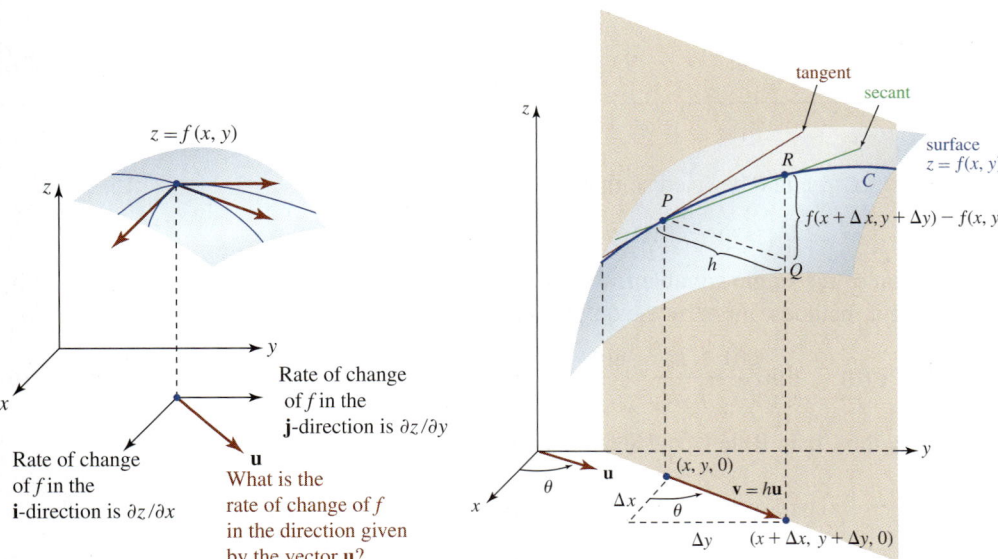

FIGURE 13.6.1 Vector $\mathbf{u}$ determines a direction

FIGURE 13.6.2 What is the slope of the tangent line to curve C at P?

From Figure 13.6.2, we see that $\Delta x = h\cos\theta$ and $\Delta y = h\sin\theta$, so that the slope of the indicated secant line through points P and R on C is

$$\frac{f(x + \Delta x, y + \Delta y) - f(x, y)}{h} = \frac{f(x + h\cos\theta, y + h\sin\theta) - f(x, y)}{h}. \qquad (3)$$

We expect the slope of the tangent at P to be the limit of (3) as $h \to 0$. This slope is the rate of change of f at P in the direction specified by the unit vector $\mathbf{u}$. This leads us to the next definition.

Definition 13.6.2 Directional Derivative

The **directional derivative** of a function $z = f(x, y)$ at (x, y) in the direction of the unit vector $\mathbf{u} = \cos\theta\,\mathbf{i} + \sin\theta\,\mathbf{j}$ is given by

$$D_{\mathbf{u}} f(x, y) = \lim_{h \to 0} \frac{f(x + h\cos\theta, y + h\sin\theta) - f(x, y)}{h}, \tag{4}$$

whenever the limit exists.

Observe that (4) is truly a generalization of (1) and (2) of Section 13.3, because:

$$\theta = 0 \text{ implies that } D_{\mathbf{i}} f(x, y) = \lim_{h \to 0} \frac{f(x + h, y) - f(x, y)}{h} = \frac{\partial z}{\partial x},$$

and

$$\theta = \frac{\pi}{2} \text{ implies that } D_{\mathbf{j}} f(x, y) = \lim_{h \to 0} \frac{f(x, y + h) - f(x, y)}{h} = \frac{\partial z}{\partial y}.$$

■ **Computing a Directional Derivative** While (4) could be used to find $D_{\mathbf{u}} f(x, y)$ for a given function, as usual we seek a more efficient procedure. The next theorem shows how the concept of the gradient of a function plays a key role in computing a directional derivative.

Theorem 13.6.1 Computing a Directional Derivative

If $z = f(x, y)$ is a differentiable function of x and y and $\mathbf{u} = \cos\theta\,\mathbf{i} + \sin\theta\,\mathbf{j}$ is a unit vector, then

$$D_{\mathbf{u}} f(x, y) = \nabla f(x, y) \cdot \mathbf{u}. \tag{5}$$

PROOF Let x, y, and θ be fixed so that

$$g(t) = f(x + t\cos\theta, y + t\sin\theta)$$

is a function of the single variable t. We wish to compare the value of $g'(0)$, which is found by two different methods. First, by the definition of a derivative,

$$g'(0) = \lim_{h \to 0} \frac{g(0 + h) - g(0)}{h} = \lim_{h \to 0} \frac{f(x + h\cos\theta, y + h\sin\theta) - f(x, y)}{h}. \tag{6}$$

Second, by the Chain Rule (1) of Section 13.5,

$$g'(t) = f_1(x + t\cos\theta, y + t\sin\theta)\frac{d}{dt}(x + t\cos\theta) + f_2(x + t\cos\theta, y + t\sin\theta)\frac{d}{dt}(y + t\sin\theta)$$

$$= f_1(x + t\cos\theta, y + t\sin\theta)\cos\theta + f_2(x + t\cos\theta, y + t\sin\theta)\sin\theta. \tag{7}$$

Here the subscripts 1 and 2 refer to the partial derivatives of $f(x + t\cos\theta, y + t\sin\theta)$ with respect to $x + t\cos\theta$ and $y + t\sin\theta$, respectively. When $t = 0$, we note that $x + t\cos\theta$ and $y + t\sin\theta$ are simply x and y, and therefore (7) becomes

$$g'(0) = f_x(x, y)\cos\theta + f_y(x, y)\sin\theta. \tag{8}$$

Comparing (4), (6), and (8) then gives

$$D_{\mathbf{u}} f(x, y) = f_x(x, y)\cos\theta + f_y(x, y)\sin\theta$$
$$= [f_x(x, y)\mathbf{i} + f_y(x, y)\mathbf{j}] \cdot (\cos\theta\,\mathbf{i} + \sin\theta\,\mathbf{j})$$
$$= \nabla f(x, y) \cdot \mathbf{u}. \qquad \blacksquare$$

EXAMPLE 3 Directional Derivative

Find the directional derivative of $f(x, y) = 2x^2y^3 + 6xy$ at $(1, 1)$ in the direction of a unit vector whose angle with the positive x-axis is $\pi/6$.

Solution Since $\partial f/\partial x = 4xy^3 + 6y$ and $\partial f/\partial y = 6x^2y^2 + 6x$ we have from (1) of Definition 13.6.1,

$$\nabla f(x, y) = (4xy^3 + 6y)\mathbf{i} + (6x^2y^2 + 6x)\mathbf{j} \quad \text{and} \quad \nabla f(1, 1) = 10\mathbf{i} + 12\mathbf{j}.$$

Now, at $\theta = \pi/6$, $\mathbf{u} = \cos\theta\mathbf{i} + \sin\theta\mathbf{j}$ becomes

$$\mathbf{u} = \frac{\sqrt{3}}{2}\mathbf{i} + \frac{1}{2}\mathbf{j}.$$

Therefore, by (5) of Theorem 13.6.1,

$$D_\mathbf{u} f(1, 1) = \nabla f(1, 1) \cdot \mathbf{u} = (10\mathbf{i} + 12\mathbf{j}) \cdot \left(\frac{1}{2}\sqrt{3}\mathbf{i} + \frac{1}{2}\mathbf{j}\right) = 5\sqrt{3} + 6. \quad\blacksquare$$

It is important that you remember that the vector $\mathbf{u}$ in Theorem 13.6.1 is a *unit* vector. If a non-unit vector $\mathbf{v}$ specifies a direction, then in order to use (5) we must normalize $\mathbf{v}$ and use $\mathbf{u} = \mathbf{v}/|\mathbf{v}|$.

EXAMPLE 4 Directional Derivative

Consider the plane that is perpendicular to the xy-plane and passes through the points $P(2, 1)$ and $Q(3, 2)$. What is the slope of the tangent line to the curve of intersection of this plane with the surface $f(x, y) = 4x^2 + y^2$ at $(2, 1, 17)$ in the direction of Q?

Solution We want $D_\mathbf{u} f(2, 1)$ in the direction given by the vector $\overrightarrow{PQ} = \mathbf{i} + \mathbf{j}$. But since $\overrightarrow{PQ}$ is not a unit vector, we form

$$\mathbf{u} = \frac{1}{|\overrightarrow{PQ}|} \overrightarrow{PQ} = \frac{1}{\sqrt{2}}\mathbf{i} + \frac{1}{\sqrt{2}}\mathbf{j}.$$

Now,

$$\nabla f(x, y) = 8x\mathbf{i} + 2y\mathbf{j} \quad \text{and} \quad \nabla f(2, 1) = 16\mathbf{i} + 2\mathbf{j}.$$

Therefore, from (5) the desired slope is

$$D_\mathbf{u} f(2, 1) = (16\mathbf{i} + 2\mathbf{j}) \cdot \left(\frac{1}{\sqrt{2}}\mathbf{i} + \frac{1}{\sqrt{2}}\mathbf{j}\right) = 9\sqrt{2}. \quad\blacksquare$$

❚ Functions of Three Variables For a function $w = f(x, y, z)$ the directional derivative is defined by

$$D_\mathbf{u} f(x, y, z) = \lim_{h \to 0} \frac{f(x + h\cos\alpha, y + h\cos\beta, z + h\cos\gamma) - f(x, y, z)}{h},$$

where α, β, and γ are the direction angles of the vector $\mathbf{u}$ measured relative to the positive x-, y- and z-axes, respectively.* But in the same manner as before, we can show that

$$D_\mathbf{u} f(x, y, z) = \nabla f(x, y, z) \cdot \mathbf{u}. \tag{9}$$

Notice, since $\mathbf{u}$ is a unit vector, it follows from (11) of Section 11.3 that

$$D_\mathbf{u} f(x, y) = \text{comp}_\mathbf{u} \nabla f(x, y) \quad \text{and} \quad D_\mathbf{u} f(x, y, z) = \text{comp}_\mathbf{u} \nabla f(x, y, z).$$

In addition, (9) reveals that

$$D_\mathbf{k} f(x, y, z) = \frac{\partial w}{\partial z}.$$

---

*Note that the numerator of (4) can be written $f(x + h\cos\alpha, y + h\cos\beta) - f(x, y)$ where $\beta = (\pi/2) - \alpha$.

EXAMPLE 5 Directional Derivative

Find the directional derivative of $f(x, y, z) = xy^2 - 4x^2y + z^2$ at $(1, -1, 2)$ in the direction of $\mathbf{v} = 6\mathbf{i} + 2\mathbf{j} + 3\mathbf{k}$.

Solution We have $\partial f/\partial x = y^2 - 8xy$, $\partial f/\partial y = 2xy - 4x^2$, and $\partial f/\partial z = 2z$ so that

$$\nabla f(x, y, z) = (y^2 - 8xy)\mathbf{i} + (2xy - 4x^2)\mathbf{j} + 2z\mathbf{k}$$
$$\nabla f(1, -1, 2) = 9\mathbf{i} - 6\mathbf{j} + 4\mathbf{k}.$$

Since $\quad |\mathbf{v}| = |6\mathbf{i} + 2\mathbf{j} + 3\mathbf{k}| = 7 \quad$ then $\quad \mathbf{u} = \dfrac{1}{|\mathbf{v}|}\mathbf{v} = \dfrac{6}{7}\mathbf{i} + \dfrac{2}{7}\mathbf{j} + \dfrac{3}{7}\mathbf{k}$

is a unit vector in the indicated direction. From (9) we obtain

$$D_{\mathbf{u}}f(1, -1, 2) = (9\mathbf{i} - 6\mathbf{j} + 4\mathbf{k}) \cdot \left(\frac{6}{7}\mathbf{i} + \frac{2}{7}\mathbf{j} + \frac{3}{7}\mathbf{k}\right) = \frac{54}{7}.$$ ∎

▍ Maximum Value of the Directional Derivative Let f represent a function of either two or three variables. Since (5) and (9) express the directional derivative as a dot product, we see from Theorem 11.3.2 that

$$D_{\mathbf{u}}f = \nabla f \cdot \mathbf{u} = |\nabla f||\mathbf{u}|\cos\phi = |\nabla f|\cos\phi, \quad (|u| = 1), \tag{10}$$

where ϕ is the angle between ∇f and $\mathbf{u}$ satisfying $0 \le \phi \le \pi$. Because $-1 \le \cos\phi \le 1$ it follows from (10) that

$$-|\nabla f| \le D_{\mathbf{u}}f \le |\nabla f|.$$

In other words:

- *The maximum value of the directional derivative is $|\nabla f|$ and it occurs when $\mathbf{u}$ has the same direction as ∇f (when $\cos\phi = 1$),* $\qquad$ (11)

and

- *The minimum value of the directional derivative is $-|\nabla f|$ and it occurs when $\mathbf{u}$ and ∇f have opposite directions (when $\cos\phi = -1$).* $\qquad$ (12)

EXAMPLE 6 Maximum Value of Directional Derivative

In Example 5 the maximum value of the directional derivative of f at $(1, -1, 2)$ is $|\nabla f(1, -1, 2)| = \sqrt{133}$. The minimum value of $D_{\mathbf{u}}f(1, -1, 2)$ is then $-\sqrt{133}$. ∎

▍ Gradient Points in Direction of Most Rapid Increase of f Put yet another way, (11) and (12) state:

- *The gradient vector ∇f points in the direction in which f increases most rapidly, whereas $-\nabla f$ points in the direction of the most rapid decrease of f.*

EXAMPLE 7 A Mathematical Model

Each year in Los Angeles there is a bicycle race up to the top of a hill by a road known to be the steepest in the city. To understand why a bicyclist, with a modicum of sanity, will zigzag up the road, let us suppose the graph of $f(x, y) = 4 - \frac{2}{3}\sqrt{x^2 + y^2}$, $0 \le z \le 4$, shown in FIGURE 13.6.3(a) is a mathematical model of the hill. The gradient of f is

$$\nabla f(x, y) = \frac{2}{3}\left[\frac{-x}{\sqrt{x^2 + y^2}}\mathbf{i} + \frac{-y}{\sqrt{x^2 + y^2}}\mathbf{j}\right] = \frac{2}{3}\frac{1}{\sqrt{x^2 + y^2}}\mathbf{r},$$

where $\mathbf{r} = -x\mathbf{i} - y\mathbf{j}$ is a vector pointing to the center of the circular base.

Thus, the steepest ascent up the hill is a straight road whose projection in the xy-plane is a radius of the circular base. Since $D_{\mathbf{u}}f = \text{comp}_{\mathbf{u}}\nabla f$, a bicyclist will zigzag, or seek a direction $\mathbf{u}$ other than ∇f, in order to reduce this component. See Figure 13.6.3(b). ∎

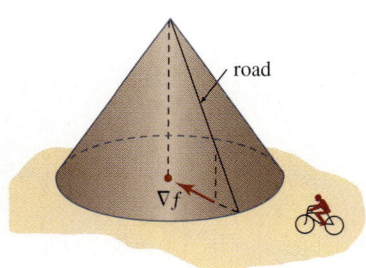

(a)

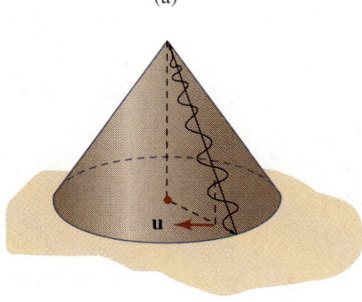

(b)

FIGURE 13.6.3 Model of a steep hill in Example 7

EXAMPLE 8　A Mathematical Model

The temperature in a rectangular box is approximated by the mathematical model $T(x, y, z) = xyz(1 - x)(2 - y)(3 - z), 0 \leq x \leq 1, 0 \leq y \leq 2, 0 \leq z \leq 3$. If a mosquito is located at $\left(\frac{1}{2}, 1, 1\right)$, in which direction should it fly to cool off as rapidly as possible?

Solution　The gradient of T is

$$\nabla T(x, y, z) = yz(2 - y)(3 - z)(1 - 2x)\mathbf{i} + xz(1 - x)(3 - z)(2 - 2y)\mathbf{j} + xy(1 - x)(2 - y)(3 - 2z)\mathbf{k}.$$

Therefore,
$$\nabla T\left(\frac{1}{2}, 1, 1\right) = \frac{1}{4}\mathbf{k}.$$

To cool off most rapidly, the mosquito should fly in the direction of $\frac{1}{4}\mathbf{k}$; that is, it should dive for the floor of the box, where the temperature is $T(x, y, 0) = 0$. ∎

Exercises 13.6　Answers to selected odd-numbered problems begin on page ANS-42.

≡ **Fundamentals**

In Problems 1–4, compute the gradient for the given function.

1. $f(x, y) = x^2 - x^3y^2 + y^4$　**2.** $f(x, y) = y - e^{-2x^2y}$

3. $F(x, y, z) = \dfrac{xy^2}{z^3}$　**4.** $G(x, y, z) = xy\cos yz$

In Problems 5–8, find the gradient of the given function at the indicated point.

5. $f(x, y) = x^2 - 4y^2$;　$(2, 4)$
6. $f(x, y) = \sqrt{x^3y - y^4}$;　$(3, 2)$
7. $f(x, y, z) = x^2z^2\sin 4y$;　$(-2, \pi/3, 1)$
8. $f(x, y, z) = \ln(x^2 + y^2 + z^2)$;　$(-4, 3, 5)$

In Problems 9 and 10, use Definition 13.6.2 to find $D_{\mathbf{u}}f(x, y)$ given that $\mathbf{u}$ makes the indicated angle with the positive x-axis.

9. $f(x, y) = x^2 + y^2$;　$\theta = 30°$
10. $f(x, y) = 3x - y^2$;　$\theta = 45°$

In Problems 11–20, find the directional derivative of the given function at the given point in the indicated direction.

11. $f(x, y) = 5x^3y^6$;　$(-1, 1), \theta = \pi/6$
12. $f(x, y) = 4x + xy^2 - 5y$;　$(3, -1), \theta = \pi/4$
13. $f(x, y) = \tan^{-1}\dfrac{y}{x}$;　$(2, -2), \mathbf{i} - 3\mathbf{j}$
14. $f(x, y) = \dfrac{xy}{x + y}$;　$(2, -1), 6\mathbf{i} + 8\mathbf{j}$
15. $f(x, y) = (xy + 1)^2$;　$(3, 2)$, in the direction of $(5, 3)$
16. $f(x, y) = x^2\tan y$;　$\left(\frac{1}{2}, \pi/3\right)$, in the direction of the negative x-axis
17. $F(x, y, z) = x^2y^2(2z + 1)^2$;　$(1, -1, 1), \langle 0, 3, 3\rangle$
18. $F(x, y, z) = \dfrac{x^2 - y^2}{z^2}$;　$(2, 4, -1), \mathbf{i} - 2\mathbf{j} + \mathbf{k}$
19. $f(x, y, z) = \sqrt{x^2y + 2y^2z}$;　$(-2, 2, 1)$, in the direction of the negative z-axis
20. $f(x, y, z) = 2x - y^2 + z^2$;　$(4, -4, 2)$, in the direction of the origin

In Problems 21 and 22, consider the plane through the points P and Q that is perpendicular to the xy-plane. Find the slope of the tangent at the indicated point to the curve of intersection of this plane and the graph of the given function in the direction of Q.

21. $f(x, y) = (x - y)^2$;　$P(4, 2), Q(0, 1)$;　$(4, 2, 4)$
22. $f(x, y) = x^3 - 5xy + y^2$;　$P(1, 1), Q(-1, 6)$;　$(1, 1, -3)$

In Problems 23–26, find a vector that gives the direction in which the given function increases most rapidly at the indicated point. Find the maximum rate.

23. $f(x, y) = e^{2x}\sin y$;　$(0, \pi/4)$　**24.** $f(x, y) = xye^{x-y}$;　$(5, 5)$
25. $f(x, y, z) = x^2 + 4xz + 2yz^2$;　$(1, 2, -1)$
26. $f(x, y, z) = xyz$;　$(3, 1, -5)$

In Problems 27–30, find a vector that gives the direction in which the given function decreases most rapidly at the indicated point. Find the minimum rate.

27. $f(x, y) = \tan(x^2 + y^2)$;　$(\sqrt{\pi/6}, \sqrt{\pi/6})$
28. $f(x, y) = x^3 - y^3$;　$(2, -2)$
29. $f(x, y, z) = \sqrt{xz}\, e^y$;　$(16, 0, 9)$
30. $f(x, y, z) = \ln\dfrac{xy}{z}$;　$\left(\frac{1}{2}, \frac{1}{6}, \frac{1}{3}\right)$

31. Find the directional derivative(s) of $f(x, y) = x + y^2$ at $(3, 4)$ in the direction of a tangent vector to the graph of $2x^2 + y^2 = 9$ at $(2, 1)$.
32. If $f(x, y) = x^2 + xy + y^2 - x$, find all points where $D_{\mathbf{u}}f(x, y)$ in the direction of $\mathbf{u} = (1/\sqrt{2})(\mathbf{i} + \mathbf{j})$ is zero.
33. Suppose $\nabla f(a, b) = 4\mathbf{i} + 3\mathbf{j}$. Find a unit vector $\mathbf{u}$ so that
　(a) $D_{\mathbf{u}}f(a, b) = 0$
　(b) $D_{\mathbf{u}}f(a, b)$ is a maximum
　(c) $D_{\mathbf{u}}f(a, b)$ is a minimum
34. Suppose $D_{\mathbf{u}}f(a, b) = 6$. What is the value of $D_{-\mathbf{u}}f(a, b)$?
35. (a) If $f(x, y) = x^3 - 3x^2y^2 + y^3$, find the directional derivative of f at a point (x, y) in the direction of $\mathbf{u} = (1/\sqrt{10})(3\mathbf{i} + \mathbf{j})$.
　(b) If $F(x, y) = D_{\mathbf{u}}f(x, y)$ in part (a), find $D_{\mathbf{u}}F(x, y)$.

36. Suppose $D_{\mathbf{u}}f(a, b) = 7$, $D_{\mathbf{v}}f(a, b) = 3$, $\mathbf{u} = \frac{5}{13}\mathbf{i} - \frac{12}{13}\mathbf{j}$, and $\mathbf{v} = \frac{5}{13}\mathbf{i} + \frac{12}{13}\mathbf{j}$. Find $\nabla f(a, b)$.

37. If $f(x, y) = x^3 - 12x + y^2 - 10y$, find all points at which $|\nabla f| = 0$.

38. If $f(x, y) = x^2 - \frac{5}{2}y^2$, then sketch the set of points in the xy-plane for which $|\nabla f| = 10$.

≡ **Applications**

39. Consider the rectangular plate shown in FIGURE 13.6.4. The temperature at a point (x, y) on the plate is given by $T(x, y) = 5 + 2x^2 + y^2$. Determine the direction an insect should take, starting at $(4, 2)$, in order to cool off as rapidly as possible.

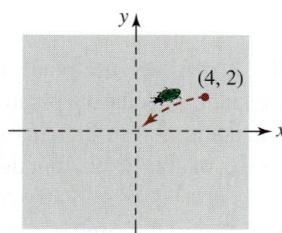

FIGURE 13.6.4   Insect on a plate in Problem 39

40. In Problem 39 observe that $(0, 0)$ is the coolest point of the plate. Find the path the cold-seeking insect, starting at $(4, 2)$, will take to the origin. If $\langle x(t), y(t) \rangle$ is the vector equation of the path, then use the fact that $-\nabla T(x, y) = \langle x'(t), y'(t) \rangle$. Why is this? [*Hint*: Review Section 8.1.]

41. The temperature T at a point (x, y) on a rectangular metal plate is given by $T(x, y) = 100 - 2x^2 - y^2$. Find the path a heat-seeking particle will take, starting at $(3, 4)$, as it

moves in the direction in which the temperature increases most rapidly.

42. The temperature T at a point (x, y, z) in space is inversely proportional to the square of the distance from (x, y, z) to the origin. It is known that $T(0, 0, 1) = 500$. Find the rate of change of T at $(2, 3, 3)$ in the direction of $(3, 1, 1)$. In which direction from $(2, 3, 3)$ does the temperature T increase most rapidly? At $(2, 3, 3)$ what is the maximum rate of change of T?

43. Consider the gravitational potential

$$U(x, y) = \frac{-Gm}{\sqrt{x^2 + y^2}},$$

where G and m are constants. Show that U increases or decreases most rapidly along a line through the origin.

≡ **Think About It**

44. Find a function f such that

$$\nabla f = (3x^2 + y^3 + ye^{xy})\mathbf{i} + (-2y^2 + 3xy^2 + xe^{xy})\mathbf{j}.$$

In Problems 45–48, assume that f and g are differentiable functions of two variables. Prove the given identity.

45. $\nabla(cf) = c\nabla f$

46. $\nabla(f + g) = \nabla f + \nabla g$

47. $\nabla(fg) = f\nabla g + g\nabla f$

48. $\nabla\left(\dfrac{f}{g}\right) = \dfrac{g\nabla f - f\nabla g}{g^2}$

49. If $\mathbf{r} = x\mathbf{i} + y\mathbf{i}$ and $r = |\mathbf{r}|$, then show $\nabla r = \mathbf{r}/r$.

50. Use Problem 49 to show that $\nabla f(r) = f'(r)\mathbf{r}/r$.

51. Let f_x, f_y, f_{xy}, f_{yx} be continuous and $\mathbf{u}$ and $\mathbf{v}$ be unit vectors. Show that $D_{\mathbf{u}}D_{\mathbf{v}}f = D_{\mathbf{v}}D_{\mathbf{u}}f$.

52. If $\mathbf{F}(x, y, z) = f_1(x, y, z)\mathbf{i} + f_2(x, y, z)\mathbf{j} + f_3(x, y, z)\mathbf{k}$, find $\nabla \times \mathbf{F}$.

13.7 Tangent Planes and Normal Lines

▮ **Introduction**   In Section 13.4 we mentioned that the three-dimensional analogue of a tangent line to a curve is a plane tangent to a surface. To obtain an equation of a tangent plane at a point on a surface we must return to the notion of the gradient of a function of either two or three variables.

▮ **Geometric Interpretation of the Gradient**   Suppose $f(x, y) = c$ is the *level curve* of the differentiable function of two variables $z = f(x, y)$ that passes through a specified point $P(x_0, y_0)$; that is, the number c is defined by $f(x_0, y_0) = c$. If this level curve is parameterized by the differentiable functions

$$x = x(t), \quad y = y(t) \qquad \text{such that} \qquad x_0 = x(t_0), \quad y_0 = y(t_0),$$

then by the Chain Rule, (1) of Section 13.5, the derivative of $f(x(t), y(t)) = c$ with respect to t is given by

$$\frac{\partial f}{\partial x}\frac{dx}{dt} + \frac{\partial f}{\partial y}\frac{dy}{dt} = 0. \tag{1}$$

By introducing the vectors

$$\nabla f(x, y) = \frac{\partial f}{\partial x}\mathbf{i} + \frac{\partial f}{\partial y}\mathbf{j} \qquad \text{and} \qquad \mathbf{r}'(t) = \frac{dx}{dt}\mathbf{i} + \frac{dy}{dt}\mathbf{j}$$

(1) can be written as the dot product $\nabla f \cdot \mathbf{r}' = 0$. Specifically, at $t = t_0$, we have

$$\nabla f(x_0, y_0) \cdot \mathbf{r}'(t_0) = 0. \tag{2}$$

Thus, if $\mathbf{r}'(t_0) \neq 0$, the vector $\nabla f(x_0, y_0)$ is orthogonal to the tangent vector $\mathbf{r}'(t_0)$ at $P(x_0, y_0)$. We interpret this to mean:

- *The gradient ∇f is perpendicular to the level curve at P.*

See **FIGURE 13.7.1**.

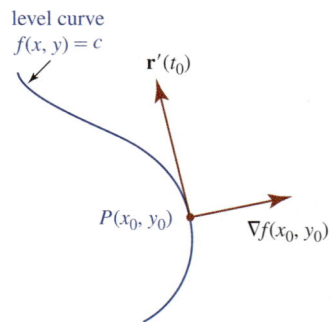

FIGURE 13.7.1 Gradient is perpendicular to a level curve

EXAMPLE 1 Gradient at a Point on a Level Curve

Find the level curve of $f(x, y) = -x^2 + y^2$ passing through $(2, 3)$. Graph the gradient at the point.

Solution Since $f(2, 3) = -4 + 9 = 5$, the level curve is the hyperbola $-x^2 + y^2 = 5$. Now,

$$\nabla f(x, y) = -2x\mathbf{i} + 2y\mathbf{j} \quad \text{and so} \quad \nabla f(2, 3) = -4\mathbf{i} + 6\mathbf{j}.$$

FIGURE 13.7.2 shows the level curve and the gradient $\nabla f(2, 3)$. ∎

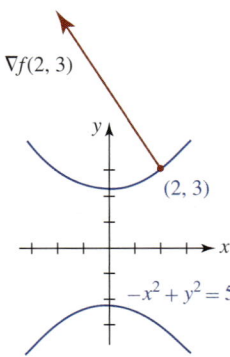

FIGURE 13.7.2 Gradient in Example 1

■ Geometric Interpretation of the Gradient—Continued Proceeding as before, let $F(x, y, z) = c$ be the *level surface* of a differentiable function of three variables $w = F(x, y, z)$ that passes through $P(x_0, y_0, z_0)$. If the differentiable functions $x = x(t)$, $y = y(t)$, $z = z(t)$ are the parametric equations of a curve C on the surface for which $x_0 = x(t_0)$, $y_0 = y(t_0)$, $z_0 = z(t_0)$, then by (3) of Section 13.5 the derivative of $F(x(t), y(t), z(t)) = c$ with respect to t is

$$\frac{\partial F}{\partial x}\frac{dx}{dt} + \frac{\partial F}{\partial y}\frac{dy}{dt} + \frac{\partial F}{\partial z}\frac{dz}{dt} = 0$$

or

$$\left(\frac{\partial F}{\partial x}\mathbf{i} + \frac{\partial F}{\partial y}\mathbf{j} + \frac{\partial F}{\partial z}\mathbf{k}\right) \cdot \left(\frac{dx}{dt}\mathbf{i} + \frac{dy}{dt}\mathbf{j} + \frac{dz}{dt}\mathbf{k}\right) = 0. \tag{3}$$

In particular, at $t = t_0$, (3) becomes

$$\nabla F(x_0, y_0, z_0) \cdot \mathbf{r}'(t_0) = 0. \tag{4}$$

Thus, (4) shows that when $\mathbf{r}'(t_0) \neq \mathbf{0}$ the vector $\nabla F(x_0, y_0, z_0)$ is orthogonal to the tangent vector $\mathbf{r}'(t_0)$. Since this argument holds for any differentiable curve through $P(x_0, y_0, z_0)$ on the surface, we conclude that:

- *The gradient ∇F is perpendicular (normal) to the level surface at P.*

See **FIGURE 13.7.3**.

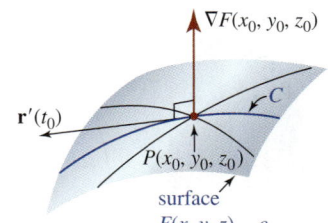

FIGURE 13.7.3 Gradient is perpendicular to a level surface

EXAMPLE 2 Gradient at a Point on a Level Surface

Find the level surface of $F(x, y, z) = x^2 + y^2 + z^2$ passing through $(1, 1, 1)$. Graph the gradient at the point.

Solution Since $F(1, 1, 1) = 3$, the level surface passing through $(1, 1, 1)$ is the sphere $x^2 + y^2 + z^2 = 3$. The gradient of the function is

$$\nabla F(x, y, z) = 2x\mathbf{i} + 2y\mathbf{j} + 2z\mathbf{k}$$

and so, at the given point,

$$\nabla F(1, 1, 1) = 2\mathbf{i} + 2\mathbf{j} + 2\mathbf{k}.$$

The level surface and $\nabla F(1, 1, 1)$ are illustrated in FIGURE 13.7.4. ∎

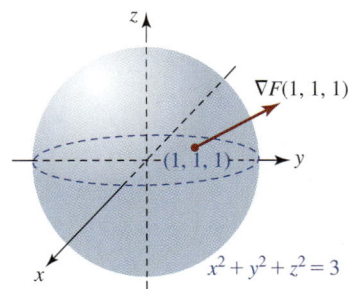

FIGURE 13.7.4 Gradient is perpendicular to a sphere in Example 2

■ Tangent Plane In earlier chapters we found equations of tangent lines to graphs of functions. In 3-space we can now solve the analogous problem of finding equations of **tangent**

planes to surfaces. We assume again that $w = F(x, y, z)$ is a differentiable function and that a surface is given by $F(x, y, z) = c$, where c is a constant.

Definition 13.7.1 Tangent Plane

Let $P(x_0, y_0, z_0)$ be a point on the graph of the level surface $F(x, y, z) = c$ where ∇F is not $\mathbf{0}$. The **tangent plane** at $P(x_0, y_0, z_0)$ is that plane through P that is perpendicular to $\nabla F(x_0, y_0, z_0)$.

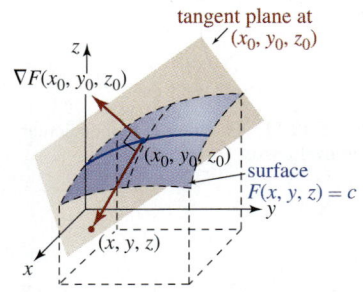

FIGURE 13.7.5 Tangent plane to a surface

Thus, if $P(x, y, z)$ and $P(x_0, y_0, z_0)$ are points on the tangent plane and $\mathbf{r} = x\mathbf{i} + y\mathbf{j} + z\mathbf{k}$ and $\mathbf{r}_0 = x_0\mathbf{i} + y_0\mathbf{j} + z_0\mathbf{k}$ are their respective position vectors, a vector equation of the tangent plane is

$$\nabla F(x_0, y_0, z_0) \cdot (\mathbf{r} - \mathbf{r}_0) = 0,$$

where $\mathbf{r} - \mathbf{r}_0 = (x - x_0)\mathbf{i} + (y - y_0)\mathbf{j} + (z - z_0)\mathbf{k}$. See FIGURE 13.7.5. We summarize this last result.

Theorem 13.7.1 Equation of Tangent Plane

Let $P(x_0, y_0, z_0)$ be a point on the graph of $F(x, y, z) = c$ where ∇F is not $\mathbf{0}$. Then an equation of the tangent plane at P is

$$F_x(x_0, y_0, z_0)(x - x_0) + F_y(x_0, y_0, z_0)(y - y_0) + F_z(x_0, y_0, z_0)(z - z_0) = 0. \tag{5}$$

EXAMPLE 3 Equation of a Tangent Plane

Find an equation of the tangent plane to the graph of the sphere $x^2 + y^2 + z^2 = 3$ at $(1, 1, 1)$.

Solution By defining $F(x, y, z) = x^2 + y^2 + z^2$, we find that the given sphere is the level surface $F(x, y, z) = F(1, 1, 1) = 3$ passing through $(1, 1, 1)$. Now,

$$F_x(x, y, z) = 2x, \quad F_y(x, y, z) = 2y, \quad \text{and} \quad F_z(x, y, z) = 2z$$

so that

$$\nabla F(x, y, z) = 2x\mathbf{i} + 2y\mathbf{j} + 2z\mathbf{k} \quad \text{and} \quad \nabla F(1, 1, 1) = 2\mathbf{i} + 2\mathbf{j} + 2\mathbf{k}.$$

It follows from (5) that an equation of the tangent plane is

$$2(x - 1) + 2(y - 1) + 2(z - 1) = 0 \quad \text{or} \quad x + y + z = 3.$$

With the aid of a CAS the tangent plane is shown in FIGURE 13.7.6. ∎

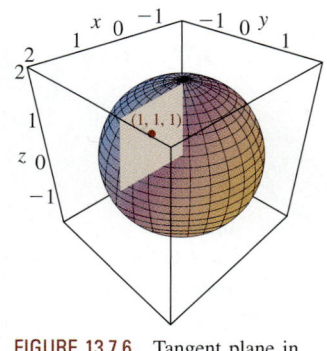

FIGURE 13.7.6 Tangent plane in Example 3

■ **Surfaces Given by $z = f(x, y)$** For a surface given explicitly by a differentiable function $z = f(x, y)$, we define $F(x, y, z) = f(x, y) - z$ or $F(x, y, z) = z - f(x, y)$. Thus a point (x_0, y_0, z_0) is on the graph of $z = f(x, y)$ if and only if it is also on the level surface $F(x, y, z) = 0$. This follows from $F(x_0, y_0, z_0) = f(x_0, y_0) - z_0 = 0$. In this case

$$F_x = f_x(x, y), \quad F_y = f_y(x, y), \quad F_z = -1$$

and so (5) becomes

$$f_x(x_0, y_0)(x - x_0) + f_y(x_0, y_0)(y - y_0) - (z - z_0) = 0$$

or

$$z = f(x_0, y_0) + f_x(x_0, y_0)(x - x_0) + f_y(x_0, y_0)(y - y_0). \tag{6}$$

A direct comparison of (6) with (7) of Section 13.4 shows that a linearization $L(x, y)$ of a function $z = f(x, y)$ that is differentiable at a point (x_0, y_0) is an equation of a tangent plane at (x_0, y_0).

EXAMPLE 4   Equation of a Tangent Plane

Find an equation of the tangent plane to the graph of the paraboloid $z = \frac{1}{2}x^2 + \frac{1}{2}y^2 + 4$ at $(1, -1, 5)$.

Solution   Define $F(x, y, z) = \frac{1}{2}x^2 + \frac{1}{2}y^2 - z + 4$ so that the level surface of F passing through the given point is $F(x, y, z) = F(1, -1, 5)$ or $F(x, y, z) = 0$. Now, $F_x = x$, $F_y = y$, and $F_z = -1$ so that

$$\nabla F(x, y, z) = x\mathbf{i} + y\mathbf{j} - \mathbf{k} \qquad \text{and} \qquad \nabla F(1, -1, 5) = \mathbf{i} - \mathbf{j} - \mathbf{k}.$$

Hence, from (5) the desired equation is

$$(x - 1) - (y + 1) - (z - 5) = 0 \qquad \text{or} \qquad -x + y + z = 3.$$

See FIGURE 13.7.7.

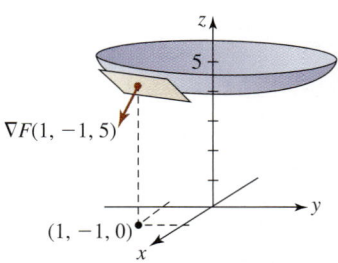

FIGURE 13.7.7   Tangent plane in Example 4

▮ Normal Line   Let $P(x_0, y_0, z_0)$ be a point on the graph of $F(x, y, z) = c$ where ∇F is not $\mathbf{0}$. The line containing $P(x_0, y_0, z_0)$ that is parallel to $\nabla F(x_0, y_0, z_0)$ is called the **normal line** to the surface at P. The normal line is perpendicular to the tangent plane to the surface at P.

EXAMPLE 5   Normal Line

Find parametric equations for the normal line to the surface in Example 4 at $(1, -1, 5)$.

Solution   A direction vector for the normal line at $(1, -1, 5)$ is $\nabla F(1, -1, 5) = \mathbf{i} - \mathbf{j} - \mathbf{k}$. It follows from (4) of Section 11.5 that parametric equations for the normal line are $x = 1 + t$, $y = -1 - t$, $z = 5 - t$.

Expressed as symmetric equations the normal line to a surface $F(x, y, z) = c$ at $P(x_0, y_0, z_0)$ is given by

$$\frac{x - x_0}{F_x(x_0, y_0, z_0)} = \frac{y - y_0}{F_y(x_0, y_0, z_0)} = \frac{z - z_0}{F_z(x_0, y_0, z_0)}.$$

In Example 5, you should verify that symmetric equations of the normal line at $(1, -1, 5)$ are

$$x - 1 = \frac{y + 1}{-1} = \frac{z - 5}{-1}.$$

∇f **NOTES FROM THE CLASSROOM**

Water flowing down a hill chooses a path in the direction of the greatest change in altitude. FIGURE 13.7.8 shows the contours, or level curves, of a hill. As shown in the figure, a stream starting at point P will take a path that is perpendicular to the contours. After reading Sections 13.7 and 13.8 you should be able to explain why.

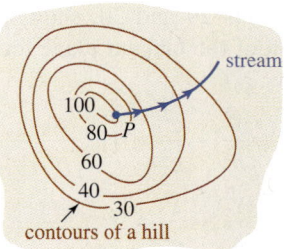

FIGURE 13.7.8   Stream flowing downhill

Exercises 13.7   Answers to selected odd-numbered problems begin on page ANS-42.

≡ Fundamentals

In Problems 1–12, sketch the level curve or surface passing through the indicated point. Sketch the gradient at the point.

1. $f(x, y) = x - 2y$; $(6, 1)$

2. $f(x, y) = \dfrac{y + 2x}{x}$; $(1, 3)$

3. $f(x, y) = y - x^2$; $(2, 5)$

4. $f(x, y) = x^2 + y^2$; $(-1, 3)$

5. $f(x, y) = \dfrac{x^2}{4} + \dfrac{y^2}{9}$; $(-2, -3)$

6. $f(x, y) = \dfrac{y^2}{x}$; $(2, 2)$

7. $f(x, y) = (x - 1)^2 - y^2$; $(1, 1)$

8. $f(x, y) = \dfrac{y - 1}{\sin x}$; $(\pi/6, \frac{3}{2})$

9. $f(x, y, z) = y + z$; $(3, 1, 1)$

10. $f(x, y, z) = x^2 + y^2 - z$; $(1, 1, 3)$

11. $F(x, y, z) = \sqrt{x^2 + y^2 + z^2}$; $(3, 4, 0)$

12. $F(x, y, z) = x^2 - y^2 + z$; $(0, -1, 1)$

In Problems 13 and 14, find the points on the given surface at which the gradient is parallel to the indicated vector.

13. $z = x^2 + y^2$; $4\mathbf{i} + \mathbf{j} + \frac{1}{2}\mathbf{k}$

14. $x^3 + y^2 + z = 15$; $27\mathbf{i} + 8\mathbf{j} + \mathbf{k}$

In Problems 15–24, find an equation of the tangent plane to the graph of the given equation at the indicated point.

15. $x^2 + y^2 + z^2 = 9$; $(-2, 2, 1)$

16. $5x^2 - y^2 + 4z^2 = 8$; $(2, 4, 1)$

17. $x^2 - y^2 - 3z^2 = 5$; $(6, 2, 3)$

18. $xy + yz + zx = 7$; $(1, -3, -5)$

19. $z = 25 - x^2 - y^2$; $(3, -4, 0)$

20. $xz = 6$; $(2, 0, 3)$

21. $z = \cos(2x + y)$; $(\pi/2, \pi/4, -1/\sqrt{2})$

22. $x^2 y^3 + 6z = 10$; $(2, 1, 1)$

23. $z = \ln(x^2 + y^2)$; $(1/\sqrt{2}, 1/\sqrt{2}, 0)$

24. $z = 8e^{-2y} \sin 4x$; $(\pi/24, 0, 4)$

In Problems 25 and 26, find the points on the given surface at which the tangent plane is parallel to the indicated plane.

25. $x^2 + y^2 + z^2 = 7$; $2x + 4y + 6z = 1$

26. $x^2 - 2y^2 - 3z^2 = 33$; $8x + 4y + 6z = 5$

27. Find points on the surface $x^2 + 4x + y^2 + z^2 - 2z = 11$ at which the tangent plane is horizontal.

28. Find points on the surface $x^2 + 3y^2 + 4z^2 - 2xy = 16$ at which the tangent plane is parallel to

    (a) the xz-plane,

    (b) the yz-plane, and

    (c) the xy-plane.

In Problems 29 and 30, show that the second equation is an equation of the tangent plane to the graph of the first equation at (x_0, y_0, z_0).

29. $\dfrac{x^2}{a^2} + \dfrac{y^2}{b^2} + \dfrac{z^2}{c^2} = 1$; $\dfrac{xx_0}{a^2} + \dfrac{yy_0}{b^2} + \dfrac{zz_0}{c^2} = 1$

30. $\dfrac{x^2}{a^2} - \dfrac{y^2}{b^2} + \dfrac{z^2}{c^2} = 1$; $\dfrac{xx_0}{a^2} - \dfrac{yy_0}{b^2} + \dfrac{zz_0}{c^2} = 1$

In Problems 31 and 32, find parametric equations for the normal line at the indicated point.

31. $x^2 + 2y^2 + z^2 = 4$; $(1, -1, 1)$

32. $z = 2x^2 - 4y^2$; $(3, -2, 2)$

In Problems 33 and 34, find symmetric equations for the normal line at the indicated point.

33. $z = 4x^2 + 9y^2 + 1$; $\left(\frac{1}{2}, \frac{1}{3}, 3\right)$

34. $x^2 + y^2 - z^2 = 0$; $(3, 4, 5)$

☰ Think About It

35. Show that every tangent plane to the graph of $z^2 = x^2 + y^2$ passes through the origin.

36. Show that the sum of the x-, y-, and z-intercepts of every tangent plane to the graph of $\sqrt{x} + \sqrt{y} + \sqrt{z} = \sqrt{a}$, $a > 0$, is the number a.

37. Show that every normal line to the graph of $x^2 + y^2 + z^2 = a^2$ passes through the origin.

38. Two surfaces are said to be **orthogonal** at a point P of intersection if their normal lines are perpendicular at P. Prove that if $\nabla F(x_0, y_0, z_0) \neq \mathbf{0}$ and $\nabla G(x_0, y_0, z_0) \neq \mathbf{0}$, then the surfaces given by $F(x, y, z) = 0$ and $G(x, y, z) = 0$ are orthogonal at $P(x_0, y_0, z_0)$ if and only if

$$F_x G_x + F_y G_y + F_z G_z = 0$$

at P.

In Problems 39 and 40, use the result of Problem 38 to show that the given surfaces are orthogonal at a point of intersection. The surfaces in Problem 39 are shown in FIGURE 13.7.9.

39. $x^2 + y^2 + z^2 = 25$; $x^2 + y^2 - z^2 = 0$

40. $x^2 - y^2 + z^2 = 4$; $z = 1/xy^2$

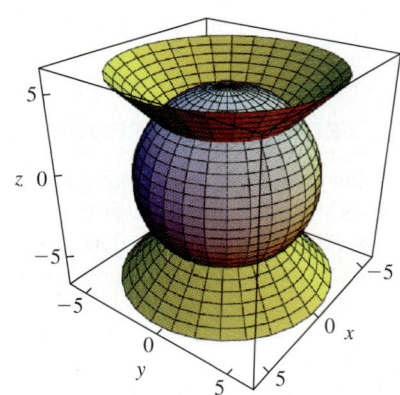

FIGURE 13.7.9 Orthogonal surfaces in Problem 39

13.8 Extrema of Multivariable Functions

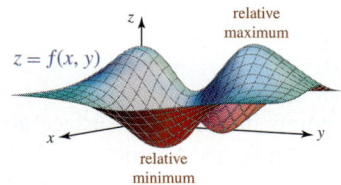

FIGURE 13.8.1 Relative extrema of f

▌ Introduction As shown in FIGURE 13.8.1 a function f of two variables can have relative maxima and relative minima. In this section we explore a way of finding these extrema. Since many of the concepts considered in this section are the three-dimensional counterparts of important definitions and theorems in Chapter 4 for functions of a single variable, a review of Sections 4.3 and 4.7 is recommended.

▌ Extrema We begin with the definition of **relative** or **local extrema** for functions of two variables x and y.

Definition 13.8.1   Relative Extrema

(*i*) A number $f(a, b)$ is a **relative maximum** of a function $z = f(x, y)$ if $f(x, y) \leq f(a, b)$
for all (x, y) in some open disk containing (a, b).

(*ii*) A number $f(a, b)$ is a **relative minimum** of a function $z = f(x, y)$ if $f(x, y) \geq f(a, b)$
for all (x, y) in some open disk containing (a, b).

Suppose for the sake of illustration that (a, b) is an interior point of a rectangular region
R at which f has a relative maximum at a point $(a, b, f(a, b))$ and, furthermore, suppose that
the first partial derivatives of f exist at (a, b). Then as seen in FIGURE 13.8.2, on the curve C_1
of intersection of the surface and the plane $x = a$, the tangent line at $(a, b, f(a, b))$ is horizontal and so its slope at the point is $f_y(a, b) = 0$. Similarly, on the curve C_2, which is the
trace of the surface in the plane $y = b$, we have $f_x(a, b) = 0$.

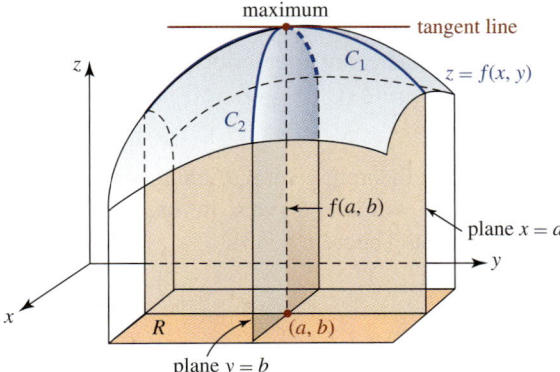

FIGURE 13.8.2   Relative maximum of a function f

Put another way, we can argue as we did in 2-space that a point on the graph of $y = f(x)$
where the *tangent line* is horizontal often leads to a relative extremum. In 3-space we can
look for a horizontal *tangent plane* to the graph of a function $z = f(x, y)$. If f has a relative
maximum or minimum at a point (a, b) and the first partials exist at the point, then an equation of the tangent plane at $(a, b, f(a, b))$ is

$$z - f(a, b) = f_x(a, b)(x - a) + f_y(a, b)(y - b). \tag{1}$$

If the plane is horizontal, its equation must be $z = $ constant, or more specifically, $z = f(a, b)$.
Using this last fact, we can conclude from (1) that we must have $f_x(a, b) = 0$ and $f_y(a, b) = 0$.
This discussion suggests the next theorem.

Theorem 13.8.1   Relative Extrema

If a function $z = f(x, y)$ has a relative extremum at a point (a, b) and if the first partial derivatives exist at this point, then

$$f_x(a, b) = 0 \quad \text{and} \quad f_y(a, b) = 0.$$

▪ **Critical Points**   In Section 4.3 we defined a **critical number** c of a function f of a single
variable x to be a number in its domain for which either $f'(c) = 0$ or $f'(c)$ does not exist. In
the definition that follows we define a **critical point** of a function f of two variables x and y.

Definition 13.8.2   Critical Points

A **critical point** of a function $z = f(x, y)$ is a point (a, b) in the domain of f for which
$f_x(a, b) = 0$ and $f_y(a, b) = 0$, or if one of its partial derivatives does not exist at the point.

The critical points correspond to points where f could *possibly* have a relative extremum. In some texts critical points are also called **stationary points**. We note in the case where the first partials exist, a critical point (a, b) is found by solving the equations

$$f_x(x, y) = 0 \quad \text{and} \quad f_y(x, y) = 0$$

simultaneously.

EXAMPLE 1 Critical Points

Find all critical points for $f(x, y) = x^3 + y^3 - 27x - 12y$.

Solution The first partial derivatives are

$$f_x(x, y) = 3x^2 - 27 \quad \text{and} \quad f_y(x, y) = 3y^2 - 12.$$

Hence, $f_x(x, y) = 0$ and $f_y(x, y) = 0$ imply that

$$x^2 = 9 \quad \text{and} \quad y^2 = 4$$

and so $x = \pm 3$, $y = \pm 2$. Thus, there are four critical points $(3, 2)$, $(-3, 2)$, $(3, -2)$, and $(-3, -2)$. ∎

▪ Second Partials Test The next theorem gives sufficient conditions for ascertaining relative extrema. The proof of the theorem will not be given. In rough terms, Theorem 13.8.2 is analogous to the Second Derivative Test (Theorem 4.7.3).

Theorem 13.8.2 Second Partials Test

Let (a, b) be a critical point of $z = f(x, y)$ and suppose f_{xx}, f_{yy}, and f_{xy} are continuous on a disk centered at (a, b). Let

$$D(x, y) = f_{xx}(x, y)f_{yy}(x, y) - [f_{xy}(x, y)]^2.$$

(*i*) If $D(a, b) > 0$ and $f_{xx}(a, b) > 0$, then $f(a, b)$ is a **relative minimum**.
(*ii*) If $D(a, b) > 0$ and $f_{xx}(a, b) < 0$, then $f(a, b)$ is a **relative maximum**.
(*iii*) If $D(a, b) < 0$, then $(a, b, f(a, b))$ is **not a relative extremum**.
(*iv*) If $D(a, b) = 0$, then the test is **inconclusive**.

Review Section 4.7 for the relationship between the second derivative and concavity.

If you are comfortable working with determinants, the function $D(x, y)$ can be written as

$$D(x, y) = \begin{vmatrix} f_{xx}(x, y) & f_{xy}(x, y) \\ f_{xy}(x, y) & f_{yy}(x, y) \end{vmatrix}.$$

EXAMPLE 2 Using the Second Partials Test

Find the extrema for $f(x, y) = 4x^2 + 2y^2 - 2xy - 10y - 2x$.

Solution The first partial derivatives are

$$f_x(x, y) = 8x - 2y - 2 \quad \text{and} \quad f_y(x, y) = 4y - 2x - 10.$$

Solving the simultaneous equations

$$8x - 2y = 2 \quad \text{and} \quad -2x + 4y = 10$$

yields the single critical point $(1, 3)$. Now,

$$f_{xx}(x, y) = 8, \quad f_{yy}(x, y) = 4, \quad f_{xy}(x, y) = -2$$

and so $D(x, y) = (8)(4) - (-2)^2 = 28$. Because $D(1, 3) > 0$ and $f_{xx}(1, 3) > 0$, it follows from part (*i*) of Theorem 13.8.2 that $f(1, 3) = -16$ is a relative minimum. ∎

EXAMPLE 3   Using the Second Partials Test

The graph of $f(x, y) = y^2 - x^2$ is the hyperbolic paraboloid given in FIGURE 13.8.3. From $f_x(x, y) = -2x$ and $f_y(x, y) = 2y$ we see that $(0, 0)$ is a critical point and that $f(0, 0) = 0$ is the only possible extremum of the function. But before using the Second Partials Test, observe

$$f(0, y) = y^2 \geq 0 \qquad \text{and} \qquad f(x, 0) = -x^2 \leq 0$$

indicates that in a neighborhood of $(0, 0)$, the points along the y-axis correspond to function values that are *greater* than or equal to $f(0, 0) = 0$ and the points along the x-axis correspond to function values that are *less* than or equal to $f(0, 0) = 0$. Hence, we can argue to the fact that $f(0, 0) = 0$ is not an extremum.

The foregoing conclusion is consistent with the results of the Second Partials Test. From $f_{xx}(x, y) = -2, f_{yy}(x, y) = 2, f_{xy}(x, y) = 0$ we see that at the critical point $(0, 0)$,

$$D(0, 0) = f_{xx}(0, 0)f_{yy}(0, 0) - [f_{xy}(0, 0)]^2$$
$$= (-2)(2) - (0)^2 = -4 < 0.$$

Hence, we conclude from part (*iii*) of Theorem 13.8.2 that $f(0, 0) = 0$ is not a relative extremum. ∎

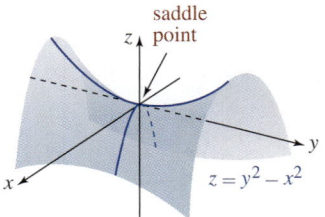

FIGURE 13.8.3   Hyperbolic paraboloid in Example 3

The point $(0, 0)$ in Example 3 is said to be a **saddle point** of the function. In general, the critical point (a, b) in case (*iii*) of Theorem 13.8.2 is a saddle point. If $D(a, b) < 0$ for a critical point (a, b), then the graph of the function f behaves essentially like the saddle-shaped hyperbolic paraboloid in a neighborhood of (a, b).

EXAMPLE 4   Saddle Point

Find the extrema for $f(x, y) = 4xy - x^2 - y^2 - 14x + 4y + 10$.

Solution   The first partial derivatives are $f_x(x, y) = 4y - 2x - 14$ and $f_y(x, y) = 4x - 2y + 4$. We then find that the only solution of the system

$$4y - 2x - 14 = 0 \qquad \text{and} \qquad 4x - 2y + 4 = 0$$

is $x = 1$ and $y = 4$; that is, $(1, 4)$ is a critical point. Now $f_{xx}(x, y) = -2, f_{yy}(x, y) = -2$, and $f_{xy}(x, y) = 4$ show that

$$D(1, 4) = (-2)(-2) - (4)^2 < 0$$

and so $f(1, 4)$ is not an extremum because $(1, 4)$ is a saddle point. The computer generated graph of f in FIGURE 13.8.4 suggests the characteristic hyperbolic-paraboloid shape in a close proximity to $(1, 4)$. ∎

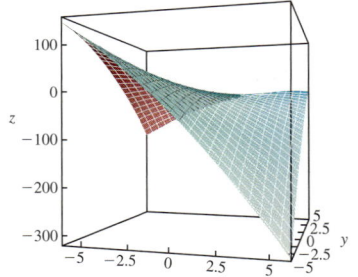

FIGURE 13.8.4   Graph of function in Example 4

EXAMPLE 5   Using the Second Partials Test

Find the extrema of $f(x, y) = x^3 + y^3 - 3x^2 - 3y^2 - 9x$.

Solution   From the partial derivatives

$$f_x(x, y) = 3x^2 - 6x - 9 = 3(x - 3)(x + 1), \qquad f_y(x, y) = 3y^2 - 6y = 3y(y - 2)$$

and the equations

$$(x - 3)(x + 1) = 0 \qquad \text{and} \qquad y(y - 2) = 0$$

we find that there are four critical points: $(3, 0), (3, 2), (-1, 0), (-1, 2)$. Since

$$f_{xx} = 6x - 6, \quad f_{yy} = 6y - 6, \quad f_{xy} = 0$$

we find $D(x, y) = 36(x - 1)(y - 1)$. The second partial derivatives test is summarized in the accompanying table.

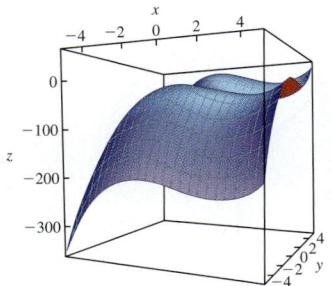

FIGURE 13.8.5 Graph of function in Example 5

Critical Point (a, b)	$D(a, b)$	$f_{xx}(a, b)$	$f(a, b)$	Conclusion
$(3, 0)$	negative	positive	-27	no extremum
$(3, 2)$	positive	positive	-31	rel. minimum
$(-1, 0)$	positive	negative	5	rel. maximum
$(-1, 2)$	negative	negative	1	no extremum

A study of the graph of f in **FIGURE 13.8.5** clearly shows the maximum and minimum. ∎

Extrema on Closed Bounded Sets Recall the Extreme Value Theorem for a function f of one variable x (Theorem 4.3.1) states that if f is continuous on a closed interval $[a, b]$, then f always possesses an absolute maximum and an absolute minimum on the interval. We also saw that these absolute extrema on $[a, b]$ occurred either at an endpoint of the interval or at a critical number c in the open interval (a, b). We next present the **Extreme Value Theorem** for a function f of two variables x and y that is continuous on a closed and bounded set R in the xy-plane.

Theorem 13.8.3 Extreme Value Theorem

A function f of two variables x and y that is continuous on a closed and bounded set R always has an **absolute maximum** and an **absolute minimum** on R.

In other words, when $x = f(x, y)$ is continuous on R, there are numbers $f(x_1, y_1)$ and $f(x_2, y_2)$ such that $f(x_1, y_1) \leq f(x, z) \leq f(x_2, y_2)$ for all (x, y) in R. The values $f(x_1, y_1)$ and $f(x_2, y_2)$ are the absolute maximum and the absolute minimum of f, respectively, on the closed set R.

Analogous to endpoint extrema, a function of two variables can have **boundary extrema**; that is, extrema on the boundary of the closed set.

Guidelines for Finding Extrema on a Closed Bounded Set R

(*i*) Find the values of f at the critical points of f in R.
(*ii*) Find all extreme values of f on the boundary of R.

The largest function value in the list of values obtained from steps (*i*) and (*ii*) is the absolute maximum of f on R; the smallest function value from this list is the absolute minimum of f on R.

EXAMPLE 6 Finding Absolute Extrema

Recall, R is called a *closed disk.*

Since $f(x, y) = 6x^2 - 8x + 2y^2 - 5$ is a polynomial function it is continuous on the closed set R defined by $x^2 + y^2 \leq 1$. Find its absolute extrema on R.

Solution We first find any critical points of f in the interior of R. From $f_x(x, y) = 12x - 8$ and $f_y(x, y) = 4y$, and

$$12x - 8 = 0, \qquad 4y = 0$$

we get the critical point $\left(\frac{2}{3}, 0\right)$. Because $\left(\frac{2}{3}\right)^2 + 0^2 < 1$ the point is in the interior of R.

In order to examine f on the boundary of the region, we represent the circle $x^2 + y^2 = 1$ by means of the parametric equations $x = \cos t$, $y = \sin t$, $0 \leq t \leq 2\pi$. Thus, on the boundary we can write f as a function of a single variable t:

$$F(t) = f(\cos t, \sin t) = 6\cos^2 t - 8\cos t + 2\sin^2 t - 5.$$

We now proceed as in Section 4.3. Differentiating F with respect to t and simplifying give

$$F'(t) = 8 \sin t (-\cos t + 1).$$

Hence, $F'(t) = 0$ implies $\sin t = 0$ or $\cos t = 1$. From these equations we find that the only critical number of F in the open interval $(0, 2\pi)$ is $t = \pi$. At this number $x = \cos \pi = -1$, $y = \sin \pi = 0$ so the corresponding point in R is $(-1, 0)$. The endpoints of the parameter-interval $[0, 2\pi]$, $t = 0$ and $t = 2\pi$, both correspond to the point $(1, 0)$ in R. From the function values

$$f\left(\frac{2}{3}, 0\right) = -\frac{23}{3}, \qquad f(-1, 0) = 9, \qquad f(1, 0) = -7$$

we see that the absolute maximum of f on R is $f(-1, 0) = 9$ and the absolute minimum is $f\left(\frac{2}{3}, 0\right) = -\frac{23}{3}$. ∎

In Example 6, we can understand what is going on by completing the square in x and rewriting the function f as

$$f(x, y) = 6\left(x - \frac{2}{3}\right)^2 + 2(y - 0)^2 - \frac{23}{3}. \tag{2}$$

From (2) it is apparent that the "vertex" of the paraboloid corresponds to the interior point $\left(\frac{2}{3}, 0\right)$ of the closed disk defined by $x^2 + y^2 \leq 1$ and that $f\left(\frac{2}{3}, 0\right) = -\frac{23}{3}$. **FIGURE 13.8.6(a)** shows one viewpoint of the graph of f; in Figure 13.8.6(b) we have superimposed the graphs of $z = 6x^2 - 8x + 2y^2 - 5$ and the cylinder defined by $x^2 + y^2 = 1$ on the same coordinate axes. In part (b) of the figure the boundary extremum $f(-1, 0) = 9$ is marked by the red point.

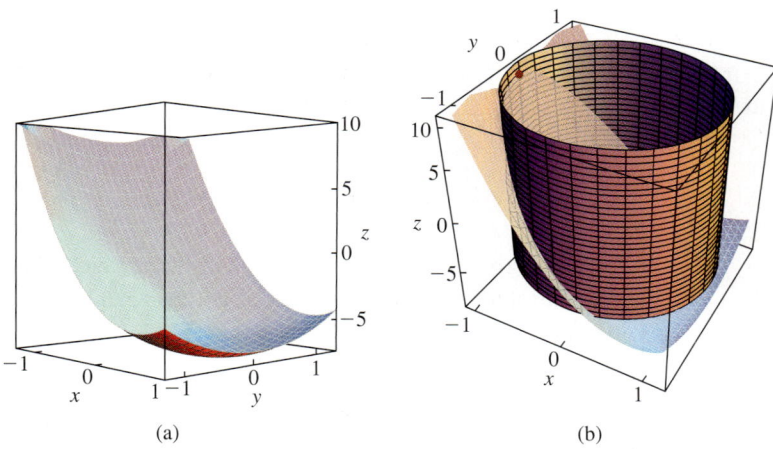

(a)   (b)

FIGURE 13.8.6 Graph of function in (a); intersection of cylinder and surface in (b)

$\dfrac{\partial z}{\partial x}$ NOTES FROM THE CLASSROOM

(*i*) The Second Partials Test has an inclusive case just like the Second Derivative Test. Recall, if c is a critical number of a function $y = f(x)$, then part (*iii*) of Theorem 4.7.3 directs us to use the First Derivative Test when $f''(c) = 0$. Unfortunately, for functions of two variables there is no convenient first derivative test to fall back on when (a, b) is a critical point for which $D(a, b) = 0$.

(*ii*) The method of solution for the system

$$f_x(x, y) = 0, \qquad f_y(x, y) = 0$$

will not always be obvious, especially when f_x and f_y are not linear. Do not be afraid to exercise your algebraic skills in the problems that follow.

Exercises 13.8 Answers to selected odd-numbered problems begin on page ANS-42.

≡ Fundamentals

In Problems 1–20, find any relative extrema of the given function.

1. $f(x, y) = x^2 + y^2 + 5$
2. $f(x, y) = 4x^2 + 8y^2$
3. $f(x, y) = -x^2 - y^2 + 8x + 6y$
4. $f(x, y) = 3x^2 + 2y^2 - 6x + 8y$
5. $f(x, y) = 5x^2 + 5y^2 + 20x - 10y + 40$
6. $f(x, y) = -4x^2 - 2y^2 - 8x + 12y + 5$
7. $f(x, y) = 4x^3 + y^3 - 12x - 3y$
8. $f(x, y) = -x^3 + 2y^3 + 27x - 24y + 3$
9. $f(x, y) = 2x^2 + 4y^2 - 2xy - 10x - 2y + 2$
10. $f(x, y) = 5x^2 + 5y^2 + 5xy - 10x - 5y + 18$
11. $f(x, y) = (2x - 5)(y - 4)$
12. $f(x, y) = (x + 5)(2y + 6)$
13. $f(x, y) = -2x^3 - 2y^3 + 6xy + 10$
14. $f(x, y) = x^3 + y^3 - 6xy + 27$
15. $f(x, y) = xy - \dfrac{2}{x} - \dfrac{4}{y} + 8$
16. $f(x, y) = -3x^2y - 3xy^2 + 36xy$
17. $f(x, y) = xe^x \sin y$
18. $f(x, y) = e^{y^2 - 3y + x^2 + 4x}$
19. $f(x, y) = \sin x + \sin y$
20. $f(x, y) = \sin xy$

21. Find three positive numbers whose sum is 21 such that their product P is a maximum. [*Hint*: Express P as a function of only two variables.]

22. Find the dimensions of a rectangular box with a volume of 1 ft³ that has a minimal surface area S.

23. Find the point on the plane $x + 2y + z = 1$ closest to the origin. [*Hint*: Consider the square of the distance.]

24. Find the least distance between the point $(2, 3, 1)$ and the plane $x + y + z = 1$.

25. Find all points on the surface $xyz = 8$ that are closest to the origin. Find the least distance.

26. Find the shortest distance between the lines whose parametric equations are
$$L_1: x = t, \ y = 4 - 2t, \ z = 1 + t,$$
$$L_2: x = 3 + 2s, \ y = 6 + 2s, \ z = 8 - 2s.$$
At what points on the lines does the minimum occur?

27. Find the maximum volume of a rectangular box with sides parallel to the coordinate planes that can be inscribed in the ellipsoid
$$\frac{x^2}{a^2} + \frac{y^2}{b^2} + \frac{z^2}{c^2} = 1, \quad a > 0, b > 0, c > 0.$$

28. The volume of an ellipsoid
$$\frac{x^2}{a^2} + \frac{y^2}{b^2} + \frac{z^2}{c^2} = 1, \quad a > 0, b > 0, c > 0$$
is $V = \frac{4}{3}\pi abc$. Show that the ellipsoid of greatest volume that satisfies $a + b + c = $ constant is a sphere.

29. The pentagon shown in FIGURE 13.8.7, formed by an isosceles triangle surmounted on a rectangle, has a fixed perimeter P. Find x, y, and θ so that the area of the pentagon is a maximum.

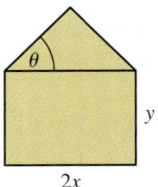

FIGURE 13.8.7 Pentagon in Problem 29

30. A 24-in. wide piece of tin is bent into a trough whose cross section is an isosceles trapezoid. See FIGURE 13.8.8. Find x and θ so that the cross-sectional area is a maximum. What is the maximum area?

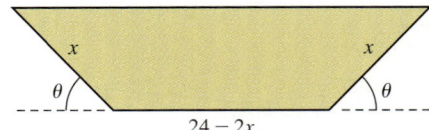

FIGURE 13.8.8 Trapezoidal cross section in Problem 30

In Problems 31–34, show that the given function has an absolute extremum but that Theorem 13.8.2 is not applicable.

31. $f(x, y) = 16 - x^{2/3} - y^{2/3}$
32. $f(x, y) = 1 - x^4 y^2$
33. $f(x, y) = 5x^2 + y^4 - 8$
34. $f(x, y) = \sqrt{x^2 + y^2}$

In Problems 35–38, find the absolute extrema of the given continuous function over the closed region R defined by $x^2 + y^2 \leq 1$.

35. $f(x, y) = x + \sqrt{3}y$
36. $f(x, y) = xy$
37. $f(x, y) = x^2 + xy + y^2$
38. $f(x, y) = -x^2 - 3y^2 + 4y + 1$

39. Find the absolute extrema of $f(x, y) = 4x - 6y$ over the closed region R defined by $\frac{1}{4}x^2 + y^2 \leq 1$.

40. Find the absolute extrema of $f(x, y) = xy - 2x - y + 6$ over the closed triangular region R with vertices $(0, 0)$, $(0, 8)$, and $(4, 0)$.

41. The function $f(x, y) = \sin xy$ is continuous on the closed rectangular region R defined by $0 \leq x \leq \pi, 0 \leq y \leq 1$.
    (a) Find the critical points in the region.
    (b) Find the points where f has an absolute extremum.
    (c) Graph the function on the rectangular region.

☰ Applications

42. A revenue function is

$$R(x, y) = x(100 - 6x) + y(192 - 4y),$$

where x and y denote the number of items of two commodities sold. Given that the corresponding cost function is

$$C(x, y) = 2x^2 + 2y^2 + 4xy - 8x + 20$$

find the maximum profit, where profit = revenue − cost.

43. A closed rectangular box is to be made so that its volume is 60 ft³. The costs of the material for the top and bottom are 10 cents per square foot and 20 cents per square foot, respectively. The cost of the sides is 2 cents per square foot. Determine the cost function $C(x, y)$, where x and y are the length and width of the box, respectively. Find the dimensions of the box that will give a minimum cost.

13.9 Method of Least Squares

▌ Introduction When performing experiments, we often tabulate data in the form of ordered pairs $(x_1, y_1), (x_2, y_2), \ldots, (x_n, y_n)$, with each x_i distinct. Given the data, it is then often desirable to be able to extrapolate or predict y from x by finding a mathematical model—that is, a function that approximates or "fits" the data. In other words, we want a function $f(x)$ such that

$$f(x_1) \approx y_1, \quad f(x_2) \approx y_2 \quad , \ldots, \quad f(x_n) \approx y_n.$$

Naturally, we do not want just any function but a function that fits the data as closely as possible. In the discussion that follows we will confine our attention to the problem of finding a linear polynomial $f(x) = mx + b$ or a straight line that "best fits" the data $(x_1, y_1), (x_2, y_2), \ldots, (x_n, y_n)$. The procedure for finding this linear function is known as **the method of least squares**.

EXAMPLE 1 Fitting a Line to Data

Consider the data $(1, 1), (2, 3), (3, 4), (4, 6), (5, 5)$ shown in FIGURE 13.9.1(a). Looking at Figure 13.9.1(b) and seeing that the line $y = x + 1$ passes through two of the data points, we might take this line as the one that best fits the data.

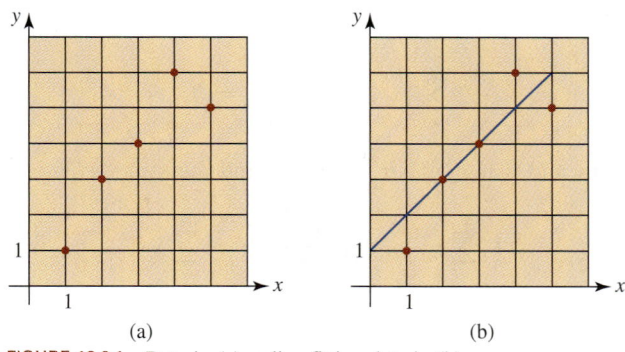

(a)   (b)

FIGURE 13.9.1 Data in (a); a line fitting data in (b)   ■

Obviously we need something better than a visual guess to determine the linear function $y = f(x)$ as in Example 1. We need a criterion that defines the concept of "best fit" or, as it is sometimes called, the "goodness of fit."

If we try to match the data points with the function $f(x) = mx + b$, then we wish to find m and b that satisfy the system of equations

$$\begin{aligned}
y_1 &= mx_1 + b \\
y_2 &= mx_2 + b \\
&\;\;\vdots \\
y_n &= mx_n + b.
\end{aligned} \tag{1}$$

Unfortunately, (1) is an *overdetermined system*; that is, the number of equations is greater than the number of unknowns. We do not expect such a system to have a solution unless, of course, the data points all lie on the same line.

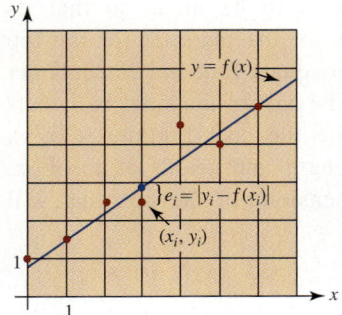

FIGURE 13.9.2 Error in approximating y_i by $f(x_i)$

■ **Least-Squares Line** If the data points are $(x_1, y_1), (x_2, y_2), \ldots, (x_n, y_n)$, then one manner of determining how well the linear function $f(x) = mx + b$ fits the data is to measure the vertical distances between the points and the graph of f:

$$e_i = |y_i - f(x_i)|, \ i = 1, 2, \ldots, n.$$

We can think of each e_i as the error in approximating the data value y_i by the function value $f(x_i)$. See **FIGURE 13.9.2**. Intuitively, the function f will fit the data well if the sum of all the e_i is a minimum. Actually, a more convenient approach to the problem is to find a linear function f so that the *sum of the squares* of all the e_i is a minimum. We define the solution of the system (1) to be those coefficients m and b that minimize the expression

$$
\begin{aligned}
E &= e_1^2 + e_2^2 + \cdots + e_n^2 \\
&= [y_1 - f(x_1)]^2 + [y_2 - f(x_2)]^2 + \cdots + [y_n - f(x_n)]^2 \\
&= [y_1 - (mx_1 + b)]^2 + [y_2 - (mx_2 + b)]^2 + \cdots + [y_n - (mx_n + b)]^2
\end{aligned}
$$

or

$$E = \sum_{i=1}^{n} [y_i - mx_i - b]^2. \tag{2}$$

The expression E is called the **sum of the square errors**. The line $y = mx + b$ that minimizes the sum of the square errors (2) is defined to be the **line of best fit** and is called the **least-squares line** or **regression line** for the data $(x_1, y_1), (x_2, y_2), \ldots, (x_n, y_n)$.

The problem remains now, how does one find m and b so that (2) is minimum? The answer can be found from the Second Partials Test, Theorem 13.8.2.

If we think of (2) as a function of two variables m and b, then to find the minimum value of E we set the first partial derivatives equal to zero:

$$\frac{\partial E}{\partial m} = 0 \quad \text{and} \quad \frac{\partial E}{\partial b} = 0.$$

The last two conditions yield in turn

$$
\begin{aligned}
-2 \sum_{i=1}^{n} x_i [y_i - mx_i - b] &= 0 \\
-2 \sum_{i=1}^{n} [y_i - mx_i - b] &= 0.
\end{aligned}
\tag{3}
$$

Expanding these sums and using $\sum_{i=1}^{n} b = nb$, we find the system (3) is the same as

$$
\begin{aligned}
\left(\sum_{i=1}^{n} x_i^2 \right) m + \left(\sum_{i=1}^{n} x_i \right) b &= \sum_{i=1}^{n} x_i y_i \\
\left(\sum_{i=1}^{n} x_i \right) m + nb &= \sum_{i=1}^{n} y_i.
\end{aligned}
\tag{4}
$$

Although we shall forego the details, the values of m and b that satisfy the system (4) yield the minimum value of E. Solving the system (4) gives

$$
m = \frac{n \sum_{i=1}^{n} x_i y_i - \sum_{i=1}^{n} x_i \sum_{i=1}^{n} y_i}{n \sum_{i=1}^{n} x_i^2 - \left(\sum_{i=1}^{n} x_i \right)^2}, \qquad b = \frac{\sum_{i=1}^{n} x_i^2 \sum_{i=1}^{n} y_i - \sum_{i=1}^{n} x_i y_i \sum_{i=1}^{n} x_i}{n \sum_{i=1}^{n} x_i^2 - \left(\sum_{i=1}^{n} x_i \right)^2}. \tag{5}
$$

EXAMPLE 2 Least-Squares Line

Find the least-squares line for the data in Example 1. Calculate the sum of the square errors E for this line and the line $y = x + 1$.

Solution From the data $(1, 1), (2, 3), (3, 4), (4, 6), (5, 5)$ we identify $x_1 = 1, x_2 = 2, x_3 = 3, x_4 = 4, x_5 = 5, y_1 = 1, y_2 = 3, y_3 = 4, y_4 = 6,$ and $y_5 = 5$. With these values and $n = 5$, we have

$$\sum_{i=1}^{5} x_i y_i = 68, \quad \sum_{i=1}^{5} x_i = 15, \quad \sum_{i=1}^{5} y_i = 19, \quad \sum_{i=1}^{5} x_i^2 = 55.$$

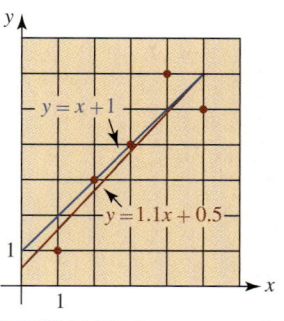

Substituting these values into the formulas in (5) yields $m = 1.1$ and $b = 0.5$. Thus, the least-squares line is $y = 1.1x + 0.5$. For this line the sum of the square errors is

$$E = [1 - f(1)]^2 + [3 - f(2)]^2 + [4 - f(3)]^2 + [6 - f(4)]^2 + [5 - f(5)]^2$$
$$= [1 - 1.6]^2 + [3 - 2.7]^2 + [4 - 3.8]^2 + [6 - 4.9]^2 + [5 - 6]^2 = 2.7.$$

For the line $y = x + 1$ that we guessed in Example 1 that also passed through two of the data points, we find the sum of the square errors is $E = 3.0$.

By way of comparison, FIGURE 13.9.3 shows the data, the line $y = x + 1$, and the least-squares line $y = 1.1x + 0.5$. ∎

FIGURE 13.9.3 Least-squares line (red) in Example 2

It is possible to generalize the least-squares technique. For example, we might want to fit the given data to a quadratic polynomial $f(x) = ax^2 + bx + c$ instead of a linear polynomial.

Exercises 13.9   Answers to selected odd-numbered problems begin on page ANS-43.

≡ Fundamentals

In Problems 1–6, find the least-squares line for the given data.

1. (2, 1), (3, 2), (4, 3), (5, 2)

2. (0, −1), (1, 3), (2, 5), (3, 7)

3. (1, 1), (2, 1.5), (3, 3), (4, 4.5), (5, 5)

4. (0, 0), (2, 1.5), (3, 3), (4, 4.5), (5, 5)

5. (0, 2), (1, 3), (2, 5), (3, 5), (4, 9), (5, 8), (6, 10)

6. (1, 2), (2, 2.5), (3, 1), (4, 1.5), (5, 2), (6, 3.2), (7, 5)

≡ Applications

7. In an experiment the correspondence given in the table was found between temperature T (in °C) and kinematic viscosity ν (in Centistokes) of an oil with a certain additive. Find the least-squares line $\nu = mT + b$. Use this line to estimate the viscosity of the oil at $T = 140$ and $T = 160$.

T	20	40	60	80	100	120
ν	220	200	180	170	150	135

8. In an experiment the correspondence given in the table was found between temperature T (in °C) and electrical resistance R (in milliohms). Find the least-squares line $R = mT + b$. Use this line to estimate the resistance at $T = 700$.

T	400	450	500	550	600	650
R	0.47	0.90	2.0	3.7	7.5	15

≡ Calculator/CAS Problems

9. **(a)** A set of data points can be approximated by a least-squares *polynomial* of degree n. Learn the syntax for the CAS you have at hand to obtain a least-squares line (linear polynomial), a least-squares quadratic, and a least-squares cubic to fit the data

$$(-5.5, 0.8), \quad (-3.3, 2.5), \quad (-1.2, 3.8),$$
$$(0.7, 5.2), \quad (2.5, 5.6), \quad (3.8, 6.5).$$

(b) Use a CAS to superimpose the plots of the data and the least-squares line obtained in part (a) on the same coordinate axes. Repeat for the plots of the data and the least-squares quadratic and then the data and the least-squares cubic.

10. Use the U.S. census data (in millions) from the year 1900 through 2000

1900	1920	1940	1960	1980	2000
75.994575	105.710620	131.669275	179.321750	226.545805	281.421906

and a least-squares line to predict the U.S. population in the year 2020.

13.10 Lagrange Multipliers

▊ **Introduction**   In Problems 21–30 of Exercises 13.8 you were asked to find the maximum or minimum of a function subject to a given side condition or constraint. The side condition was used to eliminate one of the variables in the function so that the Second Partials Test (Theorem 13.8.2) was applicable. In the present discussion, we examine another procedure for determining the so-called **constrained extrema** of a function.

Before defining that concept, let us consider an example.

EXAMPLE 1 Constrained Extremum

Determine geometrically whether the function $f(x, y) = 9 - x^2 - y^2$ has an extremum when the variables x and y are constrained by $x + y = 3$.

Solution As seen in FIGURE 13.10.1, the graph of $x + y = 3$ is a vertical plane that intersects the paraboloid given by $f(x, y) = 9 - x^2 - y^2$. It appears from the figure that the function has a *constrained maximum* for some x_1 and y_1 satisfying $0 < x_1 < 3, 0 < y_1 < 3$, and $x_1 + y_1 = 3$. The table of numerical values accompanying the figure would also seem to indicate that this new maximum is $f(1.5, 1.5) = 4.5$. Note that we cannot use numbers such as $x = 1.7$ and $y = 2.4$, since these values do not satisfy the constraint $x + y = 3$.

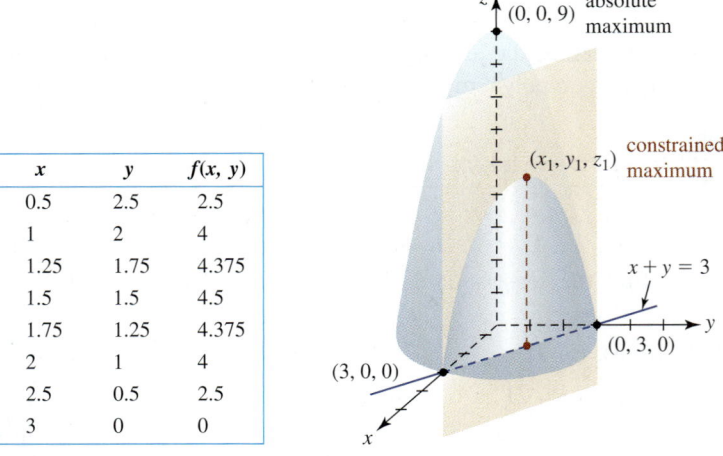

x	y	$f(x, y)$
0.5	2.5	2.5
1	2	4
1.25	1.75	4.375
1.5	1.5	4.5
1.75	1.25	4.375
2	1	4
2.5	0.5	2.5
3	0	0

FIGURE 13.10.1 Graph of function and constraint in Example 1

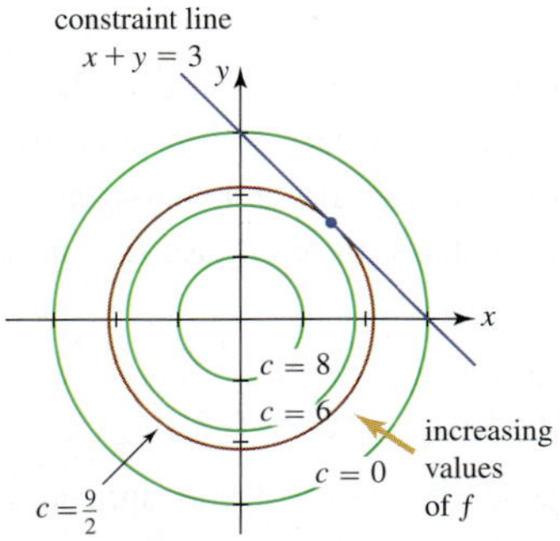

Alternatively, we can analyze Example 1 by means of level curves. As shown in FIGURE 13.10.2, increasing function values of f correspond to increasing values of c in the level curves $9 - x^2 - y^2 = c$. The maximum value of f (that is, c) subject to the constraint occurs where the level curve corresponding to $c = \frac{9}{2}$ intersects, or more precisely is tangent to, the line $x + y = 3$. By solving $x^2 + y^2 = \frac{9}{2}$ and $x + y = 3$ simultaneously, we find the point of tangency is $\left(\frac{3}{2}, \frac{3}{2}\right)$.

FIGURE 13.10.2 Level curves and constraint line

Functions of Two Variables

To generalize the foregoing discussion, suppose we wish to:

- *Find extrema of the function $z = f(x, y)$ subject to a constraint given by $g(x, y) = 0$.*

It seems plausible from FIGURE 13.10.3 that to find, say, a constrained maximum of f, we need only find the highest level curve $f(x, y) = c$ that is tangent to the graph of the constraint equation $g(x, y) = 0$. Now, recall that the gradients ∇f and ∇g are perpendicular to the curves $f(x, y) = c$ and $g(x, y) = 0$, respectively. Hence, if $\nabla g \neq \mathbf{0}$ at a point P of tangency of the curves, then ∇f and ∇g are parallel at P; that is, they lie along a common normal. Therefore, for some nonzero scalar λ (the lowercase Greek letter lambda), we must have $\nabla f = \lambda \nabla g$. We state this result in a formal fashion.

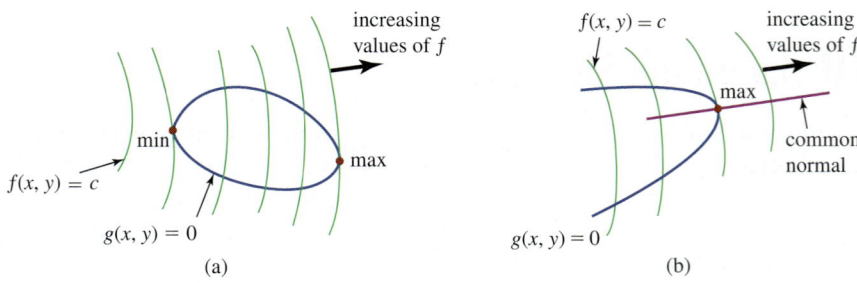

FIGURE 13.10.3 Level curves of f (green); constraint equation (blue)

Theorem 13.10.1 Lagrange's Theorem

Suppose the function $z = f(x, y)$ has an extremum at a point (x_0, y_0) on the graph of the constraint equation $g(x, y) = 0$. If f and g have continuous first partial derivatives in an open set containing the graph of the constraint equation and $\nabla g(x_0, y_0) \neq \mathbf{0}$, then there exists a real number λ such that $\nabla f(x_0, y_0) = \lambda \nabla g(x_0, y_0)$.

■ **Method of Lagrange Multipliers** The real number λ for which $\nabla f = \lambda \nabla g$ is called a **Lagrange multiplier**. After equating components, the equation $\nabla f = \lambda \nabla g$ is equivalent to

$$f_x(x, y) = \lambda g_x(x_0, y_0), \qquad f_y(x, y) = \lambda g_y(x, y).$$

If f has a constrained extremum at a point (x_0, y_0), then we have just seen there is a number λ such that

$$f_x(x_0, y_0) = \lambda g_x(x_0, y_0)$$
$$f_y(x_0, y_0) = \lambda g_y(x_0, y_0) \qquad (1)$$
$$g(x_0, y_0) = 0.$$

The equations in (1) suggest the following procedure, known as the **method of Lagrange multipliers**, for finding constrained extrema.

Guidelines for the Method of Lagrange Multipliers

(*i*) To find the extrema of $z = f(x, y)$ subject to the constraint $g(x, y) = 0$, solve the system of equations

$$f_x(x, y) = \lambda g_x(x, y)$$
$$f_y(x, y) = \lambda g_y(x, y) \qquad (2)$$
$$g(x, y) = 0.$$

(*ii*) Among the solutions (x, y, λ) of the system (2) will be the points (x_i, y_i), where f has an extremum. When f has a maximum (minimum), it will be the largest (or smallest) number in the list of function values $f(x_i, y_i)$.

EXAMPLE 2 Example 1 Revisited

Use the method of Lagrange multipliers to find the maximum of $f(x, y) = 9 - x^2 - y^2$ subject to $x + y = 3$.

Solution With $g(x, y) = x + y - 3$ and $f_x = -2x, f_y = -2y, g_x = 1, g_y = 1$ the system in (2) is

$$-2x = \lambda$$
$$-2y = \lambda$$
$$x + y - 3 = 0.$$

Equating the first and second equations gives $-2x = -2y$ or $x = y$. Substituting this result into the third equation yields $2y - 3 = 0$ or $y = \frac{3}{2}$. Thus, $x = y = \frac{3}{2}$ and the constrained maximum is $f\left(\frac{3}{2}, \frac{3}{2}\right) = \frac{9}{2}$. ■

EXAMPLE 3 Using Lagrange Multipliers

Find the extrema $f(x, y) = y^2 - 4x$ subject to $x^2 + y^2 = 9$.

Solution If we define $g(x, y) = x^2 + y^2 - 9$, then $f_x = -4, f_y = 2y, g_x = 2x$, and $g_y = 2y$. Therefore, (2) becomes

$$-4 = 2x\lambda$$
$$2y = 2y\lambda \qquad (3)$$
$$x^2 + y^2 - 9 = 0.$$

From the second of these equations, $y(1 - \lambda) = 0$, we see that either $y = 0$ or $\lambda = 1$. First, if $y = 0$, the third equation in the system gives $x^2 = 9$ or $x = \pm 3$. Hence, $(-3, 0)$ and $(3, 0)$ are solutions of the system and are points at which f *might* have an extremum. Continuing, if $\lambda = 1$, then the first equation yields $x = -2$. Substituting this value into $x^2 + y^2 - 9 = 0$ gives $y^2 = 5$ or $y = \pm\sqrt{5}$. Two more solutions of the system are $(-2, -\sqrt{5})$ and $(-2, \sqrt{5})$. From the list of function values

$$f(-3, 0) = 12, \quad f(3, 0) = -12, \quad f(-2, -\sqrt{5}) = 13, \quad \text{and} \quad f(-2, \sqrt{5}) = 13$$

we conclude that f has a constrained minimum of -12 at $(3, 0)$ and a constrained maximum of 13 at $(-2, -\sqrt{5})$ and at $(-2, \sqrt{5})$.

FIGURE 13.10.4(a) shows the graph $f(x, y) = y^2 - 4x$ intersecting the cylinder defined by the constraint equation $x^2 + y^2 = 9$. The four points we found by solving (3) lie in the xy-plane on the circle of radius 3; the three constrained extrema correspond to points $(3, 0, -12)$, $(-2, -\sqrt{5}, 13)$, and $(-2, \sqrt{5}, 13)$ in 3-space on the curve of intersection of the surface and the circular cylinder. Alternatively, Figure 13.10.4(b) shows three level curves from $y^2 - 4x = c$. Two of the level curves are tangent to the circle $x^2 + y^2 = 9$.

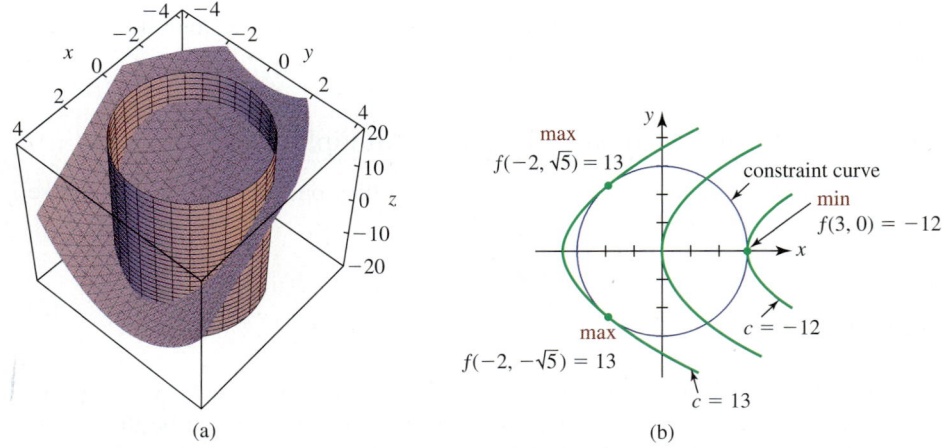

(a)   (b)

FIGURE 13.10.4 Intersection of cylinder and surface in (a); level curves of f and constraint equation in (b) ∎

In applying the method of Lagrange multipliers, we really are not very interested in finding the values of λ that satisfy the system (2). Did you notice in Example 1 that we did not even bother to find λ? In Example 3, we used the value $\lambda = 1$ to help find $x = -2$, but after that we ignored it.

EXAMPLE 4   Minimum Cost

A closed right circular cylinder will have a volume of 1000 ft³. The top and bottom of the cylinder are made of metal that costs \$2 per square foot. The lateral side is wrapped in metal costing \$2.50 per square foot. Find the minimum cost of construction.

Solution The cost function is

$$\underset{\substack{\text{cost of bottom} \\ \text{and top} \downarrow}}{} \qquad \underset{\substack{\text{cost of side} \\ \downarrow}}{}$$

$$C(r, h) = 2(2\pi r^2) + 2.5(2\pi rh)$$
$$= 4\pi r^2 + 5\pi rh.$$

Now, from the constraint $1000 = \pi r^2 h$, we can identify $g(r, h) = \pi r^2 h - 1000$, and so the first partial derivatives are $C_r = 8\pi r + 5\pi h$, $C_h = 5\pi r$, $g_r = 2\pi rh$, and $g_h = \pi r^2$. We must then solve the system

$$8\pi r + 5\pi h = 2\pi rh\lambda$$
$$5\pi r = \pi r^2\lambda \qquad (4)$$
$$\pi r^2 h - 1000 = 0.$$

By multiplying the first equation by r and the second equation by $2h$ and subtracting yield

$$8\pi r^2 - 5\pi rh = 0 \qquad \text{or} \qquad \pi r(8r - 5h) = 0.$$

Since $r = 0$ does not satisfy the constraint equation, we take $r = \frac{5}{8}h$. The constraint then gives

$$h^3 = \frac{1000 \cdot 64}{25\pi} \qquad \text{or} \qquad h = \frac{40}{\sqrt[3]{25\pi}}.$$

Thus, $r = 25/\sqrt[3]{25\pi}$ and the only solution of (4) is $\left(25/\sqrt[3]{25\pi}, 40/\sqrt[3]{25}\right)$.

The constrained minimum cost is

$$C\left(\frac{25}{\sqrt[3]{25\pi}}, \frac{40}{\sqrt[3]{25\pi}}\right) = 4\pi\left(\frac{25}{\sqrt[3]{25\pi}}\right)^2 + 5\pi\left(\frac{25}{\sqrt[3]{25\pi}}\right)\left(\frac{40}{\sqrt[3]{25\pi}}\right)$$

$$= 300\sqrt[3]{25\pi} \approx \$1284.75. \qquad \blacksquare$$

Functions of Three Variables   To find the extrema of a function of three variables $w = f(x, y, z)$ subject to the constraint $g(x, y, z) = 0$, we solve a system of four equations:

$$
\begin{aligned}
f_x(x, y, z) &= \lambda g_x(x, y, z) \\
f_y(x, y, z) &= \lambda g_y(x, y, z) \\
f_z(x, y, z) &= \lambda g_z(x, y, z) \\
g(x, y, z) &= 0.
\end{aligned}
\qquad (5)
$$

EXAMPLE 5   Function of Three Variables

Find the extrema of $f(x, y, z) = x^2 + y^2 + z^2$ subject to $2x - 2y - z = 5$.

Solution   With $g(x, y, z) = 2x - 2y - z - 5$, the system (5) is

$$
\begin{aligned}
2x &= 2\lambda \\
2y &= -2\lambda \\
2z &= -\lambda \\
2x - 2y - z - 5 &= 0.
\end{aligned}
$$

With $\lambda = x = -y = -2z$, the last equation gives $x = \frac{10}{9}$ and so $y = -\frac{10}{9}$, $z = -\frac{5}{9}$. Thus, a constrained extremum is $f\left(\frac{10}{9}, -\frac{10}{9}, -\frac{5}{9}\right) = \frac{225}{81}$. $\blacksquare$

Two Constraints   In order to optimize a function $w = f(x, y, z)$ subject to *two* constraints, $g(x, y, z) = 0$ and $h(x, y, z) = 0$, we must introduce a second Lagrange multiplier μ (the lowercase Greek letter *mu*) and solve the system

$$
\begin{aligned}
f_x(x, y, z) &= \lambda g_x(x, y, z) + \mu h_x(x, y, z) \\
f_y(x, y, z) &= \lambda g_y(x, y, z) + \mu h_y(x, y, z) \\
f_z(x, y, z) &= \lambda g_z(x, y, z) + \mu h_z(x, y, z) \\
g(x, y, z) &= 0 \\
h(x, y, z) &= 0.
\end{aligned}
\qquad (6)
$$

EXAMPLE 6   Two Constraints

Find the point on the curve C of intersection of the sphere $x^2 + y^2 + z^2 = 9$ and the plane $x - y + 3z = 6$ that is farthest from the xy-plane. Then find the point on C that is closest to the xy-plane.

Solution   FIGURE 13.10.5 suggests that there exists two such points P_1 and P_2 with nonnegative z-coordinates. The function f for which we wish to find a maximum and a minimum is simply the distance from each of these points to the xy-plane, that is, $f(x, y, z) = z$.

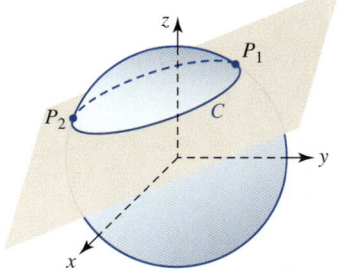

FIGURE 13.10.5   Intersection of a sphere and plane in Example 6

If we take $g(x, y, z) = x^2 + y^2 + z^2 - 9$ and $h(x, y, z) = x - y + 3z - 6$, then the system (6) is

$$0 = 2x\lambda + \mu$$
$$0 = 2y\lambda - \mu$$
$$1 = 2z\lambda + 3\mu$$
$$x^2 + y^2 + z^2 - 9 = 0$$
$$x - y + 3z - 6 = 0.$$

We add the first and second equations to obtain $2\lambda(y + x) = 0$. If $\lambda = 0$, then the first equation implies $\mu = 0$, but the third equation in the system leads to the contradiction $0 = 1$. Now if we take $y = -x$, the last two equations become

$$\begin{matrix} x^2 + x^2 + z^2 - 9 = 0 \\ x + x + 3z - 6 = 0 \end{matrix} \quad \text{or} \quad \begin{matrix} 2x^2 + z^2 = 9 \\ 2x + 3z = 6. \end{matrix}$$

By solving the last system we get

$$x = \frac{6}{11} + \frac{9}{22}\sqrt{14}, \quad z = \frac{18}{11} - \frac{3}{11}\sqrt{14}$$

and

$$x = \frac{6}{11} - \frac{9}{22}\sqrt{14}, \quad z = \frac{18}{11} + \frac{3}{11}\sqrt{14}.$$

Thus, the points on C that are farthest and closest to the xy-plane are, respectively,

$$P_1\left(\frac{6}{11} - \frac{9}{22}\sqrt{14}, -\frac{6}{11} + \frac{9}{22}\sqrt{14}, \frac{18}{11} + \frac{3}{11}\sqrt{14}\right)$$

and

$$P_2\left(\frac{6}{11} + \frac{9}{22}\sqrt{14}, -\frac{6}{11} - \frac{9}{22}\sqrt{14}, \frac{18}{11} - \frac{3}{11}\sqrt{14}\right).$$

The approximate coordinates of P_1 and P_2 are $(-0.99, 0.99, 2.66)$ and $(2.08, -2.08, 0.62)$. ■

■ **Postscript—A Bit of History** Joseph Louis Lagrange was born in 1736 as Guiseppe Lodovico Lagrangia in Turin in the Kingdom of Sardinia and died in Paris in 1813.

Lagrange was the last of his mother's eleven children and the only one to live beyond infancy. In his teens he was already a professor at the Royal Artillery School in Turin. Invited there through the efforts of Euler and D'Alembert, he spent twenty productive years at the court of Frederick the Great, until the latter's death in 1786. Thereupon Louis XVI installed him in the Louvre, where it is said he was a favorite of Marie Antoinette. He deplored the excesses of the French Revolution, yet helped the new government establish the metric system. He was the first professor of the Ècole Polytechnique, where calculus and number theory were his specialties.

Lagrange

$$\frac{\partial z}{\partial x}$$ **NOTES FROM THE CLASSROOM**

Notice in Example 5 we concluded with the vague words "a constrained extremum is." The method of Lagrange multipliers does not have a built-in indicator that flashes MAX or MIN when a single extremum is found. In addition to the graphical procedure discussed at the beginning of this section, another way of convincing oneself as to the nature of the extremum is to compare it with values obtained by calculating the given function at other points that satisfy the constraint equation. Indeed, in this manner we find that $\frac{225}{81}$ of Example 5 is actually a constrained *minimum* of the function f.

Exercises 13.10   Answers to selected odd-numbered problems begin on page ANS-43.

≡ Fundamentals

In Problems 1 and 2, sketch the graphs of the level curves of the given function f and the indicated constraint equation. Determine whether f has a constrained extremum.

1. $f(x, y) = x + 3y$, subject to $x^2 + y^2 = 1$

2. $f(x, y) = xy$, subject to $\frac{1}{2}x + y = 1$, $x \geq 0$, $y \geq 0$

In Problems 3–20, use the method of Lagrange multipliers to find the constrained extrema of the given function.

3. Problem 1

4. Problem 2

5. $f(x, y) = xy$, subject to $x^2 + y^2 = 2$

6. $f(x, y) = x^2 + y^2$, subject to $2x + y = 5$

7. $f(x, y) = 3x^2 + 3y^2 + 5$, subject to $x - y = 1$

8. $f(x, y) = 4x^2 + 2y^2 + 10$, subject to $4x^2 + y^2 = 4$

9. $f(x, y) = x^2 + y^2$, subject to $x^4 + y^4 = 1$

10. $f(x, y) = 8x^2 - 8xy + 2y^2$, subject to $x^2 + y^2 = 10$

11. $f(x, y) = x^3y$, subject to $\sqrt{x} + \sqrt{y} = 1$

12. $f(x, y) = xy^2$, subject to $x^2 + y^2 = 27$

13. $f(x, y, z) = x + 2y + z$, subject to $x^2 + y^2 + z^2 = 30$

14. $f(x, y, z) = x^2 + y^2 + z^2$, subject to $x + 2y + 3z = 4$

15. $f(x, y, z) = xyz$, subject to $x^2 + \frac{1}{4}y^2 + \frac{1}{9}z^2 = 1$,
$x > 0, y > 0, z > 0$

16. $f(x, y, z) = xyz + 5$, subject to $x^3 + y^3 + z^3 = 24$

17. $f(x, y, z) = x^3 + y^3 + z^3$, subject to $x + y + z = 1$,
$x > 0, y > 0, z > 0$

18. $f(x, y, z) = 4x^2y^2z^2$, subject to $x^2 + y^2 + z^2 = 9$,
$x > 0, y > 0, z > 0$

19. $f(x, y, z) = x^2 + y^2 + z^2$, subject to $2x + y + z = 1$,
$-x + 2y - 3z = 4$

20. $f(x, y, z) = x^2 + y^2 + z^2$, subject to $4x + z = 7$,
$z^2 = x^2 + y^2$

21. Find the maximum area of a right triangle whose perimeter is 4.

22. Find the dimensions of an open rectangular box with maximum volume if its surface area is 75 cm^3. What are the dimensions if the box is closed?

≡ Applications

23. A right cylindrical tank is surmounted by a conical cap as shown in FIGURE 13.10.6. The radius of the tank is 3 m and its total surface area is 81πm^2. Find heights x and y so that the volume of the tank is a maximum. [*Hint*: The surface area of the cone is $3\pi\sqrt{9 + y^2}$.]

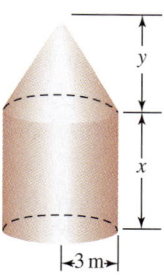

FIGURE 13.10.6   Cylinder with conical cap in Problem 23

24. In business, a utility index U is a function that gives a measure of satisfaction obtained from the purchasing of variable amounts, x and y, of two commodities that are purchased on a regular basis. If $U(x, y) = x^{1/3}y^{2/3}$ is a utility index, find its extrema subject to $x + 6y = 18$.

25. The **Haber–Bosch process*** produces ammonia by a direct union of nitrogen and hydrogen under conditions of constant pressure P and constant temperature:

$$N_2 + 3H_2 \overset{\text{catalyst}}{\rightleftharpoons} 2NH_3.$$

The partial pressures x, y, and z of hydrogen, nitrogen, and ammonia satisfy $x + y + z = P$ and the equilibrium law $z^2/xy^3 = k$, where k is a constant. The maximum amount of ammonia occurs when the maximum partial pressure of ammonia is obtained. Find the maximum value of z.

26. If a species of animal has n sources of food, the **breadth index** of its ecological niche is defined as

$$\frac{1}{x_1^2 + \cdots + x_n^2},$$

where x_i, $i = 1, 2, \ldots, n$, is the fraction of the animal's diet coming from the ith food source. For example, if a birds diet consists of 50% insects, 30% worms, and 20% seeds, the breadth index is

$$\frac{1}{(0.50)^2 + (0.30)^2 + (0.20)^2} = \frac{1}{0.25 + 0.09 + 0.04}$$

$$= \frac{1}{0.38} \approx 2.63.$$

Note that $x_1 + x_2 + \cdots + x_n = 1$ and $0 \leq x_i \leq 1$ for all i.

(a) For a species with three food sources, show that the breadth index is maximized if $x_1 = x_2 = x_3 = \frac{1}{3}$.

(b) Show that the breadth index with n sources is maximized when $x_1 = x_2 = \cdots = x_n = 1/n$.

*****Fritz Haber** (1868–1934) was a German chemist. For inventing this process, Haber won the Nobel prize in chemistry in 1918. Carl Bosch was Haber's brother-in-law and a chemical engineer who made this process practical on a large scale. Bosch won the Nobel Prize in chemistry in 1931. During World War I the German government used the Haber–Bosch process to produce large quantities of fertilizers and explosives. Haber was subsequently expelled from Germany by Adolph Hitler and died in exile.

≣ **Think About It**

27. Give a geometric interpretation of the extrema in Problem 9.

28. Give a geometric interpretation of the extrema in Problem 14.

29. Give a geometric interpretation of the extremum in Problem 19.

30. Give a geometric interpretation of the extremum in Problem 20.

31. Find the point $P(x, y)$, $x > 0$, $y > 0$, on the surface $xy^2 = 1$ that is closest to the origin. Show that the line segment from the origin to P is perpendicular to the tangent line at P.

32. Find the maximum value of $f(x, y, z) = \sqrt[3]{xyz}$ on the plane $x + y + z = k$.

33. Use the result of Problem 32 to prove the inequality

$$\sqrt[3]{xyz} \le \frac{x + y + z}{3}.$$

34. Find the point on the curve C of intersection of the cylinder $x^2 + z^2 = 1$ and the plane $x + y + 2z = 4$ that is farthest from the xz-plane. Find the point on C that is closest to the xz-plane.

Chapter 13 in Review

Answers to selected odd-numbered problems begin on page ANS-43.

A. True/False

In Problems 1–10, answer true or false.

1. If $\lim_{(x, y) \to (a, b)} f(x, y)$ has the same value for an infinite number of approaches to (a, b), then the limit exists. _____

2. The domains of the functions

$$f(x, y) = \sqrt{\ln(x^2 + y^2 - 16)} \quad \text{and} \quad g(x, y) = \ln(x^2 + y^2 - 16)$$

are the same. _____

3. The function

$$f(x, y) = \begin{cases} \dfrac{1 - \cos(x^2 + y^2)}{x^2 + y^2}, & (x, y) \ne (0, 0) \\ 0, & (x, y) = (0, 0) \end{cases}$$

is continuous at $(0, 0)$. _____

4. The function $f(x, y) = x^2 + 2xy + y^3$ is continuous everywhere. _____

5. If $\partial z / \partial x = 0$, then $z = $ constant. _____

6. If $\nabla f = \mathbf{0}$, then $f = $ constant. _____

7. ∇z is perpendicular to the graph of $z = f(x, y)$. _____

8. ∇f points in the direction in which f increases most rapidly. _____

9. If f has continuous second partial derivatives, then $f_{xy} = f_{yx}$. _____

10. If $f_x(x, y) = 0$ and $f_y(x, y) = 0$ at (a, b), then $f(a, b)$ is a relative extremum. _____

B. Fill in the Blanks

In Problems 1–12, fill in the blanks.

1. $\displaystyle\lim_{(x, y) \to (1, 1)} \frac{3x^2 + xy^2 - 3xy - 2y^3}{5x^2 - y^2} = $ _____.

2. $f(x, y) = \dfrac{xy^2 + 1}{x - y + 1}$ is continuous except at the points _____.

3. For $f(x, y) = 3x^2 + y^2$ the level curve that passes through $(2, -4)$ is _____.

4. If $p = g(\eta, \xi)$, $q = h(\eta, \xi)$, then $\dfrac{\partial}{\partial \xi} T(p, q) = $ _____.

5. If $r = g(w)$, $s = h(w)$, then $\dfrac{d}{dw} F(r, s) = $ _____.

6. If s is the distance that a body falls in time t, then the acceleration of gravity g can be obtained from $g = 2s/t^2$. Small errors Δs and Δt in the measurements of s and t, respectively, will result in an approximate error in g of _____.

7. The partial derivative $\dfrac{\partial^4 f}{\partial x\, \partial z\, \partial y^2}$ in subscript notation is _____.

8. The partial derivative f_{xyy} in ∂ notation is _____.

9. If $f(x, y) = \displaystyle\int_x^y F(t)\, dt$, then $\dfrac{\partial f}{\partial y} =$ _____ and $\dfrac{\partial f}{\partial x} =$ _____.

10. At (x_0, y_0, z_0) the function $F(x, y, z) = x + y + z$ increases most rapidly in the direction of _____.

11. If $F(x, y, z) = f(x, y)g(y)h(z)$, then $F_{xyz} =$ _____.

12. If $z = f(x, y)$ has continuous partial derivatives of any order, list all possible fourth-order partial derivatives. _____.

C. Exercises

In Problems 1–8, compute the indicated derivative.

1. $z = ye^{-x^3 y};\quad z_y$

2. $z = \ln(\cos(uv));\quad z_u$

3. $f(r, \theta) = \sqrt{r^3 + \theta^2};\quad f_{r\theta}$

4. $f(x, y) = (2x + xy^2)^2;\quad \dfrac{\partial^2 f}{\partial x^2}$

5. $z = \cosh(x^2 y^3);\quad \dfrac{\partial^2 z}{\partial y^2}$

6. $z = (e^{x^2} + e^{-y^3})^2;\quad \dfrac{\partial^3 z}{\partial x^2\, \partial y}$

7. $F(s, t, v) = s^3 t^5 v^{-4};\quad F_{stv}$

8. $w = \dfrac{xy}{z} + \dfrac{xz}{y} + \dfrac{yz}{x};\quad \dfrac{\partial^4 w}{\partial x\, \partial y^2\, \partial z}$

In Problems 9 and 10, find the gradient of the given function at the indicated point.

9. $f(x, y) = \tan^{-1}\dfrac{y}{x};\quad (1, -1)$

10. $f(x, y, z) = \dfrac{x^2 - 3y^3}{z^4};\quad (1, 2, 1)$

In Problems 11 and 12, find the directional derivative of the given function in the indicated direction.

11. $f(x, y) = x^2 y - y^2 x;\quad D_{\mathbf{u}} f$ in the direction of $2\mathbf{i} + 6\mathbf{j}$

12. $f(x, y, z) = \ln(x^2 + y^2 + z^2);\quad D_{\mathbf{u}} f$ in the direction of $-2\mathbf{i} + \mathbf{j} + 2\mathbf{k}$

In Problems 13 and 14, sketch the domain of the given function.

13. $f(x, y) = \sqrt{1 - (x + y)^2}$

14. $f(x, y) = \dfrac{1}{\ln(y - x)}$

In Problems 15 and 16, find Δz for the given function.

15. $z = 2xy - y^2$

16. $z = x^2 - 4y^2 + 7x - 9y + 10$

In Problems 17 and 18, find the total differential of the given function.

17. $z = \dfrac{x - 2y}{4x + 3y}$

18. $A = 2xy + 2yz + 2zx$

19. Find symmetric equations of the tangent line at $\left(-\sqrt{5}, 1, 3\right)$ to the trace of $z = \sqrt{x^2 + 4y^2}$ in the plane $x = -\sqrt{5}$.

20. Find the slope of the tangent line at $(2, 3, 10)$ to the curve of intersection of the surface $z = xy + x^2$ and the vertical plane that passes through $P(2, 3)$ and $Q(4, 5)$ in the direction of Q.

21. Consider the function $f(x, y) = x^2 y^4$. At $(1, 1)$ what is:

(a) the rate of change of f in the direction of $\mathbf{i}$?
(b) the rate of change of f in the direction of $\mathbf{i} - \mathbf{j}$?
(c) the rate of change of f in the direction of $\mathbf{j}$?

22. Let $w = \sqrt{x^2 + y^2 + z^2}$.

    (a) If $x = 3\sin 2t$, $y = 4\cos 2t$, $z = 5t^3$, find $\dfrac{dw}{dt}$.

    (b) If $x = 3\sin(2t/r)$, $y = 4\cos(2r/t)$, $z = 5r^3t^3$, find $\dfrac{\partial w}{\partial t}$.

23. Find an equation of the tangent plane to the graph of $z = \sin xy$ at $\left(\frac{1}{2}, \frac{2}{3}\pi, \frac{1}{2}\sqrt{3}\right)$.

24. Determine whether there are any points on the surface $z^2 + xy - 2x - y^2 = 1$ at which the tangent plane is parallel to $z = 2$.

25. Find an equation of the tangent plane to the cylinder $x^2 + y^2 = 25$ at $(3, 4, 6)$.

26. At what point is the directional derivative of $f(x, y) = x^3 + 3xy + y^3 - 3x^2$ in the direction of $\mathbf{i} + \mathbf{j}$ a minimum?

27. Find the dimensions of the rectangular box of greatest volume that is bounded in the first octant by the coordinate planes and the plane $x + 2y + z = 6$. See FIGURE 13.R.1.

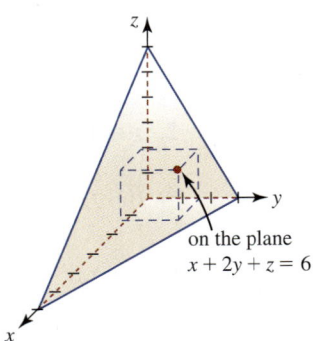

on the plane
$x + 2y + z = 6$

FIGURE 13.R.1 Box and plane in Problem 27

28. One effect of Einstein's general theory of relativity is that a massive object, such as a galaxy, can act as a "gravitational lens"; that is, if the galaxy is positioned between an observer (on Earth) and a light source (such as a quasar), then that light source appears as a ring surrounding the galaxy. If the gravitational lens is much closer to the light source than it is to the observer, then the angular radius θ of the ring (in radians) is related to the mass M of the lens and its distance D from the observer by

$$\theta = \left(\frac{GM}{c^2 D}\right)^{1/2},$$

where G is the gravitational constant and c is the speed of light. See FIGURE 13.R.2.

    (a) Solve for M in terms of θ and D.
    (b) Find the total differential of M as a function of θ and D.
    (c) If the angular radius θ can be measured with an error no greater than 2% and the distance D to the lens can be estimated with an error no greater than 10%, what is the approximate maximum percentage error in the calculation of the mass M of the lens?

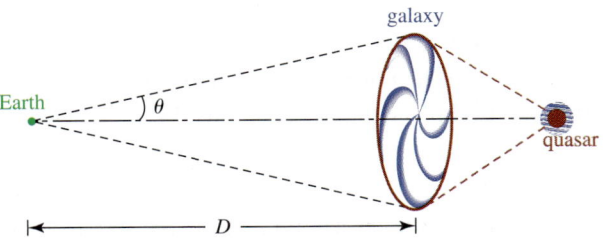

FIGURE 13.R.2 Galaxy in Problem 28

29. The velocity of the conical pendulum shown in FIGURE 13.R.3 is given by $v = r\sqrt{g/y}$, where $g = 980 \text{ cm/s}^2$. If r decreases from 20 cm to 19 cm and y increases from 25 cm to 26 cm, what is the approximate change in the velocity of the pendulum?

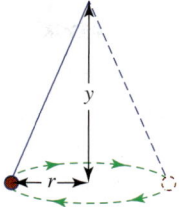

FIGURE 13.R.3   Conical pendulum in Problem 29

30. Find the directional derivative of $f(x, y) = x^2 + y^2$ at $(3, 4)$ in the direction of **(a)** $\nabla f(1, -2)$ and **(b)** $\nabla f(3, 4)$.

31. The so-called steady-state temperatures inside a circle of radius R are given by **Poisson's integral formula**

$$U(r, \theta) = \frac{1}{2\pi} \int_{-\pi}^{\pi} \frac{R^2 - r^2}{R^2 - 2rR\cos(\theta - \phi) + r^2} f(\phi) \, d\phi.$$

By formally differentiating under the integral sign, show that U satisfies the partial differential equation

$$r^2 U_{rr} + rU_r + U_{\theta\theta} = 0.$$

32. The **Cobb–Douglas production function** $z = f(x, y)$ is defined by $z = Ax^\alpha y^\beta$, where A, α, and β are constants. The value of z is called the *efficient output* for inputs x and y. Show that

$$f_x = \frac{\alpha z}{x}, \qquad f_y = \frac{\beta z}{y}, \qquad f_{xx} = \frac{\alpha(\alpha - 1)z}{x^2},$$

$$f_{yy} = \frac{\beta(\beta - 1)z}{y^2}, \qquad \text{and} \qquad f_{xy} = f_{yx} = \frac{\alpha\beta z}{xy}.$$

In Problems 33–36, suppose that $f_x(a, b) = 0, f_y(a, b) = 0$. If the given higher-order partial derivatives are evaluated at (a, b), determine, if possible, whether $f(a, b)$ is a relative extremum.

33. $f_{xx} = 4, f_{yy} = 6, f_{xy} = 5$

34. $f_{xx} = 2, f_{yy} = 7, f_{xy} = 0$

35. $f_{xx} = -5, f_{yy} = -9, f_{xy} = 6$

36. $f_{xx} = -2, f_{yy} = -8, f_{xy} = 4$

37. Express the area A of a right triangle as a function of the length L of its hypotenuse and one of its acute angles θ.

38. In FIGURE 13.R.4 express the height h of the mountain as a function of angles θ and ϕ.

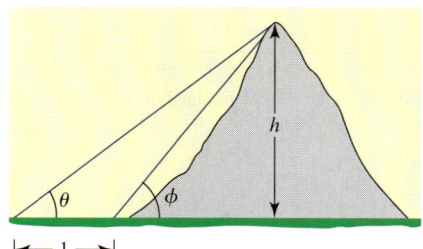

FIGURE 13.R.4   Mountain in Problem 38

39. A rectangular brick walkway shown in FIGURE 13.R.5 has a uniform width z. Express the area A of the walkway in terms of x, y, and z.

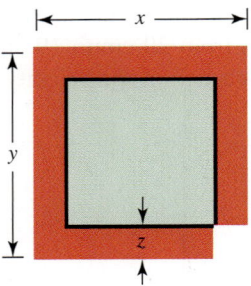

FIGURE 13.R.5 Walkway in Problem 39

40. An open box made of plastic has the shape of a rectangular parallelepiped. The outer dimensions of the box are given in FIGURE 13.R.6. If the plastic is $\frac{1}{2}$ cm thick, find the approximate volume of the plastic.

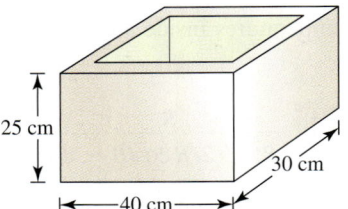

25 cm

30 cm

40 cm

FIGURE 13.R.6 Open box in Problem 40

41. A rectangular box, shown in FIGURE 13.R.7, is inscribed in the cone $z = 4 - \sqrt{x^2 + y^2}$, $0 \le z \le 4$. Express the volume V of the box in terms of x and y.

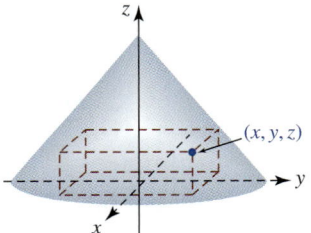

(x, y, z)

FIGURE 13.R.7 Inscribed box in Problem 41

42. The rectangular box shown in FIGURE 13.R.8 has a cover and 12 compartments. The box is made out of heavy plastic that costs 1.5 cents per square inch. Find a function giving the cost C of construction of the box.

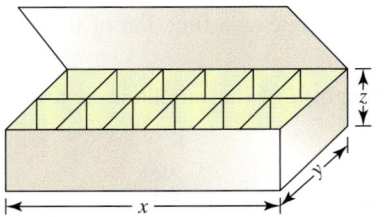

FIGURE 13.R.8 Rectangular box in Problem 42

Multiple Integrals

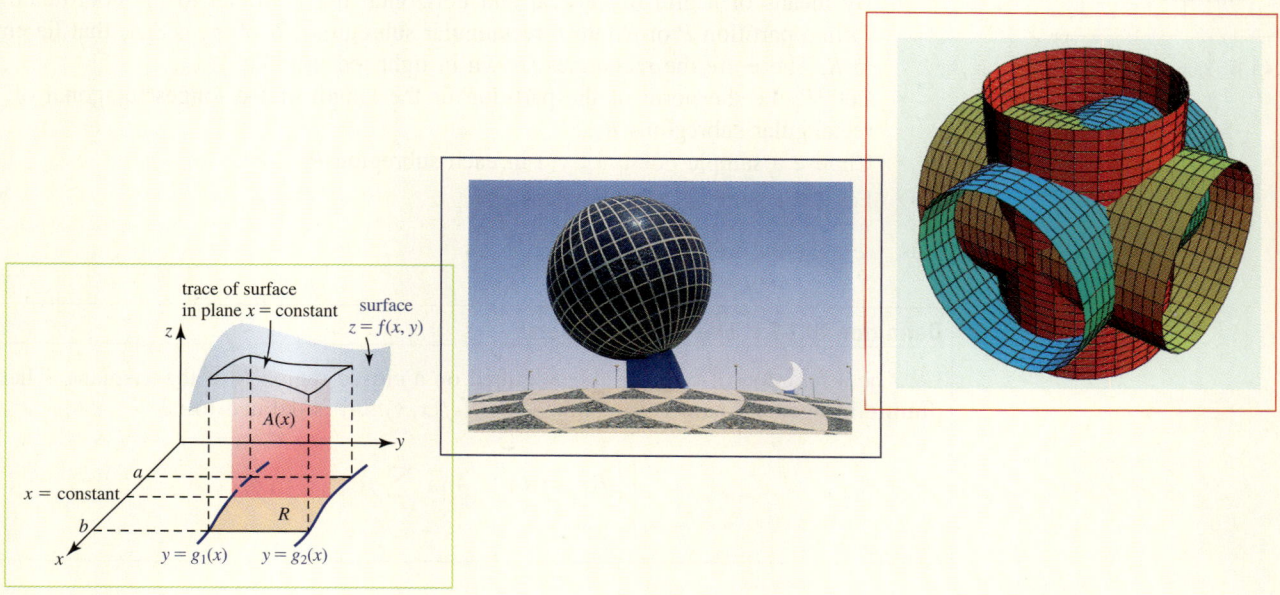

trace of surface
in plane $x = $ constant surface
$z = f(x, y)$

$A(x)$

$x = $ constant

$y = g_1(x)$ $y = g_2(x)$

R

a

b

In This Chapter   We conclude our study of the calculus of multivariable functions with the definitions and applications of the two-dimensional and three-dimensional definite integrals. These integrals are more commonly called the **double integral** and the **triple integral**, respectively.

14.1 The Double Integral

■ **Introduction**   Recall from Section 5.4 that the definition of the *definite integral* of a function of a single variable is given by a limit of a sum:

$$\int_a^b f(x)\,dx = \lim_{\|P\|\to 0}\sum_{k=1}^n f(x_k^*)\Delta x_k. \tag{1}$$

You are urged to review the steps leading to this definition on page 295. The analogous preliminary steps that lead to the concept of a *two-dimensional definite integral*, known simply as a **double integral** of a function f of two variables, are given next.

Let $z = f(x, y)$ be a function defined in a closed and bounded region R of the xy-plane. Consider the following four steps:

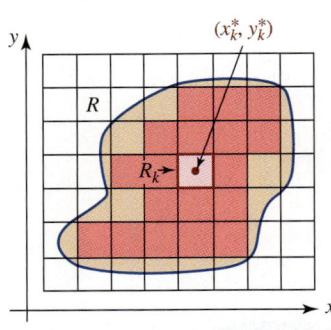

FIGURE 14.1.1   Sample point in R_k

- By means of a grid of vertical and horizontal lines parallel to the coordinate axes, form a partition P of R into n rectangular subregions R_k of areas ΔA_k that lie entirely in R. These are the rectangles shown in light red in FIGURE 14.1.1.
- Let $\|P\|$ be the norm of the partition or the length of the longest diagonal of the n rectangular subregions R_k.
- Choose a sample point (x_k^*, y_k^*) in each subregion R_k.
- Form the sum $\sum_{k=1}^n f(x_k^*, y_k^*)\Delta A_k$.

Thus, we have the following definition.

Definition 14.1.1   The Double Integral

Let f be a function of two variables defined on a closed region R of the xy-plane. Then the **double integral of f over R**, denoted by $\iint_R f(x, y)\,dA$, is defined to be

$$\iint_R f(x, y)\,dA = \lim_{\|P\|\to 0}\sum_{k=1}^n f(x_k^*, y_k^*)\Delta A_k. \tag{2}$$

If the limit in (2) exists, we say that f is **integrable over R** and that R is the **region of integration**. For a partition P of R into subregions R_k with (x_k^*, y_k^*) in R_k, a sum of the form $\sum_{k=1}^n f(x_k^*, y_k^*)\Delta A_k$ is called a **Riemann sum**. The partition of R, where the R_k lie entirely in R, is called an **inner partition** of R. The collection of shaded rectangles in the next two figures illustrate an inner partition.

Note: When f is continuous on R, the limit in (2) exists, that is, f is necessarily integrable over R.

EXAMPLE 1   Reimann Sum

Consider the region of integration R in the first quadrant bounded by the graphs of $x + y = 2$, $y = 0$, and $x = 0$. Approximate the double integral $\iint_R (5 - x - 2y)\,dA$ using a Riemann sum, the R_k shown in FIGURE 14.1.2, and the sample points (x_k^*, y_k^*) at the geometric center of each R_k.

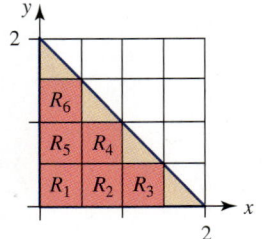

FIGURE 14.1.2   Region of integration R in Example 1

Solution   From Figure 14.1.2 we see that $\Delta A_k = \frac{1}{2}\cdot\frac{1}{2} = \frac{1}{4}$, $k = 1, 2, \ldots, 6$, and the (x_k^*, y_k^*) in the R_k for $k = 1, 2, \ldots, 6$, are in turn, $\left(\frac{1}{4}, \frac{1}{4}\right), \left(\frac{3}{4}, \frac{1}{4}\right), \left(\frac{5}{4}, \frac{1}{4}\right), \left(\frac{3}{4}, \frac{3}{4}\right), \left(\frac{1}{4}, \frac{3}{4}\right), \left(\frac{1}{4}, \frac{5}{4}\right)$. Hence, the Riemann sum is

$$\sum_{k=1}^n f(x_k^*, y_k^*)\Delta A_k = f\left(\frac{1}{4}, \frac{1}{4}\right)\frac{1}{4} + f\left(\frac{3}{4}, \frac{1}{4}\right)\frac{1}{4} + f\left(\frac{5}{4}, \frac{1}{4}\right)\frac{1}{4} + f\left(\frac{3}{4}, \frac{3}{4}\right)\frac{1}{4} + f\left(\frac{1}{4}, \frac{3}{4}\right)\frac{1}{4} + f\left(\frac{1}{4}, \frac{5}{4}\right)\frac{1}{4}$$

$$= \frac{17}{4}\cdot\frac{1}{4} + \frac{15}{4}\cdot\frac{1}{4} + \frac{13}{4}\cdot\frac{1}{4} + \frac{11}{4}\cdot\frac{1}{4} + \frac{13}{4}\cdot\frac{1}{4} + \frac{9}{4}\cdot\frac{1}{4}$$

$$= \frac{17}{16} + \frac{15}{16} + \frac{13}{16} + \frac{11}{16} + \frac{13}{16} + \frac{9}{16} = 4.875. \qquad ■$$

■ **Volume** We know that when $f(x) \geq 0$ for all x in $[a, b]$, then the definite integral (1) gives the area under the graph of f on the interval. Similarly, if $f(x, y) \geq 0$ on R, then on R_k as shown in FIGURE 14.1.3, the product $f(x_k^*, y_k^*)\Delta A_k$ can be interpreted as the volume of a rectangular parallelepiped, or prism, of height $f(x_k^*, y_k^*)$ and base of area ΔA_k. The summation of the n volumes $\sum_{k=1}^{n} f(x_k^*, y_k^*)\Delta A_k$ is an approximation to the volume V of the solid bounded between the region R and the surface $z = f(x, y)$. The limit of this sum as $\|P\| \to 0$, if it exists, will give the **volume** of this solid; that is, if f is nonnegative on R, then

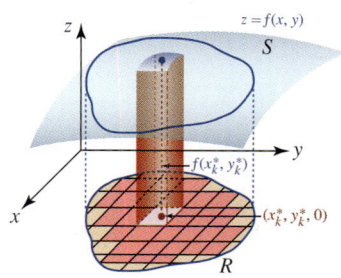

FIGURE 14.1.3 A rectangular parallelepiped is built up on each R_k

$$V = \iint_R f(x, y)\, dA. \qquad (3)$$

The parallelepipeds built up on the six R_k shown in Figure 14.1.2 are shown in FIGURE 14.1.4. Since the integrand is nonnegative on R, the value of the Riemann sum given in Example 1 represents an approximation to the volume of the solid bounded between the region R and the surface defined by the function $f(x, y) = 5 - x - 2y$.

■ **Area** When $f(x, y) = 1$ on R, then $\lim_{\|P\| \to 0} \sum_{k=1}^{n} \Delta A_k$ will simply give the **area** A of the region; that is,

$$A = \iint_R dA. \qquad (4)$$

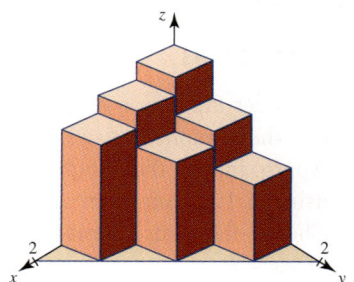

FIGURE 14.1.4 Rectangular parallelepipeds built up on each R_k in Figure 14.1.2

■ **Properties** The following properties of the double integral are similar to those of the definite integral given in Theorems 5.4.4 and 5.4.5.

Theorem 14.1.1   Properties

Let f and g be functions of two variables that are integrable over a region R of the xy-plane. Then

(i) $\displaystyle\iint_R kf(x, y)\, dA = k\iint_R f(x, y)\, dA$, where k is any constant

(ii) $\displaystyle\iint_R [f(x, y) \pm g(x, y)]\, dA = \iint_R f(x, y)\, dA \pm \iint_R g(x, y)\, dA$

(iii) $\displaystyle\iint_R f(x, y)\, dA = \iint_{R_1} f(x, y)\, dA + \iint_{R_2} f(x, y)\, dA$, where R_1 and R_2 are subregions that do not overlap and $R = R_1 \cup R_2$

(iv) $\displaystyle\iint_R f(x, y)\, dA \geq \iint_R g(x, y)\, dA$ if $f(x, y) \geq g(x, y)$ over R.

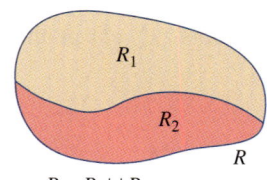

FIGURE 14.1.5 Region R is the union of two regions

Part (*iii*) of Theorem 14.1.1 is the two-dimensional equivalent of the additive interval property

$$\int_a^b f(x)\, dx = \int_a^c f(x)\, dx + \int_c^b f(x)\, dx$$

(Theorem 5.4.5). FIGURE 14.1.5 illustrates the division of a region into subregions R_1 and R_2 for which $R = R_1 \cup R_2$. The regions R_1 and R_2 can have no points in common except possibly on their common border. Furthermore, Theorem 14.1.1(*iii*) extends to any finite number of nonoverlapping subregions whose union is R. It also follows from Theorem 14.1.1(*iv*) that $\iint_R f(x, y)\, dA > 0$ whenever $f(x, y) > 0$ for all (x, y) in R.

■ **Net Signed Volume** Of course, not every double integral gives volume. For the surface $z = f(x, y)$ shown in FIGURE 14.1.6, $\iint_R f(x, y)\, dA$ is a real number but it is not volume since f is

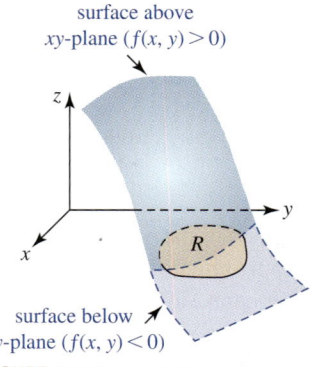

FIGURE 14.1.6 On R the surface is partly above and partly below the xy-plane

not nonnegative on R. Analogous to the concept of net signed area discussed in Section 5.4, we can interpret the double integral as the sum of the volume bounded between the graph of f and the region R whenever $f(x, y) \geq 0$ and the negative of the volume between the graph of f and the region R whenever $f(x, y) \leq 0$. In other words, $\iint_R f(x, y)\, dA$ represents a **net signed volume** between the graph of f and the xy-plane over the region R.

Exercises 14.1   Answers to selected odd-numbered problems begin on page ANS-44.

☰ Fundamentals

1. Consider the region R in the first quadrant that is bounded by the graphs of $x^2 + y^2 = 16$, $y = 0$, and $x = 0$. Approximate the double integral $\iint_R (x + 3y + 1)\, dA$ using a Riemann sum and the R_k shown in FIGURE 14.1.7. Choose the sample points (x_k^*, y_k^*) at the geometric center of each R_k.

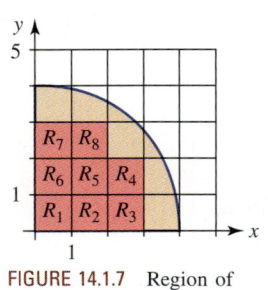

FIGURE 14.1.7   Region of integration in Problem 1

2. Consider the region R in the first quadrant bounded by the graphs of $x + y = 1$, $x + y = 3$, $y = 0$, and $x = 0$. Approximate the double integral $\iint_R (2x + 4y)\, dA$ using a Riemann sum and the R_k shown in FIGURE 14.1.8. Choose the sample points (x_k^*, y_k^*) at the upper right-hand corner of each R_k.

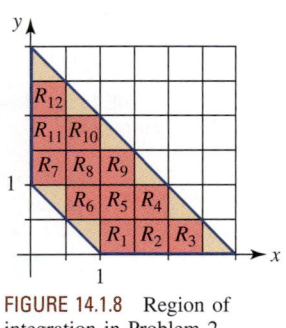

FIGURE 14.1.8   Region of integration in Problem 2

3. Consider the rectangular region R shown in FIGURE 14.1.9. Approximate the double integral $\iint_R (x + y)\, dA$ using a Riemann sum and the R_k shown in the figure. Choose the sample points (x_k^*, y_k^*) at
(a) the geometric center of each R_k and
(b) the upper left-hand corner of each R_k.

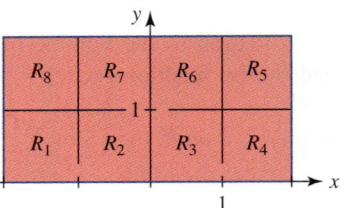

FIGURE 14.1.9   Region of integration in Problem 3

4. Consider the region R bounded by the graphs of $y = x^2$ and $y = 4$. Place a rectangular grid over R corresponding to the lines $x = -2, x = -\frac{3}{2}, x = -1, \ldots, x = 2$, and $y = 0$, $y = \frac{1}{2}, y = 1, \ldots, y = 4$. Approximate the double integral $\iint_R xy\, dA$ using a Riemann sum, where the sample points (x_k^*, y_k^*) are chosen at the lower right-hand corner of each complete rectangular R_k in R.

In Problems 5–8, evaluate $\iint_R 10\, dA$ over the given region R. Use formulas from geometry.

5.

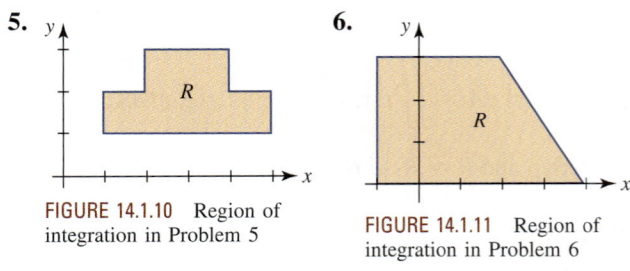

FIGURE 14.1.10   Region of integration in Problem 5

6.

FIGURE 14.1.11   Region of integration in Problem 6

7.

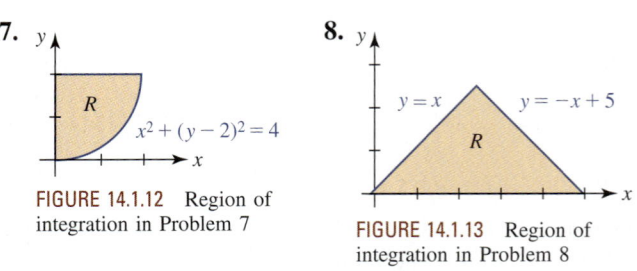

FIGURE 14.1.12   Region of integration in Problem 7

8.

FIGURE 14.1.13   Region of integration in Problem 8

9. Consider the region R bounded by the circle $(x - 3)^2 + y^2 = 9$. Does the double integral $\iint_R (x + 5y)\, dA$ represent a volume? Explain.

10. Consider the region R in the second quadrant that is bounded by the graphs of $-2x + y = 6$, $x = 0$, and $y = 0$.

Does the double integral $\iint_R (x^2 + y^2)\, dA$ represent a volume? Explain.

In Problems 11–16, suppose that $\iint_R x\, dA = 3$, $\iint_R y\, dA = 7$, and the area of R is 8. Evaluate the given double integral.

11. $\displaystyle\iint_R 10\, dA$

12. $\displaystyle\iint_R -5x\, dA$

13. $\displaystyle\iint_R (2x + 4y)\, dA$

14. $\displaystyle\iint_R (x - y)\, dA$

15. $\displaystyle\iint_R (3x + 7y + 1)\, dA$

16. $\displaystyle\iint_R y^2\, dA - \iint_R (2 + y)^2\, dA$

In Problems 17 and 18, let R_1 and R_2 be nonoverlapping regions such that $R = R_1 \cup R_2$.

17. If $\iint_{R_1} f(x, y)\, dA = 4$ and $\iint_{R_2} f(x, y)\, dA = 14$, what is the value of $\iint_R f(x, y)\, dA$?

18. Suppose $\iint_R f(x, y)\, dA = 25$ and $\iint_{R_1} f(x, y)\, dA = 30$. What is the value of $\iint_{R_2} f(x, y)\, dA$?

14.2 Iterated Integrals

▌ **Introduction**   Similar to the process of partial differentiation we can define **partial integration**. The concept of partial integration is the key to a practical method for evaluating a double integral. Since we will be using indefinite and definite integration, a review of the material in Section 5.1, Section 5.2, and Chapter 7 is strongly advised.

▌ **Partial Integration**   If $F(x, y)$ is a function such that its partial derivative with respect to y is a function f, that is $F_y(x, y) = f(x, y)$, then the **partial integral of f with respect to y** is

$$\int f(x, y)\, dy = F(x, y) + c_1(x),\tag{1}$$

where the function $c_1(x)$ plays the part of the "constant of integration." Similarly, if $F(x, y)$ is a function such that $F_x(x, y) = f(x, y)$, then the **partial integral of f with respect to x** is

$$\int f(x, y)\, dx = F(x, y) + c_2(y).\tag{2}$$

In other words, to evaluate the partial integral $\int f(x, y)\, dy$ we hold x fixed (as if it were a constant), whereas in $\int f(x, y)\, dx$ we hold y fixed.

EXAMPLE 1   Using (1) and (2)

Evaluate:

(a) $\displaystyle\int 6xy^2\, dy$ **(b)** $\displaystyle\int 6xy^2\, dx.$

Solution

(a) By holding x fixed,

$$\int 6xy^2\, dy = 6x \cdot \left(\frac{1}{3}y^3\right) + c_1(x) = 2xy^3 + c_1(x).$$

Check: $\dfrac{\partial}{\partial y}(2xy^3 + c_1(x)) = \dfrac{\partial}{\partial y}2xy^3 + \dfrac{\partial}{\partial y}c_1(x) = 2x(3y^2) + 0 = 6xy^2.$

(b) Now by holding y fixed,

$$\int 6xy^2\, dx = 6 \cdot \left(\frac{1}{2}x^2\right) \cdot y^2 + c_2(y) = 3x^2y^2 + c_2(y).$$

You should verify this result by taking its partial derivative with respect to x.   ▪

▌ **Partial Definite Integration**   When evaluating a definite integral we can dispense with the functions $c_1(y)$ and $c_2(x)$ in (1) and (2). Again, if $F(x, y)$ is a function such that $F_y(x, y) = f(x, y)$, then the **partial definite integral with respect to y** is defined to be

$$\int_{g_1(x)}^{g_2(x)} f(x, y)\, dy = F(x, y)\Big]_{g_1(x)}^{g_2(x)} = F(x, g_2(x)) - F(x, g_1(x)). \tag{3}$$

If $F(x, y)$ is a function such that $F_x(x, y) = f(x, y)$, then the **partial definite integral with respect to x** is

$$\int_{h_1(y)}^{h_2(y)} f(x, y)\, dx = F(x, y)\Big]_{h_1(y)}^{h_2(y)} = F(h_2(y), y) - F(h_1(y), y). \tag{4}$$

The functions $g_1(x)$ and $g_2(x)$ in (3) and the functions $h_1(y)$ and $h_2(y)$ in (4) are called **limits of integration**. Of course the results in (3) and (4) also hold when the limits of integration are constants.

EXAMPLE 2 Using (3) and (4)

Evaluate:

(a) $\displaystyle\int_1^2 \left(6xy^2 - 4\frac{x}{y}\right) dy$ **(b)** $\displaystyle\int_{-1}^3 \left(6xy^2 - 4\frac{x}{y}\right) dx.$

Solution
(a) It follows from (3) that

$$\int_1^2 \left(6xy^2 - 4\frac{x}{y}\right) dy = \left[2xy^3 - 4x\ln|y|\right]_1^2$$
$$= (16x - 4x\ln 2) - (2x - 4x\ln 1)$$
$$= 14x - 4x\ln 2.$$

(b) From (4),

$$\int_{-1}^3 \left(6xy^2 - 4\frac{x}{y}\right) dx = \left(3x^2y^2 - 2\frac{x^2}{y}\right)\Big]_{-1}^3$$
$$= \left(27y^2 - \frac{18}{y}\right) - \left(3y^2 - \frac{2}{y}\right)$$
$$= 24y^2 - \frac{16}{y}. \qquad \blacksquare$$

EXAMPLE 3 Using (3)

Evaluate $\displaystyle\int_{x^2}^x \sin xy\, dy.$

Solution Since we are treating x as constant, we first note that a partial integral of $\sin xy$ with respect to y is $(-\cos xy)/x$. To see this, we have by the Chain Rule,

$$\frac{\partial}{\partial y}\left(-\frac{1}{x}\cos xy\right) = -\frac{1}{x}(-\sin xy)\frac{\partial}{\partial y}xy = -\frac{1}{x}(-\sin xy)\cdot x = \sin xy.$$

Hence, by (3) the partial definite integral is

$$\int_{x^2}^x \sin xy\, dy = -\frac{\cos xy}{x}\Big]_{x^2}^x = \left(-\frac{\cos(x\cdot x)}{x}\right) - \left(-\frac{\cos(x\cdot x^2)}{x}\right) = -\frac{\cos x^2}{x} + \frac{\cos x^3}{x}. \qquad \blacksquare$$

Before proceeding we need to examine some special regions in the xy-plane.

■ **Regions of Type I and II** The region shown in FIGURE 14.2.1(a),

$$R: a \le x \le b, \quad g_1(x) \le y \le g_2(x),$$

where the boundary functions g_1 and g_2 are continuous, is called a **Type I region**. In Figure 14.2.1(b), the region

$$R: c \le y \le d, \quad h_1(y) \le x \le h_2(y),$$

where h_1 and h_2 are continuous, is called a **Type II region**.

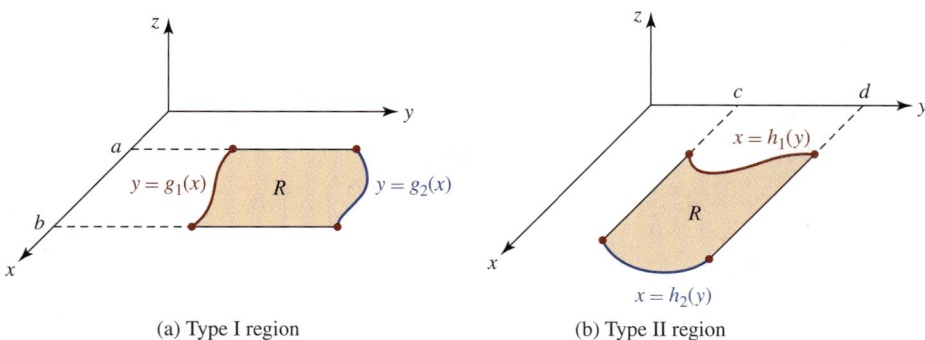

(a) Type I region (b) Type II region

FIGURE 14.2.1 Regions in the plane

■ **Iterated Integrals** Since the partial definite integral $\int_{g_1(x)}^{g_2(x)} f(x, y)\, dy$ is a function of x alone, we may in turn integrate the resulting function with respect to x. If f is continuous on a Type I region R, we define an **iterated integral of** f over the region by

$$\int_a^b \int_{g_1(x)}^{g_2(x)} f(x, y)\, dy\, dx = \int_a^b \left[\int_{g_1(x)}^{g_2(x)} f(x, y)\, dy \right] dx. \tag{5}$$

The basic idea in (5) is to carry out *repeated* or *successive* integrations. The two-step process starts with a partial definite integration that gives a function of x, which is then integrated in the usual manner from $x = a$ to $x = b$. The end result of the two integrations will be a real number. In a similar manner, we define an iterated integral of a continuous function f on a Type II region R by

$$\int_c^d \int_{h_1(y)}^{h_2(y)} f(x, y)\, dx\, dy = \int_c^d \left[\int_{h_1(y)}^{h_2(y)} f(x, y)\, dx \right] dy. \tag{6}$$

In (5) and (6), R is called the **region of integration**.

EXAMPLE 4 Iterated Integral

Evaluate the iterated integral of $f(x, y) = 2xy$ over the region shown in FIGURE 14.2.2.

Solution The region is of Type I and so by (5) we have

$$\int_{-1}^{2} \int_{x}^{x^2+1} 2xy\, dy\, dx = \int_{-1}^{2} \left[\int_{x}^{x^2+1} 2xy\, dy \right] dx = \int_{-1}^{2} xy^2 \Big]_{x}^{x^2+1} dx$$

$$= \int_{-1}^{2} [x(x^2 + 1)^2 - x^3]\, dx$$

$$= \left[\frac{1}{6}(x^2 + 1)^3 - \frac{1}{4}x^4 \right]_{-1}^{2} = \frac{63}{4}. \qquad ■$$

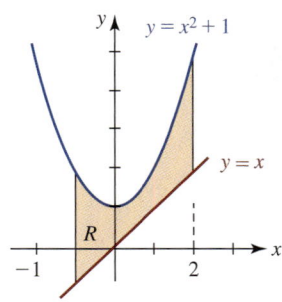

FIGURE 14.2.2 Region R in Example 4

EXAMPLE 5 Iterated Integral

Evaluate $\displaystyle\int_0^4 \int_y^{2y} (8x + e^y)\, dx\, dy$.

Solution By comparing the iterated integral to (6), we see that the region of integration is a Type II region. See FIGURE 14.2.3. We begin the successive integrations using (4):

$$\int_0^4 \int_y^{2y} (8x + e^y)\, dx\, dy = \int_0^4 \left[\int_y^{2y} (8x + e^y)\, dx \right] dy = \int_0^4 (4x^2 + xe^y) \Big]_y^{2y} dy$$

$$= \int_0^4 [(16y^2 + 2ye^y) - (4y^2 + ye^y)]\, dy$$

$$= \int_0^4 (12y^2 + ye^y)\, dy \quad \leftarrow \text{integration by parts}$$

$$= \left[4y^3 + ye^y - e^y \right]_0^4 = 257 + 3e^4 \approx 420.79. \qquad ■$$

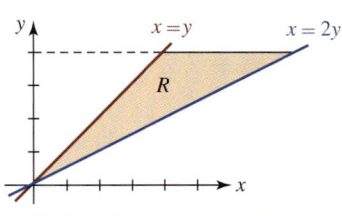

FIGURE 14.2.3 Region R in Example 5

EXAMPLE 6   Iterated Integral

Evaluate $\displaystyle\int_{-1}^{3}\int_{1}^{2}\left(6xy^2 - 4\frac{x}{y}\right) dy\,dx$.

Solution   From the result in part (a) of Example 2, we have

$$\int_{-1}^{3}\int_{1}^{2}\left(6xy^2 - 4\frac{x}{y}\right) dy\,dx = \int_{-1}^{3}\left[\int_{1}^{2}\left(6xy^2 - 4\frac{x}{y}\right) dy\right] dx$$

$$= \int_{-1}^{3}(14x - 4x\ln 2)\, dx$$

$$= (7x^2 - 2x^2\ln 2)\Big]_{-1}^{3} = 56 - 16\ln 2 \approx 44.91. \quad\blacksquare$$

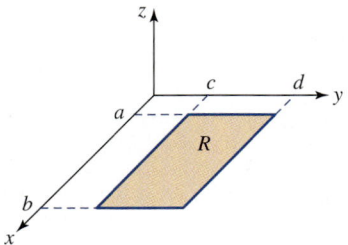

FIGURE 14.2.4   Rectangular region is both Type I and Type II

Inspection of **FIGURE 14.2.4** should convince you that a rectangular region R defined by $a \le x \le b, c \le y \le d$ is simultaneously a Type I and a Type II region. If f is continuous on R, it can be proved that

$$\int_{a}^{b}\int_{c}^{d} f(x, y)\, dy\,dx = \int_{c}^{d}\int_{a}^{b} f(x, y)\, dx\,dy. \tag{7}$$

You should verify that

$$\int_{1}^{2}\int_{-1}^{3}\left(6xy^2 - 4\frac{x}{y}\right) dx\,dy$$

gives the same result as the given iterated integral in Example 6.

A rectangular region is not the only region that can be both Type I and Type II. As in (7), if f is continuous on a region R that is simultaneously a Type I and a Type II region, then the two iterated integrals of f over R are equal. See Problems 47 and 48 in Exercises 14.2.

Exercises 14.2   Answers to selected odd-numbered problems begin on page ANS-44.

≡ Fundamentals

In Problems 1–10, evaluate the given partial integral.

1. $\displaystyle\int dy$

2. $\displaystyle\int (1 - 2y)\, dx$

3. $\displaystyle\int (6x^2y - 3x\sqrt{y})\, dx$

4. $\displaystyle\int (6x^2y - 3x\sqrt{y})\, dy$

5. $\displaystyle\int \frac{1}{x(y + 1)}\, dy$

6. $\displaystyle\int (1 + 10x - 5y^4)\, dx$

7. $\displaystyle\int (12y\cos 4x - 3\sin y)\, dx$

8. $\displaystyle\int \sec^2 3xy\, dy$

9. $\displaystyle\int \frac{y}{\sqrt{2x + 3y}}\, dx$

10. $\displaystyle\int (2x + 5y)^6\, dy$

In Problems 11–20, evaluate the given partial definite integral.

11. $\displaystyle\int_{-1}^{3}(6xy - 5e^y)\, dx$

12. $\displaystyle\int_{1}^{2}\tan xy\, dy$

13. $\displaystyle\int_{1}^{3x} x^3 e^{xy}\, dy$

14. $\displaystyle\int_{\sqrt{y}}^{y^3}(8x^3y - 4xy^2)\, dx$

15. $\displaystyle\int_{0}^{2x}\frac{xy}{x^2 + y^2}\, dy$

16. $\displaystyle\int_{x^3}^{x}e^{2y/x}\, dy$

17. $\displaystyle\int_{\tan y}^{\sec y}(2x + \cos y)\, dx$

18. $\displaystyle\int_{\sqrt{y}}^{1} y\ln x\, dx$

19. $\displaystyle\int_{x}^{\pi/2}\cos x\sin^3 y\, dy$

20. $\displaystyle\int_{1/2}^{1} y\cos^2 xy\, dx$

In Problems 21–42, evaluate the given iterated integral.

21. $\displaystyle\int_{1}^{2}\int_{-x}^{x^2}(8x - 10y + 2)\, dy\,dx$ **22.** $\displaystyle\int_{-1}^{1}\int_{0}^{y}(x + y)^2\, dx\,dy$

23. $\displaystyle\int_{0}^{\sqrt{2}}\int_{-\sqrt{2-y^2}}^{\sqrt{2-y^2}}(2x - y)\, dx\,dy$

24. $\displaystyle\int_{0}^{\pi/4}\int_{0}^{\cos x}(1 + 4y\tan^2 x)\, dy\,dx$

25. $\displaystyle\int_{0}^{\pi}\int_{y}^{3y}\cos(2x + y)\, dx\,dy$ **26.** $\displaystyle\int_{1}^{2}\int_{0}^{\sqrt{x}} 2y\sin\pi x^2\, dy\,dx$

27. $\displaystyle\int_{1}^{\ln 3}\int_{0}^{x} 6e^{x+2y}\, dy\,dx$ **28.** $\displaystyle\int_{0}^{1}\int_{0}^{2y} e^{-y^2}\, dx\,dy$

29. $\displaystyle\int_{0}^{3}\int_{x+1}^{2x+1}\frac{1}{\sqrt{y - x}}\, dy\,dx$ **30.** $\displaystyle\int_{0}^{1}\int_{0}^{y} x(y^2 - x^2)^{3/2}\, dx\,dy$

31. $\displaystyle\int_1^9 \int_0^x \frac{1}{x^2 + y^2} \, dy \, dx$

32. $\displaystyle\int_0^{1/2} \int_0^y \frac{1}{\sqrt{1 - x^2}} \, dx \, dy$

33. $\displaystyle\int_1^e \int_1^y \frac{y}{x} \, dx \, dy$

34. $\displaystyle\int_1^4 \int_1^{\sqrt{x}} 2ye^{-x} \, dy \, dx$

35. $\displaystyle\int_0^6 \int_0^{\sqrt{25 - y^2}/2} \frac{1}{\sqrt{(25 - y^2) - x^2}} \, dx \, dy$

36. $\displaystyle\int_0^2 \int_{y^2}^{\sqrt{20 - y^2}} y \, dx \, dy$

37. $\displaystyle\int_{\pi/2}^{\pi} \int_{\cos y}^{0} e^x \sin y \, dx \, dy$

38. $\displaystyle\int_0^1 \int_0^{y^{1/3}} 6x^2 \ln(y + 1) \, dx \, dy$

39. $\displaystyle\int_{\pi}^{2\pi} \int_0^x (\cos x - \sin y) \, dy \, dx$ **40.** $\displaystyle\int_1^3 \int_0^{1/x} \frac{1}{x + 1} \, dy \, dx$

41. $\displaystyle\int_{\pi/12}^{5\pi/12} \int_1^{\sqrt{2\sin 2\theta}} r \, dr \, d\theta$ **42.** $\displaystyle\int_0^{\pi/3} \int_{3\cos\theta}^{1 + \cos\theta} r \, dr \, d\theta$

In Problems 43–46, sketch the region of integration R for the given iterated integral.

43. $\displaystyle\int_0^2 \int_1^{2x+1} f(x, y) \, dy \, dx$ **44.** $\displaystyle\int_1^4 \int_{-\sqrt{y}}^{\sqrt{y}} f(x, y) \, dx \, dy$

45. $\displaystyle\int_{-1}^3 \int_0^{\sqrt{16 - y^2}} f(x, y) \, dx \, dy$ **46.** $\displaystyle\int_{-1}^2 \int_{-x^2}^{x^2+1} f(x, y) \, dy \, dx$

In Problems 47 and 48, verify by a sketch that the Type I region is the same as the Type II region. Verify that the given iterated integrals are equal.

47. Type I: $\dfrac{1}{2}x \le y \le \sqrt{x}, \quad 0 \le x \le 4$

Type II: $y^2 \le x \le 2y, \quad 0 \le y \le 2$

$$\int_0^4 \int_{x/2}^{\sqrt{x}} x^2 y \, dy \, dx = \int_0^2 \int_{y^2}^{2y} x^2 y \, dx \, dy$$

48. Type I: $-\sqrt{1 - x^2} \le y \le \sqrt{1 - x^2}, \quad 0 \le x \le 1$

Type II: $0 \le x \le \sqrt{1 - y^2}, \quad -1 \le y \le 1$

$$\int_0^1 \int_{-\sqrt{1-x^2}}^{\sqrt{1-x^2}} 2x \, dy \, dx = \int_{-1}^1 \int_0^{\sqrt{1-y^2}} 2x \, dx \, dy$$

In Problems 49–52, verify the given equality.

49. $\displaystyle\int_{-1}^2 \int_0^3 x^2 \, dy \, dx = \int_0^3 \int_{-1}^2 x^2 \, dx \, dy$

50. $\displaystyle\int_{-2}^2 \int_2^4 (2x + 4y) \, dx \, dy = \int_2^4 \int_{-2}^2 (2x + 4y) \, dy \, dx$

51. $\displaystyle\int_1^3 \int_0^{\pi} (3x^2 y - 4\sin y) \, dy \, dx = \int_0^{\pi} \int_1^3 (3x^2 y - 4\sin y) \, dx \, dy$

52. $\displaystyle\int_0^1 \int_0^2 \left(\frac{8y}{x + 1} - \frac{2x}{y^2 + 1} \right) dx \, dy =$

$$\int_0^2 \int_0^1 \left(\frac{8y}{x + 1} - \frac{2x}{y^2 + 1} \right) dy \, dx$$

☰ Think About It

53. If f and g are integrable, prove that

$$\int_c^d \int_a^b f(x)g(y) \, dx \, dy = \left(\int_a^b f(x) \, dx \right)\left(\int_c^d g(y) \, dy \right).$$

54. Use the result in Problem 53 to evaluate

$$\int_0^{\infty} \int_0^{\infty} xye^{-(2x^2 + 3y^2)} \, dx \, dy.$$

14.3 Evaluation of Double Integrals

■ Introduction   The iterated integrals of the preceding section provide the means for evaluating a double integral $\iint_R f(x, y) \, dA$ over a Type I or Type II region or a region that can be expressed as a union of a finite number of these regions. The following result is due to the Italian mathematician **Guido Fubini** (1879–1943).

Theorem 14.3.1   Fubini's Theorem

Let f be continuous on a region R.

(*i*) If R is a Type I region, then

$$\iint_R f(x, y) \, dA = \int_a^b \int_{g_1(x)}^{g_2(x)} f(x, y) \, dy \, dx. \tag{1}$$

(*ii*) If R is a Type II region, then

$$\iint_R f(x, y) \, dA = \int_c^d \int_{h_1(y)}^{h_2(y)} f(x, y) \, dx \, dy. \tag{2}$$

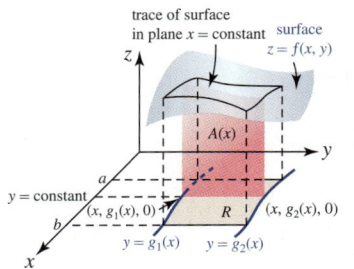

FIGURE 14.3.1 Area $A(x)$ of vertical plane is a partial definite integral of f

Theorem 14.3.1 is the double integral counterpart of Theorem 5.5.1, the Fundamental Theorem of Calculus. While Theorem 14.3.1 is difficult to prove, we can get some intuitive feeling for its significance by considering volumes. Let R be a Type I region and $z = f(x, y)$ be continuous and nonnegative on R. The area A of the vertical plane shown in FIGURE 14.3.1 is the area under the trace of the surface $z = f(x, y)$ in the plane $x = $ constant and hence is given by the partial integral

$$A(x) = \int_{g_1(x)}^{g_2(x)} f(x, y)\, dy.$$

By summing all these areas from $x = a$ to $x = b$, we obtain the volume V of the solid above R and below the surface:

$$V = \int_a^b A(x)\, dx = \int_a^b \int_{g_1(x)}^{g_2(x)} f(x, y)\, dy\, dx.$$

But as we have already seen in (3) of Section 14.1, this volume is also given by the double integral $V = \iint_R f(x, y)\, dA$. Hence,

$$V = \iint_R f(x, y)\, dA = \int_a^b \int_{g_1(x)}^{g_2(x)} f(x, y)\, dy\, dx.$$

EXAMPLE 1   Double Integral

Evaluate the double integral $\iint_R e^{x+3y}\, dA$ over the region R bounded by the graphs of $y = 1$, $y = 2$, $y = x$, and $y = -x + 5$.

Solution   As seen in FIGURE 14.3.2, R is a Type II region; hence, by (2) we integrate first with respect to x from the left boundary $x = y$ to the right boundary $x = 5 - y$:

$$\iint_R e^{x+3y}\, dA = \int_1^2 \int_y^{5-y} e^{x+3y}\, dx\, dy$$

$$= \int_1^2 e^{x+3y} \Big]_y^{5-y}\, dy$$

$$= \int_1^2 (e^{5+2y} - e^{4y})\, dy$$

$$= \left(\frac{1}{2} e^{5+2y} - \frac{1}{4} e^{4y} \right) \Big]_1^2$$

$$= \frac{1}{2} e^9 - \frac{1}{4} e^8 - \frac{1}{2} e^7 + \frac{1}{4} e^4 \approx 2771.64. \qquad \blacksquare$$

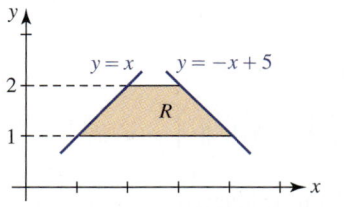

FIGURE 14.3.2   Region R in Example 1

As an aid in reducing a double integral to an iterated integral with correct limits of integration, it is useful to visualize, as suggested in the foregoing discussion, the double integral as a double summation process. Over a Type I region the iterated integral $\int_a^b \int_{g_1(x)}^{g_2(x)} f(x, y)\, dy\, dx$ is first a summation in the y-direction. Pictorially, this is indicated by the vertical arrow in FIGURE 14.3.3(a); the typical rectangle in the arrow has area $dy\, dx$. The dy placed before the dx signifies that the "volumes" $f(x, y)\, dy\, dx$ of parellepipeds built up on the rectangles are summed vertically with respect to y from the lower boundary curve $y = g_1(x)$ to the upper boundary curve $y = g_2(x)$. The dx following the dy signifies that the result of each vertical summation is then summed horizontally with respect to x from left ($x = a$) to right ($x = b$). Similar remarks hold for double integrals over regions of Type II. See Figure 14.3.3(b). Recall from (4) of Section 14.1 that when $f(x, y) = 1$, the double integral $A = \iint_R dA$ gives the area of the region. Thus, Figure 14.3.3(a) shows that $\int_a^b \int_{g_1(x)}^{g_2(x)} dy\, dx$ adds the rectangular areas vertically and then horizontally, whereas Figure 14.3.3(b) shows that $\int_c^d \int_{h_1(y)}^{h_2(y)} dx\, dy$ adds the rectangular areas horizontally and then vertically.

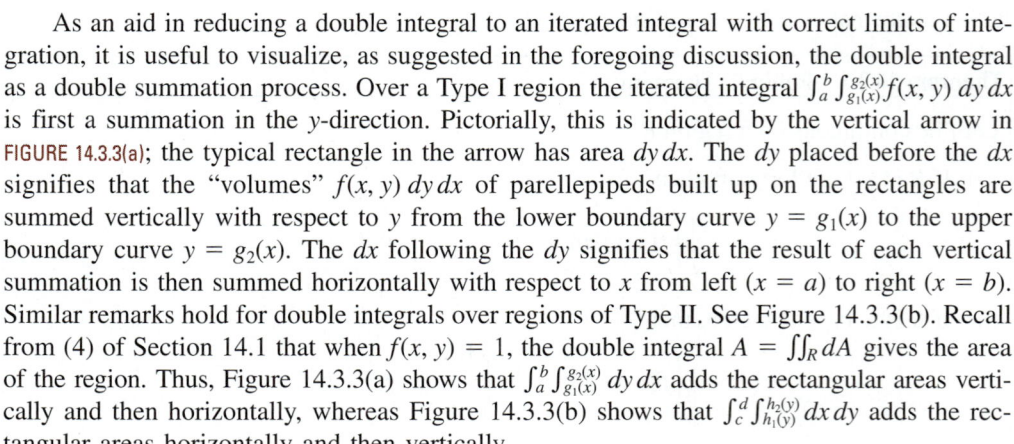

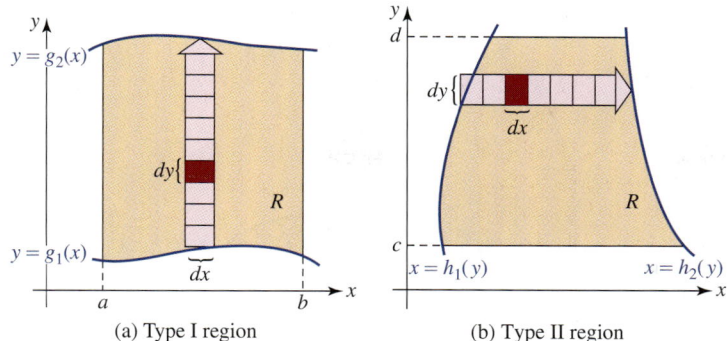

(a) Type I region   (b) Type II region

FIGURE 14.3.3 In (a) first integration is with respect to y; in (b) first integration is with respect to x

---

EXAMPLE 2 Area by Double Integration

Use a double integral to find the area of the region bounded by the graphs of $y = x^2$ and $y = 8 - x^2$.

Solution The graphs and their points of intersection are shown in FIGURE 14.3.4. Since R is evidently of Type I, we have from (1)

$$A = \iint_R dA = \int_{-2}^{2} \int_{x^2}^{8-x^2} dy\, dx$$

$$= \int_{-2}^{2} [(8 - x^2) - x^2]\, dx$$

$$= \int_{-2}^{2} (8 - 2x^2)\, dx$$

$$= \left(8x - \frac{2}{3}x^3\right)\Big]_{-2}^{2} = \frac{64}{3}.$$ ∎

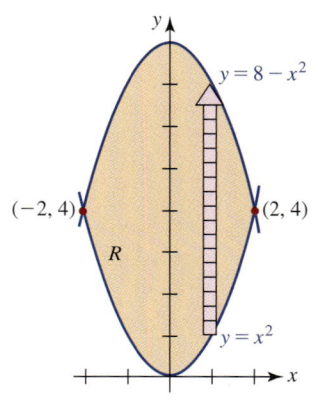

FIGURE 14.3.4 Region R in Example 2

Note: You should recognize

$$A = \iint_R dA = \int_a^b \int_{g_1(x)}^{g_2(x)} dy\, dx = \int_a^b [g_2(x) - g_1(x)]\, dx$$

as formula (3) of Section 6.2 for the area bounded between two graphs on the interval $[a, b]$.

EXAMPLE 3 Volume by Double Integration

Use a double integral to find the volume V of the solid in the first octant that is bounded by the coordinate planes and the graphs of the plane $z = 3 - x - y$ and the cylinder $x^2 + y^2 = 1$.

Solution From FIGURE 14.3.5(a) we see that the volume is given by $V = \iint_R (3 - x - y)\, dA$. Since Figure 14.3.5(b) shows that the region of integration R is a Type I region, we have from (1),

$$V = \int_0^1 \int_0^{\sqrt{1-x^2}} (3 - x - y)\, dy\, dx = \int_0^1 \left(3y - xy - \frac{1}{2}y^2\right)\Big]_0^{\sqrt{1-x^2}} dx$$

$$= \int_0^1 \left(3\sqrt{1 - x^2} - x\sqrt{1 - x^2} - \frac{1}{2} + \frac{1}{2}x^2\right) dx \quad \leftarrow \text{trig substitution}$$

$$= \left[\frac{3}{2}\sin^{-1}x + \frac{3}{2}x\sqrt{1 - x^2} + \frac{1}{3}(1 - x^2)^{3/2} - \frac{1}{2}x + \frac{1}{6}x^3\right]_0^1$$

$$= \frac{3}{4}\pi - \frac{2}{3} \approx 1.69.$$

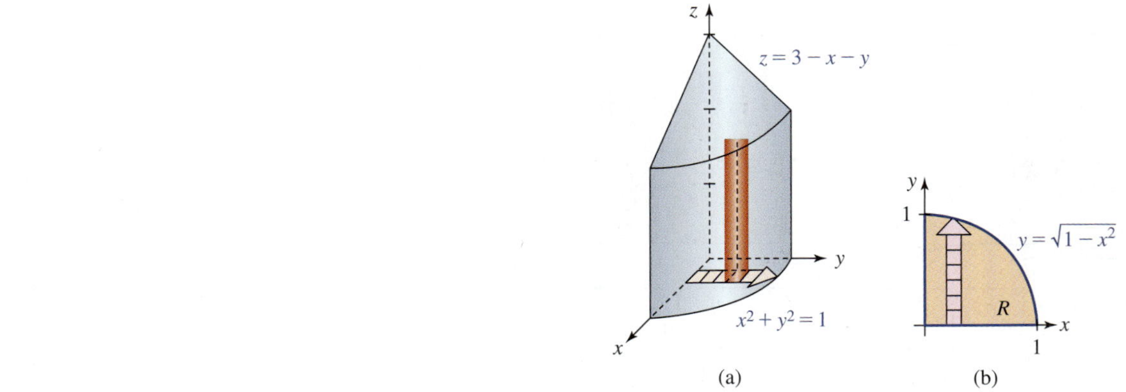

FIGURE 14.3.5 In Example 3, surface in (a); region of integration in (b)

The reduction of a double integral to either of the iterated integrals (1) or (2) depends on (a) the type of region and (b) the function itself. The next two examples illustrate each case.

EXAMPLE 4 Double Integral

Evaluate $\iint_R (x + y)\, dA$ over the region bounded by the graphs of $x = y^2$ and $y = \frac{1}{2}x - \frac{3}{2}$.

Solution The region, which is shown in FIGURE 14.3.6(a), can be written as the union $R = R_1 \cup R_2$ of two Type I regions. By solving the equation $y^2 = 2y + 3$ or $(y + 1)(y - 3) = 0$ we find that the points of intersection of the two graphs are $(1, -1)$ and $(9, 3)$. Thus, from (1) and Theorem 14.1.1(*iii*), we have

$$\iint_R (x + y)\, dA = \iint_{R_1} (x + y)\, dA + \iint_{R_2} (x + y)\, dA$$

$$= \int_0^1 \int_{-\sqrt{x}}^{\sqrt{x}} (x + y)\, dy\, dx + \int_1^9 \int_{x/2 - 3/2}^{\sqrt{x}} (x + y)\, dy\, dx$$

$$= \int_0^1 \left(xy + \frac{1}{2}y^2 \right) \Big]_{-\sqrt{x}}^{\sqrt{x}}\, dx + \int_1^9 \left(xy + \frac{1}{2}y^2 \right) \Big]_{x/2 - 3/2}^{\sqrt{x}}\, dx$$

$$= \int_0^1 2x^{3/2}\, dx + \int_1^9 \left(x^{3/2} + \frac{11}{4}x - \frac{5}{8}x^2 - \frac{9}{8} \right) dx$$

$$= \frac{4}{5}x^{5/2} \Big]_0^1 + \left(\frac{2}{5}x^{5/2} + \frac{11}{8}x^2 - \frac{5}{24}x^3 - \frac{9}{8}x \right) \Big]_1^9 \approx 46.93.$$

Alternative Solution By interpreting the region as a single Type II region, we see from Figure 14.3.6(b) that

$$\iint_R (x + y)\, dA = \int_{-1}^3 \int_{y^2}^{2y + 3} (x + y)\, dx\, dy$$

$$= \int_{-1}^3 \left(\frac{1}{2}x^2 + xy \right) \Big]_{y^2}^{2y + 3}\, dy$$

$$= \int_{-1}^3 \left(-\frac{1}{2}y^4 - y^3 + 4y^2 + 9y + \frac{9}{2} \right) dy$$

$$= \left(-\frac{1}{10}y^5 - \frac{1}{4}y^4 + \frac{4}{3}y^3 + \frac{9}{2}y^2 + \frac{9}{2}y \right) \Big]_{-1}^3 \approx 46.93.$$

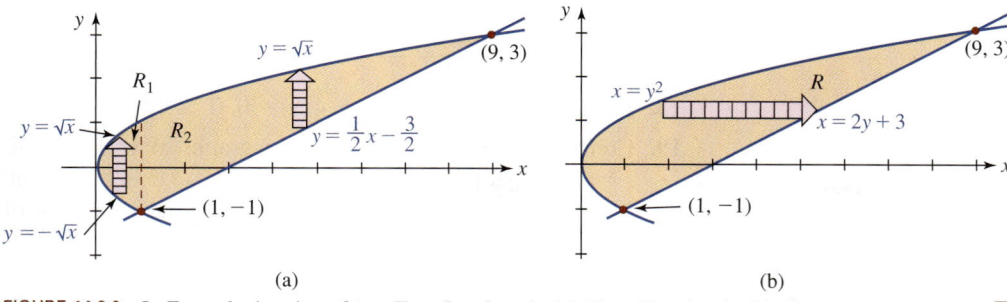

FIGURE 14.3.6 In Example 4, union of two Type I regions in (a); Type II region in (b)

Note that the answer in Example 4 does not represent the volume of the solid above R and below the plane $z = x + y$. Why not?

■ **Reversing the Order of Integration** As Example 4 illustrates, a problem may become easier when the order of integration is **changed** or **reversed**. Also, some iterated integrals that may be impossible to evaluate using one order of integration can, perhaps, be evaluated using the reverse order of integration.

EXAMPLE 5 Double Integral

Evaluate $\iint_R x e^{y^2} dA$ over the region R in the first quadrant bounded by the graphs of $y = x^2$, $x = 0$, $y = 4$.

Solution When viewed as a region of Type I, we have from **FIGURE 14.3.7(a)**, $0 \le x \le 2$, $x^2 \le y \le 4$, and so

$$\iint_R x e^{y^2} dA = \int_0^2 \int_{x^2}^4 x e^{y^2} dy \, dx.$$

The difficulty here is that the partial definite integral $\int_{x^2}^4 x e^{y^2} dy$ cannot be evaluated because e^{y^2} has no elementary-function antiderivative with respect to y. However, as we see in Figure 14.3.7(b), we can interpret the same region as a Type II region defined by $0 \le y \le 4$, $0 \le x \le \sqrt{y}$. Hence, from (2),

$$\iint_R x e^{y^2} dA = \int_0^4 \int_0^{\sqrt{y}} x e^{y^2} dx \, dy$$

$$= \int_0^4 \frac{1}{2} x^2 e^{y^2} \Big]_0^{\sqrt{y}} dy$$

$$= \int_0^4 \frac{1}{2} y e^{y^2} dy$$

$$= \frac{1}{4} e^{y^2} \Big]_0^4 = \frac{1}{4}(e^{16} - 1).$$

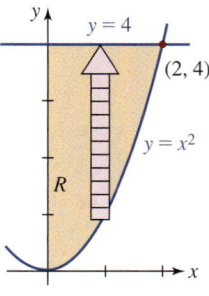

(a) Type I region

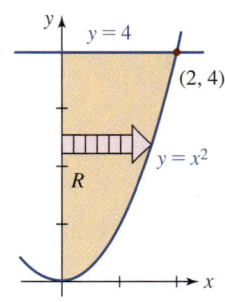

(b) Type II region
FIGURE 14.3.7 Region of integration in Example 5

$\iint_R$ **NOTES FROM THE CLASSROOM**

(*i*) As mentioned after Example 1, the double integral can be defined in terms of a double limit of a double sum such as

$$\sum_i \sum_j f(x_i^*, y_j^*) \Delta y_j \Delta x_i \qquad \text{or} \qquad \sum_j \sum_i f(x_i^*, y_j^*) \Delta x_i \Delta y_j.$$

We will not pursue the details.

(*ii*) You are encouraged to take advantage of symmetries to minimize your work when finding areas and volumes by double integration. In the case of volumes, make sure *both* the region R and the surface over the region possess corresponding symmetries. See Problem 19 in Exercises 14.3.

(*iii*) Before attempting to evaluate a double integral, *always* try to sketch an accurate picture of the region R of integration.

≡ Fundamentals

In Problems 1–10, evaluate the double integral over the region R that is bounded by the graphs of the given equations. Choose the most convenient order of integration.

1. $\displaystyle\iint_R x^3 y^2 \, dA; \quad y = x, y = 0, x = 1$

2. $\displaystyle\iint_R (x + 1) \, dA; \quad y = x, x + y = 4, x = 0$

3. $\displaystyle\iint_R (2x + 4y + 1) \, dA; \quad y = x^2, y = x^3$

4. $\displaystyle\iint_R xe^y \, dA; \quad R$ the same as in Problem 1

5. $\displaystyle\iint_R 2xy \, dA; \quad y = x^3, y = 8, x = 0$

6. $\displaystyle\iint_R \frac{x}{\sqrt{y}} \, dA; \quad y = x^2 + 1, y = 3 - x^2$

7. $\displaystyle\iint_R \frac{y}{1 + xy} \, dA; \quad y = 0, y = 1, x = 0, x = 1$

8. $\displaystyle\iint_R \sin\frac{\pi x}{y} \, dA; \quad x = y^2, x = 0, y = 1, y = 2$

9. $\displaystyle\iint_R \sqrt{x^2 + 1} \, dA; \quad x = y, x = -y, x = \sqrt{3}$

10. $\displaystyle\iint_R x \, dA; \quad y = \tan^{-1} x, y = 0, x = 1$

In Problems 11 and 12, evaluate $\iint_R (x + y) \, dA$ for the given region R.

11.

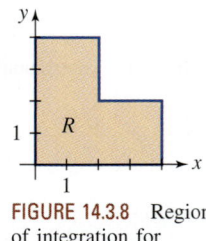

FIGURE 14.3.8 Region of integration for Problem 11

12.

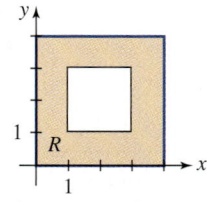

FIGURE 14.3.9 Region of integration for Problem 12

In Problems 13–18, use a double integral to find the area of the region R that is bounded by the graphs of the given equations.

13. $y = -x, y = 2x - x^2$

14. $x = y^2, x = 2 - y^2$

15. $y = e^x, y = \ln x, x = 1, x = 4$

16. $\sqrt{x} + \sqrt{y} = 2, x + y = 4$

17. $y = -2x + 3, y = x^3, x = -2$

18. $y = -x^2 + 3x, y = -2x + 4, y = 0, 0 \le x \le 2$

19. Consider the solid bounded by the graphs of $x^2 + y^2 = 4$, $z = 4 - y$, and $z = 0$ shown in FIGURE 14.3.10. Choose and evaluate the correct integral representing the volume V of the solid.

(a) $\displaystyle 4\int_0^2 \int_0^{\sqrt{4-x^2}} (4 - y) \, dy \, dx$

(b) $\displaystyle 2\int_{-2}^2 \int_0^{\sqrt{4-x^2}} (4 - y) \, dy \, dx$

(c) $\displaystyle 2\int_{-2}^2 \int_0^{\sqrt{4-y^2}} (4 - y) \, dx \, dy$

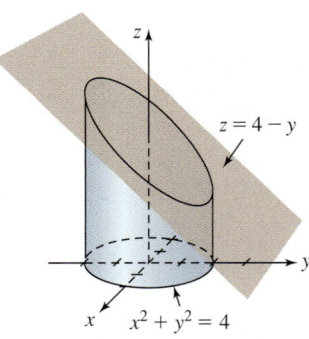

FIGURE 14.3.10 Solid in Problem 19

20. The solid bounded by the cylinders $x^2 + y^2 = r^2$ and $y^2 + z^2 = r^2$ is called a **bicylinder**. An eighth of the solid is shown in FIGURE 14.3.11. Choose and evaluate the correct integral corresponding to the volume V of the bicylinder.

(a) $\displaystyle 4\int_{-r}^r \int_{-\sqrt{r^2-x^2}}^{\sqrt{r^2-x^2}} (r^2 - y^2)^{1/2} \, dy \, dx$

(b) $\displaystyle 8\int_0^r \int_0^{\sqrt{r^2-y^2}} (r^2 - y^2)^{1/2} \, dx \, dy$

(c) $\displaystyle 8\int_0^r \int_0^{\sqrt{r^2-x^2}} (r^2 - x^2)^{1/2} \, dy \, dx$

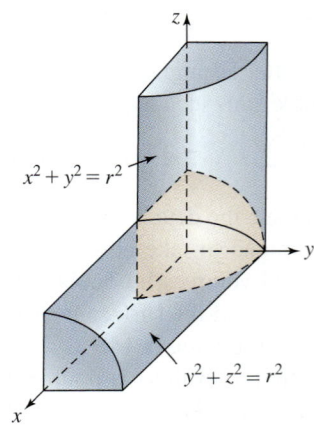

FIGURE 14.3.11 Solid in Problem 20

In Problems 21–30, find the volume of the solid bounded by the graphs of the given equations.

21. $2x + y + z = 6$, $x = 0$, $y = 0$, $z = 0$, first octant
22. $z = 4 - y^2$, $x = 3$, $x = 0$, $y = 0$, $z = 0$, first octant
23. $x^2 + y^2 = 4$, $x - y + 2z = 4$, $x = 0$, $y = 0$, $z = 0$, first octant
24. $y = x^2$, $y + z = 3$, $z = 0$
25. $z = 1 + x^2 + y^2$, $3x + y = 3$, $x = 0$, $y = 0$, $z = 0$, first octant
26. $z = x + y$, $x^2 + y^2 = 9$, $x = 0$, $y = 0$, $z = 0$, first octant
27. $yz = 6$, $x = 0$, $x = 5$, $y = 1$, $y = 6$, $z = 0$
28. $z = 4 - x^2 - \frac{1}{4}y^2$, $z = 0$
29. $z = 4 - y^2$, $x^2 + y^2 = 2x$, $z = 0$
30. $z = 1 - x^2$, $z = 1 - y^2$, $x = 0$, $y = 0$, $z = 0$, first octant

If $f_2(x, y) \geq f_1(x, y)$ for all (x, y) in a region R, then the volume of the solid bounded by the two surfaces over R is

$$V = \iint_R [f_2(x, y) - f_1(x, y)] \, dA.$$

In Problems 31–34, find the volume bounded by the graphs of the given equations.

31. $x + 2y + z = 4$, $z = x + y$, $x = 0$, $y = 0$, first octant
32. $z = x^2 + y^2$, $z = 9$
33. $z = x^2$, $z = -x + 2$, $x = 0$, $y = 0$, $y = 5$, first octant
34. $2z = 4 - x^2 - y^2$, $z = 2 - y$

In Problems 35–40, reverse the order of integration.

35. $\displaystyle\int_0^2 \int_0^{y^2} f(x, y) \, dx \, dy$ **36.** $\displaystyle\int_{-5}^5 \int_0^{\sqrt{25-y^2}} f(x, y) \, dx \, dy$

37. $\displaystyle\int_0^3 \int_1^{e^x} f(x, y) \, dy \, dx$ **38.** $\displaystyle\int_0^2 \int_{y/2}^{3-y} f(x, y) \, dx \, dy$

39. $\displaystyle\int_0^1 \int_0^{\sqrt[3]{x}} f(x, y) \, dy \, dx + \int_1^2 \int_0^{2-x} f(x, y) \, dy \, dx$

40. $\displaystyle\int_0^1 \int_0^{\sqrt{y}} f(x, y) \, dx \, dy + \int_1^2 \int_0^{\sqrt{2-y}} f(x, y) \, dx \, dy$

In Problems 41–46, evaluate the given iterated integral by reversing the order of integration.

41. $\displaystyle\int_0^1 \int_x^1 x^2 \sqrt{1 + y^4} \, dy \, dx$ **42.** $\displaystyle\int_0^1 \int_{2y}^2 e^{-y/x} \, dx \, dy$

43. $\displaystyle\int_0^2 \int_{y^2}^4 \cos \sqrt{x^3} \, dx \, dy$ **44.** $\displaystyle\int_{-1}^1 \int_{-\sqrt{1-x^2}}^{\sqrt{1-x^2}} x \sqrt{1 - x^2 - y^2} \, dy \, dx$

45. $\displaystyle\int_0^1 \int_x^1 \frac{1}{1 + y^4} \, dy \, dx$ **46.** $\displaystyle\int_0^4 \int_{\sqrt{y}}^2 \sqrt{x^3 + 1} \, dx \, dy$

The **average value** f_{ave} of a continuous function $z = f(x, y)$ over a region R in the xy-plane is defined to be

$$f_{ave} = \frac{1}{A} \iint_R f(x, y) \, dA, \tag{3}$$

where A is the area of R. In Problems 47 and 48, find f_{ave} for the given function and region R.

47. $f(x, y) = xy$; R defined by $a \leq x \leq b$, $c \leq y \leq d$
48. $f(x, y) = 9 - x^2 - 3y^2$; R bounded by the ellipse $x^2 + 3y^2 = 9$

≡ Think About It

49. From (3) we can write $\iint_R f(x, y) \, dA = f_{ave} \cdot A$, where A is the area of R. Discuss the geometric interpretation of this result in the case $f(x, y) > 0$ over R.

50. Let R be a rectangular region bounded by the lines $x = a$, $x = b$, $y = c$, and $y = d$, where $a < b$, $c < d$.
  (a) Show that

$$\iint_R \cos 2\pi(x + y) \, dA = \frac{1}{4\pi^2} (S_1 S_2 - C_1 C_2)$$

$$\iint_R \sin 2\pi(x + y) \, dA = -\frac{1}{4\pi^2} (C_1 S_2 + S_1 C_2),$$

  where

  $S_1 = \sin 2\pi b - \sin 2\pi a$, $S_2 = \sin 2\pi d - \sin 2\pi c$
  $C_1 = \cos 2\pi b - \cos 2\pi a$, $C_2 = \cos 2\pi d - \cos 2\pi c$.

  (b) Show that if at least one of the two perpendicular sides of R has an integer length, then

$$\iint_R \cos 2\pi(x + y) \, dA = 0 \quad \text{and} \quad \iint_R \sin 2\pi(x + y) \, dA = 0.$$

  (c) Conversely, show that if

$$\iint_R \cos 2\pi(x + y) \, dA = 0 \quad \text{and} \quad \iint_R \sin 2\pi(x + y) \, dA = 0,$$

  then at least one of the two perpendicular sides of R must have integer length. [*Hint*: Consider $0 = (S_1 S_2 - C_1 C_2)^2 + (C_1 S_2 + S_1 C_2)^2$.]

51. Let R be a rectangular region that has been divided into n nonoverlapping rectangular subregions $R_1, R_2, \ldots, R_n$ whose sides are all parallel to the horizontal and vertical sides of R. See FIGURE 14.3.12. Suppose that each interior rectangle has the property that one of its two perpendicular sides has integer length. Show that R has the same property. [*Hint*: Use Problem 50 and Theorem 14.1.1(*iii*).]

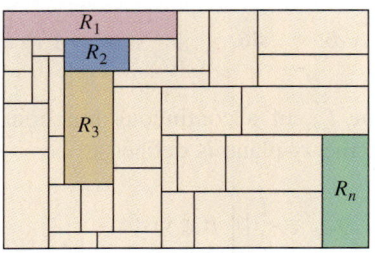

FIGURE 14.3.12   Rectangular region in Problem 51

☰ Projects

52. The solid bounded by the intersection of three cylinders $x^2 + y^2 = r^2$, $y^2 + z^2 = r^2$, and $x^2 + z^2 = r^2$ is called a **tricylinder**. See FIGURE 14.3.13. Do some Internet research and find a figure of the actual solid. Find the volume of the solid.

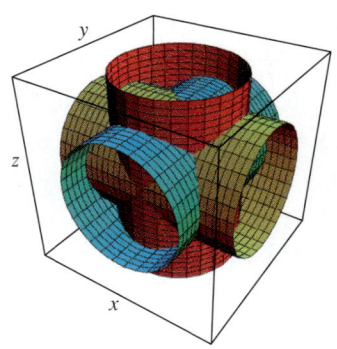

FIGURE 14.3.13   Three cylinders of the same radius intersecting at right angles in Problem 52

14.4 Center of Mass and Moments

■ **Introduction**   In Section 6.10 we saw that if ρ is a density (mass per unit area), then the mass of a two-dimensional smear of matter, or lamina, that coincides with a region bounded by the graphs of $y = f(x)$, the x-axis, and the lines $x = a$ and $x = b$ is given by

$$m = \lim_{\|P\| \to 0} \sum_{k=1}^{n} \rho \Delta A_k = \lim_{\|P\| \to 0} \sum_{k=1}^{n} \rho f(x_k^*) \Delta x_k = \int_a^b \rho f(x) \, dx. \qquad (1)$$

The density ρ in (1) can be a function of x; when $\rho =$ constant the lamina is said to be homogeneous.

We see next that if the density ρ is a function of two variables, then the mass m of a lamina is given by a double integral.

■ **Laminas with Variable Density—Center of Mass**   If a lamina corresponding to a region R in the xy-plane has a variable density $\rho(x, y)$ (units of mass per unit area), where ρ is nonnegative and continuous on R, then analogous to (1) we define its **mass** m by the double integral

$$m = \lim_{\|P\| \to 0} \sum_{k=1}^{n} \rho(x_k^*, y_k^*) \Delta A_k \quad \text{or} \quad m = \iint_R \rho(x, y) \, dA. \qquad (2)$$

As in Section 6.10, we define the coordinates of the **center of mass** of the lamina by

$$\bar{x} = \frac{M_y}{m}, \quad \bar{y} = \frac{M_x}{m}, \qquad (3)$$

where

$$M_y = \iint_R x \rho(x, y) \, dA \quad \text{and} \quad M_x = \iint_R y \rho(x, y) \, dA \qquad (4)$$

are the **moments** of the lamina about the y- and x-axes, respectively. The center of mass is the point where we consider all the mass of the lamina to be concentrated. If $\rho(x, y)$ is a constant, the lamina is said to be homogeneous and its center of mass is called the **centroid** of the lamina.

EXAMPLE 1   Center of Mass

A lamina has the shape of the region R in the first quadrant that is bounded by the graphs of $y = \sin x$ and $y = \cos x$ between $x = 0$ and $x = \pi/4$. Find its center of mass if the density is $\rho(x, y) = y$.

Solution From FIGURE 14.4.1 we see that

$$m = \iint\limits_R y\, dA = \int_0^{\pi/4}\int_{\sin x}^{\cos x} y\, dy\, dx$$

$$= \int_0^{\pi/4} \frac{1}{2}y^2 \Big]_{\sin x}^{\cos x} dx$$

$$= \frac{1}{2}\int_0^{\pi/4} (\cos^2 x - \sin^2 x)\, dx \quad \leftarrow \text{double angle formula}$$

$$= \frac{1}{2}\int_0^{\pi/4} \cos 2x\, dx$$

$$= \frac{1}{4}\sin 2x\Big]_0^{\pi/4} = \frac{1}{4}.$$

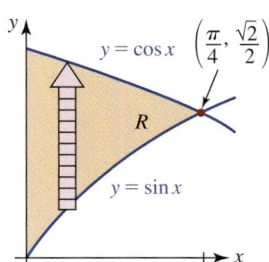

FIGURE 14.4.1 Lamina in Example 1

Now,

$$M_y = \iint\limits_R xy\, dA = \int_0^{\pi/4}\int_{\sin x}^{\cos x} xy\, dy\, dx$$

$$= \int_0^{\pi/4} \frac{1}{2}xy^2 \Big]_{\sin x}^{\cos x} dx$$

$$= \frac{1}{2}\int_0^{\pi/4} x\cos 2x\, dx \qquad \leftarrow \text{integration by parts}$$

$$= \left(\frac{1}{4}x\sin 2x + \frac{1}{8}\cos 2x\right)\Big]_0^{\pi/4} = \frac{1}{16}(\pi - 2).$$

Similarly,

$$M_x = \iint\limits_R y^2\, dA = \int_0^{\pi/4}\int_{\sin x}^{\cos x} y^2\, dy\, dx$$

$$= \frac{1}{3}\int_0^{\pi/4} (\cos^3 x - \sin^3 x)\, dx$$

$$= \frac{1}{3}\int_0^{\pi/4} [\cos x(1 - \sin^2 x) - \sin x(1 - \cos^2 x)]\, dx$$

$$= \frac{1}{3}\left(\sin x - \frac{1}{3}\sin^3 x + \cos x - \frac{1}{3}\cos^3 x\right)\Big]_0^{\pi/4} = \frac{1}{18}(5\sqrt{2} - 4).$$

Hence, from (3), the coordinates of the center of mass of the lamina are

$$\bar{x} = \frac{M_y}{m} = \frac{\frac{1}{16}(\pi - 2)}{\frac{1}{4}} = \frac{1}{4}(\pi - 2),$$

$$\bar{y} = \frac{M_x}{m} = \frac{\frac{1}{18}(5\sqrt{2} - 4)}{\frac{1}{4}} = \frac{1}{9}(10\sqrt{2} - 8).$$

Approximate coordinates of the center of mass are $(0.29, 0.68)$. ∎

EXAMPLE 2 Center of Mass

A lamina has the shape of the region R bounded by the graph of the ellipse $\frac{1}{4}x^2 + \frac{1}{16}y^2 = 1$, $0 \le y \le 4$, and $y = 0$. Find its center of mass if the density is $\rho(x, y) = |x|y$.

Solution From FIGURE 14.4.2 we see that the region is symmetric with respect to the y-axis. Furthermore, since $\rho(-x, y) = \rho(x, y)$, the density ρ is symmetric about this axis. Thus, the

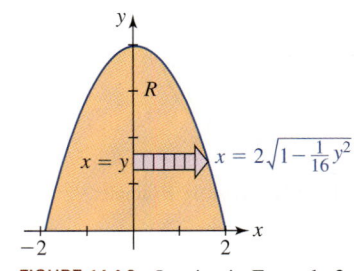

FIGURE 14.4.2 Lamina in Example 2

y-coordinate of the center of mass must lie on the axis of symmetry, and so we have $\bar{x} = 0$. Utilizing symmetry, the mass of the lamina is

$$
m = \iint_R |x|y \, dA = 2\int_0^4 \int_0^{2\sqrt{1-y^2/16}} xy \, dx \, dy
$$

$$
= \int_0^4 x^2 y \Big]_0^{2\sqrt{1-y^2/16}} \, dy
$$

$$
= 4\int_0^4 \left(y - \frac{1}{16}y^3\right) dy
$$

$$
= 4\left(\frac{1}{2}y^2 - \frac{1}{64}y^4\right)\Big]_0^4 = 16.
$$

Similarly,

$$
M_x = \iint_R |x|y^2 \, dA = 2\int_0^4 \int_0^{2\sqrt{1-y^2/16}} xy^2 \, dx \, dy = \frac{512}{15}.
$$

From (3)

$$
\bar{y} = \frac{\dfrac{512}{15}}{16} = \frac{32}{15}.
$$

The coordinates of the center of mass are $\left(0, \frac{32}{15}\right)$. ∎

Do not conclude from Example 2 that the center of mass must always lie on an axis of symmetry of a lamina. Bear in mind that the density function $\rho(x, y)$ must also be symmetric with respect to that axis.

▌ **Moments of Inertia** The integrals M_x and M_y in (4) are also called the **first moments** of a lamina about the x-axis and y-axis, respectively. The so-called **second moments** of a lamina or **moments of inertia** about the x- and y-axes are, in turn, defined by the double integrals

$$
I_x = \iint_R y^2 \rho(x, y) \, dA \qquad \text{and} \qquad I_y = \iint_R x^2 \rho(x, y) \, dA. \tag{5}
$$

A moment of inertia is the rotational equivalent of mass. For translational motion, kinetic energy is given by $K = \frac{1}{2}mv^2$, where m is mass and v is linear speed. The kinetic energy of a particle of mass m rotating at a distance r from an axis is $K = \frac{1}{2}mv^2 = \frac{1}{2}m(r\omega)^2 = \frac{1}{2}(mr^2)\omega^2 = \frac{1}{2}I\omega^2$ where $I = mr^2$ is its moment of inertia about the axis of rotation and ω is angular speed.

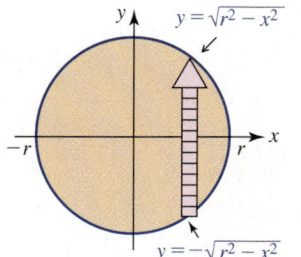

FIGURE 14.4.3 Disk in Example 3

EXAMPLE 3 Moment of Inertia

Find the moment of inertia about the y-axis of the thin homogeneous disk of mass m shown in FIGURE 14.4.3.

Solution Since the disk is homogeneous, its density is the constant $\rho(x, y) = m/\pi r^2$. Hence, from (5),

$$
I_y = \iint_R x^2 \rho(x, y) \, dA = \iint_R x^2 \left(\frac{m}{\pi r^2}\right) dA
$$

$$
= \frac{m}{\pi r^2} \int_{-r}^{r} \int_{-\sqrt{r^2-x^2}}^{\sqrt{r^2-x^2}} x^2 \, dy \, dx
$$

$$= \frac{2m}{\pi r^2} \int_{-r}^{r} x^2 \sqrt{r^2 - x^2}\, dx \quad \leftarrow \text{trig substitution}$$

$$= \frac{2mr^2}{\pi} \int_{-\pi/2}^{\pi/2} \sin^2\theta \cos^2\theta\, d\theta \quad \leftarrow \text{double angle formula}$$

$$= \frac{mr^2}{2\pi} \int_{-\pi/2}^{\pi/2} \sin^2 2\theta\, d\theta \quad \leftarrow \text{half-angle formula}$$

$$= \frac{mr^2}{4\pi} \int_{-\pi/2}^{\pi/2} (1 - \cos 4\theta)\, d\theta = \frac{1}{4} mr^2. \qquad \blacksquare$$

■ **Radius of Gyration** The radius of gyration of a lamina of mass m and the moment of inertia I about an axis is defined by

$$R_g = \sqrt{\frac{I}{m}}. \qquad (6)$$

Since (6) implies that $I = mR_g^2$, the radius of gyration is interpreted as the radial distance the lamina, considered as a point mass, can rotate about the axis without changing the rotational inertia of the body. In Example 3 the radius of gyration is $R_g = \sqrt{I_y/m} = \sqrt{(\frac{1}{4}mr^2)/m} = \frac{1}{2}r$.

Exercises 14.4 Answers to selected odd-numbered problems begin on page ANS-44.

≡ Fundamentals

In Problems 1–10, find the center of mass of the lamina that has the given shape and density.

1. $x = 0, x = 4, y = 0, y = 3$; $\rho(x, y) = xy$

2. $x = 0, y = 0, 2x + y = 4$; $\rho(x, y) = x^2$

3. $y = x, x + y = 6, y = 0$; $\rho(x, y) = 2y$

4. $y = |x|, y = 3$; $\rho(x, y) = x^2 + y^2$

5. $y = x^2, x = 1, y = 0$; $\rho(x, y) = x + y$

6. $x = y^2, x = 4$; $\rho(x, y) = y + 5$

7. $y = 1 - x^2, y = 0$; density ρ at a point P directly proportional to the distance from the x-axis

8. $y = \sin x, 0 \leq x \leq \pi, y = 0$; density ρ at a point P directly proportional to the distance from the y-axis

9. $y = e^x, x = 0, x = 1, y = 0$; $\rho(x, y) = y^3$

10. $y = \sqrt{9 - x^2}, y = 0$; $\rho(x, y) = x^2$

In Problems 11–14, find the moment of inertia about the x-axis of the lamina that has the given shape and density.

11. $x = y - y^2, x = 0$; $\rho(x, y) = 2x$

12. $y = x^2, y = \sqrt{x}$; $\rho(x, y) = x^2$

13. $y = \cos x, -\pi/2 \leq x \leq \pi/2, y = 0$;
$\rho(x, y) = k$ (constant)

14. $y = \sqrt{4 - x^2}, x = 0, y = 0$, first quadrant; $\rho(x, y) = y$

In Problems 15–18, find the moment of inertia about the y-axis of the lamina that has the given shape and density.

15. $y = x^2, x = 0, y = 4$, first quadrant; $\rho(x, y) = y$

16. $y = x^2, y = \sqrt{x}$; $\rho(x, y) = x^2$

17. $y = x, y = 0, y = 1, x = 3$; $\rho(x, y) = 4x + 3y$

18. Same R and density as in Problem 7

In Problems 19 and 20, find the radius of gyration about the indicated axis of the lamina that has the given shape and density.

19. $x = \sqrt{a^2 - y^2}, x = 0$; $\rho(x, y) = x$; y-axis

20. $x + y = a, a > 0, x = 0, y = 0$;
$\rho(x, y) = k$ (constant); x-axis

21. A lamina has the shape of the region bounded by the graph of the ellipse $x^2/a^2 + y^2/b^2 = 1$. If its density is $\rho(x, y) = 1$, find:

 (a) the moment of inertia about the x-axis of the lamina,

 (b) the moment of inertia about the y-axis of the lamina,

 (c) the radius of gyration about the x-axis [*Hint*: The area of the ellipse is πab.],

 (d) the radius of gyration about the y-axis.

22. A cross-section of an experimental airfoil is the lamina shown in **FIGURE 14.4.4**. The arc ABC is elliptical, whereas the two arcs AD and CD are parabolic. Find the moment of inertia about the x-axis of the lamina under the assumption that the density is $\rho(x, y) = 1$.

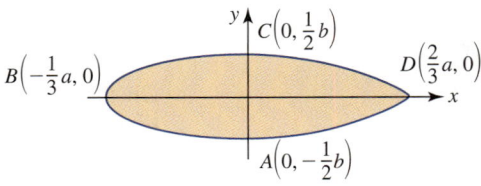

FIGURE 14.4.4 Airfoil in Problem 22

In Problems 23–26, find the polar moment of inertia I_0 of the lamina that has the given shape and density. The **polar moment of inertia** of a lamina with respect to the origin is defined to be

$$I_0 = \iint_R (x^2 + y^2)\rho(x, y)\, dA = I_x + I_y.$$

23. $x + y = a, a > 0, x = 0, y = 0;$ $\rho(x, y) = k$ (constant)

24. $y = x^2, y = \sqrt{x};$ $\rho(x, y) = x^2$ [*Hint:* See Problems 12 and 16.]

25. $x = y^2 + 2, x = 6 - y^2;$ density ρ at a point P inversely proportional to the square of the distance from the origin

26. $y = x, y = 0, y = 3, x = 4;$ $\rho(x, y) = k$ (constant)

27. Find the radius of gyration in Problem 23.

28. Show that the polar moment of inertia with respect to the origin about the center of a thin homogeneous rectangular plate of mass m, width w, and length l is $I_0 = \frac{1}{12}m(l^2 + w^2)$.

14.5 Double Integrals in Polar Coordinates

■ **Introduction** Suppose R is a region bounded by the graphs of the polar equations $r = g_1(\theta)$, $r = g_2(\theta)$ and the rays $\theta = \alpha$, $\theta = \beta$, and f is a function of r and θ that is continuous on R. In order to define the double integral of f over R, we use rays and concentric circles to partition the region into a grid of "polar rectangles" or subregions R_k. See FIGURE 14.5.1(a) and (b). The area ΔA_k of a typical subregion R_k, shown in Figure 14.5.1(c), is the difference of areas of two circular sectors:

$$\Delta A_k = \frac{1}{2}r_{k+1}^2 \Delta\theta_k - \frac{1}{2}r_k^2 \Delta\theta_k = \frac{1}{2}(r_{k+1}^2 - r_k^2)\Delta\theta_k$$

$$= \frac{1}{2}(r_{k+1} + r_k)(r_{k+1} - r_k)\Delta\theta_k = r_k^* \Delta r_k \Delta\theta_k,$$

where $\Delta r_k = r_{k+1} - r_k$ and r_k^* denotes the average radius $\frac{1}{2}(r_{k+1} + r_k)$. By

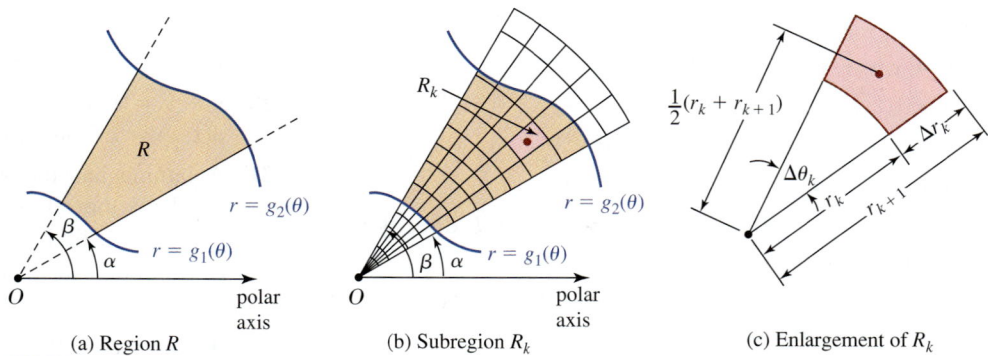

FIGURE 14.5.1 Partition of R using polar coordinates

choosing a sample point (r_k^*, θ_k^*) in each R_k, the double integral of f over R is

$$\lim_{\|P\|\to 0} \sum_{k=1}^{n} f(r_k^*, \theta_k^*) r_k^* \Delta r_k \Delta\theta_k = \iint_R f(r, \theta)\, dA.$$

The double integral is then evaluated by means of the iterated integral

$$\iint_R f(r, \theta)\, dA = \int_\alpha^\beta \int_{g_1(\theta)}^{g_2(\theta)} f(r, \theta) r\, dr\, d\theta. \tag{1}$$

On the other hand, if the region R is as given in FIGURE 14.5.2, the double integral of f over R is then

$$\iint_R f(r, \theta)\, dA = \int_a^b \int_{h_1(r)}^{h_2(r)} f(r, \theta) r\, d\theta\, dr. \tag{2}$$

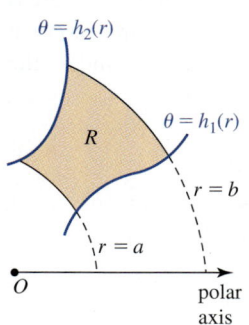

FIGURE 14.5.2 Region R of integration in (2)

EXAMPLE 1   Center of Mass

Find the center of mass of the lamina that corresponds to the region bounded by one leaf of the rose curve $r = 2\sin 2\theta$ in the first quadrant if the density at a point P in the lamina is directly proportional to the distance from the pole.

Solution   By varying θ from 0 to $\pi/2$, we obtain the graph in FIGURE 14.5.3. Now, the distance from the pole is $d(0, P) = |r|$. Hence, the density of the lamina is $\rho(r, \theta) = k|r|$, where k is a constant of proportionality. From (2) of Section 14.4, we have

$$m = \iint_R k|r|\,dA = k\int_0^{\pi/2}\int_0^{2\sin 2\theta}(r)r\,dr\,d\theta$$

$$= k\int_0^{\pi/2}\frac{1}{3}r^3\Big]_0^{2\sin 2\theta}d\theta$$

$$= \frac{8}{3}k\int_0^{\pi/2}\sin^3 2\theta\,d\theta$$

$$= \frac{8}{3}k\int_0^{\pi/2}\sin^2 2\theta\sin 2\theta\,d\theta \qquad \leftarrow \text{trig identity}$$

$$= \frac{8}{3}k\int_0^{\pi/2}(1-\cos^2 2\theta)\sin 2\theta\,d\theta$$

$$= \frac{8}{3}k\left[-\frac{1}{2}\cos 2\theta + \frac{1}{6}\cos^3 2\theta\right]_0^{\pi/2} = \frac{16}{9}k.$$

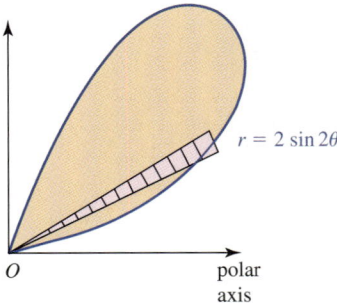

FIGURE 14.5.3   Lamina in Example 1

Since $x = r\cos\theta$, we can write the first moment $M_y = k\iint_R x|r|\,dA$ as

$$M_y = k\int_0^{\pi/2}\int_0^{2\sin 2\theta}r^3\cos\theta\,dr\,d\theta$$

$$= k\int_0^{\pi/2}\frac{1}{4}r^4\cos\theta\Big]_0^{2\sin 2\theta}d\theta$$

$$= 4k\int_0^{\pi/2}(\sin 2\theta)^4\cos\theta\,d\theta \qquad \leftarrow \text{double angle formula}$$

$$= 4k\int_0^{\pi/2}(2\sin\theta\cos\theta)^4\cos\theta\,d\theta$$

$$= 64k\int_0^{\pi/2}\sin^4\theta\cos^5\theta\,d\theta$$

$$= 64k\int_0^{\pi/2}\sin^4\theta(1-\sin^2\theta)^2\cos\theta\,d\theta$$

$$= 64k\int_0^{\pi/2}(\sin^4\theta - 2\sin^6\theta + \sin^8\theta)\cos\theta\,d\theta$$

$$= 64k\left(\frac{1}{5}\sin^5\theta - \frac{2}{7}\sin^7\theta + \frac{1}{9}\sin^9\theta\right)\Big]_0^{\pi/2} = \frac{512}{315}k.$$

Similarly, by using $y = r\sin\theta$, we find

$$M_x = k\int_0^{\pi/2}\int_0^{2\sin 2\theta}r^3\sin\theta\,dr\,d\theta = \frac{512}{315}k.$$

Here the rectangular coordinates of the center of mass are

$$\bar{x} = \bar{y} = \frac{\dfrac{512}{315}k}{\dfrac{16}{9}k} = \frac{32}{35}.$$

■

In Example 1, we could have argued to the fact that $M_x = M_y$ and hence, $\bar{x} = \bar{y}$ from the fact that the lamina and the density function are symmetric about the ray $\theta = \pi/4$.

■ **Change of Variables: Rectangular to Polar Coordinates** In some instances a double integral $\iint_R f(x, y)\, dA$ that is difficult or even impossible to evaluate using rectangular coordinates may be readily evaluated when a change of variables is used. If we assume that f is continuous on the region R and if R can be described in polar coordinates as $0 \le g_1(\theta) \le r \le g_2(\theta)$, $\alpha \le \theta \le \beta, 0 < \beta - \alpha \le 2\pi$, then

$$\iint_R f(x, y)\, dA = \int_\alpha^\beta \int_{g_1(\theta)}^{g_2(\theta)} f(r\cos\theta, r\sin\theta)r\, dr\, d\theta. \qquad (3)$$

Equation (3) is particularly useful when f contains the expression $x^2 + y^2$, since, in polar coordinates, we can now write

$$x^2 + y^2 = r^2 \qquad \text{and} \qquad \sqrt{x^2 + y^2} = r.$$

EXAMPLE 2 Change of Variables

Use polar coordinates to evaluate

$$\int_0^2 \int_x^{\sqrt{8-x^2}} \frac{1}{5 + x^2 + y^2}\, dy\, dx.$$

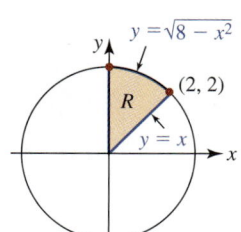

$y = \sqrt{8 - x^2}$

$(2, 2)$

R

$y = x$

FIGURE 14.5.4 Region R of integration in Example 2

Solution From $x \le y \le \sqrt{8 - x^2}, 0 \le x \le 2$, we have sketched the region R of integration in FIGURE 14.5.4. Since $x^2 + y^2 = r^2$, the polar description of the circle $x^2 + y^2 = 8$ is $r = \sqrt{8}$. Hence, in polar coordinates, the region of R is given by $0 \le r \le \sqrt{8}$, $\pi/4 \le \theta \le \pi/2$. From $1/(5 + x^2 + y^2) = 1/(5 + r^2)$ the original integral becomes

$$\int_0^2 \int_x^{\sqrt{8-x^2}} \frac{1}{5 + x^2 + y^2}\, dy\, dx = \int_{\pi/4}^{\pi/2} \int_0^{\sqrt{8}} \frac{1}{5 + r^2}r\, dr\, d\theta$$

$$= \frac{1}{2} \int_{\pi/4}^{\pi/2} \int_0^{\sqrt{8}} \frac{1}{5 + r^2}(2r\, dr)\, d\theta$$

$$= \frac{1}{2} \int_{\pi/4}^{\pi/2} \ln(5 + r^2) \Big]_0^{\sqrt{8}}\, d\theta$$

$$= \frac{1}{2}(\ln 13 - \ln 5) \int_{\pi/4}^{\pi/2}\, d\theta$$

$$= \frac{1}{2}(\ln 13 - \ln 5)\left(\frac{\pi}{2} - \frac{\pi}{4}\right) = \frac{\pi}{8} \ln \frac{13}{5}. \qquad ■$$

EXAMPLE 3 Volume

Find the volume of the solid that is under the hemisphere $z = \sqrt{1 - x^2 - y^2}$ and above the region bounded by the graph of the circle $x^2 + y^2 - y = 0$.

Solution From FIGURE 14.5.5 we see that

$$V = \iint_R \sqrt{1 - x^2 - y^2}\, dA.$$

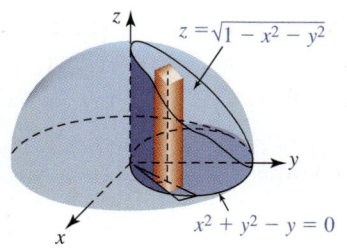

$z = \sqrt{1 - x^2 - y^2}$

$x^2 + y^2 - y = 0$

FIGURE 14.5.5 Solid within a hemisphere in Example 3

In polar coordinates the equations of the hemisphere and the circle become, respectively, $z = \sqrt{1 - r^2}$ and $r = \sin\theta$. Now, using symmetry we have

$$V = \iint_R \sqrt{1 - r^2}\, dA = 2 \int_0^{\pi/2} \int_0^{\sin\theta} (1 - r^2)^{1/2} r\, dr\, d\theta$$

$$= 2 \int_0^{\pi/2} \left[-\frac{1}{3}(1 - r^2)^{3/2} \right]_0^{\sin\theta}\, d\theta$$

$$= \frac{2}{3} \int_0^{\pi/2} \left[1 - (1 - \sin^2\theta)^{3/2} \right] d\theta$$

$$= \frac{2}{3} \int_0^{\pi/2} \left[1 - (\cos^2\theta)^{3/2} \right] d\theta$$

$$= \frac{2}{3} \int_0^{\pi/2} (1 - \cos^3\theta) \, d\theta$$

$$= \frac{2}{3} \int_0^{\pi/2} \left[1 - (1 - \sin^2\theta)\cos\theta \right] d\theta$$

$$= \frac{2}{3} \left(\theta - \sin\theta + \frac{1}{3}\sin^3\theta \right) \Big|_0^{\pi/2} = \frac{1}{3}\pi - \frac{4}{9} \approx 0.60. \qquad \blacksquare$$

■ **Area** Note that in (1) if $f(r, \theta) = 1$, then the **area** of the region R in Figure 14.5.1(a) is given by

$$A = \iint_R dA = \int_\alpha^\beta \int_{g_1(\theta)}^{g_2(\theta)} r \, dr \, d\theta. \tag{4}$$

The same observation holds for (2) and Figure 14.5.2 when $f(r, \theta) = 1$.

$\iint_R$ **NOTES FROM THE CLASSROOM**
....

You are urged to reexamine Example 3. The graph of the circle $r = \sin\theta$ is obtained by varying θ from 0 to π. However, carry out the iterated integration

$$V = \int_0^\pi \int_0^{\sin\theta} (1 - r^2)^{1/2} r \, dr \, d\theta$$

and see if you obtain the *incorrect* answer $\pi/3$. What goes wrong?

Exercises 14.5 Answers to selected odd-numbered problems begin on page ANS-44.

≡ Fundamentals

In Problems 1–4, use a double integral in polar coordinates to find the area of the region bounded by the graphs of the given polar equations.

1. $r = 3 + 3\sin\theta$ **2.** $r = 2 + \cos\theta$

3. $r = 2\sin\theta$, $r = 1$, common area

4. $r = 8\sin 4\theta$, one petal

In Problems 5–10, find the volume of the solid bounded by the graphs of the given equations.

5. One petal of $r = 5\cos 3\theta$, $z = 0$, $z = 4$

6. $x^2 + y^2 = 4$, $z = \sqrt{9 - x^2 - y^2}$, $z = 0$

7. Between $x^2 + y^2 = 1$ and $x^2 + y^2 = 9$, $z = \sqrt{16 - x^2 - y^2}$, $z = 0$

8. $z = \sqrt{x^2 + y^2}$, $x^2 + y^2 = 25$, $z = 0$

9. $r = 1 + \cos\theta$, $z = y$, $z = 0$, first octant

10. $r = \cos\theta$, $z = 2 + x^2 + y^2$, $z = 0$

In Problems 11–16, find the center of mass of the lamina that has the given shape and density.

11. $r = 1$, $r = 3$, $x = 0$, $y = 0$, first quadrant; $\rho(r, \theta) = k$ (constant)

12. $r = \cos\theta$; density ρ at a point P directly proportional to the distance from the pole

13. $y = \sqrt{3}x$, $y = 0$, $x = 3$; $\rho(r, \theta) = r^2$

14. $r = 4\cos 2\theta$, petal on the polar axis; $\rho(r, \theta) = k$ (constant)

15. Outside $r = 2$ and inside $r = 2 + 2\cos\theta$, $y = 0$, first quadrant; density ρ at a point P inversely proportional to the distance from the pole

16. $r = 2 + 2\cos\theta$, $y = 0$, first and second quadrants; $\rho(r, \theta) = k$ (constant)

In Problems 17–20, find the indicated moment of inertia of the lamina that has the given shape and density.

17. $r = a$; $\rho(r, \theta) = k$ (constant); I_x

18. $r = a$; $\rho(r, \theta) = \dfrac{1}{1 + r^4}$; I_x

19. Outside $r = a$ and inside $r = 2a\cos\theta$; density ρ at a point P inversely proportional to the cube of the distance from the pole; I_y

20. Outside $r = 1$ and inside $r = 2\sin 2\theta$, first quadrant; $\rho(r, \theta) = \sec^2\theta$; I_y

In Problems 21–24, find the **polar moment of inertia** $I_0 = \iint_R r^2\rho(r, \theta)\,dA = I_x + I_y$ of the lamina that has the given shape and density.

21. $r = a$; $\rho(r, \theta) = k$ (constant) [*Hint*: Use Problem 17 and the fact that $I_x = I_y$.]

22. $r = \theta, 0 \le \theta \le \pi, y = 0$; density ρ at a point P proportional to the distance from the pole

23. $r\theta = 1, \frac{1}{3} \le \theta \le 1, r = 1, r = 3, y = 0$; density ρ at a point P inversely proportional to the distance from the pole [*Hint*: Integrate first with respect to θ.]

24. $r = 2a\cos\theta$; $\rho(r, \theta) = k$ (constant)

In Problems 25–32, evaluate the given iterated integral by changing to polar coordinates.

25. $\displaystyle\int_{-3}^{3}\int_{0}^{\sqrt{9-x^2}} \sqrt{x^2 + y^2}\,dy\,dx$

26. $\displaystyle\int_{0}^{\sqrt{2}/2}\int_{y}^{\sqrt{1-y^2}} \frac{y^2}{\sqrt{x^2 + y^2}}\,dx\,dy$

27. $\displaystyle\int_{0}^{1}\int_{0}^{\sqrt{1-y^2}} e^{x^2 + y^2}\,dx\,dy$

28. $\displaystyle\int_{-\sqrt{\pi}}^{\sqrt{\pi}}\int_{0}^{\sqrt{\pi - x^2}} \sin(x^2 + y^2)\,dy\,dx$

29. $\displaystyle\int_{0}^{1}\int_{\sqrt{1-x^2}}^{\sqrt{4-x^2}} \frac{x^2}{x^2 + y^2}\,dy\,dx + \int_{1}^{2}\int_{0}^{\sqrt{4-x^2}} \frac{x^2}{x^2 + y^2}\,dy\,dx$

30. $\displaystyle\int_{0}^{1}\int_{0}^{\sqrt{2y - y^2}} (1 - x^2 - y^2)\,dx\,dy$

31. $\displaystyle\int_{-5}^{5}\int_{0}^{\sqrt{25 - x^2}} (4x + 3y)\,dy\,dx$

32. $\displaystyle\int_{0}^{1}\int_{0}^{\sqrt{1-y^2}} \frac{1}{1 + \sqrt{x^2 + y^2}}\,dx\,dy$

33. The improper integral $\int_{0}^{\infty} e^{-x^2}\,dx$ is important in the theory of probability, statistics, and other areas of applied mathematics. If I denotes the integral, then because the variable of integration is a dummy variable we have

$$I = \int_{0}^{\infty} e^{-x^2}\,dx \quad \text{and} \quad I = \int_{0}^{\infty} e^{-y^2}\,dy.$$

In view of Problem 53 of Exercises 14.2 we have

$$I^2 = \left(\int_{0}^{\infty} e^{-x^2}\,dx\right)\left(\int_{0}^{\infty} e^{-y^2}\,dy\right)$$

$$= \int_{0}^{\infty}\int_{0}^{\infty} e^{-(x^2 + y^2)}\,dx\,dy.$$

Use polar coordinates to evaluate the last integral. Find the value of I.

34. Evaluate $\iint_R (x + y)\,dA$ over the region shown in FIGURE 14.5.6.

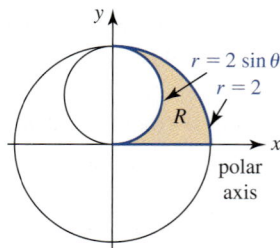

FIGURE 14.5.6 Region R in Problem 34

≡ Applications

35. The liquid hydrogen tank in the space shuttle has the form of a right circular cylinder with a semiellipsoidal cap at each end. The radius of the cylindrical part of the tank is 4.2 m. Find the volume of the tank shown in FIGURE 14.5.7.

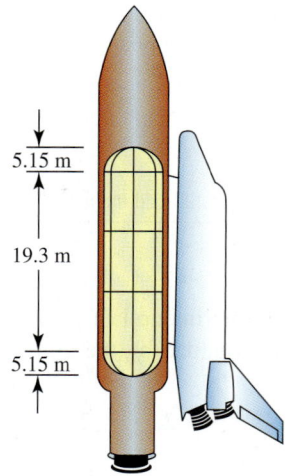

5.15 m

19.3 m

5.15 m

FIGURE 14.5.7 Space shuttle in Problem 35

36. In some studies of the spread of plant disease, the number of infections per unit area as a function of the distance from an infected source plant is described by a formula of the form

$$I(r) = a(r + c)^{-b},$$

where $I(r)$ is the number of infections per unit area at a radial distance r from the infected source plant, and a, b, and c, are (positive) parameters depending on the disease.

(a) Derive a formula for the total number of infections within a circle of radius R centered at the infected source plant; that is, evaluate $\iint_C I(r)\,dA$, where C is a circular region of radius R centered at the origin. Assume that the parameter b is not 1 or 2.

(b) Show that if $b > 2$, then the result in part (a) tends to a finite limit as $R \to \infty$.

(c) For common maize rust, the number of infections per square meter is modeled as

$$I(r) = 68.585(r + 0.248)^{-2.351},$$

where r is measured in meters. Find the total number of infections in the plane.

37. Urban population densities fall off exponentially with distance from the central business district (CBD); that is,

$$D(r) = D_0 e^{-r/d},$$

where $D(r)$ is the population density at a radial distance r from the CBD, D_0 is the density at the center, and d is a parameter.

(a) Using the formula $P = \iint_C D(r)\,dA$, find an expression for the total population living within a circular region C of radius R of the CBD.

(b) Using

$$\frac{\iint_C rD(r)\,dA}{\iint_C D(r)\,dA}$$

find an expression for the average commute (distance traveled) to the CBD for the people living within the region C.

(c) Using the results in part (a) and (b), find the total population and average commute as $R \to \infty$.

38. It is arguable that the cost, in terms of time, money, or effort, of collecting or distributing material to or from a single location is proportional to the integral $\iint_R r\,dA$, where R is the region being covered and r denotes the distance to the collection/distribution site. Suppose, for example, that a snowplow is sent to clear off a circular parking area of diameter D. Show that plowing all the snow to a single point on the perimeter is approximately 70% more costly than plowing everything to the center of the parking lot. [*Hint*: Set up the integral for each case separately, using a polar coordinate equation for the circle with the collection site at the origin.]

14.6 Surface Area

■ **Introduction**   In Section 6.5 we saw that the length of an arc of the graph of $y = f(x)$ from $x = a$ to $x = b$ was given by

$$L = \int_a^b \sqrt{1 + \left(\frac{dy}{dx}\right)^2}\,dx. \tag{1}$$

The problem in three dimensions, which is the counterpart of the arc length problem, is to find the area $A(S)$ of that portion of the surface S given by a function $z = f(x, y)$ having continuous first partial derivatives on a closed region R in the xy-plane. Such a surface S is said to be **smooth**.

■ **Building an Integral**   Suppose, as shown in FIGURE 14.6.1(a), that an inner partition P of R is formed using lines parallel to the x- and y-axes. The partition P then consists of n rectangular elements R_k of area $\Delta A_k = \Delta x_k \Delta y_k$ that lie entirely within R. Let $(x_k, y_k, 0)$ denote any point in an element R_k. As we see in Figure 14.6.1(a), by projecting the sides of R_k upward, we determine two quantities: a portion or patch S_k of the surface and a portion of T_k of a tangent plane at $(x_k, y_k, f(x_k, y_k))$. It seems reasonable to assume that when R_k is small, the area ΔT_k of T_k is approximately the same as the area ΔS_k of the patch S_k.

To find the area of T_k let us choose $(x_k, y_k, 0)$ at a corner of R_k as shown in Figure 14.6.1(b). The indicated vectors $\mathbf{u}$ and $\mathbf{v}$, which form two sides of T_k, are given by

$$\mathbf{u} = \Delta x_k \mathbf{i} + f_x(x_k, y_k)\Delta x_k \mathbf{k} \qquad \text{and} \qquad \mathbf{v} = \Delta y_k \mathbf{j} + f_y(x_k, y_k)\Delta y_k \mathbf{k},$$

where $f_x(x_k, y_k)$ and $f_y(x_k, y_k)$ are the slopes of the lines containing $\mathbf{u}$ and $\mathbf{v}$, respectively. Now from (10) of Section 11.4 we know that $\Delta T_k = |\mathbf{u} \times \mathbf{v}|$, where

$$\mathbf{u} \times \mathbf{v} = \begin{vmatrix} \mathbf{i} & \mathbf{j} & \mathbf{k} \\ \Delta x_k & 0 & f_x(x_k, y_k)\Delta x_k \\ 0 & \Delta y_k & f_y(x_k, y_k)\Delta y_k \end{vmatrix}$$

$$= [-f_x(x_k, y_k)\mathbf{i} - f_y(x_k, y_k)\mathbf{j} + \mathbf{k}]\Delta x_k \Delta y_k.$$

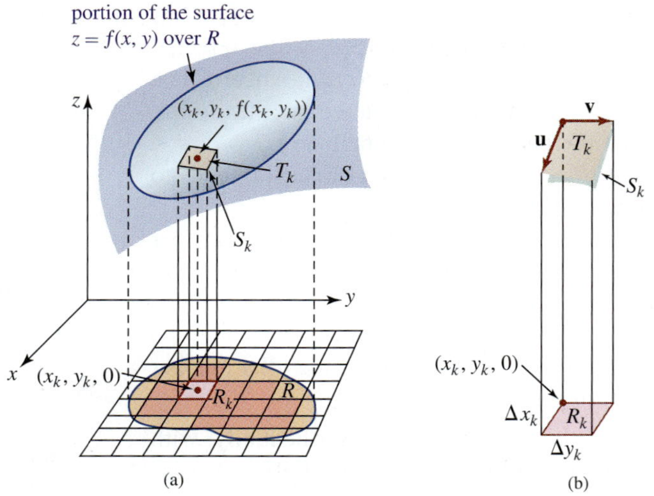

portion of the surface
$z = f(x, y)$ over R

(a)

(b)

FIGURE 14.6.1 Surface in (a); enlargements of R_k, S_k, and T_k in (b)

In other words,

$$\Delta T_k = |\mathbf{u} \times \mathbf{v}| = \sqrt{[f_x(x_k, y_k)]^2 + [f_y(x_k, y_k)]^2 + 1}\, \Delta x_k \Delta y_k.$$

Consequently, the area $A(S)$ is approximately

$$\sum_{k=1}^{n} \sqrt{1 + [f_x(x_k, y_k)]^2 + [f_y(x_k, y_k)]^2}\, \Delta x_k \Delta y_k.$$

Taking the limit of the foregoing sum as $\|P\| \to 0$ leads us to the next definition.

Definition 14.6.1   Surface Area

Let f be a function for which the first partial derivatives f_x and f_y are continuous on a closed region R. Then the **area of the surface** over R is given by

$$A(S) = \iint_R \sqrt{1 + [f_x(x, y)]^2 + [f_y(x, y)]^2}\, dA. \tag{2}$$

Note: One could have almost guessed the form of (2) by naturally extending the one-variable structure of (1) to two variables.

EXAMPLE 1   Using (2)

Find the surface area of that portion of the sphere $x^2 + y^2 + z^2 = a^2$ that is above the xy-plane and within the cylinder $x^2 + y^2 = b^2$, $0 < b < a$.

Solution   If we define $z = f(x, y)$ by $f(x, y) = \sqrt{a^2 - x^2 - y^2}$, then

$$f_x(x, y) = \frac{-x}{\sqrt{a^2 - x^2 - y^2}} \quad \text{and} \quad f_y(x, y) = \frac{-y}{\sqrt{a^2 - x^2 - y^2}}$$

and so

$$1 + [f_x(x, y)]^2 + [f_y(x, y)]^2 = \frac{a^2}{a^2 - x^2 - y^2}.$$

Hence, (2) is

$$A(S) = \iint_R \frac{a}{\sqrt{a^2 - x^2 - y^2}}\, dA,$$

where R is indicated in FIGURE 14.6.2. To evaluate this double integral, we change to polar coordinates. The circle $x^2 + y^2 = b^2$ becomes $r = b, 0 \le \theta \le 2\pi$:

$$A(S) = a \int_0^{2\pi} \int_0^b (a^2 - r^2)^{-1/2} r \, dr d\theta$$

$$= a \int_0^{2\pi} -(a^2 - r^2)^{-1/2} \Big]_0^b \, d\theta$$

$$= a(a - \sqrt{a^2 - b^2}) \int_0^{2\pi} d\theta$$

$$= 2\pi a(a - \sqrt{a^2 - b^2}).$$

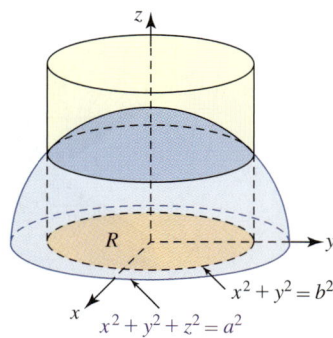

FIGURE 14.6.2 Surface in Example 1

EXAMPLE 2 Using (2)

Find the surface area of the portions of the sphere $x^2 + y^2 + z^2 = 4$ that are within the cylinder $(x - 1)^2 + y^2 = 1$.

Solution The surface area in question consists of the two darker, shaded regions of the surface (above and below the xy-plane) in FIGURE 14.6.3. As in Example 1, (2) simplifies to

$$A(S) = 2 \iint_R \frac{2}{\sqrt{4 - x^2 - y^2}} \, dA,$$

where R is the region bounded by the graph of $(x - 1)^2 + y^2 = 1$. The extra factor of 2 in the integral comes from using symmetry. Now, in polar coordinates the boundary of R is simply $r = 2\cos\theta$. Thus,

$$A(S) = 4 \int_0^\pi \int_0^{2\cos\theta} (4 - r^2)^{-1/2} r \, dr d\theta$$

$$= 8(\pi - 2) \approx 9.13.$$

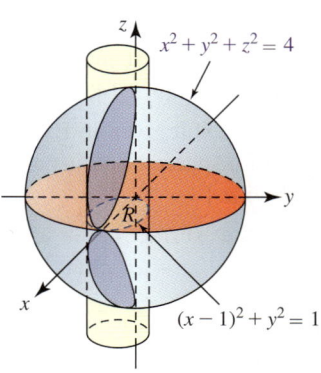

FIGURE 14.6.3 Surface in Example 2

■ **Differential of Surface Area** The function

$$dS = \sqrt{1 + [f_x(x, y)]^2 + [f_y(x, y)]^2} \, dA \qquad (3)$$

is called the **differential of the surface area**. We will use this function in Sections 15.6 and 15.9.

Exercises 14.6 Answers to selected odd-numbered problems begin on page ANS-45.

≡ Fundamentals

1. Find the surface area of that portion of the plane $2x + 3y + 4z = 12$ that is bounded by the coordinate planes in the first octant.

2. Find the surface area of that portion of the plane $2x + 3y + 4z = 12$ that is above the region in the first quadrant bounded by the graph of $r = \sin 2\theta$.

3. Find the surface area of that portion of the cylinder $x^2 + z^2 = 16$ that is above the region in the first quadrant bounded by the graphs of $x = 0, x = 2, y = 0, y = 5$.

4. Find the surface area of that portion of the paraboloid $z = x^2 + y^2$ that is below the plane $z = 2$.

5. Find the surface area of that portion of the paraboloid $z = 4 - x^2 - y^2$ that is above the xy-plane.

6. Find the surface area of the portions of the sphere $x^2 + y^2 + z^2 = 2$ that are within the cone $z^2 = x^2 + y^2$.

7. Find the surface area of that portion of the sphere $x^2 + y^2 + z^2 = 25$ that is above the region in the first quadrant bounded by the graphs of $x = 0, y = 0, 4x^2 + y^2 = 25$. [*Hint*: Integrate first with respect to x.]

8. Find the surface area of that portion of the graph of $z = x^2 - y^2$ that is in the first octant within the cylinder $x^2 + y^2 = 4$.

9. Find the surface area of the portions of the sphere $x^2 + y^2 + z^2 = a^2$ that are within the cylinder $x^2 + y^2 = ay$.

10. Find the surface area of the portions of the cone $z^2 = \frac{1}{4}(x^2 + y^2)$ that are within the cylinder $(x - 1)^2 + y^2 = 1$. See FIGURE 14.6.4.

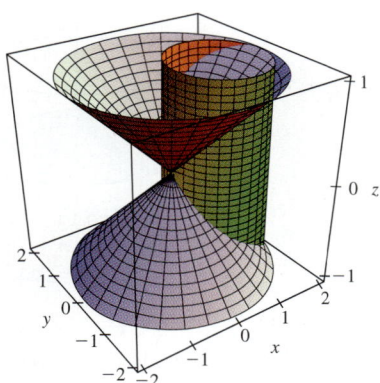

FIGURE 14.6.4 Intersecting cone and cylinder in Problem 10

11. Find the surface area of the portions of the cylinder $y^2 + z^2 = a^2$ that are within the cylinder $x^2 + y^2 = a^2$. [*Hint*: See Figure 14.3.11.]

12. Use the result given in Example 1 to prove that the surface area of a sphere of radius a is $4\pi a^2$. [*Hint*: Consider a limit as $b \to a$.]

13. Find the surface area of that portion of the sphere $x^2 + y^2 + z^2 = a^2$ that is bounded between $y = c_1$ and $y = c_2$, $0 < c_1 < c_2 < a$. [*Hint*: Use polar coordinates in the xz-plane.]

14. Show that the area found in Problem 13 is the same as the surface area of the cylinder $x^2 + z^2 = a^2$ between $y = c_1$ and $y = c_2$.

≡ Think About It

15. As shown in **FIGURE 14.6.5**, a sphere of radius 1 has its center on the surface of a sphere of radius $a > 1$. Find the surface area of that portion of the larger sphere cut out by the smaller sphere.

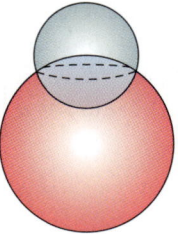

FIGURE 14.6.5 Intersecting spheres in Problem 15

16. On the surface of a globe or, more precisely, on the surface of the Earth, the boundaries of the states of Colorado and Wyoming are both "spherical rectangles." (In this problem we assume that the Earth is a perfect sphere.) Colorado is bounded by the lines of longitude 102° W and 109° W and the lines of latitude 37° N and 41° N. Wyoming is bounded by longitudes 104° W and 111° W and latitudes 41° N and 45° N. See **FIGURE 14.6.6**.

(a) Without explicitly computing their areas, determine which state is larger and explain why.

(b) By what percentage is Wyoming larger (or smaller) than Colorado? [*Hint*: Suppose the radius of the Earth is R. Project a spherical rectangle in the Northern Hemisphere that is determined by latitudes θ_1 and θ_2 and longitudes ϕ_1 and ϕ_2 onto the xy-plane.]

(c) One reference book gives the areas of the two states as $104{,}247$ mi^2 and $97{,}914$ mi^2. How does this answer compare with the answer in part (b)?

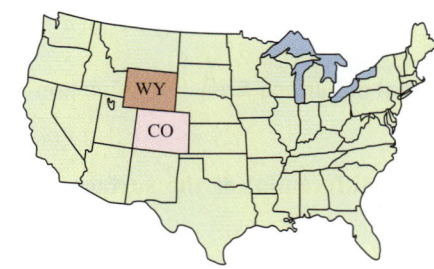

FIGURE 14.6.6 Two spherical rectangles in Problem 16

14.7 The Triple Integral

■ **Introduction** The steps leading to the definition of the *three-dimensional definite integral*, or **triple integral**, $\iiint_D f(x, y, z)\, dV$ are quite similar to those for the double integral.

Let $w = f(x, y, z)$ be defined over a closed and bounded region D of 3-space.

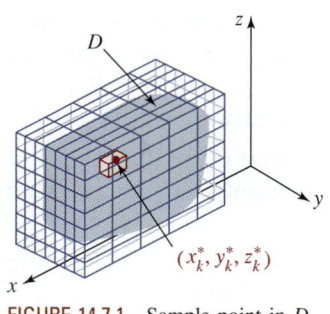

FIGURE 14.7.1 Sample point in D_k

- By means of a three-dimensional grid of vertical and horizontal planes parallel to the coordinate planes, form a partition P of D into n subregions (boxes) D_k of volumes ΔV_k that lie entirely within D. See **FIGURE 14.7.1**.
- Let $\|P\|$ be the norm of the partition or the length of the longest diagonal of the D_k.
- Choose a sample point (x_k^*, y_k^*, z_k^*) in each subregion D_k.
- Form the sum $\displaystyle\sum_{k=1}^{n} f(x_k^*, y_k^*, z_k^*)\Delta V_k$.

A sum of the form $\sum_{k=1}^{n} f(x_k^*, y_k^*, z_k^*)\Delta V_k$, where (x_k^*, y_k^*, z_k^*) is an arbitrary point within each D_k and ΔV_k denotes the volume of each D_k, is called a **Riemann sum**. The type of partition used, where all the D_k lie completely within D, is called an **inner partition** of D.

Definition 14.7.1 The Triple Integral

Let f be a function of three variables defined over a closed region D of 3-space. Then the **triple integral of f over D**, denoted by $\iiint_D f(x, y, z)\, dV$, is defined to be

$$\iiint_D f(x, y, z)\, dV = \lim_{\|P\| \to 0} \sum_{k=1}^{n} f(x_k^*, y_k^*, z_k^*)\, \Delta V_k. \tag{1}$$

As in our previous discussions on the integral, when f is continuous over D, then the limit in (1) exists; that is, f is **integrable** over D. The basic integration properties of a triple integral are the same as those of the double integral given in Theorem 14.1.1.

■ **Evaluation by Iterated Integrals** If the region D is bounded above by the graph of $z = g_2(x, y)$ and bounded below by the graph of $z = g_1(x, y)$, then it can be shown that the triple integral (1) can be expressed as a double integral of the partial integral $\int_{g_1(x, y)}^{g_2(x, y)} f(x, y, z)\, dz$; that is,

$$\iiint_D f(x, y, z)\, dV = \iint_R \left[\int_{g_1(x, y)}^{g_2(x, y)} f(x, y, z)\, dz \right] dA, \tag{2}$$

where R is the orthogonal projection of D onto the xy-plane In particular, if R is a Type I region defined by:

$$R: a \le x \le b,\ h_1(x) \le y \le h_2(x),$$

then, as shown in FIGURE 14.7.2, the triple integral of f over D can be written as an iterated integral:

$$\iiint_D f(x, y, z)\, dV = \int_a^b \int_{h_1(x)}^{h_2(x)} \int_{g_1(x, y)}^{g_2(x, y)} f(x, y, z)\, dz\, dy\, dx. \tag{3}$$

To evaluate the iterated integral in (3) we begin by evaluating the partial definite integral

$$\int_{g_1(x, y)}^{g_2(x, y)} f(x, y, z)\, dz$$

in which *both* x and y are held fixed.

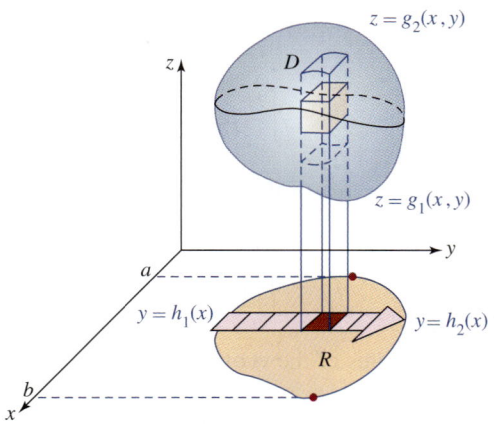

FIGURE 14.7.2 Type I region in the xy-plane

On the other hand if R is a Type II region:

$$R: c \le y \le d,\ h_1(y) \le x \le h_2(y),$$

then (2) becomes

$$\iiint_D f(x, y, z)\, dV = \int_c^d \int_{h_1(y)}^{h_2(y)} \int_{g_1(x, y)}^{g_2(x, y)} f(x, y, z)\, dz\, dx\, dy. \tag{4}$$

In a double integral there are only two possible orders of integration, $dy\,dx$ and $dx\,dy$. The triple integrals in (3) and (4) illustrate two of *six* possible orders of integration:

$$dz\,dy\,dx \qquad dz\,dx\,dy \qquad dy\,dx\,dz$$
$$dx\,dy\,dz \qquad dx\,dz\,dy \qquad dy\,dz\,dx.$$

The last two differentials tell us the coordinate plane in which the region R is located. For example, the iterated integral corresponding to the order of integration $dx\,dz\,dy$ would have the form

$$\int_c^d \int_{h_1(y)}^{h_2(y)} \int_{g_1(y,z)}^{g_2(y,z)} f(x, y, z)\, dx\, dz\, dy.$$

The geometric interpretation of this integral and the region R of integration in the yz-plane are shown in **FIGURE 14.7.3**.

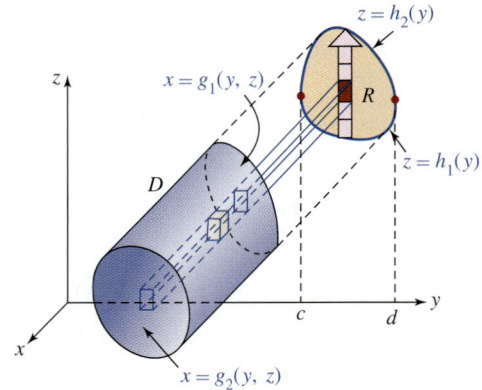

FIGURE 14.7.3 Type I region in the yz-plane

■ **Applications** A list of some of the standard applications of the triple integral follows.

Volume: If $f(x, y, z) = 1$, then the **volume** of the solid D is

$$V = \iiint_D dV.$$

Mass: If $\rho(x, y, z)$ is density (mass per unit volume), then the **mass** of the solid D is given by

$$m = \iiint_D \rho(x, y, z)\, dV.$$

First Moments: The **first moments** of the solid about the coordinate planes indicated by the subscripts are given by

$$M_{xy} = \iiint_D z\rho(x, y, z)\, dV, \quad M_{xz} = \iiint_D y\rho(x, y, z)\, dV, \quad M_{yz} = \iiint_D x\rho(x, y, z)\, dV.$$

Center of Mass: The coordinates of the **center of mass** of D are given by

$$\bar{x} = \frac{M_{yz}}{m}, \quad \bar{y} = \frac{M_{xz}}{m}, \quad \bar{z} = \frac{M_{xy}}{m}.$$

Centroid: If $\rho(x, y, z) = $ constant, the center of mass of D is called the **centroid** of the solid.

Second Moments: The **second moments**, or **moments of inertia**, of D about the coordinate axes indicated by the subscripts, are given by

$$I_x = \iiint_D (y^2 + z^2)\rho(x, y, z)\, dV, \quad I_y = \iiint_D (x^2 + z^2)\rho(x, y, z)\, dV, \quad I_z = \iiint_D (x^2 + y^2)\rho(x, y, z)\, dV.$$

Radius of Gyration: As in Section 14.4, if I is a moment of inertia of the solid about a given axis, then the **radius of gyration** is

$$R_g = \sqrt{\frac{I}{m}}.$$

EXAMPLE 1 Volume

Find the volume of the solid in the first octant bounded by the graphs of $z = 1 - y^2$, $y = 2x$, and $x = 3$.

Solution As indicated in FIGURE 14.7.4(a), the first integration with respect to z will be from 0 to $1 - y^2$. Furthermore, from Figure 14.7.4(b) we see that the projection of the solid D onto the xy-plane is a Type II region. Hence, we next integrate, with respect to x, from $y/2$ to 3. The last integration is with respect to y from 0 to 1. Thus,

$$\begin{aligned}
V = \iiint_D dV &= \int_0^1 \int_{y/2}^3 \int_0^{1-y^2} dz\, dx\, dy \\
&= \int_0^1 \int_{y/2}^3 (1 - y^2)\, dx\, dy \\
&= \int_0^1 (x - xy^2) \Big]_{y/2}^3 dy \\
&= \int_0^1 \left(3 - 3y^2 - \frac{1}{2}y + \frac{1}{2}y^3\right) dy \\
&= \left(3y - y^3 - \frac{1}{4}y^2 + \frac{1}{8}y^4\right)\Big]_0^1 = \frac{15}{8}.
\end{aligned}$$

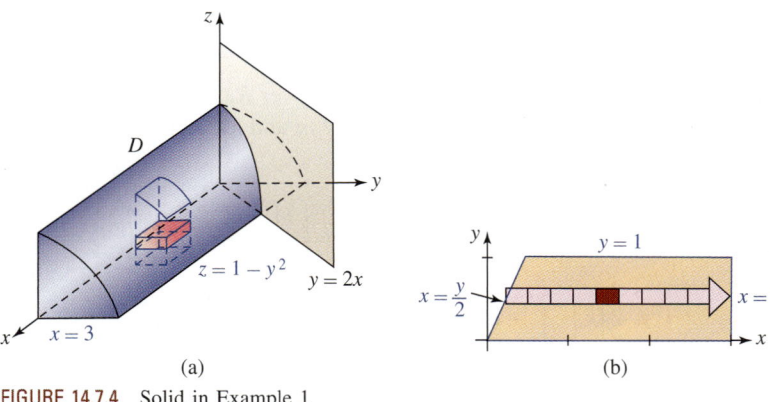

(a)   (b)

FIGURE 14.7.4 Solid in Example 1

You should observe that the volume in Example 1 could have been obtained just as easily by means of a double integral.

EXAMPLE 2 Volume

Find the triple integral that gives the volume of the solid that has the shape determined by the single-napped cone $x = \sqrt{y^2 + z^2}$ and the paraboloid $x = 6 - y^2 - z^2$.

Solution By substituting $y^2 + z^2 = x^2$ in $y^2 + z^2 = 6 - x$, we find that $x^2 = 6 - x$ or $(x + 3)(x - 2) = 0$. Thus, the two surfaces intersect in the plane $x = 2$. The projection onto the yz-plane of the curve of intersection is $y^2 + z^2 = 4$. Using symmetry and referring to FIGURE 14.7.5(a) and (b), we see that

$$V = \iiint_D dV = 4 \int_0^2 \int_0^{\sqrt{4-y^2}} \int_{\sqrt{y^2+z^2}}^{6-y^2-z^2} dx\, dz\, dy.$$

While evaluation of this integral is straightforward, it is admittedly "messy." We shall return to this integral in the next section after we have examined triple integrals in other coordinate systems.

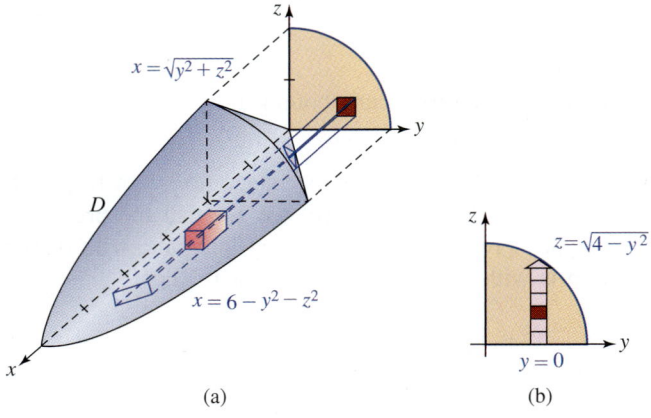

FIGURE 14.7.5 Solid in Example 2

EXAMPLE 3 Center of Mass

A solid has the shape determined by the graphs of the cylinder $|x| + |y| = 1$ and planes $z = 2$ and $z = 4$. Find its center of mass if the density is given by $\rho(x, y, z) = kz$, k a constant.

Solution The solid and its orthogonal projection onto a region R of Type I in the xy-plane are shown in **FIGURE 14.7.6(a)**. The equation $|x| + |y| = 1$ is equivalent to four lines:

$$x + y = 1, x > 0, y > 0; \qquad x - y = 1, x > 0, y < 0;$$
$$-x + y = 1, x < 0, y > 0; \qquad -x - y = 1, x < 0, y < 0.$$

Since the density function $\rho(x, y, z) = kz$ is symmetric over R, we conclude that the center of mass lies on the z-axis; that is, we need compute only m and M_{xy}. From symmetry and Figure 14.7.6(b) it follows that

$$m = 4 \int_0^1 \int_0^{1-x} \int_2^4 kz \, dz \, dy \, dx = 4k \int_0^1 \int_0^{1-x} \frac{1}{2}z^2 \Big]_2^4 dy \, dx$$

$$= 24k \int_0^1 \int_0^{1-x} dy \, dx$$

$$= 24k \int_0^1 (1 - x) \, dx$$

$$= 24k \left(x - \frac{1}{2}x^2 \right) \Big]_0^1 = 12k,$$

$$M_{xy} = 4 \int_0^1 \int_0^{1-x} \int_2^4 kz^2 \, dz \, dy \, dx = 4k \int_0^1 \int_0^{1-x} \frac{1}{3}z^3 \Big]_2^4 dy \, dx$$

$$= \frac{224}{3}k \int_0^1 \int_0^{1-x} dy \, dx = \frac{112}{3}k.$$

Hence,

$$\bar{z} = \frac{M_{xy}}{m} = \frac{\frac{112}{3}k}{12k} = \frac{28}{9}.$$

The coordinates of the center of mass are then $\left(0, 0, \frac{28}{9}\right)$.

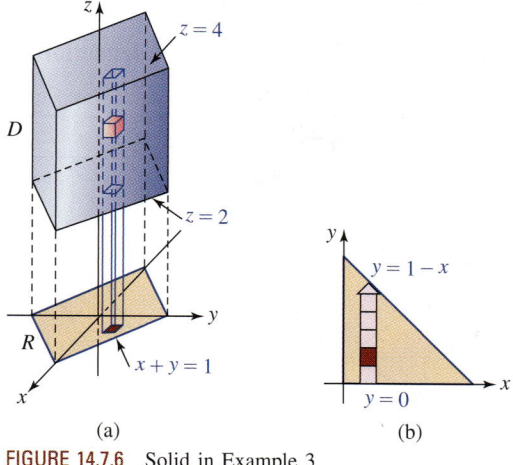

(a)   (b)

FIGURE 14.7.6   Solid in Example 3

EXAMPLE 4   Example 3 Revisited

Find the moment of inertia of the solid in Example 3 about the z-axis. Find the radius of gyration.

Solution   We know that $I_z = \iiint_D (x^2 + y^2)kz\,dV$. Using symmetry again, we can write this triple integral as

$$
I_z = 4k \int_0^1 \int_0^{1-x} \int_2^4 (x^2 + y^2)z\,dz\,dy\,dx
$$

$$
= 4k \int_0^1 \int_0^{1-x} (x^2 + y^2)\frac{1}{2}z^2 \Big]_2^4 \,dy\,dx
$$

$$
= 24k \int_0^1 \int_0^{1-x} (x^2 + y^2)\,dy\,dx
$$

$$
= 24k \int_0^1 \left(x^2 y + \frac{1}{3}y^3 \right) \Big]_0^{1-x} dx
$$

$$
= 24k \int_0^1 \left[x^2 - x^3 + \frac{1}{3}(1 - x)^3 \right] dx
$$

$$
= 24k \left[\frac{1}{3}x^3 - \frac{1}{4}x^4 - \frac{1}{12}(1 - x)^4 \right]_0^1 = 4k.
$$

From Example 3 we know that $m = 12k$ and so it follows that the radius of gyration is

$$
R_g = \sqrt{\frac{I_z}{m}} = \sqrt{\frac{4k}{12k}} = \frac{1}{3}\sqrt{3}.
$$

The last example illustrates how to change the order of integration in a triple integral.

EXAMPLE 5   Changing the Order of Integration

Change the order of integration in

$$
\int_0^6 \int_0^{4-2x/3} \int_0^{3-x/2-3y/4} f(x, y, z)\,dz\,dy\,dx
$$

to $dy\,dx\,dz$.

Solution   As seen in **FIGURE 14.7.7(a)**, the region D is the solid in the first octant bounded by the three coordinate planes and the plane $2x + 3y + 4z = 12$. Referring to Figure 14.7.7(b) and the included table, we conclude that

$$
\int_0^6 \int_0^{4-2x/3} \int_0^{3-x/2-3y/4} f(x, y, z)\,dz\,dy\,dx = \int_0^3 \int_0^{6-2z} \int_0^{4-2x/3-4z/3} f(x, y, z)\,dy\,dx\,dz.
$$

Order of Integration	First Integration	Second Integration	Third Integration
$dz\,dy\,dx$	0 to $3 - x/2 - 3y/4$	0 to $4 - 2x/3$	0 to 6
$dy\,dx\,dz$	0 to $4 - 2x/3 - 4z/3$	0 to $6 - 2z$	0 to 3

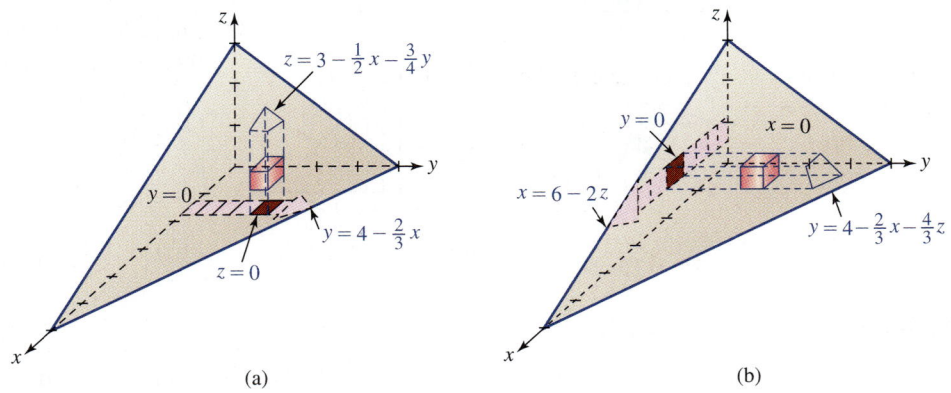

(a)   (b)

FIGURE 14.7.7 Changing integration from $dz\,dy\,dx$ to $dy\,dx\,dz$ in Example 5 ■

Exercises 14.7   Answers to selected odd-numbered problems begin on page ANS-45.

☰ Fundamentals

In Problems 1–8, evaluate the given iterated integral.

1. $\displaystyle\int_2^4 \int_{-2}^2 \int_{-1}^1 (x + y + z)\,dx\,dy\,dz$ **2.** $\displaystyle\int_1^3 \int_1^x \int_2^{xy} 24xy\,dz\,dy\,dx$

3. $\displaystyle\int_0^6 \int_0^{6-x} \int_0^{6-x-z} dy\,dz\,dx$ **4.** $\displaystyle\int_0^1 \int_0^{1-x} \int_0^{\sqrt{y}} 4x^2z^3\,dz\,dy\,dx$

5. $\displaystyle\int_0^{\pi/2} \int_0^{y^2} \int_0^y \cos\!\left(\frac{x}{y}\right) dz\,dx\,dy$ **6.** $\displaystyle\int_0^{\sqrt{2}} \int_{\sqrt{y}}^2 \int_0^{e^{x^2}} x\,dz\,dx\,dy$

7. $\displaystyle\int_0^1 \int_0^1 \int_0^{2-x^2-y^2} xye^z\,dz\,dx\,dy$

8. $\displaystyle\int_0^4 \int_0^{1/2} \int_0^{x^2} \frac{1}{\sqrt{x^2 - y^2}}\,dy\,dx\,dz$

9. Evaluate $\iiint_D z\,dV$, where D is the region in the first octant bounded by the graphs of $y = x$, $y = x - 2$, $y = 1$, $y = 3$, $z = 0$, and $z = 5$.

10. Evaluate $\iiint_D (x^2 + y^2)\,dV$, where D is the region bounded by the graphs of $y = x^2$, $z = 4 - y$, and $z = 0$.

In Problems 11 and 12, change the indicated order of integration to each of the other five orders.

11. $\displaystyle\int_0^2 \int_0^{4-2y} \int_{x+2y}^4 f(x, y, z)\,dz\,dx\,dy$

12. $\displaystyle\int_0^2 \int_0^{\sqrt{36-9x^2}/2} \int_1^3 f(x, y, z)\,dz\,dy\,dx$

In Problems 13 and 14, consider the solid given in the figure. Set up, but do not evaluate, the integrals giving the volume V of the solid using the indicated orders of integration.

13.

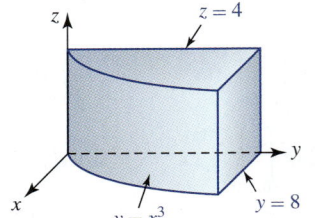

FIGURE 14.7.8 Solid in Problem 13

   (a) $dz\,dy\,dx$
   (b) $dx\,dz\,dy$
   (c) $dy\,dx\,dz$

14.

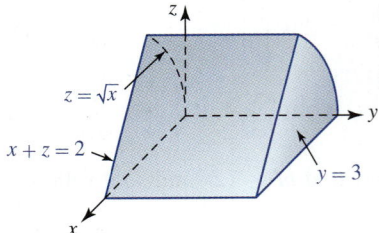

FIGURE 14.7.9 Solid in Problem 14

   (a) $dx\,dz\,dy$
   (b) $dy\,dx\,dz$
   (c) $dz\,dx\,dy$ [*Hint*: This will require two integrals.]

In Problems 15–20, sketch the region D whose volume V is given by the iterated integral.

15. $\displaystyle\int_0^4 \int_0^3 \int_0^{2-2z/3} dx\,dz\,dy$ **16.** $4\displaystyle\int_0^3 \int_0^{\sqrt{9-y^2}} \int_4^{\sqrt{25-x^2-y^2}} dz\,dx\,dy$

17. $\displaystyle\int_{-1}^1 \int_{-\sqrt{1-x^2}}^{\sqrt{1-x^2}} \int_0^5 dz\,dy\,dx$ **18.** $\displaystyle\int_0^2 \int_0^{\sqrt{4-x^2}} \int_{x^2+y^2}^4 dz\,dy\,dx$

19. $\displaystyle\int_0^2 \int_0^{2-y} \int_{-\sqrt{y}}^{\sqrt{y}} dx\,dz\,dy$ **20.** $\displaystyle\int_1^3 \int_0^{1/x} \int_0^3 dy\,dz\,dx$

In Problems 21–24, find the volume V of the solid bounded by the graphs of the given equations.

21. $x = y^2$, $4 - x = y^2$, $z = 0$, $z = 3$

22. $x^2 + y^2 = 4$, $z = x + y$, the coordinate planes, first octant

23. $y = x^2 + z^2, y = 8 - x^2 - z^2$

24. $x = 2, y = x, y = 0, z = x^2 + y^2, z = 0$

25. Find the center of mass of the solid given in Figure 14.7.8 if the density ρ at a point P is directly proportional to the distance from the xy-plane.

26. Find the centroid of the solid in Figure 14.7.9 if the density ρ is constant.

27. Find the center of mass of the solid bounded by the graphs of $x^2 + z^2 = 4, y = 0$, and $y = 3$ if the density ρ at a point P is directly proportional to the distance from the xz-plane.

28. Find the center of mass of the solid bounded by the graphs of $y = x^2, y = x, z = y + 2$, and $z = 0$ if the density ρ at a point P is directly proportional to the distance from the xy-plane.

In Problems 29 and 30, set up, but do not evaluate, the iterated integrals giving the mass m of the solid that has the given shape and density.

29. $x^2 + y^2 = 1, z + y = 8, z - 2y = 2$;
$\rho(x, y, z) = x + y + 4$

30. $x^2 + y^2 - z^2 = 1, z = -1, z = 2; \rho(x, y, z) = z^2$ [*Hint*: Do not use $dz\,dy\,dx$.]

31. Find the moment of inertia of the solid in Figure 14.7.8 about the y-axis if the density ρ is as given in Problem 25. Find the radius of gyration.

32. Find the moment of inertia of the solid in Figure 14.7.9 about the x-axis if the density ρ is constant. Find the radius of gyration.

33. Find the moment of inertia about the z-axis of the solid in the first octant that is bounded by the coordinate planes and the graph of $x + y + z = 1$ if the density ρ is constant.

34. Find the moment of inertia about the y-axis of the solid bounded by the graphs of $z = y, z = 4 - y, z = 1$, $z = 0, x = 2$, and $x = 0$ if the density ρ at a point P is directly proportional to the distance from the yz-plane.

In Problems 35 and 36, set up, but do not evaluate, the iterated integral giving the indicated moment of inertia of the solid having the given shape and density.

35. $z = \sqrt{x^2 + y^2}, z = 5$; density ρ at a point P directly proportional to the distance from the origin; I_z

36. $x^2 + z^2 = 1, y^2 + z^2 = 1$; density ρ at a point P directly proportional to the distance from the yz-plane; I_y

14.8 Triple Integrals in Other Coordinate Systems

▌ Introduction   Depending on the geometry of a region in 3-space, the evaluation of a triple integral over the region can often be made easier by utilizing a new coordinate system.

▌ Cylindrical Coordinates   The cylindrical coordinate system combines the polar description of a point in the plane with the rectangular description of the z-component of a point in space. As seen in FIGURE 14.8.1(a) the cylindrical coordinates of a point P are denoted by the ordered triple (r, θ, z). The word *cylindrical* arises from the fact that a point P in space is determined by the intersection of the planes $z = $ constant, $\theta = $ constant, with a cylinder $r = $ constant. See Figure 14.8.1(b).

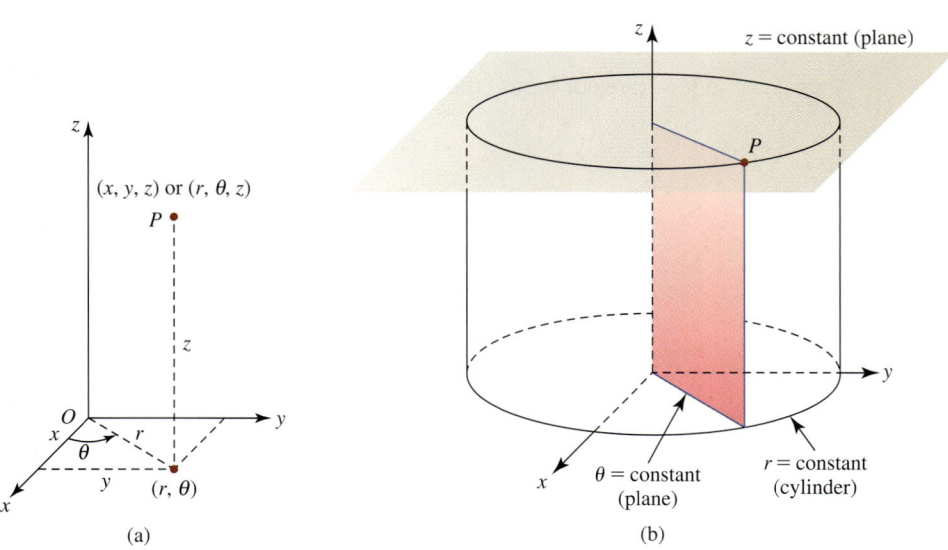

FIGURE 14.8.1 Cylindrical coordinates of a point in 3-space

■ **Cylindrical Coordinates to Rectangular Coordinates** From Figure 14.8.1(a) we also see that the rectangular coordinates (x, y, z) of a point can be obtained from the cylindrical coordinates (r, θ, z) by means of the equations

$$x = r\cos\theta, \qquad y = r\sin\theta, \qquad z = z. \tag{1}$$

EXAMPLE 1 Center of Mass

Convert $(8, \pi/3, 7)$ in cylindrical coordinates to rectangular coordinates.

Solution From (1),

$$x = 8\cos\frac{\pi}{3} = 8\left(\frac{1}{2}\right) = 4$$

$$y = 8\sin\frac{\pi}{3} = 8\left(\frac{1}{2}\sqrt{3}\right) = 4\sqrt{3}$$

$$z = 7.$$

Thus, $(8, \pi/3, 7)$ is equivalent to $(4, 4\sqrt{3}, 7)$ in rectangular coordinates. ■

■ **Rectangular Coordinates to Cylindrical Coordinates** To convert the rectangular coordinates (x, y, z) of a point to cylindrical coordinates (r, θ, z), we use

$$r^2 = x^2 + y^2, \qquad \tan\theta = \frac{y}{x}, \qquad z = z. \tag{2}$$

EXAMPLE 2 Center of Mass

Convert $\left(-\sqrt{2}, \sqrt{2}, 1\right)$ in rectangular coordinates to cylindrical coordinates.

Solution From (2) we see that

$$r^2 = \left(-\sqrt{2}\right)^2 + \left(\sqrt{2}\right)^2 = 4$$

$$\tan\theta = \frac{\sqrt{2}}{-\sqrt{2}} = -1$$

$$z = 1.$$

If we take $r = 2$, then, consistent with the fact that $x < 0$ and $y > 0$, we take $\theta = 3\pi/4$. If we use $\theta = \tan^{-1}(-1) = -\pi/4$, then we can use $r = -2$. Notice that the combinations $r = 2$, $\theta = -\pi/4$ and $r = -2$, $\theta = 3\pi/4$ are inconsistent. Consequently, $\left(-\sqrt{2}, \sqrt{2}, 1\right)$ is equivalent to $(2, 3\pi/4, 1)$ in cylindrical coordinates. See FIGURE 14.8.2. ■

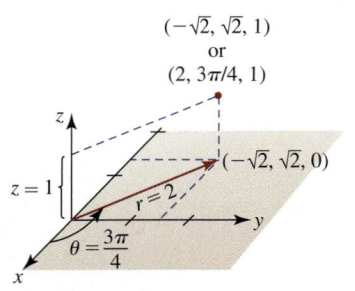

FIGURE 14.8.2. Conversion of rectangular coordinates to cylindrical coordinates in Example 2

■ **Triple Integrals in Cylindrical Coordinates** Recall from Section 14.5 that the area of a "polar rectangle" is $\Delta A = r^*\Delta r\Delta\theta$, where r^* is the average radius. From FIGURE 14.8.3(a) we see that the volume of a "cylindrical wedge" is simply

$$\Delta V = (\text{area of base}) \cdot (\text{height}) = r^*\Delta r\Delta\theta\Delta z.$$

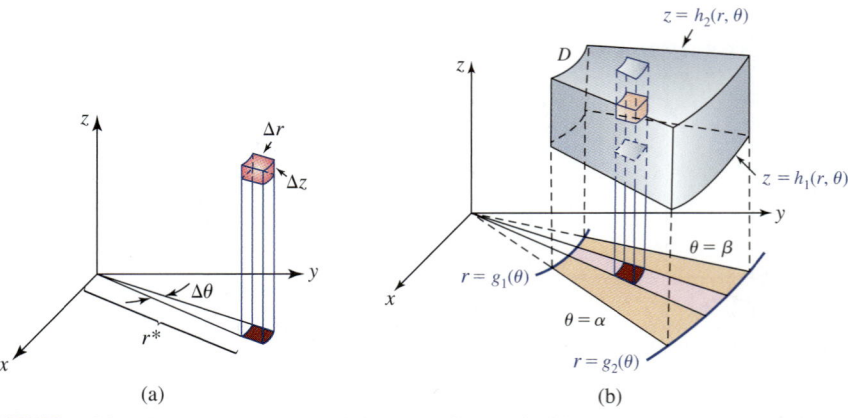

(a)   (b)

FIGURE 14.8.3 Cylindrical wedge in (a); region in 3-space in (b)

Thus, if $f(r, \theta, z)$ is a continuous function over the region D, as shown in Figure 14.8.3(b), then the triple integral of F over D is given by

$$\iiint_D f(r, \theta, z)\, dV = \iint_R \left[\int_{h_1(r, \theta)}^{h_2(r, \theta)} f(r, \theta, z)\, dz \right] dA = \int_\alpha^\beta \int_{g_1(\theta)}^{g_2(\theta)} \int_{h_1(r, \theta)}^{h_2(r, \theta)} f(r, \theta, z)\, r\, dz\, dr\, d\theta.$$

EXAMPLE 3   Center of Mass

A solid in the first octant has the shape determined by the graph of the single-napped cone $z = \sqrt{x^2 + y^2}$ and the planes $z = 1, x = 0$, and $y = 0$. Find the center of mass if the density is given by $\rho(r, \theta, z) = r$.

Solution   In view of (2), the equation of the cone is $z = r$. Hence, we see from FIGURE 14.8.4 that

$$m = \iiint_D r\, dV = \int_0^{\pi/2} \int_0^1 \int_r^1 r(r\, dz\, dr\, d\theta)$$

$$= \int_0^{\pi/2} \int_0^1 \left[r^2 z \right]_r^1 dr\, d\theta$$

$$= \int_0^{\pi/2} \int_0^1 (r^2 - r^3)\, dr\, d\theta = \frac{1}{24}\pi,$$

$$M_{xy} = \iiint_D zr\, dV = \int_0^{\pi/2} \int_0^1 \int_r^1 zr^2\, dz\, dr\, d\theta$$

$$= \int_0^{\pi/2} \int_0^1 \left[\frac{1}{2} z^2 r^2 \right]_r^1 dr\, d\theta$$

$$= \frac{1}{2} \int_0^{\pi/2} \int_0^1 (r^2 - r^4)\, dr\, d\theta = \frac{1}{30}\pi.$$

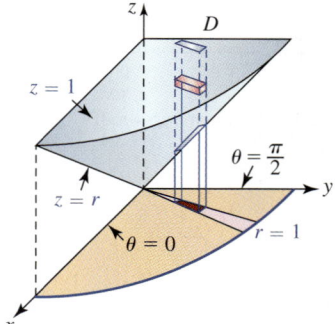

FIGURE 14.8.4   Solid in Example 3

Using $y = r\sin\theta$ and $x = r\cos\theta$, we also have

$$M_{xz} = \iiint_D r^2 \sin\theta\, dV = \int_0^{\pi/2} \int_0^1 \int_r^1 r^3 \sin\theta\, dz\, dr\, d\theta$$

$$= \int_0^{\pi/2} \int_0^1 \left[r^3 z \sin\theta \right]_r^1 dr\, d\theta$$

$$= \int_0^{\pi/2} \int_0^1 (r^3 - r^4)\sin\theta\, dr\, d\theta = \frac{1}{20},$$

$$M_{yz} = \iiint_D r^2 \cos\theta\, dV = \int_0^{\pi/2} \int_0^1 \int_r^1 r^3 \cos\theta\, dz\, dr\, d\theta = \frac{1}{20}.$$

Hence,

$$\bar{x} = \frac{M_{yz}}{m} = \frac{\dfrac{1}{20}}{\dfrac{1}{24}\pi} = \frac{6}{5\pi} \approx 0.38,$$

$$\bar{y} = \frac{M_{xz}}{m} = \frac{\dfrac{1}{20}}{\dfrac{1}{24}\pi} = \frac{6}{5\pi} \approx 0.38,$$

$$\bar{z} = \frac{M_{xy}}{m} = \frac{\dfrac{1}{30}\pi}{\dfrac{1}{24}\pi} = \frac{4}{5} = 0.8.$$

The center of mass has the approximate coordinates $(0.38, 0.38, 0.8)$.   ■

EXAMPLE 4   Center of Mass

Evaluate the volume integral

$$V = 4 \int_0^2 \int_0^{\sqrt{4-y^2}} \int_{\sqrt{y^2+z^2}}^{6-y^2-z^2} dx\,dz\,dy$$

of Example 2 in Section 14.7.

Solution   If we introduce polar coordinates in the yz-plane by $y = r\cos\theta$, $z = r\sin\theta$, then the cylindrical coordinates of a point in 3-space are (r, θ, x). The polar description of Figure 14.7.5(b) is given in FIGURE 14.8.5. Now, since $y^2 + z^2 = r^2$, we have

$$x = \sqrt{y^2 + z^2} = r \qquad \text{and} \qquad x = 6 - y^2 - z^2 = 6 - r^2.$$

Hence, the integral becomes

$$V = 4 \int_0^{\pi/2} \int_0^2 \int_r^{6-r^2} r\,dx\,dr\,d\theta = 4 \int_0^{\pi/2} \int_0^2 rx \Big]_r^{6-r^2} dr\,d\theta$$

$$= 4 \int_0^{\pi/2} \int_0^2 (6r - r^3 - r^2)\,dr\,d\theta$$

$$= 4 \int_0^{\pi/2} \left(3r^2 - \frac{1}{4}r^4 - \frac{1}{3}r^3 \right)\Big]_0^2 d\theta$$

$$= \frac{64}{3} \int_0^{\pi/2} d\theta = \frac{32}{3}\pi. \qquad \blacksquare$$

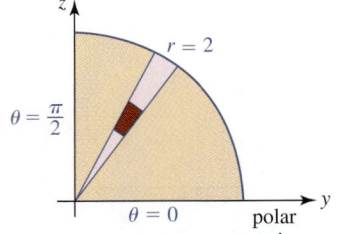

FIGURE 14.8.5   Polar version of Figure 14.7.5(b)

▌ **Spherical Coordinates**   As seen in FIGURE 14.8.6(a), the **spherical coordinates** of a point P are given by the ordered triple (ρ, ϕ, θ), where ρ is the distance from the origin to P, ϕ is the angle between the positive z-axis and the vector $\overrightarrow{OP}$, and θ is the angle measured from the positive x-axis to the vector projection $\overrightarrow{OQ}$ of $\overrightarrow{OP}$. The angle θ is the same angle as in polar and cylindrical coordinates. Figure 14.8.6(b) shows that a point P in space is determined by the intersection of a cone $\phi = $ constant, a plane $\theta = $ constant, and a sphere $\rho = $ constant; whence arises the name "spherical" coordinates.

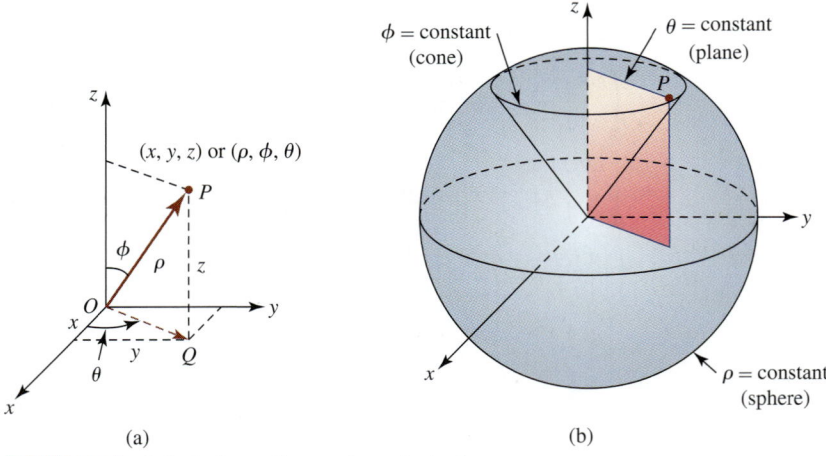

FIGURE 14.8.6   Spherical coordinates of a point in 3-space

▌ **Spherical Coordinates to Rectangular and Cylindrical Coordinates**   In order to transform from spherical coordinates (ρ, ϕ, θ) to rectangular coordinates (x, y, z), we observe from Figure 14.8.6(a) that

$$x = |\overrightarrow{OQ}|\cos\theta, \quad y = |\overrightarrow{OQ}|\sin\theta, \quad z = |\overrightarrow{OP}|\cos\phi.$$

Since $|\overrightarrow{OQ}| = \rho\sin\phi$ and $|\overrightarrow{OP}| = \rho$, the foregoing equations become

$$x = \rho\sin\phi\cos\theta, \quad y = \rho\sin\phi\sin\theta, \quad z = \rho\cos\phi. \qquad (3)$$

It is customary to take $\rho \geq 0$ and $0 \leq \phi \leq \pi$. Also, since $|\overrightarrow{OQ}| = \rho \sin\phi = r$, the formulas

$$r = \rho \sin\phi, \quad \theta = \theta, \quad z = \rho \cos\phi \qquad (4)$$

enable us to transform spherical coordinates (ρ, ϕ, θ) into cylindrical coordinates (r, θ, z). ∎

EXAMPLE 5   Center of Mass

Convert $(6, \pi/4, \pi/3)$ in spherical coordinates to
   (a) rectangular coordinates   and   **(b)** cylindrical coordinates.

Solution
   (a) Identifying $\rho = 6$, $\phi = \pi/4$, and $\theta = \pi/3$, we find from (3) that

$$x = 6\sin\frac{\pi}{4}\cos\frac{\pi}{3} = 6\left(\frac{1}{2}\sqrt{2}\right)\left(\frac{1}{2}\right) = \frac{3}{2}\sqrt{2}$$

$$y = 6\sin\frac{\pi}{4}\sin\frac{\pi}{3} = 6\left(\frac{1}{2}\sqrt{2}\right)\left(\frac{1}{2}\sqrt{3}\right) = \frac{3}{2}\sqrt{6}$$

$$z = 6\cos\frac{\pi}{4} = 6\left(\frac{1}{2}\sqrt{2}\right) = 3\sqrt{2}.$$

The rectangular coordinates of the point are $\left(\frac{3}{2}\sqrt{2}, \frac{3}{2}\sqrt{6}, 3\sqrt{2}\right)$.

   (b) From (4) we obtain

$$r = 6\sin\frac{\pi}{4} = 6\left(\frac{1}{2}\sqrt{2}\right) = 3\sqrt{2}$$

$$\theta = \frac{\pi}{3}$$

$$z = 6\cos\frac{\pi}{4} = 6\left(\frac{1}{2}\sqrt{2}\right) = 3\sqrt{2}.$$

Thus, the cylindrical coordinates of the point are $\left(3\sqrt{2}, \pi/3, 3\sqrt{2}\right)$.   ∎

■ **Rectangular Coordinates to Spherical Coordinates**   To convert rectangular coordinates (x, y, z) into spherical coordinates (ρ, ϕ, θ), we use

$$\rho^2 = x^2 + y^2 + z^2, \quad \tan\theta = \frac{y}{x}, \quad \cos\phi = \frac{z}{\sqrt{x^2 + y^2 + z^2}}. \qquad (5)$$

■ **Triple Integrals in Spherical Coordinates**   As seen in FIGURE 14.8.7, the volume of a "spherical wedge" is given by the approximation

$$\Delta V \approx \rho^2 \sin\phi \,\Delta\rho \,\Delta\phi \,\Delta\theta.$$

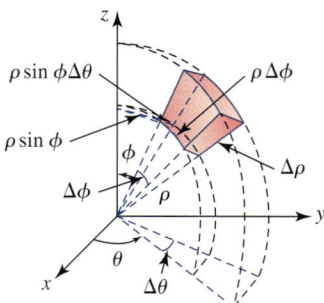

FIGURE 14.8.7   Spherical wedge

Thus, in a triple integral of a continuous spherical–coordinate function $f(\rho, \phi, \theta)$, the differential of volume dV is

$$dV = \rho^2 \sin\phi \,d\rho \,d\phi \,d\theta.$$

Hence, a typical triple integral in spherical coordinates has the form

$$\iiint\limits_{D} f(\rho, \phi, \theta)\, dV = \int_{\alpha}^{\beta} \int_{g_1(\theta)}^{g_2(\theta)} \int_{h_1(\phi,\,\theta)}^{h_2(\phi,\,\theta)} f(\rho, \phi, \theta)\, \rho^2 \sin\phi \,d\rho \,d\phi \,d\theta.$$

EXAMPLE 6   Center of Mass

Use spherical coordinates to find the volume of the solid in Example 3.

Solution   Using (3),

$$z = 1 \text{ becomes } \rho\cos\phi = 1 \text{ or } \rho = \sec\phi,$$
$$z = \sqrt{x^2 + y^2} \text{ becomes } \phi = \pi/4.$$

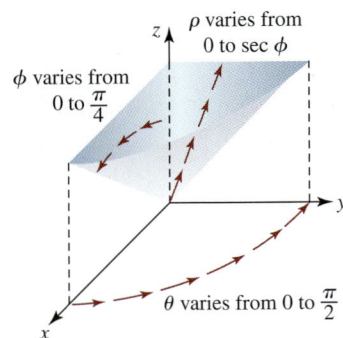

ρ varies from 0 to sec ϕ

ϕ varies from 0 to $\dfrac{\pi}{4}$

θ varies from 0 to $\dfrac{\pi}{2}$

FIGURE 14.8.8 Solid in Example 3

As indicated in FIGURE 14.8.8, $V = \displaystyle\iiint_D dV$ written as an iterated integral is

$$V = \int_0^{\pi/2}\int_0^{\pi/4}\int_0^{\sec\phi} \rho^2\sin\phi\,d\rho\,d\phi\,d\theta = \int_0^{\pi/2}\int_0^{\pi/4}\frac{1}{3}\rho^3\Big]_0^{\sec\phi}\sin\phi\,d\phi\,d\theta$$

$$= \frac{1}{3}\int_0^{\pi/2}\int_0^{\pi/4}\sec^3\phi\,\sin\phi\,d\phi\,d\theta$$

$$= \frac{1}{3}\int_0^{\pi/2}\int_0^{\pi/4}\tan\phi\,\sec^2\phi\,d\phi\,d\theta$$

$$= \frac{1}{3}\int_0^{\pi/2}\frac{1}{2}\tan^2\phi\Big]_0^{\pi/4}\,d\theta$$

$$= \frac{1}{6}\int_0^{\pi/2}d\theta = \frac{1}{12}\pi. \qquad\blacksquare$$

EXAMPLE 7   Center of Mass

Find the moment of inertia about the z-axis of the homogeneous solid bounded between the spheres $x^2 + y^2 + z^2 = a^2$ and $x^2 + y^2 + z^2 = b^2, a < b$.

> We use a different symbol for density to avoid confusion with the symbol ρ of spherical coordinates.

▶ **Solution** If $\delta(\rho, \phi, \theta) = k$ is the density, then

$$I_z = \iiint_D (x^2 + y^2)k\,dV.$$

From (3) we find $x^2 + y^2 = \rho^2\sin^2\phi$, and from the first equation in (5) we see that the equations of the spheres are simply $\rho = a$ and $\rho = b$. See FIGURE 14.8.9. Consequently, in spherical coordinates the foregoing integral becomes

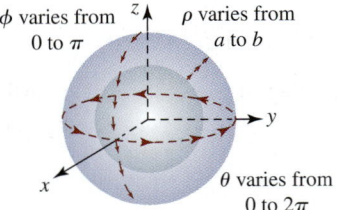

ϕ varies from 0 to π

ρ varies from a to b

θ varies from 0 to 2π

FIGURE 14.8.9 Limits of integration in Example 7

$$I_z = k\int_0^{2\pi}\int_0^{\pi}\int_a^b \rho^2\sin^2\phi(\rho^2\sin\phi\,d\rho\,d\phi\,d\theta)$$

$$= k\int_0^{2\pi}\int_0^{\pi}\int_a^b \rho^4\sin^3\phi\,d\rho\,d\phi\,d\theta$$

$$= k\int_0^{2\pi}\int_0^{\pi}\frac{1}{5}\rho^5\sin^3\phi\Big]_a^b\,d\phi\,d\theta$$

$$= \frac{1}{5}k(b^5 - a^5)\int_0^{2\pi}\int_0^{\pi}(1 - \cos^2\phi)\sin\phi\,d\phi\,d\theta$$

$$= \frac{1}{5}k(b^5 - a^5)\int_0^{2\pi}\left(-\cos\phi + \frac{1}{3}\cos^3\phi\right)\Big]_0^{\pi}\,d\theta$$

$$= \frac{4}{15}k(b^5 - a^5)\int_0^{2\pi}d\theta = \frac{8}{15}\pi k(b^5 - a^5). \qquad\blacksquare$$

$\displaystyle\iiint_D$ NOTES FROM THE CLASSROOM

Spherical coordinates are used in navigation. If we think of the Earth as a sphere of fixed radius centered at the origin, then a point P can be located by specifying two angles θ and ϕ. As shown in FIGURE 14.8.10, when ϕ is held constant the resulting curve is called a **parallel**. Fixed values of θ result in curves called **great circles**. Half of one of these great circles joining the north and south poles is called a **meridian**. The intersection of a parallel and a meridian gives the position of a point P. If $0° \leq \phi \leq 180°$ and $-180° \leq \theta \leq 180°$ the angles $90° - \phi$ and θ are said to be the **latitude** and **longitude** of P, respectively. The **prime meridian** corresponds to a longitude of $0°$. The latitude of the equator is $0°$; the latitudes of the north and south poles are, in turn, $+90°$ (or $90°$ north) and $-90°$ (or $90°$ south).

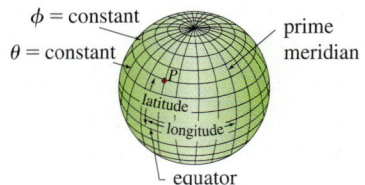

ϕ = constant

θ = constant

prime meridian

latitude

longitude

equator

FIGURE 14.8.10 Latitudes and longitudes

≡ Fundamentals

In Problems 1–6, convert the point given in cylindrical coordinates to rectangular coordinates.

1. $(10, 3\pi/4, 5)$
2. $(2, 5\pi/6, -3)$
3. $(\sqrt{3}, \pi/3, -4)$
4. $(4, 7\pi/4, 0)$
5. $(5, \pi/2, 1)$
6. $(10, 5\pi/3, 2)$

In Problems 7–12, convert the point given in rectangular coordinates to cylindrical coordinates.

7. $(1, -1, -9)$
8. $(2\sqrt{3}, 2, 17)$
9. $(-\sqrt{2}, \sqrt{6}, 2)$
10. $(1, 2, 7)$
11. $(0, -4, 0)$
12. $(\sqrt{7}, -\sqrt{7}, 3)$

In Problems 13–16, convert the given equation to cylindrical coordinates.

13. $x^2 + y^2 + z^2 = 25$
14. $x + y - z = 1$
15. $x^2 + y^2 - z^2 = 1$
16. $x^2 + z^2 = 16$

In Problems 17–20, convert the given equation to rectangular coordinates.

17. $z = r^2$
18. $z = 2r\sin\theta$
19. $r = 5\sec\theta$
20. $\theta = \pi/6$

In Problems 21–24, use a triple integral and cylindrical coordinates to find the volume of the solid that is bounded by the graphs of the given equations.

21. $x^2 + y^2 = 4, x^2 + y^2 + z^2 = 16, z = 0$
22. $z = 10 - x^2 - y^2, z = 1$
23. $z = x^2 + y^2, x^2 + y^2 = 25, z = 0$
24. $y = x^2 + z^2, 2y = x^2 + z^2 + 4$

In Problems 25–28, use a triple integral and cylindrical coordinates to find the indicated quantity.

25. the centroid of the homogeneous solid bounded by the hemisphere $z = \sqrt{a^2 - x^2 - y^2}$ and the plane $z = 0$

26. the center of mass of the solid bounded by the graphs of $y^2 + z^2 = 16, x = 0$, and $x = 5$ where the density at a point P is directly proportional to the distance from the yz-plane

27. the moment of inertia about the z-axis of the solid bounded above by the hemisphere $z = \sqrt{9 - x^2 - y^2}$ and below by the plane $z = 2$ where the density at a point P is inversely proportional to the square of the distance from the z-axis

28. the moment of inertia about the x-axis of the solid bounded by the single-napped cone $z = \sqrt{x^2 + y^2}$ and the plane $z = 1$ where the density at a point P is directly proportional to the distance from the z-axis

In Problems 29–34, convert the point given in spherical coordinates to
   (a) rectangular coordinates and
   (b) cylindrical coordinates.

29. $(\frac{2}{3}, \pi/2, \pi/6)$
30. $(5, 5\pi/4, 2\pi/3)$
31. $(8, \pi/4, 3\pi/4)$
32. $(\frac{1}{3}, 5\pi/3, \pi/6)$
33. $(4, 3\pi/4, 0)$
34. $(1, 11\pi/6, \pi)$

In Problems 35–40, convert the points given in rectangular coordinates to spherical coordinates.

35. $(-5, -5, 0)$
36. $(1, -\sqrt{3}, 1)$
37. $(\frac{1}{2}\sqrt{3}, \frac{1}{2}, 1)$
38. $(-\frac{1}{2}\sqrt{3}, 0, -\frac{1}{2})$
39. $(3, -3, 3\sqrt{2})$
40. $(1, 1, -\sqrt{6})$

In Problems 41–44, convert the given equation to spherical coordinates.

41. $x^2 + y^2 + z^2 = 64$
42. $x^2 + y^2 + z^2 = 4z$
43. $z^2 = 3x^2 + 3y^2$
44. $-x^2 - y^2 + z^2 = 1$

In Problems 45–48, convert the given equation to rectangular coordinates.

45. $\rho = 10$
46. $\phi = \pi/3$
47. $\rho = 2\sec\phi$
48. $\rho\sin^2\phi = \cos\phi$

In Problems 49–52, use a triple integral and spherical coordinates to find the volume of the solid that is bounded by the graphs of the given equations.

49. $z = \sqrt{x^2 + y^2}, x^2 + y^2 + z^2 = 9$
50. $x^2 + y^2 + z^2 = 4, y = x, y = \sqrt{3}x, z = 0$, first octant
51. $z^2 = 3x^2 + 3y^2, x = 0, y = 0, z = 2$, first octant
52. inside $x^2 + y^2 + z^2 = 1$ and outside $z^2 = x^2 + y^2$

In Problems 53–56, use a triple integral and spherical coordinates to find the indicated quantity.

53. the centroid of the homogeneous solid bounded by the single-napped cone $z = \sqrt{x^2 + y^2}$ and the sphere $x^2 + y^2 + z^2 = 2z$

54. the center of mass of the solid bounded by the hemisphere $z = \sqrt{1 - x^2 - y^2}$ and the plane $z = 0$ where the density at a point P is directly proportional to the distance from the xy-plane

55. the mass of the solid bounded above by the hemisphere $z = \sqrt{25 - x^2 - y^2}$ and below by the plane $z = 4$ where the density at a point P is inversely proportional to the distance from the origin [*Hint*: Express the upper ϕ limit of integration as an inverse cosine.]

56. the moment of inertia about the z-axis of the solid bounded by the sphere $x^2 + y^2 + z^2 = a^2$ where the density at a point P is directly proportional to the distance from the origin

14.9 Change of Variables in Multiple Integrals

■ **Introduction** In many instances it is convenient to make a substitution, or change of variable, in an integral in order to evaluate it. The idea in Theorem 5.5.3 can be rephrased as follows: If f is continuous and $x = g(u)$ has a continuous derivative and $dx = g'(u) \, du$, then

$$\int_a^b f(x) \, dx = \int_c^d f(g(u)) g'(u) \, du, \tag{1}$$

If the function g is one-to-one, then it has an inverse and so $c = g^{-1}(a)$ and $d = g^{-1}(b)$.

▶ where the y-limits of integration c and d are defined by $a = g(c)$ and $b = g(d)$. There are three things that bear emphasizing in (1). To change the variable in a definite integral we replace x where it appears in the integrand by $g(u)$, we change the interval of integration $[a, b]$ on the x-axis to the corresponding interval $[c, d]$ on the u-axis, and we replace dx by a function multiple (namely, the derivative of g) of du. If we write $J(u) = g'(u)$, then (1) has the form

$$\int_a^b f(x) \, dx = \int_c^d f(g(u)) J(u) \, du. \tag{2}$$

For example, using $x = 2 \sin \theta$, $-\pi/2 \le \theta \le \pi/2$, we have

$$\int_0^2 \overbrace{\sqrt{4 - x^2}}^{\substack{x\text{-limits}\downarrow \quad f(x)}} \, dx = \int_0^{\pi/2} \overbrace{2\cos\theta}^{\substack{\theta\text{-limits}\downarrow \quad f(2\sin\theta)}} \overbrace{(2\cos\theta)}^{J(\theta)} \, d\theta = 4 \int_0^{\pi/2} \cos^2\theta \, d\theta = \pi.$$

■ **Double Integrals** Although changing variables in a multiple integral is not as straightforward as the procedure in (1), the basic idea illustrated in (2) will carry over. To change variables in a double integral we need two equations, such as

$$x = x(u, v), \quad y = y(u, v). \tag{3}$$

To be analogous with (2), we expect that a change of variables in a double integral would take the form

$$\iint_R f(x, y) \, dA = \iint_S f(x(u, v), y(u, v)) J(u, v) \, dA', \tag{4}$$

where S is the region in the uv-plane corresponding to the region R in the xy-plane, and $J(u, v)$ is some function that depends on partial derivatives of the equations in (3). The symbol dA' on the right side of (4) represents either $du \, dv$ or $dv \, du$.

In Section 14.5 we briefly discussed how to change a double integral $\iint_R f(x, y) \, dA$ from rectangular coordinates to polar coordinates. Recall, in Example 2 of that section, the substitutions

$$x = r\cos\theta \quad \text{and} \quad y = r\sin\theta \tag{5}$$

led to

$$\int_0^2 \int_x^{\sqrt{8-x^2}} \frac{1}{5 + x^2 + y^2} \, dy \, dx = \int_{\pi/4}^{\pi/2} \int_0^{\sqrt{8}} \frac{1}{5 + r^2} r \, dr \, d\theta. \tag{6}$$

As we see in FIGURE 14.9.1, the introduction of polar coordinates changes the original region of integration R in the xy-plane to the more convenient *rectangular region* of integration S in the $r\theta$-plane. We note too that by comparing (4) with (6), we can identify $J(r, \theta) = r$ and $dA' = dr \, d\theta$.

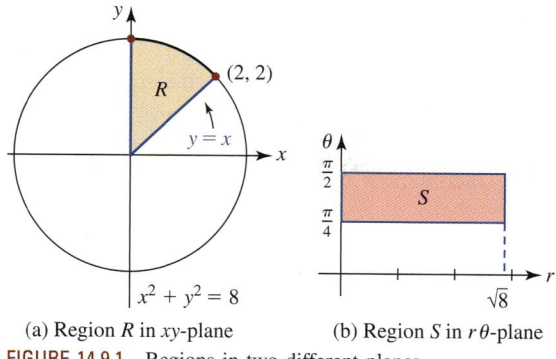

(a) Region R in xy-plane   (b) Region S in $r\theta$-plane

FIGURE 14.9.1   Regions in two different planes

The change-of-variable equations in (3) define a **transformation** or **mapping** T from the uv-plane into the xy-plane:

$$T(u, v) = (x, y).$$

A point (x_0, y_0) in the xy-plane is determined from $x_0 = x(u_0, v_0)$, $y_0 = y(u_0, v_0)$ and is said to be an **image** of (u_0, v_0), that is, $T(u_0, v_0) = (x_0, y_0)$.

EXAMPLE 1   Transformation of a Region

Find the image of the region S shown in FIGURE 14.9.2(a), under the transformation $x = u^2 + v^2$, $y = u^2 - v^2$.

Solution   We begin by finding the images of the sides of S that we have indicated by S_1, S_2, and S_3.

S_1: On the side $v = 0$ so that $x = u^2$, $y = u^2$. Eliminating u then gives $y = x$. Now, imagine moving along the boundary from $(1, 0)$ to $(2, 0)$ (that is, $1 \le u \le 2$). The equations $x = u^2$ and $y = u^2$ then indicate that x ranges from $x = 1$ to $x = 4$ and y ranges simultaneously from $y = 1$ to $y = 4$. In other words, in the xy-plane the image of S_1 is the line segment $y = x$ from $(1, 1)$ to $(4, 4)$.

S_2: On this boundary $u^2 + v^2 = 4$ and so $x = 4$. Now, as we move from the point $(2, 0)$ to $\left(\sqrt{\frac{5}{2}}, \sqrt{\frac{3}{2}}\right)$, the remaining equation $y = u^2 - v^2$ indicates that y ranges from $y = 2^2 - 0^2 = 4$ to $y = \left(\sqrt{\frac{5}{2}}\right)^2 - \left(\sqrt{\frac{3}{2}}\right)^2 = 1$. In this case the image of S_2 is the vertical line segment $x = 4$ starting at $(4, 4)$ and going down to $(4, 1)$.

S_3: Since $u^2 - v^2 = 1$, we get $y = 1$. But as we move on this boundary from $\left(\sqrt{\frac{5}{2}}, \sqrt{\frac{3}{2}}\right)$ to $(1, 0)$, the equation $x = u^2 + v^2$ indicates that x ranges from $x = 4$ to $x = 1$. The image of S_3 is the horizontal line segment $y = 1$ starting at $(4, 1)$ and ending at $(1, 1)$.

The image of S is the region R given in Figure 14.9.2(b).   ■

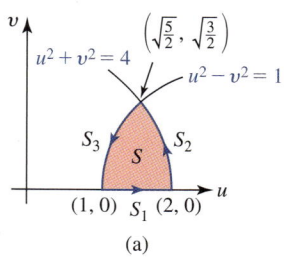

(a)

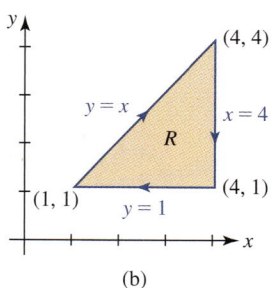

(b)

FIGURE 14.9.2   Image of S is R

Observe in Example 1 that as we traverse the boundary of S in the counterclockwise direction, the boundary of R is traversed in a clockwise manner. We say that the transformation of the boundary of S has *induced* an orientation on the boundary of R.

Although a proof of the formula for changing variables in a multiple integral is beyond the level of this text, we will give *some* of the underlying assumptions that are made about equations (3) and the regions R and S. We make the following assumptions:

- The functions $x = x(u, v)$, $y = y(u, v)$ have continuous first partial derivatives on S.
- The transformation is one-to-one.
- Each of the regions R and S consists of a piecewise smooth simple closed curve and its interior.
- The second-order determinant

$$\begin{vmatrix} \dfrac{\partial x}{\partial u} & \dfrac{\partial x}{\partial v} \\[2mm] \dfrac{\partial y}{\partial u} & \dfrac{\partial y}{\partial v} \end{vmatrix} = \dfrac{\partial x}{\partial u}\dfrac{\partial y}{\partial v} - \dfrac{\partial x}{\partial v}\dfrac{\partial y}{\partial u} \tag{7}$$

is not zero and does not change sign on S.

■ **Jacobian** A transformation T is said to be **one-to-one** if each point (x_0, y_0) in R is the image under T of a unique point (u_0, v_0) in S. Put another way, no two points in S have the same image in R. With the restrictions that $r \geq 0$ and $0 \leq \theta < 2\pi$, the equations in (5) define a one-to-one transformation from the $r\theta$-plane to the xy-plane. The determinant in (7) is called the **Jacobian determinant**, or simply **Jacobian**, of the transformation T and is the key to changing variables in a multiple integral. The Jacobian of the transformation defined by the equations in (3) is denoted by the symbol

$$\frac{\partial(x, y)}{\partial(u, v)}.$$

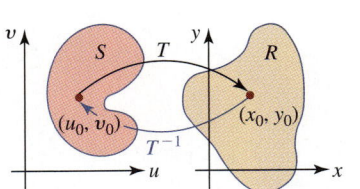

FIGURE 14.9.3 Transformations between regions

Similar to the notion of a one-to-one function introduced in Section 1.5, a one-to-one transformation T has an **inverse transformation** T^{-1} such that

$$T^{-1}(x_0, y_0) = (u_0, v_0).$$

That is, (u_0, v_0) is the image under T^{-1} of (x_0, y_0). See **FIGURE 14.9.3**. If it is possible to solve (3) for u and v in terms of x and y, then the inverse transformation is defined by the pair of equations

$$u = u(x, y), \quad v = v(x, y). \tag{8}$$

The Jacobian of the inverse transformation T^{-1} is

$$\frac{\partial(u, v)}{\partial(x, y)} = \begin{vmatrix} \dfrac{\partial u}{\partial x} & \dfrac{\partial u}{\partial y} \\ \dfrac{\partial v}{\partial x} & \dfrac{\partial v}{\partial y} \end{vmatrix} \tag{9}$$

and is related to the Jacobian (7) of the transformation T by

$$\frac{\partial(x, y)}{\partial(u, v)} \frac{\partial(u, v)}{\partial(x, y)} = 1. \tag{10}$$

EXAMPLE 2　Jacobian

The Jacobian of the transformation $x = r\cos\theta$, $y = r\sin\theta$ is

$$\frac{\partial(x, y)}{\partial(r, \theta)} = \begin{vmatrix} \dfrac{\partial x}{\partial r} & \dfrac{\partial x}{\partial \theta} \\ \dfrac{\partial y}{\partial r} & \dfrac{\partial y}{\partial \theta} \end{vmatrix} = \begin{vmatrix} \cos\theta & -r\sin\theta \\ \sin\theta & r\cos\theta \end{vmatrix} = r(\cos^2\theta + \sin^2\theta) = r. \qquad ■$$

We shall now turn our attention to the main point of this discussion: how to change variables in a multiple integral. The idea expressed in (4) is valid; the function $J(u, v)$ turns out to be the absolute value of the Jacobian; that is, $J(u, v) = |\partial(x, y)/\partial(u, v)|$. Under the assumptions made above, we have the following result for double integrals.

Theorem 14.9.1　Change of Variables in a Double Integral

If $x = x(u, v)$, $y = y(u, v)$ is a transformation that maps a region S in the uv-plane into a region R in the xy-plane and f is a function continuous on R, then

$$\iint_R f(x, y)\, dA = \iint_S f(x(u, v), y(u, v)) \left| \frac{\partial(x, y)}{\partial(u, v)} \right| dA'. \tag{11}$$

Formula (3) of Section 14.5 for changing a double integral to polar coordinates is just a special case of (11) with

$$\left| \frac{\partial(x, y)}{\partial(r, \theta)} \right| = |r| = r$$

since $r \geq 0$. In (6), then, we have $J(r, \theta) = |\partial(x, y)/\partial(r, \theta)| = r$.

A change of variables in a multiple integral can be used for either a simplification of the integrand or a simplification of the region of integration. The actual change of variables used is often inspired by the structure of the integrand $f(x, y)$ or by equations that define the region R. As a consequence, the transformation is defined by equations of the form given in (8); that is, we are dealing with the inverse transformation. The next two examples illustrate these ideas.

EXAMPLE 3 Changing Variables

Evaluate $\int_R \sin(x + 2y)\cos(x - 2y)\, dA$ over the region R shown in FIGURE 14.9.4(a).

Solution   The difficulty in evaluating this double integral is clearly the integrand. The presence of the terms $x + 2y$ and $x - 2y$ prompts us to define the change of variables

$$u = x + 2y \qquad \text{and} \qquad v = x - 2y.$$

These equations will map R onto the region S in the uv-plane. As in Example 1, we transform the sides of the region.

$S_1: y = 0$ implies $u = x$ and $v = x$ or $v = u$. As we move from $(2\pi, 0)$ to $(0, 0)$ we see that the corresponding image points in the uv-plane lie on the line segment $v = u$ from $(2\pi, 2\pi)$ to $(0, 0)$.

$S_2: x = 0$ implies $u = 2y$ and $v = -2y$ or $v = -u$. As we move from $(0, 0)$ to $(0, \pi)$, the corresponding image points in the uv-plane lie on the line segment $v = -u$ from $(0, 0)$ to $(2\pi, -2\pi)$.

$S_3: x + 2y = 2\pi$ implies $u = 2\pi$. As we move from $(0, \pi)$ to $(2\pi, 0)$, the equation $v = x - 2y$ shows that v ranges from $v = -2\pi$ to $v = 2\pi$. Thus, the image of S_3 is the vertical line segment $u = 2\pi$ starting at $(2\pi, -2\pi)$ and extending up to $(2\pi, 2\pi)$. See Figure 14.9.4(b).

Now, solving the equations $u = x + 2y$, $v = x - 2y$ for x and y in terms of u and v gives

$$x = \frac{1}{2}(u + v) \qquad \text{and} \qquad y = \frac{1}{4}(u - v).$$

Therefore,

$$\frac{\partial(x, y)}{\partial(u, v)} = \begin{vmatrix} \dfrac{\partial x}{\partial u} & \dfrac{\partial x}{\partial v} \\[2mm] \dfrac{\partial y}{\partial u} & \dfrac{\partial y}{\partial v} \end{vmatrix} = \begin{vmatrix} \dfrac{1}{2} & \dfrac{1}{2} \\[2mm] \dfrac{1}{4} & -\dfrac{1}{4} \end{vmatrix} = -\frac{1}{4}.$$

Hence, from (11) we find that

$$\iint_R \sin(x + 2y)\cos(x - 2y)\, dA = \iint_S \sin u \cos v \left| -\frac{1}{4} \right| dA'$$

$$= \frac{1}{4} \int_0^{2\pi} \int_{-u}^{u} \sin u \cos v \, dv \, du$$

$$= \frac{1}{4} \int_0^{2\pi} \sin u \sin v \bigg]_{-u}^{u} du$$

$$= \frac{1}{2} \int_0^{2\pi} \sin^2 u \, du$$

$$= \frac{1}{4} \int_0^{2\pi} (1 - \cos 2u) \, du$$

$$= \frac{1}{4} \left(u - \frac{1}{2}\sin 2u \right) \bigg]_0^{2\pi} = \frac{1}{2}\pi. \qquad \blacksquare$$

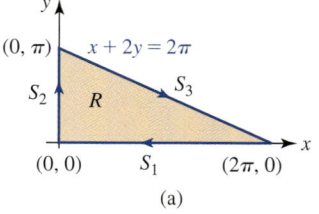

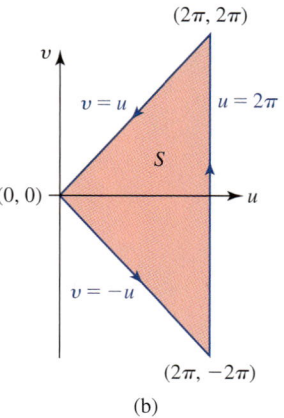

FIGURE 14.9.4 Regions R and S in Example 3

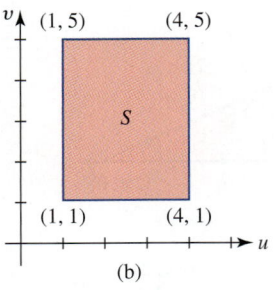

FIGURE 14.9.5 Regions R and S in Example 4

EXAMPLE 4 Changing Variables

Evaluate $\iint_R xy \, dA$ over the region R shown in FIGURE 14.9.5(a).

Solution In this case the integrand is fairly simple, but integration over the region R would be tedious, since we would have to express $\iint_R xy \, dA$ as the sum of three integrals. (Verify this.)

The equations of the boundaries of R suggest the change of variables

$$u = \frac{y}{x^2} \quad \text{and} \quad v = xy. \tag{12}$$

Obtaining the image of R is a straightforward matter in this case, since the images of the curves that make up the four boundaries are simply $u = 1$, $u = 4$, $v = 1$, and $v = 5$. In other words, the image of the region R is the rectangular region $S: 1 \le u \le 4, 1 \le v \le 5$. See Figure 14.9.5(b).

Now, instead of trying to solve the equations in (12) for x and y in terms of u and v, we can compute the Jacobian $\partial(x, y)/\partial(u, v)$ by computing $\partial(u, v)/\partial(x, y)$ and using (10). We have

$$\frac{\partial(u, v)}{\partial(x, y)} = \begin{vmatrix} \dfrac{\partial u}{\partial x} & \dfrac{\partial u}{\partial y} \\ \dfrac{\partial v}{\partial x} & \dfrac{\partial v}{\partial y} \end{vmatrix} = \begin{vmatrix} -\dfrac{2y}{x^3} & \dfrac{1}{x^2} \\ y & x \end{vmatrix} = -\frac{3y}{x^2}$$

and so from (10),

$$\frac{\partial(x, y)}{\partial(u, v)} = \frac{1}{\dfrac{\partial(u, v)}{\partial(x, y)}} = -\frac{x^2}{3y} = -\frac{1}{3u}.$$

Hence,

$$\iint_R xy \, dA = \iint_S v \left| -\frac{1}{3u} \right| dA'$$

$$= \frac{1}{3} \int_1^4 \int_1^5 \frac{v}{u} \, dv \, du$$

$$= \frac{1}{3} \int_1^4 \frac{1}{u} \left(\frac{1}{2} v^2 \right) \Big]_1^5 du$$

$$= 4 \int_1^4 \frac{1}{u} \, du = 4 \ln u \Big]_1^4 = 4 \ln 4. \qquad \blacksquare$$

Triple Integrals To change variables in a triple integral, let

$$x = x(u, v, w), \quad y = y(u, v, w), \quad z = z(u, v, w) \tag{13}$$

be a one-to-one transformation T from a region E in uvw-space to a region D in xyz-space. If the functions in (13) satisfy the three-variable counterparts of the conditions listed on page 791 and the third-order Jacobian

$$\frac{\partial(x, y, z)}{\partial(u, v, w)} = \begin{vmatrix} \dfrac{\partial x}{\partial u} & \dfrac{\partial x}{\partial v} & \dfrac{\partial x}{\partial w} \\ \dfrac{\partial y}{\partial u} & \dfrac{\partial y}{\partial v} & \dfrac{\partial y}{\partial w} \\ \dfrac{\partial z}{\partial u} & \dfrac{\partial z}{\partial v} & \dfrac{\partial z}{\partial w} \end{vmatrix}$$

is not zero and does not change signs on E, then we have the following result.

Theorem 14.9.2   Change of Variables in a Triple Integral

If $x = x(u, v, w)$, $y = y(u, v, w)$, $z = z(u, v, w)$ is a transformation that maps a region E in the uvw-space into a region D in the xyz-space and f is a function continuous on E, then

$$\iiint_D f(x, y, z)\,dV = \iiint_E f(x(u, v, w), y(u, v, w), z(u, v, w)) \left| \frac{\partial(x, y, z)}{\partial(u, v, w)} \right| dV'. \qquad (14)$$

We leave it as an exercise to show that if T is the transformation from spherical to rectangular coordinates defined by

$$x = \rho \sin\phi \cos\theta, \quad y = \rho \sin\phi \sin\theta, \quad z = \rho \cos\phi,$$

then

$$\frac{\partial(x, y, z)}{\partial(\rho, \phi, \theta)} = \rho^2 \sin\phi. \qquad (15)$$

See Problem 28 in Exercises 14.9.

■ Postscript—A Bit of History   **Carl Gustav Jacob Jacobi** (1804–1851) was born in Potsdam into a rich German family. The young Carl Gustav excelled in many areas of study, but his ability and love for intricate algebraic calculations led him to the life of an impoverished mathematician and teacher. He suffered a nervous breakdown in 1843 due to overwork. His Ph.D. dissertation was related to a topic now known to every student of calculus: partial fractions. But Jacobi's greatest contributions to mathematics were in the field of elliptic functions and number theory. He also made major contributions to the theory of determinants and to the simplification of that theory. Although Jacobi was principally a "pure" mathematician; nonetheless, every student of dynamics and quantum mechanics will recognize Jacobi's contribution to these areas through the famous Hamilton–Jacobi equations.

Jacobi

Exercises 14.9   Answers to selected odd-numbered problems begin on page ANS-45.

≡ **Fundamentals**

1. Consider a transformation T defined by $x = 4u - v$, $y = 5u + 4v$. Find the images of the points $(0, 0)$, $(0, 2)$, $(4, 0)$, and $(4, 2)$ in the uv-plane under T.

2. Consider a transformation T defined by $x = \sqrt{v - u}$, $y = v + u$. Find the images of the points $(1, 1)$,$(1, 3)$, and $\left(\sqrt{2}, 2\right)$ in the xy-plane under T^{-1}.

In Problems 3–6, find the image of the set S under the given transformation.

3. $S: 0 \le u \le 2, 0 \le v \le u$; $\quad x = 2u + v, y = u - 3v$

4. $S: -1 \le u \le 4, 1 \le v \le 5$; $\quad u = x - y, v = x + 2y$

5. $S: 0 \le u \le 1, 0 \le v \le 2$; $\quad x = u^2 - v^2, y = uv$

6. $S: 1 \le u \le 2, 1 \le v \le 2$; $\quad x = uv, y = v^2$

In Problems 7–10, find the Jacobian of the transformation T from the uv-plane into the xy-plane.

7. $x = ve^{-u}, y = ve^u$

8. $x = e^{3u}\sin v, y = e^{3u}\cos v$

9. $u = \dfrac{y}{x^2}, v = \dfrac{y^2}{x}$

10. $u = \dfrac{2x}{x^2 + y^2}, v = \dfrac{-2y}{x^2 + y^2}$

11. (a) Find the image of the region $S: 0 \le u \le 1, 0 \le v \le 1$ under the transformation $x = u - uv, y = uv$.
    (b) Explain why the transformation is not one-to-one on the boundary of S.

12. Determine where the Jacobian $\partial(x, y)/\partial(u, v)$ of the transformation in Problem 11 is zero.

In Problems 13–22, evaluate the given integral by means of the indicated change of variables.

13. $\iint_R (x + y)\,dA$, where R is the region bounded by the graphs of $x - 2y = -6, x - 2y = 6, x + y = -1, x + y = 3$; $u = x - 2y, v = x + y$

14. $\displaystyle\iint_R \frac{\cos\frac{1}{2}(x - y)}{3x + y}\,dA$, where R is the region bounded by the graphs of $y = x, y = x - \pi, y = -3x + 3, y = -3x + 6$; $u = x - y, v = 3x + y$

15. $\displaystyle\iint_R \frac{y^2}{x}\,dA$, where R is the region bounded by the graphs $y = x^2, y = \frac{1}{2}x^2, x = y^2, x = \frac{1}{2}y^2; u = x^2/y, v = y^2/x$

16. $\iint_R (x^2 + y^2)^{-3} dA$, where R is the region bounded by the circles $x^2 + y^2 = 2x$, $x^2 + y^2 = 4x$, $x^2 + y^2 = 2y$, $x^2 + y^2 = 6y$; $u = \dfrac{2x}{x^2 + y^2}$, $v = \dfrac{2y}{x^2 + y^2}$ [*Hint:* Form $u^2 + v^2$.]

17. $\iint_R (x^2 + y^2) dA$, where R is the region in the first quadrant bounded by the graphs of $x^2 - y^2 = a$, $x^2 - y^2 = b$, $2xy = c$, $2xy = d$, $0 < a < b$, $0 < c < d$; $u = x^2 - y^2$, $v = 2xy$

18. $\iint_R (x^2 + y^2) \sin xy \, dA$, where R is the region bounded by the graphs of $x^2 - y^2 = 1$, $x^2 - y^2 = 9$, $xy = 2$, $xy = -2$; $u = x^2 - y^2$, $v = xy$

19. $\displaystyle\iint_R \dfrac{x}{y + x^2} dA$, where R is the region in the first quadrant bounded by the graphs of $x = 1$, $y = x^2$, $y = 4 - x^2$; $x = \sqrt{v - u}$, $y = v + u$

20. $\iint_R y \, dA$, where R is the triangular region with vertices $(0, 0),(2, 3)$, and $(-4, 1)$; $x = 2u - 4v$, $y = 3u + v$

21. $\iint_R y^4 \, dA$, where R is the region in the first quadrant bounded by the graphs of $xy = 1$, $xy = 4$, $y = x$, $y = 4x$; $u = xy$, $v = y/x$

22. $\iiint_D (4z + 2x - 2y) \, dV$, where D is the parallelepiped $1 \le y + z \le 3$, $-1 \le -y + z \le 1$, $0 \le x - y \le 3$; $u = y + z$, $v = -y + z$, $w = x - y$

In Problems 23–26, evaluate the double integral by means of an appropriate change of variables.

23. $\displaystyle\int_0^1 \int_0^{1-x} e^{(y-x)/(y+x)} \, dy \, dx$ **24.** $\displaystyle\int_{-2}^0 \int_0^{x+2} e^{y^2 - 2xy + x^2} \, dy \, dx$

25. $\iint_R (6x + 3y) \, dA$, where R is the trapezoidal region in the first quadrant with vertices $(1, 0)$, $(4, 0)$, $(2, 4)$, and $\left(\frac{1}{2}, 1\right)$

26. $\iint_R (x + y)^4 e^{x-y} \, dA$, where R is the square region with vertices $(1, 0)$, $(0, 1)$, $(1, 2)$, and $(2, 1)$

27. Evaluate the double integral $\iint_R \left(\frac{1}{25}x^2 + \frac{1}{9}y^2\right) dA$, where R is the elliptical region whose boundary is the graph of $\frac{1}{25}x^2 + \frac{1}{9}y^2 = 1$. Use the substitution $u = \frac{1}{5}x$, $v = \frac{1}{3}y$ and polar coordinates.

28. Verify that the Jacobian of the transformation given in (14) is $\partial(x, y, z)/\partial(\rho, \phi, \theta) = \rho^2 \sin\phi$.

29. Use $V = \iiint_D dV$ and the substitutions $u = x/a$, $v = y/b$, $w = z/c$ to show that the volume of the ellipsoid $x^2/a^2 + y^2/b^2 + z^2/c^2 = 1$ is $V = \frac{4}{3}\pi abc$.

☰ Applications

30. A problem in thermodynamics is to find the work done by an ideal Carnot engine. This work is defined to be the area of the region R in the first quadrant bounded by the isothermals $xy = a$, $xy = b$, $0 < a < b$, and the adiabatics $xy^{1.4} = c$, $xy^{1.4} = d$, $0 < c < d$. Use $A = \iint_R dA$ and an appropriate substitution to find the area shown in FIGURE 14.9.6.

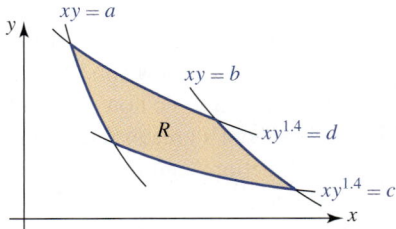

FIGURE 14.9.6 Region R for Problem 30

Chapter 14 in Review

Answers to selected odd-numbered problems begin on page ANS-46.

A. True/False

In Problems 1–6, indicate whether the given statement is true or false.

1. $\displaystyle\int_{-2}^3 \int_1^5 e^{x^2 - y} \, dx \, dy = \int_1^5 \int_{-2}^3 e^{x^2 - y} \, dy \, dx$ _____

2. If $\displaystyle\int f(x, y) \, dx = F(x, y) + c_2(y)$ is a partial integral, then $F_x(x, y) = f(x, y)$. _____

3. If I is the partial definite integral $\displaystyle\int_{g_1(x)}^{g_2(x)} f(x, y) \, dy$, then $\partial I/\partial y = 0$. _____

4. For every continuous function f, $\displaystyle\int_{-1}^1 \int_{x^2}^1 f(x, y) \, dy \, dx = 2\int_0^1 \int_{x^2}^1 f(x, y) \, dy \, dx$. _____

5. The center of mass of a lamina possessing symmetry lies on the axis of symmetry of the lamina. _____

6. In cylindrical and spherical coordinates the equation of the plane $y = x$ is the same. _____

B. Fill in the Blanks

In Problems 1–12, fill in the blanks.

1. $\displaystyle\int_{y^2+1}^{5}\left(8y^3 - \frac{5y}{x}\right) dx = $ _____ .

2. If R_1 and R_2 are nonoverlapping regions such that $R = R_1 \cup R_2$, $\iint_R f(x, y)\, dA = 10$, and $\iint_{R_2} f(x, y)\, dA = -6$, then $\iint_{R_1} f(x, y)\, dA = $ _____ .

3. $\displaystyle\int_{-a}^{a}\int_{-a}^{a} dx\, dy$ gives the area of a _____ .

4. The region bounded by the graphs of $9x^2 + y^2 = 36$, $y = -2$, $y = 5$ is a Type _____ region.

5. $\displaystyle\int_{2}^{4} f_y(x, y)\, dy = $ _____ .

6. If $\rho(x, y, z)$ is density, then the iterated integral giving the mass of the solid bounded by the ellipsoid $x^2/a^2 + y^2/b^2 + z^2/c^2 = 1$ is _____ .

7. $\displaystyle\int_{0}^{2}\int_{y^2}^{2y} f(x, y)\, dx\, dy = \int_{\rule{1em}{0.4pt}}^{\rule{1em}{0.4pt}}\int_{\rule{1em}{0.4pt}}^{\rule{1em}{0.4pt}} f(x, y)\, dy\, dx$

8. The rectangular coordinates of the point $(6, 5\pi/3, 5\pi/6)$ given in spherical coordinates are _____ .

9. The cylindrical coordinates of the point $(2, \pi/4, 2\pi/3)$ given in spherical coordinates are _____ .

10. The region R bounded by the graphs of $y = 4 - x^2$ and $y = 0$ is both Type I and Type II. Interpreted as a Type II region, $\displaystyle\iint_R f(x, y)\, dA = \int_{\rule{1em}{0.4pt}}^{\rule{1em}{0.4pt}}\int_{\rule{1em}{0.4pt}}^{\rule{1em}{0.4pt}} f(x, y)$ _____ _____ .

11. The equation of the paraboloid $z = x^2 + y^2$ in cylindrical coordinates is _____ , whereas in spherical coordinates its equation is _____ .

12. The region whose area is $A = \displaystyle\int_{0}^{\pi}\int_{0}^{\sin\theta} r\, dr\, d\theta$ is _____ .

C. Exercises

In Problems 1–14, evaluate the given integral.

1. $\displaystyle\int (12x^2 e^{-4xy} - 5x + 1)\, dy$

2. $\displaystyle\int \frac{1}{4 + 3xy}\, dx$

3. $\displaystyle\int_{y^3}^{y} y^2 \sin xy\, dx$

4. $\displaystyle\int_{1/x}^{e^x} \frac{x}{y^2}\, dy$

5. $\displaystyle\int_{0}^{2}\int_{0}^{2x} y e^{y-x}\, dy\, dx$

6. $\displaystyle\int_{0}^{4}\int_{x}^{4} \frac{1}{16 + x^2}\, dy\, dx$

7. $\displaystyle\int_{0}^{1}\int_{x}^{\sqrt{x}} \frac{\sin y}{y}\, dy\, dx$

8. $\displaystyle\int_{e}^{e^2}\int_{0}^{1/x} \ln x\, dy\, dx$

9. $\displaystyle\int_{0}^{5}\int_{0}^{\pi/2}\int_{0}^{\cos\theta} 3r^2\, dr\, d\theta\, dz$

10. $\displaystyle\int_{\pi/4}^{\pi/2}\int_{0}^{\sin z}\int_{0}^{\ln x} e^y\, dy\, dx\, dz$

11. $\displaystyle\iint_R 5\, dA$, where R is bounded by the circle $x^2 + y^2 = 64$

12. $\displaystyle\iint_R dA$, where R is bounded by the cardioid $r = 1 + \cos\theta$

13. $\displaystyle\iint_R (2x + y)\, dA$, where R is bounded by the graphs of $y = \frac{1}{2}x$, $x = y^2 + 1$, $y = 0$

14. $\displaystyle\iiint_D x\, dV$, where D is bounded by the planes $z = x + y$, $z = 6 - x - y$, $x = 0$, $y = 0$

15. Using rectangular coordinates, express

$$\iint_R \frac{1}{x^2 + y^2}\, dA$$

as an iterated integral, where R is the region in the first quadrant that is bounded by the graphs of $x^2 + y^2 = 1$, $x^2 + y^2 = 9$, $x = 0$, and $y = x$. Do not evaluate.

16. Evaluate the double integral in Problem 15 using polar coordinates.

In Problems 17 and 18, sketch the region of integration.

17. $\displaystyle\int_{-2}^{2} \int_{-x^2}^{x^2} f(x, y)\, dy\, dx$

18. $\displaystyle\int_{-1}^{1} \int_{-1}^{1} \int_{0}^{x^2 + y^2} f(x, y, z)\, dz\, dx\, dy$

19. Reverse the order of integration and evaluate

$$\int_0^1 \int_y^{\sqrt[3]{y}} \cos x^2\, dx\, dy.$$

20. Consider $\iiint_D f(x, y, z)\, dV$, where D is the region in the first octant bounded by the planes $z = 8 - 2x - y$, $z = 4$, $x = 0$, $y = 0$. Express the triple integral as six different iterated integrals.

In Problems 21 and 22, use an appropriate coordinate system to evaluate the given integral.

21. $\displaystyle\int_0^2 \int_{1/2}^1 \int_0^{\sqrt{x - x^2}} (4z + 1)\, dy\, dx\, dz$

22. $\displaystyle\int_0^1 \int_0^{\sqrt{1 - x^2}} \int_{-\sqrt{1 - x^2 - y^2}}^{\sqrt{1 - x^2 - y^2}} (x^2 + y^2 + z^2)^4\, dz\, dy\, dx$

23. Find the surface area of that portion of the graph of $z = xy$ within the cylinder $x^2 + y^2 = 1$.

24. Use a double integral to find the volume of the solid shown in FIGURE 14.R.1.

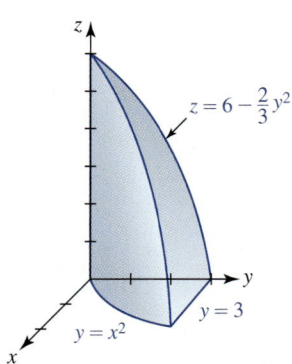

FIGURE 14.R.1   Solid in Problem 24

25. Express the volume of the solid shown in FIGURE 14.R.2 as one or more iterated integrals using the order of integration

(a) $dy\,dx$ (b) $dx\,dy$.

Choose either part (a) or part (b) to find the volume.

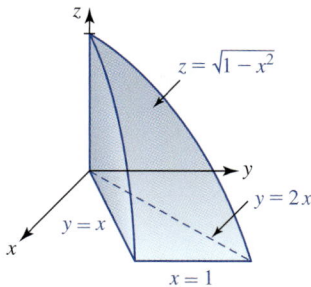

$z = \sqrt{1 - x^2}$

$y = 2x$

$y = x$

$x = 1$

FIGURE 14.R.2   Solid in Problem 25

26. A lamina has the shape of the region in the first quadrant bounded by the graphs of $y = x^2$ and $y = x^3$. Find the center of mass if the density ρ at a point P is directly proportional to the square of the distance from the origin.

27. Find the moment of inertia of the lamina described in Problem 26 about the y-axis.

28. Find the volume of the sphere $x^2 + y^2 + z^2 = a^2$ using a triple integral in

(a) rectangular coordinates,   (b) cylindrical coordinates, and (c) spherical coordinates.

29. Find the volume of the solid that is bounded between the cones $z = \sqrt{x^2 + y^2}$, $z = 3\sqrt{x^2 + y^2}$, and the plane $z = 3$.

30. Find the volume of the solid shown in FIGURE 14.R.3.

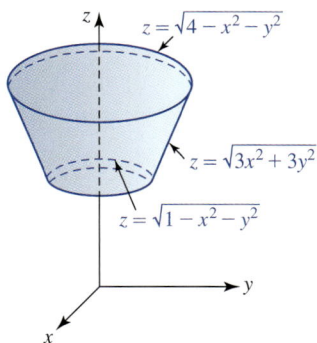

$z = \sqrt{4 - x^2 - y^2}$

$z = \sqrt{3x^2 + 3y^2}$

$z = \sqrt{1 - x^2 - y^2}$

FIGURE 14.R.3   Solid in Problem 30

31. Evaluate the integral $\iint_R (x^2 + y^2)\sqrt[3]{x^2 - y^2}\,dA$, where R is the region bounded by the graphs of $x = 0, x = 1, y = 0$, and $y = 1$ by means of the change of variables $u = 2xy, v = x^2 - y^2$.

32. Evaluate the integral

$$\iint_R \frac{1}{\sqrt{(x - y)^2 + 2(x + y) + 1}}\,dA,$$

where R is the region bounded by the graphs of $y = x, x = 2$, and $y = 0$ by means of the change of variables $x = u + uv, y = v + uv$.

Vector Integral Calculus

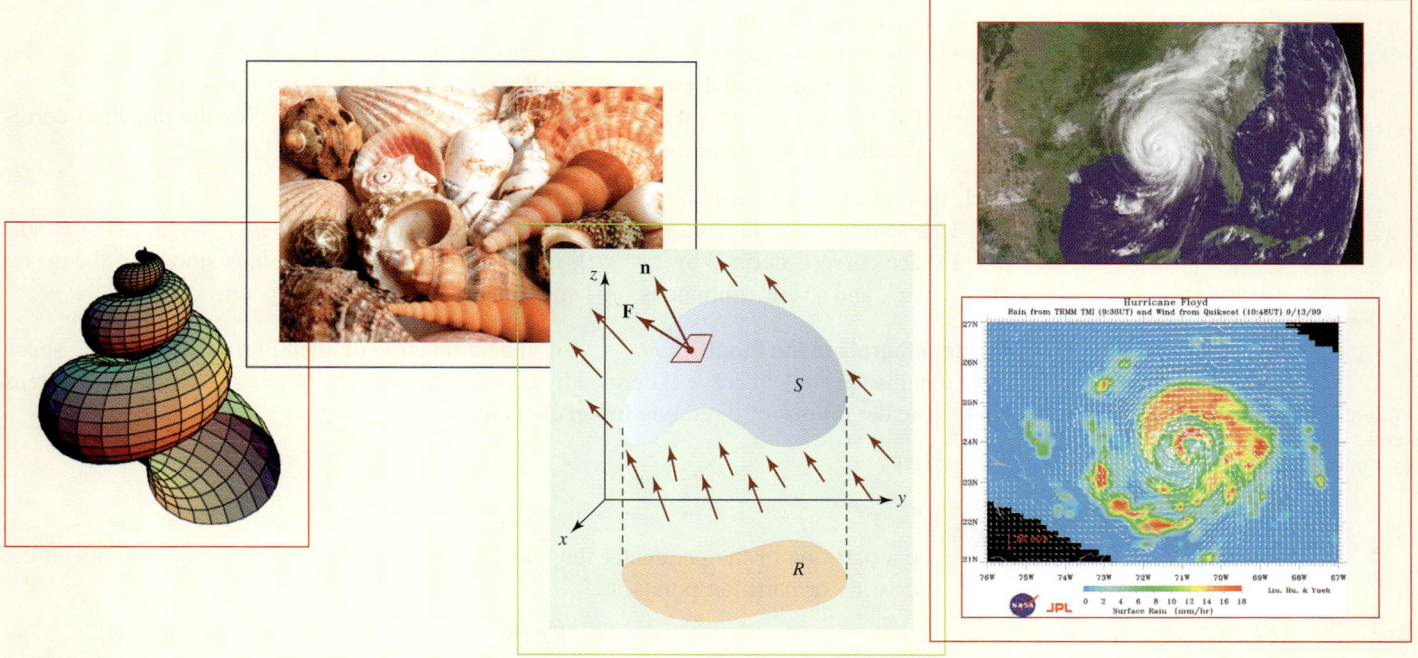

In This Chapter Up to this point in our study of calculus, we have encountered three kinds of integrals: the definite integral, the double integral, and the triple integral. In this chapter we will introduce two new kinds of integrals: line integrals and surface integrals. The development of these new concepts depends heavily on vector methods. In Section 15.2 we introduce a new kind of vector function—a function that does not define a curve but rather a field of vectors.

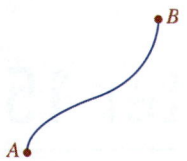

(a) Smooth curve

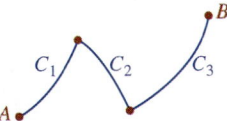

(b) Piecewise-smooth curve

(c) Closed but not simple

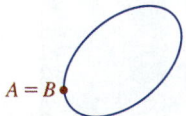

(d) Simple closed curve

FIGURE 15.1.1 Types of curves

15.1 Line Integrals

∎ **Introduction**   The notion of the definite integral $\int_a^b f(x)\,dx$—that is, *integration of a function of a single variable defined over an interval*—can be generalized to *integration of a function of several variables defined along a curve*. To this end we need to introduce some terminology about curves.

∎ **Some Terminology**   Suppose C is a curve parameterized by $x = x(t)$, $y = y(t)$, $a \le t \le b$, and A and B are the initial and terminal points $(x(a), y(a))$ and $(x(b), y(b))$, respectively. We say that:

- C is a **smooth curve** if $x'(t)$ and $y'(t)$ are continuous on the closed interval $[a, b]$ and not simultaneously zero on the open interval (a, b).
- C is a **piecewise-smooth curve** if it consists of a finite number of smooth curves $C_1, C_2, \ldots, C_n$ joined end to end; that is, $C = C_1 \cup C_2 \cup \cdots \cup C_n$.
- C is a **closed curve** if $A = B$.
- C is a **simple curve** if it does not cross itself between A and B.
- C is a **simple closed curve** if $A = B$ and the curve does not cross itself.
- If C is not a closed curve, then the **orientation** imposed on C is the direction corresponding to the increasing values of t.

Each type of curve defined above is illustrated in **FIGURE 15.1.1**.

This same terminology carries over in a natural manner to curves in 3-space. For example, a space curve C defined by $x = x(t)$, $y = y(t)$, $z = z(t)$, $a \le t \le b$, is smooth if the derivatives x', y', and z' are continuous on $[a, b]$ and not simultaneously zero on (a, b).

∎ **Line Integrals in the Plane**   Let $z = f(x, y)$ be a function defined in some region in 2-space that contains the smooth curve C defined by $x = x(t)$, $y = y(t)$, $a \le t \le b$. The following steps lead to the definitions of three **line integrals** in the plane.

An unfortunate choice of name. The ▶ term "curve integrals" would be more appropriate.

- Let
$$a = t_0 < t_1 < t_2 < \cdots < t_n = b$$
be a partition of the parameter interval $[a, b]$ and let the corresponding points on the curve C, or partition points, be
$$A = P_0, P_1, P_2, \ldots, P_n = B.$$

- The partition points $P_k = (x(t_k), y(t_k))$, $k = 0, 1, 2, \ldots, n$ divide C into n subarcs of lengths Δs_k. Let the projection of each subarc onto the x- and y-axes have lengths Δx_k and Δy_k, respectively.
- Let $\|P\|$ be the length of the longest subarc.
- Choose a sample point (x_k^*, y_k^*) on each subarc as shown in **FIGURE 15.1.2**. This point corresponds to a number t_k^* in the kth subinterval $[t_{k-1}, t_k]$ in the partition of the parameter interval $[a, b]$.
- Form the sums
$$\sum_{k=1}^{n} f(x_k^*, y_k^*)\,\Delta x_k, \qquad \sum_{k=1}^{n} f(x_k^*, y_k^*)\,\Delta y_k, \qquad \text{and} \qquad \sum_{k=1}^{n} f(x_k^*, y_k^*)\,\Delta s_k.$$

We now take the limit of these three sums as $\|P\| \to 0$. The resulting integrals are summarized next.

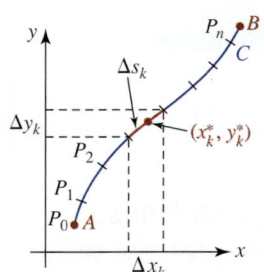

FIGURE 15.1.2 Sample point on the kth subarc

Definition 15.1.1   Line Integrals in the Plane

Let f be a function of two variables x and y defined in a region of the plane that contains a smooth curve C.

(*i*) The **line integral of f with respect to x** along C from A to B is

$$\int_C f(x, y)\,dx = \lim_{\|P\| \to 0} \sum_{k=1}^{n} f(x_k^*, y_k^*)\Delta x_k. \tag{1}$$

(continued)

(*ii*) The **line integral of *f* with respect to *y*** along *C* from *A* to *B* is

$$\int_C f(x, y)\, dy = \lim_{\|P\|\to 0} \sum_{k=1}^{n} f(x_k^*, y_k^*)\Delta y_k. \tag{2}$$

(*iii*) The **line integral of *f* with respect to arc length *s*** along *C* from *A* to *B* is

$$\int_C f(x, y)\, ds = \lim_{\|P\|\to 0} \sum_{k=1}^{n} f(x_k^*, y_k^*)\Delta s_k. \tag{3}$$

It can be proved that if $f(x, y)$ is continuous on C, then the integrals defined in (1), (2), and (3) exist. We shall assume continuity of f as a matter of course.

■ **Geometric Interpretation**   In the case of two variables, the line integral with respect to arc length $\int_C f(x, y)\, ds$ can be interpreted in a geometric manner when $f(x, y) \geq 0$ on C. In Definition 15.1.1 the symbol Δs_k represents the length of the kth subarc on the curve C. But from Figure 15.1.2 we have the approximation $\Delta s_k = \sqrt{(\Delta x_k)^2 + (\Delta y_k)^2}$. With this interpretation of Δs_k, we see from FIGURE 15.1.3(a) that the product of $f(x_k^*, y_k^*)\Delta s_k$ is the area of a vertical rectangle of height $f(x_k^*, y_k^*)$ and width Δs_k. The integral $\int_C f(x, y)\, ds$ then represents the area of one side of a "fence" or "curtain" extending from the curve C in the xy-plane up to the graph of $f(x, y)$ that corresponds to points (x, y) on C. See Figure 15.1.3(b).

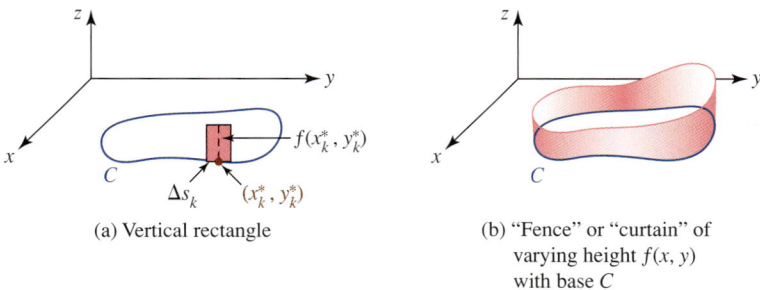

(a) Vertical rectangle

(b) "Fence" or "curtain" of
varying height $f(x, y)$
with base C

FIGURE 15.1.3   Geometric interpretation of (*iii*) of Definition 15.1.1

■ **Method of Evaluation: *C* Defined Parametrically**   The line integrals in Definition 15.1.1 can be evaluated in two ways, depending on whether the curve C is defined parametrically or by an explicit function. In either case, the basic idea is to convert the line integral to a definite integral in a single variable. If C is a smooth curve parameterized by $x = x(t), y = y(t), a \leq t \leq b$, then $dx = x'(t)\, dt, dy = y'(t)\, dt$ and so (1) and (2) become, respectively,

$$\int_C f(x, y)\, dx = \int_a^b f(x(t), y(t))x'(t)\, dt, \tag{4}$$

$$\int_C f(x, y)\, dy = \int_a^b f(x(t), y(t))y'(t)\, dt. \tag{5}$$

Furthermore, using (5) of Section 6.5 and the given parameterization, we find that $ds = \sqrt{[x'(t)]^2 + [y'(t)]^2}\, dt$. Hence, (3) can be written as

$$\int_C f(x, y)\, ds = \int_a^b f(x(t), y(t))\sqrt{[x'(t)]^2 + [y'(t)]^2}\, dt. \tag{6}$$

EXAMPLE 1   Using (4), (5), and (6)

Evaluate

(**a**) $\int_C xy^2\, dx$,   (**b**) $\int_C xy^2\, dy$,   (**c**) $\int_C xy^2\, ds$

on the quarter-circle C defined by $x = 4\cos t, y = 4\sin t, 0 \leq t \leq \pi/2$. See FIGURE 15.1.4.

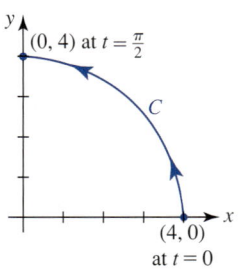

FIGURE 15.1.4   Curve C in Example 1

Solution

(a) From (4),

$$\int_C xy^2 \, dx = \int_0^{\pi/2} \overbrace{(4\cos t)}^{x}\overbrace{(16\sin^2 t)}^{y^2}\overbrace{(-4\sin t \, dt)}^{dx}$$

$$= -256\int_0^{\pi/2} \sin^3 t \cos t \, dt$$

$$= -256\left[\frac{1}{4}\sin^4 t\right]_0^{\pi/2} = -64.$$

(b) From (5),

$$\int_C xy^2 \, dy = \int_0^{\pi/2} \overbrace{(4\cos t)}^{x}\overbrace{(16\sin^2 t)}^{y^2}\overbrace{(4\cos t \, dt)}^{dy}$$

$$= 256\int_0^{\pi/2} \sin^2 t \cos^2 t \, dt \quad \leftarrow \text{use the double-angle formula for the sine}$$

$$= 256\int_0^{\pi/2} \frac{1}{4}\sin^2 2t \, dt \quad \leftarrow \text{use the half-angle formula for the sine}$$

$$= 64\int_0^{\pi/2} \frac{1}{2}(1 - \cos 4t) \, dt$$

$$= 32\left[t - \frac{1}{4}\sin 4t\right]_0^{\pi/2} = 16\pi.$$

(c) From (6),

$$\int_C xy^2 \, ds = \int_0^{\pi/2} \overbrace{(4\cos t)}^{x}\overbrace{(16\sin^2 t)}^{y^2}\overbrace{\sqrt{16(\cos^2 t + \sin^2 t)} \, dt}^{ds}$$

$$= 256\int_0^{\pi/2} \sin^2 t \cos t \, dt$$

$$= 256\left[\frac{1}{3}\sin^3 t\right]_0^{\pi/2} = \frac{256}{3}. \qquad ■$$

■ **Method of Evaluation: *C* Defined by *y* = *g*(*x*)**   If the curve *C* is defined by an explicit function $y = g(x)$, $a \leq x \leq b$, we can use x as a parameter. With $dy = g'(x) \, dx$ and $ds = \sqrt{1 + [g'(x)]^2} \, dx$, the line integrals (1), (2), and (3) become, in turn,

$$\int_C f(x, y) \, dx = \int_a^b f(x, g(x)) \, dx, \tag{7}$$

$$\int_C f(x, y) \, dy = \int_a^b f(x, g(x)) g'(x) \, dx, \tag{8}$$

$$\int_C f(x, y) \, ds = \int_a^b f(x, g(x)) \sqrt{1 + [g'(x)]^2} \, dx. \tag{9}$$

A line integral along a *piecewise-smooth* curve *C* is defined as the *sum* of the integrals over the various smooth curves whose union comprises *C*. For example, in the case of (3), if *C* is composed of smooth curves C_1 and C_2, then

$$\int_C f(x, y) \, ds = \int_{C_1} f(x, y) \, ds + \int_{C_2} f(x, y) \, ds.$$

■ **Notation**   In many applications, line integrals appear as a sum

$$\int_C P(x, y) \, dx + \int_C Q(x, y) \, dy.$$

It is common practice to write this sum without the second integral symbol as

$$\int_C P(x, y)\, dx + Q(x, y)\, dy \qquad \text{or simply} \qquad \int_C P\, dx + Q\, dy. \qquad (10)$$

EXAMPLE 2 Using (7), (8), and (10)

Evaluate $\int_C xy\, dx + x^2\, dy$, where C is given by $y = x^3$, $-1 \le x \le 2$.

Solution The curve C is illustrated in FIGURE 15.1.5 and is defined by the explicit function $y = x^3$. Hence, we can use x as the parameter. With $dy = 3x^2\, dx$, it follows from (7) and (8) that

$$\int_C xy\, dx + x^2\, dy = \int_{-1}^{2} x(\overbrace{x^3}^{y})\, dx + x^2(\overbrace{3x^2\, dx}^{dy})$$

$$= \int_{-1}^{2} 4x^4\, dx = \frac{4}{5}x^5 \Big]_{-1}^{2} = \frac{132}{5}. \qquad ■$$

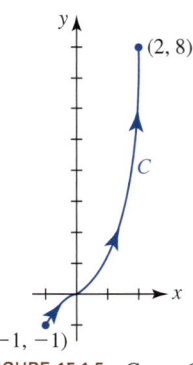

FIGURE 15.1.5 Curve C in Example 2

EXAMPLE 3 Curve C is Piecewise Defined

Evaluate $\int_C y^2\, dx - x^2\, dy$ on the closed curve C that is shown in FIGURE 15.1.6(a).

Solution Since C is piecewise smooth, we express the integral as a sum of integrals. Symbolically, we write

$$\int_C = \int_{C_1} + \int_{C_2} + \int_{C_3},$$

where C_1, C_2, and C_3 are the curves shown in Figure 15.1.6(b). On C_1, we use x as a parameter. Since $y = 0$, $dy = 0$,

$$\int_{C_1} y^2\, dx - x^2\, dy = \int_0^2 0\, dx - x^2(0) = 0.$$

On C_2, we use y as a parameter. From $x = 2$, $dx = 0$ we have

$$\int_C y^2\, dx - x^2\, dy = \int_0^4 y^2(0) - 4\, dy =$$

$$= -\int_0^4 4\, dy = -4y \Big]_0^4 = -16.$$

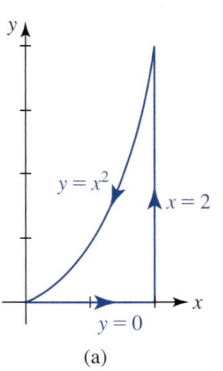

(a)

Finally, on C_3, we once again use x as a parameter. From $y = x^2$, we get $dy = 2x\, dx$ and so

$$\int_{C_3} y^2\, dx - x^2\, dy = \int_2^0 x^4\, dx - x^2(2x\, dx)$$

$$= \int_2^0 (x^4 - 2x^3)\, dx$$

$$= \left(\frac{1}{5}x^5 - \frac{1}{2}x^4 \right) \Big]_2^0 = \frac{8}{5}.$$

Hence,

$$\int_C y^2\, dx - x^2\, dy = 0 + (-16) + \frac{8}{5} = -\frac{72}{5}. \qquad ■$$

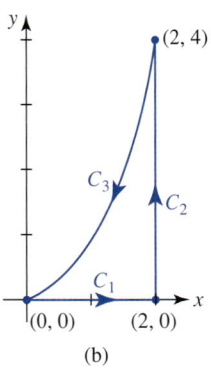

FIGURE 15.1.6 Curve C in Example 3

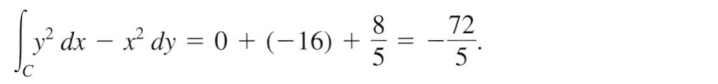

■ **Properties** It is important to note that a line integral is independent of the parameterization of the curve C provided C is given the same orientation by all sets of parametric equations defining the curve. See Problem 33 in Exercises 15.1. Recall, the positive direction of a parameterized curve C corresponds to increasing values of the parameter t.

Suppose, as shown in FIGURE 15.1.7, that the symbol $-C$ denotes the curve having the same points but the opposite orientation of C. Then it can be shown that

For ordinary definite integrals, this property is equivalent to $\int_b^a f(x)\, dx = -\int_a^b f(x)\, dx$.

$$\int_{-C} P\, dx + Q\, dy = -\int_C P\, dx + Q\, dy$$

or

$$\int_{-C} P\, dx + Q\, dy + \int_C P\, dx + Q\, dy = 0. \tag{11}$$

For example, in part (a) of Example 1 we saw that $\int_C xy^2\, dx = -64$ and so by (11) we can write $\int_{-C} xy^2\, dx = 64$.

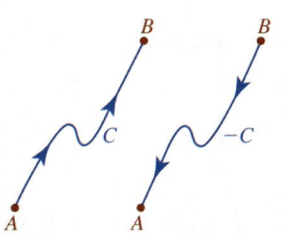

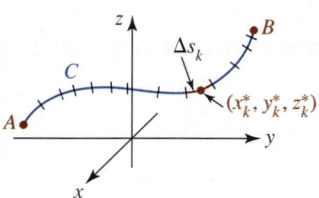

FIGURE 15.1.7 Curves C and $-C$ have opposite orientations

FIGURE 15.1.8 Sample point on the kth subarc

■ **Line Integrals in Space** Suppose C is a smooth curve in 3-space defined by the parametric equations $x = x(t), y = y(t), z = z(t), a \le t \le b$. If f is a function of three variables defined in some region of 3-space that contains C, we can define *four* line integrals along the curve:

$$\int_C f(x, y, z)\, dx, \quad \int_C f(x, y, z)\, dy, \quad \int_C f(x, y, z)\, dz, \quad \text{and} \quad \int_C f(x, y, z)\, ds.$$

The first, second, and fourth integrals are defined in a manner analogous to (1), (2), and (3) of Definition 15.1.1. For example, if C is divided into n subarcs of length Δs_k as shown in FIGURE 15.1.8, then

$$\int_C f(x, y, z)\, ds = \lim_{\|P\| \to 0} \sum_{k=1}^n f(x_k^*, y_k^*, z_k^*) \Delta s_k.$$

The new integral in the list, the **line integral of f with respect to z** along C from A to B is defined as

$$\int_C f(x, y, z)\, dz = \lim_{\|P\| \to 0} \sum_{k=1}^n f(x_k^*, y_k^*, z_k^*) \Delta z_k. \tag{12}$$

■ **Method of Evaluation** Using the parametric equations $x = x(t), y = y(t), z = z(t),$ $a \le t \le b$, we can evaluate the line integrals along the space curve C in the following manner:

$$\int_C f(x, y, z)\, dx = \int_a^b f(x(t), y(t), z(t)) x'(t)\, dt,$$

$$\int_C f(x, y, z)\, dy = \int_a^b f(x(t), y(t), z(t)) y'(t)\, dt,$$

$$\int_C f(x, y, z)\, dz = \int_a^b f(x(t), y(t), z(t)) z'(t)\, dt, \tag{13}$$

$$\int_C f(x, y, z)\, ds = \int_a^b f(x(t), y(t), z(t)) \sqrt{[x'(t)]^2 + [y'(t)]^2 + [z'(t)]^2}\, dt.$$

If C is defined by the vector function $\mathbf{r}(t) = x(t)\mathbf{i} + y(t)\mathbf{j} + z(t)\mathbf{k}$, then the last integral in (13) can be written

$$\int_C f(x, y, z)\, ds = \int_a^b f(x(t), y(t), z(t)) |\mathbf{r}'(t)|\, dt. \tag{14}$$

We will examine an integral that is analogous to (14) in Section 15.6.

As in (10), in 3-space we are often concerned with line integrals in the form of a sum:

$$\int_C P(x, y, z)\, dx + Q(x, y, z)\, dy + R(x, y, z)\, dz.$$

EXAMPLE 4 Line Integral in 3-Space

Evaluate $\int_C y\, dx + x\, dy + z\, dz$, where C is the helix $x = 2\cos t, y = 2\sin t, z = t, 0 \le t \le 2\pi$.

Solution Substituting the expressions for x, y, and z along with $dx = -2\sin t\, dt, dy = 2\cos t\, dt,$ $dz = dt$ we get

$$\int_C y\,dx + x\,dy + z\,dz = \int_0^{2\pi} \underbrace{-4\sin^2 t\,dt + 4\cos^2 t\,dt}_{4(\cos^2 t\,-\,\sin^2 t)} + t\,dt$$

$$= \int_0^{2\pi} (4\cos 2t + t)\,dt \quad \leftarrow \text{double-angle formula}$$

$$= \left(2\sin 2t + \frac{1}{2}t^2\right)\Big]_0^{2\pi} = 2\pi^2. \qquad \blacksquare$$

Exercises 15.1 Answers to selected odd-numbered problems begin on page ANS-46.

≡ Fundamentals

In Problems 1–4, evaluate $\int_C f(x, y)\,dx$, $\int_C f(x, y)\,dy$, and $\int_C f(x, y)\,ds$ on the indicated curve C.

1. $f(x, y) = 2xy$; $x = 5\cos t, y = 5\sin t, 0 \le t \le \pi/4$

2. $f(x, y) = x^3 + 2xy^2 + 2x$; $x = 2t, y = t^2, 0 \le t \le 1$

3. $f(x, y) = 3x^2 + 6y^2$; $y = 2x + 1, -1 \le x \le 0$

4. $f(x, y) = x^2/y^3$; $y = \frac{3}{2}x^{2/3}, 1 \le x \le 8$

5. Evaluate $\int_C (x^2 - y^2)\,ds$, where C is given by $x = 5\cos t$, $y = 5\sin t$, $0 \le t \le 2\pi$.

6. Evaluate $\int_C (2x + 3y)\,dy$, where C is given by $x = 3\sin 2t$, $y = 2\cos 2t$, $0 \le t \le \pi$.

In Problems 7 and 8, evaluate $\int_C f(x, y, z)\,dx$, $\int_C f(x, y, z)\,dy$, $\int_C f(x, y, z)\,dz$, and $\int_C f(x, y, z)\,ds$ on the indicated curve C.

7. $f(x, y, z) = z$; $x = \cos t, y = \sin t, z = t, 0 \le t \le \pi/2$

8. $f(x, y, z) = 4xyz$; $x = \frac{1}{3}t^3, y = t^2, z = 2t, 0 \le t \le 1$

In Problems 9–12, evaluate $\int_C (2x + y)\,dx + xy\,dy$ on the given curve C between $(-1, 2)$ and $(2, 5)$.

9. $y = x + 3$

10. $y = x^2 + 1$

11.

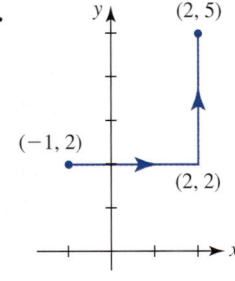

FIGURE 15.1.9 Curve in Problem 11

12.

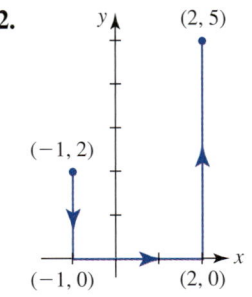

FIGURE 15.1.10 Curve in Problem 12

In Problems 13–16, evaluate $\int_C y\,dx + x\,dy$ on the given curve C between $(0, 0)$ and $(1, 1)$.

13. $y = x^2$

14. $y = x$

15. C consists of the line segments from $(0, 0)$ to $(0, 1)$ and from $(0, 1)$ to $(1, 1)$.

16. C consists of the line segments from $(0, 0)$ to $(1, 0)$ and from $(1, 0)$ to $(1, 1)$.

17. Evaluate $\int_C (6x^2 + 2y^2)\,dx + 4xy\,dy$, where C is given by $x = \sqrt{t}, y = t, 4 \le t \le 9$.

18. Evaluate $\int_C -y^2\,dx + xy\,dy$, where C is given by $x = 2t$, $y = t^3, 0 \le t \le 2$.

19. Evaluate $\int_C 2x^3y\,dx + (3x + y)\,dy$, where C is given by $x = y^2$ from $(1, -1)$ to $(1, 1)$.

20. Evaluate $\int_C 4x\,dx + 2y\,dy$, where C is given by $x = y^3 + 1$ from $(0, -1)$ to $(9, 2)$.

In Problems 21 and 22, evaluate $\int_C (x^2 + y^2)\,dx - 2xy\,dy$ on the given curve C.

21.

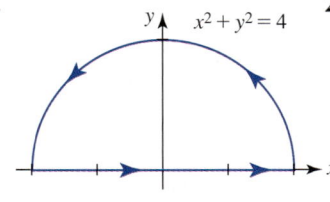

FIGURE 15.1.11 Curve in Problem 21

22.

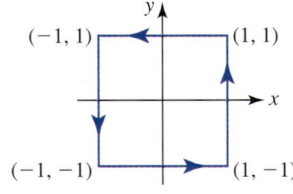

FIGURE 15.1.12 Curve in Problem 22

In Problems 23 and 24, evaluate $\int_C x^2y^3\,dx - xy^2\,dy$ on the given curve C.

23.

FIGURE 15.1.13 Curve in Problem 23

24.

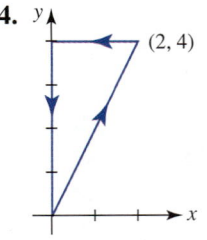

FIGURE 15.1.14 Curve in Problem 24

25. Evaluate $\int_{-C} y\,dx - x\,dy$, where C is given by $x = 2\cos t$, $y = 3\sin t$, $0 \le t \le \pi$.

26. Evaluate $\int_{-C} x^2y^3\,dx + x^3y^2\,dy$, where C is given by $y = x^4, -1 \le x \le 1$.

In Problems 27–30, evaluate $\int_C y\,dx + z\,dy + x\,dz$ on the given curve C between $(0, 0, 0)$ and $(6, 8, 5)$.

27. C consists of the line segments from $(0, 0, 0)$ to $(2, 3, 4)$ and from $(2, 3, 4)$ to $(6, 8, 5)$.

28. C defined by $\mathbf{r}(t) = 3t\mathbf{i} + t^3\mathbf{j} + \frac{5}{4}t^2\mathbf{k}, 0 \le t \le 2$

29.

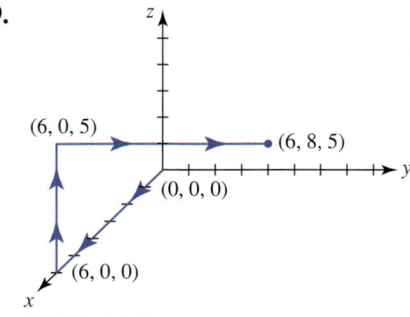

FIGURE 15.1.15   Curve in Problem 29

30.

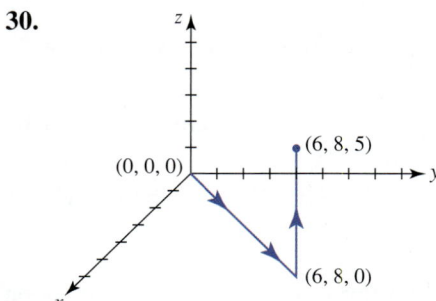

FIGURE 15.1.16   Curve in Problem 30

31. Evaluate $\int_C 10x\,dx - 2xy^2\,dy + 6xz\,dz$ where C is defined by $\mathbf{r}(t) = t\mathbf{i} + t^2\mathbf{j} + t^3\mathbf{k}, 0 \le t \le 1$.

32. Evaluate $\int_C 3x\,dx - y^2\,dy + z^2\,dz$ where $C = C_1 \cup C_2 \cup C_3$ and

C_1: the line segment from $(0, 0, 0)$ to $(1, 1, 0)$,
C_2: the line segment from $(1, 1, 0)$ to $(1, 1, 1)$,
C_3: the line segment from $(1, 1, 1)$ to $(0, 0, 0)$.

33. Verify that the line integral $\int_C y^2\,dx + xy\,dy$ has the same value on C for each of the following different parameterizations:

$$C: x = 2t + 1, y = 4t + 2, 0 \le t \le 1$$
$$C: x = t^2, y = 2t^2, 1 \le t \le \sqrt{3}$$
$$C: x = \ln t, y = 2\ln t, e \le t \le e^3.$$

34. Consider the three curves between $(0, 0)$ and $(2, 4)$.

$$C_1: x = t, y = 2t, 0 \le t \le 2$$
$$C_2: x = t, y = t^2, 0 \le t \le 2$$
$$C_3: x = 2t - 4, y = 4t - 8, 2 \le t \le 3.$$

Show that $\int_{C_1} xy\,ds = \int_{C_3} xy\,ds$, but $\int_{C_1} xy\,ds \ne \int_{C_2} xy\,ds$. Explain.

≡ Applications

35. If $\rho(x, y)$ is the density of a wire (mass per unit length), then $m = \int_C \rho(x, y)\,ds$ is the mass of the wire. Find the mass of a wire having the shape of the semicircle $x = 1 + \cos t, y = \sin t, 0 \le t \le \pi$, if the density at a point P is directly proportional to the distance from the y-axis.

36. The coordinates of the center of mass of a wire with variable density are given by

$$\bar{x} = \frac{M_y}{m}, \qquad \bar{y} = \frac{M_x}{m},$$

where

$$m = \int_C \rho(x, y)\,ds, \quad M_x = \int_C y\rho(x, y)\,ds, \quad M_y = \int_C x\rho(x, y)\,ds.$$

Find the center of mass of the wire in Problem 35.

15.2 Line Integrals of Vector Fields

▌ Introduction   The motion of wind or the flow of fluid can be described by a *velocity field* in that a vector can be assigned at each point representing the velocity of a particle at the point. See FIGURE 15.2.1(a) and (b). Notice in the velocity field superimposed on a satellite image of a hurricane in the photo in the margin that the vectors clearly show the characteristic counterclockwise rotation of winds within a low pressure area. The longer vectors near the center of

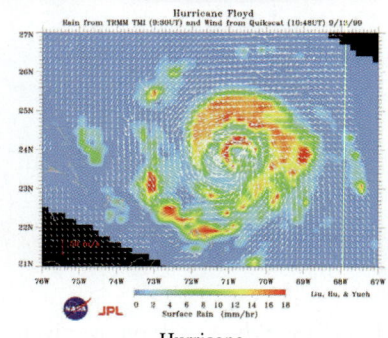

Hurricane

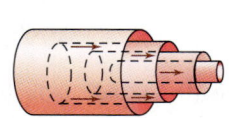

(a) Airflow around an airplane wing: $|\mathbf{v}_a| > |\mathbf{v}_b|$

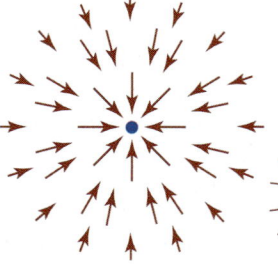

(b) Laminar flow of blood in an artery; cylindrical layers of blood flow faster near the center of the artery

(c) Inverse square force field; magnitude of the attractive force is large near the particle

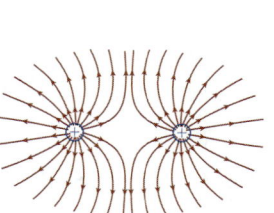

(d) Lines of force around two equal positive charges

FIGURE 15.2.1   Examples of vector fields

field indicate winds of greater velocity than those on the periphery of the field. The concept of a *force field* plays an important role in mechanics, electricity, and magnetism. See Figure 15.2.1(c) and (d). In this section we study a new vector function that describes a field of vectors, or **vector field**, in 2- or 3-space and the connection between vector fields and line integrals.

■ **Vector Fields**   A **vector field** in 2-space is a vector-valued function

$$\mathbf{F}(x, y) = P(x, y)\mathbf{i} + Q(x, y)\mathbf{j}$$

that associates a unique two-dimensional vector $\mathbf{F}(x, y)$ with each point (x, y) in a region R of the xy-plane over which the scalar component functions P and Q are defined. Similarly, a vector field in 3-space is a function

$$\mathbf{F}(x, y, z) = P(x, y, z)\mathbf{i} + Q(x, y, z)\mathbf{j} + R(x, y, z)\mathbf{k}$$

that associates a unique three-dimensional vector $\mathbf{F}(x, y, z)$ with each point (x, y, z) in a region D of 3-space with an xyz-coordinate system.

EXAMPLE 1   Vector Field in 2-Space

Graph the two-dimensional vector field $\mathbf{F}(x, y) = -y\mathbf{i} + x\mathbf{j}$.

Solution   One manner of proceeding is simply to choose points in the xy-plane and then graph the vector $\mathbf{F}$ at each point. For example, at $(1, 1)$ we would draw the vector $\mathbf{F}(1, 1) = -\mathbf{i} + \mathbf{j}$.

For the given vector field it is possible to systematically draw vectors of the same length. Observe that $|\mathbf{F}| = \sqrt{x^2 + y^2}$, and so vectors of the same length k must lie along the curve defined by $\sqrt{x^2 + y^2} = k$; that is, at any point on the circle $x^2 + y^2 = k^2$, a vector would have length k. For simplicity let us choose circles that have some points on them with integer coordinates. For example, for $k = 1$, $k = \sqrt{2}$, and $k = 2$ we have:

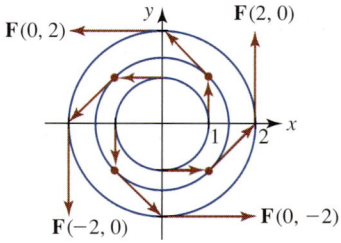

On $x^2 + y^2 = 1$: At the points $(1, 0), (0, 1), (-1, 0), (0, -1)$, the corresponding vectors $\mathbf{j}, -\mathbf{i}, -\mathbf{j}, \mathbf{i}$ have the same length 1.
On $x^2 + y^2 = 2$: At the points $(1, 1), (-1, 1), (-1, -1), (1, -1)$, the corresponding vectors $-\mathbf{i} + \mathbf{j}, -\mathbf{i} - \mathbf{j}, \mathbf{i} - \mathbf{j}, \mathbf{i} + \mathbf{j}$ have the same length $\sqrt{2}$.
On $x^2 + y^2 = 4$: At the points $(2, 0), (0, 2), (-2, 0), (0, -2)$, the corresponding vectors $2\mathbf{j}, -2\mathbf{i}, -2\mathbf{j}, 2\mathbf{i}$ have the same length 2.

The vectors at these points are shown in FIGURE 15.2.2.

FIGURE 15.2.2   Two-dimensional vector field in Example 1

In general, it is nearly impossible to sketch vector fields by hand and so we must rely on technology such as a CAS. In FIGURE 15.2.3 we have shown a computer generated version of the vector field in Example 1. Often when vectors are drawn with their correct length, the vector field becomes cluttered with overlapping vectors. See Figure 15.2.3(a). A CAS will scale the vectors in such a manner that the vectors shown have lengths proportional to their true length. See Figure 15.2.3(b). In Figure 15.2.3(c) we show the normalized version of the

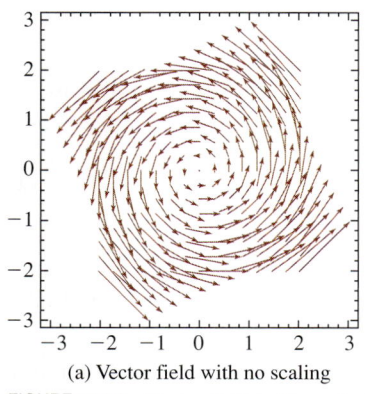

(a) Vector field with no scaling

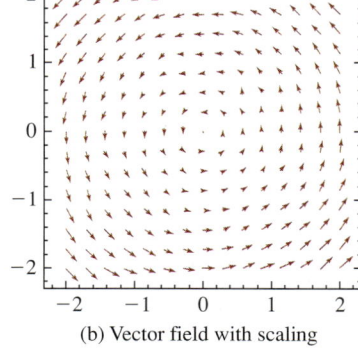

(b) Vector field with scaling

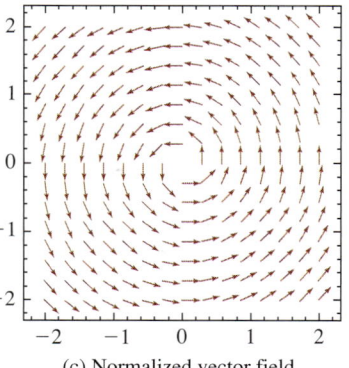

(c) Normalized vector field

FIGURE 15.2.3   Vector field in Example 1

same vector field; in other words, all the vectors have the same unit length. Note that the slight tilt in the vector field representations in Figure 15.2.3 is due to the fact that the CAS computes and plots the vector in the appropriate direction with the initial point (its tail) of the vector located at a specified point.

In **FIGURE 15.2.4** we have illustrated two vector fields in 3-space.

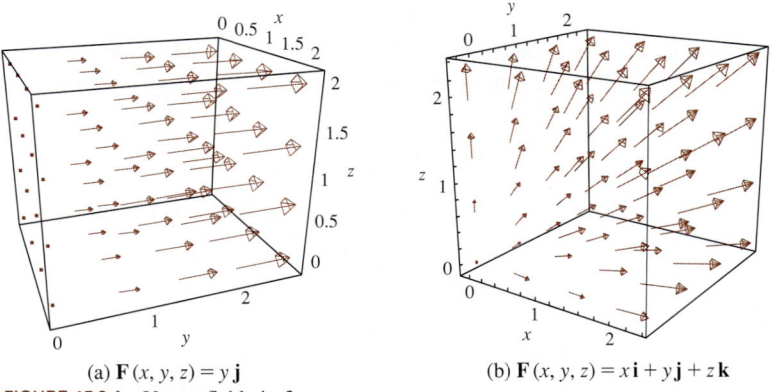

(a) $\mathbf{F}(x, y, z) = y\mathbf{j}$ (b) $\mathbf{F}(x, y, z) = x\mathbf{i} + y\mathbf{j} + z\mathbf{k}$

FIGURE 15.2.4 Vector fields in 3-space

■ **Connection with Line Integrals** We can use the concept of a vector field in 2- or 3-space to write a general line integral in a compact fashion. For example, suppose the two-dimensional vector field $\mathbf{F}(x, y) = P(x, y)\mathbf{i} + Q(x, y)\mathbf{j}$ is defined along a parametric curve $C: x = x(t)$, $y = y(t)$, $a \leq t \leq b$, and suppose the vector function $\mathbf{r}(t) = x(t)\mathbf{i} + y(t)\mathbf{j}$ is the position vector of points on C. Then the derivative of $\mathbf{r}(t)$,

$$\frac{d\mathbf{r}}{dt} = x'(t)\mathbf{i} + y'(t)\mathbf{j} = \frac{dx}{dt}\mathbf{i} + \frac{dy}{dt}\mathbf{j}$$

prompts us to define the differential of $\mathbf{r}(t)$ as

$$d\mathbf{r} = \frac{d\mathbf{r}}{dt} dt = dx\mathbf{i} + dy\mathbf{j}. \tag{1}$$

Since $\qquad\qquad \mathbf{F}(x, y) \cdot d\mathbf{r} = P(x, y)\, dx + Q(x, y)\, dy$

we can then write **a line integral of F along** C as

$$\int_C P(x, y)\, dx + Q(x, y)\, dy = \int_C \mathbf{F} \cdot d\mathbf{r}. \tag{2}$$

Similarly, for a line integral on a space curve C,

$$\int_C P(x, y, z)\, dx + Q(x, y, z)\, dy + R(x, y, z)\, dz = \int_C \mathbf{F} \cdot d\mathbf{r}, \tag{3}$$

where

$$\mathbf{F}(x, y, z) = P(x, y, z)\mathbf{i} + Q(x, y, z)\mathbf{j} + R(x, y, z)\mathbf{k} \qquad \text{and} \qquad d\mathbf{r} = dx\mathbf{i} + dy\mathbf{j} + dz\mathbf{k}.$$

If $\mathbf{r}(t) = x(t)\mathbf{i} + y(t)\mathbf{j}$, $a \leq t \leq b$, then to evaluate $\int_C \mathbf{F} \cdot d\mathbf{r}$ in (2) we define

$$\mathbf{F}(\mathbf{r}(t)) = P(x(t), y(t))\mathbf{i} + Q(x(t), y(t))\mathbf{j} \tag{4}$$

and use (1) in the form $d\mathbf{r} = \mathbf{r}'(t)\, dt$ to write

$$\int_C \mathbf{F} \cdot d\mathbf{r} = \int_a^b \mathbf{F}(\mathbf{r}(t)) \cdot \mathbf{r}'(t)\, dt. \tag{5}$$

The result in (5) extends naturally to (3) for three-dimensional vector fields defined along a space curve C given by $\mathbf{r}(t) = x(t)\mathbf{i} + y(t)\mathbf{j} + z(t)\mathbf{k}$, $a \leq t \leq b$.

EXAMPLE 2 Using (5)

Evaluate $\int_C \mathbf{F} \cdot d\mathbf{r}$ where $\mathbf{F}(x, y) = xy\mathbf{i} + y^2\mathbf{j}$ and C is defined by the vector function $\mathbf{r}(t) = e^{-t}\mathbf{i} + e^t\mathbf{j},\ -1 \le t \le 1.$

Solution From (4) we have

$$\mathbf{F}(\mathbf{r}(t)) = (e^{-t}e^t)\mathbf{i} + (e^t)^2\mathbf{j}$$
$$= \mathbf{i} + e^{2t}\mathbf{j}.$$

Since $d\mathbf{r} = \mathbf{r}'(t)\,dt = (-e^{-t}\mathbf{i} + e^t\mathbf{j})\,dt,$

$$\mathbf{F}(\mathbf{r}(t)) \cdot d\mathbf{r} = (\mathbf{i} + e^{2t}\mathbf{j}) \cdot (-e^{-t}\mathbf{i} + e^t\mathbf{j})\,dt$$
$$= (-e^{-t} + e^{3t})\,dt$$

and so from (5)

$$\int_C \mathbf{F} \cdot d\mathbf{r} = \int_{-1}^{1} \mathbf{F}(\mathbf{r}(t)) \cdot \mathbf{r}'(t)\,dt = \int_{-1}^{1} (-e^{-t} + e^{3t})\,dt$$

$$= \left(e^{-t} + \frac{1}{3}e^{3t}\right)\Big]_{-1}^{1}$$

$$= \left(e^{-1} + \frac{1}{3}e^3\right) - \left(e + \frac{1}{3}e^{-3}\right)$$

$$\approx 4.3282.$$

The vector field $\mathbf{F}$ and curve C are shown in FIGURE 15.2.5. ■

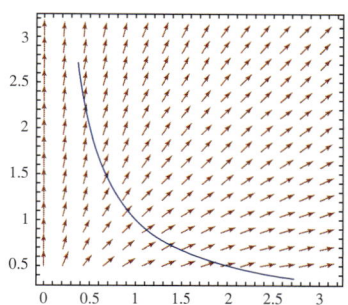

FIGURE 15.2.5 Curve and vector field in Example 2

▌ Work In Section 11.3 we saw that the work W done by a constant force $\mathbf{F}$ that causes a straight-line displacement $\mathbf{d}$ of an object is $W = \mathbf{F} \cdot \mathbf{d}$. In Section 6.8 it was shown that the work done in moving an object from $x = a$ to $x = b$ by a force $F(x)$ that varies in magnitude but not in direction is given by the definite integral $W = \int_a^b F(x)\,dx$. In general, a force field $\mathbf{F}(x, y) = P(x, y)\mathbf{i} + Q(x, y)\mathbf{j}$ acting at each point on a smooth curve $C: x = x(t),$ $y = y(t),\ a \le t \le b$, varies in both magnitude and direction. See FIGURE 15.2.6(a). If A and B are the points $(x(a), y(a))$ and $(x(b), y(b))$, respectively, we ask:

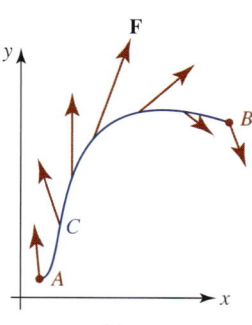

• *What is the work done by* $\mathbf{F}$ *as its point of application moves along* C *from* A *to* B?

To answer this question, suppose C is divided into n subarcs of lengths Δs_k and that (x_k^*, y_k^*) is a sample point on the kth subarc. On each subarc $\mathbf{F}(x_k^*, y_k^*)$ is a constant force. If, as shown in Figure 15.2.6(b), the length of the vector

$$\Delta\mathbf{r}_k = (x_k - x_{k-1})\mathbf{i} + (y_k - y_{k-1})\mathbf{j} = \Delta x_k\mathbf{i} + \Delta y_k\mathbf{j}$$

is an approximation to the length of the kth subarc, then the approximate work done by $\mathbf{F}$ over the subarc is

$$\big(|\mathbf{F}(x_k^*, y_k^*)|\cos\theta\big)\,|\Delta\mathbf{r}_k| = \mathbf{F}(x_k^*, y_k^*) \cdot \Delta\mathbf{r}_k$$
$$= P(x_k^*, y_k^*)\,\Delta x_k + Q(x_k^*, y_k^*)\,\Delta y_k.$$

By summing these elements of work and passing to the limit, we can naturally define the work done by $\mathbf{F}$ as the line integral of $\mathbf{F}$ along C:

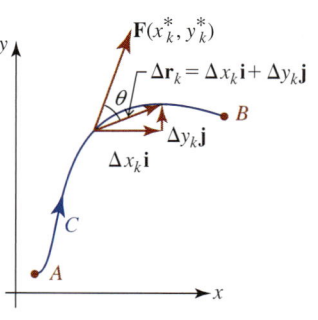

FIGURE 15.2.6 Variable force vector $\mathbf{F}$ acting along C

$$W = \int_C P(x, y)\,dx + Q(x, y)\,dy \qquad \text{or} \qquad W = \int_C \mathbf{F} \cdot d\mathbf{r}. \qquad (6)$$

In the case of a force field that acts at points on a space curve, work $\int_C \mathbf{F} \cdot d\mathbf{r}$ is defined as in (3).

Now, since

$$\frac{d\mathbf{r}}{dt} = \frac{d\mathbf{r}}{ds}\frac{ds}{dt}$$

we let $d\mathbf{r} = \mathbf{T}\, ds$, where $\mathbf{T} = d\mathbf{r}/ds$ is, as we saw in Section 12.1, a unit tangent to C. Hence,

$$W = \int_C \mathbf{F} \cdot d\mathbf{r} = \int_C \mathbf{F} \cdot \mathbf{T}\, ds = \int_C \text{comp}_\mathbf{T} \mathbf{F}\, ds. \qquad (7)$$

In other words,

- *The work done by a force $\mathbf{F}$ along a curve C is due entirely to the tangential component of $\mathbf{F}$.*

EXAMPLE 3 Work

Find the work done by
$\quad$ (a) $\mathbf{F} = x\mathbf{i} + y\mathbf{j}$ and (b) $\mathbf{F} = \frac{3}{4}\mathbf{i} + \frac{1}{2}\mathbf{j}$
along the curve C traced out by $\mathbf{r}(t) = \cos t\,\mathbf{i} + \sin t\,\mathbf{j}$ from $t = 0$ to $t = \pi$.

Solution

(a) The vector function $\mathbf{r}(t)$ gives the parametric equations $x = \cos t$, $y = \sin t$, $0 \le t \le \pi$, that we recognize as a half-circle. As seen in FIGURE 15.2.7, the force field $\mathbf{F}$ is perpendicular to C at every point. (See Problem 1 in Exercises 15.2.) Since the tangential components of $\mathbf{F}$ are zero, we expect the work done along C to be zero. To see this we use (5):

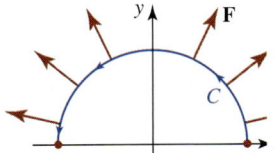

FIGURE 15.2.7 Force vector $\mathbf{F}$ acting along C in part (a) of Example 3

$$W = \int_C \mathbf{F} \cdot d\mathbf{r} = \int_C \mathbf{F}(\mathbf{r}(t)) \cdot \mathbf{r}'(t)\, dt$$

$$= \int_0^\pi (\cos t\,\mathbf{i} + \sin t\,\mathbf{j}) \cdot (-\sin t\,\mathbf{i} + \cos t\,\mathbf{j})\, dt$$

$$= \int_0^\pi (-\cos t \sin t + \sin t \cos t)\, dt = 0.$$

(b) In FIGURE 15.2.8 the vectors in gold are the projections of $\mathbf{F}$ on the unit tangent vectors. The work done by $\mathbf{F}$ is

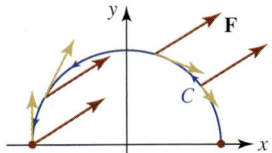

FIGURE 15.2.8 Force vector $\mathbf{F}$ acting along C in part (b) of Example 3

$$W = \int_C \mathbf{F} \cdot d\mathbf{r} = \int_C \left(\frac{3}{4}\mathbf{i} + \frac{1}{2}\mathbf{j}\right) \cdot \mathbf{r}'(t)\, dt$$

$$= \int_0^\pi \left(\frac{3}{4}\mathbf{i} + \frac{1}{2}\mathbf{j}\right) \cdot (-\sin t\,\mathbf{i} + \cos t\,\mathbf{j})\, dt$$

$$= \int_0^\pi \left(-\frac{3}{4}\sin t + \frac{1}{2}\cos t\right) dt$$

$$= \left(\frac{3}{4}\cos t + \frac{1}{2}\sin t\right)\Bigg]_0^\pi = -\frac{3}{2}.$$

The units of work depend on the units of $|\mathbf{F}|$ and on the units of distance. ∎

■ Circulation A line integral of a vector field $\mathbf{F}$ along a simple closed curve C is said to be the **circulation** of $\mathbf{F}$ around C; that is,

$$\text{circulation} = \oint_C \mathbf{F} \cdot d\mathbf{r} = \oint_C \mathbf{F} \cdot \mathbf{T}\, ds. \qquad (8)$$

In particular, if $\mathbf{F}$ is the velocity field of a fluid, then the circulation (8) is a measure of the amount by which the fluid tends to turn the curve C by rotating, or circulating, around it. For example, if $\mathbf{F}$ is perpendicular to $\mathbf{T}$ for every (x, y) on C, then $\oint_C \mathbf{F} \cdot \mathbf{T}\, ds = 0$ and the curve does not move at all. On the other hand, $\oint_C \mathbf{F} \cdot \mathbf{T}\, ds > 0$ and $\oint_C \mathbf{F} \cdot \mathbf{T}\, ds < 0$ mean that the fluid tends to rotate C in the counterclockwise and clockwise directions, respectively. See FIGURE 15.2.9.

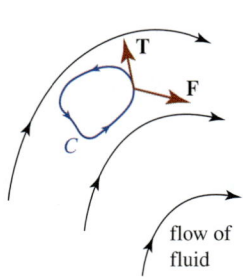

FIGURE 15.2.9 Closed curve in a velocity field

■ Gradient Vector Fields Associated with a function f of two or three variables there is a vector field. For a function of two variables $f(x, y)$, the gradient

$$\nabla f(x, y) = f_x(x, y)\mathbf{i} + f_y(x, y)\mathbf{j} \qquad (9)$$

defines a two-dimensional vector field called the **gradient field** of f. For a function of three variables $f(x, y, z)$, the three-dimensional gradient field of f is defined as

$$\nabla f(x, y, z) = f_x(x, y, z)\mathbf{i} + f_y(x, y, z)\mathbf{j} + f_z(x, y, z)\mathbf{k}. \tag{10}$$

EXAMPLE 4   Gradient Field

Find the gradient field of $f(x, y) = x^2 - y^2$.

Solution   By definition, the gradient field of f is

$$\nabla f(x, y) = \frac{\partial f}{\partial x}\mathbf{i} + \frac{\partial f}{\partial y}\mathbf{j} = 2x\mathbf{i} - 2y\mathbf{j}. \qquad \blacksquare$$

Recall from Section 13.1 curves defined by $f(x, y) = c$, for suitable c, are called the **level curves** of f. In Example 5, the level curves of f are the family of hyperbolas $x^2 - y^2 = c$, where c is a constant. With the aid of a CAS, we have superimposed in FIGURE 15.2.10 a sampling of the level curves $x^2 - y^2 = c$ (blue) and vectors in the gradient field $\nabla f(x, y) = 2x\mathbf{i} - 2y\mathbf{j}$ (red). For greater visual emphasis we have chosen to plot all the vectors in the field so that their lengths are the same. Each vector in the gradient field $\nabla f(x, y) = 2x\mathbf{i} - 2y\mathbf{j}$ is perpendicular to some level curve. In other words, if the tail or initial point of a vector coincides with a point (x, y) on a level curve, then that vector is perpendicular to the level curve at (x, y).

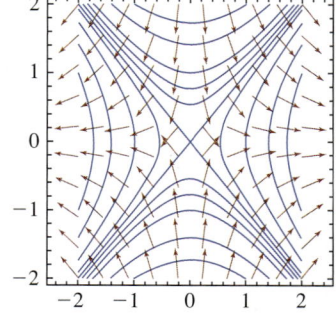

FIGURE 15.2.10   Level curves of f and gradient field of f in Example 4

■ **Conservative Vector Fields**   A vector field $\mathbf{F}$ is said to be **conservative** if $\mathbf{F}$ can be written as a gradient of a scalar function ϕ. In other words, $\mathbf{F}$ is conservative if there exists a function ϕ such that $\mathbf{F} = \nabla\phi$. The function ϕ is called a **potential function** for $\mathbf{F}$.

EXAMPLE 5   Conservative Vector Field

Show that the two-dimensional vector field $\mathbf{F}(x, y) = y\mathbf{i} + x\mathbf{j}$ is conservative.

Solution   Consider the function $\phi(x, y) = xy$. The gradient of the scalar function ϕ is

$$\nabla\phi = \frac{\partial\phi}{\partial x}\mathbf{i} + \frac{\partial\phi}{\partial y}\mathbf{j} = y\mathbf{i} + x\mathbf{j}.$$

Because $\nabla\phi = \mathbf{F}(x, y)$ we conclude that $\mathbf{F}(x, y) = y\mathbf{i} + x\mathbf{j}$ is a conservative vector field and that ϕ is a potential function for $\mathbf{F}$. The vector field is given in FIGURE 15.2.11. ■

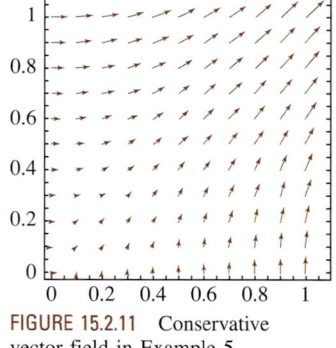

FIGURE 15.2.11   Conservative vector field in Example 5

Of course, not every vector field is a conservative field although many vector fields encountered in physics are conservative. (See Problem 51 in Exercises 15.2.) For present purposes, the importance of conservative vector fields will become evident in the next section as we continue our study of line integrals.

Exercises 15.2   Answers to selected odd-numbered problems begin on page ANS-46.

≡ Fundamentals

In Problems 1–6, graph some representative vectors in the given vector field.

1. $\mathbf{F}(x, y) = x\mathbf{i} + y\mathbf{j}$ **2.** $\mathbf{F}(x, y) = -x\mathbf{i} + y\mathbf{j}$

3. $\mathbf{F}(x, y) = y\mathbf{i} + x\mathbf{j}$ **4.** $\mathbf{F}(x, y) = x\mathbf{i} + 2y\mathbf{j}$

5. $\mathbf{F}(x, y) = y\mathbf{j}$ **6.** $\mathbf{F}(x, y) = x\mathbf{j}$

In Problems 7–10, match the given figure with one of the vector fields in (a)–(d).

    (a) $\mathbf{F}(x, y) = -3\mathbf{i} + 2\mathbf{j}$ **(b)** $\mathbf{F}(x, y) = 3\mathbf{i} + 2\mathbf{j}$

    (c) $\mathbf{F}(x, y) = 3\mathbf{i} - 2\mathbf{j}$ **(d)** $\mathbf{F}(x, y) = -3\mathbf{i} - 2\mathbf{j}$

7.

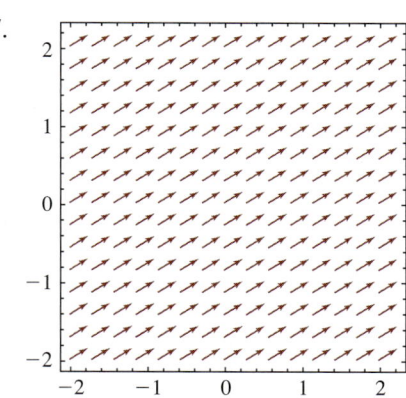

FIGURE 15.2.12   Vector field for Problem 7

8.

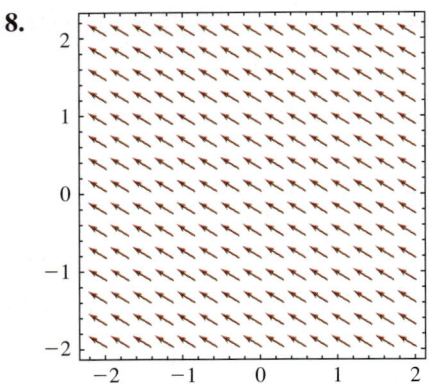

FIGURE 15.2.13 Vector field for Problem 8

9.

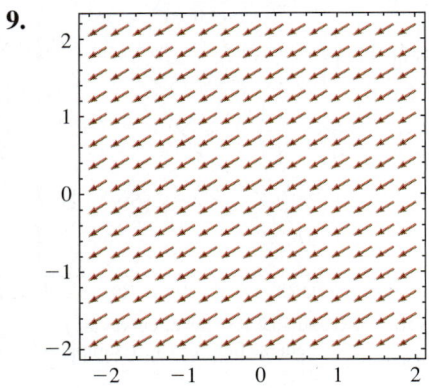

FIGURE 15.2.14 Vector field for Problem 9

10.

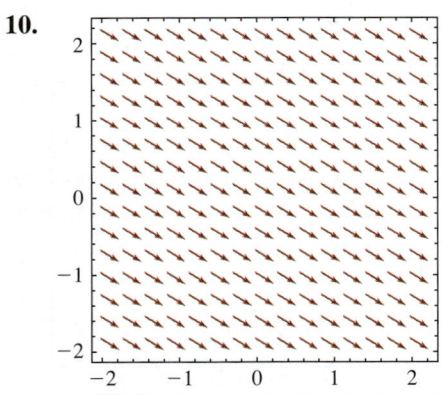

FIGURE 15.2.15 Vector field for Problem 10

In Problems 11–14, match the given figure with one of the vector fields in (a)–(d).

(a) $\mathbf{F}(x, y, z) = x\mathbf{i} + y\mathbf{j} + z\mathbf{k}$
(b) $\mathbf{F}(x, y, z) = -z\mathbf{k}$
(c) $\mathbf{F}(x, y, z) = \mathbf{i} + \mathbf{j} + z\mathbf{k}$
(d) $\mathbf{F}(x, y, z) = x\mathbf{i} + \mathbf{j} + \mathbf{k}$

11.

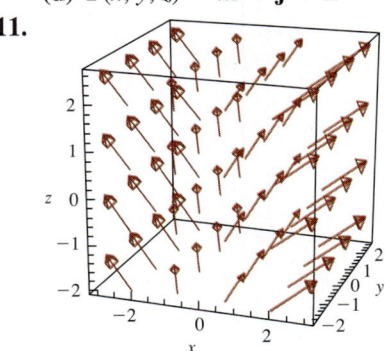

FIGURE 15.2.16 Vector field for Problem 11

12.

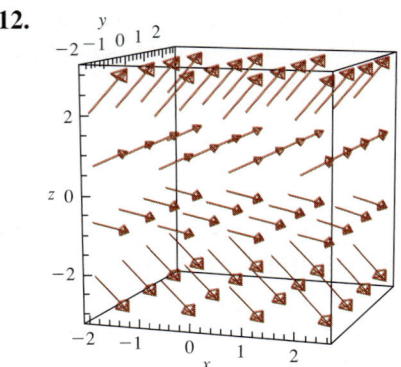

FIGURE 15.2.17 Vector field for Problem 12

13.

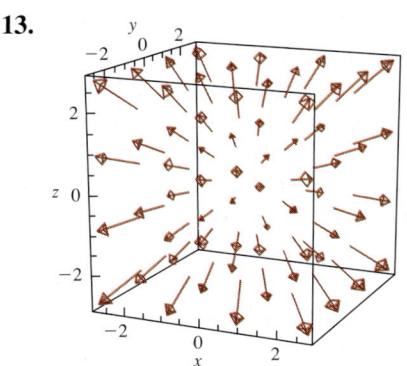

FIGURE 15.2.18 Vector field for Problem 13

14.

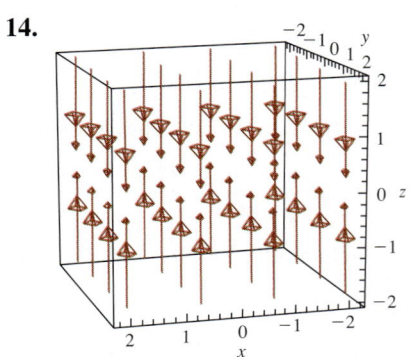

FIGURE 15.2.19 Vector field for Problem 14

In Problems 15–20, evaluate the line integral $\int_C \mathbf{F} \cdot d\mathbf{r}$.

15. $\mathbf{F}(x, y) = y^3\mathbf{i} - x^2y\mathbf{j}$; $\mathbf{r}(t) = e^{-2t}\mathbf{i} + e^t\mathbf{j}$, $0 \le t \le \ln 2$

16. $\mathbf{F}(x, y) = 2xy\mathbf{i} + x^2\mathbf{j}$; $\mathbf{r}(t) = t\mathbf{i} + t^2\mathbf{j}$, $0 \le t \le 2$

17. $\mathbf{F}(x, y) = 2x\mathbf{i} + 2y\mathbf{j}$; $\mathbf{r}(t) = (2t - 1)\mathbf{i} + (6t + 1)\mathbf{j}$, $-1 \le t \le 1$

18. $\mathbf{F}(x, y) = x^2\mathbf{i} + y\mathbf{j}$; $\mathbf{r}(t) = \cos t\mathbf{i} + \sin t\mathbf{j}$, $0 \le t \le \pi/6$

19. $\mathbf{F}(x, y, z) = -y\mathbf{i} + x\mathbf{j} + 2z\mathbf{k}$; $\mathbf{r}(t) = 2\cos t\mathbf{i} + 3\sin t\mathbf{j} + 3t\mathbf{k}, 0 \le t \le \pi$

20. $\mathbf{F}(x, y, z) = e^x\mathbf{i} + xe^{xy}\mathbf{j} + xye^{xyz}\mathbf{k}$; $\mathbf{r}(t) = t\mathbf{i} + t^2\mathbf{j} + t^3\mathbf{k}$, $0 \le t \le 1$

21. Find the work done by the force $\mathbf{F}(x, y) = y\mathbf{i} + x\mathbf{j}$ acting along $y = \ln x$ from $(1, 0)$ to $(e, 1)$.

22. Find the work done by the force $\mathbf{F}(x, y) = 2xy\mathbf{i} + 4y^2\mathbf{j}$ acting along the piecewise-smooth curve consisting of the line segments from $(-2, 2)$ to $(0, 0)$ and from $(0, 0)$ to $(2, 3)$.

23. Find the work done by the force $\mathbf{F}(x, y) = (x + 2y)\mathbf{i} + (6y - 2x)\mathbf{j}$ acting counterclockwise once around the triangle with vertices $(1, 1)$, $(3, 1)$, and $(3, 2)$.

24. Find the work done by the force $\mathbf{F}(x, y, z) = yz\mathbf{i} + xz\mathbf{j} + xy\mathbf{k}$ acting along the curve given by $\mathbf{r}(t) = t^3\mathbf{i} + t^2\mathbf{j} + t\mathbf{k}$ from $t = 1$ to $t = 3$.

25. Find the work done by a constant force $\mathbf{F}(x, y) = a\mathbf{i} + b\mathbf{j}$ acting counterclockwise once around the circle $x^2 + y^2 = 9$.

26. In an inverse square force field $\mathbf{F} = c\mathbf{r}/|\mathbf{r}|^3$, where c is a constant and $\mathbf{r} = x\mathbf{i} + y\mathbf{j} + z\mathbf{k}$, find the work done in moving a particle along the line from $(1, 1, 1)$ to $(3, 3, 3)$.

27. For the gradient vector field obtained in Example 4, find the work done by the force $\mathbf{F} = \nabla f$ acting along $\mathbf{r}(t) = 5\cos t\mathbf{i} + 5\sin t\mathbf{j}, 0 \le t \le 2\pi$.

28. For the conservative vector field in Example 5, find the work done by the force $\mathbf{F}$ acting along $\mathbf{r}(t) = 2\sin t\mathbf{i} + 10\cos t\mathbf{j}, 0 \le t \le 2\pi$.

29. A force field $\mathbf{F}(x, y)$ acts at each point on a curve C, which is the union of C_1, C_2, and C_3 shown in FIGURE 15.2.20. $|\mathbf{F}|$ is measured in pounds and distance is measured in feet using the scale given in the figure. Use the representative vectors shown to approximate the work done by $\mathbf{F}$ along C. [*Hint*: Use $W = \int_C \mathbf{F} \cdot \mathbf{T}\, ds$.]

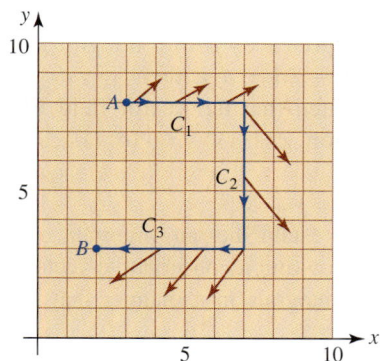

FIGURE 15.2.20 Curve C and force field $\mathbf{F}$ in Problem 29

30. Assume a smooth curve C is described by the vector function $\mathbf{r}(t)$ for $a \le t \le b$. Let acceleration, velocity, and speed be given by $\mathbf{a} = d\mathbf{v}/dt, \mathbf{v} = d\mathbf{r}/dt$, and $v = |\mathbf{v}|$, respectively. Using Newton's second law $\mathbf{F} = m\mathbf{a}$, show that, in the absence of friction, the work done by $\mathbf{F}$ in moving a particle of constant mass m from point A at $t = a$ to point B at $t = b$ is the same as the change in kinetic energy:

$$K(B) - K(A) = \frac{1}{2}m[v(b)]^2 - \frac{1}{2}m[v(a)]^2.$$

$$\left[Hint: \text{Consider } \frac{d}{dt}v^2 = \frac{d}{dt}\mathbf{v} \cdot \mathbf{v}. \right]$$

In Problems 31–36, find the gradient field of the given function f.

31. $f(x, y) = \frac{1}{6}(3x - 6y)^2$ **32.** $f(x, y) = x - y + 2x\cos 5xy$

33. $f(x, y, z) = x\tan^{-1}yz$ **34.** $f(x, y, z) = x - x^2yz^4$

35. $f(x, y, z) = y + z - xe^{-y^2}$

36. $f(x, y, z) = \ln(x^2 + 2y^4 + 3z^6)$

In Problems 37–40, match the given conservative vector field $\mathbf{F}$ with one of the potential functions in (a)–(d).

(a) $\phi(x, y) = \frac{1}{2}x^2 + \frac{1}{3}y^3 - 5$ (b) $\phi(x, y) = x^2 + \frac{1}{2}y^2$

(c) $\phi(x, y) = \frac{1}{2}x^2 + y^2 - 4$ (d) $\phi(x, y) = 2x + \frac{1}{2}y^2 + 1$

37. $\mathbf{F}(x, y) = 2x\mathbf{i} + y\mathbf{j}$ **38.** $\mathbf{F}(x, y) = x\mathbf{i} + 2y\mathbf{j}$

39. $\mathbf{F}(x, y) = 2\mathbf{i} + y\mathbf{j}$ **40.** $\mathbf{F}(x, y) = x\mathbf{i} + y^2\mathbf{j}$

In Problems 41–44, the given vector field is conservative. By trial and error, find a potential function ϕ for $\mathbf{F}$.

41. $\mathbf{F}(x, y) = \cos x\mathbf{i} + (1 - \sin y)\mathbf{j}$

42. $\mathbf{F}(x, y) = e^{-y}\mathbf{i} - xe^{-y}\mathbf{j}$

43. $\mathbf{F}(x, y, z) = \mathbf{i} + 2y\mathbf{j} - 12z^2\mathbf{k}$

44. $\mathbf{F}(x, y, z) = y^2z^3\mathbf{i} + 2xyz^3\mathbf{j} + 3xy^2z^2\mathbf{k}$

≡ Calculator/CAS Problems

In Problems 45–50, use a CAS to superimpose the plots of the gradient field of f and the level curves of f on the same set of coordinate axes.

45. $f(x, y) = x + 3y$ **46.** $f(x, y) = x - y^2$

47. $f(x, y) = \sin x\sin y$ **48.** $f(x, y) = \sin x + \sin y$

49. $f(x, y) = e^{-x}\cos y$ **50.** $f(x, y) = \cos(x + y)$

≡ Think About It

51. Every inverse square force field $\mathbf{F} = c\mathbf{r}/|\mathbf{r}|^3$, where c is a constant and $\mathbf{r} = x\mathbf{i} + y\mathbf{j} + z\mathbf{k}$, is conservative. Prove this by finding a potential function $\phi(x, y, z)$ for $\mathbf{F}$.

52. Can two different functions f and g have the same gradient field?

15.3 Independence of the Path

▌ **Introduction** In this section we will refer to a piecewise-smooth curve C between an initial point A and a terminal point B as a **path** or **path of integration**. The value of a line integral $\int_C \mathbf{F} \cdot d\mathbf{r}$ usually depends on the path of integration. In other words, if C_1 and C_2 are two different paths between the same points A and B, then in general we expect that $\int_{C_1} \mathbf{F} \cdot d\mathbf{r} \ne \int_{C_2} \mathbf{F} \cdot d\mathbf{r}$. However, there are some very important exceptions. The notion of a conservative vector field $\mathbf{F}$ plays an important role in the discussion that follows. You are urged to review this concept in Section 15.2.

Note: To avoid needless repetition we assume throughout that **F** is a continuous vector field in some region of 2- or 3-space, its component functions have continuous first partial derivatives in the region, and that the path C lies entirely in the region.

EXAMPLE 1   Path Independence

The integral $\int_C y\,dx + x\,dy$ has the same value on each path C between $(0, 0)$ and $(1, 1)$ shown in FIGURE 15.3.1. You may recall from Problems 13–16 of Exercises 15.1 that on these paths

$$\int_C y\,dx + x\,dy = 1.$$

You are also urged to verify $\int_C y\,dx + x\,dy = 1$ on the curves $y = x^3$, $y = x^4$, and $y = \sqrt{x}$ between $(0, 0)$ and between $(1, 1)$. The relevance of all this is to suggest that the integral $\int_C y\,dx + x\,dy$ does not depend on the path joining these two points. We continue this discussion in Example 2.

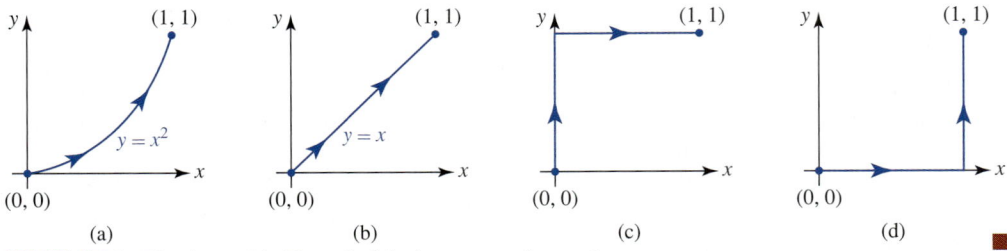

FIGURE 15.3.1   Line integral in Example 1 is the same on four paths

The integral in Example 1 can be interpreted as a line integral of a vector field **F** along a path C. If $\mathbf{F}(x, y) = y\mathbf{i} + x\mathbf{j}$ and $d\mathbf{r} = dx\mathbf{i} + dy\mathbf{j}$, then $\int_C y\,dx + x\,dy = \int_C \mathbf{F} \cdot d\mathbf{r}$. In Example 5 of Section 15.2 we demonstrated that the vector field $\mathbf{F}(x, y) = y\mathbf{i} + x\mathbf{j}$ is conservative by finding a potential function $\phi(x, y) = xy$ for **F**. Recall, this means $\mathbf{F}(x, y) = \nabla\phi$.

▌ A Fundamental Theorem   The next theorem establishes an important relationship between the value of a line integral over a path that lies within a conservative vector field. In addition, it provides a means of evaluating these line integrals in a manner that is analogous to the Fundamental Theorem of Calculus:

$$\int_a^b f'(x)\,dx = f(b) - f(a), \tag{1}$$

where $f(x)$ is an antiderivative of $f'(x)$. In the next theorem, known as the **Fundamental Theorem for Line Integrals**, the gradient of a scalar function ϕ,

$$\nabla\phi = \frac{\partial\phi}{\partial x}\mathbf{i} + \frac{\partial\phi}{\partial y}\mathbf{j},$$

plays the part of the derivative $f'(x)$ in (1).

Theorem 15.3.1   Fundamental Theorem

Suppose C is a path in an open region R of the xy-plane given by $\mathbf{r}(t) = x(t)\mathbf{i} + y(t)\mathbf{j}$, $a \le t \le b$. If $\mathbf{F}(x, y) = P(x, y)\mathbf{i} + Q(x, y)\mathbf{j}$ is a conservative vector field in R and ϕ is a potential function for **F**, then

$$\int_C \mathbf{F} \cdot d\mathbf{r} = \int_C \nabla\phi \cdot d\mathbf{r} = \phi(B) - \phi(A), \tag{2}$$

where $A = (x(a), y(a))$ and $B = (x(b), y(b))$.

PROOF   We will prove the theorem for a smooth path C. Since ϕ is a potential function for $\mathbf{F}$ we have

$$\mathbf{F} = \nabla\phi = \frac{\partial\phi}{\partial x}\mathbf{i} + \frac{\partial\phi}{\partial y}\mathbf{j}.$$

Then using $\mathbf{r}'(t) = (dx/dt)\mathbf{i} + (dy/dt)\mathbf{j}$ we can write the line integral of $\mathbf{F}$ along the path C as

$$\int_C \mathbf{F} \cdot d\mathbf{r} = \int_C \mathbf{F} \cdot \mathbf{r}'(t)\, dt$$

$$= \int_a^b \left(\frac{\partial\phi}{\partial x}\frac{dx}{dt} + \frac{\partial\phi}{\partial y}\frac{dy}{dt} \right) dt.$$

In view of the Chain Rule (Theorem 13.5.1),

$$\frac{d\phi}{dt} = \frac{\partial\phi}{\partial x}\frac{dx}{dt} + \frac{\partial\phi}{\partial y}\frac{dy}{dt}$$

and so it follows that

$$\int_C \mathbf{F} \cdot d\mathbf{r} = \int_a^b \frac{d\phi}{dt}\, dt$$

$$= \phi(x(t), y(t)) \Big]_a^b$$

$$= \phi(x(b), y(b)) - \phi(x(a), y(a))$$

$$= \phi(B) - \phi(A). \qquad \blacksquare$$

For piecewise-smooth curves, the foregoing proof must be modified by considering each smooth arc of the curve C.

■ **Path Independence**   If the value of a line integral is the same for *every* path in a region connecting the initial point A and terminal point B, then the integral is said to be **independent of the path**. In other words, a line integral $\int_C \mathbf{F} \cdot d\mathbf{r}$ of $\mathbf{F}$ along C is independent of the path if $\int_{C_1} \mathbf{F} \cdot d\mathbf{r} = \int_{C_2} \mathbf{F} \cdot d\mathbf{r}$ for any two paths C_1 and C_2 between A and B. Theorem 15.3.1 shows that if $\mathbf{F}$ is a conservative vector field in an open region in 2- or 3-space, then $\int_C \mathbf{F} \cdot d\mathbf{r}$ depends only on the initial and terminal points A and B of the path C, and not on C itself. In other words, line integrals of conservative vector fields are independent of the path. Such integrals are often written

$$\int_A^B \mathbf{F} \cdot d\mathbf{r} = \int_A^B \nabla\phi \cdot d\mathbf{r}. \qquad (3)$$

EXAMPLE 2   Example 1 Revisited

Evaluate $\int_C y\, dx + x\, dy$, where C is a path with initial point $(0, 0)$ and terminal point $(1, 1)$.

Solution   The path C shown in FIGURE 15.3.2 represents any piecewise-smooth curve with initial and terminal points $(0, 0)$ and $(1, 1)$. We have seen several times that $\mathbf{F} = y\mathbf{i} + x\mathbf{j}$ is a conservative vector field defined at each point of the xy-plane and that $\phi(x, y) = xy$ is a potential function for $\mathbf{F}$. Thus, in view of (2) of Theorem 15.3.1 and (3), we can write

$$\int_C y\, dx + x\, dy = \int_{(0,0)}^{(1,1)} \mathbf{F} \cdot d\mathbf{r} = \int_{(0,0)}^{(1,1)} \nabla\phi \cdot d\mathbf{r}$$

$$= xy \Big]_{(0,0)}^{(1,1)}$$

$$= 1 \cdot 1 - 0 \cdot 0 = 1. \qquad \blacksquare$$

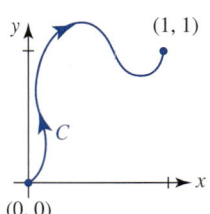

FIGURE 15.3.2   Piecewise-smooth curve in Example 2

In using the Fundamental Theorem of Calculus (1), *any* antiderivative of $f'(x)$ can be used, such as $f(x) + K$, where K is a constant. Similarly, a potential function for the vector field in Example 2 is $\phi(x, y) = xy + K$ where K is a constant. We may disregard this constant when using (2) of Theorem 15.3.1 since

$$\int_A^B \mathbf{F} \cdot d\mathbf{r} = (\phi(B) + K) - (\phi(A) + K) = \phi(B) - \phi(A).$$

Before proceeding we need to consider some special regions in the plane.

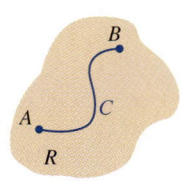

(a) Connected region R

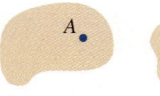

(b) R is not connected

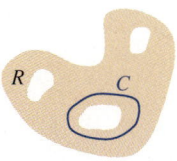

(c) Multiply-connected region R

FIGURE 15.3.3 Regions in the plane

■ **Some Terminology** We say that a region (in the plane or in space) is **connected** if every pair of points A and B in the region can be joined by a piecewise-smooth curve that lies entirely in the region. A region R in the plane is **simply connected** if it is connected and every simple closed curve C lying entirely within the region can be shrunk, or contracted, to a point without leaving R. The last condition means that if C is any simple closed curve lying entirely in R, then the region in the interior of C also lies entirely in R. Roughly put, a simply connected region has no holes in it. The region R in FIGURE 15.3.3(a) is a simply connected region. In Figure 15.3.3(b) the region R shown is not connected, or **disconnected**, since A and B cannot be joined by a piecewise-smooth curve C that lies in R. The region in Figure 15.3.3(c) is connected but not simply connected because it has three holes in it. The representative curve C in the figure surrounds one of the holes, and so cannot be shrunk to a point without leaving the region. This last region is said to be **multiply connected**.

In an open connected region R, the notions of a path independence and a conservative vector field are equivalent. This means: If **F** is conservative in R, then $\int_C \mathbf{F} \cdot d\mathbf{r}$ is independent of the path C, and conversely, if $\int_C \mathbf{F} \cdot d\mathbf{r}$ is independent of the path, then **F** is conservative.

We state this formally in the next theorem.

Theorem 15.3.2 Equivalent Concepts

In an open connected region R, $\int_C \mathbf{F} \cdot d\mathbf{r}$ is independent of the path C if and only if the vector field **F** is conservative in R.

PROOF If **F** is conservative in R, then we have already seen that $\int_C \mathbf{F} \cdot d\mathbf{r}$ is independent of the path C as a consequence of Theorem 15.3.1.

For convenience we prove the converse for a region R in the plane. Assume that $\int_C \mathbf{F} \cdot d\mathbf{r}$ is independent of the path in R and that (x_0, y_0) and (x, y) are arbitrary points in the region R. Let the function $\phi(x, y)$ be defined as

$$\phi(x, y) = \int_{(x_0, y_0)}^{(x, y)} \mathbf{F} \cdot d\mathbf{r},$$

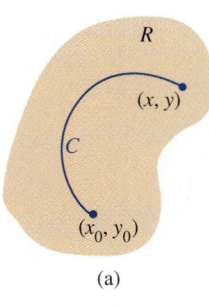

(a)

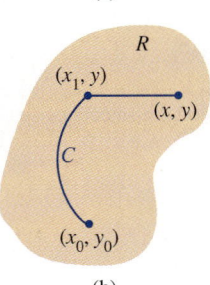

(b)

FIGURE 15.3.4 Region R in the proof of Theorem 15.3.2

where C is an arbitrary path in R from (x_0, y_0) to (x, y) and $\mathbf{F} = P\mathbf{i} + Q\mathbf{j}$. See FIGURE 15.3.4(a). Now choose a point (x_1, y), $x_1 \neq x$, so that the line segment from (x_1, y) to (x, y) is in R. See Figure 15.3.4(b). Then by path independence we can write

$$\phi(x, y) = \int_{(x_0, y_0)}^{(x_1, y)} \mathbf{F} \cdot d\mathbf{r} + \int_{(x_1, y)}^{(x, y)} \mathbf{F} \cdot d\mathbf{r}.$$

Now,

$$\frac{\partial \phi}{\partial x} = 0 + \frac{\partial}{\partial x} \int_{(x_1, y)}^{(x, y)} P\, dx + Q\, dy$$

since the first integral does not depend on x. But on the line segment between (x_1, y) and (x, y), y is constant so that $dy = 0$. Hence, $\int_{(x_1, y)}^{(x, y)} P\, dx + Q\, dy = \int_{(x_1, y)}^{(x, y)} P\, dx$. By the derivative form of the Fundamental Theorem of Calculus (Theorem 5.5.2) we then have

$$\frac{\partial \phi}{\partial x} = \frac{\partial}{\partial x} \int_{(x_1, y)}^{(x, y)} P(x, y)\, dx = P(x, y).$$

Likewise we can show that $\partial\phi/\partial y = Q(x, y)$. Hence, from

$$\nabla\phi = \frac{\partial\phi}{\partial x}\mathbf{i} + \frac{\partial\phi}{\partial y}\mathbf{j} = P\mathbf{i} + Q\mathbf{j} = \mathbf{F}(x, y)$$

we conclude that $\mathbf{F}$ is conservative.   ∎

■ Integrals Around Closed Paths   Recall from Section 15.1 that a path, or curve, C is said to be closed when its initial point A is the same as the terminal point B. If C is a parametric curve defined by a vector function $\mathbf{r}(t)$, $a \le t \le b$, then C is **closed** when $A = B$, that is, $\mathbf{r}(a) = \mathbf{r}(b)$. The next theorem is an immediate consequence of Theorem 15.3.1.

Theorem 15.3.3   Equivalent Concepts

In an open connected region R, $\int_C \mathbf{F} \cdot d\mathbf{r}$ is independent of the path if and only if $\int_C \mathbf{F} \cdot d\mathbf{r} = 0$ for every closed path C in R.

PROOF   First we show that if $\int_C \mathbf{F} \cdot d\mathbf{r}$ is independent of the path, then $\int_C \mathbf{F} \cdot d\mathbf{r} = 0$ for every closed path C in R. To see this let us suppose A and B are any two points on C and that $C = C_1 \cup C_2$, where C_1 is a path from A to B and C_2 is a path from B to A. See FIGURE 15.3.5(a). Then

$$\int_C \mathbf{F} \cdot d\mathbf{r} = \int_{C_1} \mathbf{F} \cdot d\mathbf{r} + \int_{C_2} \mathbf{F} \cdot d\mathbf{r} = \int_{C_1} \mathbf{F} \cdot d\mathbf{r} - \int_{-C_2} \mathbf{F} \cdot d\mathbf{r}, \qquad (4)$$

where $-C_2$ is now a path from A to B. Because of path independence, $\int_{C_1} \mathbf{F} \cdot d\mathbf{r} = \int_{-C_2} \mathbf{F} \cdot d\mathbf{r}$. Thus, (4) implies that $\int_C \mathbf{F} \cdot d\mathbf{r} = 0$.

Next, we prove the converse that if $\int_C \mathbf{F} \cdot d\mathbf{r} = 0$ for every closed path C in R, then $\int_C \mathbf{F} \cdot d\mathbf{r}$ is independent of the path. Let C_1 and C_2 represent any two paths from A to B and so $C = C_1 \cup (-C_2)$ is a closed path. See Figure 15.3.5(b). It follows from $\int_C \mathbf{F} \cdot d\mathbf{r} = 0$ or

$$0 = \int_C \mathbf{F} \cdot d\mathbf{r} = \int_{C_1} \mathbf{F} \cdot d\mathbf{r} + \int_{-C_2} \mathbf{F} \cdot d\mathbf{r} = \int_{C_1} \mathbf{F} \cdot d\mathbf{r} - \int_{C_2} \mathbf{F} \cdot d\mathbf{r}$$

that $\int_{C_1} \mathbf{F} \cdot d\mathbf{r} = \int_{C_2} \mathbf{F} \cdot d\mathbf{r}$. Hence, $\int_C \mathbf{F} \cdot d\mathbf{r}$ is independent of the path.   ∎

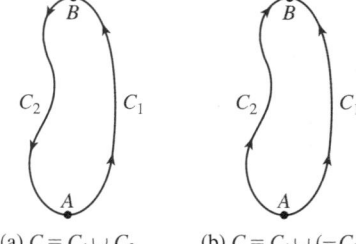

(a) $C = C_1 \cup C_2$ (b) $C = C_1 \cup (-C_2)$

FIGURE 15.3.5   Paths in the Proof of Theorem 15.3.3

Suppose $\mathbf{F}$ is a conservative vector field defined over an open connected region and C is a closed path lying entirely in the region. When the results of the preceding theorems are put together we conclude that

$$\mathbf{F} \text{ conservative} \Leftrightarrow \text{path independence} \Leftrightarrow \int_C \mathbf{F} \cdot d\mathbf{r} = 0. \qquad (5)$$

The symbol $\Leftrightarrow$ in (5) is read "equivalent to" or "if and only if."

■ Test for a Conservative Field   The implications in (5) show that if the line integral $\int_C \mathbf{F} \cdot d\mathbf{r}$ is not path independent, then the vector field is not conservative. But there is an easier way of determining whether $\mathbf{F}$ is conservative. The following theorem is a test for a conservative vector field that uses the partial derivatives of the component functions of $\mathbf{F} = P\mathbf{i} + Q\mathbf{j}$.

Theorem 15.3.4   Test for a Conservative Field

Suppose $\mathbf{F}(x, y) = P(x, y)\mathbf{i} + Q(x, y)\mathbf{j}$ is a conservative vector field in an open region R, and that P and Q are continuous and have continuous first partial derivatives in R. Then

$$\frac{\partial P}{\partial y} = \frac{\partial Q}{\partial x} \qquad (6)$$

for all (x, y) in R. Conversely, if the equality (6) holds for all (x, y) in a simply-connected region R, then $\mathbf{F} = P\mathbf{i} + Q\mathbf{j}$ is conservative in R.

PARTIAL PROOF We prove the first half of the theorem. Assume that the component functions of the conservative vector field $\mathbf{F} = P\mathbf{i} + Q\mathbf{j}$ are continuous and have continuous first partial derivatives in an open region R. Since $\mathbf{F}$ is conservative there exists a potential function ϕ such that

$$\mathbf{F} = P\mathbf{i} + Q\mathbf{j} = \nabla\phi = \frac{\partial\phi}{\partial x}\mathbf{i} + \frac{\partial\phi}{\partial y}\mathbf{j}.$$

Thus, $P = \partial\phi/\partial x$ and $Q = \partial\phi/\partial y$. Now

$$\frac{\partial P}{\partial y} = \frac{\partial}{\partial y}\left(\frac{\partial\phi}{\partial x}\right) = \frac{\partial^2\phi}{\partial y\,\partial x} \qquad \text{and} \qquad \frac{\partial Q}{\partial x} = \frac{\partial}{\partial x}\left(\frac{\partial\phi}{\partial y}\right) = \frac{\partial^2\phi}{\partial x\,\partial y}.$$

From Theorem 13.3.1 the second-order mixed partial derivatives are equal and so $\partial P/\partial y = \partial Q/\partial x$ as was to be shown. ∎

EXAMPLE 3   Using Theorem 15.3.4

The conservative vector field $\mathbf{F}(x, y) = y\mathbf{i} + x\mathbf{j}$ in Example 2 is continuous and has component functions that have continuous first partial derivatives throughout the open region R consisting of the entire xy-plane. With the identifications $P = y$ and $Q = x$ it follows from (6) of Theorem 15.3.4,

$$\frac{\partial P}{\partial y} = 1 = \frac{\partial Q}{\partial x}.$$

EXAMPLE 4   Using Theorem 15.3.4

Determine whether the vector field $\mathbf{F}(x, y) = (x^2 - 2y^3)\mathbf{i} + (x + 5y)\mathbf{j}$ is conservative.

Solution   With $P = x^2 - 2y^3$ and $Q = x + 5y$, we find

$$\frac{\partial P}{\partial y} = -6y^2 \qquad \text{and} \qquad \frac{\partial Q}{\partial x} = 1.$$

Because $\partial P/\partial y \neq \partial Q/\partial x$ for all points in the plane, it follows from Theorem 15.3.4 that $\mathbf{F}$ is not conservative. ∎

EXAMPLE 5   Using Theorem 15.3.4

Determine whether the vector field $\mathbf{F}(x, y) = -ye^{-xy}\mathbf{i} - xe^{-xy}\mathbf{j}$ is conservative.

Solution   With $P = -ye^{-xy}$ and $Q = -xe^{-xy}$, we find

$$\frac{\partial P}{\partial y} = xye^{-xy} - e^{-xy} = \frac{\partial Q}{\partial x}.$$

The components of $\mathbf{F}$ are continuous and have continuous partial derivatives. Thus, (6) holds throughout the xy-plane which is a simply-connected region. From the converse in Theorem 15.3.4 we conclude that $\mathbf{F}$ is conservative. ∎

We have one more important question to answer in this section:

- *If $\mathbf{F}$ is a conservative vector field, how does one find a potential function ϕ for $\mathbf{F}$?*   (7)

Part (b) of the next example uses partial integration. A review of Section 14.2 is recommended.

▶ In the next example we give the answer to the question posed in (7).

EXAMPLE 6   Integral That is Path Independent

(a) Show that $\int_C \mathbf{F} \cdot d\mathbf{r}$, where $\mathbf{F}(x, y) = (y^2 - 6xy + 6)\mathbf{i} + (2xy - 3x^2 - 2y)\mathbf{j}$, is independent of the path C between $(-1, 0)$ and $(3, 4)$.

(b) Find a potential function ϕ for $\mathbf{F}$.

(c) Evaluate $\displaystyle\int_{(-1,0)}^{(3,4)} \mathbf{F} \cdot d\mathbf{r}$.

Solution

(a) Identifying $P = y^2 - 6xy + 6$ and $Q = 2xy - 3x^2 - 2y$ yields

$$\frac{\partial P}{\partial y} = 2y - 6x = \frac{\partial Q}{\partial x}.$$

The vector field **F** is conservative because (6) holds throughout the xy-plane and as a consequence the integral $\int_C \mathbf{F} \cdot d\mathbf{r}$ is independent of the path between any two points A and B in the plane.

(b) Because **F** is conservative there is a potential function ϕ such that

$$\frac{\partial \phi}{\partial x} = y^2 - 6xy + 6 \qquad \text{and} \qquad \frac{\partial \phi}{\partial y} = 2xy - 3x^2 - 2y. \tag{8}$$

Employing partial integration on the first expression in (8) gives

$$\phi = \int (y^2 - 6xy + 6)\, dx = xy^2 - 3x^2y + 6x + g(y), \tag{9}$$

where $g(y)$ is the "constant" of integration. Now we take the partial derivative of (9) with respect to y and equate it to the second expression in (8):

$$\frac{\partial \phi}{\partial y} = 2xy - 3x^2 + g'(y) = 2xy - 3x^2 - 2y.$$

From the last equality we find $g'(y) = -2y$. Integrating again gives $g(y) = -y^2 + C$, where C is a constant. Thus,

$$\phi = xy^2 - 3x^2y + 6x - y^2 + C. \tag{10}$$

(c) We can now use Theorem 15.3.2 and the potential function (10) (without the constant):

$$\int_C \mathbf{F} \cdot d\mathbf{r} = \int_{(-1,0)}^{(3,4)} \mathbf{F} \cdot d\mathbf{r} = (xy^2 - 3x^2y + 6x - y^2)\Big]_{(-1,0)}^{(3,4)}$$
$$= (48 - 108 + 18 - 16) - (-6) = -52. \qquad \blacksquare$$

Note: Since the integral in Example 6 was shown to be independent of the path in part (a), we can evaluate it without finding a potential function. We can integrate along any convenient curve connecting the given points. In particular, the line $y = x + 1$ is such a curve. Using x as a parameter then gives

$$\int_C \mathbf{F} \cdot d\mathbf{r} = \int_C (y^2 - 6xy + 6)\, dx + (2xy - 3x^2 - 2y)\, dy$$

$$= \int_{-1}^{3} [(x + 1)^2 - 6x(x + 1) + 6]\, dx + [2x(x + 1) - 3x^2 - 2(x + 1)]\, dx$$

$$= \int_{-1}^{3} (-6x^2 - 4x + 5)\, dx = -52.$$

■ **Conservative Vector Fields in 3-Space**   For a three-dimensional conservative vector field

$$\mathbf{F}(x, y, z) = P(x, y, z)\mathbf{i} + Q(x, y, z)\mathbf{j} + R(x, y, z)\mathbf{k}$$

and a piecewise-smooth space curve $\mathbf{r}(t) = x(t)\mathbf{i} + y(t)\mathbf{j} + z(t)\mathbf{k}$, $a \le t \le b$, the basic form of (2) is the same:

$$\int_C \mathbf{F} \cdot d\mathbf{r} = \int_C \nabla \phi \cdot d\mathbf{r}$$
$$= \phi(B) - \phi(A) = \phi(x(b), y(b), z(b)) - \phi(x(a), y(a), z(a)). \tag{11}$$

If C is a space curve, a line integral $\int_C \mathbf{F} \cdot d\mathbf{r}$ is independent of the path whenever the three-dimensional vector field

$$\mathbf{F}(x, y, z) = P(x, y, z)\mathbf{i} + Q(x, y, z)\mathbf{j} + R(x, y, z)\mathbf{k}$$

is conservative. The three-dimensional analogue of Theorem 15.3.4 goes like this. If **F** is conservative and P, Q, and R are continuous and have continuous first partial derivatives in some open region of 3-space, then

$$\frac{\partial P}{\partial y} = \frac{\partial Q}{\partial x}, \quad \frac{\partial P}{\partial z} = \frac{\partial R}{\partial x}, \quad \frac{\partial Q}{\partial z} = \frac{\partial R}{\partial y}. \tag{12}$$

Conversely, if (12) holds throughout an appropriate region of 3-space, then **F** is conservative.

EXAMPLE 7 Integral That is Path Independent

(a) Show that the line integral

$$\int_C (y + yz)\, dx + (x + 3z^3 + xz)\, dy + (9yz^2 + xy - 1)\, dz$$

is independent of the path C between (1, 1, 1) and (2, 1, 4).

(b) Evaluate $\displaystyle\int_{(1,1,1)}^{(2,1,4)} \mathbf{F} \cdot d\mathbf{r}$.

Solution

(a) With the identifications

$$\mathbf{F}(x, y, z) = (y + yz)\mathbf{i} + (x + 3z^3 + xz)\mathbf{j} + (9yz^2 + xy - 1)\mathbf{k},$$

$$P = y + yz, \qquad Q = x + 3z^2 + xz, \qquad \text{and} \qquad R = 9yz^2 + xy - 1,$$

we see that the equalities

$$\frac{\partial P}{\partial y} = 1 + z = \frac{\partial Q}{\partial x}, \qquad \frac{\partial P}{\partial z} = y = \frac{\partial R}{\partial x}, \qquad \text{and} \qquad \frac{\partial Q}{\partial z} = 9z^2 + x = \frac{\partial R}{\partial y}$$

hold throughout 3-space. From (12) we conclude that **F** is conservative and therefore, the integral is independent of the path.

(b) The path C shown in FIGURE 15.3.6 represents any path with initial and terminal points (1, 1, 1) and (2, 1, 4). To evaluate the integral we again illustrate how to find a potential function $\phi(x, y, z)$ for **F** using partial integration.

First we know that

$$\frac{\partial \phi}{\partial x} = P, \qquad \frac{\partial \phi}{\partial y} = Q, \qquad \text{and} \qquad \frac{\partial \phi}{\partial z} = R.$$

Integrating the first of these three equations with respect to x gives

$$\phi = xy + xyz + g(y, z).$$

The derivative of this last expression with respect to y must then be equal to Q:

$$\frac{\partial \phi}{\partial y} = x + xz + \frac{\partial g}{\partial y} = x + 3z^3 + xz.$$

Hence,

$$\frac{\partial g}{\partial y} = 3z^3 \qquad \text{implies} \qquad g = 3yz^3 + h(z).$$

Consequently, $\qquad\qquad \phi = xy + xyz + 3yz^3 + h(z).$

The partial derivative of this last expression with respect to z must now be equal to the function R:

$$\frac{\partial \phi}{\partial z} = xy + 9yz^2 + h'(z) = 9yz^2 + xy - 1.$$

From this we get $h'(z) = -1$ and $h(z) = -z + K$. Disregarding K, we can write

$$\phi = xy + xyz + 3yz^3 - z. \tag{13}$$

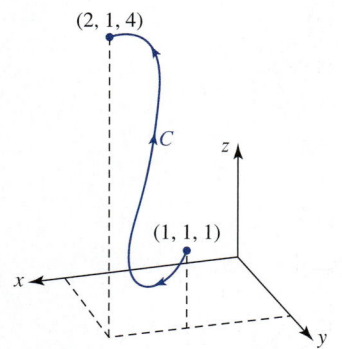

FIGURE 15.3.6 Representative path C in Example 7

Finally, from (2) and the potential function (13) we obtain

$$\int_{(1,1,1)}^{(2,1,4)} (y + yz)\, dx + (x + 3z^3 + xz)\, dy + (9yz^2 + xy - 1)\, dz$$

$$= (xy + xyz + 3yz^3 - z)\Big]_{(1,1,1)}^{(2,1,4)} = 198 - 4 = 194.\qquad\blacksquare$$

▌Conservation of Energy In a conservative force field **F**, the work done by the force on a particle moving from position A to position B is the same for all paths between these points. Moreover, the work done by the force along a closed path is *zero*. See Problem 31 in Exercises 15.3. For this reason, such a force field is also said to be **conservative**. In a conservative field **F** the *law of conservation of mechanical energy* holds: For a particle moving along a path in a conservative field,

$$kinetic\ energy + potential\ energy = constant.$$

See Problem 37 in Exercises 15.3.

$\int_C$ **NOTES FROM THE CLASSROOM**
....

A frictional force such as air resistance is *nonconservative*. Nonconservative forces are *dissipative* in that their action reduces kinetic energy without a corresponding increase in potential energy. Stated in another way, if the work done $\int_C \mathbf{F} \cdot d\mathbf{r}$ depends on the path, then **F** is nonconservative.

Exercises 15.3   Answers to selected odd-numbered problems begin on page ANS-46.

≡ **Fundamentals**

In Problems 1–10, show that the given integral is independent of the path. Evaluate in two ways:
(a) find a potential function ϕ, and
(b) integrate along any convenient path between the points.

1. $\displaystyle\int_{(0,0)}^{(2,2)} x^2\, dx + y^2\, dy$ **2.** $\displaystyle\int_{(1,1)}^{(2,4)} 2xy\, dx + x^2\, dy$

3. $\displaystyle\int_{(1,0)}^{(3,2)} (x + 2y)\, dx + (2x - y)\, dy$

4. $\displaystyle\int_{(0,0)}^{(\pi/2,\,0)} \cos x \cos y\, dx + (1 - \sin x \sin y)\, dy$

5. $\displaystyle\int_{(4,1)}^{(4,4)} \frac{-y\, dx + x\, dy}{y^2}$ on any path not crossing the x-axis

6. $\displaystyle\int_{(1,0)}^{(3,4)} \frac{x\, dx + y\, dy}{\sqrt{x^2 + y^2}}$ on any path not through the origin

7. $\displaystyle\int_{(1,2)}^{(3,6)} \mathbf{F} \cdot d\mathbf{r}, \ \mathbf{F} = (2y^2 x - 3)\mathbf{i} + (2yx^2 + 4)\mathbf{j}$

8. $\displaystyle\int_{(-1,1)}^{(0,0)} \mathbf{F} \cdot d\mathbf{r}, \ \mathbf{F} = (5x + 4y)\mathbf{i} + (4x - 8y^3)\mathbf{j}$

9. $\displaystyle\int_{(0,0)}^{(2,8)} \mathbf{F} \cdot d\mathbf{r}, \ \mathbf{F} = (y^3 + 3x^2 y)\mathbf{i} + (x^3 + 3y^2 x + 1)\mathbf{j}$

10. $\displaystyle\int_{(-2,0)}^{(1,0)} \mathbf{F} \cdot d\mathbf{r}, \ \mathbf{F} = (2x - y\sin xy - 5y^4)\mathbf{i} - (20xy^3 + x\sin xy)\mathbf{j}$

In Problems 11–18, determine whether the given vector field is a conservative field. If so, find the potential function ϕ for **F**.

11. $\mathbf{F}(x, y) = (4x^3 y^3 + 3)\mathbf{i} + (3x^4 y^2 + 1)\mathbf{j}$

12. $\mathbf{F}(x, y) = 2xy^3\mathbf{i} + 3y^2(x^2 + 1)\mathbf{j}$

13. $\mathbf{F}(x, y) = y^2 \cos xy^2\mathbf{i} - 2xy\sin xy^2\mathbf{j}$

14. $\mathbf{F}(x, y) = (x^2 + y^2 + 1)^{-2}(x\mathbf{i} + y\mathbf{j})$

15. $\mathbf{F}(x, y) = (x^3 + y)\mathbf{i} + (x + y^3)\mathbf{j}$

16. $\mathbf{F}(x, y) = 2e^{2y}\mathbf{i} + xe^{2y}\mathbf{j}$

17. $\mathbf{F}(x, y, z) = 2x\mathbf{i} + (3y^2 - z)\mathbf{j} - y\mathbf{k}$

18. $\mathbf{F}(x, y, z) = 2xy\mathbf{i} + (x^2 - ze^{-y})\mathbf{j} + (e^{-y} - 1)\mathbf{k}$.

In Problems 19 and 20, find the work done by the force $\mathbf{F}(x, y) = (2x + e^{-y})\mathbf{i} + (4y - xe^{-y})\mathbf{j}$ along the indicated curve.

19.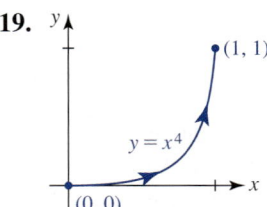

FIGURE 15.3.7 Curve in Problem 19

20.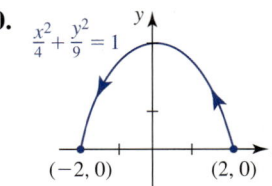

FIGURE 15.3.8 Curve in Problem 20

In Problems 21–26, show that the given integral is independent of the path. Evaluate.

21. $\displaystyle\int_{(1,1,1)}^{(2,4,8)} yz\, dx + xz\, dy + xy\, dz$

22. $\displaystyle\int_{(0,0,0)}^{(1,1,1)} 2x\,dx + 3y^2\,dy + 4z^3\,dz$

23. $\displaystyle\int_{(1,0,0)}^{(2,\pi/2,1)} (2x\sin y + e^{3z})\,dx + x^2\cos y\,dy + (3xe^{3z} + 5)\,dz$

24. $\displaystyle\int_{(1,2,1)}^{(3,4,1)} (2x + 1)\,dx + 3y^2\,dy + \frac{1}{z}\,dz$

25. $\displaystyle\int_{(1,1,\ln 3)}^{(2,2,\ln 3)} \mathbf{F}\cdot d\mathbf{r};\ \ \mathbf{F} = e^{2z}\mathbf{i} + 3y^2\mathbf{j} + 2xe^{2z}\mathbf{k}$

26. $\displaystyle\int_{(-2,3,1)}^{(0,0,0)} \mathbf{F}\cdot d\mathbf{r};\ \ \mathbf{F} = 2xz\mathbf{i} + 2yz\mathbf{j} + (x^2 + y^2)\mathbf{k}$

In Problems 27 and 28 evaluate $\int_C \mathbf{F}\cdot d\mathbf{r}$.

27. $\mathbf{F}(x, y, z) = (y - yz\sin x)\mathbf{i} + (x + z\cos x)\mathbf{j} + y\cos x\mathbf{k}$;

$\mathbf{r}(t) = 2t\mathbf{i} + (1 + \cos t)^2\mathbf{j} + 4\sin^3 t\mathbf{k},\ 0 \le t \le \pi/2$

28. $\mathbf{F}(x, y, z) = (2 - e^z)\mathbf{i} + (2y - 1)\mathbf{j} + (2 - xe^z)\mathbf{k}$;
$\mathbf{r}(t) = t\mathbf{i} + t^2\mathbf{j} + t^3\mathbf{k},\ (-1, 1, -1)$ to $(2, 4, 8)$

≡ Applications

29. The inverse square law of gravitational attraction between two masses m_1 and m_2 is given by $\mathbf{F} = -Gm_1m_2\mathbf{r}/|\mathbf{r}|^3$, where $\mathbf{r} = x\mathbf{i} + y\mathbf{j} + z\mathbf{k}$. Show that $\mathbf{F}$ is conservative. Find a potential function for $\mathbf{F}$.

30. Find the work done by the force $\mathbf{F}(x, y, z) = 8xy^3z\mathbf{i} + 12x^2y^2z\mathbf{j} + 4x^2y^3\mathbf{k}$ acting along the helix $\mathbf{r}(t) = 2\cos t\mathbf{i} + 2\sin t\mathbf{j} + t\mathbf{k}$ from $(2, 0, 0)$ to $(1, \sqrt{3}, \pi/3)$. From $(2, 0, 0)$ to $(0, 2, \pi/2)$. [*Hint*: Show that $\mathbf{F}$ is conservative.]

31. If $\mathbf{F}$ is a conservative force field, show that the work done along any simple closed path is zero.

32. A particle in the plane is attracted to the origin with a force $\mathbf{F} = |\mathbf{r}|^n\mathbf{r}$, where n is a positive integer and $\mathbf{r} = x\mathbf{i} + y\mathbf{j}$ is the position vector of the particle. Show that $\mathbf{F}$ is conservative. Find the work done in moving the particle between (x_1, y_1) and (x_2, y_2).

≡ Think About It

In Problems 33 and 34, show that the given vector field $\mathbf{F}$ is conservative. Evaluate the line integral $\int_C \mathbf{F}\cdot d\mathbf{r}$ without finding a potential function for $\mathbf{F}$.

33. $\mathbf{F}(x, y) = 2x\cos y\mathbf{i} - x^2\sin y\mathbf{j}$; C is $\mathbf{r}(t) = 2^{t-1}\mathbf{i} + \sin(\pi/t)\mathbf{j}$, $1 \le t \le 2$

34. $\mathbf{F}(x, y, z) = \sin y\mathbf{i} + x\cos y\mathbf{j} + z^2\mathbf{k}$;
C is $\mathbf{r}(t) = \sqrt{t}\mathbf{i} + t^4\mathbf{j} + te^{\sqrt{1-t}}\mathbf{k}$, $0 \le t \le 1$

35. Suppose C_1 and C_2 are two paths in an open simply-connected region that have the same initial and terminal points. If $\int_{C_1} \mathbf{F}\cdot d\mathbf{r} = \frac{3}{4}$ and $\int_{C_2} \mathbf{F}\cdot d\mathbf{r} = \frac{11}{14}$, what does this say about the vector field $\mathbf{F}$?

36. Consider the vector field

$$\mathbf{F} = -\frac{y}{x^2 + y^2}\mathbf{i} + \frac{x}{x^2 + y^2}\mathbf{j}.$$

(a) Show that $\dfrac{\partial P}{\partial y} = \dfrac{\partial Q}{\partial x}$, but demonstrate that $\mathbf{F}$ is not conservative. [*Hint*: Evaluate $\int_C \mathbf{F}\cdot d\mathbf{r}$, where $\mathbf{r}(t) = \cos t\mathbf{i} + \sin t\mathbf{j}$, $0 \le t \le 2\pi$.]

(b) Explain why this does not violate Theorem 15.3.4.

37. Suppose $\mathbf{F}$ is a conservative force field with potential function ϕ. In physics the function $p = -\phi$ is called *potential energy*. Since $\mathbf{F} = -\nabla p$, Newton's second law becomes

$$m\mathbf{r}'' = -\nabla p \qquad \text{or} \qquad m\frac{d\mathbf{v}}{dt} + \nabla p = \mathbf{0}.$$

By integrating $m\dfrac{d\mathbf{v}}{dt}\cdot\dfrac{d\mathbf{r}}{dt} + \nabla p\cdot\dfrac{d\mathbf{r}}{dt} = 0$ with respect to t, derive the law of conservation of mechanical energy: $\frac{1}{2}mv^2 + p = $ constant. [*Hint*: See Problem 30 in Exercises 15.2.]

38. Suppose C is a smooth curve between points A (at $t = a$) and B (at $t = b$) and that p is potential energy, defined in Problem 37. If $\mathbf{F}$ is a conservative force field and $K = \frac{1}{2}mv^2$ is kinetic energy, show that $p(B) + K(B) = p(A) + K(A)$.

15.4 Green's Theorem

▌ Introduction In this section we examine one of the most important theorems in vector integral calculus. We will see that this theorem relates a line integral around a piecewise-smooth simple *closed* curve with a double integral over the region bounded by the curve. We recommend that you review the terminology on page 802 of Section 15.1 and page 818 of Section 15.3.

▌ Line Integrals on Simple Closed Curves Suppose C is a piecewise-smooth simple closed curve that forms the boundary of a simply connected region R. We say the **positive orientation** around C is that direction a point on the curve must move, or the direction a person must walk, to complete a single traversal of C while keeping the region R to the left. See FIGURE 15.4.1(a). As shown in Figures 15.4.1(b) and 15.4.1(c), the *positive* and *negative* orientations correspond to *counterclockwise* and *clockwise* traversals of C, respectively.

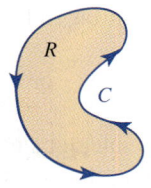 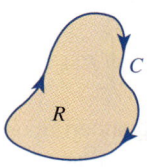

(a) Positive orientation   (b) Positive orientation   (c) Negative orientation
FIGURE 15.4.1 Orientations of simple closed curves

The next theorem is called **Green's Theorem**.

Theorem 15.4.1   Green's Theorem

Suppose that C is a piecewise-smooth simple closed curve with a positive orientation that bounds a simply connected region R. If P, Q, $\partial P/\partial y$, and $\partial Q/\partial x$ are continuous on R, then

$$\int_C P(x, y)\, dx + Q(x, y)\, dy = \iint_R \left(\frac{\partial Q}{\partial x} - \frac{\partial P}{\partial y} \right) dA. \tag{1}$$

PARTIAL PROOF   We shall prove (1) only for a region R that is simultaneously of Type I and Type II:

$$R:\quad g_1(x) \le y \le g_2(x), \quad a \le x \le b$$
$$R:\quad h_1(y) \le x \le h_2(y), \quad c \le y \le d.$$

Using FIGURE 15.4.2(a), we have

$$-\iint_R \frac{\partial P}{\partial y}\, dA = -\int_a^b \int_{g_1(x)}^{g_2(x)} \frac{\partial P}{\partial y}\, dy\, dx$$

$$= -\int_a^b [P(x, g_2(x)) - P(x, g_1(x))]\, dx$$

$$= \int_a^b P(x, g_1(x))\, dx + \int_b^a P(x, g_2(x))\, dx$$

$$= \int_C P(x, y)\, dx. \tag{2}$$

Similarly, from Figure 15.4.2(b),

$$\iint_R \frac{\partial Q}{\partial x}\, dA = \int_c^d \int_{h_1(y)}^{h_2(y)} \frac{\partial Q}{\partial x}\, dx\, dy$$

$$= \int_c^d [Q(h_2(y), y) - Q(h_1(y), y)]\, dy$$

$$= \int_c^d Q(h_2(y), y)\, dy + \int_d^c Q(h_1(y), y)\, dy$$

$$= \int_C Q(x, y)\, dy. \tag{3}$$

Adding the results in (2) and (3) yields (1). ∎

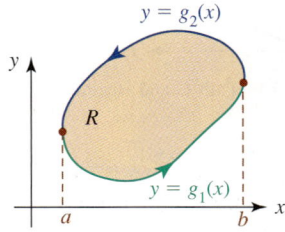

(a) R as a Type I region

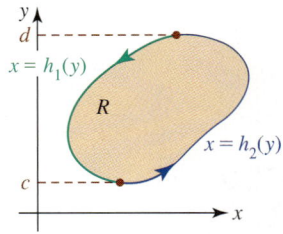

(b) R as a Type II region
FIGURE 15.4.2 Region R used in the proof in Theorem 15.4.1

Although the foregoing proof is not valid for more complicated regions, the theorem is applicable to these regions, such as that shown in FIGURE 15.4.3. The proof consists of decomposing R into a finite number of subregions to which (1) can be applied and then adding the results.

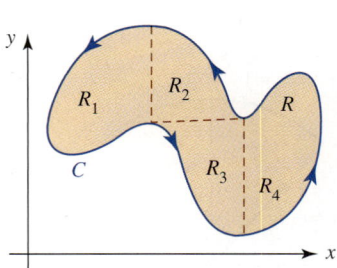

FIGURE 15.4.3 Region R decomposed into four subregions

Integration in the positive direction on simple closed curve C is often denoted by

$$\oint_C P(x, y)\, dx + Q(x, y)\, dy \quad \text{or} \quad \oint_C P(x, y)\, dx + Q(x, y)\, dy. \qquad (4)$$

The small circle superimposed on the integral sign in the first term in (4) emphasizes the fact that integration is along a closed curve; the arrow on the circle in the second term in (4) reenforces the notion that integration is along a closed curve C with a positive orientation. Although $\int_C$, $\oint_C$, and $\oint_C$ mean the same thing in this section, we will use the second integral sign for the remainder of the discussion so that you gain some familiarity with this alternative notation.

EXAMPLE 1 Using Green's Theorem

Evaluate $\oint_C (x^2 - y^2)\, dx + (2y - x)\, dy$, where C consists of the boundary of the region in the first quadrant that is bounded by the graphs of $y = x^2$ and $y = x^3$.

Solution If $P(x, y) = x^2 - y^2$ and $Q(x, y) = 2y - x$, then $\partial P/\partial y = -2y$ and $\partial Q/\partial x = -1$. From (1) and FIGURE 15.4.4 we have

$$\oint_C (x^2 - y^2)\, dx + (2y - x)\, dy = \iint_R (-1 + 2y)\, dA$$

$$= \int_0^1 \int_{x^3}^{x^2} (-1 + 2y)\, dy\, dx$$

$$= \int_0^1 (-y + y^2) \Big]_{x^3}^{x^2} dx$$

$$= \int_0^1 (-x^6 + x^4 + x^3 - x^2)\, dx = -\frac{11}{420}. \qquad \blacksquare$$

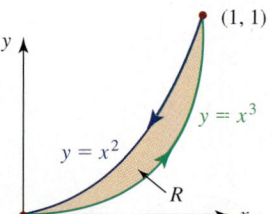

FIGURE 15.4.4 Path C and Region R in Example 1

We note that the line integral in Example 1 could have been evaluated in a straightforward manner using the variable x as a parameter. However, as you work through the next example, ponder the problem of evaluating the given line integral in the usual manner.

EXAMPLE 2 Using Green's Theorem

Evaluate $\oint_C (x^5 + 3y)\, dx + (2x - e^{y^3})\, dy$, where C is the circle $(x - 1)^2 + (y - 5)^2 = 4$.

Solution Identifying $P(x, y) = x^5 + 3y$ and $Q(x, y) = 2x - e^{y^3}$, we have $\partial P/\partial y = 3$ and $\partial Q/\partial x = 2$. Hence, (1) gives

$$\oint_C (x^5 + 3y)\, dx + (2x - e^{y^3})\, dy = \iint_R (2 - 3)\, dA = -\iint_R dA.$$

Now the double integral $\iint_R dA$ gives the area of the region R bounded by the circle of radius 2 shown in FIGURE 15.4.5. Since the area of the circle is $\pi 2^2 = 4\pi$, it follows that

$$\oint_C (x^5 + 3y)\, dx + (2x - e^{y^3})\, dy = -4\pi. \qquad \blacksquare$$

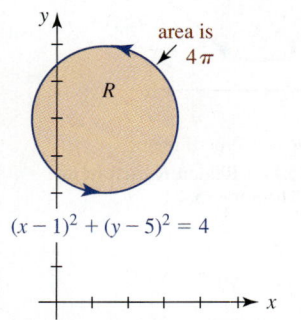

FIGURE 15.4.5 Path C and Region R in Example 2

EXAMPLE 3 Work

Find the work done by the force field $\mathbf{F} = (-16y + \sin x^2)\mathbf{i} + (4e^y + 3x^2)\mathbf{j}$ acting along the simple closed curve C shown in FIGURE 15.4.6.

Solution From (6) of Section 15.2 the work done by $\mathbf{F}$ is given by

$$W = \oint_C \mathbf{F} \cdot d\mathbf{r} = \oint_C (-16y + \sin x^2)\, dx + (4e^y + 3x^2)\, dy$$

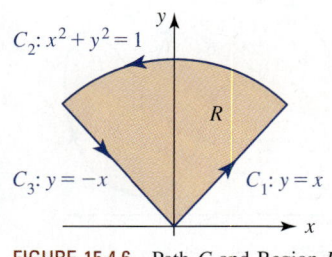

FIGURE 15.4.6 Path C and Region R in Example 3

and so by Green's Theorem,

$$W = \iint_R (6x + 16)\, dA.$$

In view of the region R the last integral is best handled in polar coordinates. In polar coordinates R is defined by $0 \le r \le 1$, $\pi/4 \le \theta \le 3\pi/4$, and so we have

$$
\begin{aligned}
W &= \int_{\pi/4}^{3\pi/4} \int_0^1 (6r\cos\theta + 16)r\, dr\, d\theta \\
&= \int_{\pi/4}^{3\pi/4} (2r^3\cos\theta + 8r^2)\Big]_0^1\, d\theta \\
&= \int_{\pi/4}^{3\pi/4} (2\cos\theta + 8)\, d\theta = 4\pi. \quad \blacksquare
\end{aligned}
$$

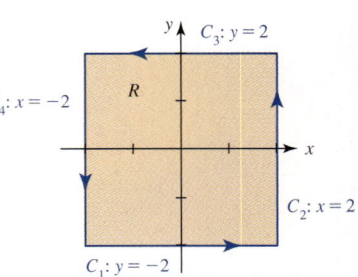

FIGURE 15.4.7 Path C and Region R in Example 4

EXAMPLE 4   Green's Theorem Not Applicable

Let C be the closed polygonal curve consisting of the four straight line segments C_1, C_2, C_3, and C_4 shown in FIGURE 15.4.7. Green's Theorem is *not* applicable to the line integral

$$\int_C \frac{-y}{x^2 + y^2}\, dx + \frac{x}{x^2 + y^2}\, dy$$

since P, Q, $\partial P/\partial y$, and $\partial Q/\partial x$ are not continuous at the origin. $\blacksquare$

■ **Green's Theorem for Multiply Connected Regions**   Green's Theorem can also be extended to a region R with holes—that is, a region that is not simply connected. Recall from Section 15.3 that a region with holes is said to be multiply connected. In FIGURE 15.4.8(a) we have shown a region R bounded by a curve C that consists of two simple closed curves C_1 and C_2; that is $C = C_1 \cup C_2$. The curve C is positively oriented, since if we traverse C_1 in a counterclockwise direction and C_2 in a clockwise direction, the region R is always to the left. If we now introduce horizontal crosscuts as shown in Figure 15.4.8(b), the region R is divided into two subregions R_1 and R_2. By applying Green's Theorem to R_1 and R_2, we obtain

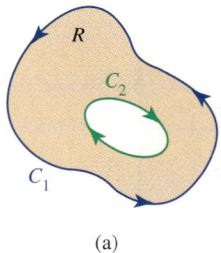

(a)

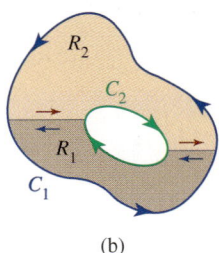

(b)

FIGURE 15.4.8 Region R with a hole

$$
\begin{aligned}
\iint_R \left(\frac{\partial Q}{\partial x} - \frac{\partial P}{\partial y}\right) dA &= \iint_{R_1} \left(\frac{\partial Q}{\partial x} - \frac{\partial P}{\partial y}\right) dA + \iint_{R_2} \left(\frac{\partial Q}{\partial x} - \frac{\partial P}{\partial y}\right) dA \\
&= \oint_{C_1} P\, dx + Q\, dy + \oint_{C_2} P\, dx + Q\, dy \qquad (5) \\
&= \oint_C P\, dx + Q\, dy.
\end{aligned}
$$

The last result follows from the fact that the line integrals on the crosscuts (paths with opposite orientations) will cancel each other. See (11) of Section 15.1.

EXAMPLE 5   Applying (5)

Evaluate $\displaystyle\oint_C \frac{-y}{x^2 + y^2}\, dx + \frac{x}{x^2 + y^2}\, dy$, where $C = C_1 \cup C_2$ is the boundary of the shaded region R shown in FIGURE 15.4.9.

Solution   Because

$$
P(x, y) = \frac{-y}{x^2 + y^2}, \quad Q(x, y) = \frac{x}{x^2 + y^2},
$$

$$
\frac{\partial P}{\partial y} = \frac{y^2 - x^2}{(x^2 + y^2)^2}, \quad \frac{\partial Q}{\partial x} = \frac{y^2 - x^2}{(x^2 + y^2)^2}
$$

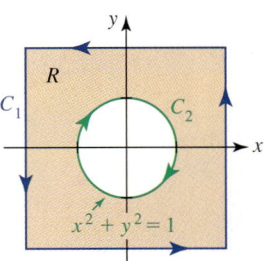

FIGURE 15.4.9 Path C and Region R in Example 5

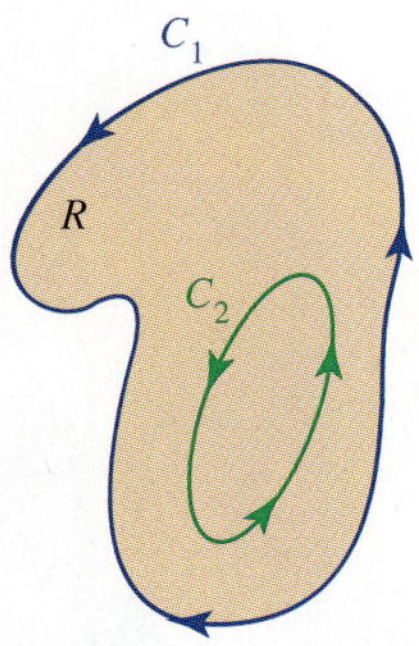

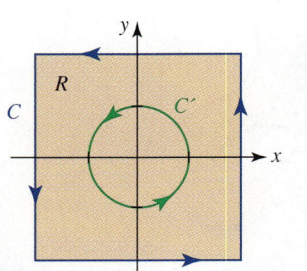

FIGURE 15.4.10 Line integral on C_1 is the same as on C_2

are continuous on the region R bounded by C, it follows from (5) that

$$\oint_C \frac{-y}{x^2 + y^2}\,dx + \frac{x}{x^2 + y^2}\,dy = \iint_R \left[\frac{y^2 - x^2}{(x^2 + y^2)^2} - \frac{y^2 - x^2}{(x^2 + y^2)^2} \right] dA = 0. \qquad \blacksquare$$

As a consequence of the discussion preceding Example 5 we can establish a result for line integrals that enables us, under certain circumstances, to replace a complicated closed path with a path that is simpler. Suppose, as shown in FIGURE 15.4.10, that C_1 and C_2 are two nonintersecting piecewise-smooth simple closed paths that have the same positive or counterclockwise orientation. Suppose further that P and Q have continuous first partial derivatives such that

$$\frac{\partial P}{\partial y} = \frac{\partial Q}{\partial x}$$

in the region R bounded between C_1 and C_2. Then from (5) above and (11) of Section 15.1 we have

$$\oint_{C_1} P\,dx + Q\,dy + \oint_{-C_2} P\,dx + Q\,dy = 0$$

or

$$\oint_{C_1} P\,dx + Q\,dy = \oint_{C_2} P\,dx + Q\,dy. \qquad (6)$$

EXAMPLE 6　Example 4 Revisited

Evaluate the line integral in Example 4.

Solution　One method of evaluating the line integral is to write

$$\oint_C = \int_{C_1} + \int_{C_2} + \int_{C_3} + \int_{C_4}$$

and then evaluate the four integrals on the line segments C_1, C_2, C_3, and C_4 shown in Figure 15.4.7. Alternatively, if we note that the circle C': $x^2 + y^2 = 1$ shown in FIGURE 15.4.11 lies entirely within C, then from Example 5 it is apparent that $P = -y/(x^2 + y^2)$ and $Q = x/(x^2 + y^2)$ have continuous first partial derivatives in the region R bounded between C and C'. Moreover,

$$\frac{\partial P}{\partial y} = \frac{y^2 - x^2}{(x^2 + y^2)^2} = \frac{\partial Q}{\partial x}$$

in R. Hence, it follows from (6) that

$$\oint_C \frac{-y}{x^2 + y^2}\,dx + \frac{x}{x^2 + y^2}\,dy = \oint_{C'} \frac{-y}{x^2 + y^2}\,dx + \frac{x}{x^2 + y^2}\,dy.$$

Using the parameterization $x = \cos t$, $y = \sin t$, $0 \le t \le 2\pi$ for C' we obtain

$$\oint_C \frac{-y}{x^2 + y^2}\,dx + \frac{x}{x^2 + y^2}\,dy = \int_0^{2\pi} [-\sin t(-\sin t) + \cos t(\cos t)]\,dt$$

$$= \int_0^{2\pi} (\sin^2 t + \cos^2 t)\,dt$$

$$= \int_0^{2\pi} dt = 2\pi. \qquad (7) \blacksquare$$

It is interesting to note that the result in (7):

$$\oint_C \frac{-y}{x^2 + y^2}\,dx + \frac{x}{x^2 + y^2}\,dy = 2\pi$$

FIGURE 15.4.11　Closed curves C and C' in Example 6

is true for every piecewise-smooth simple closed curve C with the origin in its interior. We need only choose C' to be $x^2 + y^2 = a^2$, where a is small enough so that the circle lies entirely within C.

■ **Postscript—A Bit of History** **George Green** (1793–1841) was born in Knottingham, England, of working parents. The young George left school after only four terms and at the age of 9 began work at his father's bakery. After his father's death in 1829 he used the money obtained from the sale of the bakery to pursue studies in mathematics and science. Mostly self-taught, Green produced several papers before entering Cambridge University at age 40. At his own expense he published *An Essay on the Application of Mathematical Analysis to the Theories of Electricity and Magnetism* in 1828 in which he introduced his now famous Green's Theorem. At age 45 he attained his B.A. and stayed on at Cambridge becoming a faculty member at Gonville and Caius College. Green's seminal work in mathematics, electricity, and magnetism were basically ignored after his death in 1841 but were eventually brought to the attention of the scientific and mathematical community through the efforts of William Thomson (Lord Kelvin) in 1845.

Exercises 15.4 Answers to selected odd-numbered problems begin on page ANS-47.

☰ Fundamentals

In Problems 1–4, verify Green's Theorem by evaluating both integrals.

1. $\oint_C (x - y) \, dx + xy \, dy = \iint_R (y + 1) \, dA$, where C is the triangle with vertices $(0, 0), (1, 0), (1, 3)$

2. $\oint_C 3x^2 y \, dx + (x^2 - 5y) \, dy = \iint_R (2x - 3x^2) \, dA$, where C is the rectangle with vertices $(-1, 0), (-1, 1), (1, 0), (1, 1)$

3. $\oint_C -y^2 \, dx + x^2 \, dy = \iint_R (2x + 2y) \, dA$, where C is the circle $x = 3\cos t$, $y = 3\sin t$, $0 \le t \le 2\pi$

4. $\oint_C -2y^2 \, dx + 4xy \, dy = \iint_R 8y \, dA$, where C is the boundary of the region in the first quadrant determined by the graphs of $y = 0$, $y = \sqrt{x}$, $y = -x + 2$

In Problems 5–14, use Green's Theorem to evaluate the given line integral.

5. $\oint_C 2y \, dx + 5x \, dy$, where C is the circle $(x - 1)^2 + (y + 3)^2 = 25$

6. $\oint_C (x + y^2) \, dx + (2x^2 - y) \, dy$, where C is the boundary of the region determined by the graphs of $y = x^2$, $y = 4$

7. $\oint_C (x^4 - 2y^3) \, dx + (2x^3 - y^4) \, dy$, where C is the circle $x^2 + y^2 = 4$

8. $\oint_C (x - 3y) \, dx + (4x + y) \, dy$, where C is the rectangle with vertices $(-2, 0), (3, 0), (3, 2), (-2, 2)$

9. $\oint_C 2xy \, dx + 3xy^2 \, dy$, where C is the triangle with vertices $(1, 2), (2, 2), (2, 4)$

10. $\oint_C e^{2x} \sin 2y \, dx + e^{2x} \cos 2y \, dy$, where C is the ellipse $9(x - 1)^2 + 4(y - 3)^2 = 36$

11. $\oint_C xy \, dx + x^2 \, dy$, where C is the boundary of the region determined by the graphs of $x = 0$, $x^2 + y^2 = 1$, $x \ge 0$

12. $\oint_C e^{x^2} \, dx + 2\tan^{-1} x \, dy$, where C is the triangle with vertices $(0, 0), (0, 1), (-1, 1)$

13. $\oint_C \frac{1}{3} y^3 \, dx + (xy + xy^2) \, dy$, where C is the boundary of the region in the first quadrant determined by the graphs of $y = 0$, $x = y^2$, $x = 1 - y^2$

14. $\oint_C xy^2 \, dx + 3\cos y \, dy$, where C is the boundary of the region in the first quadrant determined by the graphs of $y = x^2$, $y = x^3$

In Problems 15 and 16, evaluate the given integral on any piecewise-smooth simple closed curve C.

15. $\oint_C ay \, dx + bx \, dy$

16. $\oint_C P(x) \, dx + Q(y) \, dy$

In Problems 17 and 18, let R be the region bounded by a piecewise-smooth simple closed curve C. Prove the given result.

17. $\oint_C x \, dy = -\oint_C y \, dx = $ area of R

18. $\frac{1}{2} \oint_C -y \, dx + x \, dy = $ area of R

In Problems 19 and 20, use the results of Problems 17 and 18 to find the area of the region bounded by the given closed curve.

19. The hypocycloid $x = a\cos^3 t$, $y = a\sin^3 t$, $a > 0$, $0 \le t \le 2\pi$

20. The ellipse $x = a\cos t$, $y = b\sin t$, $a > 0$, $b > 0$, $0 \le t \le 2\pi$

21. (a) Show that
$$\int_C -y \, dx + x \, dy = x_1 y_2 - x_2 y_1,$$
where C is the line segment from the point (x_1, y_1) to (x_2, y_2).

(b) Use part (a) and Problem 18 to show that the area A of a polygon with vertices $(x_1, y_1), (x_2, y_2), \ldots, (x_n, y_n)$, labeled counterclockwise, is
$$A = \frac{1}{2}(x_1 y_2 - x_2 y_1) + \frac{1}{2}(x_2 y_3 - x_3 y_2)$$
$$+ \cdots + \frac{1}{2}(x_{n-1} y_n - x_n y_{n-1}) + \frac{1}{2}(x_n y_1 - x_1 y_n).$$

22. Use part (b) of Problem 21 to find the area of the quadrilateral with vertices $(-1, 3)$, $(1, 1)$, $(4, 2)$, and $(3, 5)$.

In Problems 23 and 24, evaluate the given line integral where $C = C_1 \cup C_2$ is the boundary of the shaded region R.

23. $\oint_C (4x^2 - y^3)\, dx + (x^3 + y^2)\, dy$

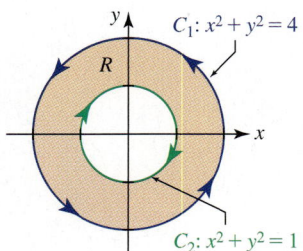

FIGURE 15.4.12   Curve C in Problem 23

24. $\oint_C (\cos x^2 - y)\, dx + \sqrt{y^2 + 1}\, dy$

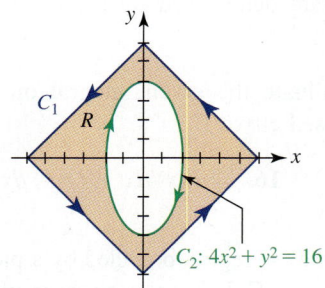

FIGURE 15.4.13   Curve C in Problem 24

In Problems 25 and 26, proceed as in Example 6 to evaluate the given line integral.

25. $\oint_C \dfrac{-y^3\, dx + xy^2\, dy}{(x^2 + y^2)^2}$, where C is the ellipse $x^2 + 4y^2 = 4$

26. $\oint_C \dfrac{-y}{(x + 1)^2 + 4y^2}\, dx + \dfrac{x + 1}{(x + 1)^2 + 4y^2}\, dy$, where C is the circle $x^2 + y^2 = 16$

In Problems 27 and 28, use Green's Theorem to evaluate the given double integral by means of a line integral. [*Hint:* Find appropriate functions P and Q.]

27. $\iint_R x^2\, dA$, where R is the region bounded by the ellipse $x^2/9 + y^2/4 = 1$

28. $\iint_R [1 - 2(y - 1)]\, dA$, where R is the region in the first quadrant bounded by the circle $x^2 + (y - 1)^2 = 1$ and $x = 0$

In Problems 29 and 30, use Green's Theorem to find the work done by the given force $\mathbf{F}$ around the closed curve in FIGURE 15.4.14.

29. $\mathbf{F} = (x - y)\mathbf{i} + (x + y)\mathbf{j}$ **30.** $\mathbf{F} = -xy^2\mathbf{i} + x^2y\mathbf{j}$

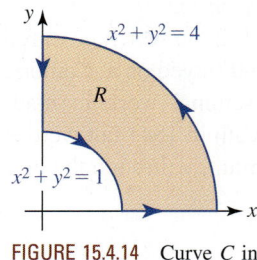

FIGURE 15.4.14   Curve C in Problems 29 and 30

☰ Applications

31. Let R be a region bounded by a piecewise-smooth simple closed curve C. Show that the coordinates of the **centroid** of the region are given by

$$\bar{x} = \frac{1}{2A} \oint_C x^2\, dy, \qquad \bar{y} = -\frac{1}{2A} \oint_C y^2\, dx.$$

32. Find the work done by the force $\mathbf{F} = -y\mathbf{i} + x\mathbf{j}$ acting along the cardioid $r = 1 + \cos\theta$.

☰ Think About It

33. Let P and Q be continuous and have continuous first partial derivatives in a simply connected region of the xy-plane. If $\int_A^B P\, dx + Q\, dy$ is independent of the path, show that $\oint_C P\, dx + Q\, dy = 0$ on every piecewise-smooth simple closed curve C in the region.

34. If f is a function of two variables that satisfies Laplace's differential equation

$$\frac{\partial^2 f}{\partial x^2} + \frac{\partial^2 f}{\partial y^2} = 0$$

in a simply connected region R, show that $\int_C f_y\, dx - f_x\, dy$ is independent of the path in R.

15.5 Parametric Surfaces and Area

❙ **Introduction**   We have seen that curves in 2-space can be defined by a function $y = f(x)$, an equation $g(x, y) = 0$, or parametrically by a set of equations $x = x(t)$, $y = y(t)$, $a \le t \le b$. A curve C described by a continuous function $y = f(x)$ can be parameterized by letting $x = t$ so that parametric equations are $x = t$, $y = f(t)$. It takes *two* variables to parameterize a surface S

in 3-space defined by a function of two variables $z = g(x, y)$. If $x = u$ and $y = v$, then parametric equations for S are $x = u$, $y = v$, $z = g(u, v)$.

■ Parametric Surfaces   In general, a set of three functions of two variables,

$$x = x(u, v), \quad y = y(u, v), \quad z = z(u, v) \tag{1}$$

are called **parametric equations**. The variables u and v are called **parameters** and the set of points (x, y, z) in 3-space defined by (1) is called a **parametric surface S**. The ordered pair (u, v) comes from a region R in the uv-plane called the **parameter domain**. The parameter domain is the 2-dimensional counterpart of the one-dimensional parameter interval for a parametric curve C. A surface S can also be described by a vector-valued function of two variables

$$\mathbf{r}(u, v) = x(u, v)\mathbf{i} + y(u, v)\mathbf{j} + z(u, v)\mathbf{k}. \tag{2}$$

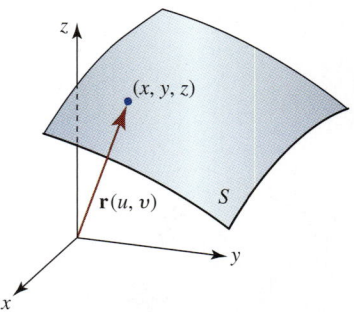

FIGURE 15.5.1   Position vector of a point on the surface S

For a given (u_0, v_0) in R the vector $\mathbf{r}(u_0, v_0)$ is the position vector of a point P on the surface S. In other words, as (u, v) varies over the region R, the surface S is traced out by the moving arrowhead of $\mathbf{r}(u, v)$. See FIGURE 15.5.1.

Parameterizations of surfaces are very important in computer graphics. Many of the complicated three-dimensional figures generated in Chapters 12, 13, and 14 were generated using a CAS and a parametric representation of a surface. For example, the seashell-like surface shown in FIGURE 15.5.2(a) was generated using *Mathematica* and the parametric equations

$$x = 2(1 - e^{u/5\pi})\cos u \cos^2(v/2),$$
$$y = 2(-1 + e^{u/5\pi})\sin u \cos^2(v/2),$$
$$z = 1 - e^{u/3\pi} - \sin v + e^{u/5\pi}\sin v,$$

over the parameter domain R in the uv-plane defined by the inequalities $0 \le u \le 8\pi$, $0 \le v \le 2\pi$. In Section 12.1 we saw that the vector function of a single variable

$$\mathbf{r}(u) = \cos u\,\mathbf{i} + \sin u\,\mathbf{j} + u\mathbf{k},$$

describes a space curve known as a circular helix that winds along the z-axis. A variation of this equation using two variables:

$$\mathbf{r}(u, v) = (3 + \sin v)\cos u\,\mathbf{i} + (3 + \sin v)\sin u\,\mathbf{j} + (u + \cos v)\mathbf{k},$$

describes what could be called a *helical tubular surface*. See Figure 15.5.2(b).

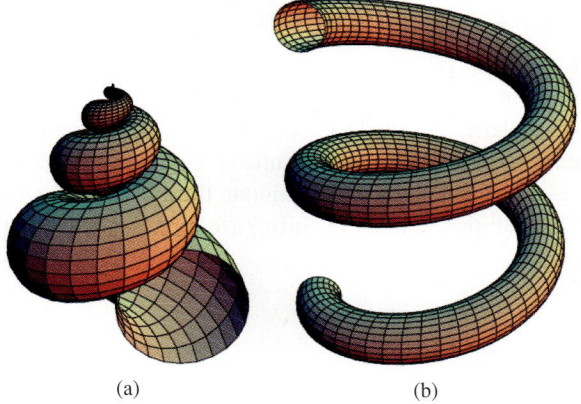

(a)   (b)
FIGURE 15.5.2   Parametric surfaces

EXAMPLE 1   Parametric Surface

Find parametric equations for the single-napped cone $z = \sqrt{x^2 + y^2}$.

Solution   If we let $x = u$ and $y = v$, the parametric surface is given by the equations $x = u$, $y = v$, $z = \sqrt{u^2 + v^2}$. Alternatively, the cone is described by the vector function $\mathbf{r}(u, v) = u\mathbf{i} + v\mathbf{i} + \sqrt{u^2 + v^2}\mathbf{k}$. ■

EXAMPLE 2 Graphs

(a) FIGURE 15.5.3(a) shows the portion of the single-napped cone $x = u, y = v, z = \sqrt{u^2 + v^2}$ in Example 1 corresponding to the parameter values $-1 \leq u \leq 1, -1 \leq v \leq 1$. The truncated sides in the figure are due to the fact that the surface intersects the display box in the figure. For example, the curve of intersection (trace) of the surface and the vertical plane $u = 1$ ($x = 1$), is one branch of a hyperbola defined by $z = \sqrt{1 + v^2} = \sqrt{1 + y^2}$. The highest points on the cone occur at the four upper corners of the display box; each of the parameter pairs $(1, 1), (1, -1), (-1, 1),$ and $(-1, -1)$ give $z = \sqrt{2}$.

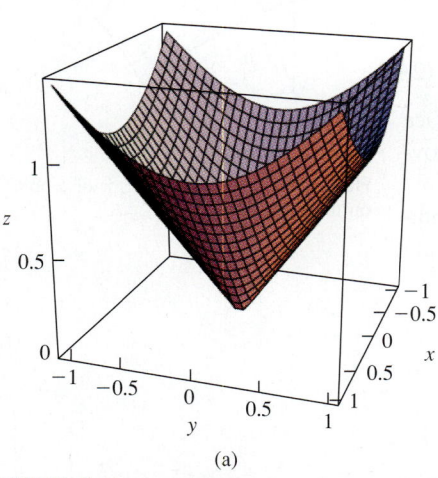

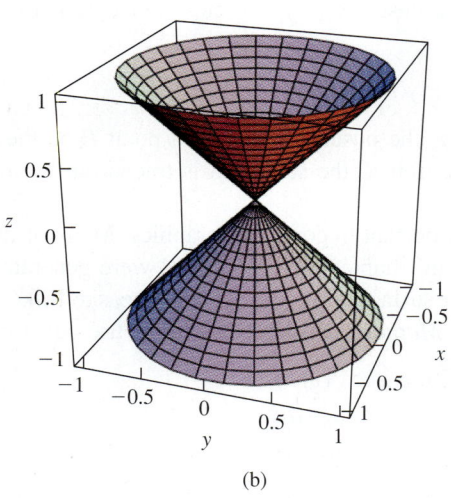

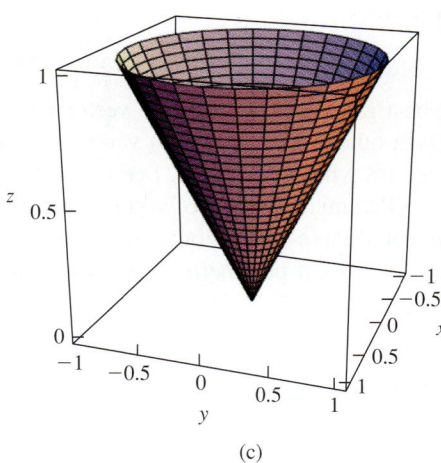

(a)

(b)

(c)

FIGURE 15.5.3 Parametric surfaces in Example 2

(b) Using polar coordinates, the parameterization

$$x = r\cos\theta, \qquad y = r\sin\theta, \qquad z = r \qquad (3)$$

also defines a cone. The graph of (3) in Figure 15.5.3(b) for $-1 \leq r \leq 1$, $0 \leq \theta \leq 2\pi$, is a double-napped cone truncated by the horizontal planes $r = -1$ ($z = -1$) and $r = 1$ ($z = 1$). By changing the parameter values to $0 \leq r \leq 1, 0 \leq \theta \leq 2\pi$ the parametric equations (3) yield the upper nappe of the cone shown in Figure 15.5.3(c). ∎

Verify this by showing that $x^2 + y^2 = z^2$

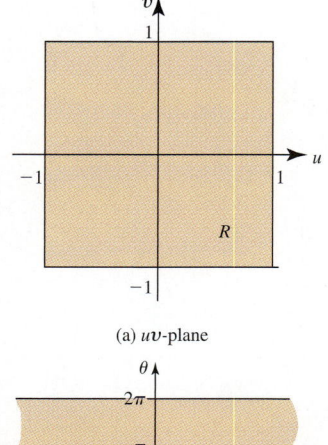

(a) uv-plane

(b) $r\theta$-plane

FIGURE 15.5.4 The parameter domain (a) is a rectangular region; the parameter domain (b) is an infinite strip

The parameter domain R defined by the inequalities in part (a) of Example 2 is a rectangular region in the uv-plane. See FIGURE 15.5.4(a). We note in passing that a parameter domain R need not be a rectangular region. If the parameter domain R in part (b) of Example 2 is defined by $-\infty \leq r \leq \infty, 0 \leq \theta \leq 2\pi$, we generate the complete double-napped cone. This set of simultaneous inequalities defines an infinite horizontal strip in the $r\theta =$ plane. See Figure 15.5.4(b).

EXAMPLE 3 Parametric Surface

Find parametric equations for the circular cylinder $y^2 + z^2 = 1$ for $3 \leq x \leq 8$.

Solution If we use $y = \cos v$ and $z = \sin v$, then it is apparent that $y^2 + z^2 = \cos^2 v + \sin^2 v = 1$. To obtain the complete lateral surface of the cylinder we use the values $0 \leq v \leq 2\pi$. We next let $x = u$, where $3 \leq u \leq 8$. Thus, parametric equations for this surface are

$$x = u, \quad y = \cos v, \quad z = \sin v, \qquad 3 \leq u \leq 8, 0 \leq v \leq 2\pi.$$

The graph of these equations over the rectangular region R defined by the inequalities $3 \leq u \leq 8, 0 \leq v \leq 2\pi$ is given in FIGURE 15.5.5. ∎

It is practically impossible to identify even well-known surfaces when given in parametric or vector form. However, in some instances a surface can be identified by eliminating the parameters.

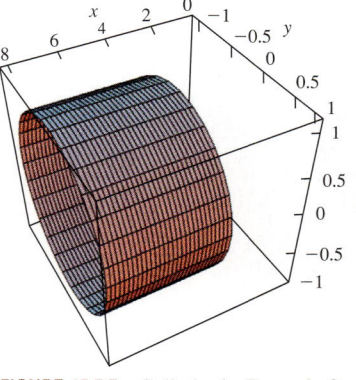

EXAMPLE 4 Eliminating the Parameters

Identify the surface with the vector function $\mathbf{r}(u, v) = (2u - v)\mathbf{i} + (u + v + 1)\mathbf{j} + u\mathbf{k}$.

Solution Parametric equations of the surface are

$$x = 2u - v, \quad y = u + v + 1, \quad z = u.$$

Adding x and y gives $x + y = 3u + 1$. Since $z = u$, we recognize $x + y = 3z + 1$ or $x + y - 3z = 1$ as an equation of a plane. ∎

FIGURE 15.5.5 Cylinder in Example 3

In Example 4, the complete plane is obtained by letting (u, v) vary over the parameter domain consisting of the entire uv-plane, that is, for $-\infty < u < \infty$, $-\infty < v < \infty$.

EXAMPLE 5 Eliminating the Parameters

The equations

$$x = a\sin\phi\cos\theta, \quad y = a\sin\phi\sin\theta, \quad z = a\cos\phi \tag{4}$$

are parametric equations of a sphere of radius $a > 0$. To see this we square the equations in (4) and add:

$$\begin{aligned} x^2 + y^2 + z^2 &= a^2\sin^2\phi\cos^2\theta + a^2\sin^2\phi\sin^2\theta + a^2\cos^2\phi \\ &= a^2\sin^2\phi(\cos^2\theta + \sin^2\theta) + a^2\cos^2\phi \\ &= a^2\sin^2\phi + a^2\cos^2\phi \\ &= a^2(\sin^2\phi + \cos^2\phi) = a^2. \end{aligned}$$

The graph of (4) for $0 \leq \phi \leq \pi, 0 \leq \theta \leq 2\pi$, is shown in **FIGURE 15.5.6**. ∎

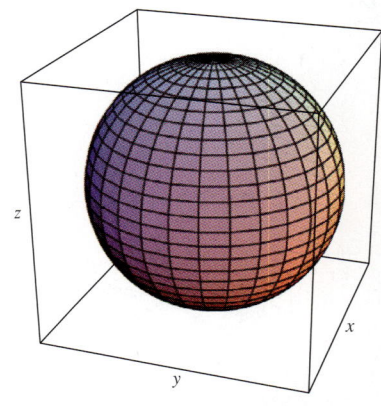

FIGURE 15.5.6 Sphere in Example 5

The parameters ϕ and θ in (4) are the polar and azimuthal angles used in spherical coordinates. You are urged to review the formulas in (3) of Section 14.8 that convert the spherical coordinates of a point to rectangular coordinates.

▌ Grid Lines The black curves that are obvious on each of the computer-generated surfaces in Figures 15.5.2, 15.5.3, 15.5.5, and 15.5.6 are called **grid lines** of the surface S. A grid line is obtained by keeping one of the parameters in either (1) or (2) constant while letting the other parameter vary. For example, if $v = v_0 = $ constant, then

$$\mathbf{r}(u, v_0) = x(u, v_0)\mathbf{i} + y(u, v_0)\mathbf{j} + z(u, v_0)\mathbf{k} \tag{5}$$

is a vector-valued function of a single variable. Consequently, (5) is an equation of a curve C_1 in 3-space that lies on the surface S traced out by $\mathbf{r}(u, v)$. Similarly, if $u = u_0 = $ constant, then

$$\mathbf{r}(u_0, v) = x(u_0, v)\mathbf{i} + y(u_0, v)\mathbf{j} + z(u_0, v)\mathbf{k} \tag{6}$$

is the vector equation of a curve C_2 on the surface S. In other words, C_1 and C_2 are grid lines of S. For a value ϕ_0 chosen from $0 \leq \phi \leq \pi$, and a value θ_0 from $0 \leq \theta \leq 2\pi$, grid lines on the sphere in Figure 15.5.6 are defined by

$$\mathbf{r}(\phi_0, \theta) = a\sin\phi_0\cos\theta\mathbf{i} + a\sin\phi_0\sin\theta\mathbf{j} + a\cos\phi_0\mathbf{k} \tag{7}$$

and $$\mathbf{r}(\phi, \theta_0) = a\sin\phi\cos\theta_0\mathbf{i} + a\sin\phi\sin\theta_0\mathbf{j} + a\cos\phi\mathbf{k}. \tag{8}$$

The vector equations in (7) and (8) are a circle and a semicircle, respectively. For $\phi_0 = $ constant the circle $\mathbf{r}(\phi_0, \theta), 0 \leq \theta \leq 2\pi$, lies on the sphere parallel to the xy-plane and is equivalent to a circle of fixed **latitude** on a world globe. For $\theta_0 = $ constant the semicircle defined by $\mathbf{r}(\phi, \theta_0), 0 \leq \phi \leq \pi$, passes through both the north pole (when $\phi = 0$ we get $(0, 0, a)$) and the south pole (when $\phi = \pi$ we get $(0, 0, -a)$) of the sphere and is called a **meridian**. On a globe a meridian corresponds to a fixed **longitude**.

■ **Tangent Plane to a Parametric Surface** For the constant parameter values $u = u_0$, $v = v_0$, the vector

$$\mathbf{r}(u_0, v_0) = x(u_0, v_0)\mathbf{i} + y(u_0, v_0)\mathbf{j} + z(u_0, v_0)\mathbf{k} = x_0\mathbf{i} + y_0\mathbf{j} + z_0\mathbf{k}$$

defines a point (x_0, y_0, z_0) on a surface S. Moreover, the vector functions of a single variable

$$\mathbf{r}(u, v_0) = x(u, v_0)\mathbf{i} + y(u, v_0)\mathbf{j} + z(u, v_0)\mathbf{k}$$
$$\mathbf{r}(u_0, v) = x(u_0, v)\mathbf{i} + y(u_0, v)\mathbf{j} + z(u_0, v)\mathbf{k}$$

define grid lines C_1 and C_2 that lie on S. Because the vector $\mathbf{r}(u_0, v_0)$ is defined by both vector functions, C_1 and C_2 intersect at (x_0, y_0, z_0). The partial derivatives of (2) with respect to u and v are defined as the vectors obtained by taking the partial derivatives of the component functions:

$$\frac{\partial \mathbf{r}}{\partial u} = \frac{\partial x}{\partial u}\mathbf{i} + \frac{\partial y}{\partial u}\mathbf{j} + \frac{\partial z}{\partial u}\mathbf{k}$$

$$\frac{\partial \mathbf{r}}{\partial v} = \frac{\partial x}{\partial v}\mathbf{i} + \frac{\partial y}{\partial v}\mathbf{j} + \frac{\partial z}{\partial v}\mathbf{k}.$$

▶ These partial derivatives are also denoted by $\mathbf{r}_u$ and $\mathbf{r}_v$.

Thus, if $\partial \mathbf{r}/\partial u \neq \mathbf{0}$ at (u_0, v_0), it represents a vector tangent to the grid line C_1 ($v = \text{constant} = v_0$) whereas $\partial \mathbf{r}/\partial v \neq \mathbf{0}$ at (u_0, v_0) is a vector that is tangent to the grid line C_2 ($u = \text{constant} = u_0$). From (2) of Section 11.4 the cross product $\partial \mathbf{r}/\partial u \times \partial \mathbf{r}/\partial v$ is defined by

$$\frac{\partial \mathbf{r}}{\partial u} \times \frac{\partial \mathbf{r}}{\partial v} = \begin{vmatrix} \mathbf{i} & \mathbf{j} & \mathbf{k} \\ \dfrac{\partial x}{\partial u} & \dfrac{\partial y}{\partial u} & \dfrac{\partial z}{\partial u} \\ \dfrac{\partial x}{\partial v} & \dfrac{\partial y}{\partial v} & \dfrac{\partial z}{\partial v} \end{vmatrix}. \tag{9}$$

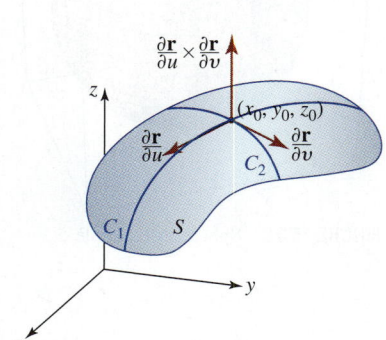

FIGURE 15.5.7 Parametric surface

The condition that the vector $\partial \mathbf{r}/\partial u \times \partial \mathbf{r}/\partial v$ is not $\mathbf{0}$ at (u_0, v_0) ensures the existence of a tangent plane at the point (x_0, y_0, z_0). Indeed, the tangent plane at $\mathbf{r}(u_0, y_0)$ or (x_0, y_0, z_0) is defined to be the plane determined by $\partial \mathbf{r}/\partial u$ and $\partial \mathbf{r}/\partial v$. Since the cross product is perpendicular to both vectors $\partial \mathbf{r}/\partial u$ and $\partial \mathbf{r}/\partial v$, the vector (9) is normal to the tangent plane to the surface S at (x_0, y_0, z_0). See FIGURE 15.5.7.

■ **Smooth Surface** Suppose S is a parametric surface whose vector equation $\mathbf{r}(u, v)$ has continuous first partial derivatives on a region R of the uv-plane. The surface S is said to be **smooth** at $\mathbf{r}(u_0, v_0)$ if the tangent vectors $\partial \mathbf{r}/\partial u$ and $\partial \mathbf{r}/\partial v$ in the u- and v-directions satisfy $\partial \mathbf{r}/\partial u \times \partial \mathbf{r}/\partial v \neq \mathbf{0}$ at (u_0, v_0). The surface S is said to be **smooth on R** if $\partial \mathbf{r}/\partial u \times \partial \mathbf{r}/\partial v \neq \mathbf{0}$ for all points (u, v) in R. In rough terms, a smooth surface has no corners, sharp points, or breaks. A **piecewise-smooth** surface S is one that can be written as $S = S_1 \cup S_2 \cup \cdots \cup S_n$, where the surfaces $S_1, S_2, \ldots, S_n$ are smooth.

EXAMPLE 6 Tangent Plane to a Parametric Surface

Find an equation of the tangent plane to the parametric surface defined by $x = u^2 + v$, $y = v$, $z = u + v^2$ at the point corresponding to $u = 3$, $v = 0$.

Solution At $u = 3$, $v = 0$, the point on the surface is $(9, 0, 3)$. If the surface is defined by the vector function $\mathbf{r}(u, v) = (u^2 + v)\mathbf{i} + v\mathbf{j} + (u + v^2)\mathbf{k}$, then

$$\frac{\partial \mathbf{r}}{\partial u} = 2u\mathbf{i} + \mathbf{k}, \quad \frac{\partial \mathbf{r}}{\partial v} = \mathbf{i} + \mathbf{j} + 2v\mathbf{k}$$

and

$$\frac{\partial \mathbf{r}}{\partial u} \times \frac{\partial \mathbf{r}}{\partial v} = \begin{vmatrix} \mathbf{i} & \mathbf{j} & \mathbf{k} \\ 2u & 0 & 1 \\ 1 & 1 & 2v \end{vmatrix} = -\mathbf{i} + (1 - 4uv)\mathbf{j} + 2u\mathbf{k}.$$

Evaluating the foregoing vector at $u = 3$, $v = 0$, gives the normal $-\mathbf{i} + \mathbf{j} + 6\mathbf{k}$ to the surface at $(9, 0, 3)$. An equation of the tangent plane at that point is

$$(-1)(x - 9) + (y - 0) + 6(z - 3) = 0 \qquad \text{or} \qquad z = \tfrac{1}{6}x - \tfrac{1}{6}y + \tfrac{3}{2}.$$

The graph of the surface and the tangent plane are given in FIGURE 15.5.8. ∎

■ **Building an Integral** We next sketch the steps leading up to an integral definition of the **area of a parametric surface**. Since the discussion is similar to that leading up to Definition 14.6.1, a review of that material is recommended. Suppose the vector function $\mathbf{r}(u, v) = x(u, v)\mathbf{i} + y(u, v)\mathbf{j} + z(u, v)\mathbf{k}$ traces out a surface S as (u, v) varies over a parameter domain R in the uv-plane. To simplify the discussion we will assume that R is a rectangular region

$$R = \{(u, v) \mid a \le u \le b, c \le v \le d\}$$

as shown in FIGURE 15.5.9(a). We use a regular partition, that is, we divide R into n rectangles each having the same width Δu and the same height Δv and let R_k denote the kth rectangular subregion. If (u_k, v_k) are the coordinates of the lower left corner of R_k, the other corners can be expressed as $(u_k + \Delta u, v_k)$, $(u_k + \Delta u, v_k + \Delta v)$, $(u_k, v_k + \Delta v)$ and so the area of R_k is $\Delta A = \Delta u \, \Delta v$. The images of the points in R_k determine a patch S_k on the surface S, where the red dot in Figure 15.5.9(b) is the point corresponding to $\mathbf{r}(u_k, v_k)$. Now two of the edges of S_k can be approximated by the vectors

$$\mathbf{r}(u_k + \Delta u, v_k) - \mathbf{r}(u_k, v_k) = \frac{\mathbf{r}(u_k + \Delta u, v_k) - \mathbf{r}(u_k, v_k)}{\Delta u} \Delta u \approx \frac{\partial \mathbf{r}}{\partial u} \Delta u$$

$$\mathbf{r}(u_k, v_k + \Delta v) - \mathbf{r}(u_k, v_k) = \frac{\mathbf{r}(u_k, v_k + \Delta v) - \mathbf{r}(u_k, v_k)}{\Delta v} \Delta u \approx \frac{\partial \mathbf{r}}{\partial v} \Delta v.$$

As seen in Figure 15.5.9(c) these vectors actually form two of the edges of a parallelogram T_k lying in the tangent plane at $\mathbf{r}(u_k, v_k)$. The area ΔT_k of

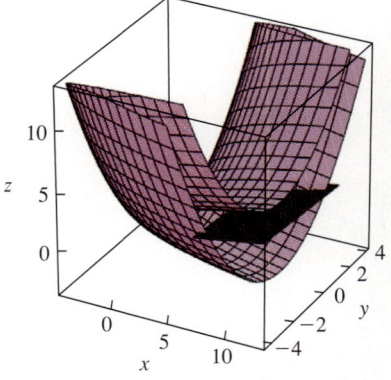

FIGURE 15.5.8 Parametric surface and tangent plane in Example 6

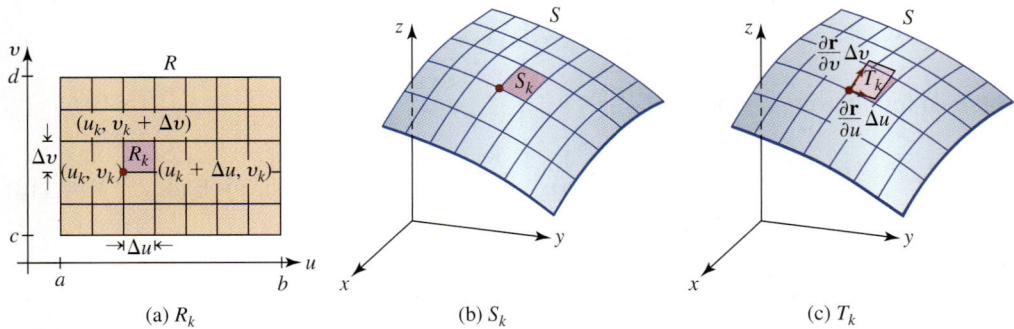

(a) R_k (b) S_k (c) T_k

FIGURE 15.5.9 Parameter domain R in (a); corresponding surface S in (b) and (c)

the parallelogram T_k approximates the area ΔS_k of S_k:

$$\Delta T_k = \left| \frac{\partial \mathbf{r}}{\partial u} \Delta u \times \frac{\partial \mathbf{r}}{\partial v} \Delta v \right| = \left| \frac{\partial \mathbf{r}}{\partial u} \times \frac{\partial \mathbf{r}}{\partial v} \right| \Delta u \, \Delta v = \left| \frac{\partial \mathbf{r}}{\partial u} \times \frac{\partial \mathbf{r}}{\partial v} \right| \Delta A \approx \Delta S_k.$$

The Riemann sum

$$\sum_{k=1}^{n} \left| \frac{\partial \mathbf{r}}{\partial u} \times \frac{\partial \mathbf{r}}{\partial v} \right| \Delta A$$

gives an approximation of the area $A(s)$ of that portion of surface S corresponding to the points in R. It is plausible then that the exact area is

$$A(S) = \lim_{n \to \infty} \sum_{k=1}^{n} \left| \frac{\partial \mathbf{r}}{\partial u} \times \frac{\partial \mathbf{r}}{\partial v} \right| \Delta A. \tag{10}$$

> **Definition 15.5.1** Area of a Surface
>
> Let S be a smooth parametric surface defined by the vector equation
>
> $$\mathbf{r}(u, v) = x(u, v)\mathbf{i} + y(u, v)\mathbf{j} + z(u, v)\mathbf{k}.$$
>
> If each point on S corresponds to exactly one point (u, v) in the parameter domain R in the uv-plane, then the area of S is
>
> $$A(S) = \iint_R \left| \frac{\partial \mathbf{r}}{\partial u} \times \frac{\partial \mathbf{r}}{\partial v} \right| dA. \tag{11}$$

As we saw in the introduction to this section, a surface described by an explicit function $z = g(x, y)$ can be parameterized by the equations $x = u, y = v, z = g(u, v)$. For this parameterization, (11) immediately reduces to

$$A(S) = \iint_R \sqrt{1 + [g_u(u, v)]^2 + [g_v(u, v)]^2} \, du \, dv$$

which is (2) of Section 14.6 with u and v playing the part of x and y.

EXAMPLE 7 Area of a Parametric Surface

Here the symbols u and v play the part of r and θ in part (b) of Example 2.

▶ Find the area of the cone $\mathbf{r} = (u\cos v)\mathbf{i} + (u\sin v)\mathbf{j} + u\mathbf{k}$, where $0 \le u \le 1, 0 \le v \le 2\pi$.

Solution The surface is an upper portion of the cone shown in Figure 15.5.3(c). First we compute

$$\frac{\partial \mathbf{r}}{\partial u} = \cos v \mathbf{i} + \sin v \mathbf{j} + \mathbf{k}$$

$$\frac{\partial \mathbf{r}}{\partial v} = -u\sin v \mathbf{i} + u\cos v \mathbf{j}$$

and then form the cross product

$$\frac{\partial \mathbf{r}}{\partial u} \times \frac{\partial \mathbf{r}}{\partial v} = \begin{vmatrix} \mathbf{i} & \mathbf{j} & \mathbf{k} \\ \cos v & \sin v & 1 \\ -u\sin v & u\cos v & 0 \end{vmatrix} = -u\cos v \mathbf{i} - u\sin v \mathbf{j} + u\mathbf{k}. \tag{12}$$

The magnitude of the vector in (12) is

$$\left| \frac{\partial \mathbf{r}}{\partial u} \times \frac{\partial \mathbf{r}}{\partial v} \right| = \sqrt{u^2\cos^2 v + u^2\sin^2 v + u^2} = \sqrt{2}u.$$

Thus, from (11) the area is

$$A(S) = \iint_R \left| \frac{\partial \mathbf{r}}{\partial u} \times \frac{\partial \mathbf{r}}{\partial v} \right| dA = \iint_R \sqrt{2}u \, du \, dv$$

$$= \sqrt{2} \int_0^{2\pi} \int_0^1 u \, du \, dv$$

$$= \sqrt{2} \int_0^{2\pi} \frac{1}{2}u^2 \Big]_0^1 dv$$

$$= \frac{1}{2}\sqrt{2} \int_0^{2\pi} dv$$

$$= \sqrt{2}\pi.$$

∎

$\iint_R$ **NOTES FROM THE CLASSROOM**

An observation about Definition 15.5.1 is in order. In Example 7 we applied (11) to find the surface area of cone defined by the vector function $\mathbf{r} = (u\cos v)\mathbf{i} + (u\sin v)\mathbf{j} + u\mathbf{k}$, even though this surface S is *not* smooth on the region R in the uv-plane defined by $0 \le u \le 1$, $0 \le v \le 2\pi$. The fact that S is not smooth should make sense since one would not expect a tangent plane to exist at the sharp point at $\mathbf{r}(0, 0)$ or $(0, 0, 0)$. We can also see this from (12), because at $u = 0, v = 0$, $\partial\mathbf{r}/\partial u \times \partial\mathbf{r}/\partial v = \mathbf{0}$ and so by definition the surface is not smooth at $\mathbf{r}(0, 0)$. The point is this: We may use (11) even though the surface S is not smooth at a finite number of points located on the boundary of the region R.

Exercises 15.5 Answers to selected odd-numbered problems begin on page ANS-47

≡ Fundamentals

In Problems 1–4, find parametric equations for the given surface.

1. the plane $4x + 3y - z = 2$

2. the plane $2x + y = 1$

3. the hyperboloid $-x^2 + y^2 - z^2 = 1$ for $y \le -1$

4. the paraboloid $z = 5 - x^2 - y^2$

In Problems 5 and 6, find a vector-valued function $\mathbf{r}(u, v)$ for the given surface.

5. the parabolic cylinder $z = 1 - y^2$ for $-2 \le x \le 2$, $-8 \le z \le 1$

6. the elliptical cylinder $x^2/4 + y^2/9 = 1$

In Problems 7–10, identify the given surface by eliminating the parameters.

7. $x = \cos u, y = \sin u, z = v$

8. $x = u, y = v, z = u^2 + v^2$

9. $\mathbf{r}(u, v) = \sin u\mathbf{i} + \sin u \cos v\mathbf{j} + \sin u \sin v\mathbf{k}$

10. $\mathbf{r}(\phi, \theta) = 2\sin\phi\cos\theta\mathbf{i} + 3\sin\phi\sin\theta\mathbf{j} + 4\cos\phi\mathbf{k}$

In Problems 11–14, use the graph to obtain the parameter domain R corresponding to the portion of the given surface. For Problems 11 and 12 see Example 3; for Problems 13 and 14 see Example 5.

11.

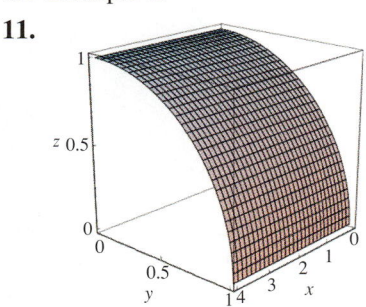

FIGURE 15.5.10 Graph for Problem 11

12.

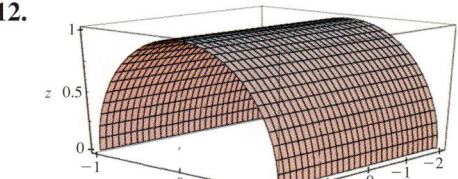

FIGURE 15.5.11 Graph for Problem 12

13.

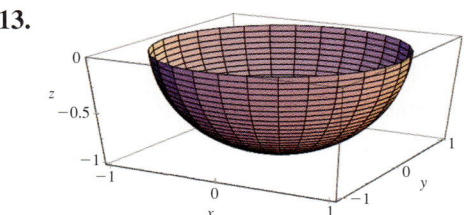

FIGURE 15.5.12 Graph for Problem 13

14.

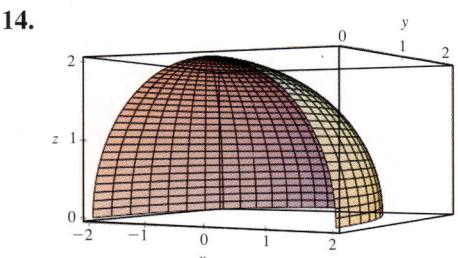

FIGURE 15.5.13 Graph for Problem 14

In Problems 15–22, find an equation of the tangent plane at the point on the surface corresponding to the given parameter values.

15. $x = 10\sin u, y = 10\cos u, z = v$; $u = \pi/6, v = 2$

16. $x = u\cos v, y = u\sin v, z = u^2 + v^2$; $u = 1, v = 0$

17. $\mathbf{r}(u, v) = (u^2 + v)\mathbf{i} + (u + v)\mathbf{j} + (u^2 - v^2)\mathbf{k}; u = 1, v = 2$

18. $\mathbf{r}(u, v) = 4u\mathbf{i} + 3u^2\cos v\mathbf{j} + 3u^2\sin v\mathbf{k}$; $u = -1, v = \pi/3$

19. $\mathbf{r}(u, v) = u\mathbf{i} + v\mathbf{j} + uv\mathbf{k}$; $u = 3, v = 3$

20. $\mathbf{r}(u, v) = u\sin v\mathbf{i} + u\cos v\mathbf{j} + u\mathbf{k}$; $u = 1, v = \pi/4$

21. $\mathbf{r}(u, v) = (u + v)\mathbf{i} + (v - u)\mathbf{j} + uv\mathbf{k}$; $\quad u = -2, v = 1$

22. $\mathbf{r}(u, v) = uv\mathbf{i} + (v + e^u)\mathbf{j} + (u + e^v)\mathbf{k}$; $\quad u = 0, v = \ln 3$

In Problems 23 and 24, find an equation of the tangent plane to the surface at the given point.

23. $x = u - v, y = 2u + 3v, z = u^2 + v^2$; $\quad (1, 7, 5)$

24. $\mathbf{r}(u, v) = v^2\mathbf{i} + (u - v)\mathbf{j} + u^2\mathbf{k}$; $\quad (1, 3, 16)$

In Problems 25–30, find the area of the given surface. If instructed, use a CAS to plot the surface.

25. the portion of the plane $\mathbf{r} = (2u - v)\mathbf{i} + (u + v + 1)\mathbf{j} + u\mathbf{k}$ for $0 \le u \le 2, -1 \le v \le 1$

26. the portion of the plane $x + y + z = 1$ inside the cylinder $x^2 + y^2 = 1$

27. the portion of $\mathbf{r}(u, v) = u\mathbf{i} + v\mathbf{j} + (u^2 + v^2)\mathbf{k}$ for $0 \le z \le 4$

28. the portion of $\mathbf{r}(r, \theta) = r\cos\theta\mathbf{i} + r\sin\theta\mathbf{j} + r\mathbf{k}$ for $0 \le r \le 2, 0 \le \theta \le 2\pi$

29. the surface $x = r\cos\theta, y = r\sin\theta, z = \theta, 0 \le r \le 2$, $0 \le \theta \le 2\pi$

30. the sphere $x = a\sin\phi\cos\theta, y = a\sin\phi\sin\theta, z = a\cos\phi$ for $0 \le \phi \le \pi, 0 \le \theta \le 2\pi$

In Problems 31–34, use Problem 30 as an aid in finding parametric equations for the indicated portion of the sphere $x^2 + y^2 + z^2 = 4$. In each case find the area of that portion of the sphere.

31. the portion of the sphere below the plane $z = 1$

32. the portion of the sphere below the plane $z = 1$ but above the plane $z = 0$

33. the portion of the sphere above the plane $z = \sqrt{2}$

34. the portion of the sphere outside the cylinder $x^2 + y^2 = 2$

35. Consider the cone given in (3) of Example 2, for $0 \le r \le 1, 0 \le \theta \le 2\pi$.
  (a) Sketch or plot using a CAS the grid lines corresponding to $r = \frac{1}{2}$ and $r = 1$. Color the curves red.
  (b) Sketch or plot using a CAS the grid lines corresponding to $\theta = \pi/2$ and $\theta = 3\pi/2$. Color the curves blue.
  (c) Superimpose the four grid lines in parts (a) and (b) on the same coordinate axes.

36. Consider the sphere given in (4) of Example 5, for $a = 2, 0 \le \phi \le \pi, 0 \le \theta \le 2\pi$.
  (a) Sketch or plot using a CAS the grid lines corresponding to $\phi = \pi/3$ and $\phi = 2\pi/3$. Color the curves red.
  (b) Sketch or plot using a CAS the grid lines corresponding to $\theta = \pi/4$ and $\theta = 5\pi/4$. Color the curves blue.
  (c) Superimpose the four grid lines in parts (a) and (b) on the same coordinate axes.

≡ Calculator/CAS Problems

In Problems 37–42, match the surface given in the figure with the plot of one of the vector-valued functions $\mathbf{r}(u, v)$ in (a)–(f).

Use a CAS and experiment with different parameter domains and viewpoints.
  (a) $\mathbf{r}(u, v) = \sin u\mathbf{i} + \sin v\mathbf{j} + \sin(u + v)\mathbf{k}$
  (b) $\mathbf{r}(u, v) = \sin^3 u\cos^3 v\mathbf{i} + \sin^3 u\sin^3 v\mathbf{j} + \cos^3 u\mathbf{k}$
  (c) $\mathbf{r}(u, v) = (u + 2\cos v)\mathbf{i} + 2\sin v\mathbf{j} + u\mathbf{k}$
  (d) $\mathbf{r}(u, v) = u\mathbf{i} + v\mathbf{j} + u^2 v^4\mathbf{k}$
  (e) $\mathbf{r}(u, v) = u\mathbf{i} + u^2\cos v\mathbf{j} + u^2\sin v\mathbf{k}$
  (f) $\mathbf{r}(u, v) = e^u\cos v\mathbf{i} + e^u\sin v\mathbf{j} + u\mathbf{k}$

37.

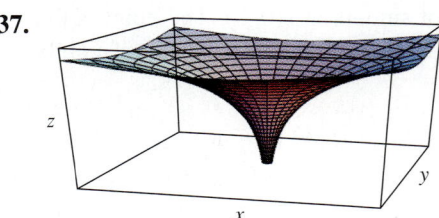

FIGURE 15.5.14   Graph for Problem 37

38.

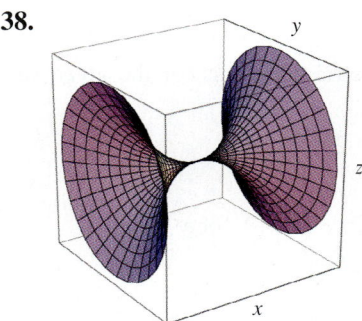

FIGURE 15.5.15   Graph for Problem 38

39

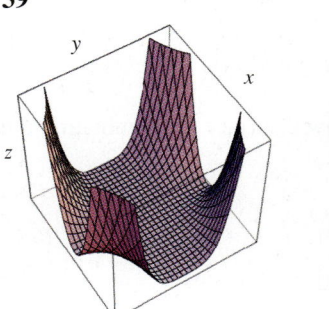

FIGURE 15.5.16   Graph for Problem 39

40.

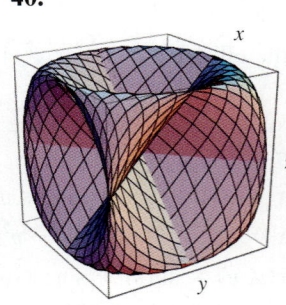

FIGURE 15.5.17   Graph for Problem 40

41.

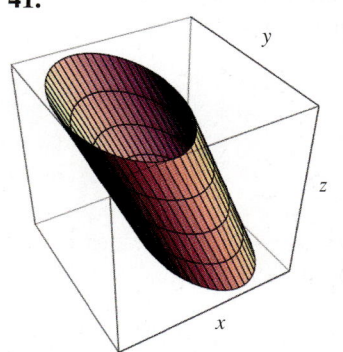

FIGURE 15.5.18   Graph for Problem 41

42.

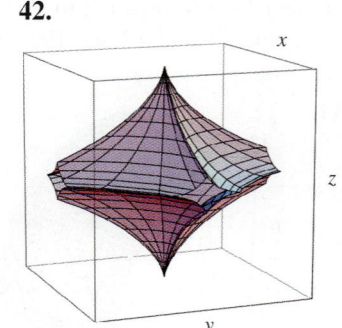

FIGURE 15.5.19   Graph for Problem 42

43. Use a CAS to plot the **torus** given by

$$\mathbf{r}(\phi, \theta) = (R - \sin\phi)\cos\theta\mathbf{i} + (R - \sin\phi)\sin\theta\mathbf{j} + \cos\phi\mathbf{k}$$

for $R = 5$ and $0 \leq \phi \leq 2\pi, 0 \leq \theta \leq 2\pi$. Experiment with different aspect ratios and viewpoints.

44. Show that for a constant $R > 1$ the surface area of the torus in Problem 43 corresponding to the given parameter domain is $A(S) = 4\pi^2 R$.

≡ Think About It

45. Find a different parameterization of the plane in Problem 1 than the one given in the answer section.

46. Find the area in Problem 11 without integration.

47. If a curve defined by $y = f(x), a \leq x \leq b$, is revolved about the x-axis, then parametric equations for the surface of revolution S are

$$x = u, y = f(u)\cos v, z = f(u)\sin v, a \leq u \leq b, 0 \leq v \leq 2\pi.$$

If f' is continuous and $f(x) \geq 0$ for all x in the interval $[a, b]$, then use (11) to show that the area of S is

$$A(S) = \int_a^b f(x)\sqrt{1 + [f'(x)]^2}\, dx.$$

See (3) of Section 6.6.

48. **(a)** Use Problem 47 to find parametric equations of the surface generated by revolving the graph of $f(x) = \sin x, -2\pi \leq x \leq 2\pi$, about the x-axis.

  (b) Use a CAS to plot the graph of the parametric surface in part (a).

  (c) Use a CAS and the formula in Problem 47 to find the area of the surface of revolution in part (a) by first finding the area of the surface corresponding to the parameter domain $0 \leq u \leq \pi, 0 \leq v \leq 2\pi$.

49. Suppose $\mathbf{r}_0 = x_0\mathbf{i} + y_0\mathbf{j} + z_0\mathbf{k}$ is the position vector of the point (x_0, y_0, z_0) and $\mathbf{v}_1$ and $\mathbf{v}_2$ are constant but nonparallel vectors. Discuss: What is the surface with the vector equation $\mathbf{r}(s, t) = \mathbf{r}_0 + s\mathbf{v}_1 + t\mathbf{v}_2$, where s and t are parameters?

50. Reread Example 5 of this section. Then find parametric equations of a sphere of radius 5 with center $(2, 3, 4)$.

15.6 Surface Integrals

■ Introduction The last kind of integral that we shall consider in this text is called a **surface integral** and involves a function f of three variables defined on a surface S.

■ Surface Integrals The steps preparatory to the definition of this integral are similar to combinations of the steps leading to the line integral, with respect to arc length, and the steps leading to the double integral. Let $w = f(x, y, z)$ be a function defined in a region of 3-space that contains a surface S, which is the graph of a function $z = g(x, y)$. Let the projection R of the surface onto the xy-plane be either a Type I or a Type II region.

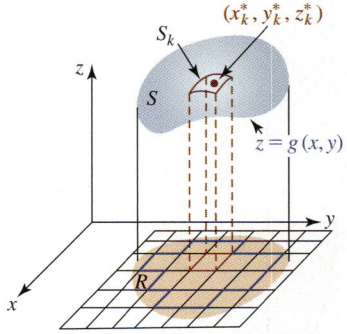

FIGURE 15.6.1 Sample point on the kth element S_k of surface

- Divide the surface S into n patches S_k with areas ΔS_k that correspond to a partition P of R into n rectangles R_k with areas ΔA_k.
- Let $\|P\|$ be the norm of the partition or the length of the longest diagonal of the R_k.
- Choose a sample point (x_k^*, y_k^*, z_k^*) on each patch S_k as shown in FIGURE 15.6.1.
- Form the sum

$$\sum_{k=1}^{n} f(x_k^*, y_k^*, z_k^*)\, \Delta S_k.$$

Definition 15.6.1 Surface Integral

Let f be a function of three variables x, y, and z defined in a region of space that contains a surface S. Then the **surface integral** of f over S is

$$\iint_S f(x, y, z)\, dS = \lim_{\|P\| \to 0} \sum_{k=1}^{n} f(x_k^*, y_k^*, z_k^*)\, \Delta S_k. \tag{1}$$

■ **Method of Evaluation** Recall from (3) of Section 14.6 that if $z = g(x, y)$ is the equation of a surface S, then the differential of the surface area is

$$dS = \sqrt{1 + [g_x(x, y)]^2 + [g_y(x, y)]^2}\, dA.$$

Thus, if f, g, g_x, and g_y are continuous throughout a region of 3-space containing S, we can evaluate (1) by means of a double integral:

$$\iint_S f(x, y, z)\, dS = \iint_R f(x, y, g(x, y))\sqrt{1 + [g_x(x, y)]^2 + [g_y(x, y)]^2}\, dA. \qquad (2)$$

Note that when $f(x, y, z) = 1$, (1) reduces to the formula for surface area (2) of Section 14.6:

$$\iint_S dS = \lim_{\|P\| \to 0} \sum_{k=1}^{n} \Delta S_k = A(S).$$

■ **Projection of S into Other Planes** If $y = g(x, z)$ is the equation of a surface S that projects onto a region R of the xz-plane, then the surface integral of f over S is given by

$$\iint_S f(x, y, z)\, dS = \iint_R f(x, g(x, z), z)\sqrt{1 + [g_x(x, z)]^2 + [g_z(x, z)]^2}\, dA. \qquad (3)$$

Similarly, if $x = g(y, z)$ is the equation of a surface S that projects onto the yz-plane, then the analogue of (3) is

$$\iint_S f(x, y, z)\, dS = \iint_R f(g(y, z), y, z)\sqrt{1 + [g_y(y, z)]^2 + [g_z(y, z)]^2}\, dA. \qquad (4)$$

■ **Mass of a Surface** Suppose $\rho(x, y, z)$ represents the density of a surface S at a point (x, y, z), or the mass per unit of surface area. Then the **mass** m of the surface is

$$m = \iint_S \rho(x, y, z)\, dS. \qquad (5)$$

EXAMPLE 1   Mass of a Surface

Find the mass of the surface of the paraboloid $z = 1 + x^2 + y^2$ in the first octant for $1 \le z \le 5$ if the density at a point P on the surface is directly proportional to the distance from the xy-plane.

Solution   The surface in question and its projection onto the xy-plane are shown in FIGURE 15.6.2. Now, since $\rho(x, y, z) = kz$, $g(x, y) = 1 + x^2 + y^2$, $g_x = 2x$, $g_y = 2y$, formulas (5) and (2) give

$$m = \iint_S kz\, dS = k\iint_R (1 + x^2 + y^2)\sqrt{1 + 4x^2 + 4y^2}\, dA.$$

By switching to polar coordinates, we obtain

$$m = k\int_0^{\pi/2}\int_0^2 (1 + r^2)\sqrt{1 + 4r^2}\, r\, dr\, d\theta$$

$$= k\int_0^{\pi/2}\int_0^2 [r(1 + 4r^2)^{1/2} + r^3(1 + 4r^2)^{1/2}]\, dr\, d\theta \quad \leftarrow \text{integration by parts}$$

$$= k\int_0^{\pi/2}\left[\frac{1}{12}(1 + 4r^2)^{3/2} + \frac{1}{12}r^2(1 + 4r^2)^{3/2} - \frac{1}{120}(1 + 4r^2)^{5/2}\right]_0^2 d\theta$$

$$= \frac{1}{2}k\pi\left[\frac{5}{12}(17)^{3/2} - \frac{1}{120}(17)^{5/2} - \frac{3}{40}\right] \approx 30.16k. \qquad \blacksquare$$

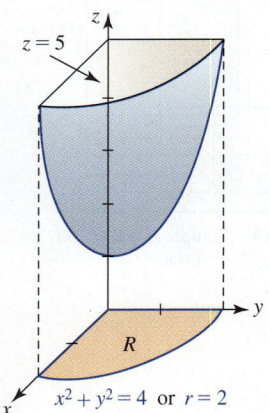

FIGURE 15.6.2   Surface in Example 1

EXAMPLE 2   Region *R* in *xz*-Plane

Evaluate $\iint_S xz^2 \, dS$, where S is that portion of the cylinder $y = 2x^2 + 1$ in the first octant bounded by $x = 0$, $x = 2$, $z = 4$, and $z = 8$.

Solution   We shall use (3) with $g(x, z) = 2x^2 + 1$ and R is the rectangular region in the *xz*-plane shown in FIGURE 15.6.3. Since $g_x(x, z) = 4x$ and $g_z(x, z) = 0$, it follows that

$$\iint_S xz^2 \, dS = \int_0^2 \int_4^8 xz^2 \sqrt{1 + 16x^2} \, dz \, dx$$

$$= \frac{1}{3} \int_0^2 z^3 x \sqrt{1 + 16x^2} \Big]_4^8 \, dx$$

$$= \frac{448}{3} \int_0^2 x(1 + 16x^2)^{1/2} \, dx = \frac{28}{9}(1 + 16x^2)^{3/2} \Big]_0^2$$

$$= \frac{28}{9}[65^{3/2} - 1] \approx 1627.3. \qquad \blacksquare$$

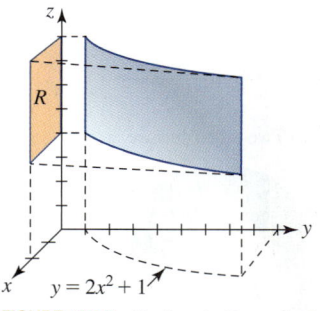

FIGURE 15.6.3   Surface in Example 2

■ **Parametric Surfaces**   If *S* is defined parametrically by the vector function

$$\mathbf{r}(u, v) = x(u, v)\mathbf{i} + y(u, v)\mathbf{j} + z(u, v)\mathbf{k},$$

where (u, v) is in the parameter domain D of the *uv*-plane and $f(x, y, z)$ is continuous on S, we have the following result.

Theorem 15.6.1   Surface Integral

Let *S* be a smooth parametric surface defined by the vector equation

$$\mathbf{r}(u, v) = x(u, v)\mathbf{i} + y(u, v)\mathbf{j} + z(u, v)\mathbf{k},$$

where (u, v) varies over the parameter region R in the *uv*-plane, and let $f(x, y, z)$ be continuous on *S*. Then

$$\iint_S f(x, y, z) \, dS = \iint_R f(x(u, v), y(u, v), z(u, v)) \left| \frac{\partial \mathbf{r}}{\partial u} \times \frac{\partial \mathbf{r}}{\partial v} \right| dA. \qquad (6)$$

Formula (6) can be thought to be the surface-integral analogue of the line integral $\int_a^b f(x(t), y(t), z(t)) |\mathbf{r}'(t)| \, dt$, (14) of Section 15.1.

EXAMPLE 3   Parametric Surface

Evaluate the surface integral $\iint_S \sqrt{1 + x^2 + y^2} \, dS$, where S is the surface defined by the vector function $\mathbf{r}(u, v) = u\cos v\mathbf{i} + u\sin v\mathbf{j} + v\mathbf{k}$, where $0 \le u \le 2$, $0 \le v \le 4\pi$.

Solution   The graph of $\mathbf{r}(u, v)$ shown in FIGURE 15.6.4 is called a *circular helicoid*. The boundary of a circular helicoid is a circular helix. See *Notes from the Classroom* in Section 10.2.

Substituting $x = u\cos v$ and $y = u\sin v$ into the integrand and simplifying gives:

$$\sqrt{1 + x^2 + y^2} = \sqrt{1 + u^2\cos^2 v + u^2\sin^2 v} = \sqrt{1 + u^2}.$$

Next,

$$\frac{\partial \mathbf{r}}{\partial u} \times \frac{\partial \mathbf{r}}{\partial v} = \begin{vmatrix} \mathbf{i} & \mathbf{j} & \mathbf{k} \\ \cos v & \sin v & 0 \\ -u\sin v & u\cos v & 1 \end{vmatrix} = \sin v\mathbf{i} - \cos v\mathbf{j} + u\mathbf{k}$$

$$\left| \frac{\partial \mathbf{r}}{\partial u} \times \frac{\partial \mathbf{r}}{\partial v} \right| = \sqrt{\sin^2 v + \cos^2 v + u^2} = \sqrt{1 + u^2}.$$

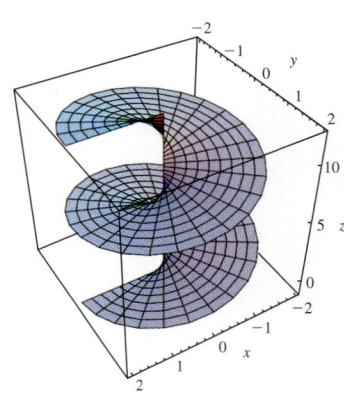

FIGURE 15.6.4   Helicoid in Example 3

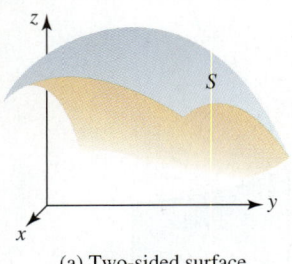

(a) Two-sided surface

(b) One-sided surface

FIGURE 15.6.5 Oriented surface (a); non-oriented surface (b)

The given integral becomes

$$\iint_S \sqrt{1 + x^2 + y^2}\, dS = \iint_R (\sqrt{1 + u^2})^2\, dA$$

$$= \int_0^{4\pi} \int_0^2 (1 + u^2)\, du\, dv$$

$$= \frac{14}{3} \int_0^{4\pi} dv$$

$$= \frac{56}{3}\pi. \qquad \blacksquare$$

■ **Oriented Surfaces** In Example 4 we shall evaluate a surface integral of a vector field. In order to do this we need to examine the concept of an **oriented surface**. In rough terms, an oriented surface S, such as that given in FIGURE 15.6.5(a), has two sides that could be painted different colors. The Möbius strip, named after the German mathematician **August Möbius** (1790–1868) and shown in Figure 15.6.5(b), is not an oriented surface and is one-sided. To construct a Möbius strip cut out a long strip of paper, give one end a half-twist, and then attach both ends by tape. A person who starts to paint the surface of a Möbius strip at a point will paint the entire surface and return to the starting point.

Specifically, we say a smooth surface S is an oriented surface if there exists a continuous unit normal function $\mathbf{n}$ defined at each point (x, y, z) on the surface. The vector field $\mathbf{n}(x, y, z)$ is called the **orientation** of S. But since a unit normal to the surface S at (x, y, z) can be either $\mathbf{n}(x, y, z)$ or $-\mathbf{n}(x, y, z)$, an oriented surface has two orientations. See FIGURE 15.6.6(a), (b), and (c). The Möbius strip shown again in Figure 15.6.6(d) is not an oriented surface because if a unit normal $\mathbf{n}$ starts at P on the surface and moves *once* around the strip on the curve C, it ends up on the opposite side of the strip at P and so points in the opposite direction. A surface S defined by $z = g(x, y)$ has an **upward orientation** (Figure 15.6.6(b)) when the unit normals are directed upward, that is, have positive $\mathbf{k}$ components, and has a **downward orientation** (Figure 15.6.6(c)) when the unit normals are directed downward, that is, have negative $\mathbf{k}$ components.

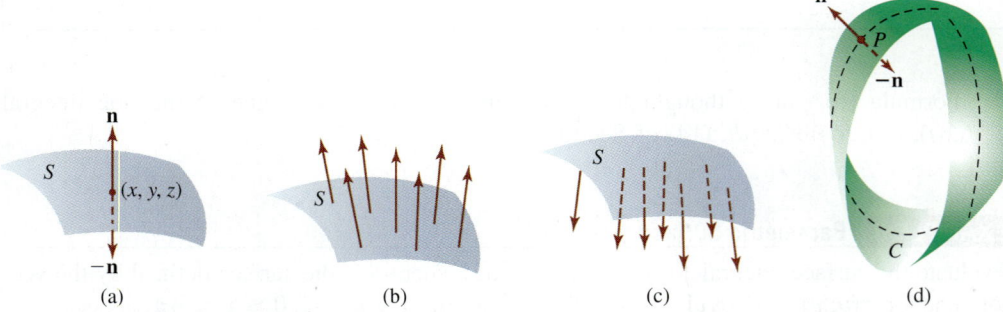

FIGURE 15.6.6 Upward orientation in (b); downward orientation in (c); no orientation in (d)

If a smooth surface S is defined implicitly by $h(x, y, z) = 0$, then recall that a unit normal to the surface is

$$\mathbf{n} = \frac{\nabla h}{|\nabla h|},$$

where $\nabla h = (\partial h/\partial x)\mathbf{i} + (\partial h/\partial y)\mathbf{j} + (\partial h/\partial z)\mathbf{k}$ is the gradient of h. If S is defined by an explicit function $z = g(x, y)$, then we can use $h(x, y, z) = z - g(x, y) = 0$ or $h(x, y, z) = g(x, y) - z = 0$ depending on the orientation of S.

As we see in the next example, the two orientations of a oriented *closed* surface are **outward** and **inward**. A **closed surface** is defined to be the boundary of a finite solid such as the surface of a sphere.

EXAMPLE 4   Region R in xz-Plane

Consider the sphere $x^2 + y^2 + z^2 = a^2$ of radius $a > 0$. If we define $h(x, y, z) = x^2 + y^2 + z^2 - a^2$, then

$$\nabla h = 2x\mathbf{i} + 2y\mathbf{j} + 2z\mathbf{k} \qquad \text{and} \qquad |\nabla h| = \sqrt{4x^2 + 4y^2 + 4z^2} = 2a.$$

Then, the two orientations of the surface are

$$\mathbf{n} = \frac{x}{a}\mathbf{i} + \frac{y}{a}\mathbf{j} + \frac{z}{a}\mathbf{k} \qquad \text{and} \qquad \mathbf{n}_1 = -\mathbf{n} = -\frac{x}{a}\mathbf{i} - \frac{y}{a}\mathbf{j} - \frac{z}{a}\mathbf{k}.$$

The vector field $\mathbf{n}$ defines an outward orientation, whereas $\mathbf{n}_1 = -\mathbf{n}$ defines an inward orientation. See FIGURE 15.6.7. ∎

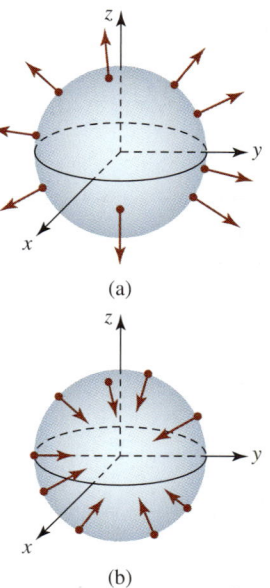

(a)

(b)

FIGURE 15.6.7 Outward orientation in (a); inward orientation in (b) in Example 4

■ **Integrals of Vector Fields**   If

$$\mathbf{F}(x, y, z) = P(x, y, z)\mathbf{i} + Q(x, y, z)\mathbf{j} + R(x, y, z)\mathbf{k}$$

is the velocity field of fluid, then, as shown in FIGURE 15.6.8(b), the volume of the fluid flowing through an element of surface area ΔS per unit time is approximated by

$$(\text{height}) - (\text{area of base}) = (\text{comp}_\mathbf{n}\mathbf{F})\Delta S = (\mathbf{F} \cdot \mathbf{n})\Delta S,$$

where $\mathbf{n}$ is a unit normal to the surface. The total volume of a fluid passing through S per unit time is called the **flux of F through S** and is given by

$$\text{flux} = \iint\limits_S (\mathbf{F} \cdot \mathbf{n})\, dS. \tag{7}$$

In the case of a closed surface S, if $\mathbf{n}$ is the outer (inner) normal, then (7) gives the volume of fluid flowing out (in) through S per unit time.

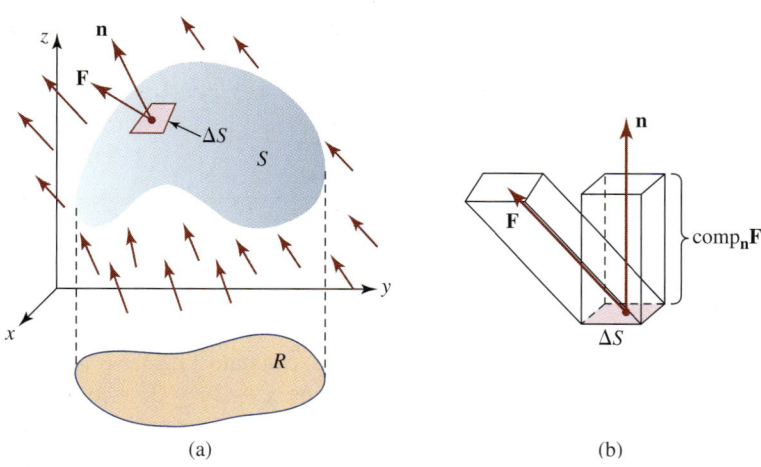

(a)   (b)

FIGURE 15.6.8   Fluid flowing through a surface

EXAMPLE 5   Flux

Let $\mathbf{F}(x, y, z) = z\mathbf{j} + z\mathbf{k}$ represent the flow of a liquid. Find the flux of $\mathbf{F}$ through the surface S given by that part of the plane $z = 6 - 3x - 2y$ in the first octant oriented upward.

Solution   The vector field and the surface are illustrated in FIGURE 15.6.9. By defining the plane by $h(x, y, z) = 3x + 2y + z - 6 = 0$, we see that a unit normal with positive $\mathbf{k}$ component is

$$\mathbf{n} = \frac{\nabla h}{|\nabla h|} = \frac{3}{\sqrt{14}}\mathbf{i} + \frac{2}{\sqrt{14}}\mathbf{j} + \frac{1}{\sqrt{14}}\mathbf{k}.$$

Because $\mathbf{F} \cdot \mathbf{n} = 3z/\sqrt{14}$ we have,

$$\text{flux} = \iint\limits_S (\mathbf{F} \cdot \mathbf{n})\, dS = \frac{1}{\sqrt{14}}\iint\limits_S 3z\, dS.$$

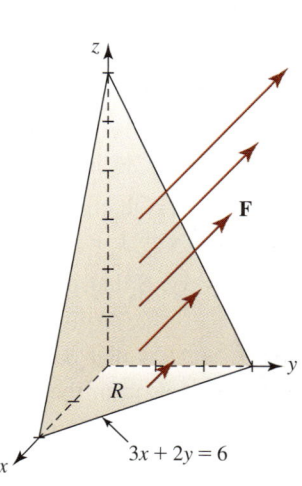

FIGURE 15.6.9   Surface in Example 5

Using the projection R of the surface onto the xy-plane shown in the figure, the last integral can be written

$$\text{flux} = \frac{1}{\sqrt{14}} \iint_R 3(6 - 3x - 2y)(\sqrt{14}\, dA)$$

$$= 3 \int_0^2 \int_0^{3 - 3x/2} (6 - 3x - 2y)\, dy\, dx = 18. \qquad \blacksquare$$

Depending on the nature of the vector field, the integral in (7) can represent other kinds of flux. For example, (7) could also give electric flux, magnetic flux, flux of heat, and so on.

$$\iint_S \quad \textbf{NOTES FROM THE CLASSROOM}$$

If the surface S is piecewise-smooth, we express a surface integral over S as the sum of the surface integrals over the various pieces of the surface. If S is given by $S = S_1 \cup \cdots \cup S_n$, where the surfaces intersect only at their boundaries, then

$$\iint_S f(x, y, z)\, dS = \iint_{S_1} f(x, y, z)\, dS + \cdots + \iint_{S_n} f(x, y, z)\, dS.$$

For example, suppose S is the oriented piecewise-smooth closed surface bounded by the paraboloid $z = x^2 + y^2$ (S_1) and the plane $z = 1$ (S_2). Then, the flux of a vector field $\mathbf{F}$ out of the surface S is

$$\iint_S \mathbf{F} \cdot \mathbf{n}\, dS = \iint_{S_1} \mathbf{F} \cdot \mathbf{n}\, dS + \iint_{S_2} \mathbf{F} \cdot \mathbf{n}\, dS,$$

where we take S_1 oriented downward and S_2 oriented upward. See FIGURE 15.6.10 and Problem 21 in Exercises 15.6.

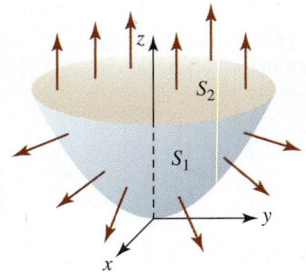

FIGURE 15.6.10   Piecewise-defined surface S

Exercises 15.6   Answers to selected odd-numbered problems begin on page ANS-47.

≡ Fundamentals

In Problems 1–10, evaluate $\iint_S f(x, y, z)\, dS$.

1. $f(x, y, z) = x$; S the portion of the cylinder $z = 2 - x^2$ in the first octant bounded by $x = 0$, $y = 0$, $y = 4$, $z = 0$

2. $f(x, y, z) = xy(9 - 4z)$; same surface S as in Problem 1

3. $f(x, y, z) = xz^3$; S the single-napped cone $z = \sqrt{x^2 + y^2}$ inside the cylinder $x^2 + y^2 = 1$

4. $f(x, y, z) = x + y + z$; S the single-napped cone $z = \sqrt{x^2 + y^2}$ between $z = 1$ and $z = 4$

5. $f(x, y, z) = (x^2 + y^2)z$; S that portion of the sphere $x^2 + y^2 + z^2 = 36$ in the first octant

6. $f(x, y, z) = z^2$; S that portion of the plane $z = x + 1$ within the cylinder $y = 1 - x^2$, $0 \le y \le 1$

7. $f(x, y, z) = xy$; S that portion of the paraboloid $2z = 4 - x^2 - y^2$ within $0 \le x \le 1$, $0 \le y \le 1$

8. $f(x, y, z) = 2z$; S that portion of the paraboloid $2z = 1 + x^2 + y^2$ in the first octant bounded by $x = 0$, $y = \sqrt{3}x$, $z = 1$

9. $f(x, y, z) = 24\sqrt{yz}$; S that portion of the cylinder $y = x^2$ in the first octant bounded by $y = 0$, $y = 4$, $z = 0$, $z = 3$

10. $f(x, y, z) = (1 + 4y^2 + 4z^2)^{1/2}$; S that portion of the paraboloid $x = 4 - y^2 - z^2$ in the first octant outside the cylinder $y^2 + z^2 = 1$

In Problems 11 and 12, evaluate $\iint_S (3z^2 + 4yz)\, dS$, where S is that portion of the plane $x + 2y + 3z = 6$ in the first octant. Use the projection of S onto the coordinate plane indicated in the given figure.

11.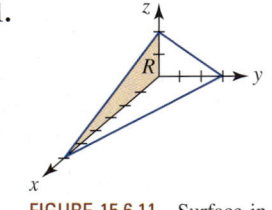

FIGURE 15.6.11   Surface in Problem 11

12.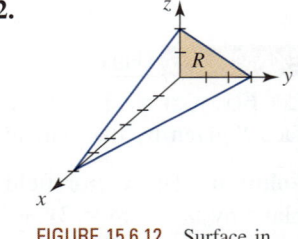

FIGURE 15.6.12   Surface in Problem 12

In Problems 13 and 14, find the mass of the given surface with the indicated density function.

13. S that portion of the plane $x + y + z = 1$ in the first octant; density at a point P directly proportional to the square of the distance from the yz-plane

14. S the hemisphere $z = \sqrt{4 - x^2 - y^2}$; $\rho(x, y, z) = |xy|$

In Problems 15–20, let **F** be a vector field. Find the flux of **F** through the given surface. Assume the surface S is oriented upward.

15. $\mathbf{F} = x\mathbf{i} + 2z\mathbf{j} + y\mathbf{k}$; S that portion of the cylinder $y^2 + z^2 = 4$ in the first octant bounded by $x = 0, x = 3$, $y = 0, z = 0$

16. $\mathbf{F} = z\mathbf{k}$; S that part of the paraboloid $z = 5 - x^2 - y^2$ inside the cylinder $x^2 + y^2 = 4$

17. $\mathbf{F} = x\mathbf{i} + y\mathbf{j} + z\mathbf{k}$; same surface S as in Problem 16

18. $\mathbf{F} = -x^3 y\mathbf{i} + yz^3\mathbf{j} + xy^3\mathbf{k}$; S that portion of the plane $z = x + 3$ in the first octant within the cylinder $x^2 + y^2 = 2x$

19. $\mathbf{F} = \frac{1}{2}x^2\mathbf{i} + \frac{1}{2}y^2\mathbf{j} + z\mathbf{k}$; S that portion of the paraboloid $z = 4 - x^2 - y^2$ for $0 \le z \le 4$

20. $\mathbf{F} = e^y\mathbf{i} + e^x\mathbf{j} + 18y\mathbf{k}$; S that portion of the plane $x + y + z = 6$ in the first octant

21. Find the flux of $\mathbf{F} = y^2\mathbf{i} + x^2\mathbf{j} + 5z\mathbf{k}$ out of the closed surface S given in Figure 15.6.10.

22. Find the flux of $\mathbf{F} = -y\mathbf{i} + x\mathbf{j} + 6z^2\mathbf{k}$ out of the closed surface S bounded by the paraboloids $z = 4 - x^2 - y^2$ and $z = x^2 + y^2$.

≡ Applications

23. Let $T(x, y, z) = x^2 + y^2 + z^2$ represent temperature and let the flow of heat be given by the vector field $\mathbf{F} = -\nabla T$. Find the flux of heat out of the sphere $x^2 + y^2 + z^2 = a^2$. [*Hint*: The surface area of a sphere of radius a is $4\pi a^2$.]

24. Find the flux of $\mathbf{F} = x\mathbf{i} + y\mathbf{j} + z\mathbf{k}$ out of the unit cube defined by $0 \le x \le 1, 0 \le y \le 1, 0 \le z \le 1$. See **FIGURE 15.6.13**. Use the fact that the flux out of the cube is the sum of the fluxes out of the sides.

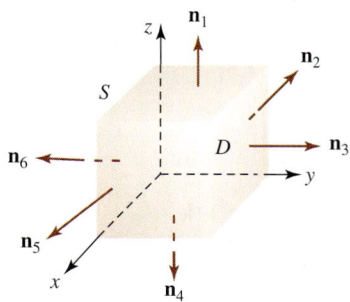

FIGURE 15.6.13 Cube in Problem 24

25. Coulomb's law states that the electric field **E** due to a point charge q at the origin is given by $\mathbf{E} = kq\mathbf{r}/|\mathbf{r}|^3$, where k is a constant and $\mathbf{r} = x\mathbf{i} + y\mathbf{j} + z\mathbf{k}$. Determine the flux out of a sphere $x^2 + y^2 + z^2 = a^2$.

26. If $\sigma(x, y, z)$ is charge density in an electrostatic field, then the total charge on a surface S is $Q = \iint_S \sigma(x, y, z)\, dS$. Find the total charge on that part of the hemisphere $z = \sqrt{16 - x^2 - y^2}$ that is inside the cylinder $x^2 + y^2 = 9$ if the charge density at a point P on the surface is directly proportional to the distance from the xy-plane.

27. The coordinates of the **centroid** of a surface are defined by

$$\bar{x} = \frac{\displaystyle\iint_S x\, dS}{A(S)}, \bar{y} = \frac{\displaystyle\iint_S y\, dS}{A(S)}, \bar{z} = \frac{\displaystyle\iint_S z\, dS}{A(S)},$$

where $A(S)$ is the area of the surface. Find the centroid of that portion of the plane $2x + 3y + z = 6$ in the first octant.

28. Use the information in Problem 27 to find the centroid of the hemisphere $z = \sqrt{a^2 - x^2 - y^2}$.

29. The **moment of inertia** of a surface S with density $\rho(x, y, z)$ at a point (x, y, z) about the z-axis is given by

$$I_z = \iint_S (x^2 + y^2)\, \rho(x, y, z)\, dS.$$

Consider the conical surface $z = 4 - \sqrt{x^2 + y^2}, 0 \le z \le 4$, with constant density k.

(a) Use Problem 27 to find the centroid of the surface.
(b) Find the moment of inertia of the surface about the z-axis.

≡ Think About It

30. Let $z = f(x, y)$ be the equation of a surface S and let **F** be the vector field $\mathbf{F}(x, y, z) = P(x, y, z)\mathbf{i} + Q(x, y, z)\mathbf{j} + R(x, y, z)\mathbf{k}$. Show that

$$\iint_S (\mathbf{F} \cdot \mathbf{n})\, dS = \iint_R \left[-P(x, y, z)\frac{\partial z}{\partial x} - Q(x, y, z)\frac{\partial z}{\partial y} + R(x, y, z) \right] dA.$$

15.7 Curl and Divergence

❚ Introduction We have seen that if a vector force field **F** is conservative, then it can be written as the gradient of a potential function ϕ:

$$\mathbf{F} = \nabla\phi = \frac{\partial\phi}{\partial x}\mathbf{i} + \frac{\partial\phi}{\partial y}\mathbf{j} + \frac{\partial\phi}{\partial z}\mathbf{k}.$$

The vector differential operator, or del operator,

$$\nabla = \mathbf{i}\frac{\partial}{\partial x} + \mathbf{j}\frac{\partial}{\partial y} + \mathbf{k}\frac{\partial}{\partial z} \tag{1}$$

that is used in the gradient can also be combined with a vector field

$$\mathbf{F}(x, y, z) = P(x, y, z)\mathbf{i} + Q(x, y, z)\mathbf{j} + R(x, y, z)\mathbf{k} \tag{2}$$

in two different ways: in one case producing another vector field and in the other producing a scalar function.

Note: We will assume throughout the following discussion that P, Q, and R have continuous partial derivatives. Throughout an appropriate region of 3-space.

▌ Curl We begin by combining the differential operator (1) with the vector field (2) to produce another vector field called the **curl** of **F**.

Definition 15.7.1 Curl of a Vector Field

The **curl** of a vector field $\mathbf{F} = P\mathbf{i} + Q\mathbf{j} + R\mathbf{k}$ is the vector field

$$\text{curl } \mathbf{F} = \left(\frac{\partial R}{\partial y} - \frac{\partial Q}{\partial z}\right)\mathbf{i} + \left(\frac{\partial P}{\partial z} - \frac{\partial R}{\partial x}\right)\mathbf{j} + \left(\frac{\partial Q}{\partial x} - \frac{\partial P}{\partial y}\right)\mathbf{k}. \tag{3}$$

There is no need to memorize the complicated components in the vector field in (3). As a matter of practicality, (3) can be interpreted as a cross product. If we interpret (1) as a vector with components $\partial/\partial x$, $\partial/\partial y$, and $\partial/\partial z$, then curl **F** can be written as a cross product of ∇ and the vector **F**:

$$\text{curl } \mathbf{F} = \nabla \times \mathbf{F} = \begin{vmatrix} \mathbf{i} & \mathbf{j} & \mathbf{k} \\ \dfrac{\partial}{\partial x} & \dfrac{\partial}{\partial y} & \dfrac{\partial}{\partial z} \\ P & Q & R \end{vmatrix}. \tag{4}$$

EXAMPLE 1 Curl of a Vector Field

If $\mathbf{F} = (x^2 y^3 - z^4)\mathbf{i} + 4x^5 y^2 z\,\mathbf{j} - y^4 z^6 \mathbf{k}$, find curl **F**.

Solution From (4),

$$\text{curl } \mathbf{F} = \nabla \times \mathbf{F} = \begin{vmatrix} \mathbf{i} & \mathbf{j} & \mathbf{k} \\ \dfrac{\partial}{\partial x} & \dfrac{\partial}{\partial y} & \dfrac{\partial}{\partial z} \\ x^2 y^3 - z^4 & 4x^5 y^2 z & -y^4 z^6 \end{vmatrix}$$

$$= \left[\frac{\partial}{\partial y}(-y^4 z^6) - \frac{\partial}{\partial z}(4x^5 y^2 z)\right]\mathbf{i} - \left[\frac{\partial}{\partial x}(-y^4 z^6) - \frac{\partial}{\partial z}(x^2 y^3 - z^4)\right]\mathbf{j}$$

$$+ \left[\frac{\partial}{\partial x}(4x^5 y^2 z) - \frac{\partial}{\partial y}(x^2 y^3 - z^4)\right]\mathbf{k}$$

$$= (-4y^3 z^6 - 4x^5 y^2)\mathbf{i} - 4z^3 \mathbf{j} + (20x^4 y^2 z - 3x^2 y^2)\mathbf{k}. \qquad \blacksquare$$

If f is a scalar function with continuous second partial derivatives, then it is easily shown that

$$\text{curl}(\text{grad } f) = \nabla \times \nabla f = \mathbf{0}. \tag{5}$$

See Problem 23 in Exercises 15.7. Since a conservative vector field **F** is a gradient field, that is, there exists a potential function ϕ such that $\mathbf{F} = \nabla\phi$, it follows from (5) that *if* **F** *is conservative, then* curl $\mathbf{F} = \mathbf{0}$.

EXAMPLE 2 A Nonconservative Vector Field

Consider the vector field $\mathbf{F} = y\mathbf{i} + z\mathbf{j} + x\mathbf{k}$. From (4),

$$\text{curl } \mathbf{F} = \begin{vmatrix} \mathbf{i} & \mathbf{j} & \mathbf{k} \\ \dfrac{\partial}{\partial x} & \dfrac{\partial}{\partial y} & \dfrac{\partial}{\partial z} \\ y & z & x \end{vmatrix} = -\mathbf{i} - \mathbf{j} - \mathbf{k}.$$

Because curl $\mathbf{F} \neq \mathbf{0}$ we can conclude that **F** is not conservative. $\qquad \blacksquare$

Under the assumption that the component function $P, Q,$ and R of a vector field $\mathbf{F}$ are continuous and have continuous partial derivatives throughout some open region D of 3-space, we can also conclude that *if* curl $\mathbf{F} = \mathbf{0}$, *then* $\mathbf{F}$ *is conservative*. We summarize these observations in the next theorem.

Theorem 15.7.1 Equivalent Concepts

Suppose $\mathbf{F} = P\mathbf{i} + Q\mathbf{j} + R\mathbf{k}$ is a vector field where $P, Q,$ and R are continuous and have continuous first partial derivatives in some open region of 3-space. The vector field $\mathbf{F}$ is conservative if and only if curl $\mathbf{F} = \mathbf{0}$.

Note that when curl $\mathbf{F} = \mathbf{0}$, then the three components of the vector must be 0. From (3) we see that this means

$$\frac{\partial R}{\partial y} = \frac{\partial Q}{\partial z}, \quad \frac{\partial P}{\partial z} = \frac{\partial R}{\partial x}, \quad \frac{\partial Q}{\partial x} = \frac{\partial P}{\partial y}.$$

Now review (12) of Section 15.3.

■ Divergence There is another combination of partial derivatives of the component functions of a vector field that occurs frequently in science and engineering. Before stating the next definition, consider the following motivation.

If $\mathbf{F}(x, y, z) = P(x, y, z)\mathbf{i} + Q(x, y, z)\mathbf{j} + R(x, y, z)\mathbf{k}$ represents the velocity field of a fluid, then as we saw in Figure 15.6.8(b) the volume of the fluid flowing through an element of surface area ΔS per unit time, that is, the flux of the vector field F through the area ΔS, is approximately

$$(\text{height}) \cdot (\text{area of base}) = (\text{comp}_\mathbf{n} \mathbf{F})\Delta S = (\mathbf{F} \cdot \mathbf{n})\Delta S, \tag{6}$$

where $\mathbf{n}$ is a unit vector normal to the surface. Now consider the rectangular parallelepiped shown in FIGURE 15.7.1. To compute the total flux of $\mathbf{F}$ through its six sides in the outward direction, we first compute the total flux out of two parallel faces. The area of face F_1 is $\Delta x \Delta z$ and its outward unit normal is $-\mathbf{j}$, and so by (6) the flux of $\mathbf{F}$ through F_1 is

$$\mathbf{F} \cdot (-\mathbf{j})\Delta x \Delta z = -Q(x, y, z)\Delta x \Delta z.$$

The flux out of face F_2, whose outward normal is $\mathbf{j}$, is given by

$$(\mathbf{F} \cdot \mathbf{j})\Delta x \Delta z = Q(x, y + \Delta y, z)\Delta x \Delta z.$$

Consequently, the total flux out of these parallel faces is

$$Q(x, y + \Delta y, z)\Delta x \Delta z + (-Q(x, y, z)\Delta x \Delta z) = [Q(x, y + \Delta y, z) - Q(x, y, z)]\Delta x \Delta z. \tag{7}$$

By multiplying (7) by $\Delta y / \Delta y$ and using the definition of a partial derivative, then for Δy close to 0,

$$\frac{[Q(x, y + \Delta y, z) - Q(x, y, z)]}{\Delta y}\Delta x \Delta y \Delta z \approx \frac{\partial Q}{\partial y}\Delta x \Delta y \Delta z.$$

Arguing in exactly the same manner, we see that the contributions to the total flux out of the parallelepiped from the two faces parallel to the yz-plane and from the two faces parallel to the xy-plane are, in turn,

$$\frac{\partial P}{\partial x}\Delta x \Delta y \Delta z \quad \text{and} \quad \frac{\partial R}{\partial z}\Delta x \Delta y \Delta z.$$

Adding the results, we see that the total flux of $\mathbf{F}$ out of the parallelepiped is approximately

$$\left(\frac{\partial P}{\partial x} + \frac{\partial Q}{\partial y} + \frac{\partial R}{\partial z}\right)\Delta x \Delta y \Delta z.$$

By dividing the last expression by $\Delta x \Delta y \Delta z$, we get the outward flux of $\mathbf{F}$ per unit volume:

$$\frac{\partial P}{\partial x} + \frac{\partial Q}{\partial y} + \frac{\partial R}{\partial z}.$$

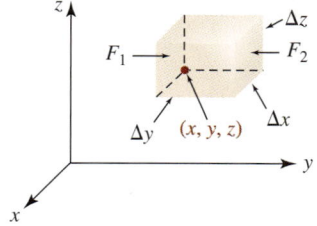

FIGURE 15.7.1 Flux through a rectangular parallelepiped

This combination of partial derivatives is a scalar function and is given the special name the **divergence** of **F**.

Definition 15.7.2 Divergence

The **divergence** of a vector field $\mathbf{F} = P\mathbf{i} + Q\mathbf{j} + R\mathbf{k}$ is the scalar function

$$\text{div } \mathbf{F} = \frac{\partial P}{\partial x} + \frac{\partial Q}{\partial y} + \frac{\partial R}{\partial z}. \tag{8}$$

The scalar function div **F** given in (8) can also be written in terms of the del operator (1) as a dot product:

$$\text{div } \mathbf{F} = \nabla \cdot \mathbf{F} = \frac{\partial}{\partial x}P(x, y, z) + \frac{\partial}{\partial y}Q(x, y, z) + \frac{\partial}{\partial z}R(x, y, z). \tag{9}$$

EXAMPLE 3 Divergence of a Vector Field

If $\mathbf{F} = xz^2\mathbf{i} + 2xy^2z\mathbf{j} - 5yz\mathbf{k}$, find div **F**.

Solution From (9),

$$\text{div } \mathbf{F} = \nabla \cdot \mathbf{F} = \frac{\partial}{\partial x}(xz^2) + \frac{\partial}{\partial y}(2xy^2z) + \frac{\partial}{\partial z}(-5yz)$$

$$= z^2 + 4xyz - 5y. \qquad \blacksquare$$

The next identity relates the notions of divergence and curl. If **F** is a vector field having continuous second partial derivatives, then

$$\text{div}(\text{curl } \mathbf{F}) = \nabla \cdot (\nabla \times \mathbf{F}) = 0. \tag{10}$$

See Problem 24 in Exercises 15.7.

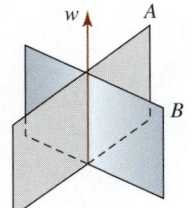

FIGURE 15.7.2 Paddle device for detecting rotation of a fluid

▌ **Physical Interpretations** The word *curl* was introduced by the Scottish mathematician and physicist **James Clerk Maxwell** (1831–1879) in his studies of electromagnetic fields. However, the curl is easily understood in connection with the flow of fluids. If a paddle device, such as shown in FIGURE 15.7.2, is inserted in a flowing fluid, then the curl of the velocity field **F** is a measure of the tendency of the fluid to turn the device about its vertical axis w. If curl $\mathbf{F} = \mathbf{0}$, then the flow of the fluid is said to be **irrotational**, which means that it is free of vortices or whirlpools that would cause the paddle to rotate. In FIGURE 15.7.3 the axis w of the paddle points straight out of the page.

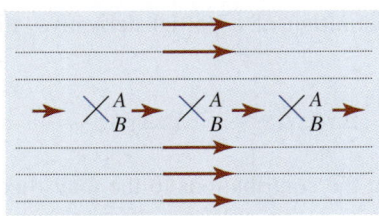

(a) Irrotational flow (b) Rotational flow

FIGURE 15.7.3 Irrotational and rotational fluid flow

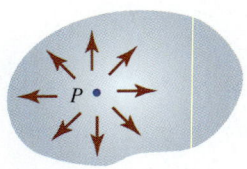

(a) Div **F**(P) > 0; P a source

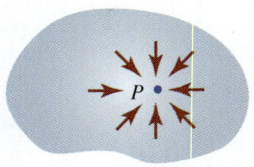

(b) Div **F**(P) < 0; P a sink

FIGURE 15.7.4 Point P is a source in (a); a sink in (b)

In the motivational discussion leading to Definition 15.7.2, we saw that the divergence of a velocity field **F** near a point $P(x, y, z)$ is the flux per unit volume. If div $\mathbf{F}(P) > 0$, then P is said to be a **source** for **F**, since there is a net outward flow of fluid near P, if div $\mathbf{F}(P) < 0$, then P is said to be a **sink** for **F**, since there is a net inward flow of fluid near P; if div $\mathbf{F}(P) = 0$, there are no sources or sinks near P. See FIGURE 15.7.4.

The divergence of a vector field has another interpretation in the context of fluid flow. A measure of the rate of change of the density of the fluid at a point is simply div **F**. In other words, div **F** is a measure of the fluid's compressibility. If $\nabla \cdot \mathbf{F} = 0$, the fluid is said to be **incompressible**. In electromagnetic theory, if $\nabla \cdot \mathbf{F} = 0$, the vector field $\mathbf{F}$ is said to be **solenoidal**.

By taking the dot product of ∇ with itself we obtain an important second-order scalar differential operator:

$$\nabla^2 = \nabla \cdot \nabla = \frac{\partial^2}{\partial x^2} + \frac{\partial^2}{\partial y^2} + \frac{\partial^2}{\partial z^2}. \tag{11}$$

When (11) is applied to a scalar function $f(x, y, z)$ the result is called the three-dimensional **Laplacian**,

$$\nabla^2 f = \frac{\partial^2 f}{\partial x^2} + \frac{\partial^2 f}{\partial y^2} + \frac{\partial^2 f}{\partial z^2} \tag{12}$$

and appears throughout applied mathematics in many partial differential equations. One of the most famous partial differential equations,

$$\frac{\partial^2 f}{\partial x^2} + \frac{\partial^2 f}{\partial y^2} + \frac{\partial^2 f}{\partial z^2} = 0, \tag{13}$$

is called **Laplace's equation** in three dimensions. Laplace's equation is often abbreviated as $\nabla^2 f = 0$. See Problems 49–54 in Exercises 13.3.

■ **Postscript—A Bit of History** **Pierre-Simon Marquis de Laplace** (1749–1827) was a noted French mathematician, physicist, and astronomer. His most famous work, the five-volume *Mécanique Céleste* (Celestial Mechanics) summarizes and extends the work of some of his famous predecessors such as Isaac Newton. Indeed, some of his enthusiastic contemporaries called Laplace the "Newton of France." Born into a poor farming family, the adult Laplace was successful in combining science and mathematics with politics. Napoleon made him a minister of the interior but later dismissed him because he "searched for subtleties everywhere and carried into administration the spirit of the infinitely small"—meaning, the infinitesimal calculus. Yet Napoleon then made him a senator. After Napoleon's abdication and the restoration of the Bourbon monarchy in 1814, Laplace was elevated to the nobility by Louis XVIII with the title of *Marquis* in 1817.

Laplace

Exercises 15.7   Answers to selected odd-numbered problems begin on page ANS-47.

≡ **Fundamentals**

In Problems 1–10, find the curl and the divergence of the given vector field.

1. $\mathbf{F}(x, y, z) = xz\mathbf{i} + yz\mathbf{j} + xy\mathbf{k}$

2. $\mathbf{F}(x, y, z) = 10yz\mathbf{i} + 2x^2z\mathbf{j} + 6x^3\mathbf{k}$

3. $\mathbf{F}(x, y, z) = 4xy\mathbf{i} + (2x^2 + 2yz)\mathbf{j} + (3z^2 + y^2)\mathbf{k}$

4. $\mathbf{F}(x, y, z) = (x - y)^3\mathbf{i} + e^{-yz}\mathbf{j} + xye^{2y}\mathbf{k}$

5. $\mathbf{F}(x, y, z) = 3x^2y\mathbf{i} + 2xz^3\mathbf{j} + y^4\mathbf{k}$

6. $\mathbf{F}(x, y, z) = 5y^3\mathbf{i} + \left(\frac{1}{2}x^3y^2 - xy\right)\mathbf{j} - (x^3yz - xz)\mathbf{k}$

7. $\mathbf{F}(x, y, z) = xe^{-z}\mathbf{i} + 4yz^2\mathbf{j} + 3ye^{-z}\mathbf{k}$

8. $\mathbf{F}(x, y, z) = yz\ln x\mathbf{i} + (2x - 3yz)\mathbf{j} + xy^2z^3\mathbf{k}$

9. $\mathbf{F}(x, y, z) = xye^x\mathbf{i} - x^3yze^z\mathbf{j} + xy^2e^y\mathbf{k}$

10. $\mathbf{F}(x, y, z) = x^2\sin yz\mathbf{i} + z\cos xz^3\mathbf{j} + ye^{5xy}\mathbf{k}$

In Problems 11–18, let $\mathbf{a}$ be a constant vector and $\mathbf{r} = x\mathbf{i} + y\mathbf{j} + z\mathbf{k}$. Verify the given identity.

11. div $\mathbf{r} = 3$

12. curl $\mathbf{r} = \mathbf{0}$

13. $(\mathbf{a} \times \nabla) \times \mathbf{r} = -2\mathbf{a}$

14. $\nabla \times (\mathbf{a} \times \mathbf{r}) = 2\mathbf{a}$

15. $\nabla \cdot (\mathbf{a} \times \mathbf{r}) = 0$

16. $\mathbf{a} \times (\nabla \times \mathbf{r}) = \mathbf{0}$

17. $\nabla \times [(\mathbf{r} \cdot \mathbf{r})\mathbf{a}] = 2(\mathbf{r} \times \mathbf{a})$

18. $\nabla \cdot [(\mathbf{r} \cdot \mathbf{r})\mathbf{a}] = 2(\mathbf{r} \cdot \mathbf{a})$

In Problems 19–26, verify the given identity. Assume continuity of all partial derivatives.

19. $\nabla \cdot (\mathbf{F} + \mathbf{G}) = \nabla \cdot \mathbf{F} + \nabla \cdot \mathbf{G}$

20. $\nabla \times (\mathbf{F} + \mathbf{G}) = \nabla \times \mathbf{F} + \nabla \times \mathbf{G}$

21. $\nabla \cdot (f\mathbf{F}) = f(\nabla \cdot \mathbf{F}) + \mathbf{F} \cdot \nabla f$

22. $\nabla \times (f\mathbf{F}) = f(\nabla \times \mathbf{F}) + (\nabla f) \times \mathbf{F}$

23. $\text{curl}(\text{grad } f) = \mathbf{0}$

24. $\text{div}(\text{curl } \mathbf{F}) = 0$

25. $\text{div}(\mathbf{F} \times \mathbf{G}) = \mathbf{G} \cdot \text{curl } \mathbf{F} - \mathbf{F} \cdot \text{curl } \mathbf{G}$

26. $\text{curl}(\text{curl } \mathbf{F} + \text{grad } f) = \text{curl}(\text{curl } \mathbf{F})$

27. Find $\text{curl}(\text{curl } \mathbf{F})$ for the vector field $\mathbf{F}(x, y, z) = xy\mathbf{i} + 4yz^2\mathbf{j} + 2xz\mathbf{k}$.

28. Suppose ∇^2 is the differential operator defined in (11). Assuming continuity of all partial derivatives, show that

$$\text{curl}(\text{curl } \mathbf{F}) = -\nabla^2\mathbf{F} + \text{grad}(\text{div } \mathbf{F}),$$

where $\nabla^2\mathbf{F} = \nabla^2(P\mathbf{i} + Q\mathbf{j} + R\mathbf{k}) = \nabla^2 P\mathbf{i} + \nabla^2 Q\mathbf{j} + \nabla^2 R\mathbf{k}$.

29. Use the identity in Problem 28 to obtain the result in Problem 27.

30. Show that $\nabla \cdot (f \nabla f) = f\nabla^2 f + |\nabla f|^2$, where $\nabla^2 f$ is the Laplacian defined in (12). [*Hint*: See Problem 21.]

Any function f with continuous second partial derivatives that satisfies Laplace's equation is said to be a **harmonic function**. In Problems 31 and 32, show that the given function f is harmonic by demonstrating f satisfies (13).

31. $f(x, y, z) = 3x^2 + 5y^2 + 4xy - 9xz - 8z^2$

32. $f(x, y, z) = \dfrac{A}{\sqrt{(x-a)^2 + (y-b)^2 + (z-c)^2}}$,

A, a, b, and c constants

Laplace's equation in two dimensions is

$$\nabla^2 f = \frac{\partial^2 f}{\partial x^2} + \frac{\partial^2 f}{\partial y^2} = 0. \qquad (14)$$

In Problems 33 and 34, show that the given function f is harmonic by demonstrating f satisfies (14).

33. $f(x, y) = \arctan\left(\dfrac{2y}{x^2 + y^2 - 1}\right)$

34. $f(x, y) = x^4 - 6x^2y^2 + y^4$

In Problems 35 and 36, assume that f and g have continuous second partial derivatives. Show that the given vector field is solenoidal. [*Hint*: See Problem 25.]

35. $\mathbf{F} = \nabla f \times \nabla g$ **36.** $\mathbf{F} = \nabla f \times (f\nabla g)$

37. If $\mathbf{F} = y^3\mathbf{i} + x^3\mathbf{j} + z^3\mathbf{k}$, find the flux of $\nabla \times \mathbf{F}$ through that portion of the ellipsoid $x^2 + y^2 + 4z^2 = 4$ in the first octant that is bounded by $y = 0$, $y = x$, $z = 0$. Assume the surface is oriented upward.

≡ Applications

38. Suppose a body rotates with a constant angular velocity $\boldsymbol{\omega}$ about an axis. If $\mathbf{r}$ is the position vector of a point P on the body measured from the origin, then the linear velocity vector $\mathbf{v}$ of rotation is $\mathbf{v} = \boldsymbol{\omega} \times \mathbf{r}$. See FIGURE 15.7.5. If $\mathbf{r} = x\mathbf{i} + y\mathbf{j} + z\mathbf{k}$ and $\boldsymbol{\omega} = \omega_1\mathbf{i} + \omega_2\mathbf{j} + \omega_3\mathbf{k}$, show that $\boldsymbol{\omega} = \frac{1}{2}\text{curl } \mathbf{v}$.

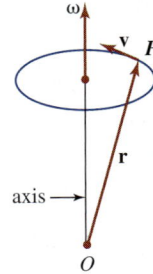

FIGURE 15.7.5 Rotating body in Problem 38

39. Let $\mathbf{r} = x\mathbf{i} + y\mathbf{j} + z\mathbf{k}$ be the position vector of a mass m_1 and let the mass m_2 be located at the origin. If the force of gravitational attraction is

$$\mathbf{F} = -\frac{Gm_1m_2}{|\mathbf{r}|^3}\mathbf{r},$$

verify that $\text{curl } \mathbf{F} = \mathbf{0}$ and $\text{div } \mathbf{F} = 0$, $\mathbf{r} \neq \mathbf{0}$.

40. The velocity vector field for the two-dimensional flow of an ideal fluid around a cylinder is given by

$$\mathbf{F}(x, y) = A\left[\left(1 - \frac{x^2 - y^2}{(x^2 + y^2)^2}\right)\mathbf{i} - \frac{2xy}{(x^2 + y^2)^2}\mathbf{j}\right],$$

for some positive constant A. See FIGURE 15.7.6.

(a) Show that when the point (x, y) is far from the origin, $\mathbf{F}(x, y) \approx A\mathbf{i}$.

(b) Show that $\mathbf{F}$ is irrotational.

(c) Show that $\mathbf{F}$ is incompressible.

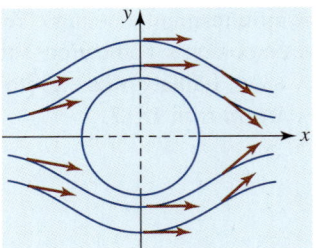

FIGURE 15.7.6 Velocity field in Problem 40

41. If $\mathbf{E} = \mathbf{E}(x, y, z, t)$ and $\mathbf{H} = \mathbf{H}(x, y, z, t)$ represent electric and magnetic fields in empty space, then Maxwell's equations are

$$\text{div } \mathbf{E} = 0, \quad \text{curl } \mathbf{E} = -\frac{1}{c}\frac{\partial \mathbf{H}}{\partial t},$$

$$\text{div } \mathbf{H} = 0, \quad \text{curl } \mathbf{H} = \frac{1}{c}\frac{\partial \mathbf{E}}{\partial t},$$

where c is the speed of light. Use the identity in Problem 28 to show that $\mathbf{E}$ and $\mathbf{H}$ satisfy

$$\nabla^2\mathbf{E} = \frac{1}{c^2}\frac{\partial^2\mathbf{E}}{\partial t^2} \quad \text{and} \quad \nabla^2\mathbf{H} = \frac{1}{c^2}\frac{\partial^2\mathbf{H}}{\partial t^2}.$$

≡ Think About It

42. Consider the vector field $\mathbf{F} = x^2yz\mathbf{i} - xy^2z\mathbf{j} + (z + 5x)\mathbf{k}$. Explain why $\mathbf{F}$ is not the curl of another vector field $\mathbf{G}$.

15.8 Stokes' Theorem

Introduction Green's Theorem of Section 15.4 can be written in two different vector forms. In this and the next section we shall generalize these forms to three dimensions.

Vector Form of Green's Theorem If $\mathbf{F}(x, y) = P(x, y)\mathbf{i} + Q(x, y)\mathbf{j}$ is a two-dimensional vector field, then

$$\text{curl } \mathbf{F} = \nabla \times \mathbf{F} = \begin{vmatrix} \mathbf{i} & \mathbf{j} & \mathbf{k} \\ \dfrac{\partial}{\partial x} & \dfrac{\partial}{\partial y} & \dfrac{\partial}{\partial z} \\ P & Q & 0 \end{vmatrix} = \left(\frac{\partial Q}{\partial x} - \frac{\partial P}{\partial y} \right)\mathbf{k}.$$

From (6) and (7) of Section 15.2, Green's Theorem

$$\oint_C P(x, y)\, dx + Q(x, y)\, dy = \iint_R \left(\frac{\partial Q}{\partial x} - \frac{\partial P}{\partial y} \right) dA$$

can be written in vector notation as

$$\oint_C \mathbf{F} \cdot d\mathbf{r} = \oint_C (\mathbf{F} \cdot \mathbf{T})\, ds = \iint_R (\text{curl } \mathbf{F}) \cdot \mathbf{k}\, dA. \tag{1}$$

That is, the line integral of the tangential component of $\mathbf{F}$ is the double integral of the normal component of curl $\mathbf{F}$.

Green's Theorem in 3-Space The vector form of Green's Theorem given in (1) relates a line integral around a piecewise-smooth simple closed curve C forming the boundary of a plane region R to a double integral over R. Green's Theorem in 3-space relates a line integral around a piecewise-smooth simple closed space curve C forming the boundary of a surface S with a surface integral over S. Suppose $z = f(x, y)$ is a continuous function whose graph is a piecewise-smooth oriented surface over a region R on the xy-plane. Let C form the boundary of S and let the projection of C onto the xy-plane form the boundary of R. The positive direction of C is induced by the orientation of the surface S; the positive direction of C corresponds to the direction a person would have to walk on C to have his or her head point in the direction of the orientation of the surface while keeping the surface to the left. See FIGURE 15.8.1. More precisely, the positive orientation of C is in accordance with the right-hand rule: If the thumb of the right hand points in the direction of the orientation of the surface, then roughly the fingers of the right hand wrap around the surface in the positive direction. Finally, let $\mathbf{T}$ be a unit tangent vector to C that points in the positive direction. The three-dimensional form of Green's Theorem, which we shall now give, is called **Stokes' Theorem**.

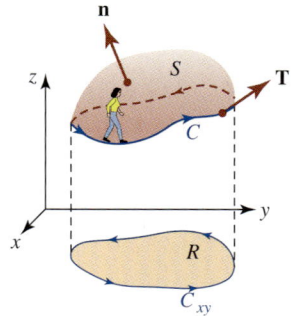

FIGURE 15.8.1. Positive direction of C

Theorem 15.8.1 Stokes' Theorem

Let S be a piecewise-smooth oriented surface bounded by a piecewise-smooth simple closed curve C. Let

$$\mathbf{F}(x, y, z) = P(x, y, z)\mathbf{i} + Q(x, y, z)\mathbf{j} + R(x, y, z)\mathbf{k}$$

be a vector field for which P, Q, and R are continuous and have continuous first partial derivatives in an open region of 3-space containing S. If C is traversed in the positive direction, then

$$\oint_C \mathbf{F} \cdot d\mathbf{r} = \oint_C (\mathbf{F} \cdot \mathbf{T})\, ds = \iint_S (\text{curl } \mathbf{F}) \cdot \mathbf{n}\, dS, \tag{2}$$

where $\mathbf{n}$ is a unit normal to S in the direction of the orientation of S.

PARTIAL PROOF Suppose the surface S is oriented upward and is defined by a function $z = g(x, y)$ that has continuous second partial derivatives. From Definition 15.7.1 we have

$$\text{curl } \mathbf{F} = \left(\frac{\partial R}{\partial y} - \frac{\partial Q}{\partial z} \right)\mathbf{i} + \left(\frac{\partial P}{\partial z} - \frac{\partial R}{\partial x} \right)\mathbf{j} + \left(\frac{\partial Q}{\partial x} - \frac{\partial P}{\partial y} \right)\mathbf{k}.$$

Furthermore, if we write $h(x, y, z) = z - g(x, y) = 0$, then

$$\mathbf{n} = \frac{\nabla g}{|\nabla g|} = \frac{-\dfrac{\partial g}{\partial x}\mathbf{i} - \dfrac{\partial g}{\partial y}\mathbf{j} + \mathbf{k}}{\sqrt{1 + \left(\dfrac{\partial g}{\partial x}\right)^2 + \left(\dfrac{\partial g}{\partial y}\right)^2}}.$$

Hence,

$$\iint_S (\text{curl } \mathbf{F}) \cdot \mathbf{n} \, dS = \iint_R \left[-\left(\frac{\partial R}{\partial y} - \frac{\partial Q}{\partial z} \right)\frac{\partial g}{\partial x} - \left(\frac{\partial P}{\partial z} - \frac{\partial R}{\partial x} \right)\frac{\partial g}{\partial y} + \left(\frac{\partial Q}{\partial x} - \frac{\partial P}{\partial y} \right) \right] dA. \quad (3)$$

Our goal now is to show that the line integral $\oint_C \mathbf{F} \cdot d\mathbf{r}$ reduces to (3).

If C_{xy} is the projection of C onto the xy-plane and has the parametric equations $x = x(t)$, $y = y(t)$, $a \leq t \leq b$, then parametric equations for C are $x = x(t), y = y(t), z = g(x(t), y(t))$, $a \leq t \leq b$. Thus,

$$\oint_C \mathbf{F} \cdot d\mathbf{r} = \int_a^b \left[P\frac{dx}{dt} + Q\frac{dy}{dt} + R\frac{dz}{dt} \right] dt$$

$$= \int_a^b \left[P\frac{dx}{dt} + Q\frac{dy}{dt} + R\left(\frac{\partial g}{\partial x}\frac{dx}{dt} + \frac{\partial g}{\partial y}\frac{dy}{dt} \right) \right] dt \quad \leftarrow \text{Chain Rule}$$

$$= \oint_{C_{xy}} \left(P + R\frac{\partial g}{\partial x} \right) dx + \left(Q + R\frac{\partial g}{\partial y} \right) dy$$

$$= \iint_R \left[\frac{\partial}{\partial x}\left(Q + R\frac{\partial g}{\partial y} \right) - \frac{\partial}{\partial y}\left(P + R\frac{\partial g}{\partial x} \right) \right] dA. \quad \leftarrow \text{Green's Theorem} \quad (4)$$

Now, by the Chain and Product Rules,

$$\frac{\partial}{\partial x}\left(Q + R\frac{\partial g}{\partial y} \right) = \frac{\partial}{\partial x}\left[Q(x, y, g(x, y)) + R(x, y, g(x, y))\frac{\partial g}{\partial y} \right]$$

$$= \frac{\partial Q}{\partial x} + \frac{\partial Q}{\partial z}\frac{\partial g}{\partial x} + R\frac{\partial^2 g}{\partial x \partial y} + \frac{\partial R}{\partial x}\frac{\partial g}{\partial y} + \frac{\partial R}{\partial z}\frac{\partial g}{\partial y}\frac{\partial g}{\partial x}. \quad (5)$$

Similarly,

$$\frac{\partial}{\partial y}\left(P + R\frac{\partial g}{\partial x} \right) = \frac{\partial P}{\partial y} + \frac{\partial P}{\partial z}\frac{\partial g}{\partial y} + R\frac{\partial^2 g}{\partial y \partial x} + \frac{\partial R}{\partial y}\frac{\partial g}{\partial x} + \frac{\partial R}{\partial z}\frac{\partial g}{\partial x}\frac{\partial g}{\partial y}. \quad (6)$$

Subtracting (6) from (5) and using the fact that $\partial^2 g/\partial x \partial y = \partial^2 g/\partial y \partial x$, we see that (4) becomes, after rearranging,

$$\iint_R \left[-\left(\frac{\partial R}{\partial y} - \frac{\partial Q}{\partial z} \right)\frac{\partial g}{\partial x} - \left(\frac{\partial P}{\partial z} - \frac{\partial R}{\partial x} \right)\frac{\partial g}{\partial y} + \left(\frac{\partial Q}{\partial x} - \frac{\partial P}{\partial y} \right) \right] dA.$$

This last expression is the same as the right side of (3), which was to be shown. ∎

EXAMPLE 1 Verifying Stokes' Theorem

Let S be the part of the cylinder $z = 1 - x^2$ for $0 \leq x \leq 1$, $-2 \leq y \leq 2$. Verify Stokes' Theorem for the vector field $\mathbf{F} = xy\mathbf{i} + yz\mathbf{j} + xz\mathbf{k}$. Assume that S is oriented upward.

Solution The surface S, the curve C (which is composed of the union of C_1, C_2, C_3, and C_4), and the region R are shown in FIGURE 15.8.2 on page 853.

The Surface Integral: For $\mathbf{F} = xy\mathbf{i} + yz\mathbf{j} + xz\mathbf{k}$ we find

$$\text{curl } \mathbf{F} = \begin{vmatrix} \mathbf{i} & \mathbf{j} & \mathbf{k} \\ \dfrac{\partial}{\partial x} & \dfrac{\partial}{\partial y} & \dfrac{\partial}{\partial z} \\ xy & yz & xz \end{vmatrix} = -y\mathbf{i} - z\mathbf{j} - x\mathbf{k}.$$

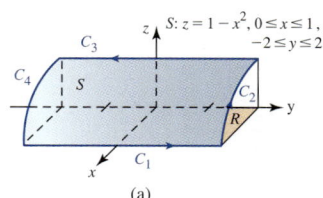

Now, if $h(x, y, z) = z + x^2 - 1 = 0$ defines the cylinder, the upper normal is

$$\mathbf{n} = \frac{\nabla h}{|\nabla h|} = \frac{2x\mathbf{i} + \mathbf{k}}{\sqrt{4x^2 + 1}}.$$

Therefore,

$$\iint_S (\text{curl } \mathbf{F} \cdot \mathbf{n})\, dS = \iint_S \frac{-2xy - x}{\sqrt{4x^2 + 1}}\, dS.$$

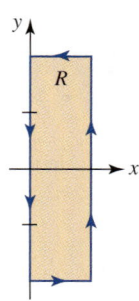

To evaluate the latter surface integral, we use (2) of Section 15.6:

$$\iint_S \frac{-2xy - x}{\sqrt{4x^2 + 1}}\, dS = \iint_R (-2xy - x)\, dA$$

(b)
FIGURE 15.8.2 Surface in Example 1

$$= \int_0^1 \int_{-2}^2 (-2xy - x)\, dy\, dx$$

$$= \int_0^1 \left[-xy^2 - xy \right]_{-2}^2 dx$$

$$= \int_0^1 (-4x)\, dx = -2. \qquad (7)$$

The Line Integral: The line integral is

$$\oint_C \mathbf{F} \cdot d\mathbf{r} = \oint_C xy\, dx + yz\, dy + xz\, dz.$$

Because C is piecewise-smooth, we write $\oint_C = \int_{C_1} + \int_{C_2} + \int_{C_3} + \int_{C_4}$.
On C_1: $x = 1$, $z = 0$, $dx = 0$, $dz = 0$, and so

$$\int_{C_1} y(0) + y(0)\, dy + 0 = 0.$$

On C_2: $y = 2$, $z = 1 - x^2$, $dy = 0$, $dz = -2x\, dx$, so

$$\int_{C_2} 2x\, dx + 2(1 - x^2)0 + x(1 - x^2)(-2x\, dx) = \int_1^0 (2x - 2x^2 + 2x^4)\, dx = -\frac{11}{15}.$$

On C_3: $x = 0$, $z = 1$, $dx = 0$, $dz = 0$, so

$$\int_{C_3} 0 + y\, dy + 0 = \int_{-2}^2 y\, dy = 0.$$

On C_4: $y = -2$, $z = 1 - x^2$, $dy = 0$, $dz = -2x\, dx$, so

$$\int_{C_4} -2x\, dx - 2(1 - x^2)0 + x(1 - x^2)(-2x\, dx) = \int_0^1 (-2x - 2x^2 + 2x^4)\, dx = -\frac{19}{15}.$$

Hence,

$$\oint_C xy\, dx + yz\, dy + xz\, dz = 0 - \frac{11}{15} + 0 - \frac{19}{15} = -2$$

which agrees with (7). ∎

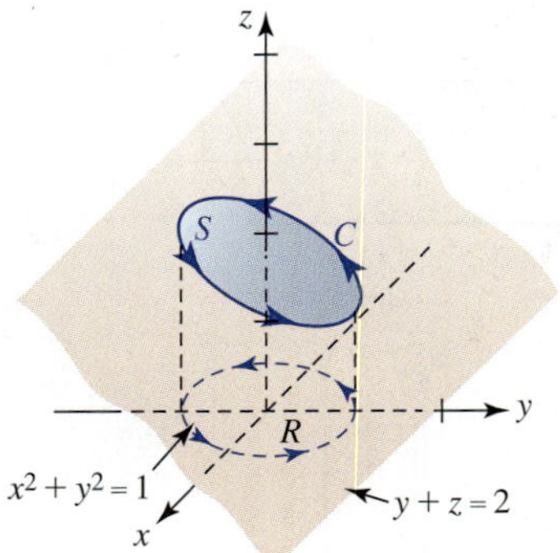

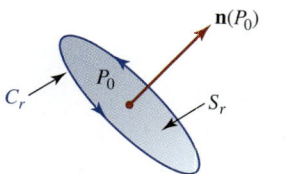

$x^2 + y^2 = 1$

$y + z = 2$

FIGURE 15.8.3 Curve C in Example 2

EXAMPLE 2 Using Stokes' Theorem

Evaluate $\oint_C z\,dx + x\,dy + y\,dz$, where C is the trace of the cylinder $x^2 + y^2 = 1$ in the plane $y + z = 2$. Orient C counterclockwise as viewed from above. See FIGURE 15.8.3.

Solution If $\mathbf{F} = z\mathbf{i} + x\mathbf{j} + y\mathbf{k}$, then

$$\text{curl } \mathbf{F} = \begin{vmatrix} \mathbf{i} & \mathbf{j} & \mathbf{k} \\ \dfrac{\partial}{\partial x} & \dfrac{\partial}{\partial y} & \dfrac{\partial}{\partial z} \\ z & x & y \end{vmatrix} = \mathbf{i} + \mathbf{j} + \mathbf{k}.$$

The given orientation of C corresponds to an upward orientation of the surface S. Thus, if $h(x, y, z) = y + z - 2 = 0$ defines the plane, then the upper normal is

$$\mathbf{n} = \frac{\nabla h}{|\nabla h|} = \frac{1}{\sqrt{2}}\mathbf{j} + \frac{1}{\sqrt{2}}\mathbf{k}.$$

Hence, from (2),

$$\oint_C \mathbf{F} \cdot d\mathbf{r} = \iint_S \left[(\mathbf{i} + \mathbf{j} + \mathbf{k}) \cdot \left(\frac{1}{\sqrt{2}}\mathbf{j} + \frac{1}{\sqrt{2}}\mathbf{k} \right) \right] dS$$

$$= \sqrt{2} \iint_S dS = \sqrt{2} \iint_R \sqrt{2}\, dA = 2\pi. \qquad \blacksquare$$

Note that if $\mathbf{F}$ is the gradient of a scalar function, then, in view of (5) of Section 15.7, (2) implies that the circulation $\oint_C \mathbf{F} \cdot d\mathbf{r}$ is zero. Conversely, it can be shown that if the circulation is zero for every simple closed curve, then $\mathbf{F}$ is the gradient of a scalar function. In other words, $\mathbf{F}$ is irrotational if and only if $\mathbf{F} = \nabla\phi$, where ϕ is the potential for $\mathbf{F}$. Equivalently, this gives the test for a conservative vector field given in Theorem 15.7.1, that is, $\mathbf{F}$ is a conservative vector field if and only if curl $\mathbf{F} = \mathbf{0}$.

■ **Physical Interpretation of Curl** In Section 15.2 we saw that if $\mathbf{F}$ is a velocity field of a fluid, then the circulation $\oint_C \mathbf{F} \cdot d\mathbf{r}$ of $\mathbf{F}$ around C is a measure of the amount by which the fluid tends to turn the curve C by circulating around it. The circulation of $\mathbf{F}$ is closely related to curl $\mathbf{F}$. To see this, suppose $P_0(x_0, y_0, z_0)$ is any point in the fluid and C_r is a small circle of radius r centered at P_0. See FIGURE 15.8.4. Then by Stokes' Theorem,

$$\oint_{C_r} \mathbf{F} \cdot d\mathbf{r} = \iint_{S_r} (\text{curl } \mathbf{F}) \cdot \mathbf{n}\, dS. \qquad (8)$$

Now, at all points $P(x, y, z)$ within the small circle C_r, if we take curl $\mathbf{F}(P) \approx$ curl $\mathbf{F}(P_0)$, then (8) gives the approximation

$$\oint_{C_r} \mathbf{F} \cdot d\mathbf{r} \approx \iint_{S_r} (\text{curl } \mathbf{F}(P_0)) \cdot \mathbf{n}(P_0)\, dS$$

$$= (\text{curl } \mathbf{F}(P_0)) \cdot \mathbf{n}(P_0) \iint_{S_r} dS$$

$$= (\text{curl } \mathbf{F}(P_0)) \cdot \mathbf{n}(P_0)\, A_r, \qquad (9)$$

where A_r is the area πr^2 of the circular surface S_r. As we let $r \to 0$, the approximation curl $\mathbf{F}(P) \approx$ curl $\mathbf{F}(P_0)$ becomes better and so (9) yields

$$(\text{curl } \mathbf{F}(P_0)) \cdot \mathbf{n}(P_0) = \lim_{r \to 0} \frac{1}{A_r} \oint_{C_r} \mathbf{F} \cdot d\mathbf{r}. \qquad (10)$$

Thus, we see that the normal component of curl $\mathbf{F}$ is the limiting value of the ratio of the circulation of $\mathbf{F}$ to the area of the circular surface. For a small but fixed value of r, we have

$$(\text{curl } \mathbf{F}(P_0)) \cdot \mathbf{n}(P_0) \approx \frac{1}{A_r} \oint_{C_r} \mathbf{F} \cdot d\mathbf{r}. \qquad (11)$$

FIGURE 15.8.4 Circle of radius r

Roughly then, curl **F** is the circulation of **F** per unit area. If curl $\mathbf{F}(P_0) \neq \mathbf{0}$, then the left-hand side of (11) is a maximum when the circle C_r is situated in a manner so that $\mathbf{n}(P_0)$ points in the same direction as curl $\mathbf{F}(P_0)$. In this case, the circulation on the right side of (11) is also a maximum. Thus, a paddle wheel inserted into the fluid at P_0 will rotate fastest when its axis points in the direction of curl $\mathbf{F}(P_0)$. See FIGURE 15.8.5. Note, too, that the paddle will not rotate if its axis is perpendicular to curl $\mathbf{F}(P_0)$.

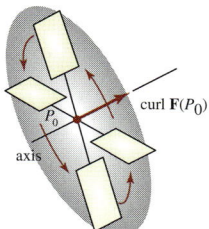

FIGURE 15.8.5 Rotating paddle wheel in a fluid

Stokes

■ **Postscript—A Bit of History** **George G. Stokes** (1819–1903) was an Irish mathematical physicist. Like George Green, Stokes was a don at Cambridge University. In 1854, Stokes posed his theorem as a problem on a prize examination for Cambridge students. It is not known whether anyone solved the problem.

$\oint_C$ NOTES FROM THE CLASSROOM

The value of the surface integral in (2) is determined solely by the integral around its boundary C. This basically means that the shape of the surface S is irrelevant. Assuming that the hypotheses of Theorem 15.8.1 are satisfied, then for two different surfaces S_1 and S_2 with the same orientation and with the same boundary C, we have

$$\oint_C \mathbf{F} \cdot d\mathbf{r} = \iint_{S_1} (\text{curl } \mathbf{F}) \cdot \mathbf{n} \, dS = \iint_{S_2} (\text{curl } \mathbf{F}) \cdot \mathbf{n} \, dS.$$

See FIGURE 15.8.6 and Problems 17 and 18 of Exercises 15.8.

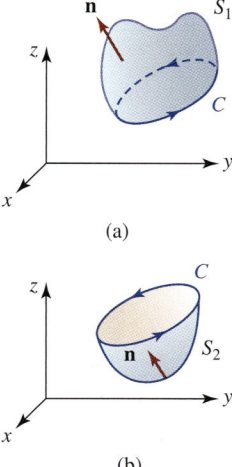

(a)

(b)

FIGURE 15.8.6 Two surfaces with the same boundary C

Exercises 15.8 Answers to selected odd-numbered problems begin on page ANS-47.

☰ Fundamentals

In Problems 1–4, verify Stokes' Theorem for the given vector field. Assume that the surface S is oriented upward.

1. $\mathbf{F} = 5y\mathbf{i} - 5x\mathbf{j} + 3\mathbf{k}$; S that portion of the plane $z = 1$ within the cylinder $x^2 + y^2 = 4$

2. $\mathbf{F} = 2z\mathbf{i} - 3x\mathbf{j} + 4y\mathbf{k}$; S that portion of the paraboloid $z = 16 - x^2 - y^2$ for $z \geq 0$

3. $\mathbf{F} = z\mathbf{i} + x\mathbf{j} + y\mathbf{k}$; S that portion of the plane $2x + y + 2z = 6$ in the first octant

4. $\mathbf{F} = x\mathbf{i} + y\mathbf{j} + z\mathbf{k}$; S that portion of the sphere $x^2 + y^2 + z^2 = 1$ for $z \geq 0$

In Problems 5–12, use Stokes' Theorem to evaluate $\oint_C \mathbf{F} \cdot d\mathbf{r}$. Assume that C is oriented counterclockwise as viewed from above.

5. $\mathbf{F} = (2z + x)\mathbf{i} + (y - z)\mathbf{j} + (x + y)\mathbf{k}$; C the triangle with vertices $(1, 0, 0)$, $(0, 1, 0)$, $(0, 0, 1)$

6. $\mathbf{F} = z^2 y \cos xy \, \mathbf{i} + z^2 x (1 + \cos xy)\mathbf{j} + 2z \sin xy \, \mathbf{k}$; C the boundary of the plane $z = 1 - y$ shown in FIGURE 15.8.7.

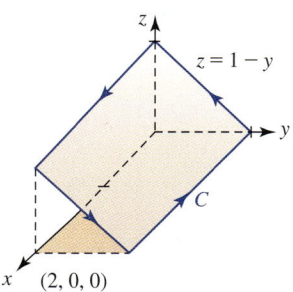

FIGURE 15.8.7 Curve in Problem 6

7. $\mathbf{F} = xy\mathbf{i} + 2yz\mathbf{j} + xz\mathbf{k}$; C the boundary given in Problem 6

8. $\mathbf{F} = (x + 2z)\mathbf{i} + (3x + y)\mathbf{j} + (2y - z)\mathbf{k}$; C the curve of intersection of the plane $x + 2y + z = 4$ with the coordinate planes

9. $\mathbf{F} = y^3\mathbf{i} - x^3\mathbf{j} + z^3\mathbf{k}$; C the trace of the cylinder $x^2 + y^2 = 1$ in the plane $x + y + z = 1$ [*Hint*: Use polar coordinates.]

10. $\mathbf{F} = x^2 y\mathbf{i} + (x + y^2)\mathbf{j} + xy^2 z\mathbf{k}$; C the boundary of the surface shown in FIGURE 15.8.8

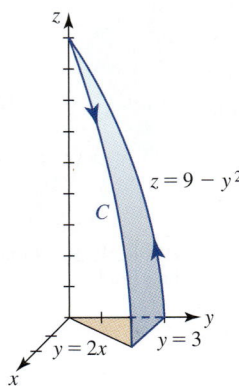

FIGURE 15.8.8 Curve in Problem 10

11. $\mathbf{F} = x\mathbf{i} + x^3y^2\mathbf{j} + z\mathbf{k}$; C the boundary of the semi-ellipsoid $z = \sqrt{4 - 4x^2 - y^2}$ in the plane $z = 0$

12. $\mathbf{F} = z\mathbf{i} + x\mathbf{j} + y\mathbf{k}$; C the curve of intersection of the cone $z = \sqrt{x^2 + y^2}$ and the sphere $x^2 + y^2 + z^2 = 1$ shown in FIGURE 15.8.9

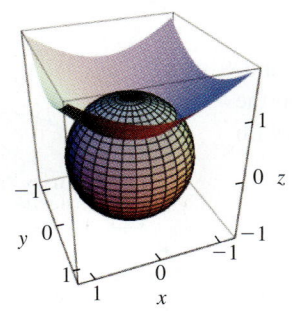

FIGURE 15.8.9 Curve in Problem 12

In Problems 13–16, use Stokes' Theorem to evaluate $\iint_S (\text{curl } \mathbf{F}) \cdot \mathbf{n} \, dS$. Assume that the surface S is oriented upward.

13. $\mathbf{F} = 6yz\mathbf{i} + 5x\mathbf{j} + yze^{x^2}\mathbf{k}$; S that portion of the paraboloid $z = \frac{1}{4}x^2 + y^2$ for $0 \le z \le 4$

14. $\mathbf{F} = y\mathbf{i} + (y - x)\mathbf{j} + z^2\mathbf{k}$; S that portion of the sphere $x^2 + y^2 + (z - 4)^2 = 25$ for $z \ge 0$

15. $\mathbf{F} = 3x^2\mathbf{i} + 8x^3y\mathbf{j} + 3x^2y\mathbf{k}$; S that portion of the plane $z = x$ that lies inside the rectangular cylinder defined by the planes $x = 0, y = 0, x = 2, y = 2$

16. $\mathbf{F} = 2xy^2z\mathbf{i} + 2x^2yz\mathbf{j} + (x^2y^2 - 6x)\mathbf{k}$; S that portion of the plane $z = y$ that lies inside the cylinder $x^2 + y^2 = 1$

17. Use Stokes' Theorem to evaluate

$$\oint_C z^2 e^{x^2} \, dx + xy^2 \, dy + \tan^{-1}y \, dz,$$

where C is the circle $x^2 + y^2 = 9$ by finding a surface S with C as its boundary and such that the orientation of C is counterclockwise as viewed from above.

18. Consider the surface integral $\iint_S (\text{curl } \mathbf{F}) \cdot \mathbf{n} \, dS$, where $\mathbf{F} = xyz\mathbf{k}$ and S is that portion of the paraboloid $z = 1 - x^2 - y^2$ for $z \ge 0$ oriented upward.

   (a) Evaluate the surface integral by the method of Section 15.6; that is, do not use Stokes' Theorem.

   (b) Evaluate the surface integral by finding a simpler surface that is oriented upward and has the same boundary as the paraboloid.

   (c) Use Stokes' Theorem to verify your result in part (b).

15.9 Divergence Theorem

▌ **Introduction** As mentioned in the introduction to Section 15.8, in this section we are going to examine another generalization of Green's Theorem. It might be worthwhile to review the first vector form of Green's Theorem in (1) of Section 15.8. This three-dimensional generalization is based on a second vector interpretation of the theorem given next.

▌ **Vector Form of Green's Theorem** Let $\mathbf{F}(x, y) = P(x, y)\mathbf{i} + Q(x, y)\mathbf{j}$ be a two-dimensional vector field and let $\mathbf{T} = (dx/ds)\mathbf{i} + (dy/ds)\mathbf{j}$ be a *unit tangent* to a simple closed plane curve C. In (1) of Section 15.8 we saw that $\oint_C (\mathbf{F} \cdot \mathbf{T}) \, ds$ can be evaluated by a double integral involving curl $\mathbf{F}$.

Similarly, if $\mathbf{n} = (dy/ds)\mathbf{i} - (dx/ds)\mathbf{j}$ is a *unit normal* to C (check $\mathbf{T} \cdot \mathbf{n}$), then $\oint_C (\mathbf{F} \cdot \mathbf{n}) \, ds$ can be expressed in terms of a double integral involving div $\mathbf{F}$. From Green's Theorem,

$$\oint_C (\mathbf{F} \cdot \mathbf{n}) \, ds = \oint_C P \, dy - Q \, dx = \iint_R \left[\frac{\partial P}{\partial x} - \left(-\frac{\partial Q}{\partial y} \right) \right] dA = \iint_R \left[\frac{\partial P}{\partial x} + \frac{\partial Q}{\partial y} \right] dA.$$

That is,

$$\oint_C (\mathbf{F} \cdot \mathbf{n}) \, ds = \iint_R \text{div } \mathbf{F} \, dA. \qquad (1)$$

▌ **Green's Theorem in 3-Space** The result in (1) is a special case of the **Divergence** or **Gauss' Theorem**. The following theorem generalizes (1) to 3-space.

Theorem 15.9.1 Divergence Theorem

Suppose that D is a bounded region in 3-space with a piecewise-smooth boundary S that is oriented outward. Let

$$\mathbf{F}(x, y, z) = P(x, y, z)\mathbf{i} + Q(x, y, z)\mathbf{j} + R(x, y, z)\mathbf{k}$$

be a vector field for which P, Q, and R are continuous and have continuous first partial derivatives in a region of 3-space containing D. Then

$$\iint_S (\mathbf{F} \cdot \mathbf{n})\, dS = \iiint_D (\text{div }\mathbf{F})\, dV, \tag{2}$$

where $\mathbf{n}$ is an outward unit normal to S.

PARTIAL PROOF We shall prove (2) for the special region D that is shown in FIGURE 15.9.1 whose surface S consists of three pieces

$$\text{(bottom) } S_1: \quad z = g_1(x, y), \quad (x, y) \text{ in } R$$
$$\text{(top) } S_2: \quad z = g_2(x, y), \quad (x, y) \text{ in } R$$
$$\text{(side) } S_3: \quad g_1(x, y) \le z \le g_2(x, y), \quad (x, y) \text{ on } C,$$

where R is the projection of D onto the xy-plane and C is the boundary of R. Since

$$\text{div }\mathbf{F} = \frac{\partial P}{\partial x} + \frac{\partial Q}{\partial y} + \frac{\partial R}{\partial z} \quad \text{and} \quad \mathbf{F} \cdot \mathbf{n} = P(\mathbf{i} \cdot \mathbf{n}) + Q(\mathbf{j} \cdot \mathbf{n}) + R(\mathbf{k} \cdot \mathbf{n})$$

we can write

$$\iint_S (\mathbf{F} \cdot \mathbf{n})\, dS = \iint_S P(\mathbf{i} \cdot \mathbf{n})\, dS + \iint_S Q(\mathbf{j} \cdot \mathbf{n})\, dS + \iint_S R(\mathbf{k} \cdot \mathbf{n})\, dS$$

and

$$\iiint_D \text{div }\mathbf{F}\, dV = \iiint_D \frac{\partial P}{\partial x}\, dV + \iiint_D \frac{\partial Q}{\partial y}\, dV + \iiint_D \frac{\partial R}{\partial z}\, dV.$$

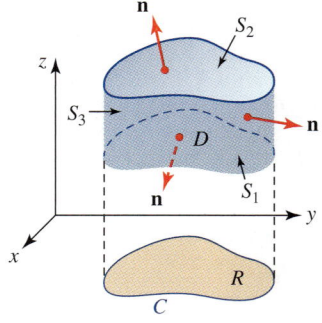

FIGURE 15.9.1 Surface used in the proof of Theorem 15.9.1

To prove (2) we need only establish that

$$\iint_S R(\mathbf{i} \cdot \mathbf{n})\, dS = \iiint_D \frac{\partial P}{\partial x}\, dV, \tag{3}$$

$$\iint_S Q(\mathbf{j} \cdot \mathbf{n})\, dS = \iiint_D \frac{\partial Q}{\partial y}\, dV, \tag{4}$$

and

$$\iint_S R(\mathbf{k} \cdot \mathbf{n})\, dS = \iiint_D \frac{\partial R}{\partial z}\, dV. \tag{5}$$

Indeed, we shall only prove (5) because the proofs of (3) and (4) follow in a similar manner. Now,

$$\iiint_D \frac{\partial R}{\partial z}\, dV = \iint_R \left[\int_{g_1(x, y)}^{g_2(x, y)} \frac{\partial R}{\partial z}\, dz \right] dA = \iint_R [R(x, y, g_2(x, y)) - R(x, y, g_1(x, y))]\, dA. \tag{6}$$

Next we write

$$\iint_S R(\mathbf{k} \cdot \mathbf{n})\, dS = \iint_{S_1} R(\mathbf{k} \cdot \mathbf{n})\, dS + \iint_{S_2} R(\mathbf{k} \cdot \mathbf{n})\, dS + \iint_{S_3} R(\mathbf{k} \cdot \mathbf{n})\, dS.$$

On S_1: Since the outward normal points downward, we describe the surface as $h(x, y, z) = g_1(x, y) - z = 0$. Thus,

$$\mathbf{n} = \frac{\nabla h}{|\nabla h|} = \frac{\dfrac{\partial g_1}{\partial x}\mathbf{i} + \dfrac{\partial g_1}{\partial y}\mathbf{j} - \mathbf{k}}{\sqrt{1 + \left(\dfrac{\partial g_1}{\partial x}\right)^2 + \left(\dfrac{\partial g_1}{\partial y}\right)^2}} \quad \text{so that} \quad \mathbf{k} \cdot \mathbf{n} = \frac{-1}{\sqrt{1 + \left(\dfrac{\partial g_1}{\partial x}\right)^2 + \left(\dfrac{\partial g_1}{\partial y}\right)^2}}.$$

From the definition of dS we then have

$$\iint_{S_1} R(\mathbf{k} \cdot \mathbf{n}) \, dS = -\iint_R R(x, y, g_1(x, y)) \, dA. \tag{7}$$

On S_2: The outward normal points upward, so we describe the surface this time as $h(x, y, z) = z - g_2(x, y) = 0$. Therefore,

$$\mathbf{n} = \frac{\nabla h}{|\nabla h|} = \frac{-\dfrac{\partial g_2}{\partial x}\mathbf{i} - \dfrac{\partial g_2}{\partial y}\mathbf{j} + \mathbf{k}}{\sqrt{1 + \left(\dfrac{\partial g_2}{\partial x}\right)^2 + \left(\dfrac{\partial g_2}{\partial y}\right)^2}} \quad \text{so that} \quad \mathbf{k} \cdot \mathbf{n} = \frac{1}{\sqrt{1 + \left(\dfrac{\partial g_2}{\partial x}\right)^2 + \left(\dfrac{\partial g_2}{\partial y}\right)^2}}.$$

From the last result we find

$$\iint_{S_2} R(\mathbf{k} \cdot \mathbf{n}) \, dS = \iint_R R(x, y, g_2(x, y)) \, dA. \tag{8}$$

On S_3: Because this side is vertical, $\mathbf{k}$ is perpendicular to $\mathbf{n}$. Consequently, $\mathbf{k} \cdot \mathbf{n} = 0$ and

$$\iint_{S_3} R(\mathbf{k} \cdot \mathbf{n}) \, dS = 0. \tag{9}$$

Finally, adding (7), (8), and (9), we get

$$\iint_R [R(x, y, g_2(x, y)) - R(x, y, g_1(x, y))] \, dA$$

which is the same as (6). ∎

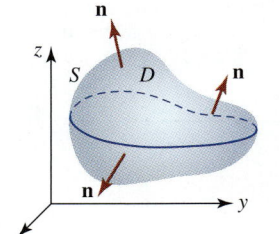

FIGURE 15.9.2 Region with no vertical side

Although we proved (2) for a special region D that has a vertical side, we note that this type of region is not required in Theorem 15.9.1. A region D with no vertical side is illustrated in FIGURE 15.9.2; a region bounded by a sphere or an ellipsoid also does not have a vertical side. The Divergence Theorem also holds for region D bounded between two closed surfaces such as the concentric spheres S_a and S_b shown in FIGURE 15.9.3; the boundary surface S of D is the union of S_a and S_b. In this case, $\iint_S (\mathbf{F} \cdot \mathbf{n}) \, dS = \iiint_D \text{div} \, \mathbf{F} \, dV$ becomes

$$\iint_{S_b} (\mathbf{F} \cdot \mathbf{n}) \, dS + \iint_{S_a} (\mathbf{F} \cdot \mathbf{n}) \, dS = \iiint_D \text{div} \, \mathbf{F} \, dV,$$

where $\mathbf{n}$ points outward from D. In other words, $\mathbf{n}$ points away from the origin on S_b but $\mathbf{n}$ points toward the origin on S_a.

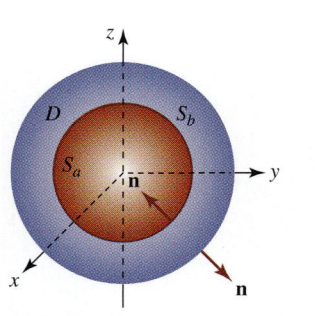

FIGURE 15.9.3 Concentric spheres

EXAMPLE 1   Verifying the Divergence Theorem

Let D be the closed region bounded by the hemisphere $x^2 + y^2 + (z - 1)^2 = 9$, $1 \le z \le 4$, and the plane $z = 1$. Verify the Divergence Theorem for the vector field $\mathbf{F} = x\mathbf{i} + y\mathbf{j} + (z - 1)\mathbf{k}$.

Solution   The closed region is shown in FIGURE 15.9.4.

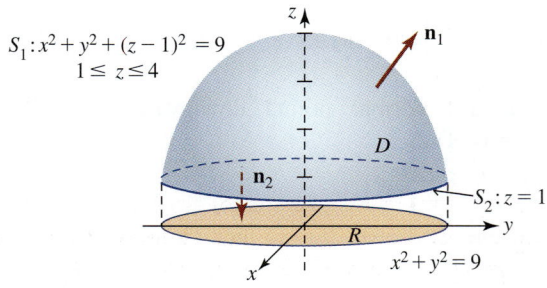

FIGURE 15.9.4 Surface in Example 1

The Triple Integral: Since $\mathbf{F} = x\mathbf{i} + y\mathbf{j} + (z - 1)\mathbf{k}$, we see div $\mathbf{F} = 3$. Hence,

$$\iiint_D \text{div } \mathbf{F} \, dV = \iiint_D 3 \, dV = 3 \iiint_D dV = 3\left[\frac{2}{3}\pi 3^3\right] = 54\pi. \tag{10}$$

In the last calculation, we used the fact that $\iiint_D dV$ gives the volume of the hemisphere.

The Surface Integral: We write $\iint_S = \iint_{S_1} + \iint_{S_2}$, where S_1 is the hemisphere and S_2 is the plane $z = 1$. If S_1 is a level surface of $h(x, y, z) = x^2 + y^2 + (z - 1)^2$, then a unit outer normal is

$$\mathbf{n} = \frac{\nabla h}{|\nabla h|} = \frac{x\mathbf{i} + y\mathbf{j} + (z - 1)\mathbf{k}}{\sqrt{x^2 + y^2 + (z - 1)^2}} = \frac{x}{3}\mathbf{i} + \frac{y}{3}\mathbf{j} + \frac{z - 1}{3}\mathbf{k}.$$

Now,

$$\mathbf{F} \cdot \mathbf{n} = \frac{x^2}{3} + \frac{y^2}{3} + \frac{(z - 1)^2}{3} = \frac{1}{3}(x^2 + y^2 + (z - 1)^2) = \frac{1}{3} \cdot 9 = 3,$$

and so with the aid of polar coordinates we get

$$\iint_{S_1} (\mathbf{F} \cdot \mathbf{n}) \, dS = \iint_R 3\left(\frac{3}{\sqrt{9 - x^2 - y^2}} \, dA\right)$$

$$= 9\int_0^{2\pi} \int_0^3 (9 - r^2)^{-1/2} r \, dr \, d\theta = 54\pi.$$

On S_2, we take $\mathbf{n} = -\mathbf{k}$ so that $\mathbf{F} \cdot \mathbf{n} = -z + 1$. But, since $z = 1$, $\iint_{S_2}(-z + 1) \, dS = 0$. Hence, we see that

$$\iint_S (\mathbf{F} \cdot \mathbf{n}) \, dS = 54\pi + 0 = 54\pi$$

agrees with (10). ∎

EXAMPLE 2 Using the Divergence Theorem

Evaluate $\iint_S(\mathbf{F} \cdot \mathbf{n}) \, dS$, where S is the unit cube defined by $0 \le x \le 1$, $0 \le y \le 1$, $0 \le z \le 1$, and $\mathbf{F} = xy\mathbf{i} + y^2z\mathbf{j} + z^3\mathbf{k}$.

Solution See Figure 15.6.13 and Problem 24 of Exercises 15.6. Rather than evaluate six surface integrals, we apply the Divergence Theorem. Since div $\mathbf{F} = \nabla \cdot \mathbf{F} = y + 2yz + 3z^2$, we have from (2),

$$\iint_S (\mathbf{F} \cdot \mathbf{n}) \, dS = \iiint_D (y + 2yz + 3z^2) \, dV$$

$$= \int_0^1 \int_0^1 \int_0^1 (y + 2yz + 3z^2) \, dx \, dy \, dz$$

$$= \int_0^1 \int_0^1 (y + 2yz + 3z^2) \, dy \, dz$$

$$= \int_0^1 \left(\frac{1}{2}y^2 + y^2z + 3yz^2\right)\Big]_0^1 \, dz$$

$$= \int_0^1 \left(\frac{1}{2} + z + 3z^2\right) dz = \left(\frac{1}{2}z + \frac{1}{2}z^2 + z^3\right)\Big]_0^1 = 2. \quad ∎$$

▮ Physical Interpretation of Divergence In Section 15.7 we saw that we could express the normal component of the curl of a vector field $\mathbf{F}$ at a point as a limit involving the circulation of $\mathbf{F}$. In view of (2) it possible to interpret the divergence of $\mathbf{F}$ at a point as a limit involving the flux of $\mathbf{F}$. Recall from (7) of Section 15.6 that the flux of the velocity field $\mathbf{F}$ of a fluid is

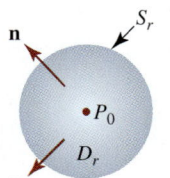

FIGURE 15.9.5 Small sphere centered at P_0

the rate of fluid flow—that is, the volume of fluid flowing through a surface per unit time. In Section 15.7 we saw that the divergence of **F** is the flux per unit volume. To reinforce this last idea, let us suppose $P_0(x_0, y_0, z_0)$ is any point in the fluid and S_r is a small sphere of radius r centered at P_0. See FIGURE 15.9.5. If D_r is the sphere S_r and its interior, then the Divergence Theorem gives

$$\iint_{S_r} (\mathbf{F} \cdot \mathbf{n})\, dS = \iiint_{D_r} \text{div } \mathbf{F}\, dV. \tag{11}$$

If we take the approximation div $\mathbf{F}(P) \approx$ div $\mathbf{F}(P_0)$ at any point $P(x, y, z)$ within the small sphere, then (11) gives

$$\iint_{S_r} (\mathbf{F} \cdot \mathbf{n})\, dS \approx \iiint_{D_r} \text{div } \mathbf{F}(P_0)\, dV$$

$$= \text{div } \mathbf{F}(P_0) \iiint_{D_r} dV$$

$$= \text{div } \mathbf{F}(P_0)\, V_r, \tag{12}$$

where V_r is the volume $\frac{4}{3}\pi r^3$ of the spherical region D_r. By letting $r \to 0$, we see from (12) that the divergence of **F** is the limiting value of the ratio of the flux of **F** to the volume of the spherical region:

$$\text{div } \mathbf{F}(P_0) = \lim_{r \to 0} \frac{1}{V_r} \iint_{S_r} (\mathbf{F} \cdot \mathbf{n})\, dS. \tag{13}$$

Hence, the divergence of **F** is flux per unit volume.

The Divergence Theorem is extremely useful in the derivation of some of the famous equations in electricity and magnetism, and hydrodynamics. In the discussion that follows we shall consider an example from the study of fluids.

■ **Continuity Equation**   At the end of Section 15.7 we mentioned that one interpretation of div **F** is a measure of the rate of change of the density of a fluid at a point. To see why this is so, let us suppose that **F** is a velocity field of a fluid and that $\rho(x, y, z, t)$ is the density of the fluid at a point $P(x, y, z)$ at time t. Let D be the closed region consisting of a sphere S and its interior. We know from Section 14.7 that the total mass m of the fluid in D is given by

$$m = \iiint_D \rho(x, y, z, t)\, dV.$$

The rate at which the mass increases in D is given by

$$\frac{dm}{dt} = \frac{d}{dt} \iiint_D \rho(x, y, z, t)\, dV = \iiint_D \frac{\partial \rho}{\partial t}\, dV. \tag{14}$$

Now from Figure 15.6.8(b) we saw that the volume of fluid flowing through an element of surface area ΔS per unit time is approximated by

$$(\mathbf{F} \cdot \mathbf{n})\, \Delta S.$$

The mass of the fluid flowing through an element of surface area ΔS per unit time is then

$$(\rho \mathbf{F} \cdot \mathbf{n})\, \Delta S.$$

If we assume that the change in mass in D is due only to the flow in and out of D, then the *volume of fluid* flowing out of D per unit time is given by (7) of Section 15.6, $\iint_S (\mathbf{F} \cdot \mathbf{n})\, dS$, whereas the *mass of the fluid* flowing out of D per unit is $\iint_S (\rho \mathbf{F} \cdot \mathbf{n})\, dS$. Hence, an alternative expression for the rate at which mass increases in D is

$$-\iint_S (\rho \mathbf{F} \cdot \mathbf{n})\, dS. \tag{15}$$

By the Divergence Theorem, (15) is the same as

$$-\iiint_D \text{div}\,(\rho\mathbf{F})\,dV. \qquad (16)$$

Equating (14) and (16) then yields

$$\iiint_D \frac{\partial \rho}{\partial t}\,dV = -\iiint_D \text{div}(\rho\mathbf{F})\,dV \qquad \text{or} \qquad \iiint_D \left(\frac{\partial \rho}{\partial t} + \text{div}(\rho\mathbf{F})\right) dV = 0.$$

Since this last result is to hold for every sphere, we obtain the **equation of continuity** for fluid flows:

$$\frac{\partial \rho}{\partial t} + \text{div}(\rho\mathbf{F}) = 0. \qquad (17)$$

On page 849 we stated that if div $\mathbf{F} = \nabla \cdot \mathbf{F} = 0$, then a fluid is incompressible. This fact follows immediately from (17). If a fluid is incompressible (such as water), then ρ is constant, so consequently $\nabla \cdot (\rho\mathbf{F}) = \rho(\nabla \cdot \mathbf{F})$. But in addition $\partial\rho/\partial t = 0$, and so (17) implies $\nabla \cdot \mathbf{F} = 0$.

▌ Postscript—A Bit of History **Johann Karl Friedrich Gauss** (1777–1855) was the first of a new breed of precise and demanding mathematicians—the "rigorists." We have seen in earlier biographical sketches that Augustin Louis Cauchy and Karl Wilhelm Weierstrass were two math-

ematicians who followed in his footsteps. Karl Friedrich Gauss, the only son of a poor gardener, was a child prodigy in mathematics. He was not yet three when he corrected his father's computation of a payroll. As an adult, Gauss often remarked that he could calculate or "reckon" before he could talk. As a college student, Gauss was torn between two loves: philology and mathematics. Although he easily mastered foreign languages, he was inspired by some original mathematical achievements as a teenager and encouraged by the mathematician Wolfgang Bolyai, so the choice between languages and mathematics was not too difficult. At the age of 20, Gauss settled on a career in mathematics. At the age of 22, he completed a book on number theory, *Disquisitiones Arithmeticae*. Published in 1801, this text was recognized as a masterpiece and

Gauss

even today remains a classic in its field. Gauss' doctoral dissertation of 1799 is also a memorable document. Using the theory of functions of a complex variable, he was the first to prove the so-called fundamental theorem of algebra: Every polynomial equation has at least one root.

Although Gauss was certainly recognized and respected as an outstanding mathematician during his lifetime, the full extent of his genius was not realized until the publication of his scientific diary in 1898, 44 years after his death. Much to the chagrin of some nineteenth-century mathematicians, the diary revealed that Gauss had foreseen, sometimes by decades, many of their discoveries or, perhaps more accurately, rediscoveries. He was oblivious to fame; his mathematical researches were often pursued, like a child playing on a beach, simply for pleasure and self-satisfaction and not for the instruction that could be given to others through publication.

On any list of "Greatest Mathematicians Who Ever Lived," Karl Friedrich Gauss must surely rank near or at the top. For his profound impact on so many branches of mathematics, Gauss is sometimes referred to as "the prince of mathematicians."

$\iint_S$ **NOTES FROM THE CLASSROOM**

Why do some objects float in water and others sink? The answer comes from **Archimedes' principle**, which states: When an object is submerged in a fluid, the fluid exerts an upward force on it, called **buoyant force**, with a magnitude that is equal to the weight of the fluid displaced. Thus, a cork has positive buoyacy or floats since the weight of the cork is less that the magnitude of the buoyant force. A submarine will attain negative buoyancy and sink by filling its ballast tanks with water, thereby making its weight greater than the magnitude of the buoyant force exerted on it. See FIGURE 15.9.6. You are asked to prove this famous theorem using the Divergence Theorem in Problem 22 of Exercises 15.9.

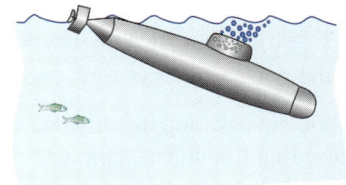

FIGURE 15.9.6 A submarine sinks when the magnitude of the buoyant force is less than the magnitude of its weight

≡ Fundamentals

In Problems 1 and 2, verify the Divergence Theorem for the given vector field.

1. $\mathbf{F} = xy\mathbf{i} + yz\mathbf{j} + xz\mathbf{k}$; D the region bounded by the unit cube defined by $0 \le x \le 1, 0 \le y \le 1, 0 \le z \le 1$

2. $\mathbf{F} = 6xy\mathbf{i} + 4yz\mathbf{j} + xe^{-y}\mathbf{k}$; D the region bounded by the three coordinate planes and the plane $x + y + z = 1$

In Problems 3–14, use the Divergence Theorem to find the outward flux $\iint_S (\mathbf{F} \cdot \mathbf{n})\, dS$ of the given vector field $\mathbf{F}$.

3. $\mathbf{F} = x^3\mathbf{i} + y^3\mathbf{j} + z^3\mathbf{k}$; D the region bounded by the sphere $x^2 + y^2 + z^2 = a^2$

4. $\mathbf{F} = 4x\mathbf{i} + y\mathbf{j} + 4z\mathbf{k}$; D the region bounded by the sphere $x^2 + y^2 + z^2 = 4$

5. $\mathbf{F} = y^2\mathbf{i} + xz^3\mathbf{j} + (z - 1)^2\mathbf{k}$; D the region bounded by the cylinder $x^2 + y^2 = 16$ and the planes $z = 1, z = 5$

6. $\mathbf{F} = x^2\mathbf{i} + 2yz\mathbf{j} + 4z^3\mathbf{k}$; D the region bounded by the parallelepiped defined by $0 \le x \le 1, 0 \le y \le 2, 0 \le z \le 3$

7. $\mathbf{F} = y^3\mathbf{i} + x^3\mathbf{j} + z^3\mathbf{k}$; D the region bounded within by $z = \sqrt{4 - x^2 - y^2}, x^2 + y^2 = 3, z = 0$

8. $\mathbf{F} = (x^2 + \sin y)\mathbf{i} + z^2\mathbf{j} + xy^3\mathbf{k}$; D the region bounded by $y = x^2, z = 9 - y, z = 0$

9. $\mathbf{F} = (x\mathbf{i} + y\mathbf{j} + z\mathbf{k})/(x^2 + y^2 + z^2)$; D the region bounded by the concentric spheres $x^2 + y^2 + z^2 = a^2$, $x^2 + y^2 + z^2 = b^2, b > a$

10. $\mathbf{F} = 2yz\mathbf{i} + x^3\mathbf{j} + xy^2\mathbf{k}$; D the region bounded by the ellipsoid $x^2/a^2 + y^2/b^2 + z^2/c^2 = 1$

11. $\mathbf{F} = 2xz\mathbf{i} + 5y^2\mathbf{j} - z^2\mathbf{k}$; D the region bounded by $z = y$, $z = 4 - y, z = 2 - \frac{1}{2}x^2, x = 0, z = 0$. See FIGURE 15.9.7.

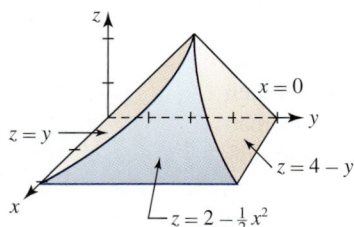

FIGURE 15.9.7   Region D in Problem 11

12. $\mathbf{F} = 15x^2y\mathbf{i} + x^2z\mathbf{j} + y^4\mathbf{k}$; D the region bounded by $x + y = 2, z = x + y, z = 3, x = 0, y = 0$

13. $\mathbf{F} = 3x^2y^2\mathbf{i} + yj - 6xy^2z\mathbf{k}$; D the region bounded by the paraboloid $z = x^2 + y^2$ and the plane $z = 2y$

14. $\mathbf{F} = xy^2\mathbf{i} + x^2y\mathbf{j} + 6(\sin x)\mathbf{k}$; D the region bounded by the cone $z = \sqrt{x^2 + y^2}$ and the planes $z = 2, z = 4$

In Problems 15 and 16, assume that S forms the boundary of a closed and bounded region D.

15. If $\mathbf{a}$ is a constant vector, show that $\iint_S (\mathbf{a} \cdot \mathbf{n})\, dS = 0$.

16. If $\mathbf{F} = P\mathbf{i} + Q\mathbf{j} + R\mathbf{k}$ and P, Q, and R have continuous second partial derivatives, show that $\iint_S (\text{curl } \mathbf{F} \cdot \mathbf{n})\, dS = 0$.

≡ Applications

17. The electric field at a point $P(x, y, z)$ due to a point charge q located at the origin is given by the inverse square field $\mathbf{E} = q\mathbf{r}/|\mathbf{r}|^3$, where $\mathbf{r} = x\mathbf{i} + y\mathbf{j} + z\mathbf{k}$.

(a) Suppose S is a closed surface, S_a is a sphere $x^2 + y^2 + z^2 = a^2$ lying completely within S, and D is the region bounded between S and S_a. See FIGURE 15.9.8. Show that the outward flux of $\mathbf{E}$ for the region D is zero.

(b) Use the result of part (a) to prove **Gauss' Law**:

$$\iint_S (\mathbf{E} \cdot \mathbf{n})\, dS = 4\pi q.$$

That is, the outward flux of the electric field $\mathbf{E}$ through *any* closed surface (for which the Divergence Theorem applies) containing the origin is $4\pi q$.

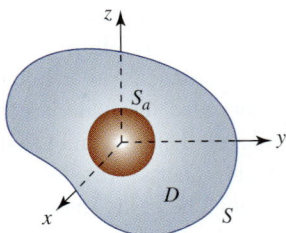

FIGURE 15.9.8   Surfaces in Problem 17

18. Suppose there is a continuous distribution of charge throughout a closed and bounded region D enclosed by a surface S. Then, the natural extension of Gauss' Law is given by

$$\iint_S (\mathbf{E} \cdot \mathbf{n})\, dS = \iiint_D 4\pi\rho\, dV,$$

where $\rho(x, y, z)$ is the charge density or charge per unit volume.

(a) Proceed as in the derivation of the continuity equation (17) to show that div $\mathbf{E} = 4\pi q$.

(b) Given that $\mathbf{E}$ is an irrotational vector field, show that the potential function ϕ for $\mathbf{E}$ satisfies Poissons equation $\nabla^2\phi = 4\pi\rho$, where $\nabla^2\phi = \nabla \cdot \nabla\phi$.

≡ Think About It

In Problems 19 and 20, assume that f and g are scalar functions with continuous second partial derivatives. Use the Divergence Theorem to establish **Green's identities**. Assume that S forms the boundary of a closed and bounded region D.

19. $\iint_S (f\nabla g) \cdot \mathbf{n}\, dS = \iiint_D (f\nabla^2 g + \nabla f \cdot \nabla g)\, dV$

20. $\iint_S (f\nabla g - g\nabla f) \cdot \mathbf{n}\, dS = \iiint_D (f\nabla^2 g - g\nabla^2 f)\, dV$

21. If f is a scalar function with continuous first partial derivatives and S forms the boundary of a closed and bounded region D, then show that

$$\iint_S f\mathbf{n}\, dS = \iiint_D \nabla f\, dV.$$

result of Problem 21 to prove Archimedes' Principle, $\mathbf{B} + \mathbf{W} = \mathbf{0}$. See FIGURE 15.9.9.

[*Hint*: Use (2) on $f\mathbf{a}$, where $\mathbf{a}$ is a constant vector, and Problem 21 in Exercises 15.7.]

22. The buoyant force on a floating object is $\mathbf{B} = -\iint_S p\mathbf{n}\, dS$, where p is the fluid pressure. The pressure p is related to the density of the fluid $\rho(x, y, z)$ by a law of hydrostatics: $\nabla p = \rho(x, y, z)\mathbf{g}$, where $\mathbf{g}$ is the constant acceleration due to gravity. If the weight of the object is $\mathbf{W} = m\mathbf{g}$, use the

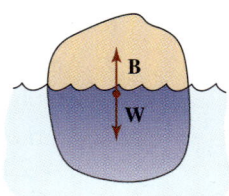

FIGURE 15.9.9 Floating object in Problem 22

Chapter 15 in Review

Answers to selected odd-numbered problems begin on page ANS-47.

A. True/False

In Problems 1–12, answer true or false. Where appropriate, assume continuity of P, Q, and their first partial derivatives.

1. The integral $\int_C (x^2 + y^2)\, dx + 2xy\, dy$, where C is given by $y = x^3$ from $(0, 0)$ to $(1, 1)$ has the same value on the curve $y = x^6$ from $(0, 0)$ to $(1, 1)$. _____

2. The value of the integral $\int_C 2xy\, dx - x^2\, dy$ between two points A and B depends on the path C. _____

3. If C_1 and C_2 are two smooth curves such that $\int_{C_1} P\, dx + Q\, dy = \int_{C_2} P\, dx + Q\, dy$, then $\int_C P\, dx + Q\, dy$ is independent of the path. _____

4. If the work $\int_C \mathbf{F} \cdot d\mathbf{r}$ depends on the curve C, then $\mathbf{F}$ is nonconservative. _____

5. Assuming continuity of all partial derivatives and $\partial P/\partial x = \partial Q/\partial y$, then $\int_C P\, dx + Q\, dy$ is independent of the path. _____

6. In a conservative force field $\mathbf{F}$, the work done by $\mathbf{F}$ around a simple closed curve is zero. _____

7. Assuming continuity of all partial derivatives, $\nabla \times \nabla f = \mathbf{0}$. _____

8. The surface integral of the normal component of the curl of a conservative vector field $\mathbf{F}$ over a surface S is equal to zero. _____

9. The work done by a force $\mathbf{F}$ along a curve C is due entirely to the tangential component of $\mathbf{F}$. _____

10. For a two-dimensional vector field $\mathbf{F}$ in the plane $z = 0$, Stokes' Theorem is the same as Green's Theorem. _____

11. If $\mathbf{F}$ is a conservative force field, then the sum of the potential and kinetic energies of an object is constant. _____

12. If $\int_C \mathbf{F} \cdot d\mathbf{r}$ is independent of the path C in an appropriate region R, then $\mathbf{F} = P\mathbf{i} + Q\mathbf{j}$ is the gradient of some function ϕ. _____

B. Fill in the Blanks

In Problems 1–10, fill in the blanks.

1. If $\phi = \dfrac{1}{\sqrt{x^2 + y^2}}$ is a potential function for a conservative force field $\mathbf{F}$, then $\mathbf{F} = $ _____.

2. If $\mathbf{F} = f(x)\mathbf{i} + g(y)\mathbf{j} + h(z)\mathbf{k}$, then curl $\mathbf{F} = $ _____.

In Problems 3–6, $\mathbf{F} = x^2 y\mathbf{i} + xy^2\mathbf{j} + 2xyz\mathbf{k}$.

3. $\nabla \cdot \mathbf{F} = $ _____

4. $\nabla \times \mathbf{F} = $ _____

5. $\nabla \cdot (\nabla \times \mathbf{F}) = $ _____

6. $\nabla(\nabla \cdot \mathbf{F}) = $ _____

7. If C is the ellipse $2(x - 10)^2 + 9(y + 13)^2 = 3$, then $\oint_C (y - 7e^{x^3})\, dx + (x + \ln \sqrt{y})\, dy = $ _____.

8. If $\mathbf{F}$ is a velocity field of a fluid for which curl $\mathbf{F} = \mathbf{0}$, then $\mathbf{F}$ is said to be _____.

9. An equation of the tangent plane to the surface $\mathbf{r}(u, v) = u\mathbf{i} + v\mathbf{j} + 2\sqrt{uv}\,\mathbf{k}$ at $u = 1, v = 4$ is _____.

10. A grid line on the surface $\mathbf{r}(u, v) = (4u + v)\mathbf{i} + (u + 2v)\mathbf{j} + (u + v)\mathbf{k}$ corresponding to $u = 2$ has the parametric equations _____.

C. Exercises

1. Evaluate $\displaystyle\int_C \frac{z^2}{x^2 + y^2}\, ds$, where C is given by $x = \cos 2t, y = \sin 2t, z = 2t, \pi \le t \le 2\pi$.

2. Evaluate $\int_C (xy + 4x)\, ds$, where C is given by $2x + y = 2$ from $(1, 0)$ to $(0, 2)$.

3. Evaluate $\int_C 3x^2 y^2\, dx + (2x^3 y - 3y^2)\, dy$, where C is given by $y = 5x^4 + 7x^2 - 14x$ from $(0, 0)$ to $(1, -2)$.

4. Evaluate $\oint_C (x^2 + y^2)\, dx + (x^2 - y^2)\, dy$, where C is the circle $x^2 + y^2 = 9$.

5. Evaluate $\int_C y \sin \pi z\, dx + x^2 e^y\, dy + 3xyz\, dz$, where C is given by $x = t, y = t^2, z = t^3$ from $(0, 0, 0)$ to $(1, 1, 1)$.

6. If $\mathbf{F} = 4y\mathbf{i} + 6x\mathbf{j}$ and C is given by $x^2 + y^2 = 1$, evaluate $\oint_C \mathbf{F} \cdot d\mathbf{r}$ in two different ways.

7. Find the work done by the force $\mathbf{F} = x \sin y\mathbf{i} + y \sin x\mathbf{j}$ acting along the line segments from $(0, 0)$ to $(\pi/2, 0)$ and from $(\pi/2, 0)$ to $(\pi/2, \pi)$.

8. Find the work done by $\mathbf{F} = \dfrac{2}{x^2 + y^2}\mathbf{i} + \dfrac{1}{x^2 + y^2}\mathbf{j}$ from $\left(-\frac{1}{2}, \frac{1}{2}\right)$ to $\left(1, \sqrt{3}\right)$ acting on the path shown in FIGURE 15.R.1.

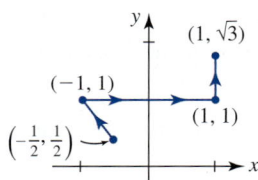

FIGURE 15.R.1 Curve C in Problem 8

In Problems 9 and 10, show that the given integral is independent of the path. Evaluate.

9. $\displaystyle\int_{(1, 1, 0)}^{(1, 1, \pi)} 2xy\, dx + (x^2 + 2yz)\, dy + (y^2 + 4)\, dz$

10. $\displaystyle\int_{(0, 0, 1)}^{(3, 2, 0)} (2x + 2ze^{2x})\, dx + (2y - 1)\, dy + e^{2x}\, dz$

11. Evaluate $\oint_C -4y\, dx + 8x\, dy$, where $C = C_1 \cup C_2$ is the boundary of the region R shown in FIGURE 15.R.2.

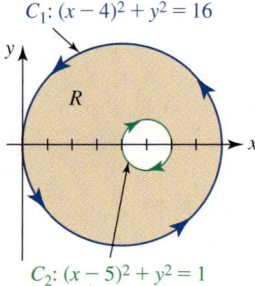

FIGURE 15.R.2 Curve C in Problem 11

12. Let C be a piecewise-smooth simple closed curve. Show that

$$\oint_C \frac{y-1}{(x-1)^2 + (y-1)^2}\,dx + \frac{1-x}{(x-1)^2 + (y-1)^2}\,dy = \begin{cases} -2\pi, & \text{if } (1, 1) \text{ is inside } C \\ 0, & \text{if } (1, 1) \text{ is outside } C. \end{cases}$$

13. Evaluate $\iint_S (z/xy)\,dS$, where S is that portion of the cylinder $z = x^2$ in the first octant that is bounded by $y = 1$, $y = 3$, $z = 1$, $z = 4$.

14. If $\mathbf{F} = \mathbf{i} + 2\mathbf{j} + 3\mathbf{k}$, find the flux of $\mathbf{F}$ through the square defined by $0 \le x \le 1$, $0 \le y \le 1$, $z = 2$.

15. Let the surface S be that portion of the cylinder $y = 2 - e^{-x}$ whose projection onto the xz-plane is a rectangular region R defined by $0 \le x \le 3$, $0 \le z \le 2$. See FIGURE 15.R.3(a). Find the flux of $\mathbf{F} = 4\mathbf{i} + (2 - y)\mathbf{j} + 9\mathbf{k}$ through the surface if S is oriented away from the xz-plane.

16. Rework Problem 15 using the region R in the yz-plane that corresponds to $0 \le x \le 3$, $0 \le z \le 2$. See Figure 15.R.3(b).

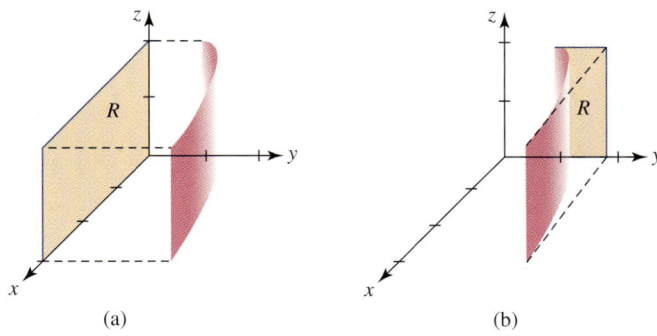

(a)   (b)

FIGURE 15.R.3   Surfaces in Problems 15 and 16

17. If $\mathbf{F} = c\nabla(1/r)$, where c is constant and $r = |\mathbf{r}|$, $\mathbf{r} = x\mathbf{i} + y\mathbf{j} + z\mathbf{k}$, find the flux of $\mathbf{F}$ through the sphere $x^2 + y^2 + z^2 = a^2$.

18. Explain why the Divergence Theorem is not applicable in Problem 17.

19. Find the flux of $\mathbf{F} = c\nabla(1/r)$, where c is constant and $r = |\mathbf{r}|$, $\mathbf{r} = x\mathbf{i} + y\mathbf{j} + z\mathbf{k}$, through any surface S that forms the boundary of a closed bounded region of space not containing the origin.

20. If $\mathbf{F} = 6x\mathbf{i} + 7z\mathbf{j} + 8y\mathbf{k}$, use Stokes' Theorem to evaluate $\iint_S (\text{curl } \mathbf{F} \cdot \mathbf{n})\,dS$, where S is that portion of the paraboloid $z = 9 - x^2 - y^2$ within the cylinder $x^2 + y^2 = 4$.

21. Use Stokes' Theorem to evaluate $\oint_C -2y\,dx + 3x\,dy + 10z\,dz$, where C is the circle $(x-1)^2 + (y-3)^2 = 25$, $z = 3$.

22. Find the work $\oint_C \mathbf{F} \cdot d\mathbf{r}$ done by the force $\mathbf{F} = x^2\mathbf{i} + y^2\mathbf{j} + z^2\mathbf{k}$ around the curve C that is formed by the intersection of the plane $z = 2 - y$ and the sphere $x^2 + y^2 + z^2 = 4z$.

23. If $\mathbf{F} = x\mathbf{i} + y\mathbf{j} + z\mathbf{k}$, use the Divergence Theorem to evaluate $\iint_S (\mathbf{F} \cdot \mathbf{n})\,dS$, where S is the surface of the region bounded by $x^2 + y^2 = 1$, $z = 0$, $z = 1$.

24. Repeat Problem 23 for $\mathbf{F} = \frac{1}{3}x^3\mathbf{i} + \frac{1}{3}y^3\mathbf{j} + \frac{1}{3}z^3\mathbf{k}$.

25. If $\mathbf{F} = (x^2 - e^y \tan^{-1}z)\mathbf{i} + (x + y)^2\mathbf{j} - (2yz + x^{10})\mathbf{k}$, use the Divergence Theorem to evaluate $\iint_S (\mathbf{F} \cdot \mathbf{n})\,dS$, where S is the surface of the region in the first octant bounded by $z = 1 - x^2$, $z = 0$, $z = 2 - y$, $y = 0$.

26. Suppose $\mathbf{F} = x\mathbf{i} + y\mathbf{j} + (z^2 + 1)\mathbf{k}$ and S is the surface of the region bounded by $x^2 + y^2 = a^2$, $z = 0$, $z = c$. Evaluate $\iint_S (\mathbf{F} \cdot \mathbf{n})\,dS$ without the aid of the Divergence Theorem. [*Hint*: The lateral surface area of the cylinder is $2\pi ac$.]

In Problems 27–30, eliminate the parameters in the set of parametric equations and obtain an equation in x, y, and z. Identify the surface.

27. $x = u\cosh v$, $\quad y = u\sinh v$, $\quad z = u^2$ 28. $x = u\cos v$, $\quad y = u\sin v$, $\quad z = u^2$

29. $\mathbf{r}(u, v) = \cos u\mathbf{i} + \cos^2 u\mathbf{j} + v\mathbf{k}$. $\qquad$ 30. $\mathbf{r}(u, v) = \cos u\cosh v\mathbf{i} + \sin u\cosh v\mathbf{j} + \sinh v\mathbf{k}$

Higher-Order Differential Equations

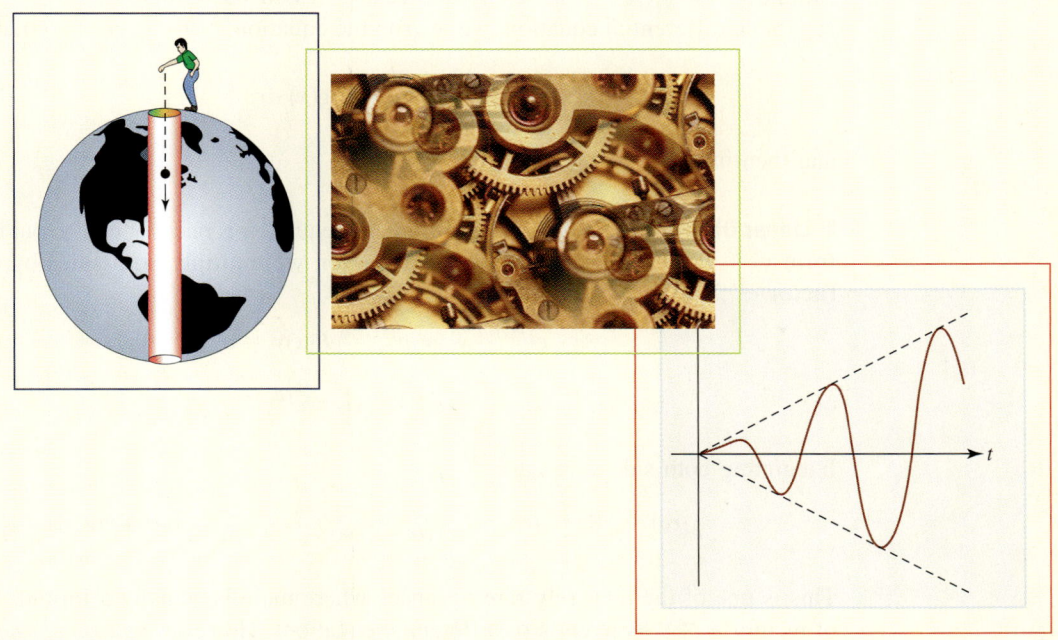

In This Chapter   In Chapter 8 we introduced two important types of first-order differential equations: separable and linear DEs. We also discussed how first-order differential equations could serve as mathematical models for various physical phenomena such as population growth, radioactive decay, and cooling of a body. In this further, and admittedly brief, discussion we shall focus our attention on an important class of second-order DEs. We will see that a mathematical model for the displacements of a mass on a vibrating spring is, except for terminology, the same as a model for the current in a series circuit containing an inductor, a resistor, and a capacitor.

16.1 Exact First-Order Equations

■ **Introduction** The notion of a first-order differential equation (DE) was introduced in Chapter 8. One of the basic problems in the study of differential equations is, How do we solve them? In Sections 8.1 and 8.2 we saw how to solve separable and linear first-order differential equations. After a brief review of these two types of equations, we examine another first-order differential equation called an **exact equation**. Since the solution method for an exact DE utilizes the differential of a function of two variables, a review of Section 13.4 is recommended.

■ **Separable DEs** Recall that a first-order differential equation $y' = F(x, y)$, is **separable** if the function $F(x, y)$ has the form $F(x, y) = g(x)f(y)$. Thus, $y' = xy/(x^2 + 1)$ is separable, because we can write

$$F(x, y) = \frac{xy}{x^2 + 1} = \frac{x}{x^2 + 1} \cdot y.$$

Similarly, $y' = xye^{x^2+y^2}$ is separable because it can be written $y' = xe^{x^2} \cdot ye^{y^2}$. To solve a separable differential equation, we rewrite the equation $dy/dx = g(x)f(y)$ in differential form

$$\frac{dy}{f(y)} = g(x)\, dx,$$

and then integrate both sides of the equation.

■ **Linear DEs** A **linear** first-order differential equation is one that can be put into the standard form $y' + P(x)y = f(x)$. To solve this equation we multiply both sides by the **integrating factor** $e^{\int P(x)\, dx}$. This gives

$$e^{\int P(x)\, dx} y' + e^{\int P(x)\, dx} P(x)y = e^{\int P(x)\, dx} f(x)$$

or

$$\frac{d}{dx}\left[e^{\int P(x)\, dx} y \right] = e^{\int P(x)\, dx} f(x). \tag{1}$$

Integrating both sides, we have

$$e^{\int P(x)\, dx} y = \int e^{\int P(x)\, dx} f(x)\, dx \qquad \text{so} \qquad y = e^{-\int P(x)\, dx} \int e^{\int P(x)\, dx} f(x)\, dx.$$

This is one of the relatively rare instances where there is actually a formula for the solution of members of a large class of differential equations. However, *you should not memorize this formula.* Rather, you should find the integrating factor and then use the equation in (1) to solve the differential equation.

■ **A Definition** We turn now to a class of first-order differential equations that are called **exact**. While the following discussion is self-contained, the main techniques for recognizing and solving an exact equation have already been covered in Section 15.3.

In addition to 13.4, you are encouraged to review Sections 14.2, 15.2, and 15.3.

The **differential** (also called the **total differential**) of a function $f(x, y)$ is

$$df = \frac{\partial f}{\partial x}\, dx + \frac{\partial f}{\partial y}\, dy. \tag{2}$$

Now consider the simple differential equation

$$y\, dx + x\, dy = 0. \tag{3}$$

This equation is both separable and linear, but it can also be solved in an alternative manner by recognizing that the left-hand side is the differential of $f(x, y) = xy$; that is, $y\, dx + x\, dy = d(xy)$. The differential equation in (3) then becomes $d(x, y) = 0$, and integrating both sides immediately yields the solution $xy = C$. In general, we want to be able to recognize when a differential form $M(x, y)\, dx + N(x, y)\, dy$ is the total differential of a function $f(x, y)$.

Definition 16.1.1 Exact Differential Equation

The differential equation $M(x, y)\, dx + N(x, y)\, dy = 0$ is **exact** in a rectangular region R of the xy-plane if there exists a function $f(x, y)$ such that

$$df = M(x, y)\, dx + N(x, y)\, dy.$$

From (2) we see that a differential equation $M(x, y)\, dx + N(x, y)\, dy = 0$ is exact if it is the same as

$$\frac{\partial f}{\partial x}\, dx + \frac{\partial f}{\partial y}\, dy = 0$$

for some function f; that is, if $M(x, y) = \dfrac{\partial f}{\partial x}$ and $N(x, y) = \dfrac{\partial f}{\partial y}$ for some function f.

EXAMPLE 1 Exact Differential

The differential equation $x^2 y^3\, dx + x^3 y^2\, dy = 0$ is exact because, when $f(x, y) = \frac{1}{3} x^3 y^3$, we have $df = x^2 y^3\, dx + x^3 y^2\, dy$. ∎

In Example 1, note that $M = x^2 y^3, N = x^3 y^2$, so

$$\frac{\partial M}{\partial y} = 3x^2 y^2 = \frac{\partial N}{\partial x}.$$

The following theorem shows that this is not a coincidence.

Theorem 16.1.1 Criterion for an Exact DE

Let $M(x, y)$ and $N(x, y)$ be continuous and have continuous partial derivatives in a rectangular region R of the xy-plane. Then a necessary and sufficient condition that

$$M(x, y)\, dx + N(x, y)\, dy = 0 \qquad (4)$$

be an exact differential equation is

$$\frac{\partial M}{\partial y} = \frac{\partial N}{\partial x}. \qquad (5)$$

Proof of Necessity We need to show that if (4) is exact, then $\partial M/\partial y = \partial N/\partial x$. By the definition of an exact differential equation, there exists a function f such that

$$M(x, y) = \frac{\partial f}{\partial x} \qquad \text{and} \qquad N(x, y) = \frac{\partial f}{\partial y}.$$

Therefore, since the first partials of M and N are continuous,

$$\frac{\partial M}{\partial y} = \frac{\partial}{\partial y}\left(\frac{\partial f}{\partial x}\right) = \frac{\partial^2 f}{\partial y\, \partial x} = \frac{\partial^2 f}{\partial x\, \partial y} = \frac{\partial}{\partial x}\left(\frac{\partial f}{\partial y}\right) = \frac{\partial N}{\partial x}. \qquad ∎$$

The sufficiency part of Theorem 16.1.1 consists of showing that there exists a function f for which $\partial f/\partial x = M(x, y)$ and $\partial f/\partial y = N(x, y)$ whenever (5) holds. The construction of f actually reflects a basic procedure for solving exact differential equations.

◀ Notice the similarity between the notions of exact differential equations and conservative vector fields, discussed in Section 15.3.

EXAMPLE 2 Solving an Exact Differential Equation

Solve $2xy\, dx + (x^2 - 1)\, dy = 0$.

Solution We first show that the equation is exact. Identifying $M(x, y) = 2xy$ and $N(x, y) = x^2 - 1$, we have

$$\frac{\partial M}{\partial y} = 2x = \frac{\partial N}{\partial x},$$

which verifies that the DE is exact. Hence, a function $f(x, y)$ exists such that

$$M(x, y) = \frac{\partial f}{\partial x} \quad \text{and} \quad N(x, y) = \frac{\partial f}{\partial y}.$$

The procedure used here for finding the function f is the same as that used in finding the potential function ϕ for a conservative vector field. See Example 6 in Section 15.3.

▶ Starting with the assumption that $\partial f / \partial x = M(x, y)$, we have

$$\frac{\partial f}{\partial x} = 2xy \quad \text{so} \quad f(x, y) = \int 2xy \, dx.$$

Using partial integration, as discussed in Section 14.2, we get $f(x, y) = x^2 y + g(y)$. Using this form for f, we have

$$\frac{\partial f}{\partial y} = x^2 + g'(y) = N(x, y) = x^2 - 1,$$

so $$g'(y) = -1 \quad \text{and} \quad g(y) = -y.$$

Hence, $f(x, y) = x^2 y - y$, and a family of solutions is $f(x, y) = C$ or

$$x^2 y - y = C. \qquad \blacksquare$$

EXAMPLE 3 An Initial-Value Problem

Solve $y(1 - x^2) y' = xy^2 - \cos x \sin x$ subject to the initial condition $y(0) = 2$.

Solution By writing the differential equation in the form

$$(\cos x \sin x - xy^2) \, dx + y(1 - x^2) \, dy = 0,$$

we identify $M = \cos x \sin x - xy^2$ and $N = y(1 - x^2)$. The equation is exact because

$$\frac{\partial M}{\partial y} = -2xy = \frac{\partial N}{\partial x}.$$

Now, starting with $\partial f / \partial y = N(x, y)$, we have

$$\frac{\partial f}{\partial y} = y(1 - x^2) \quad \leftarrow \text{use partial integration here}$$

$$f(x, y) = \tfrac{1}{2} y^2 (1 - x^2) + h(x)$$

$$\frac{\partial f}{\partial x} = -xy^2 + h'(x) = \cos x \sin x - xy^2.$$

The last equation indicates that $h'(x) = \cos x \sin x$, and so we integrate to find

$$h(x) = \int \cos x \sin x \, dx = -\int (\cos x)(-\sin x \, dx) = -\frac{1}{2} \cos^2 x.$$

Thus, the solution of the differential equation is

$$\tfrac{1}{2} y^2 (1 - x^2) - \tfrac{1}{2} \cos^2 x = C_1 \quad \text{or} \quad y^2 (1 - x^2) - \cos^2 x = C,$$

where we have replaced $2C_1$ with C. The initial condition $y = 2$ when $x = 0$ demands that $4(1) - \cos^2(0) = C$ and so $C = 3$. A solution of the problem is then

$$y^2 (1 - x^2) - \cos^2 x = 3. \qquad \blacksquare$$

Of course, not every first-order DE in the form $M(x, y) \, dx + N(x, y) \, dy = 0$ is an exact equation. For example,

$$xy \, dx + (2x^2 + 3y^2 - 20) \, dy = 0$$

is not exact. With the identifications $M = xy$ and $N = 2x^2 + 3y^2 - 20$ we see that $\partial M/\partial y = x$ and $\partial N/\partial x = 4x$. It follows from Theorem 16.1.1 that the DE is not exact because $\partial M/\partial y \neq \partial N/\partial x$. See Problem 29 in Exercises 16.1.

$\dfrac{dy}{dx}$ **NOTES FROM THE CLASSROOM** ··

In Example 2 we found the function $f(x, y)$ by first integrating $M(x, y)$ with respect to x. In Example 3 we started by integrating $N(x, y)$ with respect to y. When finding a solution of an exact differential equation, you are free to start either way; in the end it will make little difference. For example, in Example 3, you might think that by starting with $N = y(1 - x^2)$ you have avoided the need to integrate $\cos x \sin x$. As it turns out, however, this function becomes a part of $h'(x)$, and ultimately does need to be integrated.

Exercises 16.1 Answers to selected odd-numbered problems begin on page ANS-48.

≡ Fundamentals

In Problems 1–20, determine whether the given differential equation is exact. If it is exact, solve it.

1. $(2x + 4) dx + (3y - 1) dy = 0$

2. $(2x + y) dx - (x + 6y) dy = 0$

3. $(5x + 4y) dx + (4x - 8y^3) dy = 0$

4. $(\sin y - y \sin x) dx + (\cos x + x \cos y - y) dy = 0$

5. $(2xy^2 - 3) dx + (2x^2 y + 4) dy = 0$

6. $\left(2y - \dfrac{1}{x} + \cos 3x\right) \dfrac{dy}{dx} + \dfrac{y}{x^2} - 4x^3 + 3y \sin 3x = 0$

7. $(x^2 - y^2) dx + (x^2 - 2xy) dy = 0$

8. $\left(1 + \ln x + \dfrac{y}{x}\right) dx = (1 - \ln x) dy$

9. $(x - y^3 + y^2 \sin x) dx = (3xy^2 + 2y \cos x) dy$

10. $(x^3 + y^3) dx + 3xy^2 dy = 0$

11. $(y \ln y - e^{-xy}) dx + \left(\dfrac{1}{y} + x \ln y\right) dy = 0$

12. $(3x^2 y + e^y) dx + (x^3 + xe^y - 2y) dy = 0$

13. $x \dfrac{dy}{dx} = 2xe^x - y + 6x^2$

14. $\left(1 - \dfrac{3}{y} + x\right) \dfrac{dy}{dx} + y = \dfrac{3}{x} - 1$

15. $\left(x^2 y^3 - \dfrac{1}{1 + 9x^2}\right) \dfrac{dx}{dy} + x^3 y^2 = 0$

16. $(5y - 2x)y' - 2y = 0$

17. $(\tan x - \sin x \sin y) dx + \cos x \cos y \, dy = 0$

18. $(2y \sin x \cos x - y + 2y^2 e^{xy^2}) dx = (x - \sin^2 x - 4xye^{xy^2}) dy$

19. $(4t^3 y - 15t^2 - y) dt + (t^4 + 3y^2 - t) dy = 0$

20. $\left(\dfrac{1}{t} + \dfrac{1}{t^2} - \dfrac{y}{t^2 + y^2}\right) dt + \left(ye^y + \dfrac{1}{t^2 + y^2}\right) dy = 0$

In Problems 21–24, solve the given initial-value problem.

21. $(x + y)^2 dx + (2xy + x^2 - 1) dy = 0$, $y(1) = 1$

22. $(e^x + y) dx + (2 + x + ye^y) dy = 0$, $y(0) = 1$

23. $(4y + 2t - 5) dt + (6y + 4t - 1) dy = 0$, $y(-1) = 2$

24. $(y^2 \cos x - 3x^2 y - 2x) dx + (2y \sin x - x^3 + \ln y) dy = 0$, $y(0) = e$

In Problems 25 and 26, find the value of the constant k so that the given differential equation is exact.

25. $(y^3 + kxy^4 - 2x) dx + (3xy^2 + 20x^2 y^3) dy = 0$

26. $(6xy^3 + \cos y) dx + (2kx^2 y^2 - x \sin y) dy = 0$

≡ Think About It

In Problems 27 and 28, discuss how the functions $M(x, y)$ and $N(x, y)$ can be found so that each differential equation is exact. Carry out your ideas.

27. $M(x, y) dx + \left(xe^{xy} + 2xy + \dfrac{1}{x}\right) dy = 0$

28. $\left(x^{-1/2} y^{1/2} + \dfrac{x}{x^2 + y}\right) dx + N(x, y) dy = 0$

29. If the equation $M(x, y) dx + N(x, y) dy = 0$ is not exact, it is sometimes possible to find a function $\mu(x, y)$ so that $\mu(x, y)M(x, y) dx + \mu(x, y)N(x, y) dy = 0$ is exact. The function $\mu(x, y)$ is called an **integrating factor**. Find an integrating factor for

$$xy \, dx + (2x^2 + 3y^2 - 20) dy = 0$$

and then solve the DE.

30. True or False: Every separable first-order differential equation $dy/dx = g(x)h(y)$ is exact. Explain your answer.

16.2 Homogeneous Linear Equations

■ **Introduction**   A linear nth-order differential equation (DE)

$$a_n(x)\frac{d^n y}{dx^n} + a_{n-1}(x)\frac{d^{n-1}y}{dx^{n-1}} + \cdots + a_1(x)\frac{dy}{dx} + a_0(x)y = g(x)$$

is said to be **nonhomogeneous** if $g(x) \neq 0$ for some x. If $g(x) = 0$ for every x, then the differential equation is said to be **homogeneous**. In this and the following section we shall be concerned only with finding solutions of linear *second-order* differential equations with constant real coefficients:

$$ay'' + by' + cy = g(x).$$

We begin by considering the homogeneous equation

$$ay'' + by' + cy = 0. \tag{1}$$

Theorem 16.2.1   Superposition Principle

Let y_1 and y_2 be solutions of the homogeneous linear second-order differential equation (1). Then the linear combination

$$y = C_1 y_1(x) + C_2 y_2(x),$$

where C_1 and C_2 are arbitrary constants, is also a solution of the equation.

PROOF   We substitute the linear combination

$$y = C_1 y_1(x) + C_2 y_2(x)$$

into the differential equation (1), rearrange the terms, and use the fact that y_1 and y_2 are solutions of the DE. This gives

$$a(C_1 y_1'' + C_2 y_2'') + b(C_1 y_1' + C_2 y_2') + c(C_1 y_1 + C_2 y_2)$$
$$= C_1(\underbrace{ay_1'' + by_1' + cy_1}_{\text{zero}}) + C_2(\underbrace{ay_2'' + by_2' + cy_2}_{\text{zero}})$$
$$= C_1 \cdot 0 + C_2 \cdot 0 = 0. \qquad \blacksquare$$

The following results are immediate consequences of the Superposition Principle.

- *A constant multiple $y = C_1 y_1(x)$ of a solution $y_1(x)$ of a linear homogeneous differential equation is also a solution.*
- *A linear homogeneous differential equation always possesses the trivial solution $y = 0$.*

■ **Linearly Independent Functions**   Analogous with the fact that any vector in two-dimensional space can be expressed as a unique linear combination of the *linearly independent* vectors **i** and **j**, any solution of a homogeneous linear second-order differential equation can be expressed as a unique linear combination of two **linearly independent** solutions of the differential equation.

Definition 16.2.1   Linear Independence of Functions

Two functions, $y_1(x)$ and $y_2(x)$, are linearly independent if neither is a constant multiple of the other.

■ **General Solution**   Linearly independent solutions of a differential equation, y_1 and y_2, are the building blocks for all solutions of the equation. We call the two-parameter family of solutions $y = C_1 y_1(x) + C_2 y_2(x)$ the **general solution** of the differential equation.

> **Theorem 16.2.2** General Solution
>
> Let y_1 and y_2 be linearly independent solutions of the homogeneous linear second-order differential (1). Then every solution of (1) can be obtained from the general solution
>
> $$y = C_1 y_1(x) + C_2 y_2(x). \tag{2}$$

EXAMPLE 1 Linearly Independent Functions

Although $y_1 = 0$ and $y_2 = e^{2x}$ are both solutions of the differential equation $y'' + 2y' - 8y = 0$, and y_2 is not a constant multiple of y_1, y_1 and y_2 are *not* linearly independent because y_1 is a constant multiple of y_2; namely $y_1 = 0 \cdot y_2$. ∎

■ **Auxiliary Equation** The surprising fact about the differential equation in (1) is that *all* solutions either are exponential functions or are constructed out of exponential functions. If we try a solution of the form $y = e^{mx}$, then $y' = me^{mx}$ and $y'' = m^2 e^{mx}$, so that (1) becomes

$$am^2 e^{mx} + bme^{mx} + ce^{mx} = 0 \qquad \text{or} \qquad e^{mx}(am^2 + bm + c) = 0.$$

Because $e^{mx} \neq 0$ for all x, it is apparent that the only way that this exponential function can satisfy the differential equation is to choose m so that it is a root of the quadratic equation

$$am^2 + bm + c = 0.$$

This latter equation is called the **auxiliary equation** or **characteristic equation** of the differential equation (1). We shall consider three cases—namely, the solutions corresponding to distinct real roots, equal real roots, and conjugate complex roots.

CASE I: Distinct Real Roots

Under the assumption that the auxiliary equation of (1) has two unequal real roots m_1 and m_2, we find two solutions

$$y_1 = e^{m_1 x} \qquad \text{and} \qquad y_2 = e^{m_2 x}.$$

Since neither y_1 nor y_2 is a constant multiple of the other, the two solutions are linearly independent. It follows that the general solution of the DE is

$$y = C_1 e^{m_1 x} + C_2 e^{m_2 x}. \tag{3}$$

EXAMPLE 2 Distinct Real Roots of the Auxiliary Equation

Solve $2y'' - 5y' - 3y = 0$.

Solution Solving the auxiliary equation

$$2m^2 - 5m - 3 = 0 \qquad \text{or} \qquad (2m + 1)(m - 3),$$

we obtain $m_1 = -\frac{1}{2}$ and $m_2 = 3$. Hence, by (3) the general solution is

$$y = C_1 e^{-x/2} + C_2 e^{3x}. ∎$$

CASE II: Equal Real Roots

When $m_1 = m_2$, we necessarily obtain only one exponential solution $y_1 = e^{m_1 x}$. However, it is a straightforward matter of substitution into (1) to show that $y = u(x)e^{m_1 x}$ is also a solution whenever $u(x) = x$. See Problem 37 in Exercises 16.2. Then $y_1 = e^{m_1 x}$ and $y_2 = xe^{m_1 x}$ are linearly independent solutions, and the general solution is

$$y = C_1 e^{m_1 x} + C_2 x e^{m_1 x}. \tag{4}$$

EXAMPLE 3 Equal Real Roots of the Auxiliary Equation

Solve $y'' - 10y' + 25y = 0$.

Solution From the auxiliary equation $m^2 - 10m + 25 = (m - 5)^2 = 0$, we see that $m_1 = m_2 = 5$. Thus, by (4) the general solution is

$$y = C_1 e^{5x} + C_2 x e^{5x}.$$ ∎

■ Complex Numbers The last case deals with complex numbers. Recall from algebra that a number of the form $z = \alpha + i\beta$, where α and β are real numbers and $i^2 = -1$ (sometimes written $i = \sqrt{-1}$), is called a **complex number**. The complex number $\bar{z} = \alpha - i\beta$ is called the **conjugate** of z. Now, from the quadratic formula, the roots of $am^2 + bm + c = 0$ can be written

Complex numbers are reviewed in the *SRM*.

$$m_1 = \frac{-b + \sqrt{b^2 - 4ac}}{2a} \quad \text{and} \quad m_2 = \frac{-b - \sqrt{b^2 - 4ac}}{2a}.$$

When $b^2 - 4ac < 0$, the roots m_1 and m_2 are complex conjugates.

CASE III: Conjugate Complex Roots

If m_1 and m_2 are complex, then we can write

$$m_1 = \alpha + i\beta \quad \text{and} \quad m_2 = \alpha - i\beta,$$

where α and $\beta > 0$ are real numbers and $i^2 = -1$. Formally there is no difference between this case and Case I, and hence the general solution of the DE is

$$y = c_1 e^{(\alpha + i\beta)x} + c_2 e^{(\alpha - i\beta)x}. \tag{5}$$

However, in practice we would prefer to work with real functions instead of functions involving the complex number i. To do this we can rewrite (5) in a more practical form by using **Euler's formula**,

A formal derivation of Euler's formula can be obtained from the Maclaurin series $e^x = \sum_{n=0}^{\infty} x^n/n!$ by substituting $x = i\theta$, using $i^2 = -1$, $i^3 = -i$, ..., and then separating the series into real and imaginary parts. The plausibility thus established, we can adopt $\cos\theta + i\sin\theta$ as the *definition* of $e^{i\theta}$.

$$e^{i\theta} = \cos\theta + i\sin\theta,$$

where θ is any real number. From this result we can write

$$e^{i\beta x} = \cos\beta x + i\sin\beta x \quad \text{and} \quad e^{-i\beta x} = \cos\beta x - i\sin\beta x,$$

where we have used $\cos(-\beta x) = \cos\beta x$ and $\sin(-\beta x) = -\sin\beta x$. Thus, (5) becomes

$$
\begin{aligned}
y &= e^{\alpha x}(c_1 e^{i\beta x} + c_2 e^{-i\beta x}) \\
&= e^{\alpha x}[c_1(\cos\beta x + i\sin\beta x) + c_2(\cos\beta x - i\sin\beta x)] \\
&= e^{\alpha x}[(c_1 + c_2)\cos\beta x + (c_1 i - c_2 i)\sin\beta x].
\end{aligned}
$$

Since $e^{\alpha x}\cos\beta x$ and $e^{\alpha x}\sin\beta x$ are easily shown to be linearly independent solutions of the given differential equation, we can simply relabel $c_1 + c_2$ as C_1 and $c_1 i - c_2 i$ as C_2. Then we use the superposition principle to write the general solution:

$$
\begin{aligned}
y &= C_1 e^{\alpha x}\cos\beta x + C_2 e^{\alpha x}\sin\beta x \\
&= e^{\alpha x}(C_1\cos\beta x + C_2\sin\beta x).
\end{aligned}
\tag{6}
$$

When $\alpha < 0$, we call $e^{\alpha x}$ a **damping factor** because the graphs of the solution curves $\to 0$ as $x \to \infty$.

EXAMPLE 4 Complex Roots of the Auxiliary Equation

Solve $y'' + y' + y = 0$.

Solution From the quadratic formula we find that the auxiliary equation $m^2 + m + 1 = 0$ has the complex roots

$$m_1 = -\frac{1}{2} + \frac{\sqrt{3}}{2}i \quad \text{and} \quad m_2 = -\frac{1}{2} - \frac{\sqrt{3}}{2}i.$$

Identifying $\alpha = -\frac{1}{2}$ and $\beta = \frac{1}{2}\sqrt{3}$, we see from (6) that the general solution of the equation is

$$y = e^{-x/2}\left(C_1 \cos \frac{\sqrt{3}}{2}x + C_2 \sin \frac{\sqrt{3}}{2}x\right).$$ ∎

EXAMPLE 5   A Special Differential Equation

The differential equation

$$y'' + \omega^2 y = 0$$

is frequently encountered in applied mathematics. See Section 16.4. The auxiliary equation is $m^2 + \omega^2 = 0$, with roots $m_1 = \omega i$ and $m_2 = -\omega i$. It follows from (6) with $\alpha = 0$ that the general solution is

$$y = C_1 \cos \omega x + C_2 \sin \omega x.$$

FIGURE 16.2.1 shows the graph of the solution when $C_1 = -2$, $C_2 = 3$, and $\omega = 1$. If you experiment with different values for C_1, C_2, and ω, you will see that as long as C_1 and C_2 are not both 0, the solution is oscillating with a well-defined amplitude and frequency. It can be shown that this is true for any choice of C_1 and C_2 (except $C_1 = C_2 = 0$) by using trigonometry. ∎

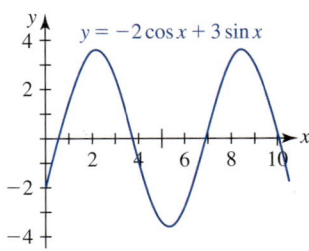

FIGURE 16.2.1 Graph of a solution in Example 5

■ **Initial-Value Problem**   The problem

$$\begin{aligned} &Solve\text{:} & ay'' + by' + cy &= g(x)\\ &Subject\ to\text{:} & y(x_0) = y_0, \quad y'(x_0) &= y_1, \end{aligned}$$

where y_0 and y_1 are arbitrary constants, is called an **initial-value problem** (**IVP**). The values y_0 and y_1 are called **initial conditions**. A solution of the problem is a function whose graph passes through (x_0, y_0) such that the slope of the tangent to the curve at that point is y_1. The next example illustrates an initial-value problem for a homogeneous equation.

EXAMPLE 6   An Initial-Value Problem

Solve $y'' - 4y' + 13y = 0$ subject to $y(0) = -1$, $y'(0) = 2$.

Solution   The roots of the auxiliary equation

$$m^2 - 4m + 13 = 0$$

are $m_1 = 2 + 3i$ and $m_2 = 2 - 3i$, so that the general solution is

$$y = e^{2x}(C_1 \cos 3x + C_2 \sin 3x).$$

The condition $y(0) = -1$ implies that

$$-1 = e^0(C_1 \cos 0 + C_2 \sin 0) = C_1,$$

from which we can write

$$y = e^{2x}(-\cos 3x + C_2 \sin 3x).$$

Differentiating this latter expression and using the second initial condition give

$$\begin{aligned} y' &= e^{2x}(3 \sin 3x + 3C_2 \cos 3x) + 2e^{2x}(-\cos 3x + C_2 \sin 3x)\\ 2 &= 3C_2 - 2, \end{aligned}$$

so that $C_2 = \frac{4}{3}$. Hence,

$$y = e^{2x}\left(-\cos 3x + \tfrac{4}{3}\sin 3x\right).$$ ∎

■ **Boundary-Value Problem**   Initial conditions for a second-order differential equation are characterized by the fact that they specify values of the solution function and its first derivative at a *single point*. By contrast, in a **boundary-value problem** (**BVP**) there are two conditions, called **boundary conditions**, that specify the values of a solution or its first derivative *at the endpoints of an interval* $[a, b]$. For example,

$$a_2 \frac{d^2y}{dx^2} + a_1 \frac{dy}{dx} + a_0 = g(x), \quad y(a) = y_0, \quad y'(b) = y_1$$

is a boundary-value problem. A solution of this problem is a function, defined on $[a, b]$, whose graph passes through the point (a, y_0) and has slope y_1 when $x = b$.

The next example shows that a boundary-value problem, unlike an initial-value problem, may have several solutions, a unique solution, or no solution at all.

EXAMPLE 7   A BVP Can Have Many, One, or No Solutions

From Example 5 we know that the general solution of the differential equation $y'' + 16y = 0$ is

$$y = C_1 \cos 4x + C_2 \sin 4x. \tag{7}$$

(a) Suppose we now wish to determine a solution of the equation that further satisfies the boundary conditions $y(0) = 0$, $y(\pi/2) = 0$. Observe that the first condition $0 = C_1 \cos 0 + C_2 \sin 0$ implies $C_1 = 0$, so that $y = C_2 \sin 4x$. But when $x = \pi/2$, $0 = C_2 \sin 2\pi$ is satisfied for any choice of C_2 since $\sin 2\pi = 0$. Hence, the boundary-value problem

$$y'' + 16y = 0, \quad y(0) = 0, \quad y(\pi/2) = 0 \tag{8}$$

has infinitely many solutions. **FIGURE 16.2.2** shows five different members of the one-parameter family $y = C_2 \sin 4x$ passing through the points $(0, 0)$ and $(\pi/2, 0)$.

(b) If the boundary-value problem in (8) is changed to

$$y'' + 16y = 0, \quad y(0) = 0, \quad y(\pi/8) = 0, \tag{9}$$

then $y(0) = 0$ still requires $C_1 = 0$ in the solution (7). But applying $y(\pi/8) = 0$ to $y = C_2 \sin 4x$ demands that $0 = C_2 \sin(\pi/2) = C_2 \cdot 1$. Hence, $y = 0$ is a solution of this new boundary-value problem. Indeed, it can be proved that $y = 0$ is the *only* solution of (9).

(c) Finally, if we change the problem to

$$y'' + 16y = 0, \quad y(0) = 0, \quad y(\pi/2) = 1, \tag{10}$$

we find again that $C_1 = 0$ from $y(0) = 0$, but that applying $y(\pi/2) = 1$ to $y = C_2 \sin 4x$ leads to the contradiction $1 = C_2 \sin 2\pi = C_2 \cdot 0 = 0$. Hence, the boundary-value problem (10) has no solution. ∎

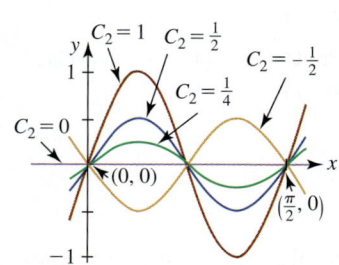

FIGURE 16.2.2 Five solutions of the BVP in part (a) of Example 7

$$\frac{d^2y}{dx^2}$$ **NOTES FROM THE CLASSROOM** ··

(i) Many of the concepts in this section can be extended to linear DEs of order three and higher with constant coefficients. For example, the auxiliary equation of

$$y''' - 4y'' + 5y' - 2y = 0$$

is $m^3 - 4m^2 + 5m - 2 = (m - 1)^2(m - 2) = 0$, and $y_1 = e^x$, $y_2 = xe^x$, $y_3 = e^{2x}$ are solutions of the differential equation. The notion of linear independence requires a more complicated definition than the one we used for two functions. See a text on differential equations.

(ii) The hyperbolic functions play an important role in the study of differential equations. Recall, these functions were introduced in Section 3.10 and have properties that are similar to the trigonometric functions. For example, the second derivatives of the hyperbolic sine and hyperbolic cosine are

$$\frac{d^2}{dx^2}(\sinh x) = \sinh x \quad \text{and} \quad \frac{d^2}{dx^2}(\cosh x) = \cosh x.$$

It then follows that $y_1 = \cosh x$ and $y_2 = \sinh x$ are solutions of the differential equation $y'' - y = 0$. Since these functions are linearly independent, the general solution of the differential equation is $y = C_1 \cosh x + C_2 \sinh x$. Another form for the general solution is easily seen to be $y = C_1 e^x + C_2 e^{-x}$. These two seemingly very different solutions are related by the definitions of the two hyperbolic functions:

$$\cosh x = \frac{e^x + e^{-x}}{2} \quad \text{and} \quad \sinh x = \frac{e^x - e^{-x}}{2}.$$

Eventually, both forms of the general solution of $y'' - y = 0$ are used in the analysis of *partial differential equations*.

◀ As the name suggests, a partial differential equation involves partial derivatives of an unknown function of several variables.

Exercises 16.2   Answers to selected odd-numbered problems begin on page ANS-48.

≡ Fundamentals

In Problems 1–20, find the general solution of the given differential equation.

1. $3y'' - y' = 0$ **2.** $2y'' + 5y' = 0$

3. $y'' - 16y = 0$ **4.** $y'' - 8y = 0$

5. $y'' + 9y = 0$ **6.** $4y'' + y = 0$

7. $y'' - 3y' + 2y = 0$ **8.** $y'' - y' - 6y = 0$

9. $\dfrac{d^2 y}{dx^2} + 8 \dfrac{dy}{dx} + 16y = 0$ **10.** $\dfrac{d^2 y}{dx^2} - 10 \dfrac{dy}{dx} + 25y = 0$

11. $y'' + 3y' - 5y = 0$ **12.** $y'' + 4y' - y = 0$

13. $12y'' - 5y' - 2y = 0$ **14.** $8y'' + 2y' - y = 0$

15. $y'' - 4y' + 5y = 0$ **16.** $2y'' - 3y' + 4y = 0$

17. $3y'' + 2y' + y = 0$ **18.** $2y'' + 2y' + y = 0$

19. $9y'' + 6y' + y = 0$ **20.** $15y'' - 16y' - 7y = 0$

In Problems 21–30, solve the given initial-value problem.

21. $y'' + 16y = 0,\ y(0) = 2,\ y'(0) = -2$

22. $y'' - y = 0,\ y(0) = y'(0) = 1$

23. $y'' + 6y' + 5y = 0,\ y(0) = 0,\ y'(0) = 3$

24. $y'' - 8y' + 17y = 0,\ y(0) = 4,\ y'(0) = -1$

25. $2y'' - 2y' + y = 0,\ y(0) = -1,\ y'(0) = 0$

26. $y'' - 2y' + y = 0,\ y(0) = 5,\ y'(0) = 10$

27. $y'' + y' + 2y = 0,\ y(0) = y'(0) = 0$

28. $4y'' - 4y' - 3y = 0,\ y(0) = 1,\ y'(0) = 5$

29. $y'' - 3y' + 2y = 0,\ y(1) = 0,\ y'(1) = 1$

30. $y'' + y = 0,\ y(\pi/3) = 0,\ y'(\pi/3) = 2$

31. The roots of an auxiliary equation are $m_1 = 4$ and $m_2 = -5$. What is the corresponding differential equation?

32. The roots of an auxiliary equation are $m_1 = 3 + i$ and $m_2 = 3 - i$. What is the corresponding differential equation?

In Problems 33–40, solve the given boundary-value problem or show that no solution exists.

33. $y'' + y = 0,\ y(0) = 0,\ y(\pi) = 0$

34. $y'' + y = 0,\ y(0) = 0,\ y(\pi) = 1$

35. $y'' + y = 0,\ y'(0) = 0,\ y'(\pi/2) = 2$

36. $y'' - y = 0,\ y(0) = 1,\ y(1) = -1$

37. $y'' - 2y' + 2y = 0,\ y(0) = 1,\ y(\pi) = -1$

38. $y'' - 2y' + 2y = 0,\ y(0) = 1,\ y(\pi/2) = 1$

39. $y'' - 4y' + 4y = 0,\ y(0) = 0,\ y(1) = 1$

40. $y'' - 4y' + 4y = 0,\ y'(0) = 1,\ y(1) = 2$

≡ Think About It

In Problems 41 and 42, find the general solution of the given third-order differential equation if it is known that y_1 is a solution.

41. $y''' - 9y'' + 25y' - 17y = 0;\quad y_1 = e^x$

42. $y''' + 6y'' + y' - 34y = 0;\quad y_1 = e^{-4x}\cos x$

In Problems 43 and 44, use the assumed solution $y = e^{mx}$ to find the auxiliary equation, roots, and general solution of the given third-order differential equation.

43. $y''' - 4y'' - 5y' = 0$ **44.** $y''' + 3y'' - 4y' - 12y = 0$

45. Consider the boundary-value problem

$$y'' + \lambda y = 0,\ y(0) = 0,\ y(1) = 0.$$

By considering the three cases $\lambda = -\alpha^2 < 0$, $\lambda = 0$, and $\lambda = \alpha^2 > 0$, find all real values of λ for which the problem possesses nonzero solutions.

≡ Projects

46. Shaft Through the Earth   Suppose a shaft is drilled through the Earth so that it passes through its center. A body with mass m is dropped into the shaft. Let the distance from the center of the Earth to the mass at time t be denoted by r. See **FIGURE 16.2.3**.

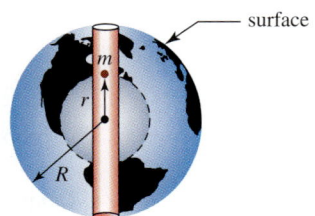

FIGURE 16.2.3 Shaft through the Earth in Problem 46

(a) Let M denote the mass of the Earth and M_r denote the mass of that portion of the Earth within a sphere of radius r. The gravitational force on m is $F = -kM_r m/r^2$, where the minus sign indicates that the force is one of attraction. Use this fact to show that

$$F = -k\frac{mM}{R^3}r.$$

[*Hint*: Assume that the Earth is homogeneous, that is, has a constant density ρ. Use mass = density $\times$ volume.]

(b) Use Newton's second law $F = ma$ and the result in part (a) to derive the differential equation

$$\frac{d^2r}{dt^2} + \omega^2 r = 0,$$

where $\omega^2 = kM/R^3 = g/R$.

(c) Solve the differential equation in part (b) if the mass m is released from rest at the surface of the Earth. Interpret your answer using $R = 3960$ mi.

16.3 Nonhomogeneous Linear Equations

▌ **Introduction**   To solve a nonhomogeneous linear differential equation

$$ay'' + by' + cy = g(x), \tag{1}$$

we must be able to do two things:

(*i*) find the general solution $y_c(x)$ of the **associated homogeneous** differential equation

$$ay'' + by' + cy = 0,$$

(*ii*) find *any* particular solution y_p of the nonhomogeneous equation (1).

As we will see, the general solution of (1) is then $y(x) = y_c(x) + y_p(x)$. In the previous section we discussed how to find $y_c(x)$; in this section we discuss two methods for finding $y_p(x)$.

▌ **Particular Solutions**   Any function y_p free of arbitrary parameters that satisfies (1) is said to be a **particular solution** of the equation.

EXAMPLE 1   A Particular Solution

We see that $y_p = x^3 - x$ is a particular solution of

$$y'' - y' + 6y = 6x^3 - 3x^2 + 1$$

by first computing $y_p' = 3x^2 - 1$ and $y_p'' = 6x$. Then, substituting into the differential equation, we have for all real numbers x

$$
\begin{aligned}
y_p'' - y_p' + 6y_p &= 6x - (3x^2 - 1) + 6(x^3 - x) \\
&= 6x^3 - 3x^2 + 1.
\end{aligned}
$$

▌ **The General Solution**   The following theorem tells us how to construct the **general solution** of (1).

Theorem 16.3.1   General Solution

Let y_p be a particular solution of the nonhomogeneous differential equation (1), and let

$$y_c(x) = C_1 y_1(x) + C_2 y_2(x)$$

be the general solution of the associated homogeneous equation.

Then the **general solution** of the nonhomogeneous equation is

$$y(x) = y_c(x) + y_p(x) = C_1 y_1(x) + C_2 y_2(x) + y_p(x). \tag{2}$$

The proof that (2) is a solution of (1) is left as an exercise. See Problem 38 in Exercises 16.3.

▌ **Complementary Function**   In Theorem 16.3.1 the solution of the associated homogeneous differential equation, $y_c(x) = C_1 y_1(x) + C_2 y_2(x)$, is called the **complementary function** of equation (1). In other words, the general solution of a nonhomogeneous linear differential equation is

$$y = complementary\ function + any\ particular\ solution.$$

■ **Undetermined Coefficients** When $g(x)$ consists of

- (*i*) a constant k,
- (*ii*) a polynomial in x,
- (*iii*) an exponential function $e^{\alpha x}$,
- (*iv*) $\sin\beta x$, $\cos\beta x$,

or finite sums and products of these functions, it is possible to find a particular solution of (1) by the **method of undetermined coefficients**. The underlying idea in this method is a conjecture, an educated guess really, about the form of y_p motivated by the distinct kinds of functions that make up $g(x)$ and its derivatives $g'(x)$, $g''(x)$, ..., $g^{(m)}(x)$.

◀ See a differential equations text for a more complete discussion of the method of undermined coefficients.

In this section we consider the special case where the n distinct functions $f_n(x)$ appearing in $g(x)$, and its derivatives, *do not* appear; that is, are not duplicated, in the complementary function y_c. Under these circumstances, a particular solution y_p having the form

$$y = A_1 f_1(x) + A_2 f_2(x) + \cdots + A_n f_n(x), \tag{3}$$

can be found. To find the specific coefficients A_k, where $k = 1, \ldots, n$, we substitute the expression in (3) into the nonhomogeneous differential equation (1). This will result in n linear algebraic equations in the n unknowns $A_1, A_2, \ldots, A_n$.

The next two examples illustrate the basic method.

EXAMPLE 2 General Solution Using Undetermined Coefficients

Solve $\dfrac{d^2y}{dx^2} + 3\dfrac{dy}{dx} + 2y = 4x^2$. \hfill (4)

Solution The complementary function is

$$y_c = C_1 e^{-x} + C_2 e^{-2x}.$$

Now, since

$$g(x) = 4x^2, \qquad g'(x) = 8x, \qquad \text{and} \qquad g''(x) = 8 \cdot 1,$$
$$\qquad\quad \underset{f_1(x)}{\uparrow} \qquad\qquad\quad \underset{f_2(x)}{\uparrow} \qquad\qquad\qquad\qquad \underset{f_3(x)}{\uparrow}$$

we seek a particular solution having the basic form

$$y_p = Ax^2 + Bx + C \cdot 1. \tag{5}$$

Differentiating (5) and substituting into the original differential equation (4) give

$$\begin{aligned} y_p'' + 3y_p' + 2y_p &= 2A + 3(2Ax + B) + 2(Ax^2 + Bx + C) \\ &= 2Ax^2 + (6A + 2B)x + (2A + 3B + 2C) \\ &= 4x^2 + 0x + 0. \end{aligned}$$

Since the last equality is supposed to be an identity, the coefficients of like powers of x must be equal:

$$\begin{aligned} 2A &= 4 \\ 6A + 2B &= 0 \\ 2A + 3B + 2C &= 0. \end{aligned}$$

Solving this system of equations then yields $A = 2$, $B = -6$, and $C = 7$. Thus, $y_p = 2x^2 - 6x + 7$ and so by (2) the general solution of the nonhomogeneous differential equation is

$$y = y_c + y_p = C_1 e^{-x} + C_2 e^{-2x} + 2x^2 - 6x + 7. \qquad ■$$

EXAMPLE 3   General Solution Using Undetermined Coefficients

Solve $y'' + 2y' + 2y = -10xe^x + 5\sin x$.

Solution   The roots of the auxiliary equation $m^2 + 2m + 2 = 0$ are $m_1 = -1 + i$ and $m_2 = -1 - i$, so

$$y_c = e^{-x}(C_1\cos x + C_2\sin x).$$

In this case,

$$g(x) = \underset{\underset{f_1(x)}{\uparrow}}{-10xe^x} + \underset{\underset{f_2(x)}{\uparrow}}{5\sin x} \qquad \text{and} \qquad g'(x) = -10xe^x - \underset{\underset{f_3(x)}{\uparrow}}{10e^x} + \underset{\underset{f_4(x)}{\uparrow}}{5\cos x}.$$

Higher-order derivatives do not generate any new functions and this suggests that a particular solution of the form

$$y_p = Axe^x + Be^x + C\sin x + D\cos x$$

can be found. Substituting y_p in the differential equation and simplifying yield

$$y_p'' + 2y_p' + 2y_p = 5Axe^x + (4A + 5B)e^x + (C - 2D)\sin x + (2C + D)\cos x$$
$$= -10xe^x + 5\sin x.$$

The corresponding system of equations is

$$5A \qquad\qquad = -10$$
$$4A + 5B = 0$$
$$C - 2D = 5$$
$$2C + D = 0,$$

so $A = -2$, $B = \frac{8}{5}$, $C = 1$, and $D = -2$. Thus, a particular solution is

$$y_p = -2xe^x + \tfrac{8}{5}e^x + \sin x - 2\cos x,$$

and the general solution is

$$y = e^{-x}(C_1\cos x + C_2\sin x) - 2xe^x + \tfrac{8}{5}e^x + \sin x - 2\cos x. \qquad \blacksquare$$

■ **Variation of Parameters**   As mentioned at the start of this discussion, the method of undetermined coefficients is limited to the case when $g(x)$ is a finite sum and product of constants, polynomials, exponentials $e^{\alpha x}$, sines, and cosines. In general, the method of undetermined coefficients will not yield a particular solution of (1) for functions such as

$$g(x) = \frac{1}{x}, \qquad g(x) = \ln x, \qquad g(x) = \tan x, \qquad \text{and} \qquad g(x) = \sin^{-1}x.$$

The method that we consider next, called **variation of parameters**, will *always* yield a particular solution y_p provided the associated homogeneous equation can be solved.

We begin our discussion of this method by putting the nonhomogenous differential equation (1) in the **standard form**

$$y'' + Py' + Qy = f(x)$$

by dividing both sides of the equation by the leading coefficient a. Next, let y_1 and y_2 be linearly independent solutions of the associated homogeneous differential equation (2); so that

$$y_1'' + Py_1' + Qy_1 = 0 \qquad \text{and} \qquad y_2'' + Py_2' + Qy_2 = 0.$$

Now we ask: Can two functions u_1 and u_2 be found so that

$$y_p = u_1(x)y_1(x) + u_2(x)y_2(x) \tag{6}$$

is a particular solution of (1)? Notice that our assumption for y_p has the same form as $y_c = C_1y_1 + C_2y_2$, but we have replaced C_1 and C_2 by the "variable parameters" u_1 and u_2. Because we seek to find two unknown functions u_1 and u_2, reason dictates that we need two equations.

Using the Product Rule to differentiate (6) twice, we get

$$y_p' = u_1 y_1' + y_1 u_1' + u_2 y_2' + y_2 u_2'$$

$$y_p'' = u_1 y_1'' + y_1' u_1' + y_1 u_1'' + u_1' y_1' + u_2 y_2'' + y_2' u_2' + y_2 u_2'' + u_2' y_2'.$$

Substituting (6) and the foregoing derivatives into the standard form of the differential equation and grouping terms then gives

$$y_p'' + P y_p' + Q y_p = u_1 \underbrace{[y_1'' + P y_1' + Q y_1]}_{\text{zero}} + u_2 \underbrace{[y_2'' + P y_2' + Q y_2]}_{\text{zero}}$$

$$+ y_1 u_1'' + u_1' y_1' + y_2 u_2'' + u_2' y_2' + P[y_1 u_1' + y_2 u_2'] + y_1' u_1' + y_2' u_2'$$

$$= \frac{d}{dx}[y_1 u_1'] + \frac{d}{dx}[y_2 u_2'] + P[y_1 u_1' + y_2 u_2'] + y_1' u_1' + y_2' u_2'$$

$$= \frac{d}{dx}[y_1 u_1' + y_2 u_2'] + P[y_1 u_1' + y_2 u_2'] + y_1' u_1' + y_2' u_2' = f(x). \tag{7}$$

At this point we make the assumption that u_1 and u_2 are functions for which $y_1 u_1' + y_2 u_2' = 0$. This assumption does not come out of the blue but is prompted by the first two terms in (7), since if $y_1 u_1' + y_2 u_2' = 0$, then (7) reduces to $y_1' u_1' + y_2' u_2' = f(x)$. We now have our desired two equations, albeit two equations for determining the derivatives u_1' and u_2'. By Cramer's Rule, the solution of the system

$$y_1 u_1' + y_2 u_2' = 0$$

$$y_1' u_1' + y_2' u_2' = f(x)$$

can be expressed in terms of 2×2 determinants:

$$u_1' = \frac{\begin{vmatrix} 0 & y_2 \\ f(x) & y_2' \end{vmatrix}}{\begin{vmatrix} y_1 & y_2 \\ y_1' & y_2' \end{vmatrix}} \quad \text{and} \quad u_2' = \frac{\begin{vmatrix} y_1 & 0 \\ y_1' & f(x) \end{vmatrix}}{\begin{vmatrix} y_1 & y_2 \\ y_1' & y_2' \end{vmatrix}}. \tag{8}$$

◀ For a review of determinants and Cramer's Rule, see the *SRM*.

The determinant $\begin{vmatrix} y_1 & y_2 \\ y_1' & y_2' \end{vmatrix}$ is called the **Wronkskian** and is usually denoted by W.

EXAMPLE 4   General Solution Using Variation of Parameters

Solve $4y'' + 36y = \csc 3x$.

Solution   To use (6) and (8) it is necessary to first write the differential equation in standard form. To this end we begin by dividing the given equation by the coefficient of y'':

$$y'' + 9y = \tfrac{1}{4}\csc 3x.$$

Since the roots of the auxiliary equation $m^2 + 9 = 0$ are $m_1 = 3i$ and $m_2 = -3i$, the complementary function is

$$y_c = C_1 \cos 3x + C_2 \sin 3x.$$

Identifying $y_1 = \cos 3x$ and $y_2 = \sin 3x$, we see that the Wronskian is

$$W = \begin{vmatrix} \cos 3x & \sin 3x \\ -3 \sin 3x & 3 \cos 3x \end{vmatrix} = 3.$$

From (8) we find

$$u_1' = \frac{\begin{vmatrix} 0 & \sin 3x \\ \tfrac{1}{4}\csc 3x & 3 \cos 3x \end{vmatrix}}{W} = \frac{(\sin 3x)(\tfrac{1}{4}\csc 3x)}{3} = -\frac{1}{12}$$

and

$$u_2' = \frac{\begin{vmatrix} \cos 3x & 0 \\ -3 \sin 3x & \tfrac{1}{4}\csc 3x \end{vmatrix}}{W} = \frac{(\cos 3x)(\tfrac{1}{4}\csc 3x)}{3} = \frac{1}{12}\frac{\cos 3x}{\sin 3x}.$$

Integrating u_1' and u_2' then yields

$$u_1 = -\tfrac{1}{12}x \qquad \text{and} \qquad u_2 = \tfrac{1}{36}\ln|\sin 3x|.$$

Therefore,

$$y_p = -\tfrac{1}{12}x\cos 3x + \tfrac{1}{36}(\sin 3x)\ln|\sin 3x|,$$

and the general solution is

$$y = y_c + y_p = C_1\cos 3x + C_2\sin 3x - \tfrac{1}{12}x\cos 3x + \tfrac{1}{36}(\sin 3x)\ln|\sin 3x|. \qquad \blacksquare$$

Constants of Integration When computing the indefinite integrals of u_1' and u_2', it is not necessary to introduce any constants. To see this, suppose a_1 and a_2 are constants introduced in the integration of u_1' and u_2'. Then the general solution $y = y_c + y_p$ becomes

$$y = \overbrace{C_1 y_1 + C_2 y_2}^{y_c} + \overbrace{(u_1 + a_1)y_1 + (u_2 + a_2)y_2}^{y_p}$$
$$= (C_1 + a_1)y_1 + (C_2 + a_2)y_2 + u_1 y_1 + u_2 y_2$$
$$= c_1 y_1 + c_2 y_2 + u_1 y_1 + u_2 y_2,$$

where $c_1 = C_1 + a_1$ and $c_2 = C_2 + a_2$ are constants.

Exercises 16.3 Answers to selected odd-numbered problems begin on page ANS-48.

≡ Fundamentals

In Problems 1–10, solve the given differential equation by undetermined coefficients.

1. $y'' - 9y = 54$

2. $2y'' - 7y' + 5y = -29$

3. $y'' + 4y' + 4y = 2x + 6$

4. $y'' - 2y' + y = x^3 + 4x$

5. $y'' + 25y = 6\sin x$

6. $y'' - 4y = 7e^{4x}$

7. $y'' - 2y' - 3y = 4e^{2x} + 2x^3$

8. $y'' + y' + y = x^2 e^x + 3$

9. $y'' - 8y' + 25y = e^{3x} - 6\cos 2x$

10. $y'' - 5y' + 4y = 2\sinh 3x$

In Problems 11 and 12, solve the given differential equation by undetermined coefficients subject to the initial conditions $y(0) = 1$ and $y'(0) = 0$.

11. $y'' - 64y = 16$

12. $y'' + 5y' - 6y = 10e^{2x}$

In Problems 13–32, solve the given differential equation by variation of parameters.

13. $y'' + y = \sec x$

14. $y'' + y = \tan x$

15. $y'' + y = \sin x$

16. $y'' + y = \sec x \tan x$

17. $y'' + y = \cos^2 x$

18. $y'' + y = \sec^2 x$

19. $y'' - y = \cosh x$

20. $y'' - y = \sinh 2x$

21. $y'' - 4y = e^{2x}/x$

22. $y'' - 9y = 9xe^{-3x}$

23. $y'' + 3y' + 2y = 1/(1 + e^x)$

24. $y'' - 3y' + 2y = e^{3x}/(1 + e^x)$

25. $y'' + 3y' + 2y = \sin e^x$

26. $y'' - 2y' + y = e^x \arctan x$

27. $y'' - 2y' + y = e^x/(1 + x^2)$

28. $y'' - 2y' + 2y = e^x \sec x$

29. $y'' + 2y' + y = e^{-x}\ln x$

30. $y'' + 10y' + 25y = e^{-10x}/x^2$

31. $4y'' - 4y' + y = 8e^{-x} + x$

32. $4y'' - 4y' + y = e^{x/2}\sqrt{1 - x^2}$

In Problems 33 and 34, solve the given differential equation by variation of parameters subject to the initial conditions $y(0) = 1$ and $y'(0) = 0$.

33. $y'' - y = xe^x$

34. $2y'' + y' - y = x + 1$

35. Given that $y_1 = x$ and $y_2 = x\ln x$ are linearly independent solutions of $x^2 y'' - xy' + y = 0$, use variation of parameters to solve $x^2 y'' - xy' + y = 4x\ln x$ for $x > 0$.

36. Given that $y_1 = x^2$ and $y_2 = x^3$ are linearly independent solutions of $x^2 y'' - 4xy' + 6y = 0$, use variation of parameters to solve $x^2 y'' - 4xy' + 6y = 1/x$ for $x > 0$.

≡ Applications

37. Since phosphate is often the limiting nutrient for algae growth in lakes, it is important for the management of water quality to be able to predict phosphate input into lakes. One source is from the sediment in the lake bed. A mathematical model that describes phosphate concentration in lake bed sediment is the differential equation

$$\frac{d^2 C}{dx^2} = \frac{C(x) - C(\infty)}{\lambda^2},$$

where $C(x)$ is the phosphate concentration at depth x from the surface of the sediment, $C(\infty)$ is the equilibrium concentration at "infinite" depth, that is, $C(\infty) = \lim_{x\to\infty} C(x)$, and $\lambda > 0$ is a "yardstick of thickness" parameter involving the porosity of the sediment, the diffusion coefficient of the phosphate ion, and an adsorption rate constant. Solve this differential equation subject to the initial condition $C(0) = 0$.

≡ **Think About It**

38. If y_c and y_p are the complementary function and particular solution, respectively, of the nonhomogeneous differential equation (1), prove that $y = y_c + y_p$ is a solution of (1).

39. (a) Show that a particular solution of the form $y_p = Ae^x$ cannot be found for the differential equation $y'' + 2y' - 3y = 10e^x$.
   (b) Find a particular solution of the DE in part (a) of the form $y_p = Axe^x$.
   (c) Find the general solution of the DE in part (a).

40. Use undetermined coefficients and the ideas in Problem 39 to find the general solution of $y'' - y = e^{-x} - e^x$.

16.4 Mathematical Models

■ **Introduction** An important branch of mathematics involves the study of *dynamical systems*—roughly systems that change or evolve with time. More precisely, a dynamical system consists of a set of time-dependent variables, called *state variables*, together with a rule that enables us to determine the state of the system (this may be a past, present, or future state) in terms of a state prescribed at some time t_0.

In this section we first concentrate on a mathematical model of one such dynamical system—a spring/mass system—whose state (position x and velocity dx/dt of the mass) at any future time $t > 0$ depends on initial conditions $x(0) = x_0$ and $x'(0) = x_1$. The initial conditions represent the state of the mass at time $t = 0$. We then introduce an analogous system that can be used to model electrical circuits.

■ **Hooke's Law** Suppose, as in FIGURE 16.4.1(b), a mass m_1 is attached to a flexible spring suspended from a rigid support. When m_1 is replaced with a different mass m_2, the amount of stretch, or elongation of the spring, will of course be different.

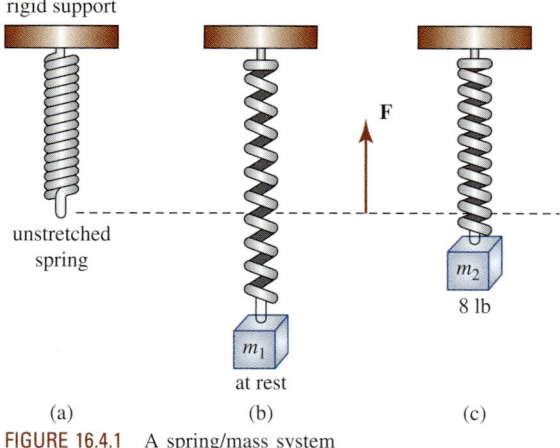

rigid support

unstretched spring

F

m_1
at rest

m_2
8 lb

(a)   (b)   (c)

FIGURE 16.4.1 A spring/mass system

By Hooke's law, the spring itself exerts a restoring force F opposite to the direction of elongation and proportional to the amount of elongation s. Simply stated, $F = ks$, where k is a constant of proportionality. Although masses with different weights stretch a spring by different amounts, the spring is essentially characterized by the number k. For example, if a mass weighing 10 lb stretches a spring $\frac{1}{2}$ ft, then

$$10 = k \cdot \tfrac{1}{2} \qquad \text{implies} \qquad k = 20 \text{ lb/ft.}$$

Necessarily then, a mass weighing 8 lb stretches the same spring $s = \frac{8}{20} = \frac{2}{5}$ ft.

Newton's Second Law After a mass m is attached to a spring, it will stretch the spring by an amount s and attain an **equilibrium position** at which its weight W is balanced by the restoring force ks. Recall that weight is defined by $W = mg$, where the mass m is measured in slugs, kilograms, or grams, and $g = 32$ ft/s², 9.8 m/s², or 980 cm/s², respectively. As indicated in FIGURE 16.4.2(b), the condition of equilibrium is $mg = ks$ or $mg - ks = 0$.

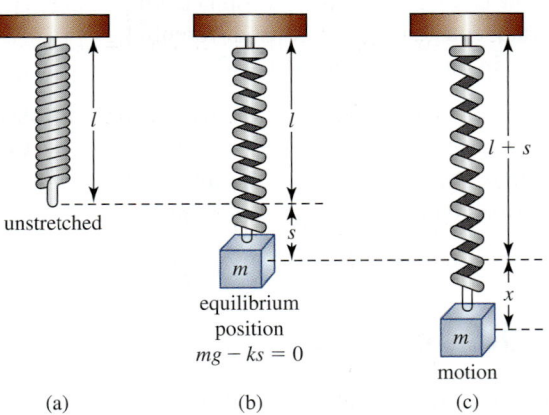

unstretched

equilibrium position
$mg - ks = 0$

motion

(a) (b) (c)

FIGURE 16.4.2 A spring/mass system in equilibrium and in motion

If the mass is then displaced by an amount x from its equilibrium position and released, the net force F in this dynamic case is given by **Newton's second law of motion**, $F = ma$, where a is the acceleration d^2x/dt^2. Assuming that there are no retarding forces acting on the system and assuming that the mass vibrates free of other external influencing forces—**free motion**—we can equate F to the resultant force of the weight and the restoring force:

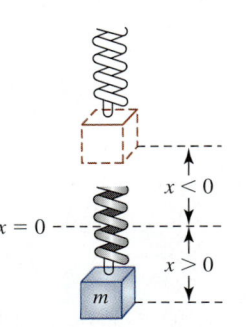

FIGURE 16.4.3 Negative and positive displacements

$$m\frac{d^2x}{dt^2} = -k(s + x) + mg = -kx + \underbrace{mg - ks}_{\text{zero}} = -kx. \tag{1}$$

The negative sign in (1) indicates that the restoring force of the spring acts opposite to the direction of motion. Furthermore, we shall adopt the convention that displacements measured *below* the equilibrium position are positive. See FIGURE 16.4.3.

Free Undamped Motion By dividing (1) by the mass m we obtain the second-order differential equation

$$\frac{d^2x}{dt^2} + \frac{k}{m}x = 0 \qquad \text{or} \qquad \frac{d^2x}{dt^2} + \omega^2 x = 0, \tag{2}$$

where $\omega^2 = k/m$. Equation (2) is said to describe **simple harmonic motion**, or **free undamped motion**. There are two obvious initial conditions associated with this differential equation:

$$x(0) = s_0, \qquad \left.\frac{dx}{dt}\right|_{t=0} = v_0, \tag{3}$$

representing the amount of initial displacement and the initial velocity, respectively. For example, if $s_0 > 0$, $v_0 < 0$, the mass would start from a point *below* the equilibrium position with an imparted *upward* velocity. If $s_0 < 0$, $v_0 = 0$, the mass would be released from rest from a point $|s_0|$ units *above* the equilibrium position, and so on.

The Solution and the Equation of Motion To solve (2) we note that the solutions of the auxiliary equation $m^2 + \omega^2 = 0$ are the complex numbers

$$m_1 = \omega i, \qquad m_2 = -\omega i.$$

Thus, from (6) of Section 16.2, we find the general solution of the equation to be

$$x(t) = C_1 \cos \omega t + C_2 \sin \omega t. \tag{4}$$

The **period** of free vibrations described by (4) is $T = 2\pi/\omega$ and the **frequency** is $f = 1/T = \omega/2\pi$. Finally, when the initial conditions (3) are used to determine the constants

C_1 and C_2 in (4), we say that the resulting particular solution is the **equation of motion** of the mass.

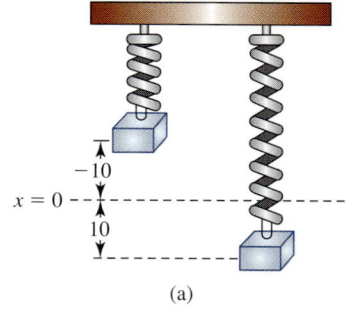

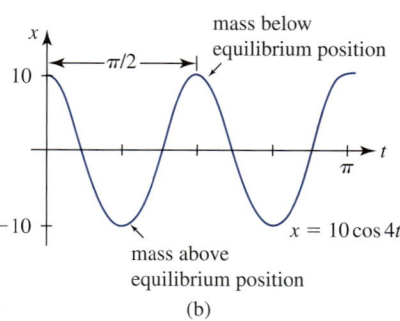

EXAMPLE 1 A System Describing Simple Harmonic Motion

Solve and interpret the initial-value problem

$$\frac{d^2x}{dt^2} + 16x = 0, \qquad x(0) = 10, x'(0) = 0.$$

Solution The problem is equivalent to pulling a mass on a spring down 10 units below the equilibrium position, holding it until $t = 0$, and then releasing it from rest. Applying the initial conditions to the solution

$$x(t) = C_1 \cos 4t + C_2 \sin 4t$$

gives

$$x(0) = 10 = C_1 \cdot 1 + C_2 \cdot 0,$$

so that $C_1 = 10$, and hence,

$$x(t) = 10 \cos 4t + C_2 \sin 4t$$
$$x'(t) = -40 \sin 4t + 4C_2 \cos 4t$$
$$x'(0) = 0 = 4C_2 \cdot 1.$$

The last equation implies that $C_2 = 0$, so the equation of motion is $x(t) = 10 \cos 4t$.

The solution clearly shows that once the system is set in motion, it stays in motion with the mass bouncing back and forth 10 units on either side of the equilibrium position $x = 0$. As shown in FIGURE 16.4.4(b), the period of oscillation is $2\pi/4 = \pi/2$ s. ■

FIGURE 16.4.4 Simple harmonic motion of a spring/mass system in Example 1

■ Free Damped Motion The discussion of simple harmonic motion in Example 1 is somewhat unrealistic. Unless the mass is suspended in a perfect vacuum, there will be at least a resisting force due to the surrounding medium. For example, as FIGURE 16.4.5 shows, the mass m could be suspended in a viscous medium or connected to a dashpot damping device. In the study of mechanics, damping forces acting on a body are considered to be proportional to a power of the instantaneous velocity. In particular, we shall assume that this force is given by a constant multiple of dx/dt. Thus, when no other external forces are impressed on the system, it follows from Newton's second law that

$$m\frac{d^2x}{dt^2} = -kx - \beta\frac{dx}{dt}, \qquad (5)$$

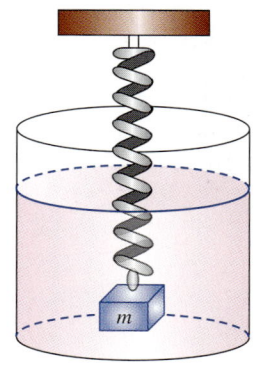

where β is a positive *damping constant* and the negative sign is a consequence of the fact that the damping force acts in a direction opposite to the motion. When we divide (5) by the mass m, the differential equation of **free damped motion** is

$$\frac{d^2x}{dt^2} + \frac{\beta}{m}\frac{dx}{dt} + \frac{k}{m}x = 0 \qquad \text{or} \qquad \frac{d^2x}{dt^2} + 2\lambda\frac{dx}{dt} + \omega^2 x = 0, \qquad (6)$$

where $2\lambda = \beta/m$ and $\omega^2 = k/m$. The symbol 2λ is used only for algebraic convenience, since the auxiliary equation is $m^2 + 2\lambda m + \omega^2 = 0$ and the corresponding roots are then

$$m_1 = -\lambda + \sqrt{\lambda^2 - \omega^2}, \quad m_2 = -\lambda - \sqrt{\lambda^2 - \omega^2}.$$

When $\lambda^2 - \omega^2 \neq 0$ the solution of the differential equation has the form

$$x(t) = e^{-\lambda t}\left(C_1 e^{\sqrt{\lambda^2 - \omega^2}\,t} + C_2 e^{-\sqrt{\lambda^2 - \omega^2}\,t}\right), \qquad (7)$$

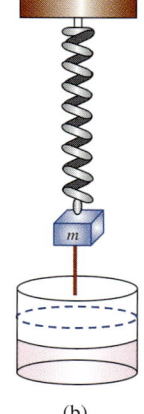

and we see that each solution will contain the **damping factor** $e^{-\lambda t}, \lambda > 0$. (As shown below, this is also the case when $\lambda^2 - \omega^2 = 0$.) Thus, displacements of the mass will become negligible as time increases. We now consider the three possible cases determined by the algebraic sign of $\lambda^2 - \omega^2$.

FIGURE 16.4.5 A spring/mass system with damped harmonic motion

CASE I: $\lambda^2 - \omega^2 > 0$ Here, the roots of the auxiliary equation are real and unequal and the system is said to be **overdamped**. It is easily shown that when $\lambda^2 - \omega^2 > 0, \beta^2 > 4km$, so

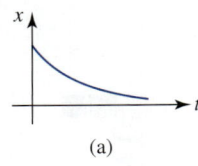

(a)

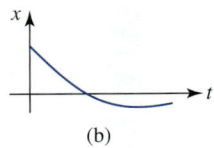

(b)

FIGURE 16.4.6 Overdamped motion of a spring/mass system

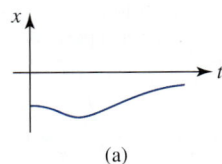

(a)

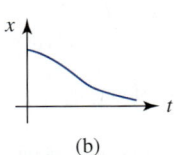

(b)

FIGURE 16.4.7 Critically damped motion of a spring/mass system

FIGURE 16.4.8 Underdamped motion of a spring/mass

that the damping constant β is large when compared to the spring constant k. The corresponding equation of motion is given by (7):

$$x(t) = e^{-\lambda t}\left(C_1 e^{\sqrt{\lambda^2 - \omega^2}\, t} + C_2 e^{-\sqrt{\lambda^2 - \omega^2}\, t}\right).$$

Two possible graphs of $x(t)$ are shown in FIGURE 16.4.6, illustrating the fact that the motion of the mass is nonoscillatory and quickly moves toward the equilibrium position.

CASE II: $\lambda^2 - \omega^2 = 0$ Here $m_1 = m_2 = -\lambda$ and the system is said to be **critically damped**, since any slight decrease in the damping force would result in oscillatory motion. The general solution of (6) is

$$x(t) = C_1 e^{m_1 t} + C_2 t e^{m_1 t}$$

or
$$x(t) = e^{-\lambda t}(C_1 + C_2 t). \tag{8}$$

Some graphs of typical motion are given in FIGURE 16.4.7. Notice that the motion is quite similar to that of an overdamped system. It is also apparent from (8) that the mass can pass through the equilibrium position at most one time.

CASE III: $\lambda^2 - \omega^2 < 0$ In this case we have $\beta^2 < 4km$, so the damping constant is small compared with the spring constant k, and the system is said to be **underdamped**. The roots m_1 and m_2 are now complex numbers:

$$m_1 = -\lambda + \sqrt{\omega^2 - \lambda^2}\, i, \qquad m_2 = -\lambda - \sqrt{\omega^2 - \lambda^2}\, i,$$

so the equation of motion given in (7) can be written as

$$x(t) = e^{-\lambda t}\left(C_1 \cos \sqrt{\omega^2 - \lambda^2}\, t + C_2 \sin \sqrt{\omega^2 - \lambda^2}\, t\right). \tag{9}$$

As indicated in FIGURE 16.4.8, the motion described by (9) is oscillatory, but because of the coefficient $e^{-\lambda t}$ we see that the amplitudes of vibration $\to 0$ as $t \to \infty$.

EXAMPLE 2 A System with Critically Damped Motion

A mass weighing 8 lb stretches a spring 2 ft. Assuming a damping force numerically equal to two times the instantaneous velocity acts on the system, determine the equation of motion if the mass is released from the equilibrium position with an upward velocity of 3 ft/s. Determine the type of damping exhibited by the system and graph the equation of motion.

Solution From Hooke's law we have

$$8 = k \cdot 2 \qquad \text{so} \qquad k = 4 \text{ lb/ft},$$

and from $m = W/g$,

$$m = \frac{8}{32} = \frac{1}{4} \text{slug}.$$

Since the damping constant is $\beta = 2$, the differential equation of motion is

$$\frac{1}{4}\frac{d^2 x}{dt^2} = -4x - 2\frac{dx}{dt} \qquad \text{or} \qquad \frac{d^2 x}{dt^2} + 8\frac{dx}{dt} + 16x = 0.$$

Since the mass is released from the equilibrium position with an upward velocity of 3 ft/s, the initial conditions are

$$x(0) = 0, \qquad \frac{dx}{dt}\bigg|_{t=0} = -3.$$

Since the auxiliary equation of the differential equation is

$$m^2 + 8m + 16 = (m + 4)^2 = 0,$$

we have $m_1 = m_2 = -4$, and the system is critically damped. The general solution of the differential equation is

$$x(t) = C_1 e^{-4t} + C_2 t e^{-4t}.$$

The initial condition $x(0) = 0$ immediately implies that $C_1 = 0$, whereas using $x'(0) = -3$ gives $C_2 = -3$. Thus, the equation of motion is

$$x(t) = -3te^{-4t}.$$

To graph $x(t)$ we find the time at which $x'(t) = 0$:

$$x'(t) = -3(-4te^{-4t} + e^{-4t}) = -3e^{-4t}(1 - 4t).$$

Clearly, $x'(t) = 0$ when $t = \frac{1}{4}$, and the corresponding displacement is

$$x\left(\tfrac{1}{4}\right) = -\tfrac{3}{4}e^{-1} = -0.276 \text{ ft}.$$

As shown in FIGURE 16.4.9, we interpret this value to mean that the weight reaches a maximum height of 0.276 ft above the equilibrium position. ∎

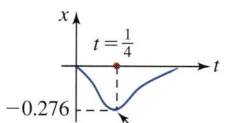

maximum height above equilibrium position

FIGURE 16.4.9 Graph of equation of motion in Example 2

EXAMPLE 3 A System with Underdamped Motion

A mass weighing 16 lb is attached to a 5-ft-long spring. At equilibrium the spring measures 8.2 ft. If the mass is pushed up and released from rest at a point 2 ft above the equilibrium position, find the displacements $x(t)$ if it is further known that the surrounding medium offers a resistance numerically equal to the instantaneous velocity. Determine the type of damping exhibited by the system.

Solution The elongation of the spring after the weight is attached is $8.2 - 5 = 3.2$ ft, so it follows from Hooke's law that

$$16 = k \cdot (3.2) \qquad \text{and} \qquad k = 5 \text{ lb/ft}.$$

In addition,

$$m = \frac{16}{32} = \frac{1}{2} \text{ slug} \qquad \text{and} \qquad \beta = 1,$$

so that the differential equation is given by

$$\frac{1}{2}\frac{d^2x}{dt^2} = -5x - \frac{dx}{dt} \qquad \text{or} \qquad \frac{d^2x}{dt^2} + 2\frac{dx}{dt} + 10x = 0.$$

This latter equation is solved subject to the initial conditions

$$x(0) = -2, \qquad \left.\frac{dx}{dt}\right|_{t=0} = 0.$$

Proceeding, we find that the roots of $m^2 + 2m + 10 = 0$ are $m_1 = -1 + 3i$ and $m_2 = -1 - 3i$, which then implies the system is underdamped and

$$x(t) = e^{-t}(C_1 \cos 3t + C_2 \sin 3t).$$

Now

$$x(0) = -2 = C_1$$
$$x(t) = e^{-t}(-2 \cos 3t + C_2 \sin 3t)$$
$$x'(t) = e^{-t}(6 \sin 3t + 3C_2 \cos 3t) - e^{-t}(-2 \cos 3t + C_2 \sin 3t)$$
$$x'(0) = 0 = 3C_2 + 2,$$

which gives $C_2 = -\frac{2}{3}$. Thus, the equation of motion is

$$x(t) = e^{-t}\left(-2 \cos 3t - \tfrac{2}{3} \sin 3t\right). \qquad ∎$$

▪ Forced Motion Suppose we now take into consideration an external force $f(t)$ acting on a vibrating mass on a spring. For example, $f(t)$ could represent a driving force causing an oscillatory vertical motion of the support of the spring. See FIGURE 16.4.10. The inclusion of $f(t)$ in the formulation of Newton's second law gives

$$m\frac{d^2x}{dt^2} = -kx - \beta\frac{dx}{dt} + f(t),$$

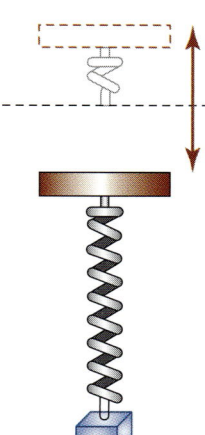

FIGURE 16.4.10 A forced spring/mass system

so,

$$\frac{d^2x}{dt^2} + \frac{\beta}{m}\frac{dx}{dt} + \frac{k}{m}x = \frac{f(t)}{m} \quad \text{or} \quad \frac{d^2x}{dt^2} + 2\lambda\frac{dx}{dt} + \omega^2 x = F(t), \tag{10}$$

where $F(t) = f(t)/m$ and, as before, $2\lambda = \beta/m$, and $\omega^2 = k/m$. To solve the latter nonhomogeneous equation, we can employ either the method of undetermined coefficients or variation of parameters.

The next example illustrates undamped forced motion.

EXAMPLE 4   A System with Forced Motion

Solve the initial-value problem

$$\frac{d^2x}{dt^2} + \omega^2 x = F_0 \sin \gamma t, \qquad F_0 = \text{constant},$$

$$x(0) = 0, \qquad \frac{dx}{dt}\bigg|_{t=0} = 0.$$

Solution   The complementary function is $x_c(t) = C_1 \cos \omega t + C_2 \sin \omega t$. To obtain a particular solution we require that $\gamma \neq \omega$ and use the method of undetermined coefficients. Then, assuming $x_p = A \cos \gamma t + B \sin \gamma t$, we have

$$x_p' = -A\gamma \sin \gamma t + B\gamma \cos \gamma t$$
$$x_p'' = -A\gamma^2 \cos \gamma t - B\gamma^2 \sin \gamma t$$
$$x_p'' + \omega^2 x_p = A(\omega^2 - \gamma^2) \cos \gamma t + B(\omega^2 - \gamma^2) \sin \gamma t$$
$$= F_0 \sin \gamma t.$$

It follows that

$$A = 0 \quad \text{and} \quad B = \frac{F_0}{\omega^2 - \gamma^2}.$$

Therefore,

$$x_p(t) = \frac{F_0}{\omega^2 - \gamma^2} \sin \gamma t.$$

Applying the given initial conditions to the general solution

$$x(t) = C_1 \cos \omega t + C_2 \sin \omega t + \frac{F_0}{\omega^2 - \gamma^2} \sin \gamma t$$

yields $C_1 = 0$ and $C_2 = -\gamma F_0/\omega(\omega^2 - \gamma^2)$. Thus, the equation of motion of the forced system is

$$x(t) = \frac{F_0}{\omega(\omega^2 - \gamma^2)} (-\gamma \sin \omega t + \omega \sin \gamma t), \quad \gamma \neq \omega. \tag{11} \ \blacksquare$$

∎ Pure Resonance   Although (11) is not defined for $\gamma = \omega$, it is interesting to observe that its limiting value as $\gamma \to \omega$ can be obtained by applying L'Hôpital's Rule. This limiting process is analogous to "tuning in" the frequency of the driving force $\gamma/2\pi$ to the frequency of free vibrations $\omega/2\pi$. Intuitively, we expect that over a length of time we should be able to substantially increase the amplitudes of vibration. For $\gamma = \omega$, we define the solution to be

$$x(t) = \lim_{\gamma \to \omega} F_0 \frac{-\gamma \sin \omega t + \omega \sin \gamma t}{\omega(\omega^2 - \gamma^2)}$$

$$= F_0 \lim_{\gamma \to \omega} \frac{\dfrac{d}{d\gamma}[-\gamma \sin \omega t + \omega \sin \gamma t]}{\dfrac{d}{d\gamma}[\omega^3 - \omega\gamma^2]} \qquad \leftarrow \text{ by L'Hôpital's Rule}$$

$$= F_0 \lim_{\gamma \to \omega} \frac{-\sin \omega t + \omega t \cos \gamma t}{-2\omega\gamma}$$

$$= \frac{F_0}{2\omega^2}(\sin\omega t - \omega t \cos\omega t)$$

$$= \frac{F_0}{2\omega^2}\sin\omega t - \frac{F_0}{2\omega}t\cos\omega t. \tag{12}$$

As suspected, when $t \to \infty$ the displacements become large; in fact, $|x(t)| \to \infty$. The phenomenon we have just described is known as **pure resonance**. The graph in FIGURE 16.4.11 displays typical motion in this case.

It should be noted that there is no actual need to use a limiting process on (11) to obtain the solution for $\gamma = \omega$. Alternatively, (12) follows by solving the initial-value problem

$$\frac{d^2x}{dt^2} + \omega^2 x = F_0 \sin\omega t, \quad x(0) = 0, \quad \frac{dx}{dt}\bigg|_{t=0} = 0$$

directly by conventional methods.

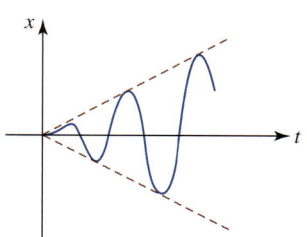

FIGURE 16.4.11   Pure resonance

■ **Series Circuit Analogue**   In conclusion, we examine another physical system that can be modeled by a linear second-order differential equation similar to the DE of a forced spring/mass system with damping:

$$m\frac{d^2x}{dt^2} + \beta\frac{dx}{dt} + kx = f(t).$$

If $i(t)$ denotes current in the **LRC-series electrical circuit** shown in FIGURE 16.4.12(a), then the voltage drops across the inductor, resistor, and capacitor are as shown in Figure 16.4.12(b). By Kirchhoff's second law, the sum of these voltages equals the voltage $E(t)$ impressed on the circuit; that is,

$$L\frac{di}{dt} + Ri + \frac{1}{C}q = E(t). \tag{13}$$

Since the charge $q(t)$ on the capacitor is related to the current $i(t)$ by $i = dq/dt$, (13) becomes the linear second-order differential equation

$$L\frac{d^2q}{dt^2} + R\frac{dq}{dt} + \frac{1}{C}q = E(t). \tag{14}$$

Except for the interpretation of the symbols used in (10) and (14), there is no difference between the mathematics of vibrating springs and simple series circuits. Even much of the terminology is the same. If $E(t) = 0$, the electrical vibrations of the circuit are said to be **free**. Since the auxiliary equation for (14) is $Lm^2 + Rm + 1/C = 0$, there will be three forms of the solution when $R \neq 0$, depending on the value of the discriminant $R^2 - 4L/C$. We say that the circuit is

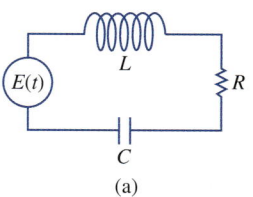

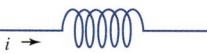

Inductor
Inductance L: henrys (h)
voltage drop across: $L\dfrac{di}{dt}$

Resistor
Resistance R: ohms (Ω)
voltage drop across: iR

Capacitor
Capacitance C: farads (f)
voltage drop across: $\dfrac{1}{C}q$

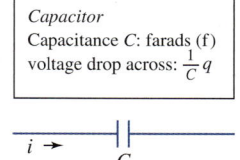

FIGURE 16.4.12   An *LRC*-series electrical circuit with voltage drops

overdamped if	$R^2 - 4L/C > 0$,	
critically damped if	$R^2 - 4L/C = 0$,	
and **underdamped if**	$R^2 - 4L/C < 0$.	

In each of these three cases, the general solution of (14) contains the factor $e^{-Rt/2L}$ and so the charge $q(t) \to 0$ as $t \to \infty$. In the underdamped case, when at least one of the initial charge and initial current are not zero, the charge on the capacitor oscillates as it decays. That is, the capacitor is charging and discharging as $t \to \infty$.

◀ Remember that current i is the derivative of the charge q so this is the same as saying that one of the initial conditions is nonzero.

EXAMPLE 5   A Series Circuit

Find the charge $q(t)$ on the capacitor in an *LRC*-series circuit when $L = 0.25$ henry (h), $R = 10$ ohms (Ω), $C = 0.001$ farad (f), $E(t) = 200\sin 40t$ volts (V), $q(0) = 3$ coulombs (C), and $i(0) = 0$ amperes (A).

Solution   Since $1/C = 1000$, equation (14) becomes

$$\tfrac{1}{4}q'' + 10q' + 1000q = 200\sin 40t$$

or

$$q'' + 40q' + 4000q = 800\sin 40t.$$

The circuit is underdamped because the roots of the auxiliary equation are the complex numbers $m_1 = -20 + 60i$ and $m_2 = -20 - 60i$. Thus, the complementary function of the differential equation is $q_c(t) = e^{-20t}(c_1 \cos 60t + c_2 \sin 60t)$. To find a particular solution q_p we use undetermined coefficients and assume a solution of the form $q_p = A \sin 40t + B \cos 40t$. Substituting this expression into the differential equation we find $A = \frac{3}{13}$ and $B = -\frac{2}{13}$. Thus, the charge on the capacitor is

$$q(t) = e^{-20t}(c_1 \cos 60t + c_2 \sin 60t) + \tfrac{3}{13} \sin 40t - \tfrac{2}{13} \cos 40t.$$

Applying the initial conditions $q(0) = 3$ and $i(0) = q'(0) = 0$, we get $c_1 = \frac{41}{13}$ and $c_2 = \frac{35}{39}$, so

$$q(t) = e^{-20t}\left(\tfrac{41}{13} \cos 60t + \tfrac{35}{36} \sin 60t\right) + \tfrac{3}{13} \sin 40t - \tfrac{2}{13} \cos 40t. \qquad \blacksquare$$

$$\dfrac{d^2x}{dt^2}$$ **NOTES FROM THE CLASSROOM**

(*i*) Acoustic vibrations can be as destructive as large mechanical vibrations. In television commercials, jazz singers have inflicted destruction on the lowly wine glass. Sounds from organs and piccolos have been known to crack windows.

As the horns blew, the people began to shout. When they heard the signal horn, they raised a tremendous shout. The wall collapsed. . . . (Joshua 6:20)

Did the power of acoustic resonance cause the walls of Jericho to tumble down? This is the conjecture of some contemporary scholars.

(*ii*) The phenomenon of resonance is not always destructive. For example, it is resonance of an electrical circuit that enables a radio to be tuned to a specific station.

Shattering effect of acoustic resonance

Exercises 16.4   Answers to selected odd-numbered problems begin on page ANS-48.

≡ Fundamentals

In Problems 1 and 2, state in words a physical interpretation of the given initial-value problem.

1. $\frac{4}{32}x'' + 3x = 0$; $x(0) = -3, x'(0) = -2$

2. $\frac{1}{16}x'' + 4x = 0$; $x(0) = 0.7, x'(0) = 0$

3. A mass weighing 8 lb attached to a spring exhibits simple harmonic motion. Determine the equation of motion if the spring constant is 1 lb/ft and if the mass is released 6 in. below the equilibrium position with a downward velocity of $\frac{3}{2}$ ft/s.

4. A mass weighing 24 lb attached to a spring exhibits simple harmonic motion. When placed on the spring, the mass stretches the spring 4 in. Find the equation of motion if the mass is released from rest from a point 3 in. above the equilibrium position.

5. A force of 400 N stretches a spring 2 m. A mass of 50 kg attached to the spring exhibits simple harmonic motion. Find the equation of motion if the mass is released from the equilibrium position with an upward velocity of 10 m/s.

6. A mass weighing 2 lb attached to a spring exhibits simple harmonic motion. At $t = 0$ the mass is released from a point 8 in. below the equilibrium position with an upward velocity of $\frac{4}{3}$ ft/s. If the spring constant is $k = 4$ lb/ft, find the equation of motion.

In Problems 7 and 8, state in words a physical interpretation of the given initial-value problem.

7. $\frac{1}{16}x'' + 2x' + x = 0$; $x(0) = 0, \left.\dfrac{dx}{dt}\right|_{t=0} = -1.5$

8. $\frac{16}{32}x'' + x' + 2x = 0$; $x(0) = -2, x'(0) = 1$

9. A mass weighing 4 lb attached to a spring exhibits free damped motion. The spring constant is 2 lb/ft and the medium offers a resistance to the motion of the mass numerically equal to the instantaneous velocity. If the mass is released from a point 1 ft above the equilibrium position with a downward velocity of 8 ft/s, determine the time that the mass passes through the equilibrium position. Find the time at which the mass attains its maximum displacement from the equilibrium position. What is the position of the mass at this instant?

10. A mass of 40 g stretches a spring 10 cm. A damping device imparts a resistance to motion numerically equal to 560 times the instantaneous velocity. Find the equation of free motion if the mass is released from the equilibrium position with a downward velocity of 2 cm/s.

11. After a mass weighing 10 lb is attached to a 5-ft spring, the spring measures 7 ft. The mass is removed and replaced with another mass weighing 8 lb and then the entire system is placed in a medium offering a resistance numerically equal to the instantaneous velocity. Find the equation of motion if the mass is released $\frac{1}{2}$ ft below the equilibrium position with a downward velocity of 1 ft/s.

12. A mass weighing 24 lb stretches a spring 4 ft. The subsequent motion takes place in a medium offering a resistance numerically equal to β ($\beta > 0$) times the instantaneous velocity. If the mass starts from the equilibrium position with an upward velocity of 2 ft/s, show that if $\beta > 3\sqrt{2}$, the equation of motion is

$$x(t) = \frac{-3}{\sqrt{\beta^2 - 18}} e^{-2\beta t/3} \sinh\frac{2}{3}\sqrt{\beta^2 - 18}\,t.$$

13. A mass weighing 10 lb attached to a spring stretches it 2 ft. The system is then set in motion in a medium that offers a resistance numerically equal to β ($\beta > 0$) times the instantaneous velocity. Determine the values of β so that the motion is

    (a) overdamped,
    (b) critically damped, and
    (c) underdamped.

14. A mass of 1 slug when attached to a spring stretches it 2 ft and then comes to rest in the equilibrium position. Starting at $t = 0$, an external force equal to $f(t) = 8 \sin 4t$ is applied to the system. Find the equation of motion if the surrounding medium offers a damping force numerically equal to 8 times the instantaneous velocity.

15. Solve Problem 14 when $f(t) = e^{-t} \sin 4t$. Analyze the displacements for $t \to \infty$.

16. Solve and interpret the initial-value problem:

$$\frac{d^2x}{dt^2} + 9x = 5 \sin 3t, \quad x(0) = 2, \quad \frac{dx}{dt}\bigg|_{t=0} = 0.$$

17. Find the charge on the capacitor in an *LRC*-series circuit at $t = 0.01$ s when $L = 0.05$ h, $R = 2\Omega$, $C = 0.01$ f, $E(t) = 0$ V, $q(0) = 5$ C, and $i(0) = 0$ A. Determine the first time at which the charge on the capacitor is equal to 0.

18. Find the charge on the capacitor in an *LRC*-series circuit when $L = \frac{1}{4}$ h, $R = 20\Omega$, $C = \frac{1}{300}$ f, $E(t) = 0$ V, $q(0) = 4$ C, and $i(0) = 0$ A. Is the charge on the capacitor ever equal to zero?

In Problems 19 and 20, find the charge on the capacitor and the current in the given *LRC*-series circuit. Find the maximum charge on the capacitor.

19. $L = \frac{5}{3}$ h, $R = 10\,\Omega$, $C = \frac{1}{30}$ f, $E(t) = 300$ V, $q(0) = 0$ C, $i(0) = 0$ A

20. $L = 1$ h, $R = 100\Omega$, $C = 0.0004$ f, $E(t) = 30$ V, $q(0) = 0$ C, $i(0) = 2$ A

≡ Projects

21. Beats When the frequency of a periodic impressed force is exactly the same as the frequency of free undamped vibration, then a spring/mass system is in a state of pure resonance. In this state, the displacements of the mass grow without bound as $t \to \infty$. But when the frequency $\gamma/2\pi$ of a periodic driving function is close to the frequency $\omega/2\pi$ of free vibrations, the mass undergoes complicated but bounded oscillations known as **beats**.

    (a) To examine this phenomenon, use undetermined coefficients to show that the solution of the initial-value problem

$$\frac{d^2x}{dt^2} + \omega^2 x = F_0 \cos\gamma t, \quad x(0) = 0, \quad x'(0) = 0$$

    is $x(t) = \dfrac{F_0}{\omega^2 - \gamma^2}(\cos\gamma t - \cos\omega t), \quad \gamma \neq \omega.$

    (b) Suppose $\omega = 2$ and $F_0 = \frac{1}{2}$. Use a graphing utility to graph, on separate coordinate axes, solution curves corresponding to $\gamma = 1$, $\gamma = 1.5$, $\gamma = 1.75$, and $\gamma = 1.9$. Use $0 \leq t \leq 60$ and $-3 \leq x \leq 3$.

    (c) Use a trigonometric identity to show that the solution in part (a) can be written as the product

$$x(t) = \frac{2F_0}{\gamma^2 - \omega^2} \sin\frac{1}{2}(\gamma - \omega)t \sin\frac{1}{2}(\gamma + \omega)t, \quad \gamma \neq \omega.$$

    (d) If $\varepsilon = \frac{1}{2}(\gamma - \omega)$, show that when ε is small, the solution in part (c) is approximately

$$x(t) = \frac{F_0}{2\varepsilon\gamma} \sin\varepsilon t \sin\gamma t.$$

    (e) Use a graphing utility to graph the function in part (d) with $\omega = 2$, $F_0 = \frac{1}{2}$, and $\gamma = 1.75$. Then graph $\pm(F_0/2\varepsilon\gamma) \sin\varepsilon t$ on the same set of coordinate axes. The graphs of $\pm(F_0/2\varepsilon\gamma) \sin\varepsilon t$ are called an *envelope* for the graph of $x(t)$. Compare the graph of $x(t)$ obtained in this manner with the third graph in part (b).

16.5 Power Series Solutions

▌ Introduction Some homogeneous linear second-order differential equations with *variable coefficients* can be solved by using power series. The procedure consists of assuming a solution of the form

$$y = c_0 + c_1 x + c_2 x^2 + c_3 x^3 + \cdots = \sum_{n=0}^{\infty} c_n x^n,$$

differentiating

$$y' = c_1 + 2c_2x + 3c_3x^2 + \cdots = \sum_{n=1}^{\infty} nc_nx^{n-1}, \tag{1}$$

$$y'' = 2c_2 + 6c_3x + \cdots = \sum_{n=2}^{\infty} n(n-1)c_nx^{n-2}, \tag{2}$$

and substituting the results into the differential equation with the expectation of determining a recurrence relation that will yield the coefficients c_n. To do this it is important that you become adept at simplifying the sum of two or more power series, each series expressed in sigma notation, to an expression with a single Σ. As the next example illustrates, combining two or more summations as a single summation often requires a reindexing, that is, a shift in the index of summation. To add two series written in sigma notation, it is necessary that

- *both summation indices start with the same number, and*
- *the powers of x in each series be in "phase," that is, if one series starts with, say, x to the first power, then we want the other series to start with the same power.*

EXAMPLE 1 Series Solution of a Differential Equation

Find a power series solution of $y'' - 2xy = 0$.

Solution Substituting $y = \sum_{n=0}^{\infty} c_nx^n$ into the differential equation and using (2) we have

$$y'' - 2xy = \sum_{n=2}^{\infty} n(n-1)c_nx^{n-2} - 2x\sum_{n=0}^{\infty} c_nx^n$$

$$= \sum_{n=2}^{\infty} n(n-1)c_nx^{n-2} - \sum_{n=0}^{\infty} 2c_nx^{n+1} = 0.$$

In each series, we now substitute k for the exponent on x. In the first series we use $k = n - 2$, and in the second series, $k = n + 1$. Thus

$$\sum_{n=2}^{\infty} n(n-1)c_nx^{n-2} - \sum_{n=0}^{\infty} 2c_nx^{n+1} = \overset{\underset{k=n-2}{\downarrow}}{\sum_{k=0}^{\infty} (k+2)(k+1)c_{k+2}x^k} - \overset{\underset{k=n+1}{\downarrow}}{\sum_{k=1}^{\infty} 2c_{k-1}x^k}.$$

$$\underset{\text{when } n=2,\, k=0}{\uparrow} \qquad \underset{\text{when } n=0,\, k=1}{\uparrow}$$

So that both series start with $k = 1$, we write the first term of the first series outside of the sigma notation and then combine the two series:

$$y'' - 2xy = 2c_2 + \sum_{k=1}^{\infty} (k+2)(k+1)c_{k+2}x^k - \sum_{k=1}^{\infty} 2c_{k-1}x^k$$

$$= 2c_2 + \sum_{k=1}^{\infty} [(k+2)(k+1)c_{k+2} - 2c_{k-1}]x^k = 0.$$

Corresponding coefficients of equal power series are themselves equal. ▶ Since the last equality is an identity, the coefficient of each power of x must be zero. That is,

$$2c_2 = 0 \quad \text{and} \quad (k+2)(k+1)c_{k+2} - 2c_{k-1} = 0. \tag{3}$$

Since $(k+1)(k+2) \neq 0$ for all values of k, we can solve (3) for c_{k+2} in terms of c_{k-1}:

$$c_{k+2} = \frac{2c_{k-1}}{(k+2)(k+1)}, \qquad k = 1, 2, 3, \ldots. \tag{4}$$

Now $2c_2 = 0$ obviously indicates that $c_2 = 0$. But the expression in (4), called a **recurrence relation**, determines the remaining c_k in such a manner that we can choose a certain subset of these coefficients to be *nonzero*. By letting k take on the indicated successive integers, (4) generates consecutive coefficients of the assumed solution one at a time:

$$c_3 = \frac{2c_0}{3 \cdot 2}$$

$$c_4 = \frac{2c_1}{4 \cdot 3}$$

$$c_5 = \frac{2c_2}{5 \cdot 4} = 0 \qquad \leftarrow c_2 = 0$$

$$c_6 = \frac{2c_3}{6 \cdot 5} = \frac{2^2}{6 \cdot 5 \cdot 3 \cdot 2} c_0$$

$$c_7 = \frac{2c_4}{7 \cdot 6} = \frac{2^2}{7 \cdot 6 \cdot 4 \cdot 3} c_1$$

$$c_8 = \frac{2c_5}{8 \cdot 7} = 0 \qquad \leftarrow c_5 = 0$$

$$c_9 = \frac{2c_6}{9 \cdot 8} = \frac{2^3}{9 \cdot 8 \cdot 6 \cdot 5 \cdot 3 \cdot 2} c_0$$

$$c_{10} = \frac{2c_7}{10 \cdot 9} = \frac{2^3}{10 \cdot 9 \cdot 7 \cdot 6 \cdot 4 \cdot 3} c_1$$

$$c_{11} = \frac{2c_8}{11 \cdot 10} = 0, \qquad \leftarrow c_8 = 0$$

and so on. It should be apparent that both c_0 and c_1 are arbitrary. Now

$$y = c_0 + c_1 x + c_2 x^2 + c_3 x^3 + c_4 x^4 + c_5 x^5 + c_6 x^6 + c_7 x^7 + c_8 x^8$$
$$+ c_9 x^9 + c_{10} x^{10} + c_{11} x^{11} + \cdots$$

$$= c_0 + c_1 x + 0 + \frac{2}{3 \cdot 2} c_0 x^3 + \frac{2}{4 \cdot 3} c_1 x^4 + 0 + \frac{2^2}{6 \cdot 5 \cdot 3 \cdot 2} c_0 x^6$$

$$+ \frac{2^2}{7 \cdot 6 \cdot 4 \cdot 3} c_1 x^7 + 0 + \frac{2^3}{9 \cdot 8 \cdot 6 \cdot 5 \cdot 3 \cdot 2} c_0 x^9$$

$$+ \frac{2^3}{10 \cdot 9 \cdot 7 \cdot 6 \cdot 4 \cdot 3} c_1 x^{10} + 0 + \cdots$$

$$= c_0 \left[1 + \frac{2}{3 \cdot 2} x^3 + \frac{2^2}{6 \cdot 5 \cdot 3 \cdot 2} x^6 + \frac{2^3}{9 \cdot 8 \cdot 6 \cdot 5 \cdot 3 \cdot 2} x^9 + \cdots \right]$$

$$+ c_1 \left[x + \frac{2}{4 \cdot 3} x^4 + \frac{2^2}{7 \cdot 6 \cdot 4 \cdot 3} x^7 + \frac{2^3}{10 \cdot 9 \cdot 7 \cdot 6 \cdot 4 \cdot 3} x^{10} + \cdots \right]$$

$$= c_0 y_1(x) + c_1 y_2(x). \qquad \blacksquare$$

A power series will represent a solution of the differential equation on some interval of convergence. Since the pattern of coefficients in Example 1 is clear, we can write the solutions in terms of summation notation. By using the properties of the factorial we have

$$y_1(x) = 1 + \sum_{k=1}^{\infty} \frac{2^k [1 \cdot 4 \cdot 7 \cdots (3k - 2)]}{(3k)!} x^{3k} \tag{5}$$

and

$$y_2(x) = x + \sum_{k=1}^{\infty} \frac{2^k [2 \cdot 5 \cdot 8 \cdots (3k - 1)]}{(3k + 1)!} x^{3k+1}. \tag{6}$$

The Ratio Test can be used on the forms in (5) and (6) to show that each series converges on the interval $(-\infty, \infty)$.

EXAMPLE 2 Series Solution of a Differential Equation

Find the power series solution of $(x^2 + 1)y'' + xy' - y = 0$.

Solution The assumption $y = \sum_{n=0}^{\infty} c_n x^n$ leads to

$$(x^2 + 1) \sum_{n=2}^{\infty} n(n-1)c_n x^{n-2} + x \sum_{n=1}^{\infty} nc_n x^{n-1} - \sum_{n=0}^{\infty} c_n x^n$$

$$= \underbrace{\sum_{n=2}^{\infty} n(n-1)c_n x^n}_{k=n} + \underbrace{\sum_{n=2}^{\infty} n(n-1)c_n x^{n-2}}_{k=n-2} + \underbrace{\sum_{n=1}^{\infty} nc_n x^n}_{k=n} - \underbrace{\sum_{n=0}^{\infty} c_n x^n}_{k=n}$$

$$= \sum_{k=2}^{\infty} k(k-1)c_k x^k + \sum_{k=0}^{\infty} (k+2)(k+1)c_{k+2} x^k + \sum_{k=1}^{\infty} kc_k x^k - \sum_{k=0}^{\infty} c_k x^k$$

$$= 2 \cdot 1 c_2 + 3 \cdot 2 c_3 x + c_1 x - c_0 - c_1 x$$

$$+ \sum_{k=2}^{\infty} [k(k-1)c_k + (k+2)(k+1)c_{k+2} + kc_k - c_k]x^k$$

$$= 2c_2 - c_0 + 6c_3 x$$

$$+ \sum_{k=2}^{\infty} [(k+1)(k-1)c_k + (k+2)(k+1)c_{k+2}]x^k = 0.$$

Thus, we must have

$$2c_2 - c_0 = 0$$
$$c_3 = 0$$
$$(k+1)(k-1)c_k + (k+2)(k+1)c_{k+2} = 0.$$

The foregoing equations yield $c_2 = \frac{1}{2}c_0$, $c_3 = 0$, and the recurrence relation

$$c_{k+2} = -\frac{k-1}{k+2}c_k, \quad k = 2, 3, 4, \ldots.$$

By letting k take on the values 2, 3, 4, . . . the last formula gives

$$c_4 = -\frac{1}{4}c_2 = -\frac{1}{2 \cdot 4}c_0 = -\frac{1}{2^2 2!}c_0 \qquad \leftarrow c_2 = \frac{1}{2}c_0$$

$$c_5 = -\frac{2}{5}c_3 = 0 \qquad \leftarrow c_3 = 0$$

$$c_6 = -\frac{3}{6}c_4 = \frac{3}{2 \cdot 4 \cdot 6}c_0 = \frac{1 \cdot 3}{2^3 3!}c_0$$

$$c_7 = -\frac{4}{7}c_5 = 0 \qquad \leftarrow c_5 = 0$$

$$c_8 = -\frac{5}{8}c_6 = -\frac{3 \cdot 5}{2 \cdot 4 \cdot 6 \cdot 8}c_0 = -\frac{1 \cdot 3 \cdot 5}{2^4 4!}c_0$$

$$c_9 = -\frac{6}{9}c_7 = 0 \qquad \leftarrow c_7 = 0$$

$$c_{10} = -\frac{7}{10}c_8 = \frac{3 \cdot 5 \cdot 7}{2 \cdot 4 \cdot 6 \cdot 8 \cdot 10}c_0 = \frac{1 \cdot 3 \cdot 5 \cdot 7}{2^5 5!}c_0,$$

and so on. Therefore,

$$y = c_0 + c_1 x + c_2 x^2 + c_3 x^3 + c_4 x^4 + c_5 x^5 + c_6 x^6 + c_7 x^7 + c_8 x^8 + \cdots$$

$$= c_1 x + c_0 \left[1 + \frac{1}{2}x^2 - \frac{1}{2^2 2!}x^4 + \frac{1 \cdot 3}{2^3 3!}x^6 - \frac{1 \cdot 3 \cdot 5}{2^4 4!}x^8 + \frac{1 \cdot 3 \cdot 5 \cdot 7}{2^5 5!}x^{10} + \cdots \right]$$

$$= c_0 y_1(x) + c_1 y_2(x).$$

Two solutions of the differential equation are

$$y_1(x) = 1 + \frac{1}{2}x^2 + \sum_{n=2}^{\infty} (-1)^{n-1} \frac{1 \cdot 3 \cdot 5 \cdots (2n-3)}{2^n n!} x^{2n}$$

and

$$y_2(x) = x.$$

≡ Fundamentals

In Problems 1–18, find power series solutions of the given differential equation.

1. $y'' + y = 0$

2. $y'' - y = 0$

3. $y'' = y'$

4. $2y'' + y' = 0$

5. $y'' = xy$

6. $y'' + x^2 y = 0$

7. $y'' - 2xy' + y = 0$

8. $y'' - xy' + 2y = 0$

9. $y'' + x^2 y' + xy = 0$

10. $y'' + 2xy' + 2y = 0$

11. $(x - 1)y'' + y' = 0$

12. $(x + 2)y'' + xy' - y = 0$

13. $(x^2 - 1)y'' + 4xy' + 2y = 0$

14. $(x^2 + 1)y'' - 6y = 0$

15. $(x^2 + 2)y'' + 3xy' - y = 0$

16. $(x^2 - 1)y'' + xy' - y = 0$

17. $y'' - (x + 1)y' - y = 0$

18. $y'' - xy' - (x + 2)y = 0$

In Problems 19 and 20, use the power series method to solve the given differential equation subject to the indicated initial conditions.

19. $(x - 1)y'' - xy' + y = 0; \quad y(0) = -2, y'(0) = 6$

20. $y'' - 2xy' + 8y = 0; \quad y(0) = 3, y'(0) = 0$

Chapter 16 in Review

Answers to selected odd-numbered problems begin on page ANS-49.

A. True/False

In Problems 1–8, indicate whether the given statement is true or false.

1. If y_1 is a solution of $ay'' + by' + cy = 0$, a, b, c constants, then $C_1 y_1$ is also a solution for every real number C_1. _____

2. A general solution of $y'' - y = 0$ is $y = C_1 \cosh x + C_2 \sinh x$. _____

3. $y_1 = e^x$ and $y_2 = 0$ are linearly independent solutions of the differential equation $y'' - y' = 0$. _____

4. The differential equation $y'' - y' = 10$ possesses a constant particular solution $y_p = A$. _____

5. The differential equation $y'' - y' = 0$ possesses infinitely many constant solutions. _____

6. The first-order differential equation $2xy\, dx = (x^2 - e^{-y})\, dy$ is exact. _____

7. Undamped and unforced motion of a mass on a spring is called simple harmonic motion. _____

8. Pure resonance cannot occur when damping is present. _____

B. Fill in the Blanks

In Problems 1–5, fill in the blanks.

1. A solution of the initial-value problem $y'' + 9y = 0$, $y(0) = 0, y'(0) = 0$ is _____.

2. A solution of the boundary-value problem $y'' - y' = 0$, $y(0) = 1, y(1) = 0$ is _____.

3. If a mass weighing 10 lb stretches a spring 2.5 ft, then a mass weighing 32 lb will stretch the same spring _____ ft.

4. If the auxiliary equation $am^2 + bm + c = 0$ for a homogeneous second-order DE possesses the solutions $m_1 = m_2 = -7$, then the general solution of the differential equation is _____.

5. Without solving, the form of a particular solution of $y'' + 6y' + 9y = 5x^2 - 3xe^{2x}$ is $y_p =$ _____.

C. Exercises

In Problems 1 and 2, determine whether the given differential equation is exact. If exact, solve.

1. $2x \cos y^3\, dx = (1 + 3x^2 y^2 \sin y^3)\, dy$

2. $(3x^2 + 2y^3)\, dx + y^2(6x + 1)\, dy = 0$

In Problems 3 and 4, solve the given initial-value problem.

3. $\frac{1}{2}xy^{-4} dx + (3y^{-3} - x^2y^{-5}) dy = 0, y(1) = 1$

4. $(y^2+y\sin x) dx + \left(2xy - \cos x - \dfrac{1}{1+y^2}\right) dy = 0, y(0) = 1$

In Problems 5–10, find the general solution of the given differential equation.

5. $y'' - 2y' - 2y = 0$ **6.** $y'' - 8y = 0$ **7.** $y'' - 3y' - 10y = 0$

8. $4y'' + 20y' + 25y = 0$ **9.** $9y'' + y = 0$ **10.** $2y'' - 5y' = 0$

In Problems 11 and 12, solve the given initial-value problem.

11. $y'' + 36y = 0, y(\pi/2) = 24, y'(\pi/2) = -18$

12. $y'' + 4y' + 4y = 0, y(0) = -2, y'(0) = 0$

In Problems 13 and 14, solve each differential equation by the method of undetermined coefficients.

13. $y'' - y' - 12y = (x + 1)e^{2x}$ **14.** $y'' + 4y = 16x^2$

In Problems 15 and 16, solve each differential equation by the method of variation of parameters.

15. $y'' - 2y' + 2y = e^x \tan x$ **16.** $y'' - y = 2e^x/(e^x + e^{-x})$

In Problems 17 and 18, solve the given initial-value problem.

17. $y'' + y = \sec^3 x, y(0) = 1, y'(0) = \frac{1}{2}$ **18.** $y'' + 2y' + 2y = 1, y(0) = 0, y'(0) = 1$

In Problems 19 and 20, find power series solutions of the given differential equation.

19. $y'' + xy = 0$ **20.** $(x - 1)y'' + 3y = 0$

21. A spring with constant $k = 2$ is suspended in a liquid that offers a damping force numerically equal to 4 times the instantaneous velocity. If a mass m is suspended from the spring, determine the values of m for which the subsequent free motion is nonoscillatory.

22. Find a particular solution for $\dfrac{d^2x}{dt^2} + 2\lambda\dfrac{dx}{dt} + \omega^2 x = A$, where A is a constant force.

23. A mass weighing 4 lb is suspended from a spring whose constant is 3 lb/ft. The entire system is immersed in a fluid offering a damping force numerically equal to the instantaneous velocity. Beginning at $t = 0$, an external force equal to $f(t) = e^{-t}$ is impressed on the system. Determine the equation of motion if the mass is released from rest at a point 2 ft below the equilibrium position.

24. A mass weighing W lb stretches one spring $\frac{1}{2}$ ft and stretches a different spring $\frac{1}{4}$ ft. If the two springs are attached in series, the effective spring constant k of the system is given by $1/k = 1/k_1 + 1/k_2$. The mass is then attached to the double spring, as shown in FIGURE 16.R.1. Assume that the motion is free and that there is no damping force present.

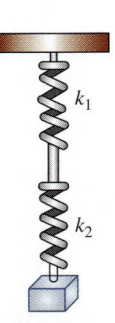

FIGURE 16.R.1 Attached springs in Problem 24

(a) Determine the equation of motion if the mass is released at a point 1 ft below the equation position with a downward velocity of $\frac{2}{3}$ ft/s.

(b) Show that the maximum speed of the mass is $\frac{2}{3}\sqrt{3g + 1}$.

25. The vertical motion of a mass attached to a spring is described by the initial-value problem

$$\frac{1}{4}\frac{d^2x}{dt^2} + \frac{dx}{dt} + x = 0, \quad x(0) = 4, x'(0) = 2.$$

Determine the maximum vertical displacement.

Proofs of Selected Theorems

■ **Section 2.2**

PROOF OF THEOREM 2.2.1(i): Let $\varepsilon > 0$ be given. To prove (i) we must find a $\delta > 0$ so that

$$|c - c| < \varepsilon \qquad \text{whenever} \qquad 0 < |x - a| < \delta.$$

Since $|c - c| = 0$, the preceding line is equivalent to

$$\varepsilon > 0 \qquad \text{whenever} \qquad 0 < |x - a| < \delta.$$

The last statement is always true for any choice of $\delta > 0$. ■

PROOF OF THEOREM 2.2.3(i): Let $\varepsilon > 0$ be given. To prove (i) we must find a $\delta > 0$ so that

$$|f(x) + g(x) - L_1 - L_2| < \varepsilon \qquad \text{whenever} \qquad 0 < |x - a| < \delta.$$

Since $\lim\limits_{x \to a} f(x) = L_1$ and $\lim\limits_{x \to a} g(x) = L_2$, we know there exist numbers $\delta_1 > 0$ and $\delta_2 > 0$ for which

$$|f(x) - L_1| < \frac{\varepsilon}{2} \qquad \text{whenever} \qquad 0 < |x - a| < \delta_1, \tag{1}$$

and

$$|g(x) - L_2| < \frac{\varepsilon}{2} \qquad \text{whenever} \qquad 0 < |x - a| < \delta_2. \tag{2}$$

Now, if we choose δ to be the smallest number in the set of positive numbers $\{\delta_1, \delta_2\}$, then (1) and (2) *both* hold and so

$$\begin{aligned}
|f(x) + g(x) - L_1 - L_2| &= |f(x) - L_1 + g(x) - L_2| \\
&\leq |f(x) - L_1| + |g(x) - L_2| \\
&< \frac{\varepsilon}{2} + \frac{\varepsilon}{2} = \varepsilon,
\end{aligned}$$

whenever $0 < |x - a| < \delta$. ■

PROOF OF THEOREM 2.2.3(ii): By the triangle inequality,

$$\begin{aligned}
|f(x)g(x) - L_1L_2| &= |f(x)g(x) - f(x)L_2 + f(x)L_2 - L_1L_2| \\
&\leq |f(x)g(x) - f(x)L_2| + |f(x)L_2 - L_1L_2| \\
&= |f(x)||g(x) - L_2| + |L_2||f(x) - L_1| \\
&\leq |f(x)||g(x) - L_2| + (1 + |L_2|)|f(x) - L_1|.
\end{aligned} \tag{3}$$

Since $\lim\limits_{x \to a} f(x) = L_1$ and $\lim\limits_{x \to a} g(x) = L_2$, we know there exist numbers $\delta_1 > 0, \delta_2 > 0, \delta_3 > 0$ such that $|f(x) - L_1| < 1$ or

$$|f(x)| < 1 + |L_1| \qquad \text{whenever} \qquad 0 < |x - a| < \delta_1, \tag{4}$$

$$|f(x) - L_1| < \frac{\varepsilon/2}{1 + |L_2|} \qquad \text{whenever} \qquad 0 < |x - a| < \delta_2, \tag{5}$$

and

$$|g(x) - L_2| < \frac{\varepsilon/2}{1 + |L_1|} \qquad \text{whenever} \qquad 0 < |x - a| < \delta_3. \tag{6}$$

Hence, if we choose δ to be the smallest number in the set of positive numbers $\{\delta_1, \delta_2, \delta_3\}$, then we have from (3), (4), (5), and (6),

$$|f(x)g(x) - L_1L_2| < (1 + |L_1|) \cdot \frac{\varepsilon/2}{1 + |L_1|} + (1 + |L_2|) \cdot \frac{\varepsilon/2}{1 + |L_2|} = \frac{\varepsilon}{2} + \frac{\varepsilon}{2} = \varepsilon. \quad \blacksquare$$

PROOF OF THEOREM 2.2.3(iii): We will first prove that

$$\lim_{x \to a} \frac{1}{g(x)} = \frac{1}{L_2}, \quad L_2 \neq 0.$$

Consider

$$\left| \frac{1}{g(x)} - \frac{1}{L_2} \right| = \frac{|g(x) - L_2|}{|L_2||g(x)|}. \tag{7}$$

Since $\lim_{x \to a} g(x) = L_2$, there exists a $\delta_1 > 0$ such that

$$|g(x) - L_2| < \frac{|L_2|}{2} \quad \text{whenever} \quad 0 < |x - a| < \delta_1.$$

For these values of x, the inequality

$$|L_2| = |g(x) - (g(x) - L_2)| \leq |g(x)| + |g(x) - L_2| < |g(x)| + \frac{|L_2|}{2}$$

gives

$$|g(x)| > \frac{|L_2|}{2} \quad \text{and} \quad \frac{1}{|g(x)|} < \frac{2}{|L_2|}.$$

Thus, from (7),

$$\left| \frac{1}{g(x)} - \frac{1}{L_2} \right| < \frac{2}{|L_2|^2}|g(x) - L_2|. \tag{8}$$

Now for $\varepsilon > 0$ there exists a $\delta_2 > 0$ such that

$$|g(x) - L_2| < \frac{|L_2|^2}{2}\varepsilon \quad \text{whenever} \quad 0 < |x - a| < \delta_2.$$

By choosing δ to be the smallest number in the set of positive numbers $\{\delta_1, \delta_2\}$, it follows from (8) that

$$\left| \frac{1}{g(x)} - \frac{1}{L_2} \right| < \varepsilon \quad \text{whenever} \quad 0 < |x - a| < \delta.$$

We conclude the proof using Theorem 2.2.3(ii):

$$\lim_{x \to a} \frac{f(x)}{g(x)} = \lim_{x \to a} \frac{1}{g(x)} \cdot f(x) = \lim_{x \to a} \frac{1}{g(x)} \cdot \lim_{x \to a} f(x) = \frac{L_1}{L_2}. \quad \blacksquare$$

■ **Section 2.3**

PROOF OF THEOREM 2.3.3: To prove the theorem we must find a $\delta > 0$ so that

$$|f(g(x)) - f(L)| < \varepsilon \quad \text{whenever} \quad 0 < |x - a| < \delta.$$

To this end we first use the fact that f is continuous at L, in other words, $\lim_{u \to L} f(u) = f(L)$. This means for a given $\varepsilon > 0$ there exists a $\delta_1 > 0$ such that

$$|f(u) - f(L)| < \varepsilon \quad \text{whenever} \quad |u - L| < \delta_1.$$

Now if $u = g(x)$, then the last line is

$$|f(g(x)) - f(L)| < \varepsilon \quad \text{whenever} \quad |g(x) - L| < \delta_1.$$

Also from the assumption that $\lim_{x \to a} g(x) = L$, we know there exists a $\delta > 0$ such that

$$|g(x) - L| < \delta_1 \quad \text{whenever} \quad 0 < |x - a| < \delta.$$

We now combine the last two results. That is, whenever $0 < |x - a| < \delta$, then $|g(x) - L| < \delta_1$; but whenever $|g(x) - L| < \delta_1$, then necessarily $|f(g(x)) - f(L)| < \varepsilon$. $\quad \blacksquare$

∎ Section 2.4

PROOF OF THEOREM 2.4.1:   We assume that $g(x) \leq f(x) \leq h(x)$ for all x in an open interval that contains the number a (except possibly at a itself) and that $\lim_{x \to a} g(x) = L$ and $\lim_{x \to a} h(x) = L$. Let $\varepsilon > 0$ be given. Then there exist numbers $\delta_1 > 0$ and $\delta_2 > 0$ such that $|g(x) - L| < \varepsilon$ whenever $0 < |x - a| < \delta_1$ and $|h(x) - L| < \varepsilon$ whenever $0 < |x - a| < \delta_2$. That is,

$$L - \varepsilon < g(x) < L + \varepsilon \qquad \text{whenever} \qquad 0 < |x - a| < \delta_1$$
$$L - \varepsilon < h(x) < L + \varepsilon \qquad \text{whenever} \qquad 0 < |x - a| < \delta_2.$$

Also there must exist $\delta_3 > 0$ such that

$$g(x) \leq f(x) \leq h(x) \qquad \text{whenever} \qquad 0 < |x - a| < \delta_3.$$

If δ is taken to be the smallest number in the set of positive numbers $\{\delta_1, \delta_2, \delta_3\}$, then for $0 < |x - a| < \delta$ we have

$$L - \varepsilon < g(x) \leq f(x) \leq h(x) < L + \varepsilon$$

or equivalently $|f(x) - L| < \varepsilon$. This means $\lim_{x \to a} f(x) = L$.   ∎

∎ Section 9.10

PROOF OF THEOREM 9.10.2   Let x be a fixed number in the interval $(a - r, a + r)$ and let the difference between $f(x)$ and the nth degree Taylor polynomial of f at a be denoted by

$$R_n(x) = f(x) - P_n(x).$$

For any t in the interval $[a, x]$ we define

$$F(t) = f(x) - f(t) - \frac{f'(t)}{1!}(x - t) - \frac{f''(t)}{2!}(x - t)^2 - \cdots - \frac{f^{(n)}(t)}{n!}(x - t)^n - \frac{R_n(x)}{(x - a)^{n+1}}(x - t)^{n+1}. \qquad (9)$$

With x held constant we differentiate F with respect to t using the Product and Power Rules:

$$F'(t) = -f'(t) + \left[f'(t) - \frac{f''(t)}{1!}(x - t) \right] + \left[\frac{f''(t)}{1!}(x - t) - \frac{f'''(t)}{2!}(x - t)^2 \right] + \cdots$$
$$+ \left[\frac{f^{(n)}(t)}{(n - 1)!}(x - t)^{n-1} - \frac{f^{(n+1)}(t)}{n!}(x - t)^n \right] + \frac{R_n(x)(n + 1)}{(x - a)^{n+1}}(x - t)^n,$$

for all t in the open interval (a, x). Since the last sum telescopes, we obtain

$$F'(t) = -\frac{f^{(n+1)}(t)}{n!}(x - t)^n + \frac{R_n(x)(n + 1)}{(x - a)^{n+1}}(x - t)^n. \qquad (10)$$

Now it is evident from (9) that F is continuous on $[a, x]$ and that

$$F(x) = f(x) - f(x) - 0 - \cdots - 0 = 0.$$

Furthermore, $\qquad\qquad F(a) = f(x) - P_n(x) - R_n(x) = 0.$

Thus, $F(t)$ satisfies the hypotheses of Rolle's Theorem (Theorem 4.4.1) on $[a, x]$ and so there exists a number c between a and x for which $F'(c) = 0$. From (10) we obtain

$$R_n(x) = \frac{f^{(n+1)}(c)}{(n + 1)!}(x - a)^{n+1}.$$ ∎

Review of Algebra

Integers

$\{\ldots, -4, -3, -2, -1, 0, 1, 2, 3, 4, \ldots\}$

Positive Integers (Natural Numbers)

$\{1, 2, 3, 4, 5, \ldots\}$

Nonnegative Integers (Whole Numbers)

$\{0, 1, 2, 3, 4, 5, \ldots\}$

Rational Numbers

A rational number is a number of the form p/q, where p and $q \neq 0$ are integers

Irrational Numbers

An irrational number is a number that cannot be written in the form p/q, where p and $q \neq 0$ are integers

Real Numbers

The set R of real numbers is the union of the sets of rational and irrational numbers

Laws of Exponents

$$a^m a^n = a^{m+n}, \quad \frac{a^m}{a^n} = a^{m-n}$$

$$(a^m)^n = a^{mn}, \quad (ab)^n = a^n b^n$$

$$\left(\frac{a}{b}\right)^n = \frac{a^n}{b^n}, \quad a^0 = 1, a \neq 0$$

Negative Exponent

$$a^{-n} = \frac{1}{a^n}, \ n > 0$$

Radical

$$a^{1/n} = \sqrt[n]{a}, \ n > 0 \text{ an integer}$$

Rational Exponents and Radicals

$$a^{m/n} = \left(a^m\right)^{1/n} = \left(a^{1/n}\right)^m$$

$$a^{m/n} = \sqrt[n]{a^m} = \left(\sqrt[n]{a}\right)^m$$

$$\sqrt[n]{ab} = \sqrt[n]{a}\sqrt[n]{b}$$

$$\sqrt[n]{\frac{a}{b}} = \frac{\sqrt[n]{a}}{\sqrt[n]{b}}$$

Quadratic Formula

Roots of a quadratic equation $ax^2 + bx + c = 0, a \neq 0$, are

$$x = \frac{-b \pm \sqrt{b^2 - 4ac}}{2a}$$

Binomial Expansions

$(a + b)^2 = a^2 + 2ab + b^2$

$(a + b)^3 = a^3 + 3a^2b + 3ab^2 + b^3$

$(a + b)^4 = a^4 + 4a^3b + 6a^2b^2 + 4ab^3 + b^4$

$(a + b)^5 = a^5 + 5a^4b + 10a^3b^2 + 10a^2b^3 + 5ab^4 + b^5$

Pascal's Triangle

Coefficients in the expansion of $(a + b)^n$ follow the pattern:

```
         1
       1   1
      1  2  1
    1  3   3  1
  1  4   6   4  1
         ⋮
```

Each number in the interior of this array is the sum of the two numbers directly above it:

```
  1   4   6   4   1
1   5   10  10  5   1
```

The last row are the coefficients in the expansion of $(a + b)^5$.

Factorization Formulas

$a^2 - b^2 = (a - b)(a + b)$

$a^3 - b^3 = (a - b)(a^2 + ab + b^2)$

$a^3 + b^3 = (a + b)(a^2 - ab + b^2)$

$a^4 - b^4 = (a - b)(a + b)(a^2 + b^2)$

Definition of Absolute Value

$$|a| = \begin{cases} a & \text{if } a \text{ is nonnegative } (a \geq 0) \\ -a & \text{if } a \text{ is negative } (a < 0) \end{cases}$$

Properties of Inequalities

If $a > b$ and $b > c$, then $a > c$.

If $a < b$, then $a + c < b + c$.

If $a < b$, then $ac < bc$ for $c > 0$.

If $a < b$, then $ac > bc$ for $c < 0$.

Formulas from Geometry

Area A, Circumference C, Volume V, Surface Area S

RECTANGLE	PARALLELOGRAM	TRAPEZOID
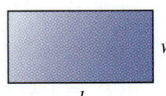		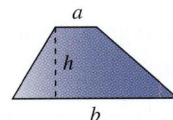
$A = lw, \ C = 2l + 2w$	$A = bh$	$A = \frac{1}{2}(a + b)h$

RIGHT TRIANGLE	TRIANGLE	EQUILATERAL TRIANGLE
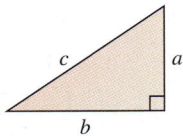	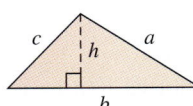	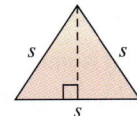
Pythagorean Theorem: $c^2 = a^2 + b^2$	$A = \frac{1}{2}bh, \ C = a + b + c$	$h = \frac{\sqrt{3}}{2}s, \ A = \frac{\sqrt{3}}{4}s^2$

CIRCLE	CIRCULAR RING	CIRCULAR SECTOR
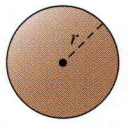	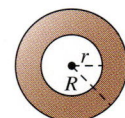	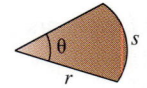
$A = \pi r^2, \ C = 2\pi r$	$A = \pi(R^2 - r^2)$	$A = \frac{1}{2}r^2\theta, \ s = r\theta$

ELLIPSE	ELLIPSOID	SPHERE
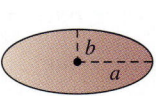	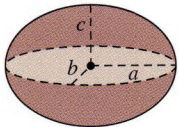	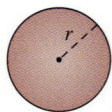
$A = \pi ab$	$V = \frac{4}{3}\pi abc$	$V = \frac{4}{3}\pi r^3, \ S = 4\pi r^2$

RIGHT CYLINDER	RIGHT CIRCULAR CYLINDER	RECTANGULAR PARALLELEPIPED
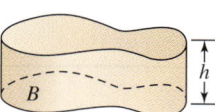	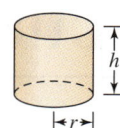	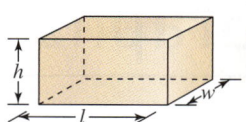
$V = Bh$, B area of base	$V = \pi r^2 h$, $S = 2\pi rh$ (lateral side)	$V = lwh$, $S = 2(hl + lw + hw)$
CONE	RIGHT CIRCULAR CONE	FRUSTUM OF A CONE
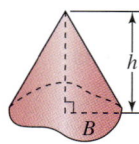	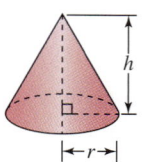	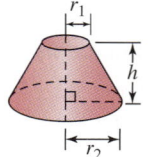
$V = \frac{1}{3}Bh$, B area of base	$V = \frac{1}{3}\pi r^2 h$, $S = \pi r\sqrt{r^2 + h^2}$	$V = \frac{1}{3}\pi h(r_1^2 + r_1 r_2 + r_2^2)$

Graphs and Functions

To Find Intercepts

y-intercepts: Set $x = 0$ in an equation and solve for y
x-intercepts: Set $y = 0$ in an equation and solve for x

Polynomial Functions

$$f(x) = a_n x^n + a_{n-1} x^{n-1} + \cdots + a_1 x + a_0,$$

where n is a nonnegative integer

Linear Function

$f(x) = ax + b, a \neq 0$

Graph of a linear function is a straight line.

Equation forms of lines:
  Point-Slope: $y - x_0 = m(x - x_0)$,
  Slope-Intercept: $y = mx + b$,

where m is slope

Quadratic Function

$$f(x) = ax^2 + bx + c, a \neq 0$$

Graph of a quadratic function is a parabola.

Vertex (h, k) of a Parabola

Complete the square in x for $f(x) = ax^2 + bx + c$ to obtain $f(x) = a(x - h)^2 + k$. Alternatively, compute the coordinates

$$\left(-\frac{b}{2a}, f\left(-\frac{b}{2a} \right) \right).$$

Even and Odd Functions

Even: $f(-x) = f(x)$;   Symmetry of graph: y-axis
Odd: $f(-x) = -f(x)$;   Symmetry of graph: origin

Rigid Transformations

Graph of $y = f(x)$ for $c > 0$:
$y = f(x) + c$,   shifted up c units
$y = f(x) - c$,   shifted down c units
$y = f(x + c)$,   shifted left c units
$y = f(x - c)$,   shifted right c units
$y = f(-x)$,   reflection in y-axis
$y = -f(x)$,   reflection in x-axis

Rational Function

$$f(x) = \frac{p(x)}{q(x)} = \frac{a_n x^n + \cdots + a_1 x + a_0}{b_m x^m + \cdots + b_1 x + b_0},$$

where $p(x)$ and $q(x)$ are polynomial functions

Asymptotes

If the polynomial functions $p(x)$ and $q(x)$ have no common factors, then the graph of a rational function

$$f(x) = \frac{p(x)}{q(x)} = \frac{a_n x^n + \cdots + a_1 x + a_0}{b_m x^m + \cdots + b_1 x + b_0}$$

has a

  Vertical asymptote:
    $x = a$ when $q(a) = 0$,
  Horizontal asymptote:
    $y = a_n/b_m$ when $n = m$, and $y = 0$ when $n < m$,
  Slant asymptote:
    $y = ax + b$ when $n = m + 1$.

The graph has no horizontal asymptote when $n > m$. A slant asymptote is found by division.

Power Function

$f(x) = x^n$,

where n is any real number

Review of Trigonometry

Unit Circle Definition of Sine and Cosine

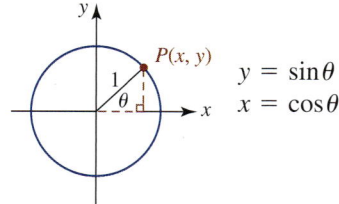

$$y = \sin\theta$$
$$x = \cos\theta$$

Other Trigonometric Functions

$$\tan\theta = \frac{y}{x} = \frac{\sin\theta}{\cos\theta}, \quad \cot\theta = \frac{x}{y} = \frac{\cos\theta}{\sin\theta}$$

$$\sec\theta = \frac{1}{x} = \frac{1}{\cos\theta}, \quad \csc\theta = \frac{1}{y} = \frac{1}{\sin\theta}$$

Conversion Formulas

$$1 \text{ degree} = \frac{\pi}{180} \text{ radians}$$

$$1 \text{ radian} = \frac{180}{\pi} \text{ degrees}$$

Right Triangle Definition of Sine and Cosine

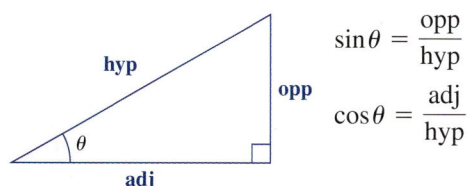

$$\sin\theta = \frac{\text{opp}}{\text{hyp}}$$

$$\cos\theta = \frac{\text{adj}}{\text{hyp}}$$

Other Trigonometric Functions

$$\tan\theta = \frac{\text{opp}}{\text{adj}}, \quad \cot\theta = \frac{\text{adj}}{\text{opp}}$$

$$\sec\theta = \frac{\text{hyp}}{\text{adj}}, \quad \csc\theta = \frac{\text{hyp}}{\text{opp}}$$

Signs of Sine and Cosine

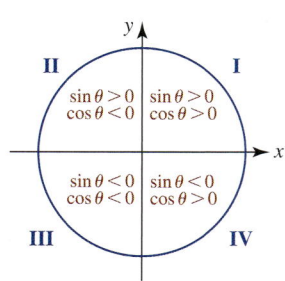

Values of Sine and Cosine for Special Angles

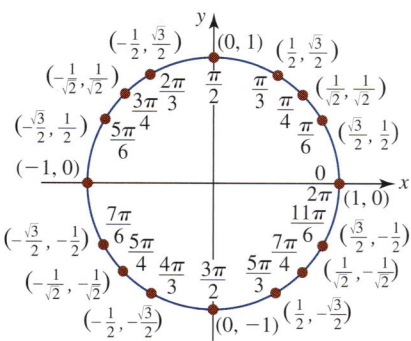

Bounds for Sine and Cosine Functions

$$-1 \leq \sin x \leq 1 \quad \text{and} \quad -1 \leq \cos x \leq 1$$

Periodicity of Trigonometric Functions

$$\sin(x + 2\pi) = \sin x, \quad \cos(x + 2\pi) = \cos x$$
$$\sec(x + 2\pi) = \sec x, \quad \csc(x + 2\pi) = \csc x$$
$$\tan(x + \pi) = \tan x, \quad \cot(x + \pi) = \cot x$$

Cofunction Identities

$$\sin\left(\frac{\pi}{2} - x\right) = \cos x$$

$$\cos\left(\frac{\pi}{2} - x\right) = \sin x$$

$$\tan\left(\frac{\pi}{2} - x\right) = \cot x$$

Pythagorean Identities

$$\sin^2 x + \cos^2 x = 1$$
$$1 + \tan^2 x = \sec^2 x$$
$$1 + \cot^2 x = \csc^2 x$$

Even/Odd Identities

Even	Odd
$\cos(-x) = \cos x$	$\sin(-x) = -\sin x$
$\sec(-x) = \sec x$	$\csc(-x) = -\csc x$
	$\tan(-x) = -\tan x$
	$\cot(-x) = -\cot x$

RESOURCE PAGES

Sum Formulas

$\sin(x_1 + x_2) = \sin x_1 \cos x_2 + \cos x_1 \sin x_2$

$\cos(x_1 + x_2) = \cos x_1 \cos x_2 - \sin x_1 \sin x_2$

$\tan(x_1 + x_2) = \dfrac{\tan x_1 + \tan x_2}{1 - \tan x_1 \tan x_2}$

Difference Formulas

$\sin(x_1 - x_2) = \sin x_1 \cos x_2 - \cos x_1 \sin x_2$

$\cos(x_1 - x_2) = \cos x_1 \cos x_2 + \sin x_1 \sin x_2$

$\tan(x_1 - x_2) = \dfrac{\tan x_1 - \tan x_2}{1 + \tan x_1 \tan x_2}$

Double-Angle Formulas

$\sin 2x = 2 \sin x \cos x$

$\cos 2x = \cos^2 x - \sin^2 x$

Alternative Double-Angle Formulas for Cosine

$\cos 2x = 1 - 2 \sin^2 x$

$\cos 2x = 2 \cos^2 x - 1$

Half-Angle Formulas as Used in Calculus

$\sin^2 x = \frac{1}{2}(1 - \cos 2x)$

$\cos^2 x = \frac{1}{2}(1 + \cos 2x)$

Law of Sines

$$\frac{\sin \alpha}{a} = \frac{\sin \beta}{b} = \frac{\sin \gamma}{c}$$

Law of Cosines

$$a^2 = b^2 + c^2 - 2bc \cos \alpha$$
$$b^2 = a^2 + c^2 - 2ac \cos \beta$$
$$c^2 = a^2 + b^2 - 2ab \cos \gamma$$

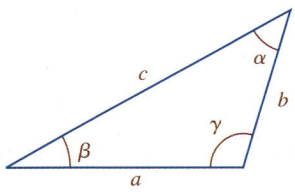

Inverse Trigonometric Functions

$y = \sin^{-1} x$ if and only if $x = \sin y$, $\quad -\pi/2 \le y \le \pi/2$

$y = \cos^{-1} x$ if and only if $x = \cos y$, $\quad 0 \le y \le \pi$

$y = \tan^{-1} x$ if and only if $x = \tan y$, $\quad -\pi/2 < y < \pi/2$

Cycles for Sine, Cosine, and Tangent

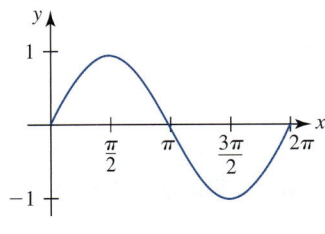

sine

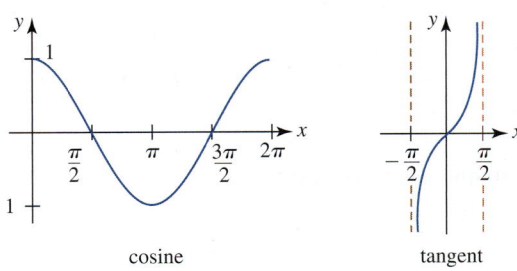

cosine   tangent

Exponential and Logarithmic Functions

The Number e

$e = 2.718281828459...$

Definitions of the Number e

$$e = \lim_{x \to \infty}\left(1 + \frac{1}{x}\right)^x$$

$$e = \lim_{h \to 0}(1 + h)^{1/h}$$

Exponential Function

$f(x) = b^x, \; b > 0, b \neq 1$

Natural Exponential Function

$f(x) = e^x$

Logarithmic Function

$f(x) = \log_b x, \quad x > 0$

where $y = \log_b x$ is equivalent to $x = b^y$

Natural Logarithmic Function

$f(x) = \log_e x = \ln x, \quad x > 0$

where $y = \ln x$ is equivalent to $x = e^y$

Laws of Logarithms

$\log_b MN = \log_b M + \log_b N$

$\log_b \dfrac{M}{N} = \log_b M - \log_b N$

$\log_b M^c = c\log_b M$

Properties of Logarithms

$\log_b b = 1, \qquad \log_b 1 = 0$

$\log_b b^x = x, \qquad b^{\log_b x} = x$

Change from Base b to Base e

$$\log_b x = \frac{\ln x}{\ln b}$$

Hyperbolic Functions

$$\sinh x = \frac{e^x - e^{-x}}{2}, \quad \cosh x = \frac{e^x + e^{-x}}{2}$$

$$\tanh x = \frac{\sinh x}{\cosh x}, \quad \coth x = \frac{\cosh x}{\sinh x}$$

$$\operatorname{sech} x = \frac{1}{\cosh x}, \quad \operatorname{csch} x = \frac{1}{\sinh x}$$

Inverse Hyperbolic Functions as Logarithms

$\sinh^{-1}x = \ln\left(x + \sqrt{x^2 + 1}\right)$

$\cosh^{-1}x = \ln\left(x + \sqrt{x^2 - 1}\right), \; x \geq 1$

$\tanh^{-1}x = \dfrac{1}{2}\ln\left(\dfrac{1 + x}{1 - x}\right), \; |x| < 1$

$\coth^{-1}x = \dfrac{1}{2}\ln\left(\dfrac{x + 1}{x - 1}\right), \; |x| > 1$

$\operatorname{sech}^{-1}x = \ln\left(\dfrac{1 + \sqrt{1 - x^2}}{x}\right), \; 0 < x \leq 1$

$\operatorname{csch}^{-1}x = \ln\left(\dfrac{1}{x} + \dfrac{\sqrt{1 + x^2}}{|x|}\right), \; x \neq 0$

Even/Odd Identities

Even	Odd
$\cosh(-x) = \cosh x$	$\sinh(-x) = -\sinh x$

Additional Identities

$\cosh^2 x - \sinh^2 x = 1$

$1 - \tanh^2 x = \operatorname{sech}^2 x$

$\coth^2 x - 1 = \operatorname{csch}^2 x$

$\sinh(x_1 \pm x_2) = \sinh x_1 \cosh x_2 \pm \cosh x_1 \sinh x_2$

$\cosh(x_1 \pm x_2) = \cosh x_1 \cosh x_2 \pm \sinh x_1 \sinh x_2$

$\sinh 2x = 2\sinh x \cosh x$

$\cosh 2x = \cosh^2 x + \sinh^2 x$

$\sinh^2 x = \frac{1}{2}(-1 + \cosh 2x)$

$\cosh^2 x = \frac{1}{2}(1 + \cosh 2x)$

Differentiation

Rules

1. **Constant:** $\dfrac{d}{dx}c = 0$

2. **Constant Multiple:** $\dfrac{d}{dx}cf(x) = cf'(x)$

3. **Sum:** $\dfrac{d}{dx}[f(x) \pm g(x)] = f'(x) \pm g'(x)$

4. **Product:** $\dfrac{d}{dx}f(x)g(x) = f(x)g'(x) + g(x)f'(x)$

5. **Quotient:** $\dfrac{d}{dx}\dfrac{f(x)}{g(x)} = \dfrac{g(x)f'(x) - f(x)g'(x)}{[g(x)]^2}$

6. **Chain:** $\dfrac{d}{dx}f(g(x)) = f'(g(x))g'(x)$

7. **Power:** $\dfrac{d}{dx}x^n = nx^{n-1}$

8. **Power:** $\dfrac{d}{dx}[g(x)]^n = n[g(x)]^{n-1}g'(x)$

Functions

Trigonometric:

9. $\dfrac{d}{dx}\sin x = \cos x$

10. $\dfrac{d}{dx}\cos x = -\sin x$

11. $\dfrac{d}{dx}\tan x = \sec^2 x$

12. $\dfrac{d}{dx}\cot x = -\csc^2 x$

13. $\dfrac{d}{dx}\sec x = \sec x \tan x$

14. $\dfrac{d}{dx}\csc x = -\csc x \cot x$

Inverse trigonometric:

15. $\dfrac{d}{dx}\sin^{-1}x = \dfrac{1}{\sqrt{1-x^2}}$

16. $\dfrac{d}{dx}\cos^{-1}x = -\dfrac{1}{\sqrt{1-x^2}}$

17. $\dfrac{d}{dx}\tan^{-1}x = \dfrac{1}{1+x^2}$

18. $\dfrac{d}{dx}\cot^{-1}x = -\dfrac{1}{1+x^2}$

19. $\dfrac{d}{dx}\sec^{-1}x = \dfrac{1}{|x|\sqrt{x^2-1}}$

20. $\dfrac{d}{dx}\csc^{-1}x = -\dfrac{1}{|x|\sqrt{x^2-1}}$

Hyperbolic:

21. $\dfrac{d}{dx}\sinh x = \cosh x$

22. $\dfrac{d}{dx}\cosh x = \sinh x$

23. $\dfrac{d}{dx}\tanh x = \operatorname{sech}^2 x$

24. $\dfrac{d}{dx}\coth x = -\operatorname{csch}^2 x$

25. $\dfrac{d}{dx}\operatorname{sech} x = -\operatorname{sech} x \tanh x$

26. $\dfrac{d}{dx}\operatorname{csch} x = -\operatorname{csch} x \coth x$

Inverse hyperbolic:

27. $\dfrac{d}{dx}\sinh^{-1}x = \dfrac{1}{\sqrt{x^2+1}}$

28. $\dfrac{d}{dx}\cosh^{-1}x = \dfrac{1}{\sqrt{x^2-1}}$

29. $\dfrac{d}{dx}\tanh^{-1}x = \dfrac{1}{1-x^2}$

30. $\dfrac{d}{dx}\coth^{-1}x = \dfrac{1}{1-x^2}$

31. $\dfrac{d}{dx}\operatorname{sech}^{-1}x = -\dfrac{1}{x\sqrt{1-x^2}}$

32. $\dfrac{d}{dx}\operatorname{csch}^{-1}x = -\dfrac{1}{|x|\sqrt{x^2+1}}$

Exponential:

33. $\dfrac{d}{dx}e^x = e^x$

34. $\dfrac{d}{dx}b^x = b^x(\ln b)$

Logarithmic:

35. $\dfrac{d}{dx}\ln|x| = \dfrac{1}{x}$

36. $\dfrac{d}{dx}\log_b x = \dfrac{1}{x(\ln b)}$

Integration Formulas

Basic Forms

1. $\displaystyle\int u\,dv = uv - \int v\,du$

2. $\displaystyle\int u^n\,du = \frac{1}{n+1}u^{n+1} + C,\, n \neq -1$

3. $\displaystyle\int \frac{du}{u} = \ln|u| + C$ **4.** $\displaystyle\int e^u\,du = e^u + C$

5. $\displaystyle\int a^u\,du = \frac{1}{\ln a}a^u + C$ **6.** $\displaystyle\int \sin u\,du = -\cos u + C$

7. $\displaystyle\int \cos u\,du = \sin u + C$ **8.** $\displaystyle\int \sec^2 u\,du = \tan u + C$

9. $\displaystyle\int \csc^2 u\,du = -\cot u + C$

10. $\displaystyle\int \sec u\tan u\,du = \sec u + C$

11. $\displaystyle\int \csc u\cot u\,du = -\csc u + C$

12. $\displaystyle\int \tan u\,du = -\ln|\cos u| + C$

13. $\displaystyle\int \cot u\,du = \ln|\sin u| + C$

14. $\displaystyle\int \sec u\,du = \ln|\sec u + \tan u| + C$

15. $\displaystyle\int \csc u\,du = \ln|\csc u - \cot u| + C$

16. $\displaystyle\int \frac{du}{\sqrt{a^2 - u^2}} = \sin^{-1}\frac{u}{a} + C$

17. $\displaystyle\int \frac{du}{a^2 + u^2} = \frac{1}{a}\tan^{-1}\frac{u}{a} + C$

18. $\displaystyle\int \frac{du}{u\sqrt{u^2 - a^2}} = \frac{1}{a}\sec^{-1}\left|\frac{u}{a}\right| + C$

19. $\displaystyle\int \frac{du}{a^2 - u^2} = \frac{1}{2a}\ln\left|\frac{u + a}{u - a}\right| + C$

20. $\displaystyle\int \frac{du}{u^2 - a^2} = \frac{1}{2a}\ln\left|\frac{u - a}{u + a}\right| + C$

Forms Involving $\sqrt{a^2 + u^2}$

21. $\displaystyle\int \sqrt{a^2 + u^2}\,du = \frac{u}{2}\sqrt{a^2 + u^2} + \frac{a^2}{2}\ln\left|u + \sqrt{a^2 + u^2}\right| + C$

22. $\displaystyle\int u^2\sqrt{a^2 + u^2}\,du = \frac{u}{8}(a^2 + 2u^2)\sqrt{a^2 + u^2}$
$$- \frac{a^4}{8}\ln\left|u + \sqrt{a^2 + u^2}\right| + C$$

23. $\displaystyle\int \frac{\sqrt{a^2 + u^2}}{u}\,du = \sqrt{a^2 + u^2} - a\ln\left|\frac{a + \sqrt{a^2 + u^2}}{u}\right| + C$

24. $\displaystyle\int \frac{\sqrt{a^2 + u^2}}{u^2}\,du = -\frac{\sqrt{a^2 + u^2}}{u} + \ln\left|u + \sqrt{a^2 + u^2}\right| + C$

25. $\displaystyle\int \frac{du}{\sqrt{a^2 + u^2}} = \ln\left|u + \sqrt{a^2 + u^2}\right| + C$

26. $\displaystyle\int \frac{u^2\,du}{\sqrt{a^2 + u^2}} = \frac{u}{2}\sqrt{a^2 + u^2} - \frac{a^2}{2}\ln\left|u + \sqrt{a^2 + u^2}\right| + C$

27. $\displaystyle\int \frac{du}{u\sqrt{a^2 + u^2}} = -\frac{1}{a}\ln\left|\frac{\sqrt{a^2 + u^2} + a}{u}\right| + C$

28. $\displaystyle\int \frac{du}{u^2\sqrt{a^2 + u^2}} = -\frac{\sqrt{a^2 + u^2}}{a^2 u} + C$

29. $\displaystyle\int \frac{du}{(a^2 + u^2)^{3/2}} = \frac{u}{a^2\sqrt{a^2 + u^2}} + C$

Form Involving $\sqrt{a^2 - u^2}$

30. $\displaystyle\int \sqrt{a^2 - u^2}\,du = \frac{u}{2}\sqrt{a^2 - u^2} + \frac{a^2}{2}\sin^{-1}\frac{u}{a} + C$

31. $\displaystyle\int u^2\sqrt{a^2 - u^2}\,du = \frac{u}{8}(2u^2 - a^2)\sqrt{a^2 - u^2}$
$$+ \frac{a^4}{8}\sin^{-1}\frac{u}{a} + C$$

32. $\displaystyle\int \frac{\sqrt{a^2 - u^2}}{u}\,du = \sqrt{a^2 - u^2} - a\ln\left|\frac{a + \sqrt{a^2 - u^2}}{u}\right| + C$

33. $\displaystyle\int \frac{\sqrt{a^2 - u^2}}{u^2}\,du = -\frac{1}{u}\sqrt{a^2 - u^2} - \sin^{-1}\frac{u}{a} + C$

34. $\displaystyle\int \frac{u^2\,du}{\sqrt{a^2 - u^2}} = -\frac{u}{2}\sqrt{a^2 - u^2} + \frac{a^2}{2}\sin^{-1}\frac{u}{a} + C$

35. $\displaystyle\int \frac{du}{u\sqrt{a^2 - u^2}} = -\frac{1}{a}\ln\left|\frac{a + \sqrt{a^2 - u^2}}{u}\right| + C$

36. $\displaystyle\int \frac{du}{u^2\sqrt{a^2 - u^2}} = -\frac{1}{a^2 u}\sqrt{a^2 - u^2} + C$

37. $\displaystyle\int (a^2 - u^2)^{3/2}\, du = -\frac{u}{8}(2u^2 - 5a^2)\sqrt{a^2 - u^2}$
$$+ \frac{3a^4}{8}\sin^{-1}\frac{u}{a} + C$$

38. $\displaystyle\int \frac{du}{(a^2 - u^2)^{3/2}} = \frac{u}{a^2\sqrt{a^2 - u^2}} + C$

Forms Involving $\sqrt{u^2 - a^2}$

39. $\displaystyle\int \sqrt{u^2 - a^2}\, du = \frac{u}{2}\sqrt{u^2 - a^2}$
$$- \frac{a^2}{2}\ln|u + \sqrt{u^2 - a^2}| + C$$

40. $\displaystyle\int u^2\sqrt{u^2 - a^2}\, du = \frac{u}{8}(2u^2 - a^2)\sqrt{u^2 - a^2}$
$$- \frac{a^4}{8}\ln|u + \sqrt{u^2 - a^2}| + C$$

41. $\displaystyle\int \frac{\sqrt{u^2 - a^2}}{u}\, du = \sqrt{u^2 - a^2} - a\cos^{-1}\frac{a}{u} + C$

42. $\displaystyle\int \frac{\sqrt{u^2 - a^2}}{u^2}\, du = -\frac{\sqrt{u^2 - a^2}}{u}$
$$+ \ln|u + \sqrt{u^2 - a^2}| + C$$

43. $\displaystyle\int \frac{du}{\sqrt{u^2 - a^2}} = \ln|u + \sqrt{u^2 - a^2}| + C$

44. $\displaystyle\int \frac{u^2\, du}{\sqrt{u^2 - a^2}} = \frac{u}{2}\sqrt{u^2 - a^2}$
$$+ \frac{a^2}{2}\ln|u + \sqrt{u^2 - a^2}| + C$$

45. $\displaystyle\int \frac{du}{u^2\sqrt{u^2 - a^2}} = \frac{\sqrt{u^2 - a^2}}{a^2 u} + C$

46. $\displaystyle\int \frac{du}{(u^2 - a^2)^{3/2}} = -\frac{u}{a^2\sqrt{u^2 - a^2}} + C$

Forms Involving $a + bu$

47. $\displaystyle\int \frac{u\, du}{a + bu} = \frac{1}{b^2}(a + bu - a\ln|a + bu|) + C$

48. $\displaystyle\int \frac{u^2\, du}{a + bu} = \frac{1}{2b^3}[(a + bu)^2 - 4a(a + bu)$
$$+ 2a^2\ln|a + bu|] + C$$

49. $\displaystyle\int \frac{du}{u(a + bu)} = \frac{1}{a}\ln\left|\frac{u}{a + bu}\right| + C$

50. $\displaystyle\int \frac{du}{u^2(a + bu)} = -\frac{1}{au} + \frac{b}{a^2}\ln\left|\frac{a + bu}{u}\right| + C$

51. $\displaystyle\int \frac{u\, du}{(a + bu)^2} = \frac{a}{b^2(a + bu)} + \frac{1}{b^2}\ln|a + bu| + C$

52. $\displaystyle\int \frac{du}{u(a + bu)^2} = \frac{1}{a(a + bu)} - \frac{1}{a^2}\ln\left|\frac{a + bu}{u}\right| + C$

53. $\displaystyle\int \frac{u^2\, du}{(a + bu)^2} = \frac{1}{b^3}\left(a + bu - \frac{a^2}{a + bu} - 2a\ln|a + bu|\right) + C$

54. $\displaystyle\int u\sqrt{a + bu}\, du = \frac{2}{15b^2}(3bu - 2a)(a + bu)^{3/2} + C$

55. $\displaystyle\int \frac{u\, du}{\sqrt{a + bu}} = \frac{2}{3b^2}(bu - 2a)\sqrt{a + bu} + C$

56. $\displaystyle\int \frac{u^2\, du}{\sqrt{a + bu}} = \frac{2}{15b^3}(8a^2 + 3b^2u^2 - 4abu)\sqrt{a + bu} + C$

57. $\displaystyle\int \frac{du}{u\sqrt{a + bu}} = \frac{1}{\sqrt{a}}\ln\left|\frac{\sqrt{a + bu} - \sqrt{a}}{\sqrt{a + bu} + \sqrt{a}}\right| + C,\ \text{if } a > 0$
$$= \frac{2}{\sqrt{-a}}\tan^{-1}\sqrt{\frac{a + bu}{-a}} + C,\ \text{if } a < 0$$

58. $\displaystyle\int \frac{\sqrt{a + bu}}{u}\, du = 2\sqrt{a + bu} + a\int \frac{du}{u\sqrt{a + bu}}$

59. $\displaystyle\int \frac{\sqrt{a + bu}}{u^2}\, du = -\frac{\sqrt{a + bu}}{u} + \frac{b}{2}\int \frac{du}{u\sqrt{a + bu}}$

60. $\displaystyle\int u^2\sqrt{a + bu}\, du = \frac{2u^n(a + bu)^{3/2}}{b(2n + 3)}$
$$- \frac{2na}{b(2n + 3)}\int u^{n-1}\sqrt{a + bu}\, du$$

61. $\displaystyle\int \frac{u^n\, du}{\sqrt{a + bu}} = \frac{2u^n\sqrt{a + bu}}{b(2n + 1)} - \frac{2na}{b(2n + 1)}\int \frac{u^{n-1}\, du}{\sqrt{a + bu}}$

62. $\displaystyle\int \frac{du}{u^n\sqrt{a + bu}} = -\frac{\sqrt{a + bu}}{a(n - 1)u^{n-1}}$
$$- \frac{b(2n - 3)}{2a(n - 1)}\int \frac{du}{u^{n-1}\sqrt{a + bu}}$$

Trigonometric Forms

63. $\displaystyle\int \sin^2 u\, du = \frac{1}{2}u - \frac{1}{4}\sin 2u + C$

64. $\displaystyle\int \cos^2 u\, du = \frac{1}{2}u + \frac{1}{4}\sin 2u + C$

65. $\displaystyle\int \tan^2 u\, du = \tan u - u + C$

66. $\displaystyle\int \cot^2 u\, du = -\cot u - u + C$

67. $\displaystyle\int \sin^3 u\, du = -\frac{1}{3}(2 + \sin^2 u)\cos u + C$

68. $\displaystyle\int \cos^3 u\, du = \frac{1}{3}(2 + \cos^2 u)\sin u + C$

69. $\displaystyle\int \tan^3 u\, du = \frac{1}{2}\tan^2 u + \ln|\cos u| + C$

70. $\displaystyle\int \cot^2 u\, du = -\frac{1}{2}\cot^2 u - \ln|\sin u| + C$

71. $\displaystyle\int \sec^3 u\, du = \frac{1}{2}\sec u\tan u + \frac{1}{2}\ln|\sec u + \tan u| + C$

72. $\int \csc^3 u \, du = -\frac{1}{2}\csc u \cot u + \frac{1}{2}\ln|\csc u - \cot u| + C$

73. $\int \sin^n u \, du = -\frac{1}{n}\sin^{n-1}u\cos u + \frac{n-1}{n}\int \sin^{n-2}u \, du$

74. $\int \cos^n u \, du = \frac{1}{n}\cos^{n-1}u\sin u + \frac{n-1}{n}\int \cos^{n-2}u \, du$

75. $\int \tan^n u \, du = \frac{1}{n-1}\tan^{n-1}u - \int \tan^{n-2}u \, du$

76. $\int \cot^n u \, du = \frac{-1}{n-1}\cot^{n-1}u - \int \cot^{n-2}u \, du$

77. $\int \sec^n u \, du = \frac{1}{n-1}\tan u \sec^{n-2}u + \frac{n-2}{n-1}\int \sec^{n-2}u \, du$

78. $\int \csc^n u \, du = \frac{-1}{n-1}\cot u \csc^{n-2}u + \frac{n-2}{n-1}\int \csc^{n-2}u \, du$

79. $\int \sin au \sin bu \, du = \frac{\sin(a-b)u}{2(a-b)} - \frac{\sin(a+b)u}{2(a+b)} + C$

80. $\int \cos au \cos bu \, du = \frac{\sin(a-b)u}{2(a-b)} + \frac{\sin(a+b)u}{2(a+b)} + C$

81. $\int \sin au \cos bu \, du = -\frac{\cos(a-b)u}{2(a-b)} - \frac{\cos(a+b)u}{2(a+b)} + C$

82. $\int u \sin u \, du = \sin u - u\cos u + C$

83. $\int u \cos u \, du = \cos u + u\sin u + C$

84. $\int u^n \sin u \, du = -u^n\cos u + n\int u^{n-1}\cos u \, du$

85. $\int u^n \cos u \, du = u^n\sin u - n\int u^{n-1}\sin u \, du$

86. $\int \sin^n u \cos^m u \, du = -\frac{\sin^{n-1}u\cos^{m+1}u}{n+m}$
$$+ \frac{n-1}{n+m}\int \sin^{n-1}u\cos^m u \, du$$
$$= \frac{\sin^{n+1}u\cos^{m-1}u}{n+m}$$
$$+ \frac{m-1}{n+m}\int \sin^n u\cos^{m-2}u \, du$$

87. $\int \frac{du}{1-\sin au} = \frac{1}{a}\tan\left(\frac{\pi}{4} + \frac{au}{2}\right) + C$

88. $\int \frac{du}{1+\sin au} = -\frac{1}{a}\tan\left(\frac{\pi}{4} - \frac{au}{2}\right) + C$

89. $\int \frac{u\,du}{1-\sin au} = \frac{u}{a}\tan\left(\frac{\pi}{4} + \frac{au}{2}\right)$
$$+ \frac{2}{a^2}\ln\left|\sin\left(\frac{\pi}{4} - \frac{au}{2}\right)\right| + C$$

Inverse Trigonometric Forms

90. $\int \sin^{-1}u \, du = u\sin^{-1}u + \sqrt{1-u^2} + C$

91. $\int \cos^{-1}u \, du = u\cos^{-1}u - \sqrt{1-u^2} + C$

92. $\int \tan^{-1}u \, du = u\tan^{-1}u - \frac{1}{2}\ln(1+u^2) + C$

93. $\int u\sin^{-1}u \, du = \frac{2u^2-1}{4}\sin^{-1}u + \frac{u\sqrt{1-u^2}}{4} + C$

94. $\int u\cos^{-1}u \, du = \frac{2u^2-1}{4}\cos^{-1}u - \frac{u\sqrt{1-u^2}}{4} + C$

95. $\int u\tan^{-1}u \, du = \frac{u^2+1}{2}\tan^{-1}u - \frac{u}{2} + C$

96. $\int u^n\sin^{-1}u \, du = \frac{1}{n+1}\left[u^{n+1}\sin^{-1}u\right.$
$$\left. - \int \frac{u^{n+1}\,du}{\sqrt{1-u^2}}\right], \quad n \neq -1$$

97. $\int u^n\cos^{-1}u \, du = \frac{1}{n+1}\left[u^{n+1}\cos^{-1}u\right.$
$$\left. + \int \frac{u^{n+1}\,du}{\sqrt{1-u^2}}\right], \quad n \neq -1$$

98. $\int u^n\tan^{-1}u \, du = \frac{1}{n+1}\left[u^{n+1}\tan^{-1}u\right.$
$$\left. - \int \frac{u^{n+1}\,du}{1+u^2}\right], \quad n \neq -1$$

Exponential and Logarithmic Forms

99. $\int ue^{au} \, du = \frac{1}{a^2}(au-1)e^{au} + C$

100. $\int u^n e^{au} \, du = \frac{1}{a}u^n e^{au} - \frac{n}{a}\int u^{n-1}e^{au} \, du$

101. $\int e^{au}\sin bu \, du = \frac{e^{au}}{a^2+b^2}(a\sin bu - b\cos bu) + C$

102. $\int e^{au}\cos bu \, du = \frac{e^{au}}{a^2+b^2}(a\cos bu + b\sin bu) + C$

103. $\int \ln u \, du = u\ln u - u + C$

104. $\int \frac{1}{u\ln u}\, du = \ln|\ln u| + C$

105. $\int u^n\ln u \, du = \frac{u^{n+1}}{(n+1)^2}[(n+1)\ln u - 1] + C$

106. $\int u^m\ln^n u \, du = \frac{u^{m+1}\ln^n u}{m+1}$
$$- \frac{n}{m+1}\int u^m\ln^{n-1}u \, du, \quad m \neq -1$$

107. $\int \ln(u^2 + a^2)\, du = u\ln(u^2 + a^2) - 2u + 2a\tan^{-1}\dfrac{u}{a} + C$

108. $\int \ln|u^2 - a^2|\, du = u\ln|u^2 - a^2| - 2u + a\ln\left|\dfrac{u+a}{u-a}\right| + C$

109. $\int \dfrac{du}{a + be^u} = \dfrac{u}{a} - \dfrac{1}{a}\ln|a + be^u| + C$

Hyperbolic Forms

110. $\int \sinh u\, du = \cosh u + C$

111. $\int \cosh u\, du = \sinh u + C$

112. $\int \tanh u\, du = \ln(\cosh u) + C$

113. $\int \coth u\, du = \ln|\sinh u| + C$

114. $\int \operatorname{sech} u\, du = \tan^{-1}(\sinh u) + C$

115. $\int \operatorname{csch} u\, du = \ln|\tanh\tfrac{1}{2}u| + C$

116. $\int \operatorname{sech}^2 u\, du = \tanh u + C$

117. $\int \operatorname{csch}^2 u\, du = -\coth u + C$

118. $\int \operatorname{sech} u\tanh u\, du = -\operatorname{sech} u + C$

119. $\int \operatorname{csch} u\coth u\, du = -\operatorname{csch} u + C$

Forms Involving $\sqrt{2au - u^2}$

120. $\int \sqrt{2au - u^2}\, du = \dfrac{u - a}{2}\sqrt{2au - u^2}$
$\qquad + \dfrac{a^2}{2}\cos^{-1}\left(\dfrac{a-u}{a}\right) + C$

121. $\int u\sqrt{2au - u^2}\, du = \dfrac{2u^2 - au - 3a^2}{6}\sqrt{2au - u^2}$
$\qquad + \dfrac{a^3}{2}\cos^{-1}\left(\dfrac{a-u}{a}\right) + C$

122. $\int \dfrac{\sqrt{2au - u^2}}{u}\, du = \sqrt{2au - u^2} + a\cos^{-1}\left(\dfrac{a-u}{a}\right) + C$

123. $\int \dfrac{\sqrt{2au - u^2}}{u^2}\, du = -\dfrac{2\sqrt{2au - u^2}}{u} - \cos^{-1}\left(\dfrac{a-u}{a}\right) + C$

124. $\int \dfrac{du}{\sqrt{2au - u^2}} = \cos^{-1}\left(\dfrac{a-u}{a}\right) + C$

125. $\int \dfrac{u\, du}{\sqrt{2au - u^2}} = -\sqrt{2au - u^2} + a\cos^{-1}\left(\dfrac{a-u}{a}\right) + C$

126. $\int \dfrac{u^2\, du}{\sqrt{2au - u^2}} = -\dfrac{(u + 3a)}{2}\sqrt{2au - u^2}$
$\qquad + \dfrac{3a^2}{2}\cos^{-1}\left(\dfrac{a-u}{a}\right) + C$

127. $\int \dfrac{du}{u\sqrt{2ua - u^2}} = -\dfrac{\sqrt{2au - u^2}}{au} + C$

Some Definite Integrals

128. $\displaystyle\int_0^{\pi/2} \sin^{2n}x\, dx = \int_0^{\pi/2} \cos^{2n}x\, dx$
$\qquad = \dfrac{\pi}{2}\dfrac{1 \cdot 3 \cdot 5 \cdots (2n-1)}{2 \cdot 4 \cdot 6 \cdots 2n},\ n = 1, 2, 3, \ldots$

129. $\displaystyle\int_0^{\pi/2} \sin^{2n+1}x\, dx = \int_0^{\pi/2} \cos^{2n+1}x\, dx$
$\qquad = \dfrac{2 \cdot 4 \cdot 6 \cdots 2n}{1 \cdot 3 \cdot 5 \cdots (2n+1)},\ n = 1, 2, 3, \ldots$

Answers to Test Yourself

1. false

2. true

3. false

4. true

5. 12

6. -243

7. $\dfrac{3x^3 + 8x}{\sqrt{x^2 + 4}}$

8. $2\left(x + \frac{3}{2}\right)^2 + \frac{1}{2}$

9. (a) $0, 7$ (b) $-1 + \sqrt{6}, -1 - \sqrt{6}$
    (c) 1 (d) 1

10. (a) $(5x + 1)(2x - 3)$ (b) $x^2(x + 3)(x - 5)$
    (c) $(x - 3)(x^2 + 3x + 9)$ (d) $(x - 2)(x + 2)(x^2 + 4)$

11. false

12. false

13. true

14. $6; \ -6$

15. $-a + 5$

16. (a), (b), (d), (e), (g), (h), (i), (l)

17. (*i*) (d);   (*ii*) (c),   (*iii*) (a);   (*iv*) (b)

18. (a) $-2 < x < 2$;   (b) $|x| < 2$

19.

20. $(-\infty, -2)\cup\left(\frac{8}{3}, \infty\right)$

21. $(-\infty, -5]\cup[3, \infty)$

22. $(-\infty, -2)\cup[0, 1]$

23. fourth

24. $(5, -7)$

25. $-12; 9$

26. (a) $(1, -5)$ (b) $(-1, 5)$ (c) $(-1, -5)$

27. $(-2, 0), (0, -4), (0, 4)$

28. second and fourth

29. $x = 6$ or $x = -4$

30. $x^2 + y^2 = 25$

31. $d(P_1, P_2) + d(P_2, P_3) = d(P_1, P_3)$

32. (c)

33. false

34. -27

35. 8

36. $\dfrac{2}{3}; (-9, 0); (0, 6)$

37. $y = -5x + 3$

38. $y = 2x - 14$

39. $y = -\dfrac{1}{3}x + 3$

40. $y = -\dfrac{5}{8}x$

41. $x - \sqrt{3}y + 4\sqrt{3} - 7 = 0$

42. (*i*) (g); (*ii*) (e); (*iii*) (h); (*iv*) (a); (*v*) (b); (*vi*) (f);
    (*vii*) (d); (*viii*) (c)

43. false

44. false

45. $4\pi/3$

46. 15

47. 0.23

48. $\cos t = -\dfrac{2\sqrt{2}}{3}$

49. $\sin\theta = \frac{3}{5}; \ \cos\theta = \frac{4}{5}; \ \tan\theta = \frac{3}{4}; \ \cot\theta = \frac{4}{3}; \ \sec\theta = \frac{5}{4};$
    $\csc\theta = \frac{5}{3}$

50. $b = 10\tan\theta, c = 10\sec\theta$

51. $k = 10\ln 5$

52. $4 = 64^{1/3}$

53. $\log_b 125$

54. approximately 2.3347

55. 1000

56. true

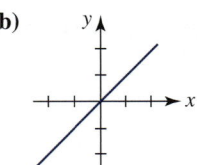
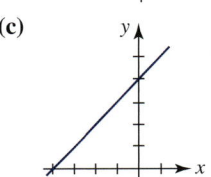
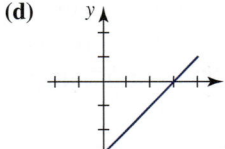
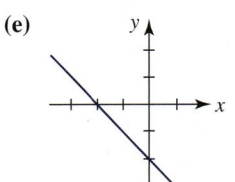
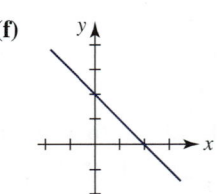
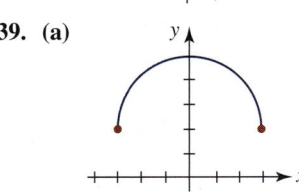
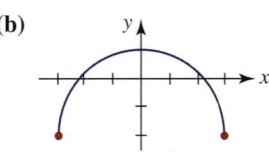
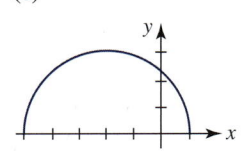
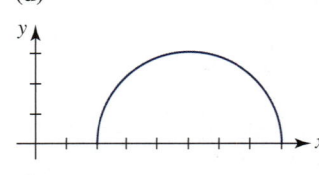
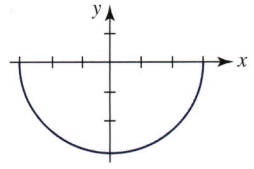
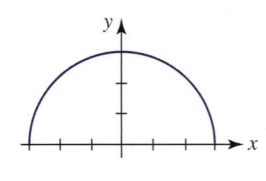

Answers to Selected Odd-Numbered Problems

Exercises 1.1, Page 8

1. 24; 2; 8; 35, **3.** 0; 1; 2; $\sqrt{6}$

5. $-\dfrac{3}{2}$; 0; $\dfrac{3}{2}$; $\sqrt{2}$

7. $-2x^2 + 3x$; $-8a^2 + 6a$; $-2a^4 + 3a^2$; $-50x^2 - 15x$; $-8a^2 - 2a + 1$; $-2x^2 - 4xh - 2h^2 + 3x + 3h$

9. $-2, 2$ **11.** $\left[\frac{1}{2}, \infty\right)$

13. $(-\infty, 1)$ **15.** $\{x \mid x \neq 0, x \neq 3\}$

17. $\{x \mid x \neq 5\}$ **19.** $(-\infty, \infty)$

21. $[-5, 5]$ **23.** $(-\infty, 0] \cup [5, \infty)$

25. $(-2, 3]$ **27.** not a function

29. function

31. domain: $[-4, 4]$; range: $[0, 5]$

33. domain: $[1, 9]$; range: $[1, 6]$

35. $(8, 0), (0, -4)$ **37.** $\left(\frac{3}{2}, 0\right), \left(\frac{5}{2}, 0\right), (0, 15)$

39. $(-1, 0), (2, 0), (0, 0)$ **41.** $\left(0, -\frac{1}{4}\right)$

43. $(-2, 0), (2, 0), (0, 3)$

45. 0; -3.4; 0.3; 2; 3.8; 2.9; $(0, 2)$

47. 3.6; 2; 3.3; 4.1; 2; -4.1; $(-3.2, 0), (2.3, 0), (3.8, 0)$

49. $f_1(x) = \sqrt{x + 5}, f_2(x) = -\sqrt{x + 5}$; $[-5, \infty)$

51. (a) 2; 6; 120; 5040 (c) 5; 42 (d) $(n + 1)(n + 2)(n + 3)$

Exercises 1.2, Page 18

1. $-2x + 13$; $6x - 3$; $-8x^2 - 4x + 40$; $\dfrac{2x + 5}{-4x + 8}, x \neq 2$

3. $\dfrac{x^2 + x + 1}{x(x + 1)}$; $\dfrac{x^2 - x - 1}{x(x + 1)}$; $\dfrac{1}{x + 1}$; $\dfrac{x^2}{x + 1}, x \neq 0, x \neq -1$

5. $2x^2 + 5x - 7$; $-x + 1$; $x^4 + 5x^3 - x^2 - 17x + 12$; $\dfrac{x + 3}{x + 4}, x \neq 1, x \neq -4$

7. the interval $[1, 2]$ **9.** the interval $[1, 2)$

11. $3x + 16$; $3x + 4$ **13.** $x^6 + 2x^5 + x^4$; $x^6 + x^4$

15. $\dfrac{3x + 3}{x}$; $\dfrac{3}{3 + x}$ **17.** $(-\infty, -1] \cup [1, \infty)$

19. $\left[-\sqrt{5}, \sqrt{5}\right]$ **21.** $128x^9$; $\dfrac{1}{4x^9}$

23. $36x^2 - 36x + 15$ **25.** $-2x + 9$

27. $f(x) = 2x^2 - x, g(x) = x^2$ **29.** $(-2, 3), (3, -2)$

31. $(-8, 1), (-3, -4)$ **33.** $(-6, 2), (-1, -3)$

35. $(2, 1), (-3, -4)$

37. (a) (b) (c) (d) (e) (f)

39. (a) (b) (c) (d) (e) (f)

41. (a)
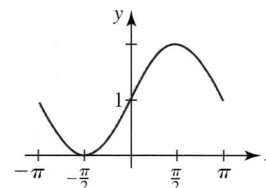

(b)
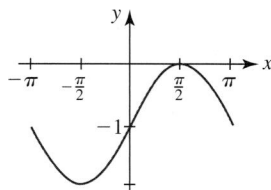

(c)
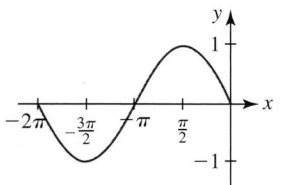

(d)
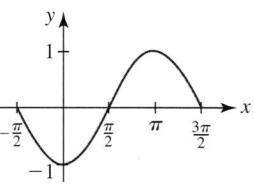

(e)
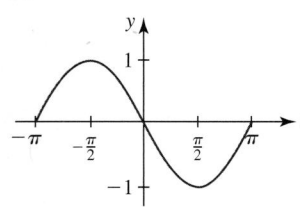

(f)
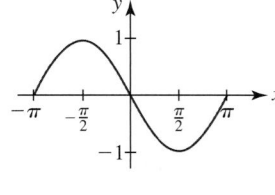

(g)
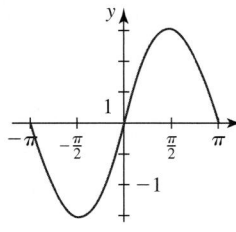

(h)
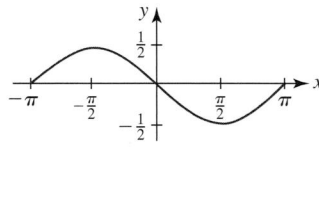

43. $y = (x-1)^3 + 5$

45. $y = -(x+7)^4$

47.
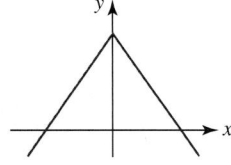

49. $10, 8, -1, 2, 0$

51.
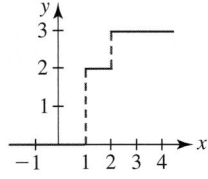

53. $y = 2 - 3U(x-2) + U(x-3)$

55.
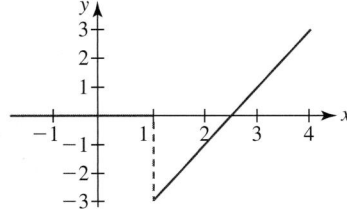

Exercises 1.3, Page 28

1. $y = \frac{2}{3}x + \frac{4}{3}$

3. $y = 2$

5. $y = -x + 3$

7. $\frac{3}{4}$; $(-4, 0), (0, 3)$;
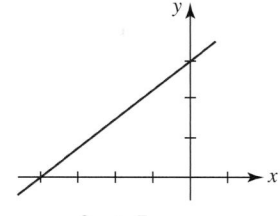

9. $\frac{2}{3}$; $\left(\frac{9}{2}, 0\right), (0, -3)$;
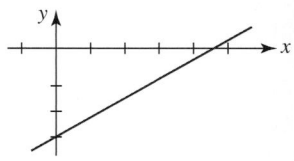

11. $y = -2x + 7$

13. $y = -3x - 2$

15. $y = -4x + 11$

17. $f(x) = \frac{1}{2}x + \frac{11}{2}$

19. $y = x + 3$

21. (a) $(0, 0), (-5, 0)$
(b) $y = \left(x + \frac{5}{2}\right)^2 - \frac{25}{4}$
(c) $\left(-\frac{5}{2}, -\frac{25}{4}\right)$; $x = -\frac{5}{2}$
(d)
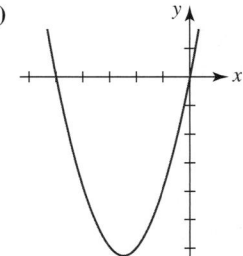

(e) $\left[-\frac{25}{4}, \infty\right)$
(f) $\left[-\frac{5}{2}, \infty\right)$; $\left(-\infty, -\frac{5}{2}\right]$

23. (a) $(-1, 0), (3, 0), (0, 3)$
(b) $y = -(x-1)^2 + 4$
(c) $(1, 4)$; $x = 1$
(d)
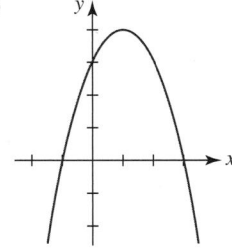

(e) $(-\infty, 4]$
(f) $(-\infty, 1]$; $[1, \infty)$

25. (a) $(1, 0), (2, 0), (0, 2)$
(b) $y = \left(x - \frac{3}{2}\right)^2 - \frac{1}{4}$
(c) $\left(\frac{3}{2}, -\frac{1}{4}\right)$; $x = \frac{3}{2}$
(d)
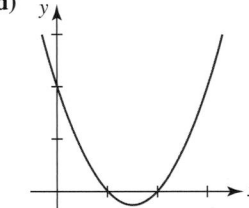

(e) $\left[-\frac{1}{4}, \infty\right)$
(f) $\left[\frac{3}{2}, \infty\right)$; $\left(-\infty, \frac{3}{2}\right]$

27. graph is shifted horizontally 10 units to the right

29. graph is compressed vertically, followed by a reflection in the x-axis, followed by a horizontal shift of 4 units to the left, followed by a vertical shift of 9 units upward

31. graph is shifted horizontally 6 units to the left, followed by a vertical shift of 4 units downward

33.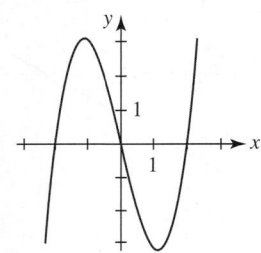

35.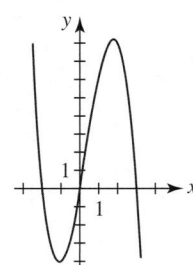

37.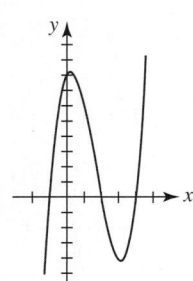

39.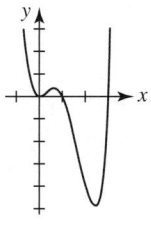

41.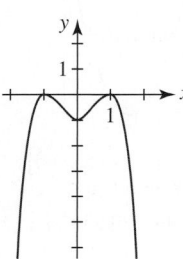

43. (f)

45. (e)

47. (b)

49. asymptotes: $x = -\frac{3}{2}, y = 2$;   intercepts: $\left(\frac{9}{4}, 0\right), (0, -3)$;

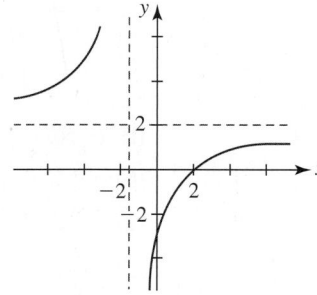

51. asymptotes: $x = 1, y = 0$;   intercepts: $(0, 1)$;

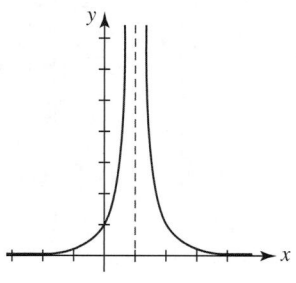

53. asymptotes: $x = -1, x = 1, y = 0$;   intercepts: $(0, 0)$;

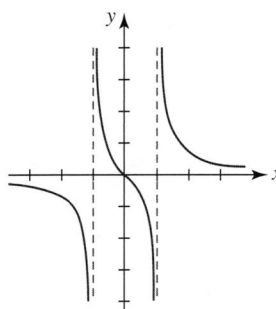

55. asymptotes: $x = 0, y = -1$;   intercepts: $(-1, 0), (1, 0)$;

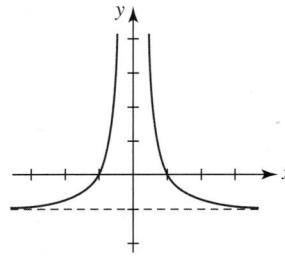

57. asymptotes: $x = 0, y = x$;   intercepts: $(-3, 0), (3, 0)$;

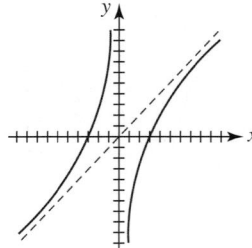

59. asymptotes: $x = -2, y = x - 2$;   intercepts: $(0, 0)$;

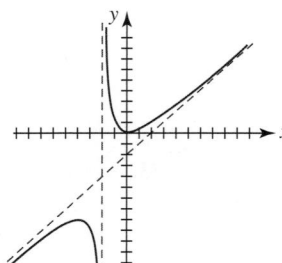

61. asymptotes: $x = 1, y = x - 1$;   intercepts: $(-1, 0), (3, 0), (0, 3)$;

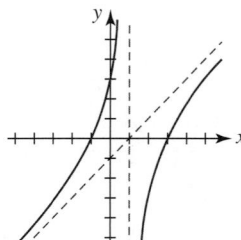

63. -1 is in the range of f, but 2 is not in the range of f

65. $T_F = \dfrac{9}{5}T_C + 32$

67. 1680;   approximately 35.3 years

69. $t = 0$ and $t = 6$;

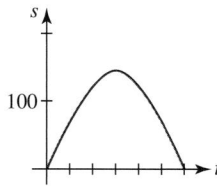

Exercises 1.4, Page 35

1.
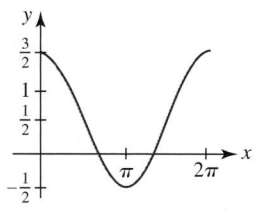

3.
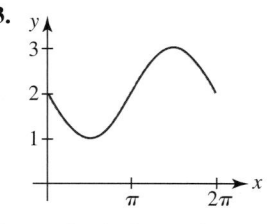

5.
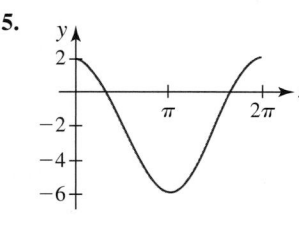

7. amplitude: 4; period: 2;
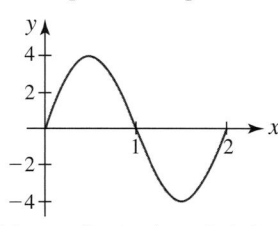

9. amplitude: 3; period: 1;
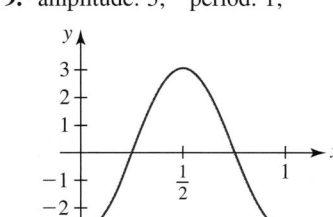

11. amplitude: 4; period: 2π;
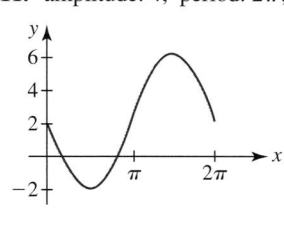

13. amplitude: 1; period: 3π;
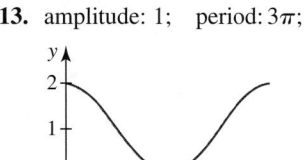

15. $y = -3\sin x$

17. $y = 1 - 3\cos x$

19. $y = 3\sin 2x$

21. $y = \dfrac{1}{2}\cos \pi x$

23. $y = -\sin \pi x$

25. amplitude: 1; period: 2π; phase shift: $\pi/6$;
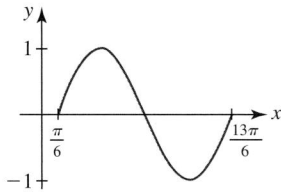

27. amplitude: 1; period: 2π; phase shift: $\pi/4$;
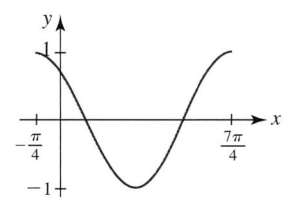

29. amplitude: 4; period: π; phase shift: $3\pi/4$;
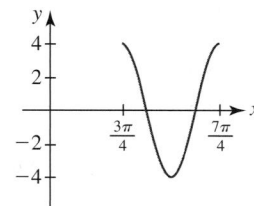

31. amplitude: 3; period: 4π; phase shift: $2\pi/3$;
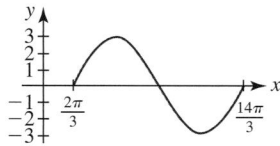

33. amplitude: 4; period: 6; phase shift: 1;
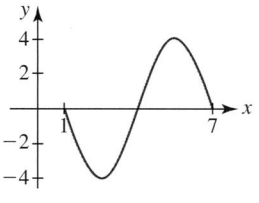

35. $y = 5\sin\left(\pi x - \dfrac{\pi}{2}\right)$

37. $(\pi/2, 0)$; $(\pi/2 + 2n\pi, 0)$, where n is an integer

39. $(n, 0)$, where n is an integer

41. $((2n + 1)\pi, 0)$, where n is an integer

43. $(\pi/4 + n\pi, 0)$, where n is an integer

45. period: 1; x-intercepts: $(n, 0)$, where n is an integer; asymptotes: $x = \dfrac{1}{2}(2n + 1)$, where n is an integer;
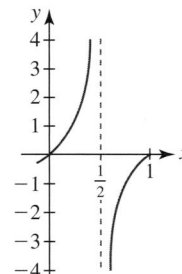

47. period: $\dfrac{\pi}{2}$; x-intercepts: $\left(\dfrac{1}{4}(2n + 1)\pi, 0\right)$, where n is an integer; asymptotes: $x = n\pi/2$, where n is an integer;
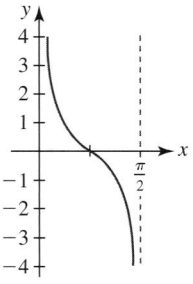

49. period: 2π; x-intercepts: $(\pi/2 + 2n\pi, 0)$, where n is an integer; asymptotes: $x = 3\pi/2 + 2n\pi$, where n is an integer;

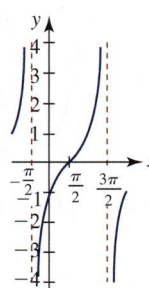

51. period: 1; x-intercepts: $\left(\frac{1}{4} + n, 0\right)$, where n is an integer; asymptotes: $x = n$, where n is an integer;

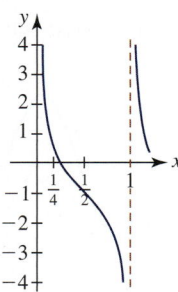

53. period: 2; asymptotes: $x = n$, where n is an integer;

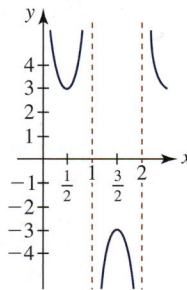

55. period: $2\pi/3$; asymptotes: $x = n\pi/3$, where n is an integer;

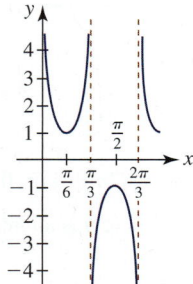

57.

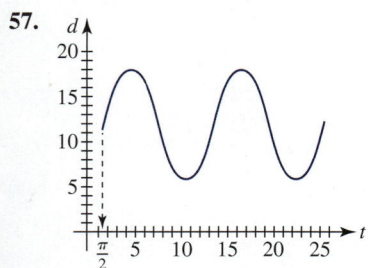

59. **(a)** 978.0309 cm/s² **(b)** 983.21642 cm/s²
(c) 980.61796 cm/s²

Exercises 1.5, Page 46

1. because $f(0) = 1$ and $f(5) = 1$ **3.** not one-to-one
5. one-to-one   **7.** one-to-one

9. $f^{-1}(x) = \sqrt[3]{\dfrac{x-7}{3}}$ **11.** $f^{-1}(x) = \dfrac{2-x}{1-x}$

15. domain: $[0, \infty)$;   range: $[-2, \infty)$

17. domain: $(-\infty, 0) \cup (0, \infty)$;   range: $(-\infty, -3) \cup (-3, \infty)$
19. $(20, 2)$ **21.** $x = 12$

23.

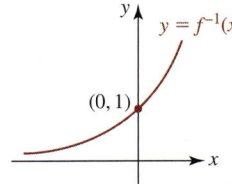

25.

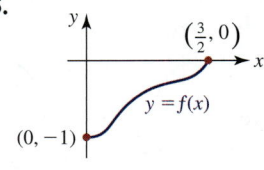

27. $f(x) = (5 - 2x)^2, x \geq \frac{5}{2}$; $f^{-1}(x) = \frac{1}{2}(5 - \sqrt{x})$

29. $f(x) = x^2 + 2x + 4, x \geq -1$; $f^{-1}(x) = -1 + \sqrt{x-3}$

33. $3\pi/4$ **35.** $\pi/4$

37. $3\pi/4$ **39.** $-\pi/3$

41. $\dfrac{4}{5}$ **43.** 2

45. $4\sqrt{2}/9$ **47.** $\sqrt{3}(2 + \sqrt{10})/9$

49. $\sqrt{1 - x^2}$ **51.** $\sqrt{1 + x^2}$

57. $\cos t = \sqrt{5}/5, \tan t = -2, \cot t = -\dfrac{1}{2}, \sec t = \sqrt{5},$
$\csc t = -\sqrt{5}/2$

63. **(a)** $\pi/4$ **(b)** 0.942 radian $\approx 53.97°$

Exercises 1.6, Page 53

1. $(0, 1); y = 0;$ **3.** $(0, -1); y = 0;$

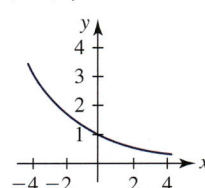   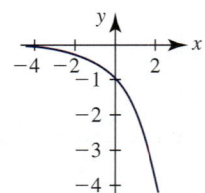

5. $(0, -4); y = -5;$ **7.** $f(x) = 6^x$

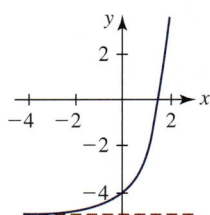

9. $f(x) = e^{-2x}$ **11.** $x > 4$

13. $x < 2$ **15.**

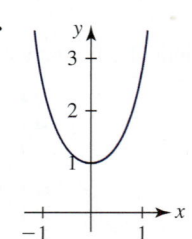

17.
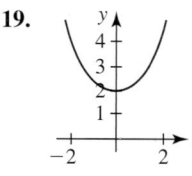

19.
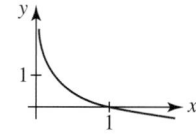

21.
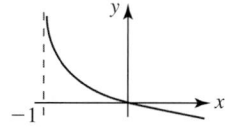

23. $-\dfrac{1}{2} = \log_4\dfrac{1}{2}$

25. $4 = \log_{10} 10{,}000$

27. $2^7 = 128$

29. $\left(\sqrt{3}\right)^8 = 81$

31. $f(x) = \log_7 x$

33. e

35. 36

37. $\dfrac{1}{7}$

39. $(0, \infty);\quad (1, 0);\quad x = 0;$
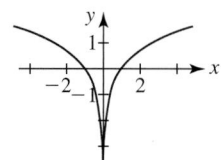

41. $(-1, \infty);\quad (0, 0);\quad x = -1;$

43. the interval $(-3, 3)$

45. $(-1, 0), (1, 0);\quad x = 0;$

47. $\ln(x^2 - 2)$

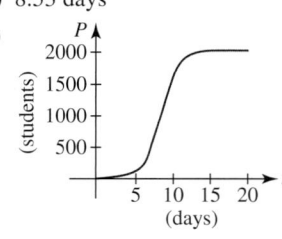

49. 0

51. $10\ln x + \dfrac{1}{2}\ln(x^2 + 5) - \dfrac{1}{3}\ln(8x^3 + 2)$

53. $5\ln(x^3 - 3) + 8\ln(x^4 + 3x^2 + 1) - \dfrac{1}{2}\ln x - 9\ln(7x + 5)$

55. $\log_6 51 = \dfrac{\ln 51}{\ln 6} \approx 2.1944$

57. $-5 + \dfrac{\ln 9}{\ln 2} \approx 1.8301$

59. $\dfrac{1 + \ln 2}{-1 + \ln 5} \approx 2.7782$

61. 3

63. (a) $P(t) = P_0 e^{0.3466t}$ (b) $5.66P_0$ (c) 8.64 h

65. (a) 82 (b) 8.53 days (c) 2000 (d)

Exercises 1.7, Page 59

1. $S(x) = x + \dfrac{50}{x};\quad (0, \infty)$

3. $S(x) = 3x^2 - 4x + 2;\quad [0, 1]$

5. $A(x) = 100x - x^2;\quad [0, 100]$

7. $A(x) = 2x - \dfrac{1}{2}x^2;\quad [0, 4]$

9. $d(x) = \sqrt{2x^2 + 8};\quad (-\infty, \infty)$

11. $P(A) = 4\sqrt{A};\quad (0, \infty)$

13. $d(C) = C/\pi;\quad (0, \infty)$

15. $A(h) = \dfrac{1}{\sqrt{3}}h^2;\quad (0, \infty)$

17. $A(x) = \dfrac{1}{4\pi}x^2;\quad (0, \infty)$

19. $C(x) = 8x + \dfrac{3200}{x};\quad (0, \infty)$

21. $S(w) = 3w^2 + \dfrac{1200}{w};\quad (0, \infty)$

23. $d(t) = 20\sqrt{13t^2 + 8t + 4};\quad (0, \infty)$

25. $V(h) = \begin{cases} 120h^2, & 0 \le h < 5 \\ 1200h - 3000, & 5 \le h \le 8 \end{cases};\quad [0, 8]$

27. $h(\theta) = 300\tan\theta;\quad (0, \pi/2)$

29. $L(\theta) = 3\csc\theta + 4\sec\theta;\quad (0, \pi/2)$

31. $\theta(x) = \tan^{-1}(1/x) - \tan^{-1}(1/2x);\quad (0, \infty)$

Chapter 1 in Review, Page 61

A. 1. false **3.** true

5. false **7.** true

9. false **11.** true

13. true **15.** true

17. true **19.** true

B. 1. $[-2, 0) \cup (0, \infty)$ **3.** $(-8, 6)$

5. $(1, 0);\quad (0, 0), (5, 0)$ **7.** $\left(0, -\dfrac{4}{5}\right)$

9. 6 **11.** 0

13. $(3, 5)$ **15.** $\log_3 5 = \dfrac{\ln 5}{\ln 3}$

17. $\dfrac{1}{9}$ **19.** $y = \ln x$

C. 1. (a) 3 (b) 0 (c) -2 (d) 0 (e) 2.5
  (f) 2 (g) 1 (h) 0 (i) 3 (j) 4

3. 1 and 8 are in the range; 5 is not in the range

5. $-3x^2 + 4x - 3xh - h^2 + 2h - 1$

7. (f) **9.** (d)

11. (h) **13.** (c)

15. (b) **17.** $\dfrac{3^{1-h} - 3}{h}$

19. (a) ab (b) b/a (c) $1/b$

21. $f(x) = 5e^{\left(-\frac{1}{6}\ln 5\right)x} = 5e^{-0.2682x}$ **23.** $f(x) = 5 + \left(\frac{1}{2}\right)^x$

25. (b) **27.** (d)

29. (c)

31. (a) $V = 6l^3$ (b) $V = \dfrac{2}{9}w^3$ (c) $V = \dfrac{3}{4}h^3$

33. $V(\theta) = 360 + 75\cot\theta$

35. $A(\phi) = 100\cos\phi + 50\sin 2\phi$ **37.** $V(x) = 2\sqrt{3}(1 - x^2)$

Exercises 2.1, Page 72

1. 8

3. does not exist

5. 2

7. does not exist

9. 0

11. 3

13. 0

15. (a) 1 (b) -1 (c) 2 (d) does not exist

17. (a) 2 (b) -1 (c) -1 (d) -1

19. correct

21. $\displaystyle\lim_{x\to 1^-}\sqrt{1 - x} = 0$

23. $\displaystyle\lim_{x\to 0^+}\lfloor x\rfloor = 0$

25. correct

27. $\displaystyle\lim_{x\to 3^-}\sqrt{9 - x^2} = 0$

29. (a) -1 (b) 0 (c) -3 (d) -2 (e) 0 (f) 1

35. does not exist

37. $-\dfrac{1}{4}$

39. -2

41. -3

43. 0

45. $\dfrac{1}{3}$

47. $\dfrac{1}{4}$

49. 5

Exercises 2.2, Page 80

1. 15

3. -12

5. 4

7. 4

9. $-\dfrac{8}{5}$

11. 14

13. $\dfrac{28}{9}$

15. -1

17. $\sqrt{7}$

19. does not exist

21. -10

23. 3

25. 60

27. 14

29. $\dfrac{1}{5}$

31. $-\dfrac{1}{8}$

33. 3

35. does not exist

37. 2

39. $\dfrac{128}{3}$

41. -2

43. $a^2 - 2ab + b^2$

45. 16

47. $-1/x^2$

49. $\dfrac{1}{2}$

51. $\dfrac{1}{5}$

53. 32

55. $\dfrac{1}{2}$

57. does not exist

59. $8a$

Exercises 2.3, Page 86

1. none

3. 3 and 6

5. $n\pi/2, n = 0, \pm 1, \pm 2, \ldots$

7. 2

9. none

11. e^{-2}

13. (a) continuous (b) continuous

15. (a) continuous (b) continuous

17. (a) not continuous (b) not continuous

19. (a) continuous (b) not continuous

21. (a) not continuous (b) not continuous

23. (a) not continuous (b) continuous

25. $m = 4$

27. $m = 1; n = 3$

29. discontinuous at $n/2$, where n is an integer;

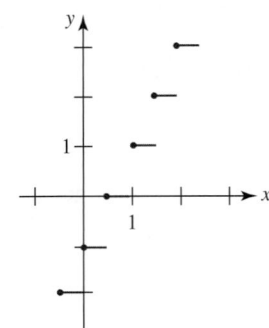

31. define $f(9) = 6$

33. $\dfrac{\sqrt{3}}{2}$

35. 0

37. 1

39. 1

41. $-\pi/6$

43. $(-3, \infty)$

45. $c = 4$

47. $c = 0, c = \pm\sqrt{2}$

55. $-1.22, -0.64, 1.34$

57. 2.21

59. 0.78

Exercises 2.4, Page 93

1. $\dfrac{3}{2}$

3. 0

5. 1

7. 4

9. 0

11. 36

13. $\dfrac{1}{2}$

15. does not exist

17. 3

19. $\dfrac{3}{7}$

21. 0

23. -4

25. 4

27. $\dfrac{1}{2}$

29. 5

31. $\dfrac{1}{6}$

33. 8

35. $\sqrt{2}$

37. $\dfrac{\sqrt{2}}{2}$

43. 3

Exercises 2.5, Page 102

1. $-\infty$

3. ∞

5. ∞

7. ∞

9. $\dfrac{1}{4}$

11. 5

13. $-\dfrac{1}{4}$

15. $\dfrac{5}{2}$

17. $\dfrac{1}{\sqrt{2}}$

19. 0

21. 1

23. $-\pi/6$

25. $-4; 4$

27. $-\dfrac{2}{\sqrt{3}}; \dfrac{2}{\sqrt{3}}$

29. $-1; 1$

31. $-1; 1$

33. VA: none;   HA: $y = 0$;

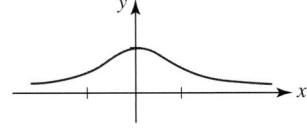

35. VA: $x = -1$;   HA: none;

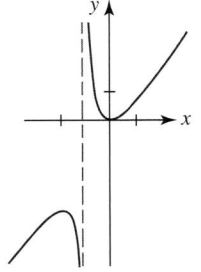

37. VA: $x = 0, x = 2$;   HA: $y = 0$;

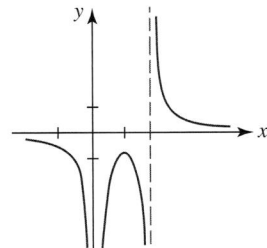

39. VA: $x = 1$;   HA: $y = 1$;

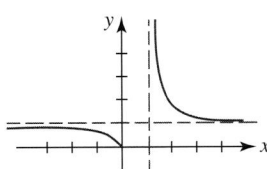

41. VA: none;   HA: $y = -1, y = 1$;

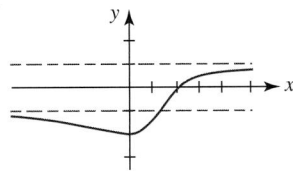

43. (a) 2   **(b)** $-\infty$ **(c)** 0   **(d)** 2
45. (a) $-\infty$ **(b)** -1 **(c)** ∞ **(d)** 0
51. 3

Exercises 2.6, Page 110

1. choose $\delta = \varepsilon$
3. choose $\delta = \varepsilon$
5. choose $\delta = \varepsilon$
7. choose $\delta = \varepsilon/3$
9. choose $\delta = 2\varepsilon$
11. choose $\delta = \varepsilon$
13. choose $\delta = \varepsilon/8$
15. choose $\delta = \sqrt{\varepsilon}$

17. choose $\delta = \varepsilon^2/5$
19. choose $\delta = \varepsilon/2$
21. choose $\delta = \min\{1, \varepsilon/7\}$
23. choose $\delta = \sqrt{\varepsilon}$
25. choose $\delta = \sqrt{a\varepsilon}$
31. choose $N = 7/(4\varepsilon)$
33. choose $N = -30/\varepsilon$

Exercises 2.7, Page 116

1. -4.5;

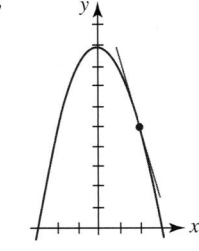

3. 7;

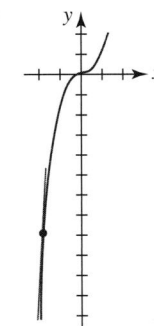

5. $\dfrac{3\sqrt{3} - 6}{\pi}$;

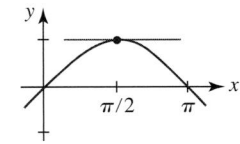

7. $m_{\tan} = 6$; $y = 6x - 15$

9. $m_{\tan} = -1$; $y = -x - 1$
11. $m_{\tan} = -23$; $y = -23x + 32$
13. $m_{\tan} = -\dfrac{1}{2}$; $y = -\dfrac{1}{2}x - 1$
15. $m_{\tan} = 2$; $y = 2x + 1$
17. $m_{\tan} = \dfrac{1}{4}$; $y = \dfrac{1}{4}x + 1$
19. $m_{\tan} = \dfrac{\sqrt{3}}{2}$; $y = \dfrac{\sqrt{3}}{2}x - \dfrac{\sqrt{3}\pi}{12} + \dfrac{1}{2}$
21. not a tangent line   **23.** $y = x - 2$; $(0, -2)$
25. $m_{\tan} = -2x + 6$; $(3, 10)$
27. $m_{\tan} = 3x^2 - 3$; $(-1, 2), (1, -2)$
29. 58 mi/h   **31.** 3.8 h
33. -14
35. (a) -4.9 m/s   **(b)** 5 s   **(c)** -49 m/s
37. (a) 448 ft;   960 ft;   1008 ft;   960 ft
(b) 144 ft/s   **(d)** 16 s   **(e)** $-32t + 256$
(f) -256 ft/s   **(g)** 1024 ft

Chapter 2 in Review, Page 118

A. 1. true   **3.** false
5. false   **7.** true
9. false   **11.** false
13. true   **15.** true
17. false   **19.** true
21. false

B. 1. 4   **3.** $-\dfrac{1}{5}$
5. 0   **7.** ∞

9. 1

13. $-\infty$

17. 10

21. 9

C. 5. (a), (e), (f), (h)

9. (b), (c), (d), (e), (f)

11.

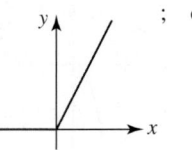 ; continuous everywhere

13. $(-\infty, -1), (-1, 0), (0, 1), (1, \infty)$

15. $(-\infty, -\sqrt{5}), (\sqrt{5}, \infty)$

17. $\dfrac{1}{6}$

21. $y = 8x - 6$

11. 3^-

15. -2

19. continuous

7. (c), (h)

19. $y = 4x + 24$

23. $y = \dfrac{1}{2}x + \dfrac{3}{2}$

Exercises 3.1, Page 128

1. 0

5. $6x$

9. $2x + 2$

13. $-3x^2 + 30x - 1$

17. $5/(x + 4)^2$

21. $y = -x - 4$

25. $(-4, -6)$

29. $x; \left(3, \dfrac{7}{2}\right)$

33. $f_+'(2) = 2$ but $f_-'(2) = -1$

37. $3a^2 - 8a$

41. $y = \dfrac{1}{2}x + 3;$ $f(-3) = \dfrac{3}{2};$ $f'(-3) = \dfrac{1}{2}$

43.

47.

51. (b)

3. -3

7. $-2x + 4$

11. $3x^2 + 1$

15. $-2/(x + 1)^2$

19. $-1/(2x^{3/2})$

23. $y = 2x - 2$

27. $(1, -2), (-1, 2)$

31. $-3x^2;$ $(2, -4), (-2, 12)$

35. $20a$

39. $4/(3 - a)^2$

45.

49. (e)

53. (a)

Exercises 3.2, Page 136

1. 0

5. $14x - 4$

9. $x^4 - 12x^3 + 18x$

13. $6x^5 + 40x^3 + 50x$

17. $192u^2$

21. $y = 6x + 3$

25. $(4, -11)$

3. $9x^8$

7. $2x^{-1/2} + 4x^{-5/3}$

11. $20x^4 - 20x^3 - 18x^2$

15. $16 + 4/\sqrt{x}$

19. $-1/r^2 - 2/r^3 - 3/r^4 - 4/r^5$

23. $y = \dfrac{1}{4}x + 5$

27. $(3, -25), (-1, 7)$

29. $y = \dfrac{1}{4}x - \dfrac{7}{2}$

33. -2

37. $60/x^4$

41. $(-4, \infty), (-\infty, -4)$

45. $(1, \infty), (-\infty, 1)$

51. $\left(\dfrac{1}{4}, -\dfrac{3}{16}\right)$

55. $S = 4\pi r^2$

31. $x = 4$

35. 32

39. $1440x^2 + 120x$

43. $(-4, 48)$

49. $(2, 8)$

53. $y = -7x$

57. -15 N

Exercises 3.3, Page 142

1. $5x^4 - 9x^2 + 4x - 28$

5. $-20x/(x^2 + 1)^2$

9. $72x - 12$

13. $(x^2 + 2x)/(2x^2 + x + 1)^2$

17. $(6x^2 + 8x - 3)/(3x + 2)^2$

19. $(2x^3 + 8x^2 - 6x - 8)/(x + 3)^2$

21. $y = -4x + 1$

25. $(0, 24), (\sqrt{5}, -1), (-\sqrt{5}, -1)$

27. $(0, 0), \left(-1, \dfrac{1}{2}\right), \left(1, \dfrac{1}{2}\right)$

31. $(-4, 0), (-6, 2)$

35. -28

39. -30

43. $(x^2 f''(x) - 2xf'(x) + 2f(x))/x^3$

3. $8x^{-7/3} - 4x^{-5/6} + 12^{1/2}$

7. $-17/(5 - 2x)^2$

11. $(2x^5 + x^2 - 40x - 12)/x^4$

15. $18x^2 + 22x + 6$

23. $y = 7x - 1$

29. $\left(3, \dfrac{3}{2}\right), \left(-5, \dfrac{1}{2}\right)$

33. $k = -21$

37. $\dfrac{11}{3}$

41. $\dfrac{13}{2}$

45. $f'(x) > 0$ on $(-\infty, 0) \cup (0, 1);$ $f'(x) < 0$ on $(1, 2) \cup (2, \infty)$

47. $f'(x) > 0$ on $\left(-\infty, \dfrac{5}{8}\right);$ $f'(x) < 0$ on $\left(\dfrac{5}{8}, \infty\right)$

49. $-16 \, km_1m_2$

51. $-\dfrac{RT}{(V - b)^2} + \dfrac{2a}{V^3}$

Exercises 3.4, Page 147

1. $2x + \sin x$

5. $x \cos x + \sin x$

9. $x^2 \sec x \tan x + 2x \sec x + \sec^2 x$

11. 0

15. $\dfrac{-x \csc^2 x - \csc^2 x - \cot x}{(x + 1)^2}$

19. $\dfrac{1}{1 + \cos x}$

21. $x^4 \sin x \sec^2 x + x^4 \sin x + 4x^3 \sin x \tan x$

23. $y = -\dfrac{\sqrt{3}}{2}x + \dfrac{1}{2} + \dfrac{\sqrt{3}\pi}{6}$

27. $\pi/6, 5\pi/6$

31. $y = 2x - \dfrac{\sqrt{3}}{2} - \dfrac{8\pi}{3}$

35. $2(\cos^2 x - \sin^2 x) = 2\cos 2x$

39. $\dfrac{-x^2 \sin x - 2x \cos x + 2 \sin x}{x^3}$

41. $\csc x \cot^2 x + \csc^3 x$

45. $-\dfrac{160}{3};$ as the angle of elevation increases, the length s of the shadow decreases

3. $7 \cos x - \sec^2 x$

7. $(x^3 - 2) \sec^2 x + 3x^2 \tan x$

13. $\cos x$

17. $\dfrac{-2x^2 \sec^2 x + 4x \tan x + 2x}{(1 + 2\tan x)^2}$

25. $y = \dfrac{2}{3}x + \dfrac{2}{\sqrt{3}} - \dfrac{\pi}{9}$

29. $\pi/2$

33. $y = x - 2\pi$

37. $2 \cos x - x \sin x$

53. not differentiable at 0, $\pm\pi$, $\pm 2\pi$, …

55. (b) $-\dfrac{14(0.2\cos\theta - \sin\theta)}{(0.2\sin\theta + \cos\theta)^2}$ **(c)** 0.1974 radians

   (d) approximately 13.7281

   (e) The minimum force required to pull the sled is about 13.73 lb when θ is approximately 0.1974 radians or 11.31°.

Exercises 3.5, Page 155

1. $-150(-3x)^{29}$ **3.** $200(2x^2 + x)^{199}(4x + 1)$

5. $-4(x^3 - 2x^2 + 7)^{-5}(3x^2 - 2x)$

7. $-2(3x - 1)^3(-2x + 9)^4(27x - 59)$

9. $\dfrac{\cos\sqrt{2x}}{\sqrt{2x}}$ **11.** $\dfrac{2x}{\sqrt{x^2 - 1}(x^2 + 1)^{3/2}}$

13. $10(1 + 6x(x^2 - 4)^2)(x + (x^2 - 4)^3)^9$

15. $\dfrac{5x^{14} + 9x^{13} + 13x^{12}}{(x^2 + x + 1)^5}$ **17.** $\pi\cos(\pi x + 1)$

19. $15\sin^2 5x\cos 5x$ **21.** $-3x^5\sin x^3 + 3x^2\cos x^3$

23. $10(2 + x\sin 3x)^9(3x\cos 3x + \sin 3x)$

25. $-x^{-2}\sec^2(1/x)$

27. $-3\sin 2x\sin 3x + 2\cos 2x\cos 3x$

29. $5(\sec 4x + \tan 2x)^4(4\sec 4x\tan 4x + 2\sec^2 2x)$

31. $2\cos 2x\cos(\sin 2x)$

33. $-(2x + 5)^{-1/2}\cos\sqrt{2x + 5}\,\sin(\sin\sqrt{2x + 5})$

35. $24x\sin^2(4x^2 - 1)\cos(4x^2 - 1)$

37. $360x^2(1 + x^3)^3(1 + (1 + x^3)^4)^4(1 + (1 + (1 + x^3)^4)^5)^5$

39. -54 **41.** -7

43. $y = -8x - 3$ **45.** $y = 6x - 1 - \dfrac{3\pi}{2}$

47. $y = \dfrac{\sqrt{6}}{4} + \dfrac{12}{\pi(2\sqrt{2} + 3\sqrt{6})}\left(x - \dfrac{1}{2}\right)$

49. $-\pi^3\cos\pi x$ **51.** $-125x\cos 5x - 75\sin 5x$

53. $(\sqrt{3}/3, 3\sqrt{3}/16), (-\sqrt{3}/3, -3\sqrt{3}/16)$; no

55. $\dfrac{1}{18}$

57. If $0 \le \theta \le \pi$, then $\theta = \pi/4$ or $\theta = 3\pi/4$.

59. $dr/dt = 5/(8\pi)$ in/min

Exercises 3.6, Page 160

1. $4x^2y^3\dfrac{dy}{dx} + 2xy^4$ **3.** $-2y\sin y^2\dfrac{dy}{dx}$

5. $\dfrac{1}{2y - 2}$ **7.** $\dfrac{2x - y^2}{2xy}$

9. $\dfrac{2x}{3 - \sin y}$ **11.** $\dfrac{4x - 3x^2y^2}{2x^3y - 2y}$

13. $\dfrac{x^2 - 4x(x^2 + y^2)^5}{y^2 + 4y(x^2 + y^2)^5}$ **15.** $\dfrac{2x^4y^4 + 3y^{10} - 6x^9y}{6xy^9 - 3x^{10}}$

17. $\dfrac{1 - x}{y + 4}$ **19.** $\dfrac{3}{2y(x + 2)^2}$

21. $\dfrac{\cos(x + y) - y}{x - \cos(x + y)}$ **23.** $\cos y\cot y$

25. $\dfrac{\cos 2\theta}{r}$ **27.** $-\dfrac{2}{5}$

29. $-\dfrac{1}{3}$ and $-\dfrac{2}{3}$ **31.** $y = \dfrac{8}{3}x + \dfrac{22}{3}$

33. $y = \dfrac{1}{2}x - \dfrac{1}{2} + \dfrac{\pi}{4}$ **35.** $(1, 2), (-1, -2)$

37. $(-\sqrt{5}, 2\sqrt{5}), (\sqrt{5}, -2\sqrt{5})$

39. $(8, 4)$ **41.** $\dfrac{y^3 - 2x^2}{y^5}$

43. $\dfrac{-25}{y^3}$ **45.** $\dfrac{-\sin y}{(1 - \cos y)^3}$

47. $\dfrac{-2}{(y - x)^3}$ **49.** $\dfrac{2x - 1}{2\sqrt{x^2 - x}}, -\dfrac{2x - 1}{2\sqrt{x^2 - x}}$

51. $\dfrac{-2x - 3}{x^4}$ **53.** $y = 1 - \sqrt{x - 2}$

55. $y = \begin{cases} \sqrt{4 - x^2}, & -2 \le x < 0 \\ -\sqrt{4 - x^2}, & 0 \le x < 2 \end{cases}$

57. $\dfrac{dy}{dt} = -\dfrac{x}{y}\dfrac{dx}{dt}$

59. (a) $y = -x + 3$ **(b)** $(\sqrt[3]{2}, \sqrt[3]{4})$

65. (b) $\dfrac{4(252 - x^2)}{(x^2 + 252)^2 + 16x^2}$ **(c)** $x = 6\sqrt{7} \approx 15.87$ ft

Exercises 3.7, Page 167

1. $f'(x) > 0$ for all x shows that f is increasing on $(-\infty, \infty)$. It follows from Theorem 3.7.3 that f is one-to-one.

3. $f(0) = 0, f(1) = 0$ implies f is not one-to-one.

5. $\dfrac{2}{3}$ **7.** $(f^{-1})'(x) = -1/(x - 2)^2$

9. $(5, 3)$; $y = \dfrac{1}{10}x + \dfrac{5}{2}$ **11.** $(8, 1)$; $y = \dfrac{1}{60}x + \dfrac{13}{15}$

13. $\dfrac{5}{\sqrt{1 - (5x - 1)^2}}$ **15.** $\dfrac{-8}{4 + x^2}$

17. $\dfrac{1}{1 + x} + \dfrac{\tan^{-1}\sqrt{x}}{\sqrt{x}}$ **19.** $\dfrac{2(\cos^{-1}2x + \sin^{-1}2x)}{\sqrt{1 - 4x^2}(\cos^{-1}2x)^2}$

21. $\dfrac{-2x}{(1 + x^4)(\tan^{-1}x^2)^2}$ **23.** $\dfrac{2 - x}{\sqrt{1 - x^2}} + \cos^{-1}x$

25. $3\left(x^2 - 9\tan^{-1}\dfrac{x}{3}\right)^2\left(2x - \dfrac{27}{9 + x^2}\right)$

27. $\dfrac{1}{t^2 + 1}$ **29.** $\dfrac{-4\sin 4x}{|\sin 4x|}$

31. $\dfrac{2x\sec^2(\sin^{-1}x^2)}{\sqrt{1 - x^4}}$ **33.** $\dfrac{2x(1 + y^2)}{1 - 2y - 2y^3}$

35. $\sin^{-1}x + \cos^{-1}x = $ constant **37.** $\sqrt{3}/3$

39. $y = \dfrac{2 + \pi}{4}x - \dfrac{1}{2}$ **41.** $(5\pi/6, 4), (7\pi/6, 6)$

Exercises 3.8, Page 171

1. $-e^{-x}$ **3.** $\dfrac{e^{\sqrt{x}}}{2\sqrt{x}}$

5. $5^{2x}(2\ln 5)$ **7.** $x^2e^{4x}(3 + 4x)$

9. $\dfrac{-e^{-2x}(2x + 1)}{x^2}$ **11.** $-\dfrac{5}{2}(1 + e^{-5x})^{-1/2}e^{-5x}$

13. $-\dfrac{e^{x/2} - e^{-x/2}}{(e^{x/2} + e^{-x/2})^2}$ **15.** $8e^{8x}$

17. $3e^{3x-3}$ **19.** $\dfrac{1}{3}x^{-2/3}e^{x^{1/3}} + \dfrac{1}{3}e^{x/3}$

21. $\sec^2 e^x - e^{-x}\tan e^x$

23. $\dfrac{e^{x\sqrt{x^2+1}}(2x^2+1)}{\sqrt{x^2+1}}$

37. $\dfrac{2}{x^3}$

39. $\dfrac{2-2\ln|x|}{x^2}$

25. $2xe^{x^2}e^{e^{x^2}}$

27. $y = 4x + 4$

29. $(\ln 3, 3)$

31. $x = \pi/4 + n\pi,\ n = 0, \pm 1, 2, \ldots$

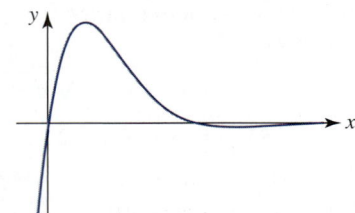

33. $4e^{x^2}(2x^3 + 3x)$

35. $4e^{2x}\cos e^{2x} - 4e^{4x}\sin e^{2x}$

41. $\dfrac{e^{x+y}}{1 - e^{x+y}}$

43. $\dfrac{-ye^{xy}\sin e^{xy}}{1 + xe^{xy}\sin e^{xy}}$

45. $\dfrac{-y^2 + ye^{x/y}}{2y^3 + xe^{x/y}}$

43. $\dfrac{y}{2xy^2 - x}$

45. $\dfrac{y - xy}{2xy^2 + x}$

47. $\dfrac{2x - x^2y - y^3}{x^3 + xy^2 - 2y}$

49. $x^{\sin x}\left[\dfrac{\sin x}{x} + (\cos x)\ln x\right]$

51. $x(x-1)^x\left[\dfrac{1}{x} + \dfrac{x}{x-1} + \ln(x-1)\right]$

53. $\dfrac{\sqrt{(2x+1)(3x+2)}}{4x+3}\left[\dfrac{1}{2x+1} + \dfrac{3/2}{3x+2} - \dfrac{4}{4x+3}\right]$

55. $\dfrac{(x^3-1)^5(x^4+3x^3)^4}{(7x+5)^9}\left[\dfrac{15x^2}{x^3-1} + \dfrac{16x^3+36x^2}{x^4+3x^3} - \dfrac{63}{7x+5}\right]$

57. $y = 3x - 2$

59. $(e^{-1}, e^{-e^{-1}})$;

47. (a)

(b) $f'(x) = \begin{cases} e^x, & x > 0 \\ -e^{-x}, & x < 0 \end{cases}$

(c)

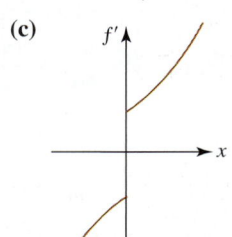

(d) no

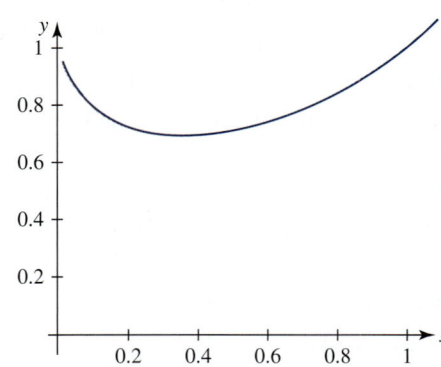

65. (b) one interval is $(\pi, 2\pi)$ **67.** $4 - 4\ln 4 \approx -1.55$

Exercises 3.10, Page 185

1. $\cosh x = \sqrt{5}/2,\ \tanh x = -\sqrt{5}/5,\ \coth x = -\sqrt{5},$
$\operatorname{sech} x = 2\sqrt{5}/5,\ \operatorname{csch} x = -2$

3. $10\sinh 10x$

5. $\dfrac{1}{2}x^{-1/2}\operatorname{sech}^2\sqrt{x}$

7. $-6(3x-1)\operatorname{sech}(3x-1)^2\tanh(3x-1)^2$

9. $-3\sinh 3x\,\operatorname{csch}^2(\cosh 3x)$

11. $3\sinh 2x\sinh 3x + 2\cosh 2x\cosh 3x$

13. $2x^2\sinh x^2 + \cosh x^2$ **15.** $3\sinh^2 x\cosh x$

17. $\dfrac{2}{3}(x - \cosh x)^{-1/3}(1 - \sinh x)$ **19.** $4\tanh 4x$

21. $\dfrac{e^x + 1}{(1 + \cosh x)^2}$ **23.** $e^{\sinh t}\cosh t$

25. $\dfrac{\cos t + \cos t\sinh 2t - 2\sin t\cosh 2t}{(1 + \sinh 2t)^2}$

27. $y = 3x$

29. $(0, -2), (-2, 2\cosh 2 - 4\sinh 2), (2, 2\cosh 2 - 4\sinh 2)$

31. $-2\operatorname{sech}^2 x\tanh x$ **35.** $\dfrac{3}{\sqrt{9x^2+1}}$

37. $\dfrac{-2x}{1 - (1 - x^2)^2}$ **39.** $\sec x$

41. $\dfrac{3x^3}{\sqrt{x^6+1}} + \sinh^{-1}x^3$ **43.** $-\dfrac{1}{x^2\sqrt{1-x^2}} - \dfrac{\operatorname{sech}^{-1}x}{x^2}$

45. $\dfrac{-1}{x\sqrt{1 - x^2}\operatorname{sech}^{-1}x}$ **47.** $\dfrac{3}{\sqrt{\cosh^{-1}6x}\sqrt{36x^2-1}}$

49. (b) $v_{\text{ter}} = \sqrt{mg/k}$ **(c)** 56 m/s

49. (b) $P = 0, P = 2$ **(c)**

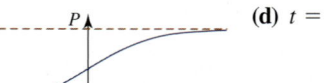

(d) $t = 0$

61. $f'(0) = 0$

Exercises 3.9, Page 177

1. $\dfrac{10}{x}$

3. $\dfrac{1}{2x}$

5. $\dfrac{4x^3 + 6x}{x^4 + 3x^2 + 1}$

7. $3x + 6x\ln x$

9. $\dfrac{1 - \ln x}{x^2}$

11. $\dfrac{1}{x(x+1)}$

13. $\tan x$

15. $\dfrac{-1}{x(\ln x)^2}$

17. $\dfrac{1 + \ln x}{x\ln x}$

19. $\dfrac{1}{4x\sqrt{\ln\sqrt{x}}}$

21. $\dfrac{2}{t} + \dfrac{2t}{t^2+2}$

23. $\dfrac{1}{x+1} + \dfrac{1}{x+2} - \dfrac{1}{x+3}$

25. $y = x - 1$

27. 4

29. -8

31. (e, e^{-1})

33. $\dfrac{1}{\sqrt{x^2-1}}$

35. $\sec x$

Chapter 3 in Review, Page 186

A. 1. false   **3.** false

  5. true   **7.** true

  9. true   **11.** true

13. false   **15.** true

17. false   **19.** true

B. 1. 0   **3.** $-\dfrac{1}{4}$

  5. $y = -\dfrac{5}{4}x - \dfrac{3}{2}$ **7.** -3

  9. 23

11. $-16F'(\sin 4x)\sin 4x + 16F''(\sin 4x)\cos^2 x$

13. $a = 6$; $b = -9$ **15.** $(1, 5)$

17. $\dfrac{1}{x(\ln 10)}$ **19.** catenary

C. 1. $0.08x^{-0.9}$

  3. $10(t + \sqrt{t^2 + 1})^9(1 + t(t^2 + 1)^{-1/2})$

  5. $x^2(x^4 + 16)^{1/4}(x^3 + 8)^{-2/3} + x^3(x^4 + 16)^{-3/4}(x^3 + 8)^{1/3}$

  7. $-\dfrac{16x\sin 4x + 4\sin 4x + 4\cos 4x}{(4x + 1)^2}$

  9. $10x^3 \sin 5x \cos 5x + 3x^2 \sin^2 5x$

11. $\dfrac{-3}{|x|\sqrt{x^2 - 9}}$ **13.** $\dfrac{1}{(\cot^{-1} x)^2(1 + x^2)}$

15. $\dfrac{-4x^2}{\sqrt{1 - x^2}}$ **17.** $-xe^{-x}$

19. $7x^6 + 7^x(\ln 7) + 7e^{7x}$ **21.** $\dfrac{1}{x} + \dfrac{2}{4x - 1}$

23. $\dfrac{1}{\sqrt{(\sin^{-1} x)^2 + 1}\sqrt{1 - x^2}}$

25. $e^{x\cosh^{-1} x}\left[\dfrac{x^2}{\sqrt{x^2 - 1}} + x\cosh^{-1} x + 1\right]$

27. $3x^2 e^{x^3}\cosh e^{x^3}$ **29.** $\dfrac{405}{8\sqrt{1 + 3x}}$

31. $\dfrac{120}{t^6}$ **33.** $4e^{\sin 2x}(\cos^2 2x - \sin 2x)$

35. $\dfrac{4}{x + 5} - \dfrac{3}{2 - x} - \dfrac{10}{x + 8} - \dfrac{2}{6x + 4}$

37. $\dfrac{1}{4}$ **39.** $\dfrac{e^x - y^2}{2xy + e^y}$

41. $y = \dfrac{1}{3}x - \dfrac{2}{27}, y = \dfrac{1}{3}x + \dfrac{2}{27}$ **43.** $y = 6x - 9, y = -6x - 9$

45. $(4, 2)$ **47.** $0, 2\pi/3, \pi, 4\pi/3, 2\pi$

53. (a) $(2, 0), (2, -1), (2, 1)$ **(b)** $4, -2, -2$

55. $y = \sqrt{3}x - \dfrac{\sqrt{3}}{2}, y = -\sqrt{3}x + \dfrac{\sqrt{3}}{2}$

Exercises 4.1, Page 195

  1. $-1, 19$; $-2, 18$; $2, 18$; $8, 8$

  3. $18, 6$; $-23, 1$; $23, 1$; $18, -6$

  5. $-\dfrac{15}{4}, 0$; $17, 2$; $17, 2$; $-128, -2$

  7. $1, \dfrac{1}{2}$; $1 - \pi, 1$; $\pi - 1, 1$; $0, \pi^2$

  9. (a) $-6, 6$ **(b)** $-8, 8$

11. (a) $-6\sqrt{2}, 6\sqrt{2}$ **(b)** 15 **(c)** $-4, 8$

13. slowing down on the time intervals $(-\infty, -3), (0, 3)$; speeding up on the time intervals $(-3, 0), (3, \infty)$

15. $v(t) = 2t, a(t) = 2$; slowing down on the time interval $(-1, 0)$; speeding up on the time interval$(0, 3)$;

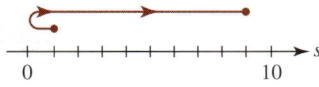

17. $v(t) = 2t - 4, a(t) = 2$; slowing down on the time interval $(-1, 2)$; speeding up on the time interval $(2, 5)$;

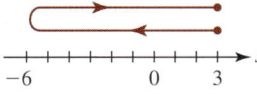

19. $v(t) = 6t^2 - 12t, a(t) = 12t - 12$; slowing down on the time intervals $(-2, 0), (1, 2)$; speeding up on the time intervals $(0, 1), (2, 3)$;

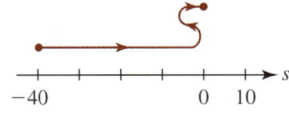

21. $v(t) = 12t^3 - 24t^2, a(t) = 36t^2 - 48t$;

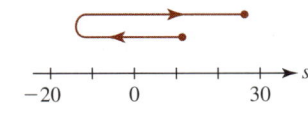

23. $v(t) = 1 - 2t^{-1/2}, a(t) = t^{-3/2}$;

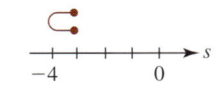

25. $v(t) = \dfrac{\pi}{2}\cos\dfrac{\pi}{2}t, a(t) = -\left(\dfrac{\pi}{2}\right)^2\sin\dfrac{\pi}{2}t$;

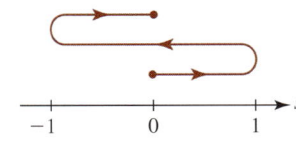

27. $v(t) = e^{-t}(-t^3 + 3t^2), a(t) = e^{-t}(t^3 - 6t^2 + 6t)$;

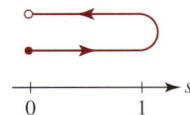

29.

positive	negative
zero	zero
positive	positive
positive	negative
negative	negative
negative	positive

; slowing down on the time intervals $(a, b), (d, e), (f, g)$; speeding up on the time intervals $(c, d), (e, f)$

31. (a) $v > 0$ on $\left[0, \frac{3}{2}\right)$, $v < 0$ on $\left(\frac{3}{2}, \frac{1}{4}(6 + \sqrt{42})\right]$

    (b) 42 ft

33. $64\sqrt{2}$ ft/s; 16 ft/s²

35. $-8\sqrt{\pi}$ ft/s; the y-coordinate is decreasing

Exercises 4.2, Page 200

1. $\dfrac{dV}{dt} = 3x^2 \dfrac{dx}{dt}$

3. $8\sqrt{3}$ cm²/h

5. $\dfrac{4}{3}$ in/h

7. $\dfrac{dx}{dt} = s\cos\theta \dfrac{d\theta}{dt} + \sin\theta \dfrac{ds}{dt}$

9. -6 or 6

11. $\dfrac{4}{9}$ cm²/h

13. (a) 1 ft/s

(b) 4 ft/s

15. $-\dfrac{1}{\sqrt{2}}$ ft/min

19. 17 knots

21. $-\dfrac{5}{4}$ ft/s

23. 15 rad/h

25. -360 mi/h

27. $\dfrac{8\pi}{9}$ km/min

29. (a) $500\sqrt{3}$ mi/h

(b) 500 mi/h

31. $\dfrac{5}{32\pi}$ m/min

33. (a) $-\dfrac{1}{4\pi}$ ft/min

(b) $-\dfrac{1}{12\pi}$ ft/min

(c) approximately -0.0124 ft/min

35. (a) $\dfrac{\sqrt{3}}{10}$ ft/min

(c) $\dfrac{165\sqrt{3}}{4} \approx 71.45$ min; 0.035 ft/min

39. $-\dfrac{1}{3}$ in²/min

41. 668.7 ft/min

43. $\dfrac{dR}{dt} = \dfrac{R^2}{R_1^2}\dfrac{dR_1}{dt} + \dfrac{R^2}{R_2^2}\dfrac{dR_2}{dt}$

45. (a) increases (b) approximately 2.8% per day

47. (a) 24,000 kg km/h² (b) 2,023,100 kg km/h²

Exercises 4.3, Page 209

1. (a) abs. max. $f(2) = -2$, abs. min. $f(-1) = -5$

(b) abs. max. $f(7) = 3$, abs. min. $f(3) = -1$

(c) no extrema

(d) abs. max. $f(4) = 0$, abs. min. $f(1) = -3$

3. (a) abs. max. $f(4) = 0$, abs. min. $f(2) = -4$

(b) abs. max. $f(1) = f(3) = -3$, abs. min. $f(2) = -4$

(c) abs. min. $f(2) = -4$

(d) abs. max. $f(5) = 5$

5. (a) no extrema

(b) abs. max. $f(\pi/4) = 1$, abs. min. $f(-\pi/4) = -1$

(c) abs. max. $f(\pi/3) = \sqrt{3}$, abs. min. $f(0) = 0$

(d) no extrema

7. $\dfrac{3}{2}$

9. $-1, 6$

11. $\dfrac{4}{3}, 2$

13. 1

15. $\dfrac{3}{4}$

17. $-2, -\dfrac{11}{7}, 1$

19. $2n\pi$, n an integer

21. 2

23. abs. max. $f(3) = 9$, abs. min. $f(1) = 5$

25. abs. max. $f(8) = 4$, abs. min. $f(0) = 0$

27. abs. max. $f(0) = 2$, abs. min. $f(-3) = -79$

29. abs. max. $f(3) = 8$, abs. min. $f(-4) = -125$

31. abs. max. $f(2) = 16$, abs. min. $f(0) = f(1) = 0$

33. abs. max. $f(\pi/6) = f(5\pi/6) = f(7\pi/6) = f(11\pi/6) = \frac{3}{2}$,

abs. min. $f(\pi/2) = f(3\pi/2) = -3$

35. abs. max. $f(\pi/8) = f(3\pi/8) = f(5\pi/8) = f(7\pi/8) = 5$,

abs. min. $f(0) = f(\pi/4) = f(\pi/2) = f(3\pi/4) = f(\pi) = 3$

37. endpoint abs. max. $f(3) = 3$, rel. max. $f(0) = 0$,

abs. min. $f(-1) = f(1) = -1$

39. (a) c_1, c_3, c_4, c_{10}

(b) $c_2, c_5, c_6, c_7, c_8, c_9$

(c) abs. min. $f(c_7)$, endpoint abs. max. $f(b)$

(d) rel. max. $f(c_3), f(c_5), f(c_9)$, rel. min. $f(c_2), f(c_4), f(c_7), f(c_{10})$

41. (a) $s(t) \geq 0$ only for $0 \leq t \leq 20$ (b) $s(10) = 1600$

53. (b) $0, \pi/3, \pi, 5\pi/3, 2\pi$

(c) abs. max. $f(\pi) = 3$, abs. min. $f(\pi/3) = f(5\pi/3) = -\frac{3}{2}$

Exercises 4.4, Page 215

1. $c = 0$

3. $f(-3) = 0$ but $f(-2) \neq f(-3)$

5. $c = -\dfrac{2}{3}$

7. $c = -\pi/2, \pi/2,$ or $3\pi/2$

9. f is not differentiable on the interval

11. $f(a) \neq 0$ and $f(b) = 0$, so $f(a) \neq f(b)$

13. $c = 3$

15. $c = \sqrt{13}$

17. f is not continuous on the interval

19. $c = \dfrac{9}{4}$

21. $c = 1 - \sqrt{6}$

23. f is not continuous on $[a, b]$

25. f increasing on $[0, \infty)$; f decreasing on $(-\infty, 0]$

27. f increasing on $[-3, \infty)$; f decreasing on $(-\infty, -3]$

29. f increasing on $(-\infty, 0]$ and $[2, \infty)$; f decreasing on $[0, 2]$

31. f increasing on $[3, \infty)$; f decreasing on $(-\infty, 0]$ and $[0, 3]$

33. f decreasing on $(-\infty, 0]$ and $[0, \infty)$

35. f increasing on $(-\infty, -1]$ and $[1, \infty)$; f decreasing on $[-1, 0)$ and $(0, 1]$

37. f increasing on $[-2, 2]$; f decreasing on $[-2\sqrt{2}, -2]$ and $[2, 2\sqrt{2}]$

39. f increasing on $(-\infty, 0]$; f decreasing on $[0, \infty)$

41. f increasing on $(-\infty, 1]$ and $[3, \infty)$; f decreasing on $[1, 3]$

43. f increasing on $[-\pi/2 + 2n\pi, \pi/2 + 2n\pi]$; f decreasing on $[\pi/2 + 2n\pi, 3\pi/2 + 2n\pi]$, where n is an integer

45. f increasing on $[0, \infty)$; f decreasing on $(-\infty, 0]$

47. f is increasing on $(-\infty, \infty)$

49. If the motorist travels at the speed limit, he will have gone no more than 65 mi.

61. $c \approx 0.3451$ radian

Exercises 4.5, Page 222

1. 0

3. 2

5. $\dfrac{2}{3}$

7. 10

9. -6

11. $\dfrac{1}{2}$

13. $\dfrac{7}{5}$

15. $\dfrac{1}{6}$

17. does not exist

19. $\dfrac{1}{2}$

21. $2e^4$

23. 0

25. $\dfrac{1}{3}$

27. ∞

29. -2

31. $-\dfrac{1}{8}$

33. -1

35. does not exist

37. $\dfrac{1}{9}$

39. 3

41. $\infty - \infty$; $-\dfrac{1}{2}$

43. $0 \cdot \infty$; 1

45. 0^0; 1

47. $\infty - \infty$; 0

49. $\infty - \infty$; $\dfrac{1}{24}$

51. $0 \cdot \infty$; $\dfrac{1}{4}$

53. ∞^0; 1

55. 1^∞; e^3

57. 0^0; 1

59. denominator is $0 \cdot \infty$; $\dfrac{1}{4}$

61. $\infty - \infty$; $\dfrac{1}{5}$

63. $0 \cdot \infty$; 0

65. $0 \cdot \infty$; 1

67. $0 \cdot \infty$; 5

69. $\infty - \infty$; does not exist

71. 1^∞; $e^{-1/3}$

73. 0^0; 1

75. $\dfrac{1}{2}$

79. 0

81. **(a)** $A(\theta) = 25\,\dfrac{\theta - \frac{1}{2}\sin 2\theta}{\theta^2}$ **(b)** 0 **(c)** $\dfrac{50}{3}$

83. **(b)** $p_1 v_1 \ln(v_2/v_1)$

Exercises 4.6, Page 228

1. rel. max. $f(1) = 2$;

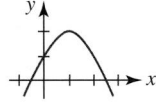

3. rel. max. $f(-1) = 2$, rel. min. $f(1) = -2$;

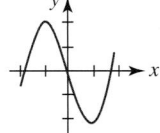

5. rel. max. $f\left(\frac{2}{3}\right) = \frac{32}{27}$, rel. min. $f(2) = 0$;

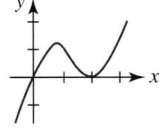

7. no extrema;

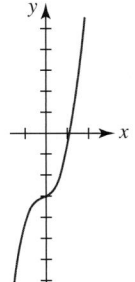

9. rel. min. $f(-1) = -3$;

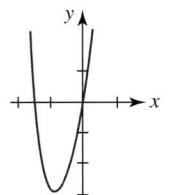

11. rel. min. $f(0) = 0$;

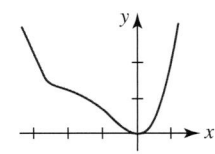

13. rel. max. $f(0) = f(3) = 0$, rel. min. $f\left(\frac{3}{2}\right) = -\frac{81}{16}$;

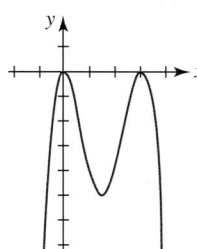

15. rel. max. $f(0) = 0$, rel. min. $f(1) = -1$;

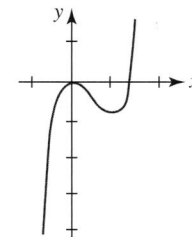

17. rel. max. $f(-3) = -6$, rel. min. $f(1) = 2$;

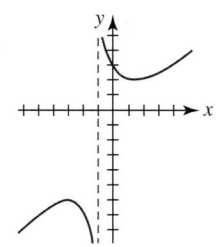

19. rel. max. $f\left(\sqrt{3}\right) = \dfrac{2\sqrt{3}}{9}$, rel. min. $f\left(-\sqrt{3}\right) = -\dfrac{2\sqrt{3}}{9}$;

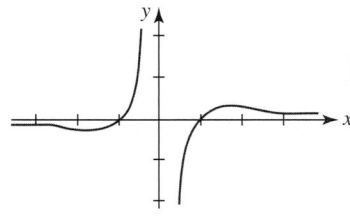
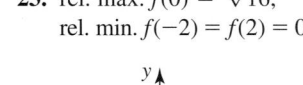

21. rel. max. $f(0) = 10$;

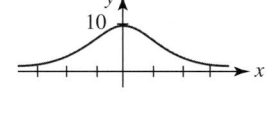

23. rel. max. $f(0) = \sqrt[3]{16}$, rel. min. $f(-2) = f(2) = 0$;

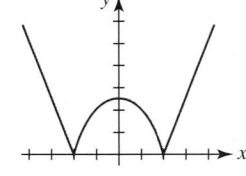

25. rel. max. $f\left(\frac{\sqrt{2}}{2}\right) = \frac{1}{2}$, rel. min. $f\left(-\frac{\sqrt{2}}{2}\right) = -\frac{1}{2}$;

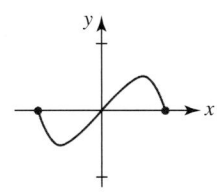

27. rel. max. $f(-8) = 16$, rel. min. $f(8) = -16$;

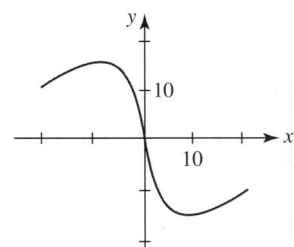

29. rel. min. $f(2) \approx -8.64$;

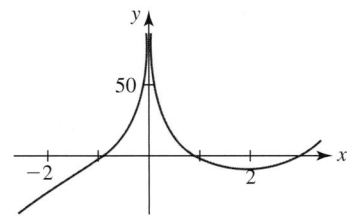

31. rel. min. $f(-3) = 0$, rel. max. $f(-1) = 4e$;

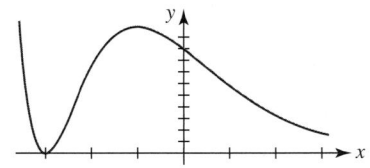

33.

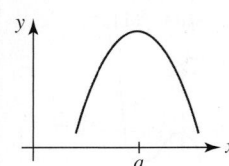

35.

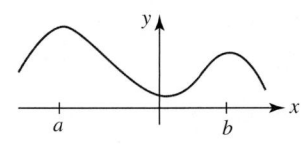

37.

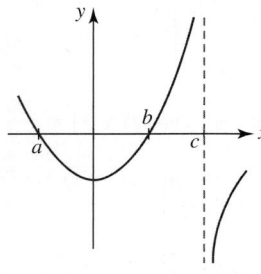

43. rel. min. $f'(-2) = -13$

45. **(a)** $(n\pi, \pi/2 + n\pi), (\pi/2 + n\pi, \pi + n\pi)$, n an integer
   (b) $n\pi/2$, n an integer; rel. max. is $f(-\pi/2) = f(\pi/2) = \cdots 1$,
   rel. min. is $f(0) = f(\pi) = \cdots 0$
   (c)

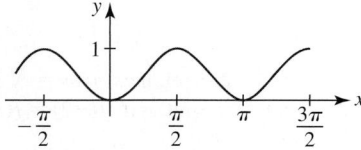

Exercises 4.7, Page 233

1. concave down on $(-\infty, \infty)$

3 concave up on $(-\infty, 2)$;   concave down on $(2, \infty)$

5. concave up on $(-\infty, 2)$ and $(4, \infty)$;   concave down on $(2, 4)$

7. concave up on $(-\infty, 0)$;   concave down on $(0, \infty)$

9. concave up on $(0, \infty)$;   concave down on $(-\infty, 0)$

11. concave up on $(-\infty, -1)$ and $(1, \infty)$;   concave down on $(-1, 1)$

13. approximate answers: f' increasing on $(-2, 2)$; f' decreasing on $(-\infty, -2)$ and $(2, \infty)$

15. approximate answers: f' increasing on $(-\infty, -1)$ and $(3, \infty)$; f' decreasing on $(-1, 3)$

19. $(-\sqrt{2}, -21 - \sqrt{2}), (\sqrt{2}, -21 + \sqrt{2})$

21. $(n\pi, 0)$, n an integer

23. $(n\pi, n\pi)$, n an integer

25. $(2, 2 + 2e^{-2})$

27. rel. max. $f\left(\frac{5}{2}\right) = 0$;

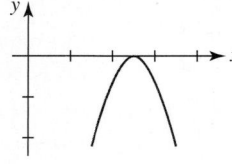

29. point of inflection: $(-1, 0)$;

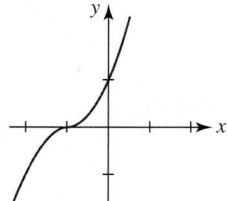

31. rel. max. $f(-1) = 4$, rel. min. $f(1) = -4$; points of inflection:
   $(0, 0), \left(-\frac{\sqrt{2}}{2}, \frac{7\sqrt{2}}{4}\right), \left(\frac{\sqrt{2}}{2}, -\frac{7\sqrt{2}}{4}\right)$;

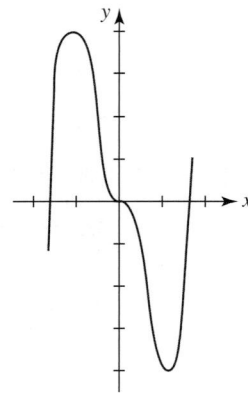

33. rel. max. $f(\sqrt{2}) = \frac{\sqrt{2}}{4}$, rel. min. $f(-\sqrt{2}) = -\frac{\sqrt{2}}{4}$;
   points of inflection: $(0, 0), \left(-\sqrt{6}, -\frac{\sqrt{6}}{8}\right), \left(\sqrt{6}, \frac{\sqrt{6}}{8}\right)$;

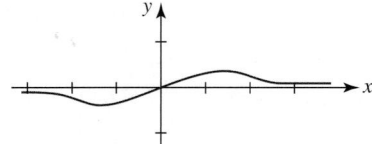

35. rel. max. $f(0) = 3$;

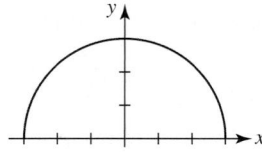

37. rel. min. $f\left(-\frac{1}{4}\right) = -3/4^{4/3}$;
   points of inflection: $(0, 0), (1/2, 3/2^{4/3})$;

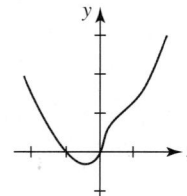

39. rel. max. $f(2\pi/3) = f(4\pi/3) = 1$,
   rel. min. $f(\pi/3) = f(\pi) = f(5\pi/3) = -1$;
   points of inflection: $(\pi/6, 0), (\pi/2, 0), (5\pi/6, 0), (7\pi/6, 0)$,
   $(9\pi/6, 0), (11\pi/6, 0)$;

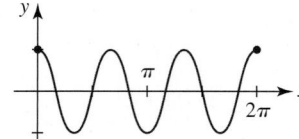

41. rel. max. $f(\pi/4) = \sqrt{2}$, rel. max. $f(5\pi/4) = -\sqrt{2}$;
   points of inflection: $(3\pi/4, 0), (7\pi/4, 0)$;

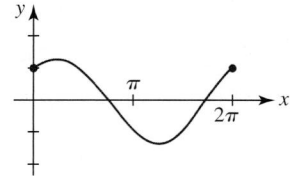

43. rel. max. $f(e) = e$;

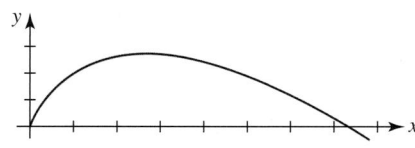

45. rel. max. $f(\pi/4) = \frac{1}{2}$ **47.** rel. min. $f(\pi) = 0$

Exercises 4.8, Page 240

1. 30 and 30 **3.** $\frac{1}{2}$

5. $\frac{1}{3}$ and $\frac{2}{3}$ **7.** $(2, 2\sqrt{3}), (2, -2\sqrt{3}); (0, 0)$

9. $\left(\frac{4}{3}, -\frac{128}{27}\right)$ **11.** base $\frac{3}{2}$, height 1

13. (4, 0) and (0, 8) **15.** 750 ft by 750 ft

17. 2000 m by 1000 m

19. yard should be rectangle 40 ft long and 20 ft wide

21. base 40 cm by 40 cm, height 20 cm

23. base $\frac{80}{3}$ cm by $\frac{80}{3}$ cm, height $\frac{20}{3}$ cm; max. vol. $\frac{128,000}{27}$ cm^3

25. height $\frac{15}{2}$ cm, width 15 cm

27. 10 ft from the flag pole on the right in Figure 4.8.19

29. radius of circular portion $10/(4 + \pi)$ m, width $20/(4 + \pi)$ m, height of rectangular portion $10/(4 + \pi)$ m

31. $L \approx 20.81$ ft **33.** radius 16/3, height 4

35. radius $\sqrt[3]{16/\pi}$, height $2\sqrt[3]{16/\pi}$

37. fly to the point 17.75 km from the nest

39. minimum cost when $x = \dfrac{4}{\sqrt{3}}$

41. $r = \sqrt[3]{9}, h = 2\sqrt[3]{9}$

43. minimum length when $x = 6.375$ in.

45. square with length of side $(a + b)/\sqrt{2}$

47. length of cross section $\sqrt{3}d/3$, width of cross section $\sqrt{6}d/3$

49. $\frac{50}{11}$ m from bulb with illumincance I_1

53. $-\dfrac{1}{8}$

55. (a) $w_0L^4/384EI$

(b)
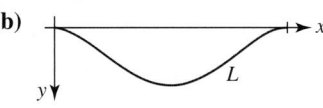

65. She should swim from A to a point B about 3.18 mi down the beach from the point on the beach closest to A, and then proceed directly to C.

67. (a) $L = x + 2\sqrt{4 + (4 - x)^2}$

(c) $x = 4 - \dfrac{2}{3}\sqrt{3}$

(d) $L = x + \sqrt{1 + (4 - x)^2} + \sqrt{4 + (4 - x)^2}$

(f) $x \approx 3.1955$

Exercises 4.9, Page 252

1. $L(x) = 3 + \dfrac{1}{6}(x - 9)$ **3.** $L(x) = 1 + 2\left(x - \dfrac{\pi}{4}\right)$

5. $L(x) = x - 1$ **7.** $L(x) = 2 + \dfrac{1}{4}(x - 3)$

17. 0.98 **19.** 11.6

21. 0.7 **23.** 0.96

25. 16 **27.** 0.325

29. 0.4 **31.** $\dfrac{1}{2} + \dfrac{\sqrt{3}\pi}{120} \approx 0.5453$

33. $L(x) = 4 + 2(x - 1)$; 4.08

35. $\Delta y = 2x\,\Delta x + (\Delta x)^2$; $dy = 2x\,dx$

37. $\Delta y = 2(x + 1)\Delta x + (\Delta x)^2$; $dy = 2(x + 1)\,dx$

39. $\Delta y = -\dfrac{\Delta x}{x(x + \Delta x)}$; $dy = -\dfrac{1}{x^2}\,dx$

41. $\Delta y = \cos x \sin \Delta x + \sin x\,(\cos \Delta x - 1)$; $dy = \cos x\,dx$

43.

x	Δx	Δy	dy	$\Delta y - dy$
2	1	25	20	5
2	0.5	11.25	10	1.25
2	0.1	2.05	2	0.05
2	0.01	0.2005	0.2	0.0005

45. (a) 1.11 (b) -2.9 **47.** (a) 9π cm^2 (b) 8π cm^2

49. exact volume is $\Delta V = \frac{4}{3}\pi(3r^2t + 3rt^2 + t^3)$; approximate volume is $dV = 4\pi r^2 t$, where $t = \Delta r$; $(0.1024)\pi$ in^3

51. ± 6 cm^2; ± 0.06; $\pm 6\%$ **55.** 2048 ft; 160 ft

57. (a) minimum at the equator ($\theta = 0°$); maximum at the North Pole ($\theta = 90°$ N)

(b) 981.9169 cm/s^2 (c) 0.07856 cm/s^2

59. 0.0102 s

Exercises 4.10, Page 257

1. one real root **3.** no real roots

5. one real root **7.** 3.1623

9. 1.5874 **11.** 0.6823

13. ± 1.1414 **15.** 0, 0.8767

17. 2.4981 **19.** 1.6560 ft

21. 0.7297 **23.** (b) 0.0915 ft

25. (b) 0.33711, 44.494 (c) 44.497

27. 1.8955 radians **29.** 1.0000, -1.2494, -2.6638

31. (d) 1.4645

Chapter 4 in Review, Page 260

A. 1. false **3.** false

5. true **7.** false

9. true **11.** true

13. true **15.** false

17. true **19.** false

B. 1. the velocity function **3.** $y = \tan^{-1}x$

5. 0 **7.** 2

9. $2x\Delta x - \Delta x + (\Delta x)^2$

C. 1. abs. max. $f(-3) = 348$, abs. min. $f(4) = -86$

3. abs. max. $f(3) = \frac{9}{7}$, abs. min. $f(0) = 0$

7. max. vel. $v(2) = 12$, max. speed $|v(-1)| = |v(5)| = 15$;

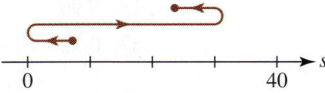

9. (b) $a, b, (a + b)/2$

11. rel. max. $f(-3) = 81$,
rel. min. $f(2) = -44$;

13. rel. max. $f(0) = 2$,
rel. min. $f(1) = 0$;

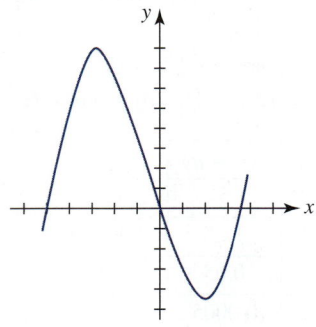

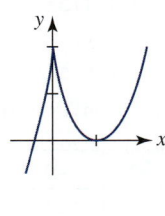

15. rel. min. $f(0) = 0$, points of inflection: $(-3, 27), (-1, 11)$

17. point of inflection: $(3, 10)$ **19.** (c), (d)

21. (c), (d), (e)   **23.** (c)

25. $(a + b + c)/3$ **27.** 32 in²/min

31. $y = \frac{1}{2}h$; max. distance is h

33. $x = 195$ ft, $y = 390$ ft; 57,037.5 ft²

39. $8\sqrt{3}\pi/9$ **41.** -2

43. 1   **45.** e^{-1}

47. $-\infty$ **49.** 1.6751

Exercises 5.1, Page 274

1. $3x + C$ **3.** $\frac{1}{6}x^6 + C$

5. $\frac{3}{2}x^{2/3} + C$ **7.** $t - \frac{25}{12}t^{0.48} + C$

9. $x^3 + x^2 - x + C$ **11.** $\frac{2}{7}x^{7/2} - \frac{4}{3}x^{3/2} + C$

13. $\frac{16}{3}x^3 + 4x^2 + x + C$

15. $16w^4 - 16w^3 + 6w^2 - w + C$

17. $\ln|r| + 10r^{-1} - 2r^{-2} + C$

19. $-\frac{1}{2}x^{-2} + \frac{1}{3}x^{-3} - \frac{1}{4}x^{-4} + C$

21. $-4\cos x - x - 2x^{-4} + C$ **23.** $-\cot x + \csc x + C$

25. $-2\cot x + 3x + C$ **27.** $4x^2 + x - 9e^x + C$

29. $x^2 - x + 5\tan^{-1}x + C$ **31.** $\tan x - x + C$

41. $x^2 - 4x + 5$ **43.** $2x^3 + 9x + C$

45. $-x^{-1} + C$ **47.** $x - x^2 - \cos x + C$

49. $y = x^2 - x + 1$

51. $f'(x) = x^2 + C_1$; $f(x) = \frac{1}{3}x^3 + C_1x + C_2$

53. $f(x) = x^4 + x^2 - 3x + 2$ **55.** G

57. $y = \frac{\omega^2}{2g}x^2$

Exercises 5.2, Page 285

1. $-\frac{1}{6}(1 - 4x)^{3/2} + C$ **3.** $-\frac{1}{10}(5x + 1)^{-2} + C$

5. $\frac{1}{3}(x^2 + 4)^{3/2} + C$ **7.** $\frac{1}{18}\sin^6 3x + C$

9. $\frac{1}{6}\tan^3 2x + C$ **11.** $-\frac{1}{4}\cos 4x + C$

13. $\frac{1}{3}(2t)^{3/2} - \frac{1}{6}\sin 6t + C$ **15.** $-\frac{1}{2}\cos x^2 + C$

17. $\frac{1}{3}\tan x^3 + C$ **19.** $-2\csc\sqrt{x} + C$

21. $\frac{1}{7}\ln|7x + 3| + C$ **23.** $\frac{1}{2}\ln(x^2 + 1) + C$

25. $x - \ln|x + 1| + C$ **27.** $\ln|\ln x| + C$

29. $-\cos(\ln x) + C$ **31.** $\frac{1}{10}e^{10x} + C$

33. $-\frac{1}{6}e^{-2x^3} + C$ **35.** $-2e^{-\sqrt{x}} + C$

37. $\ln(e^x + e^{-x}) + C$ **39.** $\sin^{-1}\left(\frac{x}{\sqrt{5}}\right) + C$

41. $\frac{1}{5}\tan^{-1}5x + C$ **43.** $\tan^{-1}e^x + C$

45. $-2\sqrt{1 - x^2} - 3\sin^{-1}x + C$

47. $\frac{1}{2}(\tan^{-1}x)^2 + C$ **49.** $-\frac{1}{5}\ln|\cos 5x| + C$

51. $\frac{1}{2}x - \frac{1}{4}\sin 2x + C$ **53.** $\frac{1}{2}x + \frac{1}{16}\sin 8x + C$

55. $11x + 12\cos x - \sin 2x + C$ **57.** $-\frac{3}{4}(1 - x)^{4/3} + C$

59. $y = x + 2\cos 3x + 1 - \pi$

63. (b) $\frac{1}{2}\pi\sqrt{L/g}$ **(c)** $2\pi\sqrt{L/g}$

Exercises 5.3, Page 293

1. $3 + 6 + 9 + 12 + 15$ **3.** $\frac{2}{1} + \frac{2^2}{2} + \frac{2^3}{3} + \frac{2^4}{4}$

5. $-\frac{1}{7} + \frac{1}{9} - \frac{1}{11} + \frac{1}{13} - \frac{1}{15} + \frac{1}{17} - \frac{1}{19} + \frac{1}{21} - \frac{1}{23} + \frac{1}{25}$

7. $(2^2 - 4) + (3^2 - 6) + (4^2 - 8) + (5^2 - 10)$

9. $-1 + 1 - 1 + 1 - 1$ **11.** $\sum_{k=1}^{7}(2k + 1)$

13. $\sum_{k=0}^{12}(3k + 1)$ **15.** $\sum_{k=1}^{5}\frac{(-1)^{k+1}}{k}$

17. $\sum_{k=1}^{8}6$ **19.** $\sum_{k=1}^{4}\frac{(-1)^{k+1}}{k^2}\cos\frac{k\pi}{p}x$

21. 420   **23.** 65

25. 109   **27.** 3069

29. 18   **31.** 28

33. $\frac{8}{3}$ **35.** $\frac{4}{3}$

37. $\frac{16}{3}$ **39.** $\frac{1}{4}$

41. $\frac{25}{2}$ **43.** $\frac{77}{60}; \frac{25}{12}$

45. 9

47.
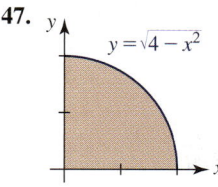
$y = \sqrt{4 - x^2}$

Exercises 5.4, Page 303

1. $\dfrac{33}{2}$; 1

3. $\dfrac{189}{256}$; $\dfrac{3}{4}$

5. $\dfrac{1}{4}(3 - \sqrt{2})\pi$; π

7. 5

9. $\displaystyle\int_{-2}^{4} \sqrt{9 + x^2}\,dx$

11. $\displaystyle\int_{0}^{2} (1 + x)\,dx$

13. -4

15. $\dfrac{5}{6}$

17. $-\dfrac{3}{4}$

21. 4

23. 12

25. -3

27. 40

29. $-\dfrac{28}{3}$

31. -32

33. $\dfrac{28}{3}$

35. 36

37. 0

39. 2.5

41. 11

43. **(a)** -2.5 **(b)** 3.9 **(c)** -1.2 **(d)** 1.4 **(e)** 2.7 **(f)** 0.2

45.
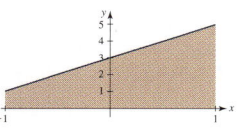

47.
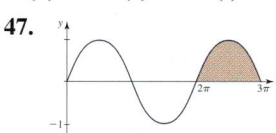

49. 18

51. $\dfrac{9}{4}\pi$

53.
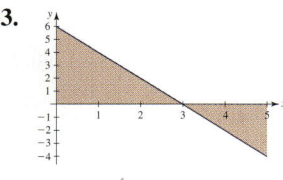

55.
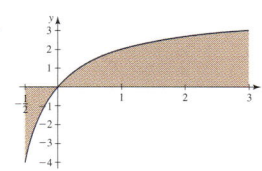

57. 15

59. $-\dfrac{\pi}{2}$

61. -2

63. $\dfrac{5}{2}$

69. $\geq$

Exercises 5.5, Page 313

1. 4

3. 12

5. 46

7. 1

9. $-\dfrac{1}{3} - \dfrac{\sqrt{2}}{6}$

11. $\dfrac{2}{3}$

13. $e - e^{-1}$

15. $-\dfrac{2}{3}$

17. $-\dfrac{28}{3}$

19. $\dfrac{8}{3}$

21. $\dfrac{\pi}{12}$

23. $\dfrac{128}{3}$

25. 1

27. $\dfrac{65}{4}$

29. $\sqrt{6} - \sqrt{3}$

31. $\dfrac{1}{2}$

33. 1

35. $\dfrac{2}{3}$

37. $\dfrac{4\pi + 6}{(\pi + 2)(\pi + 3)}$

39. $\dfrac{3}{8} + \dfrac{1}{4\pi}$

41. $\dfrac{1}{2}\ln\dfrac{11}{3}$

43. xe^x

45. $(3t^2 - 2t)^6$

47. $6\sqrt{24x + 5}$

49. $\dfrac{2x}{x^6 + 1} - \dfrac{3}{27x^3 + 1}$

53. **(a)** 0 **(b)** $\ln 3$ **(c)** $\dfrac{2}{3}$ **(d)** $-\dfrac{4}{9}$

55. $\dfrac{19}{6}$

57. 9

59. $\dfrac{38}{3}$

61. 5

63. 22

65. 4

67. $\dfrac{1}{6}(1 + \ln 2)^6$

69. $\dfrac{1}{2}\ln\left(\dfrac{2}{1 + e^{-2}}\right)$

Chapter 5 in Review, Page 316

A. 1. false

3. true

5. true

7. true

9. false

11. true

13. false

15. true

B. 1. $f(x)$

3. $\dfrac{\ln x}{x}$

5. $-f(g(x))g'(x)$

7. $\displaystyle\sum_{k=1}^{5} \dfrac{k}{2k + 1}$

9. $\displaystyle\int_{5}^{17}$

11. $\dfrac{5}{2}$

13. $\displaystyle\int_{0}^{4} \sqrt{x}\,dx$; $\dfrac{16}{3}$

15. $2 + e^{-1} - e$; $e - e^{-1}$

C. 1. -6

3. $\dfrac{1}{505}(5t + 1)^{101} + C$

5. $\dfrac{1}{2}$

7. 0

9. $-\dfrac{1}{56}\cot^7 8x + C$

11. $\dfrac{1}{40}(4x^2 - 16x + 7)^5 + C$

13. $\dfrac{1}{2}(x^3 + 3x - 16)^{2/3} + C$

15. $\dfrac{1}{2}\ln 2$

17. $\dfrac{\pi}{6}$

19. $-\dfrac{1}{10}\ln|\cos 10x| + C$

21. 5

23. $\dfrac{11}{2}$

25. 0

27. $\dfrac{2}{3\sqrt{3}}\pi$

29. $\dfrac{1}{2}$

31. 156 lb; approximately 20 min

33. $\dfrac{51}{4}$

Exercises 6.1, Page 323

1. $s(t) = 6t - 7$

3. $s(t) = \dfrac{1}{3}t^3 - 2t^2 + 15$

5. $s(t) = -\dfrac{5}{2}\sin(4t + \pi/6) + \dfrac{5}{2}$

7. $v(t) = -5t + 9;\quad s(t) = -\dfrac{5}{2}t^2 + 9t - \dfrac{9}{2}$

9. $v(t) = t^3 - 2t^2 + 5t - 3;\ s(t) = \dfrac{1}{4}t^4 - \dfrac{2}{3}t^3 + \dfrac{5}{2}t^2 - 3t + 10$

11. $v(t) = \dfrac{21}{4}t^{4/3} - t - 26;\quad s(t) = \dfrac{9}{4}t^{7/3} - \dfrac{1}{2}t^2 - 26t - 48$

13. 17 cm   **15.** 34 cm

17. 24 cm   **19.** $\dfrac{1}{30}$ mi = 176 ft

21. 256 ft   **23.** 30.625 m

25. 400 ft;   6 s   **27.** -80 ft/s

Exercises 6.2, Page 331

1. $\dfrac{4}{3}$ **3.** $\dfrac{81}{4}$

5. $\dfrac{9}{2}$ **7.** $\dfrac{11}{2}$

9. $\dfrac{11}{4}$ **11.** $\dfrac{11}{6}$

13. 2   **15.** $\dfrac{3}{4}(2^{4/3} + 3^{4/3})$

17. 4   **19.** 2π

21. $\dfrac{7}{3}$ **23.** $\dfrac{27}{2}$

25. $\dfrac{32}{3}$ **27.** $\dfrac{81}{4}$

29. 4   **31.** $\dfrac{10}{3}$

33. $\dfrac{64}{3}$ **35.** $\dfrac{128}{5}$

37. $\dfrac{118}{3}$ **39.** 22

41. $\dfrac{9}{2}$ **43.** $\dfrac{8}{3}$

45. 8   **47.** $2\sqrt{2} - 2$

49. $4\sqrt{3} - 4\pi/3$ **53.** $7 + 3\ln\dfrac{3}{4} \approx 6.1370$

55. $9\pi/4$ **57.** $4 + 2\pi$

59. πab **61.** $\dfrac{52}{3}$

63. $A = \displaystyle\int_0^{\ln\frac{3}{2}}(e^x - 1)\,dx + \int_{\ln\frac{3}{2}}^{\ln 2}(2 - e^x)\,dx,$

$A = \displaystyle\int_1^2\left[\ln y - \ln\dfrac{1}{2}(y+1)\right]dy;\qquad \ln\dfrac{32}{27} \approx 0.1699$

Exercises 6.3, Page 338

1. $\dfrac{256\sqrt{3}}{3}$ **3.** 128

5. $10\pi/3$ **7.** 9

9. $\pi/2$ **11.** $4\pi/5$

13. $\pi/6$ **15.** $1296\pi/5$

17. $\pi/2$ **19.** $32\pi/5$

21. 32π **23.** $7\pi/3$

25. $256\pi/15$ **27.** $3\pi/5$

29. 36π **31.** $500\pi/3$

33. $16\pi/105$ **35.** $\pi\left(2e^{-1} - \dfrac{1}{2}e^{-2} - \dfrac{1}{2}\right)$

37. π^2 **39.** $\dfrac{1}{4}(4\pi - \pi^2)$

Exercises 6.4, Page 344

1. $4\pi/5$ **3.** $\pi/6$

5. $8\pi/15$ **7.** $250\pi/3$

9. $36\sqrt{3}\pi/5$ **11.** $3\pi/2$

13. 16π **15.** $8\pi/5$

17. $21\pi/10$ **19.** $\pi/6$

21. $243\pi/10$ **23.** 4π

25. $625\pi/6$ **27.** $248\pi/15$

29. $\dfrac{1}{2}(\pi^2 - 2\pi)$ **31.** $\dfrac{1}{3}\pi r^2 h$

33. $\dfrac{4}{3}\pi r^3$ **35.** $\dfrac{4}{3}\pi ab^2$

37. $V = \pi r^2 h - \dfrac{\pi\omega^2 r^4}{4g}$

Exercises 6.5, Page 347

1. $2\sqrt{2}$ **3.** $\dfrac{1}{27}(13^{3/2} - 8) \approx 1.4397$

5. 45   **7.** $\dfrac{10}{3}$

9. $\dfrac{4685}{288} \approx 16.2674$ **11.** 9

13. $\displaystyle\int_{-1}^{3}\sqrt{1 + 4x^2}\,dx$ **15.** $\displaystyle\int_0^{\pi}\sqrt{1 + \cos^2 x}\,dx$

17. $\dfrac{1}{27}(40^{3/2} - 8) \approx 9.0734$ **19.** **(b)** 6

21. $\pi/2$

Exercises 6.6, Page 350

1. $208\pi/3$ **3.** $\dfrac{\pi}{27}(10^{3/2} - 1) \approx 3.5631$

5. $\dfrac{\pi}{6}(37^{3/2} - 1) \approx 117.3187$ **7.** $100\sqrt{5}\pi$

9. $253\pi/20$

11. **(a)** $(\pi r/6h^2)\left[(r^2 + 4h^2)^{3/2} - r^3\right]$
     (b) approximately $0.99\% < 1\%$

13. $20\sqrt{2}\pi$

Exercises 6.7, Page 354

1. -4

3. $\dfrac{34}{3}$

5. 3

7. 0

9. 2

11. $\dfrac{61}{9}$

13. 24

15. $\dfrac{1}{12}$

17. 0

19. $3\sqrt{3}/\pi$

21. $-1 + \dfrac{2\sqrt{3}}{3} \approx 0.1547$

23. 12

25. $103°$

29. $2kt_1/3$

Exercises 6.8, Page 360

1. 3300 ft-lb

3. $\dfrac{2}{5}$ ft

5. (a) 10 joules   **(b)** 27.5 joules

7. (a) 7.5 ft-lb   **(b)** 37.5 ft-lb

9. 453.1×10^8 joules

11. 127,030.9 ft-lb

13. 45,741.6 ft-lb

15. 57,408 ft-lb

17. 64,000 ft-lb

19. (a) 5200 ft-lb   **(b)** 6256.25 ft-lb

21. $3k/4$, where k is constant of proportionality

Exercises 6.9, Page 365

1. (a) 196,000 N/m^2; $4,900,000\pi$ N
   (b) 196,000 N/m^2; $784,000\pi$ N
   (c) 196,000 N/m^2; $19,600,000\pi$ N

3. (a) 499.2 lb/ft^2;   244,640 lb   **(b)** 59,904 lb;   29,952 lb

5. 129.59 lb

7. 1280 lb

9. 3660.8 lb

11. 13,977.6 lb

13. 9984π lb

15. 5990.4 lb

Exercises 6.10, Page 372

1. $-\dfrac{2}{7}$

3. $-\dfrac{13}{30}$

5. 1

7. $\dfrac{115}{36}$

9. $\dfrac{4}{7}$

11. $\dfrac{19}{15}$

13. $\dfrac{11}{10}$

15. $\dfrac{15}{2}$

17. $\bar{x} = -\dfrac{2}{7}, \bar{y} = \dfrac{17}{7}$

19. $\bar{x} = \dfrac{17}{11}, \bar{y} = -\dfrac{20}{11}$

21. $\bar{x} = \dfrac{10}{9}, \bar{y} = \dfrac{28}{9}$

23. $\bar{x} = \dfrac{3}{4}, \bar{y} = \dfrac{3}{10}$

25. $\bar{x} = \dfrac{12}{5}, \bar{y} = \dfrac{54}{7}$

27. $\bar{x} = \dfrac{93}{35}, \bar{y} = \dfrac{45}{56}$

29. $\bar{x} = \dfrac{1}{2}, \bar{y} = \dfrac{8}{5}$

31. $\bar{x} = \dfrac{16}{35}, \bar{y} = \dfrac{16}{35}$

33. $\bar{x} = \dfrac{3}{2}, \bar{y} = \dfrac{121}{540}$

35. $\bar{x} = -\dfrac{7}{10}, \bar{y} = \dfrac{7}{8}$

37. $\bar{x} = 0, \bar{y} = 2$

39. $\bar{x} = 0, \bar{y} = \dfrac{1}{8}(\pi + 8)$

Chapter 6 in Review, Page 373

A. 1. false

3. true

5. true

7. true

9. true

11. false

B. 1. joule

3. 2500 ft-lb

5. 6

7. smooth

C. 1. $-\displaystyle\int_0^a f(x)\, dx$

3. $\displaystyle\int_0^a \left[f(x) - \dfrac{f(a)}{a}x \right] dx$

5. $-\displaystyle\int_a^b 2f(x)\, dx + \int_b^c 2f(x)\, dx$

7. $\displaystyle\int_b^c [a - f(y)]\, dy + \int_c^d [f(y) - a]\, dy$

9. $\dfrac{1}{4}a^2 + b^2$

11. $\bar{x} = \dfrac{\displaystyle\int_0^2 x\,[f(x) - g(x)]\, dx}{\displaystyle\int_0^2 [f(x) - g(x)]\, dx}, \bar{y} = \dfrac{\dfrac{1}{2}\displaystyle\int_0^2 ([f(x)]^2 - [g(x)]^2)\, dx}{\displaystyle\int_0^2 [f(x) - g(x)]\, dx}$

13. $2\pi \displaystyle\int_0^2 x[f(x) - g(x)]\, dx$

15. $2\pi \displaystyle\int_0^2 (2 - x)[f(x) - g(x)]\, dx$

17. $\dfrac{5}{2}$

19. (a) 4   **(b)** π

21. $\dfrac{315\sqrt{41}}{16}\pi$ ft$^2 \approx 396.03$ ft^2

23. $\dfrac{256}{45}$

25. 37.5 joules

27. 624,000 ft-lb

29. 2040 ft-lb

31. 691,612.83 ft-lb

33. $\dfrac{1}{27}(40^{3/2} - 8) \approx 9.07$

35. 17,066.7 N

37. $\frac{3}{4}$ m from the left along the 1-m bar and $\frac{6}{5}$ m from the left along the 2-m bar

Exercises 7.1, Page 382

1. $-\dfrac{5^{-5x}}{5\ln 5} + C$

3. $-2\cos\sqrt{1+x} + C$

5. $-\dfrac{1}{4}\sqrt{25 - 4x^2} + C$

7. $\dfrac{1}{5}\sec^{-1}\left|\dfrac{2}{5}x\right| + C$

9. $\dfrac{1}{10}\tan^{-1}\left(\dfrac{2}{5}x\right) + C$

11. $\dfrac{1}{20}\ln\left|\dfrac{2x-5}{2x+5}\right| + C$

13. $\dfrac{1}{10}\ln|\sin 10x| + C$

15. $(3 - 5t)^{-1.2} + C$

17. $\dfrac{1}{3}\ln|\sec 3x + \tan 3x| + C$

19. $\dfrac{1}{2}(\sin^{-1}x)^2 + C$

21. $-\tan^{-1}(\cos x) + C$

23. $\dfrac{1}{4}\tanh x^4 + C$

25. $\dfrac{1}{2}\sec 2x + C$

27. $\csc(\cos x) + C$

29. $\dfrac{1}{3}(1 + \tan x)^3 + C$

31. $\dfrac{1}{2}\ln(1 + e^{2x}) + C$

Exercises 7.2, Page 385

1. $\frac{1}{5}(x+1)^5 - \frac{1}{4}(x+1)^4 + C$

3. $\frac{4}{5}(x-5)^{5/2} + \frac{22}{3}(x-5)^{3/2} + C$

5. $\frac{2}{3}(x-1)^{3/2} + 2(x-1)^{1/2} + C$

7. $\frac{2}{9}(3x-4)^{1/2} - \frac{26}{9}(3x-4)^{-1/2} + C$

9. $2\sqrt{x} - 2\tan^{-1}\sqrt{x} + C$

11. $(\sqrt{t}+1)^2 - 10(\sqrt{t}+1) + 8\ln(\sqrt{t}+1) + C$

13. $\frac{3}{10}(x^2+1)^{5/3} - \frac{3}{4}(x^2+1)^{2/3} + C$

15. $-\frac{1}{x-1} - \frac{1}{(x-1)^2} - \frac{1}{3(x-1)^3} + C$

17. $2\sqrt{e^x - 1} - 2\tan^{-1}\sqrt{e^x - 1} + C$

19. $\frac{4}{5}(1 - \sqrt{v})^{5/2} - \frac{4}{3}(1 - \sqrt{v})^{3/2} + C$

21. $\frac{4}{3}(1 + \sqrt{t})^{3/2} + C$

23. $\ln(x^2 + 2x + 5) + \frac{5}{2}\tan^{-1}\left(\frac{x+1}{2}\right) + C$

25. $-2\sqrt{16 - 6x - x^2} - \sin^{-1}\left(\frac{x+3}{5}\right) + C$

27. $2x^{1/2} + 3x^{1/3} + 6x^{1/6} + 6\ln|x^{1/6} - 1| + C$

29. $\frac{506}{375}$ **31.** $6 + 20\ln\frac{11}{14}$

33. $\frac{177}{2}$ **35.** $\frac{1}{1326}$

37. $3 + 3\ln\frac{2}{3}$ **39.** $\frac{1}{168}$

43. $-\frac{3}{2} + 3\ln 2$ **45.** $\frac{32\pi}{3} - 4\pi\ln 3$

47. $\frac{232}{15}$

Exercises 7.3, Page 392

1. $\frac{2}{3}x(x+3)^{3/2} - \frac{4}{15}(x+3)^{5/2} + C$

3. $x\ln 4x - x + C$ **5.** $\frac{1}{2}x^2\ln 2x - \frac{1}{4}x^2 + C$

7. $-x^{-1}\ln x - x^{-1} + C$ **9.** $t(\ln t)^2 - 2t\ln t + 2t + C$

11. $x\sin^{-1}x + \sqrt{1 - x^2} + C$ **13.** $\frac{1}{3}xe^{3x} - \frac{1}{9}e^{3x} + C$

15. $-\frac{1}{4}x^3e^{-4x} - \frac{3}{16}x^2e^{-4x} - \frac{3}{32}xe^{-4x} - \frac{3}{128}e^{-4x} + C$

17. $\frac{1}{2}x^2e^{x^2} - \frac{1}{2}e^{x^2} + C$ **19.** $\frac{1}{8}t\sin 8t + \frac{1}{64}\cos 8t + C$

21. $-x^2\cos x + 2x\sin x + 2\cos x + C$

23. $\frac{1}{3}x^3\sin 3x + \frac{1}{3}x^2\cos 3x - \frac{2}{9}x\sin 3x - \frac{2}{27}\cos 3x + C$

25. $\frac{1}{17}e^x(\sin 4x - 4\cos 4x) + C$ **27.** $\frac{1}{5}e^{-2\theta}(\sin\theta - 2\cos\theta) + C$

29. $\theta\sec\theta - \ln|\sec\theta + \tan\theta| + C$

31. $\frac{1}{3}\cos x\cos 2x + \frac{2}{3}\sin x\sin 2x + C$

33. $\frac{1}{3}x^2(x^2+4)^{3/2} - \frac{2}{15}(x^2+4)^{5/2} + C$

35. $\frac{1}{2}x\sin(\ln x) - \frac{1}{2}x\cos(\ln x) + C$

37. $-\frac{1}{2}\csc x\cot x + \frac{1}{2}\ln|\csc x - \cot x| + C$

39. $x\tan x + \ln|\cos x| + C$ **41.** $\frac{3}{2}\ln 3$

43. $-12e^{-2} + 8e^{-1}$ **45.** $\frac{\pi}{4} - \frac{1}{2}\ln 2$

47. $3\ln 3 + e^{-1}$

49. $5\pi(\ln 5)^2 - 10\pi\ln 5 + 8\pi$

51. $2\pi^2$ **53.** $\frac{\pi}{4} - \frac{1}{2}\ln 2$

55. $v(t) = -te^{-t} - e^{-t} + 2;\quad s(t) = te^{-t} + 2e^{-t} + 2t - 3$

57. $(124.8)\cdot\frac{8(\pi-2)}{\pi^2} \approx 115.48\,\text{lb}$

59. $4\tan^{-1}2 - \pi/2 - \ln\frac{5}{2}$

61. $-2\sqrt{x+2}\cos\sqrt{x+2} + 2\sin\sqrt{x+2} + C$

67. $-\frac{1}{3}\sin^2 x\cos x - \frac{2}{3}\cos x + C$

69. $\frac{1}{30}\cos^2 10x\sin 10x + \frac{1}{15}\sin 10x + C$

73. $\frac{35\pi}{256}$ **83. (b)** $\frac{17\pi}{4}$

Exercises 7.4, Page 398

1. $\frac{2}{3}(\sin x)^{3/2} + C$ **3.** $\sin x - \frac{1}{3}\sin^3 x + C$

5. $-\cos t + \frac{2}{3}\cos^3 t - \frac{1}{5}\cos^5 t + C$ **7.** $\frac{1}{4}\sin^4 x - \frac{1}{6}\sin^6 x + C$

9. $\frac{3}{8}t - \frac{1}{4}\sin 2t + \frac{1}{32}\sin 4t + C$

11. $\frac{1}{16}x - \frac{1}{64}\sin 4x + \frac{1}{48}\sin^3 2x + C$

13. $\frac{3}{128}x - \frac{1}{128}\sin 4x + \frac{1}{1024}\sin 8x + C$

15. $\frac{1}{8}\tan^4 2t + \frac{1}{12}\tan^6 2t + C$

17. $\frac{1}{4}\tan x\sec^3 x - \frac{1}{8}\sec x\tan x - \frac{1}{8}\ln|\sec x + \tan x| + C$

19. $\frac{2}{3}(\sec x)^{3/2} + 2(\sec x)^{-1/2} + C$ **21.** $\frac{1}{7}\sec^7 x - \frac{1}{5}\sec^5 x + C$

23. $\frac{1}{4}\tan x\sec^3 x + \frac{3}{8}\sec x\tan x + \frac{3}{8}\ln|\sec x + \tan x| + C$

25. $\ln|\sin x| + \frac{1}{2}\cos^2 x + C$

27. $-\frac{1}{11}\cot^{11}x - \frac{1}{13}\cot^{13}x + C$

29. $\frac{1}{7\tan^7(1-t)} + \frac{1}{5\tan^5(1-t)} + C$

31. $\frac{1}{2}\sec x\tan x + 2\sec x + \frac{1}{2}\ln|\sec x + \tan x| + C$

33. $\frac{1}{3}\tan^3 x - \tan x + x + C$ **35.** $-\frac{1}{2}\csc^2 t - \ln|\sin t| + C$

37. $\frac{1}{5}\tan^5 x - \frac{1}{3}\tan^3 x + C$ **39.** $-\frac{1}{2}\cos x^2 + \frac{1}{6}\cos^3 x^2 + C$

41. $\dfrac{25\sqrt{2}}{168}$ **43.** 0

45. $\dfrac{3}{4}$ **47.** $-\frac{1}{6}\cos 3x + \frac{1}{2}\cos x + C$

49. $\frac{1}{4}\sin 2x - \frac{1}{12}\sin 6x + C$ **51.** $\dfrac{5}{12}$

55. $\dfrac{16\pi}{3}$ **57.** $\dfrac{5\sqrt{2}}{3}$

Exercises 7.5, Page 405

1. $-\sin^{-1}x - \dfrac{\sqrt{1-x^2}}{x} + C$ **3.** $\ln\left|\dfrac{x+\sqrt{x^2-36}}{6}\right| + C$

5. $\frac{1}{3}(x^2+7)^{3/2} + C$

7. $-\frac{1}{3}(1-x^2)^{3/2} + \frac{1}{5}(1-x^2)^{5/2} + C$

9. $-\dfrac{x}{4\sqrt{x^2-4}} + C$

11. $\frac{1}{2}x\sqrt{x^2+4} + 2\ln x\left|\dfrac{\sqrt{x^2+4}+x}{2}\right| + C$

13. $\sin^{-1}\left(\dfrac{x}{5}\right) + C$

15. $\frac{1}{4}\ln\left|\dfrac{4-\sqrt{16-x^2}}{x}\right| + C$

17. $\ln\left|\dfrac{\sqrt{x^2+1}-1}{x}\right| + C$ **19.** $-\dfrac{(1-x^2)^{3/2}}{3x^3} + C$

21. $\dfrac{x}{\sqrt{9-x^2}} - \sin^{-1}\left(\dfrac{x}{3}\right) + C$

23. $\frac{1}{2}\tan^{-1}x + \dfrac{x}{2(1+x^2)} + C$

25. $\dfrac{x}{16\sqrt{4+x^2}} - \dfrac{x^3}{48(4+x^2)^{3/2}} + C$

27. $\ln\left|\dfrac{\sqrt{x^2+2x+10}+x+1}{3}\right| + C$

29. $\frac{1}{16}\tan^{-1}\left(\dfrac{x+3}{2}\right) + \dfrac{x+3}{8(x^2+6x+13)} + C$

31. $\dfrac{-5x-1}{9\sqrt{5-4x-x^2}} + C$ **33.** $\ln(x^2+4x+13) + C$

35. $x - 4\tan^{-1}\left(\dfrac{x}{4}\right) + C$

37. $\frac{9}{2}\sin^{-1}\left(\dfrac{x-3}{3}\right) + \frac{1}{2}(x-3)\sqrt{9-(x-3)^2} + C$

39. $\dfrac{2\pi}{3} + \sqrt{3}$ **41.** $\dfrac{\sqrt{2}}{50}$

43. $2\sqrt{3} - \dfrac{172}{81}$

45. $\frac{1}{3}x^3\sin^{-1}x + \frac{1}{3}\sqrt{1-x^2} - \frac{1}{9}(1-x^2)^{3/2} + C$

47. $\dfrac{1}{\sqrt{3}}\ln\left(\dfrac{\sqrt{2}-1}{2-\sqrt{3}}\right)$ **51.** $\dfrac{\pi\sqrt{3}}{9}\left(\sqrt{3} - 1 - \dfrac{\pi}{12}\right)$

53. $12\pi\sqrt{2} - 4\pi\ln(\sqrt{2}+1)$

55. $2 - \sqrt{2} - \ln(\sqrt{6} - \sqrt{3})$

57. **(b)** $y = -10\ln\left(\dfrac{10-\sqrt{100-x^2}}{x}\right) - \sqrt{100-x^2}$

59. $15.6\pi \approx 49.01$ lb

Exercises 7.6, Page 413

1. $\dfrac{A}{x} + \dfrac{B}{x+1}$

3. $\dfrac{A}{x-1} + \dfrac{B}{x+2} + \dfrac{C}{(x+2)^2} + \dfrac{D}{(x+2)^3}$

5. $\dfrac{A}{x} + \dfrac{B}{x^2} + \dfrac{C}{x^3} + \dfrac{Dx+E}{x^2+3}$

7. $\dfrac{Ax+B}{x^2+9} + \dfrac{Cx+D}{(x^2+9)^2}$

9. $-\frac{1}{2}\ln|x| + \frac{1}{2}\ln|x-2| + C$

11. $-2\ln|x| + \frac{5}{2}\ln|2x-1| + C$

13. $\frac{5}{8}\ln|x-4| + \frac{3}{8}\ln|x+4| + C$

15. $-\frac{1}{6}\ln|2x+1| + \frac{2}{3}\ln|x+2| + C$

17. $6\ln|x| - \frac{7}{2}\ln|x+1| - \frac{3}{2}\ln|x-1| + C$

19. $\frac{1}{2}\ln|x+1| - \ln|x+2| + \frac{1}{2}\ln|x+3| + C$

21. $-2\ln|t| - t^{-1} + 6\ln|t-1| + C$

23. $\ln|x| - \ln|x+1| + (x+1)^{-1} + C$

25. $-2(x+1)^{-1} + \frac{3}{2}(x+1)^{-2} + C$

27. $-\frac{1}{32}\ln|x+1| - \frac{1}{16}(x+1)^{-1} + \frac{1}{32}\ln|x+5|$
$\qquad - \frac{1}{16}(x+5)^{-1} + C$

29. $-\frac{19}{16}\ln|x| - \frac{19}{8}x^{-1} + \frac{11}{8}x^{-2} - \frac{3}{2}x^{-3} + \frac{35}{16}\ln|x+2| + C$

31. $-\ln|x| + \frac{1}{2}\ln(x^2+1) + \tan^{-1}x + C$

33. $\frac{1}{2}(x+1)^{-1} + \frac{1}{2}\tan^{-1}x + C$

35. $\frac{1}{3}\tan^{-1}x - \frac{1}{6}\tan^{-1}\left(\dfrac{x}{2}\right) + C$

37. $\frac{1}{3}\ln|x-1| - \frac{1}{6}\ln|x^2+x+1| - \dfrac{1}{\sqrt{3}}\tan^{-1}\left(\dfrac{2x+1}{\sqrt{3}}\right) + C$

39. $5\ln|x+1| - \ln(x^2+2x+2) - 7\tan^{-1}(x+1) + C$

41. $\dfrac{1}{2(x^2+4)} + \frac{1}{2}\tan^{-1}\left(\dfrac{x}{2}\right) + C$

43. $\frac{1}{2}\ln(x^2+4) - \frac{11}{16}\tan^{-1}\left(\dfrac{x}{2}\right) + \dfrac{5x+12}{8(x^2+4)} + C$

45. $\frac{1}{3}x^3 - x^2 + 6x - 10\ln|x+1| - 8(x+1)^{-1} + C$

47. $-\dfrac{1}{2}\ln 3$

49. $2\ln\dfrac{5}{3}-\dfrac{14}{15}$

51. $\dfrac{1}{6}\ln\dfrac{8}{3}+\dfrac{1}{3\sqrt{2}}\tan^{-1}\left(\dfrac{1}{\sqrt{2}}\right)+C$ **53.** 0

55. $\dfrac{1}{4}\ln\left|\dfrac{1+\sqrt{1-x^2}}{1-\sqrt{1-x^2}}\right|-\dfrac{\sqrt{1-x^2}}{2x^2}+C$

57. $3(x+1)^{1/3}+\ln|(x+1)^{1/3}-1|-$

$\dfrac{1}{2}\ln|(x+1)^{2/3}+(x+1)^{1/3}+1|-\sqrt{3}\tan^{-1}\left(\dfrac{2(x+1)^{1/3}+1}{\sqrt{3}}\right)+C$

59. $\dfrac{1}{4}\ln\dfrac{15}{7}\approx 0.191$

61. $7\ln 2-8\ln 3+3\ln 4\approx 0.222$

63. $8\pi\ln\dfrac{2}{3}+\dfrac{11\pi}{3}\approx 1.329$

65. $8\pi\ln 2-4\pi\approx 4.854$

Exercises 7.7, Page 421

1. $\dfrac{1}{81}$

3. diverges

5. $\dfrac{1}{2}e^6$

7. diverges

9. $\dfrac{1}{2}$

11. 0

13. $-\dfrac{1}{18}$

15. $3e^{-2}$

17. 1

19. $\dfrac{\pi}{2}$

21. $\dfrac{1}{2}$

23. 4

25. $\ln 2$

27. $\dfrac{1}{4}\ln\dfrac{7}{3}$

29. $\dfrac{1}{21}$

31. diverges

33. 100

35. $2\sqrt{2}$

37. diverges

39. 6

41. $-\dfrac{1}{4}$

43. diverges

45. diverges

47. $-\dfrac{4}{3}$

49. $\dfrac{\pi}{4}$

51. $\dfrac{\pi}{2}$

53. $\dfrac{\pi}{6}$

55. $\dfrac{1}{6}$

57. 2

59. 8

61. $\dfrac{1}{2}\ln 2$

63. 2.86×10^{10} joules

65. $\dfrac{1}{s}, s>0$

67. $\dfrac{1}{s-1}, s>1$

69. $\dfrac{1}{s^2+1}, s>0$

71. $\dfrac{e^{-s}}{s}, s>0$

Exercises 7.8, Page 430

1. 78; $M_3=77.25$

3. 22; $T_3=22.5$

5. 1.7564; 1.8667

7. 1.1475; 1.1484

9. 0.4393; 0.4228

11. 0.4470; 0.4900

13. $\dfrac{26}{3}$; $S_4=8.6611$

15. 1.6222

17. 0.7854

19. 0.4339

21. 11.1053

23. $n\geq 8$

25. 1.11

27. Simpson's Rule: $n\geq 26$; Trapezoidal Rule: $n\geq 366$

29. Trapezoidal Rule gives 1.10

31. For $n=2$ and $n=4$, the Midpoint Rule gives the exact value of the integral: 36

33. (a) $\dfrac{2}{3}$ (b) $M_8=\dfrac{21}{22}$ (c) $T_8=\dfrac{11}{16}$

(d) $E_8=\frac{1}{96}$ for Midpoint Rule and $E_8=\frac{1}{48}$ for Trapezoidal Rule. The error for the Midpoint Rule is one-half the error for the Trapezoidal Rule.

37. 7.0667

39. approximately 4975 gal

41. 41.4028

43. (b) 1.2460

45. 1.4804

47. 14.9772

Chapter 7 in Review, Page 433

A. 1. true

3. true

5. true

7. false

9. false

11. true

13. true

15. false

17. true

19. false

B. 1. $\dfrac{1}{5}$

3. $\sqrt{\pi}$

5. $\ln\sqrt{2}$

C. 1. $2\sqrt{x}-18\ln(\sqrt{x}+9)+C$

3. $(x^2+4)^{1/2}+C$

5. $\dfrac{3}{256}\tan^{-1}\left(\dfrac{x}{2}\right)+\dfrac{x}{32(x^2+4)}+\dfrac{x}{32(x^2+4)^2}$

$-\dfrac{x^3}{128(x^2+4)^2}+C$

7. $x-\dfrac{4}{x}+C$

9. $\dfrac{1}{2}\ln(x^2+4)-\dfrac{5}{2}\tan^{-1}\left(\dfrac{x}{2}\right)+C$

11. $\dfrac{1}{10}(\ln x)^{10}+C$

13. $\dfrac{1}{2}t^2\sin^{-1}t-\dfrac{1}{4}\sin^{-1}t+\dfrac{1}{4}t\sqrt{1-t^2}+C$

15. $\dfrac{1}{5}(x+1)^5-\dfrac{3}{4}(x+1)^4+C$

17. $x\ln(x^2+4)-2x+4\tan^{-1}\left(\dfrac{x}{2}\right)+C$

19. $-\dfrac{2}{125}\ln|x|-\dfrac{1}{25}x^{-1}+\dfrac{2}{125}\ln|x+5|-\dfrac{1}{25}(x+5)^{-1}+C$

21. $-\dfrac{1}{12}\ln|x+3|-\dfrac{1}{2}(x+3)^{-1}+\dfrac{1}{12}\ln|x-3|+C$

23. $\tan t - t + C$

25. $\frac{1}{13}\tan^{13}t + \frac{1}{11}\tan^{11}t + C$

27. $y\sin y + \cos y + C$

29. $\sin t - \frac{1}{5}\sin^5 t + C$

31. $\frac{1}{6}(1 + e^w)^6 + C$

33. $-\frac{1}{8}\csc^2 4x - \frac{1}{4}\ln|\sin 4x| + C$

35. $\frac{1}{4}$

37. $\sec x - \tan x + x + C$

39. $\frac{5}{2}\ln 2 - \frac{3}{2}\ln 3$

41. $\frac{1}{10}e^x(\cos 3x + 3\sin 3x) + C$

43. $\frac{1}{2}t\cos(\ln t) + \frac{1}{2}t\sin(\ln t) + C$

45. $2\sqrt{x}\sin\sqrt{x} + 2\cos\sqrt{x} + C$

47. $-\frac{2}{3}\cos^3 x + C$

49. $\frac{1}{2}(x + 1)\sqrt{x^2 + 2x + 5} + 2\ln\left|\frac{\sqrt{x^2 + 2x + 5} + x + 1}{2}\right| + C$

51. $\frac{1}{7}\tan^7 x - \frac{2}{5}\sec^5 x + \frac{1}{3}\sec^3 x + C$

53. $\frac{1}{4}t^4 - \frac{1}{2}t^2 + \frac{1}{2}\ln(1 + t^2) + C$

55. $\frac{5}{2}\ln(x^2 + 1) + \tan^{-1}x - \frac{1}{2}(x^2 + 1)^{-1} + C$

57. $\frac{1}{4}x^2 - \frac{1}{4}x\sin 2x - \frac{1}{8}\cos 2x + C$

59. $2(\sin x)e^{\sin x} - 2e^{\sin x} + C$

61. $\sqrt{6} - 2$

63. $t\sinh^{-1}t - \sqrt{t^2 + 1} + C$

65. $\ln\frac{3}{2}$

67. $\frac{1}{39}\tan^{13}3u + \frac{1}{45}\tan^{15}3u + C$

69. $3\tan x + \sec x + C$

71. $\frac{1}{2}x^2(1 + \ln x)^2 - \frac{1}{2}x^2(1 + \ln x) + \frac{1}{4}x^2 + C$

73. $e^{e^x} + C$

75. $t^2 - \ln(1 + e^{t^2}) + C$

77. $\frac{1}{5}\sin^{-1}(5x + 2) + C$

79. $(\sin x)\ln|\sin x| - \sin x + C$

81. $\frac{3}{2}\sqrt[3]{9}$

83. 0

85. diverges

87. 0

89. diverges

91. $2 - 2e^{-1}$

95. $\frac{1}{2}$

97. $\frac{2}{3}$

99. (a) 2π (b) the areas are infinite

101. 126 joules

Exercises 8.1, Page 444

1. $y = -\frac{1}{5}\cos 5x + C$

3. $y^{-2} = 2x^{-1} + C$

5. $y + y^2 + \frac{1}{3}y^3 = x + x^2 + \frac{1}{3}x^3 + C$

7. $\cos y = x^{-1} - 5x + C$

9. $y = Cx^4$

11. $-3e^{-2y} = 2e^{3x} + C$

13. $\frac{1}{3}x^3\ln x - \frac{1}{9}x^3 = \frac{1}{2}y^2 + 2y + \ln|y| + C$

15. $\ln|N| = te^{t+2} - e^{t+2} - t + C$

17. $P = \dfrac{5}{1 + Ce^{-5t}}$

19. $(y + 3)^5 e^x = C(x + 4)^5 e^y$

21. $y^3 = -3x^{-1} + 30$

23. $x = \tan\left(4t - \frac{3}{4}\pi\right)$

25. $y = \dfrac{e^{-(1+1/x)}}{x}$

27. $y = \frac{1}{2}x + \frac{\sqrt{3}}{2}\sqrt{1 - x^2}$

29. $y = 3$

31. $y = -4, y = 5$

33. (a) $y = \dfrac{1}{1 + Cx}$ (b) $y = 0$ (c) $y = \dfrac{1}{1 + 2x}$

Exercises 8.2, Page 448

1. $y = Ce^{4x}$

3. $y = \frac{1}{10} + Ce^{-5x}$

5. $y = \frac{1}{4}e^{3t} + Ce^{-t}$

7. $y = \frac{1}{3} + Ce^{-x^3}$

9. $y = \dfrac{\ln x}{x} + \dfrac{C}{x}$

11. $y = \dfrac{C}{1 + e^x}$

13. $y = -x\cos x + Cx$

15. $y = \sin x + C\cos x$

17. $y = \sin x + C\csc x$

19. $y = \frac{5}{3}(x + 2)^{-1} + C(x + 2)^{-4}$

21. $y = \dfrac{e^x}{2x^2} + C\dfrac{e^{-x}}{x^2}$

23. $y = -x - 1 - 3e^x$

25. $y = \dfrac{e^x + 2 - e}{x}$

27. $y = 2x^2 - \frac{49}{5}x$

29. $(t + 1)x = t\ln t - t + 21$

31. $i = \frac{E}{R} + \left(i_0 - \frac{E}{R}\right)e^{-Rt/L}$

33. (a) $y = e^{x^2}[1 + \sqrt{\pi}\,\mathrm{erf}(x)]$ (b) $y(2) = 150.92;$

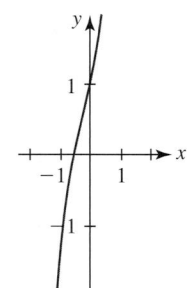

Exercises 8.3, Page 455

1. 7.9 years; 10 years

3. 760

5. approximately 11 h

7. 136.5 h

9. $0.00098 I_0$

11. 15,600 years

13. $36.67°$;   approximately 3.06 min

15. $A(t) = 200 - 170e^{-t/50}$

17. $A(t) = 1000 - 1000e^{-t/100}$

19. 100 min

21. $s(t) = \frac{mg}{k}t - \frac{m}{k}\left(v_0 - \frac{mg}{k}\right)e^{-kt/m} + \frac{mv_0}{k} - \frac{m^2 g}{k^2}$

23. $X(t) = \frac{A}{B} - \frac{A}{B}e^{-Bt};$ $X(t) \to \frac{A}{B}$ as $t \to \infty$; $t = (\ln 2)/B$

25. $E(t) = E_0 e^{-(t-t_1)/RC}$

27. $i(t) = \frac{3}{5} - \frac{3}{5}e^{-500t};$ $i(t) \to \frac{3}{5}$ as $t \to \infty$

31. 276

Exercises 8.4, Page 465

13. 0 is asymptotically stable, 3 is unstable

15. 2 is semi-stable

17. -2 is unstable, 0 is semi-stable; 2 is asymptotically stable

19. -1 is asymptotically stable, 0 is unstable

21.

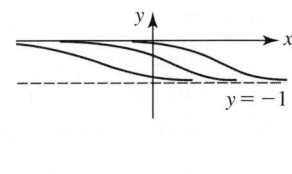

(a)

(b)

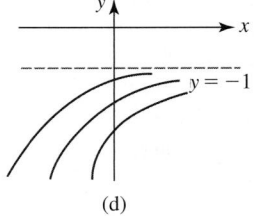

(c)

(d)

25. mg/k

27. $i = E/R$ is an equilibrium solution and E/R is asymptotically stable

Exercises 8.5, Page 470

1. $y_2 = 2.9800$, $y_4 = 3.1151$

3. $y_{10} = 2.5937$, $y_{20} = 2.6533$; $y = e^x$

5. $y_5 = 0.4198$, $y_{10} = 0.4124$

7. $y_5 = 0.5639$, $y_{10} = 0.5565$

9. $y_5 = 1.2194$, $y_{10} = 1.2696$

Chapter 8 in Review, Page 471

A. 1. true   **3.** true

B. 1. $y = x - 3x^2 + 4e^{3x} + C$ **3.** e^{-x}

   5. half life   **7.** $dP/dt = 0.16P$, $P(0) = P_0$

C. 1. $y = C\csc x$ **3.** $y = -\dfrac{1}{4}t + Ct^5$

   5. $y = \dfrac{1}{4} + C(x^2 + 4)^{-4}$ **7.** $y = \sin(x^2 + C)$

   9. $y = xe^{3x} - e^{3x} - \dfrac{1}{2}x^2e^{2x} + Ce^{2x}$

   11. $P(t) = 1000e^{0.05t}$

   13. $y = \dfrac{1}{25}t^{-1} + \dfrac{1}{25}t^4(-1 + 5\ln t)$ **15.** $y = \dfrac{6}{5e^{-2x} - 3}$

   17. $y = \tan(x - 7\pi/12)$ **19.** $y = \dfrac{1}{2(1 + x^4)}$

   21. $3y^4 = 4x^2 + 48$

   25. (a) $A(t) = \dfrac{k_1 M}{k_1 + k_2}(1 - e^{-(k_1 + k_2)t})$

      (b) $A \to \dfrac{k_1 M}{k_1 + k_2}$ as $t \to \infty$, the material will never be completely memorized

(c)

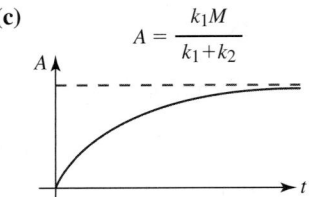

$$A = \frac{k_1 M}{k_1 + k_2}$$

27. (a) $P(t) = P_0 e^{k\sin t}$ **(b)**

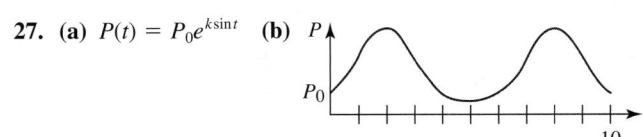

31. 1.3214

Exercises 9.1, Page 483

1. $\dfrac{1}{3}, \dfrac{1}{5}, \dfrac{1}{7}, \dfrac{1}{9}, \ldots$ **3.** $-1, \dfrac{1}{2}, -\dfrac{1}{3}, \dfrac{1}{4}, \ldots$

5. $10, 100, 1000, 10000, \ldots$ **7.** $2, 4, 12, 48, \ldots$

9. $1, 1 + \dfrac{1}{2}, 1 + \dfrac{1}{2} + \dfrac{1}{3}, 1 + \dfrac{1}{2} + \dfrac{1}{3} + \dfrac{1}{4}, \ldots$

15. 0 **17.** 0

19. $\dfrac{1}{2}$ **21.** sequence diverges

23. sequence diverges   **25.** 0

27. 0 **29.** sequence diverges

31. 0 **33.** $\dfrac{5}{7}$

35. 1 **37.** 6

39. 1 **41.** 1

43. $\ln\dfrac{4}{3}$ **45.** 0

47. $\left\{\dfrac{2n}{2n - 1}\right\}$, converges to 1

49. $\{(-1)^{n+1}(2n + 1)\}$, diverges   **51.** $\left\{\dfrac{2}{3^{n-1}}\right\}$, converges to 0

53. $-\dfrac{1}{2}, -\dfrac{1}{4}, -\dfrac{1}{8}, -\dfrac{1}{16}, \ldots$ **55.** $3, 1, \dfrac{1}{3}, \dfrac{1}{3}, \ldots$

57. 8 **59.** $a_{n+1} = \dfrac{5}{n + 1}a_n$, $a_1 = 5$

61. converges to 0 **63.** converges to 0

67. $\dfrac{40}{9}$ ft; $15\left(\dfrac{2}{3}\right)^n$ ft

69. $15, 18, 18.6, 18.72, 18.744, 18.7488, \ldots$

71. 32

Exercises 9.2, Page 489

1. increasing   **3.** not monotonic

5. increasing   **7.** nonincreasing

9. increasing   **11.** not monotonic

13. bounded and increasing   **15.** bounded and increasing

17. bounded and decreasing   **19.** bounded and decreasing

21. bounded and increasing   **23.** bounded and decreasing

25. 10 **27.** 7

Exercises 9.3, Page 498

1. $3 + \dfrac{5}{2} + \dfrac{7}{3} + +\dfrac{9}{4} + \cdots$

3. $\dfrac{1}{2} - \dfrac{1}{6} + \dfrac{1}{12} - \dfrac{1}{20} + \cdots$

5. $1 + 2 + \dfrac{3}{2} + \dfrac{2}{3} + \cdots$

7. $2 + \dfrac{8}{3} + \dfrac{16}{5} + \dfrac{128}{35} + \cdots$

9. $-\dfrac{1}{7} + \dfrac{1}{9} - \dfrac{1}{11} + \dfrac{1}{13} - \cdots$

11. 1

13. $\dfrac{1}{2}$

15. $\dfrac{15}{4}$

17. $\dfrac{2}{3}$

19. diverges

21. 9000

23. diverges

25. $\dfrac{2}{9}$

27. $\dfrac{61}{99}$

29. $\dfrac{1313}{999}$

31. $\dfrac{17}{6}$

43. $-2 < x < 2$

45. $-2 < x < 0$

47. 75 ft

49. $\dfrac{N_0}{1 - s};\quad 1000$

51. 18.75 mg

Exercises 9.4, Page 503

1. converges

3. converges

5. diverges

7. converges

9. converges

11. converges

13. diverges

15. converges

17. converges

19. diverges

21. converges

23. diverges

25. converges

27. converges

29. converges

31. diverges

33. converges

35. converges for $p > 1$, diverges for $p \le 1$

Exercises 9.5, Page 507

1. converges

3. diverges

5. diverges

7. diverges

9. converges

11. converges

13. converges

15. diverges

17. converges

19. converges

21. converges

23. converges

25. diverges

27. converges

29. diverges

31. diverges

33. converges

35. diverges

37. converges

39. diverges

Exercises 9.6, Page 511

1. converges

3. diverges

5. converges

7. diverges

9. converges

11. converges

13. converges

15. diverges

17. converges

19. diverges

21. converges

23. converges

25. diverges

27. converges

29. diverges

31. converges

33. converges for $0 \le p < 1$

35. converges for all real values of p

39. use the Ratio Test

Exercises 9.7, Page 517

1. converges

3. diverges

5. converges

7. converges

9. converges

11. converges

13. diverges

15. conditionally convergent

17. absolutely convergent

19. absolutely convergent

21. absolutely convergent

23. divergent

25. conditionally convergent

27. divergent

29. conditionally convergent

31. absolutely convergent

33. divergent

35. 0.84147

37. 5

39. 0.9492

41. less than $\frac{1}{101} \approx 0.009901$

43. the series contains mixed algebraic signs but the signs do not alternate;   converges

45. the algebraic signs do not alternate; converges

47. $a_{k+1} \le a_k$ is not satisfied for k sufficiently large. The sequence of partial sums $\{S_{2n}\}$ is the same as the sequence of partial sums for the harmonic series. This implies that the series diverges.

49. diverges

51. converges

Exercises 9.8, Page 522

1. $(-1, 1];\quad 1$

3. $\left[-\frac{1}{2}, \frac{1}{2}\right);\quad \frac{1}{2}$

5. $[2, 4];\quad 1$

7. $(-5, 15);\quad 10$

9. $\{0\};\quad 0$

11. $\left[0, \frac{2}{3}\right];\quad \frac{1}{3}$

13. $[-1, 1);\quad 1$

15. $(-16, 2);\quad 9$

17. $\left(-\frac{75}{32}, \frac{75}{32}\right);\quad \frac{75}{32}$

19. $\left[\frac{2}{3}, \frac{4}{3}\right];\quad \frac{1}{3}$

21. $(-\infty, \infty);\quad \infty$

23. $(-3, N);\quad 3$

25. $(-\infty, \infty);\quad \infty$

27. $\left(-\frac{15}{4}, -\frac{9}{4}\right);\quad \frac{3}{4}$

29. 4

31. $x > 1$ or $x < -1$

33. $x < -\frac{1}{2}$

35. $-2 < x < 2$

37. $x < 0$

39. $0 \le x < \pi/3, 2\pi/3 < x < 4\pi/3, 5\pi/3 < x \le 2\pi$

41. (a) $(-\infty, \infty)$

Exercises 9.9, Page 528

1. $\displaystyle\sum_{k=0}^{\infty} \dfrac{x^k}{3^{k+1}};\quad (-3, 3)$

3. $\displaystyle\sum_{k=0}^{\infty} (-1)^k 2^k x^k;\quad \left(-\frac{1}{2}, \frac{1}{2}\right)$

5. $\sum_{k=0}^{\infty} (-1)^k x^{2k};$ $(-1, 1)$

7. $\sum_{k=0}^{\infty} \frac{(-1)^k}{4^{k+1}} x^{2k};$ $(-2, 2)$

9. $\sum_{k=1}^{\infty} \frac{k}{3^{k+1}} x^{k-1};$ $(-3, 3)$

11. $\sum_{k=2}^{\infty} \frac{(-1)^k k(k-1)2^{k-3}}{5^{k+1}} x^{k-2};$ $\left(-\frac{5}{2}, \frac{5}{2}\right)$

13. $\sum_{k=1}^{\infty} (-1)^{k+1} k x^{2k-1};$ $(-1, 1)$

15. $\sum_{k=0}^{\infty} \frac{(-1)^k}{2k+1} x^{2k+1};$ $[-1, 1]$

17. $\sum_{k=0}^{\infty} \frac{(-1)^k}{k+1} x^{2k+2};$ $[-1, 1]$

19. $\ln 4 + \sum_{k=0}^{\infty} \frac{(-1)^k}{(k+1)4^{k+1}} x^{k+1};$ $(-4, 4]$

21. $1 + \frac{3}{2} \sum_{k=1}^{\infty} (-1)^k (2x)^k;$ $\left(-\frac{1}{2}, \frac{1}{2}\right)$

23. $\frac{1}{2} \sum_{k=2}^{\infty} (-1)^k k(k-1) x^k;$ $(-1, 1)$

25. $\sum_{k=0}^{\infty} \frac{(-1)^k}{k+1} x^{2k+3};$ $[-1, 1]$

27. $\sum_{k=0}^{\infty} \frac{(-1)^k}{(2k+1)(2k+2)} x^{2k+2};$ $[-1, 1]$

29. $\sum_{k=0}^{\infty} \frac{(-1)^{k+1}}{5^{k+1}} (x-6)^k;$ $(1, 11)$

31. $-1 + 2 \sum_{k=0}^{\infty} (-1)^k (x+1)^{k+1};$ $(-2, 0)$

33. $\sum_{k=1}^{\infty} \left[\frac{(-1)^k}{4^k} - \frac{1}{3^k} \right] x^k;$ $(-3, 3)$

35. $\frac{1}{2} + \frac{3}{4}x + \frac{7}{8}x^2 + \frac{15}{16}x^3 + \cdots$ **37.** $(-3, 3]$

39. 0.0953 **41.** 0.4854

43. 0.0088

Exercises 9.10, Page 539

1. $\sum_{k=0}^{\infty} \frac{x^k}{2^{k+1}}$

3. $\sum_{k=0}^{\infty} \frac{(-1)^k}{k+1} x^{k+1}$

5. $\sum_{k=0}^{\infty} \frac{(-1)^k}{(2k+1)!} x^{2k+1}$

7. $\sum_{k=0}^{\infty} \frac{x^k}{k!}$

9. $\sum_{k=0}^{\infty} \frac{x^{2k+1}}{(2k+1)!}$

11. $x + \frac{1}{3}x^3 + \frac{2}{15}x^5 + \frac{17}{315}x^7 + \cdots$

13. $\sum_{k=0}^{\infty} \frac{(-1)^k}{5^{k+1}} (x-4)^k$

15. $\sum_{k=0}^{\infty} (-1)^k (x-1)^k$

17. $\frac{\sqrt{2}}{2} + \frac{\sqrt{2}}{2}\left(x - \frac{\pi}{4}\right) - \frac{\sqrt{2}}{2 \cdot 2!}\left(x - \frac{\pi}{4}\right)^2 - \frac{\sqrt{2}}{2 \cdot 3!}\left(x - \frac{\pi}{4}\right)^3 + \cdots$

19. $\frac{1}{2} - \frac{\sqrt{3}}{2}\left(x - \frac{\pi}{3}\right) - \frac{1}{2 \cdot 2!}\left(x - \frac{\pi}{3}\right)^2 + \frac{\sqrt{3}}{2 \cdot 3!}\left(x - \frac{\pi}{3}\right)^3 + \cdots$

21. $\sum_{k=0}^{\infty} \frac{e}{k!} (x-1)^k$

23. $\ln 2 + \sum_{k=1}^{\infty} \frac{(-1)^{k+1}}{k2^k} (x-2)^k$ **25.** $\sum_{k=0}^{\infty} \frac{(-1)^k}{k!} x^{2k}$

27. $\sum_{k=0}^{\infty} \frac{(-1)^k}{(2k)!} x^{2k+1}$ **29.** $\sum_{k=1}^{\infty} \frac{-1}{k} x^k$

31. $1 + x^2 + \frac{2}{3}x^4 + \frac{17}{45}x^6 + \cdots$ **33.** 6

35. $\sum_{k=0}^{\infty} \frac{x^{2k}}{(2k)!}$

37. $1 + 2x + \frac{5}{2}x^2 + \frac{8}{3}x^3 + \frac{65}{24}x^4 + \cdots$

39. $1 + x + x^2 + \frac{2}{3}x^3 + \frac{1}{2}x^4 + \cdots$

43. $\frac{\pi}{4}$ **45.** -1

47. 0.71934; four decimal places

49. 1.34983; four decimal places

55. (c) $y = 7.92$ in. **(d)** $y = 7.92000021$ in.

Exercises 9.11, Page 543

1. $1 + \frac{1}{3}x - \frac{1 \cdot 2}{3^2 \cdot 2!}x^2 + \frac{1 \cdot 2 \cdot 5}{3^3 \cdot 3!}x^3 - \cdots;$ 1

3. $3 - \frac{3}{2 \cdot 9}x - \frac{3 \cdot 1}{2^2 \cdot 2! \cdot 9^2}x^2 - \frac{3 \cdot 1 \cdot 3}{2^3 \cdot 3! \cdot 9^3}x^3 - \cdots;$ 9

5. $1 - \frac{1}{2}x^2 + \frac{1 \cdot 3}{2^2 \cdot 2!}x^4 - \frac{1 \cdot 3 \cdot 5}{2^3 \cdot 3!}x^6 + \cdots;$ 1

7. $8 + \frac{8 \cdot 3}{2 \cdot 4}x + \frac{8 \cdot 3 \cdot 1}{2^2 \cdot 2! \cdot 4^2}x^2 - \frac{8 \cdot 3 \cdot 1}{2^3 \cdot 3! \cdot 4^3}x^3 + \cdots;$ 4

9. $\frac{1}{4}x - \frac{2}{4 \cdot 2}x^2 + \frac{2 \cdot 3}{4 \cdot 2! \cdot 2^2}x^3 - \frac{2 \cdot 3 \cdot 4}{4 \cdot 3! \cdot 2^3}x^4 + \cdots;$ 2

11. $|S_2 - S| < a_3 = \frac{1}{9}x^2$

13. $x + \sum_{k=1}^{\infty} \frac{1 \cdot 3 \cdot 5 \cdots (2k-1)}{2^k k!(2k+1)} x^{2k+1}$

17. $P_0(x) = 1, P_1(x) = x, P_2(x) = \frac{1}{2}(3x^2 - 1)$

19. $\sqrt{2} + \frac{\sqrt{2}}{2^2}(x-1) - \frac{\sqrt{2}}{2^4 \cdot 2!}(x-1)^2 + \frac{\sqrt{2} \cdot 1 \cdot 3}{2^6 \cdot 3!}(x-1)^3 - \cdots$

Chapter 9 in Review, Page 544

A. 1. false **3.** false

5. true **7.** false

9. true **11.** false

13. true **15.** false

17. true **19.** false

21. false **23.** false

25. false **27.** true

29. true

B. 1. 20; 9; $\frac{4}{5}$; 16 **3.** 4

5. $n/9$; $22/9$

7. e^x

9. $x < -5$ or $x > 5$

11. $(-1, 1]$

C. 1. converges

3. converges

5. converges

7. diverges

9. diverges

11. converges

13. $\dfrac{61,004}{201}$

15. $\left[-\frac{1}{3}, \frac{1}{3}\right]$

17. $\{-5\}$

19. $\dfrac{4}{3}$

21. $\dfrac{1}{\alpha - 1}$

23. $1 - \dfrac{1}{3}x^5 + \dfrac{2}{9}x^{10} - \cdots$

25. $x - \dfrac{2}{3}x^3 + \dfrac{2}{15}x^5 - \cdots$

27. $\displaystyle\sum_{k=0}^{\infty} \dfrac{(-1)^{k+1}}{(2k+1)!}(x - \pi/2)^{2k+1}$

29. \$6 million

Exercises 10.1, Page 558

1. vertex: $(0, 0)$; focus: $(1, 0)$; directrix: $x = -1$; axis: $y = 0$;

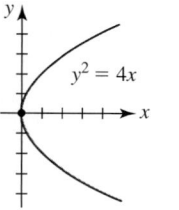

$y^2 = 4x$

3. vertex: $(0, 0)$; focus: $(0, -4)$; directrix: $y = 4$; axis: $x = 0$;

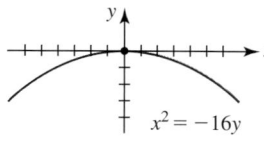

$x^2 = -16y$

5. vertex: $(0, 1)$; focus: $(4, 1)$; directrix: $x = -4$; axis: $y = 1$;

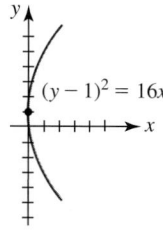

$(y - 1)^2 = 16x$

7. vertex: $(-5, -1)$; focus: $(-5, -2)$; directrix: $y = 0$; axis: $x = -5$;

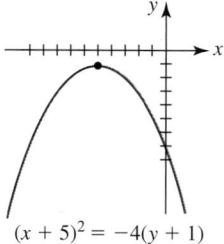

$(x + 5)^2 = -4(y + 1)$

9. vertex: $(-5, -6)$; focus: $(-4, -6)$; directrix: $x = -6$; axis: $y = -6$;

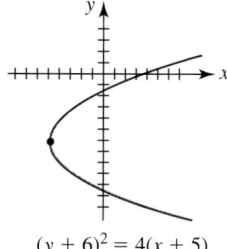

$(y + 6)^2 = 4(x + 5)$

11. vertex: $\left(-\frac{5}{2}, -1\right)$; focus: $\left(-\frac{5}{2}, -\frac{15}{16}\right)$; directrix: $y = -\frac{17}{16}$; axis: $x = -\frac{5}{2}$;

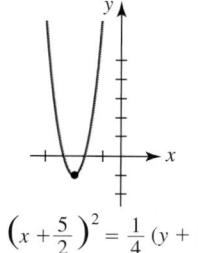

$\left(x + \frac{5}{2}\right)^2 = \frac{1}{4}(y + 1)$

13. vertex: $(3, 4)$; focus: $\left(\frac{5}{2}, 4\right)$; directrix: $x = \frac{7}{2}$; axis: $y = 4$;

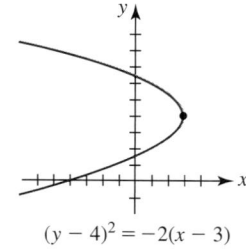

$(y - 4)^2 = -2(x - 3)$

15. $x^2 = 28y$

17. $y^2 = 10x$

19. $(y + 7)^2 = 12(x + 2)$

21. $x^2 = \dfrac{1}{2}y$

23. $(3, 0), (0, -2), (0, -6)$

25. center: $(0, 0)$; foci: $\left(0, \pm\sqrt{15}\right)$; vertices: $(0, \pm 4)$; minor axis endpoints: $(\pm 1, 0)$; eccentricity: $\frac{\sqrt{15}}{4}$;

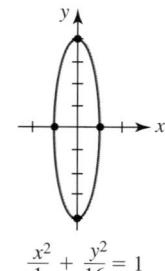

$\dfrac{x^2}{1} + \dfrac{y^2}{16} = 1$

27. center: $(0, 0)$; foci: $\left(\pm\sqrt{7}, 0\right)$; vertices: $(\pm 4, 0)$; minor axis endpoints: $(0, \pm 3)$; eccentricity: $\frac{\sqrt{7}}{4}$

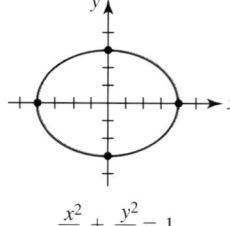

$\dfrac{x^2}{16} + \dfrac{y^2}{9} = 1$

29. center: $(1, 3)$; foci: $\left(1 \pm \sqrt{13}, 3\right)$; vertices: $(-6, 3), (8, 3)$; minor axis endpoints: $(1, -3), (1, 9)$; eccentricity: $\frac{\sqrt{13}}{7}$;

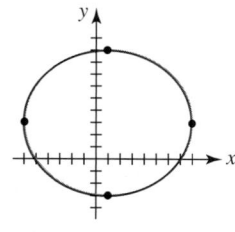

$\dfrac{(x - 1)^2}{49} + \dfrac{(y - 3)^2}{36} = 1$

31. center: $(-5, -2)$;   foci: $\left(-5, -2 \pm \sqrt{15}\right)$;   vertices: $(-5, -6)$, $(-5, 2)$;   minor axis endpoints: $(-6, -2)$, $(-4, -2)$; eccentricity: $\frac{\sqrt{15}}{4}$;

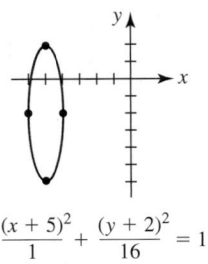

$$\frac{(x + 5)^2}{1} + \frac{(y + 2)^2}{16} = 1$$

33. center: $\left(0, -\frac{1}{2}\right)$;   foci: $\left(0, -\frac{1}{2} \pm \sqrt{3}\right)$; vertices: $\left(0, -\frac{5}{2}\right)$, $\left(0, \frac{3}{2}\right)$; minor axis endpoints: $\left(-1, -\frac{1}{2}\right)$, $\left(1, -\frac{1}{2}\right)$; eccentricity: $\frac{\sqrt{3}}{2}$;

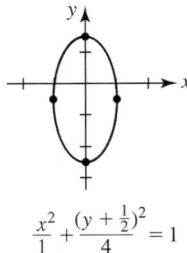

$$\frac{x^2}{1} + \frac{\left(y + \frac{1}{2}\right)^2}{4} = 1$$

35. center: $(2, -1)$;   foci: $(2, -5)$, $(2, 3)$;   vertices: $(2, -6)$, $(2, 4)$;   minor axis endpoints: $(-1, -1)$, $(5, -1)$;   eccentricity: $\frac{4}{5}$;

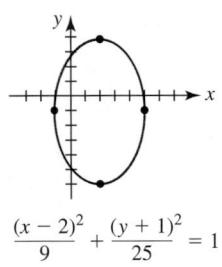

$$\frac{(x - 2)^2}{9} + \frac{(y + 1)^2}{25} = 1$$

37. center: $(0, -3)$; foci: $\left(\pm\sqrt{6}, -3\right)$; vertices: $(-3, -3)$, $(3, -3)$;   minor axis endpoints: $\left(0, -3 \pm \sqrt{3}\right)$; eccentricity: $\frac{\sqrt{6}}{3}$;

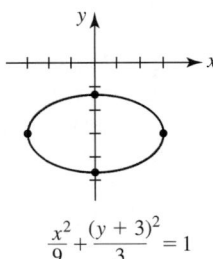

$$\frac{x^2}{9} + \frac{(y + 3)^2}{3} = 1$$

39. $\dfrac{x^2}{25} + \dfrac{y^2}{16} = 1$

41. $\dfrac{(x - 1)^2}{16} + \dfrac{(y + 3)^2}{4} = 1$

43. $\dfrac{x^2}{11} + \dfrac{y^2}{9} = 1$

45. $\dfrac{x^2}{3} + \dfrac{y^2}{12} = 1$

47. $\dfrac{(x - 1)^2}{7} + \dfrac{(y - 3)^2}{16} = 1$

49. center: $(0,0)$;   foci: $\left(\pm\sqrt{41}, 0\right)$;   vertices: $(\pm 4, 0)$; asymptotes: $y = \pm\frac{5}{4}x$;   eccentricity: $\frac{\sqrt{41}}{4}$;

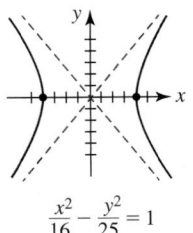

$$\frac{x^2}{16} - \frac{y^2}{25} = 1$$

51. center: $(0, 0)$;   foci: $\left(0, \pm 2\sqrt{6}\right)$;   vertices: $\left(0, \pm 2\sqrt{5}\right)$; asymptotes: $y = \pm\sqrt{5}x$;   eccentricity: $\sqrt{\frac{6}{5}}$;

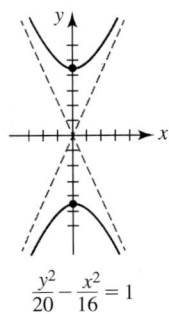

$$\frac{y^2}{20} - \frac{x^2}{16} = 1$$

53. center: $(5, -1)$;   foci: $\left(5 \pm \sqrt{53}, -1\right)$; vertices: $(3, -1)$, $(7, -1)$; asymptotes: $y = -1 \pm \frac{7}{2}(x - 5)$;   eccentricity: $\frac{\sqrt{53}}{2}$;

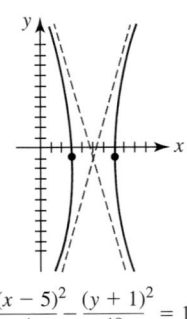

$$\frac{(x - 5)^2}{4} - \frac{(y + 1)^2}{49} = 1$$

55. center: $(0, 4)$;   foci: $\left(0, 4 \pm \sqrt{37}\right)$;   vertices: $(0, -2)$, $(0, 10)$; asymptotes: $y = 4 \pm 6x$;   eccentricity: $\frac{\sqrt{37}}{6}$;

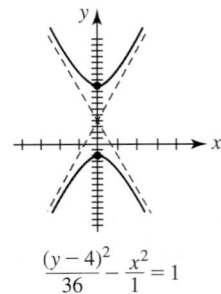

$$\frac{(y - 4)^2}{36} - \frac{x^2}{1} = 1$$

57. center: (3, 1); foci: $\left(3 \pm \sqrt{30}, 1\right)$; vertices: $\left(3 \pm \sqrt{5}, 1\right)$; asymptotes: $y = 1 \pm \sqrt{5}\,(x - 3)$; eccentricity: $\sqrt{6}$;

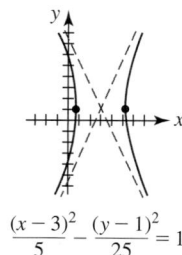

$$\frac{(x-3)^2}{5} - \frac{(y-1)^2}{25} = 1$$

59. center: (2, 1); foci: $\left(2 \pm \sqrt{11}, 1\right)$; vertices: $\left(2 \pm \sqrt{6}, 1\right)$; asymptotes: $y = 1 \pm \sqrt{\frac{5}{6}}(x - 2)$; eccentricity: $\sqrt{\frac{11}{6}}$;

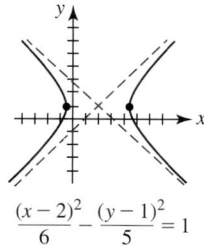

$$\frac{(x-2)^2}{6} - \frac{(y-1)^2}{5} = 1$$

61. center: (1, 3); foci: $\left(1, 3 \pm \frac{\sqrt{5}}{2}\right)$; vertices: (1, 2), (1, 4); asymptotes: $y = 3 \pm 2(x - 1)$; eccentricity: $\frac{\sqrt{5}}{2}$;

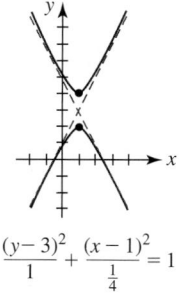

$$\frac{(y-3)^2}{1} + \frac{(x-1)^2}{\frac{1}{4}} = 1$$

63. $\dfrac{y^2}{4} - \dfrac{x^2}{12} = 1$

65. $\dfrac{(y + 3)^2}{4} - \dfrac{(x - 1)^2}{5} = 1$

67. $(y - 3)^2 - \dfrac{(x + 1)^2}{4} = 1$

69. $(y - 4)^2 - \dfrac{(x - 2)^2}{4} = 1$

71. at the focus 6 in. from the vertex

73. 76.5625 ft

75. 12.65 m from the point on the ground directly beneath the end of the pipe

77. least distance is 28.5 million mi; greatest distance is 43.5 million mi

79. approximately 0.97   **81.** 12 ft

Exercises 10.2, Page 564

1.

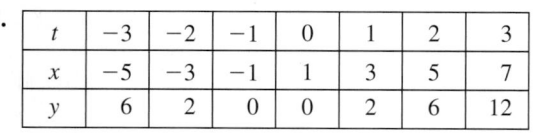

t	-3	-2	-1	0	1	2	3
x	-5	-3	-1	1	3	5	7
y	6	2	0	0	2	6	12

3.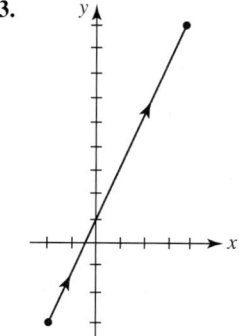

5.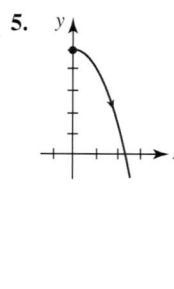

7.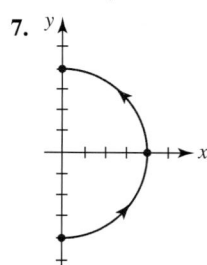

9.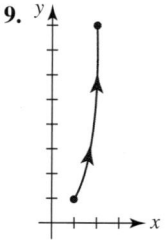

11. $y = x^2 + 3x - 1, x \ge 0$

13. $x = -1 + 2y^2, -1 \le x \le 0$ **15.** $y = \ln x, x > 0$

17.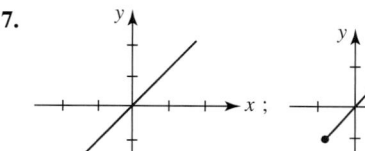

19.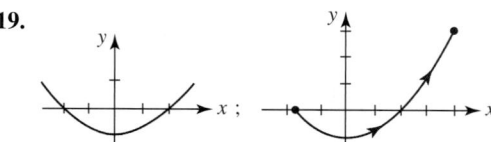

21.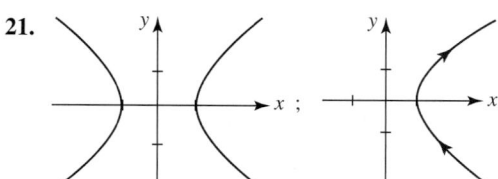

23.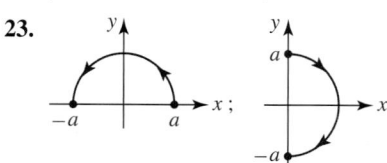

25.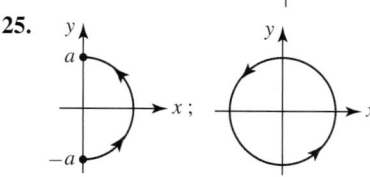

27.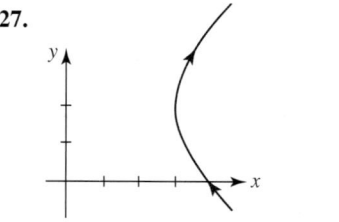

29. yes   **31.** no

33. no

35. $x = \pm\sqrt{r^2 - L^2 \sin^2\phi}$, $y = L\sin\phi$

37. $x = a(\cos\theta + \theta\sin\theta)$, $y = a(\sin\theta - \theta\cos\theta)$

39. (b)   (c) $x^{2/3} + y^{2/3} = b^{2/3}$

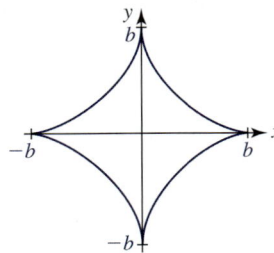

41. (b)

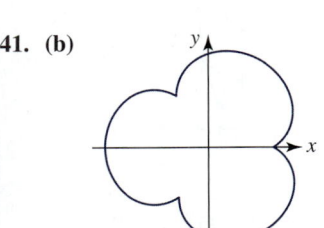

Exercises 10.3, Page 572

1. $\dfrac{3}{5}$ **3.** 24

5. -1 **7.** $y = -2x - 1$

9. $y = \dfrac{4}{3}x + \dfrac{4}{3}$ **11.** $\dfrac{\sqrt{3}}{4}$

13. $y = 3x - 7$

15. horizontal tangent at $(0, 0)$, vertical tangent at $\left(-\frac{2}{3\sqrt{3}}, \frac{1}{3}\right)$ and $\left(\frac{2}{3\sqrt{3}}, \frac{1}{3}\right)$;

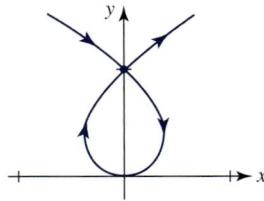

17. horizontal tangents at $(-1, 0)$ and $(1, -4)$, no vertical tangents;

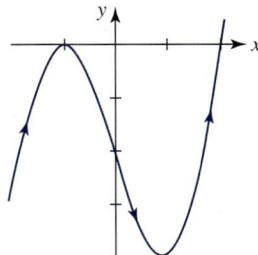

19. $3t$; $1/(2t)$; $-1/(12t^3)$

21. $-2e^{3t} - 3e^{4t}$; $6e^{4t} + 12e^{5t}$; $-24e^{5t} - 60e^{6t}$

23. concave up for $0 < t < 2$, concave down for $t < 0$ and $t > 2$

25. $\dfrac{104}{3}$ **27.** $\sqrt{2}(e^\pi - 1)$

29. $\dfrac{3}{2}|b|$

31. (a) -0.6551 (b) $-5.9991, 1.0446, 9.7361$

Exercises 10.4, Page 576

1.

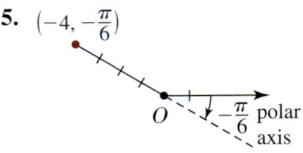

3.

5. $\left(-4, -\dfrac{\pi}{6}\right)$

7. (a) $(2, -5\pi/4)$ (b) $(2, 11\pi/4)$
      (c) $(-2, 7\pi/4)$ (d) $(-2, -\pi/4)$

9. (a) $(4, -5\pi/3)$ (b) $(4, 7\pi/3)$
      (c) $(-4, 4\pi/3)$ (d) $(-4, -2\pi/3)$

11. (a) $(1, -11\pi/6)$ (b) $(1, 13\pi/6)$
      (c) $(-1, 7\pi/6)$ (d) $(-1, -5\pi/6)$

13. $\left(-\frac{1}{4}, \frac{\sqrt{3}}{4}\right)$ **15.** $(-3, 3\sqrt{3})$

17. $(-2\sqrt{2}, -2\sqrt{2})$

19. (a) $(2\sqrt{2}, -3\pi/4)$ (b) $(-2\sqrt{2}, \pi/4)$

21. (a) $(2, -\pi/3)$ (b) $(-2, 2\pi/3)$

23. (a) $(7, 0)$ (b) $(-7, \pi)$

25.

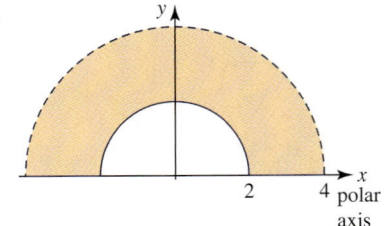

27.   **29.**

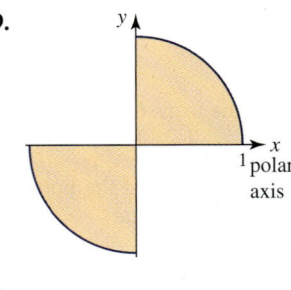

31. $r = 5\csc\theta$ **33.** $\theta = \tan^{-1}7$

35. $r = 2/(1 + \cos\theta)$ **37.** $r = 6$

39. $r = 1 - \cos\theta$ **41.** $x = 2$

43. $(x^2 + y^2)^3 = 144x^2y^2$ **45.** $(x^2 + y^2)^2 = 8xy$

47. $x^2 + y^2 + 5y = 0$ **49.** $8x^2 - 12x - y^2 + 4 = 0$

51. $3x + 8y = 5$

Exercises 10.5, Page 583

1. circle;

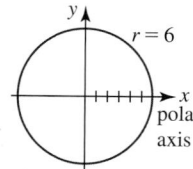

3. line through origin;

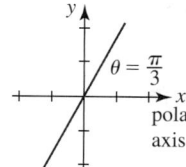

5. spiral;

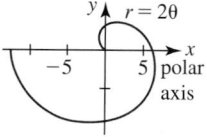

7. cardioid;

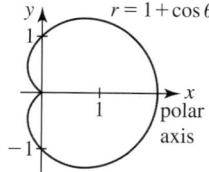

9. cardioid;

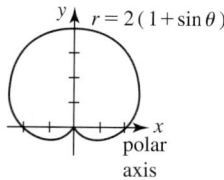

11. limaçon with an interior loop;

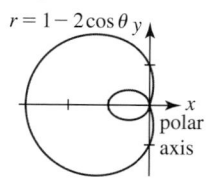

13. dimpled limaçon;

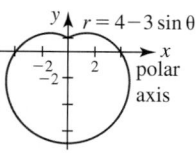

15. convex limaçon;

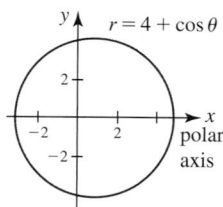

17. rose curve;

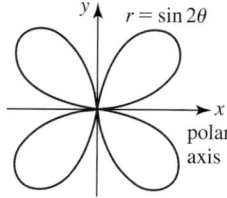

19. rose curve;

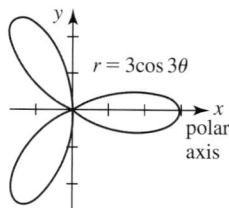

21. rose curve;

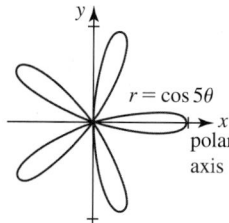

23. circle with center on x-axis;

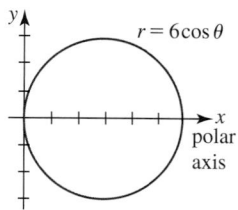

25. circle with center on y-axis;

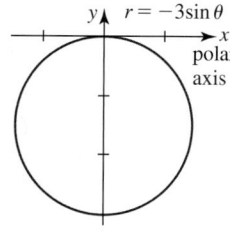

27. lemniscate;

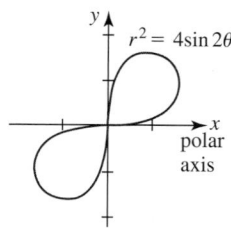

29. lemniscate;

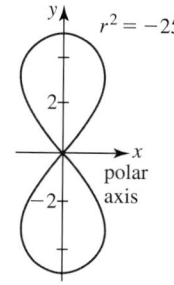

31.

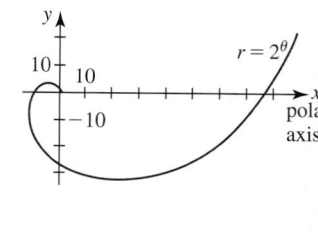

33. $r = \dfrac{5}{2}$

35. $r = 4 - 3\cos\theta$

37. $r = 2\cos 4\theta$

39. $(2, \pi/6), (2, 5\pi/6)$

41. $(1, \pi/2), (1, 3\pi/2)$, origin

43. $(3, \pi/12), (3, 5\pi/12), (3, 13\pi/12), (3, 17\pi/12), (3, -\pi/12),$
$(3, -5\pi/12), (3, -13\pi/12), (3, -17\pi/12)$

45. $(0, 0), \left(\frac{\sqrt{3}}{2}, \pi/3\right), \left(\frac{\sqrt{3}}{2}, 2\pi/3\right)$ **49.** (d)

51. (b)

Exercises 10.6, Page 590

1. $-2/\pi$

3. $\dfrac{\sqrt{3} - 2}{2\sqrt{3} - 1}$

5. $\sqrt{3}$

7. horizontal tangent at $(3, \pi/3)$ and $(3, 5\pi/3)$, vertical tangent at $(4, 0), (1, 2\pi/3)$, and $(1, 4\pi/3)$

9. $y = \dfrac{1}{\sqrt{3}}x + \dfrac{8}{\sqrt{3}}, y = -\dfrac{1}{\sqrt{3}}x - \dfrac{8}{\sqrt{3}}$

11. $\theta = 0$

13. $\theta = 5\pi/4, \theta = 7\pi/4$

15. $\theta = \pi/10, \theta = 3\pi/10, \theta = \pi/2, \theta = 7\pi/10, \theta = 9\pi/10$

17. π

19. 24π

21. 11π

23. $\dfrac{9}{2}\pi$

25. $\dfrac{9}{4}\pi^3$

27. $\dfrac{1}{4}(e^{2\pi} - 1)$

29. $\dfrac{1}{8}(4 - \pi)$

31. $\pi - \dfrac{3\sqrt{3}}{2}$

33. $\dfrac{1}{6}(2\pi + 3\sqrt{3})$

35. $\pi + 6\sqrt{3}$

37. $18\sqrt{3} - 4\pi$

39. 6π

41. $\sqrt{5}(e^2 - 1)$

43. 24

Exercises 10.7, Page 596

1. $e = 1$; parabola;

3. $e = \dfrac{1}{4}$; ellipse;

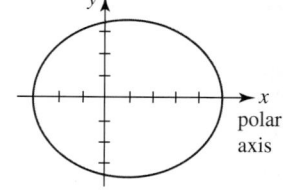

ANSWERS TO SELECTED ODD-NUMBERED PROBLEMS, CHAPTER 10

5. $e = 2$; hyperbola

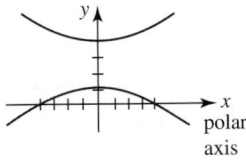

7. $e = 2$; hyperbola

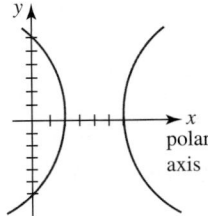

9. $e = \dfrac{4}{5}$; ellipse;

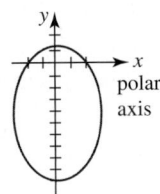

11. $e = 2$; $\dfrac{(y-4)^2}{4} - \dfrac{x^2}{12} = 1$

13. $e = \dfrac{2}{3}$; $\dfrac{\left(x - \dfrac{24}{5}\right)^2}{\dfrac{1296}{25}} + \dfrac{y^2}{\dfrac{144}{5}} = 1$

15. $r = \dfrac{3}{1 + \cos\theta}$

17. $r = \dfrac{4}{3 - 2\sin\theta}$

19. $r = \dfrac{12}{1 + 2\cos\theta}$

21. $r = \dfrac{3}{1 + \cos(\theta + 2\pi/3)}$

23. $r = \dfrac{3}{1 - \sin\theta}$

25. $r = \dfrac{1}{1 - \cos\theta}$

27. $r = \dfrac{1}{2 - 2\sin\theta}$

29. vertex: $(2, \pi/4)$

31. vertices: $(10, \pi/3)$ and $\left(\dfrac{10}{3}, 4\pi/3\right)$

33. $r_p = 8000$ km

35. $r = \dfrac{1.495 \times 10^8}{1 - 0.0167\cos\theta}$

Chapter 10 in Review, Page 597

A. **1.** true **3.** true

5. true **7.** false

9. true **11.** true

13. false **15.** true

17. true **19.** false

21. true **23.** true

25. false

B. **1.** $\left(0, \frac{1}{8}\right)$ **3.** $(0, -3)$

5. $y = -5$ **7.** $(-10, -2)$

9. $(2, -1), (6, -1)$ **11.** $(4, -3)$

13. $\left(0, \sqrt{5}\right), \left(0, -\sqrt{5}\right)$ **15.** line through origin

17. circle through origin **19.** $\theta = 0, \theta = \pi/3, \theta = 2\pi/3$

21. $(0, 0), (5, 3\pi/2)$

C. **1.** $y = -\dfrac{\sqrt{3}}{3}x + \dfrac{\sqrt{3}\pi}{9}$ **3.** $(8, -26)$

5. **(b)** $x = \sin t, y = \sin 2t, 0 \le t \le 2\pi$

  (c) $\left(\frac{\sqrt{2}}{2}, 1\right), \left(\frac{\sqrt{2}}{2}, -1\right), \left(-\frac{\sqrt{2}}{2}, 1\right), \left(-\frac{\sqrt{2}}{2}, -1\right)$

  (d)

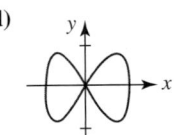

7. $5\pi/4$

9. **(a)** $x + y = 2\sqrt{2}$

  (b) $r = 2\sqrt{2}/(\cos\theta + \sin\theta)$

11. $x^2 + y^2 = x + y$

13. $r^2 = 5\csc 2\theta$

15. $r = 1/(1 - \cos\theta)$

17. $r = 3\sin 10\theta$

19. $\dfrac{y^2}{100} - \dfrac{x^2}{36} = 1$

21. $x = \dfrac{3at}{1 + t^3}, y = \dfrac{3at^2}{1 + t^3}$

23. **(a)** $r = \dfrac{3a\cos\theta\sin\theta}{\cos^3\theta + \sin^3\theta}$

  (b) $\dfrac{3}{2}a^2$

25. $\pi - \dfrac{3\sqrt{3}}{2}$

27. **(a)** $r = 2\cos(\theta - \pi/4)$

  (b) $x^2 + y^2 = \sqrt{2}x + \sqrt{2}y$

29. 10^8 m; 9×10^8 m

Exercises 11.1, Page 606

1. (a) $6\mathbf{i} + 12\mathbf{j}$ (b) $\mathbf{i} + 8\mathbf{j}$ (c) $3\mathbf{i}$ (d) $\sqrt{65}$ (e) 3
3. (a) $\langle 12, 0 \rangle$ (b) $\langle 4, -5 \rangle$ (c) $\langle 4, 5 \rangle$ (d) $\sqrt{41}$ (e) $\sqrt{41}$
5. (a) $-9\mathbf{i} + 6\mathbf{j}$ (b) $-3\mathbf{i} + 9\mathbf{j}$ (c) $-3\mathbf{i} - 5\mathbf{j}$
   (d) $3\sqrt{10}$ (e) $\sqrt{34}$
7. (a) $-6\mathbf{i} + 27\mathbf{j}$ (b) $\mathbf{0}$ (c) $-4\mathbf{i} + 18\mathbf{j}$ (d) 0 (e) $2\sqrt{85}$
9. (a) $\langle 6, -14 \rangle$ (b) $\langle 2, 4 \rangle$
11. (a) $10\mathbf{i} - 12\mathbf{j}$ (b) $12\mathbf{i} - 17\mathbf{j}$
13. (a) $\langle 20, 52 \rangle$ (b) $\langle -2, 0 \rangle$
15. $2\mathbf{i} + 5\mathbf{j}$

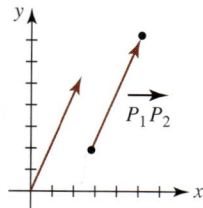

17. $2\mathbf{i} + 2\mathbf{j}$

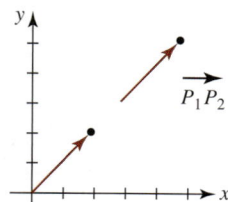

19. $(1, 18)$
21. (a), (b), (c), (e), (f)
23. $\langle 6, 15 \rangle$
25. (a) $\left\langle \frac{1}{\sqrt{2}}, \frac{1}{\sqrt{2}} \right\rangle$ (b) $\left\langle -\frac{1}{\sqrt{2}}, -\frac{1}{\sqrt{2}} \right\rangle$
27. (a) $\langle 0, -1 \rangle$ (b) $\langle 0, 1 \rangle$
29. $\left\langle \frac{5}{13}, \frac{12}{13} \right\rangle$
31. $\dfrac{6}{\sqrt{58}}\mathbf{i} + \dfrac{14}{\sqrt{58}}\mathbf{j}$
33. $\left\langle -3, -\frac{15}{2} \right\rangle$
35.

3b − a
3b
a

37. $-(\mathbf{a} + \mathbf{b})$
41. $\mathbf{a} = \dfrac{5}{2}\mathbf{b} - \dfrac{1}{2}\mathbf{c}$
43. $\pm\dfrac{1}{\sqrt{2}}(\mathbf{i} + \mathbf{j})$
45. (a) $|\mathbf{a} + \mathbf{b}| \le |\mathbf{a}| + |\mathbf{b}|$
    (b) when $P_1, P_2,$ and P_3 are collinear and P_2 lies between P_1 and P_3

47. (b) approximately $31°$
49. approximately 153 lb

Exercises 11.2, Page 612

1, 3, 5.

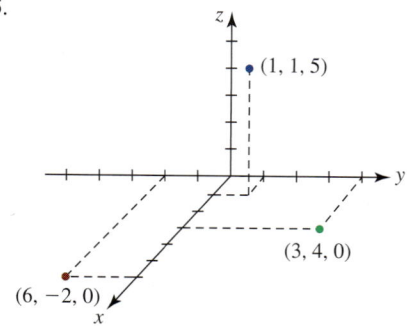

7. The set $\{(x, y, 5)\,|\,x, y \text{ real numbers}\}$ is a plane perpendicular to the z-axis, 5 units above the xy-plane.
9. The set $\{(2, 3, z)\,|\,z \text{ a real number}\}$ is a line perpendicular to the xy-plane at $(2, 3, 0)$.
11. $(2, 0, 0), (2, 5, 0), (2, 0, 8), (2, 5, 8), (0, 5, 0), (0, 5, 8),$ $(0, 0, 8), (0, 0, 0)$
13. (a) $(-2, 5, 0), (-2, 0, 4), (0, 5, 4)$
    (b) $(-2, 5, -2)$ (c) $(3, 5, 4)$
15. The union of the coordinate planes
17. The point $(-1, 2, -3)$
19. The union of the planes $z = 5$ and $z = -5$
21. $\sqrt{70}$
23. (a) 7 (b) 5
25. right triangle
27. isosceles triangle
29. collinear
31. not collinear
33. 6 or -2
35. $\left(4, \frac{1}{2}, \frac{3}{2}\right)$
37. $(-4, -11, 10)$
39. $\langle -3, -6, 1 \rangle$
41. $\langle 2, 1, 1 \rangle$
43.

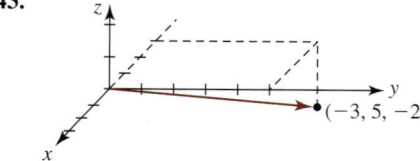

45.

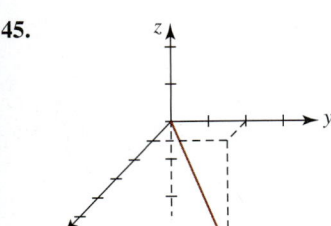

$(1, 2, -3)$

47. xy-plane

49. positive z-axis, xz-plane, yz-plane

51. $\langle 2, 4, 12 \rangle$

53. $\langle -11, -41, -49 \rangle$

55. $\sqrt{139}$

57. 6

59. $\langle -\frac{2}{3}, \frac{1}{3}, -\frac{2}{3} \rangle$

61. $4\mathbf{i} - 4\mathbf{j} + 4\mathbf{k}$

Exercises 11.3, Page 620

1. 12

3. -16

5. 48

7. 29

9. 25

11. $\langle -\frac{2}{5}, \frac{4}{5}, 2 \rangle$

13. $25\sqrt{2}$

15. -3

17. 1.11 radians or 63.43°

19. 1.89 radians or 108.4°

21. (a) and (f), (c) and (d), (b) and (e)

23. $\langle \frac{4}{9}, -\frac{1}{3}, 1 \rangle$

27. $\cos\alpha = \dfrac{1}{\sqrt{14}}, \cos\beta = \dfrac{2}{\sqrt{14}}, \cos\gamma = \dfrac{3}{\sqrt{14}};$

$\alpha = 74.5°, \beta = 57.69°, \gamma = 36.7°$

29. $\cos\alpha = \dfrac{1}{2}, \cos\beta = 0, \cos\gamma = -\dfrac{\sqrt{3}}{2};$

$\alpha = 60°, \beta = 90°, \gamma = 150°$

31. 0.9553 radian or 57.74°; 0.6155 radian or 35.26°

33. $\dfrac{5}{7}$

35. $-\dfrac{6}{\sqrt{11}}$

37. $\dfrac{72}{\sqrt{109}}$

39. (a) $-\dfrac{21}{5}\mathbf{i} + \dfrac{28}{5}\mathbf{j}$ (b) $-\dfrac{4}{5}\mathbf{i} - \dfrac{3}{5}\mathbf{j}$

41. (a) $\langle -\frac{12}{7}, \frac{6}{7}, \frac{4}{7} \rangle$ (b) $\langle \frac{5}{7}, -\frac{20}{7}, \frac{45}{7} \rangle$

43. $\dfrac{72}{25}\mathbf{i} + \dfrac{96}{25}\mathbf{j}$

45. 1000 ft-lb

47. 45 N-m

49. $\dfrac{78}{5}$ ft-lb

Exercises 11.4, Page 628

1. $-5\mathbf{i} - 5\mathbf{j} + 3\mathbf{k}$

3. $\langle -12, -2, 6 \rangle$

5. $-5\mathbf{i} + 5\mathbf{k}$

7. $\langle -3, 2, 3 \rangle$

9. 0

11. $6\mathbf{i} + 14\mathbf{j} + 4\mathbf{k}$

13. $-3\mathbf{i} - 2\mathbf{j} - 5\mathbf{k}$, or any nonzero multiple of this vector

17. (a) $\mathbf{j} - \mathbf{k}; -\mathbf{i} + \mathbf{j} + \mathbf{k}$

19. $2\mathbf{k}$

21. $\mathbf{i} + 2\mathbf{j}$

23. $-24\mathbf{k}$

25. $5\mathbf{i} - 5\mathbf{j} - \mathbf{k}$

27. 0

29. $\sqrt{41}$

31. $-\mathbf{j}$

33. 0

35. 6

37. $12\mathbf{i} - 9\mathbf{j} + 18\mathbf{k}$

39. $-4\mathbf{i} + 3\mathbf{j} - 6\mathbf{k}$

41. $-21\mathbf{i} + 16\mathbf{j} + 22\mathbf{k}$

43. -10

45. (b) 14

47. $\dfrac{1}{2}$

49. $\dfrac{7}{2}$

51. 10

53. The vectors are coplanar.

55. The points are coplanar.

57. (a) 32 (b) 30° from the positive x-axis in the direction of the negative y-axis (c) $16\sqrt{3}\mathbf{i} - 16\mathbf{j}$

Exercises 11.5, Page 633

1. $\langle x, y, z \rangle = \langle 4, 6, -7 \rangle + t\langle 3, \frac{1}{2}, -\frac{3}{2} \rangle$

3. $\langle x, y, z \rangle = t\langle 5, 9, 4 \rangle$

5. $\langle x, y, z \rangle = \langle 1, 2, 1 \rangle + t\langle 2, 3, -3 \rangle$

7. $\langle x, y, z \rangle = \langle \frac{1}{2}, -\frac{1}{2}, 1 \rangle + t\langle -2, 3, -\frac{3}{2} \rangle$

9. $\langle x, y, z \rangle = \langle 1, 1, -1 \rangle + t\langle 5, 0, 0 \rangle$

11. $x = 2 + 4t, y = 3 - 4t, z = 5 + 3t$

13. $x = 1 + 2t, y = -2t, z = -7t$

15. $x = 4 + 10t, y = \dfrac{1}{2} + \dfrac{3}{4}t, z = \dfrac{1}{3} + \dfrac{1}{6}t$

17. $\dfrac{x - 1}{9} = \dfrac{y - 4}{10} = \dfrac{z + 9}{7}$

19. $\dfrac{x + 7}{11} = \dfrac{z - 5}{-4}, y = 2$ **21.** $x = 5, \dfrac{y - 10}{9} = \dfrac{z + 2}{12}$

23. $x = 6 + 2t, y = 4 - 3t, z = -2 + 6t$

25. $x = 2 + t, y = -2, z = 15$

27. Both lines pass through the origin and have parallel direction vectors.

29. (a) $t = -5$ (b) $s = 12$

31. $(0, 5, 15), (5, 0, \frac{15}{2}), (10, -5, 0)$

33. $(2, 3, -5)$

35. The lines do not intersect.

37. yes

39. $x = 2 + 4t, y = 5 - 6t, z = 9 - 6t, 0 \le t \le 1$

41. 40.37°

43. $x = 4 - 6t, y = 1 + 3t, z = 6 + 3t$

45. The lines are not parallel and do not intersect.

Exercises 11.6, Page 638

1. $2x - 3y + 4z = 19$

3. $5x - 3z = 51$

5. $6x + 8y - 4z = 11$

7. $5x - 3y + z = 2$

9. $3x - 4y + z = 0$

11. The points are collinear.

13. $x + y - 4z = 25$

15. $z = 12$

17. $-3x + y + 10z = 18$

19. $9x - 7y + 5z = 17$

21. $6x - 2y + z = 12$

23. perpendicular: (a) and (d), (b) and (c), (d) and (f), (b) and (e); parallel: (a) and (f), (c) and (e)

25. (c), (d)

27. $x = 2 + t, y = \dfrac{1}{2} - t, z = t$

29. $x = \frac{1}{2} - \frac{1}{2}t, y = \frac{1}{2} - \frac{3}{2}t, z = t$

31. $(-5, 5, 9)$ **33.** $(1, 2, -5)$

35. $x = 5 + t, y = 6 + 3t, z = -12 + t$

37. $3x - y - 2z = 10$ **39.**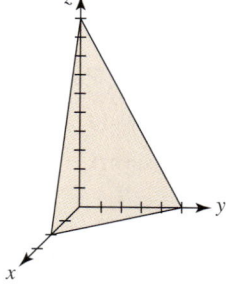

41. 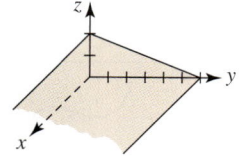   **43.**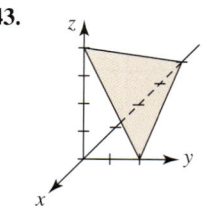

47. $\frac{3}{\sqrt{11}}$ **49.** $107.98°$

Exercises 11.7, Page 642

1. 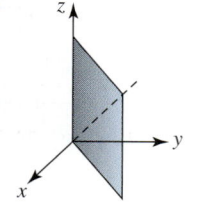   **3.**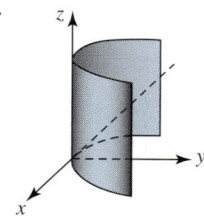

5. 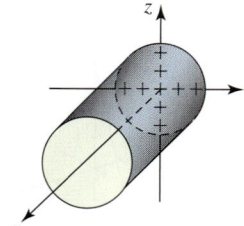   **7.**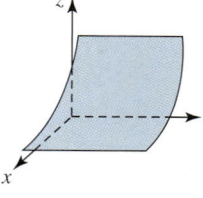

9. 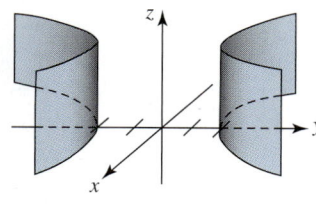   **11.**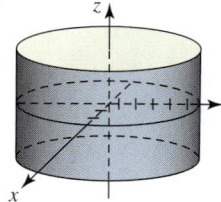

13. 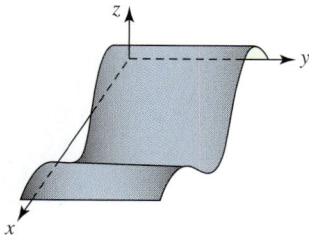   **15.**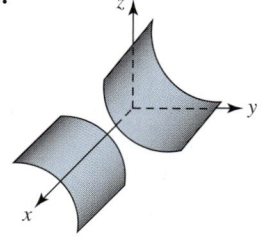

17. 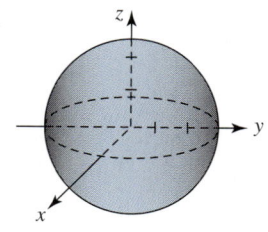   **19.**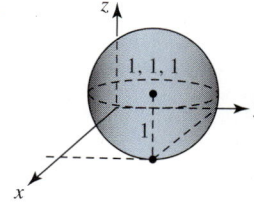

21. center $(-4, 3, 2)$; radius 6   **23.** center $(0, 0, 8)$; radius 8

25. $(x + 1)^2 + (y - 4)^2 + (z - 6)^2 = 3$

27. $(x - 1)^2 + (y - 1)^2 + (z - 4)^2 = 16$

29. $x^2 + (y - 4)^2 + z^2 = 4$ or $x^2 + (y - 8)^2 + z^2 = 4$

31. $(x - 1)^2 + (y - 4)^2 + (z - 2)^2 = 90$

33. all points on the upper half of the sphere $x^2 + y^2 + (z - 1)^2 = 4$ (upper hemisphere)

35. all points on and outside of the sphere $x^2 + y^2 + z^2 = 1$

37. all points on and between concentric spheres of radius 1 and radius 3 centered at the origin

Exercises 11.8, Page 649

1. paraboloid; 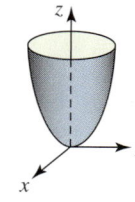   **3.** ellipsoid;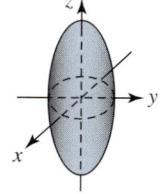

5. hyperboloid of one sheet; 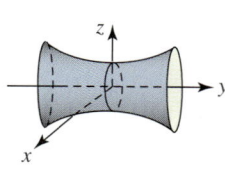   **7.** elliptic cone;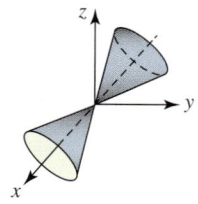

9. hyperbolic paraboloid; 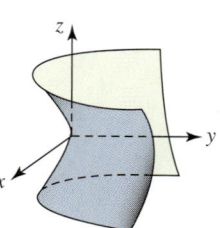   **11.** hyperboloid of two sheets;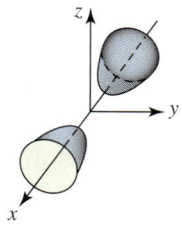

13. elliptic paraboloid; 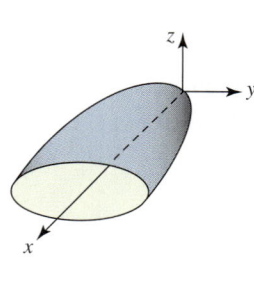   **15.**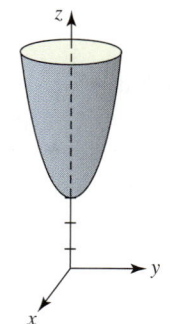

17.

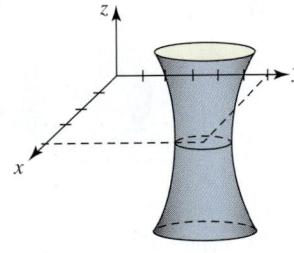

19. one possibility is $y^2 + z^2 = 1$; z-axis

21. one possibility is $y = e^{x^2}$; y-axis

23. $y^2 = 4(x^2 + z^2)$

25. $y^2 + z^2 = (9 - x^2)^2, x \geq 0$

27. $x^2 - y^2 - z^2 = 4$

29. $z = \ln\sqrt{x^2 + y^2}$

31. The surfaces in Problems 1, 4, 6, 10, and 14 are surfaces of revolution about the z-axis. The surface in Problem 2 is a surface of revolution about the y-axis. The surface in Problem 11 is surface of revolution about the x-axis.

33.

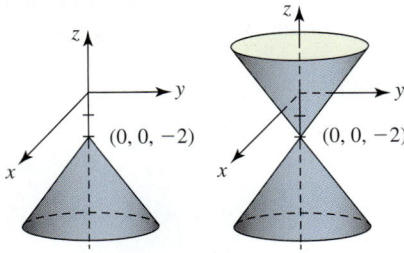

35. (a) area of a cross section is $\pi ab(c - z)$ (b) $\frac{1}{2}\pi abc^2$

37. $(2, -2, 6), (-2, 4, 3)$

Chapter 11 in Review, Page 650

A. 1. true

3. false

5. true

7. true

9. true

11. true

13. true

15. false

17. false

19. true

B. 1. $9\mathbf{i} + 2\mathbf{j} + 2\mathbf{k}$

3. $5\mathbf{i}$

5. 14

7. 26

9. $-6\mathbf{i} + \mathbf{j} - 7\mathbf{k}$

11. $(4, 7, 5)$

13. $(5, 6, 3)$

15. $-36\sqrt{2}$

17. $(12, 0, 0), (0, -8, 0), (0, 0, 6)$ **19.** $\dfrac{3\sqrt{10}}{2}$

21. 2

23. ellipsoid

C. 1. $\dfrac{1}{\sqrt{11}}(\mathbf{i} - \mathbf{j} - 3\mathbf{k})$

3. 2

5. $\left\langle \frac{16}{5}, \frac{12}{5}, 0 \right\rangle$

7. elliptic cylinder

9. hyperboloid of two sheets   **11.** hyperbolic paraboloid

13. $x^2 - y^2 + z^2 = 1$, hyperboloid of one sheet; $x^2 - y^2 - z^2 = 1$, hyperboloid of two sheets

15. (a) sphere   (b) plane

17. $\dfrac{x - 7}{4} = \dfrac{y - 3}{-2} = \dfrac{z + 5}{6}$

19. The direction vectors are orthogonal and the point of intersection is $(3, -3, 0)$.

21. $14x - 5y - 3z = 0$

23. $-6x - 3y + 4z = 5$

27. (a) $-qvB\mathbf{k}$

(b) $\mathbf{v} = \dfrac{1}{m|\mathbf{r}|^2}(\mathbf{L} \times \mathbf{r})$

29. approximately 192.4 N-m

Exercises 12.1, Page 659

1. $(-\infty, -3] \cup [3, \infty)$ **3.** $[-1, 1]$

5. $\mathbf{r}(t) = \sin\pi t\,\mathbf{i} + \cos\pi t\,\mathbf{j} - \cos^2\pi t\,\mathbf{k}$

7. $\mathbf{r}(t) = e^{-t}\mathbf{i} + e^{2t}\mathbf{j} + e^{3t}\mathbf{k}$ **9.** $x = t^2, y = \sin t, z = \cos t$

11. $x = \ln t, y = 1 + t, z = t^3$

13.

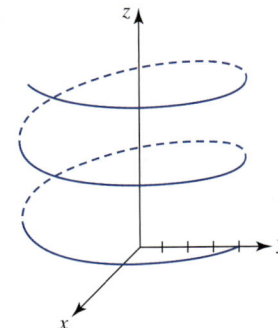

15.

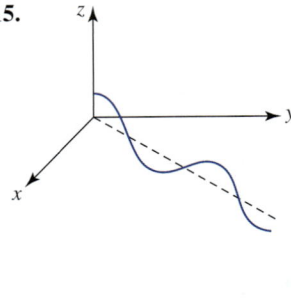

17.

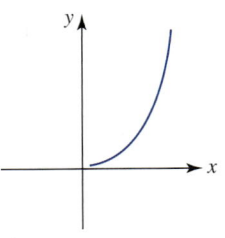

19.

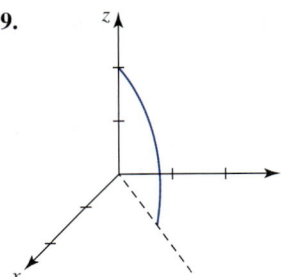

21.

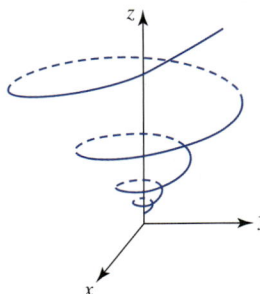

23.

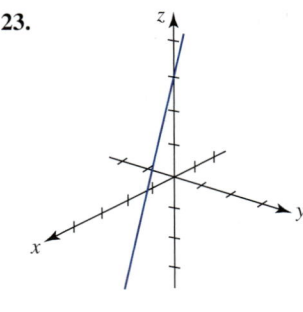

25. $\mathbf{r}(t) = (1 - t)\langle 4, 0 \rangle + t\langle 0, 3 \rangle$,   **27.** $\mathbf{r}(t) = t\mathbf{i} + t\mathbf{j} + 2t^2\mathbf{k}$;
$0 \leq t \leq 1$;

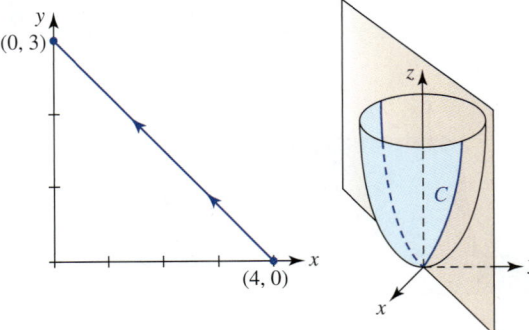

29. $\mathbf{r}(t) = 3\cos t\,\mathbf{i} + 3\sin t\,\mathbf{j} + 9\sin^2 t\,\mathbf{k}$;

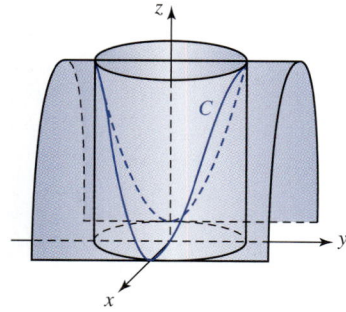

31. $\mathbf{r}(t) = t\mathbf{i} + t\mathbf{j} + (1 - 2t)\mathbf{k}$;

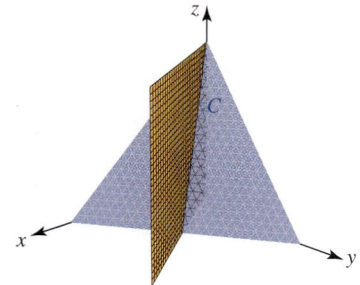

33. (b)   **35. (d)**

43. (a)

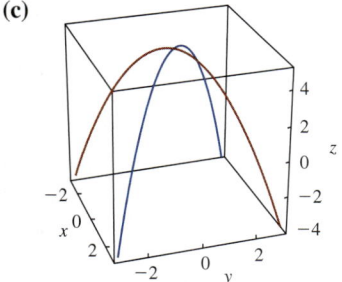

(b) $\mathbf{r}_1(t) = t\mathbf{i} + t\mathbf{j} + (4 - t^2)\mathbf{k}$, $\mathbf{r}_2(t) = t\mathbf{i} - t\mathbf{j} + (4 - t^2)\mathbf{k}$

(c)

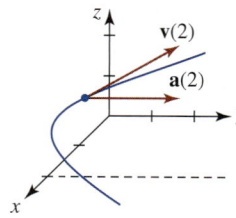

Exercises 12.2, Page 667

1. $8\mathbf{i} + 16\mathbf{j} + 32\mathbf{k}$ **3.** $\langle 2, 2, 2 \rangle$

5. $2\mathbf{i} + 23\mathbf{j} + 17\mathbf{k}$ **7.** discontinuous

9. $3\mathbf{i} + 8\mathbf{j} + 9\mathbf{k}$; $3\mathbf{i} + 8.4\mathbf{j} + 9.5\mathbf{k}$

11. $(1/t)\mathbf{i} - (1/t^2)\mathbf{j}$; $-(1/t^2)\mathbf{i} + (2/t^3)\mathbf{j}$

13. $\langle 2te^{2t} + e^{2t}, 3t^2, 8t - 1 \rangle$; $\langle 4te^{2t} + 4e^{2t}, 6t, 8 \rangle$

15.

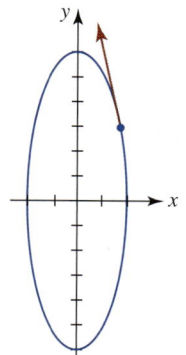

17.

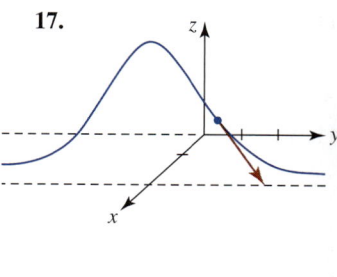

19. $x = 2 + t,\ y = 2 + 2t,\ z = \dfrac{8}{3} + 4t$

21. $\dfrac{1}{\sqrt{6}}\mathbf{i} + \dfrac{2}{\sqrt{6}}\mathbf{j} - \dfrac{1}{\sqrt{6}}\mathbf{k}$; $x = \dfrac{1}{\sqrt{6}}t,\ y = \dfrac{2}{\sqrt{6}}t,\ z = -\dfrac{1}{\sqrt{6}}t$

23. $\mathbf{r}(t) = \left\langle \frac{1}{2} - \frac{\sqrt{3}}{2}t,\ \frac{\sqrt{3}}{2} + \frac{1}{2}t,\ \pi/3 + t \right\rangle$

25. $\mathbf{r}(t) \times \mathbf{r}''(t)$ **27.** $\mathbf{r}(t) \cdot (\mathbf{r}'(t) \times \mathbf{r}'''(t))$

29. $2\mathbf{r}_1'(2t) - (1/t^2)\mathbf{r}_2'(1/t)$ **31.** $\dfrac{3}{2}\mathbf{i} + 9\mathbf{j} + 15\mathbf{k}$

33. $(te^t - e^t)\mathbf{i} + \dfrac{1}{2}e^{-2t}\mathbf{j} + \dfrac{1}{2}e^{t^2}\mathbf{k} + \mathbf{C}$

35. $(6t + 1)\mathbf{i} + (3t^2 - 2)\mathbf{j} + (t^3 + 1)\mathbf{k}$

37. $(2t^3 - 6t + 6)\mathbf{i} + (7t - 4t^{3/2} - 3)\mathbf{j} + (t^2 - 2t)\mathbf{k}$

39. $2\pi\sqrt{a^2 + c^2}$ **41.** $\sqrt{6}(e^{3\pi} - 1)$

43. $\mathbf{r}(s) = 9\cos(s/9)\mathbf{i} + 9\sin(s/9)\mathbf{j}$

45. $\mathbf{r}(s) = \left(1 + \frac{2}{\sqrt{29}}s\right)\mathbf{i} + \left(5 - \frac{3}{\sqrt{29}}s\right)\mathbf{j} + \left(2 + \frac{4}{\sqrt{29}}s\right)\mathbf{k}$

Exercises 12.3, Page 671

1. Speed is $\sqrt{5}$;   **3.** Speed is 2;

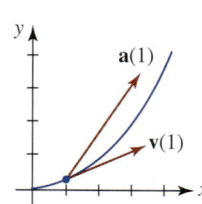

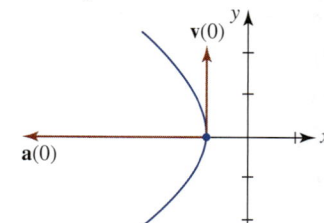

5. Speed is $\sqrt{5}$;   **7.** Speed is $\sqrt{14}$;

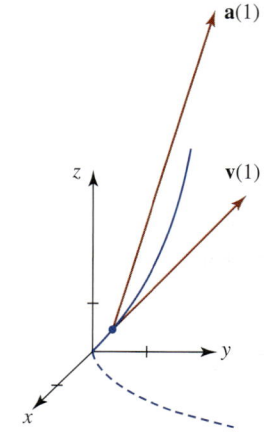

9. (a) $(0, 0, 0)$ and $(25, 115, 0)$

(b) $\mathbf{v}(0) = -2\mathbf{i} - 5\mathbf{k}$, $\mathbf{a}(0) = 2\mathbf{i} + 2\mathbf{k}$;

$\mathbf{v}(5) = 10\mathbf{i} + 73\mathbf{j} + 5\mathbf{k}$, $\mathbf{a}(5) = 2\mathbf{i} + 30\mathbf{j} + 2\mathbf{k}$

11. (a) $\mathbf{r}(t) = 240\sqrt{3}\,t\mathbf{i} + (-16t^2 + 240t)\mathbf{j}$;

$x(t) = 240\sqrt{3}\,t$, $y(t) = -16t^2 + 240t$

(b) 900 ft

(c) approximately 6235 ft

(d) 480 ft/s

13. 72.11 ft/s   **15.** 97.98 ft/s

19. Assume that (x_0, y_0) are the coordinates of the center of the target at time $t = 0$. Then $\mathbf{r}_p = \mathbf{r}_t$ when $t = x_0/(v_0\cos\theta) = y_0/(v_0\sin\theta)$. This implies $\tan\theta = y_0/x_0$. In other words, aim directly at the target at $t = 0$.

21. approximately 191.33 lb

25. $\mathbf{r}(t) = k_1 e^{2t^3}\mathbf{i} + \dfrac{1}{2t^2 + k_2}\mathbf{j} + (k_3 e^{t^2} - 1)\mathbf{k}$

27. Since $\mathbf{F}$ is directed along $\mathbf{r}$, we must have $\mathbf{F} = c\mathbf{r}$ for some constant c. Hence, $\boldsymbol{\tau} = \mathbf{r} \times (c\mathbf{r}) = c(\mathbf{r} \times \mathbf{r}) = \mathbf{0}$. If $\boldsymbol{\tau} = \mathbf{0}$, then $d\mathbf{L}/dt = \mathbf{0}$. This implies that $\mathbf{L}$ is a constant.

Exercises 12.4, Page 678

1. $\mathbf{T} = \dfrac{1}{\sqrt{5}}(-\sin t\mathbf{i} + \cos t\mathbf{j} + 2\mathbf{k})$

3. $\mathbf{T} = (a^2 + b^2)^{-1/2}(-a\sin t\mathbf{i} + a\cos t\mathbf{j} + c\mathbf{k})$;

$\mathbf{N} = -\cos t\mathbf{i} - \sin t\mathbf{j}$;

$\mathbf{B} = (a^2 + b^2)^{-1/2}(c\sin t\mathbf{i} - c\cos t\mathbf{j} + a\mathbf{k})$; $\kappa = a/(a^2 + c^2)$

5. (a) $3\sqrt{2}x - 3\sqrt{2}y + 4z = 3\pi$

(b) $-4\sqrt{2}x + 4\sqrt{2}y + 12z = 9\pi$

(c) $x + y = 2\sqrt{2}$

7. $a_T = 4t/\sqrt{1 + 4t^2}$; $a_N = 2/\sqrt{1 + 4t^2}$

9. $a_T = 2\sqrt{6}$; $a_N = 0$, $t > 0$

11. $a_T = 2t/\sqrt{1 + t^2}$; $a_N = 2/\sqrt{1 + t^2}$

13. $a_T = 0$; $a_N = 5$

15. $a_T = -\sqrt{3}e^{-t}$; $a_N = 0$

17. $\kappa = \dfrac{\sqrt{b^2c^2\sin^2 t + a^2c^2\cos^2 t + a^2b^2}}{(a^2\sin^2 t + b^2\cos^2 t + c^2)^{3/2}}$

23. $\kappa = 2$, $\rho = \frac{1}{2}$; $\kappa = 2/\sqrt{125} \approx 0.18$, $\rho = \sqrt{125}/2 \approx 5.59$; the curve is sharper at $(0, 0)$

25. 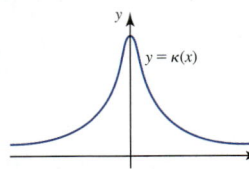 ; for large values of $|x|$ the graph of $y = x^2$ behaves like a straight line since $\kappa(x) \to 0$.

Chapter 12 in Review, Page 679

A. 1. true   **3.** true   **5.** true   **7.** true   **9.** false

B. 1. $y = 4$ **3.** $\langle 1, 2, 1 \rangle$

5. $\dfrac{\sqrt{2}}{6}$ **7.** $\left\langle -\frac{1}{\sqrt{2}}, 0, \frac{1}{\sqrt{2}} \right\rangle$

9. $3x + 6y + 3z = 10$

C. 1. $\sqrt{2}\pi$

3. $x = -27 - 18t$, $y = 8 + t$, $z = 1 + t$

5.

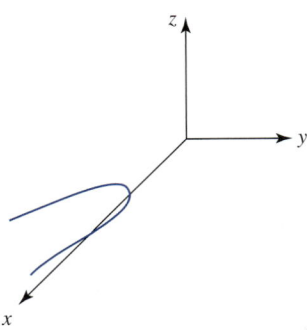

7. $-t^2\sin t + 2t\cos t - 2\sin t\cos t + 8t^3 e^{2t} + 12t^2 e^{2t}$

9. $(t + 1)\mathbf{i} + \left(\dfrac{1}{m}t^2 + t + 1\right)\mathbf{j} + t\mathbf{k}$;

$x = t + 1$, $y = \dfrac{1}{m}t^2 + t + 1$, $z = t$

11. $\mathbf{v}(1) = 6\mathbf{i} + \mathbf{j} + 2\mathbf{k}$, $\mathbf{v}(4) = 6\mathbf{i} + \mathbf{j} + 8\mathbf{k}$,

$\mathbf{a}(1) = 2\mathbf{k}$, $\mathbf{a}(4) = 2\mathbf{k}$

13. $\mathbf{i} + 4\mathbf{j} + (3\pi/4)\mathbf{k}$

15. $\mathbf{T} = \dfrac{1}{\sqrt{2}}(\tanh 1\mathbf{i} + \mathbf{j} + \operatorname{sech} 1\mathbf{k})$;

$\mathbf{N} = \operatorname{sech} 1\mathbf{i} - \tanh 1\mathbf{k}$;

$\mathbf{B} = \dfrac{1}{\sqrt{2}}(-\tanh 1\mathbf{i} + \mathbf{j} - \operatorname{sech} 1\mathbf{k})$;

$\kappa = \dfrac{1}{2}\operatorname{sech}^2 1$

Exercises 13.1, Page 686

1. $\{(x, y)\,|\,(x, y) \neq (0, 0)\}$ **3.** $\{(x, y)\,|\,y \neq -x^2\}$

5. $\{(s, t)\,|\,s, t \text{ any real numbers}\}$

7. $\{(r, s)\,|\,r \text{ any real number},\,|s| \geq 1\}$

9. $\{(u, v, w)\,|\,u^2 + v^2 + w^2 \geq 16\}$

11. (c)   **13. (b)**

15. (d)   **17. (f)**

19.   **21.**

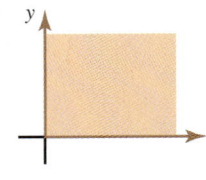   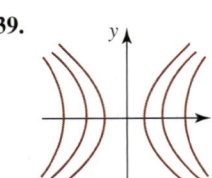

23. $\{z\,|\,z \geq 10\}$ **25.** $\{w\,|\,-1 \leq w \leq 1\}$

27. $10, -2$ **29.** $4, 4$

31. plane through the origin perpendicular to the xz-plane

33. upper nappe of a circular cone

35. upper half of an ellipsoid

37.   **39.**

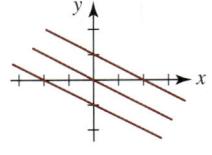

41.

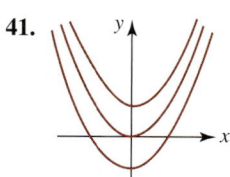

43. elliptic cylinders

45. ellipsoids

47.

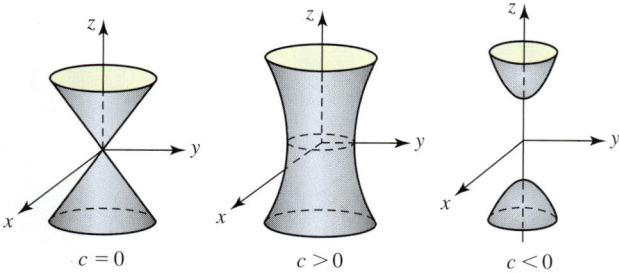

$c = 0$ $c > 0$ $c < 0$

49. P

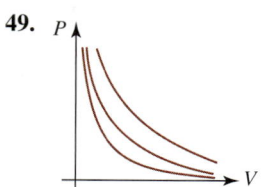

51. $C(r, h) = 2.8\pi r^2 + 4.6\pi rh$

53. $V = \dfrac{11}{9}\pi r^2 h$

55. $15,600 \text{ cm}^2$

Exercises 13.2, Page 694

1. 26

3. does not exist

5. 1

7. does not exist

9. 108

11. 1

13. $\dfrac{1}{3}$

15. $-\dfrac{1}{3}$

17. 360

19. does not exist

21. -3

23. 0

25. 0

27. 0

29. 0

31. $\{(x, y)\,|\,x \geq 0 \text{ and } y \geq -x\}$

33. $\{(x, y)\,|\,y \neq 0 \text{ and } x/y \neq (2n + 1)\pi/2, n = 0, \pm 1, \pm 2, \dots\}$

35. **(a)** continuous   **(b)** not continuous   **(c)** not continuous

37. f is continuous at $(0, 0)$.

Exercises 13.3, Page 701

1. $\partial z/\partial x = 7, \ \partial z/\partial y = 16y$

3. $\partial z/\partial x = 6xy + 4y^2, \ \partial z/\partial y = 3x^2 + 8xy$

5. $\partial z/\partial x = 2x - y^2, \ \partial z/\partial y = -2xy + 20y^4$

7. $\partial z/\partial x = 20x^3y^3 - 2xy^6 + 30x^4, \ \partial z/\partial y = 15x^4y^2 - 6x^2y^5 - 4$

9. $\partial z/\partial x = 2x^{-1/2}/(3y^2 + 1), \ \partial z/\partial y = -24y\sqrt{x}/(3y^2 + 1)^2$

11. $\partial z/\partial x = -3x^2(x^3 - y^2)^{-2}, \ \partial z/\partial y = 2y(x^3 - y^2)^{-2}$

13. $\partial z/\partial x = -10\cos 5x\sin 5x, \ \partial z/\partial y = 10\sin 5y \cos 5y$

15. $f_x = (3x^3y + 1)e^{x^3y}, \ f_y = x^4 e^{x^3y}$

17. $f_x = 7y/(x + 2y)^2, \ f_y = -7x/(x + 2y)^2$

19. $g_u = 8u/(4u^2 + 5v^3), \ g_v = 15v^2/(4u^2 + 5v^3)$

21. $w_x = x^{-1/2}y, \ w_y = 2\sqrt{x} - (y/z)e^{y/z} - e^{y/z}, \ w_z = (y^2/z^2)e^{y/z}$

23. $F_u = 2uw^2 - v^3 - vwt^2\sin(ut^2), \ F_v = -3uv^2 + w\cos(ut^2),$
$F_x = 128x^7t^4, \ F_t = -2uvwt\sin(ut^2) + 64x^8t^3$

25. -16

27. $x = -1, y = 4 + t, z = -24 + 2t$

29. -2

31. $\partial^2 z/\partial x^2 = y^2 e^{xy}$

33. $f_{xy} = 20xy - 6y^2$

35. $w_{tuv} = 18uv^2t^2$

37. $F_{r\theta r} = -2e^{r^2}(2r^2 + 1)\sin\theta$ **39.** $-60x^3y^2 + 8y$

41. $-48uvt^2$

43. $\partial z/\partial x = -x/z, \ \partial z/\partial y = -y/z$

45. $\partial z/\partial u = (vz - 2uv^3)/(2z - uv), \ \partial z/\partial v = (uz - 3u^2v^2)/(2z - uv)$

47. $A_x = y\sin\theta, \ A_y = x\sin\theta, \ A_\theta = xy\cos\theta$

59. **(a)** $\dfrac{\partial u}{\partial t} = \begin{cases} -gx/a, & 0 \leq x \leq at \\ -gt, & x > at \end{cases}$;   for $x > at$ the motion is
that of a freely falling body

(b) $\dfrac{\partial u}{\partial x} = \begin{cases} (-g/a^2)(at - x), & 0 \leq x \leq at \\ 0, & x > at \end{cases}$;   for $x > at$ the
string is horizontal

Exercises 13.4, Page 709

1. $L(x, y) = 2 - 2(x - 1) + 6(y - 1)$

3. $L(x, y) = 136 + \dfrac{353}{17}(x - 8) + \dfrac{120}{17}(y - 15)$

5. $L(x, y) = \ln 2 - (x + 1) + \dfrac{3}{2}(y - 1)$

7. 13.0907   **9.** 61.44

11. $dz = 2x\sin 4y\, dx + 4x^2\cos 4y\, dy$

13. $dz = 2x(2x^2 - 4y^3)^{-1/2}\, dx - 6y^2(2x^2 - 4y^3)^{-1/2}\, dy$

15. $df = 7t(s + 3t)^{-2}\, ds - 7s(s + 3t)^{-2}\, dt$

17. $dw = 2xy^4z^{-5}\, dx + 4x^2y^3z^{-5}\, dy - 5x^2y^4z^{-6}\, dz$

19. $dF = 3r^2\, dr - 2s^{-3}\, ds - 2t^{-1/2}\, dt$

21. $dw = du/u + dv/v - ds/s - dt/t$

23. $\Delta z = 0.2, \ dz = 0.2$

25. $\Delta z = -0.79, \ dz = -0.8$

27. $\varepsilon_1 = 5\Delta x, \ \varepsilon_2 = -\Delta x$

29. $\varepsilon_1 = y^2\Delta x + 4xy\Delta y + 2y\Delta x\Delta y,$
$\varepsilon_2 = x^2\Delta y + 2x\Delta x\Delta y + (\Delta x)^2\Delta y$

31. 0.9%   **33.** $-mg(0.009)$;   decreases

35. 15%   **37.** 4.9%

Exercises 13.5, Page 716

1. $\dfrac{dz}{dt} = \dfrac{4xt - 4yt^{-3}}{x^2 + y^2}$ **3.** $\left.\dfrac{dz}{dt}\right|_{t=\pi} = -2$

5. $\dfrac{dp}{du} = \dfrac{2u}{2s + t} + \dfrac{4r}{u^3(2s + t)^2} - \dfrac{r}{2\sqrt{u}(2s + t)^2}$

7. $\partial z/\partial u = 3u^2y^2e^{xy^2} + 2xye^{xy^2}, \ \partial z/\partial v = -4vxye^{xy^2}$

9. $\partial z/\partial u = 16u^3 - 40y(2u - v), \ \partial z/\partial v = -96v^2 + 20y(2u - v)$

11. $\partial w/\partial t = -3u(u^2 + v^2)^{1/2}e^{-t}\sin\theta - 3v(u^2 + v^2)^{1/2}e^{-t}\cos\theta,$
$\partial w/\partial\theta = 3u(u^2 + v^2)^{1/2}e^{-t}\cos\theta - 3v(u^2 + v^2)^{1/2}e^{-t}\sin\theta$

13. $\partial R/\partial u = s^2t^4e^{v^2} - 4rst^4uve^{-u^2} + 8rs^2t^3uv^2e^{u^2v^2},$
$\partial R/\partial v = 2s^2t^4uve^{v^2} + 2rst^4e^{-u^2} + 8rs^2t^3u^2ve^{u^2v^2}$

15. $\dfrac{\partial w}{\partial t} = \dfrac{xu}{\sqrt{x^2 + y^2}\,(rs + tu)} + \dfrac{y\cosh rs}{u\sqrt{x^2 + y^2}}$,

$\dfrac{\partial w}{\partial r} = \dfrac{xs}{\sqrt{x^2 + y^2}\,(rs + tu)} + \dfrac{sty\sinh rs}{u\sqrt{x^2 + y^2}}$,

$\dfrac{\partial w}{\partial u} = \dfrac{xt}{\sqrt{x^2 + y^2}\,(rs + tu)} - \dfrac{ty\cosh rs}{u^2\sqrt{x^2 + y^2}}$

17. $dy/dx = (4xy^2 - 3x^2)/(1 - 4x^2y)$

19. $dy/dx = y\cos xy/(1 - x\cos xy)$

21. $\partial z/\partial x = x/z,\ \partial z/\partial y = y/z$

23. $\partial z/\partial x = (2x + y^2z^3)/(10z - 3xy^2z^2)$,
$\partial z/\partial y = (2xyz^3 - 2y)/(10z - 3xy^2z^2)$

33. 5.31 cm²/s   **35.** 0.5976 in²/yr

39. (a) approximately 380 cycles per second   **(b)** decreasing

Exercises 13.6, Page 723

1. $(2x - 3x^2y^2)\mathbf{i} + (-2x^3y + 4y^3)\mathbf{j}$

3. $(y^2/z^3)\mathbf{i} + (2xy/z^3)\mathbf{j} - (3xy^2/z^4)\mathbf{k}$

5. $4\mathbf{i} - 32\mathbf{j}$ **7.** $2\sqrt{3}\mathbf{i} - 8\mathbf{j} - 4\sqrt{3}\mathbf{k}$

9. $\sqrt{3}x + y$ **11.** $\dfrac{15}{2}(\sqrt{3} - 2)$

13. $-\dfrac{1}{2\sqrt{10}}$ **15.** $\dfrac{98}{\sqrt{5}}$

17. $-3\sqrt{2}$ **19.** -1

21. $-\dfrac{12}{\sqrt{17}}$ **23.** $\sqrt{2}\mathbf{i} + \dfrac{1}{\sqrt{2}}\mathbf{j};\ \sqrt{\dfrac{5}{2}}$

25. $-2\mathbf{i} + 2\mathbf{j} - 4\mathbf{k};\ 2\sqrt{6}$

27. $-8\sqrt{\pi/6}\,\mathbf{i} - 8\sqrt{\pi/6}\,\mathbf{j};\ -8\sqrt{\pi/3}$

29. $-\dfrac{3}{8}\mathbf{i} - 12\mathbf{j} - \dfrac{2}{3}\mathbf{k};\ -\dfrac{\sqrt{83{,}281}}{24}$

31. $\pm\dfrac{31}{\sqrt{17}}$

33. (a) $\mathbf{u} = \dfrac{3}{5}\mathbf{i} - \dfrac{4}{5}\mathbf{j}$ **(b)** $\mathbf{u} = \dfrac{4}{5}\mathbf{i} + \dfrac{3}{5}\mathbf{j}$ **(c)** $\mathbf{u} = -\dfrac{4}{5}\mathbf{i} - \dfrac{3}{5}\mathbf{j}$

35. (a) $D_{\mathbf{u}}f = \dfrac{1}{\sqrt{10}}(9x^2 + 3y^2 - 18xy^2 - 6x^2y)$

(b) $D_{\mathbf{u}}F = \dfrac{1}{5}(-3x^2 - 27y^2 + 27x + 3y - 36xy)$

37. $(2, 5), (-2, 5)$ **39.** $-16\mathbf{i} - 4\mathbf{j}$

41. $x = 3e^{-4t}, y = 4e^{-2t}$ or $16x = 3y^2, y \geq 0$

Exercises 13.7, Page 727

1.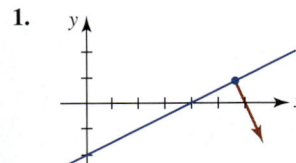

3.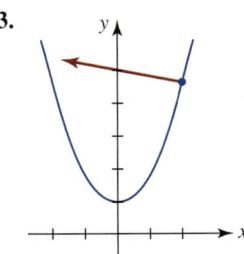

5.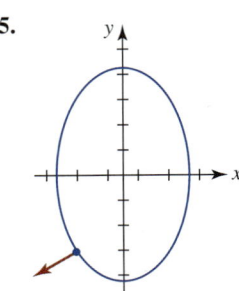

7.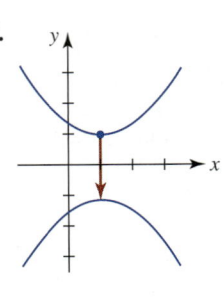

9.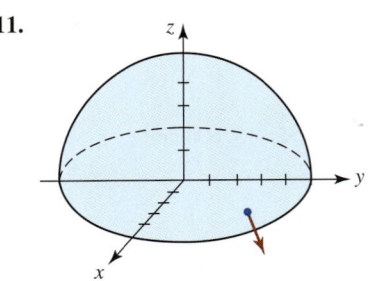

11.

13. $(-4, -1, 17)$ **15.** $-2x + 2y + z = 9$

17. $6x - 2y - 9z = 5$ **19.** $6x - 8y + z = 50$

21. $2x + y - \sqrt{2}z = 1 + \dfrac{5}{4}\pi$ **23.** $\sqrt{2}x + \sqrt{2}y - z = 2$

25. $\left(\dfrac{1}{\sqrt{2}}, \sqrt{2}, \dfrac{3}{\sqrt{2}}\right), \left(-\dfrac{1}{\sqrt{2}}, -\sqrt{2}, -\dfrac{3}{\sqrt{2}}\right)$,

27. $(-2, 0, 5), (-2, 0, -3)$

31. $x = 1 + 2t, y = -1 - 4t, z = 1 + 2t$

33. $\dfrac{x - \frac{1}{2}}{4} = \dfrac{y - \frac{1}{3}}{6} = \dfrac{z - 3}{-1}$

Exercises 13.8, Page 734

1. rel. min. $f(0, 0) = 5$

3. rel. max. $f(4, 3) = 25$

5. rel. min. $f(-2, 1) = 15$

7. rel. max. $f(-1, -1) = 10$;   rel. min. $f(1, 1) = -10$

9. rel. min. $f(3, 1) = -14$

11. no extrema

13. rel. max. $f(1, 1) = 12$

15. rel. min. $f(-1, -2) = 14$

17. rel. max. $f(-1, (2n + 1)\pi/2) = e^{-1}, n$ odd;
rel. min. $f(-1, (2n + 1)\pi/2) = -e^{-1}, n$ even

19. rel. max. $f((2m + 1)\pi/2, (2n + 1)\pi/2) = 2, m$ and n even;
rel. min. $f((2m + 1)\pi/2, (2n + 1)\pi/2) = -2, m$ and n odd

21. $x = 7, y = 7, z = 7$

23. $\left(\frac{1}{6}, \frac{1}{3}, \frac{1}{6}\right)$

25. $(2, 2, 2), (2, -2, -2), (-2, 2, -2), (-2, -2, 2)$; at these points least distance is $2\sqrt{3}$

27. $\frac{8}{9}\sqrt{3}abc$

29. $x = P/(4 + 2\sqrt{3}), y = P(\sqrt{3} - 1)/(2\sqrt{3}), \theta = 30°$

31. abs. max. $f(0, 0) = 16$ **33.** abs. min. $f(0, 0) = -8$

35. abs. max. $f\left(\frac{1}{2}, \frac{\sqrt{3}}{2}\right) = 2$;   abs. min. $f\left(-\frac{1}{2}, -\frac{\sqrt{3}}{2}\right) = -2$

37. abs. max. $f\left(\frac{\sqrt{2}}{2}, \frac{\sqrt{2}}{2}\right) = f\left(-\frac{\sqrt{2}}{2}, -\frac{\sqrt{2}}{2}\right) = \frac{3}{2}$;
abs. min. $f(0, 0) = 0$

39. abs. max. $f\left(\frac{8}{5}, -\frac{3}{5}\right) = 10$;   abs. min. $f\left(-\frac{8}{5}, \frac{3}{5}\right) = -10$

41. (a) $(0, 0)$ and all points $(x, 2\pi/x)$ for $0 < x \le \pi$
(b) abs. max. $f(x, \pi/2x) = 1, 0 < x \le \pi$;
abs. min. $f(0, 0) = f(0, y) = f(x, 0) = f(\pi, 1) = 0$
(c)

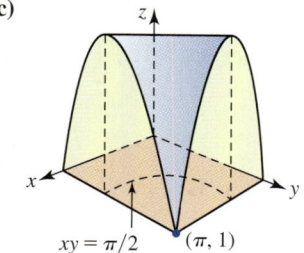

$xy = \pi/2 \quad (\pi, 1)$

43. $x = 2, y = 2, z = 15$

Exercises 13.9, Page 737

1. $y = 0.4x + 0.6$

3. $y = 1.1x - 0.3$

5. $y = 1.3571x + 1.9286$

7. $v = -0.8357T + 234.333$; $117.335, 100.621$

9. (a) $y = 0.5996x + 4.3665$;
$y = -0.0232x^2 + 0.5618x + 4.5942$;
$y = 0.00079x^3 - 0.0212x^2 + 0.5498x + 4.5840$

Exercises 13.10, Page 743

1.

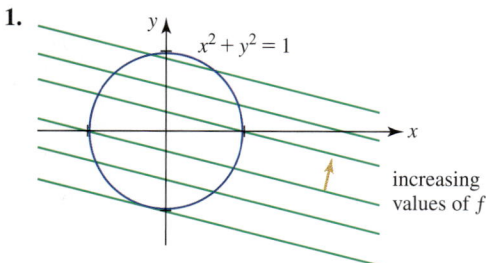

$x^2 + y^2 = 1$

increasing values of f

f appears to have a constrained maximum and a constrained minimum

3. max. $f\left(\frac{1}{\sqrt{10}}, \frac{3}{\sqrt{10}}\right) = \sqrt{10}$;
min. $f\left(-\frac{1}{\sqrt{10}}, -\frac{3}{\sqrt{10}}\right) = -\sqrt{10}$

5. max. $f(1, 1) = f(-1, -1) = 1$;
min. $f(1, -1) = f(-1, 1) = -1$

7. min. $f\left(\frac{1}{2}, -\frac{1}{2}\right) = \frac{13}{2}$
min. $f(0, 1) = f(0, -1) = f(1, 0) = f(-1, 0) = 1$

9. max. $f(1/\sqrt[4]{2}, 1/\sqrt[4]{2}) = f(-1/\sqrt[4]{2}, -1/\sqrt[4]{2}) = f(1/\sqrt[4]{2}), -1/\sqrt[4]{2}) = f(-1/\sqrt[4]{2}, 1/\sqrt[4]{2}) = \sqrt{2}$;
min. $f(0, 1) = f(0, -1) = f(1, 0) = f(-1, 0) = 1$

11. max. $f\left(\frac{9}{16}, \frac{1}{16}\right) = \frac{729}{65,536}$;   min. $f(0, 1) = f(1, 0) = 0$

13. max. $f(\sqrt{5}, 2\sqrt{5}, \sqrt{5}) = 6\sqrt{5}$;
min. $f(-\sqrt{5}, -2\sqrt{5}, -\sqrt{5}) = -6\sqrt{5}$

15. max. $f\left(\frac{1}{\sqrt{3}}, \frac{2}{\sqrt{3}}, \sqrt{3}\right) = \frac{2}{\sqrt{3}}$

17. min. $f\left(\frac{1}{3}, \frac{1}{3}, \frac{1}{3}\right) = \frac{1}{9}$

19. min. $f\left(\frac{1}{3}, \frac{16}{15}, -\frac{11}{15}\right) = \frac{134}{75}$

21. max. $A\left(\frac{4}{2 + \sqrt{2}}, \frac{4}{2 + \sqrt{2}}\right) = \frac{4}{3 + 2\sqrt{2}}$

23. $x = 12 - \frac{9}{2\sqrt{5}}$ in, $y = \frac{6}{\sqrt{5}}$ in

25. $z = P + \frac{4}{\sqrt{27k}}(2 - \sqrt{4 + P\sqrt{27k}})$

Chapter 13 in Review, Page 744

A. 1. false   **3.** true

5. false   **7.** false

9. true

B. 1. $-\frac{1}{4}$ **3.** $3x^2 + y^2 = 28$

5. $\frac{\partial F}{\partial r}g'(w) + \frac{\partial F}{\partial s}h'(w)$ **7.** f_{yyzx}

9. $F(y); -F(x)$

11. $f_x(x, y)g'(y)h'(z) + f_{xy}(x, y)g(y)h'(z)$

C. 1. $e^{-x^3y}(-x^3y + 1)$ **3.** $-\frac{3}{2}r^2\theta(r^3 + \theta^2)^{-3/2}$

5. $6x^2y\sinh(x^2y^3) + 9x^4y^4\cosh(x^2y^3)$

7. $-60s^2t^4v^{-5}$ **9.** $\frac{1}{2}\mathbf{i} + \frac{1}{2}\mathbf{j}$

11. $\frac{1}{\sqrt{10}}(3x^2 - y^2 - 4xy)$ **13.**

y

x

15. $2x\Delta y + 2y\Delta x + 2\Delta x\Delta y - 2y\Delta y - (\Delta y)^2$

17. $dz = 11y\,dx/(4x + 3y)^2 - 11x\,dy/(4x + 3y)^2$

19. $x = -\sqrt{5}, \frac{z - 3}{4} = \frac{y - 1}{3}$

21. (a) 2   **(b)** $-\sqrt{2}$ **(c)** 4

23. $4\pi x + 3y - 12z = 4\pi - 6\sqrt{3}$

25. $3x + 4y = 25$

27. $x = 2, y = 1, z = 2$

29. approximately -8.77 cm/s

33. not an extremum

35. relative maximum

37. $A = \dfrac{1}{2}L^2\cos\theta\sin\theta$

39. $A = 2xz + 2yz - 5z^2$

41. $V = 16xy - 4xy\sqrt{x^2 + y^2}$

Exercises 14.1, Page 752

1. 52

3. (a) 8   (b) 8

5. 60

7. 10π

9. No. The integrand $f(x, y) = x + 5y$ is not nonnegative over the region R.

11. 80

13. 34

15. 66

17. 18

Exercises 14.2, Page 756

1. $y + c_1(x)$

3. $2x^3y - \dfrac{3}{2}x^2\sqrt{y} + c_2(y)$

5. $\dfrac{\ln|y + 1|}{x} + c_1(x)$

7. $3y\sin 4x - 3x\sin y + c_2(y)$

9. $y(2x + 3y)^{1/2} + c_2(y)$

11. $24y - 20e^y$

13. $x^2e^{3x^2} - x^2e^x$

15. $\dfrac{1}{2}x\ln 5$

17. $2 - \sin y$

19. $\cos^2 x - \dfrac{1}{3}\cos^4 x$

21. 37

23. $-\dfrac{4\sqrt{2}}{3}$

25. $-\dfrac{4}{21}$

27. $18 - e^3 + 3e$

29. $\dfrac{10}{3}$

31. $\dfrac{\pi}{4}\ln 9$

33. $\dfrac{1}{4}e^2 + \dfrac{1}{4}$

35. π

37. e^{-1}

39. $2 - \pi$

41. $\dfrac{1}{6}(3\sqrt{3} - \pi)$

43.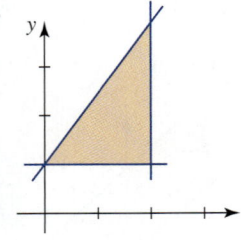

45.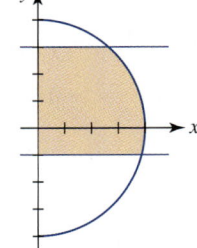

47. Both integrals equal $\dfrac{32}{5}$.

49. Both integrals equal 9.

51. Both integrals equal $13\pi^2 - 16$.

Exercises 14.3, Page 762

1. $\dfrac{1}{21}$

3. $\dfrac{25}{84}$

5. 96

7. $2\ln 2 - 1$

9. $\dfrac{14}{3}$

11. 40

13. $\dfrac{9}{2}$

15. $e^4 - e + 3 - 4\ln 4$

17. $\dfrac{63}{4}$

19. The volume is 16π.

21. 18

23. 2π

25. 4

27. $30\ln 6$

29. $\dfrac{15\pi}{4}$

31. $\dfrac{16}{9}$

33. $\dfrac{35}{6}$

35. $\displaystyle\int_0^4\int_{\sqrt{x}}^2 f(x, y)\, dy\, dx$

37. $\displaystyle\int_1^{e^3}\int_{\ln y}^3 f(x, y)\, dx\, dy$

39. $\displaystyle\int_0^1\int_{y^3}^{2-y} f(x, y)\, dx\, dy$

41. $\dfrac{1}{18}(2\sqrt{2} - 1)$

43. $\dfrac{2}{3}\sin 8$

45. $\dfrac{\pi}{8}$

47. $\dfrac{a + b}{2} \cdot \dfrac{c + d}{2}$

Exercises 14.4, Page 767

1. $\bar{x} = \dfrac{8}{3}, \bar{y} = 2$

3. $\bar{x} = 3, \bar{y} = \dfrac{3}{2}$

5. $\bar{x} = \dfrac{17}{21}, \bar{y} = \dfrac{55}{147}$

7. $\bar{x} = 0, \bar{y} = \dfrac{4}{7}$

9. $\bar{x} = \dfrac{3e^4 + 1}{4(e^4 - 1)}, \bar{y} = \dfrac{16(e^5 - 1)}{25(e^4 - 1)}$

11. $\dfrac{1}{105}$

13. $\dfrac{4}{9}k$

15. $\dfrac{256}{21}$

17. $\dfrac{941}{10}$

19. $\dfrac{\sqrt{10}}{5}a$

21. (a) $\dfrac{1}{4}ab^3\pi$ (b) $\dfrac{1}{4}a^3b\pi$ (c) $\dfrac{1}{2}b$ (d) $\dfrac{1}{2}a$

23. $\dfrac{1}{6}ka^4$

25. $\dfrac{16\sqrt{2}}{3}k$

27. $\dfrac{1}{\sqrt{3}}a$

Exercises 14.5, Page 771

1. $\dfrac{27}{2}\pi$

3. $\dfrac{1}{6}(4\pi - 3\sqrt{3})$

5. $\dfrac{25}{3}\pi$

7. $\dfrac{2}{3}\pi(15\sqrt{15} - 7\sqrt{7})$

9. $\dfrac{5}{4}$

11. $\bar{x} = \dfrac{13}{3\pi}, \bar{y} = \dfrac{13}{3\pi}$

13. $\bar{x} = \dfrac{12}{5}, \bar{y} = \dfrac{3\sqrt{3}}{2}$

15. $\bar{x} = \dfrac{1}{6}(4 + 3\pi), \bar{y} = \dfrac{4}{3}$

17. $\dfrac{1}{4}\pi a^4 k$

19. $\dfrac{1}{12}ak(15\sqrt{3} - 4\pi)$

21. $\dfrac{1}{2}\pi a^4 k$

23. $4k$

25. 9π

27. $\dfrac{1}{4}\pi(e - 1)$

29. $\dfrac{3}{8}\pi$

31. 250

33. $\frac{1}{2}\sqrt{\pi}$

35. approximately 1450 m³

37. (a) $2\pi dD_0[d - (R + d)e^{-R/d}]$

(b) $\dfrac{2d^2 - (R^2 + 2dR + 2d^2)e^{-R/d}}{d - (R + d)e^{-R/d}}$

(c) $2\pi d^2 D_0,\ 2d$

Exercises 14.6, Page 775

1. $3\sqrt{29}$

3. $\dfrac{10}{3}\pi$

5. $\dfrac{1}{6}\pi(17\sqrt{17} - 1)$

7. $\dfrac{25}{6}\pi$

9. $2a^2(\pi - 2)$

11. $8a^2$

13. $2\pi a(c_2 - c_1)$

Exercises 14.7, Page 782

1. 48

3. 36

5. $\pi - 2$

7. $\dfrac{1}{4}e^2 - \dfrac{1}{2}e$

9. 50

11. $\displaystyle\int_0^4\int_0^{2-(x/2)}\int_{x+2y}^4 f(x, y, z)\, dz\, dy\, dx,\quad \int_0^2\int_{2y}^4\int_0^{z-2y} f(x, y, z)\, dx\, dz\, dy,$

$\displaystyle\int_0^4\int_0^{z/2}\int_0^{z-2y} f(x, y, z)\, dx\, dy\, dz,\quad \int_0^4\int_x^4\int_0^{(z-x)/2} f(x, y, z)\, dy\, dz\, dx,$

$\displaystyle\int_0^4\int_0^z\int_0^{(z-x)/2} f(x, y, z)\, dy\, dx\, dz$

13. (a) $\displaystyle\int_0^2\int_{x^3}^8\int_0^4 dz\, dy\, dx$ (b) $\displaystyle\int_0^8\int_0^4\int_0^{\sqrt[3]{y}} dx\, dz\, dy$

(c) $\displaystyle\int_0^4\int_0^2\int_{x^3}^8 dy\, dx\, dz$

15.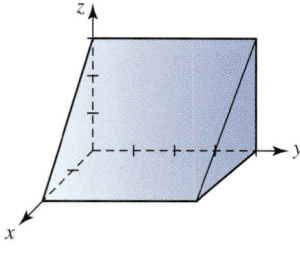

17.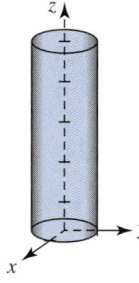

19.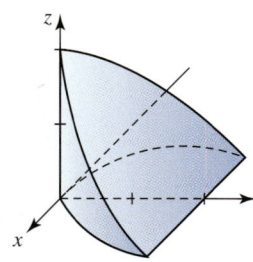

21. $16\sqrt{2}$

23. 16π

25. $\bar{x} = \dfrac{4}{5}, \bar{y} = \dfrac{32}{7}, \bar{z} = \dfrac{8}{3}$

27. $\bar{x} = 0, \bar{y} = 2, \bar{z} = 0$

29. $\displaystyle\int_{-1}^1\int_{-\sqrt{1-x^2}}^{\sqrt{1-x^2}}\int_{2y+2}^{8-y} (x + y + 4)\, dz\, dy\, dx$

31. $\dfrac{2560}{3}k, \dfrac{4}{3}\sqrt{5}$

33. $\dfrac{1}{30}k$

35. $k\displaystyle\int_{-5}^5\int_{-\sqrt{25-x^2}}^{\sqrt{25-x^2}}\int_{\sqrt{x^2+y^2}}^5 (x^2 + y^2)\sqrt{x^2 + y^2 + z^2}\, dz\, dy\, dx$

Exercises 14.8, Page 789

1. $(-5\sqrt{2}, 5\sqrt{2}, 5)$

3. $\left(\dfrac{\sqrt{3}}{2}, \dfrac{3}{2}, -4\right)$

5. $(0, 5, 1)$

7. $(\sqrt{2}, -\pi/4, -9)$

9. $(2\sqrt{2}, 2\pi/3, 2)$

11. $(4, -\pi/2, 0)$

13. $r^2 + z^2 = 25$

15. $r^2 - z^2 = 1$

17. $z = x^2 + y^2$

19. $x = 5$

21. $\dfrac{2}{3}\pi(64 - 24\sqrt{3})$

23. $\dfrac{625}{2}\pi$

25. $\left(0, 0, \tfrac{3}{8}a\right)$

27. $\dfrac{8}{3}\pi k$

29. (a) $\left(\dfrac{\sqrt{3}}{3}, \dfrac{1}{3}, 0\right)$ (b) $\left(\dfrac{2}{3}, \pi/6, 0\right)$

31. (a) $(-4, 4, 4\sqrt{2})$ (b) $(4\sqrt{2}, 3\pi/4, 4\sqrt{2})$

33. (a) $(2\sqrt{2}, 0, -2\sqrt{2})$ (b) $(2\sqrt{2}, 0, -2\sqrt{2})$

35. $(5\sqrt{2}, \pi/2, 5\pi/4)$

37. $(\sqrt{2}, \pi/4, \pi/6)$

39. $(6, \pi/4, -\pi/4)$

41. $\rho = 8$

43. $\phi = \pi/6, \phi = 5\pi/6$

45. $x^2 + y^2 + z^2 = 100$

47. $z = 2$

49. $9\pi(2 - \sqrt{2})$

51. $\dfrac{2}{9}\pi$

53. $\left(0, 0, \tfrac{7}{6}\right)$

55. πk

Exercises 14.9, Page 795

1. $(0, 0), (-2, 8), (16, 20), (14, 28)$

3.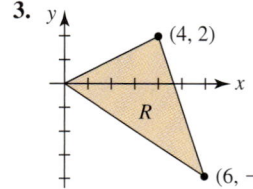

5.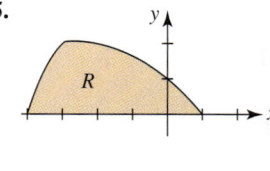

7. $-2v$

9. $-\dfrac{1}{3u^2}$

11. (a) 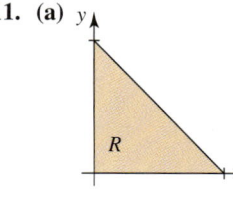 (b) $(0, 0)$ is the image of every point on the boundary $u = 0$.

13. 16

15. $\dfrac{1}{2}$

17. $\dfrac{1}{4}(b - a)(d - c)$

19. $\dfrac{1}{2}(1 - \ln 2)$

21. $\dfrac{315}{4}$

23. $\dfrac{1}{4}(e - e^{-1})$

25. 126

27. $\dfrac{15}{2}\pi$

Chapter 14 in Review, Page 796

A. **1.** true   **3.** true

5. false

B. **1.** $32y^3 - 8y^5 + 5y\ln(y^2 + 1) - 5y\ln 5$

3. square region   **5.** $f(x, 4) - f(x, 2)$

7. $\displaystyle\int_0^4 \int_{x/2}^{\sqrt{x}} f(x, y)\, dy\, dx$ **9.** $(\sqrt{2}, 2\pi/3, \sqrt{2})$

11. $z = r^2$; $\rho = \csc\phi\cot\phi$

C. **1.** $-3xe^{-4xy} - 5xy + y + c_1(x)$

3. $-y\cos y^2 + y\cos y^4$ **5.** $e^2 - e^{-2} + 4$

7. $1 - \sin 1$ **9.** $\dfrac{10}{3}$

11. 320π **13.** $\dfrac{37}{60}$

15. $\displaystyle\int_0^{1/\sqrt{2}} \int_{\sqrt{1-x^2}}^{\sqrt{9-x^2}} \dfrac{1}{x^2 + y^2}\, dy\, dx + \int_{1/\sqrt{2}}^{3/\sqrt{2}} \int_x^{\sqrt{9-x^2}} \dfrac{1}{x^2 + y^2}\, dy\, dx$

17.

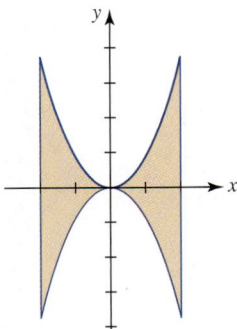

19. $\frac{1}{2}(1 - \cos 1)$

21. $\dfrac{5}{8}\pi$ **23.** $\dfrac{2}{3}\pi(2\sqrt{2} - 1)$

25. (a) $\displaystyle\int_0^1 \int_x^{2x} \sqrt{1 - x^2}\, dy\, dx$

(b) $\displaystyle\int_0^1 \int_{y/2}^y \sqrt{1 - x^2}\, dx\, dy + \int_1^2 \int_{y/2}^1 \sqrt{1 - x^2}\, dx\, dy$

(c) $\dfrac{1}{3}$

27. $\dfrac{41}{1512}k$ **29.** 8π

31. 0

Exercises 15.1, Page 807

1. $-\dfrac{125\sqrt{2}}{6}$; $\dfrac{125}{6}(4 - \sqrt{2})$; $\dfrac{125}{2}$

3. 3;   6; $3\sqrt{5}$ **5.** 0

7. -1; $\dfrac{1}{2}(\pi - 2)$; $\dfrac{1}{8}\pi^2$; $\dfrac{1}{8}\pi^2\sqrt{2}$

9. 21   **11.** 30

13. 1   **15.** 1

17. 460   **19.** $\dfrac{26}{9}$

21. $-\dfrac{64}{3}$ **23.** $-\dfrac{8}{3}$

25. 6π **27.** $\dfrac{123}{2}$

29. 70   **31.** 7

33. On each curve the line integral has the value $\dfrac{208}{3}$.

35. $k\pi$

Exercises 15.2, Page 813

1.

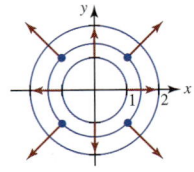

3.

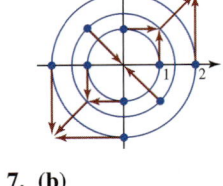

5.

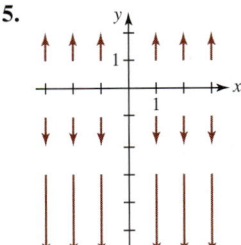

7. (b)

9. (d)   **11.** (d)

13. (a)   **15.** $-\dfrac{19}{8}$

17. 16   **19.** $9\pi^2 + 6\pi$

21. e **23.** -4

25. 0   **27.** 0

29. approximately 21.5 lb

31. $\nabla f = (3x - 6y)\mathbf{i} + (12y - 6x)\mathbf{j}$

33. $\nabla f = \tan^{-1} yz\,\mathbf{i} + \dfrac{xz}{1 + y^2 z^2}\mathbf{j} + \dfrac{xy}{1 + y^2 z^2}\mathbf{k}$

35. $\nabla f = -e^{-y^2}\mathbf{i} + (1 + 2xye^{-y^2})\mathbf{j} + \mathbf{k}$

37. (b)   **39.** (d)

41. $\phi(x, y) = y + \cos y + \sin x$ **43.** $\phi(x, y, z) = x + y^2 - 4z^3$

Exercises 15.3, Page 823

1. $\dfrac{16}{3}$ **3.** 14

5. 3   **7.** 330

9. 1096   **11.** $\phi = x^4 y^3 + 3x + y + K$

13. not a conservative field   **15.** $\phi = \dfrac{1}{4}x^4 + xy + \dfrac{1}{4}y^4 + K$

17. $\phi(x, y, z) = x^2 + y^3 - yz + K$ **19.** $3 + e^{-1}$

21. 63   **23.** $8 + 2e^3$

25. 16

27. $\pi - 4$

29. $\phi = (Gm_1m_2)/|\mathbf{r}|$

Exercises 15.4, Page 829

1. 3

3. 0

5. 75π

7. 48π

9. $\dfrac{56}{3}$

11. $\dfrac{2}{3}$

13. $\dfrac{1}{8}$

15. $(b - a) \times$ (area of region bounded by C)

19. $\dfrac{3}{8}a^2\pi$

23. $\dfrac{45}{2}\pi$

25. π

27. $\dfrac{27}{2}\pi$

29. $\dfrac{3}{2}\pi$

Exercises 15.5, Page 837

1. $x = u, y = v, z = 4u + 3v - 2$

3. $x = u, y = -\sqrt{1 + u^2 + v^2}, z = v$

5. $\mathbf{r}(u, v) = u\mathbf{i} + v\mathbf{j} + (1 - v^2)\mathbf{k}, -2 \le u \le 2, -3 \le v \le 3$

7. $x^2 + y^2 = 1$, circular cylinder

9. $x^2 = y^2 + z^2$, portion of a circular cone

11. parameter domain defined by $0 \le u \le 4, 0 \le v \le \pi/2$

13. parameter domain defined by $0 \le \theta \le 2\pi, \pi/2 \le \phi \le \pi$

15. $x + \sqrt{3}y = 20$

17. $-6x + 10y + z = 9$

19. $3x + 3y - z = 9$

21. $x + 3y + 2z = 4$

23. $8x + 6x - 5z = 25$

25. $4\sqrt{11}$

27. $\frac{1}{6}\pi(17\sqrt{17} - 1)$

29. $2\sqrt{5}\pi + \pi\ln(2 + \sqrt{5})$

31. $x = 2\sin\phi\cos\theta, y = 2\sin\phi\sin\theta, z = 2\cos\phi, \pi/3 \le \phi \le \pi,$ $0 \le \theta \le 2\pi; \quad 12\pi$

33. $x = 2\sin\phi\cos\theta, y = 2\sin\phi\sin\theta, z = 2\cos\phi, 0 \le \phi \le \pi/4,$ $0 \le \theta \le 2\pi; \quad 4\pi(2 - \sqrt{2})$

35.

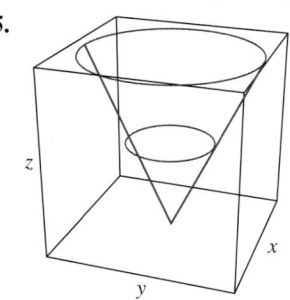

Exercises 15.6, Page 844

1. $\dfrac{26}{3}$

3. 0

5. 972π

7. $\dfrac{1}{15}(3^{5/2} - 2^{7/2} + 1)$

9. $9(17^{3/2} - 1)$

11. $12\sqrt{14}$

13. $\dfrac{\sqrt{3}}{12}k$

15. 18

17. 28π

19. 8π

21. $\dfrac{5}{2}\pi$

23. $-8\pi a^3$

25. $4\pi kq$

27. $\left(1, \frac{2}{3}, 2\right)$

29. (a) $\left(0, 0, \frac{4}{3}\right)$ (b) $128\sqrt{2}\pi k$

Exercises 15.7, Page 849

1. $(x - y)\mathbf{i} + (x + y)\mathbf{j}; \quad 2z$ **3.** 0; $4y + 8z$

5. $(4y^3 - 6xz^2)\mathbf{i} + (2z^3 - 3x^2)\mathbf{k}; \quad 6xy$

7. $(3e^{-z} - 8yz)\mathbf{i} - xe^{-z}\mathbf{j}; \quad e^{-z} + 4z^2 - 3ye^{-z}$

9. $(xy^2e^y + 2xye^y + x^3yze^z + x^3ye^z)\mathbf{i} - y^2e^y\mathbf{j} + (-3x^2yze^z - xe^x)\mathbf{k};$ $xye^x + ye^x - x^3ze^z$

27. $2\mathbf{i} + (1 - 8y)\mathbf{j} + 8z\mathbf{k}$ **37.** 6

Exercises 15.8, Page 855

1. -40π

3. $\dfrac{45}{2}$

5. $\dfrac{3}{2}$

7. -3

9. $-\dfrac{3}{2}\pi$

11. π

13. -152π

15. 112

17. take the surface to be $z = 0$; $\dfrac{81}{4}\pi$

Exercises 15.9, Page 862

1. $\dfrac{3}{2}$

3. $\dfrac{12}{5}a^5\pi$

5. 256π

7. $\dfrac{62}{5}\pi$

9. $4\pi(b - a)$

11. 128

13. $\dfrac{1}{2}\pi$

Chapter 15 in Review, Page 863

A. 1. true

3. false

5. false

7. true

9. true

11. true

B. 1. $\nabla\phi = -\dfrac{x}{(x^2 + y^2)^{3/2}}\mathbf{i} - \dfrac{y}{(x^2 + y^2)^{3/2}}\mathbf{j}$

3. $6xy$

5. 0

7. 0

9. $4x + y - 2z = 0$

C. 1. $\dfrac{56}{3}\sqrt{2}\pi^3$

3. 12

5. $2 + \dfrac{2}{3\pi}$

7. $\frac{1}{2}\pi^2$

9. 5π

11. 180π

13. $\frac{1}{12}(\ln 3)(17^{3/2} - 5^{3/2})$

15. $6(e^{-3} - 1)$

17. $-4\pi c$

19. 0

21. 125π

23. 3π

25. $\dfrac{5}{3}$

27. $z = x^2 - y^2$; hyperbolic paraboloid

29. $y = x^2$; parabolic cylinder

Exercises 16.1, Page 871

1. $x^2 + 4x + \frac{3}{2}y^2 - y = C$

3. $\frac{5}{2}x^2 + 4xy - 2y^4 = C$

5. $x^2y^2 - 3x + 4y = C$

7. not exact

9. $xy^3 + y^2\cos x - \frac{1}{2}x^2 = C$

11. not exact

13. $xy - 2xe^x + 2e^x - 2x^3 = C$

15. $x^3y^3 - \tan^{-1}(3x) = C$

17. $-\ln|\cos x| + \cos x \sin y = C$

19. $t^4y - 5t^3 - ty + y^3 = C$

21. $xy^2 + x^2y - y + \frac{1}{3}x^3 = \frac{4}{3}$

23. $4ty + t^2 - 5t + 3y^2 - y = 8$

25. $k = 10$

Exercises 16.2, Page 877

1. $y = C_1 + C_2e^{x/3}$

3. $y = C_1e^{-4x} + C_2e^{4x}$

5. $y = C_1\cos 3x + C_2\sin 3x$

7. $y = C_1e^x + C_2e^{2x}$

9. $y = C_1e^{-4x} + C_2xe^{-4x}$

11. $y = C_1e^{(-3/2+\sqrt{29}/2)x} + C_2e^{(-3/2-\sqrt{29}/2)x}$

13. $y = C_1e^{2x/3} + C_2e^{-x/4}$

15. $y = e^{2x}(C_1\cos x + C_2\sin x)$

17. $y = e^{-x/3}\left(C_1\cos\dfrac{\sqrt{2}}{3}x + C_2\sin\dfrac{\sqrt{2}}{3}x\right)$

19. $y = C_1e^{-x/3} + C_2xe^{-x/3}$

21. $y = 2\cos 4x - \frac{1}{2}\sin 4x$

23. $y = -\dfrac{3}{4}e^{-5x} + \dfrac{3}{4}e^{-x}$

25. $y = e^{x/2}\left(-\cos\frac{1}{2}x + \sin\frac{1}{2}x\right)$

27. $y = 0$

29. $y = e^{2(x-1)} - e^{x-1}$

31. $y'' + y' - 20y = 0$

33. $y = C_2\sin x$

35. $y = -2\cos x$

37. no solution

39. $y = xe^{2(x-1)}$

Exercises 16.3, Page 882

1. $y = C_1e^{-3x} + C_2e^{3x} - 6$

3. $y = C_1e^{-2x} + C_2xe^{-2x} + \frac{1}{2}x + 1$

5. $y = C_1\cos 5x + C_2\sin 5x + \frac{1}{4}\sin x$

7. $y = C_1e^{-x} + C_2e^{3x} - \dfrac{4}{3}e^{2x} - \dfrac{2}{3}x^3 + \dfrac{4}{3}x^2 - \dfrac{28}{9}x + \dfrac{80}{27}$

9. $y = e^{4x}(C_1\cos 3x + C_2\sin 3x) + \dfrac{1}{10}e^{3x} - \dfrac{126}{697}\cos 2x + \dfrac{96}{697}\sin 2x$

11. $y = \dfrac{5}{8}e^{-8x} + \dfrac{5}{8}e^{8x} - \dfrac{1}{4}$

13. $y = C_1\cos x + C_2\sin x + x\sin x + \cos x \ln|\cos x|$

15. $y = C_1\cos x + C_2\sin x + \dfrac{1}{2}\sin x - \dfrac{1}{2}x\cos x$

$\quad = C_1\cos x + C_3\sin x - \dfrac{1}{2}x\cos x$

17. $y = C_1\cos x + C_2\sin x + \dfrac{1}{2} - \dfrac{1}{6}\cos 2x$

19. $y = C_1e^x + C_2e^{-x} + \dfrac{1}{4}xe^x - \dfrac{1}{4}xe^{-x} = C_1e^x + C_2e^{-x} + \dfrac{1}{2}x\sinh x$

21. $y = C_1e^{-2x} + C_2e^{2x} + \frac{1}{4}e^{2x}\ln|x| - \frac{1}{4}e^{-2x}\displaystyle\int_{x_0}^{x}\frac{e^{4t}}{t}\,dt, \ x_0 > 0$

23. $y = C_1e^{-2x} + C_2e^{-x} + (e^{-2x} + e^{-x})\ln(1 + e^x)$

25. $y = C_1e^{-2x} + C_2e^{-x} - e^{-2x}\sin e^x$

27. $y = C_1e^x + C_2xe^x - \frac{1}{2}e^x\ln(1 + x^2) + xe^x\tan^{-1}x$

29. $y = C_1e^{-x} + C_2xe^{-x} + \frac{1}{2}x^2e^{-x}\ln x - \frac{3}{4}x^2e^{-x}$

31. $y = C_1e^{x/2} + C_2xe^{x/2} + \frac{8}{9}e^{-x} + x + 4$

33. $y = \dfrac{3}{8}e^{-x} + \dfrac{5}{8}e^x + \dfrac{1}{4}x^2e^x - \dfrac{1}{4}xe^x$

35. $y = C_1x + C_2x\ln x + \dfrac{2}{3}x(\ln x)^3$

37. $C(x) = C(\infty)(1 - e^{-x/\lambda})$

Exercises 16.4, Page 890

1. A mass weighing 4 lb $\left(\frac{1}{8}\text{ slug}\right)$ attached to a spring is released from a point 3 units above the equilibrium position with an initial upward velocity of 2 ft/s. The spring constant is 3 lb/ft.

3. $x(t) = \dfrac{1}{2}\cos 2t + \dfrac{3}{4}\sin 2t$

5. $x(t) = -5\sin 2t$

7. A mass weighing 2 lb $\left(\frac{1}{16}\text{ slug}\right)$ is attached to a spring whose constant is 1 lb/ft. The system is damped with a resisting force numerically equal to 2 times the instantaneous velocity. The mass starts from the equilibrium position with an upward velocity of 1.5 ft/s.

9. $\dfrac{1}{4}$ s, $\dfrac{1}{2}$ s, $x\left(\frac{1}{2}\right) = e^{-2} \approx 0.14$

11. $x(t) = \frac{1}{2}e^{-2t}(\cos 4t + \sin 4t)$

13. (a) $\beta > \dfrac{5}{2}$ (b) $\beta = \dfrac{5}{2}$ (c) $0 < \beta < \dfrac{5}{2}$

15. $x(t) = \dfrac{1}{625}e^{-4t}(24 + 100t) - \dfrac{1}{625}e^{-t}(24\cos 4t + 7\sin 4t)$;

$\quad x(t) \to 0$ as $t \to \infty$.

17. 4.568 C;   0.0509 s

19. $q(t) = 10 - 10e^{-3t}(\cos 3t + \sin 3t)$; $i(t) = 60e^{-3t}\sin 3t$;

$\quad 10.432$ C

Exercises 16.5, Page 895

1. $y_1(x) = c_0\displaystyle\sum_{n=0}^{\infty}\dfrac{(-1)^n}{(2n)!}x^{2n}$, $y_2(x) = c_1\displaystyle\sum_{n=0}^{\infty}\dfrac{(-1)^n}{(2n+1)!}x^{2n+1}$

3. $y_1(x) = c_0$, $y_2(x) = c_1\displaystyle\sum_{n=1}^{\infty}\dfrac{1}{n!}x^n$

5. $y_1(x) = c_0\left[1 + \dfrac{1}{3\cdot 2}x^3 + \dfrac{1}{6\cdot 5\cdot 3\cdot 2}x^6\right.$

$\quad\quad \left. + \dfrac{1}{9\cdot 8\cdot 6\cdot 5\cdot 3\cdot 2}x^9 + \cdots\right]$,

$\quad y_2(x) = c_1\left[x + \dfrac{1}{4\cdot 3}x^4 + \dfrac{1}{7\cdot 6\cdot 4\cdot 3}x^7\right.$

$\quad\quad \left. + \dfrac{1}{10\cdot 9\cdot 7\cdot 6\cdot 4\cdot 3}x^{10} + \cdots\right]$

7. $y_1(x) = c_0\left[1 - \dfrac{1}{2!}x^2 - \dfrac{3}{4!}x^4 - \dfrac{7\cdot 3}{6!}x^6 - \cdots\right]$,

$\quad y_2(x) = c_1\left[x + \dfrac{1}{3!}x^3 + \dfrac{5}{5!}x^5 + \dfrac{9\cdot 5}{7!}x^7 + \cdots\right]$

9. $y_1(x) = c_0\left[1 - \dfrac{1}{3!}x^3 + \dfrac{4^2}{6!}x^6 - \dfrac{7^2 \cdot 4^2}{9!}x^9 + \cdots\right]$,

$y_2(x) = c_1\left[x - \dfrac{2^2}{4!}x^4 + \dfrac{5^2 \cdot 2^2}{7!}x^7 - \dfrac{8^2 \cdot 5^2 \cdot 2^2}{10!}x^{10} + \cdots\right]$

11. $y_1(x) = c_0, \ y_2(x) = c_1\displaystyle\sum_{n=1}^{\infty}\dfrac{1}{n}x^n$

13. $y_1(x) = c_0\displaystyle\sum_{n=0}^{\infty}x^{2n}, \ y_2(x) = c_1\displaystyle\sum_{n=0}^{\infty}x^{2n+1}$

15. $y_1(x) = c_0\left[1 + \dfrac{1}{4}x^2 - \dfrac{7}{4 \cdot 4!}x^4 + \dfrac{23 \cdot 7}{8 \cdot 6!}x^6 - \cdots\right]$,

$y_2(x) = c_1\left[x - \dfrac{1}{6}x^3 + \dfrac{14}{2 \cdot 5!}x^5 - \dfrac{34 \cdot 14}{4 \cdot 7!}x^7 + \cdots\right]$

17. $y_1(x) = c_0\left[1 + \dfrac{1}{2}x^2 + \dfrac{1}{6}x^3 + \dfrac{1}{6}x^4 + \cdots\right]$,

$y_2(x) = c_1\left[x + \dfrac{1}{2}x^2 + \dfrac{1}{2}x^3 + \dfrac{1}{4}x^4 + \cdots\right]$

19. $y = 6x - 2\left[1 + \dfrac{1}{2!}x^2 + \dfrac{1}{3!}x^3 + \dfrac{1}{4!}x^4 + \cdots\right]$

Chapter 16 in Review, Page 895

A. 1. true

3. false

5. true

7. true

B. 1. $y = 0$

3. 8 ft

5. $y_p = Ax^2 + Bx + C + Dxe^{2x} + Ee^{2x}$

C. 1. $x^2\cos y^3 - y = C$

3. $\dfrac{1}{4}x^2y^{-4} - \dfrac{3}{2}y^{-2} = -\dfrac{5}{4}$

5. $y = C_1e^{(1-\sqrt{3})x} + C_2e^{(1+\sqrt{3})x}$

7. $y = C_1e^{-2x} + C_2e^{5x}$

9. $y = C_1\cos\dfrac{1}{3}x + C_2\sin\dfrac{1}{3}x$

11. $y = -24\cos 6x + 3\sin 6x$

13. $y = C_1e^{-3x} + C_2e^{4x} - \dfrac{1}{10}xe^{2x} - \dfrac{13}{100}e^{2x}$

15. $y = e^x(C_1\cos x + C_2\sin x) - e^x\cos x\ln|\sec x + \tan x|$

17. $y = \dfrac{1}{2}\sin x + \dfrac{1}{2}\cos x + \dfrac{1}{2}\sec x$

19. $y_1(x) = c_0\left[1 - \dfrac{1}{3 \cdot 2}x^3 + \dfrac{1}{6 \cdot 5 \cdot 3 \cdot 2}x^6\right.$

$\left. - \dfrac{1}{9 \cdot 8 \cdot 6 \cdot 5 \cdot 3 \cdot 2}x^9 + \cdots\right]$,

$y_2(x) = c_1\left[x - \dfrac{1}{4 \cdot 3}x^4 + \dfrac{1}{7 \cdot 6 \cdot 4 \cdot 3}x^7\right.$

$\left. - \dfrac{1}{10 \cdot 9 \cdot 7 \cdot 6 \cdot 4 \cdot 3}x^{10} + \cdots\right]$

21. $0 < m \le 2$

23. $x(t) = e^{-4t}\left(\dfrac{26}{17}\cos 2\sqrt{2}t + \dfrac{28}{17}\sqrt{2}\sin 2\sqrt{2}t\right) + \dfrac{8}{17}e^{-t}$

25. $x\left(\dfrac{1}{10}\right) = 5e^{-0.2} \approx 4.0937$

INDEX

INDEX

INDEX

INDEX

M

INDEX

Reflecting property:
  of an ellipse, 556
  of a parabola, 556
Reflection of a graph, 15, 39
Region:
  bounded, 689
  closed, 689
  connected, 818
  disconnected, 818
  of integration, 750, 755
  multiply connected, 818
  open, 689
  polar, 768
  rectangular, 756
  simply connected, 818
  of Type I, 754–755
  of Type II, 754–755
  unbounded, 689
Regression line, 736
Regular partition, 290
Related rates, 196
Relative error, 248, 469
Relative extrema:
  definition of, 206, 729
  First Derivative Test for, 224
  Second Derivative Test for, 231
Removable discontinuity, 84
Repeated zero, 25
Repeller, 465
Restricted domains, 40
Resultant force, 603, 672
Return of the plague, 423
Reversing the limits of integration, 299
Reversing the order of integration, 299, 761, 778
Review:
  of algebra, RP-1
  of differentiation rules and formulas, RP-8
  of exponential and logarithmic functions, RP-7
  of geometric formulas, RP-2, RP-3
  of graphs of functions, RP-4
  of integration formulas, RP-9
  of trigonometry, RP-4, RP-5
Revolution, solids of, 334, 340
Riemann, Georg Friedrich Bernhard, 295, 303
Riemann sum:
  for a definite integral, 295
  for a double integral, 750
  for a triple integral, 776
Rhombus, 620
Right cylinder, 333
Right endpoint, 291
Right-hand derivative, 125
Right-hand limit, 69, 107
Right-handed coordinate system, 609
Right-hand rule, 609, 625
Rigid transformation, 14
Rise of a line, 21
Robotic arm, 710–711
Roll of spacecraft, 614
Rolle, Michel, 214
Rolle's Theorem, 210
Root of an equation, 5,
Root Test, 510, 517

Rose curves, 580
Rotation of a polar graph, 582, 595
Rotation of a vector field, 848
Rule of correspondence, 2
Rule of 70, 546
Rules for differentials, 251
Rules of differentiation:
  chain, 151
  constant function, 131
  constant multiple, 132
  difference, 132
  sum, 132
  power, 130–131, 149
  product, 139
  quotient, 140
Ruling of a cylinder, 641
Run of a line, 21

S

Saddle point, 731
Sample point, 295, 750, 776, 802
Sandwiching Theorem, 89
Scalar, 602
Scalar multiple of a vector, 602, 604, 611
Scalar product, 614
Scalar triple product, 626
Seashell surface, 831
Secant function:
  derivative of, 146, 147, 153
  domain of, 31
  graph of, 32
  period of, 32
Secant line, 111
Second derivative:
  ordinary, 135
  partial, 699
Second Derivative Test, 231
Second moments, 766, 778
Second-order differential equation:
  auxiliary equation of, 873
  definition of, 440,
  general solution of, 873
  homogeneous, 872
  nonhomogeneous, 872
Second Partials Test, 730
Semicircles, 6
Semistable critical point, 465
Separable differential equation, 441
Sequence(s):
  of absolute values, 482
  bounded, 487
  bounded above, 487
  bounded below, 487
  constant, 478
  convergent, 476–477
  decreasing, 486
  definition of, 476
  divergent, 477
  divergent to infinity, 478
  divergent to negative infinity, 478
  divergent by oscillation, 478
  Fibonacci, 485, 511

INDEX

INDEX

Photo Credits

Chapter 9

Opener (middle left) © E.A. Janes/age footstock; **Opener (middle right)** © Suzanne Tucker/ShutterStock, Inc.; **page 494** Courtesy of Michael Maggs; **page 499** © Andrjuss/ShutterStock, Inc.

Chapter 10

Opener (middle) Courtesy of The Observatories of the Carnegie Institution of Washington; **page 556 (left)** © Dennis Donohue/ShutterStock, Inc.; **page 556 (right)** Courtesy of the Palomar Observatory, California Institute of Technology; **page 557** © Architect of the Capital; **page 564** © Jan Kliciak/ShutterStock, Inc.; **page 577** © Mircea Bezergheanu/ShutterStock, Inc.; **page 578** © Jgroup/Dreamstime.com; **page 595** Courtesy of Mariner 10, Astrogeology Team, and USGS

Chapter 11

Page 649 (left) © Martin Mette/ShutterStock, Inc.; **page 649 (middle)** Courtesy of The North Carolina State Archives; **page 649 (right)** © Marcie Fowler — Shining Hope Images/ShutterStock, Inc.

Chapter 12

Opener (middle) © Corbis/age footstock

Chapter 13

Opener (middle) © Brand X Pictures/age footstock

Chapter 14

Opener (middle) © GraphEast/age footstock

Chapter 15

Opener (left) © Netfalls/ShutterStock, Inc.; **Opener (top right)** Courtesy of NOAA; **Opener (bottom right)** Image by Liu, Hu, and Yeuh, NASA Jet Propulsion Laboratory; **page 808 (top)** Courtesy of NOAA; **page 808 (bottom)** Image by Liu, Hu, and Yeuh, NASA Jet Propulsion Laboratory

Chapter 16

Opener (middle) © Nancy A. Thiele/ShutterStock, Inc.; **page 890** © Kameel4u/Dreamstime.com

Unless otherwise indicated, all photographs and illustrations are under copyright of Jones and Bartlett Publishers, LLC. Portrait of mathematicians were illustrated by Diana Coe.

Therapy Thieves

*How to Save Mental Health Care
From Its Providers*

FRANCIS A. MARTIN, PHD

OXFORD
UNIVERSITY PRESS

OXFORD
UNIVERSITY PRESS

Oxford University Press is a department of the University of Oxford. It furthers
the University's objective of excellence in research, scholarship, and education
by publishing worldwide. Oxford is a registered trade mark of Oxford University
Press in the UK and certain other countries.

Published in the United States of America by Oxford University Press
198 Madison Avenue, New York, NY 10016, United States of America.

© Oxford University Press 2020

CIP data is on file at the Library of Congress
ISBN 978-0-19-751678-2

9 8 7 6 5 4 3 2 1
Printed by Sheridan Books, Inc., United States of America

For her courageous curiosity, persistent generosity, disciplined determination, loving spirit, loyal devotion, boundless friendship, and so much more, this book is dedicated to Martha, my wife.

CONTENTS

Just like many other counselors and therapists, I understand much about my professional world because of those with whom I share it. Some of them easily come to mind—Harvey, Jan, Cris, Dick, Daniella, and Jennifer. These individuals and others have reassured me of the positive future for our profession that I wish to see. Indeed, during my entire career, I have prized my peers who could teach me how to become a better clinician and teacher and how to fill other roles in our profession. The huge body of research that confirms the effectiveness of psychotherapy has also been reassuring. Nothing in this book is intended to be a refutation of that research. Recently, though, the reassurance from both my peers and the research has darkened under threatening clouds over the practice of counseling and psychotherapy and our collective future. One of the largest clouds has appeared in the form of unusual therapies. As I explain later, unusual therapies are promoted by large numbers of practicing counselors and psychotherapists; these include (to name only 20 of the thousands that I have located) source therapy, hypnoanalytic therapy, Indo-Tibetan Buddhism, love first interventions, restorative justice therapy, alchemical psychotherapy, laser allergy treatment, RoHun purification, bright light therapy, Shifaa soul counseling, Jin Shin Jyutsu, happiness work, parapsychology, channeling, Himalayan singing bowls therapy, matrix synergetics, interLife exploration, psicoastrology, quantum medicine, and life principles of the Holy Spirit.

No doubt, some of the thousands of unusual approaches to counseling and psychotherapy are harmful. However, although harm or potential harm to clients should compel others to seek to eliminate these approaches, I emphasize in this book that the large volume of approaches is itself the shape and substance of the threatening clouds over mental health care. While some of the approaches to counseling and psychotherapy may be harmful to clients and to the profession of those who advertise them, they should not as a group be categorized as harmful merely because they are new to those who see them in this book or because they appear to be weird. Also, as this book will show, there are other threatening clouds that should compel action by professionally accountable counselors and psychotherapists. Without such action, the viability of counseling and psychotherapy will continue to decline.

When I began to find out about these unusual approaches to counseling and psychotherapy, I believed that I was one of the very few, if not the only one, who saw them as threats. I later realized that I was correct in seeing the threats, but incorrect in thinking I was the only one who saw them. Some of the publications of others who have written about aberrant therapies will be included later in this discussion.

With no deliberate intention of writing this book, I gathered information from the websites where practicing counselors and psychotherapists advertise their services. An explanation of how this began appears in Chapter 3, "21,758: The Scariest Number in the Psychotherapy World." The outcome of gathering this information from websites surprised me. No, "surprise" is an understatement. As I gathered information, I was shocked and disappointed in us—practicing counselors and psychotherapists. The question that emerged is one that I could not have imagined, intended, or even expected when I began to gather this information. Nevertheless, for me, it was unavoidable: "Why are we—practicing counselors and psychotherapists—committed to failure?" Even now, after having lived with this question for a while, it evokes an aura of unreality.

Why are we committed to failure? As important as this question is, it does not stand alone as a necessary question for counselors and psychotherapists. A further question was, "Why do we tolerate the widespread and obviously inappropriate practices of counseling and psychotherapy?" Several other questions followed that I could not readily answer, and some among them particularly offered challenges for me: "Where did we go wrong in our attempts to train counselors and psychotherapists?" "Where did we go wrong in clinical supervision?" "What's wrong with licensing?" "Does licensing really matter, if very large numbers of licensed, practicing counselors and psychotherapists offer services that their respective scopes of practice do not allow?" "With apparent neglect of professional accountability among counselors and psychotherapists, have we failed to regulate ourselves as mental health care providers?" "Do professional ethics really matter to those who practice counseling and psychotherapy?" "Do counselors and psychotherapists possess a minimally relevant understanding of what they are doing?" "Are competent counselors and psychotherapists a declining proportion of those who practice, as the research concludes?"

As a long-serving mental health care provider, I have felt a strong and enduring commitment to our profession. Based on a considerable volume of research and my long and active participation in our profession as a clinician and trainer of clinicians, I know that counseling and psychotherapy are effective and have confidence that they can be delivered in professionally accountable ways. However, based on a large volume of evidence that contradicts my commitment and confidence, I felt constrained to try to answer the unavoidable question, "Why are counselors and psychotherapists committed to failure?" More than explaining why we have been committed to failure, I try to acknowledge our collective failure and propose ways to move our profession toward a viable future.

In his article, "Fringe Psychotherapies: The Public at Risk," Beyerstein (2001) offers useful insights in his answer to the question, "How did this state of affairs come about?" His explanation asserts that psychotherapies "were founded on ill-conceived assumptions" (p. 70) and that providers of psychotherapies have largely sustained these assumptions in their work. He concludes, "As long as people refuse to think critically and to put psychotherapy methods to hard-nosed empirical tests, bogus treatments will continue to flood the market" (p. 77). His conclusion is not likely to evoke strong arguments against it. Significantly, the question is not whether readers of Beyerstein's argument agree with him, but whether such arguments have an impact on those who practice counseling and psychotherapy.

"Why are we committed to failure?" is an important question, but it is hardly sufficient. It merely acknowledges that we have a problem. Much more than this is needed. Therefore, I tried to answer the other questions that necessity mandated after problems were identified. I asked, "If we are committed to failure, how may we change our outlook and practices so that we are committed to something else? And, what should the 'something else' be?" In a word, the "something else" became "reform." To emphasize the significance of needed reform of counseling and psychotherapy, I concluded that, if we do not seek and achieve major reform, our profession is likely to suffer decline and, possibly, extinction. However, instead of framing our failure as disease and death, I determined that "thievery among us" was a better conceptualization of where we are. This is a way to say that numerous practicing counselors and psychotherapies have "stolen" mental health care. They have stolen its integrity, its future, and its viability. Likewise, in response to their thievery, the rest of us should acknowledge our inadvertent and often enthusiastic abetting of their thievery and attempt to rescue counseling and psychotherapy from the thieves. Also, I should say that the conclusion that our profession is in peril is not new to me or to many other observers of our profession. Evidence for this appears throughout the book.

With almost all of them being licensed professionals, practicing counselors and psychotherapists have stolen counseling and psychotherapy. They have taken it into their private places where the curative power of sunlight fails to penetrate. By their neglect of their captive, except for feeding it empty promises, they allow psychotherapy to atrophy, lose intellectual and ethical muscle, and struggle to survive. In their private places, they practice as if the normative standards of personal integrity have lost appeal to them. They threaten the survival of counseling and psychotherapy. Worse, often under the guise of good intentions and personal commitments—both of which fall seriously below normative professional standards—they often harm their clients.

As dramatic as these statements may appear, the evidence that clearly indicates the need for immediate CPR—counseling and psychotherapy reform—is indisputable. Further, the evidence is not new. Generally unrecognized, the evidence is more abundant than it has ever been.

Quite clearly, the purpose of this book is to advance the cause of reforming counseling and psychotherapy. The need for the book and for reform comes from obvious facts. Among the facts, some of the palpably salient ones are these:

Competence among counselors and therapists continues to decline.

Counselors and therapists do not know what counseling and therapy is.

Counseling and therapy operate almost totally without regulations and/ or oversight. On a massive scale, counselors and therapists "just make up stuff" and call it therapy, compromising their clinical integrity and often harming their clients. In their approaches to counseling and psychotherapy, they are more focused on expressing themselves than on delivering professionally accountable services. There is more unethical behavior in these aberrant approaches to therapy than all other unethical behaviors combined.

Licensing of counselors and therapists is, at best, chaotic.

Training and preparation for counselors and therapists appear to have little positive impact on the practice of counseling and psychotherapy.

Clinical supervision appears to have only a marginally positive impact on the possibility of developing effective clinicians.

Unlicensed counselors and therapists advertise their services with no hesitation, let alone fear of censure or reprimand or termination of their inadequate services.

Codes of professional ethics have proliferated, causing unnecessary confusion and conflict.

Credentialing, especially certifications, among counselors and psychotherapists is much more a sign of profiteering and self-aggrandizement than of professional accountability and clinical effectiveness.

While the mental health care–related professions are practice-based, the maintenance of their future has largely become based in academic institutions, with the consequence of separating the practice of counseling and psychotherapy from research and evidence that many academic professionals hold with justifiable pride.

Even if these assertions characterize the status of mental health care services in minimally accurate ways, they pose serious and long-term challenges for professionally accountable counselors and psychotherapists. At least, they call for reform of counseling and psychotherapy. This book adds to the call for such reform. To clarify and advance the call for reform, this book offers clear evidence to support the assertions listed previously. The evidence comes in several forms, with the largest and likely most persuasive derived from a review of tens of thousands of websites where practicing counselors and therapists advertise their professional services. For now, with more than 1,200 approaches to counseling and psychotherapy included in this book, suffice it to say that a much larger list of approaches to therapy than this book contains has been compiled. One of the obvious conclusions is that the world of counseling and psychotherapy contains an indefinitely large number of these approaches, with a huge proportion of them lacking credibility.

Among the tempting responses to the huge numbers of unusual approaches to counseling and psychotherapy, one to which I opted not to succumb was to excoriate those who advertised them. While excoriation was tempting, I concluded that it was more of a distraction than a potentially valuable contribution to the purpose of the book and to desperately needed reforms in counseling and psychotherapy. For one thing, making individuals targets of analysis would not likely help to move mental health care in a good direction. For another thing, if this book included an analysis that led to the exposure of, say, 5,000 obviously inappropriate approaches to counseling and psychotherapy, such analysis would not have examined an indefinitely larger number of inappropriate approaches.

The therapy thieves who have stolen counseling and psychotherapy include others who have crept into mental health care, as if they know what to do with what they have taken. These thieves include legislators who propose and sometimes enact legislation that seriously compromises the rightful prerogatives of counselors and psychotherapists. One of these is reported in detail in Chapter 7, "Counseling and Psychotherapy Reform (CPR): What We Can Do Together." These thieves also include religious zealots who conclude that all of their world, including counseling and psychotherapy, should comply with their passionately held and often bigoted ideologies. Another category of therapy thieves includes numerous and often very large social service agencies that allege to offer counseling and psychotherapy for their clients. Sadly, their respective personnel are often untrained and transparently ill-prepared to provide the services that they allege to provide. The importance of these and other thieves is that they "reform" counseling and psychotherapy. This is to say that they make counseling and psychotherapy into something that professionally accountable counselors and psychotherapists hardly recognize and should not accept. Thus, the question is not whether counseling and psychotherapy will be reformed. The question is who will initiate and accomplish reform.

Chapter 1, "Kismetic Psychotherapy," is my attempt to offer a plausible, integrated approach to counseling and psychotherapy. While kismetic psychotherapy may evoke uncomfortable responses from well-informed and practicing counselors and psychotherapists, they are encouraged to read the chapter carefully and to raise questions about it, reflect on ways that this innovative approach to therapy may be integrated into their respective practices, and heed the insights that emerge in the chapter. However, as much as readers may respond in a very positive manner to the chapter, they must keep in mind that kismetic psychotherapy is a fantasy—a serious fantasy. It is an approach to therapy that I created only to illustrate how easily anyone can create an approach to psychotherapy and offer it as if it were a real approach to psychotherapy.

Chapter 2, "The Long Conversation About Counseling and Psychotherapy," briefly discusses several nagging issues with which clinical practice is currently suffering. These issues include the decline of competence among practicing counselors and psychotherapists and doubts about the sufficiency of preparation programs. As such, the discussion aligns this book with the long and large conversation that has endured from the beginning of psychotherapy.

Chapter 3, "21,758: The Scariest Number in the Psychotherapy World," introduces information gathered from thousands of websites where practicing counselors and therapists advertise their services. Mainly, it reports about their approaches to counseling and psychotherapy. The number, 21,758, refers to the number of approaches to counseling and psychotherapy that had been identified at the time the chapter was written. Significantly, as large and as shocking as this number may be, it continues to rise as additional websites are reviewed.

Chapter 4, "Observations About Multiple Therapeutic Approaches," reports on the characteristics of the numerous therapeutic approaches that were located on the websites of practicing counselors and psychotherapists. Among several observations, it says that technology has become a dominating influence in mental health care, that psychotherapists continue to turn away from evidence and toward ideology-defined and ideology-driven approaches to their work, and that psychotherapy has become highly and excessively individualized. The chapter includes several other observations.

Chapter 5, "Interpretation of Multiple Therapeutic Approaches," offers "a case conceptualization" of multiple approaches to counseling and psychotherapy. It interprets the status of clinical training, professional competence, and professional identity, among other approaches. It raises questions about the condition of professional ethics among practicing counselors and psychotherapists. As such, it attempts to give meaning and understanding to the current clinical environment, generally characterized as chaotic and cacophonous. In the end, this helps to explain the need for and the aims of reform in counseling and psychotherapy.

Chapter 6, "What Is Psychotherapy?" provides a definition of psychotherapy. The need for a definition of psychotherapy emerges from the lack of one that captures the existential immediacy of the work of counselors and psychotherapists. As such, extremely more important than an efficient and terse entry in a dictionary, we need a definition that sets and maintains the boundaries around the "territory" in which professionally accountable providers deliver their service. Such a definition also suggests that numerous practices that counselors and psychotherapists believe to be approaches to their work lack credibility because they lie outside the boundaries of counseling and psychotherapy.

Chapter 7, "Counseling and Psychotherapy Reform (CPR): What We Can Do Together," identifies several specific reforms that are needed. Among them, given a very large number of codes of ethics, a needed reform is that the number of codes of ethics be reduced. Another is that, given that the practice of counseling and psychotherapy is harmfully underregulated, regulatory initiatives are urgently needed. My colleague, Janet Turner, PhD, generously agreed to coauthor this chapter. She delivered wise and useful guidance toward ensuring that the chapter fulfilled its intended purposes. As an experienced and acutely attuned observer of and participant in mental health care, she has been an energetic supporter of reform. This chapter aims to offer needed macro-reforms, the large-scale changes that must be made if counseling and psychotherapy are to survive and thrive.

Chapter 8, "Counseling and Psychotherapy Reform (CPR): What Each Counselor and Psychotherapist Can Do," describes several practices that may lead to the reform of counseling and psychotherapy and that do not require the individual's involvement in large political or professional movements. It answers the question about what individual counselors and psychotherapist can do, mostly, with or without others to elevate their clinical effectiveness. To illustrate, this chapter offers several recommendations that individuals can pursue, including finding a mentor, studying clinical performance, and developing a network of clinically useful resources.

Finally, Chapter 9, "Guiding Principles of Mental Health Care Reform," recommends several principles of reform, such as seeking cohesion among mental health care professional associations. Simply stated, the importance of this principle is that it contains the belief that reform must be a broadly supported goal for practicing counselors and psychotherapists, instead of an exclusively or mostly individual goal or the mission of, say, social workers and not psychologists.

Throughout the book, I provide lists of approaches to therapy, many of which are aberrant approaches; some of them are likely harmful to the clients who receive treatments based on them. The total number of approaches to therapy in this book comes to more than 1,200. However, as the accrued evidence shows, the number of approaches to therapy is an indefinitely larger one. At the time of writing this book, I have identified 21,758.

Some of the approaches to therapy in this book, such as cognitive behavior therapy or person-centered therapy, are readily identified and widely accepted, but most of them are neither readily recognized nor widely accepted. When readers see the approaches to counseling and psychotherapy that are new to them, they are encouraged to be cautious about their conclusions. From his article, "Psychological Treatments to Avoid," Thompson's advice may be useful for readers who see approaches to counseling and psychotherapy that are new and possibly shocking to their clinical sensitivities. Thompson (2010) wrote,

> Although some therapeutic approaches have clearly demonstrated that they can be dangerous, such as rebirthing, others are thought to be only potentially harmful. Eventually some of the currently unsupported therapies may be shown to be safe and effective. Psychotherapists should be open minded and willing to consider new approaches when their worth is demonstrated. However, those who use unsupported therapies have the burden of proof that they are safe and effective, and until such evidence is provided, they are probably best avoided. (p. 41)

Clearly, given the large number of approaches to counseling and psychotherapy, judgmental responses to them lack professionalism, just as judgmental responses to clients lack professionalism. In this book, the primary response to unusual approaches to therapy is intended to be more invitational than confrontational. This is to say that individuals who advertise approaches to therapy that most of us do not recognize should invite dialogue about their approaches, with the singular aim of trying to answer a couple of simple questions: "How may we make our profession better?" and "Does my identified approach to counseling and psychotherapy help to improve the profession?" Currently, the lack of dialogue sustains a stormy professional environment that likely eventuates in profoundly negative outcomes for the practice of counseling and psychotherapy. To survive the storm, major reforms are needed.

Lacking reform, counseling and psychotherapy will continue to decline. Its practitioners will continue to devolve into declining relevance and effectiveness. Its practitioners will continue to express themselves as the important parties to a less and less coherent and professionally accountable activity. Its leaders and spokespersons will continue to speak into the vacuum where they wish their followers were. External influencers, such as legislators, insurance companies, and others, will continue to pull counseling and psychotherapy into their preferred way of managing mental health care, failing to recognize that it is health care. Prospective clients will continue to embrace many forms of magic as sources of hoped-for healing and, thereby, sustain the ongoing decline in professionally accountable counseling and psychotherapy (Miller & Hubble, 2011). And, much like the lung cancer patient who continues to smoke as he feels the anguish of his impending death, counselors and psychotherapists, "gasping for breath," will wonder, "How could I have been so damned stupid?" Or, in the words of this introduction, "How could I have been so clearly committed to failure?" While we enjoyed its considerable benefits, how could my professional friends and colleagues not recognize the dangers to our profession?

Kismetic Psychotherapy

INTRODUCTION

This chapter provides an introduction to kismetic psychotherapy. It includes a definition of kismetic psychotherapy as a conceptual framework for application as a therapeutic approach to a diverse mental health clientele. It interprets kismetic psychotherapy as a foundation for practicing psychotherapy, including a description of currently understood principles of kismetic psychotherapy, the role of the therapist, and several types of kismetic practices in psychotherapy. The expectation is that the introduction of kismetic psychotherapy will become an important resource for practicing therapists who wish to improve their understanding of clinical work and their services to their clients. Among other important emphases, kismetic psychotherapy focuses on adaptation. Therefore, the reader is encouraged to adapt—to invent plausible, sustainable, responses to kismetic psychotherapy and to consciously pursue such responses with professional discipline and high energy.

As an adaptable profession, mental health care must continuously change. The reader is asked to engage with kismetic psychotherapy with the expectation that she/he will adapt by integrating this new approach to psychotherapy into her/his clinical practice. The singularly important caution is that the reader should recognize that kismetic psychotherapy is a fantasy. It is an entirely plausible fantasy, but a fantasy nevertheless.

Many of the important principles in kismetic psychotherapy already define much of the practice of psychotherapy. This is to say that some of the principles of kismetic psychotherapy are already familiar ones to mental health clinicians, even if they are integrated and configured in new ways that may enhance the range of effectiveness for many practicing mental health clinicians. This merely illustrates that almost all psychotherapists recognize that they share common beliefs and practices, as many writers have emphasized (e.g., Kazdin, 2007; Stevens, Hynan, & Allen, 2000; Wampold, 2015).

To narrow the focus on kismetic psychotherapy, in uniquely personal ways, every mental health clinician knows that she/he was born female or male. Among many other facts that may describe her or him, being born female or male determines many consequent thoughts, roles, ambitions, limitations, opportunities, and so on, without necessarily being narrowly and fixedly deterministic. In other words, most mental health care providers intuitively and objectively know that the biology of being born female or male and the social roles associated with being female or male bear heavy influence on how individuals function as female or male. Moreover, all psychotherapists know this about themselves as female or male individuals. This is not a condition that they created for themselves. Nor, once recognized, as a fact of their existence, they almost inevitably did little to change it, except in small and usually socially constructed, cosmetic ways. This is a kismetic psychotherapist's understanding of gender, briefly stated.

Instead of arriving with both obvious and immutable features and acquiring others, each human person emerges into the world and adapts to it, with commonly identifiable and generally acknowledged constraints and possibilities. Along with its constraints and possibilities, it is this adaptation that partially but significantly expresses kismetic principles. Stated in this manner, the basic premises of kismetic psychotherapy likely appear to be somewhat obtuse. To make it more translucent, illustrations of some of its basic principles may be helpful. Therefore, a brief discussion of two luminaries in the history of psychotherapy may illustrate the importance and necessity of adaptation that almost universally flows from one's personal history into the work of being a therapist. The luminaries are George Kelly, a primary founder of cognitive therapy, and Carl Rogers, the founder of person-centered therapy.

GEORGE ALEXANDER KELLY

Few individuals epitomize the emergence and development of a profession more than George Kelly. In spirit and character, Kelly is much like David Love. Love personifies the development and emergence of modern geology. In *Rising from the Plains*, Love's biographer artfully integrates Love's personal history with the geological history of the Great Plains and the mountains in the western United States where Love was born and raised (McPhee, 1986). Love's immense curiosity about rocks and where they came from drove him to understand the then-nascent discipline of geology, a foreshadowing influence in geology that Love could not likely have imagined. From his home schooling and life as nature's child, Love timidly took the challenge of formal education, eventually completing a PhD at Yale University. And, as "they" say, the rest is history. He became one of the preeminent geologists—if not the preeminent geologist—who shaped the profession of geology.

Similarly, George Kelly was born and raised on the Great Plains in Kansas, with educated parents who sought to teach him at home, mostly. From the many documents that describe Kelly's professional work, the view of him as a man whose formation came outside conventional ways of thinking about psychology and psychotherapy is easy to form (e.g., Fransella, 1995; Kelly, 1955; Neimeyer, 2000; Peyser, 1994). Indeed, his idea of "constructive alternativism" serves not only as a way of understanding important aspects of human functioning but also as a way to picture Kelly's way of functioning—namely, seeing alternate ways of viewing what others saw and, thereby, forming new and useful ways of understanding human functioning.

Kelly (1905–1967) may be the most influential and, proportionately, least known of the truly major figures in the development of psychotherapy. Despite the lack of apparent acclaim, when carefully examined, his influence is undeniable. Like many other third- or fourth-generation psychotherapists, Kelly trained in psychoanalytic concepts and practices, completing his PhD at Iowa State University in 1930. So, when he began to teach at Fort Hays State University, he conscientiously taught and practiced as a psychoanalytic therapist (Doorey, 2014). Along with his teaching and very limited practice, though, he felt the call of conscience to deliver psychotherapy to needy families in rural Kansas during the Great Depression. With support from the state legislature, Kelly's traveling clinic had funding.

Kelly and several of his students conducted therapy sessions for farmers whose lives were seriously challenged by life in the infamous Dust Bowl. As psychoanalytic therapists, though, they did what psychoanalytic therapists do. As Boeree (2014) reports about Kelly,

> He had these folks lie down on a couch, free associate, and tell him their dreams. When he saw resistances or symbols of sexual and aggressive needs, he would patiently convey his impressions to them. It was surprising, he thought, how readily these relatively unsophisticated people took to these explanations of their problems. Surely, given their culture, the standard Freudian interpretations should seem terribly bizarre? Apparently, they placed their faith in him, the professional. Kelly himself, however, wasn't so sure about these standard Freudian explanations. He found them a bit farfetched at times, not quite appropriate to the lives of Kansan farm families.

As farfetched as psychoanalytic concepts and practices may have seemed to Kelly, he sought, nevertheless, to help his desperately needy clientele. The farfetched quality of his work, though, seems to have both discouraged and inspired him, "So, as time went by, he noticed that his interpretations of dreams and such were becoming increasingly unorthodox. In fact, he began 'making up' explanations!" (Boeree, 2014). Yes, because Kelly did not know what else to do, he began to make up stuff to tell his clients, with some of his so-called "interpretations" being much more bizarre than psychoanalytic interpretations. He reflected that he sometimes made up explanations for his clients' problems that were entertaining, bizarre, and quite unfounded, but that had a positive impact on the progress of his clients. "His clients listened as carefully as before, believed in him as much as ever, and improved at the same slow but steady pace" (Boeree, 2014). Apparently, the progress of his clients surprised and confused Kelly.

If many psychotherapists can imagine just making up stuff to tell their clients and seeing their clients improve, what could they reasonably conclude from this? Perhaps Kelly's unusual management of his therapeutic contacts with his clients could be construed as "the mother lode of paradoxical interventions," insofar as he told them stories that might in some farfetched way account for their difficulties and that required his clients to reject the stories and claim their own way of construing their problems. For Kelly, though, an explanation like this seemed too easy. Instead of this, "It began to occur to him that what truly mattered to these people was that they had an explanation of their difficulties, that they had a way of understanding them. What mattered was that the 'chaos' of their lives developed some order" (Boeree, 2014). Eventually, he recognized that he saw cognitive therapy in the process of being developed. Preceding this recognition, though, Kelly "discovered that, while just about any order and understanding that came from an authority was accepted gladly, order and understanding that came out of their own lives, their own culture, was even better" (Boeree, 2014). This is to say that Kelly believed that by giving his clients an alternate way of understanding—a different way of thinking about—their difficulties, he could help them to improve. At its heart, this is Kelly's constructive alternativism and the essence of what would become cognitive therapy.

The importance of Kelly's discoveries is that they occurred during the 1930s, long before several, more prominent cognitive therapists, including Aaron Beck and Albert Ellis, became prominent. In fact, Beck, Ellis, and others have correctly and graciously acknowledged Kelly as an important source of their evolving into the ranks of cognitive therapists (e.g., Leahy, 1996; Rosner, 2012). As Kelland (2017) affirms, "Albert Ellis and Aaron Beck are best known for developing therapeutic techniques that are based on a cognitive perspective of personality and behavior. Although they are not known for developing actual theories of personality, their clinical approaches are based on underlying theoretical perspectives, which shed light on how they view the nature of personality." Making a necessary connection with Kelly, Kelland continues, "Thus, their influential work is naturally connected to that of Kelly, whose theory of personality was entirely cognitive . . ." (Kelland, 2017). Beyond the timing of his discoveries, though, Kelly ingeniously and openly allowed his clinical experiences to influence him to change in fundamental ways. This experiential dynamic appears to be a feature of virtually all psychotherapists who have directed or redirected the profession of psychotherapy. The reality of Kelly's influence—just like all other luminaries in psychotherapy's history—reveals uniquely personal dimensions of this influence. He allowed the phenomenology of human existence to inform his view of human experiencing, but he also allowed his personal history to influence his views of psychotherapy. During a time when psychoanalysis was the primary—almost exclusive—way of conducting therapy and behaviorism was the primary way of researching and teaching human behavior among psychologists—a largely academic profession at the time—Kelly did something different, although he was not totally alone in such a quest. Still, just as Love became something akin to an intellectual sire for modern geology, Kelly became something similar for psychotherapy.

In other words, although Kelly—or any other luminary in the history of psychotherapy—should not be idealized, he seemed to be uniquely able to learn from his experiences as a therapist. He developed but also personified constructive alternativism. In its fuller expression, constructive alternativism may be recognized in Kelly's classic, *The Psychology of Personal Constructs* (1955). In personifying constructive alternativism, Kelly expressed the heritage of a child of freedom from the constraints of normative schooling. This meant that he thought and acted differently from his peers, while maintaining membership in their professional community. Moreover, as a likely inadvertent compatriot with other luminaries, his history impinged on his peers in ways that changed theirs.

Virtually all of those who have studied Kelly's ideas and biography understand that this somewhat fate-based view of his influence seems obvious (Boeree, 1997, 2006; Fransella, 2008; Kenny, 1984; Neimeyer, 2000; Peyser, 1994; Thompson, 1968). As a child whose parents sought economic security by emigrating in a covered wagon to Colorado, failed in this pursuit, returned to Kansas to resume farming, and became his primary teachers during his formative years, Kelly's biography may be seen in his psychology of personal constructs. As Kenny (1984) concludes, "Kelly undoubtedly demonstrated the spirit of the great American pioneers with the publication of his major work at the age of fifty, when he more figuratively acted in the pioneering tradition by pushing back the boundaries of psychology as it was known and accepted at the time." In the same piece, Kenny adds, "The conditions which challenged Kelly to re-construct his own psychological outlook were formed partly out of the American depression of 1931 and the subsequent devastation of the Kansas 'dust-bowl,' together with the exigencies of his clinical practice where he found that people needing to change themselves were not significantly helped by either behaviorism or psychoanalysis."

One of the challenges of reconstruing psychotherapy, in the modes of Freud, Kelly, Rogers, and others, from the frame of their respective biographies is that it inevitably necessitates both acceptance and rejection. Among other reasons that could be explored, but that will wait for another day, the personal, existential reality is that Freud's or Kelly's or Roger's biography cannot match, blend with, or complement the biographies of the complex array of psychotherapists who study them. As if it is in some way an objective process of selecting psychoanalysis, person-centered therapy, solution-focused therapy, cognitive therapy, or another of the many possible selections, some therapists will readily identify with and adopt a theory, and even adore the ideas of a theorist, while others will have responses that are extremely different from theirs, confusing both. The evidence that supports a theoretical orientation alone, no matter what the strength of that evidence, cannot explain an individual psychotherapist's adoption of one theory or another. Further, the evidence and the biography of the inventor of a treatment orientation cannot explain a mental health clinician's decision to accept and practice a specific orientation. Kismetic thinking, though, can and does explain this.

Also, even a cursory reading of *The Psychology of Personal Constructs* evokes thoughts about the author. Why would Kelly construct his theory as he did? Considering their high intellectual capacity and their attraction to the profession of psychotherapy, why would Rogers (discussed later) and Kelly not arrive at essentially the same ideas? What may be the source(s) of their ideas? To what degree are Kelly's views of psychotherapy products of his personal history? The answers to these questions, along with many others that could be raised, are not likely to be obvious with the precision that Kelly himself sought as he composed his views. Nevertheless, with no aim of high-definition precision or convincing others of its rightness, Kelly's personal story is recapitulated in his views of psychotherapy, in major ways. Suffice it to say here that Kelly displayed a strong proclivity for technical issues and that he constructed his understanding of human personality accordingly, concluding, among other things, that each human being is "a scientist."

CARL RANSOM ROGERS

The preeminent cognitive therapist, Aaron Beck, grew up with a depressed mother. Before Beck was born, his mother and father suffered the loss of their children in an influenza epidemic. Apparently, from the time Beck entered the family as a newborn, he received special protection, making him very special—alive and teeming with potential. Many years into his professional work, Beck observed that he finally realized that his professional preoccupation with depression came from his family history–based need to help his depressed mother. Clearly, for psychotherapy community and for numerous clients who have received help from therapists who displayed his influence, Aaron Beck has been one of the most important sources of change in the practice of psychotherapy (Beck, 1996; Beck & Beck, 2013; Famous Psychologists, 2018; Encyclopedia.com, 2018a, 2018b; Psychologist Anywhere Anytime, 2018a).

Similarly, Carl Rogers invented an approach to psychotherapy that expressed who he was and what he most needed from others. This may be arguable for many psychotherapists. However, Rogers himself would very likely have sided with the view that his approach to therapy was, indeed, an expression of who he was. According to Martin, "Just as he [Rogers] would view himself as a 'lone wolf' professionally, he saw himself in youth as a 'pretty solitary boy' and went all through high school with only two dates" (Martin, 1973, p. 2). More than these somewhat terse characterizations, though, Rogers' description (Rogers, 1961, p. 5) of his early years provides several indicators of his experience of himself as lonely and needy of warmth and caring that his family of origin lacked. Some of these indicators include the following descriptions:

- "A very strict and uncompromising religious and ethical atmosphere"
- "A worship of the virtue of hard work"
- His parents were "very controlling of our [Carl and his five siblings] behavior."
- "Even carbonated beverages had a sinful aroma."
- "We had good times together within the family, but we did not mix."

Apparently, Rogers kept his need for warm and caring relationships neatly contained, but barely, as he progressed through the University of Wisconsin. He began his studies there as an aspirant toward a career in agriculture. Later, recognizing that his views on many aspects of life were becoming "much more liberal in religion and politics due to my exposure to a wide spectrum of opinions" (Rogers, 1967, p. 350), Rogers redirected his studies toward a career in ministry. He wrote that his "intellectual horizon was incredibly stretched through this period" (Rogers, 1967), suggesting that his historically rigid "self" may have begun to evolve toward new expressions that he could not then see or anticipate.

Rogers' commitment to a career in ministry took him to Union Theological Seminary in New York City "because it was the most liberal in the country and an intellectual leader in religious work" (Rogers, 1967, p. 353). He began his studies at the Seminary, but after a while began to take courses at Columbia University, eventually leaving the Seminary to pursue a new career direction. At Columbia, he entered the psychology program and completed his PhD in 1928. His evolution toward seeing his need for warm and caring relationships still had a long way to go. After the birth of his first child, David, he stated that he and his wife, Helen, "endeavored to raise him 'by the book' of Watsonian behaviorism" (Rogers, 1967, p. 356), a retrospective recognition of his apparent lack of awareness of possible directions his career would later take. This recognition, though, captures Rogers' almost inevitable discovery of ways to express himself in the practice of psychotherapy.

Eventually, and mainly through clinical experiences, Rogers discovered ways to express himself in an authentic manner in the practice of psychotherapy. While there may be numerous indicators of his evolution toward his authentic presence during the practice of therapy, few of them are as clearly and thoroughly articulated as in his book, *Counseling and Psychotherapy* (Rogers, 1942). Still consolidating his thinking about how therapy works, his claim for a newer approach in 1942 served as a precursor to his fully developed ideas in person-centered therapy. He wrote,

> In the first place, it [the newer approach to therapy] relies much more heavily on the individual drive toward health, growth, and adjustment. Therapy is not a matter of doing something to the individual, or of inducing him to do something about himself. It is instead a matter of freeing him for normal growth and development, of removing obstacles so that he can move forward.
>
> In the second place, this newer therapy places greater stress upon the emotional elements, the feeling aspects of the situation, than upon intellectual aspects.
>
> In the third place, this newer therapy places greater stress upon the immediate situation than upon the individual's past.
>
> [Finally,] . . . this approach lays stress upon the therapeutic relationship itself as a growth experience. (p. 29)

Rogers effectively articulated his belief that psychotherapy will and should flow from a relationship of importance to the client—and, in this discussion, more importantly, to the therapist. Through client-centered therapy and, later, its relabeled title, person-centered therapy, Rogers gave the world, including the psychotherapy

community, the clearest and most profound definition of the therapeutic relationship. In the intellectual community at large and virtually unanimously within the
therapeutic community, his definition stands as strongly now as when he first
presented it many years ago. It includes the following conditions that therapists
should provide for their clients (Rogers, 1951):

- Authenticity or genuineness
- Unconditional positive regard
- Accurate empathic understanding

In short, Carl Rogers found ways to express himself in the practice of psychotherapy, transitioning from "a solitary boy" who was constrained by a strict religious heritage and overcontrolling parents to a man whose life, spirit, intellectual
offerings, and influence epitomize the best of healthy relationships. Pursuing
relationships of which he was largely deprived early in his life, he gained understanding and fulfillment in relationships that also redirected the development of
psychotherapy. With authenticity, unconditional positive regard, and accurate
empathic understanding, these relationships are more attached than distant, more
meaningful than utilitarian, more durable than weak, more adaptable than closed,
more inclusive than judgmental, and more authentic than expedient. Moreover,
they are more therapeutic than psychotherapy could be without them. Together,
these features of Rogers' approach to psychotherapy stand with almost universal
approval almost 70 years since he first presented them.

   In the context of a discussion about kismetic psychotherapy, the looming question here is whether Rogers was fated to invent person-centered therapy. The
short answer is no. He was no more fated than any other psychotherapist. Quite
understandably, he could have succumbed to the demands of his family—namely,
that he maintain their rigid, unaffected style of relating. Or, he could have opted
for any number of other courses for his life, including addiction to a substance
that could have neutralized the pain of his loneliness. Or, he could have sustained his once-committed interest in becoming a minister. Conversely, because
of his personal history, he was extremely unlikely to become a devotee of cognitive therapy or of behavioral therapy. Instead, however, again in the context of
this discussion, a more nuanced answer than "no" is that Rogers, like the rest of
psychotherapists, utilized his unique, personal history in ways that he deemed
would meet his uniquely personal needs. And, to the good fortune of the professional mental health care community, he accomplished this.

   Clearly, Rogers transcended the mission of meeting his needs. Much more
than meeting his needs, as described earlier, with vigorous intellectual discipline,
he investigated and researched psychotherapy. So, although he may have found
ways to meet his unique, personal needs through service as a psychotherapist, his
gifts to the profession of psychotherapy stand on their own, just as Aaron Beck's
contributions to the profession of psychotherapy stand on their own. More generally, given the scientific muscularity of his work—work for which he received
numerous awards—Rogers raised questions about whether behavioral scientists

"will use their power to dehumanize and depersonalize the person, or whether they will use their power to free the person, to develop the creativity of the person, and to facilitate their self-directed process of becoming" (Martin, 1973, p. 65). "While recognizing the enormous complexity of the choice of the use of psychological knowledge, Rogers clearly chose the latter" (Martin, 1973, p. 63). Rogers aimed for personal power, creativity, freedom, and self-directed becoming for others—just as he needed them, sought them, and chose them for himself.

DEFINITION OF KISMETIC PSYCHOTHERAPY

"Kismetic" derives from a commonly used word, *kismet*. Typically, the usual translation of kismet is "fate" or "destiny." Placing "kismet" in the context of kismetic psychotherapy, the meaning of "fate" or "fated-ness" expands considerably, while maintaining the essential concept of fate. In its expanded and elucidated presentation, "kismetic" refers to the fate-suffused circumstances of life that in many meaningful and practical ways set directions in the lives of individuals, as illustrated by the reports about George Kelly and Carl Rogers, discussed previously.

With particular reference to psychotherapy, "kismetic" maintains, on the one hand, a historical and concurrent fated-ness of the individual and, on the other hand, a future and malleable fated-ness. This is largely a time-related measure of kismetic thinking. As such, kismetic psychotherapy holds that a person's past, instead of a mere historical record such as a somewhat pedantic biographer may pen, contains a vast array of features over which the person had no control, that frequently if not always shaped the individuals uniqueness as a psychological being, and that continue to shape the individual's behavior, if unquestioned and unaltered. Indeed, questioning and altering, sometimes eliciting daring and painful adaptations, catch the essence of kismetic psychotherapy.

To expand kismetic thinking, these comments should be noted: Virtually all mental health clinicians recognize the feature of fated-ness in their clients and in themselves. For example, they recognize that a child who is born and raised with professional persons—parents—who are also nurturing, dedicated, supportive, disciplined, financially secure, resourceful, and energetically focused on their child's welfare, will likely fare much better than the child whose single parent struggles in isolated poverty and works three part-time jobs. Which child is likely to be mentally healthy? Which child is likely to pursue higher education and a professional career? The belief in fated-ness, not fixed determinism, comes easily when relevant facts are taken into account. This is not deterministic, with the implication of choice-less-ness, as much as it is a recognition that all human beings are informed by their environment, relational histories, various social and natural events, circumstances, and biology. Specifically, for example, being reared in Tempe, Arizona, or Portland, Maine surely influences a child's view of the world, while not rigidly determining that view.

What should an individual be, when he is born male? Should he be gregarious? Should he be athletic? Or, intellectual? Or, gay or straight or something else?

Obviously, being born male or female is not a self-selected option for the one being born. After being born, say, as male, the options will come, along with biologically and socially constructed constraints and opportunities. At least partially, this illustrates the kismetic quality of human experiencing. In addition, examples of the somewhat kismetic features of human experiencing include the following ones, keeping in mind that the deterministic quality of these examples should not be construed as universally applied to all persons who may have these features:

- Being born female or male
- Developing during childhood in, say, Tunisia or Canada or Japan or Hawaii
- Being of a certain race, such as Caucasian or African
- Developing in a one-parent and impoverished household
- Developing with a physical disability, such as spina bifida

Each of these features—or two or more, taken together—shapes a person's sense of relevance, of importance, of place in the world, and more. The significance of these features, along with others, is that each one establishes a large measure of an individual's identity and fated-ness. Then, necessarily, the matter of the quality of this fated-ness must be addressed. This matter is addressed in later discussions of various types of kismetic issues and their treatment in psychotherapy.

Maybe, a brief discussion of what kismetic psychotherapy *is not* could be helpful, to distinguish it from some common and commonly, if not universally, accepted ways of viewing personal mental health issues.

Kismetic psychotherapy rejects mindless determinism, although some aspects of being human are clearly determined, such as birth, specific genetic features, aging, and death. Between birth and death, individuals adapt. Kismetic psychotherapy presses the question of how adaptation occurs.

Unlike classical psychoanalysis, kismetic beliefs do not rest on the necessity of the unconscious as a critically important feature of psychological functioning. While biological sex may come with birth and shape an individual's consciousness of self and relating with the world, sexual identity and its fullest implications are hardly unconscious features of human functioning, even when they involve intense struggle toward clarification.

Unlike much of cognitive therapy, along with various versions of it, kismetic beliefs do not hold that cognitive processes dominate human functioning. Just as Aaron Beck invested decades of his professional life to researching and exploring noncognitive approaches to psychotherapy before discovering his "truth," many therapists understand, albeit tacitly, that neither cognitive therapy, psychoanalysis, narrative therapy, behavioral therapy, person-centered therapy, emotion-focused therapy, nor any other orientation to psychotherapy can justly lay claim to holding the singularly superior approach to therapy. However, with specific reference to cognitive therapy, kismetic thinking conceives it as another sense, comparable to sight, tactility, hearing, and smell. As such, cognitive functioning may not be as governable by individuals as cognitive theorists wish to believe.

Unlike person-centered and other humanistic approaches to psychotherapy, kismetic psychotherapy concludes that psychological freedom cannot be as free as probably previously overly constrained individual converts to the humanistic ideology wish to believe. The previous report about Carl Rogers illustrates this very well. The difference between kismetic psychotherapy and humanistic approaches to psychotherapy is one of shades of meaning in which the understandable desire for freedom as the humanistic therapists describe it is conditioned by many obvious constraints on freedom that almost everyone else acknowledges. Nevertheless, despite the exaggerated emphasis on freedom among humanistic therapists, kismetic psychotherapists place a strong and ever-present emphasis on readapting—or overcoming personal problems—so as to gain as many healthy options as possible. This is kismetic freedom, as paradoxical a this may seem. Both approaches believe in the possibility of significant change toward health as an outcome of psychotherapy. (A presentation of the psychologically healthy individual is provided later. Kismetic psychotherapists hope to assist individuals with their emergent adaptations as healthy.)

Unlike behavioral psychotherapies, kismetic psychotherapy holds a much more encompassing view of human potential. For example, generally, behavioral understandings of human functioning lack a capacity to make minimally appropriate sense of creativity and innovation. Kismetic psychotherapy endorses creativity and innovation as important features of being human, if not universally expressed ones. More than endorsing them as features of human experiencing, kismetic thinking holds that creativity, innovation, and similar activities may be genetic endowments in the human species, much like vision, hunger, and movement. Further, whether it is creativity or innovation or similar concepts, adaptation is a crucial and indispensable feature of being human. Kismetic therapists invent ways to channel adaptation toward health for their clients.

These ways of differentiating kismetic psychotherapy from other general orientations to psychotherapy are much too stark. As generalizations, they provide only sketches of some of the essential differences. Some of the principles of kismetic psychotherapy, presented next, add details that assert relatively autonomous views that do not require differentiation from other major orientations to psychotherapy. Such differentiations are likely better left to historians of psychotherapy and critics who may provide objectivity that those inside "the frame" of kismetic psychotherapy may not have.

EMERGING PRINCIPLES OF KISMETIC PSYCHOTHERAPY

Following the previous discussion of George Kelly and Carl Rogers, the first-mentioned principle of kismetic psychotherapy is not the most important one, but a nicely illustrative one. It is a transition from the discussion of Kelly and Rogers to the other, more basic, principles. It is that *therapy is biography*. Kelly's history defined and forecast much of what he would become as a psychotherapist. Rogers' history defined and forecast much of what he would become as a psychotherapist. The view of kismetic psychotherapy is that the history of the therapist defines and forecasts much of what the psychotherapist will become. This does not preclude the possibility that individual psychotherapists will/can transcend the limitations of their respective histories. Personal history expresses itself, somehow. To conclude that personality is merely a recapitulation of one's history, as if dynamic and changing relationships, personal choices, and random large and small events do not concurrently alter expressions of personality, would deny too much of the reality of being human. Furthermore, therapists may be seriously limited, if not impaired, by their histories. Quite separate from what may define or forecast the emergence of a therapist, all individuals may be largely defined and forecast with regard to what they may become.

More fundamentally, a necessary principle of kismetic psychotherapy is that, more than unconscious forces or drives toward inevitable health when the conditions are "right," *adaptation moves individuals toward increasing confirmation of their character as their respective personal histories have defined them.* Adaptation is energized by needs, some of which have arisen in the unique personal, relational environment of family. This describes both Kelly and Rogers, along with virtually everyone else. In kismetic psychotherapy, adaptation may move an individual toward increasing confirmations of an individual's character in ways that are transparently healthy, such as happened with Kelly and Rogers. Or, adaptation may move individuals toward poor functioning, such as overly anxious or depressed or addictive or self-harming functioning. Health is a way of adapting, but so is pathology. As such, neither mental health nor mental pathology is inevitable. Generally, both are adaptations. The kismetic psychotherapist seeks to facilitate client development toward less suffering and, more hopefully, toward health.

Followed by 11 corollaries, Kelly (1955) worded his fundamental postulate as follows: "A person's processes are psychologically channelized by the ways in which he anticipates events" (Vol. 1, p. 46). Capturing his view of the individual as scientist, Kelly concluded that human beings construct a view of their future and experiment with possibilities and, therefore, make decisions about how to get into the future by anticipating what may happen—testing their view. Kismetic psychotherapy takes a view that is similar to Kelly's. The primary difference is that anticipation and testing are important features of individual functioning, but that these features express a rolling, ongoing confirmation of an individual's history and the view of her/himself that continues to evolve from this history. Kelly could not have opted to be a girl who grew up in New York City. He was a boy who grew up in rural Kansas. So, the principle that kismetic psychotherapy holds is that *anticipation and adaptation are a future orientation, including decision-making about how to get to the future, emerging from a personal history that shapes all future anticipating and decision-making.*

Adding to Kelly's idea about the individual as scientist, kismetic psychotherapy incorporates the idea that individuals truly experiment with possibilities for their future. For example, even in colloquial terms, a common view is that "Michael has been experimenting with drugs." This indicates that "Michael" anticipates a future for himself, including decision-making about how to get to his future, that emerges from his personal history. He may be confirming himself as a "loser." Further, as scientist-person, Michael tests his view of himself as a loser each time he experiments with drugs. Obviously, addictions are far more complex than this illustration may suggest. Significantly, the opposite of addictions as experimenting is important, too. Instead of addictions, Michael could have tested his loser identity by achieving desired, positive goals, such as starting and maintaining a successful tree-trimming business or volunteering to deliver meals to home-bound elderly persons.

The concept and practice of reconciliation encapsulate another principle of kismetic psychotherapy. Mentally healthy individuals know this. But, what exactly do they know? In short, healthy individuals know that their mental health "flows," instead of being consolidated into an achievement that is maintained as something achieved or awarded, like a trophy. Mentally healthy individuals know that their mental health is vulnerable. After all, today's mental health may dissolve into maddening chaos following an unfortunate series of life-changing events. First, the automobile accident takes away a not-yet-paid-for car. Then, the slip on a patch of black ice fractures the right femur. Then, the second of three children, 9-year-old Samuel, is diagnosed with lung cancer in an advanced stage. Then, the oldest child, 15-year-old Glorietta, announces that she is pregnant. Then, the exhausted parent tries to get a night's sleep after a day like this one. The emphasis on reconciliation in kismetic psychotherapy means that historical fated-ness, concurrent fated-ness—as illustrated in the circumstances just described—and individuals' handling of these circumstances challenge individuals to renew their commitment to resolving the inherent and sometimes overwhelming conflict between despair because of what they cannot control and friendship with their capacity to move or try to move their life toward well-being. Reconciliation is ongoing adaptation to one's past, complex decision-making about how to function in the present moment, and the envisioning of a useable future. It is this renewal of commitment to resolving life's inherent conflict between despair and friendship with life that distinguishes those who get knocked down and stay down and those who get knocked down and get up. This is resignation to fate or accommodating it in constructive and effective ways or transcending it.

Also, reconciliation in kismetic psychotherapy points to openness to one's own historical experiences and ongoing experiencing, as practicable and feasible as possible and in a moment-to-moment manner. In this way of thinking, openness refers to openness to hopeful and life-affirming features of one's own history as well as to despairing and life-denying aspects of one's history.

A clear marker of the potential for reconciliation is curiosity. Heightened curiosity about one's self and one's primary relationships signals movement toward mental/emotional health. Most mental health care providers understand that those who lack curiosity—or whose curiosity has been thwarted—generally function poorly. Kismetic psychotherapists, then, look for the evolution of curiosity in their clients. And, as curiosity evolves—with an emphasis on heightening self-awareness from curiosity about one's self—self-examination becomes increasingly purposeful. While the purposefulness in therapy should be client-defined and not therapist-defined, the kismetic psychotherapist anticipates that the client's purpose will be seen in a clear and demonstrated interest in becoming as healthy as possible. In this, the inherent challenge for all psychotherapists, but especially kismetic psychotherapists, is the emergent immediacy of knowing what human mental-emotional-relational health is and a clinically plausible and realistic judgment about the capacity of a client to achieve such health. Analogously, when a physician confronts a patient with a sprained ankle or lung cancer, she/he makes a determination about the patient's potential for restoration to health.

Many experienced mental health clinicians have treated parents who have abused their children. Even when the abuse is obvious and undeniable or even witnessed by accountable and reliable reporters, including mental health clinicians, police officers, attorneys, and others, offending parents often readily avow love for their children. Often mental health clinicians are aghast on hearing these parents declare their love for their children. How could this be? These parents severely abuse their children and declare their love for their children? They often reveal guilt about committing the abuse and state that their children do not deserve such treatment. Observing this may be the easier part of helping these disturbed parents. The harder part comes when these parents begin to wonder about themselves and, with visible anguish, ask hard and larger unanswerable questions: "Why in hell do I want to hurt my kids?" "How could I be so goddamn stupid?" As an objective case study in which observers have no accountability or other interest, these parents hold relatively little meaning for psychotherapists. As living organisms whose grimaced, anguished faces and flooding tears emit almost immeasurable pain and suffering only a few feet away from the psychotherapist, these parents have enormous meaning.

Although this is not a universal observation, the observing mental health clinicians often come to a common conclusion about the abusing parent. Namely, *unmet relational needs energize behavior, but so do met relational needs.* In the case of the abusing parents, unmet relational needs—usually from a long history of relational deficits in their history—of the abusers sustain what they know to be true about their lives. Usually, the abusing parents know that life is onerous, frustrating, lacking meaningful connections with other human beings, and even pervasively hostile. In contrast, if a parent's relational needs are met, abuse is extremely unlikely to happen.

The practice of psychotherapy will be discussed later. However, just now, in the context of a discussion of the concept of unmet relational needs, the implication for practicing psychotherapy should be noted. Specifically, when treating almost any client, including the abusing parent, *the psychotherapist looks for disclosures of the client's self-identified relational needs. Further, the more strongly and clearly these needs are declared, the healthier the client is becoming.* The process may be tormenting for these clients, but because relational needs seek expression and expand into a gathering repertoire of relational skills through therapy, clients somewhat paradoxically welcome the torment.

These are some of the principles of kismetic psychotherapy. Following a large body of clinical evidence, gathering data and guidance from relevant research, and engaging in serious contemplation about how to practice psychotherapy, kismetic psychotherapy adds to its practice many learnings from the accrued wisdom of clinicians that have emerged during the first 100 years or so of the practice of psychotherapy. For example, from the earliest days of the emergence of psychoanalysis, humane treatment of clients—"do no harm"—has been suffused with the thinking of almost all psychotherapists. However, not doing harm hardly suffices as therapeutic. Being therapeutic requires a clinician to have a goal of mental health and an ambition of helping to move the client toward mental health—the healthy person.

THE HEALTHY PERSON

Most theories of psychotherapy place understandable emphasis on pathology, disorder, poor functioning, or other problem-focused behaviors. Clearly, this is important. After all, treating individuals who epitomize mental and relational health is not likely to benefit anyone. While the emphasis on pathology—or whatever comparable term is used—is understandable, it often establishes a clinical environment in which psychotherapists and clients lack ideas about how to understand and recognize healthy functioning. Analogously, clinical supervisors generally lack clear understanding of their fundamental reason for conducting clinical supervision, helping to create a highly effective psychotherapist (e.g., American Psychological Association, 2012; Crits-Christoph et al., 1991; Jennings & Skovolt, 1999, 2004; Miller, 2013; Miller & Hubble, 2011; Mullenbach & Skovholt, 2004; Skovholt, Jennings, & Mullenbach, 2004). Including its careful attention to clients—or to clinical supervisees—*kismetic psychotherapy seeks to maintain focus on human health.* In what ways may this be seen? What is human mental health?

Regardless of the limitations or possibilities that may define any particular theory of psychotherapy, the obligation to articulate a view of healthy functioning should be unavoidable. This represents another principle of kismetic psychotherapy to the extent that it serves as an important source of guidance for kismetic clinicians. In the interest of limitations that come with containing a plausible but manageable volume of information about the healthy person, it should be said that much more could be written about the healthy person than appears here. The characteristics provided here are more illustrative than complete.

The healthy person carries an ever-present awareness of her/his freedom to choose. In most social contexts in which having a "side"—a team, a religion, a race, and so on—seems to be a necessity, the healthy person is relatively "sideless," while maintaining a clear moral steadiness that transcends sides. This means that the competitivism that characterizes much of American culture, insofar as it promotes division, disunity, hostility, hierarchy, and other generally unwanted sources of hurtful conflict, evokes reactions of conscience in the healthy person. Instead of competitivism, the healthy person seeks to fulfill life-affirming and meaningful behaviors. Specifically, for example, through generosity, the healthy person places emphasis on ensuring that others have needed resources, particularly those resources that others cannot provide for themselves. Typically, the healthy person maintains awareness of the pressure to take sides, to align with those who can bring benefits to her/him, but nevertheless chooses life-affirming actions. Drawing both admiration and enmity, the healthy person stands in contrast with an American culture that is largely narcissistic, self-promoting, and even greedy, a culture in which the "have-nots" have persistently increased as a proportion of the population for almost 4 decades.

The healthy person has a heightened and effective sense of proportionality. This characteristic of healthy individuals may be expressed in numerous ways. Here are some of them:

> They feel good about themselves, but not with exaggeration or denial of problems. They take justifiable pride in good or excellent relationships or in work or other activities, but know that pride is much more about future expectations than self-satisfaction from achievements.
> They are searching, curious persons who know that there are answers to many of life's pressing questions and that the answers are almost always tentative.
> They understand that adversity comes with living and that this fact only partially describes human experiencing, insofar as companionship, good food, good sleep, and other "feel good" realities come with life, too. So, more likely than not, they understand that adversity or its opposite is momentary and that something significant happens after either one.
> It is this cognizance of "something" that permits the healthy person to maintain openness to possibilities.

The healthy person demonstrates cognitive complexity. For concrete, either-or thinkers, this feature of the healthy person may be confusing. In short, this feature says that the healthy person shows high tolerance for ambiguity, likes puzzles in many forms, and invites new experiences, including discovering new acquaintances. This is not to suggest that healthy persons reside only in communities of intellectuals. Instead, healthy persons may be found in many kinds of communities, including families, universities, religious groups, various occupations, all income levels, and more. Indeed, an important indicator of complexity is the individual's ability to receive and understand the views of others, especially when their views conflict with one's own.

Whether they are more or less expressive with their feelings than others, *healthy persons are considerably more fluid with their feelings,* commonly demonstrating a capacity for expressing a wider range of feelings than others. They embody the antithesis of alexithymia (Sifneos, 1973). In their relations with others, they show a readiness to be receptive to a wide variety of expressions of emotions. Generally, this equips them to attunement to their own feelings and, therefore, to effective self-care.

Healthy persons possess strong and positive relational skills. The idea of strong relational skills is often associated with leading political figures. While the difference between healthy persons and political figures cannot be narrowed or statically set, it is one of health, more than obvious effectiveness. The healthy person relates in a manner that affirms and endorses health in others, while politicians often seek benefits for themselves, even denying benefits for those whom they seek to influence. Typically, healthy persons emerge from a history of relationships in which they have been positively valued. In addition, they almost universally believe that strong, mutually affirming relationships form the basis for effective living. Sometimes as a stated philosophy of life and sometimes with simple intuition, they know that their significance, their mattering, comes from relationships and not so much in academic degrees, awards won, or other common markers of success, or in the number of high-profile names that may be "dropped" so as to gain vicarious value.

THE PRACTICE OF KISMETIC PSYCHOTHERAPY

Neither this material nor any other book can report and explain everything about the practice of psychotherapy. The need to economize is clear in any attempt to report about such things. Taking limitations into account, the information here about the practice of kismetic psychotherapy is summarized within the following categories: the role of the therapist, the process of psychotherapy, and the practice of psychotherapy.

The psychotherapist carries several important functions that define the role of the therapist. Lacking "answers" that, in almost all client situations, are the prerogatives and properties of the client, the therapist is a co-discoverer and, to a significant degree, a co-adventurer. As such, some of the major functions of the therapist are described next.

The therapist offers persistent demonstrations of curiosity about the client. The need for this assumes that the therapist cannot know all clinically relevant information that resides in the history and mind of the client. Further, while most reports from the client accurately represent factually correct and clinically relevant information, the common conclusion about memory as reconstruction prompts the necessity for curiosity. In other words, the therapist asks questions, instead of giving answers, believing that the clients' discovered answers will much more likely be owned and followed by them, mainly because the clients experience and manage life within their own frame of discretion and accountability. This is to say that the clients live life for themselves, not for the psychotherapist or the psychotherapist's welfare or needs. Kismetic psychotherapists hold great respect for this reality.

The therapist delivers multiple actions of disciplined spontaneity. Allowing for the vagaries of psychological and relational life that clients inevitably bring to therapy, psychotherapists must demonstrate the attitudes and actions that they seek to develop in their clients—mainly, change toward healthier ways of functioning. As co-discoverers, living through complex processes of change, therapists deliver discipline in clinical judgment and other actions, but also maintain a readiness to innovate in ways that cannot be planned—spontaneity. By definition, spontaneity is not scripted, even while it is accompanied by professional clinical judgment. The importance of the practice of disciplined spontaneity or disciplined improvisation has endured for many years in psychotherapy. Increasingly, it is an empirically validated practice. For example, Crane et al. (2015) confirm this, in their article, "Disciplined Improvisation: Characteristics of Inquiry in Mindfulness-Based Teaching." Cited by Kindler (2010), Nachmanovitch captures the essence and importance of disciplined improvisation in therapy.

> Improvisation is the normal mode of human communication. Whether the context is therapy, art, love-making, parenting or any number of other arenas in which we interact, we do not write down what we are about to say before we say it. We simply say and do, prompted by the incalculable mixture of conscious and unconscious influences that shape the self at that time and place in our development. (p. 223)

The therapist focuses on change, always mindful of encouraging and affirming of changes toward health in clients. Being mindful of change requires that mental health clinicians develop clear parameters and markers of changes toward health so that these may be recognized when they occur. Most therapists give attention to change as a necessary feature of psychotherapy, but they almost universally preclude the possibility of change by placing a higher priority on other aspects of therapy, such as building and consolidating a therapeutic alliance with the client. Instead of this, the kismetic psychotherapist contracts with clients from the beginning of therapy, with central agreement that the client identifies and claims desired goals and that the therapist facilitates dialogue through which these goals maintain a high level of visibility during sessions. The high visibility of client-defined goals rests on careful problem definition and mutual recognition between the therapist and the client of that definition.

The therapist maintains a professional, clinical environment for clients. A professional environment includes many features. It includes many elements that most therapists usually know very well, such as accurate empathic awareness, thoughtful assessment of the client's presenting issues, appropriate comfort with client pain and struggle, and many more. In addition, the therapist insists on coming to a clear understanding with the client about the benefits and risks of therapy, including but surely not limited to the following:

- Elevated stress from making needed changes
- Clarity about payment
- Agreement about the length and frequency of sessions
- Information about the therapist's role, qualifications, and expectations
- Information about the risks and positive potential of participating in psychotherapy
- Conditions under which therapy may be preemptively terminated
- Limitations around the clinical relationship

The therapist integrates relevant research into the process of therapy, understanding that each therapeutic encounter may pose clinical challenges for which little or no relevant research is available. When such research is available, though, the kismetic psychotherapist asks how it may be integrated into the process of therapy. The importance of this is that, pursued in intellectually honest and professionally accountable ways, therapy is much more likely to benefit the client than to indulge the therapist. The long history of psychotherapy confirms that, lacking information from research, clinical expertise is more often than not a somewhat whimsical notion (e.g., Garfield & Bergin, 1986; Miller & Hubble, 2011, Rowan, 1992). Unfortunately, the "strayward" thinking of this may be easily seen in the self-interested advocacy of many so-called therapies, such as the purposeful pursuit of God's purpose, Pranic healing, horse-informed archetypical therapy, ear candling, bouldering, or any of thousands of others.

The therapist functions as an informed analyst of human interaction, with emphasis on unmet relational needs. The kismetic psychotherapist exerts energetic attempts to identify unmet needs, to be sure, but also the multiple, surrogate, and usually noneffectual ways that clients seek to meet their relational needs. As abstracted from real human experience, it is truly real and not so abstract. Indeed, the interest in identifying such behaviors is usually easily satisfied. For example, the smart, haughty, somewhat smug, and easily disliked 52-year-old male client unintentionally broadcasts his desperate need for affection, but also his intense fear of exposure to the risk of failure in exploring his need for affection. With this quite limited profile of the client, even relatively unskilled therapists would likely and effectively assist this client in discovering his unmet need and setting reachable goals that stem from this need, knowing that important and enduring change may be long in coming and difficult to achieve. This approach defies a common research-minded clinician or academic researcher who may be inclined to rely too heavily on isolated variables that are used to explain client behavior (Frank & Frank, 1991).

ADDITIONAL, PRACTICAL CONSIDERATIONS IN KISMETIC PSYCHOTHERAPY

Every experienced therapist likely knows that some aspects of psychotherapy involve subtlety beyond ordinary and convenient ways of expressing what they mean and how they work. This is life, it seems. It is much like dedicated and healthy parents who conscientiously invest in their child's well-being. No matter how truly good the parents are, their wish for the child's inexorable and ideal way of being is gnawingly known to be neither inexorable nor ideal. Similarly, just because "the client is there," the therapist conscientiously invests in the client's well-being, knowing that movement toward health is neither inexorable nor ideal. This conscientious investment requires all therapists—kismetic or not—to make judgments on incomplete information and to act on behalf of clients, utilizing the best of tactical thinking in the moment. This last point, then, refers to several expressions of tactical thinking. The kismetic psychotherapist makes decisions to act on behalf of clients in many of the following ways, knowing that these ways are always increasing in number:

- Receives clients, making immediate observations and reaching conclusions about the appropriateness of therapy with "this particular client." This is a gate-keeping function, a function that many therapists believe is not theirs.
- Carefully and as thoroughly as is practicable, explains therapy to the prospective client
- Describes kismetic psychotherapy for the client
- Evaluates/assesses the client's presentation of problems, keeping in mind that the evaluation/assessment may be necessarily incomplete, if an alternate kind of assessment, such as a neurological examination, may be needed
- Consults with the client about the best possible site and kind of therapy that may be available for the client
- Within the context of identifiable constraints, readily affirms the client's capacity and willingness to change
- Utilizes questions, much more than giving answers or solving problems for the client. The kismetic psychotherapist refuses to tell clients what they can know for themselves independently. Exceptions to this involve times during which the kismetic therapist practices psychoeducation, temporarily taking the role of a teacher.
- Remains alert to any client movement away from engaging in therapy

- Coordinates treatment with other health care providers and possibly others who may be needed in relation to treatment, such as parole officers or teachers, and actively communicates with them
- Periodically interprets therapy for clients
- Invites/allows the client to define problems and set the pace of the psychotherapy work, with both the client and the therapist being mindful of needed change toward clinically relevant goals
- Keeps a clear and steady focus on clinically relevant issues
- Ensures that treatment goals are realistic
- Uncovers and defines resistance to treatment, with emphasis on re-examining the role of the therapist and, to a lesser degree, avoiding client-blaming, and giving attention to the therapist's accountability. With therapist accountability, the client's resistance to treatment becomes another clinically relevant issue. Without it, therapy ceases.
- Grants permission for the client to feel her/his feelings and experience her/his needs and to express them both
- Encourages clients to utilize healthy psychological defenses, instead of merely exposing self-defeating, dysfunctional defenses
- Reframes the client's argumentation, avoidance, confrontation, self-harm, and other dysfunctions as energetic, invested ways of vainly trying to meet relational needs that may be met in other ways when the client is ready
- Maintains congruence in the clinical relationship. The therapist is not likely effective in the role of therapist without congruence. In addition, though, the therapist serves a model of good mental health.
- Facilitates increasingly healthy communications between the client and significant others
- Frees the client—especially at the end of successful therapy—releasing the client to fledge as a healthy person and taking expressions of pleasure and pride in the client's productive struggles and accomplishments
- In deference to what should be common clinical knowledge and to a specific legal principle, psychotherapists hold special accountability to recognize and manage transference and countertransference issues.

As interpreted here, the role of the therapist is incomplete. No doubt, some readers will shrug, "So, what's new with this?" The goal here is not necessarily simply to be "new." Further, if practicing mental health care providers recognize established value in the role of the therapist as it is presented here, this is acknowledged as being a good thing. This is considerably better than this presentation having no resonance with practicing psychotherapists. Quite beyond resonance or the lack of it, though, a more basic intention is to give a professionally accurate and accountable picture of the role of a kismetic psychotherapist. Adding to the role of the kismetic psychotherapist, the following discussion of the practice of psycho-therapy builds on what has been written.

A PARTIAL COLLECTION OF PRACTICAL THERAPEUTIC ACTIONS IN KISMETIC PSYCHOTHERAPY

With regard to specific clinical techniques in kismetic psychotherapy, it should be noted that this approach to clinical work readily incorporates useful, reliable, and commonly supported techniques from other therapeutic orientations. To men-tion only a small number of these borrowed techniques, the following ones may be listed:

- Exaggeration, a cognitive therapy technique
- Scaling, a solution-focused technique
- Externalizing the problem, a narrative therapy technique
- Shaping behavior through reinforcement, a behavioral therapy technique
- A-B-C technique, from rational emotive behavioral therapy

As noted earlier, kismetic psychotherapists define themselves as co-discoverers with their clients. In addition, it was noted that questioning and adapting are major themes and techniques in kismetic psychotherapy. Thinking about therapy in these terms almost inevitably produces techniques. Without trying to provide a complete list of kismetic techniques, here are some of them.

Explore the Story

The practice of exploring the story usually allows clients to reveal salient events from their respective histories—as these events, relationships, self-definitions, and more are experienced in the moment. Just as "therapy is biography," client needs are biography, too. Or, more precisely, the client's accrued strategies for meeting her/his needs constitutes her/his relational biography. This is the relevance of the reports about George Kelly and Carl Rogers presented earlier in this chapter.

How may a kismetic psychotherapist explore the client's story? First, most mental health clinicians recognize that clients sometimes tell their stories without prompting from their therapists. For these clients, the story comes easily and in a manner that both the therapist and the client can make productive use of. Clearly, this is a good thing. However, when clients' stories are not told without prompting and evocation, therapists need to invite them to reveal the living-ness of their stories, more as expressions of their functioning character than a recitation of per-sonal history. Typically, clients become engaged in revealing their stories through sensitive questioning. Some of these questions are illustrated later. For now, the client's story reveals themes or fated ways of thinking, acting, and feeling. As these are revealed, an adult client may divulge a great deal in saying, "Mom was a lousy cook." While this may have no precise meaning, it may provide opportunity for useful discoveries, such as revealing operative assumptions like "Women are supposed to cook," or "Mom failed to nurture me as I very much needed her to nurture me," or "She was a lousy cook, but a wonderful, loving, invested mother." These are likely to be valued and useful themes during the process of therapy.

Raise Essential, Clarifying Questions

During the process of conducting therapy, questions naturally arise, no matter what theoretical orientation may support a therapist. However, quite beyond normative clinical inquiry, kismetic psychotherapists usually, but not routinely, intend to raise a category of questions that may be called "fate-generating questions." In the end, the therapist hopes to enable the clients to reshape their destiny. These are some of the questions that kismetic therapists are likely to consider during the process of conducting therapy:

What are the circumstances to which you need to adapt?

How do you currently function that would help you to make any necessary adaptations?

What is your personal story/history? And, what thoughts, events, persons, commitments, and so forth from your story have an influencing impact on your current problem(s)?

Are there specific challenges with which you believe you are stuck? These may be fate-based or kismetic issues. Everyone has kismetic issues.

How has your fated-ness prepared you to function well?

How has your fated-ness prepared you to function not so well?

What parts of functioning well would you like to develop beyond your current level of effectiveness?

What is it about your past, your present, and your future that you need to reconcile? If these things were reconciled in the way they should be, how would your life be different from what it is?

Establish Anticipations

The last question is intended to engage clients in creating a very detailed picture of their respective futures. Clearly, this involves trying to establish a sketch of the future to which a client may be able to commit. However, a sketch is merely a beginning. If kismetic psychotherapy works well, the client will return to the sketch, adding details, almost as if she/he is anticipating a physical relocation to a new life—with its work and income, its home, its relationships, the home's furnishings, the age and model of a car, and so on. This picture—an ever-increasing reality in the mind of the client—rises in clarity in terms of sights, colors, people, and every other conceivable aspect of life. Usually, as clients establish anticipations of their future, they are very much more likely to conclude that such a future is possible. As a consequence of such a conclusion, they move toward becoming the healthy, productive person they want to be.

A desirable feature of establishing anticipations is that clients elevate into awareness as much of a view of their healthy functioning as possible. In short, the therapist and the client seek to gain as much clarity as possible about the healthiest possible person that the client is capable of becoming. This capability varies extremely. For one client, removing painful problems may suffice. For another client, healthy functioning becomes much more aspirational or somewhat ideal healthy functioning.

Introduce Preclusive Interventions

These types of interventions assist clients in moving in two but complementary directions at the same time. One direction involves enabling a client to frame her/his current situation with extraordinary clarity and confidence so as reasonably to predict consequences that are likely to appear because of the client's current functioning in this situation. "As reasonably as you can, what do you believe happens to your relationship with your children, if you continue to yell at them?" With as much clarity and certainty as possible, the answer to this question will be established. Thus, the other direction is the preclusive one. It asks a different question, but one that naturally grows from the client's same situation. "As reasonably as you can, what do you believe is your best motivation for yelling at your children?" While the question may be raised/reworded in many different ways, it is intended to assist the client in claiming her/his best motivation for what may be her/his unwanted behavior. Therefore, as the client claims her/his best motivation for acting toward her/his children, she/he precludes yelling at them. In the end, the client may engage in appropriate self-regulation in relating with her/his children, by saying something like this: "I find it impossible to yell at my kids because I love them so much." Or, "My love for my children stops me from yelling at them." This is preclusive action.

Utilize Occlusive Desensitization

While occlusive desensitization may be utilized with a wide array of client issues/problems, it is most easily illustrated with reference to treating psychological trauma. Experienced mental health care providers know that clients who present with psychological trauma give a picture of heightened distress and reactivity, among other things. In extreme cases, these clients become almost totally nonfunctional. Whether the client is nonfunctional or seriously appears to be living in a manner that approaches nonfunctionality, occlusive desensitization may be a useful intervention.

In medical—not psychological—treatment, an occlusive dressing refers to an airtight and watertight bandage that is used when a patient has sustained a physical injury. Instead of a physical injury, though, mental health care providers treat psychological injury. The goal of occlusive desensitization is to assist clients in constructing both an internal and external environment in which they are in a "stress-tight" and "stimulus-tight" protective "bandage." With reference to the external environment, a kismetic psychotherapist works with the client toward removing as many external demands as possible. Among other things, with the clear consent of the client, this means limiting contact with others so as to lower the potential for inappropriately reactive responses to others. (i.e., "Who calms you?" "Around whom are you the most comfortable?" Alternately, "Who alarms you?" "Who frustrates you?" "Who angers you?"). With reference to the internal environment, kismetic psychotherapists seek to equip their clients with as many stress-lowering actions as possible. Usually, this means that the therapist and the client experiment with numerous stress reduction techniques so as to identify ones that work most effectively for the client. Together, they establish as many comforting, safe, predictable, and largely controllable "bandages" as possible. The "bandages" are both personal and relational, both internal and external, and both fixed-fated and malleable.

Integrate Proprioception

Analogous to an individual's body functions, a client's psychological and relational functions form a whole. Clearly, despite many individuals' attempts to isolate, deny, or avoid them, as if they are not parts of their existence, their painful memories, physical scars, chemical dependence, and harmful and disappointing relationships shape and largely determine their current functioning—in a comprehensive way. Just as one part of an individual's body affects other parts, one part of an individual's psychological functioning affects every other part. Despite this natural reality, for example, a father somewhat thoughtlessly overconsumes alcohol and acknowledges little impact of his drinking on his children. This is fragmentation of awareness. With the expectation that this fragmentation may be largely eliminated, integrated proprioception comes from involving clients in a process of identifying the "parts" of functioning and establishing efficacious links among the parts. The emphasis here is on making connections—largely internal connections.

Typically, initiating techniques that facilitate integrating proprioception comes with identifying other techniques and, significantly, the therapist's clear conceptualization of the client's unmet personal and relational needs. This is a way to say that, in isolation, integrating proprioception may be more of a nuisance than productive or may be more overwhelming than useful for clients. Nevertheless, as a technique or a set of related techniques, integrating proprioception flows from a single concept: How is "A" related to "B"? For example, the kismetic psychotherapist may say to the client, "You indicate that you drink too much and that you are fatigued most of the time. How is your drinking related to your fatigue?" Or, "You report that your wife expects and needs more intimacy with you and that you drink a lot. How is your responsiveness to your wife related to your drinking?" Often, these questions facilitate a process through which clients integrate their fragmented awareness of the many "parts" of their lives. As the strength of these connections develops and this integration occurs, clients usually experience themselves more fully, including their pain and their potential for moving away from their pain in healthy ways. Utilizing the example of the husband who lacks intimacy with his wife, if he makes enough connections between his excessive drinking and his lack of intimacy, he may acknowledge that he fears intimacy, his drinking keeps him safe from intimacy, and, despite his fear, wishes to seek intimacy with his wife.

A few kismetic ideas about clients may be useful here. By definition, the role of the client lacks the clarity of the therapist's role. After all, the therapist is the one who knows and understands the work of therapists and, to a lesser degree of precision, the work of clients. Nevertheless, the work of the client is, indeed, work. With emphasis on self-imposed and therapist-supported demands on the client, effective therapy is the work of the client much more than the work of the therapist. Allowing for this, the question persists, "What is the role of the client?" The following features of the role of the client may help therapists to reconsider their conceptions of clients.

Most commonly, clients receive therapy because they desire the benefits that therapy may provide—namely, movement toward specific, recorded, and agreed upon goals. Without expressed and clear desires for change, effective therapy fails to happen. Experienced therapists know that many clients seek relief from suffering more than they seek clear and life-affirming goals. They know, too, that effective therapy usually engages clients in defining life-affirming goals. In other words, a necessary feature of clients is that they bring a clinically relevant need to therapy.

As a co-discoverer with the therapist, the client finds ways to engage in a process of change. This is a feature of therapy, but it is also an important function of a client. Surely, the therapist brings curiosity about the client. This is understood as a feature of professional functioning. The therapist, though, must place positive value on the client's capacity for discovery, too, or nurture it into existence when the client lacks it. In a summative way, the client's engagement in the adventure of change—as frightening as this may be, at times—provides the energy and direction of therapy, along with, if not more important than, the therapist's engagement. At least, engagement involved helping the client answer the question, "What do you want from therapy?"

Explore Kismetic Confirmation

This is as simple as it is daunting. Kismetic confirmation involves clients in a process of identifying the features of their existence that are more or less immutable—out of the control of the client. By identifying these features, such as age, sex, death of significant other (e.g., father, child, or grandmother), the client can reframe them as sources of confidence in the certainties on which they may rely and about which they do not need to worry. Permanence, in other words, can be a source of security, particularly during times of heightened personal distress. The idea of kismetic confirmation is likely a familiar one with most psychotherapists. It nicely overlaps with Niebuhr's often cited sagacity, "O God, give us the serenity to accept what cannot be changed, the courage to change what can be changed, and the wisdom to know the one from the other" (Niebuhr, 1950).

VARIATIONS AND UNANSWERED QUESTIONS IN KISMETICISM

If the long evolution of psychotherapy poses any important learning about psychotherapy, it is that psychotherapy remains largely unsettled and even conflicted, continuing to evolve. This says what virtually all psychotherapists know very well. Namely, psychotherapists disagree about many things. This fact characterizes kismetic psychotherapist, too. However, reporting careful and evidentiary arguments among kismetic psychotherapists almost immediately poses barriers. The singularly most important barrier comes from the newness of kismetic psychotherapy. This is to ask whether the approach of kismetic psychotherapy is too new to warrant serious debate. Nevertheless, some disagreements among kismetic therapists have emerged. Therefore, to add useful detail to these disagreements and to add focus to the kismetic view of things, the following disagreements among kismetic psychotherapists provide intellectual texture to kismetic psychotherapy.

New, old, or evolving? Some kismetic psychotherapists have concluded that they do not offer a new "brand" of therapy, but instead offer a new integration of several ideas and practices of therapy. They add that some of what they believe to be true about their clinical work is very similar to beliefs that were held by Freud and Ferenczi, among others, while they acknowledge that they generously borrow from other approaches to therapy. However, distinct from this view, other kismetic psychotherapists argue that virtually all of their therapy is new. A third view is held by those who conclude that kismetic psychotherapy is largely new. They assert that much of what constitutes kismetic psychotherapy has yet to be discovered. In other words, it is evolving. They take an integrative and progressive, futuristic view of kismetic psychotherapy.

Constructionist or scientists? Among kismetic psychotherapists, debate often erupts with regard to the question about whether their work grows within their unique construction of ideas about therapy—constructionists—or emerges from their research about specific kismetic interventions. Is kismetic psychotherapy a product of constructionism, science, or both? As this question continues to be answered, it will clarify the essential nature of kismetic psychotherapy and its place among many other therapies.

Solitary or relational? Similar to almost every other approach to therapy, the question about with whom it works best inevitably arises. This is a question among kismetic psychotherapists. Is it best suited for treatment of individuals or for relationships? Kismetic psychotherapists who argue that it is best suited for treating individuals emphasize that each person's history—or collection of fated features—is unique and complicated to a degree that only individuals should be treated. Obviously, kismetic psychotherapists who argue an almost opposite point of view emphasize that almost everything of significance in every individual's life involves relationships with others, both in the individual's history and in the individual's current functioning.

Needs and drives or contextual influences? While no one, including kismetic psychotherapists, denies the influence of needs and drives or contextual factors, the debates about the relative significance of these influences appears to have emerged with human self-consciousness. While the terms "nature" or "nurture" distort the issue, they capture the essence of the debate. Kismetic therapists do not wholly discount either nature or nurture. However, most of them emphasize con-textual influences as ones that give each individual her/his character/personality, understood in the most encompassing manner. After all, needs and drives may be universal, but contexts are not. Therefore, accurately and in clinically appropriate ways, to comprehend each client's needs in the most relevant manner requires un-derstanding the client's contextual history.

CONCLUSION

In response to every innovative and new approach to psychotherapy, questions and challenges must be raised. Surely, this is a necessary response to kismetic psychotherapy.

Based on this cursory presentation of kismetic psychotherapy, how may it be evaluated? What specific contributions does this approach to psychotherapy make? On a continuum of isolative to integrative, where does this approach stand? What is the evidentiary basis for this approach to psychotherapy? What is the future of kismetic psychotherapy? Where may interested clinicians receive training in kismetic psychotherapy? What degree of influence, if any, will kismetic psychotherapy have in the larger world of counseling and psychotherapy?

Obviously, the attempts to answer these and more questions will lead to addi-tional questions. However, to preclude the slightest effort toward answering these questions, readers should know that kismetic psychotherapy is bogus. There is no such thing as kismetic psychotherapy. It is just made up. The presentation of kismetic psychotherapy, though, is not a useless exercise. Instead, it indicates how easily a new approach to therapy may be invented and perpetrated as if it is a viable approach to psychotherapy. One of the major themes of this book is that massive numbers of practicing counselors and psychotherapists, along with many of the preparation programs from which they graduated, "just make up stuff."

Worse, though—much worse—many thousands of practicing therapists just make up conspicuously inappropriate "therapies" that likely pose harm to their clients. They practice them with little apparent regard for the needs of clients and the welfare of the profession of which they are members—including counselors, psychologists, social workers, and marital and family therapists. Chapter 2 addresses this as a critically important issue in mental health care. In advance of going there, it should be said that postulating an ever-expanding variety of sometimes sophisticated, sometimes transparently ego-driven, sometimes intellectual sophistry, and sometimes merely strange approaches to therapy, along with likely harmful treatments of clients and patently irrational views of human mental health issues and psychotherapy, has not helped much and will continue not to help much.

The Long Conversation About Counseling and Psychotherapy

In virtually any context, when the broad topic of counseling and psychotherapy is raised among their providers, inevitable differences arise. Among providers, there are very few settled answers to important questions. Probably, the disputes about basic questions began as soon as Sigmund Freud shared his thoughts about psychoanalysis with someone else. Indeed, soon after Freud began to promote his ideas about psychotherapy, Carl Jung and others began to promote their versions of psychotherapy, disagreeing with Freud, their mentor. Today, setting aside disputes within the history of psychoanalysis, virtually all practicing counselors and psychotherapists recognize that "settled answers" is not a concept that they associate with their clinical work. Reinforcing the lack of answers, Wampold concluded, "Decades of psychotherapy research have failed to find a scintilla of evidence that any specific ingredient is necessary for therapeutic change" (cited by Miller & Hubble, 2017). Wampold's reference to the idea that there is no "specific ingredient" that "is necessary for therapeutic change" adds emphasis to the lack of settled answers to counselors' and psychotherapists' important questions. For example, while the questions may be too numerous to present here, surely, some of them include one such as these: What is psychotherapy? How may highly effective therapists be defined and located? What causes mental and emotional disturbances? What characterizes the mentally/emotionally healthy individual? What are the psychological differences between women and men?

Or, present a clinical vignette to group of experienced therapists and ask them to render a diagnosis and offer recommendations for treatment. Predictably, members of the group will engage in friendly disputes about whether the client in the vignette is suffering with anxiety, borderline personality disorder, dysthymia, adjustment demands from stressful circumstances, or something else. They will likely express appreciation for the subtle differentiations of various diagnoses among them, opting to ensure one another that their differences are useful and desirable products of their thoughtful analyses of the facts. Then, following their several speculations about what the client's problem may be, they will offer their treatment recommendations. Some will speak with the certainty that confuses others, provoking insecurity among those who hold much less certain recommendations. Some will reveal the confusion from their doubt about a therapist who speaks with certainty. Others will describe an array of possible interventions but avow that the interventions must remain tentative, pending the acquisition of needed information. Others will defer opinions about interventions and allege that interventions are nonrelevant until the therapist knows what the client wishes to achieve through therapy. Others in the group will acknowledge the importance of forming an appropriately trusting and bonded therapeutic relationship before specific treatment recommendations can be made. And, others will place specific conditions, such as a "need to develop a clear conceptualization of the client's issues" or "a responsible consideration of whether my skill set matches what the client needs," on their reluctance to offer recommendations. In the end, the participants in the group will have openly confirmed that they have differences among them and that inwardly and personally the differences confirm each one's correctness.

Even when the context involves only one therapist, differences could arise. After all, each practicing counselor and psychotherapist understands that formulating a clinically relevant view of a client involves ambiguity and usually conflict. The practitioner may wonder, "Is this client irrationally angry or is he right about how thoroughly his mother misunderstands him?" Or, "Should I utilize a paradoxical intervention, listen carefully and gather more information, or something else?" Or, more fundamentally, "What's wrong with my preferred way of helping clients, since much of what I do seems not to be working well?" Or, sustaining a mental debate, "I can't believe what my client's previous therapist tried to do with my client; her approach is just wrong."

In the context of this book, readers will see any number of approaches to counseling and psychotherapy and feel the urge to argue about them. For example, if the list of approaches to counseling and psychotherapy in Table 2.1 were submitted to a group of licensed mental health care providers, would any of them become a cause for debate? If any of them caused debate, what would guide the debate?

As an aside, readers should note that lists of unusual approaches to counseling and psychotherapy appear throughout this book. The purpose of each list depends on the context in which it is presented. Regardless of the context, though, the lists are important because collectively they include more than 1,200 approaches to counseling and psychotherapy and represent a much larger number of approaches. As this book confirms, the number of approaches to counseling and psychotherapy is an indefinitely long one, depending largely on the frequently exercised option of individual counselors and psychotherapists to invent their own approaches or to adopt approaches that accommodate their individualistic desires.

Whether the focus of attention is the therapist who practices alone or a survey of the history of counseling and psychotherapy, debate is and has been an important

Table 2.1 UNUSUAL APPROACHES TO COUNSELING AND PSYCHOTHERAPY

Usui shiki ryoho	Talent development counseling
Supportive touch	Visual coding displacement therapy
Character education	Christ conscious counseling
Deconstructionism	Buddhist and yogic philosophy
Equine ecosomatics	Mandala process work
Moral injury therapy	Secular treatment
NIASZIIH healing	Sanctuary counseling
Photo therapy	Zyto biotechonology scans
Postural integration	Pranic healing procedures
Psycho-organic analysis	Therapeutic taro and reiki
SOZO attainment therapy	Shambala meditation
Reiki womb therapy	Radix intensive therapy

feature of counseling and psychotherapy. From the earliest days of modern psychotherapy, beginning with Sigmund Freud, approaches to counseling and psychotherapy have been debated. Perhaps someone can find an exception to this, but the probability is that there has never been an approach to counseling and psychotherapy that has not received criticism within the clinical community. Every approach has been denounced by someone. Almost as soon as Sigmund Freud published his book on hysteria in 1895, several critics arose, including Freud himself, whose views about hysteria and the then-ascending method of hypnosis that was used to treat it dramatically changed.

Whereas Freud had enthusiastically participated in developing ideas about hysteria and hypnosis, along with Breuer, Charcot, and others, he remained unsettled about what was happening with his patients. As Sharpe and Faulkner (2008) observe, "Freud came to hypothesize that hypnosis, by momentarily suspending the patients' resistances to recalling these painful memories, also prevented the doctor from confronting these resistances" (p. 3). Guided by his puzzlement about his patients, he challenged the assumptions that he shared with Charcot, Breuer, and others:

> When the patient, through talking, followed associations in her memory, she was able to recover the forgotten event, which led to the cure. Freud eventually gave up the process of using hypnotism for the use of a technique he came to call "free association," in which the patient was encouraged to put aside all inhibitions and follow her associations, which would eventually, even without hypnosis, lead to the recovery of unconscious memory. (Robbins, no date)

Freud's search for ways to understand resistances led him to psychoanalysis. And, inexorably, his promotion of psychoanalysis led to conflict with others, including those who were intellectually, personally, and clinically closest to him, such as Carl Jung (Donn, 2011). As unsettled as they were for Freud and his contemporaries, the unsettled questions have remained as visible features of counseling and psychotherapy.

More recently, when cognitive behavioral therapy (CBT) emerged, it rapidly became the dominant approach to counseling and psychotherapy because of its consistent record of confirmatory scientific research. With its success, though, came serious problems, including schisms among CBT advocates. Now, CBT shows internal incompatibilities, with the emergence of its clinical descendants, such as dialectical behavior therapy (DBT), mindfulness-based stress reduction (MBSR), mindfulness-based cognitive therapy (MBCT), and acceptance and commitment therapy (ACT), among others. Among CBT advocates, a common reference is to "third-wave" CBT. The "camps" within CBT politely debate, if not argue, among themselves. No doubt, therapists who adopt the new offspring of CBT conclude that therapy has been improved.

In this book, we report much more than we argue. The arguments for and against numerous specific approaches to therapy would be easy, as our collective history clearly demonstrates, but the initiating and settling of debates about thousands of unusual approaches to counseling and psychotherapy is not what this book intends to accomplish. More than advancing arguments about specific approaches to counseling and psychotherapy, this book argues that the massive proliferation of approaches, along with other deficiencies in the profession of mental health care, continues to have a corrosive and possibly irreversible adverse consequences.

Our collective history confirms the importance of debate about approaches to therapy. This is normative. The debate should continue, insofar as it makes mental health care better than it is. Also, the accrued volume of critiques of aberrant approaches to therapy adds confirmation to the kind of "debate" that this book encompasses. Some of the lively issues of concern and critique are discussed in the next section, "The Conversation."

THE CONVERSATION

Something like staring at a map of the United States that has no place names on it, the next priority is to place information on the map. The information comes, first, from active participants in the long conversation about the condition of mental health care and, then, from practicing counselors' and psychotherapists' websites (Chapter 3). The information in Chapters 2 and 3 assumes that many who read them may not know about the long conversation or the tens of thousands of orientations to counseling and psychotherapy. As information is added to the "map," the picture of counseling and psychotherapy gets clearer than it is now. And, possibly, with a map that allows mental health care providers to see the landscapes in which they function, they may develop ways to change what they see.

Based mostly on the widespread pervasiveness of unusual approaches to counseling and psychotherapy, along with other factors, this book calls for major reforms in counseling and psychotherapy. (Major reforms are discussed in Chapter 7.) Leading to a discussion of needed reform, the present chapter shares the view of O'Donohue and Lilienfeld (2013) that "we are deeply concerned about the large number of professional psychological services sold and delivered that are not evidence-based" (p. xi). In words that should evoke alarm among those who provide professionally accountable psychological services, they continue, "This is a preventable tragedy that often harms the very people we ought to be helping" (p. xi). The preventable tragedy flows from the massive proliferation of aberrant approaches to counseling and psychotherapy, but it also flows from other deleterious influences. The comments by O'Donohue and Lilienfeld suggest that a conversation about problematic issues in counseling and psychotherapy has been evolving for many years. Indeed, this is the case. The issues that have emerged in this conversation should occupy more time and space than the present discussion. Recognizing such limitations, the discussion here refers only to the following topics: pseudoscience in counseling and psychotherapy, declining competence among counselors and psychotherapists, the inadequacy of preparation programs, the chaos in licensing mental health care professionals, the increasing visibility of ideology-driven approaches to counseling and psychotherapy, the need to account for the social world of which counseling and psychotherapy are parts, and the expanding influence of technology on the practice of counseling and psychotherapy. This chapter is intended more to introduce these topics than to be a thorough discussion of them. As the book develops, many details are added to this introduction.

PSEUDOSCIENCE IN COUNSELING AND PSYCHOTHERAPY

Instead of standing alone, the judgments in this book join a large conversation about the proliferation of approaches to counseling and psychotherapy. Other writers have offered similar judgments about many of the aberrant approaches to therapy that appear in this book, although none of them presents or describes the huge volume of unusual approaches as this book does. Regardless of the number of unusual approaches that they identify, they call attention to them, based on their conspicuous concern about the adverse consequences for clients, mental health care providers, and others. For example, Singer and Lalich (1996) wrote a book, *Crazy Therapies: What Are They? Do They Work?*, apparently attempting to warn prospective clients about the dangers inherent in receiving mental health care from therapists who practice one of the crazy therapies about which they wrote. Thyer and Pignotti (2015) included several portentous approaches in their book, *Science and Pseudoscience in Social Work Practice*. In their article, "Should Social Workers Be Engaged in These Practices?" Holden and Barker (2018) listed more than 400 aberrant approaches to therapy. In 2005, Lilienfeld, Fowler, Lohr, and Lynn addressed "Pseudoscience, nonscience, and nonsense in clinical psychology: Dangers and remedies." In the articles, "Psychological Treatments That Cause Harm" (Lilienfeld, 2007) and "Why Ineffective Psychotherapies Appear to Work: A Taxonomy of Causes of Spurious Therapeutic Effectiveness" (Lilienfeld, Ritschel, Lynn, Cautin, & Latzman, 2014), the authors analyzed the deleterious impact of several approaches to therapy. Similarly, Lee and Hunsley (2015) justly lament the unwanted challenge of separating science from pseudoscience in clinical practice and argue for ways to do this. Thompson (2010) reported several lists of "psychological treatments to avoid." And Beyerstein artfully laments the proliferation of "fringe psychotherapies" that put "the public at risk," concluding,

> It continues to amaze me that many people who demand extensive, impartial evaluations of automobiles or televisions before making a purchase will put themselves in the hands of psychotherapists with little or no prior investigation of their credentials, theoretical orientations, professional affiliations, or their records of successfully helping their clients in the past. (p. 77)

Beyerstein acknowledges that consumers of mental health care services generally demand little of their providers. However, his larger theme is that mental health care professionals have too long relied on "intuition and folk wisdom" (p. 70), instead of science. Altogether, these several analyses of pseudoscience in the practice of counseling and psychotherapy should have caused widespread alarm among mental health care providers, but their analyses appear to have had little impact on the practice of counseling and psychotherapy.

In contrast to those who discuss pseudoscience in psychotherapy, McNamee (2009) raises important challenges to traditional assessments of therapy and ethics that guide therapy. In her article, "Postmodern Psychotherapeutic Ethics: Relational Responsibility in Practice," McNamee effectively captures emerging relativistic themes and practices in psychotherapy that make easy judgments about aberrant approaches to therapy a lot less easy. She suggests that perhaps a welcoming attitude toward multiplicity (diversity?) and a broadly celebrated emphasis on postmodern approaches to psychotherapy may inadvertently promote highly individualized inventions of aberrant approaches to psychotherapy, without regard for normative professional ethics. With constructionism's long-standing emphasis on rejecting the normative authority of professional mental health care providers, McNamee concludes, "The uncertainty that is associated with a constructionist philosophical stance is one that invites multiplicity and thereby invites therapists and clients alike to question their assumptions and explore alternative resources for personal, relational, and social transformation" (p. 70). As appealing as this uncertainty may be and as inviting as the call to creativity may be, it has consequences. As McNamee adds, "We could call this generative uncertainty, a term that I believe echoes Wittgenstein's notion about a de-contextualized ethic when he claims, 'It is clear that ethics cannot be formulated'" (p. 70).

A constructionist philosophical stance may, indeed, invite multiplicity of thought, accompanied by elevated tolerance of differences among counselors and psychotherapists. More generally, postmodernism invites a multiplicity of approaches to counseling and psychotherapy. Thus, insofar as McNamee's and other similar views may be influential, they may be inviting counselors and psychotherapists to approach their work with excessive relativism and the belief that they can create unique and sometimes inappropriate approaches to their clinical work. However, more than blaming anyone for the apparent decline in the quality of mental health care, the larger discussion of relevant issues begs the question, "What should we do to make needed corrections in the practice of mental health care?" In the context of this question, do agreed upon standards have a place in making corrections or, as McNamee avows, "what is therapeutic (and thus ethical) remains open and indeterminate, just like conversation" (p. 64). Further, she raises the challenge of how mental health care professionals bring "diverse and competing moral orders" and "disparate ideas and practices into" the dialogue about what counseling and psychotherapy may be (p. 64). Suffice it to say that, at this time, the clinical community of about 800,000 providers has no way to define its work, establish meaningful boundaries around its work, and build minimal cohesion or consensus about what their work is. The sufficiency of these characterizations may be confirmed, when approaches to clinical work include guided afterlife repair, cinematherapy, kabbalistic healing, popular culture–based therapy, tantric models, yogamotion, soulful hypnosis, animal-love therapy, therapeutic psychogenics, the love of Jesus Christ, Tal Ben-Shahar coaching model, faith Christian counseling, integrative massage therapy, yoga unchained, dancing mindfulness, psycho-organic analysis, seven spiritual practices, postural integration, past life emotional blueprint readings, and the mandala of santosha, to name only 20 approaches that represent competing moral orders, disparate ideas, and a wide variety of practices.

Obviously, along with basic questions about what psychotherapy is (e.g., ACT vs. nouthetic methods, CBT vs. nature photo therapy, or elixer qigong vs. systematic desensitization), pseudoscience in counseling and psychotherapy has received significant, but not necessarily consequential, attention. Moreover, the mere presence of pseudoscientific approaches to psychotherapy evokes questions about what psychotherapy is. Is psychotherapy a scientific discipline? Is psychotherapy an adjunct to entrepreneurial pursuits, with hopeful expectations that the normative vicissitudes of life can somehow be redirected into joy, happiness, hope, or even ecstasy—or, in another word, hedonistic fulfillment and profit? Is psychotherapy another "competing moral order" through which reformation of individuals and society at large should be expected? Is psychotherapy an enlarged altruism that encompasses the need fulfillment of the ones who provide it to others, enabling them vicariously to live satisfying lives? The questions are not new among psychotherapists. Because of competing moral orders and other disparate ideas about what psychotherapy is, virtually every approach to counseling and psychotherapy has been denounced by someone.

The very long debate about why psychotherapy works or whether it works at all should be acknowledged. A snapshot of this debate is Wampold's article (2013), "The Good, the Bad, and the Ugly: A 50-Year Perspective on the Outcome Problem." Wampold (2013) nicely captures the conflicting perspectives of two luminaries in the history of psychotherapy, Eysenck and Strupp. Another example of the long debate about what psychotherapy is appears in Kirschenbaum and Henderson (1989), *Carl Rogers-Dialogues: Conversations with Martin Buber, Paul*

Tillich, B. F. Skinner, Gregory Bateson, Michael Polanyi, Rollo May, and Others.
The debate between Rogers and Skinner is a particularly poignant one because
it captures the tension-filled, friendly, and powerful thoughts of two of the most
influential figures in the history of mental health care. The necessary features
of counseling and psychotherapy will likely continue to be matters of disagreement among counselors and psychotherapists. Additional examples of this debate appear later in this book. The debate that began with Freud's puzzlement
about resistances has continued to the present, affecting the newest, most naïve
practicing therapists. Currently, the debate is being shaped, at least partially, by
the challenge of competence or incompetence among practicing counselors and
psychotherapists, as well as other pressing challenges, as the ensuing discussion
describes.

THE CHALLENGE OF COMPETENCE

Pseudoscience among practicing counselors and psychotherapists and the debate
about the necessary features of therapy are accompanied by other important issues. Another one is the decline of competence among practicing counselors and
psychotherapists. Recently, for example, the research by Goldberg et al. (2016)
epitomizes the burgeoning concern about the condition of competence among
practicing counselors and psychotherapists. The title of their article tersely states
their challenge, "Do Psychotherapists Improve With Time and Experience?
A Longitudinal Analysis of Outcomes in a Clinical Setting." Their conclusion may
be an unsettling one for trainers of counselors and psychotherapists, including
instructors in preparation programs and clinical supervisors. They wrote, "At the
same time, the present analyses show that, in the aggregate, therapists did not
improve with more experience. . . . Indeed, results suggest that therapists on the
whole became less effective over time, although the magnitude of the deterioration was extremely small" (p. 7). Later, additional contributions to the discussion
of professional competence are presented. For now, augmenting the ongoing concern about the decline of clinical competence, others have examined the question
of whether expertise in psychotherapy (e.g., Hill, Hoffman, Kivlighan, Spiegel, &
Gelso, 2017; Hill, Spiegel et al., 2017; Tracey, Wampold, Goodyear, & Lichtenberg,
2015; Tracey, Wampold, Lichtenberg, & Goodyear, 2014). No doubt, the status of
expertise in psychotherapy will remain an unanswered question. This book adds
another dimension to the decline of competence. It is the massive proliferation of
unusual approaches to counseling and psychotherapy.

THE SUFFICIENCY OF PREPARATION PROGRAMS

Another issue is the question of whether preparation for clinical work is effective, including discussions later in this book. The need to ask questions about
whether preparation programs are effective is an indication of concern about

the adequacy of training for clinical practice. Among the numerous expressions of this concern, research by Hill et al. (2015), "Is Training Effective? A Study of Counseling Psychology Doctoral Trainees in a Psychodynamic/Interpersonal Training Clinic," characterizes much of the current situation. Their conclusions offer compelling challenges for those who train counselors and psychotherapists On the one hand, they conclude, "Over their time [12 to 42 months] in the clinic, trainees were able to form stronger working alliances (as rated by both clients and therapists) and stronger real relationships (as rated by clients), indicating that as therapists progressed in their externship and gained experience, they were better able to form relationships with clients" (p. 198). "On the other hand, trainees did not change in terms of therapist- and client-rated session quality, therapist-rated real relationship, engagement of clients after intake or eighth session, or reductions in client's symptomatic distress" (p. 198). The conclusion that carefully monitored psychotherapy trainees did not change in terms of effectuating "reductions in client's symptomatic distress" shines light on the need to question the efficacy of clinical training. In the context of this book, this conclusion adds significance to findings reported here. Surely, studies of clinical training will continue, as they should, but the confounding question in this book, "Why are we committed to failure?" comes more from finding many thousands of aberrant therapies advertised by practicing counselors and psychotherapists than from examinations of carefully monitored graduate students. In other words, if practicing clinicians advertise thousands of aberrant approaches to their clinical work, is it reasonable to consider the possibility that clinical practice may be substantially unrelated to clinical training? As important, but seemingly as bizarre as this question may be, the reasonable possibility that it describes the current condition of clinical training adds to the necessary call for major reforms in counseling and psychotherapy, including major reforms in preparation programs.

Studies such as the one that Hill et al. (2015) led, along with the questions that arise from the proliferation of unusual approaches to counseling and psychotherapy, add to the causes for concern about the sufficiency of preparation programs. A brief study of preparation programs for mental health care providers reveals that, in pursuit of a license to practice, they may complete a programs that hold extremely different requirements. For example, one program may requires a low of 28 graduate semester hours, including an internship, while another may require 110 graduate semester hours, not including an internship. Such disparity may be located in a single university, such as Loyola University of Chicago (Loyola University Chicago, 2019a, 2019b). At Loyola, an aspiring counselor/psychotherapist may complete 28 graduate semester hours in social work or 110 graduate semester hours in counseling psychology. The first option requires approximately 1 year and the other approximately 6 years of preparation for a license, while both licensees will likely provide many of the same kinds of services. The observation about this difference intends only to be an observation, not a judgment about the quality of the programs, the number of graduate semester hours, or the competence of licensees from Loyola University. Instead, along with several other observations that appear throughout this book, this observation intends only to confirm the extraordinary variability in preparation programs. The importance of nonjudgment about Loyola University's programs is important because its programs represent a differential in requirements that is virtually universal, not unique to Loyola University. Allowing for nonjudgment, a question remains. Are counselors and psychotherapists ill-prepared for their work because they receive too little training, or are others unduly burdened by receiving far more training than they need? Or, is there another possibility that adequately accounts for the difference in requirements between clinical social workers and counseling psychologists?

The Educational Testing Service (ETS) has reported that applicants for admission to graduate programs in counseling and psychotherapy achieve scores (i.e., GRE) that place them at or near the bottom of scores for all applicants to graduate programs (ETS, 2017a). Among applicants for admission to psychology programs, counselor education programs, social work programs, and marriage and family therapy programs, prospective psychology graduate students achieve the highest scores and prospective social work graduate students achieve the lowest scores (ETS, 2017a). Also, according to ETS (2017b), among the IQ rankings for 57 college majors, students in physics and astronomy achieve the highest scores, while psychology students rank 39th and social work students rank 57th. These data should not be taken as conclusive or definitive descriptions of the capacity or potential of students in preparation programs that lead to licensing for mental health care professionals. Nevertheless, they should invite leaders in preparation programs to consider whether they are recruiting and admitting candidates who may be capable of delivering effective clinical services to their clients.

THE PRESUMED NECESSITY OF LICENSING

Almost universally, government-approved licensing of mental health care providers stands on the belief that the presumed regulation that comes from licensing protects those who receive services (Carnahan & Junger, 2015). This belief has been an essential feature of arguments for licensing by psychologists, counselors, social workers, and others. Another common belief about licensing was succinctly stated by Gross (1986) many years ago: "The generally stated purpose for licensing and the primary justification for this use of the police power of the state is to ensure quality in services offered to the public" (p. 1). These beliefs are increasingly doubted and challenged. Others, including Hogan (1983) and Smith (2011), have challenged the value of licensing, with Smith (2011) speculating that licensing of psychotherapists may be "evidence of societal regression." Licensing of psychotherapists may fulfill the expectations with which it has become associated, but this book raises serious objections to the assumption that licensing protects those who receive services from licensed mental health care providers or ensures quality of services offered to the public. The discussions around the relevance of licensing should account for the ongoing decline of competence among licensees, the extraordinary variability in the requirements for licensing, the conspicuous and deleterious profusion of approaches to mental health care, the lack of regulation of the practice of counseling and psychotherapy, the rising visibility of ideology-driven approaches to counseling and psychotherapy, the unprofessional intrusion into mental health care by legislators, the ongoing issuance of numerous dubious certifications, and the increasing number of unlicensed "therapists," among other signs of ongoing deterioration of the practice of counseling and psychotherapy. These conditions receive attention later in this book. For now, the current condition of licensing and the practice of counseling and psychotherapy appear to be ones in which counselors and psychotherapists emphatically want public sanction of their professional prerogatives through licensing, but with equal emphasis resist regulations and defy minimal standards of professional accountability.

THE IDEOLOGY-DRIVEN CHARACTER OF COUNSELING AND PSYCHOTHERAPY

The already cited works of Holden and Barker, Thyer and Pignotti, Lilienfeld, and others characterize another cause for concern about the current status of counseling and psychotherapy. It is the continuing surge of ideology-driven approaches to counseling and psychotherapy. While an argument can be made that the practice of counseling and psychotherapy was ideology-driven from its earliest days, beginning with Freud's psychoanalysis, the nature of the ideology is different from the major ideologies that have historically shaped the practice of mental health care (i.e., psychoanalysis, behaviorism, humanism, cognitivism, and postmodernism). Currently, the ideology-driven practice of counseling and psychotherapy may be described as ideological, to be sure, but also as highly individualized. The already cited sources and several yet to be cited confirm this, but the material presented in this book adds considerable detail to this reality. For example, among the large number of approaches to counseling and psychotherapy to which this book frequently refers, almost 1,000 titles for some form of Christian counseling and therapy have been located. These are titles, not therapists. The number of therapists who advertise some form of "Christian" therapy is much larger than 1,000. The importance of reporting about 1,000 "Christian" approaches to counseling and psychotherapy is not to express an assessment of these approaches as "good" or "bad," but to indicate that, along with thousands of other highly individualized approaches, they define counseling and psychotherapy as poorly regulated, increasingly ideological, and highly individualized. For illustration, Table 2.2 includes some of the Christian-related approaches.

To be clear, "Christian" ideologies among practicing counselors and psychotherapists are a relatively small number of ideology-driven approaches. Almost any interest that an individual counselor or therapy may have has likely become an approach to counseling and psychotherapy. Such interests may include a contemplative lifestyle, playing the piano, martial arts, praying, knitting, gardening, or any of thousands of others. These approaches to counseling and

Table 2.2 A PARTIAL LIST OF CHRISTIAN APPROACHES TO COUNSELING
AND PSYCHOTHERAPY

Child-centered (Christian) play therapy	Christian activities therapy
Christian archetypical psychology	Christian Buddhism
Christian cognitive behavioral therapy	Christian control mastery therapy
Christian creationist approach	Christian family therapy
Christian Gospel-based therapy	Christian healing process
Christian hypnotherapy	Christian identity counseling
Christian individual therapy	Christian marriage counseling
Christian mysticism	Christian systems theory
Christian sensitive counseling	Christian discipleship counseling
Christ conscious counseling	Christian caregiving model

psychotherapy further illustrate the imperative that professionally accountable mental health care providers define what counseling and psychotherapy are and commensurately what they are not. Further, defining what counseling and psychotherapy are should be followed by establishing standards by which approaches to counseling and psychotherapy may be utilized.

THE EXIGENT REALITY OF SOCIAL DYNAMICS

Perhaps those who provide professional mental health care and those who train them to provide it should account for social dynamics that may impinge on the texture of practice. For example, Gray (2011) discusses "the decline of play and the rise of psychopathology in children and adolescents" (p. 443). With persuasive documentation of his assertion, Gray correlates the decline of play—"free play" through which children gain social and self-management skills—with a commensurate rise in narcissism. If Gray's assertion is correct, how does this influence the practice of psychotherapy? Should counselors and psychotherapists attempt to ensure that children have opportunities to experience free play? Are counselors and therapists increasingly narcissistic? If so, does their narcissism contribute to the massive proliferation of individualized approaches to counseling and psychotherapy? Similar to the matter of play with regard to social dynamics, how do evolving attitudes about sexuality impinge on the practice of counseling and psychotherapy? To heighten awareness of evolving attitudes about sexuality, watching a romantic movie from each of the most recent five decades will reveal stunning changes in attitudes toward sexuality. While the human experience of sexuality is not necessarily well-represented in movies, it is represented in many different ways in the lives of clients who receive mental health care and those who provide their care. This observation prompted Heinemann, Atallah, and Rosenbaum (2016) to conclude, "Clinicians should be aware and attempt to gain an understanding of the different cultural beliefs with which their clients may present. These principles affect all aspects of sexuality including beliefs regarding an appropriate partner or partners, appropriate age of marriage, appropriate sexual behaviors as well as how they should be approached by the clinician" (p. 148). As members of their social world and just as their clients have attitudes and values about almost everything, counselors and psychotherapists have attitudes and values about almost everything, too. Whether the topic is play, sexuality, synagogues, firearms, political parties, racial identity, or many other features of their social world, counselors and psychotherapists are influenced by the attitudes and values inherent in the social world that surrounds them.

THE RISING IMPORTANCE OF TECHNOLOGY

Still others have tried to assess the impact of technology on their larger social world, but more specifically on the practice of counseling and psychotherapy. A common question and challenge for mental health care providers revolve around their use of social media. Should counselors and psychotherapists have a social media presence? If they have such a presence, what should they publish? Should they reveal personal information, such as marital status, names of their children, their personal interests, and so on? What should they advertise about themselves as helping professionals? Should they communicate with their clients through social media? If so, what are the clinical or ethical implications and consequences of such communication? What are the specific ethical and legal issues involved in using e-therapy, telehealth, telepsychology, or e-health?

An attorney who publishes opinions about many aspects of counseling and psychotherapy says that clients may waive client privilege when they publish information about their therapeutic experience on social media (Wheeler, 2019). Should an informed consent document establish the boundaries around contact with clients on social media? Quite apart from what counselors and psychotherapists may do with social media, should they advise their clients about how they may put themselves at risk when they disclose information about the therapy that they receive?

Separate from counselors' and psychotherapists' utilization of technology, they are acutely aware of some of the mental health consequences of technology for their clients. Have they visited with frustrated parents who report that their child has become "impossible to manage" because of the child's immersion in video games and commensurate neglect of schooling, bathing, and other normative activities? Surely, they have. Have they heard the strained, stressful concern about difficult experiences with online dating? Surely, they have. Have they received client couples in which one of the members complains about how the other member of the couple has become obsessed with online pornography? Surely, they have. Specific clinical needs will likely continue to arise from the technology-related activities of clients.

A relatively recent entry into the clinical lexicon is electronic health records (EHRs). Surely, the Health Information Technology for Economic and Clinical

Health (HITECH) Act of 2009 gave energy and focus to the common use of the term. More than a term, though, EHRs have become widespread and, in many clinical situations, mandated. Recognizing that EHRs often evoke a somewhat odious reaction from those who are mandated to use them, Menachemi and Collum (2011) sought to describe the potential benefits of EHRs. They assert that the "benefits of EHRs include clinical outcomes (e.g., financial and operational benefits), and social outcomes (e.g., improved ability to conduct research, improved population health, reduced costs)" (p. 47) In their concluding comments, Menachemi and Collum state, "Many of the benefits [of EHRs] accrue to patients and society overall" (p. 52). However, they caution about the currently awkward balance between benefits and drawbacks in the ambitious movement toward EHRs, reporting that several factors elevate the awkwardness. These factors include financial incentives and disincentives, organizational desires for EHRs and the formidable expense of systems for EHRs, the gaps in EHRs that come from their ubiquitous use in Medicare and Medicaid and other programs, and more. They emphasize that Medicare and Medicaid offer financial resources for the development of EHRs, while alternate health care provider systems generally do not provide such resources. They suggest that technological problems and challenges will continue to pose barriers to the best utilization of EHRs (p. 53). Menachemi and Collum close their article with this comment:

> Nationwide implementation of EHRs is a necessary, although not sufficient, part in transforming the US health care system for the better. EHR adoption must be considered one of many approaches that diversify our focus on quality improvement and cost reduction. The current major legislative and political support for EHRs represents the greatest investment in health information technologies in US history. Over time, providers and researchers will be eager to quantify the returns that are expected from these investments. (p. 53)

Important changes in the world of mental health care, including frequent and ongoing contributions from discoveries in neuroscience and inventions of empirically validated treatment interventions, continue to reshape the work of mental health care providers. As ubiquitous and powerful, the influence of technology has become another force that continues to reshape mental health care.

Clearly, technology provides opportunities for counselors and psychotherapists to present themselves to prospective clients. They advertise their services online in a virtually countless number of ways. They offer an extreme variety of beliefs, manners of introducing themselves, and approaches to counseling and psychotherapy. Indeed, without the massive proliferation of advertised approaches to counseling and psychotherapy, this book would not exist. One of the sources of alarm about the current condition of mental health care comes from gathering information from the websites of practicing counselors and psychotherapists, as the accrued evidence reported here makes clear.

COMPLEMENTARY AND ALTERNATIVE MEDICINE, ADAPTATION, AND CAUTION

Complementary and alternative medicine (CAM) is a broadly encompassing term that includes many practices in mental health care. In this section "adaptation" refers to one of the obvious facts in the long history of mental health care. Specifically, mental health care has made numerous adaptations and will continue to need to make numerous adaptations. "Caution" refers to the need to exercise caution and discretion, with regard to assessing the clinical relevance of alternative approaches to counseling and psychotherapy. The heading, "Complementary and Alternative Medicine, Adaptation, and Caution," intends to introduce judgments about the current condition of counseling and psychotherapy and to encourage readers to consider an expansive view of this condition.

The observations and conclusions presented in this book are clear. For example, one of the observations is that preparation programs for counselors and psychotherapists are demonstrably inadequate. Another is that licensing for mental health care providers is chaotic and largely nonrelevant to effective practice. Still another is that the regulation of professional practice is virtually nonexistent. These observations, along with the evidentiary base for them, come later in the book. For now, based on such observations, this discussion attempts to establish a conceptual context in which the evidence may be understood and utilized.

Given the somewhat bleak view of the current situation and the call for major reforms of counseling and psychotherapy, what about "adaptation?" How does adaptation play a role in understanding the current situation? In short, the position of this book is that adaptation has been a significant feature of the collective history of counseling and psychotherapy and that it will be a significant feature of the future. The question is much more one of which adaptations will occur than whether they will occur. This takes the discussion to "CAM."

According to the National Center for Complementary and Integrative Health (NCCIH, 2019a), CAM includes "treatments that are used along with standard medical treatments but are not considered to be standard treatments." Usually, the concept of CAM encompasses treatments for problems that many counselors and psychotherapists would recognize as problems that they or their peers may treat. The importance of CAM in this discussion is that many treatments identified as CAMs may be unfamiliar to many mental health clinicians but are commonly utilized. For example, the NCCIH (2019b) reports, "The 2017 National Health Interview Survey (NHIS) found that the use of yoga by U.S. adults increased significantly from 2012 (from 9.5 percent in 2012 to 14.3 percent in 2017). The percentage of U.S. children who used yoga more than doubled during this time (from 3.1 percent in 2012 to 8.4 percent in 2017)." Clearly, yoga is commonly utilized. However, are yoga and, by implication, many other CAMs utilized by counselors and therapists on the basis of good evidence to support their practice? The short answer is a qualified "yes."

The qualified "yes" requires explanation. The explanation involves several parts. First, an apparent assumption is that CAMs will be utilized. The issue, then, is not one of whether CAMs will be utilized, but one of whether they will be utilized in a professionally accountable manner, including professional recognition of the CAM. Natwick (2018) represents this assumption in her article, "Counselors Are Doing What Now? Exploring the Ethics of Complementary Methods." She says, "Counselors are finding ways to combine traditional talk therapy with a vast number of other practices, specialties, methods and techniques" (p. 10). She adds that these alternative practices include "yoga, acupuncture, reiki, hiking, aromatherapy, equine therapy, nutrition, neurofeedback and mindfulness" (p. 9). More than assuming that complementary methods will be utilized, Natwick provides ethical guidelines for those who wish to utilize them, including the recommendation that "counselors should consult their state's scope of practice laws and regulations" (p. 9). She adds that counselors should consider whether the practice is evidence based, whether they are competent to provide the treatment, whether they maintain the dignity and welfare of their clients, and whether they practice appropriate informed consent (pp. 9–10). These are important guidelines because they direct counselors toward professionally accountable service.

Second, CAM interventions for almost any human problem may be easily located. With reference to mood disorders, post-traumatic stress disorder (PTSD), multiple forms of spirituality, anxiety, and many other needs that their clients present, counselors and psychotherapists can find complementary and alternative interventions. To refer to only a small sampling of affirmative support for various CAMs, the following may be sources that warrant review. Cook-Cattone (2015) advocates for mindfulness and yoga. Pradhan, Kluewer, Makani, and Parikh (2016) support various alternative approaches to treatment of PTSD. Lee (2009) advocates for an integrative mind–body–spirit approach to treatment. Tusaie (2013) offers alternative interventions for mood disorders. Buggio, Barbara, Facchin, Frattaruolo, Aimi, and Berlanda (2017) advance the conclusion that the psychological and sex-related problems associated with endometriosis can be effectively treated with alternative methods. And, Sadiq (2007) describes alternative treatments for attention deficit/hyperactivity disorder. Clearly, clinicians have applied their preferred CAM to the clinical needs of their clients.

Third, CAM interventions receive considerable endorsement as a comprehensive orientation to counseling and psychotherapy. For example, more generally than the specific treatments referred to in the previous paragraph, Lumadue et al. (2005) somewhat enthusiastically advocate for alternative and complementary therapies, including the need for training in counselor preparation programs. Similarly, in his book, *Transforming Clinical Practice Using the MindBody Approach: A Radical Integration,* Broom (2013) argues for the revolution of counseling and psychotherapy by integrating alternative treatments into clinical practice. Somewhat more committedly than the others named here, Lees and Tovey (2012) argue stridently for the cause of complementary and alternative treatments in counseling and psychotherapy. They write, "Fundamentally, we are arguing that the potential of the therapy profession to have an impact on healthcare systems can be enhanced by learning from, entering into dialogue with and even establishing alliances with CAM practitioners in view of their shared characteristics" (p. 78). Finally, the considerable body of research-based evidence that supports various CAM interventions should be consulted and referenced by those who wish to practice any of them, while remaining open to questions about the inherent ambiguity of the methods used, such as mindfulness, meditation, yoga, or other approaches. The evidence may be easily found, but no matter how convincing the evidence is—for CAM interventions or for commonly accepted interventions, such as CBT—the skill of the clinician may or may not ensure effectiveness in the delivery of treatment.

Fourth, quite apart from advocacy of CAM interventions, valuable guidance is available for those who wish to understand and evaluate such interventions. At its website, the NCCIH (2019a) says, "The mission of NCCIH is to define, through rigorous scientific investigation, the usefulness and safety of complementary and integrative health interventions and their roles in improving health and health care." Through dissemination of information, the NCCIH Clearinghouse serves this mission. The Clearinghouse "is the public's point of contact for scientifically based information on complementary and integrative health interventions and for information about NCCIH" (NCCIH, 2019b). Further, "The NCCIH Clearinghouse provides information on complementary health approaches and NCCIH, including publications and searches of Federal databases of scientific and medical literature" (NCCIH, 2019a). Obviously, the Clearinghouse is a useful resource for anyone who needs information about CAM interventions. In addition to being an important source of information about CAM interventions, the NCCIH has completed research on many CAMs and continues to provide funding for such research. The NCCIH is a complex operation that cannot be adequately described here, but it should be said that, because of its mission and research on CAMs, it provides a considerably large repository of information about many different treatment possibilities, with clear evidence that supports many of them, including many that may not be well known, and clear evidence that rejects others.

Therefore, with regard to CAMs and the inexorable adaptation and evolution of counseling and psychotherapy, great caution is needed. Granted that, while the specific, identified approaches to counseling and psychotherapy are wildly out of control, observers of any particular approach should exercise discretion and caution with regard to assessing the clinical relevance of the approach. Wholesale, blanketed condemnation of alternative approaches to counseling and psychotherapy is likely to be counterproductive, at least, because it alienates those who may be making important and positive contributions to mental health care. Also, such condemnation may prevent clients from receiving needed treatments. Discretion and caution are advised, for the reasons cited previously.

CONCLUSION

The long conversation about the status and future of modern counseling and psychotherapy that began more than a century ago will continue. The conversation has included numerous serious and thoughtful contributions. It has included controversy and unwanted but necessary adaptations. Inevitably, the conversation has been shaped by external forces, such as major economic turmoil, world wars, political disturbances, and more. Obviously, the conversation has also been shaped by innovators within the community of counselors and psychotherapists, including luminaries such as Aaron Beck and Carl Rogers. However, until the current time, counselors and psychotherapists have not performed in such ways that erode and potentially destroy the profession of which they are members. This is what appears to be happening now, as the following chapters report. The potential for destruction, though, should not be the closing of the history of counseling and psychotherapy. Instead of closing their history in ignominious cowering, counselors and psychotherapists may want to advance the cause of their clients by claiming or reclaiming their profession through major and urgently needed reforms.

21,758

The Scariest Number in the Psychotherapy World

This chapter reports information gathered from the websites of practicing counselors and psychotherapists. Mostly, it is an account of what was found on the websites. It confirms that practicing counselors and therapists advertise approaches that compromise the integrity of counseling and psychotherapy and sometimes likely cause harm to their clients. Many of the clinical approaches raise doubts about professional accountability among practicing counselors and psychotherapists. In addition to confirming the proliferation of advertised approaches to counseling and psychotherapy, this chapter also forms the basis for observations about the place of this proliferation in the larger picture of mental health care (see Chapters 4 and 5). As such, it briefly prompts concern about the apparent marginality of preparation programs, the somewhat nonrelevant status of professional licensing, the ongoing decline of competence among counselors and therapists, and the continuing surge of ideology-driven approaches to counseling and psychotherapy, among others. These issues receive attention in subsequent chapters.

INFORMATION FROM THERAPISTS' WEBSITES

Chapter 2 briefly discussed several causes of concern about the current condition of counseling and psychotherapy. It identified several discussants who confirm the causes for concern. Generally, the discussants have addressed specific potentially harmful treatments, along with other issues (e.g., declining competence, poor preparation programs, ideology-driven approaches to mental health care). They are presumed to be important contributions to needed reforms in counseling and psychotherapy. Adding to their contributions, the hope on which this book rests is that it may add impetus to movement toward reform. In advance of presenting the data from the websites of practicing counselors and psychotherapists, it should be said that Chapters 4 and 5 offer discussions about the meaning of these approaches.

This chapter presents several tables, which include many numbers and lists of mostly unusual approaches to counseling and psychotherapy. They come from four rounds of gathering information from practicing therapists' websites. Because of my concern about seeing odd and somewhat disturbing information on the website of one of my clinical supervisees, I decided to look at a large number of websites. The initial gathering located information on 600 websites. Dissatisfied with what I found, I looked at another round of 600. Based on my growing concern about what I found on these 1,200 websites, I decided to look at the websites of licensed professional counselors. Because I am a licensed professional counselor, I committed to compare the websites of licensees who completed accredited preparation programs and licensees who completed unaccredited preparation programs. This added 200 websites on which this chapter is based, for a new total of 1,400 websites. Finally, I gathered information from four cities that were approximately the same size, small enough that all available websites could be located and large enough to offer a useful number of websites. The four cities included Santa Rosa, CA, Chattanooga, TN, Providence, RI, and Sioux Falls, SD. The number of websites in these four cities was 622. The final total of websites from which information was gathered for this chapter was 2,022.

To provide useful information, I composed a narrative that explains how I proceeded and illustrated the narrative with charts and lists. The total number of charts and lists is 26. Combined, the charts and lists provide a detailed and factual picture of the information from the websites of 2,022 practicing therapists. It is this picture that, as a matter of professional conscience, required me to prepare this book.

Also, this chapter represents a partial acknowledgment of the fact that I was quite unaware of how therapists really work and what they allege to offer to their clients, although I have been an active participant in the therapy world for many years. In quite general ways, it tracks the evolution of my awareness of how therapists really work, but it also introduces others to the world that I discovered, a world that alarmed me and one that will likely alarm others. The number, 21,758, in the title of this chapter and the alarm that comes with it will likely make solid and formidable sense to those who read this chapter.

PROVOCATIVE DISCOVERIES

Note to the reader: The goal of this chapter is to provide information. The information is likely new to you. After you have read this chapter, you may be worried, disturbed, upset, and alarmed by the condition of our shared profession. If you are not concerned, I will be surprised. If you are a professionally accountable mental health care provider—professional counselor, psychologist, social worker, marriage and family therapist, or other—and if you are not seriously concerned by the number 21,758, I hope that you will remain open to the call for needed reforms of counseling and psychotherapy. Your profession is in peril. Without reform, the threats to your profession will continue and will likely cause adversities for you. To make sense of these statements, information is needed.

Approximately 8 years ago, I provided clinical supervision for a postgraduate-degreed individual for the purpose of her pursuit of a license in our state. She had completed her degree in May of that year. Our supervision began shortly after her graduation. The events described here emerged in October. Mostly, supervision went smoothly. We met regularly. We discussed her clients, all of whom she saw at the agency where we met. However, just as a passing conversation, one of my supervisee's friends, an employee at the agency where my supervisee worked, mentioned that my supervisee had a private practice. This was news to me. Before the next meeting with my supervisee, the friend mentioned also that my supervisee had a website. This, too, was news to me.

Curious about my supervisee's website, I checked it out. The information on my supervisee's website disturbed me. I found that she identified 28 therapeutic orientations and 68 clinical specialties, such as weight loss therapy, psychosis treatment, attachment issues, and so on. As I perused the website, I felt confusion. How could my supervisee have a private practice without my knowledge? Why had she not disclosed the fact of her private practice and her website to me? Moreover, how could anyone, especially a novice therapist, claim to have proficiency in 28 therapeutic orientations? And, how could she reasonably claim to have 68 clinical specialties? Worse, she did not indicate that she was not licensed. She did not indicate that she was "pre-licensed." And, adding to reasons for concern, she indicated that she had been in practice for 9 years, a length of time that, if correct, meant that she began to practice around the time she entered her undergraduate degree program.

At our next supervision meeting, my supervisee and I had the proverbial heart-to-heart talk. She acknowledged that she had acted as a clinician who promised more than she was capable of delivering. She ended her private practice and closed her website. Nevertheless, the memory of her website continued to surface, as I provided clinical supervision for others. None of the others posed challenges that were similar to the one who ignited my curiosity about what clinicians offer on their websites, although some of them presumed to offer clinical services for which they were not yet ready. Quite clearly, back then, my belief and assumption were that my supervisee gave an aberrant picture of someone in her situation and could not be representative of new clinicians or mental health care providers in general.

So, to confirm my belief and assumption, I decided to conduct a somewhat informal but careful study of several other websites of practicing mental health clinicians. I began with the belief that, if I were to gather information from a large number of websites, I would find that counselors and psychotherapists offer services that almost all of them would regard as professional and professionally accountable and that they agreed about far more than they disagreed. Therefore, the plan I pursued was fairly simple. Online, I visited 12 cities and took a random sample of 50 therapists from each city, giving me a total of 600 therapists' websites. From the information that I gathered, two facts appeared to be very clear. One was that the consensus among these therapists—a consensus that I sought to confirm—was there. Among other things, they agreed about several approaches to their work; approximately 75% of them identified cognitive behavioral therapy (CBT) as one of their therapeutic orientations.

This fact caused low-level but clear smugness. I was right, after all. However, other facts soon shattered my smugness and left me chaffed and challenged. Among these 600 therapists, a total of 489 different therapeutic approaches to therapy were advertised. Counting them was an afterthought that followed counting the therapeutic approaches about which they agreed. Nevertheless, taking the number seriously, I preferred to conclude that 489 was a mistake. How was it possible for 600 therapists, including counselors, social workers, psychologists, marriage and family therapists, and a small number of unlicensed mental health care providers, to offer such a number of therapeutic approaches? Obviously, while several of the therapeutic approaches were ones that they seemed to have in common, the huge majority of their therapeutic approaches showed an extreme lack of consensus.

Initially, I opted to look at the number, 489, and find a path toward reasoning it away. Surely, I thought, a lack of consensus may not seem so bad. After all, each of us who is a professional mental health care provider needs to find his/her own way to operate as a therapist. Naturally, then, as each of us finds a way to think about and deliver therapy, we will select numerous different ways. This thought brought echoes of hearing, "How am I supposed to choose a theory for therapy?" or "What do you think about this or that approach to therapy?" or "As you anticipate entering the role of a therapist, you must give serious thought to selecting your theoretical orientation." Also, as a small rationalization, I thought that the variety and complexity of therapeutic approaches reflected the immense complexity of being human. If a group of therapists planned a conference that included sessions for each of the numerous human problems that therapists treat, the number of sessions would be virtually limitless. Or, if, say, a group of 12 long-experienced therapists sat around a conference table and began to list the variety of human problems that they have treated, the list may rise into the thousands. So, why not have 489 therapeutic approaches for 600 therapists?

However, the information I gathered broke the credulity of my attempts to reason away the number. The idea of variability, instead of consensus, is good. Along with almost everyone in a mental health care profession, I believe in diversity, including alternate ways of thinking about and conducting therapy. The challenges to my wish to pretend that I was seeing necessary and honorable diversity came in two forms. One was that among these 600 therapists, more than one-third of them advertised an approach that none of the other 599 offered. The other was that some of the therapeutic orientations shocked me. Just to illustrate this, each of 237 advertised therapeutic orientations came from one individual. Yes, among these 600 therapists, 39.5% of the therapeutic orientations were solo orientations—an approach to therapy that only one individual advertised. This did not look like the consensus that I had believed I would find.

Another fact—the shocking one—became much more captivating than the mere fact of the solo orientations. It was that, among the solo orientations, many

Table 3.1 UNUSUAL APPROACHES TO THERAPY I

Ancient Hebraic dream interpretation	Astrology
Bach flower therapy	Pray therapy
Past life soul retrieval therapy	Mud therapy
Clairvoyance	ZenCare
Mormon/Latter-Day Saint therapy	Theophostic prayer therapy

seemed to be truly weird. Table 3.1 lists the therapeutic orientations that were advertised.

These so-called therapies are ones that I did not recognize. Mostly, I concluded that the process of gathering information must have been flawed. Surely, licensed mental health care providers would not offer these therapies or several others that I found. So, I decided to gather information from another 600 therapists. And, because the first group of 600 therapists came from urban areas, I decided to gather information from licensees whose advertising came from rural areas or small cities, believing that these areas may provide a correcting ballast, if this were needed. A correcting ballast is not what I found, as the numbers in Table 3.2 indicate.

The numbers evoked uncertainty about how to think about the information that had been gathered in these two samples of therapeutic approaches of therapists in urban and rural/small-city areas. While the population of sampled therapists was the same—600 in each group—inconsistencies arose. Mainly, therapists in urban areas advertised more therapeutic approaches than those in the other group, but offered fewer solo orientations. Conversely, the rural and small-city therapists advertised fewer therapeutic approaches but a much larger number of solo approaches. Another feature of these data is that the combined total number of therapeutic approaches—489 plus 452 equals 941—may be misleading. Because of overlapping items among these approaches, such as CBT, the blended total comes to 893. Along with other numbers, the blended total raised challenging questions, too. Are there really this many approaches to counseling and psycho-therapy? If there are this many approaches, how may this be explained? How do therapists invent or discover these approaches?

Within the numbers, a consensus about CBT appeared, again. Among the second group of 600 therapists, 484 (81%) identified CBT as one of their

Table 3.2 URBAN AND RURAL THERAPISTS' ADVERTISING

	Urban ($N = 600$)	Rural/Small City ($N = 600$)	Blended Total
Number of different orientations	489	452	893
Solo orientations	237	312	543
Number of orientations per therapist	8.4	6.3	7.8

Table 3.3 UNUSUAL APPROACHES TO THERAPY II

Anthetic therapy	Aroma freedom technique
Biblical counseling	BSFF—be set free fast
Circus arts therapy	DBDDD treatment
Discipleship counseling	Emotional freedom technique
Leveraging theory	Polarity realization therapy
Tapas acupressure technique	Zanker therapy

approaches to therapy. In addition, a majority of therapists identified family and systems approaches (317, or 53%) and solution-focused therapy (310, or 52%). None of the other approaches to therapy came close to being offered by a majority of therapists in this group, but several familiar approaches predictably appeared on the list. They included relational therapies (30%), motivational interviewing (20%), play therapy (21%), and psychodynamic therapies (26%). However, the unusual approaches—or, no doubt, some would conclude, "inappropriate" approaches—arose again. With each copied and pasted, the new, unusual approaches included those given in in Table 3.3, although the list of unusual therapeutic approaches is much longer than this.

Quite apart from this short list of unusual approaches to therapy, others were simply disturbing and alarming as therapeutic approaches that would be advertised by licensed mental health care providers. For example, past life regression therapy, or, similarly, Biblical existential therapy, plant spirit medicine, or past life regression and soulful hypnosis, is an approach to therapy that most therapists would likely reject. In addition, most therapists would probably wonder whether the approaches to therapy in Table 3.4 are viable, professional, and appropriate.

Later, this book discusses the lack of regulation of practicing counselors and therapists. For now, the lack of regulation may be indicated by saying that the approaches to therapy that have been identified so far in this chapter are easily located. After all, they are widely advertised. At least, they strongly suggest that regulation of the practice of counseling and psychotherapy may be seriously deficient.

The two rounds of information gathering led to tentative conclusions that were not entirely compatible. For example, one was that practicing therapists

Table 3.4 UNUSUAL APPROACHES TO THERAPY III

Fairy tale model	Universal values counseling
Power animal retrieval	Shamanic healing
Bible counseling	Vibrator therapy
Ancient Christian wisdom	Comedy therapy
After death communications	Conjoint therapy (God-client-therapist)
Soul oriented therapy	Diamond approach
Essential oil healing	Trager psychophysical movement
Contemplative Tibetan Buddhist therapy	Eco-spiritual therapy

who advertised their approaches to therapy reflected a consensus around several therapeutic approaches, such as CBT. In contrast, another conclusion held that practicing therapists who advertise their approaches to therapy reflected an extreme lack of consensus about treatment approaches, as indicated by the large number of "solo approaches." Additional observations and conclusions will be presented later. For now, though, a third conclusion, one that was tentative, became a nagging one, as I considered the information that I had gathered from 1,200 practicing therapists. It was that therapists "just make up stuff," and that they do this on a very large scale. Naturally, this thought endured and became truly a nagging one. And, maybe naturally, too, I converted the thought into one that prompted me to want to validate my work as a therapist and preparer of therapists.

Therefore, as a licensed professional counselor, I decided to gather information from the websites of counselors, but in such a way that I could distinguish the licensees who completed accredited preparation programs from those who completed unaccredited preparation programs. Surely, I predicted, those who completed accredited programs would advertise only therapeutic approaches that represented the accrued wisdom of the profession and the best evidence-supported therapies about which virtually everyone would agree. If my prediction could be confirmed, my positive view of therapists would be at least partially restored. In contrast with the two previous rounds of gathering information, I decided that the new study must be systematic and accountable to all of my peers, especially all other licensed professional counselors. My thought was that, of course, they would want to know what I found and see it as credible.

ACCREDITED VERSUS UNACCREDITED PROGRAMS

The accrediting agency for professional counseling preparation programs is the Council for the Accreditation of Counseling and Related Educational Programs (CACREP, 2017). Because of its importance to professional counseling, my new study became a comparison of counselor-selected treatment approaches from graduates of CACREP-accredited training programs and graduates of unaccredited training programs. The information for this study comes from gathering licensed counselors' advertised, self-selected treatment approaches, with their respective websites being the singular source of information. The study sought to compare the treatment approaches of licensees who completed CACREP-accredited programs and licensees who completed unaccredited programs. The need for this study came from the simple observation that many of the unusual and, possibly, aberrant or even harmful approaches to therapy discovered in two previous attempts to gather information about the therapeutic approaches of practicing therapists were practiced by licensed professional counselors. For me, the necessary question became "How may these unusual treatment approaches among professional counselors be clarified and understood, if in fact they are there?" The information already gathered did not help to answer this question.

On reflection, this question evolved into a different question: "What, if any, differences may there be between the treatment approaches of licensed professional counselors who completed CACREP-accredited programs and licensed professional counselors who completed unaccredited programs?" This is the question that I sought to answer, in at least a partial manner. Clearly, another approach to gathering information was needed. Therefore, to answer this question, carefully defined procedures, as described next, were followed.

Procedures. CACREP accredits counseling-related graduate programs that seek its professional endorsement through accreditation. As a creation of the American Counseling Association, since 1981 CACREP has sought "to promote the professional competence of counseling and related practitioners through the development of preparation standards; the encouragement of excellence in program development; and the accreditation of professional preparation programs" (CACREP, 2017).

Here, the term "unaccredited" refers to counseling-related graduate programs that hold no endorsements from specialized accrediting agencies or professional associations. My use of this term does not offer a judgment about the quality of these programs. It merely identifies the absence of accreditation at the time information was gathered about them.

Gathering information for this report involved several procedures, with each one adding assurance that the report could be a valid one. The procedures included the following:.

1. To be selected for this analysis, professional counselors needed to advertise their respective practices in states where they are licensed to practice. The information in this study comes from their online advertising.
2. States were selected randomly. The selection was simple. A map of the continental United States was divided into quadrants, with each part having approximately equal populations. This was intended to ensure that every region of the continental United States would be represented. These selected states were Colorado, Florida, Idaho, Iowa, New Mexico, North Carolina, Ohio, Pennsylvania, Texas, and Utah.
3. After selecting the states, the next step was to select locations within each state. Almost always, the search began with a major city. However, with the inherent challenges of identifying licensed counselors who met specific criteria (see the criteria listed under procedure 6), a single location within a state did not prove to be sufficient.
4. To gather a sampling of information about counselors, an Internet search using the term "psychotherapists [name of city or state]" located an almost uncountable number of therapists. Nevertheless, whatever the number, the websites of licensed counselors were available. Far more sites were visited than could be utilized for this report because a huge majority of sites visited did not meet criteria.

5. Ten licensed professional counselors who completed CACREP-accredited program were selected from each state. Likewise, 10 licensed professional counselors who completed unaccredited programs were selected from each state. After 10 licensed counselors in each of these groups was located, providing 20 from each state and a total of 200 from all of the selected states, the search was stopped.

6. As each randomly sampled licensed counselor's information was found, it needed to meet the following additional criteria:
   - The counselor must be licensed.
   - The counselor must hold no other license. Commonly, but not typically, licensed counselors hold more than one license. Selecting individuals with only one license ensured that the individuals selected would not be confused with those whose primary licensing identification may be uncertain.
   - The counselor must have identified the institution from which the licensing-qualifying degree was earned (i.e., university, college, seminary, institute, or other). If more than one institution was named by the counselor, the counselor's information was not collected. The name of the institution and the date of graduating from it provided ways to determine whether the individual completed a CACREP-accredited program or an unaccredited program.
   - If the counselor completed a program that was accredited by another accrediting agency, such as the Commission on Accreditation for Marriage and Family Therapy Education (2018), the Council on Social Work Education (2018), the American Psychological Association Commission on Accreditation (2018), or another, similar accrediting agency, the counselor's information was not gathered.
   - In states where licensing for professional counselors involves two or more tiers of licensing, the highest level of licensing was selected.

7. The information collected did not include identifying data, such as a counselor's name or address or website address or license number. Only the counselor's degree-granting institution, date of graduation, name of license, and treatment approaches were gathered. This ensures that inappropriate attention is not focused on the individual or inadvertently offered explicit or implicit judgment about any individual.

8. The name of licenses for professional counselors appears in many forms (e.g., LPC, LPC-MH, LCPC, LCPCC, LMHC). Regardless of the particular title of the license, the licenses selected for this study had to be unambiguously professional counseling licenses.

General Finding. The general finding from this study is that the 100 licensees of CACREP-accredited programs identified 197 different treatment approaches, while licensees from unaccredited programs identified 258 treatment approaches. When overlapping treatment approaches were identified, the combined total of treatment approaches dropped from 455 to 294 (see the details given later).

Table 3.5 provides a summative account of the information gathered about licensees from CACREP-accredited program graduates and licensees from unaccredited programs. The account requires a definition of terms. "Undifferentiated approaches" refers merely to an arithmetic total, meaning that all approaches were added to the total even when they were duplicates. If a counselor identified, say, 10 treatment approaches, even when other counselors identified some of the same approaches, all 10 of them were added to this total. This number is accompanied by the average number, in parentheses, of treatment approaches for counselors in both of these groups. The information indicates that the licensees from CACREP-accredited programs offered 11.2 treatment approaches, while the licensees from unaccredited programs offered 10.3 treatment approaches.

"Differentiated totals" refers to a count of each different treatment approach, meaning that each treatment approach counted only once, no matter how many times it was named by licensees. For example, CBT was identified as a treatment approach by 87 licensees from CACREP-accredited programs and by 75 licensees from unaccredited programs. In the aggregated total for each group, CBT counted only once, making it an unduplicated item on the list.

The aggregated total for each of these groups gives a view of an interesting reversal. The number of total approaches from CACREP-accredited licensees exceeds the number of total approaches from unaccredited licensees, but the number of differentiated approaches from unaccredited licensees exceeds the number of approaches from CACREP-accredited licensees. This is similar to the difference between urban therapists and rural/small-city therapists reported earlier.

"Approach identified by one counselor" or "solo approaches" refers to approaches that are identified by only one counselor. For example, past life regression therapy was identified as a treatment approach by only one counselor. From the licensees who completed CACREP-accredited programs, each of the 139 approaches was identified by only one counselor. From the licensees who completed unaccredited programs, the number of approaches that only one counselor identified was 187. This is another reversal.

"Approach identified by two counselors" refers to a treatment approach that only two counselors listed on their respective websites. For example, "dream interpretation" as an approach to treatment was identified by only two licensees.

Table 3.5 General Findings About the Number of Self-Selected Treatment Approaches

	Unaccredited	CACREP-Accredited	Blended Totals
Undifferentiated approaches	1,032 (10.3)	1,123 (11.2)	
Differentiated totals	258	197	294
Solo approaches	187	139	222
Approach identified by two counselors	20	14	31

Similarities and Dissimilarities. Within these numbers, similarities and dissimilarities in self-selected treatment approaches among licensed professional counselors who completed either CACREP-accredited programs or unaccredited programs definitely emerge in this study. Tables 3.6 and 3.7 identify several of the major similarities and major differences between these two groups.

At least, these charts indicate that there are both large similarities and dissimilarities between licensees from CACREP-accredited programs and licensees from unaccredited programs. Additional differences, though, may be significant in addressing the question with which this article began, "What, if any, differences are there between the treatment approaches of licensed professional counselors who are graduates of CACREP-accredited programs and graduates of unaccredited programs?"

Consensus Issues. As indicated previously, the general finding is that the 100 licensees from CACREP-accredited programs identified 197 different treatment approaches, while the licensees from unaccredited programs identified 258 different treatment approaches. These numbers may appear to be worrisome, especially if they are extrapolated in a manner that concludes that, say, 10,000 practicing counselors and psychotherapists advertise 20,000 therapeutic approaches. Such an extrapolation would be incorrect, as indicated later.

Table 3.6 THE 10 TREATMENT APPROACHES MOST CLOSELY HELD IN COMMON BETWEEN LICENSEES FROM CACREP-ACCREDITED PROGRAMS AND LICENSEES FROM UNACCREDITED PROGRAMS

Treatment Approach	Number Held by Licensees From Unaccredited Programs	Number Held by Licensees From CACREP-Accredited Programs
1. Life coaching	13	13
2. Art therapy	11	12
3. Career counseling	8	7
4. Mindfulness-based cognitive therapy	40	42
5. Psychodynamic therapy	29	27
6. Coping skills	15	13
7. Existential therapy	27	31
8. Gestalt therapy	15	19
9. Cognitive behavioral therapy	74	87
10. Eclectic theoretical approach	44	34

Note: Between these two groups of licensed counselors, numerous treatment approaches appeared in low and similar numbers but because of the low numbers did not rank in the top 10 approaches held in common. For example, neurofeedback was identified by three licensees from the unaccredited group and by four from the CACREP-accredited group. Sex therapy was identified by three individuals from the unaccredited group and by four from the CACREP-accredited group.

Table 3.7 THE 10 TREATMENT APPROACHES LEAST CLOSELY HELD IN COMMON
BETWEEN LICENSEES FROM CACREP-ACCREDITED PROGRAMS AND LICENSEES
FROM UNACCREDITED PROGRAMS

Treatment Approach	Number Held by Licensees From Unaccredited Programs	Number Held by Licensees From CACREP Programs
1. Internal family systems	8	0
2. Emotion focused therapy	4	28
3. Expressive arts	6	18
4. Somatic therapy	4	9
5. Play therapy	14	28
6. Spirituality	19	12
7. Relational	21	32
8. Family systems orientation to therapy	26	47
9. Narrative	14	25
10. Dialectical behavioral therapy	15	26

Note: As indicated, both groups of licensees identified approaches in low numbers. Based merely on the size of difference between the two, the difference may appear to be significant. However, because these approaches were reported in low numbers, they are not included in the list of 10 least closely held in common. For example, meditation was identified by four in the CACREP group of licenses and none in the unaccredited group, and person-centered therapy was identified by four in the CACREP group and none in the unaccredited group. The four-to-one difference is significant, but because the numbers are so low, these approaches do not appear in the table.

Characterizing either of these groups as having a consensus about treatment approaches would be incorrect. Nevertheless, within the group of licensees that completed CACREP-accredited programs, the level of disagreement about treatment approaches is lower than the level of disagreement among licensees who completed unaccredited programs. The apparent difference in the level of disagreement arises within the number of different treatment approaches: 197 for CACREP-accredited licensees and 258 for licensees from unaccredited programs. Another potentially significant set of numbers indicates the volume of treatment approaches that only one or two individuals identified. Table 3.8 provides this information.

Among the approaches that were identified by only one counselor, some appeared on the list of those identified by CACREP-accredited licensees as well as unaccredited licensees. Twenty identical matches appeared—or a combined total of 40 treatment approaches. These included Child Counseling, Christian Counseling, and Clinical Sexology. In addition, five clusters of obviously overlapping approaches appeared. The largest of these was a cluster of 13 that included "mindfulness" in some form. The number of approaches that appeared in the clusters is 26.

Table 3.8 TREATMENT APPROACHES IDENTIFIED BY ONLY ONE OR TWO INDIVIDUALS
WITHIN THE GROUP OF LICENSEES FROM CACREP-ACCREDITED PROGRAMS AND
THE GROUP OF LICENSEES FROM UNACCREDITED PROGRAMS

	CACREP-Accredited	Unaccredited
Approach identified by one counselor	139	187
Approach identified by two counselors	14	20

After the matches and obviously overlapping approaches are added together, the resulting number of approaches identified by only one or two individuals changes. Among the CACREP-accredited licensees, 117 approaches were identified by only one or two individuals; 105 of these were solo approaches and 12 were identified by two individuals. Among the unaccredited licensees, 177 approaches were identified by one or two individuals; 158 of these were solo approaches and 19 were identified by two individuals. To be clear, each of these 253 treatment approaches was an approach that only one or two individuals out of 200 licensed professional counselors advertised. This reveals a high level of individualized approaches to treatment. Collectively, licensees from CACREP-accredited programs identified 40% of the solo treatment approaches that only one individual advertised. Licensees from unaccredited programs advertised 60% of the solo approaches that only one individual advertised. Licensees from unaccredited programs advertised a much larger number of solo approaches to therapy.

Separately and together, CACREP-accredited licensees and unaccredited licensees appear to express a considerable lack of consensus about treatment approaches. Clearly, within both groups, the number of different treatment approaches points away from consensus about treatment approaches and, possibly, about what treatment is. Some possible meanings of the lack of consensus are addressed later. For now, a major observation arises from these numbers: Approaches to counseling and psychotherapy are highly individualized.

Gender Differences. Quite apart from the apparent lack of consensus about treatment approaches, other kinds of difference emerge. These include gender, the type of degree-granting institutions from which these two groups earned graduate degrees, and the appropriateness of their respective treatment orientations.

Initially, the question about gender differences arose because of an interest in knowing whether the number of treatment approaches for females may be different from those for males. There appeared to be virtually no differences in the number or kind of treatment approaches identified by females and males. However, an unexpected difference appeared in the ratio of females to males between the two groups of licensees. Table 3.9 provides this information.

With a sample size of 200 licensed counselors, with 100 in each group, conclusions about gender differences should be tentative ones, at best. Based on this acknowledgment, gender differences in this sample suggest that females are more likely than males to complete CACREP-accredited programs. Also, the numbers beg the question about consensus-building among licensees from CACREP-accredited programs and whether this may be a product of the differential in the

Table 3.9 GENDER DIFFERENCES AMONG LICENSEES FROM CACREP-
ACCREDITED PROGRAMS AND LICENSEES FROM UNACCREDITED PROGRAMS

State	CACREP-Accredited		Unaccredited	
	Females	Males	Females	Males
Colorado	8	2	8	2
Florida	10	-0-	9	1
Idaho	7	3	9	1
Iowa	8	2	8	2
New Mexico	9	1	4	6
North Carolina	10	-0-	8	2
Ohio	9	1	7	3
Pennsylvania	9	1	9	1
Texas	9	1	7	3
Utah	7	3	7	3
Total	86	14	76	24

populations of men and women from CACREP-accredited programs. However, as indicated later, alternate explanations of gender differences may be reasonably offered.

The Degree-Granting Institutions. A review of the types of institutions from which individuals earned their licensing-based degrees adds another category of differences between licensees who completed CACREP-accredited programs and licensees who completed unaccredited programs. Specifically, as Table 3.10 indicates, licensees who completed CACREP-accredited program are much more likely than licensees who completed unaccredited programs to have earned their graduate degrees at public or state-supported institutions. Licensees who completed unaccredited programs are much more likely to have completed degrees at private, religion-oriented, theological institutions, and private independent institutions.

For clarity, the difference between "private, religion-oriented" and "theological" institutions represents the nature of the institution and the training program that a licensee completed. Licensees in both categories completed degrees at private, religion-oriented schools. For example, some of the licensees who completed CACREP-accredited programs earned degrees at Duquesne University (2018) or Mississippi College (2018). Both of these institutions have histories of affiliation with their religious groups—Catholic and Baptist, respectively. However, their

Table 3.10 TYPES OF INSTITUTIONS/PROGRAMS THAT LICENSEES COMPLETED

	CACREP-Accredited	Unaccredited
Public institutions	69	40
Private religious institutions	11	22
Theological institutions	6	12
For-profit institutions	10	7
Private independent institution	4	19

training programs in mental health counseling are not features of a theology-based curriculum. Likewise, individuals in both categories of licensing completed degrees at theological institutions. For example, licensees who completed degrees at Denver Seminary (2018) appear on both lists of licensees, with the distinction that some of them completed degrees long before its counseling program received accreditation from CACREP and some of them completed their degrees after it became accredited. Accredited or not, a theological school such as the Denver Seminary would presumably make theological inquiry an important feature of graduate education.

The difference in the type of institutions between the licensees who completed a CACREP-accredited program or an unaccredited program likely helps to explain some of the differences in treatment approaches. For example, the licensees from CACREP-accredited programs identify 12 specifically religious treatment approaches, while the licensees from unaccredited programs identify 32 specifically religious approaches to treatment.

To maintain equity in presenting this difference between these two groups, the same numbers of examples of religion-oriented approaches from each group are listed here. As examples of their religion-oriented treatment approaches, licensees from CACREP-accredited programs included those listed in Table 3.11.

Similarly, licensees who completed unaccredited programs included those listed in Table 3.12.

In addition, theological schools usually have largely male populations of students and graduates. This very likely contributes to the gender differences between licensees from CACREP-accredited programs and licensees from unaccredited programs, as represented in Table 3.9.

Outlying, if Not Inappropriate Approaches to Therapy. In the discussion at the beginning of this chapter, the number of therapists—two groups with 600 in each group—was much larger than the number in this discussion of licensees who are graduates of CACREP-accredited and unaccredited preparation programs. The earlier discussion included several categories of licensees, including professional counselors, social workers, psychologists, marital and family therapists, and others. Tentatively, it suggested that many mental health care providers advertised approaches to therapy that they "just made up," probably based primarily on their personal interests and not the accrued wisdom of their profession or evidence from research. The information from 1,200 licensed professionals and a small number of unlicensed providers indicated that 893 treatment approaches were advertised by these providers. Five hundred forty-three of these approaches

Table 3.11 EXAMPLES OF RELIGION-ORIENTED THERAPIES AMONG LICENSES
FROM CACREP-ACCREDITED PREPARATION PROGRAMS

Christian counseling	Living the 12 steps spirituality relapse
Buddhist spirituality	Psychological principles with Biblical teaching
Christian	Spiritual-based counseling
Spiritual direction	

Table 3.12 Examples of Religion-Oriented Therapies Among Licensees
From Unaccredited Preparation Programs

Holistic-Christian perspective	Biblical foundations
Biblical truth	Depth (soul) psychology
Earth-based religious	Healing light of God's grace and truth
The truth of God's word	

were solo approaches, with each one of them being advertised by only one therapist. Some of these approaches to treatment were deemed to be inappropriate, although I did not conclude this by examining their inappropriateness in a systematic way. Nevertheless, the tentative conclusion about their inappropriateness evoked my interest in looking specifically at differences among licensed professional counselors with regard to their appropriateness as well as whether some of them may have been harmful to clients. The inappropriateness of an approach to treatment is easier to determine than its harmfulness or potential for harmfulness. Obviously, many approaches named in this book are unusual, but nothing here concludes or confirms that clients have been harmed by these approaches.

The question about the appropriateness of counselors' approaches to therapy, therefore, invited an attempt to answer it. In seeking to answer this question, criteria for judging approaches to therapy were needed. The following criteria were followed to determine which of the self-selected treatment approaches by counselors in this study were inappropriate.

1. To conclude that a therapeutic approach may be inappropriate, a list of approaches was distributed to 10 licensed professional counselors. One half of the counselors completed CACREP-accredited programs, and the other half, including me, completed unaccredited programs. Also, two of these licensed professional counselors had Christian-based theological training as a feature of their professional histories. One of the counselors avowed commitment to practicing Buddhism.
2. Except for me, the counselors who received the list did not know whether items on the list came from licensees from CACREP-accredited programs or licensees from unaccredited programs.
3. All therapeutic approaches that were identified by 10 or more licensees who were either from CACREP-accredited programs or unaccredited programs were eliminated from any further consideration about their inappropriateness. The operative assumption is that, when 10% of licensed counselors or more offer an approach to psychotherapy, this represents substantial support and, therefore, credibility for an approach. These approaches included CBT, solution-focused brief therapy, and marriage and family therapy, along with others.
4. The list did not include approaches to therapy that are easily identified as "okay," insofar as they have been broadly accepted, even if they are practiced by a small minority of counselors in this study. These

approaches included person-centered therapy, career counseling, and twelve-step orientations, along with others.

5. Several approaches were excluded from the list of possibly inappropriate approaches to therapy because they are virtually impossible to evaluate. These included alternative approaches, progressive blend of approaches, collaborative methods, conflict resolution, inclusive holistic methods, and traditional approaches. These approaches simply cannot be evaluated.

6. Some approaches were excluded from the list of possibly inappropriate approaches to therapy because they addressed a specific problem and, because of this, indicated little about an approach to therapy. These included gambling counseling, life transitions therapy, work-related stress, and anger management. Clearly, these are important clinical issues, but they would not usually be regarded as approaches to therapy.

7. The list included all approaches, except those that were eliminated by the previous criteria.

8. If a treatment approach was deemed to be inappropriate, it needed to be characterized as having one or more of the following qualities: ethically questionable, lack of clinical relevance, personal (i.e., nonclinical) expression of the licensee, weak and inadequate concept, potentially harmful to clients, too ambiguous, "unknowable," "just made up stuff," or other indicators of nonvalidity. The aim of using these qualities is not to affix a label to them as "good" or "bad."

None of the determinations was made only on the basis of a judge not preferring or disagreeing with a treatment approach, even while disagreements and subjectivity are features of these determinations. Expectedly, among the 10 counselors who offered judgments about the appropriateness of the treatment approaches in this study, serious disagreements arose. Opinions for and against them arose in relation to Jungian archetypes, moral reconation therapy, and eye movement desensitization and reprocessing (EMDR), among others. While any of us may disagree with the clinical value of these therapeutic approaches and others, we usually approve of others' practice of them.

The task given to the 10 counselors was simple: Make a discriminating judgment about whether an approach to therapy is one that you regard as appropriate or inappropriate for licensed professional counselors to advertise and to practice. The counselors received a copy of the criteria. The responses from the counselors were assessed with the added criterion that only those that received seven of 10 (70%) responses as being inappropriate were included in the list of approaches that were deemed to be "probably inappropriate" for counselors to offer and to practice. When 10 of 10 counselors judged that an approach was inappropriate, the approach was labeled "inappropriate." To be clear, if a therapeutic approach was placed in the category of "inappropriate," the approach must have received this judgment in unanimity—10 of 10 counselors. Table 3.13 includes asterisks that indicate the treatment approaches that received unanimous votes of "inappropriate."

To be sure, none of the counselors who participated in this study wished merely to make judgments about the professionalism of other licensed counselors, their competence, or even whether they should be conducting psychotherapy with the approaches that appear in Table 3.13. More than anything else, making these judgments should be understood as an invitation to a conversation about the status of professional counseling and mental health care generally. The added significance of these judgments, if there is one, appears in the concluding section of this chapter, "Closing Brief Analysis."

In advance of the presentation of inappropriate approaches to counseling and psychotherapy, acknowledgment of the absence of explanations for judging each of these approaches as inappropriate is needed. This is to say that this discussion does not provide an analysis of each of these approaches to therapy. Such an analysis would likely take this discussion well beyond its intended purposes and would require multiple volumes. Also, those of us who participated in this process wish to say that another panel of judges may include approaches that do not appear on this list and may exclude some of the ones that appear on the list. Therefore, while many of these judgments would likely be confirmed by almost all licensed professional counselors, some would not, so the judgments presented here come with confidence as well as caution. Most of them can be easily supported, but we recognize that they are subject to modification. Also, given the apparent individualistic nature of the advertised treatment approaches, another study of this kind would surely produce valuable and useful results that would contribute positively to the discussion.

Another note of caution should be raised. As time passes, the assessment of these approaches may evolve in ways that lead to different conclusions. For example, somatic experiencing was a new approach to therapy for some of the reviewers. Would these reviewers reach a different conclusion about Somatic Experiencing if they studied it or received training in it?

Regarding these approaches to therapy, explanatory notes about the list may be helpful.

1. This is possibly the most important explanatory note. Quite simply, while judgments about approaches to clinical work are made, they are intended to be invitations to discussions much more than final, definitive statements about the inadequacy of these approaches. At least, the approaches listed in Table 3.13 should lead to discussions about them, with reference to the future of counseling and psychotherapy.
2. The asterisk (*) indicates treatment approaches that were deemed to be "inappropriate" by 10 of the 10 panel members who judged the approaches.
3. The others were deemed to be "probably inappropriate" by seven, eight, or nine members of the panel and proved to pose a wide range of challenges. Here are some examples: Seven of 10 licensed counselors agreed with the presumed inherent values in "GLBT-friendly" counseling but concluded that it is not an approach to psychotherapy. It may be quite appropriate for a therapist to promote this feature of the services

Table 3.13 INAPPROPRIATE TREATMENT APPROACHES ADVERTISED BY LICENSED
PROFESSIONAL COUNSELORS

Treatment Approach	Unaccredited or CACREP-Accredited
Alchemy*	Unaccredited
Aquamassage*	Unaccredited
Arbinger facilitation	CACREP
Aromatouch (massage)*	Unaccredited
Biblical foundations*	Unaccredited
Biblical truth*	Unaccredited
Buddhist studies*	CACREP
Chakra balancing*	Unaccredited
Chiropractic care*	Unaccredited
Christian counseling	CACREP
Christian	CACREP
Comic books in therapy	CACREP
Contemplative and humanistic psychology	Unaccredited
Contemplative	Unaccredited
Counseling that works	CACREP
Craniosacral therapy*	Unaccredited
Depth (soul) psychology*	Unaccredited
Dream interpretation	Unaccredited
Dream tending*	CACREP
Dream work/nightmares	CACREP
Earth-based religious*	Unaccredited
Embodied experiencing	CACREP
Emotional freedom techniques*	Unaccredited
Essential oils*	Unaccredited
Feline-assisted therapy*	CACREP
GLBT-friendly	Unaccredited
Healing light of God's grace and truth*	Unaccredited
Heart centered hypnotherapy	Unaccredited
Heathen	Unaccredited
Hermetic philosophy*	Unaccredited
Holistic healing approaches*	Unaccredited
Holistic, Christian perspective	Unaccredited
Hormone and amino acid therapy*	CACREP
Hormone balancing and coaching*	CACREP
Hypnotic regression*	Unaccredited
Interconnection and spirituality	Unaccredited
Jungian dream interpretation	Unaccredited
Kalish practice*	CACREP
Kundalini yoga*	Unaccredited
Law of attraction*	Unaccredited
Light touch therapies*	Unaccredited
Mind–body bridging	CACREP
Neurolinguistic*	Unaccredited
Neuropsychiatry*	CACREP

Table 3.13 CONTINUED

Treatment Approach	Unaccredited or CACREP-Accredited
Neuropsychotherapy	CACREP
Neurotherapy*	Unaccredited
New regression therapy*	Unaccredited
Nutrition counseling	CACREP
Past life regression therapy*	Unaccredited
Psychological principles with biblical teaching*	CACREP
Purpose focused approach	Unaccredited
Reiki energy healing*	Unaccredited
Shadow integration*	Unaccredited
Shamanism*	Unaccredited
Somatic experiencing (SE)	Unaccredited
Spiritual based counseling	CACREP
Spiritual direction	CACREP
Spiritual direction/seeking*	Unaccredited
The power of prayer	Unaccredited
Thera-P	Unaccredited
Time travel*	CACREP
Truth of God's word*	Unaccredited
Video games in therapy	CACREP
Vipassana or insight meditation	CACREP
Vitamin/nutrition consulting*	Unaccredited
Whatever works!*	CACREP
Wicca*	Unaccredited
Yogic techniques and chakra balancing*	Unaccredited

* Indicates treatment approaches that were deemed to be "inappropriate" by 10 of the 10 panel members who judged the approaches.

that are offered and to acknowledge that it is not an approach to therapy. The power of prayer received a similar split decision. No one argued against the power of prayer, but they concluded that it is not an approach to psychotherapy. This kind of recognition applied to several other items on the list, including Biblical truth and vipassana or insight meditation. These beliefs and practices may be quite good, even admirable, but are they clinically appropriate?

4. Maintaining focus on the fact that all of the therapists in this study are licensed professional counselors, it is important to emphasize that the less-than-unanimous votes about the appropriateness of the treatment approaches affirm interest in having discussions about what counseling/psychotherapy is and are not intended to establish boundaries of division among licensed counselors. Unsolicited, all members of the panel wished to emphasize their commitment to diversity among counselors. I suspect that most of us, including counselors, social workers, psychologists, and

others, wish to protect our clinical and professional prerogatives with regard to advancing the interests of our clients. In addition, the panel of professional counselors concluded that some of the approaches on the list are ambiguously associated with practices that are not particularly clinical, even if they are otherwise valuable and useful, such as spiritual direction, purpose focused approach, and contemplative. Ambivalent feelings about some of the items on the list should not obscure the fact that several of them are clearly inappropriate. If nothing else, these approaches should be carefully scrutinized.

5. An unexpected view emerged in response to this exercise, although the view fell short of a consensus. Nevertheless, the view might forecast changes in the list if the exercise were repeated. Some of the counselors who judged these approaches may have encountered some of the approaches for the first time and, because of inexperience with them, judged them to be inappropriate. For example, although somatic experiencing was judged to be inappropriate, some who made this judgment expressed caution about their judgment and their interest in knowing more about Somatic Experiencing. Also, when they reviewed Somatic Experiencing and became aware of multiple research reports (e.g., Brom et al., 2017; Warner et al., 2014) that support the clinical efficacy of somatic experiencing, those who expressed caution about their judgment later indicated that their earlier judgment was inadequate. However, to preserve the integrity of the study, their judgment has been retained.

6. Even when judgments about some therapeutic approaches were unanimous, they came with thoughtfulness and sometimes serious disagreement.

7. Despite ambiguities and disagreements, the panel concluded that most of the items on this list would be judged as "inappropriate" by a preponderant majority of mental health care providers. For example, these items would include alchemy, past life regression therapy, and chiropractic care, among others.

8. All members of the panel offered approval of Christian counseling. Despite this, the emergent view was that much too often Christian counseling becomes a way to practice the therapist's religion and not serve the clinically relevant needs of clients. Disagreement about this may have been stronger than disagreement about any other item on the list and included the observations that (a) many who allege to practice Christian counseling do not indicate that they have any specific training or credentials that would qualify them to provide this kind of counseling and (b) a religion-based therapy of one kind (e.g., Christian) necessarily promotes the inclusion of other religion-based therapies, such as ones that emerge from Islam, Judaism, Hinduism, or others. In the end, the panel concluded that religious orientations should not be clinical orientations to therapy.

9. Some of the judgments may lack awareness of what a therapeutic approach is.
10. A review of the judgments may not account for certain demand characteristics that the panel may have inadvertently considered. This is to say, when the request to the members of the panel involved making judgments about the appropriateness or inappropriateness of the therapeutic orientations, did they feel pressed to judge items on the list as inappropriate?

Table 3.14 presents a numerical comparison of "probably inappropriate" and "inappropriate" treatment approaches that licensed counselors advertised. "Probably inappropriate" treatment approaches were those that received less-than-unanimous votes of disapproval from 10 licensed counselors.

The comparisons in Table 3.14 appear to indicate that both licensees from unaccredited programs and licensees from CACREP-accredited programs advertise a significant volume of probably inappropriate and inappropriate approaches to treatment. Licensees from CACREP-accredited programs advertise 35% of the total of 68 items that appear on this list, and licensees from unaccredited programs advertise the other 65%. Possibly, a slightly mitigating factor in considering this volume of treatment approaches that may not be appropriate is that some combinations of approaches on the list are advertised by one person. This means that the 74 items on the list do not indicate that 74 licensed counselors advertised them. Instead, because of combined approaches, the number of licensed counselors whose approaches to treatment may be deemed to be probably inappropriate or inappropriate falls from 68 to 51, or 25.5%, of the total number of licensees in this study. This means that approximately 10% of licensees from CACREP-accredited programs offer inappropriate treatment approaches and approximately 20% of licensees from unaccredited programs offer inappropriate approaches.

Another possible source of mitigation of the significance of the number of approaches that may be inappropriate is the number of approaches that are deemed to be appropriate. The total number of differentiated treatment approaches is 294, while the total number of possible approaches that are inappropriate is 68. This means that 77% of the total treatment approaches were not on the list of inappropriate approaches.

So, what have we discovered about the therapeutic approaches of CACREP-accredited licensees and unaccredited licensees? The following are some thoughts about this.

Table 3.14 Numerical Representation of Inappropriate Treatment Approaches Offered by Licensed Professional Counselors

	Unaccredited	CACREP-Accredited	Total
Probably inappropriate approaches	12	14	26
Inappropriate approaches	32	10	42
Total	44	24	

Accredited Versus Unaccredited Wrap-Up. Gathering information from 200 therapists hardly lends itself to reaching conclusions that characterize either professional counselors or other mental health care providers. Agreement on a conclusion likely will require many more studies of this kind, possibly energizing ideas and actions about how to eliminate conspicuously inappropriate approaches to therapy and affirm and support approaches to therapy. This investigation discovered that 200 licensed professional counselors offered a total of 294 different treatment approaches through their advertising. This is approximately one and one-half unique treatment approaches for each counselor. Based on the information from this chapter so far, from 1,400 therapists who advertise online, therapists of all kinds who are professionally accountable should wonder what the larger world of therapy may be advertising. Information from another 622) counselors and therapists appears later in this chapter.

The information gathered for this study may be useful as a cause of discussion among mental health care providers. Observations and conclusions about the information here are offered later. For now, questions may be more important than conclusions. The following questions could be some of the necessary ones.

What do the apparently aberrant therapies say about practicing therapists?

How widely spread are these aberrant therapies?

Insofar as these aberrant therapies are "out there," how may they be regulated or eliminated?

To whom, if anyone, are the advertisers of aberrant therapies accountable?

Has the specifically religious expression among therapists overwhelmed clinical judgment and professional clinical services to clients?

Has preparation of prospective mental health providers failed to prepare them for professionally accountable professional services?

With regard to licensed professional counselors, are unaccredited programs producing graduates and licensees who contaminate the larger clinical world with their inappropriate clinical offerings? Alternately, given that licensees from both accredited and unaccredited programs advertise unusual therapeutic approaches, what may be the implications for their respective programs and the importance of accreditation?

How much alike are licensed professional counselors and other types of licensees, including social workers, psychologists, and marriage and family therapists, with regard to offering inappropriate therapeutic approaches?

Are unusual therapeutic approaches the outcome of professional socialization or inappropriate indoctrination into a treatment orientation?

Licensed professional counselors lack consensus, let alone agreement, about approaches to counseling and psychotherapy. This may point to a more fundamental observation. Licensed professional counselors, along with other mental health care providers, may not know what counseling and psychotherapy are.

The apparent proliferation of treatment approaches suggests this. The options appear to be limited only by what a therapist says therapy is. For some, it may be an expression of religion. For others, it may be health care. For others, it may be an implementation of important personal values. For others, it may be an art form that is limited only by the therapist's imagination. Additional possible explanations of the extraordinary proliferation of approaches to counseling and psychotherapy will likely be contributed by others. For now, the questions may be more important than the answers. The pivotal question is this: Is psychotherapy whatever a therapist says it is, whatever interests a therapist may have, a means of self-discovery for the therapist, or something else? Along with a large body of supporting evidence, the question prompted the necessity of writing Chapter 6.

Along with other types of mental health care licensees, licensed professional counselors proliferate treatment approaches. Among other possible outcomes of this proliferation, therapists, clients, and potential clients may be confused and uncertain about what mental health care providers do and may be capable of doing. This may cause erosion in the credibility of mental health care providers and the services they offer. For some observers, it may indicate self-indulgent affirmation of the therapists' clinical prowess.

Licensees from both CACREP-accredited and unaccredited training programs sometimes offer approaches to counseling that may evoke suspicion, if not rejection, from their peers. Further, some of these approaches may be harmful to clients. Readers may want to review the list of treatment approaches that were deemed to be inappropriate—especially the ones that received unanimous rejection—and investigate which ones may be harmful. Further investigation is needed.

Combined, approximately one in five licensees from CACREP-accredited programs and licensees from unaccredited programs compromise the integrity and credibility of counseling and psychotherapy. Their self-selected treatment approaches appear in many instances to stem much more from a need to express their personal interests than from professional accountability toward assisting their clients. Taken from this investigation, this may be exemplified through the approaches in Table 3.15, among others.

Table 3.15 UNUSUAL APPROACHES TO THERAPY BY LICENSEES FROM CACREP-ACCREDITED PREPARATION PROGRAMS AND LICENSEES FROM UNACCREDITED PREPARATION PROGRAMS

Alchemy	Aromatouch (massage)
Comic books in therapy	Craniosacral therapy
Earth-based religious	Essential oils
Feline-assisted therapy	Kundalini yoga
Light touch therapies	Past life regression therapy
Shamanism	Spiritual direction
The power of prayer	Time traveling
Truth of God's word	Wicca

Licensees from CACREP-accredited programs appear to be oriented some-what more toward offering professionally accountable approaches to treatment. Combining the numbers from both groups, approximately 20% of professional counselors appear to offer inappropriate therapeutic approaches. Even if the list of inappropriate treatments were half of what it is, it would still be shocking and dis-appointing. It would very likely cause concern from professionally accountable, li-censed professional counselors and almost all other mental health care providers. And even if just 10% of licensed professional counselors were offering inappro-priate approaches to counseling, this would be cause for serious concern. Therefore, licensed professional counselors need to consider the potential consequences of the treatment approach selections of some of their peers. Analogously, if 10% of pharmacists were known to dispense inappropriate medications, the responses from their peers and the patients who needed medications would be swift and definitive. They would soon be stopped.

Following the examination of professional counselors' approaches to therapy, my curiosity turned to other mental health care providers, including mar-riage and family therapists, social workers, and psychologists. This is where I went next.

FOUR CITIES: SIOUX FALLS, CHATTANOOGA, PROVIDENCE, AND SANTA ROSA

In 2015 and 2016, I gathered information from the websites of all therapists who advertised their clinical services online in the Chattanooga, Tennessee area. At the time, my search identified 67 therapists who advertised online. In early 2018, I de-cided to locate some of the therapists whose advertising I had found almost 2 years earlier. Some no longer had a website that I could locate. Others had slightly mod-ified their websites. Most important, though, I found many more websites than I had found almost 2 years earlier. Intrigued by the dramatic increase—doubling in 2 years—in the number of therapists who advertised online, I wondered how cities whose populations were similar in size to Chattanooga would compare with regard to the online advertising by therapists. Also, by expanding a search beyond Chattanooga, I hoped that I would have a sampling of therapists that would be large enough to support credible conclusions about therapists' advertising.

Several cities have populations that came close to the population of Chattanooga. For the purpose of geographical distribution, I somewhat arbitrarily identified three more cities where I would try to locate all therapists who advertised online. The cities, their populations, and the number of therapists appear in Table 3.16.

The populations of these cities are close in number, but the number of therapists who advertised online varied considerably, ranging from 50 in Sioux Falls to 258 in Santa Rosa. This difference may suggest that Sioux Falls is grossly underserved or that Santa Rosa is grossly overserved. However, accounting for this difference may be easier than this. Sioux Falls is a somewhat geographically isolated city, located in the central, northern plains. In contrast, Santa Rosa is located in close

Table 3.16 POPULATIONS OF THE FOUR CITIES STUDIED*

City	Population	Number of Therapists
Sioux Falls, South Dakota	174,360	50
Chattanooga, Tennessee	177,571	133
Providence, Rhode Island	179,219	181
Santa Rosa, California	175,155	258

* Population data taken from Wikipedia, 2018.

proximity to major metropolitan areas, including San Francisco and Sacramento, California.

One of the primary goals of gathering information about therapists' advertising in the four cities was to locate and report the comparative numbers of advertisers who hold mental health care–related licenses. This, too, brought unexpected results, as Table 3.17 indicates.

Comparing cities, the difference between the high and low percentages for type of license was larger than expected. For licensed marriage and family therapists

Table 3.17 TYPES OF LICENSES

License	City			
	SFSD	CHTN	PRRI	SRCA
LMFT	10% (5)	12% (16)	3% (5)	62% (159)
LPC	68% (34)	34% (45)	23% (39)	2% (4)
PSY	6% (3)	10% (13)	20% (37)	21% (55)
LCSW	14% (7)	17% (22)	50% (91)	7% (18)
Other*	2% (1)	18% (24)	5% (9)	6% (14)
None†	2% (1)	10% (13)	0% (0)	2% (8)

* Individuals who do not hold a license but who practice under supervision by an individual who holds a qualifying license in her/his respective state. These individuals practice as de facto licensees insofar as their respective supervisor's license gives them defined legal status in their respective states.

† Most, if not all, states have individuals who practice without a license. The information gathered for this study indicates that, among the 622 therapists who advertised online, 21 individuals (3%) appear to hold no license. In our earlier and other reviews of therapists' websites, the percentage of unlicensed therapists who advertise online was higher—usually about 9%.

CHTN, Chattanooga, Tennessee; LCSW, licensed clinical social worker; LMFT, licensed marriage and family therapists; LPC, licensed professional counselor; PRRI, Providence, Rhode Island; PSY, psychologist; SFSD, Sioux Falls, South Dakota; SRCA, Santa Rosa, California.

Note: In the states in which these cities are located, the titles of these licenses sometimes vary. For example, the primary license for professional counselors in South Dakota is LPC-MH, while in Tennessee, it is LPC-MHSP.

Table 3.18 THERAPISTS' GENDER BY CITY STUDIED

	SFSD	CHTN	PRRI	SRCA
Female	86%	74%	77%	76%
Male	14%	26%	23%	24%

CHTN, Chattanooga, Tennessee; PRRI, Providence, Rhode Island; SFSD, Sioux Falls, South Dakota; SRCA, Santa Rosa, California.

(LMFTs), the high was 62% in Santa Rosa and the low was 3% in Providence. Similarly, for licensed professional counselors (LPCs), the high was 68% in Sioux Falls and the low was 2% in Santa Rosa. For licensed clinical social workers (LCSWs), the high was 50% in Providence and the low was 7% in Santa Rosa. For psychologists, the high was 21% in Santa Rosa and the low was 6% in Sioux Falls. At 18%, Chattanooga had elevated numbers of pre-licensed therapists and unlicensed therapists compared with the other cities.

Along with many others, I have observed that gender plays a significant role in the identity, work, and populations of counselors and therapists. Bringing this observation into this study as a question about the gender differences among therapists' populations in these cities seemed to be a necessary task, whether or not it produced useful results. The populations of men and women among these four cities appear to be fairly consistent, except for Sioux Falls, which has a considerably higher proportion of female therapists, as Table 3.18 indicates.

Proceeding toward an analysis of the therapeutic orientations of therapists who advertised in these four cities, a numerical picture of their approaches appears in Table 3.19. Among other things, the numbers say that the 622 therapists who advertised their approaches to therapy identified an undifferentiated total of 7,017 approaches to therapy. "Undifferentiated" refers to the all-inclusive number of approaches. For example, if 37 therapists in Sioux Falls named CBT, as they did, and 154 therapists in Santa Rosa named CBT, as they did, the undifferentiated total would be 191 who named CBT as an approach to therapy. In contrast, the "differentiated" total of CBT would be one. Therefore, as the numbers in the second row indicate, the therapists in these four cities advertised a total of 1,128 differentiated approaches to therapy. However, when the lists of therapeutic

Table 3.19 NUMBER OF THERAPEUTIC ORIENTATIONS ADVERTISED

	SFSD	CHTN	PRRI	SRCA	Totals or Blended
Undifferentiated total	634	1,372	1,748	3,263	7,017
Differentiated total	108	247	281	492	1,128 or 871
Solo orientations	56	126	208	367	757 or 743
Mean (average) of approaches	8.8	10.3	9.7	12.6	11.3

CHTN, Chattanooga, Tennessee; PRRI, Providence, Rhode Island; SFSD, Sioux Falls, South Dakota; SRCA, Santa Rosa, California.

approaches are combined so that duplications are eliminated, the total number of different approaches to therapy becomes 871.

Possibly, a more surprising outcome of counting therapeutic approaches is that the number of solo approaches constitutes the preponderant majority. When the differentiated approaches from all four cities are blended, the number of approaches is 871. When the solo approaches are blended, the number is 743. Solo approaches (meaning that each of these was listed by only one therapist) constitute 85% of the total number of different approaches. As Table 3.19 suggests, this appears to indicate that therapists generally advertise highly individualized approaches to therapy. This is consistent with the information from three reports earlier in this chapter.

The numbers in Table 3.19 may provoke thoughts about what therapists in these four cities advertise on their respective websites. Such thoughts may or may not be confirmed with details about what these therapists actually advertised. Table 3.20 offers some of the details. Specifically, it identifies the 15 most commonly advertised approaches to therapy from each of the four cities, ranking them from one to 15 for each city. No doubt, CBT stands alone as the most commonly

Table 3.20 THE 15 MOST COMMONLY IDENTIFIED TREATMENT APPROACHES IN EACH OF THE FOUR CITIES

CBT	76%	CHTN
CBT	75%	PRRI
CBT	74%	SFSD
CBT	60%	SRCA
Family systems therapy	58%	SFSD
SFBT	58%	SFSD
Family and marital counseling	56%	SFSD
Psychodynamic therapy	55%	SRCA
Family systems therapy	50%	CHTN
Humanistic therapy	50%	SRCA
Trauma-focused therapy	48%	SFSD
MBCT	47%	SRCA
Family and marital counseling	46%	SRCA
Psychodynamic therapy	45%	PRRI
Family and marital counseling	45%	CHTN
MBCT	44%	PRRI
Eclectic approach	42%	PRRI
Interpersonal therapy	42%	PRRI
Interpersonal therapy	42%	CHTN
Attachment-based therapy	41%	SRCA
SFBT	41%	CHTN
Trauma-focused therapy	41%	CHTN
Relational therapy	38%	SRCA
Strength-based therapy	36%	SFSD
Eclectic approach	36%	CHTN

(*continued*)

Table 3.20 CONTINUED

Relational therapy	36%	CHTN
MBCT	35%	CHTN
Interpersonal therapy	34%	SFSD
Eclectic approach	34%	SRCA
Family systems therapy	34%	SRCA
Family and marital counseling	33%	PRRI
Family systems therapy	33%	PRRI
DBT	32%	SFSD
Trauma-focused therapy	32%	SRCA
Interpersonal therapy	32%	SRCA
MBCT	30%	SFSD
Attachment-based therapy	30%	SFSD
Humanistic therapy	30%	PRRI
EMDR	30%	CHTN
EFT	29%	CHTN
Eclectic approach	28%	SFSD
EFT	28%	SFSD
Person-centered therapy	28%	SFSD
Somatic therapy	28%	SRCA
SFBT	28%	PRRI
Existential therapy	27%	SRCA
Motivational interviewing	27%	PRRI
Relational therapy	27%	PRRI
Attachment-based therapy	27%	CHTN
Psychodynamic therapy	27%	CHTN
EMDR	26%	SFSD
Motivational interviewing	26%	SFSD
Trauma-focused therapy	26%	PRRI
EMDR	26%	SRCA
Strength-based therapy	25%	PRRI
Person-centered therapy	25%	CHTN
EFT	24%	SRCA
Integrative therapy	23%	CHTN
DBT	21%	PRRI
Coaching	20%	PRRI

CBT, cognitive behavioral therapy; CHTN, Chattanooga, Tennessee; DBT, dialectical behavioral therapy; EFT, emotionally focused therapy; EMDR, eye movement desensitization and reprocessing; MBCT, mindfulness-based cognitive therapy; PRRI, Providence, Rhode Island; SFBT, solution-focused brief therapy; SFSD, Sioux Falls, South Dakota; SRCA, Santa Rosa, California.

advertised approach. Following this, the rankings appear to have relatively little in common.

When the 15 most commonly advertised approaches to therapy among these four cities are blended into a single list of the most commonly advertised therapies, differences among the four cities appear. Therapists in Santa Rosa and Chattanooga advertised 14 of 15 of the therapies on the list. In contrast, therapists in Providence and Sioux Falls advertised 11 of 15. Does this suggest that Santa Rosa and Chattanooga stand more in the mainstream of therapy than the other two cities? Alternately, individually, with 85% of the 622 therapists in these four cities advertising therapeutic approaches that none of the other 621 advertised, do the numbers in these charts indicate that in the aggregate, practicing therapists have little in common? Table 3.21 may provide partial answers to this question.

Making sense of or assigning meaning to these numbers may be premature, based on what has been presented so far. A good next step in trying to make sense

Table 3.21 THE 15 MOST COMMONLY ADVERTISED TREATMENT APPROACHES FOR ALL THERAPISTS IN FOUR CITIES (*N* = 622)

Approach	Rank	Number	Proportion of All Therapists	Cities Identifying
Cognitive behavioral therapy	1	428	69%	SFSD, PRRI, CHTN, SRCA
Family and marital counseling	2	266	43%	SFSD, PRRI, CHTN, SRCA
Psychodynamic therapy	3	260	42%	PRRI, CHTN, SRCA
Mindfulness-based cognitive therapy	4	259	42%	SFSD, PRRI, CHTN, SRCA
Family systems therapy	5	244	39%	SFSD, PRRI, CHTN, SRCA
Interpersonal therapy	6	231	37%	SFSD, PRRI, CHTN, SRCA
Eclectic approach	7	229	37%	SFSD, PRRI, CHTN, SRCA
Trauma-focused therapy	8	208	33%	SFSD, PRRI, CHTN, SRCA
Relational therapy	9	196	32%	PRRI, CHTN, SRCA
Humanistic therapy	10	184	30%	PRRI, SRCA
Attachment-based therapy	11	158	25%	SFSD, CHTN, SRCA
Solution-focused brief therapy	12	133	21%	SFSD, PRRI, CHTN
Eye movement desensitization and reprocessing	13	121	20%	SFSD, CHTN, SRCA
Emotionally focused therapy	14	116	19%	SFSD, CHTN, SRCA
Somatic therapy	15	74	12%	SRCA

CHTN, Chattanooga, Tennessee; PRRI, Providence, Rhode Island; SFSD, Sioux Falls, South Dakota; SRCA, Santa Rosa, California.

of these numbers is to look at the solo approaches that therapists in these four cities advertised.

Unusual Approaches to Counseling and Psychotherapy. In all four cities, the number of "solo" approaches to therapy was higher than the most commonly identified approach to therapy, as indicated in Table 3.22. Most of the solo approaches are unusual, and the table serves merely to confirm the number of solo approaches.

In each of these cities, significantly more mental health care providers advertised a solo approach to therapy than those who advertised the most agreed upon approaches to counseling and psychotherapy. In 756 instances, an individual provider advertised an approach to therapy that none of the other 621 advertised. This appears to indicate a gross lack of consensus about what a psychotherapeutic approach may be and what psychotherapy is. Further, this appears to indicate that each provider of mental health care can and does offer a highly individualized understanding of and approach to psychotherapy. Also, the solo approaches appear to be evidence for a serious lack of regulation of the practice of psychotherapy.

While many of the solo approaches to therapy may easily be regarded as "inappropriate," they are not the exclusive source of inappropriate approaches. Depending on an individual's view of what therapy is or should be, an assessment of it will be made, accordingly. For the purpose of this discussion, when referring to "inappropriate," the question is whether an approach to therapy demonstrates professional accountability. As noted earlier in this discussion, several factors may indicate a lack of professional accountability. Specifically, if a treatment approach is deemed to be inappropriate, it needs to be characterized as having one or more of the following qualities: ethically questionable, lack of clinical relevance, personal (i.e., nonclinical) expression of the licensee, weak and inadequate concept, potentially harmful to clients, too ambiguous, "unknowable," "just made up stuff," or other indicators of nonvalidity. These criteria imply standards for therapeutic approaches that may not be unusual or inappropriate. The essential question is this: "Is this approach to therapy a professionally accountable one?" This question aims to emphasize the importance of professional ethics, clinical judgment, and reasonably confident support among therapists because the approach has an evidentiary base and reasonable consensus among professionals who utilize the approach as well as among those who do not. This is to say that an approach to

Table 3.22 SOLO APPROACHES TO COUNSELING AND PSYCHOTHERAPY

City	Most Commonly Identified Therapy: Behavioral Therapy	Number of Solo Approaches	Number of Providers
Sioux Falls	37 (74%)	56 (112%)	50
Providence	136 (75.1%)	208 (114.9%)	181
Chattanooga	101 (75.9%)	126 (94.7%)	133
Santa Rosa	154 (59.6%)	367 (142.2%)	258
Total	528	757	622

therapy should be very carefully scrutinized if it represents the interests of only one therapist.

Because I am a licensed professional counselor, I felt reasonably assured that I could and should offer an assessment of those who hold the same license. However, as a matter of simple respect for those who hold different licenses, I have decided to avoid characterizing other licensees, except in general ways. Further, I have decided to avoid characterizing each of the four cities in this study, except for some features that boldly stand out as needing comments. Some of these comments appear later in this discussion and in the next chapter. For now, my interest is to present a list of so-called approaches to therapy that appear in the information gathered from the online advertising in these four cities.

Quite arbitrarily, I decided that, if I could identify 70 inappropriate approaches to therapy, this would represent slightly more than 10% of all the therapists in these four cities. The inappropriate approaches appear in Table 3.23. Based on the standard of professional accountability, more than 10% of all therapists in these four cities advertise approaches that are inappropriate. As the notes in Table 3.23 indicate, the number of therapists who offer questionable, if not inappropriate, approaches clearly exceeds 10%. Also, although this will be discussed later, the accrued information from therapists' websites indicate that aberrant approaches to therapy pose formidable problems for therapists and, possibly, threats to their continued existence.

Although I am offering judgments about these so-called therapeutic approaches, other writers have offered similar judgments about many of them. Earlier, this chapter included information from Thyer and Pignotti (2015), Holden and Barker (2018), Lilienfeld et al. (2005, 2007, 2014), Lee and Hunsley (2015), Thompson (2010), and Beyerstein (2001). Along with material in this book, these authors may elevate confidence in the judgments about aberrant therapies. Collectively, they help to establish a body of scholarship that defines a major problem for counselors and psychotherapists. Readers are encouraged to pursue other sources that may help them to assess various approaches to therapy, including the National Center for Complementary and Integrative Health (NCCIH, 2019a, 2019b), as discussed earlier in Chapter 2.

In the information gathered for this chapter, numerous acronyms appear, such as AMST, HAES (probably, "health at every size"), DNMS (probably, "developmental needs meeting strategies"), FCT, RECT, CBT, and many others. Assuming to know to what an acronym refers may pose risks for the assumer. Usually, for example, most clinicians assume that "CBT" refers to cognitive behavioral therapy. When a therapist uses only the acronym, though, it may refer to something else, such as "Christian biblical therapy." In the information gathered, therapists sometimes connect acronyms with recognized therapies, such as "gold-standard EFT." Such a reference leaves open the question of what "EFT" is. When the acronym, "EFT," alone is given, it may be emotion focused therapy or emotional freedom technique. These two EFTs are extremely different approaches to therapy. Another example of this is "EMDR tapping." EMDR and tapping are not usually the same thing, but some therapists connect them anyway. Another ambiguous acronym

Table 3.23 INAPPROPRIATE "THERAPIES" IN THE FOUR CITIES

7 Path self hypnosis

Affect-centered therapy

Alternating bilateral stimulation

Ancient Hebraic dream interpretation

Ancient traditions

AQAL therapy

Aromatherapy

Body intelligence

Buddhism*

Character analytic approach

Christ-centered approach†

Classical Chinese medicine

Cognitive mind stimulation therapy

Comprehensive energy psychology

Core energetics (mind/body)‡

Cutting-edge laser coaching

Diamond approach/Diamond breathwork

Earth-based spirituality

Energetic ecstatic practice

Emotional freedom technique/tapping

Energy tapping/energetic‡

Enneagram soul based

First argument technique

Gracious Christian counseling

Heart-centered and sacred soul work

Herbal medicine

Hindu practices

Holistic and spiritual psychotherapies

Holistic Christian

HPN neurofeedback

Image de-construction protocol (IPT)

Image transformation therapy

Imaginal psychology

Integral

Integrative psychology and spirituality

Integrative restoration

Interpersonal and social rhythm therapy

Interpersonal neurobiology

Intersectionality

Intersubjective

iRest

Japanese acupressure

Jungian psychology and theology

Kundalini yoga

Marschak interaction method (MIM)

Matrix reimprinting for trauma

Meridian tapping

Moral recognition

Native American practices

Nature-based therapy

Neurolinguistic therapy

Nondual philosophy

Past life regression

Psycho-shamanic counseling

Psychospiritual inquiry

Reichian orgonomic (3)

Relational neurobiology

Sacred imagination

Self-inquiry counseling

Shamanic hypnotherapy processes (8)

Soul retrieval counseling§

Spiritual cognitive therapy (SCT)

Spiritual formation/direction

Spiritual psychology

Tapas energy psychology

Techniques from Sandra Ingerman

Temperament analysis profile

Trauma-informed hypnotherapy

Voice dialogue

Zen meditation/Zen philosophy/Zen-
vipassana/ZenCare¶

* With a variety of descriptions, several titles for approaches that therapists had connected with Buddhism were advertised. Twenty references to some form of Buddhism were found. Placing Buddhism on this list should not diminish its importance as a philosophy of life or meaningful personal practices. Instead of challenging Buddhism, its placement on the list merely questions whether it is a professionally accountable approach to psychotherapy.

† With a variety of names, 57 therapists advertised "Christian" approaches to therapy. From this gathering of information and others, I had found almost 500 titles for Christian-related counseling and psychotherapy at the time of this writing.

Table 3.23 CONTINUED

‡ Numerous references to energy-related therapies were found in the information gathered. Thirty-nine therapists advertised them.

§ "Soul retrieval counseling" is not a therapy. However, the therapists who advertised a therapy that included a reference to "soul" surely included some who would reject soul retrieval counseling. This study located 12 different approaches that included the word "soul," including "soul work (inner child work)," "soul-centered psychotherapy," and "soul-heart-mind-body perspective."

¶ Some form of Zen was found among 10 providers.

is "NLP." Commonly, this stands for "neurolinguistic programming." However, it is sometimes also used to stand for "new life psychotherapy." Finally, "ACT" may stand for "acceptance and commitment therapy" or "advanced Christian therapy," or for several other names of approaches to counseling and psychotherapy.

Among the therapeutic approaches gathered, several simply lack definition or professional orientation. Apparently, therapists assert that they use these approaches but fail to connect them with anything about which other clinicians may inform themselves. Like many other approaches to therapy, these seem to be ones that therapists "just make up." Table 3.24 contains some of these.

Therapists offer services, but the services are not ones that fit within the scope of counseling and psychotherapy (e.g., business coaching, dance instruction, and nutrition and closely related services). Other approaches, such as documentary arts therapy, appear to represent only the highly individualized interests of a therapist.

Timeline Incongruence. A review of the reports about length of service among the 622 mental health care providers in four cities produced peculiar and quite unexpected discoveries. Clearly, the discoveries that provoked my interest in gathering the information from four cities included the expectation that a review of therapists' advertised services would result in finding truly inappropriate approaches to psychotherapy. While I hoped for failure in the pursuit of inappropriate approaches to psychotherapy, a large number and variety of such approaches were easily found. However, unexpected discoveries came, too. Among them, the information gathering revealed glaring discrepancies in the timelines that providers reported on their websites. The following examples from 2018 indicate how easily these

Table 3.24 THERAPEUTIC APPROACHES THAT LACK DEFINITION

Contemporary alternative treatments	Gentle martial arts movements
Grace therapy	Integration of faith/spiritual perspective
Martial arts	Mental fitness coaching
Neuroscience and biochemistry techniques	Spiritual practice
Unconditional care model	Western psychology
Wisdom-based therapy	Women's psychology

Table 3.25 TIMELINE DISCREPANCIES

Years in Practice	Year Graduated	Year Information Located
10	2015	2018
10+	2009	2018
12	2017	2018
15	2009	2018
16	2012	2018
16+	2009	2018
18	2010	2018
19	2010	2018
24	2015	2018
30+	2011	2018
6	2017	2018
8	2015	2018

discrepancies may be found. A licensed mental health care provider reported on her webpage that she graduated from her graduate program in 2016 and had been in practice for 6 years in 2018. Another reported that she graduated in 2013 and had been in practice for more than 15 years. Another reported that she graduated in 2016 and had been in practice 12 years. Easily located examples of timeline incongruence appear in Table 3.25.

These timeline incongruities come from the analysis of websites of counselors and psychotherapists in four cities. At a glance, these may appear to be isolated and unusual reports. Instead, timeline incongruities appear frequently in the advertising from practicing therapists. An analysis of these incongruities appears in Chapter 4. For now, suffice it to say that approximately 15% of practicing therapists include timeline incongruities in their online advertising.

AGGREGATED NUMBERS

The aggregated numbers from these studies appear in Table 3.26: Several numbers in this table beg for attention. Perhaps the loudest voices rise from two numbers: 2,022 therapists identified 2,267 approaches to counseling and psychotherapy.

The interpretation of the aggregated numbers may be stated in many different ways. One way is to suggest that mental health care providers demonstrate their commitment to diversity and multiculturalism. Another is that mental health care providers have developed a large variety of responses to a large variety of clients' problems. After all, each provider needs to find approaches that work best for him/her. However, no matter whether one or another of these interpretations is adopted, the interpreter must account for the fact that a surprising number of these approaches appear to betray counselors' and psychotherapists' commitment to diversity and fail to demonstrate credible understanding of counseling, psychotherapy, and the mental health needs of clients. More than a commitment to diversity or to the needs of clients, the aggregated numbers represent the

Table 3.26 AGGREGATED NUMBERS

	Number of Counselors	Differentiated Approaches	Solo Approaches	Blended Solo Approaches
Urban	600	489	237	
Rural	600	452	312	543
	1,200	941	549	543
CACREP-accredited	100	197	139	
Unaccredited	100	258	187	
	200	455	326	323
Four cities	622	871	757	743
Total	2,022	2,267	1,632	1,609
Blended total*		2,267*		1,593†

* Blending the differentiated approaches so that duplications in this number were eliminated pressed the question of what a duplication is. For example, when one therapist offers "Christian counseling" and another therapist offers "Faith-based Christian counseling," are these approaches duplicates? Are "reality therapy" and "reality-based psychotherapy" the same?

† When all solo approaches are combined, those that appear more than once are removed from the list of solo approaches, reducing the total number of solo approaches.

collective failure of mental health care providers to seek and find an agreed upon way of thinking about and conducting counseling and psychotherapy. Again, this strongly implies that mental health care providers may not know what counseling and psychotherapy are. Also, the extraordinary variety and number of unusual approaches to counseling and psychotherapy indicate a serious lack of regulation of the practice of counseling and psychotherapy. Among other observations, the numbers strongly suggest that approaches to counseling and psychotherapy are increasingly personal expressions of individual counselors and psychotherapists.

CLOSING BRIEF ANALYSIS

This brief analysis provides a transition to the next chapter. In Chapter 4, the importance of this chapter becomes clearer. For now, suffice it to say that the number 21,758 with which this chapter began is the number of therapeutic approaches that I have gathered from therapists' websites so far. If it is a threatening number, it is threatening because (a) the number is indefinitely larger than 21,758, (b) it reveals massive disarray in clinical work, (c) it points to likely potential harm to clients by a significant number of therapists, (d) among other questions, it raises the critically important question about whether there is such a thing as counseling or psychotherapy, and (e) it threatens the future of counseling and psychotherapy. Moreover, it strongly suggests that many therapists "just make up stuff" and self-interestedly utilize therapy for their benefit, regardless of whether clients gain from this or not. The number confirms that therapists demonstrate declining competence in their clinical work, as several researchers have reported and as discussed elsewhere in this book.

Thus, 21,758 is the scariest number in the therapy world. Much like a thief who burgles your home with you in it, the number provides a sketch of therapy thieves whose thefts come from the work that they allege to share with the rest of us. Insofar as this is correct, it places an urgent demand on us who live in the real therapy world as therapists.

Somewhat naïvely, when I started this endeavor, I would never have guessed that anyone could discover more than 21,000 approaches to therapy. Now, I recognize that the number is an infinitely larger one because I have gathered information from more than 60,000 websites and have necessarily concluded that, if I were to gather treatment approaches from another 30,000 practicing therapists' websites, I would discover another disturbingly large number of approaches. Pointing toward this eventuality, the fact that 2,022 therapists reported in this chapter identified 2,267 approaches to therapy suggests that the number 21,758 is a modestly low one and that numerous additional approaches to therapy are yet to be located.

If anyone wishes to locate unusual and harmful approaches to therapy, a simple activity will very likely fulfill this wish. An Internet search of, say, 100 therapists' websites will likely produce several conspicuously inappropriate approaches to therapy.

Observations About Multiple Therapeutic Approaches

The preceding chapter presented a record of a large number of and often inappropriate approaches to counseling and psychotherapy. It included only 200 unusual approaches to counseling and psychotherapy but also indicated that the number of unusual approaches to therapy is much larger than 200. Going forward, this chapter and Chapter 5 discuss the implications of these unusual approaches and, when the discussion allows, add some not yet identified and unusual approaches. This chapter discusses some of the characteristics of these unusual approaches to therapy. Chapter 5 describes some of the implications and consequences of these approaches.

Along with other features of contemporary counseling and psychotherapy, the record of massive numbers of inappropriate approaches to counseling and psychotherapy emphasizes the necessity of referring to "therapy thieves." As signs of theft, the extremely large number of approaches to counseling and psychotherapy requires the conclusion that many practicing therapists have "stolen" counseling and psychotherapy. However, the massive number of approaches to therapy alone should not be understood as the only indicator of theft. Thieves also appear under several other guises. They appear in the form of inadequate licensing for counselors and psychotherapists that allows inadequately trained individuals to receive licenses to practice. They come in the form of a nearly complete lack of regulations of clinical practice that permit inappropriate "therapies" to be massively advertised. They come in the form of multiple codes of ethics that are largely ignored and unenforced, with the predictable result of a countless number of inappropriate therapies being perpetrated on unsuspecting clients, among other results. They come in the form of widespread, wholesale, and conspicuous rejection of relevant evidence for the practice of therapy. Therapy thieves take many roles and receive considerable support among those who abet their thievery. They come in many guises.

Probably, the most salient conclusion from the presentation of the multiple therapeutic approaches provided in Chapter 3 is that counseling and psychotherapy must be reformed. The presentation of kismetic psychotherapy in the first chapter reinforces the necessity of reform in the practice of therapy because it makes clear that adding another therapy to the thousands that others have "just made up" can be easily done. As evidence will show, the questions are not really whether the practice of therapy will be reformed or changed. The increasingly conspicuous question is who will do the reforming. Another question is whether reform may already be too late.

Advocating for reform is the goal of this book. However, before advancing ways to reform counseling and psychotherapy, establishing the case for reform is needed. The presentation of kismetic psychotherapy and the report about multiple approaches to therapy are not a sufficient basis for advocating for reform. Toward making such a case, this chapter offers additional observations about the multiple approaches to therapy and discusses their implications for the practice of counseling and psychotherapy. With a brief discussion of each one, the observations and implications include the following topics:

- Large numbers of therapeutic approaches
- Technology as a dominating influence
- Extraordinary and irreconcilable polarities
- Timeline incongruence
- Excessive claims
- Lack of a license to practice
- Ideology-defined/ideology-driven approaches
- Suffusion of religion with psychotherapy
- Highly individualized approaches
- Proliferation of degrees and titles of degrees
- Proliferation of certifications

LARGE NUMBERS

The most obvious observation about the multiple treatment approaches in Chapter 3 is that the number of these approaches is very large. At this time and to the best of my knowledge, the report about these numbers is entirely new, with emphasis on the fact that the number of approaches to counseling and psychotherapy extremely exceeds any previous estimates. And, although the number of therapeutic approaches is quite large, it continues to grow rapidly.

Before extensive advertising through the Internet, knowledge of the number of treatment approaches was mostly out of reach of researchers. If they had been located, the effort to find them would have been a virtually impossible one. It would have involved postal mail, telephone surveys, or person-to-person interviews. All of these means of gathering information would not likely have eventuated in the discovery of more than 20,000 approaches to counseling and psychotherapy and the necessary conclusion that the number of approaches is indefinitely larger than the one reported in Chapter 3.

The large number of treatment approaches does not speak to the meaning of the number. It is large. This fact alone compels attention. The number requires interpretation, though. As numbers that are newly discovered, the meaning of them cannot yet be understood. Toward such understanding, this and the remaining chapters take on this task.

TECHNOLOGY AS A DOMINATING INFLUENCE

Another obvious observation is that technology continues to reshape and redefine humanity's view of itself and of counseling and psychotherapy. Alone in her/his office or home, a therapist can advance ideas to almost anyone in the world. This includes the potential for advertising.

In 2012, *Counseling Today* published an article (Shallcross, 2012), "What the Future Holds for the Counseling Profession," in which 17 leaders among counselors gave their views on this topic. Almost without exception, all of them referred to the predictable and large influence of technology on the profession of counseling. Herlihy stated, "Technology is changing our world at an astonishing pace" (Shallcross, 2012). Who could doubt this? Naturally, neither in 2012 nor today, accurate predictions about the future of mental health care are unlikely. However, regardless of predictions, the omnipresence and influence of technology is undeniable. The question is not so much whether presence and influence exist, but what kinds of presence and influence exist. How is psychotherapy being defined by therapists who advertise online? More narrowly, how is psychotherapy being defined by therapists who advertise aberrant approaches to therapy? The idea that they may be shaping the image and influence of psychotherapy in a good way is hardly a viable one.

Also, in the same article, Erford, a recent past president of the American Counseling Association, stated, "There are over 400 published counseling theories, but the outcome literature only supports use of a small fraction of these helping approaches and only for limited developmental and clinical applications" (Shallcross, 2012). His reference to "400 published counseling theories," although it is a commonly cited number, appears to be a strange one, given the information in this book. Understandably, Erford confirms the information that many mental health care leaders have understood to be true. More than this, though, the importance of his comment is that it implies a serious warning. He says that while there are more than 400 published counseling theories, "the outcome literature supports" only a "small fraction of these helping approaches and only for limited developmental and clinical applications." If his judgment is roughly correct—a judgment that most informed individuals would endorse—it leads to a scandalous observation about the current number of approaches to therapy that have no basis in "the outcome literature." The observation may be simply stated: Numerous aberrant therapies are broadly advertised but have little more than the whims of the advertising therapist to support them. Again, with the massive proliferation, including advertising, of aberrant therapies, how is psychotherapy being redefined? Is psychotherapy being made more or less viable as a result of this misuse of technology? If psychotherapy is being made less viable, what should be done about this? With more than 20,000 approaches to counseling and psychotherapy gathered from the websites of practicing counselors and psychotherapists, perhaps another important question is whether technology not only makes some investigations of mental health care easier but also exposes the pervasive inadequacy of counseling and psychotherapy.

EXTRAORDINARY AND IRRECONCILABLE POLARITIES

The advertising by therapists indicates that they share common interests. While the numbers cannot reveal altruism, compassion, sincerity, and clinical skill, among other personal and professional assets, a presumption that most counselors and psychotherapists possess these assets may be gracious and needed, not to mention that such a presumption nicely accommodates the value of diversity. They take immense good will and good intentions to their clients. However, clearly many of them express their good will and good intentions in ways that betray the profession of which they are members and offer approaches to therapy that may cause harm to their clients.

Extending this impression, the numbers reveal extraordinary and irreconcilable polarities among therapists. The polarities may be easily illustrated. On the one hand, a substantial majority of therapists advertise that they provide cognitive behavioral therapy (CBT). On the other hand, they advertise that they offer a huge variety of approaches to therapy that are quite incompatible with CBT, including Christian cognitive behavioral therapy and Zumba fitness. Another illustration of polarity is shown by the emphasis on evidence-based approaches to therapy on one end and ancient love pattern analysis, four temperaments model, and vegetotherapy on the other. Interpreting the extraordinary and irreconcilable polarities comes later in this book, but for now, they suggest incoherence in the delivery of therapy, the lack of cohesion among therapists, and serious challenges for professionally accountable therapists.

TIMELINE INCONGRUENCE REVISITED

In Chapter 3, a review of the reports about length of service among the 622 mental health care providers in four cities produced peculiar and quite unexpected discoveries. As noted in the preceding chapter, the discoveries that provoked the interest in gathering the information included the expectation that a review of therapists' advertised services would result in finding truly inappropriate approaches to psychotherapy. While hoping for failure of the pursuit of inappropriate approaches to psychotherapy, a large number and variety of such approaches were easily found. They included acutonics and Tibetan bowls, applied Godly wisdom, virtues orientation, auriculotherapy (ear acupuncture), two step revolution, absolute happiness, psycho-shamanic healing, bioaquatic exploration, scriptural principles therapy, WAR, pychopomp, superhero therapy, bootcamp therapy, clairvoyance, ceramics arts therapy, and many others. The expectation of finding unusual, if not inappropriate, approaches to therapy was, with sadness, fulfilled. The unanticipated discovery was that in four cities counselors and psychotherapists who advertised their services online appeared to distort, if not intentionally misrepresent, their years of providing professional services. At a glance, these may appear to be isolated and unusual reports. However, before concluding that they are unusual, additional facts should be considered. For one thing, the previous examples demonstrate much more than typographical errors or other incidental mistakes. Surely, given the volume of information from more than 600 individuals who advertise their mental health care services, mistakes of this kind would almost inevitably be found. These significant discrepancies appeared too frequently to be understood as mere slips or typographical entries. The gross discrepancies of this kind were made 79 times, or by 13% of the therapists who advertised in the four cities.

Another fact is that, among the 622 providers whose advertising was gathered, 95 provided too little information for an observation about the timeline to be made. Some provided no information about how long they had been in practice. Others did not indicate when they completed their graduate programs. And others provided no information about when they completed their graduate program or how long they had been practicing. Therefore, among the 527—(i.e., 622 minus 95)—who provided information that was sufficient for such an observation, 79 (15%) of them provided discrepant, incongruent information.

Also, a larger but undetermined number of providers reported information that is likely also significantly incongruent. For example, if a provider reported that he completed his graduate program in 2010 and had been in practice for 9 years in 2018, he was not judged to have reported incongruent information. Such a report may or may not be slightly exaggerated. Surely, it is not grossly incongruent. Therefore, small discrepancies such as this were not included among the 79 providers who advertised incongruent information. To be counted as timeline incongruent, an advertising counselor or psychotherapist needed to have provided at least a 6-year incongruity.

The discrepancies in the timelines reported by these therapists appear to be lapses in judgment and, possibly, failures in professional ethical behavior. Their lapse in judgment is not limited to these 95 individuals. In the process of gathering information from counselors' and psychotherapists' websites, I did not consider timeline incongruities. The timeline incongruities were noticed only after reviews of approximately 10,000 websites had been completed. After they were noticed as a fairly consistent feature of online advertising, information was gathered more consistently than it had been. Thus, among the larger number of approximately 52,200 therapists who advertise their services online, the proportion of exaggerators is the same, 15%. This means that among the 52,200 counselors and psychotherapists who advertise their services, approximately 7,830 presented timeline incongruities and, thereby, demonstrated a similar lapse in judgment.

However, to characterize the incongruities as lapses in judgment may be premature, or even incorrect, insofar as it explains almost nothing.

EXCESSIVE CLAIMS

Mental health care providers presented narratives about their services on their websites. A majority of the narratives would likely evoke reassured responses from most readers. Also, as instruments of marketing, providers should within reason and professionalism be allowed to advertise their services for prospective clients and to do this in a manner that is positive and hopeful. In other words, while anyone may review another's narrative and conclude that it includes exaggerated claims, such conclusions should be reached with great caution. After all, what exactly should anyone say about her/his professional services? How much self-promotion is too much? We have concerns about this but declined to gather examples of narratives that clearly went too far and, therefore, may have misled prospective clients. Suffice it to say that a therapist who advertises that she/he provides clinical services that facilitate joyful and pain-free change is likely making a serious mistake. While many of these types of assurances were found, they were extraordinarily difficult to quantify and were not, therefore, systematically gathered in the review of websites. In contrast, a therapist who advertises that she/he will provide clinical services in such ways that aim to facilitate hope is likely advertising appropriately.

However, regardless of the content of the narratives that mental health care providers offered to their prospectively consuming public, some of them offered services that they may not be appropriately prepared to deliver. Some of these are isolated types of services. For example, a licensed professional counselor indicated that he provides "medication management." Another, a nonpsychiatrist, indicated that he practices "neuropsychiatry." This isolated type of an inappropriate offering may be a mere mistake that could be easily corrected. Quite beyond an isolated type of inappropriate service, others are not so isolated.

A common and apparently exaggerated claim is that mental health care providers advertise that they as individuals offer one of more than 800 approaches to "Christian counseling." Their claims are exaggerated in at least two ways. One is that Christian counseling could be almost anything and, therefore, almost nothing. For example, what could possibly be the connection between Christian cognitive behavioral therapy, Christian Buddhism, Christian feminism, Christian Biblical principles therapy, and Christian meditation and contemplative counseling? "Counseling" and "therapy" have become such extremely common terms that refer to almost every conceivable problem that arises for human beings and almost every conceivable way of thinking about and treating these problems. Similarly, just like "counseling" and "therapy," "Christian" seems to rely almost exclusively on the whims of the one who offers it. This form of distortion, if not exaggeration or deception, is adjectival. Individual counselors and psychotherapists arbitrarily place their preferred adjectives on their preferred approach to counseling and psychotherapy. However, the adjectives include many more than "Christian." With similar arbitrariness, they frequently include "holistic," "collaborative," "spiritual," "integrative," and others.

The second form of exaggeration among those who advertise Christian counseling is that a large majority appear to have no specific training that would qualify them to provide it. In the early stages of reviewing mental health providers' websites, no account of this phenomenon was considered. After several thousand websites had been reviewed and this repeated phenomenon arose, a systematic accounting of it was conducted. In short, most of those who advertised Christian counseling completed their training at state colleges and universities. The large majority—78%—of those who identified the institution where they completed their training completed it in state colleges and universities. When they advertised that they offered Christian counseling or a similar offering, they did not indicate that they had completed relevant training or held relevant credentials that would qualify them to offer what they advertised. The appearance is that they made exaggerated claims. This kind of exaggerated claim was also observed on websites that advertised other specifically religious theory orientations. As the discussion indicates in the following paragraphs, exaggerated claims involve much more than offering some sort of faith-based counseling.

Among the 622 mental health care providers, 103 advertised "play therapy" as a service. Only 8% of them indicated that they hold a certification in play therapy, as a Registered Play Therapist (RPT) or the equivalent. This appears to involve excessive claims about the services that they offer. Similarly, 66 of them advertised "sex therapy" as one of their services. However, only 9% of them indicated that they have a certification in sex therapy, such as the certification from the American Association of Sexuality Educators, Counselors and Therapists (2018) or an equivalent organization. Another much larger number advertised specifically religion-based, religion-associated, soul-related, or spiritual approaches. A very small proportion of these therapists referred to relevant training, an academic program, certification, or another indicator of preparation or recognized competence that would support such advertising. These advertising therapists appear to have concluded that they could practice some sort of religion-oriented or spirituality-related therapy based almost exclusively on their personal interest in it and not on professional preparation.

Professional discretion requires a generous interpretation of these facts. An individual who practices play therapy, sex therapy, or some form of religion-related therapy is not required to hold a certification in any of these practices. Allowing for this, the apparent lack of credentials that would indicate sufficient preparation in a specialization like sex therapy necessarily raises the question of whether those who advertise sex therapy as a professional specialization are sufficiently prepared to deliver sex therapy. Having a strong personal interest in sex cannot qualify a therapist to provide sex therapy. This reasoning aptly relates to almost every other concern with which clients need professional help. Nevertheless, the excessive claims by these therapists appear to be lapses in judgment and, probably, failures in professional ethical behavior.

LACK OF A LICENSE TO PRACTICE

Another unexpected observation from the current review of websites of mental health care providers who advertise in Sioux Falls, Chattanooga, Providence, and Santa Rosa is that some of those who advertise lack a license to practice. Those who practice without a license may easily be placed into two groups. One group includes individuals who practice without a license but who practice under the license of a qualified clinical supervisor. Although members of this group do not hold a license, they practice as if they hold a license, insofar as their practice is legally and ethically sanctioned as a consequence of working under another professional's license. This group includes 38 providers or 6% of the total of 622.

The other group includes individuals who lack a license and who do not practice under the license of another provider. This group includes 22 so-called psychotherapists (3.5%) of the total of 622. (As an aside, what percentage of unlicensed cardiologists, radiologists, pediatricians, physical therapists, or other health care providers would be acceptable?) This percentage is significantly lower than the percentage of unlicensed individuals in previous reviews of websites that we have completed. The earlier reviews indicate that the proportion is about 9%. If the review of therapists' advertising in four cities indicates a decline in the percentage of unlicensed providers, this will be a positive change. Whether the percentage of unlicensed providers is a low of 3.5% or a high of 12%, the fact that they offer—or allege to offer—services that appear to be very much like those of licensed professionals should cause alarm among licensed professionals.

If at least 3.5% of all therapists are practicing without a license and this percentage is generalized and applied to the 60,200 therapists from whose websites information was gathered, the result is that approximately 2,107 "therapists" are practicing without a license. Many of the 2,107 have little or no training in counseling and psychotherapy. Likely, this reflects the failure of qualified mental health care professionals, including all licensees, to regulate their profession.

With reference to therapists who practice without a license, at least two additional categories of providers should be mentioned. One are the providers who get their "license" from nonstate agencies, such as the National Christian Counselors Association (NCCA), Pastoral Medical Association (PMA), and International Ministry Institute (IMI). These agencies advertise that they provide advanced degrees and issue licenses. These agencies are discussed in Chapter 6. For now, suffice it to say that practicing counselors and psychotherapists may advertise that they are licensed, but fail to disclose that their licenses were issued by non-state entities.

IDEOLOGY-DEFINED/IDEOLOGY-DRIVEN APPROACHES

This discussion opens the proverbial can of worms. It aims to describe certain features of counseling and psychotherapy, without necessarily categorizing them as good or bad. Many varieties of ideological approaches to therapy may be easily located online on therapists' websites.

Among many other kinds of information, the data about the four cities (i.e., Sioux Falls, Santa Rosa, Chattanooga, and Providence) included information about the mental health care providers' interest in treating issues of spirituality. In these four cities, 286 (46%) of the providers identified "spirituality" as a treatment issue. There was no other treatment issue that they identified as much as they identified spirituality. As an indicator of apparent interest in spirituality in these four cities, this percentage may be misleading. The range of interest among the four cities went from a low of 28% in Providence to a high of 61% in Chattanooga. The other two cities came in at 32% in Sioux Falls and 53% in Santa Rosa. These numbers should not be taken to conclude much about these four cities. However, they invite curiosity about what the numbers may say about how the therapists in these four cities view psychotherapy.

Very little information in the data gathered from these four cities helps to define how spirituality may be understood in these four cities or, more broadly, in the whole of mental health care. Despite this limitation, several of the providers offered important clues about how spirituality and, therefore, psychotherapy may be defined. For example, from the websites on which the four cities review is based, a therapist stated, "I borrow practices from several spiritual traditions, Native American, Buddhist, and Hindu." Allowing for the possibility that a single therapist has mastered Native American, Buddhist, and Hindu spiritual traditions and has abilities that allow her/him to integrate these traditions into psychotherapy, what may be the therapist's implied understanding of psychotherapy? Unfortunately, the story of spirituality in psychotherapy gets much more complicated than this. The complexity may be easily illustrated by comments that therapists posted on their websites:

"My understandings of hope, healing, humans, and the world are rooted in Christianity."

"I offer a Christ centered approach and integrate that fully in the work I do with my clients."

Another indicates that she offers "a soul-focused alternative to psychotherapy," with emphasis on "Hakomi, Psychosynthesis, and Spiritual Guidance" and "Transformational Process Coaching and Soul Therapy."

"I have personally found and witnessed with my clients that all pain can be an opportunity to move towards wholeness and deeper into the abundant life Jesus offers us in him."

"With my clients' . . . help and working from a Gospel-centered worldview, I have learned how to support them in successfully navigating their own life challenges and relationships in the pursuit of real meaning and true joy."

Another therapist proffered the view that her mission is "to bring clients into a loving relationship with Jesus Christ."

Obviously, these approaches to spirituality involve conflicts. The extraordinary differences among these therapists include several approaches to spirituality, including Native American, Buddhist, and Hindu spiritual traditions, Christianity, a Christ-centered approach, hakomi, psychosynthesis, transformational process coaching and soul therapy, a Jesus-oriented approach, and a Gospel-centered worldview, among others. All of these approaches to therapy come from either Sioux Falls, Chattanooga, Providence, or Santa Rosa. However, as daunting a picture of spirituality in psychotherapy as this may be, the list is not complete, based on this four cities review. Among the 622 therapists, others emphasized various religious and spiritual orientations, as indicated in Table 4.1.

From earlier searches, we can locate more than 900 approaches that therapists offer as forms of Christian counseling and psychotherapy. Only to illustrate some of the possibilities, we found theistic therapy/Christian values therapy, Christian Buddhism, Bible-based Christian counseling, and systems approach/Christian principles. This is not a complete list of the spirituality-related approaches to psychotherapy that we found in our review of the advertising of 622 therapists in four cities. The many variations of spirituality-related and religion-related approaches to therapy necessitate the question, "Can psychotherapy be all of the things that these providers say that it is and still be psychotherapy?" Then, adding almost another 1,000 treatment approaches from the four cities and approximately another 8,000 from information gathered from providers in other locations, the question about how psychotherapy may be defined and delivered becomes even more important. What is psychotherapy, or is there such a thing as psychotherapy, if it can be expressed in 10,000 or 20,000—and counting—different ways? If counseling and psychotherapy can emerge from ideology-defined and ideology-driven approaches, can it be reasonably and professionally supported?

Are there other ideology-defined and ideology-driven approaches? Cleary, the answer to this question is an emphatic "Yes." The answer invites an important question, though. "What is an ideology-defined or ideology-driven approach to

Table 4.1 VARIOUS RELIGIOUS AND SPIRITUAL ORIENTATIONS

12-step philosophy of spirituality	Buddhist Vipassana
Christian experiential therapy	Core energetics
Diamond approach	Divine nature of each soul/client
Earth-based spirituality	Energy psychology
Gracious Christian counseling	Integrative restoration
Jungian theology	Kundalini yoga
Past life regression	Psychospiritual inquiry
Psychospiritual integration	Religious cognitive emotional therapy
Sacred imagination	Shamanic spirituality
Soul retrieval therapy	Soul-based psychotherapy
SoulTalk/SoulCare	Spirit care counseling
Spiritual and somatic approach	Spiritual cognitive therapy (SCT)
Spiritual direction	The Enneagram

therapy?" Or, "If an approach to therapy is not based on ideology, what could or should it be based on, instead of ideology?" Identifying an approach to therapy as based on ideology may appear to denigrate it. This is not the intent. Instead, if an approach is based on ideology—and, usually, ideology alone—it likely lacks professional accountability and an evidentiary base. As a personal reflection on this matter, I find this observation to be both necessary and discomfiting. It is necessary because it is correct and true, and it is discomfiting because I wish that it was not necessary and true. The therapeutic approaches listed in Table 4.2 are based primarily in ideology.

The accrued wisdom of mental health care, professional accountability, and evidentiary support has been identified in this discussion as important to an approach to therapy. All of these features infer another important feature: scrutiny. An approach to therapy should have been subjected to scrutiny by the profession of which it is a part. If any of these approaches, such as psycho-shamanic therapy, are rejected by the professional community, they should not be continued. Obviously, offering this standard leaves too much unsaid about how a professional community may organize for, evaluate, and accept or reject approaches to therapy. Chapter 6 provides details about how this standard may be implemented. In the approaches listed earlier, such scrutiny is generally lacking. To preserve a useable future for psychotherapy, carefully scrutinizing approaches to therapy should become a feature of reforming psychotherapy. Without a means of approving approaches to therapy, the number of approaches will continue to rise, perpetuating many that are conspicuously inappropriate and likely harmful.

For many years, professionally accountable counselors and psychotherapists have urged one another to establish an evidentiary base for their clinical decision-making. For now, suffice it to observe that numerous mental health care providers demonstrate a conspicuous lack of interest in evidence and accountability, when they offer and provide stoicism philosophical perspective, spiritual/biblical stance, sound weaving, Satatove consciousness living, principle-based therapy, mytho-poetic therapy, or numerous other so-called therapies that express only their individualized, nonclinical, and potentially harmful services. Insofar as this observation is correct, an ensuing and necessary conclusion is that it betrays the profession of which these providers are members but also poses serious consequences for all providers and their clients. The consequences include the likelihood that all counselors and psychotherapists continue to lose credibility.

Table 4.2 VARIOUS IDEOLOGY-DRIVEN APPROACHES

Contemplative Buddhist therapy	Core energetic therapy
Feminist therapy	Gestalt therapy
IMAGO dialogue	Integral therapy
LGBTQ-affirmative therapy	Past life soul retrieval therapy
Psycho-shamanic therapy	Reichian biophysical therapy
Schema therapy	Vedic astrology

This is one of the unfortunate and undesirable outcomes of ideology-driven approaches to psychotherapy.

SUFFUSION OF RELIGION WITH PSYCHOTHERAPY

Note that nothing here challenges the religious views of individuals or groups. As I indicated elsewhere, I have written and published religious materials. As a person of faith, I support other persons of faith, recognizing that a faith orientation is not synonymous with a clinical orientation. Moreover, I aspire to protect the personal faith orientations of others, including those whose orientation contradicts or conflicts with or understands faith very differently from mine.

A significant but narrower observation about ideology-driven psychotherapy is that religion has become synonymous with psychotherapy for a rapidly increasing proportion of therapists. First, how may the suffusion of religion with psychotherapy be demonstrated? It may be demonstrated in at least two ways. One is the frequently reported title that numerous mental health care providers give to their approach to therapy. Some of these have already been identified in this book. For clarity, though, some of the titles are listed in Table 4.3.

As these titles indicate, several kinds of religion may be expressed as therapy. Most of the titles refer to some sort of Christian counseling and psychotherapy. So far, I have located almost 900 titles for Christian counseling and psychotherapy. Clearly, the titles provide indicators of the suffusion of religion with psychotherapy. The other way that this suffusion may be indicated involves citations from therapists' websites where they advertise their respective approaches to therapy. In the citations, the therapists make little or no distinction between their religion and their therapy. From thousands of these that have been gathered, the list below includes only 15, taken from licensed providers. The rationale for citing them here should be clear. They are cited for the purpose of illustrating the suffusion of religion with counseling and psychotherapy. They are not cited for the purpose of belittling or rejecting their faith or their religious point of view.

Table 4.3 ADDITIONAL IDEOLOGY-DRIVEN APPROACHES

Christian principles and psychology	God's saving work
Multicultural Christian counseling	Life reformation Christian counseling
Bible counseling	Bible-based counseling
Chassidic spirituality	Christian Buddhism
Christian cognitive behavioral therapy	Christian counseling
Faith-based therapy	Immanuel prayer process
Kabbalistic healing	Mormon/Latter-Day Saint therapy
Prayer counseling	Scriptural principles therapy
Theophostic prayer counseling	Word of God counseling
God-image issues counseling	Embodiment of the Divine

1. "I combine sound biblical principles with proven, professional counseling practices to help you discover the hope, healing and growth that's possible through Jesus Christ."

2. "My studies at _____ and _____ focused not only on the clinical aspects of counseling, but the Bible, understanding God's Word and the firm foundation it creates."

3. "I specialize in Christian counseling, where the principles of the Bible are integrated into therapy. In session sometimes we talk about specific Bible passages and pray in session, sometimes we don't. It really depends on what we are working on and how comfortable you are with Christian spirituality."

4. "Interwoven through all of my therapy is a Christian world view."

5. "I am a licensed professional counselor who is certified in Cognitive Behavioral Therapy through the NACBT. I am a born again Christian who strongly believes in the power of Jesus Christ and His saving grace. Through treatment, prayer and the word of God, all can overcome addiction."

6. "I am unapologetically Christian. I can work with anyone. I believe that true peace comes from a relationship with Jesus Christ."

7. "I am a Christian counselor who integrates faith into the counseling process while providing mental health services. I do not impose my views, instead I look to scripture to bring the truth of God's Word to a hurting client. In my counseling, I offer practical, biblical help, striving to keep central the transforming truth of the Gospel. I will guide you to emotional and spiritual growth as you respond biblically to your struggles, whatever they may be."

8. "Having spent decades in Christian ministry, I can offer counseling based on spiritual sensitivity and biblical principles for growth and transformation. You can gain clarity in how to move forward based on Godly wisdom. You can learn how to establish worth based on identity in Christ and relationship with God."

9. "I strive to integrate faith-based or Biblical truth with proven methods of psychological healing in order that you find physical, emotional, and spiritual wholeness."

10. "Maladaptive patterns of thoughts and actions are explored from a Biblical perspective. Though I'm a Christian, I don't intend to impose any beliefs upon you. I respect your right to your beliefs and hope you respect mine. Please understand that my views may reflect those of Biblical principles."

11. "I believe freedom comes when one replaces the lies believed about self and experiences with the Truth of God's Word and identity in Christ."

12. "Steve's goals are to help couples transform their relationships and to experience everything that God intended for their marriage. Often this includes addressing anxiety, depression, addiction, and career issues. Steve utilizes both Biblical principles as well as counseling theories and

tools to help couples begin to work on their marriages in completely new ways."

13. "My passion and commitment is to help individuals and families navigate through difficult times in their life using Christian principals [*sic*]."

14. "I became a counselor to walk alongside people—believers and non-believers alike—as they confront any issues or situations that may be preventing them from living out the plan God has for them."

15. "I am a seasoned professional and licensed therapist with a passion to empower individuals, couples, and families. I strive to assist clients in their journey to find fulfillment, joy, and peace of mind through services that promote personal and Christian spiritual growth, emotional healing, improved relationships, and purposeful living."

These excerpts from therapists' websites clearly illustrate the suffusion of religion with counseling and psychotherapy but do not necessarily define the scope and size of the suffusion. Thus, the necessary question is, "How big is the suffusion of religion and psychotherapy?" The answer takes more than one form. One form is that strongly religion-oriented institutions have adopted therapist preparation programs. For example, when the licensing for mental health care providers was established in all states, with rare exceptions—for the major types of mental health care providers by the end of the 1990s, including licensing for social workers, counselors, psychologists, and marriage and family therapists—relatively few of the institutions that prepared therapists were strongly religion-oriented. Now, the number is large and increasing as a proportion of institutions that prepare therapists. In their report about counselor training programs in the State of Tennessee, Martin and Cannon (2010) documented the changing nature of counselor preparation programs.

Another form of increase has been the extraordinary rise in strongly religion-oriented mental health care associations, such as the American Association of Christian Counselors (AACC). The AACC represents the suffusion of religion and psychotherapy in at least two ways. One is that it advances the belief that "[t]he Scriptures, both Old and New testaments, are the inspired, inerrant and trustworthy Word of God, the complete revelation of His will for the salvation of human beings, and the final authority for all matters about which it speaks" (AACC, 2018). This is to report that the source of authority for mental health care is "the Scriptures" and not other forms of evidence, such as outcomes of clinical work or research reports about clinical work. The other indicator of suffusion is that the AACC makes no distinction between professionally trained and untrained counselors. Its mission statement says, "It is our intention to equip clinical, pastoral, and lay care-givers with biblical truth and psychosocial insights that minister to hurting persons and helps them move to personal wholeness, interpersonal competence, mental stability, and spiritual maturity" (AACC, 2018).

Similarly, the National Association of Christian Social Workers (NACSW) clearly identifies with the suffusion of religion and psychotherapy. Its view is that "[t]here need not be a distinction between effective social work practice and practice which embodies Christian values" (NACSW, 2018). Likewise, the Christian Association for Psychological Studies (CAPS) holds the view that the association is "a fellowship" whose members "are united in their commitment to Christ and to professional excellence" (CAPS, 2018). In addition to the AACC, NACSW, and CAPS, the Society for Christian Psychology (SCP) espouses views that confirm its commitment to suffusing psychotherapy and religion:

> Christian psychology began in the Scriptures of the Hebrews and early Christians. Later, Christian thinkers and ministers throughout the ensuing centuries developed many understandings of human beings, using the Bible as a canon or standard for reflection. As a result, the history of Christian thought contains countless works of psychological import that offer the Christian community a rich treasure of insights, themes, and foundational assumptions upon which to ground the project of a Christian psychology. (SCP, 2018)

Along with many other mental health care associations that could be added to those already identified, these alone establish that the suffusion of religion and psychotherapy has occurred, although it is far from a universally supported occurrence. And, when these associations are added to the ideology-driven, theological schools and seminaries that allege to prepare candidates for licensing, the suffusion becomes even more solidly established. In addition to the ones that have already been described, the New Orleans Baptist Theological Seminary (NOBTS) prepares individuals for licensing as mental health care providers, as indicated in its description of its degree programs. Among its several programs that purport to equip students for careers in counseling and psychotherapy, the Clinical Mental Health Counseling program is designed "[t]o serve as a Christian counselor in a church-based ministry, social service agency, community-based ministry, or other Christian ministries. This degree fulfills the academic requirements in most states for Licensed Professional Counselor (LPC) or its equivalent" (NOBTS, 2018a, pp. 80–83).

As a theological institution, NOBTS holds theological views that it seeks to perpetuate through its training programs. Its views have major and obvious implications for the practice of counseling and psychotherapy by its graduates. Statements from NOBTS make this clear. One is its view of the Bible. Drawing from the Baptist Faith and Message, a statement of beliefs to which all Southern Baptist institutions and programs must subscribe, NOBTS offers this statement about the Bible:

> It is a perfect treasure of divine instruction. It has God for its author, salvation for its end, and truth, without any mixture of error, for its matter. Therefore, all Scripture is totally true and trustworthy. It reveals the principles by which God judges us, and therefore is, and will remain to the end of the world, the true center of Christian union, and the supreme standard

by which all human conduct, creeds, and religious opinions should be tried. (NOBTS, 2018b, p. 10)

From the same document, NOBTS states its view of marriage and the marital relationship:

> Marriage is the uniting of one man and one woman in covenant commitment for a lifetime. It is God's unique gift to reveal the union between Christ and His church and to provide for the man and the woman in marriage the framework for intimate companionship, the channel of sexual expression according to biblical standards, and the means for procreation of the human race. The husband and wife are of equal worth before God, since both are created in God's image. The marriage relationship models the way God relates to His people. A husband is to love his wife as Christ loved the church. He has the God-given responsibility to provide for, to protect, and to lead his family. A wife is to submit herself graciously to the servant leadership of her husband even as the church willingly submits to the headship of Christ. She, being in the image of God as is her husband and thus equal to him, has the God-given responsibility to respect her husband and to serve as his helper in managing the household and nurturing the next generation. (NOBTS, 2018b, pp. 13–14)

Before entering the licensure track at NOBTS, students must complete Basic Ministerial Competency Component courses (NOBTS, 2018a). The courses systematically indoctrinate individuals toward views that the NOBTS is entirely free and justified in holding. The question is whether these views contribute to mental health development among clients who may be served by licensees who complete their training at NOBTS or whether these clients may not be helped at all or may be harmed. Similarly, do these views compromise the integrity of other mental health care professionals? These questions necessarily emerges from the program description:

> The NOBTS Counseling program provides training and supervised experience in evidence-based counseling methods to help people deal with life issues in a biblically sound way and prepares students for licensure as professional counselors or marriage and family therapists. We stand on the principle that true healing comes from God. We offer a program that equips students to be instruments for the Lord, guiding people to biblical solutions to life's problems. (NOBTS, 2018a)

While questions about the preparation programs at NOBTS necessarily arise, this discussion does not aim to answer them. Attempts to answer questions about ideology-driven preparation programs will come later. For now, suffice it to observe that the suffusion of religion and psychotherapy has occurred and that NOBTS illustrates this.

At this time, the picture of licensed mental health care providers who practice some form of religious, ideology-driven therapy is an incomplete one. However, relevant facts help to create a sketch of the situation. Based on the review of approximately 60,200 websites where therapists advertise their approaches to therapy, many of them advertise some form of Christian-related or other religion-driven counseling and psychotherapy. Oddly, among the self-identified Christian counselors, a small minority indicate that they have specific training for such services. Approximately 82% of them do not indicate that they have theological training, affiliation with a theological school, a related certification, membership in a Christian-related clinical organization, or other professional preparation for advertising and delivering Christian-related professional services. Based on this observation, the question is, "On what basis do these licensed mental health care providers offer their expertise in Christian counseling?"

Another indicator of the suffusion of counseling and psychotherapy with religious practices is the emergence of independent, nonstate, usually for-profit agencies that advertise their offerings of degrees, certifications, and licenses. For example, the National Christian Counselors Association and the Pastoral Medical Association advertise such offerings on their websites. They are discussed in Chapter 6.

Within the relatively short time that we have been reviewing therapists' website advertising, the proportion who advertise some form of religion-based counseling and psychotherapy has steadily increased. Here, the reference to "religion-based" needs clarification. It is intended to refer only to treatment approaches that are conspicuously religious. Many therapists advertise that their therapy may express a "Buddhist philosophy" or that they encourage their clients to practice "Zen" or a related approach. These are not counted as religion-based.

When we first gathered information from therapists' websites in 2010 and 2011, the visibility of religion-based aberrant therapies was nearly nonexistent. At the time, we did not count them. Later, during 2014, we gathered advertising from 600 therapists' websites. Ten of them—or less than 2%—advertised an approach that was clearly religion-based. In 2018, we gathered advertising from 622 therapists' websites from four cities. The number of overtly religion-based treatment approaches was 55, or almost 9%. Recently, from reviews of additional websites, the percentage of specifically religion-based therapies rose to 14%. While the increases from 2% to 9% to 14% invite additional careful study, they appear to be significant indicators of major changes in counseling and psychotherapy. If these numbers indicate a trend, they suggest that the practice of therapy and the practice of religion are becoming increasingly synonymous. Also, the historical respectful professional distance between psychotherapy and religion has disappeared. The consequences of this disappearance need to be explored.

In what ways may psychotherapy be redefined by its suffusion with religion? What may be the consequences of this suffusion? The answers likely frustrate and disappoint professionally accountable counselors and psychotherapists. The suffusion of religion and psychotherapy blurs both religion and psychotherapy so that neither may be recognized as having integrity. It promotes the view that psychotherapy has little more than personal biases to support it. It exposes the vulnerability of psychotherapists to a burgeoning reaction to them as cultic enclaves. Whether intended or not, those who suffuse their religion with counseling and psychotherapy create and maintain polytheistic babble or, alternately, theological psychobabble. Either way, it is babble.

Numerous well-trained and appropriately credentialed individuals express their faith commitments through counseling and psychotherapy. Probably, many of them also volunteer to help feed hungry families, provide comfort and support for terminally ill individuals, or deliver other kinds of needed services. Typically, their services are expressions of faith, and they do not regard the food they share as "Christian" food.

LACK OF REGULATION

The observation about the lack of regulation of mental health care providers may seem to be self-evident. It may be easily missed, though. After all, compared with the status of mental health care providers of a few decades ago, licensing is now virtually universal. And, along with licensing, professional associations, such as the American Psychological Association, American Counseling Association, and the National Association of Social Workers, have stable memberships and political visibility. These associations and other groups, such as the American Association of Sexuality Educators, Counselors and Therapists, have their codes of ethics as sources of regulation. With all of this regulation and more, how may the lack of regulation show itself? The lack of regulation of mental health care providers is indeed self-evident, as revealed by the question: Adding another 30 dubious approaches to therapy (see Table 4.4) to the long list of those already identified in

Table 4.4 APPROACHES THAT INDICATE A LACK OF REGULATION

5-element theory	Access consciousness
Alchemical imagination	Amma therapy
An-Ra energy healing	Ancient wisdom of Ayurveda therapy
Angelic channeling	Aquatic and yoga therapies
Attachment core pattern therapy	Balametrics
Bible life coaching	Capoeira
Clairvoyant psychotherapy	Collaborative stage model (CSM)
Color puncture	Cynthia Bourgeault approach
Divine integration healing	Eastern coaching
Financial social work	Feminine power coaching
Kalla treatment	Pyramid model
Qabalistic pathworking	Radical aliveness
Raindrop and aroma therapy techniques	Rising star healing
ROAR (radically orgasmically alive reality)	Rohun purification process
Sacred card readings	Therapeutic Yeshiva

this book, who or what regulates the practice of counseling and psychotherapy, if counselors and therapists may advertise and practice these approaches?

These 30 approaches to counseling and psychotherapy appear in advertising by licensed mental health care providers. So far, this book has identified approximately 350 approaches to therapy that fall below a normative standard of care that mental health care providers should be understood to provide. Three hundred fifty unusual approaches to counseling and psychotherapy may evoke concern, if not alarm. However, this number of such approaches is a small fraction of the unusual therapies that we have found. Therefore, another observation that emerges from the evidence is that the practice of counseling and psychotherapy is almost entirely unregulated.

HIGHLY INDIVIDUALIZED APPROACHES

A review of the websites of therapists reveals odd approaches to therapy. As the material in Chapter 3 makes clear, solo approaches to therapy are much more the norm than the exception. On a massive scale, psychotherapists define psychotherapy in highly individualistic ways, as if their individual practice is, at best, only tangentially related to a larger profession.

In what ways would the professional clinical community be expected to understand and support something like energetic Asian systems, contemplative Tibetan Buddhist psychotherapy, clear belief process, rapid prayer intervention counseling, induced communications after death, knitting therapy, or most of the thousands of others that are easily located online? Probably, the answer to the question is that the professional community cannot understand, accept, support, or reject an approach to therapy about which almost every member of the community knows nothing. When the professional community knows nothing about an approach to therapy because it is a highly individualistic one, professionally accountable mental health care suffers a loss of a coherent identity, fails to exist in a meaningful way, and feels little or no pain as its credibility declines, a gradual and steady process that may be akin to the proverbial "death by a thousand cuts." Professionally accountable counselors and psychotherapists must consider whether the massive number of highly individualized approaches to therapy may precipitate the inevitable loss of credibility, if not death, of psychotherapy.

Before members of the professional community pursue and punish or aggressively modify these highly individualized approaches to therapy, they should consider whether the community itself has encouraged and in a de facto manner endorsed these approaches. For one thing, almost from the beginning of psychotherapy, the commitment to inventing approaches to therapy, especially ideologically driven approaches, expressed itself. For example, Freud did not discover an approach to therapy; he invented it. His successors in the psychoanalytic movement likewise invented approaches to therapy. Largely just like Freud, their approaches came from their individualistic interests and little more than this. And, today, students in graduate programs are often persuaded to adopt the approach of their instructors or to put on their professional plates the cafeteria offerings of multiple therapeutic approaches that appeal to them. Sometimes, the advice from instructors is cringe-worthy. "Find an approach with which you are comfortable, something that fits your personality," for example. The historical habit of inventing individualized approaches continues.

Most members of the psychotherapeutic community, including me, believe in diversity as a professional commitment, but also as a personal value. Through our sincere and entirely right desire to promote diversity, we have embodied discipline toward advancing diversity and welcoming individuals, including colleagues, clients, and others, with reference to "age, culture, disability, ethnicity, race, religion/spirituality, gender, gender identity, sexual orientation, marital/partnership status, language preference, socioeconomic status, immigration status, or any basis proscribed by law" (American Counseling Association, 2014, p. 9). All major mental health care provider associations have committed to diversity. However, they may also have inadvertently promoted others' consuming, self-serving therapies. How else may a homophobic "Christian" approach to therapy be explained? Or, how may punching bag therapy or circus arts therapy be explained? Is the conscientious and welcoming advancement of diversity a professional virtue of which some will take undue and excessive advantage? Is this a necessary risk of advocating for diversity? Is the advocacy of diversity an invitation for some therapists to invent their individualized and sometimes harmful approaches to therapy? Is the collective and personal commitment to diversity among counselors and psychotherapists a source of vulnerability through which they are victimized by their peers, who express their betrayal of their profession though self-serving and highly individualized therapies?

Highly individualized approaches to therapy appear to be mere expressions of the personal interests of the therapists who offer them. If a therapist has an interest in bouldering, the therapist offers bouldering. If a therapist has an interest in martial arts, the therapist offers martial arts therapy. If a therapist has an interest in silent hiking, the therapist offers silent hiking therapy. The list of personal interests is a very long one. It includes numerous activities, not to mention a virtually limitless number of idea-based and ideological approaches. A partial list of such activities appears in Table 4.5.

Interestingly, each of these appears in multiple forms. For example, play therapy may appear as humanistic play therapy or Adlerian play therapy, along with many other options. Similarly, equine-assisted therapy may appear as horse-informed archetypical therapy or experiential equine therapy, along with many others. And, to confuse what may be play therapy and horse therapy, there is horse play therapy.

Almost any human activity in which nonclinicians have been involved has been brought into the practice of psychotherapy. Typically, these so-called therapies express the interests of individual therapists much more than they represent responsible clinical judgment or the needs of clients. Commonly, on their websites, therapists often report about the significance of the activity in their lives. Usually, with flourishes and embellishments, their reports go something like this: "I know firsthand how powerful movies can be in facilitating personal growth because

Table 4.5 SOME ACTIVITIES THAT APPEAR IN CLINICAL PRACTICE

Animal-feeding therapy	Ballet therapy
Bible reading	Biking therapy
Boxing therapy	Chess therapy
Ceramics arts therapy	Circus arts therapy
Climbing therapy	Clown therapy
Cooking therapy	Cycling therapy
Dance therapy	Digital art therapy
Dowsing therapy	Drama therapy
Drawing therapy	Editing therapy
Equine-assisted therapy	Feline-assisted therapy
Film–video arts therapy	Firing-range therapy
Fishing	Flower-arranging therapy (floratherapy)
Fly-fishing therapy	Gardening therapy
Golf therapy	Gymnastics psychology
Hiking therapy	Hockey therapy
Journaling	Juggling therapy
Labyrinth-walking therapy	Kinetic counseling
Knitting therapy	LEGO therapy
Lifting therapy	Magnet therapy
Massage therapy	Mazes therapy
Movies therapy	Music therapy
Origami therapy	Painting therapy
Pet massage therapy	Photography therapy
Physical exercise counseling	Piano therapy
Play therapy	Poetry therapy
Pray therapy	Punching-bag therapy
Puppet therapy	Riding therapy
Riding llamas therapy	Rowing therapy
Running therapy	Sailing therapy
Sand play	Sand-tray therapy
Sculpting therapy	Singing therapy
Storytelling	Swimming therapy
Theater counseling	Trail cycling
Vibrator therapy	Video production therapy
Vocal arts therapy	Walking therapy
Wire-sculpting therapy	Yoga therapy

there are some that have meant so much to me. And, because of this, I have integrated movies into my practice of therapy." More broadly, the activities listed earlier raise the basic question, again, "What are counseling and psychotherapy?"

As a personal aside, I should say that, based on good evidence for using them, I support and believe in multiple forms of expressive therapies. However, I reject the arbitrary and thoughtless utilization of any of the therapies listed previously, when the therapy primarily expresses the interests and needs of the therapist. A therapist's interest alone in any of these activities cannot be a justification for utilizing it in therapy, no matter how strongly the therapist likes the activity or personally finds that the activity is therapeutic. Only to illustrate this idea, no matter how sincerely a therapist loves to cook, bringing clients into a kitchen to cook is not necessarily a good thing. As another example, just because a therapist experiences the physicality of lawn-mowing as therapeutic does not mean the therapist should expect clients to mow lawns as a therapeutic activity. The possibilities for including activities in which individual therapists indulge is virtually an endless list.

PROLIFERATION OF DEGREES AND DEGREE TITLES

The number of titles of academic degrees that practicing counselors and psychotherapists report may be larger than imaginable. Currently, based on review of their websites, they identify at least 53 degree titles on which their licenses were issued (see Table 4.6).

Although this list is not complete, it causes questions and concerns to be raised. Does the proliferation of degrees—titles, sizes, emphases—cause confusion for clients and prospective clients, but also for mental health care providers? Does the proliferation of degrees generate a massive and ongoing redefinition of counseling and psychotherapy? Does the proliferation of degrees accelerate the deterioration of counseling and psychotherapy, simply because it denies any reasonable likelihood of appropriate professional accountability? What, if anything, may a doctoral degree program at the South Florida Bible College and Theological Seminary have in common with a doctoral degree program at the University of Missouri? One says that its vision is "to provide a quality Christian education under the Christian tenets provided by the Holy Bible" (South Florida Bible College and Theological Seminary, 2018). The other says that its mission is "to train counseling psychology leaders who will transform local, national, and international communities by reducing social inequities through scholarship, practice, and advocacy. Rooted in a scientist-practitioner model, students develop these competencies within a culturally diverse learning community supported by strong, individualized mentorship with faculty" (University of Missouri, 2018). One appears to live in a closed, sect-like, cultic world, while the other appears to live in an open, diversity-oriented, exploratory, and world that is subject to scrutiny.

PROLIFERATION OF PROFESSIONAL CERTIFICATIONS

Professional certifications are discussed in Chapter 5. Suffice it to say that they, too, continue to proliferate. Holden and Barker (2018), for example, list 58 unusual, if not maniacally bizarre, professional certifications, no doubt causing readers to laugh about how comedically silly some therapists are and to cry about how

Table 4.6 A Partial List of Degree Titles of Practicing Counselors and Psychotherapists

Doctor of Philosophy	Doctor of Education
Doctor of Psychology	Doctor of Osteopathy
Doctor of Philosophy in Christian Counseling	Doctor of Biblical Counseling
Doctor of Medicine	Doctor of Science
Doctor of Theology	Doctor of Ministry
Doctor of Professional Counseling	Doctor of Social Work
Doctor of Marriage and Family Therapy	Doctor of Divinity
Doctor of Psychoanalysis	Master of Science in Mental Health
Doctor of Public Health	Doctor of Pastoral Science
Doctor of Pastoral Theology	Doctor of Pastoral Psychology
Doctor of Christian Counseling	Doctor of Professional Counseling
Master of Arts	Education Specialist
Master of Science in Clinical Rehabilitation Counseling	Master of Mental Health Counseling
Master of Science	Master of Science in Education
Master of Science in Mental Health Counseling	Master of Social Work
Master of Science in Professional Counseling,	Master of Arts in Psychology
Master of Science in Psychology	Master of Science in Social Work
Master of Marriage and Family Therapy	Master of Education
Master of Counseling in Marriage and Family Therapy	Master of Social Science
Master of Arts in Professional Counseling	Master of Science in Education
Master of Human Services	Master of Arts in Human Service
Master of Counseling	Master of Divinity
Master of Professional Counseling	Master of Professional Studies
Master of Public Health	Master of Science in Nursing
Master of Science in Rehabilitation Counseling	Education Specialist in Counseling
Master of Arts in Rehabilitation Counseling	Master of Pastoral Counseling
Master of Science in Rehabilitation and Mental Health Counseling	

tragically silly some therapist are. Reviews of websites indicate that the number of professional certifications, just like the number of therapeutic approaches, is indefinitely large. So far, we have located more than 1,800 certifications that appear on websites of practicing counselors and psychotherapists. The implications of this may be quite large and widespread. One of the implications is that the numerous aberrant certifications held by licensed providers contaminate and spoil professionally accountable certifications. Another, more serious, implication is that clients and prospective clients are seriously misled and placed at risk by unethical providers.

CONCLUSION

This chapter answers the question, "What was found on the websites of practicing counselors and psychotherapists?" The answers to this question came in the form of observations. No matter what the responses to these observations may be, the answers are as incomplete as they are important. Therefore, the next necessary task is to address the matter of the meaning of these observations. This is the focus of Chapter 5.

Interpretation of Multiple Therapeutic Approaches

The preceding two chapters presented a clear record of a large number of and often inappropriate approaches to counseling and psychotherapy and relevant observations about the large number. This chapter advances the discussion in the direction of attempting to make sense of the large number of approaches to therapy, including the large number of inappropriate approaches to therapy. It follows a presentation of several observations about multiple approaches to therapy, including the following:

- A very large number of therapies are advertised and presumably practiced.
- The use and misuse of technology has become a dominant influence in counseling and psychotherapy.
- Therapists advertise approaches to therapy that reveal extraordinary and irreconcilable polarities among therapists.
- Therapists advertise their services in ways that include glaring incongruities.
- In their advertising, therapists make excessive and unreliable claims.
- A small but omnipresent number of "therapists" who advertise their services lack a license to practice.
- Therapists' advertising reveals massive ideology-defined and ideology-driven approaches to therapy.
- Therapists advertise approaches to therapy that reveal the suffusion of religion and psychotherapy.
- Based on the extraordinary variety and mostly individualized approaches to therapy, the practice of counseling and psychotherapy is almost entirely unregulated.
- Because therapists advertise a large number of "solo" approaches to therapy, they apparently practice highly individualized approaches to therapy.

This chapter seeks to interpret the information that has so far been provided in this book. It extends the discussion beyond facts and observations and asks the terse but poignant question, "So, what?" Or, "What does it matter that these facts and observations currently characterize counseling and psychotherapy?" The answers to this question will almost inevitably take many shapes. For example, an easy answer is that, contrary to common expectations, counselors and psychotherapists are largely avoidant of, if not hostile toward, science and empirical evidence. Another easy conclusion is that counselors and psychotherapists struggle for significance, satisfying their interest in being right and not so much their interest in being clinically effective. Conclusions such as these may be useful or even correct speculations, but they also may succumb to the professional clinical traps that they are seeking to characterize. Instead, critical analyses and evidence are needed as predicates for conclusions.

The pursuit of answers to these questions almost inevitably leads to the conclusion that major reforms in counseling and psychotherapy are needed. Before advocating for these reforms, though, a solid evidentiary case for them should be made. Building on critical analyses and evidence, this chapter gives meaning and understanding to the facts and observations that have so far been presented. The extension of the discussion gives attention to the following topics:

Clinical training
Clinical development, competence, and professional ethics
Evidence-based approaches
Professional identity
Scope of practice
Advertising professional services
Clinical supervision
External threats
Chaos and cacophony

CLINICAL TRAINING

Training of prospective mental health care licensees needs reform. Such reform should include universally agreed upon standards for training. The extreme lack of agreed upon standards for training is presumed to contribute to the proliferation of treatment approaches, including inappropriate and harmful ones, although the evidence for this is readily available. When needed and specific reforms are discussed later in Chapter 7, the case is easily made that clinical training is wildly inconsistent. At this time and in all states, the acquisition of a license may be based on a low of about 30 graduate semester hours or a high of about 140 semester hours. Without any necessary judgment about which is correct or better, two programs at Loyola University in Chicago illustrate this inconsistency. In one program that leads to a license, the prospective licensee may complete 28 graduate semester hours, including an internship, while the prospective licensee in another program will need to complete 110 graduate semester hours. One license is for social work (Loyola University Chicago, 2019a), and the other is for counseling psychology (Loyola University Chicago, 2019b). The differential between social work preparation programs and counseling psychology programs is not unique to Loyola University. It is a nationwide differential. Despite this huge differential in their respective preparation programs, the licensee is authorized by her/his respective state to provide clinical services that are much like the clinical services that every other state-licensed individual provides. This exposes an extreme lack of agreed upon standards for preparing counselors and psychotherapists. The lack of standards is much worse than the difference in the number of graduate semester indicates.

A comparison of one preparation program with another indicates added reasons for concern about the lack of standards for training counselors and psychotherapists. For example, an online visit to Naropa University results in the University's transparent commitment to "the fundamentals of mind training, especially the practice of sitting meditation, so that inner development and outer knowledge go hand in hand" (Naropa University, 2018). Elaborating, with its emphasis on contemplative education, the University states, "Naropa University has offered contemplative education to students for more than thirty-five years. Drawn from ancient Eastern teaching philosophies and Greek educational traditions, contemplative education at Naropa provides students a different way of knowing by combining rigorous liberal arts training with disciplined training of the heart" (Naropa University, 2018). The point of a reference to Naropa University or other training programs is not so much to evaluate it, but rather to identify it as an outlier that necessarily evokes the conclusion that clinical training for therapists lacks agreed upon standards, sizes, and methods.

Another example is the Reformed Theological Seminary. At its website (Reformed Theological Seminary, 2018), the Seminary says that "[t]he mission of Reformed Theological Seminary is to serve the Church by preparing its leaders, through a program of graduate theological education, based upon the authority of the inerrant Word of God, and committed to the Reformed Faith." Giving detail to its mission statement, the Seminary says,

> Since the Bible is absolutely and finally authoritative as the inerrant Word of God, it is the basis for the total curriculum. Students are equipped with the necessary skills to understand and teach the Scriptures, developing above all a burning desire to know and do the will of God as revealed in the Old and New Testaments, for the essence of Reformed theology is a willingness constantly to conform all of life to the Word of God. Our primary distinctives are a commitment to historic Reformed theology and the Bible as God's inerrant Word. (Reformed Theological Seminary, 2018)

Much of the Seminary's graduate curriculum in its counselor preparation programs is Bible-centered. Much of its curriculum is Bible study that is not related to mental health care, but its graduates receive licenses to practice. While the Seminary may not be as well-known as, say, the University of Tennessee, Idaho State University, or other major universities that offer training for several types of prospective licensees, it produces more candidates for licensing than either one of them. As an outlier in its approach to preparing mental health care clinicians, it also betrays any assumption that clinical training for therapists results from agreed upon standards.

For comparison, another example is the California Institute for Integral Studies (CIIS). CIIS defines its mission in the following manner:

> California Institute of Integral Studies (CIIS) is an accredited university that strives to embody spirit, intellect, and wisdom in service to individuals, communities, and the earth. CIIS expands the boundaries of traditional degree programs with transdisciplinary, cross-cultural, and applied studies utilizing face-to-face, hybrid, and online pedagogical approaches. Offering a personal learning environment and supportive community, CIIS provides an excellent multifaceted education for people committed to transforming themselves, others, and the world. (CIIS, 2018)

For many who are not familiar with CIIS or "integral studies," the natural question is "What is integral therapy?" In his book, *A Guide to Integral Therapy: Complexity, Integration, and Spirituality in Practice*, Forman (2010) introduces integral therapy. He contextualizes integral therapy by positioning it as an emerging integrative therapy:

> Integral Psychotherapy represents this next, integrated stage in therapeutic orientation. Grounded in the work of theoretical psychologist and philosopher Ken Wilber, Integral Psychotherapy organizes the key insights and interventions of pharmacological, psychodynamic, cognitive, behavioral, humanistic, existential, feminist, multicultural, somatic, and transpersonal approaches to psychotherapy. (p. 2)

At a glance, Forman appears to suggest that integral therapy integrates almost every major aspect of psychological functioning into a unified approach to psychotherapy. However, he explains,

> As we will see, the Integral approach does not simply melt all of these orientations into one or seek some grand unifying common factor. Instead, it takes a meta-theoretical perspective, giving general guidelines as to when each of these therapies is most appropriate for use with a client, allowing each approach to retain its individual flavor and utility. It is because it facilitates this organization of complete systems of therapy that the Integral approach can be so useful in helping the therapist confront human psychological complexity. (p. 2)

No doubt, practicing therapists see integral therapy as having appeal because it predicts its usefulness in confronting immense human complexity. Forman extends the importance of integral therapy far beyond this appeal. He avows, "Therapists who employ this comprehensive, multi-perspectival approach will gain confidence, strengthen their client work, deepen multicultural and spiritual understanding, and improve their interactions with colleagues of different specialties and orientations" (p. 2). To achieve such aspirations, Forman states that a critically important strength of integral therapy is that "it strongly emphasizes the therapist's personal development." "Specifically," he writes, "it brings the understanding of the therapist's role into line with constructivist–developmental theory" (p. 3). Taking CIIS and Forman seriously, a necessary question is whether their approach to preparing mental health care clinicians is normative or something else.

The point of briefly describing Naropa University, the Reformed Theological Seminary, and CIIS has almost nothing to do with offering a critique of their approach to therapy. Instead, it is to acknowledge an obvious reality that much of the training for counselors and therapists comes from philosophical orientations, not evidence for what works in counseling and psychotherapy or even a consensus about a philosophical orientation about what counseling and psychotherapy should be. This is to say that each of these institutions begins with a set of beliefs about what life should be and, therefore, seek to press mental health care into the shape of these beliefs. They are not alone in such an endeavor. The same observation may be made about Adler University, the Psychoanalytic Institute of New York, and many others. For example, according to information on its website, Duquesne University emphasizes that it has been "[i]nternationally recognized for over three decades as a center for existential phenomenology" (Duquesne University, 2018). Additional examples of degree programs that begin with a set of beliefs comes from the Association for the Development of the Person Centered Approach (ADPCA). The ADPCA publishes a directory of graduate degree programs, including related courses and individuals with this specialization (ADPCA, 2018). Similarly, the same observation may correctly be made about many different approaches to counseling and therapy. Among the many examples, various 12-step approaches for substance abusers, hypnotherapy, feminist therapy, Gestalt therapy, and various attachment-related therapies could easily be added to an extremely long list of therapeutic approaches. Again, the emphasis here is not an assessment of each of these approaches to clinical work or to educational programs. Instead, it is to report that, if a preparation program is constructed to provide training in, say, Gestalt therapy, it is not constructed to provide training in solution-focused therapy. If it is built around psychoanalysis, it not built around behavioral therapy. If it built around contemplative therapy, it is not built around emotion-focused therapy. The implicative conclusion is that many preparation programs and the licensees who completed them provide clinical services based on arbitrariness much more than agreed upon standards of preparation and clinical practice.

This observation becomes critically important in trying to make sense of the current status of counseling and psychotherapy. A necessary interpretation of this observation is that training for counselors and therapists often perpetuates the belief that an ephemeral philosophical approach to therapy—usually, a highly individual choice, but also institutional commitment—has become far more important than broadly understood professional accountability and professional discipline or a profession that is based on credible evidence or a consensus about what the profession is. Surely, the long-standing belief that individual therapists can "just make up stuff" continues to be tacitly supported by these institutions. More fundamentally, though, it confronts mental health care professionals with the dreaded question of whether there is such a thing as psychotherapy. In other words, if psychotherapy can be any one approach or a combination of approaches arbitrarily selected from tens of thousands of approaches, is there such a thing as psychotherapy, both explicitly and tacitly? Analogously, if food can be sparkling sand, mud from a river bottom, French fries, ink, venison, or cardboard cutouts of vegetables, can anyone know what food is?

In addition, the critically necessary question is whether clinical training as it currently operates prepares individuals to function minimally well as clinicians. The necessity of the question arises from the evidence presented in this book, along with evidence from others, such as Spengler and Pilipis (2015), Hill et al. (2015), Goldberg et al. (2016), and others whose work has been cited in this book. More tersely, professionally accountable mental health care providers, including those who prepare other providers, "Where have we gone wrong in clinical training and why do we continue to go this way?" Or, framed in another way, if licensed mental health care providers advertise tens of thousands of approaches to their work, what implication, if any, may this have for the way they were prepared for their work in their preparation programs?

CLINICAL DEVELOPMENT, COMPETENCE, AND PROFESSIONAL ETHICS

Continued professional development and clinical competence should be agreed upon standards for practicing therapists. They aren't. Instead, professional development generally does not occur, and clinical competence continues to decline. A review of Chapter 2 strongly suggests that those who practice counseling and psychotherapy advertise numerous therapeutic approaches that fall below minimal standards of professionalism and a normative standard of professional care and likely reveals incompetence. This observation evokes the question, "Are counselors and psychotherapists committed to incompetence?" If not, why have they so clearly invited the pervasive, obvious, and capacious expression of therapies that have no basis in clinical or professional reality? Beyond observation, the judgment here is that incompetence has become an expanding feature of the practice of counseling and psychotherapy and one that is tacitly and broadly endorsed. However, this book should not be taken as the sole indicator of the failure of therapists to develop their clinical competence or to maintain their competence.

In 2011, Brown acknowledged "the meager literature on the measurement of therapist competence" (Brown, 2011, p. 8). The situation has not substantially improved since she offered her acknowledgment. Allowing for this limitation, obvious facts should not be ignored. Among these facts, as this book reports, on a large scale, counselors and therapists "just make up stuff." In addition, as a general observation, Tracey et al. (2014) state, "There is no demonstration of accuracy and skill that is associated with experience as a therapist" (p. 218), concluding that "little is known about what differentiates the more effective therapist from others" (p. 225). The echoes from their observation continue to reverberate around mental health care professions. Their observation, though, may be regarded only as an observation. The conclusions of Spengler and Pilipis (2015) cannot be so comfortably regarded. In their article, "A Comprehensive Meta-reanalysis of the Robustness of the Experience-Accuracy Effect in Clinical Judgment," among other conclusions, they write, "It is of no surprise that scholars have concluded that experience is not associated with improvements in accuracy. . . . The results of this experience-accuracy clinical judgment meta-analysis raise significant questions regarding the benefits of education, the time invested, and clinical experience for counseling and other psychologists" (p. 373). Their conclusion stands close to Haderlie's. "Although it is intuitive that the judgments made by mental-health clinicians become increasingly accurate as they gain clinical experience, research has demonstrated only minimal effects of experience on clinical judgment" (Haderlie, 2011, p. iii). Another recent and informative article is, "Do Psychotherapists Improve With Time and Experience? A Longitudinal Analysis of Outcomes in a Clinical Setting" by Goldberg, Miller, Nielsen, Rousmaniere, Whipple, Hoyt, and Wampold (2016). They conclude, "the present analyses show that, in the aggregate, therapists did not improve with more experience, operationalized as either time or number of cases" (p. 7). Their study indicates that in the aggregate therapists tend toward a slight decline in competence over the span of their practicing. Others, including Webb, Derubeis, and Barber (2010), raise questions about the place of competence in the effective functioning of counselors and psychotherapists.

Generally, with regard to competence, including clinical judgment, the conclusions about counselors and therapists should be discomfiting to them. "Despite considerable evidence that psychotherapists are not alert to treatment failure (e.g., Hannan et al., 2005; Hatfield, et al., 2010), and strong evidence that clinical judgments are usually inferior to actuarial methods (Garb, 2005), therapists' confidence in their clinical judgment alone stands as a barrier to implementation of monitoring and feedback systems" (Boswell, Kraus, Miller, & Lambert, 2015, p. 13). Confirming this, Walfish et al. (2012) have consistently found that counselors and psychotherapists hold exaggeratedly positive views of their work. The consensus among researchers and observers of counselors' and therapists' competence is that it is generally weak. Most agree that it is also declining. Obviously, accounting for the current condition of counseling and psychotherapy involves multiple and complex factors. Suffice it to say here that some of the salient factors include poor preparation programs, inadequate licensing, marginal clinical supervision, emphasis on self-expression as therapy, chaos in ethics, and poor to nonexistent regulation of professional practice.

Imel et al. (2015) conclude that incompetent therapists may not be eliminated from practicing, mainly because "these data [about low-performing therapists] simply do not exist." On behalf of clients, they argue,

> Given the state of the science in psychotherapy on therapist variability in pa-
> tient response, modern medical records technology, and ongoing pressures
> for outcome-based accountability in health care, this state of affairs is unac-
> ceptable and unsustainable. The integrity of psychotherapy as a professional
> activity, and the well-being of patients who trust us with their care requires
> that we begin the difficult work of determining how to hold therapists ac-
> countable for their performance with patients. This must be done in a
> way that protects therapists from wrongful action, but that also protects
> patients from harm (or the illusory expectation of benefit). (Imel et al., 2015,
> pp. 334–335)

The voices of concern about the lack of expertise and competence in counseling and psychotherapy continue to rise in volume and concern. Among others, the group that includes Tracey, Wampold, Goodyear, and Lichtenberg has been writing about this for several years. Perhaps their summary comment aptly describes the status of expertise and competence in counseling and psychotherapy: "We are concerned about the lack of empirical demonstration of expertise in the profession as a whole" (Tracey et al., 2015). Recently, they extend their concern and sharpen their focus when they assert that "psychotherapy expertise should mean superior outcomes and demonstrable improvement over time" (Goodyear et al., 2017, p. 54). With considerable detail about some of the parameters within which the cause of expertise in psychotherapy may be advanced, Hill et al. (2017) tersely conclude, "We need better research on expertise" (p. 33).

The discussion about expertise in psychotherapy will continue. It should continue. The compelling need for expanded and ongoing research stems from the simplest of obligations that all psychotherapists carry. It is to provide the most effective clinical service of which they are capable. At this time, the professional mental health care community cannot assure members of a consuming public that they can receive minimally competent services from counselors and psychotherapists. Lacking a capacity to offer such assurance, research on expertise and competence is obviously needed.

However, the emphasis on research of expertise may not be relevant in the larger world of the practice of psychotherapy. As this book clearly shows, the practice world exists outside the academic research bubble in which discussions about and research on expertise occur. Quite apart from these discussions and research, reform of counseling and psychotherapy must happen. Surely, all types of reform will include and express the outcomes of research on expertise and other important features of counseling and psychotherapy, but they must also include significant changes in preparation programs, extensive revision in codes of ethics, radical changes in licensing, major and intensive profession-based regulation of the practice of counseling and psychotherapy—and more. These reforms and others are discussed in detail later in Chapters 7 and 8.

If anyone, including me, can locate tens of thousands of therapeutic approaches, as they are advertised by practicing therapists, the call for reforms should be regarded as an urgent one. The question is not so much whether reform is needed, but whether reforms may be too late.

Based on identifying an indefinitely large number of aberrant approaches to therapy, to call for reform that is urgent would appear to be self-evident. An added rationale for needed reform has long been embedded in the codes of ethics of the major mental health care provider associations, which all speak to the necessity of providing treatment within the boundaries of professional competence. These include the codes of ethics of the American Counseling Association (ACA), American Psychological Association (APA), National Association of Social Workers (NASW), and American Association for Marriage and Family Therapy (AAMFT), to name a few. Without exception, the codes emphasize the importance of providing professional services that meet the standards of the profession for which the ethics have been developed.

As an orientation to its emphasis on "professional competence and integrity," the AAMFT (2018) asserts, "Marriage and family therapists maintain high standards of professional competence and integrity." Further, the AAMFT holds its members to several competence-related standards, including these:

3.1 Maintenance of Competency. Marriage and family therapists pursue knowledge of new developments and maintain their competence in marriage and family therapy through education, training, and/or supervised experience.

3.2 Knowledge of Regulatory Standards. Marriage and family therapists pursue appropriate consultation and training to ensure adequate knowledge of and adherence to applicable laws, ethics, and professional standards.

3.6 Development of New Skills. While developing new skills in specialty areas, marriage and family therapists take steps to ensure the competence of their work and to protect clients from possible harm. Marriage and family therapists practice in specialty areas new to them only after appropriate education, training, and/or supervised experience.

3.10 Scope of Competence. Marriage and family therapists do not diagnose, treat, or advise on problems outside the recognized boundaries of their competencies. (AAMFT, 2018)

The ethical standards of the AAMFT stand in solidarity with the other major mental health care provider associations and include language that is much like the language from the ACA, APA, and NASW. The following excerpts from these associations make this clear.

> C.2. Professional Competence. C.2.a. Boundaries of Competence. Counselors practice only within the boundaries of their competence, based on their education, training, supervised experience, state and national professional credentials, and appropriate professional experience. Whereas multicultural counseling competency is required across all counseling specialties, counselors gain knowledge, personal awareness, sensitivity, dispositions, and skills pertinent to being a culturally competent counselor in working with a diverse client population.

> C.2.b. New Specialty Areas of Practice. Counselors practice in specialty areas new to them only after appropriate education, training, and supervised experience. While developing skills in new specialty areas, counselors take steps to ensure the competence of their work and protect others from possible harm. (ACA, 2014, p. 8)

> 2.01 Boundaries of Competence. (a) Psychologists provide services, teach, and conduct research with populations and in areas only within the boundaries of their competence, based on their education, training, supervised experience, consultation, study, or professional experience. (b) Where scientific or professional knowledge in the discipline of psychology establishes that an understanding of factors associated with age, gender, gender identity, race, ethnicity, culture, national origin, religion, sexual orientation, disability, language, or socioeconomic status is essential for effective implementation of their services or research, psychologists have or obtain the training, experience, consultation, or supervision necessary to ensure the competence of their services, or they make appropriate referrals, except as provided in Standard 2.02, Providing Services in Emergencies. (c) Psychologists planning to provide services, teach, or conduct research involving populations, areas, techniques, or technologies new to them undertake relevant education, training, supervised experience, consultation, or study.

> 2.03 Maintaining Competence. Psychologists undertake ongoing efforts to develop and maintain their competence.

> 2.04 Bases for Scientific and Professional Judgments. Psychologists' work is based upon established scientific and professional knowledge of the discipline. (APA, 2017)

> 1.04 Competence. (a) Social workers should provide services and represent themselves as competent only within the boundaries of their education, training, license, certification, consultation received, supervised experience, or other relevant professional experience. (b) Social workers should provide services in substantive areas or use intervention techniques or approaches that are new to them only after engaging in appropriate study, training, consultation, and supervision from people who are competent in those interventions or techniques. (c) When generally recognized standards do

not exist with respect to an emerging area of practice, social workers should exercise careful judgment and take responsible steps (including appropriate education, research, training, consultation, and supervision) to ensure the competence of their work and to protect clients from harm. (d) Social workers who use technology in the provision of social work services should ensure that they have the necessary knowledge and skills to provide such services in a competent manner. This includes an understanding of the special communication challenges when using technology and the ability to implement strategies to address these challenges. (NASW, 2017)

All licensed mental health care professionals should become considerably more alert so that they may identify and assist their peers toward maintaining understanding of and fidelity to professional ethics. The need for this comes from the observation that many treatment approaches that licensed professionals offer are self-evidently inappropriate and may be harmful to clients. Each licensed professional needs to ask, "What is the professional and ethically accountable way to ensure that counselors and psychotherapists provide clinically appropriate services to their clients?" Or, "How may we effectively monitor one another so that we deliver competent services?" Or, arguing that the current "state of affairs is unacceptable and unsustainable," Imel et al. assert that the care of patients "requires that we begin the difficult work of determining how to hold therapists accountable for their performance with patients" (2015, pp. 334–335).

Currently, licensed counselors and therapists advertise and presumably practice approaches to therapy in such ways that reveal not only blatant indifference to competent clinical services but also the lack of regulation of their practices. Later, this book addresses the challenges of regulatory accountability for mental health care providers. Whether regulatory accountability becomes a feature of professional practice will likely remain an open question until the inertia around the need for it changes. Between now and then, each professional provider should be asking, "How may I be better than I am?" And, at least, the answer to this question should include consultation among those who wish to improve their services to clients, assessment of one's clinical work, readiness to welcome others who can assess their clinical work, aggressive reading about areas in which competence needs to be improved, participation in professional development events, and much more. The International Center for Clinical Excellence (ICCE), among others, has initiated ideas and practices that hold promises for clinicians who wish to improve their work, with particular emphasis on feedback-informed practice and deliberate practice (ICCE, 2018).

The discussion of professional competence in mental health care should include comments about the status of competence. In short, as many researchers and observers have confirmed, the doubt about competence among licensed mental health care providers continues to arouse disagreements, at least, and widely divergent projections about the status of competence and the collective future of counseling and psychotherapy (Boswell et al., 2015; Chow et al., 2015; Fairburn & Cooper, 2011; Goodyear, Wampold, Tracey, & Lichtenberg, 2017; Hill,

Hoffman et al., 2017; Hill, Sharon et al., 2017; Imel et al., 2015; Martin & Cannon, 2010; Miller, 2013; Overholser & Fine, 1990; Sburiati, Lyneham, & Schniering, 2012; Spengler & Pilipis, 2015; Tao, Owen, Pace, & Imel, 2015; Tracey et al., 2014, 2015). Probably, the doubt about the status of competence among counselors and psychotherapists is seriously—and possibly terminally—exacerbated by those who advertise and practice grossly inappropriate approaches to therapy.

The status of competence among mental health care providers may be an understandably confounding one. On the one side, the evidence that supports the effectiveness of counseling and psychotherapy is clear and positive. For example, the APA (2012) endorsed a resolution, "Recognition of Psychotherapy Effectiveness," with considerable documentation to support it. On the other side, as this discussion indicates, the evidence that supports the lack of competence among counselors and therapists is also considerable. The gap between these two bodies of evidence appears to be that, on the first side, evidence comes from an academic research bubble and, on the second side, evidence comes from a clinical practice research bubble. Neither bubble serves clinicians or clients well. However, choosing to believe one side or the other is not likely to be helpful or productive. And, as research continues, perspective-taking is necessary. In receiving his Distinguished Psychologist Award from the APA, somewhat wryly, Mahrer urged such perspective-taking. He commented, "Set our scientific measure-makers on the task, and they can prove the existence of schizophrenia, devil possession, elves and goblins, witches and warlocks. No problem. We have scientific measures" (Mahrer, 1999, p. 1150).

To advance perspective-taking, this discussion should refer to the fact that many therapists who advertise do so in ways of which, allowing for many kinds of understandable differences among them, most of their peers would approve. For example, a therapist who works primarily with children may commonly provide play therapy, while another therapist who works primarily with troubled adult couples may commonly provide emotion-focused therapy, and another therapist who works primarily with overanxious adults may commonly provide cognitive behavioral therapy. None of these three therapists is likely to consider the other two as providing inappropriate therapy. While the measure of such advertising and approval is not a feature of this discussion, based on a review of 60,200 websites where therapists advertise their services, approximately 50% to 55% of their websites—an admittedly rough estimate—are ones about which few, if any, serious questions or objections will be raised. Without doubt, though, the volume of ethical lapses and violations in advertising and practice exceeds the number of all other ethical lapses and violations. The lapses and violations include several that this book discusses elsewhere (e.g., promoting years of clinical experience that the individual misrepresents and offering specific clinical services for which the individual has no apparent training). The volume of ethical lapses and violations alone should provoke counselors' and psychotherapists' collective interest in reforming counseling and psychotherapy.

EVIDENCE-BASED APPROACHES

For decades, counselors and psychotherapists have recognized that their credibility depends on a few critically important assets, including professionally accountable competence, attaining markers of professional status (e.g., relevant academic degrees, membership in professional associations, professional certifications, standards of professional ethics, licensing), and having solid evidentiary support for their mental health care. During these decades, leaders of professional associations carried the awareness of the comparatively high credibility of other health care professions, particularly physicians. They understood that credibility comes from good will and good evidence and that neither good will nor good evidence alone was sufficient. The information gathered for this book, however, necessarily evokes the question, "Is therapy evidence-based?" The answer to this simple question is much too clear. Psychotherapy is not evidence-based. How could psychotherapy be seriously considered as an evidence-based health care service, when large numbers of therapists advertise treatment approaches that include alchemy, agnostic therapeutic framework, the truth of God's word, feline-assisted therapy, akashic record revelation readings, aroma freedom technique, Bible insights based therapy, and numerous other that have already been named in this book? The lack of evidentiary support for much of the practice of psychotherapy strongly suggests that a major task for professional therapists is to define, locate, and utilize the best evidentiary support for their work that they can find or, dreadfully, to work toward the termination of a dysfunctional profession. Alternatively, they may also work toward removing approaches to therapy that clearly lack evidence to support them and, possibly, the therapists who practice aberrant approaches to therapy that have no evidentiary support.

Along with hundreds of other approaches to counseling and psychotherapy identified in this book, what evidence, if any, may account for the following approaches to counseling and psychotherapy? This question may be extended to include added questions about whether any of the approaches in Table 5.1 express professional accountability and whether they may undermine counseling and psychotherapy, owing to the conspicuous lack of evidence to support them.

The lack of evidentiary support, along with the pervasive rejection of the accrued wisdom of professionally accountable therapists, places the future of counseling and psychotherapy in doubt. As this book makes clear in various discussions, some leaders among counselors and psychotherapists have begun to speak of the extinction of psychotherapy.

To turn clinical practice away from perpetuating a profession that lacks evidentiary support, many useful actions may be taken that will turn mental health care in a better direction. For now, at least, to offer a modicum of encouragement, a suggestion for practicing therapists is that they do what good research recommends, if they wish to improve their competence. One of the best actions they can take is to consult with other professionals. They may initiate ongoing relationships with other practicing therapists. Through these ongoing relationships, with emphasis on

Table 5.1 APPROACHES THAT EVOKE QUESTIONS ABOUT PROFESSIONAL
ACCOUNTABILITY

Advanced past life therapy	Amazon warrior anger therapy
Astrotherapy	Ayurveda
Bionomic psychotherapy	BLACK love
Bohemian approach	Carpal tunnel hypnotherapy
Celtic healing	Character analytic vegetotherapy
Clinical thermographor	Conscious enlightenment therapy
Contemporary Christian dream interpretation	Divorce discernment counseling
Dr. Margaret Paul/inner bonding	Duality reality therapy
Eastern-Afrikan and Western	Eduardo Duran's work
Energetic empathic processes	Energetic healing modalities
GAP counseling	Gemstone therapy
Goddess psychology	Healing love energy direction
Ho'oponopono	Huna healing
Ichthyotherapy	Integrated synergistic approach
Integrative mystical states of consciousness	Integrity model perspective
Light touch therapies	LSD psychotherapy
Male responsibility therapy	Meridian psychotherapy
Moon circle	Morita therapy
Nature-based archetypal depth psychology	Recasting therapy
Reiki energy work/essential oils	Rejection therapy

diversity of age, sex, clinical interests, experience, and other variables, they may establish and utilize a network of trustable and committed individuals, perhaps with six to eight others who can provide supportive, candid, and expert feedback. They can explore specific clinical interventions that work well, explaining how they work and how to make them better. They can give attention to their competence and how to overcome their clinical weaknesses. They can locate and share evidence-based materials, such as *Interventions for Disruptive Behavior Disorders: Evidence-Based Practices* (Substance Abuse and Mental Health Services Administration [SAMSHA], 2011). Also, they may visit the ICCE website to get ideas about ICCE's work on feedback-informed therapy and deliberate practice (ICCE, 2018). Similarly, they can get other evidence-based materials from SAMHSA, such as *Addressing the Specific Behavioral Health Needs of Men* (SAMHSA, 2013) or *Improving Cultural Competence* (SAMHSA, 2014) They may study the materials and share their gains with others. The important point here is that there are effective and inexpensive ways to develop and practice evidence-based treatments.

PROFESSIONAL IDENTITY

This book does not present a detailed or comprehensive study of professional identity among counselors and psychotherapists. Nevertheless, the information gathered from 62,200 websites of practicing counselors and psychotherapists warrants the observation that professional identity among them is, at best, fragmented and, at worse, irreparably damaged. Whether the status of the current professional identity of counselors and psychotherapists may be characterized as fragmented, damaged, or something else, the condition of professional identity is surely aggravated by several obvious features of the profession. The list below contains some of them:

- A casual search for titles of licenses for mental health care providers discloses that as many as 100 can be found. They are listed in Appendix B.
- As many as 80 codes of ethics for mental health care providers may be easily located. In Chapter 6, when proposals for reforming mental health care are discussed, a list of many of these codes of ethics is included.
- Graduate degree programs that prepare or allege to prepare individuals to function as psychotherapists range from requiring a low of 30 semester hours for social workers to a high of 140 semester hours for counseling psychologists. Despite the extreme difference in the size of their graduate programs, they allege to prepare social workers and counseling psychologists to deliver services that appear to be very much alike.
- Preparation programs encompass ideologies that stand in blatant contrast, if not contradiction, to one another. For example, a program that expresses a research-oriented and scholarly approach to training its students stands apart from a Christian, fundamentalist theological school that expresses its homophobic and misogynist beliefs. With wildly various philosophies of counseling and psychotherapy, preparation programs appear in many different kinds of institutions. Some appear in theological schools that represent almost all Christian denominations and almost all other major religions, with each one expressing its self-identified and self-limiting ideology. They appear in state-sponsored universities, with many of them building their respective programs around a selected theoretical orientation, such as Adlerian, existential, or solution-focused therapy, among others. They appear in private universities that may be formed around any of a large number of theoretical orientations, such as integral, contemplative, or other therapies. They appear in for-profit institutions, with an extraordinary variety of approaches, including professionally accountable preparation of their students for exploitive and mercenary programs.
- In addition, as the preceding chapter makes clear, practicing therapists adopt a therapeutic orientation or combinations of therapeutic orientations or "just make up" their own therapeutic orientations in a virtually limitlessly large number. To be clear, individual therapists adopt and deliver approaches to therapy that come in extremely large numbers, with many of them posing extraordinary challenges to intellectual honesty and clinical efficacy and credulity.

- Antecedent to receiving a license to practice therapy, post-degree internships or residencies cover a range from 1,000 clock hours to 6,000 clock hours.
- Counselors and psychotherapists may hold certifications in almost any area in which they have an interest. If they have an interest in career counseling, clinical supervision, addictions treatment, play therapy, art therapy, dialectical behavior therapy, sex therapy, distance counseling, animal-assisted therapy, rehabilitation counseling, martial arts therapy, or numerous other interests, they may seek and receive a certification for it. For example, among the American Association of Sexuality Educators, Counselors and Therapists (AASECT), the American Association of Christian Counselors (AACC), and the National Board for Certified Counselors (NBCC, including its Center for Credentialing and Education, [CCE]), there are 20 or more certifications (AACC, 2018; AASECT, 2018; CCE, 2018; NBCC, 2018). A low-intensity moment of an Internet search will quickly reveal that a licensed therapist advertises that she is a "Certified T'ai Chi Chih" practitioner. The importance of this is that the multiple certifications—there are hundreds, if not thousands of them—contribute to the fragmentation of counseling and psychotherapy, mainly, because broadly agreed upon standards for certifications do not exist. The review of websites of practicing counselors and psychotherapists located more than 1,800 certifications. There is little or no oversight of certifications or of the agencies and associations that grant them. Adding complication and deficiency to complexity and disparity with regard to certifications, as indicated earlier in this chapter, the overwhelming majority of therapists who practice a specialization for which certification is available do not have it.
- Copied and pasted from therapists' websites, the 20 certifications listed in Box 5.1 may be easily located. Each one appears on the website of a licensed mental health care provider.

The idea that too many mental health care providers hold too many nonrelevant, if not patently harmful, certifications may come as a surprise to most licensed mental health care providers. They should read a recent article by Holden and Barker (2018), who list more than 50 certifications in their article, "Should Social Workers Be Engaged in These Practices?" Among the certifications on their list (pp. 3–4), they include the ones in Box 5.2.

Including the certifications on their list, Holden and Barker identify more than 400 approaches to therapy about which they raise serious doubt. In addition, speaking from the context of social work education, they wonder about the future of mental health care, "if the micro-credentialing fad takes off" (2018, p.12). They conclude their article with a compelling question: "We began by asking if social workers should be engaged in these practices. If your answer is no, then what will you do about it?" (p. 12).

Box 5.1

A PARTIAL LIST OF CERTIFICATIONS FROM CLINICIANS' WEBSITES

Anger Management Specialist Certification
Certified Alcohol and Drug Counselor
Certified EMDR Therapist or Certified EMDR Teacher
Certified Family Trauma Professional (CFTP)
Certified Gottman Method Couples Therapy
Certified Group Practitioner
Certified Holistic Health Coach
Certified in Art Therapy
Certified in Classical Homeopathy
Certified in Compassionate Bereavement Care
Certified in Jungian Psychotherapy
Certified Level 2 Reiki Energy Work
Certified Personal Fitness Trainer
Certified Practitioner of Psychodrama
Certified Sex Offender Treatment Specialist
Certified Yoga Therapy Teacher
Dance/Movement Therapy Certification
National Board Certified Clinical Hypnotherapist
Nationally Certified School Psychologist
Senior Certified Gottman Therapist

These observations alone suggest that psychotherapy may not exist in a form that may justify a minimally credible definition. For now, though, the concern of whether psychotherapists may have a coherent professional identity is the emphasis. Obviously, professional identity cannot be established by a single source, such as a university preparation program, a professional association, holding a license, or affiliation with a theoretical orientation to psychotherapy. Also, with regard to an individual's professional identity, unique and individual values, decisions, and experiences contribute to shaping it. This discussion includes concern about each therapist's professional identity, but the larger concern is whether psychotherapists may have a coherent professional identity. Without question, such an identity does not exist. Worse, the extreme variability among therapists, including the factors listed previously, not to mention massive and blatant defiance of professionalism among practicing therapists, means not only that psychotherapists may lack the potential for a minimally appropriate professional identity but also that psychotherapy itself may lack definition and a useful and useable future.

My interest is in reporting about the current condition of counseling and psychotherapy and not in assigning blame. The current condition evolved over the

Box 5.2

SOME OF THE CERTIFICATIONS FROM HOLDEN AND BARKER*

Certificate in Totem Pole Imagery Process
Certified Cuddle Party Facilitator
Certified Eden Energy Medicine Advanced Practitioner
Certified Flower and Gem Essence Therapist
Certified in BrainWorking Recursive Technique (BWRT)
Certified Pranic Healing® Practitioner
Certified Reference Point Practitioner
Certified Soul Entrainment®
Certified Theta Healer
Certified in White Time Healing
Certified Zero Balancing Practitioner

*Holden, G., & Barker, K. (2018). Should social workers be engaged in these practices? *Journal of Evidence-informed Social Work, 15*(1), 1–13.

span of more than 100 years during which mental health care professionals have expressed their concern for human suffering of virtually every kind that human experience and imagination can create. Their responses have likely included many forms of effective and widely supported treatments, but treatments by licensed providers have also included feng shui consulting, business coaching, shamanic healing, vibrator therapy, compassionate de-possession, courage beyond therapy, direct therapy, divine relationship counseling, equality counseling, and five se-quence relational drawing process, along with thousands of other approaches to counseling and psychotherapy. The current condition of the practice of mental health care is that it has become a device through which an individual therapist expresses "therapy" through anything that appeals to her/him. As such, the prac-tice of counseling and psychotherapy has become a means of self-expression for counselors and therapists, much more than a clinically oriented and profession-ally accountable service. Generally, it continues to lose coherence and meaning as a profession and, possibly, to devolve toward extinction.

"Therapy" has come to express the unique, personal identity of therapists. For decades, mental health care professionals have emphasized the necessity of "pro-fessional identity" and almost always equate this with personal identity. This has been our mistake. Much more than personal identity that expresses itself in specif-ically clinical ways, the emphasis should be placed on identification with the pro-fession. The difference is that one is not clinically and professionally accountable and the other one is. Questions about professional identity will likely continue. For now, suffice it to say that, from the historical emphasis on expressing per-sonal identity through professional activities, therapy has become almost every-thing and, therefore, has come to be less relevant and obviously less professional. If a researcher can use his/her computer to examine the online advertising and relatively easily locate more than 20,000 approaches to therapy, therapy appears to be flooded and drowning in a narcissistic tsunami.

If a minimally coherent professional identity within the psychotherapeutic community existed, discussions of this kind would not likely be held. Anticipating a discussion of needed reforms in counseling and psychotherapy, the focus of attention on professional identity should be expected to shift away from individual identity that shapes clinical practice and toward the identity of the profession that shapes the individual.

SCOPE OF PRACTICE

"Scope of practice" refers to the definition of professional functioning and accountability that attends a license that is granted by a state. Among states, the definitions of professional functioning and accountability appear to be fairly consistent, even when obvious variations are taken into account, whether the license is issued to a social worker, psychologist, professional counselor, or marriage and family therapist. Drawn randomly from each of four states, with a scope of practice for social workers, another for professional counselors, another for psychologists, and another for marriage and family therapists, the following excerpts from scope of practice statements illustrate their similarities.

Social worker in Ohio: An independent social worker may provide "counseling, psychosocial interventions, and social psychotherapy," including "psychosocial assessment: intervention planning, psychosocial intervention, and social psychotherapy, which includes the diagnosis and treatment of mental and emotional disorders and counseling." In addition, a social worker may provide "program assessment, planning, and development, program implementation and evaluation"; "organizational assessment, planning and development, intervention, accountability, and supervision"; and "specialized problem-oriented assessment, specialized project or case-oriented planning, specialized intervention, [and] evaluation of consultation activities" (Licensed Social Worker Scope of Practice in Ohio, 2018). The heart of the scope of practice for social workers in Ohio appears to be the authorization to diagnose and treat mental disorders and related problems, according to established standards of care within the profession of social work.

Professional counselor in Michigan: The State of Michigan authorizes licensed professional counselors to assess, test, and evaluate an "individual, family and group" to conduct "counseling and psychotherapy" and to diagnose and treat "mental and emotional disorders." Further, licensed professional counselors conduct counseling and psychotherapy according to "counseling principles, methods or procedures," referring to "a developmental approach that systematically assists an individual through the application of any of the following procedures: evaluation and appraisal techniques, exploring alternative solutions, and developing and providing a counseling plan for mental and emotional development" (Licensed Professional Counselors Scope of Practice in Michigan, 2018). Based on "counseling principles, methods and procedures," licensed professional counselors may diagnose and treat mental disorders and related problems.

Psychologists in New Hampshire: Comparable to many other states, licensed psychologists in New Hampshire receive authorization within the following scope of practice:

> "Psychology services" means the observation, description, evaluation, interpretation, diagnosis, and modification of human behavior by the application of psychological and systems principles, methods, and procedures for the purpose of preventing or eliminating symptomatic, maladapted, or undesirable behavior and of enhancing interpersonal relationships, work and life adjustments, personal effectiveness, behavioral health, and mental health, as well as the diagnosis and treatment of the psychological and social aspects of physical illness, accident, injury, or disability.

Toward implementing psychological services according to "psychological and systems principles, methods and procedures" and extending the scope of practice statement, New Hampshire adds that licensed psychologists may diagnose and treat mental and emotional disorders, with specific reference to the current edition of the *Diagnostic and Statistical Manual of Mental Disorders*. The heart of this scope of practice appears to be the authorization to diagnose and treat mental disorders and related problems within the boundaries of normative psychological, clinical practices (Licensed Psychologists Scope of Practice in New Hampshire, 2018).

Licensed marriage and family therapists in Wyoming: According to law in Wyoming, licensed marriage and family therapy refers to "the rendering of professional marital and family therapy services and treatment to individuals, family groups and marital pairs, singly or in groups." Further, clarifying the scope of accountability, the law says, "Marital and family therapy includes but is not limited to the diagnosis and treatment, including psychotherapy, of nervous, emotional, and mental disorders, whether cognitive, affective or behavioral, within the context of marital and family systems. Marital and family therapy involves the professional application of psychotherapeutic and family systems theories and techniques in the delivery of services to individuals, marital pairs and families for the purpose of treating such diagnosed nervous and mental disorders" (Licensed Marriage and Family Therapists Scope of Practice in Wyoming, 2018). Allowing for the emphasis on "family systems theories" and other marriage- and family-related issues, the heart of the scope of practice for marriage and family therapists in Wyoming appears to authorize marriage and family therapists to diagnose and treat mental disorders and related problems.

Clearly, the scopes of practice for these major types of licenses are similar. In addition, and quite typically, the laws, regulations, and policies proscribe adherence to professional ethics. Viewed as a singular professional orientation, all licensed mental health care providers should practice within their respective scope of practice, within the boundaries of their competence, and according to professional ethics. Viewed in this somewhat succinct and general manner, questions about many of the so-called treatment approaches necessarily arise. Does any particular approach to therapy fit within a licensee's scope of practice? Can the therapist demonstrate competence in the practice of any particular therapeutic approach? Can the therapist assure a client that the practice of any particular approach to therapy comports with a normative standard of care of her/his profession? And, is the therapeutic approach an ethical one?

The professionally accountable constraints of the therapist's scope of practice, competence, commitment to a normative standard of care, and professional ethics pose challenges for those who wish to advertise approaches to therapy that distort, if not self-consciously reject or destroy, these constraints. Nevertheless, with these constraints in view, how may a therapist respond to them, if she/he advertises and practices any of the approaches to therapy in Table 5.2, along with thousands of others that could easily be added to the list?

All professionally accountable counselors and psychotherapists must maintain their rightful autonomy to practice their profession. Normative practice often involves sensitive, ambiguous, complex, demanding, and often immensely consequential clinical judgments. For these judgments, autonomy is necessary. Likewise, among therapists, disagreements about clinical judgments are also necessary. Collectively, therapists simply are not capable of unanimity of clinical judgment. Moreover, they would likely suffer failure if they succumbed to the belief that they should have unanimity of judgment among them. Diversity and discretion come with the proverbial "territory" of counseling and psychotherapy.

However, diversity and discretion must operate within the boundaries that are prescribed by laws and regulations, normative standards of practice, demonstrated clinical competence, and professional ethics. The so-called therapeutic approaches

Table 5.2 APPROACHES THAT MAY OR MAY NOT COMPLY WITH SCOPE OF
PRACTICE LAWS AND REGULATIONS

Akashic record readings	Ancient wisdom
Bible lifestyle counseling	Buddhism and the wisdom traditions
Catholic coaching	Chakra energy
Christian mind/body work—splanka	Classic hatha therapy
CMC power counseling	Cognitive bias modification
Contemplative Tibetan psychology	Eastern spiritual wisdom
Gunborg Palme's method	Hassidic spirituality
Hypnobirthing	Intuitive readings
Light laser therapy	Matrix reimprinting
Mayan healing	Resnick model
Sandra Ingerman model	Shamanic energy healing
Soteriological counseling	Spiritual root system
Thought field therapy	Uzazu coaching
Whitehawk process practice	Word of God universal truths
Young living raindrop technique	Zentangle teaching

named earlier raise doubt with regard to professional accountability. Clearly, some of these approaches seriously compromise the integrity of counseling and psychotherapy. Some of them may perpetrate harm. So, adapting the question that Holden and Barker (2018) raise, "What shall we do about this?" Their question appears to be a necessary one for the counseling and psychotherapy profession. While several varieties of failure (e.g., breach of confidentiality and violation of boundaries) to comply with their respective scopes of practice are evident, all of the failures combined are smaller than the failures in advertising and the practice of unusual approaches to counseling and psychotherapy. This observation alone should provoke counselors' and psychotherapists' collective interest in reforming counseling and psychotherapy.

ADVERTISING PROFESSIONAL SERVICES

When presenting their services to the public, licensed mental health care professionals should express their messages in language that is reasonably accessible to prospective clients, current clients, and others whose views and attitudes toward counseling and psychotherapy may be influenced by advertising. In other words, they represent much more than their personal interests. They represent themselves, of course, but they also represent all of mental health care and speak to a consuming public, including insurance companies, attorneys, government officials, and others whose views of mental health care hold significance. The failure to advertise in a professionally accountable manner occurs on a large scale and in very public ways. Consider the following examples of approaches advertised by licensed mental health care providers: whatever WORKS!, queer theory, inner humanism, the Amen method, aromatouch (massage), chakra balancing, neuropsychiatry, the truth of God's word, craniosacral therapy, Peruvian Inkan tradition, freedom counseling, young living essential oils (YLEOS), image-bearer orientation, fulfillment-oriented therapy, critical mixed race theories, tarot cards, and cellular support protocol. Quite reasonably, most prospective clients and almost all qualified professional counselors and therapists would see these as quirky, misguided, and unprofessional. After all, they are quirky, misguided, unprofessional, and likely unethical.

Similarly, how may licensed mental health care professionals expect their prospective clients accurately to interpret their approaches to therapy and to make reasonably good sense of therapy generally? The language of therapy should understandably and necessarily be very familiar to those who hold licenses to practice. This familiarity, though, may inadvertently dim their awareness of how little most clients understand about professional services. For example, if a professional speaks with a client about cognitive behavior therapy, reality therapy, internal family systems, or solution-focused therapy, what would they expect a client to understand from these references? Mental health care professionals bear the burden of effectively translating these and many other approaches to therapy to their clients. Moreover, when they advertise their approaches to therapy, counselors and psychotherapists should follow the dictum, "Say it in ways that they cannot possibly misunderstand it." To achieve the intention of this dictum, they may wish to seek and follow guidance from their respective state's regulations and laws and from professional ethics.

With regard to professional ethics and public presentations, the ACA nicely expresses needed guidance. For example, if a counselor or psychotherapist wishes to develop a new area of specialization, the *ACA Code of Ethics* says, "Counselors practice in specialty areas new to them only after appropriate education, training, and supervised experience. While developing skills in new specialty areas, counselors take steps to ensure the competence of their work and protect others from possible harm" (ACA, 2014, p. 8). With reference to advertising clinical services, the *Code of Ethics* says, "When advertising or otherwise representing their services to the public, counselors identify their credentials in an accurate manner that is not false, misleading, deceptive, or fraudulent" (p. 9). And, with regard to a therapist's credentials, the *Code of Ethics* says. "Counselors claim or imply only professional qualifications actually completed and correct any known misrepresentations of their qualifications by others" (p. 9). Although many counselors and psychotherapists are not obligated to comply with the *ACA Code of Ethics*, the guidance inherent in the code may be a valuable resource for them.

CLINICAL SUPERVISION

Clinical supervisors carry legal and ethical accountability for the clinical work of their supervisees, especially for post-degree, prospective licensees (Martin & Turner, 2020). As these supervisors review studies like the ones reported in this book, they must consider the potential consequences of providing clinical supervision for a prospective licensee who may practice, say, past life regression, alchemy, or massage, or any of the other, inappropriate approaches to therapy. Do legally and ethically accountable clinical supervisors want to be accountable for, say, neuropsychiatry, rover therapy, alchemy, lymph drainage massage, iridology, tuina massage, lanktonian psychotherapy, Torah perspective and principled living, recovery dynamics, M.E.T.A. attachment treatment, or seven challenges? If a clinical supervisee advertises any of these therapeutic approaches, her/his clinical supervisor is accountable for them. Insofar as these and numerous other aberrant "therapies" are advertised by supervisees and former supervisees, clinical supervisors may need to ask themselves where they went wrong.

Clinical supervisors also carry responsibility for what is commonly referred to as "gatekeeping." This is to say that they often determine who enters the profession of counseling and psychotherapy. Given the massive volume of truly unusual, if not blatantly unethical, approaches to therapy, clinical supervisors may need to examine their work and wonder whether they may be contributing to the collective failure to maintain diligence and protection of the "gate." Further, they may need to examine the outcomes of their clinical supervision if any of their former supervisees who are now independently licensed practice therapies that lie beyond their competence or that may be found on lists of inappropriate therapies.

EXTERNAL THREATS

In 1996, the Supreme Court of the United States decided to uphold client privilege for those who receive counseling or psychotherapy from licensed mental health care providers. The details of the case, *Jaffee v. Redmond* (1996) tell a compelling story that included a police officer who shot and killed Ricky Allen, a young man who aggressively threatened another man with a knife. The police officer, Ms. Redmond, who shot and killed Allen, sought therapy with a licensed social worker. Allen's family, the Jaffees, sued Redmond, claiming "wrongful death." As the suit evolved, the attorney for the family requested the clinical records from the social worker, Karen Beyer. Beyer refused to provide the clinical records, claiming that the records were protected by client privilege (Appeals Court, 2017). Eventually, the case was heard by the Supreme Court. The Court decided in favor of Redmond. The decision of the court extended privilege to clients of all appropriately state-credentialed mental health care providers in the country. As a result, the legal doctrine of client privilege became the law of the land. Justly, many mental health care providers celebrated the Court's decision.

However, in a discussion of therapy thieves who have stolen the integrity and reputation of mental health care, the dissenting opinions in 1996 have importance. The Court's decision had two dissenters, Associate Justice Scalia and Chief Justice Rehnquist. In response to the majority's view that client privilege served "the public good," the minority posed several arguments to the contrary. Scalia argued that deciding in favor of client privilege in this case should not be extended to cover all licensed counselors and therapists. He argued that there is no real evidence to support the belief that client privilege is needed so that clients are increasingly likely to seek and receive help. In addition to his written and reasoned legal views, Scalia spoke somewhat sarcastically about the majority's opinion (Scalia, 2018).

Scalia stated that social workers are to psychotherapy as legal aides are to attorneys (*Jaffee v. Redmond*, 1996, 2018). They are not real therapists, in his view. Moreover, he said that there is no basis for distinguishing "psychotherapists from others in society in whom people place valuable confidences." He added that "for most of history, men and women have worked out their difficulties by talking to . . . parents, siblings, best friends, and bartenders—none of whom was awarded a privilege against testifying in court" (*Jaffee v. Redmond*, 2018). Strongly implicit in Scalia's opinion is the conclusion that counseling and psychotherapy merely pretend to be health care. After all, if an individual really needs to talk with someone, he/she can do what people have always done: speak with a spouse, a sibling, a priest or other religious leader, friends, or other trusted confidants (*Jaffee v. Redmond*, 1996, 2018).

The story of *Jaffee v. Redmond* becomes a part of this discussion, with the reasoning that, if the US Supreme Court were deciding the case today, the outcome may be a very different one. To address the issue to colleagues who are mental health care professionals, how would you decide the case today, knowing that large numbers of your peers advertise and practice many different kinds of therapy that may be aptly described as "goofy," "weird," "specifically religious," "mere expressions of personal quirks," or some other characterization that defies professional accountability. To add only 30 unusual therapeutic approaches to those already identified in this book, if you were sitting on a judge's bench, how might you respond to the argument for client privilege based on the practice of the "therapies" listed in Table 5.3?

The invitation to consider the question of how a judge may react to the aberrant therapies listed in the table comes from an imagined but realistic threat. The threat is entirely plausible, but its likelihood is not predictable. However, immediate threats to the existence of counseling and psychotherapy are indeed real. Recently, for example, "an Arizona legislator [proposed legislation] to repeal the Board of Behavioral Health Examiners board and nine other boards" (National Board for Certified Counselors, 2018). If the legislation had succeeded, it would have eliminated the state's oversight and regulation of all of the helping professions and, among other consequences, terminate current licensing for mental health care providers. The legislation in Arizona failed. In Tennessee, enacted legislation and proposed legislation have seriously compromised mental health care. Often regarded as authorizing therapists in Tennessee to discriminate against others, particularly sexual minorities, the situation receives detailed attention in Chapter 7. A recent governor of Iowa sought to eliminate licensing for mental health care providers (Branstad, 2017). He failed.

Table 5.3 LIST OF THERAPIES FOR DISCUSSION ABOUT JUDGES' AND JURIES' POSSIBLE RESPONSES

Adyashanti approach	Anderson+Anderson model
Corporate psychoanalysis	Crystal healing
Deconstructionist orientation	DeTUR method
Divination	Enhanced/advanced leadership coaching
EVOX Voice Mapping therapy	Fundamental scriptural principles
Harmonic medicine	Harner Shamanic counseling
Holy Fire Karuna reiki	Induced after-death communication
ISIS practice	Light energy art therapy
Limited natural sleep	Mental feng shui
Naturopathic ministry	Peace circles
Prayer	Psychopomp work
Soul coaching	Soul psychiatry [lpc]
Splankna energy therapy	Street yoga
Thomas Hubl orientation	Genuine wisdom
Yuen method	Zensight energy healing

In Texas, a 10-year-long lawsuit brought by the Texas Medical Association (TMA) against the Texas State Board of Examiners of Marriage and Family Therapy (TSBEMFT) provides another example of an external threat. The substance of the lawsuit is that it challenges whether Licensed Marital and Family Therapists (LMFTs) in Texas can continue to diagnose mental disorders and practice independently, based on the position that LMFTs in Texas practice medicine without a license. The TMA prevailed in two Texas courts (Appeals Court, 2017), followed by an appeal to the Texas Supreme Court by the TSBEMFT.

Following a loss in the Texas Court of Appeals, Dr. Peter Bradley wrote to members of the Texas Association of Marriage and Family Therapists (TAMFT) to say, "If the current ruling holds, LMFT's (and potentially LPC's, Psychologists and Social Workers) may lose the right to render a diagnosis according to the DSM" (Bradley, 2015). In his message, Bradley added, "When reviewing the statutes governing LPC's and Psychologists, the language is similar enough that the current court's decision could place their license in harm's way." After a lengthy legal contest, the Texas Supreme Court issued its judgment in February 2017. On its website, the TAMFT reported the Court's judgment in the following manner:

> Last week TAMFT held its annual conference. This is always a time for learning, laughter and lively surprises. None was ever met with more excitement than the news of the Texas State Supreme Court unanimous ruling in favor Marriage and Family Therapists ability to diagnose. We heard the good news on Friday, February 24th and are just now able to share our thoughts with our members.
>
> The court concluded that the ability of MD's to diagnose "does not preclude MFT's from making diagnostic assessments of emotional, mental and behavioral problems." This ruling will no doubt be felt across the entire country as we breathe a sigh of relief after 7 long years of litigation. Without this decision, MFT's in Texas were on the verge of losing their jobs and programs across the 19 Universities in Texas who have MFT programs seeing a decline in attendance.
>
> This is a wonderful victory not only for those of us in the field but also for Texas Families and Individuals now and for years to come. We are grateful for the court's opinion. TAMFT appreciates their deliberate and thoughtful assessment of the situation. (Texas Association of Marriage and Family Therapists, 2017)

Naturally, the question of the relevance of this lawsuit and other threats arises. In short, the relevance is that mental health care providers correctly see that their personal and collective well-being has been placed in serious jeopardy. Further, looking at these threats prospectively, the relevance is that, if any of these threats become active in the future, the long list of aberrant approaches to therapy, along with other weaknesses, emboldens those making threats and increases the likelihood of losses for counselors and psychotherapists.

Are there other external threats to counseling and psychotherapy? Maybe or probably. In addition to the threats that may come from a stained reputation, the potential for unwelcome legislative reforms, and the kind of lawsuits discussed previously, are there are other external threats? The short answer is "yes." As explained later, some of them are charges of insurance fraud and class action lawsuits.

Another external threat is that licensed mental health care providers may be subject to charges of insurance fraud. This may be framed in the form of a question for providers to consider: If you were managing audits of payments for a health insurance company and discovered that various mental health care providers, including licensed professional counselors, psychologists, social workers, and others, were making claims and collecting payment from your insurance company for Sogyal Rinpoche, vegetotherapy, word of God counseling, unitive psychotherapy, or other obviously questionable approaches to treatment, would you want your company's money back? If you could not retrieve your company's money without a lawsuit, would you initiate a lawsuit? If you located, say, just 1,000 therapists who received insurance payments for aberrant approaches to therapy, would you claim insurance fraud against all of the therapists whose claims were paid by your company? If the number of those who practiced these aberrant approaches to therapy was, say, 60,000 or more, would you initiate a lawsuit against all of the providers who practice one of these approaches to psychotherapy? Also, quite apart from the matter of insurance fraud, how many professionally accountable mental health care providers want to share the liability insurance risk pool with those who conspicuously practice unethically? How many wish to share the liability insurance risk pool with a provider who practices symphonic healing gong or soul tracking?

Another external threat is the potential for class action lawsuits against individuals or various groups of mental health care providers. For context, "A 'class action' lawsuit is one in which a group of people with the same or similar injuries caused by the same product or action sue the defendant as a group" (FindLaw, 2018). The potential for class action lawsuits may be actualized because some unknown number of thousands of psychotherapists practice inappropriate approaches to therapy and harm clients. Some of the clients initiate lawsuits because they were harmed. This is not unusual. However, instead of suing only the offenders, these clients could unite as a class of complainants and sue all providers whose clients know about a complaint, received the harmful treatment, and join the class of complainants. In such a complaint, all providers who practiced a harmful approach to therapy may become a defendant in the case. Presumably, these providers knew or should have known that the therapy they practiced was harmful. Further, after complainants have formed a class, they could sue a group of providers who practice "under the same roof," even if most of the providers in the group have not in fact practiced one or more of the aberrant therapies that caused harm. Even if the "innocent" members of the group have not practiced one or more of the aberrant therapies, they did nothing to stop the harm about which they surely knew and may have conspired to allow the practice of such therapies.

CHAOS AND CACOPHONY

Can counselors and psychotherapists speak understandably among themselves? Can the practitioner of, say, tabletop therapy speak understandably to
the practitioner of Stephen Porges's safe and sound protocol? Or, can the advocate of Christian counseling theory hold a sensible and effective dialogue with
the advocate of therapy for religious abuse? Can a practitioner of food-mood
wellness have anything in common with a practitioner of cognition-oriented
theology? What does a proponent of the good life model have in common
with the proponent of interactive metronome training? These questions suggest that the volume of unusual and aberrant "therapies" is so large that a
common understanding of what counseling and psychotherapy may be cannot
be achieved without major reforms in counseling and psychotherapy. To emphasize the lack of common understanding, and to continue to add to the
list of unusual therapies, the same question may be reasonably raised with
regard to those who practice any of the therapies listed in Table 5.4. Do the
counselors and psychotherapists who practice one of these therapies have anything in common with those who practice the other approaches to therapy
listed there?

From these questions, an observation about the current status and critique of
counseling and psychotherapy becomes somewhat obvious. It is as simple as it is
worrisome. At this time, the large body of counselors and therapists hold virtually
nothing in common with each other. They have no shared language. They have no
shared standards through which individuals become counselors and therapists.
They have no shared ethics. They have no shared understanding of who they are
and what they do. They—we, all of us—reek of chaos and cacophony.

Table 5.4 A WIDE VARIETY OF CLINICAL APPROACHES

CAMt Christian auditory meditation therapy	Co-active coaching Gestalt
Crafting vision	Eclectic version of therapy
Family financial responsibility counseling	Geek therapy
Higher brain living	Kink aware therapy
LOCUS/CAFAS	Medical and past life hypnotherapy
MICPS	Moshe Feldenkrais approach
Muscle testing	Mysticism
Naikan	Nanj therapy
Native Mesoamerican healing	Orgone therapy
PERMA model	Personality trait theories
Pre- and post-natal massage	Professional transformational conversation
Psychic medium	Psycho-organic analysis
Yuen method	Quantum healing
Radix neo-Reichian approach	Reason-based approach
Sleep hygiene	Youth coaching

Chaos and cacophony call for reform of counseling and psychotherapy. Without reform, counselors and therapists must ask whether they have any good reasons to believe that they have a profession that is worthy of their continued involvement. Worse, they must wonder whether they may be involved in a profession that harms others.

The results—real practical, economic, and political—of chaos and cacophony are disagreements among providers, highly individualized approaches to therapy, and a very large number of likely harmful approaches to therapy. The potential results may be tersely summed in these questions: Given the large numbers of aberrant therapies and the extreme lack of cohesion among counselors and therapists, if you were deciding whether to sustain counseling and psychotherapy as a profession, would you sustain the profession? Or, would you seek to reform counseling and psychotherapy? If you would seek reform, what would reform look like?

CONCLUDING COMMENTS

The current condition of counseling and psychotherapy is declining, if not poor. Surely, the mass of details encompassed in the profession make conclusions or general characterizations hazardous, at best. Nevertheless, generally, the evidence says that the profession is threatened and, possibly, desperately needs life support. For me, the "life support" must come through needed reforms. Without major and pervasive reform, the thieves may have taken the profession, held it captive, and killed it before life-saving reform could save it.

As dramatic as these comments may appear, they call more for understanding than rejection, more for dialogue than acquiescence, more for change than maintenance, more for discomfort than assurance that everything will be okay, more for long-term endurance than momentary venting, and more for action than passive monitoring. And, insofar as the call is an urgent one, it is one that begs immediate and sustained action. Again, the question is not whether reform will happen, but who will take it on and whether it may be too late.

What Is Psychotherapy?

This chapter offers several approaches to counseling and psychotherapy that the author has not been able to locate anywhere. They are shared here, with the hope that readers may help to refine these approaches and, maybe, contribute innovative therapies of their own. Then, the chapter discusses and proposes ways to define what psychotherapy is.

> *Seasonal Secession Therapy or Migration Therapy.* This approach to psychotherapy assumes that in the northern parts of the United States the weather conditions are cold, snowy, and generally uncomfortable during the winter months and that the southern parts of the United States are hot, sticky, and generally uncomfortable during the summer months. Therefore, based on this assumption, those who live in northern states and who need psychotherapy during the winter months will need to relocate to a southern state for needed services. Likewise, those who live in southern states and who need psychotherapy during the summer months will need to relocate to a northern state. This ensures that everyone—therapists and clients—will be comfortable in the therapeutic environment.
>
> *Cold Heart Collaborative Brief Therapy.* This therapy helps individuals to invent ways to escape from unloving, cold-hearted lovers and to select a way to escape. Usually, during therapy sessions, Paul Simon's song, *Fifty Ways to Lose Your Lover*, plays in the background.
>
> *Asynchronous Melody Therapy (Reprised).* This therapy rests on a simple concept. Individuals like some songs more than others. Likewise, almost everyone manages some parts of their lives well and other parts poorly. Therapy consists of combining these two facts. The client receives a long selection of songs that she/he seriously dislikes—provided on CDs or less expensive digital formats—and instructions for listening carefully to each song. Following exposure to the most disliked songs, the client moves to a selection of songs that she/he likes a little more than the seriously disliked ones. Eventually, over a period of time of approximately a year, as the client listens from seriously disliked music to ecstatically pleasurable music, she/he adapts from distressing and disturbing aspects

of her/his life to increasingly positive experience of listening to music. As an aside, the benefits of this approach to therapists are just incredible. The therapists can relieve many, many clients of their problems and their money, merely by distributing music. Just imagine the income from having 1,000 clients each week who pay only $50 a week.

Christian, Christian, Christian, Christian Therapy (CCCCT). This therapy may just as well be called "Hindu, Hindu, Hindu, Hindu, Hindu Therapy" or "Methodist, Methodist, Methodist, Methodist Therapy." The idea is really very simple, and its outcomes are profoundly effective. In advance of receiving therapy or surely from the first moments of receiving therapy, the client is vigorously assured that his/her deeply held religious views are absolutely correct and that life is everything that he/she needs it to be. By conditioning clients to believe that their religion is the singular source of ultimate well-being and fulfillment, they do not have to change anything. Typically, Christian, Christian, Christian, Christian Therapy—or any of its subparts, such as "Jehovah's Witnesses, Jehovah's Witnesses, Jehovah's Witnesses, Jehovah's Witnesses Therapy"— is brief. From a professional well-being point of view, the potential for income security may be clearly stated. After a therapist "gets inside" of a religious group by dogmatically and enthusiastically affirming its beliefs—no matter how skewed or bigoted or shallow its beliefs are—the members of the group will just keep coming. Naturally, a therapist should take good care to affirm the religion that the client needs affirmed. Just imagine the embarrassment and loss of income when the therapist says, "Baptist, Baptist, Baptist, Baptist," and the client says, "But I'm Jewish."

Past Life Money Retrieval (PLMR). PLMR—or sometimes referred to as "Palmer"—draws from ancient practices but has been modified to accommodate contemporary values. Whereas in the ancient past PLMR required human sacrifice, usually crucifixion or beheading, as a prerequisite for the successful retrieval of money from a past life, the contemporary practice of PLMR requires only that an economically impoverished individual or family consign one of its members to the tiring job of being a permanent blood donor. The proceeds cover an agreed upon portion of the therapist's fees. Only individuals who receive certification from the International Institute for the Abundant Accrual of Earthly Riches Through Past Life Money Retrieval, Inc. (IIAAERPLMRI) hold authorization to practice PLMR. Practitioners of PLMR continue to be amazed when clients take the power of suggestion about retrieving money from a past life and commit to pay for therapeutic services long beyond any reasonable expectation that they will actually receive money. The power of suggestion and the power of promise are indeed powerful, calling on the highest level of thespian skills of which PLMR therapists are capable. Still, with the kind of money that PLMR therapists take home, they would be foolish to discontinue their extraordinarily lucrative practice of PLMR.

NOW, THE REAL BEGINNING

The "therapies" described earlier are not real therapies. However, as easily invented as they are—just like kismetic psychotherapy with which this book began—they make a statement about the current condition of mental health care. At least, they evoke questions about possible ways that psychotherapy may be formulated. If these six therapies are not real therapies, as most therapists would readily acknowledge, how should counselors and therapists think about what psychotherapy is, instead of what it is not? What is psychotherapy? This is the question that this chapter is designed to answer. The need to define psychotherapy imposes a demand that far exceeds the satisfaction of composing an entry for a dictionary. Instead of a proper dictionary entry, the definition of psychotherapy that counselors and psychotherapists need is one that defines their lives and adequately expresses the meaningfulness of what they do. Surely, for those who live the life of a counselor and psychotherapist, their definition of psychotherapy encapsulates their lives. They feel the reality of their training, their anguished progress from observing the distant goal of being a psychotherapist to being a psychotherapist, their chiseled and often affectionate professional biography that their clients unknowingly caused to take shape as it did, their anticipation of living a live full of relational adventures, their fulfilled and reassuring self-discipline, their embodiment of service, and so much more. The best definition, then, is not so much intellectual clarification, as good as such clarification may be, but, at the risk of being more poetic than this discussion calls for, it is the fresh and renewed taste of the most nourishing, delicious, shared, and sustaining food that an individual has ever had, recognizing that her/his hunger may be morally satisfying only when the food belongs to all who need it. Therapy and the health that comes from it belong to everyone. Viewed this way, psychotherapy becomes the process through which individual therapists identify with and become members of the large human struggle for significance in their corporate and sometimes confused need for health.

The expectation is that the critical importance of the question, "What is psychotherapy?" and its answer will expand and add details to the implied definition of psychotherapy that has already been developed throughout this book. More than this, though, it will stand as a heartfelt affirmation of those who fulfill the nobler mandates of their lives as psychotherapists.

As noted earlier, at this time, the professional mental health care community cannot assure members of a consuming public that they can receive minimally competent services from counselors and psychotherapists. As discussed elsewhere in this book, several researchers have concluded that mental health care providers do not generally improve with experience, but instead they decline in competence. And they conclude that, at this time, the procedures for eliminating or improving competence among practicing providers do not exist. To be clear, if, say, only 30% of practicing therapists cannot provide minimally competent services, the other 70% cannot assure a consuming public that they can receive minimally competent services.

The problem of incompetence spreads well beyond those who provide clinical services. Beginning with the incapacity of assuring others that mental health care providers are competent, related others are clearly involved. Specifically, along with practicing clinicians, if those who prepare clinicians for their work cannot assure a consuming public that they can receive minimally competent services, they should ask whether there are others whom they cannot assure. Can they assure legislators that they are competent? If they cannot offer such an assurance to legislators, what should legislators do about this? Can they assure government agencies, such as the Department of Veterans Affairs, that they are competent? If they cannot offer such an assurance to government agencies, what should government agencies do about this? Can they assure licensing boards that they are competent? If they cannot offer such an assurance to licensing boards, what should licensing boards do about this? Can they assure judges and juries that they are competent? If they cannot offer such an assurance to judges and juries, what should judges and juries do about this? Currently, lacking a sufficient basis for assuring a consuming public and others that they are competent and that they are not misleading or doing harm to others, what must counselors and psychotherapists do? What should their professional associations do?

Aiming toward recommending specific reforms in counseling and psychotherapy and basing this discussion on the evidence already presented, counselors and therapists must answer a simple question, "What is psychotherapy?" Without an answer to this question, and an answer to which almost all counselors and therapists agree, the prolific practice of "just making up stuff" and practicing various ideology-driven and sometimes harmful approaches will continue. Merely agreeing about what psychotherapy is, though, is a beginning of a long process, not the end. Such an agreement begins the process of assuring that psychotherapists can determine what is therapy and what is not therapy. They can determine that their work is what they say that it is. They can determine how to prepare individuals to become psychotherapists. They can determine how psychotherapy is to be regulated, including regulation by states and other jurisdictions and by professional ethics. They can design preparation programs that express the agreed upon definition of psychotherapy. The hope on which this chapter, along with the rest of the book, rests is that it will begin an expanding dialogue among psychotherapists and that the dialogue will eventuate in several specific reforms.

Without making such a beginning and without needed reforms, professionally accountable psychotherapists must reckon with necessary and uncomfortable questions. What degree of insensitivity to the needs of clients would allow aberrant therapies and harmful therapists to continue? What degree of compromise of conscience would allow these things to continue? And, somewhat less worse, why would counselors and therapists allow themselves to participate in a profession that would inevitably harm them—their reputations and their incomes, at least? As cited earlier, "The integrity of psychotherapy as a professional activity, and the well-being of patients who trust us with their care requires that we begin the difficult work of determining how to hold therapists accountable for their performance with patients" (Imel et al., 2015, pp. 334–335). Holding counselors and therapists accountable for their performance, for the most part, simply is not done. While this book is not the sole source, it surely adds evidence that supports the conclusion that the mental health care community has failed to monitor itself and to hold its members accountable.

The protection of the integrity of psychotherapy assumes that psychotherapy is a known and understood activity. As Mahrer (1999) stated, though, "Maybe, there is no field of psychotherapy" (p. 1148). He refers to multiple types of training programs for psychotherapists, multiple titles for degrees, and many titles for psychotherapists, among other sources of his questioning about whether there is such a thing as a field of psychotherapy. While his explanation may be different from the explanations in this book, Mahrer and this book have a common view that if psychotherapy is everything, it is nothing. Illustratively, drawing from a virtually limitless list of so-called therapies that are advertised by licensed mental health care providers, how may the therapies in Table 6.1 define psychotherapy?

So far, this book has identified approximately 650 approaches to psychotherapy, including a small number of commonly accepted approaches and a much larger number of unusual and aberrant approaches. Many of the unusual approaches to therapy likely are also harmful to clients on whom they are imposed. Appendix A includes the complete list of therapies that are referenced in this book.

Even a casual reading of the list of approaches to therapy that are included in this book evokes responses. Among the responses, surely, is the response that says no one knows what psychotherapy is. Another likely response is that psychotherapy is in jeopardy. And, another is that psychotherapy desperately needs

Table 6.1 VARIOUS EXPRESSIONS AND CONCEPTIONS OF PSYCHOTHERAPY

Advanced progression counseling	Advanced spiritual psychology
Brian Weiss/past life regression therapy	Chinese and ayurvedic medicine
HALO light therapy	Harmonial stress management system
Hedy and Yumi imago therapy	Hypnosis: age and past lives regression
IPN/PRN expertise	Iridology
John of God crystal healing bed	Life SOULutions therapy
Maya womb healing	Muscle testing
Music and aroma therapy	Muslim therapy
Physical fitness/health	Popular psychology
Quantum techniques	RAAD model for interventions
Sistahpeaceful life-style changes	Social location therapy
Sociotherapy	Soul wisdom therapy
Spiraculturally based	Spirit-in-nature flower essences
Spiritual divination	Spiritual emergence seekers
Sufi psychology	Surrogacy communication (TM)
The 11 principles of transformation	Transformation/transcendence techniques
Trigger point therapy	Ubuntu philosophy
Unconventional therapy	Watchwaitwonder
Wildish work	Wisdom recovery
Womin-affirmative	Yandara yoga

reform. The responses to and discussion of the approaches poses a vast array of possible additional responses. All of this makes a discussion of this kind even more formidable. While metaphors may invite discussion, they can also be a means of escape from intellectually accountable discussion. Allowing for this caveat, these responses may be framed by a metaphorical question: Can psychotherapists share a home that lacks a foundation, reeks of a gas leak, and houses both lung-pure health pursuers and dedicated chain-smokers? Presumably, such a home cannot forever be occupied without significant changes—reforms. The first step in reform is to ensure that the home has a solid foundation. In this book, the foundation begins with a clear conceptualization of what psychotherapy is not and proceeds to define what psychotherapy is.

WHAT PSYCHOTHERAPY IS NOT

Psychotherapy is not ideology-driven. Currently, the practice of psychotherapy appears to be driven more by ideology that is unrelated to the appropriate practice of psychotherapy than it has ever been. A substantial body of evidence supports this observation, including the evidence compiled for this book. Another illustration of this appears in the following paragraphs.

The duplicity of intellect and professionalism—becoming a part of a profession while simultaneously denying and possibly harming its integrity—may be seen in many institutions that allege to prepare individuals for professional roles in counseling and psychotherapy. Another example of this is the Biblical Theological Seminary (BTS), in Hatfield, Pennsylvania. The Seminary's view of its role is that it is first a "missional" theological school or an evangelical school. It encourages prospective students: "Our focus on missional theology and missional training will help prepare you for ministry no matter where you serve." Further, "where you serve" includes licensing as a professional counselor. As the Seminary says, the Master of Arts degree "is designed for students who wish to pursue licensure in the Commonwealth of PA." It indicates that students in the program will "learn effective evidence-based counseling skills from faculty who are practicing clinical counselors." Virtually nothing in its curriculum indicates that the Seminary is prepared to uphold "evidence-based counseling skills." Instead, upholding its evangelical, biblical tradition, it states that "BTS prepares them to skillfully apply the grace and truth of the Gospel first to their lives and then to those of their counselees" (BTS, 2018).

Along with many other institutions, BTS is duplicitous in its approach to counseling and psychotherapy. It utilizes counseling and psychotherapy as devices for narrowly religious expression. Surely, while there may be many reasons to disagree with BTS's religious views, none of those reasons should cause its views to be banned or even criticized. Its supporters should possess their rightful prerogative to believe what they wish to believe and to practice their faith as they see fit. However, just as they would never conclude that their religious views would qualify them to practice, say, chemical engineering or dermatologic medicine, they should not succumb to the easy and inappropriate rationale that their religious views somehow authorize them to practice counseling and psychotherapy. On the other side of this proverbial "coin," they are entirely within their religious prerogatives to reject any or all aspects of counseling and psychotherapy. After all, rejecting counseling and psychotherapy is what they have tacitly done. On behalf of intellectual honesty, basic human ethics of right and wrong, and respect for an evidentiary base for psychotherapy, neither BTS nor any other school of its kind should, on the one hand, claim to be "evidence-based" in its approach to psychotherapy and, on the other hand, deliver an approach to psychotherapy that applies "the truth of the Gospel" to clients. This is very much like claiming that they will seek medical advice from, say, a cardiologist and accept cardiac treatment as it was practiced in the first century or the way it may be practiced by, say, a potato farmer.

As an important aside, I wish to acknowledge that I am a person of faith. A small expression of this is my recent book (2019), *Notes to Grieving Friends: What to Say and Do When Their Loss Challenges Your Faith*. As a mental health care provider, I conclude that my faith may motivate me to provide care, but that it cannot define or prescribe care. For me, just as there is no "Muslim MRI," "Christian cardiac surgery," or "Jewish aspirin," there is no specifically religious psychotherapy. Additional analysis of this conclusion appears in the next chapter.

To agree with the assertion that psychotherapy should not be ideology-driven poses difficulties for therapists who reject the approach of the BTS, while showing positive enthusiasm for humanistic, feminist, psychoanalytic, Gestalt, or any of many other ideology-driven approaches to therapy. After all that may be said in their favor, many approaches to therapy are ideology-driven. Along with the approach of ideologies that clearly operate outside of the mainstream of psychotherapy, such as the approach of the BTS, humanism, feminism, psychoanalysis, or Gestalt therapy is no less ideology-driven. Among other implications of neutralizing the adverse influence of ideologies, psychotherapists must review how they train other psychotherapists. Should a curriculum include a course or two or three on historical and commonly accepted ideologies, such as the ones named in this paragraph? Or, instead, should they include formidable demands on their students for compliance with treatment that stands solidly on the best evidence for treating, say, post-traumatic stress disorder, marital schism, or whatever the clinical need may be? In other words, should preparation for the practice of psychotherapy shift the emphasis away from traditional theories to a coherent body of evidence that researchers have been establishing for decades?

Psychotherapy is not a uniquely personal expression. In Chapter 4, a list of therapies that very likely emerged only from individual interests included ballet therapy, chess therapy, editing therapy, firing-range therapy, knitting therapy, origami therapy, sailing therapy, vocal arts therapy, juggling therapy, video production therapy, gymnastics psychology, and other personal interests of therapists. Following a review of thousands of websites where counselors and psychotherapists advertise their approaches to therapy, an obvious and necessary observation is that their practice of psychotherapy is a highly individualized pursuit. More than expressing the profession of which they are part, therapists express themselves. More than integrating solidly good evidence into their practice of therapy, they express themselves. More than responding to the specific clinical needs of their clients, they express themselves. And, more than doing clinical good, they express themselves and, no doubt, sometimes help their clients.

Probably, all of the interests in the previous paragraph become compelling and enjoyable ones for many individuals. Insofar as these interests engage therapists as individuals, they should be encouraged or, at least, tolerated. As expressions of self-interest, the activities have understandable importance. Still, as clinical interventions, the fact of self-interest is, at most, a marginal consideration. Therefore, more about this should be said. The therapists who demonstrate conspicuous and dubious self-interest in their practice of therapy are a largely unknown number or percentage of therapists. A review of the advertising on their websites does not allow this number or percentage to be precise. For example, if origami therapy may be judged to be merely a highly individualized expression of the therapist's interests, by what criteria is this determined? Likewise, what about music therapy or some of its specific manifestations? How about vocal arts therapy or singing therapy or listening to music? Whether the precise number or percentage of therapists is known, it is more or less clear that, more than providing appropriate clinical services for their clients, many of them place a higher priority on their own needs.

Therapists who express their uniquely personal interests, such as chess therapy or ballet therapy, serve themselves more than they serve the interests of their clients. However, those who place a high priority on their personal interests join others who appear to be ill-prepared for the work of psychotherapy. Unlicensed "therapists" who advertise online should be added to this number. They represent another 9% of those who advertise through websites. Another significant number of therapists who currently deliver some sort of counseling and psychotherapy is the large number of untrained providers who work in social service agencies. They are largely hidden from the professional psychotherapy community. Because they are untrained, generally they hold no memberships in professional associations. They are not alumni of professional preparation programs. Many of the agencies that employ them are licensed by their respective states or are contracted with their states to deliver counseling and psychotherapy services. We estimate that these untrained counselors or therapists constitute another 20 to 30% of all counseling and psychotherapy services. Another category of providers who appear to lack relevant training and competence are those who are "licensed" by private agencies, such as the National Christian Counselors Association, International Ministry Institute, and Pastoral Medical Association. These agencies advertise that they offer nonstate licenses and degrees that have no credible accreditation status. Taking these categories of individuals who advertise their services, a rough estimate is that they provide 40 to 60% of all mental health care services.

Psychotherapy is not "everything." Recognizing that most are inappropriate, more than 600 approaches to therapy have been identified in this book, so far, and more will be added (see Appendix A). Based on the number and extreme variety of approaches to therapy that can be easily located at the websites where practicing therapists advertise their approaches to therapy, a clear and unambiguous reality is that collectively therapists try to treat every conceivable human problem. They not only treat every conceivable human problem, but they also invent every conceivable way of thinking about how to conduct therapy and every conceivable method for doing therapy. For the purpose of affirming this reality—and keeping in mind that there are thousands more similarly peculiar approaches to therapy than those advertised by licensed mental health care providers—Table 6.2 adds another 30 unusual approaches to therapy.

What do counselors and psychotherapists not treat? The short answer is almost nothing. Worse, they treat almost every conceivable human problem, but they provide treatments that confuse and likely harm many of their clients. They treat excessive anxiety, but they also balance chakras. They help with marital schisms, but they also help with past life soul retrieval. They facilitate self-management skills with impulsive children, but they also engage them in Bible study. They treat anguished depression, but they also engulf naïve clients with energy alchemy. They offer support for suffering during overwhelming changes in the lives of their clients, but they also offer ancient Chinese medicine. They help to create hope for an overly angry mother, but they also impose faith-based ABIDE debriefing and renewal. They help the suffering perfectionist to lower his intensity to a manageable level of comfort, but they also confuse him with advanced progression counseling. They listen to their clients with finely tuned empathic awareness, but they also practice the nine rites of munay-ki. And, all of these treatments, along with thousands more, come from licensed mental health care providers. So, what do counselors and psychotherapists not treat?

Obviously, counselors and psychotherapists treat almost everything, as this book indicates. Worse, lacking scrutiny and oversight of their services, they

Table 6.2 ADDED UNUSUAL APPROACHES TO COUNSELING AND PSYCHOTHERAPY

9 Rites of munay-ki	African and Native American rituals
Arno temperament profiling	Biblical nauthetic principle
Biofield tuning	Chakra realignment
Chinese herbal medicine	Christian dream interpretation
Energy alchemy	Existential personalistic anthropology
Formational prayer counseling	Grief yoga
Horticultural therapy	Intuitive healing
Jin shin do bodymind acupressure	Karuna reiki practice
LEGO-based social skills	Marconics no-touch healing
Muzen mindful creative	Neurocise methods
Radical compassion	Reichian reiki
Secret garden healing	Splankna (Christian-based)
Strategic whole-life assessment	Three-to-one couples therapy
Traditional bhakti music	Vibrational sound therapy
Wisdom of God's word	Yoga and mudra exercises

sometimes provide treatment that confuses and harms clients. They do this on a massive scale. Analogously, collectively, counselors and psychotherapists appear to be like, say, a civil engineer who designs attitudes toward sunsets that occur in the month of August. Or, they appear to be like a plumber who installs fingernails. Or, they appear to be like a pediatrician who prescribes only ancient Chinese peanut butter as medication for the children she/he treats, regardless of the medical needs of the child. Or, they appear to be like a roofing contractor who installs guided imagination roofing materials. Collectively, licensed counselors and therapists appear to be as dysfunctional as their clients, if not more.

Why do professionally accountable counselors and psychotherapists allow these disturbing approaches to therapy to continue unchallenged? Why do they allow themselves to be compromised by their peers who lack professional accountability? What may be done to regulate the practice of counseling and psychotherapy so that these inappropriate approaches to therapy are eliminated? How do counselors and psychotherapists reasonably conceive of some of these bizarre approaches to therapy? And, why do counselors and psychotherapists believe that they can treat any and all problems that beset the human condition?

Psychotherapy cannot address all human problems, including the ones about which individual therapists may feel deeply and strongly. Apart from their role as therapists, individual therapists may invest much of what they have, including their money, time, and influence, toward addressing a social issue, such as criminal justice reform, emergency financial assistance for sufferers from tornado damage, education reform, or the nagging challenge of feeding hungry children and their families. None of these issues, though, is one that psychotherapy can or should directly address. Of course, therapists should treat sufferers and victims of these social inequities and maladies. And, as opportunities arise, they should advocate for social change toward eliminating these maladies. Analogously, an oncologist should treat cancer patients, but should not confound treatment with engaging in advocacy that, say, seeks to eliminate the products of a manufacturer whose products are carcinogenic. In the practice of psychotherapy, affirming the overanxious client who suffers from her employer's misogynist practices should not be confused with political activism that seeks to eliminate employment-related misogyny.

Psychotherapy is not "just made up stuff." Although the total is not precise, huge numbers of practicing counselors and psychotherapists make up approaches to psychotherapy. Insofar as large numbers of counselors and psychotherapists offer approaches to therapy that have no basis in clinical reality, they steal therapy from professionally accountable therapists. Moreover, many of them harm their clients through the practice of aberrant therapies. Such random and nonclinical inventions lack profession accountability, compromising the integrity of those who are professionally accountable. This book includes numerous examples of highly individualistic "just made up stuff" that is offered as approaches to counseling and psychotherapy.

Psychotherapy is not academic. If those who deliver preparation programs and provide clinical supervision for prospective counselors and psychotherapists had complied with high clinical standards, along with high academic standards, would the mental health care community be suffering its current multiple embarrassments? Surely, in his article about some of the major embarrassments of psychotherapy, Mahrer's (1999) prescient description of the condition of psychotherapy makes this clear. He refers to therapists' attempts to cure "unreal fictions" (p. 1150). This likely includes numerous therapists' attempts to cure lost souls from previous lifetimes or to repair damaged and ill-defined "self-esteem" and other such unreal fictions. Despite widespread and commonly accepted belief in graduate programs that prepare prospective counselors and psychotherapists, neither their instructors nor the graduates can competently define numerous basic terms, such as "psychotherapy" or "mental health." Or, as Mahrer wryly observes, "Psychotherapy is a pseudoscience of non-existing unrealities, measured with rigorous precision" (p. 1150). He continues, "Set our scientific measure-makers to the task, and they can prove the existence of schizophrenia, devil possession, elves and goblins, witches and warlocks. No problem. We have scientific measures" (p. 1150). Professionally accountable mental health care providers must wonder whether their profession has arrived at a time when it is losing credibility. If the loss of credibility aptly describes the current condition of the practice of psychotherapy, how has it sustained the illusion of a profession? Probably, it has done this because of its connection with reputable academic institutions and not because of its clinical efficacy or credulity. Psychotherapy has maintained the mirage of accountability much more by being academically smart than by having appropriate clinical prowess. Through the visibility of is connection with academic institutions, psychotherapy has falsely concluded that academic achievement almost inevitably produces clinical proficiency and effectiveness. The evidence for general clinical proficiency and effectiveness is lacking. The evidence for academic proficiency and effectiveness is indisputably abundant. In the context of the information in this book, the issue may be framed in the following manner: If preparation programs were working well, effectively preparing counselors and psychotherapists, would tens of thousands of their graduates and licensees advertise so many unusual and conspicuously inappropriate approaches to their work?

The things that psychotherapy is not may be indefinitely long. The ones presented previously may be sufficient, for now. Also, for those who try to observe the condition of counseling and psychotherapy, these emphases may appear to be self-evident. However, if they are taken seriously, they hold important implications for the future of counseling and psychotherapy. The implications, however, build on a clear view of what psychotherapy is more than what it is not.

WHAT PSYCHOTHERAPY IS

Psychotherapy is about health. Counseling and psychotherapy emerged from social service and social sciences. Beginning in universities as a tool for studying human beings, *psychology* slowly morphed into counseling and clinical psychology. Beginning in schools and guidance programs, *professional counseling* morphed into professional mental health counseling. Beginning mostly as government employees, *social welfare* workers morphed into clinical social workers. Beginning mostly as *pastors and other religious leaders,* the membership of the American Association for Marriage and Family Counselors morphed into increasingly sophisticated therapists and became AAMFT. Although they came from widely different histories, these four groups of mental health care providers have demonstrated their interest in and concern about providing services for their respective clients. However, a historical failure among practicing counselors and psychotherapists has been that they have not defined or sought clarity about what mental health is. As Hightower (1988) and others have noted, "Psychologists [along with the others] have not devoted enough of their energies to the study of psychological health" (p. 527). He adds, "We have too seldom asked how the healthy person lives" (p. 527). Instead of asking and answering this basic question, counselors and psychotherapists have succumbed to their personal needs for individual and professional fulfillment, likely assuming that they may be models of mental health and relational well-being. In clinically practical terms, how would, say, goal-setting in therapy be different if practicing clinicians had clear, identifiable, useful, and sustainable information about mentally healthy individuals? Moreover, instead of presuming that members of the mental health care community are themselves mentally health, how would their clinical services improve if mental health were defined and necessitated in its members?

Among others, Carl Rogers (1961, 1963), Eric Erikson (1950), Gordon Allport 1955, 1961), and Abraham Maslow (1970), have sought to develop and encourage others to develop sophisticated conceptualizations of mentally healthy individuals. Beyond their historically important efforts, more is needed. All practicing mental health care providers need and should possess clinically useful definitions and criteria for measuring mentally healthy individuals. While this would not immediately convert them into highly effective providers, it would likely move mental health care providers toward this necessary aspiration. Simply stated, knowing what mental health is increases the possibility that providers and clients alike can achieve it.

Psychotherapy is health care. To separate psychotherapy from health care and make it merely an expression of one's personal interests or the interests of bringing individuals into a loving relationship with the clinician's preferred god denies that psychotherapy is health care. As health care, psychotherapy gives primary attention to helping others to establish and maintain health—mental health.

Defining psychotherapy as health care carries important consequences. One is that it aligns the work of counseling and psychotherapy with the larger health care community. As such, it shares the principles, practices, and general outlook of the larger health care community. Specifically, psychotherapy holds itself to the standards of health care that make it compatible with the standards of health care of, say, cardiologists or oncologists. Suffice it to say that, if cardiologists instructed their patients to massage grape jelly into their kneecaps, the advice would likely

cause cardiologists to lose their licenses to practice and perhaps suffer other adverse consequences. In mental health care, though, when conspicuously bad
approaches to therapy appear frequently in the advertising and practices of large
numbers of counselors and therapists, nothing of consequence happens, except for
the erosion of their credibility, the potential severe compromising of the profession of which they are members, and the elevated risk of harm to their clients. In
contrast to the standards of the larger health care community, when therapists advertise one or more of the following so-called therapies listed in Table 6.3, do they
move mental health care toward higher standards or lower standards? Do they
enhance the credibility of mental health care or erode it? Do they make mental
health care a more integrated health care service or a more fragmented one? Do
they establish circumstances that lead to a secure professional future or preclude
the possibility of a secure professional future? Or, speaking directly to counselors
and therapists, would you encourage a member of your family to receive needed
therapy from a therapist who advertised one or more of the therapies in Table 6.3?

The conclusions here are unpleasant ones. Promotions such as these readily
ensure that psychotherapy will forever stand outside of the intellectual honesty
and professional accountability of the larger health care community. They justify
rising skepticism about mental health care's integrity and ethical viability. They
encourage removal from the larger health care community. They continue to
erode the potential for a vibrant future for mental health care.

Instead of self-promotion, the fundamentally important criterion by which
mental health care should be judged is whether it is effective in alleviating suffering
and restoring health. To comply with this criterion, all therapists should view
their work in terms of disciplined, objective, profession-based, carefully measured outcomes. Conversely, they should reject highly individualized expressions
of their uniquely personal and likely aberrant approaches to counseling and

Table 6.3 APPROACHES TO THERAPY THAT MAY NOT BE HEALTHY ONES

3-D energetics	Afterglow curriculum
Anatomic therapy	Aquatic therapy
Aura therapy	Biblical Christian sex therapy
Christian theoretical foundation	Enneagram/Myers Briggs
Human givens trauma rewind	HUSO pro
Indigenous healing practices	Integrative Biblical principals [sic]
Interpersonal hypnotherapy	Jungian-focused spiritual direction
Magnetic field therapy	MAT core-12
Medifast coaching	Mesotherapy
New life therapy	Nouthetic therapy
Phage therapy	Self healing
Sogyal rinpoche	Soul care Christian counseling
Soul healing love	Spartan SGX coaching
Spritual [sic] disciplines	Therapuetic [sic] fly fishing
Toltec wisdom	Walking the middle path

psychotherapy, no matter how strongly they feel about such interests. Holding a strong interest in fly fishing, Kabbalistic healing, Biblical principles, tarot, Wicca, Buddhist practices, ancient Chinese martial arts, the sanctity of marriage, or any of thousands of other personal interests cannot be a sufficient basis for practicing this interest as if it is therapeutic. The work of counselors and psychotherapists must be health care, or it has no basis on which to exist.

Psychotherapy is mental health care. For broad consideration, take this expanded statement: Mental health care is a professionally accountable service that, in a very large variety of ways, treats individuals, couples, and families in response to relationship problems, environmental stressors, and mental illnesses. To help make sense of this definition, this discussion briefly explains each part.

1. "Mental health care" refers to many different kinds of helping services in response to a virtually limitless array of needs. The services may include divorce mediation, parent–child counseling, medication management, anger management, substance abuse treatment, eating-related problems, fear of the dark, psychological assessment, or many other mental health needs with which individuals, couples, and families need assistance. The range of possible mental health problems is similar to the range of possible physical health problems. To illustrate this, a list of common physical health problems could easily be composed. For example, the items in Table 6.4 could appear on such a list.

   These physical health problems may be quickly listed. Also, they easily show that these kinds of problem cover a very wide range of possibilities. The same thing may be said of problems for which mental health care is needed.

2. "A professionally accountable service" refers to a high level of competence, as defined by the mental health care profession and ways of helping (e.g., American Counseling Association, American Psychological Association, National Association of Social Workers, and American Association for Marriage and Family Therapy). Most mental health care providers can and should be able to tell their clients what "a professionally accountable service" means. The following list includes several specific indicators of a professionally accountable service.

   - A license to practice in the state where mental health care services are provided
   - A demonstrable knowledge of professional ethics

Table 6.4 COMMON PHYSICAL HEALTH PROBLEMS

Failing eyesight	Breast cancer	Elevated blood sugar
Displaced vertebrae	High blood pressure	Sprained ankle
Stomach ulcers	Appendicitis	Acute sun burn
Morbid obesity	Irregular heartbeat	Dementia
Gout	A bad head cold	Rapid, excessive weight loss

- A demonstrable knowledge of laws and regulations that cover their practice
- A readiness and ability to explain—to clients and to peers—their treatments, including their specialties and the evidence from research on which their treatments are based
- A clear focus on the clinically relevant needs of clients for treatment and not a focus on persuading clients that they need help, nor on the providers' uniquely personal needs to express themselves, including interests in religion, personal practices, sex, or many others
- Active participation in the professional association that represents their profession
- Profession-determined practices of seeking consultation from other professionals
- Demonstrated knowledge of and commitment to normative standards of professional mental health care and ongoing development of skills, competence, and accountability
- Ongoing pursuit of professional competence, as the profession defines it

3. "Treats individuals, couples, families, groups, and others" recognizes that that mental health needs come in multiple forms and that treatment commensurately comes in multiple forms, too. Adding to the ones that are already acknowledged, mental health care sometimes involves coworkers or other who are not family. For example, a much-loved coworker may be tragically killed in an automobile accident. Her/his associates may need help in coping with their loss. Mostly, though, mental health care involves individuals and their families. Also, treatment sometimes occurs in settings, such as a psychiatric hospital, where a group of clients receives treatment.

4. "Relationship problems" recognizes that almost all clients bring relationship problems to therapy. Indeed, disturbances in the primary relationships of those who seek help is a common reason for seeking help. In this category of need, marital and divorce issues are common, along with parent–child problems. However, relationship problems are also commonly associated with depression, substance abuse, and other mental dysfunctions.

5. "Stress" refers to the "wear and tear" of life. Along with everyone else, counselors and psychotherapists sometimes work too hard for too long and get too tired, leaving them to wonder about whether to maintain their commitment to their profession. Or, threatening circumstances, such as the collapse of the economy in 2008, cause elevated levels of wear and tear. Some stress is considerably more serious than tiredness at the end of the day or threats from economic circumstances.

It may come with serious, possibly terminal, illness or injury. It may come with divorce or death of a spouse. Stress is normal, but extreme stress is not. Mental health care responds to both normal stress and extreme stress.

6. "Environmental stressors" recognizes that stress results from normal and abnormal responses to stressful events that are largely external to the individual and family. They include many different kinds of events, such as the unexpected death of the owner of the family's business, multiple losses from a tornado, sudden death of an especially loved nephew, irreparable damages from a flood, a high school band bus accident that killed a child's best friends, a house fire, a sexual assault at a summer camp, and so on. And, while environmental stressors are not universally shared, most individuals understand that they may worry or feel stressed by many common stressors, including traffic congestion, violent weather, victimization from scammers, school shootings, earthquakes, and many others.

7. "Mental illnesses" refers to intensely personal suffering that almost all mental health care providers readily recognize and is one of the major needs to which they respond. Such suffering may include major depression, excessive anxiety, various psychoses, and so on. Many mental health care providers prefer not to use the term "mental illness." Some of them prefer the term "mental disorder." Others use several other terms, including "mental problem," "mental–emotional problem," "problems in living," or others. Similar to the language of individuals who are not mental health care providers, many different terms are used to refer to "mental illness." No matter what words are used to refer to it, "mental illness" is a very common reference. And, the most frequently cited definition of "mental illness" comes from the fifth edition of the *Diagnostic and Statistical Manual of Mental Disorders*—or the *DSM-5*—a publication of the American Psychiatric Association (2013):

> A mental disorder is a syndrome characterized by clinically signifi-cant disturbance in an individual's cognition, emotion regulation, or behavior that reflects a dysfunction in the psychological, biological, or developmental processes underlying mental functioning. Mental disorders are usually associated with significant distress in social, oc-cupational, or other important activities. An expectable or culturally approved response to a common stressor or loss, such as the death of a loved one, is not a mental disorder. Socially deviant behavior (e.g., political, religious, or sexual) and conflicts that are primarily be-tween the individual and society are not mental disorders unless the deviance or conflict results from a dysfunction in the individual, as described above. (p. 20)

As informative and useful as it may be, this definition of mental illness likely fails to satisfy everyone who may want such a definition, including counselors and psychotherapists. Whether this definition satisfies everyone, it should be placed in the setting where it is often needed: psychotherapy.

Typically, when they meet with clients, counselors and psychotherapists con-front the existential immediacy in their work and understandably seldom reflect on the kind of definitional presentation of their work as in this discussion. To define psychotherapy as mental health care invites, if not requires, some reflection on this existential immediacy. Mainly, on a day-to-day basis, mental health care becomes the "territory" in which counselors and psychotherapists provide their services. As such, the work is paradoxically enormous and confined, huge and general, but also individual and personal. As a statement about what mental health care is, the paradox may establish as much doubt as clarity and confirmation. The idea is, though, that the existential immediacy reflects the demands with which mental health care providers are confronted and to which they attempt to respond. However, it reflects also the complexity of human life and the virtually limitless ways that something may go wrong with human life. Probably, too, it reflects the relative immaturity of mental health care and the exigent necessity of trying to be good enough to alleviate suffering from and about almost everything. Within the context of attempting to be as disciplined, objective, and professionally accountable as they can be, counselors and therapists provide services for the following examples:

- An adult couple in which one member wants "out"
- A tantrum-plagued 11-year-old child who throws dangerous objects at others
- A young engineer who recognizes that she has become dependent on pain medications
- A middle-aged man who was gang-raped
- A 16-year-old girl who cuts herself almost every day
- A recently widowed woman whose depression is expressed in suicidal thoughts
- A new retiree who is trying to figure out how to enjoy life
- A perfectionistic physician who harshly judges everything that he does
- Parents who see their 12-year-old daughter as having an excessive interest in sex
- A licensed therapist who concludes that the life of a therapist is not for him
- An extremely successful musician whose success isolates her from others
- A husband and father who feels guilty about relocating his family to a new city, knowing that his wife and child will have to make huge and difficult adjustments

In the proverbial "final analysis," a discussion of what psychotherapy is may appear to be a detached, objective, and perhaps even defensive attempt to define it. However, practicing and professionally accountable counselors and therapists understand that, no matter what the analysis concludes, they must try to determine what will help the person in front of them. This is a noble, professional, personal, and indispensable value of what they do. As a personal note, my hope is that this book will help to serve their interests, protect them, and preserve their life-saving and life-sustaining potential.

Psychotherapy is professionally accountable. Most counselors and psycho-therapists probably read this feature of a definition of psychotherapy and reasonably place it in the context of professional ethics, laws and regulations, and normative standards of care. This is the way it should be. However, as this discussion strongly suggests, psychotherapy cannot and should not primarily serve the interests and needs of counselors and psychotherapists. It has referred to "a narcissistic tsunami" as a way to characterize much of what licensed providers advertise at their respective websites. As such, despite presumably knowing about professional ethics, laws and regulations, and normative standards of care, they demonstrate clear lack of accountability. A review of the so-called therapeutic approaches identified in this book makes this apparent. However, to further establish clarity about the lack of professional accountability among many practicing counselors and psychotherapists, readers of this book are encouraged to consider the approaches to therapy listed in Table 6.5.

The therapists who practice any of these therapies may be able to explain and justify them to the satisfaction of the professional community of which they are a part or the licensing board to which they are accountable. However, the likelihood is that almost all practicing and licensed counselors and psychotherapists do not recognize any of these therapies as ones about which they know anything. And, similarly, the likelihood is that members of licensing boards do not recognize any of these therapies. So, with regard to those who practice these therapies, to whom are they accountable?

The volume of relevant data, including information in this book, should convince readers to conclude that professional accountability among practicing counselors and psychotherapists is lacking among many providers. However, the evidence also prompts the conclusion that many counselors and psychotherapists

Table 6.5 Approaches to Mental Health Care That Raise Questions About Professional Accountability

Aikido	Circle sanctuary
Cowboy Gestalt therapy	Critical mixed race theories
Chrystal BOWL SOUND THERAPY	Divine's symbolic communication
Eden ENERGY MEDICINE	God designed counseling
Indian Saint Sri Kaleshwar	Inner healing prayer
Mysticism	Neurotransmitter restoration therapy
Nourishing choices	Ontological hermeneutics
Radix therapy	RoHun transpersonal (soul) psychology
Running the bars/access bars	SHEN therapy
Sophia analysis	Soul alignment
Soul detective protocols	Spiritual self-schema therapy
Spiritually fluent psychotherapy	Techniques based on Dr. Dan Allender
Transformational anthropology	Trim life healing
Usui reiki healing	Virotherapy
Voice dialogue therapy	Young living essential oils (YLEOS)

are conscientious and professionally accountable as individuals. Yes, numerous practicing professionals have a sincere interest in sex but refuse to leap to the conclusion that their interest in sex qualifies them to practice sex therapy. Or, numerous practicing professionals love children and playing with children but refuse to leap to the conclusion that their love of children and play qualifies them to offer play therapy. Or, numerous practicing professionals hold enduring and meaningful religious faith but refuse to leap to the conclusion that their faith is a sufficient basis on which to offer faith-based therapy. They demonstrate professional accountability. They will claim only the expertise that their profession acknowledges and that they are competent to provide. They will offer treatment approaches that make clear and broadly acceptable clinical sense. They will practice within the boundaries of professional competence, their respective licensing scope of practice, and professional ethics. Further, when in doubt, they will consult with their qualified peers.

The discussions in this book include references to numerous approaches to psychotherapy. Many of them have become synonyms of psychotherapy and, because of this, appear to define psychotherapy. Unfortunately, their presence itself defies and denies the likelihood of achieving a useable definition of psychotherapy. They serve as wedges that split psychotherapy into ever smaller pieces and, therefore, prohibit the effective understanding of psychotherapy and a useable future for psychotherapy. For example, when the proponents of psychoanalysis, Christian counseling, integral therapy, or another approach to therapy presume to conclude that their view of psychotherapy is the best and often exclusive view and "territory" of psychotherapy, they have contaminated, if not poisoned, both their view and psychotherapy generally. If the proponents of any specific therapeutic modality—widely accepted or not—presume to operate as if their approach adequately gathers almost every important feature and function of psychotherapy, their calcified thinking will have arrested and perhaps executed the potential of psychotherapy. Insofar as they do this, they embody the lack of professional accountability. Instead of professional accountability, they engage in another delusional circularity. "I am good at conducting therapy because I have an approach that I like, and my approach to conducting therapy is good because I provide it."

As noted previously, professional accountability among psychotherapists involves several necessary features. Among these features, a license to practice in the state where mental health care services are provided, a demonstrable knowledge of professional ethics, a demonstrable knowledge of laws and regulations that cover their practice, and others have already been listed. Somewhat more specifically, this discussion provides added details about professional accountability in the clinical lives of practicing counselors and psychotherapists.

Professional accountability begins with the high intention and personal capacity of being professionally accountable. Without this, professional accountability is not likely to happen. As intention and capacity, professional accountability is largely an internal phenomenon, although the expression of such intentionality should be seen in quite specific ways. This discussion includes most, if not all, of them.

Beyond intentionality and capacity, the question for practicing counselors and psychotherapist is, "Can you give a broadly endorsed account of your professional development and your clinical judgments, actions, and relationships?" The question assumes that the individual is accountable to something. In Chapters 7 and 8, which discuss ways to reform of counseling and psychotherapy, some of the possibilities are presented. At least, the approach to professional accountability in the future must operate on the foundational commitment to ensuring that it cannot happen in isolation.

Professional accountability means conforming to the expectations inherent in empirically validated clinical practices that may be seen in research literature and best practices statements about effective, outcomes-supported clinical service. Allowing for the ambiguous character of clinical service, it should, nevertheless, stand on the best available evidence for it. In contrast with the chaotic and cacophonous condition of mental health care, the future of professional accountability may require regular retesting of providers' knowledge, examination of their skills, and assessment of the outcomes of their clinical work, their attitudes toward their clinical work, and their competence in completing necessary tasks (e.g., maintaining current informed consent documents and appropriate documentation of clients' progress) associated with their clinical work. Such regular retesting would likely help to support counselors' and psychotherapists' commitment to client welfare and professional effectiveness, based on their profession-endorsed standards.

Professionally accountable counselors and psychotherapists represent their profession by offering only clinically tested approaches to therapy and delivering these approaches in ways that support their profession. To view this issue in a real-world context, consider the example of a cancer patient who seeks treatment resources online. The patient visits the websites of oncologists who treat his particular cancer. Because he recently received the diagnosis of prostate cancer, he begins his search for a physician who treats prostate cancer. He finds more websites of physicians who advertise their services than he could ever need. Among the many physicians' websites he visits, he finds information that he copies and pastes, including the following:

1. Real Men's Clinic welcomes you to our online home. We affirm your courage in taking your first step toward getting the best help you deserve. We support Real Men because we are Real Christians. You may be totally assured to know that our treatment provides you with Biblical Christian Prostate Cancer Therapy (BCPCT).
2. Hey, man, c'mon in. At Prostate Pros, we know how to move you out of any worry you have about your prostate cancer. Relax. Your worries are about to end. Past life health retrieval will take you into a life of health that you can recover, with our help. It is our unique blend of spiritual intuition, mystical encounters with wraithlike beings, and a heated environment of our salt cave. When you are ready for prostate health, just call Prostate Pros. We're here for you, man.

3. If you have been diagnosed with prostate cancer, you may feel overwhelmed and worried. We understand how you feel. We take your worry and fear about having prostate cancer very seriously. We promise to provide a safe and supportive home-like treatment center where you will soon feel like you are part of a large and caring family. We rely on your trust and confidence in us. In addition to our supportive manner of relating with you, we will also treat the cancer-enemy within you and fight as hard as we possibly can. We will use every weapon that the best of scientific evidence has given us. For you, the best weapon against your cancer may be extra-virgin olive oil gently massaged into your scalp behind your ears or apricot pits ground to a powder and mixed with honey and tenderly wrapped around your feet, or it may be stone-ground peanut butter, unless you have a peanut butter allergy, in which case we use sunflower butter. The peanut butter or sunflower butter may be eaten or merely chewed until it is expectorated.

4. Thank you for visiting our clinic. We are dedicated to providing our patients with the best, most effective medical services. We have the experience, training, dedication, and heart to help you or your loved one with the issues you are facing. Whether you are struggling with a new prostate cancer diagnosis or need help navigating a transitional time in your life, we are here to help. We have experience working with a wide range of individuals, including adults, adolescents, and children who are prostate cancer patients. We utilize a holistic wellness model that looks at the whole person to help you become the happiest, most successful version of yourself. Our model engages you in multiple forms of effective prostate cancer therapies, including anatomic therapy, aquatic therapy, aura therapy, indigenous healing practices, integrative wholistic wellness principals, interpersonal hypnotherapy, magnetic field therapy, new prostate life therapy, prostate healing love, Spanish OXQBDH coaching, and many others. With all of the treatment approaches at our disposal, we will find a cure for you.

The absurdity of these bogus advertisements makes the point. Lacking profession-ally accountable endorsements of their respective clinical approaches, numerous counselors and psychotherapists have betrayed their clients and their profession by offering numerous approaches to therapy that, at least, are simply inappro-priate and, at worst, are extremely harmful for their clients. In contrast to the profligate madness among counselors and psychotherapists, physicians who treat prostate cancer clearly understand that surgery is an option, along with watchful waiting or active surveillance, radiation therapy, cryotherapy (cryosurgery), hor-mone therapy, chemotherapy, vaccine treatment, or bone-directed treatment (American Cancer Society, 2018). Allowing for no compromise of their profes-sionalism, physicians who treat prostate cancer would not be able to practice if their preferred approach deviated from the options endorsed by their profession. If they treated prostate cancer with extra-virgin olive oil, apricot pits mixed with

Table 6.6 How Professionally Accountable Are These Approaches
to Counseling and Psychotherapy?

3 Principles/single paradigm approach	Akoma healing
Ancient African wisdom	Ancient Chinese martial arts
Ancient wisdom wellness practices	Archetypal art therapy
Blue brain training	Chakra balancing flow orientation
Chinese alternative medicine	Christ-centered identity discovery process
Diamond heart approach	EFT unwinding
Freedom adaptation techniques	Harmonic sound therapy
Heart assisted therapy	HMS holistic therapy
Holodynamic therapy	Hypnosis/past-life regression therapy
John Pierrakos' core energetics	Journey of the soul
Light touch healing	May cause miracles practice
Oneness blessing giver	Past life soul retrieval
Philosophy tools	Pure energy integration
Shadow work–women's empowerment	Sound therapy (crystal bowl)
Spirit junkie masterclass	Sufism
Svaroopa yoga embodyment [*sic*] therapy	Traditional Native American talking circle

honey, or the other imagined treatments in the four preceding paragraphs, they would very likely be sued successfully by their patients, lose their license to practice, and possibly be criminally prosecuted. What are the consequences, though, when counselors and psychotherapists practice any of the "therapies" in Table 6.6, not to mention more than 600 unusual therapies that have already been named in this book? Table 6.6 contains another 30 unusual approaches to counseling and psychotherapy. Are they demonstrations of professional accountability?

Psychotherapy is a profession. The chaos and cacophony in counseling and psychotherapy are readily and easily confirmed. Such confirmation raises questions about what psychotherapy is, with specific emphasis on doubt about whether psychotherapy is a profession. Assuming that it is a profession, the assumption requires clarity about what a profession is. If a profession cannot be defined by what it is or what its members have in common or an identifiable purpose and mission or more, is there a profession? In this discussion, the assumption is that psychotherapy—or whatever synonym is used to refer to it—is the profession. Further, this chapter has defined what psychotherapy is and what it is not. To continue toward refined observations about what psychotherapy is, the discussion turns toward briefly defining what a profession is. Accordingly, a profession meets the following criteria:

> Professions have an identifiable body of knowledge. If a body of knowledge defines a profession, what knowledge do counselors and psychotherapists share? More narrowly, is it a profession if the counselors and psychotherapists provide Chinese five elements and emotion-focused therapy? Or, Eastern body work and cognitive behavioral therapy? Image-bearer counseling and acceptance and commitment therapy? Body wisdom

dialogue and dreamwork and rational emotive behavioral therapy? Sufi healing circle and applied behavioral analysis? Clearly, a minimally appropriate, useful, and professional understanding of psychotherapy excludes some of these approaches from belonging to the profession of psychotherapy, although they frequently appear with many others that are also alien to the profession.

A profession has a specific kind of expertise. Currently, the practice of counseling and psychotherapy includes numerous, highly individualized and just made up nonsense, along with including numerous agreed upon and clinically effective approaches to counseling and psychotherapy. The highly individualized approaches raise doubt about whether the practice of counseling and psychotherapy has an identifiable and specific kind of expertise. What options may characterize such expertise? Astrological consultation or career counseling? sat nam rasayan healing or trauma-focused cognitive behavioral therapy? mystic aroma therapy or child trends lifecourse model? Christian behavioral analysis or systematic desensitization? Chakra balancing flow orientation or dialectical behavior therapy? Quantum healing hypnosis or motivational interviewing? Soul level positive psychology or community reinforcement approach? Schoolhouse yoga or assertive community treatment? Kunga yoga or illness management and recovery? Or which of tens of thousands of other options? These paired options represent counselors and psychotherapists who likely stand in conflict, recognizing that each side is not likely to select treatment approaches that the other advertises. With tens of thousands of such conflicts among mental health care providers, the inevitable questions persist: "Does expertise in psychotherapy have any relevant meaning?" "Is there such a thing as a specific kind of expertise that may be called 'psychotherapy?'" Necessarily, these questions point to the disappointing answer, "No." At this time, psychotherapists cannot be described as having a specific kind of expertise.

A profession has practices and procedures that members of the profession and the recipients of its services recognize and accept as features of their profession. Do counselors and psychotherapists recognize and accept massage therapy, acupuncture, converting clients to a religion, or many other practices and procedures identified in this book as features of their profession? Based on the evidence presented in this book, the likely conclusion is that psychotherapy lacks identifiable practices and procedures that psychotherapists recognize and accept or that their clients recognize and accept.

A profession has a set of rules that guide its members. Clearly, psychotherapy suffers from the pseudoscientific approaches of many of its providers, along with other long-standing, apparently habituated practices (discussed in Chapter 2). The practice of psychotherapy is largely unregulated. It lacks agreed upon rules that guide its members. Further, its rules, as represented in codes of ethics, suggest that multiple codes of ethics (see Chapter 7) may impair the capacity of professional psychotherapy to regulate itself.

A profession has a broadly recognized capacity to manage itself. Does psychotherapy have a broadly recognized capacity to manage itself? The idea of answering this question affirmatively and supporting such an affirmation poses challenges. The idea must account for the fact that, instead of agreed upon approaches to the practice of psychotherapy, a virtually limitless number of approaches may be located. At the time of writing this, we have identified 21,758 approaches to psychotherapy. The idea must account for the fact that many of these approaches represent no more than the whims of their providers. It must account for the fact that numerous individuals practice psychotherapy without a license and many without training. It must account for numerous licensees whose licenses were issued by nonstate agencies and have no professionally accountable reason to exist. It must account for the fact that numerous certifications that presumably attest to a specialization of some kind lack professional accountability (Holden & Barker, 2018). With these kinds of evidence, along with much more, the lack of a capacity of psychotherapy as a profession to manage itself at this time is clear. To illustrate the challenge of affirming the idea that psychotherapy has a capacity to manage itself, a so-called psychotherapist may practice spiritual warriorship, or another psychotherapist may practice mindfulness-based cognitive therapy. With regard to certifications, an individual may earn a certificate in rational emotive behavioral therapy through the Albert Ellis Institute (2018). In contrast, an individual may earn a certification in soul entrainment, issued by an apparently untrained and unlicensed individual, Karen Paolino Correia (2018) at Create Heaven Soul Entrainment Training. A certification in Soul Entrainment is advertised by licensed mental health care providers, such as a licensed social worker, Kim Conway (2018). Alternately, a certification may be earned in Traditional Tibetan Medicine from the Tibetan Medication Training Center (2019). These facts hardly support the capacity of psychotherapy to manage itself as a profession.

Finally, a profession has commonly understood and utilized standards and steps through which individuals change from being nonmembers of the profession to being members of the profession. As noted earlier, a prospective psychotherapist may complete as few as 28 graduate semester hours as a basis for receiving a license to practice or as many as 120 graduate semester hours as a basis for receiving a license to practice. Moreover, an unknown but large number of individuals receive questionable licenses, such as Licensed Clinical Christian Counselor from the National Christian Counselors Association (NCAA). Apparently, those who hold this license usually have little or no professional training in providing psychotherapy. The so-called license may be issued to individuals who have not completed high school. Advocating for the NCCA, the Above and Beyond Counseling Academy notes, "Individuals who do not wish to pursue a degree may engage in a field of study that results in receiving a Christian Counseling license only. They will be licensed to counsel by the National Christian Counselors Association (NCCA License). A High School Diploma or GED equivalent, as a minimum, is required" (Above and Beyond Counseling Academy, 2019). Alternately, apparently without appropriate professional training, individuals may receive degrees and licenses from the American Association of Christian Therapists (2019). Taken together, the number of nonstate licenses is large but unknown. Nevertheless, individuals who receive these licenses may be added to those who hold no license to practice counseling and psychotherapy, but who practice anyway and advertise their services online, often through "respectable" promotion services. Then, all of these individuals may be added to those who provide counseling and psychotherapy though social service agencies for which no license is required. Many individuals—maybe, enough psychotherapy pretenders to constitute about half of all psychotherapy services—practice psychotherapy and conspicuously lack minimal qualifications for such a practice. This raises serious doubt about whether psychotherapy as a profession has commonly understood and utilized standards and steps through which individuals change from being nonmembers of the profession to being members of the profession.

CONCLUSION

Professionally accountable individuals who practice psychotherapy face formidable threats to their professional survival. Despite this, they face a future that may be accessible to them if they work conscientiously toward establishing their profession as truly professional. This chapter may be a good opening for a discussion about how establishing such professional accountability may occur.

More than a discussion starter, though, this chapter may serve as another resource for helping to articulate a view about what counseling and psychotherapy reform may look like. This is the task of Chapter 7.

Counseling and Psychotherapy Reform (CPR)

What We Can Do Together

FRANCIS A. MARTIN AND JANET P. TURNER ■

INTRODUCTION

This chapter describes and proposes reforms of counseling and psychotherapy that have emerged from a long and careful review of the current condition of the profession of counseling and psychotherapy. With emphasis, it draws from the accrued evidence that has already been presented in this book. And, as indicated in several parts of this book, the conclusion is that the professional mental health care community should energetically pursue needed reforms of their profession— or others will—if they expect their professional community to survive.

No doubt, the call for reform elicits discomfort from those who have not previously considered the need for it. This is understandable and expected. This chapter may elevate discomfort, or it may reassure readers that those who call for reform have done this, only after the evidence for it has been diligently gathered. In anticipation of presenting the evidence and following considerable evidence that has already been presented, the call for reform of counseling and psychotherapy encompasses several important issues. Surely, legislative initiatives, governmental agencies, insurance companies, and other entities external to counseling and psychotherapy pose some of the threats. Mostly, though, the threats to counseling and psychotherapy are internal. Based on the internal threats, four general reforms are advanced. First, codes of professional ethics need revision and integration so that, instead of having a very large number of them, the number is reduced to one. Second, the practice of counseling and psychotherapy must be regulated so that the massive numbers of quite unusual and potentially harmful approaches to therapy can be eliminated. Third, preparation programs need to immediately pursue significant enhancements so that they produce appropriated effective clinicians. And fourth, to replace the currently chaotic approach to licensing, state-approved licensing will reduce the profligate number of licenses to only one or two in each state. None of these reforms is likely to be completed easily or soon. However, as this book has noted several times already, the question is not whether reform will come. The question is who will bring the reform. Further, an operative and evidence-supported premise is that reform is happening, with some of its specific expressions being ones that do not favor positive outcomes for counselors and psychotherapists.

In the briefest form, some of the answers to the question of why reform is needed include the following:

- Poor counselor and psychotherapist preparation programs
- Contradiction and incoherence among mental health care provider preparation programs
- Incompetence among substantial numbers of mental health care providers
- Declining competence among mental health care providers
- Lack of regulation of the practice of mental health care services
- Massive proliferation of aberrant approaches to counseling and psychotherapy, adding to the precipitous decline of counseling and psychotherapy
- Extraordinary and unnecessary difficulties, including confusion and rising potential harm, imposed on those who need mental health care
- Proliferation of unlicensed and otherwise unqualified "counselors" and "psychotherapists"
- Proliferation of types of licenses, with each of them closely resembling the others
- Proliferation of codes of ethics

To add important information to the need for concern, if not alarm, this chapter discusses external attempts to reform counseling and psychotherapy. Earlier, the book referred to a legislative initiative in Arizona (American Counseling Association [ACA], 2018) that aims to eliminate licensing for mental health care providers and a recent governor of Iowa (Branstad, 2017) who sought and failed to eliminate licensing for professional counselors, social workers, and marriage and family therapists. The legislation in Arizona failed. The governor failed, too. So far, the initiatives to eliminate licensing for mental health care providers have failed. However, several initiatives to change or "reform" mental health care have succeeded.

We have seen several attempts in the legislature of the state of Tennessee to change the character of mental health care. These attempts have been successful at times and unsuccessful at other times. This chapter offers a response to a recent legislative initiative in Tennessee. The aim of this response only incidentally is to call attention to legislation in Tennessee. More important, the aim is to make clear that many individuals who know little about mental health care have sought to reform it and will continue to seek ways to reform it. The underlying assumption is that

reform of counseling and psychotherapy will come. The question is not whether reform will come; the question is who will succeed in reforming mental health care.
Naturally, another inevitable question is what specific reforms will come.

The recent initiative in Tennessee affects only licensed professional counselors,
although other licensees under the same licensing board may also be affected. As a
major feature of this response, this chapter seeks to expand the already established
base of evidence for the conclusion that counseling and psychotherapy reform
(CPR) is urgently needed. While this chapter comes as a response to recent events
in Tennessee that target professional counselors, it expands the frame of reference
to include all mental health care providers, not just professional counselors, and
to the need for reform in all states, not just Tennessee. Further, it reports that,
along with several prominent representatives of counseling and psychotherapy,
counseling and psychotherapy professions are in serious jeopardy. The jeopardy
comes because of long-standing and evolving failures among counselors and
psychotherapists and not simply because of threatening legislation. Finally, the
chapter offers specific recommendations that are necessary reforms if counseling
and psychotherapy are to survive as mental health care professions.

Our view is that, while none of the legislative attempts to eliminate licensing for
mental health care have succeeded, sooner or later one of the attempts will likely
succeed if mental health care professionals fail to manage their professions as they
have demonstrably done. Concerned counselors and psychotherapists should not
take recent legislative attempts as only external manipulations and assaults on
their profession. To be sure, they are manipulations and assaults. They are more
than this, though. They are responses to invitations from mental health care
providers for these attempts. Concerned counselors and therapists need only to
consider the multiple, religion-based and ideology-driven approaches to psychotherapy that have been identified in this book. The counselors and therapists who
wish to advance these approaches as religious-political initiatives generally wish
also to secure legal protection of their almost always skewed and inappropriate
but also sometimes often bigoted and harmful approaches to mental health care.
Unfortunately, their legislative representatives hear their wishes and sometimes
respond affirmatively to them. When legislators hear the wishes of counselors
and psychotherapists who advocate for reshaping counseling and psychotherapy
into religious-political activities and do not hear from professionally accountable
counselors and psychotherapists, what should they do?

TENNESSEE LEGISLATION

State Senator Jack Johnson proposed Senate Bill 1 for the 2017 legislative session
in Tennessee. According to LegisScan, a legislative reporting agency,

> As introduced, prohibits the Board for Professional Counselors, Marital
> and Family Therapists, and Clinical Pastoral Therapists from adopting any

rule that incorporates by reference a national association's code of ethics, including, but not limited to, the American Counseling Association Code of Ethics; revises other provisions related to allowing counselors to not counsel when doing so conflicts with beliefs.—Amends TCA Title 4; Title 49 and Title 63. (LegisScan, 2017, Bill Title section)

In short, the bill eliminates the *ACA Code of Ethics* as enforceable by the licensing board for professional counselors in Tennessee and limits the licensing board to professional ethics that it writes/creates for licensed professional counselors in Tennessee. Also, the bill allows for the possibility that codes of ethics for other licensees under the same board may be adversely affected as well. These include marriage and family therapists and pastoral therapists. In news releases, Senator Johnson stated that he does not want a group outside the state to manage/control the standards of ethics for professional counselors in Tennessee.

In addition, with reference to his doubts about the ethics of termination and referral (Bartlett, 2016), Senator Johnson apparently has concerns about a specific standard in the *ACA Code of Ethics*. It is the following one:

A.11.b. Values Within Termination and Referral
Counselors refrain from referring prospective and current clients based solely on the counselor's personally held values, attitudes, beliefs, and behaviors. Counselors respect the diversity of clients and seek training in areas in which they are at risk of imposing their values onto clients, especially when the counselor's values are inconsistent with the client's goals or are discriminatory in nature. (ACA, 2014, §A.11.b.)

Senator Johnson indicates, too, that his preparation of his proposed bill did not include consultation with anyone else (e.g., Tennessee Counseling Association or other mental health care providers) (Humphrey, 2017). Regardless of the specific content of his legislative initiative, he believes that professional counselors ought to operate differently in Tennessee as a result of his legislation.

Johnson's proposed legislation allows "counselors to refer a client to another counselor when the counselor's 'personally held beliefs' conflict with that of the client" (Bartlett, 2016, para. 4). In an announcement to the media, "Johnson says the legislation 'protects the rights of counselors' by being able to refer the client to another therapist when 'the goals, outcomes or behaviors for which they are seeking counseling are a violation of his or her [the counselor's] sincerely held beliefs'" (Bartlett, 2016, para. 6). Apparently, Johnson's proposed legislation rejects the idea that a counselor should "refrain from referring prospective and current clients based solely on the counselor's personally held values, attitudes, beliefs, and behaviors," with reference to the *ACA Code of Ethics*. Johnson advances this idea with reference to only counselors. He seems not to know that the real and useful meaning of this standard has relatively little to do with a counselor's values, attitudes, beliefs, and behaviors. It merely gives attention to some of the important guidelines for "Termination and Referral" (ACA, 2014, §A.11.). And, for the protection of clients and counselors, the simple significance of the emphasis on "values, attitudes, beliefs, and behaviors" is that appropriate clinical judgment prompts referral and/or termination, not the specifically nonclinical judgments of the counselor. This is protective of both clients and counselors because it clarifies useful boundaries between the two, simply by emphasizing the obligation of counselors to make needed decisions about clients that come from solid clinical judgment.

To clarify, a primary care physician (PCP) would not likely refer a man to receive services from a gynecologist, no matter how extremely well-qualified the gynecologist may be or no matter whether the two physicians may be close friends. To do so would violate the normative standards of good clinical judgment. The PCP may place a high positive value on the friend, a gynecologist, but referral of a man for gynecological services would lack solid professional judgment. In other words, the importance of clinical judgment is the understandable emphasis in A.11.b.

So, A.11.b should be read more carefully than Johnson appears to have read it. "Solely" is a critically important word in the standard. The standard presumes that a counselor's personally held values, attitudes, beliefs, and behaviors are always operative in clinical work. An example is that mental health care providers place a high value on mental health and work hard to eliminate self-destructive behaviors, such as heroin abuse. Placing a high value on mental health is a personal value as well as a professional value. The standard, then, intends only to protect clients and counselors from capriciousness and arbitrariness when they conclude necessary judgments about their clients.

Similarly, to clarify, among all health care providers, the transparently clear and premier value of client/patient welfare should not be abrogated. So, the simple meaning of the standard, A.11.b., is that this value should almost always be upheld, instead of upholding the subjective, personal values of the counselor or psychotherapist. Or, to state the matter more simply, counseling and psychotherapy exist to help clients more than to help counselors and psychotherapists.

Despite vigorous protests to Johnson's views among counselors (Freedom for All Americans, 2017; Gervin, 2016; Humphrey, 2016; Meyer, 2016, 2017), his rejection of the *ACA Code of Ethics* likely makes more sense than counselors wish to acknowledge. With particular emphasis, if his rejection of the *ACA Code of Ethics* (2014) on standard A.11.b. has importance, it requires counselors individually and collectively to face some difficult questions. Specifically, if counselors or other mental health care providers should not refrain from referring prospective and current clients based solely on personally held values, attitudes, beliefs, and behaviors, they should raise the questions, "Which values, attitudes, beliefs, and behaviors can be reasons for referral of a client? If the personally held values of the counselor should not be the basis for a referral, what other values, if any, should be the basis for a referral?"

Without an answer to these questions, Johnson's objection to A.11.b. or the rejection of Johnson's legislative proposal seems to hold little or no meaning of significance. Counselors, though, have answers to this difficult question. They should make referrals that are based on solid clinical judgment and reasonable conclusions about what actions may be taken that best protect and promote client welfare.

Lacking a minimally reasonable rationale for objecting to A.11.b., Johnson's objection appears merely to be a negative one. Analogously, this is to say that he refuses to paint his living room the color blue. This is not a problem. However, if he refuses to paint his living room the color blue, what other color may he want to paint his living room? Merely objecting to blue simply leaves his living room unpainted. Raising objections to the *ACA Code of Ethics* is not necessarily a bad thing, but to have no reasonable basis for such objections is a bad thing. Further, having nothing with which to replace the *ACA Code of Ethics* appears to be an attempt to meet a need that does not exist.

Almost any practicing counselor could offer numerous hypothetical challenges to Johnson's objection to the *ACA Code of Ethics*. However, practicing mental health care providers know that they are talking about real human beings, not hypotheticals. They are real citizens of Tennessee and every other state in the country about whom clinical judgments must be made. Johnson's rejection of the *ACA Code of Ethics* seems not to take this into account. His view appears to be one of protecting counselors, not clients—protecting one at the expense of the other. At this time, nothing in the *ACA Code of Ethics* compromises or denies counselors' "values, attitudes, beliefs, and behaviors." Moreover, practicing counselors know that almost all clients contradict the values, attitudes, beliefs, and behaviors of counselors. Such contradictions are normative, as seen in the following examples:

- A man in his late 30s carries a heavy load of unresolved anger. Unfortunately, he displays his anger toward his two children. While he has never hit them or physically harmed them in any way, he has caused them to cower in perpetual fear.
- A 42-year-old woman fears that she is losing her appeal as a woman. Unfortunately, she chooses to confirm her appeal by pursuing as many extramarital "affirmations" as she can find.
- A midlife couple persist in demanding that their adult children maintain them as partners in the adult children's marriages, including having their names on checking accounts and titles to automobiles. The couple is confused by their children's hostility.

Hundreds, if not thousands, of additional illustrations could be reported, here. For now, suffice it to say that these individuals seriously contradict the values, attitudes, beliefs, and behaviors of counselors. Despite this, counselors provide treatment for them, knowing that this contradiction is not a sufficient clinical basis for referring them or terminating therapy with them. Further, these clients represent the much larger clientele who receive services from professional counselors and other mental health care providers. They are not exceptional. They are normative.

Truly, a sensitive issue is how professional mental health care providers make distinctions among clients whose problems may reasonably be treated and problems that may not be treated. For example, American culture emphasizes independence in relationships, whereas some other cultures emphasize interdependence. So, from the view of American therapists who treat individuals from other cultures, counselors wish to respect the other culture, while also affirming their own. For both the counselor and the client, this may cause confusion. In very real ways, this confusion may be seen in a family that brings its 16-year-old son for treatment because the parents have concluded that their son has rejected and dishonored the family. The counselor, though, concludes that their son has expressed his needed independence from the family and that no rejection or dishonor has occurred. So, should this child and/or his family be treated? And, if he and/or the family is/are treated, what should be the reasonable, clinically appropriate outcome of therapy? This story emphasizes the need for solid clinical judgment and not a counselor-centric need to affirm the personal values, attitudes, beliefs, and behaviors of counselors. Also, it emphasizes therapists' obligation to understand and implement the principles of diversity in their clinical and other professional practices.

MAJOR UNINTENDED CONSEQUENCES

Eliminating the *ACA Code of Ethics* or similar attempted changes forecasts major unintended consequences. For example, among those who have tracked professional ethics, relevant case law, and liability insurance for mental health care professionals, we conclude that a likely unintended consequence of Johnson's efforts will be a *rise in liability insurance premiums* for those who practice in Tennessee. While the *ACA Code of Ethics* currently stands as the most commonly recognized standard of professional behavior for licensed professional counselors, its absence will cause uncertainty among liability insurance carriers. Quite simply, uncertainty—an inability to utilize a relevant and predictable plan for coverage—means that premiums for liability will very likely increase. After all, if an insurance carrier faces uncertainty about the normative standards by which their insured professionals practice, how do they set their premiums? In short, they set premiums that will predictably cover their expenses and generate a profit. This means that, if the *ACA Code of Ethics* is eliminated, licensed professional counselors should plan to pay more for their liability insurance than they currently pay.

Another unintended consequence of Johnson's effort to eliminate the *ACA Code of Ethics*—without relevant and compensatory actions—is the likely *denial of licenses in other states*, when licensees in Tennessee relocate to other states or when new graduates from training programs in Tennessee apply for licensing in other states. This is merely a way to say that the *ACA Code of Ethics* constitutes a critically important and national normative standard of care and that its absence will cause major doubts about the quality of licensing and training in Tennessee. Professionally accountable mental health care providers have worked for many decades to establish the standards of professional accountability. To take these standards away surely lowers these standards, but with Johnson's legislation lowers them only in Tennessee.

Generally, if legislation such as Johnson's succeeds, *the quality of licensing in Tennessee will continue to fall behind* the quality of licensing in other states. Compared with other states, the license for professional counselors in Tennessee is strong but is growing weaker. The easily available facts on which this observation stands are far too many and complex to provide in detail here. Suffice it to say that, compared with many other states, several counselor preparation programs in Tennessee are notoriously deficient. Also, the long, post-degree, supervised clinical experience for prospective licensees is ill-defined and lacks relevant standards. The current qualifications for acceptable clinical supervisors seriously lag behind other states. These conditions and others mean that graduates of training programs in Tennessee and those who are already licensed will face unnecessary barriers in getting a license in many other states. The emphasis here should be clear: Weaknesses in Tennessee's current approach to clinical training and licensing, particularly for professional counselors, already pose unnecessary challenges for counselors who wish to relocate to some other states, and Johnson's legislation elevates the difficulty of these challenges. However, this emphasis should not obscure the fact that Tennessee has several counselor preparation programs and leaders that have shaped and will continue to shape the positive growth of mental health care in Tennessee and nationally.

Another unintended consequence is that *colleges and universities in Tennessee will suffer impairment in their ability to recruit outstanding students from other states.* Prospective students in bordering states (i.e., Kentucky, Virginia, North Carolina, South Carolina, Georgia, Alabama, Mississippi, Arkansas, and Missouri) must be apprised of the limitations of preparing for a career in Tennessee because the possibility of or requirement of compliance with a Tennessee-specific code of ethics poses insurmountable limitations. The probability that other states will produce their own state-specific codes of ethics is extremely low, at best. And, if other states produce their own state-specific ethics, interstate educational opportunities are likely to decline.

Clearly, the mental health care community has already spread the message that training and regulation of mental health care in Tennessee continues to slide into nonrelevance and oppressive political interference. One indicator of this is that, following legislation in 2016, the ACA relocated its annual convention from Nashville to San Francisco because of a new law in Tennessee. Among professional counselors and other major mental health care groups, the new law back then was viewed as an assault on counseling and psychotherapy. Numerous published statements through the ACA confirm that negative messages about Tennessee continue to spread. To cite only a few examples: A feature article in *Counseling Today* characterizes the 2016 legislation in Tennessee as a "license to deny service" to clients with whom counselors disagree (Meyers, 2016). An editorial by the CEO of the ACA describes the 2016 legislation in Tennessee as "a dangerous precedent" (Yep, 2016, para. 2). And, a recent president of the ACA wrote,

> This past spring, the American Counseling Association Governing Council made the wrenching decision to move our 2017 national conference from Nashville to San Francisco. That decision was the result of a bill passed by the Tennessee Legislature and signed into law by the state's governor, that is discriminatory in nature, targeting the LGBTQ community and challenging the 2014 *ACA Code of Ethics*. (Roland, 2016, para. 3)

Extending the difficulty of recruiting desirable students from other states, another unintended consequence is that in Tennessee *academic/clinical programs that train professional counselors and other mental health professionals, such as school psychologists, marriage and family therapists, school counselors, and others, will have to revise their curricula so that they are state-specific and not profession-specific.* Moreover, as a matter of practicing professionally accountable informed consent with prospective students, they will be obligated to inform prospective students that these respective training programs have been compromised by political interference. In addition, professional accountability requires counselors and psychotherapists to recognize that political interference has emerged because of their (our) failure to define and promote professionally accountable standards of training and practice.

More than merely revising their curricula, the elimination of the *ACA Code of Ethics* for licensees *places training programs in the position of having to teach the* ACA Code of Ethics *anyway, along with a Tennessee-specific code of ethics.* Their appeal is likely to drop precipitously because the Tennessee-specific code of ethics would likely make these programs professionally unwanted outliers among those who may wish to teach in these programs or become students in them. And, possibly, these programs will have to teach two ethics courses: one that prepares their students to live in a professional world in which the *ACA Code of Ethics* is commonly recognized as their code of ethics and another that prepares them for a Tennessee-specific world.

If the *ACA Code of Ethics* is eliminated in Tennessee or elsewhere, licensing boards, training programs, professional associations, practicing mental health care providers, and others will have begun a long and frustrating game of "ethics whack-a-mole." The *ACA Code of Ethics* will be eliminated, but then the game will require politicians and others to "whack" every other code of ethics, knowing that a very similar one is like to "pop up."

With each maverick, ill-informed deletion of another code of ethics, mental health providers and others will, again, have to persuade other licensing boards in other states that Tennessee has a viable license. This is already a challenge. And, they will have to convince liability insurance providers that licensees in Tennessee practice ethically. This will likely require several years. In addition, as indicated later, licensed mental health care providers will have to initiate numerous measures toward regulating the practice of mental health care in Tennessee and elsewhere. Currently, the professional practice of mental health care providers in Tennessee and elsewhere is already poorly regulated, as this book has so far clearly described. And, predictably, academic/clinical programs will suffer consequences if the *ACA Code of Ethics* is eliminated, as noted previously.

To be clear, the recent legislative initiative to eliminate the *ACA Code of Ethics* in Tennessee failed. However, the initiative was not the first of its kind and is not likely to be the final one.

MENTAL HEALTH CARE AS HEALTH CARE

Real feelings of awkwardness sometimes swirl around how to think about mental health care. And, an individual's way of thinking about mental health care turns the discussion in one of many directions.

One direction is that, much like an artist, some individuals regard mental health care as a means of personal expression. If counselors and psychotherapists desire to engage in a pleasurable hobby, such as fly fishing, martial arts, cooking, or thousands of others that we have located online, the hobby almost magically becomes "therapeutic." These individuals may—at the risk of taking this too lightly—seek to express their passion for their personal interests through counseling and psychotherapy. They seek more to express themselves than to engage their clients in healing and recovery. This approach to mental health care usually means that therapists are largely satisfied with their work but usually provide demonstrably ineffective mental health care. Their self-expressions are individualistic, quirky, and unhelpful—and sometimes very harmful. This conclusion has been confirmed in reporting about the multiple therapeutic approaches that have been named in this book. Somewhat unique personal expressions are suffused with counseling and psychotherapy with increasing frequency. The examples of this are much too numerous to list here, although many of them appear throughout this book. Suffice it to offer only a small number of examples: ancient instruments for healing, clay sculpture and pottery, cross-cultural music in healing, enneagram in the narrative tradition, MARI and mandala drawing, money success-planning, music/sound healing, nature-based rites of passage, noetic balancing, picture framing, TERRA essential oils, and therapeutic singing sessions.

Another common view of mental health care is that it conveys the beliefs of the caregiver to the care receiver. This kind of care comes from individuals who may, for example, practice one or more of the therapies in Table 7.1 that are individualized, ideology-driven "therapies," many of which are specifically religious.

This very short list exemplifies the many thousands of unusual so-called therapies that are practiced by licensed mental health providers. Each one of these strongly suggests that the therapist who practices these seeks to convey her/his beliefs to her/his clients. Many of these beliefs are specifically religious. While the conveyance of beliefs is not necessarily a religious practice, it is often the approach of those who wish to convey their beliefs to their care receivers. Sadly, sometimes these beliefs are political ones that represent, say, specific political agendas, with regard to such matters as sexual orientation and abortion-related issues. The

Table 7.1 BELIEF-BASED APPROACHES TO COUNSELING AND PSYCHOTHERAPY

African psychology	Ancient healing modalities
Anthroposophical counseling	Avesa quantum healing
Aztec healing	Bibliotherapy = Bible counseling
Biorhythm therapy	Birthing from within
Body-soul integration	Buddhist recovery models
Buteyko method	Cellular release therapy
Ceremonial healing	Christ consciousness counseling
Christian ACT	Christian auditory meditation therapy
Contemplative evocative approach	Depth psychotherapy/Christian tradition
Depth soul work and discovery	Exploration of the soul
Healing wisdom	Integrative soul coaching
Mystical meditation	Pure feeling awareness
Repentance therapy	RUH healing
Sacred contracts	Sacred/spiritual approaches
Samatha and vipassana meditation	Scale-walking
Scientology	Secular Buddhism
Soul entrainment therapy	Soul restoration
Spiritual healing ceremonies	Sufi healing circle
The divine feminine	The hero's journey
Visceral massage	Watsu (water shiatsu)

capricious and often covert attempts to convey religious beliefs are the actions that the *ACA Code of Ethics* seeks to prevent.

Clearly, another common view of mental health care is that it is health care. This is the view of mental health care in this book. As health care, mental health care offers the skills of clinical experts and applies these skills to a vast array of mental, emotional, and behavioral problems. These experts provide mental health care for an almost limitless variety of individuals and their mental health needs. Whether they live in severe poverty or high luxury, these individuals suffer with many different problems (see Table 7.2). The small sample of problems provided in the table merely illustrates the need to define the work of mental health care providers as comparable to almost all other health care providers.

Table 7.2 COMMON REASONS FOR GETTING MENTAL HEALTH CARE

Morbid obesity	Excessive anxiety
Various chemical addictions	Marital discord
Self-harming children	Depressed/suicidal individuals
Partner violence	Sexual dysfunction
Eating-related problems	Adjustment to physical disability
Psychological trauma	Poorly managed anger
Dysfunctional/harmful parenting	Overcontrolling executives
Irrational fear of driving a car	Excessive dependency on others
Loss of a spouse (death)	Career-related frustrations
"Workaholism"	Family illness and medical demands

Mental health care providers expect that their clients will engage in behaviors, such as some of those listed in the table, with which they disagree. For example, a mental health care provider would never agree with a client who wishes uses heroin. Further, the client who uses heroin likely presents behaviors that contradict the sincerely held religious beliefs of almost all practicing therapists. Similarly, mental health care providers never approve of or agree with compulsive sexual behaviors (i.e., sex addiction), knowing that it poses serious risks to the client and others.

Among other things, this means that therapists who treat individuals and families with these problems do this with the knowledge that they will likely disagree with their clients, such as the angry and harmful parent. Practicing therapists know very well the probability that the preponderance of their clients will present problems and behaviors that are incompatible with the therapists' sincerely held religious beliefs and other personal values. Generally, just because it is commonly expected and a common feature of their work, this is not a problem. Just as the oncologist seriously disagrees with the committed smoker who suffers with lung cancer but provides treatment for the patient anyway, mental health care providers commonly treat individuals and families with whom they seriously disagree. Or, just as the emergency department physician disagrees with the motorcyclist who drove at excessive speeds while intoxicated and suffered serious injuries in a preventable crash treats the injured patient, mental health care providers treat individuals with whom they seriously disagree.

Also, under professional ethics and within the boundaries of solid competent clinical judgment, therapists almost always have the prerogative of treating a client or not.

TOO MANY CODES OF ETHICS?

The recent legislative initiative in Tennessee aims to eliminate the *ACA Code of Ethics* as a code with which licensed professional counselors in the state are now required to comply. However, licensed professional counselors will be constrained by other codes of ethics that enshrine the standards to which Johnson's legislation objects. This is to say that, even if the *ACA Code of Ethics* is eliminated, several other codes of ethics will continue to require compliance by professional counselors and other licensees to standards to which Johnson objects. This neutralizes the impact of what he has tried to do.

For one thing, some licensees hold membership in other associations or hold certifications that have their own codes of ethics. Among the many associations in which licensed professional counselors have membership and, therefore, have obligations to comply with the ethics of the association, 16 are listed in Box 7.1.

More specifically, all of these associations, as well as others, have codes of ethics that closely parallel the standards and values, resulting in their approximate alignment with the *ACA Code of Ethics*. For example, from the *ACA Code of Ethics*:

Box 7.1

MAJOR PROFESSIONAL ASSOCIATIONS WITH CODES OF ETHICS

American Art Therapy Association (2013)
American Association of Christian Counselors (2014)
American Association for Marriage and Family Therapy (2018a)
American Association of Pastoral Counselors (2012)
American Association of Sexuality Educators, Counselors, and
Therapists (2014)
American Counseling Association (2014)
American Group Psychotherapy Association (2002)
American Mental Health Counselors Association (2015)
American Psychiatric Association (2013)
American Psychological Association (2017)
American School Counselor Association (2016)
Association for Addiction Professionals (2016)
International Expressive Arts Therapy Association (n.d.)
National Association of Social Workers (2017)
National Career Development Association (2015)
United States Association for Body Psychotherapy (USABP, 2007).

C.5. Nondiscrimination
    Counselors do not condone or engage in discrimination against prospective or current clients, students, employees, supervisees, or research participants based on age, culture, disability, ethnicity, race, religion/spirituality, gender, gender identity, sexual orientation, marital/partnership status, language preference, socioeconomic status, immigration status, or any basis proscribed by law. (ACA, 2014, p. 9)

Similarly, the National Career Development Association (NDCA) *Code of Ethics* includes the following statement:

NCDA opposes discrimination against any individual based on age, culture, disability, ethnicity, race, religion/spirituality, creed, gender, gender identity and expression, sexual orientation, marital/partnership status, language preference, socioeconomic status, or any other characteristics not specifically relevant to job performance. (2015, Contents section, Nondiscrimination statement, para. 2)

Aligned with ACA and NCDA, all of the professionally accountable professional associations make clear that being open-minded and welcoming differences within a multicultural world should be sustained. Also, the similarities in their codes of ethics suggests that they may be able to contain their ethical standards in a single code of ethics.

In addition, numerous certifications for mental health care providers come with the constraints of ethics. Some of the common certifications are listed in Box 7.2, but this list of 26 is an illustrative one and not an exhaustive one.

Among professional counselors in Tennessee and elsewhere, all of these certifications and more are held. Some of them are held by individuals who hold other kinds of mental health care licenses in Tennessee and elsewhere, including social workers, psychologists, pastoral therapists, marriage and family therapists, and others. For example, certified play therapists may hold any of these licenses. Many more certifications than these may be readily located. So far, we have located approximately 1,800 certifications. The point is that Johnson's or others'

Box 7.2

Professional Certifications With Codes of Ethics

AAMFT Approved Supervisor (American Association for Marriage and Family Therapy 2018b)

Approved Clinical Supervisor through the National Board for Certified Counselors' Center for Credentialing and Education (2017)

Board Certified Music Therapist (2014)

Certified Career Adjustment Associates (2017)

Certified Eating Disorders Specialist (2003)

Certified Group Therapist (n.d.)

Certified Rehabilitation Counselor (2017)

Certified Sex Therapist (CST, 2014)

Certified Vocational Evaluation Specialists (2009)

Certified Work Adjustment Specialists (2017)

Master Addictions Counselor (2016)

National Certified Addiction Counselor (Level I, 2016)

National Certified Addiction Counselor (Level II, 2016)

National Certified Adolescent Addiction Counselor (2016)

National Certified Clinical Mental Health Counselor (2016)

National Certified Counselor (2016)

National Certified School Counselor (2016)

National Clinical Supervision Endorsement (2016)

National Endorsed Student Assistance Professional (2016)

National Endorsed Co-occurring Disorders Professional (2016)

National Peer Recovery Support Specialist (2016)

Nicotine Dependence Specialist (2016)

Professional Certified Coach (2015)

Registered Expressive Arts Therapist (n.d.)

Registered Play Therapist (2015)

School-Based Registered Play Therapist (2015)

---

similar efforts may eliminate the *ACA Code of Ethics*, but the numerous licensees will be held to virtually the same standards of ethics through other codes of ethics.

These multiple codes of ethics evoke the question, "Are all of these codes of ethics really needed?" An answer to this question appears later in this chapter. Suffice it to say that efforts to modify professional ethics through political micro-management, such as Johnson's, will likely accomplish almost nothing, except to accelerate the ongoing decline of counseling and psychotherapy. In addition, such efforts may lead to unintended consequences that will frustrate attempts toward reforming mental health care with regard to professional ethics.

The most conspicuous virtue of having the *ACA Code of Ethics* for professional counselors is that it stands as the most commonly identified and utilized code of ethics among counselors. Additionally, without challenge, it is extraordinarily client-supportive and broadly accepted as a necessary code of mental health care ethics. From its beginning 50 years ago, the *ACA Code of Ethics* has grown into a major source of guidance for counselors and psychotherapists around the world. And, despite Johnson's obviously incorrect view that counselors and psychotherapists are receiving an oppressive dictation of ethics from "a special interest group in D.C." (Humphrey, 2016, para. 2), or "a private organization" (Bartlett, 2016, para. 33), the *ACA Code of Ethics* has emerged from practicing counselors in every state, including Tennessee. Further, the *Code of Ethics* has been adopted by every state counseling association, without direction, let alone coercion, by anyone outside of each state. The members of the Tennessee Counseling Association (TCA) have adopted the *ACA Code of Ethics*.

Clearly, in response to proposed legislation that would eliminate the *ACA Code of Ethics*, numerous individuals and relevant professional associations voiced strong objections to the legislation. Despite earlier successful legislative actions, this proposal failed.

As noted earlier in this book, the recently proposed and failed legislation in Tennessee is one of several legislative and administrative initiatives that propose significant changes in counseling and psychotherapy. Thus, the vestigial, ongoing importance of the proposed legislation should not be missed: Reform of counseling and psychotherapy is happening. Further, as this book concluded earlier, there are more ethical violations by counselors and psychotherapists who offer aberrant and often harmful approaches than all of the other ethical violations taken together. Along with other urgently and long overdue reforms, such as reform in licensing, regulation of clinical practice, and others, reform in professional ethics is needed, too.

Toward reform of professional ethics and given that there are multiple codes of ethics, a beginning question is, "With which code of ethics should mental health care providers be expected to comply?" The fact of multiple codes of ethics poses problems. One problem is that an individual mental health care provider may hold membership in more than one professional association and hold more than one professional certification, knowing that each professional association and certifying agency has its own code of ethics. Another common problem is that within a clinical practice or agency, supervisors and supervisees who may be licensed, say, as professional counselors and psychologists, are expected to comply with different codes of ethics that are sometimes in conflict. A similar kind of problem is that, following the completion of their degrees, prospective licensed counselors sometimes receive clinical supervision from, say, a clinical social worker, with each of them having to comply with a different code of ethics. With which code should they be expected to comply? More important than the confusion among mental health care providers, another problem is the question about which code of mental health care ethics, among only the 42 identified in this chapter, best serves the needs and interests of those who receive mental health care? Asked differently, which of their codes of ethics takes priority over the other one when conflicts in the codes of ethics arise? Or, in the context of legislative initiatives, such as Senator Johnson's, which code or codes of ethics may be seriously compromised or eliminated when legislators wish to manipulate mental health care ethics for political purposes and when mental health care providers lack even superficial cohesion around a code of ethics?

REFORM OF PROFESSIONAL ETHICS

To eliminate confusion, conflict, doubt, and excessive duplication, all mental health care providers need to establish a broadly agreed upon code of ethics for all appropriately credentialed mental health are providers and follow it. The details of executing a plan to establish one code of ethics for all mental health care providers in Tennessee and elsewhere may seem to be a daunting task. However, if mental health care providers are expected to "follow the law" with regard to professional ethics, having only one code for all providers may be one of the few logical and effective ways to achieve this. Such a code would eliminate huge ambiguity for providers in Tennessee and elsewhere, insofar as all of them would have one guide to follow instead of multiple guides. Currently and unfortunately, there are multiple codes of ethics for mental health care providers.

Just imagine the chaos—accidents, injuries, and deaths—that would inevitably occur if Tennessee or any other state had multiple sets of regulations for licensed drivers.

What would driving be like if adult females—or licensed psychologists—over 40 years of age had their own driving regulations?

Or, what would driving be like if drivers of American-made vehicles had their own driving regulations and this rule applied only to licensed professional counselors?

Or, what would driving be like if very tall individuals who are licensed clinical social workers had their own driving regulations?

Or, what would driving be like if members of a political party or members of the American Association of Christian Counselors had their own driving regulations?

Or—just one more—what would driving be like if those who hold a
graduate degree and licensed marriage and family therapists had their
own driving regulations?

The point is that, just as multiple sets of driving regulations are untenable and
wrong, having multiple codes of ethics is untenable and wrong, too. As whimsical as they appear to be, these illustrations provide a reasonably accurate characterization of the mostly unregulated and highly individualistic approach to
counseling and psychotherapy by those who provide these services. To blunt the
deleterious impact of multiple codes of ethics, not to mention a virtually limitless
number of professional certifications, a single code of ethics should be established.
With a single code of ethics, the most important beneficiary would be the client.
This probability challenges all mental health care providers to pursue one code
of ethics. However, as formidable as this challenge may be, other challenges may
be more daunting. Additional kinds of reform are similarly and urgently needed.

In addition, one code of ethics for all mental health care providers would ensure that efforts, such as Senator Johnson's, to target a code of ethics for elimination would be almost impossible, based on the support of a single code of ethics
by all licensed providers. Moreover, in the context of the world of practice, such
as a mental health center or a private practice group, all licensed providers would
have clarity about their professional ethics.

Instead of trying to eliminate the *ACA Code of Ethics*, other actions may be
compelling and useful ones, if not critical to the survival of counseling and psychotherapy. Or, to frame possible, plausible actions in a different way, "Whether
the *ACA Code of Ethics* is eliminated or not, what may the mental health care
community (i.e., licensed counselors, psychologists, social workers, and others)
do to achieve positive, productive, enduring, and reasonable goals on behalf of
the professional community, but moreover on behalf of those who receive mental
health care?" As noted several times in this book, if the professional community fails to reform itself, others will. This is one of the implications of Senator
Johnson's legislative initiatives and the governmental initiatives in several other
states. However, long before Johnson initiated his uninformed attempts to reform
professional ethics, individuals within the psychotherapy community advanced
the need to reform professional ethics. In 1998, cited by Grohol (1998/2004),
Goodrich offered several proposals for improving psychotherapy, including the
idea that there should be "a consistent set of functional and ethical regulations
for all forms [all licenses] of psychotherapists." The difference between Goodrich's
proposal and the proposals in this book is that while both hope to improve psychotherapy, this book adds the very real and easily documented threat to the survival of psychotherapy as a reason for reform.

Whether Johnson or anyone else outside of the professional mental health care
community pursues reform of mental health care, mental health care professionals
should. So, if working toward having only one code of professional ethics is a
needed reform, what else may mental health care providers do to ensure that

major, current deficiencies and difficulties no longer exist? In response to this question, the remainder of this chapter discusses additional needed reforms.

REGULATION OF PRACTICE

The following story is a composite of real stories. It reflects real events but is not a report about a real individual. Living approximately halfway between Chattanooga and Nashville, a woman in rural Tennessee sees her 17-year-old daughter suffering with an ill-defined disturbance. The mother is not a therapist, but she believes that her daughter needs help. She begins to review the websites of mental health care providers in her area. Finding very few therapists in her area, she expands her online search to Chattanooga and Nashville. As she does this, she, too, begins to suffer. Her suffering comes from finding disturbing information on the websites of licensed mental health care providers. The more she looks, the more disturbed she becomes.

Finding these disturbing things, she contacts friends and asks for help. Naturally, her friends ask for details about what she may have found on the websites of licensed mental health care providers. She reports that, along with a confusing mass of information, she found specific treatment offerings that disturbed her. She shared the following information with her friends.

She found several kinds of licensees, including LPC-MHSPs, pre-licensed but supervised therapists, LPCs, psychologists, LCSWs, and LMFTs. From these mental health care providers, she gathered a staggeringly large number of treatment approaches—approximately 600, from about 200 therapists. However, her disturbance did not come from the number of treatment orientations. It came from the kinds of treatment orientations that she found. Just to avoid overwhelming her friends with her lengthy list of "crazy therapies"—her word for them—she reduced her list to just "the top 50," including those listed in Table 7.3.

Disturbed by the strange offerings by these "therapists," the mother asked university faculty about them. In clear, definitive, and conclusive terms, she received confirmation of her concern. The faculty observed that almost all of these so-called therapies were utter nonsense. They added that those who practice these therapies very likely provide services that fall outside of the scope of practice that their respective licenses authorize them to practice. Unequivocally, they stated that these practices are unethical.

In the end, almost any observer of the licensed mental health care providers in Tennessee and elsewhere must conclude that the practice of mental health care is almost entirely unregulated, as strongly implied by the list of "therapies" in the table. In addition, as this book confirms, the number of therapeutic approaches is indefinitely large. So far, almost 22,000 approaches have been located at websites where counselors and psychotherapists advertise their services. Approximately 18,500 of these approaches lack credibility, likely fail to comply with professional ethics, and harm clients. Among other important observations from gathering these approaches, a necessary one is that, after an individual receives a license

Table 7.3 A CLIENT'S LIST OF "CRAZY" THERAPIES

Alternative medicine	Biodynamic craniosacral therapy
Body psychotherapy and breathwork	Bohmian dialogue
Bonny method	Circle process (groups)
Comfort (therapy) dog	Compassion power
Contemplative psychotherapy (Buddhist)	Continuing bond with the deceased
Core shamanic practices	CSEFEL (pyramid model)
Dream analysis	Encouraging coaching
Eastern and Western [J. Krishnamurti]	Energy medicine
Etiotropic trauma management	ETM TRT SHOM theory
Existential philosophy	Exposure therapy
Feminist	Guided SELF-INQUIRY
Herbalist/herbalism	Hollistic [*sic*]
Image-bearer	Impact therapy approach
Interfaith/nonsectarian	Intuitive therapy
IPP	Journey towards rebirth
Maslovian motivational therapy	Medicinal aromatherapy
Meditation	Multi-systems
Neurobiology	Nonduality
PSYCH-K	Peripheral and neurobiofeedback techniques
Prolonged exposure	Psychodrama
Reiki	Relational life therapy
RRT	Somatic
Soultalk [*sic*]	Spiritual and energy therapy
Survivor-fighter-thriver model of care	Taylor Johnson analysis
Veg*n lifestyle	Yoga

Note: Our interest in providing this list is merely to provide a report and not necessarily to offer an assessment of each of these approaches to therapy. Instead of assessing them, we report them as sources of confusion among clients or prospective clients. Indeed, given what we know about feminist therapy or relational life therapy, we conclude that it is an acceptable approach to therapy, albeit an ideological one.

to practice, the effective regulation of practice almost completely disappears. Whether practicing in truly professional ways or not, therapists can "just make up stuff" and call it "therapy" if they wish. Table 7.3 and other lists provided in this book illustrate this all too well.

The lack of professionally accountable regulation of mental health care providers imposes a specific obligation and burden on those who are professionally accountable. With urgency, *they must pursue plausible and effective regulation of the practice of counseling and psychotherapy.*

The urgency of regulating counseling and psychotherapy is clear. Multiple kinds of evidence for effective regulation have been presented. To reiterate just a few of them, the evidence includes the following kinds: Research shows that competence among counselors and psychotherapists continues to decline. Inappropriate credentialing, such as a large number of bizarre certifications, continues to prolif-erate. Counselors and psychotherapists advertise and presumably practice many thousands of transparently poor, if not harmful, treatment approaches.

For emphasis, our review of 60,200 websites of licensed counselors and psychotherapists concludes that approximately 27% of them practice therapies like the unusual ones listed throughout this book. And, an almost limitless number of unusual therapies can be added to the list. So far, from our review of websites where counselors and psychotherapists advertise their approaches to therapy, we have identified almost 22,000 unusual therapies. Again, for emphasis on the size of this problem, some additional unusual approaches appear in Table 7.4.

Taken by itself, identifying unusual therapeutic approaches confirms the need for serious, urgent, and pervasive reform of counseling and psychotherapy. It calls for regulation of the practice of counseling and psychotherapy. The administra-tion of such regulation and the agreed upon procedures for it exceed the call for regulation in this book. However, if such regulatory oversight were in place, it could demonstrate flexible and adaptable approaches to regulation. For example, a professional association in a state could receive requests from a licensee with regard to the licensee's interest in adopting a particular therapy. Through the asso-ciation, the licensee's broadly supported peers could make a determination about which therapies could be delivered and which could not be delivered. The same could be achieved through a state-administered oversight board. Regardless of the specific mechanism for oversight, almost any regulation beyond the current

Table 7.4 ANOTHER LIST OF UNUSUAL APPROACHES TO
COUNSELING AND PSYCHOTHERAPY

Alaska wisdom recovery	AMATI project
Auricular acupuncture (NADA protocol)	Bert Hellinger approach
Biblical Buddhism	BTB feng shui
Buddhist Jungian psychology	Chinese five elements
Eastern bodywork	Embodied Gestalt approaches
Embodied relating	Emerging wisdom
German play-therapy	Guided dream journeying
Interactive metronome	Japanese psychology approaches
Jungian and fairytale study	Katherine Woodward Thomas approach
Kripalu yoga	Kundalini arousal
Multi-generational inclusion	MuZen mindful creative
Myofascial release	Neural-based emotional release
Oriental medicine	Peaceful living therapy
Personalized theory	Reflexology
Regressionism	Reiki stress reduction and relaxation
Reunification therapy	Ridhwan approach
Rosen method bodywork	Shamanic therapy
Sound mnemonics	Subtle energy enhancement
Sustainable happiness	TARA approach
Three pinciples [sic] therapy	Total body modification

professional environment, in which numerous clinicians "just make up stuff" and call it therapy, could be achieved with relative ease and timeliness. Naturally, procedures for enforcement would be needed, too. However, all of this would have a fairly narrow justification: Clients should receive the best care of which licensees are capable, and in receiving the best care, they should receive care that the licensee's profession endorses. If the licensee's profession does not endorse a particular therapy, such as Biblical love model, past life soul retrieval, acupressure with pebbles, community herbalism, or ecstatic dance, the therapy ought to be prohibited and become the target of specific action against the provider who offers it.

The effective regulation of professionally accountable counseling and psychotherapy may cause the termination or upgrading of identified incompetent mental health care providers. The idea of terminating the work and career of a colleague is, at least, a disturbing one. Great caution should guide any judgments about mental health care providers as incompetent. This discussion embraces such caution. However, when licensees practice any of a very large number of unethical and nonclinical approaches, they harm their clients and the profession of which they are members. Specifically, when a licensee practices such therapies as power animal retrieval, past life soul regression, astrological consultations, or others that fall extremely below normative professional standards of mental health care, the professional community should stop them or require them to upgrade their services so that they practice within their respective scope of practice and professional accountability. If they continue to provide their deficient services, they harm their clients and the professional community. In such cases, the professional work of the individual should be terminated.

In addition, the proliferation of unlicensed individuals who allege to provide counseling and psychotherapy should be stopped. Based on our review of thousands of websites of counselors and psychotherapists who advertise their services—services that mostly look like they are authentically professional—our estimate is that approximately 9% of them are not licensed. In the information gathered from four cities and reported in Chapter 3, the proportion of unlicensed therapists who advertised online was 3.2%, or 20 of 622. If these apparently unqualified providers of mental health care in fact deliver adequate or superior services, all of the licenses and certifications for counselors and psychotherapists will not matter. On the contrary, if training and credentialing matter, especially with regard to the welfare of clients, unlicensed individuals should be stopped. To be clear, "unlicensed" does not refer to unlicensed practitioners who practice as prelicensed individuals under the supervision of a licensed mental health care provider. They are licensed de facto.

Additionally, another category of proliferation of unlicensed individuals is those who are employed in agencies. These employees provide "counseling and psychotherapy" in agencies that, because they are usually contracted with their respective states, are often not required or expected to utilize licensed mental health care professionals. Based on reviewing several agencies, we conservatively estimate that these individuals constitute 10 to 15% of the total number of mental health care providers. This contributes to questionable, if not poor, services for clients. Combined with the private-practice individuals in the preceding paragraph, a good estimate of the unlicensed and mostly untrained counselors and psychotherapists is approximately 20 to 30% of the total number of mental health care providers. The individuals who provide services through state-contracted agencies presumably achieve a great deal of good outcomes for their clients. They should not be stopped. However, as marginally qualified individuals, they should be eliminated by attrition and replaced by appropriately trained, credentialed, accountable, and effective professional providers.

There is a category of "licensed" providers whose practices appear to be considerably more insidious than the ones already discussed. These are the "licensees" who receive their "degrees" and "licenses" from institutions that have no accredited degree programs and no professionally recognized authority to grant licenses. They include the National Christian Counselors Association (NCCA, 2019), the Pastoral Medical Association (PMA), and the Ministry International Institute (2019). The NCCA says, "We train, certify and license" (NCCA, 2019). The PMA (2019) says its primary purpose is "[p]roviding a constitutionally protected path for practice (license), for spiritually-minded professionals who wish to offer natural, ecclesiastical based physical, mental and spiritual health services." These organizations issue diplomas and licenses. They make no claims that their degree programs and their licenses meet the standards or requirements of state-endorsed or other government-endorsed or generally accepted accreditation agencies. However, their degree holders or licensees sometimes advertise in such ways that their lack of professionalism is not made clear to consumers of their services. At this time, we have no solid basis for estimating the number of these providers, although we can observe that practitioners advertise their services online and sometimes fail to define their degrees and licenses accurately These licensees appear again later in the discussion of licensing.

In contrast with any other category of licensed health care providers, would this lack of training for unlicensed individuals or agency-employed providers and lack of credentialing be tolerated? Would anyone wish to receive medications from a pharmacist who had no training and no professional credentials? Would anyone wish to receive treatment from a cardiologist who had no training and no professional credentials? Would anyone wish to take his/her child to a pediatrician who had no training and no professional credentials? Despite the predictable and correct answer to these questions, untrained and unqualified mental health care providers are tolerated on a very large scale. What, if anything, should be done about this? Obviously, the procedures through which this situation is changed will take some time to develop, but why would professionally accountable counselors and psychotherapists not try to develop them? The emphasis here is that the conspicuous lack of professionally accountable regulation should be corrected. Later, in the discussion of licensing reform, the misleading advertising of licenses that fall extremely below minimal standards become another contributor to the urgency of reform of licensing for mental health care providers.

PREPARATION PROGRAMS

The information presented in this book calls attention to an uncomfortable question: "Where have we gone wrong in our preparation programs for counselors and psychotherapists?" If there are contributions in this book, one of them is the exposure of thousands of aberrant approaches to counseling and psychotherapy. However, in contrast to this exposure, the idea that preparation programs that ostensibly equip individuals to begin their service as counselors and psychotherapists are inadequate is not newly identified in this book. Indeed, the view has been emerging for at least 20 years. An early example of the idea came from Alvin Mahrer (1999) in his article, "Embarrassing Problems in the Field of Psychotherapy." Among other embarrassments, Mahrer wrote, "Psychotherapists are licensed, registered, certified, and accredited without having to demonstrate that they can do psychotherapy" (p. 1151). Worse, he stated, "A program to train a person to do what psychotherapists do takes about two or three days" (p. 1152). The apparent meaning of this is that the demands for effectiveness, competence, and professional accountability are low. More pointedly, training to deliver counseling and psychotherapy may have little impact on those who receive training.

Between the publication of Mahrer's article and recent years, several researchers have raised doubts about the impact of preparation programs for counselors and psychotherapists. Based on gathering a massive volume of data in their "comprehensive meta-re-analysis" of competence and experience in psychotherapy, Spengler and Pilipis (2015) report that experience improves clinical decision-making only marginally. Their analysis of data prompted them to assert, "We strongly urge graduate counseling and psychology to seriously reconsider their approach to training clinical assessment, judgment, and decision-making" (p. 373). Erekson, Janis, Bailey, Cattani, and Pedersen (2017) investigated client outcomes for graduate students, interns, postdoctoral trainees, and licensed professionals. According to their analysis of the data, they concluded that "a potentially radical implication of these study results—that existing structures for training psychologists, including formal graduate training, internship, postdoctoral appointments, and licensure, do not appear to be connected to improvement in client outcomes or therapy efficiency" (p. 522). Similarly, Hill et al. (2015) studied psychology doctoral trainees, based on their earlier conclusion that "there is only tentative evidence that graduate training is effective" (p. 184), and further concluded that their then current study produced mixed and additionally tentative indicators of the effectiveness of training. Recently, the Society for the Advancement of Psychotherapy (2018) has promoted its view that "graduate programs must take greater responsibility for students who are not achieving appropriate outcomes and may not be suitable as therapists." It adds,

> This can be a touchy subject. Yet, just like any other profession, students will vary in performance and some may not have the potential to be effective therapists. It is very difficult to accurately assess who will be inappropriate

for clinical work during the graduate school application process. Therefore, *once it is clear a student is not capable of providing appropriate clinical care, we should take it upon ourselves to ensure (with empathy and care) that they seek alternative career paths.* Educators should ask themselves, "Would I want my partner/family member seeing this student?" (Society for the Advancement of Psychotherapy, 2018, emphasis as in the original)

At this time, researchers and other careful and objective observers clearly and justifiably conclude that preparation programs for counselors and psychotherapists lack sufficiency. This necessitates the question, "What may be done to improve preparation for counseling and psychotherapy?" In response to this question, it should be said that this book is not large enough to provide a comprehensive and detailed answer to this question. However, it can offer some recommendations for improving preparation programs, such as the ones that follow. Readers are encouraged to regard these recommendations as a call to action toward urgently needed reform of counseling and psychotherapy and to expand these recommendations with their own improvements and added recommendations.

Among major professional groups, agreed upon standards of clinical performance should be established. The major groups include the ACA, American Psychological Association (APA), National Association of Social Workers (NASW), and the American Association for Marriage and Family Therapy (AAMFT). The need for this is clear. Among other indicators of this need, as noted earlier in this book, a license to practice as a counselor or psychotherapist may be based on a graduate degree program that requires only 28 graduate semester hours or one that requires as many as 130 semester hours. As another indicator, very little evidence shows that professional training of any size results in improved client outcomes. With the emphases in this book (e.g., highly individualized approaches to therapy and tens of thousands of very unusual approaches to therapy), another indicator of the need for agreed upon standards of clinical performance is that there are none. Further, as this chapter strongly suggests, either counselors and psychotherapists reform their profession, or others—namely, political and religious leaders—will. Establishing agreed upon standards of clinical performance is a good place to begin. Without them, the mental health care community may have little to guard itself against predictable assaults, including assaults from within the mental health care community.

Among major preparation groups, agreed upon standards of clinical education should be established. The essential rationale for having preparation programs for counselors and psychotherapists is that those who complete such programs can deliver competent and professionally accountable clinical services. Without this, preparation programs have no reason to exist. While the ongoing decrial of the sometimes tragic failure of preparation program may come easily, the extremely more important challenge is to create programs that produce effective clinicians. Currently, the obvious lack of agreement about standards for such programs allows programs that are sometimes misogynistic, sometimes homophobic, sometimes captive of conspicuously nonclinical ideologies, sometimes provided by clearly unqualified faculty, and so on. Instead of this condition, professionally accountable mental health care associations and their members should establish clinically relevant standards for preparation programs.

"Good psychotherapists strive for self-awareness, appreciate their professional limitations, and monitor the quality of their services" (Knapp et al., 2017, p. 170). With this commonly, if not universally, held conclusion, *psychotherapy for candidates who wish to become counselors and psychotherapists should become a mandate for these candidates.* The need for psychotherapy for prospective counselors and psychotherapists stems from two concerns. One is that the candidate should be sufficiently and competently aware of personal issues that may enhance or frustrate her/his effective delivery of clinical services. The other is that faculty and clinical supervisors should know about the candidate's personal issues so that they can endorse the individual or not, based on their assessment of these personal issues. The obligation for practicing counselors and psychotherapists to possess a high level of self-awareness is neither new nor lacking support. Mandated psychotherapy for prospective counselors and psychotherapists is a way to facilitate self-awareness.

Graduate preparation programs should become much more clinical than they are. This implies that they should be much less academic than they are, but this need not be the case. Both clinical preparation and academic preparation can be sustained and mutually supportive. One should not supplant the other. Instead, because the more important criterion for success of preparation is that licensed counselors and psychotherapists achieve positive outcomes for their clients, this should be the necessary condition for completing a preparation program, regardless of the intellectual and/or academic strength of individuals in preparation programs. Hill et al. (2015) confirm the importance of clinical experience: "Clinical experiences, both in the graduate program and in the clinic, were perceived as influential to growth. Specifically in the clinic, therapists liked that they had learned how to conduct intakes, assess client strengths and pathology, negotiate and collect fees, establish and maintain boundaries, trust their instincts, be more open and authentic, and manage countertransference reactions" (p. 199). Implicit in their report, Hill and her colleagues affirm the recognition that, after all, equipping their students to provide effective clinical services is the reason and purpose for preparation programs.

Naturally, the recommendation for preparation programs to be much more clinical than they are begs the question of what may be different if such a recommendation were implemented. Preparation programs for counselors and psychotherapists should avoid the comfort and security that come with those who function as "academic ruminants," to use Nietzsche's somewhat derisive term (Howard, 2000, p. 280). Analogously, preparing "clinical ruminants," those who have absorbed many of the right and good ideas about counseling and psychotherapy but whose clinical activity more closely resembles cattle who wander from one tasty bit of information to another and spontaneously defecate at will, forebodes poor clinical outcomes. Instead, such programs should persist in engaging students in the existential immediacy of participating in clinically related and relevant activity. This is not an argument for or against existential thinking. It is an emphasis on clinical activity, along with clinically relevant intellectual development.

If a new model for a preparation program appeared today, what might be its important features? If it achieved the "right" level of emphasis on clinically relevant activity, perhaps it would involve beginning graduate students in clinical activity, such as participating in a therapy group as clients. Concurrent with this participation, students would also study group psychotherapy and, through these two activities, gain experience with therapy, gain knowledge about such experience, and integrate the two. The group would be led by advanced students in the program and by clinically skilled faculty. From the initiation into clinical work through being a client in a therapy group, students would become involved in progressively accountable clinical activity until they begin independently to treat clients. The progression from being outside the professional role to living inside the professional role would develop according to established standards for endorsing each student's acquired clinical competence and intellectual understanding of counseling and psychotherapy. The student would advance through her/his program according to her/his successful demonstration of established competencies and criteria for assessing these competencies.

Place emphasis on the best available evidence for clinical decision-making. This emphasis carries specific and impactful implications for preparing counselors and psychotherapists. One of those is that preparation should avoid—or, more precisely, escape from—ideology-based and ideology-driven approaches to mental health care, including existential, psychoanalysis, feminist therapy, and other long-taught ideologies, but also numerous others including the many multiples of "Christian" and other specifically religious approaches. Instead of these various oppositions to evidence, preparation programs should ask a simple but daunting question, "What is the best available evidence for teaching counseling and psychotherapy and for conducting counseling and psychotherapy?" Surely, an overanxious individual may choose to pray to God for assistance in "settling my nerves." Prayer, though, is not an evidence-based approach to counseling and psychotherapy. [As an aside, I, Francis Martin, may assuage the concern from advocates of prayer by reporting that I have written books on prayer and related experiences.] With reference to using evidence for specific treatment interventions, would, say, ophthalmologists use their own spit to treat eye problems? To elaborate, what is the best available evidence for treating eye-related problems, such as glaucoma or blindness? According to the Bible,

They came to Bethsaida, and some people brought a blind man and begged Jesus to touch him. He took the blind man by the hand and led him outside the village. When he had spit on the man's eyes and put his hands on him, Jesus asked, "Do you see anything?" He looked up and said, "I see people; they look like trees walking around." Once more Jesus put his hands on the man's eyes. Then his eyes were opened, his sight was restored, and he saw everything clearly. Jesus sent him home, saying, "Don't even go into[a] the village." (Mark 8:22–26, New International Version)

In the case of blindness or other eye-related problems, should the suffering person want to consult with an ophthalmologist or seek a faith healer who will spit in her eyes? Of course, normative good sense and a serious pursuit of effective treatment will lead to the ophthalmologist. Nothing about consulting with an ophthalmologist denies or contradicts an individual's personal faith as much as a naïve and anti-intellectual dependence on "evidence" that has no basis in reality. Worse than a patient who seeks treatment that involves spit in the eye is the appropriately trained and duly licensed ophthalmologist who uses spit in the eyes of patients. Unfortunately, an ophthalmologist who uses such a practice would bear a disappointing resemblance to the treatments of numerous appropriately trained and duly licensed mental health care providers.

Educators in preparation programs should seek to implement the best of evidence that supports their teaching. Teaching in higher education, including graduate education, has been studied for decades. Relatively little study has been given to creating the best approaches to teaching mental health care for prospective counselors and psychotherapists. Nevertheless, evidence-based approaches to teaching in mental health care programs are available. Readers may want to explore ways to improve their teaching through these resources. For example, they may wish to locate and study the following resources:

Feinstein, R., Heiman, N., & Yager, J. (2015). Common factors affecting psychotherapy outcomes: Some implications for teaching psychotherapy. *Journal of Psychiatric Practice, 21*(3), 180–189.

Mackrill, T., & Iwakabe, S. (2013). Making a case for case studies in psychotherapy training: A small step towards establishing an empirical basis for psychotherapy training. *Counselling Psychology Quarterly, 26*(3–4), 250–266.

Malott, K. M., Hall, K. H, Sheely-Moore, A., Krell, M. M., & Cardaciotto, L. A. (2014). Evidence-based teaching in higher education: Application to counselor education. *Counselor Education and Supervision, 53,* 294–305.

McCauliff, G., & Eriksen, K. (2011). *Handbook of counselor preparation: Constructivist, developmental, and experiential approaches.* Thousand Oaks, CA: Sage.

Osberg, T. M. (1997). Teaching psychotherapy outcome research methodology using a research-based checklist. *Teaching of Psychology, 24*(4), 271–274.

Wheeler, K., & Delaney, K. (2008). Challenges and realities of teaching psychotherapy: A survey of psychiatric-mental health nursing graduate programs. *Perspectives in Psychiatric Care, 44*(2), 72–80.

These resources will likely not be universally applicable nor relevant. However, whether or not these particular resources prove to be useful for all instructors who seeks to improve their teaching, the emphasis here is larger than these resources. Those who teach psychotherapy should provide exemplary clinical functioning as well as exemplary stewardship of their students' needs, anxiety, and goals through superior, evidence-based teaching. Just as instructors in clinical programs prepare their students to integrate the best evidence for their clinical work, the instructors themselves should conduct their teaching according to the best evidence for teaching prospective counselors and psychotherapists. Lacking such an approach, instructors may perpetuate the likelihood of the demise of the program in which they teach and the profession of which they are members.

Preparation programs for counselors and psychotherapists should *integrate specialists in clinical supervision,* instead of operating on the assumption that instructors possess superior clinical supervision skills based on their clinically related accomplishments, including licensing. Surely, degrees, clinical work, and licensing are necessary markers of professionalism and preparation for providing clinical supervision, but they seriously lack sufficiency. For clinicians who have little or no training in clinical supervision but seek to provide it, several important constructs and practices may seem foreign. For example, what is their understanding of "vicarious liability?" Or, what is their understanding of "parallel process?" To which specific case law decisions can they refer that provide guidance for their clinical supervision?

As a complementary recommendation to integrating specialists in clinical supervision into preparation programs, mental health care preparation programs of all kinds should create specific standards and possibly a credential for all clinical supervisors. Analogous to developing a specialization in, say, play therapy, sex therapy, or acceptance and commitment therapy, those who develop a specialization in clinical supervision should possess special knowledge and skills. Insofar as this specialization is pursued, it should include the pursuer's ability to answer several basic questions about clinical supervision, such as the following:

What is clinical supervision?
What is a philosophy of clinical supervision?
What are counseling and psychotherapy?
What do I believe are the necessary outcomes of clinical supervision?
What is the role of the supervisor?
What is the role of the supervisee?
How do I define the good clinical supervisor?
How do I conceptualize the good or excellent counselor or therapist?
Why is my philosophy of clinical supervision a necessary feature of my
    supervision?
With which primary model of clinical supervision do I most closely
    identify?
What are the ethically sensitive issues that I am likely to encounter in
    conducting clinical supervision?

What is informed consent in supervision and why would I want to
   include it?
What do my supervisees need to know about me?
What do I need to ensure that my supervisees know about the process and
   procedures of clinical supervision?
What should I include in a supervision contract?
How will I evaluate my supervisees?

These questions are indicators of a much larger body of questions for which competent clinical supervisors will want answers. The volume of valuable materials
that can help clinical supervisors or prospective clinical supervisors to answer
these questions is quite large. One of those is an edited work by Watkins and
Milne (2014), *The Wiley International Handbook of Clinical Supervision*. In this
Handbook, Watkins and Milne include several chapters from which specialists in
clinical supervision could readily receive assistance. One is "Toward an Evidence-
Based Approach to Clinical Supervision" by Milne (pp. 38–60) and another is "On
the Education of Clinical Supervisors" by Watkins and Wang (pp. 177–203). Also,
Martin and Turner (in press) have prepared a book, *Clinical Supervision in the
Real World*, that will assist clinical supervisors in their pursuit of a professionally
accountable specialization. The previous questions are adaptations of material in
Martin's and Turner's book, *Clinical Supervision in the Real World* (2020).They
build on the earlier work by Martin and Cannon (2010b).

*Preparation programs for counselors and psychotherapists should integrate models
of highly effective mental health clinicians.* The importance of knowing the characteristics of highly effective mental health clinicians is that it gives instructors
and their preparation programs needed goals that they should seek to achieve.
Specifically, preparation programs should equip individuals to provide highly effective mental health care services. The models that may be adopted by a preparation program allow for considerable flexibility and experimentation. For example,
if the classic study by Jennings and Skovolt (1999) is compared with the study
by Six (2014), a plausible conclusion, among many possible conclusions, is that
a preparation program has considerable latitude to shape a clinical curriculum
that its faculty deems to be appropriate. Toward integrating models of highly effective clinicians, preparation programs will also want to include evidence-based
methodologies for developing such clinicians. One of these methodologies is "deliberate practice." Readers may wish to refer to the article by Chow et al. (2015)
and the book by Rousimaniere (2017). Quite apart from integrating this model or
other models into the curriculum of a preparation program, practicing clinicians
will gain ideas and methods for improving their effectiveness when they utilize
models of highly effective clinicians.

Another important feature of integrating models of highly effective mental
health clinicians is that it enhances the ability of program leaders to articulate
and implement specific, encompassing, and clinically relevant learning objectives.
Likewise, the importance of integrating models of effective practice, such as "deliberate practice," is that program leaders also have specific methodologies for

assisting their students in their professional development toward being effective clinicians. Thus, models of effective practice development may also be translated into specific and clinically relevant learning objectives.

Preparation programs should ensure that faculty hold competencies that are expected of the individuals whom they are training. Clinical faculty competence should be placed in the context of the primary reason for the existence of the preparation program in which they teach. If the primary reason is to prepare scholar-researchers, faculty whom the program employs should be demonstrably competent scholar-researchers. Obviously, the same rationale should be seen in programs that seek to prepare mental health care clinicians. However, allowing for diverse missions among preparation programs, each program should establish specific and measurable standards with which faculty performance is measured.

While long-standing commitments within the mental health care community do not necessarily translate into effective counseling and psychotherapy, they should continue to stand. Specifically, allowing for ambiguities that pose challenges to their effective implementation, preparation programs *should maintain their emphases on openness and diversity, social reform, and social justice.* All of the codes of ethic from professionally accountable associations emphasize the importance of the inherent values and necessity of diversity, social reform, and social justice. Conversely, mental health care stands against harmful prejudice, mindless bigotry, and injustice in its many forms. It decries racism, sexism, classism, and other harmful "isms."

What may be the professionally accountable and clinically relevant impact of these recommendations or others that may be needed? The intended impact is that preparation programs will equip their completers to begin their clinical work with the reasonable expectation that they will provide effective clinical services. As an aim for preparation programs, this is as right as it is necessary.

However, as right and necessary as this aim may be, it leads to consequences that many individuals and preparation programs will almost inevitably reject, if it is done well. If needed reform is pursued and executed well, it will likely terminate or require serious upgrades of deficient training programs. Terminating deficient programs should not be the primary goal of reform, but it may be an inevitable outcome. At this time, there may be no way to generalize about deficient preparation programs. Allowing for this constraint, perhaps illustrations of deficient preparation programs will help.

One is a state-supported university in Tennessee that offers a "mental health counseling program" that ostensibly prepares its graduates for a license from the state. It has almost never had a licensed mental health care provider on its faculty and thus is deficient. Another is a theological school in Pennsylvania that prepares its graduates for licensing as mental health care providers. Through its "missional character," it prepares its students and graduates to pursue licensing and, with their licenses, "to skillfully apply the grace and truth of the Gospel first to their lives and then to those of their counselees" (Biblical Theological Seminary, 2018). Its missional character is entirely within their rightful prerogatives, but to equate their prerogative with counseling and psychotherapy lacks intellectual honesty and professional accountability. Advocating and living within a contemplative education community, another institution that prepares its students and graduates for licensing also prepares them "to be compassionate, skilled, and knowledgeable professional counselors by drawing on the insights of the world wisdom traditions, experiential self-reflection, and contemporary empirical findings in order to work inclusively with diverse populations" (Naropa University, 2018). Analogous to the Biblical Theological Seminary's exaggerated emphasis on a Gospel-centric approach to counseling and psychotherapy, Naropa demonstrates an exaggerated emphasis on a contemplation-centric approach to counseling and psychotherapy. It offers a specialization in "Wilderness Therapy" and describes it as "a concentration that integrates clinical and theoretical course work in counseling psychology with contemplative practice" (Naropa University, 2018). We should make clear that we have no argument with the Gospel or a contemplative lifestyle. These are good and worthwhile views of life. However, neither should be the basis of nor the primary approach to counseling and psychotherapy. More than a commitment to the welfare of clients, the Biblical Theological Seminary and Naropa University demonstrate a commitment to their respective ideologies.

In Tennessee, the details about grossly deficient programs that allege to train mental health care providers come in complex and large volumes (Martin & Cannon, 2010a). To avoid undue complexity here, suffice it to say that among the approximately 60 programs that train mental health professionals for licensing, several are demonstrably and grossly poor ones. Locating such programs in Tennessee or elsewhere poses little challenge. The fact that professionally accountable counselors and psychotherapists have done little or nothing about them is the problem.

LICENSING FOR MENTAL HEALTH CARE PROVIDERS

The need to reform licensing for mental health care providers is not new. In 1998, cited by Grohol (1998/2004), Goodrich reported, "A recent multi-year study commissioned by the State Legislature found that there was no evidence that licensing provided any degree of consumer protection overall." Even earlier, in 1983, Hogan asserted, "It is absolutely essential to recognize that licensing laws are not meant to ensure a high level of professional competence, only that a practitioner is not likely to harm the public" (Hogan, 1983, p. 134). Further, based on his analysis of the effectiveness of licensing, Hogan offers several proposals.

First, any definition of practice regulating a health care profession should be narrowly and precisely drawn, in sharp contrast to the broad and vague definitions now in force. . . .

Second, standards and criteria to determine who is qualified to practice should be competence-based and related to actual performance. Academic credentials should only be used if they pass this test.

Third, alternative paths to certification should be kept open, including apprenticeships, proficiency examinations, educational equivalency exams, or certification by appropriate accredited organizations.

Fourth, a high degree of emphasis should be placed on the training of competent paraprofessionals.

Fifth, the design and administration of regulatory policies and programs should include a balanced representation of appropriate constituencies. Normally, these would involve the public, professionals, government officials, clients, and any other affected parties.

Finally, licensing laws should only restrict the use of certain titles, not the right of a person to practice. (Hogan, 1983, pp. 133–134)

The question of whether Hogan's proposals have currency with the practice of psychotherapy more than 35 years after he wrote them may cause readers to miss the significance of what he wrote, which is that the need for reforming licensing in fundamental ways has been around for many years and has not yet occurred. What must be added, though, is that the need for reform is more critically urgent than it was in 1983. In 1983, the goal of reform was refinement of clinical services. Today, the goal of reform is survival of clinical services and the profession that delivers them.

Individuals and families who need mental health care often confront a truly mystifying array of licenses for mental health care providers. Should they seek help from a Licensed Clinical Social Worker (LCSW), a Licensed Psychological Examiner (LPE), a Licensed Master Social Worker (LMSW), a Licensed Professional Counselor (LPC), a Senior Psychological Examiner (SPE), a Licensed Psychologist (LP), a Temporary Licensed Professional Counselor (TLPC-MHSP), a Licensed Marriage and Family Therapist (LMFT), a Licensed Nurse Practitioner (LNP), a Licensed Professional Counselor with Mental Health Service Provider designation (LPC-MHSP), a Licensed Psychologist with Health Service Provider designation (LP-HSP), a Licensed Advanced Practice Social Worker (LAPSW), or a Licensed Alcohol and Drug Abuse Counselor LADAC)? These are the designated mental health licenses in the state of Tennessee. In Appendix B, the reader can find more than 100 titles of licenses.

So, to which one should an individual or family go for help? Often, consumers/clients see only the acronyms of the professional license, leaving them to see confounding items, such as those in Table 7.5, from only Tennessee.

The individuals who hold one of these licenses receive authorization from the state of Tennessee to provide services that are almost identical to all of the services that holders of the other licenses receive authorization from their state to provide,

Table 7.5 Types of Mental Health Care Licenses in Tennessee

LAPCSW	LCSW	LPE	LMSW	LPC
SPE	LP	TLPC-MHSP	LMFT	LNP
LPC-MHSP	LP-HSP	LADAC	MD	APRN

with only a couple of exceptions, including MDs and nurses. So, why would anyone want to maintain this confusing array of licenses? Worse, from the view of those who need help, why would this confusion be imposed on them? Why not integrate all of these licenses into one license? Or, alternately, why not integrate these licenses into as few licenses as possible? Maybe, all of the nonmedical licenses could be integrated into one or two. One could be a temporary license, and the other could be a permanent one.

More than two decades ago, Loretta Bradley, a national leader among professional counselors, wrote, "It is apparent that the counseling profession is facing a dilemma with regard to certification and licensure. As specialties have flourished at both the state and national levels, the title distinctions have created confusion and a lack of unity" (Bradley, 1995, p. 185). Clearly, the condition of certification and licensing among counselors is more confusing and chaotic than it was more than two decades ago. Further, when certification and licensing for professional counselors are placed in the larger mental health care provider picture with psychologists, social workers, and others, the scene of chaos, cacophony, and confusion becomes even more intense and formidably threatening for counseling and psychotherapy. Thus, again, our recommendation is that, following the establishment of agreed upon standards for preparation programs and professional competence, all licenses should be reduced to one or two.

This proposal should not exclude anyone who is currently licensed from continuing to hold his/her license. Instead, either all should retain their current license or receive "grandfathering" endorsements for the new, encompassing license. The proposal intends to help, not hurt, licensed mental health care providers and to increase the viability of their useable future.

However, reform of licensing should also seek to eliminate nonclinicians who allege to be licensed. For example, as already noted, the NCCA (2019) website says, "We train, certify, and license Christain [sic] counselors." The NCCA promises its prospective candidates for licensing that they will very likely have extraordinary success with their clients, claiming that "Counseling With Proven Success: The Arno Profile System (A.P.S.) is an easy-to-learn counseling technique that has achieved a success rate of over 90 percent" (NCCA, 2018). With their "license," the licensees may advertise that they are licensed. With or without credible degrees or training or any other credentialing, these individuals advertise services that look very much like services that well-trained and credentialed licensees— those with state-issued licenses—advertise. Recognizing that NCCA is merely one of many groups that provides licensing for counselors and other mental health–related titled providers, reform of counseling and psychotherapy should seek to eliminate the possibility that misleading promotion of licenses that have little or no credible reason to exist can be advertised.

To illustrate the insidiousness of licensing from conspicuously nonclinical groups, the Abundant Life Christian Counseling Services and Training Center says that it offers "Certification, Licensure and Degree Training Programs" and that it is a "Certified Academic Institution through the National Christian Counselors Association. The N.C.C.A. is recognized as being one-of-the best, most-effective and the largest training and licensing agencies/associations, for Christians, in the world." And, it boasts, "As a partner with N.C.C.A., the tuition remains low because of offering the studies online and eliminating unneeded class subjects such ash [sic] History, Science, Math, etc." (Abundant Life Christian Counseling Services and Training Center, 2018). Clearly, the NCCA and related organizations constitute merely the tip of the proverbial iceberg. In the previous discussion about the need for reform in the regulation of counseling and psychotherapy, PMA and the Ministry International Institute (2019) were identified as organizations that offer degree programs and licenses.

This book has already provided information about the massive proliferation of certifications for counselors and psychotherapists. The report from Holden and Barker (2018) includes a list of 58 truly bizarre certifications for licensed clinical social workers. Independently, we have located approximately 1,800 certifications for mental health care professionals. Compatible with Holden and Barker's assessment of the certifications that they list, a majority of the certifications on our list are truly strange. Therefore, we conclude that reforming counseling and psychotherapy should entail reform of certifying mental health care professionals, including broadly agreed upon standards for issuing certifications and limiting their issuance to professionally accountable agencies, colleges, universities, and institutes that receive endorsement from major mental health care associations. Without such limitations, certifications, along with licensing, will continue to erode the credibility and effectiveness of mental health care.

The long and complicated history of each of the major types of mental health care licenses deserves considerable understanding and respect, especially for many mental health care professionals who have been a part of this history for almost all of their careers and for many years. Allowing for this, despite the hugely important, personal investments in their respective professional histories and the professional identities that come from these histories, change is needed. We have composed these recommendations in good faith and hopefulness about our future in the mental health care provider community, but we recognize the uncertainty and stress that come with our recommendations. Uncertainty and stress, though, have not prevented extraordinary advances and benefits achieved by brave, thoughtful, creative, and persistent leaders, almost all of which we could not have imagined. This suggests that, just as Sigmund Freud, Carl Rogers, George Kelly, Aaron Beck, and many others invested their considerable personal and professional assets in our potential, we can do this, too. Despite the heartfelt admiration that counselors, psychologists, and social workers should have for their histories, mental health care should not be imbedded in silo-like histories that are no longer relevant, any more than, say, medicine should be imbedded in the practice of barbering, as it was many years ago. Moreover, the respective histories of counseling, psychology, social work, and marriage and family therapy can be enhanced and honored by creating one, substantial, clinically based, and truly professional license for all mental health care providers, only because their clients will be better served than they are now.

Also, with regard to initiating professionally accountable regulation, reforming professional ethics, revising preparation programs, and undertaking other needed reforms, all major professional associations should work together. One group alone, such as the ACA, APA, NASW, or AAMFT, cannot achieve effective reform. Specifically with regard to the need to reform licensing for mental health care providers, the need for reform has historically been based on the importance of providing competent services for clients and improving clinical services. These priorities continue to be reasons for reform. Today, however, the call for reform of licensing is based on the added need to ensure the survival of clinical services.

AN UNRESOLVABLE TENSION

During his long and distinguished career in psychotherapy and, more broadly, as an influence on multiple aspects of the way life is perceived, Carl Rogers spoke and wrote eloquently about psychotherapy and its potential. No doubt, Rogers raised the level of understanding of psychotherapy and its effectiveness. The process of making his many and enduring contributions, though, did not come with sudden, expected, or realized easiness. Instead, he frequently identified the necessary tension between developing disciplined, high standards for psychotherapists, while simultaneously advocating for freedom, creativity, and originality. When he asked the question, "Where are we going in clinical psychology?" Rogers (1951) affirmed the need for training, high standards, competence, and credentialing, but he also warned about the potential for stifling freedom and originality and for maintaining the status quo. Looking forward, this is a necessary tension that we should welcome. How may we advance the cause of the clients we serve and establish and maintain high standards of ethics, practice, licensing, and other features of our profession, according to the best standards we can create, while simultaneously affirming and engaging in the adventure and immense personal fulfillment that come with service to our clients' best interests? Our belief is that this is what we do collectively when we are effective in our work as truly professionally accountable counselors and psychotherapists. We engage with our clients with focused and disciplined objectivity and analysis of their problems, while sensitively relating to them with supportive understanding and empathic awareness.

In a not-so-subtle manner, counseling and psychotherapy reform has been referred to in this book as "CPR." A more common meaning of these letters is "cardiopulmonary resuscitation." When the more common meaning refers to an activity, the urgency of resuscitating an individual whose heart is failing is quite clear. CPR for counseling and psychotherapy is not quite as clear, but it is nevertheless urgent. This chapter began with a list of some of the causes of this urgency. Our aim here is to provide evidence and a rationale for necessary changes in counseling and psychotherapy, but also to invite all counselors and psychotherapists to engage in the dialogue that leads to productive and durable changes—CPR!

Scott Miller founded and directs the International Center for Clinical Excellence. Through his research and advocacy for excellence in psychotherapy,

Miller has become a visible and important influence on psychotherapy. In his 2013 address, *The Evolution of Psychotherapy: An Oxymoron*, he provided details about the perilous condition of psychotherapy and the possible extinction of psychotherapy. Yes, he spoke about the possibility of the extinction of psychotherapy. If this is a possibility, it adds to the urgency of CPR. [His address may be easily found and heard by conducting an Internet search to locate his YouTube presentation (Miller, 2013).] In addition, throughout this book, we have included other voices that have raised alarm about the condition of psychotherapy, as well as reporting our research. Collectively, the voices continue to get louder. Together, they call for CPR.

Many of us who have provided mental health care services dread that our clinical independence may be seriously compromised if reform is effectuated. This dread is understandable. Because of this, reform should seek to maintain and improve the delicate balance between rigorous scientific affirmations of our work and the unique human factors that appear each time we meet with a client. The tension between these two features of our work have always been a part of it. Because both are necessary, they should always be features of our work.

CONCLUDING THOUGHTS

This chapter opened with a list of symptoms of the compromised health of counseling and psychotherapy. Following the list of symptoms, the chapter introduced the misguided legislative initiatives in the state of Tennessee. Members of the professional community of counselors and psychotherapists may have understandably wondered why a discussion of the initiatives in Tennessee would be prominently featured in a chapter on reforming counseling and psychotherapy. Briefly stated, the discussion of legislative initiatives in Tennessee was intended to identify another symptom of the compromised health of mental health care and to call attention to the fact that counseling and psychotherapy are being reformed, whether counselors and psychotherapists know about it or not, let alone whether they agree with it or not.

Also, the discussion of the legislative initiatives in Tennessee serves to alert others in the mental health care community to the need to consider how they would want to improve their profession. With regard to improving our profession, we recommended several courses of action, accompanied by brief explanations of the recommendations, based partially on our strong belief that mental health care is health care. At the end of our research and analysis of huge volumes of information, our conclusion is that counseling and psychotherapy reform (CPR) is urgently needed. Nearing the conclusion of this chapter, we should say that the urgency of needed reform brings no satisfaction to us. Instead of satisfaction, we feel deep concern about the status of our profession, broadly defined, to include all mental health care providers.

To be clear and certain, CPR is necessary. Based on this broad conclusion, the following views are affirmed.

One, as discussed earlier, counseling and psychotherapy are health care. This should be a matter of serious and sustained advocacy by counselors, psychologists, social workers, and others. While the specific activities among counselors, psychologists, social workers, and others may be seen in a very large number of places, styles, and personal beliefs, they share the broad mission of helping others to heal and to recover from problems. So, whether the named health care providers are physical therapists, pharmacists, pediatricians, oncologists, or psychotherapists (i.e., social workers, counselors, psychologists, or others), they stand together as health care providers.

The delivery of mental health care necessarily involves ambiguity. If Tennessee's Senator Johnson or others who share his views wish to protect the religious prerogatives of a minority of mental health care providers, they should do this, while simultaneously recognizing that mental health care is not essentially a religious activity. It is not an activity that can possibly be validated on the basis of "sincerely held religious beliefs." When counseling and psychotherapy receive their validation by religious belief, they are no longer health care; they are instead means of religious expression.

Surely, most politicians and mental health care providers recognize, as we do, that their religious and spiritual values truly matter in their work. However, while these values matter and may provide important guidance, they are also limiting. For example, Christian grace may require some counselors and psychotherapists to demonstrate heightened and sensitive compassion toward those who suffer. Such compassion should not, however, evolve into the necessity of identifying a broken leg as "Christian," depression as "Muslim," breast cancer as "Methodist," pain management as "Anglican," marital conflict as "Mormon," psychological trauma as "Buddhist," or an oppositional defiant disorder as "Lutheran." More generally, counseling and psychotherapy should not be conflated with "being Muslim," "being Christian," or "being Buddhist" or with other religious or spiritual experiences. Such thinking betrays both those who suffer with mental-emotional challenges and the religious-spiritual experience of Christians, Muslims, Methodists, Mormons, Buddhists, Lutherans, and many others who are not mentioned here.

To frame this issue another way, do politicians plan to make specific religious standards requirements for holding a political office, such as State Senator in Tennessee? If not, why not? Maybe, because of sincerely held religious beliefs, they will make clear to their constituents that only those who agree with their sincerely held religious beliefs may vote for them. After all, given what Senator Johnson, or others like him, is saying about how he wishes to regulate professional counselors and other health care providers, he should accept only the votes of those who agree with his religious beliefs. If he solicits votes from his constituents, knowing that some of them disagree with his political views, should he take a clear and ethical stand and refuse to accept their votes? If he does not do this, is he unethical according to his "reasoning"?

Two, a nationally and internationally recognized code of ethics for mental health care providers is necessary. Analogously, without one, say, a pediatrician may refuse to provide urgently needed care for Senator Johnson's children—or children of others like him—because the pediatrician sincerely holds political beliefs that diverge from Senator Johnson's. The issue at hand is not Senator Johnson or his children. The compelling issue or question is what guides the clinical decision-making of counselors and psychotherapists. An agreed upon code of ethics would settle or help to settle this issue.

Toward encouraging reform that benefits those who need mental health care, the various mental health–related associations should immediately pursue reform of professional ethics so that, instead of subjecting providers and their clients to multiple codes of ethics, they comply with only one code of professional ethics.

Three, toward encouraging reform that benefits those who need mental health care, the various mental health–related associations should immediately pursue reform of regulations of the practice of mental health care services and the providers who deliver these services. The aim of such reform is, first, to provide the best possible services for clients. Beyond this, such reform protects mental health providers by, at least, requiring them to eliminate harmful and likely harmful approaches to counseling and psychotherapy.

Four, toward ensuring a viable future for clinical preparation programs, academic institutions should immediately begin to reform their approaches to training counselors and psychotherapists. Such reform should develop along several important lines, including establishing agreed upon standards for effective clinical functioning, developing specializations in clinical supervision, and giving a higher priority to training effective clinicians.

Five, toward encouraging reform that benefits those who need mental health care, the various mental health–related associations should immediately pursue reform of licensing for mental health care providers so that only one kind of license is issued by a state. Instead of issuing the huge array of confusing licenses, as illustrated previously, a state would issue only one kind of license for all mental health care providers. Obviously, this would necessitate extensive rewriting of current laws and regulations for counselors, social workers, and others.

As an adjunct to encouraging reform that benefits those who need mental health care, the various mental health–related professional associations should immediately pursue either eliminating grossly deficient mental health provider preparation programs or seriously upgrading them. This, too, would require new laws and regulations. However, this would obviously serve the interests of those who need mental health care.

Reform of mental health care has long been an urgent need. In a foreboding, menacing, and enduring way, the singular failure to initiate reform sits squarely on the shoulders of mental health care providers, including counselors, social workers, psychologists, and marriage and family therapists, not state legislators, governors, or others outside of the professional community. As long as they neglect their commitment to their clientele, they will have acquiesced to others, such as legislators and governors and those who agree with them, who will reform their work for them.

Whether the specific and formidable challenge is deficient training programs, unregulated practice, burgeoning numbers of untrained and unlicensed providers, strange legislative initiatives, the proliferation of licenses, the proliferation of codes of ethics, or others, mental health care providers must face that challenge and immediately initiate reform. If not in their own survival interests, they should pursue reform on behalf of their clients.

In the context of emerging legislation and the other factors that appear to hold the potential of dramatically reshaping mental health care and the work of those who deliver it, we call for leaders of professional associations to facilitate cohesion among the various mental health care associations. These associations include the ACA, APA, NASW, AAMFT, and others. Without cohesion, without one voice that carries a singularly powerful influence on those who wish to control us without our permission, mental health care is likely to be reshaped in ways that do not serve the interests and needs of clients or counselors and psychotherapists.

Our belief is that reform is inevitable. For us, the reforms that we propose are neither palatable nor welcome, but they are necessary. The question is not whether reform is needed or whether it will occur. The questions are, "Who will initiate reform?" and "Who will achieve reform?" Whether we—mental health care professionals—desire it or not, reform will come. If we do not take the opportunity and obligation to get it done, others will. This is the significance of Senator Johnson's proposed legislation.

In their article, "Should Social Workers Be Engaged in These Practices?" Holden and Barker (2018) report more than 400 very unusual approaches to psychotherapy, with many of them appearing to be harmful. At the end of their article, they raise the critically important question that this chapter wishes to raise with every counselor and psychotherapist: "We began by asking if social workers should be engaged in these practices. If your answer is no, then what will you do about it?" (p. 12). Following the presentation of many more than 450 very unusual approaches to counseling and psychotherapy in this book, with many of them likely causing harm to clients, our question is obvious: "What will you do about it?" This hugely implicative question may be emphasized by its counterpoint: "What happens to you if you do nothing about it?"

Counseling and Psychotherapy Reform (CPR)

What Each Counselor and Psychotherapist Can Do

The mandate for reformation is clear. Mental health care is in trouble. If it has a truly professional and useable future, mental health care providers for whom professional accountability matters must take action to establish the likelihood of having such a future.

All counselors and psychotherapists should understand a couple of facts. Anyone can do therapy. Anyone who does therapy can do anything that he or she wishes to do. These two facts represent the massive failure of mental health care providers to demonstrate professional accountability. Lacking accountability, the question is not whether reform of mental health care will come, but who will bring it.

Alarming facts about the condition of mental health care are easily located and easily understood. The need for reform is clear, based on these facts. Despite the clarity of the facts and the need for reform, the overwhelming majority of counselors and psychotherapists and those who train counselors and psychotherapists effectively tune out and dismiss the facts and the need for reform. A couple of anecdotal illustrations and evidence from research may help to establish the reality of these observations.

Recently, along with others, I helped to present some of the research that appears in this book. The setting was a conference that brought attendees from three states. In addition to the charts and observations, our team offered recommendations for major reforms in counseling and psychotherapy, including the ones that appeared in Chapter 7. During the presentation, the question and answer discussions, and in talks following the presentation, no one asked, "What can we do about this?" or "What should we do first?" or "What can we do to help?" Nothing. The audience of approximately 120 individuals may have had an "off" day. Alternately, the presenters may have delivered their information poorly. Also recently, along with others, I received an invitation to present at a university-based conference. Again, we presented our facts, had wonderful, supportive responses, but again no one indicated an interest in needed reforms. They express alarm that some counselors and psychotherapists provided very unusual approaches to their work but seemed to conclude that, if wrong-doing was happening among therapists, others were doing it. The single exception came from a graduate student who at the time worked in a psychiatric hospital as a counselor. She lamented her recognition, "We are taught that everybody should be doing the right thing and that theories are the ones we should try to follow, but this madness that you have talked about scares me. It makes me wonder about our profession." She did not indicate an interest in the need for the reform of counseling and psychotherapy.

Research suggests that counselors and psychotherapists do not see the need for reform. In short, why would reform be needed, if all practicing counselors and psychotherapists are above average in their clinical performance? While this question may appear to reveal an absurdity that lacks relevant support, the evidence suggests that it is, indeed, correct. Walfish, McAllister, and Lambert (2012) report that "[c]linicians rated their skills to be above average compared to other clinicians with similar credentials. On average, they viewed their skills to be at the 80th percentile" (p. 3). They continue,

On average, clinicians believed that 77.01% (SD = 12.63) of their clients improved as a result of being in psychotherapy with them, with 3.66% (SD = 4.91) deteriorating. Nearly two-thirds (58.4%) of the clinicians believed that 80% or more of their clients improved as a result of being in psychotherapy with them. This included 21.2% believing that 90% or more of their clients improved as a result of psychotherapy. (p. 3)

They observe, "Based on the data from the current sample of psychotherapists, all rated their effectiveness as above average when compared to other psychotherapists in the same discipline" (p. 3). Finally, they state that "no psychotherapist self-rated his own skill as below average compared to other psychotherapists" (p. 3).

In their longitudinal study of practicing psychotherapists, Goldberg et al. (2016) raise the question, "Do psychotherapists improve with time and experience?" Their answer is that "the present analyses show that in the aggregate, therapists did not improve with more experience, operationalized as either time or number of cases. Indeed, results suggest that therapists on the whole became slightly less effective over time" (p. 7). These two studies strongly suggest that counselors and psychotherapists believe that they are extremely better than they are. This appears to be confirmed in "a comprehensive meta-re-analysis of competence" among psychotherapists by Spengler and Pilipis (2015). Additional confirmation comes from studies by Imel, Baldwin, Sheng, and Atkins (2015). They conclude that the most important barrier to removing low-performing therapists is that low-performing therapists do not and generally cannot acknowledge that they are low-performing.

Including the information in this book, the evidence for low performance among counselors and psychotherapists stands as irrefutable. Looking beyond the evidence, though, the challenge for the future of counseling and psychotherapy is whether individual counselors and psychotherapists will actively seek to improve their performance. As a practice-based profession, its future depends on the quality of the performance of those who practice it. It is this future that this book intends to influence.

In the introduction to this book, I raised a question, "Why are we—practicing counselors and psychotherapists—committed to failure?" Between the introduction and this chapter, the case against us has been made. It has been made by the accrual of several kinds of evidence and in substantial quantities. The particular kind of failure to which counselors and psychotherapists have been committed is one of theft or, maybe, hostage-taking. We have tacitly allowed numerous practicing counselors and psychotherapists to steal our profession. We have tacitly allowed numerous major social service agencies to steal our profession. We have tacitly allowed numerous academic institutions to steal our profession. We have tacitly allowed numerous inventors of approaches to counseling and psychotherapy to steal our profession. We have allowed numerous unlicensed individuals to steal our profession. Whatever the particular image that may best describe our failure, we have allowed our profession, counseling and psychotherapy, to be stolen by numerous individuals who express themselves through their therapy and only marginally, if at all, deliver professional services for their clients.

And, now, following a review of the evidence, we should ask, "What is the condition of counseling and psychotherapy?" The condition is alarming. Starkly put, gasping for breath and appearing to be dying, mental health care sweats aggressively, while its providers and others paradoxically administer poisonous treatments and believe that they are performing highly effectively.

If members of any other profession performed as poorly as mental health care providers perform in their profession, they would exist only in the memory of their grandchildren. They would be a quaint footnote to an era of social reform that simply petered out decades ago, owing to the pervasive neglect of advancing beyond self-interest, predatory initiatives, outrageous ideologies, shallow good intentions, and other self-defeating characteristics. Framed differently, we—counselors and psychotherapists—do not know who we are or what we are doing. We have become almost everything to everyone and, by virtue of being everything to everyone, we have declined in effectiveness and relevance.

Overly dramatic? Attention-seeking? Inaccurate? Unduly pessimistic? Judgmental? Foolish? Histrionic? Greedy? Self-interested? Dishonest? Possibly, the statements about the condition of counseling and psychotherapy could be all of these. With the evidence presented in this book so far, however, attempts to dismiss the evidence with these distorted attributions merely perpetuate the decline of counseling and psychotherapy. The challenge and obligation for all mental health clinicians is one of answering the question about how they may contribute to reform of their profession, even if they wish to argue against the evidence here or to establish evidence of their own.

Challenging attributions, such as "judgmental," "pessimistic," or "histrionic," comes much too easily. So far, this book has presented approximately 900 quite unusual approaches to mental health care. This number is disturbing. However, a virtually limitless number of unusual approaches are currently advertised by duly credentialed counselors and psychotherapists. To add to the already established fact of numerous therapists "just making up stuff," another list of their aberrant, so-called therapies may be a good way to continue the discussion in this chapter. Again, readers should keep in mind that these approaches come from licensed counselors and psychotherapists. Table 8.1 lists another 60 unusual approaches that have not previously appeared in this book.

Are these thousands of unusual approaches to counseling and psychotherapy a problem? Obviously, the conclusion in this book is an emphatic, loud, urgent, and compelling "yes!" Taken separately, each of the unusual approaches may not be much of a threat to the health and future of mental health care. Each one, though, requires some serious reflection on its importance. While such a conclusion may not apply to all of the unusual approaches to mental health care, each one may be imagined as a 22-caliber short slug. "It can hardly be felt in your pocket; it's so small and light. But it can kill you, when it slams into your chest and penetrates

Table 8.1 ADDITIONAL UNUSUAL APPROACHES TO COUNSELING AND PSYCHOTHERAPY

Asian inspiration	Bagua
Chinese medical science	CIMBS
Contemplative-realizational	Earth sangha meditation
Decolonized interventions and indigenous philosophy	Eastern methods of healing
Eastern and Western mystical traditions	Eastern mindfulness philosophies
Eastern Wisdom traditions	Expressive analysis
Eye movement desensitization reinstallation	Feng Shui coaching/consulting
Executive-business-entrepreneur coaching	Focusing oriented art therapy
Havening techniques	Heartbeat method
Helper model	Hold sacred space
Hypnosis/EFT/TFT/energy psychotherapies	Inner bonding
Indigenous focusing oriented therapy	Inner radiance
Karol McBride's 5-step model	Ketamine-assisted psychotherapy
Life between life spiritual regression	Music and sound therapies
Metaphysics—science of the mind/new thought	Neuro-affective touch for trauma
Nightmare resolution therapy (NRT)	Nondual Kabbalistic healing
Open studio process	Organic mind energy
Own therapies	Plant medicine integration
Private and public standing counseling and coaching	Point approach
Psychedelic integration	QTPOC conversations
Resonance repatterning	Sacred feminine guide
Self-relations–generative trance	Shiatsu and martial arts
Sorenson method	Soul travel
Sound and vibrational therapy	Spiritual counseling and divination
Taoist/tantric practices	Traditional Tibetan medicine
Transpersonal coaching	Traya chakra psychotherapy
Twelve-step weddings	Unique psychotherapy
Universal life counseling	Visual journaling
Vitalistic paradigms	Wild feminine soul psychotherapy
Zen and vipassana meditation practice	Zen dharma

your heart. No, it ain't much, but it can kill you." The point is that the accrued weight and impact of these thousands of unusual approaches to mental health care are killing mental health care.

Each one of the thousands of aberrant approaches to counseling and psychotherapy usually stands with a large number of others. For example, the practice of past life soul retrieval very likely disturbs most counselors and psychotherapists, as it should. If only one individual offered past life soul retrieval, it would likely come and go, with little impact on anyone except the unfortunate clients on whom it is applied. In gathering information from the websites of practicing counselors and psychotherapists, hundreds of practitioners of past life soul retrieval were located. Also, for example, the information from websites gleaned more than 900 titles for some sort of Christian-related counseling, including such as Immanuel prayer process, pray therapy, Christian lifestyle counseling, Christ-centered professional counseling, and a possibly misleading one, "CBT"—Christian-based therapy. These five, though, do not even scratch the proverbial surface; they merely expose more surface. To emphasize this point, a partial list of specifically Christian approaches that are advertised as Bible-related approaches to counseling and psychotherapy appears in Table 8.2.

Table 8.2 SELECTED CHRISTIAN APPROACHES TO COUNSELING AND PSYCHOTHERAPY

Bible-based counseling	Bible believers hope-filled, inspired counseling
Bible believing therapy (BBT)	Bible counseling
Bible reading	Bible studies coaching
Bible study	Bible-based perspective
Bible-based therapy	Bible-based supportive psychotherapy
Bible-centered eclectic approach	Biblical
Biblical and life coaching	Biblical and universal faith resourcing
Biblical base	Biblical coaching
Biblical counseling	Biblical Christian counseling and soul care
Biblical counseling and guidance	Biblically based salvation oriented counseling
Biblical discipleship counseling	Biblical counseling/complementary methodologies
Biblical foundations	Biblical framework
Biblical framework counseling	Biblical instuction [*sic*]
Biblical integration	Biblical integrative therapy
Biblical knowledge and prayer	Biblical life coaching
Biblical nauthetic principle	Biblical perspective
Biblical practices	Biblical principles
Biblical principles counseling	Biblical principles in God's word
Biblical recasting therapy	Biblical truth
Biblical values	Biblical truths and sound counseling psychology
Biblical view	Biblical world view
Biblical/Christian	Biblical-Christian perspective
Biblically based counseling	Biblically based perspective
Biblically Christian counseling	Biblically based soul care/Christian counseling
Biblically-based enlightenment	Biblically-based principles

These 50 titles of various Christian counseling, Bible-related approaches come from websites where licensed mental health care providers advertise their counseling services. They are a small sampling of the much larger number of specifically Christian, Bible-related approaches. In this book, I am not approving or disapproving of the titles or the attendant practices. Instead, while serious objections to these approaches may be easily offered, the larger question is what collectively they say about how their practitioners understand—or, more likely, misunderstand—counseling and psychotherapy. More broadly than anything that may have a Biblical, Christian, or other religious perspective, these approaches define counseling and psychotherapy as a religious practice and not as mental health care. The obvious and mistaken belief is that their religious views are superior to clinical views and, critically important, superior to the clinically relevant needs of their clients. If the proponents of these approaches can do this, others whose religious outlooks are different can do this, too. In fact, they do. The list of alternate religious perspectives is simply too large to include here. Nevertheless, for the purpose of illustration and in contrast to the 50 Bible-centered, Christian approaches in Table 8.2, Table 8.3 contains another 50 religious approaches.

Table 8.3 ADDITIONAL SPECIFICALLY RELIGIOUS APPROACHES TO COUNSELING AND PSYCHOTHERAPY

Mormon therapy	Theophostic prayer therapy
Catholic coaching	Chassidic spirituality
Earth-based religious	Eastern mystical traditions
Enneagram soul based	Entheogenic shamanism
Frum therapy	God designed counseling
Muslim psychotherapy	Torah-based therapy
99 Divine qualities	Christian cognitive behavioral therapy
Latter Day Saints counseling	Jewish spirituality
John of God crystal healing bed	Kabbalistic healing
Life SOULutions therapy	Life between life spiritual regression
Magical SOULutions	Muslim healing
Muslim psychotherapy	Mysticism
Nondual Kabbalistic healing	Principles of soul healing love
Qabalistic pathworking	Reincarnation therapy
Scriptural values counseling	Shamanic healing practices
Shambhala Buddhist approach	Soteriological counseling
Soul coaching	Soul expansion shamanic mentoring
Soul support coaching	Soul travel
Spiritual cognitive therapy (SCT)	Spiritual root system
Splankna (Christian based)	Tao of self-realization
Taoist/tantric practices	Temperament analysis profile
The 11 principles of transformation	Torah-based counseling and therapy
Torah life coaching	Wicca
Wild feminine soul psychotherapy	Wisdom from the Holy Spirit
Wisdom of God's word	Word of God counseling

These, too, come from websites where counselors and psychotherapists advertise their preferred approaches to counseling and psychotherapy.

So far, this chapter has introduced more than 150 approaches to counseling and psychotherapy. A majority of them express specifically religious views. Collectively, the views lack coherence. They stand as reminders of the chaos and cacophony that characterize mental health care and, worse, the likelihood that, if these multiple views persist in mental health care, the profession suffers, if not dies. At least, among the questions that should be raised, readers should ask, "Which religion or expression of religion is best suited for effective counseling and psychotherapy?" Is it Wicca? Is it Torah-based therapy? Is it Bible counseling? Is it 99 divine qualities (Muslim)? Or, is it one of the other thousands of specifically religious approaches to counseling and psychotherapy? No matter what the answer to this question may be, all of these ideologically driven approaches to therapy have emerged from religious systems and the individuals who affiliate with them. Extending these systems into counseling and psychotherapy, individual counselors and psychotherapists decided to offer each one of these religious approaches as a health care service that is not religious. As individuals, they have betrayed the interests of their clients. They have betrayed the interests of the profession in which they are members.

Just as individuals have decided to offer these approaches, they can and should decide not to offer these approaches. With a virtually countless number of specifically religious approaches to counseling and psychotherapy, a necessary question is, "Whose religion offers the best counseling and psychotherapy?" Limited only to the large number of religious approaches identified in this chapter, the inevitable answer is that religious orientations offer little of value to clients of counselors and psychotherapists. However, the more basic and necessary question is, "How may individual counselors and psychotherapists contribute to and professionalize the profession of which they are members?" Or, more tersely, "Where do we go from here?"

Addressed to my clinical colleagues and those who help to train them, the inescapable question for us is, "If we expect psychotherapy to be better than it is, how should we be better than we are?" Or, "What must we do to improve our clinical performance?" Regardless of whether the improvements are personal and relatively small or major profession-encompassing, all mental health care providers can engage in improving their professional performance. Sooner or later, this must become the mandate of our profession. Without such a mandate, every provider may be contributing to the decline of counseling and psychotherapy. Or, in the language of this book, we all become therapy thieves, presuming to take what is not ours, claiming it as personal property, and remodeling it to suit our personal tastes. This is the problem. Counselors and psychotherapists have so extremely personalized their approach to their work that they have transformed their work into something that commonly is no longer counseling and psychotherapy. While this chapter places emphasis on individuals, it should not be understood as releasing academic institutions, licensing boards, social service agencies, and others from the burden of their deleterious impact on the practice of counseling and psychotherapy and their culpability in perpetuating harm or potential harm to clients who receive mental health care.

Counselors and psychotherapists develop in many different ways. At this time, proponents of their development have advocated for several models of development. So far, the models of development of counselors and psychotherapists have emphasized the development of beginners and/or supervisees, not licensed therapists. Clearly, placing a major emphasis on helping beginners to develop their various competencies is critically important for them, their clients, and the profession of which they are members. However, one of the problems with placing the almost exclusive emphasis on beginners is that it may inadvertently affirm the belief that practicing licensed therapists no longer need to develop. Further, research such as Spengler's and Pilipis' (2015) has established that practicing therapists generally fail to develop and maintain their clinical competence. The emphasis in this chapter, then, is placed on the need for and ways that duly credentialed counselors and psychotherapists may develop and maintain their competence. Moreover, as a way to sustain the profession of which they are members, the chapter answers the question about how they may contribute to the urgently needed reforms in psychotherapy. Obviously, this should be matter of self-interest among practicing counselors and psychotherapists, but it should also possibly ignite momentum toward reforms that may save their profession from extinction.

As already stated, "If you expect psychotherapy to be better than it is, you must ask how you need to be better than you are." The mandate for improvement and professionally accountable mental health care is obvious. The ways to fulfill such a mandate, though, are not so obvious. Just as clinical work is immensely complex, the professional development of counselors and psychotherapists is similarly complex. Allowing for these complexities, this chapter offers recommendations of ways that individual development may contribute to the desperately needed reform of counseling and psychotherapy. Specifically, the ways include the following ones.

HOW TO IMPROVE CLINICAL EFFECTIVENESS

Perhaps a good place to begin is to *look for examples of clinical excellence, study them, and adopt and adapt their practices.* A likely helpful feature of this search is to locate research about examples of excellence. Once found, the colleagues who deliver excellent services and the research about excellent providers will very likely provide useful clinical resources, actions that contribute to improvement, encouragement toward achieving excellence in clinical service, contacts with other clinicians, and potential for developing a culture of excellence. Surely, a feature of such a search is to locate research reports, including many that are cited in this book, as sources of information about what highly effective therapists do and who they are. Using these research reports, practicing counselors and psychotherapists may want to identify locally accessible peers who provide excellent clinical services and seek to learn from them.

Establish an individual plan for change. For the sake of clients, practicing clinicians should assume that their clinical performance can be better than it is. Quite likely, none of the suggestions for individual reform will be feasible for everyone. This recognition emphasizes the need for all of us to develop an individual plan for improvement. An individual's plan may include some or all of the suggestions presented next. It will likely include features that add to them. Whatever form the plan may take, it should be energized by the recognition that the welfare of clients is at stake. Moreover, because reform influences real individuals who are doing real things in real places with real people, the necessity of reform is no abstraction. It is a personal obligation of every practicing counselor and psychotherapist and every instructor or clinical supervisor who trains them. Further, as noted several times already, the question is not whether reform will occur, but who will initiate reform and who will control it. The encouragement here is that all of us should develop an individual plan for improvement.

Establish professional development goals. By the time in their careers that counselors and psychotherapists receive their licenses to practice, they have already established important goals and reached them. Their goals include those of applying for entry into a graduate program, with the belief that they qualify for such a program and with the expectation that they will complete such a program, and applying for and receiving a license to practice. Then, with their licenses in hand, they continue to practice in an extraordinary variety of treatment environments. So, the idea of setting goals for themselves as professional persons is a familiar one.

In the context of the need to reform mental health care, the importance of setting goals may already be clear. However, to ensure that it is clear, the need for reform is, at least, to save the future of counseling and psychotherapy. And, to the extent to which individuals achieve reform in their respective practices of psychotherapy, this need will be partially met. Further, to meet this need, establishing goals will surely help. The recommendation about setting goals is an easy one to make. Taking the recommendation and fulfilling the intentionality of it are something else.

Goals may take many forms. The list of possibilities is a very long one. Instead of a list that may include numerous possibilities, the list of possible goals presented here includes some with an emphasis on the existential immediacy of interacting with clients and peers, others with an emphasis on specific actions that may or may not involve other people, and still others with an emphasis on specific ways to examine a counselor's clinical work with other counselors. More important, some of them emphasize goals as standards of performance.

- I would like to have more confidence in knowing that I am helping my clients.
- During the next 2 months, I will meet with two or three of my peers for the purpose of discussing my clinical work so that I may improve what I do.

- To improve my clinical expertise and effectiveness, I will seek consultation with those who can challenge my intellectual, emotional, and relational functioning.
- When I attend professional meetings in my area, I will look for a struggling therapist and do all I can to become a resource for her/him.
- I will read at least one book on neuroscience and mental health this year.
- I will study my performance as a therapist and identify weaknesses that I will work to turn into strengths.
- If I could raise my hopefulness about some of the most challenging clients I work with, I would achieve a lot for them and for me.
- To ensure that my state's legislature supports the best interests of clients, I will examine my legislators' websites at least once a month. Compatible with this, I will follow the legislative concerns that arise from my professional association.
- Just for the purpose of getting outside the bubble of my practice, I will dedicate time to being with other therapists.
- For randomly selected clients, I will examine my clinical rationale/justification for every major aspect of what I have done with them, including diagnosis and ongoing clinical decision-making.
- I will develop my own assessment plan so that I can ensure that clients and peers can assess the effectiveness of what I do in counseling.

Find a mentor. This recommendation may be as important as it is risky. First, on the side of risk, as reported in this book several times already, time and experience are not generally signs of improved or superior clinical skill and effectiveness. Therefore, when a counselor and psychotherapist looks for a mentor, she/he should look for qualities more than length of service as a counselor and psychotherapist. This recognition points to the importance of locating a mentor who may be capable of helping others to improve their clinical skills and effectiveness. A prospective mentee may want to ask some of the following questions: In what areas of clinical work does the prospective mentor have expertise? Does the prospective mentor have good information about her/his clinical effectiveness? Is the prospective mentor prepared to share her/his clinical cases as a means of disclosing her/his clinical effectiveness and commitment to her/his professional growth? Does the prospective mentor receive mentoring?

Study your clinical performance. A well-researched approach to studying one's clinical performance is feedback-informed treatment (FIT). One of the progenitors of FIT, Scott Miller, says that FIT "is a pan-theoretical approach for evaluating and improving the quality and effectiveness of behavioral health services." He adds, "FIT involves routinely and formally soliciting feedback from clients regarding the therapeutic alliance and outcome of care and using the resulting information to inform and tailor service delivery" (Miller, as cited in Nylund & Filippelli, 2018). The importance of FIT is that it is solidly evidence-based, practical, and readily available. Many useful explanations of FIT and helpful forms are available online as well as in the following references:

Bertolino, B., & Miller, S. D. (2012). *Feedback-informed clinical work: The basics.* Chicago, IL: International Center for Clinical Excellence.

Goldberg, S. B., Babins-Wagner, R., Rousmaniere, T., Berzins, S., Hoyt, W. T., Whipple, J. L . . . Wampold, B. E. (2016). Creating a climate for therapist improvement: A case study of an agency focused on outcomes and deliberate practice. *Psychotherapy, 53*(3), 367–375.

Low, D. C., Miller, S. D., & Squire, B. (2012). *The Outcome Rating Scales (ORS) & Session Rating Scales (SRS): Feedback informed treatment in child and adolescent mental health services (CAMHS).* Retrieved from http://www.aft.org.uk/SpringboardWebApp/userfiles/aft/file/Events/2012/David%20Low%20paper%20for%20CYP-IAPT.pdf

McMahan, E. H. (2014). Supervision, a non-elusive component of deliberate practice toward expertise. *American Psychologist, 69*(7), 712–713.

Miller, S. D., & Donahey, K. M. (2012). Feedback informed treatment (FIT): Improving the outcome of sex therapy one person at a time. In P. Kleinplatz (Ed.), *New directions in sex therapy: Innovations and alternatives* (pp. 195–211) New York, NY: Routledge.

Rousmanaiere, T. (2017). *Deliberate practice for psychotherapists: A guide to improving clinical effectiveness.* New York, NY: Routledge.

Yates, C. M., Holmes, C. M., Coe Smith, J. C., & Nielson, T. (2018). The benefits of implementing a feedback informed treatment system within counselor education curriculum. *Professional Counselor, 5*(1), 22–32.

Develop a self-assessment instrument. Few clinicians regard themselves as researchers, let alone creators of assessment instruments. This kind of self-regard is quite understandable. Despite this, all clinicians evaluate their work, although almost always it is informal and unsystematic. The idea here is that clinicians can elevate their informal and unsystematic evaluation of their work and make it somewhat formal and systematic. The question is whether counselors and psychotherapists examine their work and gain from the experience.

Counselors and psychotherapists may want to consider including the following items in their personal and professional self-assessment instrument. They should feel quite free and flexible about creating questions and issues of their own. The questions and issues in the instrument presented here were somewhat arbitrarily limited to 20 items. They are intended to involve a clinician in assessing several aspect of her/his clinical work.

To utilize the following instrument, simply indicate in the blank space to the left of each question the degree to which the issue is "strongly correct" about you (1), "strongly incorrect" about you (10), or a number in between 1 and 10.

_____ I enjoy delivering counseling services.
_____ I have a clearly articulated approach to counseling/therapy.
_____ I take care of myself in ways that I encourage my clients to care for themselves.

_____ I am alert to my potential for burnout.

_____ I find that the practice of psychotherapy is highly rewarding and gratifying.

_____ My sense of accomplishment continues to decline in my clinical work.

_____ I comfortably seek consultation with my peers.

_____ I have clear goals for my personal and professional growth and development.

_____ My clinical work continues to be intellectually stimulating.

_____ I review my clinical effectiveness with my clients, when this is feasible.

_____ I somewhat eagerly welcome every day of clinical work.

_____ I have recently reviewed my clinical files to find areas of possible improvement.

_____ I have tried to compose my ideas about what a good therapist is.

_____ I have a strong sense of belonging to my professional community.

_____ I frequently locate and develop new interests in my clinical work.

_____ I have enjoyable activities (e.g., friends and hobbies) unrelated to my work.

_____ I seriously engage in activities that will improve my clinical effectiveness.

_____ I can make a list of my clinical strengths and weaknesses.

_____ My family (e.g., partners, children, intimate friends) know that I find appropriate fulfillment in my clinical work.

_____ I seriously engage in activities that will improve my profession.

After counselors and psychotherapists create and utilize their individual self-assessment instrument, they may wish to consider the possibility of sharing it with other counselors and psychotherapists. Also, they may wish to engage others in creating and utilizing an assessment instrument. This may contribute to building a clinical community.

Review your model of therapy. Given that numerous models of counseling and psychotherapy have been readily available for many decades and that they continue to increase in number, counselors and psychotherapists almost inevitably select one or more as their orientation to their clinical work. Or, they somewhat naturally integrate and blend models into their preferred way of conducting counseling and psychotherapy. Whether the model is a self-evident and self-consciously utilized approach or one that is not so easily described because it has grown from the experience of conducting clinical work, it is "there" mainly because counselors and psychotherapists carry major assumptions and beliefs into their clinical work. The following questions might help clinicians to examine their model of therapy:

- What are my personal beliefs that may influence my work with clients? Detailing this question, what do I believe about parenting? Or, what

do I believe about how men and women should relate with each other?
What do I believe about a faith-orientation to my life?

- How do I influence others to meet my needs? This question stands on
  the assumption that others do not come to us fully prepared to respond
  to our needs. We do something to influence them so that they can
  respond to us and possibly provide what we need. How do my ways of
  getting my needs met influence my clinical work?
- What is my model of therapy? How did I choose this one or combination
  of therapies? If my preferred approaches disappeared, what therapy
  would I choose to use and why would I choose it?
- For which kinds of client issues is my preferred model most effective?
  With which kind of clients is my preferred model least effective?
- Which model of therapy would I never or almost never consider using?
  Why? What does this way of thinking about this model say about my
  understanding of counseling and psychotherapy?
- In only one sentence, what is my definition of my approach to counseling
  and psychotherapy?
- If one of my clients asked me how I make sense of my approach to
  my work, how would I respond so that my client could not possibly
  misunderstand my approach?

Start a workbook-journal. With the goal of ensuring that you can deliver
effective clinical services, consider creating your own workbook. The idea is
a simple but demanding one. Simply stated, counselors and psychotherapists
who wish to ensure that they can deliver effective clinical services will prepare
questions that they will answer. After they answer the questions, they may utilize them as a source of ongoing reflection and growth. For example, if a counselor or psychotherapist includes the question, "To which client populations
am I drawn?" the answer itself may be useful just because it has been recorded.
However, it may continue to be useful when it is reexamined over a period of
months or years.

A workbook-journal could include almost anything that a counselor or psychotherapist deems to be needed. The questions that may be included are virtually
limitless in number and kind. For illustration, here are some possible questions
for a workbook-journal:

- Who helps me to develop as a clinician? How does she/he or they do this
  for me?
- How do I define good therapy? How do I assess my effectiveness?
- To which clients am I drawn? What is it about me that draws me to
  them? What is it about them that draws me to them?
- When I review my professional development history, what persons
  or factors or events shaped my development? What have been major
  turning points in my professional history? What kinds of change have

been more difficult for me? What kinds of changes have I eagerly
pursued?
- How do I confirm and affirm growth in my clients?
- What is my understanding of mental health? What does a mentally
healthy person look like? What do I conclude about my mental health?
- What are my ideas about encouraging and coaching a beginning
therapist to excel in her/his work as a therapist?
- What is the most enjoyable feature of my work as a therapist today? If
I record my most enjoyable moments from each day, say, for 6 months,
what will I learn about what brings me enjoyment?
- How do I best confront my clients who need confrontation?
- What inhibits my professional development?
- What are the most important lessons I have learned in my clinical work?
What have these lessons given me that have made my life better that
I would have been?
- What are the 10 most appealing features of my clinical workday?
- How do I handle my best clinical successes?
- How do I handle my worst clinical failures?
- What are the clinical issues that appeal most to me? Why?
- How do I affirm my clinical strengths?
- The list of terms that I use to express who I am as a therapist includes the
following 10.
- What resources—books, mentors, income from clients, articles— do I or
may I utilize, to help me clarify an approach to my clinical work?
- Knowing what I now know about clinical work, the picture of clinical
work that I would give myself if I were beginning my career now would
include the following insights.

Many more questions than these could be raised. The list of questions for any individual would likely be a more valuable asset if its development were shared by
trusted peers.

A variation on the theme of starting a workbook is to start a reflective and
information-based clinical journal. A journal may be as simple as recording your
impressions of your clinical service with selected clients. Alternately, a journal
may involve searching reflections on why you failed or succeeded with selected
clients. The potential benefits of keeping a professional journal may cover a
wide range of professional functioning. A likely benefit is that a journal helps to
clarify and establish effective clinical interventions or helps to eliminate ineffective interventions. A journal may help to identify unanswered questions that a
counselor or therapist wants to explore with peers and/or supervisors. Of course,
keeping a professional journal may help counselors and psychotherapists to see
ways of working with a specific client that improve clinical services for the client.
And, almost inevitably, a journal not only affirms clinical assets but also helps to
bring focus to specific needs for professional development. For example, when a

clinician records in a journal something like, "I think I do as much to avoid my borderline client as I do to help," this may be an indication of a need for training on how to work with clients who suffer with borderline personality disorder.

Change the big picture. Just as each counselor and psychotherapist contributes to the large social experiment of which she/his is a part, she/he will influence the large social experiment by giving attention to it, asking how it may be improved, and initiating actions that may help to change its direction. Or, to borrow an often cited rallying quip from the civil rights movement, "Keep your eyes on the prize." With regard to mental health care, instead of the current environment, the prize is a professional environment in which the means of professional accountability are clear so that those who need mental health care can receive truly professional care. Ultimately, the significance of the reform of mental health care is that, if it is achieved and sustained, client welfare can be protected. The big picture may be changed when each practicing clinician seeks to improve her/his clinical effectiveness, but it may also be changed when each practicing clinician seeks to improve the entire body of mental health care providers.

Develop a network of shared resources. Practicing clinicians know that clinical work can be isolating. For many of them, the question about how they may break out of their isolation arises. The suggestion here is that they develop a network of shared resources. The resources may include ones that, say, five or six individuals wish to share. If one of the five or six individuals has a professional specialization in play therapy or trauma-focused cognitive behavioral therapy, she/he may share resources (e.g., articles, manuals, books, DVDs, training events) with the others. Similarly, if one of them knows about a place to receive training in a specialization, she/he may share this, too. Or, if one of them knows effective techniques for working with a particular insurance company or a government official, she/he may share this, too. Most practicing clinicians possess resources that can bring real benefits to their peers.

To refer to only one very good example of important resources that may be shared, the American Foundation for Suicide Prevention (AFSP, 2019) offers for relatively inexpensive sale several DVDs. They include "Living With Bi-polar Disorder" and "Depression and Bi-polar Awareness." The DVDs and other materials may be found at AFSP's website, https://afsp.org. In addition to sharing resources, DVDs such as those from AFSP may be viewed in informal groups and discussed, with the aim of improving clinical competence.

Seek counseling. One of the small gestures of support for mental health care providers is that they seek help from one another. However, more important than showing support for the profession, seeking counseling serves to affirm one's belief in her/his professional self and develop a clearer, more substantial base of honesty and confidence in clinical work. In other words, healthy clinicians help their clients in healthy ways.

Enjoy the work of counseling and psychotherapy. Clearly, life involves adversity. Clinicians carry this knowledge into every clinical encounter. Further, they know this from their own lives. As persons and as counselors and therapists, they know adversity. In their own families, they likely know a kinsperson who drinks excessively. They know the woman who dreads the next breast cancer treatment. They know another one who smokes cigarettes. They know the man who fears his unreliable heart. They know another one who seems unable to maintain emotional regulation. This is the stuff of life. However, these same clinicians also know that many of their clients and many of their kinspersons demonstrate heroism and inspire them. And, just as clinicians seek to equip their clients with increasing, life-affirming skills and attitudes, they, too, can build within themselves increasing capacity for life-affirming skills and attitudes. Amid the chaos, disappointments, losses, and other adversities, they can look for joy. They can look for joy in the heroic struggles of their clients and of their suffering kinspersons. And, moreover, they can find it in themselves, as clinicians, when they examine their own heroism that they bring to every client, every day. For practicing clinicians, the practice of counseling and psychotherapy can bring joy.

Enjoy your life! Providing mental health care for clients energizes counselors and psychotherapists and gives them meaning in their lives. This should be recognized and affirmed. However, if providing mental health care stands as the primary or sole source of meaning in the life of a counselor or psychotherapist, the therapist has committed a serious misjudgment. Instead of centering life on providing professional services, they should find other centers. Just as they help their clients to pursue healthy living, they themselves should be ever conscious of their need to pursue healthy living and to enjoy life. So, to state this in a declarative and personal way, enjoy your life! Invest your time, effort, and love in others and allow them to invest their time, effort, and love in you. Your life is quite worth living, apart from your professional life. Do what brings you joy. Conversely, reject the possibility that you will get through your week and discover that you are tired and emotionally spent, with little to anticipate that brings you joy. If, at this moment, your review of your life apart from professional service contains little that holds your attention and brings you joy, take an inventory. Name those persons and activities that can bring you joy. Make specific plans for being with these people and doing these things. In short, take good care of yourself and those who love you. This will improve your delivery of mental health care. It will make your life better than it is. Enjoy your life!

CONCLUSION

Older counselors and psychotherapists have adapted to numerous changes in their work. They endured the change from one edition of the *DSM* to another. They adapted to the emergence of problems with which they had little or no experience, such as sex addiction, prescription drug abuse, and many others. They adapted to new approaches to counseling and psychotherapy, such as emotion-focused therapy and acceptance and commitment therapy. They adapted to regulatory changes, such as HIPAA and the Affordable Care Act. Adaptation is inherent to the profession. Indeed, it is adaptation to healthier ways of living that clinicians wish to see in their clients. More than merely adapting to changing circumstances, though, mental health care providers seek to make life better for others and themselves. And, when they are effective, they may characterize their work as "good therapy."

Good therapy is the product of intentional action. It does not come naturally. It comes from discipline and commitment. Its value is that it fulfills the purpose for which counseling and psychotherapy exist. Namely, it makes life better for those who receive it. This chapter is based on the simple premise that all counselors and psychotherapists can engage in activities that enable them to make life better for their clients. While the list of activities could be much longer, it includes the following:

Look for examples of clinical excellence, study them, and adopt and adapt
    their practices. Establish an individual plan for change.
Establish professional development goals.
Find a mentor.
Study your clinical performance.
Develop a self-assessment instrument.
Review your model of therapy.
Start a workbook or journal.
Change the big picture.
Develop a network of shared resources.
Seek therapy.
Enjoy your work.
Enjoy your life!

Guiding Principles of Mental Health Care Reform

This brief chapter's intents are to be aspirational and to offer encouragement. When we who are counselors and psychotherapists conscientiously seek to do the right things, we will likely achieve good results. However, like most good things, such as raising healthy children, the good results almost always involve risks, losses, pain, and other adversities, even when the good results seem inevitably to come. Surely, counseling and psychotherapy reform (CPR) involves risks and, moreover, potentially necessary and good results.

As we seek to reform ourselves, what principles should guide us? None of us, including me, has the last word about the principles that should guide us as we seek reform. The principles depend on good will. If nothing else, mental health care providers almost always possess good will. This chapter encourages all of us to express our individual and collective good will, a sincere interest in making things better for everyone, toward mental health care reform.

THE PRINCIPLES OF REFORM

The accrued evidence about mental health care providers says that they—or, we—do not get better over time. Generally, therapists get comfortable with what they do and conclude that they are good at what they do. A review of the article by Spengler and Pilipis (2015) makes this clear. Another recent and informative article also makes this clear. It is "Do Psychotherapists Improve With Time and Experience? A Longitudinal Analysis of Outcomes in a Clinical Setting" by Goldberg, Miller, Nielsen, Rousmaniere, Whipple, Hoyt, and Wampold (2016). They conclude that "the present analyses show that, in the aggregate, therapists did not improve with more experience, operationalized as either time or number of cases." The importance of these two sources is not so much that they confirm a general weakness in the practice of mental health care, but that they should urge providers to be better than they are. Thus, *the first principle is that individually—and consequently, collectively—each provider should ask, "How may I be better than I am in my delivery of mental health care?"* In addition, the accrued evidence in this book makes clear that counseling and psychotherapy need improvement and reform.

Do counseling and psychotherapy need reform? Yes. However, while the need for reform is clear, the need to pursue reform in life-affirming and productive ways is also clear. Addressing this need, a necessary question is, "What principles should guide reform?" Here are some ideas with which our collective discussion of principles may begin.

Client welfare is the preeminent priority. Helping clients is the primary reason for the existence of counselors and psychotherapists. Further, client welfare should be the preeminent priority, not the needs of a subgroup among counselors and psychotherapists.

Among the priorities of client welfare, such as ethical therapist behavior or informed consent for clients, clinical service should be first. This is simply a way to advance the reason for the existence of counselors and psychotherapists. They exist to provide counseling and psychotherapy, broadly defined. With client welfare as the reason for their existence, all providers should demonstrate their commitment to providing highly effective clinical services.

Within the priority of client welfare, the other category is maximum access to mental health care. Without access to it, there is no mental health care. No matter how good counseling and psychotherapy may be, access to it translates into providing it. Collectively, mental health care providers should actively advance the cause of access to professional services.

Cohesion is needed. Reform and improvement must begin somewhere. In CPR's beginning, it may be limited to books like this, a session at a conference, a discussion among clinical peers, advocacy with a state legislator, or other relatively small activities. Inevitably, though, if counseling and psychotherapy are to be saved, cohesion among counselors and therapists will occur. Instead of merely a good idea, though, cohesion and cohesive action will be needed from coordinated and agreed upon strategies from the American Counseling Association, the American Psychological Association, and the National Association of Social Workers. However, smaller professional associations, such as the American Mental Health Counselors Association and the American Association for Marriage and Family Therapy, will also need to share in such strategies. Cohesion must be achieved because if one of these groups neither suffers the risks of reform nor receives its benefits, reform is unlikely to happen.

Each of our major mental health care associations and types of licensing emerged from their unique histories. Their histories should be respected and admired. Whether the category of licensing is social work, professional counseling, or psychology, all have a history of significant achievements and contributions to mental health care. Nevertheless, we may be at a time in our collective history when we should "grow up" and separate from our respective histories. More than rejecting our respective histories, cohesively seeking reform is a significant concession to a hopeful and useable future.

If mistrust and wariness among mental health care providers inhibits cohesion, they should be felt and expressed within professional groups and associations and not among them. To be clear, social workers should be mistrusting and wary of social workers who practice "angelic channeling" (Holden & Barker, 2018) or any of a large number of other inappropriate approaches to counseling and psychotherapy. Professional counselors should be mistrusting and wary of professional counselors who practice chiropractic care, or neuropsychiatry, or any of a large number of other inappropriate approaches to counseling and psychotherapy. Psychologists should react in very similar ways to psychologists who practice Vedic astrology or past life soul retrieval. And, marriage and family therapists should react similarly to their peers who practice aromatherapy or power animal retrieval. After all, the illicit practices are the ones that inhibit cohesion and corrode the image and practice of counseling and psychotherapy.

The possibility of achieving useful cohesion may need to begin with presumed confidence in others with whom cohesion is needed. Presumed confidence means that social workers should begin to work toward cohesion with the belief that professional counselors share interest in the welfare of social workers and that professional counselors hold a reciprocal view. Likewise, all of the other major professional associations begin with the same presumed confidence so that psychologists have confidence in marriage and family therapists and marriage and family therapists have confidence in psychologists. In other words, presumed confidence should be the characteristic feature of the beginning of cohesion and, therefore, the achievement of important reforms of counseling and psychotherapy. Without cohesion, reform will not likely happen. Worse, without cohesion, the decline of counseling and psychotherapy will continue.

Adaptation is needed. All mental health care providers have seen individuals and families who maintain nonadaptive rigidity and self-defeating ways of relating to one another. In contrast, providers have engaged in serious adaptation and understood that it is a necessary feature of healthy living. Expanding the scope of healthy living, if the profession of mental health care itself is to become healthy, it must adapt. Several specific kinds of adaptation are discussed in an earlier chapter. They include, for example, major changes in licensing and regulation of mental health care providers. Crudely stated, "We either adapt or we die." If death does not come from failure to reform mental health care, surely, adverse actions against providers will come. Among the adverse actions, providers will become objects of increasing ethics complaints, denial of third party payments, suffer legislative constraints that other health care providers do not suffer, and, of course, suffer increasing defeats in legal actions against them.

Aspirational thinking and action are needed. In the context of delivering clinical services, clinicians often recognize that clients will truly suffer when they seek to make their lives better. The process of change means, at least, that clients will relinquish important thoughts and behaviors that they have held as necessary and valuable during most of their lives. The process of change toward CPR requires counselors and psychotherapists to relinquish important thoughts and behaviors that have been seen as necessary and valuable to them. One of them, at least, in defiance of a normative standard of care, is that clinicians have been free to "just make up stuff" and call it "therapy," when their "stuff" is clearly not therapy or therapeutic. This is well-documented in this book. Despite the difficulty of relinquishing such freedom, counselors and therapists must think in aspirational terms, just as they want their clients to do. For clinicians, this means that beyond the inconvenience, frustration, confusion, and new forms of necessary adaptation, reform leads to improvement. A useable and positive future can be achieved through reform. Aspirational thinking and action are needed. The proverbial "brighter day" may not be merely proverbial. It may become reality for counselors and psychotherapists if they work for it.

The interests of all qualified clinicians should be positively valued. To the degree that the interests of qualified clinicians are served by reform, reform is more likely than not to happen. Indeed, no matter how qualified clinicians arrived at the present time, their histories should be respected, along with their individual potential for effective clinical service.

Reform should be as democratic as possible. While leadership is needed for reform, it will have major consequences for approximately 800,000 state-licensed individuals. A democratic approach is needed so that as many licensees as possible may participate in reform if they wish to do so.

A long view of change is needed. Because of the long, somewhat chaotic and complicated but rich history of counseling and psychotherapy, short-term and easy changes are not likely to occur. A long view of reform, based on the commitment to professional accountability and the assurance for almost all mental health care providers that their interests will be protected, is much more likely to produce agreement about broadly based and desirable change. Possibly, for example, the long view of reform takes a short-term view that all of the major health care provider associations will seek to establish agreed upon standards about how to admit candidates to their respective preparation programs. This could be a relatively small step toward working together to achieve comprehensive reform.

Toward achieving successful reform that makes professional life better than it is for all providers, *information should be openly and freely shared.* Understandably, complex and demanding agendas and priorities among professional groups, such as the National Association of Social Workers and the American Psychological Association, or between regional or state groups can easily prevent urgently needed reform. An antidote for this is a clear and enduring commitment to openness about agendas and priorities, but also resources that may be useful toward successful reform of counseling and psychotherapy.

Finally, while principles for reform may be added to the ones identified here, the highest priority placed on client welfare should be paired with *a collective and demonstrated commitment to the core values in the major codes of professional ethics.* With this commitment, reform affirms more than contradicts counselors and psychotherapists and delivers more of importance to them than it takes away. With such a commitment in place, urgently needed reform can happen.

SUMMARY OF GUIDING PRINCIPLES

Affirming the significance of each counselor and psychotherapist, the over-arching principle that guides the others is that individually—and consequently, collectively—each provider should ask, "How may I be better in my delivery of mental health care?" Based on this principle, the following ones may guide reform of counseling and psychotherapy:

1. The first principle is that individually—and consequently, collectively—each provider should ask, "How may I be better in my delivery of mental health care?"
2. Client welfare is the preeminent priority.
3. Cohesion is needed
4. Adaptation is needed.
5. Aspirational thinking and action are needed.
6. The interests of all qualified clinicians should be positively valued.
7. Reform should be as democratic as possible.
8. A long view of change is needed.
9. Information should be openly and freely shared.
10. A collective and demonstrated commitment to the core values in the major codes of professional ethics.

FINALLY, A PERSONAL NOTE

The review of numerous websites of practicing therapists brought information that I was not prepared to see. Nevertheless, as I saw things that fueled my concern about our profession, I felt an increasing burden about how to communicate my concern to others. The burden led to this book. Now, at the end of the book, I still wonder how best to communicate concern about the condition of our profession. So, before closing the book, I wish to share another short list of unusual approaches to counseling and psychotherapy. Each item in Table 9.1 comes from the websites of licensed, practicing therapists.

In addition to the unusual therapies in Table 9.1, the number of therapies that begin with the letter "A" is 1,508, with 188 containing the adjective "advanced," as in "advanced coherence therapy." And, as previously reported, this book includes more than 1,200 approaches to counseling and psychotherapy. Also as previously reported, at the time of writing Chapter 3, 21,758 approaches to therapy had been gathered.

When I began to review the websites of practicing therapists, I expected to find what I wanted to find. I expected to see that my peers advertised professionally accountable information, including their narratives about what they offer, the skills that they may justly claim, approaches to therapy that most of us would

Table 9.1 Unusual Approaches to Therapy, From the Letter "A"

Access bars	Acu-energy flow
Ad astra therapy	Addo recovery
Afro-Brazilian dance	Agnostic/Socratic questioning
AiKi training	Alternative esoteric spiritual approaches
Amazonian ayahuasca shamanism	Ambient music clinical environment
Anapanasati	Ancestral mapping
Ancient personality typing	Ancient Peruvian healing
Angel therapy and energy healing	Angelic healing fire
Animal spirit archetypal approach	Animal-focused wisdom
Anodyne coaching	Anointed emotional processing
Anuttarayoga tantra	Approved Reflexology Practice
Archangelic light	Arcturian healing
Arts-based premarital counseling	Ashtanga yoga
Assisi pattern dream analysis	Astrocartography
Atheist theoretical constructs	Audio visual entrainment
Aura balance and clearing	Ayahuasca healing

readily recognize, conspicuous attempts to be honest about themselves, and occasional candor about the risks of receiving therapy. My expectations declined as I searched. Instead of fulfilling my expectations, my concern and disappointment rose, sort of like a fever that comes on slowly but surely with a bout of influenza. And, over time, I sought to make sense of what I found. Mostly, I could not. I was left with questions instead of answers. Here are some of them:

- Why do licensed therapists report that they have many more years of clinical experience than the evidence would justify?
- Why do licensed therapists get and advertise certifications that obviously lack credibility?
- Why do licensed therapists promote their respective ideologies, with clear implications for close-mindedness and sometimes bigotry?
- Why do licensed therapists hold inflated positive views of themselves?
- Why do licensed therapists tolerate unlicensed and other transparently unqualified individuals who provide clinical services?
- Why do licensed therapists boast about their "unique" approaches to therapy and emphasize their "passion" for their selected interests?
- Why do licensed therapists advertise approaches to therapy in which they have no apparent preparation or skill?
- Why do licensed therapists invent or adopt highly individualized approaches to counseling and psychotherapy?
- Why do licensed therapists promote unusual and sometimes truly bizarre approaches to therapy?

Surely, along with others, I can easily speculate about the answers to these questions. For example, I suspect that many practicing therapists express a largely unrecognized hunger for significance. I suspect that some of my peers dread serious engagement with the ambiguous but real demands of intellectual pursuits and inadvertently succumb to anti-intellectualism. I suspect that many wish to be acknowledged as somehow special and display this as they align their professional work with their presumed expectations of their respective god. I suspect that many inadvertently conceal their fear of exposure as being deficient and inadequate, hiding themselves in the cloister of certainty that attends rigid religious convictions or rigid clinical convictions, accommodating the community of which they are members. I suspect that many of my peers correctly abhor the conspicuous ethical lapses among us, such as breaches of confidentiality or violation of the personal space of their clients, while simultaneously failing to comply with minimally appropriate professional standards when they select or invent a highly personalized and peculiar approach to therapy. While many more speculations than these could be added, all of them may be wrong or, possibly, right. Whether these speculations are right or wrong, one conclusion about the massive proliferation appears to be correct. It is that counseling and psychotherapy have been corrupted, similar to a virus that destroys the operations of a computer.

The question with which this book began is, "Why are we—counselors and psychotherapists—committed to failure?" Now, more than 100,000 words later, the question that most challenges me is simple. With all of the evidence that clearly indicates that counseling and psychotherapy are troubled and declining, why are we doing almost nothing to save our profession? However, because I refuse to believe that we will do nothing, another question is, "To save our profession, what shall we do next?" While we may grimace or argue among ourselves in response to our collective failures, it is our failures that now give us the opportunity to actualize the promise of gathering relevant information, assessing the situation in which we find ourselves, construing a "diagnosis," setting appropriate professional goals, and working hard to achieve our goals. This is the invitation that we extend to our clients. It is the invitation that we should extend to one another. Just as Freud, Beck, Rogers, Linehan, and many others have felt the sting of their failures and gave their talent and anguish to the task of making us better, we can make us better, too.

Therapeutic Approaches to Counseling
and Psychotherapy Identified in This Book

More than 1,000 therapeutic approaches have been identified in this book. They are gathered here for the convenience of the reader who wishes to have such a list.

While the list may appear to be formidably long, it is merely representative of a much larger number of therapeutic approaches that are advertised and, presumably, practiced by counselors and therapists. The list that this research has so far gathered includes almost 15,000 approaches to therapy.

12 Step philosophy of spirituality
3 Principles/single paradigm approach
3-D energetics
5 element theory
7 Path self hypnosis
9 Rites of munay-ki
99 Divine qualities [Muslim]
ABIDE debriefing and renewal
Absolute happiness therapy
Acceptance and commitment therapy
Access bars
Access consciousness
ACT
Acu-energy flow
Acupressure with pebbles
Acutonics and Tibetan bowls
Ad astra therapy
Addictions treatment
Addo recovery
Adlerian
Adlerian play therapy
Advanced Christian therapy
Advanced Eden energy medicine
Advanced past life therapy

Advanced progression counseling
Advanced spiritual psychology
Adyashanti approach
Affect centered therapy
African and Native American rituals
African psychology
Afro-Brazilian dance
After death communications
Afterglow curriculum
Agnostic therapeutic framework
Agnostic/Socratic questioning
AiKi training
Aikido
Akashic record readings
Akashic record revelation readings
Alaska wisdom recovery
Alchemical imagination
Alchemical psychotherapy
Alchemy
Alternating bilateral stimulation
Alternative approaches
Alternative esoteric spiritual approaches
Alternative medicine
AMATI project
Amazon warrior anger therapy
Amazonian ayahuasca shamanism
Ambient music clinical environment
Amma therapy
AMST
An-ra energy healing
Anapanasati
Anatomic therapy
Ancestral mapping
Ancient African wisdom
Ancient Chinese martial arts
Ancient Chinese medicine
Ancient Christian wisdom
Ancient healing modalities
Ancient Hebraic dream interpretation
Ancient instruments for healing
Ancient love pattern analysis
Ancient personality typing
Ancient Peruvian healing
Ancient traditions

Ancient wisdom
Ancient wisdom of ayurveda therapy
Ancient wisdom wellness practices
Anderson + Anderson model
Angel therapy and energy healing
Angelic channeling
Angelic healing fire
Anger management
Animal-assisted therapy
Animal feeding therapy
Animal spirit archetypal approach
Animal-focused wisdom
Anodyne coaching
Anointed emotional processing
Anthetic therapy
Anthroposophical counseling
Anuttarayoga tantra
Applied behavioral analysis
Applied Godly wisdom
Approved reflexology practice
AQAL therapy
Aquamassage
Aquatic and yoga therapies
Aquatic therapy
Arbinger facilitation
Archangelic light
Archetypal art therapy
Arcturian healing
Arno temperament profiling
Aroma freedom technique
Aromatherapy
Aromatouch (massage)
Art therapy
Arts-based pre-marital counseling
Ashtanga yoga
Asian inspiration
Assertive community treatment
Assisi pattern dream analysis
Astrocartography
Astrological consultation
Astrology
Astrotherapy
Atheist theoretical constructs
Attachment core pattern therapy

Attachment-based
Attachment-related therapies
Audio visual entrainment
Aura balance and clearing
Aura therapy
Auricular acupuncture (NADA protocol)
Auriculotherapy (ear acupuncture)
Avesa quantum healing
Ayahuasca healing
Ayurveda
Aztec healing
Bach flower therapy
Bagua
Balametrics
Balance chakras
Ballet therapy
Behavioral therapy
Bert Hellinger approach
Bible-based counseling
Bible believers hope-filled, inspired counseling
Bible believing therapy (BBT)
Bible counseling
Bible insights–based therapy
Bible life coaching
Bible lifestyle counseling
Bible reading
Bible studies coaching
Bible study
Bible-based Christian counseling
Bible-based counseling
Bible-based perspective
Bible-based supportive psychotherapy
Bible-based therapy
Bible-centered eclectic approach
Biblical
Biblical and universal faith resourcing
Biblical and life coaching
Biblical base
Biblical Buddhism
Biblical Christian counseling and soul care
Biblical Christian sex therapy
Biblical coaching
Biblical counseling
Biblical counseling and guidance

Biblical counseling/complementary methodologies
Biblical discipleship counseling
Biblical existential therapy
Biblical foundations
Biblical framework
Biblical framework counseling
Biblical instuction [*sic*]
Biblical integration
Biblical integrative therapy
Biblical knowledge and prayer
Biblical life coaching
Biblical love model
Biblical nauthetic principle
Biblical perspective
Biblical practices
Biblical principles counseling
Biblical principles in God's word
Biblical recasting therapy
Biblical truth
Biblical truths and sound counseling psychology
Biblical values
Biblical view
Biblical world view
Biblical-Christian perspective
Biblical/Christian
Biblically based counseling
Biblically based perspective
Biblically based salvation oriented counseling
Biblically based soul care/Christian counseling
Biblically Christian counseling
Biblically-based enlightenment
Biblically-based principles
Bibliotherapy = Bible counseling
Biking therapy
Bioaquatic exploration
Biodynamic craniosacral therapy
Biofield tuning
Bionomic psychotherapy
Biorythm therapy
Birthing from within
BLACK love
Blue brain training
Body intelligence
Body psychotherapy and breathwork

Body wisdom dialogue and dreamwork
Body-soul integration
Bohemian approach
Bohmian dialogue
Bonny method
Bootcamp therapy
Bouldering
Boxing therapy
Brainworking recursive technique (BWRT)
Brian Weiss/past life regression therapy
Bright light therapy
BSFF—be set free fast
BTB feng shui
Buddhism
Buddhism and the wisdom traditions
Buddhist and yogic philosophy
Buddhist Jungian psychology
Buddhist philosophy
Buddhist practices
Buddhist recovery models
Buddhist spirituality
Buddhist studies
Buddhist vipassana
Business coaching
Buteyko method
CAMt Christian auditory meditation therapy
Capoeira
Career counseling
Carpal tunnel hypnotherapy
Catholic coaching
CBT
Cellular release therapy
Cellular support protocol
Celtic healing
Ceramics arts therapy
Ceremonial healing
Chakra balancing
Chakra balancing flow orientation
Chakra energy
Chakra realignment
Channeling
Character analytic approach
Character analytic vegetotherapy
Character education

Chassidic spirituality
Chess therapy
Child counseling
Child trends lifecourse model
Child-centered (Christian) play therapy
Chinese and ayurvedic medicine
Chinese alternative medicine
Chinese five elements
Chinese herbal medicine
Chinese medical science
Chiropractic care
Christ centered approach
Christ conscious counseling
Christ consciousness counseling
Christ-centered 12-step approach
Christ-centered approach
Christ-centered identity discovery process
Christ-centered professional counseling
Christian
Christian ACT
Christian activities therapy
Christian archetypical psychology
Christian auditory meditation therapy
Christian-based therapy
Christian behavioral analysis
Christian Biblical principles therapy
Christian Biblical therapy (CBT)
Christian Buddhism
Christian caregiving model
Christian cognitive behavioral therapy
Christian control mastery therapy
Christian counseling
Christian counseling theory
Christian creationist approach
Christian discipleship counseling
Christian dream interpretation
Christian experiential therapy
Christian family therapy
Christian feminism
Christian Gospel-based therapy
Christian healing process
Christian hypnotherapy
Christian identity counseling
Christian individual therapy

Christian lifestyle counseling
Christian marriage counseling
Christian meditation and contemplative counseling
Christian mind/body work—splanka
Christian mysticism
Christian principles and psychology
Christian sensitive counseling
Christian systems theory
Christian theoretical foundation
Christian-related counseling and psychotherapy
Christian-related professional services
Christianity
Chrystal bowl sound therapy
CIMBS
Circle process (groups)
Circle sanctuary
Circus arts therapy
Clairvoyance
Clairvoyant psychotherapy
Classic hatha therapy
Classical Chinese medicine
Classical homeopathy
Clay sculpture and pottery
Clear belief process
Climbing therapy
Clinical sexology
Clinical thermographor
Clown therapy
CMC power counseling
Co-active coaching Gestalt
Coaching
Cognition-oriented theology
Cognitive
Cognitive behavior therapy
Cognitive behavioral (CBT)
Cognitive behavioral therapy
Cognitive bias modification
Cognitive mind stimulation therapy
Collaborative methods
Collaborative stage model (CSM)
Color puncture
Comedy therapy
Comfort (therapy) dog
Comic books in therapy

Community herbalism
Community reinforcement approach (CRA)
Compassion power
Compassionate de-possession
Comprehensive energy psychology
Comprehensive, multi-perspectival approach
Conflict resolution
Conjoint therapy (God-client-therapist)
Conscious enlightenment therapy
Contemplative
Contemplative and humanistic psychology
Contemplative Buddhist therapy
Contemplative evocative approach
Contemplative psychotherapy (Buddhist)
Contemplative therapy
Contemplative Tibetan Buddhist therapy
Contemplative Tibetan psychology
Contemplative-realizational
Contemporary alternative treatments
Contemporary Christian dream interpretation
Continuing bond with the deceased
Cooking therapy
Coping skills
Core energetic therapy
Core energetics
Core energetics (mind/body)
Core shamanic practices
Corporate psychoanalysis
Counseling that works
Courage beyond therapy
Cowboy Gestalt therapy
Crafting vision
Craniosacral therapy
Create heaven soul entrainment training
Critical mixed race theories
Cross-cultural music in healing
Crystal healing
CSEFEL (pyramid Model)
Cuddle party facilitator
Cutting-edge laser coaching
Cycling therapy
Cynthia Bourgeault approach
Dance therapy
DBDDD treatment

Decolonized interventions and indigenous philosophy
Deconstructionism
Deconstructionist orientation
Depth (soul) psychology
Depth psychotherapy/Christian tradition
Depth soul work and discovery
DeTUR method
Developmental needs meeting strategies
Dialectical behavioral therapy (DBT)
Diamond approach
Diamond approach/diamond breathwork
Diamond heart approach
Digital art therapy
Direct therapy
Discipleship counseling
Distance counseling
Divination
Divine integration healing
Divine nature of each soul/client
Divine relationship counseling
Divine's symbolic communication
Divorce discernment counseling
DNMS
Dowsing therapy
Dr. Margaret Paul/inner bonding
Drama therapy
Drawing therapy
Dream analysis
Dream interpretation
Dream tending
Dream work/nightmares
Duality reality therapy
Ear candling
Earth sangha meditation
Earth-based religious
Earth-based spirituality
Eastern and Western [J. Krishnamurti]
Eastern and Western mystical traditions
Eastern body work cognitive behavioral therapy
Eastern bodywork
Eastern coaching
Eastern methods of healing
Eastern mindfulness philosophies
Eastern mystical traditions

Eastern spiritual wisdom
Eastern wisdom traditions
Eastern-Afrikan and Western
Eclectic
Eclectic theoretical approach
Eclectic version of therapy
Eco-spiritual therapy
Ecstatic dance
Ecstatic energetic practice
Eden energy medicine
Editing therapy
Eduardo Duran's work
EFT
EFT unwinding
Embodied experiencing
Embodied Gestalt approaches
Embodied relating
Embodiment of the divine
EMDR
EMDR tapping
Emerging wisdom
Emotion focused therapy
Emotional freedom technique
Emotional freedom technique/tapping
Emotionally focused therapy (EFT)
Encouraging coaching
Energetic Asian systems
Energetic empathic processes
Energetic healing modalities
Energy alchemy
Energy medicine
Energy psychology
Energy tapping/energetic
Energy-related therapies
Enhanced/advanced leadership coaching
Enneagram in the narrative tradition
Enneagram soul based
Enneagram/Myers Briggs
Entheogenic shamanism
Equality counseling
Equine ecosomatics
Equine-assisted therapy
Essential oil healing
Essential oils

Essential oils
Etiotropic trauma management
ETM TRT SHOM theory
EVOX voicemapping therapy
Executive-business-entrepreneur coaching
Existential
Existential personalistic anthropology
Existential phenomenology
Existential philosophy
Experiential equine therapy
Exploration of the soul
Exposure therapy
Expressive analysis
Expressive arts
Eye movement desensitization reinstallation
Fairy tale model
Faith-based Christian counseling
Faith-based therapy
Family/marital
Family and marital counseling
Family financial responsibility counseling
Family systems
Family systems orientation to therapy
Family systems therapy
FCT
Feedback informed treatment (FIT)
Feline-assisted therapy
Feminine power coaching
Feminism
Feminist therapy
Feng shui coaching/consulting
Feng shui consulting
Film-video arts therapy
Financial social work
Firing range therapy
First argument technique
Fishing therapy
FIT
Five sequence relational drawing process
Flower and gem essence therapy
Flower arranging therapy (floratherapy)
Fly-fishing therapy
Focusing oriented art therapy
Food-mood wellness

Formational prayer counseling
Four temperaments model
Freedom adaptation techniques
Freedom counseling
Frum therapy
Fulfillment-oriented therapy
Fundamental scriptural principles
Gambling counseling
GAP counseling
Gardening therapy
Geek therapy
Gemstone therapy
Gentle martial arts movements
Genuine wisdom
German play-therapy
Gestalt therapy
GLBT-friendly
God designed counseling
God-image issues counseling
Goddess psychology
Gods saving work
Gold standard EFT
Golf therapy
Gospel-centered worldview
Grace therapy
Gracious Christian counseling
Grief yoga
Guided dream journeying
Guided self-inquiry
Gunborg Palme's method
Gymnastics psychology
HAES
Hakomi
HALO light therapy
Happiness work
Harmonial stress management system
Harmonic medicine
Harmonic sound therapy
Harner shamanic counseling
Hassidic spirituality
Havening techniques
Healing light of God's grace and truth
Healing love energy direction
Healing wisdom

Health at every size
Heart assisted therapy
Heart centered hypnotherapy
Heart-centered and sacred soul work
Heartbeat method
Heathen
Hedy and Yumi imago therapy
Helper model
Herbal medicine
Herbalist/herbalism
Hermetic philosophy
Higher brain living
Hiking therapy
Himalayan singing bowls therapy
Hindu practices
HMS holistic therapy
Ho'oponopono
Hockey therapy
Hold sacred space
Holistic and spiritual psychotherapies
Holistic Christian
Holistic healing approaches
Holistic-Christian perspective
Hollistic (*sic*)
Holodynamic therapy
Holy fire karuna reiki
Hormone and amino acid therapy
Hormone balancing and coaching
Horse play therapy
Horse-informed archetypical therapy
Horticultural therapy
HPN neurofeedback
Human givens trauma rewind
Humanism
Humanistic
Humanistic play therapy
Huna healing
HUSO pro
Hypnoanalytic therapy
Hypnobirthing
Hypnosis: age and past lives regression
Hypnosis/EFT/TFT/energy psychotherapies
Hypnosis/past-life regression therapy
Hypnotherapy

Hypnotic regression
Ichthyotherapy
Illness management and recovery
Image de-construction protocol (IPT)
Image transformation therapy
Image-bearer counseling
Image-bearer orientation
Imaginal psychology
IMAGO dialogue
Immanuel prayer process
Impact therapy approach
Inclusive holistic methods
Indian Saint Sri Kaleshwar
Indigenous focusing oriented therapy
Indigenous healing practices
Indo-Tibetan Buddhism
Induced after-death communication
Induced communications after death
Inner bonding
Inner healing prayer
Inner humanism
Inner radiance
Integral therapy
Integrated synergistic approach
Integration of faith/spiritual perspective
Integrative
Integrative Biblical principals [sic]
Integrative mystical states of consciousness
Integrative psychology and spirituality
Integrative restoration
Integrative soul coaching
Integrity model perspective
Interactive metronome therapy
Interactive metronome training
Interconnection and spirituality
Interfaith/nonsectarian
Interlife exploration
Internal family systems
Interpersonal and social rhythm therapy
Interpersonal hypnotherapy
Interpersonal neurobiology
Interpersonal therapy
Intersectionality
Intersubjective

Intuitive healing
Intuitive readings
Intuitive therapy
IPN/PRN expertise
IPP
iRest
Iridology
ISIS practice
Japanese acupressure
Japanese psychology approaches
Jesus-oriented approach
Jewish spirituality
Jin shin do bodymind acupressure
Jin shin jyutsu
John of God crystal healing bed
John Pierrakos' core energetics
Journaling
Journey of the soul
Journey towards rebirth
Juggling therapy
Jungian and fairytale study
Jungian archetypes
Jungian dream interpretation
Jungian psychology and theology
Jungian theology
Jungian-focused spiritual direction
Kabbalistic healing
Kalish practice
Kalla treatment
Karol McBride's 5 step model
Karuna reiki practice
Katherine Woodward Thomas approach
Ketamine-assisted psychotherapy
Kinetic counseling
Kink aware therapy
Knitting therapy
Kripalu yoga
Kundalini arousal
Kundalini yoga
Kunga yoga
Labyrinth walking therapy
Lanktonian psychotherapy
Laser allergy treatment
Latter Day Saints counseling

Law of attraction
LEGO-based social skills
LEGO therapy
Leveraging theory
LGBTQ-affirmative therapy
Life between life spiritual regression
Life coaching
Life principles of the Holy Spirit
Life reformation Christian counseling
Life SOULutions therapy
Life transitions therapy
Lifting therapy
Light energy art therapy
Light laser therapy
Light touch healing
Light touch therapies
Limited natural sleep
Living the twelve steps spirituality relapse
LOCUS/CAFAS
Love first interventions
LSD psychotherapy
Lymph drainage massage
Lanktonian psychotherapy
M.E.T.A. attachment treatment
Magical SOULutions
Magnet therapy
Magnetic field therapy
Male responsibility therapy
Mandala process work
Marconics no-touch healing
MARI and mandala drawing
Marriage and family therapy
Marschak interaction method (MIM)
Martial arts therapy
Maslovian motivational therapy
Massage therapy
MAT core-12
Matrix reimprinting
Matrix reimprinting for trauma
Matrix synergetics
May cause miracles practice
Maya womb healing
Mayan healing
Mazes therapy

Medical and past life hypnotherapy
Medicinal aromatherapy
Medifast coaching
Meditation
Mental feng shui
Mental fitness coaching
Meridian psychotherapy
Meridian tapping
Mesotherapy
Metaphysics—science of the mind/new thought
MICPS
Mind-body bridging
Mindfulness
Mindfulness-based cognitive therapy
Money success-planning
Moon circle
Moral injury therapy
Moral recognition
Moral reconation therapy
Morita therapy
Mormon therapy
Mormon/LDS therapy
Moshe Feldenkrais approach
Motivational interviewing
Movies therapy
Mud therapy
Multi-generational inclusion
Multi-systems
Multicultural
Multicultural Christian counseling
Muscle testing
Music and aroma therapy
Music and sound therapies
Music therapy
Music/sound healing
Muslim healing
Muslim therapy
Muzen mindful creative
Myofascial release
Mystic aroma therapy (MAT)
Mystical meditation
Mysticism
Mytho-poetic therapy
Naikan

Nanj therapy
Narrative
Native American practices
Native American rituals
Native American, Buddhist, and Hindu spiritual traditions
Native Mesoamerican healing
Nature-based rites of passage
Nature-based therapy
Nature-based archetypal depth psychology
Naturopathic ministry
Neural-based emotional release
Neuro-affective touch for trauma
Neuro-linguistic
Neuro-linguistic programming
Neuro-psychotherapy
Neurobiology
Neurocise methods
Neurofeedback
Neuropsychiatry
Neuroscience and biochemistry techniques
Neurotherapy
Neurotransmitter restoration therapy
New life psychotherapy
New life therapy
New regression therapy
NIASZIIH healing
Nightmare resolution therapy (NRT)
Nine rites of munay-ki
NLP
Noetic balancing
Nondual kabbalistic healing
Nondual philosophy
Nonduality
Nourishing choices
Nouthetic therapy
Nutrition counseling
Oneness blessing giver
Ontological hermeneutics
Open studio process
Organic mind energy
Orgone therapy
Oriental medicine
Origami therapy
Own therapies

Painting therapy
Para-psychology
Past life regression therapy
Past life soul retrieval therapy
Peace circles
Peaceful living therapy
Peripheral and neurobiofeedback techniques
PERMA model
Person-centered therapy
Personal fitness training
Personality trait theories
Personalized theory
Peruvian Inkan tradition
Pet massage therapy
Phage therapy
Pharmacological
Philosophical orientations
Philosophy tools
Photo therapy
Photography therapy
Physical exercise counseling
Physical fitness/health
Piano therapy
Picture framing
Plant medicine integration
Plant spirit medicine
Play therapy
Poetry therapy
Point approach
Polarity realization therapy
Popular psychology
Postural integration
Power animal retrieval
Pranic healing
Pranic healing procedures
Pray therapy
Prayer
Prayer counseling
Pre- and post-natal massage
Principle-based therapy
Principles of soul healing love
Private and public standing counseling and coaching
Professional transformational conversation
Progressive blend of approaches

Prolonged exposure
Psicoastrology
PSYCH-K
Psychedelic integration
Psychic medium
Psycho-organic analysis
Psycho-shamanic healing
Psycho-shamanic therapy
Psycho-spiritual inquiry
Psycho-spiritual integration
Psychoanalysis
Psychodrama
Psychodynamic
Psychological principles with biblical teaching
Psychopomp
Psychopomp work
Psychosynthesis
Punching bag therapy
Puppet therapy
Pure energy integration
Pure feeling awareness
Purpose focused approach
Pyramid model
Qabalistic pathworking
QTPOC conversations
Quantum healing
Quantum healing hypnosis
Quantum medicine
Quantum techniques
Queer theory
RAAD model for interventions
Radical aliveness
Radical compassion
Radix intensive therapy
Radix neo-Reichian approach
Radix therapy
Raindrop and aroma therapy techniques
Rapid prayer intervention counseling
Rational emotive behavioral therapy
Reality therapy
Reality-based psychotherapy
Reason-based approach
Recasting therapy
Recovery dynamics

RECT
Reference point practitioner
Reflexology
Regressionism
Rehabilitation counseling
Reichian biophysical therapy
Reichian orgonomic
Reichian reiki
Reiki
Reiki energy healing
Reiki energy work/essential oils
Reiki stress reduction and relaxation
Reiki womb therapy
Reincarnation therapy
Rejection therapy
Relational
Relational life therapy
Relational neurobiology
Religious cognitive–emotional therapy
Repentance therapy
Resnick model
Resonance repatterning
Restorative justice therapy
Reunification therapy
Ridhwan approach
Riding llamas therapy
Riding therapy
Rising star healing
ROAR (radically orgasmically alive reality)
RoHun purification
Rohun purification process
RoHun transpersonal (soul) psychology
Rosen method bodywork
Rover therapy
Rowing therapy
RRT
RUH healing
Running the bars/access bars
Running therapy
Sacred card readings
Sacred contracts
Sacred feminine guide
Sacred imagination
Sacred/spiritual approaches

Sailing therapy
Samatha and vipassana meditation
Sanctuary counseling
Sand play
Sand tray therapy
Sandra Ingerman model
Sat nam rasayan healing
Satatove consciousness living
Scale-walking
Schema therapy
Scientology
Scriptural principles therapy
Scriptural values counseling
Sculpting therapy
Secret garden healing
Secular Buddhism
Secular treatment
Self healing
Self-inquiry counseling
Self-relations-generative trance
Seven challenges
Sex therapy
Shadow integration
Shadow work—women's empowerment
Shamanic energy healing
Shamanic healing
Shamanic healing practices
Shamanic hypnotherapy processes (8)
Shamanic spirituality
Shamanic therapy
Shamanism
Shambala meditation
Shambhala Buddhist approach
SHEN therapy
Shiatsu and martial arts
Shifaa soul counseling
Silent hiking
Singing therapy
Sistahpeaceful life-style changes
Sleep hygiene
Social location therapy
Sociotherapy
Sogyal rinpoche
Solution focused brief (SFBT)

Solution-focused therapy
Somatic experiencing
Somatic experiencing (SE)
Somatic therapy
Sophia analysis
Sorenson method
Soteriological counseling
Soul alignment
Soul care Christian counseling
Soul coaching
Soul detective protocols
Soul entrainment therapy
Soul expansion shamanic mentoring
Soul healing love
Soul level positive psychology
Soul oriented therapy
Soul psychiatry [lpc]
Soul restoration
Soul retrieval therapy
Soul support coaching
Soul tracking
Soul travel
Soul wisdom therapy
Soul work (inner child work)
Soul-based psychotherapy
Soul-centered psychotherapy
Soul-heart-mind-body perspective
SoulTalk
SoulTalk/SoulCare
Sound and vibrational therapy
Sound mnemonics
Sound therapy (crystal bowl)
Sound weaving
Source Therapy
SOZO attainment therapy
Spartan SGX coaching
Spiraculturally based
Spirit care counseling
Spirit junkie masterclass
Spirit-in-nature flower essences
Spiritual and somatic approach
Spiritual and energy THERAPY
Spiritual-based counseling
Spiritual cognitive therapy (SCT)

Spiritual counseling and divination
Spiritual direction
Spiritual DIRECTION/seeking
Spiritual divination
Spiritual emergence seekers
Spiritual formation/direction
Spiritual healing ceremonies
Spiritual practice
Spiritual psychology
Spiritual root system
Spiritual self-schema therapy
Spiritual warriorship
Spiritual/Biblical stance
Spirituality
Spiritually fluent psychotherapy
Splankna (Christian-based)
Splankna energy therapy
Spritual [sic] disciplines
Stephen Porges's safe and sound protocol
Stoicism philosophical perspective
Storytelling
Strategic whole-life assessment
Street yoga
Strength-based
Subtle energy enhancement
Sufi healing circle
Sufi psychology
Sufism
Superhero therapy
Supportive touch
Surrogacy communication (TM)
Survivor-fighter-thriver model of care
Sustainable happiness
Svaroopa yoga embodyment [sic] therapy
Swimming therapy
Symphonic healing gong
Systematic desensitization
Systems approach/Christian principles
Tabletop therapy
Talent development counseling
Tao of self-realization
Taoist/tantric practices
Tapas acupressure technique
Tapas energy psychology

TARA approach
Tarot cards
Taylor Johnson analysis
Techniques based on Dr. Dan Allender
Techniques from Sandra Ingerman
Temperament analysis profile
TERRA essential oils
The 11 principles of transformation
The Amen method
The Arno profile system (A.P.S.)
The divine feminine
The enneagram
The good life model
The hero's journey
The power of prayer
The purposeful pursuit of God's purpose
The truth of God's word
The truth of the Gospel
Theater counseling
Theistic therapy/Christian values therapy
Theophostic prayer therapy
Thera-P
Therapeutic singing sessions
Therapeutic taro and reiki
Therapeutic yeshiva
Therapuetic [sic] fly fishing
Therapy for religious abuse
Third-wave cognitive behavioral therapy
Thomas Hubl orientation
Thought field therapy
Three pinciples [sic] therapy
Three-to-one couples therapy
Time travel
Time traveling
Toltec wisdom
Torah-based counseling and therapy
Torah-based therapy
Torah life coaching
Torah perspective and principled living
Total body modification
Totem pole imagery process
Traditional approaches
Traditional bhakti music
Traditional Native American talking circle

Traditional Tibetan medicine
Trager psychophysical movement
Trail cycling
Transformation/transcendence techniques
Transformational anthropology
Transformational process coaching and soul therapy
Transpersonal approaches to psychotherapy
Transpersonal coaching
Trauma focused
Trauma informed hypnotherapy
Trauma-focused cognitive behavioral therapy
Traya chakra psychotherapy
Trigger point therapy
Trim life healing
Truth of God's word
Tuina massage
Twelve-step approaches
Twelve-step orientations
Twelve-step weddings
Two step revolution
Ubuntu philosophy
Unconditional care model
Unconventional therapy
Unique psychotherapy
Unitive psychotherapy
Universal life counseling
Universal values counseling
Usui reiki healing
Usui shiki ryoho
Uzazu coaching
Vedic astrology
Veg*n lifestyle
Vegetotherapy
Vibrational sound therapy
Vibrator therapy
Video games in therapy
Video production therapy
Vipassana or insight meditation
Virotherapy
Virtues orientation
Visceral massage
Visual coding displacement therapy
Visual journaling
Vitalistic paradigms

Vitamin/nutrition consulting
Vocal arts therapy
Voice dialogue therapy
Walking the middle path
Walking therapy
WAR
Wat-su (water shiatsu)
Watchwaitwonder
Western psychology
Whatever works!
White time healing
Whitehawk process practice
Wicca
Wild feminine soul psychotherapy
Wilderness therapy
Wildish work
Wire sculpting therapy
Wisdom from the Holy Spirit
Wisdom of God's word
Wisdom recovery
Wisdom–based therapy
Women's psychology
Womin-affirmative
Word of God counseling
Word of God universal truths
Work-related stress
Yandara yoga
Yoga and mudra exercises
Yoga therapy
Yogic techniques and chakra balancing
Young living essential oils (YLEOS)
Young living raindrop technique
Youth coaching
Yuen method
Zanker therapy
Zen
Zen and vipassana meditation practice
Zen dharma
Zen meditation/Zen philosophy
Zen-vipassana
Zencare
Zensight energy healing
Zentangle teaching
Zero balancing practice
Zyto biotechonology scans

Titles of Licenses for Mental Health Care Providers

Agency Affiliated Counselor
Associate Clinical Mental Health Counselor
Associate Clinical Mental Health Counselor Extern
Associate Licensed Counselor
Certified Advanced Practice Social Worker
Certified Counselor
Certified Independent Practice Social Worker
Certified Master Social Worker
Certified Mental Health Practitioner
Certified Professional Counselor
Certified Psychological Assistant
Certified Social Work Manager
Certified Social Worker
Certified Social Worker Manager
Certified Social Worker—Clinical
Certified Social Worker—Independent Practice
Clinical Social Work Associate
Licensed Advanced Macro Social Worker
Licensed Advanced Practice Social Worker
Licensed Advanced Social Worker
Licensed Alcohol and Drug Counselor
Licensed Art Therapist
Licensed Associate Clinical Social Worker
Licensed Associate Counselor
Licensed Associate Counselor of Mental Health
Licensed Associate Marriage and Family Therapist
Licensed Associate Professional Counselor
Licensed Baccalaureate Social Worker (LBSW)
Licensed Bachelor's Social Worker (LBSW)
Licensed Certified Social Worker
Licensed Certified Social Worker—Clinical

Licensed Clinical Counselor Intern
Licensed Clinical Marriage and Family Therapist
Licensed Clinical Mental Health Counselor
Licensed Clinical Professional Counselor
Licensed Clinical Social Worker
Licensed Clinical Social Worker—Private Independent Practice
Licensed Creative Arts Therapist
Licensed Dance/Movement Therapist
Licensed Graduate Professional Counselor
Licensed Graduate Social Worker (LGSW)
Licensed Independent Clinical Social Worker
Licensed Independent Marriage and Family Therapist
Licensed Independent Mental Health Practitioner
Licensed Independent Social Worker
Licensed Independent Social Worker—Advanced Practice
Licensed Independent Social Worker—Clinical Practice
Licensed Marriage and Family Therapist
Licensed Marriage and Family Therapist Associate
Licensed Marriage, Family, and Child Counselor
Licensed Master Social Worker
Licensed Master Social Worker—Advanced Practice
Licensed Master Social Worker—Clinical
Licensed Master Social Worker—Macro
Licensed Master Social Worker—Advanced Practice
Licensed Master Social Worker—Clinical Conditional
Licensed Master Social Worker—Independent
Licensed Master's Psychologist
Licensed Master's Social Worker—Clinical Designation
Licensed Master's Social Worker—Macro Designation
Licensed Mental Health Counselor
Licensed Mental Health Practitioner
Licensed Professional Clinical Counselor
Licensed Professional Counselor Associate
Licensed Professional Counselor of Mental Health
Licensed Professional Counselor Trainee
Licensed Professional Counselor With Appraisal Privileges
Licensed Professional Counselor/Intern
Licensed Professional Counselor/Mental Health Service Provider
Licensed Professional Music Therapist
Licensed Psychoanalyst
Licensed Psychological Assistant
Licensed Psychological Associate
Licensed Psychological Examiner
Licensed Psychologist

Licensed Psychologist Associate
Licensed School Counselor
Licensed School Psychologist
Licensed Senior Psychological Examiner
Licensed Sex Offender Treatment Provider
Licensed Social Service Worker
Licensed Social Work Associate
Licensed Social Work Associate and Advanced
Licensed Social Work Associate and Independent Clinical
Licensed Social Worker
Licensed Social Worker Associate
Licensed Social Worker—Administration
Licensed Specialist Clinical Social Worker
Limited License Bachelor Social Worker
Limited License Master Social Worker
Limited License Psychology
Marriage and Family Therapist—Intern
Marriage and Family Therapy Intern
Private Independent Practice (Certificate)
Provisional Licensed Clinical Social Worker
Provisional Licensed Professional Counselor
Provisional Marriage and Family Therapist
Provisional Professional Counselor
Provisionally Licensed Counselor
Provisionally Licensed Marital and Family Therapist
Registered Counselor Intern
Supervised Marital and Family Therapist
Licensed Professional Art Therapist
Licensed Clinical Professional Art Therapist
Licensed Certified Art Therapist
Licensed Professional Counselor With Specialty Designation in Art Therapy
Registered Art Therapist With a License to Practice Psychotherapy

REFERENCES

INTRODUCTION

Beyerstein, B. L. (2001). Fringe psychotherapies: The public at risk. *Scientific Review of Alternative Medicine, 5*(2), 70–79.

Miller, S., & Hubble, M. (2011). The road to mastery. *Psychotherapy Networker, 35*(2), 22–60.

Thompson, T. C. (2010). Psychological treatments to avoid. *Alabama Counseling Association Journal, 36*(1), 39–48.

CHAPTER 1

American Psychological Association. (2012). *Recognition of psychotherapy effectiveness.* Washington, DC: Author. Retrieved from http://www.apa.org/about/policy/resolution-psychotherapy.aspx

Beck, A. (1996). The past and the future of cognitive therapy. *Journal of Psychotherapy Practice and Research, 6*(4), 276–284.

Beck, A., & Beck, J. (2013). *A conversation with Aaron Beck.* Retrieved from https://www.youtube.com/user/BeckInstitute

Boeree, C. G. (1997/2006). *Personality theories.* Retrieved from http://www.social-psychology.de/do/pt_intro.pdf

Boeree, C. G. (2014). *George Kelly.* Retrieved from http://www.social-psychology.de/do/pt_kelly.pdf

Brom, D., Stokar, Y., Lawi, C., Nuriel-Porat, V., Ziv, Y., Lerner, K., & Ross, G. (2017). Somatic Experiencing for Posttraumatic Stress Disorder: A Randomized Controlled Outcome Study. *Journal of Traumatic Stress, 30*(3), 304–312.

Crane, R. S., Stanley, S., Rooney, M. W., Bartley, T., Cooper, L., & Mardula. J. (2015). Disciplined improvisation: Characteristics of inquiry in mindfulness-based teaching. *Mindfulness, 6*(5), 1104–1114.

Crits-Christoph, P., Baranackie, K., Kurcias, J., Beck, A., Carroll, K., Perry, K., . . . Zitrin, C. (1991). Meta-analysis of therapist effects in psychotherapy outcome studies. *Psychotherapy Research, 1*, 81–91.

Doorey, M. (2014). *George Alexander Kelly.* Retrieved from http://psychology.jrank.org/pages/357/George-Alexander-Kelly.html

Encyclopedia.com. (2018a). *Aaron Temkin Beck.* Retrieved from https://www.encyclopedia.com/psychology/arts-construction-medicine-science-and-technology-magazines/beck-aaron-temkin

Encyclopedia.com. (2018b). *Rogers, Carl (1902–1987).* Retrieved from https://www.encyclopedia.com/people/medicine/psychology-and-psychiatry-biographies/carl-rogers

Famous   Psychologists.   (2018).   *Aaron   Beck*.   Retrieved   from   http://www.
    famouspsychologists.org/aaron-beck/

Frank, J. D., & Frank, J. B. (1991). *Persuasion and healing: A comparative study of psycho-
    therapy* (3rd ed.). Baltimore, MD: Johns Hopkins University Press.

Fransella, F. (1995). *George Kelly*. London, UK: Sage.

Fransella, F. (2008). Book review: *George Kelly: The psychology of personal constructs by
    Trevor Butt*. Retrieved from http://www.pcp-net.org/journal/pctp08/fransella08.html

Garfield, S. L., & Bergin, A. E. (Eds.). (1986). *Handbook of psychotherapy and behaviour
    change*. New York, NY: Wiley.

Jennings, L., & Skovholt, T. M. (1999). The cognitive, emotional, and relational charac-
    teristics of master therapists. *Journal of Counseling Psychology, 46*(1), 3–11.

Jennings, L., & Skovholt, T. M. (2004). The cognitive, emotional, and relational char-
    acteristics of master therapists. In T. M. Skovholt & L. Jennings (Eds.), *Master
    therapists: Exploring expertise in therapy and counseling* (pp. 31–52). Boston,
    MA: Allyn & Bacon.

Kazdin, A. E. (2007). Mediators and mechanisms of change in psychotherapy research.
    Annual Review of Clinical Psychology, 3, 1–27.

Kelland, M. (2017). *Personality theory*. Retrieved from https://www.oercommons.org/
    authoring/22859-personality-theory/12/view

Kelly, G. A. (1991). *The psychology of personal constructs*. London, UK: Routledge (orig-
    inally published in 1955).

Kenny, V. (1984). *An introduction to the personal construct psychology of George A. Kelly*.
    Retrieved from http://www.oikos.org/vincpcp.htm

Kindler, A. (2010). Spontaneity and improvisation in psychoanalysis. *Psychoanalytic
    Inquiry, 30*(3), 222–234.

Leahy, R. L. (1996). *Cognitive therapy: Basic principles and applications*. Oxford,
    UK: Rowman and Littlefield.

Martin, F. A. (1973). *A study of Carl Rogers' philosophy of persons* (Unpublished disserta-
    tion). Louisville, KY: The Southern Baptist Theological Seminary.

McPhee, J. (1986). *Rising from the plains*. New York, NY: Farrar, Straus & Giroux.

Miller, S. (2013). *The evolution of psychotherapy*. Presented at the annual meeting of the
    Evolution of Psychotherapy Conference. Retrieved from https://www.youtube.com/
    watch?v=pI8Hww1xjK4

Miller, S., & Hubble, M. (2011). The road to mastery. *Psychotherapy Networker,
    35*(2), 22–60.

Mullenbach, M., & Skovholt, T. M. (2004). Emotional wellness and professional resiliency
    of master therapists. In T. M. Skovholt & L. Jennings (Eds.), *Master therapists: Exploring
    expertise in therapy and counseling* (pp. 77–106). Boston, MA: Allyn & Bacon.

Niebuhr, R. (1950). *The A.A. Grapevine*. January 1950, pp. 6–7.

Neimeyer, R. A. (2000). Kelly, George Alexander. In A. E. Kazdin (Ed.), *2000 Encyclopedia
    of psychology* (pp. 805–807). Vol. 2. Washington, DC: American Psychological
    Association

Peyser, C. S. (1994). Kelly, George A. In R. J. Corsini (Ed.), *Encyclopedia of psychology*
    (2nd ed., Vol. 4, pp. 353–354). New York, NY: John Wiley & Sons.

Psychologist Anywhere Anytime. (2018a). *Famous psychologists: Aaron Beck*. Retrieved
    from   http://www.psychologistanywhereanytime.com/famous_psychologist_and_
    psychologists/psychologist_famous_aaron_beck.htm

Psychologist Anywhere Anytime. (2018b). *Famous psychologists: Carl Rogers.* Retrieved from http://www.psychologistanywhereanytime.com/famous_psychologist_and_psychologists/psychologist_famous_carl_rogers.htm

Rogers, C. R. (1942). *Counseling and psychotherapy.* Boston, MA: Houghton Mifflin.

Rogers, C. R. (1951). *Client-centered therapy.* Boston, MA: Houghton Mifflin.

Rogers, C. R. (1961). *On becoming a person: A therapist's view of personality.* Boston, MA: Houghton Mifflin.

Rogers, C. R. (1967). Carl R. Rogers. In E. G. Boring & G. Lindzey. *A history of psychology in autobiography* (Vol. 5, pp. 341–384). New York, NY: Appleton-Century-Crofts.

Rosner, R. I. (2012). Aaron T. Beck's drawings and the psychoanalytic origin story of cognitive therapy. *History of Psychology, 15*(1), 1–18.

Rowan, J. (1992). In a response to, Mair, K. The myth of therapist expertise. In W. Dryden & C. Feltham (Eds.), *Psychotherapy and its discontents* (pp. 160–165). Buckingham, UK: Open University Press.

Sburiati, E. S., Lyneham, H. J., & Schniering, C. A. (2012). A model of therapist competencies for the empirically supported interpersonal psychotherapy for adolescent depression. *Clinical Child and Family Psychology, 15*(2), 93–112.

Sifneos, P. E. (1973). The prevalence of "alexithymic" characteristics in psychosomatic patients. *Psychotherapy and Psychosomatics, 22*(2-6), 255–262.

Skovholt, T. M., Jennings, L., & Mullenbach, M. (2004). Portrait of the master therapist: The highly-functioning self. In T. M. Skovholt & L. Jennings (Eds.), *Master therapists: Exploring expertise in therapy and counseling* (pp. 125–146). Boston, MA: Allyn & Bacon.

Stevens, S. E., Hynan, M. T., & Allen, M. (2000). A meta-analysis of common factor and specific treatment effects across the outcome domains of the phase model of psychotherapy. *Clinical Psychology: Science and Practice, 7*(3), 273–290.

Thompson, G. G. (1968). George Alexander Kelly (1905–1967). *Journal of General Psychology, 79,* 19–24.

Wampold, B. E. (2015). How important are the common factors in psychotherapy? An update. *World Psychiatry, 14*(3), 270–177.

CHAPTER 2

Beyerstein, B. L. (2001). Fringe psychotherapies: The public at risk. *Scientific Review of Alternative Medicine, 5*(2), 70–79.

Broom, B. (2013). *Transforming clinical practice using the MindBody approach: A radical integration.* London, UK: Routledge.

Buggio, L. Barbara, G., Facchin, F., Frattaruolo, M. P., Aimi, G., & Berlanda, N. (2017). Self-management and psychological-sexological interventions in patients with endometriosis: Strategies, outcomes, and integration into clinical care. *International Journal of Women's Health, 9,* 281–293.

Carnahan, B., & Jungers, C. (2015). Understanding how counselors are regulated. *Counseling Today.* Retrieved from https://ct.counseling.org/2015/11/understanding-how-counselors-are-regulated/

Cook-Cattone, C. P. (2015). *Mindfulness and yoga for self-regulation: A primer for mental health professionals.* New York, NY: Springer.

Donn, L. (2011). *Freud and Jung: Years of friendship, years of loss.* East Burke, VT: Adin Louis.

Educational Testing Service. (2017a). *GRE: General test interpretive data by graduate major field*. Retrieved from https://www.ets.org/s/gre/pdf/gre_table4_extended.pdf

Educational Testing Service [as cited by The Tab]. (2017b). *Experts have worked out which majors have the highest IQ*. Retrieved from https://thetab.com/us/2017/04/10/which-major-has-highest-iq-64811

Goldberg, S. B., Babins-Wagner, R., Rousmaniere, T., Berzins, S., Hoyt, W. T., Whipple, J. L., Miller, S. D., & Wampold, B. E. (2016). Creating a climate for therapist improvement: A case study of an agency focused on outcomes and deliberate practice. *Psychotherapy, 53*(3), 367–375.

Gray, P. (2011). The decline of play and the rise of psychopathology in children and adolescents. *American Journal of Play, 3*(4), 443–463.

Gross, S. J. (1986). Professional licensure and quality: The evidence. *Cato Institute Policy Analysis No. 79*. Retrieved from https://object.cato.org/pubs/pas/pa079.pdf

Heinemann, J., Atallah, S., & Rosenbaum, T. (2016). The impact of culture and ethnicity on sexuality and sexual function. *Current Sexual Health Reports, 8*(3), 144–150.

Hill, C. E., Baumann, E., Shafran, N., Gupta, S., Morrison, A., Rojas, A. E., . . . Gelso, C. J. (2015). Is training effective? A study of counseling psychology doctoral trainees in a psychodynamic/interpersonal training clinic. *Journal of Counseling Psychology, 62*(2), 184–201.

Hill, C. E., Hoffman, M. A., Kivlighan, D. M., Spiegel, S. B., & Gelso, C. J. (2017). Therapist expertise: The debate continues. *Counseling Psychologist, 45*(1), 99–112.

Hill, C. E., Spiegel, S. B., Hoffman, M. A., Kivlighan, D. M., & Gelso, C. J. (2017). Therapist expertise in psychotherapy revisited. *Counseling Psychologist, 45*(1), 1–47.

Hogan, D. B. (1983). The effectiveness of licensing: History, evidence, and recommendations. *Law and Human Behavior, 7*(2/3), 117–138.

Holden, G., & Barker, K. (2018). Should social workers be engaged in these practices? *Journal of Evidence-informed Social Work, 15*(1), 1–13.

Kirschenbaum, H., & Henderson, V. L. (Eds.) (1989). *Carl Rogers—dialogues: Conversations with Martin Buber, Paul Tillich, B. F. Skinner, Gregory Bateson, Michael Polanyi, Rollo May, and others*. Boston, MA: Houghton Mifflin.

Lee, C. M., & Hunsley, J. (2015). Evidence-based practice: Separating science from pseudoscience. *Canadian Journal of Psychiatry, 60*(12), 534–540.

Lee, M. Y. (2009). *Integrative mind-body-spirit social work: An empirically based approach to assessment and treatment*. Oxford, UK: Oxford University Press.

Lees, J., & Tovey, P. (2012). Counselling and psychotherapy, complementary medicine and the future of healthcare. *British Journal of Guidance and Counseling, 40*(1), 67–81.

Lilienfeld, S. O. (2007). Psychological treatments that cause harm. *Perspectives on Psychological Science, 2*(1), 53–70.

Lilienfeld, S. O., Fowler, K. A., Lohr, J. M., & Lynn, S. J. (2005). Pseudoscience, nonscience, and nonsense in clinical psychology: Dangers and remedies. In R. H. Wright & N. A. Cummings (Eds.), *Destructive trends in mental health: The well-intentioned path to harm* (pp. 187–218). New York, NY: Routledge.

Lilienfeld, S. O., Ritschel, L. A., Lynn, S. J., Cautin, R. L., & Latzman, R. D. (2014). Why ineffective psychotherapies appear to work: A taxonomy of causes of spurious therapeutic effectiveness. *Perspectives on Psychological Science, 9*(4), 355–387.

Loyola University Chicago. (2019a). Retrieved from http://www.luc.edu/socialwork/academics/graduate/msw-advanced-standing/

Loyola University Chicago. (2019b). Retrieved from http://www.luc.edu/education/graduate/doctoraldegrees/counseling-psychology/curriculum/

Lumadue, C. A., Munk, M., & Wooten, H. R. (2005). Inclusion of alternative and complementary therapies in CACREP training programs: A survey. *Journal of Creativity in Mental Health. 1*(1), 7–19.

McNamee, S. (2009). Postmodern psychotherapeutic ethics: relational responsibility in practice. *Human Systems: The Journal of Therapy, Consultation & Training, 20*(1), 57–71.

Menachemi, N., & Collum, T. H. (2011). Benefits and drawbacks of electronic health record systems. *Risk Management and Healthcare Policy, 4,* 47–55.

Miller, S., & Hubble, M. (2017). How psychotherapy lost its magick: The art of healing in an age of science. *Psychotherapy Networker.* Retrieved from https://www.psychotherapynetworker.org/magazine/article/1077/how-psychotherapy-lost-its-magick

National Center for Complementary and Integrative Health (NCCIH). (2019a). *NCCIH facts-at-a-glance and mission.* Retrieved from https://nccih.nih.gov/about/ataglance

National Center for Complementary and Integrative Health (NCCIH). (2019b*). NCCIH clearinghouse.* Retrieved from https://nccih.nih.gov/health/clearinghouse

Natwick, J. (2018). Counselors are finding ways to combine traditional talk therapy with a vast number of other practices, specialties, methods and techniques. *Counseling Today, September,* 9–11.

O'Donohue, W., & Lilienfeld, S. O. (Eds.). (2013). *Case studies in clinical psychological science: Bridging the gap from science to practice.* New York, NY: Oxford University Press.

Pradhan, B., Kluewer D. J., Makani, R., & Parikh, T. (2016). Nonconventional interventions for chronic post traumatic stress disorder: Ketamine, repetitive transcranial magnetic stimulation (rTMS), and alternative approaches. *Journal of Trauma Dissociation, 17*(1), 35–54.

Robbins, B. D. (no date). *Backup of: A brief history of psychoanalytic thought—and related theories of human existence.* Retrieved from http://www5.csudh.edu/dearhabermas/selfshape01bk.htm

Sadiq, A. J. (2007). Attention-deficit/hyperactivity disorder and integrative approaches. *Psychiatric Annals, 37*(9), 630–638.

Sharpe, M., & Faulkner, J. (2008). *Understanding psychoanalysis.* Stocksfield, UK: Routledge.

Singer, M., & Lalich, J. (1996). *Crazy therapies: What are they? Do they work?* San Francisco, CA: Josey-Bass.

Smith, J. B. (2011). Licensing of psychotherapists in the United States: Evidence of societal regression? *Transactional Analysis Journal, 41,* 139–146.

Thompson, T. C. (2010). Psychological treatments to avoid. *Alabama Counseling Association Journal, 36*(1), 39–48.

Thyer, B. A., & Pignotti, M. G. (2015). *Science and pseudoscience in social work practice.* New York, NY: Springer.

Tracey, T. J. G., Wampold, B. E., Goodyear, R. K., & Lichtenberg, J. W. (2015). Improving expertise in psychotherapy. *Psychotherapy Bulletin, 50*(1), 7–13.

Tracey, T. J. G., Wampold, B. E., Lichtenberg, J. W., & Goodyear, R. K. (2014). Expertise in psychotherapy: An elusive goal? *American Psychologist, 69*(3), 218–229.

Tusaie, K. R. (2013). *Integrative management of disordered mood.* New York, NY: Springer.

Wampold, B. E. (2013). The good, the bad, and the ugly: A 50-year perspective on the outcome problem. *Psychotherapy, 50*(1), 16–24.

Wheeler, A. M. (2019). Social media and privileged communication. *Counseling Today, February,* 13.

CHAPTER 3

American Psychological Association Commission on Accreditation. (2018). Retrieved from https://coaportal.apa.org/login

Beyerstein, B. L. (2001). Fringe psychotherapies: The public at risk. *Scientific Review of Alternative Medicine, 5*(2), 70–79.

Commission on Accreditation for Marriage and Family Therapy Education. (2018). Retrieved from https://www.coamfte.org

Council for the Accreditation of Counseling and Related Educational Programs. (2017). Retrieved from http://www.cacrep.org/about-cacrep/vision-mission-and-core-values/

Council on Social Work Education. (2018). Retrieved from https://www.socialworkdegreecenter.com/what-is-the-council-on-social-work-education-cswe/

Denver Seminary. (2018). *Denver seminary.* Retrieved from https://denverseminary.edu

Duquesne University. (2018). Department of Psychology, Duquesne University. Retrieved from http://www.duq.edu/academics/schools/liberal-arts/academic-programs/psychology

Holden, G., & Barker, K. (2018). Should social workers be engaged in these practices? *Journal of Evidence-Informed Social Work, 15*(1), 1–13.

Lee, C. M., & Hunsley, J. (2015). Evidence-based practice: Separating science from pseudoscience. *Canadian Journal of Psychiatry, 60*(12), 534–540.

Lilienfeld, S. O. (2007). Psychological treatments that cause harm. *Perspectives on Psychological Science, 2*(1), 53–70.

Lilienfeld, S. O., Fowler, K. A., Lohr, J. M., & Lynn, S. J. (2005). Pseudoscience, nonscience, and nonsense in clinical psychology: Dangers and remedies. In R. H. Wright & N. A. Cummings (Eds.), *Destructive trends in mental health: The well-intentioned path to harm* (pp. 187–218). New York, NY: Routledge.

Lilienfeld, S. O., Ritschel, L. A., Lynn, S. J., Cautin, R. L., & Latzman, R. D. (2014). Why ineffective psychotherapies appear to work: A taxonomy of causes of spurious therapeutic effectiveness. *Perspectives on Psychological Science, 9*(4), 355–387.

Mississippi College. (2018). *Mississippi College.* Retrieved from https://www.mc.edu

National Center for Complementary and Integrative Health (NCCIH). (2019a). *NCCIH facts-at-a-glance and mission.* Retrieved from https://nccih.nih.gov/about/ataglance

National Center for Complementary and Integrative Health (NCCIH). (2019b). *NCCIH clearinghouse.* Retrieved from https://nccih.nih.gov/health/clearinghouse

Thompson, T. C. (2010). Psychological treatments to avoid. *Alabama Counseling Association Journal, 36*(1), 39–48.

Thyer, B. A., & Pignotti, M. G. (2015). *Science and pseudoscience in social work practice.* New York, NY: Springer.

Warner, E., et al. (2014). The Body Can Change the Score: Empirical Support for Somatic Regulation in the Treatment of Traumatized Adolescents. *Journal of Child and Adolescent Trauma.* Retrieved from: Retrieved from: http://www.traumacenter.org/products/pdf_files/Body_Change_Score_W0001.pdf

Wikipedia. (2018). *List of United States cities by population.* Retrieved from https://en.wikipedia.org/wiki/List_of_United_States_cities_by_population

Chapter 4

American Association of Christian Counselors. (2018). *About [the American Association of Christian Counselors]*. Retrieved from https://www.aacc.net/about/

American Association of Sexuality Educators, Counselors, and Therapists. (2014). *American Association of Sexuality Educators, Counselors and Therapists: Code of ethics and conduct for AASECT certified members*. Retrieved from https://www.aasect.org/sites/default/files/documents/Code%20of%20Ethics%20and%20Conduct_0.pdf

American Counseling Association (ACA). (2014). *Code of ethics*. Alexandria, VA: Author.

Christian Association for Psychological Studies. (2018). Retrieved from https://www.caps.net

Holden, G., & Barker, K. (2018). Should social workers be engaged in these practices? *Journal of Evidence-Informed Social Work, 15*(1), 1–13.

Martin, F. A., & Cannon, W. C. (2010). A profession in peril. *Counseling Today*, May.

National Association of Christian Social Workers. (2018). Retrieved from https://www.nacsw.org/AudioConf/042902Handouts.htm

New Orleans Baptist Theological Seminary. (2018a). *Graduate catalog*. Retrieved from http://www.nobts.edu/counseling/masters-degrees.html

New Orleans Baptist Theological Seminary. (2018b). *Graduate catalog*. Retrieved from http://www.nobts.edu/_resources/pdf/academics/GraduateCatalog.pdf#page=80

Shallcross, L. (2012). What the future holds for the counseling profession. *Counseling Today, March*. Retrieved from https://ct.counseling.org/2012/03/what-the-future-holds-for-the-counseling-profession/

Society for Christian Psychology. (2018). Retrieved from http://www.christianpsych.org/wp_scp/about-the-society/

South Florida Bible College and Theological Seminary. (2018). Retrieved from https://www.sfbc.edu/mission-philosophy/

University of Missouri. (2018). University of Missouri, Department of Educational, School and Counseling Psychology. Retrieved from https://education.missouri.edu/educational-school-counseling-psychology/degrees-programs/counseling-psychology-program/

Chapter 5

American Association of Christian Counselors. (2018). *Four credentials now offered by IBCC*. Retrieved from https://www.aacc.net/2006/09/20/four-credentials-now-offered-by-ibcc/

American Association for Marriage and Family Therapists. (2018). *Code of ethics*. Retrieved from http://www.aamft.org/iMIS15/AAMFT/Content/Legal_Ethics/Code_of_Ethics.aspx

American Association of Sexuality Educators, Counselors and Therapists. (2018). *Certification overview*. Retrieved from https://www.aasect.org/aasect-certification

American Counseling Association. (2014). *ACA code of ethics*. Alexandria, VA: Author.

American Psychological Association. (2012). *Recognition of psychotherapy effectiveness*. APA: Washington, DC: Author. Retrieved from http://www.apa.org/about/policy/resolution-psychotherapy.aspx

American Psychological Association. (2017). *Ethical principles of psychologists and code of conduct*. Washington, DC: Author.

Appeals Court. (2017). Retrieved from http://cases.justia.com/texas/third-court-of-appeals/2014-03-13-00077-cv.pdf?ts=1416566645

Association for the Development of the Person Centered Approach. (2018). *Directory of institutions in the United States and Canada.* Retrieved from http://www.adpca.org/content/directory-institutions-united-states-and-canada

Boswell, J. F., Kraus, D. R., Miller, S. D., & Lambert, M. J. (2015). Implementing routine outcome monitoring in clinical practice: Benefits, challenges, and solutions. *Psychotherapy Research, 25*(1), 6–19.

Bradley, P. Retrieved from http://www.tamft.org/#!LMFT%20License%20at%20Risk!/c1e20/D617DEF9-7C19- 4B0B-9FAA-706FAEC65448

Branstad, T. (2017). *Branstad urges reform to professional licensing requirements despite legislative pushback.* Retrieved from https://www.desmoinesregister.com/story/news/politics/2017/03/06/branstad-urges-reform-professional-licensing-requirements-despite-legislative-pushback/98802600/

Brown, R. C. C. (2011). *The development of common factor therapist competence scale for youth psychotherapy.* Retrieved from https://scholarscompass.vcu.edu/cgi/viewcontent.cgi?article=3632&context=etd

California Institute for Integral Studies. (2018). *About CIIS.* Retrieved from https://www.ciis.edu/about-ciis

Center for Credentialing and Education. (2018). *Credentialing.* Retrieved from http://www.cce-global.org

Chow, D., Miller, S., Seidel, J. A., Kane, R. T., Thornton, J. A., & Andrews, W. P. (2015). The role of deliberate practice in the development of highly effective therapists. *Psychotherapist, 52*(3), 337–345.

Duquesne University. (2018). *Department of Psychology, Duquesne University.* Retrieved from http://www.duq.edu/academics/schools/liberal-arts/academic-programs/psychology

Fairburn, C. G., & Cooper, Z. (2011). Therapist competence, therapy quality, and therapist training. *Journal of Behavior Research and Therapy, 49*(6–7), 373–388.

FindLaw. (2018). Retrieved from http://litigation.findlaw.com/legal-system/class-action-cases.html

Forman, M. D. (2010). *A guide to integral therapy: Complexity, integration, and spirituality in practice.* Albany, NY: State University of New York Press.

Goldberg, S. B., Miller, S. D., Nielsen, S. L., Rousmaniere, T., Whipple, J., Hoyt, W. T., & Wampold, B. E. (2016). Do psychotherapists improve with time and experience? A longitudinal analysis of outcomes in a clinical setting. *Journal of Counseling Psychology, 63*(1), 1–11.

Goodyear, R., Wampold, B., Tracey, T., & Lichtenberg, J. (2017). Psychotherapy expertise should mean superior outcomes and demonstrable improvement over time. *Counseling Psychologist, 45*(1), 54–65.

Haderlie, M. M. (2011). *Enhancing therapists' clinical judgments of client progress subsequent to objective feedback.* Retrieved from https://digitalscholarship.unlv.edu/cgi/viewcontent.cgi?referer=https://www.google.com/&httpsredir=1&article=2225&context=thesesdissertations

Hill C. E., Baumann E., Shafran N., Gupta S., Morrison A., Rojas A. E., . . . Gelso C. J. (2015). Is training effective? A study of counseling psychology doctoral trainees in a psychodynamic/interpersonal training clinic. *Journal of Counseling Psychology, 62*(2), 184–201.

Hill, C. E., Hoffman, M. A., Kivlighan, D. M., Spiegel, S. B., & Gelso, C. J. (2017). Therapist expertise: The debate continues. *Counseling Psychologist, 45*(1), 99–112.

Hill, C. E., Spiegel S. B., Hoffman, M. A., Kivlighan, D. M., & Gelso, C. J. (2017). Therapist expertise in psychotherapy revisited. *Counseling Psychologist, 45*(1), 1–47.

Holden, G., & Barker, K. (2018). Should social workers be engaged in these practices? *Journal of Evidence-Informed Social Work, 15*(1), 1–13.

Imel, Z. E., Sheng, E., Baldwin, S. A., & Atkins, D. C. (2015). Removing very low-performing therapists: A simulation of performance-based retention in psycho-therapy. *Psychotherapy, 52*(3), 329–336.

International Center for Clinical Excellence. (2018). *Feedback informed practice.* Retrieved from https://www.centerforclinicalexcellence.com

Jaffee v. Redmond, 518 U.S. 1 (1996). *Justia: U. S. Supreme Court.* Retrieved from https://supreme.justia.com/cases/federal/us/518/1/case.html

Jaffee v. Redmond. (2018). Retrieved from https://en.wikipedia.org/wiki/Jaffee_v._Redmond

Licensed Marriage and Family Therapists Scope of Practice in Wyoming. (2018). *2010 Wyoming Statutes: Title 33—Professions And Occupations: Chapter 38—Professional Counselors, Marriage and Family Therapists, Social Workers and Chemical Dependency Specialists.* Retrieved from https://law.justia.com/codes/wyoming/2010/Title33/chapter38.html

Licensed Professional Counselors Scope of Practice in Michigan. (2018). *Licensed Professional Counselor.* Retrieved from http://www.michigancounselingassociation.com/uploads/2/6/3/4/2634297/2010_lpc_quick_fact_sheet.pdf

Licensed Psychologists Scope of Practice in New Hampshire. (2018). *Board of Psychologists.* Retrieved from http://www.gencourt.state.nh.us/rsa/html/XXX/329-B/329-B-2.htm

Licensed Social Worker Scope of Practice in Ohio. (2018). *Chapter 4757-21 scope of practice for social workers.* Retrieved from http://codes.ohio.gov/oac/4757-21

Loyola University Chicago. (2019a). Retrieved from http://www.luc.edu/socialwork/academics/graduate/msw-advanced-standing/

Loyola University Chicago. (2019b). Retrieved from http://www.luc.edu/education/graduate/doctoraldegrees/counseling-psychology/curriculum/

Mahrer, A. (1999). Embarrassing problems in the field of psychotherapy. *Journal of Clinical Psychology, 55*(9), 1147–1156.

Martin, F. A., & Cannon, W. C. (2010). A profession in peril. *Counseling Today, 52*(5), 50–53.

Martin, F. A., & Turner, J. P. (2020). *Clinical Supervision in the Real World: A Practical Guide to Ethics, Legal Issues, and Personal Development.* New York: Routledge.

Miller, S. (2013). *The evolution of psychotherapy.* Presented at the annual meeting of the Evolution of Psychotherapy Conference. Retrieved from https://www.youtube.com/watch?v=pI8Hww1xjK4https://www.youtube.com/watch?v=pI8Hww1xjK4

Naropa University. (2018). *Contemplative practice.* Retrieved from http://www.naropa.edu/the-naropa-experience/contemplative-practice/index.php

National Association of Social Workers. (2017). *NASW code of ethics (guide to the everyday professional conduct of social workers)*. Washington, DC: Author.

National Board for Certified Counselors. (2018). *Deregulation?* Retrieved from http://nbcc.informz.net/NBCC/pages/February_Deregulation?_zs=JSCkD1&_zmi=jkTa

Overholser, J. C., & Fine, M. A. (1990). Defining the boundaries of professional competence: Managing subtle cases of clinical incompetence. *Professional Psychology: Research and Practice, 21*(6), 462–469.

Reformed Theological Seminary. (2018). *Beliefs.* Retrieved from https://www.rts.edu/site/about/beliefs.aspx

Sburiati, E. S., Lyneham, H. J., & Schniering, C. A. (2012). A model of therapist competencies for the empirically supported interpersonal psychotherapy for adolescent depression. *Clinical Child and Family Psychology, 15*(2), 93–112.

Scalia, A. (2018). *Dissenting opinion in Jaffee v. Redmond.* Retrieved from http://nsuworks.nova.edu/cgi/viewcontent.cgi?article=1466&context=nlr

Spengler, P. M., & Pilipis, L. A. (2015). A comprehensive meta-reanalysis of the robustness of the experience-accuracy effect in clinical judgment. *Journal of Counseling Psychology, 62*(3), 360–378.

Substance Abuse and Mental Health Services Administration. (2011). *Interventions for disruptive behavior disorders: Evidence-based and promising practices.* HHS Pub. No. SMA-11-4634, Rockville, MD: Center for Mental Health Services, Substance Abuse and Mental Health Services Administration, US Department of Health and Human Services.

Substance Abuse and Mental Health Services Administration. (2013). *Addressing the specific behavioral health needs of men.* Treatment Improvement Protocol (TIP) Series 56. HHS Publication No. (SMA) 13–4736. Rockville, MD: Substance Abuse and Mental Health Services Administration.

Substance Abuse and Mental Health Services Administration. (2014). *Improving cultural competence.* Treatment Improvement Protocol (TIP) Series No. 59. HHS Publication No. (SMA) 14-4849. Rockville, MD: Substance Abuse and Mental Health Services Administration.

Tao, K. W., Owen, J., Pace, B. T., & Imel, Z. E. (2015). A meta-analysis of multicultural competencies and psychotherapy process and outcomes. *Journal of Counseling Psychology, 62*(3), 337–350.

Texas Association of Marriage and Family Therapists. (2017). *Thrilling victory!* Retrieved from https://www.tamft.org/single-post/2017/02/27/THRILLING-VICTORY

Tracey, T. J. G., Wampold, B. E., Goodyear, R. K., & Lichtenberg. (2015). Improving expertise in psychotherapy. *Psychotherapy Bulletin, 50*(1), 7–13.

Tracey, T. J. G., Wampold, B. E., Lichtenberg, J. W., & Goodyear, R. K. (2014). Expertise in psychotherapy: An elusive goal? *American Psychologist, 69*(3), 218–229. Retrieved from http://dx.doi.org/10.1037/a0035099

Walfish, S., McAllister, B., & Lambert, M. J. (2012). An investigation of self-assessment bias in mental health providers. *Psychological Reports, 110,* 639–644.

Webb, C. A., Derubeis, R. J., & Barber, J. P. (2010). Therapist adherence/competence and treatment outcome: A meta-analytic review. *Journal of Consulting and Clinical Psychology, 78,* 200–211

CHAPTER 6

Above and Beyond Counseling Academy. (2019). *Christian counseling degrees*. Retrieved from https://academy.aandbcounseling.com/ncca-counseling-license-only/

Albert Ellis Institute. (2018). *Trainings*. Retrieved from http://albertellis.org/professional-rebt-cbt/trainings/

Allport, G. (1955). *Becoming: Basic considerations for a psychology of personality*. New Haven, CT: Yale University Press.

Allport, G. (1961). *Pattern and growth in personality*. New York, NY: Holt, Rinehart & Winston.

American Association of Christian Therapists. (2019). *American Association of Christian Therapists catalog of certification and license*. Retrieved from http://www.aact1995.com/sitebuildercontent/sitebuilderfiles/aactecatalog01112013poboxpriceeditedee.pdf

American Cancer Society. (2018). *Treatments for prostate cancer*. Retrieved from https://www.cancer.org/cancer/prostate-cancer/treating.html

American Psychiatric Association. (2013). *Diagnostic and statistical manual of mental disorders* (5th ed.). Arlington, VA: American Psychiatric Publishing.

American Psychiatric Association. (2018). *What is psychotherapy?* Retrieved from https://www.psychiatry.org/patients-families/psychotherapy

Biblical Theological Seminary. (2018). *Our mission, vision, and core values*. Retrieved from http://www.biblical.edu/about-bts/about-bts/mission-vision-core-values

Conway, K. (2018). *Kim Conway*. Retrieved from https://www.psychologytoday.com/us/therapists/kim-conway-andover-ma/303227

Correia, K. P. (2018). *Soul Entrainment® training and certification*. Retrieved from http://createheaven.com/soul-entrainment-training/

Erikson, E. H. (1950). Growth and crises of the "healthy personality." In M. J. E. Senn (Ed.), *Symposium on the healthy personality* (pp. 91–146). Oxford, UK: Josiah Macy, Jr. Foundation.

Hightower, E. (1988). Four illustrations of healthy personality: A prescription for living the good life. *Journal of Clinical Psychology, 44*(4), 527–535.

Holden, G., & Barker, K. (2018). Should social workers be engaged in these practices? *Journal of Evidence-informed Social Work, 15*(1), 1–13.

Imel, Z. E., Sheng, E., Baldwin, S. A., & Atkins, D. C. (2015). Removing very low-performing therapists: A simulation of performance-based retention in psychotherapy. *Psychotherapy, 52*(3), 329–336.

Mahrer, A. (1999). Embarrassing problems in the field of psychotherapy. *Journal of Clinical Psychology, 55*(9), 1147–1156.

Martin, F. A. (2020). *Notes to Grieving Friends: What to Say and Do When Their Loss Challenges Your Faith*. Eugene, Oregon: Resource Publications.

Maslow, A. (1970). *Motivation and personality*. New York, NY: Harper & Row.

Rogers, C. R. (1961). *On becoming a person: A therapist's view of personality*. Boston, MA: Houghton Mifflin.

Rogers, C. R. (1963). Actualizing tendency in relation to motives and to consciousness. In M. R. Jones (Ed.), *Nebraska symposium on motivation*. Lincoln, NE: University of Nebraska Press.

Tibetan Medical Training Center. (2019). *Tibetan medicine training*. Retrieved from https://www.tibetanmedicine-edu.org/index.php/tibetan-medicine-course

CHAPTER 7

Abundant Life Christian Counseling Services and Training Center. (2018). *Training: Here is God's pathway for training to set the captives free!* Retrieved from https://abundantlife4me.org/training/#degree+training

American Art Therapy Association. (2013, December). *Ethical principles for art therapists.* Retrieved from http://arttherapy.org/aata-ethics/

American Association of Christian Counselors. (2014). *AACC code of ethics.* Retrieved from http://aacc.net/files/AACC%20Code%20of%20Ethics%20%20Master%20Document.pdf

American Association of Marriage and Family Therapists. (2018a). *Code of ethics.* Retrieved from http://www.aamft.org/iMIS15/AAMFT/Content/Legal_Ethics/Code_of_Ethics.aspx

American Association of Marriage and Family Therapists. (2018b). *Responsibilities and guidelines for AAMFT approved supervisors and supervisor candidates.* Retrieved from https://www.aamft.org/AAMFT/supervision/Responsibilities.aspx

American Association of Pastoral Counselors. (2012, April). *AAPC code of ethics.* Retrieved from http://www.aapc.org/Default.aspx?ssid=74&NavPTypeId=116American

American Association of Sexuality Educators, Counselors, and Therapists. (2014). *Code of ethics and conduct for AASECT certified members.* Retrieved from https://www.aasect.org/sites/default/files/documents/Code%20of%20Ethics%20and%20Conduct_0.pdf

American Counseling Association. (2014). *ACA code of ethics.* Alexandria, VA: Author.

American Counseling Association. (2018). *ACA takes action against state-based legislation threatening licensure for counselors.* Retrieved from https://www.counseling.org/news/aca-blogs/aca-government-affairs-blog/aca-government-affairs-blog/2018/01/29/aca-takes-action-to-block-bill-that-could-eliminate-licensure-for-counselors-in-arizona

American Group Psychotherapy Association. (2002, February). *AGPA and IBCGP guidelines for ethics.* Retrieved from http://www.agpa.org/home/practice-resources/ethics-in-group-therapy

American Mental Health Counselors Association. (2015). *AMHCA code of ethics.* Retrieved from http://www.amhca.org/?page=codeofethics

American Psychiatric Association. (2013). *The principles of medical ethics: With annotations especially applicable to psychiatry.* Retrieved from file:///C:/Users/User%201/Downloads/principles-medical-ethics.pdf

American Psychological Association. (2017). *Ethical principles of psychologists and code of conduct.* Washington, DC: Author.

American School Counselor Association. (2016). *ASCA ethical standards for school counselors.* Retrieved from https://www.schoolcounselor.org/asca/media/asca/Ethics/EthicalStandards2016.pdf

Association for Addiction Professionals. (2016, October 9). *NAADAC/NCC AP code of ethics.* Retrieved from http://www.naadac.org/assets/1959/naadac-nccap-code-ofethics11-04-16.pdf

Bartlett, K. (2016, December 6). Sen. Johnson proposes counseling board to devise new code of ethics: TCA deeply saddened. *The Williamson Herald.* Retrieved from http://www.williamsonherald.com/news/article_86cbeeaa-bc0b-11e6-831d-f3a475989882.html

Biblical Theological Seminary. (2018). *Our mission, vision, and core values.* Retrieved from http://www.biblical.edu/about-bts/about-bts/mission-vision-core-values

Board Certified Music Therapist. (2011).*CBMT code of professional practice.* Retrieved from http://www.cbmt.org/wp-content/uploads/2019/09/CBMT_Code-of-Professional-Practice_2015.pdf

Bradley, L. J. (1995). Certification and licensure issues. *Journal of Counseling and Development, 74*(November/December), 185–185.

Center for Credentialing and Education. (2016). *Approved clinical supervisor code of ethics.* Retrieved from http://www.cce-global.org/Assets/Ethics/ACScodeofethics.pdf

Center for Credentialing and Education. (2018). *Credentialing.* Retrieved from http://www.cce-global.org

Certified Career Adjustment Associates. (2017, January 1). *Code of ethics.* Retrieved from https://www.crccertification.com/code-of-ethics-4

Certified Eating Disorders Specialist. (2003). *International association of eating disorders professionals foundation certification manual, Appendix I: Ethical principles of psychologists and code of conduct.* Retrieved from http://www.iaedp.com/MH-Cert-Manual-in-PDF.pdf

Certified Group Therapist. (n.d.). *Ethics in group therapy: AGPA and NRCGP guidelines for ethics.* Retrieved from http://www.groupsinc.org/certification/ethics-in-group-therapy/

Certified Rehabilitation Counselor. (2017). *Code of professional ethics for rehabilitation counselors.* Retrieved from https://www.crccertification.com/code-of-ethics-4

Certified Sex Therapist. (2014). *American Association of Sexuality Educators, Counselors and Therapists: Code of ethics and conduct for AASECT certified members.* Retrieved from https://www.aasect.org/sites/default/files/documents/Code%20of%20Ethics%20and%20Conduct_0.pdf

Certified Vocational Evaluation Specialists. (2009, April). *Code of professional ethics for vocational evaluation specialist, work adjustment specialist, and career assessment associates.* Retrieved from https://www.crccertification.com/filebin/pdf/VE/VECodeofEthics.pdf

Certified Work Adjustment Specialists. (2009, April). *Code of professional ethics for vocational evaluation specialist, work adjustment specialist, and career assessment associates.* Retrieved from https://www.crccertification.com/filebin/pdf/VE/VECodeofEthics.pdf

Chow, D., Miller, S., Seidel, J. A., Kane, R. T., Thornton, J. A., & Andrews, W. P. (2015). The role of deliberate practice in the development of highly effective therapists. *Psychotherapist, 52*(3), 337–345.

Erekson, D. M., Janis, R., Bailey, R. J., Cattani, K., & Pedersen, T. R. (2017). A longitudinal investigation of the impact of psychotherapist training: Does training improve client outcomes? *Journal of Counseling Psychology, 64*(5), 514–524.

Feinstein, R., Heiman, N., & Yager, J. (2015). Common factors affecting psychotherapy outcomes: Some implications forteaching psychotherapy. *Journal of Psychiatric Practice, 21*(3), 180–189.

Freedom for All Americans. (2017). *Tennessee Senate bill 1.* Retrieved from http://www.freedomforallamericans.org/tennessee-senate-bill-1/

Gervin, C. W. (2016, December 5). One anti-gay counseling bill wasn't enough. Now There's a second one. *Nashville Scene.* Retrieved from http://www.nashvillescene. com/news/pith-in-the-wind/article/20845882/one-antigay-counseling-bill-wasnt-enough-now-theres-a-second-one

Grohol, J. (1998/2004). *Why don't current psychotherapy licensing regulations work?* Retrieved on from https://psychcentral.com/archives/licensing.htm

Hill, C. E., Baumann, E., Shafran, N., Gupta, S., Morrison, A., Rojas, A. E., . . . Gelso C. J. (2015). Is training effective? A study of counseling psychology doctoral trainees in a psychodynamic/interpersonal training clinic. *Journal of Counseling Psychology, 62*(2), 184–201.

Hogan, D. B. (1983). The effectiveness of licensing: History, evidence, and recommendations. *Law and Human Behavior, 7*(2/3), 117–138.

Holden, G., & Barker, K. (2018). Should social workers be engaged in these practices? *Journal of Evidence-Informed Social Work, 15*(1), 1–13.

Howard, A. (2000). *Philosophy for counselling and psychotherapy: Pythagoras to postmodernism.* London, UK: McMillan Press.

Humphrey, T. (2016, December 6). Bill mandates new code of ethics for TN counselors. *The Tennessee Journal's Humphrey on the hill.* Retrieved from http://humphreyonthehill. tnjournal.net/bill-calls-new-code-ethics-tn-counselors/

International Expressive Arts Therapy Association. (n.d.). *Code of ethics: Ethical guidelines.* Retrieved from http://www.ieata.org/

Jennings, L., & Skovholt, T. M. (1999). The cognitive, emotional, and relational characteristics of master therapists. *Journal of Counseling Psychology, 46*(1), 3–11.

Knapp, S., Gottlieb, M. C., & Handelsman, M. M. (2017). Self-awareness questions for effective psychotherapists: Helping good psychotherapists become even better. *Practice Innovations, 2*(4), 163–172.

LegisScan. (2017). *Bill Text: TN SB0001 | 2017-2018 | 110th General Assembly | Draft.* Retrieved from https://legiscan.com/TN/text/SB0001/2017

Mackrill, T., & Iwakabe, S. (2013).Making a case for case studies in psychotherapy training: A small step towards establishing an empirical basis for psychotherapy training, *Counselling Psychology Quarterly, 26*(3–4), 250–266.

Mahrer, A. (1999). Embarrassing problems in the field of psychotherapy. *Journal of Clinical Psychology, 55*(9), 1147–1156.

Malott, K. M., Hall, K. H, Sheely-Moore, A., Krell, M. M., & Cardaciotto, L. A. (2014). Evidence-based teaching in higher education: Application to counselor education. *Counselor Education & Supervision, 53*(4), 294–305.

Martin, F. A., & Cannon, W. C. (2010a). A profession in peril. *Counseling Today, 52*(5), 52–53.

Martin, F. A., & Cannon, W. C. (2010b). *The necessity of a philosophy of clinical supervision.* Retrieved from http://counselingoutfitters.com/vistas/vistas10/Article_45.pdf

Martin, F., & Turner, J. (2020). *Clinical supervision in the real world: A Practical Guide to Ethics, Legal Issues, and Personal Development.* New York: Routledge.

Master Addictions Counselor. (2016). *NAADAC/NCC AP code of ethics.* Retrieved from http://www.naadac.org/code-of-ethics

McCauliff, G., & Eriksen, K. (2011). *Handbook of counselor preparation: Constructivist, developmental, and experiential approaches.* Thousand Oaks, CA: Sage.

Meyer, H. (2016, December 7) Counseling debate continues with new Tennessee bill. *The Tennessean.* Retrieved from http://www.tennessean.com/story/news/politics/2016/12/07/counselingdebatecontinues-new-tennessee-bill/95039534/

Meyer, H. (2017, January 5). Local group lists 5 reasons it opposes counseling bill. *The Tennessean.* Retrieved from http://www.tennessean.com/story/news/religion/2017/01/05/anothertherapyorganization-opposes-counseling-bill/96172612/

Meyers, L. (2016, June 27). License to deny service. *Counseling Today.* Retrieved from http://ct.counseling.org/2016/06/license-deny-services/

Miller, S. (2013). *The evolution of psychotherapy.* Presented at the annual meeting of the Evolution of Psychotherapy Conference. Retrieved from https://www.youtube.com/watch?v=pI8HwwlxjK4

Milne, D. (2014). Toward and evidence-based approach to clinical supervision. In C. E. Watkins & D. Milne. (2014). *The Wiley international handbook of clinical supervision* (pp. 38–60). Malden, MA: John Wiley and Sons.

Ministry International Institute. (2019). Ministry International Institute. Retrieved from https://ministryinternational.tv/mii-institute/

Naropa University. (2018). *Naropa University.* Retrieved from http://www.naropa.edu/the-naropa-experience/contemplative-practice/index.php

National Association of Social Workers. (2017). *NASW code of ethics.* Washington, DC: Author.

National Career Development Association. (2015). *NCDA code of ethics.* Retrieved from http://www.ncda.org/aws/NCDA/asset_manager/get_file/3395

National Certified Addiction Counselor, Levels I or II. (2016). *NAADAC/NCC AP code of ethics.* Retrieved from http://www.naadac.org/code-of-ethics

National Certified Adolescent Addiction Counselor. (2016). *NAADAC/NCC AP code of ethics.* Retrieved from http://www.naadac.org/code-of-ethics

National Certified Clinical Mental Health Counselor. (2016). *National board for certified counselors (NBCC) code of ethics.* Retrieved from http://www.nbcc.org/Certification/Ethics

National Certified Counselor. (2016). *National board for certified counselors (NBCC) code of ethics.* Retrieved from http://www.nbcc.org/Certification/Ethics

National Certified School Counselor. (2016). *National board for certified counselors (NBCC) code of ethics.* Retrieved from http://www.nbcc.org/Certification/Ethics

National Christian Counselors Association. (2018). *New member information.* Retrieved from https://www.ncca.org/newmember.html

National Christian Counselors Association. (2019). *National Christian Counselors Association.* Retrieved from https://www.ncca.org/default.html

National Clinical Supervision Endorsement. (2016). *NAADAC/NCC AP code of ethics.* Retrieved from http://www.naadac.org/code-of-ethics

National Endorsed Co-Occurring Disorders Professional. (2016). *NAADAC/NCC AP code of ethics.* Retrieved from http://www.naadac.org/code-of-ethics

National Endorsed Student Assistance Professional. (2016). *NAADAC/NCC AP code of ethics.* Retrieved from http://www.naadac.org/code-of-ethics

National Peer Recovery Support Specialist. (2016). *NAADAC/NCC AP code of ethics.* Retrieved from http://www.naadac.org/code-of-ethics

Nicotine Dependence Specialist. (2016). *NAADAC/NCC AP code of ethics.* Retrieved from http://www.naadac.org/code-of-ethics

Osberg, T. M. (1997). Teaching psychotherapy outcome research methodology using a research-based checklist. *Teaching of Psychology, 24*(4), 271–274.

Pastoral Medical Association. (2019). *About us.* Retrieved from https://www.pmai.us/about-us

Professional Certified Coach. (2015). *International coach federation code of ethics.* Retrieved from http://coachfederation.org/about/ethics.aspx?ItemNumber=854

Registered Expressive Arts Therapist. (n.d.). *Code of ethics for registered expressive arts therapist.* Retrieved from http://www.ieata.org/downloads/reat_ethics.pdf

Registered Play Therapist. (2016, September). *Play therapy best practices: Clinical, professional and ethical issues.* Retrieved from http://c.ymcdn.com/sites/www.a4pt.org/resource/resmgr/publications/Best_Practices__-_Sept_2016.pdf?hhSearchTerms=%22code+and+ethics%22

Rogers, C. R. (1951). *Client-centered therapy.* Boston: Houghton Mifflin.

Roland, C. B. (2016, December 22). 2017: Counselors stepping up to step forward. *Counseling Today.* Retrieved from http://ct.counseling.org/2016/12/2017-counselors-stepping-step-forward/

Rousmanaiere, T. (2017). *Deliberate practice for psychotherapists: A guide to improving clinical effectiveness.* New York, NY: Routledge.

School-Based Registered Play Therapist. (2015). *Play therapy best practices: Clinical, professional and ethical issues.* Retrieved from http://c.ymcdn.com/sites/www.a4pt.org/resource/resmgr/publications/Best_Practices__-_Sept_2016.pdf?hhSearchTerms=%22code+and+ethics%22

Six, T. (2014). *Characteristics of highly effective therapists.* Unpublished master's thesis, Gent University. Retrieved from https://lib.ugent.be/fulltxt/RUG01/002/166/096/RUG01-002166096_2014_0001_AC.pdf

Society for the Advancement of Psychotherapy. (2018). Retrieved from http://societyforpsychotherapy.org/10-ways-to-improve-psychotherapy-outcome/

Spengler, P. M., & Pilipis, L. A. (2015). A comprehensive meta-reanalysis of the robustness of the experience-accuracy effect in clinical judgment. *Journal of Counseling Psychology, 62*(3), 360–378.

United States Association for Body Psychotherapy. (2013). *Ethical guidelines.* Retrieved from http://usabp.org/wp-content/uploads/2013/12/USABPethics2.pdf

Watkins, C. E., & Milne, D. (Eds.) (2014). *The Wiley International Handbook of Clinical Supervision.* Malden, MA: John Wiley and Sons.

Watkins, C. E., & Wang, C. D. C. (2014). On the education of clinical supervisors. In C. E. Watkins & D. Milne, D. *The Wiley international handbook of clinical supervision* (pp. 177–203). Malden, MA: John Wiley and Sons.

Wheeler, K., & Delaney, K. (2008). Challenges and realities of teaching psychotherapy: A survey of psychiatric-mental health nursing graduate programs. *Perspectives in Psychiatric Care, 44*(2), 72–80.

Yep, R. (2016, April 28). CEO's message: A dangerous precedent. *Counseling Today.* Retrieved from http://ct.counseling.org/2016/04/ceos-message-a-dangerous-precedent/

CHAPTER 8

American Foundation for Suicide Prevention. (2019). Retrieved from https://afsp.org

Bertolino, B., & Miller, S. D. (2012). *Feedback-informed clinical work: The basics.* Chicago, IL: The International Center for Clinical Excellence.

Goldberg, S. B., Babins-Wagner, R., Rousmaniere, T., Berzins, S., Hoyt, W. T., Whipple, J. L., ... Wampold, B. E. (2016). Creating a climate for therapist improvement: A case study of an agency focused on outcomes and deliberate practice. *Psychotherapy, 53*(3), 367–375.

Goldberg, S. B., Miller, S. D., Nielsen, S. L., Rousmaniere, T., Whipple, J., Hoyt, W. T., & Wampold, B. E. (2016). Do psychotherapists improve with time and experience? A longitudinal analysis of outcomes in a clinical setting. *Journal of Counseling Psychology, 63*(1), 1–11.

Imel, Z. E., Baldwin, S. A., Sheng, E., & Atkins, D. C. (2015). Removing very low-performing therapists: A simulation of performance-based retention in psychotherapy. *Psychotherapy, 52*(3), 329–336.

Low, D. C., Miller, S. D., & Squire, B. (2012). *The Outcome Rating Scales (ORS) and Session Rating Scales (SRS): Feedback informed treatment in child and adolescent mental health services (CAMHS)*. Retrieved from http://www.aft.org.uk/SpringboardWebApp/userfiles/aft/file/Events/2012/David%20Low%20paper%20for%20CYP-IAPT.pdf

McMahan, E. H. (2014). Supervision: A non-elusive component of deliberate practice toward expertise. *American Psychologist, 69*(3).

Miller, S. D., & Donahey, K. M. (2012). Feedback informed treatment (FIT): Improving the outcome of sex therapy one person at a time. In P. Kleinplatz (Ed.), *New directions in sex therapy: Innovations and alternatives* (pp. 195–211). New York, NY: Routledge.

Nylund, D., & Filippelli, A. (2018). Feedback informed treatment. Retrieved from https://www.cibhs.org/sites/main/files/file-attachments/thurs_1030_garden_fillipi_nylund_fit_powerpoint_jan_2016.pdf

Rousmanaiere, T. (2017). *Deliberate practice for psychotherapists: A guide to improving clinical effectiveness.* New York, NY: Routledge.

Spengler, P. M., & Pilipis, L. A. (2015). A comprehensive meta-reanalysis of the robustness of the experience-accuracy effect in clinical judgment. *Journal of Counseling Psychology, 62*(3), 360–378.

Walfish, S., McAllister, B., & Lambert, M. J. (2012). An investigation of self-assessment bias in mental health providers. *Psychological Reports, 110,* 639–644.

Yates, C. M., Holmes, C. M., Coe Smith, J. C., & Nielsen, T. (2018). The benefits of implementing a feedback informed treatment system within counselor education curriculum. *Professional Counselor, 5*(1), 22–32.

Chapter 9

Goldberg, S. B., Miller, S. D., Nielsen, S. L., Rousmaniere, T., Whipple, J., Hoyt, W. T., & Wampold, B. E. (2016). Do psychotherapists improve with time and experience? A longitudinal analysis of outcomes in a clinical setting. *Journal of Counseling Psychology, 63*(1), 1–11.

Holden, G., & Barker, K. (2018). Should social workers be engaged in these practices? *Journal of Evidence-informed Social Work, 15*(1), 1–13.

Spengler, P. M., & Pilipis, L. A. (2015). A comprehensive meta-reanalysis of the robustness of the experience-accuracy effect in clinical judgment. *Journal of Counseling Psychology, 62*(3), 360–378.

Tables and boxes are indicated by *t* and *b* following the page number

For the benefit of digital users, indexed terms that span two pages (e.g., 52–53) may, on occasion, appear on only one of those pages.